Edited by
Louis W. Chang

TOXICOLOGY OF METALS

Associate Editors
László Magos
Tsuguyoshi Suzuki

Foreword by Thomas W. Clarkson

LEWIS PUBLISHERS
Boca Raton   New York   London   Tokyo

TOXICOLOGY
OF
METALS

Acquiring Editor:	Ken McCombs
Editorial Assistant	Susan Alfieri
Project Editor:	Carrie L. Unger
Marketing Manager:	Greg Daurelle
Direct Marketing Manager:	Arline Massey
Cover design:	Shayna Murry
PrePress:	Carlos Esser
Manufacturing:	Sheri Schwartz

Library of Congress Cataloging-in-Publication Data

Toxicology of metals / edited by Louis W. Chang; associate editors László Magos, Tsuguyoshi Suzuki
    p.   cm.
    Includes bibliographical references and index.
    ISBN 1-56670-803-6 (alk. paper)
    1. Metals—toxicity   2. Metals—toxicology
 I. Chang, Louis W.   II. Magos, L.   III. Suzuki, Tsuguyoshi 1932– .
 [DNLM: 1. Metals—toxicity.   QV 290 T7548 1996]
RA1231.M52T695   1996
615.9′253—dc20   95-42586
                                                                                                                                                   CIP

    This book contains information obtained from authentic and highly regarded sources. Reprinted material is quoted with permission, and sources are indicated. A wide variety of references are listed. Reasonable efforts have been made to publish reliable data and information, but the author and the publisher cannot assume responsibility for the validity of all materials or for the consequences of their use.

    Neither this book nor any part may be reproduced or transmitted in any form or by any means, electronic or mechanical, including photocopying, microfilming, and recording, or by any information storage or retrieval system, without prior permission in writing from the publisher.

    All rights reserved. Authorization to photocopy items for internal or personal use, or the personal or internal use of specific clients, may be granted by CRC Press, Inc., provided that $.50 per page photocopied is paid directly to Copyright Clearance Center, 27 Congress Street, Salem, MA 01970 USA. The fee code for users of the Transactional Reporting Service is ISBN 1-56670-803-6/96/$0.00+$.50. The fee is subject to change without notice. For organizations that have been granted a photocopy license by the CCC, a separate system of payment has been arranged.

    The consent of CRC Press does not extend to copying for general distribution, for promotion, for creating new works, or for resale. Specific permission must be obtained from CRC Press for such copying.

    Direct all inquiries to CRC Press, Inc., 2000 Corporate Blvd., N.W., Boca Raton, Florida 33431.

© 1996 by CRC Press, Inc.
Lewis Publishers is an imprint of CRC Press

No claim to original U.S. Government works
International Standard Book Number 1-56670-803-6
Library of Congress Card Number   95-42586
Printed in the United States of America   1   2   3   4   5   6   7   8   9   0
Printed on acid-free paper

Dedication

This book is dedicated to ...

MY TEACHERS, who have paved the way for me to travel

MY COLLEAGUES, who have provided me with support, encouragement, and comfort throughout my professional journey

MY STUDENTS, who will carry on the torch and continue the journey with renewed youth and vigor

Louis W. Chang

Preface

Metals are perhaps one of the earliest medicines and poisons known to humankind. Indeed, mercury, for example, has been used for therapeutic purposes since the dawn of human civilization, yet mercury is also one of the most potent toxins known to man. In short, humankind has always been fascinated and fearful towards this "two-edged sword" which can be neither created nor destroyed.

Because of this, there is a fascination regarding the immense benefits that metals can offer (indeed, human lives are heavily dependant on the utilization of metals — from the nutrients we eat to the utilities we use daily; from the machines that we operate to the weapons that we build; from agricultural chemicals to medications, etc.). From the dark fears of the potent but incidious toxic potentials of metals, humankind has engaged in metal research since the dawn of sciences; toxicology may be defined as the study of the adverse effects of chemicals and the biomolecular mechanisms underlying such adverse effects. The chemical actions and health effects of metals have become one of the prime subjects of interest since the birth of both pharmacology and toxicology.

Because of this unyielding interest in metals, metal research represents one of the most rapidly developing fields in toxicology. Numerous research articles as well as textbooks have been devoted to this subject. When I was asked to consider putting together a new volume on the toxicology of metals, I was perplexed with the thought about what I could do differently than those who had written and edited similar publications in the past. How was I going to present to the readers something that was more than simply a "review" of what we had already done, while also including our current achievements as well as the issues and concepts that are important to the future development of metal toxicology? Most importantly, what topics and authors should be included to assure the quality and authority of the volume?

With much deliberation, I have decided that it will take a large volume to adequately encompass all the major topics and interests in metal toxicology. Furthermore, in order to assure quality and authority, the volume should draw contributions from the best scientists internationally rather than (like many previous publications) being limited to local or regional experts. I further recognized that a monumental and ambitious effort as such would require a true "team" effort starting from the initial days of planning.

I am most fortunate and grateful that Prof. László Magos of the U.K. and Prof. Tsuguyoshi Suzuki of Japan were willing to serve as my Associate Editors. Their scholarly experiences have truly been an inspiration to me throughout the organization of these volumes. I have also retained Prof. Thomas Clarkson, Prof. Robert Goyer, Prof. George Cherian, Prof. Ernest Foulkes, Prof. Arthur Furst, Prof. Lars Friberg, Prof. Gunnar Nordberg, Prof. Kazuo Suzuki, and Prof. Felicia Chen-Wu to be on the Advisory Board for this volume. Their encouragement as well as their constructive criticism throughout the years certainly have helped to shape this book. Their contributions are well appreciated by the editors.

A textbook can only be as good as the subject matters covered (i.e., the contents) and the authors involved (i.e., the contributors). As the editor (i.e., the organizer) of this book, it has become my responsibility to determine the contents of the volume as well as who would be the best contributors for each subject matter. Metal toxicology has become a broad and multidisciplinary science. It is practically impossible for any *one* person to be familiar with all areas of metal toxicology (e.g., risk assessment, carcinogenesis, neurotoxicology, immunotoxicity, etc.). With the assistance and advice of my Associate Editors and Advisory Board, I have designated four sections for Part 1 and five sections for Part 2 according to the subject matters that I believed to be of major importance in metal toxicology. One or two top experts in each of these subject matters were then identified and invited to serve as Section Heads for these sections. The Editors, Advisory Board members, and Section Heads then worked together to identify the "contents" (i.e., chapter topics) for individual sections and the best possible scientists to be invited as chapter contributors. The general guidelines for contributor selections were (1) the contributor(s) must be true expert(s) in the field of metal toxicology and must be currently active in research and in the subject matter covered in the chapter, and (2) if several experts of equal recognitions on the same subject matter were identified, we preferred to select the one(s) who had **not** recently or previously published a review article or chapter on a similar topic. I believe that these guidelines would help to assure us that **current** information would to be included in each chapter and that a fresh approach to each chapter would be taken (rather than an "update" of a previously published review). I believe that our current "team" approach (Editor, Associate Editors, Advisory Board, Section Heads, and internationally selected expert-contributors) allowed us to share ideas and expertise to create and produce the most comprehensive and authoritative texts of scholarly quality that we had all hoped for.

Part 1 on General and Clinical Aspects, Risk Assessment, and Genotoxicity of Metals consists of four sections: Monitoring Environmental and Human Assessments; Bioresponses and Reactivities of Metals; Carcinogenesis and Genotoxicity of Metals; and Clinical Aspects of Metal Toxicology. I am most indebted to Dr. Anna M. Fan (California Environmental Protection Agency), Prof. Max Costa (New York University Medical Center), Dr. Lionel A. Poirier (National Center for Toxicological Research), Prof. Kazuo Suzuki (Chiba University, Japan), and Prof. Tsuguyoshi Suzuki (University of Tokyo, Japan), who all agreed to serve as Section Heads for this part. Their assistance in identifying the chapter topics as well as their suggestions for expert contributors for each chapter were most helpful. Their well recognized expertise in their respective areas of metal toxicology was animportant assets to us in the planning and design of each section in order to assure the highest quality and authority possible.

Part 2 is devoted to Target Organ Toxicology. While many previously published texts organize the health effects in accordance to metals (e.g., neurotoxicity of mercury, neurotoxicity of lead, etc.), we preferred to organize the chapters in accordance to concepts of importance. Multiple metals will be discussed together within each conceptual confine so that "metals" can be viewed as a whole and comparative bases can be made between them (e.g., Comparative Neurobehavioral Toxicology of Heavy Metals, Oxygen Generation as a Bases for Neurotoxicity by Metals, etc.) As metal toxicology has evolved rapidly from the classical studies on "health effects" of metals to the current interests on the biomolecular bases of metal toxicity, I believe that our current approach in organizing these volumes will meet the needs and expectations of our readers.

Part 2 is divided into five sections: Neurotoxicology of Metals, Renal Toxicology of Metals, Immunomodulation by Metals, Effects of Metals on Other Organs, and Reproductive and Developmental Toxicology of Metals. Each of these sections is headed by an internationally recognized expert of the field to assure quality as well as authority excellence in each section in this volume. I would like to take the opportunity to thank Prof. Bruce Fowler (University of Maryland), Prof. Judith T. Zelikoff (New York University Medical Center), Dr. Donald Gardner (ManTech, Inc.), and Dr. John Rogers (U.S. Environmental Protection Agency) for their assistance in the organizations of these sections.

This volume of text consists of 72 chapters involving over 120 expert-contributors from over 10 countries. Each of the contributors was selected authorities from the various fields of metal toxicology. I want to thank them for their commitment to this project. Despite their own busy schedules, they have come through for us in producing a text that we can all be proud of for years to come. Without their sense of commitment and professionalism, this project can never be completed as planned.

As we are at the twilight of the 20th century, we can look back and be proud of all of the accomplishments and achievements that we have made in the last decades in the field of metal toxicology. We can also look forward to the dawning of the 21st century where new challenges await us. I hope that this volume of *Toxicology of Metals* serves not just as a record of our past accomplishments, but also as a stepping stone for those who are interested in metal toxicology to go forth boldly to meet those new challenges awaiting for us. Many contributors in this book have devoted decades of their lives to the development of metal toxicology. Through this volume, we are also passing the torch to the new generation of metal scientists, enabling them to carry this field of toxicology forth with pride and commitment.

Louis W. Chang
Editor

The Editors

Louis W. Chang, Ph.D., was born in Hong Kong, China in 1944. After completing his high school education at St. Jospeh's College in Hong Kong, China, he moved to Brazil with his parents for a year before moving to the U.S. for his college education. Prof. Chang received his B.A. from the University of Massachusetts-Amherst in 1966, his M.S. in Anatomy and Histochemistry from Tufts University School of Medicine in 1969, and his Ph.D. in Pathology from the University of Wisconsin-Madison School of Medicine in 1972. Prof. Chang also received his education and training in neurocytology from Harvard Medical School and *in vitro* and biochemical neurotoxicology from the Brain Research Institute at the University of California-Los ANgeles School of Medicine.

Currently, Dr. Chang is a Professor in the Departments of Pathology, Pharmacology and Toxicology at the University of Arkansas for Medical Sciences College of Medicine. He has also served as Program Director in Toxicology at UAMS between 1988-1990. A this time, he is the Graduate Program Director in Pathology at UAMS. Aside from being an author of over 200 scientific publications (mainly in metal toxicology and neurotoxicology), Prof. Chang is also responsible for the following important publications: *Principles of Neurotoxicology* (Marcel Dekker, Inc.), *Handbook of Neurotoxicology* (Marcel Dekker, Inc.), *Neurotoxicology: Approaches and Methods* (Academic Press, Inc.), *Toxicology and Risk Assessment* (Marcel Dekker, Inc.), and *Toxicology of Metals* (Lewis Publishers/CRC Press, Inc.). Other new textbooks currently in preparation include *Developmental Toxicology* (Academic Press, Inc.) and *Neurobehavioral Toxicology* (Academic Press, Inc.).

Prof. Chang is an active member of the Society of Toxicology, the Society of Neurosciences, and the American Association of Neuropathologists, among others. He has also served on the editorial boards of numerous scientific journals, including *Neurotoxicology, Toxicology and Applied Pharmacology, Fundamentals and Applied Toxicology, Behaviorial Neurotoxicology and Teratology, Biomedical and Environmental Sciences,* to name a few. Prof. Chang is a frequent invited speaker in international symposia and conferences in the areas of environmental health, metal toxicology, neurotoxicology, pharmaceutical evaluations, risk assessments, and developmental toxicology. He is well recognized internationally and hold numerous international academic appointments such as Visiting and Honored Professor at the Beijing Medical University, the Nanjing Medical University, the Chinese Academy of Preventive Medicine, The Institute of Occupational Medicine in Beijing, and the National Institute for Control of Pharmaceutical and Biological Products in China. Prof. Chang has organized numerous symposia, workshops, and conferences in China, Taiwan, Hong Kong, Korea, Japan, Singapore, and Brazil and has received numerous recognitions and awards both nationally and internationally.

Prof. Chang has recently been elected as a Fellow of the Academy of Toxicological Sciences and is also board certified as a Diplomate of the American Board of Forensic Examiners.

László Magos, M.D., obtained his medical degree from the University of Zagreb, Zagreb, Hungary in 1948; his diploma in clinical pathology from the Board of Medical Specialization, Budapest, Hungary in 1954; a degree for research on Raynaud's phenomenon of occupational origins from the Hungarian Scientific Academy in 1957; and a diploma of the University of Pavia for recognition of research in the field of toxicology in 1959. In addition, he was elected as a Fellow of the Royal College of Pathologists in 1982.

After his university years, Dr. Magos worked in the Department of Hygiene and Epidemiology of his university, but in 1950 he moved to the Institute of Occupational Health, Budapest, Hungary, where his interests increasingly turned to chemical hazards, mainly compounds which react with hemoglobin.

Foreword

The field of metal toxicology is one of the broadest areas of toxicology. The metals account for over 75% of all the elements in the periodic table, not to mention the recent transuranic metals. The chemistry of metals which underlies their toxic action is also diverse. The toxic outcomes cover virtually every adverse effect from the cellular to the whole animal level. Several human inheritable diseases are related to imbalances in metal metabolism.

As one of the oldest areas of study, metal toxicology is also one of the most rapidly developing fields. This volume records many of the recent advances in our understanding of the mechanisms of toxic action. In the past two decades, we have seen exciting discoveries at every level of metal interaction with living cells. Metals act as mutagens and carcinogens, modifying the structure of DNA or interfering with the transcription processes. The "toxic" metals can displace "essential" metals acting as cofactors of enzymes or supporting key structures in the cytoskeleton. Membrane transport proteins can be targets for metal action. At the same time, certain transport proteins can serve as means of introducing metals into the cell. For example, the oxyanions of metals, such as arsenates and chromates, sufficiently mimic their endogenous counterparts, phosphates and sulfates, that they are transported into the cell on proteins intended for the latter.

New insights have been gained into how the cell protects itself. Perhaps the most dramatic story is the discovery of the metallothioneins. This family of small molecular weight proteins, rich in thiol groups, has been shown to protect all types of cells from the toxic action of many metals, including Ag, Cd, Cu, Hg, and Zn. Indeed, the study of metal interaction with the genes coding for the metallothioneins advanced our understanding of the processes involved in gene regulation.

Chapters in this volume also address key public health issues associated with human exposure to toxic metals. Advances have been made in the field of risk assessment through studies on the epidemiology of toxic metals. By far the most celebrated story is that of childhood exposure to lead making us aware of the subtle but insidious damage that can be inflicted on the developing nervous system. Recent studies on other human populations have led to establishing key dose–response relationships for such metals as As, Cd, Hg, and Pb. These quantitative relationships between dose and risk of adverse effect are the key to modern risk assessment procedures.

The present volume on toxicology of metals fills an important gap in the literature. Nowhere else has such a broad and exhaustive account of metal toxicology been attempted. This volume promises to be the key reference source on the toxicology of metals for many years to come.

Thomas W. Clarkson
Rochester, New York

From 1963 and on, at which time he joined the MRC Toxicology Unit, his main interest became and remained the toxicology of mercurials.

Dr. Magos was a World Health Organization (WHO) consultant in Iraq during the methylmercury epidemic of 1972 as well as on heavy metal monitoring in India in 1977. He also worked for the Group of Experts of the Scientific Aspects of Marine Pollution (GESAMP) from 1978 to 1986.

He has written or contributed to several documents for the WHO, The Mediterranean Action Plan, IARA, and UNESCO, as well as national and international organizations. He retired from the MRC Toxicology Unit in 1986, but remains active as a consultant and as the European Editor of the *JAT*.

Prof. Tsuguyoshi Suzuki, Ph.D., was born in Tokyo, Japan in 1932. After graduating from the Faculty of Medicine at the University of Tokyo in 1955, he completed his Ph.D. in the Medical Sciences at the University of Tokyo Graduate School in 1960. Since then, he has dedicated his efforts to a variety of occupational diseases, especially those related to heavy metal poisoning.

As a professor and researcher in the Departments of Public Health and Toxicology at the University of Tokyo, Prof. Suzuki has devoted his life to solving problems related to human–environmental interactions. Recently, he advocated the use of multi-element analysis and stable isotope analysis as tools to dissect the environmental system. Currently, Prof. Suzuki is serving as the Director General of the National Institute for Environmental Studies at the Environmental Agency of Japan. He is also serving as the Emeritus Professor at the University of Tokyo.

Contributors

N. Joan Abbott, Ph.D.
Physiology Group
Biomedical Sciences Division
King's College London
London, England

Carol R. Angle, M.D.
Department of Pediatrics
University of Nebraska Medical Center
Omaha, Nebraska

Michael Aschner, Ph.D.
Department of Physiology and Pharmacology
Bowman Gray School of Medicine
Winston-Salem, North Carolina

Gerald Audesirk, Ph.D.
Department of Biology
University of Colorado at Denver
Denver, Colorado

Teresa Audesirk, Ph.D.
Department of Biology
University of Colorado at Denver
Denver, Colorado

Delon W. Barfuss, Ph.D.
Department of Biology
Georgia State University
Atlanta, Georgia

David A. Basketter, B.Sc.
Unilever Environmental Safety Laboratory
Bedford, England

Frances W.J. Beck, Ph.D.
Department of Internal Medicine
University Health Center
Detroit, Michigan

Janet M. Benson, Ph.D.
Inhalation Toxicology Research Institute
Albuquerque, New Mexico

Maryka H. Bhattacharyya, Ph.D.
Center for Mechanistic Biology and Biotechnology
Argonne National Laboratory
Argonne, Illinois

Pierluigi E. Bigazzi, M.D.
Department of Pathology
University of Connecticut Health Center
Farmington, Connecticut

Stephen Bondy, Ph.D.
Community and Environmental Medicine
University of California/Irvine
Irvine, California

Darlene H. Bowser, Ph.D.
Institute of Environmental Medicine
New York University Medical Center
Tuxedo, New York

Michael W. B. Bradbury, Ph.D.
Physiology Group
King's College London
London, England

Mariano E. Cebrían, Ph.D.
Department of Pharmacology and Toxicology
CINVESTAV-IPN
Mexico City, Mexico

Andrew C. Chang, Ph.D.
Department of Soil and Environmental Sciences
University of California
Riverside, California

Louis W. Chang, Ph.D.
Departments of Pharmacology, Pathology, and Toxicology
University of Arkansas for Medical Sciences
Little Rock, Arkansas

Mitchell D. Cohen, Ph.D.
Institute of Environmental Medicine
New York University Medical Center
New York, New York

Jacinto Corbella, Ph.D.
Faculty of Medicine
University of Barcelona
Barcelona, Spain

Deborah A. Cory-Slechta, Ph.D.
Department of Neurobiology and Anatomy
University of Rochester
School of Medicine and Dentistry
Rochester, New York

Max Costa, Ph.D.
Institute of Environmental Medicine
New York University Medical Center
Tuxedo, New York

George P. Daston, Ph.D.
Miami Valley Laboratories
The Procter and Gamble Company
Cincinnati, Ohio

J. Michael Davis, Ph.D.
National Center for Health and Environmental Assessment
U.S. Environmental Protection Agency
Research Triangle Park, North Carolina

Frederik A. de Wolff, M.Sc., M.A., Ph.D.
Department of Human Toxicology
Academic Medical Center
University of Amsterdam
Amsterdam, The Netherlands

P. Michael Dews, Ph.D.
Department of Molecular Microbiology and Immunology
University of Southern California School of Medicine
USC/Norris Comprehensive Cancer Center
Los Angeles, California

P. Markus Dey, Ph.D.
Neurotoxicology Laboratories
Rutgers College of Pharmacy
Piscataway, New Jersey

Jose L. Domingo, Ph.D.
Laboratory of Toxicology and Environmental Health
Rovira i Virgili University
Reus, Spain

Raviprahash R. Dugyala, Ph.D.
Department of Physiology and Pharmacology
University of Georgia
College of Veterinary Medicine
Athens, Georgia

Carol J. Eisenmann, Ph.D.
Science International, Inc.
Alexandria, Virginia

Hassan A.N. El-Fawal, Ph.D.
Institute of Environmental Medicine
New York University Medical Center
Tuxedo, New York

Robert W. Elias, Ph.D.
National Center for Health and Environmental Assessment
U.S. Environmental Protection Agency
Research Triangle Park, North Carolina

Douglas Parker Evans, Ph.D.
Department of Molecular Microbiology and Immunology
University of Southern California School of Medicine
USC/Norris Comprehensive Cancer Center
Los Angeles, California

Jerry H. Exon, Ph.D.
Department of Food Science and Toxicology
University of Idaho
Moscow, Idaho

Anna M. Fan, Ph.D.
Office of Environmental Health Hazard Assessment
California Environmental Protection Agency
Berkeley, California

Glenn G. Fletcher, M.Sc.
Department of Biochemistry
McMaster University Health Science Center
Hamilton, Ontario, Canada

Ernest C. Foulkes, D.Phil.
Department of Environmental Health
University of Cincinnati Medical Center
Cincinnati, Ohio

Bruce A. Fowler, Ph.D.
Program in Toxicology
University of Maryland
Baltimore, Maryland

P.E. Fraser, Ph.D.
Tanz Neuroscience Building
University of Toronto
Toronto, Ontario, Canada

Gonzalo G. García-Vargas, Ph.D.
Department of Pharmacology and Toxicology
CINVESTAV-IPN
Mexico City, Mexico

Donald E. Gardner, Ph.D.
National Health and Environmental Effects
  Research Lab
U.S. Environmental Protection Agency
Research Triangle Park, North Carolina

Lars Gerhardsson, Ph.D.
Department of Occupational Health and
  Environmental Medicine
University Hospital
Lund, Sweden

Peter L. Goering, Ph.D.
Center for Devices and Radiological Health
Food and Drug Administration
Rockville, Maryland

Ronald D. Graff, Ph.D.
Department of Biochemistry
University of North Carolina at Chapel Hill
  School of Medicine
Chapel Hill, North Carolina

Rikuzo Hamada, Ph.D.
Division of Neurology
Minamikyushu-Chuo National Hospital
Kagashima, Japan

Kathleen Hendrix, Ph.D.
Department of Food Science and Technology
University of Idaho
Moscow, Idaho

Naohide Inoue, M.D., Ph.D.
Department of Hygiene
Kyushu University
Faculty of Medicine
Fukuoka, Japan

Richard M. Jacobs, Ph.D.
U.S. Food and Drug Administration
San Francisco District
Alameda, California

E. Jaikaran, Ph.D.
Tanz Neuroscience Building
University of Toronto
Toronto, Ontario, Canada

Elizabeth Jeffery, Ph.D.
Institute for Environmental Study
University of Illinois
Urbana, Illinois

Kenneth Jenkins, Ph.D.
Molecular Ecology Institute
California State University
Long Beach, California

Sam Kacew, Ph.D.
Department of Pharmacology
University of Ottawa
Ottawa, Ontario, Canada

Kazimierz S. Kasprzak, Ph.D.
Laboratory of Comparative Carcinogenesis
National Institutes of Health/FCRDC
Frederick, Maryland

Carl L. Keen, Ph.D.
Department of Nutrition
University of California/Davis
Davis, California

Julian P. Keogh, Ph.D.
Institute for Toxicology
Medical University of Lubeck
Lubeck, Germany

Teruhiko Kido, Ph.D.
Department of Hygiene
Chiba University School of Medicine
Chiba, Japan

Ian Kimber, Ph.D.
Zeneca Central Toxicology Laboratory
Cheshire, England

Harold K. Kimelberg, Ph.D.
Division of Neurosurgery
Albany Medical College
Albany, New York

Catherine B. Klein. Ph.D.
Department of Environmental Medicine
New York University Medical Center
Tuxedo, New York

Leslie M. Klevay, M.D., S.D.
U.S. Department of Agriculture/ARS
Grand Forks Research Center
Grands Forks, North Dakota

Hiroko Kodama, M.D.
Department of Pediatrics
Teikyo University School of Medicine
Tokyo, Japan

Yukinori Kusaka, Ph.D.
Department of Environmental Health
Fukui Medical Center
Fukui, Japan

Joseph R. Landolph, Ph.D.
Department of Molecular Microbiology and
  Immunology
University of Southern California School of
  Medicine
USC/Norris Comprehensive Cancer Center
Los Angeles, California

Lawrence H. Lash, Ph.D.
Department of Pharmacology
Wayne State University
School of Medicine
Detroit, Michigan

Cindy P. Lawler, Ph.D.
Department of Psychiatry and Curriculum in
  Toxicology
Brain Development Research Center
University of North Carolina at Chapel Hill
  School of Medicine
Chapel Hill, North Carolina

Lois D. Lehman-McKeeman, Ph.D.
Miami Valley Laboratories
The Procter and Gamble Company
Cincinnati, Ohio

John J. Lemasters, M.D., Ph.D.
Department of Cell Biology and Anatomy
University of North Carolina at Chapel Hill
  School of Medicine
Chapel Hill, North Carolina

László Magos , Ph.D.
BIBRA Toxicology International
Surrey, England

Neil A. Littlefield, Ph.D.
Division of Nutritional Toxicology
National Center for Toxicology Research
Jefferson, Arkansas

W.J. Lukiw, Ph.D.
Tanz Neuroscience Building
University of Toronto
Toronto, Ontario, Canada

Neil A. Littlefield, Ph.D.
Division of Nutritional Toxicology
National Center for Toxicology Research
Jefferson, Arkansas

W.J. Lukiw, Ph.D.
Tanz Neuroscience Building
University of Toronto
Toronto, Ontario, Canada

Richard B. Mailman, Ph.D.
Department of Child Development
Brain and Development Research Center
University of North Carolina
  School of Medicine
Chapel Hill, North Carolina

Yuji Makita, M.D., Ph.D.
Department of Hygiene
Faculty of Medicine
Kyushu University
Fukuoka, Japan

Edward J. Massaro, Ph.D.
Perinatal Toxicology
U.S. Environmental Protection Agency
Health Effects Research Laboratory
Research Triangle Park, North Carolina

Mechelle Mayleben, M.A.
Curriculum in Neurobiology
University of North Carolina at Chapel Hill
  School of Medicine
Chapel Hill, North Carolina

D.R. McLachlan, Ph.D.
Tanz Neuroscience Building
University of Toronto
Toronto, Ontario, Canada

Fiona McNeill, Ph.D.
Program in Toxicology
University of Maryland
Baltimore, Maryland

Toshio Narahashi, Ph.D.
Department of Molecular Pharmacology and
  Biological Chemistry
Northwestern University Medical Center
Chicago, Illinois

Herbert L. Needleman, M.D.
Department of Psychiatry
University of Pittsburgh School of Medicine
Pittsburgh, Pennsylvania

Evert Nieboer, Ph.D.
Department of Biochemistry
McMaster University Health Science Center
Hamilton, Ontario, Canada

Anna-Liisa Nieminen, Ph.D.
Department of Anatomy
Case Western Reserve University
  School of Medicine
Cleveland, Ohio

Koji Nogawa, Ph.D.
Department of Hygiene
Chiba University School of Medicine
Chiba, Japan

Monica Nordberg, Ph.D.
Institute for Environmental Medicine
Karolinska Institute
Stockholm, Sweden

Magnus Nylander, Ph.D.
Institute for Applied Biochemistry
Clinical Research Center
Karolinska Institute
Stockholm, Sweden

Mitsuhiro Osame, Ph.D.
The 3rd Department of Internal Medicine
Faculty of Medicine
Kagoshima University
Kagoshima, Japan

Laurent Ozbun, B.S.
Department of Molecular Microbiology and
  Immunology
University of Southern California School of
  Medicine
USC/Norris Comprehensive Cancer Center
Los Angeles, California

Jozef M. Pacyna, Ph.D.
Norwegian Institute for Air Research
Kjeller, Norway

Albert L. Page, Ph.D.
Department of Soil and Environmental Sciences
University of California
Riverside, California

Lionel A. Poirier, Ph.D.
National Center for Toxicological Research
Division of Nutritional Toxicology
Jefferson, Arkansas

Ananda S. Prasad, Ph.D.
Department of Internal Medicine
University Health Center
Detroit, Michigan

Kenneth R. Reuhl, Ph.D.
Neurotoxicology Laboratories
Rutgers College of Pharmacy
Piscataway, New Jersey

Deborah C. Rice, Ph.D.
Toxicology Research Division
Health Protection Branch/Health Canada
Ottawa, Ontario, Canada

John M. Rogers, Ph.D.
Perinatal Toxicology
National Health and Environmental Effects
  Research Laboratory
U.S. Environmental Protection Agency
Research Triangle Park, North Carolina

Ignacio A. Romero, Ph.D.
Physiology Group
Biomedical Sciences Division
King's College London
London, England

Toby G. Rossman, Ph.D.
Kaplan Cancer Center
New York University Medical Center
New York, New York

Brenda M. Sanders, Ph.D.
Molecular Ecology Institute
California State University
Long Beach, California

Mary Jane K. Selgrade, Ph.D.
National Health and Environmental Effects
  Research Lab
U.S. Environmental Protection Agency
Research Triangle Park, North Carolina

Raghubir P. Sharma, D.V.M, Ph.D.
Department of Physiology and Pharmacology
University of Georgia
College of Veterinary Medicine
Athens, Georgia

Ellen K. Silbergeld, Ph.D.
Epidemiology and Preventive Medicine
University of Maryland Medical School
Baltimore, Maryland

Claus-Peter Siegers, Ph.D.
Institute for Toxicology
Medical University of Lubeck
Lubeck, Germany

Staffan Skerfving, Ph.D.
Department of Occupational and Environmental Medicine
University Hospital
Lund, Sweden

Anja Slikkerveer, M.D., Ph.D.
Toxicology Laboratory
University Hospital Leidan
Leidan, The Netherlands

Donald R. Smith, Ph.D.
Environmental Toxicology
Applied Sciences
University of California
Santa Cruz, California

Ralph J. Smialowicz, Ph.D.
Environmental Research Center
U.S. Environmental Protection Agency
Research Triangle Park, North Carolina

Lee S.F. Soderberg, Ph.D.
Department of Microbiology and Immunology
University of Arkansas for Medical Sciences
Little Rock, Arkansas

John R. J. Sorenson, Ph.D.
Department of Medicinal Chemistry
University of Arkansas Medical Sciences
Little Rock, Arkansas

Elizabeth H. South, Ph.D.
Department of Food Science and Technology
University of Idaho
Moscow, Idaho

Katherine Squibb, Ph.D.
Program in Toxicology
University of Maryland
Baltimore, Maryland

Kazuo T. Suzuki, Ph.D.
Faculty of Pharmaceutical Sciences
Chiba University
Chiba, Japan

Tsuguyoshi Suzuki, M.D., Ph.D.
National Institute for Environmental Studies
Ibaraki, Japan

M. Anthony Verity, M.D.
Department of Pathology and Laboratory Medicine
Division of Neuropathology
UCLA Medical Center
Los Angeles, California

Gijsbert B. van der Voet, Ph.D.
Toxicology Laboratory
University Hospital Leiden
Leiden, The Netherlands

Michael P. Waalkes, Ph.D.
Laboratory of Comparative Carcinogenesis
National Cancer Institute/FCRDC
Frederick, Maryland

Zaolin Wang, Ph.D.
Department of Environmental Medicine
New York University Medical Center
Tuxedo, New York

Jan A. Weiner, Ph.D.
Institute for Applied Biochemistry
Clinical Research Center
Karolinska Institute
Stockholm, Sweden

James S. Woods, Ph.D.
Battelle Seattle Research Center
Seattle, Washington

Hiroshi Yamauchi, Ph.D.
Department of Public Health
St. Marianna University School of Medicine
Kawasaki, Japan

Rudolfs K. Zalups, Ph.D.
Division of Basic Medical Sciences
Mercer University School of Medicine
Macon, Georgia

Judith T. Zelikoff, Ph.D.
Institute of Environmental Medicine
New York University Medical Center
Tuxedo, New York

Wei Zhang, Ph.D.
Division of Environmental Sciences
Columbia University
New York, New York

Table of Contents

PART 1: GENERAL AND CLINICAL ASPECTS, RISK ASSESSMENT, AND GENOTOXICITY OF METALS

Section I: Monitoring Environmental and Health Risk Assessments
Section Head: Anna M. Fan

Overview
An Introduction to Monitoring and Environmental and
Human Risk Assessment of Metals 5
Anna M. Fan

Chapter 1
Monitoring and Assessment of Metal Contaminants in the Air 9
Jozef M. Pacyna

Chapter 2
Assessment of Ecological and Health Effects of Soil-Borne Trace Elements and Metals 29
Andrew C. Chang and Albert L. Page

Chapter 3
Assessment of Metals in Drinking Water With Specific References to
Lead, Copper, Arsenic, and Selenium 39
Anna M. Fan

Chapter 4
Risk Assessment of Metals 55
J. Michael Davis and Robert W. Elias

Chapter 5
Techniques Employed for the Assessment of Metals in Biological Systems 69
Richard M. Jacob

Chapter 6
Concepts on Biological Markers and Biomonitoring for Metal Toxicity 81
Lars Gerhardsson and Staffan Skerfving

Section II: Bioresponses and Reactivities in Metal Toxicity
Section Head: Louis W. Chang

Overview
An Introduction to Bioresponses and Reactivities of Metals 111
Louis W. Chang

Chapter 7
Determinants of Reactivity in Metal Toxicology 113
Evert Nieboer and Glenn G. Fletcher

Chapter 8
Metals and Biological Membranes 133
Ernest C. Foulkes

Chapter 9
Interactions Between Glutathione and Mercury in the Kidney, Liver, and Blood 145
Rudolfs K. Zalups and Lawrence H. Lash

Chapter 10
The Role of General and Metal-Specific Cellular Responses in Protection and
Repair of Metal-Induced Damage: Stress Proteins and Metallothioneins 165
Brenda M. Sanders, Peter L. Goering, and Kenneth Jenkins

Chapter 11
Metal-Metal Interactions 189
Ananda S. Prasad and Frances W. J. Beck

Section III: Carcinogenesis and Genotoxicity of Metals
Section Heads: Max Costa and Lionel A. Poirier

Part A: Carcinogenicity and Genotoxicity of Specific Metals
Overview
An Introduction to Metal Toxicity and Carcinogenecity of Metals 203
Max Costa

Chapter 12
Carcinogenicity and Genotoxicity of Chromium 205
Catherine B. Klein

Chapter 13
The Carcinogenicity of Arsenic 221
Zaolin Wang and Toby G. Rossman

Chapter 14
Cadmium Carcinogenicity and Genotoxicity 231
Michael P. Waalkes and R. Rita Misra

Chapter 15
Mechanisms of Nickel Genotoxicity and Carcinogenicity 245
Max Costa

Chapter 16
Carcinogenicity and Genotoxicity of Lead, Beryllium, and Other Metals 253
Mitchell D. Cohen, Darlene H. Bowser, and Max Costa

Part B: Selected Concepts in Metal Carcinogenesis

Overview
An Introduction to Selected Concepts in Metal Carcinogenesis 287
Lionel A. Poirier

Chapter 17
Metal Interactions in Chemical Carcinogenesis 289
Lionel A. Poirier and Neil A. Littlefield

Chapter 18
Oxidative DNA Damage in Metal-Induced Carcinogenesis 299
Kazimierz S. Kasprzak

Chapter 19
Metal-Induced Gene Expression and Neoplastic Transformation 321
Joseph R. Landolph, Michael Dews, Laurent Ozbun, and Douglas Parker Evans

Section IV: Clinical Aspects of Metal Toxicity
Section Heads: Kazuo T. Suzuki and Tsuguyoshi Suzuki

Overview
An Introduction to Clinical Aspects of Toxicity 333
Kazuo T. Suzuki and Tsuguyoshi Suzuki

Chapter 20
Minamata Disease and Other Mercury Syndromes 337
Rikuzo Hamada and Mitsuhiro Osame

Chapter 21
Itai-Itai Disease and Health Effects of Cadmium 353
Koji Nogawa and Teruhiko Kido

Chapter 22
Genetic Disorders of Copper Metabolism 371
Hiroko Kodama

Chapter 23
Alzheimer's Disease and Other Aluminum-Associated Health Conditions 387
D.R. McLachlan, P.E. Fraser, E. Jaikaran, and W.J. Lukiw

Chapter 24
Current Status of Childhood Lead Exposure at Low Dose 405
Herbert L. Needleman

Chapter 25
Neurological Aspects in Human Exposure to Manganese 415
Naohide Inoue and Yuji Makita

Chapter 26
Health Effects of Arsenic 423
Gonzalo G. García-Vargas and Mariano E. Cebrían

Chapter 27
Toxicity of Bismuth and Its Compounds 439
Anja Slikkerveer and Frederik A. de Wolff

Chapter 28
Human Exposure to Lithium, Thallium, Antimony, Gold, and Platinum 455
Gijsbert B. van der Voet and Frederik A. de Wolff

Chapter 29
Cobalt and Nickel Induced Hard Metal Asthma 461
Yukinori Kusaka

Chapter 30
Aspects on Health Risks of Mercury from Dental Amalgams 469
Jan A. Weiner and Magnus Nylander

Chapter 31
Chelation Therapies for Metal Intoxication 487
Carol R. Angle

PART 2: TARGET ORGAN TOXICOLOGY

Section V: Neurotoxicology of Metals
Section Head: Louis W. Chang

Overview
An Introduction to Neurotoxicology of Metals 509
Louis W. Chang

Chapter 32
Toxico-Neurology and Neuropathology Induced by Metals 511
Louis W. Chang

Chapter 33
Comparative Neurobehavioral Toxicology of Heavy Metals 537
Deborah A. Cory-Slechta

Chapter 34
The Blood-Brain Barrier in Normal CNS and in Metal-Induced Neurotoxicity 561
Ignacio A. Romero, N. Joan Abbott, and Michael W.B. Bradbury

Chapter 35
Astrocytes: Potential Modulators of Heavy Metal-Induced Neurotoxicity 587
Michael Aschner and Harold K. Kimelberg

Chapter 36
Choroid Plexus and Metal Toxicity 609
Wei Zhang

Chapter 37
Effects of Toxic Metals on Neurotransmitters 627
Richard B. Mailman, Mechelle Mayleben, and Cindy P. Lawler

Chapter 38
Cytoskeletal Toxicity of Heavy Metals 639
Ronald D. Graff and Kenneth R. Reuhl

Chapter 39
Lead Neurotoxicity: Concordance of Human and Animal Research 659
Deborah Rice and Ellen Silbergeld

Chapter 40
Effects of Metals on Ion Channels 677
Toshio Narahashi

Chapter 41
Oxygen Generation as a Basis for Neurotoxicity by Metals 699
Stephen C. Bondy

Chapter 42
Disruption of Protein Synthesis as Mechanistic Basis of Metal Neurotoxicity 707
M. Anthony Verity

Section VI: Renal Toxicology of Metals
Section Head: Bruce A. Fowler

Overview
An Introduction to Metal-Induced Nephrotoxicity 719
Bruce A. Fowler

Chapter 43
The Nephropathology of Metals 721
Bruce A. Fowler

Chapter 44
Roles of Metal-Binding Proteins in Mechanisms of Nephrotoxicity of Metals 731
Katherine S. Squibb

Chapter 45
In Vivo Measurement and Speciation of Nephrotoxic Metals 737
Donald R. Smith and Fiona McNeill

Chapter 46
Metal-Induced Alterations in Renal Gene Expression 751
Hiroshi Yamauchi and Bruce A. Fowler

Chapter 47
Biomarkers of Metal-Induced Nephrotoxicity 759
Bruce A. Fowler and Monica Nordberg

Chapter 48
Study on the Transport and Toxicity of Metals Along the Nephron by
Means of the Isolated Perfused Tubule Technique 765
Rudolfs K. Zalups and Delon W. Barfuss

Section VII: Immunomodulation by Metals
Section Heads: Judith T. Zelikoff and Donald E. Gardner

Overview
An Introduction to Immunomodulation by Metals 783
Donald E. Gardner and Judith T. Zelikoff

Chapter 49
Effects of Metals on Cell-Mediated Immunity and Biological Response Modulators 785
Raghubir P. Sharma and Raviprakash R. Dugyala

Chapter 50
Effects of Metals on the Humoral Immune Response 797
Jerry H. Exon, Elizabeth H. South, and Kathleen Hendrix

Chapter 51
Metal-Induced Alterations in Innate Immunity 811
Judith T. Zelikoff and R.J. Smialowicz

Chapter 52
Contact Hypersensitivity to Metals 827
Ian Kimber and David A. Basketter

Chapter 53
Autoimmunity Induced by Metals 835
Pierluigi E. Bigazzi

Chapter 54
Altered Host Defenses and Resistance to Respiratory Infections
Following Exposure to Airborne Metals 853
Mary Jane K. Selgrade and Donald E. Gardner

Chapter 55
Concepts of Immunological Biomarkers of Metal Toxicity 861
Hassan A. N. El-Fawal

Chapter 56
Modulation of Immune Functions by Metallic Compounds 871
Lee S. F. Soderberg, John R. J. Sorenson, and Louis W. Chang

Section VIII: Effects of Metals On Other Organ Systems
Section Head: Louis W. Chang

Overview
An Introduction to Metal Effects on Other Organ Systems 885
Louis W. Chang

Chapter 57
Hepatic Injury by Metal Accumulation 887
Anna-Liisa Nieminen and John J. Lemasters

Chapter 58
Influences on the Gastrointestinal System by Essential and Toxic Metals 901
Julian P. Keogh and Claus-Peter Siegers

Chapter 59
Copper and Other Chemical Elements That Affect the Cardiovascular System 921
Leslie M. Klevay

Chapter 60
Respiratory Toxicology of Metals 929
Janet M. Benson and Judith T. Zelikoff

Chapter 61
Effects of Metals on the Hematopoietic System and Heme Metabolism 939
James S. Woods

Chapter 62
Bone Metabolism: Effects of Essential and Toxic Trace Metals 959
Maryka H. Bhattacharyya, Elizabeth Jeffery, and Ellen K. Silbergeld

Section IX: Reproductive and Developmental Toxicology of Metals
Section Head: John M. Rogers

Overview
An Introduction to Developmental and Reproductive Toxicology of Metals 975
John M. Rogers

Chapter 63
Teratogenic Effects of Essential Trace Metals: Deficiencies and Excesses 977
Carl L. Keen

Chapter 64
Placental Transport, Metabolism, and Toxicology of Metals 1003
Carol J. Eisenmann and Richard K. Miller

Chapter 65
The Development of Cadmium and Arsenic With Notes on Lead 1027
John M. Rogers

Chapter 66
The Developmental Cytotoxicity of Mercurials 1047
Edward J. Massaro

Chapter 67
Developmental and Reproductive Effects of Aluminum, Uranium, and Vanadium 1083
Jacinto Corbella and Jose L. Domingo

Chapter 68
Cell Adhesion Molecules in Metal Neurotoxicity 1097
Kenneth R. Reuhl and P. Markus Dey

Chapter 69
The Effects of Metals on Neurite Development 1111
Teresa Audesirk and Gerald Audesirk

Chapter 70
Mammary Heavy Metal Content: Contribution of Lactational Exposure to
Toxicity in Suckling Infants 1129
Sam Kacew

Chapter 71
Constitutive and Induced Metallothionein Expression in Development 1139
George P. Daston and Lois D. Lehman-McKeeman

Chapter 72
Developmental and Reproductive Effects of Metal Chelators... 1153
Jose L. Domingo

Epilogue 1171
László Magos

Index.. 1177

PART 1
General and Clinical Aspects, Risk Assessment, and Genotoxicity of Metals

Section I
Monitoring Environmental and Health Risk Assessments

Overview

An Introduction to Monitoring and Environmental and Human Risk Assessment of Metals

Anna M. Fan

Metals are commonly found in the environment. Their presence could be due to natural occurrence or as a result of anthropogenic activities. The geographical distribution and concentrations found in different environmental media such as air, water, or soil are important factors contributing to the extent of human or ecosystem exposures. The actual exposures may be reflected in biologic media such as blood and urine, and in some cases biomarkers have been developed for the study of these chemicals in biologic systems. This section discusses environmental, ecological, and human health assessments, laboratory methodologies, biomonitoring, and the interrelationships among these aspects.

Adequate environmental assessment and biomonitoring are dependent on the availability of state of the art methodologies in analyzing environmental and biological levels in the samples. The accuracy, reliability, quality control, variability, and detection/quantitation limits of such methodologies have a profound effect on the predictive value of the use of the resultant analytical data for ecological or human risk assessment. The primary influence is on the exposure parameters. The appropriate methodology may be different for different types of environmental or biological samples and the kind of metal or its compounds to be analyzed. Composite vs. individual sampling may be conducted, and different estimation methods may be used for analyzing results below the detection limit.

Although it is well known that different chemical forms of the same chemical can have different toxicities, and there is a need for speciation of metals in various media, the analytical methodology used may not readily distinguish these forms. Environmental measurements of arsenic in water, selenium in food or water, and methylmercury in fish are often reported as the total arsenic, selenium, or mercury, respectively. Compliance monitoring will require specification to measure the target substance. Examples are total lead and total nickel as the monitoring target for the regulated substances lead and nickel subsulfide, respectively. The measurement of hexavalent chromium requires reliable sampling and analytical methods in order to distinguish this carcinogenic form from the noncarcinogenic ones (trivalent and elemental chromium).

Environmental monitoring can be detection monitoring or compliance monitoring. Besides being used for exposure assessment, environmental data are also used for establishment of priorities for intervention and site mitigation efforts. False negative or false positive values can result from inadequate monitoring design which should involve the balancing of such values. Inadequate monitoring may lead to failure to reduce uncertainty about human exposure and the associated risk estimate.

The use of biological markers is one way to reduce the uncertainty of human exposure. Biological monitoring has the advantage of measuring integrated exposure from all sources. It is intended to measure a chemical, its metabolite, or a reversible biochemical marker or other effects of exposure. However, it is more invasive and often requires medical professionals and other precautionary measures. Biological exposure indices (BEIs) have been established for occupational exposures by the American Conference of Governmental Industrial Hygienists (ACGIH), but the general population may be exposed to chemical forms that are different from those encountered (e.g., selenium in food vs. selenium dioxide) or absent (e.g., methylmercury in fish vs. mercury vapor) in the workplace. For selenium, human blood levels reflect past exposure and urine levels recent exposure. For methylmercury, human hair levels can be used to project blood levels in the past, and blood levels can be used to provide a perspective on exposure. For lead, human blood levels are used as an indicator of exposure and for a determination of the health concern level. These can be used as exposure indices and provide a perspective on the health status of individuals. The same biological values may have different health implications for humans compared to fish and wildlife.

Risk assessment of metals generally follows the same methodology commonly used by toxicologists, which is to identify the toxic properties and the dose-response relationship, followed by assessment of exposure and risk characterization. Inherent in this four-step process are the following aspects that deserve special considerations.

Chemicals are evaluated as having certain toxic characteristics in animals or humans, depending on the availability of data or evidence. Such classification can be exemplified by that presented by the International Agency for Research in Cancer (IARC) on their evaluation of carcinogens. The chemicals may also be identified as having certain toxic properties, either first seen in animals and then confirmed in humans, or vice versa. One metal, arsenic, stands out as a unique chemical in this regard: repeated attempts have been unsuccessful in identifying this as an animal carcinogen in experimental studies, although human epidemiological studies have demonstrated its carcinogenicity in humans by producing cancers of the skin, lung, bladder, and liver.

The chemical forms of metal are also related to the chemical's toxicity. For arsenic, the inorganic form is believed to be the carcinogenic form, while the organic forms are not. Similarly, differences in chemical forms also contribute to the differences in toxicity of various metals such as chromium, cadmium, and nickel. Therefore, knowledge of the specific form of metal involved is important for risk assessment.

One assumption in exposure assessment is that the effects from exposure to a chemical by a certain route can be extrapolated to another route, with the use of appropriate conversion factors. In the case of metals, several compounds have produced local tumors at the injection site (e.g., fibrosarcomas by cadmium), but not by other routes, thus raising the question whether such tumors can be extrapolated to other exposure situations, such as exposure by the ingestion route.

Some metals are ubiquitous in the environment, are present in the diet of the general population, and thereby have multiple sources of exposure. In particular, certain metals such as manganese, chromium, selenium, and zinc have nutritional values, thus requiring a special approach in their risk assessment. Results of risk assessment that may lead to levels indicative of toxicity or regulatory limits below the nutritional need levels could be misleading and have undesirable consequences. This potential nutritional deficiency deserves particular attention in the risk characterization.

For environmental applications, risk assessment is used as a basis for determining acceptable levels of exposure for toxic chemicals in order to prevent unnecessarily excessive exposure. In the case of lead, more recent epidemiological data point to a concern level lower than that previously estimated. The recent National Human and Nutritional Estimates Survey (NHANES) shows that in the U.S. population, the blood lead levels have generally decreased as compared to the earlier study, primarily due to the phase-out of leaded gasoline. However, there is still a segment of the population that has levels approaching or exceeding the level of concern, thus any lead intake would cause an incremental increase of the blood lead level and would not be advisable.

Methylmercury in fish is of particular concern to pregnant women and children. Risk assessment for methylmercury is still struggling with the issue of identifying how much more sensitive a pregnant woman or fetus is than a nonpregnant adult. It is also noteworthy that methylmercury in fish is often referred to or reported as mercury. That toxicity from eating contaminated fish is primarily due to methylmercury, and not mercury, should be clearly identified.

Metal-metal interaction has been studied extensively; however, the available information has not been adequate for use in quantitative risk assessment.

Certain properties (Kow's) of metals have also led to the suggestion that the Tholman model not be used for these metals, but only for the more lipophilic organic chemicals. This model is used for evaluating the bioaccumulation/biomagnification of chemicals in the aquatic food chain. Thus, there are more limitations on the overall risk assessment for metals based on the ability to estimate related exposure parameters for both ecological and human assessment.

Overall, an integration of the use of environmental monitoring, biomonitoring, up-to-date laboratory methodologies, and an understanding of the special properties of metals would be necessary in order to gain a full perspective regarding the presence of metals in our environment and their effects on the ecological system and human health and welfare.

Chapter 1

Monitoring and Assessment of Metal Contaminants in the Air

Jozef M. Pacyna

I. INTRODUCTION

Recent studies on the behavior of trace metals in the environment have concluded that many of these compounds create serious problems due to their toxicity and bioaccumulation in various environmental compartments. Several scientific questions have been posed in order to improve our understanding of the nature and the extent of these problems. The major questions can be grouped with respect to:

- Identification of trace metals causing adverse environmental and health effects
- Sources and fluxes of those compounds responsible for their origin in the environment
- Major pathways of trace metals through the environmental compartments studied on local, regional, and global scales
- Environmental and health effects and the degree of public health concern associated with various trace metals
- Technical and other possibilities to reduce the levels of trace metals in the environment

Preliminary responses to some of the above-mentioned topics are summarized in Table 1, after discussions during the Dahlem Conference on Changing Metal Cycles and Human Health in 1983 (Nriagu, 1984) and modified in a view of results from the latest studies. Considerable effort in these studies has been devoted to the development of schemes to identify and rank those trace metals which pose a threat to human health or to the environment (Fisher et al., 1992). New information on the above-mentioned subjects is discussed in this chapter on the basis of monitoring and assessment of trace metals.

General conclusions from the recent studies on the atmospheric trace metals are that:

- Several of these compounds are ubiquitous in various raw materials and industrial products.
- Many trace metals evaporate from raw materials during the production of industrial goods and during the consumption of fuels, ores, etc., incineration processes, and application of various chemicals.
- Releases to other environmental compartments (e.g., spills to water bodies, landfills, sewage lagoons, holding ponds) may result in volatilization and entrainment of several trace metals.
- While emitted to the atmosphere, many trace metals are subject to long-range transport within air masses and migration through the ecosystem, which causes perturbations of their geochemical cycles not only on a local scale but also on regional and even global scales, as seen from the Arctic data.
- Long-range transport may also occur through interactions of pollutants with other environmental compartments, e.g., through ocean currents.

Table 1 Perturbations of the Geochemical Cycles of Trace Metals by Society

Element	Scale of perturbation[a] Global	Regional	Local	Most diagnostic environments[b]	Mobility[c]	Health concern[d]	Critical pathway[e]
Pb	+	+	+	A, Sd, I, W, H, So	v, a	+	F, A[h]
V	+	+, c	+	A	i	(+)	A?
As	+	+	+	A, Sd, So, W	v, s, a	+	A, W
Sn	+	+	+	A, Sd, W	v, a	+[f]	F
Zn	+	+	+	A, Sd, W, So	v, s	E	F
Cd	+	+	+	A, Sd, So, W	v, s	+	F
Hg	+	+	+	A, Sd, Fish, So	v, a	+[f]	F, (A)
Sb	+	+	+	A, Sd	v, s	(+)	F, W, A?
Cu	+	+	+	A, Sd, W, So	v, s	E	F?
Ag	+	+	+	A, Sd, W	(v)	(+)	?
Se	+	(+)	+	A	v, s, a	E	F
Ge	?	+	+	A, So, W?	v, s, a	(+)[f]	?
Ni	–	+	+	A, Sd	–	E	F, W, A?
Cr	–	+	+	A, Sd, W, Gw	s, v[g]	E	W, F
B	–	(+)	+	A, Sd, Gw	v, s	E	W
K	–	(+)	+	A	s	E	F
Pt	?	?	+	A, Sd	s	(+)	?
Pd	?	?	+	Sd	s	(+)	?
Mo	?	?	+	A, W, So, Sd	s	E	F, W
Tl	?	?	+	Em, So	v, s	(+)	A, F?
In	?	?	+	A, So, Em	v	(+)	?
Bi	?	?	+	A, So, Em	v	(+)	?
Be	?	?	+	A, So, Em	–	(+)	A
Ga	?	?	+	Em	v	(+)	?
Te	?	?	(+)	So	v, a?	(+)	?
Mn	–	c, +	+	A, Sd, W	r	E	A
Fe	–	c	+	A, Sd, W	r	E	F, A, W
Al	–	c	+	A, Sd	–	(+)	W?
Si	–	c	+	A	–	(E)	–
Ti	–	c	+	A, Sd, So	–	–	–
Co	–	c	(+)	Sd	r	E	F?
Na	–	c, +	+	W, A, So	s	E	F, W
Mg, Ca	–	c	+	A, Sd, Em	(s)	E	
Ba	–	c	+	A, Sd, Em	–	(+)	F, W
U	–	c	+	A, So, Gw	s	(+)	A, W
Th	–	c	+	A, So, Gw	–	(+)	A
Zr	–	c	+	A	–	–	–
Y	–	c	(+)	Sd	–	–	–

[a] + = significant perturbation; (+) = possible perturbation; (–) = enriched relative to crustal abundances, but the enrichment may not be anthropogenic; – = no perturbation; ? = not enough information; c = enhanced due to mobilization of crustal materials (soil, dust).

[b] A = air; Sd = sediments (coastal, lake); So = soils; I = ice cores; W = surface waters; Gw = groundwaters; H = humans; Em = emission studies (only listed when little geochemical information is available).

[c] v = volatile; s = soluble; r = soluble only under reducing conditions; a = mobile as alkylated organometallic species; – = not mobile.

[d] + = toxic in excess; (+) = toxic, but little data available; E = essential, but toxic in excess.

[e] F = food; W = water; A = air; – = no significant exposure likely.

[f] Organometallic forms only.

[g] Hexavalent form volatile and toxic, trivalent form essential.

[h] Exposure through hand-to-mouth activity is critical for lead in children.

[i] Enriched relative to crustal abundance from fuel oil combustion (vanadium porphyrins).

Note: Data from Nriagu, 1984 and Pacyna and Winchester, 1990.

- Deposition of trace metals in the emission source area and during their atmospheric transport has reached values which for some of the compounds in certain regions had exceeded the maximum permissible values.

Based on the above conclusions, it is clear that international cooperation is needed to reduce the levels of many trace metals released into the environment and particularly to the atmosphere in order to limit their migration through the other environmental media and to remediate hot spots of contaminants. The UN Economic Commission for Europe (ECE) has organized a task force on heavy metals to provide by 1994 the basis for a possible protocol on reductions of their releases into the atmosphere.

This chapter presents information on the current state of our knowledge on fluxes, atmospheric transport, and deposition of trace metals based on data collected during the UN ECE task force on heavy metals (e.g., Pacyna, 1993). It aims at the substantiation of the above-presented conclusions by reviewing research within major international programs such as the Paris Convention for the Prevention of Marine Pollution (PARCOM) program ATMOS, the Baltic Sea Joint Comprehensive Environmental Action Programme (HELCOM), the Arctic Monitoring and Assessment Programme (AMAP), and various U.S.-Canadian programs in the Great Waters region, research activities within international organizations, particularly the UN Economic Commission for Europe (UN ECE) and the Organization for Economic Cooperation and Development (OECD), and studies carried out at a number of national institutions and authorities in Europe and North America.

II. EMISSIONS OF TRACE METALS INTO THE ATMOSPHERE

High-temperature processes, such as coal and oil combustion in electric power plants and heat and industrial plants, roasting and smelting of ores in nonferrous metal smelters, melting operations in ferrous foundries, refuse incineration, and kiln operations in cement plants emit various trace metals into the atmosphere. The amount of these emissions depends on:

- The contamination of fossil fuels and other raw materials; content of trace metals frequently reported for major types of coal burned in Europe is presented in Figure 1.

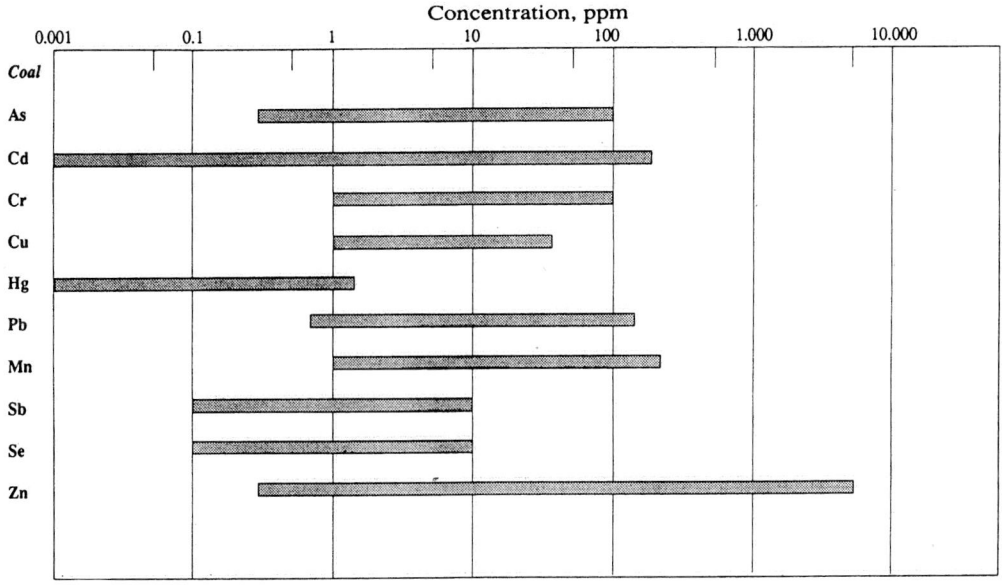

Figure 1 Range of concentrations of trace metals in coal burned in Europe.

- Physicochemical properties of trace metals, affecting their behavior during the industrial processes; for example, affecting the evaporation of metals from coal in the combustion furnace and then condensation on fine particles in exhaust gases.
- The technology of the industrial processes; an example of emission factors for various types of combustion furnaces is given in Table 2.
- The type and efficiency of control equipment; the penetration rate for various trace metals passing through electrostatic precipitators (ESPs) and Venturi scrubbers are presented in Table 3.

Table 2 Emission Factors of Trace Metals From Coal-Fired Industrial, Commercial, and Residential Units (Grams of Trace Element/Tons of Coal Burned)

Element	Industrial boilers			Commercial and residential units
	Cyclone	Stoker	Pulverized	
As	0.34	1.68	1.32	0.59
Cd	0.11	0.52	0.42	0.18
Hg	0.01	0.06	0.03	—
Pb	1.23	7.70	4.54	2.71
Zn	1.72	11.42	6.54	4.03

Table 3 Penetration (%) of Elements Contained in Particles Emitted From a Venturi Wet Scrubber and an ESP-Equipped Coal-Fired Generating Unit

Element	Penetration range		Element	Penetration range	
	ESP	Venturi		ESP	Venturi
As	4.3–11.5	2.5–7.5	Sb	3.1–7.7	3.0–6.6
Be	0.5–0.9		Se	3.8–8.1	10–21
Cd	3.3–8.8		V	1.6–3.7	0.5–1.1
Co	1.2–3.2	0.06–2.1	W	3.1–7.2	1.7–3.5
Cr	1.2–12.1	0.6–36	Zn	2.3–6.3	0.3–8.5
Mo	1.8–6.8	0.9–2.2	Zr	0.5–1.6	0.05–0.14
Mn	0.3–1.6	0.07–4.6			
Pb	2.2–5.5				

Note: ESP, electrostatic precipitator.

Data from Ondor et al., 1979 and Pacyna, 1989.

Detailed discussion on to what extent the above parameters affect trace metal emissions is presented by Pacyna (1989). The major conclusion was that contamination of raw materials and the type and efficiency of control equipment are the most important parameters.

The first quantitative world-wide estimate of the annual industrial input of 16 trace metals into the air, soil, and water has been published by Nriagu and Pacyna (1988). The summary of this estimate for arsenic, cadmium, mercury, lead, and zinc is presented in Table 4. The study illustrated that emissions from various human activities significantly alter the cycling of many trace metals in the environment on a global scale, not just in New York City or in a local community sewage treatment plant (Weisburd, 1988).

Table 4 Global Emissions of Trace Metals to the Atmosphere in 1983 (in 10^3 Tons)

	Source category	As	Cd	Hg	Pb	Zn
1.	Fossil fuel combustion	0.4–3.7	0.2–1.1	0.6–3.5	2.7–18.4	3.1–23.4
2.	Gasoline combustion	—	—	—	248.0	—
3.	Nonferrous metal industry	9.6–15.1	2.6–8.2	0–0.2	30.0–69.6	51.0–93.8
4.	Other anthropogenic sources	2.0–6.8	0.3–2.7	0.3–2.5	8.0–40.0	15.9–76.3
A.	Total anthropogenic emissions	12.0–25.6	3.1–12.0	0.9–6.2	288.7–376.0	70.0–193.5
	(Mean values)	18.8	7.6	3.6	332.0	132.0
B.	Natural emissions	7.8	1.0	6.0	19.0	4.0

The above study used information collected during the estimation of atmospheric emissions of trace metals in Europe and North America. Recently, the global emission inventory of lead was updated for 1989. The estimates show that these emissions range from 150,000 to 209,000 t, with 62% of the lead from gasoline combustion, followed by nonferrous metal production with 26% (Pacyna et al., 1993). The upper range of these estimates is presented in Table 5.

One of the first attempts to estimate atmospheric emissions of trace metals from anthropogenic sources in Europe was completed at the beginning of the 1980s (Pacyna, 1984). This work included information

Table 5 Global Emission Inventory for Lead in 1989 — Maximum Scenario (in Tons)

Continent	Gasoline combustion	Fossil fuel combustion	Nonferrous metal production	Waste incineration	Cement production	Steel and iron	Total
Africa	12,316	628	4,421		85	86	17,536
Asia	44,006	4,209	21,323	207	867	3,713	74,325
Australia	2,000	411	2,775	62	12	118	5,378
Europe	47,579	3,477	13,041	537	641	4,278	69,553
North America	14,192	993	7,613	2,498	177	1,316	26,789
South America	8,796	195	5,422		98	555	15,066
Total	128,889	9,913	54,595	3,304	1,880	10,066	208,647

on emissions of 16 compounds. Previous works have dealt with either a single metal or a given source category. Most of the effort was on inventorying emissions of lead and cadmium. Emissions during the combustion of fuels in coal-fired power plants and the combustion of gasoline were studied most frequently.

The above-mentioned European survey was then updated, completed, emission gridded, and applied in long-range transport models to study deposition processes, particularly within PARCOM. The 1992 version of this emission inventory has been presented by Axenfeld et al. (1992) and was limited to five components: arsenic, cadmium, mercury, lead, and zinc. The summary of these estimates is presented in Table 6. The following conclusions can be drawn from this work:

- Pyrometallurgical processes in the primary nonferrous metal industries are the major sources of atmospheric arsenic, cadmium, and zinc in Europe. Combustion of coal in utility and industrial boilers is the major source of anthropogenic mercury, and combustion of leaded gasoline is still the major source of lead. Little information is available on emission of trace metals from various diffuse sources.
- The emission estimates were based on emission factors calculated for all source categories considered. It was recommended that emissions from major point sources, such as electric power plants, central heat plants, and primary smelters, should be measured and not calculated on the basis of emission factors.
- More national data on emission and emission factors are necessary in order to improve the quality of emission inventories in Europe. This is specially important for source categories related to the various applications of metals and incineration of wastes. PARCOM has been actively involved in collecting this information; however, the results obtain so far are not fully satisfactory. Only a few countries in Europe report trace metal emissions and emission factors and a first manual of emission factors has been made mostly on the experience gained in the Netherlands (van der Most and Veldt, 1992).
- Verification of the European emission estimates through their application in dispersion models and comparison of modeled atmospheric concentrations with measurements has indicated an accuracy of emission estimates within a factor of 2.

Table 6 Trace Element Emissions in Europe (in Tons/Year)

No.	Category	As (1982)	Cd (1982)[a]	Hg (1987)	Pb (1985)[a]	Zn (1982)
1.	Fuel combustion in utility boilers	330	125	189	1,300	1,510
2.	Fuel combustion in industrial, commercial, and residential units	380	145	216	1,600	1,780
3.	Gasoline combustion	—	—	—	64,000	—
4.	Nonferrous metal industry	3,660	730	29	13,040	26,700
5.	Iron and steel production	230	53	2	3,900	9,410
6.	Waste incineration	10	37	35	540	650
7.	Other sources	360	30	255	1,120	4,540
	Total	4,970	1,120	726	85,500	44,590

[a] The 1990 emission of Pb in Europe was estimated between 32,200 and 54,150 t.
The 1990 emission of Cd in Europe was estimated between 270 and 1,950 t (678 t as average value).

Special emphasis was placed on estimating emissions of atmospheric mercury. It was concluded that information on total mercury emissions is not sufficient because of differences in atmospheric behavior of the various physical and chemical species of the compound and their different toxic effects. Therefore, an estimate of emissions of gaseous mercury and mercury on particles has been made focusing on elemental and oxidized forms of the compound. The results are presented in Figure 2.

Natural mercury emissions result from various processes, including the off-gassing of mercury-ladden rock, and volatilization of mercury from soils and vegetation as well as water bodies and the ocean. Mercury enters the atmosphere as a result of forest fires, volcanic activity, and other biomass burning. Elevated ambient temperatures tend to increase the rate of mercury loss from soils. Very few direct measurements of the flux of mercury, by species, have been made to date under a wide enough set of meteorological conditions to allow for extrapolation of the old data to the present day. Emissions from natural sources are difficult to distinguish from secondary emissions and diffusive reemissions from anthropogenic sources. These include reemissions of previously deposited mercury, emissions resulting

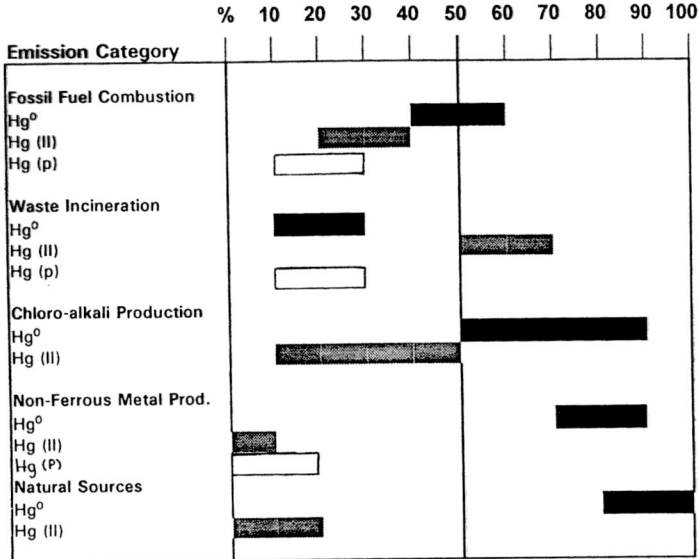

Figure 2 Contributions of various forms of mercury to the emissions from major source categories in Europe in 1987 (in %).

from discharge into water bodies, and from contaminated soils. Hence, it is more appropriate to differentiate between preindustrial and postindustrial diffuse sources (e.g., Lindqvist, 1991).

A spatial distribution of trace metal emission estimates in Europe is also available within the grid system of 150 by 150 km. A geographical location of sources was used to spatially distribute point source emissions and the population density and the road density for area source emissions. A spatial distribution of the 1989 cadmium emissions in Europe is shown in Figure 3.

Pursuant to the requirements of the 1990 Clean Air Act Amendment, an interim toxic emission inventory has been developed for the continental United States. Preliminary results of this work include geographical distribution and source type analysis (e.g., Benjey and Coventry, 1992) and are based on the 1985 National Acid Precipitation Program (NAPAP) inventory. This inventory is now under revision. Older emission estimates of arsenic, cadmium, and lead from major source categories in the U.S. are presented in Table 7 (after Voldner and Smith, 1989). The U.S. Environmental Protection Agency (U.S. EPA) and the Great Lakes Protection Fund are currently funding an air toxics emission inventory project being developed by the eight Great Lakes states through the Great Lakes Commission (e.g., Radian, 1992). The Air Clearinghouse for Inventories and Emission Factors (AIR CHIEF) has provided information on estimating air emissions of criteria and toxic pollutants from selected sources. The U.S. EPA has launched a number of projects on locating and estimating air emissions from various sources of trace metals. As much as 302 t of mercury were emitted into the air in 1990 with combustion of coal and wastes contributing more than 30% each (U.S. EPA, 1993).

Environment Canada has carried out several projects on emission inventory for several trace metals. Lead, mercury, cadmium, and arsenic have been given a priority in these projects. The contributions of atmospheric emissions from major source categories in 1982 are shown in Table 8 on the basis of data from Jacques (1987). Emission estimates for mercury and lead have been revised to account for major changes in consumer patterns.

The North American emission estimates for trace metals have generally been performed in connection with the contamination of the Great Lakes. Identification of emission sources in this region and their characterization with respect to the atmospheric emissions have been carried out for some time in both the U.S. and Canada, as well as by the International Joint Commission (IJC). Several of the conclusions from these studies are similar to the conclusions reached during the estimation of emissions in Europe, particularly with respect to the contributions from major source categories to the total emissions of trace metals and the group of compounds studied. An important additional conclusion drawn from emission estimates and source characterization in the Great Lakes region is that the spatial distribution of trace metal emissions in North America is not uniform with respect to the major sources as it is in Europe.

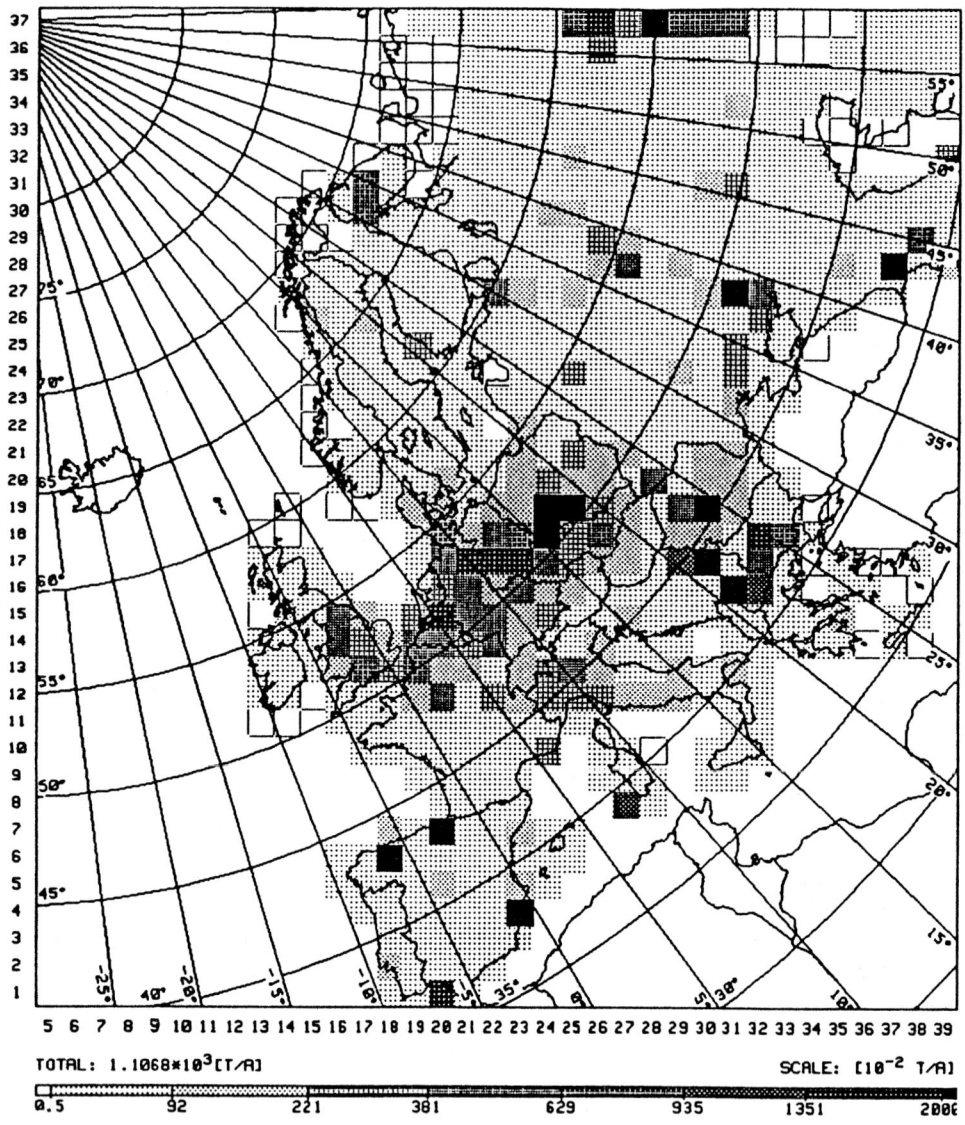

Figure 3 Spatial distribution of cadmium emissions in Europe in 1989 within the EMEP grid of 150 × 150 km.

Table 7 Emission Estimates for Atmospheric Arsenic, Cadmium, and Lead in the United States (in Tons/Year)

No.	Source	Arsenic	Cadmium	Lead
1.	Fossil fuel combustion	1,410	122	805
2.	Industrial processes	383	68	2,300
3.	Metal application	536	1	—
4.	Solid waste disposal	3	20	2 300
	Total	2,332	211	5,405

Note: After Voldner, E. and Smith, L., 1989.

Emissions data currently reported by the U.S. EPA and the IJC revealed that major source regions are located in the eastern part of the continent (e.g., a review by Keeler et al., 1993).

Table 8 Atmospheric Emissions of Selected Trace Metals From Major Source Categories in Canada in 1982 (in Tons)

Source	Arsenic	Cadmium	Chromium	Copper	Mercury	Nickel
Fuel combustion/ stationary sources	19	53	39	10	8	232
Industrial process	451	260	20	1,672	16	599
Transportation	—	4	10	4	—	13
Solid waste incineration	1	5	—	2	2	1
Miscellaneous sources	—	—	—	—	5	—
Total	471	322	69	1,688	31	845

Note: After Jacques, 1987.

III. ATMOSPHERIC TRANSPORT

The results of long-term measurements on pollution of the North Sea and Baltic Sea waters led to the conclusion that as much as 50% of lead and mercury and between 30 and 50% of arsenic, cadmium, chromium, copper, nickel, and zinc enter these waters through atmospheric deposition. Atmospheric mercury is primarily in the vapor phase with more than 80% of the compound in an elemental form (as summarized by Pacyna, 1992).

Using the "best available" data, quantitative estimates have recently been made on the fluxes of many toxic contaminants into the Great Lakes (e.g., Strachan and Eisenreich, 1988). The general conclusion can be drawn on the basis of this and other studies (e.g., those reviewed by ICF, 1992, for the U.S. EPA) that between 35 and 50% of the annual load of the toxic compounds enter the Great Lakes waters through atmospheric deposition. Concerning lead, it was concluded that the atmospheric input was responsible for 35 to 47% of all inputs of the element to the Great Lakes during the period from 1977 to 1981.

Recent studies of the origin of the Arctic air pollution have concluded that between 3 and 14% of the total emissions of arsenic, cadmium, lead, zinc, vanadium, and antimony in all of Eurasia is deposited in the Arctic (Akeredolu et al., 1994).

The above examples of studies on the origin of trace metals in various parts of the globe come to the same conclusion that trace metals emitted into the atmosphere can be transported a long distance and that atmospheric deposition is an important pathway for contamination of the terrestrial and aquatic ecosystems by these compounds world-wide. More soluble species can be stored in the oceans for a very long time and transported according to the oceanic circulation. Therefore, an ecosystem approach rather than atmospheric transport alone should be considered.

A. LONG-RANGE TRANSPORT MODELS

There have been several applications of various long-range transport models to study the origin of atmospheric trace metals observed at remote locations, such as the northern parts of Scandinavia and the Arctic, as well as the degree to which the atmospheric pathway contributes to the contamination of certain regions by these pollutants, including the North Sea, the Baltic Sea, the Great Lakes, and larger regions such as Europe and eastern North America.

The following conclusions have been reached from the application of a simple Lagrangian model to study the long-range transport of trace metals to remote sites in Scandinavia (e.g., Pacyna et al., 1989) and the Norwegian Arctic:

- The emission estimates for As, Sb, V, and Zn from anthropogenic sources in Europe (as applied in the model) can be related to the concentration measurements in Scandinavia with the help of the trajectory model, and the agreement between the measured and modeled concentrations was within a factor of two. The accuracy of emission estimates was found to be the most sensitive parameter affecting the model performance.
- Trace metals are removed from the atmosphere by dry and wet deposition processes. In general, dry deposition appears to be more important for the removal of pollutants close to the emission source, while wet deposition is more important to remove pollutants "en route".
- A range of dry deposition velocities has been suggested as shown in Figure 4. Fixed values of the washout ratio used in the model have been a simplification of the problem and contributed to a higher inaccuracy of the modeling.

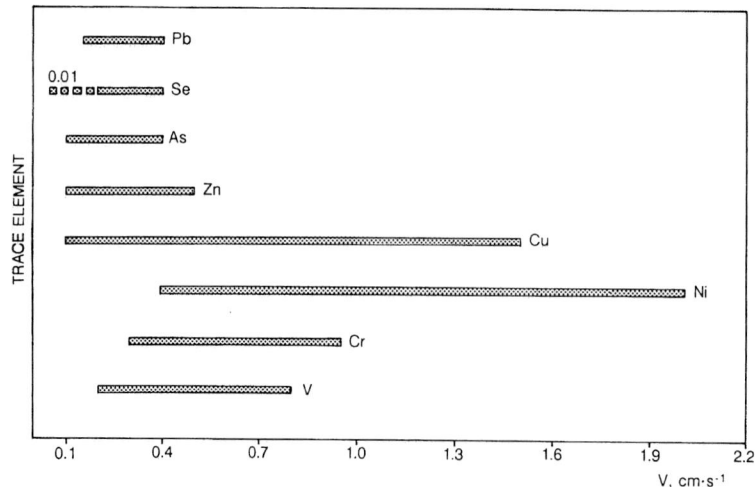

Figure 4 Range of dry deposition velocities for several trace metals.

- Mixing height for pollutants within the air masses also affects the performance of the model, but far less than the emission estimates and the wet and dry deposition processes.

The above conclusions were confirmed through the application of a climatological-type model (Alcamo et al., 1992) to study the transport of trace metals over Europe (Trace Toxic Air Concentrations in Europe — TRACE Model). The additional conclusions from the application of TRACE were that:

- The mean residence time of the mass of As, Cd, and Zn aerosol in Europe's atmosphere is estimated to be 27 h, and for Pb 64 h.
- Wet deposition of As and other metals dominates dry deposition throughout most of Europe except in some high-emission and outlying dry areas.

Two major approaches to model deposition of trace metals to the North Sea have been made in the Netherlands (the so-called TREND model: van Jaarsveld and Onderdelinden, 1991) and Germany (e.g., Petersen et al., 1989). The following conclusions have been drawn from the application of TREND:

- Model studies give much lower deposition values than those based on measurements due to the uncertain quality of emission data and, particularly, the completeness of emission source categories inventoried. The best results were obtained for Pb.
- Agreement between the model estimates and measurements was again within a factor of two for most of the metals studied.

The general conclusion was that a combined model and monitoring approach is the most useful one for the assessment of atmospheric deposition of trace metals to large surface waters.

The German application of the UN ECE European Monitoring and Evaluation Programme (EMEP) model was to study the transport of trace metals to the North Sea and the Baltic Sea. Many of the conclusions reached during the performance of the model were similar to those from TREND, particularly concerning the agreement between model estimates and measurements. However, one should admit that the emission input data to both models were also similar. Some additional conclusions were that:

- Wet deposition is the dominating removal process.
- Refined meteorological input data from a limited-area weather prediction model do not significantly change the spatial distribution of the total annual deposition of trace metals to the studied seas.

The heavy metals model (HMET) has been developed at the Meteorological Synthesizing Centre-West (MSC-W) of EMEP. It is an Eulerian model operating within a 150 by 150 km grid which uses a pseudospectral advection scheme. The model is used to study the transport and wet and dry deposition of As, Cd, Pb, Zn, and Hg. The modeled depositions of Cd, As, and Pb agree generally within a factor of two with measurements, while Zn is significantly underestimated.

Mercury has been given special attention in recent modeling estimations of long-range transport of toxic compounds within air masses. It has been realized that recent progress in the understanding of mercury chemistry and biogeochemistry, as well as development of mercury emission data bases, would make it possible to attempt to model the atmospheric transport of the compound, its chemical transformations in the atmosphere, and the fluxes of mercury to and from the terrestrial and aquatic ecosystems. The predominant form of mercury in the air is elemental mercury, which is not very water soluble. Wet deposition will occur after the elemental mercury has been converted to more water-soluble forms that can be subject to wet removal. Dry deposition of mercury may also occur, although the mechanisms leading to this process are not well understood. The oxidation of elemental mercury to more water-soluble mercury salts, followed by wet deposition or condensation, remains the most probable pathway for the removal of mercury from the air (e.g., Munthe, 1993).

An EMEP-type Lagrangian trajectory model was applied in Germany, and simultaneously an Eulerian grid model (ADOM) was used in Canada to study the mercury transport in Europe and eastern North America, respectively (e.g., Petersen et al., 1990). A preliminary conclusion from this study is that even with a simplified mercury chemistry scheme in a comprehensive Eulerian model and with a linear chemistry in the Lagrangian model, an agreement of a factor below 2 has been obtained when comparing the model and measured concentrations of the compound in the air and precipitation samples. This conclusion holds an interesting promise, particularly in a view of a new chemical module for the liquid-phase chemistry of mercury, which has been recently developed. This new module takes into account the complexity of oxidation, reduction, and adsorption processes in rain and cloudwater.

Very recently, Petersen (1993) summarized information on the modeling of trace metals in Europe. The models, their attributes, and the main area of application are presented in Table 9 (after Petersen, 1993). Intercomparison of results from three models (TREND, HHLRT, and EMEP) calculating deposition of pollutants to the North Sea and comparison with data from seven measurement sites around the sea indicate that:

- The models perform well in predicting total annual inputs.
- Predicted monthly values may show significant spatial and temporal differences mainly due to errors associated with information on precipitation fields over the sea.

A long-range transport model has also been applied to study the origin of trace metals in the Great Waters area. Clark (1992) has utilized the RELMAP model to calculate air concentrations and deposition of several compounds. The model results were used to assess to what extent the emissions from sources outside the study area can contribute to the contamination of the Great Waters area. Significant contribution was suggested for lead and cadmium.

More than 30% of the deposition of trace metals, except mercury, to Lake Michigan is estimated to originate from sources more than 200 km away. The same fraction of the deposition to Lake Superior is calculated to arrive from sources located 500 to 1000 km away. It must be admitted, however, that due to inaccurate emission inventories the predicted deposition values for Lake Michigan and Lake Superior seem to be much higher than the results of measurements. A summary of the modeling activity in North America is presented in Table 10.

B. APPLICATION OF SOURCE APPORTIONMENT TECHNIQUES

Statistical techniques have been developed which use information on the chemical composition of aerosols to study contribution of sources or even source regions to the contamination at a given receptor (e.g., Stevens et al., 1980; Dzubay et al., 1988; Keeler and Samson, 1989; Gordon, 1991; Henry, 1991). The applicability of multivariate techniques for resolving sources and source regions for aerosols measured at several remote locations far from major emission regions has been tested with the use of the absolute principal component analysis (APCA) and the chemical mass balance (CMB) methods. The APCA method determines the composition of the major source components, such as coal combustion-related, crustal, or sea salt-related which contributed to the measured concentrations at the receptors. In the past, the APCA method was applied to total suspended particulate concentrations (particles measured in both the fine and coarse fractions). APCA has been utilized to study the origin of the Arctic aerosol. The results of these studies can be summarized as follows. The anthropogenic component contained many toxic compounds; however, some crustal material was also found. A second component containing the crustal elements was observed with proportions similar to those found in the average crustal rock. However, most of the elements in the soil component were highly enriched and this fact suggests that the second component also contained some material from anthropogenic sources. The composition of the third component found in the Arctic aerosol was fairly similar to that of bulk seawater, indicating

Table 9 A Comparative Summary of Models Applied For Trace Metals in Europe

Model	Type	Institution	Application		Output	
			Input		Receptor	
			Meteorological input	Calculated substances	area	Time scale
EMEP	Lagrangian one-layer trajectory model, sulfur version modified for inert particle-associated substances	GKSS Research Centre Geesthacht, Germany	Six hourly from the Norwegian Numerical Weather Prediction Model and from the European monitoring network	Pb, Cd, Zn, As	North Sea Baltic Sea	Monthly and annual averages
	Sulfur version modified for reactive mercury species in gaseous and particulate form			Hg		
TRACE (Trace toxic Air Concentrations in Europe)	Improved climatological-type model	IIASA (International Institute for Applied System Analysis) Laxenburg, Austria	Improved climatological input	Pb, Cd, Zn, As	Europe	Annual averages
GESIMA (Geesthacht Simulation Model of the Atmosphere)	High-resolution nonhydrostatic mesoscale model	GKSS-Research Centre Geesthacht, Germany	Numerical weather prediction models, e.g., "Deutschland" model of the German weather service	Pb	North Sea coastal area in Germany	Days
EMEP (European Monitoring and Evaluation Programme)	Lagrangian one-layer trajectory model, sulfur model modified for inert particle-associated substances	NILU (Norsk Instituttt for LUftforskning) Lillestrøm, Norway	Six hourly from the Norwegian Numerical Weather Prediction Model	V, Mn, Cd, Pb, As, Sb, Se	Birkenes, Southern Norway	Daily averages

Model	Description	Institute	Meteorological input	Contaminants	Region	Output
TREND	Statistical approach of a Gaussian plume model and a trajectory model	RIVM (Rijksinstituut voor Volksgesondheit en Milieuhygiene) Bilthoven, The Netherlands	Long-term averages from the Dutch meteorological office	As, Sb, Cd, Cr, Cu, Ni, Pb, Zn	North Sea	Annual averages
OPS (Operational Priority Substances)	More operational version of TREND	TNO (Netherlands organization for applied scientific research) Delft, the Netherlands		23 contaminants Cd, Cr, Cu, Pb, Hg, Ni, Zn	North Sea Dutch Wadden Sea	
				As above plus 6 additional pesticides	River Rhine drainage basin	
HHLRT (Hamburg Long Range Transport Model)	Three-dimensional stochastic model	Meteorological Institute University of Hamburg Germany	Six-hourly analyses of the European Centre for Medium Weather Forecast (ECMWF)	Pd, Cd	North Sea	Monthly averages

Note: Adapted from Petersen, 1993.

Table 10 A Comparative Summary of Models Applied For Trace Metals in North America

Model	Type	Institution	Application			
			Input		Output	
			Meteorological input	Calculated substances	Receptor area	Time scale
ASTRAP (Advanced Statistical Trajectory Air Pollution)	Statistical Lagrangian, trajectory, 9 vertical layers	Argonne National Laboratory, USA; Environment Canada	Six hourly analysis from Canadian Meteorological Center	Pb	North America — Great Lakes	Seasonal
RELMAP	Lagrangian one-layer trajectory	NOAA/US EPA	12 hourly wind	Pb, Cd Toxaphene As, Cd, Cr, Pb, Ni	Lake Michigan	Daily
AES-LRT	Lagrangian one-layer trajectory	Environment Canada	Six hourly analysis from Canadian Meteorological Center	As, Cd, Cr, Pb, Ni	Lake Superior	Daily
					Arctic	Daily
ADOM	Eulerian, 12 vertical layers	Ontario Ministry of the Environment, Environment Canada	Six hourly analysis from Canadian Meteorological Center processed through PBL model	Hg	Eastern North America	Hourly

that this component is essentially sea salt. The three-component solution for the Arctic winter aerosol was confirmed by several studies (e.g., Barrie et al., 1992).

Further improvement of this receptor modeling method was obtained by applying APCA to aerosol elemental concentration measurements in separate particle size fractions (e.g., Li and Winchester, 1990). The results of this application of APCA gives the basis for interpretation of coupled chemical reactions and physical processes in remote locations, as well as giving information concerning atmospheric aging processes and, therefore, the history of the aerosols.

Principal component analysis has also been applied to assess the origin of the remote aerosols on the basis of results from automated microanalysis of individual particles. For example, Anderson et al. (1992) employed a modified scanning electron microscope (SEM) analysis of particles to study the origin of the Arctic aerosol. They concluded that the collected aerosol contained a mixture of altered and unaltered particles types from a variety of sources such as coal combustion, sea salt, and crustal dust. Two important findings were that (1) the Arctic aerosol has pollution products from human activity even in a normal period of spring (no episodes of long-range transport of air pollutants), and (2) many of the apparent pollutant particles in the fine fraction, S-rich species and perhaps Br-rich species, are of natural origins.

One important limitation of the PCA methods is, however, that the results do not allow one to obtain a fine resolution of the contributions from various distant source regions to the chemical composition of the remote aerosol. To attempt this resolution, CMB source apportionment must be performed, using either a set of emission source profiles or a set of elemental signatures. The CMB source apportionment proved to be a good technique to assess contributions from major source regions but was often limited to provide fine resolution of the contributions from more defined areas due to the high collinearity of signatures or source profiles used. For example, the use of three European signatures in the Lowenthal and Rahn (1985) studies was supposed to provide a fine resolution of the contributions from various parts of Europe to the contamination of the Arctic aerosol. This attempt failed due to high collinearity of the European signatures. Therefore, both the PCA and CMB methods have limitations when it comes to fine resolutions of emission source regions contributing to the remote aerosol.

IV. MONITORING OF TRACE METALS IN THE AIR

Long-range transport models based on emission inventories are used to assess the origin of trace metals measured at various receptors. Their results provide the evidence that these compounds can be transported within air masses over long distances. It is then interesting to quantify this transport in terms of amounts of the compounds being removed from the air by wet and dry deposition processes and reaching the terrestrial and aquatic ecosystems, as well as return fluxes from these ecosystems to the air. In other words, it is important to assess so-called net fluxes of atmospheric trace metals in order to study their pathways through various environmental media, and human intake and exposure.

The assessment of net fluxes of trace metals has been carried out in various regions of the globe using two major methods based on measurements and calculations with the help of various statistical techniques. The most studied regions include the North Sea and the Baltic Sea, the Great Waters, and the Arctic. There are also other networks routinely collecting air and precipitation data for trace metals, e.g., the AEROCE network at island sites in the North Atlantic Ocean and a number of sites in the Pacific and Indian Oceans.

A. NORTH SEA

Major studies on the deposition of trace metals to the North Sea have been carried out within PARCOM (ATMOS, 1992). Both measurements and estimates are used to assess the deposition. Various methods are applied to assess the wet deposition on the basis of measurements with arithmetic averaging of wet deposition over all stations, a fixed value of precipitation amounts at open sea (taking into account that precipitation rates at coastal stations are highly variable), and with the correction factor being the ratio of average deposition at a given station to the average deposition over the total area.

The results of the deposition assessment for several trace metals to the North Sea in 1991 are presented in Table 11.

B. BALTIC SEA

Deposition estimates are carried out on the basis of measurements and long-range transport models. Two methods are used to assess wet deposition on the basis of measurements. The first experimental method relies exclusively on measurement data on concentrations in precipitation and the precipitation

Table 11 Estimates of Annual Atmospheric Inputs to the North Sea Based on Measurements During 1991. Calculated for an Area of 525,000 km² (in Tons)

	Cd	As	Cr	Cu	Ni	Pb	Zn
Wet Deposition	23	75	85	572	168	1,147	2,648
Dry Deposition	4	8	9	38	31	94	451
Total	27	83	94	610	199	1,241	3,099

recorded at various coastal stations. The method presupposes that this precipitation is also representative for the open sea and that it is a crude approximation. The results of this method must therefore be viewed with some caution.

The second, hybrid estimation method relies on pollution measurements and both measured and calculated precipitation amounts. Model calculations are considered more reliable for estimating precipitation over the open sea than extrapolation of actual coastal measurements.

Average wet deposition values for lead of between 965 and 1285 t per year were calculated for the period from 1986 through 1990. No data were available for other trace metals. The following conclusions were drawn:

- The decreasing trend of Pb concentrations in precipitation stopped in 1988, remaining stable during the last few years.
- There is a clear tendency for higher concentrations in the southern parts of the Baltic Sea.

Model calculations of lead deposition have also been carried out with the help of a Lagrangian trajectory model based on the EMEP model structure (e.g., Grassl et al., 1989). A total deposition of 1400 t of the compound per year has been calculated. The following conclusions have been reached:

- About 70% of the total input of lead to the Baltic Sea was caused by the emissions in countries around the sea and the rest was due to long-range transport from other regions in Europe.
- Wet deposition contributed over 60% to the total deposition of the compound in 1985.

A similar version of the model was used to calculate the deposition of mercury, indicating that about 12 t of the compound was deposited in 1985 (Petersen et al., 1990).

C. GREAT WATERS IN NORTH AMERICA

The Great Water area includes the region of Great Lakes, Lake Champlain, and Chesapeake Bay. An excellent review of the current understanding of atmospheric deposition processes for trace metals and other pollutants in the Great Waters area, particularly with a view to constructing defensible mass balances of these compounds, has been prepared for the U.S. EPA (Baker et al., 1993). Specific questions addressed in this work have dealt with:

- The degree of our understanding of processes leading to atmospheric deposition of the studied compounds
- The availability of methods estimating deposition rates accurate enough to base regulatory decisions upon them
- The current understanding as well as availability of accurate and sufficient data required to estimate relative (input-output) atmospheric loadings of the studied compounds to the studied area
- Further work needed in order to improve our ability to address the issue of atmospheric loadings of the studied pollutants in the Great Waters area

The estimated atmospheric deposition of mercury, lead, cadmium, and arsenic to the Great Lakes is presented in Table 12 (Eisenreich and Strachan, 1992).

An interesting subject related to the budget studies is to assess the role of atmospheric deposition in mass balances in terms of deposited fluxes, including information on estimation uncertainties. The following conclusions were drawn concerning trace metals:

- Atmospheric input of cadmium to Delaware Bay was estimated to be at least 10% of the total input, and in other less urbanized estuaries would probably be more.

Table 12 Estimated Atmospheric Deposition of Trace Metals to the Great Lakes in Tons Per Year

Lake	As	Cd	Hg	Pb
Superior				
Wet deposition	12.5	9.4	1.3	62.4
Dry deposition	4.6	2.8	0.9	4.7
Total	17.1	12.2	2.2	67.1
Michigan				
Wet deposition	9.1	6.9	0.9	12.8
Dry deposition	3.3	2.0	0.7	13.1
Total	12.4	8.9	1.6	25.9
Huron				
Wet deposition	9.1	6.8	0.9	9.1
Dry deposition	3.4	2.0	0.7	1.4
Total	12.5	8.8	1.6	10.5
Erie				
Wet deposition	4.3	3.2	0.4	90.7
Dry deposition	3.4	2.0	0.3	5.8
Total	7.7	5.2	0.7	96.5
Ontario				
Wet deposition	3.5	2.6	0.4	43.2
Dry deposition	1.1	0.7	0.2	4.4
Total	4.6	3.3	0.6	47.6

Note: Data after Eisenreich and Strachan, 1992.

- Atmospheric deposition dominates the flux of mercury to lacustrine systems and the open ocean, and it appears that modest increases in atmospheric mercury loading could lead directly to enhanced levels of mercury in biota.

D. THE ARCTIC

Atmospheric deposition of trace metals and other pollutants in the Arctic has been a subject of studies focusing on the:

- Anthropogenic origin and magnitude of these pollutants arriving within air masses from very distant source regions,
- Perturbations of geochemical cycles of trace metals by their anthropogenic emissions,
- Historical evolution of environmental contamination by the studied compounds.

The following conclusions have been drawn from the above-mentioned studies concerning the origin of Arctic air pollution:

- Emissions from anthropogenic sources are the main contributors to the pollution phenomenon called arctic haze.
- Emissions from sources in Eurasia dominate the pollution measured in the Arctic during winter, as can be expected from meteorological studies of transport paths of air masses to the region. During summer, transport of pollutants to the Arctic from sources in Europe seems to be more important and a North American source contribution is also more evident.

Although there have been a number of successful studies carried out to explain the origin of the Arctic air pollution using information on the concentrations of trace metals, a quantitative assessment of a portion of the pollution load to be deposited in the area has been less evaluated. A very interesting review of research and assessment of sources, occurrence, and pathways of Arctic contaminants has been prepared by Barrie et al. (1992).

The importance of ecosystem cycling and the role of the atmosphere and ocean currents in the transport of pollutants to the Arctic have been emphasized in various recent studies. The results of these studies show that a multimedia approach should be applied to explain the behavior of various pollutants, including trace metals, during their transport to the region.

More information on the deposition of trace metals in the Arctic can be obtained from studies by Davidson (1991), who confirmed that atmospheric chemistry over the ice sheet differs greatly from the patterns observed at the coastal sites. This observation is very important when assessing the magnitude of atmospheric deposition of trace metals on the basis of measurements.

Results of studies on the Arctic air contamination have been used to assess to what extent the anthropogenic emissions have altered the geochemical cycles of trace metals (e.g., Pacyna and Winchester, 1990). Using the enrichment factor method, the authors concluded that the major parameters affecting these perturbations are

- The quantity of trace metal emissions in the source regions affecting the air mass transport to the Arctic
- The solubility of various chemical species of trace metals and thus wet deposition
- The size of particles carrying trace metals and thus dry deposition

The following trace metals were found to have their geochemical cycles perturbated the most: Ag, As, Cd, Pb, Sb, Se, and Zn.

Several studies have been carried out to explain the increasing trend of trace metal deposition in the earth. The following can be concluded from these efforts:

- The Greenland and Antarctic polar ice sheets are an archive of gas and particle air chemistry over at least 200,000 years (e.g., Wolff, 1990).
- Present concentrations of lead in Greenland snow are 200 times greater than the natural value (e.g., Murozumi et al., 1969) but decreased by a factor of 7.5 since the 1960s (e.g., Boutron et al., 1991) due to reduced consumption of leaded gasoline in many countries of the Northern Hemisphere.
- There has been significant progress made in developing the instrumental methods to trace the origin and changes in deposition of trace metals as observed on the basis of concentrations in snow and ice (e.g., Rosman et al., 1993).

V. FINAL REMARKS

Significant progress has been made during the past two decades to obtain information on emission sources and source fluxes for trace metals, their transport within air masses, and atmospheric deposition to various surfaces of the aquatic and terrestrial ecosystems. This progress accounts for both the development of assessment tools and presentation of quantitative evaluations.

While much has been done, several questions within the above-mentioned study areas of atmospheric trace metals still need to be addressed. Concerning emissions, more information is needed on the releases of various chemical and physical forms of some trace metals, particularly mercury. Further work is needed to improve the quality of information presented in current guidebooks of emission factors and to develop other methods for emission estimates, e.g., models relating the amount of emissions to the meteorological, technical, physical, and chemical conditions during the emission-generating processes. The accuracy of emission estimates has been poorly evaluated. Techniques and procedures to evaluate emission inventories as part of the validation process need to be further developed. Of particular interest is the development of emission factor validation procedures. It is believed that the reliability of emission data presented in Europe and North America decreases in the following order:

$$Pb > Hg > Cd > \text{rest of trace metals}$$

Application of deterministic long-range transport models, in both episodic investigations with detailed physical and chemical processes included, and long-term deposition or concentration estimates with highly parameterized processes proved useful to study the transport of trace metals. An agreement between modeled and measured concentrations and depositions within a factor of two should be regarded as good considering the inaccuracy of the input data to the models, namely the emission and meteorological data.

The uncertainty of the emission data seems to be the most important factor contributing to the accuracy of model estimates. Further improvement is needed in determining the atmospheric lifetimes of trace metals, their deposition velocities over various surfaces and scavenging within and below clouds, and gas exchange processes with water, land, and vegetation.

Concerning receptor modeling, it should be noted that the techniques currently used for sampling high-temperature combustion sources are inappropriate for aerosol receptor modeling applications since they do not provide size-fractionated data and techniques which can provide this information are urgently needed.

ACKNOWLEDGMENTS

This chapter is based on a working document on emissions, atmospheric transport, and deposition of heavy metals and persistent organic pollutants prepared by the author for discussion at the First Workshop on Emissions and Modelling of Atmospheric Transport of Persistent Organic Pollutants and Heavy Metals organized in accordance with the work plan of the UN ECE Task Force on Persistent Organic Pollutants by the U.S. Environmental Protection Agency in Durham, NC, 6–7 May, 1993. The author would like to acknowledge the cooperation of Dr. Eva Voldner and Dr. Terry Bidleman, both of Environment Canada, Dr. Jerry Keeler of the University of Michigan, and Mr. Gary Evans of the U.S. EPA in collecting information for the working paper used in the chapter.

REFERENCES

Akeredolu, F., Barrie, L. A., Olson, M. P., Oikawa, K. K., Pacyna, J. M., and Keeler, G. J., *Atmos. Environ.*, 28, 1557–1572, 1994.

Alcamo, J., Bartnicki, J., Olendrzynski, K., and Pacyna, J. M., *Atmos. Environ.*, 26A, 3355–3369, 1992.

Anderson, J. R., Buseck, P. R., Saucy, D. A., and Pacyna, J. M., *Atmos. Environ.*, 26A, 1747–1762, 1992.

ATMOS, Paris Convention for the Prevention of Marine Pollution, London, November 9–12, 1992.

Axenfeld, F., Munch, J., Pacyna, J. M., Duiser, J. A., and Veldt, C., Umweltforschunsplan des Bundesministers fur Umwelt Naturschutz und Reaktorsicherheit, Luftreinhaltung, Forschungsbericht 104 02 588, Dornier GmbH Report, Friedrichshafen, Germany, 1992.

Baker, J. E., Church, T. M., Eisenreich, S. J., Fitzgerald, W. F., and Scudlark, J. R., A report for the U.S. EPA, University of Maryland, Solomons, MD, 1993.

Barrie, L. A., Gregor, D., Hargrave, B., Lake, R., Muir, D., Shearer, R., Tracey, B., and Bidleman T. F., *Sci. Total Environ.*, 122, 1–174, 1992.

Benjey, W. G. and Coventry, D. H., Presented at Int. Symp. Measurement of Toxic and Related Air Pollutants, Durham, NC, May 3–8, 1992.

Boutron, C. F., Gorlach, U., Candelone, J. P., Bolshov, M. A., and Delmas, R. J., *Nature*, 353, 153–156, 1991.

Clark, T. L., Presented at the Joint Workshop on Mercury and Multimedia Risk Assessment, Durham, NC, October 22, 1992.

Davidson, C. I., Presented at ACS Natl. Meet., Atlanta, GA, April 14–19, 1991, 1991.

Dzubay, T. G., Stevens, R. K., Gordon, G. E., Olmez, I., and Sheffield, A. E., *Environ. Sci. Technol.*, 22, 1132–1141, 1988.

Eisenreich, S. J. and Strachan, W. M. J., Report from Workshop in Burlington, Ontario, Jan. 31-Feb. 2, 1992, sponsored by the Great Lakes Protection Fund and Environment Canada, 1992.

Fisher, G., Andren, A. W., and Wittman, S., Report submitted to the International Joint Commission Virtual Elimination Task Force, Windsor, Ontario, 1992.

Gordon, G. E., *Environ. Sci. Technol.*, 11, 1822–1828, 1991.

Grassl, H., Eppel, D., Petersen, G., Schneider, B., Weber, H., Gandrass, J., Reinhardt, K. H., Wodarg, D., and Fliess, J., GKSS-Forschungszentrum Geesthacht GmbH, GKSS Rept. 89/E/8, Geesthacht, Germany, 1989.

Henry, R. C., In: *Receptor Modeling for Air Quality Management*, Hopke, P. K., Ed., Elsevier, New York, pp. 117–147, 1991.

ICF, ICF Inc., Fairfax, VA, 1992.

van Jaarsveld, J. A. and Onderdelinden, D., Report No. 228603009, RIVM, the Netherlands, 1991.

Jacques, A. P., Environmental Analysis Branch, Conservation and Protection, Environment Canada, Ottawa, 1987.

Keeler, G. J., Pacyna, J. M., Bidleman, T. F., and Nriagu, J. O., A report for the U.S. EPA, University of Michigan, Ann Arbor, MI, 1993.

Keeler, G. J. and Samson, P. J., *Environ. Sci. Technol.*, 23, 1358–1364, 1989.

Li, S.-M. and Winchester, J., *J. Geophys. Res.*, 95, D2, 1797–1810, 1990.

Lindqvist, O., *Mercury in the Swedish Environment*, Vol. 55, Kluwer Academic, Dordrecht, the Netherlands, 1991.

Lowenthal, D. H. and Rahn, K. A., *Atmos. Environ.*, 19, 2011–2024, 1985.

van der Most, P. F. J. and Veldt, C., TNO Report 92–235, Apeldoorn, the Netherlands, 1992.

Munthe, J., Proc. First Workshop on Emissions and Modelling of Atmospheric Transport of Persistent Organic Pollutants and Heavy Metals, Durham, NC, 6–7 May, 1993, 1993.

Murozumi, M., Chow, T. J., and Patterson, C. C., *Geochim. Cosmochim. Acta*, 33, 1247–94, 1969.

Nriagu, J. O., Ed., *Changing Metal Cycles and Human Health*, Dahlem Konferenzen, Springer-Verlag, Berlin, 1984.

Nriagu, J. O. and Pacyna, J. M., *Nature*, 333, 134–139, 1988.

Ondov, J. M., Ragaini, R. C., and Bierman, A. H., *Environ. Sci. Technol.*, 13, 588–601, 1979.

Pacyna, J. M., *Atmos. Environ.*, 18, 41–50, 1984.

Pacyna, J. M., In *Control and Fate of Atmospheric Trace Metals*, Pacyna, J.M. and Ottar, B., Eds., Kluwer Academic, Dordrecht, the Netherlands, 15, 1989.

Pacyna, J. M., Final Report NILU OR 46/92, Norwegian Institute for Air Research, Lillestrøm, Norway, 1992.

Pacyna, J. M., Proc. First Workshop on Emissions and Modelling of Atmospheric Transport of Persistent Organic Pollutants and Heavy Metals, Durham, NC, 6–7 May, 1993.

Pacyna, J. M. and Winchester, J. W., *Palaeogeogr. Palaeoclimatol. Palaeoecol.* (Global and Planetary Change Sect.), 82, 149–157, 1990.
Pacyna, J. M., Bartonova, A., Cornille, Ph., and Maenhaut, W., *Atmos. Environ.,* 23, 107–114, 1989.
Pacyna, J. M., Shin, B. D., and Pacyna, P., A report for the Atmospheric Environment Service, Environment Canada, Ottawa, 1993.
Petersen, G., Weber, H., and Grassl, H., In *Control and Fate of Atmospheric Trace Metals,* Pacyna, J.M. and Ottar, B., Eds., Kluwer Academic, Dordrecht, the Netherlands, 57–84, 1989.
Petersen, G., Schneider, B., Eppel, D., Grassl, H., Iverfeldt, A., Misra, P. K., Bloxam, R., Wong, S., Schroeder, W. H., Voldner, E., and Pacyna, J. M., GKSS-Forschunszentrum Geesthacht GmbH, GKSS Rept. 90/E/24. Geesthacht, Germany, 1990.
Petersen, G., Paper prepared for the UN ECE Task Force on POPs, Durham, NC, 6–7 May, 1993.
Radian, Software Technical Specification for the Emission Factor Information System, Draft Report, Radian Corporation, Research Triangle Park, NC, 1992.
Rosman, K. J. R., Chisholm, W., Boutron, C. F., Candelone, J. P., and Gorlach, U., *Nature,* 1993.
Stevens, R. K., Dzubay, T. G., Shaw, R. W., McClenory, W. A., Lewis, C. W., and Wilson, W. E., *Environ. Sci. Technol.,* 14, 14911–14918, 1980.
Strachan, W. M. J. and Eisenreich, S. J., International Joint Commission, Windsor, Canada, 1988.
U.S. EPA, EPA-454/R-93–023, U.S. Environmental Protection Agency, Research Triangle Park, NC, 1993.
Voldner, E. and Smith, L., Presented at the Workshop on Great Lakes Atmospheric Deposition, October 29–31, 1986, 1989.
Weisburd, S., *Science News,* 133, 309.
Wolff, E. W., *Antarct. Sci.,* 2, 189–205, 1988.

Chapter 2

Assessment of Ecological and Health Effects of Soil-Borne Trace Elements and Metals

Andrew C. Chang and Albert L. Page

I. INTRODUCTION

Soils are derived from geological material which has undergone thousands and sometimes millions of years of weathering processes. All soils are heterogeneous porous media consisting of mineral fragments, organic matter, biota, water, and air. Five interplaying factors — parent material, climate, organisms, topography, and time — are important to soil genesis and to determining the compositions and properties of soils, resulting in a divergence of soils around the world (Jenny, 1941). They are an integral part of the terrestrial environment and form interfaces with the atmosphere, hydrosphere, biosphere, and lithosphere. The nature of soil is constantly evolving in response to inputs of matter from natural and anthropogenic sources and to interactions among soil constituents. Substances entering the soil may be mineralized and assimilated through the chemical-biochemical cycling processes in soils, and also through these processes the chemical, physical, and biological conditions of the soil may be maintained at a steady state. An element, however, will accumulate in the soil if its input exceeds the rate of its biochemical cycling.

Elemental compositions of biota are influenced in many respects by the geochemical nature of their habitat reflecting in general a pattern of elemental transfer from the terrestrial to the biotic environment (Hamilton, 1987; Li, 1984). Food, fiber, and many products essential to sustain human life come directly or indirectly from plants which are supported by soils. Through food grown on soils, humans obtain energy and essential elements. Also, through food grown on soils, humans may be exposed to harmful substances that are absorbed by plants. Both the presence of potentially toxic substances and the absence of essential elements in the soil could result in disorders. Since ancient times, the trace element deposition onto soils has been the cause of many human ailments (Anonymous, 1989). For example, millions of people inhabiting the iodine-deficient soils in eastern Africa are still susceptible to goiter today (Tebeb, 1993; Jaffiol et al., 1992); the cause of both Keshan and Kaschin-Beck diseases in China is related to Se concentrations of soils (Yang et al., 1980; Hou et al., 1980), and the etiology of Itai-Itai disease is attributed to consumption of rice grown on soils contaminated by Cd-containing wastes (Friberg et al., 1971).

Elements of environmental and toxicological importance are numerous. According to Nieboer and Richardson (1980), the toxic potential of these elements may be classified in terms of their chemical and biological reactivities into categories of Class A, Class B, and Borderline Class (Table 1). Ions belonging to Class B are expected to be more toxic than ions in the Borderline category which, in turn,

Table 1 Elements of Environmental and Toxicological Importance

Category	Ion
Class A	Li^+, Na^+, K^+, Cs^+, Ca^{2+}, Ba^{2+}, Sr^{2+}, Mg^{2+}, Be^{2+}, La^{2+}, Y^{3+} Gd^{3+}, Lu^{3+}, Sc^{3+}, and Al^{3+}
Class B	Ag^+, Au^+, Tl^+, Cu^+, Hg^{2+}, Pd^{2+}, Bi^{2+}, Tl^{2+}, and Pb^{2+}
Borderline	Mn^{2+}, V^{2+}, Ti^{2+}, Zn^{2+}, Ni^{2+}, Fe^{2+}, Co^{2+}, Cd^{2+}, Cu^{+2}, Sn^{2+}, Cr^{3+}, Fe^{3+}, Ga^{3+}, As^{3+}, Sn^{3+}, Pb^{4+}, In^{4+}, and Sb^{4+}

Note: Derived from Nieboer and Richardson (1980).

are more toxic than the ions in Class A (Ahrland et al., 1958; Nieboer and Richardson, 1980). Metal ion-induced toxicity usually is the result of chemical and/or biochemical reactions that block the essential functional groups of biomolecules, displace essential metal ions in biomolecules, and modify the active configuration of biomolecules (Ochiai, 1977). The ions in the Class B group and to a lesser extent the ions in the Borderline group are more likely to take part in reactions resulting in biotoxicities because of their greater inclination to form cation/ligand complexes. Under many circumstances, elements may bioaccumulate through food chain transfer and affect organisms not directly exposed to the toxic substances. Because of the tendency for soil reactions to attenuate the toxic effects of trace elements, cases of trace element-related toxicities are relatively rare.

Problems of the deteriorating environmental quality, which was brought about when ancient nomadic tribes first settled into villages and started utilizing fire, cultivating land, and producing wastes, have become increasingly acute because of the exponential population growth and global industrialization. Millions of chemicals are being used world-wide as manufacturing stock and consumer products (Piemental and Coonrod, 1987). During the course of manufacturing, industrial processing, and product consumption, these chemicals may be released deliberately or inadvertently into the environment. In the U.S. alone more than 10 billion kilograms of potentially harmful chemicals are released into the environment each year (U.S. Environmental Protection Agency, 1993a). Most of those released onto land are the result of waste disposal. Judging from compositions of wastes, many elements of environmental toxicological importance (Table 1) are routinely released into the environment. Around the world the physical, chemical, and biological integrity of soils have never been so stressed by anthropogenic inputs of pollutants in the recorded history of mankind. Environmental contamination of air, water, and soil has become a potential threat to the safety of food. The soil-borne trace elements are particularly important, because food chain transfer is by far the most significant route of human exposure to trace elements (McKone and Ryan, 1989; Galal-Gorchev, 1991; Prasad, 1993). Therefore, trace elements deposition in soils and biogeochemical interactions between those present in soils and the vegetative cover are controlling factors in human exposure to toxic elements in the environment.

II. OCCURRENCE OF TRACE ELEMENTS IN SOILS AND PLANTS

The earth's crust is composed primarily of minerals of Si, Al, Fe, Ca, Na, Mg, Ti, P, S, and C. In igneous rocks, trace elements usually exist as minor constituents, and they are not readily soluble and biologically reactive. Due to reactions such as dissolution, hydration, hydrolysis, oxidation, and reduction, a part of the trace elements in primary minerals may be solubilized and readsorbed by the secondary minerals and organic matter during the course of weathering. As a result, the concentration of trace elements in weathered minerals are often higher than in the parent material. Metal concentrations of soils around the world have been extensively reported and they vary considerably from location to location, reflecting differences in the soil genesis processes (Table 2).

In soils, trace elements are distributed throughout the solid matrix in forms ranging from water soluble, exchangeable, sorbed and organically bound, occluded in oxides and other secondary minerals, precipitated, and remained as a part of the primary mineral lattice. The trace elements in soils may be chemically speciated by sequentially extracting the element associated or bound to particular soil phases by selective extractants of different strengths (Tessier et al., 1979). The chemical forms obtained from the sequential extraction, however, are functionally defined and do not always reflect the true distribution of trace metals in soils. It is, nevertheless, the only practical way to understand the trace elements present in the solid phase of soils.

Regardless of their origins, trace metals react rapidly with soil components to form sparingly soluble organic and inorganic compounds. Studies indicate that trace elements in the environment exhibit strong affinity for a solid phase such as hydrous oxides of iron and manganese, organic matter and biota, clays,

Table 2 Concentrations of Trace Elements in Soils[a]

Element	Alloway, 1990	Kabata-Pendias and Pendias, 1992	Page, 1974
Ag	0.01–8	—	0.1–40
As	0.1–50	0.07–197	—
Au	0.001–0.02	—	—
Cd	0.01–2.4	0.01–2.53	0.01–7
Co	1–40	0.1–122	—
Cr	5–1500	1–1100	5–3000
Cu	2–250	1–323	2–100
Hg	0.01–0.3	0.0014–5.8	—
Mn	20–10000	7–8423	100–4000
Mo	0.2–5	0.2–17	0.2–5
Ni	2–1000	0.2–450	10–1000
Pb	2–300	1.5–286	2–200
Sb	0.05–260	0.05–4	—
Se	0.01–2	0.005–4	0.1–2.0
Sn	1–200	—	—
Tl	0.03–10	—	—
U	0.7–9	—	—
V	3–500	0.7–500	20–500
W	0.5–83	—	—
Zn	10–300	3–770	10–300

[a] Concentrations expressed in mg kg^{-1} dry weight.

and minerals of carbonate, phosphate, sulfide, and hydroxide (Kabata-Pendias and Pendias, 1992; Alloway, 1990). For this reason, soils are sinks for trace elements and under ordinary conditions, only a very small fraction of the total trace elements in the soil is readily soluble or exchangeable at any given time. Trace metals such as Cu, Co, Pb, Zn, Mn, Cd, and many other divalent species exhibit rather high affinities for organic matter in the soil. Trace metal-organic matter complexes are difficult to characterize because their chemical structures and chemical properties are not well defined. It is generally known that positively charged trace metal ions bind principally with carboxyl, phenolic, and amide functional groups in the soil organic matter, and negatively charged trace element oxyanions are adsorbed by oxides and hydroxides of Al, Fe, and Mn. In situations where solubility of trace elements exceed solubility products of inorganic compounds known or suspected to occur in soils, it is thought that the trace element existed in the soil solution as a soluble organic complex. However, a significant portion of trace elements in soils are present in inorganic forms, and they are chemically stable and may remain essentially unchanged over a long period of time. Lindsay (1972) has shown that concentration of Cu, Zn, and Mn in many soil solutions are less than those predicted from the solubility of hydroxides and carbonates of these elements. Similar observations have been made in natural water systems. Although no discrete solid phase has been identified, it is believed that trace elements frequently are specifically adsorbed by iron and manganese oxides and/or are complexed by organic matter in the soils (Alloway, 1990).

The trace elements Mn, Ba, Cu, Zn, Ni, Cd, Co, and Pb, if not complexed with organic matter, most probably exist in soil solutions predominantly as divalent cations. In aqueous solutions, many other inorganic complex ions, molecules, and ion pairs of these elements are known to occur in equilibrium with the divalent form. These complex ions have the general formula of $M_aX_b^{(2a-b)}$ where M is the divalent cation and X is the complexing anion. The complexes can be cationic, neutral, or anionic. The ionic speciation in soil solution may be computed if concentrations of soluble species and thermodynamic stability constants of the chemical equilibria are known. Several computer programs have been developed to model the chemical equilibria in soil solutions (Sposito and Mattigod, 1980; Parker et al., 1992).

The reactivity and biological availability of trace elements in the terrestrial environment are determined by chemical equilibria between the metals in solution and the metals associated with the solid phases (Sposito, 1981). As plants absorb only the trace elements in the solution phase, the adsorption and desorption kinetics and diffusion of trace elements in the soil solution are as equally important parameters as their solubilities in controlling plant uptake (Brummer, 1986). In addition to the soil chemical properties, the geometry and absorbing capacity of root systems, solute diffusion, and tortuosity of the growing media all affect the rates and amounts of trace metal ions absorbed by plants (Barber,

1984). If plants are grown under the same environmental conditions, the above-mentioned physical and biophysical parameters would be the same, and the activities of metal ions in the soil solution would determine their availability to plants (Sparks, 1984; Halvorson and Lindsay, 1977; Bingham et al., 1992; Minnich et al., 1987; Checkai et al., 1987; Xian, 1989; Bell et al., 1991)

Trace elements indigenous to the soils seldom are bioaccumulated in the plant tissue as their solubilities are low and the kinetics of their solubilization from the solid phase is slow. With growing industrial and agricultural activities world-wide, trace metal concentrations of the soils have been steadily increasing (Page and Sposito, 1985). The plant uptake of potentially toxic trace elements from contaminated soils, however, would be significantly higher than those grown in the uncontaminated sites (Table 3).

Table 3 Concentration of Trace Elements of Food Plants Grown on Uncontaminated and Contaminated Soils[a]

Element	Plant	Uncontaminated soils (mg kg^{-1} dry weight)	Contaminated soils (mg kg^{-1} dry weight)
As	Leafy vegetables	0.02–1.5	—
	Root/tuber vegetables	0.03–0.2	0.26–1.1
	Grain/cereal	0.003–0.4	1.2
Cd	Leafy vegetables	0.05–0.66	5.2–45
	Root/tuber vegetables	0.03–0.24	1.7–3.7
	Grain/cereal	0.01–0.21	0.72–4.2
Cu	Leafy vegetables	2.9–8.1	64
	Root/tuber vegetables	4–8.4	—
	Grain/cereal	1.1–7.8	21
Hg	Leafy vegetables	0.007–0.008	—
	Root/tuber vegetables	0.006–0.009	0.5–0.13
	Grain/cereal	0.003–0.005	—
Zn	Leafy vegetables	24–73	213–1300
	Root/tuber vegetables	10–45	74–458
	Grain/cereal	5–37	132–194

[a] Data extracted from Kabata-Pendias and Pendias, 1992.

III. ENVIRONMENTAL EXPOSURE ASSESSMENT

In the soil, toxic pollutants are subject to various chemical and biochemical degradation mechanisms and may be transported by several media (Tinsley, 1979). To assess the environmental exposure to trace elements, one must account for their fate and transport in the environment. Because it is difficult to obtain comprehensive environmental measurement data for exposure analysis, model simulations are an acceptable and normal part of the exposure assessment process (U.S. Environmental Protection Agency, 1986). Models, however, do not always describe the interactive and time-dependent behavior of a pollutant in the environment in its entirety.

Two approaches may be used to evaluate and predict pollutant distribution and transport in the ecosystem (Jones et al., 1991; Jackson and Smith, 1987). Some models divide the environmental exposure route, through which pollutants undergo transformation and reach human subjects, into broad and general compartments, and assume concentration of pollutants between the environmental compartments along an exposure route are in a steady-state equilibrium (equilibrium models). Mathematically, the partition of a chemical between two adjoining environmental compartments is described by a linear transfer coefficient:

$$C_a = f_{ab} \times C_b \tag{1}$$

where C_a and C_b are pollutant concentrations in compartments a and b, respectively, and f_{ab} is the pollutant transfer coefficient. The mathematical relationship defined by Equation 1 is empirical because parameters describing processes and mechanisms of all the reactions are aggregated into one transfer coefficient which must be determined on a case by case basis. The other types of models are designed to account for spatial- and temporal-dependent behavior of a pollutant in the environment (dynamic models).

$$\frac{\partial C_i(t)}{\partial t} = \sum_{i=1}^{n} \mu_{ij} C_j(t) + D_i(t) \quad (2)$$

$$i = 1 \ldots\ldots n,$$

where $C_i(t)$ is the concentration in spatial compartment i at time t; $C_j(t)$ is the concentration in spatial compartment j at time t; μ_{ij} is the rate constant for transfer from spatial compartment j to i, if $i \neq j$; μ_{ij} is the pollutant degradation rate constant in compartment i, if $j = i$; D_i is the pollutant influx into compartment i; and n is the number of spatial compartments (Jackson and Smith, 1987). In principle, the spatial and temporal changes of a pollutant with respect to all of the degradation processes may be represented. Consequently, dynamic models are structured to approximate the kinetics in a real system and are applicable to all situations (e.g., acute, intermittent, and continuous exposures). They are, however, impractical to use, because spatial-, temporal-, and concentration-dependent model rate parameters are numerous and are difficult, if not impossible, to define.

As pollutant exposure assessment in its most comprehensive manner, the U.S. Environmental Protection Agency (1993b) developed and employed 14 pollutant transport pathways to track the pollutants released from land disposal of sewage sludge (Table 4). The pathways identified in this document described almost all the possible environmental transport routes for pollutants released through land application of wastes, with receptors of pollutants ranging from child and adult humans, livestock animals, plants, soil biota, and predators of soil biota. Mathematical models based on linear transfer between compartments were constructed to describe the transfer of pollutants through each pathway.

Table 4 Hypothetical Pollutant Transport Routes for Land Application of Sewage Sludge Considered by U.S. Environmental Protection Agency (1993b)

Pathway	Pollutant exposure route
1	Sludge-soil-plant-human
2	Sludge-soil-plant-home gardener
3	Sludge-soil-child (ingesting soil)
4	Sludge-soil-plant-animal-human
5	Sludge-soil-animal-human
6	Sludge-soil-plant-animal
7	Sludge-soil-animal
8	Sludge-soil-plant
9	Sludge-soil-soil biota
10	Sludge-soil-soil biota-predator of soil biota
11	Sludge-soil-airborne particulate-human
12	Sludge-soil-contaminated surface water-fish-human
13	Sludge-soil-air-human
14	Sludge-soil-groundwater-human

Typical dynamic models describing the pollutant transport from soil to plants may be depicted by the schematic diagrams in Figure 1. Each link in the diagram must be represented by a mathematical expression of Equation 2 and must have the rate constant precisely defined. To track a pollutant according to the mechanistic processes of chemical and biochemical transformations from soil through roots and various plant parts to the site of toxicity in the human body requires that rate constants in all of the equations are properly defined, and then the equations are solved simultaneously. Realistically, this type of model has only a limited utility because of the difficulties in defining model parameters and in validating model predictions (Jackson and Smith, 1987). McCarty and MacKay (1993) reviewed models for ecotoxicological risk assessment, and they are convinced that "there is now a well-derived capability of calculating concentrations, or body residues, in organisms from (pollutant) loading data." The "critical body residue" method they advocated takes the form of a multimedia equilibrium model in which pollutant inputs from various sources are integrated. Although the examples they presented were all for aquatic systems, the principles should be applicable to other parts of the ecosystem including the soils, if data are available.

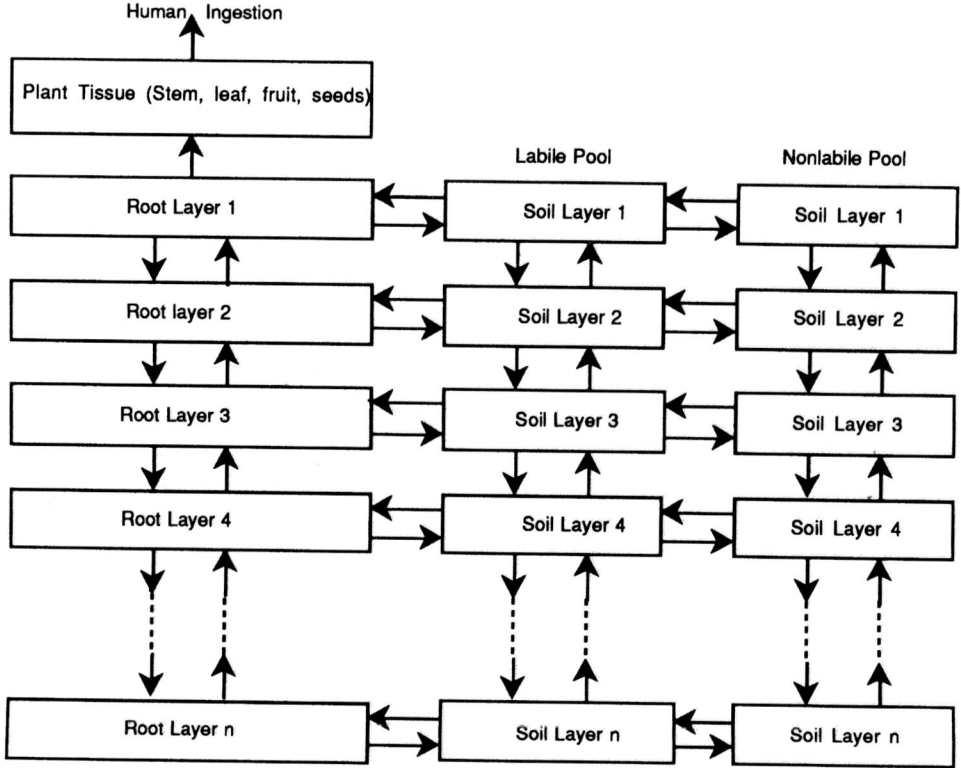

Figure 1 Flow diagram describing the transfer of polutants from soil to plants.

IV. LINEAR PARTITION COEFFICIENT FOR TRACE ELEMENTS

For trace element exposures via the human food chain, the transfer from soil to plant is a key step. In theory, plant uptake of trace elements from soil is a function of the activity of free ions at the surface of the randomly dispersed root system of the plant (Corey et al., 1987) and can be expressed in a mathematical form. Modeling plant uptake of trace elements from the soil, however, is a complicated problem. As the trace element concentrations at the root surface are depleted due to absorption, they must be replenished by trace elements carried in the mass flow and diffusing through the soil matrix. These processes are affected by the conditions of the soil and the plant. It has been proposed that plant uptake of trace elements may be described by the following equation:

$$U = C_{fi} \times b \times \left[1 - e^{\frac{-2\pi\alpha A_1 r_0 L_v t}{b\left(1 + \frac{\alpha A_0 r_0}{D_1 \theta F} \ln \frac{r_h}{1.65 r_0}\right)}} \right] \qquad (3)$$

where

C_{fi} = initial soil solution concentration of trace element (mol cm^{-3})

b = buffer power of soil (e.g., change in total concentration of trace element in labile form (mol cm^{-3} soil) per unit change of concentration of trace element in dissolved form (mol cm^{-3} soil solution)

A_1 = fractional area of soil solution (cm^2 water cm^{-2} soil)

θ = volumetric water content (cm^3 water cm^{-3} soil)

D_1 = diffusion coefficient in soil solution (cm^2 sec^{-1})

U = uptake of trace element per unit volume of soil in time t (mol cm^{-3} soil)

α = root absorbing power (e.g., uptake flux density [mol cm^{-2} root sec^{-1}] per unit concentration in soil solution mol [cm^{-3}])
t = time (sec)
r_h = half-distance between roots (cm)
r_o = root radius (cm)
L_v = root (cm root cm^{-3} soil)
F = conductivity factor (cm^2 soil cm^{-2} water)
π = 3.1416

Many of the parameters included in the above equation, however, cannot be realistically determined, and the model's ability to predict plant uptake of pollutants from soil has not yet been verified by experimental data (Mullins and Sommers, 1985). Perwak et al. (1986) have reviewed methods for assessing environmental pathways of food contamination. They concluded that very little experimentally based quantitative information is available to evaluate pollutant transfer from soil to plants.

Because of the inherent difficulties in considering all of the mechanistic processes and incorporating the dynamic reaction kinetics in simulating the plant uptake of pollutants from soils, the equilibrium approach is customarily employed to track the transfer of pollutants along the exposure pathways of the human food chain (U.S. Environmental Protection Agency, 1993b; McKone and Ryan, 1989). Data from the literature indicate that soil-borne trace elements do not bioaccumulate in the plant tissue, and the soil-to-plant transfer coefficients are invariably less than unity. It is, however, problematic to obtain representative soil-to-plant transfer coefficients (f_{ab} in Equation 1). Chang et al. (1993) reviewed approximately 300 references and concluded that the plant tissue concentrations of Cr, Cu, Ni, and Zn cannot be related to respective amounts of Cr, Cu, Ni, and Zn in the soil, even if the soil pH and soil texture are considered. In determining the soil-to-plant pollutant transfer coefficients, the U.S. Environmental Protection Agency (1993b) consulted data in over 2000 references in the published literature and discovered serious gaps in the availability of data, especially for elements such as Se, Mo, and Hg. After examining the data statistically, McKone and Ryan (1989) concluded that much of the uncertainty in calculating human exposure to toxic chemicals via the food chain is attributable to the uncertainty in "biotransfer coefficients". The outcomes computed by the mean or median values were lower than those obtained by a Monte Carlo simulation process by a factor of 2.

McKone and Ryan (1989) described fruit, vegetable, and grain contamination by pollutants in the soil as

$$F_{si} = \left(\ell_i \times f_i \times K_{sp}\right) \times C_{fd} \tag{4}$$

where

F_{si} = pathway exposure factor from soil to crop (vegetable, fruit, grain, root/tuber) (kg/day)
ℓ_i = human intake of food (kg/day fresh mass) (fruit, vegetable, grain, root/tuber)
f_i = fraction of target population's fruit (vegetables grain and root/tuber) foods that comes from the contaminated source
K_{sp} = plant/soil transfer coefficient (mg/kg DW of pollutant in plant tissue per mg/kg DW of pollutant in soil)
C_{fd} = the fresh weight to dry weight conversion factor. C_{fd} = 0.9, 0.05, 0.2, and 0.05 for grain, vegetable, root/tuber, and fruit foodstuffs, respectively.

The daily average ingestion exposure, E (mg/day), may then be determined by:

$$E = \left(\Sigma F_{si}\right) C_s \tag{5}$$

where C_s is the pollutant concentration in soil (mg/kg).

Similar mathematical relationships may be developed to estimate exposures from ingestion of meat and dairy products. Soil concentrations and transfer coefficients of selected trace elements derived from data compiled by Kabata-Pendias and Pendias (1984) and Fergusson (1990) are presented in Table 5. These data reflect the trace element levels of uncontaminated soil from various regions around the world.

Table 5 Concentrations of Trace Elements in Soils and Plant Tissue[a]

Element	Soil (mg kg^{-1}) Range	Soil (mg kg^{-1}) Average	Transfer coefficient Grain	Transfer coefficient Vegetable	Transfer coefficient Root/tuber	Transfer coefficient Fruit
As	0.1	40	6	0.004	0.037	0.004	0.003
Cd	0.01	2	0.35	0.036	0.223	0.008	0.09
Pb	<1	300	19	0.002	0.0016	2E-5	14E-5
Hg	0.01	0.5	0.06	0.085	0.009	0.002	0.009
Se	0.01	1.2	0.4	0.002	0.015	0.042	0.021

[a] Data derived from Kabata-Pendias and Pendias (1984) and Fergusson (1990).

V. ANALYSIS OF DIETS

The food consumption patterns, which must be used along with the transfer coefficients to determine the pollutant exposure levels, vary considerably from region to region (Ryan et al., 1982; Galal-Gorchev, 1991). There are significant differences in terms of the total amounts consumed as well as types of food consumed. Even if the total dietary intake of the regions can be normalized, there are still significant differences in regional preference of food groups whose K_{sp} are not always known. The Food and Agriculture Organization (FAO) recommends that the average food consumption data in FAO Food Balance Sheets (global diet) be used to estimate human exposure to pollutants via food consumption (World Health Organization, 1989). The food consumption described in the global diet does not resemble any specific diet (Table 6), but rather a generic aggregation of diets from various regions. In the global diet, grain/cereal, vegetables, root/tuber, and fruit account for 76% of the total daily food consumption. The potential for pollutants to be transferred to humans through other food groups (dairy and animal products, oil/fat/shortening, sugar/honey, etc.) is small (Galal-Gorchev, 1991; McKone and Ryan, 1989).

Table 6 Regional Food Consumption Patterns and Global Diet[a]

Food group	Middle Eastern	Far Eastern	African	Latin Amer.	European	Global diet
Cereals	432	452	320	254	226	405
Roots/tubers	62	108	321	159	242	288
Pulses	21	27	18	21	9.3	23
Sugar/honey	95	50	43	104	105	58
Nuts/oil seeds	4.3	18	15	19	12	18
Vegetable oil/fat	38	15	24	26	48	38
Stimulants	8.0	1.5	0.5	5.3	14	8.1
Spices	2.3	2.0	1.6	0.3	0.3	1.4
Vegetables	193	168	74	124	298	194
Fish/seafood	13	32	32	40	46	39
Eggs	14	13	3.6	12	38	38
Fruit	220	94	85	288	213	235
Milk and products	132	33	42	168	338	55
Meat/offals	72	47	32	78	221	111
Animal oil/fat	0.5	1.5	0.3	5.0	10	5.3
Other	4.3	—	0.5	7.0	2.0	3.2
Total	1311	1062	1012	1311	1822	1520

[a] All expressed in grams person^{-1} day^{-1}.

VI. SAMPLE CALCULATION ON HUMAN EXPOSURE TO TRACE ELEMENTS

Based on the data in Tables 5 and 6 and assuming exposure to trace elements resulting primarily from consumption of grain, vegetable, root/tuber, and fruit, Equations 4 and 5 may be used to estimate the potential human exposure to trace elements (Table 7). Based on this estimation, it appears that if the soil is not contaminated, human exposure to trace elements through food consumption is significantly below the provisional tolerable intake published by the Joint FAO/WHO Expert Committee on Food Additives (World Health Organization, 1989) and is not substantially affected by the diets. In fact, the computations suggest that uptake of Se by the food groups selected is not sufficient to satisfy human nutritional requirements for Se, which is reported to be about 25 µg day^{-1}. Although the estimates in general are all within the range of exposures calculated by other means, the accuracy of these estimates is not known. The exposure may be 50 to 400 times greater if the maximum concentrations in Table 6 are used in the calculation. For example, daily dietary intake of As, Cd, Pb, Hg, and Se by adults have been reported to be 10–130, 10–120, 5–1700, ≈30, and <10–5000 µg day^{-1} (Fergusson, 1990; Anke, 1986).

Table 7 Estimated Human Exposure to Selected Trace Elements Via Food Intake

Diet	Exposure (µg day^{-1})				
	As	Cd	Pb	Hg	Se
Middle Eastern	12	6	17	2	0.7
Far Eastern	12	6	17	2	0.8
African	9	4	12	1	1
Latin American	7	4	11	1	0.8
European	9	4	10	1	1
Global	12	6	16	2	1

VII. CONCLUSION

We have demonstrated the relationship between soil and human health with mathematical expressions that link the trace elements in the soil to concentrations in plants, and methodologies for calculating human exposure. Available data, however, can provide only a rough approximation of global human exposure to trace elements. Human exposure to potentially hazardous trace elements through food grown on uncontaminated soils does not appear to be significant. If the cropland soils are polluted by trace metals, the human dietary intake of trace elements will increase. The extent of this increment, however, can only be evaluated by carefully tracking the pollutant transfer through the food chain. A thorough investigation on the food chain transfer of trace elements relies on the availability of accurate soil-to-plant transfer coefficients.

ACKNOWLEDGMENT

A portion of the material in this chapter was presented at the 15th World Congress of Soil Science, Acapulco, Mexico, July 10–17, 1994.

REFERENCES

Ahrland, S., Chatt, J., and Davis, N. R., *Q. Rev. Chem. Soc.,* 12, 265–276, 1958.
Alloway, B., *Heavy Metals in Soils,* John Wiley & Sons, New York, 339 pp., 1990.
Anke, M., Arsenic. In *Trace Elements in Human and Animal Nutrition,* Vol. 2 5th ed., Mertz, W., Ed., Academic Press, New York, 347, 1986.
Anon., *The Atlas of Endemic Diseases and Their Environments in the People's Republic of China,* Science Press, Beijing, People's Republic of China, 194 pp., 1989.
Barber, S. A., *Soil Nutrient Availability,* John Wiley & Sons, New York, 1984.
Bell, P. F., Chaney, R. L. and Angle, J. S., *Plant Soil,* 130, 51–62, 1992.
Bingham, F. T., Sposito, G., and Strong, J. E., *J. Environ. Qual.,* 13, 160–165, 1984.
Brummer, G. W., In *The Importance of Chemical Speciation in Environmental Process,* Bernhard, M., Brinkman, F. E., and Sadler, P., Eds., Springer-Verlag, Berlin, 186, 1986.
Chang, A. C., Granato, T. C., and Page, A. L., *J. Environ. Qual.,* 21, 521–536, 1993.
Checkai, R. T., Corey, R. B., and Helmke, P. A., *Plant Soil,* 99, 335–345, 1987.
Corey, R. B., King, L. D., Lue-Hing, C., Fanning, D. S., Street, J. J., and Walker, J. M., In *Land Application of Sludge,* Page, A.L., Logan, T.J., and Ryan, J.A., Eds., Lewis Publishers, Chelsea, MI, 25, 1987.
Friberg, L., Pescator, M., and Nordberg, G., *Cadmium in the Environment,* Chemical Rubber Co. Press, Cleveland, Ohio, 1971.
Fergusson, J. E., *The Heavy Elements: Chemistry, Environmental Impact and Health Effects,* Pergamon Press, Oxford, 614 pp., 1990.

Galal-Gorchev, H., *Food Additives Contaminants,* 8, 793–806, 1991.

Halvorson, A. D. and Lindsay, W. L., *Soil Sci. Soc. Am. J.,* 41, 531–534, 1977.

Hamilton, E. I., In *Pollutant Transport and Fate in Ecosystems,* Spec. Pub. No. 6, Coughtrey, P. J., Mattin, M. H., and Unsworth, M. H., Eds., British Ecological Society, Blackwell Scientific, Oxford, 414 pp., 1987.

Hou, S. F., Zhu, Z. Y., and Tan, J. A., *Acta Geogr. Sin.,* 39, 75, 1980.

Jaffiol, H. N., Perezi, N., Baylet, R., and Baldet, L., *Bull. Acad. Natl. Med.,* 176(N4), 557–567, 1992.

Jackson, D. and Smith, A. D., In *Pollutant Transport and Fate in Ecosystems,* Spec. Publ. No. 6, Coughtrey, P. J., Mattin, M. H., and Unsworth, M. H., Eds. British Ecological Society, Blackwell Scientific, Oxford, 414 pp., 1987.

Jenny, H., *The Factors of Soil Formation,* McGraw-Hill, New York, 281 pp., 1941.

Jones, K. C., Keating, T., Diage, P., and Chang, A. C., *J. Environ. Qual.,* 20, 317–329, 1991.

Kabata-Pendias, A. and Pendias, H., *Trace Elements in Soils and Plants,* CRC Press, Boca Raton, FL, 365, 1992.

Kabata-Pendias, A. and Pendias, H., *Trace Elements in Soils and Plants,* CRC Press, Boca Raton, FL, 315, 1984.

Li, Y. H., *Schweiz. Z. Hydrol.,* 46, 1–8, 1984.

Lindsay, W. L., *Chemical Equilibria in Soils,* Wiley-Interscience, New York, 449 pp., 1979.

McCarty, L. S. and MacKay, D., *Environ. Sci. Technol.,* 27(9), 1719–1728, 1993.

McKone, T. E. and Ryan, P. R., *Environ. Sci. Technol.,* 23, 1154–1163, 1989.

Minnich, M. M., McBride, M. B., and Chaney, R. L., *Soil Sci. Soc. Am. J.,* 51, 573–578, 1987.

Mullins, G. L. and Sommers, L. E., *Plant Soil,* 96, 153–164, 1985.

Nieboer, E. and Richardson, D. H., *Environ. Pollut. Ser. B,* 1, 3–26, 1980.

Ochiai, E., *Bioorganic Chemistry: An Introduction,* Allyn and Bacon, Boston, MA, 515 pp., 1977.

Page, A. L., Fate and Effects of Trace Elements in Sewage Sludge When Applied to Agricultural Land: A Literature Review Study, U. S. Environmental Protection Agency, Cincinnati, OH, 105 pp., 1974.

Page, A. L. and Sposito, G., In H. Sigel (Ed.), *Metal Ions in Biological Systems,* Vol. 18, Marcel Dekker, New York, 287, 1985.

Parker, D. R., Norvell, W. A., and Chaney, R. L., *Soil Chemical Equilibria and Reaction Models,* Lippert, R.H. et al., Eds., Soil Science Society of America, Madison, WI, 1993.

Piemental, G. C. and Coonrod, J. A., *Opportunities in Chemistry: Today and Tomorrow,* National Academy Press, Washington, D.C., 244 pp., 1987.

Perwak, J. H., Ong, J. H., and Whelan, R., EPA560/5–85/008, U.S. Environmental Protection Agency, Washington, D.C., 1986.

Prasad, A., Ed., *Progress in Clinical and Biological Research,* Wiley-Liss, New York, 1993.

Ryan, J. A., Pahren, H. R., and Lucas, J. B., *Environ. Res.,* 28, 251–302, 1982.

Sparks, D. L., *Soil Sci. Soc. Am. J.,* 48, 415–418, 1984.

Sposito, G., *The Thermodynamics of Soil Solutions,* Oxford University Press, New York, 223 pp., 1981.

Sposito, G. and Mattigod, S. V., Kearney Foundation of Soil Science, University of California, Berkely, 1980.

Tebeb, H. N., *Am. J. Clin. Nutr.,* 57, S315–S316, 1993.

Tessier, A., Campbell, C. P. G., and Bisson, M., *Anal. Chem.,* 51, 844–851, 1979.

Tinsley, I. J., *Chemical Concepts in Pollutant Behavior,* Wiley-Interscience, New York, 265 pp., 1979.

U.S. Environmental Protection Agency, *Federal Register,* 51, 34042–34054, 1986.

U.S. Environmental Protection Agency, *EPA Environmental News,* April 12, 1989

U.S. Environmental Protection Agency, Toxic Chemical Release Inventory CD-Rom Database, U.S. Environmental Protection Agency, Washington D.C., 1993a.

U.S. Environmental Protection Agency, *Federal Register,* 58, 9248–9415, 1993b.

World Health Organization, WHO Offset Publ. No. 87, World Health Organization, Geneva, Switzerland, 102 pp., 1989.

Xian, X., *Plant Soil,* 113, 257–265, 1989.

Yang, G. Q., Yin, T. A., Sun, S. Z., Wang, H. Z., and You, D. Q., *Chin. J. Prev. Med.,* 14, 14–25, 1980.

Chapter 3

Assessment of Metals in Drinking Water With Specific References to Lead, Copper, Arsenic, and Selenium

Anna M. Fan

I. INTRODUCTION

The occurrence of metals as chemical contaminants in drinking water and the potential human exposure to these chemicals with the associated toxicological impacts have been receiving increasing attention. This attention is partly due to the availability of new data indicating the need to reevaluate the health effects and to revise the existing drinking water standards for certain metals, and partly due to the finding of these chemicals at concentrations exceeding the current standards.

The toxicological evaluation of chemical contaminants has involved two major program activities: one is the development of permissible levels or water quality standards for chemicals in drinking water in order to ensure the safety of the drinking water supply for the general public. The other is the investigation of chemical contamination situations which are chemical and location specific, and which often involve a more restrictive geographic location, human subpopulation, and other factors which affect the extent of contamination, risk assessment, and risk management options that follow. The present discussion will focus on a description of the methodology for regulatory standards development for drinking water, followed with an update of the status of reevaluation of four metals of current public health interest, namely lead, copper, arsenic, and selenium. It will also briefly point out selected situations of water contamination associated with these chemicals.

II. DRINKING WATER REGULATIONS IN THE UNITED STATES

In the U.S. the Environmental Protection Agency (EPA) has national authority in carrying out the provisions of the Safe Drinking Water Act (SDWA, PL 93–523), which was passed in Congress in 1974 and has since been revised several times. The National Interim Primary Drinking Water Regulations (NIPDWR), effective in 1977, included ten inorganic constituents, including metals. The 1986 amendment requires EPA to establish regulations for 83 drinking water contaminants. Since then, EPA has promulgated National Primary Drinking Water Regulations (NPDWR) for volatile organic chemicals and fluoride and maximum contaminant level goals (MCLGs) for a group of inorganic and organic chemicals, pesticides, and microbial contaminants. To improve the existing regulations, the reauthorization of the

SDWA needs to improve funding and provide flexibility for state programs to meet federal regulations, and reconsider the numerical objectives of developing 25 standards every 3 years.

The various drinking water standards and criteria terminology describe the different bases for the supporting documentation and different regulatory impacts associated with each of those terms. An understanding of these terms as described below will help to interpret and use the standards and criteria appropriately.

The Maximum Contaminant Levels (MCLs) are the maximum permissible levels of chemical contaminants in water which enter the distribution system of a public water system. (MCLs for bacteria and trihalomethanes are measured in the distribution system.) These are legally enforceable standards which must be met by all public drinking water systems to which they apply. These are generally derived with consideration of the animal and human health effects data, pharmacokinetics and exposure parameters, along with balancing the technologic and economic concerns that are directly related to the use of water for domestic supplies. No long-term adverse health effects are anticipated with daily ingestion at these levels, and they allow an adequate margin of safety after a comprehensive data evaluation and risk assessment. The derivation of the MCLs involves a comprehensive risk assessment, and the adoption of MCLs goes through a rigorous regulatory process involving public notification.

The Maximum Contaminant Level Goals (MCLGs) are maximum levels of contaminants in drinking water for lifetime consumption, which are nonenforceable health goals and are strictly health based, and do not include a technical feasibility or economic evaluation. For example, the MCLG for a carcinogen is zero, but it is neither technically or economically feasible to set a zero value and, therefore, a MCL is set and enforced. A full risk assessment is also conducted.

The MCLs can be classified as primary or secondary. The primary MCLs are health based, while the secondary MCLs are based on aesthetic effects and are maintained to protect public welfare and to assure a supply of pure, wholesome, and potable water. They are applied at the point of delivery to the consumer and generally involve protection of the taste, odor, or appearance of drinking water (e.g., toluene). The state may adopt the federal MCLs or establish independent ones under its own program that can either be the same or more stringent than federal MCLs. Federal secondary MCLs are nonenforceable. However, state secondary MCLs are enforceable for all new systems and new sources developed by existing systems.

The Action Levels (ALs) are interim guidance levels that are health based. They have not gone through the thorough risk assessment or vigorous regulatory process, but they do take into account analytical detection levels. Often ALs are established to provide guidance to drinking water programs before adequate data, time, or resources are available to develop a MCL, especially when the chemicals involved are often detected during water monitoring. They may trigger mitigation action on the part of a water purveyor. Public notification is not required when an AL is exceeded, but may be recommended. An AL is replaced by an MCL once the latter is promulgated and becomes final.

Health Advisories (HA) are guidance levels developed by EPA for 1-day, 10-day, or longer-term exposures for children as well as for longer-term or lifetime exposure for adults. Discrepancies between longer term and lifetime values may occur due to the Agency's conservative policies, especially with regard to carcinogenicity, relative source contribution, and less than lifetime exposures in chronic toxicity testing. The interplay of these factors is further described below. These HAs are not enforceable.

A. DRINKING WATER REGULATIONS IN CALIFORNIA

In California, the Department of Health Services (DHS) has obtained "primacy" or the authority to develop and administer the state's drinking water program. As a requirement, DHS must adopt water standards that are the same or more stringent than EPA within the time period provided by the regulations. The Department has been promulgating regulations under the California Safe Drinking Water Act, many of which are more stringent than the federal regulations. The Division of Water and Environmental Management is responsible for adopting the regulations and the monitoring and enforcement activities. It also evaluates the technical feasibility of water treatment methods and calculates the cost to utilities and consumers of different treatment options. The Pesticide and Environmental Toxicology Section (PETS) of the Office of Environmental Health Hazard Assessment, California Environmental Protection Agency, is responsible for the risk assessment functions, which evaluate the health effects data on the chemicals and recommends health based values for establishing water standards for California. The chemicals being evaluated or for which risk assessment is performed include pesticides, organics, inorganics, and industrial solvents. The procedure involved in establishing drinking water standards is shown in Figure 1.

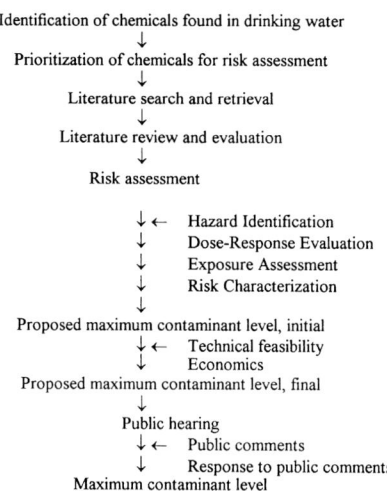

Figure 1 Procedure for development of chemical drinking water standards.

The 1989 revision to the California Safe Drinking Water Act (AB 21) requires an updated review and evaluation of the California MCLs every 5 years. In addition, it requires the development of recommended public health levels (RPHLs) that are health based only without the consideration of technical and economic factors. Some of these are likely to be the same as the MCLs, especially for the noncarcinogens. Chemicals that are carcinogens may have lower values for RPHLs compared to the MCLs, the latter of which has technical and economic limitations. This is due to the general finding that carcinogenicity is the most sensitive endpoint and, thereby, a permissible exposure limit that is adequately protective for carcinogenicity would also be protective of other toxicological endpoints. Health based negligible risk standards would lead to nonzero MCLGs for some carcinogens. For potent carcinogens (e.g., arsenic), however, the permissible limits to be established may be influenced by the sensitivity of the analytical methodology (i.e., detection limit or technical feasibility) and the cost for implementation of a stricter standard (i.e., economics). The law is designated to improve water quality as technical and economic factors become more favorable.

III. RISK ASSESSMENT OF WATER CONTAMINANTS

The considerations for assessment and management of health risks from carcinogenic and noncarcinogenic chemicals found in drinking water have been described by the U.S. EPA (1987, 1990) and the National Academy of Sciences (1986, 1987). The scientific process for risk assessment used for the setting of permissible levels for chemicals in drinking water have recently been described (Lam et al., 1994) and further shown in Figures 2 and 3.

Two approaches are used for risk assessment of chemicals in drinking water: one for noncarcinogens (e.g., selenium, lead — some lead compounds are carcinogenic in animals — and mercury) (see Figure 2) and one for carcinogens (e.g., arsenic) (see Figure 3). For risk assessment of all chemicals, the entire toxicological data base is evaluated. This would include the evaluation of the epidemiological data and experimental dose-response data on acute toxicity, subacute toxicity, chronic toxicity (e.g., systemic effects, target organ toxicity), reproductive toxicity, teratogenicity, genotoxicity, immunotoxicity, neurotoxicity, and carcinogenicity. For noncarcinogens, data from chronic studies and an uncertainty factor approach are used based on a threshold phenomenon for the critical health endpoint of concern. Subchronic studies may be used with the application of an additional uncertainty factor. Pharmacokinetic data, when available and when determined to be adequate, are used for dose adjustment. For carcinogens assumed not to have a threshold in the absence of convincing data, a commonly used approach is mathematical modeling for quantitative risk assessment. Various models are used for carcinogenic risk assessment, a common one being the linearized multistage model. A potency factor for tumor induction is developed based mostly on animal data extrapolated to humans, taking into consideration the animal-to-human extrapolation, exposure pattern, dose, and body weight to surface area conversions. Water consumption is assumed to be 2 l per day for an adult. A 20% relative source contribution (RSC) from

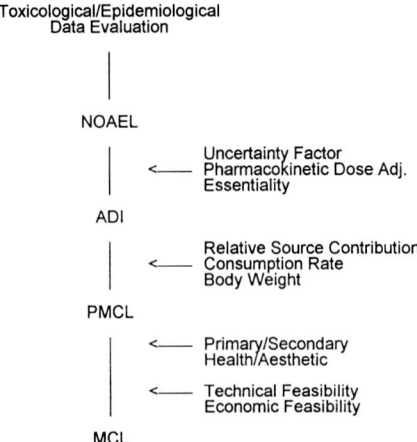

Figure 2 Development of water standards for noncarcinogens.

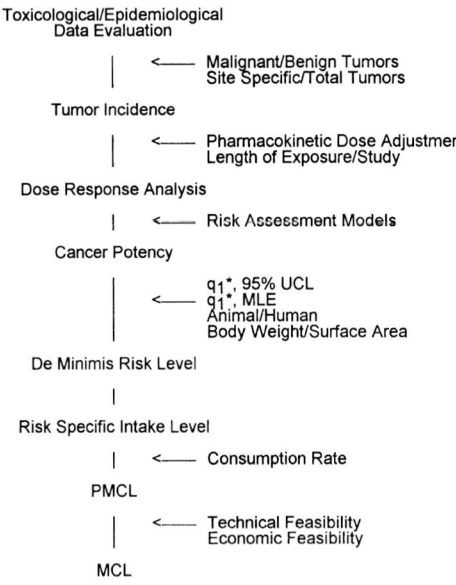

Figure 3 Development of water standards for carcinogens.

drinking water is often assumed, unless these are adequate data to indicate a more precise value. Diet is often a major source of overall exposure, such as in the case of lead and selenium. A full risk assessment is performed for each chemical for drinking water standard development.

For noncarcinogens, the first step in the risk assessment process, hazard identification, would allow an identification of one or more health effect endpoints of concern. A no-observable-adverse-effect level (NOAEL) or lowest-observable-adverse-effect level (LOAEL) is then derived for the critical endpoint as a reference exposure value for comparison to an individual's total exposure from all sources to the chemical. The reference value sets a limit which should not be exceeded by an individual for the purpose of health protection. Application of the uncertainty factor and RSC would provide a calculation of the permissible level in water as follows:

$$\frac{(\text{NOAEL/LOAEL, mg/kg-d}) (70\text{-kg}) (\text{RSC})}{(\text{UF}) (W, 1/d)}$$

The NOAEL or LOAEL values are in milligrams per kilogram per day. The standard adult human body weight is 70 kg. RSC is the relative source contribution, usually assumed to be 20% or 0.2, if adequate data on other sources of human exposure such as diet are absent but can reasonably be believed to contribute to major exposure. The RSC can be 100% if water is believed to be the likely sole source. UF is the overall uncertainty factor. It is usually the product of tenfold uncertainty factors reflecting the quality of data on which the NOAEL or LOAEL is based. The lesser the quality, the greater is the factor to account for the uncertainty. Uncertainty factors are also applied to account for interspecies (animal to human) and intraspecies (individual human sensitivity) differences. However, deviations from the use of the tenfold factors are sometimes seen, as in the cases of selenium, lead, copper, and arsenic (see below). Guidelines for selection of uncertainty factors can be found in the reports of Dourson and Stara (1983) and the U.S. EPA (1990). W is the daily water consumption rate of 2 l/d for drinking water only (oral) for an adult. In establishing permissible levels of metals (which are essential trace elements) in drinking water, often the nutritional essentiality is one factor to be considered, such as in the case of selenium and copper. For arsenic, it is believed to have some nutritional essentiality but this has not been clearly established.

For carcinogens, risk assessment guidelines were developed by the U.S. EPA in 1986. Since then, the methodology has undergone considerable debate regarding issues such as the use of data from combined tumor types and the use of which ratio for animal-to-human body surface conversion. The strength of carcinogenic activity of a chemical, expressed as carcinogenic potency, is first determined. The human carcinogenic potency for a chemical can be estimated directly from human epidemiological data or from experimental animal data. However, human data are rarely available or adequate for cancer risk assessment. Therefore, animal data are often used to derive the human potency estimate, using a mathematical model (e.g., the linearized multistage model) fitted to the animal tumor incidence data from carcinogenicity studies in rodents. This default approach uses the 95% upper-bound estimate (95% Upper Confidence Limit, UCL) of the low-dose slope obtained from animal dose-response data, assuming linearity at very low doses. In fitting the linearized multistage model to the animal cancer data, experimental data on dosing regimen are converted to lifetime averaged daily doses. When available and appropriate, other tissue dosimetry, which include pharmacokinetic behavior, are also considered. In cases of a shorter than lifetime study duration, the potency values are corrected for intercurrent mortality by the factor $(L/Le)^3$ where Le is the duration of the experiment and L is the life span of the animal, usually 102 weeks is used for rats and mice. To extrapolate animal potency data to humans, a conversion based on the surface area or body weight ratio is used as follows:

$$q_1^*(H) = q_1^*(A)(W_H/W_A)^{1/3}$$

where $q_1^*(H)$ is the human potency in (mg/kg-d)$^{-1}$, $q_1^*(A)$ is the animal potency in (mg/kg-d)$^{-1}$, W_H is the standard adult human body weight in kg (e.g., 70 kg), and W_A is the animal (e.g., rat or mouse) weight in kg.

The adjustment for species differences is made using the cube root of the body weight ratio based on the assumption that the metabolic rate is proportional to the 2/3 power of body weight (Crump, 1984). U.S. EPA is currently proposing the use of a cross-species scaling factor for carcinogen risk assessment based on the equivalence of mg/kg$^{3/4}$/day (EPA, 1992) as part of the efforts in achieving consistency in risk assessment among all agencies. In the carcinogenicity assessment of metals, one issue being debated is whether a positive response in one route (e.g., inhalation or intramuscular injection) should be extrapolated to another route (e.g., oral), as in the case of chromium and cadmium. For lead, although some of its compounds are evaluated to be carcinogenic, this endpoint has not been used to establish exposure criteria.

Using the potency determined from above, the permissible level in drinking water for a carcinogen can be derived as follows:

$$\frac{(\text{Risk})(70\ \text{kg})(\text{RSC})}{[q_1^*(H), (\text{mg/kg-d})^{-1}(W, \text{l/d})]}$$

Risk is the theoretical de minimus (or insignificant) individual excess lifetime cancer risk of 10^{-6}; 70 kg is adult body weight; RSC is the relative source contribution often, but not always, assumed to be 1.0; $q_1^*(H)$ is the human potency in mg/kg-d^{-1}, and W is the water consumption rate in liters per day or liter equivalents per day if additional exposure routes other than drinking water are considered.

IV. HEALTH EFFECTS OF METALS AND DRINKING WATER STANDARDS

The national drinking water standards, or MCLs, established by the U.S. EPA (1993) for selected metals are shown in Table 1. The health effects associated with exposure to these chemicals are also presented. Selected findings reported on metals in drinking water in California and the estimated exposed population are shown in Table 2. These are chemicals found to exceed the drinking water standards. A knowledge of the size of the population exposed would help to provide a better perspective in the risk characterization process. Biological monitoring is often used to evaluate human exposure to metals. For this purpose the reference values for metals in biological media are shown in Table 3. Blood and urine are the two common biological media used for analysis. The values are often affected by dietary intakes and one medium may be more reflective of current or past exposure than the other, depending on the specific metal.

A. LEAD

Recent concern relating to lead is the belief that there is no threshold for the toxicity of lead, as more recent data indicate effects at exposure levels below those previously established.

The toxicology of lead in children and adults has been described extensively and more recently by ATSDR (1988) and CDC (1991). Lead causes various health effects on virtually every system in the body, and the severity is dependent on the level of exposure and individual sensitivity. Children are at higher risk for lead exposure than adults because they have more hand-to-mouth activity and absorb lead at a higher level. At low levels of exposure, lead produces developmental/behavioral effects in children, primarily seen as decrements in cognition. High blood pressure, or hypertension, was observed in adult males. Increasing exposures lead to kidney damage, anemia, encephalopathy, and death.

The indicator of human exposure and the associated concern exposure level to lead has been based on the level of lead in blood related to lead intake rather than a direct estimation of a specific intake level. The blood level considered to indicate lead toxicity has progressively shifted downward as more information becomes available on childhood exposure and related responses (CDC 1991). An evaluation of 24 major cross-sectional studies strongly support the hypothesis of an inverse relationship between children's intelligence quotient (IQ) scores and lead burden (Needleman and Gatsonis, 1990). Based on these studies correlating blood lead levels with decreased IQ in children, the Centers for Disease Control (CDC, 1991) has identified a blood level of ≥ 10 µg/dl to be the concern level, equivalent to a lowest-observed-adverse-effect level. The implications of varying blood lead levels in children are as follows:

1. ≥ 9 µg/dl, not lead poisoned
2. 10–14 µg/dl, need frequent rescreening
3. 15–19 µg/dl, need nutritional intervention and more frequent screening
4. 20–44 µg/dl, need medical evaluation and maybe pharmacologic treatment of lead poisoning
5. 45–69 µg/dl, need medical intervention, including chelation therapy
6. >70 µg/dl, need immediate medical management.

The epidemiology data on children also show that the blood lead level increases by 0.16 µg/dl blood for every additional microgram per day of lead intake. This relationship in blood vs. intake level has been used to predict blood lead levels with increasing dietary intake (Ryu et al., 1983, Mahaffey, 1985). The blood lead level of 10 µg/dl corresponds to a daily intake of 62.5 µg for children. Lead-based paint in old homes is the most common source of lead exposure for children. The average dietary lead intake for a 2-year-old child has decreased from 30 µg/dl in 1982 to about 5 µg/dl in the period of 1986 through 1988 (CDC, 1991). Drinking water contributes from 5 to 50% of total lead exposure in children (EPA, 1991). Contamination of drinking water with lead usually occurs in the distribution system. The 1986 Safe Drinking Water Act Amendment banned the use of lead in public drinking water distribution systems and limited the lead content of brass used for plumbing to 8%. Recently, EPA eliminated the MCL of 50 µg/l for lead and established a MCLG of zero and an action level of 15 µg/dl at the tap.

Since lead intake has no beneficial value, any level of lead intake will increase the health risk from lead toxicity and add on to the baseline level of blood lead which, in adults, could be close to or exceeding the concern level of ≤ 10 µg/dl. All U.S. children are at risk for lead poisoning. Based on a 1984 estimate, it is believed that 17% of all U.S. children younger than 6 years old (between 3 and 4 million) had blood levels above 15 µg/dl (ATSDR, 1988). Recent data from the National Health and Nutrition Examination Survey (NHANES) showed that blood lead levels fell dramatically for all groups in the U.S. population between the study periods of 1976 to 1980 (NHANES II) and 1988 to 1991 (NHANES III, Phase 1)

Assessment of Metals in Drinking Water

Table 1 Drinking Water Standards and Health Advisories for Selected Metals

Chemical	Standards		Health advisories							Potential health effects
	MCLG (mg/L)	MCL (mg/L)	10 kg Child				70 kg Adult			
			One day (mg/L)	Ten day (mg/L)	Longer term (mg/L)	Longer term (mg/L)	RfD (mg/kg/day)	DWEL (mg/L)	Lifetime (mg/L)	
Aluminum	—	—	—	—	—	—	—	—	—	—
Antimony	0.006	0.006	0.015	0.015	0.015	0.015	0.0004	0.015	0.003	Decreased longevity, blood effects, gastrointestinal effects (h), liver and kidney effects (a)
Arsenic	—	0.05	—	—	—	—	—	—	—	Skin (hyperpigmentation, keratosis). Vascular complications. Neurotoxicity, liver injury (h). Reproductive/developmental effects (a)
Barium	2	2	—	—	—	—	—	—	2	Blood pressure effects
Beryllium	0.004	0.004	30	30	4	20	0.005	0.2	—	Contact dermatitis, pulmonary effects (h), skeletal effects, genotoxicity (a)
Boron	—	—	4	0.9	0.9	3	0.9	3	0.6	Testicular atrophy, spermatogenic arrest (a)
Cadmium	0.005	0.005	0.04	0.04	0.005	0.02	0.0005	0.02	0.005	Pulmonary and renal tubular effects, skeletal changes associated with effects on calcium metabolism (h), reproductive/teratogenic effects, effect on myocardium (a)
Chromium	0.1	0.1	1	1	0.2	0.8	0.005	0.2	0.1	Renal tubular necrosis (h), genotoxicity (a)
Copper	1.3	1.3 (AL)	—	—	—	—	—	—	—	Gastrointestinal effects (h), genotoxicity (a)
Lead (at tap)	zero	0.015 (AL)	—	—	—	—	—	—	—	Children — neurological, neurobehavioral, developmental effects Adult — peripheral and/or chronic neuropathy, hypertension also gastrointestinal, hematological, and renal effects (h), genotoxicity (a)
Manganese	0.2	—	—	—	—	—	0.14/0.005	—	—	Central nervous system effect (h). Biochemical changes in brain (a)
Mercury	0.002	0.002	—	—	—	0.002	0.0003	0.01	0.002	Kidney, central nervous system effects (h), genotoxicity (a)
Molybdenum	—	—	0.08	0.08	0.01	0.05	0.005	0.2	0.04	—
Nickel	0.1	0.1	1	1	0.5	1.7	0.02	0.6	0.1	Liver effects contact dermatitis (h), reproductive effects, genotoxicity (a)

Table 1 (continued)

| Chemical | Standards | | Health advisories | | | | | | | Potential health effects |
| | MCLG (mg/L) | MCL (mg/L) | 10 kg Child | | | 70 kg Adult | | | | |
			One day (mg/L)	Ten day (mg/L)	Longer term (mg/L)	Longer term (mg/L)	RfD (mg/kg/day)	DWEL (mg/L)	Lifetime (mg/L)	
Selenium	0.05	0.05	—	—	—	—	0.005	—	—	Nail changes, hair loss, skin lesions, nervous system effects (h), reproductive effects, genotoxicity (a)
Silver	—	—	0.2	0.2	0.2	0.2	0.005	0.2	0.1	Argyria (medically benign but permanent skin discoloration)
Strontium	—	—	25	25	25	90	0.6	90	17	—
Thallium	0.0005	0.002	0.007	0.007	0.007	0.02	0.00007	0.002	0.00004	Kidney, liver, brain, intestine effects
Vanadium	—	—	—	—	—	—	—	—	—	—
Zinc	—	—	6	6	3	12	0.3	11	2	Gastrointestinal distress, diarrhea (h), genotoxicity (a)

Notes: MCLG = Maximum Contaminant Level Goal.
MCL = Maximum Contaminant Level.
RfD = Reference dose, estimated doses that is likely to be without appreciable risk of deleterious effects during a lifetime.
DWEL = Drinking Water Equivalent Level.
AL = Action Level.

Potential health effects: h = based on human data; a = based on animal data, except for genotoxicity which is based on *in vitro* assays.
Standards and advisories are based on information from EPA (1992). Health effects information is based on literature review.

(Brody et al., 1994; Pirkle et al., 1994). The decline is attributed to the removal of leaded gasoline and lead-soldered cans. The geometric mean blood lead level declined from 12.8 to 2.8 µg/dl for all persons and from 15 to 3.6 µg/dl for children to 5 years. The percentage of all persons with a blood lead level higher than

Table 2 Populations Exposed to Chemicals Exceeding MCL in Active Drinking Water Sources in California

Chemical	A Active sources sampled in systems > MCL	B Active sources > MCL	C Total population of affected systems	D Potentially exposed population
Mercury	279	25	17,945,528	1,608,022
Cadmium	251	33	3,774,689	496,274
Chromium (total)	190	15	3,904,527	308,252
Selenium	328	70	1,084,744	231,500
Aluminum	235	37	962,672	151,570
Arsenic	239	40	621,041	103,940
Lead	204	15	489,225	35,972
Silver	28	4	210,423	30,060
Barium	13	4	4,500	1,292

Note: D = (B/A) × C

From Storm, D., *Water Contamination and Health*, Wang, R., Ed., Marcel Dekker, New York. With permission.

Table 3 Human Reference Values for Biological Monitoring

	Normal Range	
Chemical	Blood	Urine
Aluminum	<2.5–7 µg/L serum	
Antimony		<1–1 µg/L 0.5–2.6 µg/24 h
Arsenic		10–300 µg/L
Beryllium		2 ng/g fresh tissue
Cadmium	<1–4 µg/L	
Cobalt		0.5–7 µg/L 0.2–16 µg/24 h
Lead	7–22 µg/100 g	10–40 µg/L (could be up to 190 µg/L)
Nickel	0.28–0.68 µg/dL	4.5 µg/L
Mercury (Total and methyl mercury)	No cited value. Working range is 0.2–100 ng.	No cited value. Working range is 0.2–100 ng.
Selenium	18.2 µg/dL	34 µg/L

Note: Tabulated based on information from Kneip and Crable, Ed. 1988. *Methods for Biological Monitoring. A Manual for assessing Human Exposure to Hazardous Substances.* American Public Health Association, Washington D.C. Units are based on original units used in the reference.

the concern level declined from 77.8 to 4.3%, and the percentage for children 1 to 5 years of age declined from 88.2 to 8.9%. However, there is still a significant number of individuals with levels higher than the 10 µg/dl blood lead level, particularly in certain racial and ethnic groups.

The MCLG of zero reflects the goal to prevent any additional exposure to lead as technology becomes favorable. The action level of 15 µg/l will provide a blood level of 2.4 µg/dl in children at 1 l/day of water consumption. For adults, a similar concern level of 10 µg/dl is estimated for hypertension in males. Based on a correlation of 0.04 µg lead per deciliter of blood per additional 1 µg/day of lead intake for adults (Bolger, 1987), the action level would provide a blood lead level of 1.2 µg/dl in an adult at 2 l/day of water consumption. Alternately, the estimation of a drinking water concentration leading to a given steady-state blood lead level based on the use of clearance concepts is described by Bois et al. (1989).

B. COPPER

Copper is an essential element in humans (EPA, 1987). It can bind to proteins, such as cytochrome oxidase, and it is essential for many enzyme functions. Copper deficiency is associated with reduced hemoglobin formation, reduced elastin formation, teratogenesis, and abnormal amino oxidase activity, and it affects carbohydrate metabolism and catecholamine biosynthesis. On the other hand, toxicity can result from excessive exposure. Ingestion of copper has produced the typical acute symptomology of copper poisoning ranging from gastrointestinal (GI) disturbances, headache, dizziness and metallic taste in the mouth to death. Respiratory collapse, hemolytic anemia, hemoglobinemia, hepatic and renal failure, and GI bleeding have been reported. Chronic toxicity is characterized by studies of individuals with

Wilson's disease, an inborn error of metabolism found in 1 in 200,000 individuals, which results in copper accumulation in the liver, brain, kidney, and cornea. Hepatolenticular degeneration results from an excess retention of hepatic copper and impaired biliary copper excretion.

Children may be particularly susceptible to the toxicity of copper. Normal newborns can have up to eight times the levels of hepatic copper concentration than adults. Serious cases involving hepatic and renal necrosis, coma, and death were reported in Indian Childhood Cirrhosis (ICC), a condition affecting children under 5 years of age in India. Also, 13% of the American Negro male population have red blood cell glucose-6-phosphate dehydrogenase (G-6-PD) deficiency and may be at increased risk to the toxic effects of copper. Acute hemolytic anemia in hemodialysis patients has been attributed to excess copper in the analysis fluid. Individuals having occupational exposure to copper may be at additional risk from copper toxicity.

The human homeostatic mechanisms can prevent toxicity from the wide normal variations in copper by storing copper, preventing excessive absorption from the gut, utilizing the metal in cellular function, and excreting it through the bile or urine. Intoxication in humans mostly arises from accidental poisoning or suicide attempts, although poisoning from drinking water is uncommon. Chronic effects are not likely because the acute toxicity limits its intake, and copper does not tend to bioaccumulate in normal individuals. The lowest level of exposure reported to cause acute toxicity was a single oral dose of 5.3 mg. copper. Childhood cirrhosis occurred in infants and young children ingesting contaminated milk, which provided copper at 1 mg/kg body weight.

There are no data on carcinogenicity of copper in humans, and evidence of carcinogenicity in experimental animals is inadequate. Mutagenicity studies have reported mixed results, and thus no evaluation can be made of the mutagenicity of copper.

The current MCLG for copper (EPA, 1991) (see below) is based on the symptoms of nausea, vomiting, diarrhea, abdominal cramps, dizziness, and headaches resulting from poisoning at a cocktail party at the lowest dose of 5.3 mg (Wyllie, 1957). An uncertainty factor of 2 and a water consumption rate of 2 l/day were used to derive a level of 1.3 mg/l. Recurrent episodes of gastrointestinal problems reported in three out of four family members, including two children, ingesting water containing 7.8 mg/l of copper (Spitalny et al., 1984) would suggest a lower permissible level for children when adjusted for body weight and when a tenfold uncertainty factor is used.

Recently, "Blue Water" containing copper at 5 ppm or higher has been reported in several residential areas in California. The higher copper levels had been attributed at different times to the copper flux used in the water pipes, underground electromagnetic fields, and certain microorganisms. A clear determination for the cause of the elevated copper levels has not been made.

The final Lead and Copper Rule was promulgated by the EPA in June 1991 (56 FR 26460), with corrections published in July 1991 and June 1992. The effective date for monitoring was July 7, 1991. The remaining regulations, including action levels and treatment requirements, became effective in December 1992. For monitoring, the first-flush water samples from consumers' taps are to be analyzed. If more than 10% of these samples contain greater than the AL of 0.015 mg/l for lead or 1.3 mg/l for copper, three required actions must initially be taken. These requirements include corrosion control treatment, source water treatment, and public education. Then, if a system continues to exceed the lead action level, lead service lines will have to be replaced. The regulations also eliminate the MCL of 0.05 mg/l for lead and the secondary MCL of 1.0 mg/l for copper. Instead, the MCLGs of 0 and 1.3 mg/l were set for lead and copper, respectively.

C. ARSENIC

The health effects of arsenic have been described in several major documents (EPA, 1984; 1988). A review and risk assessment for the development of the California drinking water standard is reported by Brown and Fan (1994) and further summarized below. An oral intake of 10 mg of arsenic may be fatal to humans. The general health effects are vascular disorders, dermal effects, reproductive effects, and neurological disturbances. Subacute and chronic exposures affect the gastrointestinal tract, circulatory system, skin, liver, kidneys, nervous system, and heart. Adverse dermal effects such as hyperkeratosis, hyperpigmentation, and depigmentation are attributed to high arsenic levels in drinking water. In southwest Taiwan, Blackfoot disease is an endemic peripheral vascular disorder associated with consumption of arsenic in well water, although a causal relationship is being disputed.

Experimentally, arsenic compounds are fetotoxic and teratogenic in mice, rats, and hamsters. Common developmental effects seen include malformations of the brain, urogenital organs, skeleton, ear, as well as small or missing eyes. Generally, these are only seen at dose levels that also result in maternal toxicity.

Deficiencies in study design, reporting, and documentation have made it difficult to determine with confidence the effect or no-effect levels for developmental endpoints. In humans, conclusive evidence of human reproductive or developmental toxicity following arsenic exposure is lacking.

When tested for genotoxicity, arsenic compounds inhibited DNA repair and induced chromosome aberrations and sister chromatid exchanges. The compounds are usually negative in routine *in vitro* tests for mutagenicity. Recent findings suggested that inorganic arsenic may induce gene amplification in mammalian cells and may possibly affect the later stages of carcinogenesis.

When tested in animals, inorganic or organic arsenic compounds failed to exhibit significant carcinogenic activity. Knoth (1966) reported adenomas of the skin, lung, peritoneum, and lymph nodes in mice exposed for 5 months. Ishinishi et al. (1983) and Pershagen et al. (1984) reported an increased but low incidence of lung tumors in hamsters given arsenic trioxide by intratracheal instillation. The applicability of these studies to human environmental exposure is questionable. In humans, studies in Taiwan (Tseng, 1977; Tseng et al., 1968), South America (Zaldivar, 1974, Borgono et al., 1987), India (Cebrian et al., 1983), and Mexico (Chakraborty and Saha, 1987) supported an association between ingestion of arsenic in drinking water and the development of skin cancer and related disorders. Similar findings were not seen in studies conducted in the U.S. in Oregon (Morton et al., 1976), Utah (Southwick et al., 1984), Alaska (Harrington et al., 1978), California (Goldsmith et al., 1972), and Nevada (Vig et al., 1984).

The negative responses seen in animal carcinogenicity studies and gene mutation assays have generated suggestions that arsenic might act as a co-carcinogen or tumor promoter, rather than as a direct carcinogen. The possibility of a threshold has been discussed by some investigators (EPA, 1984; Marcus and Rispin, 1988; Stohrer, 1991; Petito and Beck, 1991), but it has also been argued that human data do not support the methylation threshold hypothesis (Hopenhayn-Rich et al., 1993). Recently, evaluation of dose-response data on occupational exposure to arsenic and lung cancer (Hertz-Picciotto and Smith, 1993) suggested that the use of linear models applied to epidemiological data may result in an underestimation of the true risk at lower exposures.

The recent increased interest in arsenic is due to the finding of an association with fatal internal cancers such as cancers of the bladder, lung, liver, and kidney based on data from Taiwan (Chen et al., 1986, 1988; Chen and Wang, 1990). Early investigations showed only an association of arsenic ingestion with generally nonfatal skin cancers and Blackfoot disease (Tseng et al., 1968; Tseng, 1977). On the basis of the EPA risk estimates, the current drinking water standard (Maximum Contaminant Level, or MCL) of 50 ppb would present a lifetime (skin) cancer risk level of greater than 1:1000 (2.5×10^{-3}). The recent findings on internal cancers at similar risk levels (Smith et al., 1992; Bates et al., 1992) have generated an increased concern regarding human exposure to arsenic in drinking water. A reduction in the drinking water standard is currently being considered at the federal level.

In 1988, EPA estimated the human cancer potency of arsenic to be 2×10^{-3}/mg/kg-d with a 10^{-6} lifetime skin cancer risk at 20 ppt based on 2 l per day water consumption. A potency estimate of 5.3×10^{-3}/mg/kg-d was derived by Smith et al. (1992) associated with a lifetime skin cancer risk of 8×10^{-3} at the MCL of 50 ppb and a risk of 1×10^{-6} at 2 ppt. Potency and risk estimates were higher for other internal cancers. Currently, efforts are being made to closely examine and validate the well sampling and dietary exposure data on arsenic obtained in Taiwan to determine the adequacy for use in exposure and risk assessment for carcinogenicity.

Arsenic occurs primarily in groundwater. In California, it is found frequently in the groundwater in the Central Valley and, therefore, affects water systems in that region which are mostly small, having less than 200 service connections, and serving about 600 persons or less (Fan et al., 1993). Some large systems (more than 200 service connections) are also affected, but these mostly have small distributions serving approximately 1500 persons or less. The monitoring data on arsenic in large water systems updated through April 1993 showed that most of the wells (710 among a total of 741 in 48 counties) have detectable levels of arsenic between 5 and 50 ppb; 24 wells had levels in the >50–100 ppb range, and 7 had levels >10–20 ppb ranges. In most counties, the well water contains arsenic within the >20–50 ppb range. Kings county has the highest number of levels exceeding the MCL. Data on arsenic levels in small water system sources is limited. A telephone survey (1991) of selected counties showed arsenic levels exceeding the MCL in San Benito, San Joaquin, Sacramento, Sonoma, Kern, and Kings counties. A reduction in the MCL, as is being considered currently, will be more health protective, but will have a major cost impact in affected areas.

D. SELENIUM

The toxicology and nutritional value of selenium have been described in several documents (EPA, 1989; Combs and Combs, 1986). Early data were primarily available on the effects on farm animals because selenium is added to the diet of poultry and livestock. Recent interest focuses on the identification of a reference value for human toxicity evaluation following reports of the following (Fan et al., 1988; Fan and Kizer, 1990): (1) human intoxication in China and U.S., (2) beneficial uses for Keshan and Kaschin-Beck diseases and after parenteral alimentation, and (3) environmental contamination in California (agricultural drainage water, fish, ducks). More recent studies provided information on deriving the maximal intake levels with no signs of toxicity.

The critical studies which provide quantitative data in relation to selenium toxicity are summarized as follows:

1. Thickened but fragile nails and characteristic dermal garlic odor were reported in an individual (in China) who consumed selenium tablets at 1 mg Se/day (1000 µg) for 2 years (Yang et al., 1983).
2. In a study of three separate geographical areas in China with low, medium, and high selenium levels in soil and food, the average dietary selenium intakes for 400 individuals were determined. Persistent clinical signs of selenosis were seen in a sensitive subpopulation at an intake of 1261 µg/day, and no signs of significant biochemical change or selenosis were found at 853 µg/day (or 1050 based on 70-kg body wt) (Yang et al., 1989).
3. In a study of selected geographical areas in South Dakota and Wyoming with high selenium in the soil and reported cases of livestock selenosis, physical examinations were conducted for 142 volunteers, including blood, urine, toenail, and food analyses. There were no effects on liver function and clinical chemistry or selenosis at intakes of up to 724 µg/day (Longnecker et al., 1991).
4. An investigation of endemic selenium intoxication in China involving about 200 individuals showed that estimated dietary ingestion of 3.2–6 mg/d (3200–6000 µg/day) was associated with hair and nail loss, skin lesions, tooth decay, and neurologic abnormalities (Yang et al., 1983).
5. An investigation of selenium intoxication involving 13 individuals in the U.S. following ingestion of improperly manufactured dietary supplement showed that ingestion of tablets containing about 4 to 31 mg of total selenium resulted in nausea, vomiting, nail changes, hair loss, fatigue, irritability, and paresthesia (FDA, 1984; CDC, 1984; Helzsouer et al., 1985).

On the basis of the above, chronic symptoms of toxicity are not likely to occur until the daily intake of selenium approaches 1000 µg/day. The transition state appears to lie between the range of 724 (U.S.) to 1050 µg/day (China). This association between selenium intake and the appearance of signs and symptoms of toxicity is further summarized in Figure 4. These values do not include an uncertainty factor which should be applied in establishing a permissible intake level not anticipated to have an effect. For this purpose, the U.S. Environmental Protection Agency uses a factor of 3 to derive a reference dose of 350 µg/day (1050/3). The same information data base is used by the National Academy of Sciences (1989) to recommend a recommended dietary allowance of 0.87 µg/kg, or 70 and 50 µg/day for males and females, respectively.

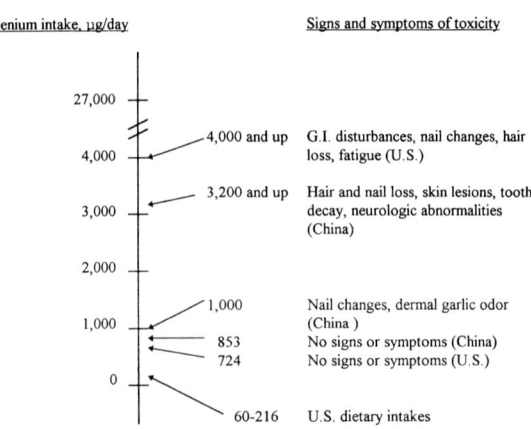

Figure 4 Selenium intake and associated signs and symptoms of toxicity.

Health effects from short-term dietary intake of selenium is not known. Gastrointestinal disturbances from eating nuts from a seleniferous area in Venezuela had been reported but quantitative data on the doses are not available. Toxicity from ingestion was reported in the supplement intoxication episode in the U.S. (see above) at the lowest single intake level of 4 mg/day. The lower limit for toxicity is not known. In animals, the lowest acute dose reported to have an effect (edema) was 2 mg/kg (2000 µg/kg) given by gavage to rabbits (140 mg/70-kg person). LD_{50} values varying from 1 to 8.6 mg/kg/d have also been reported in five different gavage studies in rats, mice, rabbits, and guinea pigs, but human data on the supplement intoxication episode did not show any lethality at up to 27 mg/day (27,000 µg/day).

The daily intake level leading to toxicity can be affected by the chemical form of selenium in the dietary items, which can vary in toxicity and bioavailability. For example, selenite has been shown to be more toxic than selenomethionine. Studies using exudative diathesis or pancreatic fibrosis in chicks indicated a superior availability of selenium from plant over animal source. Using plasma glutathione peroxidase activity to correlate selenium intake in Se-depleted chicks, selenite was found to be most available followed by selenomethionine, fish meal, corn meal, and soy bean meal. Human studies have shown selenomethionine, selenium-rich wheat, and selenium yeast to be more bioavailable than selenite. Often, measurements of selenium in dietary items are made or reported based on total selenium without the distinction made for different chemical forms. The quantitative information for identifying selenium intakes and the associated human toxicity is often, though not always, based on data reported on total selenium, not specific chemical forms. Data on water analysis are also mostly reported as total selenium.

Selenium administered in drinking water has produced reproductive toxicity in rats and mice and reproductive toxicity and teratogenicity in avian species. Both mutagenic and antimutagenic activities in experimental systems have been reported. Selenium sulfide was observed to be carcinogenic in rats and female mice by gavage, but the results cannot be extrapolated to other inorganic compounds (selenite or selenate). Other carcinogenicity studies have been deficient in experimental design and reporting and do not allow for a clear interpretation of the data.

Based on the recent studies in China and U.S. described above, EPA (1992) has revised its MCL for selenium from 10 µg/l to 50 µg/l. This is based on a reference dose of 5 µg/kg/day (350 µg/day), a dietary intake of 120 µg selenium, a water consumption rate of 2 l/day, and an uncertainty factor of 3. The health effects endpoints are those related to selenosis characterized by nail changes and hair loss. Individuals taking selenium supplementation will have a higher total intake.

V. SUMMARY

The considerations and procedures for the risk assessment of chemicals for the purpose of establishing permissible levels in drinking water are described. Case samples are presented for four metals, namely lead, copper, arsenic, and selenium. The updated toxicology and health basis for the establishment of the drinking water standards, including the supporting information for a revision of such standards, are discussed. The occurrence of water contaminants will result in higher health risks to some individuals than to others or higher than de minimis risk levels at exposures permissible under the present drinking water regulations. The challenge for health authorities is to maintain public health protection with programmatic support and efficient water regulations. Both the state and federal programs have worked to maintain credibility and public confidence in providing a safe and high quality public drinking water supply to the nation's population. The federal government is looked upon for providing leadership and guidance while flexibility at the state level is maintained to address specific local needs.

REFERENCES

Agency for Toxic Substances and Disease Registry (ATSDR), The nature and extent of lead poisoning in children in the United States: A report to the Congress. Atlanta, Georgia, 1988.

Bates, M. N., Smith, A. H., and Hopenhayn-Rich, C., *Am. J. Epidemiol.*, 135, 462–476, 1992.

Bois, F. et al., *Am. J. Public Health*, 79, 827, 1989.

Bolger, M., "Revision of Action Level for Lead (PB) in Large Ceramic Hollowware", memo to Ms. Evelyn Osman, Division of Regulatory Guidance, U.S. Food and Drug Administration, Washington, D.C., 1987.

Borgono, J. M., Vincent, P., Venturino, H., and Infante, A., *Environ. Health Perspect.*, 19, 326–334, 1987.

Brody, D., Pirkle, J., Kramer, R., Flegal, K., Matte, T., Gunter, E., and Paschal, D., *J. Am. Med. Assoc.*, 272, 277–283, 1994.

Brown, J. and Fan, A., *J. Hazardous Mater.*, 39, 149–159, 1994.

CDC (Centers for Disease Control), Selenium Intoxication. Centers for Disease Control MMWR 33:157, Atlanta, GA, 1984.

Cebrian, M. E., Albores, A., Aguilar, M., and Blakely, E., *Hum. Toxicol.*, 2, 121–133, 1983.

Centers for Disease Control (CDC), Preventing Lead Poisoning in Young Children, U.S. Department of Health and Human Services, Public Health Service, Washington, D.C., 1991.

Chakraborty, A. K. and Saha, K. C., *Indian J. Med. Res.*, 85, 326–334, 1987.

Chen, C.-J. and Wang, C.-J., *Cancer Res.*, 50, 5470–5474, 1990.

Chen, C.-J., Chuang, Y.-C., Lin, T.-M., et al., *Cancer Res.*, 45, 5895–5899, 1985.

Chen, C.-J., Chuang, Y.-C., You, S.-L., Lin, T.-M., and Wu, T.-M., *Br. J. Cancer,* 53, 399–405, 1986.

Chen, C.-J., Kuo, T.-L., and Wu, M.-M., *Lancet*, 1, 414, 1988.

Combs, G. F., Jr. and Combs, S. B., *The Role of Selenium in Nutrition,* Academic Press, San Diego, CA, 1986.

Crump, K. S., *J. Environ. Pathol. Toxicol. Oncol.*, 5, 339–348, 1983.

Dourson, M. L. and Stara, J. F., *Regul. Toxicol. Pharmacol.*, 3, 224–238, 1983.

Fan, A. M. and Kizer, K. W., *WJ Med.*, 153, 160–167, 1990.

Fan, A. M., Brown, J., Milea, A., and Spath, D., Arsenic in California drinking water: Toxicology, monitoring and regulatory issues. Presented at the First Int. Conf. Arsenic, New Orleans, July 17–18, 1993, 1993.

Fan, A. M., Neutra, R., Book, S. A., and Epstein, D., *J. Environ. Toxicol.*, 1988.

FDA (Food and Drug Administration), *Food Drug Bull.*, 14, 19, 1994.

Goldsmith, J. R., Deane, M., Thom, J., and Gentry, G., *Water Res.*, 6, 1133–1136, 1972.

Harrington, J. M., Middaugh, J. P., Morse, D. L., and Housworth, J., *Am. J. Epidemiol.*, 108, 377–385, 1978.

Helzlsouer, K., Jacobs, R., and Morris, S., *Fed. Proc.*, 44, 1670, 1985.

Hertz-Picciotto, I. and Smith, A. H. *Scand. J. Work Environ. Health*, 19, 217–226, 1993.

Hopenhayn-Rich, C., Smith, A. H., and Goeden, H. M., *Environ. Res.*, 60, 161–177, 1993.

Ishinishi, N., Yamamoto, A., Hisanaga, A., and Inamasu, T., *Cancer Lett.*, 21, 141–147, 1989.

Knoth, W., *Arch. Klin. Exp. Dermatol.*, 227:228–234, 1966.

Lam, R. H. F., Brown, J. P., Fan, A. M., and Milea, A., *J. Hazardous Mater.*, 39, 173–192, 1996.

Longnecker, M. P., Taylor, P. R., Levander, O. A., et al., *Am. J. Clin. Nutr.*, 53, 1288–1294, 1991.

Mahaffey, K., Dietary and Environmental Lead: Human Health Effects, pp. 1–459, 1985.

Marcus, W. L. and Rispin, A. S. in: *Advances in Modern Environmental Toxicology: Risk Assessment and Risk Management of Industrial and Environmental Chemicals,* Cothern, C.R. and Mehlman, M.A., Eds., Princeton Publishing, Princeton, NJ, 1988, 133.

Morton, W., Starr, G., Pohl, D., et al., *Cancer,* 37, 2523–2532, 1976.

National Academy of Sciences (NAS), in *Drinking Water and Health,* Vol. 6, National Academy Press, Washington, D.C., pp 250–293, 1986.

National Academy of Sciences (NAS), in *Drinking Water and Health,* Vol. 1, National Academy Press, Washington, D.C., pp 19–62, 1977.

National Academy of Sciences (NAS), in *Drinking Water and Health,* Vol. 8, National Academy Press, Washington, D.C., 1987.

National Academy of Sciences (NAS), *Recommended Dietary Allowances,* 10th ed., National Academy Press, Washington, D.C., 217, 1989.

Needleman, H. L. and Gatsonis, C. A., *J. Am. Med. Assoc.*, 263, 673–678, 1990.

Pershagen, G., Nordberg, G., and Bjorklund, N. E., *Environ. Res.*, 34, 227–241, 1984.

Petito, C. T. and Beck, B. D., in: *Trace Substances in Environmental Health,* Vol. XXIV, Hemphhill, D.D., Ed., University of Missouri, Columbia, 1991, pp. 143–176.

Pirkle, J., Brody, D., Gunter, E., Kramer, R., Paschal, D., Flegal, K., and Matte, T., *J. Am. Med. Assoc.*, 272, 284–291, 1994.

Ryu, J. R., Zeigler, E., Nelson, S., and Fomon, S., *Am. J. Dis. Child.*, 137, 886–891, 1983.

Smith, A. H., Hopenhagen-Rich, C., Bates, M., Goeden, H., Hertz, I., Allen, H., Hood, R., Kosnett, M., and Smith, M., *Environ. Health Perspect.*, 97, 259–267, 1992.

Southwick, J. W., Western, A. E., Beck, M. M., et al., in: *Arsenic: Industrial, Biomedical, Environmental Perspectives,* Lederer, W. H. and Fensterheim, R. J., Eds., Van Nostrand Reinhold Co., New York, pp. 210–225.

Spitalyny, K. C., Brondum, J., Vogf, R. L., Sargent, H. E., and Kappel, S., *Pediatrics,* 74, 1103–1106, 1984.

Stöhrer, G., *Arch. Toxicol.*, 65, 525–531, 1991.

Storm, D., *Water Contamination and Health,* Wang, R., Ed., Marcel Dekker, New York, 1994.

Tseng, W. P., *Environ. Health Perspect.*, 19, 109–119, 1977.

Tseng, W. P., Chu, H. M., How, S. W., Fong, J. M., Lin, C. S., and Yeh, S., *J. Natl. Cancer Inst.*, 40, 453–463, 1968.

U.S. Environmental Protection Agency (U.S. EPA), The Risk Assessment Guidelines of 1986. U.S. EPA, Office of Health and Environmental Assessment, Washington, D.C., 1987

U.S. Environmental Protection Agency (U.S. EPA), Workshops on Assessment and Management of Drinking Water Contamination, U.S. EPA/Office of Water, EPA/600/M-86/026, Washington, D.C., 1987.

U.S. Environmental Protection Agency (U.S. EPA), Risk Assessment, Management and Communication of Drinking Water Contamination, U.S. EPA/Office of Water, EPA/625/4–89/024, Washington, D.C., 1990.

U.S. Environmental Protection Agency (U.S. EPA), Maximum Contaminant Level Goals and National Primary Drinking Water Regulations for Lead and Copper; Final Rule, 40 CFR Parts 141 and 142, *Federal Register,* 56, 26460–26564, 1991.

U.S. Environmental Protection Agency (U.S. EPA), Maximum Contaminant Level Goals and National Primary Drinking Water Regulation for Lead and Copper; Final Rule. 40 CFR Parts 141 and 142, *Federal Register,* 56: 26460, 1991.

U.S. Environmental Protection Agency (U.S. EPA), Integrated Risk Information Service, U.S. Environmental Protection Agency, Washington, D.C., 1992.

U.S. Environmental Protection Agency (U.S. EPA), Special Report on Ingested Arsenic: Skin Cancer; Nutritional Essentiality, EPA/625/3–87/013, Washington, D.C., 1984.

U.S. Environmental Protection Agency (U.S. EPA), Drinking Water Criteria Document for Copper, U.S. EPA, Cincinnati, OH, 1987.

U.S. Environmental Protection Agency (U.S. EPA), Health Assessment Document for Inorganic Arsenic, EPA-600/8–83–021F, Environmental Criteria and Assessment Office, Washington, D.C., 1988.

U.S. Environmental Protection Agency (U.S. EPA), Drinking Water Regulations and Health Advisories, Office of Water, Washington, D.C., 1993.

U.S. Environmental Protection Agency, Arsenic, Inorganic. Iris on line file (last revised 1/1/92), U.S. EPA, Washington, D.C., 1992.

Vig, B. K., Figeroa, M. L., Cornforth, M. N., and Jenkins, S. H., *Am. J. Ind. Med.,* 6, 325–338, 1984.

Wyllie, J., *Am. J. Public Health,* 47, 617, 1957.

Yang, G., Wang, S., Zhou, R., and Sun, S., *Am. J. Clin. Nutr.,* 37, 872–881, 1983.

Yang, G., Yin, S., Zhou, R., Gu, L., et al., *J. Trace Elem. Electrolytes Health Dis.,* 3(2), 77–87, 1989a.

Yang, G., Yin, S., Zhou, R., Gu, L., et al., *J. Trace Elem. Electrolytes Health Dis.,* 3(2), 123–130, 1989b.

Zaldivar, R., *Beitr. Pathol.,* 151, 384–400, 1974.

Chapter 4

Risk Assessment of Metals

J. Michael Davis and Robert W. Elias

ABSTRACT

Risk assessments of two neurotoxic metals, lead and manganese, have been conducted by the U.S. Environmental Protection Agency in connection with the use of these metals in gasoline. Although some features of the two risk assessments are fundamentally similar, others are rather different, in large part due to differences in the respective databases of information for the two metals. Some of the key features of these assessments are highlighted to illustrate approaches used to assess the potential public health risks of metals.

I. INTRODUCTION

The process for characterizing the potential health risks of neurotoxicants is essentially the same as that for substances that primarily affect target organs or systems other than the nervous system. Basically, it involves the integration of health effects information with exposure information. As described by the well-known schematic of risk assessment by the National Research Council (1983), the health effects aspect of risk assessment consists of two steps: hazard identification and dose-response analysis. Hazard identification establishes the qualitative types of toxic effects that a substance can induce; for example, is it neurotoxic, immunotoxic, carcinogenic, or a fetal toxicant? Dose-response analysis defines the quantitative relationship between dose (or exposure) and response (or effect). The health effects assessment is then coupled with an assessment of human exposures to the substance in question. The integration of hazard identification and dose-response analysis with exposure assessment constitutes health risk characterization.

The risk assessment process depends fundamentally on the amount and quality of information available. As the amount and quality of information increase, the precision of the risk characterization increases, along with the degree of confidence appropriate to the characterization. If relatively little information is available on one or more aspects of the health effects and exposure assessments, uncertainties in the risk characterization increase. To illustrate how differences in the databases of information for a neurotoxic metal affect the risk characterization of the neurotoxicant, the assessments of two metals, lead and manganese, are discussed here.

The potential health risks of lead and manganese have been the subjects of separate assessments by the U.S. Environmental Protection Agency (EPA) primarily because of their use, or requested use, in gasoline. Organic compounds of lead, viz., tetraethyl lead (TEL) and tetramethyl lead (TML), and of manganese, viz., methylcyclopentadienyl manganese tricarbonyl (MMT), boost the octane rating of gasoline. Lead has the longer history of such usage in the U.S., dating to the 1920s; manganese has been allowed to some extent in leaded U.S. gasoline since the 1970s. With leaded gasoline being the predominant source of lead in the ambient air, EPA assessed the health risks of lead on various occasions over the past several years (e.g., U.S. Environmental Protection Agency, 1977, 1986, 1990). However, with the phasedown of lead in gasoline, the importance of this source of lead exposure has declined in relation to other sources (e.g., old leaded paint); meanwhile, requests to allow the expanded use of MMT in gasoline, specifically in unleaded gasoline, have been repeatedly submitted to EPA. In responding to the latest of a series of such requests, EPA denied the petitioner's request, citing concerns about potential public health risks and the lack of adequate information to evaluate such potential health risks quantitatively (Federal Register, 1994).

The EPA assessments of lead and manganese offer a study in contrasts in several respects. Although neither an in-depth nor comprehensive examination of the assessments is possible here, some of the more significant issues and considerations involved in the EPA assessments of these two neurotoxic metals are highlighted. Following separate summaries of the lead and manganese assessments, specific points of comparison are discussed.

II. RISK ASSESSMENT OF LEAD

Information on the nature of some of lead's effects on humans has existed for centuries, but a more complete, scientific understanding of the extent of these effects has only come about over the past few decades. Although lead affects multiple organs and systems of the body, its effects on the nervous system, especially the developing nervous system, have been of particular interest and concern. The developmental neurotoxicity of lead has been the focus, albeit not necessarily the exclusive focus, of several risk assessments by EPA and other agencies. In these assessments, the critical and most contentious issue has not been whether lead is neurotoxic to young children, for clearly it is. Rather, the issue under debate has been the level of exposure at which lead causes neurotoxic effects. Thus, the characterization of the dose-response relationship has received the most attention in the assessments of the health risks of lead.

Based on the collective evidence from numerous epidemiological and experimental studies of lead, including the findings of several independent prospective studies of neurobehavioral development in populations of children having rather typical levels of lead exposure, EPA (U.S. Environmental Protection Agency, 1986, 1990) concluded that a blood lead level of 10 µg/dl constituted a level of concern for adverse health effects such as delayed neurobehavioral development in children. The evidence underlying this conclusion has been described in more detail elsewhere (e.g., U.S. Environmental Protection Agency, 1986, 1990; Davis, 1990), but some key elements may be recounted here.

First, it should be noted that the most widely accepted indicator of lead exposure for dose-response characterization is the concentration of lead in whole blood. Although blood lead is an imperfect indicator of target tissue lead burden, it is the best available measure of low level lead exposure for risk assessment/management purposes, and it integrates exposure via all routes — inhalation, ingestion, and dermal. Several cross-sectional epidemiological studies of populations of lead-exposed children dating from the 1970s indicated that an association existed between contemporaneous blood lead levels and intelligence quotient (IQ) test performance. An evaluation of these studies by EPA (U.S. Environmental Protection Agency, 1986) concluded that about a 5-point IQ deficit existed at blood lead levels averaging around 50 to 70 µg/dl, about a 4-point deficit existed around 30–50 µg/dl, and about a 1- to 2-point deficit existed around 15 to 30 µg/dl. This judgment of the dose-response relationship was subjected to extensive review by the scientific community and was examined and approved by the Clean Air Scientific Advisory Committee, an independent body of expert scientists. Nevertheless, the possibility that this association was confounded by other variables could not be altogether eliminated with the available evidence at that time.

With the advent of long-term prospective studies of populations of lead-exposed children and methodological improvements in the measurement of blood lead concentrations, many of the confounding issues that plagued earlier cross-sectional studies were better controlled. For example, by periodically measuring blood lead levels, beginning pre- or perinatally and continuing at least through early childhood, the pattern and extent of a child's lead exposure could be much better characterized than was possible with a single measurement around age 5 to 8 years, as was often the case in the cross-sectional studies. Evaluation of the findings from several independent prospective studies (Table 1) led EPA to conclude

Table 1 Summary of Selected Cognitive Effects From Prospective Studies of Early Developmental Lead Exposure

Study	Blood lead measure Mean (range/±S.D.), μg/dl	Time/source	N	Endpoint	Effect	P Value
Boston (Bellinger et al., 1987a,b, 1989)	7.4 (±5.5)	Delivery/cord	201	6-mo MDI	−4.3 pts. at PbB ≥10 μg/dl	0.095
	"	"	199	12-mo MDI	−5.8 "	0.020
	"	"	187	18-mo MDI	−6.7 "	0.049
	"	"	182	24-mo MDI	−7.8 "	0.006
	6.8 (±6.3)	24-mo/child	170	57-mo GCI	−6.9 "	0.040
Cincinnati (Dietrich et al., 1987, 1988, 1990, 1991, 1992, 1993)	6.3 (±4.5)	Delivery/cord	96	3-mo MDI	−6.0 pts. per 10 μg/dl	0.02
	8.0 (±3.7)	Prenatal/mother	266	"	−3.4 "	0.05
	"	"	249	6-mo MDI	−8.4 " (males)	≤0.01
	4.6 (±2.8)	10-d/child	283	"	−7.3 " (lower SES)	≤0.03
	"	"	257	12-mo MDI	−6.2 "	0.04
	"	Prenatal/mother	237	24-mo MDI	+5.1 "	0.022
	"	10-d/child	270	"	−0.2 "	0.948
			247	4-yr K-ABC (MPC)	−6.3 "	≤0.01
			258	" (SIM)	−5.0 "	≤0.05
	14.1 (7.3)	4-yr/child	259	5-yr K-ABC (SIM)	−2.0 "	≤0.05
	<12 (<12)	6-yr/child	231	6-yr WISC-R (FS)	−3.3 "	≤0.05
	"	"	231	" (P)	−5.2 "	≤0.01
Cleveland (Ernhart et al., 1986, 1987; Wolf et al., 1985)	5.8 (2.6–14.7)	Delivery/cord	132	>1-d Soft Signs	−? at PbB <15 μg/dl sig. assoc. with Soft Signs	0.008
				12-mo MDI		?
	6.5 (2.7–11.8)	Delivery/mother	119	6-mo KID	−? at PbB <12 μg/dl	0.002
	"	"	127	6-mo MDI	−? "	<0.05
	"	"	145	12-mo MDI	−? "	N.S.
	"	"	142	24-mo MDI	+? "	N.S.
	"	"	138	36-mo S-B IQ	+? "	N.S.
	"	"	134	58-mo WPPSI	−? "	N.S.
Port Pirie (McMichael et al., 1986, 1988; Wigg et al., 1988; Baghurst et al., 1992)	14.4 g.m. (?)	6-mo/child	575	24-mo MDI	−1.6 pts. per 10 μg/dl	0.07
	21.2 g.m. (5–57)	24-mo/child	534	48-mo GCI	−3.1 "	0.04
	17.6 g.m. (?)	0–4 yr avg/child	463	48-mo GCI	−3.8 "	0.04
	17.4 g.m. (?)	0–3 yr avg/child	≤494	7.5-yr WISC-R	−1.2 "	0.04

Table 1 (continued)

Study	Blood lead measure		N	Endpoint	Effect	P Value
	Mean (range/±S.D.), µg/dl	Time/source				
Sydney (Cooney et al., 1989a,b)	9.1 g.m. (3–28)	Delivery/mother	235	6-mo MDI	?	>0.25
	8.1 g.m. (1–36)	Delivery/cord		12-mo MDI	?	>0.30
				24-mo MDI	?	>0.70
	~12.5 g.m. (?)	Prior & current/child	207	36-mo GCI	?	>0.70
				48-mo GCI	?	0.14
Yugoslavia (Wasserman et al., 1992)	14.4 (±10.4)	Delivery/cord	348	24-mo MDI	−0.8 pts. per 10 µg/dl	0.12
	15.1 (±11.8)	6-mo/child	291	"	−0.6 "	0.34
	18.3 (±13.8)	12-mo/child	268	"	−0.4 "	0.17
	22.0 (±16.7)	18-mo/child	273	"	−0.4 "	0.16
	24.3 (±17.0)	24-mo/child	312	"	−1.3 "	0.03

Note: ? not reported; GCI, McCarthy General Cognitive Index; g.m., geometric mean; K-ABC, Kaufman Assessment Battery for Children (MPC = Mental Processing Composite; SIM = Simultaneous Processing); KID, Kent Infant Development Scale; MDI, Bayley Mental Development Index; N.S., not significant at $p < 0.05$, two-tailed; PbB, blood lead level; S-B IQ, Stanford-Binet Intelligence Quotient; Soft Signs, Graham-Rosenblith Neurological Soft Signs Scale; WISC-R, Wechsler Intelligence Scale for Children, Revised (FS = Full Scale; P = Performance); WPPSI, Wechsler Preschool and Primary Scales of Intelligence.

that the level of concern for neurotoxicity in developing children was 10 to 15 µg/dl, and possibly lower (U.S. Environmental Protection Agency, 1986, 1990).

Extracting this conclusion from the collective evidence available at the time required judgments about the results from various studies and how they related to each other. For example, in the "Boston Study" (Bellinger et al., 1986, 1987a,b), infants were categorized according to their umbilical cord blood lead levels into three groups: low (<3 µg/dl), medium (6 to 7 µg/dl), and high (10 to 25 µg/dl). During the first 2 years of life, children in the high-lead group had significantly lower scores on the mental development index (MDI) of the Bayley Scales of Infant Development. Follow-up analyses of the data (Bellinger, 1989) indicated that the effect observed in the high-lead group was not limited to the highest exposed individuals in that group, but was rather evenly distributed between individuals with blood lead levels above and below 15 µg/dl. Thus, the effects were evident at blood lead levels at least as low as 10 µg/dl.

In initial reports of the "Cincinnati Study" (Dietrich et al., 1987), lower scores on the Bayley MDI were found to be mediated in part by lead-related reductions in birth weight and gestational age. A related report (Bornschein et al., 1989) presented the results of a threshold analysis for birth-weight effects in the Cincinnati study population and noted a significant decrease in birth weight between blood lead groups of 7 to 12 µg/dl and 13 to 18 µg/dl. This finding suggested that the LOAEL for the effect of lead on Bayley MDI performance could be as low as 7 to 18 µg/dl.

Reports of the "Cleveland Study" (Ernhart et al., 1986, 1987; Wolf et al., 1985) provided some additional indirect evidence of lead-induced impairment of Bayley MDI performance, but did not provide any information that would allow discernment of a threshold or breakpoint in the dose-response curve. Nevertheless, the findings were consistent with a level of concern of "10 to 15 µg/dl, and possibly lower" because the average blood lead concentration in the Cleveland study was around 6 µg/dl and the highest blood lead level was less than 15 µg/dl.

Initial results from the "Port Pirie (South Australia) Study" (Wigg et al., 1988; McMichael et al., 1988) indicated that the average blood lead level of the children in that study was about 14 µg/dl at the time in postnatal development when the association with poorer Bayley MDI performance was strongest. Subsequent reports from these and other prospective studies in Sydney (Cooney et al., 1989a,b) and the former Yugoslavia (Wasserman et al., 1992) have supplemented the findings that figured into EPA's 1986 assessment (U.S. Environmental Protection Agency, 1986). Although many of the results from the prospective studies were not necessarily statistically significant, and a few results were even in the opposite direction from those expected, the overall pattern of findings, particularly when one considers the limited statistical power of most studies to detect such effects, clearly pointed to an effect of lead on cognitive development at blood lead levels around 10 to 15 µg/dl, and possibly lower.

Note that the judgment regarding a level of concern at blood lead levels around 10 µg/dl does not imply a biological threshold for lead effects, which almost certainly occur at even lower levels. Rather, the statement "reflects scientific judgment for purposes of protecting public health" in the context of providing a foundation for regulatory or policy positions (Davis, 1990). As such, this conclusion was first stated by EPA in 1986 (U.S. Environmental Protection Agency, 1986) and reaffirmed in 1990 (U.S. Environmental Protection Agency, 1990), and has been supported by the statements of other agencies and organizations, including the Agency for Toxic Substances and Disease Registry (1988), the Centers for Disease Control (1991), and the World Health Organization (1995).

The assessment of population exposures to lead has been the subject of several notable studies (see U.S. Environmental Protection Agency, 1986 for review). With respect to the use of TEL or TML in gasoline, the primary concern about lead has been the widespread exposure of the general population to the inorganic lead emitted from the tailpipes of large numbers of cars rather than the more limited exposure of a relatively small number of individuals to the fuel itself, either by inhalation or dermal contact. In the 1970s, the second National Health and Nutrition Examination Survey (NHANES II) provided a valuable characterization of blood lead levels in the U.S. by age, race, socioeconomic status, and other variables. One major finding of the NHANES II study was the documentation of a rather dramatic decline in blood lead levels that correlated closely with a decline in the use of leaded gasoline (Annest and Mahaffey, 1984). This evidence supported regulatory actions by EPA to reduce the lead content of leaded gasoline (Federal Register, 1985).

In addition to population surveys, the biokinetics of lead exposure have received considerable attention. For example, EPA developed an integrated exposure uptake biokinetic (IEUBK) model for predicting the distribution of blood lead levels in populations of children based on exposures to lead from various environmental sources and through various pathways (U.S. Environmental Protection Agency,

1994b,c,d). Historically, atmospheric lead has made a major contribution to total lead exposure in children, both by direct inhalation of lead-bearing particles and by ingestion of dust particles that deposit on surfaces that children typically contact through hand-to-mouth activities. Although the reduction of lead in gasoline has resulted in a major reduction in atmospheric lead, other sources, such as lead emitted from smelters, power plants, and incinerators, also contribute to the inhalation and dust ingestion pathways of exposure. Other sources of lead, such as lead-based paint, continue to represent a major component of lead in both soil and household dust.

As a risk assessment tool, the IEUBK model for lead in children provides a means of performing rapid calculations of a complex series of equations that predict the uptake, distribution, and elimination of lead by children. The model does not by itself reduce the uncertainty about the calculation or the variability within the population. However, by predicting children's blood lead levels at any stage of development through the first 6 years, the model assists the risk assessor in accounting for past exposure more systematically than can usually be done with a simple equation. By this means, it has been possible to refine regulatory or abatement efforts to address the specific sources that make the largest contributions to lead exposure on a population by population basis.

III. RISK ASSESSMENT OF MANGANESE

As noted above, most of the attention devoted by EPA to manganese health risks has been in the context of its use in a gasoline additive known as MMT (U.S. Environmental Protection Agency, 1994a). Although MMT itself is toxic, the main focus of EPA's health risk assessment has been on the inhalation of Mn_3O_4, the primary manganese combustion product of MMT in unleaded gasoline. However, the limited health effects information specifically on Mn_3O_4 has necessitated consideration of data for inorganic manganese compounds in general.

The toxicity of manganese is related to the route of exposure. Ingested manganese is relatively low in toxicity. Few cases of frank poisoning due to manganese ingestion have been reported. Indeed, manganese is considered a nutritionally essential element (although actual cases of deficiency in humans appear to be quite rare because of its abundance in nature). Manganese is a constituent of various enzymes and thus is very important to certain physiological processes, as illustrated by the role of superoxide dismutase in mitochondrial oxygen radical metabolism (Gavin et al., 1992). Inhaled manganese, on the other hand, has been known since the early 1800s to be neurotoxic (as well as toxic to the respiratory and reproductive systems). Manganism is characterized by various psychiatric and movement disorders, with some general resemblance to Parkinson's disease in terms of impairments in the fine control of some movements, lack of facial expression, and involvement of underlying neuroanatomical and neurochemical factors. The greater toxicity of inhaled manganese may owe to the fact that inhaled manganese first passes the brain, whereas ingested manganese first passes the liver, which has a high capacity to metabolize and eliminate ingested manganese. Although much remains to be learned about the biokinetics of manganese, especially in relation to its passage across the blood-brain barrier, this difference between inhaled and ingested manganese may help explain why blood manganese levels have not proven to be very well correlated with outcome measures in studies of manganese health effects.

As part of EPA's health risk assessment of manganese, an inhalation reference concentration (RfC) was derived for manganese (IRIS, 1993). The RfC is defined as an estimate (with uncertainty spanning about an order of magnitude) of a continuous inhalation exposure level for the human population (including sensitive subpopulations) that is likely to be without appreciable risk of deleterious noncancer effects during a lifetime (U.S. Environmental Protection Agency, 1994e). The RfC is derived by dividing a dosimetrically adjusted no-observed-adverse-effect level (NOAEL) or, lacking a NOAEL, a lowest-observed-adverse-effect level (LOAEL) by numerical uncertainty factors for specific limitations in the available health effects data. (An oral reference dose [RfD] for manganese has also been developed by EPA [IRIS, 1994], but the relevance of oral exposure to the MMT issue is limited, as explained above.)

In deriving the manganese RfC, the most useful available data came from epidemiological studies of worker populations (Roels et al., 1987, 1992; Mergler et al., 1994; Iregren, 1990). All of these studies indicated that measures of psychomotor function (e.g., eye-hand coordination, hand steadiness, speed of movement, reaction time) were significantly worse in manganese-exposed workers than in control workers. Moreover, neurobehavioral function appeared to be the most sensitive endpoint, i.e., the LOAELs for neurobehavioral function were lower than those for other endpoints such as reproductive or respiratory function. Thus, in terms of the qualitative nature of the health hazard posed by inhaled manganese, the evidence has rather consistently pointed to neurotoxicity.

As in the case of lead, the dose-response characterization of manganese neurotoxicity has been a more complicated endeavor. Because none of the epidemiological studies in question reported a NOAEL, derivation of the RfC involved identifying a LOAEL from the study by Roels et al. (1992). The geometric mean of the integrated respirable Mn dust concentration (793 $\mu g/m^3$ × exposure years) was first divided by the average period of worker exposure (5.3 years) to eliminate time (in years) from the time-weighted average, thereby yielding a LOAEL of 150 $\mu g/m^3$ of Mn. This workplace-based LOAEL was then adjusted for differences between occupational and general population exposures (8 vs. 24 h/day, 5 vs. 7 days/week, 10 vs. 20 m^3/day air breathed). The resulting adjusted LOAEL of 50 $\mu g/m^3$ of Mn was divided by a total uncertainty factor of 1000, which included a factor of 10 for the lack of adequate information on potentially susceptible subpopulations (e.g., children, the elderly) and a composite factor of 10 for inadequate information on reproductive or developmental effects, on the effects of chronic exposure, and on possible differences in toxicity of different manganese compounds. In addition, a factor of 10 was included for the lack of a NOAEL, yielding a total uncertainty factor of 1000. The result was an RfC of 0.05 $\mu g/m^3$ Mn.

In more extensive analyses of the data from the study by Roels et al. (1992), EPA (U.S. Environmental Protection Agency, 1994a) later examined various statistical approaches to assessing the dose-response relationship. These included "benchmark dose" analyses of the type described by Crump (1984) and related Bayesian analyses of the type described by Jarabek and Hasselblad (1991). These approaches use a mathematical model to estimate the effective dose or concentration at which a specified percentage (usually 5 or 10%) of a population would show an adverse effect. In performing and interpreting these analyses, several issues arose, including questions regarding selection of a mathematical dose-response model (e.g., linear vs. nonlinear), a benchmark level (e.g., 5% vs. 10% effective concentration), a severity factor (is a 10% benchmark level for the neurotoxic effects in question equivalent to a NOAEL or a minimal-severity LOAEL?), and the use of continuous vs. dichotomous data. These issues have been discussed in greater detail in the EPA assessment (U.S. Environmental Protection Agency, 1994a). After considering the merits and weaknesses of various alternative approaches, EPA concluded that the leading candidate alternative estimates of an RfC for manganese were in the range of 0.09 to 0.2 $\mu g/m^3$.

Given that MMT is not already in widespread use in the U.S., the exposure assessment component of EPA's assessment of MMT had to make projections about manganese exposure levels that would result from its addition to unleaded gasoline. This exposure assessment focused on personal exposure levels because of the disparity that typically exists between ambient pollutant concentrations and personal exposure levels (i.e., in the space immediately around a person during typical daily activities). Fortuitously, a major study of personal exposures to particles, including manganese, was conducted in Riverside, CA in the fall of 1990, at which time MMT was still being used in leaded gasoline (Pellizzari et al., 1992). From this and other information bearing on the automotive contribution to environmental manganese levels, EPA was able to model a distribution of personal exposure levels to particulate manganese that would be predicted to result from the wider use of MMT in unleaded gasoline (U.S. Environmental Protection Agency, 1994a). Although the predictions involved various uncertainties, the resulting distribution indicated that on the order of 5 to 10% (depending on the methods and assumptions used to derive the estimates) of the Riverside population would have long-term personal exposure levels around 0.1 $\mu g/m^3$ or higher. A lack of data for other times of the year made it impossible to estimate exposure levels for any period but the fall, albeit essentially for a lifetime of fall seasons (Wallace et al., 1994).

If the Riverside projections are considered representative of the Los Angeles metropolitan area (with a population of more than 14 million people), hundreds of thousands of persons would be predicted to have manganese exposure levels exceeding 0.1 $\mu g/m^3$ in that locale alone. The fact that there is overlap between RfC estimates (0.09–0.2 $\mu g/m^3$) and personal exposure estimates does not necessarily indicate a public health risk would exist, because the RfC is defined merely as a protective level against health risk. The method used to derive an RfC, namely dividing a NOAEL or LOAEL by a total uncertainty factor, which may be several orders of magnitude in size, does not enable one to make any estimation about health risks at exposures above the RfC but below the NOAEL or LOAEL. On the other hand, it would be obviously incorrect to conclude that a risk probably does not exist at chronic exposure levels exceeding the RfC.

IV. COMPARISON OF LEAD AND MANGANESE RISK ASSESSMENTS

In terms of the NRC/NAS (1983) risk assessment paradigm, both lead and manganese are well identified as neurotoxicity hazards, although the fact that both metals have toxic effects on other organ systems should not be overlooked. As neurotoxicants, however, they differ somewhat in the nature of their effects. Lead seems to be a more potent toxicant to the developing nervous system than to the mature CNS (Davis and Grant, 1992), whereas manganese seems to be more toxic to elderly persons than to younger persons (Kawamura et al., 1941). However, this characterization must be qualified by noting the lack of research on manganese neurotoxicity during early development. Lead is best known for its effects on cognitive development, whereas manganese is best known for its effects on motor function. Here again, there may be some circularity in this description, in that the difference in the reported effects of the two metals may merely reflect the dominant focus and methodological strengths or weaknesses of the efforts that have been made thus far to investigate the neurotoxicity of the two metals, for it is clear that both cognitive and motor functions can be impaired by lead as well as manganese.

It is also of interest to note differences in the methods used to measure neurotoxicity in these two cases. The assessment of the developmental neurotoxicity of lead has relied heavily on standardized measures of neurobehavioral development, such as the Bayley MDI, the McCarthy General Cognitive Index, and other scales for which normative data have been well established. In the case of the neurobehavioral effects of manganese on adult workers, the tests may not have had norms established or have been as widely used as the developmental scales used in lead research, but they provide objective measures of motor function. The comparable results from several independent studies of manganese-exposed workers (Roels et al., 1987, 1992; Mergler et al., 1994; Iregren, 1990) provide evidence that the tests are valid and sensitive methods for assessing neurobehavioral function (e.g., eye-hand coordination, hand steadiness, and visual reaction time) in relation to manganese exposure. However, in neither the case of lead nor that of manganese should it be assumed that the most sensitive methods for detecting neurotoxicity have necessarily been employed to date. It remains to be seen whether some other technique might provide even greater sensitivity and specificity than the methods employed thus far.

The dose-response analyses for lead and manganese provide a number of interesting comparisons. One contrast lies in the fact that a level of concern could be identified directly from the collective findings of several studies of the health effects of lead, whereas a modeling approach was applied to the data available from a single study (Roels et al., 1992) in the case of manganese. This is not to say that the manganese assessment was based on a single study, for other studies did in fact provide important supporting evidence. Nevertheless, the substantially greater database for lead allows a rather direct approach to identifying an overall LOAEL for lead developmental neurotoxicity. In the case of manganese neurotoxicity, the modeling approach allows one to interpolate a boundary between the NOAEL and LOAEL on the dose-response curve, but the multiple models and assumptions involved in this approach make the determination of a NOAEL or LOAEL less self-evident than in the case of lead.

Despite the differences in the bases for and approaches to dose-response characterizations of lead and manganese, the two assessments share a basic similarity in that a judgment is made about the point on the dose-response curve at which adverse health effects begin. This type of judgment requires careful evaluation of all the available data and consideration of what constitutes an adverse effect. It should be noted that such judgments can come into the assessment process at various stages. For example, a judgment often is made initially by a researcher in categorizing the responses of subjects as normal or abnormal. Although a statistical criterion may be used to dichotomize data (e.g., responses more than two standard deviations from the mean might be termed "abnormal"), the distinction between such categories of responses is fundamentally a value judgment.

Even if nondichotomized (i.e., continuous) data are available and used in an assessment, it is typically the risk assessor who decides where the line between adverse and nonadverse health effects occurs. To illustrate, some risk assessors might consider a 10% benchmark level equivalent to a NOAEL, whereas others might consider the same level a minimal-severity LOAEL. In either case, a judgment is made about whether or not an adverse effect is present. Implicit in this judgment is a consideration of the nature of the endpoint itself. For example, a subtle reversible effect (e.g., headache) would be judged less adverse (and thus possibly allowed to occur in a larger percentage of the population) than a frank irreversible effect (e.g., paralysis). However, even the same endpoint may be judged differently in different situations. Minor hand tremor can be caused by caffeine as well as by manganese. The tremor induced by caffeine is generally accepted because it is recognized as an easily reversible condition, whereas the tremor associated with manganese exposure is considered an adverse health effect because it may not be reversible and may presage other adverse neurobehavioral effects as well.

A conclusion that a given blood lead level constitutes a level of concern also involves a judgment about the adverseness of health effects that occur above or below that level. In the case of lower IQ scores at blood lead levels around 10 µg/dl, the decrement in IQ appears to average around 1 to 4 points. Some have questioned whether an effect of this size is meaningful because it would have little or no significance in a clinical evaluation of a given child. When viewed in population terms, however, even a small decline in IQ can have significant public health implications. Figure 1 illustrates how a normal distribution of IQ scores might be related to blood lead levels across a range of exposures. Note the decrease in the size of the upper tail (IQ > 120) and increase in the size of the lower tail (IQ < 80) of the distribution as blood lead level increases. For example, a decline of 4 points in the mean IQ in conjunction with an increase in average blood lead levels from 10 to 20 µg/dl implies an increase of about 50% in the number of children scoring at least one standard deviation below the mean. Conversely, the number of children scoring more than one standard deviation above the mean is diminished by 50% as well. Although the effects of shifts in average IQ may be more apparent by focusing on the tails of the distributions, it is the effect of lead on the entire distribution of IQ scores that is ultimately of concern.

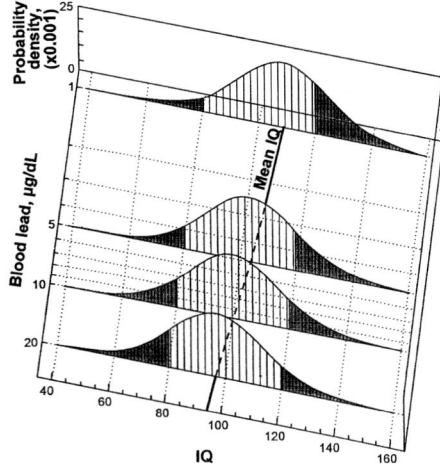

Figure 1 Normal distributions of IQ scores with standard deviations of 16 and means ranging from 108 (at an average blood lead level of 1 µg/dl) to 96 (at an average blood lead level of 20 µg/dl).

From the standpoint of exposure assessment, several points of comparison are noteworthy. A fundamental similarity in the assessments of lead and manganese is the focus on inhalation exposure to the combustion emissions of inorganic compounds of lead and of manganese. Although the organometallic additives to gasoline are themselves neurotoxic, their greater import for public health lies in the widespread dispersion of their combustion products throughout the ambient atmosphere. This is not to ignore the possible importance of localized or point sources, which in some cases may overshadow ambient air exposure. Rather, it serves to illustrate that the most direct source or route of exposure may not be the most significant to the population as a whole.

The importance of the route of exposure is also underscored by the greater toxicity of inhaled vs. ingested manganese. As a result, different sources or routes of exposure may have quite different health risk implications. Whereas lead-contaminated soil and dust are quite significant sources of lead exposure for young children, manganese-contaminated soil and dust were not considered particularly significant concerns in EPA's decision on MMT (Federal Register, 1994), partly because of the relatively low potential for significant accumulation of manganese in soil and dust as a result of using MMT in gasoline, and partly because of the relatively low oral toxicity of manganese. The differential bioavailability of different forms of lead has been the subject of a growing area of research in recent years (LaVelle et al., 1991; U.S. Environmental Protection Agency, 1994c), particularly as regulatory and remediation efforts become more focused on specific sources and pathways of lead exposure. However, too few studies have been devoted to examining the bioavailability of different compounds of manganese to allow EPA to assess adequately the relative toxicity of Mn_3O_4 compared to other manganese oxides or to manganese compounds in general.

Another key difference between lead and manganese is the lack of a bioindicator of exposure for manganese comparable to that of blood lead concentration, which provides a useful measure of integrated exposure resulting from various sources, pathways, and routes of uptake. Because blood lead level is commonly used by researchers both as the metric of dose-response relationships for lead health effects and as the metric of lead exposure levels, it facilitates the integration of exposure and dose-response information in arriving at a health risk characterization. However, when a specific source of lead exposure is the focus of a health risk assessment, the relative contributions of different sources of exposure must be distinguished.

In this respect, EPA's assessment of manganese exposure (U.S. Environmental Protection Agency, 1994a) is no different from EPA's assessment of lead exposure for describing air quality criteria (U.S. Environmental Protection Agency, 1986). Nevertheless, the approaches used to distinguish the contributions of different sources of exposure were quite different for lead and manganese. The lead exposure assessment was built upon several studies of different populations in which lead concentrations in various environmental media were measured and related to blood lead levels through statistical techniques such as regression analysis. In some cases, methods such as lead isotope ratio profiles of gasoline and blood samples have been used to isolate the automotive contribution to blood lead levels. For the manganese exposure assessment, the automotive contribution to the personal exposure levels of a single population had to be estimated, largely based on the assumption that fine-mode (≤ 2.5 μm) manganese particles in the Riverside, CA area originated from automotive emissions rather than crustal material. Because the key health effects study in the manganese assessment measured particulate manganese in the respirable mode (≤ 5 μm), additional assumptions and inferences were made so the health effects and exposure information could be better compared and integrated into the risk characterization (see U.S. Environmental Protection Agency, 1994a). These and various other inferences and assumptions involved in the manganese exposure assessment contributed to the uncertainties in the resulting risk characterization.

In the absence of information, uncertainty increases. The current manganese RfC of 0.05 μg/m³ was previously verified based on a total uncertainty factor of 1000. However, total uncertainty factors of 100 to 300 were used in obtaining more recent (not yet officially verified) alternative manganese RfC estimates of 0.09 to 0.2 μg/m³. The difference in uncertainty factors for the verified and alternative RfC estimates reflects a judgment about the information used in the various derivations. In the case of the RfC of 0.05 μg/m³, the total uncertainty factor of 1000 includes a factor of 10 for using a LOAEL in lieu of a NOAEL. In the case of the alternative analyses, the total uncertainty factor was either 100 or 300, depending on whether a 5 or 10% benchmark level was used in deriving the RfC estimates. A 5% benchmark was treated essentially as a NOAEL, whereas a 10% benchmark level was treated as a LOAEL of minimal severity. Thus, instead of including an uncertainty factor of 10 for using a LOAEL instead of a NOAEL, either a factor of 1 was used for the surrogate NOAEL (5% benchmark level), or a factor of 3 (i.e., approximately one-half of 10 on a log scale) was used for the minimal severity LOAEL (10% benchmark level). Although the resulting RfC estimates were slightly higher than the verified RfC of 0.05 μg/m³, the difference is not meaningful in view of the inherent order-of-magnitude imprecision of the RfC. However, even the higher values are well below the current National Ambient Air Quality Standard (NAAQS) for lead, 1.5 μg/m³. The difference between a NAAQS of 1.5 μg/m³ for lead and an RfC of 0.05 (or even 0.2) μg/m³ for manganese does not imply that manganese is much more toxic than lead. Rather, it reflects a difference in the data available, and therefore the methods used, to derive these values.

The health effects of lead have been extensively investigated in susceptible populations such as children and pregnant women. The health effects of manganese have primarily been investigated in populations of healthy, adult male workers, rather than in the general population or in potentially susceptible subpopulations such as children and the elderly. To make an extrapolation about the health effects that might occur in the general population, including susceptible subpopulations, an uncertainty factor of 10 is typically used in deriving RfCs. In the case of the manganese RfC, such a factor was especially warranted in view of some clinical evidence indicating that the elderly are more sensitive to the toxic effects of manganese (Kawamura et al., 1941).

Several studies of lead-exposed populations such as children or adults (primarily male workers) have followed subjects over extended periods of time or have investigated subjects after long periods of past lead exposure. By contrast, no long-term studies of manganese-exposed children have been conducted, and even in studies of manganese-exposed workers the periods of exposure have been considerably less than an occupational lifetime. For example, the mean duration of manganese exposure was only 5.3 years (range: 0.2 to 17.7 years) in the study by Roels et al. (1992), and the longest average exposure in any of the recent key studies (Mergler et al., 1994) was only 16.7 years. Therefore, to extrapolate from findings obtained after these relatively limited periods of exposure to a 70-year lifetime, an uncertainty factor was judged appropriate as part of a composite factor of 10 for database limitations. In addition to the uncertainty of extrapolating from less than chronic exposure to a lifetime exposure, other uncertainties included in the composite factor reflected the lack of adequate data on reproductive toxicity and on possible differences in the toxicity of different chemical forms of manganese. Although some work has indicated the existence of reproductive effects (e.g., Gennart et al., 1992) and differences in toxicity of different manganese compounds (e.g., Komura and Sakamoto, 1991, 1992, 1993), not enough information existed in these areas to eliminate or substantially reduce concerns about such uncertainties.

Because the health effects of lead have been subjected to much more investigation than those of manganese, much less uncertainty is associated with the identification of a level of concern. Consequently, exceeding the level of concern for lead implies an increased level of public health risk, whereas exceeding the manganese RfC does not necessarily imply increased health risk. Even so, it is not possible to infer, solely on the basis of a blood lead measurement above 10 µg/dl, the actual consequences that any given child would experience, for the evidence on which the level of concern is based pertains to populations of children, not individual children. Thus, the level of concern for lead health effects is a matter of public health concern in the collective sense, not in the sense that each individual is necessarily adversely affected.

A few years prior to the recent EPA assessment of MMT, EPA issued a report (U.S. Environmental Protection Agency, 1991) on the information that would be needed to reduce the uncertainties involved in assessing the potential health risks of manganese associated with MMT usage in gasoline. Until the information identified in that report (U.S. Environmental Protection Agency, 1991) is obtained and the scientific uncertainties surrounding MMT usage are reduced, the assessments of health risks for lead and manganese are likely to continue to be rather different, and the RfC for manganese is likely to remain about an order of magnitude below the NAAQS for lead.

ACKNOWLEDGMENTS

We thank M. Barrier, D. Fennell, S. Lassiter, W. Lloyd, C. Thomas, and Information Organizers, Inc. for assistance in the preparation of this paper. We also thank A. Jarabek and M. Stanton for reviewing the draft manuscript. The U.S. Government has the right to retain a nonexclusive royalty-free license in and to any copyright covering this chapter.

REFERENCES

Agency for Toxic Substances and Disease Registry, The nature and extent of lead poisoning in children in the United States: a report to Congress. Atlanta, GA: U.S. Department of Health and Human Services, Public Health Service. Available from: NTIS, Springfield, VA; PB89–100184/XAB, 1988.

Annest, J. L. and Mahaffey, K., Blood lead levels for persons ages 6 months–74 years: United States, 1976–80. U.S. Department of Health and Human Services, Public Health Service, National Center for Health Statistics; DHHS publ. (PHS) 84–1683, Washington, D.C., 1984.

Baghurst, P. A., McMichael, A. J., Wigg, N. R., Vimpani, G. V., Robertson, E. F., Roberts, R. J., and Tong, S.-L., *N. Engl. J. Med.*, 327, 1279–1284, 1992.

Bellinger, D., in *Research in Infant Assessment,* Paul, N. W. and Golia, S. R., Eds., Proc. Symp. Association for Children and Adults with Learning Disabilities (ACLD), Inc., Miami Beach, FL. Birth Defects Orig. Artic. Ser. 25(6), 73–97, 1989.

Bellinger, D., Leviton, A., Needleman, H. L., Waternaux, C., and Rabinowitz, M., *Neurobehav. Toxicol. Teratol.,* 8, 151–161, 1986.

Bellinger, D., Sloman, J., Leviton, A., Waternaux, C., Needleman, H., and Rabinowitz, M., in: Lindberg, S. E. and Hutchinson, T. C., Eds., *Heavy Metals in the Environment,* Vol. 1, CEP Consultants, Ltd., Edinburgh, U.K., 49, 1987a.

Bellinger, D., Leviton, A., Waternaux, C., Needleman, H., and Rabinowitz, M., *N. Engl. J. Med.,* 316: 1037–1043, 1987b.

Bornschein, R. L., Grote, J., Mitchell, T., Succop, P. A., Dietrich, K. N., Krafft, K. M., and Hammond, P. B., in Smith, M. A., Grant, L. D., and Sors, A. I., Eds., *Lead Exposure and Child Development: An International Assessment,* Kluwer Academic, Norwell, MA, 307, 1989.

Centers for Disease Control, Preventing lead poisoning in young children: a statement by the Centers for Disease Control — October 1991, U.S. Department of Health and Human Services, Public Health Service, Atlanta, GA, 1991.

Cooney, G. H., Bell, A., McBride, W., and Carter, C., *Neurotoxicol. Teratol.,* 11, 95–104, 1989a.

Cooney, G. H., Bell, A., McBride, W., and Carter, C., *Dev. Med. Child Neurol.*, 31, 640–649, 1989b.

Crump, K. S., *Fundam. Appl. Toxicol.*, 4, 854–871, 1984.

Davis, J. M.,*Neurotoxicology*, 11, 285–292, 1990.

Davis, J. M. and Grant, L. D., in Guzelian, P. S., Henry, C. J., and Olin, S. S., Eds., *Similarities and Differences Between Children and Adults: Implications for Risk Assessment*, International Life Sciences Institute Press, Washington, D.C., 150, 1992.

Dietrich, K. N., Krafft, K. M., Bornschein, R. L., Hammond, P. B., Berger, O., Succop, P. A., and Bier, M., *Pediatrics*, 80, 721–730, 1987.

Dietrich, K. N., Krafft, K. M., Shukla, R., Bornschein, R. L., Berger, O., Succop, P. A., and Hammond, P. B., Developmental patterns of neurobehavioral deficit and recovery in infancy: response to fetal and early postnatal lead exposure [unpublished manuscript], University of Cincinnati, College of Medicine. Available for inspection at: U.S. Environmental Protection Agency, Central Docket Section, ECAO-CD-81–2 IIA.C.13, Washington, D.C., 1988.

Dietrich, K. N., Succop, P. A., Bornschein, R. L., Krafft, K. M., Berger, O., Hammond, P. B., and Buncher, C. R., *Environ. Health Perspect.*, 89, 13–19, 1990.

Dietrich, K. N., Succop, P. A., Berger, O. G., Hammond, P. B., and Bornschein, R. L., *Neurotoxicol. Teratol.*, 13, 203–211, 1991.

Dietrich, K. N., Succop, P. A., Berger, O. G., and Keith, R. W., *Neurotoxicol. Teratol.*, 14, 51–56, 1992.

Dietrich, K. N., Berger, O. G., Succop, P. A., Hammond, P. B., and Bornschein, R. L., *Neurotoxicol. Teratol.*, 15, 37–44, 1993.

Ernhart, C. B., Wolf, A. W., Kennard, M. J., Erhard, P., Filipovich, H. F., and Sokol, R. J., *Arch. Environ. Health*, 41, 287–291, 1986.

Ernhart, C. B., Morrow-Tlucak, M., Marler, M. R., and Wolf, A. W., *Neurotoxicol. Teratol.*, 9, 259–270, 1987.

Federal Register, Regulation of fuel and fuel additives; gasoline lead content, F. R., (March 7) 50, 9386–9408, 1985.

Federal Register, Fuels and fuel additives; waiver decision/circuit court remand, F. R., (August 17) 59, 42227–42247, 1994.

Gavin, C. E., Gunter, K. K., and Gunter, T. E., *Toxicol. Appl. Pharmacol.*, 115, 1–5.

Gennart, J. P., Buchet, J. P., Roels, H., Ghyselen, P., Ceulemans, E., and Lauwerys, R., *Am. J. Epidemiol.*, 135, 1208–1219, 1992.

IRIS, Integrated Risk Information System [Database], Inhalation Reference for Manganese, 9/23/93, U.S. Environmental Protection Agency, Office of Health and Environmental Assessment, Environmental Criteria and Assessment Office, Washington, D.C. Available online from: TOXNET, National Library of Medicine, Rockville, MD, 1993.

IRIS, Integrated Risk Information System [Database], Oral Reference for Manganese, 4/1/94, U.S. Environmental Protection Agency, Office of Health and Environmental Assessment, Environmental Criteria and Assessment Office, Washington, D.C. Available online from: TOXNET, National Library of Medicine, Bethesda, MD, 1994.

Iregren, A., *Neurotoxicol. Teratol.*, 12, 673–675, 1990.

Jarabek, A. M. and Hasselblad, V., Inhalation reference concentration methodology: impact of dosimetric adjustments and future directions using the confidence profile method. Paper 91–173.3, presented at: 84th annu. meet. Air and Waste Management Association, June, Vancouver, B.C., Canada, 1991.

Kawamura, R., Ikuta, H., Fukuzumi, S., Yamada, R., Tsubaki, S., Kodama, T., and Kurata, S., *Kitasato Arch. Exp. Med.*, 18, 145–169, 1941.

Komura, J. and Sakamoto, M., *Bull. Environ. Contam. Toxicol.*, 46, 921–928, 1991.

Komura, J. and Sakamoto, M., *Environ. Res.*, 57, 34–44, 1992.

Komura, J. and Sakamoto, M., *Toxicol. Lett.*, 66, 287–294, 1993.

LaVelle, J. M., Poppenga, R. H., Thacker, B. J., Giesy, J. P., Weis, C., Othoudt, R., and Vandervoort, C., *Chem. Speciation Bioavailability*, 3, 105–111, 1991.

McMichael, A. J., Vimpani, G. V., Robertson, E. F., Baghurst, P. A., and Clark, P. D., *J. Epidemiol. Commun. Health*, 40, 18–25, 1986.

McMichael, A. J., Baghurst, P. A., Wigg, N. R., Vimpani, G. V., Robertson, E. F., and Roberts, R. J., *N. Engl. J. Med.*, 319, 468–475, 1988.

Mergler, D., Huel, G., Bowler, R., Iregren, A., Belanger, S., Baldwin, M., Tardif, R., Smargiassi, A., and Martin, L., *Environ. Res.*, 64, 151–180, 1994.

National Research Council, Risk Assessment in the Federal Government: Managing the Process, National Academy Press, Washington, D.C., 1983.

Pellizzari, E. D., Thomas, K. W., Clayton, C. A., Whitmore, R. W., Shores, R. C., Zelon, H. S., and Perritt, R. L., Particle total exposure assessment methodology (PTEAM): Riverside, California pilot study, Vol. I. U.S. Environmental Protection Agency, Atmospheric Research and Exposure Assessment Laboratory, Report no. EPA/600/R-93/050, Washington, D.C., 1992.

Roels, H., Lauwerys, R., Buchet, J.-P., Genet, P., Sarhan, M. J., Hanotiau, I., de Fays, M., Bernard, A., and Stanescu, D., *Am. J. Ind. Med.*, 11, 307–327, 1987.

Roels, H. A., Ghyselen, P., Buchet, J. P., Ceulemans, E., and Lauwerys, R. R., *Br. J. Ind. Med.*, 49, 25–34, 1992.

U.S. Environmental Protection Agency, Air quality criteria for lead, Health Effects Research Laboratory, Criteria and Special Studies Office, Report no. EPA-600/8-77-017. Available from: NTIS, Springfield, VA, 1977.

U.S. Environmental Protection Agency, Air quality criteria for lead, Office of Health and Environmental Assessment, Environmental Criteria and Assessment Office, Report no. EPA-600/8-83/028aF-dF. 4v. Available from: NTIS, Springfield, VA, 1986.

U.S. Environmental Protection Agency, Air quality criteria for lead: supplement to the 1986 addendum, Office of Health and Environmental Assessment, Environmental Criteria and Assessment Office, Report no. EPA/600/8–89/049F. Available from: NTIS, Springfield, VA, 1990.

U.S. Environmental Protection Agency, Information needed to improve the risk characterization of manganese tetraoxide (Mn_3O_4) and methylcyclopentadienyl manganese tricarbonyl (MMT), Office of Research and Development, Washington, D.C. 1991.

U.S. Environmental Protection Agency, Reevaluation of inhalation health risks associated with methylcyclopentadienyl manganese tricarbonyl (MMT) in gasoline [final]. Office of Research and Development; Report no. EPA/600/R-94/062, Washington, D.C., 1994a.

U.S. Environmental Protection Agency, Integrated exposure uptake biokinetic model for lead in children (IEUBK) version 0.99d. Office of Solid Waste and Emergency Response. Available from: NTIS, Springfield, VA, 1994b.

U.S. Environmental Protection Agency, Guidance manual for the integrated exposure uptake biokinetic model for lead in children, Office of Emergency and Remedial Response, Report no. EPA/540/R-93/081. Available from: NTIS, Springfield, VA, 1994c.

U.S. Environmental Protection Agency, Technical support document: parameters and equations used in integrated exposure uptake biokinetic model for lead in children (v 0.99d). Office of Solid Waste and Emergency Response, Report no. EPA/540/R-94/040. Available from: NTIS, Springfield, VA, 1994d.

U.S. Environmental Protection Agency, Methods for derivation of inhalation reference concentrations and application of inhalation dosimetry [draft final], Office of Research and Development, Report no. EPA/600/8–88/066F, Research Triangle Park, NC, 1994e.

Wallace, L. A., Duan, N., and Ziegenfus, R., Risk Anal., 14, 75–85, 1994.

Wasserman, G., Graziano, J. H., Factor-Litvak, R., Popovac, D., Morina, N., Musabegovic, A., Vrenezi, N., Capuni-Paracka, S., Lekic, V., Preteni-Redjepi, E., Hadzialjevic, S., Slavkovich, V., Kline, J., Shrout, P., and Stein, Z., J. Pediatr., 121, 695–703, 1992.

Wigg, N. R., Vimpani, G. V., McMichael, A. J., Baghurst, P. A., Robertson, E. F., and Roberts, R. J., , J. Epidemiol. Commun. Health, 42, 213–219, 1988.

Wolf, A. W., Ernhart, C. B., and White, C. S., in Lekkas, T. D., Ed., Heavy Metals in the Environment, Vol. 2, CEP Consultants, Ltd., Edinburgh, U.K., 153, 1985.

World Health Organization, Inorganic Lead Environmental Health Criteria 165, International Programme on Chemical Safety, WHO, Geneva, Switzerland.

Chapter 5

Techniques Employed for the Assessment of Metals in Biological Systems

Richard M. Jacobs

I. INTRODUCTION

In order to determine exposure to or intoxication from a metal, it is usually necessary to conduct some type of metal analysis of either a tissue and/or fluid and/or the exposed substance. This analysis may be qualitative, semiquantitative, or quantitative in nature. The analytical objective may be directed toward the determination of the total metal content or it may be directed toward the determination of a distinct chemical species of that metal or even the particular isotopic form, especially when that form of the metal may have particular toxicological significance, or when there may be multiple forms of the metal of varying toxicological consequences in that tissue.

There are approximately 30 elements that are practical toxicants[1] (Table 1). Some are biologically essential nutrients.[2] Others may have no known beneficial effects or functions. While metals are neither created nor destroyed in biology, they may be significantly altered chemically through metabolism and other chemical reactions. In tissues these metals may be of various physical and chemical forms: free uncompounded forms or compounded forms either organic or inorganic. In this chapter the term metal refers to the elements of practical toxicological significance in any chemical form. A majority of the analytical techniques discussed will be those that determine the total tissue content without regard to chemical form, since most analytical techniques employed require "destructive" means of sample preparation or can not discern the various forms of metals.

In principle, many analytical approaches yield acceptable results for an individual metal. In practice, each has its particular advantages, disadvantages, or practicality for a given metal, tissue, and toxicological assessment. In principle and practice, every methodological approach should be validated for each metal and each tissue type. An appropriate quality assurance program should be established to substantiate the accuracy and precision of the data generated.

The choice of sample type and quantity is most often limited by accessibility in the living. Often the choice is limited to blood and its components, gastric contents, feces, urine, hair, and other accessible but nonvital tissues and fluids. Often the accessible tissues may not be those that most accurately reflect the exposure to the metal. Likewise, the accessible tissue may not be collected at a time when it could be of greatest diagnostic significance. Even with an analytically acceptable result, the result, by itself, may be of limited value without corresponding biological signs, symptoms, indices of adverse effects, and the availability of normal tissue levels for comparison.

For most nonessential metals, even though true "normal" levels may not be readily available for many tissues and fluids, for most of these metals there is thought to be a large difference between those levels

Table 1 Metals of Practical Toxicological Significance

Biologically Essential Metals[a]		
Cobalt (Co)	Copper (Cu)	Chromium (Cr)
Iron (Fe)	Magnesium (Mg)	Manganese (Mn)
Molybdenum (Mo)	Selenium (Se)	Zinc (Zn)

Metals With No Established Biological Functions		
Aluminum (Al)	Antimony (Sb)	Arsenic (As)
Barium (Ba)	Beryllium (Be)	Bismuth (Bi)
Cadmium (Cd)	Gallium (Ga)	Germanium (Ge)
Gold (Au)	Indium (In)	Lead (Pb)
Lithium (Li)	Mercury (Hg)	Nickel (Ni)
Platinum (Pt)	Silver (Ag)	Strontium (Sr)
Tellurium (Te)	Thallium (Tl)	Tin (Sn)
Titanium (Ti)	Vanadium (V)	Uranium (U).

[a] Data from Reference 2.

associated with normal exposure and those that result in toxicity. A notable exception may be lead. Because the form(s) of a metal in the tissue may be an important factor in relating a given value with a potential toxicological effect, it may be important to determine the form(s) of the element. For example, methyl mercury is significantly more toxic than elemental mercury and perhaps some other forms of inorganic mercury. Therefore, in instances where methyl mercury is the suspected toxicant, methyl mercury should be the analytical objective rather than total mercury, where inorganic forms of mercury may also be present.

The form may be particularly important for metals that are biologically essential, since typically there are numerous forms of those metals in tissues and body fluids. Here too, there may be a lack of "normal" values and therefore a lack of clear demarcation of a level that indicates toxicity. However, for most biologically essential metals, a relatively wide range between a level that yields "good" nutrition and a level that is toxic usually exists. However, for some of these metals it may be important to distinguish the natural forms, i.e., those present in metalloenzymes, etc., from the "toxic" form(s).

The purpose of this chapter is to acquaint the reader with the methodological approaches and techniques that yield accurate and precise results for various metals in biological tissues. The primary focus will be toward modern multielemental techniques that will yield results either sequentially or simultaneously in the amounts of tissue typically available. These methods will largely depend on destructive, rather than nondestructive methods of sample preparation, and therefore are able to determine the total tissue content of the metal, rather than that of a particular form.

II. SAMPLE SELECTION, COLLECTION, STORAGE, AND PRESERVATION

In living animals, including humans, the choice of sample is usually limited to relatively small quantities of a relatively limited number of body tissues, fluids, and excretions, e.g., blood and its components (cells, plasma, serum), urine, feces, hair, nails, etc. Under special circumstances other tissues, e.g., liver, bone, cerebrospinal fluid, etc., may be available for analysis through biopsy. A greater variety of tissues are only available through post-mortem examination.

Although metals are neither created nor destroyed during sample processing and analysis, preserving the integrity of the sample with regard to mass (including water content), the analyte's physical distribution within the sample, analyte content, and analyte form (in case analyte speciation may be required) are essential to accurate quantitative results that can be compared to "normal" results from nonexposed or nonintoxicated individuals. For values for tissues and body fluids reported on a fresh weight basis, the moisture content of the samples must be safeguarded through refrigeration or freezing, depending on the length of storage prior to analysis. Special care must be observed when thawing refrigerated and frozen samples so that the analyte of interest remains homogeneously distributed within the sample. Some fluids and materials such as urine and feces may need to be preserved with strong mineral acid or other antibacterial agents in order to preserve analyte form or for safety considerations. Dry weights may be preferred for some types of samples, e.g., feces, where the moisture content may be difficult to control and the homogeneity of the sample can not be expected, or where the live moisture content has not been maintained. Regardless of the preservation technique used, the technique must not contribute significant amounts of analyte to the sample.

The means of sample collection and storage may be crucial for a valid, accurate result. For example, for collection of blood for Pb analysis, the collection device must be specially prepared in order to exclude either exogenous Pb from the device or Pb from the anticoagulant. Blood obtained from "finger sticks" can be significantly contaminated by Pb from the skin, especially considering the relatively small volume of this type of sample. The suitability of any collection device should be evaluated prior to use. If necessary, the actual amount of potential contamination should be determined through analysis. Ideally, the analyte under study should not be a component of either the device nor the sample storage container. Stainless steel implements such as needless, scalpel blades, etc. may significantly contribute to the analyte content, e.g., Cr, V, or Ni to blood plasma and serum. While one expects to find markedly higher amounts of metals in the samples obtained from intoxicated individuals, the inadvertent addition of metals to the normal control samples may raise questions regarding the integrity of the results in general.

Many types of suitable sample storage containers are readily available for use. These include various forms of plastics and glass. If glass is to be used as a storage container or in sample preparation, it must be meticulously prepared. It should be initially cleaned in detergent and then soaked in a relatively strong mineral acid solution. Commonly employed acid soaking mixtures include 20% (v/v) hydrochloric or nitric acids, either individually or sequentially, over a several-day period. After soaking, the glassware must be rinsed thoroughly with deionized water of suitable purity,[3,4] then dried in a metal-free environment. Glassware, as well as all wares used for samples storage, must be maintained in a metal-free environment prior to use. It is also advisable that glassware used for sample preparation and analysis should be dedicated to metals analysis use only, in order to limit cross contamination from other laboratory activities.

By far, the most popular type of storage container for samples to undergo metals analysis are composed of some form of plastic. Various forms of polyethylene and fluorohydrocarbon-based plastics are most suitable for sample storage and sample analysis. Many commercial polyethylene bags can be used without concern of contamination. Some vessels composed of PTFE-based plastics are being used effectively in sample mineralization procedures.

III. CONTAMINATION CONTROL AND SAMPLE PREPARATION

A. CONTAMINATION CONTROL

Analysts must recognize the particular demands of trace metal analysis in regard to contamination control, especially for those determination techniques that utilize small samples and measure picogram amounts of analytes. Scrupulous cleanliness of the laboratory environment, the "tools" of the analysis, and the reagents are essential to the production of precise and accurate metals data. The quality of the reagents, especially the mineral acids, must meet appropriate requirements of purity. Typically, distilled or double distilled grades of acids are needed to assure low reagent levels of the analyte.

Because most analysts handle a wide array of sample types of varying metal composition, the potential for severe contamination always exists. Extreme care should be taken to eliminate contamination caused by touching surfaces of containers or vessels that may come in contact with the sample. Vinyl gloves should be worn while handling the sample, its vessel, or other potential vehicles of contamination.

Mineralization steps pose the greatest opportunity for sample contamination through the environment, the ashing vessels, and through the addition of reagents employed in the ashing steps. The exposure of the sample to environmental contaminants is typically greatest during this phase of the analysis. When possible and feasible, sample preparative steps should be carried out in a HEPA-filtered, "clean-air" environment. Samples should be covered or protected at all times to limit particulate contamination.

Regardless of the sample environment during the various stages of analysis, the contaminant contribution to the metal from all sources needs to be monitored and routinely determined. The total contaminant contribution should be accurately determined and subtracted from the sample analyte value, if significant. An excellent source of information regarding most of the aspects of contamination has been prepared by Zief and Mitchell.[5]

B. SAMPLE PREPARATION TECHNIQUES

The degree and means of sample pretreatment required prior to the determinative step is dependent on several factors, including the tissue to be analyzed, the metal of interest, and the determination technique employed. Determination techniques may be classified as destructive or nondestructive. By definition, destructive techniques require, or employ, means that destroy the native condition of the sample. Although nondestructive techniques do not consume the sample, these techniques may effect changes in the physical nature of the sample. Rains[6] has reviewed sample preparation techniques of various types.

Most often, determination techniques are chosen that require either partial or complete destruction of the sample matrix. This sample treatment may take place either *in vitro* or *in situ*. Samples may require varying degrees of preparation, from simple dilution, dissolution, or complete mineralization. The analysis of some liquid samples types such as urine, blood, serum, etc. for some metals may proceed with only minimal sample pretreatment *in vitro*. For example, a number of metals can be determined accurately by flame atomic absorption spectrophotometry of serum samples after simple dilution. For some other determinative techniques, e.g., induced coupled plasma atomic emission, and graphite furnace atomic absorption spectrophotometry, while not requiring complete sample mineralization prior to sample introduction, they afford mineralization *in situ*, i.e., in the plasma or in the graphite "furnace".

Destructive means of sample preparation are employed for a number of reasons, including the formation of a liquid sample with similar physical characteristics to the standards, the removal of interfering matrix substances (organic material), to reduce sample complexity, and to convert the metal in the sample to an equivalent chemical form as in the determination standards. When the total metal content of the sample is the objective of the analysis, rather than a specific form of the metal, it is usually desirable to carry out mineralization procedures to completion, i.e., to remove all organic materials and convert the metal to a single form. However, the final chemical form of the metal must be consistent with the requirements of the determination technique. Hydride generation atomic absorption spectrophotometry requires a particular valence state of the metal in order to form a hydride. For example, Se must be in the selenite form, rather than the more oxidized form of selenate, in order to form a hydride.

The most commonly employed techniques used for mineralization are those that either require high-temperature thermal ashing, also known as "dry ashing" techniques, or those that utilize strong oxidizing conditions such as those provided by concentrated mineral acids and strong chemical oxidants, e.g., nitric acid, sulfuric acid, perchloric acid, hydrogen peroxide, etc. This type of procedure is generally referred to as "wet ashing".

Each has its advantages and disadvantages. Wet ashing procedures are generally more rapid, but require more attention and the use of more reagents. Dry ashing procedures are slow, often requiring days to complete. While they may proceed without persistent attention, unexpected and often unseen behavior may lead to analyte loss, or worse, cross contamination. Analyte losses during mineralization can be a significant problem for either dry or wet ashing procedures. During dry ashing, mechanical losses due to foaming or spattering not only can cause significant sample analyte losses, but also pose potential cross-contamination problems. Because some metals have chemical forms that are relatively low-boiling, significant losses through analyte volatilization can be expected if ashing temperatures are not rigorously controlled. Pb, Ge, and Hg suffer considerable losses during thermal ashing in the presence of excess chloride. The use of ashing aids that boost the volatilization temperature of the metal of interest is often necessary to prevent these losses.

Even under wet ashing conditions significant losses of some analytes can occur. The strong reducing conditions provided by charring may cause losses of metals such as Se and As, and perhaps others. This condition can occur when sulfuric acid is a component of the digestion mixture and the oxidation of organic material is incomplete and the primary oxidizing acid, e.g., nitric acid, is expended.

Microwave digestion is a relatively new wet ashing technique. It is considered an ideal mineralization technique, because of its ability to rapidly ash samples with a minimum use of acid, while limiting the potential exposure to contaminants.[6] Sample preparation utilizing microwave digestion may be complete in minutes, compared to hours or days required for wet ashing and dry ashing techniques. Modern microwave digestion apparatus utilize closed vessels capable of withstanding relatively high temperatures and pressures. The sample vessels are typically fabricated from noncontaminating PTFE. Sensors that gauge the internal pressure in the vessel control the microwave energy transmitted to the sample container. The elevated pressures and superheated acids resulting from these design features dramatically increase the rate of sample dissolution and mineralization. These features are especially important for samples with high levels of fat. Moreover, because the vessels are in a "closed" configuration during ashing, these devices offer the additional advantage of protection from loss of sample analyte through volatilization, a severe problem for some analytes in some sample preparation procedures. For determinative techniques that don't require complete sample mineralization, this apparatus can be utilized for sample dissolution, a relatively short procedure.

It should be noted that perchloric acid, a powerful oxidizing acid at high temperatures, should not be utilized in these systems due to the "closed" nature of the apparatus and the hazards created by the potential for chemical decomposition of PTFE aided by perchloric acid.

IV. DETERMINATION TECHNIQUES OF CHOICE

Numerous types of analytical methods employing a variety of determinative techniques are available to the analytical chemist for determining these analytes in biological tissues. In practice, however, a relative few are most often effectively employed in these analyses. The choice of technique is predicated on many factors, e.g., selectivity, sensitivity, ease of analysis, multi-element capabilities, lack of interferences, practicality, commercial availability of instrumentation, etc.

Quantitative techniques rather than qualitative techniques are most often selected for metals analyses. These techniques include, atomic spectroscopy, polarography, electrochemical, X-ray fluorescence, "wet chemistry" methods and assorted "hyphenated" techniques. The wet chemistry and hyphenated techniques are beyond the scope of this chapter.

Techniques based on a form of atomic spectroscopy are the most widely employed for the analysis of metals, in general. Of these, the ones most employed are based on atomic absorption spectroscopy (AAS) and atomic emission spectroscopy (AES). Jackson and Mahmood[7] and Jackson and Qiao[8] have recently reviewed the literature with regard to AAS techniques. Hill et al.[9,10] reviewed the instrumentation associated with AAS techniques.

A. ATOMIC ABSORPTION SPECTROSCOPY

The AAS techniques that have commercially available instrumentation and that are suitable for these types of analyses include: Flame-AAS (FAAS), Graphite Furnace-AAS (GFAAS), and Hydride Generation-AAS (HGAAS). Although, in theory at least, AAS techniques have the ability to perform simultaneous multielemental analysis, the present commercially available instruments are largely capable only of sequential multielemental analyses.

The principal difference in the various AAS techniques is in the means and form of sample presentation and atomization. In FAAS, the sample presentation is usually by continuous aspiration in liquid form, most often after the sample has been completely mineralized and dissolved in a suitable solvent, usually dilute acid. The sample is usually aspirated to the flame as an aerosol. In GFAAS, a small aliquot, rather than a continuous flow of sample, is deposited in a graphite "furnace". However, unlike FAAS where the sample has usually been completely mineralized by pretreatment, in GFAAS the sample can be introduced following dissolution, dilution, or other simple pretreatment. Sample desolvation and mineralization is completed *in situ* in the graphite furnace. In GFAAS some type of matrix modifier is used[11] in order to protect the analyte from premature volatilization prior to vaporization. In GFAAS the analyte is volatilized into the light path at high temperature. In the case of HGAAS only a few elements can be practically determined, based on their ability to form volatile hydrides (As, Bi, Sb, Se, Te) or based on their high volatility (Hg).[12] In HGAAS, the sample analyte is ultimately determined in gaseous form from an aliquot of a completely mineralized sample.

Some instrumental detection limits[13] of AAS techniques are shown in Table 2. Because these particular limits were determined optimally for a particular group of commercial instruments, actual method detection limits (and actual method quantitation limits) for a given analyte and instrument may be markedly different (usually greater) than those recited here. The primary value to the reader is in gauging the relative detention limits of the AAS methods.

Of the three AAS techniques, GFAAS generally has the greatest sensitivity except for those elements that can be analyzed by HGAAS. However, GFAAS has a relatively short analytical working range and the slowest sample throughput of these techniques. Most modern GFAAS instruments are equipped to operate in an unattended fashion and have automated sample dilution when the sample analyte value falls outside the range of the standards or the analytical working range. Delves and Shuttler[14] have recently reviewed the analysis of biological fluids and tissues by GFAAS.

AAS techniques for many of the analytes can suffer significant interferences due to the sample matrix and from the physical and chemical conditions encountered in the light path. These interferences can be due to background absorption (nonatomic absorption), vaporization interferences, ionization effects, or from spectral interferences. Dulude has recently reviewed these interferences.[15] In general, some type of interference should always be assumed, except under the most simple analyte-matrix conditions — ones that are seldom encountered in the analysis of biological samples. Most interferences can be

Table 2 Optimal Instrumental Atomic Spectroscopy Detection Limits[a] (µg/l)

Element	FAAS	HGAAS	GFAAS	ICP	ICP-MS
Ag	1.5	—	0.05	1.5	0.003
Al	45	—	0.3	6	0.006
As	150	0.03	0.5	30	0.006
Au	9	—	0.4	6	0.001
Ba	15	—	0.9	0.15	0.002
Be	1.5	—	0.02	0.09	0.03
Bi	30	0.03	0.6	30	0.0005
Cd	0.8	—	0.02	1.5	0.003
Co	9	—	0.4	3	0.0009
Cr	3	—	0.08	3	0.02
Cu	1.5	—	0.25	1.5	0.003
Fe	5	—	0.3	1.5	0.4
Ga	75	—	—	15	0.001
Ge	300	—	—	15	0.003
Hg	300	0.009	1.5	30	0.004
In	30	—	—	45	<0.0005
Li	0.8	—	0.15	1.5	0.03
Mg	0.15	—	0.01	0.15	0.007
Mn	1.5	—	0.09	0.6	0.002
Mo	45	—	0.2	7.5	0.003
Ni	6	—	0.8	6	0.005
Pb	15	—	0.15	30	0.001
Sb	45	0.15	0.4	90	0.001
Se	100	0.03	0.7	90	0.06
Sn	150	—	0.5	60	0.002
Sr	3	—	0.06	0.075	0.0008
Te	30	0.03	1	75	0.01
Ti	75	—	0.9	0.75	0.006
Tl	15	—	0.4	60	0.0005
U	15000	—	—	15	<0.0005
V	60	—	0.3	3	0.002
Zn	1.5	—	0.3	1.5	0.003

Note: Determined in dilute aqueous solution. These concentrations could be distinguished from zero 95 to 98% of the time. Individual method detection (and quantitation) limits may be several orders of magnitude higher depending on several factors, e.g., instrument used, sample matrix, operator skill, etc.

[a] Data from Reference 13.

overcome or minimized through techniques such as matrix matching, using matrix modifiers, background correction techniques (deuterium background correction in FAAS and GFAAS, by background correction using the Zeeman effect in GFAAS, or other instrumentally employed techniques).[7,16] Many of the physical and chemical interferences encountered in FAAS and GFAAS can be effectively overcome by employing the method of standard additions.

B. ATOMIC EMISSION SPECTROSCOPY

The most commonly employed AES techniques for the determination of metals in biological tissues are the argon plasma emission techniques, i.e., induced coupled plasma (ICP), direct current coupled plasma (DCP), and induced coupled plasma-mass spectrometry (ICP-MS). These techniques have gained in popularity as the ease of use has improved and the practicability of the instrumentation has improved. However, the cost for most of these instruments still remains substantially more than that for the AAS instrument of comparable or superior analytical sensitivity. Unlike commercially available AAS instrumentation, ICP instrumentation is available in various designs and configurations that allow simultaneous or sequential multielement analysis. Some available ICP instrumentation can detect all of the metals of interest and can quantify many at the levels usually encountered in metal toxicity.

Unlike FAAS, where the dissolved sample is commonly entrained through the combustion mixture of acetylene-air, in ICP the dissolved sample is entrained through an argon plasma, maintained by a radio-frequency field and ionized argon gas. The sample experiences extremely high temperatures in the plasma, 5500–8000 K. These temperatures promote complete atomization of the sample. These instruments are available in two different configurations, i.e., one for simultaneous multielemental determinations and one for sequential multielemental determinations. Analyte emission lines are separated either by a polychromater-based or monochromater-based optical system.

Whereas ICP has comparable detection limits to FAAS, the range of signal linearity in ICP is far superior to that of FAAS or any of the AAS techniques. For some analytes, linearity of response in ICP is more than five orders of magnitude. Except for ICP-MS, GFAAS has detection limits that are far superior to any of the AES techniques (see Table 2).

Because of the extremely high temperature of the plasma, compared to that achieved in AAS techniques, AES techniques suffer inherently fewer interference problems, in general. At these high temperatures fewer chemical interferences are experienced than in AAS techniques. Physical differences in the sample matrix and spectral interferences are more important interference problems in ICP. Spectral interferences may be especially important when trace element levels are being determined in the presence of macro amounts of other elements that may have similar emission lines. Matching the elemental composition of the standards with the sample may resolve this type of interference. Typically, the resolution of spectral interferences may be more efficiently achieved with simultaneous instruments than with sequential ones. Physical differences in the sample matrix, e.g., viscosity, compared to determination standards that result in different efficiencies of the analyte reaching the plasma, will also result in inaccurate sample determinations unless effective countermeasures are employed.[17] Physical interferences can often be corrected by using internal standards.[18] Background correction in ICP is accomplished by several instrumental means, usually involving some "off-line" measurement. Dawson and Snook recently reviewed this subject.[19]

ICP-Mass Spectrometry (ICP-MS) employs a quadrapole mass spectrometer as a measurement device rather than a mono- or polychromator-based system.[20] For ICP-MS the ICP part of the instrument is used to generate singly charged ions, not analyte line emissions. Individual mass values, i.e., isotopes of metals, are determined in ICP-MS. This unique capability may be exploited for determinations where a particular isotope of a metal is the analytical objective, for example, after neutron activation or stable isotope enrichment.[21] Multielement analysis of biological samples has been reported by Vaughn and Horlick.[22]

Compared to the other AS techniques, ICP-MS has by far the lowest limits of detection. Because of these low detection limits, small samples can be analyzed even after considerable dilution. This sensitivity also implies the need for extreme care in sample preparation in order to exclude environmental contamination. Because of the remarkable sensitivity of ICP-MS, sample "memory" can be an important consequence when samples have widely disparate levels of analytes. Lower, but detectable, levels of analytes may be observed in subsequent determinations. Longer "wash-out" times may be required on these occasions.

ICP-MS suffers from the same matrix-based, physical interferences as ICP. However, the spectral interferences seen in ICP are not encountered in ICP-MS, since the detection is based on mass and not on emitted light. Correction of physical interferences that result from different efficiencies of analytes in the sample and the standards reaching the plasma can be corrected either by the use of internal standards or an alternate means of sample presentation, such as ultrasonic nebulization. Interferences that result in mass overlap with the analyte may be corrected by choosing an alternate mass value for the analyte, or by using a higher mass resolution.

V. LESS-OFTEN EMPLOYED TECHNIQUES

A. ELECTROMETRIC TECHNIQUES

Polarography, voltametry, ion selective potentiometry, and related techniques have been utilized in determining a number of the toxic metals. However, these techniques, for the most part, are not generally applicable for many of the toxic metals, either individually or in systematic multimetal analyses of tissues and body fluids. A comprehensive review of the methodology and applicability of electrochemistry to metals and other analytes appeared recently.[23]

In theory, the most applicable voltametric methods are those that employ anodic stripping, since most of these metals are thought to be in cationic states after mineralization. Notable exceptions are Se and

As. Anodic stripping voltametric (ASV) techniques have demonstrated great sensitivity for some of the toxic metals such as Cu, Zn, Cd, Pb and can easily determine these metals quantitatively in the low parts per billion range in tissue and other biological matrices, such as food. However, while ASV methods are direct in nature, highly sensitive, and can be effectively employed, they demand a level of expertise of the analyst and a level of care in operation that discourages their use except where single or a few analytes are determined repetitively in routine samples, e.g., Pb in blood, where the instrumentation is largely automated. In virtually all ASV analyses the sample must first undergo complete mineralization prior to analysis. Moreover, to overcome interferences from the sample matrix, a method of standard additions or some equivalent technique must usually be employed to insure quantitative results.

With the improvement of the sensitivity and selectivity of the new generation of GFAAS instrumentation equipped with superior background correction systems, such as those based on principles by Zeeman or others, and the development of practical ICP-MS instrumentation, the advantage of superior sensitivity once offered by the ASV techniques has become less of an advantage.

B. X-RAY SPECTROMETRY TECHNIQUES (XRS)

Analytical techniques based on XRS, e.g., X-ray fluorescence (XRF), electron probe microanalysis (EPMA), and particle-induced X-ray emission analysis (PIXE), have been applied to the analysis of some toxic metals in some biological samples. Although these techniques have not, in general, enjoyed the popularity of those based on AS techniques, under some circumstances they may offer unique advantages. Their application to metals analysis in general, and to biological samples has recently been reviewed.[24]

XRS techniques are based on the detection and measurement of X-rays emitted from metal atoms when outer shell electrons are captured by a vacant inner electron shell(s) (primarily the K, L, and M shells). This capture occurs when some or all of these inner shell electrons have been displaced by some form of irradiation or particle bombardment, e.g., high-energy photons (from an X-ray or radioisotopic source), electrons, protons, etc. Depending on the energy of the excitation source, an individual metal may emit relatively few X-rays of characteristic energy. Quantitation of the emitted X-rays may be achieved by employing means similar to those employed in gamma spectrometry of radioisotopes. Electronic energy discrimination and multichannel analysis permit simultaneous, multimetal analysis by these techniques.

These techniques may yield qualitative or quantitative results for metals in biological samples, depending on a number of factors and circumstances. The degree of accuracy of the result is largely dependent on the technique employed, sample preparation, sample matrix, the metal, the level of occurrence of the metal, and the degree of validation employed. XRF techniques can be quantitative if uniform, thin films of the samples are prepared. The analysis of each sample type must be validated for each for the metals to be quantitatively determined. Ideally, certified materials should be prepared in an identical fashion to the samples for this purpose. Likewise, quantitation standards must be similarly prepared in order to eliminate the effects of the sample matrix on the determination. Because EPMA and PIXE utilize electrons and protons for excitation, penetration into the sample is limited to only relatively short distances. The results for these two techniques are usually considered qualitative in nature, without employing extraordinary means validation. Because the primary use for EPMA techniques is to reveal the metal composition of only limited areas, e.g., subcellular areas, the analytical information is expected to be qualitative in nature.

XRS techniques such as XRF and PIXE may be ideal for hair and nails.[25,26] Semiquantitative analysis of hair for numerous toxic metals may be performed rapidly by XRF and PIXE, with little sample preparation except to remove extraneous contaminants. Some toxic metals tend to accumulate in hair. Hair appears to be a means of determining exposure to some metals, at least for chronic exposures, or acute exposures after some period of time has elapsed between exposure and sampling. Sequential analysis performed along the length of a hair theoretically can reveal various aspects of the exposure, e.g., the onset, cessation, and period of exposure to the metal. XRF employing radioisotopic sources are being used for the in vivo determination of the toxic metal content of bone.[27]

C. NUCLEAR TECHNIQUES

Nuclear techniques (NT), such as neutron activation analysis (NAA), coupled with typical radiometric counting or ICP-MS offer quantitative, simultaneous, multielemental capability for many of the toxic metals. NTs employ neutrons, charged particles, or photons in order to induce radionuclide formation within the sample. Depending on the means, the intensity, the duration of the activation step, and the elemental content of the sample, the metal of interest achieves a level of radioactivity that can be determined and the metal concentration computed. Excellent, up-to-date reviews of NT and their application are available.[28,29]

The applicability of these techniques to specific metals depends on several factors, e.g., the activation characteristics of the metal, its concentration in the sample, other elemental constituents of the sample, the means of radionuclide measurement, etc. Smith et al.[30] has compared the detection by ICP-MS with state of the art gamma spectrometry systems. The limits of detection can exceed those of methods that employ AS techniques.[31]

Although these techniques have some advantages, they are not routinely utilized for toxic metal analysis, primarily due to the lack of equipment accessible to the analyst. The activation step usually employs a nuclear reactor, or an accelerator. An additional drawback is that they produce radionuclides and employ instrumentation that requires more care and oversight than other metal analytical techniques employ. Because of the lack of convenience and the time-consuming nature of these analyses, these techniques are more often employed for systematic studies of large numbers of samples, such as environmental surveys.

While somewhat inconvenient and time consuming, NAA offers some distinct advantages apart from its capability as a sensitive, simultaneous, multielemental analytical tool. Because the analysis of solid samples can proceed without sample dissolution, fewer blank corrections are needed and the potential for sample contamination can be substantially lower than that for other techniques that require destructive sample preparations.

VI. QUALITY CONTROL AND QUALITY ASSURANCE CONSIDERATIONS (QA/QC)

Metal analyses, especially those for trace levels, have numerous procedural and methodological pitfalls that can lead to serious failures both in accuracy and precision. The literature is replete with significant errors in trace metal content of tissues and other materials. Only through properly designed and executed QA/QC procedures can many of these problems be detected and, hopefully, controlled. A sound, well-designed QA/QC program should prevent inadvertent use of imprecise and/or inaccurate data. Equally important, meaningful (QC/QA) procedures lend credibility and confidence to the use of all analytical data. A general but comprehensive approach to QA/QC for chemical analyses is offered by Taylor and others.[32-34]

Modern standards of professional conduct require that the credibility of all data be established in relationship to their intended use. Many organizations and institutions provide standards of conduct for quality assurance.[35-38] While these standards do not explicitly address the analysis and reporting of metals results, they do provide a framework of standard practices that applies to metals analyses. The level of data credibility (proof of accuracy and precision) required needs to be framed in relationship to the use of the data, e.g., research, medical treatment, activities intended for regulatory decisions, those intended for forensic evidence destined for the courtroom, etc. Perhaps, in most cases of metal toxicity, the analytical levels encountered in indicator tissues are either at a markedly increased level than those of normal tissues, or the form of the metal or the metal itself (or both) are essentially nondetectable in the indicator tissues of the normal, unexposed individual.

Analytical error, or bias, can originate from numerous sources, e.g., procedural contamination, procedural analyte losses, instrumental interferences, etc. Regardless of the origin(s), in principle, biases should be determined for each metal-matrix combination. Methods and procedures should be appropriately selected or modified in order to eliminate or limit the significant effects of biases on the analytical result where possible.

The method of choice must be able to detect and accurately quantify the metal of interest with a degree of precision necessary to meet the goals of the activity. In general, it is better to choose a method that can determine the metal of interest at a fair margin above the limit of quantitation for the method and sample size, than one that operates near the limit of quantitation for the sample size. Instrumental sensitivity (limit of detection and limit of quantitation) should be determined and should be near those specified for that instrument. The limit of quantitation for the overall method should likewise be determined.

Once selected, the method of choice should be validated using suitable reference materials, i.e., certified reference material or standard reference material. It is most desirable that the reference material should be of similar or identical matrix as the tissue sample. Ideally, the form of the metal in the sample (if known) and that of the reference material should be identical or equivalent. Unfortunately, suitable certified or standard reference materials are not always available for method validation. Under these conditions, a "spiked" material can be used as a surrogate. The metal of interest in similar or identical form can be added to the "blank" sample of the same tissue and carried through the entire procedure. This material can be used to essentially achieve method validation.

Typically, it is recommended that analysis of duplicates be conducted at a frequency of 10% or more. Each run or batch of samples should also contain a reference material of some type, preferably certified. These routine analyses will provide an ongoing basis for determining operational performance.

Reference materials are available from a number of sources. A computerized data base that lists reference materials available world-wide can be obtained from the National Institute of Standards and Technology[39] (also see Appendix A).

VII. CONCLUDING REMARKS

Medicine, nutrition, the environment, and technical needs are continuing to fuel rapid and highly competitive development of commercial instrumentation for metals analysis. Recent developments in instrumentation has favored instruments with simultaneous, multielemental capabilities, such as ICP and ICP-MS. While ICP-MS, with its enviable detection capabilities, still remains a relatively expensive tool for metals analyses, the use of axially configured torches, like those used in ICP-MS, have improved the detection limits for many metals by ICP. These new-generation ICP systems coupled with new signal detection systems, such as charge-injection device solid-state detectors or segmented-array charge-coupled device detectors have greatly improved their utility, and reduced their operational complexity. The abandonment of the traditional optical bench and its array of photomultiplier tubes, one of which was required for each element determined in simultaneous ICP, has dramatically lessened the size of most of these instruments and reduced their cost.

Apart from improvements in instrumental sensitivity, selectivity, etc., considerable progress has been made for most instruments in regard to utility of use. This has been especially important for the GFAAS instruments. Fully automated sampling systems, improved background correction capabilities, and computer-assisted functions have facilitated (unattended) sample analysis.

There continues to be a need to develop techniques and companion commercial instrumentation that will allow for the safe, accurate, and precise determination of toxic metals, such as lead, *in vivo*. This is especially important for lead and other toxic metals that accumulate in relatively inaccessible tissues of the body.

There is likewise a critical need to better understand the metabolism of many of the toxic metals at various levels of intake. Because the biological form(s) of toxic metals may be important, there needs to be means of speciating metals. Improvements in separation technology appear to be key in this area.

While some of the recent developments in instrumentation are focused on techniques that require little or no sample pretreatment, there still remains a need for automated or semiautomated means of sample preparation (mineralization) that is rapid, requires small amounts of reagents, and subjects the sample to little potential contamination or analyte loss. The first-generation microwave digestion apparatus have met that need for many sample types and metals.

REFERENCES

1. Goyer, R. A., Toxic effects of metals, in *Casarett and Doull's Toxicology,* 3rd ed., Klaassen, C. D., Amdur, M. O., and Doull, J., Eds., Macmillan, New York, 1986, 582.
2. NRC Subcommittee on the Tenth Edition of the RDAs, Food and Nutrition Board, Commission on Life Sciences, National Research Council, Recommended Dietary Allowances, 10th ed., National Academy Press, Washington, D.C., 1986, 174.
3. ASTM Water, Part 31, *ASTM Standards,* American Society of Testing and Materials, Philadelphia, PA, 20.
4. Environmental Protection Agency Handbook for Analytical Quality Control in Water and Wastewater Laboratories, EPA-600/4–79–019, Cincinnati, OH, 1979.
5. Zief, M. and Mitchell, J. W., *Contamination Control in Trace Element Analysis,* John Wiley & Sons, New York.
6. Rains, T. C., Application of atomic absorption spectrometry to the analysis of foods, *Analytical Spectroscopy Library,* Vol. 5, Haswell, S. J., Ed., Elsevier, Amsterdam, chap. 4d, 1991.
7. Jackson, K. W. and Mahmood, T. M., *Anal. Chem.,* 66, 252R–279R, 1994.
8. Jackson, K. W. and Qioa, H., *Anal. Chem.,* 64, 50R–66R, 1992.
9. Hill, S. J., Dawson, J. B., Price, W. J., Shuttler, I. L., and Tyson, J. F., *J. Anal. At. Spectrom.,* 7, 215R–277R, 1992.
10. Hill, S. J., Dawson, J. B., Price, W. J., Shuttler, I. L., and Tyson, J. F., *J. Anal. At. Spectrom.,* 8, 197R–237R, 1993.

11. Sommer, L., Komarek, J., and Burns, D. T., *Pure Appl. Chem.,* 64, 213–226, 1992.
12. Campbell, A. D., *Pure Appl. Chem.,* 64, 227–244, 1994.
13. Anon, *Guide to Techniques and Applications of Atomic Spectroscopy,* Tech. Bull. no. L-655F, Perkins-Elmer Corp., Norwalk, CT, 1993.
14. Delves, H. T. and Shuttler, I. L., *Analytical Spectroscopy Library,* Vol. 5, Atomic Absorption Spectrometry. Theory, Design, and Applications, Haswell, S.J., Ed., Elsevier, Amsterdam, 381, 1991.
15. Dulude, G., In *Advances in Atomic Spectrometry,* Vol. 1, Sneddon, J., Ed., JAI Press, Greenwich, CT, 126, 1992.
16. Rossi, G., in *Applications of Zeeman Graphite Furnace Atomic Absorption Spectrometry in the Chemical Laboratory and in Toxicology,* Minola, C. and Carroll, S., Eds., Pergamon Press, Oxford, 3, 1992.
17. Thomas, R. J. and Anderau, *At. Spectrosc.,* 10, 2, 1989.
18. Myers, S. A. and Tracy, D. H., *Spectrochim. Acta,* 38B, 9, 1983.
19. Dawson, J. B. and Snook, R. D., *J. Anal. Spectrosc.,* 18, 517–537, 1993.
20. Douglas, D. and Houk, R. S., *J. Anal. At. Spectrosc.,* 8, 18, 1985.
21. Ting, B. and Janghorbani, M., *Mikrochim. Acta,* 3, 315, 1989.
22. Vaughn, M. and Horlick, G., *Clin. Chem.,* 37, 2, 1991.
23. Ryan, M. D., Bowden, E. F., and Chambers, J. Q., *Anal. Chem.,* 66, 360R–427R, 1994.
24. Torok, S. B. and Van Grieken, R. E., *Anal. Chem.,* 66, 186R–206R, 1994.
25. Valkovic, V., Jaksic, M., Watt, F., Grime, G. W., Wells, J., and Hopewell, J. W., *Nucl. Instrum. Methods Phys. Res.,* B75, 173–176, 1993.
26. Horino, Y. Mokuno, Y., and Fujii, K., *Nucl. Instrum. Methods Phys. Res.,* B75, 535–538, 1993.
27. Todd, A. C., McNeill, F. E., and Fowler, B. A., *Environ. Res.,* 59, 326–335, 1992.
28. Ehmann, W. D., Robertson, J. D., and Yates, S. W., *Anal. Chem.,* 66, 229R–251R, 1994.
29. Ehmann, W. D., Robertson, J. D., and Yates, S. W., *Anal. Chem.,* 64, 1R–22R, 1992.
30. Smith, M. R., Wyse, E. J., and Koppenaal, D. W., *J. Radioanal. Nucl. Chem.,* 160(2), 341–354, 1992.
31. Vandecasteele, C., *Mikrochim. Acta,* 2, 370–389, 1991.
32. Taylor, J. K., *Quality Assurance of Chemical Measurements,* CRC Press, Boca Raton, FL, 1987.
33. Keith, L. H., Libby, R. A., Crummett, W., Taylor, J. K., Deegan, J., and Wentler, G., *Anal. Chem.,* 55, 2210–2218, 1983.
34. Taylor, J. K., *J. Test. Eval.,* 11, 355–357, 1983.
35. ACIL, *Quality Control System Requirements for a Testing and Inspection Laboratory,* American Council of Independent Laboratories (ACIL), Washington, D.C.
36. Code of Federal Regulations, Title 21, Foods and Drugs, Chap. 1, FDA, D part 38, GLPs for Non-Clinical Laboratory Studies, U.S. Government Printing Office, Washington, D.C.
37. Code of Federal Regulations, Title 42, Public Health, Chap. 1, Public Service, Part 74, Clinical Laboratories, U.S. Government Printing Office, Washington, D.C.
38. Federal Register Vol. 57, No. 40, Subpart P, Quality Assurance for Moderate or High Complexity, or Both (Para 493.1701), U.S. Government Printing Office, Washington, D.C., 1992.
39. COMAR (COde MAteriaux Reference), Certified Reference Materials Data Base, Version 6.0, NIST Special Database 15, National Institute of Standards and Technology, Gaithersburg, MD, 1994.

APPENDIX

CONTACTS FOR THE STANDARD REFERENCE DATA PROGRAM

In the U.S. Only

Phoebe Fagan
National Institute of Standards and technology
Standard Reference Data Program
Building 221, Room A323
Gaithersburg, MD 20899, U.S.
Internet: fagan@enh.nist.gov
Phone: (301) 975–2213
FAX: (301) 926–0416

Outside the U.S.

Dr. Alain Marschal
Chef du Department
References et Essais Chimiques
1, Rue Gaston Boissier 75015 Paris, FRANCE
Telex: 202 31 9 F
Phone: (33) (1) 40.43.37.50
FAX: (33) (1) 40.43.37.37

Chapter 6

Concepts on Biological Markers and Biomonitoring for Metal Toxicity

Lars Gerhardsson and Staffan Skerfving

I. INTRODUCTION

Exposure to toxic metals is often assessed by measurements in air, foods, and/or water, and sometimes also in soil and other materials, i.e., environmental monitoring. Although this is valuable in many cases, e.g., for identification of sources of exposure and checking of effects and measures to reduce them, such exposure estimates have many weaknesses.

Thus, it is often only possible to assess the levels in a limited number of samples obtained during a short time period. Also, air samples only reflect exposure via inhalation and foods only orally. This often does not cover all routes of exposure. Thus, there is often combined inhalation and oral intake and, in addition, skin uptake. Further, inhalation exposure is affected not only by the air level, but also by the pulmonary ventilation, which varies widely between individuals. Moreover, the choice of foods and other behavior (such as mouthing in infants, which is important for exposure from soil) differ.

Some of these problems may be overcome by employment of biological monitoring, i.e., the use of biomarkers for assessment of exposure to metals. The biomarkers take into account exposure from different sources (air, foods, water, soil, etc.) and through various routes (inhalation, gastrointestinal, and skin). Also, they may reflect exposure during different time periods. In some cases, the biomarker may reflect exposure during long periods, up to decades. In others, it only indicates the most recent uptake.

Mostly, the content of the metal itself is analyzed in biological samples. For a long time, the possibilities of biological monitoring were restricted by the limited performance of the analytical techniques. However, the situation was dramatically improved by the introduction a few decades ago of the neutron activation and atomic absorption techniques. Recently, new possibilities have been offered by the inductively coupled plasma/mass spectrometry technique, which allows analysis of low concentration of small samples, and determination of many elements simultaneously.

Some of the elements, e.g., lead and mercury, exist in different chemical forms with varying toxicity. In those cases it may not be sufficient to analyze the total content of the element. Instead, a speciation of the chemical entity may be necessary (Schütz et al., 1994).

In most cases, blood, urine, or hair are used as the index media for biological monitoring. With blood, whole blood, plasma, or erythrocytes may be employed. For some of the elements, this choice may be important. Thus, some of the metals, e.g., lead, methylmercury, and cadmium, display considerable variation of concentrations between plasma and erythrocytes. Further, the distribution of different chemical species may vary. Also, the turnover rates are substantial in some cases.

In the case of urine, the concentration varies with the urinary flow, which also varies considerably. One way of handling this is to relate the excretion to time, e.g., 24 h. However, this is not always suitable, since there is often an incomplete sampling and, in particular in connection with occupational exposure, a risk of contamination at the work place. Thus, spot samples are often used. In that case, there is a need for adjustment for degree of dilution. This is particularly important when the urinary element concentration is related to an effect parameter also determined in the urine, e.g., proteinuria. The adjustment may be made by recalculation of the concentration to a chosen urinary density (usually 1.020) or osmolarity. Alternatively, the element concentration may be related to the urinary creatinine content, which is particularly useful in cases with glucosuria. However, the creatinine adjustment does not fully compensate for the urinary flow. This is a problem when the urine is very dilute (density <1.010 or about <0.3 g creatinine/l) or concentrated (>1.030 or about >2.5 g/l). Samples outside these limits are mostly disregarded. Further, for some of the elements there is a diurnal variation of the excretion. Thus, sampling should preferably be performed on samples obtained at a certain time of the day, usually in the morning.

For some elements, especially lead and aluminum, and to some extent also mercury, mobilization tests have been widely used for assessment of the relevant body burden. In such cases, a chelating agent is administered parenterally or orally, and the urinary excretion of the element is measured during a specified time period, usually 24 h or less. Since the procedure increased the urinary element concentrations considerably, such tests were particularly important earlier, when the analytical methods had limited performance. These tests are not without problems, e.g., because of redistribution of the element to sensitive organs. Further, modern analytical techniques are, in most cases, sensitive enough for accurate determination of the elements without chelation. Moreover, the information obtained from samples from subjects who have not received chelating agent is as good as that with chelation. Thus, it is likely that the mobilization tests will lose importance in the future.

In the past decade, new possibilities for biological monitoring of elements have been developed. Thus, neutron activation and X-ray fluorescence (XRF) techniques may be used for assessment *in vivo* of contents of, in particular, cadmium in the kidney and liver, and by XRF, lead in bone (Ahlgren et al., 1976; Scott and Chettle, 1986; Hu et al., 1989; Nilsson and Skerfving, 1993). These methods give valuable information on long-term exposure and, in the case of cadmium in kidney, on the concentration in the critical organ, i.e., the organ first suffering damage at chronic long-term exposure. The equipment is still rather complicated, and thus only available at a limited number of institutions. However, mobile units have been developed, and may be used in field studies.

To interpret the information on element concentrations in index media, there is a need for knowledge on the relationship between exposure and levels as well as patterns of distribution, in particular to the critical organ, and excretion. This information is often described in metabolic models, which may be classical compartment models or physiologically based ones. For some of the elements, subclinical biochemical effects of the exposure have been widely used for biological monitoring of exposure. For example, effects on heme synthesis have been widely used for assessment of lead exposure. Monitoring may be made of enzyme inhibition, or of changes of metabolite patterns in urine or blood induced by such. Such measurements are often both simple from an analytical point of view and sensitive. However, the specificity is sometimes low, as the enzyme activity and metabolite patterns may be affected by other elements and also by other factors.

Here, the term biological monitoring will be employed for the use of biomarkers for assessment of the exposure to elements, i.e., of internal dose, body burden, and or/concentration in the target tissue (critical organ; Elinder et al., 1994). As a rule, such biomarkers will also give information about the risk of toxic effects in the exposed individual/group. Biomarkers may also be used for health monitoring/surveillance, i.e., as bioindicators of effects, early health effects, and/or overt health impairment. Biomarkers of such adverse effects will not be discussed in any detail here, although a few examples will be mentioned, in particular the effects on the kidney and heme synthesis. However, there are, of course, many more biomarkers and examinations which are sometimes used for surveillance of health (WHO, 1993) which will not be discussed here.

In this chapter, only the most important metals are described. There are many more metals. For these, more comprehensive handbooks should be consulted (e.g., Friberg et al., 1986; Clarkson et al., 1988a).

II. CADMIUM

There is a wealth of information on the toxicity of cadmium. Several extensive reviews have been published recently (Friberg et al., 1985; Nordberg and Nordberg, 1988; WHO, 1992; Nordberg, 1993; IARC, 1994). To limit the length of this treatise, these reviews will be referred to for further references.

A. EXPOSURE

Humans are exposed to cadmium in occupational settings by inhalation and through contaminated foods, and tobacco. Heavy exposure may occur in the mining and smelting of cadmium (and zinc, since zinc ores often contain cadmium as an impurity), manufacture or use of cadmium pigments, alloys (including welding and soldering), surface coatings, stabilizers (of polyvinyl chloride), manufacture of alkaline accumulators, and in handling and recycling of scrap containing cadmium (WHO, 1992). The use of cadmium has now been restricted in several countries.

As to the general environment, there is fairly extensive exposure to cadmium through smoking, since tobacco contains cadmium, in general corresponding to 1–2 µg per cigarette (Svensson et al., 1987; WHO, 1992; Bensryd et al., 1994). Further, there is oral exposure through water and foods. In Sweden, the daily intake through foods is about 10 µg (Schütz, 1979). In certain areas the exposure may be extremely high, in particular from rice exposed to cadmium-containing irrigation water contaminated by zinc mines (WHO, 1992).

B. METABOLISM

1. Absorption

At inhalation exposure, absorption is dependent upon the particle sizes and their solubility. Uptakes ranging from 0.1–35% of the inhaled amounts have been reported (Nordberg and Nordberg, 1988). Some of the particles are cleared from the respiratory tract and swallowed. Thus, the total uptake, including the gastrointestinal contribution, may range from 5–37%. Of the cadmium inhaled at tobacco smoking, about 10% is absorbed.

The cadmium absorption from the gastrointestinal tract is usually about 5% (Nordberg and Nordberg, 1988). However, it varies considerably. Thus, subjects with iron deficiency may absorb as much as about 20%. Absorption of cadmium through the skin is extremely low.

2. Distribution

Cadmium is absorbed by the blood plasma. There, it is bound to albumin and other high molecular weight proteins (Nordberg and Nordberg, 1998). The level of cadmium in plasma is extremely low. Cadmium penetrates the blood cells, and the erythrocyte concentration is much higher than in plasma.

The cadmium in plasma is taken up by the liver (Nordberg and Nordberg, 1988). There, it induces synthesis of metallothionein, a low molecular weight protein containing a high proportion of cystein, the sulfhydryl groups of which bind cadmium efficiently. A limited fraction of the metallothionein-cadmium complex from the liver reenters the circulation.

Mainly because of its low molecular weight, the cadmium-metallothionein in plasma is effectively filtered through the renal glomeruli and is reabsorbed in the proximal tubuli, where it gradually accumulates (Nordberg and Nordberg, 1988). Thus, at chronic exposure, the kidney contains a major part of the body burden of cadmium.

The metallothionein–cadmium complex is continuously broken down in lysosomes of the tubuli cells (Nordberg and Nordberg, 1988). The free cadmium induces renal synthesis of metallothionein, which again binds cadmium, which is then degraded again. Thus, at any time, only a fraction of the cadmium is not bound. The bound cadmium is not toxic, while the free metal ion may exert toxicity. Such occurs when the accumulation is so high that the synthesis capacity is exceeded. Some cadmium from the kidneys is continuously released into the blood stream. Smaller proportions of cadmium is distributed to other organs, e.g., the pancreas and testis. Cadmium passes the blood-brain barrier only to a very limited extent.

There is no biotransformation of cadmium.

3. Elimination

As mentioned above, the metallothionein–complex is filtered through the renal glomerula, the main fraction of which is reabsorbed in the proximal tubuli and accumulated in these cells. However, there is shedding of these cells into the urine, which is the major route of excretion (Nordberg and Nordberg, 1988). Cadmium is also, to a minor extent, excreted via bile and pancreatic juice into feces. Small amount of cadmium are incorporated into hair.

Before the formation of the placenta, cadmium reaches the embryo. Later cadmium is deposited in the placenta, and there is limited passage of cadmium into the fetus (Nordberg and Nordberg, 1988). Moreover, there is only minimal excretion in milk.

4. Metabolic Models

Cadmium models have been proposed (Kjellström and Nordberg, 1985; WHO, 1992). The main parts of the body burden are contained in the liver and kidneys. The elimination of cadmium from the kidney is very slow, with a biological half-time of several decades. There is a relative increase of the renal proportion with increasing time of exposure (up to one-third to half the body burden). Within the kidney the highest level is found in the cortex, the concentration of which is 1.25 times higher than the whole kidney. Elimination from the kidney is very slow (see below) and governs the total body excretion.

C. HEALTH EFFECTS

1. Mechanisms of Toxicity

Rather little is known about the basic mechanism behind cadmium toxicity. However, it is well established that cadmium binds to the sulfydryl group of proteins. If this occurs on an enzyme, its function may be inhibited, which may result in toxic effects. However, there are also other possible mechanisms (Nordberg and Nordberg, 1988).

2. Lung Disease

Welding or soldering with materials containing cadmium may cause exposure to cadmium fumes, which may, within hours, induce severe lung damage (metal fume fever and pneumonitis, with pulmonary edema), and may later result in lung fibrosis. Moreover, lower (but still fairly considerable) inhalation exposure during a long time period may be associated with chronic obstructive lung disease (Nordberg and Nordberg, 1988).

3. Gastrointestinal Tract

Intake of food or water containing very high amounts (milligrams) of cadmium may cause acute abdominal disorder (Nordberg, 1993).

4. Kidney Disease

The kidney is the critical organ at long-term exposure to cadmium (Friberg et al., 1985). The proximal tubuli are mainly affected, although glomerular effects have also been described. The damage results in tubular proteinuria, with dominance of low-molecular weight proteins, i.e., β_2-microglobulin, retinol binding protein (RBP), and alpha$_1$-microglobulin (protein HC), as a result of a decrease of reabsorption in the proximal tubuli of plasma proteins filtered through the glomeruli. For the same reason, there is excretion of glucose and amino acids. Also, there is urinary excretion of tubular enzymes, e.g., N-acetyl-β-glucosaminidase (NAG).

Application of the NAA technique (see Section II.D) in studies of cadmium workers has been useful in estimating the critical concentration in kidney cortex, i.e., the concentration at which tubular dysfunction occurs (Roels et al., 1981; Ellis et al., 1984). Tubular damage occurs in about 10% of workers who have a cadmium concentration of about 200 µg/g in the renal cortex.

The U-Cd that corresponds to a renal concentration of 200 µg/g is about 10 µg/g creatinine (corresponding to about 10 µmol/mol creatinine, 10 µg/l, and 15 µg/24 h; WHO, 1992). Recent studies of the general population have indicated that a fraction may have slight tubular proteinuria already at U-Cd of 2–4 µg/24 h, and that diabetics may be particularly sensitive (Buchet et al., 1990). This U-Cd may correspond to a kidney cadmium concentration as low as 50 µg/g. Further, at approximate steady state, an U-Cd of 10 µg/g corresponds to a B-Cd of about 100 nmol/l (10 µg/l).

5. Genotoxicity

Cadmium causes genotoxic effects in a variety of types of eukaryotic cells, including human. Further, cadmium and its compounds are carcinogenic in humans, as are cadmium compounds in animals (IARC, 1994). Thus, cadmium and its compounds were classified as "carcinogenic to humans (Group 1; IARC, 1994)".

D. BIOLOGICAL MONITORING

1. Blood Cadmium Levels

B-Cd increases after onset of cadmium exposure. Thus, blood cadmium is often used for biological monitoring. It has many advantages; samples are easy to obtain and the analysis is uncomplicated, but there are some limitations.

After a decrease of cadmium exposure there is a decrease of B-Cd. This decay has at least two components: one relatively fast one, which has a half-time of about a couple of months (Welinder et al., 1977; Järup et al., 1983). Later, a very slow elimination phase dominates, which reflects elimination from the kidney (and the rest of the body).

Thus, B-Cd reflects both recent exposure and the body burden resulting from earlier uptake. After a strong increase of exposure, the relative contribution of the ongoing uptake is important. At steady state, which occurs only after decades, the B-Cd mainly reflects the body burden, which is also the case a few months after a sharp decrease of exposure (Welinder et al., 1977).

When whole-blood cadmium concentration is analyzed, it is mainly the erythrocyte content that is assessed. Theoretically, serum/plasma levels would be more suitable, as they may reflect the metabolically most active pool. However, the cadmium levels in plasma/serum are so low that problems with contamination and analysis are difficult to overcome. Metallothionein-bound cadmium in plasma has been proposed as an index of exposure (Nordberg et al., 1982), but is still not available for routine use (Nordberg and Nordberg, 1988).

The B-Cd in subjects without particular cadmium exposure varies between different regions and between smokers and nonsmokers. In Sweden, the present level in adult never-smokers without occupational exposure is about 1.7 nmol/l (0.18 µg/l), in ex-smokers 2.3 (0.25 µg/l), and in current smokers 12 (1.3 µg/l) nmol/l (Svensson et al., 1987; Bensryd et al., 1994). The levels in children are lower (Willers et al., 1988). In many countries it is considerably higher (Nordberg and Nordberg, 1988). There is an association between lead levels in maternal and cord blood (though the latter is somewhat lower).

The relationship between various kinds of renal toxic effects and B-Cd has been discussed above (see Section II.C).

2. Cadmium Concentration in Urine

U-Cd has been fairly widely used as an index of cadmium load, since the excretion in urine is generally proportional to the kidney content, and thus to the body burden (Nordberg and Nordberg, 1988).

However, there is a considerable interindividual variation in the relationship between U-Cd and the kidney cadmium content. One important factor behind this is variations in the dilution of the urine. This can be handled by adjustment for the creatinine content or density, or by sampling urine for 24-h periods, which, however, is often not possible, and is connected with a considerable risk of contamination and sampling error (see Section I). Moreover, when the renal cortex cadmium concentration is sufficiently high to have caused tubular damage, the relative U-Cd increases.

The U-Cd in subjects without occupational exposure varies considerably between different areas, and between smokers and nonsmokers (Nordberg and Nordberg, 1988). In nonsmokers, it is generally below 1 µg/g creatinine (corresponding to about 1 µmol/mol creatinine, 1 µg/l, or 1.5 µg/24 h). In smokers it is several times higher (Christoffersson et al., 1987a; Nilsson et al., 1995), as it may be in nonsmokers in areas with a high intake through foods.

The relationship between renal toxic effects and U-Cd has been discussed above (see Section II.C).

3. Cadmium Concentrations in Kidney and Liver

As stated above, cadmium accumulates in these organs. During the 1970s and 1980s, several groups used NAA techniques for the measurement of cadmium in liver and total kidney (Vartsky et al., 1977; Roels et al., 1981; Ellis et al., 1984; Mason et al., 1988; Morgan et al., 1990; Chettle and Ellis, 1992). The method is based on the fact that neutrons are absorbed by cadmium, causing emission of gamma rays of characteristic energy. Suitably calibrated measurement of these gamma rays enables the amount of cadmium in the organ to be determined.

Later, it became possible to determine cadmium *in vivo* in kidney cortex by means of XRF techniques, in which either a ^{241}Am source or a partly plane-polarized X-ray spectrum is used for excitation (Ahlgren and Mattsson, 1981; Christoffersson and Mattsson, 1983; Christoffersson et al., 1987a; Nilsson et al., 1990). The XRF technique has a significantly lower effective dose equivalent than the NAA technique, the detection limit is better or comparable, and the measurements are restricted to the toxicologically relevant part of the kidney, the cortex. The duration of measurement is 30 min. A drawback of the XRF technique is that the detection limit is strongly dependent upon the distance between the skin and the kidney, which must be determined and adjusted for by an ultrasonic positioning of the kidney. The detection limit is sufficient to measure levels in subjects in the general population not subject to any particular cadmium exposure. The precision is about 20% (Christoffersson et al., 1987a). The cadmium levels in kidney cortex in Swedish nonsmokers is about 6 µg/g, and in smokers 26 µg/g (Nilsson et al., 1995), which is in agreement with the levels found in kidneys obtained at autopsy. Subjects with occupational exposure to cadmium had much higher levels (up to 320 µg/g; Christoffersson et al., 1987a).

The relationship between renal toxic effects and kidney cadmium has been discussed above (see Section II.C).

4. Other Indices

In health surveillance of cadmium-exposed populations, other indices are often used, e.g., serum creatinine levels in plasma and/or excretion of proteins or enzymes in urine as indices of renal effects. However, such effects must be considered to be adverse (see Section II.C) and are thus not optimal biomarkers of exposure. Here, only one comment will be made: β_2-microglobulin is presently the most widely used index for proximal tubular damage. However, this protein is very unstable in acid urine, which already occurs in the bladder. Thus, there is a risk of underestimating the excretion if the subject is not pretreated with sodium bicarbonate hours before sampling. RBP and alpha$_1$-microglobulin (Grubb, 1992) are far more stable, and thus more suitable for health surveillance of cadmium-exposed subjects.

E. CONCLUSIONS

Biological monitoring of cadmium exposure has several advantages over external exposure assessment. Mainly, concentrations of cadmium in blood (B-Cd) or urine (U-Cd) are employed. B-Cd changes fairly rapidly with changes of exposure, and is thus used as an index of recent exposure even if it also reflects the body burden, which has a very slow turnover. U-Cd reflects the cadmium level in the kidney, which is the critical organ at long-term exposure. During the past decade, techniques for *in vivo* determination by NAA or XRF of cadmium in kidney have become available. The renal cadmium level reflects the long-term exposure and the risk of kidney damage.

III. LEAD

The information on the toxicity of lead is enormous. Several comprehensive reviews of the toxicology of lead have been published recently (U.S. EPA, 1986; Skerfving, 1988, 1993). To limit the length of this treatise, these reviews will be referred to for further references.

Exposure occurs to inorganic lead and lead compounds, as well a to organic lead compounds. From a practical point of view, exposure to inorganic lead is of greatest interest in both the general and occupational environments. Thus, this review will focus on inorganic lead.

IV. INORGANIC LEAD

A. EXPOSURE

Humans are exposed to inorganic lead in the occupational setting by inhalation, and through contaminated foods and tobacco. Heavy exposure may occur in lead smelters, metal scrapping, spray painting, and storage battery manufacturing (U.S. EPA, 1986; Skerfving, 1993). Less severe exposure occurs in many settings, e.g., in indoor shooting ranges (Svensson et al., 1992a).

In the general environment there is oral exposure through foods (from the foods as such, or through contamination from soldered tin cans or lead-glazed or lead-painted pottery). In Sweden, the intake via foods is about 30 µg/day (Schütz, 1979). Even crystal glass may cause exposure. In some areas, there is considerable exposure through plumbosolvent drinking water, contaminated from lead pipes (Mushak et al., 1989). Also, children may suffer considerable exposure through intake of lead-containing paint (now prohibited in most countries) flakes, dust, and soil (Mushak et al., 1989). Moreover, alcoholic beverages cause exposure (Elinder et al., 1988), particularly "moonshine whiskey".

Cigarette smoking causes inhalation exposure to lead (Svensson et al., 1987). Furthermore, there is inhalation exposure from automobile exhausts and industrial emissions (Schütz et al., 1984, 1989). In children, exposure to environmental tobacco smoke is associated with higher blood lead levels (Andrén et al., 1988; Willers et al., 1988), though probably not because of lead exposure through the smoke.

In most countries there has, in the last 15 years, been a dramatic decrease in exposure (as reflected by blood-lead levels, B-Pbs; Skerfving et al., 1986; Schütz et al., 1989; Strömberg et al., 1995), mainly because of the reduction of alkyl lead additions to gasoline. However, in several areas lead exposure in the general population, and in particular among pregnant women and infants, is still a considerable public health problem (Mushak et al., 1989).

B. METABOLISM

1. Absorption

At inhalation exposure, 10–60% of the particles with a size in the range 0.01 to 5 µm are deposited in the alveolar region of the respiratory tract (U.S. EPA, 1986; Skerfving, 1993). Larger particles are mainly deposited in the nose, mouth, and upper part of the airways. A major fraction of this is cleared and swallowed.

Lead is absorbed from the gastrointestinal tract (U.S. EPA, 1986; Skerfving, 1993). In radiotracer experiments in fasting subjects, the absorption ranged between 37 and 70%. Of soluble lead salts taken with meals, a lower fraction was absorbed, between 4 and 21%. The absorption of lead salts through the skin is generally low.

2. Distribution

Lead is absorbed to the blood plasma. Lead rapidly equilibrates between plasma and the extracellular fluid. More slowly, but within minutes, lead is transferred from plasma to blood cells. The turnover of lead in plasma is very rapid; the half-life after intravenous injection in humans was about a minute (Campbell et al., 1984).

Among the soft tissues, lead is distributed to the bone marrow, the liver, and the kidney (Barry, 1975; Skerfving et al., 1983). Lead does, to some extent, pass the blood-brain barrier into the nervous system. Such passage is probably higher in infants than in adults.

A large proportion of the absorbed lead is transferred into the skeleton (Gusserow, 1861). The skeleton contains about 90% of the body burden (Barry, 1975). There are at least two pools of lead in the skeleton. One is in trabecular bone (spongy bone; Schütz et al., 1987a, b), the other in cortical bone (compact; Christoffersson et al., 1984, 1986; Somervaille et al., 1989; Nilsson et al., 1991; Tell et al., 1992; Gerhardsson et al., 1992, 1993). The trabecular bone constitutes about 10% of the marrow-free bone mass, the rest is in the cortical bone (O'Flaherty, 1991a). Lead is also incorporated in teeth (Skerfving, 1988, 1993).

Lead is continuously mobilized from the skeleton, which causes a considerable "endogenous" lead exposure (Christoffersson et al., 1984, 1986, 1987b; Schütz et al., 1987a, 1987b, 1987c; Nilsson et al., 1991).

3. Elimination

There is glomerular filtration of lead in the kidney, probably followed by partial tubular reabsorption. Lead is also excreted in bile and pancreatic juice, and appears in feces (U.S. EPA, 1986; Skerfving, 1993). Lead is also, to some extent, excreted in sweat, seminal fluid, hair, and nails (U.S. EPA, 1986; Skerfving, 1988, 1993).

Lead is deposited in the placenta (U.S. EPA, 1986; Skerfving, 1993). There is also a transplacental passage of lead to the fetus. Also, there is lead excretion in the milk. The levels in human milk are much lower than in whole blood, but higher than in plasma (Skerfving, 1993).

4. Metabolic Model

A metabolic model for inorganic lead should at least comprise plasma, blood cell, soft tissue, trabecular bone, and cortical bone pools (Skerfving et al., 1985, 1987; Christoffersson et al., 1987b; Bert et al., 1989; O'Flaherty, 1991b; Rabinowitz, 1991; Skerfving et al., 1993).

C. HEALTH EFFECTS

1. Mechanisms of Toxicity

Rather little is known about the basic mechanism behind lead toxicity (plumbism or saturnism). However, it is well established that lead binds to the sulfydryl groups of proteins. If this occurs on an enzyme, its function may be inhibited, which may result in toxic effects. Also, lead alters calcium-mediated cellular processes.

2. Nervous System

Exposure to inorganic lead may cause symptoms and signs from both the peripheral and central (CNS) nervous systems (U.S. EPA, 1986; Skerfving, 1993). This is due to demyelinization, axonal degeneration, and possibly also presynaptic block.

The peripheral nervous system damage causes paralysis, as well as pain in the extremities. Several studies have shown that chronic lead exposure reduces nerve conduction velocity in peripheral nerves in adult subjects without clinical symptoms or signs of disease (U.S. EPA, 1986; Skerfving, 1993). In some studies, such effects have been recorded in subjects with B-Pbs as low as 1.5–2.0 µmol/l (30 to 40 µg/dl). The significance of such neurophysiological changes is not clear.

In humans, especially in children, lead exposure may cause encephalopathy, with ataxia, coma, and convulsions (U.S. EPA, 1986). In subjects without obvious clinical signs of encephalopathia, subjective and nonspecific symptoms (e.g., fatigue, impaired concentration, loss of memory, insomnia, anxiety, and irritability) may occur, as well as impaired performance in psychometric tests. In such tests, minor effects, mainly of visual intelligence and visual-motor coordination, as well as changes of somatosensory-, visual-, and auditory-evoked potentials have been recorded in lead workers with B-Pbs as low as 2.0 to 2.5 µmol/l (Skerfving, 1988, 1993). These CNS effects are, at least in some cases, partially reversible.

3. Blood and Blood-Forming Organs

Lead has an inhibitory effect on steps in the chain of reactions that lead to the formation of heme (ALAD and ferrochelatase = heme synthetase; cf. U.S. EPA, 1986). Lead also inhibits the activity of the enzyme pyrimidine-5-nucleotidase (P5N) in red cells.

Heavy lead exposure is associated with reticulocytosis and occurrence of stippled erythrocytes in peripheral blood (U.S. EPA, 1986), possibly mediated through the effect of P5N. Also, the life span of circulating erythrocytes becomes shortened, probably because of inhibition of the Na^+, K^+-ATPase, possibly also of the erythrocyte P5N, and changes of the membrane proteins. This may lead to anemia, which is normocytic and sideroblastic.

Inhibition of ALAD starts at B-Pbs of about 0.5 µmol/l (Schütz and Skerfving, 1976), and is complete at about 3 µmol/l. Inhibition of P5N occurs at similar levels. At a B-Pb of 1.5 µmol/l, there is an increase of ZPP in a considerable fraction of the population (Schütz and Haeger-Aronsen, 1978; Skerfving et al., 1993). Increases of ALA and CP in urine starts at higher levels (Schütz and Skerfving, 1976). These effects cannot, in themselves, be considered detrimental to health. Anemia ("lead pallor") occurs at high B-Pbs (about 3 µmol/l; U.S. EPA, 1986).

4. Kidneys

Lead exposure may cause kidney damage. In acute lead toxicity, there is proximal tubular damage which may result in a reversible Fanconi syndrome-like condition (with aminoaciduria, glucosuria, and hyperphosphaturia).

Further, the tubular damage may cause leakage of enzymes (e.g., the lysosomal NAG) from the cells into the urine (Skerfving, 1988, 1993). It is unclear whether slight enzymuria is an adverse effect. It seems to be reversible. Such effects seem to occur at B-Pbs in the range 1.5 to 2.0 µmol/l (Skerfving, 1990; Gerhardsson et al., 1992).

After heavy exposure for years, interstitial nephritis, with interstitial fibrosis, tubular atrophy, and arteriosclerotic changes may occur (granular contracted kidneys). Functionally, there is a decrease of renal plasma flow with a reduction of the glomerular filtration rate, resulting in azotemia (increase of blood urea nitrogen = BUN and serum creatinine) and an increase of the tubular reabsorption of uric acid, resulting in an increase of serum levels of uric acid (eventually hyperuricemia), which is probably a cause of gout with arthritis (saturnine gouty arthritis; Skerfving, 1988, 1993). Such changes seem to occur at B-Pbs of 2.5–3.0 µmol/l, or higher. However, there is a remarkable lack of a dose-response relationship.

5. Gastrointestinal Tract

Lead exposure may result in the precipitation of dark-bluish lead sulfide in the gingiva ("lead line"; "Burtoninan line"). Lead also affects the gastrointestinal tract, causing diarrhea, epigastric pain, nausea, indigestion, loss of appetite, and colic (Skerfving, 1988, 1993). Such symptoms and signs usually occur at B-Pbs higher than 3.5 µmol/l.

6. Cardiovascular System

In several studies of the general populations, there were associations between blood pressure and B-Pb (Skerfving, 1993). However, all reported associations were weak, and a causal relationship has not been established since there are several possible confounding factors, as well as a possibility of reverse causation (Skerfving, 1993).

There was an increase of the systolic and diastolic blood pressure by 1 to 2 mm Hg for each doubling of the B-Pb. Thus, there are indications of a leveling off of the effects as B-Pb increased. This may be the reason why the results in recent studies of lead workers have not been consistent (Skerfving, 1993). In earlier studies of more heavily exposed workers, the observed blood pressure effects may depend on kidney damage with secondary hypertension. There was no increase of cardio- or cerebrovascular deaths among smelter workers (Gerhardsson et al., 1986), while glass workers had increased risks (Wingren and Englander, 1990).

7. Genotoxicity

Lead acetate and lead subacetate caused kidney and brain tumors and lead phosphate caused kidney tumors in rodents following oral or parenteral administration (IARC, 1987). However, the doses were high and caused gross morphological changes in the kidney.

Epidemiological studies of workers exposed to lead do not support a carcinogenic effect in humans (IARC, 1987; Gerhardsson et al., 1986). Still, on basis of the animal data, IARC (1987) classified lead as "possibly carcinogenic to humans (Group 2B)".

8. Reproductive Effects

There are data from lead-poisoned males that may indicate lead effects on the testis and spermatogenesis (Skerfving, 1993). In a recent study of lead workers, there were associations between sperm vitality, motility, and morphology, on the one hand, and B-Pb (average 2 µmol/l) on the other. Possibly, such effects are mediated through interference with the endocrine system (Gustafson et al., 1989).

As early as 1860, it was reported that severely lead-poisoned pregnant women were likely to abort, while those less severely intoxicated were more likely to deliver stillborn babies (Paul, 1860). Recent studies of women much less exposed to lead have given varying results (Mushak et al., 1989). In some studies, but not all, there have been indications of stillbirth, preterm delivery, reductions of gestational age and birth weight, and sudden infant death syndrome (Skerfving, 1993). Prospective (during pregnancy and onwards) studies in the U.S. and Australia have shown lead-exposure-associated developmental effects on cognitive abilities and behavioral functions in the infants who were followed for years after birth (Mushak et al., 1989). In summary, the response rates of such effects in the child was dependent on the B-Pbs in the pregnant woman, the newborn infant, and/or the young child. Taken together, the data indicate that slight effects may already be present at B-Pbs in the range 0.50 to 0.75 µmol/l, perhaps even lower. These results are in general agreement with a long series of cross-sectional studies in the U.S. and Europe on the relationship between neurotoxic effects, on the one hand, and B-Pb or teeth-Pb on the other (Mushak et al., 1989; Skerfving, 1993). However, all studies do not reveal any lead-associated effects. It should be stressed that there are several methodological problems. In a metaanalysis of eight European cross-sectional studies of children, covering B-Pbs in the range 0.2 to 2.9 µmol/l, there were weak negative associations between psychometric intelligence and visual-motor integration, on the one hand, and B-Pb on the other, but the explained variance was low (Winneke et al., 1989).

D. BIOLOGICAL MONITORING OF EXPOSURE

1. Blood Lead Levels

This is the prevailing index of lead exposure and risk (U.S. EPA, 1986; Skerfving 1988, 1993). It has many advantages; samples are easy to obtain and the analysis is uncomplicated, but there are several limitations which must be carefully considered when interpreting a B-Pb result.

One of these problems is the nonlinear relationship between lead exposure/uptake and B-Pb, both at inhalation and gastrointestinal exposure (U.S. EPA, 1986). Further, there is a nonlinear relationship between B-Pb, on the one hand, and levels in other media, such as serum (De Silva, 1981), urine (U-Pb; Skerfving et al., 1985, Schütz et al., 1987b; Tell et al., 1992), and milk (Oskarsson et al., 1992) on the other. Moreover, there are nonlinear relationships between B-Pb, on the one hand, and different metabolic/toxic effects, such as on the heme (Schütz and Skerfving, 1976; Schütz and Hager-Aronsen, 1978; Skerfving et al., 1993) and nucleotide (Skerfving, 1988, 1993) synthesis on the other. The probable explanation of the nonlinear behavior of B-Pb is a saturation of erythrocytes. A consequence of the nonlinearity is that B-Pb is a much more sensitive index of changes of exposure at low exposure intensities than at high ones (Schütz et al., 1989). There was a remarkable variation of both U-Pb (Skerfving et al., 1993) and blood concentration of ZPP (Skerfving et al., 1993) on the one hand, and B-Pb on the other. Theoretically, serum/plasma levels would be more suitable. However, the levels are so low, that problems with contamination and analysis would be insurmountable.

The B-Pb in subjects without particular lead exposure varies between different regions. In Sweden, the present level in children without particular exposure is about 30 µg/l (0.15 µmol/l; Schütz et al., 1989; Strömberg et al., 1995). In many countries, it is considerably higher (U.S. EPA, 1986; Skerfving, 1988, 1993). There is an association between lead levels in maternal and cord blood (though the latter is somewhat lower; U.S. EPA, 1986; Skerfving, 1988, 1993). The relationships between various kinds of toxic effects, on the one hand, and B-Pb on the other, have been discussed above in connection with each effect (Section IV.C).

After the end of occupational lead exposure, there is a decrease of B-Pb (Christoffersson et al., 1986; Schütz et al., 1987c; Nilsson et al., 1991). The decay pattern may be described by a three-compartment exponential model: a fast pool with a half-time of one month, an intermediate pool with a half-time of about a year (which probably reflects the trabecular bone; Schütz et al., 1987a), and a slow pool having a half-time of 13 years (which reflects the cortical bone, see below). There is a pronounced interindividual variation in kinetics, which would mean a different B-Pb and body burden at the same exposure (Schütz et al., 1987c).

2. Lead Concentration in Urine

This has been fairly widely used as an index of lead exposure. There is a nonlinear relationship between lead levels in blood and urine, with a relative increase of U-Pb with increasing B-Pb (Skerfving et al., 1985, 1993). This is probably due to the above-discussed saturation of the blood cells. However, there seems to be a linear relationship between plasma lead and U-Pb (Manton and Cook, 1984).

Another option is U-Pb concentration. However, this would also involve problems as the risk of contamination is higher than for B-Pb, and the available information on the relationship between U-Pb and exposure/effects is much more limited.

3. Lead Concentrations in Calcified Tissues

As stated above, lead accumulates in calcified tissues. Levels in shed deciduous teeth have been rather widely used to assess lead exposure during the fetal period and infancy, especially in studies of central nervous system effects (Needleman et al., 1979; U.S. EPA, 1986; Skerfving, 1988, 1993).

In the last two decades, other possibilities for monitoring lead in calcified tissues have occurred: determinations *in vivo* of lead in bone (bone-Pb) by X-ray fluorescence (XRF) techniques. A ^{57}Co source has been used to excite the lead, and measure the characteristic K X-rays of lead (Ahlgren et al., 1976; Christoffersson et al., 1984, 1986; Schütz et al., 1987a, 1987b, 1987c; Somervaille et al., 1989; Nilsson et al., 1991; Price et al., 1992; Skerfving and Nilsson, 1992; Tell et al., 1992; Nilsson and Skerfving, 1993; Skerfving et al., 1993).

In addition, ^{109}Cd has been used for measurements of lead in the tibia and calcaneus (Batuman et al., 1989; Hu et al., 1989; Somervaille et al., 1989; Sokas et al., 1990; Erkkilä et al., 1992; Tell et al., 1992; Gerhardsson et al., 1992, 1993; Nilsson and Skerfving, 1993; Skerfving et al., 1993). While these K techniques measure lead up to a considerable depth in the bone, L techniques mainly assess superficially located lead (Rosen et al., 1991). Thus, results obtained by the two techniques may not be directly comparable.

The bone-Pb increased with rising time of employment, from a few up to 100 µg/g, or more (Somervaille et al., 1989; Nilsson et al., 1991; Gerhardsson et al., 1993). Thus, bone-Pb is a valuable index of long-term exposure. There are significant associations between the levels at the different bone sites (Somervaille et al., 1989; Gerhardsson et al., 1993; Nilsson and Skerfving, 1993; Skerfving et al., 1993). The lead mobilized from the skeleton causes a considerable "endogenous" lead exposure (Christoffersson et al., 1984, 1986; Schütz et al., 1987a, 1987c; Nilsson et al., 1991; Erkkilä et al., 1992).

After the end of employment there is a decrease of finger-bone lead (Christoffersson et al., 1987b; Nilsson et al., 1991). The trabecular bone pool has a faster turnover than the cortical one (Schütz et al., 1987a; Somervaille et al., 1989; Erkkilä et al., 1992). The lead in these compartments seems to have half-lives of about a year and decades, respectively (Nilsson et al., 1991; Erkkilä et al., 1992).

The skeletal lead should be a useful index of risk, especially for chronic effects of lead exposure. Little research has yet been done in this promising field. In a recent study of smelter workers, there was no association between renal function, on the one hand, and tibia or calcaneus lead concentrations on the other (Gerhardsson et al., 1992). However, in a study of lead workers with a higher lead exposure, there were indications of slight effects on the tubuli in one population of smelter workers (Skerfving, 1990).

4. Lead Concentration in Other Media

The lead content of cerebrospinal fluid is very low (Manton and Cook, 1984). Thus, because of risk of contamination and analytical difficulties, it has not been used for biological monitoring.

5. Mobilization Tests

Mobilization tests have been widely employed to estimate the body burden. A chelating agent is first given, and the resulting excretion of lead in the urine is determined (Schütz et al., 1987b; Skerfving, 1988, 1993; Tell et al., 1992). The subject is given penicillamine orally (Schütz et al., 1987b) or calcium disodium edetate (EDTA) intravenously (Tell et al., 1992), and the excretion of lead into the urine is measured (chelated lead). There is close associations between chelated lead, on the other hand, and B-Pb as well as urinary lead levels before chelation, on the other (Schütz et al., 1987b; Tell et al., 1992). However, while the relationship with B-Pb was nonlinear, with relatively less increase of U-Pb with rising B-Pb (which is probably explained by the above-mentioned saturation of B-Pb), the association with U-Pb seemed to be linear (Skerfving et al., 1993).

There is also an association, though less close, between the chelated lead and the lead level in trabecular bone (Schütz et al., 1987b), but far less so between chelated lead, on the one hand, and levels in either finger, tibia, or calcaneus on the other (Tell et al., 1992). Thus, cortical bone lead, which makes up the major part of the body burden, is only to a very limited degree available for chelation. Hence, chelatable lead makes up a minor part of the total body burden (Aono and Araki, 1986). Thus, chelated lead is mainly a reflection of the soft tissue pools and a rapid bone compartment, but it does not indicate the compact bone lead, which constitutes the major part of the body burden. Thus, determinations of skeletal lead has a separate meaning: an index of long-term exposure. Further, as an index of the soft tissue pool, which probably reflects the metabolically active and thus toxicologically interesting pool, the chelated lead does not, at least in adults, offer a clearcut advantage over B-Pb or over U-Pb before chelation.

6. Disturbances of Heme Metabolism

Heme has often been used for biological monitoring (U.S. EPA, 1986; Skerfving, 1988, 1993). As stated above, lead has an inhibitory effect on ALAD (U.S. EPA, 1986). Inhibition of ALAD activity in whole blood or erythrocytes has been used as an index of lead exposure. The inhibition starts at very low B-Pbs (see above, Section IV.C). A major limitation is that a complete inhibition occurs at about 2 μmol/l (Skerfving, 1988, 1993).

The ALAD inhibition leads to an accumulation of ALA and coproporphyrin (CP) in blood plasma, which in turn causes an excretion of these metabolites in the urine. The urinary concentrations have been widely used earlier as an index of lead exposure. Increases of ALA and CP in urine starts at fairly high exposure levels (Schütz and Skerfving, 1976; see above, Section IV.C). ALA and CP in urine have now generally been abandoned as indices of lead exposure (Skerfving, 1988, 1993).

Also, the inhibition of ferrochelatase causes an accumulation of protoporphyrins (ZPP) in the red blood cells. The level of ZPP may be determined by simple fluorimeters, and has been fairly widely used, especially in field studies, as an index of lead exposure (Skerfving, 1988, 1993). However, an increase starts only at a fairly significant exposure level (see above, Section IV.C), and an iron deficiency causes an interfering increase (Schütz and Haeger-Aronsen, 1976; Skerfving et al., 1993).

7. Other Indices

In health surveillance of lead-exposed populations other indices are often used, e.g., serum creatinine levels in plasma and/or protein excretion in urine as indices of renal effects, and blood hemoglobin concentrations as an index of effects on the bone marrow. However, such effects must be considered to be adverse (see Section IV.C) and are thus not optimal biomarkers of exposure.

E. CONCLUSIONS

Biological monitoring of lead exposure has several advantages over external exposure assessment. Traditionally, levels in blood (B-Pb) have been widely employed. However, a problem with B-Pb is a saturation of the erythrocyte lead concentration, which causes a nonlinear relationship between B-Pb and uptake, and also between metabolic/toxic effects and B-Pb. During the last decades, techniques for *in vivo* determination by XRF of lead in fingerbone, tibia, or calcaneus have become available. The bone lead levels reflect long-term exposure, and should become a valuable tool in epidemiological studies, especially of chronic effects. Mobilization tests have been widely used for biological monitoring of lead. They seem to mainly reflect the lead content of the soft tissues, and are not an index of the total body burden, the major part of which is in the skeleton.

V. ORGANIC LEAD

A. EXPOSURE

The alkyl lead compounds, tetraethyl and tetramethyl lead, are widely used as antiknock agents in gasoline (up to about 1 g lead per liter) although their use has decreased dramatically in many countries during the last two decades (Strömberg et al., 1995). The alkyl lead compounds have high vapor pressures. Thus, there is occupational exposure in workers in factories producing alkyl lead, oil refineries, tanker ships, and gasoline filling stations (especially when cleaning tanks). Also, exposure occurs while sniffing gasoline.

During combustion of gasoline in the engine, there is decomposition of most of the alkyl lead into inorganic lead; only a minor fraction of alkyl lead is emitted in the exhausts. Thus, the exposure of the general population to alkyl lead is low (Grandjean and Nielsen, 1979; Tsuchiya, 1986).

B. METABOLISM

The metabolism of alkyl lead is very different from that of inorganic lead (Grandjean and Nielsen, 1979). There is a high degree of absorption at inhalation exposure, it is absorbed through the skin and in the gastrointestinal tract, it efficiently passes the blood-brain barrier, it occurs in the blood only to a limited extent, it is biotransformed through dealkylation into trialkyl lead which is rapidly excreted in the urine.

C. HEALTH EFFECTS

Alkyl lead poisoning is generally acute. The brain is the critical organ, with symptoms of irritability, headache, convulsion, delirium, and coma.

D. BIOLOGICAL MONITORING

Biological monitoring of exposure is mainly performed through analysis of total lead levels in urine, preferably adjusted for creatinine or density. The urinary concentration of lead in subjects without high exposure is about 10 µg/l (Tsuchiya, 1986) or lower. Poisoning is generally associated with urinary lead concentrations of 300 µg/l or more; slight poisoning occurs with about 150 µg/l (Grandjean and Nielsen, 1979).

The concentration of lead in blood is not a good index of exposure to alkyl lead, though there is usually some increase; neither is bone lead a good index, though gasoline sniffers had increased levels (Eastwell et al., 1983), probably as a result of biotransformation into inorganic lead, which is incorporated in the skeleton (see above).

VI. INORGANIC MERCURY

A. EXPOSURE

Degassing from the earth's crust is a major natural source of environmental mercury pollution. Metallic mercury pollution is also produced from mining operations, smelters, combustion of fossil fuel, and refining of gold. It is used in large quantities in the chloralkali industry during the production of chlorine and caustic soda through the electrolysis of brine (sodium chloride solution). It is also used in the electrical industry, in the production of control instruments such as thermometers and barometers, fluorescent tubes, and alkaline batteries, and in dentistry for amalgam fillings (WHO, 1991a). Occupational exposure occurs in workers in all these settings. The concentration of inorganic mercury in food is low. The daily intake with food and water is very low as compared to organic mercury, and mostly less than 1 µg/day. For occupationally unexposed humans, mercury vapor released from amalgam fillings is the dominant source of inorganic mercury (Clarkson et al., 1988b; Molin et al., 1990, 1991; Åkesson et al., 1991; Svensson et al., 1992b). The absorbed amount has been estimated to range from 3–18 μg/day by use of data from different studies (Clarkson et al., 1988b).

B. METABOLISM

Approximately 80% of inhaled mercury vapor is absorbed in the lungs. The absorption of metallic mercury (Hg^0) is low in the gastrointestinal tract: less than 1%. It is somewhat higher (<10%; WHO, 1991a) for divalent mercuric mercury (Hg^{2+}).

Dissolved elemental mercury is transported in blood and distributed to different organs, e.g., the brain, kidney, thyroid gland, pituitary gland, liver, ovaries, pancreas, prostate, and testes. It easily passes the placental and brain barriers, as well as the erythrocyte membrane. After oxidation in red blood cells and in tissues to divalent ionic mercury, which takes place within minutes, it can no longer penetrate these barriers (WHO, 1991a). After exposure to mercury vapor, the mercury levels in plasma and erythrocytes are approximately equal (WHO, 1976). The kidney is the dominant tissue depot of inorganic mercury in humans, both after exposure to Hg^0 and Hg^{2+}. Excretion is via urine and feces (WHO, 1991a). The urinary route is dominant at high exposure levels (Clarkson et al., 1988c).

The estimated biological half-time in whole body after inhalation of mercury vapor is a couple of months. In blood, the first phase of elimination has a half-time of about 2–4 days (Barregård et al., 1992). The second phase has a longer half-time: approximately 2–6 weeks (WHO, 1991; Barregård et al., 1992; Sällsten et al., 1993) and about 2 months in urine (Sällsten et al., 1994a).

C. HEALTH EFFECTS

Exposure to high levels of mercury vapor can cause acute irritation of the respiratory tract and interstitial pneumonitis. Ingestion of mercuric compounds may lead to gastroenteritis and acute tubular necrosis.

Critical organs after exposure to inorganic mercury are the central nervous system and the kidney (WHO, 1976; WHO, 1991a). Long-term exposure to mercury concentrations in air at 100 μg/m^3 or higher may lead to mercurialism, characterized by gingivitis, hypersalivation, metal taste in the mouth, and symptoms from the central nervous system (erethism, irritability, excitability, excessive shyness, and insomnia). A fine tremor, initially involving the hands, is often reported (WHO, 1991a). Studies in mercury-exposed workers have shown decreased peripheral nerve conduction velocity. Mercuric mercury mainly affects the kidney, producing an increased excretion of lysosomal enzymes (Barregård et al., 1988) and proteinuria. Glomerular damage leading to albuminuria and occasionally to nephrotic syndrome (Kazantzis et al., 1962) may appear in more severe cases, for example among highly exposed chloralkali workers.

WHO (1991a) concluded that urinary mercury levels of about 100 μg/g creatinine (corresponding to air concentrations >80 μg/m^3) increase the risk for neurological symptoms of mercurial intoxication (tremor, erethism), as well as for adverse kidney effects such as proteinuria. Less severe toxic effects may appear in the interval 30–100 μg Hg/g creatinine corresponding to air mercury levels between 25 and 80 μg/m^3. In some studies, tremor, recorded electrophysiologically, has been observed at low urinary concentrations (25–35 μg/g creatinine), while other investigations have been negative. At these concentration ranges, low molecular weight proteinuria and microalbuminuria also have been reported in some studies.

D. BIOLOGICAL MONITORING

Monitoring of inorganic mercury exposure through the determination of total mercury in body fluids is interfered with by simultaneous exposure to organic (methyl) mercury. However, due to the different metabolic patterns, the mercury concentration in plasma and urine mainly reflects the exposure to inorganic mercury, while the concentration in erythrocytes mirrors the exposure to organic mercury. Different speciation techniques can be used to differentiate between organic and inorganic mercury compounds, e.g., selective reduction of inorganic mercury, selective extraction, steam distillation, ion exchange, gas chromatography, and liquid chromatography (Schütz et al., 1994).

One method of speciation is to determine the concentrations of total mercury and of inorganic mercury in blood and tissue samples by, e.g., cold vapor atomic absorption spectrophotometry. The difference between these two measurements will be an estimate of the amount of organic mercury. Inorganic mercury in biological samples can also be determined by selective reduction of Hg^0, without affecting alkyl mercury compounds present (Schütz et al., 1994; Bergdahl et al., 1995).

Reported concentrations in blood in occupationally unexposed humans with a low to moderate fish consumption is about 3 to 4 µg/l (Barregård, 1993). The level in plasma is around 1.5 to 2 µg/l (7.5 to 10 nmol/l; Barregård et al., 1991; Åkesson et al., 1991; Svensson et al., 1992b). The urinary excretion probably reflects the kidney accumulation of inorganic mercury. The average concentration in morning urine samples from occupationally unexposed subjects in Sweden is approximately 2 to 3 µg/g creatinine (1 to 2 nmol/mmol creatinine; Langworth et al., 1991; Åkesson et al., 1991; Barregård, 1993). Levels higher than 10 µg/l (approximately 5 nmol/mmol creatinine) are seldom found. In subjects with occupational exposure, levels in the range from 20 to 100 µg/l (approximately 10 to 50 nmol/mmol creatinine) are commonly reported (Langworth et al., 1991; Barregård et al., 1992). An association between the concentrations in urine (Langworth et al., 1991) and brain (Nylander et al., 1987) on the one hand, and the number of amalgam fillings or amalgam surfaces on the other, has been reported. For nonoccupationally exposed individuals, a moderate number of amalgam fillings will increase the brain mercury concentration by about 10 µg/kg. The corresponding increase in the kidney is estimated to be about 300 to 400 µg/kg (WHO, 1991a). However, the individual variation is large. Normal concentrations in kidney in subjects without particular exposure is ranging from 20 to 800 µg/kg (Nylander et al., 1987; Schütz et al., 1994).

In occupationally unexposed subjects, the lowest reported brain levels are about 10 µg/kg (Nylander et al., 1987). At these levels, inorganic mercury is probably the main fraction. In humans with intense and long-term occupational exposure to mercury vapor, or with a high fish intake from polluted waters, brain concentrations may exceed 1 mg/kg (Schütz et al., 1994). Long-term inhalational exposure to elemental mercury vapor, corresponding to a daily absorption of about 5 to 10 µg/day, will give a mercury excretion in urine of about 5 µg/l and an average mercury concentration in the occipital lobe cortex and kidney of approximately 10 and 500 µg/kg, respectively (WHO, 1991a).

On a group basis, mercury levels in blood and urine reflect recent exposure and also long-term exposure, if the exposure by air, food and water is relatively constant. In a study by Smith et al (1970), evaluated by WHO (1976), long-term time-weighted occupational exposure to an average air mercury concentration of 50 µg/m^3 was associated, on a group basis, with blood mercury levels of approximately 35 µg/l, and with urinary levels of about 150 µg/l. However, the ratio of urine-to-air concentrations was later reevaluated by WHO in 1980 and found to be closer to 2.0 to 2.5 instead of 3.0. In further studies, using personal sampling, the ratio between urinary mercury (µg/l or µg/g creatinine) and mercury in air (µg/m^3) has, as a rule, been about 1 to 2. Thus, it has been indicated that a urine mercury level of 50 µg/g creatinine is associated with a blood mercury level of about 15 to 20 µg/l, which corresponds to air mercury levels of about 40 µg/m^3 (Roels et al., 1987a). However, concomitant exposure to methylmercury must be taken into consideration.

Mobilization tests with chelating agents, e.g., DMPS (2,3-dimercaptopropane-1-sulfonate) and DMSA (2,3-dimercaptosuccinic acid) to increase the excretion of mercury, have been used in workers with current mercury exposure and in humans with and without amalgam fillings (Roels et al., 1991; Molin et al., 1991; Sällsten et al., 1994a). However, the additional mercury which is excreted after a single dose of DMPS is about the same as is normally excreted during a week. Accordingly, a mobilization test with a single dose of a chelating agent gives no supplementary information on the size of the mercury pool in the kidneys (Sällsten et al., 1994a).

Other media, such as feces, expired air, and cerebrospinal fluid (SCF), are seldom used for biological monitoring (Sällsten, 1994). In a cross-sectional study, slightly increased levels of CSF-Hg were observed in chloralkali workers — 1.1 nmol/l compared to 0.3 nmol/l in referents. The concentration of SCF-Hg was about 2% of the level of plasma-Hg in the exposed group (Sällsten et al., 1994b).

Recently, mercury concentrations in kidney have been determined by an X-ray fluorescence technique in a cross-sectional study of chloralkali workers (Börjesson et al., 1995). The detection limit was approximately 26 µg/g wet weight (mean; SD 12 to 45 µg/g), varying with the kidney depth. The method is noninvasive and gives a low radiation dose. However, the technique could be further improved by a more precise localization of the organs investigated, more efficient detectors, better methods for curve fitting and, in some cases, also by longer measurement times (Börjesson et al., 1995).

E. CONCLUSION

Kidney and brain are target organs after long-term exposure to mercury vapor. Determinations of mercury in plasma and urine reflects the current exposure to mercury vapor. Repeated sampling may be used for risk assessment. Fish consumption must be considered if mercury levels in whole blood are determined, as such food items can have high concentrations of organic mercury. Speciation techniques can be used to determine the concentration of inorganic mercury in biological samples. Classical signs of mercurialism have been reported at air exposure levels of about 100 µg/m^3. At urinary levels of 100 µg Hg/g creatinine, the probability of developing neurological signs of mercury intoxication and proteinuria is high. Subclinical effects have been reported at exposure levels corresponding to 30 to 100 µg Hg/g creatinine in urine. For the time being, there is no available indicator medium that reflects long-term accumulation of inorganic mercury in brain. Data are not sufficient for constructing a dose-response curve or for estimating a no-observed-effect-level.

The kidney is the critical organ after long-term exposure to mercuric mercury. Plasma and urine determinations may be used to evaluate the exposure. Urinary levels reflect the kidney burden of mercury and may assist in risk assessment.

VII. ORGANIC MERCURY

A. EXPOSURE

Organic mercury compounds, such as alkyl (e.g., methyl and ethyl), alkoxy (mainly methoxyethyl), and aryl (mainly phenylmercury) have been used for seed treatment and as fungicides and algaecides (WHO, 1990). Occupational exposure occurs in industries producing disinfectants, fungicides, and insecticides.

Environmental exposure to methylmercury originates from consumption of fish and other seafood from mercury-polluted waters. In aquatic environments inorganic mercury is methylated both through microbial action or nonenzymical reactions into monomethylmercury, which is then enriched in the marine food chain. Fish from nonpolluted water areas may contain 0.01 to 0.05 mg/kg (Schütz et al., 1994). However, in polluted areas, mercury concentrations may exceed 1 mg/kg. The estimated daily human intake from all sources, in subjects with low intake of noncontaminated fish, is about 2.4 µg, with a daily uptake of approximately 2.3 µg (WHO, 1990).

B. METABOLISM

After peroral intake, methylmercury is readily absorbed from the gastrointestinal tract (>90%; Clarkson et al., 1988c; WHO, 1990). It is transported in the blood and evenly distributed to all tissues within 4 days (Kershaw et al., 1980). In blood, the major part is bound to hemoglobin in the red cells, and the remainder to plasma proteins. The ratio between the levels in erythrocytes and in plasma is usually about 20:1. Species differences exist. Methylmercury penetrates the blood-brain and placental barriers and accumulates in the brain and in the fetus. In the brain methylmercury is demethylated to inorganic mercury. The ratio between methylmercury and inorganic mercury depends on the exposure time, and the time since the cessation of exposure.

Methylmercury is secreted in bile and has an enterohepatic circulation (WHO, 1976). In the gut, a fraction of the methylmercury is demethylated by the intestinal microflora, and most of this inorganic mercury is excreted in the feces. Excretion in urine is limited. Methylmercury is accumulated in hair, which can be used for retrospective evaluation of exposure and for calculation of the biological half-time (WHO, 1990).

The daily elimination is about 1% of the body burden. The estimated biological half-time in blood and the whole body, described by a single compartment model, is about 40 to 70 days (Clarkson 1988c, WHO 1990).

Phenylmercury and other organic mercury compounds are fairly rapidly metabolised to inorganic mercury, and will not be treated separately.

C. HEALTH EFFECTS

The nervous system is the critical organ after exposure to methylmercury. The fetus is more sensitive than adults (WHO 1990). Sensory, visual, and auditory cerebral functions, as well as cerebellar (coordination), are most commonly affected. Early symptoms include paraesthesia in the region around the mouth and with the hands. Later, at higher exposure levels, concentric constriction of the visual field, deafness, and ataxia (including dysarthria) may follow (WHO 1976, WHO 1990). Clonic seizures, coma, and death may appear. If the intoxication is less severe, partial functional recovery is possible, depending on the compensatory mechanisms of the central nervous system. Also, the peripheral nervous system may be affected at high doses. Methylmercury has been classified as an animal carcinogen (Group 2B) by IARC (1993).

Methylmercury poisoning is characterized by a latency period, which can last up to several months. In infants exposed to high concentrations of methylmercury from their mothers during pregnancy, cerebral palsy may follow. The syndrome is characterized by microcephaly, hyperreflexia, and gross motor and mental impairment, sometimes associated with blindness and deafness.

D. BIOLOGICAL MONITORING

Reference values for total mercury in non-exposed populations without high fish consumption are as follows: mean concentration in whole blood about 8 µg/l, in hair about 2 µg/g, and in urine about 4 µg/l (WHO 1990). In regions with a high daily fish consumption, these levels will be considerably higher. The intake of about 200 µg Hg/day will lead to blood mercury concentrations of about 200 µg/l and hair levels of 50 µg/g.

Gas chromatography (GC) has been the most frequently used analytical technique for determination of organic mercury compounds (Brunmark et al., 1992; Schütz et al., 1994). Early GC-ECD methods (electron capture detector) often involved extraction of MeHg chloride into benzene. Later, less toxic solvents, such as toluene, have been used (Schütz et al., 1994).

Mercury concentrations in blood cells mainly reflect the exposure to methylmercury. Thus, the total mercury level in blood in non-occupationally exposed individuals is affected by fish consumption. After a long-term and fairly constant exposure to methylmercury in food (mainly fish), there is a linear relationship between the daily intake of methylmercury and the concentration of mercury in blood (Åkesson et al., 1991; Svensson et al., 1992). Mercury in brain is also at constant exposure situations related to the blood mercury levels (5 to 10:1).

Methylmercury is even more strongly accumulated in hair (blood to hair ratio 1:250). Hair strand analyses may be used to estimate the exposure during different growth periods of the hair strands. One example is the possibility for estimating maternal exposure to methylmercury during different periods of pregnancy. Children born by mothers who had hair mercury levels between 70 to 640 µg/g during pregnancy showed a 30% increased risk of psychomotor function disturbances and other neurological disorders (Marsh et al., 1981; Marsh et al., 1987). Studies of populations exposed to methylmercury by ingestion of treated seed in Iraq indicate that maternal hair mercury levels between 10 and 20 µg/g will increase the risk for negative health effects in the children by about 5% (WHO, 1990). Similar risk figures have been calculated for adults, with high fish consumption giving blood methylmercury levels of about 200 µg/l, which corresponds to concentrations of about 50 µg/g in hair (WHO, 1990). This group will show a 5% increased risk of neurological disturbances.

E. CONCLUSION

The fetus is more sensitive to methylmercury than adults, and the nervous system is the critical organ. Determination of mercury in blood, especially in the erythrocytes, can be used for evaluation of the exposure to methylmercury. Such determinations can also be used to estimate the brain levels and for risk assessment, as the relationship between blood and brain concentrations is well established. There is also a well-established relationship between blood and hair levels of methylmercury. Thus, hair concentrations can be used to estimate past exposure during specific time intervals of the growth of the hair strands, and also for risk assessment.

VIII. ALUMINUM

A. EXPOSURE

Aluminum is an abundant metal, comprising about 8% of the earth's crust. In spite of this, the level in drinking water and most biological materials is low. The daily intake of aluminum from food has been estimated to about 7 mg. However in patients taking antacids containing aluminum it can be considerably higher, up to approximately 3 g. Occupational exposure takes place at aluminum smelters and aluminum welding operations, and in the aluminum industry.

B. METABOLISM

The information about uptake, transport, distribution, and elimination of aluminum is limited. The absorption in the gastrointestinal tract is low (Savory and Wills, 1988), probably less than 1%. However, simultaneous intake of citric acid markedly increase the absorption from the gut. In connection with occupational exposure, there is also a small uptake of inhaled aluminum particles in the lungs, probably also less than 1%.

Aluminum is widely distributed to several organs, e.g., lung, liver, bone, muscles, and brain (Skalsky and Carchman, 1983). The highest levels have been found in lung tissue. Values exceeding 100 mg/kg have been reported in adults (Elinder and Sjögren, 1986). Urinary excretion is fairly rapid. The short-term biological half-time after heavy exposure is estimated to be about 8 h (Sjögren et al., 1985).

C. HEALTH EFFECTS

Inhalation of aluminum powder may cause lung problems, e.g., pneumoconiosis (aluminosis). Some aluminum compounds such as aluminum fluoride may have an irritative effect on the epithelium in the respiratory tract, leading to obstructive lung disease.

As urine is the major elimination route, there is a risk for retention and whole-body accumulation in uremic patients. This may cause systemic effects such as dialysis encephalopathy or dialysis dementia (Alfrey et al., 1976), which is characterized by impaired memory functions, dementia, aphasia, ataxia, convulsions, and characteristic EEG changes. Death may follow if the syndrome is untreated. Other symptoms related to aluminum toxicity include dialysis osteodystrophy (Walker et al., 1982), microcytic anemia, and metastatic extraskeletal calcification.

D. BIOLOGICAL MONITORING

The abundance of the metal complicates analytical determinations as the risk for contamination is high. Thus, reported normal concentrations in blood and plasma may vary from <2 to about 40 µg/l. Normal values in urine in healthy, nonoccupationally exposed individuals with a normal kidney function is usually less than 3 µg/l (Sjögren et al., 1983a). In highly exposed aluminum welders, the urinary concentration can be up to 100 times the normal levels. For these workers, the urinary concentration correlates with both current air aluminum concentrations as well as with the number of exposed years (Sjögren et al., 1985, 1988).

Infusion of desferrioxamine (DFO) has been used as a diagnostic and well-tolerated test of aluminum-related osteodystrophy. In a study of 54 patients on hemodialysis, an intravenous injection of a standard dose of DFO gave an increase in plasma aluminum concentration of 534 ± 260 (SD) µg/l in patients with aluminum-related bone disease compared to 214 ± 92 µg/l in control patients without aluminum-related bone disease. In this investigation (Milliner et al., 1984), an increase in the plasma aluminum level of more than 200 µg/l identified 35 of 37 patients with aluminum-related osteodystrophy, giving a sensitivity of 94% and a specificity of 50%.

E. CONCLUSION

In order to avoid dialysis encephalopathy, plasma concentrations should be less than 200 µg/l. Less severe symptoms may, however, appear in the interval 50–100 µg/l. In aluminum welders, slightly increased blood aluminum concentrations have been reported. However, these concentrations are far below the levels noted in uremic patients.

Critical organs for aluminum toxicity are the brain and bone. Due to the increased risk for dialysis encephalopathy, aluminum in blood should be determined in dialysis patients. For certain groups of occupationally exposed individuals, e.g., aluminum workers, determinations of aluminum in urine can be used for exposure assessment. However, so far it is not possible to use these analyses for risk assessment.

IX. ARSENIC

A. EXPOSURE

Both trivalent and pentavalent forms of inorganic arsenic are found in the environment. Inorganic arsenic, mainly in the form of arsenic trioxide, is produced as a byproduct in the smelting of copper and lead ores. Several inorganic arsenic compounds are used in pesticides, e.g., calcium and lead arsenate. Gallium arsenide has been used in electronic devices. Certain fish species and crustacea contain high levels of organic arsenic, mainly in the form of arsenobetaine (Vahter, 1988).

The average daily intake of inorganic arsenic in the general population is about 10 to 50 µg in Europe and the U.S. and 100 µg or more in Japan. The average daily intake of organic arsenic from seafood is about 15 to 20 µg, but can be considerably higher, up to about 1 mg, if fish consumption is very high.

B. METABOLISM

About 70–90% of ingested inorganic arsenic is readily absorbed from the gastrointestinal tract (Pomroy et al., 1980) and then distributed to different organs. Organic arsenic in the form of arsenobetaine is almost completely absorbed. Soluble arsenic compounds, e.g., arsenic trioxide, are rapidly absorbed from the lungs, while slowly soluble compounds such as calcium arsenate and lead arsenate have a longer retention time.

After exposure to inorganic arsenic, the longest retention times are found in the skin, gastrointestinal tract, epididymis, thyroid, and skeleton. The highest tissue concentrations in humans have been observed in hair, nails, skin, and lungs (Liebscher and Smith, 1968). Most inorganic arsenic is methylated in the body, mainly in the liver, and then excreted into the urine in the form of about 10–20% inorganic arsenic, 10–15% monomethylarsonic acid (MMA), and 60–80% dimethylarsinic acid (DMA; Vahter, 1988). With increasing doses, the methylated fraction decreases. About 50–60% of an oral dose of inorganic arsenic is excreted within a week. About 70–75% of ingested arsenobetaine is excreted into the urine within a week without being metabolized (Vahter, 1988).

C. HEALTH EFFECTS

Acute intoxication from inorganic arsenic may cause irritation of the gastrointestinal tract (vomiting and diarrhea) and the respiratory tract (rhinitis, pharyngitis, and bronchitis; WHO, 1981a). The skin is the critical organ after long-term oral exposure. Inorganic arsenic may cause lesions of the skin and mucous membranes, leading to melanosis, hyperpigmentation, depigmentation, and hyperkeratosis. Skin cancer may follow. Long-term exposure may also affect the peripheral nervous system and the cardiovascular system (e.g., Blackfoot disease and Raynaud's syndrome) as well as the hematopoietic system (anemia and leukopenia). In epidemiological studies an increased risk of lung cancer have been observed among workers exposed to arsenic (WHO, 1981a).

D. BIOLOGICAL MONITORING

Arsenic concentrations in blood in nonoccupationally exposed individuals is usually in the range 1 to 4 µg/l (Bencko and Symon, 1977). Urinary concentrations of inorganic arsenic and its metabolites are normally less than 10 µg/l in European countries, somewhat higher in the U.S., and highest in Japan. After short-term exposure to inorganic or organic arsenic, the blood concentration increases rapidly. However, due to the rapid elimination via urine, the blood concentration is only elevated for a short time-period. In a situation with a long-term and fairly stable exposure leading to steady-state, blood arsenic will reflect the magnitude of exposure on a group basis. However, blood arsenic determinations may grossly overestimate the exposure to inorganic arsenic if the intake of additional arsenobetaine is high. So far, no routine methods are available for the determination of different arsenic metabolites in blood.

On a group basis, total arsenic in urine reflects the intake of inorganic arsenic if the intake of arsenobetaine from seafood is limited. Urinary concentrations of total arsenic in the range 50 to 100 µg/l indicate a high intake of seafood arsenic or increased exposure to inorganic arsenic from food or air (inhalation). The exposure can be further clarified by complementary separate determinations of inorganic arsenic and its metabolites MMA and DMA in urine.

In studies of smelter workers, a linear relationship between work-site air arsenic concentrations and the excretion of inorganic arsenic and its metabolites in urine has been observed; arsenic in urine (µg As/g creatinine) = 29 (µg As/g creatinine) + 2 × arsenic in air (µg As/m^3) (Vahter et al., 1986). In a recently performed study at a chemical factory (Offergelt et al., 1992), significant correlations (log scales) were found between airborne time-weighted average exposure to arsenic trioxide and the inorganic arsenic metabolites in urine collected immediately after the shift, or just before the next shift.

Contrary to organic arsenic and methylated metabolites, the inorganic form is accumulated in hair where the trivalent form, particularly, is bound to sulfhydryl groups of keratin. Normal concentrations are usually less than 1 mg As/kg (Liebscher and Smith, 1968). As the concentration of inorganic arsenic in the root is in equilibrium with the concentration in blood, analyses of different parts of hair strands will reflect the exposure to inorganic arsenic at certain time periods. However, the evaluation is hampered by the risk of external contamination from, e.g., air, water, dust, shampoos, and soaps. Thus, on a group basis, hair analyses can be used to get a rough estimate of the exposure in situations with a continuous and fairly stable uptake via food or drinking water.

E. CONCLUSION

Because of the rapid clearance of inorganic arsenic and its metabolites from blood, and the influence of arsenobetaine, this medium is of limited use for exposure evaluations. Determinations of inorganic arsenic and its metabolites MMA and DMA in urine may be used to estimate the exposure to inorganic arsenic. For soluble inorganic arsenic compounds, repeated measurements of these biological indices may be used for risk assessment after oral or inhalational intake. However, for poorly absorbed arsenic compounds, e.g., calcium arsenate, lead arsenate, and gallium arsenide, the urinary levels will only mirror the amount absorbed, and not the total intake. The high risk for external contamination limits the use of hair and nail analyses for biological monitoring.

X. COBALT

A. EXPOSURE

Cobalt is an essential metal for humans and is part of the enzyme cyanocobalamin (vitamin B_{12}). It is a relatively rare metal used in the production of hard metals. The advantages of cobalt alloys are high melting points, strength, and resistance to oxidation. Tungsten and carbon, as well as titanium and tantalum, are other metals involved in the production of hard metals. Addition of cobalt chloride to beer has been used to improve the quality of the froth in Canada, the U.S., and Belgium. The daily intake from food is about 5 to 45 µg/day.

B. METABOLISM

After oral intake, the absorption in the gastrointestinal tract of soluble cobalt compounds is about 30 to 40%. The knowledge about pulmonary absorption after inhalation is limited. In humans, the highest tissue concentrations have been found in the liver and the kidneys (Alessio and Dell'Orto, 1988). A major portion of absorbed cobalt is excreted into the urine within days. However, a proportion of cobalt is retained, with a biological half-time in the magnitude of one year.

C. HEALTH EFFECTS

Critical organs after cobalt exposure include the skin, heart, and the respiratory tract. Skin exposure may lead to allergic contact dermatitis, characterized by erythematous papules. Endemic outbreaks of cardiomyopathy have been reported among heavy beer consumers, following addition of cobalt chloride to beer as a foam stabilizer (Bonenfant et al., 1967). Fatalities were not uncommon. Cobalt fortification of beer is, however, no longer in use.

Even at fairly low air cobalt concentrations, inhalation may lead to irritation of the mucous membranes in the respiratory tract. Lung function test may reveal signs of obstructive lung disease. Inhalation of cobalt in hard metal processing may cause hard metal lung disease, which is a severe type of pneumoconiosis. The disease progresses from chronic airway obstruction to interstitial lung fibrosis, causing increasingly severe dyspnea (Alessio and Dell'Orto, 1988).

D. BIOLOGICAL MONITORING

The concentrations of cobalt in blood and urine from nonoccupationally exposed individuals are usually in the range 0.1 to 2 µg/l. The increase of cobalt levels in these media is proportional to the exposure (Ichikawa et al., 1985; Stebbins et al., 1992), being most pronounced in urine. The excretion in urine shows a two-phase elimination pattern (Alexandersson and Lidums, 1979). The first phase of elimination is rapid and reflects the exposure during a working shift. Furthermore, there is an increase of the urinary excretion from Monday to Friday during a working week. Thus, the difference in urinary cobalt concentrations before and after a shift mirrors the daily exposure. Similarly, the increase in urinary levels from Monday to Friday reflects the cumulative week's exposure (Pellet et al., 1984; Scansetti et al., 1985).

E. CONCLUSION

Determinations of cobalt, particularly in urine, can be used to assess exposure by inhalation in occupationally exposed workers. However, the current knowledge is not sufficient for risk assessment.

XI. CHROMIUM

A. EXPOSURE

Principal industrial users of chromium compounds are the metallurgical processors of ferrochromium and stainless steel. These compounds are also used for electroplating, pigment production, and tanning. Other sources of chromium in air and water include the burning of fossil fuels and waste incineration (WHO, 1988). Handling of chromium in cement is an important source of dermal exposure.

Chromium has several oxidation states, however, only the zero-, di-, tri-, and hexavalent oxidation states are of biological importance. The trivalent form is an essential nutrient for humans of importance for the maintenance of normal glucose metabolism. Reported intakes from food and water are usually in the range 50 to 200 µg/day (WHO, 1988).

B. METABOLISM

The gastrointestinal absorption of trivalent chromium is low, less than 1% (Aitio et al., 1984). The absorption for hexavalent chromium is somewhat higher, approximately 3 to 6%. Data about inhalational exposure in humans are sparse. Hexavalent chromium compounds readily cross the cell membranes via anion transport systems. Inside the cell it is reduced to trivalent chromium, e.g., by thiols such as glutathione and cysteine, and cytochrome P-450. After absorption, trivalent chromium is transported mainly bound to transferrin in plasma, while hexavalent chromium is partly taken up by erythrocytes. In a human autopsy study, Schroeder et al. (1962) reported chromium concentrations of 15.6 mg/kg ash in lung, aorta 9.1 mg/kg, pancreas 6.5 mg/kg, heart 3.8 mg/kg, kidney 2.1 mg/kg, and liver 1.8 mg/kg. After the cessation of occupational exposure, chromium may be retained in the lungs for several years, indicating a long biological half-time (WHO, 1988). After intravenous injection, most of the trivalent chromium is excreted in urine, while part of the similarly injected hexavalent sodium chromate is excreted in feces (Aitio et al., 1988).

C. HEALTH EFFECTS

Hexavalent chromium compounds can cause irritation of mucous membranes and skin, leading, e.g., to skin ulcers, nasal septum perforations, and chronic bronchitis (WHO, 1988). Chrome eczema may follow handling of hexavalent chromium in cement. Excessive peroral intake may lead to tubular necrosis of the kidneys. Hexavalent chromium compounds, especially those of low solubility, have caused lung cancer in a number of animal experiments (WHO, 1988). The development of lung cancer has also been observed in epidemiological studies of workers in chromate production, chrome pigment production, and the chromium plating industry (IARC, 1990). Data are not conclusive for other types of chromium exposures, e.g., ferrochromium production and stainless steel welding.

D. BIOLOGICAL MONITORING

The risk for external contamination must be considered when analyzing chromium in blood and urine. One important source of error is the leakage of chromium into blood when using stainless steel needles for blood sampling. This factor may lead to a gross overestimation of chromium concentrations in blood in occupationally unexposed populations. Reported normal concentrations in plasma or serum in unexposed individuals is approximately 0.1 µg/l. Urinary concentrations are usually in the range 0.1 to 0.5 µg/l (Aitio et al., 1988).

Analyses in blood and urine have mainly been used for biological monitoring. Linear relationships have been reported between air chromium concentrations and urinary excretions in studies of chrome platers (Lindberg and Vesterberg, 1983) and stainless steel welders (Sjögren et al., 1983b; Welinder et al., 1983). However, the urinary excretion in chromate workers is considerably higher than in stainless steel welders at the same air chromium concentration. This may partly be explained by differences in particle size and bioavailability of the aerosols.

Long-term exposure to hexavalent chromium acid mist giving an urinary excretion lower than 5 µg/l indicates air exposure levels lower than 2 µg/m^3. Higher chromium exposure may lead to nasal mucous irritation and ulcers (Lindberg and Hedenstierna, 1983).

Determinations in plasma and erythrocytes can be used to differentiate between different chromium species, as trivalent chromium is mainly transported in plasma, while hexavalent chromium is partly found in erythrocytes (Aitio et al., 1988).

E. CONCLUSION

Because of established quantitative relationships between chromium exposure and chromium excretion in urine for chrome platers and stainless steel welders, biological monitoring can be used for exposure assessment in these groups. Determinations of chromium in urine can partly be used for risk assessment in chrome platers, as negative health effects on the nasal mucousa and respiratory tract has not been observed below a U-Cr of 5 µg/l.

XII. MANGANESE

A. EXPOSURE

Manganese is an essential element for humans. It is part of the enzyme mitochondrial SOD (superoxide dismutase) in rats and is essential for many animal species for the formation of bone and connective tissue, and for the metabolism of carbohydrate and lipids. The daily requirement for humans is about 2 to 3 mg (WHO, 1981b). Important uses include the production of steel, nonferrous alloys, and dry cell batteries. It is also used by the chemical industry. Considerably exposure may take place in manganese-containing ores. However, most iron ores contain manganese as well. Organic manganese compounds have been used as fungicides and as antiknocking agents (MMT = methylcyclopentadienyl manganese tricarbonyl) in gasoline (Oberdoerster and Cherian, 1988).

B. METABOLISM

Homeostatic mechanisms regulate the absorption of manganese after oral intake. Normally, about 3% of the ingested amount is absorbed in the gastrointestinal tract (EPA, 1984), but the uptake is increased by iron deficiency. In occupational settings, inhalation is the predominant route of exposure for exposed workers. Manganese is retained in the lungs after inhalation exposure to MnO_2, with a half-time of approximately 2 to 3 months.

After absorption, manganese is transported by blood proteins such as transferrin and alpha$_2$-macroglobulin (Scheuhammer and Cherian, 1985) and distributed to different organs. The highest concentrations are found in the liver, kidney, the endocrine system, and in the small and large intestines. The uptake in brain is slow. However, local concentrations in basal ganglia can be high. The major part is excreted via bile into the feces. Only small amounts are excreted via urine (<1 µg/l), sweat, hair, and nails. The organic form, MMT, is excreted in equal proportions in feces and urine.

C. HEALTH EFFECTS

Lung and brain are critical organs after manganese exposure. Exposure to manganese may cause irritation of the respiratory airways and dyspnea. Abnormalities in lung function tests have been observed. Increased incidences of pneumonia and bronchitis have been reported among populations living near manganese-emitting industries.

Long-term inhalation exposure may cause brain accumulation due to the slow elimination from this organ. An irreversible neurological disorder, resembling both Parkinson's disease and dystonia have been reported in miners exposed to MnO_2. The syndrome is characterized by gait disorder, postural instability, tremor, and a mask-like expressionless face (Oberdoerster and Cherian, 1988). Symptoms usually appear after one to two years of exposure and progresses rapidly. The underlying mechanism is unknown, but disturbance of the cathecholamine system has been suggested.

D. BIOLOGICAL MONITORING

Normal concentrations of manganese in blood range from 7 to 12 µg/l and in urine from 1 to 8 µg/l (Oberdoerster and Cherian, 1988). Due to the short biological half-time in blood of less than 5 min (WHO, 1981b), blood concentrations will be elevated shortly after manganese exposure. The homeostatic mechanisms involved in trying to keep tissue concentrations on a constant level will make it more difficult to detect increased blood levels. Because of the very low excretion in urine, this medium is not a good indicator of exposure. However, on a group basis, manganese in urine to some extent seem to reflect recent exposure (Roels et al., 1987b). After exposure to an organic form of manganese, e.g., MMT, a linear increase in the urinary excretion has been observed. Thus, urinary determinations are more suitable for exposure assessment for this compound since about 50% of the absorbed amount is excreted through the kidneys. Analyses of manganese in hair are not so well documented that they can be considered for exposure assessment.

E. CONCLUSION

No suitable indicator media are available for biological monitoring of inorganic manganese. Thus, biological monitoring can not be used for risk assessment in this context.

XIII. NICKEL

A. EXPOSURE

Nickel is a ubiquitous metal. The concentration in the earth's crust is about 0.008%. It is mainly used for production of stainless steel and other nickel alloys. Important emission sources of nickel in air include, e.g., combustion of coal and oil for heat and power generation, incineration of waste and sewage sludge, nickel mining and production, steel manufacture, and electroplating (WHO, 1991b). The daily intake of nickel from food is 100 to 300 µg/day in most countries. Concentrations in tap water are usually below 10 µg/l. Reported levels of atmospheric nickel range from about 5 to 35 ng/m^3 (Bennett, 1984), depending on grade of urbanization. Such exposure levels can give an inhalational uptake of 0.1 to 0.7 µg/day (WHO, 1991b). Considerably higher levels have been observed in the working environment, from a few micrograms to a few milligrams per cubic meter.

B. METABOLISM

Gastrointestinal absorption is low, less than 1% after food intake (Sunderman et al., 1989). The remainder is eliminated in feces. About 30% of inhaled nickel is deposited in the lung, and approximately 20% of inhaled nickel is absorbed from the respiratory tract (Sunderman, 1988). After absorption, nickel is transported in plasma, mainly bound to albumin and to ultrafiltrable ligands (e.g., amino acids), and then distributed and accumulated in bone, kidney, liver, and to some extent in lung. A large fraction is bound to nickeloplasmin, a macroglobulin which, however, does not have an important role in the extracellular transport of nickel. In electroplaters, the estimated biological half-time in plasma is approximately 20–34 h, as compared to 17 to 39 h in urine (Tossavainen et al., 1980). Nickel carbonyl can pass the blood-brain barrier and accumulate in the brain. Urine is the dominant excretion pathway, which is probably best described by a two-compartment model. Small amounts are also excreted in bile, sweat, as well as being exhaled. The excretion in feces reflects the unabsorbed part from dietary intake and the biliary and tracheally cleared nickel. The whole-body content in an occupationally unexposed individual of 70 kg is about 0.5 mg (Bennett, 1984).

C. HEALTH EFFECTS

Critical organs for nickel exposure in humans are the respiratory system, especially the nasal cavities and sinuses, the immune system, and the skin. Nickel carbonyl is the most acutely toxic nickel compound. Poisoning can lead to headache, vertigo, nausea, vomiting, nephrotoxic effects, and severe pneumonia. Pulmonary fibrosis may follow. Chronic irritative effects observed in nickel refinery and nickel plating workers include rhinitis, sinusitis, perforations of the nasal septum, and bronchial asthma (WHO, 1991b). Nickel contact dermatitis is not uncommon in the general population (up to 10% in females and 1% in males; WHO, 1991b). It has also frequently been observed in a number of occupations in which the workers have been exposed to soluble nickel compounds. Long-term exposure to nickel compounds with low solubility, e.g., nickel subsulfide and nickel oxide, give a high risk of lung and nasal malignancies, which has been observed in a number of epidemiological studies (Peto et al., 1984; WHO, 1991b). High risks have been reported in nickel refinery workers, and in workers involved in processes with exposure to soluble nickel, e.g., nickel sulfate (Magnus et al., 1982), often combined with some exposure to nickel oxide (IARC, 1990; WHO, 1991b). Nickel compounds have been classified as carcinogenic to humans (Group 1) and metallic nickel as possibly carcinogenic to humans (Group 2B; IARC, 1990).

D. BIOLOGICAL MONITORING

Serum and urine samples can be used for biological monitoring. However, special precautions must be undertaken to avoid contamination. Normal levels in unexposed populations of adults are about 0.05 to 1.1 µg/l in serum, and somewhat higher in urine, 0.5–4.0 mg/g creatinine (WHO, 1991b). Feces is probably the most reliable indicator of oral exposure to nickel (Hassler et al., 1983). Expired air may be used for monitoring, following inhalational exposure to nickel carbonyl (Sunderman et al., 1968). Exposed workers display higher concentrations in serum and urine than unexposed populations. However, the levels in these indicator media are weakly correlated to air concentrations because of a large individual variation (Sunderman, 1988). The absorption in workers is to a great extent dependent on the solubility of the nickel compounds inhaled, soluble nickel compounds having a much faster clearance. Nickel concentrations in serum mainly reflects recent exposure because of the short biological half-time in this compartment (Sunderman, 1988). At steady-state conditions, only determinations in serum will reflect long-term exposure. The excretion in urine may reflect more extended exposure. However, for a proper evaluation the accumulation and elimination rates in the kidney must be considered.

E. CONCLUSION

Determinations of nickel in serum and urine can be used for biological monitoring. However, they do not give a good picture of past exposures, and they can not be used for risk assessment as current knowledge is not sufficient to relate nickel concentrations in these indicator media to specific adverse health effects.

ACKNOWLEDGMENTS

Some of the studies quoted in this review were supported by the Swedish Work Environment Fund, the National Swedish Environment Protection Board, and the Medical Faculty, Lund University.

REFERENCES

Ahlgren, L., Lidén, K., Mattsson, S., and Tejning, S., *Scand. J. Work Environ. Health,* 2, 82–86, 1976.
Ahlgren, L. and Mattsson, S., *Phys. Med. Biol.,* 26, 19–26, 1981.
Aitio, A., Jarvisalo, J., Kiilunen, M., Tossavainen, A., and Vaittinen, P., *Int. Arch. Occup. Environ. Health,* 54, 241–249, 1984.
Aitio, A., Jarvisalo, J., Kiilunen, M., Kalliomaki, P.-L., and Kalliomaki, K., in *Biological Monitoring of Toxic Metals,* Clarkson, T.W., Friberg, L., Nordberg, G.F., and Sager, P.R., Eds., Rochester Ser. Environ. Toxicity, Plenum Press, New York, 369, 1988.
Alessio, L, and Dell'Orto. A., in *Biological Monitoring of Toxic Metals,* Clarkson, T.W., Friberg, L., Nordberg, G.F., and Sager, P.R., Eds., Rochester Ser. Environ. Toxicity, Plenum Press, New York, 407, 1988.
Alexandersson, R. and Lidums, V., *Arb. och Hälsa,* 8, 1–23, 1979 (in Swedish, English summary).
Alfrey, A. C., LeGendre, G. R., and Kaehny, W. D., *N. Engl. J. Med.,* 294, 184–188, 1976.
Andrén, P., Schütz, A., Vahter, M., Attewell, R., Johansson, L, Willers, S., and Skerfving, S., *Sci. Total Environ.,* 77, 25–34, 1988.
Aono, H. and Araki, S., *Ind. Health,* 24, 129–138, 1986.
Barregård, L., Hultberg, B., Schütz, A., and Sällsten, G., *Int. Arch. Occup. Environ. Health,* 61, 65–69, 1988.
Barregård, L., Högstedt, B., Schütz, A., Karlsson, A., Sällsten, G., and Thiringer, G., *Scand. J. Work Environ. Health,* 17, 263–268, 1991.
Barregård, L., Sällsten, G., Schütz, A., Attewell, R., Skerfving, S., and Järvholm, B., *Arch. Environ. Health,* 47, 176–184, 1992.
Barregård, L., *Scand. J. Work Environ. Health,* 19 (Suppl. 1), 45–49, 1993.
Barry, P. S. I., *Br. J. Ind. Med.,* 32, 119–139, 1975.
Batuman, V., Wedeen, R., Bogden, J. D., Balestra, D. J., Jones, K., and Schidlovsky, G., *Environ. Res.,* 48, 70–5, 1989.
Bencko, V. and Symon, K., *Environ. Res.,* 13, 378–385, 1977.
Bennett, B. G., in *Nickel in the Human Environment,* Proc. Joint Symp., Lyon, 8–11 March, 1983, IARC Sci. Publ. No. 53, International Agency for Research on Cancer, Lyon, France, 487, 1984.
Bensryd, I., Rylander, L., Högstedt, B., Aprea, P., Bratt, I., Fåhraéus, C., Holmén, A., Karlsson, A., Nilsson, A., Svensson, B.-L., Schütz, A., Thomassen, Y., and Skerfving, S., *Sci. Total Environ.,* 145, 81–102, 1994.
Bergdahl, I. A., Schütz A., and Hansson, G.-Å., *Analyst,* 120, 1205–1209, 1995.
Bert, J. L., van Dusen, L. J., and Grace, J. R., *Environ. Res.,* 48, 117–27, 1989.
Bonenfant, J. L., Miller, G., and Roy, P. E., *Can. Med. Assoc. J.,* 97, 910–916, 1967.
Brunmark, P., Skarping, G., Schütz, A., *J. Chromatogr.,* 573, 35–41, 1992.
Buchet, J. P., Lauwerys, R., Roels, H., Bernard, A., Braux, P., Claeys, F., Ducoffre, G., de Plaen, P., Staessen, J., Amery, A., et al., Renal effects of cadmium body burden of the general population, *Lancet,* 336, 699–702, 1990.
Börjesson, J., Barregård, L., Sällsten, G., Schütz, A., Jonson, R., Alpsten, M., and Mattsson, S., *Phys. Med. Biol.,* 40, 413–426, 1995.
Campbell, B. C., Meredith, P. A., Moore, M. R., and Watson W. S., *Toxicol. Lett.,* 21, 231–235, 1984.
Chettle, D. R. and Ellis, K. J., *Am. J. Ind. Med.,* 22, 117–124, 1992.
Christoffersson, J. O. and Mattsson, S., *Phys. Med. Biol.,* 28 1135–1144, 1983.
Christoffersson, J. O., Schütz, A., Ahlgren, L., Haeger-Aronsen, B., Mattsson, S., and Skerfving, S., *Am. J. Ind. Med.,* 6, 447–457, 1984.
Christoffersson, J. O., Schütz, A., Skerfving, S., Ahlgren, L., and Mattson, S., *Arch. Environ. Health,* 41, 312–318, 1986.
Christoffersson, J. O., Welinder, H., Spång, G., Mattsson, S., and Skerfving, S., *Environ. Res.,* 42, 489–499, 1987a.

Christoffersson, J. O., Schütz, A., Skerfving, S., Ahlgren, L., and Mattsson, S., in In vivo *Body Composition Studies*, Ellis, K.J., Yasumura, S., and Morgan, W.D., Eds., Bocardo Press, Oxford, 334, 1987b.

Clarkson, T. W., Friberg, L., Nordberg, G.F., and Sager, P.R., Eds., *Biological Monitoring of Toxic Metals*, Plenum Press, New York, 1988a.

Clarkson, T. W., Friberg, L., Hursh, J. B., Nylander, M., in *Biological Monitoring of Toxic Metals*, Clarkson, T.W., Friberg, L., Nordberg, G.F., and Sager, P.R., Eds., Plenum Press, New York, 247, 1988b.

Clarkson, T. W., Hursh, J. B., Sager, P. R., and Syversen, T. L. M., in *Biological Monitoring of Toxic Metals*, Clarkson, T. W., Friberg, L., Nordberg, G.F., and Sager, P. R., Eds., Plenum Press, New York, 199, 1988c.

De Silva, P. E., *Br. J. Ind. Med.*, 38, 209–217, 1981.

Eastwell, H. D., Thomas, B. J., and Thomas, B. W., *Lancet*, ii, 524–525, 1983.

Elinder, C.-G. and Sjögren, B., in *Handbook on the Toxicology of Metals*, Vol. II, Friberg, L., Nordberg, G.F., and Vouk, V.B., Eds., Elsevier, New York, 1, 1986.

Elinder, C.-G., Lind, B., Nilsson, B., and Oskarsson, A., *Food Add. Contam.*, 5, 641–644, 1988.

Elinder, C. G., Friberg, L., Kjellström, T., Nordberg, G., and Oberdoerster, G., Biological Monitoring of Metals. Int. Prog. Chem. Safety, WHO/EHG/94.2, World Health Organization, Geneva, 1994.

Ellis, K. J., Yuen, K., Yasumura, S., and Cohn, S. H., *Environ. Res.*, 33, 216–226, 1984.

EPA, Health Assessment Document for Manganese, Final Rep. 1–1 to 10–77, U.S. Environmental Protection Agency, Washington, D.C., 1984.

Erkkilä, J., Armstrong, R., Riihimäki, V., Chettle, D. R., Paakari, A., Scott, M., Somervaille, L., Starck, J., and Aitio, A., *Br. J. Ind. Med.*, 49, 631–644, 1992.

Friberg, L., Elinder, C. G., Kjellström, T., and Nordberg, G. F., Eds., *Cadmium and Health*. Vol. 1 and 2, CRC Press, Boca Raton, FL, 1985.

Friberg, L., Nordberg, G. F., and Vouk, V. B., Eds., *Handbook on the Toxicology of Metals*, Elsevier, Amsterdam, 1986.

Gerhardsson, L., Lundström, N. G., Nordberg, G. F., and Wall, S., *Br. J. Ind. Med.*, 43, 707–712, 1986.

Gerhardsson, L., Chettle, D. R., Englyst, V., Nordberg, G. F., Nyhlin, H., Scott, M. C., Todd, A. C., and Vesterberg, O., *Br. J. Ind. Med.*, 49, 186–192, 1992.

Gerhardsson, L., Attewell, R., Chettle, D. R., Englyst, V., Lundström, N. G., Nordberg, G. F., Nyhlin, H., Scott, M. C., and Todd, A. C., *Arch. Environ. Health*, 48, 147–156, 1993.

Grandjean, P. and Nielsen, T., *Res. Rev.*, 72, 97–148, 1979.

Grubb, A., *Nephrology*, 38, 20–27, 1992.

Gusserow, A., *Virchows Arch. Pathol.*, 21, 443–452, 1861.

Gustafson, Å., Hedner, P., Schütz., A., and Skerfving, S., *Int. Arch. Occup. Environ. Health*, 61, 277–281, 1989.

Hassler, E., Lind, B., Nilsson, B., and Piscator, M., *Ann. Clin. Lab. Sci.*, 13, 217–224, 1983.

Hu, H., Milder, F. L., and Burger, D. E., *Environ. Res.*, 49, 295–317, 1989.

IARC, Monographs on the Evaluation of Carcinogenic Risks to Humans, overall evaluations of carcinogenicity: an updating of IARC Monographs Vol. 1 to 42. Suppl. 7. International Agency for Research on Cancer, Lyon, 230–232, 1987.

IARC, Chromium, nickel and welding, in IARC Monographs on the Evaluation of Carcinogenic Risks to Humans, Vol. 49, International Agency for Research on Cancer, Lyon, 1990.

IARC, Mercury and mercury compounds. In IARC Monographs on the Evaluation of Carcinogenic Risks to Humans, Vol. 58, International Agency for Research on Cancer, 239–345, 1993.

IARC, Cadmium and cadmium compounds, in IARC Monographs on the Evaluation of Carcinogenic Risks to Humans, Vol. 58. Beryllium, cadmium, mercury, and exposures in the glass manufacturing industry, International Agency for Research on Cancer, Lyon, 119–237, 1994.

Ichikawa, Y., Kusaka, Y., and Goto, S., *Int. Arch. Occup. Environ. Health*, 55, 269–276, 1985.

Järup, L., Rogenfelt, A., Elinder, C. G., Nogawa, K., and Kjellström, T., *Scand. J. Work Environ. Health*, 9, 327–331, 1983.

Kazantzis, G., Schiller, K. F. R., Asscher, A. W., and Drew, R. G., *Q. J. Med.*, 31, 403–418, 1962.

Kershaw, T. G., Clarkson, T. W., and Dhahir, P. H., *Arch. Environ. Health*, 35, 28–36, 1980.

Kjellström, T. and Nordberg, G. F., in *Cadmium and Health*, Vol. 1, Friberg, L., Elinder, C.G., Kjellström, T., and Nordberg, G.F., Eds., CRC Press, Boca Raton, FL, 179–197, 1985.

Langworth, S., Elinder, C. G., Göthe, C.-J., and Vesterberg, O., *Int. Arch. Occup. Environ. Health*, 63, 161–167, 1991.

Liebscher, K. and Smith, H., *Arch. Environ. Health*, 17, 881–890, 1968.

Lindberg, E. and Vesterberg, O., *Scand. J. Work Environ. Health*, 9, 333–340, 1983.

Lindberg, E. and Hedenstierna, G., *Arch. Environ. Health*, 38, 367–374, 1983.

Magnus, K., Andersen, A., Hogetveit, A. C., *Int. J. Cancer*, 30, 681–685, 1982.

Manton, W. I. and Cook, J.-D., *Br. J. Ind. Med.*, 41, 313–319, 1984.

Marsh, D. O., Myers, G. J., Clarkson, T. W., Amin-Zaki, L., Tikriti, S., Majeed, M. A., and Dabbagh, A. R., *Clin. Toxicol.*, 18, 1311–18, 1981.

Marsh, D. O., Clarkson, T. W., Cox, C., Myers, G. J., Amin-Zaki, L., and Al-Tikriti, S., *Arch. Neurol.*, 44, 1017–1022, 1987.

Mason, H. J., Davison, A. G., Wright, A. L., Guthrie, C. J. G., Fayers, P. M., Venables, K. M., Smith, N. J., Chettle, D. R., Franklin, D. M., Scott, M. C., Holden, H., Gompertz, D., and Newman-Taylor, A. J., *Br. J. Ind. Med.*, 45, 793–802, 1988.

Milliner, D. S., Nebeker, H. G., Ott, S. M., Andress, D. L., Sherrard, D. J., Alfrey, A. C., Spatopolsky, E. A., and Coburn, J. W., *Am. Int. Med.*, 101, 775–780, 1984.

Molin, M., Bergman, M., Marklund, S. L., Schütz, A., and Skerfving, S., *Acta Odontol. Scand.*, 48, 189–202, 1990.

Molin, M., Schütz, A., Skerfving, S., and Sällsten, G., *Int. Arch. Occup. Environ. Health*, 63, 187–192, 1991.

Morgan, W. D., Ryde, S. J. S., Jones, S. J., Wyatt, R. M., Hainsworth, I. R., Cobbold, S. S., Evans, C. J., and Braithwait, R. A., *Biol. Trace Elem. Res.*, 26–27, 407–414, 1990.

Mushak, P., Davis, J. M., Crocetti, A. F., and Grant, L. D., *Environ. Res.*, 50, 11–36, 1989.

Needleman, H. L., Gunnoe, C., Leviton, A., Reed, R., Peresie, H., Maher, C., and Barrett, P., *N. Engl. J. Med.*, 300, 689–695, 1979.

Nilsson, U., Ahlgren, L., Christoffersson, J. O., and Mattsson, S., in *Advances in In Vivo Body Composition Studies*, Yasumura, S., Harrison, J. E., McNeill, K. G., Woodhead, A. D. and Diemanian, S. F., Eds., Plenum Press, New York, 297, 1990.

Nilsson, U., Attewell, R., Christoffersson, J. O., Schütz, A., Ahlgren, L., Skerfving, S., and Mattsson, S., *Pharmacol. Toxicol.*, 68, 477–484, 1991.

Nilsson, U. and Skerfving, S., *Scand. J. Work Environ. Health*, 19 (Suppl. 1), 54–58, 1993.

Nilsson, U., Schütz, A., Skerfving, S., and Mattsson, S., *Int. Arch. Occup. Environ. Health*, 67, 405–411, 1995.

Nordberg, G. F., Garvey, J. S., and Chang, C. C., *Environ. Res.*, 28, 179–182, 1982.

Nordberg, G. F. and Nordberg, M., in *Biological Monitoring of Toxic Metals*, Clarkson, T. W., Friberg, L., Nordberg, G. F., and Sager, P.R., Eds., Plenum Press, New York, 151, 1988.

Nordberg, G., *Arb. Hälsa*, 1–123, 1993.

Nylander, M., Friberg, L., and Lind, B., *Swed. Dent. J.*, 11, 179–187, 1987.

Oberdoerster, G. and Cherian, G., in *Biological Monitoring of Toxic Metals*, Clarkson, T. W., Friberg, L., Nordberg, G. F., and Saget, P.R., Eds., Rochester Ser. Environ. Toxicity, Plenum Press, New York, 283–301, 1988.

Offergelt, J. A., Roels, H., Buchet, J. P., Boeckx, M., and Lauwerys, R., *Br. J. Ind. Med.*, 49, 387–393, 1992.

O'Flaherty, E. J., *Toxicol. Appl. Pharmacol.*, 111, 313–331, 1991a.

O'Flaherty, E. J., *Toxicol. Appl. Pharmacol.*, 111, 332–341, 1991b.

Oskarsson, A., Jorhem, L., Sundberg, J., Nilsson, N. G., and Albanus, L., *Sci. Total Environ.*, 111, 83–94, 1992.

Paul, C., *Arch. Gen. Med.*, 43, 513–533, 1860.

Pellet, F., Perdrix, A., Vincent, M., and Mallion, J. M., *Arch. Mal. Prof.*, 45, 81–85, 1984.

Peto, J., Cuckle, H., Doll, R., Hermon, C., and Morgan, L. G., Respiratory cancer mortality of Welsh nickel refinery workers, in *Nickel in the Human Environment*, IARC Sci. Publ. 53, Proc. Joint Symp. Lyon, 8–11 March, 1983, International Agency for Research on Cancer, Lyon, 37, 1984.

Pomroy, C., Charbonneau, S. M., McCullough, R. S., and Tam, G. K. H., *Toxicol. Appl. Pharmacol.*, 53, 550–556, 1980.

Price, J., Grudzinski, A. W., Craswell, P. W., and Thomas, B. J., *Arch. Environ. Health*, 47, 256–262, 1992.

Rabinowitz, M. B., *Environ. Health Perspect.*, 91, 33–37, 1991.

Roels, H. A., Lauwerys, R. R., Buchet, J. P., Bernard, A., Chettle, D. R., Harvey, T. C., and Al-Haddad, I. K., *Environ. Res.*, 26, 217–240, 1981.

Roels, H., Abdeladim, S., Ceulemans, E., and Lauwerys, R., *Ann. Occup. Hyg.*, 31, 135–145, 1987a.

Roels, H., Lauwerys, R., Genet, P., Sarhan, M. J., de Fays, M., Hanotiau, I., and Buchet, J.-P., *Am. J. Ind. Med.*, 11, 297–305, 1987b.

Roels, H. A., Boeckx, M., Ceulemans, E., and Lauwerys, R. R., *Br. J. Ind. Med.*, 48, 247–253, 1991.

Rosen, J. F., Markowitz, M. E., Bijur, P. E., Jenks, S. T., Wielopolski, L., Kalef-Ezra, J. A., and Slatkin, D. N., *Environ. Health Perspect.*, 91, 57–62, 1991.

Savory, J. and Wills, M. R., in *Biological Monitoring of Toxic Metals*, Clarkson, T.W., Friberg, L., Nordberg, G.F., and Sager, P.R., Eds., Rochester Ser. Environ. Toxicity, Plenum Press, New York, 323, 1988.

Scansetti, G., Lamon, S., Talarico, S., Botta, G. C., Spinelli, P., Sulotto, F., and Fantoni, F., *Int. Arch. Occup. Environ. Health*, 57, 19–26, 1985.

Scheuhammer, A. M. and Cherian, M. G., *Biochim. Biophys. Acta*, 840, 163–169, 1985.

Schroeder, H. A., Balassa, J. J., and Tipton, I. H., *J. Chronic Dis.*, 15, 941–964, 1962.

Schütz, A. and Skerfving, S., *Scand. J. Work Environ. Health*, 3, 176–184, 1976.

Schütz, A. and Haeger-Aronsen, B., *Läkartidningen*, 75, 3427–3430, 1978 (in Swedish with English summary).

Schütz, A., *Scand. J. Gastroenterol.*, 14, 223–235, 1979.

Schütz, A., Ranstam, J., Skerfving, S., and Tejning, S., *Ambio*, 13, 115–117, 1984.

Schütz, A., Skerfving, S., Christoffersson, J. O., Ahlgren, L., and Mattson, S., *Arch. Environ. Health*, 42, 340–346, 1987a.

Schütz, A., Skerfving, S., Christoffersson, J. O., and Tell, I., *Sci. Total Environ.*, 61, 201–209, 1987b.

Schütz, A., Skerfving, S., Ranstam, J., Gullberg, B., and Christoffersson, J. O., *Scand. J. Work Environ. Health*, 13, 221–231, 1987c.

Schütz, A., Attewell, R., and Skerfving, S., *Arch. Environ. Health*, 44, 391–394, 1989.

Schütz, A., Skarping, G., and Skerfving, S., in *Trace Element Analysis in Biological Specimens, Techniques and Instrumentation in Analytical Chemistry*, Vol. 15, Herber, R. F. M. and Stoeppler, M., Eds., Elsevier, Amsterdam, 403, 1994.

Scott, M. C. and Chettle, D. R., *Scand. J. Work Environ. Health*, 12, 81–96, 1986.

Sjögren, B., Lundberg, I., and Lidums, V., *Br. J. Ind. Med.*, 40, 301–304, 1983a.

Sjögren, B., Hedström, L., and Ulfvarson, U., *Int. Arch. Occup. Environ. Health*, 51, 347–354, 1983b.

Sjögren, B., Lidums, V., Håkansson, M., and Hedström, L., *Scand. J. Work Environ. Health*, 11, 39–43, 1985.

Sjögren, B., Elinder, C.-G., Lidums, V., and Chang, G., *Int. Arch. Occup. Environ. Health*, 60, 77–79, 1988.

Skalsky, H. L. and Carchman, R. A., *J. Am. Coll. Toxicol.,* 2, 405–423, 1983.
Skerfving, S., Ahlgren, L., Christoffersson, J. O., Haeger-Aronsen, B., Mattsson, S., and Schütz, A., *Arh. Hig. Rada Toksikol.,* 34, 277–286, 1983.
Skerfving, S., Ahlgren, L., Christoffersson, J. O., Haeger-Aronsen, B., Mattson, S., Schütz, A., and Lindberg, G., *Nutr. Res.,* Suppl. 1, 601–607, 1985.
Skerfving, S., Schütz, A., and Ranstam, J., *Sci. Total Environ.,* 58, 225–229, 1986.
Skerfving, S., Christoffersson, J. O., Schütz, A., Welinder, H., Spång, G., Ahlgren, L., and Mattsson, S., *Biol. Trace Elem. Res.,* 13, 241–251, 1987.
Skerfving, S., in *Biological Monitoring of Toxic Metals,* Clarkson, T.W., Friberg, L., Nordberg, G.F., and Sager, P.R., Eds., Plenum Press, New York, 169, 1988.
Skerfving, S., in *Trace Elements in Clinical Medicine,* H. Tomita, Ed., Springer Verlag, Tokyo, 479, 1990.
Skerfving, S. and Nilsson, U., *Toxicol. Lett.,* 64/65, 17–24, 1992.
Skerfving, S., *Arb. och Hälsa,* 125–138, 1993.
Skerfving, S., Nilsson, U., Schütz, A., and Gerhardsson, L., *Scand. J. Work Environ. Health,* 19 (Suppl. 1), 59–64, 1993.
Smith, R. G., Vorwald, A. J., Patil, L. S., and Mooney, T. F., Jr., *Am. Ind. Hyg. Assoc. J.,* 31, 687–700, 1970.
Sokas, R. K., Besarab, A., McDiarmid, M. A., Shapiro, I. M., and Bloch, P., *Arch. Environ. Health,* 45, 268–272, 1990.
Somervaille, L. J., Nilsson, U., Chettle, D. R., Tell, I., Scott, M. C., Schütz, A., Mattsson, S., and Skerfving, S., *Phys. Med. Biol.,* 34, 1833–1845, 1989.
Stebbins, A. I., Horstman, S. W., Daniell, W. E., and Atallah, R., *Am. Ind. Hyg. Assoc. J.,* 53, 186–192, 1992.
Strömberg, U., Schütz, A., and Skerfving, S., *Occup. Environ. Med.,* 52, 764–769, 1995.
Sunderman, F. W., Jr., Roszel, N. O., and Clark, R. J., *Arch. Environ. Health,* 16, 836–843, 1968.
Sunderman, F. W., Jr., in *Biological Monitoring of Toxic Metals,* Clarkson, T.W., Friberg, L., Nordberg, G.F., and Sager, P.R., Eds., Rochester Ser. Environ. Toxicity, Plenum Press, New York, 265, 1988.
Sunderman, F. W., Jr., Hopfer, S. M., Sweeney, K. C., Marcus, A. H., Most, B. M., and Creason, J., *Proc. Soc. Exp. Biol. Med.,* 191, 5–11, 1989.
Svensson, B. G., Björnham, Å., Schütz, A., Lettevall, U., Nilsson, A., and Skerfving, S., *Sci. Total Environ.,* 67, 101–1015, 1987.
Svensson, B. G., Schütz, A., Nilsson, A., and Skerfving, S., *Int. Arch. Occup. Environ. Health,* 64, 219–221, 1992a.
Svensson, B.-G., Schütz, A., Nilsson, A., Åkesson, I., Åkesson, B., and Skerfving, S., *Sci. Total Environ.,* 126, 61–74, 1992b.
Sällsten, G., Barregård, L., and Schütz, A., *Br. J. Ind. Med.,* 50, 814–821, 1993.
Sällsten, G., Occupational Exposure to Inorganic Mercury. Exposure Assessment and Elimination Kinetics, Academic dissertation, Department of Internal Medicine, Göteborg University, Sweden, 1994.
Sällsten, G., Barregård, L., and Schütz, A., *Occup. Environ. Med.,* 51, 337–342, 1994a.
Sällsten, G., Barregård, L., Wikkelsö, C., and Schütz, A., *Environ. Res.,* 65, 195–206, 1994b.
Tell, I., Somervaille, L. J., Nilsson, U., Bensryd, I., Schütz, A., Chettle, D. R., Scott, M. C., and Skerfving, S., *Scand. J. Work Environ. Health,* 18, 113–119, 1992.
Tossavainen, A., Nurminen, M., Mutanen, P., Tola, S., *Br. J. Ind. Med.,* 37, 285–291, 1980.
Tsuchiya, K., in *Handbook on the Toxicology of Metals,* Vol. II. Friberg, L., Nordberg, G.F., and Vouk, V.B., Eds., Elsevier, Amsterdam, 1986, 298, 1986.
U.S. EPA, Air quality criteria for lead. EPA-600/8–83/028aF, Vol I-IV. Environmental Protection Agency, Environmental Criteria and Assessment Office, Research Triangle Park, NC, 1986.
Vahter, M., Friberg, L., Rahnster, B., Nygren, Å., and Norlinder, P., *Int. Arch. Occup. Environ. Health,* 57, 79–91, 1986.
Vahter, M. E., in *Biological Monitoring of Toxic Metals,* Clarkson, T.W., Friberg, L., Nordberg, G.F., and Sager, P.R., Eds., Rochester Ser. Environ. Toxicity, Plenum Press, New York, 303, 1988.
Vartsky, D., Ellis, K. J., Chen, N. S., and Cohn, S. H., *Phys. Med. Biol.,* 22, 1085–1096, 1977.
Walker, G. S., Aaron, J. E., Peacock, M., Robinson, P. J. A., and Davison, A. M., *Kidney Int.,* 21, 411–415, 1982.
Welinder, H., Skerfving, S., and Henrikssen, O., *Br. J. Ind. Med.,* 34, 221–228, 1977.
Welinder, H., Littorin, M., Gullberg, B., and Skerfving, S., *Scand. J. Work Environ. Health,* 9, 397–403, 1983.
WHO, Mercury, Environmental Health Criteria 1, 1–131. World, Health Organization, Geneva, 1976.
WHO, Recommended health-based limits in occupational exposure to heavy metals. Report of a WHO Study Group, WHO Tech. Rep. Ser. No. 647, World Health Organization, Geneva, 1980.
WHO, Arsenic, Environmental Health Criteria 18, World Health Organization, Geneva, 1981a.
WHO, Manganese, IPCS Int. Prog. Chemical Safety, Environmental Health Criteria 17, World Health Organization, Geneva, 1981b.
WHO, Chromium. Environmental Health Criteria 61, 1–197. World Health Organization, Geneva, 1988.
WHO, Methylmercury, Environmental Health Criteria 101, 1–144. World Health Organization. Geneva, 1990.
WHO, Inorganic mercury, Environmental Health Criteria 118, 1–168. World Health Organization, Geneva, 1991a.
WHO, Nickel, IPCS Int. Prog. Chemical Safety, Environmental Health Criteria 108, 1–383, World Health Organization, Geneva, 1991b.
WHO, Cadmium, Environmental Health Criteria 134, Int. Prog. Chemical Safety, World Health Organization, Geneva, 1992.
WHO, Biomarkers and risk assessment: concepts and principles, Int. Prog. Chemical Safety, World Health Organization, Geneva, 1993.
Willers, S., Schütz, A., Attewell, R., and Skerfving, S., *Scand. J. Work Environ. Health,* 14, 385–389, 1988.

Wingren, G. and Englander, V., *Int. Arch. Occup. Environ. Health,* 62, 253–257, 1990.
Winneke, G., Collet, W., Krämer, U., Brockhaus, A., Ewert, T., and Krause, C., in *Lead Exposure and Child Development: An International Assessment,* Smith, M.J., Grant, L.D., and Sors, A.I., Eds., Kluwer Academic Publishers, Lancaster, MA, 260, 1989.
Åkesson, I., Schütz, A., Attewell, R., Skerfving, S., and Glantz, P.-O., *Arch. Environ. Health,* 46, 102–109, 1991.

Section II
Biorespones and Reactivities in Metal Toxicity

Overview

An Introduction to Bioresponses and Reactivities of Metals

Louis W. Chang

The "toxicity" of a chemical, metal included, is frequently influenced by the reactivities of the chemical within the biological system and the various responses to that chemical by the biological system. In most traditional texts of metal toxicology, the basic "metabolism" (absorption, distribution, and excretion) of metals is usually presented. While readers can seek such well established information in practically all standard textbooks of metal toxicology, comprehensive information on the "actions" and "reactions" of metals in the biological system still need to be focused and reviewed. This section, through a series of chapters, will discuss the various issues and concepts on the "actions" and "reactions" of metals in cells and tissues.

The general coordination chemistry of metal ions is usually included in the biological chemistry of metals. In addition to this information, Prof. Nieboer and his co-author will discuss the determinants of reactivity in metal toxicology. Various factors that may be responsible for or may alter the toxicity of metals will be presented and discussed in the leading chapter of this section.

Because many biological membranes are sulfhydral (–SH), rich, biological membranes have been considered to be the primary targets sites in metal toxicity. Metals bound to membranes are likely to modify membrane integrity and function, inhibiting the various transport systems of the membranes. In the chapter by Professor E.C. Foulkes, the actions of metals on the biological membranes are reviewed and discussed. Knowledge of how metals react with or cross cell membranes is fundamental to the understanding of their toxic actions. We are certain that the readers will appreciate the selection and focus in this subject.

Metals have a special affinity toward sulfhydral (–SH) rich components in the biological system. Asides from the biological membranes, other –SH rich proteins (thiol proteins) are also affected. Increasing interest and attention have been casted recently on the glutathione (GSH), which is one of the predominant intracellular non-protein thiols in the biological system. A chapter in this section is devoted to the presentation and discussion on the significance of glutathione system on metal toxicity. In this chapter, Prof. Zalups and Dr. Lash provide detailed examples with mercury interactions with glutathione in various organs and tissues.

One of the most important and current interests on metal toxicology is the "protective reaction" to metals by the biological system. In the chapter by Prof. Sanders, Goering, and Jenkins, the mechanisms by which organisms protect themselves from metal-induced damage at cellular and molecular levels are presented and discussed. The two primary proteins of concern, involved with protection against metal toxicity, are the stress proteins and metallothioneins. Each of these aspects is discussed in perspective regarding to their function in cell physiology, the molecular bases of their protective mechanisms, and the interrelationships between these two responses. Specific metals will also be used to exemplify these concepts.

Last but not least, the metal-metal interactions in the biological system is discussed by Professor Prasad and Professor Beck in the final chapter of this section. The influence and modulation of metal toxicity by another metal or trace element are well known phenomenon in metal toxicology. This well-known, yet poorly understood, issue represents a much needed area for future research. As many metal toxicities, such as lead, mercury, and cadmium, can be "lessened" by nutritional elements such as calcium, selenium, and zinc, the presence or absence of these "protective" elements should be considered in future risk assessment evaluations for toxic metals.

It is our hope that through these five chapters in this section, the reader will acquire the basic concepts on how cells react or respond to metals and the various "protective" strategies that may be important to the biological system as a result of metal exposures. This information will be important for the understanding of toxic actions and reactions induced by metals in various organs as presented in Part 2 of this volume.

Chapter 7

Determinants of Reactivity in Metal Toxicology

Evert Nieboer and Glenn G. Fletcher

I. INTRODUCTION

A review of the general coordination chemistry of metal ions is usually included in textbooks on the biological chemistry of metals (e.g., da Silva and Williams, 1991) or handbooks on their toxicology (e.g., Vouk, 1986; Martin, 1988). It is the goal of this article to take a more comprehensive perspective since some properties in addition to those displayed in aqueous media are relevant to toxic responses. This broader approach is consistent with the definition of speciation of the elements formulated in 1991 and embraced in 1994, respectively, at the First and Second International Symposium on Speciation of Elements in Toxicology and in Environmental and Biological Sciences held at Loen, Norway (Nieboer, 1992a; Nieboer and Thomassen, 1995):

> Speciation is the occurrence of an element in separate, identifiable forms (i.e., chemical, physical or morphological state).

The determinants of toxicity proposed and reviewed are summarized in Table 1. In the citing of examples, compounds of the metalloids, namely arsenic (As), antimony (Sb), selenium (Se), and tellurium (Te), will also be considered since these elements exhibit metallic properties albeit to different degrees.

II. PHYSICAL STATE

A. SOLIDS

Most metallic compounds are solids. In the occupational setting, workers are often exposed to particulates such as pyrometallurgical intermediates. Nickel oxides and sulfides, as well as metallic nickel, are examples in the nickel-producing industry (Doll, 1990). As discussed in a subsequent section, particle size determines where in the respiratory tract inhaled particles are deposited and thus can exert their toxic effects (Gibson et al., 1987; Vincent, 1993). Evidence that water solubility and surface properties are important determinants of reactivity will also be reviewed.

B. LIQUIDS

Not many metals or metal compounds are liquids at standard conditions (25°C and 101.3 kPa). Liquids sustain measurable vapor pressures above them which permit exposure by inhalation. Mercury metal is a liquid at room temperature and atmospheric pressure (mp, –39°C; bp, 357°C), while gallium metal melts at 30°C (bp, 2403°C) and cesium metal at 28.5°C (bp, 669°C). Nickel tetracarbonyl [$Ni(CO)_4$] is a liquid (mp, –25°C; bp, 43°C), is used in the industrial refining of nickel (Nieboer, 1992b), and is one

Table 1 Determinants of Metal Toxicity

Physical State	Atomic Properties
Solids	Ion Size
Liquids	Geometry
Vapors and gases	Oxidation state
	Electronegativity

Reactivity
Ionic and covalent bonding tendencies
Donor-atom preference
Complex formation and stability
Kinetic aspects
Radical formation
Solubility
Particle size of solid compounds
Physicochemical properties of solid surfaces

Biological phenomena
Compartmentalization
Respiratory tract clearance
Bioavailability
Biological residence time
Resistance

of the most toxic metal compounds known (Morgan, 1992; Sunderman, 1992; Shi, 1994a,b). Chloride derivatives of germanium ($GeCl_4$, mp, $-49.5°C$; bp, $84°C$) and arsenic ($AsCl_3$, mp, $-8°C$; bp, $130°C$) are also liquids (Stokinger, 1981; Lide, 1990). Further, organometallic complexes of lead and mercury are liquids at standard conditions: tetraethyllead, $Pb(CH_2CH_3)_4$ (mp, $-137°C$; decomposes $\approx 200°C$; bp, $91°C$ at 2.5 kPa); tetramethyllead, $Pb(CH_3)_4$ (mp, $-28°C$; bp, $110°C$ or $6°C$ at 1.3 kPa); and dimethylmercury, $Hg(CH_3)_2$ (bp, $96°C$).

C. VAPORS AND GASES

If not contained the liquids described above release vapors, which in most cases are extremely toxic (Stokinger, 1981; Friberg et al., 1986). Mercury and nickel tetracarbonyl are prime examples of such industrial liquids. Mercury vapor causes bronchiolitis, interstitial pneumonitis, renal damage, and neurotoxic effects (ATSDR, 1989; Rowens et al., 1991); nickel tetracarbonyl induces serious lesions in the lung and brain, and if the exposure is acute, can be fatal (Nieboer et al., 1988; Nieboer and Fletcher, 1995a). A number of solid compounds sublime and can thus also sustain significant vapor pressures. Arsenic metal (subl, $613°C$), ethylmercuric chloride (mp, $193°C$; subl $> 40°C$) and antimony trioxide (mp, $650°C$; subl $1550°C$) are examples (Stokinger, 1981; Lide, 1990). Cadmium metal (mp, $321°C$; bp, $767°C$) and lead metal (mp, $328°C$; bp, $1740°C$) have relative low melting points and are readily vaporized during industrial processes such as refining, welding, and machining, and thus pose an additional hazard compared to other metals.

A limited number of metal compounds are gases at room temperature and pressure. The hydrides of germanium (GeH_4, germane), antimony (SbH_3, stibine), arsenic (AsH_3, arsine), selenium (H_2Se, hydrogen selenide) and tellurium (H_2Te, hydrogen telluride), as well as the electrical insulator selenium hexafluoride (SeF_6), are examples (Stokinger, 1981). Unless extremely water soluble, gases penetrate deep into the lungs where they can cause tissue damage, as well as being absorbed in significant amounts into the bloodstream because of the large alveolar surface area (Davies, 1985). Although water solubility protects the lungs, soluble gases are nevertheless rapidly cleared from the nasal and bronchial epithelium to the bloodstream.

III. ATOMIC PROPERTIES

A. INTRODUCTORY COMMENTS

Nonisomorphous replacement of naturally occurring metal ions is an important toxicologic principle (Nieboer et al., 1984b; Nieboer and Sanford, 1985). Such substitution often results in complexes of different stabilities and spatial orientations, which have the potential of inducing detrimental conformational changes in enzymes, structural proteins, DNA, or membranes. Metal-ion cavities in biomolecules frequently match the size and preferred geometry of specific cations (Williams, 1971; da Silva and Williams, 1991). Biological systems also depend on the ability of cations to accept or give up electrons and thus oxidation/reduction properties are of crucial importance (Williams, 1973, 1981a). Further, the nature of the bonding in complexes with biomolecules — whether it is largely ionic or has a significant covalent contribution — is also an important factor. The fundamental property of electronegativity constitutes the ability of an atom or ion to attract electrons in a molecule, or, more pertinent, in a metal complex (Nieboer and Richardson, 1980; Huheey et al., 1993).

The relevance to the interpretation of metal biochemistry and toxicity of ion size, geometry, oxidation state and electronegativity is described in this section.

B. ION SIZE

A perusal of the effective ionic radii compiled in Table 2 illustrates that ion size depends on the coordination number or geometry, the oxidation state, and the electron spin state (the latter applies to transition metal ions, and refers to whether the d electrons assume an unpaired or paired configuration due to the ligand field) (Cotton and Wilkinson, 1988).

Ion size appears to have played a central role in the evolutionary design of metal-ion specificity. This is readily illustrated. Ion channels can achieve specificity by matching the *hydrated* radius of the ion of interest. Potassium channels of nerves display the selectivity of $Li^+ < Na^+ <<< K^+ > Rb^+ > Cs^+$, while calcium channels often have the preference $Mg^{2+} << Ca^{2+} > Sr^{2+} > Ba^{2+}$ (da Silva and Williams, 1991). By contrast, the *ionic* radius determines the selectivity when complex formation is involved. Polyfunctional biomolecular ligands often provide cavities that exactly match the preferred metal ion. It is for this reason that the bacterial antibiotics nonactin and valinomycin have the selectivity $Li^+ << Na^+ < K^+ > Rb^+ > Cs^+$ (da Silva and Williams, 1991). Similarly, the cavity in hemoglobin and myoglobin exactly matches the ionic radius of Fe^{2+}, specifically, the low-spin form in which the six d electrons are spin-paired.

By virtue of its comparable size to Ca^{2+}, Cd^{2+} readily replaces this essential cation (e.g., Nieboer and Richardson, 1980; Avery and Tobin, 1993), may exhibit comparable binding parameters (e.g., Reid and McDonald, 1991), and often induces inhibition or toxicity (e.g., Evans and Weingarten, 1990). Similarly, Mg^{2+} and Ni^{2+} have comparable radii, and indeed both isomorphous and nonisomorphous replacements of the former ion are well established (Wetterhahn-Jennette, 1981; Nieboer et al., 1984b). In animal studies of nickel carcinogenesis, co-administration of magnesium and nickel compounds has been shown to have an ameliorating effect on the carcinogenic potency of nickel compounds (Kasprzak, 1992; Kasprzak and Rodriguez, 1992). A direct competition between critical binding sites (e.g., on DNA) or cellular uptake is suspected.

C. GEOMETRY

Metal-ion centers in proteins and enzymes have unique features which may require specific spatial orientations of the ligand-attachment sites. The three-dimensional structure of the protein can impose a specific geometry upon an ion. Five-coordinated, square-pyramidal structures occur for Fe^{2+} in deoxyhemoglobin providing an "open site" for the incoming molecular oxygen (O_2) molecule (Huheey et al., 1993). Activated four- and five-coordinated Zn^{2+} centers are found in carbonic anhydrase and carboxypeptidase, respectively, exhibiting an activated water molecule poised for hydrolytic action or which is ready to serve as a convenient leaving group (da Silva and Williams, 1991). Not surprisingly, and as illustrated in Table 3, very few ions that are inserted into enzymes to replace Zn^{2+} result in a retention of catalytic activity. A mismatch in ion size likely contributes to the distortion of the active site and the inactivation by Cd^{2+}, Hg^{2+} and Pb^{2+}. By contrast, the esterase activity of carboxypeptidase appears to have less demanding requirements for size and spatial orientation as it is activated to some extent by all the metal ions tested, except Cu^{2+}. The latter ion has a strong propensity to form square planar complexes, while octahedral and tetrahedral geometries are common for the remaining divalent ions listed in Table 3 (Cotton and Wilkinson, 1988).

A second type of metal binding center in proteins or enzymes tends to impose a specific geometry on the protein, thereby inducing an allosteric event with regulatory potential. Calcium trigger-proteins, such as calmodulins, ATPase-dependent transporters of H^+, Na^+, K^+, Ca^{2+}, or Cu^{2+}, and Mg^{2+} in phosphoglucomutase are examples (Williams, 1971; da Silva and Williams, 1991; Bull and Cox, 1994). Rapid exchange of the metal ion usually occurs. It is clear from the data in Table 3 for phosphoglucomutase,

Table 2 Effective Ionic Radii of Selected Metal Ions[a] (in Picometers, [10^{-12} Meter])

Dependence on Coordination Number (CN)

r(Ca^{2+})	CN	r(Zn^{2+})	CN	r(Cd^{2+})	CN	r(Pb^{2+})	CN	r(Hg^{2+})	CN
100	6	60	4	78	4	98	4(PY)	69	2
106	7	68	5	87	5	119	6	96	4
112	8	74	6	95	6	123	7	102	6
118	9	90	8	103	7	129	8		
123	10			110	8				

Dependence on Electron Spin State (SP) (CN = 6)

r(Fe^{2+})	SP	r(Fe^{3+})	SP	r(Mn^{2+})	SP	r(Co^{2+})	SP
61	LS	54	LS	67	LS	65	LS
78	HS	65	HS	83	HS	75	HS

Dependence on Oxidation State (OX) (CN = 6; High Spin, Unless Specified Otherwise)

M	r(M^{n+})	OX	M	r(M^{n+})	OX
Cu	77	+1	Ni	69	+2
	73	+2		56	+3 (LS)
Mn	83	+2		60	+3
	65	+3		48	+4 (LS)
	53	+4	Pb	119	+2
Hg	119	+1		78	+4
	102	+2	Tl	150	+1
				89	+3

Radii for Selected Essential Metal Ions (CN = 6; High Spin) | Radii for Selected Toxic Metal Ions (CN = 6; High Spin)

M^{n+}	r(M^{n+})	M^{n+}	r(M^{n+})	M^{n+}	r(M^{n+})	M^{n+}	r(M^{n+})
Na$^+$	102	Co^{3+}	61	Li$^+$	76	Cd^{2+}	95
K$^+$	138	Fe^{2+}	78	Cs$^+$	167	Hg^{2+}	102
Mg^{2+}	72	Fe^{3+}	65	Be^{2+}	45	Pb^{2+}	119
Ca^{2+}	100	Cu^{2+}	73	Sr^{2+}	118	[Cr^{3+}][c]	62
Mn^{2+}	83	Zn^{2+}	74	Al^{3+}	54	Ni^{2+}	69
				Ln^{3+}	86–103[b]	Co^{2+}	75

[a] Shannon and Prewitt "traditional" ionic radii based on r(O^{2-}) = 140 pm (CN = 6), from Shannon, 1976; r = radius; M = metal; M^{n+} = metal ion with charge n+; LS = low spin; HS = high spin; PY = square pyramidal; unless specified otherwise, for CN = 4 the geometry is tetrahedral.

[b] Ln = lanthanide series; values given are for La^{3+} (103 pm) and Lu^{3+} (86 pm).

[c] There is some evidence that chromium is an essential metal in humans.

which has an absolute requirement for Mg^{2+}, that replacement studies show high selectivity. Mg^{2+} has a high demand for octahedral symmetry, and the only ions with comparable size to Mg^{2+} (r = 72 ppm, Table 2) and that can satisfy this geometric requirement are Ni^{2+} (r = 69 pm) and Co^{2+} (r = 75 pm).

D. OXIDATION STATE

As illustrated in Table 2, the size of a metal ion is dependent on its oxidation state, with ionic radii usually decreasing with increasing oxidation state. In addition, preferred geometries vary with oxidation state. For example, Cu$^+$ has a preference for tetrahedral geometry, while for Cu^{2+} the tetrahedral, square planar and octahedral configurations are often not sharply distinguished because of inherent spatial distortions (Cotton and Wilkinson, 1988; da Silva and Williams, 1991). Similarly, Co^{2+} prefers tetrahedral or octahedral sites, while Co^{3+} has a preference for the octahedral arrangement.

The metals Mn, Fe, and Cu have variable oxidation states and are crucial in the production of high energy electrons in photosynthesis and respiration. Proteins containing these metals are also central to

Table 3 Noniosmorphic Replacement of Essential Metal Ions in Enzymes

Metal Ion	Bovine Carboxypeptidase[a]			Rabbit Phosphoglucomutase[b]	
	Relative(%) Peptidase Activity[c]	Relative(%) Esterase Activity[d]	Apparent Stability Constant[e] (log K)	Relative(%) Catalytic Activity[f]	Apparent Stability Constant[g] (log K)
Zn^{2+}	100	100	10.5	0.3	11.4
Co^{2+}	95	96	7.0	15.0	—
Ni^{2+}	36	87	8.2	60.0	—
Mn^{2+}	11	35	5.6	5.0	7.3
Cu^{2+}	0	0	10.6	—	—
Hg^{2+}	0	117	21.0	—	—
Cd^{2+}	0	152	10.8	0.8	—
Pb^{2+}	0	52	—	—	—
Ca^{2+}	—	—	—	<0.5	—
Mg^{2+}	—	—	—	100.0	5.0

[a] From Coleman and Vallee (1961).
[b] From Ray (1969).
[c] At 0°C and pH 7.5, with benzoyl-glycyl-L-phenylalanine as substrate.
[d] At 25°C and pH 7.5 with hippuryl-dl-β-phenyllactate as substrate.
[e] For pH 8.0 and 4°C.
[f] At 30°C, in the presence of saturating levels of both glucose-1-phosphate and glucose-1,6-diphosphate.
[g] At pH 8.5 and 25°C.

processes that generate "energy for living" in the form of ATP; they, including Mo proteins, are involved in electron transfer (e.g., cytochromes) and in the transport and biological use/control of "active oxygen" as illustrated by oxygen carriers (e.g., hemoglobin), oxygenases (e.g., P-450s), oxidases/reductases (e.g., cytochrome c oxidase, dehydrogenases), superoxide dismutase (scavenging of superoxide anion), and catalase (scavenging of hydrogen peroxide) (Nieboer and Sanford, 1985; da Silva and Williams, 1991). Consequently, most anabolic and catabolic processes depend on enzymes or proteins with metal centers possessing oxidation/reduction capabilities. Oxidation state can also dominate the toxicity of elements. For example, the toxic effects of mercuric salts are quite distinct from those induced by mercury vapor, and chromium(III) compounds tend to be considerably less toxic than chromates.

E. ELECTRONEGATIVITY

The energy of the empty valence orbital of a metal ion is often taken as a measure of its ability to accept electrons and thus to form covalent bonds. Orbital energy is related to electronegativity [$\chi(M)$] which, in simple terms, may be defined as the electron attracting capability of an atom or ion in a molecule. Pauling was the first to calculate χ by comparing the bond energies between unlike atoms, E (A-B), with that of the homoatomic bonds E (A-A) and E (B-B) (Allred, 1961; Huheey et al., 1993). From the compilation provided in Table 4, it is clear that χ depends on the position in the Periodic Table and on the oxidation state, increasing with ion charge (Z). For comparison, the χ value for N is 3.04 (trivalent), $\chi(O) = 3.44$ (divalent), $\chi(S) = 2.58$ (divalent), $\chi(F) = 3.98$ (monovalent), and $\chi(Cl) = 3.16$ (monovalent). For small to moderate differences in electronegativity between bond partners (<1.8), the ionic character of a bond increases steadily with the magnitude of the electronegativity difference. Consequently, ions such as H^+ ($\chi = 2.20$), Cu^{2+} (2.02), or Hg^{2+} (2.00) tend to form bonds with sulfur donor atoms that have a higher degree of covalent character than do, for example, Mn^{2+} ($\chi = 1.55$), Zn^{2+} (1.65), or Pb^{2+} (1.87). Electronegativity is therefore an important atomic property as it relates to the degree of ionic and covalent bonding in metal ion-donor atom interactions.

IV. REACTIVITY

A. IONIC AND COVALENT BONDING TENDENCIES

A useful measure employed in classifying metal ions is the index $(\chi_m)^2 r$, with χ_m the Pauling electronegativity, and r the ionic radius corresponding to the most common coordination number. $(\chi_m)^2 r$ is a quotient that compares valence orbital energy with ionic energy (Nieboer and Richardson, 1980)

Table 4 Pauling Electronegativities of Selected Metal Ions[a]

M^{n+}	$\chi(M^{n+})$	M^{n+}	$\chi(M^{n+})$	M^{n+}	$\chi(M^{n+})$	M^{n+}	$\chi(M^{n+})$
H^+	2.20	Ln^{3+}	1.10–1.27[b]	Pd^{2+}	2.20	Tl^{3+}	2.04
Li^+	0.98	Sc^{3+}	1.36	Pt^{2+}	2.28	Sn^{2+}	1.80
Na^+	0.93	V^{2+}	1.63	Ag^+	1.93	Sn^{4+}	1.96
K^+	0.82	Cr^{2+}	1.66	Au^+	2.54	Pb^{2+}	1.87
Rb^+	0.82	Mn^{2+}	1.55	Zn^{2+}	1.65	Pb^{4+}	2.33
Cs^+	0.79	Fe^{2+}	1.83	Cd^{2+}	1.69	As^{3+}	2.18
Be^{2+}	1.57	Fe^{3+}	1.96	Hg^{2+}	2.00	Sb^{3+}	2.05
Mg^{2+}	1.31	Co^{2+}	1.88	Al^{3+}	1.61	Bi^{3+}	2.02
Ca^{2+}	1.00	Ni^{2+}	1.91	Ga^{3+}	1.81		
Sr^{2+}	0.95	Cu^+	1.90	In^{3+}	1.78		
Ba^{2+}	0.89	Cu^{2+}	2.02	Tl^+	1.62		

[a] From Allred (1961) in units of (energy)$^{1/2}$.
[b] The values listed correspond to Ln^{3+} (1.10) and Lu^{3+} (1.27).

and is therefore considered to be a measure of the relative ability of metal ions to participate in covalent interactions compared to ionic interactions. Comparable indices have been devised by others (Williams and Hale, 1966; Turner and Whitfield, 1983; also see da Silva and Williams, 1991). By contrast, the quantity Z^2/r correlates successfully with interactions that are known to be highly ionic such as the hydration of cations and anions as measured by hydration energies (Phillips and Williams, 1965; Turner et al., 1981; Kaiser, 1980; Huheey et al., 1993). By plotting the covalent index $(\chi_m)^2 r$ vs. the ionic index Z^2/r, Nieboer and Richardson (1980) effected a division of metal and metalloid ions into the traditional groupings of *Class A* (also referred to as class (a) or "hard" acids), *Borderline* and *Class B* (class (b) or "soft" acids; Pearson, 1963; Williams and Hale, 1966; da Silva and Williams, 1991; Huheey et al., 1993). The original figure is reproduced in Figure 1. The interactions of *Class A* metal ions are largely ionic, while significant covalent contributions to the interaction energy can occur for *Class B* ions; *Borderline* ions exhibit intermediate bonding tendencies. With reference to Figure 1, for *Class A* metal ions $(\chi_m)^2 r$ has values less than 1.75; for *Borderline* ions, values are greater than 1.75, but less than 3.0 (Pb^{2+} being the only exception); *Class B* ions have an index value greater than 3.0. As explained in the next section, this classification may be linked to donor-atom preferences.

B. DONOR-ATOM PREFERENCES

As described in detail elsewhere (Pearson, 1963; Williams and Hale, 1966; Nieboer and Richardson, 1980; Huheey et al., 1993), the separation of metal ions into the three classes was based on empirical thermodynamic data, namely trends in the magnitude of equilibrium constants that describe the formation of metal ion/ligand complexes. Nieboer and Richardson (1980) extended this classification to biological systems by examining crystal structures of proteins and polynucleotides to elucidate the type of donor atoms metal ions bind to. On this basis, *Class A* metal ions may be designated as oxygen-seeking, *Class B* metal ions as nitrogen/sulfur seeking, and the *Borderline* category as ambivalent or intermediate by displaying comparable affinity for all three types of donor-atom sites. Although the original survey was done in 1979, crystallographic and spectroscopic data published subsequently corroborate these conclusions as illustrated by the examples portrayed in da Silva and Williams (1991) and Huheey et al. (1993). In spite of the preference of *Class B* ions for nitrogen/sulphur ligands, both *Class B* and *Borderline* ions form complexes with oxygen ligands which are more stable than those of *Class A* ions of comparable size and charge. This general phenomenon is independent of the type of ligand and is the basis for the much quoted Irving-Williams order of complex stability (e.g., Martin, 1988; da Silva and Williams, 1991; Huheey et al., 1993). Presumably, this feature suggests that in addition to the largely ionic interaction observed for *Class A* ions, *Borderline* and *Class B* ions have significant covalent contributions to the overall interaction energy.

Nieboer and Richardson (1980) demonstrated the usefulness of the concept of donor-atom preference in the understanding of relative toxicities of metal and metalloid ions in a range of organisms (i.e., toxicity sequences). More recent applications include the interpretation of: metal-ion interactions with algae (Crist et al., 1988); ionoregulation in fish (McDonald et al., 1989); metal speciation in soil leachates (Duffy et al., 1989); inhibition of plant enzymes (van Assche and Clijsters, 1990); renal toxicity in rats (Templeton and Chaitu, 1990); assimilation of metals in marine copepods (Reinfelder and Fisher, 1991), phytotoxicity in wheat seedlings (Taylor et al., 1992); and metal adsorption by fungi (Avery and Tobin,

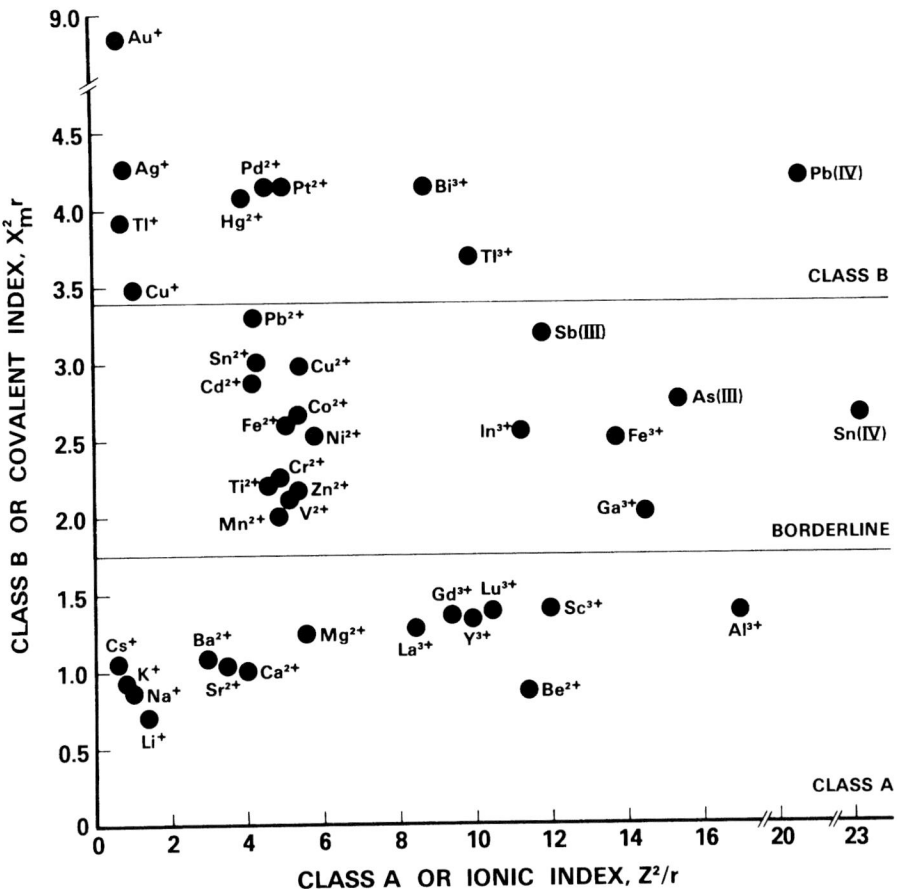

Figure 1 A separation of metal ions and metalloid ions (As(III) and Sb(III)) into three categories: *Class A*, *Borderline*, and *Class B*. The *Class B* index $\chi_m^2 r$ is plotted for each ion against the *Class A* index Z^2/r. In these expressions, χ_m is the metal-ion electronegativity, r its ionic radius, and Z its formal charge. The Allred values (1961) of Pauling's electronegativity and the crystal "IR" ionic radii (in angstrom units), corresponding to six (octahedral) coordination and compiled by Shannon and Prewitt (1969, 1970), were used to calculate these indices. In those few cases where the "IR" values were not available, the "Ahrens" ionic radii were abstracted from the same sources. An ionic radius of 0.94 Å, corresponding to a coordination number of four, was selected for Pb as the resulting $\chi_m^2 r$ value was more commensurate with the known solution coordination chemistry of Pb^{2+} (Nieboer & McBryde, 1973). Oxidation states given by Roman numerals imply that simple cations do not exist even in acidic aqueous solutions. (From Niebor, E. and Richardson, D. H. S., *Environ. Pollut.* (Series B), 1:3–26, 1980. With permission.)

1993). As pointed out by Niebor and Richardson (1980) and more recently by Goering (1993) in an extensive review of the molecular mechanisms of lead toxicity, the dogma that Pb toxicity is mediated through its interaction with sulfhydryl groups is an oversimplification. Because of its *Borderline* character, Pb^{2+} forms stable complexes with O, N, and S donors. This explains its interference with the biochemistries of the *Class A* ion Ca^{2+} and the *Borderline* ion Zn^{2+}. Similarly, Cd^{2+}/Ca^{2+} competition is an important toxicological principle (Nieboer and Richardson, 1980; Lag and Helgeland, 1987; Reid and McDonald, 1991; Evans and Weingarten, 1990); Cd^{2+} and Ca^{2+} have nearly identical ionic radii (see Table 2). The prediction that Al^{3+} is toxic to fish because it blocks external *Class A* sites such as on gills has also been confirmed (e.g., Reid et al., 1991; Wilkinson et al., 1990, 1993). Recently and as illustrated in Figure 2, a strong correlation was observed between the (metal ion)/(poly d (G C)) mole ratio required for conformational transition midpoints of synthetic DNA and the covalent index $(\chi_m)^2 r$. (Poly d(G-C) is a synthetic polynucleotide with alternating guanosine (G) and cytosine (C) bases.) This relationship was independent of the type of conformational transitions observed (whether monophasic or biphasic) or the specific polynucleotide conformation generated. It is clear from the data in Figure 2 that metal ions with considerable *Class B* character induce conformational changes in DNA more readily (i.e., at lower concentrations) than *Class A* ions. The observed trend is interpreted to indicate that affinity for the nitrogen centers on the nucleotide bases is the determining factor. The most likely site of metal-nitrogen interactions in poly d(G-C) is the N-7 site of guanine (Rossetto and Nieboer, 1994).

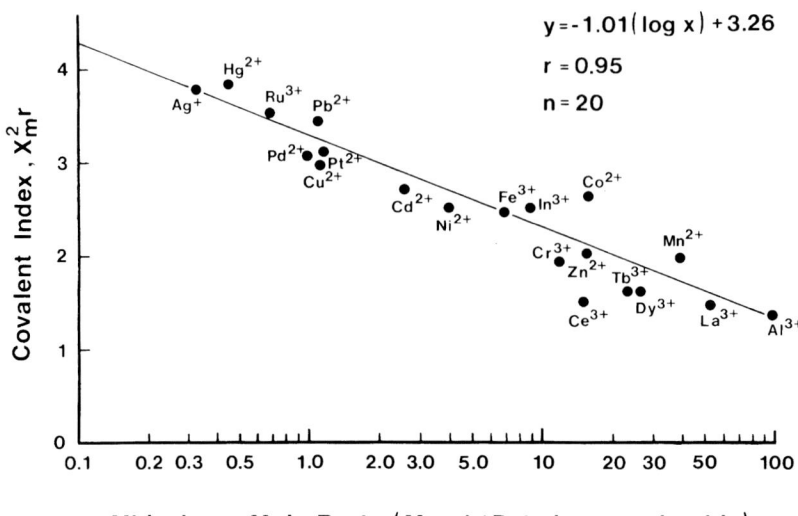

Figure 2 Correlation between the covalent index $\chi_m^2 r$ and phase transition mid-points. For the covalent index, χ_m is the Pauling electronegativity (Allred, 1961) and r is the effective ionic radius for the most common metal-ion coordination number (i.e., 9, lanthanides; 4, Hg(II), Pd(II), Pb(II), Pt(II) and Ag(I); and 6 for all other metal ions) (Shannon, 1976). A correction factor was applied to the mole ratios for Cr(III), Pt(II), Pd(II), and In(III) to compensate for competing reactions involving the anion of the metal salt. The use of uncorrected concentration data has little effect on the slope and intercept, but reduced the correlation coefficient somewhat (r = 0.92). (From Rossetto, F. E. and Niebor, E., *J. Inorg. Biochem.*, 54:167–186, 1994. With permission.)

C. COMPLEX FORMATION AND STABILITY

Natural waters, extracellular and intracellular fluids, as well as soils and sediments, contain an array of complexing agents or ligands which favor complex formation.

$$M + iL \rightleftarrows ML_i \qquad \beta_i = \frac{[ML_i]}{[M][L]^i} \tag{1}$$

$$M + iHL \rightleftarrows ML_i + iH \qquad \kappa_i = \frac{[ML_i][H]^i}{[M][LH]^i} \tag{2}$$

In expressions 1 and 2, M represents the metal ion; L, the unprotonated ligand; HL, the protonated ligand; ML_i, the metal complex with i ligands bound to it; β_i, the stoichiometric stability constant and is the product of the stepwise stoichiometric stability (formation) constants K_i, 1 to i ($\beta_i = K_1 K_2 K_3 \ldots K_i$). κ_i is the overall stoichiometric proton-displacement constant. Formal charges on M, L, and ML are omitted for convenience; square brackets denote concentrations, and thus β_i and κ_i are valid only in a solution of a particular composition, since the activity coefficients of the various species are not specified (Rossotti and Rossotti, 1961; Butler, 1964; Cotton and Wilkinson, 1988; Evans, 1989). Equation 2 is important since the proton directly competes with metal ions for binding sites, and thus pH regulates complex formation. Further, metal ions undergo hydrolysis (Baes and Mesmer, 1976). The ligand L can be polyprotic with more than one donor atom available for attachment to the metal ion to form a chelate (see Nieboer and Richardson, 1980, for the terminology used in coordination chemistry). Polynuclear complexes are also common such as for Al^{3+} (Nieboer et al., 1995).

Generally speaking, it is clear from a perusal of stability constants (Smith and Martell, 1976) that for fixed values of the covalent index $(\chi_m)^2 r$ in Figure 1 complex stability with nearly all ligands increases with the magnitude of the ionic index; conversely, for fixed values the ionic index Z^2/r complex stability increases with the covalent index. Consequently, endogenous ions such as Mg^{2+}, Ca^{2+}, Mn^{2+}, and Zn^{2+} are readily displaced by ions above them in Figure 1 (e.g., Pb^{2+} and Cd^{2+}); similarly Be^{2+}, La^{3+}, and Al^{3+} which are located to the right of Mg^{2+} and Ca^{2+}, can displace them. As already indicated, nonisomorphous replacement of naturally occurring metal ions is an important toxicological principle.

When the ligand concentration in a compartment or the availability of binding sites is not limiting, it can be deduced from Equation 1 that the amount of complex formed is determined by the product of the free-ion concentration [M] and the equilibrium constant β_i. This is referred to as the free-ion activity model of reactivity (Morel, 1983; da Silva and Williams, 1991). In many situations, this paradigm accounts for the observed trends in interactions with toxic responses of aquatic organisms (Campbell, 1995). The underlying assumption is that a metal complex forms on the cell surface as a prerequisite to a response or internalization, with M determining the extent of surface complexation. However, exceptions have been noted. For example, aluminum toxicity in juvenile salmon responds to a weighted function of both $[Al^{3+}]$ and $[AlF^{2+}]$ when F^- is present (Wilkinson et al., 1990, 1993). The formation of a ternary gill-surface complex appears responsible. Further, Clarkson (1993) and his colleagues have demonstrated that the CH_3Hg^+ complex of cysteine structurally resembles the amino acid methionine and can compete with the carrier-mediated uptake of the latter. It also appears that both Hg^{2+} and CH_3Hg^+ can be transported across cell membranes as complexes of glutathione using the latter's transport system. Williams (1981b) not only suggested such transport involving endogenous ligands, but further elaborated that M might diffuse across the cell membrane under the influence of favorable "steady state" complex-formation conditions inside the cell. Membranes are known to have highly selective channels with gates that bypass the need for complex formation. *In vitro* experiments with human erythrocytes and lymphocytes have demonstrated the validity of the diffusion model for Ni^{2+} (Nieboer et al., 1984a). Similarly, it has been shown that ligands that render metal complexes lipophilic enhance metal-ion uptake (Menon and Nieboer, 1986). The inherent importance of metal complexes in human physiology is also readily illustrated. At the level of the filtering unit of the kidney, the glomerulus, metal complexes of albumin, and other proteins with relative molecular masses exceeding 60,000 are not filtered from the plasma, while complexes of smaller proteins or amino acids are. It is also known that the gastrointestinal uptake of Al^{3+} is facilitated by citrate (Nieboer et al., 1995).

Chelating agents are used in the treatment of acute metal poisoning. Their use in the management of lead poisoning in children is quite common in the U.S. Versenate (calcium disodium ethylenediaminetetraacetic acid) is an example of an intraveneously administered drug, while succimer (*meso*-2,3-dimercaptosuccinic acid) is given orally (Chisolm, 1990; CDC, 1991). All such drugs have potential side effects, are not specific for lead, and thus must be used with caution.

As already described in earlier sections, metal centers in biomolecules have numerous roles. One important consequence of complex formation is the ability to alter the oxidation/reduction potentials of metal centers. The Fe^{3+}/Fe^{2+} redox couple has the potential range 0 to –500 mV in proteins with iron-sulfur proteins and –400 to +400 mV in heme-containing proteins (da Silva and Williams, 1991).

D. KINETIC ASPECTS

Thermodynamic arguments such as those based on the magnitudes of equilibrium constants can be used to judge whether a reaction takes place, but can make no prediction about the time frame. An enormous range of time scales is involved in inorganic reaction kinetics, as is illustrated in Figure 3 for water exchange at aqua cations. The mean residence time of a water molecule in the primary hydration sphere of cations varies from near the diffusion-controlled limit of about 1 ns for Pb^{2+}, Cu^{2+}, as well as for Na^+, K^+ (which would fall in between Li^+ and Cs^+ in Figure 3), and Hg^{2+} (not shown), to values between 1 ns and 1 µs for Ca^{2+}, Mn^{2+}, Zn^{2+}, and Cd^{2+}, to times between 1 µs and 1 ms for Mg^{2+}, Ni^{2+}, and Fe^{3+}, to close to 1s for Al^{3+} and near 1 day for Cr^{3+} (Cotton and Wilkinson, 1988; Burgess, 1988, 1992; Huheey et al., 1993). Simple ligand substitutions take place on a time scale about 10-fold more slowly than the water-exchange rates and are nearly independent of the ligand (Cotton and Wilkinson, 1988). Successive replacement of water molecules leads to steadily increasing rate constants for water exchange, while attachment of polydentate anionic ligands slows the replacement of the remaining water molecules (e.g., 10- to 100-fold) (Burgess, 1992). For trivalent ions such as Fe^{3+}, Al^{3+}, and Cr^{3+}, coordination of the hydroxide ion significantly labilizes the remaining water molecules (1000-fold is typical). This improved lability has important toxicological consequences. Not surprisingly, complex

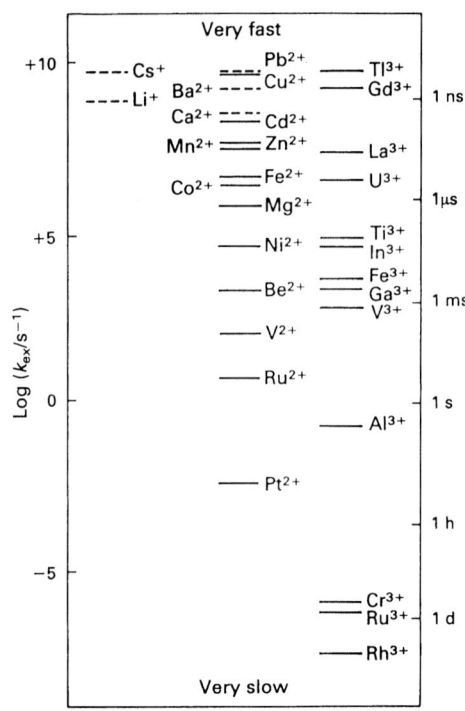

Figure 3 Rate constants for water exchange and mean residence times for primary hydration-shell water molecules for 2+ and 3+ aqua-metal cations at 298.2°K. The dashed line denotes that the available kinetic data are approximate. (From Burgess, J., *Analyst*, 117:605–611, 1992. With permission.)

formation with polydentate ligands is considerably slower than with monodentate ligands (e.g., up to 10^5-fold for Cu^{2+}) (Burgess, 1992). Kinetic regulation of biological processes involving metal-ion complex formation/dissociation as triggers is thus possible. Kinetic aspects may well delineate the biological roles of Ca^{2+} and Mg^{2+}. In fact, it appears that the presence of Mg^{2+} in the chlorin ring of chlorophyll is the result of kinetic trapping rather than thermodynamic control (da Silva and Williams, 1991). Conversely, phaeophytization of chlorophyll in plants by ambient sulfur dioxide is a recognized toxic response and involves the replacement of Mg^{2+} by protons and presumably reflects the thermodynamic instability (Puckett et al., 1973).

Langford and Cook (1995) point out that soil and water systems are dynamic systems far from equilibrium. For example, measurements of the binding of Cu^{2+} to colloidal hydrous oxide particles identified rapidly exchanging surface sites, while sites in the interior of the colloids appeared to release Cu^{2+} much more slowly presumably because of diffusion requirements. Consequently, sites that appeared thermodynamically equivalent were different from a kinetic perspective.

E. RADICAL FORMATION

Free radicals are chemical entities with unpaired electrons capable of damaging tissues. Molecular oxygen (O_2) and its reduced derivatives (superoxide anion, O_2^-; hydrogen peroxide, H_2O_2, which is not a radical species; and the hydroxyl radical, OH·) are the species of greatest concern (Gutteridge, 1995). These species when in excess can damage proteins, DNA, lipids, lipoproteins, and thus tissues (Rusting, 1992; Heinecke, 1995). What is often not mentioned but which follows directly from our earlier discussions under Oxidation State is that aerobic systems are inherently dependent on these active oxygen species: namely, as electron acceptors and reactive metabolites in many essential anabolic, catabolic and immunological processes (Stryer, 1995; Kehrer, 1993; Gelvan and Saltman, 1995). Metalloproteins (mainly of Mn, Fe, Cu, Mo) are central to these functions. In addition, organic radicals generated in reactions involving vitamin B_{12}-dependent enzymes permit unique molecular rearrangements and syntheses (e.g., methionine synthesis, rearrangement of diols) (da Silva and Williams, 1991). Fortunately, defenses against oxidative damage are intrinsic to biological systems. The enzymes superoxide dismutase and catalase are very effective scavengers of superoxide anion and hydrogen peroxide, respectively.

Further low-molecular radical scavengers abound such as vitamin C and uric acid (Kehrer, 1993). Systems that repair damaged proteins, lipids, and DNA are also in place (Rusting, 1992).

Uncontrolled generation of reactive oxygen species and derivatives such as lipid peroxides and alkoxy radicals are damaging to tissues and have been linked to disease. Inherited defects exist that increase the risk of radical tissue injury such as iron overload (e.g., thalassemia) and Lou Gehrig's disease (possibly linked to a defective superoxide dismutase gene) (Halliwell and Gutteridge, 1984; Marx, 1993; Deng et al., 1993; McNamara and Fridovich, 1993). Further, perturbation of active oxygen metabolism by environmental/occupational exposures to substances capable of inducing unusual fluxes of radicals is hypothesized. The latter is illustrated for chromium(VI) in Figure 4 in relation to its genotoxicity in the context of carcinogenesis. Chromium(VI) compounds are potent human carcinogens (Yassi and Nieboer, 1988; de Flora et al., 1995). Under physiologic conditions, certain nickel compounds can also participate in active oxygen biochemistry (Nieboer et al., 1989; Kasprzak, 1991; Bal et al., 1994). Interestingly, nickel compounds encountered in the nickel refining industry are also human carcinogens (Doll, 1990; Nieboer and Fletcher, 1995b). The direct interaction of Cr^{3+} and Ni^{2+} with DNA is also an important factor in current models of chromium or nickel carcinogenesis (Costa, 1995; de Flora et al., 1995).

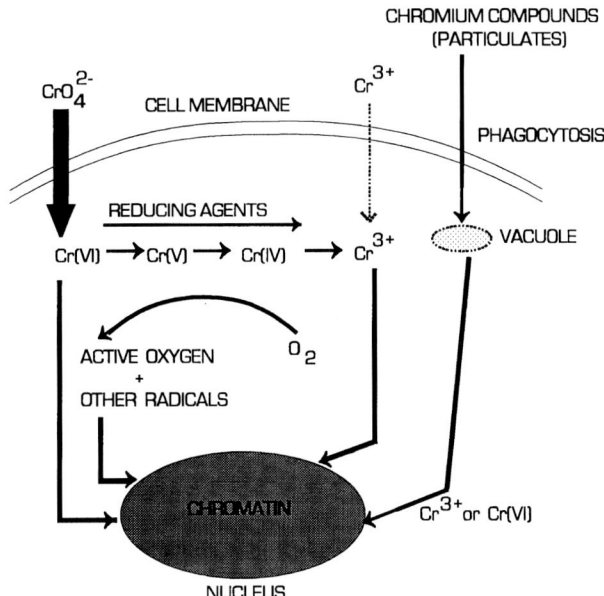

Figure 4 Postulated radical production associated with the intracellular reduction of chromium(VI). This reduction is favored since, generally speaking, biological systems provide a reducing environment. At physiological pH (i.e., 7.4), the chromate ion is not protonated (Nieboer and Jusys, 1988) and Cr^{3+} is actually present as cationic hydroxy derivatives (Baes and Mesmer, 1976). The schematic depicts that CrO_4^{2-} is readily taken up by cells, while the trivalent cation, or related cationic species, is not. This is consistent with the general trend that anions cross cell membranes more readily than cations. The uptake of particulate compounds by phagocytosis is discussed in a subsequent section of the text. In terms of genotoxicity, both radical damage of DNA and Cr(III)-DNA complex formation are assumed to contribute. For a detailed discussion of the scientific basis for the model shown, refer to Nieboer and Jusys (1988), Nieboer and Shaw (1988), and de Flora et al. (1995).

Oxidative damage has been suggested as a determinant in a number of diseases (Halliwell and Gutteridge, 1984; Kehrer, 1993). Suspected radical-mediated processes or disorders include aging (Rusting, 1992); cardiovascular disease (Minotti et al., 1995; Heinecke, 1995); cancer (Feig and Loeb, 1995); inflammation and rheumatoid arthritis; oxygen toxicity; and as already indicated, Lou Gehrig's disease, more specifically amyotrophic lateral sclerosis (ALS), and iron-overload diseases.

F. SOLUBILITY

Reactions between solids and natural waters are paramount in the biogeochemical distribution and circulation of metals (Stumm and Morgan, 1981; Millward, 1995). Airborne particulate emissions from natural (e.g., volcanic), industrial (e.g., pyrometallurgical refining), and municipal (e.g., incinerator) sources are not only responsible for the contamination of air, water, and soils, but also for the long-distance (global) transport of metals (Nriagu and Pacyna 1988; Nriagu, 1989; Jervis et al., 1995). In biology, the hard structures such as skeletons, shells, bones, and teeth involve the systematic deposition of mineral phases either extracellularly or intracellularly (Pautard and Williams, 1982; Williams, 1984; da Silva and Williams, 1991). Further, solubility is an important determinant in metal toxicology. In aquatic systems, release of metal ions from sediments is an important factor, while detoxication in plants or mammalian systems may involve deposition of toxic metals into compartments. In plants, phytochelatins are involved in vacuolar deposition of metals (Ernst, 1977; Steffens, 1990), while deposits of protein complexes of lead as inclusion bodies have been identified in animal and human kidney cells (Moore et al., 1973; Fowler and DuVal, 1991; Fowler, 1992). In humans, about 95% of the body burden of lead is located in the bones, although this deposit is not inert and is suspected of having some toxicologic impact on bone cells (Chettle et al., 1991; Rabinowitz, 1991; Pounds et al., 1991). Gastrointestinal uptake of metals is very dependent on the solubility of metal complexes. A number of natural ligands such as phytates and polyphenols inhibit uptake of metal ions by precipitation while others enhance it, presumably by ensuring solubility or directly affecting the uptake pathway as discussed for citrate and Al^{3+} (Lönnerdal and Sandström, 1995). Finally, lipid solubility has a drastic effect on uptake, distribution, route of excretion, and the nature of toxicity in the body. Lipid solubility promotes accumulation in lipid-rich compartments such as the brain. Mercury and lead compounds provide good illustrations. In terms of the central nervous system, a primary target of CH_3Hg^+ is neuromotor function, while for mercury vapor the effect is more neuropsychiatric in nature (IPCS, 1990, 1991a; Halbach, 1995). Similar demarcations are established for the relative neurotoxicities of organolead (e.g., triethyl-lead, $(CH_3)_3Pb^+$) and inorganic lead compounds (Grandjean and Grandjean, 1984; Verity, 1990; Skerfving, 1995). Interestingly, Nieboer and Richardson (1980) point out with reference to Figure 1 that ions with intermediate values of $(\chi_m)^2 r$ and concurrent large Z^2/r values or ions with large values of $(\chi_m)^2 r$ and concurrent intermediate Z^2/r values form water-stable organometallic cations, such as methyl derivatives of Pb(IV), Tl(III), Sn(IV), Hg(II), and As(III); they form an outer arc across the top of Figure 1. Water-resistant alkyl derivatives of Au(III) and Pt(IV) are also known.

G. PARTICLE SIZE OF SOLID COMPOUNDS

Submicron (<1 μm in diameter) mineral colloid matter, such as oxyhydroxides of iron, are strongly implicated in modulating the bioavailability and mode of dispersion of both metal contaminants and nutrients (Leppard, 1992; Ledin et al., 1995). In human exposures by inhalation, three size-dependent particulate fractions are defined, which determine where deposition occurs in the respiratory tract to exert a response (Vincent, 1993, 1994a,b). The *inhalable fraction* (diameter ≤100 μm) is the fraction of ambient airborne particles that enters the body through the nose and/or mouth during breathing. It is relevant to health effects anywhere in the respiratory tract such as rhinitis, nasal cancer, or systemic effects. The *thoracic fraction* (≤20 μm) is the inhaled particle component which penetrates into the lung (i.e., the whole region below the larynx) and is relevant for asthma, bronchitis, or lung cancer. The *respirable fraction* (≤10 μm) constitutes the inhaled particles that penetrate to the alveolar region of the lung (i.e., includes the respiratory bronchioles, the alveolar ducts and sacs) and is pertinent to the development of such chronic diseases as pneumoconiosis or emphysema. This particle classification is especially relevant to the industries employing pyrometallurgical refining techniques such as in the nickel-producing industry (Doll, 1990). Lung and nasal cancers and asthma have been reported (Nieboer et al., 1988; Nieboer and Fletcher, 1995a,b). Separate threshold limit values (TLVs) for workplace exposures to water-soluble and water-insoluble metal compounds are often promulgated (Fairfax and Blotzer, 1994). Particulate air pollution of diameters ≤10 μm (the PM_{10} fraction) appears to be linked to increased mortality from lung cancer, cardiopulmonary disease, and other respiratory causes. These effects were quite strongly correlated with suspended sulfates, and contributions from metals are also suspected (Utell and Samet, 1993; Dockery et al., 1993).

At the cellular level, particle size is very important and relates to uptake by phagocytosis and subsequent toxicity. Animal cells have diameters of 10 to 30 μm and thus for particles to be endocytosed ("eaten" or phagocytosed), they must be considerably smaller (<10 μm in diameter). Costa et al. (1981) have shown for Chinese hamster ovary (CHO) cells that the mean particle size of crystalline αNiS correlated inversely with uptake and cytotoxicity as measured by colony formation in the range 1 to 6 μm. In cell culture, many mammalian cells are phagocytes, while in multicellular organisms the "professional" phagocytes are macrophages and neutrophils and few other cells are able to ingest large particles (Alberts et al., 1994). Little is known whether, for example, epithelial cells lining the human lung have such facility.

H. PHYSICOCHEMICAL PROPERTIES OF SOLID SURFACES

In vitro cell culture and experimental animal studies with particulate nickel compounds clearly have demonstrated that surface properties have a strong influence on biological activity. Because solid metal compounds have exposed ions or anion sites on their surfaces with reduced coordination number, hydrolysis or chemisorption results in surface-active functional groups such as hydroxyl groups (Stumm and Morgan, 1981; Nieboer et al., 1984b). Sunderman et al. (1987) examined six nickel oxides that had been sintered at temperatures from <650°C (the parent material) to 1045°C. X-ray diffraction and chemical analysis were consistent with bunsenite (NiO) as the major crystalline phase in all six test compounds. Dissolution half-times in rat serum and renal cytosol were around 1 year for the parent compound and 10-fold longer for the remaining five; the parent oxide also had a significantly greater surface area per unit weight. The parent material was the most biologically active in four tests (phagocytosis and cell transformation, but not cell survival, in cell culture tests; erythropoietic stimulation and renal pathology in a rat intra-renal injection assay).

Protein adsorption data provide direct evidence that physical properties of solid nickel compounds might determine their biological activity (Nieboer et al., 1986). Protein adsorption capacities determined with human serum albumin at pH 7.4 for the non-crystalline, amorphous NiS and colloidal $Ni(OH)_2$ are significantly higher than those for the crystalline nickel compounds and the metal powder. Slopes of protein absorption curves yielded the sequence (in $\mu g\ mg^{-1}$): colloidal $Ni(OH)_2$ (570 ± 10) > amorphous NiS (300) >>> NiO (8.0 ± 0.5) > Ni powder (4.3 ± 0.4) > α-NiS and β-NiS (3.4 ± 0.2) > dried $Ni(OH)_2$ (2.9 ± 0.1) > α-Ni_3S_2 (2.2 ± 0.4). The diameter of the particles was mostly <10 μm. Ranking by the ability to lyse human red blood cells roughly paralleled this order. The non-crystalline compounds are the least carcinogenic in animals (colloidal $Ni(OH)_2$ is non-carcinogenic and amorphous NiS weakly so). For rats, Sunderman (1984) established the carcinogenicity sequence: α-Ni_3S_2, β-NiS > NiO >> Ni powder >>> amorphous NiS (material of median particle size of <2 μm was injected intramuscularly). Dried, crystalline $Ni(OH)_2$ is also known to be carcinogenic (Kasprzak et al., 1983). A review of the available laboratory studies (see Sunderman, 1984; Sunderman et al., 1987; Costa et al., 1981; Abbrachio et al., 1982; Nieboer et al., 1984b,1988; Coogan et al., 1989; IARC, 1990; Fletcher et al., 1994) suggests that surface passivity of solids, namely smooth exterior, crystallinity, low surface charge, low surface activity with respect to protein absorption and cell lysing ability, and moderate solubility appear to be a predisposition to experimental carcinogenicity (Nieboer et al., 1986). These properties parallel the ability of particles to be phagocytized, which constitutes an effective uptake mechanism (Sen and Costa, 1986). Interestingly, exposure to nickel oxides and sulfides figure prominently in the epidemiology of occupational nickel-related lung and nasal cancer.

It is clear from the above deliberations that the physicochemical properties of metal compounds used in laboratory toxicologic assessments must be well characterized for the experimental results to be interpreted meaningfully.

V. BIOLOGICAL PHENOMENA

A. COMPARTMENTALIZATION

It is a well-established fact that the exclusion of Ca^{2+} from cells is the basis for the intracellular responses to small inward fluxes of this ion that lead to such complicated processes as muscle contraction and excocytosis of, e.g., neurotransmitters (da Silva and Williams, 1991). The source of the Ca^{2+} can be extracellular, but in muscle cells the release is largely from internal stores (vesicles). Not surprisingly, intracellular compartmentalization of toxic metals is also important. This is illustrated in a recent study by the authors (Fletcher et al., 1994) in which the uptake, cytoxicity, and mutagenicity of three water-soluble nickel salts (chloride, sulfate, and acetate), three water-insoluble nickel compounds (oxides and sulphides) of particle diameter ≤5 μm, and two relatively insoluble nickel compounds (carbonate and hydroxide) of the same particle size were evaluated in AS52 CHO cells. LC_{50} values, corresponding to the concentration at which the survival by colony forming ability of exposed cells is 50% of that of nonexposed control cells, ranged from 2 to 130 μg Ni/mL for particulate nickel compounds and from 45 to 60 μg Ni/mL for the water-soluble salts. While the administered LC_{50} doses differed by a factor of 75 for the 11 nickel compounds, the cytosolic and nuclear concentrations of dissolved (non-particulate) nickel differed only by about a factor of 10. The most toxic compounds exhibited the highest nuclear and cytosolic levels. Interestingly, the particulate compounds delivered dissolved nickel to both compartments, while the water-soluble salts loaded the cytosol but very little nickel reached the nucleus. Presumably this reflects the different modes of uptake, respectively phagocytosis (particles) and diffusion (water-soluble salts). While the cytosolic nickel concentrations for the 11 compounds did not correlate with the LC_{50} values, the nuclear levels did ($p < 0.05$). All the compounds tested were weakly mutagenic. Mutant profile analyses involving a specific gene locus were compound specific and suggested different mutagenic mechanisms for the small number of water-soluble and particulate compounds examined in this part of the study. These results were interpreted to reflect the different intracellular compartmentalization patterns of dissolved nickel mentioned (Rossetto et al., 1994).

Organ specific toxicity reflects compartmentalization. Since inhalation is the main route of occupational exposure for most metal compounds, the nasal passages and lungs are the tissues most often affected by cancer. Analysis of lung and nasal tissues of deceased nickel refinery workers has confirmed considerable residual deposits (presumably particulate and interstitial) (Andersen and Svenes, 1989).

Cadmium is known to accumulate in the kidneys by virtue of its association with the Cd-scavaging protein metallothionein, and this organ is the major target for this toxic metal (IPCS, 1992; Fowler, 1992). By contrast, lead is distributed throughout the body before being deposited in the bone. Not surprisingly, it is a systemic poison. When the exposures are high and chronic, lead inclusion bodies appear in the kidney and significant perturbations of renal function occur (Fowler and Duval, 1991; Skerfving et al, 1995). Recent research suggests that certain lead-binding proteins act as transcription factors, which appear to regulate lead uptake by cells and some of the toxic responses (Fowler, 1992; Goering, 1993). However, compartmentalization does not necessarily result in toxicity. For example, in patients with chronic renal failure who were exposed to acute levels of aluminum, only the central nervous system and bone were affected, even though high levels of aluminum occurred in other organs (Nieboer et al, 1995).

It is clear from the above examples that the target organ receiving the exposure insult is highly susceptible to injury or disease. The respiratory tract, skin, and eyes are examples. Organ specific toxicity can also be the consequence of preferential compartmentalization after absorption into the blood compartment. Both uptake and toxicity may be linked to unique metabolic processes, induced changes in gene expression, and organ- or organelle-specific susceptibility. Similarity in chemistry between a toxic ion and an essential metal ion may also be an important determinant, as occurs in the bone compartment for Pb^{2+} and Ca^{2+}.

B. RESPIRATORY TRACT CLEARANCE

The epithelial lining of the bronchi have single-celled exocrine glands that secrete mucus, as well as ciliated cells. Submucosal glands of the bronchial lining also produce mucus to blanket the bronchial epithelium. The cilia rhythmically move the mucus blanket covering the bronchial epithelium toward the trachea and pharynx, where it can be swallowed or expectorated by coughing (Guyton, 1986; McCance and Huether, 1994). Most of the nasal passages clear foreign particles and microorganisms similarly. The alveolar region of the lung does not share this mucociliary clearance ability, but alveolar macrophages ingest foreign material that reaches the alveoli, permitting removal through the lymphatics. Clearance of metal-containing materials from nasal and lung tissue appears to be slow.

C. BIOAVAILABILITY

Bioavailability may be defined as the "extent to which a substance to which the body is exposed (by ingestion, inhalation, injection, or skin contact) reaches the systemic circulation, and the rate at which this occurs" (Duffus, 1993). The extent a toxic agent reaches the site of action constitutes a more pertinent definition. Gastrointestinal uptake has already been discussed in the *Solubility* section. As indicated, it is very dependent on the ligands available in the food (also see Johnson, 1995; Serfaty-Lacrosnière et al., 1995). Further, the presence of other metals is known to be important, presumably due to competition at the level of the absorptive cells (Mahaffey, 1990). Chemical extraction protocols have been devised to assess the bioavailability in soils. The extractability of, for example, Cd is then correlated with some index of toxicity (e.g., Krishnamurti et al., 1995).

In terms of bioavailability to other organs, release of metal in the form of a cation or anion from objects that come into contact with the skin is essential to the development of allergic contact dermatitis. Examples are nickel released from jewelry and chromate from cement (Menné and Nieboer, 1989). For exposures by inhalation, the nature of the toxic substance is crucial. Gases such as nickel tetracarbonyl, carbon monoxide, hydrogen cyanide, or arsine are nearly completely bioavailable. Exposures to aqueous aerosols such as nickel(II) salts or potassium dichromate are accompanied by significant absorption, perhaps involving a short delay (hours). By contrast, most particulate metal compounds of low water-solubility are not readily absorbed, as indicated by estimates of biological residence times (see material which follows). It is well established by magnetopneumography that chronic deposits of lung-retained contaminants occur in workers exposed to metal-containing particles such as welders (Kalliomäki et al, 1981, 1983). Further, as already indicated, chemical analysis of lung tissue of deceased nickel refinery workers tells the same story. Presumably, solubility in biological fluids of the deposited material, as well as the surface properties discussed earlier that determine biological processes such as uptake and removal by macrophages, regulate the rate of clearance. Interestingly, workers exposed primarily to aqueous aerosols of nickel chloride and sulfate in an electroplating department had significantly less ($p < 0.01$) nickel in autopsied lung tissue than those employed primarily in a roasting and smelting department (Andersen and Svenes, 1989).

D. BIOLOGICAL RESIDENCE TIME

Metals can be eliminated by way of urinary, biliary, pancreatic, and intestinal excretion (Gregus and Klaassen, 1995; Rosenberg, 1995). The urinary pathway is the best defined route for most metals. Less is known with certainty about the remaining excretion routes in humans since the available data are plagued by analytical uncertainty related to inadvertent contamination during collection, handling, storage, and analysis of samples. For instance, compare the evidence for biliary excretion of nickel reviewed by Nieboer et al. (1992a) and Gregus and Klaassen (1995). For most metals, concentrations in the blood compartments (whole blood, plasma, or serum) and urine reflect exposure. Consequently, these parameters may be employed to assess biological residence or elimination half-times, as well as occupational or accidental exposures (e.g., Sunderman et al, 1988; Friberg and Elinder, 1993). When exposures are to water-soluble compounds or vapor, the turnover for many metals is relatively rapid with half-times ($t_{1/2}$) in terms of hours (e.g., Al; Nieboer et al, 1995), 1 to 3 days (e.g., Ni, rapid phase for Hg vapor, chromate; IPCS, 1991a; Nieboer and Jusys, 1988; Sunderman et al., 1989; Nieboer et al., 1992a), weeks to month (e.g., Pb, slow phase for Hg vapor; IPCS, 1990; Skerfving et al, 1995), or months (e.g., methylmercury; IPCS, 1990). By contrast, release of Pb from bone occurs with $t_{1/2}$ of 10 to 30 years (Rabinowitz, 1991), which is of comparable magnitude to that of the turnover of Cd in the body (IPCS, 1992). Both Cd and Pb may therefore be considered persistent and as already described, toxic pollutants in the context of human exposure. Clearance of inhaled particulates is also relatively slow. For Al welders, $t_{1/2}$ values between 20 and 400 days have been reported (Nieboer et al, 1995) for a slow release compartment in the absence of current exposure. Similarly, for retired nickel refinery workers, $t_{1/2}$ for the removal of nickel from nasal mucosa tissue or its appearance in urine was estimated as 3.5 years (Torjussen and Andersen, 1979; Boysen et al, 1982). Further, the clearance rate of welding fumes from the lungs of mild steel welders has been estimated as 20% per year (i.e., a $t_{1/2}$ of 3.5 years) by the magnetic method mentioned earlier (Kalliomäki et al, 1981, 1982).

It is clear that in the absence of suitable clearance pathways for inhaled metal-containing mineral dust, the risk of chronic diseases such as cancer and pneumoconiosis increases considerably. Short-term toxic effects of metals may also be exacerbated by unfavorable clearance rates. Chelating agents are therefore administered therapeutically to speed up the removal of toxic metals from the blood and soft tissue compartments (CDC, 1991).

E. METAL RESISTANCE

Toxicity can be modulated by the development of tolerance or resistance. This is readily illustrated for nickel since a number of terrestrial and aquatic organisms can accumulate enormous quantities of this metal without apparent ill effect (IPCS, 1991b). As discussed under *Solubility*, vacuolar deposition as metal complexes is one mechanism at the cellular level (Steffens, 1990). Other cellular defense mechanisms are possible: cellular export, blocked entry, extracellular scavenging, and/or conversion to nontoxic forms (e.g., as mercury vapor), development of target tolerance, and repair of incurred damage (Walsh et al, 1988; Ono, 1988). In bacteria, such resistance to toxic metals occurs primarily by regulation through plasmid-encoded processes. Hg, Cd, As, Cr, and Cu resistance are examples (Silver and Walderhaug, 1992). By contrast, bacterial uptake of essential nutrients (e.g., phosphate, sulfate, Mn, Ni, Zn, and in part Cu) usually are chromosomally encoded. This is also true for Cu uptake in yeast (Welch et al., 1989). Presumably, sensitivity to toxic metals would be conferred by gene mutation (Welch et al., 1989). Interestingly, mutations in two closely related genes which encode proteins involved in copper transport are responsible for either Menkes disease (copper deficiency) or Wilson's disease (copper toxicity due to excess) (Bull and Cox, 1994).

Immunological tolerance has been recognized for some time in humans in the context of allergic contact dermatitis (Menné and Nieboer, 1989). Because of the Human Genome Project, which has as its goal to map and sequence the entire genome, inherited susceptibility to environmental insults is a topic of considerable interest for toxicologists and environmental health professionals (Calabrese, 1986; Forrester and Wolf, 1990; Nieboer et al., 1992b). Genetic susceptibility to asthma is a very recent example (Postma et al, 1995). The hypothesis is that such inherited sensitivity will exacerbate the effect of chemical respiratory allergens, which include metal compounds such as of nickel (e.g., Nieboer and Fletcher, 1995a).

VI. CONCLUDING REMARKS

An understanding of the factors that determine toxicological responses to metals and their compounds requires an integrated, holistic approach. Not only are basic atomic/molecular properties, and factors that determine chemical reactivity important, but anatomical/biological structures, parameters and processes and genetic factors clearly modulate the toxic outcome. Obviously, interdisciplinary research efforts are required to enhance our understanding of metal toxicology further. It is hoped that the content of this article will stimulate such collaborative efforts.

ACKNOWLEDGMENTS

I (E.N.) acknowledge with considerable gratitude the collaboration of many colleagues. Special tributes are extended to Professor R.J.P. Williams, F.R.S., Inorganic Chemistry Laboratory, Oxford University for opening the door to Inorganic Biochemistry; to Dr. D.H.S. Richardson, now Dean of Science, Saint Mary's University, Halifax, Canada for an exciting partnership in environmental research that lasted a decade; and to Dr. Albert Cecutti, Director, Occupational Health and Hygiene, Falconbridge Limited, Toronto, Canada for nudging me toward the challenges of Occupational Toxicology. Of course, the contributions of my graduate students should not be overlooked as we have liberally quoted their research results in the preparation of this article. I am indebted to them.

REFERENCES

Abbracchio, M. P., Heck, J. D. and Costa, M., *Carcinogenesis*, 3, 175–180, 1982.
Alberts, B., Bray, D., Lewis, J., Raff, M., Roberts, K. and Watson, J. D., *Molecular Biology of the Cell, Third Ed.*, Garland, New York, 1994.
Allred, A. L., *J. Inorg. Nucl. Chem.*, 17, 215–221, 1961.
Andersen, I. and Svenes, K. B., *Int. Arch. Occup. Environ. Health*, 61, 289–295, 1989.
van Assche, F. and Clijsters, H., *Plant Cell Environ.*, 13, 195–206, 1990.
ATSDR, *Toxicological Profile for Mercury*, Agency for Toxic Substances and Disease Registry, U.S. Public Health Service, Atlanta, GA, 1989.
Avery, S. V. and Tobin, J. M., *Appl. Environ. Microbiol.*, 59, 2851–2856, 1993.
Baes, C. F. and Mesmer, R. F., *The Hydrolysis of Cations*, John Wiley & Sons, New York, 1976.
Bal, W., Djuran, M. I., Margerum, D. W., Gray, E. T. Jr., Mazid, M. A., Tom, R. T., Nieboer, E., and Sadler, P. J., *J. Chem. Soc., Chem. Commun.*, 1889–1890, 1994.
Boysen, M., Solberg, L. A., Andersen, I., Høgetveit, A. C. and Torjussen, W., *Scand. J. Work Environ. Health*, 8, 283–289, 1982.
Bull, P. C. and Cox, D. W., *TIG*, 10, 246–252, 1994.
Burgess, J., *Ions in Solution*, Ellis Horwood, Chichester, 1988.
Burgess, J., *Analyst*, 117, 605–611, 1992.
Butler, J. N., *Ionic Equilibrium. A Mathematical Approach*. Addison-Wesley, Reading, PA, 1964.
Calabrese, E. J., *J. Occup. Med.*, 28, 1096–1102, 1986.
Campbell, P. G. C., in *Metal Speciation and Bioavailability*, A. Tessier, and D. R. Turner, Eds. John Wiley, New York, 45–102, 1995.
CDC, *Preventing Lead Poisoning in Young Children*, Centers for Disease Control, Public Health Service, U.S. Department of Health and Human Services, Atlanta, GA, 1991.
Chettle, D. R., Scott, M. C., and Somervaille, L. J., *Environ. Health Perspect.*, 91, 49–55, 1991.
Chisolm, J. J., Jr., *Environ. Health Perspect.*, 89, 67–74, 1990.
Clarkson, T. W., *Annu. Rev. Pharmacol. Toxicol.*, 32, 545–571, 1993.

Coleman, J. E. and Vallee, B. L., *J. Biol. Chem.*, 236, 2244–2249, 1961.
Coogan, T. P., Latta, D. M., Snow, E. T., and Costa, M., *CRC Crit. Rev. Toxicol.*, 19, 341–384, 1989.
Costa, M., in *Handbook of Metal-Ligand Interactions in Biological Fluids: Bioinorganic Medicine Vol. 2*, G. Berthon, Ed., Marcel Dekker, New York, 1003–1008, 1995.
Costa, M., Abbracchio, M. P., and Simmons-Hansen, J., *Toxicol. Appl. Pharmacol.*, 60, 313–323, 1981.
Cotton, F. A. and Wilkinson, G., *Advanced Inorganic Chemistry. A Comprehensive Text*, 5th ed., John Wiley & Sons, New York, 1988.
Crist, R. H., Oberholser, K., Schwartz, D., Marzoff, J., Ryder, D., and Crist, D. R., *Environ. Sci. Technol.*, 22, 755–760, 1988.
daSilva, J. J. R. F. and Williams, R. J. P., *The Biological Chemistry of the Elements*, Clarendon Press, Oxford, 1991.
Davies, C. N., *Ann. Occup. Hyg.*, 29, 13–25, 1985.
De Flora, S. Camoirano, A., Bagnasco, M., and Zanacchi, P., in *Handbook of Metal-Ligand Interactions in Biological Fluids: Bioinorganic Medicine Vol. 2*, G. Berthon, Ed., Marcel Dekker, New York, 1020–1036, 1995.
Deng, H.-X., Hentati, A., Tainer, J. A., Iqbal, Z., Cayabyab, A., Hung, W.-Y., Getzoff, E. D., Hu, P., Herzfeldt, B., Roos, R. P., Warner, C., Deng, G., Soriano, E., Smyth, C., Parge, H. E., Ahmed, A., Roses, A. D., Hallewell, R. A., Pericak-Vance, M. A., and Siddique, T., *Science*, 261, 1047–1051, 1993.
Dockery, D. W., Pope III, C. A., Xu, X., Spengler, J. D., Ware, J. H., Fay, M. E., Ferris, B. G., Jr., and Speizer, F. E., *New Engl. J. Med.*, 329, 1753–1759, 1993.
Doll, R., *Scand. J. Work Environ. Health*, 16, 1–82, 1990.
Duffus, J. H., *Pure Appl. Chem.*, 65, 2003–2122, 1993.
Duffy, S. J., Hay, G. W., Micklethwaite, R. K., and Vanloon, G. W., *Sci. Total Environ.*, 87/88, 189–197, 1989.
Ernst, W. H. O., in *Symposium Proceedings. Intenational Conference on Heavy Metals in the Environment, Toronto, 27-31 October, 1975*, 2(1), 121–136, 1977.
Evans, L. J., *Environ. Sci. Technol.*, 23, 1046–1056, 1989.
Evans, D. H. and Weingarten, K., *Toxicology*, 61, 275–281, 1990.
Fairfax, R. and Blotzer, M., *Appl. Occup. Environ. Hyg.*, 9, 683–686, 1994.
Feig, D. I. and Loeb, L. A., in *Handbook of Metal-Ligand Interactions in Biological Fluids: Bioinorganic Medicine Vol. 2*, G. Berthon, Ed., Marcel Dekker, New York, 995–1002, 1995.
Fletcher, G. G., Rossetto, F. E., Turnbull, J. D., and Nieboer, E., *Environ. Health Perspect.*, 102 (Suppl 3), 69–79, 1994.
Forrester, L. M. and Wolf, C. R., in *The Metabolic and Molecular Basis of Acquired Disease, Volume One*, R. D. Cohen, B. Lewis, K. G. M. M. Alberti, and A. M. Denman, Eds., Ballière Tindall, London, 3–18, 1990.
Fowler, B. A., *Environ. Health Perspect.*, 100, 57–63, 1992.
Fowler, B. A. and DuVal, G., *Environ. Health Perspect.*, 91, 77–80, 1991.
Friberg, L. and Elinder, C.-G., *Scand. J. Work Environ. Health*, 19 (Suppl 1), 7–13, 1993.
Friberg, L., Nordberg, F. G., and Vouk, V. B., *Handbook on the Toxicology of Metals*, 2nd ed., Elsevier, Amsterdam, 1986.
Gelvan, D. and Saltman, P., in *Handbook of Metal-Ligand Interactions in Biological Fluids: Bioinorganic Medicine Vol. 2*, G. Berthon, Ed., Marcel Dekker, New York, 976–985, 1995.
Gibson, H., Vincent, J. H., and Mark, D., *Ann. Occup. Hyg.*, 31, 463–479, 1987.
Goering, P. L., *Neurotoxicology*, 14, 45–60, 1993.
Grandjean, P. and Grandjean, E. C., *Biological Effects of Organolead Compounds*, CRC Press, Boca Raton, FL, 1984.
Gregus, Z. and Klaassen, C. D., in *Handbook of Metal-Ligand Interactions in Biological Fluids: Bioinorganic Medicine Vol. 1*, G. Berthon, Ed., Marcel Dekker, New York, 445–460, 1995.
Gutteridge, J. M. C., in *Handbook of Metal-Ligand Interactions in Biological Fluids: Bioinorganic Chemistzy, Vol. 2*, G. Berthon, Ed., Marcel Dekker, New York, 848–856, 1995.
Guyton, A. C., *Textbook of Medical Physiology, Seventh Ed.*, W. B. Saunders, Philadelphia, 53–54, 474–478, 1986.
Halbach, S., in *Handbook of Metal-Ligand Interactions in Biological Fluids: Bioinorganic Medicine Vol. 2*, G. Berthon, Ed., Marcel Dekker, New York, 749–754, 1995.
Halliwell, B. and Gutteridge, J.M. C., *Biochem. J.*, 219, 1–14, 1984.
Heinecke, J. W., in *Handbook of Metal-Ligand Interactions in Biological Fluids: Bioinorganic Medicine Vol. 2*, G. Berthon, Ed., Marcel Dekker, New York, 986–994, 1995.
Huheey, J. E., Keiter, E. A., and Keiter, R. L., *Inorganic Chemistry. Principles of Structure and Reactivity*, 4th ed. Harper Collins, New York, 1993.
IARC, in *IARC Monographs on the Evaluation of Carcinogenic Risks to Humans. Volume 49, Chromium, Nickel and Welding*, International Agency for Research on Cancer, Lyon, France, 257–445, 1990.
IPCS, *Environmental Health Criteria 101: Methylmercury*, International Programme on Chemical Safety, World Health Organization, Geneva, 1990.
IPCS, *Environmental Health Criteria 118: Inorganic Mercury*, International Programme on Chemical Safety, World Health Organization, Geneva, 1991a.
IPCS, *Environmental Health Criteria 108: Nickel*, International Programme on Chemical Safety, World Health Organization, Geneva, 1991b.
IPCS, *Environmental Health Criteria 134: Cadmium*, International Programme on Chemical Safety, World Health Organization, Geneva, 1992.
Jervis, R. E., Krishnan, S. S., Ko, M. M., Vela, L. D., Pringle, T. G., Chan, A. C., and Xing, L., *Analyst*, 120, 651–657, 1995.

Johnson, P. E., in *Handbook of Metal-Ligand Interactions in Biological Fluids: Bioinorganic Medicine Vol. 1*, G. Berthon, Ed., Marcel Dekker, New York, 346–350, 1995.
Kaiser, K. L. E., *Can. J. Fish. Aquat. Sci.*, 37, 211–218, 1980.
Kalliomäki, P.-L., Kalliomaki, K., Rahkonen, E., and Aittoniemi, K., *Ann. Occup. Hyg.*, 27, 449–452, 1983.
Kalliomäki, P.-L., Rahkonen, E., Vaaranen, V., Kalliomaki, K. and Aittoniemi, K., *Int. Arch. Occup. Environ. Health*, 49, 67–75, 1981.
Kasprzak, K.S., *Chem. Res. Toxicol.*, 4, 604–615, 1991.
Kasprzak, K. S., in *Nickel and Human Health. Current Perspectives*, E. Nieboer and J. O. Nriagu, Eds., John Wiley & Sons, New York, 387–420, 1992.
Kasprzak, K. S., Gabryel, P. and Jarczewska, K., *Carinogenesis*, 4, 275–279, 1983.
Kasprzak, K. S. and Rodriguez, R. E., in *Nickel and Human Health. Current Perspectives*, E. Nieboer and J. O. Nriagu, Eds., John Wiley & Sons, New York, 545–559, 1992.
Kehrer, J. P., *Crit. Rev. Toxicol.*, 23, 21–48, 1993.
Krishnamurti, G. S. R., Huang, P. M., Van Rees, K. C. J., Kozak, L. M., and Rostad, H. P. W., *Analyst*, 120, 659-665, 1995.
Lag, M. and Helgeland, K., *Pharmacol. Toxicol.*, 60, 318–320, 1987.
Langford, C. H. and Cook, R. L., *Analyst*, 120, 591–596, 1995.
Ledin, A., Karlsson, S., Düker, A., and Allard, B., *Analyst*, 120, 603–608, 1995.
Leppard, G. G., *Analyst*, 117, 595–603, 1992.
Lide, D. R., Ed-in-Chief, *Handbook of Chemistry and Physics*, 71st ed. CRC Press, Boca Raton, FL, 1990.
Lönnerdal, B. and Sandström, B., in *Handbook of Metal-Ligand Interactions in Biological Fluids: Bioinorganic Medicine Vol. 1*, G. Berthon, Ed., Marcel Dekker, NewYork, 331–337, 1995.
Mahaffey, K. R., *Environ. Health Perspect.*, 89, 75–89, 1990, 1990.
Martin, R. B., in *Handbook on the Toxicity of Inorganic Compounds*, H. G. Seiler, H. Sigel, and A. Sigel, Eds., Marcel Dekker, New York, 9–25, 1988.
Marx, J., *Science*, 261, 986, 1993.
McCance, K. L. and Huether, S. E., in *Pathophysiology: The Biologic Basis for Disease in Adults and Children*, Mosby, St. Louis, MO, 1122–1129, 1994.
McDonald, D. G., Reader, J. P., and Dalziel, T. R. K., in *Acid Toxicity and Aquatic Animals*, R. Morris, E. W. Taylor, D. J. A. Brown, and J. A. Brown, Eds., Cambridge University Press, Cambridge, UK, 221–242, 1989.
Menné, T. and Nieboer, E., *Endeavour*, 13, 117–122, 1989.
Menon, C. R. and Nieboer, E., *J. Inorg. Biochem.*, 28, 217–225, 1986.
Millward, G. E., *Analyst*, 120, 609–614, 1995.
Minotti, G., Mordente, A., and Cavaliere, A. F., in *Handbook of Metal-Ligand Interactions in Biological Fluids: Bioinorganic Medicine Vol. 2*, G. Berthon, Ed., Marcel Dekker, New York, 962–975, 1995.
Moore, J. F., Goyer, R. A., and Wilson, M., *Lab. Invest.*, 29, 488–494, 1973.
Morel, F. M. N., *Principles of Aquatic Chemistry*, John Wiley & Sons, New York, 1983.
Morgan, L. G., in *Nickel and Human Health. Current Perspectives*, E. Nieboer and J. O. Nriagu, Eds., John Wiley & Sons, New York, 261–271, 1992.
Nieboer, E., *Analyst*, 117, 550, 1992a (Special March 1992 Issue 551–691).
Nieboer, E., in *Nickel and Human Health. Current Perspectives*, E. Nieboer and J. O. Nriagu, Eds., John Wiley & Sons, New York, 37–47, 1992b.
Nieboer, E. and McBryde, W. A. E., *Can. J. Chem.*, 51, 2512–2524, 1973.
Nieboer, E. and Richardson, D. H. S., *Environ. Pollut. Ser. B*, 1, 3–26, 1980.
Nieboer, E. and Sanford, W. E., in *Reviews in Biochemical Toxicology, Vol. 7*, E. Hodgson, J. R. Bend and R. M. Philpot, Eds., Elsevier Science, New York, 205–245, 19.
Nieboer, E. and Shaw, S. L., in *Chromium in the Natural & Human Environments Advances in Environmental Science and Technology*, Vol. 20, J. O. Nriagu and E. Nieboer, Eds., John Wiley & Sons, New York, 399–441, 19.
Nieboer, E. and Fletcher, G. G., in *Handbook of Metal-Ligand Interactions in Biological Fluids. Bioinorganic Medicine, Vol. 2*, G. Berthon, Ed., Marcel Dekker, New York, 709–715, 1995a.
Nieboer, E. and Fletcher, G. G., in *Handbook of Metal-Ligand Interactions in Biological Fluids: Bioinorganic Medicine Vol. 2*, G. Berthon, Ed., Marcel Dekker, New York, 1014–1019, 1995b.
Nieboer, E. and Jusys, A. A., in *Chromium in the Natural & Human Environments, Advances in Environmental Science and Technology, Vol. 2*, J. O. Nriagu and E. Nieboer, Eds., John Wiley & Sons, New York, 21–79, 1995c.
Nieboer, E. and Thomassen, Y., Speciation of Elements in Toxicology and in Environmental and Biological Sciences, Second Intemational Symposium, June 15–18, 1994, Loen, Norway. *Analyst*, 120, 30N. (Special March 1995 Issue, 583-763.)
Nieboer, E., Stafford, A. R., Evans, S. L., and Dolovich, J., in *Nickel in the Human Environment*, IARC Sc. Publ. No. 53, F. W. Sunderman, Jr., Ed-in-Chief, International Agency for Research on Cancer, Lyon, France, 321–331, 1984a.
Nieboer, E., Maxwell, R. I., and Stafford, A. R., in *Nickel in the Environment*, F. W. Sunderman, Ed-in-Chief, International Agency for Research on Cancer, Lyon, France, 439–458, 1984b.
Nieboer, E., Maxwell, R. I., Rossetto, F. E., Stafford, A. R., and Stetsko, P. I., in *Frontiers in Bioinorganic Chemistry*, A. V. Xavier, Ed., VCH Verlagsgesellschaft mbH, Weinheim, Federal Republic of Germany, 142–151, 1986.
Nieboer, E., Tom, R. T., and Sanford, W. E., in *Metal Ions in Biological Systems, Vol. 23, Nickel and its Role in Biology*, H. Sigel and A. Sigel, Eds., Marcel Dekker, New York, 91–121, 1988.

Nieboer, E., Rossetto, F. E., and Menon, C. R., in *Metal Ions in Biological Systems, Vol. 23, Nickel and its Role in Biology*, H. Sigel and A. Sigel, Eds., Marcel Dekker, New York, 359–402, 1988.
Nieboer, E., Tom, R. T., and Rossetto, F. E., *Biol. Trace Elem. Res.*, 21, 23–33, 1989.
Nieboer, E., Sanford, W. E., and Stace, B. C., in *Nickel and Human Health: Current Perspectives (Advances in Environmental Science and Technology, Vol. 25)*, E. Nieboer and J. O. Nriagu, Ed., John Wiley & Sons, New York, 49–68, 1992a.
Nieboer, E., Rossetto, F. E., and Turnbull, J. D., *Toxicol. Lett.*, 64/65, 25–32, 1992b.
Nieboer, E., Gibson, B. L., Oxman, A. D., and Kramer, J. R., *Environ. Rev.*, 3, 29-81, 1995.
Nriagu, J. O., *Nature*, 338, 47–49, 1989.
Nriagu, J. O. and Pacyna, J. M., *Nature*, 333, 134–139, 1988.
Ono, B.-I., in *Chromium in the Natural & Human Environments Advances in Environmental Science and Technology, Vol. 20*, J. O. Nriagu and E. Nieboer, Eds., John Wiley & Sons, New York, 351–368, 1988.
Pautard, F. G. E. and Williams, R. J. P., *Chem. Brit.*, 18, 14–16, 1982.
Pearson, R. G., *J. Am. Chem. Soc.*, 85, 3533–3539, 1963.
Phillips, C. S. G. and Williams, R. J. P., *Inorganic Chemistry, Vol. 1*, Clarendon Press, Oxford, 1965.
Postma, D. S., Bleecker, E. R., Amelung, P. J., Holroyd, K. J., Xu, J., Panhuysen, C. I. M., Meyers, D. A., and Levitt, R. C., *New Engl. J. Med.*, 333, 894–900, 1995.
Pounds, J. G., Long, G. J., and Rosen, J. F., *Environ. Health Perspect.*, 91, 17–32, 1991.
Puckett, K. J., Nieboer, E., Flora, W. P., and Richardson, D. H. S., *New Phytol.*, 72, 141–154, 1973.
Rabinowitz, M. B., *Environ. Health Perspect.*, 91, 33–37, 1991.
Ray, W. J., *J. Biol. Chem.*, 244, 3740–3747, 1969.
Reid, S. D. and McDonald, D. G., *Can. J. Fish. Aquat. Sci.*, 48, 1061–1068, 1991.
Reid, S. D., McDonald, D. G., and Rhem, R. R., *Can. J. Fish. Aquat. Sci.*, 48, 1996–2005, 1991.
Reinfelder, J. R. and Fisher, N. S., *Science*, 251, 794–796, 1991.
Rosenberg, D. W., in *Handbook of Metal-Ligand Interactions in Biological Fluids: Bioinorganic Medicine Vol. 1*, G. Berthon, Ed., Marcel Dekker, New York, 461–466, 1995.
Rossetto, F. E. and Nieboer, E., *J. Inorg. Biochem.*, 54, 167–186, 1994.
Rossetto, F. E., Turnbull, J. D., and Nieboer, E., *Sci. Total Environ.*, 148, 201–206, 1994.
Rossotti, F. J. C. and Rossotti, H., *The Determination of Stability Constants*, McGraw Hill, New York, 1961.
Rowens, B., Guerrero-Betancourt, D., Gotlieb, C. A., Boyes, R. J., and Eichenhorn, M.-S., *Chest*, 99, 185–190, 1991.
Rusting, R. L., *Sci. Am.*, 267(Dec), 130–141, 1992.
Sen, P. and Costa, M., *Toxicol. Appl. Pharmacol.*, 84, 278–285, 1986.
Serfaty-Lacrosnière, C., Rosenberg, I. H., and Wood, R. J., in *Handbook of Metal-Ligand Interactions in Biological Fluids: Bioinorganic Medicine Vol. 1*, G. Berthon, Ed., Marcel Dekker, New York, 322–330, 1995.
Shannon, R. D., *Acta Cryst.*, A32, 751–767, 1976.
Shannon, R. D. and Prewitt, C. T., *Acta Crystallogr.*, B25, 925–946, 1969.
Shannon, R. D. and Prewitt, C. T., *Acta Crystallogr.*, B26, 1046–1048, 1970.
Shi, Z.-C., *Sci. Total Environ.*, 148, 293–298, 1994a.
Shi, Z.-C., *Sci. Total Environ.*, 148, 299–301, 1994b.
Silver, S. and Walderhaug, M., *Microbiol. Rev.*, 56, 195–228, 1992.
Skerfving, S., Gerhardsson, L., Schütz, A., and Svensson, B.-G., in *Handbook of Metal-Ligand Interactions in Biological Fluids: Bioinorganic Medicine Vol. 2*, G. Berthon, Ed., Marcel Dekker, New York, 755–765, 1995.
Smith, R. M. and Marten, A. E., *Critical Stability Constants*, Plenum Press, New York, 1976.
Steffens, J. C., *Annu. Rev. Plant Physiol. Plant Mol. Biol.*, 41, 553–575, 1990.
Stokinger, H. E., in *Patty's Industrial Hygiene and Toxicology*, 3rd revised ed., G. D. Clayton and F. E. Clayton, Eds., Vol. 2A, Toxicology, John Wiley & Sons, New York, 1493–2060, 1981.
Stryer, L., *Biochemistry, 4th Ed.*, W. H. Freeman, New York, 1995.
Stumm, W. and Morgan, J. J., *Aquatic Chemistry: An Introduction Emphasizing Chemical Equilibria in Natural Waters, 2nd Ed.*, John Wiley & Sons, New York, 1981.
Sunderman, F. W., Jr., in *Nickel in the Human Environment*, IARC Sci. Publ. No. 53, F. W. Sunderman Jr., Ed-in-Chief, International Agency for Research on Cancer, Lyon, France, 127–142, 1984.
Sunderman, F. W., Jr., in *Nickel and Human Health. Current Perspectives*, E. Nieboer and J. O. Nriagu, Eds., John Wiley & Sons, New York, 281–293, 1992.
Sunderman, F. W., Jr., Hopfer, S. M., Knight, J. A., McCully, K. S., Cecutti, A. G., Thornhill, P. G., Conway, K., Miller, C., Patierno, S. R., and Costa, M., *Carcinogenesis*, 8, 305–313, 1987.
Sunderman, F. W., Jr., Dingle, B., Hopfer, S. M., and Swift, T., *Am. J. Ind. Med.*, 14, 257–266, 1988.
Sunderman, F. W., Jr., Hopfer, S. M., Sweeney, K. R., Marcus, A. H., Most, B. M., and Creason, J., *Proc. Soc. Exp. Biol. Med.*, 191, 5–11, 1989.
Taylor, G. J., Stadt, K. J., and Dale, M. R. T., *Environ. Exp. Biol.*, 32, 281–293, 1992.
Templeton, D. M. and Chaitu, N., *Toxicology*, 61, 119–133, 1990.
Torjussen, W. and Andersen, I., *Ann. Clin. Lab. Sci.*, 9, 289–298, 1979.
Turner, D. R. and Whitfield, M., *Ecol. Bull. (Stockholm)*, 35, 9–37, 1983.
Turner, D. R., Whitfield, M., and Dickson, A. G., *Geochim. Cosmochim. Acta*, 45, 855–881, 1981.
Utell, M. J. and Samet, J. M., *Am. Rev. Respir. Dis.*, 147, 1334–1335, 1993.

Verity, M. A., *Environ. Health Perspect.*, 89, 43–48, 1990.
Vincent, J. H., *Appl. Occup. Environ. Hyg.*, 8, 233–238, 1993.
Vincent, J. H., *Analyst*, 119, 13–18, 1994a.
Vincent, J. H., *Analyst*, 119, 19–25, 1994b.
Vouk, V., in *Handbook on the Toxicology of Metals, Volume 1*, L. Friberg, G.F. Nordberg and V.B. Vouk, Eds., Elsevier, Amsterdam, 14–35, 1986.
Walsh, C. T., Distefano, M. D., Moore, M. J., Shewchuk, L. M., and Verdine, G. L., *FASEB J.*, 2, 124–130, 1988.
Welch, J., Fogel, S. Buchman, C., and Karin, M., *EMBO J.*, 8, 255–260, 1989.
Wetterhahn-Jennette, K., *Environ. Health Perspect.*, 40, 233–252, 1981.
Wilkinson, K. J., Campbell, P. G. C., and Couture, P., *Can. J. Fish. Aquat. Sci.*, 47, 1446–1452, 1990.
Wilkinson, K. J., Bertsch, P. M., Jagoe, C. H., and Campbell, P. G. C., *Environ. Sci. Technol.*, 27, 1132–1138, 1993.
Williams, R. J. P., *Inorg. Chim. Acta Rev.*, 5, 137–155, 1971.
Williams, R. J. P., *Trans. Biochem. Soc.*, 1, 1–29, 1973.
Williams, R. J. P., *Proc. R. Soc. Lond.*, B213, 361–397, 1981a.
Williams, R. J. P., *Phil. Trans. R. Soc. Lond.*, B294, 57–74, 1981b.
Williams, R. J. P., *Phil. Trans. R. Soc. Lond.*, B 304, 411–424, 1984.
Williams, R. J. P. and Hale, J. D., *Struct. Bond. (Berlin)*, 1, 249–281, 1966.
Yassi, A. and Nieboer, E., in *Chromium in the Natural and Human Environments (Advances in Environmental Science and Technology, Vol. 20)*, J. O. Nriagu and E. Nieboer, Eds., John Wiley & Sons, New York, 443–495, 1988.

Chapter 8

Metals and Biological Membranes

Ernest C. Foulkes

I. INTRODUCTION

Heavy metals in solution are highly reactive and readily combine with many biological molecules, including constituents of cell membranes such as proteins and phospholipids. Reaction with proteins is determined by

1. Electric charge: a cationic metal is attracted to proteins by their usual electronegativity (Perkins, 1981), with metals binding to carboxy and other anionic groups.
2. Presence of chelating structures such as SH groups, imidazole residues, etc.
3. The tertiary structure of proteins, as illustrated by the specific complex formation of Cd with the cysteine clusters in metallothionein (see, e.g., Elinder and Nordberg, 1985).

Interaction between metals and phospholipids involves primarily phosphate residues.

Several predictions can be made about heavy metals on the basis of their general reactivity in biological systems. They are obviously not likely to be able to enter cells by ionic diffusion without reacting with cell membranes; this is true even for experiments *in vitro* where cells are exposed to inorganic metal compounds. *In vivo* the metals, except perhaps very transiently, will always be found as complexes with a great variety of ligands. These include diffusible molecules like glutathione, cysteine and other amino acids, metabolites including citric acid and others, inorganic anions as in the case of uranyl bicarbonate, etc. Macromolecules such as those constituting the cell membranes compete for the metals with the diffusible ligands.

Metals bound to membranes are likely to modify membrane function, for instance, inhibiting transmembrane transport of electrolytes, sugars, amino acids, and other solutes. This may result directly from reaction of the toxic metal with transport systems, but could also reflect an indirect effect related to alterations in membrane properties (see Section III). In addition to effects exerted at the membrane, metals may also indirectly alter membrane function by interfering with the supply of metabolic energy and with other vital cell processes.

This chapter focuses on the reactions of heavy metals with plasma (or outer) membranes from (mostly) animal cells, as well as on the alterations in membrane function resulting directly or indirectly from such interactions. Not surprisingly, subcellular membrane structures including mitochondrial or nuclear membranes, cytoplasmic reticulum, etc. also appear to react with metals. Thus, it can be calculated from the work of Zhang and Lindup (1993) that the water content of mitochondria isolated from slices of rat renal cortex which had been exposed to cisplatin is increased above control values by one third; swollen mitochondria, of course, are commonly observed in toxic states. While cisplatin treatment in these studies presumably altered the permeability of subcellular membranes, little is known of the mechanisms of these metal-membrane interactions, and they will not be further considered here.

The main emphasis here is placed on toxic and nonessential metals (TNEM) like Cd, Hg, and Pb. However, because this topic covers a very broad field, no attempt was made to cover the subject exhaustively. Instead, suitable references were selected to illustrate the overall significance of cell membranes to metal toxicology. The review first shortly summarizes the major relevant properties of these membranes (Section II). Their reaction with metals may directly involve enzymes or transporters, or it may alter function as a result of less specific changes in general membrane properties (see Section III). Toxic metals may also act primarily on basic intracellular processes, as pointed out above, and thereby indirectly alter membrane function; this is discussed in Section IV. Transient reaction of a metal with the membrane as the first step in cellular metal uptake is considered in Section V. Section VI deals with cellular retention of metals and their ultimate extrusion across the cell membranes.

II. NATURE AND FUNCTION OF MEMBRANES

Detailed discussion of the chemical composition and the variety of functions of cell membranes lies beyond the scope of this chapter. Briefly, however, the basic structure of the plasma membrane enclosing the cellular cytoplasm is a bilayer of lipids and proteins. The structure is pierced by polar pores, whose contribution to metal movement, however, remains uncertain (Foulkes and Bergman, 1993). The presence of transmembrane proteins, complex carbohydrates, sialic acid, phospholipids and other compounds all contribute to the complexity of the structure and usually give it a net electronegative charge. Relatively high charge densities on cell membranes have been reported. For instance, Hanck and Sheets (1992) calculated a value of $0.72 \times 10^{14}/cm^2$ for the density of anionic sites on canine cardiac Purkinje cells. Heavy metals can therefore bind to membranes by electrostatic interaction, as illustrated for instance in the NMR study of Tacnet et al. (1991) on the binding of Cd and Zn to intestinal brush border phosphatidylinositol and phosphatidylserine. Alternatively, metals may be chelated at the membrane by a variety of reactive sites including sulfhydryl, imidazole, and other residues (see, e.g., Rothstein, 1959).

Electrostatic interaction with fixed anionic charges is, of course, not relevant for metals when these exist in solution in the form of anionic complexes. One such metal is mercury: 44% of mercuric ion in the presence of chloride was reported to be present as $HgCl_4^{2-}$ and 28% as $HgCl_3^-$ (Clarkson and Cross, 1961). A net negative charge on the membrane will tend to decrease its ability to react with metals present in anionic complexes. Such an explanation was offered for the observation that neutralization of fixed anionic charges with La^{3+} depresses binding of Cd to the jejunal brush border (Foulkes, 1985), but increases that of Hg (Foulkes and Bergman, 1993). However, transmembrane movement of Hg into mucosal cells does not appear to utilize anion channels, as judged from its insensitivity to the channel inhibitor diisocyanatostilbene (DIDS). Contrary findings have been reported with erythrocytes where uptake of several heavy metals is inhibited by DIDS (Simons, 1986; Lou et al., 1991).

Active membrane function is associated with a variety of receptors, binding sites, membrane-bound enzymes, and solute transport mechanisms. As a result, agonists such as hormones and antagonists and inhibitors including heavy metals in some cases appear to regulate function directly at the membrane. For instance, Wada et al. (1991) reported that cadmium acts directly on endothelin receptors to inhibit binding of this peptide. However, the once widely held view that the membrane is the primary site of action of many of these agents (see, e.g., Kinter and Pritchard, 1977) is now known to be too simple (cf. Section IV). Instead, the membrane reaction frequently serves only as signal triggering a cascade of subsequent events; these may ultimately and only indirectly lead to functional alterations at the membrane.

III. METAL BINDING AND ITS EFFECTS ON MEMBRANE STRUCTURE AND FUNCTION

As discussed in the previous section, heavy metals readily bind to cell membranes. If the binding sites form part of some specific carrier, an immediate inhibition of a particular function would be expected. The metal may, however, also bind at some distal site on the membrane, and alter function indirectly as a consequence of nonspecific changes in membrane properties. This could, for instance, involve lipid peroxidation, as reviewed by Christie and Costa (1984), or result from redistribution of electric charges. Other changes in physicochemical characteristics of the membrane may also occur, as illustrated by metal effects on the mechanical and osmotic fragility of erythrocytes (see, e.g., Valley and Ullmer, 1972).

An interesting illustration of these concepts is provided by the work of Jourd'heuil et al. (1993). These authors reported that exposure of membrane vesicles from the intestinal brush border to Fe^{2+} in the presence of ascorbic acid inhibits glucose transport by two different mechanisms: one third of the inhibition was attributed to changes in the rigidity of the membrane, and could be reversed upon restoring its normal fluidity; the remaining two thirds of the inhibition were believed to represent direct effects on glucose carriers. Another possible mechanism whereby a metal could alter membrane transport of certain solutes would be through an effect on passive membrane permeability. This, in turn, might lead to dissipation of the solute gradients essential for various coupled transport processes.

Some binding sites on the membrane are easily accessible to, e.g., external chelators, while others lie relatively deeply buried in the membrane structure, or may even be located on the inside of the membrane. Such an explanation has been suggested for the observation that only a portion of the Cd accumulated by purified membrane vesicles from the jejunal brush border can subsequently be removed by EDTA (Bevan and Foulkes, 1989). The suggestion is based on the further finding that osmotic shrinkage of the vesicles does not influence Cd uptake, so that the metal must be accumulated in bound rather than osmotically free form. In absence of cell constituents not related to the membrane, the vesicular accumulation of Cd can thus be equated to binding of the metal to the membrane, in part at a site not readily accessible to EDTA. It is only after destruction of the vesicles that all the accumulated metal can be extracted with the chelator.

In intact cells, on the other hand, internalized metals readily react with cytoplasmic constituents and subcellular organelles. By analogy with vesicle preparations, the portion of Cd or Hg accumulated by cells in a form freely accessible to suitable extracellular and nonpenetrating chelators presumably consists of metal bound to the outside of the membrane. The remaining portion of heavy metals in intact cells, as in the case of vesicles, is readily extracted by these chelators only after destruction of the cell membrane (Foulkes, 1988). These findings provide the basis for the operational definition of two metal compartments, one externally bound to the membrane, and the second metal sequestered deep in the membrane, or bound to the inside of the membrane, or accumulated in the cell. The nature of the binding and subsequent internalization process is discussed in Section V.

The ability of an external chelator to remove metals from the membrane depends on its ability to reach the bound metals and on the relative metal affinities of the chelator and of the membrane binding sites. Thus, while EDTA removes Cd from the membrane, it cannot dissociate membrane-bound Hg. Dimercaptosuccinate (DMSA), on the other hand, has been used to separate externally bound from internalized Hg (Foulkes and Bergman, 1993). Short exposure of Hg-containing cells to this agent in the cold distinguishes between a readily removable Hg pool whose filling with extracellular Hg is independent of temperature, and a second, more DMSA-resistant compartment, metal uptake into which is significantly faster at 37°C than in the cold (see Section V).

The role of the membrane barrier in restricting access of a heavy metal to some of its potential sites of action is illustrated by the report that large nonpenetrating mercurials may be inactive, while the time lag before smaller molecules begin to affect function can in certain cases be abolished by membrane disruption (Kinter and Pritchard, 1977). The proposed explanation invoked metal binding to critical sites deep within the membrane. An alternative explanation for time lags would be to assume an indirect metal action in this instance, a possibility further considered in Section IV.

Rothstein and Hayes (1956) had already concluded from their work on metabolic inhibition by externally bound metals that these metals exert direct effects on membrane function. They showed that metal cations, especially UO_2^{2+}, are firmly bound to external binding sites on yeast cell membranes; provisionally these binding sites were identified as carboxy or phosphate groups. Mercury also rapidly reacts with the rat diaphragm, where it initially localizes at the cell membrane (Demis and Rothstein, 1955).

Many examples of direct metal action on membrane function have been described in the literature. For instance, mercury *in vitro* appears to release the α-subunit of Na,K,-ATPase from its membrane anchor (Imesch et al., 1992). The results of Sellinger et al. (1991) also point to a direct and specific toxic action of Hg on solute transport at the membrane. These authors worked with membrane vesicles from hepatocytes of *Raja erinacea* and found that under conditions where Hg exerts no effect on Na gradients and on the membrane permeability to Na, it strongly inhibits isotopic exchange of amino acids across the membranes. Preston and Chen (1989) similarly observed that low concentrations of Hg inhibit the active accumulation of taurine in the nucleated red cells of the marine polychaete *Glycera dibranchiata,* without altering the membrane potential, fluxes of Na and K across the membrane, or the maintenance of cell volume. The effect could be reversed by treatment with various thiol compounds such as dithiothreitol, presumably by removing the metal from its site of action (Preston et al., 1991). Grosso and De Sousa (1993) reported that 1 mM $HgCl_2$ reverses the high water permeability induced in toad skin by vasopressin. The authors also referred to earlier reports on the effects of mercury on water permeability of membranes from other tissues; the action of the metal in all these instances is believed to represent a direct effect on membranes.

Metal inhibition of a membrane enzyme on the "cytoplasmic" side was described by Anner and Moosmayer (1992) in their work on the Hg inhibition of purified Na,K-ATPase in reconstituted liposomes. Activity of this transport ATPase, well known to be sensitive to mercurials *in vitro,* was monitored by transport of ^{86}Rb; at low concentration (10 μM), Hg^{2+} inhibited Rb transport only at the cytoplasmic side. It is important to reiterate, however, that many proteins react with heavy metals *in vitro,* so that metal sensitivity of purified enzymes cannot be taken as proof that they are specific metal targets *in vivo.*

Another apparently direct toxic effect of a heavy metal on membrane function was described by Verbost et al. (1987). Nanomolar concentrations of Cd were found to competitively inhibit ATP-driven Ca transport in basolateral membrane vesicles from rat duodenum. The reaction appears to involve Ca-binding sites on the enzyme itself (Verbost et al., 1988). A final example of presumably direct membrane action of a metal is the dose-dependent Cd inhibition of 5'-nucleotidase in the liver of Cd-treated rats (Morselt, 1991). The effect of metals on this outwardly facing enzyme (ectoenzyme) has also been studied *in vitro:* Stefanovic et al. (1976) reported that a nominal level of 1 mM Cd or Hg reduces enzyme activity in cultured glioma cells by only about one third. However, the significance of this finding is obscured by the use of complex and relatively alkaline media in which the concentration of ionized metal can be expected to have been much lower than 1 mM.

Other membrane-associated enzymes have also been reported to be metal sensitive. Thus, Aggarwal and Niroomand-Rad (1983) described the inactivation of various phosphatases on the plasma membrane of ascites tumor cells by cisplatin. The inhibition develops relatively slowly, over a period of 30 to 60 min. Even if we accept the likelihood that the action is exerted directly on the membrane, it is difficult to decide whether it represents a direct but slow interaction between the enzymes and the metal complex. Conceivably, the Pt compound here acts on the membrane but not directly with the enzyme molecules (see Section IV). Metal action in such instances, though initiated at the membrane, would therefore represent an indirect effect on the specific membrane function. Actions of metals indirectly exerted on the membrane following a primary lesion elsewhere in the cell are further considered in the next section.

Transient binding of metals to membranes during their uptake into cells has been studied, especially in epithelial cells of the jejunal mucosa, and is further discussed in Section V. Only a fraction of metal which has reacted with the membrane may, however, subsequently reach the cell interior. This was described for uranyl ions on the yeast membrane (Rothstein and Hayes, 1956), and is true also of Cd bound to the apical membrane of jejunal epithelium (Foulkes, 1988). The Cd studies are illustrated in Figure 1. In these experiments, everted sacs of rat jejunum had been shortly exposed to 20 μM $CdCl_2$, and then rinsed in cold saline; next the tissues were maintained for 60 s in saline, either in the cold or at 37°C. After cold saline treatment, 69% of tissue Cd could be rapidly removed by EDTA. In contrast, at 37°C, a significant portion of the EDTA-sensitive Cd was transferred into a chelator-resistant pool, so that only 45% of tissue Cd remained accessible to EDTA.

The nature of the two Cd compartments on the membrane, i.e., that which readily traverses the membrane and that which does not, remains unclear; no kinetic evidence for compartmentation of membrane-bound Cd could be found at the brush border of intact epithelium (Foulkes, 1991a). On the other hand, Scatchard plots revealed the presence of two compartments of externally bound Zn in membrane vesicles from pig jejunum (Tacnet et al., 1991a, see Section V). It is not clear, however, whether such compartmentation results from inhomogeneity of the vesicle preparations.

Unlike in the mucosa of the rat jejunum, no membrane-bound and EDTA-sensitive fraction of Cd could be demonstrated in rabbit kidney *in vivo* (Foulkes and Blanck, 1990). This may well be due, however, to the relatively poor time resolution of experiments with whole animals, as contrasted to the fact that in everted sacs rapid processes can readily be followed over intervals as short as 5 s.

Where two or more EDTA-sensitive metal compartments exist on the membrane, they might differ not only in their ability to subsequently cross into the cells, but also in their effect on various membrane functions. As a minimum, binding of polyvalent metals will alter the electrical charge distribution on the membrane, and could thereby influence reactions of cells with their environment.

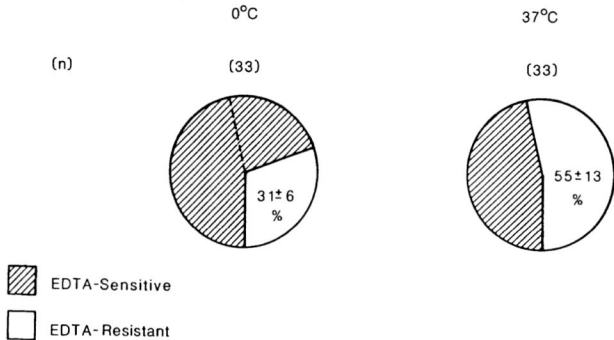

Figure 1 Internalization of membrane-bound Cd. Everted sacs of rat jejunum had been exposed to ^{109}Cd for 10 s at 37°C, then rapidly washed with cold saline and further incubated in saline for 60 s at the stated temperature. They were next extracted with 5 mM EDTA-saline in the cold for 60 s; finally, radioactivities of extracts and tissues were determined. (Reprinted from Foulkes, E.C., *Toxicology*, 52, 263–272, 1988. With permission.)

IV. DIRECT VS. INDIRECT MEMBRANE EFFECTS OF METALS

In spite of the evidence discussed in Section III that metals may exert toxic effects directly on the cell membrane, this is clearly not always the immediate site of metal action. Instead, the reaction of an agent at the membrane frequently constitutes only the first step in a series of events (signal transduction), both at the membrane and within cells. Changes ultimately observed in membrane function following cellular contact with, e.g., metals, may therefore result indirectly from actions elsewhere in the cells.

It may be difficult at times to distinguish direct from indirect metal actions on membrane function. An example is the report of Takada and Hayashi (1980) that Cd amplifies the short circuit current in frog skin. The toxicological significance of this finding is obscured by the fact that high Cd concentration in SO_4-Ringer's solution was used. Hillyard and Gonick (1976) attributed the Cd stimulation to an increased permeability of apical membranes. The likelihood of indirect effects on membrane function may be suspected when they can be observed only after a significant time lag following exposure of the target cells. Delays preceding appearance of overt effects following administration of metals *in vivo* may, of course, arise from the time required for the metal to reach a critical concentration at its site of action. Where such an explanation is not appropriate, as for instance when luminal Cd inhibits amino acid absorption from perfused jejunum only after a significant delay (Foulkes and Blanck, 1991), the time lag is normally not compatible with a direct action of metals on a membrane carrier system. This point was previously emphasized, e.g., by Yuile (1973). An indirect toxic effect is also suggested when it can be prevented by inhibitors of protein synthesis like oligomycin, or when, as in the case of the nephrotoxic action of Hg, the immune system can be shown to be involved.

Hypothetically, indirect effects on membrane function could result from interference with intracellular processes such as the production of available metabolic energy. Inhibition of the energy supply would then indirectly reduce the active transport capacity of the cell membrane. Even more indirectly, and speculatively, a primary effect on energy metabolism could conceivably result in changes of cell volume and thereby in membrane geometry and function; stretch effects on membrane function are well known. Similarly, reaction of a metal with the cytoskeleton (e.g., Mills and Ferm, 1989) could alter the spatial structure of the membrane.

Differences between direct and indirect actions of metals on membrane function may be illustrated by a comparison of effects of Cd and of Hg on amino acid transport, both in renal and in jejunal epithelium. In the intact rabbit, Cd inhibits aspartate uptake across the basolateral cell membrane of the renal tubule only after as long as 12 h following intraarterial injection (Foulkes and Blanck, 1990). In contrast, equivalent doses of Hg strongly inhibit amino acid uptake within minutes of administration (Foulkes, 1991). Similarly, in perfused rat jejunum, addition of 0.5 mM Hg immediately suppressed absorption of the amino acid analogue cycloleucine, whereas the effects of the same concentration of Cd required 1 h to develop. In another study of the cadmium inhibition of basolateral solute fluxes in the rabbit kidney (Foulkes and Blanck, 1991), the metal was found to inhibit active uptake as much as the passive backflux for both cycloleucine and *p*-aminohippurate. Such an unexpected finding is difficult to reconcile with direct transport inhibition, but could be explained on the assumption that here Cd modifies the properties of the cell membrane in such a way that solute movement becomes indirectly and perhaps nonspecifically depressed (see also Section III).

Reference has already been made to the fact that metal sensitivity of an isolated enzyme *in vitro* should not be taken as evidence that this enzyme necessarily constitutes the site of toxic action of the metal in the intact tissue. Thus, the toxicological significance of the observation that membrane Na,K,-ATPase is inhibited by mercurials *in vitro* (e.g., Nechay et al., 1967) remains somewhat uncertain. Membrane action of metals in some cases probably does not involve direct effects on solute transport systems at all. For instance, Cd has been shown to interact with membrane receptors for atrial natriuretic peptide (Giridhar et al., 1992), and could in this indirect fashion greatly influence normal physiological homeostasis.

V. METAL-MEMBRANE INTERACTION DURING METAL UPTAKE

Cellular uptake of heavy metals generally involves their interaction with the cell membrane. If the metal is present in the extracellular space in tightly complexed form, it may not be able to react with the membrane and may therefore be excluded from the cell. For instance, Cd-EDTA is handled by the kidney like an extracellular marker (Foulkes, 1974), filtered and excreted as is insulin; the same result had previously been reported for Cr-EDTA (Stacy and Thornburn, 1966). The Cd complex is also entirely inert in the intestine (Foulkes, 1991). In contrast, if binding sites on the membrane possess a sufficiently high affinity for the metal, they can compete for it with the chelator. This is illustrated by the fact that Hg-EDTA, unlike Cd-EDTA, readily reacts with renal tubular epithelium and with the intestinal mucosa (Foulkes, 1991). EDTA does not interfere with renal Hg uptake, presumably because it does not possess as high an affinity for the metal as do Hg-binding sites on the epithelial cell membranes. This constitutes, of course, a corollary of the inability of EDTA to remove Hg from membranes, as already discussed in Section III.

Many probable exceptions, however, may be cited from the proposed requirement that a polyvalent metal must be able to react with membranes before it can penetrate cells. An obvious instance is the uptake of relatively lipid-soluble metal complexes, such as, in part, methylmercury (see below). Another exception applies to metals bound very tightly in a nonreactive form in compounds which are transported by more or less specific membrane carrier mechanisms. Examples here are the tubular secretion of the mercurial diuretic chlormerodrin in the dog kidney (Borghgraef et al., 1956), and the presence in the proximal renal tubule of a transport system specific for anionic proteins such as Cd-metallothionein and myoglobin (Foulkes, 1978).

Similarly, transmembrane movement of anionic metal compounds through anion channels, as in the case of the DIDS-sensitive metal uptake in erythrocytes described by Simons (1986) (see Section II), presumably does not require binding of the metal to the membrane. This is likely to be true also of anionic compounds of hexavalent Cr, whose translocation across membranes resembles that of, e.g., phosphate (Alexander et al., 1982). In this regard, Cr^{6+} differs from Cr^{3+}; the latter reacts with the membrane charges involved in Cd uptake (Foulkes, 1985). Both the suggested role of glutathione in renal Hg uptake (Tanaka et al., 1990), and the apparent transport of methylmercury-cysteine by an amino acid carrier system in the intestine (Hirayama, 1975) and at the blood-brain barrier (Kerper et al., 1992), support the conclusion that some transmembrane movement of metals may proceed independent of direct metal-membrane interactions.

Exploration of the role of glutathione and other low molecular weight thiols in transmembrane metal transport *in vivo* is complicated by the fact that a large portion of their action undoubtedly is involved simply in preventing metal sequestration by plasma protein (Foulkes, 1974). Binding of the metals to diffusible ligands permits the metals in blood plasma to remain diffusible and filterable, and able to approach the renal and other cell membranes. What this hypothesis does not explain is why in the case of Cd-cysteine, for instance, it is specifically renal rather than, e.g., hepatic cells which accumulate Cd *in vivo*. Why is the major portion of the body burden of Cd and Hg found in the renal cortex? A hypothesis which has not yet been tested is that membranes of renal cells differ from hepatocyte membranes in some important property relevant to metal transport. Among evidence for the existence of specific membrane mechanisms involved in metal uptake by target organs is the aforementioned presence of a transport mechanism for metallothionein and other anionic proteins in the brush border of the renal tubule (Foulkes, 1978). Activity of such a system, however, could explain preferential Cd-metallothionein uptake by the kidney only if the liver and other organs do not possess a similar capacity.

In the kidney, Cd uptake in presence of excess mercaptoethanol is inhibited by Zn (Foulkes and Blanck, 1990). It is not intuitatively obvious how Zn could interfere with passive diffusion of a somewhat lipid-soluble complex. Instead, this observation suggests again that mercaptoethanol simply serves here to permit the metal to diffuse to the membrane; consequently, membrane binding sites can compete for the metal with the monothiol. The action of Zn here may resemble that seen in the jejunal mucosa, where Zn depresses apical Cd uptake by neutralizing fixed anionic charges on the membrane (Foulkes, 1985).

Uptake of metals has been studied in a variety of cell systems *in vitro,* including erythrocytes (e.g., Garty et al., 1986; Frame and Milanick, 1991), hepatocytes (e.g., Blazka and Shaikh, 1991), cells from Chinese hamster ovaries (Corrigan and Huang, 1981), kidney (Templeton, 1990), pituitary (Hinkle et al., 1987), etc. The variety of mechanisms proposed suggests that uptake mechanisms in various cell types may not all be identical. Presumably, however, in many of these cases the metal reacts with the membrane as a required first step in its uptake. This interaction between metals and membranes has been studied convincingly with purified membrane vesicles from various cell types. For instance, Victery et al. (1984) reported that Pb binds to purified brush border membrane vesicles from rat kidney; the relationship of this lead-binding to absorption of the metal from the tubular lumen remains uncertain. Bevan and Foulkes (1989) noted the similarities between Cd uptake by intact rat jejunum and by brush border preparations. Cd uptake in the two systems is temperature dependent and leads to accumulation in both an EDTA-sensitive and an EDTA-insensitive pool; the amount of Cd externally bound to the vesicles, i.e., in the EDTA-sensitive pool, could be correlated with inhibition of glucose transport.

Essential metals like Zn also react with membranes. It is not clear whether the reaction seen with Zn, for instance, forms part of the specific and homeostatically controlled Zn uptake mechanism, or whether it represents an additional and nonspecific reaction similar to that of the toxic and nonessential heavy metals. Like Ca^{2+}, La^{3+}, and other polyvalent cations, Zn noncompetitively inhibits Cd uptake by jejunal mucosal cells (Foulkes, 1985; see also below), presumably by neutralization of fixed charges on the membrane. Binding of Zn to membranes was observed in vesicle preparations from the intestinal mucosa by Menard and Cousins (1983). Tacnet et al. (1991a) described the presence of two Zn compartments in brush border vesicles from pig jejunum: one was chelator-sensitive and probably corresponds to externally bound metal; a second compartment contained Zn in a form not accessible to chelation and may be assumed to represent internalized metal.

A great advantage of vesicles for the study of metal-membrane interactions derives from the detailed control over metal concentration and speciation in the suspending medium. This advantage is lost in the study of metal uptake in blood-perfused organs. *In vivo,* the difficulty can be avoided by studying metal uptake from the lumen of microperfused renal tubules, as reported for instance for Cd by Felley-Bosco and Diezi (1987) and for Hg by Zalups et al. (1991). Useful results have also been obtained by following the process of metal absorption from the perfused intestinal lumen; this is described in more detail below. It is interesting to see to what extent the processes in these two systems are similar. In both instances, the reaction of Cd with cell membranes, followed by its subsequent transfer into the cells, is diffusion-limited and depressed by chelators like cysteine and similar compounds.

The process of intestinal metal absorption is particularly relevant to a discussion of metal-membrane interactions because in adult animals the metals move along transmembrane and transcellular rather than intercellular diffusion pathways (Foulkes and McMullen, 1987). Metal absorption has frequently been reviewed in the literature (see, e.g., Foulkes, 1984). The preferred transcellular absorption can be inferred from the observations that cytoplasmic metallothionein completely traps low concentrations of Cd removed from the intestinal lumen (Foulkes and McMullen, 1986), and that absorbed Cd passes through the total tissue store of the metal (Foulkes and McMullen, 1987). To assume that Cd does not follow transcellular pathways during absorption would have to imply that significant extracellular pools of metallothionein and Cd must be present in the tissue; this seems unlikely.

A model was proposed for the interaction between Cd and apical brush border membranes during intestinal absorption (Foulkes, 1988). This process, in brief, is believed to involve the following events. The metal first binds to the membrane by apparently largely electrostatic interactions, the so-called step 1A of absorption. Presence of significant numbers of fixed anionic charges on the cell membrane is well known (e.g. Hanck and Sheets, 1992; see Section II). This binding step is relatively rapid and independent of temperature (Foulkes, 1991a). The reaction is noncompetitively inhibited by all polyvalent cations tested, probably as a result of charge neutralization (Foulkes, 1985). Even the apparent saturability of the process may be attributed to charge effects, as it cannot be observed if the total divalent cation concentration is maintained constant by addition of a second heavy metal (Foulkes, 1991a). In contrast to the cation sensitivity of Cd uptake, step 1A of Hg apparently involves the metal in an anionic form, so that its reaction with the membrane is accelerated when other cations reduce the net negative charge on the membrane (Foulkes and Bergman, 1993).

As illustrated in Figure 1, portions of bound metal may then cross the membrane by a temperature-sensitive process (internalization) not requiring a supply of metabolic energy (step 1B) (Foulkes, 1991a; Foulkes and Bergman, 1993). In the cold, accordingly, step 1B is relatively slow, permitting the transient accumulation of, e.g., Hg or Cd in the chelation-sensitive compartment on the membrane. The characteristics of step 1B suggested that membrane fluidity might well play a crucial role, a hypothesis further supported by the finding that Cd uptake by purified membrane vesicles is accelerated by n-butanol (Bevan and Foulkes, 1989).

Note that this scheme can account for the apparent saturability, the metal inhibition, and the temperature dependence of the transfer of Cd and Hg across the brush border membrane in rat jejunum; it probably applies also to other toxic and nonessential heavy metals (TNEM) (Foulkes and Bergman, 1993). There is no need for making the biologically meaningless assumption that specific carrier systems have evolved to mediate TNEM absorption. Instead, relatively nonspecific physicochemical properties of the cell membrane (charge and fluidity) suffice to account for TNEM uptake. The concept of homeostasis is not applicable to uptake of TNEM, and the proposed model does not allow for it. Absence of homeostatic control over intestinal transport of TNEM was confirmed by Cotzias et al. (1961), who found that Cd absorption from the gut is independent of its body burden. Similarly, absorption of Pb is not affected by the body burden of this metal in the rat (Conrad and Barton, 1978).

The finding of significant similarities between the mechanisms responsible for Cd and Hg uptake further supports the hypothesis that TNEM move across the jejunal brush border by processes based on nonspecific membrane properties. It is interesting in this connection that Ni is also taken up in a manner closely resembling that of Cd (Foulkes and McMullen, 1986a).

The proposed transfer mechanism of heavy metals across the jejunal brush border in the mature rat is not fully applicable to immature animals. The high fractional absorption and retention of Cd in the neonatal gut was previously noted, for instance, by Sasser and Jarboe (1977). Comparisons of intestinal function in mature and immature animals cannot always be readily interpreted: the calculated rates of these processes are influenced by differences in such variables as diet, tissue thickness, density of microvilli, etc. Comparative rates of absorption therefore need to be expressed on the basis of surface area rather than weight or length of tissue. For this purpose, advantage was taken of the fact that the rate of ethanol absorption is directly related to surface area available for absorption, while being independent of surface composition (Foulkes et al., 1991). The relative intrinsic permeability of jejunal membranes accordingly can be defined by the ratios of unidirectional transmembrane fluxes of a solute to that of ethanol. Table 1 shows in arbitrary units the permeability of immature and mature jejunum to metals, an amino acid (cycloleucine), and to urea. Permeability to Cd dropped around 40% with maturation, that for Hg by 54%, whereas that for cycloleucine did not appreciably change. At the same time, the intrinsic permeability to urea, a measure of the presence of polar pores in the membrane (Macey and Farmer, 1970), dropped by 58%.

Table 1 Intrinsic Permeabilities of Brush Border in Mature and Immature Rat Jejunum

		Weanlings	Mature rats	Δ(%)
A	Cadmium	3.4	1.9	−44
	Cycloleucine	1.5	1.6	+7
B	Cadmium	0.94	0.60	−36
	Mercury	0.85	0.39	−54
	Urea	0.43	0.18	−58

Notes: Permeabilities are expressed by the ratio of solute to ethanol fluxes.
  A: Tied-off segments, Foulkes et al. 1991.
  B: Perfused segments, Foulkes and Bergman, 1993.

The fall in metal permeability with age may be related to the loss in membrane fluidity (Israel et al., 1987) and/or the reduction in polar diffusion channels, possibly including intercellular pathways (Foulkes and Bergman, 1993). The same factors might well explain the higher intrinsic metal permeability of alveolar than of mature jejunal epithelium, as reflected in the greater fractional absorption of heavy metals from the alveolar space than from the intestine (see, e.g., Nordberg et al., 1985).

The reaction of essential metals with cell membranes is likely to be more complex than that of the nonessential toxic heavy metals. Indeed, specific and physiologically controlled mechanisms are well known to be responsible for uptake of essential metals like Ca and Zn. In some systems, the toxic metals appear to react with uptake mechanisms for essential metals. For instance, Cd competes with Ca for voltage-gated Ca channels in secretory cells from the pituitary (Hinkle et al., 1987). Concentrations of Cd as low as 0.3 μM block Ca channel currents in single cells of guinea pig taenia caeci (Lang and Paul, 1991).

However, in rat jejunum no evidence could be found for reactions between toxic heavy metals and homeostatically controlled transport mechanisms for absorption of Ca and Zn. The two essential metals do inhibit Cd uptake, but the inhibition is noncompetitive and therefore more adequately explained as above, in terms of neutralization of fixed anionic membrane charges. Moreover, the inhibition of Cd uptake by La^{3+} kinetically resembles that caused by Ca and Zn (Foulkes, 1985), and no suggestion has ever been made that La movement is mediated by specific mechanisms. Additional evidence against a role of Zn carrier mechanisms in Cd uptake is provided by the observation that the increased Zn uptake from the lumen of the perfused jejunum in Zn-deficient rats is not accompanied by increased Cd uptake (Foulkes and Voner, 1981). It must be added that this conclusion, based on experiments with the proximal portion of the jejunum, was not confirmed by Hoadley and Cousins (1985), using the whole length of the doubly perfused rat small intestine. In that preparation an inverse relationship between Zn nutrition and Cd absorption was reported.

VI. CELLULAR TRAPPING AND EXTRUSION OF HEAVY METALS

Comparison of the kinetics of Cd, Ni, and Zn transferred across the mucosal epithelium in rat jejunum showed that the transfer of these metals from the intestinal lumen to the portal vein is delayed in direct correlation with their affinity for metallothionein (Foulkes and McMullen, 1987); metallothionein is the well-known low molecular weight metal binding protein present in the cytoplasm of most cells. Trapping of metals in the mucosal cells provides an important means of protecting the body against toxic metals, given the relatively short half-life of mucosal cells and the consequent loss of their contents from the body upon desquamation. The role of metallothionein in controlling Cd absorption across the intestinal barrier was directly demonstrated in experiments in which movement of the metal from the cell into the body (cellular Cd extrusion) could be reduced by increasing the metallothionein content of the cells (Foulkes and McMullen, 1986). By contrast, mucosal retention of Hg was little affected under these conditions. This could be predicted from the fact that, unlike for Cd, the high affinity of Hg for thiol groups is not specific for metallothionein (Foulkes and Bergman, 1993).

Metal extrusion from the mucosal cell across the basolateral cell membrane has been labeled step 2 of the absorption process. Relatively little is known of its nature, but results available suggest that it consists of passive transmembrane diffusion of complexes with low molecular weight ligands such as glutathione (GSH) (Foulkes, 1993). Because the affinity of Cd for monothiols like GSH is much less than that for metallothionein, Cd is effectively trapped in cells. In contrast, the affinity of Hg for monothiols more closely approaches that for metallothionein, and Hg is accordingly more readily mobilized than Cd. These findings can also offer an explanation for the much longer half-life of Cd in renal cortex than of Hg. The proposed role of metallothionein in renal Cd accumulation is fully compatible with the short half-life of the protein moiety (Cherian and Shaikh, 1975); presumably, Cd set free and/or Zn displaced from Zn-metallothionein by Cd continue to induce metallothionein synthesis, thus maintaining an elevated level of the protein in the cytoplasm.

VII. CONCLUSIONS

Knowledge of how heavy metals react with and cross cell membranes is fundamental to the understanding of their toxic action. While some membrane-bound metal remains relatively inert, bound metals may directly affect membrane structure and function. Alternatively, transient binding could represent a step in metal uptake into cells. Unless internalized metals are sequestered in inactive form, as is probably the case with the Cd complex of endogenous metallothionein, they are likely to exert widespread toxic effects. This, in turn, could indirectly alter function at the membrane, for instance when inhibition of energy metabolism depresses active solute transport across the membrane or where membrane geometry is altered following changes in cell volume or interference with the cytoskeleton.

Extrusion of intracellular metals, for instance during the process of intestinal absorption, may not involve specific reactions with cell membranes but appears to consist of passive diffusion of low molecular weight complexes with glutathione and similar compounds. Cellular metal uptake and extrusion assume special significance in light of the conclusion that absorption of heavy metals in the mature rat jejunum follows transcellular diffusion pathways. In summary, a satisfactory description of the effects of heavy metals in the body therefore demands an understanding both of their direct and indirect (or specific and nonspecific) structural and functional effects on membranes, as well as of the role of cell membranes in metal toxicokinetics.

REFERENCES

Aggarwal, S. K. and Niroomand-Rad, I., *J. Histochem. Cytochem.*, 31, 307–317, 1983.
Alexander, J., Aaseth, J., and Norseth, T., *Toxicology*, 24, 115–122, 1982.
Anner, B. M. and Moosmayer, M., *Am. J. Physiol.*, 262, F843–F848, 1992.
Bernard, A., Lauwerys, R., and Amor, A., *Arch. Toxicol.*, 66, 272–278, 1992.
Bevan, C. and Foulkes, E. C., *Toxicology*, 54, 297–309, 1989.
Blazka, M. E. and Shaikh, Z. A., *Toxicol. Appl. Pharmacol.*, 110, 355–363, 1991.
Borghgraef, R. R. M., Kessler, R. H., and Pitts, R. F., *J. Clin. Invest.*, 35, 1055–1066, 1956.
Cherian, M. G. and Shaikh, Z. A., *Biochem. Biophys. Res. Commun.*, 65, 863–869, 1975.
Christie, N. T. and Costa, M., *Biol. Trace Elem. Res.*, 6, 139–158, 1984.
Clarkson, T. W. and Cross, A., Studies on the Action of Mercuric Chloride on Intestinal Absorption, AEC Report 588, University of Rochester, Rochester, New York, 1961.
Conrad, M. E. and Barton, J. C., *Gastroenterology*, 74, 731–740, 1978.
Corrigan, A. J. and Huang, P. C., *Biol. Trace Elem. Res.*, 3, 197–216, 1981.
Cotzias, G. C., Borg, D. C., and Selleck, B., *Am. J. Physiol.*, 201, 927–930, 1961.
Demis, D. J. and Rothstein, A., *Am. J. Physiol.*, 180, 566–574, 1955.
Elinder, C.-G. and Nordberg, M., in *Cadmium and Health*, Friberg, F.L. et al., Eds., CRC Press, Boca Raton, FL, 65, 1985.
Felley-Bosco, E. and Diezi, J., *Toxicol. Appl. Pharmacol.*, 91, 204–211, 1987.
Foulkes, E. C., *Am. J. Physiol.*, 227, 1356–1360, 1974.
Foulkes, E. C., *Proc. Soc. Exp. Biol. Med.*, 159, 321–323, 1978.
Foulkes, E. C., in *Pharmacology of Intestinal Permeation*, Vol. 70, Csaky, T.Z., Ed., *Handbook of Experimental Pharmacology*, Springer-Verlag, Heidelberg, 543, 1984.
Foulkes, E. C., *Toxicology*, 37, 117–125, 1985.
Foulkes, E. C., *Toxicology*, 52, 263–272, 1988.
Foulkes, E. C., *Toxicology*, 69, 177–185, 1991.
Foulkes, E. C., *Toxicology*, 70, 261–270, 1991a.
Foulkes, E. C., *Life Sci.*, 52, 1617–1620, 1993.
Foulkes, E. C. and Bergman, D., *Toxicol. Appl. Pharmacol.*, 120, 89–95, 1993.
Foulkes, E. C. and Blanck, S., *Toxicol. Appl. Pharmacol.*, 102, 464–473, 1990.
Foulkes, E. C. and Blanck, S., *Toxicol. Appl. Pharmacol.*, 108, 150–156, 1991.
Foulkes, E. C. and McMullen, D. M., *Toxicology*, 38, 285–291, 1986.
Foulkes, E. C. and McMullen, D., *Toxicology*, 38, 35–42, 1986a.
Foulkes, E. C. and McMullen, D. M., *Am. J. Physiol.*, 253, G134–G138, 1987.
Foulkes, E. C., Mort, T., and Buncher, R., *Proc. Soc. Exp. Biol. Med.*, 197, 477–481, 1991.
Foulkes, E. C. and Voner, C., *Toxicology*, 22, 115–122, 1981.
Frame, M. D. S. and Milanick, M. A., *Am. J. Physiol.*, 261, C467–C475, 1991.
Garty, M., Bracken, W. M., and Klaassen, C. D., *Toxicology*, 42, 111–119, 1986.
Grosso, A. and De Sousa B. C., *J. Physiol.*, 468, 741–752, 1993.
Giridhar, J., Rathinaveln, A., and Isom, G. E., *Toxicology*, 75, 133–143, 1992.
Hanck, D. A. and Sheets, M. F., *J. Physiol.*, 454, 267–298, 1992.
Hillyard, S. D. and Gonick, H. C., *J. Membr. Biol.*, 26, 109–119, 1976.
Hinkle, P. M., Kinsella, P. A., and Osterhoudt, K. C., *J. Biol. Chem.*, 262, 16333–16337, 1987.
Hirayama, K., *Kumamoto Med. J.*, 28, 151–163, 1975.
Hoadley, J. E. and Cousins, R. J., *Proc. Soc. Exp. Biol. Med.*, 180, 296–302, 1985.
Imesch, E., Moosmayer, M., and Anner, B. M., *Am. J. Physiol.*, 262, 837–842, 1992.
Israel, E. J., Pang, K. Y., Harmutz, P. R., and Walker, W. A., *Am. J. Physiol.*, 252, G762–G767, 1987.
Jourd'heuil, D., Vaamanen, P., and Meddings, J. B., *Am. J. Physiol.*, 264, G1009–G1015, 1993.
Kinter, W. B. and Pritchard, J. B., in *Reactions to Environmental Agents*, Handbook of Env. Physiol., Sect. 9, American Physiological Society, Bethesda, MD, 536, 1977.

Kostial, K., Simonovic, I., and Pisonic, M., *Nature,* 233, 564, 1971.
Lang, R. J. and Paul, R. J., *J. Physiol.,* 433, 1–24, 1991.
Lou, M., Garay, R., and Alda, J. O., *J. Physiol.,* 443, 123–136, 1991.
Macey, R. I. and Farmer, R. E. L., *Biochim. Biophys. Acta,* 211, 104–106, 1970.
Menard, M. P. and Cousins, R. J., *J. Nutr.,* 113, 1434–1442, 1983.
Mills, J. W. and Ferm, V. H., *Toxicol. Appl. Pharmacol.,* 101, 245–254, 1989.
Morselt, A. F. W., *Toxicology,* 70, 5–37, 1991.
Nechay, B. R., Palmer, R. F., Chinoy, D. A., and Posey, V. A., *J. Pharmacol. Exp. Ther.,* 157, 599–617, 1967.
Nordberg, G. F., Kjellstrom, T., and Nordberg, M., in *Cadmium and Health,* Friberg, L., et al., Eds., CRC Press, Boca Raton, FL, 103, 1985.
Perkins, D. H., *Biochem. J.,* 80, 668–672, 1980.
Preston, R. L. and Chen, C. W., *Bull. Environ. Contam. Toxicol.,* 42, 620–627, 1989.
Preston, R. L., Janssen, S. J., Lu, S., McQuade, K. L., and Beal, L., *Bull. M. Desert Isl. Biol. Lab.,* 30, 72–74, 1991.
Rothstein, A., , *Fed. Proc.,* 18, 1026–1035, 1959.
Rothstein, A. and Hayes, A. D., *Arch. Biochem. Biophys.,* 63, 87–99, 1956.
Sahagian, B. M., Harding-Barlow, I., and Perry, H. M., *J. Nutr.,* 93, 291–300, 1967.
Sasser, L. B. and Jarboe, G. E., *Toxicol. Appl. Pharmacol.,* 41, 423–431, 1977.
Sellinger, M., Ballatori, N., and Boyer, J. L., *Toxicol. Appl. Pharmacol.,* 107, 369–376, 1991.
Simons, T. J. B., *J. Physiol.,* 378, 287–312, 1986.
Stacy, B. D. and Thornburn, G. D., *Science,* 152, 1076–1077, 1966.
Stefanovic, V., Mandel, P., and Rosenberg, A., *J. Biol. Chem.,* 251, 3900–3905, 1976.
Tacnet, F., Ripoche, P., Roux, M., and Neumann, J. M., *Eur. Biophys. J.,* 19, 317–322, 1991.
Tacnet, F., Watkins, D. W., and Ripoche, P., *Biochim. Biophys. Acta,* 1063, 51–59, 1991a.
Takada, M. and Hayashi, H., *Jpn. J. Physiol.,* 30, 257–269, 1980.
Tanaka, T., Naganuma, A., and Imura, N., *Toxicology,* 60, 187–198, 1990.
Templeton, D. M., *J. Biol. Chem.,* 265, 21764–21770, 1990.
Valley, B. L. and Ullmer, D. D., *Ann. Rev. Biochem.,* 41, 91–128, 1972.
Verbost, P. M., Senden, M. H. M. N., and van Os, C. H., *Biochim. Biophys. Acta,* 902, 247–252, 1987.
Verbost, P. M., Flik, G., Lok, R. A. C., and Wendelaar Bonga, S. E., *J. Membr. Biol.,* 102, 97–104, 1988.
Victery, W., Miller, C. R., and Fowler, B. A., *J. Pharmacol. Exp. Ther.,* 231, 589–596, 1984.
Wada, K., Fujii, Y., Watanabe, H., Satoh, M., and Furuichi, Y., *FEBS Lett.,* 285, 71–74, 1991.
Yuile, C. L., in *Handbook of Experimental Pharmacology,* Vol. 36, Hodge, H. C. et al., Eds., Springer-Verlag, Berlin, 165, 1973.
Zalups, R. K., Robinson, M. K., and Barfuss, D. W., *J. Am. Soc. Nephrol.,* 2, 866–878, 1991.
Zhang, J.-G. and Lindup, W. E., *Biochem. Pharmacol.,* 45, 2215–2222, 1993.

Chapter 9

Interactions Between Glutathione and Mercury in the Kidney, Liver, and Blood

Rudolfs K. Zalups and Lawrence H. Lash

I. INTRODUCTION

Mercury exists in biological systems in multiple forms, including elemental mercury (Hg^0), inorganic mercurous (Hg^+) and mercuric (Hg^{2+}) compounds, and organic mercuric compounds (R-Hg^+ or R-Hg-R; where R represents an organic ligand). Mercury remains a major environmental contaminant that has well-characterized toxic effects. Although the kidneys are the major sites of accumulation for most forms of mercury, particularly the inorganic forms of mercury, organic mercurials have patterns of distribution in the body leading to neurotoxicity as well as nephrotoxicity. Elemental mercury is readily oxidized in erythrocytes and tissues to the inorganic mercuric ion, so that the fate of elemental mercury vapor is very similar to that of inorganic mercury (Magos and Clarkson, 1977; Magos et al., 1978). Moreover, elemental mercury must be oxidized to inorganic mercury in order for it to interact with biological molecules. The oxidation step appears to occur very soon after absorption since the tissue distributions of elemental and inorganic mercury are very similar (Clarkson, 1972). The biological disposition of, and effects of, inorganic and organic forms of mercury are largely determined by their high affinity for sulfhydryl groups, in particular glutathione (GSH), which is one of the predominant intracellular non-protein thiols. To gain an understanding of the pharmacokinetics and pharmacodynamics of mercurous and mercuric compounds, one must consider the interactions between the different forms of mercury and thiols, especially GSH.

This chapter will focus on interactions between inorganic and organic forms of mercury and GSH, and how they pertain to the translocation of mercury between tissues, to the renal and hepatic processing of mercury, and to the nephrotoxicity of mercury. After a brief consideration of the chemical reactivity of mercury towards GSH, which underlies the biological reactivity of mercury, disposition of mercury and thiols in erythrocytes and plasma, in liver, and finally in the primary target organ, the kidneys, will be discussed. Significant advances have been made recently in our understanding of the use of GSH and other thiol-containing compounds as therapeutic agents to protect cells from damage due to a variety of toxicants including mercury, thus adding a new dimension to the consideration of the interaction of mercury with thiols.

Humans can be exposed to heavy metals, such as mercury, in food and water or as an environmental contaminant in soil and air. In the first situation, gastrointestinal absorption is the limiting factor that determines interaction of external mercury with the internal milieu of the organism. In the second situation, pulmonary and dermal absorption enable external mercury to gain access to the body. Once mercury gains access to the circulatory system, the three primary tissues or organ systems that determine

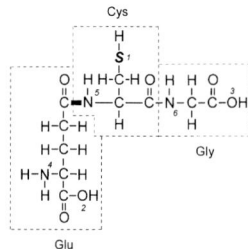

Figure 1 Structure of glutathione. Potential nucleophilic binding sites for ions of mercury and other metals are numbered 1 through 6. Note: in biological systems, the sulfhydryl group on the cysteinyl residue of glutathione is the primary site where mercuric ions bind.

the status of mercury are the blood, liver, and kidneys. In each of these compartments, GSH status appears to be an important determinant in the disposition of mercury. This chapter will describe the current knowledge of the role of GSH in determining the metabolism and toxicity of mercury in the blood, liver, and kidneys. Certain aspects of the biological functions of GSH as they may relate to the mechanism by which mercury acts will be presented. These include functions of GSH as a nucleophile in conjugation reactions and as a reductant in maintaining redox homeostasis and protecting cells from oxidative injury.

II. CHEMICAL INTERACTIONS BETWEEN GSH AND MERCURY

GSH is a tripeptide containing glutamyl, cysteinyl, and glycyl amino acid residues. One of the unique features of this peptide is that the amino nitrogen of the cysteinyl residue (which is the central residue of the molecule) is bonded to the γ-carbon of the glutamyl residue. The formation of this peptide bond is catalyzed by the enzyme γ-glutamylcysteine synthetase, which is the rate-limiting enzyme involved in the synthesis of GSH. Once the γ-glutamylcysteine bond is formed, it is resistant to most protease activity and requires the enzyme γ-glutamyltransferase to cleave it.

Due to the presence of the free sulfhydryl group on the cysteinyl residue, GSH has a great propensity for forming complexes with ions of metals that have strong electrophilic characteristics, such as mercury. An understanding of how these complexes form and their biological activities is necessary for an understanding of the toxicity of mercury.

In a recent review of glutathione-metal complexes, Rabenstein (1989) points out that GSH can act as a polydentate ligand in reactions with metal ions. The potential binding sites include two carboxylate oxygens, one amino nitrogen, one sulfhydryl group, and two amide groups. These binding sites are depicted numerically in a two-dimensional representation of a molecule of GSH in Figure 1. Each of the three amino acid residues of GSH contains two of the potential sites where metal ions can bind. The binding of metals to these six sites is greatly dependent on pH, among other factors.

In biological systems, however, the sulfhydryl group on the cysteinyl residue of GSH is perhaps one of the most important sites when it comes to the formation of complexes between metals and GSH. A number of metal ions including cadmium (Cd(II)), lead (Pb(II)), and mercury (Hg(II)) form bonds with the sulfhydryl group of GSH. When GSH is in a 2:1 ratio with any of these metals, there is a tendency for coordinate covalent bond formation between the sulfhydryl group of two molecules of GSH and one ion of the respective metal. Rabenstein (1989) points out that the tendency for linear II coordination is particularly strong between inorganic mercuric ions and GSH. Figure 2 shows potential complexes of mercury with GSH, GSH and cysteine, and GSH and a protein sulfhydryl group. Rabenstein also notes that linear, two-coordinate bonding also occurs with GSH in the presence of methyl mercuric chloride ($CH_3Hg(II)$) in a 1:1 ratio. These bonds appear to be quite stable *in vitro* since negligible dissociation has been detected throughout a pH range of 0–14.

III. GSH AND MERCURY IN PLASMA AND ERYTHROCYTES

A. INTERORGAN TRANSLOCATION OF MERCURY AND MERCURY COMPLEXES

After exposure to a heavy metal such as mercury much of the dose of the metal eventually enters the systemic circulation. The rate of entry is dependent on a number of factors, including the ionic and

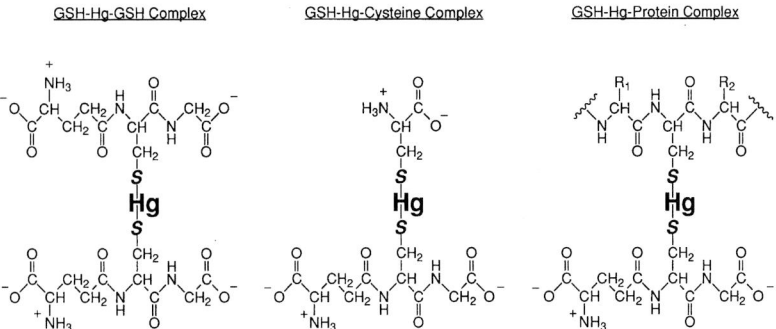

Figure 2 Potential complexes of the inorganic mercuric ion and glutathione (GSH). Inorganic mercury tends to form linear II coordinate complexes with thiols, such as GSH, cysteine, and cysteinyl residues of proteins. Each of these complexes may be found in the plasma. Epithelial transport of mercury-thiol complexes may be a means by which mercuric ions enter into target cells in the body. Note: there is only one site where GSH or other thiols can bind on organic mercuric ions.

molecular species and the route of exposure of the metal. Once the metal enters into the blood, it can either become bound to, or enter into, the cellular components of blood (mainly erythrocytes), or it can bind onto some ligand in the plasma. The rate of abstraction or elimination of a metal from the blood and the distribution of a metal between the two fractions of blood depends greatly on the chemical nature of the metal and the time after and type of exposure to the metal. For example, Zalups (1993) demonstrated recently that only about 43% of the dose of mercury remained in the blood of rats 1 min after they received an intravenous nontoxic dose of mercuric chloride, indicating rapid abstraction of mercury from the blood. Of the mercury remaining in the blood, 50% was in the plasma and 50% was associated with the cellular components of blood. Zalups (1993) also showed that by the end of 3 h following injection, less than 10% of the dose of mercury was present in the blood, and of that, only 30% was in the plasma. Much of the reduction in the content of mercury in the blood was presumably due to abstraction and uptake by the kidneys, since 50% of the administered dose was present in the total renal mass 3 h after injection. These findings exemplify the fact that the rate of entry of a heavy metal into the blood and the manner in which it distributes between the cellular and plasma fractions of blood play an important role in the manner in which the metal is subsequently handled in the kidneys. Metals that tend to remain longer in the plasma fraction of blood appear to be more likely to be handled to a greater extent in the kidneys.

In spite of the focus on handling of mercury by the kidneys because of the nephropathy induced by mercury, the liver and intestines are the primary organ involved in the early elimination of mercury, with fecal excretion greatly exceeding the urinary excretion (Zalups et al., 1987; Zalups et al., 1992). Hence, translocation of mercury, presumably in the form of various mercury complexes, to the liver and intestines is quantitatively the major route of mercury interorgan flux.

B. GSH AND OTHER THIOLS IN PLASMA

While in the plasma of blood, metals tend to bind to albumin as well as other plasma proteins, especially those containing free sulfhydryl groups. For example, there are data showing that a large percentage of the mercury in the plasma after exposure to inorganic mercury is bound to albumin (Cember et al., 1968; Friedman, 1957; Mussini, 1958). Albumin is by far the most abundant protein in plasma (3.5 to 5.0 g/dl), and it possesses a single free sulfhydryl group located on the terminal cysteinyl residue of the molecule with which metal ions can bind (Jocelyn, 1972; Brown and Shockley, 1982). Other potential ligands for heavy metal ions in plasma that contain a free sulfhydryl group are cysteine and GSH. Both of these ligands are present in the plasma of blood at concentrations approximating 10 μM (Lash and Jones, 1985). Although these concentrations are considerably less than those for albumin, they are high enough to consider GSH and cysteine as important ligands in plasma to which heavy metal ions bind. An even larger plasma pool of GSH and cysteine is available in the form of low molecular weight protein or mixed disulfides (Lash and Jones, 1985). Although the concentration of glutathione disulfide (GSSG) in plasma is only about 1 μM or less, the plasma does contain cystine and mixed disulfides of GSH and cysteine with protein at concentrations of 50 to 150 μM.

C. GSH IN ERYTHROCYTES AND MERCURY SEQUESTRATION

Besides plasma, erythrocytes comprise a major portion of the volume of blood, and thus represent a significant site for sequestration of mercury. Mature red blood cells contain GSH at concentrations of approximately 2 mM (Beutler and Dale, 1989). This pool of GSH has a half-life of approximately 4 days. Since mature erythrocytes lack γ-glutamyltransferase, this suggests that efflux of GSH in some form is responsible for GSH turnover. GSH-equivalents in erythrocytes are transported by an ATP-dependent system across the membrane in an outward direction into plasma as either GSSG or as GSH S-conjugates (Beutler, 1983). GSH, as an intact tripeptide, is not transported in either direction across erythrocyte plasma membranes. Because the normal intracellular concentration of GSSG in erythrocytes is only 5 μM, transport of GSSG can only be detected when an oxidative stress is imposed. Additional studies, however, have demonstrated that ATP-dependent GSSG efflux occurs continuously under normal physiological conditions, and that this efflux can account quantitatively for turnover of the pool of glutathione. Studies with inside-out vesicles prepared from erythrocyte membranes showed the presence of two transport systems, a high-affinity (K_m = 0.1 mM) and a low-affinity (K_m = 7.1 mM) system (Beutler and Dale, 1989).

The rationale for having an energy-dependent efflux system for GSSG was not apparent until it was realized that this transport system could also catalyze efflux of GSH S-conjugates. Since mature erythrocytes have the capability to synthesize GSH *de novo* and have significant activity levels of GSH peroxidase and GSH S-transferase, erythrocytes can be viewed as a first line of defense against blood-borne oxidants and reactive electrophiles. Once GSH in erythrocytes acts on these potential toxicants, the cells transport GSSG or the thioether conjugate into plasma for translocation to other tissues, particularly the liver, kidneys, and intestines, for further metabolism or excretion. Complexes of GSH with inorganic or organic mercury may be handled by erythrocytes in a manner similar to that of thioether conjugates. GSH-mercury conjugates can be viewed as transport forms of mercury that enable translocation to other tissues.

Additional research is necessary to confirm this suggestive mechanism of mercury transport in erythrocytes.

IV. GSH, MERCURY, AND THE LIVER

A. UPTAKE FROM PLASMA

In considering the mechanism of hepatic uptake of inorganic and organic forms of mercury, at least two points must be kept in mind. First, the great affinity between cations of mercury and sulfhydryl groups suggests that virtually all of the mercury that is delivered to the liver through the portal circulation will be in the form of complexes with proteins such as albumin or with low molecular weight thiols such as cysteine or GSH. Second, the general assumption that organic forms of mercury (such as methylmercury) are lipid soluble and, therefore, easily diffuse across biological membranes, is incorrect, as simple diffusion plays only a minimal role in the transport of methylmercury across biological membranes (Ballatori, 1991). Thus, it appears that specific transport carriers for mercury-containing complexes are present on the sinusoidal plasma membranes that mediate uptake of mercury into the hepatocyte. Figures 3 and 4 summarize some possible mechanisms involved in the absorptive uptake and subsequent excretion into bile of inorganic mercury (Figure 3) and methylmercury (Figure 4) in hepatocytes. The mechanistic schemes presented in these figures focus only on transport when either the inorganic mercuric or methylmercuric ion is part of a complex containing GSH or some degradative product of GSH (such as cysteinylglycine or cysteine). These representations are not meant to present all possible mechanisms involved in the transport of inorganic or organic forms of mercury in hepatocytes. They simply serve to present some of the possible mechanisms involving GSH.

Uptake of both inorganic mercury and methylmercury by isolated rat hepatocytes is linearly related to their concentration in the extracellular medium and does not appear to show kinetic saturability at subtoxic doses (Ballatori, 1991). However, the presence of serum proteins in the extracellular buffer or increases in the intracellular content of the metal-binding protein metallothionein alter the uptake of inorganic mercury. Little else is known about the mechanism of mercury uptake across the hepatic sinusoidal plasma membrane. Additional research is needed to understand how different forms of inorganic and organic mercury are taken up by the liver and, in the absence of endocytosis, to determine how mercuric ions bound to proteins gain access into hepatocytes.

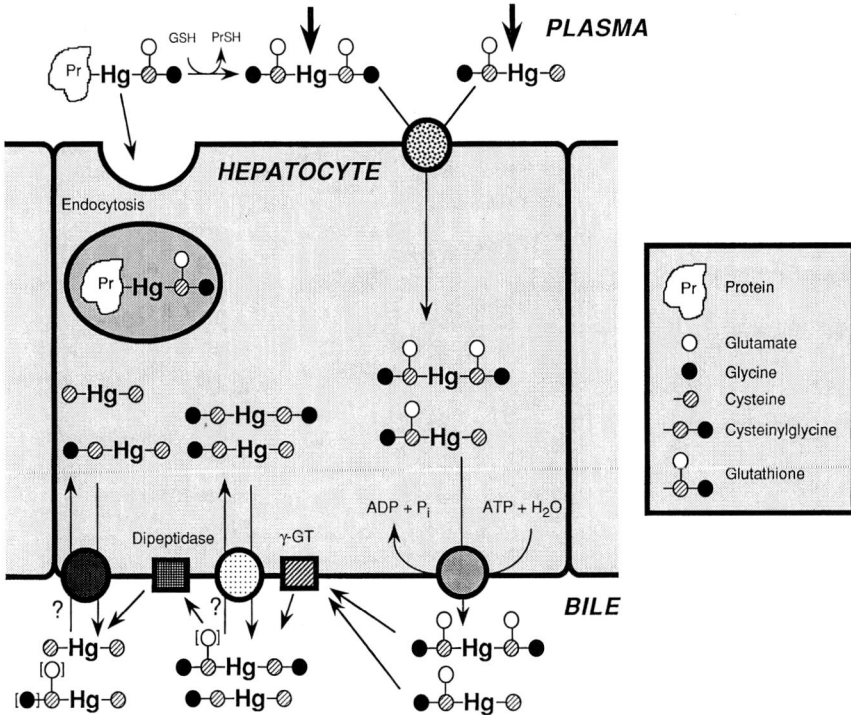

Figure 3 Scheme of transport of inorganic mercury in hepatocytes that involve the cotransport of inorganic mercury with glutathione (GSH) or a degradative product of GSH. The scheme shows possible routes of cellular entry and exit for the mercuric ion when bound to GSH or a catabolic product of GSH. Three possible complexes of inorganic mercury and GSH can be presented to the sinusoidal membrane by the plasma. These complexes include Protein:Hg:GSH, GSH:Hg:GSH, and Cysteine:Hg:GSH. The scheme presents the following possible mechanisms: (1) the complex involving protein (predominantly albumin) can be taken up by the hepatocyte by a form of endocytosis, or the protein on the complex may be exchanged for a molecule by GSH in the plasma; (2) both the GSH:Hg:GSH and the Cysteine:Hg:GSH complexes can be transported into the heptocyte from the plasma by the GSH and/or GSH S-conjugate transport system on the sinusoidal plasma membrane of the hepatocyte; (3) these complexes can be secreted by the heptocyte into the bile by a ATP-dependent GSSG and GSH S-conjugate transport system on the canalicular plasma membrane; the canalicular or biliary epithelial membrane has both γ-glutamytransferase (γ-GT) and a dipeptidase for degrading GSH to cysteinylglycine and cysteine; (4) the complexes of mercury and the degradative products of GSH catabolism can be reabsorbed by a poorly defined biliary-hepatic recycling mechanism.

Note: mechanisms other than the ones presented in the above-mentioned scheme may also be involved in the hepatocellular transport of inorganic mercury. The purpose of the scheme is to relate the possible role of GSH in the transport of inorganic mercury in hepatocytes. In addition, while parts of the scheme were generated from currently accepted mechanisms involved in the heptocellular transport of inorganic mercury, other parts were generated from logical possibilities that need to be explored and tested.

B. INTRACELLULAR FATE OF MERCURY

Once inside the hepatocyte, mercury has numerous potential ligands with which to bind, including metallothionein (MT), other sulfhydryl-containing proteins, and low molecular weight thiols such as cysteine and GSH. Inorganic mercury has a high affinity for MT, which may act as a buffer for the cell. Binding to MT plays only a minor role in the intracellular sequestration of methylmercury, because the binding affinity of methylmercury for MT is much lower than that of inorganic mercury for MT. Because of the predominance of GSH in the cell and the existence of specific plasma membrane transport systems, GSH appears to play a central role in the hepatic disposition of both organic and inorganic mercury. In hepatocytes, intracellular GSH is present at a concentration of approximately 5 mM, while cysteine is generally found at a concentration of less than 0.2 mM. It is probable, therefore, that mercury is presented to the active site of transport proteins on the canalicular membrane, with at least one GSH molecule bound to each mercuric ion.

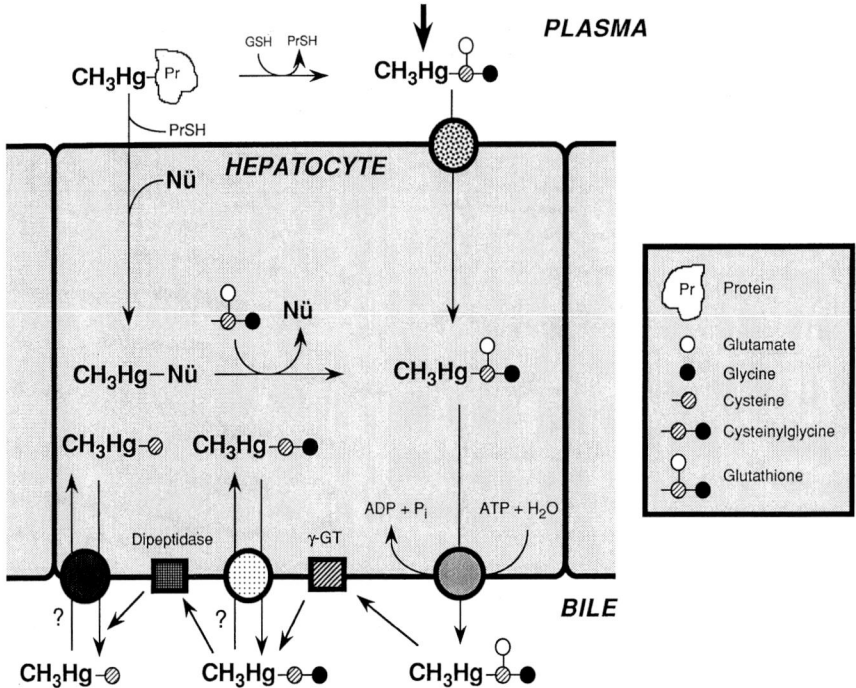

Figure 4 Scheme of transport of methylmercury in hepatocytes that involve the cotransport of methylmercury with glutathione (GSH) or a degradative product of GSH. The scheme shows possible routes of cellular entry and exit for the methylmercuric ion when bound to GSH or a catabolic product of GSH. Two possible complexes of methylmercury are shown to be presented to the sinusoidal membrane by the plasma. These complexes include CH_3Hg:Protein and CH_3Hg:GSH. The scheme presents the following possible mechanisms: (1) hepatocellular uptake of methylmercury may involve the delivery of methylmercury to the sinusoidal membrane by a protein ligand; once at the plasma membrane, the methylmercuric ion may dissociate from the protein and enter into the plasma membrane of the hepatocyte by a lipophilic mechanism involving substitution with a cellular nucleophile (Nü); (2) the CH_3Hg:GSH complex may enter the hepatocyte by a lipophilic mechanism and/or may involve a GSH/GSH S-conjugate transporter; (3) once the CH_3Hg:GSH complex has entered into the hepatocyte, it may be transported into the bile across the canalicular membrane by an ATP-dependent transporter; the canalicular or biliary epithelial membrane has both γ-glutamytransferase (γ-GT) and a dipeptidase for degrading GSH to cysteinylglycine and cysteine; (4) the complexes of mercury and the degradative products of GSH catabolism can be reabsorbed by a poorly defined biliary-hepatic recycling mechanism.
Note: mechanisms other than the ones presented in the above-mentioned scheme may also be involved in the hepatocellular transport of methylmercury. The purpose of the scheme is to relate the potential role of GSH in the transport of methylmercury in hepatocytes. While parts of the scheme were generated from currently accepted mechanisms involved in the heptocellular transport of methylmercury, other parts were generated from logical possibilities that need to be explored and tested.

Although the liver is a primary site of the excretion of mercury, little hepatotoxicity is generally observed *in vivo*. This suggests that the mercury that is taken up by hepatocytes for transport across the sinusoidal plasma membrane does not interact significantly with hepatocellular thiols or macromolecules. Rather, the mercury that is taken up by the hepatocyte must be efficiently delivered into the bile across the canalicular plasma membrane for excretion into the intestine.

C. EFFLUX OF GSH, MERCURY, AND MERCURY COMPLEXES INTO PLASMA AND BILE

Transport of inorganic and organic forms of mercury across the canalicular plasma membrane into bile has been studied much more thoroughly than transport across the sinusoidal plasma membrane. GSH plays a major role in the biliary secretion of a number of essential and toxic metals besides inorganic and methylmercury, including cadmium, copper, and zinc (Ballatori and Clarkson, 1983, 1984a,b). ATP-dependent transport systems for GSH, GSSG, and GSH S-conjugates have been identified and characterized on the canalicular membrane, and these system(s) mediate efflux from the hepatocyte into bile (see Figures 3 and 4). Complexes of inorganic and organic mercury with GSH are apparently transported as GSH S-conjugates into bile. Once in the bile, γ-glutamyltransferase and dipeptidases catalyze extensive metabolism of GSH or GSH S-conjugates to produce the corresponding amino acids or cysteine S-conjugates (Ballatori et al., 1986, 1988). Consequently, complexes of mercury with cysteine will normally be rapidly formed in the bile. These complexes may be recycled back to the liver by a biliary-hepatic recycling mechanism (Dutczak and Ballatori, 1992) or they may be delivered to the intestines for eventual excretion or reabsorption and return to the liver (i.e., enterohepatic circulation) and other organs. Exposure of the kidneys to these mercury complexes could also occur by reabsorption into hepatocytes and efflux across the sinusoidal membrane into plasma (i.e., hepatic-renal circulation).

An important consideration in evaluating the interorgan metabolism and transport of mercury-GSH complexes is species-dependent differences in the activities of key enzymes in the metabolic pathway for GSH, particularly in the liver and kidney (Hinchman and Ballatori, 1990). Hinchman and Ballatori (1990) determined the renal/hepatic ratio of γ-glutamyltransferase to be 875 in the rat, 413 in the mouse, 100 in the rabbit, and between 15 and 20 in the guinea pig, pig, and human. This approximate 60-fold range in activity ratios suggests that the form of the mercury complex presented to the renal circulation may differ significantly among various species.

V. GSH, MERCURY, AND THE KIDNEY

A. GENERAL PRINCIPLES PERTAINING TO THE DELIVERY OF MERCURY AND OTHER METALS TO RENAL TUBULAR EPITHELIAL CELLS

Mechanisms involved in the renal tubular uptake and transport of mercury as well as other heavy metals are not well understood. In order for mercury or other heavy metal ions to be taken up by specific renal tubular epithelial cells, they must first be delivered to the luminal and/or basolateral membrane(s) of the epithelial cells. For the case where luminal binding and/or uptake of a metal takes place, one of two events must occur. The metal must either pass through the glomerular filtration barrier, or it must be secreted into the lumen by the epithelial cells of a particular segment of the nephron, which can be the same segment of the nephron or a segment more proximal to where the metal interacts with the luminal plasma membrane. Basolateral binding and/or uptake of a metal would require the metal to be delivered to the basolateral membrane from peritubular capillaries. Both luminal and basolateral interactions with metals along the nephron are greatly dependent on the fraction of the metal in blood that is present in the plasma and the size and chemical characteristics of organic complexes formed with the metal in plasma.

B. CURRENT KNOWLEDGE CONCERNING THE RENAL TUBULAR UPTAKE AND TRANSPORT OF MERCURY

Of all the organs in the body of a mammal, the kidneys are the primary site for accumulation of mercury after exposure to inorganic or elemental mercury (Adam, 1951; Ashe et al., 1953; Berlin and Gibson, 1963; Cherian and Clarkson, 1976; Friberg, 1956, 1959; Hahn et al., 1989, 1990; Rothstein and Hayes, 1960; Swensson and Ulfarson, 1968; Zalups, 1991a,b,c, 1993; Zalups and Barfuss, 1990; Zalups and Diamond, 1987a,b). Organic forms of mercury also accumulate in the kidneys of mammals to a significant degree (Friberg, 1959; Magos and Butler, 1976; Magos et al., 1981, 1985; McNeil et al., 1988; Norseth and Clarkson, 1970a,b; Prickett et al., 1950; Zalups et al., 1992), but to a lesser extent than inorganic or elemental forms of mercury.

Findings from a number of studies show that inorganic mercury accumulates primarily in the renal cortex and outer stripe of the outer medulla (Bergstrand et al., 1959; Friberg et al., 1957; Taugner et al., 1966; Zalups, 1991a,b,c, 1993; Zalups and Barfuss, 1990; Zalups and Cherian, 1992a,b; Zalups and Lash, 1990; Zalups et al., 1987, 1988). Intrarenal distribution of mercury after exposure to organic forms of mercury appears to be similar to that of inorganic mercury (Berlin, 1963; Berlin and Ullberg, 1963a,b). Data from histochemical, autoradiographic, and microdissected tubule studies indicate that inorganic mercury is taken up primarily by the three segments of the proximal tubule (Hultman et al., 1985; Hultman and Enestrom, 1986; Magos et al., 1985; Rodier et al., 1988; Taugner et al., 1966; Zalups 1991a,b; Zalups and Barfuss, 1990). Histochemical evidence also indicates that mercury accumulates primarily along the proximal tubule after exposure to organic forms of mercury (Magos et al., 1985; Rodier et al., 1988). It is currently unknown whether there is any significant uptake and/or transport of inorganic or organic forms of mercury along other segments of the nephron or collecting duct. If the data gathered to date are reflecting that mercury is handled mainly or exclusively along the proximal tubule then one might ask, what are some of the unique characteristics of the segments of the proximal tubule that could explain the avid uptake, transport, and accumulation of mercury along the these tubular segments?

There are both morphological and physiological differences between the segments of the proximal tubule and other segments of the nephron and collecting duct. Reabsorption of filtered protein occurs almost exclusively along the proximal tubule. The proximal tubule is the primary site for the transport of organic anions and cations. In addition, the epithelial cells along the proximal tubule contain on their luminal membrane large amounts of the enzyme γ-glutamyltransferase, which plays an important role in GSH metabolism in and along the proximal tubule. Other differences also exist. In order to better understand the mechanisms involved in the uptake and transport of mercury, it would seem logical to pursue or test hypotheses generated from the unique structural, physiological, and biochemical aspects of the epithelial cells along the proximal tubule.

One such hypothesis has been generated and evaluated in recent years. This hypothesis states that some mercury is cotransported with albumin or other filtered proteins during the endocytotic uptake of albumin. Since albumin is one of the principal ligands in plasma to which mercurous or mercuric ions bind, and since some ultrafiltration of mercury (Zalups and Barfuss, 1990) and albumin (Dirks et al., 1964; Landwegr et al., 1977; Maack, 1992; Oken and Flamebaum, 1971) occurs at the glomerulus, it is likely that some filtered mercury is, or becomes, bound to albumin as it enters the ultrafiltrate. Moreover, since albumin is reabsorbed as an intact molecule by adsorptive endocytosis along segments of the proximal tubule (Bourdeau et al., 1972; Clapp et al., 1988; Maack, 1992), it would seem that the cotransport of mercury with albumin would be a reasonable possibility.

In one histochemical study, Hultman et al. (1985) showed that presumed deposits of inorganic mercury appeared in the lumen of proximal tubules within 30 s after injection of mercuric chloride. By 90 s after injection, these deposits appeared to be initially associated with the brush border and then with the apical vesicular system of the proximal tubular cells, which is involved in endocytosis. By 30 min after injection, deposits presumed to contain mercury were found in secondary lysosomes. The pathway of uptake and processing of these deposits is consistent with the uptake and degradation of reabsorbed protein. In an earlier study, Madsen (1980) demonstrated in rats made proteinuric that mercury excreted in the urine after injection of inorganic mercury was largely bound to albumin. Moreover, this investigator showed that a fraction of the mercury taken up by the kidneys was associated with lysosomes, which is consistent with the notion that some inorganic mercury is taken up by endocytosis.

In a recent study (Zalups and Barfuss, 1993a), however, where rats were injected simultaneously with radiolabeled mercury and radiolabeled albumin, it was not possible to demonstrate that all of the mercury that was taken up by the kidney after exposure to inorganic mercury was accomplished exclusively by cotransport with albumin. Although these findings were not inconsistent with the notion that some of the mercury taken up by proximal tubular cells is as a result of cotransport with albumin during the endocytosis of filtered albumin, they tended to indicate that additional mechanism(s) are also involved in the uptake of mercury along the proximal tubule.

C. EXTRACELLULAR GSH AND THE RENAL TUBULAR UPTAKE OF MERCURY

Both cysteine and GSH are in sufficient concentrations in the blood to allow for the formation of complexes between either of these ligands and inorganic or organic mercuric ions. Unlike albumin, both GSH and cysteine are filtered freely at the glomerulus and are handled and reabsorbed avidly along the proximal tubule (Silbernagl, 1992). Complexes of mercuric ions bound to cysteine or GSH are also small enough so that they should filter freely at the glomerulus.

A scheme of potential mechanisms involved in the uptake of inorganic mercury and methylmercury across the luminal plasma membrane of proximal tubular epithelial cells when the respective form of mercury is in complexes with GSH or degradative products of GSH is depicted in Figures 4 and 5, respectively. It is assumed that inorganic mercury and methylmercury form complexes with GSH in the blood and/or ultrafiltrate.

One potential mechanism, depicted in Figure 4, is that complexes of inorganic mercury and GSH formed in the blood or ultrafiltrate may be taken up across the luminal membrane by direct transport mechanisms. This seems to be unlikely since there is no evidence to date of GSH being reabsorbed across the luminal membrane as an intact tripeptide, although efflux of GSH from the proximal tubular cell into the lumen is well documented (Griffith and Meister, 1979). A more plausible mechanism involves the reabsorption of GSH and mercury after the catalytic cleavage of the γ-glutamylcysteine bound by γ-glutamyltransferase, which releases glutamate from GSH for reabsorption by the sodium-dependent amino acid transport system. Further processing may be required by the dipeptidase on the luminal

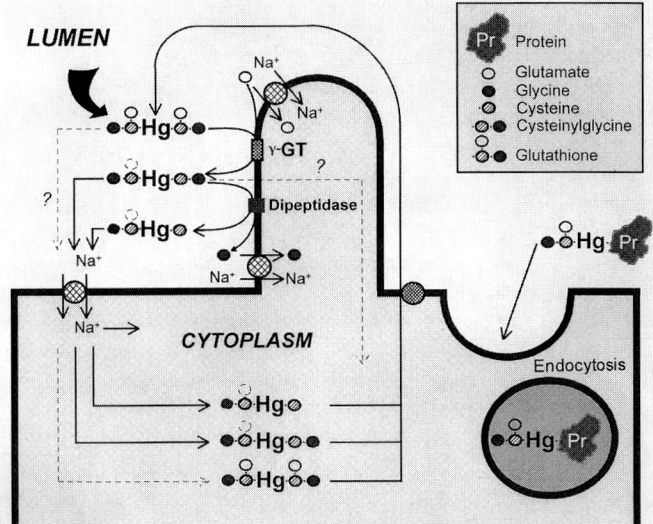

Figure 5 Scheme of transport of inorganic mercury at the brush border membrane of proximal tubular epithelial cells involving cotransport of the inorganic mercuric ion with glutathione (GSH) or a degradative product of GSH. The scheme shows that complexes of GSH:Hg:GSH, GSH:Hg:Cysteine, and GSH:Hg:Protein can be delivered to the luminal membrane of proximal tubular epithelial cells after ultrafiltration. The GSH:Hg:Protein complex may be taken up into the cell during the endocytosis of filtered proteins. Complexes containing GSH and inorganic mercury may be taken up directly across the luminal membrane by some tripeptide transport system. This seems unlikely since no evidence of reabsorptive transport of glutathione as a tripeptide has been demonstrated. More likely, mechanisms involve the catabolism of GSH by the γ-glutamyltransferase (which cleaves the γ-glutamyl-cysteine bond) and/or subsequent catabolism of cysteinylglycine by a dipeptidase on the luminal membrane. Inorganic mercury bonded to either cysteinylglycine or cysteine may be taken up across the luminal membrane by the sodium-dependent amino acid or dipeptide transport systems. There is evidence that GSH:Hg complexes may be secreted into the lumen by some transport mechanism and, once in the lumen, the GSH on the complexes can be acted on by γ-glutamyltransferase in a GSH recycling pathway. In the scheme, dashed lines on the amino acid residues indicate that the residue may or may not be present.

Note: mechanisms other than the ones presented in the above-mentioned scheme may also be involved in the proximal tubular reabsorptive or secretory transport of inorganic mercury. The purpose of the scheme is to relate the potential role of GSH in the transport of inorganic mercury across the luminal membrane of proximal tubular epithelial cells. While parts of the scheme were generated from currently accepted mechanisms involved in the transport of inorganic mercury across the luminal membrane of proximal tubular epithelial cells, other parts were generated from logical possibilities that need to be explored and tested.

membrane that is responsible for hydrolysis of the cysteinylglycine bond (Silbernagl, 1992), which allows both cysteine and glycine to be reabsorbed by the sodium-dependent amino acid transport system. If action by both γ-glutamyltransferase and the dipeptidase are necessary for the uptake of inorganic mercury, then it may be the transporter of cysteine that translocates inorganic mercury into the cytoplasm during the reabsorption of cysteine. It is possible, however, that mercury may be taken up with cysteinylglycine since there is evidence for dipeptide transport at the luminal membrane of proximal tubular epithelial cells (Barfuss et al., 1988; Silbernagl, 1992).

There is some experimental evidence indicating that γ-glutamyltransferase plays an important role in the proximal tubular reabsorption of mercury. This evidence comes from studies showing that inhibition of the γ-glutamyltransferase with acivicin prior to exposure to either inorganic mercury (Tanaka et al., 1990) or methylmercury (Naganuma et al., 1988) causes a decrease in the renal burden of mercury.

The schemes depicted in Figure 4 and in Figure 5 also show that complexes of GSH and inorganic mercury or methylmercury in the cytoplasm of proximal tubular cells may be secreted into the tubular lumen, where they can be acted upon in the same manner as complexes of GSH and mercury that are filtered or are formed in the ultrafiltrate. Tanaka-Kagawa et al. (1993) provide some very recent *in vivo* evidence from mice supporting the hypothesis that some mercury present in the cytoplasm of renal epithelial cells along the proximal tubule may enter into the lumen by a secretory pathway as a GSH-mercury complex. Once in the lumen, the mercury is reabsorbed back into the proximal tubular epithelial cells by a mechanism requiring the activity of γ-glutamyltransferase. They showed that inhibition of γ-glutamyltransferase with acivicin prior to treatment with inorganic mercury or methylmercury caused the urinary excretion of mercury to increase and the renal accumulation of mercury to decrease. Additional evidence supporting this hypothesis comes from previous studies in which it was shown that the urinary excretion of mercury and GSH increased in rats (Berndt et al., 1985) and mice (Tanaka et al., 1990) when they were pretreated with acivicin.

Another possibility, depicted in Figure 4, is that some inorganic mercury bound to both GSH and a filtered plasma protein, such as albumin, is taken up by an endocytotic mechanism.

Recent data from both mice (Tanaka et al., 1990) and rats (Zalups and Barfuss, 1994) indicate that exogenous GSH causes an increase in the renal tubular uptake of mercury. In one study, Zalups and Barfuss (1994) demonstrated that when a nontoxic dose of inorganic mercury and GSH (in a 1:2 ratio) was coadministered intravenously to rats, the accumulation of inorganic mercury in the kidneys, specifically in the cortex and outer stripe of the outer medulla, increased significantly during the first hour after injection. During this hour, the amount of mercury in the blood decreased by about one half and more mercury was in the plasma than in the cellular components of blood. These findings tend to indicate that the mercury was injected as a complex with GSH and that many of these complexes remained in the blood for at least 1 h after injection. These findings are consistent with $in\ vitro$ ^{13}C-NMR data showing that when mercury is in a 1:2 ratio with GSH, a linear II coordinate covalent bond forms between each inorganic mercuric ion and two molecules of GSH (Fuhr and Rabenstein, 1973; Rabenstein, 1978, 1989). Zalups and Barfuss also have some preliminary (unpublished) data indicating that the nephropathy induced by inorganic mercury is made more severe when GSH is coadministered with mercury, presumably as a result of enhanced entry of mercury into the pars recta segments of the proximal tubule. All these findings, put together, tend to indicate a link between exogenous or extracellular GSH and the uptake of mercury along the proximal tubule.

In a couple of $in\ vitro$ renal systems, addition or modification of extracellular GSH greatly modifies the toxicity and cellular uptake of both inorganic mercury and methylmercury in epithelial cells of proximal tubules. In suspensions of segments of proximal tubules from rabbits, the cellular release of lactate dehydrogenase and accumulation of inorganic mercury were greatly decreased when GSH was present in the bathing medium in a ratio of 4:1 with inorganic mercury (Zalups et al., 1993). A decrease in uptake and toxicity of mercury was also observed when cysteine or bovine serum albumin was present in the medium in a ratio of 4:1 with inorganic mercury.

Using the isolated perfused tubule technique, Zalups et al. (1991) showed that addition of GSH to a perfusate containing inorganic mercury at a ratio of 4:1 provided complete protection to isolated perfused S1, S2, and S3 segments of the proximal tubule of the rabbit from the toxic effects of mercury. The protective effect of GSH appeared to be linked to reduced luminal uptake of mercury (Figure 6). The greatest reduction in the flux of mercury occurred in the S1 segment and the least amount of reduction of flux occurred in the S3 segment. Thus, compared with inorganic mercury alone, the entry of mercury across the luminal membrane when in a complex with GSH is substantially lower.

In another study, Zalups and Barfuss (1993b) showed that addition of GSH to a perfusate containing methylmercury at a ratio of 4:1 greatly reduced the disappearance flux of mercury from the lumen and the appearance flux into the bath in S2 and S3 segments of the proximal tubule (Figure 7). It is interesting that in the S3 segment of the proximal tubule, the addition of the GSH to the perfusate actually caused cellular injury and necrosis to be exacerbated. This enhanced severity of injury presumably was due to enhanced accumulation of mercury, which appeared to be related to the fact that the appearance flux of mercury in the bath was reduced to a greater extent than the disappearance flux of mercury from the lumen (Figure 6). It should be pointed out that comparing these $in\ vitro$ findings to the $in\ vivo$ findings discussed earlier should be done with caution. In the living organism, inorganic or organic mercuric ions are almost never delivered to the proximal tubular epithelium in a state where they are not bound to some organic ligand.

There is some $in\ vivo$ evidence from rabbits indicating that secretion of mercury may occur (Foulkes, 1974). Until substantial proof of the contrary is provided, one must consider the possibility that some uptake of mercury may occur at the basolateral membrane of proximal tubules, and that some of this uptake is linked to uptake of GSH. Figures 8 and 9 depict schemes for possible mechanisms involved in the uptake of inorganic mercury and methylmercury when the respective mercuric ion is in some complex with GSH. It is possible that complexes of GSH and inorganic or organic mercury are taken up across the basolateral membrane, since there is evidence that GSH can be transported across the basolateral membrane as an intact tripeptide in a manner that is dependent on the activity of the sodium pump (Lash and Jones, 1983, 1984). Another possibility is that some mercury is delivered to a site or

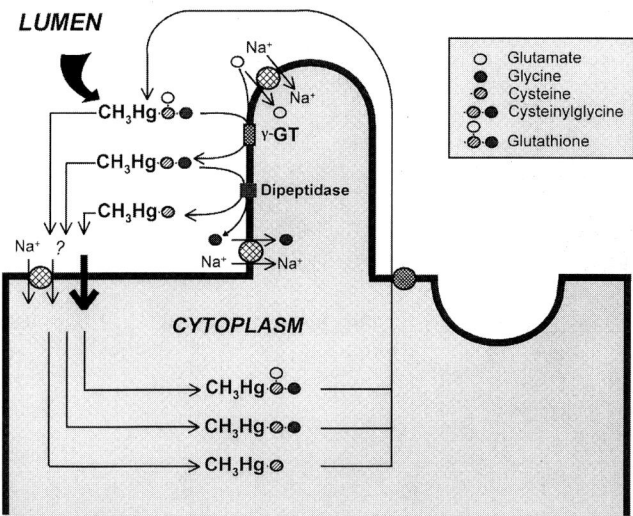

Figure 6 Scheme of transport of methylmercury at the brush border membrane of proximal tubular epithelial cells involving cotransport of the methylmercuric ion with glutathione (GSH) or a degradative product of GSH. The scheme shows that complexes of CH_3Hg:GSH can be delivered to the luminal membrane of proximal tubular epithelial cells after ultrafiltration. Complexes containing GSH and methylmercury may be taken up directly across the luminal membrane by some mechanism involving facilitated diffusion or transport of the lipophilic methylmercuric ion. It seems unlikely that a transport mechanism would involve reabsorptive uptake of GSH as a tripeptide, since luminal uptake of GSH as a tripeptide has not been well documented. More likely, mechanisms involve the catabolism of GSH by the γ-glutamyltransferase (which cleaves the γ-glutamylcysteine bond) and/or subsequent catabolism of cysteinylglycine by a dipeptidase on the luminal membrane. Methylmercury bonded to either cysteinylglycine or cysteine may be taken up across the luminal membrane by the sodium-dependent amino acid or dipeptide transport systems. These conjugates may also enter the cell by a mechanism relating to the lipophilic nature of methylmercury. There is evidence that CH_3Hg:GSH complexes may be secreted into the lumen by some transport mechanism and, once in the lumen, the GSH on the complexes can be acted on by γ-glutamyltransferase in a GSH recycling pathway.
Note: mechanisms other than the ones presented in the above-mentioned scheme may also be involved in the proximal tubular reabsorptive or secretory transport of methylmercury. The purpose of the scheme is to relate the potential role of GSH in the transport of methylmercury across the luminal membrane of proximal tubular epithelial cells. While parts of the scheme were generated from currently accepted mechanisms involved in the transport of methylmercury across the luminal membrane of proximal tubular epithelial cells, other parts were generated from logical possibilities that need to be explored and tested.

receptor on the basolateral plasma membrane (that is nucleophilic) via a complex with GSH or a plasma protein. Once the mercury comes in contact with the site on the basolateral membrane, it could bind to the site and remain outside the cell, or the mercury could be internalized by some translocation process. Methylmercury-GSH complexes may move more readily across the basolateral membrane due to the lipophilic nature of methylmercury.

D. INTRACELLULAR GSH AND THE ACCUMULATION AND TOXICITY OF MERCURY IN THE KIDNEY

Modulation of intracellular GSH in renal epithelial cells greatly influences the metabolism and toxicity of mercuric compounds. Both diethyl maleate and buthionine sulfoximine are frequently used to reduce intracellular content of GSH in the kidney. Diethyl maleate reduces intracellular GSH by forming conjugates with GSH and buthionine sulfoximine depletes intracellular GSH by inhibiting the synthesis of GSH. Johnson (1982) and Berndt and colleagues (Berndt et al., 1985; Baggett and Berndt, 1986) reported that depletion of renal intracellular GSH or thiols with diethyl maleate prior to treatment with mercuric chloride resulted in a decrease in the renal uptake and accumulation of inorganic mercury in rats. Moreover, Berndt and colleagues reported that the nephropathy induced by mercuric chloride was made more severe in rats when GSH was depleted prior to treatment with mercuric chloride. Tanaka-Kagawa et al. (1993), however, were recently unable to detect any changes in the renal accumulation of mercury in mice treated with either inorganic mercury or methylmercury after the animals were

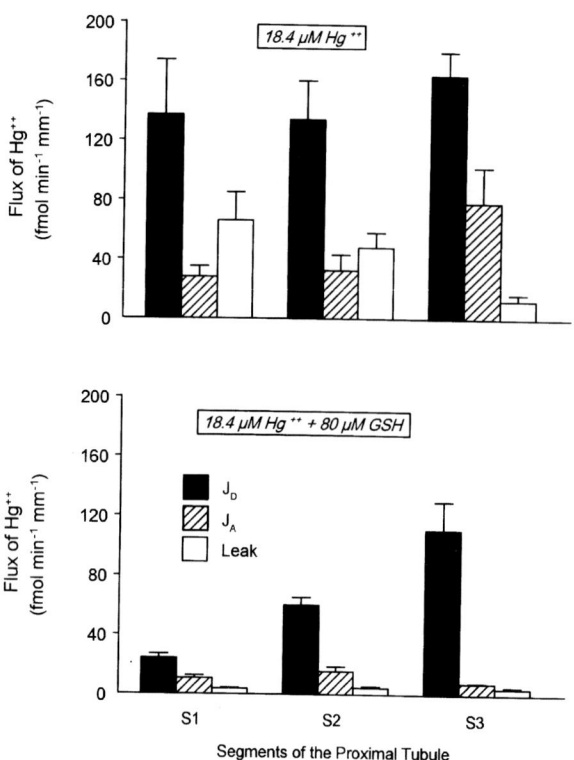

Figure 7 Influence of glutathione (GSH) on the lumen-to-bath transport of inorganic mercury (Hg^{2+}) in isolated perfused S1, S2, and S3 segments of the proximal tubule of the rabbit. Addition of 80 μM GSH to a perfusate containing 18.4 μM $^{203}Hg^{2+}$ caused the rate of disappearance of $^{203}Hg^{2+}$ from the lumen (J_D), the rate of appearance of $^{203}Hg^{2+}$ in the bath (J_A) and the leak of the volume marker ^{3}H-L-glucose to decrease significantly. In addition to causing a decrease in the lumen-to-bath transport of inorganic mercury, the addition of GSH to the perfusate provided complete protection to the epithelial cells of all three segments of the proximal tubule from the toxic effects of 18.4 μM inorganic mercury. The perfusate used in the experiments was a buffered electrolyte solution (280 mOsmol/kg) containing no organic ligands with which inorganic mercury could bind. The perfusion experiments lasted between 45 and 60 min. Values represent means ± SE.

pretreated with buthionine sulfoximine and acivicin, which is a potent inhibitor of γ-glutamyltransferase. To make the matter somewhat confusing, Girardi and Elias (1991) reported that the renal accumulation of inorganic mercury increased in mice pretreated with diethyl maleate. The reasons for these discrepancies are not clear at present.

As stated earlier, increased urinary excretion of both inorganic mercury and GSH have been observed in rats (Berndt et al., 1985) and mice (Tanaka et al., 1990) pretreated with acivicin. In another study with mice (Tanaka-Kagawa et al., 1993), inhibition of γ-glutamyltransferase prior to treatment with inorganic mercury or methylmercury caused the urinary excretion of mercury to increase and the renal accumulation of mercury to decrease. These data again indicate that the brush border enzyme γ-glutamyltransferase is probably involved in the renal uptake and accumulation of mercury.

Tanaka et al. (1990) also showed recently that the renal accumulation of mercury decreased in mice when intracellular GSH in hepatocytes was depleted specifically with 1,2-dichloro-4-nitrobenzene prior to treatment with inorganic mercury. They observed that the depletion of hepatocellular GSH caused a decrease in the severity of the nephropathy induced by inorganic mercury. These data seem to indicate the presence of a hepato-renal shunt that delivers mercury to the kidney as a GSH-mercury conjugate or complex that is first formed in the liver. These data also indicate that the proximal tubular uptake of mercury is linked to the activity of the γ-glutamyltransferase on the luminal (brush border) membrane.

Incubation of isolated proximal tubular cells, from both normal and uninephrectomized rats, with GSH has been shown by Lash and Zalups (1992) to provide protection from the toxic effects of subsequent exposure to mercuric chloride. Figure 11 shows that preincubation of isolated proximal tubular cells

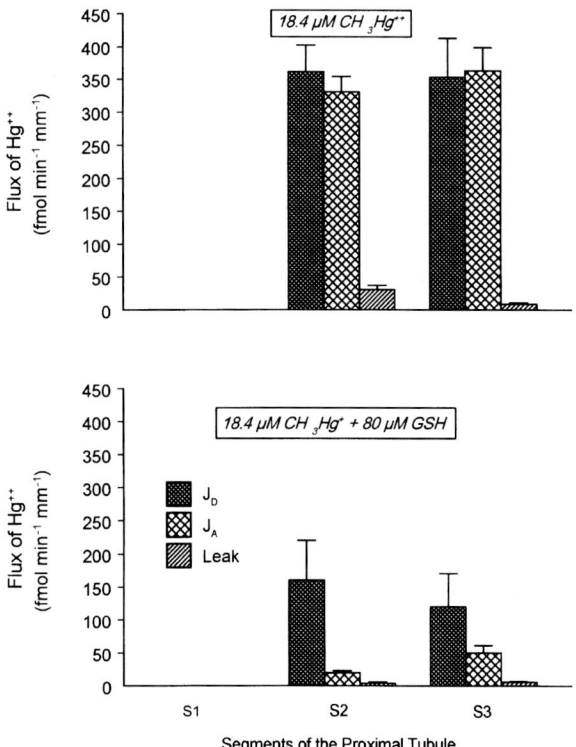

Figure 8 Influence of glutathione (GSH) on the lumen-to-bath transport of methylmercury (CH$_3$Hg$^+$) in isolated perfused S2 and S3 segments of the proximal tubule of the rabbit. Rates of transport of CH$_3^{203}$Hg$^+$ could not be determined in the S1 segments of the proximal tubule due to the toxic effects of 18.4 μM CH$_3^{203}$Hg$^+$ on the tubular epithelium. Addition of 80 μM GSH to a perfusate containing 18.4 μM CH$_3^{203}$Hg$^+$ caused the rate of disappearance of CH$_3^{203}$Hg$^+$ from the lumen (J$_D$), the rate of appearance of CH$_3^{203}$Hg$^+$ in the bath (J$_A$), and the leak of the volume marker ^{3}H-L-glucose to decrease significantly. Addition of GSH to the perfusate did not noticably affect the toxicity of methylmercury in the S2 segments of the proximal tubule. However, addition of the GSH to the perfusate seemed to enhance the toxic effects of methylmercury in the S3 segments. It should be mentioned that the perfusate used in the experiments was a buffered electrolyte solution (280 mOsmol/kg) containing no other organic ligands with which the methylmercury could bind. The perfusion experiments lasted about 45 and 60 min. Values represent means ± SE.

with GSH provided concentration-dependent protection from injury induced by mercuric chloride, as assessed by decreases in total intracellular activity of lactate dehydrogenase. Preincubation with a twofold higher concentration of GSH (500 μM) was required to provide complete protection from the cytotoxic effects of 250 μM mercuric chloride. This concentration presumably increased the probability for linear II coordinate complexes forming between two molecules of GSH and each mercuric ion, inside and/or outside the cells, which in turn decreased the probability of the mercuric ions interacting with vital binding sites on or in the proximal tubular epithelial cells.

As a final point, Girardi and Elias (1991) reported recently that treatment of mice with N-acetylcysteine (which was used to increase renal cellular contents of GSH) prior to injection with inorganic mercury, caused renal and hepatic accumulation of mercury to decrease. One might have expected to see the opposite, since increasing the intracellular concentration of GSH should have provided more ligands with which mercuric ions could bind. These findings are in contrast to what is observed when the concentration of MT is increased in renal epithelial cells. Induction of the synthesis of MT in renal tubular epithelial cells with zinc, prior to treatment with mercuric compounds, causes the renal concentration of mercury to increase (Fukino et al., 1984; Zalups and Cherian, 1992a,b). Decreased renal injury induced by mercury has also been observed in association with the induction of synthesis of MT. It should be pointed out, however, that MT is primarily confined to the intracellular environment. The findings of Girardi and Elias (1991) lead one to conclude that the renal cellular disposition of mercury-

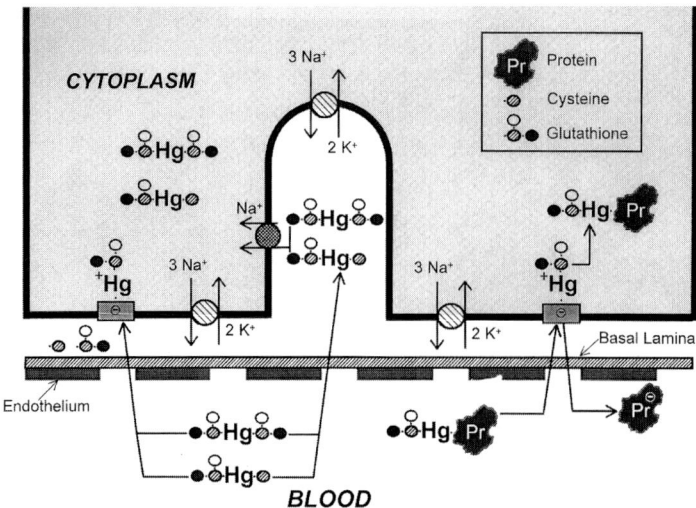

Figure 9 Scheme of transport of inorganic mercury at the basolateral membrane of proximal tubular epithelial cells involving cotransport of the inorganic mercuric ion with glutathione (GSH). The scheme shows that complexes of GSH:Hg:GSH, GSH:Hg:Cysteine, and GSH:Hg:Protein can be delivered to the basolateral membrane of proximal tubular epithelial cells by the blood. Any of the three complexes may come in contact with the basolateral membrane and allow one of the electrophilic binding sites of the mercuric ion to bind to a nucleophilic site on the plasma membrane, which could result in the internalization of the mercuric-GSH complex or even the mercuric ion by itself. The internalized ion could then potentially bind to some cellular protein or nonprotein thiol. Inorganic mercury may also enter proximal tubular epithelial cells at the basolateral membrane in complexes of GSH:Hg:GSH or GSH:Hg:Cysteine, which may be taken up across the basolateral membrane by the sodium-dependent GSH transporter. Transport of GSH as a tripeptide has been demonstrated to occur across the basolateral membrane.

Note: Mechanisms other than the ones presented in the above-mentioned scheme may also be involved in the proximal tubular basolateral transport of inorganic mercury. The purpose of the scheme is to relate the potential role of GSH in the transport of inorganic mercury across the basolateral membrane of proximal tubular epithelial cells. While parts of the scheme were generated from currently accepted mechanisms involved in the transport of inorganic mercury across the luminal membrane of proximal tubular epithelial cells, other parts were generated from logical possibilities that need to be explored and tested.

containing compounds must be regulated by a more complex set of factors than the intracellular availability of thiol ligands on GSH.

E. INFLUENCE OF MERCURY ON RENAL CELLULAR METABOLISM OF GSH

Not only can GSH alter the renal metabolism of mercury, but mercury can alter the renal metabolism of GSH. Both inorganic and organic mercuric compounds cause alterations in the intracellular metabolism of GSH in the kidney. Most of these alterations have been detected after short-term single treatments, and are concentration dependent. It has been demonstrated in several *in vivo* and *in vitro* renal studies that intracellular concentrations of GSH in renal epithelial cells increase after treatment with relatively low toxic doses or subtoxic doses of either methylmercury (Woods et al., 1992) or inorganic mercury (Aleo et al., 1987; Chung et al., 1982; Fukino et al., 1986; Siegers et al., 1987; Zalups and Lash, 1990; Zalups and Veltman, 1988). At higher, nephrotoxic doses of inorganic mercury, however, the renal content of GSH decreases (Addya et al., 1984; Aleo et al., 1987, 1992; Fukino et al., 1984, 1986; Girardi and Elias, 1991; Gstraunthaler et al., 1983; Lash and Zalups, 1992; Zalups and Lash, 1990; Zalups and Veltman, 1988), presumably due primarily to extensive cellular and tubular necrosis.

Figure 12 shows the concentration dependence of the effect of treatment with inorganic mercury on renal GSH. These data were obtained from the kidneys of male Sprague-Dawley rats given intravenous doses (0.0, 0.5, 2.0, or 3.0 µmol/kg body weight) of mercuric chloride 24 h prior to making determinations of the renal concentrations of GSH. The 0.5-µmol/kg dose produced no demonstrable renal injury, the 2.0-µmol/kg dose produced a moderate amount of cellular necrosis in the pars recta segments of proximal tubules, and the 3.0-µmol/kg dose of mercuric chloride produced very severe necrosis in pars recta segments of the proximal tubule. The 0.5- and 2.0-µmol/kg doses of mercuric chloride induced significant

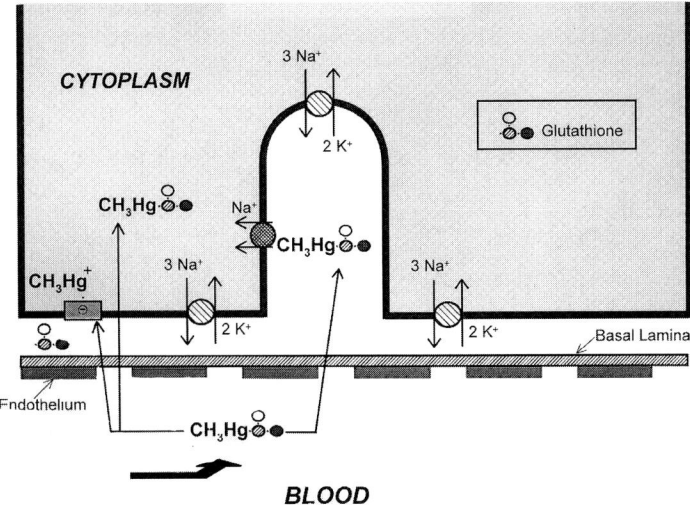

Figure 10 Scheme of transport of methylmercury at the basolateral membrane of proximal tubular epithelial cells involving cotransport of the methylmercuric ion with glutathione (GSH). The scheme shows that complexes of $CH_3Hg:GSH$ can be delivered to the basolateral membrane of proximal tubular epithelial cells by the blood. The methylmercury in these complexes may gain entry into the proximal tubular epithelial cells by one of three potential pathways. The first pathway involves direct entry by some lipophilic transport system. The second pathway involves the transport of the $CH_3Hg:GSH$ complex by the the sodium-dependent GSH transporter on the basolateral membrane. The third pathway involves nucleophilic substitution at a binding site on the basolateral membrane, following by internalization of the methylmercuric ion.

Note: Mechanisms other than the ones presented in the above-mentioned scheme may also be involved in the proximal tubular basolateral transport of methylmercury. The purpose of the scheme is to relate the potential role of GSH in the transport of methylmercury across the basolateral membrane of proximal tubular epithelial cells. While parts of the scheme were generated from currently accepted mechanisms involved in the transport of methylmercury across the luminal membrane of proximal tubular epithelial cells, other parts were generated from logical possibilities that need to be explored and tested.

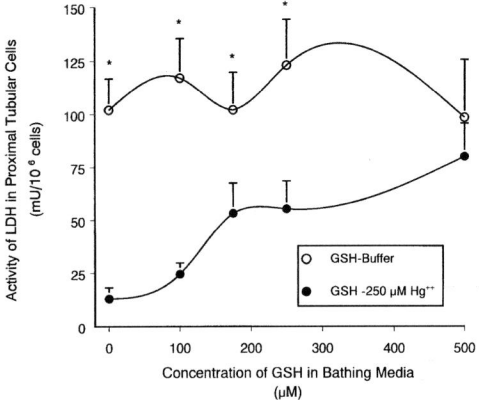

Figure 11 Protective effect of glutathione GSH on the cytotoxic effects of mercuric chloride in isolated proximal tubular epithelial cells from the rat. Proximal tubular cells ($2–3 \times 10^6$ cells/ml) were isolated from the kidneys of rats by collagenase perfusion and digestion followed by Percoll density-gradient centrifugation. Cells were preincubated for 15 min with the indicated concentration of GSH under an atmosphere of 95% O_2/5% CO_2, resuspended in fresh buffer without GSH, and then incubated for 1 h with either buffer or 250 μM $HgCl_2$. Lactate dehydrogenase (LDH) activity (mU/10^6 cells) was measured as NADH oxidation in the presence of NADH, pyruvate, and Triton X-100®. Results are presented as means ± SE for measurements from three to four separate cell preparations;* significantly different ($p < .05$) from the corresponding samples incubated with buffer.

increases in the concentration of GSH in the kidneys, which were due to increases in the concentration of GSH in the renal cortex or outer stripe of the outer medulla. The greatest increase in the concentration of GSH was detected in the renal outer stripe of the outer medulla of the rats treated with the 2.0-μmol/kg dose of mercuric chloride, where the concentration of GSH increased by 85%. At the highest dose of mercuric chloride, the renal concentration of GSH did not change significantly. The decrease in the concentration of GSH relative to that at the 2.0-μmol/kg dose can be explained by the extensive necrosis that occurred along pars recta segments of the proximal tubule in the cortex and outer stripe of the outer medulla. Extensive necrosis would cause some depletion in the total renal content of GSH.

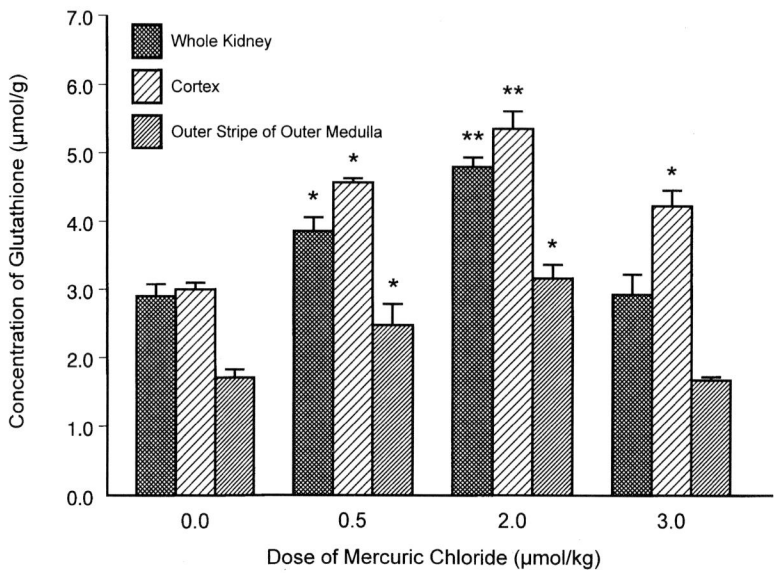

Figure 12 Concentration of glutathione (GSH) in renal tissue of rats 24 h after the rats were treated with a 0.0, 0.5, 2.0, or 3.0 μmol/kg dose of mercuric chloride. Concentrations of GSH (expressed as μmol/g tissue) were determined by high performance liquid chromatography in samples of whole kidney, cortex, and outer stripe of the outer medulla. Animals were administered $HgCl_2$ (in 2.0 ml/kg 0.9%, w/v, sodium chloride) by injection into the femoral vein during anesthesia induced by ether. Values are means ± SE for 12 animals (0.0 μmol $HgCl_2$/kg) or for 6 animals (all other doses);* significantly different ($p < .05$) from the corresponding mean for the rats given the 0.0 μmol/kg dose of mercuric chloride;** significantly different ($p < .05$) from the corresponding means for the animals given the 0.0 μmol/kg dose of mercuric chloride and the animals given the 0.5 μmol/kg dose of mercuric chloride.

Since the cellular content of GSH is under feedback control, the increases in the concentrations and content of GSH that occur after treatment with nontoxic or low toxic doses of mercuric chloride may be due to mercury inducing the synthesis of the rate-limiting enzyme in the synthesis of GSH, γ-glutamylcysteine synthetase. Lash and Zalups (1994) tested this hypothesis recently by measuring the activity of γ-glutamylcysteine synthetase in freshly isolated renal proximal tubular and distal tubular epithelial cells from control rats and from rats treated with a nontoxic, 0.5-μmol/kg dose of mercuric chloride 24 h prior to cell isolation. Both proximal tubular and distal tubular cells from rats treated with the nontoxic dose of inorganic mercury exhibited significant increases in the activity of γ-glutamylcysteine synthetase (Figure 13). Woods et al. (1992) have also provided evidence that mercury can induce synthesis of γ-glutamylcysteine synthetase in the kidney. They showed that mRNA for γ-glutamylcysteine synthetase increased significantly in the kidneys of male Fischer 344 rats that were treated chronically with methylmercury hydroxide. Thus, there is evidence that subtoxic doses of both inorganic and organic mercury can induce synthesis of GSH in the kidney.

In addition to causing alterations in the synthesis and activity of the rate-limiting enzyme involved in the biosynthesis of GSH in the kidneys, inorganic mercury also alters the activity of GSH-dependent drug metabolism enzymes in a dose-dependent manner. Lash and Zalups (1994) detected increases in the activities of GSSG reductase and GSH peroxidase in isolated epithelial cells from both proximal and distal tubular regions of the kidneys from rats treated with a subtoxic 0.5 μmol/kg dose of mercuric chloride 24 h prior to the isolation of the cells. In contrast, *in vivo* data from Addya et al. (1984) show that the activities of GSSG reductase and GSH peroxidase decreased in the kidneys of rats that were

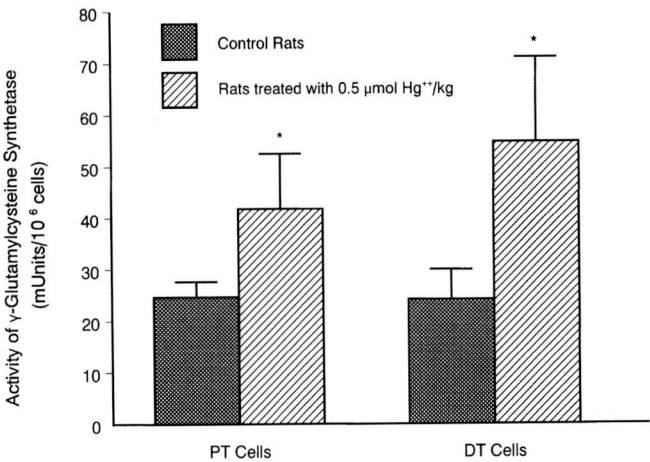

Figure 13 Activity of γ-glutamylcysteine synthetase (mUnits/10^6 cells) in proximal tubular and distal tubular cells isolated from the kidneys of rats treated or not treated with a single 0.5 μmol/kg subtoxic dose of mercuric chloride 24 h prior to the isolation of cells. The mercuric chloride was administered intraperitoneally in 0.9% aqueous sodium chloride (2.0 ml/kg body weight). Proximal tubular (PT) and distal tubular (DT) cells were isolated from the kidneys of rats by collagenase perfusion and digestion followed by Percoll density-gradient centrifugation. The activity of γ-glutamylcysteine synthetase was measured with L-α-aminobutyrate and L-glutamate as substrates by a pyridine nucleotide-linked spectrophotometric assay. As can be seen, *in vivo* treatment with mercuric chloride caused the activity of γ-glutamylcysteine synthetase to increase in both PT and DT cells;* significantly different ($p < .05$) from the mean for the cells from control rats not treated with mercuric chloride.

treated chronically with a relatively high dose of mercuric chloride. Chung et al. (1982), Gstraunthaler et al. (1983) and Fukino et al. (1984) have also detected significant decreases in renal GSSG reductase activity after administration of high nephrotoxic doses of mercuric chloride. These findings are really not surprising because of the severe tubular injury induced by the highly toxic doses of mercuric chloride.

More work clearly needs to be performed to better elucidate the role of mercury, and other metal ions on the intracellular regulation of the metabolism of GSH and other thiols. Studies directed at this task should avoid using large toxic doses of metal, for the data are made ambiguous by renal cellular injury. To better study the effect of mercury and other metals on GSH metabolism, nontoxic or subtoxic doses of metal should be used.

VI. CONCLUSIONS

Various forms of mercury are present in the environment and workplace, which pose a risk to human health. The disposition and effects of both the inorganic and organic forms of mercury in mammals are largely determined by reactions with molecules containing free sulfhydryl groups. This chapter has focused mainly on the interactions between inorganic and organic forms of mercury and GSH. The kidneys, liver, and blood play a major role in the handling and disposition of mercury in the body of mammals.

Based on the findings and proposed models presented in this chapter, it is very clear that further studies are needed to better characterize the mechanisms involved in the transport of mercury in various compartments of the body. Modulation of various aspects of transport, with particular focus on interactions with GSH, may provide useful information on the mechanisms involved in the transport and toxicity of the various forms of mercury in target organs and tissues.

REFERENCES

Adam, K. R., *Br. J. Pharmacol.,* 6, 483–491, 1951.

Addya, A., Chakravarti, K., Basu, A., Santra, M., Haldar, S., and Chatterjee, G. C., *Acta Vitaminol. Enzymol.,* 6, 103–107, 1984.

Aleo, M. D., Taub, M. L., Olson, J. R., Nickerson, P. A., and Kostyniak, P. J., in In Vitro *Toxicology: Approaches to Validation,* Goldberg, A.M., Ed., Mary Ann Liebert, Inc., New York, 211, 1987.

Aleo, M. D., Taub, M. L., and Kostyniak, P. J., *Toxicol. Appl. Pharmacol.,* 112, 310–317, 1992.
Ashe, W. F., Largent, E. J., Dutra, F. R., Hubbard, D. M., and Blackstone, M., *Arch. Ind. Hyg.,* 7, 19–43, 1953.
Baggett, J. McC. and Berndt, W. O., *Toxicol. Appl. Pharmacol.,* 83, 556–562, 1986.
Ballatori, N., *Drug Metab. Rev.,* 23, 83–132, 1991.
Ballatori, N. and Clarkson, T. W., *Am. J. Physiol.,* 244, G435–G441, 1983.
Ballatori, N. and Clarkson, T. W., *Biochem. Pharmacol.,* 33, 1087–1092, 1984a.
Ballatori, N. and Clarkson, T. W., *Biochem. Pharmacol.,* 33, 1093–1098, 1984b.
Ballatori, N., Jacob, R., and Boyer, J. L., *J. Biol. Chem.,* 261, 7860–7865, 1986.
Ballatori, N., Jacob, R., Barrett, C., and Boyer, J. L., *Am. J. Physiol.,* 254, G1–G7, 1988.
Barfuss, D. W., Ganapathy, V., and Leibach, F. H., *Am. J. Physiol.,* 255, F177–F181, 1988.
Bergstrand, A. Friberg, L., Mendel, L., and Odeblad, E., *J. Ultrastruct. Res.,* 3, 238–239, 1959.
Berlin, M., *Arch. Environ. Health,* 6, 610–616, 1963.
Berlin, M. and Gibson, S., *Arch. Environ. Health,* 6, 617–625, 1963.
Berlin, M. and Ullberg, S., *Arch. Environ. Health,* 6, 582–601, 1963a.
Berlin, M. and Ullberg, S., *Arch. Environ. Health,* 6, 602–609, 1963b.
Berndt, W. O., Baggett, J. McC., Blacker, A., and Houser, M., *Fundam. Appl. Toxicol.,* 5, 832–839, 1985.
Beutler, E., in *Functions of Glutathione: Biochemical, Physiological, Toxicological, and Clinical Aspects,* Larsson, A., Orrenius, S., Holmgren, A., and Mannervik, B., Eds., Raven Press, New York, 65, 1983.
Beutler, E. and Dale, G. L., in *Glutathione: Chemical, Biochemical and Medical Aspects,* Vol. 3, Dolphin, D., Auramovic, O., and Poulson, R., Eds., John Wiley & Sons, New York, 292, 1989.
Bourdeau, J. E., Carone, F. A., and Ganote, C. E., *J. Cell Biol.,* 54, 382–398, 1972.
Brown, J. R. and Shockley, P., in *Lipid-Protein Interactions,* Vol. 1, Jost P.C. and Griffith, O.H., Eds., John Wiley & Sons, New York, 25, 1982.
Cember, H., Gallagher, P., and Faulkner, A., *Am. Ind. Hyg. Assoc. J.,* 29, 233–237, 1968.
Cherian, M. G. and Clarkson, T. W., *Chem.-Biol. Interact.,* 12, 109–120, 1976.
Chung, A.-S., Maines, M. D., and Reynolds, W. A., *Biochem. Pharmacol.,* 31, 3093–3100, 1982.
Clapp, W. L., Park, C. H., Madsen, K. M., and Tisher, C. C., *Lab. Invest.,* 58, 549–558, 1988.
Clarkson, T. W., *Annu. Rev. Pharmacol. Toxicol.,* 12, 375–406, 1972.
Dirks, J. H., Clapp, J. R., and Berliner, R. W., *J. Clin. Invest.,* 45, 916–921, 1964.
Dutczak, W. J. and Ballatori, N., *J. Pharmacol. Exp. Ther.,* 262, 619–623, 1992.
Foulkes, E. C., *Am. J. Physiol.,* 227, 1356–1360, 1974.
Friberg, L., *Acta Pharmacol. Toxicol.,* 12, 411–427, 1956.
Friberg, L., *AMA Arch. Ind. Health,* 20, 42–49, 1959.
Friberg, L., Odeblad, E., and Forssman, S., *AMA Arch. Ind. Health,* 16, 163–168, 1957.
Friedman, H. L., *Ann. N.Y. Acad. Sci.,* 65, 461–470, 1957.
Fuhr, B. J. and Rabenstein, D. L., *J. Am. Chem. Soc.,* 95, 6944–6950, 1973.
Fukino, H., Hirai, M., Hsueh, Y. M., and Yamane, Y., *Toxicol. Appl. Pharmacol.,* 73, 395–401, 1984.
Fukino, H., Hirai, M., Hsueh, Y. M., Moriyasu, S., and Yamane, Y., *J. Toxicol. Environ. Health,* 19, 75–89, 1986.
Girardi, G. and Elias, M. M., *Toxicology,* 67, 155–164, 1991.
Griffith, O. W. and Meister, A., *Proc. Natl. Acad. Sci. U.S.A.,* 76, 268–272, 1979.
Gstraunthaler, G., Pfaller, W., and Kotanko, P., *Biochem. Pharmacol.,* 32, 2969–2972, 1983.
Hahn, L. J., Kloiber, R., Vimy, M. J., Takahashi, Y., and Lorscheider, F. L., *FASEB J.,* 3, 2641–2646, 1989.
Hahn, L. J., Kloiber, R., Leininger, R. W., Vimy, M. J., and Lorscheider, F. L., *FASEB J.,* 4, 3256–3260, 1990.
Hinchman, C. A. and Ballatori, N., *Biochem. Pharmacol.,* 40, 1131–1135, 1990.
Hultman, P. and Enestrom, S., *Br. J. Exp. Pathol.,* 67, 493–503, 1986.
Hultman, P., Enestrom, S., and von Schenck, H., *Virchows Arch. B,* 49, 209–224, 1985.
Jocelyn, P. C., *Biochemistry of the SH Group,* Academic Press, London.
Johnson, D. R., *J. Toxicol. Environ. Health,* 9, 119–126, 1982.
Landwegr, D. M., Carvalho, J. S., and Oken, D. E., *Kidney Int.,* 11, 9–17, 1977.
Lash, L. H. and Jones, D. P., *Biochem. Biophys. Res. Commun.,* 112, 55–60, 1983.
Lash, L. H. and Jones, D. P., Renal glutathione transport: *J. Biol. Chem.,* 259, 14508–14514, 1984.
Lash, L. H. and Jones, D. P., *Arch. Biochem. Biophys.,* 240, 583–592, 1985.
Lash, L. H. and Zalups, R. K., *J. Pharmacol. Exp. Ther.,* 261, 819–829, 1992.
Lash, L. H. and Zalups, R. K., *Arch. Biochem. Biophys.,* in Press.
Maack, T., Renal handling of proteins and polypeptides. In *Handbook of Physiology,* Vol. II, Sect. 8, Windhager, E.E., Ed., Oxford University Press, New York, 2039, 1992.
Madsen, K. M., *Kidney Int.,* 18, 445–453, 1980.
Magos, L. and Butler, W. H., *Arch. Toxicol.,* 35, 25–39, 1976.
Magos, L. and Clarkson, T. W., in *Handbook of Physiology,* Sect. 9, Lee, D.H.K., Ed., American Physiological Society, Bethesda, MD, 503, 1977.
Magos, L., Halbach, S., and Clarkson, T. W., *Biochem. Pharmacol.,* 27, 1327–1331, 1978.
Magos, L., Peristianis, G. G., Clarkson, T. W., Brown, A., Preston, S., and Snowden, R. T., *Arch. Toxicol.,* 48, 11–20, 1981.
Magos, L., Brown, A. W., Sparrow, S., Bailey, E., Snowden, R. T., and Skipp, W. R., *Arch. Toxicol.,* 57, 260–267, 1985.

McNeil, S. I., Bhatnagar, M. K., and Turner, C. J., Combined toxicity of ethanol and methylmercury in rat, *Toxicology,* 53, 345–363, 1988.
Mussini, E., *Boll. Soc. Ital. Biol. Sper.,* 34, 1588–1590, 1958.
Naganuma, A., Oda-Urano, N., Tanaka, T., and Imura, N., *Biochem. Pharmacol.,* 37, 291–296, 1988.
Norseth, T. and Clarkson, T. W., *Arch. Environ. Health,* 21, 717–727, 1970a.
Norseth, T. and Clarkson, T. W., *Biochem. Pharmacol.,* 19, 2775–2783, 1970b.
Oken, D. E. and Flamebaum, W., *J. Clin. Invest.,* 50, 1498–1505, 1971.
Prickett, C. S., Laug, E. P., and Kunze, F. M., *Proc. Soc. Exp. Biol. Med.,* 73, 585–588, 1950.
Rabenstein, D. L., *Acc. Chem. Res.,* 11, 100–107, 1978.
Rabenstein, D. L., in *Glutathione: Chemical, Biochemical and Medical Aspects,* Vol. 3, Dolphin, D., Auramovic, O., and Poulson, R., Eds., John Wiley & Sons, New York, 147, 1989.
Rodier, P. M., Kates, B., and Simons, R., *Toxicol. Appl. Pharmacol.,* 57, 235–245, 1988.
Rothstein, A. and Hayes, A. D., *J. Pharmacol. Exp. Ther.,* 130, 166–176, 1960.
Silbernagl, S., in *Handbook of Physiology,* Vol. II, Sect. 8, Windhager, E.E., Ed., Oxford University Press, New York, 1938, 1992.
Siegers, C.-P., Schenke, M., and Younes, M., *J. Toxicol. Environ. Health,* 22, 141–148, 1987.
Swensson, A. and Ulfvarson, U., *Acta Pharmacol. Toxicol.,* 26, 273–283, 1968.
Tanaka, T., Naganuma, A., and Imura, N., *Toxicology,* 60, 187–198, 1990.
Tanaka-Kagawa, T., Naganuma, A., and Imura, N., *J. Pharmacol. Exp. Ther.,* 264, 776–782, 1993.
Taugner, R., Winkel, K., and Iravani, J., *Virchows Arch. Pathol. Anat. Physiol.,* 340, 369–383, 1966.
Woods, J. S., Davis, H. A., and Baer, R. P., *Arch. Biochem. Biophys.,* 296, 350–353, 1992.
Zalups, R. K., *J. Pharmacol. Methods,* 26, 89–104, 1991a.
Zalups, R. K., *Exp. Mol. Pathol.,* 54, 10–21, 1991b.
Zalups, R. K., *J. Toxicol. Environ. Health,* 33, 213–228, 1991c.
Zalups, R. K., *Toxicology,* 79, 215–228, 1993.
Zalups, R. K. and Barfuss, D. W., *Toxicol. Appl. Pharmacol.,* 106, 245–253, 1990.
Zalups, R. K. and Barfuss, D. W., *J. Toxicol. Environ. Health,* 40, 77–103, 1993a.
Zalups, R. K. and Barfuss, D. W., *Toxicol. Appl. Pharmacol.,* 121, 176–185, 1993b.
Zalups, R. K. and Barfuss, D. W., *Toxicologist,* in Press.
Zalups, R. K. and Cherian, M. G., *Toxicology,* 71, 83–102, 1992a.
Zalups, R. K. and Cherian, M. G., *Toxicology,* 71, 103–117, 1992b.
Zalups, R. K. and Diamond, G. L., *Bull. Environ. Contam. Toxicol.,* 38, 67–72, 1987a.
Zalups, R. K. and Diamond, G. L., *Virchows Arch. B,* 53, 336–346, 1987b.
Zalups, R. K. and Lash, L. H., *J. Pharmacol. Exp. Ther.,* 254, 962–970, 1990.
Zalups, R. K. and Veltman, J. C., *Life Sci.,* 42, 2171–2176, 1988.
Zalups, R. L., Klotzbach, J. M., and Diamond, G. L., *Toxicol. Appl. Pharmacol.,* 89, 226–236, 1987.
Zalups, R. K., Cox, C., and Diamond, G. L., in *Biological Monitoring of Toxic Metals,* Clarkson, T.W., Friberg, L., Nordberg, G.F., and Sager, P.R., Eds., Plenum Press, New York, 1988.
Zalups, R. K., Robinson, M. K., and Barfuss, D. W., *J. Am. Soc. Nephrol.,* 2, 866–878, 1991.
Zalups, R. K., Barfuss, D. W., and Kostyniak, P. J., *Toxicol. Appl. Pharmacol.,* 115, 174–182, 1992.
Zalups, R. K., Knutson, K. L., and Schnellman, R. G., *Toxicol. Appl. Pharmacol.,* 119, 221–227, 1993.

Chapter 10

The Role of General and Metal-Specific Cellular Responses in Protection and Repair of Metal-Induced Damage: Stress Proteins and Metallothioneins

Brenda M. Sanders, Peter L. Goering, and Kenneth Jenkins

ABSTRACT

This chapter discusses our current knowledge of the mechanisms by which organisms protect themselves from metal-induced damage at the molecular and cellular levels. Initially we discuss the mechanisms of metal toxicity, both direct and indirect, and the nature of the potential molecular damage. We then focus on two cellular strategies in which exposure to metals induces the synthesis of proteins that act to protect and repair macromolecules from damage. One system, the cellular stress response, functions to repair and protect macromolecules from many types of environmentally induced damage, including metal toxicity. It acts to provide cellular protection through the protection and repair of proteins and protein complexes. In addition, another strategy which entails the metal-specific induction of metallothionein synthesis is also often brought online to alter metal metabolism in a specific manner. Each of these protective strategies is discussed from the perspective of what is known regarding their function in cell physiology under both normal and stressful conditions, the ways in which each system provides protection from metal toxicity, and interrelationships between the two responses. In addition, the induction of the cellular stress response and metallothionein by specific metals is reviewed.

I. INTRODUCTION: GENERAL VS. METAL-SPECIFIC RESPONSES TO METALS

Although organisms encounter stress often in their environment, our knowledge of how they adapt to it is quite limited at the molecular and cellular levels. It is now apparent that cells have evolved a number of lines of defense to protect themselves from environmental adversity. Several of these strategies involve inducible systems in which exposure to stressful conditions induces the synthesis of proteins which act to protect and repair macromolecules from stress-induced damage. General stress responses involve molecular interactions which function in concert to repair and protect macromolecules from environmentally induced damage and maintain metabolic and genetic integrity. In this chapter we will focus on the epigenetic response, often referred to as the cellular stress response or the heat-shock response. The cellular stress response is induced by a broad range of stress conditions and protects against stressor-induced damage to proteins and protein complexes. Protection of DNA from metal-induced damage will be addressed in another chapter. In addition, stressor-specific systems such as the induced synthesis of metallothionein act to alter metal metabolism in a specific manner. In this chapter we will describe both general and metal-specific responses which contribute to the protection of cells from metal toxicity.

We discuss what is currently known about the induction of the stress response elicited by metals and metalloids, with emphasis on their role in stress physiology and their applications in toxicology. We also review the role of the metal-binding ligand metallothionein in metal metabolism and detoxification. We summarize our current understanding of the mechanisms which regulate expression of the stress protein and metallothionein genes. The biochemical functions of these inducible proteins in cells under normal and stressful conditions are also reviewed to provide an integrated and mechanistic understanding of their role in repair and protection from metal-induced damage. More extensive reviews which cover these topics can be found in Gething and Sambrook (1992); Rothman (1989); Nover (1991); Morimoto et al., (1994a); Mason and Jenkins (1995); and Klaassen and Suzuki (1991). The chapter also includes a discussion of the relationship between the induced synthesis of stress proteins and cellular damage, and the mechanisms by which stress proteins and metallothionein appear to protect cells and organisms from metal toxicity.

II. MECHANISMS OF METAL-INDUCED TOXICITY

As with heat-shock treatment, proteins are major targets of damage from metals (Goering, 1993; Pohl, 1993; Sugiyama, 1994; Hojima et al., 1994). Loss of protein function is usually a consequence of protein modifications involving both direct and indirect interactions with metals. Metal-protein interactions which result in toxicity can arise through several mechanisms. For example, metals can block functional sites through binding to sulfhydryl groups which are part of the catalytic or binding domains. Mercury and arsenite inactivate a number of enzymes through this mechanism (Vallee and Ulmer, 1972; Massey et al., 1962; Peters, 1955). Alternatively, one metal may displace another metal which is essential for biological activity (Dixon and Webb, 1967; Coleman, 1967). Since these essential metals often act to stabilize the conformation of a metalloprotein or as electron donors/acceptors, their metal-protein interaction is highly specific and metal substitutions usually result in loss of biological activity (Vallee and Ulmer, 1972; Wood and Wang, 1983). Displacements will most likely occur when the concentration of the competitive metal is elevated and the essential metal is deficient. Factors such as geometry, size, and coordination chemistry also come into play. The inactivation of the Zn-requiring enzyme, δ-aminolevulinic acid dehydratase (ALAD) by lead is an example of this latter mechanism since it occurs through the direct displacement of zinc by lead from a SH-mediated binding site (Goering, 1993).

Metals can also alter protein conformation by covalently binding to sulfhydryl groups or creating protein adducts through modification of side chains leading to changes in protein shape and activity (Pohl, 1993; Bruschi et al., 1993). Such alterations are often the result of the generation of free radicals by metals (Held and Biaglow, 1993; Yoshida et al., 1993). Genotoxic effects of some metals and metalloids, including arsenic, copper, and cadmium, result from their modification of proteins that participate in DNA repair (Sunderman, 1984; Yager and Wiencke, 1993). For example, a number of metals and metalloids, which include arsenic, cadmium, silver, copper, nickel, chromium, manganese, cobalt, and lead have been shown to alter DNA polymerase (Christie and Costa, 1983; Nocentini, 1987).

As is discussed in detail in a subsequent section (III.C), which addresses the induction of the stress response by environmental perturbation, the induced synthesis of stress proteins by metals probably occurs as a result of metal-induced protein damage (Ananthan et al., 1986; Hightower, 1991). Both direct and indirect induction pathways may be involved, all acting through the same transcription factor. Metal induction of the stress response that is mediated through oxidative damage is expressed quite differently from that of heat-induced activation and is most likely the result of differences in cellular damage (Keyse and Tyrrell, 1987; Bruce et al., 1993). The oxidative pathway may be a primary mechanism of induction of the response for metals such as cadmium, mercury, nickel, arsenite, copper, lead, and iron, which induce oxygen free radicals or promote formation of lipid peroxides (Stacey and Klaassen, 1981; Amoruso et al., 1982; Halliwell and Gutteridge, 1984; Christie and Costa, 1984; Kasprzak, 1991; Donati et al., 1991). An additional mechanism may involve depletion of intracellular thiol pools, since several sulfhydryl reactive agents (e.g., diamide, mercury, cadmium, diethylmaleate, and iodoacetamide) increase synthesis of stress proteins. Further, specific induction of the subset of stress proteins called glucose regulated proteins (grps) by metals such as cadmium (Goering et al., 1993b) and lead (Shelton et al., 1986) may be related to the capacity of these agents to disrupt cellular calcium homeostasis (Dwyer et al., 1991; Vig and Nath, 1991), since grps are induced by other agents which alter calcium homeostasis (Nover, 1991).

III. THE CELLULAR STRESS RESPONSE

The cellular stress response is involved in limiting protein damage as a result of exposure to a wide variety of environmental stressors, both natural and anthropogenic (for reviews see Lindquist, 1986; Nover, 1991; Sanders, 1993). This cellular stress response is induced as a result of a number of normal and abnormal conditions that affect homeostatic processes, which include changes in physiological state involving hormones and growth factors, pathophysiological conditions, and environmental stressors (Morimoto et al., 1994a). The environmental stressors that can elicit the response encompass a wide variety of physical variables (e.g., anoxia, elevated temperatures, UV radiation), as well as a wide range of chemicals, which include metals and metalloids.

A major feature of the response is the rapid synthesis of proteins, referred to as stress proteins, in response to environmental stress. These proteins are highly conserved in evolution and play similar roles in organisms from bacteria to humans. However, the specific characterization of the stress response including the number and size of the proteins induced and the extent of induction, is dependent upon a number of factors, which include: (1) the species examined, (2) the specific stressor used to induce the response, (3) chemical speciation of the toxicant, (4) the exposure protocol, (5) the tissue or cell type, and (6) the developmental stage of the organism (Nover, 1991; Shuman and Przybyla, 1988; Lai et al., 1993).

Implicit in what is known regarding the protection and repair functions of stress proteins is the premise that they may play an important role in protecting organisms from metal-induced damage. In addition, the induction of stress proteins by metals and their subsequent accumulation provide information on the mechanisms of metal toxicity and can be used to help identify vulnerable target organelles and tissues.

The stress proteins which make up the heat shock protein (hsp) families of 90, 70, 60, and 16 to 24 kDa and ubiquitin are the most studied of the heat-inducible proteins. The term stress protein, however, is often used to denote the broader class of stress-inducible proteins of which the heat-shock proteins are a subset. Although nomenclature for these proteins is not consistent in the literature, in this chapter we will use the terms "stress90" and "stress70" for proteins in the 90-kDa and 70-kDa families, respectively, and the 60-kDa family will be referred to as chaperonin60 (cpn60). Members of these major stress protein families are found in a number of cellular compartments and the specific terminology for each protein is dependent upon both their cellular location and the species in which they are found (see Sanders, 1993). Since many stress proteins are constitutively expressed it should be noted that the term stress proteins does not imply that they are present only under stressful conditions (Craig et al., 1983).

Numerous studies have demonstrated that stress proteins protect cells and organisms from environmentally induced stress (Parsell and Lindquist, 1994; Morimoto et al., 1994a). Studies comparing the cellular stress response in closely related species which inhabit different environments demonstrate differences in the response that can be related to their thermal resistance, suggesting that stress proteins have an adaptive role (Bosch et al., 1988; Sanders et al., 1991a; White et al., 1994). There is also evidence that the stress response is induced by chemicals at concentrations found in polluted environments (Sanders and Martin, 1994).

A. MECHANISMS OF STRESS PROTEIN INDUCTION

Heat-inducible genes include a conserved sequence referred to as the heat-shock element (HSE) in their upstream regulatory region. Activation of the gene is initiated by a family of regulatory proteins, called heat-shock factors (HSF), which bind to this element (Morimoto et al., 1994b). There are at least two primary classes of HSFs; HSF1 binds to the HSE in response to heat shock, oxidative stress, and metals, whereas HSF2 responds to physiological factors such as developmental stage, hormones, and spermatogenesis. Although we do not precisely know how these adverse environmental conditions activate HSF1, a number of studies suggest that heat-shock and other stressors which cause an increase in damaged or abnormal proteins activate it in some manner which involves trimerization and a series of phosphorylation reactions (Wu et al., 1994). Stress70 appears to modulate the interactions between the HSF and the HSE, possibly through binding to the HSF and regulating induction of the stress response.

Numerous studies on the regulation of the cellular stress response support the premise that the induced synthesis of stress proteins is related to protein damage. Injection of denatured, but not native, protein results in transcription of heat-shock genes in *Xenopus* oocytes (Ananthan et al., 1986). Amino acid analogs that create abnormal proteins induce stress protein synthesis (Hightower, 1980). Also, biochemical conditions which alter protein conformation, such as addition of glycerol and other polyhydroxyl alcohols and D_2O, inhibit the stress response induced by heat shock but not by sodium arsenite, as expected (Edington et al., 1989; Mosser et al., 1990). However, the most intriguing aspect of this regulation of induction of the stress proteins is that denatured proteins appear to be both the signal that activates transcription of the stress protein genes and the substrate for the proteins themselves.

B. STRESS PROTEINS MAINTAIN PROTEIN INTEGRITY

A number of recent reviews describe the role of stress proteins under normal conditions; several of the major stress proteins are present at low levels and function as molecular chaperones to facilitate the folding, assembly, and distribution of newly synthesized proteins (Craig et al., 1994; Hightower et al., 1994; Frydman and Hartl, 1994). Under conditions of environmental stress, they are involved in the protection and repair of vulnerable protein targets (Rothman, 1989; Gething and Sambrook, 1992). They also vector damaged proteins to the lysosomal and ubiquitin protein degradation pathways (Dice et al., 1994; Schlesinger, 1993). In essence, the cellular stress response entails the orchestrated induction of key proteins that form the basis for the cell's protein repair and recycling systems.

The way in which these proteins act as catalysts of protein folding and repair is best understood for members of the stress70 and cpn60 families. These are illustrated in Figure 1A. Stress70 is a large multigene family with members in a number of subcellular compartments including the cytoplasm, mitochondria, and endoplasmic reticulum (Hartl et al., 1992; Craig et al., 1994). Under normal conditions they prevent incorrect folding of newly synthesized peptides by binding to the growing peptide chain and maintaining it in a loosely folded state until synthesis is complete (Figure 1A: a) (Beckmann et al., 1990; Gething and Sambrook, 1992). A number of other accessory proteins (e.g., hsp40) appear to interact with stress70 and participate in various aspects of this process (Hightower et al., 1994). Stress70 disassociation, an ATP-dependent process, occurs as the protein proceeds down its appropriate folding pathway to reach its correct three-dimensional shape (Figure 1A: b,h). In addition, proteins which need to be distributed to other subcellular compartments (e.g., mitochondria, endoplasmic reticulum) are maintained in an unfolded state and escorted to that destination for translocation (Figure 1A: c) (Marshall et al., 1990; Phillips and Silhavy, 1990; Nover 1991; Sanders et al., 1992). Once inside the organelle the target protein interacts with another member of the stress70 family which performs similar folding functions (Figure 1: d,e) (Chirico et al., 1988; Craig, 1990; Hartl and Neupert, 1990).

Additional folding functions and assembly are carried out by the cpn60 family which is found in eubacteria, mitochondria, plastids and, most recently, in the nucleus. Chaperonin60 assembles into large "double donut" shaped complexes that direct higher-level folding and the assembly of subunits into complexes through an ATP-dependent process (Figure 1A: f,g) (McMullin and Hallberg, 1988; Ellis, 1990; Buchner et al., 1991). A functional homolog, TCP1, is present in the cytoplasm, where it facilitates folding of actin and tubulin and perhaps other proteins (Figure 1A: i,j). Although this group of chaperonins is weakly related to the cpn60 group it is not considered a stress protein because its synthesis is not induced by stress.

Stress90 appears to have a more subtle folding role. It interacts with target proteins, including enzymes (Brugge et al., 1981), transcription factors (Twomey et al., 1993), cytoskeletal proteins (Koyasu et al., 1986, 1989), and steroid hormone receptors (Brunt et al., 1990). The binding of stress90 results in a transitional complex which can regulate the activity of the protein and participate in the later stages of the folding pathway (Gething and Sambrook, 1992).

A more diverse group of stress proteins, referred to as the small or low molecular weight stress proteins (sHSP), in the size range of approximately 20 to 30 kDa, are induced by a variety of environ-

The Role of General and Metal-Specific Cellular Responses

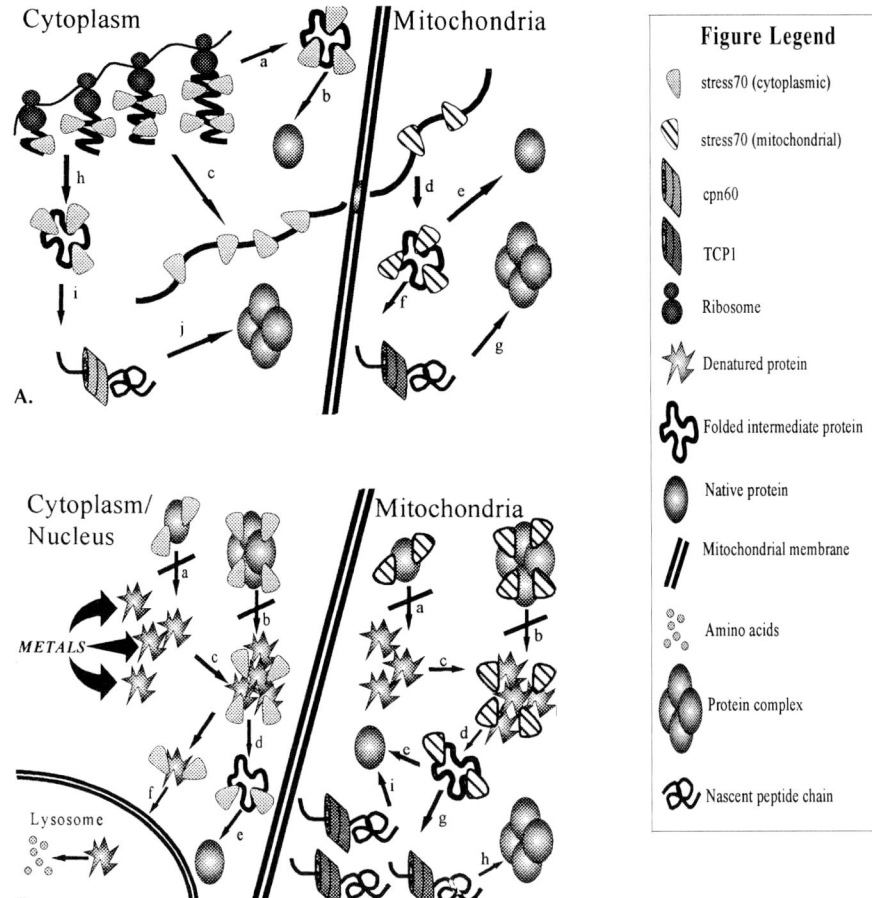

Figure 1A Cellular functions of stress70 and cpn60 under normal conditions. Stress70 prevents incorrect folding of newly synthesized peptides by binding to the growing peptide chain and maintaining it in a loosely folded state until synthesis is complete (a). Stress70 disassociation, an ATP-dependent process, occurs as the protein proceeds down its appropriate folding pathway to reach its correct three-dimensional shape (b). Proteins which need to be distributed to other subcellular compartments are maintained in an unfolded state and escorted to that destination for translocation (c). Once inside the organelle, the target protein interacts with another member of the stress70 family which performs similar folding functions (d and e). Chaperonins assemble into large "double donut"-shaped complexes that act to direct the higher-level folding and the assembly of subunits into complexes (f and g). A functional homolog, TCP1, is present in the cytoplasm where it facilitates folding of actin and tubulin, and perhaps other proteins (i and j).

Figure 1B The related roles of stress70 and cpn60 under conditions of metal-induced stress. In response to metals stress70 increases to protect the cell from metal-induced damage by binding to vulnerable proteins preventing denaturation (a and b) and the formation of insoluble aggregates (c). Stress70 also can break up existing aggregates and repair proteins to complete biological activity (d and e), and vector irreparably damaged proteins to the lysosome for degradation (f). Metals also stimulate cpn60 synthesis and it takes on related roles in protection and repair by preventing protein aggregation of misfolded proteins and facilitating renaturation and assembly into complexes (g, h and i).

mental factors. They are also under hormonal and developmental regulation (Ciocca et al., 1993). The sHSPs are homologous to the lens protein, α-crystallin. In humans, hsp27/28 and αB crystallin, which share a similar structure and promoter region, are the major sHSPs (Kato et al., 1993). It has been suggested that hsp27 also performs chaperon functions in protein folding since this protein has been shown to prevent aggregation and facilitate refolding of damaged proteins (Horwitz, 1992; Jakob et al., 1993; Ciocca et al., 1993; Arrigo and Landry, 1994). However, there is still much to be determined to understand their specific role in protein folding and repair.

In response to environmentally stressful conditions, including exposure to elevated trace metals and metalloids, stress proteins take on related roles in the protection and repair of proteins and protein complexes in various cellular compartments that are targets of stress-induced damage (Figure 1B). Stress70 protects preexisting proteins from damage (Figure 1B: a,b). Stress70 can also repair damaged proteins and dissolve existing aggregates to restore proteins to complete biological activity (Figure 1B: c,d,e) (Skowyra et al., 1990; Gaitanaris et al., 1990; Ellis, 1990; Schroder et al., 1993). In addition to its protective role in the cytoplasm and mitochondria, it migrates to the nucleus and nucleolus where it binds to nuclear proteins, preribosomes, and other protein complexes and protects them from potential damage (Lindquist, 1986; Ellis 1990; Pelham, 1990; Gething and Sambrook, 1992). Proteins which are damaged beyond repair are escorted by stress70 to the lysosome or the ubiquitin pathway for breakdown (Figure 1B: f). Another, lesser known stress protein, hsp 110, which associates with the nucleolus upon heat shock, may also facilitate the breakup of protein aggregates (Sanchez and Lindquist, 1990). Exposure to adverse environmental conditions also induces cpn60 synthesis and it takes on additional roles in protection and repair (Figure 1B: g,h,i) (Gething and Sambrook, 1992). It also prevents aggregation of incompletely folded proteins (Cheng et al., 1989; Ostermann et al., 1989; Martin et al., 1992).

C. THE RECYCLING OF DAMAGED PROTEINS

Proteins that are beyond repair are recycled through the major cytosolic and lysosomal degradation pathways which complement the resolubilization and stabilization activities of chaperonin, stress70, and other chaperones (Figure 2). An increase of stress-induced damage to proteins induces the synthesis of proteins which participate in protein protection and turnover through three discrete pathways: (1) increasing levels of chaperones which, in addition to their protection and repair functions, vector irreparably damaged proteins to the lysosome for degradation via several breakdown pathways and to several cytoplasmic degradation pathways; (2) increasing levels of ubiquitin, which increases the capacity of this cytoplasmic pathway to breakdown proteins; and (3) increasing levels of a group of proteases which selectively degrade nonnative proteins (Leonhardt et al., 1993; Goldberg and Guillou 1994; Chiang et al., 1989; Dice et al., 1994). Each pathway appears to be highly selective in terms of the substrate accepted for degradation. For example, certain cytoplasmic proteins have been shown to contain specific peptide sequences which bind stress70 and subsequently are targeted to the lysosomes for breakdown.

Ubiquitin is a small molecular weight (7 kDa) protein involved in the nonlysosomal degradation of intracellular proteins (Schlesinger and Hershko, 1988; Schlesinger, 1990). Small multigene families code for both the constitutively expressed and inducible forms of ubiquitin. Under stressful conditions ubiquitin acts as the cofactor for cytoplasmic degradation of abnormal proteins by covalently binding to proteins and tagging them for degradation. Thus, an increase in ubiquitin increases the capacity for turnover of severely damaged proteins along this nonlysosomal pathway.

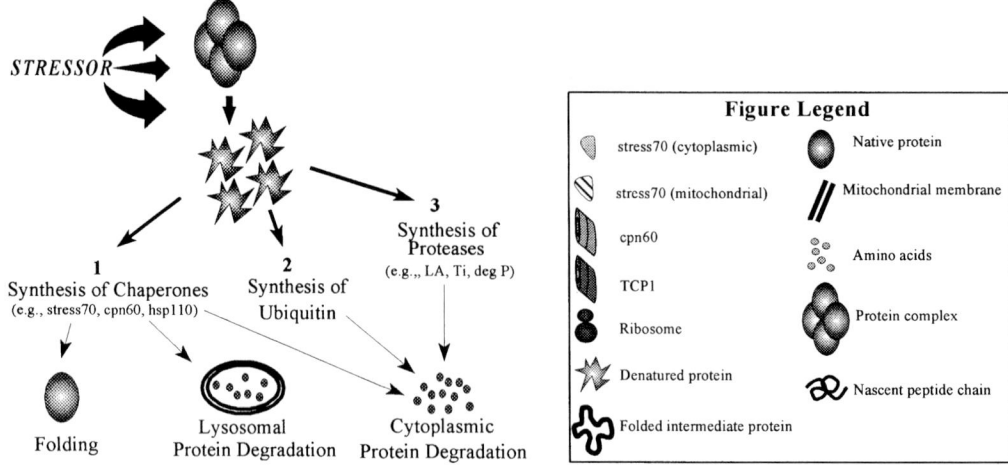

Figure 2 Overview of the function of the cellular stress response in maintaining protein homeostasis. See text for discussion.

D. ACQUIRED THERMOTOLERANCE

At both the cellular and the organismal levels induction of synthesis of stress proteins correlates with acquired tolerance, in which exposure to a mildly stressful condition increases the ability to survive a subsequent severe stressful condition that would otherwise be lethal (Lindquist, 1986; Lindquist and Craig, 1988; Black and Subjeck, 1991; Welte et al., 1993). Acquired thermotolerance in which heat shock is used as the stressor has been studied most extensively and has been demonstrated in organisms from diverse phyla (Boon-Niermeijer et al., 1986; Lindquist, 1986; Mosser et al., 1987; Landry et al., 1982, 1989; Hahn and Li, 1990; Sanders, 1993). Other stressors that induce stress proteins, such as arsenite, cadmium, and ethanol, also induce thermotolerance (Li, 1983; Kapoor and Sveenivasan, 1988; Nover, 1991; Hahn and Li, 1990; Burgman et al., 1993). At the cellular level, acquired tolerance is related to protection of such universal processes as protein synthesis and RNA processing and other more specific cellular functions in differentiated tissue (Black and Subjeck, 1991; Brown et al., 1992).

Gerner and Schneider (1975) were the first to demonstrate that heat shock induces a transient state of heat resistance in mammalian cells. Subsequent research demonstrated that a heat shock sufficient to induce stress protein synthesis confers thermotolerance in both cellular and embryonic systems (Li and Laszlo, 1985; Mirkes, 1987; Johnston and Kucey, 1988; Riabowol et al., 1988). Thermotolerance is attributable to the induction and expression of stress proteins. Correlations between the ability to synthesize stress proteins and the expression of thermotolerance have been observed in diverse species (Landry et al., 1982; Heikkila et al., 1986; Welch and Mizzen, 1988; Mosser and Bols, 1988). A number of studies, have demonstrated that the kinetics of thermotolerance induction and decay correlate with stress protein synthesis and degradation (Landry et al., 1982; Li and Werb, 1982; Lavoie et al., 1993). In addition, yeast mutants which are unable to synthesize stress proteins are highly sensitive to elevated temperatures, whereas mutants that constitutively express the proteins are resistant to damage from high temperatures (Krause et al., 1987; Chretien and Landry, 1988). Rodent cells which express a cloned gene for human stress70 become thermally resistant (Li et al., 1992). Studies using site-directed mutagenesis also demonstrate that stress proteins are required for thermotolerance (Landry et al., 1989; Sanchez and Lindquist, 1990).

There is also increasing evidence at the organismal level to suggest that stress proteins play a role in environmental adaptation (Sanders, 1993). Species adapted to environmental extremes exhibit a number of unique features of the stress response, including a higher basal expression of stress proteins (Kee and Nobel, 1986; Koban et al., 1991; Ulmasov et al., 1993), enhanced ability to induce the response (Loomis and Wheeler, 1980; Bosch et al., 1988), and the presence of a greater number of isoforms of the major stress proteins groups (Berger and Woodward, 1983; Landry et al., 1989; Sanders et al., 1991a, 1992; White et al., 1994).

E. SUBCELLULAR LOCALIZATION OF STRESS PROTEINS IN RESPONSE TO STRESS

The role of stress70 and cpn60 in repair of environmentally induced damage is further substantiated by their cellular localization and distribution in response to stress. Lindquist (1986) was the first to show that cytoplasmic stress70 moves into the nucleus in response to heat shock, where it interacts with the nucleolus, the site of ribosomal assembly, and returns to the cytoplasm during recovery. More recently, Hattori et al. (1993) demonstrated that its smaller companion, hsp40, responds to heat shock in a similar manner. It appears that cpn60 also localizes in the nucleus in a heat-inducible manner (Sanders et al., 1994b).

The intracellular localization of stress proteins differs when other stressors besides heat shock are used to elicit the response. To illustrate this point, a series of micrographs of a fish cell line are presented in Figures 3 and 4, in which stress70 and cpn60 are visualized by indirect immunofluorescence after a heat-shock treatment or exposure to elevated copper. Stress70 is normally present at low levels (Figure 3A: Sanders et al., 1994b); however, 2 h after a 30-min heat-shock treatment of 38°C it increases in the perinuclear region of the cytoplasm and in the nucleus, where it associates with the nucleolus. Stress70 disperses throughout the cytoplasm and forms a pattern of concentric rings in the nucleus 2 h later (Figure 3A), and within 24 h after the heat-shock treatment its abundance and distribution returns to normal (Figure 3A).

Chaperonin60 is also normally present at low levels under normal culturing conditions (Figure 3B). Its abundance increases within the mitochondria 2 to 6 h after the same heat-shock treatment. Then, 12 to 16 h after heat shock, it can be visualized in the nucleus in association with the nucleolus and as discrete foci (Figure 3B: f–g). The nuclear localization of cpn60 in these foci is similar to the distribution of the snRNP-rich organelles, referred to as coiled bodies, which are believed to be involved

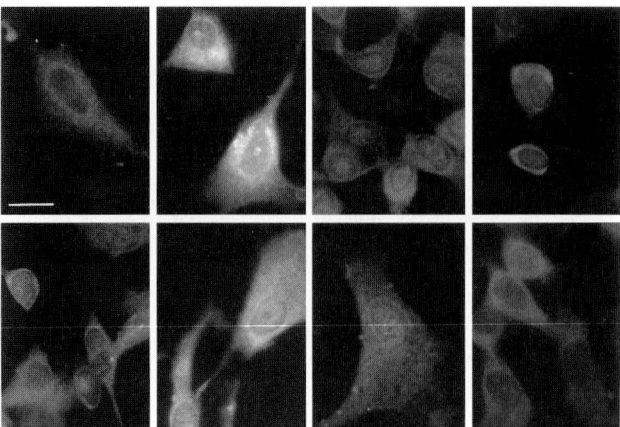

Figure 3A Indirect immunofluorescence micrographs of fish epithelial cells probed with anti-stress70 pAb and detected by antibody conjugated with FITC. Top row: (left to right) cells cultured at 24°C for 2 h; 4 h; 6 h; bottom row: (left to right) 8 h; 12 h; 16 h; and 24 h after heat shock. Scale bar equals 10 µm. (From Sanders, B.M., Nguyen, J., Douglass, T.G., and Miller, S., *Biochem. J., 297,* 21–25, 1994. With permission.)

in pre-mRNA splicing (Zhang et al., 1992; Lamond and Carmo-Fonseca, 1993; Matunis et al., 1993). They have also been shown to be disrupted by heat shock (Welch and Mizzen, 1988).

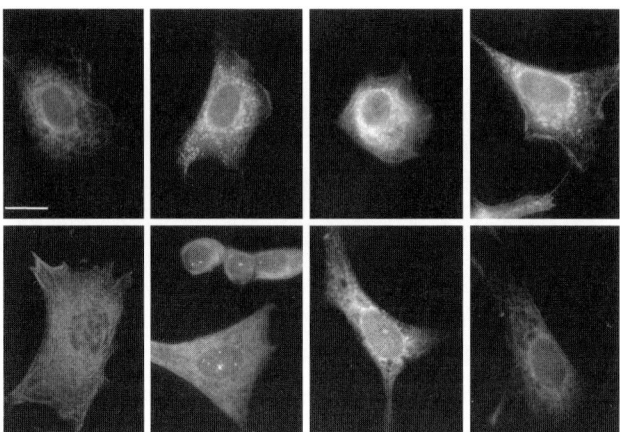

Figure 3B Indirect immunofluorescence micrographs of fish epithelial cells probed with anti-cpn60 pAb and detected by antibody conjugated with FITC. Top row: (left to right) cells cultured at 24°C for 2 h; 4 h; 6 h; bottom row: (left to right) 8 h; 12 h; 16 h; and 24 h after heat shock. Scale bar equals 10 µm. (From Sanders, B.M., Nguyen, J., Douglass, T.G., and Miller, S., *Biochem. J., 297,* 21–25, 1994. With permission.)

It is apparent that these two major stress proteins interact differently with target proteins within the cell. The translocation of stress70 into the nucleus is more rapid than cpn60. Also, cpn60 localizes as highly discrete foci, whereas, stress70 does not. Further, although both proteins interacted with the nucleolus, the nucleolar association of cpn60 occurs well after stress70 has migrated from the nucleolus. This sequential interaction is also observed during the folding of multimeric proteins in the mitochondria where the mitochondrial member of the stress70 family facilitates early folding and passes the partly folded intermediate on to cpn60 for subsequent folding and assembly (Manning-Krieg et al., 1991; Hartl et al., 1992). That these stress proteins have distinct interactions in the nucleus suggest that they play different roles in facilitating repair of nuclear structures.

Subcellular localization of stress proteins after exposure to a range of copper concentrations differs from that observed with heat-shock treatments (Figure 4). Cytoplasmic stress70 increases with increasing copper concentrations and is dispersed relatively uniformly, with staining most intense around the perinuclear region at 750 µM copper (Figure 4A: d). Notably, stress70 is absent in the nucleus and the

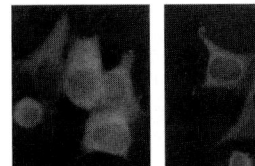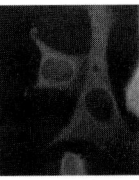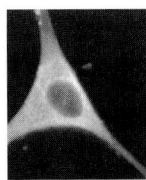

Figure 4A Indirect immunofluorescence micrographs of fathead minnow epithelial cells probed with anti-stress70 pAb and detected by antibody conjugated with FITC. Top row: (left to right) cells cultured for 24 h at 24°C in 0, 250, 500, 750 μM copper. Scale bar equals 10 μm. (From Sanders, B.M., Nguyen, J., Martin, L.S., Howe, S.R., and Coventry, S., *Comp. Biochem. Physiol.*, 112C, 335–343, 1995. With permission.)

nucleolus. Chaperonin60 in the mitochondria did not appear to increase appreciably with copper exposure, but staining of mitochondria became more distinct at the higher concentrations (Figure 4B: a through d) (Sanders et al., 1994a). As is observed with stress70, cpn60 is absent in the nucleus at all copper concentrations. These observations suggest that, in contrast to the damage resulting from heat-shock treatments, the cellular targets of copper-induced damage are not nuclear, but cytoplasmic and mitochondrial. Similar differences in subcellular localization of stress proteins have been observed in mammalian cells for other stressors including arsenite (Welch and Feramisco, 1984; Vincent and Tanguay, 1982). Kato et al. (1993) suggest that a similar relationship may occur with hsp28. From a toxicological perspective this association of stress proteins with protein complexes particularly sensitive to stress-induced damage may prove helpful to identifying subcellular sites of toxicity for various chemicals and other stressors.

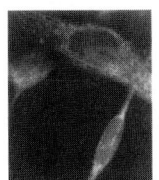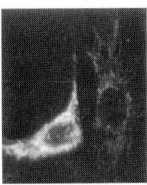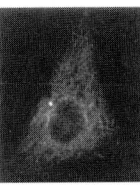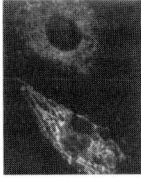

Figure 4B Indirect immunofluorescence micrographs of fathead minnow epithelial cells probed with anti-cpn60 pAb and detected by antibody conjugated with FITC. Bottom: (left to right) cells cultured for 24 h at 24°C in 0, 250, 500, 750 μM copper. Scale bar equals 10 μm. (From Sanders, B.M., Nguyen, J., Martin, L.S., Howe, S.R., and Coventry, S., *Comp. Biochem. Physiol.*, 112C, 335–343, 1995. With permission.)

F. INDUCTION OF THE STRESS RESPONSE AMONG TISSUES IS STRESSOR SPECIFIC

Induction of stress protein synthesis in response to environmental stress is also highly tissue specific. In response to heat-shock treatment the temperature range of induction and the extent and time course for induction of each stress protein appears to depend upon tissue type (Dyer et al., 1991; Sanders, 1993; Wilkinson and Pollard, 1993; Arrigo and Landry, 1994). This tissue specificity is probably a result of several mechanisms, including differences in gene expression and cellular metabolism among specialized cell types, and the extent of tissue level damage. In light of our current understanding of the regulation and function of the stress response, its intensity and the relative concentrations of stress proteins among the various tissues should be greatest in tissues which are most vulnerable to a particular environmental stressor. For chemical stressors the factors determining potential damage would include: (1) the distribution of the chemical among tissues; (2) the ability of each tissue type to detoxify the contaminant and minimize damage; and (3) molecular mechanisms of chemical toxicity. As a consequence of the interplay between the many physiological processes involved, the relative concentrations of stress proteins among tissues differ significantly with the stressor. Such differences might be particularly useful in identifying target tissues and evaluating the extent of damage.

Studies at the organismal level in mammals, fish, and invertebrates lend further support to the notion that stress proteins associate with damaged tissues. Within hours after an acute dosing of rats with cadmium, the synthesis of 70-, 90-, and 110-kDa proteins is induced in the liver, a primary target organ for this metal, but not in the kidney (Goering et al., 1993b). Further, induced synthesis of these proteins occurs prior to detection of cell injury. Two of these proteins, hsp72 and grp94, are elevated for up to 24 h after the exposure. When cadmium is complexed with cysteine, which targets cadmium to the kidney, a primary target organ for chronic exposure to cadmium, stress proteins are induced within 4 h

(Goering et al., 1993c). Further, induction of the response occurs at cadmium concentrations that are lower than those in which renal injury is detected. Mercury also induces the synthesis of stress proteins in rats *in vivo* in a tissue-, dose-, and time-dependent manner (Goering et al., 1992). Proteins of 43, 70, 90, and 110 kDa are induced in the kidney, but not the liver from mercury-injected rats. Changes in renal protein synthesis is rapid (2 to 4 h) and occurs prior to overt renal injury.

In fathead minnows the route of exposure dictates the tissue level stress response to arsenite, a hydrophilic protein denaturant in which all tissues are potential targets (Dyer et al., 1993). The gill which is in direct contact with arsenite in the water column is more sensitive than tissues not directly exposed. In contrast, the tissue level response upon exposure to two hydrophobic pesticides, diazinon and lindane, is less dependent upon the mode of exposure and is targeted to specific tissues. Diazinon is a cholinesterase inhibitor, and lindane inhibits the GABA-activated chloride channel in neurons. The primary targets for induction of the stress response after exposure to these pesticides are the tissues with the greatest innervation — brain and muscle tissue — and expression of the stress response is lowest in the gill. Similar tissue-specific differences are observed in the bivalve *Mytilus edulis* exposed to copper, another hydrophilic compound which acts to damage proteins (Sanders et al., 1994a). Upon exposure to copper in seawater, the accumulation of cpn60 is more than an order of magnitude higher in the gill, a primary organ of copper accumulation, than the mantle. Further, the specific isoforms of stress70 which are induced differ between these two tissues.

Perhaps the most dramatic evidence of the tissue specific induction of stress proteins by different types of stressors, including metals, is observed in the patterns observed in the transgenic nematode *Caenorhabditis elegans* derived from the fusion of the hsp16 gene promoter to the *Escherichia coli lacZ* reporter gene (Stringham and Candido, 1994). In this system, induction of the stress response results in an increased transcription and translation of the enzyme β-galactosidase, which stains a dark blue upon exposure to substrate (Figure 5). Exposure to sodium arsenite results in the diffuse expression of β-galactosidase throughout most of the somatic tissues, a pattern similar to that observed with heat-shock induction. Induction of the stress response with mercury stains only in the gut, cadmium stains most of the cells in the pharynx, and copper stains only the neurons and muscles of the anterior end of the pharynx. Lead induces staining in only one muscle cell in the posterior bulb of the pharynx, and zinc induction stains hydrodermal cells of the cuticle. As a comparison, the herbicide paraquat induces staining in the intestinal cells.

The tissue-specific patterns of expression of the stress response by different metals and other contaminants observed in these diverse species and the association of stress proteins with known target tissues serves to reinforce the notion that induction of the stress response reflects tissue level damage. Accordingly, differences in the levels of accumulation of stress proteins among tissues may prove useful in identifying target tissues that are particularly vulnerable to metal-induced damage and the specific routes of exposure of the metal (Sanders et al., 1991b; Goering et al., 1993a).

G. THE PROTECTIVE ROLE OF THE CELLULAR STRESS RESPONSE

Under specific circumstances the prior induction of stress proteins by diverse stressors, including heat shock and elevated metals, affords protection for cells and organisms from stressor-induced injury. Studies have demonstrated a strong correlation between the induced synthesis and translocation of stress proteins and the development of this protection. The molecular mechanisms for this tolerance relate to those underlying acquired thermotolerances discussed previously in this chapter (Section III.D) and involve the inherent role of stress proteins in protecting and repairing target proteins from damage.

Thermal stress may confer chemical tolerance (for review see Nover, 1991). Renfro et al. (1993) showed that a mild heat shock can inhibit the reduction of transport capacity of renal proximal tubules caused by exposure to the herbicide 2,4-D, and that exposure to elevated zinc can protect tissue function. Some metals readily confer thermotolerance, but heat shock generally does not confer tolerance to the effects of metals. Sodium arsenite confers protection against a variety of physiologic stressors and confers thermotolerance (Wiegant et al., 1987; Lee and Dewey, 1987, 1988; Lee and Hahn, 1988; Kampinga et al., 1992; Burgman et al., 1993). Although a few exceptions have been reported, heat-shock treatment does not generally confer tolerance to cadmium in cultured cells (Cervera, 1985; Ciavarra and Simeone, 1990a,b). Cadmium exposure confers protection from cadmium toxicity, whereas pretreatment with zinc or cadmium generally does not confer thermotolerance in cultured cells (Cervera, 1985) and sea urchin embryos (Roccheri et al., 1988). An exception has been reported; cadmium treatment induced thermotolerance in *Neurospora crassa* (Kapoor and Sveenivasan, 1998).

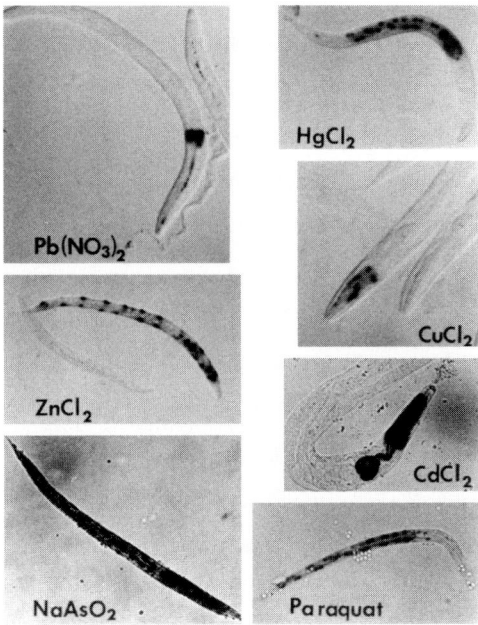

Figure 5 Tissue distribution of β-galactosidase activity in the *C. elegans* transgenic PC72 exposed to various chemical stressors as detected by X-gal staining (see text for details). Larvae were exposed to each stressor for 24 h. Concentrations: Pb(NO$_3$)$_2$, 30 μM; HgCl$_2$, 75 μM; CdCl$_2$, 100 μM; CuCl$_2$, 100 μM; NaAsO$_2$, 100 μM; ZnCl$_2$, 7 μM; paraquat, 1 μM. (From Stringham, E.G. and Candido, E.P.M., *Environ. Toxicol. Chem.*, 13, 1211–1220, 1994. With permission.)

Additional evidence reinforces the notion that stress proteins are involved in the synergism and acquisition of tolerance observed when organisms are exposed to diverse stressors. Further, this protection tends to be general in nature and independent of the specific stressor which induces the initial response. There is a direct correlation between levels of stress proteins and the degree of myocardial protection from ischemia-induced injury when the stress response is induced by heat-shock treatment (Hutter et al., 1994), sublethal ischemia (Kuzuya et al., 1993) or if stress70 is overexpressed through transfection (Mestril et al., 1994). Stress70 induced by prior ischemia also protects against experimental stroke (Simon et al., 1993) and its overexpression in dorsal ganglia protects them from heat damage (Uney et al., 1993).

In light of the known functions of stress proteins, a simple underlying theme appears to emerge in terms of the ability of a particular stressor challenge to provide protection from a subsequent stress condition, which may explain the apparent contradictions reported in various studies. In general, each stress protein probably plays a series of distinct molecular roles in protection and repair. However, the kinetics of contaminant uptake and distribution, target organ toxicity, and the nature of the stressor-induced damage will dictate both the extent and tissue level localization of stressor-induced damage for each exposure regime. In order for a preexposure treatment to provide protection from a subsequent stressor, including metal toxicity, it seems logical to assume that it must result in an increased localization of the relevant stress proteins in the target tissues for that particular stressor. In addition, stress proteins do not operate in isolation, but in concert with other stressor-specific cellular protective mechanisms in a multi-tiered system to protect cells from toxic injury. For metals, such mechanisms include the detoxication roles played by such metal-binding proteins as metallothionein and glutathione, a topic discussed in the next section.

IV. METAL-SPECIFIC CELLULAR RESPONSES

As we have seen in the previous sections, the cellular stress response focuses on maintaining the integrity of protein structure and function regardless of the nature of the stressor. In contrast, metal-specific cellular responses provide protection through a strategy of limiting nonspecific binding of metals to cellular macromolecules, including proteins. Control of nonspecific binding appears to be accomplished through the regulation of metal compartmentalization and speciation within the cell and is mediated through a wide variety of intracellular ligands which serve as "sinks" for both essential and nonessential metals (Mason and Jenkins, 1995). These ligands include high-affinity metal-binding proteins such as metallothionein, lysosomes, and mineralized and organic-based concretions. For the purposes of this chapter we will focus on metallothionein, the most well characterized of these ligands.

A. MAJOR CHARACTERISTICS OF METALLOTHIONEIN

Metallothioneins (MTs) are sulfhydryl-rich, low molecular weight metal-binding proteins that have been implicated in the metabolism and detoxication of a number of borderline and Class B metals (Karin, 1985). MTs were originally isolated from equine kidney cortex (Margoshes and Vallee, 1957; Kagi and Vallee, 1960) and similar proteins have been identified from such widely related species as mammals (Lee et al., 1977; Kwohn et al., 1986), reptiles (Hamer, 1986), amphibians (Suzuki and Takanaka, 1983; Suzuki and Ebihara, 1984), fish (Noël-Lambot et al., 1978; Hamilton and Mehrle, 1986), and various invertebrates (Olafson et al., 1979; Lerch et al., 1982; Brouwer et al., 1984; Engel and Brouwer, 1987; Brouwer et al., 1989; Roesijadi et al., 1989). The wide phylogenetic distribution of these proteins implies that they play an important role in cellular function.

Comparisons of MTs from different species have shown some diversity and variation which has caused them to be divided into three classes based upon their mode of synthesis and primary structure (Fowler et al., 1987). Class I MTs have a primary structure similar to that of the equine renal protein and are characterized by Cys-X-Cys, Cys-X-X-Cys, and Cys-Cys motifs. They have been found in all vertebrates and several invertebrate phyla studied to date. Class II MT proteins, which have been isolated in sea urchins, nematodes, and yeast, show a different distribution of cysteine residues and appear to be only distantly related. Class III are nontranscriptionally derived oligopeptides such as phytochelatons. Phytochelatons are enzymatically produced metal-binding ligands which have been isolated from plants and contain γ-glutamyl-cysteinyl motifs in their structure. This chapter focuses on the prevalent Class I MTs.

The Class I MTs in mammals contain approximately 61 amino acids and have a molecular weight in the range of 6 to 7 kDa. Typically, these proteins contain approximately 30% cysteine residues and lack aromatic amino acids. The sulfhydryl groups of the cysteine residues have been implicated in the coordination binding of a number of metals, including cadmium, zinc, copper, silver, and mercury (Hamer, 1986). The relative affinity of the protein for the metal is $Hg^{2+} > Ag^{1+} > Cu^{1+} > Cd^{2+} > Zn^{2+}$ (Hamer, 1986). Studies of MT structure indicate that the metals are ligated by multiple cysteine thiolate side chains in two metal-binding domains (Otvos and Armitage, 1980; Furey et al., 1986). The specific coordination and the number of atoms of metal complexed varies with each metal. Cadmium metallothionein (Cd-MT) shows tetrahedral coordination forming two unequal metal-binding domains, with the N-terminal domain containing 3 cadmium atoms ligated to 9 cysteine residues and the second domain containing 4 metal atoms in association with 11 cysteines (Otvos and Armitage, 1980; Otvos et al., 1985). Zinc (Zn-MT) and cadmium/zinc (Zn/Cd-MT) forms of the protein show tetrahedral coordination geometry like that of the fully cadmium-saturated protein. Stoichiometric studies on the binding of copper and silver to the protein show differences from the Cd/Zn form. In the case of Cu^+, the protein generally binds approximately 12 atoms/mol (Cu_{12}-MT) (Nielson et al., 1985; Bremner, 1987). In mammals, the metal in the Cu_{12} form is bound in either a tetragonal (Freedman et al., 1986) or trigonal (Nielson et al., 1985) coordination, while in the yeast *Saccharomyces* coordination is trigonal (George et al., 1988). In general, Ag^+ shows a comparable stoichiometry and coordination to Cu^+. Normally the protein binds 12 g-atom of metal in a trigonal geometry (Nielson et al., 1985), although under certain conditions the protein can bind approximately 17 to 18 g-atom of silver *in vitro* (Scheuhammer and Cherian, 1986), indicating a possible metal:thiol ratio of 1:1 and disruption of the metal-binding cluster domains.

Most species studied carry multiple copies of the metallothionein gene. Initial studies indicated two major isoforms termed MT-I and MT-II, which can be readily separated by ion exchange chromatography (Klauser et al., 1983). Data for several mammalian species indicate that MT-I and MT-II are regulated in a coordinated fashion, expressed in most cell types, and appear to be functionally equivalent (Hamer, 1986). Other members of the MT gene family, however, show different patterns of expression; two other members of the MT gene family, MT-III and MT-IV, have been identified in humans and mice (Palmiter et al., 1993). Expression for these isoforms suggests differences in function between cell types since MT-III appears to be limited to the brain and MT-IV to stratified squamous epithelial cells.

B. REGULATION OF METALLOTHIONEIN SYNTHESIS

The synthesis of MT can be induced by a wide variety of metals, including cadmium, copper, mercury, and zinc (Hamer, 1986; Roesijadi and Fellingham, 1987; Zafarullah et al., 1993). This metal inducible system is of particular importance in metal detoxication since it provides an efficient feedback mechanism for controlling the concentrations of a high-affinity biological ligand which, in turn, controls the speciation of selected metals within the cell.

Induction of MT synthesis occurs at the level of transcription and is regulated by specific CIS-acting sequences, termed metal regulatory elements (MREs), which are located in the 5' flanking region of the MT gene (Hamer, 1986). TRANS-acting transcription factors (TF) have also been identified in several species (Hamer, 1986; Palmiter, 1987; Thiele, 1988; Fürst et al., 1988). The interactions between the TF and MRE have been studied in some detail in the yeast *Saccharomyces cerevisae*. The CUP 1 gene in *S. cerevisae* encodes a small copper-binding metallothionein of the type II class which protects the cell against copper toxicity. Normally, the gene is transcribed at low basal levels, but increased transcription can be induced by copper. Both basal and induced transcription require the eukaryotic transcription factor TFIID together with a functional TATA binding site. Induction by Cu(I) is caused by the binding of the metal to the 24-kDa ACE 1 transcriptional activator protein (Fürst et al., 1988), the product of the ACE 1 gene which has been sequenced (Thiele, 1988; Fürst et al., 1988). Spectroscopic studies indicate that the metal is bound to the ACE 1 protein as a Cu-cysteinyl thiolate cluster, possibly as a Cu_8S_{12} cluster (Fürst et al., 1988). Binding causes a conformational change in the amino-terminal domain which allows the subsequent recognition and binding of the protein to multiple sites in the activator region upstream of the MT structural gene. Induction is caused by the interaction of the c-terminal domain of the protein with TFDII and additional factors which stimulates the formation of an activated transcription factor. Although the TF and MREs of higher eukaryotes are different than those of yeast, the mechanisms of induction should be conceptually similar.

The specific interactions between the various metals and the transcription factor are central to the regulation of MT synthesis and the ability of MT to function in both metal metabolism and detoxication (Mason and Jenkins, 1995). Although a number of metals and metalloids appear capable of inducing MT synthesis *in vivo* (Hamer, 1986; Roesijadi and Fellingham, 1987), it is not clear if each of these metals induces MT synthesis through the direct interaction with the transcription factor or by acting indirectly through the competitive displacement of a metal such as zinc from MT, which would in turn induce MT synthesis.

From a conceptual standpoint, there are several avenues of exploration which would determine if MT induction is direct or indirect. First, for each metal, the relative binding constant of the metal for the transcription factor should reflect the potential sensitivity of the MT induction system to the concentrations of the metal within the cell. The coordination of the binding of the metal and its affect on the final conformation of the metal-transcription factor complex are also critical factors to be considered. These factors determine the ability of the transcription factor to bind properly with the DNA sequence of the MRE and with proteins required for the initiation of MT transcription. Therefore, the affinity of each metal-transcription factor complex for the MRE base sequence should provide the most direct measure of the ability of a specific metal to induce MT synthesis. For those metals which form metal-transcription factor complexes with the appropriate conformation, the binding constant would provide information on the relative range of metal concentrations over which the MT gene is capable of responding to metal challenges.

Although transcription factors have been isolated from a number of species no data are available regarding the metal-binding constants or the affinities of various metal-transcription factor complexes in higher eukaryotes which would allow us to distinguish direct from indirect induction of MT. However, given the breadth of metals and their differing coordination chemistries, it is difficult to envision a TF complex capable of interacting with each of these metals and attaining a shape capable of interacting with the regulatory element. In higher eukaryotes, MT is directly induced by zinc and has been implicated in zinc metabolism (Imbra and Karin, 1987). If the TF evolved to respond primarily to zinc, then the induction of MT by metals such as Hg(II), Cu(I), Cd(II), and Ag(I) could be explained entirely by their ability to displace zinc from MT and other molecules and thus increase zinc ion activity within the cell. It follows that the development of a zinc-dependent transcription factor in higher eukaryotic cells may have provided the opportunity for MT to detoxify metals such as cadmium and mercury which readily bind to the MT molecule.

A number of other factors also regulate MT gene expression in higher eukaryotes, including glucocorticoid hormones, growth factors, protein kinase C activators, various cytokines, catecholamines, and interferons (Kägi and Schaffer, 1988; Imbra and Karin, 1987). Although regulatory elements for several of these factors, such as glucocorticoids, have been identified in the 5' flanking region of mammalian MT genes, the specific mechanisms of induction of most of these factors have not been clearly defined (Hamer, 1986; Kägi and Schaffer, 1988; Karin, 1985; Palmiter et al., 1993).

C. ROLE OF METALLOTHIONEIN IN INTRACELLULAR METAL METABOLISM

Biochemical and physiological studies have provided important insights into the biological role of MT. Initial research was devoted largely to the cadmium inducible form of the protein and its role in cadmium sequestration and detoxification. More recently, however, a variety of studies have been undertaken to identify the potential roles of MT in the physiological buffering of cellular copper and zinc during normal cellular functioning. *In vitro* studies have shown that zinc can be donated from Zn-thionein for the activation of zinc-dependent enzymes (Li et al., 1980; Winge and Miklossy, 1982). Moreover, Cu-MT acts as a donor of Cu^{+1} for the synthesis of hemocyanin during molting in the lobster *Homarus americanus* (Brouwer et al., 1986). Only specific isoforms of the protein appear to be able to facilitate the transfer and restore the oxygen binding capacity of apohemocyanin. The donation and activation of apohemocyanin by Cu-MT is facilitated by glutathione, which has a specific binding site on the MT molecule (Brouwer et al., 1993). Transfer of the metal is synchronized to the molt cycle of the crab and detailed analyses have shown a complex inverse relationship between Cu-MT and Zn-MT that is apparently tied to the anabolism and catabolism of the blood protein (Engel and Brouwer, 1991). These data indicate that MT may play an important physiological role in the buffering of essential metals during the molt cycle. This system appears capable of sequestering these essential metals when their concentrations exceed cellular requirements, and in so doing limits the nonspecific binding of these metals to other biomolecules (Mason and Jenkins, 1994). Once sequestered, the metals can be donated to apometalloenzymes as required. In this respect it may be suggested that one of the principal roles of MT in copper and zinc metabolism is to extend the homeostatic range for these metals by acting as an inducible metal buffer. In support of this, it has been shown that yeast mutants which lack a functional MT gene are only capable of growing at a single copper concentration, while yeast with functional MT genes have a fairly broad homeostatic range for copper (Hamer, 1986).

In addition to elevated metal concentrations, various physiological factors are capable of inducing MT gene expression which suggests that this protein plays several important roles in cellular physiology. As an example, the induction of MTs by serum growth factors and activators of protein kinase C lends credence to the notion that MTs play a role in regulating zinc metabolism during cell proliferation (Imbra and Karin, 1987). Since a number of zinc-containing enzymes, including the RNA and DNA polymerases, participate in cell division, MT could provide an intracellular sink for zinc which could be accessed when needed during the cell cycle. Other studies have suggested that MT plays an important role in copper metabolism in developing systems (Kägi and Kojima, 1987; Masters, et al., 1994).

Further, other factors which have been implicated in MT induction, such as interleukins 1 and 6 and tumor necrosis factor α, are released from activated macrophages in tissues experiencing inflammation. These observations have led to the suggestion that MT may be involved in detoxifying excess free radical species generated during the inflammation process (Hamer, 1986; Masters et al., 1994), a function consistent with the high cysteine content of this protein (Vallee, 1987). In contrast, other recent studies contradict this notion and raise questions regarding the importance of MT in the metabolism of essential metals and the general physiology in higher organisms (Masters et al., 1994). In these studies, MT-I and MT-II were inactivated in embryonic stem cells of mice. Mice which were homozygous for inactivated MT-I and MT-II genes showed no adverse effects on reproduction or development, indicating that these members of the MT multigene family are not essential for normal growth and differentiation.

D. EVIDENCE FOR A ROLE FOR METALLOTHIONEIN IN METAL TOLERANCE

In addition to participating in metal metabolism, metallothioneins also appear to play a central role in protecting cells from potential damage as a result of exposure to elevated concentrations of nonessential and essential metals (Karin et al., 1980; Waalkes and Goering, 1990). Perhaps the clearest indication of the role of this protein in metal detoxication came from studies of metal-resistant mammalian cell lines which had been selected for by exposure to progressively higher concentrations of Cd^{2+} (Durnam and Palmiter, 1987). Almost all of these metal-resistant cell lines carry additional copies of the MT gene (Gick and McCarty, 1982; Crawford et al., 1985). The degree of amplification of the MT gene relative to control cells ranged from 2-fold to 75-fold. Moreover, cells containing the highest number of MT genes generally show the highest rate of MT synthesis and the greatest resistance to increased cadmium concentrations. Comparable metal resistance conferred by the duplication of the MT gene has also been demonstrated in intact organisms.

The amplification process can also increase the rapidity of the response, which may also be important in conferring tolerance. Cadmium resistance in Cd^{r}20F4 strains of Chinese hamster ovary (CHO) cells has been shown to correlate with gene amplification, but in this instance resistance is not due to increased production of the protein but to an ability to synthesize MT more rapidly (Enger et al., 1984). There is also evidence in other cell lines that the constitutive levels of proteins may also be important in determining resistance (Enger et al., 1984). It would appear, therefore, that the kinetics of induction as well as the levels of induction are important in the detoxication process.

Additional evidence for metallothionein being involved in cadmium resistance has been provided by studying the molecular basis for the genetic reversion of cadmium-sensitive cell lines to cadmium-resistant ones. Restriction enzyme analyses using Msp I and Hpa II demonstrated that the patterns of methylation for a CCGG sequence in the region of the MT-1 gene of CHO cells are related to cadmium tolerance (Hildebrand et al., 1982). Cadmium-sensitive cells show a high degree of methylation at the internal C (i.e., –CC$_{meth}$GG–), while resistant phenotypes selected either by culturing in 5-azacytidine (a nonmethylatable derivative of cytidine) or Cd^{2+} display demethylation at six sites relative to the sensitive cells. This switch in methylation and increased resistance correlates with a concomitant reactivation in inducibility of both major isometallothioneins, MT-I and MT-II (Hildebrand et al., 1982).

The studies of mice with inactivated MT-I and MT-II genes mentioned in the previous section have also proved useful in evaluating the role of MTs in metal tolerance in multicellular organisms (Masters et al., 1994). In these studies normal mice and the so-called MT-null mice were injected with 10 μmol of cadmium per kilogram per day. Normal mice tolerated this dose with no observed effects while the MT-null mice died within 2 to 3 days of the initial injection. Histopathological examinations of the MT-null mice revealed that death had occurred as a result of hepatocellular injury. In normal mice, cadmium in the liver was associated with MTs, while in the MT-null mice cadmium was associated with other cellular proteins. All of these results provide strong support for the role of MT in metal tolerance and detoxication.

E. RELATIONSHIPS BETWEEN METAL METABOLISM AND DETOXIFICATION

The apparent role of MT in the metabolism of essential metals and in metal detoxification are not mutually exclusive. Indeed, the transient sequestration of copper and zinc by this high-affinity protein provides a nonreactive reservoir of the metals for apometalloenzymes. This mechanism represents a transient form of detoxification which is essential for optimizing the availability of toxic metals as components of biologically important molecules. The available data suggest that MT can act as a metal buffer system for copper and zinc. Thus, MT provides a nontoxic reservoir of these metals for a number of potentially rate-limiting copper and zinc metalloenzymes which modulate many important cellular processes. Further, MT limits the nonspecific binding of nonessential metals and, thereby, reduces their potential toxicity. It is interesting to note that the relative affinities of metals for MT based on *in vitro* studies (i.e., Hg^{2+} > Ag^{1+} > Cu^{1+} > Cd^{2+} > Zn^{2+}) provide an indirect mechanism for induction of MT via zinc displacement and concomitantly allow these more toxic metals to be sequestered while the less toxic zinc is released.

V. METALS AND METALLOIDS AND INDUCTION OF GENERAL AND STRESSOR-SPECIFIC PROTEINS

Recent studies have examined a wide variety of metals to determine if they induce stress proteins. Fischbach et al. (1993) used a transgenic mouse cell line with the human growth hormone gene under control of the human stress70 promoter to examine induction of stress70 by 31 metals and metalloids. Of these metals, 15 induced the promoter: As(III), Zn(II), Hg(II), Cd(II), Pd(II), Ag(I), and Cu(II) were strong inducers; Se(IV), Ni(II), and As(V) were intermediate inducers; and Cr(III), Mn(II), Co(II), Tl(I), and La(II) were weak inducers. For most metals, induction occurred in a dose-response manner, however, for Pd(II), Se(IV), Ni(II), Mn(II), and Tl(I) the intensity of the response decreased at the highest concentrations, perhaps as a result of impairment of the ability of the cell to mount a stress response. In all cases, induction of stress70 was observed before cytotoxicity and the relative toxicity among metals appeared to correlate with other toxicity data, including that from LD$_{50}$ studies (Williams et al., 1982; Tan et al., 1984). These rankings also correlated well with metal-induced genotoxicity endpoints such as mutagenesis, chromosomal aberrations, and DNA repair (Christie and Costa, 1983; Fischbach et al.,

1993), which suggests that stress70 may play a role in protection of DNA repair mechanisms (Williams et al., 1989; Yager and Wiencke, 1993). Gallium and indium are also inducers of the stress response (Aoki, 1993).

If induction of stress proteins is related to stress-induced damage to specific proteins and protein complexes, then the induction of the stress response by a particular metal should relate to factors that affect metal-protein interactions such as its mode of uptake, chemistry (which dictates the potential for metal-induced toxicity), subcellular distribution, and tissue level accumulation. In this section we will explore this hypothesis by examining the stress response induced by a number of metals and metalloids within the context of these factors.

A. GENERAL

When compared to the stress response induced by heat-shock treatment, three basic patterns of metal-induced protein synthesis are observed: (1) the same proteins are induced by both heat shock and metals, (2) the proteins induced by metals represent a subset of the proteins induced by heat, and (3) specific proteins are induced by metals or metalloids which are not heat inducible (de Jong et al., 1986; Cohen et al., 1991; Honda et al., 1992; Miller and Qureshi, 1992; for review, see Goering and Fisher, 1995).

In mammals the stress proteins stress70 and stress90 are induced by heat and many metals. Differences in the stress response which appear dependent upon the stressor include: (1) the rate of induced synthesis and the levels of accumulation; (2) duration and persistence of the response; (3) differences in the induced synthesis of various stress70 isoforms; and (4) the relative accumulation of cpn60 and stress70 (Nover, 1991). For example, Bauman et al. (1993) reported induction of both stress70 and stress90 in rat hepatocytes by heat shock and a variety of metals. The total amount of protein induced and the kinetics of induction varies with specific metals; heat treatment results in the greatest increase in stress70 and stress90. Cadmium produces a fivefold elevation in stress70 and a twofold increase in stress90; whereas arsenite results in a sixfold rise in stress70 and a twofold increase in stress90, and nickel causes a twofold increase in stress70 and only a 30% increase in stress90.

Since both the rate at which the response is induced and the extent to which stress proteins accumulate should be a measure of the severity of damage, each of these characteristics would be dependent upon such factors as mode of uptake and level of exposure. Fischbach et al. (1993) showed that metals which are strong inducers of the stress response induce levels of stress proteins equivalent to that observed with heat shock. However, the onset of the response with chemicals appears to be consistently slower than that with heat shock treatment, i.e., hours vs. minutes (Nover, 1991). In addition, induction of specific families of stress proteins tend to be highly stressor specific and may reflect the nature of the damage caused by different types of stressors. For example, Kato et al. (1993) demonstrated induction of hsp28 and αB crystallin in human glioma cells by both heat-shock treatment and arsenite. However, the relative increase of the two proteins is about twofold for arsenite induction, whereas αB crystallin is induced to a greater extent than hsp28 upon heat shock, i.e., fivefold vs. twofold. Other stressors such as cadmium, aluminum, zinc, ethanol, caffeine, and nicotine do not induce the two proteins.

There are a number of proteins which are induced by metals, but not other stressors such as heat-shock treatment. Exposure of human or murine melanoma cells to cadmium, copper, zinc, and other thiol-reactive agents elicits the synthesis of 100-, 90-, 72-, and 32/34-kDa proteins (Caltabiano et al., 1986). In contrast, heat and calcium ionophore induce the synthesis of the major stress proteins, but not the synthesis of the 32/34-kDa proteins identified as heme oxygenase. Proteins, in particular metallothionein and other as yet unidentified proteins, are also sensitive to specific metals, but are generally not heat inducible (Rodenhiser et al., 1986; Sauk et al., 1988; Kohane et al., 1990; Lee et al., 1991; Delpino et al., 1992). These latter proteins are probably involved in specific pathways of metal detoxification or in the protection and repair of metabolic pathways that are particularly vulnerable to metal-induced damage.

B. ARSENIC

The majority of literature on metal-induced stress proteins results from studies employing the arsenic oxides, sodium arsenate and sodium arsenite. Although these compounds are very effective inducers of the stress response, arsenite has been shown to be the most potent inducer (Bournias-Vardiabasis et al., 1990; Bauman et al., 1993). In some instances arsenite induces the same proteins as heat-shock treatment, whereas in others, there are reported differences in the stress proteins produced by these two stressors (de Jong et al., 1986; Bournias-Vardiabasis et al., 1990; Cohen et al., 1991; Honda et al., 1992; Mirkes and Cornel, 1992).

Many of the differences observed in the stress response elicited by heat shock and arsenite hold true for other metals as well. Amaral et al., (1988) showed that arsenite does not inhibit synthesis of constitutive, non-stress proteins as dramatically as does heat shock. Several studies have reported differences in the kinetics of synthesis and cellular redistribution of specific stress proteins following treatment with either heat-shock treatment or arsenite (Darasch et al., 1988; Lee et al., 1991). Although transcripts of these proteins may accumulate at the same rate after arsenite or heat treatments, decay of the stress protein message and the accumulation of stress proteins appear to be protracted after arsenite exposure (Shuman and Przybyla, 1988). It has been suggested that stress protein mRNA induced by arsenite may be more stable than the mRNA transcript induced by heat shock; however, such differences could also be explained by differences in the interactions of each stressor with protein targets (Amaral et al., 1988).

C. CADMIUM

Cadmium has also been one of the most studied inducers of stress proteins. Mitochondrial cpn60 is induced in a dose-response manner which parallels cadmium accumulation in mitochondria and cadmium-induced dysfunction (Hiranuma et al., 1993). Cadmium induces stress protein synthesis in primary hepatocyte cultures and *Drosophila* cells prior to cytotoxicity (Courgeon et al., 1984; Bauman et al., 1993; Goering et al., 1993d). Tissue specificity has been observed *in vitro* in studies of the stress response induced by cadmium. Misra et al. (1989) demonstrated enhanced synthesis of stress70 in rainbow trout hepatoma cells and chinook salmon embryonic cells exposed to heat shock and zinc, whereas only the embryonic cells are also responsive to cadmium. Cadmium also induces synthesis of metallothionein in the hepatoma cells but not in the embryonic cells. Exposure of neuroblastoma cells to cadmium results in increased expression of mRNAs for stress70, stress90, heme oxygenase, metallothionein, N-myc, and the multidrug-resistance gene MDR1, also referred to as P-glycoprotein (Murakami et al., 1991). Cadmium also induces the synthesis of the 72- and 90-kDa proteins in human keratinocytes, skin fibroblasts, and human epithelial tumor cells 3 h after exposure (Edwards et al., 1991).

D. MERCURY

Mercury elicits the synthesis of stress proteins *in vitro* and *in vivo*. Bournias-Vardiabasis et al. (1990) reported that the suite of proteins induced in *Drosophila* embryonic cells are identical after heat shock or mercury treatment. In contrast, yeast cells *(Candida albicans)* exposed to mercury induces the synthesis of three stress proteins, only one of which corresponds in molecular mass to a major stress protein (Zeuthen and Howard, 1989). Goering et al. (1992) demonstrated mercury-induced stress proteins in rat kidney within 4 h of exposure, and prior to the onset of histopathological damage.

E. COPPER

A recent study of a copper-inducible protein in HeLa cells supports the premise that induction of stress proteins is related to cell damage (Hatayama et al., 1991). Incubation with 100 μM cupric sulfate induced the synthesis of MT with no observed changes in stress70 synthesis or cell growth. In contrast, 200 μM cupric sulfate induces stress70 and also results in inhibition of cell growth. This concentration-dependent difference in induction of metallothionein and stress protein synthesis suggests that the induction of metallothionein is more sensitive to copper and that metallothionein may provide protection from metal toxicity by sequestering the copper. At higher copper concentrations where growth inhibition is observed, the stress response is induced.

A similar relationship is observed *in vivo* between copper induction of the stress protein response and scope for growth (SFG), a common organismal stress index based on bioenergetic changes (Sanders et al., 1991b). The SFG index for mussels exposed to 32 and 100 µg/l copper was significantly lower than controls, a condition indicative of growth inhibition. The accumulation of cpn60 in mantle tissue was significantly higher at 3.2 µg/l copper and increased with increasing copper exposure. Consequently, significant increases in chaperonin were observed at a copper concentration which was an order of magnitude lower than organismal stress as measured by SFG, supporting the view that increased accumulation of stress proteins may be a useful early warning for higher-level effects.

The accumulation of stress proteins in response to copper appears to be tissue specific in *Mytilus* (Sanders et al., 1994a). Differences in the stress response are observed between the gill and mantle tissue of mussels exposed to elevated copper concentrations. Chaperonin60 concentrations are more than an order of magnitude higher in the gill, a target organ for this metal, than in the mantle. Further, two additional isoforms of stress70 are induced in gill, but not mantle tissue. This study provides further evidence that the physiological processes involved in contaminant uptake, distribution, and detoxification may affect the tissue level expression of the stress response in multicellular organisms.

F. ZINC

Whelan and Hightower (1985) demonstrated that even the low level of zinc introduced into cell culture inadvertently through contaminated media induces the stress response in chicken embryo cells. Exposure of *Drosophila* embryonic cells to zinc induces a stress response pattern similar to that elicited by "classical" teratogens (Bournias-Vardiabasis et al., 1990). Treatment of sea urchin embryos with zinc induces the same stress proteins as those observed in heat-treated embryos (Roccheri et al., 1988). Unlike heat shock, zinc treatment did not inhibit overall protein synthesis or induce thermotolerance in these zinc-treated sea urchin embryos. As with copper, metallothionein is induced at lower zinc concentrations than stress70 (Hatayama et al., 1992). Incubation of HeLa cells with 100 µM zinc sulfate induces the synthesis of metallothionein with no observed changes in stress70 synthesis or cell growth, whereas 200 µM zinc sulfate induces stress70 and inhibition of growth.

G. LEAD

Interestingly, lead does not induce the classic heat-inducible stress proteins, including stress70. Instead, exposure of primary rat kidney epithelial cells and rat fibroblasts to lead glutamate induced synthesis of two members of the stress70 gene family that are not heat inducible, called glucose regulated proteins (grps), and the 32-kDa heme oxygenase (Shelton et al., 1986). It has been suggested that the induction of grps by lead might result from its disruption of calcium homeostasis, or from its binding to sulfhydryl group targets.

H. IRON

Iron is an ineffective inducer of stress70 and stress90 in primary cultures of rat hepatocytes (Bauman et al., 1993), but induces expression of ubiquitin and heme oxygenase (Kantengwa and Polla, 1993; Uney et al., 1993). Treatment of mouse neuroblastoma cells with iron increases mRNA levels of stress70 and ubiquitin (Uney et al., 1993). The response is blocked by the chelating agents alpha-tocopherol or desferroxamine, demonstrating that the induction of these transcripts is mediated by the presence of iron. It has been suggested that induction of the stress response by iron is mediated through the generation of free radicals and subsequent cell injury (Cajone and Bernelli-Zazzera, 1988). Oxygen free radicals generated in the presence of hemoglobin-released iron are also believed to initiate stress protein induction by human monocytes-macrophages during erythrophagocytosis (Clerget and Polla, 1990; Donati et al., 1991).

VI. CONCLUSIONS

Cells respond to exposure to elevated metal concentrations through both general and specific strategies which function in concert to provide protection from metal toxicity. The general cellular stress response provides a strategy to protect proteins and protein complexes from a wide variety of environmental stressors, including metals. In contrast, the induction of metallothionein is a quite specific and direct response to an increase in available metal. This strategy directly limits nonspecific binding of metals to vulnerable macromolecules, including but not limited to proteins, by sequestering the metal and lowering its availability. An understanding of the roles that each of these cellular responses play in minimizing metal toxicity provides additional insight into the mechanisms of metal toxicity and may prove useful in assessing the extent of metal-induced damage.

ACKNOWLEDGMENTS

The authors extend appreciation to C. Taylor for help in manuscript preparation and L. Martin for figure preparation. This effort was supported in part by grants from the AFOSR/NL (93-NL-119) to BMS and the U.E. EPA (R-818378–01–0) to KDJ. The opinions expressed in this chapter do not reflect official policy of the U.S. Food and Drug Administration.

REFERENCES

Amaral, M. D., Galego, L., and Rodriguez-Pousada, C., *Eur. J. Biochem.*, 171, 463–470, 1988.
Amoruso, M. A., Witz, G., and Goldstein, B. D., *Toxicol. Lett.*, 10, 133–138, 1982.
Ananthan, J., Goldberg, A. L., and Voellmy, R., *Science*, 232, 522–524, 1986.
Aoki, Y., *Jpn. J. Toxicol. Environ. Health*, 39, 55–57, 1993.
Arrigo, A. P. and Landry, J., in *The Biology of Heat Shock Proteins and Molecular Chaperones*, Morimoto, R. I., Tissieres, A., and Georgopoulos, C., Eds., Cold Spring Harbor Laboratory Press, Cold Spring Harbor, NY, 335, 1994.
Bauman, J. W., Liu, J., and Klaassen, C. D., *Fundam. Appl. Toxicol.*, 21, 15–22, 1993.
Beckmann, R. B., Mizzen, L. A., and Welch, W. J., *Science*, 248, 850–853, 1990.
Berger, H. M. and Woodward, M. P., *Exp. Cell Res.*, 147, 437–442, 1983.
Black, A. R. and Subjeck, J. R., *Methods Achiev. Exp. Pathol.*, 15, 126–166, 1991.
Boon-Niermeijer, E. K., van der Scheur, H., and Tuyl, M., *Int. J. Hyperthermia*, 2, 93–105, 1986.
Bosch, T. C. G., Krylow, S. M., Bode, H. R., and Steele, R. E., *Proc. Natl. Acad. Sci. U.S.A.*, 85, 7927–7931, 1988.
Bournias-Vardiabasis, N., Buzin, C., and Flores, J., *Exp. Cell Res.*, 189, 177–182, 1990.
Bremner, I., *Prog.* 11, 1–37, 1987.
Brouwer, M., Hoexum-Brouwer, T., and Cashon, R. E., *Biochem. J.*, 294, 219–225, 1993.
Brouwer, M., Hoexum-Brouwer, T. M., and Engel, D. W., *Mar. Environ. Res.*, 14, 71–88, 1984.
Brouwer, M., Whaling, P., and Engel, D. W., *Environ. Health Perspect.* 65, 93–100, 1986.
Brouwer, M., Winge, D. R., and Gray, W. R., *J. Inorg. Biochem.*, 35, 289–303, 1989.
Brown, M. A., Upender, R. P., Hightower, L. E., and Renfro, J. L., *Proc. Natl. Acad. Sci. U.S.A.*, 89, 3246–3250, 1992.
Bruce, J. L., Price, B. D., Coleman, C. N., and Calderwood, S. K., *Cancer Res.*, 53, 12–15, 1993.
Brugge, J. S., Erikson, E., and Erikson, R. L., *Cell*, 25, 363–372, 1981.
Brunt, S. A., Riehl, R., and Silver, J. C., *Mol. Cell. Biol.*, 10, 273–281, 1990.
Bruschi, S. A., West, K. A., Crabb, J. W., Gupta, R. S., and Stevens, J. L., *J. Biol. Chem.*, 268, 23157–23161, 1993.
Buchner, J., Schmidt, M., Fuchs, M., Jaenicke, R., Rudolph, R., Schmid, F. X., and Kiefhaber, T., *Biochemistry*, 30, 1586–1591, 1991.
Burgman, P. W., Kampinga, H. H., and Konings, A. W., *Int. J. Hyperthermia*, 9, 151–162, 1993.
Cajone, F. and Bernelli-Zazzera, A., *Chem.-Biol. Interact.*, 65, 235–246, 1988.
Caltabiano, M. M., Koestler, T. P., Poste, G., and Greig, R. G., *J. Biol. Chem.*, 261, 13381–13386, 1986.
Cervera, J., *Cell. Biol. Int. Rep.*, 9, 131–141, 1985.
Cheng, M. Y., Hartl, F.-U., Martin, J., Pollock, R. A., Kalousek, F., Neupert, W., Hallberg, E. M., Hallberg, R. L., and Horwich, A. L., *Nature*, 337, 620–624, 1989.
Chiang, H.-L., Terlecky, S. R., Plant, C. P., and Dice, J. F., *Science*, 246, 382–385, 1989.
Chirico, W. J., Waters, M. G., and Blobel, G., *Nature*, 333, 805–810, 1988.
Chretien, P. and Landry, J., *J. Cell. Physiol.*, 137, 157–166, 1988.
Christie, N. T. and Costa, M., *Biol. Trace Elem. Res.*, 5, 55–71, 1983.
Christie, N. T. and Costa, M., *Biol. Trace Elem. Res.*, 6, 139–158, 1984.
Ciavarra, R. P. and Simeone, A., *Cell. Immunol.*, 129, 363–376, 1990a.
Ciavarra, R. P. and Simeone, A., *Cell. Immunol.*, 131, 11–26, 1990b.
Ciocca, D. R., Oesterreich, S., Chamness, G. C., McGuire, W. L., and Fuqua, S. A. W., *J. Natl., Cancer Inst.*, 85, 1558–1570, 1993.
Clerget, M. and Polla, B. S., *Proc. Natl. Acad. Sci. U.S.A.*, 87, 1081–1085, 1990.
Cohen, D. S., Palmer, E., Welch, W. J., and Sheppard, D., *Am. J. Respir. Cell. Mol. Biol.*, 5, 133–143, 1991.
Coleman, J. E., *Nature*, 214, 193–194, 1967.
Courgeon, A. M., Maisonhaute, C., and Best-Belpomme, M., *Exp. Cell Res.*, 153, 515–521, 1984.
Craig, E. A., in *Stress Proteins in Biology and Medicine*, Morimoto, R. I., Tissieres, A., and Georgopoulos, C., Eds., Cold Spring Harbor Laboratory Press, Cold Spring Harbor, NY, 279, 1990.
Craig, E. A., Baxter, B. K., Becker, J., Halladay, J., and Ziegelhoffer, T., in *The Biology of Heat Shock Proteins and Molecular Chaperones*, Morimoto, R. I., Tissieres, A., and Georgopoulos, C., Eds., Cold Spring Harbor Laboratory Press, Cold Spring Harbor, NY, 31, 1994.
Craig, E. A., Ingolia, T. D., and Manseau, L. J., *Dev. Biol.*, 99, 418–426, 1983.
Craig, E., Kang, P. J., and Boorstein, W., in *Heat Shock*, Maresca, B. and Lindquist, S., Eds., Springer-Verlag, Berlin, 77, 1993.
Crawford, B. D., Enger, M. D., Griffith, B. B., Griffith, J. K., Hanners, J. L., Longmire, J. L., Munk, A. C., Stallings, R. L., Tesmer, J. G., Walters, R. A., and Hildebrand, C. E., *Mol. Cell. Biol.*, 5, 320–329, 1985.
Darasch, S., Mosser, D. D., Bols, N. C., and Heikkila, J. J., *Biochem. Cell Biol.*, 66, 862–870, 1988.
de Jong, W. W., Hoekman, W. A., Mulders, J. W. M., and Bloemendal, H., *J. Cell Biol.*, 102, 104–111, 1986.
Delpino, A., Spinsanti, P., Mattei, E., Mileo, A. M., Vismara, D., and Ferrini, U., *Melanoma Res.*, 2, 369–375, 1992.

Dice, J. F., Agarraberes, F., Kirven-Brooks, M., Terlecky, L. J., and Terlecky, S. R., in *The Biology of Heat Shock Proteins and Molecular Chaperones*, Morimoto, R. I., Tissieres, A., and Georgopoulos, C., Eds., Cold Spring Harbor Laboratory Press, Cold Spring Harbor, NY, 137, 1994.

Dixon, M. and Webb, E. C., *Enzymes*, Longmans, London, 1967.

Donati, Y. R., Kantengwa, S., and Polla, B. S., *Pathobiology*, 59, 156–161, 1991.

Durnam, D. M. and Palmiter, R. D., *Experientia Suppl.*, 52, 457–463, 1987.

Dwyer, S. D., Zhuang, Y., and Smith, J. B., *Exp. Cell Res.*, 192, 22–31, 1991.

Dyer, S. D., Brooks, G. L., Dickson, K. L., Sanders, B. A., and Zimmerman, E. G., *Environ. Toxicol. Chem.*, 12, 1–12, 1993.

Dyer, S. D., Dickson, K. L., Zimmerman, E. G., and Sanders, B. M., *Can. J. Zool.*, 69, 2021–2027, 1991.

Edington, B. V., Whelan, S. A., and Hightower, L. E., *J. Cell. Physiol.*, 139, 219–228, 1989.

Edwards, M. J., Marks, R., Dykes, P. J., Merrett, V. R., Morgan, H. E., and O'Donovan, M. R., *J. Invest. Dermatol.*, 96, 392–396, 1991.

Ellis, R. J., *Semin. Cell Biol.*, 1, 1–9, 1990.

Engel, D. W. and Brouwer, M., *Biol. Bull.*, 172, 239–251, 1987.

Engel, D. W. and Brouwer, M., *Biol. Bull.*, 180, 1–6, 1991.

Enger, M. D., Hildebrand, C. E., Griffith, J. K., and Walters, R. E., in *Metabolism of Trace Metals in Man: Genetic Implications*, Rennert, O. M. and Chan, W.-Y., Eds., CRC Press, Boca Raton, FL, 7, 1984.

Fischbach, M., Sabbioni, E., and Bromley, P., *Cell Biol. Toxicol.*, 9, 177–188, 1993.

Fowler, B. A., Hildebrand, C. E., Kojima, Y., and Webb, M., *Experientia Suppl.*, 52, 19–22, 1987.

Freedman, J. H., Powers, L., and Peisach, J., S *Biochemistry*, 25, 2342–2349, 1986.

Frydman, J. and Hartl, F.-U., in *The Biology of Heat Shock Proteins and Molecular Chaperones*, Morimoto, R. I., Tissieres, A., and Georgopoulos, C., Eds., Cold Spring Harbor Laboratory Press, Cold Spring Harbor, NY, 251, 1994.

Furey, W. F., Robbins, A. H., Clancy, L. L., Winge, D. R., Wang, B. C., and Stout, C. D., *Science*, 231, 704–710, 1986.

Fürst, P., Hu, S., Hackett, R., and Hamer, D., *Cell*, 55, 705–717, 1988.

Gaitanaris, G. A., Papavassiliou, A. G., Rubock, P., Silverstein, S. J., and Gottesman, M. E., *Cell*, 61, 1013–1020, 1990.

George, G. N., Byrd, J., and Winge, D. R., *J. Biol. Chem.*, 263, 8199–8203, 1988.

Gerner, E. W. and Schneider, M. J., *Nature*, 256, 500–502, 1975.

Gething, M. J. and Sambrook, J., *Nature*, 355, 33–45, 1992.

Gick, G. G. and McCarty, K. S., *J. Biol. Chem.*, 257, 9049–9053, 1982.

Goering, P. L., *Neurotoxicology*, 14, 45–60, 1993.

Goering, P. L. and Fisher, B. R., in *Handbook of Experimental Pharmacology: Toxicology of Metals. Biochemical Aspects*, Goyer, R. A. and Cherian, M. G., Eds., Springer-Verlag, New York, 229, 1995.

Goering, P. L., Fisher, B. R., Chaudhary, P. P., and Dick, C. A., *Toxicol. Appl. Pharmacol.*, 113, 184–191, 1992.

Goering, P. L., Fisher, B. R., Kimmel, C. A., and Kimmel, G. L., in *Use of Biomarkers in Assessing Health and Environmental Impacts of Chemical Pollutants*, Travis, C. C., Ed., Plenum Press, New York, 95, 1993a.

Goering, P. L., Fisher, B. R., and Kish, C. L., *Toxicol. Appl. Pharmacol.*, 122, 139–148, 1993b.

Goering, P. L., Kish, C. L., and Fisher, B. R., *Toxicology*, 85, 25–39, 1993c.

Goering, P. L., Kish, C. L., and Dick, S. E., *Toxicologist*, 13, 160, 1993d.

Goldberg, M. E. and Guillou, Y., *Protein Sci.*, 3, 883–887, 1994.

Hahn, G. M. and Li, G. C., in *Stress Proteins in Biology and Medicine*, Morimoto, R. I., Tissieres, A., and Georgopoulos, C., Eds., Cold Spring Harbor Press, Cold Spring Harbor, NY, 79, 1990.

Halliwell, B. and Gutteridge, J. M. C., *Biochem. J.*, 219, 1–14, 1984.

Hamer, D. H., *Ann. Rev. Biochem.*, 55, 913–951, 1986.

Hamilton, S. J. and Mehrle, P. M., *Trans. Am. Fish Soc.*, 115, 596–609, 1986.

Hartl, F.-U., Martin, J., and Neupert, W., *Ann. Rev. Biophys. Biomol. Struct.*, 21, 293–322, 1992.

Hartl, F.-U. and Neupert, W., *Science*, 247, 930–938, 1990.

Hatayama, T., Tsukimi, Y., Wakatsuki, T., Kitamura, T., and Imahara, H., *J. Biochem.*, 110, 726–731, 1991.

Hatayama, T., Tsukimi, Y., Wakatsuki, T., Kitamura, T., and Imahara, H., *Mol. Cell. Biochem.*, 112, 143–153, 1992.

Hattori, H., Kaneda, T., Lokeshwar, B., Laszlo, A., and Ohtsuka, K., *J. Cell Sci.*, 104, 629–638, 1993.

Heikkila, J. J., Browder, L. W., Gedamu, L., Nickells, R. W., and Schultz, G. A., *Can. J. Genet. Cytol.*, 28, 1093–1105, 1986.

Held, K. D. and Biaglow, J. E., *Radiat. Res.*, 134, 375–382, 1993.

Hightower, L. E., *J. Cell. Physiol.*, 102, 407–427, 1980.

Hightower, L. E., *Cell*, 66, 1–20, 1991.

Hightower, L. E., Sadis, S. E., and Takenaka, I. M., in *The Biology of Heat Shock Proteins and Molecular Chaperones*, Morimoto, R. I., Tissieres, A., and Georgopoulos, C., Eds., Cold Spring Harbor Laboratory Press, Cold Spring Harbor, NY, 179, 1994.

Hildebrand, C. E., Griffith, J. K., Tobey, R. A., Walters, R. A., and Enger, M. D., *PFS*, 1, 280, 1982.

Hiranuma, K., Hirata, K., Abe, T., Hirano, T., Matsuno, K., Hirano, H., Suzuki, K., and Higashi, K., *Biochem. Biophys. Res. Commun.*, 194, 531–536, 1993.

Hojima, Y., Behta, B., Romanic, A. M., and Prockop, D. J., *Matrix Biol.*, 14, 113–120, 1994.

Honda, K., Hatayama, T., Takahashi, K., and Yukioka, M., *Teratogen. Carcinogen. Mutagen.*, 11, 235–244, 1992.

Horwitz, J., *Proc. Natl. Acad. Sci. U.S.A.*, 89, 10449–10453, 1992.

Hutter, M. M., Sievers, R. E., Barbosa, V., and Wolfe, C. L., *Circulation*, 89, 355–360, 1994.

Imbra, R. J. and Karin, M., *Mol. Cell. Biol.,* 7, 1358–1363, 1987.
Jakob, U., Gaestel, M., Engel, K., and Buchner, J., *J. Biol. Chem.,* 268, 1517–1520, 1993.
Johnston, R. N. and Kucey, B. L., *Science,* 242, 1551–1554, 1988.
Kägi, J. H. R. and Kojima, Y., *Experientia Suppl.,* 52, 35–61, 1987.
Kägi, J. H. R. and Schaffer, A., *Biochemistry,* 27, 8509–8515, 1988.
Kägi, J. H. R. and Vallee, B. L., *J. Biol. Chem.,* 235, 3460–3465, 1960.
Kampinga, H. H., Brunsting, J. F., and Konings, A. W., *J. Cell. Physiol.,* 150, 406–416, 1992.
Kantengwa, S. and Polla, B. S., *Infect. Immunol.,* 61, 1281–1287, 1993.
Kapoor, M. and Sveenivasan, G. M., *Biochem. Biophys. Res. Commun.,* 156, 1097–1102, 1988.
Karin, M., *Cell,* 41, 9–10, 1985.
Karin, M., Anderson, R. D., Slater, E., Smith, K., and Herschman, H. R., *Nature,* 286, 295–297, 1980.
Kasprzak, K. S., *Chem. Res. Toxicol.,* 4, 604–615, 1991.
Kato, S., Hirano, A., Kato, M., Herz, F., and Ohama, E., *Neuropathol. Appl. Neurobiol.,* 19, 436–442, 1993.
Kee, S. C. and Nobel, P. S., *Plant Physiol.,* 80, 596–598, 1986.
Keyse, S. M. and Tyrrell, R. M., *J. Biol. Chem.,* 262, 14821–14825, 1987.
Klaassen, C. D. and Suzuki, K. T., *Metallothionein in Biology and Medicine,* CRC Press, Boca Raton, FL, 1991.
Klauser, S., Kägi, J. H. R., and Wilson, K. J., *Biochem. J.,* 209, 71–80, 1983.
Koban, M., Yup, A. A., Agellon, L. B., and Powers, D. A., *Mol. Mar. Biol. Biotechnol.,* 1, 1–17, 1991.
Kohane, D. S., Sarzani, R., Schwartz, J. H., Chobanian, A. V., and Brecher, P., *Am. J. Physiol.,* 258, H1699–H1705, 1990.
Koyasu, S., Nishida, E., Kadowaki, T., Matsuzaki, F., Iida, K., Harada, F., Kasuga, M., Sakai, H., and Yahara, I., *Proc. Natl. Acad. Sci. U.S.A.,* 83, 8054–8058, 1986.
Koyasu, S., Nishida, E., Miyata, Y., Sakai, H., and Yahara, I., *J. Biol. Chem.,* 264, 15083–15087, 1989.
Krause, K. W., Good, P. J., and Hallberg, R. L., *Proc. Natl. Acad. Sci. U.S.A.,* 84, 383–387, 1987.
Kuzuya, T., Hoshida, S., Yamashita, N., Fuji, H., Oe, H., Hori, M., Kamada, T., and Tada, M., *Circ. Res.,* 72, 1293–1299, 1983.
Kwohn, Y.-T., Yamazaki, S., Okubo, A., Toshimura, E., Tatsukawa, R., and Toda, S., *Agric. Biol. Chem.,* 52, 837–841, 1986.
Lai, Y., Shen, C., Cheng, T., Hou, M., and Lee, W., *J. Cell. Biochem.,* 51, 369–379, 1993.
Lamond, A. I. and Carmo-Fonseca, M., *Trend Cell Biol.,* 3, 198–203, 1993.
Landry, J., Bernier, D., Chretien, P., Nicole, L. M., Tanguay, R. M., and Marceau, N., *Cancer Res.,* 42, 2457–2461, 1982.
Landry, J., Chretien, P., Lambert, H., Hickey, E., and Weber, L., *J. Cell Biol.,* 109, 7–15, 1989.
Lavoie, J. N., Hickey, E., Weber, L. A., and Landry, J., *J. Biol. Chem.,* 268, 24210–24214, 1993.
Lee, K. J. and Hahn, G. M., *J. Cell. Physiol.,* 136, 411–420, 1988.
Lee, S. S., Mate, B. R., von der Trenck, K. T., Rimerman, R. A., and Bühler, D. R., *Comp. Biochem. Physiol.,* 57C, 45–53, 1977.
Lee, Y. J., Curetty, L., and Corry, P. M., *J. Cell. Physiol.,* 149, 77–87, 1991.
Lee, Y. J. and Dewey, W. C., *J. Cell. Physiol.,* 132, 41–48, 1987.
Lee, Y. J. and Dewey, W. C., *J. Cell. Physiol.,* 135, 397–406, 1988.
Leonhardt, S. A., Fearon, K., Danese, P. N., and Mason, T. L., *Mol. Cell. Biol.,* 13, 6304–6313, 1993.
Lerch, K., Ammer, D., and Olafson, R. W., *J. Biol. Chem.,* 257, 2420–2426, 1982.
Li, G. C. and Werb, Z., *Proc. Natl. Acad. Sci. U.S.A.,* 79, 3218–3222, 1982.
Li, G. C., *J. Cell. Physiol.,* 115, 116–122, 1983.
Li, G. C. and Laszlo, A., *J. Cell. Physiol.,* 122, 91–97, 1985.
Li, G. C., Li, L., Liu, R. Y., Rehman, M., and Lee, W. M. F., *Proc. Natl. Acad. Sci. U.S.A.,* 89, 2036–2040, 1992.
Li, T. Y., Kraker, A. J., Shaw, C. F., and Petering, D. H., *Proc. Natl. Acad. Sci. U.S.A.,* 77, 6334–6338, 1980.
Lindquist, S., *Ann. Rev. Biochem.,* 55, 1151–1191, 1986.
Lindquist, S. and Craig, E. A., *Ann. Rev. Genet.,* 22, 631–677, 1988.
Loomis, W. F. and Wheeler, S., *Dev. Biol.,* 79, 399–408, 1980.
Manning-Krieg, U. C., Scherer, P. E., and Schatz, G., *EMBO J.,* 10, 3273–3280, 1991.
Margoshes, M. and Vallee, B. L., *J. Am. Chem. Soc.,* 79, 4813–4814, 1957.
Marshall, J. S., DeRocher, A. E., Keegstra, K., and Vierling, E., *Proc. Natl. Acad. Sci. U.S.A.,* 87, 374–378, 1990.
Martin, J., Horwich, A. L., and Hartl, F.-U., *Science,* 258, 995–998, 1992.
Mason, A. Z. and Jenkins, K. D., in *Interactions Between Trace Metals and Aquatic Organisms,* Tessier, A. and Turner, D., Eds., Lewis Publishers, Chelsea, MI.
Massey, V., Hofmann, T., and Palmer, J., *J. Biol. Chem.,* 237, 3820–3828, 1962.
Masters, B. A., Kelly, E. J., Quaife, C. J., Brinster, R. L., and Palmiter, R. D., *Proc. Natl. Acad. Sci. U.S.A.,* 91, 584–588, 1994.
Matunis, E. L., Matunis, M. J., and Dreyfuss, G., *J. Cell Biol.,* 121, 219–228, 1993.
McMullin, T. W. and Hallberg, R. L., *Mol. Cell. Biol.,* 8, 371–380, 1988.
Mestril, R., Chi, S. H., Sayen, M. R., and Dillmann, W. H., *Biochem. J.,* 298, 561–569, 1994.
Miller, L. and Qureshi, M. A., *Poult. Sci.,* 71, 988–998, 1992.
Mirkes, P. E., *Dev. Biol.,* 119, 115–122, 1987.
Mirkes, P. E. and Cornel, L., *Teratology,* 46, 251–259, 1992.
Misra, S., Zafarullah, M., Price-Haughey, J., and Gedamu, L., *Biochem. Biophys. Acta,* 1007, 325–333, 1989.
Morimoto, R. I., Tissieres, A., and Georgopoulous, C., *The Biology of Heat Shock Proteins and Molecular Chaperones,* Cold Spring Harbor Laboratory Press, Cold Spring Harbor, NY, 1994a.

Morimoto, R. I., Jurivich, D. A., Kroeger, P. E., Mathur, S. K., Murphy, S. P., Nakai, A., Sarge, K., Abravaya, K., and Sistonen, L. T., in *The Biology of Heat Shock Proteins and Molecular Chaperones,* Morimoto, R. I., Tissieres, A., and Georgopoulos, C., Eds., Cold Spring Harbor Laboratory Press, Cold Spring Harbor, NY, 417, 1994b.
Mosser, D. D. and Bols, N. C., *J. Comp. Physiol. B.,* 158, 457–467, 1988.
Mosser, D. D., Kotzbauer, P. T., Sarge, K. D., and Morimoto, R. I.,, *Proc. Natl. Acad. Sci. U.S.A.,* 87, 3748–3752, 1990.
Mosser, D. D., van Oostrom, J., and Bols, N. C., *J. Cell. Physiol.,* 132, 155–160, 1987.
Murakami, T., Ohmori, H., Katoh, T., Abe, T., and Higashi, K., *Sangyo Ika Daigaku Zasshi,* 13, 271–278, 1991.
Nielson, K. B., Atkin, C. L., and Winge, D. R.,, *J. Biol. Chem.,* 260, 5342–5350, 1985.
Nocentini, S., *Nucleic Acids Res.,* 15, 4211–4225, 1987.
Nöel-Lambot, F., Gerday, C., and Disteche, A., *Comp. Biochem. Physiol.,* C61, 177–187, 1978.
Nover, L., *The Heat Shock Response,* CRC Press, Boca Raton, FL, 1991.
Olafson, R. W., Abel, K., and Sim, R. G., *Biochem. Biophys. Res. Commun.,* 89, 36–43, 1979.
Ostermann, J., Horwich, A. L., Neupert, W., and Hartl, F.-U *Nature,* 341, 125–130, 1989.
Otvos, J. D. and Armitage, I. M., *Proc. Natl. Acad. Sci. U.S.A.,* 77, 7094–7098, 1980.
Otvos, J. D., Engeseth, H. R., and Wehrli, S., *Biochemistry,* 24, 6735–6740, 1985.
Palmiter, R., *Experientia Suppl.,* 52, 63–80, 1987.
Palmiter, R. D., Sandgren, E. P., Koeller, D. M., and Brinster, R. L., *Mol. Cell. Biol.,* 13, 5266–5275, 1993.
Parsell, D. A. and Lindquist, S., H in *The Biology of Heat Shock Proteins and Molecular Chaperones,* Morimoto, R. I., Tissieres, A., and Georgopoulos, C., Eds., Cold Spring Harbor Laboratory Press, Cold Spring Harbor, NY, 457, 1994.
Pelham, H. R. B., in *Stress Proteins in Biology and Medicine,* Morimoto, R. I., Tissieres, A., and Georgopoulos, C., Eds., Cold Spring Harbor Laboratory Press, Cold Spring Harbor, NY, 287, 1990.
Peters, R. A., *Bull. Johns Hopkins Hosp.,* 97, 1–20, 1955.
Phillips, G. J. and Silhavy, T. J., *Nature,* 344, 882–884, 1990.
Pohl, L. R., *Chem. Res. Toxicol.,* 6, 786–793, 1993.
Renfro, J. L., Brown, M. A., Parker, S. L., and Hightower, L. E., *J. Pharmacol. Exp. Ther.,* 265(2), 992, 1993.
Riabowol, K. T., Mizzen, L. A., and Welch, W. J., *Science,* 242, 433–436, 1988.
Roccheri, M. C., La Rosa, M., Ferraro, M. G., Cantone, M., Casino, D., Giudice, G., and Sconzo, G., *Cell Differ.,* 24, 209–214, 1988.
Rodenhiser, D. I., Jung, J. H., and Atkinson, B. G., *Can. J. Genet. Cytol.,* 28, 1115–1124, 1986.
Roesijadi, G. and Fellingham, G. W., *Can. J. Fish Aquat. Sci.,* 44, 680–684, 1987.
Roesijadi, G., Kielland, S. L., and Klerks, P. L., *Arch. Biochem. Biophys.,* 273, 403–413, 1989.
Rothman, J. E., *Cell,* 59, 591–601, 1989.
Sanchez, Y. and Lindquist, S. L., *Science,* 248, 1112–1115, 1990.
Sanders, B. M., *Crit. Rev. Toxicol.,* 23(1), 49–75, 1993.
Sanders, B. M., Hope, C., Pascoe, V. M., and Martin, L. S., *Physiol. Zool.,* 64, 1471–1489, 1991a.
Sanders, B. M., Martin, L. S., Nelson, W. G., Phelps, D. K., and Welch, W., *Mar. Environ. Res.,* 31, 81–97, 1991b.
Sanders, B. M., Pascoe, V. M., Nakagawa, P. A., and Martin, L. S., *Mol. Mar. Biol. Biotechnol.,* 1, 147–154, 1992.
Sanders, B. M. and Martin, L. S., *Comp. Biochem. Physiol.,* 109:3, 295–307, 1994.
Sanders, B. M., Martin, L. S., Howe, S. R., Nelson, W. G., Hegre, E. S., and Phelps, D. K., *Toxicol. Appl. Pharmacol.,* 125, 206–213, 1994a.
Sanders, B. M., Nguyen, J., Douglass, T. G., and Miller, S., *Biochem. J.,* 297, 21–25, 1994b.
Sanders, B. M., Nguyen, J., Martin, L. S., Howe, S. R., and Coventry, S., *Comp. Biochem. Physiol.,* 112C, 335–343, 1995.
Sanders, S. L., Whitfield, K. M., Vogel, J. P., Rose, M. D., and Schekman, R. W., *Cell,* 69, 353–365, 1992.
Sauk, J. J., Norris, K., Foster, R., Moehring, J., and Somerman, M. J., *J. Oral Pathol. Med.,* 17, 496–499, 1988.
Scheuhammer, A. M. and Cherian, M. G., *Toxicol. Appl. Pharmacol.,* 82, 417–425, 1986.
Schlesinger, M. J., *J. Biol. Chem.,* 265, 12111–12114, 1990.
Schlesinger, M. J., in *Stress Proteins — Induction and Function,* Schlesinger, M. J., Santoro, M. G., and Garaci, E., Eds., Springer-Verlag, Berlin, 81, 1993.
Schlesinger, M. and Hershko, A., *The Ubiquitin System,* Cold Spring Harbor Press, Cold Spring Harbor, NY, 1988.
Schroder, H., Langer, T., Hartl, F. U., and Bukau, B., *EMBO J.,* 12, 4137–4144, 1993.
Shelton, K. R., Todd, J. M., and Egle, P. M., The induction of stress-related proteins by lead, *J. Biol. Chem.,* 261, 1935–1940, 1986.
Shuman, J. and Przybyla, A., *DNA,* 7, 475–482, 1988.
Simon, R. P., Niiro, M., and Gwinn, R., *Neurosci. Lett.,* 163, 135–137, 1993.
Skowyra, D., Georgopoulos, C., and Zylicz, W., *Cell,* 62, 939–944, 1990.
Stacey, N. H. and Klaassen, C. D., *J. Toxicol. Environ. Health,* 7, 139–147, 1981.
Stringham, E. G. and Candido, E. P. M., *Environ. Toxicol. Chem.,* 13, 1211–1220, 1994.
Sugiyama, M., *Cell Biol. Toxicol.,* 10, 1–22, 1994.
Sunderman, F. W., *Recent Advances in Metal Carcinogenesis,* Second International Symposium Clinical Chemistry-Chemical Toxicology of Metals, pp. 93–122, 1984.
Suzuki, K. T. and Ebihara, Y., *Comp. Biochem. Physiol.,* 78C, 35–38, 1984.
Suzuki, K. T. and Takanaka, Y., *Comp. Biochem. Physiol.,* 74C, 311–317, 1983.

Tan, E. L., Williams, M. W., Schenley, R. L., Perdue, S. W., Hayden, T. L., Turner, J. E., and Hsie, A. W. A., *Toxicol. Appl. Pharmacol.*, 74, 330–336, 1984.

Thiele, D. J., *Mol. Cell. Biol.*, 8, 2745–2752, 1988.

Twomey, B. M., Dhillon, V. B., McCallum, S., Isenberg, D. A., and Latchman, D. S., *J. Autoimmun.*, 6, 495–506, 1993.

Ulmasov, H. A., Karaev, K. K., Lyashko, V. N., and Evgenev, M. B., *Comp. Biochem. Physiol.*, 106, 867–872, 1993.

Uney, J. B., Anderson, B. H., and Thomas, S. M., *J. Neurochem.*, 60, 659–665, 1993.

Vallee, B. L., in *Metallothionein II*, Kägi, J.H.R. and Kojima, Y., Eds., Birkhäuser Verlag, Basel, 5, 1987.

Vallee, B. L. and Ulmer, D. D., *Ann. Rev. Biochem.*, 41, 91–128, 1972.

Vig, P. J. and Nath, R., *Biochem. Int.*, 23, 927–934, 1991.

Vincent, M. and Tanguay, R. M., *J. Mol. Biol.*, 162, 365–378, 1982.

Waalkes, M. P. and Goering, P. L., *Chem. Res. Toxicol.*, 3, 281–288, 1990.

Welch, W. J. and Feramisco, J. R., *J. Biol. Chem.*, 259, 4501–4510, 1984.

Welch, W. J. and Mizzen, L. A., *J. Cell Biol.*, 106, 1117–1130, 1988.

Welte, M. A., Tetrault, J. M., Dellavalle, R. P., and Lindquist, S. L., *Curr. Biol.*, 3, 842–853, 1993.

Whelan, S. A. and Hightower, L. E., *J. Cell. Physiol.*, 122, 205–209, 1985.

White, C. N., Hightower, L. E., and Schultz, R. J., *Mol. Biol. Evol.*, 11, 106–119, 1994.

Wiegant, F. A., Van Bergen En Henegouwen, P. M., Van Dongen, G., and Linnemans, W. A., *Cancer Res.*, 47, 1674–1680, 1987.

Wilkinson, J. M. and Pollard, I., *Anat. Rec.*, 237, 453–457, 1993.

Williams, K. J., Landgraf, B. E., Whiting, N. L., and Zurlo, J., *Cancer Res.*, 49, 2735–2742, 1989.

Williams, M. W., Hoeschele, J. D., Turner, J. E., Bruce Jacobson, K., Christie, N. T., Platon, C. L., Smith, L. H., Witschi, H. R., and Lee, E. H., *Toxicol. Appl. Pharmacol.*, 63, 461–469, 1982.

Winge, D. R. and Miklossy, K.-A., *J. Biol. Chem.*, 257, 3471–3476, 1982.

Wood, J. M. and Wang, H. K., *Environ. Sci. Technol.*, 17, 582a–590a, 1983.

Wu, C., Clos, J., Giorgi, G., Haroun, R. I., Kim, S. J., Rabindran, S. K., Westwood, J. T., Wisniewski, J., and Yim, G., in *The Biology of Heat Shock Proteins and Molecular Chaperones*, Morimoto, R. I., Tissieres, A., and Georgopoulos, C., Eds., Cold Spring Harbor Laboratory Press, Cold Spring Harbor, NY, 395, 1994.

Yager, J. W. and Wiencke, J. K., *Environ. Health Perspect.*, 101, 79–82, 1993.

Yoshida, Y., Furuta, S., and Niki, E., *Biochim. Biophys. Acta*, 1210, 81–88, 1993.

Zafarullah, M., Su, S. M., and Gedamu, L., *Exp. Cell Res.*, 208, 371–377, 1993.

Zeuthen, M. L. and Howard, D. H., *J. Gen. Microbiol.*, 135, 2509–2518, 1989.

Zhang, M., Zamore, P. D., Carmo-Fonseca, M., Lamond, A. I., and Green, M. R., *Proc. Natl. Acad. Sci. U.S.A.*, 89, 8769–8773, 1992.

Chapter 11

Metal-Metal Interactions

Ananda S. Prasad and Frances W.J. Beck

ABSTRACT

Biological interactions between trace elements are common. It has been hypothesized that those elements whose physical and chemical properties are similar will act antagonistically to each other biologically. In humans, inorganic iron, when added to test solutions of zinc salts in Fe/Zn ratios of 2.23, significantly lowered zinc absorption. The effect of iron on zinc absorption in humans is largely dependent upon the nature and extent of zinc complex formation with food in the intestine. If large iron supplements are ingested in the absence of food, it is probable that iron could affect absorption of zinc adversely. Iron storage may be affected by a high level of zinc supplementation. High levels of zinc ingestion lead to copper deficiency in humans, and this is the basis for the treatment of Wilson's disease (a copper-storage disorder) with zinc supplementation. The metallothionein (MT) level in the intestine is directly related to the zinc status. After high doses of zinc supplementation, the MT level in the intestine increases. Inasmuch as MT has a higher affinity for copper, the increased intestinal MT binds and traps copper which is then excreted in the feces when the cells are sloughed off. MT is an important metal-binding protein, and one other important function of the protein is believed to be cellular detoxification of metals. Zinc protects against cadmium toxicity. From the above, it is clear that metal-metal interactions are of great physiological importance, and knowledge in this area may provide therapeutic insights of metal toxicity.

I. INTRODUCTION

Many biological interactions between trace elements are known to occur. As examples, high levels of dietary manganese decrease hemoglobin synthesis and iron stores, and this effect is reversible with the administration of supplemental iron. High levels of zinc in the diet produces an anemia which is corrected by copper supplementation. Another example is the observation that selenium toxicity could be overcome by the addition of arsenic to the drinking water. Cadmium in the diet decreases growth and increases hock and feather abnormalities in turkey poults, and these symptoms are correctable by zinc supplementation. High levels of tungstate inhibit growth and xanthine oxidase activity in chicks and rats, and these effects are reversed by the addition of molybdenum in the diet. The above examples led Hill and Matrone (1970) to hypothesize that "those elements whose physical and chemical properties are similar will act antagonistically to each other biologically."

The orbitals of an element exist in an orderly arrangement. By convention these are called, in order of increasing distance from the nucleus, 1s, 2s, 2p, 3s, 3p, 3d, 4s, 4p, 4d, 4f, 5s, 5p, 5d, 5f, 5g, and so forth. Each orbital may have a maximum of two electrons and these must be of opposite spin status.

Copper exists in either the monovalent or divalent form, and sum of its biological importance is related to its ability to oscillate between the cuprous and cupric state, i.e., the ability to accept and donate electrons. The 1s, 2s, 2p, 3s, and 3p orbital of the cuprous ions are also filled. The cupric ion has one less electron and it happens because of energy considerations that one 3d electron is promoted to a 4p orbital, leaving a 3d orbital as well as the 4s orbital without electrons and available for bond formation.

The Zn^{2+}, Cd^{2+}, and Hg^{2+} ions all have the same electronic structure of the valence shell as the cuprous ion, while the Ag^{2+} ion has the same structure as the cupric ion.

The experimental studies of Hill and Matrone (1970) showed that zinc acts as a copper antagonist. Copper increased the weights of the chicks and cadmium significantly decreased them. There was a significant zinc-cadmium interaction, indicating that in the presence of zinc cadmium was much less effective in reducing growth.

Silver at 100 ppm reduces growth in the copper-deficient but not in the control chicks (Hill and Matrone, 1970). Mortality was increased in the copper-deficient chick when silver was administered at a 50-ppm level, but not in those groups receiving copper; thus, in these experiments, silver acted as copper antagonist.

Thus, the experimental data reported by Hill and Matrone (1970) showed that zinc, cadmium, and silver acted as copper antagonists and that cadmium acted as a zinc antagonist, consistent with the hypothesis stated earlier.

In this chapter, metal-metal interactions of biological significance will be presented.

II. INTERACTION OF ZINC AND IRON

It has been suggested that an antagonistic effect of iron on zinc absorption may have a detrimental effect on zinc nutrition, particularly in population groups that receive iron supplements routinely (Flanagan and Valberg, 1988; Solomons, 1988). Both iron and zinc belong to the first transition series of the periodic table and they share an identical outer electronic configuration with manganese, cobalt, and nickel. Both zinc and iron are essential for growth and development and are similar in amounts ingested, and absorbed amounts represent only a fraction of the total amount present in the body (Flanagan and Valberg, 1988). In spite of these similarities, body homeostasis mechanisms of iron and zinc differ. Iron, once absorbed, is retained tenaciously in the body, whereas zinc is both absorbed and endogenously secreted by the intestines (Flanagan and Valberg, 1988; Matseshe et al., 1980; Lee et al., 1989).

In rodents, the capacity of the intestine to absorb iron is greatly enhanced by feeding a low-iron diet and, in iron-deficient rats and mice, the oral absorption of zinc is also increased (Forth and Rummel, 1973). It is probable that in animals with a high capacity to absorb iron, a significant portion of zinc absorption occurs via the iron-absorbing mechanism, and transport of zinc is inhibited partly by iron (Pollack et al., 1965; Hamilton et al., 1979).

The absorption of iron and zinc are not completely similar. Iron absorption is restricted to the duodenum, whereas zinc is absorbed throughout the upper small intestine (Forth and Rummel, 1973; Flanagan et al., 1983; Lee et al., 1989). It appears that zinc is less inhibitory of iron absorption than iron is of zinc absorption. Finally, in mice with sex-linked anemia (sla), the genetic lesion affects the absorption of iron but not zinc (Flanagan et al., 1984). These observations suggest that the absorption of zinc is not entirely restricted to the iron pathway.

In humans, inorganic iron added to test solutions of zinc salts in Fe/Zn ratios significantly lowered zinc absorption (Solomons and Jacob, 1981; Solomons et al., 1983; Valberg et al., 1984; Sandstrom et al., 1985). Zinc and iron interaction in humans may not be to the same extent as in rodents, and this may explain why higher Fe/Zn ratios were required to inhibit zinc absorption in the human studies (Flanagan and Valberg, 1988).

In contrast to the above studies, in three studies in which iron was given with food, no effect of iron on zinc absorption was observed. Oysters providing about 54 mg of zinc were consumed with 100 mg ferrous iron; absorption of zinc was unaffected (Solomons and Jacob, 1981). In another study, turkey meat containing 4 mg zinc was fed to volunteers with either 17 or 34 mg ferric iron without any effect on zinc absorption (Valberg et al., 1984). Finally, when ferrous iron at an Fe/Zn ratio was added to a composite meal containing 2.6 mg zinc, the absorption of zinc was not altered (Sandstrom et al., 1985).

Thus, it appears that human zinc absorption is largely dependent upon the nature and extent of zinc complex formation with food in the intestine, and normally the influence of iron on zinc absorption may not be detrimental. Under unusual circumstances, however, if large iron supplements are ingested in the absence of food, it is probable that iron could affect absorption of zinc adversely.

According to Solomons (1988), however, an iron-zinc interaction in human subjects is of nutritional importance. He cites several observations supporting this conclusion. The first one is a study from Iran, where Mahloudji et al. (1975) reported that growth of iron-deficient Iranian schoolboys improved when they were supplemented with only 20 mg iron daily, in comparison to when a combination of 20 mg iron and 20 mg zinc was used, suggesting that oral zinc may have decreased the absorption of iron in these children. An additional interpretation could be that the effect of zinc supplementation on growth in these children was not observed because iron may have adversely affected zinc absorption. The ^{59}Fe absorption studies of Aggett et al. (1983) lend support to the view that iron and zinc mutually inhibit absorption.

The second evidence of interaction between iron and zinc in humans comes from the study of Prasad et al. (1978) of four volunteers who were fed for several months on a semisynthetic soy protein-based zinc-deficient diet containing 3.5 mg zinc per day. The fall in plasma zinc was more pronounced in two subjects who received 130 mg iron per day (Fe/Zn ratio 37:1) in comparison to the other two who received 20.3 mg iron per day (Fe/Zn ratio 8:1), suggesting that iron excess may have significantly contributed to zinc deficiency in the first two subjects.

In another study, the growth rates of healthy, middle-class white infants in Denver, Colorado were recorded when they received 1.8 mg zinc per liter of infant formula vs. those receiving 5.8 mg zinc per liter of infant formula (Solomons, 1988). More zinc was required for optimal growth of the infants when the formula contained 12 mg iron per liter.

A final, population-level support for an antagonistic iron-zinc interaction comes from observation of pregnant women receiving different levels of prenatal mineral supplements. Hambidge et al. (1983) observed an inverse relationship between the dosages of iron supplement and the plasma zinc concentration in the first and third trimesters of pregnancy. Campbell-Brown et al. (1985) observed that three women who were supplemented with 100 mg or more iron daily had the lowest plasma zinc levels relative to other pregnant women who were supplemented with less iron. Breskin et al. (1983) showed that pregnant women who were supplemented with 30 mg of iron daily had significantly lower plasma zinc concentrations in comparison to those who received either no iron supplement or less than 30 mg per day. In contrast to the above, some investigators found no effect of iron supplementation (160 mg iron daily) on plasma zinc levels in pregnant women (Solomons, 1988).

The absorption of trace element involves intraluminal ligands, mucosal binding sites or channels, intracellular transport proteins, intracellular carrier proteins, basolateral energetic mechanisms, and circulating carrier proteins in the bloodstream (Solomons, 1988). The simultaneous presence of two ions with similar electronic configurations might lead to competition for movement at one or more of these specific levels.

A direct competition of iron and zinc in the intestine has been studied by several investigators using sample solutions of minerals and analyzing their effects on zinc uptake in the plasma (Abu-Hamdan et al., 1984; Aggett et al., 1983; Payton et al., 1982; Sandstrom et al., 1985; Solomons, 1983; Solomons and Jacob, 1981; Solomons et al., 1983a,b; Valberg et al., 1982). Consistent with the electronic configuration hypothesis, ferrous iron produced a greater inhibition of zinc uptake than did ferric iron (Hill and Matrone, 1970).

Iron storage is affected by a high level of zinc supplementation. Zinc interferes with iron uptake by the liver and storage of iron as ferritin is decreased (Settlemire and Matrone, 1967b). Decreased iron content in the liver and kidneys in response to excessive zinc supplementation has been observed by several investigators (Cox and Harris, 1960; Magee and Matrone, 1960; Kang et al., 1977; Hamilton et al., 1979). Since transferrin in plasma transports both iron and zinc, the interaction of iron and zinc may reflect competition at the transport level.

An increase in iron and a decrease in zinc concentration in various organs such as the liver, bone, pancreas, and testes have been reported in zinc-deficient animals relative to pair-fed controls (Moses and Parker, 1964; Prasad et al., 1967; Prasad et al., 1969a,b). These changes are reversible following zinc supplementation. Alterations in iron absorption may be responsible for changes in the iron concentration of organs in zinc-deficient conditions.

Effects on rat tissue concentrations of zinc, iron, copper, and magnesium of feeding various amounts of zinc was reported by Kang et al. (1977). The rats were fed the following three levels of zinc for four weeks. Group A: this group of rats were fed *ad libitum* a zinc-deficient diet (1.3 μg of zinc per gram); Group B: the second group of rats received the same diet as group A, but were supplemented with 55 μg of zinc per gram; and Group C: the third group of rats, received the same diet as the first group but were supplemented with 550 μg zinc per gram. The control rats (groups B and C) were pair-fed, and they received each day an amount of diet equal to the diet intake of their pair-mates in group A.

The zinc-supplemented rats showed increased levels of zinc in the blood, heart, kidneys, and liver, and a marked decrease of iron in the kidney and liver. Other elements showed no change. Thus, the changes in iron concentrations in tissues of rats were also observed when zinc was administered in normal or subtoxic but excess amounts in the diet.

III. ZINC AND COPPER

The interaction between zinc and copper is mutually antagonistic. In zinc-deficient animals, copper concentration in the liver and bone is increased (Moses and Parker, 1964; Prasad et al., 1969a; Petering et al., 1971; Burch et al., 1975; Roth and Kirchgessner, 1977). Milk excretion of copper is increased in zinc-deficient cows (Kirchgessner et al., 1982). When the dietary zinc is increased excessively, copper concentration in the liver, heart, and serum, and the activities of copper enzymes such as ceruloplasmin and cytochrome oxidase, are decreased (Duncan et al., 1953; Van Reen, 1953; Cox and Harris, 1960). These changes are reversed by increasing the dietary copper, thus decreasing the zinc-copper ratio. Similarly, zinc is protective against copper toxicity (Suttle and Mills, 1966a,b; Bremner et al., 1976).

Copper and zinc inhibit the intestinal absorption of each other. Decreased zinc absorption in the presence of excess copper is observed only in rats administered an adequate level of zinc, since in zinc-deficient rats zinc absorption is not inhibited by copper (Evans et al., 1974; Schwarz and Kirchgessner, 1973; Schwarz and Kirchgessner, 1974a,b). In zinc deficiency, absorption of both zinc and copper is improved. Whereas absorption of copper is increased in copper deficiency, zinc absorption is not affected.

The metallothionein (MT) level in the intestine is directly related to the zinc status. After high doses of zinc supplementation, the MT level in the intestine increases (Richards and Cousins, 1975). Inasmuch as MT has a higher affinity for copper, the increased intestinal MT binds and traps copper which is then excreted in the feces when the cells are sloughed off (Richards and Cousins, 1976). No such relationship between MT and reduced Zn absorption after excess copper exposure has been reported (Hall et al., 1979).

A. ZINC AND COPPER INTERACTIONS IN HUMANS

We observed hypocupremia and hypoceruloplasminemia in an adult with sickle cell anemia (SCA) who received zinc (25 mg elemental zinc every 4 h) as an antisickling agent for two years (Prasad et al., 1978a). The hypocupremia was associated with microcytosis and neutropenia. Administration of copper resulted in an increase in the size of the red blood cells and neutrophil counts. We have since observed hypoceruloplasminemia in several other SCA patients who were receiving excess zinc as therapy. This complication was correctable with copper administration.

These observations led us to use zinc for reduction of copper burden in patients with Wilson's disease. Wilson's disease is an autosomal recessive inborn error involving low excretion of copper by the liver, which leads to excessive accumulation of copper in the liver, brain, kidneys, and other tissues and, if untreated, becomes fatal. Penicillamine, a copper chelator, is currently used for treatment of Wilson's disease. In about 30% of the patients, penicillamine therapy results in acute sensitivity reactions including skin eruptions, fever, eosinophilia, leukopenia, thrombocytopenia, and lymphadenopathy, which occur within two weeks of treatment. If the drug is withdrawn for a few days, therapy frequently can be restarted, usually at a lower dose. In 10% of the cases, however, intolerance to penicillamine is so great that the drug cannot be taken at all (Hirschman and Isselbacher, 1965). We therefore studied the efficacy of zinc treatment in Wilson's disease.

Initially, we treated five patients with Wilson's disease with zinc (Brewer et al., 1983). We administered 25 mg of elemental zinc (as acetate) every 4 h during the day plus a 50 mg dose at bedtime. In studies on our first two patients with Wilson's disease, we assumed that direct competition between zinc and copper at the intestinal level would result in decreased absorption of zinc almost immediately. This effect was not observed. Copper excretion increased after 21 days of zinc therapy in the first patient and after 6 days of zinc therapy in the second patient. We hypothesized that a buildup of zinc in body tissues, especially in the intestine, must occur before copper absorption is inhibited.

In our next experiment, we pretreated five patients with zinc for 3 weeks before the study. Penicillamine was discontinued 1 week prior to admission. The patients were then maintained on zinc therapy alone. We induced a negative or neutral copper balance in all five subjects who were receiving no therapy other than zinc.

A significant amount of copper is excreted into the gastrointestinal tract (endogenous excretion of copper). Whereas biliary copper is not reabsorbed to a large extent, the copper in saliva and gastric juices, totaling 1.5 mg/d, is normally reabsorbed. Reabsorption of nonbiliary copper as well as dietary copper are therefore targets for intestinal MT, which could bind copper and prevent its absorption. This hypothesis is consistent with our results, inasmuch as pretreatment with zinc most likely resulted in induction of MT synthesis in the intestine and inasmuch as MT has greater affinity for binding copper we were able to induce a negative copper balance by zinc therapy.

We validated the copper balance data in Wilson's disease patients by measurement of ^{64}Cu uptake in blood following oral ingestion of 0.5 mCi ^{64}Cu acetate in 40 ml of cow's milk (Hill et al., 1986). The mean plasma peak of ^{64}Cu uptake in nine Wilson's disease patients on D-penicillamine, Trien, or no medication was $6.04 \pm 2.74\%$, in contrast to patients on zinc therapy in whom the mean peak of ^{64}Cu was $0.79 \pm 1.05\%$. These results showed that prevention of copper intake into blood by zinc therapy in Wilson's disease patients can be evaluated by ^{64}Cu studies, and that a peak uptake of less than 1% occurs in patients with neutral or negative copper balance.

Our studies also showed that the 24 h urine copper decreases following zinc therapy, (Brewer et al., 1987a). In penicillamine-treated Wilson's disease patients, urinary copper is high because of the increased amount of mobilizable copper load in the body available for excretion. In the absence of penicillamine therapy (as is the case with zinc therapy), the quantity of copper in the urine becomes a reflection of the level of excess copper. With zinc therapy we see a fairly rapid decline of urinary copper, which returns to the normal range (69 ± 24 µg/d).

The range of nonceruloplasmin plasma copper (non-cp Cu) of normal subjects is approximately 10 to 20 µg/dl. Non-cp Cu is increased in untreated Wilson's disease patients, and it is presumably the copper that causes toxicity in Wilson's disease (Brewer et al., 1987a). In zinc-treated patients, non-cp Cu returns to normal levels.

In another study, 12 patients with Wilson's disease, most of whom had received intensive treatment with penicillamine, were given zinc therapy as the sole treatment for copper control (Brewer et al., 1987b). Liver copper was measured in serial biopsies during a 12- to 20-month follow-up period. Mean ($\pm SD$) baseline copper concentration was 255 ± 194 µg/g dry wt whereas after therapy it was 239 ± 185 µg/g dry wt. No subject showed hepatic reaccumulation of copper during zinc therapy. Copper balance, 24 h urinary copper, and non-cp Cu indicated good copper control during zinc therapy. Hepatic zinc concentration increased two- to threefold over baseline values, but no toxicity was seen. Hepatic zinc concentration appeared to reach a plateau after 12 to 18 months of zinc therapy. These results showed that oral zinc as the sole maintenance therapy in Wilson's disease prevents hepatic reaccumulation of copper. On the basis of the size of the positive copper balances in patients who have received no treatment, and making some assumptions about skin surface losses and the proportion of accumulated copper stored in the liver, we estimate an average yearly accumulation of 150 µg/g of dry weight in the liver in untreated patients. All patients on zinc therapy remained clinically stable, and none showed worsening or progression of neurologic disease or speech abnormality.

Our recent data indicate that a more simplified regimen such as 50 mg zinc three times a day is equally effective in controlling copper balance in patients with Wilson's disease and that administration of zinc six times a day at 4-h intervals was not necessary (Brewer et al., 1987a,b).

B. ZINC AND COPPER INTERRELATIONSHIP IN OTHER CONDITIONS

A decreased plasma zinc and increased plasma copper concentration have been reported in pregnancy, women on oral contraceptives, acute infection, malignancy, cardiovascular disease, renal disease, schizophrenia, and certain endocrine diseases such as acromegaly and Addison's disease. The exact mechanism of the reciprocal relationship between zinc and copper in the above conditions is not known.

IV. MANGANESE AND ZINC

Manganese deficiency in experimental animals results in a decline in zinc levels in the liver, kidney, spleen, and bone (Heiseke and Kirchgessner, 1978). Zinc retention is increased in response to diets supplemented with increased levels of manganese (Grace, 1973; Ivan and Grieve, 1975; Jarvinen and Ahlstrom, 1975).

In short-term studies, manganese levels in the liver, small intestine, heart, and bone decreased when zinc deficiency was induced in experimental animals (Schwarz and Kirchgessner, 1980). In other studies, however, no effect or even a positive effect of zinc depletion on manganese accumulation in certain organs was reported (Prasad et al., 1967, 1969a; Roth and Kirchgessner, 1979). It is possible that zinc and manganese interact at the site of intermediary metabolism, leading to an increased requirement for manganese when zinc nutrition is suboptimal, and conversely for zinc when manganese nutrition is inadequate. There is no evidence that manganese absorption is affected due to zinc deficiency or vice versa.

Two groups of male Sprague-Dawley rats were treated intraperitoneally for 30 d with either 3.0 mg Mn/kg or an equal volume of 0.9% NaCl. Liver, kidney, pancreas, duodenum, spleen, testes, lungs, brain, skeletal muscle, bone, and blood were analyzed for Mn, Mg, Zn, Fe, and Cu. Mn increased in all tissues except liver due to treatment. Bone and pancreas revealed the largest increases. In blood, increased Mn levels were accounted for by increase in the erythrocyte fraction. Subcellularly, all fractions revealed increases in Mn content due to treatment. Mn exposure was accompanied by decreased Zn levels in plasma and bone, decreased Mg levels in heart and bone, increased pancreatic Fe concentration, and increased Cu concentration in plasma and several tissues.

V. METALLOTHIONEIN

Metallothionein (MT) is an important metal-binding protein which occurs in varying amounts in a wide range of tissues. The liver, kidneys, intestine, and pancreas contain relatively high amounts of MT. Synthesis of MT is induced by Zn and Cu. Other nonessential elements such as Cd also induce its synthesis. A variety of stress factors stimulate MT synthesis in liver. The turnover rate of MT in tissues is high and depends to a large extent on its metal content. One important function of this protein is believed to be cellular detoxification of Cu, Zn, and other metals. MT also appears to participate in metabolic interactions between zinc and copper. MT occurs in small amounts in blood and urine, and its assay may be useful in assessment of trace element status.

VI. CADMIUM, LEAD, AND OTHER METALS

Zinc interacts negatively with cadmium. The excretion of cadmium increases when zinc is administered and, in general, zinc protects against cadmium toxicity (Ahokas et al., 1980; Lucis et al., 1972; Schroeder et al., 1970; Stowe, 1976). High ratios of cadmium to zinc in the kidneys have been reported in both rats and humans (Schroeder et al., 1970). In rats, zinc displaces cadmium from the kidneys and decreases blood pressure.

Marginal zinc deficiency in the rat leads to a greater accumulation of lead in rat pups whose mothers have been administered lead during lactation (Ashrafi and Fosmire, 1985). The clinical relevance of this observation in humans is not known.

Jamall and Roque (1990) demonstrated that feeding 50 ppm cadmium to rats results in cadmium accumulation in the eyes after 7 weeks. Rats fed the basal diet and given 100 ppm cadmium via their feed for 6 weeks exhibited a 69% reduction in the activity of the selenoenzyme, glutathione peroxidase, in the eye. Iron levels increased by 30% in rats fed a low-selenium diet and decreased by 40% in rats fed a selenium-supplemented diet, compared to animals fed identical levels of selenium without cadmium. Ocular copper levels were increased only in rats fed the low-selenium diet and treated with cadmium. Ocular zinc levels were not significantly affected by dietary cadmium or selenium.

In one study, newly hatched white Pekin ducklings were fed a commercial starter mash adequate in selenium and vitamin E, either alone or with supplements of silver (Ag), copper (Cu), cobalt (Co), tellurium (Te), cadmium (Cd), zinc (Zn) or vanadium (V) (Van Vleet et al., 1981). The ducklings fed Ag, Cu, Co, Te, Cd, and Zn frequently developed lesions characteristic of Se-E deficiency. This consisted of necrosis of skeletal and cardiac muscle and of smooth muscle of the gizzard and intestine. Complete protection from these lesions was provided by vitamin E (200 IU/kg alpha-tocopherol acetate) and Se (2 mg/kg, as selenite). Ducklings fed Ag were protected by supplements of vitamin E, and partial protection was achieved by Se addition. The ducklings fed excessive Zn developed pancreatic necrosis and fibrosis that was not prevented by vitamin E or Se.

The protective effect of Cu, Zn, or Co against Pb toxicity in rats was investigated by Flora et al. (1982). Administration of essential trace elements, together with Pb, decreased the hepatic and renal uptake of lead and reduced the Pb-induced inhibition of blood delta-aminolevulinic acid dehydratase activity. The hepatic uptake of Zn or Co was increased in animals administered Pb and Zn, or Pb and Co, respectively.

Cu and Zn are known to induce synthesis of hepatic MT. Thus, the protective effect of Cu and Zn against Pb may be related to MT induction in the liver.

Relatively small dietary changes in young Japanese quails markedly affected tissue levels of cadmium, and low intake of zinc significantly increased the risk to dietary cadmium exposure.

In one study, effects of a chronic low dose of cadmium (50 ppm additional cadmium in water) added to an otherwise normal diet on the liver and kidney accumulation of zinc, copper, iron, manganese, and chromium in the mouse at different times after the cessation of cadmium ingestion were reported. Zinc concentration in the liver did not change, but there were decreases in copper and iron concentrations with cadmium ingestion. Similar results have been reported earlier by other investigators (Suzuki et al., 1983; Bunn and Matrone, 1966). Cadmium is known to be antagonistic to copper and iron absorption (Hill and Matrone, 1974; Weigel et al., 1984).

In one study, human subjects were either fed mixed diets containing 0.11 mg tin daily (control diet) or containing 49.67 mg tin daily (test diet) for 40 days. The level of tin in the control diet was typical of the levels of tin found in diets that contain only fresh and frozen foods; the level of tin in the test diet was typical of the amount of tin in diets that contain two cups of certain canned foods. Subjects fed the test diet lost significantly more zinc in their feces and less zinc in their urine. The fecal and urinary losses of copper, iron, manganese, and magnesium were not significantly affected by the dietary treatments.

Biological interactions between selenium and other elements (arsenic, mercury, cadmium, and copper) occur such that selenium is rendered much less toxic than when it is present alone. The presence of selenium also reduces the toxicity of mercury and cadmium. It has been shown that the reaction products of selenium with mercury and cadmium are less toxic than an equal amount of selenium fed alone to chicks. The presence of arsenic shifts the excretion of selenium to the bile. There is no conclusive evidence that the presence of other elements reduces the absorption or retention of selenium. It is likely that some of the interactions are caused by the formation of a compound by selenium and other elements which has less affinity for active groups on biologically active compounds.

VII. CALCIUM AND ZINC

At the intestinal level, calcium and zinc have an antagonistic relationship. In experimental animals, if the intake of calcium is high the absorption of zinc is decreased and vice versa (Hanson et al., 1958). In pigs, parakeratosis is partly a result of high calcium intake in the presence of a low intake of zinc (Hanson et al., 1958).

The exacerbation of an essential fatty acid deficiency by zinc deficiency was shown to be enhanced by a high calcium intake. In rats, fetal malformations induced by zinc deficiency were also dependent on calcium intake, in that when calcium intake was low, the malformations did not occur, whereas the malformations were increased when the intake of calcium was high (Hurley and Tao, 1972).

Zinc interacts with calcium at the red cell membrane by suppressing calmodulin, a calcium regulating protein (Baudier et al., 1983). This property of zinc has been utilized to suppress formation of irreversible sickle cells (ISCs) in which the intracellular concentration of calcium is known to rise and then calcium binds to hemoglobin and the membrane, forming ISCs. Oral administration of zinc to SCA patients is known to decrease the number of ISCs *in vivo* (Brewer, 1980; Brewer et al., 1979).

VIII. MAGNESIUM AND ZINC

Very little is known concerning the interaction of magnesium with zinc. A mild deficiency of magnesium in the rat is associated with decreased levels of zinc in lower femur and whole carcass (Kubena et al., 1985). Pyridoxal phosphokinase enzyme is catalyzed by both magnesium and zinc (McCormick et al., 1961).

IX. SODIUM AND ZINC

Zinc (0.1 to 0.9 µm) infusion into the jejunum decreases sodium absorption in humans (Steinhardt and Adibi, 1984; Lee et al., 1989). Low dietary intake of sodium is associated with increased urinary zinc excretion, suggesting that a sodium-dependent mechanism may control renal tubular reabsorption of zinc (Matustik et al., 1982). The sodium/potassium content of muscle is increased in zinc-deficient rats, and there is no increase in the aldosterone secretion in sodium-depleted rats as a result of zinc deficiency.

ACKNOWLEDGMENTS

This work was supported in part by the National Institutes of Health/National Institute of Diabetes and Digestive and Kidney Diseases Grant No. DK-31410, Food and Drug Administration Grant No. FDA-U-000457, NIH/National Cancer Institute Grant No. CA 43838, and Labcatal Laboratories.

REFERENCES

Abu-Hamdan, D. K., Mahajan, S. K., Migdal, S. D., Prasad, A. S., and McDonald, F. D., *J. Am. Coll. Nutr.*, 3, 283–284, 1984.
Aggett, P. J., Crofton, R. W., Khin, C., Gvozdanovic, S., and Gvozdanovic, D., in *Zinc Deficiency in Human Subjects*, Prasad, A.S., Cavder, A.O., Brewer, G.J., and Aggett, P.J., Eds., Alan R. Liss, New York, 117, 1983.
Ahokas, R. A., Dilts, P. V., and Lahaye, E. B., *Am. J. Obstet. Gynecol.*, 136, 216–221, 1980.
Ashrafi, M. H. and Fosmire, G. J., *J. Nutr.*, 115, 334–346, 1985.
Baudier, J., Haglid, K., Haiech, J., and Gerard, D., *Biochem. Biophys. Res. Commun.*, 114, 1138–1146, 1983.
Bremner, I., *Progr. Food Nutr. Sci.*, 11, 1–37, 1987.
Bremner, I., Young, B. W., and Mills, C. F., *Br. J. Nutr.*, 36, 551–561, 1976.
Breskin, M. W., Worthington-Roberts, B. S., Knopp, R. H., Brown, Z., Plovie, B., Mottet, N. K., and Mills, J. L., *Am. J. Clin. Nutr.*, 38, 943–953, 1983.
Brewer, G. J., *Am. J. Hematol.*, 8, 231–248, 1980.
Brewer, G. J., Aster, J. C., Knutsen, C. A., and Kruckberg, W. D., *Am. J. Hematol.*, 7, 53–60, 1979.
Brewer, G. J., Hill, G. M., Prasad, A. S., Cossack, Z. T., and Rabbani, P., *Ann. Intern. Med.*, 99, 314–320, 1983.
Brewer, G. J., Hill, G. M., Dick, R. D., Nostrant, T. T., Sams, J. S., Wells, J. J., and Prasad, A. S., *J. Lab. Clin. Med.*, 109, 526–531, 1987a.
Brewer, G. J., Hill, G. M., Prasad, A. S., and Dick, R., *Proc. Soc. Exp. Biol. Med.*, 184, 446–455, 1987b.
Burch, R. E., Williams, R. V., Hahn, H. K. J., Jetton, M. M., and Sullivan, J. F., *Clin. Chem.*, 21, 568–577, 1975.
Campbell-Brown, M., Ward, R. J., Haines, A. P., North, W. R. S., Abraham, R., and McFayden, I. R., *Br. J. Obstet. Gynecol.*, 92, 975–985, 1985.
Cox, D. H. and Harris, D. L., *J. Nutr.*, 70, 514–520, 1960.
Duncan, G. D., Gray, L. F., and Daniel, L. J., *Proc. Soc. Exp. Biol. Med.*, 83, 625–627, 1953.
Evans, G. W., Grace, C. I., and Hahn, C., *Bioinorg. Chem.*, 3, 115–120, 1974.
Flanagan, P. R. and Valberg, L. S., in *Essential and Toxic Trace Elements in Human Health and Disease*, Prasad, A.S., Ed., Alan R. Liss, New York, 501, 1988.
Flanagan, P. R., Haist, J., and Valberg, L. S., *J. Nutr.*, 113, 962–972, 1983.
Flanagan, P. R., Haist, J., MacKenzie, I., and Valberg, L. S., *Can. J. Physiol. Pharmacol.*, 62, 1124–1128, 1984.
Flora, S. J. S., Jain, V. K., Behari, J. R., and Taudon, S. K., *Toxicol. Lett.*, 13, 51–56, 1982.
Forth, W. and Rummel, W., *Physiol. Rev.*, 63, 724–792, 1973.
Fox, M. R. S., Jacobs, R. M., Jones, A. O. L., and Fry, B. E., Jr., *Environ. Health Perspect.*, 28, 107–114, 1979.
Friel, J. K., Borgman, R. F., and Chandra, R. K., *Bull. Environ. Contam. Toxicol.*, 38, 588–593, 1987.
Grace, N. D., *N.Z. J. Agric. Res.*, 16, 177, 1973.
Hall, A. C., Young, B. W., and Bremner, I., *J. Inorg. Biochem.*, 11, 57–66, 1979.
Hambidge, K. M., Krebs, N. F., Jacobs, M. A., Favier, A., Guyette, L., and Ickle, D. N., *Am. J. Clin. Nutr.*, 37, 429–442, 1983.
Hamilton, R. P., Fox, M. R. S., Fry, B. E., Jr., Jones, A. O. L., and Jacobs, R. M., *J. Food Sci.*, 44, 738–741, 1979.
Hanson, L. J., Sorenson, D. K., and Kernkamp, H. C. H., *Am. J. Vet. Res.*, 18, 921–930, 1958.
Heiseke, D. and Kirchgessner, M., *Zentralbl. Veterinaermed. Reihe A.*, 25, 307–311, 1978.
Hill, C. H., *Fed. Proc.*, 34, 2096–2100, 1975.
Hill, C. H. and Matrone, G., *Fed. Proc.*, 29, 1474–1481, 1970.
Hill, G. M., Brewer, G. J., Juni, J. E., Prasad, A. S., and Dick, R. D., *Am. J. Med. Sci.*, 29, 344–349, 1986.
Hirschman, S. Z. and Isselbacher, K. J., *Ann. Intern. Med.*, 62, 1297–1300, 1965.
Hurley, L. S. and Tao, S. H., *Am. J. Physiol.*, 222, 322–325, 1972.
Ivan, M. and Grieve, C. M., *J. Dairy Sci.*, 58, 410–415, 1975.
Jamall, I. S. and Roque, H., *Biol. Trace Elem. Res.*, 23, 55–63, 1990.
Jarvinen, R. and Ahlstrom, A., *Med. Biol.*, 53, 93–99, 1975.
Johnson, M. A., Baier, M. J., and Greger, J. L., *Am. J. Clin. Nutr.*, 13, 1332–1338, 1982.
Kang, H. K., Harvey, P. W., Valentine, J. L., and Swendseid, M. E., *Clin. Chem.*, 23, 1834–1837, 1977.
Kirchgessner, M., Schwarz, F. J., and Schnegg, A., in *Clinical, Biochemical, and Nutritional Aspects of Trace Elements* Prasad, A.S., Ed., Alan R. Liss, New York, 477, 1982.
Kubena, K. S., Landmann, W. A., Young, C. R., and Carpenter, Z. L., *Nutr. Res.*, 5, 317–328, 1985.

Lee, H. H., Prasad, A. S., Brewer, G. J., and Owyang, C., *Am. J. Physiol.,* 256, G87–G91, 1989.
Lucis, O. J., Lucis, R., and Shaikh, Z. A. *Arch. Environ. Health,* 25, 14–22, 1972.
Magee, A. C. and Matrone, G., *J. Nutr.,* 72, 233–242, 1960.
Mahloudji, M., Reinhold, J. G., Haghasenass, M., Ronaghy, H. A., Fox, M. R., and Halsted, J. A., *Am. J. Clin. Nutr.,* 28, 721–725, 1975.
Matseshe, J. W., Phillips, S. F., Malageldad, J. R., and McCall, J. T., *Am. J. Clin. Nutr.,* 33, 1946–1953, 1980.
Matustik, M. C., Chausner, A. B., and Meyer, W. J., *J. Am. Coll. Nutr.,* 1, 331–336, 1982.
McCormick, D. B., Gregroy, M. E., and Snell, E. E., *J. Biol. Chem.,* 236, 2076–2084, 1961.
Moses, H. A. and Parker, H. E., *Fed. Proc.,* 23, 132, 1964.
Neary, J. T. and Divan, W. F., *J. Biol. Chem.,* 245, 5585–5593, 1970.
Payton, K. B., Flanagan, P. R., Stinson, E. A., Chrodiker, D. R., Chamberlain, M. J., and Valberg, L. S.,*Gastroenterology,* 83, 1264–1270, 1982.
Petering, H. G., Johnson, M. A., and Horwitz, J. P., *Arch. Environ. Health,* 23, 93–101, 1971.
Pollack, S., George, J. N., Reba, R. C., Kaufman, R. M., and Crosby, W. J., *J. Clin. Invest.,* 44, 1470–1473, 1965.
Prasad, A. S., Oberleas, D., Wolf, P., and Horwitz, J. P., *J. Clin. Invest.,* 46, 549–557, 1967.
Prasad, A. S., Oberleas, D., Wolf, P., Horwitz, J. P., Miller, E. R., and Luecke, R. W., *Am. J. Clin. Nutr.,* 22, 628–637, 1969.
Prasad, A. S., Oberleas, D., Wolf, P., and Horwitz, J. P., *J. Lab. Clin. Med.,* 73, 486–494, 1969.
Prasad, A. S., Brewer, G. J., Schoomaker, E. B., and Rabbani, P., *J. Am. Med. Assoc.,* 240, 2166–2168, 1978a.
Prasad, A. S., Rabbani, P., Abassi, A., Bowersox, E., and Spivey-Fox, M. R. S., *Ann. Intern. Med.,* 89, 483–490, 1978b.
Richards, M. P. and Cousins, R. J., *Biochem. Biophys. Res. Commun.,* 64, 1215–1223, 1975.
Richards, M. P. and Cousins, R. J., *J. Nutr.,* 106, 1591–1599, 1976.
Roth, H. P. and Kirchgessner, M., *Zentralbl. Veterinaermed. Reihe. A.,* 24, 177–188, 1977.
Roth, H. P. and Kirchgessner, M., *Z. Tierphysiol. Tierernaehr. Futtermittelkd.,* 42, 277–286, 1979.
Sandstrom, B., Davidson, L., Cederblad, A., and Lonnerdal, B., *J. Nutr.,* 115, 411–414, 1985.
Schroeder, H. A., Baker, J. T., Hansen, N. M., Size, J. G., and Wise, R. A., *Arch. Environ. Health,* 21, 609–614, 1970.
Schwarz, F. J. and Kirchgessner, M., *Z. Tierphysiol. Tierernaehr. Futtermittelkd.,* 31, 91–98, 1973.
Schwarz, F. J. and Kirchgessner, M., *Int. J. Vitam. Nutr. Res.,* 44, 258–266, 1974a.
Schwarz, F. J. and Kirchgessner, M., *Int. J. Vitam. Nutr. Res.,* 44, 116–126, 1974b.
Schwarz, F. J. and Kirchgessner, M., *Z. Tierphysiol. Tierernaehr. Futtermittelkd.,* 43, 272–282, 1980.
Settlemire, C. T. and Matrone, G., *J. Nutr.,* 92, 153–158, 1967b.
Solomons, N. W., in *Essential and Toxic Trace Elements in Human Health and Disease,* Prasad, A.S., Ed., Alan R. Liss, New York, 509, 1988.
Solomons, N. W. and Jacob, R. A., *Am. J. Clin. Nutr.,* 34, 475–482, 1981.
Solomons, N. W., Pineda, O., Viteri, F., and Sandstead, H. H., *J. Nutr.,* 113, 337–349, 1983a.
Solomons, N. W., Marchini, J. S., Duarte-Favaro, R. M., Vannuchi, H., and Dutra de Oliveira, J. E., *Am. J. Clin. Nutr.,* 37, 566–571, 1983b.
Steinhardt, H. J. and Adibi, S. A., *Am. J. Physiol.,* 247, G176–G182, 1984.
Stowe, H. D., *J. Toxicol. Environ. Health,* 2, 45–53, 1976.
Suttle, N. F. and Mills, C. F., *Br. J. Nutr.,* 20, 135–148, 1966a.
Suttle, N. F. and Mills, C. F., *Br. J. Nutr.,* 20, 149–161, 1966b.
Valberg, L. S., Flanagan, P. R., and Chamberlain, M. J., *Am. J. Clin. Nutr.,* 40, 536–541, 1984.
Van Reen, R., *Arch. Biochem. Biophys.,* 46, 337–344, 1953.
Van Vleet, J. F., Boon, G. D., and Ferrans, V. J., *Am. J. Vet. Res.,* 42, 1206–1217, 1981.

Section III
Carcinogenesis and Genotoxicity of Metals

PART A

Carcinogenesis and Genotoxicity of Specific Metals

Overview

An Introduction to Metal Toxicity and Carcinogenicity of Metals

Max Costa

In this section the genotoxicity and carcinogenicity of nickel, cadmium, chromium, and arsenic are addressed in considerable detail by devoting an entire chapter to each of these areas. The chapters are written by experts in their respective fields.

Three of these chapters address mechanisms involved in the genotoxicity and carcinogenicity, because it is from an understanding of these mechanisms that we obtain a better handle on how these metals are carcinogenic or genotoxic. The concept that emerges from these studies is the understanding that each metal has a unique mechanism of action and thus it is inappropriate and misleading to discuss metal carcinogenesis under one umbrella. For example, the carcinogenicity of chromium with a hexavalent oxidation which is converted to a trivalent form of chromium in cells that forms adducts with DNA and protein is quite different from nickel which undergoes oxidation but does not directly bind tightly to DNA or protein. It does not form strong bonds with DNA and protein as does chromium. Perhaps the common way that nickel, chromium, and several other metals act in cells is by production of oxygen radicals; however, there is little evidence that arsenic and cadmium work by these mechanisms. There is evidence that inhibition of Ni carcinogenesis by Mg^{2+} does not alternate oxidative stress.

The final chapter presents the genotoxicity and carcinogenicity of other metals considered in less detail such as lead, beryllium, zinc, titanium, etc. A synopsis is presented in these areas as not much information is available from the literature, though at least the reader will have some knowledge of what information exists regarding the carcinogenicity and genotoxicity of these metals.

Chapter 12

Carcinogenicity and Genotoxicity of Chromium

Catherine B. Klein

I. INTRODUCTION

Chromium was first suspected to be a human carcinogen in the late nineteenth century when epidemiological studies linked nasal tumors in Scottish chrome workers with chromium exposure (Newman, 1890). Since then, other chromium industries, including leather tanning, chrome plating, and stainless steel production, have been implicated as potential sources of human occupational exposures to chromium. The environmental contamination that results from industrial chromium production, use, and disposal has also been recognized as a potential source of nonoccupational human exposure to chromium. The mechanisms by which chromium, an essential metal in small quantities, is a carcinogen are not yet completely known, but we have come closer to understanding some of the processes involved in chromium carcinogenesis. In this chapter, an updated review of the literature on chromium carcinogenicity and genotoxicity will be concisely summarized, and many of the newer experimental findings will be briefly described.

II. CHROMIUM CARCINOGENESIS IN HUMANS AND ANIMALS

Whereas it is well established that chromium is essential to life and must be supplemented as a trace element in the diets of humans and animals (Doisy et al., 1976), the reasons for this are not fully understood. Chromium dietary deficiencies have been clinically correlated with abnormalities of glucose metabolism and insulin function (Nath et al., 1979), the metabolism of cholesterol (Mossop, 1983) and triglycerides, and with impairments of growth and immunologic and reproductive functions (reviewed in Cohen et al., 1993). Despite its metabolic essentiality, chromium is a recognized human allergen, as well as a human and animal carcinogen that most frequently yields respiratory cancers (IARC, 1980; Hayes, 1988; IARC, 1990; Langard, 1990). Bronchial carcinomas, lung cancers, and nasal tumors predominate (Leonard and Lawreys, 1980; Langard, 1990). In this section, the literature pertaining to the epidemiology of human chromium carcinogenesis will be considered separately from the experimental data for chromium-induced tumors in animal models.

A. HUMAN EPIDEMIOLOGY

The first established case of chromium-induced cancer was diagnosed in a "chrome worker" in Scotland (Newman, 1890). Since then, numerous case histories of suspected chromium carcinogenesis in occupationally exposed workers have originated from industrial nations worldwide (summarized in Sunderman, 1976; IARC, 1990; Squibb and Snow, 1993; Cohen et al., 1993). The cancers described in these reports include those of the oral and nasal cavities, larynx, lungs (Machle and Gregorious, 1948;

Baetjer, 1950; Enterline, 1974), as well as those involving the digestive system (i.e., esophagus, stomach) (Langard and Norseth, 1979; Hayes et al., 1989; Langard et al., 1990), bladder (Claude et al., 1988), kidneys and prostate (Langard et al., 1980). The most conclusive epidemiological data have been derived from the studies of chromate production workers, for whom a significant carcinogenic risk is believed to be correlated with the inhalation of chromium particulates and dusts (Ohsaki et al., 1978; Aldersen et al., 1981; Satoh et al., 1981; Korallus et al., 1982; DeMarco et al., 1988). Measurable concentrations of chromium have been found in the lungs and other tissues of some of these workers (Tsuneta et al., 1978; Hyodo et al., 1980). For chromate workers, the lung cancer risks were at one time estimated to be as much as 10 to 20% higher than for the general population (reviewed in Sunderman, 1976), although modern occupational safety procedures and regulated industry standards have significantly reduced those risks (Squibb and Snow, 1993).

In addition to chromate manufacturing, other industries also resulted in occupational human exposure to chromium. Cancer incidences have been studied among workers involved in chromium pigment manufacturing (Langard and Norseth, 1975; Davies, 1978; Davies, 1979; Langard and Norseth, 1979; Frentzel-Beyme, 1983; Hayes et al., 1989; Davies, 1991), and significant correlations were noted, especially among workers exposed to $ZnCrO_4$ pigments (Dalager et al., 1980; Haguenoer et al., 1981; Sheffet et al., 1982; Davies, 1984). Electroplating was another primary source of occupational chromium exposure (Royle, 1975a,b; Waterhouse, 1975; Okubo and Tsuchiya 1977, 1979; Silverstein et al., 1981; Franchini et al., 1983; Sorahan et al., 1987) prior to the implementation of industrial safety protocols. A comprehensive evaluation of modern (1980 to 1989) industrial exposure to chromium as determined by air, blood, and urine measurements was recently completed by the Finnish Institute of Occupational Health (Kiilunen, 1994). This study shows that the highest levels of modern-day chromium exposure result from occupations involving metal grinding, plasma cutting, sheet metal working, and welding. High levels of exposure to Cr(VI) compounds, however, were also specifically found for dye workers and painters.

Although welders may suffer chromium exposure and increased risks of cancer, the epidemiological data are not as strong as those for chromate production workers (IARC, 1990). Although some welder studies do not show specific cancer risks (Becker et al., 1985), an increased risk for lung cancer has been noted among stainless steel welders (Sjogren et al., 1987). Arc welders, who are exposed to chromium as well as to nickel in the welding fumes generated by the alkaline-coated chromium/nickel alloy electrodes (Becker et al., 1985) suffer more frequent cancers than argon welders. This may be because argon produces a protective inert gas barrier surrounding the uncoated chromium/nickel electrodes, and these welders are believed to be less exposed to electrode-generated chromium-laden fumes. Stainless steel welders may also be exposed to chromium in the fumes generated by the stainless steel welding material, which can contain up to 10 to 20% chromium and 0.5 to 2% nickel (Sjogren et al., 1983, 1987; IARC, 1990). In contrast, mild steel welding would be less hazardous due to the lower chromium and nickel concentrations in the mild steel alloys (Bonde and Christensen, 1991).

Humans can also be exposed to chromium pollution in the environment, primarily via air or soil contamination. Soil contamination can occur in locations with either a previous industrial history, or that may have been chromium waste dump sites (Paustenbach et al., 1991; Sheehan et al., 1991). In the U.S. chromium is believed to be a substantial contaminant in about one third of the Environmental Protection Agency's identified Superfund sites (ATSDR, 1989). In studies of polluted regions of the Ruhr valley in Germany, high chromium levels in the air were found to be associated with an increased incidence of lung cancer (Kollmeier et al., 1987). However, in studies of individuals residing in residential areas of New Jersey that are burdened with large tracts of chromium-contaminated soils, an increased risk for chromium-associated cancer or other illnesses has not yet been found (Paustenbach et al., 1991; Sheehan et al., 1991). Food and water can also be other potential sources of chromium exposure and, in fact, naturally existing chromium is the 21st most abundant element in the earth's crust (Squibb and Snow, 1993). Food chain sources of chromium exposure are not likely to be toxic or carcinogenic, but rather should serve to satisfy essential dietary chromium requirements.

The epidemiology literature attempts to unravel the variables that are associated with human chromium exposure and subsequent cancer development. Among these potential variables are chromium chemistry (valency), place of employment, duration of exposure, and chromium concentrations in assorted tissues. The carcinogenic risk associated with human exposure to chromium seems to be most directly correlated with particular chromium valance states. The greatest observed risks are associated with exposures to Cr(VI) compounds such as chromates, which primarily occurs during chromate production, chrome pigment manufacturing and use, plating, welding, and spray painting (IARC, 1980,

1987). In 1990, an IARC committee again reviewed the available data on chromium exposure and human risk and concluded that there was sufficient evidence of a human lung cancer risk associated with occupational exposure to chromium for chromate and chromate pigment production workers, as well as for chromium platers (IARC, 1990). At that time, however, there was still insufficient data on lung or other cancer risks associated with occupational chromium exposures incurred during chromium mining, steel and ferrochromium alloy production, and during the handling, cutting, and grinding of chromium alloys (Axelsson et al., 1980; Langard et al., 1980, 1990). These occupations primarily involve exposures to chromium metal (Cr), chromium ore ($FeOCr_2O_3$), Cr(III) compounds, and other intermediary chromium valance states (Norseth, 1981) which are generally less hazardous due to their extremely poor cellular uptake (Gray and Sterling, 1950; Beyersmann and Koster, 1987; Alcedo and Wetterhahn, 1990) and thus, limited bioavailability (see Section III.A).

The types of cancers that have been correlated with occupational chromate exposure are varied. In one of the early studies of lung cancer in over 100 chromate workers, squamous cell carcinomas, adenocarcinomas, and anaplastic type cancers were diagnosed (Hueper, 1966). In a later study where the incidence of lung cancer in chromium workers was more than fifteen times that of the general population, the predominant cancers were squamous cell and small-cell carcinomas (Abe et al., 1982; Tsuneta, 1982). Histopathology often shows hyperplastic changes and metaplasia in the bronchial epithelium of chromate workers (Mikami, 1982; Uyama et al., 1989). Thus, diagnostic surveillance of occupational chromate-exposed populations by bronchoscopy and chest X-rays may lead to the early detection and treatment of lung cancer.

In addition to lung cancers, upper respiratory tract and oral cavity tumors including pharyngeal and esophageal cancers have also been reported among chromate industry workers. Nasal and sinonasal tumors have been associated with exposures to fumes from welding and soldering iron as well as from chromium contaminants in wood dusts (Hernberg et al., 1983 a,b). Several studies reported an increased risk for the development of rare sinonasal cancer among workers employed in primary chromium production and chromium pigment production (reviewed in Cohen et al., 1993). Cancers other than those associated with the respiratory tract have also been associated with chromium exposure, although their incidences are usually less frequent (IARC, 1987). In several cancer studies of chromium workers, including chromium electroplaters, pigment production workers, and ferrochromium plant employees, gastrointestinal tumors were suggested to be correlated with chromate exposure (Hayes et al., 1989; Langard and Norseth, 1979; Langard et al., 1990). Urological cancers have also been found in some of these workers (Langard et al., 1980; Claude et al., 1988).

In the past, the duration of employment in the chromate industry was a contributing factor for subsequent cancer development and lung cancer death (reviewed in Cohen et al., 1993). Thus, the early human epidemiology data suggested a dose-response relationship for carcinogenesis among chromate workers. The latency period for lung cancer development in these populations was estimated to be 15 to 27 years although a somewhat shorter latency was suggested specifically for chromate pigment workers (Hayes, 1980). Employment duration was also found to play a role in the type of cancer that developed. For example, small-cell lung carcinomas developed in workers with significantly shorter employment histories than those workers who developed squamous cell carcinomas (Abe et al., 1982; Tsuneta, 1982). However, length of exposure was not the only determinant for the development of this type of cancer. The workers who developed small-cell carcinomas were those whose primary employment was in early phases of chromate production where they were heavily exposed to Cr(VI) dusts. In contrast, many of the workers who developed squamous cell carcinomas worked in subsequent production phases where they were exposed to more refined Cr(VI) products and reduced levels of Cr(III) dusts. Employment duration may no longer present a significant occupational cancer risk; several recent studies show that the chromium-induced lung cancer incidence has been significantly reduced by mandatory plant modifications that have resulted in significantly cleaner work environments (Hill and Ferguson, 1979; Alderson et al., 1981; Davies, 1991; Squibb and Snow, 1993; Kiilunen, 1994). Indeed, a recent study of chromate workers in Japan did not show any correlation between employment length (0.1 to 40 years) and chromium exposure as determined by urinary chromium measurements (Nagaya et al., 1994).

Measurements of chromium concentrations in human lung and other tissues have been performed in an attempt to establish a causal relationship between chromium exposure and ensuing cancer in occupationally exposed populations (Kollmeier et al., 1987; Hyodo et al., 1980; Tsuneta et al., 1980). Chromium can accumulate in the lung tissue during the course of a lifetime (Kishi et al., 1987), and deposition in the lungs primarily occurs in the upper rather than lower lobes (Tsuneta et al., 1980). Several cases of presumed chromium-induced lung cancer have been positively correlated with increased chromium concentrations in lung tissue (Kim et al., 1985; Kishi et al., 1987). In one case, a chromate production worker who died of lung cancer after nearly three decades of occupational exposure had almost 100 times the normal level of chromium in his lung tissue (Kishi et al., 1987). In contrast, the chromium levels in nonrespiratory tissues did not differ in exposed workers vs. controls in another study of chromate factory workers (Tsuneta, 1982). Increased levels of blood and urinary chromium have also been detected in several studies of stainless steel welders, suggesting that inspired welding fumes can release chromium into the bloodstream (Sjogren et al., 1983; IARC, 1990; Bonde and Christensen, 1991).

Since chromium can be nephrotoxic, several recent studies have focused on investigating urinary chromium levels as predictors of renal damage following occupational exposure to low levels of chromium. A study of more than 150 male chromate platers in Japan suggested a positive correlation between age-adjusted total urinary protein and urinary chromium levels in the chromium workers; however, age-adjusted total urinary protein levels did not differ between the worker and control populations (Nagaya et al., 1994). There was also no overt evidence of renal dysfunction among these platers, who had a mean employment duration of 12.6 years. Although no apparent excess mortality has been associated with kidney disease in chromium industry workers, a recent reported case of chronic interstitial nephropathy in a stainless steel plasma cutter was accompanied by significantly elevated blood and urinary chromium concentrations (Petersen et al., 1994).

B. CHROMIUM CARCINOGENESIS IN ANIMAL MODELS

Animal bioassays have provided a wealth of evidence showing that chromium(VI), but not chromium(III), compounds are carcinogenic. The primary information derived from these studies, in addition to determining the importance of chromium valence, is that chromium-induced tumors originate mainly at the site of exposure. The reason for the lack of chromium-induced tumors at distant tissue sites is likely to be dependent on chromium chemistry, since the reduction of Cr(VI) to Cr(III) in the blood or other body fluids would yield chromium ions that are not bioavailable for uptake by distant cells. Animal studies have also confirmed that respiratory (lung and nasal) tumors are the predominant type in animals, as in humans. In fact, similar types of cancer, such as lung adenomas and adenocarcinomas, have been found in both chromium-exposed humans and experimental animals. Although some of the most pertinent historical animal data will be summarized here, the reader is also directed to other more comprehensive reviews of the literature on this subject (Sunderman, 1986; Langard, 1988; IARC, 1980, 1990; Cohen et al., 1990; Squibb and Snow, 1993).

Industrial environments are typically contaminated with an assortment of trivalent and hexavalent chromium compounds. Chromium metal has a valence of 0, while chromium acetate, carbonate, phosphate, and chromic oxide (Cr_2O_3) have a valence of +3. The chromate and dichromate salts have a chromium valence of +6, as does chromic acid (CrO_3). Although both trivalent and hexavalent chromium exist in nature, the trivalent form is the most prevalent both in the environment and in biological tissues (Mertz, 1969). Hexavalent chromium is a strong oxidizing agent and is rapidly reduced to Cr(III) *in vitro* and *in vivo*. The earliest animal studies attempted to demonstrate which of the chromium species were the most carcinogenic. To date, the overwhelming majority of experimental evidence shows that exposure to metallic or trivalent chromium does not induce tumors in animals (rodents, guinea pigs, and rabbits) regardless of the administration protocol (i.e., inhalation, instillation, implantation, or by intramuscular, intravenous, intraperitoneal, subcutaneous, or intrapleural injection) (Hueper, 1955; Hueper and Payne, 1962; Schroeder et al., 1964; Ivankovic and Preussmann, 1975; Sunderman, 1976; Shimkin et al., 1977; Sunderman, 1984; Langard, 1988). An exception to this is one study in which chromic(III) oxide instillation in rats yielded a reported increase in lung sarcomas (Dvizhkov and Fedorova, 1967). However, control data were not reported and the conclusions were therefore deemed to be weak (IARC, 1980, 1990).

Chromium compound solubility, and route of administration of these compounds, are important variables in the animal carcinogenesis studies. Intramuscular implantation of insoluble chromate compounds and subcutaneous or intrapleural injections of various chromate compounds produce injection-site sarcomas (reviewed in IARC, 1987; Langard, 1988). Metastases are only rarely noted, and are likely to arise from other components of the chromate salts such as lead (Furst et al., 1976). However, injection or implantation-site sarcomas are less well induced by soluble chromates (Hueper and Payne, 1962; Levy and Venitt, 1986; IARC, 1987). In the majority of animal experiments, the most effective tumor-inducing chromium compounds are the slightly soluble hexavalent salts, such as calcium and lead chromate. By far the most successful administration protocol for inducing respiratory tumors in animals involves tracheal or intrabronchial implantation, as demonstrated by an abundance of studies (reviewed

Carcinogenicity and Genotoxicity of Chromium

in IARC, 1980, 1990). It has been suggested that animal studies employing mice and rats may be more informative than those employing hamsters, guinea pigs, or rabbits with regard to the type of lung cancers that develop following chromium exposure (Cohen et al., 1993).

Historically, inhalation exposure to chromium compounds has not yielded tumors in most animal bioassays. Although the majority of early animal chromate inhalation studies did not demonstrate an increase in lung tumors (Baetjer et al., 1959; Hueper and Payne, 1962; Steffee and Baetjer, 1965; IARC, 1980; Langard, 1988), occasional experiments suggested that this route of exposure could be effective (Nettesheim et al., 1971). Another study suggested that exposure to inhaled chromate dusts could reduce the age at which tumors appeared, even if it did not increase the tumor incidence (Baetjer et al., 1959). In more recent experiments, primary lung tumors have been found to develop in rats administered sodium dichromate aerosols (Glaser et al., 1986). In inhalation studies in which mice were exposed to calcium chromate dusts or chromic acid aerosols, the development of lung adenomas and carcinomas is consistent with human epidemiological data (Steinhoff et al., 1986; Adachi et al., 1986; Adachi, 1987; Adachi and Takemoto, 1987). However, the statistical significance for induced tumor incidence is marginal even in these studies. It is surprising that experimental animal studies have not confirmed the putative epidemiological correlation between human occupational carcinogenesis and inhalation of chromium-containing dusts or fumes.

Whereas oral ingestion of excessive doses of chromium is toxic and potentially lethal to humans (Leonard and Lauwreys, 1980; Hamilton and Wetterhahn, 1987; Michie et al., 1991), experimental studies have shown that oral exposure to chromium compounds does not yield tumor formation in test animals. In drinking studies in which rats and mice were dosed with chromic acetate in their drinking water for their lifetime, tumor development at all examined organ sites did not differ from control animals (Schroeder et al., 1964, 1965). A recent short-term drinking study of mice and rats orally exposed to potassium chromate confirmed that tissue accumulation of chromium following gastrointestinal tract absorption was much lower than that following intraperitoneal injection (Kargacin et al., 1993). However species differences in tissue accumulation were observed for the two different dosing routes, and may signify species differences in the reduction of Cr(VI) to Cr(III). In a feeding study, ingestion of solid food spiked with high doses of Cr_2O_3 (1 to 3 g/kg/day) did not yield tumors (Ivankovic and Preussmann, 1975), however, exposure to Cr(III) compounds is not generally carcinogenic by other routes of exposure either.

III. CHROMIUM GENOTOXICITY

Genotoxicity can be defined as perturbation of the genome, either by disruption of the DNA itself, or by interference with its proper division and distribution in mitosis. In the remainder of this chapter, chromium genotoxicity will be discussed in several sections that focus on chromium-DNA interactions *in vitro*, chromium mutagenesis in bacterial and mammalian cell systems, and chromium effects on gene expression and chromosomal integrity. Since the genotoxic effects of chromium compounds have recently been comprehensively reviewed by several authors (Alcedo and Wetterhahn, 1990; DeFlora et al., 1990; Snow, 1992; Cohen et al., 1993; Rossman, 1993; Squibb and Snow, 1993), this chapter will concentrate on summarizing some of the most recent findings. The discussions of DNA effects *in vitro*, mutagenesis, and gene expression, will be focused on defining the mechanisms by which chromium compounds are carcinogenic. The section on chromosomal effects will conclude with a brief discussion of human population biomonitoring.

A. BIOAVAILABILITY AND INTRACELLULAR CHEMISTRY

As noted in the human epidemiology studies and animal carcinogenesis bioassays, chromium chemistry is an important determinant of its genotoxicity. Hexavalent chromium exists as an oxyanion at physiological pH and is believed to be transported into living cells via the sulfate anion transport system (Standeven and Wetterhahn, 1989). Within a short period of time, hexavalent chromium becomes reduced to the trivalent form which is highly stable, less prone to ligand substitution, and thus often described in the literature as "kinetically inert". Since trivalent chromium cannot be readily transported into or out of living cells (Kitagawa et al., 1982), it has historically been measured as an indicator of erythrocyte half-life. Intermediate chromium oxidation states, such as Cr(V) and Cr(IV), are formed during the process of intracellular reduction (reviewed in Standeven and Wetterhahn, 1989). Accumulating evidence suggests that these chromium intermediates are not free ions, but rather exist as an assortment of often short-lived organic ligands. There are several dominant biochemical pathways for Cr(VI) reduction, and an assortment of potentially genotoxic intermediates and ligands can be formed (reviewed in Standeven and Wetterhahn, 1991a). As a result of this complex biochemistry, our understanding of chromium genotoxicity in biological systems has been difficult to unravel.

B. CHROMIUM INTERACTIONS WITH DNA

The simplest systems in which to study chromium-DNA effects are those which are not complicated by cellular intricacy. Thus *in vitro* experiments with pure DNA have been instrumental in beginning to dissect the mechanisms involved in chromium biochemistry, genotoxicity, and carcinogenicity (reviewed in Cohen et al., 1990, 1993). It is now well established that hexavalent chromium compounds produce DNA strand breaks (Robison et al., 1982; Sugiyama et al., 1986a,b; Snyder, 1988; Sugiyama et al., 1989a,b), DNA-DNA and DNA-protein cross-links (reviewed in Costa, 1991), as well as modified nucleotides including oxidized base damage (8-hydroxyguanine) *in vitro* and *in vivo* (Aiyar et al., 1989, 1991; Faux et al., 1992). Chromium is also reported to complex free amino acids such as cysteine and tyrosine to DNA (Lin et al., 1992; Salnikow et al., 1992). In addition, depurination of DNA by chromium(VI) yields residual alkali-labile abasic (AP) sites that can be mutagenic (Schaaper et al., 1987; Liebross and Wetterhahn, 1990). However, chromium(VI) does not react with isolated DNA *in vitro* in the absence of reducing agents (Tsapakos and Wetterhahn, 1983).

As noted throughout this chapter, trivalent chromium does not readily induce genotoxic effects due to its poor uptake into living cells. However, Cr(III) does react *in vitro* with DNA (reviewed in Snow, 1991) and can in fact come into contact with cellular DNA *in vivo* (Cupo and Wetterhahn, 1985). It has been shown *in vitro* that Cr(III) binds to isolated nuclei (Koster and Beyersmann, 1985), and interacts with nucleotides and nucleic acids (Wolf et al., 1989; Kortenkamp and O'Brien, 1991). Although different mechanisms may be involved, chromium(III), like hexavalent chromium, can also produce DNA-protein cross-links (Snow and Xu, 1989). In studies of DNA polymerase activity, it was found that Cr(III) can alter the fidelity and kinetics of DNA replication such that processivity is increased and fidelity is decreased (Snow and Xu, 1989, 1991). Recent studies show that Cr(III)-induced DNA-DNA cross-links can block DNA replication (Bridgewater et al., 1994). In addition, *in vitro* bypass of permanganate lesions is enhanced when chromium(III) is present (Snow, 1994), suggesting that chromium can potentiate the mutagenicity of endogenous or carcinogen-induced DNA oxidation. Certain chromium(III) complexes have been shown to be mutagenic in bacteria, and may act via redox cycling of the chromium ions (Sugden et al., 1992). Chromium(III) complexes with glutathione, ascorbate, and cysteine have also been found to cause DNA-DNA cross-links *in vitro* (Gulanowski et al., 1994), however, these complexes were not found to be toxic or genotoxic in preliminary studies with bacteria and additional studies are needed to define the genotoxicity of chromium(III) complexes in cellular systems.

Neither hexavalent nor trivalent chromium, however, is believed to be the biologically active ultimate carcinogen. Rather, it is the reactive intermediates of chromium(VI) reduction that are thought to be the most significant genotoxins. Numerous cellular reductants including, but not limited to cytochrome P_{450}, NADH, glutathione (GSH), ascorbate (vitamin C), cysteine, NADPH:quinone oxidoreductase (DT diaphorase), and hydrogen peroxide (H_2O_2) are capable of reducing chromate *in vitro* (reviewed in DeFlora and Wetterhahn, 1989; Aiyar et al., 1992; Cohen et al., 1993; Snow, 1993). The focus of many recent studies has pointed to ascorbate and sulfhydryl compounds, including glutathione and cysteine, as the biologically relevant mediators of chromium reduction *in vivo* (Standeven and Wetterhahn, 1991b, 1992). Both glutathione and ascorbate are the most prevalent intracellular reductants in many mammalian tissues, including the lung (Suzuki, 1990) which is a primary target tissue for chromium-exposed populations. Although glutathione serves to protect cells against overt chromium toxicity (Standeven and Wetterhahn, 1991b), several studies demonstrate that Cr(V)-glutathione complexes may also be genotoxic, as evidenced by DNA breakage *in vitro* and the formation of DNA adducts and chromium-glutathione DNA cross-links (Kortenkamp et al., 1990; Borges and Wetterhahn, 1989; Aiyar et al., 1991). Recent evidence suggests that glutathione may be important in the intracellular formation of Cr(V) intermediates, but not Cr(III) (Sugiyama and Tsuzuki, 1994). Whether chromium-glutathione adducts or DNA cross-links play a direct role in mutagenesis or altered gene expression has yet to be directly demonstrated.

Accumulating evidence suggests that oxidative processes may also play a role in chromium genotoxicity (reviewed in Klein et al., 1991; Standeven and Wetterhahn, 1991; Snow, 1991). Chromium reduction intermediates, such as Cr(V)-complexes, or end products like Cr(III), may be genotoxic by mechanisms that ultimately involve reactive hydroxyl or thionyl radicals. These radicals can be generated during physiological chromate reduction, as shown by the spin trapping of activated oxygen species following reactions of chromate with ascorbate or glutathione (Jones et al., 1991; Lefebrve and Pezerat, 1992). Chromate reduction via hydrogen peroxide *in vitro* has been shown to produce hydroxyl radicals via what is now believed to be a Cr(VI)-mediated Fenton-like reaction (Shi and Dalal, 1990; Shi et al., 1994). More recently, cysteinyl radicals were also demonstrated following reactions of Cr(VI) with cysteine (Shi et al., 1994). However, it is not known whether hydroxyl-, cysteinyl-, or thionyl-mediated metabolic pathways are biologically relevant (Standeven and Wetterhahn, 1991a) or important in chromium carcinogenesis. Numerous *in vitro* studies with antioxidant and free radical scavengers have produced an abundance of confusing results. Ascorbic acid, for example, increases Cr(VI) cytotoxicity and DNA-protein cross-link yield in Chinese hamster V79 cells, but decreases the formation of alkali-labile DNA damage (Sugiyama et al., 1991a). Vitamin B_2 mitigates chromate toxicity in Chinese hamster V79 cells (Sugiyama et al., 1989c), but at low doses it enhances mutations and chromosomal aberrations with no effect upon cytotoxicity (Sugiyama et al., 1992). Orthophenanthroline (OP), which can chelate metal ions including iron and Cr(III), can also reduce the yield of Cr(V) and hydrogen peroxide-mediated DNA strand breaks and alkali-labile sites in V79 cells (Sugiyama et al., 1993a); however, studies with hydrogen peroxide-resistant CHO cells show that chromium(V)-induced strand breaks do not appear to be responsible for chromium cytotoxicity (Sugiyama et al., 1993b). Thus, the biological significance of chromium metabolism-generated reactive radicals remains to be clearly defined.

Whereas the *in vitro* studies summarized above have been useful in identifying potentially genotoxic chromium reduction intermediates and reactive ligands, it has become increasingly evident that further research is needed to determine the biological relevance of this information. On the other hand, traditional *in vitro* endpoints can be measured in cultured cells, as well as *in vivo*, and may be exploited for biomonitoring chromium-exposed human populations. For example, DNA strand break production, and the rapid repair of these lesions, has been studied with low doses of chromate (<40 μM) in cultured rat hepatocytes and human lymphocytes (Gao et al., 1992, 1993). In studies of Wistar rats treated with intratracheal instillations of chromate, DNA strand breaks could be demonstrated *in vivo* in peripheral lymphocytes (Gao et al., 1992). In this study, preliminary data also suggested that fluorescent analysis of DNA unwinding in peripheral lymphocytes could possibly be a sensitive measure of occupational exposure of chromium workers. Alternatively, a sensitive new assay to detect low levels of metal-induced DNA-protein cross-links (Zhitkovitch and Costa, 1992) has recently been validated and applied to a study of chromium- and nickel-exposed welders with very promising results (Costa et al., 1993).

C. MUTAGENICITY OF CHROMIUM COMPOUNDS

Chromium compounds are mutagenic in most bacterial and mammalian assays (reviewed in DeFlora et al., 1990; Snow, 1992; Cohen et al., 1993; Rossman, 1993). In fact, as summarized by IARC (1990), hexavalent chromium compounds are mutagenic in more genotoxicity assays than any other carcinogen/mutagen. The reader will note that chromium valency, solubility, and bioavailability are important modulators of chromium activity in the short-term biological test systems, as in the whole animal and human carcinogenesis studies discussed earlier. The types of mutations that are observed, as well as the possible mechanisms involved in their production, are also discussed.

Historically, bacterial assays provided the initial evidence that chromium compounds were mutagenic and, hence, genotoxic (Venitt and Levy, 1974; Nakamuro et al., 1978; Nestmann et al., 1979; Leonard and Lauwreys, 1980). Chromate- or dichromate-induced base substitution mutations were detected in *Escherichia coli (trp* locus), *Bacillus subtilis (rec* locus), and *Salmonella typhimurium (his* locus). In the Ames *Salmonella* strains, base substitution mutations at A-T were more common than those at G-C sequences, and some frame shift mutations were also detected (Petrilli and DeFlora, 1978; Nestmann et al., 1979). The oxidation-sensitive Ames strain TA102, which detects reversion at an A-T site, was particularly sensitive to chromate (Bennicelli et al., 1983). The *Salmonella* strains that are SOS proficient by the presence of the pKM101 plasmid were found to be more sensitive to chromate than those that do not have SOS capabilities (Bennicelli et al., 1983). Hexavalent chromium also induces SOS responsive genes such as *recA, umuC,* and *sifA,* in *E. coli* (Llagostera et al., 1986). Trivalent chromium salts, such as $CrCl_3$ or $CrK(SO_4)_2$, were not mutagenic in bacteria (Nishioka, 1975; Warren et al., 1981). However, when Cr(III) was complexed with organic ligands such as 2,2'-bipyridyl or 1,10-phenanthroline Cr(III), mutagenic activity was observed in *Salmonella* (Warren et al., 1981; Sugden et al., 1990). Again, these experiments show that Cr(III) uptake by living cells is limited, but can be enhanced by the formation of chromium complexes with organic ligands.

In higher organisms such as the yeast *Saccharomyces cerevisiae,* chromate produces mitochondrial *petit* mutants (Henderson, 1989). In mammalian cell assays, chromium mutagenesis has been studied at several genetic markers that differ in their capacity to detect certain types of mutations. The X-linked hypoxanthine phosphoribosyl transferase gene *(hprt),* for example, detects a broad spectrum of mutations including base substitutions, frame shifts, and some deletions. However, large multilocus deletions involving *hprt* and flanking essential genes may not yield viable mutants (reviewed in DeMarini et al., 1989; Klein et al., 1994a). In contrast, the autosomal (non-X-linked) thymidine kinase (tk) locus can also detect all types of mutations, including multilocus deletions (DeMarini et al., 1989). Hexavalent chromium produces a moderate mutagenic response (6-thioguanine resistance, $6TG^R$) at *hprt* in mammalian cells, with the strongest response elicited by exposure to the less soluble compounds such as calcium chromate (Bianchi et al., 1983; Elias et al., 1986; Celotti et al., 1987; Patierno and Landolph, 1989; Biedermann and Landolph, 1990). In human fibroblasts, calcium chromate was also more mutagenic at *hprt* than other more soluble chromate salts (Biedermann and Landolph, 1990). However, for very insoluble lead chromate particles $6TG^R$ mutagenesis could not be demonstrated in CHO cells (Patierno et al., 1988), despite the highly clastogenic nature of this compound (Wise et al., 1992). Interestingly, insoluble trivalent chromic oxide was found to be somewhat mutagenic at *hprt* in Chinese hamster cells (Elias et al., 1986).

In contrast to the *hprt* studies, slightly soluble calcium chromate is much less mutagenic than the soluble potassium or sodium chromates and dichromates at the deletion-sensitive *tk* locus in mouse lymphoma cells (Oberly et al., 1982). This may suggest that the soluble chromates tend to primarily induce deletions, while less soluble calcium chromate may yield mostly base substitutions. In support of this, calcium chromate (Patierno and Landolph, 1989), but not potassium chromate (Klein et al., 1994b), has been shown to induce ouabain resistance, which can arise only by a limited subset of base substitutions in the essential $Na^+/K^+ATPase$ gene (Muriel et al., 1987). Also, chromate produces an unexpected frequency of deletions of the *gpt* transgene in the Chinese hamster G12 cell line (Klein and Rossman, 1990; Klein et al., 1994a,b) and is even more mutagenic in another *gpt* transgenic cell line (G10) that is both highly sensitive to oxidative mutagenesis and prone to transgene deletions (Klein et al., 1994a).

Chromium compounds are often mutagenic at a very narrow dose range (reviewed in Cohen et al., 1990, 1993). This may be due to persistent toxicity after the initial cell treatments because of the inability of reduced Cr(III) to exit the cells (Kitagawa et al., 1982; Standeven and Wetterhahn, 1989). Accumulations of high levels of intracellular and intranuclear chromium have been demonstrated for Chinese hamster V79 cells (Shelmeyer et al., 1990) and residual chromium toxicity has been observed a week after chromate treatment of V79 and V79-derived G12 cells (Klein et al., 1992). This effect is also reflected in the chromate mutagenesis curves, which decrease at the highest doses in these and other experiments (Cohen et al., 1993; Klein et al., 1994b). In other studies, sodium chromate was shown to inhibit CHO cell growth for up to 8 days after treatment, and this delay was accompanied by significant DNA synthesis inhibition for up to 4 days (Blackenship et al., 1994). The cells that did not survive the chromate treatment were found to undergo apoptosis. Vitamin E can reduce both chromate cytotoxicity and *hprt* mutagenicity in Chinese hamster V79 cells (Sugiyama et al., 1989b, 1991b). Thus, oxidative processes (perhaps induced by the intracellular reduction of chromium) are thought to be involved in chromium genotoxicity in mammalian cells. However, as discussed above, the complexities of chromium metabolism and the genotoxic effects of these processes are still not well defined in biological systems.

Despite several recent studies to determine the types of mutations that are caused by chromate in mammalian cells, the mutation spectra is still not clear. In one study, chromium(IV) oxide predominantly produced T $\rightarrow$ A and T $\rightarrow$ G transversions in 18/27 individually isolated and sequenced *hrpt* mutants of CHO cells (Yang et al., 1992). The mutations, which occur throughout the entire gene, are believed to involve T residues on the nontranscribed strand. Several mutants exhibited DNA base changes at two adjacent bases, and one mutant showed four base changes. In addition, numerous splicing mutations, and single base insertions or deletions were also identified. Larger deletions may have also been induced, but they could not be examined since the mutated *hprt* sequence could not be PCR amplified in many (up to 20%) of the mutants. T $\rightarrow$ A transversions were also noted in 4/5 potassium chromate and 3/3 lead chromate mutants in the same study. In contrast, the analysis of *hrpt* mutants of human TK6 cells reveals a different spectrum which primarily involves G:C base pairs (Chen and Thilly, 1994). The system developed by them to identify recurrent *hprt* mutations utilizes batch-selected chromium-treated cell populations whose altered DNA ("mutants") can be differentiated from wild-type sequences on denaturing gradient gels. The study identified four distinct chromium mutagenesis hot spots within exon 3 of the human *hprt* gene. Each of these hotspots was estimated to occur in 2 to 4.5% of the total $6TG^R$ mutant population. Although there is some site-specific overlap between the observed chromate hotspots and those for hydrogen peroxide and X-rays, there are distinct differences in the mutagenic events that have occurred at these sites. Whereas chromate produced G:C → A:T at hotspot 243, hydrogen peroxide yields G:C → C:G at the same site. At hotspot 247, chromate generated A:T → T:A, whereas X-rays delete this A:T base pair. Thus, if chromium mutagenesis does indeed occur via reactive oxygen intermediates, there are likely to be distinct differences in the free radical chemistry generated by chromium, X-rays, and hydrogen peroxide. Information derived from shuttle vectors such as pZ189 may also be useful in determining the spectrum and mechanisms of chromate mutagenesis in mammalian cells (Liu and Dixon, 1994).

D. CHROMIUM EFFECTS ON GENE EXPRESSION

Experimental evidence shows that chromium can alter gene expression, perhaps by the presence of persistent chromium-DNA adducts or chromium-generated cross-links. Chromium was first found to suppress the expression of inducible genes such as 5-aminolevulinate synthase, cytochrome P_{450}, and metal-responsive metallothionein in chick embryo liver without altering the expression of housekeeping genes β-actin or albumin (Hamilton and Wetterhahn, 1989; Wetterhahn and Hamilton, 1989). Expression of a hormone-responsive inducible gene, phosphoenolpyruvate carboxykinase (PEPCK), has also been shown to be inhibited by chromate (McCaffrey et al., 1994). However, whether specific regulatory promoter sequences may be targets of chromium action, or whether the promotors of inducible genes are generally more open in their chromatin structure and are therefore susceptible to carcinogen attack, has yet to be determined (McCaffrey and Hamilton, 1994). For genes that exhibit both constitutive and inducible levels of expression, chromium can reduce the inducible expression without affecting the constitutive levels of gene product. This has been demonstrated for the gene encoding glucose-regulated protein (GRP78) in a study in which the presence of DNA-protein cross-links, and perhaps DNA-adducts, were likely to be responsible for the reduced tunicamycin-inducible expression of GRP78 (Manning et al., 1992). In further studies, chromate-induced DNA-protein cross-links were found to be preferentially associated with nuclear matrix DNA (Xu et al., 1994). Since the nuclear matrix is a subcellular location where the DNA associates with an assortment of replication, repair, and transcription proteins, chromate-induced DNA-protein cross-links could effectively hinder all these essential processes *in vivo*. It is therefore not surprising that the repair of chromate-induced DNA damage was also preferentially found in nuclear matrix-associated DNA fractions (Xu et al., 1994). The overall picture that is emerging from these studies shows that inducible gene expression can be inhibited by hexavalent chromium. It is also becoming clear that DNA-damaging agents and growth factors can alter the expression of many genes through common routes of action (Blattner et al., 1994). It will be interesting to determine whether the expression of any oncogenes or tumor suppressor genes, such as p53 for example, can also be affected by carcinogenic chromium.

E. CHROMOSOMAL EFFECTS

Chromium induces chromosome damage in nonmammalian species (e.g., yeast, plants, and insects) as well as in mammalian or human cells (reviewed in Leonard and Lauwerys, 1980; DeFlora et al., 1990). Chromium is highly clastogenic, which means that it breaks the chromosomes and produces microscopically visible aberrations such as unrejoined chromatid gaps and larger breaks. Clastogenic agents also tend to produce sister chromatid exchanges between homologous sequences in replicated, but not yet separated, mitotic chromosomes (Majone and Levis, 1979; Bianchi et al., 1983; Sharma and Talukder, 1987). At low chromium doses (<1.0 µg/ml $K_2Cr_2O_7$), chromatid gaps were the predominant visible aberrations (Tsuda and Kato, 1976). As in the animal carcinogenesis experiments, the soluble hexavalent chromium compounds yielded higher dose-dependent increases in chromosome damage compared to trivalent chromium in cultured or primary rodent cells (reviewed in Leonard and Lauwreys, 1980; Leonard, 1988). Chromium compounds are also clastogenic to cultured human lymphocytes (Nakamuro et al., 1978), and several studies suggest that human cells may be more sensitive than rodent cells to chromium damage. It has been shown that chromate concentrations below those needed to produce DNA strand breaks can enhance the incidence of SCE in CHO cells (Sen and Costa, 1986). These studies also show that chromium-induced chromosome damage occurs randomly throughout the genome, and is not concentrated within heterochromatin as observed for nickel damage (Sen and Costa, 1986; Sen et al., 1987).

For less soluble hexavalent chromium compounds, such as lead chromate ($PbCrO_4$), chromosome damage and SCEs were detected in cultured human lymphocytes but not in Chinese hamster ovary cells in a study in which the lead chromate was apparently solubilized to some degree (Douglas et al., 1980).

However, highly insoluble lead chromate particles did induce chromosomal damage in both Chinese hamster ovary cells and human foreskin fibroblasts (Wise et al., 1992) when the cells were exposed for much longer durations (24 h) and the lead chromate was maintained as insoluble particles. This allowed phagocytic uptake of lead chromate particles by the cultured cells, as was previously demonstrated by electron microscopy (Patierno et al., 1988). Other particulate chromium compounds, such as insoluble crystalline chromium oxide (Cr_2O_3) have also been reported to induce SCE in cultured Chinese hamster cells (IARC, 1987). Whether intracellular dissolution of Cr ions from the particles yields the observed chromosome damage, or whether the phagocytic process itself produces oxidative toxicity (Klein et al., 1991), has not yet been resolved. In either case, reactive oxygen intermediates may be involved. The antioxidant vitamin E has been shown to reduce both chromate-induced DNA strand breaks and chromosomal aberrations in Chinese hamster V79 cells (Sugiyama et al., 1987, 1991b). Another antioxidant, ascorbate, has been shown to eliminate the clastogenicity of phagocytized lead chromate particles without inhibiting their cellular internalization (Wise et al., 1993).

Whereas chromium(III) chloride ($CrCl_3$) induced some measurable increases in sister chromatid exchanges in cultured mammalian cells (Levis and Majone, 1981; Ohno et al., 1982; Bianchi et al., 1983; Elias et al., 1983), this was not true in all studies (Uyeki and Nishio, 1983). As shown in detail by Elias et al. (1983) and summarized by IARC (1987), it seems that the effective concentrations of chromium(III) that are needed to yield significant levels of chromosomal aberrations are several hundred to one thousand times higher than those required for Cr(VI). Thus the reduced bioavailability of Cr(III) to living cells does not permit sufficient accumulation of intracellular chromium to produce measurable chromosomal damage.

In dividing cells, the proper segregation of mitotic chromosomes ensures that each daughter cell receives a full complement of genetic material. Abnormal segregation can arise by damage to the mitotic spindle apparatus or to the kinetichore structure of the mitotic chromosome itself. Chromosomal nondysjunction leads to aneuploidy, in which the daughter cells receive more or less than the normal chromosome content. Aneuploidy can be readily detected *in vivo* or *in vitro* by examining interphase cells for the presence of micronuclei which contain whole chromosomes or chromosome fragments that may subsequently be lost from the dividing cell. Hexavalent chromium disrupts chromosome segregation, yielding aneuploidy and micronucleus formation in yeast, *Drosophila,* rodents, and humans (IARC, 1987; DeFlora et al., 1990). In contrast, trivalent chromium does not induce micronuclei *in vivo*, as studied in rodents (IARC, 1987). In rats and guinea pigs, dietary vitamin C but not vitamin E was found to reduce the levels of chromate-induced micronuclei (Chorvatovicova et al., 1991).

In cytogenetic studies of chromate workers, increased levels of SCE and chromosome aberrations in peripheral lymphocytes have been noted (Bigaliev et al., 1977; Stella et al., 1982). Early studies of workers exposed to chromium during stainless steel welding, however, did not show increases in aberrations, SCE, or micronuclei (IARC, 1987). More recently, chromium particle-laden welding fumes were found to produce SCEs in cultured mammalian cells (Koshi, 1979; de Raat and Bakker, 1988), and investigations of electric welders show significant correlations between SCE frequency, DNA breakage levels, and chromium concentrations in the urine (Popp et al., 1991). Similarly, significant increases in chromatid breaks and aberrations have been found for manual metal stainless steel arc welders (Jelmert et al., 1994). Micronucleus assays were recently validated for monitoring human urban populations, and preliminary studies suggest that these assays may be useful in evaluating metal genotoxicity (Berces et al., 1993; Koteles et al., 1993). Thus, cytogenetic surveillance of individuals who are exposed to chromium by occupation or environmental contamination is a biomonitoring tool that is still relevant to the study of human populations and can provide exposure assessment data that may be useful in validating newer biomarkers, such as DNA-protein cross-links (Costa et al., 1993). However, the application of modern cytogenetics technology, including fluorescent kinetichore staining of micronuclei (Eastmond and Tucker, 1989) or aberration studies utilizing specific fluorescent chromosome probes, are likely to be more informative than traditional SCE and aberration analyses.

IV. CONCLUSIONS

It is now well established that chromium is a human carcinogen. In recent years, the evaluation of human health hazards associated with chromium exposure has shifted from the occupational to the environmental arena. This is the result of drastically improved working conditions, tightly controlled disposal regulations, and the ongoing discovery of chromium-laden environmental dump sites. Hexavalent chromium compounds are the most hazardous, because the nuances of chromium chemistry do not allow significant cellular uptake of trivalent chromium. *In vitro* experiments suggest that once Cr(III) is produced inside cells by chemical reduction, or is taken up by alternative routes such as phagocytosis, it may be genotoxic by interferring with DNA replication, repair, or transcription. Whether this actually occurs in living cells has yet to be demonstrated. Complex pathways of chromium metabolism produce an assortment of reduced chromium ligands and reactive radical byproducts. It is not yet clear which of these intermediates are ultimate carcinogens *in vivo*, however, glutathione- and perhaps cysteine-mediated DNA-protein cross-links may turn out to be important lesions. It will be interesting to see whether chromium-induced cross-links can cause gene deletions, or otherwise alter the expression of genes involved in the development of cancer. In conclusion, several readily measured endpoints such as DNA-protein cross-links and chromosomal damage can be studied in human populations that may have been exposed to chromium. It is hoped that biomarkers like these can serve not only to identify exposed individuals and monitor their health, but can also be used to study preventative therapies as well.

ACKNOWLEDGMENTS

The preparation of this manuscript was supported by the National Institute of Environmental Health Sciences Grant ES00260. The author thanks Dr. Elizabeth Snow for thoughtful discussions, and Dr. Max Costa and Carrie Zinaman for critical reading of the manuscript.

REFERENCES

Abe, S., Ohsaki, Y., Kimura, K., Tsuneta, Y., Mikami, H., and Murao, M., *Cancer,* 49, 783–787, 1982.
Adachi, S., *Jpn. J. Ind. Health,* 29, 17–33, 1987.
Adachi, S. and Takemoto, K., *Jpn. J. Ind. Health,* 29, 345–357, 1987.
Adachi, S., Yoshimura, H., Katayama, H., and Takemoto, K., *Jpn. J. Ind. Health,* 28, 283–287, 1986.
Aiyar, J., Borges, K. M., Floyd, R. A., and Wetterhahn, K. E., *Toxicol. Environ. Chem.,* 22, 135–148, 1989.
Aiyar, J., Berkovits, H. J., Floyd, R. A., and Wetterhahn, K. E., *Environ. Health Perspect.,* 92, 53–62, 1991.
Aiyar, J., DeFlora, S., and Wetterhahn, K. E., *Carcinogenesis,* 13, 1159–1166, 1992.
Alcedo, J. A. and Wetterhahn, K. E., *Int. Rev. Exp. Pathol.,* 31, 85–108, 1990.
Aldersen, M. R., Rattan, N. S., and Bidstrup, L., *Br. J. Ind. Med.,* 38, 117–124, 1981.
ATSDR Agency for Toxic Substances and Disease Registry, Toxicological Profile for Chromium, ATSDR/RP-88/10, Public Health Service Center for Disease Control, Atlanta, GA, 1989.
Axelsson, G., Rylander, R., and Schmidt, A., *Br. J. Ind. Med.,* 37, 121–127, 1980.
Baetjer, A. M., *Arch. Ind. Hyg. Occup. Med.,* 2, 505–516, 1950.
Baetjer, A. M., Lowney, J. F., Steffee, H., and Budacz, V., *A.M.A. Arch. Ind. Health,* 20, 124–135, 1959.
Becker, N., Claude, J., and Frentzel-Beyme, R., *Scand. J. Work Environ. Health,* 11, 75–82, 1985.
Bennicelli, C., Camoirano, A., Petruzzelli, S., Zanacchi, P., and DeFlora, S., *Mutat. Res.,* 122, 1–5, 1983.
Berces, J., Otos, M., Szirmai, S., Crane-Uruena, C., and Koteles, G. J., *Environ. Health Perspect.,* 101 (Suppl. 3), 1993.
Beyersmann, D. and Koster, A., *Toxicol. Environ. Chem.,* 14, 11–22, 1987.
Bianchi, V., Celotti, L., Lanfranchini, G., Majone, F., Marin, G., Motaldi, A., Sponza, G., Tamino, G., Venier, P., Zantedeschi, A., and Levis, A. G., *Mutat. Res.,* 117, 279–300, 1983.
Biedermann, K. A. and Landolph, J. R., *Cancer Res.,* 50, 7835–7842, 1990.
Bigaliev, A. B., Turebaev, M. N., Bigalieva, R. K., and Elemesova, M. S., *Genetika,* 13, 545, 1977.
Blackenship, L. J., Manning, F. C. R., Orenstein, J. M., and Patierno, S. R., *Toxicol. Appl. Pharmacol.,* 126, 75–83, 1994.
Blattner, Ch., Knebel, A., Radler-Pohl, A., Sachsenmaier, Ch., Herrlich, P., and Rhamsdorf, H. J., *Environ. Mol. Mutagen.,* 24, 3–10, 1994.
Bonde, J. P. and Christensen, J. M., *Arch. Environ. Health,* 46, 225–229, 1991.
Borges, K. M. and Wetterhan, K. E., *Carcinogenesis,* 10, 2165–2168, 1989.
Bridgewater, L. C., Manning, F. C. R., Woo, E. S., and Patierno, S. R., *Mol. Carcinogenesis,* 9, 122–133, 1994.
Celotti, L., Furlan, D., Seccati, L., and Levis, A. G., *Mutat. Res.,* 190, 35–39, 1981.
Chen, J. and Thilly, W. G., *Mutat. Res.,* 323, 21–27, 1994.
Chorvatovicova, D., Ginter, E., Kosinova, A., and Zloch, Z., *Mutat. Res.,* 262, 41–46, 1991.
Claude, J. C., Frentzel-Beyme, R. R., and Kunze, E., *Int. J. Cancer,* 41, 371–379, 1988.
Cohen, M., Latta, D., Coogan, T., and Costa, M., in *Biological Effects of Heavy Metals,* Vol. 2, Foulkes, E., Ed., CRC Press, Boca Raton, FL, 19, 1990.
Cohen, M. D., Kargacın, B., Klein, C. B., and Costa, M., *Crit. Rev. Toxicol.,* 23, 255–281, 1993.
Costa, M., *Environ. Health Perspect.,* 92, 45–52, 1991.
Costa, M., Zhitkovich, A., and Toniolo, P., *Cancer Res.,* 53, 460–463, 1993.
Cupo, D. Y. and Wetterhahn, K. E., *Cancer Res.,* 45, 1146–1151, 1985.

Dalager, N. A., Mason, T. J., Fraumeni, J. F., Hoover, R., and Payne, W. W., Cancer mortality among workers exposed to zinc chromate paints, *J. Occup. Med.,* 22, 25–29, 1980.
Davies, J. M., *Lancet,* 1, 384, 1978.
Davies, J. M., *J. Oil. Colour Chem. Assoc.,* 62, 157–163, 1979.
Davies, J. M., *Br. J. Ind. Med.,* 41, 158–169, 1984.
Davies, J. M., *Br. J. Ind. Med.,* 48, 299–313, 1991.
de Raat, P. W. and Bakker, G. L., *Ann. Occup. Hyg.,* 32, 191–202, 1988.
DeFlora, S. and Wetterhahn, K. E., *Life Chem. Rep.,* 7, 169–244, 1989.
DeFlora, S., Bagnasco, M., Serra, D., and Zanacchi, P., *Mutat. Res.,* 238, 99–172, 1990.
DeMarco, R., Bernardinelli, L., and Mangione, M. P., *Med. Lav.,* 79, 368–376, 1988.
DeMarini, D. M., Brockman, H. E., de Serres, F. J., Evans, H. H., Stankowski, L. F., Jr., and Hsie, A. W. *Mutat. Res.,* 220, 11–29, 1989.
Doisy, R. J., Streeten, D. H. P., Freiberg, J. M., and Schneider, A. J., in *Trace Elements in Human Health and Disease,* Vol. 2, Prasad, A. S., Ed., Academic Press, New York, 79, 1976.
Douglas, G. R., Bell, R. D. L., Grant, C. E., Wytsma, J. M., and Bora, K. C., *Mutat. Res.,* 77, 157–163, 1980.
Dvizhkov, P. P. and Fedorova, V. I., *Vopr. Onkol.,* 13, 57–62, 1967.
Eastmond, D. A. and Tucker, J. D., *Environ. Mol. Mutagenet.,* 13, 34–43, 1989.
Elias, Z., Schneider, O., Aubry, F., Daniere, M., and Poirot, O., *Carcinogenesis,* 4, 605–611, 1983.
Elias, Z., Poirot, O., Schneider, O., Daniere, M. C., Terzetti, F., Guedenet, J. C., and Cavelier, C., *Mutat. Res.,* 169, 159–170, 1986.
Enterline, P. E., *J. Occup. Med.,* 16, 523–526, 1974.
Faux, S. P., Gao, M., Chipman, J. K., and Levy, L. S. *Carcinogenesis,* 13, 1667–1669, 1992.
Franchini, I., Magnani, F., and Mutti, A., *Scand. J. Work Environ. Health,* 9, 247–252, 1983.
Frentzel-Beyme, R., *J. Cancer Res. Clin. Oncol.,* 105, 183–188, 1983.
Furst, A., Schlauder, M., and Sasmore, D. P., *Cancer Res.,* 36, 1779–1783, 1976.
Gao, M., Binks, S. P., Chipman, J. K., and Levy, L. S., *Toxicology,* 77, 171–180, 1992.
Gao, M., Binks, S. P., Chipman, J. K., Levy, L. S., Braithwaite, R. A., and Brown, S. S., *Hum. Exp. Toxicol.,* 11, 77–82, 1993.
Glaser, U., Hochrainer, D., Kloppel, H., and Oldiges, H., *Toxicology,* 42, 219–232, 1986.
Gray, S. J. and Sterling, K., *J. Clin. Invest.,* 29, 1604–1613, 1950.
Gulanowski, B., Cieslak-Golonka, M., Szyba, K., and Urban, J., *Biometals,* 7, 177–184, 1994.
Hagaya, T., Ishikawa, N., Hata, H., Takahashi, A., Yoshida, I., and Okamoto, Y., *Arch. Toxicol.,* 68, 322–324, 1994.
Haguenoer, J. M., Dubois, G., Frimat, P., Cantineau, A., Lefrancois, H., and Furon, D., in *Prevention of Occupational Cancer International Symposium, Occupational Safety and Health Series No. 46,* International Labor Office, Geneva, 168, 1981.
Hamilton, J. W. and Wetterhahn, K. E., in *Handbook on Toxicity of Inorganic Compounds,* Seiler, H. G. and Sigel, H., Eds., Marcel Dekker, New York, 239, 1987.
Hamilton, J. W. and Wetterhahn, K. E., *Mol. Carcinogenesis,* 2, 274–286, 1989.
Hayes, R. B., in *Reviews in Cancer Epidemiology,* Vol. 1, Lilienfeld, A. M., Ed., Elsevier, New York, 293, 1980.
Hayes, R. B., *Sci. Total Environ.,* 71, 331–339, 1988.
Hayes, R. B., Sheffet, A., and Spirtas, R., *Am. J. Ind. Med.,* 16, 127–133, 1989.
Henderson, G., *Biol. Met.,* 2, 83–88, 1989.
Hernberg, S., Collan, Y., Degerth, R., Englund, A., Engzell, U., Kuosma, E., Mutanen, P., Norlinder, H., Hansen, H. S., Schultz-Larsen, K., Sogaard, H., and Westerholm, P., *Scand. J. Work Environ. Health,* 9, 208–213, 1983.
Hernberg, S., Westerholm, P., Schultz-Larsen, K., Degerth, R., Kuosma, E., Englund, A., Engzell, U., Hansen, H. S., and Mutanen, P., *Scand. J. Work Environ. Health,* 9, 315–326, 1983.
Hill, W. H. and Ferguson, W. S., *J. Occup. Med.,* 21, 103–106, 1979.
Hueper, W. C., *J. Natl. Cancer Inst.,* 16, 447–469, 1955.
Hueper, W. C., *Occupational and Environmental Cancers of the Respiratory System,* Hueper, W. C., Ed., Springer Verlag, New York, 66, 1966.
Hueper, W. C. and Payne, W. W., *Arch. Environ. Health,* 5, 445–462, 1962.
Hyodo, K., Suzuki, S., Furuya, N., and Meshizuka, K., *Int. Arch. Occup. Environ. Health,* 46, 141–150, 1980.
IARC, IARC Monograph on the Evaluation of the Carcinogenic Risk to Humans, Vol. 23. Some Metals and Metallic Compounds, International Agency for Research on Cancer, Lyon, France, 1980.
IARC, IARC Monograph on the Evaluation of the Carcinogenic Risk to Humans, Suppl. 7. Overall Evaluations of Carcinogenicity: An Updating of IARC Monographs Volumes 1 to 42, International Agency for Research on Cancer, Lyon, France, 1987.
IARC, Monograph on the Evaluation of Carcinogenic Risk to Humans. Chromium, Nickel and Welding, Vol. 48, International Agency for Research on Cancer, Lyon, France, 1990.
Ivankovic, S. and Preussmann, R.,, *Food Cosmet. Toxicol.,* 13, 347–351, 1975.
Jelmert, O., Hansteen, I.-L., and Langard, S., *Mutat. Res.,* 320, 223–233, 1994.
Jones, P., Kortenkamp, A., O'Brien, P., Wang, G., and Yang, G., *Arch. Biochem. Biophys.,* 286, 652–655, 1991.
Kargacin, B., Squibb, K. S., Cosentino, S., Zhitkovitch, A., and Costa, M., *Biol. Trace Elem. Res.,* 36, 307–318, 1993.
Kiilunen, M., *Ann. Occup. Hyg.,* 38, 171–187, 1994.

Kim, S., Iwai, Y., Fujino, M., Furumoto, M., Sumino, K., and Miyasaki, K., *Acta Pathol. Jpn.*, 35, 643–654, 1985.
Kishi, R., Tarumi, T., Uchino, E., and Miyake, H., *Am. J. Ind. Med.*, 11, 67–74, 1987.
Kitagawa, K. S., Seki, H., Katemani, F., and Sakurai, H., *Chem.-Biol. Interact.*, 40, 265–274, 1982.
Klein, C. B. and Rossman, T. G., *Environ. Mol. Mutagenet.*, 16, 1–12, 1990.
Klein, C. B., Frenkel, K., and Costa, M., *Chem. Res. Toxicol.*, 4, 592–604, 1991.
Klein, C. B., Su, L. X., and Snow, E. T., *Environ. Mol. Mutagen.*, 19, 29a, 1992.
Klein, C. B., Su, L. X., Rossman, T. G., and Snow, E. T., *Mutat. Res.*, 304, 217–228, 1994a.
Klein, C. B., Su, L., Kargacin, B., Cosentino, S., Snow, E. T., and Costa, M., *Environ. Health Perspect.*, 102 (Suppl. 4), 1994b.
Kollmeier, H., Seemann, J. W., Muller, K. M., Rothe, G., Wittig, P., and Schejbal, V. B. *Am. J. Ind. Med.*, 11, 659–669, 1987.
Korallus, U., Lange, H. J., Neiss, A., Wustefeld, E., and Zwingers, T., *Arbeitsmed. Sozialmed. Praeventivmed.*, 17, 159–167, 1982.
Kortenkamp, A. and O'Brien, P., *Carcinogenesis*, 12, 921–926, 1991.
Kortenkamp, A., Oetken, G., and Beyersmann, D., *Mutat. Res.*, 232, 155–161, 1990.
Koshi, K., *Ind. Health*, 17, 339–349, 1979.
Koster, A. and Beyersmann, D., *Toxicol. Environ. Chem.*, 10, 307–313, 1985.
Koteles, G. J., Bojtor, I., Szirmai, S., Berces, J., and Otos, M., *Mutat. Res.*, 319, 267–271, 1993.
Langard, S., *Sci. Total Environ.*, 71, 341–350, 1988.
Langard, S., *Am. J. Ind. Med.*, 17, 189–215, 1990.
Langard, S. and Norseth, T., *Br. J. Ind. Med.*, 32, 62–65, 1975.
Langard, S. and Norseth, T., *Arh. Hig. Rada. Toksikol.*, 30, 301–304, 1979.
Langard, S., Andersen, A., and Gylseth, B., *Br. J. Ind. Med.*, 37, 114–120, 1980.
Langard, S., Andersen, A., and Ravnestad, J., *Br. J. Ind. Med.*, 47, 14–19, 1990.
Lefebrve, Y. and Pezerat, H., *Chem. Res. Toxicol.*, 5, 461–463, 1992.
Leonard, A., *Mutat. Res.*, 198, 321–326, 1988.
Leonard, A. and Lauwreys, R. R., *Mutat. Res.*, 76, 227–239, 1980.
Levis, A. G. and Majone, F., *Br. J. Cancer*, 44, 219–236, 1981.
Levy, L. S. and Venitt, S., *Carcinogenesis*, 7, 831–835, 1986.
Liebross, R. H. and Wetterhahn, K. E., *Chem. Res. Toxicol.*, 3, 401–403, 1990.
Lin, X., Zhuang, Z., and Costa, M., *Carcinogenesis*, 13, 1763–1768, 1992.
Liu, S. and Dixon, K., *Environ. Mol. Mutagenet.*, 23 (Suppl. 23), 39, 1994.
Llagostera, M., Gasrrido, S., Guerrero, R., and Barbé, J., *Environ. Mutagen.*, 8, 571–577, 1986.
Machle, W. and Gregorious, F., *Public Health Rep.*, 63, 1114–1127, 1948.
Majone, F. and Levis, A. G., *Mutat. Res.*, 67, 231–238, 1979.
Manning, F. C. R., Xu, J., and Patierno, S. R., *Mol. Carcinogenesis*, 6, 270–279, 1992.
McCaffrey, J. and Hamilton, J. W., *Environ. Mol. Mutagen.*, 23, 164–170, 1994.
McCaffrey, J., Wolf, C. M., and Hamilton, J. W., *Mol. Carcinogenesis*, 10, 1994.
Mertz, W., Chromium occurrence and function in biological systems, *Physiol. Rev.*, 49, 163–239, 1969.
Michie, C. A., Hayhurst, M., Knobel, G. J., Stokol, J. M., and Hensley, B., *Hum. Exp. Toxicol.*, 10, 129–131, 1991.
Mikami, H., *Hokkaido Igaku Zasshi*, 57, 537–549, 1982.
Mossop, R. T., *Cent. Afr. J. Med.*, 29, 80–82, 1983.
Muriel, W. J., Cole, J., and Lehmann, A. R., *Mutagenesis*, 2, 383–389, 1987.
Nagaya, T., Ishikawa, N., Hata, H., Takahashi, A., Yoshida, I., and Okamoto, Y., *Arch. Toxicol.*, 68, 322–324, 1994.
Nakamuro, K., Yoshikawa, K., Sayato, Y., and Kurata, H., *Mutat. Res.*, 58, 175–181, 1978.
Nath, R., Minocha, J., Lyall, V., Sunder, S., Kumar, V., Kapoor, S., and Dhar, K. L., in *Chromium in Nutrition and Metabolism* Shapcott, D. and Hubert, J., Eds., Elsevier, Amsterdam, 213, 1979.
Nestmann, E. R., Matula, T. I., Douglas, G. R., Bora, K. C., and Kowbel, D. J., *Mutat. Res.*, 66, 357–365, 1979.
Nettesheim, P., Hanna, M. G., Doherty, D. G., Newell, R. F., and Hellman, A., *J. Natl. Cancer Inst.*, 47, 1129–1144, 1971.
Newman, D., *Glasgow Med. J.*, 33, 469–470, 1890.
Nishioka, H., *Mutat. Res.*, 31, 185–189, 1975.
Norseth, T., *Environ. Health Perspect.*, 40, 121–130, 1981.
Oberly, T. J., Piper, C. E., and McDonald, D. S., *J. Toxicol. Environ. Health*, 9, 367–376, 1982.
Ohno, H., Hanaoka, F., and Yamada, M., *Mutat. Res.*, 104, 141–145, 1982.
Ohsaki, Y., Abe, S., Kimura, K., Tsuneta, Y., Mikami, H., and Murao, M., *Thorax*, 33, 372–374, 1978.
Okubo, T. and Tsuchiya, K., *Keio J. Med.*, 26, 171–177, 1977.
Okubo, T. and Tsuchiya, K., *Scand. J. Work Environ. Health*, 13, 179, 1979.
Patierno, S. R. and Landolph, J. R., *Biol. Trace Elem. Res.*, 21, 469–474, 1989.
Patierno, S. R., Bahn, D., and Landolph, J. R., *Cancer Res.*, 48, 5280–5288, 1988.
Paustenbach, D. J., Meyer, D. M., Sheehan, P. J., and Lau, V., *Toxicol. Ind. Health*, 7, 159–196, 1991.
Payne, W. W., *A.M.A. Arch. Ind. Health*, 21, 530–535, 1960.
Peterson, R., Mikkelsen, S., and Thomsen, O. F., *Occup. Environ. Med.*, 51, 259–261, 1994.
Petrilli, F. L. and DeFlora, S., *Mutat. Res.*, 54, 139–147, 1978.
Petrilli, F. L. and DeFlora, S., *Mutat. Res.*, 58, 167–173, 1979.

Popp, W., Vahrenholz, C., Schmieding, W., Krewet, E., and Norpoth, K., *Int. Arch. Occup. Environ. Health,* 63, 115–120, 1991.
Robison, S. H., Cantoni, O., and Costa, M., *Carcinogenesis,* 3, 657–662, 1982.
Rossman, T. G., in *Handbook of Experimental Pharmacology. Toxicology of Metals — Biochemical Aspects,* Goyer, R. A. and Cherian, M. G., Eds., Springer-Verlag, New York, 1993.
Royle, H., *Environ. Res.,* 10, 39–53, 1975a.
Royle, H., *Environ. Res.,* 10, 141–163, 1975b.
Salnikow, K., Zhitkovich, A., and Costa, M., *Carcinogenesis,* 13, 2341–2346, 1992.
Satoh, K., Fukuda, Y., Torii, K., and Katsuno, N., *J. Occup. Med.,* 23, 835–838, 1981.
Schaaper, R. M., Koplitz, R. M., Tkeshelashvili, L. K., and Loeb, L. A., *Mutat. Res.,* 177, 179–188, 1987.
Schroeder, H. A., Balassa, J. J., and Vinton, W. H., *J. Nutr.,* 83, 239–250, 1964.
Schroeder, H. A., Balassa, J. J., and Vinton, W. H., *J. Nutr.,* 86, 51–66, 1965.
Sen, P. and Costa, M., *Carcinogenesis,* 7, 1527–1533, 1986.
Sen, P., Conway, K., and Costa, M., *Cancer Res.,* 47, 2142–2147, 1987.
Sharma, A. and Talukder, G., *Environ. Mutagen.,* 9, 191–226, 1987.
Sheehan, P. J., Meyer, D. M., Sauer, M. M., and Paustenbach, D. J., *J. Toxicol. Environ. Health,* 32, 161–201, 1991.
Sheffet, A., Thind, I., Miller, A. M., and Louria, D. B., *Arch. Environ. Health,* 37, 44–52, 1982.
Shelmeyer, U., Hectenberg, S., Klyszcz, H., and Beyersmann, D., *Arch. Toxicol.,* 64, 506–508, 1990.
Shi, X. and Dalal, N. S., *Arch. Biochem. Biophys.,* 277, 342–350, 1990.
Shi, X., Dong, Z., Dalal, N. S., and Gannett, P. M., *Biochim. Biophys. Acta,* 1226, 65–72, 1994.
Shimkin, M. B., Stoner, G. D., and Theiss, J. C., *Adv. Exp. Med. Biol.,* 91, 85–91, 1977.
Silverstein, M., Mirrer, F., Kotelchuck, D., Silverstein, B., and Benett, M., *Scand. J. Work Environ. Health,* 7 (Suppl. 4), 156–165, 1981.
Sjogren, B., Hedstrom, L., and Ulfrarson, U., *Int. Arch. Occup. Environ. Health,* 51, 347–354, 1983.
Sjogren, B., Gustavsson, A., and Hedstrom, L., *Scand. J. Work Environ. Health,* 13, 247–251, 1987.
Snow, E. T., *Environ. Health Perspect.,* 92, 75–81, 1991.
Snow, E. T., *Pharmacol. Ther.,* 53, 31–65, 1992.
Snow, E. T., *Environ. Health Perspect.,* 102 (Suppl. 4), 1994.
Snow, E. T. and Xu, L. S., *Biol. Trace Elem. Res.,* 21, 61–71, 1989.
Snow, E. T. and Xu, L. S., *Biochemistry,* 30, 11238–11245, 1991.
Snyder, R. D., *Mutat. Res.,* 193, 237–244, 1988.
Sorahan, T., Burges, D. C. L., and Waterhouse, J. A. H., *Br. J. Ind. Med.,* 44, 250–258, 1987.
Squibb, K. S. and Snow, E. T., in *Handbook of Hazardous Materials,* Corn, M., Ed., Academic Press, New York, 127, 1993.
Standeven, A. M. and Wetterhahn, K. E., Chromium(VI) toxicity: uptake, reduction, and DNA damage, *J. Am. Coll. Toxicol.,* 8, 1275–1283, 1989.
Standeven, A. M. and Wetterhahn, K. E., *Chem. Res. Toxicol.,* 4, 616–625, 1991a.
Standeven, A. M. and Wetterhahn, K. E., *Pharmacol. Toxicol.,* 68, 469–476, 1991b.
Standeven, A. M. and Wetterhahn, K. E., *Carcinogenesis,* 13, 1319–1324, 1992.
Steffee, C. H. and Baetjer, A. M., *Arch. Environ. Health,* 11, 66–75, 1965.
Steinhoff, D., Gad, S. C., Hatfield, K., and Mohr, U., *Exp. Pathol.,* 30, 129–141, 1986.
Stella, M., Montaldi, A., Rossi, R., Rossi, G., and Levis, A. G., *Mutat. Res.,* 101, 151–164, 1982.
Sugden, K., Burris, R. B., and Rogers, S. J., *Mutat. Res.,* 244, 239–244, 1990.
Sugden, K. D., Geer, R. D., and Rogers, S. J., *Biochemistry,* 31, 626–631, 1992.
Sugiyama, M. and Tsuzuki, K., *FEBS Lett.,* 341, 273276, 1994.
Sugiyama, M., Patierno, S. R., Cantoni, O., and Costa, M., *Mol. Pharmacol.,* 29, 606–613, 1986a.
Sugiyama, M., Wang, X. W., and Costa, M., *Cancer Res.,* 46, 4547–4551, 1986b.
Sugiyama, M., Ando, A., Furuno, H., Furlong, N. B., Hidaka, T., and Ogura, R., *Cancer Lett.,* 38, 1–7, 1987.
Sugiyama, M., Ando, A., and Ogura, R., *Biochem. Biophys. Res. Commun.,* 159, 1080–1085, 1989a.
Sugiyama, M., Ando, A., and Ogura, R., *Carcinogenesis,* 10, 737–741, 1989b.
Sugiyama, M., Ando, A., Nakao, K., Ueta, H., Hidaka, T., and Ogura, R., *Cancer Res.,* 49, 6180–6184, 1989c.
Sugiyama, M., Tsuzuki, K., and Ogura, R., *J. Biol. Chem.,* 266, 3383–3386, 1991a.
Sugiyama, M., Lin, X., and Costa, M., *Mutat. Res.,* 260, 19–23, 1991b.
Sugiyama, M., Tsuzuki, K., Lin, X., and Costa, M., *Mutat. Res.,* 283, 211–214, 1992.
Sugiyama, M., Tsuzuki, K., and Haramaki, N., *Arch. Biochem. Biophys.,* 305, 261–266, 1993a.
Sugiyama, M., Tsuzuki, K., and Haramaki, N., *Mutat. Res.,* 299, 95–102, 1993b.
Sunderman, F. W., Jr., *Prev. Med.,* 5, 279–294, 1976.
Sunderman, F. W., Jr., *Ann. Clin. Lab. Sci.,* 14, 93–122, 1984.
Sunderman, F. W., Jr., *Prev. Med.,* 5, 279–294, 1986.
Suzuki, Y., *Ind. Health,* 228, 9–19, 1990.
Tsapakos, M. J. and Wetterhahn, K. J., *Chem. Biol. Interact.,* 46, 265–277, 1983.
Tsuda, H. and Kato, K., *Gann,* 67, 469–470, 1976.
Tsuneta, Y., *Hokkaido J. Med. Sci.,* 57, 175–187, 1982.
Tsuneta, Y., Mikami, H., Kimura, K., Abe, S., Osaki, Y., and Murao, M., *Haigan,* 18, 341–348, 1978.

Tsuneta, Y., Ohsaki, Y., Kimura, K., Mikami, H., Abe, S., and Murao, M., *Thorax,* 35, 294–297, 1980.
Uyama, T., Monden, Y., Tsuyuguchi, M., Harada, K., Kimura, S., and Taniki, T., *J. Surg. Oncol.,* 41, 213–218, 1989.
Uyeki, E. M. and Nishio, A., *J. Toxicol. Environ. Health,* 11, 227–235, 1983.
Venitt, S. and Levy, L. S., *Nature (London),* 250, 493–495, 1974.
Warren, G., Schultz, P., Bancroft, D., Bennett, K., Abbott, E. H., and Rogers, S., *Mutat. Res.,* 90, 111–118, 1981.
Waterhouse, J. A. H., *Br. J. Cancer,* 32, 262, 1975.
Wetterhahn, K. E. and Hamilton, J. W., *Sci. Total Environ.,* 86, 113–129, 1989.
Wise, J. P., Leonard, J. C., and Patierno, S. R., *Mutat. Res.,* 278, 69–79, 1992.
Wise, J. P., Orenstein, J. M., and Patierno, S. R., *Carcinogenesis,* 14, 429–434, 1993.
Wolf, T., Kasemann, R., and Ottenwalder, H., *Arch. Toxicol.,* Suppl. 13, 48–51, 1989.
Xu, J., Manning, C. R., and Patierno, S. R., *Carcinogenesis,* 15, 1443–1450, 1994.
Yang, J. L., Hsieh, Y. C., Wu, C. W., and Lee, T. C., *Carcinogenesis,* 13, 2053–2057, 1992.
Zhitkovich, A. and Costa, M., *Carcinogenesis,* 13, 1485–1489, 1992.

Chapter 13

The Carcinogenicity of Arsenic

Zaolin Wang and Toby G. Rossman

I. ARSENIC: CHARACTERISTICS AND HUMAN EXPOSURE

Arsenic (atomic number 33, atomic mass 74.9126) belongs to subgroup V.A of the periodic table, situated below phosphorus and above antimony. Arsenic has a number of valence states and can exist in both cationic and anionic forms. Arsenate (As^{5+}) is probably the most common environmental form of inorganic arsenic. It is less toxic than arsenite (As^{3+}) and many of the organoarsenicals. Arsenic compounds are used in the preparation of insecticides, herbicides (weedkillers for railroad and telephone posts), fungicides, rodenticides, desiccants to facilitate mechanical cotton harvesting, and wood preservatives. Other uses of arsenic are in the glass industry and in the smelting of other metals. With the rapid development of the electronic industry, arsenic exposure is likely to increase because gallium arsenide is widely used in semiconductors and integrated circuits. Arsenic compounds are also used as veterinary medicines and are still in use in human medicine in Africa.

Arsenic is mobile within the environment and can leach from soil or rocks into hot spring waters. The drinking water supply has been found to be contaminated by arsenic in many countries including the U.S. Fresh water was estimated to contain about 1 µg/l (IARC, 1980). Mineral waters and hot springs may contain up to 50 and 300 times more arsenic than fresh water, respectively. Waters in some regions contain more than 1000 µg/l arsenic. Seawater also contains levels of arsenic ranging from 0.09 to 24 µg/l (Ishinishi et al., 1986; Léonard, 1991).

Environmental arsenic contamination occurs mainly from industrial processes such as smelting of other metals, application of arsenical pesticides and herbicides, and power generation from coal or geothermal sources. The burning of coal and smelting of metals are major sources of arsenic in air. Airborne arsenic can be transported over long distances in the atmosphere before being precipitated.

Humans may also be exposed to arsenic in food. Arsenic can accumulate in various vegetables, grains, fruits, seafood, meats, and tobacco products. Use of arsenical pesticides may increase arsenic concentration in some species of plants. Marine species of fish or shellfish usually contain higher concentrations of arsenic than freshwater species. The average total human intake of arsenic is about 90 µg/day (Lisella et al., 1972), but can vary depending mainly on the amount of seafood consumed. Exposure can also occur through ingestion of arsenic-containing drinking water or industrial dust. As a natural toxicant, arsenic cannot be destroyed and only a part of human exposure to arsenic is under our control. The average arsenic intake of 45 µg/day from food and 10 µg/day from drinking water may be unavoidable (Stöhrer, 1991).

Arsenic compounds can be absorbed through the GI tract, lungs, and skin. After absorption, arsenic is transported to other parts of the body and distributed in various organs. The highest concentration of arsenic in humans is in hair and nails, followed by skin and lungs (Aposhian, 1989). Arsenate (As^{5+}) is reduced *in vivo* to arsenite (As^{3+}), and it is the later (trivalent) form that is then active in animal tissues

(Ginsburg, 1965; Ginsburg and Lotspcich, 1964; Gordon and Quastel, 1948; Lander et al., 1975; Peoples, 1964; Winkler, 1962). However, arsenite has also been reported to be oxidized to arsenate in the detoxication of arsenic (Bencko et al., 1976; Léonard, 1991). The reduction of As^{5+} to As^{3+} appears to be rapid in all species studied (Vahter and Marafante, 1985). As^{3+} is then methylated via a nonenzymatic mechanism to monomethylarsonic acid (MMA), which is further methylated enzymatically to dimethylarsinic acid (DMA) (Petito and Beck, 1991). Excess As^{3+} can inhibit the first nonenzymatic step and saturate the second enzymatic step (Buchet and Lauwerys, 1985). In mammals, reduction takes place in the blood and methylation primarily in the liver (Marafante and Vahter, 1984; Marafante et al., 1985; Buchet and Lauwerys, 1985; Vahter and Marafante, 1987). Humans may be more sensitive to arsenic than other species because arsenic methylation in humans is less efficient than it is in other species (Vahter and Marafante, 1985).

Excretion of arsenic is mainly through urine. Pentavalent arsenic is excreted more rapidly than trivalent, and organic faster than inorganic. The urine arsenic concentration is thought to be the best indicator of current or of recent past exposure to arsenic (Baker et al., 1977; Milham and Strong, 1974; Harrington et al., 1978; Landrigan, 1992). In contrast, blood determination is found to have little practical value because of the short half-life of arsenic in blood (Landrigan, 1992). The normal average blood and urinary levels are about 100 and 15 µg/l, respectively (Fowler, 1977). The urinary concentration of inorganic arsenic plus MMA and DMA is probably a more accurate biological marker of the absorbed dose of inorganic arsenic than the total urinary arsenic concentration (Hopenhayn-Rich et al., 1993) because it excludes other arsenic compounds, in particular arsenobetaine, which is abundant in some fish and shellfish (Vahter, 1983). Arsenobetaine is thought to be nontoxic and is rapidly excreted in the urine without metabolism (U.S. EPA, 1988; Crecelius, 1977). Hair arsenic determination is a useful semiquantitative indicator of past exposure to arsenic (Baker et al., 1977; Landrigan, 1992).

Occupational exposure to arsenic is of main concern for workers involved in the processing of copper, gold, lead, and antimony ores, producing and using arsenicals and arsenic-containing pesticides, burning of arsenic-containing coal in power plants, and treating wood with arsenic preservatives and working with such treated wood. Although acute arsenic poisoning was reported in the past, chronic exposure is now of more concern than acute exposure mainly because of the carcinogenic effects of chronic exposure.

II. EVIDENCE OF ARSENIC CARCINOGENICITY

Arsenic has long been thought to contribute to the incidence of human cancer. The major evidence for arsenic as a human carcinogen came from studies of lung cancer in arsenic ore smelters and skin cancer in people exposed to arsenic-containing drinking water or, therapeutically, to Fowler's solution (potassium arsenite).

As early as 1820, Ayrton Paris described the high incidence of scrotal carcinoma among copper smelter workers in Cornwall, England as a consequence of exposure to arsenic fumes. In 1879, it was suggested that inhaled arsenic might be the cause of high rates of lung cancer in German miners (Neubauer, 1947). A few years later, Hutchinson (1887, 1888) reported skin carcinomas in patients who had taken long-term arsenical medications. Studies of smelter workers, insecticide manufacturing workers, and sheep dip workers in many countries around the world have all found an association between occupational arsenic exposure and lung cancer mortality. Table 1 summarizes findings from some recent and historical studies relating occupational arsenic exposure and lung cancer. Matanoski et al. (1981) reported an excess risk of lung cancer in a population residing near a pesticide manufacturing plant. The association between skin cancer and arsenic ingestion was confirmed by studies in Taiwan, Chile, Argentina, and Mexico (Tseng et al., 1968; Tseng, 1977; Borgono and Greiber, 1971; Borgono et al., 1977; Bergoglio, 1964; Cebrian et al., 1983), where drinking water supplies are contaminated by arsenic. A high risk of skin cancer was also found in persons treated with arsenic-containing medicines (Sommers and McManus, 1953). Results showing no excess skin cancer among U.S. residents using drinking water contaminated by arsenic (Morton et al., 1976; Southwick et al., 1981) appear to be of inadequate statistical power because of small sample size and therefore are not necessarily inconsistent with the positive results mentioned above.

Although it has been more than 100 years since the carcinogenic properties of arsenic were first suggested, it is still uncertain whether inorganic arsenic causes cancers other than skin and lung. A follow-up study of the population living in the area of Taiwan, where a high prevalence of skin cancer was associated with arsenic contamination in the water supply, also found significantly elevated standard

Table 1 Occupational Studies of Arsenic Exposure and Lung Cancer

Type of Industry	Lung Cancer Risk Ratio[a] Range (Average)	
	Overall	Highly exposed subgroup
Sheep dip manufacturing	4.55 (one study)	
Smelting and/or mining	1.65–11.89 (3.61)	2.91–25 (10.79)
Insecticide manufacturing	1.68–3.45 (2.46)	3.13–16.67 (8.93)

[a] Standardized mortality ratio for most studies; odds ratio, risk ratio, or proportional mortality ratio in other studies. Data from Hertz-Picciotto et al., 1992.

mortality ratios for cancers of the bladder, lung, liver, kidney, and colon (Tseng, 1977). Like skin cancer, the incidence of bladder, liver, and lung cancer cases in this study showed a dose-related increase with arsenic exposure. Similar results were reported by Chen et al. (1988, 1992) and Smith et al. (1992). Recently, Bates et al. (1992) analyzed all published epidemiological and clinical studies concerning inorganic arsenic ingestion and internal cancers, and found that many studies were uninformative due to low statistical power or potential bias, either in collection or analysis of data. The studies from Taiwan (Chen et al., 1985, 1986, 1988) and Japan (Tsuda et al., 1989), involving exposure to arsenic in drinking water, were found to be the most informative. These studies strongly suggest that ingestion of inorganic arsenic is associated with cancers in the bladder, kidney, lung, liver, and possibly other sites. It was suggested that in the future, the debate will center on whether arsenic ingestion increases the risk of internal cancers (Bates et al., 1992).

Arsenite is the most likely active carcinogenic species. Tinwell et al. (1991) treated mice with a single i.p. dose of Fowler's solution (arsenic trioxide dissolved in potassium bicarbonate), sodium arsenite, and potassium arsenite and with a single oral dose of the natural ore orpiment (principally As_2S_3). Bone marrow micronuclei were induced only in mice treated with Fowler's solution, sodium arsenite, and potassium arsenite. The blood arsenic concentration in orpiment-treated mice (which did not have micronuclei) reached 300 to 900 ng/ml in 24 h, a much higher level than that in arsenite-treated mice, which was only 100 ng/ml or lower. In addition, both arsenite and arsenate were able to transform Syrian hamster embryo cells, but arsenite was more than tenfold more potent than arsenate (Barrett et al., 1989).

III. ARSENIC CARCINOGENICITY DOSE-RESPONSE RELATIONSHIP

Recently, Hertz-Picciotto and Smith (1993) evaluated all published reports which relate lung cancer risk to quantified occupational arsenic exposure. Relative risks from six studies (Enerline et al., 1987; Lee-Feldstein, 1986; Jarup et al., 1989, 1991; Taylor et al., 1989; Ott et al., 1974) in three countries (U.S., Sweden, and China) were plotted against cumulative arsenic exposure. With one exception, the suggestion of a declining slope for the risk of lung cancer at higher cumulative exposure was observed. The dose-response curves appear to rise more steeply at low as compared with high exposures, suggesting that the use of linear models applied to occupational epidemiological data may result in underestimation of the true risk at lower exposures.

On the other hand, a review by Stöhrer (1991) discusses a possible threshold dose for arsenic carcinogenicity, which had also been postulated previously (U.S. EPA, 1988; Marcus and Rispin, 1988). Stöhrer points out that data from the 100 years since the discovery of arsenical cancer indicate a sharp drop-off in cancer cases at daily intakes below 1 mg/day. The best dose-response data came from studies in the southwest of Taiwan, where drinking water from artesian wells has been contaminated by arsenic for 45 years. Among 40,000 people living in this area, more than 7000 developed arsenical diseases including hyperpigmentation and keratosis (especially on the palms and soles), Blackfoot disease, and skin cancers (Tseng et al., 1968; Yeh, 1973). No arsenical cancer or other arsenical lesions were observed in a control group of 7500 people exposed to less than 19 µg/l arsenic in the drinking water (plus about 15 µg/day inorganic arsenic from food). Stöhrer concluded that: (1) more than 400 µg arsenic daily is required to cause arsenical disease; (2) skin cancers, internal cancers, and noncancerous effects of arsenic all have the same threshold dose and probably result from the same primary interaction; (3) hyperpigmentation can therefore serve as a sensitive indicator of arsenic exposure and of future cancer risk; (4) synergistic interactions change the log-normal dose response slope but leave unchanged the intercept or the threshold dose.

According to studies by Petito and Beck (1991), the mechanism for the threshold is due to the metabolism and detoxification of inorganic arsenic via methylation to less toxic metabolites, which are excreted from the body more efficiently than inorganic arsenic. When arsenic exposures are high enough to exceed the methylation capacity, excess inorganic arsenic can be retained in the body for longer periods of time, resulting in contact with target tissues and leading to adverse health effects. The methylation process for arsenic begins to be affected at a daily intake rate of approximately 200–250 µg/day. When the daily intake reaches 500 µg/day, the methylation capacity is limited by saturation of the enzymatic conversion of MMA to DMA and by the inhibition of nonenzymatic conversion of As^{3+} to MMA by excess As^{3+}. However, the methylation threshold hypothesis has recently been challenged (Hopenhayn-Rich et al., 1993). All available human data which measure the urinary metabolites of inorganic arsenic (MMA, DMA, and inorganic arsenic) in different populations with various exposure levels were analyzed. It was concluded that regardless of the exposure level, the level of unmethylated inorganic arsenic is approximately 20% of the absorbed dose (measured as inorganic arsenic + MMA + DMA). Thus, other threshold mechanisms must be involved in the toxic effect of inorganic arsenic.

IV. THE INADEQUACY OF AN ANIMAL MODEL FOR ARSENIC CARCINOGENESIS

Early attempts to induce carcinomas in animals by arsenic either failed or provided very limited significance (Leicht and Kennaway, 1922). The review of arsenic carcinogenicity by IARC (1980) lists four different species (mouse, rat, dog, and rabbit) given various arsenic compounds (arsenic trioxide; sodium arsenite; sodium arsenate; dimethylarsenic acid; arsanilic acid; lead arsenate; copper ore containing 3.95% arsenic; flue dust containing 10% arsenic; Bordeaux mixture containing calcium arsenate, copper sulfate, and calcium hydroxide; metallic arsenic; and arsenic in combination with chemical carcinogens) by different routes of exposure. There was no consistent demonstration of arsenic carcinogenicity in these studies.

However, a few reports of successful arsenic carcinogenesis assays exist. As discussed by Bencko (1977, 1987), arsenic-induced benign and malignant teratomas in rat embryos were reported as early as 1927. In this protocol, rat embryos are transplanted into the peritoneal cavity of rats exposed to arsenic via drinking water. Similar results were observed in mice after diaplacental and postnatal application of arsenic (Oswald and Goerttler, 1971) and in mice treated transplacentally and neonatally with arsenic trioxide (Bencko, 1987). It was suggested that embryonic cells are unusually sensitive to arsenic. When skin application and oral administration were used in an attempt to induce tumors in white mice, carcinoma was observed only in mice treated by repeated, long-term skin application of a solution of 18% sodium arsenate (Currie, 1947). Treatment of BD rats intratracheally with a single 0.1-ml instillation of a vineyard pesticide containing calcium arsenate resulted in the death of 10 out of 25 rats. Of 15 surviving rats with a normal lifespan, 9 developed lung carcinoma. However, most of the tumors were rather small and could only be detected by serial cuts, and parts of the lung were severely damaged (Ivankovic et al., 1979). Similar positive observations in hamsters were also reported (Ishinishi et al., 1983; Yamamoto et al., 1987). Pershagen et al. (1984) applied a carrier dust (charcoal carbon) for longer lung retention of arsenic and other chemicals to mimic the situation encountered in smelter workroom air. The authors concluded that arsenic trioxide could induce lung carcinomas. Stomach cancers could also be induced by using implantation of arsenic-containing capsules (Katsnel'son et al., 1986).

Despite the positive results described above, it must be remembered that very high doses of arsenic compounds are required for tumor induction. Unlike laboratory animals, humans are always exposed to a mixture of toxicants for long periods of time, a situation which is difficult recreate in a laboratory setting. Since most of the animal experiments provided negative results, and some even reported a decrease in tumor induction by arsenic (Boutwell, 1963; Milner, 1969), it seems reasonable to suggest that arsenic might act as a cocarcinogen, or tumor promoter, rather than an initiating direct carcinogen with a no-threshold effect.

V. ARSENIC AND MUTATION

Most of the studies on the genotoxic activity of arsenic compounds have yielded negative results for gene mutations but positive results for chromosomal damage. Reversion assays in *Salmonella typhimurium* failed to detect mutations induced by sodium arsenite ($NaAsO_2$), sodium arsenate ($AsHNa_2O_4$), methane arsonic acid (CH_5AsO_3), or monosodium methane arsonate (CH_4AsNaO_3) (Löfroth and Ames, 1978; Andersen et al., 1972). In *Escherichia coli,* negative results were obtained with sodium arsenite, sodium arsenate, octyl ammonium methyl arsonate ($CH_3(CH_2)_7NH_4AsCH_3O_3$), and dodecyl ammonium methyl arsonate ($CH_3(CH_2)_{11}NH_4AsCH_3O_3$) (Ficsor and Poccolo, 1972; Rossman et al., 1977, 1980; Léonard and Lauwerys, 1980; Léonard, 1984). Positive results were obtained in some rec-assays with *Bacillus subtilis* (Nishioka, 1975; Nishioka and Takagi, 1975; Shirasu, 1976; Kada et al., 1980). However, positive results in the rec-assay indicate only that the compound is more toxic to DNA repair deficient cells (suggestive of DNA damage). It is not a mutagenic endpoint.

As in bacterial cells, arsenic compounds generally fail to induce mutations in Chinese hamster cells, at least under conditions of high survival (Rossman et al., 1980; Lee et al., 1985b). Since the lack of mutagenicity by arsenite in bacteria or in Chinese hamster cells does not rule out the possibility that arsenite might cause large deletions (which are either unselectable or lethal events in these systems), the mutagenicity of arsenite was also assayed in a transgenic line, G12 — a pSV2*gpt* transformed Chinese hamster V79 cell which is able to detect deletions in addition to point mutations at the transgenic *gpt* locus (Klein and Rossman, 1990). Arsenite was not significantly mutagenic in these cells (Li and Rossman, 1991). On the other hand, Yamanaka et al. (1989a) found that dimethylarsine, a volatile metabolite of DMA, one of the main metabolites of inorganic arsenic in mammals, is mutagenic in *Escherichia coli*. They also found that oxygen was required to induce the mutations, suggesting that the reaction product(s) between dimethylarsine and oxygen are mutagenic. The same laboratory found that DMA induced lung-specific DNA damage in mice and rats via the dimethylarsenic peroxyl radical and other active oxygen species produced by the metabolism of DMA (Yamanaka et al., 1989b, 1990). Evidence for a cellular protective response to the oxidative damage was seen (Yamanaka et al., 1991). However, it is not clear whether these metabolites would be formed in significant amounts *in vivo* after exposure to the usual environmental arsenic compounds. Although other positive mutagenesis results were claimed in *E. coli* (Nishioka, 1975) and in L5178Y TK +/− mouse lymphoma cells (Harrington-Brock, 1993), these studies used very high toxic concentrations of arsenic compounds and these results have not been reproduced by other laboratories (Rossman et al., 1980).

In vitro experiments with many inorganic and organic arsenic compounds showed that they were powerful clastogens in many cell types (Nakamuro and Sayato, 1981; King and Lunford, 1950; Larramendy et al., 1981; Oppenheim and Fishbein, 1965; Paton and Allison, 1972; Wan et al., 1982; Lee et al., 1985b; Sweins, 1983; Nordenson et al., 1981; Lee et al., 1986). Trivalent arsenic compounds are usually more efficient in inducing chromosomal aberrations than pentavalent arsenic, though both show a linear dose-response relationship (Nakamuro and Sayato, 1981). Although clastogenic effects were obtained in one animal study (Sram, 1976), other studies in animals (Poma et al., 1981) or humans (Burgdorf et al., 1977; Vig et al., 1984) gave negative results. A much smaller data base is available on micronucleus formation. Both positive (Deknudt et al., 1986; Tinwell et al., 1991) and negative results (De Brabander et al., 1976; Poma et al., 1981) were reported in animals treated with sodium arsenite, potassium arsenate, and the natural ore orpiment (principally As_2S_3), and more data are available from *in vivo* assays than from *in vitro* assays.

Sister chromatid exchanges (SCEs) require a much lower inducing dose of arsenic compounds compared with chromosome aberrations. Like chromosome aberrations, *in vitro* studies of SCE induction by arsenic compounds gave positive results (Ohno et al., 1982; Wan et al., 1982; Lee et al., 1985b; Wen et al., 1981; Zanzoni and Jung, 1980) with a few exceptions (Larramendy et al., 1981; Crossen, 1983). Both positive and negative observations were reported for *in vivo* SCE induction (Burgdorf et al., 1977; Vig et al., 1984). The discrepancy between the findings from *in vitro* and *in vivo* data for chromosome aberrations and SCEs might be due to the difference in doses tolerated for *in vitro* and *in vivo* assays. *In vitro* systems allow the use of higher concentrations of the test chemical whereas *in vivo* experiments require doses compatible with animal survival. Since SCE is the more sensitive assay, it was expected that lower doses would suffice for *in vivo* SCE induction. However, since consistent positive SCE results were not seen *in vivo*, dose differences used for *in vivo* and *in vitro* assays could not account for the discrepancy between these two systems. Other mechanisms, such as pharmacodynamics and pharmacokinetics of the test compounds, might be involved.

VI. OTHER GENETIC EFFECTS

Although low concentrations of arsenic compounds failed to induce gene mutations in either bacterial or mammalian cells, arsenite was found to be comutagenic with UV in *E. coli* (Rossman, 1981) and with UV, methyl methanesulfonate (MMS), and methyl nitrosourea (MNU) in Chinese hamster cells (Lee et al., 1985b; Li and Rossman, 1989a, 1991; Yang et al., 1992). The comutagenicity of arsenite and its lack of DNA-damaging ability suggest that it might modify the repair of DNA lesions induced by other agents. There is some evidence that arsenic compounds can interfere with a variety of DNA repair processes, including inhibition of pyrimidine dimer removal in human SF34 cells after UV irradiation (Okui and Fujimara, 1986), blockage of the repair of DNA damage induced by X-rays and UV (Snyder et al., 1989), inhibition of postreplication repair of UV-induced damage (Lee-Chen et al., 1992), inhibition of completion of repair of MNU-induced DNA damage (Li and Rossman, 1989a), and inhibition of mammalian DNA ligase II activity (Li and Rossman, 1989b). The theory that arsenic compounds inhibit DNA repair is also supported by the findings that arsenic compounds potentiate X-ray- and UV-induced chromosomal damage in peripheral human lymphocytes and fibroblasts (Jha et al., 1992), alter the mutational spectrum (but not the strand bias) of UV-irradiated Chinese hamster ovary cells (Yang et al., 1992) and synergistically enhance chromosome aberrations induced by diepoxybutane, a DNA cross-linking agent (Wiencke and Yager, 1992). Taken together, these results support the concept that arsenic compounds may act as cocarcinogens, enhancing the carcinogenicity of mutagenic carcinogens.

Arsenite is a very good inducer of gene expression, especially of heat shock proteins (Li, 1983; Burden, 1986; Deaton et al., 1990). The expression of the multidrug resistance (MDR1) gene (Chin et al., 1990), heme oxygenase (Keyse and Tyrrell, 1989), and *c-fos* oncogene (Gubits, 1988) is also increased by arsenite. Administration of arsenic compounds also induce metallothionein gene expression *in vivo* but not *in vitro* (Kreppel et al., 1992; Albores et al., 1992; Maitani et al., 1987; Koropatnick et al., 1991), suggesting an indirect induction *in vivo*. Arsenic compounds induce gene expression in organisms spanning the entire eukaryotic kingdom, including plants (Rice et al., 1985). Arsenite is also an inducer of gene amplification in rodent and human cells (Lee et al., 1988; Rossman and Wolosin, 1992). Arsenic compounds have been shown to induce cell transformation in Syrian hamster embryo cells (DiPaolo and Casto, 1979; Lee et al., 1985a; Barrett et al., 1989).

VII. POSSIBLE MECHANISM FOR THE CARCINOGENICITY OF ARSENIC

Nongenotoxic indirect carcinogens have a variety of stress-related effects that ultimately induce or affect the expression of genes controlling proliferation and differentiation. Effective intracellular concentrations of the inducing agent are required for the inducible expression of the target genes. This is consistent with the concept of a threshold of biological action in the field of classical toxicology. Data from dose-response determinations of indirect carcinogens in animal studies support a threshold model (Schröter et al., 1987; Wolff et al., 1987). Arsenic shares great similarities with other known "nongenotoxic" carcinogens and cocarcinogens with respect to threshold of dose response, induction of gene expression, stress-related effects, and lack of a mutagenic effect in gene mutation assays. Trivalent arsenic (the likely carcinogenic form) may activate nuclear oncogenes, perhaps via its ability to bind to vicinal dithiols within a protein, or to bridge two thiols between two proteins, as shown in Figure 1. The number of proteins with vicinal dithiols is relatively small, but this structure is common in the zinc fingers found in DNA binding proteins and transcription factors, and in some DNA repair proteins (Berg, 1990). Because arsenite inhibits many steps in DNA repair (discussed above) a second mode of action for arsenite might involve blocking the repair of DNA damage induced by genotoxic carcinogens.

Among all established human carcinogens, arsenic is the only one for which there is no good animal model. One possible explanation for the difficulty of inducing arsenic cancer in laboratory animals might be associated with arsenic tolerance. Inducible arsenic tolerance was reported in Chinese hamster cells and their arsenite-resistant variants (Wang and Rossman, 1993; Lee et al., 1989). Inducible arsenite tolerance was not observed in a number of human cell lines, and human cells are much more sensitive to arsenite than are rodent cells (Wang et al., unpublished). Most of the animal carcinogenicity experiments were performed using rodent species which can easily induce the arsenic-tolerant state. It is possible that the failure in many rodent bioassays was caused by such inducible tolerance mechanisms. The few positive results in carcinogenicity assays were obtained if arsenic was applied to the embryo or neonates (Oswald and Goerttler, 1971; Bencko, 1987), which appear to be very sensitive to arsenic

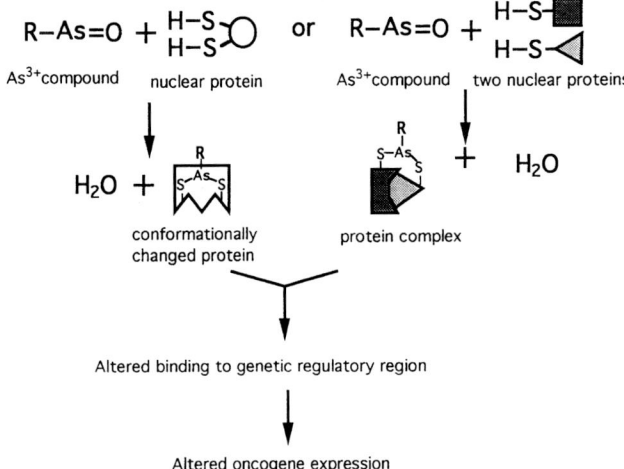

Figure 1 Scheme for one of the possible mechanisms of carcinogenesis by trivalent arsenic compounds.

and may not be able to induce tolerance. Positive results were also observed in a few experiments where very high doses were used, a situation in which the tolerant state may not be able to develop.

In summary, arsenite (the most likely active carcinogenic species) acts as a nongenotoxic indirect carcinogen probably via induction of oncogene expression and inhibition of DNA repair. The failure in most rodent bioassays might be due to the induction of an arsenic-tolerant state in rodents. It would be of interest to determine whether arsenic compounds act as cocarcinogens in animal bioassays.

REFERENCES

Albores, A., Koropatnick, J., Cherian, M. G., and Zelazowski, A. J.,*Chem.-Biol. Interact.*, 85, 127–140, 1992.
Andersen, K. J., Leighty, E. G., and Takahashi, M. T., *J. Agric. Food Chem.*, 20, 649–666, 1972.
Aposhian, H. V., *Rev. Biochem. Toxicol.*, 10, 265–299, 1989.
Baker, E. L., Jr., Hayes, C. G., Landrigan, P. J., Handke, J. L., Leger, R. T., Housworth, W. L., and Harrington, J. M., *Am. J. Epidemiol.*, 106, 261–273, 1977.
Barrett, J. C., Lamb, P. W., Wang, T. C., and Lee, T. C., *Biol. Trace Elem. Res.*, 21, 421–429, 1989.
Bates, M. N., Smith, A. H., and Hopenhayn-Rich, C., *Am. J. Epidemiol.*, 135, 462–476, 1992.
Bencko, V., *Environ. Health Perspect.*, 19, 179–182, 1977.
Bencko, V., in *Advances in Modern Environmental Toxicology*, Vol. XI. Genotoxic and Carcinogenic Metals: Environmental and Occupational Occurrence and Exposure, Fishbein, L., Furst, A., and Mehlman, M.A., Eds., Princeton Scientific, Princeton, NJ, 1, 1987.
Bencko, V., Benes, B., and Cikrt, M., *Arch. Toxicol.*, 36, 159–162, 1976.
Berg, J., *Annu. Rev. Biophys. Chem.*, 19, 405–421, 1990.
Bergoglio, R. M., *Prensa Med. Argent.*, 51, 994–998, 1964.
Borgono, J. M. and Greiber, R., in *Trace Substances in Environmental Health-V*, Proc. 5th Annu. Conf. June 29–July 1, Hemphill, D.C., Ed., University of Missouri, Columbia, MO, 13, 1971.
Borgono, J. M., Vicent, P., Venturino, H., and Infante, A., *Environ. Health Perspect.*, 19, 103–105, 1977.
Boutwell, R. K., *J. Agric. Food Chem.*, 11, 381–385, 1963.
Buchet, J. P. and Lauwerys, R., *Arch. Toxicol.*, 57, 125–129, 1985.
Burden, R. H., *Biochem. J.*, 240, 313–324, 1986.
Burgdorf, W., Kurvink, K., and Cervenka, J., *Mamm. Chromatogr. Newsl.*, 18, 12, 1977.
Cebrian, M. E., Albores, A., Aguilar, M., and Blakely, E., *Hum. Toxicol.*, 2, 121–133, 1983.
Chen, C.-J., Chuang, S.-L., You, T.-M., Lin, T.-M., and Wu, H.-Y., *Br. J. Cancer*, 53, 399–405, 1986.
Chen, C.-J., Chuang, Y.-C., Lin, T.-M., and Wu, H.-Y., *Cancer Res.*, 45, 5895–5899, 1985.
Chen, C.-J., Kuo, T.-L., and Wu, M.-M., *Lancet*, 1, 414–415, 1988.
Chen, C.-J., Chen, C. W., Wu, M.-M., and Kuo, T.-L., *Br. J. Cancer*, 66, 888–892, 1992.
Chin, K. V., Tanaka, S., Darlington, G., Pastan, I., and Gothesman, M. M., *J. Biol. Chem.*, 265, 221–226, 1990.
Crecelius, E. A., *Environ. Health Perspect.*, 19, 147–150, 1977.
Crossen, P. E., *Mutat. Res.*, 119, 415–419, 1983.
Currie, N. A., *Br. Med. Bull.*, 4, 402, 1947.
De Brabander, M., Van De Veire, R., Aerts, F., Geuens, S., and Hoebeke, J., *J. Natl. Cancer Inst.*, 56, 357–363, 1976.

Deaton, M. A., Bowman, P. D., Jones, G. P., and Powanda, M. C., *Fundam. Appl. Toxicol.*, 14, 471–476, 1990.
Deknudt, G. H., Leonard, A., Jenar-Du Buisson, G., and Delavignette, E., *Mutagenesis*, 1, 33–34, 1986.
DiPaolo, J. A. and Casto, B. C., *Cancer Res.*, 39, 1008–1313, 1979.
Ficsor, G. and Nii Lo Piccolo, G. M., *EMS Newsl.*, 6, 6–8, 1972.
Fowler, B. A., in *Toxicology of Trace Elements*, Goyer, R.A. and Mehlman, M.S., Eds., John Wiley & Sons, New York, 79, 1977.
Ginsburg, J. M. and Lotspcich, W. W., *Am. J. Physiol.*, 205, 707–714, 1964.
Ginsburg, J. M., *Am. J. Physiol.*, 208, 832–840, 1965.
Gordon, J. J. and Quastel, J. H., *Biochem. J.*, 42, 337–350, 1948.
Gubits, R. M., *Oncogene*, 3, 163–168, 1988.
Harrington, J. M., Middaugh, J. P., Morse, D. L., and Housworth, J. A., *Am. J. Epidemiol.*, 108, 377–385, 1978.
Harrington-Brock, K., Smith, T. W., Doerr, C. L., and Moore, M. M., *Environ. Mol. Mutagenet.*, 21 (Suppl. 22), 7, 1993.
Hertz-Picciotto, I. and Smith, A. H., *Scand. J. Work Environ. Health*, 19, 217–226, 1993.
Hertz-Picciotto, I., Smith, A. H., Holtzman, D., Lipsett, M., and Alexeeff, G., *Epidemiology*, 3, 23–31, 1992.
Hopenhayn-Rich, C., Smith, A. H., and Goeden, H. M., *Environ. Res.*, 60, 161–177, 1993.
Hutchinson, J., *Br. Med. J.*, 2, 1280–1281, 1887.
Hutchinson, J., *Trans. Pathol. Soc. (London)*, 39, 352–363, 1888.
IARC, IARC Monographs on the Evaluation of Carcinogenic Risk of Chemicals to Man, Vol. 23. Some Metals and Metallic Compounds, World Health Organization, Lyon, France, 1980.
Ishinishi, N., Tsuchiya, K., Vahter, M., and Fowler, B. A., in *Handbook on the Toxicology of Metals*, 2nd ed., Vol. II, Friberg, L., Nordberg, G. F., and Vouk, V. B., Eds., Elsevier, Amsterdam, 43, 1986.
Ishinishi, N., Yamamoto, A., Hisanaga, A., and Inamasu, T., *Cancer Lett.*, 21, 141–147, 1983.
Ivankovic, S., Eisenbrand, G., and Preussmann, R., *Int. J. Cancer*, 34, 786–788, 1979.
Jarup, L., Pershagen, G., and Wall, S., *Am. J. Ind. Med.*, 15, 31–41, 1989.
Jha, A. N., Noditi, M., Nilsson, R., and Natarajan, A. T., *Mutat. Res.*, 284, 215–221, 1992.
Kada, T., Hirano, K., and Shirasu, Y., in *Chemical Mutagens: Principles and Methods for Their Detection*, Vol. 6, de Serres, F.J. and Hollaender, A. Eds., Plenum Press, New York, 149, 1980.
Katsnel'son, B. A., Neizvestnova, E. M., and Blokhin, V. A., *Vopr. Onkol.*, 32, 68–73, 1986.
Keyse, S. M. and Tyrrell, R. M., *Proc. Natl. Acad. Sci. U.S.A.*, 85, 99–103, 1989.
King, H. and Lunford, R. J., *J. Chem. Soc.*, 8, 2086–2088, 1950.
Klein, C. B. and Rossman, T. G., *Environ. Mol. Mutagenet.*, 16, 1–12, 1990.
Koropatnick, J., Zelazowski, A. J., Cebrian, M. E., Vesely, S., and Cherian, M. G., *Toxicologist*, 1, 76, 1991.
Kreppel, H., Bauman, J. W., Liu, J., McKim, J. M., Jr., and Klaassen, C. D., *Fundam. Applied Toxicol.*, 20, 184–189, 1992.
Landrigan, P. J., in *Environmental and Occupational Medicine*, 2nd ed., Rom, W. N., Ed., Little, Brown, Boston, 773, 1992.
Lander, J. J., Stanely, R. J., Summer, H. W., Boswell, D. C., and Aach, R. D., *Gastroenterology*, 68, 1582–1586, 1975.
Larramendy, M. L., Popescu, N. C., and DiPaolo, J., *Environ. Mutagenet.*, 3, 597–606, 1981.
Lee, T.-C., Oshimura, M., and Barrett, J. C., *Carcinogenesis*, 6, 1421–1426, 1985a.
Lee, T.-C., Huang, R. Y., and Jan, K. Y., *Mutat. Res.*, 148, 83–89, 1985b.
Lee, T.-C., Lee, K. C., Tseng, Y. J., Huang, R. Y., and Jan, K. Y., *Environ. Mol. Mutagenet.*, 8, 119–128, 1986.
Lee, T.-C., Tanaka, N., Lamb, P. W., Gillmer, T. M., and Barrett, C. J., *Science*, 241, 79–81, 1988.
Lee, T.-C., Wei, M. L., Chang, W. J., Ho, I. C., Lo, J. F., Jan, K. Y., and Huang, H., *In Vitro Cell. Develop. Biol.*, 25, 442–448, 1989.
Lee-Chen, S. F., Yu, C. T., and Jan, K. Y., *Mutagenesis*, 7, 51–55, 1992.
Lee-Feldstein, A., *J. Occup. Med.*, 28, 296–302, 1986.
Leicht, A. and Kennaway, E. L., , *Br. J. Med.*, 2, 1107–1108, 1922.
Léonard, A., *Toxicol. Environ. Chem.*, 7, 241–250, 1984.
Léonard, A., in *Metals and Their Compounds in the Environment*, Merian, E., Ed., VCH Publishers, New York, 751, 1991.
Léonard, A. and Lauwerys, R. R., *Mutat. Res.*, 75, 49–62, 1980.
Li, G. C., *J. Cell. Physiol.*, 115, 116–122, 1983.
Li, J.-H and Rossman, T. G., *Biol. Trace Elem. Res.*, 21, 373–381, 1989a.
Li, J.-H. and Rossman, T. G., *Mol. Toxicol.*, 2, 1–9, 1989b.
Li, J.-H. and Rossman, T. G., *Biol. Metals*, 4, 197–200, 1991.
Lisella, F. S., Long, K. R., and Scott, H. G., *J. Environ. Health*, 34, 511–518, 1972.
Löfroth, G. and Ames, B. N., *Mutat. Res.*, 53, 65–66, 1978.
Maitani, T., Staio, N., Abe, M., Uchiyama, S., and Saito, Y., *Toxicol. Lett.*, 39, 63–70, 1987.
Marafant, E. and Vahter, M., *Chem. Biol. Interact.*, 50, 49–57, 1984.
Marafant, E., Vahter, M., and Envoll, J., *Chem. Biol. Interact.*, 56, 225–238, 1985.
Marcus, W. L. and Rispin, A. S., in *Advances in Modern Environmental Toxicology: Risk Assessment and Risk Management of Industrial and Environmental Chemicals*, Cothern, C.R. and Mehlman, M.A., Eds., Princeton Publishing, Princeton, NJ, 133, 1988.
Matanoski, G., Landau, E., Tonascia, J., Lazar, C., Elliot, E., McEnroe, W., and King, K., *Environ. Res.*, 25, 8–28, 1981.
Milham, S., Jr. and Strong, T., *Environ. Res.*, 7, 176–182, 1974.
Milner, J. E., *Arch. Environ. Health*, 18, 7–11, 1969.

Morton, W., Starr, G., Pohl, D., Stoner, J., Wagner, S., and Weswig, P., *Cancer,* 37, 2523–2532, 1976.
Nakamuro, K. and Sayato, Y., *Mutat. Res.,* 88, 73–80, 1981.
Neubauer, O. *Br. J. Cancer,* 1, 192–251, 1947.
Nishioka, H., *Mutat. Res.,* 31, 185–189, 1975.
Nishioka, H. and Takagi, K., *Jpn. J. Genet.,* 50, 485–486, 1975.
Nordenson, I., Sweins, A., and Beckman, L., *Scand. J. Work Environ. Health,* 7, 277–281, 1981.
Ohno, H., Hanaoka, F., and Yamada, M., *Mutat. Res.,* 104, 141–145, 1982.
Okui, T. and Fujiwara, Y., *Mutat. Res.,* 172, 69–76, 1986.
Openheim, J. P. and Fishbein, W. N., *Cancer Res.,* 25, 980–985, 1965.
Oswald, H. and Goerttler, K., *Verh. Dtsch. Ges. Pathol.,* 55, 289–293, 1971.
Paton, G. R. and Allison, A. C., *Mutat. Res.,* 16, 332–336, 1972.
Peoples, S. A., *Ann. N.Y. Acad. Sci.,* 111, 644–649, 1964.
Pershagen, G., Nordberg, G., and Bjorklund, N. E., *Environ. Res.,* 34, 227–241, 1984.
Petito, C. T. and Beck, B. D., in *Trace Substances in Environmental Health,* Vol. XXIV, Hemphill, D.D., Ed., University of Missouri, Columbia, MO, 143, 1991.
Poma, K., Degraeve, N., Kirsch-Volders, M., and Susanne, C., *Experientia,* 37, 129–130, 1981.
Rice, J. D., Nikdel, S., and Purvis, A. C., *Proc. Fla. State Hortic. Soc.,* 98, 224–228, 1985.
Rossman, T. G., Meyn, M. S., and Troll, W., *Environ. Health Perspect.,* 19, 229–233, 1977.
Rossman, T. G., Stone, D., Molina, M., and Troll, W., *Environ. Mutagenet.,* 2, 371–379, 1980.
Rossman, T. G., *Mutat. Res.,* 91, 207–211, 1981.
Rossman, T. G. and Wolosin, D., *Mol. Carcinogenet.,* 6, 203–213, 1992.
Schröter, C., Parzefall, W., Schröter, H., and Schulte-Hermann, R., *Cancer Res.,* 47, 80–88, 1987.
Shirasu, Y., Moriya, M., Kato, K., Furuchashi, A., and Kada, T., *Mutat. Res.,* 40, 19–30, 1976.
Smith, A. H., Hopenhayn-Rich, C., Bates, M. N., Goeden, H. M., Hertz-Picciotto, I. H., Duggan, I., Wood, R., Kosnett, M., and Smith, M. T., *Environ. Health Perspect.,* 97, 259–267, 1992.
Snyder, R. D., Davis, G. F., and Lachmann, P., *Biol. Trace Elem. Res.,* 21, 389–398, 1989.
Sommers, S. C. and McManus, R. G., *Cancer,* 6, 347–359, 1953.
Southwick, J., Western, A., Beck, M., Whitley, T., Isaacs, R., Petajan, J., and Hansen, C., Community health associated with arsenic in drinking water in Millard County, Utah, Health Effects Research Laboratory, EPA-600/1–81–064, Cincinnati, OH, 1981.
Sram, R. J., *Mutat. Res.,* 41, 25–42, 1976.
Stöhrer, G., *Arch. Toxicol.,* 65, 525–531, 1991.
Sweins, A., *Hereditas,* 98, 249–252, 1983.
Taylor, P. R., Qiao, Y.-L., Schatzkin, A., Yao, S.-X., Lubin, J., Mao, B.-L., Rao, J.-Y., McAdams, M., Xuan, X. Z., and Li, I.-Y., *Br. J. Ind. Med.,* 46, 881–886, 1989.
Tinwell, H., Stephens, S. C., and Ashby, J., *Environ. Health Perspect.,* 95, 205–210, 1991.
Tseng, W. P. *Environ. Health Perspect.,* 19, 109–119, 1977.
Tseng, W. P., Chu, H. M., How, S. W., Fong, J. M., Lin, C. S., and Yeh, S., *J. Natl. Cancer Inst.,* 40, 453–463, 1968.
Tsuda, T., Nagira, T., Yamamoto, M., et al., *Sangyo Ika Daigaku Zasshi,* 11, 289–301, 1989.
U.S. EPA, Special report on ingested inorganic arsenic: skin cancer; Nutritional Essentiality (EPA/625/3–87/013). Risk Assessment Forum, U.S. Environmental Protection Agency, Washington, D.C., 1988.
Vahter, M., in *Biological and Environmental Effects of Arsenic,* Fowler, B.A., Ed., Elsevier, Amsterdam, 171, 1983.
Vahter, M. and Marafante, E., *Arch. Toxicol.,* 57, 119–124, 1985.
Vahter, M. and Marafante, E., *Toxicol. Lett.,* 37, 41–46, 1987.
Vig, B. K., Figueroa, M. L., Cornforth, M. N., and Jenkins, S. H., *Am. J. Ind. Med.,* 6, 325–338, 1984.
Wan, B., Christian, R. T., and Sookup, S. W., *Environ. Mutagenet.,* 4, 493–498, 1982.
Wang, Z. and Rossman, T. G., *Toxicol. Appl. Pharmacol.,* 118, 80–86, 1993.
Wen, W.-N., Lieu, T.-L., Chang, H.-J., Wuu, S. W., Yau, M.-L., and Jan, K. Y., *Hum. Genet.,* 59, 201–203, 1981.
Wiencke, J. K. and Yager, J. W., *Environ. Mol. Mutagenet.,* 19, 195–200, 1992.
Winkler, W. O., *J. Assoc. Anal. Chem.,* 45, 80–91, 1962.
Wolff, G. L., Roberts, D. W., Morissey, D. W., Greenman, D. L., Allen, R. R., Campbell, W. L., Bergman, H., Nesnow, S., and Frith, C. H., *Carcinogenesis,* 8, 1889–1897, 1987.
Yamamoto, A., Hisanaga, A., and Ishinishi, N., *Int. J. Cancer,* 40, 220–223, 1987.
Yamanaka, K., Ohba, H., Hasegawa, A., Sawamura, R., and Okada, S., *Chem. Pharm. Bull.,* 37, 2753–2756, 1989a.
Yamanaka, K., Hasegawa, A., Sawamura, R., and Okada, S., *Biochem. Biophys. Res. Commun.,* 165, 43–50, 1989b.
Yamanaka, K., Hoshino, M., Okamoto, M., Sawamura, R., Hasegawa, A., and Okada, S., *Biochem. Biophys. Res. Commun.,* 168, 58–64, 1990.
Yamanaka, K., Hasegawa, A., Sawamura, R., and Okada, S., *Toxicol. Appl. Pharmacol.,* 108, 205–213, 1991.
Yang, J.-L., Chen, M.-F., Wu, C.-W., and Lee, T.-C., *Mol. Mutagenet.,* 20, 156–164, 1992.
Yeh, H., *Hum. Pathol.,* 4, 469–485, 1973.
Zanzoni, F. and Jung, E. G., *Arch. Dermatol. Res.,* 267, 91–95, 1980.

Chapter 14

Cadmium Carcinogenicity and Genotoxicity

Michael P. Waalkes and R. Rita Misra

I. INTRODUCTION

Cadmium is a toxic transition metal of continuing occupational and environmental concern. Several reviews are available concerning various aspects of cadmium toxicity (IARC, 1976, 1993; Friberg et al., 1986a,b; Kazantzis, 1987; Waalkes and Oberdörster, 1990; Waalkes et al., 1992a). Cadmium has recently been designated a human carcinogen (IARC, 1993) and is clearly a potent animal carcinogen (IARC, 1976, 1994; Oberdörster, 1986; Waalkes and Oberdörster, 1990; Waalkes et al., 1992a). Occupational exposure to cadmium has been linked to lung cancers in humans, while at other sites, such as the prostate, linkage has not been definitively established (IARC, 1993; Waalkes and Oberdörster, 1990; Waalkes et al. 1992a). Several studies in rodents have shown that chronic inhalation of cadmium causes pulmonary carcinomas (Takenaka et al., 1983), in clear support of human data. Cadmium can also be a prostatic carcinogen after systemic or direct exposure (Waalkes et al., 1988a, 1989; Waalkes and Rehm, 1992; Hoffman et al., 1985a,b, 1988). Other target tissues of cadmium carcinogenesis in animals include injection sites (Heath et al., 1962; Kazantzis, 1963; Haddow et al., 1964), the testes (Gunn and Gould, 1970), and the hematopoietic system (Waalkes et al., 1992b, 1994; Waalkes and Rehm, 1992, 1994a,b). Certain treatments modify cadmium carcinogenicity, including administration of zinc, which prevents cadmium-induced injection site and testicular tumors while facilitating prostatic tumor formation (Gunn et al., 1963, 1964; Waalkes et al., 1989). Diets deficient in zinc increase the progression of testicular tumors (Waalkes et al., 1991a) but reduce the progression of prostatic tumors (Waalkes and Rehm, 1992). There are definite species- and strain-related differences in sensitivity to cadmium carcinogenicity (Heinrich et al., 1989; Waalkes et al., 1991b, 1994; Waalkes and Rehm, 1994a,b). The potential mechanism or mechanisms of cadmium carcinogenesis as yet are unknown but may well be target tissue specific.

Cadmium has an extremely long biological half-life which essentially makes the metal a cumulative toxin. To date there are no proven effective treatments for chronic cadmium intoxication. Cadmium accumulates primarily in the liver and kidney where it is bound to metallothionein, a low molecular weight metal binding protein thought to detoxify the metal through high-affinity sequestration (Waalkes and Goering, 1990). The toxic effects of cadmium often stem from interference with various zinc-mediated metabolic processes, and zinc treatments frequently reduce or abolish the effects of cadmium (Goering and Klaassen, 1984a; Gunn and Gould, 1970). There are several sources of human exposure to this toxic metal, including employment in primary metal industries and consumption of tobacco products (IARC, 1993; Friberg et al., 1986a,b).

II. PATHOBIOLOGIC CHARACTERISTICS OF CADMIUM

The metabolism of cadmium has several unique facets. The absorption of cadmium shows marked route dependency (Friberg et al., 1986a,b; Kazantzis, 1987). Absorption from the gastrointestinal tract is only ~5% of the ingested dose of cadmium although various dietary factors can elevate the amount absorbed (Friberg et al., 1986a,b; Kazantzis, 1987; Shaikh and Smith, 1980). In contrast, absorption of cadmium from the respiratory system is very high and as much as 90% of soluble cadmium compounds deposited in the deep lung are absorbed (Oberdörster, 1986). Cadmium does not accumulate to high levels in the lung and most of what is absorbed is distributed to other tissues (Oberdörster, 1986; Oberdörster and Kordel, 1981). Once absorbed, cadmium is rapidly cleared from blood into various tissues and has a high volume of distribution (Klaassen, 1981).

Hepatic and renal cadmium usually make up the bulk of the total body burden (Klaassen, 1981). The deposition of cadmium in the liver and kidney may be due to the ability of these organs to produce large amounts of metallothionein, a metal-binding protein with high affinity for cadmium (Klaassen, 1981; Waalkes and Klaassen, 1985). The biological half-life of cadmium is approximately 25 to 30 years in humans and cadmium is only very slowly eliminated from the body (Friberg et al., 1986a,b). This long residence time may well enhance the probability of neoplastic transformation (Waalkes et al., 1992a).

In high-dose acute exposures to cadmium, the tissue of first contact is most highly affected. For instance the lung is the critical organ after inhalation, while the gastric tract is the tissue most affected after ingestion (Waalkes et al., 1992c). With chronic exposure, the kidney is the critical organ for noncarcinogenic effects regardless of route (Friberg et al., 1986a,b; Foulkes, 1990; Waalkes et al., 1992c). Chronic respiratory disease has been associated with occupational exposure to cadmium fumes and dust (Friberg et al., 1986a,b; Kazantzis, 1987).

III. CHEMISTRY AND USES OF CADMIUM

Cadmium has a variety of industrial uses (see IARC, 1994; Waalkes et al., 1992a,c for review). Primary uses of cadmium include battery manufacture and metal coating by electroplating. Cadmium compounds are also used as pigments in paints, enamels, and plastics. Cadmium stearate finds use as a stabilizer for plastics. Cadmium can be also found in silver solder and welding electrodes.

Several excellent articles are available on the chemistry of cadmium (Aylett, 1979; Martell, 1981; Jacobson and Turner, 1980). The fact that zinc clearly plays a major role in the toxicity of cadmium may well be due to the similar chemical nature of cadmium and zinc and their common interactions within living systems. This similar chemistry, combined with the greater affinity of cadmium for various bioligands, probably allows cadmium to displace zinc in many biological processes. Cadmium intoxication can also affect calcium homeostasis and has been associated with the debilitative disease called "Itai-Itai", first observed in a group of Japanese women exposed to cadmium and characterized by severe osteoporosis (Nomiyama, 1980).

IV. HUMAN EXPOSURE TO CADMIUM

Increased industrial usage of cadmium has caused an increase in cadmium production and concomitant rise in contamination of soil, air, and water. The primary sources of cadmium contamination within the environment include mining, refining, and smelting operations. Additional sources include fossil fuel consumption, municipal waste incineration, agricultural use of sewage sludge and phosphate fertilizers, and other uses such as batteries, alloys, paints, and plastics (Aylett, 1979; Friberg et al., 1986a; Chmielnicka and Cherian, 1986). Less than 5% of the cadmium used in the U.S. is ever recycled (Willard, 1986) and in the essential absence of recycling, the continued usage of the metal has resulted in an increase in abundance of cadmium within the biosphere over the last several decades (Chmielnicka and Cherian, 1986).

It is estimated that ~1.5 million U.S. workers are potentially exposed to cadmium at the work place (NIOSH, 1984). Cadmium dusts are generated by mechanical processes whereas cadmium fumes can be generated by heating during metal smelting and processing or during electroplating. Industries utilizing large quantities of cadmium, such as in the production of nickel-cadmium batteries, can also represent sources of high exposure. In addition, welders can face high cadmium exposure levels. Improper use of silver- or cadmium-based solders can also be a significant source of exposure.

Inhaled tobacco smoke is another important source of human exposure to cadmium due to the ready absorption of cadmium by tobacco plants. It is estimated that smoking two packs per day can result in a doubling of the lifetime body burden of cadmium (Lewis et al., 1972; Friberg et al., 1986a). Clearly, this source of exposure to cadmium has important implications due to the association of cadmium with cancers of the lung.

Cadmium enters the food chain and concentrates within organisms because of its long biologic half-life. In individuals who do not smoke tobacco or work in cadmium-related industries, contamination of the food chain is the main source of exposure (Chmielnicka and Cherian, 1986). In general, meat byproducts, including liver and kidney, can be significant sources of cadmium, and in certain cases shellfish and other seafood are important sources (IARC, 1976; Friberg et al., 1986a; Chmielnicka and Cherian, 1986). Exposures to cadmium directly from drinking water and from the air are, in most cases, considered minor (Friberg et al., 1986a; Chmielnicka and Cherian, 1986). Industrialized areas provide the greatest chance of exposure to cadmium, whereas in rural areas exposure can be can be quite low.

V. CADMIUM CARCINOGENESIS

A. EPIDEMIOLOGY

The International Agency for Research on Cancer (IARC, 1993) has concluded that there is adequate evidence that cadmium is a human carcinogen based on all available data. The designation of cadmium as a Group 1 (human) carcinogen was prompted primarily by continued findings of an association between occupational cadmium exposure and lung cancer (Lemen et al., 1976; Thun et al., 1985; Stayner et al., 1992; IARC, 1993), as well as very strong rodent data which included the lung as a target site (IARC, 1993). Compounding factors in pulmonary carcinogenesis, such as exposures to other human carcinogens including arsenic, have now been adequately accounted for (Stayner et al., 1992; IARC, 1993). Thus the lung is clearly the most well-established site of human carcinogenesis from cadmium exposure (Thun et al., 1985; Leman et al., 1976; Elinder et al., 1985; Stayner et al., 1992).

In some studies, occupational or environmental cadmium exposure has also been associated with development of cancers of the prostate, kidney, liver, hematopoietic system, and stomach (Kipling and Waterhouse, 1967; Leman et al., 1976; Bako et al., 1982; Abd Elghany et al., 1990; Campbell et al., 1990; Kazantzis et al., 1988; Kolonel, 1976; Berg and Burbank, 1972). The linkage of cadmium exposure with human neoplasia of any specific site other than the lung has, however, not been absolutely established (IARC, 1976, 1993; Kazantzis, 1987; Waalkes and Oberdörster, 1990; Waalkes et al., 1992a). For instance, the role of cadmium in human prostatic cancer continues to be a matter of some controversy (Doll, 1992). Associations have been seen between prostatic cancer and occupational cadmium exposure (Kipling and Waterhouse, 1967; Kjellstrom et al., 1979; Lemen et al., 1976; Abd Elghany et al., 1990; West et al., 1991) and presumed environmental cadmium exposure (Bako et al., 1982). Many studies, however, show no association between prostatic cancer and cadmium (e.g., Kazantzis et al., 1988; Kazantzis and Blanks, 1992). Prostatic cancer has an extremely complex etiology which has proved very difficult to precisely define and this may make association with any one factor very difficult to discern (Piscator, 1981).

For the associations between cadmium and renal (Kolonel, 1976), hepatic (Campbell et al., 1990), gastric (Kazantzis, et al., 1988), or hematopoietic (Berg and Burbank, 1972) cancers, only a single report exists for each. In the absence of further verification these sites must be considered tentative at best. Clearly, further epidemiological and experimental work is necessary to determine the precise sites and exact carcinogenic risk of cadmium to humans.

B. ANIMAL STUDIES
1. Early Studies in Rodents

The earliest suspicion that cadmium might be carcinogenic came from the work of Haddow et al. (1961), who injected ferritin (s.c. or i.m.) which had been prepared from rat liver by cadmium precipitation and subsequently found malignant tumors at the site of injection in rats and mice. It was unclear at the time if cadmium was the causative agent, but it was indeed suspected (Haddow et al., 1961). This result prompted further studies and cadmium has now been recognized as a potent carcinogen in rodents for over 30 years (Heath et al., 1962). The carcinogenic potential of this metal was first established at repository-type injection sites, such as i.m. or s.c., where it forms sarcomas (Heath et al., 1962; Kazantzis, 1963; Haddow et al., 1964). Early studies also showed that cadmium is a very effective testicular tumorigen, with a single dose giving rise to a high incidence of Leydig cell tumors (Gunn et al., 1963,

1964; Roe et al., 1964). Initial mechanistic studies detected the profound ability of zinc treatments to prevent the carcinogenic effects of cadmium at both the injection site and within the testes (Gunn et al., 1963, 1964).

2. Pulmonary Tumors

Several studies have established that inhaled cadmium is a potent pulmonary carcinogen in the rat, a fact clearly supportive of its potential as a human pulmonary carcinogen (IARC, 1993). In the first such study, rats that had been chronically exposed by inhalation to cadmium chloride aerosols at doses up to 50 µg/m^3 for 16 months showed more than a 70% incidence of pulmonary carcinoma over 26 months (Takenaka et al., 1983). This effect of cadmium chloride was clearly dose related (Takenaka et al., 1983). Several other forms of cadmium, including forms perhaps more relevant to human exposures such as cadmium oxide, have subsequently been shown to be effective pulmonary carcinogens after chronic inhalation in rats (Oldiges et al., 1989; Glaser et al., 1990). Cadmium will also induce malignant pulmonary tumors after either continuous (Takenaka et al., 1983) or discontinuous (Oldiges et al., 1989) inhalation. As seen with the several other sites of cadmium carcinogenesis, zinc can prevent or reduce cadmium-induced lung tumors (Oldiges et al., 1989; Oberdörster and Cox, 1990).

In contrast to the activity in rat lung, inhaled cadmium does not appear to be an effective pulmonary carcinogen in the mouse or hamster (Heinrich et al., 1989). In a study on the chronic inhalation of several doses of various cadmium compounds ($CdCl_2$, CdO, CdS, $CdSO_4$), neither the mouse nor hamster showed overt neoplastic changes in the lung, with the possible exception of the mouse following inhalation of cadmium oxide fumes where a dose-related increase appeared to occur (Heinrich et al., 1989). All cadmium compounds did induce significant increases in bronchiolar-alveolar hyperplasia in both mice (Heinrich et al., 1989) and hamsters (Heinrich et al., 1989; Aufderheide et al., 1990). A variable baseline of pulmonary tumor incidence was a complicating factor in this study, at least in mice (Heinrich et al., 1989). Thus the rat seems to be one of the most sensitive rodent species to the pulmonary carcinogenic effects of inhaled cadmium.

The results with inhaled cadmium have been essentially duplicated with intratracheal instillations of several salts of cadmium in the rat, at least in one study (Pott et al., 1987). Pulmonary adenocarcinoma incidence was clearly elevated in rats after repeated intratracheal instillations of cadmium chloride, sulfide and oxide (Pott et al., 1987). On the other hand, Sanders and Mahaffey (1984) did not detect any pulmonary tumors after single or multiple intratracheal instillations of cadmium oxide. No studies using intratracheal instillation are available in the mouse or hamster.

Systemic cadmium exposure can have a mixed effect on pulmonary tumor incidence in rodents. In a recent study, pulmonary tumors in NFS mice were induced by a single s.c. exposure to cadmium but not by multiple exposures (Waalkes and Rehm, 1994a). Similarly, multiple i.p. injections of cadmium failed to modify the incidence or multiplicity of lung adenomas in strain A/Strong mice (Stoner et al., 1976). Other studies indicate that chronic oral cadmium exposure in the drinking water significantly reduces the incidence of lung tumors in Swiss (Schroeder et al., 1964) and B6C3F1 (Waalkes et al., 1991c; 1993) mice. Thus the capacity of cadmium to induce pulmonary tumors after systemic exposure is apparently related to various factors, perhaps including route of exposure and total dose, which are as yet not completely defined.

3. Prostatic Tumors

Prostatic cancer is an important and deadly human malignant disease of essentially undefined etiology. Several recent studies indicate that cadmium, given by various routes, can induce tumors of the prostate in rats (Waalkes et al., 1988a, 1989a; Waalkes and Rehm, 1992; Hoffman et al., 1985a,b, 1988). The ability of cadmium to induce prostate cancer appears to be highly dose dependent and also dependent on the effects of the metal on other tissues such as the testes.

An analysis of the carcinogenicity of a single s.c. injection of cadmium chloride in rats over two years, using a wide range of doses, showed prostatic tumor incidence was elevated only at doses of cadmium below the threshold for significant testicular toxicity (~5.0 µmol/kg; Waalkes et al., 1988a). At these lower doses a clear dose-related increase in prostatic tumors occurred (Waalkes et al., 1988a). Tumors were mainly adenomas, and were exclusively of the ventral lobe. Multiplicity of tumorous foci of the prostate showed a similar dose-relationship, and was elevated only at doses of cadmium ≤2.5 µmol/kg. On the other hand, multiplicity of preneoplastic, proliferative (hyperplastic) foci was increasingly elevated throughout most of the cadmium dosage range (0 to 20 µmol/kg), indicating that cadmium, at doses >2.5 µmol/kg, induced initiating events, but promotional factors (i.e., androgens) were not present at sufficient levels at these higher doses. Testicular production of androgens is essential for the growth and maintenance of the prostate and prostate tumors are frequently testosterone dependent, at least in the early stages (Coffey and Issacs, 1981). In rodents, testosterone alone will increase the incidence of prostatic carcinoma (Nobel, 1977) and androgens clearly can enhance the effectiveness of other prostatic carcinogens, such as N-nitroso-N-methylurea, N-nitrobis(2-oxopropyl)amine, and 3,2′-dimethyl-4-aminobiphenyl (Pour and Stepan, 1987; Bosland et al., 1983; Shirai et al., 1988). Thus the testicular toxicity of cadmium probably was responsible for the lack of prostatic tumorigenicity at cadmium doses resulting in such toxicity.

The effects of zinc pretreatment on cadmium carcinogenicity also indicate the carcinogenic potential for cadmium in the prostate. When zinc is given at doses sufficient to prevent cadmium-induced chronic degeneration in the testes, cadmium-induced prostatic tumors occurred (Waalkes et al., 1989). Doses of zinc that were ineffective in preventing the cadmium-induced testicular toxicity did not result in an elevation of prostatic tumors (Waalkes et al., 1989). Cadmium, when given i.m. at doses that did not result in chronic testicular degeneration, also induced an elevated incidence of prostatic tumors in this study (Waalkes et al., 1989), again indicating a dependence of tumor formation on appropriate testicular function. Prostatic tumors occurred only in the ventral prostate in this study as well (Waalkes et al., 1989).

Oral cadmium exposure also induces proliferative lesions of the rat ventral prostate (Waalkes and Rehm, 1992). When rats were fed cadmium mixed with diets adequate or marginally deficient in zinc (Waalkes and Rehm, 1992) an increase in the overall incidence of prostatic proliferative lesions (focal atypical hyperplasia and adenomas) occurred in rats fed cadmium (25–200 ppm) in zinc-adequate diets compared to the rats fed cadmium in zinc-deficient diets or in controls. The lower incidence of prostatic lesions in zinc-deficient rats was associated with a marked increase in prostatic atrophy, an indication of poor androgen support of the prostate. Prostatic proliferative lesions again occurred exclusively in the ventral lobe of the prostate. Zinc deficiency has a marked suppressive effect on testicular function and reduces its ability to support sex-accessory tissues such as the prostate (Prasad, 1982). Thus, it is evident that testicular function is important in cadmium induction of prostatic tumors, whether cadmium is given orally (Waalkes and Rehm, 1992) or parenterally (Waalkes et al., 1988a, 1989).

Direct injection of cadmium into the rat ventral prostate has also been shown to produce malignant prostatic adenocarcinomas. Hoffman et al. (1985a,b) gave rats single injections of cadmium chloride into the right lobe of the ventral prostate and the first case of invasive prostatic carcinoma was detected only 56 days after the injection. Eventually 5 cases of invasive carcinoma occurred in the 100 rats examined (Hoffman et al., 1985a,b). Other proliferative lesions resulting from cadmium exposure included 11 cases of carcinoma *in situ,* 29 cases of atypical hyperplasia, and 38 of simple hyperplasia (Hoffman et al., 1985a,b). In 20 control rats, 5 had simple hyperplasia (25%) and 1 rat had atypical hyperplasia (5%). Hoffman et al. (1988) also studied the effects of repeated (two or three) injections of cadmium into the rat ventral prostate and over 9 months. Two prostatic carcinomas occurred in eight rats (25%) in the group receiving two injections, while of those treated with three injections fully 60% (9/15) had prostatic adenocarcinoma.

Cadmium can also enhance the appearance of prostatic tumors in rats. Shirai et al., (1993) recently found in rats that a combination of cadmium and 3,2′-dimethyl-4-aminophenyl given i.m. acted synergistically to induce prostatic carcinomas. Again, these tumors occurred exclusively in the ventral lobe of the prostate.

Thus, chronic studies in rats establish that the rat ventral prostate is a target site for cadmium induction of neoplasia. It is noteworthy that cadmium induces the full spectrum of prostatic proliferative lesions in rats, including invasive adenocarcinomas. The finding of prostatic tumors in cadmium-treated rats clearly supports a possible role in human prostatic neoplasia.

4. Testicular Tumors

The rodent testes are extremely sensitive to cadmium. Relatively low doses of cadmium can rapidly induce severe testicular hemorrhagic necrosis, despite the fact that very little cadmium actually reaches the testes (Gunn and Gould, 1970). After the initial toxic lesion from a parenteral cadmium treatment, a high incidence of interstitial cell (i.e., Leydig cell) tumors is observed in rats (Gunn et al., 1963, 1964; Roe et al., 1964; Waalkes et al., 1988a, 1989, 1991a). Recent work also indicates that oral cadmium exposure can result in interstitial cell tumors in the rat testes (Waalkes and Rehm, 1992). The development of testicular tumors in rats is thought to be at least partially related to the chronic degenerative effects of cadmium in this tissue (Gunn et al., 1963, 1964; Waalkes et al., 1989), though they can occur in the absence of such degenerative lesions (Waalkes and Rehm, 1992).

Only one study has reported the occurrence of cadmium-induced interstitial cell tumors in mice (Gunn et al., 1963) although some strains show proliferative (hyperplastic) lesions within the testes (Roe et al., 1964; Waalkes et al., 1994). Hamsters also do not develop interstitial tumors but do show a high incidence of testicular hyperplasia with cadmium treatment (Waalkes et al., 1994). Thus, as is the case with the lung, the rat appears to be the most sensitive rodent species for induction of tumors of the testes. Rare testicular tumors have been associated with systemic cadmium exposure in rats or mice including seminomas, rete testes adenocarcinomas, mixed Sertoli-Leydig cell tumors, and leiomyosarcoma (Rehm and Waalkes, 1988; Boorman et al., 1987; Waalkes and Rehm, 1994a) although not in sufficiently high numbers to draw definitive conclusions about cadmium causation. Cadmium directly injected into the testes also induces teratomas (Guthrie, 1964).

5. Injection Site Sarcomas

Malignant tumors develop at the site of repository-type injections of cadmium (Heath et al., 1962; Kazantzis, 1963; Haddow et al., 1964; Poirier et al., 1983; Waalkes et al., 1988a, 1989). Cadmium induces tumors when injected s.c., i.m., or subperiosteally (Gunn et al., 1967) and the tumors produced at these sites are typically fibrosarcomas. Injection site sarcomas induced by cadmium appear to be strictly related to total accumulated dosage at the site, and several studies have shown dose-related increases in sarcoma formation with cadmium exposure (Waalkes et al., 1988a, 1989). The formation of injection site sarcomas is a common occurrence with many metals and could be related, in part, to a solid state phenomenon where malignancies form during encapsulation of a chronically irritating implant, and are due to the physical rather than chemical nature of the implant (Waalkes and Oberdörster, 1990). Cadmium at the s.c. injection sites do form a distinct calcified area. However, several studies argue against an exclusively solid-state mechanism. Repeated injections cause a marked increase in the rate of distant metastases of cadmium-induced injection site tumors, indicating a dose-dependent modification of cellular characteristics (Waalkes et al., 1988b). Furthermore, zinc can markedly reduce cadmium induction of injection site tumors (Gunn et al., 1963, 1964), even when given by a totally different route (Waalkes et al., 1989), indicating antagonism of the effects of cadmium by zinc as a mechanism for reduction of carcinogenesis. The strain of rats or mice has a pronounced effect on the final incidence and latency of cadmium induction of injection site sarcomas (see below; Waalkes et al., 1988b, 1991b, 1994; Waalkes and Rehm, 1994a,b), indicating a genetic basis of susceptibility. In fact mice are generally much less sensitive than rats to cadmium induction of injection site sarcoma (Waalkes et al., 1988a,b, 1989, 1991b, 1994; Waalkes and Rehm, 1994a,b), including mouse strains that are normally sensitive to solid-state carcinogenesis in the subcutis (Brand et al., 1977). Thus a solid-state mechanism for cadmium carcinogenicity at the site of injection does not fully explain the occurrence of these tumors.

6. Hematopoietic Tumors

Cadmium can affect tumors of the hematopoietic system in rodents. Oral exposure to cadmium in mice infected with lymphocytic leukemia virus increased death from leukemia by over 30% (Blakley, 1986). This was attributed to cadmium-impaired immunosurveillance allowing emergence of the virus. In rats, oral cadmium can induce a dose-related six-fold increase of the incidence of leukemia (Waalkes et al., 1992b; Waalkes and Rehm, 1992). Dose-related increases in lymphoma have been shown to be induced by s.c. injections of cadmium in certain strains of mice (Waalkes et al., 1994; Waalkes and Rehm, 1994b). In contrast, a single high-dose subcutaneous injection of cadmium (30 µmol/kg) markedly decreased the spontaneous incidence of LGL leukemia in Fischer rats (Waalkes et al., 1991b). Hence, cadmium appears to have a possible role in both the induction and suppression of tumors of the hematopoietic system.

7. Tumors At Other Sites

Cadmium has a mixed effect on tumors of the pancreas in rats. The rat pancreas, depending on strain, often shows a fair incidence of spontaneous tumors. Cadmium, when given over a wide range of s.c. dosages, can cause a dose-related reduction of endocrine and exocrine pancreatic tumors in Wistar rats (Waalkes et al., 1988a). In contrast, when cadmium is given s.c. concurrently with calcium, an elevated incidence of islet cell tumors of the rat pancreas has been shown (Poirier et al., 1983). Thus cadmium has variable effects on the incidence of tumors of the rat pancreas, depending at least in part on conditions of exposure. Recent work indicates that multiple s.c. injections of cadmium in rats results in marked transdifferentiation of pancreatic cells into hepatocytes (Konishi et al., 1990). The role of these metaplastic lesions in the formation or prevention of pancreatic neoplasia by cadmium is unknown.

Induction of tumors of the adrenals has been reported with s.c. cadmium treatment in hamsters in one study (Waalkes et al., 1994) and the incidence of proliferative lesions of the adrenal cortex approached 55% in these animals. There is other evidence of cadmium toxicity in the adrenal cortex of rodents. For instance, in rats cadmium stimulates adrenal DNA synthesis concurrently with hypertrophy of the gland (Nishiyama and Nakamura, 1984). However, the relevance of the finding of proliferative lesions in the adrenal cortex of hamsters, induced by cadmium, will require confirmation and further study.

Pott et al. (1987) gave rats i.p. injections of cadmium oxide or cadmium sulfide and induced malignant peritoneal cavity tumors. Of the rats given cadmium oxide, 3/47 had peritoneal cavity tumors, while in rats given cadmium sulfide 54/81 had such tumors. Tumors were sarcomas, mesothelioma, or carcinoma (Pott et al., 1987).

Sanders and Mahaffey (1984) gave groups of male rats single or multiple intratracheal instillations of cadmium oxide and found significant increases in mammary gland fibroadenomas associated with the treatment. Other studies have not found the mammary gland to be a target of cadmium.

8. Metal-Metal Interactions in Cadmium Carcinogenesis

Zinc can clearly have an important impact on cadmium carcinogenesis. In several tissues, including the lung, testes, and at the injection site, zinc treatment ameliorates the effects of cadmium (Gunn et al., 1963, 1964; Waalkes et al., 1989; Oldiges et al., 1989). Calcium and magnesium, are relatively ineffective in reducing the carcinogenic effects of cadmium compared to zinc (Poirier et al., 1983). This selective antagonism by zinc of the carcinogenic effects of cadmium at so many different target sites could point to a basic mechanism of cadmium carcinogenesis. Other metals or metalloids which have shown the capacity to reduce the acute toxic effects of cadmium, such as selenium (Wahba et al., 1993), have not been studied for inhibition of carcinogenesis. In contrast to inhibition in some tissues, zinc treatment can actually facilitate cadmium carcinogenesis in the prostate, probably through protection of testes and the consequent maintenance of androgen support (Waalkes et al., 1989). Dietary zinc deficiency clearly reduces the carcinogenic effects of cadmium in the prostate, since prostatic proliferative lesions induced by oral exposure to cadmium in rats are reduced with reduced zinc intake (Waalkes and Rehm, 1992). Zinc deficiency induces atrophy of the prostate, which again may be due to a reduction in testicular androgen secretion (Waalkes and Rehm, 1992). A diet deficient in zinc enhances the progression of testicular lesions and increases the incidence of injection site sarcomas induced by cadmium in rats (Waalkes et al., 1991a). Thus zinc can either facilitate or inhibit cadmium carcinogenesis, depending on the tissue in question.

9. Synergism and Antagonism of Carcinogenesis By Cadmium

Cadmium can both antagonize and enhance the carcinogenic effects of organic carcinogens. For instance, a synergistic increase in lung tumors occurred in rats exposed to combined cadmium, N-nitrosoheptamethyleneimine, and crocidolite asbestos fibers (Harrison and Heath, 1986). Chronic oral cadmium also enhanced the appearance of preneoplastic renal dysplasias in rats that previously received N-ethyl-N-hydroxyethylnitrosamine (Kurokawa et al., 1985), while the incidence of diethylnitrosamine (DEN)-induced hepatic and renal tumors was markedly enhanced in rats by cadmium injections given soon after DEN (Wade et al., 1987). Shirai et al. (1993) have also recently found that a combination of cadmium and 3,2'-dimethyl-4-aminophenyl acts synergistically to induce prostatic tumors in the rat.

In marked contrast, chronic oral treatment with cadmium starting two weeks after DEN exposure markedly reduced liver and lung tumor incidence in the mouse (Waalkes et al., 1991c). Kurokawa et al. (1989) similarly tested the effects of cadmium in the liver of rats treated with DEN and found that cadmium in the drinking water also significantly reduced the incidence of hepatocellular carcinomas. This suppression of liver and lung tumors by cadmium can occur even when the metal is given after tumor formation and may have therapeutic potential (Waalkes et al., 1993). Oral cadmium exposure also suppressed spontaneously occurring liver tumors in mice (Schroeder et al., 1964; Waalkes et al., 1991c), indicating suppression is dependent on the tumor type rather than the precise carcinogen. It now appears that liver tumors may have unusually low levels of metallothionein, which could render them highly sensitive to cadmium cytotoxicity (Waalkes et al., 1993).

Another study indicates that cadmium can suppress radiation-induced tumors. When Schmahl et al. (1986) assessed the transplacental effects of cadmium in rats on induction of pituitary tumors by radioisotopic strontium they found that cadmium treatment produced a much lower incidence of pituitary adenomas.

Clearly cadmium can have a pronounced, and sometimes contrary, effect on carcinogenicity of other compounds, and the impact of such interactions can be quite dramatic but are as yet poorly defined. Chronic animal experiments of combined exposures to multiple carcinogenic metals, such as would be seen most frequently with human exposures, have unfortunately not been performed with cadmium. The role of cadmium as a promoter or cocarcinogen is as yet poorly defined and deserves further attention.

10. Molecular Basis of Tolerance or Sensitivity to Cadmium Carcinogenicity

It is quite clear that metallothionein plays an important role in the tolerance to cadmium in many cases. Cadmium has a very high binding affinity for metallothionein and it is thought that cadmium bound to metallothionein within the cell is relatively inert and therefore nontoxic (Waalkes and Goering, 1990; Goering and Klaassen, 1984a,b). The expression of the metallothionein gene is stimulated by both cadmium and zinc in most tissues (Waalkes and Goering, 1990; Waalkes and Klaassen, 1985). It thus seems a fair conclusion to presume that resistance or sensitivity to cadmium carcinogenesis would in some way involve expression of metallothionein. Accumulating evidence from several laboratories indicates that several target sites of cadmium carcinogenesis may actually be deficient in metallothionein (Deagan and Whanger, 1985; Waalkes et al., 1984, 1988c,d, 1992d; Waalkes and Perantoni, 1986, 1989; Ohata et al., 1988; Kaur et al., 1993) or that the metallothionein gene is not sensitive to metal stimulation (Wahba et al., 1993; Coogan et al., 1994a; Waalkes et al., 1992d). This appears to be the case with the rat ventral prostate (Coogan et al., 1994b; Waalkes et al., 1992d; Tohyama et al., 1993), the specific lobe where cadmium induces tumors (Waalkes et al., 1988a, 1989; Waalkes and Rehm, 1992; Hoffman et al., 1988; Shirai et al., 1993). The ventral prostatic metallothionein gene shows minimal basal expression compared to liver or dorsal prostate, which are both typically nontarget sites of cadmium carcinogenesis (Coogan et al., 1994a; Waalkes et al., 1992d). Immunohistochemical evidence also indicates minimal expression of metallothionein in the rat ventral lobe (Tohyama et al., 1993). Perhaps more importantly, the metallothionein gene is not stimulated by cadmium in the ventral prostate, unlike the gene in liver or dorsal prostate (Coogan et al., 1994b; Waalkes et al., 1992d). Metallothionein expression has been shown to reduce cadmium genotoxicity *in vitro* (Coogan et al., 1992, 1994b) while cells poorly expressing the metallothionein gene are quite sensitive to the genotoxic effects of the metal (Hochadel et al., 1994). Thus, the poor expression of metallothionein may play an important role in the sensitivity of certain tissues to cadmium carcinogenesis. Clearly, further research in this area may lead to important findings on the mechanism of cadmium carcinogenicity.

11. Effects of Species and Strain on Cadmium Carcinogenicity

Cadmium has clear differential effects depending on species and strain. Differing responses and susceptibility of various rodent species to inhaled cadmium carcinogenicity are quite evident as rats are vulnerable, whereas mice and hamsters are generally resistant. Clear strain-dependent differences in incidence and latency of cadmium-induced injection site sarcomas in rats have occurred (Waalkes et al., 1991b) and again mice and hamsters appear to be typically less sensitive to these tumors (Waalkes and Rehm, 1994a,b; Waalkes et al., 1994a). Differences in distribution of cadmium probably would have little bearing on these particular tumors, indicating a genetic basis for susceptibility. There is clear evidence that sensitivity to acute toxic effects of cadmium has a genetic basis in mice (Gunn et al., 1965; Taylor et al., 1973; Chellman et al., 1984, 1985; Waalkes et al., 1988d, Kershaw and Klaassen, 1991), and the potential for cadmium carcinogenesis on the whole is lower in acutely resistant strains than in acutely susceptible strains (Waalkes and Rehm, 1994a,b; Waalkes et al., 1994a). Further work is warranted to determine the genetic basis of sensitivity to cadmium carcinogenesis and such information could provide the opportunity to determine molecular mechanisms.

VI. GENOTOXICITY OF CADMIUM

Over the last 15 years the genotoxic effects of cadmium have been studied in a wide variety of experimental systems. Unfortunately, no consensus on how cadmium induces genetic damage has yet emerged. At least three different hypotheses regarding this metal's mode of action currently exist: (1) cadmium may interact directly with chromatin to induce strand breakage, cross-linking, or conformational changes in DNA; (2) cadmium may act indirectly, by inhibiting various proteins involved in DNA repair; (3) cadmium may act by catalyzing cellular redox reactions whose byproducts subsequently produce strand breaks, cross-links, or covalent adducts in DNA. Key findings from studies of cadmium-induced genotoxicity are discussed below.

In the early 1980s, investigators showed that cadmium ions could bind to bases and phosphate groups in purified DNA (Jacobson and Turner, 1980; Waalkes and Poirier, 1985). However, experiments conducted more recently indicate that additional factors, such as metal-protein complex formation, may be required in order to produce significant levels of DNA damage *in vivo* (Müller et al., 1991; Rossman et al., 1992).

In 1983, Ochi et al. were the first investigators to demonstrate that soluble cadmium could induce DNA damage in eukaryotic cells. Using the alkaline elution technique, they showed that $CdCl_2$ exposure led to increased production of single-strand breaks in the DNA of V79 hamster cells (Ochi et al., 1983). Furthermore, marked reductions in the induction of strand breakage were observed when $CdCl_2$ was administered under anaerobic conditions, or when superoxide dismutase (SOD) was added to the culture medium. These results suggested that an active oxygen species may be ultimately responsible for the metal's genotoxic effects. Two years later, investigators in the same laboratory showed that the incidence of chromosomal aberrations was dramatically reduced in V79 cells that received catalase, D-mannitol, or butylated hydroxytoluene, prior to cadmium exposure (Ochi and Ohsawa, 1985). However, in contrast to the results from the earlier work, pretreatment with SOD offered no protection against cadmium genotoxicity (Ochi and Ohsawa, 1985). Using a human fibroblast line, Snyder (1988) later confirmed that cadmium-induced chromosomal damage could be offset by pretreatment with catalase, but not SOD. Taken together, these results supported the theory that cadmium acts by stimulating the production of hydrogen peroxide which, in turn, forms highly reactive hydroxyl radicals in the presence of intracellular iron or copper.

The apparent discrepancies between the results obtained from different experiments with cadmium and SOD (Ochi et al., 1983; Ochi and Ohsawa, 1985) suggest that cadmium-induced production of single-strand DNA breaks and chromosomal aberrations may occur by different mechanisms. Alternatively, these discrepancies may simply reflect dose-related differences in the type of DNA damage produced; higher doses of cadmium were used in the earlier study (Ochi et al., 1983). It is well known that high doses of $CdCl_2$ ($>2 \times 10^{-5}$ M) can stimulate lipid peroxidation (Stacey et al., 1980), or decrease the levels of various antioxidants in cells (Ochi et al., 1987). Such effects could, in turn, alter the profile of cadmium-induced oxidative damage in target tissues. One caveat for all of the aforementioned studies is that DNA damage was measured using doses of cadmium that were extremely toxic to cells. It is therefore quite conceivable that other biologically relevant lesions, (e.g., conformationally altered chromatin, DNA-protein cross-links, or specific DNA adducts) may have been completely overlooked in such experiments.

Another way to evaluate the genotoxicity of suspected carcinogens is by monitoring their ability to transform cells. Two different studies of cadmium-induced cell transformation have been conducted thus far (Rivedal and Sanner, 1981; Terracio and Nachtigal, 1986). In the first study, Rivedal and Sanner (1981) demonstrated that cadmium could initiate and promote the morphological transformation of cultured hamster embryo cells; in this study, the metal's mode of action depended upon the timing of cadmium administration. This dual activity of cadmium was later confirmed in a study where a single, prolonged exposure to $CdCl_2$ resulted in the complete transformation of rat ventral prostate cells (Terracio and Nachtigal, 1986). This is a highly significant finding as the rat ventral prostate is a known target for cadmium carcinogenicity (Waalkes et al., 1988a, 1989; Waalkes and Rehm, 1992; Hoffman et al., 1985a,b, 1988; Shirai et al., 1993). Eventually, these same transformed cell lines proved capable of forming metastatic tumors when injected into newborn rats (Terracio and Nachtigal, 1988). Exactly how cadmium achieves transformation, however, awaits further investigation.

Surprisingly few studies have been conducted on the mutagenicity of cadmium in eukaryotic cells. Existing evidence suggests, however, that cadmium is mutagenic at concentrations well below those used to induce direct damage in DNA. For example, dose-related increases in mutation frequency have been observed in the HPRT locus of V79 cells exposed to $1-3 \times 10^{-6}$ M $CdCl_2$ (Ochi and Ohsawa, 1983) while studies monitoring strand breakage in the same cell line were performed at $CdCl_2$ doses that were more than ten times higher. Limited insight into the mechanisms underlying cadmium mutagenicity has also been provided by experiments where $CdCl_2$ proved to have an additive, rather than synergistic, effect on the mutagenicity of activated benzo(*a*)pyrene (Ochi and Ohsawa, 1983). Such results suggest that cadmium mutagenesis proceeds by a mechanism that is substantially different from that of bulky DNA adducts.

More recently, Biggart and Murphy (1988) characterized the mutational spectrum of cadmium in rat kidney cells containing a stably integrated reporter gene. By analyzing the mutant proteins and RNA transcripts collected from five revertant colonies, these investigators discovered that extremely low doses of cadmium can induce deletions which may vary from 1 to 300 base pairs in size (Biggart and Murphy, 1988). Although the number of mutants analyzed was extremely limited, such results provided indirect evidence that cadmium may interact with proteins involved in repair or recombination of the genetic material. Additional evidence for this so-called "epigenetic" activity of cadmium has been provided by a number of other laboratories as well. For example, Scicchitano and Pegg (1987) had previously reported that cadmium had a strong inhibitory effect on the ability of O^6-methylguanine methyl transferase to remove promutagenic adducts from purified DNA. In other experiments, Hartwig and Beyersmann (1989) demonstrated that $CdCl_2$ exposure could reduce the incidence of UV-induced chromosomal breaks; according to their experimental scheme, such breaks were directly associated with the repair of UV-induced DNA damage. Most recently, Yamada et al. (1993) showed that cadmium exposure can lead to an increased incidence of induced chromosomal aberrations, depending upon the repair capacity of the host cell line.

The first study aimed at elucidating the exact mechanism(s) underlying cadmium mutagenesis dates back to 1976. In that study, 31 different metal salts were ranked on their ability to alter the fidelity of DNA synthesis *in vitro* (Sirover and Loeb, 1976). Of the metals tested, all ten known carcinogens (including cadmium) reduced the accuracy of the DNA polymerase. In comparison, all 17 noncarcinogenic metals had no effect on the polymerase, and the three "possible carcinogens" produced mixed results. In subsequent work, Nocentini (1987) used a variety of experimental techniques to demonstrate that micromolar concentrations of cadmium could inhibit replication and transcription of the DNA template in normal and UV-damaged mammalian cells. In this same study, Nocentini (1987) also discovered that zinc could counteract such effects if it was administered at concentrations that were five to ten times higher than the concentration of cadmium. This is noteworthy given the ability of zinc to inhibit the chronic carcinogenic effects of cadmium in the rat or mouse (Gunn et al., 1963, 1964; Waalkes et al., 1989). Such findings further support the position that cadmium may produce genetic damage by interfering with normal DNA metabolism.

It is widely believed that proteins involved in regulating eukaryotic gene expression are composed of specific amino acid "motifs" which allow the protein to recognize particular regions of DNA. One such motif is the zinc-finger loop. Examples of nuclear proteins known to contain this motif include oncogene products, transcription coupling factors, and enzymes involved in DNA repair (Sunderman and Barber, 1988; Tanaka et al., 1990; Rhodes and Klug, 1993). It is quite conceivable that cadmium may exert genotoxicity by substituting for zinc in these proteins thus forming coordination complexes with thiol-sulfur and imidazole-nitrogen atoms, thereby altering the conformation of the protein's active site, or via free-radical formation, damaging specific regions of the genome that are critical for proper replication and/or differentiation of cells (Sunderman and Barber, 1988).

It has recently been demonstrated that extremely low doses of cadmium can stimulate DNA replication (Lohmann and Beyersmann, 1993) and/or prevent apoptosis (Von Zglinicki et al., 1992). In light of these results and the experimental data collected thus far, it is tempting to speculate that low doses of cadmium may achieve genotoxicity by compromising the cell's ability to cope with spontaneously occurring DNA damage. At high doses, however, cadmium may act by damaging DNA directly, or by stimulating the production of reactive intermediates which subsequently attack the genetic material. In any case, it is likely that the biological outcome of cadmium exposure at the level of the whole animal will ultimately be determined by the balance of such processes within particular target cells. For instance, Coogan et al. (1992, 1994a) have shown that induction of metallothionein can lead to a marked reduction in the incidence of cadmium-induced DNA strand breakage. In other experiments, R2C cells, a rat testicular Leydig cell line in which the metallothionein gene appears to be quiescent and nonresponsive to metal induction stimuli, proved to be extremely sensitive to cadmium-induced DNA strand breakage at levels of cadmium that were clearly noncytotoxic (Hochadel et al., 1994). Metallothionein is a good example of one cell-specific factor which may play an important role in modulating the genotoxic effects of cadmium.

VII. SUMMARY

The carcinogenic effects of cadmium have now been clearly established in humans and in experimental animals, but further epidemiological efforts are warranted in order to determine more precisely the risks and target tissues in humans. The mechanism or mechanisms of cadmium carcinogenesis remain unknown but there is a pronounced target site specificity in rodents. This pronounced site specificity for cadmium may be related to target cell-specific factors, such as poor expression of the metallothionein gene. As yet, no consensus on the molecular events associated with cadmium-induced genetic damage has emerged, and this too is an area deserving further effort.

REFERENCES

Abd Elghany, N., Schumacher, M. C., Slattery, M. L., West, D. W., and Lee, J. S., *Epidemiology,* 1, 107–115, 1990.
Aufderheide, M., Mohr, U., Thiedemann, K.-U., and Heinrich, U., *Toxicol. Environ. Chem.,* 27, 173–180, 1990.
Aylett, B. J., in *The Chemistry, Biochemistry and Biology of Cadmium,* Webb, M., Ed., Elsevier, New York, 1979.
Bako, G., Smith, E. S. O., Hanson, J., and Dewar, R., *Can. J. Public Health,* 73, 92–96, 1982.
Berg, J. W. and Burbank, F., *Ann. N.Y. Acad. Sci.,* 199, 249–264, 1972.
Biggart, N. W. and Murphy, E. C., Jr., *Mutat. Res.,* 198, 115–129, 1988.
Blakley, B. R., *J. Appl. Toxicol.,* 6, 425–429, 1986.
Boorman, G., Rehm, S., Waalkes, M. P., Elwell, M. R., and Eustis, S. L., in *Pathology of Laboratory Animals,* Vol. 5, Genital System, Jones, T. C., Mohr, U., and Hunt, R. D., Eds., Springer-Verlag, New York, 192, 1987.
Bosland, M. C., Prinsen, M. K., and Kroes, R., *Cancer Lett.,* 18, 69–78, 1983.
Brand, I., Buoen, L. C., and Brand, K. G., *J. Natl. Cancer Inst.,* 58, 1443–1447, 1977.
Campbell, T. C., Chen, J., Liu, C., Li, J., and Parpia, B., *Cancer Res.,* 50, 6882–6893, 1990.
Chellman, G. J., Shaikh, Z. A., and Baggs, R. B., *Toxicology,* 30, 157–169, 1984.
Chellman, G. J., Shaikh, Z. A., Baggs, R. B., and Diamond, G. L., *Toxicol. Appl. Pharmacol.,* 79, 511–523, 1985.
Chmielnicka, J. and Cherian, M. G., *Biol. Trace Elem. Res.,* 10, 243–262, 1986.
Coffey, D. S. and Isaacs, J. T., *Urology,* Suppl. 3, 17–24, 1981.
Coogan, T. P., Bare, R. M., Bjornson, E. J., and Waalkes, M. P., *J. Toxicol. Environ. Health,* 41, 129–141, 1994a.
Coogan, T. P., Bare, R. M., and Waalkes, M. P., *Toxicol. Appl. Pharmacol.,* 113, 227–233, 1992.
Coogan, T. P., Shiraishi, N., and Waalkes, M. P., *Environ. Health Perspect.,* 1994b.
Deagen, J. T. and Whanger, P. D., *Biochem. J.,* 231, 279–285, 1985.
Doll, R., in *Cadmium in the Human Environment: Toxicity and Carcinogenicity,* Nordberg, G.F., Alessio, L., and Herber, R.F.M., Eds., IARC Sci. Publ., International Agency For Research on Cancer, Lyon, France, 1992.
Elinder, G.-C., Kjellstrom, T., Hogstedt, C., Andersson, K., and Spang, G., *Br. J. Ind. Med.,* 42, 651–655, 1985.
Foulkes, E. C., *Crit. Rev. Toxicol.,* 20, 327, 1990.
Friberg, L., Elinder, C.-G., Kjellström, T., and Nordberg, G. F., *Cadmium and Health: A Toxicological and Epidemiological Appraisal,* Vol. I and II, CRC Press, Boca Raton, FL, 1986a.
Friberg, L., Kjellström, T., and Nordberg, G. F., in *Handbook of the Toxicology of Metals,* 2nd ed., Vol. II, Friberg, L., Nordberg, G.F., and Vouk, V., Eds., Elsevier, Amsterdam, 130, 1986b.
Glaser, U., Hochrainer, D., Otto, F. J., and Oldiges, H., *Chem. Environ. Toxicol.,* 27, 153–162, 1990.
Goering, P. L. and Klaassen, C. D., *Toxicol. Appl. Pharmacol.,* 74, 308–313, 1984a.
Goering, P. L. and Klaassen, C. D., *J. Toxicol. Environ. Health,* 14, 803–812, 1984b.
Goering, P. L. and Klaassen, C. D., *Toxicol. Appl. Pharmacol.,* 74, 299–307, 1984.
Gunn, S. A. and Gould, T. C., in *The Testes, Influencing Factors,* Vol. III, Johnson, A.D., Gomes, W.R., and Vandemark, N.L., Eds., Academic Press, New York, 377, 1970.
Gunn, S. A., Gould, T. C., and Anderson, W. A. D., *J. Natl. Cancer Inst.,* 31, 745–753, 1963.
Gunn, S. A., Gould, T. C., and Anderson, W. A. D., *Proc. Soc. Exp. Biol. Med.,* 115, 653–657, 1964.
Gunn, S. A., Gould, T. C., and Anderson, W. A. D., *Arch. Pathol.,* 83, 493–499, 1967.
Gunn, S. A., Gould, T. C., and Anderson, W. A. D., *J. Reprod. Fertil.,* 10, 273–275, 1965.
Guthrie, J., *Br. J. Cancer,* 18, 255–260, 1964.
Haddow, A., Dukes, C. E., and Mitchley, B. C. V., *Annu. Rep. Br. Empire Cancer Campaign,* 37, 74–76, 1961.
Haddow, A., Roe, F. J. C., Dukes, C. E., and Mitchley, B. C. V., *Br. J. Cancer,* 18, 667–673, 1964.
Harrison, P. T. C. and Heath, J. C., *Carcinogenesis,* 7, 1903–1908, 1986.
Hartwig, A. and Beyersmann, D., *Biol. Trace Elem. Res.,* 21, 359–365, 1989.
Heath, J. C., Daniel, I. R., Dingle, J. T., and Webb, M., *Nature,* 193, 592–593, 1962.
Heinrich, U., Peters, L., Ernst, H., Rittinghausen, S., Dasenbrock, C., and König, H., *Exp. Pathol.,* 37, 253–258, 1989.
Hochadel, J. F., Shiraishi, N., Coogan, T. P., Koropatnick, J., and Waalkes, M. P., *The Toxicologist,* 14, 1994.
Hoffmann, L., Putzke, H.-P., Simonn, C., Gase, P., Russbült, R., Kampehl, H.-J., Erdmann, T., and Huckstorf, C., *Z. Ges. Hyg.,* 31, 224–227, 1985a.
Hoffman, L., Putzke, H.-P., Bendel, L., Erdmann, T., and Huckstorf, C., *J. Cancer Res. Clin. Oncol.,* 114, 273–278, 1988.
Hoffmann, L., Putzke, H.-P., Kampehl, H.-J., Russbült, R., Gase, P., Simonn, C., Erdmann, T., and Huckstorf, C., *J. Cancer Res. Clin. Oncol.,* 109, 193–199, 1985b.
IARC, Monographs: Vol. 11, Cadmium, Nickel, Some Epoxides, Miscellaneous Industrial Chemicals and General Considerations on Volatile Anesthetics, International Agency for Research on Cancer, Lyon, France, 39, 1976.
IARC, Monographs: Vol. 58, Beryllium Cadmium, Mercury, Beryllium and in the Glass Manufacturing Industry, International Agency for Research on Cancer, Lyon, France, 1994.
Jacobson, K. B. and Turner, J. E., *Toxicology,* 16, 1–37, 1980.
Kaur, G., Nath, R., and Gupta, G. S., *J. Trace Elem. Exp. Med.,* 6, 1–13, 1993.

Kazantzis, G., , in *Advances in Modern Toxicology,* Vol. XI, Fishbein, L., Furst, A., and Mehlman, M.A., Eds., Princeton Scientific, Princeton, NJ, 127, 1987.
Kazantzis, G., *Nature,* 198, 1213–1214, 1963.
Kazantzis, G. and Blanks, R. G., in *7th International Cadmium Conference,* Cook, M. E., Hiscock, S. A., Morrow, H., and Volpe, R. A., Eds., Cadmium Association, London, 150, 1992.
Kazantzis, G., Lam, T.-H., and Sullivan, K. R., *Scand. J. Environ. Health,* 14, 220–223, 1988.
Kershaw, W. C. and Klaassen, C. D., *Chem. Biol. Interact.,* 78, 269–282, 1991.
Kipling, M. D. and Waterhouse, J. A. H., *Lancet,* i, 730–731, 1976.
Klaassen, C. D., *Fundam. Appl. Toxicol.,* 1, 353–357, 1981.
Kolonel, L. N., *Cancer,* 37, 1782–1787, 1976.
Konishi, N., Ward, J. M., and Waalkes, M. P., *Toxicol. Appl. Pharmacol.,* 104, 149–156, 1990.
Kurokawa, Y., Matsushima, M., Imazawa, T., Takamura, N., Takahashi, M., and Hayashi, Y., *J. Am. Coll. Toxicol.,* 4, 321–330, 1985.
Kurokawa, Y., Takahashi, M., Maekawa, A., and Hayashi, Y., *J. Am. Coll. Toxicol.,* 8, 1235–1239, 1989.
Lemen, R. A., Lee, J. S., Wagoner, J. K., and Blejer, H. P., *Ann. N.Y. Acad. Sci.,* 271, 273–279, 1976.
Lewis, G. P., Jusko, W. J., Coughlin, L. L., and Hartz, S., *Lancet,* 1, 291, 1972.
Lohmann, R. D. and Beyersmann, D., *Biochem. Biophys. Res. Commun.,* 190, 1097–1103, 1993.
Martell, A. E., *Environ. Health Perspect.,* 40, 207–226, 1981.
Muller, T., Schuckelt, R., and Jaenicke, L., *Arch. Toxicol.,* 65, 20–26, 1991.
NIOSH, Current intelligence bulletin 42: Cadmium, DHHS [NIOSH] publ. no. 84–116, National Institute of Occupational Safety and Health, Cincinnati, OH, 1984.
Nishiyama, S. and Nakamura, K., *Toxicol. Appl. Pharmacol.,* 74, 337–344, 1984.
Nobel, R. L., *Cancer Res.,* 37, 1929–1933, 1977.
Nocentini, S., *Nucl. Acids Res.,* 15, 4211–4225, 1987.
Nomiyama, K., *Sci. Total Environ.,* 14, 199–232, 1980.
Oberdörster, G., *Scand. J. Work Environ. Health,* 12, 523–537, 1986.
Oberdörster, G. and Cox, C., *Chem. Environ. Toxicol.,* 27, 181–195, 1990.
Oberdöster, G. and Kordel, W., in *Proceedings of the International Conference on Heavy Metals in the Environment,* CEP Consultants, Edinburgh, 505, 1981.
Ochi, T. and Ohsawa, M., *Mutat. Res.,* 111, 69–78, 1983.
Ochi, T. and Ohsawa, M., *Mutat. Res.,* 143, 137–142, 1985.
Ochi, T., Ishiguro, T., and Ohsawa, M., *Mutat. Res.,* 122, 169–175, 1983.
Ochi, T., Takahashi, K., and Ohsawa, M., *Mutat. Res.,* 180, 257–266, 1987.
Ohata, H., Seki, Y., and Imamiya, S., *Arch. Environ. Contam. Toxicol.,* 41, 195–200, 1988.
Oldiges, H., Hochrainer, D., and Glaser, U., *Chem. Environ. Toxicol.,* 19, 217–222, 1989.
Piscator, M., *Environ. Health Perspect.,* 40, 107–120, 1981.
Poirier, L. A., Kasprzak, K. S., Hoover, K., and Wenk, M. L., *Cancer Res.,* 43, 4575–4581, 1983.
Pour, P. P. and Stepan, K., *Cancer Res.,* 47, 5699–5706, 1987.
Pott, F., Ziem, U., Reiffer, F. J., Huth, F., Ernst, H., and Mohr, U., *Exp. Pathol.,* 32, 129–152, 1987.
Prasad, A. S., in *Clinical, Biochemical, and Nutritional Aspects of Trace Elements,* Prasad, A.S., Ed., Alan R. Liss, New York, 1982.
Rehm, S. and Waalkes, M. P., *Vet. Pathol.,* 25, 163–166, 1988.
Rhodes, D. and Klug, A., *Sci. Am.,* 56–65, 1993.
Rivedal, E. and Sanner, T., *Cancer Res.,* 41, 2950–2953, 1981.
Roe, F. J. C., Dukes, C. E., Cameron, K. I., Pugh, R. C. B., and Mitchley, B. C. V., *Br. J. Cancer,* 18, 674–681, 1964.
Rossman, T. G., Roy, N. K., and Lin, W., in *Cadmium in the Human Environment: Toxicity and Carcinogenicity.* Norberg, G.F., Herber, R.F.M., and Allessio, L., Eds., International Agency for Research on Cancer, Lyon, France, 367, 1992.
Sanders, C. L. and Mahaffey, J. A., *Environ. Res.,* 33, 227–233, 1984.
Schroeder, H. A., Balassa, J. J., and Vinton, W. H., Jr., *J. Nutr.,* 83, 239–250, 1964.
Schmahl, W., Kollmer, W. E., and Berg, D., *J. Trace Elem. Exp. Med.,* 3, 14–18, 1986.
Scicchitano, D. A. and Pegg, A. E., *Mutat. Res.,* 192, 207–210, 1987.
Sirover, M. A. and Loeb, L. A., *Science,* 194, 1434–1436, 1976.
Shaikh, Z. A. and Smith, J. C., in *Mechanisms of Toxicity and Hazard Evaluation,* Holmstedt, B., Lauwerys, R., Mecier, M., and Roberfroid, M., Eds., Elsevier, Amsterdam, 569, 1980.
Shirai, T., Iwasaki, S., Masui, T., Mori, T., Kato, T., and Ito, N., *Jpn. J. Cancer Res.,* 84, 1023–1030, 1993.
Shirai, T., Tagawa, Y., Taguchi, O., Ikawa, E., Mutai, M., Fukushima, S., and Ito, N., *Jpn. J. Cancer Res.,* 79, 1293–1296, 1993.
Snyder, R. D., , *Mutat. Res.,* 193, 237–246, 1988.
Stacey, N. H., Cantilena, L. R., Jr., and Klassen, C. D., *Toxicol. Appl. Pharmacol.,* 53, 470–480, 1980.
Stayner, L., Smith, R., Thun, M., Schorr, T., and Lemen, R., *Ann. Epidemiol.,* 2, 177–194, 1992.
Stoner, G. D., Shimkin, M. B., Troxell, M. C., Thompson, T. L., and Terry, L. S., *Cancer Res.,* 36, 1744–1747, 1976.
Sunderman, F. W., Jr. and Barber, A. M., *Ann. Clin. Lab. Sci.,* 18, 267–288, 1988.
Takenaka, S., Oldiges, H., König, H., Hochrainer, D., and Oberdörster, G., *J. Natl. Cancer Inst.,* 70, 367–373, 1983.

Tanaka, K., Miura, N., Satokata, I., Miyamuto, M. C., Satoh, Y., Kondo, S., Yasui, A., Okayama, H., and Okada, Y., *Nature*, 348, 73–76, 1990.
Taylor, B. A., Heiniger, H. J., and Meier, H., *Proc. Soc. Exp. Biol. Med.*, 143, 629–633, 1973.
Terracio, L. and Nachtigal, M., *Arch. Toxicol.*, 58, 141–151, 1986.
Terracio, L. and Nachtigal, M., *Arch. Toxicol.*, 61, 450–456, 1988.
Thun, M. J., Schnorr, T. M., Smith, A. B., Halperin, W. E., and Lemen, R. A., *J. Natl. Cancer Inst.*, 74, 325–333, 1985.
Tohyama, C., Suzuki, J. S., Homma, N., Nishimura, N., and Nishimura, H., in *Metallothionein III*, Suzuki, K. T., Imura, N., and Kimura, M., Eds., Birkhäuser Verlag, Basel, 443, 1993.
Von Zglinicki, T., Edwall, C., Ostlund, E., Lind, B., Nordberg, M., Ringertz, N. R., and Wroblewski, J., *J. Cell Sci.*, 103, 1073–1081, 1992.
Waalkes, M. P., Chernoff, S. B., and Klaassen, C. D., *Biochem. J.*, 220, 811–819, 1984.
Waalkes, M. P., Coogan, T. P., and Barter, R. A., *Crit. Rev. Toxicol.*, 22, 175–201, 1992a.
Waalkes, M. P., Diwan, B. A., Bare, R. M., Ward, J. M., Weghorst, C., and Rice, J. M. *Toxicol. Appl. Pharmacol.*, 110, 327–335, 1991c.
Waalkes, M. P., Diwan, B. A., Weghorst, C. M., Ward, J. M., Rice, J. M., Cherian, M. G., and Goyer, R., *J. Pharmacol. Exp. Ther.*, 266, 1656–1663, 1993.
Waalkes, M. P., Diwan, B. A., Rehm, S., Ward, J. M., Rice, J. M., Moussa, M., Cherian, M. G., and Goyer, R. M., *The Toxicologist*, 14, 1994.
Waalkes, M. P. and Goering, P. L., *Chem. Res. Toxicol.*, 3, 281–288, 1990.
Waalkes, M. P. and Klaassen, C. D., *Fundam. Appl. Toxicol.*, 5, 473–477, 1985.
Waalkes, M. P., Kovatch, R., and Rehm, S., *Toxicol. Appl. Pharmacol.*, 108, 448–456, 1991a.
Waalkes, M. P. and Poirier, L. P., *Biochem. Pharmacol.*, 81, 250–257, 1985.
Waalkes, M. P. and Oberdörster, G., in *Biological Effects of Heavy Metals*, Vol. II, Foulkes, E.C., Ed., CRC Press, Boca Raton, FL, 129, 1990.
Waalkes, M. P. and Perantoni, A., *J. Biol. Chem.*, 261, 13097–13103, 1986.
Waalkes, M. P. and Perantoni, A., *Toxicol. Appl. Pharmacol.*, 101, 83–94, 1989.
Waalkes, M. P., Perantoni, A., Bhave, M. R., and Rehm, S., *Toxicol. Appl. Pharmacol.*, 93, 47–61, 1988d.
Waalkes, M. P., Perantoni, A., and Palmer, A. E., *Biochem. J.*, 256, 131–137, 1988c.
Waalkes, M. P. and Rehm, S., *Toxic Subst. J.*, 1994b.
Waalkes, M. P. and Rehm, S., *Fundam. Appl. Toxicol.*, 19, 512–520, 1992.
Waalkes, M. P. and Rehm, S., *Fundam. Appl. Toxicol.*, 1994a.
Waalkes, M. P., Rehm, S., Perantoni, A., and Coogan, T. P., in *Cadmium in the Human Environment: Toxicity and Carcinogenicity*, Nordberg, G. F., Alessio, L., and Herber, R. F. M., Eds., IARC Sci. Publ., International Agency for Research on Cancer, Lyon, France, 390, 1992d.
Waalkes, M. P., Rehm, S., Riggs, C. W., Bare, R. M., Devor, D. E., Poirier, L. A., Wenk, M. L., Henneman, J. R., and Balaschak, M. S., *Cancer Res.*, 48, 4656–4663, 1988a.
Waalkes, M. P., Rehm, S., Riggs, C. W., Bare, R. M., Devor, D. E., Poirier, L. A., Wenk, M. L., and Henneman, J. R., *Cancer Res.*, 49, 4282–4288, 1989.
Waalkes, M. P., Rehm, S., Sass, B., Konishi, N., and Ward, J. M., *Environ. Res.*, 55, 40–50, 1991b.
Waalkes, M. P., Rehm, S., Sass, B., Kovatch, R., and Ward, J. M., *Toxic Subst. J.*, 13, 15–28, 1994.
Waalkes, M. P., Rehm, S., Sass, B., and Ward, J. M., in *Cadmium in the Human Environment: Toxicity and Carcinogenicity*, Nordberg, G. F., Alessio, L., and Herber, R. F. M., Eds., IARC Sci. Pub., International Agency for Research on Cancer, Lyon, France, 401, 1992b.
Waalkes, M. P., Wahba, Z. Z., and Rodriguez, R. E., in *Hazardous Materials Toxicology; Clinical Principles of Environmental Health*, Sullivan, J.B. and Krieger, K.R., Eds., Williams & Wilkins, Baltimore, 845, 1992c.
Waalkes, M. P., Ward, J. M., and Konishi, N., *Proc. Am. Assoc. Cancer Res.*, 29, 132, 1988b.
Wade, G. G., Mandel, R., and Ryser, H. J.-P., *Cancer Res.*, 47, 6606–6613, 1987.
Wahba, Z. Z., Coogan, T. P., Rhodes, S. W., and Waalkes, M. P., Protective effects of selenium on cadmium toxicity in rats. Role of altered toxicokinetics and metallothionein, *J. Toxicol. Environ. Health*, 38, 171–182, 1993.
Wahba, Z. Z., Miller, M. S., and Waalkes, M. P., *Hum. Exp. Toxicol.*, 12, 1–3, 1993.
West, D. W., Slattery, M. L., Robison, L. M., French, T. K., and Mahoney, A. W., *Cancer Causes Control*, 2, 85–94, 1991.
Willard, R. E., Assessment of cadmium exposure and toxicity risk in an American vegetarian population, Project Summary, Doc. No. EPA/600/S1–85/009, Environmental Protection Agency, Washington, D.C., 1986.
Yamada, H., Miyahara, T., and Sasaki, Y. F., *Mutat. Res.*, 302, 137–145, 1993.

Chapter

Mechanisms of Nickel Genotoxicity and Carcinogenicity

Max Costa

I. INTRODUCTION

The coordination chemistry of nickel encompasses many different geometries, coordination numbers, and oxidation states.[1] Nickel complexes with oxidation states ranging from –1 to +4 and geometries of all major structural types are included.[1] The most common nickel oxidation state is Ni(II) $3d^8$ and much work has been done on nickel coordination and complexes involving the Ni(II) species.[1] The atomic number of nickel is 28 and the atomic weight is 58.71. Stable isotopes and their relative abundances are ^{58}Ni (68%), ^{60}Ni (26%), ^{61}Ni (1%), ^{62}Ni (4%), and ^{64}Ni (1%).[1] ^{61}Ni is an important isotope for biophysical investigation since it has a nuclear spin of 3 → 2 and a nuclear magnetic moment of –0.748 BM. The most abundant isotopes, ^{58}Ni and ^{60}Ni, have no nuclear spin.[1] The isotope ^{63}Ni is beta-emitting (0.067 MeV) with a half-life of 92 years, making it a very useful isotope for biological studies. Solid nickel is a hard silver-white metal with high electrical and thermal conductivity. Nickel represents 0.018% of the earth's crust, which is more than twice that of cobalt and 500 times less than that of iron. In nature, nickel is mainly complexed with sulfur, arsenic, and antimony.[1] The main source of nickel is pyrrhotite, which contains approximately 3–5% nickel.[1]

Nickel is one of the few transition metals which can combine with carbon monoxide at atmospheric pressure.[1] Nickel carbonyl was one of the first metal carbonyls to be discovered. The toxicity of nickel carbonyl is greater than that of hydrogen cyanide.[1] Many forms of metallic nickel are resistant to oxidation by air and water at ordinary temperatures, and this is why nickel coatings by electroplating are important. Ni^{+2} is considered to be a borderline metal ion in terms of hardness and softness because it binds to both types of ligands.[1] Examples of hard metal ions include Na^+, $Cr(VI)$, Mn^{7+}, Fe^{3+}, and Co^{2+}, which bind to ligands such as NH_3 and Fl^-. Hard metal ligands are smaller and less polarized than those of the soft metals. Soft metal ions include Cu^+, Pt^{2+}, Hg^{2+}, and they bind to ligands such as RS^-, CO, and CN^-.[1]

In 1915, nickel was discovered to stimulate the growth of certain plants.[2] Since that time, nickel has been clearly implicated as important in a number of enzymes in bacteria including dehydrogenases, carbon monoxide dehydrogenases, and methylcholinase reductase.[2] Nickel has also been shown to be important in the activity of the enzyme urease, the first enzyme ever crystallized.[2] Clearly, nickel also has an important role as an essential element in higher organisms, however, the function of nickel in these organisms is not known.[2]

The most significant toxicological effect of nickel is its ability to cause human cancer, which was first described in 1939 by an excess of lung and nasal cancer among workers in a nickel refining company in Clydack, Wales.[3] Nickel is primarily a hazard in the occupational setting, and nonoccupational environmental exposures to nickel and its compounds are rare.[3] Ambient nickel concentrations are 0.008 µg/m^3. A major atmospheric contribution of nickel comes from the combustion of fossil fuels that contain nickel sulfides and oxides.[3]

Occupational exposure to nickel compounds is dependent upon industrial processing.[3] Aerosols of dissolved nickel salts are the types of exposure found in the electroplating and electrolysis areas of nickel refineries. Nickel concentrations that are as high as 0.2 mg/m^3 have been reported in these occupational environments.[3] Ambient exposure levels exceeding 1 mg/m^3 have been reported in pyrometallurgical refining processes relating to exposure to nickel oxides, sulfides, and nickel powder. The route of nickel uptake is by inhalation and ingestion, but dermal penetration of soluble Ni(II) is important in contact dermatitis.[3] In the occupational setting inhalation is the primary route of exposure to nickel, while in the general population ingestion is considered to be the most important route of exposure. Dietary intake of nickel averages 150 µg, but the values depend on age and country of residence.[3] Dietary intakes derive from eating foods that are rich in nickel, such as cocoa, soy products, nuts, oats, buckwheat, oatmeal, bitter chocolate, and dried legumes. Nickel may also be leached from kitchen utensils containing this metal.[3]

The most potent carcinogenic nickel compounds in humans are thought to be the crystalline nickel subsulfides and sulfides, although exposures to certain nickel oxides is also deemed to be a severe carcinogenic hazard.[3] Some epidemiological data suggest that exposure to soluble nickel salts in the electrolytic areas of nickel refineries may have also contributed to a higher incidence of lung and nasal cancers.[3] The epidemiological data suggesting that water-soluble salts are carcinogenic are not consistent with animal studies showing that water-soluble nickel salts were generally incapable of inducing cancers.[3] However, water-insoluble salts such as crystalline nickel sulfide and subsulfide are extremely potent carcinogens in rodents, inducing essentially 100% incidence of cancers at virtually any site of their administration including intramuscular, intratracheal, as well as other sites.[3,4] Many routes of exposure are not relevant to the human situation, however, inhalation and multiple intratracheal exposures to nickel subsulfide have been positive in inducing cancer in both rats and mice.[3] Exhaustive reviews on nickel carcinogenesis in experimental animals have been compiled, and these will not be duplicated in this chapter.[3,4] This chapter attempts to discuss possible mechanisms of nickel carcinogenesis.

II. CHEMICAL CONSIDERATIONS IN RELATIONSHIP TO NICKEL GENOTOXICITY AND CARCINOGENICITY

The major oxidation state of nickel in most biologically relevant systems is Ni^{2+}. There has been much controversy in the last few years as to whether Ni^{2+} can undergo oxidation to Ni^{3+} in biological systems and participate in Fenton chemistry.[1] These studies are based largely on the observation that when Ni^{2+} was reacted with histidine, H_2O_2, and dG, there was increased oxidation of deoxyguanine to 8-hydroxyguanine.[5,6] However, the detection of Ni^{3+} in these systems has been difficult. Therefore, there is controversy as to what oxidative intermediates are derived from Ni^{2+}. Some believe that hydroxyl radicals are formed subsequent to Ni^{3+} formation,[6] while others propose that a Ni^{4+} hydroperoxy complex is involved in the oxidation of DNA.[5] Certain biological ligands can lower the oxidation potential of Ni^{2+} and allow it to be oxidized by very strong oxidants, such as H_2O_2. Generally, the concentration of H_2O_2 is not high inside the cell, however, treatment of cells with nickel compounds shifts the balance of reductants/oxidants toward oxidants, and H_2O_2 levels are higher. Normally, the oxidation potential of Ni^{2+} is 1.09 V, but following its binding to various peptide ligands, such as gly-gly-His or tetraglycine, the oxidation potential is lowered from 1.09 V to 0.79 to 0.94 V, depending upon the peptide ligand.[7] This may not seem a substantial change in potential, however it is significant in biological systems. Metals such as iron and copper readily exhibit Fenton chemistry, but Ni^{2+} Fenton chemistry is dependent on nickel binding to ligands that lower its oxidation potential to levels that allow oxidants such as H_2O_2 to oxidize Ni^{2+}.

A. UPTAKE OF NICKEL COMPOUNDS

The carcinogenic potency of all nickel compounds is directly related to their ability to enter cells.[8] Water-soluble nickel salts do not readily enter cells. Therefore, these compounds are generally not carcinogenic in animals and, to a large extent, have not been considered potent human carcinogens, although recent studies have suggested an increase in cancer in nickel refinery areas where exposure to water-soluble nickel salts occurs.[3] Many carcinogenic nickel compounds are those that are intermediates in the process of nickel refining.[3] Smelting of nickel results in the production of a nickel iron sulfide matte that contains a high concentration of the potent carcinogen, crystalline nickel subsulfide. Individuals employed in the crushing operations of nickel refining were exposed to nickel subsulfide dusts and developed a high incidence of nasal and pharyngeal cancers.[3]

Exposure of nonphagocytic cells to crystalline nickel subsulfide resulted in active phagocytosis of nickel sulfide particles.[8,9] In experimental animals, the crystalline nickel subsulfide particles were highly carcinogenic but amorphous nickel sulfide was not.[4] If cells in tissue culture were exposed to amorphous nickel sulfide particles, the particles were not phagocytized.[9] The potent carcinogenic activity of the nickel subsulfide particles is due to their ability to enter cells by phagocytosis.[9] Water-soluble nickel salts produce genetic effects *in vitro* because their delivery and exposure can be controlled, but *in vivo* water-soluble nickel compounds are rapidly removed from the body. Thus, extracellular exposure does not occur for a sufficiently long time to achieve high intracellular concentrations of Ni^{2+}. Once the particles enter cells they can be dissolved. Note that not all nickel-containing particles have to completely dissolve inside the cell for the concentrations of solubilized nickel to reach high levels. Even a small percentage of particle dissolution can greatly increase the levels of soluble Ni^{2+} inside cells, as shown in Table 1.[10] Additionally, it must be remembered that the entry of the particle itself can be very destructive due to a surface active chemistry that can produce oxygen radicals inside cells which have genetic consequences.[5] This is particularly true for the black form of nickel oxide that is calcined at low temperatures and exhibits potent carcinogenicity, but inside the cells would not dissolve as readily as nickel sulfide compounds.[11] Intracellular particles are contained in vacuoles and aggregate around the nucleus and, with time, lysosomes fuse with the vacuoles and hydrogen peroxide is generated.[12,13] Recent studies from this laboratory have shown that in cells treated with crystalline NiS or Ni_3S_2, H_2O_2 levels increase in the nucleus. In contrast, water-soluble nickel compounds can increase H_2O_2 levels in the cytosol, but not in the nucleus.

Table 1 Potential Intracellular Concentration of a Phagocytized Crystalline NiS Particle[a]

Mean particle diameter used in calculation (μm)	Approximate NiS cellular concentration[b] *(M)*
1.45	0.25
4.00	4.75

[a] Cell volume was determined in CHO cells with a Coulter counter-particle size analyzer and log range expander.
[b] Cell volume 393.5 μm³; density of NiS, 5.5 g/cm³; particles assumed spherical.

If water-soluble nickel salts should enter cells, they are likely to be bound to ligands that are not readily exchanged. This is in contrast to nickel ions dissolving from a particle inside the cell that will probably undergo substantial ligand exchange reactions, permitting the formation of stable complexes involving important biological molecules such as nuclear transcription factors. Water soluble salts also compete with Mg^{2+} for uptake, and this requires mM levels of Ni^2.

B. INTERACTION OF Ni^{2+} WITH PROTEINS AND OXIDATIVE DAMAGE AS A MECHANISM OF GENOTOXICITY

Since nickel ions accumulate inside cells, they bind to various peptides and amino acids and this binding lowers their oxidation potential.[7] Ni^{2+} can be oxidized and generates oxygen radicals that produce damage to cells.[5,6] A summary of the genotoxic effects of nickel compounds is provided in Table 2. Nickel is very active at oxidizing protein, as detected by carbonyl formation.[14] Various nickel-binding proteins that have a higher affinity for Ni^{2+} compared to other proteins have been described in the literature.[7] This is particularly interesting with regard to the selective damage nickel produces in the heterochromatin regions of mouse and Chinese hamster cells.[9] The distribution of heterochromatin is different in these two species. Very little is known about heterochromatin function (salient features are shown in Table 3), but there is evidence from using a variety of endpoints (Table 4) that nickel selectively damages heterochromatin.[9]

Most discussions of the interaction of nickel with cellular ligands have involved its binding to proteins and peptides. This does not mean that Ni^{2+} cannot bind to DNA.[15] The most common complexes involving Ni^{2+} are square planer or octahedral, meaning that nickel must coordinate with electronegative groups

Table 2 Summary of Genotoxic Effects of Crystalline NiS, $Ni_3S_2 > Ni^{2+}$

1.	Very small quantity of DNA single-strand breaks, DNA-protein cross-links
2.	↑ SCE
	↑ Chromosomal aberrations (heterochromatin)
3.	Oxidized DNA bases (i.e., 8 OH G, 8 OH A)
	Very small amount
	Most studies done *in vitro* with H_2O_2

Table 3 Salient Features of Heterochromatin

1.	Remains highly condensed through the cell cycle
2.	Stains positively by the C-band technique
3.	Is enriched in highly repeated or satellite DNA sequences
4.	Replicates during late S phase
5.	Chromosomal distribution differs between species
6.	Has been thought not to be transcribed
7.	Functions unknown

Table 4 Evidence Showing Selective Interactions of Ni^{2+} with Heterochromatin

1.	More $^{63}Ni^{2+}$ binding to heterochromatin in intact mammalian cells (Mg^{2+} insoluble fraction)
2.	More cross-linking of protein in heterochromatin by NiS and $NiCl_2$ in intact mammalian cells
3.	Ni^{2+} inhibits specific protein binding (DNA gel retardation assay) to heterochromatin DNA (mouse satellite DNA sequence)
4.	$NiCl_2$ and NiS selectively damage heterochromatin in a number of species: a. Chinese hamster b. Mouse c. Human

such as N, S, or O. Ni^{2+} can interact at the N7 position of guanine with high affinity and with phosphate groups if the DNA can assume a structure that accommodates common Ni^{2+} complexes, such as octahedral or square planer geometry. The binding of Ni^{2+} to guanine may affect the ability of the base-pair cytosine to be methylated by DNA 5′ cytosine methylase. Gene promoter regions are generally rich in poly dG sequences and exhibit quite dramatic and unique conformational changes that are required to regulate gene expression. Thus, nickel could potentially have a direct effect on the DNA in the promoter region. This subject needs to be investigated further.

Recent studies have demonstrated that heterochromatic proteins, when reacted with Ni^{2+}, enhance the oxidation of deoxyguanine and bind more nickel compared to euchromatin. Additionally, the proteins found in heterochromatin are also more readily oxidized by Ni^{2+}. We believe that the high concentration of histone H-1 in heterochromatin may be the reason for the selective effects of Ni^{2+}, since nickel interacts avidly with H-1. The oxidized proteins and amino acids can react with DNA. Figure 1 illustrates a potential reaction mechanism of an oxidized amino acid where the carbonyl was the result of nickel-induced oxidative damage. This carbonyl is shown reacting with the N1 or N2 position of guanine or the exocyclic nitrogens on DNA bases. These are among the most likely sites for reaction of the carbonyl.[14] While there are many oxidation products of amino acids, formation of carbonyl is one of the more reactive chemicals that could cause a DNA-protein cross-link. Figure 2 illustrates how Ni^{2+} may oxidize any amino acid to a reactive carbonyl. This figure also describes the oxidation products of amino acids that have been reported in the literature.

Cysteine, histidine, and tyrosine have been shown to be increased in their cross-linking to DNA following treatment of cells with nickel compounds.[14] In further support of the role of oxidative damage in nickel-induced genotoxicity, vitamin E suppressed nickel-induced chromosomal aberrations, and the addition of H_2O_2 with $NiCl_2$ greatly increased DNA strand breaks beyond the effect of $NiCl_2$ alone.[9] Nickel compounds have also been shown to induce lipid peroxidation[16] which could, in turn, yield malonyldialdehyde that reacts with DNA.

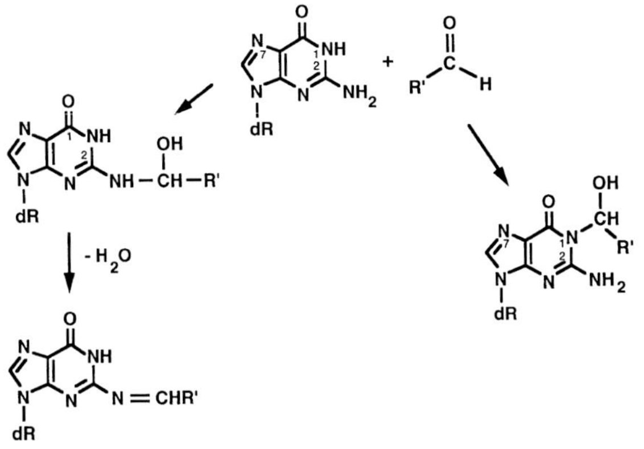

Figure 1 Example of site of reaction of oxidized amino acid reactive carbonyl with guanine.

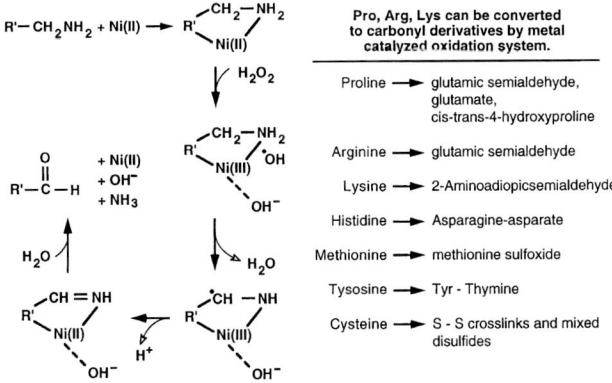

Figure 2 Oxidation of amino acids by Ni(II).

Nickel produces a low level of genetic damage. In particular, when bound to heterochromatin, it oxidized protein and DNA bases that produced localized structural changes. Nickel-binding proteins may exist at other sites in addition to heterochromatin, such as the promoter regions that are active in protein-DNA interactions. Nickel binding to these proteins lowers its oxidation potential, resulting in protein oxidation and altering the normal function of that protein. In fact, water-soluble nickel salts have been shown to affect protein-DNA interactions by using a satellite DNA sequence that is found in abundance in the heterochromatin region of mouse cells.[17] Previous studies have shown that histone H-1 was a primary protein that was cross-linked to DNA in heterochromatin[18] and we have recently observed that ^{63}Ni avidly binds to histone H-1. Recent studies have shown that the selective damage that nickel produced in heterochromatin was probably due to the abundance of H-1 in heterochromatin compared with euchromatin. However, inhibition of metal carcinogenesis by Mg^{2+} does not alternate nickel-induced oxidative stress.[32]

Recent studies have shown that Ni^{2+} can inactivate the transcription of a number of genes, presumably by inducing hypermethylation of the promoter.[14] The mechanisms involved remain very unclear, however, the ability of nickel to turn off the transcription of selected genes if a tumor suppressor or a senescence gene is inactivated no doubt is very important in its carcinogenic action.[20] It is interesting that nickel is synergistic at many cancer, mutagenic, or genotoxic endpoints where exposure occurs together with many other types of carcinogens, such as UV light, X-rays, alkylating agents, and benzopyrene, suggesting that nickel has a unique mechanism of action that is different from the effects of these other agents.

An interesting system that may model the interaction of nickel involves its effects in various transgenic cell lines where the endogenous *hpgrt* gene has been inactivated and a bacterial *gpt* gene under SV 40 promoter control has been inserted at different locations within the genome.[2] When the *gpt* gene was inserted in a heterochromatic region of chromosome 1, it was found that the cell readily became resistant to 6-thioguanine (6-TG) following nickel treatment.[2] These resistant cells did not exhibit structural mutations in the gene, however, the transcription of the bacterial *gpt* gene was turned off by what is thought to be DNA cytosine hypermethylation. If the same gene was inserted at another site not near heterochromatin, little 6-TG resistance developed following nickel treatment. It should be noted that the g12 cells that became resistant to 6-TG following nickel treatment were in fact less responsive to other mutagens, such as UV light, X-rays, and bleomycin. Thus, the insertion of this gene at that site did not induce a "hypermutable" state. The position of the gene in the chromosome was critical as to whether its transcription was affected by Ni^{2+}. The effects of nickel in this transcription were inherited, making this important in carcinogenesis.

The method by which nickel affects DNA methylation is not understood, but there are many studies in the literature that refer to nickel as a cocarcinogen, tumor promoter, and other names that suggest it has an epigenetic effect. Many investigators point to nickel compounds as tumor promoters and not initiators of carcinogenesis. An effect on DNA methylation would certainly be consistent with Ni^{2+} exhibiting tumor promoting activity. The extent of genetic lesions produced by nickel and its mutagenic response are inconsistent with its potent carcinogenic activity, pointing to the probable role of nickel in tumor promotion rather than initiation. The most selective damage that nickel produced was in heterochromatin, a genetically inactive region. Heterochromatic DNA is known to be heavily methylated and very condensed with proteins. Perhaps the ability of nickel to interact with proteins found in abundance in heterochromatin, such as H-1, and their association with highly methylated DNA, caused the structure to decondense. Neighboring hypermethylation in genes in the same chromosome may be produced as a result of the cellular compensatory response.

A most interesting effect of Ni^{2+} is the inactivation it produced in an X chromosome senescence gene by hypermethylation.[20] This indicated that since nickel interacted with heterochromatin and resulted in selective decondensation of the X chromosome that contained a large amount of heterochromatin, it produced neighboring effects on genes in the same chromosome but located in euchromatin.[22] Since heterochromatin was decondensed by Ni^{2+} the cell probably was trying to recondense the heterochromatin by producing a hypermethylation state. This compensatory response caused other genes (such as the senescence gene) that were under methylation control to become hypermethylated and inactivated. This provided the best evidence yet of a link between Ni^{2+} effects on heterochromatin and changes in methylation of genes at other genetically active sites in the same chromosome.

REFERENCES

1. Coyle, C. L. and Stiefel, E. I., in *The Bioinorganic Chemistry of Nickel*, Lancaster, J.R., Jr., Ed., VCH Publishers, New York, chap. 1, 1988.
2. Ankel-Fuchs, D. and Thauer, R. K., in *The Bioinorganic Chemistry of Nickel*, Lancaster, J.R., Jr., Ed., VCH Publishers, New York, chap. 5, 1988.
3. IARC Monograph on the Evaluation of Carcinogenic Risk to Humans, *Chromium, Nickel, Welding*, Vol. 49, International Agency for Research on Cancer, Lyon, France, 677, 1990.
4. Coogan, T. P., Latta, D. M., and Costa, M., *Crit. Rev. Toxicol.*, 19, 341–384, 1989.
5. Klein, C. B., Frenkel, K., and Costa, M., *Chem. Res. Toxicol.*, 4, 592–604, 1991.
6. Kasprzak, K. S., *Chem. Res. Toxicol.*, 4, 604–615, 1991.
7. Margerum, D. W. and Anliker, S. L., in *The Bioinorganic Chemistry of Nickel*, Lancaster, J.R., Jr., Ed., VCH Publishers, New York, chap. 2, 1988.
8. Costa, M. and Mollenhauer, H. H., *Science*, 209, 515–517, 1980.
9. Costa, M., *Annu. Rev. Pharmacol. Toxicol.*, 31, 321–337, 1991.
10. Abbracchio, M. P., Simmons-Hansen, J. S., and Costa, M., *J. Toxicol. Environ. Health*, 9, 663–676, 1982.
11. Sunderman, F. W., Jr., Hopfer, S. M., Knight, J. A., McCully, K. S., Cecutti, A. G., Thornhill, P. G., Conway, K., Miller, C., Patierno, S. R., and Costa, M., *Carcinogenesis*, 8, 305–313, 1987.
12. Evans, R. M., Davies, P. J. A., and Costa, M., *Cancer Res.*, 42, 2729–2735, 1982.
13. Huang, X., Frenkel, K., Klein, C. B., and Costa, M., *Toxicol. Appl. Pharmacol.*, 120, 29–36, 1993.
14. Zhuang, Z., Huang, X., and Costa, M., *Toxicol. Appl. Pharmacol.*, 1994.
15. Rossetto, F. E. and Nieboer, E., *J. Inorg. Biochem.*, 1994.
16. Sunderman, F. W., Jr., Marzork, A., Hopper, S. M., Zaharen, O., and Reid, M. C., *Ann. Clin. Lab. Sci.*, 15, 229–236, 1985.
17. Imbra, R. J., Latta, D. M., and Costa, M., *Toxicol. Environ. Chem.*, 22, 167–179, 1989.
18. Patierno, S. R., Sugiyama, M., Basilion, J. B., and Costa, M., *Cancer Res.*, 45, 5787–5794, 1985.
19. Salnikow, K., Cosentino, S., Klein, C. B., and Costa, M., *Mol. Cell Biol.*, 14, 851–858, 1994.
20. Klein, C. B., Conway, K., Wang, X. W., Bhamra, R. K., Lin, X., Cohen, M. D., Annab, L., Barrett, J. C., and Costa, M., *Science*, 251, 796–799, 1991.
21. Kargacin, B., Klein, C. B., and Costa, M., *Mutat. Res.*, 300, 63–72, 1993.

22. Wang, X.-W., Lin, X., Klein, C. B., Bhamra, R. K., Lee, Y.-W., and Costa, M., *Carcinogenesis,* 13, 555–561, 1992.
23. Huang, X., Zhuang, Z., Frankel, K., Klein, C. B., and Costa, M., *Environ. Health Perspect.*, 102, 281–284, 1994.
24. Lee, Y. W., Klein, C. B., Kargacim, B., Salmikow, K., Kitahara, J., Dowjar, K., Zhitkovich, A., Christie, N. T., and Costa, M., *Molec. Cell Biol.*, 15, 2547–2557, 1995.

Chapter 16

Carcinogenicity and Genotoxicity of Lead, Beryllium, and Other Metals

Mitchell D. Cohen, Darlene H. Bowser, and Max Costa

I. INTRODUCTION

Aluminum, beryllium, cobalt, copper, germanium, iron, lead, manganese, molybdenum, scandium, selenium, silicon, thorium, tin, titanium, vanadium, and zinc have been evaluated for their carcinogenic and genotoxic activity. Palladium, tellurium, and thallium have also been considered, but in less detail.

II. ALUMINUM (Al)

There is increasing evidence for an association between employment in the aluminum processing industry and the development of several types of cancers, including those of the lung, bladder, and pancreas, as well as lymphomas and reticulosarcomas (Gibbs and Horowitz, 1979; IARC, 1984; Abramson et al., 1989; Spinelli et al., 1991; Ronneberg and Langmark, 1992). However, the suspect vapors that workers are exposed to during processing are more often a noxious mixture of aluminum particulates (including aluminum metal, alumina [Al_2O_3], and aluminum fluoride [AlF_3]) and relatively high concentrations of polycyclic aromatic hydrocarbons, including benzo(α)pyrene. There is no epidemiological evidence supporting that aluminum may be a human carcinogen.

There is also a lack of animal data to demonstrate a link between exposure to aluminum metal and/or its derivatives and tumor induction. In studies where thin pieces of aluminum foil were implanted subcutaneously or intramuscularly into rats, localized myosarcomas were induced (O'Hara and Brown, 1967). However, depending on the geometry of the implants (i.e., <4 μM) tumors were not inducible even after intrapleural or intraperitoneal injection (Stanton, 1974; Pigott and Ishmael, 1981). The positive results with foil are questionable in light of the fact that any substance in the form of a solid flat plate has the capacity to induce sarcomas when implanted subcutaneously (Bischoff and Bryson, 1964).

While it was shown that subcutaneous injections of aluminum dextran into rodents were able to induce sarcomas (Haddow and Horning, 1960), the administration of aluminum metal powder, aluminum hydroxide (Al[OH]$_3$), Al_2O_3, aluminum phthalocyanine, or aluminum phosphate (AlPO$_4$) by various routes to rabbits, rats, mice, and guinea pigs failed to induce cancer (Haddow and Horning, 1960; Furst and Haro, 1969; Shabik and Hartwell, 1969; Furst, 1971a). Oddly enough, one study indicated that rather than inducing cancers in exposed rodents, the aluminum dextran mixture was actually antitumorigenic (cited in Leonard and Gerber, 1988).

The antitumorigenic activity of aluminum compounds has been demonstrated in several model systems. When rats were administered aluminum nitrate ($Al[NO_3]_3$), the growth rate of intraperitoneally transplanted Walker 256 carcinosarcomas was reduced (Hart and Adamson, 1971; Adamson et al., 1975).

Aluminum salts also are able to reduce the formation of tumors *in situ* induced by other carcinogens. Subcutaneous injections of aluminum chloride ($AlCl_3$) into mice prior to treatment with dimethyl nitrosamine caused a reduction in the number of resultant nodules (Yamane and Ohtawa, 1979). However, in dust form, when Al_2O_3 is inhaled simultaneously with benzo(α)pyrene, there is an increase in the incidence of bronchotracheal tumors in exposed mice above that due to the hydrocarbon alone (Stenback et al., 1976; Knizhnikov et al., 1989). Thus, a true anticarcinogenic effect for aluminum is highly dependent upon the route of exposure as well as its physical state.

Although it is unclear whether aluminum compounds are carcinogenic, they do display clastogenic activity in several higher organisms. Chromosomal aberrations (including, gaps, breaks, and failure of pairing) were produced in the gametes of grasshoppers fed $AlCl_3$ (Manna and Parida, 1965). Similar aberrations were observed in the bone marrow cells of mice injected with $AlCl_3$ solutions (Manna and Das, 1972). A chronic oral administration of aluminum sulfate ($Al_2[SO_4]_3$) to mice resulted in spindle damage in the blood cells of mice and also chromosomal irregularities (reviewed in Sharma and Talukder, 1987). These changes in spindle functionality may help in understanding the mechanisms of the effect of aluminum on cell division of mammalian cells (Gelfant, 1963).

Aluminum compounds are negative in tests for mutagenicity (reviewed in Leonard and Gerber, 1988). Aluminum metal and $AlCl_3$ failed to induce reverse mutations in *Salmonella typhimurium* strain TA102 (Marzin and Phi, 1985), had no effect in the SOS chromotest with *Escherichia coli* strains P937 and P935 (Olivier and Marzin, 1987), and in the *Bacillus subtilis* (strains H17 and M45) rec-assay (Nishioka, 1975); similar lack of mutagenicity was evident in the rec-assay with $Al_2[SO_4]_3$ and Al_2O_3 (Kada et al., 1980a,b; Kanematsu et al., 1980). In Syrian hamster embryo (SHE) cells, both $AlCl_3$ and $Al_2[SO_4]_3$ failed to cause morphological transformations or to enhance the transformations induced by simian adenovirus SA7 (Casto et al., 1979; DiPaolo and Casto, 1979). Similarly, $AlCl_3$ did not induce forward mutations in the mouse L5178Y lymphoma line (Oberly et al., 1982).

Even in test systems utilizing purified DNA metabolizing enzymes, $Al_2[SO_4]_3$ had no effect upon the accuracy of DNA synthesis, although overall replication was reduced (Sirover and Loeb, 1976). Similarly, $AlCl_3$ did not depurinate isolated calf thymus DNA (Schaaper et al., 1987), nor was it able to modify the biological activity of $\phi\chi 174$ DNA transfected into *E. coli* spheroblasts (Miller and Levine, 1974).

III. BERYLLIUM (Be)

The carcinogenicity of beryllium has been well established in a number of animal models (reviewed in Kuschner, 1981; Reeves, 1986; Leonard and Lauwerys, 1987; Cohen et al., 1990; ATSDR, 1992). One of the earliest reports of beryllium-induced cancers was osteosarcomas induced following intravenous injections of beryllium sulfate ($BeSO_4$) into rabbits (Gardner and Heslington, 1946). Other early studies using different beryllium salts (i.e., BeO, $ZnBeSiO_3$, $BeSiO_3$, $BeHPO_4$) and similar exposure protocols have also demonstrated the carcinogenic nature of this metal (Barnes et al., 1950; Dutra and Largent, 1950; Sissons, 1950; Araki et al., 1954; Janes et al., 1954). Dermal application of beryllium salts have thus far not yielded tumorigenic responses in test animals (ATSDR, 1992). Nor has beryllium been shown to cause cancers following its ingestion by animals in their diets or drinking water, although one study demonstrated (in a non-dose-related manner) an increased incidence of reticulum cell sarcomas in rats fed $BeSO_4$ for two years (Morgareidge et al., 1975) while another study indicated a general increase in tumor formation over the same period (Schroeder and Mitchener, 1975a,b).

Although beryllium is not carcinogenic (or only weakly so) by these latter routes, it is a carcinogen in both rodent and primate animal models following intratracheal instillation or inhalation (reviewed in Kuschner, 1981; ATSDR, 1992). In a wide number of studies, exposure to BeO, $BeSO_4$, or $BeHPO_4$ resulted in the formation of lung carcinomas that were both transplantable and metastatic (Dutra et al., 1951; Schepers et al., 1957, Vorwald and Reeves, 1959; Schepers, 1961; Reeves and Vorwald, 1967; reviewed in Reeves, 1986); other cancers, such as bronchial alveolar cell tumors, adenomas, adenocarcinomas, epidermoid tumors, and primary cancers of the bronchiole, have been detected as well. The carcinogenic properties of inhaled beryllium salts were not related to their solubility (unlike lead or chromate salts) nor to their retention within the lungs proper (Reeves, 1986).

While beryllium is considered an "established" carcinogen in animals, the IARC working group designates the metal as a "suspect" or "probable" carcinogen in humans (IARC, 1980a), although the

National Toxicology Program does consider it a "certain carcinogen" (NTP, 1989). Epidemiological evidence for an increased risk of cancer in humans occupationally exposed to beryllium is uncertain, with a few early reports suggesting limited or no evidence of carcinogenicity (Hardy et al., 1967; Mancuso and El-Attar, 1969; Mancuso, 1970; Shapley, 1977; Stoeckle et al., 1979). However, more recent studies suggest that there appears to be an excess of lung cancers in workers exposed to beryllium on the job (Infante et al., 1980; Mancuso, 1980; Smith and Suzuki, 1980; Wagoner et al., 1980), although the statistical approaches used by the authors have been criticized (Kuschner, 1981; Smith, 1981; Kazantzis and Lilly, 1986; Reeves, 1986).

No studies are available regarding the genotoxicity of beryllium in humans following oral, dermal, or inhalation exposure. Mice given $BeSO_4$ by gavage exhibited no increase in the incidence of micronuclei in polychromatic erythrocytes up to 72 h after exposure (Ashby et al., 1990).

Both positive and negative clastogenic effects of beryllium compounds have been observed in mammalian cell systems. Cultured human fibroblasts, leukocytes, or Chinese hamster ovary (CHO) cells, exposed to $BeSO_4$ for either short (up to 24 h) or long (several weeks) periods failed to display any significant levels of chromosomal damage (Paton and Allison, 1972; Brooks et al., 1989). Similarly, in a cultured macrophage cell line (P338D1) and in human lymphocytes, $BeSO_4$ failed to induce significant increases in the levels of sister chromatid exchanges above control (Andersen, 1983).

Larramendy observed increased sister chromatid exchanges and chromosomal aberrations in SHE cells or cultured human leukocytes exposed to $BeSO_4$ (Larramendy et al., 1981). Increased levels of somatic mutations at the hypoxanthine-guanine phosphoribosyl transferase locus (Hsie et al., 1978, 1979; Miyaki et al., 1979), as well as enhanced morphological transformations in the presence/absence of viral agents, have been reported with Be compounds (Casto et al., 1976, 1979; Pienta et al., 1977; DiPaolo and Casto, 1979; Dunkel et al., 1981).

In nonmammalian mutation assay systems, beryllium compounds are negative. This may be due to a problem of uptake or other factors that generally make most bacterial systems unresponsive to metals. In the Ames reverse mutation assay, $BeSO_4$, as well as $Be(NO_3)_2$ proved to be nonmutagenic in at least six different *S. typhimurium* tester strains (TA98, TA100, TA1535, TA1536, TA1537, and TA1538) (Simmon, 1979a; Dunkel et al., 1981; Tso and Fung, 1981; Arlauskas et al., 1985). A similar lack of mutagenicity was also observed for $BeSO_4$ and $BeCl_2$ in both the *B. subtilis* rec-assay (Nishioka, 1975; Kanematsu et al., 1980) as well as in the *Sacharomyces cerevisiae* D3 mitotic recombination assay (Simmon, 1979b).

In forward mutations in the lacI gene of *E. coli*, $BeCl_2$ increased the levels of *amber* and *ochre* mutants significantly (Zakour and Glickman, 1984). This same compound also induced reverse mutations in a tester system utilizing *Photobacterium fischeri* (Ulitzur and Barak, 1988). It should be noted that there are complications with the chemistry of Be compounds that are often not considered. For example, at physiological pH, water-soluble Be salts tend to form polymers.

In an assay of DNA replication fidelity, it was determined that $BeCl_2$ could enhance the base substitution error frequency of avian myeloblastosis virus DNA polymerase (Sirover and Loeb, 1976). These inducible errors might play a role in the reduction in DNA synthesis in regenerating rat liver following host treatment with beryllium (Witschi, 1968), although similar effects upon RNA synthesis were not observed (Marcotte and Witschi, 1972). Contradictory results regarding the effects of beryllium on several RNA/DNA-associated processes and their enzymes (re: transcription and translation) in cultured cells or using isolated enzyme systems have been reported (reviewed in Cohen et al., 1990). Thus, while the carcinogenicity of beryllium is fairly well established, the underlying mechanisms are not clear.

IV. COBALT (Co)

The carcinogenicity of cobalt in mammals has been demonstrated (reviewed in Kazantzis, 1981; Jensen and Tüchsen, 1990; Steinhoff and Mohr, 1991) primarily when cobalt compounds were administered to animals by injection as opposed to ingestion or inhalation (Domingo, 1989). The earliest studies found that intrafemoral injections of metallic cobalt into rabbits produced peritoneal metastases and injection-site sarcomas after a period of 3–6 years (Vollmann, 1938; Schinz and Uehlinger, 1942). Similar studies utilizing soluble cobalt powder or cobalt chloride ($CoCl_2$) administered in a variety of different ways to hooded rats showed that a significant number of malignant rhabdomyosarcomas and fibrosarcomas were found at the intramuscular injection site, while increased numbers of sarcomas, induced by thoracic injection, were found proximal to the heart (Heath, 1954; Heath and Daniel, 1962).

Subcutaneous injections of $CoCl_2$ also caused an increase of fibrosarcomas in rats (Shabaan et al., 1977), while intraperitoneal and intramuscular injections of cobalt nitrate ($Co[NO_3]_2$) induced fibrosarcomas in rabbits (Thomas and Thiery, 1953). Intraperitoneal and intramuscular injections of cobalt (II) naphthenate in rabbits, and its intravenous administration to mice, induced significantly greater numbers of cancers in both animals (Nowak, 1961, 1966). Cobalt oxide (CoO) was shown to produce metastisizing tumors in rats by intramuscular administration over a period of one year (Gilman, 1962; Gilman and Ruckerbauer, 1962) while this same agent was only weakly carcinogenic when administered by intermittent intratracheal instillations (Steinhoff and Mohr, 1991). These investigators also determined that insoluble cobalt compounds (such as cobalt sulfide [CoS]) often displayed a greater carcinogenic potential in some rodent models than did water-soluble cobalt.

There was no evidence of tumor formation in mice exposed to identical intramuscular injections of CoO as rats, albeit over a longer period of time (Gilman and Ruckerbauer, 1962). Intrarenal injection of CoS also was not found to induce tumors in rats (Jasmin and Riopello, 1976). Respirable aerosols of CoO produced no carcinogenic effects in Syrian golden hamsters (Wehner et al., 1977). Mice injected intraperitoneally with cobalt acetate ($[CH_3COO]_2Co$) showed no pulmonary tumors (Stoner et al., 1976; Shimkin et al., 1977).

In contrast to metals such as cadmium, nickel, and arsenic, there has been no significant demonstration of cobalt-induced cancers in humans (Kazantzis, 1981; Elinder and Friberg, 1986; Jensen and Tüchsen, 1990). This lack of positive data is mainly due to the paucity of epidemiological studies of cancer incidence in cobalt-exposed individuals. Medical therapy with $CoCl_2$, including oral treatment and implants, was considered to have caused malignant neoplasms in a number of patients (McDougall, 1956; Dodion et al., 1982; Penman and Ring, 1984; Swann, 1984; Weber, 1986); however, this number was subsequently considered negligible and coincidental as compared to the large number of patients which were treated overall (Hughes et al., 1987). There has been recent evidence to support the hypothesis that workers exposed to cobalt in the hard-metal industry may have a greater propensity for developing lung cancer (Lundgren and Ohman, 1954; Bech et al., 1962; Alexandersson and Hogstedt, 1987, 1988; Hogstedt and Alexandersson, 1987; Hogstedt, 1988). However, the evidence is still not sufficient enough to suggest a valid relationship, as many of the epidemiological studies on cobalt carcinogenesis are confounded by other metals (nickel and arsenic) and have a limited population size (Jensen and Tüchsen, 1990).

There have been relatively few *in vivo* investigations that explore the genotoxic effects of cobalt. In Syrian hamsters, intraperitoneal administration of $CoCl_2$ caused aneuploidy in bone marrow cells and testes (Farah, 1983), while ingestion of sublethal doses of $CoCl_2$ in the diet/drinking water, or as a single peroral dose of cobalt blue paints, by mice gave rise to increased levels of chromosomal aberrations in white blood cells (Nehez et al., 1982; Palit et al., 1991). Enhanced early embryonic losses were detected in mice exposed to $CoCl_2$ (Pedigo et al., 1988); however, since this dominant lethal assay is an indirect measure of genotoxicity, it remains to be seen whether cobalt compounds can cause heritable damage *in vivo* (Beyersmann and Hartwig, 1992).

Most prokaryotic mutagenicity studies have failed to demonstrate that $CoCl_2$ and the majority of cobalt agents are mutagens. As with other metallic ions, factors contributing to this inactivity are precipitation of metal-phosphate chelates in the bacterial media, a low rate of metal uptake, or interference with indirect mechanisms of interaction between the genetic material and the metal (Beyersmann and Hartwig, 1992). Cobalt chloride was inactive in the λ prophage induction assay (Rossman et al., 1984) and the results were variable in the *B. subtilis* rec-assay (Kanematsu et al., 1980). Conversely, $CoCl_2$ did induce gene conversions and mitochondrial mutations in *S. cerevisae* (Prazmo et al., 1975; Egilsson et al., 1979; Fukunaga et al., 1982; Kharab and Singh, 1987). Results with *S. typhimurium* and *E. coli* have similarly been conflicting (Rossman et al., 1984; Arlauskas et al., 1985; Ogawa et al., 1986, 1988; Pagano and Zeiger, 1992).

The genotoxicity of soluble cobalt has also been examined in mammalian *in vitro* assays. A majority of the studies yielded positive results with cultured mammalian cells (i.e., CHO cells and human diploid fibroblasts), including DNA strand breaks, induction of DNA cross-links, formation of micronuclei, and increased levels of sister chromatid exchanges (Hamilton-Koch et al., 1986; reviewed in Beyersmann and Hartwig, 1992). DNA strand breaks in CHO cells were observed (Robison et al., 1982) upon treatment with CoS, and the crystalline form was more active than the amorphous form in the induction of transformation of SHE cells (Costa et al., 1982a). There was also an increased induction of mutations by $CoCl_2$ at the *hprt* locus in V79 Chinese hamster lung cells (Miyaki et al., 1979; Hartwig et al., 1990) and enhanced SA7 adenovirus transformation of SHE cells treated with $(CH_3COO)_2Co$ (Casto et al., 1979).

In some instances, cobalt has been shown to have antimutagenic effects. Cultured Chinese V79 lung cells incubated in the presence of $CoCl_2$ after γ-ray irradiation had a marked reduction of 8-azaguanine-resistant mutation formation (Yokoiyama et al., 1990). This modulation of mutagenicity by cobalt in the mammalian system has been shown to be variable and may be due to effects upon DNA repair systems. Similarly, several soluble cobalt compounds, when employed at nontoxic levels, modulate the effect of other genotoxic agents (i.e., MNNG, caffeine) in bacterial systems by apparently reducing mutations (Kada and Kanematsu, 1978; Inoue et al., 1981; Kada, 1984; Clarke and Shankel, 1988). Using purified enzymes, divalent cobalt ions have been shown to substitute for Mg^{2+} ions in the activation of DNA synthesis, but reduced the fidelity of DNA synthesis (Sirover and Loeb, 1976). However, this phenomenon only occurred when cobalt levels were very high and so it is unclear as to whether these effects might be relevant *in vivo*. Clearly, further studies are necessary to determine the mutagenic potential of cobalt, as well as its mechanism of genotoxicity in mammalian systems.

V. COPPER (Cu)

Copper is necessary in the diet as a trace element and is crucial as a cofactor in enzymes associated with oxidative metabolism. Due to its proposed role in formation of oxygen free radicals in tissue and DNA damage, it is not unreasonable to assume that copper bioavailability may be an important consideration in neoplastic formation.

Epidemiologic evidence, such as a high incidence of cancer among coppersmiths, suggested a primary carcinogenic role for copper ions (Agnese et al., 1959). In nonindustrial settings, a higher incidence of human stomach cancers was detected in regions where soil Zn:Cu ratios were dramatically altered (Stocks and Davies, 1960). However, while treatment of laboratory rodents with organocopper complexes caused an increase in the incidence of tumors in bone and lung (Stanton, 1967), similar studies using animals exposed to various copper salts by several different routes failed to give rise to an increased incidence of tumors above levels observed in untreated control animals (Gilman and Ruckerbauer, 1962). Similarly, repeated injections of mice with cupric acetate ($Cu[CH_3COO]_3$) failed to give rise to tumors at either the sites of injection or distally (i.e., lung). Therefore, a direct carcinogenic effect of copper remains doubtful.

Most studies examining the effects of copper carcinogenicity in animals have almost exclusively focused upon the role of supplemental dietary copper. It was shown that the addition of 0.25% $Cu(CH_3COO)_3$ as a dietary supplement inhibited the effects of ethionine in inducing hepatomas in rats (Kamamoto et al., 1973; Yamane et al., 1976). Similar protective effects were demonstrated in rats fed copper and subsequently treated with dimethylaminobenzene dyes (Sharpless, 1946; Clayton et al., 1953). Trace amounts of copper can also impact upon the activity of exogenous cancer-fighting agents; copper treatment greatly accentuated the potency of the drug 3-oxy-2-oxybutyraldehyde bis-(thiosemicarbazone) against Walker 256 tumor cells in rats bearing tumors (Petering et al., 1967; Van Geissen et al., 1973).

However, supplemental copper has proven to be less effective in preventing a variety of other tumors such as dimethylnitrosamine-induced renal and lung carcinomas, 2-aminofluorene-induced epithelial carcinomas, as well as tumors of the lungs, skin, intestine, pancreas, and muscles induced by acetylaminofluorine (Fare and Howell, 1964; Carleton and Price, 1973).

Other means by which copper ions can affect carcinogenesis may be related to promotional and progression events. Copper-dependent angiogenesis in relation to tumor formation has been examined. It has been shown that tissue levels of copper increase prior to vascularization (Ziche et al., 1982), and that the metal ions complex with fibroblast growth factor (Shing, 1988), heparin, and ceruloplasm (Raju et al., 1984), rendering these agents angiogenic. Corneal implants of human brain tumor cells into copper-deficient rabbits displayed decreased neovascularization and tumor growth as compared with the subsequent tumors found in copper-sufficient controls (Alpern-Ehran and Brem, 1985). Similar effects upon tumor growth and angiogenesis in brain neoplasms were observed in dietary copper-deprived rats and rabbits (Brem et al., 1990a,b). However, in other studies it was shown that an induced copper deficiency had no effect on tumor growth and neovascularization in rat striated muscle (Schuschke et al., 1992a,b) or in tumor microcirculation itself (Johnson and Saari, 1991).

Rather than examining the epidemiological effects of copper exposure, most human studies have been concerned with serum copper levels as a clinical biomarker in carcinogenesis. Patients diagnosed with a variety of malignant melanomas and Hodgkin's disease have elevated serum copper levels (as compared with the control populations) which decreased significantly after successful chemo- or radiotherapy (Tessmer et al., 1972, 1973; Hrgovcic et al., 1973). Increased copper levels in serum have also been found in other forms of cancer including squamous cell carcinoma of the larynx (DeJorge et al., 1966), bronchogenic carcinoma (Tani and Kokkola, 1972), cervical (Oldfield et al., 1971), bladder (Albert et al., 1972), and breast cancers (DeJorge et al., 1965). This increase in serum copper concentrations has also been observed prior to the onset of tumor growth in mice implanted with Maloney murine rhabdomyosarcoma cells, after subcutaneous implantation of Morris hepatoma cells into rats, and in marmosets infected with *Herpes saimiri* (Melendez et al., 1970).

Useful as the clinical evidence of copper levels as a tumor biomarker might be, investigations suggest that these tests cannot be considered as an effective parameter in screening for cancer risk. The results show that the positive relationship between serum copper levels and cancer is relevant only among cases that were diagnosed within four years of blood collection, and as more time elapsed, the significance became weaker (Coates et al., 1989).

Metal-induced oxygen radicals, including those caused by copper ions, are known to cause DNA damage. Copper binds with phosphate on nucleotides and nucleic acids (Sharma and Talukder, 1987) and these complexes, in the presence of hydrogen peroxide or other activated oxygen species, can produce point mutations (Wetterhahn et al., 1992). The types of mutations observed experimentally with copper ions are often clustered within the DNA molecules, and are predominantly single base substitutions. The actual mutagenic processes for copper also appears to be less sensitive to oxygen radical scavengers than is observed with some other known mutagenic metals (Tkeshelashvili et al., 1991).

The effects of copper on base incorporation by DNA polymerases and upon RNA synthesis rates have been studied. The potential for the copper salts cuprous chloride ($CuCl_2$) and cuprous acetate $Cu(CH_3COO)_2$ to act as putative mutagens and carcinogens has been shown by their ability to alter the fidelity of DNA synthesis *in vitro* (Sirover and Loeb, 1976). The inhibition of human DNA polymerase β by divalent copper ions is due to a direct interaction of the metal ion with the enzyme (Popenoe and Schmaeler, 1977). Overall RNA synthesis and the rate of initiation of RNA synthesis *in vivo*, using *E. coli* RNA polymerase with calf thymus DNA and T4 DNA templates, were reduced by incubation with $CuCl_2$ (Hoffman and Niyogi, 1977). The presence of Cu^{2+} ions was shown to strongly inhibit the efficiency of transcription of a poly(dA-dT) template by *E. coli* polymerase, and also increased the misincorporation of CMP into the poly(rA-rU) product (Niyogi and Feldman, 1981). While copper at the concentrations tested was shown to decrease *E. coli* polymerase I activity, the presence of the divalent Cu^{2+} ions did not affect DNA replication fidelity (Miyaki et al., 1977).

It has been reported that the presence of divalent copper ions increased the mutagenic potential of thymine hydroperoxidase (Thomas et al., 1976). Similarly, the addition of glycine-Cu^{2+} complexes to ascorbate-treated CHO cells enhanced both mitotic inhibition and frequency of chromosomal aberrations due to ascorbate alone (Stich et al., 1979). This "synergistic" effect was also apparent in studies of the induction of chromasomal aberrations in CHO cells by isoniazid (Whiting et al., 1980a) and the enhancement of the mutagenicity of creatinine following its acidification and gentle heating (Ishizawa et al., 1978). However, the presence of copper ions themselves can have an effect upon mammalian cells, as shown by the enhancement of the viral transformation of SHE cells by SA7 simian adenovirus after pretreatment with copper sulfide (CuS_2) or sulfate ($CuSO_4$) (Casto et al., 1979).

In the majority of commonly employed short-term tests of the mutagenic potential for copper compounds, most results have been negative. There was no effect on recombination or production of disomic and/or diploid spores in *S. cerevisiae* yeast after treatment with $CuSO_4$ (Sora et al., 1986). Similarly, the induction of reverse mutations and mitotic gene conversions in a different strain of this yeast was not observed with this same agent (Singh, 1983). Negative effects of copper were also observed in the *B. subtilis* rec-assay (Nishioka, 1975; Kanematsu et al., 1980), although a much earlier study concluded that copper ions directly inhibited the transformation of this organism (Weed, 1963). To date, only one study has demonstrated any positive direct mutagenic effect from a copper compound (Demerec et al., 1951).

VI. GERMANIUM (Ge)

Germanium compounds have gained more notoriety for their anticarcinogenic/antitumorigenic effects than for their cancer-causing potentials. In life-term feeding studies, mice that received sodium germanate developed significantly fewer spontaneous tumors than did control animals (Kanisawa and Schroeder, 1967).

Numerous inorganic and organogermanium compounds have been tested for their anticarcinogenic effects in animals. Germanium oxide antagonized the genotoxic/carcinogenic effects of simultaneously administered cadmium in mice (Han et al., 1992). This antagonistic effect was also apparent when germanium in either an inorganic form, a synthesized organogermanate complex, or a natural organogermanium compound was administered to rats that received the carcinogen 1,2-dimethylhydrazine (Jao et al., 1990).

Germanium also has intrinsic antitumorigenic effects. The sera from mice treated with carboxyethylgermanium sesquioxide (Ge-132) were found to exhibit antitumor activity against both allogeneic and syngeneic ascites tumors in different strain hosts (Suzuki et al., 1985). Treatment of rodents with the organogermanium agent PCAGeS was shown to block the normal pulmonary metastases that arose from implanted Lewis lung carcinomas (Sato et al., 1988). Several novel metallocene (i.e., germanocene) agents have also been shown to act as antitumorigenic agents in hosts bearing ascites tumors (Kopf-Maier et al., 1988).

While the mechanisms by which germanium acts to block the carcinogenic activities of known cancer-causing agents is not established, such effects are thought to arise from immunopotentiation (Mizushima et al., 1980; Arimori et al., 1981). Several organogermanium compounds are potent inducers of interferons in treated hosts, as well as inducers of natural killer cells and macrophages (Tanaka et al., 1984; Aso et al., 1985). The precise mechanisms by which germanium compounds cause immunomodulation remains undefined.

It should be noted that while germanium agents do display anticarcinogenic/antitumorigenic activities, they also have well-defined, potentially lethal side effects. These include overt irritation and localized inflammation, neurotoxicity, and nephrotoxicity (Nagata et al., 1985; Vouk, 1986; Arts et al., 1990; van der Spoel et al., 1990; Obara et al., 1991; Schauss, 1991), and as such, their use in chemoprevention is not warranted.

Germanium has antimutagenic/anticlastogenic effects in many commonly used assays of genotoxicity. Germanium oxide prevented the induction of chromosomal aberrations, micronuclei formation, sister chromatid exchanges, and spermhead abnormalities inducible by cadmium in mouse testes (Han et al., 1992). This compound also antagonized Trp-P-2-induced mutagenesis in *S. typhimurium* strains TA98 and TA1538 (Kada et al., 1984), possibly through antagonism of the SOS repair function within the cells. The organogermanium agent Ge-132 was shown to be antimutagenic towards γ-ray-induced mutagenesis in *E. coli* B/r WP$_2$ trp$^-$, most likely via an antagonizing effect against the error proneness of the DNA replicating enzyme (Mochizuki and Kada, 1982).

VII. IRON (Fe)

The carcinogenicity of iron in mammals is unclear. Occupational exposure to iron compounds, mainly oxides, is common in mining, iron and steel foundry work, and in arc welding. Epidemiological studies have shown that an excess mortality from lung cancer has been observed in iron-ore miners from a number of countries, such as those from the Lorraine basin in France (Braun et al., 1960; Monlibert and Roubille, 1960; Pham et al., 1991, 1992), Minnesota (Hueper, 1966; Lawler et al., 1985), British Cumbria (Kinlen and Willows, 1988), and the Kiruna region of Sweden (Jorgensen, 1973, 1984; Radford and Renard, 1984). An increased incidence of lung cancer has also been reported in iron and steel foundry workers from Sheffield, England (McLaughlin and Harding, 1956). More recently, French epidemiologists have found an increased mortality of lung and stomach cancer in Lorraine iron miners (Chau et al., 1993). It was concluded that the risk of lung cancer in these miners was mainly due to dust, diesel engine soot, and other airborne carcinogens (including radon daughters), with the stomach cancers arising from postinhalation reflux and swallowing of inhaled toxicants (Meyer et al., 1980). A study of stomach cancer showed that many of the subjects were excessively exposed to iron dust, and that the risk of cancer formation increased with concentration and duration of the exposures (Kraus et al., 1957).

While human evidence is limited, the carcinogenicity of metallic iron and iron oxide (Fe$_2$O$_3$) in experimental animals was negative, since only benign nonmetastisizing tumors appeared (Muller and Erhardt, 1957; Sunderman, 1971; Stockinger, 1984). However, repeated intramuscular, subcutaneous, or parenteral administration of large doses of select iron-carbohydrate complexes (i.e., iron(III) dextran or dextrin, and saccharated iron oxide) resulted in increased formation of metastisizing sarcomas in rats, mice, hamster, and rabbits (Richmond, 1959; Fielding, 1962; Haddow et al., 1964; Langvad, 1968). The iron complex is essential for the sarcomatous response, since neither the carbohydrate injected alone (Richmond, 1959; Haddow and Horning, 1960) nor inorganic iron compounds induced sarcomas at the injection site (Haddow et al., 1961). Still, not all organoiron compounds of this type (i.e., iron sorbitol and citrate) were shown to be carcinogenic (Lundin, 1961; Fielding, 1962; Roe and Haddow, 1965). Even as a soluble noncarbohydrate salt (i.e., ferric acetate), iron was unable to induce tumors in experimental animals (Stoner et al., 1976; Shimkin et al., 1977).

Intratracheally-instilled iron oxide acts synergistically when given together with benzo(α)pyrene or diethylnitrosamine. Hamsters receiving the postcarcinogen instillation of iron displayed an increased lung tumor incidence in comparison to animals receiving the carcinogens alone (Saffioti et al., 1972). It has been suggested that ferric oxides serve as carcinogenic cofactors by either causing a retarded clearance of inhaled carcinogens or by inducing cytophysiological changes which make respiratory tract cells more "susceptible" when exposed to carcinogens. More recently, reports have suggested that oxygen radicals formed by iron ions may be important mediators of pulmonary damage from asbestos (Lund and Aust, 1991; reviewed in Kamp et al., 1992), natural and synthetic mineral fibers (Nejjari et al., 1993), and Fe^{2+}-containing dusts (Costa et al., 1989; Huang et al., 1993).

A major health concern has been the suggestion that iron supplementation for nonanemic adults may increase the risk for cancer, while low body iron stores may be in some ways protective (Selby and Friedman, 1988; Stevens et al., 1988). Clinical studies have shown that iron promotes cancer cell growth (Laskey et al., 1988; Weinberg, 1992). This hypothesis was examined in a methyl nitrosourea-induced experimental model for breast cancer (Thompson et al., 1991). It was shown that excess dietary iron in female rats was more prominent than iron deficiency in the modification of the progress of mammary carcinogenesis. However, low dietary iron increased the incidence of urethane-induced lung adenomas in mice by 86% while iron overload did not influence adenoma development (Omara and Blakley, 1993).

The genotoxic effects of iron compounds appear to be dependent upon several factors, primarily the physical nature of the iron agent itself. Ferrous chloride ($FeCl_2$) and ferrous sulfate ($FeSO_4$) have not been found to enhance the transformation frequency of SHE cells with simian adenovirus SA7 (Casto et al., 1979; DiPaolo and Casto, 1979). More recent studies have shown that $FeSO_4$ in aqueous solution is cytotoxic, but again not directly transforming in SHE cells (Huang, 1991). In contrast, when $FeSO_4$ is deposited on a particle, and then phagocytized by these cells, the released $FeSO_4$ gave rise to cellular transformation. It has been suggested that water-soluble divalent ferrous ion was difficult to be incorporated into cells and, when internalized, any free iron was carried and stored under high surveillance (Huang, 1991). In contrast, phagocytosis of a particle containing $FeSO_4$ delivered a large quantity of free divalent ion which was then able to produce reactive oxygen species (ROS) to attack DNA, ultimately giving rise to transformed cells (reviewed in Reizenstein, 1991; Ryan and Aust, 1992).

It has also been shown that several classes of iron minerals (i.e., biotite, hematite, nemalite, and pyrite) mediate the formation of 7,8-dihydro-8-oxo-2'-deoxyguanosine (8-oxo-dG) and -deoxyadenosine *in vitro* (Berger et al., 1993). The 8-oxo-dG in the DNA template was found to cause polymerase-α to miscode incorporation of nucleotides in the replicated strands and induce G $\to$ T and A $\to$ C base substitutions (Cheng et al., 1992). In a study of the effect of mineral particles containing Fe^{2+} on primary cultures of rabbit tracheal epithelial cells, it was shown that among the tested particles (i.e., chrysotile, nemalite, and hematite), the most cytostatic after 24 h of treatment was the one that had more Fe^{2+} available on the surface (nemalite) and which produced the most ROS (Guilianelli et al., 1993).

This potential to induce ROS that can directly damage DNA should make iron an excellent clastogen. However, data on the clastogenicity of iron compounds are scarce. Chronic treatment of mice with iron resulted in a progressive decrease in the mitotic frequency of circulating blood cells in a dose-dependent, but not time-dependent, manner. Spindle disturbances formed the great majority of damage, possibly as a result of the affinity of iron for sulfur ligands in proteins; at higher concentrations, chromatin matter became increasingly affected (Dhir et al., 1985). Still, the mere presence of iron greatly enhances the clastogenic effects of other known agents (i.e., isoniazid, ascorbate) (Stich et al., 1979; Whiting et al., 1980a), lending greater credence to the belief that iron acts more like a cocarcinogen than a primary one.

Studies of the mutagenicity of iron compounds in yeast and bacteria have been contradictory. While $FeCl_2$ and $FeCl_3$ failed to induce responses in the *B. subtilis* rec-assay (Nishioka, 1975; Kanematsu et al., 1980) or $FeSO_4$ in *S. typhimurium* TA102 (Marzin and Phi, 1985), other studies using modified reaction conditions found the latter to yield positive mutagenic responses in strain TA97 (Brusick et al., 1976; Pagano and Zeiger, 1992). Similarly, the effects from iron varied among the differing strains of *S. cerevisiae* yeast. In strain DS13, $FeSO_4$ had no effect during meiotic cell division, while in strain D7, the same compound induced gene conversions and reverse mutations at two separate loci (Singh, 1983). However, in this same set of studies, FeCl$_3$ was completely inactive at all endpoints tested.

VIII. LEAD (Pb)

Lead is one of the most commonly encountered and utilized metals. Its widespread industrial usage and continual release into the environment as an exhaust emission, has made lead one of the most studied metals for carcinogenic and genotoxic effects.

While the epidemiological evidence to implicate lead as a human carcinogen is not conclusive, the findings are suggestive. Significantly increased rates of lung cancer, as well as cases of kidney and stomach cancer, have been documented in occupationally-exposed battery, smelter, and pigment plant workers, as well as plumbers and pipefitters exposed to lead solder fumes (Cooper, 1976; Baker et al., 1980; Kang et al., 1980; Sheffet et al., 1982; Selevan et al., 1984; Cantor et al., 1986; all reviewed in: IARC, 1980b; Kazantzis, 1981; Schlag, 1987; and Cohen et al., 1990). While these and other studies demonstrated the increased occurrence of cancers (primarily respiratory or gastrointestinal varieties), none has yet been able to establish a dose-response relationship between lead and cancer incidence. Most of the results are confounded by factors such as smoking as well as exposures to other carcinogens (i.e., chromium or arsenic); thus, IARC declared that in the absence of concrete human data, it is reasonable to regard lead compounds (both organic and inorganic) as if they presented a carcinogenic risk to humans (IARC, 1980b).

In contrast to the epidemiological data, there is substantial evidence for lead-induced carcinogenesis in experimental animals. Lead acetate (Pb[OOCCH$_3$]$_2$) and basic lead acetate (Pb[OOCCH$_3$]$_2$ · Pb[OH]$_2$) are carcinogenic when fed to rats and mice, with tumors arising most commonly in the kidneys, although tumors in the testes, lungs, pituitary, prostate, and adrenal glands have also been documented (Boyland et al., 1962; van Esch et al., 1962; Schroeder et al., 1965; Oyasu et al., 1970; Koller et al., 1985; Blakely, 1987). Other agents, such as lead powder, lead nitrate, and some other highly soluble lead salts, failed to induce significant increases in the incidence of tumors in treated rodents (Schroeder et al., 1964, 1970; Furst et al., 1976). Studies with dogs, hamsters, rabbits, and monkeys have not yielded strong positive results, but many of these were of shorter duration than the rodent studies, in effect, not allowing sufficient time for tumor formation (van Esch and Kroes, 1969; Azar et al., 1973).

Cancers have also been induced in rodents using other routes of exposure. Application of lead naphthenate to the skin of mice induced kidney neoplasias, but oddly no skin tumors (Baldwin et al., 1964). Inhalation/intratracheal instillation of lead oxide (alone and in combination with benzo(α)pyrene) resulted in increased lung adenomas only in the hamsters given the mixture, lending credence to the claim that lead might act as a cocarcinogen in exposed hosts (Kobayashi and Okamoto, 1974). Subcutaneous/intramuscular injections of organolead or inorganic lead salts have also been shown to cause increased numbers of tumors. Injections of mice with tetraethyl lead resulted in the formation of malignant lymphomas (Epstein and Mantel, 1968), while repeated injections of rats with lead phosphate resulted in increased numbers of renal cancers (Zollinger, 1953; Balo et al., 1965; Roe et al., 1965). Similar injections of rats with free lead powder did not cause increases in renal tumors, and only gave rise to fibrosarcomas at the site of injection (Furst et al., 1976). Repeated intraperitoneal injections of rats and mice with lead subacetate have been shown to increase the incidence of renal adenomas/carcinomas and lung adenomas, respectively (Coogan et al., 1972; Stoner et al., 1976; Shimkin et al., 1977).

As with mammalian carcinogenicity, the genotoxicity of lead compounds has been extensively reviewed (Leonard et al., 1972, 1984; Damstra, 1977; Gerber et al., 1980; IARC, 1980b; Cohen et al., 1990). Many studies have analyzed the clastogenic effects of lead agents on human and animal blood cells. Cultured lymphocytes from exposed hosts or naive cells cultured with lead salts demonstrated increased levels of chromosomal aberrations, including chromatid exchanges, translocations, gaps, rings, polycentric chromosomes, and dispirilizations (Ahlberg et al., 1972; Bauchinger and Schmid, 1972; Beek and Obe, 1974; Deknudt et al., 1977a,b; Deknudt and Deminatti, 1978; Nordenson et al., 1978; Deknudt and Gerber, 1979; Forni et al., 1980; Columbano et al., 1983; Al-Hakkak et al., 1986). Similar clastogenic results were obtained with *Drosophila* and CHO cells (Ahlberg et al., 1972; Bauchinger and Schmid, 1972), with increased achromatic lesions rather than outright aberrations or mitotic abnormalities occurring. The majority of studies analyzing the effect of lead upon sister chromatid exchanges have indicated that lead ions are in some cases positive (Andersen, 1983; Grandjean et al., 1983). There are also many other studies which did not demonstrate a clastogenicity for lead compounds using similar assays as described in the studies cited above (Schmid et al., 1972; Beek and Obe, 1975; Bauchinger et al., 1977; Horvat et al., 1981; Maki-Paakkanen et al., 1981; Costa et al., 1982b; Dalpra et al., 1983).

Although the clastogenicity studies suggest a potential for positive genotoxicity for lead, *in vitro* studies with both bacterial and mammalian systems have failed to demonstrate any significant mutagenic activity. Both lead acetate and chloride were inactive in the *B. subtilis* rec-assay (Nishioka, 1975; Kada et al., 1980a,b), as well as with the *Salmonella*/microsome test or with plate/fluctuation tests with *E. coli* WP_2 (Nestmann et al., 1979). Lead acetate was also found inactive in assays for point mutations (Rosenkranz and Poirier, 1979), DNA-modifying effects in the *E. coli polA$^+$/polA$^-$* assay, mitotic recombination in the *S. cerevisiae* D3 system (Simmon, 1979b), and in the gene conversion/reverse mutation assay using *S. cerevisiae* D7 (Singh, 1983). The only studies that have demonstrated even a weak effect by lead ions were those that examined the induction of mutants in the L5178Y mouse lymphoma cell line (Amacher and Paillet, 1980; Oberly et al., 1982), the formation of disomic and diploid *S. cerevisiae* spores (Sora et al., 1986), the changes in fidelity of DNA replication, and alterations in rates of RNA synthesis using purified polymerases (Sirover and Loeb, 1976; Hoffman and Niyogi, 1977; Niyogi et al., 1981).

Clearly, while the carcinogenic and clastogenic effects of lead are little in doubt, improved assay systems are required to better define the mutagenic potential of both inorganic and organic lead agents.

IX. MANGANESE (Mn)

Manganese is an essential element in all mammals, and is considered not to be carcinogenic. However, although there is a paucity of epidemiologic data to confirm or refute its carcinogenic potential in humans, some studies have indicated a strong relationship between the increased incidences of colorectal and digestive cancers and exposure to workplace atmospheres containing several metals including manganese (Brown et al., 1984; Wingren and Axelson, 1987; Andersson et al., 1990).

Studies of the carcinogenicity of manganese compounds in animal models are similarly few in number. Chronic subcutaneous and intraperitoneal injections of mice with manganese chloride ($MnCl_2$) increased the frequency of lymphosarcomas, as well as distal mammary adenocarcinomas and leukemias (DiPaolo, 1964). In rats and mice injected intramuscularly with manganous acetylacetonate, a similar increase in injection site fibrosarcomas was evidenced; similar studies with pure manganese powder or manganese dioxide failed to elicit tumor formation (Furst, 1978). In studies with mice injected intraperitoneally with manganous sulfate ($MnSO_4$), there was again an increased incidence of distal (i.e., lung carcinomas) tumors (Stoner et al., 1976; Shimkin et al., 1977). As in the early studies, the lack of a localized site of deposition (as occurs in intramuscular exposures) suggests that eventual deposition of the manganese ions occurs in preferential tissues; however, no studies of manganese deposition and subsequent tumor formation following intravenous injections are available.

It has been demonstrated that manganese compounds display antitumorigenic/anticarcinogenic effects. This may be the result of an increased cell capacity to generate manganese-containing superoxide dismutase to counter the effects of carcinogens which act through activation of cellular oxygen (Zhong et al., 1990) or via direct competition with DNA-binding agents/DNA-enzyme modifiers to ameliorate their carcinogenic potentials (Yamane and Sakai, 1973; Sunderman, 1977, 1989a; Kasprzak et al., 1986; Francis et al., 1988). It has also been shown that manganese compounds can directly stimulate those components of the host immune system which are essential to an early and full antitumor cell response (i.e., interferons and natural killer cells) (Smialowicz et al., 1984, 1985, 1987; Judde et al., 1987).

Although the question of the carcinogenicity of manganese compounds is still open, both a divalent ($MnCl_2$) and heptavalent form ($KMnO_4$) of the metal are known to be strong clastogens both *in vivo* and *in vitro*. In cultured mammalian cells treated with these agents, there was an increase in the levels of chromosomal breaks, fragments, and exchanges (Umeda and Nishimura, 1979; Skreb et al., 1980). Similar exposures also caused true DNA strand breaks in CHO cells and human diploid fibroblasts (Hamilton-Koch et al., 1986; Snyder, 1988). *In vitro,* manganese ions also enhance the incidence of chromosomal aberrations induced by other agents, such as ascorbate and isoniazid (Stich et al., 1979; Whiting et al., 1980a). Even following short- and long-term ingestion of $KMnO_4$ and $MnSO_4$ by mice, the incidence of bone marrow cell chromosomal aberrations, strand breaks, and micronuclei, as well as that of spermhead abnormalities, was increased in a dose-dependent fashion (Joardar and Sharma, 1990).

In conjunction with this ability to act as clastogens, it is not surprising that manganese ions can also induce both forward and point mutations in cultured mammalian cells, such as the mouse lymphoma line and CHO cells (Hsie, 1978; Nishimura and Umeda, 1978; Hsie et al., 1979; Miyaki et al., 1979;

Oberly et al., 1982), and they can enhance the transformation of cultured SHE cells by the simian adenovirus SA7 (Casto et al., 1979).

In yeast, manganese compounds (primarily the divalent $MnCl_2$ and $MnSO_4$) have been shown to induce genetic damage. In *S. cerevisiae* DIS13, the sulfate induced an increase in the numbers of disomic and diploid meiotic products (Sora et al., 1986). In the D7 strain, this compound also increased the incidence of gene conversions and reverse mutations at two different loci (Singh, 1983, 1984). Similar results have been obtained using these compounds and measurements of gene conversions and/or intergenic recombinations in several heteroallelic and homoallelic diploid, as well as haploid, yeast strains (Putrament et al., 1975, 1977; Baranowska et al., 1977).

Manganese ions have been shown to be mutagenic, or to act as anti- or comutagens in several prokaryotic systems, with the effects being strongly dependent upon species, cell stage, and treatment conditions (Auerbach, 1976; Flessel, 1981). Manganese was capable of inducing mutations in *S. typhimurium* and *E. coli*, depending on strain and incubation conditions (i.e., studies performed as suspension, fluctuation, or plate assays) (Demerec and Hanson, 1951; Arlauskas et al., 1985; Marzin and Phi, 1985; Mortelmans et al., 1986; DeMeo et al., 1991; Pagano and Zeiger, 1992). Results in the *B. subtilis* rec-assay have been conflicting, although $KMnO_4$ was consistently negative in both studies and in both the H17 and M45 strains (Nishioka, 1975; Kanematsu et al., 1980). In one particular study of effects of manganese on the *E. coli lacI* gene locus (Zakour and Glickman, 1984), it was shown that $MnCl_2$-induced mutations were not the nonsense type, but included primarily A:T → G:C transitions, frameshift, addition, and deletion events. Manganese has also been shown to induce mutations in the *rII* gene of bacteriophage T4 (Orgel and Orgel, 1965; Ripley, 1975) and to induce λ prophage in *E. coli* $WP2_s$ (Rossman et al., 1984).

Using purified enzymes associated with DNA metabolism, it has been shown that divalent manganese ions activate DNA polymerases from bacteriophage T4, *E. coli*, and avian myeloblastosis virus; however, in each case, there was a substantial decrease in the fidelity of replication (Hall and Lehman, 1968; Sirover and Loeb, 1976; Kunkel and Loeb, 1979). It was demonstrated that a major effect from divalent manganese is its ability to act as a substitute for divalent magnesium during polymerase reactions, but that this results in an increased frequency of base misincorporations (Goodman et al., 1983; Beckman et al., 1985), as well as an altered accuracy of base selection and specificity of hydrolysis by the polymerase (El-Deiry et al., 1984). One study has even suggested that the manganese ion associated with the DNA polymerase and resulted in the incorporation of ribonucleotides rather than deoxyribonucleotides into the DNA (cited in Rasmuson, 1985).

It is clear that manganese is a potent mutagen and clastogen *in vivo* and *in vitro*. However, the lack of clearer evidence for the carcinogenic effect of this metal, in any of its multiple valence states, remains a major problem both from a biochemical and from an epidemiological standpoint.

X. MOLYBDENUM (Mo)

Unlike many other metals whose compounds are nearly either all carcinogenic or all noncarcinogenic, molybdenum and its compounds represent a dichotomy. While inorganic molybdenum agents are considered noncarcinogenic, the absence of molybdenum in human and animal diets has been correlated with increases of gastric and esophageal cancers due to other etiologic agents (Schwartz, 1975; Tuyns et al., 1978; Li, 1982; Kibblewhite et al., 1984; Khojasteh and Kraybill, 1988; Wang and Tang, 1988; Barch, 1989). The mechanism by which molybdenum is thought to act in cancer prevention is through the inactivation of the phase I P_{450}-dependent monooxygenase systems (Bompart, 1990). The diminution of the activity of these systems prevents the bioactivation of numerous carcinogens, including members of the benzopyrene family, N-nitroso compounds, and polycyclic mycotoxins encountered in the workplace and in the diet (van Rensburg et al., 1985; Bogden et al., 1986; Wei et al., 1987). Still, a few studies have shown that repeated exposures of experimental animals to solutions of molybdenum trioxide (MoO_3) gave rise to significant increases in the numbers of lung tumors (Stoner et al., 1976; Shimkin et al., 1977).

Molybdenum is active in select short-term genotoxicity assays and studies indicate that molybdenum exposure correlated with significant increases in chromosomal aberrations and sister chromatid exchanges in the lymphocytes of workers and animals and the sperm of mammals and *Drosophila* (Babain et al., 1980; Bobyleva et al., 1991; Chopikashvili et al., 1991; Bobyleva et al., 1993).

In bacterial systems, the majority of inorganic molybdenum compounds failed to induce or gave rise to weakly positive mutagenic responses in the *B. subtilis* rec-assay (Nishioka, 1975; Kanematsu et al.,

1980) or in the *S. cerevisiae* gene conversion/reverse mutation assay (Singh, 1983). However, the use of the polyanionic agent ammonium heptamolybdate [$(NH_4)_6Mo_7O_{24}$] yielded positive results. Subsequent testing also revealed that this compound was also capable of inducing reversions of three *E. coli* strains, i.e., WP_2 (*uvr*A$^+$, *rec*A$^+$), WP_2 uvrA (*uvr*A$^-$, *rec*A$^+$), and CM_{571} (*uvr*A$^+$, *rec*A$^-$) (Nishioka, 1975).

XI. SCANDIUM (Sc)

Although scandium is detectable in all experiments on determination of trace element contents from molecular to tissue levels, in humans as well as animal models (Andronikashvili and Mosulishvili, 1980), information regarding both its carcinogenicity and genotoxicity is lacking.

There are reports that scandium is more associated with the DNA and RNA from tumors than from normal tissue and, within tumors, its association with DNA is higher than with RNA (Andronikashvili et al., 1974). It has also been shown than scandium ions are readily bound to several metalloproteins both *in vivo* and *in vitro* (Ford-Hutchinson and Perkins, 1971, 1972; Schroeder and Mitchener, 1971). The carcinogenic/genotoxic implications of these observations have as yet not been described.

The only reports suggesting possible carcinogenicity for scandium are based upon the formation of skin and lung granulomas following ingestion or inhalation of scandium salts (Haley, 1965), or the induction of liver neoplasms following life-term feeding of soluble scandium salts to mice in their drinking water (Luckey and Venugopal, 1977).

Genotoxicity studies on scandium are completely lacking. A few studies indicate that scandium ions are active competitors with trivalent iron in the iron/enterochelin system of certain Gram-negative bacteria, including *Klebsiella pneumoniae* and *E. coli* (Rogers et al., 1980, 1982). The ultimate effect from scandium in these organisms is an inhibition of RNA and DNA synthesis, primarily during the initiation of chromosomal replication (Plaha and Rogers, 1983; Plaha et al., 1984).

XII. SELENIUM (Se)

As with several other trace elements discussed here, selenium has been shown to display both carcinogenic and anticarcinogenic properties in animals and humans. In an early study, groups of rats were fed daily diets supplemented with either seleniferous corn or wheat, or a selenium supplement of mixed inorganic selenide (Nelson et al., 1943). Of the 73 animals sacrificed after 3–18 months, most developed cirrhotic livers, but no tumors or adenomatoid hyperplasia were observed. Of the 53 animals surviving to 2 years, 21% developed liver cell adenomas or low-grade carcinomas (without metastases) but only in the cirrhotic livers, whereas unexposed rats had low rates of spontaneous tumors. In another study, rats fed bis-4-acetaminophenyl selenium dihydroxide for four months developed multiple hepatic adenomatous hyperplasia (Seifter et al., 1946). Selenium diethyldithiocarbamate given by gavage to infant rats resulted in tenfold and fivefold increases, respectively, in the incidence of hepatomas and lymphomas in adults as compared with their controls (Innes et al., 1969). Neoplastic growths have also been observed in rats fed moderate protein diets supplemented with selenium salts (Tscherkes et al., 1963).

Selenium sulfide (SeS_2) has also been shown to cause an increase in the incidence of primary liver tumors when administered to rats (NCI, 1980). Neoplastic nodules appeared in the liver of groups given a high dosage (i.e., 15 mg/kg BW) by gavage for 5 days a week over a period of 2 years. A similar study was performed with mice, except the highest dose was 100 mg/kg selenium sulfide; an increased incidence of hepatocellular carcinomas as well as primary lung tumors were observed in high-dose females and marginally in high-dose males (NCI, 1980).

However, not all animal studies have demonstrated a positive correlation between selenium intake and tumor incidence. Often, results have varied depending upon the particular selenium agent employed as well as the host model. Mice fed a diet supplemented with either sodium selenite (Na_2SeO_3) or sodium selenate (Na_2SeO_4) in their drinking water throughout their lifetime did not develop a significant number of tumors as compared with unsupplemented controls (Schroeder and Mitchener, 1972; Jacobs and Frost, 1981). In contrast, in a similar study with rats, a significant increase in a large variety of tumors became apparent over the life span of the exposed hosts (Schroeder and Mitchener, 1971). This latter set of results conflicts with those performed a few years later in which no neoplasms were observed in rats fed semipurified diets containing large amounts of selenium (>2 mg/kg BW), with most of the animals dying within the first 100 days (Harr et al., 1967).

As a result of the early positive carcinogenicity studies in rodents, selenium was included as a carcinogen in the 1958 Delaney Clause (Griffin, 1979). However, it was later reported that some of these early animal studies were deficient in experimental design and in the interpretation of the observed lesions (Shapiro, 1973; Fan, 1990). Upon review of the latter conflicting animal data, IARC determined that the evidence regarding the carcinogenicity of selenium is inadequate for its evaluation (IARC, 1987a).

Epidemiological evidence also points out conflicting results for a role of selenium in carcinogenesis. There was proposed to be an inverse relationship between cancer death rates and selenium levels in forage crops and blood samples, as seen in data taken from surveys throughout various parts of the U.S. (Shamberger and Frost, 1969; Shamberger and Willis, 1971; Allaway, 1972, 1978; Shamberger et al., 1976; Clark, 1985). These studies concluded that some specific types of cancers (i.e., stomach, colon, liver, pancreas) occurred more frequently in individuals with a low selenium blood level or in areas classified as having a low selenium forage content. However, the high mortality from these same and other forms of cancer appeared in populations having high selenium levels and living in environmentally-high selenium content areas. In China, case studies demonstrated a relation between a high incidence of liver cancer and low selenium blood serum levels (Yu et al., 1985). The authors concluded that although selenium deficiency was not a cause of primary liver cancer, a low-selenium diet might decrease the ability of the body to respond to cancer-causing stress. The age-adjusted mortality rates for colon and breast cancer were found to be significantly lower in Finland than the U.S. despite the well-documented low dietary selenium intakes of their populations (Levander, 1986), and data gathered from 27 countries (Schrauzer et al., 1977) failed to produce any significant or direct correlations between liver cancer and dietary selenium intake.

To further complicate the understanding of the roles of selenium as a possible carcinogen, several studies have demonstrated that high levels of selenium administered under a variety of circumstances can reduce the tumorigenic potential of other known carcinogens and can alter the rates of spontaneous tumor formation in models prone to cancer. Skin tumor formation in mice following topical application of 7,12-dimethylbenzanthrene, croton oil, or benzo(α)pyrene was reduced by concomitant topical application of Na_2SeO_3 (Shamberger and Rudolph, 1966; Riley, 1968; Shamberger, 1970). In a second study of similar design, groups of animals that were dermally treated with either of the three chemical compounds and simultaneously fed high dietary supplements of selenium (>1.0 mg/kg) as sodium selenite had 32% fewer papillomas than did members of selenite-deprived groups (Shamberger, 1970). Sprague-Dawley rats fed selenite in their drinking water and subsequently fed diets containing 2-acetylaminofluorene had 50% fewer liver tumors than did control rats (Marshall et al., 1978). Daily dietary selenium supplements fed to rats resulted in a reduced incidence of 3'-methyl-4-dimethylaminoazobenzene-induced liver tumors (Griffin and Jacobs, 1977) and a 47% decrease in dimethylhydrazine-induced colon tumors (Jacobs et al., 1977).

In other studies, the protective effect of supplemental dietary selenium in reducing hepatocarcinomas was shown in rats given daily oral doses of aflatoxin B_1 (Newberne and Connor, 1974; Grant et al., 1977; Milks et al., 1985). Similar protective effects against UVR-induced carcinogenesis were observed in nude mice fed high doses of Na_2SeO_3 (Overvad et al., 1985). In other studies, it was shown that pharmacological doses of selenium inhibited the growth of transplantable tumors, prevented the production of the carcinogenic metabolite N-hydroxyacetylaminofluorene (Thompson, 1984; Milner, 1985) and, in some cases, when given with different dietary fat levels, decreased the incidence of palpable tumors in rats treated with dimethylbenzathracene (Ip and Sinha, 1981).

Spontaneous mouse mammary tumors in C_3H/St mice given selenium as selenium dioxide (SeO_2) in their drinking water had a rate of tumor formation 72% less than in mice deprived of selenium (Schrauzer and Ishmael, 1974). In two further studies (Schrauzer, 1976; Schrauzer et al., 1978), similar comparative tumor profiles were observed when various concentrations of either SeO_2 or Na_2SeO_3 were provided in drinking water.

Mutagenicity studies of chromosomal damage due to selenium compounds have disclosed that the form of selenium tested greatly influences the degree of mutagenicity. The results also show that tetravalent selenium is more genotoxic than the hexavalent form in several different *in vitro* test systems. The tissue culture inhibition dose ($TCID_{50}$) of 20 µg/ml for potassium selenate (K_2SeO_4) did not increase the rates of sister chromatid exchange (SCE) in Don Chinese hamster cells (Ohno et al., 1982), nor was it mutagenic at the CHO cell HGPRT locus (Hsie et al., 1979). The SCE frequency was increased further in both Don Chinese hamster cells and human whole-blood cells when the salt was switched from selenate to potassium selenite (K_2SeO_3) (Ray and Altenburg, 1980; Haruki, 1983). While selenite was the most potent inducer of SCE in Chinese hamster V79 lung cells when an S9 activating system was present, selenide was most damaging in its absence; no changes were observed with selenate-treated cells in either the presence or absence of S9 (Sirianni and Huang, 1983). Chromosomal aberrations and mitotic inhibition activity was shown to increase fivefold with selenite treatment of human fibroblasts incubated with the S9 mouse liver fraction (Lo et al., 1978) and comparable results were found when xeroderma pigmentosum fibroblasts were exposed to selenite. The DNA-damaging effects of tetravalent selenium upon human leukocyte and diploid fibroblast cultures were observed to be more effective than hexavalent forms (Nakamuro et al., 1976; Hamilton-Koch et al., 1986). These effects were not mitigated by the use of active oxygen radical scavengers (such as mannitol, catalase, or superoxide dismutase) and, in fact, were exacerbated by the presence of reduced glutathione (Whiting et al., 1980b; Ray, 1984; Snyder, 1988). Still, not all studies using selenite/selenate were able to produce this clastogenic effect (Paton and Allison, 1972).

In vivo, tetravalent selenium has been shown to be clastogenic in both hamsters and rats. Intravenous injection of selenite increased the incidence of chromosomal aberrations in both Chinese hamster and rat bone marrow cells (Norppa et al., 1980; Newton and Lilly, 1986). The levels of spindle breakage and overt chromosomal abnormalities in the bone marrow cells after oral administration of sodium selenite were initially high, but leveled off by three weeks after termination of either the chronic or acute exposures (Das, 1982; Mitra, 1984).

In mutagenicity studies with different strains of *S. typhimurium* (i.e., TA100 and TA1537), both Na_2SeO_3 and Na_2SeO_4 were able to induce base-pair substitution-type mutations; in TA98, Na_2SeO_3 was ineffective (Lofroth and Ames, 1978; Jansson et al., 1978; Noda et al., 1979). Mutagenicity results with *E. coli* have thus far proven negative (Kanematsu et al., 1980). Both agents, as well as SeO_2, have yielded positive mutagenicities in the *B. subtilis* rec-assay (Noda et al., 1979; Kanematsu et al., 1980) although negative results were obtained with the potassium salts of these same two compounds (Nishioka, 1975). In studies using purified DNA-metabolizing enzymes, assays of DNA replication fidelity found that selenium had no effects with respect to the misincorporation of nucleotides into daughter DNA strands synthesized using microbial DNA polymerase-α (Zakour et al., 1981a,b).

While DNA damage by selenium agents in prokaryotes was primarily relegated to tetravalent compounds (Nakamuro et al., 1976), as in the cancer studies, selenium can also act to mitigate the DNA-damaging effects of other known mutagens. At subcytotoxic concentrations, the presence of Na_2SeO_3 contributed to a reduction in the levels of chromosomal breakage inducible by malonaldehyde, β-propiolactone, acetylaminofluorene, *N*-hydroxy-acetylaminofluorene, *N*-hydroxy-aminofluorene, *N*-methyl-*N'*-nitro-*N*-nitrosoguanidine, and *N*-acetoxy-2-acetylaminofluorene in several *S. typhimurium* tester strains (Jacobs, 1977; Rosin and Stich, 1979; Shamberger et al., 1979). This same compound has also been shown to suppress the rates of spontaneous mutagenesis in different strains of *S. cerevisiae* (Rosin, 1981).

XIII. SILICON (Si)

The vast majority of studies of the carcinogenicity of silicon deal primarily with two of the most abundant mineral derivatives of the element, i.e., silica and asbestos. Asbestos encompasses a wide group of minerals that are distinctive in their particle size, chemical composition, crystalline structure, and biologic activity. The asbestos fibers (primarily chrysolite and crocidolite) are polymeric chains of silicon dioxide (SiO_2) linked through iron or magnesium ions, although other known carcinogenic metals (i.e., nickel, chromium, or cadmium) may also be present. As there are many other texts that deal with asbestos in great detail, the remainder of this section will focus primarily upon silica and other silicon derivatives.

The three most common forms of crystalline silica are quartz, tridymite, and cristobolite. As with asbestos, the particles are all multimers of SiO_2, but utilize oxygen groups to form their cross-linkages, much in the manner of chromium during olation. The presence of silanol (SiOH), silica-based free radicals, the degree of surface charge and area, and the capacity to promote hydrogen peroxide formation in aqueous environments are all important to the carcinogenic/toxic activity of silicas (Shi et al., 1989; Holland, 1990). After deposition, the silica particles display membranolytic activity and an overt cytotoxic effect against erythrocytes, macrophages, and perhaps polymorphonuclear leukocytes (Shi et al., 1989). The macrophage death, the release of cytoplasmic enzymes, and the enhanced presence of macrophage-derived fibrogenesis factor (MFF) result in fibroblast stimulation and enhanced localized formation of collagen (Heppleston, 1984). Although this mechanism helps clarify how silica-induced fibroplasia might arise (now referred to as silicosis), the linkage to subsequent neoplasia is still lacking.

While the epidemiologic evidence suggesting a carcinogenic potential for silica is vast (Goldsmith et al., 1982; Heppleston, 1985; Forastiere et al., 1986, 1989; Hessel et al., 1986; Lynge et al., 1986; IARC, 1987b, 1990; Mastrangelo et al., 1988; McDonald, 1989; Siemiatycki et al., 1989; Spivac, 1990; Amandus et al., 1991; Carta et al., 1991; Chia et al., 1991; Hnizdo and Sluis-Cremer, 1991), the IARC working group, as of 1987, declared that evidence for the carcinogenicity of crystalline silica in man was "limited" (IARC, 1987b). However, this same group has stated that there was "sufficient" evidence for establishing the carcinogenicity of the same agents in experimental animal models.

In studies using totally nonrepresentative routes of silica exposure, implantation of silica (i.e., quartz) into rabbit lung parenchyma or the direct intrapleural instillation of Min-U-Sil quartz (as well as alkaline-washed quartz into rats) gave rise to significant numbers of lung adenocarcinomas and malignant lymphoma histiocytic type (MLHT) cancers, respectively (along with a few thymomas and lymphosarcomas in the latter study) (Kahlau, 1961; Wagner, 1970; Wagner, 1976; Wagner and Wagner, 1972).

Using intratracheally-instilled silica particles, rats implanted with Min-U-Sil or novaculite again developed significantly greater numbers of lung adenocarcinomas and squamous cell (epidermoid) carcinomas (Groth et al., 1986; Holland et al., 1986). In nearly all cases, the tumors appeared to have arisen from, or were associated with, scar tissue that formed during the silica-derived fibroplasia.

Similar tumorigenic profiles have been obtained following long-term (1.5–2 years) inhalation studies using rats. In a 2-year whole-body exposure study, all of the tumors that developed were characterized as of the epidermoid (squamous) type (Dagle et al., 1986). In the slightly shorter nose-only exposure study, the majority of malignant carcinomas were adenocarcinomas originating from type II alveolar cells (although increased numbers of benign adenomas were also apparent) (Holland et al., 1986). In a large number of exposed rats, irrespective of whether tumor formation occurred, there were also increases in the thickening of the interalveolar septa, metaplasia of the alveolar epithelium, lipoproteinosis, granulomatous lymphadenopathy, and discrete nodular and interstitial fibrosis.

Its should be noted here that the choice of experimental animal model has proven critical in the studies of the carcinogenic character of silica. When mice or hamsters have been employed in similar intratracheal instillation or inhalation studies, tumorigenic responses were not achieved (Holland et al., 1983; Renne et al., 1985; Niemeier et al., 1986; Wilson et al., 1986). While particle size and penetration might play a significant role for the lack of response in mice, the major difference in the hamster model appears to be in the cellular chemistry/function of resident macrophages.

Because of its crystalline nature, silica is not readily testable for genotoxicity in most nonmammalian systems such as the *S. typhimurium*-utilizing Ames assay. Kanematsu et al. (1980) did use the SiO_2 monomer in their *B. Subtilis* rec-assay, but the results were negative. As with most inorganic particles, the use of mammalian test systems have provided more practical results. In SHE cells exposed to Min-U-Sil or quartz, there were increased levels of binuclei and micronuclei, but no concurrent increases in tetraploid cells (Hesterberg et al., 1986). In studies using the small particle tridymite, the incidence of sister chromatid exchanges in cultured human lymphocytes was enhanced (Pairon et al., 1990). However, other studies have failed to demonstrate similar genotoxic effects for certain forms of quartz with respect to mammalian cell chromosomal aberrations, cellular transformations, or inducible metabolic cooperation (Chamberlain, 1983; Oshimura et al., 1984).

As with the cancer studies, little information is available about the genotoxicity of other silicon compounds (i.e., SiN_4, Si_3N_4, Na_2SiF_6, Na_2SiO_3). Thus, a comprehensive statement with regards to the carcinogenicity and/or genotoxicity of silicon cannot yet be formulated.

XIV. THORIUM (Th)

Thorium, the naturally radioactive α-particle-emitting second element of the actinide series, is atypical in the context of the metals discussed in this chapter. Unlike the other metals, carcinogenic thorium was purposely introduced into thousands of humans worldwide as opposed to being internalized as a consequence of occupational or environmental contamination.

The agent Thorotrast ($^{242}ThO_2$) was used widely between 1928 and 1954 (and as long as into the 1960s worldwide) as a diagnostic agent for systemic radiological studies of cerebral angiography, limb arteriography, hepatosplenography, aortography, and portography, as well as locally for bladder, sinus, and fistula visualization (Abbatt, 1979). However, once injected into the body, thorium remains forever, as practically no elimination occurs (except for exhalation of radioactive thoron daughter gas) (reviewed in Tavares et al., 1979; Fisenne et al., 1985). Principally, the thorium accumulates in the liver, spleen, lymph nodes, and bone marrow, thereby allowing long-lasting contact with a great diversity of blood-borne cells as well as with those of the resident tissue.

The carcinogenicity of ThO_2 was first demonstrated in animals as early as 1934 (Roussy et al., 1934). Most of the reported tumors that were observed in exposed animals were spindle cell carcinomas and osteosarcomas (Selbie, 1936; Guimaraes et al., 1955, 1956; Swarm et al., 1962). Overwhelmingly, however, the majority of the data confirming the carcinogenicity of thorium arose from the epidemiological studies and clinical reports that examined Thorotrast-exposed patients. The types of cancers that have been reported in exposed patients cover a wide spectrum — granulomas, hemangioendotheliomas, hepatomas, tumors of the rectum and colon, stomach and duodenum, esophagus and larynx, uterus, bladder, Fallopian tubes, breast, bile duct, nasal sinuses, and brain (reviewed in Abbatt, 1979; Farid, 1982; Mohr et al., 1984; Kahn, 1985; Wargotz et al., 1988). While most of these cancers appeared in random patients, the most commonly encountered cancers amongst all groups were hepatic angiosarcomas (MacMahon et al., 1947; Falk et al., 1979; Telles et al., 1979; Kojiro et al., 1985), skeletal sarcomas (reviewed in Mays and Spiess, 1979), leukemias (Faber, 1978; da Silva Horta et al., 1978; van Kaick et al., 1978; Hirose et al., 1991; Andersson et al., 1993), and thorotrastomas (injection-site granulomas) (general review of all induced cancers in Stover, 1983).

While some humans have been exposed to thorium as the result of medical procedures, workers in thorium-producing or processing industries are also exposed to this metal, albeit in less pure forms and at substantially lower concentrations per exposure. However, the latter more than match the doses given some patients by undergoing repeated exposures over a number of years. In these individuals, the incidence of lung-associated cancers, as well as skeletal- and blood-related cancers, are significantly higher than those observed in nonexposed populations (Polednak et al., 1983; Stover, 1983; Solli et al., 1985).

There is little information on the genotoxic effects of thorium compounds in nonmammalian systems. When used in the *B. Subtilis* rec-assay, thorium tetrachloride ($ThCl_4$) failed to display any mutagenic effects (Nishioka, 1975). Most thorium agents have not been tested by investigators in the other studies routinely cited throughout this chapter.

However, Thorotrast patients have been studied extensively for the incidence of thorium-dependent clastogenic effects (Buckton et al., 1967; Kemmer et al., 1971; Woodliff et al., 1972; Texeira-Pinto and Azevedo e Silva, 1979; Steinstrasser, 1981; Hoegerman and Cummins, 1983). In large numbers of cases, very high frequencies of dicentric and ring chromosomes, indicators of radiation-induced damage, were detected in the lymphocytes of these patients. In occupationally-exposed workers, however, the frequencies were much lower than in Thorotrast patients, although still much higher than in nonexposed controls (Costa-Ribeiro et al., 1975). Some patients have also been shown to have increased *in vivo* somatic mutation frequencies in these same cells (i.e., at the T-cell receptor genetic locus) (Umeki et al., 1991; Kyoizumi et al., 1992). Studies of large populations of Thorotrast patients and thorium industry workers are still in progress, and it is anticipated that further descriptive information regarding thorium genotoxicity will be forthcoming.

XV. TIN (Sn)

Tin is encountered in both inorganic and organic forms of its stable divalent and trivalent ions. As such, each class (as well as valence) can give rise to different genotoxic outcomes in the commonly used assay systems, and can also have differing carcinogenic potentials.

Although there is strong epidemiological evidence for increased development of bronchopulmonary cancers in tin-mine workers around the world (Fox et al., 1981; Hodgson and Jones, 1990; Chen et al., 1992), it is unclear as to whether tin metal or any derivatives were the underlying causative agents. Animal studies using industrial dust from tin mines indicated that the dust was capable of inducing significant numbers of carcinomas in exposed animals (Huang, 1990). However, as the test dust was a mixture of 15 nonradioactive inorganic elements, conclusive proof of the carcinogenicity of tin was not provided.

In studies using rodents exposed to various inorganic tin preparations, the data regarding tumor formation is conflicting. When tin(II) oleate or sodium chlorostannate was administered to mice in their diet and/or drinking water over a 12-month period, no increases in the incidence of lymphomas, hepatomas, or adenomas was demonstrated (Walters and Roe, 1965). However, when rats were provided a similar regimen, the incidence of sarcomas and carcinomas was increased (Roe et al., 1965). In life-term feeding studies, when mice and rats were provided with stannous chloride ($SnCl_2$) in their water/diets, the incidence of cancers was again unaffected (Stoner et al., 1976, Kanisawa and Schroeder, 1967, 1969).

Other routes of exposure to inorganic tin have been employed to assess carcinogenic potential. Implantation of tinfoil was found to give rise to localized tumors in rats, but the effect was dependent upon disc thickness (Alexander and Horning, 1958). Intrathoracic injection of tin needles into mice caused increases in the incidence of nodular hyperplasia, giant-cell granulomas, and enhanced angiogenic capillary formation, but no overt increases in local or general tumor incidence (Bischoff and Bryson, 1976). Intraperitoneal injections of $SnCl_2$ into mice also failed to yield increased lung or abdominal cancers (Shimkin et al., 1977).

Organotin compounds, primarily in the forms of di-, tri-, or tetraalkyl tin complexes, are widely used in industry and farming, but little is known about their carcinogenic potential. In chronic feeding studies, mice fed low levels (<0.5 mg/kg BW) of triphenyltin acetate for 18 months had no increases in the formation of tumors (Innes et al., 1969). Rats fed low to moderate (up to 12 mg/kg BW) doses of tricyclohexyltin hydroxide for 2 years had increases in the number of tumors, but their patterns and distributions were fairly random and did not exhibit a dose-dependent relationship (WHO, 1971, 1980). In a 2-year feeding study, rats fed high doses (50 mg/kg BW) of bis(tri-*n*-butyltin)oxide (TBTO) demonstrated increased numbers of endocrine gland tumors. Those sites primarily affected were the anterior pituitary, the adrenals (increased numbers of pheochromocytomas), the parathyroid, and the pancreas (exocrine adenocarcinomas, but no alterations in islet cells) (Wester et al., 1990). The authors contend that they cannot be certain whether the TBTO was the cause of the neoplastic changes or whether TBTO-linked changes in host metabolic, endocrine, or immune profiles might have contributed to the increased incidence of cancers above background.

Compared with the results from the carcinogenicity studies, more is known regarding the genotoxicity of these tin agents. In mammalian cells, $SnCl_2$ produced extensive DNA damage in treated CHO cells, while $SnCl_4$ had no effect (McClean et al., 1983a). The authors noted that the divalent tin ions produced about 200 times more detectable DNA damage on a equimolar basis than does hexavalent chromium. Similar results have also been observed in treated human white blood cells (McClean et al., 1983b). These studies also indicated that although most of the DNA damage was repaired promptly, long-lived effects that arise from tin particle aggregate dissolution continue to hamper DNA biochemical processes even after source removal. These long-term effects become manifested in cell cycle delays and reduced DNA replication after mitogenic stimulation (Gulati et al., 1989).

The mechanisms underlying tin-induced DNA damage are somewhat similar to those for iron or chromium. The borderline soft acid nature of divalent tin makes it a good electron donor, thereby allowing extensive reductive bond cleavage in the DNA to occur. The tin(II)/tin(III) ionization potential is similar to that for iron(II)/iron(III), and so tin(II) might catalyze the conversion of cellular peroxides to DNA-damaging hydroxyl radicals. This particular mechanism might be the underlying basis for the clastogenicity of divalent tin in CHO cells (Gulati et al., 1989), although this phenomenon is not consistent (Ohno et al., 1982). Lastly, divalent tin can coordinate many atoms (e.g., S, O, Se, N, P, and halogens) and so might lead to the formation of direct adducts with the DNA, or the formation of DNA-DNA and DNA-protein cross-links.

In nonmammalian test systems, such as the *Drosophila* wing spot test (Tripathy et al., 1990), the *B. subtilis* rec-assay (Nishioka, 1975; Kanematsu et al., 1980), and the *S. cerevisiae* reverse mutation/mitotic gene conversion assay (Singh, 1983), all inorganic tin compounds yielded negative mutagenicities.

In studies of the genotoxicity of tri-*n*-butyltin compounds, no chromosomal aberrations in *Mytilus edulis* mussels were noted, and only a weak response was found in the fluctuation test with *S. typhimurium* strain TA100 (Dixon and Prosser, 1986; Davis et al., 1987). However, strongly positive genotoxicities were reported for these agents in the rec-assay and a modified Ames assay, but not in an SOS chromotest system (Hamasaki et al., 1992, 1993). Mono- and di-*n*-butyltin compounds, as well as various mono- and multimethyltin compounds, also displayed mutagenicities in the same assay systems, but were also positive in the SOS test. The methyltin agents also caused nondisjunction and/or loss of sex chromosomes in *Drosophila* (Ramel and Magnusson, 1979). The butyltin and methyltin agents are also now known to cause increases in micronuclei formation and chromosomal aberrations in human lymphocytes (Ghosh et al., 1990, 1991) and rat bone marrow cells (Mazaev and Slepnina, 1973). Similar genotoxicity results have yet to be demonstrated with phenyltin compounds in any of the commonly employed assay systems.

XVI. TITANIUM (Ti)

Few studies have been carried out to determine the carcinogenicity of titanium metal or any of its inorganic/organic derivatives. Epidemiological surveys of workers exposed to titanium-bearing atmospheres (primarily containing titanium oxychloride [$TiOCl_2$], titanium dioxide [TiO_2] particles, or titanium tetrachloride [$TiCl_4$] vapor) indicate that the risk from developing lung (or distal) cancers (as well as other fatal respiratory diseases) was no greater than for reference nonexposed workers (Uragoda and Pinto, 1972; Garabrant et al., 1987; Chen et al., 1988; Fayerweather et al., 1992). The predominant toxicities are localized increases in lung connective tissue (subpleural fibrosis) and titanium deposition along the epithelia and on the lung surface proper (Elo et al., 1972). Titanium-containing alloys used as surgical implants have not been associated with cancers following long-term contact with tissues (Kazantzis, 1981; Sunderman, 1989b), although a soft tissue sarcoma arising in a patient implanted with a titanium-covered pulse generator has been reported (Fraedrich et al., 1984).

While studies with humans failed to demonstrate any carcinogenic potential for titanium compounds, studies with laboratory animals have, in some cases, yielded positive results. Intramuscular injections of rats with pure titanium powder over a six-month period gave rise to a significant increase in the numbers of fibrosarcomas and lymphosarcomas (Furst, 1971b). Similar results, in addition to an increased incidence of hepatomas and splenic lymphomas, were obtained with rats when the sandwich titanocene compound was intramuscularly injected (Furst and Haro, 1969). Via inhalation, rats exposed to very high concentrations of TiO_2 (250 mg/m^3, as can arise in poorly-ventilated industrial settings) developed bronchoalveolar adenomas and cystic keratinizing squamous cell carcinomas; exposures to lower levels (i.e., 10 or 50 mg/m^3) did not yield compound-related tumor formation (Lee et al., 1985). While similar cancer profiles were observed in inhalation studies with $TiCl_4$ that used a 2-year exposure to low to moderate concentrations of titanium (0.1–10 mg/m^3), the incidence of alveolar cell hyperplasia, cholesterol granuloma, and nasal glandular epithelial hyperplasia were also increased at the higher levels tested (Lee et al., 1986). In combination with a known carcinogen (e.g., benzo(α)pyrene), TiO_2 significantly increased the carcinogenic potential of the polycyclic hydrocarbon (Stenback et al., 1976).

Still, not all animal studies have indicated that titanium is carcinogenic. Neither the intratracheal administration of lead titanate to guinea pigs, nor the life-term administration of titanium salts via the diet/drinking water of rats and mice gave rise to greater numbers of tumors than observed in control animals (Schroeder et al., 1964; Steffee and Baetjer, 1965; NCI, 1978). On the basis of most available animal and human data, titanium is considered to belong to that group of metals with a low carcinogenicity (WHO, 1982).

The proof for the mutagenic/genotoxic activity of titanium compounds is even scarcer than that for its carcinogenicity. The overwhelming majority of commonly cited studies (Kada et al., 1980a,b; Kanematsu et al., 1980; Ohno et al., 1982; Singh, 1983; Ivett et al., 1989; Tripathy et al., 1990) indicate no genotoxic effects in the specific bacterial/insect/mammalian system utilized. Even studies using novel test systems failed to demonstrate positive genotoxic effects from titanium (Li, 1986; Kodama, 1989). Only three references have been located which support some marginal genotoxic potential for titanium compounds. Titanium metal and TiO_2 enhance human colon cancer cell growth and surface marker expression (Moyer et al., 1991), increase the incidence of stick chromosomes and anaphase bridges in *Allium cepa* root cells (Levan, 1945), and induce (when present as a titanium-hexachloroethane mixture) reversions in *S. typhimurium* strains TA98 and TA100 in the absence of metabolic (S9) activation (Karlsson et al., 1991).

XVII. VANADIUM (V)

Although vanadium compounds are among the most abundant in the natural environment, and are readily encountered within occupational settings, very little is known regarding the carcinogenicity of this metal in both human and animal models (NIOSH, 1985; WHO, 1988; ATSDR, 1990).

Following life-term feeding studies of rodents with tetravalent vanadium, there was inconclusive evidence for carcinogenicity (Kanisawa and Schroeder, 1967; Schroeder and Balassa, 1967; Schroeder et al., 1970; Schroeder and Mitchener, 1975b). This is likely the result of the low levels of uptake of vanadium, as is the case with many known carcinogenic metals which do not display carcinogenic potentials. Using an intraperitoneal exposure route, repeated exposures of mice to vanadium(III) 2,4-pentanedione also failed to induce significant numbers of lung adenomas (Stoner et al., 1976; Shimkin et al., 1977).

The more likely route for vanadium intoxication is via inhalation of metal-containing dusts or particles. Recent studies using mice and rats exposed to atmospheres of vanadium pentoxide (V_2O_5) indicated a dose-related increase in the incidence of pulmonary, as well as sino-nasal, epithelial hyperplasia and metaplasia (NIH, 1993). Epidemiological studies have demonstrated that both acute and chronic exposure to moderate-to-high levels of vanadium (pentoxide or vanadate) dusts/fumes in the workplace resulted in increased incidences of noncancerous pulmonary diseases (boilermakers' bronchitis) (reviewed in Faulkner-Hudson, 1964; McTurk et al., 1966; Roschin, 1967; reviewed in Waters, 1977; Levy et al., 1984), increased localized fibrotic foci and lung weights, as well as an enhanced incidence of lung cancers initiated by other agents (Stocks, 1960; Hickey et al., 1967; Hopkins et al., 1977; Kivuoloto, 1980; Rhoads and Sanders, 1985; Zychlinksi, 1980; Kowalska, 1989).

Apparently, vanadium, like several of the other metals discussed in this chapter, might not act as a direct carcinogen, but rather exert secondary toxicities (i.e., immunosuppressive effects) within exposed hosts, thereby allowing initiated cancers to be expressed into neoplasms. In addition, vanadium, in certain forms, can also act as an anticarcinogen. If vanadium is given in the form of vanadocene to hosts bearing Erlich ascites or liver tumors, the metal (as with germanium, molybdenum, and titanium) can be cancerostatic (Kopf-Maier and Kopf, 1979; Kopf-Maier et al., 1980). As a dietary supplement, the metal can block cancer induction by other carcinogenic compounds (Thompson et al., 1984; Kingsnorth et al., 1986).

Vanadium has been shown to be a mutagen and a clastogen in numerous mammalian and prokaryote assay systems. The majority of these *in vitro* studies demonstrate positive genotoxic effects for vanadium in test systems using bacteria, yeast, and cultured mouse and human cells for endpoints such as recombination repair, gene mutation, and/or DNA synthesis.

Vanadium compounds, such as NH_4VO_3, V_2O_5, and $VOCl_2$, were able to induce revertants in *B. subtilis* rec⁻ mutants (Kanematsu et al., 1980), but could not induce a similar effect in *S. typhimurium* or in *E. coli* strains WP_2 and B/r WP_2 (Kanematsu and Kada, 1978; Kada et al., 1980a,b). However, subsequent studies have confounded the issue of vanadium mutagenicity. The compound NH_4VO_3 was found to be mutagenic in the *S. typhimurium* tester strain 1537 in a modified incorporation assay, as well as in a fluctuation test with TA100 (Arlauskas et al., 1985). Insoluble V_2O_5 was able to induce reverse mutations with *E. coli* WP_2, WP_2 uvr A, and Cm-981, but was still unable to induce frameshift mutations in *E. coli* ND160 or MR102, nor was it mutagenic with *Salmonella* strains TA1535, TA1537, TA100, or TA98 in standard reversion assays (Sun, 1987).

In yeast, pentavalent vanadium compounds were capable of increasing both the convertant and revertant mutation frequencies of mitotic gene conversions and reverse point mutations, respectively, in *S. cerevisiae* D7, and giving rise to aneuploidy in strain D61M (Bronzetti et al., 1990; Galli et al., 1991). In addition, tetravalent $VOSO_4$ was able to increase the formation of diploid spores in yeast (Sora et al., 1986).

In cultured mammalian cells, pentavalent vanadium has been shown to inhibit or enhance DNA synthesis, depending on the concentration of metal present in the culture medium (Hori and Oka, 1980; Carpenter, 1981; Jackson and Linskens, 1982; Sabbioni et al., 1983; Smith, 1983). Vanadium has also been able to cause increases in the formation of DNA-protein cross-links in both CHO and human leukemic MOLT4 cells (Cohen et al., 1992). Other forms of genetic damage, such as DNA strand breaks, have been demonstrated in vanadium-treated human leukocytes (Birnboim, 1988). Chromosomal aberrations and sister chromatid exchanges were documented in studies using CHO cells treated with NH_4VO_3, $VOSO_4$, or V_2O_3 (Owusu-Yaw et al., 1990). However, as with the prokaryote assays, other investigations using similar test systems did not obtain similar results (McLean et al., 1982; Roldan and Altamirano, 1992).

The antitubulin nature of many vanadium agents (Cande and Wolniak, 1978; Wang and Choppin, 1981; Smith, 1983) has been thought to underlie the observations that the treatment of human lymphocytes with vanadate ions results in increased numbers of polyploid cells without concurrent increases in mitotic indices (Bracken and Sharma, 1985; Sharma and Talukder, 1987; Roldan and Altamirano, 1992). Vanadium also has the capacity to alter the colony formation capacity of cultured human tumor cell lines (Hanauske et al., 1987; Sheu et al., 1990), and can affect the transformation of cultured cells by itself or after transfection with intact viral DNA or specific oncogenes (Klarland, 1985; Dessureault and Weber, 1990; Kowalski et al., 1992).

As a result of all of these conflicting results among and within both the prokaryotic and eukaryotic test systems, no firm conclusions with respect to vanadium genotoxicity can as yet be drawn.

XVIII. ZINC (Zn)

Zinc exists as a divalent ion in all of its major inorganic and organozinc derivatives. Although most experimental data have demonstrated that zinc deficiency is by far more crucial to human and animal health than is zinc toxicity, there is still evidence to support the hypothesis that overexposure to zinc can result in deleterious health effects.

Although it has been shown that cancerous tissues often have higher concentrations of zinc than do normal tissues (Tupper et al., 1955; Mulay et al., 1971), the role of zinc as a carcinogenic agent remains controversial. While little epidemiological evidence regarding the carcinogenicity of zinc in humans is available, its potential as a carcinogen has been demonstrated in several animal models. In the first experiment of metal carcinogenesis, direct injections of zinc chloride ($ZnCl_2$) into rooster testes caused the formation of testicular teratomas (Michalowsky, 1926). Similar experiments using other fowl and mammalian models revealed similar results (Bagg, 1936; Riviere et al., 1960; Sunderman, 1971; Guthrie and Guthrie, 1974). Similarly, intratracheal instillation or intrapleural injection of zinc powder also resulted in an increased incidence of testicular seminomas, as well as localized lung reticulosarcomas (Dvizhkov and Potapova, 1970). However, using other routes of exposure, zinc salts failed to produce local or distal tumors in animals after either subcutaneous (Gunn et al., 1964, 1967), intraperitoneal (Stoner et al., 1976; Shimkin et al., 1977), or intramuscular injections (Heath et al., 1962).

The ingestion of zinc salts has also been shown to induce increased tumor formation, as well as to enhance the formation of tumors induced by other carcinogens, in the early segments of the gastrointestinal tract (Chahovitch, 1955; Stocks and Davies, 1964; McGlashan, 1972; Bespalov et al., 1990). The presence of $ZnCl_2$ in drinking water has been correlated with an increased incidence of mammary carcinomas of rodents (Halme, 1961). Conversely, both the ingestion of zinc as well as an induced deficiency have been shown to inhibit the development of transplanted tumors or those initiated *in situ* (DeWys et al., 1970; McQuity et al., 1970; Poswillo and Cohen, 1971; DeWys and Poires, 1972).

Previously considered to be one of the less noxious metals, zinc has been shown to be clastogenic both *in vivo* and *in vitro* (reviewed in Sharma and Talukder, 1987). Over a wide range of concentrations tested, both acute and chronic repeated injections of mice with $ZnCl_2$ induced chromosomal aberrations in bone marrow cells (Gupta et al., 1991). The incidence of translocations and dicentric and ring chromosomes was found to be increased; this effect was even more pronounced if the animals were simultaneously calcium deficient (Deknudt and Gerber, 1979). A similar pattern of inducible clastogenicity was observed in rats following the chronic inhalation of aerosols of zinc salts (Voroshilin et al., 1978).

When normal stimulated cultured human lymphocytes were pretreated with various soluble zinc salts, there was an increase in the numbers of cells with chromosomal fragmentation, diploidy, dicentrism, and chromatid gaps and breaks (Deknudt, 1978; Deknudt and Deminatti, 1978). A later study using a similar treatment regimen indicated that zinc induced a dose-dependent decrease in lymphocytic mitotic indices (Saksoong and Campiranon, 1983). It has been suggested that the structural chromosomal aberrations observed in the peripheral lymphocytes of zinc industry workers who were simultaneously exposed to cadmium and lead were influenced by a zinc-related synergistic effect and not by either cadmium and/or lead alone (Bauchinger and Rohr, 1976).

The current classification of zinc as a mutagen is ambiguous and inconclusive. In the Ames test for the induction of *S. typhimurium* reversions, $ZnCl_2$ was positive in promoting frame-shift mutations (Kalinina and Polukhina, 1977; Polukhina et al., 1977). However, in the *B. subtilis* rec-assay (Nishioka, 1975; Kada et al., 1980; Kanematsu et al., 1980), the *E. coli* WP2 system (Shirasu et al., 1976), as well as in the *S. cerveisiae* D7 gene conversion/reverse mutation assay (Singh, 1983), both $ZnCl_2$ and $ZnSO_4$ were only weakly active or totally nonmutagenic.

In higher organisms, zinc salts were again capable of inducing mutagenic effects. The numbers of dominant lethals and sex-linked mutations in the offspring of *Drosophila* were increased by treatment of adult flies with $ZnCl_2$ (Carpenter and Ray, 1969). In cultured Syrian hamster embryo cells, zinc nitrate ($Zn[NO_3]_2$) and zinc sulfate ($ZnSO_4$) were able to enhance the incidence of viral transformations, although direct transforming activity or the enhancement of transformations induced by other chemical agents were not affected by $ZnCl_2$ (Casto et al., 1979; DiPaolo and Casto, 1979; Riverdal and Sanner, 1981; Sirover, 1981).

ACKNOWLEDGMENT

The authors would like to thank Dr. Xi Huang for his assistance in the preparation of portions of this chapter.

REFERENCES

Abbatt, J. D., *Environ. Res.,* 18, 6–12, 1979.
Abramson, M. J., Wlodarczyk, J. H., Saunders, N. A., and Hensley, M. J., *Am. Rev. Respir. Dis.,* 139, 1042–1057, 1979.
Adamson, R. H., Canellos, G. P., and Sieber, S. M., *Cancer Chemother. Rep. Part 1,* 59, 599–610, 1975.
Agnese, T., Veris, B., and Stantolini, B., *Ig. Mod.,* 52, 149–160, 1959.
Ahlberg, J. C., Ramel, C., and Wachmeister, C., *Ambio,* 1, 29–31, 1972.
Al-Hakkak, Z. S., Hamamy, H. A., Murad, A. M. B., and Hussain, A. F., *Mutat. Res.,* 171, 53–60, 1986.
Albert, L., Hienzsch, E., Arndt, J., and Kriester, A., *Z. Urol.,* 8, 561–566, 1972.
Alexander, P. and Horning, H. S., *CIBA Foundation on Carcinogenesis: Mechanisms of Action,* Little, Brown, Boston, 1958.
Alexandersson, R. and Hogstedt, C., *Nord. Arbetmiljö Mötet,* 25, 27–33, 1987.
Alexandersson, R. and Hogstedt, C., *Arbetsmiljöfondens Projekt No. 82–0846,* Slutapport, Stockholm, 1988.
Allaway, W. H., *Ann. N.Y. Acad. Sci.,* 199, 17–25, 1972.
Allaway, W. H., in *Trace Substances in Environmental Health,* Vol. XII, Hemphill, D. D., Ed., University of Missouri Press, Columbia, 3, 1978.
Alpern-Ehran, H. and Brem, S., *Surg. Forum.,* 36, 498–500, 1985.
Amacher, D. E. and Paillet, S. C., *Mutat. Res.,* 78, 279–288, 1980.
Amandus, H. E., Shy, C., Wing, S., Blair, A., and Heineman, E. F., *Am. J. Ind. Med.,* 20, 57–70, 1991.
Andersen, O., *Environ. Health Perspect.,* 47, 239–253, 1983.
Andersson, L., Wingren, G., and Axelson, O., *Int. Arch. Occup. Environ. Health,* 62, 249–252, 1990.
Andersson, M., Carstensen, B., and Visfeldt, J., *Radiat. Res.,* 134, 224–233, 1993.
Andronikashvili, E. L. and Mosulishvili, L. M., in *Metal Ions in Biological Systems,* Vol. 10, Sigel, H., Ed., Marcel Dekker, New York, 167, 1980.
Andronikashvili, E. L., Mosulishvili, L. M., Belokobiliski, A. I., Tevzieva, T. K., and Efremova, E. Y., *Cancer Res.,* 34, 271–274, 1974.
Araki, M. S., Okada, S., and Fujita, M., *Gann,* 45, 449–451, 1954.
Arimori, S., Watanabe, K., Yoshida, M., and Nagao, T., in *Immunomodulation by Microbial Products and Related Synthetic Compounds,* Yamamura, Y., Ed., Elsevier, Amsterdam, 498, 1981.
Arlauskas, A., Baker, R. S. U., Bonin, A. M., Tandon, R. K., Crisp, P. T., and Ellis, J., *Environ. Res.,* 36, 379–388, 1985.
Arts, J. H. E., Reuzel, P. G. J., Falke, H. E., and Beems, R. B., *Food Chem. Toxicol.,* 28, 571–579, 1990.
Ashby, J., Ishidate, J., Stoner, G. D., Morgan, M. A., Ratpan, F., and Callander, R. D., *Mutat. Res.,* 240, 217–225, 1990.
Aso, H., Suzuki, F., Yamaguchi, T., Hayashi, Y., Ebina, T., and Ishida, N., *Microbiol. Immunol.,* 29, 65–74, 1985.
ATSDR, Toxicological Profile for Vanadium and Compounds, Agency for Toxic Substances and Disease Registry, U.S. Department of Health and Human Services, Public Health Service, Atlanta, GA, 1990.
ATSDR, Toxicological Profile for Beryllium, Agency for Toxic Substances and Disease Registry, U.S. Department of Health and Human Services, Public Health Service, Atlanta, GA, 1992.
Auerbach, C., *Mutation Research; Problems, Results, and Perspectives,* Auerbach, C., Ed., Chapman and Hall, London, 1976.
Azar, A., Trochiowicz, H. J., and Maxfield, M. E., Review of lead studies in animals carried out at Haskell Laboratory. Two-year feeding study and response to hemorrhage study, in Environmental Health Aspects of Lead: Proc. Int. Symp. Barth, D., Berlin, A., Engel, R., Recht, P., and Smeets, J., Eds., Commission of the European Communities Directorate General for Dissemination of Knowledge Center for Information and Documentation, Amsterdam, 199, 1973.
Babain, E. A., Bagramian, S. B., and Pogosian, A. S., *Gig. Tr. Prof. Zabol.,* 9, 33–36, 1980.
Bagg, H. J., *Am. J. Cancer,* 26, 69, 1936.
Baker, E. L., Goyer, R. A., Fowler, B. A., Khettry, U., Bernard, D. B., Adler, S., White, R. D., Babayan, R., and Feldman, R. G., *Am. J. Ind. Med.,* 1, 139–148, 1980.
Baldwin, R. W., Cunningham, G. J., and Pratt, D., *Br. J. Cancer,* 18, 503–507, 1964.
Balo, J., Batjai, A., and Szende, B., *Mag. Onkol.,* 9, 144–151, 1965.
Baranowska, H., Ejchart, A., and Putrament, A., *Mutat. Res.,* 42, 343–348, 1977.
Barch, D. H., *J. Am. Coll. Nutr.,* 8, 99–107, 1989.
Barnes, J. M., Denz, F. A., and Sissons, H. A., *Br. J. Cancer,* 4, 212–222, 1950.
Bauchinger, M. and Rohr, G., *Mutat. Res.,* 38, 102–103, 1976.
Bauchinger, M. and Schmid, E., *Mutat. Res.,* 14, 95–100, 1972.
Bauchinger, M., Dresp, J., Schmid, E., Englert, N., and Krause, C., *Mutat. Res.,* 56, 75–80, 1977.

Bech, A. O., Kipling, M. D., and Heather, J. C., *Br. J. Ind. Med.,* 19, 239–252, 1962.
Beckman, R. A., Mildvan, A. S., and Loeb, L. A., *Biochemistry,* 24, 5810–5817, 1985.
Beek, B. and Obe, G., *Experientia,* 30, 1006–1007, 1974.
Beek, B. and Obe, G., *Humangenetik,* 29, 127–134, 1975.
Berger, M. de Hazen, M., Nejjari, A., Fournier, J., Guignard, J., Pezerat, H., and Cadet, J., *Carcinogenesis,* 14, 41–46, 1993.
Bespalov, V. G., Troian, D. N., Petrov, A. S., and Aleksandrov, V. A., *Vopr. Onkol.,* 36, 559–563, 1990.
Beyersmann, D. and Hartwig, A., *Toxicol. Appl. Pharmacol.,* 115, 137–145, 1992.
Birnboim, H. C., *Biochem. Cell Biol.,* 66, 374–381, 1988.
Bischoff, F. and Bryson, G., *Proc. Exp. Tumor Res.,* 5, 85–133, 1964.
Bischoff, F. and Bryson, G., *Res. Commun. Chem. Pathol. Pharmacol.,* 15, 331–340, 1976.
Blakely, B. R., *J. Appl. Toxicol.,* 7, 167–174, 1987.
Bobyleva, L. A., Chopikashvili, L. V., Alekhina, N. I., and Zasukhina, G. D., *Tsitol. Genet.,* 25, 18–23, 1991.
Bobyleva, L. A., Chopikashvili, L. V., Alekhina, N. I., and Zasukhina, G. D., *Genetika,* 29, 430–434, 1993.
Bogden, J. D., Chung, H. R., Kemp, F. W., Holding, K., Bruening, K. S., and Naveh, Y., *J. Nutr.,* 116, 2432–2442, 1986.
Bompart, G., *J. Toxicol. Clin. Exp.,*10, 95–104, 1990.
Boyland, E., Dukes, C. E., Grover, P. L., and Mitchley, B. C., *Br. J. Cancer,* 16, 283–288, 1962.
Bracken, W. M. and Sharma, R. P., *Biochem. Pharmacol.,* 34, 2465–2470, 1985.
Braun, P., Guillerm, J., Pierson, B., Lacoste, J., and Sadoul, P., *Rev. Med. Nancy,* 85, 702–708, 1960.
Brem, S. B., Tsanaclis, A. M. C., and ZagZag, D., *Neurosurgery,* 26, 391–396, 1990a.
Brem, S. B., ZagZag, D., Tsanaclis, A. M. C., Gately, S., Elkouby, M. P., and Brein, S. E., *Am. J. Pathol.,* 137, 1121–1142, 1990b.
Bronzetti, G., Morichetti, E., Della Croce, C., Del Carratore, R., Giromini, L., and Galli, A., *Mutagenesis,* 5, 293–295, 1990.
Brooks, A. L., Griffith, W. C., Johnson, N. F., Finch, G. L., and Cuddihy, R. G., *Radiat. Res.,* 120, 494–507, 1989.
Brown, L. M., Pottern, L. M., and Blot, W. J., *Environ. Res.,* 34, 250–261, 1984.
Brusick, D., Gletten, F., Jaganrath, D. R., and Weekes, U.., *Mutat. Res.,* 38, 386–387, 1976.
Buckton, K. E., Langlands, A. O., and Woodcock, G. E., *Int. J. Radiat. Biol.,* 112, 565–577, 1967.
Cande, W. Z. and Wolniak, S. M., *J. Cell Biol.,* 79, 573–580, 1978.
Cantor, K. P., Santag, J. M., and Held, M. F., *Am. J. Ind. Med.,* 10, 73–89, 1986.
Carleton, W. W. and Price, P. S., *Food Cosmet. Toxicol.,* 11, 827–840, 1973.
Carpenter, G., *Biochem. Biophys. Res. Commun.,* 102, 1115–1121, 1981.
Carpenter, J. M. and Ray, J. H., *Ann. Zool.,* 9, 1121–1126, 1969.
Carta, P., Cocco, P. L., and Casula, D., *Br. J. Ind. Med.,* 48, 122–129, 1991.
Casto, B. C., Meyer, J. D., and DiPaolo, J. A., *Cancer Res.,* 39, 193–198, 1979.
Casto, B. C., Pieczynski, W. J., Nelson, R. L., and DiPaolo, J., *Proc. Am. Assoc. Cancer Res.,* 17, 12, 1976.
Chahovitch, X., *Glas. Srp. Akad. Nauka Umet. Od. Med. Nauka,* 215, 143–146, 1955.
Chamberlain, M., *Environ. Health Perspect.,* 51, 5–9, 1983.
Chau, N., Benamghar, L., Pham, Q. T., Teculescu, D., Rebstock, E., and Mur, J. M., *Br. J. Ind. Med.,* 50, 1017–1031, 1993.
Chen, J. L. and Fayerweather, W. E., *J. Occup. Med.,* 30, 937–942, 1988.
Chen, J., McLaughlin, J. K., Zhang, J. Y., Sone, B. J., Luo, J., Chen, R., Dosemeci, M., Rexing, S. H., Wu, Z., Hearl, F. J., McCawley, M. A., and Blot, W. J., *J. Occup. Med.,* 34, 311–316, 1992.
Cheng, K. C., Cahill, D. S., Kasai, H., Nishimura, S., Loeb, L. A., *J. Biol. Chem.,* 267, 166–172, 1992.
Chia, S. E., Chia, K. S., Phoon, W. H., and Lee, H. P., *Scand. J. Work Environ. Health,* 17, 170–174, 1991.
Chopikashvili, L. V., Bobyleva, L. A., and Zolotareva, G. N., *Tsitol. Genet.,* 25, 45–49, 1991.
Clark, L. C., *Fed. Proc.,* 44, 2584–2589, 1985.
Clarke, C. H. and Shankel, D. M., *Mutat. Res.,* 202, 19–23, 1988.
Clayton, C. C., King, J. H., and Spain, J. D., *Fed. Proc.,* 12, 190, 1953.
Coates, R. J., Weiss, N. S., Daling, J. R., Rettmer, R. L., and Warnick, G. R., *Cancer Res.,* 49, 4353–4356, 1989.
Cohen, M. D., Klein, C. B., and Costa, M., *Mutat. Res.,* 269, 1412–1418, 1992.
Cohen, M. D., Latta, D., Coogan, T., and Costa, M., in *Biological Effects of Heavy Metals,* Vol. II, Foulkes, E. C., Ed., CRC Press, Boca Raton, FL, 19, 1990.
Columbano, A., Ledda, G. M., Sirigu, P., Perra, T., and Pani, P., *Am. J. Pathol.,* 110, 83–88, 1983.
Coogan, P., Stein, L., Hsu, G., and Hass, G., *Lab. Invest.,* 26, 473, 1983.
Cooper, W. C., *Ann. N.Y. Acad. Sci.,* 271, 250–259, 1976.
Costa, D., Guignard, J., Zalma, R., and Pezerat, H., *Toxicol. Ind. Health,* 5, 1061–1078, 1989.
Costa, M., Cantoni, O., DeMars, M., and Swartzendruber, D. E., *Res. Commun. Chem. Pathol.,* 38, 405–419, 1982b.
Costa, M., Heck, J. D., and Robison, S. H., *Cancer Res.,* 42, 2757–2763, 1982a.
Costa-Ribeiro, C., Barcinski, M. A., Figueiredo, N., Penna Franca, N., Lobai, N., and Krieger, H., *Health Phys.,* 28, 225–231, 1975.
da Silva Horta, J., da Silva Horta, M. E., da Motta, L. C., and Tavares, M. H., *Health Phys.,* 35, 137–151, 1978.
Dagle, G. E., Wehner, A. P., Clark, M. L., and Buschbom, R. L., in *Silica, Silicosis, and Cancer,* Goldsmith, D. F., Winn, D. M., and Shy, C. M., Eds., Praeger Press, New York, 255, 1986.
Dalpra, L., Tibiletti, M. G., Nocera, G., Giulotto, P., Auriti, L., Carnelli, V., and Simoni, G., *Mutat. Res.,* 120, 249–256, 1983.
Damstra, T., *Environ. Health Perspect.,* 19, 297–307, 1977.

Das, S. K., Cytological and Cytochemical Effects of Metallic Salts on Mammalian Systems, Doctoral Thesis, University of Calcutta, 1982.
Davis, A., Barale, R., Brun, G., Foster, R., Gunther, T., Hautefeuille, H., van der Heiden, C. A., Knapp, A. G., Krowke, R., Kuroki, T., Lopreino, N., Malaveille, C., Merker, H. J., Monaco, M., Mosesso, P., Neubert, D., Norppa, H., Sorsa, M., Vogel, E., Voogd, C. E., Umeda, M., and Bartsch, H., *Mutat. Res.*, 188, 65–95, 1987.
DeJorge, F. B., Goes, J. S., Guedes, J. L., and DeUlhoa Cintra, A. B., *Clin. Chim. Acta*, 12, 403–406, 1965.
DeJorge, F. B., Paiva, L., Mion, P., and DeNova, R., *Acta Otolaryngol.*, 62, 454–458, 1966.
Deknudt, G., *Mutat. Res.*, 53, 176, 1978.
Deknudt, G. and Deminatti, M., *Toxicology*, 10, 67–75, 1978.
Deknudt, G. and Gerber, G. B., *Mutat. Res.*, 68, 163–168, 1979.
Deknudt, G., Colle, A., and Gerber, G. B., *Mutat. Res.*, 454, 77–83, 1977a.
Deknudt, G., Manuel, Y., and Gerber, G. B., *J. Toxicol. Environ. Health*, 3, 885–891, 1977b.
DeMeo, M., Laget, M., Castegnaro, M., and Dumenil, G., *Mutat. Res.*, 260, 295–306, 1991.
Demerec, M. and Hanson, J., *Cold Spring Harbor Symp. Qual. Biol.*, 16, 215–228, 1951.
Demerec, M., Bertani, G., and Flint, J., *Am. Nat.*, 85, 119–136, 1951.
Dessureault, J. and Weber, J. M., *J. Cell. Biochem.*, 43, 293–296, 1990.
DeWys, W. D. and Poires, W., *J. Natl. Cancer Inst.*, 48, 375–381, 1972.
DeWys, W. D., Poires, W., Richter, M. C., and Strain, W. H., *Proc. Soc. Exp. Biol. Med.*, 135, 17–22, 1970.
Dhir, H., Sharma, A., and Talukder, G., *J. Cytol. Genet.*, 20, 36–45, 1985.
DiPaolo, J. A., *Fed. Proc.*, 23, 393, 1964.
DiPaolo, J. A. and Casto, B. C., *Cancer Res.*, 39, 1008–1013, 1979.
Dixon, D. R. and Prosser, H., *Aquat. Toxicol.*, 8, 185–195, 1986.
Dodion, P., Putz, P., Amiri-Lamraski, M. H., Efira, A., de Martelaere, E., and Heimann, R., *Histopathology*, 6, 807–814, 1982.
Domingo, J. L., *Res. Environ. Contam. Toxicol.*, 108, 105–132, 1989.
Dunkel, V. C., Pienta, R. J., Sivak, A., and Traul, K. A., *J. Natl. Cancer Inst.*, 67, 1303–1315, 1981.
Dutra, F. R. and Largent, E. J., *Am. J. Pathol.*, 26, 197–204, 1950.
Dutra, F. R., Largent, E. J., and Roth, J., *Arch. Pathol.*, 51, 473–479, 1951.
Dvizhkov, P. P. and Potapova, I. M., *Int. J. Vitalst. Zivilisat.*, 2, 23–29, 1970.
Egilsson, V., Evans, I. H., and Wilkie, D., *Mol. Gen. Genet.*, 174, 39–46, 1979.
El-Deiry, W. S., Downey, K. M., and So, A., *Proc. Natl. Acad. Sci. U.S.A.*, 81, 7378–7382, 1984.
Elinder, C. G. and Friberg, L., in *Toxicology of Metals*, Vol. 2, 2nd ed., Friberg, L., Nordberg, G. F., and Vouk, V., Eds., Elsevier Amsterdam, 211, 1986.
Elo, R., Maatta, K., Uksila, E., and Arstila, A. U., *Arch. Pathol.*, 94, 417–424, 1972.
Epstein, S. S. and Mantel, N., *Experientia*, 24, 580–581, 1968.
Faber, M., *Health Phys.*, 35, 153–158, 1978.
Falk, H., Telles, N. C., Ishak, K. G., Thomas, L. B., and Popper, H., *Environ. Res.*, 18, 65–73, 1979.
Fan, A. M., *J. Toxicol. Sci.*, 15 (Suppl. 4), 162–175, 1990.
Farah, S. B., *Rev. Brasil. Genet.*, VI, 433–442, 1983.
Fare, G. and Howell, J. S., *Cancer Res.*, 24, 1279–1284, 1964.
Farid, I., *J. Am. Med. Assoc.*, 248, 550, 1982.
Faulkner-Hudson, T. G., *Vanadium: Toxicology and Biological Significance*, Browning, E., Ed., Elsevier, New York, 1964.
Fayerweather, W. E., Karns, M. E., Gilby, P. G., and Chen, J. L., *J. Occup. Med.*, 34, 164–169, 1992.
Fielding, J., *Br. Med. J.*, 1, 1800–1804, 1962.
Fisenne, J. M., Demoleas, P. N., and Harley, N. H., *Strahlentherapie Sonderb.*, 80, 151–156, 1985.
Flessel, C. P., in *Inorganic and Nutritional Aspects of Cancer*, Schrauzer, G. N., Ed., Plenum Press, New York, 117, 1981.
Forastiere, F., Lagorio, S., Michelozzi, P., Cavariani, F., Arca, M., Borgia, P., Perucci, C., and Axelson, O., *Am. J. Ind. Med.*, 10, 363–370, 1986.
Forastiere, F., Lagorio, S., Michelozzi, P., Perucci, C., and Axelson, O., *Br. J. Ind. Med.*, 46, 877–880, 1989.
Ford-Hutchinson, A. W. and Perkins, D. J., *Eur. J. Biochem.*, 21, 55–59, 1971.
Ford-Hutchinson, A. W. and Perkins, D. J., *Radiat. Res.*, 51, 244–248, 1972.
Forni, A., Sciame, A., Bertazzi, P. A., and Alessio, L., *Arch. Environ. Health*, 35, 139–146, 1980.
Fox, A. J., Goldblatt, P., and Kinlen, L. J., *Br. J. Ind. Med.*, 38, 378–380, 1981.
Fraedrich, G., Kracht, J., Scheld, H. H., Jundt, G., and Munch, J., *Thorac. Cardiovasc. Surg.*, 32, 67–69, 1984.
Francis, A. R., Shetty, T. K., and Bhattacharya, R. K., *Mutat. Res.*, 199, 85–93, 1988.
Fukunaga, M., Kurachi, Y., and Mizuguchi, Y., *Chem. Pharm. Bull.*, 30, 3017–3019, 1982.
Furst, A., *Cancer Geol. Soc. Am. Mem.*, 123, 109–114, 1971a.
Furst, A., in *Environmental Geochemistry in Health and Disease*, Cannon, H. L., and Hopps, H. C., Eds., Geological Society of America, Boulder, CO, 109, 1971b.
Furst, A., *J. Natl. Cancer Inst.*, 60, 1171–1173, 1978.
Furst, A. and Haro, R. T., *Prog. Exp. Tumor Res.*, 12, 102–133, 1969.
Furst, A., Schlauder, M., and Sasmore, D. P., *Cancer Res.*, 36, 1779–1783, 1976.
Galli, A., Vellosi, R., Florio, R., Della Croce, C., Del Carrotore, R., Morichetti, E., Giromini, L., Rosellini, D., and Bronzetti, G., *Teratogen. Carcinogen. Mutagen.*, 11, 175–183, 1991.

Garabrant, D. H., Fine, L. J., Oliver, C., Bernstein, L., and Peters, J. M., *Scand. J. Work Environ. Health,* 13, 47–51, 1987.
Gardner, L. U. and Hestington, H. F., *Fed. Proc.,* 5, 21, 1946.
Gelfant, S., in *International Review of Cytology,* Bourne, G. H. and Danielli, J. F., Eds., Academic Press, New York, 1963.
Gerber, G., Leonard, A., and Jacquet, P., *Mutat. Res.,* 76, 115–141, 1980.
Ghosh, B. B., Talukder, G., and Sharma, A., *Mutat. Res.,* 245, 33–39, 1990.
Ghosh, B. B., Talukder, G., and Sharma, A., *Mech. Ageing Dev.,* 57, 125–137, 1991.
Gibbs, G. W. and Horowitz, I., *J. Occup. Med.,* 21, 347–353, 1979.
Gilman, J. P. W., *Cancer Res.,* 22, 158–162, 1962.
Gilman, J. P. W. and Ruckerbauer, G. M., *Cancer Res.,* 22, 152–157, 1962.
Goldsmith, D. F., Guidotti, T. L., and Johnston, D. R., *Am. J. Ind. Med.,* 3, 423–440, 1982.
Goodman, M. F., Keener, S., and Guidotti, S., *J. Biol. Chem.,* 258, 3469–3475, 1983.
Grandjean, P., Wulf, H., and Niebuhr, E., *Environ. Res.,* 32, 199–204, 1983.
Grant, K. E., Connor, M. W., and Newberne, P. M., *Toxicol. Appl. Pharmacol.,* 41, 166, 1977.
Griffin, A. C., *Adv. Cancer Res.,* 29, 419–442, 1979.
Griffin, A. C. and Jacobs, M. M., *Cancer Lett.,* 3, 177–181, 1977.
Groth, D. H., Stetler, L. E., Platek, S. F., Lal, J. B., and Burg, J. R., in *Silica, Silicosis, and Cancer,* Goldsmith, D. F., Winn, D. M., and Shy, C. M., Eds., Praeger Press, New York, 243, 1986.
Guilianelli, C., Baeza-Squiban, A., Boisvieux-Ulrich, E., Houcine, O., Zalma, R., Guennou, C., Pezerat, H., and Marano, F., *Environ. Health Perspect.,* 101, 436–442, 1993.
Guimaraes, J. P. and Lamerton, L. F., *Br. J. Cancer,* 10, 527–532, 1956.
Guimaraes, J. P., Lamerton, L. F., and Christensen, W. R., *Br. J. Cancer,* 9, 253–267, 1955.
Gulati, D. K., Witt, K., Anderson, B., Zeiger, E., and Shelby, M. D., *Environ. Mol. Mutagen.,* 13, 133–145, 1989.
Gunn, S. A., Gould, T. C., and Anderson, W. A. D., *Arch. Pathol.,* 83, 493–499, 1967.
Gunn, S. A., Thelma, C., and Anderson, W. A., *Proc. Soc. Exp. Biol. Med.,* 115, 653, 1964.
Gupta, T., Talukder, G., and Sharma, A., *Biol. Trace Elem. Res.,* 30, 95–101, 1991.
Guthrie, J. and Guthrie, O., *Cancer Res.,* 34, 2612–2614, 1974.
Haddow, A. and Horning, E. S., *J. Natl. Cancer Inst.,* 24, 109–147, 1960.
Haddow, A., Dukes, C. E., and Mitchley, B. C. V., *Rep. Br. Emp. Cancer Camp.,* 39, 74–79, 1961.
Haddow, A., Roe, F. J. C., and Mitchley, B. C. V., *Br. Med. J.,* 1, 1593–1594, 1964.
Haley, T. J., *J. Pharm. Sci.,* 54, 663–670, 1965.
Hall, Z. W. and Lehman, I. R., *J. Mol. Biol.,* 36, 321–333, 1968.
Halme, E., *Vitalst. Zivilisat.,* 6, 59–61, 1961.
Hamasaki, T., Sato, T., Nagase, H., and Kito, H., *Mutat. Res.,* 280, 195–203, 1992.
Hamasaki, T., Sato, T., Nagase, H., and Kito, H., *Mutat. Res.,* 300, 265–271, 1993.
Hamilton-Koch, W., Snyder, R. D., and Lavelle, J. M., *Chem.-Biol. Interact.,* 59, 17–28, 1986.
Han, C., Wu, Y., Yin, Y., and Shen, M., *Food Chem. Toxicol.,* 30, 521–524, 1992.
Hanauske, U., Hanauske, A. R., and Marshall, M. H., *Int. J. Cell Cloning,* 5, 170–178, 1987.
Hardy, H. L., Robe, E. W., and Loren, S., *J. Occup. Med.,* 9, 271–276, 1967.
Harr, J. R., Bone, J. F., Tinsley, I. J., Weswig, P. H., and Yamamoto, R. S., in *Selenium in Biomedicine* Mith, O. H., Oldfield, J. E., and Weswig, P. H., Eds., AVI Publishing, Westport, CT, 153, 1967.
Hart, M. M. and Adamson, R. H., *Proc. Natl. Acad. Sci. U.S.A.,* 68, 1623–1626, 1971.
Hartwig, A., Kasten, U., Boakye-Dankwa, K., Schlepegrell, R., and Beyersmann, D., *Toxicol. Environ. Chem.,* 28, 205–215, 1990.
Haruki, K., *Osaka-shi Igakki Zasshi,* 32, 193–205, 1983.
Heath, J. C. and Daniel, M. R., *Br. J. Cancer,* 16, 473–478, 1962.
Heath, J. C., Daniel, M. R., and Webb, M., *Nature,* 193, 592–593, 1962.
Heath, J. C., *Nature,* 173, 822–823, 1954.
Heath, J. C., *Br. J. Cancer,* 10, 668–673, 1956.
Heppleston, A. G., *Environ. Health Perspect.,* 55, 111–127, 1984.
Heppleston, A. G., *Am. J. Ind. Med.,* 7, 285–294, 1985.
Hessel, P. A., Sluis-Cremer, G. K., and Hnizdo, E., *Am. J. Ind. Med.,* 10, 57–62, 1986.
Hesterberg, T. W., Oshimura, M., Brody, A. R., and Barrett, J. C., in *Silica, Silicosis, and Cancer* Goldsmith, D. F., Winn, D. M., and Shy, C. M., Eds., Praeger Press, New York, 177, 1986.
Hickey, R. J., Schoff, E. P., and Clelland, R. C., *Arch. Environ. Health,* 15, 728–739, 1967.
Hirose, Y., Konda, S., Sasaki, K., Konishi, F., and Takazakura, E., *Jpn. J. Med.,* 30, 43–46, 1991.
Hnizdo, E. and Sluis-Cremer, G. K., *Br. J. Ind. Med.,* 48, 53–60, 1991.
Hodgson, J. T. and Jones, R. D., *Br. J. Ind. Med.,* 47, 665–676, 1990.
Hoegerman, S. F. and Cummins, H. T., *Health Phys.,* 44, 365–371, 1983.
Hoffman, D. J. and Niyogi, S. K., *Science,* 198, 513–514, 1977.
Hogstedt, C., *Scand. J. Work Environ. Health,* 14 (Suppl. 1), 17–20, 1988.
Hogstedt, C. and Alexandersson, R., *Scand. J. Work Environ. Health,* 13, 177–178, 1987.
Holland, L. M., *Reg. Toxicol. Pharmacol.,* 12, 224–237, 1990.

Holland, L. M., Gonzales, M., Wilson, J. S., and Tillery, M. I., in *Health Issues Related to Metal and Nonmetallic Mining,* Wagner, W. L., Rom, W. N., and Merchant, J. A., Eds., Butterworth, Boston, 485, 1983.

Holland, L. M., Wilson, J. S., Tillery, M. I., and Smith, D. M., in *Silica, Silicosis, and Cancer* Goldsmith, D. F., Winn, D. M., and Shy, C. M., Eds., Praeger Press, New York, 267, 1986.

Hopkins, L. L., Cannon, H. L., Niesch, A. T., Welch, R. M., and Nielsen, F. H., in *Geochemistry and the Environment,* Vol. II, National Academy of Sciences, Washington, D.C., 93, 1977.

Hori, C. and Oka, T., *Biochem. Biophys. Acta,* 610, 235–240, 1980.

Horvat, D., Racic, J., and Rozgaz, R., *Arh. Hig. Rada. Toksikol.,* 32, 147–156, 1981.

Hrgovcic, M., Tessmer, C. F., Thomas, F. B., Ong, P. S., Gamble, J. F., and Shullenberger, C. C., *Cancer,* 32, 1512–1524, 1973.

Hsie, A. W., *Environ. Sci. Res.,* 15, 295–315, 1978.

Hsie, A. W., Johnson, N. P., Couch, D. B., San Sebastian, J. R., O'Neill, J. P., Hoeschele, J. D., Rahn, R. O., and Forbes, N. L., in *Trace Metals in Health and Disease,* Raven Press, New York, 55, 1979.

Hsie, A. W., O'Neill, J. P., San Sebastian, J. R., Couch, D. B., Brimer, P. A., Sun, W. N. C., Fuscoe, J. C., Forbes, N. L., Machanoff, R., Riddle, J. C., and Hsie, M. H., *Chin. Cancer J.,* 12, 417–420, 1990.

Huang, X., Properties of Coals Capable of Playing a Role in the Mechanisms of Emphysema, Stomach Cancer, and Pneumoconiosis, Doctoral Thesis, University of Paris VII, 1991.

Huang, X., Laurent, P. A., Zalma, R., and Pezerat, H., *Chem. Res. Toxicol.,* 6, 452–458, 1993.

Hueper, W. C., *Occupational and Environmental Cancers of the Respiratory System,* Hueper, W. C., Ed., Springer-Verlag, Berlin, 93, 1966.

Hughes, A. W., Sherlock, D. A., Hamblen, D. L., and Reid, R., *J. Bone J. Surg.,* 69, 470–472, 1987.

IARC, Monographs on the Evaluation of Carcinogenic Risk of Chemicals to Humans, Vol. 23, Some Metals and Metallic Compounds, International Agency for Research on Cancer, IARC Sci. Publ., Lyon, France, 325, 1980a.

IARC, Monographs on the Evaluation of Carcinogenic Risk of Chemicals to Humans, Vol. 23, Some Metals and Metallic Compounds, International Agency for Research on Cancer, IARC Sci. Publ., Lyon, France, 143, 1980b.

IARC, Monographs on the Evaluation of the Carcinogenic Risk of Chemicals to Humans, Vol. 34, Polynuclear Aromatic Compounds, Part 3, Industrial Exposures in Aluminum Production, Coal Gasification, Coke Production, and Iron and Steel Founding, International Agency for Research on Cancer, IARC Sci. Publ., Lyon, France, 37, 1984.

IARC, Monographs on the Evaluation of the Carcinogenic Risk of Chemicals to Humans, Suppl. 6, International Agency for Research on Cancer, IARC Sci. Publ., Lyon, France, 132, 1987a.

IARC, Monographs on the Evaluation of the Carcinogenic Risk of Chemicals to Humans, Vol. 42, International Agency for Research on Cancer, IARC Sci. Publ., Lyon, France, 111, 1987b.

IARC, Occupational Exposure to Silica and Cancer Risk, Simonato, L., Fletcher, A. C., Saracci, R., and Thomas, T. L., Eds., IARC Sci. Publ. No. 97, International Agency for Research on Cancer, Lyon, France, 1990.

Infante, P. F., Wagoner, J. K., and Sprince, N. L., *Environ Res.,* 21, 35–43, 1980.

Innes, J. R. M., Ulland, B. M., Valerio, M. G., Petrucelli, L., Fishbein, L., Hart, E. R., Pallotta, A. J., Bates, R. R., Falk, H. L., Gart, J. J., Klein, M., Mitchell, I., and Peters, J., *J. Natl. Cancer Inst.,* 42, 1101–1114, 1969.

Inoue, T., Ohta, Y., Sadaie, Y., and Kada, T., *Mutat. Res.,* 91, 41–45, 1981.

Ip, C. and Sinha, D. K., *Cancer Res.,* 41, 31–34, 1981.

Ishizawa, M., Endo, T., and Endo, H., *Mutat. Res.,* 54, 214, 1978.

Ivett, J. L., Brown, B. M., Rodgers, C., Anderson, B. E., Resnick, M. A., and Zeiger, E., *Environ. Mol. Mutagen.,* 14, 165–187, 1989.

Jackson, J. F. and Linskens, H. F., *Mol. Gen. Genet.,* 187, 112, 1982.

Jacobs, M. M., *Cancer,* 40, 2557–2564, 1977.

Jacobs, M. M. and Frost, C., *J. Toxicol. Environ. Health,* 8, 587–598, 1981.

Jacobs, M. M., Matney, T. S., and Griffin, A. C., *Cancer Lett.,* 2, 319–322, 1977.

Janes, J. M., Higgins, G. M., and Herrick, J. F., *J. Bone J. Surg.,* 36B, 543–552, 1954.

Jansson, B., Jacobs, M. M., and Griffin, A. C., in *Inorganic and Nutritional Aspects of Cancer,* Schrauzer, G. N., Ed., Plenum Press, New York, 305, 1978.

Jao, S. W., Lee, W., and Ho, Y. S., *Dis. Colon Rectum,* 33, 99–104, 1990.

Jasmin, G. and Riopello, J. L., *Lab. Invest.,* 35, 71–78, 1976.

Jensen, A. A. and Tüchsen, F., *Crit. Rev. Toxicol.,* 20, 427–437, 1990.

Joardar, M. and Sharma, A., *Mutat. Res.,* 240, 159–163, 1990.

Johnson, W. T. and Saari, J. T., *Nutr. Res.,* 11, 1403–1414, 1991.

Jorgensen, H. S., *Work Environ. Health,* 10, 126–133, 1973.

Jorgensen, H. S., *Ann. Acad. Med. Singapore,* 13, 371–377, 1984.

Judde, J. G., Breillout, F., Clemenceau, C., Poupon, M. F., and Jasmin, C., *J. Natl. Cancer Inst.,* 78, 1185–1190, 1987.

Kada, T., in *Problems of Threshold in Chemical Mutagenesis,* Tazima, Y., Ed., Environmental Mutagen Society of Japan, Mishima, 73, 1984.

Kada, T. and Kanematsu, N., *Proc. Jpn. Acad.,* 54, 234–237, 1978.

Kada, T., Hirano, K., and Shirasu, Y., *Chem. Mutat.,* 6, 149–173, 1980a.

Kada, T., Hirano, K., and Shirasu, Y., in *Chemical Mutagens: Principles and Methods for Their Detection,* deSerres F. J. and Hollander, A., Eds., Plenum Press, New York, 149, 1980b.

Kada, T., Mochizuki, H., and Miyao, K., *Mutat. Res.,* 125, 145–151, 1984.

Kahlau, G., *Frankfurt Z. Pathol.,* 71, 3–13, 1961.
Kahn, A. A., *Am. J. Gastroenterol.,* 80, 699–703, 1985.
Kalinina, L. M. and Polukhina, G. H., *Mutat. Res.,* 46, 223–224, 1977.
Kamamoto, Y., Makimura, S., Sugihara, S., Hiasa, Y., Arai, M., and Ito, K., *Cancer Res.,* 33, 1129–1133, 1973.
Kamp, D. W., Graceffa, P., Pryor, W. A., and Weitzman, S. A., *Free Rad. Biol. Med.,* 12, 293–315, 1992.
Kanematsu, N. and Kada, T., *Mutat. Res.,* 53, 207–208, 1978.
Kanematsu, N., Hara, M., and Kada, T., *Mutat. Res.,* 77, 109–116, 1980.
Kang, H. K., Infante, P. F., and Carra, J. S., *Science,* 207, 935–936, 1980.
Kanisawa, M. and Schroeder, H. A., *Cancer Res.,* 27, 1192–1195, 1967.
Kanisawa, M. and Schroeder, H. A., *Cancer Res.,* 29, 892–895, 1969.
Karlsson, N., Fangmark, I., Haggqvist, I., Karlsson, B., Rittfeldt, L., and Marchner, H., *Mutat. Res.,* 260, 39–46, 1991.
Kasprzak, K. S., Waalkes, M. P., and Poirier, L. A., *Toxicol. Appl. Pharmacol.,* 82, 336–343, 1986.
Kazantzis, G., *Environ. Health Perspect.,* 40, 143–161, 1981.
Kazantzis, G. and Lilly, L. J., in *Handbook on the Toxicology of Metals,* Vol. I, Friberg, L., Norberg, G. F., and Vouk, V. B., Eds., Elsevier Amsterdam, 319, 1986.
Kemmer, W., Muth, H., Tranekjer, F., and Edelmann, L., *Biophysik,* 7, 343–351, 1971.
Kharab, P. and Singh, I., *Indian J. Exp. Biol.,* 25, 141–142, 1987.
Khojasteh, A. and Kraybill, W. G., *South. Med. J.,* 81, 878–882, 1988.
Kibblewhite, M. G., van Rensburg, S. J., Laker, M. C., and Rose, E. F., *Environ. Res.,* 33, 270–278, 1984.
Kingsnorth, A. N., LaMuraglia, G. M., Ross, J. S., and Malt, R. A., *Br. J. Cancer,* 53, 683–686, 1986.
Kinlen, L. J. and Willows, A. N., *Br. J. Ind. Med.,* 45, 219–224, 1988.
Kivuoloto, M., *Br. J. Ind. Med.,* 37, 363–366, 1980.
Klarland, J. K., *Cell,* 41, 707–717, 1985.
Knizhnikov, V. A., Komleva, V. A., Nifatov, A. P., Novoselova, G. P., and Molokanov, A. A., *Gig. Sanit.,* 9, 18–21, 1989.
Kobayashi, N. and Okamoto, T., *J. Natl. Cancer Inst.,* 52, 1605–1610, 1974.
Kodama, T., *J. Stomatol. Soc. Jpn.,* 56, 263–288, 1989.
Kojiro, M., Nakashima, T., Ito, Y., Ikezaki, H., Mori, T., and Kido, C., *Arch. Pathol. Lab. Med.,* 109, 853–857, 1985.
Koller, L. D., Kerkvliet, N. I., and Exon, J. H., *Toxicol. Pathol.,* 13, 50–57, 1985.
Kopf-Maier, P. and Kopf, H., *Z. Naturforsch.,* 34B, 805–807, 1979.
Kopf-Maier, P., Hesse, B., and Kopf, H., *J. Cancer Res. Clin. Oncol.,* 96, 43–51, 1980.
Kopf-Maier, P., Janiak, C., and Schumann, H., *J. Cancer Res. Clin. Oncol.,* 114, 502–506, 1988.
Kowalska, M., *Toxicol. Lett.,* 47, 185–190, 1989.
Kowalski, L. A., Tsang, S. S., and Davison, A. J., *Cancer Lett.,* 64, 83–90, 1992.
Kraus, A. S., Levin, M. L., and Gerhardt, P. R., *Am. J. Public Health,* 47, 961–970, 1957.
Kunkel, T. A. and Loeb, L. A., *J. Biol. Chem.,* 254, 5718–5725, 1979.
Kuschner, M., *Environ. Health Perspect.,* 40, 101–105, 1981.
Kyoizumi, S., Umeki, S., Akiyama, M., Hirai, Y., Kusonoki, Y., Nakamura, N., Endoh, K., Konishi, J., Sasaki, M. S., and Mori, T., *Mutat. Res.,* 265, 173–180, 1992.
Langvad, E., *Int. J. Cancer,* 3, 415–423, 1968.
Larramendy, M. L., Papescu, N. C., and DiPaolo, J. C., *Environ. Mutagen.,* 3, 596–606, 1981.
Laskey, J., Wess, I., Schulman, H. M., and Ponka, P., *Exp. Cell Res.,* 176, 87–95, 1988.
Lawler, A. B., Mandel, J. S., Schuman, L. M., and Lubin, J. H., *J. Occup. Med.,* 27, 507–517, 1985.
Lee, K. P., Kelly, D. P., Schneider, P. W., and Trochimowicz, H. J., *Toxicol. Appl. Pharmacol.,* 83, 30–45, 1986.
Lee, K. P., Trochimowicz, H. J., and Reinhardt, C. F., *Toxicol. Appl. Pharmacol.,* 79, 179–192, 1985.
Leonard, A. and Gerber, G. B., *Mutat. Res.,* 196, 247–257, 1988.
Leonard, A. and Lauwerys, R., *Mutat. Res.,* 186, 35–42, 1987.
Leonard, A., Gerber, G. B., Jacquet, P., and Lawreys, P. P., in *Mutagenicity, Carcinogenicity, and Teratogenicity of Industrial Pollutants,* Kirsch-Volders, M., Ed., Plenum Press, New York, 84, 1984.
Leonard, A., Linden, B., and Gerber, G. B., in *Proc. Int. Symp. Environmental Health Aspects of Lead,* pp. 303–309, 1972.
Levan, A., *Nature,* 156, 751–752, 1945.
Levander, O. A., in *Trace Elements,* Vol. 2, Mertz, M., Ed., Academic Press, New York, 209, 1986.
Levy, B. S., Hoffman, L., and Gottsegen, S., *J. Occup. Med.,* 26, 567–570, 1984.
Li, A. P., *Food Chem. Toxicol.,* 24, 527–534, 1986.
Li, J. Y., *Natl. Cancer Inst. Monogr.,* 62, 113–120, 1982.
Lo, L. W., Koropatnick, J., and Stich, H. F., *Mutat. Res.,* 49, 305–312, 1978.
Lofroth, G. and Ames, B. N., *Mutat. Res.,* 53, 65–66, 1978.
Luckey, T. D. and Venugopal, B., in *Metal Toxicity in Mammals,* Vol. 1, Luckey, T. D. and Venugopal, B., Eds., Plenum Press, New York, 137, 1977.
Lund, L. G. and Aust, A. E., *Biofactors,* 3, 83–89, 1991.
Lundin, P. M., *Br. J. Cancer,* 15, 838–847, 1961.
Lynge, E., Kurppa, K., Kristofersen, L., Malker, H., and Sauli, H., *J. Natl. Cancer Inst.,* 77, 883–889, 1986.
MacMahon, H. E., Murphy, A. S., and Mates, M. I., *Am. J. Pathol.,* 23, 585–612, 1947.
Maki-Paakkanen, J., Sorsa, M., and Vaino, H., *Hereditas,* 94, 269–276, 1981.

Mancuso, T. F., *Environ. Res.,* 3, 251–275, 1970.
Mancuso, T. F., *Environ. Res.,* 21, 48–55, 1980.
Mancuso, T. F. and El-Attar, A. A., *J. Occup. Med.,* 11, 422–434, 1969.
Manna, G. and Das, R. K., *Nucleus,* 15, 180–186, 1972.
Manna, G. and Parida, B. B., *Naturwissenschaften,* 52, 647–648, 1965.
Marcotte, J. and Witschi, H. P., *Chem. Pathol. Pharmacol.,* 3, 97–104, 1972.
Marshall, M. V., Jacobs, M. M., and Griffin, A. C., *Proc. Am. Assoc. Cancer Res.,* 19, 75, 1978.
Marzin, D. R. and Phi, H. V., *Mutat. Res.,* 155, 49–51, 1985.
Mastrangelo, G., Zambon, P., Simonato, L., and Rizzi, P., *Int. Arch. Occup. Environ. Health,* 60, 299–302, 1988.
Mays, C. W. and Spiess, H., *Environ. Res.,* 18, 88–93, 1979.
Mazaev, V. T. and Slepnina, T. G., *Gig. Sanit.,* 8, 10–13, 1973.
McClean, J. R. N., Birnboim, H. C., Pontefact, R., and Kaplan, J. G., *Chem.-Biol. Interact.,* 46, 189–200, 1983b.
McClean, J. R. N., Blakely, D. H., Douglas, G. R., and Kaplan, J. G., *Mutat. Res.,* 119, 195–201, 1983a.
McDonald, J. C., *Br. J. Ind. Med.,* 46, 289–291, 1989.
McDougall, A., *J. Bone J. Surg.,* 38, 709–713, 1956.
McGlashan, N. D., *Lancet,* 1, 578, 1972.
McLaughlin, A. I. G. and Harding, H. E., *Arch. Ind. Health,* 14, 350–378, 1956.
McLean, J. R., McWilliams, R. S., Kaplan, J. G., and Birnboim, H. C., in *Chemical Mutagenesis, Human Population Monitoring, and Genetic Risk Assessment,* Bora, K. C., Douglas, G. R., and Nestmann, E. R., Eds., Elsevier, New York, 137, 1982.
McQuity, J. T., DeWys, W. D., Monaco, L., Strain, W. H., Rob, O. G., Apgar, J., and Poires, W., *Cancer Res.,* 30, 1381–1390, 1970.
McTurk, L. C., Huis, C. H., and Eckhardt, R. E., *Ind. Med. Surg.,* 25, 29–36, 1966.
Melendez, L. V., Hunt, R. D., Daniel, M. D., Fraser, C. E. O., Garcia, F. G., and Williamson, M. E., *Int. J. Cancer,* 6, 431–435, 1970.
Meyer, M. B., Luk, G. D., Sotelo, J. M., Cohen, B. H., and Menkes, H. A., *Am. Rev. Respir. Dis.,* 121, 887–892, 1980.
Michalowsky, I., *Zentralbl. Allg. Pathol. Pathol. Anat.,* 38, 585, 1926.
Milks, M. M., Wilt, S. R., Ali, I. I., and Couri, D., *Fundam. Appl. Toxicol.,* 5, 320–326, 1985.
Miller, C. A. and Levine, E. M., *J. Neurochem.,* 22, 751–758, 1974.
Milner, J. A., *Fed. Proc.,* 44, 2568–2572, 1985.
Mitra, A., Antagonistic and Synergistic Effects of Some Metals of the Group II on Genetic Systems. Doctoral Thesis, University of Calcutta, 1984.
Miyaki, M. N., Akamatsu, N., Ono, T., and Koyama, H., *Mutat. Res.,* 68, 259–263, 1979.
Miyaki, M., Murata, I., Osabe, M., and Ono, T., *Biochem. Biophys. Res. Commun.,* 77, 854–860, 1977.
Mizushima, Y., Shoji, Y., and Kaneko, K., *Int. Arch. Allergy Appl. Immunol.,* 63, 338–339, 1980.
Mochizuki, H. and Kada, T., *Int. J. Radiat. Biol.,* 42, 653–659, 1982.
Mohr, C. P., Kaup, F. G., Gossmann, H. H., and Teschke, R., *Z. Gastroenterol.,* 22, 153–157, 1984.
Monlibert, L. and Roubille, R., *J. Fr. Med. Chir. Thorac.,* 14, 435–442, 1960.
Morgareidge, K., Cox, G. E., and Bailey, D. E., *Chronic Feeding Studies With Beryllium Sulfate in Rats: Evaluation of Carcinogenic Potential,* Food and Drug Research Laboratories, Inc., Aluminum Company of America, Pittsburgh, PA, 1975.
Mortelmans, K., Haworth, S., Lawlor, T., Speck, W., Tainer, B., and Zeiger, E., *Environ. Mutagen.,* 8 (Suppl. 7), 1–119, 1986.
Moyer, M. P., Leifur, R. L., Cantu, A. A., Hestilow, K. L., Egan, J. W., and Windeler, A. S., *Proc. Annu. Meet. Am. Assoc. Cancer Res.,* 32, A143, 1991.
Mulay, I. L., Roy, R., Knox, B. E., Suhr, N. H., and Delaney, W. E., *J. Natl. Cancer Inst.,* 47, 1–13, 1971.
Muller, E. and Erhardt, W., *Z. Krebsforsch.,* 6, 65–77, 1957.
Nagata, N., Yoneyama, T., Yanagida, K., Ushio, K., Yanagihara, S., Matsubara, O., and Eishi, Y., *J. Toxicol. Sci.,* 10, 333–341, 1985.
Nakamuro, K., Yoshikawa, K., Sayato, Y., Kurata, H., Tonomura, M., and Tonomura, A., *Mutat. Res.,* 40, 177–184, 1976.
NCI, National Cancer Institute. *Federal Register,* 43, 54299, 1978.
NCI Bioassay of Selenium Sulfide for Possible Carcinogenicity (Gavage Study), U.S. Department of Health and Human Services, Public Health Service, Publ. No. NIH 80–1750, Bethesda, MD, 1980.
Nehez, M., Berencsi, G., Freye, H. A., Mazzag, E., Scheifler, H., and Selypes, A., *Mutat. Res.,* 97, 206–207, 1982.
Nejjari, A., Fournier, J., Pezerat, H., and Leanderson, P., *Br. J. Ind. Med.,* 50, 501–504, 1993.
Nelson, A. A., Fitzhugh, O. G., and Calvery, H. O., *Cancer Res.,* 3, 230–236, 1943.
Nestmann, E. R., Matula, T. I., Douglas, G. R., Bora, K. C., and Kowbel, D. J., *Mutat. Res.,* 66, 357–365, 1979.
Newberne, P. M. and Connor, M. W., in *Trace Substances in Environmental Health,* Vol. VIII, Hemphill, D.D., Ed., University of Missouri Press, Columbia, MO, 323, 1974.
Newton, M. F. and Lilly, L. J., *Mutat. Res.,* 169, 61–69, 1986.
Niemeier, R. W., Mulligan, L. T., and Rowland, J., in *Silica, Silicosis, and Cancer,* Goldsmith, D. F., Winn, D. M., and Shy, C. M., Eds., Praeger Press, New York, 215, 1986.
NIH, Preliminary Report of the 90-Day Subchronic Inhalation Toxicity Study of Vanadium Pentoxide in F344 Rats and $B_6C_3F_1$ Mice, National Institutes of Health, Research Triangle Park, NC, 1993.

NIOSH, National Institutes for Occupational Safety and Health Pocket Guide to Chemical Hazards, 5th ed., U.S. Department of Health and Human Services, Washington, D.C., 234, 1985.
Nishimura, M. and Umeda, M., *Mutat. Res.,* 54, 246–247, 1979.
Nishioka, H., *Mutat. Res.,* 31, 185–189, 1975.
Niyogi, S. K. and Feldman, R. P., *Nucl. Acids Res.,* 9, 2615–2627, 1981.
Niyogi, S. K., Feldman, R. P., and Hoffman, D. J., *Toxicology,* 22, 9–23, 1981.
Noda, M., Takano, T., and Sakurai, H., *Mutat. Res.,* 66, 175–179, 1979.
Nordenson, I., Beckman, G., Beckman, L., and Nordstrom, S., *Hereditas,* 88, 263–267, 1978.
Norppa, H., Westermarck, T., Laasonen, M., Knuutila, K., and Knuutila, S., *Hereditas,* 93, 93–105, 1980.
Nowak, H. F., *Rocz. Akad. Med. Bialymstoku.,* 7, 323–327, 1961.
Nowak, H. F., *Arch. Immunol. Ther. Exp.,* 14, 774–780, 1966.
NTP, National Toxicology Program Fifth Annual Report on Carcinogens, Public Health Service, U.S. Department of Health and Human Services, Research Triange Park, NC, 1989.
O'Hara, R. W. and Brown, J. M., *J. Natl. Cancer Inst.,* 38, 947–957, 1967.
Obara, K., Saito, T., Sato, H., Yamakage, K., Watanabe, T., Kakizawa, M., Tsukamoto, T., Kobayash, K., Hngo, M., and Yoshinaga, K., *Jpn. J. Med.,* 30, 67–72, 1991.
Oberly, T. J., Piper, C. E., and McDonald, D. S., *J. Toxicol. Environ. Health,* 9, 367–376, 1982.
Ogawa, H. I., Liu, S. Y., Sakata, K., Niyitani, Y., Tsuruta, S., and Kato, Y., *Mutat. Res.,* 204, 117–121, 1988.
Ogawa, H. I., Sakata, K., Inouye, T., Jyosui, S., Niyitani, Y., Kakimoto, K., Morishita, M., Tsuruta, S., and Kato, Y., *Mutat. Res.,* 172, 97–104, 1986.
Ohno, H., Hanaoka, F., and Yamada, M., *Mutat. Res.,* 104, 141–145, 1982.
Oldfield, J. R., Allaway, W. H., Draper, H. H., Frost, D. V., Jensen, L. S., Scott, M. L., and Wright, P. I., *Selenium in Nutrition,* Oldfield, J. R., Allaway, W. H., Draper, H. H., Frost, D. V., Jensen, L. S., Scott, M. L., and Wright, P. I., Eds., National Academy of Sciences. Washington, D.C., 1971.
Olivier, P. and Marzin, D., *Mutat. Res.,* 189, 263–269, 1987.
Omara, F. O. and Blakley, B. R., *Can. J. Vet. Res.,* 57, 209–211, 1993.
Orgel, A. and Orgel, L. E., *J. Mol. Biol.,* 14, 453–457, 1965.
Oshimura, M., Hestenberg, T. W., Tsutsui, T., and Barrett, J. C., *Cancer Res.,* 44, 5017–5022, 1984.
Overvad, K., Thorling, E. B., Bjerring, P., and Ebbesen, P., *Cancer Lett.,* 27, 163–170, 1985.
Owusu-Yaw, J., Cohen, M. D., Fernando, S. Y., and Wei, C. I., *Toxicol. Lett.,* 50, 327–336, 1990.
Oyasu, R., Battifora, H. A., Clasen, R. A., McDonald, J. H., and Hass, G. M., *Cancer Res.,* 30, 1248–1261, 1970.
Pagano, D. A. and Zeiger, E., *Environ. Mol. Mutagen.,* 19, 139–146, 1992.
Pairon, J. C., Jaurand, M. C., Kheuang, L., Janson, X., Brochard, P., and Bignon, J., *Br. J. Ind. Med.,* 47, 110–115, 1990.
Palit, S., Sharma, A., and Talukder, G., *Biol. Trace Elem. Res.,* 29, 139–145, 1991.
Paton, G. R. and Allison, A. C., *Mutat. Res.,* 16, 332–336, 1972.
Pedigo, N. G., George, W. V., and Anderson, M. B., *Reprod. Toxicol.,* 2, 45–53, 1988.
Penman, H. G. and Ring, P. A., *J. Bone J. Surg.,* 66, 632–634, 1984.
Petering, H. G., Buskirk, H. H., and Crim, J. A., *Cancer Res.,* 27, 1115–1121, 1967.
Pham, Q. T., Chau, N., Patris, A., Trombert, B., Henquel, J. C., Geny, M., and Teculescu, D., *Cancer Detect. Prevent.,* 15, 449–454, 1991.
Pham, Q. T., Teculescu, D., Bruant, A., Chau, N., Viaggi, M. N., and Rebstock, E., *Eur. J. Epidemiol.,* 8, 594–600, 1992.
Pienta, R. J., Pailey, J. A., and Lebherz, W. B., *Int. J. Cancer,* 19, 642–655, 1977.
Pigott, R. and Ishmael, J., *Toxicol. Lett.,* 8, 153–163, 1981.
Plaha, D. S. and Rogers, H. J., *Biochim. Biophys. Acta,* 760, 246–255, 1983.
Plaha, D. S., Rogers, H. J., and Williams, G. W., *J. Antibiot. Tokyo,* 37, 588–595, 1984.
Polednak, A. P., Stehney, A. F., and Lucas, H. F., *Health Phys.,* 44, 239–251, 1983.
Polukhina, G. H., Kalinina, L. M., and Lukasheva, L. I., *Genetika,* 13, 1491–1494, 1977.
Popenoe, E. A. and Schmaeler, M. A., *Arch. Biochem. Biophys.,* 196, 109–120, 1979.
Poswillo, D. E. and Cohen, B., *Nature,* 231, 447–448, 1971.
Prazmo, W., Balbin, E., Baranowska, H., Ejchart, A., and Putrament, A., *Genet. Res. Cambridge,* 26, 21–29, 1975.
Putrament, A., Baranowska, H., Ejchart, A., and Jachymczyk, W., *Mol. Gen. Genet.,* 151, 69–76, 1977.
Putrament, A., Baranowska, H., Ejchart, A., and Prazmo, W., *J. Gen. Microbiol.,* 62, 265–270, 1975.
Radford, E. P. and Renard, K. G., *N. Engl. J. Med.,* 310, 1485–1494, 1984.
Raju, K. S., Alessandri, G., and Gullino, P. M., *Cancer Res.,* 44, 1579–1584, 1984.
Ramel, C. and Magnusson, J., *Environ. Health Perspect.,* 31, 59–66, 1979.
Rasmuson, A., *Mutat. Res.,* 157, 157–162, 1985.
Ray, J. H., *Mutat. Res.,* 141, 49–53, 1984.
Ray, J. H. and Altenburg, L. C., *Mutat. Res.,* 78, 261–266, 1980.
Reeves, A. L., in *Handbook on the Toxicology of Metals,* Vol. II, Friberg, L., Norberg, G. F., and Vouk, V. B., Eds., Elsevier, New York, 95, 1986.
Reeves, A. L. and Vorwald, A. J., *Cancer Res.,* 27, 446–451, 1967.
Reizenstein, P., *Med. Oncol. Tumor Pharmacother.,* 8, 229–233, 1991.
Renne, R. A., Eldridge, S. R., Lewis, T. R., and Stevens, D. L., *Toxicol. Pathol.,* 13, 306–314, 1985.

Rhoads, K. and Sanders, C. L., *Environ. Res.*, 36, 359–378, 1985.
Richmond, H. G., *Br. Med. J.*, 1, 947–949, 1959.
Riley, J. F., *Experientia*, 24, 1237–1238, 1968.
Ripley, L. S., *Mol. Gen. Genet.*, 141, 23–40, 1975.
Riverdal, E. and Sanner, T., *Cancer Res.*, 41, 2950–2953, 1981.
Riviere, M. R., Chouroulenkov, I., and Guerin, M., *Bull. Assoc. Fr. Etude Cancer*, 47, 55–87, 1960.
Robison, S. H., Cantoni, O., and Costa, M., *Carcinogenesis*, 3, 657–662, 1982.
Roe, F. J. C., Boyland, E., Dukes, C. E., and Mitchley, B. C., *Br. J. Cancer*, 19, 860–866, 1965.
Roe, F. J. and Haddow, A., *Br. J. Cancer*, 19, 855–859, 1965.
Roe, F. J., Boyland, E., and Millican, K., *Food Cosmet. Toxicol.*, 3, 277–280, 1965.
Rogers, H. J., Synge, C., and Woods, V. E., *Antimicrob. Agents Chemother.*, 18, 63–68, 1980.
Rogers, H. J., Woods, V. E., and Synge, C., *J. Gen. Microbiol.*, 128, 2389–2394, 1982.
Roldan, R. E. and Altamirano, L. M. A., *Mutat. Res.*, 245, 61–65, 1992.
Ronneberg, A. and Langmark, F., *Am. J. Ind. Med.*, 22, 573–590, 1992.
Roschin, I. V., *Gig. Sanit.*, 32, 26–32, 1967.
Rosenkranz, H. S. and Poirier, L. A., *J. Natl. Cancer Inst.*, 62, 873–892, 1979.
Rosin, M. P., *Cancer Lett.*, 13, 7–14, 1981.
Rosin, M. P. and Stich, H. F., *Int. J. Cancer*, 23, 722–727, 1979.
Rossman, T. G., Molina, M., and Meyer, L. W., *Environ. Mutagen.*, 6, 59–69, 1984.
Roussy, G., Oberling, C., and Guerin, M., *Bull. Acad. Natl. Med. Paris*, 112, 809–816, 1934.
Ryan, T. P. and Aust, S., *Crit. Rev. Toxicol.*, 22, 119–141, 1992.
Sabbioni, E., Clerici, L., and Brazzelli, A., *J. Toxicol. Environ. Health*, 12, 737–748, 1983 .
Saffiotti, U., Montesano, R., Sellakumar, A. R., and Kaufman, D. G., *J. Natl. Cancer Inst.*, 49, 1199–1204, 1972.
Saksoong, P. and Campiranon, A., Genotoxicity of heavy metals. I. Zinc. Abst. Proc. XV Int. Congr. Genetics, New Delhi, p. 324, 1983.
Sato, I., Nishimura, T., Kakimoto, N., Suzuki, H., and Tanaka, N., *J. Biol. Response Modif.*, 7, 1–5, 1988.
Schaaper, R. M., Koplitz, R. M., Tkeshelashvilli, L. K., and Loeb, L. A., *Mutat. Res.*, 177, 179–188, 1987.
Schauss, A. G., *Biol. Trace Elem. Res.*, 29, 267–280, 1991.
Schepers, G. W. H., *Prog. Exp. Tumor Res.*, 2, 203–204, 1961.
Schepers, G. W. H., Durkan, T. M., Delehant, A. B., and Creedon, F. T., *Arch. Ind. Health*, 15, 32–58, 1957.
Schinz, H. R. and Uehlinger, E., *Z. Krebsforsch.*, 52, 425–437, 1942.
Schlag, R. D., in *Genotoxic and Carcinogenic Metals: Environmental and Occupational Occurrence and Exposure, Advances in Modern Environmental Toxicology*, Vol. XI, Fishbein, L., Furst, A., and Mehlman, M. A., Eds., Princeton Scientific, Princeton, NJ, 211, 1987.
Schmid, E., Bauchinger, M., Pietruck, S., and Hall, G., *Mutat. Res.*, 16, 401–406, 1972.
Schrauzer, G. N., *Med. Hypoth.*, 2, 39–49, 1976.
Schrauzer, G. N. and Ishmael, D., *Ann. Clin. Lab. Sci.*, 4, 441–447, 1974.
Schrauzer, G. N., White, D. A., and Schneider, C. J., *Bioinorg. Chem.*, 7, 23–34, 1977.
Schrauzer, G. N., White, D. A., and Schneider, C. J., in *Trace Element Metabolism in Man and Animals*. Vol. III, Kirchgessner, M., Ed., Workshop on Animal Nutrition Research, Freising-Weihenstephan, Germany, 387, 1978.
Schroeder, H. A., Mitchener, M., and Nason, A. P., *J. Nutr.*, 100, 59–68, 1970.
Schroeder, H. A. and Balassa, J. J., *J. Nutr.*, 92, 245–252, 1967.
Schroeder, H. A. and Mitchener, M., *J. Nutr.*, 101, 1431–1437, 1971.
Schroeder, H. A. and Mitchener, M., *J. Nutr.*, 101, 1531–1537, 1971.
Schroeder, H. A. and Mitchener, M., *Arch. Environ. Health*, 24, 66–71, 1972.
Schroeder, H. A. and Mitchener, M., *J. Nutr.*, 105, 421–427, 1975a.
Schroeder, H. A. and Mitchener, M., *J. Nutr.*, 105, 452–458, 1975b.
Schroeder, H. A., Balassa, J. J., and Vinton, J. H., *J. Nutr.*, 83, 239–250, 1964.
Schroeder, H. A., Balassa, J. J., and Vinton, J. H., *J. Nutr.*, 86, 51–66, 1965.
Schroeder, H. A., Mitchener, M., and Nason, A. P., *J. Nutr.*, 100, 59–68, 1970.
Schuschke, D. A., Reed, M. W. R., Saari, J. T., and Miller, F. N., *J. Nutr.*, 122, 1547–1552, 1992a.
Schuschke, D. A., Reed, M. W. R., Saari, J. T., Olson, D. M., Ackerman, D. M., and Miller, F. N., *Br. J. Cancer*, 66, 1059–1064, 1992b.
Schwartz, M. K., *Cancer Res.*, 35, 3481–3487, 1975.
Seifter, J., Ehrich, W. E., Hudyma, G., and Mueller, G., *Science*, 103, 762, 1946.
Selbie, F. R., *Lancet*, 2, 847–848, 1936.
Selby, J. V. and Friedman, G. D., *Int. J. Cancer*, 41, 677–682, 1988.
Selevan, S. G., Landrigan, P. J., Stern, F. B., and Jones, J. H., *Am. J. Epidemiol.*, 122, 673–683, 1984.
Shabaan, A. A., Marks, V., Lancaster, M. C., and Dufeu, G. N., *Lab. Anim.*, 11, 43–46, 1977.
Shabik, P. and Hartwell, L. L., Survey on compounds which have been tested for carcinogenic activity, Suppl. 2, Publ. No. 149, U.S. Public Health Service, Bethesda, MD, 3, 1969.
Shamberger, R. J., *J. Natl. Cancer Inst.*, 44, 931–936, 1970.
Shamberger, R. J. and Frost, D. V., *Can. Med. Assoc. J.*, 100, 682, 1969.

Shamberger, R. J. and Rudolph, G., *Experientia*, 22, 116, 1966.
Shamberger, R. J. and Willis, C. E., *Crit. Rev. Clin. Lab. Sci.*, 2, 211–221, 1971.
Shamberger, R. J., Cortlett, C. L., Beaman, K. D., and Kasten, B. L., *Mutat. Res.*, 66, 349–355, 1979.
Shamberger, R. J., Tytko, S. A., and Willis, C. E., *Arch. Environ. Health*, 31, 231–235, 1976.
Shapiro, J. R., *Ann. N.Y. Acad. Sci.*, 192, 215–219, 1973.
Shapley, D., *Science*, 198, 898–901, 1977.
Sharma, A. and Talukder, G., *Environ. Mutagen.*, 9, 191–226, 1987.
Sharpless, G. R., *Fed. Proc.*, 5, 239–240, 1946.
Sheffet, A., Thind, I., Miller, A. M., and Louria, D. B., *Arch. Environ. Health*, 37, 44–52, 1982.
Sheu, C. W., Rodriguez, I., and Lee, K. W., *Environ. Mol. Mutagen.*, 15 (Suppl. 17), 55, 1990.
Shi, X., Dalal, N. S., Hu, X. N., and Vallyathan, V., *J. Toxicol. Environ. Health*, 27, 435–454, 1989.
Shimkin, M. B., Stoner, G. D., and Theiss, J. C., *Adv. Exp. Med. Biol.*, 91, 85–91, 1977.
Shing, Y., *J. Biol. Chem.*, 263, 9059–9062, 1988.
Shirasu, Y., Moriya, M., Kato, K., Furuhashi, A., and Kada, T., *Mutat. Res.*, 40, 19–30, 1976.
Siemiatycki, J., Dewar, R., Lakhani, R., Nadon, L., Richardson, L., and Gerin, M., *Am. J. Ind. Med.*, 16, 547–567, 1989.
Simmon, V. F., *J. Natl. Cancer Inst.*, 63, 893–899, 1979a.
Simmon, V. F., *J. Natl. Cancer Inst.*, 62, 901–909, 1979b.
Singh, I., *Mutat. Res.*, 117, 149–152, 1983.
Singh, I., *Mutat. Res.*, 137, 47–49, 1984.
Sirianni, S. R. and Huang, C. C., *Cancer Lett.*, 18, 109–116, 1983.
Sirover, M. A., *Environ. Health Perspect.*, 40, 162–172, 1981.
Sirover, M. A. and Loeb, L. A., *Science*, 194, 1434–1436, 1976.
Sissons, H. A., *Acta Unio Int. Contra Cancrum*, 7, 171, 1950.
Skreb, Y., Racic, J., and Hors, N., *Mutat. Res.*, 74, 241, 1980.
Smialowicz, R. J., Luebke, R. W., Rogers, R. R., Riddle, M. M., and Rowe, D. G., *Immunopharmacology*, 9, 1–11, 1985.
Smialowicz, R. J., Rogers, R. R., Riddle, M. M., Luebke, R. W., Fogelson, L. D., and Rowe, D. G., *J. Toxicol. Environ. Health*, 20, 67–80, 1987.
Smialowicz, R. J., Rogers, R. R., Riddle, M. M., Luebke, R. W., Rowe, D. G., and Garner, R. J., *J. Immunopharmacol.*, 6, 1–23, 1984.
Smith, A. B. and Suzuki, Y., *Environ. Res.*, 21, 10–14, 1980.
Smith, J. B., *Proc. Natl. Acad. Sci. U.S.A.*, 80, 6162–6166, 1983.
Smith, R. J., *Science*, 211, 556–557, 1981.
Snyder, R. D., *Mutat. Res.*, 193, 237–246, 1988.
Solli, H. M., Andersen, A., Stranden, E., and Langard, S., *Scand. J. Work Environ. Health*, 11, 7–13, 1985.
Sora, S., Carbone, M. L., Pacciarini, M., and Magni, G. E., *Mutagenesis*, 1, 21–28, 1986.
Spinelli, J. J., Band, P. R., Svirchev, L. M., and Gallagher, R. P., *Occup. Med.*, 33, 1150–1155, 1991.
Spivack, S. D., *Lancet*, 335, 854–855, 1990.
Stanton, M. F., *Cancer Res.*, 28, 1000–1006, 1967.
Stanton, M. F., *J. Natl. Cancer Inst.*, 52, 633–634, 1974.
Steffee, C. H. and Baetjer, A. M., *Arch. Environ. Health*, 11, 66–75, 1965.
Steinhoff, D. and Mohr, U., *Exp. Pathol.*, 41, 169–174, 1991.
Steinstrasser, A., *Radiat. Environ. Biophys.*, 19, 1–15, 1981.
Stenback, F., Rowland, J., and Sellakumar, A., *Oncology Basel*, 33, 29–34, 1976.
Stevens, R. G., Jones, D. Y., Micozzi, M. S., and Taylor, P. R., *N. Engl. J. Med.*, 319, 1047–1052, 1988.
Stich, H. F., Wei, L., and Whiting, R. F., *Cancer Res.*, 39, 4145–4151, 1979.
Stockinger, H. E., *Am. Ind. Hyg. Assoc. J.*, 45, 127–133, 1984.
Stocks, P., *Br. J. Cancer*, 14, 397–418, 1960.
Stocks, P. and Davies, R. I., *Br. J. Cancer*, 14, 8–22, 1960.
Stocks, P. and Davies, R. I., *Br. J. Cancer*, 18, 14–24, 1964.
Stoeckle, J. D., Hardy, H. L., and Weber, A. L., *Am. J. Med.*, 46, 545–561, 1979.
Stoner, G. D., Shimkin, M. B., Troxell, M. C., Thompson, T. L., and Terry, L. S., *Cancer Res.*, 36, 1744–1747, 1976.
Stover, B. J., *Health Phys.*, 44, 253–257, 1983.
Sun, M., *Toxicity of Vanadium and its Environmental Health Standard* Sun, M., Ed., West China University Medical Sciences, Chengdu, China, 1987.
Sunderman, F. W., *Food Cosmet. Toxicol.*, 9, 105–120, 1971.
Sunderman, F. W., in *Advances in Modern Toxicology*, Vol. 2, Goyer R.A. and Mehlman, M. A., Eds., Hemisphere, Washington, D.C., 257, 1977.
Sunderman, F. W., *Scand. J. Work Environ. Health*, 15, 1–12, 1989a.
Sunderman, F. W., *Fundam. Appl. Toxicol.*, 13, 205–216, 1989b.
Suzuki, F., Brutkiewicz, R. R., and Pollard, R. B., *Br. J. Cancer*, 52, 757–763, 1985.
Swann, M., *J. Bone J. Surg.*, 66, 629–631, 1984.
Swarm, R. L., Miller, E., and Michelitch, H. J., *Pathol. Microbiol.*, 25, 27, 1962.

Tanaka, N., Ohida, J., Ono, M., Yoshiwara, H., Beika, T., Terasawa, A., Yamada, J., Morioka, S., Mannami, T., and Orita, K., *Gan Kagaku Ryoho,* 11, 1303–1306, 1984.
Tani, P. and Kokkola, K., *Scand. J. Respir. Dis.,* 80, 121–128, 1972.
Tavares, M. H., Saragoca, A., Oliveira, E. A., Rocha Oliveira, M. P., and da Silva Horta, J., *Environ. Res.,* 18, 173–177, 1979.
Telles, N. C., Thomas, L. B., Popper, H., Ishak, K. G., and Falk, H., *Environ. Res.,* 18, 74–87, 1979.
Tessmer, C. F., Hrgovcic, M., Brown, B. W., Wilbur, J., and Thomas, F. B., *Cancer,* 29, 173–179, 1973.
Tessmer, C. F., Hrgovcic, M., Thomas, F. B., Fuller, L. M., and Castro J. R., *Ther. Radiat.,* 106, 635–639, 1972.
Texeira-Pinto, A. A. and Azevedo e Silva, M. C., *Environ. Res.,* 18, 225–230, 1979.
Thomas, H. F., Herriott, R. M., Hahn, B. S., and Wang, S. Y., *Nature,* 259, 341–342, 1976.
Thomas, J. A. and Thiery, J. P., *C.R. Acad. Sci. Paris,* 236, 1387–1392, 1953.
Thompson, H. J., *J. Agric. Food Chem.,* 32, 422–425, 1984.
Thompson, H. J., Chasteen, N. D., and Meeker, L. D., *Carcinogenesis,* 5, 849–851, 1984.
Thompson, H. J., Kennedy, K., Witt, M., and Juzefyk, J., *Carcinogenesis,* 12, 111–114, 1991.
Tkeshelashvili, L. K., McBride, T. J., Spence, K., and Loeb, L. A., *J. Biol. Chem.,* 266, 6401–6406, 1991.
Tripathy, N. K., Wurgler, F. E., and Frei, H., *Mutat. Res.,* 242, 169–180, 1990.
Tscherkes, L. A., Volgarev, M. N., and Aptekar, S. G., *Acta Unio Int. Contra Cancrum.,* 19, 632–633, 1963.
Tso, W. W. and Fung, W. P., *Toxicol. Lett.,* 8, 195–200, 1981.
Tupper, R., Watts, R. W. E., and Normall, A., *Biochem. J.,* 59, 264–268, 1955.
Tuyns, A. J., Pequignot, G., and Jensen, O. M., *J. Biolumin. Chemilumin.,* 2, 95–99, 1988.
Umeda, M. and Nishimura, M., *Mutat. Res.,* 67, 221–229, 1979.
Umeki, S., Kyoizumi, S., Kusonoki, Y., Nakamura, N., Sasaki, M., Mori, T., Ishikawa, Y., Cologne, J. B., and Akiyama, M., *Jpn. J. Cancer Res.,* 82, 1349–1353, 1991.
Uragoda, C. G. and Pinto, M. R. M., *Med. J. Aust.,* 1, 167–169, 1972.
Van Geissen, G. J., Crim, J. A., Petering, D. H., and Petering, H. G., *J. Natl. Cancer Inst.,* 51, 139–146, 1973.
van der Spoel, J. I., Stricker, B. H., Esseveld, M. R., and Schipper, M. E. I., *Lancet,* 336, 117, 1990.
van Esch, G. J. and Kroes, R., *Br. J. Cancer,* 23, 765–771, 1969.
van Esch, G. J., van Genderen, H., and Vink, H. H., *Br. J. Cancer,* 16, 289–297, 1962.
van Kaick, G., Lorenz, D., Muth, H., and Kaul, A., *Health Phys.,* 35, 127–136, 1978.
van Rensburg, S. J., Hall, J. M., and du Bruyn, D. B., *J. Natl. Cancer Inst.,* 75, 561–566, 1985.
Vollmann, J., *Schweiz. Z. Allg. Pathol. Bakteriol.,* 1, 440–443, 1938.
Voroshilin, S., Plotko, E. G., Fink, T. V., and Nikiforova, V. Y., *Tsitol. Genet.,* 12, 241–243, 1978.
Vorwald, A. J. and Reeves, A. L., *Arch. Ind. Health,* 19, 190–199, 1959.
Vouk, V. B., in *Handbook on the Toxicology of Metals,* 2nd ed., Friberg, L., Nordberg, G. F., and Vouk, V. B., Eds., Elsevier, Amsterdam, 225, 1986.
Wagner, J. C., in *Morphology of Experimental Respiratory Carcinogenesis,* Nettesheim, P., Hanna, M. G., and Deatherage, J. W., Eds., U.S. Atomic Energy Commission, Oak Ridge, TN, 347, 1970.
Wagner, M. M. F., *J. Natl. Cancer Inst.,* 57, 509–518, 1976.
Wagner, M. M. F. and Wagner, J. C., *J. Natl. Cancer Inst.,* 49, 81–91, 1972.
Wagoner, J. K., Infante, P. F., and Bayliss, D. L., *Environ. Res.,* 21, 15–34, 1980.
Walters, M. and Roe, F. J., *Food Cosmet. Toxicol.,* 3, 271–276, 1965.
Wang, E. and Choppin, P. W., *Proc. Natl. Acad. Sci. U.S.A.,* 78, 2363–2367, 1981.
Wang, G. T. and Tang, Z. D., *Cancer,* 62, 58–66, 1988.
Wargotz, E. S., Sidawy, M. K., and Jannotta, F. S., *Cancer,* 62, 58–66, 1988.
Waters, M. D., Toxicology of vanadium, *Adv. Mod. Toxicol.,* 2, 147–189, 1977.
Weber, P. C., *J. Bone J. Surg.,* 68, 824–826, 1986.
Weed, L. L., *J. Bacteriol.,* 85, 1003–1010, 1963.
Wehner, A. P., Busch, R. H., Olson, R. J., and Craig, D. K., *Am. Ind. Hyg. Assoc. J.,* 38, 338–346, 1977.
Wei, H. J., Luo, X. M., and Yang, X. P., *Chin. Cancer J.,* 9, 204–207, 1987.
Weinberg, E. D., *Biol. Trace Elem. Res.,* 34, 123–140, 1992.
Wester, P. W., Krajnc, E. I., van Leeuwen, F. X. R., Loeber, J. G., van der Heijden, C. A., Vaessen, H. A., and Helleman, P. W., *Food Chem. Toxicol.,* 28, 179–196, 1990.
Wetterhahn, K. E., Demple, B., Kulesz-Martin, M., and Copeland, E. S., *Cancer Res.,* 52, 4058–4063, 1992.
Whiting, R. F., Wei, L., and Stich, H. F., *Biochem. Pharmacol.,* 29, 842–845, 1980a.
Whiting, R. F., Wei, L., and Stich, H. F., *Mutat. Res.,* 78, 159–169, 1980b.
WHO, 1970 Evaluations of Some Pesticide Residues in Food, Food and Agricultural Organization, World Health Organization, Rome, 527, 1971.
WHO, Environmental Health Criteria 15. Tin and Organotin Compounds — A Preliminary Review, World Health Organization, Geneva, 77, 1980.
WHO, Environmental Health Criteria 24. Titanium, World Health Organization, Geneva, 42, 1982.
WHO, Environmental Health Criteria 81. Vanadium, World Health Organization, Geneva, 1988.
Wilson, T., Scheuchenzuber, W. J., Eskew, M. L., and Zankower, A., *Environ. Res.,* 39, 331–344, 1986.
Wingren, G. and Axelson, O., *Scand. J. Work Environ. Health,* 13, 412–416, 1987.
Witschi, H. P., *Lab. Invest.,* 19, 67–70, 1968.

Woodliff, H. J., Cohen, G., and Gallon, W., *Med. J. Aust.*, 2, 768–771, 1972.
Yamane, Y. and Ohtawa, M., *Gann,* 70, 147–152, 1979.
Yamane, Y. and Sakai, K., *Gann,* 64, 563, 1973.
Yamane, Y., Sakai, K., and Kojima, S., *Gann,* 67, 295–302, 1976.
Yokoiyama, A., Kada, T., and Kuroda, Y., *Mutat. Res.,* 245, 99–105, 1990.
Yu, S., Chu, Y., Gong, X., and Hou, C., *Ecol. Trace Elem. Res.,* 7, 21–23, 1985.
Zakour, R. A. and Glickman, B. W., *Mutat. Res.,* 126, 9–18, 1984.
Zakour, R. A., Kunkel, I. A., and Loeb, L. A., *Environ. Health Perspect.,* 40, 197–205, 1981a.
Zakour, R. A., Tkeshelashvili, L. K., Shearman, C. W., Koplitz, R. M., and Loeb, L. A., *J. Cancer Res. Clin. Oncol.,* 99, 187–196, 1981b.
Zhong, Z., Troll, W., Koenig, K. L., and Frenkel, K., *Cancer Res.,* 50, 7564–7570, 1990.
Ziche, M., Jones, J., and Gullino, P. M., *J. Natl. Cancer Inst.,* 69, 475–482, 1982.
Zollinger, H. U., *Virchows Arch. Pathol. Anat. Physiol.,* 326, 694–710, 1953.
Zychlinski, L., *Bromatol. Chem. Toksykol.,* 8, 195–199, 1980.

PART B
Selected Concepts on Metal Carcinogenesis

Overview

An Introduction to Selected Concepts in Metal Carcinogenesis

Lionel A. Poirier

The previous sections of this work have been concerned with the carcinogenic effects and mechanisms of action of specific metals. The present section deals with aspects of metal carcinogenesis which appear to be generally applicable to several different examples. As a class, the metal carcinogens have been much less widely investigated than the organic carcinogens. Because of their ubiquitous presence in the environment, their known activity in humans, and their great range of activity, they constitute an important class of agents requiring further investigation. While in many respects the biological effects of the well-investigated carcinogens such as nickel, chromium, and cadmium, resemble those of the more classical organic carcinogens, the mechanisms of action of even these metals remain relatively ill defined. The relationship of the inorganic carcinogens to the organic initiators, promoters, or cocarcinogens remains to be established. Most of the metal carcinogens are transition elements. As such, they possess multiple valences and readily participate in redox reactions, both chemically and *in vivo*. As indicated in Part A of Section III, this feature enables them to augment the oxidative stress resulting in oxidative damage to biological systems. On the other hand, the metal carcinogens exhibit many features resembling those of organic carcinogens. Some act directly on DNA, proteins, or other macromolecules. Some alter hormonal regulation and metabolism. Thus, recent studies have shown the capacity of metal carcinogens to cause oxidative damage to DNA, to inhibit repair, to produce strand breaks both *in vivo* and in cell culture, and to cause chromosomal alterations. These metals have induced mutations and abnormal gene expression, as well as abnormal cellular differentiation. While both the toxic and the physiologically essential metals have been shown to alter markedly the activities of organic carcinogens, the effects of metal/metal interactions in carcinogenesis have been most striking.

The present section deals with three topics of general interest in metal carcinogenesis. The first, by Poirier and Littlefield, describes the interactive effects of metals on carcinogenesis by other agents. This chapter illustrates that despite the wide differences between the metabolism and initial molecular targets of the organic and inorganic carcinogens, recent studies have shown an increasing relatedness between them in their effects on parameters closely associated with the oncogenic process. The second, by Kasprzak, presents a strong case implicating oxidative damage to DNA as a major mechanism of carcinogenesis by the metals. He sifts through the maze of complex oxidative damage in cells to show that similar molecular genetic effects can be produced by several metal carcinogens. Lastly, the chapter by Landolph et al. describes *in vitro* model systems by which the cell-transforming and mutagenic activity of metals can be investigated. In some respects, these *in vitro* assays may provide a more appropriate model of human carcinogenesis by metals than do the whole-animal rodent assays. In sum, these chapters provide a broad view of the oncogenic effects of several metals.

Chapter 17

Metal Interactions in Chemical Carcinogenesis

Lionel A. Poirier and Neil A. Littlefield

I. INTRODUCTION

The probable involvement of metals in tumorigenesis has been known since the 1920s and 1930s, when it was shown that the direct injection of Zn into the testes of fowl would result in teratoma formation (Michalowsky, 1926; Falin and Gromzewa, 1939). The similar formation of teratomas by the direct injection of Zn into the testes of rodents has subsequently been demonstrated (Riviere et al., 1960; Guthrie and Guthrie, 1974). The first interactive study of a metal in carcinogenesis showed that Zn administration inhibited testicular carcinoma formation via Cd in rats (Gunn et al., 1963). Battifora et al. (1968) showed that while Mg deficiency per se was sufficient to induce lymphomas in rats, the formation of these tumors was unaffected by the simultaneous feeding of 2-acetylaminofluorene (AAF). Since then, an increasing number of metal-carcinogen interactions have been examined both *in vivo* and *in vitro*, principally with metal carcinogens. To date, most of the interactions investigated have focused on the biological effects of the metals in chemical carcinogenesis by other agents. Relatively few studies have been reported describing the mechanisms by which the metals exert their effects in carcinogenesis. However, the delineation of the carcinogenic process into the stages of tumor initiation, tumor promotion, progression, and tumor development for organic carcinogens has been well described, particularly for those metabolized to reactive electrophiles (Harris, 1991; Kadlubar, 1994). In comparison, the mechanisms and processes by which the carcinogenic metals exert their activities are not well understood, particularly the means by which metals interfere with or enhance the tumorigenic process. For the purposes of this chapter, the amphoteric elements, Se and As, will be included among the metals.

II. BIOLOGY AND TUMOR FORMATION

A. ZINC

Abnormalities in Zn nutriture have long been associated with human cancer. Extensive epidemiological evidence has shown an association between long-term Zn intake and the development of esophageal cancer in humans (Barch and Iannaccone, 1986; Nelson, 1990). Serum Zn levels are generally low in tumor-bearing patients and almost all tumors studied have low Zn levels. In experimental animals, the administration of Zn has generally led to decreases in the carcinogenic activity of both inorganic and organic carcinogens. The administration of Zn-deficient diets to rats has led to the increased formation of esophageal cancer by methylbenzylnitrosamine, and of forestomach carcinoma by both dimethylnitrosamine and a dietary combination of methylbenzylamine and sodium nitrite (Barch and Iannaccone, 1986). Consistent with the tumor-enhancing activity of a deficiency of dietary Zn, the administration of excess Zn, either by injection or in the diet, inhibited tumor formation by DMBA in rats and hamsters, and by 3-methylcholanthrene on the skin of mice (Kasprzak and Waalkes, 1986).

Treatment of rats with Zn inhibited the tumorigenicity of Cd, both at the injection site and in the testes (Kasprzak and Waalkes, 1986). Similarly, treatment with Zn has inhibited the tumorigenicity of other carcinogenic metals. Zn injection has also been shown to decrease the tumorigenicity of Ni subsulfide injected s.c. (Kasprzak and Waalkes, 1986; Kasprzak et al., 1988). In the latter study, the time to tumor formation was affected by Zn administration. It is interesting to note the mutual interactions of Zn and Cd on the pancreas of rats since these two metals are well-known biochemical antagonists (Kasprzak and Waalkes, 1986). Cd treatment has been shown to cause pancreatic islet cell tumors (Poirier et al., 1983) as well as a hepatic transdifferentiation of pancreatic acinar cells (Konishi et al., 1990).

Other experimental carcinogenesis studies have demonstrated a tumor-enhancing effect by Zn salts. Oral tumor formation by 4-nitroquinoline-N-oxide was decreased in Zn-deficient rats, but increased in animals treated with supplementary Zn (Barch and Iannaccone, 1986; Kasprzak and Waalkes, 1986). Similarly, excess Zn enhanced the tumorigenicity of N-ethyl-N-nitrosourea towards the brain of rats while a deficiency of this metal decreased the skin tumorigenicity of 3-methylcholanthrene (Barch and Iannaccone, 1986; Kasprzak and Waalkes, 1986). In line with its cocarcinogenic effect, Zn was shown to offset the protection afforded by Se against Cd carcinogenesis towards the testes of rats (Mason and Young, 1967). Zn excess is associated with increased cancer risk in many organs in humans (Nelson, 1990).

B. MAGNESIUM

Mg is another metal which has shown a number of associations with tumor development in humans and experimental animals (Durlach et al., 1986; Hass et al., 1989). In regions with low Mg content in the soil and water, there has long been an association between low levels of Mg in the serum of humans and animals and an increased incidence of neoplasms, particularly of lympho-proliferative tissues, as well as of decreased immunocompetence (Kasprzak and Waalkes, 1986). Consistent with this observation in humans were early studies showing that Mg-deficient diets produced thymomas in rats (Bois, 1964; Jasmin, 1968). The leukemogenic effect of Mg deficiency was noted in animals treated with AAF (Battifora et al. 1968), even though no significant interaction between the Mg deficiency and the AAF feeding was noted. In addition, the intratracheal administration of Mg oxide inhibited lung tumor formation by BP and DEN (Kasprzak and Waalkes, 1986).

In line with the previously described negative association between Mg and tumor development in animals treated with either no carcinogen or an organic carcinogen, are the chemoprotective effects of Mg against inorganic carcinogens. Thus, Mg treatment inhibited renal tumor formation by Ni in the rat (Kasprzak et al., 1994). Relatively low doses of Mg completely suppressed the tumorigenic effects of Ni and of Pb salts towards the lungs of strain A mice (Poirier et al., 1984). Mg injected along with Cd or Ni salts at the injection site of the carcinogenic metal inhibited sarcoma formation by each of these metals (Kasprzak et al., 1987; Kasprzak and Poirier, 1985a; Poirier et al., 1983). On the other hand, Mg added to the diet had no significant effect on the formation of testicular carcinomas or injection site sarcomas in rats receiving an i.m. dose of Cd (Poirier et al., 1983). It appears that Mg deficiency tends to enhance while Mg supplementation tends to suppress tumor formation by both organic and inorganic agents (Hass et al., 1981; Kasprzak and Waalkes, 1986).

In related studies, the presence of an active tumor tended to decrease the Mg content of the nontumor tissue in the host animal (Waalkes and Poirier, 1984) while severe Mg deficiency retarded the growth rates of several transplantable tumors in rats (Mills et al., 1984; Kasprzak and Waalkes, 1986).

C. CALCIUM

Ca and its salts have received a great deal of attention as potential chemopreventive agents, but they also act as prospective cocarcinogens. Epidemiological evidence indicates a decreased risk of developing cancer, particularly of the colon, with a high dietary intake of Ca (Nelson, 1990; Garland et al., 1991; Peters et al., 1991; Scalmati et al., 1992). Cancer-bearing humans or animals have sufficient changes in their Ca homeostasis to produce hypercalcemia, a frequent accompaniment to cancer development (Kasprzak and Waalkes, 1986). Also, virtually all tumors studied have exhibited marked increases in their Ca-dependent calmodulin levels (Kasprzak and Waalkes, 1986). In a number of experimental systems high dietary Ca has been shown to inhibit the formation of colon tumors in rodents (Newmark and Lipkin, 1991; Pence and Budding, 1988; Scalmati et al., 1992; Wargovich et al., 1991). High levels of dietary Ca have suppressed the markers of preneoplasia such as cell proliferation and increased levels of both ornithine decarboxylase and thymidine kinase in the colons of rats treated with azoxymethane (Scalmati et al., 1992). Dietary Ca inhibits colon tumor formation initiated by AOM and promoted by cholic acid (Pence et al., 1994). The administration of high levels of Ca salts in the diet led to decreased tumor formation in the colons of mice promoted with fatty acids and bioacids (Kasprzak and Waalkes, 1986). Dietary Ca exerts its effect both on tumor multiplicity and on tumor incidence (Scalmati et al., 1992). Ca administered in the postinitiation phase of carcinogenesis also inhibited tumor formation in the glandular stomachs of rats initiated with MNNG (Nishikawa et al., 1994).

Like Mg, Ca administered i.p. to strain A mice, along with the lung tumorigens Ni and Pb, markedly suppressed the tumorigenic activity of these metal ions towards lung (Poirier et al., 1984). In the same study, Ca administration **increased** the formation of lung adenomas compared to the control animals (Poirier et al., 1984). In a similar vein, the administration of high dietary levels of Ca to rats receiving the carcinogen Pb enhanced both the multiplicity of renal tumors as well as increased the renal toxicity of the Pb in rats (Kasprzak et al., 1985a). Previous studies have shown that Ca could increase tumor formation by DMBA when administered along with the tumor promoter TPA or with retinoic acid applied to the cheek pouches of hamsters to enhance tumor progression (Kasprzak and Waalkes, 1986). Thus, at least three separate studies show that, unlike their general chemopreventive effects in the colon, Ca salts may in fact enhance carcinogenesis in other organs. In other studies, Ca exhibited no significant effect on carcinogenesis. Ca, either administered in the diet or s.c., proximally had no significant effect on the induction of Cd injection-site sarcomas or testicular carcinomas (Poirier et al., 1983). Also, Ca administered in the diet or s.c. as a prospective cocarcinogen had no significant effect on the injection-site sarcomas produced by Ni (Kasprzak et al., 1985b). Finally, consistent with the antagonistic effects noted in these investigations between the physiological and the carcinogenic divalent metals, it was shown that Cd treatment decreased the toxicity of Ca in the hearts of rats (Nishiyama et al., 1990).

D. IRON AND COPPER

The excess accumulation of Fe in humans is associated with an increased risk of cancer (Nelson, 1990; Nelson et al., 1992; Whittaker et al., 1994). Hereditary hemochromatosis appears to produce liver cirrhosis and carcinomas in humans (Nelson et al., 1992; Whittaker et al., 1994). Similarly, the body Fe stores are positively associated with the formation of colon polyps and with increased colon carcinoma risk in humans (Nelson, 1990; Nelson et al., 1992).

Fe appears to increase the development of cancer in a number of experimental systems. Fe-dextran administered s.c. to rats, rabbits, and mice has resulted in the formation of sarcomas at the injection site (IARC, 1987c). BP bound to Fe oxide dust particles and administered by intratracheal instillation to hamsters resulted in squamous-cell and anaplastic carcinomas (IARC, 1987b). In other studies, the topical administration of Fe oxide to hamsters enhanced lung and nasal cavity tumor formation by DEN and dimethylnitrosamine, respectively (IARC, 1987b). Dietary supplementation with Fe results in enhanced colorectal carcinoma formation in DMH-treated rats and mice. Kasprzak et al. (1994) have shown that Fe accelerates Ni-induced carcinogenesis in the rat kidney. Finally, dietary overload of Fe produced hepatic hyperplasia, pancreatic degeneration, and other toxic lesions in rats (Whittaker et al., 1994; Wolff et al., 1994).

In general, the effects of Cu salts in carcinogenesis have not been widely investigated. However, dietary Cu has been shown to inhibit the hepatocarcinogenic activity of an aminoazo dye in rats (Fare and Howell, 1964). Further, investigations have shown that CuDIPS, a chemical equivalent of superoxide dismutase, inhibits skin tumor promotion in mice by TPA (cf. review by Badawi [1990]). Studies showing the possible effects of Cu and of Fe in carcinogenesis have been examined in Long-Evans Cinnamon (LEC) rats. These animals develop a high spontaneous incidence of kidney and liver cancer (Izumi et al., 1994; Kato et al., 1994); abnormal Cu metabolism has been associated with the hepatitis and liver cancer seen in these animals (Izumi et al., 1994). Fe deficiency delays the onset of hepatic lesions in these animals while Fe accumulation in their livers appears to hasten hepatitis (Kato et al., 1994). This strain may serve as an excellent model to study the possible common mechanisms of actions of Fe and Cu, possibly via oxidative damage to DNA.

E. ARSENIC

The IARC has determined that there is sufficient evidence to conclude that As is a human carcinogen (IARC, 1980; IARC, 1987a). Exposure of humans to As has been associated with an excess of cancers of the skin and lung, as well as with occasional hepatic angiosarcomas. While most animal studies have found As to be inactive as a carcinogen, inorganic As compounds have produced lung tumors in mice, respiratory tract tumors in hamsters, and stomach adenocarcinomas in rats; sodium arsenite has also enhanced the formation of DEN-induced kidney tumors in mice (IARC, 1987a). Further, BD IX male rats, treated with a Bordeaux mixture, develop lung tumors (IARC, 1980). Still, the large majority of animal studies on a number of As compounds have not shown evidence for the carcinogenicity of As. However, in cell culture studies with the mouse embryo cell line BALB/3T3 clone A31–1-1, both the trivalent As (sodium arsenite) and the pentavalent As (sodium arsenate) showed transformation activities (Bertolero et al., 1987; Saffiotti and Bertolero, 1989). Thus, As does seem to exhibit carcinogenic potential even with animal systems.

F. SELENIUM

Se compounds have been widely investigated as chemopreventive agents against a number of well-known chemical carcinogens. The epidemiological evidence available indicates that Se intake is negatively associated with the formation of colon cancer in humans (Nelson, 1990). The epidemiological evidence also suggests that Se is a protective agent in humans for cancers of the lung, ovary, prostate, rectum, intestine, and against leukemia (cf. review by Medina, 1986b). Se also protects against a variety of carcinogen- and virus-induced tumors in experimental animals (Ip, 1986; Medina, 1986a; Yu and Lu, 1992). Among the experimental animal models in which Se has shown significant protective effects are the following: mammary tumor formation in mice and rats by N-nitroso-N-methylurea, DMBA, adenovirus type 9, and MMTV (Medina 1986b; Nelson, 1990); colon cancers in rats caused by DMH, azoxymethane, and BOP (Medina, 1986b; Nelson, 1990; Reddy et al., 1987) liver cancer in rats by dimethylaminoazobenzene and AAF (Medina, 1986b); and skin tumor formation in mice (Milner, 1986). In the latter studies, Se inhibited both the initiation and the promotion stages of carcinogenesis (Milner, 1986). A series of organic and inorganic Se compounds were investigated for their relative chemopreventive activities in the standard DMBA-induced mammary tumor model using female Sprague-Dawley rats (Ip and Ganther, 1992). The two most effective chemopreventive derivatives were selenobetaine and Se-methylselenocysteine, both of which could give rise quite readily to methylselenol (CH_3-SeH) (Ip and Ganther, 1992). Corresponding to the protective effects of supplementary Se, the administration of Se-deficient diets to rats results in liver and pancreatic necrosis as well as cardiomyopathy (Medina, 1986b). In addition to its chemopreventive effects against experimentally induced tumor formation, Se inhibits the growth of established tumor lines (Yu and Lu, 1992). However, Se has also counteracted the protection afforded by Zn against Cd carcinogenesis towards the testes of rats (Mason and Young, 1967).

G. COBALT AND CADMIUM

These two metals are both regarded as human and animal carcinogens (IARC, 1991; Waalkes, 1994). While the inhibition of Cd carcinogenesis by physiologically essential divalent metals has been described above, such studies have been published describing their interactions with organic carcinogens. Co, administered by intratracheal instillations, was co-carcinogenic with similarly administered BP towards the lungs of rats; it did not show such activity in DEN-treated hamsters (IARC, 1991). Cd, administered to B6C3F1 mice during: (1) chronic DEN treatment; (2) promotion by sodium barbital; (3) after tumor promotion with barbital; (4) as an initiator with barbital as a promoter; and (5) with no further treatment, inhibited liver tumor formation in all groups investigated (Waalkes, 1994). Cd appears to act as an antineoplastic drug on a tumor cell population lacking the protective protein metallothionein (MT).

III. BIOCHEMICAL AND MOLECULAR MECHANISMS

A. GENERAL ASPECTS

This section is concerned with the prospective mechanisms by which the metals may exert their modulatory effects during the carcinogenic process. In general, our knowledge of the processes occurring during chemical carcinogenesis has been obtained using organic compounds.

These steps have been extensively discussed in recent reviews (Harris, 1991; Kadlubar, 1994; Yuspa, 1994). A simplified version of these processes is provided in Table 1.

To date, with the exception of Se, little evidence has been obtained showing the modulatory effects of metals on each of the steps shown in Table 1. In large measure this is because most of the investigations on the effects of metals in the carcinogenic process have examined such effects on the activities of inorganic carcinogens and not on organic carcinogens. The above scheme was developed almost exclusively from studies using organic compounds. The areas explored in metal effects in carcinogenesis have

Table 1 Outline of the Major Steps in Cancer Causation By Chemicals

Carcinogen metabolism
    Activation to reactive, electrophilic intermediates
        Inactivation of the carcinogen or its initial metabolite
    By Phase I enzymes
    By Phase II enzymes
    Adduct formation
        Can be blocked by providing alternate noncritical targets
Initiation
    A consequence of adduct formation
    Result of DNA damage, gene mutation, abnormal DNA repair
Promotion and progression
    Associated with increased cell replication
    Intermediate causation by secondary messengers
    Frequently accompanied by oxidative damage and
        abnormal control of gene expression
        Tumor suppressor genes
        Gene hypomethylation
        Altered cell differentiation

largely examined competition between metals and oxidative damage. However, the studies with Se have largely paralleled the carcinogenesis models developed with the organic carcinogens.

B. CARCINOGEN METABOLISM

Studies on the alterations in carcinogen metabolism have largely focused on the effect of the modulatory metal on the accumulation of the carcinogenic metal in its target tissue. Thus, the multiple injection of Mg acetate with the lung tumorigen Ni acetate decreased the uptake of Ni in the lungs of strain A mice (Kasprzak and Waalkes, 1986; Rodriguez and Kasprzak, 1989). Similarly, multiple injections of Mg acetate at the s.c. injection site at which Cd had been previously administered decreased the accumulation of the carcinogen at the injection site in rats (Kasprzak and Poirier, 1985b). In each of these cases, Mg inhibited the activity of the carcinogenic metal (see above). On the other hand, the injection of Mg carbonate at the same i.m. site at which the carcinogen Ni subsulfide had previously been administered had no effect on the concentration of Ni in the target tissue even though the Mg salt inhibited the sarcomagenic activity of the Ni (Kasprzak and Waalkes, 1986). In addition, the chronic feeding of Ca acetate in the diets of rats fed a carcinogenic dose of Pb subacetate led to decreased levels of Pb in the kidney, but enhanced tumor formation by Pb in this organ (Kasprzak and Waalkes, 1986). Also, while Ca acetate inhibited the tumorigenicity of Ni towards the lungs of strain A mice, it actually increased the accumulation of Ni in this organ (Kasprzak and Waalkes, 1986). In the same set of investigations, the chronic feeding of Ca or Mg salts or their administration at sites distant from the carcinogen injection site had no effect on the sarcomagenic activities of Ni and Cd (Kasprzak and Waalkes, 1986). Thus, the best correlation between protection and altered carcinogen metabolism was provided by Mg administered simultaneously at the same site as the metal carcinogen.

In other studies, Se was found to shift the metabolism of the carcinogens BP and AAF from activation to inactivation pathways (Medina, 1986b), and Zn was found to increase the detoxification of the carcinogen methylbenzylnitrosamine to benzaldehyde both *in vitro* and *in vivo* (Barch and Iannaccone, 1986).

C. ADDUCT FORMATION

To a greater or lesser extent in *in vitro* studies, virtually all of the physiologically essential divalent metals inhibit the binding of the carcinogenic divalent metals to DNA. Ca salts were shown to inhibit the binding of Ni and Cd to DNA (Kasprzak et al., 1986; Waalkes and Poirier, 1985). Mg, which stabilizes the DNA structure by binding to specific phosphate sites (Anastassopoulou et al., 1992), inhibits the binding to DNA of the carcinogens Ni and Cd, as well as the physiological metal Zn (Kasprzak and Waalkes, 1986; Kasprzak et al., 1986; Waalkes and Poirier, 1984). Mg also inhibited the reactions of the carcinogens vinyl chloride and BP diol epoxides with DNA *in vitro* (Kasprzak and Waalkes, 1986). Like Mg, Zn antagonizes the binding of Cd and Ni cations to DNA *in vitro* (Kasprzak et al., 1986;

Waalkes and Poirier, 1984). The inhibition by Zn and Mg of DNA binding by Cd and Ni may be a partial explanation for the chemoprotective effects of these physiological cations *in vivo*.

The effects of Se on binding by organic carcinogens to the DNA of target tissue has been extensively studied *in vivo* (for reviews cf. Ip, 1986; Medina, 1986b; Milner, 1986). In general, while the effects of Se on chemical carcinogenesis are often quite marked, its effects on adduct formation are relatively slight. Thus, Se did not alter the DNA binding of DMBA in the skin of mice (Milner, 1986) and had little effect on the DNA binding by AAF in the liver and by DMH in the colon of rats (Medina, 1986b). There is evidence that Se proteins can bind heavy metals *in vivo* (Burk and Hill, 1993). The provision of alternate nucleophilic targets for possible attack by reactive intermediates can also be seen in the case of the protective metals. For example, Zn induces the protein MT, which strongly bind such heavy metals as Cd and inhibits their toxicities (Kasprzak and Waalkes, 1986; Sunderman et al., 1994; Lazo et al., 1994); the development of jaundice in aging female LEC rats fed a high-Cu diet is seen only after the hepatic MT is saturated (Suzuki et al., 1994).

D. INITIATION, MUTAGENESIS, AND DNA REPAIR

While there are very few studies on the modulation by metals of the process of initiation, the essential metals Se, Mg, and Ca have exhibited protective effects against the DNA damage caused by organic and inorganic carcinogens. Thus, Se was found to increase the repair of the DNA lesions caused by BOP in the colon and by AAF in the livers of rats (Medina, 1986b) and to inhibit the mutagenicity of malondialdehyde, β-propiolactone, MNNG, and *N*-acetoxy-AAF in the Ames assay (Medina, 1986b; Rosin and Stich, 1979). In chemical studies Mg inhibited the formation of DNA damage by Ni and Cd and of single strand-breaks by Cd and Hg (Littlefield and Hass, 1994; Littlefield et al., 1991, 1994). Increasing the extracellular concentration of Mg similarly reduced the formation of DNA single-strand breaks, DNA-protein cross-links and sister chromatid exchanges induced by Ni in cells (Conway et al., 1987). Feeding high dietary levels of Ca to rats receiving DMH suppressed mutations in the K-*ras* protooncogene observed in the developing colon tumors (Rozen, 1991). Consistent with an oxidative mechanism of metal carcinogenesis are the observations that Ni-induced renal carcinomas have GGT to GTT transversions in codon 12 of the K-*ras* oncogene (Higinbotham et al., 1992; Kasprzak, 1994) and that Fe bound to crocidolite increases single-strand breaks in phage DNA (Hardy and Aust, 1994).

E. PROMOTION AND PROGRESSION

Relatively few studies have been conducted on the possible impact of metals on the tumor promotion and progression stages of carcinogenesis. Studies which have examined the biological and biochemical aspects commonly associated with these stages of carcinogenesis have tended to focus more upon oxidant stress and abnormal gene expression than upon cell replication and second messengers. In the DMBA rat mammary tumor model *in vivo*, Se inhibited tumor formation principally in the postinitiation stages of carcinogenesis; such inhibition was associated with decreased cell proliferation (Medina, 1986a,b). The increase in Ni-induced cell replication in the lungs of strain A mice, which are susceptible to tumor formation by this metal, was decreased by a chemopreventive dose of Mg, but enhanced by the corresponding treatment with Ca (Kasprzak and Poirier, 1985a; Kasprzak and Waalkes, 1986). Zn, which often exhibits chemoprotective effects against tumor formation, appeared to have mixed effects on cell replication: it enhanced the process of wound healing, but decreased DNA synthesis in a transplanted hepatoma and in CHO cells in culture (Kasprzak and Waalkes, 1986). Despite the known role of Ca in the regulation of protein kinase C, whose activity is enhanced during tumor promotion, an effect of exogenous Ca upon this pathway yet needs to be demonstrated (Kasprzak and Waalkes, 1986; Merrill and Schroeder, 1993). However, increasing the concentrations of Ca in normal and H-*ras*-transfected rat kidney cells with the drug sulofenur resulted in induction of the *fos* and *jun* proteins (Gu et al., 1994).

F. OXIDATIVE DAMAGE TO DNA

Oxidative damage *in vivo* and *in vitro* has received considerable attention in recent years as a mechanism of heavy metal toxicity (Fields et al., 1994; Gutteridge, 1990; Kasprzak, 1994). Fe and Ni are two of the heavy metals whose roles in producing oxidative damage, both *in vivo* and *in vitro*, have been most widely investigated (Eborn and Aust, 1994; Enright et al., 1992; Fields et al., 1992; Higinbotham et al., 1992; Kasprzak, 1994; Mello-Filho and Meneghini, 1991). Fe causes lipid peroxidation and free radical formation *in vivo* (Fields et al., 1992) and causes double-strand breaks in DNA through the hydroxyl radical (Enright et al., 1992; Fields et al., 1992; Mello-Filho and Meneghini, 1991). The chelation of Fe prevents the oxidative damage to DNA caused by this metal (Mello-Filho and Meneghini, 1991). As indicated above, Ni-induced renal tumors have K-*ras* mutations consistent with the formation of 8-hydroxyguanine *in vivo*, i.e., a G to T transversion (Higinbotham et al., 1992; Kasprzak, 1994). Mg has been shown to inhibit the oxidative damage produced by Ni, both *in vitro* and in cells (Littlefield et al., 1991). In modulating the toxicity of reactive oxygen species, Cu acts as a double-edged sword: this ion is necessary for the activity of Cu superoxide dismutase, and dietary Cu deficiency decreases the hepatic levels of the protective enzymes, catalase, superoxide dismustase, and glutathione peroxidase (Fields et al., 1994). Similarly, Cu deficiency may result in hepatic Fe overload in rats fed fructose (Fields et al., 1992). On the other hand, as is the case for Fe^{+2}, the aerobic incubation of Cu^{+1} and Cu^{+2} in the presence of phage DNA can result in the G to T transversions considered to be characteristic for base mispairing by 8-oxodeoxyguanosine (Kasprzak, 1994). Two other physiological metals, Ca and Zn, have been shown to modulate oxidative damage — Ca by modifying the oxidation of DNA by H_2O_2 in a chemical system and Zn by serving as a cofactor for Zn-dependent superoxide dismutase (Davies and Chipman, 1994; Nelson, 1990). Se also appears to play a major role in the inhibition of oxidative stress. The major effect of this element appears to be mediated through its presence in the enzyme glutathione peroxidase (Burk and Hill, 1993; Medina, 1986b; Sunde, 1990). The major function of this enzyme is to minimize the accumulation of lipid peroxides.

G. ABNORMAL GENE EXPRESSION

Altered control of gene expression is one of the most frequent changes seen in tumor promotion (Yuspa, 1994). It is generally accompanied by changes in DNA or gene hypomethylation and altered cell differentiation (Fearon and Vogelstein, 1990; Yuspa, 1994). Metals alter each of these processes. The normal differentiation of mouse skin keratinocyte *in vitro* is highly dependent upon the Ca concentration in the medium (Yuspa, 1994). In studies conducted *in vivo* the administration of a Cu-deficient diet to rats resulted in the transdifferentiation of pancreatic acinar cells to hepatocytes; a similar effect has been observed in the pancreata of Cd-treated rats (Konishi et al., 1990). The transdifferentiation of pancreatic acinar cells to hepatocytes was first reported by Scarpelli and Rao (1981) and subsequently by Hoover and Poirier (1986). In each of these instances, the animals had been treated with methyl-deficient diets. Thus, a link could be made between the effects of metals and of methyl deprivation on differentiation *in vivo*. Three other sets of studies have established links between methyl insufficiency and metal toxicity. In the first, Wallwork and Duerre (1985) showed that the livers of Zn-deficient rats were also methyl deficient. The LEC rats, which as described previously are very sensitive to the hepatotoxicity of Fe and Cu and develop a high spontaneous incidence of liver carcinomas, have low levels of 5-methyldeoxycytidine in their hepatic DNA (Suzuki et al., 1991). Finally, two agents which induce DNA hypomethylation, ethionine and azacytidine, were shown to induce MT in the livers of rats (Waalkes et al. 1985; Waalkes and Poirier, 1985).

IV. SUMMARY AND CONCLUSIONS

In the last 10 years a great deal of knowledge has accumulated showing the influence of metals on the process of carcinogenesis. Many of these studies have focused on metal/metal interactions, while a large portion have also examined the impact of inorganic agents on carcinogenesis by the classical chemical carcinogens and tumor promoters. The mechanisms underlying the effects of the metals have also been examined. Among the major findings of these investigations are the following. The physiologically essential metals, particularly Mg and Zn, often counteract the activities of the carcinogenic heavy metals. This protection appears to result from competition between the metals from the metabolic down to the molecular level. Another mechanism is the induction of protective proteins serving as alternate targets to DNA for the heavy metal ions. Both Zn and Ca have exerted significant protection against classical organic carcinogens by mechanisms which are as yet not well understood. While Se has inhibited the carcinogenic process at each of its successive stages, a major mechanism underlying its activity is its role as an inhibitor of oxidant stress. DNA damage caused by the presence of metal carcinogens, particularly via oxidation, is frequently observed *in vivo*, in cell culture, and in model chemical systems. Under some circumstances such damage can be inhibited by the physiologic metals. Finally, biochemical and biological effects similar to those noted in methyl-deficient animals can also be observed in animals fed feed deficient in the essential metals Cu or Zn, or treated with Cd.

ABBREVIATIONS

AAF, 2-acetylaminofluorene
BP, benzo[α]pyrene
BOP, *bis*-(2-oxopropyl)-nitrosamine
DEN, diethylnitrosamine
DMBA, 9,10-dimethylbenz[a]anthracene
DMH, dimethylhydrazine
MT, metallothionein
MNNG, *N*-methyl-*N'*-nitro-*N*-nitrosoguanidine
TPA, 12-*o*-tetra-decanoylphorbol-13-acetate

REFERENCES

Anastassopoulou, J., Polissiou, M., Manfait, M., and Theophanides, Th., in *Metal Ions in Biology and Medicine,* Vol. 2, Anastassopoulou, J., Collery, Ph., Etienne, J.C., and Theophanides, Th., Eds., John Libbey Eurotext, Paris, 26, 1992.
Badawi, A. M., in *Metal Ions in Biology and Medicine,* Vol. 1, Collery, Ph., Poirier, L.A., Manfait, M., and Etienne, J.C., Eds., John Libbey Eurotext, Paris, 349, 1990.
Barch, D. H. and Iannaccone, P. M., in *Essential Nutrients in Carcinogenesis,* Poirier, L.A., Newberne, P.M., and Pariza, M.W., Eds., Plenum Press, New York, 517, 1986.
Battifora, H. A., McCreary, P. A., Hahneman, B. M., Laing, G. H., and Hass, G. M., *Arch. Pathol.,* 86, 610–620, 1968.
Bertolero, F., Pozzi, G., Sabbioni, E., and Saffiotti, U., *Carcinogenesis,* 8, 803–808, 1987.
Bois, P., Tumor of the thymus in Mg-deficient rats, *Nature,* 204, 1316, 1964.
Burk, R. F. and Hill, K. E., in *Annual Review of Nutrition,* Olson, R.E., Bier, D.M., and McCormick, D.B., Eds., Annual Reviews, Inc., Palo Alto, CA, 65, 1993.
Conway, K., Wang, X, Xu, L., and Costa, M., *Carcinogenesis,* 8, 1115–1121, 1987.
Davies, J. and Chipman, J. K., *Tox. Vitro,* 8, 29–36, 1994.
Durlach, J., Bara, M., Guiet-Bara, A., and Collery, P., *Anticancer Res.,* 6, 1353–1362, 1986.
Eborn, S. K. and Aust, A. E., *Proc. Am. Assoc. Cancer Res.,* 35, 154, 1994.
Enright, H. U., Miller, W. J., and Hebbel, R. P., *Nucl. Acids Res.,* 20, 3341–3346, 1992.
Falin, L. I. and Gromzewa, K. E., *Am. J. Cancer,* 36, 233, 1939.
Fare, G. and Howell, J., *Cancer Res.,* 24, 1279–1283, 1964.
Fearon, E. R. and Vogelstein, B., *Cell,* 61, 759–767, 1990.
Fields, M., Lewis, C. G., Antholine, W. E., Burns, W. A., and Lure, M. D., in *Metal Ions in Biology and Medicine,* Vol. 2, Anastassopoulou, J., Collery, Ph., Etienne, J.C., and Theophanides, Th., Eds., John Libbey Eurotext, Paris, 408, 1992.
Fields, M., Lure, M. D., and Lewis, C. G., in *Metal Ions in Biology and Medicine,* Vol. 3, Collery, Ph., Poirier, L.A., Littlefield, N.A., and Etienne, J.C., Eds., John Libbey Eurotext, Paris, 467, 1994.
Garland, C. F., Garland, F. C., and Gorham, E. D., in *Calcium, Vitamin D, and Prevention of Colon Cancer,* Lipkin, M., Newmark, H.L., and Kelloff, G., Eds., CRC Press, Boca Raton, FL, 81, 1991.
Gu, H., Merriman, R. L., Berezesky, I. K., Boder, G. B., and Trump, B. F., *Proc. Am. Assoc. Cancer Res.,* 35, 222, 1994.
Gunn, S. A., Gould, T. C., and Anderson, W. A. D., *J. Natl. Cancer Inst.,* 31, 745–759, 1963.
Guthrie, J. and Guthrie, O. A., *Cancer Res.,* 34, 2612, 1974.
Gutteridge, J. M. C., in *Metal Ions in Biology and Medicine,* Vol. 1, Collery, Ph., Poirier, L.A., Manfait, M., and Etienne, J.C., Eds., John Libbey Eurotext, Paris, 75, 1990.
Hardy, J. A. and Aust, A. E., *Proc. Am. Assoc. Cancer Res.,* 35(920), 154, 1994.
Harris, C. C., *Cancer Res.,* Suppl. 51, 5023s–5044s, 1991.
Hass, G. M., Laing, G. H., Galt, R. M., and McCreary, P. A., *Mg Bull.,* 3, 217–228, 1981.
Hass, G. M., Galt, R. M., Laing, G. H., Coogan, P. S., Maganini, R. O., and Friese, J. A., *Mg Bull.,* 8, 45–55, 1989.
Higinbotham, K. G., Rice, J. M., Bhalchandra, A. D., Kasprzak, K. S., Reed, C. D., and Perantoni, A. O., *Cancer Res.,* 52, 4747–4751, 1992.
Hoover, K. L. and Poirier, L. A., *Am. Inst. Nutr.,* 1569–1575, 1986.
IARC Monographs, Arsenic and Arsenic Compounds, Vol. 23, International Agency for Research on Cancer, Lyon, France, 39, 1980.
IARC Monographs, Arsenic and Arsenic Compounds, Suppl. 7, International Agency for Research on Cancer, Lyon, France, 100, 1987a.
IARC Monographs, Haematite and Ferric Oxide, Suppl. 7, International Agency for Research on Cancer, Lyon, France, 216, 1987b.
IARC Monographs, Iron-Dextran Complex, Suppl. 7, International Agency for Research on Cancer, Lyon, France, 226, 1987c.
IARC Monographs, Cobalt and Cobalt Compounds, Vol. 52, International Agency for Research on Cancer, Lyon, France, 363, 1991.

Ip, C., in *Essential Nutrients in Carcinogenesis,* Poirier, L.A., Newberne, P.M., and Pariza, M.W., Eds., Plenum Press, New York, 431, 1986.

Ip, C. and Ganther, H. E., in *Cancer Chemoprevention,* Wattenberg, L., Lipkin, M., Boone, C.W., and Kelloff, G.J., Eds., CRC Press, Boca Raton, FL, 479, 1992.

Izumi, K., Kitaura, K., Chone, Y., Suzuki, Y., and Matsumoto, K., *Proc. Am. Assoc. Cancer Res.,* 35, 161, 1994.

Jasmin, G., in *Dans Endocrine Aspects of Disease Processes,* Jasmin, G., Ed., Warren H. Green Publishers, St. Louis, 356, 1968.

Kadlubar, F. F., in *DNA Adducts: Identification and Biological Significance,* Hemminki, K., Dipple, A., Shuker, D.E.G., Kadlubar, F.F., Segerbäck, D., and Bartsch, H., Eds., IARC Sci. Publ. 125, International Agency for Research on Cancer, Lyon, France, 199, 1994.

Kasprzak, K. S., in *Metal Ions in Biology and Medicine,* Vol. 3, Collery, Ph., Poirier, L.A., Littlefield, N.A., and Etienne, J.C., Eds., John Libbey Eurotext, Paris, 37, 1994.

Kasprzak, K. S., Diwan, B. A., and Rice, J. M., *Toxicology,* 90, 129–140, 1994.

Kasprzak, K. S., Hoover, K. L., and Poirier, L. A., *Carcinogenesis,* 6, 279–282, 1985, 1985a.

Kasprzak, K. S., Kovatch, R. M., and Poirier, L. A., *Toxicology,* 52, 253–262, 1988.

Kasprzak, K. S. and Poirier, L. A., *Carcinogenesis,* 6, 1819–1821, 1985a.

Kasprzak, K. S. and Poirier, L. A., *Toxicology,* 34, 221–230, 1985b.

Kasprzak, K. S., Quander, R. V., and Poirier, L. A., *Carcinogenesis,* 8, 1161–1166, 1985b.

Kasprzak, K. S., Waalkes, M. P., and Poirier, L. A., *Biol. Trace Elem. Res.,* 13, 253–273, 1987.

Kasprzak, K. S. and Waalkes, M. P., in *Essential Nutrients in Carcinogenesis,* Poirier, L.A., Newberne, P.M., and Pariza, M.W., Eds., Plenum Press, New York, 497, 1986.

Kasprzak, K. S., Waalkes, M. P., and Poirier, L. A., *Toxicol. Appl. Pharmacol.,* 82, 236–343, 1986.

Kato, J., Kohgo, Y., Sugawara, N., Kobune, M., and Niitsu, Y., *Proc. Am. Assoc. Cancer Res.,* 35(777), 130, 1994.

Konishi, N., Ward, J. M., and Waalkes, M. P., *Toxicol. Appl. Pharmacol.,* 104(1), 149–156, 1990.

Lazo, J. S., Schwarz, M. A., and Pitt, B. R., in *Metal Ions in Biology and Medicine,* Vol. 3, Collery, Ph., Poirier, L.A., Littlefield, N., and Etienne, J.C., Eds., John Libbey Eurotext, Paris, 15, 1994.

Littlefield, N. A. and Hass, B. S., in *Metal Ions in Biology and Medicine,* Collery, Ph., Poirier, L.A., Littlefield, N.A., and Etienne, J.C., Eds., John Libbey Eurotext, Paris, 507, 1994.

Littlefield, N. A., Fullerton, F. R., and Poirier, L. A., *Chem.-Biol. Interact.,* 79, 217–228, 1991.

Littlefield, N. A., Hass, B. S., James, S. J., and Poirier, L. A., *Cell Biol. Toxicol.,* 10, 127–135, 1994.

Mason, K. E. and Young, J. O., in *Selenium in Biomedicine,* Muth, O.H., Oldfield, J.E., and Weswig, P.H., Eds., AVI Publishing, Westport, CT, 383, 1967.

Medina, D., in *Essential Nutrients in Carcinogenesis,* Poirier, L.A., Newberne, P.M., and Pariza, M.W., Eds., Plenum Press, New York, 465, 1986a.

Medina, D., in *Diet, Nutrition, and Cancer: A Critical Evaluation,* Vol. II, Reddy, B.S. and Cohen, L.A., Eds., CRC Press, Boca Raton, FL, 23, 1986b.

Mello-Filho, A. C. and Meneghini, R., *Mutat. Res.,* 251, 109–113, 1991.

Merrill, A. H., Jr. and Schroeder, J. J., in *Annual Review of Nutrition,* Vol. 13, Olson, R.E., Bier, D.M., and McCormick, D.B., Eds., Annual Reviews Inc., Palo Alto, CA, 1993.

Michalowsky, I., *Centralbl. Allg. Pathol. Pathol. Anat.,* 38, 585, 1926.

Mills, B. J., Broghamer, W. L., Higgens, P. J., and Lindeman, R. D., *J. Nutr.,* 114, 739–745, 1984.

Milner, J. A., in *Essential Nutrients in Carcinogenesis,* Poirier, L.A., Newberne, P.M., and Pariza, M.W., Eds., Plenum Press, New York, 449, 1986.

Nelson, R. L., in *Metal Ions in Biology and Medicine,* Vol. 1, Collery, Ph., Poirier, L.A., Manfait, M., and Etienne, J.C., Eds., John Libbey Eurotext, Paris, 35, 1990.

Nelson, R. L., Davis, F., Bowen, P., and Kikendal, J. W., in *Metal Ions in Biology and Medicine,* Vol. 2, Anastassopoulou, J., Collery, Ph., Etienne, J.C., and Theophanides, Th., Eds., John Libbey Eurotext, Paris, 358, 1992.

Newmark, H. L. and Lipkin, M., in *Calcium, Vitamin D, and Prevention of Colon Cancer,* Lipkin, M., Newmark, H.L., and Kelloff, G., Eds., CRC Press, Boca Raton, FL, 145, 1991.

Nishikawa, A., Furukawa, F., and Takahashi, M., *Proc. Am. Assoc. Cancer Res.,* 35, 141, 1994.

Nishiyama, S., Saito, N., Konishi, Y., Abe, Y., and Kusumi, K., Cardiotoxicity in Mg-deficient rats fed cadmium, *J. Nutr. Sci. Vitaminol.,* 36, 33–44, 1990.

Pence, B. C. and Budding, F., *Carcinogenesis,* 9, 187, 1988.

Pence, B. C., Dunn, D. M., Zhao, A., Landers, M., and Wargovich, M. J., *Proc. Am. Assoc. Cancer Res.,* 35, 624, 1994.

Peters, R. K., Mack, T. M., Garabrant, D. H., Homa, D. M., and Pike, M. C., in *Calcium, Vitamin D, and Prevention of Colon Cancer,* Lipkin, M., Newmark, H.L., and Kelloff, G., Eds., CRC Press, Boca Raton, FL, 113, 1991.

Poirier, L. A., Kasprzak, K. S., Hoover, K. L., and Wenk, M. L., *Cancer Res.,* 43, 4575–4581, 1983.

Poirier, L. A., Theiss, J. C., Arnold, L. J., and Shimkin, M. B., *Cancer Res.,* 44, 1520–1522, 1984.

Reddy, B. S., Sugle, S., Maruyama, H., El-Bayoumy, K., and Marra, P., *Cancer Res.,* 47, 5901–5904, 1987.

Riviere, M. A., Chouroulinkov, I., and Guerin, M., *Bull. Assoc. Fr. Etude Cancer,* 47, 55, 1960.

Rodriguez, R. E. and Kasprzak, K. S., *J. Am. Coll. Toxicol.,* 1265–1269, 1989.

Rozen, P., in *Calcium, Vitamin D, and Prevention of Colon Cancer,* Lipkin, M., Newmark, H.L., and Kelloff, G., Eds., CRC Press, Boca Raton, FL, 1991.

Rosin, M. P. and Stich, H. F., *Int. J. Cancer,* 23, 722, 1979.
Saffiotti, U. and Bertolero, F., *Biol. Trace Elem. Res.,* 21, 475–482, 1989.
Scalmati, A., Lipkin, M., and Newmark, H., in *Cancer Chemoprevention,* Wattenberg, L., Lipkin, M., Boone, C.W., and Kelloff, G.J., Eds., CRC Press, Boca Raton, FL, 249, 1992.
Scarpelli, D. G. and Rao, M. S., *Proc. Natl. Acad. Sci. U.S.A.,* 78, 2577–2581, 1981.
Sunde, R. A., in *Annual Review of Nutrition 10,* Annual Reviews, Inc., Palo Alto, CA, 451, 1990.
Sunderman, F. W., Jr., Plowman, M. C., Slaisova, O., Grbac-Ivankovic, S., Foglia, L., and Crivello, J. F., in *Metal Ions in Biology and Medicine,* Vol. 3, Collery, Ph., Poirier, L.A., Littlefield, N.A., and Etienne, J.C., Eds., John Libbey Eurotext, Paris, 17, 1994.
Suzuki, K. T., Kanno, S., Ogra, Y., Misawa, S., and Aoki, Y., in *Metal Ions in Biology and Medicine,* Vol. 3, Collery, Ph., Poirier, L.A., Littlefield, N.A., and Etienne, J.C., Eds., John Libbey Eurotext, Paris, 187, 1994.
Suzuki, K., Sugiyama, T., Ookawara, T., Kurosawa, T., and Taniguchi, N., *Biochem. Int.,* 23, 9, 1991.
Waalkes, M. P., in *Metal Ions in Biology and Medicine,* Vol. 3, Collery, Ph., Poirier, L.A., Littlefield, N.A., and Etienne, J.C., Eds., John Libbey Eurotext, Paris, 137, 1994.
Waalkes, M. P. and Poirier, L. A., *Toxicol. Appl. Pharmacol.,* 75, 539–546, 1984.
Waalkes, M. P. and Poirier, L. A., *Toxicol. Appl. Pharmacol.,* 79, 47–53, 1985.
Waalkes, M. P., *Toxicol. Lett.,* 26, 133–138, 1985.
Wallwork, J. C. and Duerre, J. A., *J. Nutr.,* 115, 252–262, 1985.
Wargovich, M. J., Baer, A., and Levin, B., in *Calcium, Vitamin D, and Prevention of Colon Cancer,* Lipkin, M., Newmark, H.L., and Kelloff, G., Eds., CRC Press, Boca Raton, FL, 267, 1991.
Whittaker, P., Robl, M. G., and Dunkel, V. C., *Proc. Am. Assoc. Cancer Res.,* 35(614), 103, 1994.
Wolff, G. L., Whittaker, P., and Dunkel, V. C., *Proc. Am. Assoc. Cancer Res.,* 35(616), 103, 1994.
Yu, S.-Y. and Lu, X.-P., in *Metal Ions in Biology and Medicine,* Vol. 2, Anastassopoulou, J., Collery, Ph., Etienne, J.C., and Theophanides, Th., Eds., John Libbey Eurotext, Paris, 151, 1992.
Yuspa, S. H., The pathogenesis of squamous cell cancer: lessons learned from studies of skin carcinogenesis, Thirty-third G.H.A. Clowes Memorial Award Lecture, *Cancer Res.,* 54, 1178–1189, 1994.

Chapter 18

Oxidative DNA Damage in Metal-Induced Carcinogenesis

Kazimierz S. Kasprzak

I. INTRODUCTION

Several metals have been found to be carcinogenic to humans and/or animals (Tables 1 and 2). However, the underlying mechanisms remain unclear. It is believed that neoplastic transformation of cells results from a heritable alteration in the genetic code. If this belief is correct, any molecule that can bind with constituents of cell nuclei may affect the genetic code. It is obvious that DNA, having an abundance of phosphate anions and nitrogen and oxygen donor groups, is an ideal binding partner for metals (as cations). Nuclear proteins bind metals as well. The metal is essential for the proper structure and function of some of these proteins such as "zinc fingers" (Berg, 1986; Sunderman and Barber, 1988), DNA polymerases (Leonard, 1986), or certain DNA repair enzymes (Mo et al., 1992; O'Connor et al., 1993). Therefore, it is not surprising that *in vivo* exposure to toxic metals results in their association with nuclear components (Bryan, 1981). However, whether or not a metal cation is capable of changing the genetic code by substituting for magnesium, the native DNA counterion, and thus altering the conformation of the DNA double helix (Sunderman, 1989) remains an open question. Several carcinogenic metal cations were tested *in vitro* for their effect on DNA replication and found to markedly decrease replication fidelity (Loeb and Mildvan, 1981; Sunderman, 1984, 1986a), but the relatively high concentrations needed to produce a significant number of errors *in vitro* would hardly be attainable *in vivo* without killing the cells.

Strand scission, depurination, cross-linking, and base modifications are the major DNA lesions formed after exposure of experimental animals and cultured cells to carcinogenic metals (Hamilton-Koch et al., 1986; Kasprzak et al., 1992a, 1994; Misra et al., 1993; Oleinick et al., 1987; Sunderman, 1986a; Wetterhahn et al., 1989). This DNA damage must result from disruption of the normal pattern of covalent chemical bonds in the chromatin in addition to any conformational changes produced by ionic substitution of the carcinogenic metal for magnesium. Hence, not only the direct conformational effects of metal binding, but also some other obviously indirect effects of metals on nuclear chromatin must be considered. Interestingly, the damage caused in nuclear chromatin by carcinogenic metals can also be produced by oxygen radicals and/or other free radical species generated in aqueous media by ionizing radiation (Ames, 1989; Angelov et al., 1991; Breimer, 1990; Dizdaroglu, 1991; Dizdaroglu and Gajewski, 1990; Oleinick et al., 1987). This striking similarity provokes formulation of the hypothesis that metal carcinogenicity is mediated by free radicals. To test this hypothesis, we must find answers to the following questions:

1. Does the chemistry of carcinogenic metals allow for production of radical or other highly reactive species from molecular oxygen and/or other molecules under biologically relevant conditions?
2. Do carcinogenic metals sustain production of such species in living cells?
3. Do the metals assist in the attack of active oxygen species on DNA?
4. Do the metals and oxygen radicals produce the same fingerprint mutations?

Table 1 Metal Exposures Associated with Human Cancer

Metal	Exposure/route	Tumor location	Ref.
Ni	Pyrometallurgy, hydrometallurgy, and electroplating of nickel/inhalation	Lung and sinonasal cancer	IARC, 1990
Cr	Chromate and chromate pigment production, chromium plating industries/inhalation	Lung and sinonasal cancer	IARC, 1990
As	Production and use of arsenic trioxide and its derivatives/inhalation, skin and oral exposure	Lung, skin and gastro-intestinal cancers; precancerous dermal keratoses	Sunderman, 1984, 1986a; Leonard, 1985; Leonard and Lauwerys, 1980.
Cd	Production and use/inhalation	Lung	IARC, 1994; Waalkes and Oberdoerster, 1990

Note: Limited evidence suggests that exposure to certain Fe and Be derivatives increases risk of cancer in humans (IARC, 1973, 1987b, 1994; Skilleter, 1985; Sunderman, 1984).

Table 2 Metal Derivatives Producing Tumors in Experimental Animals

Metal	Derivatives/exposure route	Tumor location	Ref.
Be	BeO, BeHPO$_4$/inhalation, i.v., i.os.	Lung carcinomas, bone sarcomas	Sunderman, 1984; Skilleter, 1985
Cd	Cd(II) acetate, CdCl$_2$, CdS/s.c., i.m., i.ts.	Local sarcomas, gonadal adenomas	Sunderman, 1984; Waalkes and Oberdorster, 1990
Co	Co^0, CoCl$_2$, CoS, CoO/i.m., s.c., i.os.	Local sarcomas	IARC, 1987a; Leonard and Lauwerys, 1990; Taylor, 1990; Sunderman, 1984
Cr	CaCrO$_4$, PbCrO$_4$, CrCl$_3$/inhalation, s.c., i.m., p.e.	Lung tumors, local sarcomas	Anderson et al., 1994; IARC, 1990; Sunderman, 1984
Fe	Fe(III) dextran, Fe(III)-NTA/s.c., i.m., i.p.	Local sarcomas, renal carcinomas	IARC, 1973; Li et al., 1987; Okada et al., 1983; Sunderman, 1984
Ni	Ni^0, Ni$_3$S$_2$, crst. NiS, Ni(II) acetate, Ni(CO)$_4$/inhalation, s.c., i.m., i.v., i.ren., i.oc., t.p.	Local sarcomas, lung renal carcinomas	IARC, 1990; Kasprzak, 1987; Sunderman, 1984
Pb	Pb(II)acetate/oral, s.c., i.p.	Renal adenomas and carcinomas	Kasprzak et al., 1985a; Schlag, 1987; Sunderman, 1984
Pt	cis-Pt(II)/s.c., i.p., t.p.	Lung adenomas, thymus lymphomas, local sarcomas; skin tumors after TPA promotion	Diwan et al., 1993; Kazantzis, 1981; Sunderman, 1984

Note: Also, Al(III)dextran, Mn(II)acetylacetonate, and Ti(IV)dicyclopentadiene induced local sarcomas after s.c. or i.m. injections; NaAsO$_2$ given transplacentally and postnatally induce leukemias/lymphomas in mice; CuCl$_2$ and ZnSO$_4$ induced testicular tumors after local injection (Kazantzis, 1981; Sunderman, 1984).

Abbreviations: i.v., intravenous; i.p., intraperitoneal; s.c., subcutaneous; i.m., intramuscular; i.os., intraosseous; i.ren., intrarenal; i.oc., intraocular; i.ts., intratesticular; t.p., transplacental; p.e., preconception i.p. exposure of male mice, lung tumors in the progeny; Co^0, metallic cobalt powder; Ni^0, metallic nickel powder; acet, acetate; NTA, nitrilotriacetate; cis-Pt(II), cis-dichlorodiammineplatinum; TPA, 12-O-tetradecanoylphorbol-13-acetate.

The aim of this chapter is to answer those questions based on a critical review of chemical and biological data on the mechanisms of metal-induced carcinogenesis. We do not attempt to present a complete list of relevant publications; those can be found elsewhere (Aust et al., 1993; Cadet et al., 1991, 1993; Frenkel, 1992; Janssen et al., 1993). We shall first consider the basic redox biochemistry of transition metals and the types of metal-induced oxidative damage to certain biomolecules, and then evaluate the significance of that damage to carcinogenesis. Although various target biomolecules and carcinogenic metals are regarded, special attention is paid to DNA as the primary target of carcinogens, and to nickel and chromium, two metals that are well documented as carcinogenic to humans.

II. BASIC REDOX BIOCHEMISTRY OF METAL CARCINOGENS

For the purpose of this presentation, let us focus on metal-mediated production of only those oxygen species that are capable of attacking cellular genetic material. More general data on the role of metals in driving free radical reactions are available elsewhere (Sawyer, 1987; Spiro, 1980). The most important mechanisms of oxygen activation by transition metals involve Fenton/Haber-Weiss chemistry and autoxidation. The substrates are molecular oxygen (O_2) and two common metabolic products, hydrogen peroxide (H_2O_2) and the superoxide anion radical ($O_2^{-\cdot}$) (Halliwell and Gutteridge, 1986; Imlay and Linn, 1988), which do not react with DNA (Blakely et al., 1990; Halliwell and Aruoma, 1991). Equation 1 illustrates the conversion of H_2O_2 into a powerful DNA-damaging hydroxyl radical ($\cdot OH$) via the oxidation of a metal cation:

$$M^{n+} + H_2O_2 \rightarrow M^{(n+1)+} + OH^- + \cdot OH \quad \text{(Fenton)} \tag{1}$$

$O_2^{-\cdot}$ provides an important function by reducing the metal via a one-electron process and thus recycling it back to Reaction 1:

$$M^{(n+1)+} + O_2^{-\cdot} \rightarrow M^{n+} + O_2 \tag{2}$$

The balance of those two reactions is

$$H_2O_2 + O_2^{-\cdot} \rightarrow O_2 + OH^- + \cdot OH \quad \text{(Haber-Weiss)} \tag{3}$$

Two oxidation states of the metal cation (M^{n+} and $M^{(n+1)+}$) form a catalytic electron transfer (redox) couple. In the absence of chelators, the above reactions are driven by Cu(I), Fe(II), Co(II), Ti(III), and Cr(VI) ions (Gutteridge, 1985; Shi and Dalal, 1992; Walling, 1975). Some other metal ions become reactive after proper chelation (see below). H_2O_2 and $O_2^{-\cdot}$ needed for those reactions may originate not only from cell metabolism but also from oxidation of some metal ions with O_2 (autoxidation) (Sawyer, 1987; Spiro, 1980). The latter is illustrated below:

$$M^{n+} + O_2 \rightarrow M^{(n+1)+} + O_2^{-\cdot} \tag{4}$$

$$M^{n+} + O_2^{-\cdot} + 2H^+ \rightarrow M^{(n+1)+} + H_2O_2 \tag{5}$$

The yield of $O_2^{-\cdot}$ and/or H_2O_2 in the above reactions depends on the conditions and is influenced by the acidity of the solution and by accompanying anions and chelators (Sawyer, 1987; Spiro, 1980).

Transition metal compounds interacting with H_2O_2 produce not only free $\cdot OH$ radical but also other strong oxidants, such as singlet oxygen ($^1O_2$) (Kawanishi et al., 1989a), and metal-oxo and -peroxo species, all capable of damaging DNA and proteins (Ito et al., 1992; Kawanishi and Yamamoto, 1991; Kawanishi et al., 1986; Rush et al., 1990; Stadtman and Berlett, 1991; Wink et al., 1994; Yamamoto and Kawanishi, 1989, 1991, 1992, Yamamoto et al., 1989, 1993a). The formation of metal oxo (Equation 6) and -peroxo (Equation 7) species is suggested as an alternative to the classical Fenton reaction (reviewed by Wink et al., 1994):

$$M^{n+} + H_2O_2 \rightarrow [MO]^{n+} + H_2O \tag{6}$$

$$M^{n+} + H_2O_2 \rightarrow [MO_2]^{(n-2)+} + 2H^+ \tag{7}$$

In Reactions 6 and 7, the metal cation changes its oxidation state n into n+2 or n+1, respectively. Such metal-associated oxidants acting at sites of metal binding to DNA allow for a better explanation of the observed site specificity of metal-mediated oxidative DNA damage than free (i.e., "diffusible") hydroxyl radicals (Ito et al., 1992; Kawanishi and Yamamoto, 1991; Kawanishi et al., 1986; Mouret et al., 1991; Rush et al., 1990; Stadtman and Berlett, 1991; Wink et al., 1994; Yamamoto and Kawanishi, 1989, 1991, 1992; Yamamoto et al., 1989, 1993a).

Although it is the metal cation itself that drives electron transfer in the above reactions, the ligand with which the cation is complexed has a profound effect on that transfer. It may either enhance or inhibit the reactions. For example, autoxidation of Fe(II) to Fe(III) is enhanced by ethylenediaminetetraacetic acid (EDTA) and nitrilotriacetic acid (NTA) (Hamazaki et al., 1989), but inhibited by o-phenanthroline (Nassi-Calo et al., 1989) and deferoxamine (DFO) (Loeb et al., 1988). NTA enables Fe(III) to react with H_2O_2 and produce ·OH (Inoue and Kawanishi, 1987). DFO inhibits free radical generation and dG hydroxylation mediated by Cr(V) (Shi et al., 1992b). Ni(II) aquo-cation reacts with neither O_2 nor H_2O_2. Reactivity of Ni(II) towards H_2O_2 is provided, however, by several bioligands, such as oligopeptides or proteins (Table 3). Likewise, the inorganic Co(II) cation, which is stable under air, becomes sensitive to oxidation with atmospheric O_2 in the presence of organic ligands (Moorhouse et al., 1985). Quite often, the chelator itself becomes the first target for oxidation damage mediated by its "own" metal. The resulting products may interact with other molecules, including proteins and DNA (Kasprzak and Bare, 1989).

A biologically important role for carcinogenic transition metals is catalysis of decomposition of organic peroxides (Akman et al., 1992; Cadet et al., 1993; Halliwell and Gutteridge, 1984; Sunderman, 1986b; Tofigh and Frenkel, 1989). The peroxides (e.g., lipid peroxides and nucleoside hydroperoxides produced by ionizing radiation) are degraded through chain radical reactions with the formation of transient radicals, including ·OH. For example, degradation of 5-hydroperoxy-methyl-2'-deoxyuridine (a dT oxidation product) to 5-hydroxymethyl-2'-deoxyuridine (5-OHMe-dU) and 5-formyl-2'-deoxyuridine proceeds rapidly with Cu(I), Cu(II), Fe(II), and Sn(II); slowly with Co(II) and Ni(II); but not at all in the presence of Fe(III), Mn(II), Mn(III), Al(III), Sn(IV), or Ca(II) (Tofigh and Frenkel, 1989). Owing to those reactions, 5-hydroxyperoxymethyl-2'-deoxyuridine in DNA appears to be capable of oxidizing neighboring bases (Patel et al., 1992). Benzoyl peroxide, a commercial oxidant widely used for industrial, household, and medicinal purposes, was found to decompose and produce promutagenic DNA damage in the presence of Cu(I) (Akman et al., 1992). Chelating agents, EDTA, diethylenetriamine-pentaacetic acid (DTPA), DFO, and proteins may either inhibit or enhance degradation (Frenkel and Tofigh, 1989; Tofigh and Frenkel, 1989).

Certain transition metal sulfides display high genotoxicity and carcinogenicity relative to other derivatives of the same metal (Kargacin et al., 1993; Kasprzak, 1978a,b; Lee et al., 1993; Lin et al., 1991). A possible reason for this is the potential of both the metal cation and the sulfide anion (S^{2-}) to undergo autoxidation and thus activate O_2 (Kasprzak, 1978a,b; Kasprzak and Sunderman, 1977; Lin et al., 1991; Shi et al., 1994). One product of S^{2-} autoxidation, the sulfite anion (SO_3^{2-}), has been demonstrated to produce genotoxic effects by deamination of DNA bases (Singer and Grunberger, 1983) and degradation of nuclear proteins (Ito and Kawanishi, 1991). SO_3^{2-} is able to activate more O_2 during further autoxidation to the sulfate (SO_4^{2-}) and thus promote DNA base oxidation, another genotoxic effect. For example, nickel subsulfide (Ni_3S_2) in aerobic solution caused oxidation of 2'-deoxyguanosine (dG) to 8-oxo-2'-deoxyguanosine (8-oxo-dG) (Kasprzak and Hernandez, 1989) and deamination of 5-methyl-2'-deoxycytidine to thymidine (Kasprzak et al., 1991b). Ni_3S_2-rich nickel mattes, to which nickel refinery workers are exposed, were found to catalyze oxidation by O_2 of the formate to the $CO_2^{-\cdot}$ radical (Costa et al., 1989a). Oxygen-activating capacity was observed for particles of various sulfides (Costa et al., 1989a,b,c; Zalma et al., 1989) and other classes of insoluble metal compounds, including asbestos (Berger et al., 1993; Costa et al., 1989a,b,c; Vallyathan et al., 1992; Zalma et al., 1989).

Because of the high carcinogenic potency of nickel and chromium, the redox biochemistry of these two metals is worthy of more attention. This subject has been discussed in more detail in several publications (Lancaster, 1988; Magos, 1991; Snow, 1992; Wetterhahn et al. 1989). Shortly, the biological

Table 3 Reactivity of Ni(II) Complexes with Peptides and Proteins Towards Oxygen Species

Ni(II) complex with	Substrate	Active oxygen species produced/ligand degradation	Ref.
Gly_4, Ala_4, Gly_3NH_2 Gly-Gly-His	O_2	Organic hydroperoxides, $O_2^-\cdot$(?) /yes, mainly decarboxylation	Bossu et al., 1978 Nieboer et al., 1986
Gly_2, Gly_3, Gly_4, Gly_5	H_2O_2	NiO^{2+}(?), oxygen- and carbon-centered radicals/possible	Bossu et al., 1978 Cotelle et al., 1992 Inoue and Kawanishi, 1989
Gly-His	H_2O_2	·OH/unknown	Inoue and Kawanishi, 1992
Gly-Gly-His	H_2O_2	·OH, $O_2^-\cdot$, $^1O_2$/possible breakdown of the imidazole ring	Cotelle et al., 1992 Inoue and Kawanishi, 1989 Nieboer et al., 1989 Torreilles and Guerin, 1990
	Organic peroxides	·OH, alkyl and alkoxyl radicals/possible	Shi et al., 1992a Torreilles and Guerin, 1990
	$O_2^-\cdot$	H_2O_2/possible; the peptide acts as radical scavenger	Nieboer et al., 1984, 1989
Gly_4	$O_2^-\cdot$	H_2O_2, ·OH/unknown	Cotelle et al., 1992
Gly-His-Lys	H_2O_2	None	Cotelle et al., 1992
	$O_2^-\cdot$	H_2O_2/unknown	Cotelle et al., 1992
Carnosine[a]	H_2O_2	·OH/unknown	Cotelle et al., 1992
	$O_2^-\cdot$	H_2O_2, ·OH/unknown	Cotelle et al., 1992
Carnosine, anserine, homocarnosine, GSH	Organic peroxides	·OH, alkyl and alkoxyl radicals/unknown	Datta et al., 1993 Shi et al., 1992a
Gly-Gly-His-Gly	H_2O_2	·OH, $O_2^-\cdot$/unknown	Torreilles and Guerin, 1990 Nieboer et al., 1989
	$O_2^-\cdot$	H_2O_2/unknown	Nieboer et al., 1989
Ala-Gly-Gly-His, Gly-His-Ala	H_2O_2	·OH/unknown	Torreilles and Guerin, 1990
Asp-Ala-His-Lys	H_2O_2	·OH, $O_2^-\cdot$/unknown	Cotelle et al., 1992
	$O_2^-\cdot$	H_2O_2, ·OH/unknown	Cotelle et al., 1992
Gly_2, Gly-Phe-Ala	H_2O_2	None	Torreilles and Guerin, 1990
Val^5-Angiotensin-II-Asp^1-β-amide	H_2O_2	Unknown/yes, degradation to smaller peptides	Curtius et al., 1968
Polymyxin B Albumin	H_2O_2	·OH, NiO^{2+}(?)/unknown	Cotelle et al., 1992; Nieboer et al., 1984, 1986
	$O_2^-\cdot$	H_2O_2, ·OH/possible; albumin acts as radical scavenger	Cotelle et al., 1992; Nieboer et al., 1984, 1989

[a] Carnosine, β-Ala-His; anserine, γ-aminobutyryl-His; homocarnosine, β-Ala-3-methyl-His; GSH, reduced glutathione (γ-Glu-Cys-Gly).

oxidative effects of nickel depend on its ability to form an electron transfer couple Ni(III)/Ni(II) (compare Reactions 1 and 2) when complexed by certain oligopeptides (reviewed in Kasprzak, 1991; Klein et al., 1991; Margerum and Anliker, 1988). Examples of relevant nickel-mediated reactions and the resulting free radical and other chemically active products are listed in Table 3. It must be remembered that Ni(II) complexation with oligopeptides and certain other organic ligands is necessary to make this cation reactive with H_2O_2 (Bossu et al., 1978). Ni(II)Gly-Gly-His and Ni(II)Gly-Gly-His-Gly, besides converting H_2O_2 into ·OH, also produce $O_2^-\cdot$ (Bossu et al., 1978; Shi et al., 1993b). For Ni(II) complexes with some histidine (His)-containing peptides or human serum albumin reacting with H_2O_2, Nieboer et al.

(1986) postulated the formation of the nickel-oxo cation NiO^{2+}. Such a metal-oxo species would be a strong site-specific oxidizing agent. Site specificity of DNA oxidation in the presence of Ni(II) and Cu(II) complexes has been described by several authors (Kawanishi et al., 1989b; Li and Rokita, 1991; Mack and Dervan, 1992). Neither Ni(II) alone, nor the histidyl peptides alone, exhibit any reactivity towards H_2O_2 (Nieboer et al., 1986). Cu(II), Mn(II), Zn(II), and Cd(II) complexes with the peptides are inactive (Nieboer et al., 1986). An important result of the reactions of certain Ni(II) complexes with O_2 or H_2O_2 is the production of not only ·OH (or NiO^{2+}), but also other oxygen-, carbon-, and, perhaps, sulfur-centered radicals originating from the ligands. All those species may attack neighboring molecules (Bossu et al., 1978; Torreilles and Guerin, 1990).

The redox biochemistry of chromium is richer and considers several oxidation states: Cr(VI), Cr(V), Cr(IV), Cr(III), and possibly Cr(II). Chromium is most potent as a carcinogen in the form of chromate and dichromate anions [Cr(VI)] which, unlike the Cr(III) cation, are easily taken up by cells (IARC, 1990). The biochemistry of chromium relevant to carcinogenesis has been studied in great detail (Aiyar et al., 1991; Borges et al., 1991; Dillon et al., 1993; Faux et al., 1992; Hamilton and Wetterhahn, 1989; Hneihen et al., 1993; IARC, 1990; Kawanishi et al., 1986, 1989a; Salnikow et al., 1992; Shi and Dalal, 1990a,b,c; Shi et al., 1993c; Standeven and Wetterhahn, 1991a,b, 1992). The results have led to formulation of an "uptake-reduction" model for Cr(VI) carcinogenesis (Wetterhahn et al., 1989). In brief, under physiological conditions, Cr(VI) does not react with DNA. It is, however, reduced intracellularly by H_2O_2, glutathione (GSH) reductase, carbohydrates, ascorbic acid, GSH, and other molecules to more reactive products, including Cr(V), Cr(IV), and Cr(III). The byproducts of the reduction, $^1O_2$ and oxygen- and sulfur-centered radicals (e.g., ·OH, lipid hydroperoxide-derived, and GSH-thiyl radicals), can damage DNA. Hence, the ability of Cr(VI) to attack DNA depends on cellular redox systems (Goodgame and Joy, 1986; Shi and Dalal, 1989, 1990b,c; Shi et al., 1993c; Sugiyama et al., 1991; Wetterhahn et al., 1989) and the "ultimate" carcinogenic species in Cr(VI)-induced carcinogenesis may be ·OH and other radical species (Shi et al., 1993c; Shi and Dalal, 1989, 1990b).

There is growing evidence that oxidative DNA damage may also be involved in cobalt-induced toxicity and carcinogenesis (Beyersmann and Hartwig, 1992; Kasprzak et al., 1994; Nackerdien et al., 1991; Shi et al., 1993a). That evidence is strongly supported by the redox biochemistry of cobalt (Hanna et al., 1992; Kadiiska et al., 1989; Shi et al., 1993a). This metal, in the form of Co(II), complexes with some ligands, e.g., GSH and L-cysteine (Cys), and has been found to generate ·OH and other oxygen- and carbon-centered radicals from model lipid peroxides at physiological pH (Shi et al., 1993a). Likewise, NADH, GSH, and anserine rendered Co(II) reactive with H_2O_2 to produce ·OH (Kadiiska et al., 1989; Shi et al., 1993a).

In contrast, evidence that carcinogenic metals other than nickel, chromium, or cobalt act through oxidative damage, although sometimes quite suggestive (e.g., for cadmium [Frenkel et al., 1994]), remains only conjectural.

III. METAL-MEDIATED OXIDATIVE DNA DAMAGE AND ITS MUTAGENICITY

It is important for this discussion to stress the fact that $O_2^{-·}$ and H_2O_2, which along with O_2 are the major substrates for oxygen activation to ·OH and metal-associated oxidants (e.g., NiO^{2+}), are produced metabolically within a mammalian cell, including the nucleus (Bartoli et al., 1977; Peskin and Shlyahova, 1986; Szatrowski and Nathan, 1991). The pool of substrates is further enriched by H_2O_2 and lipid peroxides generated by cells in response to metal-induced stress (Huang et al., 1993). Hence, the reactions leading to oxygen activation can proceed *in vivo* around and inside the cell nucleus (Huang et al., 1993; Kadiiska et al., 1992).

It has been known from radiobiology that ·OH reacts with all components of nuclear chromatin. It can modify DNA bases and deoxyribose and produce DNA-protein cross-links (Dizdaroglu, 1992; Oleinick et al., 1987; von Sonntag, 1987). As well, it can cause DNA depurination and strand scission (von Sonntag, 1987). Besides nuclear DNA, an important target for metals is also mitochondrial DNA (Rossi and Wetterhahn, 1989; Rossi et al., 1988). The techniques of detection and quantitation of DNA damage have recently been described by several authors (Berger et al., 1990; Cadet and Weinfeld, 1993; Dizdaroglu, 1991; Frenkel et al., 1991; Halliwell and Dizdaroglu, 1992).

A. DNA BASE DAMAGE

The pattern of chemical changes produced by ·OH in the base moiety of DNA (Figure 1) is so characteristic that it can be used for identification of ·OH attack (Dizdaroglu, 1991, 1992; Halliwell and

Aruoma, 1991; Teoule and Cadet, 1978; von Sonntag, 1987). Such a pattern has been found in human K562 cell chromatin exposed to H_2O_2 plus Ni(II) or Co(II) (Nackerdien et al., 1991), or in murine HyHEL-10 cell chromatin exposed to H_2O_2 plus Cu(II) or Fe(III) (Dizdaroglu et al., 1991). Ni(II) inflicted damage even when H_2O_2 was not added, an indication that Ni(II) bound to chromatin was able to activate ambient O_2 (Nackerdien et al., 1991). Production of 8-oxo-dG in isolated DNA by active oxygen arising from the reactions of Cr(VI) with H_2O_2 (Wetterhahn et al., 1989) or GSH reductase (Shi and Dalal, 1989) was also observed.

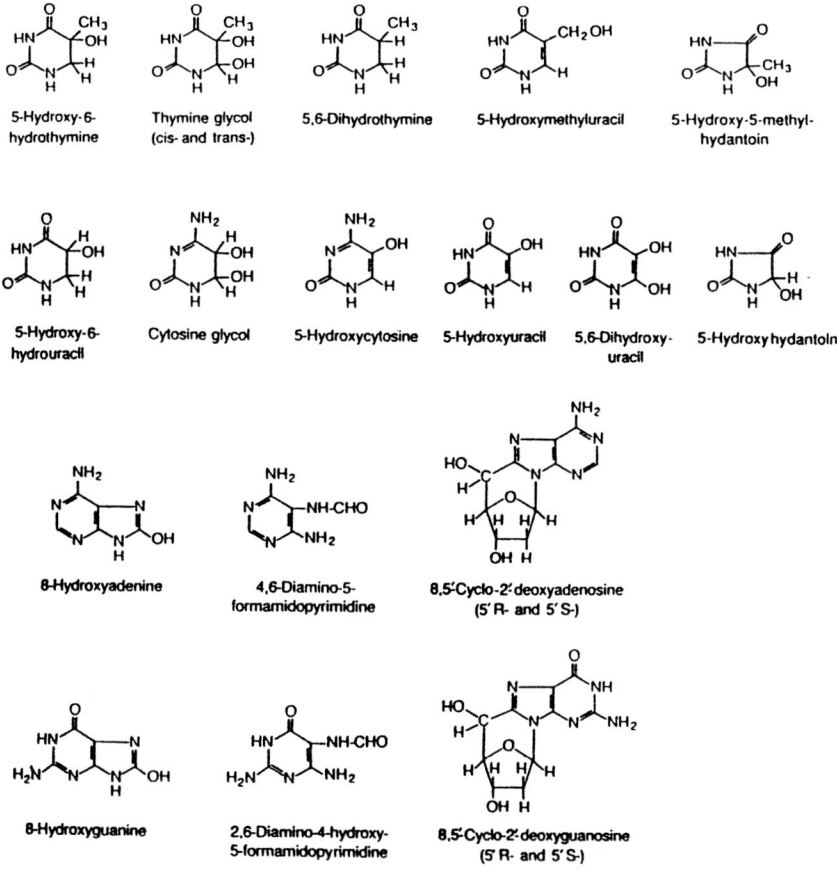

Figure 1 Structures of free radical-induced products of pyrimidines and purines in DNA which were identified by the use of gas chromatography/mass spectrometry technique. (According to Dizdaroglu, M., *Free Rad. Biol. Med.*, 10, 225, 1991. With permission.)

Very importantly, increased amounts of oxidatively damaged DNA bases were found in organs of Fischer rats (Kasprzak et al., 1990, 1992a; Misra et al., 1993) and BALB/c mice (Kasprzak et al., 1991a) following *in vivo* exposure to Ni(II). Moreover, the distribution of damaged bases in the rat kidney, which is a target organ for nickel carcinogenesis (Kasprzak et al., 1990), differed significantly from that in liver, a nontarget organ (Kasprzak et al., 1992a). In Ni(II)-treated mice, renal 8-oxo-dG levels were increased only in the BALB/c strain, which had low GSH and GSH peroxidase levels compared to two other strains investigated, the B6C3F1 and C3H (Kasprzak et al., 1991a; Misra et al., 1991; Rodriguez et al., 1991). Data pertinent to carcinogenic metal-mediated modifications of DNA bases *in vitro* and *in vivo* are presented in Table 4. As can be seen in this table, elevated contents of the modified bases have been found thus far in DNA exposed to nickel, chromium, cobalt, and iron carcinogens.

B. CROSS-LINKING

The most common nonlethal effect of metals on nuclear chromatin observed *in vitro* and *in vivo* is DNA-protein cross-linking (Borges and Wetterhahn, 1989; Chang et al., 1993; Wedrychowski et al.,

Table 4 Mediation of DNA Base Modifications by Carcinogenic Metal Derivatives *In Vivo* and *In Vitro*

Metal derivative	Experimental system	Result	Ref.
$NiSO_4$, $CdSO_4$, $Ni(OH)_2$, and $Cd(OH)_2$	Workers in an alkaline battery plant	Elevation of 5-OHMe-Ura in the blood	Frenkel et al., 1994
Ni(II) acetate	A single i.p. injection to Fischer rats	Elevated renal 8-oxo-dG 16–48 h post injection	Kasprzak et al., 1990
	A single i.p. injection to BALB/c, B6C3F1, C3H, and C57BL mice	Elevated renal 8-oxo-dG up to 48 h post injection in BALB/c mice only	Kasprzak et al., 1991a
	Exposure of NIH 3T3 and NRK-52 cells for 24 h	Variable (±25%) Ni(II) concentration- and time-related changes in 8-oxo-dG	Kasprzak et al., 1991a
	I.p. injection to pregnant Fischer rats	Increased levels of several damaged DNA bases in kidneys and livers of mothers and fetuses	Kasprzak et al., 1992
Ni(II)His$_2$ complex	I.v. injection to Fischer rats	Increased levels of 8-OH-G, FapyGua, 8-OH-Ade, and Cyt glycol in renal DNA	Misra et al., 1993
Ni(II) complexes with carnosine, homocarnosine, and anserine.[a]	Incubation of dG, DNA, or nucleohistone with the complexes plus H_2O_2	Enhanced production of 8-oxo-dG in free dG, but not in DNA and nucleohistone	Datta et al., 1993
$NiCl_2$	Exposure of nuclear chromatin from K562 cells to Ni(II) under air	Increase in several DNA base products typical for ·OH attack	Nackerdien et al., 1991
$NiCl_2$, $CoCl_2$	Exposure of nuclear chromatin from K562 cells to Ni(II) or Co(II) plus H_2O_2	Increase in 11 modified DNA bases to a greater extent than without H_2O_2; Co(II) ≫ Ni(II)	Nackerdien et al., 1991
Co(II) acetate	I.p. injection to Fischer rats	Increased amounts of several DNA base products in the kidney, liver, and lung	Kasprzak et al., 1994
$Na_2Cr_2O_7$	Incubation of calf thymus DNA with the dichromate plus GSH and H_2O_2	Increased formation of 8-oxo-dG	Wetterhahn et al., 1989
$CuSO_4$, $FeCl_3$	Exposure of nuclear chromatin of murine SP-2/0 cells to the metals plus H_2O_2	Increase in 10 DNA base products typical for ·OH attack. Chelation with EDTA or NTA suppressed Cu(II) effect but enhanced Fe(III) effect	Dizdaroglu et al., 1991
Fe(II)EDTA	Incubation of calf thymus DNA with Fe(II)EDTA and stimulated neutrophils	Increase in DNA base products typical for ·OH attack	Jackson et al., 1989
Fe(III)EDTA	Incubation of calf thymus DNA and dG with Fe(III)EDTA plus polyphenols and H_2O_2	Increased formation of 8-oxo-dG	Kasai and Nishimura, 1984c
$FeCl_3$, Fe(III)EDTA	Incubation of calf thymus DNA with Fe(III) and O_2^--producing hypoxanthine/xanthine oxidase system	Increase in DNA base products typical for ·OH attack; enhancement of this effect by EDTA	Aruoma et al., 1989

Table 4 (continued)

Metal derivative	Experimental system	Result	Ref.
Fe(III)NTA	Incubation of DNA with the iron complex plus H_2O_2	DNA base damage; mostly G and T	Inoue and Kawanishi, 1987
	A single i.p. injection to Wistar rats	Increase in renal 8-oxo-dG up to 24 h post injection	Umemura et al., 1990a,-b, 1991
Fe-rich asbestos	Incubation of calf thymus DNA with asbestos of various types for up to 20 h	Increase in 8-oxo-dG in proportion to Fe contents	Kasai and Nishimura, 1984b

Note: Abbreviations: i.p., intraperitoneal; 8-oxo-dG, 8-hydroxy-2′-deoxyguanosine; dG, 2′-deoxyguanosine; EDTA, ethylenediaminetetraacetate; GSH, reduced glutathione (γ-Glu-Cys-Gly).

1986a; Wetterhahn et al., 1989). Morphologic aberrations of chromosomes resulting from this type of damage were observed, for example, in lymphocytes of workers exposed to nickel and chromium compounds (Deng et al., 1988; Sunderman, 1989). In cultured Chinese hamster ovary cells, chromosomal damage by Ni(II) was predominantly localized in the protein-rich, heterochromatic region of the X chromosome (Sen and Costa, 1985).

Test tube experiments have revealed that DNA-protein cross-links may differ greatly in strength. If the cross-links are formed by polyvalent metal cations bridging DNA to the protein, the bonds are weak and dissociate easily when the ionic strength of the solution is increased (Patierno et al., 1987; Wedrychowski et al., 1986b) or upon addition of metal chelators (Borges and Wetterhahn, 1989; Wedrychowski, 1986a). However, there is a class of metal-induced cross-links in which DNA and protein(s) are bound together with strong covalent bonds involving no metal bridges (Lesko et al., 1982). Formation of these bonds is characteristic for free radical attack on chromatin and is likely to involve ·OH and/or other reactive oxygen intermediates (Dizdaroglu, 1991; Lesko et al., 1982; Misra et al., 1993; Oleinick et al., 1987). A good example of covalent cross-linking is that produced in isolated nuclear chromatin by H_2O_2 plus Fe(II)EDTA (Lesko et al., 1982). H_2O_2 alone, or Fe(II)EDTA alone, does not cause such cross-linking. Further support for active oxygen involvement in the formation of DNA-protein cross-links comes from *in vivo* experiment in which this effect mediated by Ni(II) was greatly enhanced by His, known to complex Ni(II) and promote its oxidative capability (Datta et al., 1992). Likewise, Gly_4 facilitated DNA-protein cross-linking by Ni(II) in isolated calf thymus nucleohistone and protein-protein cross-linking among histones (Kasprzak and Bare, 1989).

As mentioned earlier, carcinogenic Cr(VI) reacts with DNA only after, or in the course of, metabolic reduction of the metal to lower oxidation states. Oxygen- and sulfur-centered radicals derived from the reductants may attack DNA and become cross-linked to its molecule. The resulting adducts with GSH (Goodgame and Joy, 1986; Wetterhahn et al., 1989), Cys (Borges and Wetterhahn, 1989), and proteins of different molecular weights (Borges and Wetterhahn, 1989; Miller and Costa, 1989) were identified. Unfortunately, little is known about the influence of specific metals on the type and extent of the covalent cross-links and the exact chemistry of their formation (Costa et al., 1993; Dizdaroglu, 1992; Salnikow et al., 1992; Zhitkovich and Costa, 1992). Cross-links between thymine and tyrosine were identified in isolated chromatin treated with H_2O_2 in the presence of Fe(III) or Cu(II) (Dizdaroglu, 1992). DNA-protein cross-links produced in calf thymus nucleohistone by ·OH generated by ionizing radiation involved formation of covalent bonds mainly between the pyrimidine bases and glycine, alanine, valine, leucine, isoloeucine, threonine, lysine, and tyrosine (Dizdaroglu, 1991).

Intrastrand cross-linking between two adjacent bases in DNA is also possible. Formation of this type of cross-link between two cytosines or two thymines is characteristic for DNA damage produced by free radicals generated by γ and UV radiation (Dizdaroglu and Simic, 1984a,b). Such cross-links are believed to result in tandem double mutations that are produced by DNA following exposure to radiation or to metals (Dizdaroglu and Simic, 1984a,b; Tkeshelashvili et al., 1993).

C. CLEAVAGE

Kawanishi and co-workers (1989b) found that Ni(II) facilitates DNA cleavage by H_2O_2 in a site-specific way characteristic for the action of an active nickel-oxygen complex rather than ·OH or $^1O_2$.

The most sensitive sites appear to be at the C, T, and G residues, and rarely at the A residues. This pattern is surprising, especially in view of the "hole migration" phenomenon (Candeias and Steenken, 1993) and stronger interaction of Ni(II) with the purines (mainly guanine) than with the pyrimidines (Datta et al., 1991). The results of Kawanishi et al. (1989b) might indicate that binding of Ni(II) by the DNA molecule as a whole does not follow affinity rules established for separate nucleotides. Perhaps the difference results from the participation of other ligands such as proteins, which may form ternary (mixed ligand) complexes with Ni(II) and DNA. This would explain the site specificity of the DNA-protein cross-linking observed (Dizdaroglu, 1992; Salnikov et al., 1992). DNA cleavage mediation by Ni(II) complexes and ligand effects on the selectivity of DNA oxidation with various oxidants, including O_2 and H_2O_2, were studied in detail by Mack and Dervan (1992) and Burrows et al. (Cheng et al., 1993; Li and Rokita, 1991; Muller et al., 1992; Wey et al., 1993).

D. DEPURINATION

A more general and potentially mutagenic effect that results from the exposure of DNA to carcinogenic metals is depurination. For example, Cr(VI) and Cu(II) were found to release guanine (G) and adenine (A), respectively, and Ni(II) appeared to release A (the product was not definitely identified) from the DNA molecule (Shaaper et al., 1987). Schaaper et al. (1987) speculate that the mechanism of depurination involves oxygen free radicals. Depurination occurs concurrently with DNA strand scission and both effects can result from ·OH attack on the DNA sugar moiety (Dizdaroglu, 1991). Modified sugars that remain in the DNA backbone constitute alkali-labile sites that are frequently observed in DNA from metal-treated cells (Dizdaroglu, 1991; Hamilton-Koch, 1986; Sugiyama et al., 1991). Thus, the presence of such sites may also indicate possible depurination. A great enhancement by Fe(II) and Ni(II) of *in vitro* depurination was observed in dG treated with an oxidizing mixture of H_2O_2 + ascorbate (Kasprzak and Hernandez, 1989; Littlefield et al., 1991).

E. MUTAGENIC SPECTRUM OF THE DAMAGE

The significance of the results described above to carcinogenesis relies on growing evidence that at least some types of oxidative DNA damage are potentially mutagenic. The mutation spectra produced by agents associated with generation of oxygen radicals, including H_2O_2, Fe(II), Cu(I), and Cu(II), were established by several authors (Loeb et al., 1991; McBride et al., 1991; Moraes et al., 1990; Tkeshelashvili et al., 1991). However, the underlying types of the initial DNA lesions were not identified. The mutations induced by the metals were predominantly single base substitutions which clustered characteristically for each agent. The most frequent mutations observed for single-stranded M13mp2 phage DNA after aerobic incubation with Fe(II) were G → C transversions followed by C → T transitions, and G → T transversions (McBride et al., 1991). With Cu(I) and Cu(II), the mutations were predominantly C → T transitions followed by G → T transversions (Tkeshelashvili et al., 1991). The relative low frequency of the G → T transversions, considered to be characteristic for base mispairing by 8-oxo-dG (see later), is intriguing since chemical models of Fe(II)- or Fe(III)-mediated DNA oxidation (Dizdaroglu et al., 1991; Kasai and Nishimura, 1984a,b,c) and *in vivo* experiments with Fe(III)NTA (Umemura et al., 1990a,b) point at 8-oxo-dG as the most profuse lesion produced in DNA by active oxygen species. The Fe(III)NTA complex is strongly carcinogenic to the rat and mouse kidney (Li et al., 1987; Okada et al., 1983). Umemura et al. (1990a,b) found a significant increase of 8-oxo-dG production in renal DNA of male Wistar rats given a single i.p. injection of Fe(III)NTA, but not Na(I)NTA or Fe(III) chloride.

Reid and Loeb (1993) established that tandem double CC → TT mutations, known to occur via UV damage to DNA, can also be produced by treatments generating active oxygen species, e.g., by Fe(II), Cu(I), and Cu(II) plus O_2 and/or H_2O_2. In a follow-up study, Tkeshelashvili et al. (1993) found such mutations in bacteria transfected with Ni(II)Gly-Gly-His + H_2O_2-treated DNA. The authors speculated that the mutations occurred because of base mispairing at sites of cytosine dimers produced by ·OH (Dizdaroglu and Simic, 1984a); however, those dimers were not assayed in this particular experiment.

Thus far, the DNA damage/mutation relationship has been firmly established only for mutations produced by the most frequently assayed base lesion, 8-oxo-dG, present in DNA templates (Cheng et al., 1992; Kamiya et al., 1992; Kuchino et al., 1987; Moriya et al., 1991; Shibutani et al., 1991; Wood et al., 1990, 1992) or in the free nucleotide pool (Cheng et al., 1992). Despite some controversies concerning mutations at sites adjacent to 8-oxoguanine (8-oxo-G) in a DNA template (compare Kamiya et al., 1992 and Kuchino et al., 1987 vs. Cheng et al., 1992; Moriya et al., 1991; Shibutani et al., 1991; and Wood et al., 1990, 1992), it has been accepted that in addition to coding for C, 8-oxo-G may also code for A. The latter results in a G → T transversion mutation (Cheng et al., 1992; Moriya et al., 1991;

Shibutani et al., 1991; Wood et al., 1990, 1992). The same damaged base used as a substrate for DNA synthesis in the form of the corresponding deoxynucleoside triphosphate, 8-oxo-dGTP, is again mispaired with A, causing A → C substitutions (Cheng et al., 1992). The G → T transversion mutation was found by our laboratory to occur exclusively in the K-*ras* oncogene isolated from nickel subsulfide (Ni_3S_2)-induced kidney tumors in rats (Higinbotham et al., 1992). The latency of tumors was shortened and the incidence of codon 12 mutations in K-*ras* was greatly increased by the addition of metallic iron powder to Ni_3S_2. Since the iron powder did not produce any tumors by itself and did not affect the final tumor incidence by Ni_3S_2, the enhancement of Ni_3S_2 carcinogenicity by iron was most likely due to assistance in tumor progression. The latter is consistent with *ras* involvement in tumor latency (Higinbotham et al., 1992). The K-*ras* mutation is not mandatory for nickel carcinogenesis since most of the tumors in the Ni_3S_2-only group did not contain transforming mutations in K-*ras*.

Besides 8-oxo-dG, one more product of DNA base oxidation, 5-OHMe-dU, is also potentially mutagenic (Bilimoria and Gupta, 1986; Shirname-More et al., 1987). Thymidine glycol is a suspected promutagen (Akman et al., 1991; Basu et al., 1989; Hayes et al., 1988). According to Wood et al. (1992), in *E. coli* with normal DNA repair capabilities, 8-oxo-2'-deoxyadenosine (8-oxo-dA) is at least an order of magnitude less mutagenic than 8-oxo-dG. The latter is likely to cause mutations by escaping from a proofreading function, whereas misincorporation at the 8-oxo-dA lesion may be corrected (Shibutani et al., 1991). Mutagenicity of other ·OH-modified nucleobases is still debated or unexplored (Akman et al., 1991, 1993; Ames, 1989; Breimer, 1990; Wood et al., 1992).

Possible promutagenic effects of metal-induced DNA strand scission and depurination are not fully recognized. Strand scissions have been found to be mutagenic by causing deletions (Runger and Kraemer, 1989) and/or insertions (Chiocca et al., 1991). For example, treatment of cultured 6m2 cells (rat renal cells with a defective MuSVts110 v-*mos* gene) with Ni(II) chloride resulted in mutation consisting of duplication of a 70-base-long DNA strand, an indication of strand breakage followed by faulty repair, or erratic DNA replication (Chiocca et al., 1991). Depurination of the G site may result in G → T transversions (McBride et al., 1991).

IV. OTHER RELEVANT EFFECTS
A. LIPID PEROXIDATION ENHANCEMENT

Transition metals promote lipid peroxidation in living cells (Andersen and Andersen, 1989; Athar et al., 1987a; Aust and Svingen, 1982; Buettner, 1993; Dix and Aikens, 1993; Knight and Voorhees, 1990; Sole et al., 1990; Sunderman, 1986b) and thus provide a source for genotoxic active oxygen species (Shi et al., 1993a,b,c). Enhancement of lipid peroxidation *in vitro* and *in vivo* was observed for Cd(II), Co(II), Cu(II), Hg(II), Ni(II), Pb(II), Sn(II), V(V) (Sunderman, 1986b) and Fe(III) (Hartwig et al., 1993; Okada, 1987; Umemura et al., 1990a) compounds. Lipid peroxidation products were found to mediate the formation of 8-oxo-dG (Park and Floyd, 1992) and strand breaks (Hartwig et al., 1993) in DNA. Another indication of a possible causal relationship between peroxidation and DNA damage is that metal-induced lipid peroxides and the concurrent chromosomal aberrations *in vivo* and *in vitro* could be inhibited by antioxidants, e.g., GSH, selenium, or vitamin E (Sugiyama et al., 1991; Sunderman, 1986b; Umemura et al., 1991). Lipid peroxidation by Ni(II) in mice was found to be high in strains (e.g., BALB/c) that are low in GSH and GSH peroxidase compared to other strains (e.g., B6C3F1 or C3H) (Kasprzak 1991a; Misra et al., 1991; Rodriguez et al., 1991).

B. STIMULATION OF INFLAMMATION

Metal carcinogens, especially those acting locally, trigger an inflammatory response which results in infiltration of the affected tissue by phagocytes with all their excretory oxidative weapons. Activated neutrophils are known to damage nucleobases in isolated DNA (Frenkel, 1989; Jackson et al., 1989; Reid and Loeb, 1992). Although it seems unlikely that phagocyte-derived active oxygen products can diffuse far enough to attack DNA in neighboring cells, there is experimental evidence that at least one such product does, in fact, achieve that goal. Frenkel and Chrzan (1987) incubated HeLa cells with stimulated polymorphonuclear leukocytes and found increased amounts of 5-OHMe-dU (a dT oxidation product) present in DNA of the cells. Formation of this derivative could be prevented with catalase, indicating that H_2O_2 was the diffusible agent responsible for dT oxidation. The authors speculate that oxidation of thymine and other DNA bases by H_2O_2 was catalyzed by DNA-bound iron. Phagocytes are also known to cause bacterial cell mutations. Thus, His-requiring mutants of *Salmonella typhimurium* TA100 were reverted into His-independence following incubation with human blood leukocytes. Heat-killed leukocytes or leukocytes from a patient with chronic granulomatous disease ($O_2^{-\cdot}$ production defect) were inactive (Weitzman and Stossel, 1981). It is worth noting that some carcinogenic metal sulfide particles were found to stimulate polymorphonuclear leukocytes to produce H_2O_2, whereas noncarcinogenic sulfides suppressed H_2O_2 production (Zhong et al., 1990).

C. EFFECT ON CELLULAR OXYGEN-HANDLING SYSTEMS

In addition to direct oxygen activation, carcinogenic metals may also sustain oxidative damage indirectly through inhibition of antioxidant cellular defenses or enhancement of physiological oxygen activation systems (Table 5). For example, in the liver and kidney of rats injected with Ni(II), a transient decrease in catalase and GSH peroxidase activity concurrent with elevated lipid peroxidation was observed during the first day post injection (Misra et al., 1990). Catalase and GSH peroxidase are also known to be inhibited by Ni(II) *in vitro* (Rodriguez and Kasprzak, 1992; Rodriguez et al., 1990). Superoxide dismutase appears to be less sensitive to Ni(II) inhibition (Misra et al., 1990). However, great variability of the effects of Ni(II) on catalase, GSH peroxidase, GSH reductase, GSH-S-transferase, or superoxide dismutase in different species, strains, and tissues does not allow more general conclusions to be drawn as to the possible significance of antioxidant enzyme inhibition to oxidative damage by this metal.

Table 5 Effects of Carcinogenic Metal Derivatives on Cellular Oxygen Activation and Deactivation Systems

Metal derivative	Effect	Ref.
Ni(II) acetate, NiCl$_2$	Inhibition of isolated CAT and GSH-Px, but enhancement of MPO in a Ni(II) concentration-related manner	Rodriguez and Kasprzak, 1989, 1992; Rodriguez et al., 1990
	Transient inhibition of CAT activity in red blood cells, liver, and kidney of Ni(II)-treated rats	Misra et al., 1990; Rodriguez et al., 1990
	Transient inhibition of CAT, GSH-Px, and GSSG-R in the kidney and liver, but not in muscle of Ni(II)-treated rats; no effect on SOD in kidney and liver, but decrease in skeletal muscle	Misra et al., 1990
	Inhibition of CAT, SOD, and GSH in livers of 4 mouse strains, but variable effects on GSH-Px, GSSG-R, and GST in the same mice. No consistent patterns of Ni(II) effect on the same enzymes in kidneys	Misra et al., 1991; Rodriguez et al., 1991
	Inhibition of hepatic GSH-Px activity, but increase in GSH, GSSG-R, and GST in Ni(II)-treated rats	Athar et al., 1987a,b
	Depletion of hepatic GSH in 8–12 wk old CBA mice but not in younger mice	Andersen and Andersen, 1989; Shukla et al., 1987
Cd(II) acetate	Complete inhibition of SOD activity in bovine blood and in rat brain *in vitro* and strong inhibition in various parts of the rat brain *in vivo*	
CdCl$_2$	Possible stimulation of cell membrane-bound pyridine nucleotide oxidase in human granulocytes and rat alveolar macrophages	Amoruso et al., 1982
Pb(II) acetate	Depletion of hepatic GSH-Px, SOD, and nonprotein SH groups and plasma vitamin E in mice treated with Pb(II) and bacterial endotoxin	Hermes-Lima et al., 1991
NaAsO$_2$	Arsenite affects reactivity of reduced XO with O_2 by enhancing the O_2 and/or $O_2^{-\cdot}$ reactivity of the reduced Mo-center of the enzyme	Nordenson and Beckman, 1991; Steward et al., 1985

Note: Abbreviations: CAT, catalase; GSH, reduced glutathione; GSH-Px, glutathione peroxidase; GSSG-R, glutathione reductase; GST, glutathione-S-transferase; MPO, myeloperoxidase; SOD, superoxide dismutase; SH, sulfhydryl; XO, xanthine oxidase.

Unlike Ni(II), Cd(II) appears to be a powerful and apparently specific inhibitor of superoxide dismutase, thus promoting metabolic $O_2^{-\cdot}$ build-up in affected tissues (Amoruso et al., 1982; Ochi, 1983; Shukla et al., 1987). At the same time, there are indications that Cd(II) may enhance production of $O_2^{-\cdot}$ through stimulation of pyridine nucleotide oxidase (Amoruso et al., 1982; Ochi, 1983). These may be significant in carcinogenesis owing to the DNA-nicking activity of $O_2^{-\cdot}$ (Peskin and Shlyahova, 1986). Likewise, arsenite has the potential to aggravate oxidative damage by a mechanism involving xanthine oxidase because it can bind to the active molybdenum center of this enzyme (Steward et al., 1985). This binding enhances reactivity of the enzyme with O_2 and makes it also reactive with $O_2^{-\cdot}$, one of its products. In effect, arsenite partially diverts the activity of xanthine oxidase towards production of relatively more H_2O_2 than does As(III)-free enzyme. Cantoni et al. (1984) found that the addition of superoxide dismutase or catalase to cells treated with Hg(II) chloride markedly reduced the extent of DNA single-strand breaks without decreasing Hg(II) uptake. Pb(II) provides an interesting example whereby the ultimate oxidative cell damage is only remotely controlled by the primary insult, the Pb(II) poisoning. Pb(II) causes accumulation of δ-aminolevulinic acid (a heme precursor) which undergoes autoxidation accompanied by formation of active species, such as H_2O_2 and $O_2^{-\cdot}$. These products may then react with cellular Fe(II)/Fe(III) to form ·OH (Hermes-Lima et al., 1991).

Enhancement of oxygen-activating enzymes, or the mimicking of such enzymes, can contribute to oxidative damage by two more transition metals, Ni(II) and Cu(II). Ni(II) was found to augment the activity of myeloperoxidase (Rodriguez and Kasprzak, 1989), whereas Cu(II) was found to mimic that enzyme (Frenkel et al., 1986a). The metals increased or elicited production of the highly reactive hypohalite anion from sodium chloride and H_2O_2. However, since both effects were observed *in vitro*, their relevance to genotoxicity of Ni(II) and Cu(II) in animals remains to be established.

Carcinogenic metal cations also have the ability to bind to cellular antioxidants such as ascorbate, Cys, His, GSH, and others, and modify their reactions with oxygen species to produce free radicals (Datta et al., 1992; Kasprzak and Hernandez, 1989; Kasprzak et al., 1992b; Shi et al., 1993b). For example, Ni(II) was found to elicit generation of a variety of free radicals, including ·OH, from Cys oxidation with O_2 or cumene or *t*-butyl hydroperoxides (Shi et al., 1993b) and from GSH oxidation with the same hydroperoxides (Shi et al., 1992a). The effect of Ni(II) on Cys oxidation was not inhibited as might be expected, but augmented by His (Shi et al., 1993b). Cys and His are both major low molecular weight tissue carriers for Ni(II) and other transition metals (Jones et al., 1980; May et al., 1977). Ni(II) complex(es) with His enhanced *in vitro* oxidation of the G moiety in free dG and in DNA (Datta et al., 1992) and *in vivo* oxidation of several DNA bases in the rat kidney (Misra et al., 1993). However, the presumed histidyl oligopeptide antioxidants, carnosine, homocarnosine, and anserine, enhanced Ni(II)-mediated *in vitro* oxidation with H_2O_2 only of free dG, but not DNA- or nucleohistone-bound dG (Datta et al., 1993). Similar enhancement by His of dG oxidation with H_2O_2 was observed for Fe(III), Cr(III), and Cu(II), but not Co(II) (Datta et al., 1994). In the case of DNA-bound dG, however, the enhancement was noticed only for Cu(II). The effect of other metals remained unchanged (Cr(III), Co(II)), or was inhibited (Fe(III)) by His (Datta et al., 1994).

D. INHIBITION OF DNA REPAIR

Oxidative DNA damage can be prevented and/or repaired in living cells by various mechanisms (Bessho et al., 1993; Beyersmann and Hartwig, 1992; Chung et al., 1991a,b; Demple and Amabile-Cuevas, 1991; Grollman and Moriya, 1993; Kasai et al., 1986; Klein et al., 1992; Vos and Wauthier, 1991; Yamamoto et al., 1992). The cellular protection against 8-oxo-dG mutagenicity includes several enzymes capable of removing 8-oxo-2'-deoxyguanosine triphosphate (8-oxo-dGTP) from the deoxynucleotide pool and 8-oxo-dG from the DNA molecule. Data on those enzymes have been recently reviewed by Grollman and Moriya (1993). It is noteworthy that the activity of at least two enzymes engaged in the repair depends on essential metals: that of the specific triphosphatase (MutT protein) which removes 8-oxo-dGTP from the nucleotide pool (Akiyama et al., 1989; Grollman and Moriya, 1993; Maki and Sekiguchi, 1992; Mo et al., 1992) depends on Mg(II) (Mo et al., 1992), while that of formamidopyrimidine DNA glycosylase (Fpg protein) which removes 8-oxo-dG and some other purine base products from DNA (Boiteux et al., 1992) depends on Zn(II) (Fpg is a zinc finger protein) (O'Connor et al., 1993). Therefore, we may suspect that both enzymes should be sensitive to inhibition by other divalent cations. However, despite results indicating that Zn(II) in the Fpg protein can be replaced by Cu(II), Cd(II), and Hg(II) (O'Connor, 1993), and the possibility that Mg(II) in the MutT protein can be displaced by other divalent cations, e.g., Ni(II) is known to compete with Mg(II) in several other systems (Hartwig et al., 1992; Kasprzak, 1990; Kasprzak et al., 1985c; Littlefield and Poirier, 1992; Luo et al., 1993), the effect of the nonessential metals on the function of both proteins has not yet been determined.

Transition metal cations have long been known to reduce fidelity of *in vitro* DNA replication (Sirover and Loeb, 1976). It is not surprising, therefore, that the repair of damaged DNA in some metal-exposed cells is found to be impaired. Thus, Ni(II) and Co(II) inhibited repair of DNA strand breaks in cultured mammalian cells after UV- or X-irradiation (Hartwig and Beyersmann, 1989; Hartwig et al., 1991, 1992), apparently by affecting the polymerization and/or ligation steps (Hartwig et al., 1992). Lee-Chen et al. (1993) observed Ni(II) inhibition only at the ligation step. The difference might depend on the cell lines used in both studies. In the case of Ni(II), the DNA strand-breaks repair inhibition could be prevented by Mg(II). This indicated that Ni(II) affected the repair enzymes (DNA polymerases and ligases) and/or the DNA substrate by substituting for Mg(II) (Hartwig et al., 1992). Co(II) was found to suppress repair of the UV-generated pyrimidine dimer lesion by blocking its excision step (Hartwig et al., 1991). Other metal compounds, Pb(II), Cr(VI), Fe(III), and Sn(II) had no detectable effects on dimer removal (Snyder et al., 1989). There are indications that UV-induced DNA repair may be obstructed also by Cr(VI), Cd(II) (Hartwig and Beyersmann, 1989), and As(III) (Lee-Chen et al., 1991).

V. CONCLUSION

Current hypotheses on the mechanisms of metal carcinogenesis are based on the assumption of metal cations binding by cell nuclei (Sunderman, 1984, 1986a). Such binding should result in damaging effects observed *in vitro* and/or *in vivo*, including conformational changes of DNA and nuclear proteins, strand breakage and depurination of the DNA molecule, cross-linking of chromatin components, and DNA base modification. Those effects may, in turn, lead to mutations due to erratic repair and/or replication of the damaged DNA template; metal-affected enzymes controlling these processes may also contribute to increased error frequency. The same effects may also cause abnormal gene expression owing to exposure of segments of the genome that are normally repressed, or to erratic transcription and/or translation processes. The latter may be caused by metals bound to nuclear RNA and gene regulatory proteins, e.g., the "zinc fingers" (Berg, 1986; Sunderman and Barber, 1988). More details on the above are presented in reviews by Sunderman (1984; 1986a; 1989) and Costa (1991). However, can metal binding alone (i.e., DNA-metal "adduct" formation) really result in all that damage?

There is no doubt, at the present time, that following *in vitro* or *in vivo* exposures, transition metals do reach the cell nucleus and are retained there (Berg, 1986; Bryan, 1981; Costa, 1991; Leonard, 1986; Sunderman and Barber, 1988). However, since the nucleus has to compete for the metal with a plethora of other cellular components and the binding is further suppressed by competition with native cations, especially Mg(II) (Kasprzak et al., 1985c), the amounts of transition metals found in the nuclei following nonlethal exposures seem to be too low to produce the coarse effects observed in test tubes, such as the B-DNA to Z-DNA transformation, depurination, and/or strand scission [reviewed by Sunderman (1989)]. DNA strand scission in cultured cells was observed only for lethally high exposures to nickel carcinogens, but not for nontoxic exposures (Swierenga and McLean, 1985). Direct evidence for the occurrence of discrete DNA conformation changes in living cells is not available. Nonetheless, it seems possible that such effects occur. Concerning metal-caused depurination, we may suspect that it really happens *in vivo* from the nature of the resulting mutations: the G $\rightarrow$ T transversion mutation observed in renal tumors (Higinbotham et al., 1992), originally linked to 8-oxo-dG (see above), could equally well result from depurination of the G site.

Although development of conformational changes in DNA and nuclear proteins and depurination of DNA in metal-exposed cells can result directly from "metal adduct" formation (and subsequent loosening of glycosidic bonds) and the observed variety and differing strengths of the cross-links can partially be due to metal-bridged protein-DNA adducts formation, the DNA base damage cannot be explained by the "metal adduct" concept itself because the damaged bases found *in vitro* and *in vivo* are not adducts of metals but oxidation products. However, a common explanation for all those effects may be provided assuming that the gene-damaging action of chromatin-bound metal is of predominantly indirect (catalytic) nature. This assumption is consistent with the chemistry of metals in question and the results of numerous experiments in both *in vivo* and in cell-free systems.

Based on the data reviewed above, we can offer the following answers to the questions asked in the Introduction. The answer to question (1) is positive: most of the carcinogenic metals belong to the category of transition elements and thus have rich coordination and redox chemistry. This factor, among others, allows those metals to react with and activate oxygen species under physiological conditions. The most important examples include O_2 conversion to $O_2^{-\cdot}$ and/or H_2O_2 during autoxidation, or H_2O_2 activation to $\cdot OH$ through Fenton chemistry. Formation of other strong metal-associated oxidants such as metal-oxo and metal-peroxo complexes is also observed.

A positive answer to question (2) comes from evidence based on identification of increased amounts of oxygen activation products, such as H_2O_2 and lipid peroxides, in metal-exposed cells. Along with O_2

and metabolic H_2O_2, these products may be utilized for generation of the ultimate oxidants mentioned above.

Data supporting a positive answer to question (3) include inhibition by metals of cellular antioxidant and DNA repair systems and the site specificity of metal-directed oxidative attacks on DNA. Consequently, in both *in vitro* and *in vivo* experiments, exposure to carcinogenic transition metals results in increased production of damaged DNA bases and DNA-protein cross-linking, typical for ·OH and metal-associated oxidants' attack on chromatin.

The answer to question (4) is also positive, however, with some reservation. The significance and specificity of cross-links observed in chromatin with respect to certain types of mutation and carcinogenesis remain to be defined. It is believed that persistent cross-links may impair functions of the nuclear matrix, especially during replication and transcription, and thus introduce genetic and epigenetic alterations into the affected cells (Oleinick et al., 1987). The bond-breaking effects, DNA depurination and strand scission, are promutagenic events (Chiocca et al., 1991; McBride et al., 1991; Runger and Kraemer, 1989), but they cannot be ascribed solely to the catalytic facet of metal action on DNA. Their contribution to metal-related carcinogenesis needs further elucidation.

The best evidence supporting the hypothesis of the oxidative nature of metal-induced genotoxic damage is provided by the wide spectrum of nucleobase products typical for the active oxygen attack on DNA, found in cultured cells and animals exposed to metal carcinogens (Table 4). Some of those base products have been identified as promutagens (Akman et al., 1991; Basu et al., 1989; Cheng et al., 1992; Kamiya et al., 1992; Moriya et al., 1991; Shibutani et al., 1991; Shirname-More et al., 1987; Wood et al., 1990, 1992). However, there are certain gaps in the literature that should make us cautious in jumping to conclusions. First, base modification studies by metals suffer from a lack of experiments in cultured cells. Indeed, 5-OHMe-dU, a promutagen, was found in HeLa cells exposed to H_2O_2 (originating from phagocytes), but not in cells exposed to a metal; involvement of iron in production of this derivative was only postulated by inference (Frenkel and Chrzan, 1987; Frenkel et al., 1986b). Our own attempts to reveal the impact of nontoxic Ni(II) concentrations on 8-oxo-dG production in NIH 3T3, NRK-52, and V79 cells indicate a rather weak and inconsistent effect (Kasprzak et al., 1991; unpublished data). Secondly, the *in vivo* evidence available thus far for oxidative damage to DNA bases by carcinogenic metals (Kasprzak et al., 1990, 1991, 1992a, 1994; Misra et al., 1993; Umemura et al., 1990a,b) is not accompanied by evidence of gene mutation(s) gathered in the same bioassay. Thus, Ni(II) acetate, known to initiate renal cortical epithelial tumors by systemic injection, was found to increase levels of promutagenic 8-oxo-dG in kidneys over the first days after treatment (Kasprzak et al., 1990). Unfortunately, the tumors from this experiment were not tested for activated oncogenes. On the other hand, rat renal mesenchymal tumors induced with another nickel carcinogen, Ni_3S_2 (alone or plus metallic iron powder), were tested for oncogene activation (Higinbotham et al., 1992), but the same kidneys had not been analyzed earlier for 8-oxo-dG. The Ni_3S_2-induced tumors were found to contain K-*ras* oncogene activated at codon 12 exclusively by the G → T transversion mutation (Higinbotham et al., 1992), typical for 8-oxo-dG (Cheng et al., 1992; Wood et al., 1992). Although the data suggest a causative association between that mutation and 8-oxo-dG, we still lack a direct confirmation that this is really true in a single experiment. It is noteworthy that the magnitude of Ni(II) effects on 8-oxo-dG production in renal DNA may be species- and strain-dependent (Kasprzak et al., 1990, 1991a, 1992a). However, no correlation has yet been made between this phenomenon and the corresponding susceptibility of different animals and strains to renal carcinogenesis by Ni(II). The most recent strong support for the discussed hypothesis comes from *in vitro* experiments of Tkeshelashvili et al. (1993). They revealed that mutagenic effects of metal-mediated DNA damage were identical with those produced by radiolytically generated oxygen radicals. However, specific DNA lesions responsible for particular mutagenic effects were not identified. Also, the damage was inflicted on pure DNA in a test tube and thus did not reflect the full complexity of multimolecular interactions and the mutagenicity of all products arising from such interactions in whole-cell exposure models.

In addition to the well-established carcinogenic metals, chromium and nickel, a less notorious carcinogen, iron, has thus far provided the most convincing experimental data relative to the role of active oxygen in metal carcinogenicity and acute toxicity. Moreover, owing to its relative abundance and high capacity to activate oxygen, iron displaced from its natural stores by a toxic insult is often thought to be the ultimate carcinogen (Halliwell and Gutteridge, 1986; Kon, 1978; Loeb et al., 1988; Puppo and Halliwell, 1988). Copper, considered to be noncarcinogenic, may be yet another candidate for this position (Akman et al., 1993; Dizdaroglu et al., 1991; Suzuki et al., 1993; Yamamoto et al., 1993b). This situation is paradoxical: if oxygen activation plays a crucial role in metal carcinogenesis, why are the highly redox-active metals, iron and copper (Dizdaroglu et al., 1991), not the strongest carcinogens when administered to animals? There are two possible answers to this riddle. The first answer must obviously consider bioavailability. It is believed that tight physiological control (chelation and compartmentalization) over iron and copper, two physiological metals, makes them inaccessible for adventitious reactions with oxygen species, e.g., H_2O_2. Indeed, in contrast to the case of nickel and chromium, many natural tissue ligands prevent participation of both metals in radical reactions (Gutteridge, 1990). Binding of iron by transferrin, lactoferrin, hemoglobin, myoglobin, and storage of this metal in the Fe(III) form, which is less dangerous than Fe(II), and binding of copper by ceruloplasmin and albumin greatly reduce the danger of uncontrolled redox reactions by these two metals (Gutteridge, 1990; Hanna and Mason, 1992). This condition may be reversed only by introduction of excessive amounts of metal cation plus an exogenous chelator which can overwhelm the control functions of endogenous ligands, deliver the metal into target cells, and sustain its redox activity. In the case of Fe(III), all these functions are apparently provided by NTA (Inoue and Kawanishi, 1987; Umemura et al., 1991) and, to some extent, also by dextran (Sunderman, 1984). Also, NTA and dextran may attenuate the inflammatory response to Fe(III) and prevent the "overkill" effect, discussed below. There are no known iron complexes other than Fe(III)NTA capable of providing all these functions at the same time and thus rendering this metal carcinogenic after systemic administration (Hartwig et al., 1993). Fe(III) dextran is carcinogenic only locally (Sunderman, 1984). Thus far, we do not know of any copper complex of this type.

Another answer to our puzzle may depend on the "overkill" effect. If it happens that iron or copper overwhelms chelation control (e.g., following parenteral injection), its powerful redox catalysis in conjunction with the inflammatory response (active oxygen substrates) will result not in injury, but in death of the target cells. In concordance with the above presumption, Tkeshelashvili et al. (1993) noticed that the ratio of increased mutagenesis to loss of survival of cells transfected with DNA exposed to metal-generated reactive oxygen species was greater for nickel than for iron or copper. Severe inflammatory/necrotizing effects of Fe^0 or $FeCl_3$ were observed in contrast to the local cell immunosuppressive effects of Ni_3S_2 in rat skeletal muscle (Kasprzak et al., 1985b, 1987). The addition of mycobacterial antigen to Ni_3S_2, which greatly enhances phagocytic infiltration into the target tissue, prevents Ni_3S_2 carcinogenesis (Kasprzak and Ward, 1991). Likewise, prevention of the immunosuppressive action of Ni(II) by Mg(II) inhibits Ni_3S_2 carcinogenesis (Kasprzak et al., 1987). These findings obviously undermine a popular conviction of the possible villainous role of phagocytes in carcinogenesis. They are, however, consistent with the "overkill" concept: the phagocytes have been designed to destroy cells, not just to tamper with their genetic code.

The overall picture which emerges from our discussion portrays carcinogenic metal derivatives as multipotent reagents. They are capable of interacting with almost any cell constituent, including physiological metals (Kasprzak, 1990), and causing a plethora of damaging effects in the mammalian genome. The effects may be direct, due to metal binding-related conformational and functional distortions of biomolecules, or indirect, due to a variety of structural modifications of biomolecules caused by reactive oxygen species arising in metal-catalyzed reactions. For uptake- and toxicity-related reasons, the catalytic effects of metals seem to be more important for carcinogenesis than the direct effects. Published data supporting the concept of the crucial role of oxidative damage in metal carcinogenesis are particularly strong for two of the most powerful human metal carcinogens, nickel and chromium. However, without excluding the contribution of other effects, oxidative damage seems to be slowly taking the leading role in explaining mechanisms of cancer causation and acute toxicity by other metals as well.

ACKNOWLEDGMENTS

The author is grateful to Drs. K.A. Canella, L. Keefer, and J.M. Rice for valuable critical comments on this chapter and to Ms. K. Breeze for editorial help.

ABBREVIATIONS AND COMMON NAMES

A, adenine
8-oxo-A, 7,8-dihydro-8-oxoadenine (syn.: 8-OH-A, 8-hydroxyadenine)
dA, 2′-deoxyadenosine
8-oxo-dA, 7,8-dihydro-8-oxo-2′-deoxyadenosine (syn.: 8-OH-dA, 8-hydroxy-2′-deoxyadenosine)
C, cytosine

Cyt glycol, cytosine glycol
G, guanine
8-oxo-G, 7,8-dihydro-8-oxoguanine (syn.: 8-OH-G, 8-hydroxyguanine)
dG, 2′-deoxyguanosine
8-oxo-dG, 7,8-dihydro-8-oxo-2′-deoxyguanosine (syn.: 8-OH-dG, 8-hydroxy-2′-deoxyguanosine)
FapyGua, 2,6,-diamino-4-hydroxy-5-formamidopyrimidine
T, thymine
dT, thymidine
U, uracil
dU, 2′-deoxyuridine
5-OHMe-dU, 5-hydroxymethyl-2′-deoxyuridine
EDTA, ethylenediaminetetraacetic acid
NTA, nitrilotriacetic acid
DFO, deferoxamine (syn.: desferoxamine, desferrioxamine)
DTPA, diethylenetriaminepentaacetic acid (syn.: DETAPAC)
Asp, L-aspartic acid
Ala, L-alanine
Phe, L-phenylalanine
Cys, L-cysteine
Gly, glycine
Gly_2–Gly_5, diglycine to pentaglycine
His, L-histidine
Lys, L-lysine
O_2^-·, superoxide anion radical
·OH, hydroxyl radical
$^1O_2$, singlet oxygen
GSH, glutathione, γ-glutamylcysteinylglycine
carnosine, β-alanyl-L-histidine
homocarnosine, γ-aminobutyryl-L-histidine
anserine, β-alanyl-3-methyl-L-histidine.

REFERENCES

Akiyama, M., Maki, H., Sekiguchi, M., and Horiuchi, T., *Proc. Natl. Acad. Sci. U.S.A.,* 86, 3949–3952, 1989.
Akman, S. A., Doroshow, J. H., and Kensler, T. W., *Carcinogenesis,* 13, 1783–1787, 1992.
Akman, S. A., Forrest, G. P., Doroshow, J. H., and Dizdaroglu, M., *Mutat. Res.,* 261, 123–130, 1991.
Akman, S. A., Kensler, T. W., Doroshow, J. H., and Dizdaroglu, M., *Carcinogenesis,* 14, 1971–1974, 1993.
Ames, B. N., *Environ. Mol. Mutagen.,* 14 (Suppl. 16), 66–77, 1989.
Amoruso, M. A., Witz, G., and Goldstein, B. D., *Toxicol. Lett.,* 10, 133–138, 1982.
Andersen, H. R. and Andersen, O., *Biol. Trace Elem. Res.,* 21, 255–261, 1989.
Anderson, L. K., Kasprzak, K. S., and Rice, J. M., in *Male-Mediated Developmental Toxicity,* Olshan, A. and Madison, D., Eds., Plenum Press, New York, 129, 1994.
Angelov, D., Berger, M., Cadet, J., Getoff, N., Keskinova, E., and Solar, S., *Radiat. Phys. Chem.,* 37, 717–727, 1991.
Aruoma, O. I., Halliwell, B., and Dizdaroglu, M., *J. Biol. Chem.,* 264, 13024–13028, 1989.
Athar, M., Hasan, S. K., and Srivastava, R. C., *Biochem. Biophys. Res. Commun.,* 147, 1276–1281, 1987a.
Athar, M., Hasan, S. K., and Srivastava, R. C., *Res. Commun. Chem. Pathol. Pharmacol.,* 57, 421–424, 1987b.
Aust, S. D. and Svingen, B. A., in *Free Radicals in Biology,* Vol. 5, Academic Press, New York, 1, 1982.
Aust, S. D., Chignell, C. F., Bray, T. M., Kalyanaraman, B., and Mason, R. P., *Toxicol. Appl. Pharmacol.,* 120, 168–178, 1993.
Ayiar, J., Berkovits, H. J., Floyd, R. A., and Wetterhahn, K. E., *Environ. Health Perspect.,* 92, 53–62, 1991.
Bartoli, G. M., Galeotti, T., and Azzi, A., *Biochem. Biophys. Acta,* 497, 622–626, 1977.
Basu, A. K., Loechler, E. L., Leadon, S. A., and Essigman, J. M., *Proc. Natl. Acad. Sci. U.S.A.,* 84, 7677–7681, 1989.
Berg, J. M., *Science,* 232, 485–487, 1986.
Berger, M., Anselmino, C., Mouret, J. F., and Cadet, J., *J. Liquid Chromatogr.,* 13, 929–940, 1990.
Berger, M., de Hazen, M., Nejjari, A., Fournier, J., Guignard, J., Pezerat, H., and Cadet, J., *Carcinogenesis,* 14, 41–46, 1993.
Bessho, T., Roy, R., Yamamoto, K., Kadai, H., Nishimura, S., Tano, K., and Mitra, S., *Proc. Natl. Acad. Sci. U.S.A.,* 90, 8901–8904, 1993.
Beyersmann, D. and Hartwig, A., *Toxicol. Appl. Pharmacol.,* 115, 137–145, 1992.
Bilimoria, M. H. and Gupta, S. V., *Mutat. Res.,* 169, 123–127, 1986.
Blakely, W. F., Fuciarelli, A. F., Wegher, B. J., and Dizdaroglu, M., *Radiat. Res.,* 121, 338–343, 1990.

Boiteux, S., Gajewski, E., Laval, J., and Dizdaroglu, M., *Biochemistry,* 31, 106–110, 1992.
Borges, K. M. and Wetterhahn, K. E., *Carcinogenesis,* 10, 2165–2168, 1989.
Borges, K. M., Boswell, J. S., Liebross, R. H., and Wetterhahn, K. E., *Carcinogenesis,* 12, 551–561, 1991.
Bossu, F. P., Paniago, E. B., Margerum, D. W., Kirksey, S. T., and Kurtz, J. L., *Inorg. Chem.,* 17, 1034–1042, 1978.
Breimer, L. H., *Mol. Carcinogen.,* 3, 188–197, 1990.
Bryan, S. E., in *Metal Ions in Genetic Information Transfer,* Eichhorn, G.L. and Marzilli, L.G., Eds., Elsevier, New York, 87, 1981.
Buettner, G. R., *Arch. Biochem. Biophys.,* 300, 535–543, 1993.
Cadet, J. and Weinfeld, M., *Analyt. Chem.,* 65, 675A–682A, 1993.
Cadet, J., Berger, M., Buchko, G. W., Incardona, M. F., Morin, B., Raoul, S., Ravanat, J. L., and Wagner, J. R., *Lipids,* in press.
Cadet, J., Berger, M., Decarroz, C., Mourret, J. F., van Lier, J. E., and Wagner, R. J., *J. Chem. Phys.,* 88, 1021–1042, 1991.
Candeias, L. P. and Steenken, S., *J. Am. Chem. Soc.,* 115, 2437–2440, 1993.
Cantoni, O., Christie, N. T., Swann, A., Drath, D. B., and Costa, M., *Mol. Pharmacol.,* 26, 360–368, 1984.
Chang, J., Watson, W., Randerath, E., and Randerath, K., *Mutat. Res.,* 291, 147–159, 1993.
Cheng, C. C., Rokita, S. E., and Burrows, C. J., *Angew. Chem. Int. Ed. Engl.,* 32, 277–278, 1993.
Cheng, K. C., Cahill, D. S., Kasai, H., Nishimura, S., and Loeb, L. A., *J. Biol. Chem.,* 267, 166–172, 1992.
Chiocca, S. M., Sterner, D. A., Biggart, N. W., and Murphy, E. C., Jr., *Mol. Carcinogen.,* 4, 61–71, 1991.
Chung, M. H., Kasai, H., Jones, D. S., Inoue, H., Ishikawa, H., Ohtsuka, E., and Nishimura, S., *Mutat. Res.,* 254, 1–12, 1991a.
Chung, M. H., Kim, H. K., Ohtsuka, E., Kasai, H., Yamamoto, F., and Nishimura, S., *Biochem. Biophys. Res. Commun.,* 178, 1472–1478, 1991b.
Costa, M., *Annu. Rev. Pharmacol. Toxicol.,* 31, 321–327, 1991.
Costa, D., Guignard, J., and Pezerat, H., *Toxicol. Ind. Health,* 5, 1079–1097, 1989a.
Costa, D., Guignard, J., and Pezerat, H., *NATO ASI Ser.,* H30, 189–196, 1989b.
Costa, D., Guignard, J., Zalma, R., and Pezerat, H., *Toxicol. Ind. Health,* 5, 1061–1078, 1989c.
Costa, M., Zhitkovich, A., and Toniolo, P., *Cancer Res.,* 53, 460–463, 1993.
Cotelle, N., Tremolieres, E., Bernier, J. L., Catteau, J. P., and Henichart, J. P., *J. Inorg. Chem.,* 46, 7–15, 1992.
Curtius, H. C., Anders, P., Erlenmeyer, H., and Sigel, H., *Helv. Chem. Acta,* 51, 896–899, 1968.
Datta, A. K., Misra, M., North, S. L., and Kasprzak, K. S., *Carcinogenesis,* 13, 283–287, 1992.
Datta, A. K., North, S. L., and Kasprzak, K. S., *Toxicologist,* 14, in press, 1994.
Datta, A. K., Riggs, C. W., Fivash, M. J., and Kasprzak, K. S., *Chem. Biol. Interact.,* 79, 323–334, 1991.
Datta, A. K., Shi, X., and Kasprzak, K. S., *Carcinogenesis,* 14, 417–422, 1993.
Demple, B. and Amabile-Cuevas, C. F., *Cell,* 67, 837–839, 1991.
Deng, C., Lee, H. H., Xian, H., Yao, M., Huang, J., and Ou, B., *J. Trace Elem. Exp. Med.,* 1, 57–62, 1988.
Dillon, C. T., Lay, P. A., Bonin, A. M., Dixon, N. E., Collins, T. J., and Kostka, K. L., *Carcinogenesis,* 14, 1875–1880, 1993.
Diwan, B. A., Anderson, L. M., Rehm, S., and Rice, J. M., *Cancer Res.,* 53, 3874–3876, 1993.
Dix, T. A. and Aikens, J., *Chem. Res. Toxicol.,* 6, 2–18, 1993.
Dizdaroglu, M., *Free Rad. Biol. Med.,* 10, 225–242, 1991.
Dizdaroglu, M., *Mutat. Res.,* 275, 331–342, 1992.
Dizdaroglu, M. and Gajewski, E., *Methods Enzymol.,* 186, 530–544, 1990.
Dizdaroglu, M. and Simic, M. G., *Radiat. Res.,* 100, 41–46, 1984a.
Dizdaroglu, M. and Simic, M. G., *Int. J. Radiat. Biol.,* 46, 241–246, 1984b.
Dizdaroglu, M., Rao, G., Halliwell, B., and Gajewski, E., *Arch. Biochem. Biophys.,* 285, 317–324, 1991.
Faux, S. P., Gao, M., Chipman, J. K., and Levy, L. S., *Carcinogenesis,* 13, 1667–1669, 1992.
Frenkel, K., *Environ. Health Perspect.,* 81, 45–54, 1989.
Frenkel, K., *Pharm. Ther.,* 53, 127–166, 1992.
Frenkel, K. and Chrzan, K., in *Anticarcinogenesis and Radiation Protection,* Cerutti, P.A., Nygaard, O.F., and Simic, M.G., Eds., Plenum Press, New York, 97, 1987.
Frenkel, K. and Tofigh, S., *Biol. Trace Elem. Res.,* 21, 351–357, 1989.
Frenkel, K., Blum, F., and Troll, W., *J. Cell. Biochem.,* 30, 181–193, 1986a.
Frenkel, K., Chrzan, K., Troll, W., Teebor, G. W., and Steinberg, J. J., *Cancer Res.,* 46, 5533–5540, 1986b.
Frenkel, K., Karkoszka, J., Cohen, B., Baranski, B., Jakubowski, M., Cosma, G., Taioli, M., and Toniolo, P., *Environ. Health Perspect.,* 102(Suppl), 221–225, 1994.
Frenkel, K., Zhong, Z., Wei, H., Karkoszka, J., Patel, U., Rashid, K., Georgescu, M., and Solomon, J. J., *Analyt. Biochem.,* 196, 126–136, 1991.
Goodgame, D. M. L. and Joy, A. M., *J. Inorg. Biochem.,* 26, 219–224, 1986.
Grollman, A. P. and Moriya, M., *Trends Genet.,* 9, 246–249, 1993.
Gutteridge, J. M. C., *FEBS Lett.,* 185, 19–23, 1985.
Gutteridge, J. M. C., in *Metal Ions in Biology and Medicine,* Collery, P., Poirier, L.A., Manfait, M., and Etienne, J.C., Eds., John Libbey Eurotext, Paris, 75, 1990.
Halliwell, B. and Aruoma, O. I., *FEBS Lett.,* 281, 9–91, 1991.
Halliwell, B. and Dizdaroglu, M., *Free Rad. Res. Commun.,* 16, 75–87, 1992.
Halliwell, B. and Gutteridge, J. M. C., *Biochem. J.,* 219, 1–14, 1984.
Halliwell, B. and Gutteridge, J. M. C., *Arch. Biochem. Biophys.,* 246, 501–514, 1986.

Hamazaki, S., Okada, S., Li, J. L., Toyokuni, S., and Midorikawa, O., *Arch. Biochem. Biophys.,* 272, 10–17, 1989.
Hamilton, J. W. and Wetterhahn, K. E., *Mol. Carcinogen.,* 2, 274–286, 1989.
Hamilton-Koch, W., Snyder, R. D., and Lavelle, J. M., *Chem.-Biol. Interact.,* 59, 17–28, 1986.
Hanna, P. M. and Mason, R. P., *Arch. Biochem. Biophys.,* 295, 205–213, 1992.
Hanna, P. M., Kadiiska, M. B., and Mason, R. P., *Chem. Res. Toxicol.,* 5, 109–115, 1992.
Hartwig, A. and Beyersmann, D., *Biol. Trace Elem. Res.,* 21, 359–365, 1989.
Hartwig, A., Klyszcz-Nasko, H., Schleppegrell, R., and Beyersmann, D., *Carcinogenesis,* 14, 107–112, 1993.
Hartwig, A., Schleppegrell, R., and Beyersmann, D., in *Metal Compounds in Environment and Life,* Vol. 4, Merian, E. and Haerdi, W., Eds., Science Reviews Inc., Wilmington, DE, 475, 1992.
Hartwig, A., Snyder, R. D., Schleppegrell, R., and Beyersmann, D., *Mutat. Res.,* 248, 177–185, 1991.
Hayes, R. C., Petrullo, L. A., Huang, H., Wallace, S. S., and LeClerc, J. E., *J. Mol. Biol.,* 201, 239–246, 1988.
Hermes-Lima, M., Valle, V. G. R., Vercesi, A. E., and Bechara, E. J. H., *Biochem. Biophys. Acta,* 1056, 57–63, 1991.
Higinbotham, K. G., Rice, J. M., Diwan, B. A., Kasprzak, K. S., Reed, C. D., and Perantoni, A., *Cancer Res.,* 52, 4747–4751, 1992.
Hneihen, A. S., Standeven, A. M., and Wetterhahn, K. E., *Carcinogenesis,* 14, 1795–1803, 1993.
Huang, X., Frenkel, K., Klein, C. B., and Costa, M., *Toxicol. Appl. Pharmacol.,* 120, 29–36, 1993.
IARC, Some inorganic and organometallic compounds. In Monographs on the Evaluation of Carcinogenic Risk of Chemicals to Man, Vol. 2, International Agency for Research on Cancer, Lyon, France, 1973.
IARC, Chlorinated drinking water; chlorinated by-products; some other halogenated compounds; cobalt and cobalt compounds. In Monographs on the Evaluation of Carcinogenic Risk to Humans, Vol. 52, International Agency for Research on Cancer, Lyon, France, 363, 1987a.
IARC, Overall evaluations of carcinogenicity. An updating of IARC Monographs Vol. 1 to 42. In Monographs on the Evaluation of Carcinogenic Risk to Humans, Suppl. 7, International Agency for Research on Cancer, Lyon, France, 1987b.
IARC, Chromium, nickel and welding. In Monographs on the Evaluation of Carcinogenic Risk to Humans, International Agency for Research on Cancer, Lyon, France, 1990.
IARC, Cadmium, mercury, beryllium and the glass industry. In Monographs on the Evaluation of Carcinogenic Risk to Humans, Vol 58, International Agency for Research on Cancer, Lyon, France, 1994.
Imlay, J. A. and Linn, S., *Science,* 240, 1302–1309, 1988.
Inoue, S. and Kawanishi, S., *Cancer Res.,* 47, 6522–6527, 1987.
Inoue, S. and Kawanishi, S., *Biochem. Biophys. Res. Commun.,* 159, 445–451, 1989.
Ito, K. and Kawanishi, S., *Biochem. Biophys. Res. Commun.,* 176, 1306–1312, 1991.
Ito, K., Yamamoto, K., and Kawanishi, S., *Biochemistry,* 31, 11606–11613, 1992.
Jackson, J. H., Gajewski, E., Schraufstatter, I., Hyslop, P. A., Fuciarelli, A. F., Cochrane, C. G., and Dizdaroglu, M., *J. Clin. Invest.,* 84, 1644–1649, 1989.
Janssen, Y. M. W., Van Houten, B. V., Borm, P. J. A., and Mossman, B. T., *Lab. Invest.,* 69, 261–274, 1993.
Jones, D. C., May, P. M., and Williams, D. R., in *Nickel Toxicology,* Brown, S.S. and Sunderman, F.W., Jr., Eds., Academic Press, New York, 73, 1980.
Kadiiska, M. B., Hanna, P. M., Hernandez, L., and Mason, R. P., *Mol. Pharmacol.,* 42, 723–729, 1992.
Kadiiska, M. B., Maples, K. R., and Mason, R. P., *Arch. Biochem. Biophys.,* 275, 98–111, 1989.
Kamiya, H., Miura, K., Ishikawa, H., Inoue, H., Nishimura, S., and Ohtsuka, E., *Cancer Res.,* 52, 3483–3485, 1992.
Kargacin, B., Klein, C. B., and Costa, M., *Mutat. Res.,* 300, 63–72, 1993.
Kasai, H. and Nishimura, S., *Nucl. Acid Res.,* 12, 2137–2145, 1984a.
Kasai, H. and Nishimura, S., *Gann,* 75, 841–844, 1984b.
Kasai, H. and Nishimura, S., *Gann,* 75, 565–566, 1984c.
Kasai, H., Crain, P. F., Kuchino, Y., Nishimura, S., Ootsuyama, A., and Tanooka, H., *Carcinogenesis,* 7, 1849–1851, 1986.
Kasprzak, K. S., *Metabolic Problems of Carcinogenic Nickel Compounds,* Technical University Press, Poznan, Poland, 1978a.
Kasprzak, K. S., *Nickel Subsulfide — Ni_3S_2: Chemistry, Applications, Carcinogenicity,* Technical University Press, Poznan, Poland, 1978b.
Kasprzak, K. S., in *Advances in Modern Environmental Toxicology,* Vol. XI, Fishbein, L., Furst, A., and Mehlman, M.A., Eds., Princeton Scientific Publishing, Princeton, NJ, 145, 1987.
Kasprzak, K. S., in *Biological Effects of Heavy Metals,* Vol. II, Foulkes, E. C., Ed., CRC Press, Boca Raton, FL, 173, 1990.
Kasprzak, K. S., *Chem. Res. Toxicol.,* 4, 604–615, 1991.
Kasprzak, K. S. and Bare, R. M., *Carcinogenesis,* 10, 621–624, 1989.
Kasprzak, K. S. and Hernandez, L., *Cancer Res.,* 49, 5964–5968, 1989.
Kasprzak, K. S. and Sunderman, F. W., Jr., *Res. Commun. Chem. Pathol. Pharmacol.,* 16, 95–108, 1977.
Kasprzak, K. S. and Ward, J. M., *Toxicology,* 67, 97–105, 1991.
Kasprzak, K. S., Hoover, K. L., and Poirier, L. A., *Carcinogenesis,* 6, 279–282, 1985a.
Kasprzak, K. S., Quander, R. V., and Poirier, L. A., *Carcinogenesis,* 6, 1161–1166, 1985b.
Kasprzak, K. S., Waalkes, M. P., and Poirier, L. A., *Toxicol. Appl. Pharmacol.,* 82, 336–343, 1985c.
Kasprzak, K. S., Ward, J. M., Poirier, L. A., Reichardt, D. A., Denn, A. C., III, and Reynolds, C. W., *Carcinogenesis,* 8, 1005–1011, 1987.

Kasprzak, K. S., Diwan, B. A., Konishi, N, Misra, M., and Rice, J. M., *Carcinogenesis,* 11, 647–652, 1990.
Kasprzak, K. S., Misra, M., Rodriguez, R. E., and North, S. L., *Toxicologist,* 11, 233, 1991a.
Kasprzak, K. S., North, S. L., and Keefer, L. K., *Proc. Am. Assoc. Cancer Res.,* 32, 108, 1991b.
Kasprzak, K. S., Diwan, B. A., Rice, J. M., Misra, M., Riggs, C. W., Olinski, R., and Dizdaroglu, M., *Chem. Res. Toxicol.,* 5, 809–815, 1992a.
Kasprzak, K. S., North, S. L., and Hernandez, L., *Chem. Biol. Interact.,* 84, 11–19, 1992b.
Kasprzak, K. S., Zastawny, T. H., North, S. L., Riggs, C. W., Diwan, B. A., Rice, J. M., and Dizdaroglu, M., *Chem. Res. Toxicol.,* 7, 329–335, 1994.
Kawanishi, S. and Yamamoto, K., *Biochemistry,* 30, 3069–3075, 1991.
Kawanishi, S., Inoue, S., and Sano, S., *J. Biol. Chem.,* 261, 5952–5958, 1986.
Kawanishi, S., Inoue, S., and Yamamoto, K., *Biol. Trace Elem. Res.,* 21, 367–372, 1989a.
Kawanishi, S., Inoue, S., and Yamamoto, K., *Carcinogenesis,* 12, 2231–2235, 1989b.
Kazantzis, G., *Environ. Health Perspect.,* 40, 143–162, 1981.
Klein, C. B., Frenkel, K., and Costa, M., *Chem. Res. Toxicol.,* 4, 592–604, 1991.
Klein, J. C., Bleeker, M. J., Saris, C. P., Roelen, H. C. P. F., Brugghe, H. F., van den Elst, H., van der Marel, G. A., van Boom, J. H., Westra, J. G., Kriek, E., and Berns, A. J. M., *Nucl. Acids Res.,* 20, 4437–4443, 1992.
Knight, J. A. and Voorhees, R. P., *Ann. Clin. Lab. Sci.,* 20, 347–352, 1990.
Kon, S. H., *Med. Hypotheses,* 4, 445–463, 1978.
Kuchino, Y., Mori, F., Kasai, H., Inoue, H., Iwai, S., Miura, K., Ohtsuka, E., and Nishimura, S., *Nature,* 327, 77–79,1987.
Lancaster, J. R., Jr., Ed., *The Bioinorganic Chemistry of Nickel,* VCH Publishers, New York, 1988.
Lee, Y. W., Pons, C., Tummolo, D. M., Klein, C. B., Rossman, T. G., and Christie, N. T., *Environ. Mol. Mutagen.,* 21, 365–371, 1993.
Lee-Chen, S. F., Yu, C. T., and Jan, K. Y., *Mutagenesis,* 7, 51–55, 1991.
Lee-Chen, S. F., Wang, M. C., Yu, C. T., Wu, D. R., and Jan, K. Y., *Biol. Trace Elem. Res.,* 37, 39–50, 1993.
Leonard, A., *Curr. Top. Environ. Toxic Chem.,* 8, 443–452, 1985.
Leonard, A., in *Metal Ions in Biological Systems,* Vol. 20, Sigel, H., Ed., Marcel Dekker, New York, 229, 1986.
Leonard, A. and Lauwerys, R. R., *Mutat. Res.,* 75, 49–62, 1980.
Leonard, A. and Lauwerys, R., *Mutat. Res.,* 239, 17–27, 1990.
Lesko, S. A., Drocourt, J.-L., and Yang, S.-U., *Biochemistry,* 21, 5010–5015, 1982.
Li, J.-L., Okada, S., Hamazaki, S., Ebina, Y., and Midorikawa, S., *Cancer Res.,* 47, 1867–1869, 1987.
Li, T. and Rokita, S. E., *J. Am. Chem. Soc.,* 113, 7771–7773, 1991.
Lin, X., Sugiyama, M., and Costa, M., *Mutat. Res.,* 260, 159–164, 1991.
Littlefield, N. A., Fullerton, F. R., and Poirier, L. A., *Chem. Biol. Interact.,* 79, 217–228, 1991.
Littlefield, N. A. and Poirier, L. A., in *Metal Ions in Biology and Medicine,* Vol. 2, Anastassopoulou, J., Collery, P., Etienne, J.C., and Teophanides, T., Eds., John Libbey Eurotext, Paris, 157, 1992.
Loeb, L. A. and Mildvan, A. S., in *Metal Ions in Genetic Information Transfer,* Eichhorn, G.L. and Marzilli, L.G., Eds., Elsevier, New York, 125, 1981.
Loeb, L. A., James, E. A., Waltersdorph, A. M., and Klebanoff, S. J., *Proc. Natl. Acad. Sci. U.S.A.,* 85, 3918–3922, 1988.
Loeb, L. A., McBride, T. J., Reid, T. M., and Cheng, K. C., in *New Horizons in Molecular Toxicology,* Probst, G.S., Vodicnik, M.J., and Dorato, M.A., Eds., FASEB Publ., Bethesda, MD, 35, 1991.
Luo, S. Q., Plowman, M. C., Hopfer, S. M., and Sunderman, F. W., Jr., *Ann. Clin. Lab. Sci.,* 23, 121–129, 1993.
Mack, D. P. and Dervan, P. B., *Biochemistry,* 31, 9399–9405, 1992.
Magos, L., *Environ. Health Perspect.,* 95, 157–189, 1991.
Maki, H. and Sekiguchi, M., *Nature,* 355, 273–275, 1992.
Margerum, D. W. and Anliker, S. L., in *The Bioinorganic Chemistry of Nickel,* Lancaster, J.R., Jr., Ed., VCH Publishers, New York, 29, 1988.
May, P. M., Linder, P. W., and Williams, D. R., *J. Chem. Soc. Dalton Trans.,* 588, 1977.
McBride, T. J., Preston, B. D., and Loeb, L. A., *Biochemistry,* 30, 207–213, 1991.
Miller, C. A., III and Costa, M., *Carcinogenesis,* 10, 667–672, 1989.
Misra, M., Rodriguez, R. E., and Kasprzak, K. S., *Toxicology,* 64, 1–17, 1990.
Misra, M., Rodriguez, R. E., North, S. L., and Kasprzak, K. S., *Toxicol. Lett.,* 59, 121–133, 1991.
Misra, M., Olinski, R., Dizdaroglu, M., and Kasprzak, K. S., *Chem. Res. Toxicol.,* 6, 33–37, 1993.
Mo, J. Y., Maki, H., and Sekiguchi, M., *Proc. Natl. Acad. Sci. U.S.A.,* 89, 11021–11025, 1992.
Moorhouse, C. P., Halliwell, B., Grootveld, M., and Gutteridge, J. M. C., *Biochem. Biophys. Acta,* 843, 261–268, 1985.
Moraes, E. C., Keyse, S. M., and Tyrrell, R. M., *Carcinogenesis,* 11, 283–293, 1990.
Moriya, M., Ou, C., Bodepudi, C., Johnson, F., Takeshita, M., and Grollman, A. P., *Mutat. Res.,* 254, 281–288, 1991.
Mouret, J. F., Berger, M., Anselmino, C., Polverelli, M., and Cadet, J., *J. Chem. Phys.,* 88, 1053–1061, 1991.
Muller, J. G., Chen, X., Dadiz, A. C., Rokita, S. E., and Burrows, C. J., *J. Am. Chem. Soc.,* 114, 6407–6411, 1992.
Nackerdien, Z., Kasprzak, K. S., Rao, G., Halliwell, B., and Dizdaroglu, M., *Cancer Res.,* 51, 5837–5842, 1991.
Nassi-Calo, L., Mello-Filho, A. C., and Meneghini, R., *Carcinogenesis,* 10, 1055–1057, 1989.
Nieboer, E., Stetsko, P. I., and Hin, P. Y., *Ann. Clin. Lab. Sci.,* 14, 409, 1984.
Nieboer, E., Maxwell, R. I., Rosetto, F. E., Stafford, A. R., and Stetsko, P. I., in *Frontiers in Bioinorganic Chemistry,* Xavier, A.V., Ed., VCH Publishers, Heidelberg, 142, 1986.

Nieboer, E., Tom, R. T., and Rosetto, F. E., *Biol. Trace Elem. Res.,* 21, 23–33, 1989.
Nordenson, I. and Beckman, L., *Hum. Hered.,* 41, 71–73, 1991.
Ochi, T., Ishiguro, T., and Ohsawa, M., *Mutat. Res.,* 122, 169–175, 1983.
O'Connor, T. R., Graves, R. J., de Murcia, G., Castaing, B., and Laval, J., *J. Biol. Chem.,* 268, 9063–9070, 1993.
Okada, S., Hamazaki, S., Ebina, Y., Fujioka, M., and Midorikawa, O., in *Structure and Function of Iron Storage and Transport Proteins,* Urushizaki, I., Aisen, P., Litowsky, I., and Drysdale, J.W., Eds., Elsevier, New York, 473, 1983.
Okada, S., Hamazaki, S., Ebina, Y., Li, J. L., and Midorikawa, O., *Biochim. Biophys. Acta,* 922, 28–33, 1987.
Oleinick, N. L., Chiu, S., Ramakrishnan, N., and Xue, L., *Br. J. Cancer,* 55 (Suppl. VIII), 135–140, 1987.
Park, J. W. and Floyd, R. A., *Free Rad. Biol. Med.,* 12, 245–250, 1992.
Patel, U., Bhimani, R., and Frenkel, K., *Mutat. Res.,* 283, 145–156, 1992.
Patierno, S. R., Sygiyama, M., and Costa, M., *J. Biochem. Toxicol.,* 2, 13–23, 1987.
Peskin, A. V. and Shlyahova, L., *FEBS Lett.,* 194, 317–321, 1986.
Puppo, A. and Halliwell, B., *Biochem. J.,* 249, 185–190, 1988.
Reid, T. M. and Loeb, L. A., *Cancer Res.,* 52, 1082–1086, 1992.
Reid, T. M. and Loeb, L. A., *Proc. Natl. Acad. Sci. U.S.A.,* 90, 3904–3907, 1993.
Rodriguez, R. E. and Kasprzak, K. S., *Proc. Am. Assoc. Cancer Res.,* 30, 204, 1989.
Rodriguez, R. E. and Kasprzak, K. S., in *Nickel and Human Health: Current Perspectives,* Nieboer, E. and Nriagu, J.O., Eds., John Wiley & Sons, New York, 375, 1992.
Rodriguez, R. E., Misra, M., and Kasprzak, K. S., *Toxicology,* 63, 45–52, 1990.
Rodriguez, R. E., Misra, M., North, S. L., and Kasprzak, K. S., *Toxicol. Lett.,* 57, 269–281, 1991.
Rossi, S. C. and Wetterhahn, K. E., *Carcinogenesis,* 10, 913–920, 1989.
Rossi, S. C., Gorman, N., and Wetterhahn, K. E., *Chem. Res. Toxicol.,* 1, 101–107, 1988.
Runger, T. M. and Kraemer, K. H., *EMBO J.,* 8, 1419–1425, 1989.
Rush, J. D., Maskos, Z., and Koppenol, W. H., *FEBS Lett.,* 261, 121–123, 1990.
Salnikow, K., Zhitkovich, A., and Costa, M., *Carcinogenesis,* 13, 2341–2346, 1992.
Sawyer, D. T., in *Oxygen Complexes and Oxygen Activation by Transition Metals,* Martell, A.E. and Sawyer, D.T., Eds., Plenum Press, New York, 131, 1987.
Schaaper, R. M., Koplitz, R. M., Tkeshelashvili, L. K., and Loeb, L. A., *Mutat. Res.,* 177, 179–188, 1987.
Schlag, R. D., in *Advances in Modern Environmental Toxicology,* Vol. XI, Fishbein, L., Furst, A., and Mehlman, M.A., Eds., Princeton Scientific, Princeton, NJ, 211, 1987.
Sen, P. and Costa, M., *Cancer Res.,* 45, 2320–2325, 1985.
Shi, X. and Dalal, N. S., *Biochem. Biophys. Res. Commun.,* 163, 627–634, 1989.
Shi, X. and Dalal, N. S., *Arch. Biochem. Biophys.,* 281, 90–95, 1990a.
Shi, X. and Dalal, N. S., *Arch. Biochem. Biophys.,* 277, 342–350, 1990b.
Shi, X. and Dalal, N. S., *J. Inorg. Biochem.,* 40, 1–12, 1990c.
Shi, S. and Dalal, N. S., *Arch. Biochem. Biophys.,* 292, 323–327, 1992.
Shi, X., Dalal, N. S., and Kasprzak, K. S., *Arch. Biochem. Biophys.,* 299, 154–162, 1992a.
Shi, X., Sun, X., Gannett, P. M., and Dalal, N. S., *Arch. Biochem. Biophys.,* 293, 281–286, 1992b.
Shi, X., Dalal, N. S., and Kasprzak, K. S., *Chem. Res. Toxicol.,* 6, 277–283, 1993.
Shi, X., Dalal, N. S., and Kasprzak, K. S., *J. Inorg. Biochem.,* 50, 211–225, 1993b.
Shi, X., Dalal, N. S., and Kasprzak, K. S., *Arch. Biochem. Biophys.,* 302, 294–299, 1993c.
Shi, X., Dalal, N. S., and Kasprzak, K. S., *Environ. Health Perspect.,* 102 (Suppl. 3), 91–96, 1994.
Shibutani, S., Takeshita, M., and Grollman, A. P., *Nature,* 349, 431–434, 1991.
Shirname-More, L., Rossman, T. G., Troll, W., Teebor, G. W., and Frenkel, K., *Mutat. Res.,* 178, 177–186, 1987.
Shukla, G. S., Hussain, T., and Chandra, S. V., *Life Sci.,* 41, 2215–2221, 1987.
Singer, B. and Grunberger, D., *Molecular Biology of Mutagens and Carcinogens,* Plenum Press, New York, 1983.
Sirover, M. A. and Loeb, L. A., *Science,* 194, 1434–1436, 1976.
Skilleter, D. N., in *Carcinogenic and Mutagenic Metal Compounds,* Merian, E., Frei, R.W., Hardi, W., and Schlatter, C., Eds., Gordon and Breach, New York, 371, 1985.
Snow, E. T., *Pharm. Ther.,* 53, 31–65, 1992.
Snyder, R. D., Davis, G. F., and Lachmann, P. J., *Biol. Trace Elem. Res.,* 21, 389–398, 1989.
Sole, J., Huguet, J., Arola, L., and Romeu, A., *Bull. Environ. Contam. Toxicol.,* 44, 686–691, 1990.
Spiro, T. G., Ed., *Metal Ion Activation of Dioxygen,* John Wiley & Sons, New York, 1980.
Stadtman, E. R. and Berlett, B. S., *J. Biol. Chem.,* 266, 17201–17211, 1991.
Standeven, A. M. and Wetterhahn, K. E., *Chem. Res. Toxicol.,* 4, 616–625, 1991a.
Standeven, A. M. and Wetterhahn, K. E., *Pharmacol. Toxicol.,* 68, 469–476, 1991b.
Standeven, A. M. and Wetterhahn, K. E., *Carcinogenesis,* 13, 1319–1324, 1992.
Steward, R. C., Hille, R., and Massey, V., *J. Biol. Chem.,* 260, 8892–8904, 1985.
Sugiyama, M., Tsuzuki, K., and Ogura, R., *J. Biol. Chem.,* 266, 3383–3386, 1991.
Sunderman, F. W., Jr., *Ann. Clin. Lab. Sci.,* 14, 93–122, 1984.
Sunderman, F. W., Jr., Carcinogenicity and Mutagenicity of Some Metals and Their Compounds, Vol. 71, IARC Sci. Publ., International Agency for Research on Cancer, Lyon, France, 17, 1986a.
Sunderman, F. W., Jr., *Acta Pharmacol. Toxicol.,* 59 (Suppl. VII), 248–255, 1986b.

Sunderman, F. W., Jr., *Scand. J. Work Environ. Health*, 15, 1–12, 1989.
Sunderman, F. W., Jr. and Barber, A. M., *Ann. Clin. Lab. Sci.*, 18, 267–288, 1988.
Suzuki, K., Miyazawa, N., Nakata, T., Seo, H. G., Sugiyama, T., and Taniguchi, N., *Carcinogenesis*, 14, 1881–1884, 1993.
Swierenga, S. S. H. and McLean, J. R., in *Progress in Nickel Toxicology,* Brown, S.S. and Sunderman, F.W., Jr., Eds., Blackwell Scientific, Oxford, 1985, 101.
Sygiyama, M., Lin, X., and Costa, M., *Mutat. Res.*, 260, 19–23, 1991.
Szatrowski, T. P. and Nathan, C. F., *Cancer Res.*, 51, 794–798, 1991.
Taylor, A., in *Biological Effects of Heavy Metals: Metal Carcinogenesis,* Vol. II, Foulkes, E.C., Ed., CRC Press, Boca Raton, FL, 159, 1990.
Teoule, R. and Cadet, J., in *Effects of Ionizing Radiation on DNA,* Hutterman, J., Kohnlein, W., Teoule, R., and Bertinchamps, J., Eds., Springer-Verlag, New York, 171, 1978.
Tkeshelashvili, L. K., McBride, T. J., Spence, K., and Loeb, L. A., *J. Biol. Chem.*, 266, 6401–6406, 1991.
Tkeshelashvili, L. K., Reid, T. M., McBride, T. J., and Loeb, L. A., *Cancer Res.*, 53, 4172–4174, 1993.
Tofigh, S. and Frenkel, K., *Free Rad. Biol. Med.*, 7, 131–143, 1989.
Torreilles, J. and Guerin, M.-C., *FEBS Lett.*, 272, 58–60, 1990.
Umemura, T., Sai, K., Takagi, A., Hasegawa, R., and Kurokawa, Y., *Cancer Lett.*, 54, 95–100, 1990a.
Umemura, T., Sai, K., Takagi, A., Hasegawa, R., and Kurokawa, Y., *Carcinogenesis*, 11, 345–347, 1990b.
Umemura, T., Sai, K., Takagi, A., Hasegawa, R., and Kurokawa, Y., *Cancer Lett.*, 58, 49–56, 1991.
Vallyathan, V., Mega, J. F., Shi, X., and Dalal, N. S., *Am. J. Respir. Cell Mol. Biol.*, 6, 404–413, 1992.
von Sonntag, C., *The Chemical Basis of Radiation Biology,* Taylor and Francis, London, 1987.
Vos, J. M. H. and Wauthier, E. L., *Mol. Cell. Biol.*, 11, 2245–2252, 1991.
Waalkes, M. P. and Oberdorster, G., in *Biological Effects of Heavy Metals: Metal Carcinogenesis,* Vol. II, Foulkes, E.C., Ed., CRC Press, Boca Raton, FL, 129, 1990.
Walling, C., *Acc. Chem. Res.*, 8, 125–131, 1975.
Wedrychowski, A., Schmidt, W. N., and Hnilica, L. S., *J. Biol. Chem.*, 261, 3370–3376, 1986a.
Wedrychowski, A., Schmidt, W. N., and Hnilica, L. S., *Arch. Biochem. Biophys.*, 251, 397–402, 1986b.
Weitzman, S. A. and Stossel, T. P., *Science*, 212, 546–547, 1981.
Wetterhahn, K. E., Hamilton, J. W., Aiyar, J., Borges, K. M., and Floyd, R., *Biol. Trace Elem. Res.*, 21, 405–411, 1989.
Wey, S. J., O'Connor, K. J., and Burrows, C. J., *Tetrahedron Lett.*, 34, 1905–1908, 1993.
Wink, D. A., Wink, C. B., Nims, R. W., and Ford, P. C., *Environ. Health Perspect.*, 102 (Suppl. 3), 11–15, 1994
Wood, M. L., Dizdaroglu, M., Gajewski, E., and Essigmann, J. M., *Biochemistry*, 29, 7024–7032, 1990.
Wood, M. L., Esteve, A., Morningstar, M. L., Kuziemko, G. M., and Essigmann, J. M., *Nucl. Acids Res.* 20, 6023–6032, 1992.
Yamamoto, K. and Kawanishi, S., *J. Biol. Chem.*, 264, 15435–15440, 1989.
Yamamoto, K. and Kawanishi, S., *J. Biol. Chem.*, 266, 1509–1515, 1991.
Yamamoto, K. and Kawanishi, S., *Chem. Res. Toxicol.*, 5, 440–446, 1992.
Yamamoto, K., Inoue, S., Yamazaki, A., Yoshinaga, T., and Kawanishi, S., *Chem. Res. Toxicol.*, 2, 234–239, 1989.
Yamamoto, F., Kasai, H., Bessho, T., Chung, M. H., Inoue, H., Ohtsuka, E., Hori, T., and Nishimura, S., *Jpn. J. Cancer Res.*, 83, 351–357, 1992.
Yamamoto, K., Inoue, S., and Kawanishi, S., *Carcinogenesis*, 14, 1397–1401, 1993a.
Yamamoto, F., Kasai, H., Togashi, Y., Takeichi, N., Hori, T., and Nishimura, S., *Jpn. J. Cancer Res.*, 84, 508–511, 1993b.
Zalma, R., Guignard, J., and Pezerat, H., *NATO ASI Ser.*, H30, 257–264, 1989.
Zhitkovich, A. and Costa, M., *Carcinogenesis*, 13, 1485–1489, 1992.
Zhong, Z., Troll, W., Koenig, K. L., and Frenkel, K., *Cancer Res.*, 50, 7564–7570, 1990.

Chapter 19

Metal-Induced Gene Expression and Neoplastic Transformation

Joseph R. Landolph, P. Michael Dews, Laurent Ozbun, and Douglas Parker Evans

ABSTRACT

Carcinogenic arsenic and nickel compounds and lead chromate induced morphological and neoplastic transformation but no mutation to ouabain resistance in 10T1/2 mouse embryo cells; lead chromate also did not induce mutation to ouabain or 6-thioguanine resistance in Chinese hamster ovary cells. Therefore, the mechanism of arsenic-, nickel-, and lead chromate-induced morphological and neoplastic transformation was likely not due to the specific type of base substitution mutations measured in assays for mutation to ouabain resistance and, for lead chromate, also not due to this type of base substitution mutation or to frameshift mutations. Preliminary data from our laboratory indicate that there are increases in steady-state levels of c-*myc* RNA in arsenic-, nickel-, and chromium-transformed cell lines and increased c-*myc* protein levels in lead chromate-transformed cell lines. This is due to an increased stability of c-*myc* RNA in the cases of arsenic- and lead chromate-induced transformed cell lines.

We also showed that carcinogenic nickel, chromium, and arsenic compounds and *N*-methyl-*N'*-nitro-*N*-nitrosoguanidine (MNNG) induced stable anchorage independence (AI) in diploid human fibroblasts but not focus formation or immortality. Nickel subsulfide and lead chromate induced AI but not mutation to 6-thioguanine resistance. Induction of AI by metal salts in human fibroblasts was likely not caused by the type of base substitution or frameshift mutations measured in this assay. In contrast, MNNG facilely induced dose-dependent AI, mutation to ouabain resistance, and mutation to 6-thioguanine resistance over the same concentration ranges. Therefore, MNNG likely induced AI by inducing base substitution or frameshift mutations of the type measured in these mutation assays. Utilizing inhibitors to probe the molecular mechanisms of metal-induced AI in human fibroblasts, we found that dexamethasone, aspirin, and salicylic acid inhibited nickel subsulfide and MNNG-induced AI in diploid human fibroblasts. This suggests that nickel subsulfide and MNNG-induced arachidonic acid metabolism and consequent oxygen radical generation off this pathway plays a role in induction of AI.

We therefore propose that carcinogenic nickel compounds activate arachidonic acid metabolism, consequent oxygen radical generation, and also generate oxygen radicals by acting as redox active agents, and that these oxygen radicals react with DNA to cause mutational activation of specific protooncogenes to oncogenes and mutational inactivation or deletions in specific tumor suppressor genes. Reduction of chromium(VI) intracellularly to Cr(V) or other species which generate oxygen radicals, and also to Cr(III), followed by reaction of these species with DNA, also causes activation of specific oncogenes and inactivation of specific tumor suppressor genes — the latter likely through the mechanism of chromosome breakage. Arsenite causes chromosome breaks. We propose that arsenic, nickel, and chromium compounds specifically cause small deletions or mutations in the 5' or 3' regulatory regions of the c-*myc* and other protooncogenes, resulting in stabilization of c-*myc* RNA and higher steady-state levels of c-*myc* RNA and protein. We also postulate that nickel-induced oxygen radical generation, Cr(V) ions or oxygen radicals generated by chromium, and arsenite-induced DNA strand breaks/chromosome breaks induce inactivating mutations or deletions in tumor suppressor genes. We therefore postulate that arsenic-, chromium-, and nickel compound-induced neoplastic transformation proceeds through a combination of activation of c-*myc* and/or other specific protooncogenes and inactivation of one or more specific tumor suppressor genes. These cumulative activations of oncogenes and inactivations of tumor suppressor genes are proposed to be the overall mechanism of arsenic-, nickel-, and chromium-induced carcinogenesis, with an emphasis being placed on the likelihood that these metalloid (arsenic)/metal carcinogens (nickel/chromium) cause deletional types of mutations to accomplish carcinogenesis. Our preliminary evidence in support of this hypothesis is that transformed 10T1/2 cell lines induced by lead chromate have higher steady-state levels of c-*myc* RNA due to an enhanced stability of c-*myc* RNA, and also higher steady-state levels of c-*myc* protein.

I. INTRODUCTION

Exposure of humans to arsenic, nickel, or chromium compounds in occupational settings correlates with increased frequencies of skin, lung, esophageal, and nasal carcinomas (reviewed in Hernberg, 1977; Norseth, 1977; IARC, 1980; IARC, 1982; IARC, 1983; Peters et al., 1986; Yu et al., 1988; Landolph, 1989, 1990, 1994; Christie et al., 1990; Cohen et al., 1990; Smith et al., 1992). Nickel and insoluble hexavalent chromium compounds are also carcinogenic in animal bioassays (Ottolenghi et al., 1974; Sunderman and Maenza, 1976; Sunderman, 1989; reviewed in Sunderman, 1978, 1989; Peters et al., 1986; Smith et al., 1992; Landolph, 1989, 1990, 1994a). Epidemiological evidence suggests that occupational exposure to arsenic compounds correlates with increased skin and respiratory cancer in humans, but there has not yet been a study showing the carcinogenicity of arsenic compounds in animals (reviewed in IARC, 1983b; Peters et al., 1986; Landolph, 1989, 1990, 1994; Smith et al., 1992). The findings that nickel, hexavalent chromium, and arsenic compounds induce (Costa et al., 1979; DiPaolo and Casto, 1979; Costa and Mollenauer, 1980; Saxholm et al., 1981; Heck and Costa, 1982; Patierno et al., 1988; Miura et al., 1989; reviewed in Landolph, 1989, 1990, 1994) or promote (Rivedal and Sanner, 1981) morphological transformation of cultured rodent cells and induce anchorage-independence in diploid human fibroblasts (Biedermann and Landolph, 1987, 1990) have extended our search for the molecular mechanisms of metal/metalloid carcinogenesis to the cellular level (reviewed in Landolph, 1989, 1990, 1994; Cohen et al., 1990; Christie et al., 1990).

Substantial effort has gone into studying the molecular mechanisms of metal carcinogenesis, which is just beginning to become understood. Metal compounds readily induce chromosomal aberrations in cultured mammalian cells (Nishimura and Umeda, 1978; Lavramendy et al., 1981). However, most carcinogenic metal compounds, except for hexavalent chromium compounds and platinum complexes, are inactive or only weakly active in bacterial and mammalian cell mutation assays (reviewed in Heck and Costa, 1983; Landolph, 1989, 1990, 1994). *In vitro* mammalian and bacterial mutagenesis assays available in the past did not detect DNA alterations caused by carcinogenic metalloid/metal salts (reviewed in Heck and Costa, 1983; Landolph, 1989, 1990, 1994). Recent evidence indicates that metal compounds, particularly carcinogenic nickel compounds, induced deletions in a gpt gene transfected into mammalian V79 cells (Kargacin et al., 1993; Lee et al., 1993). Lee et al. (1993) observed mutation at the transfected gpt locus in G12 V79 Chinese hamster cells and showed that the gpt gene was amplifiable by PCR analysis, indicating that the mutated gene had no large deletions. Interestingly, Reid et al. showed that metal compound-induced oxygen radicals caused mutations (Reid et al., 1994).

In this chapter, we review studies from our laboratory and other laboratories on induction of morphological transformation in C3H/10T1/2 Cl 8 (10T1/2) mouse embryo cells, anchorage-independence in diploid human fibroblasts by carcinogenic metal and metalloid (arsenic) salts, and hypothesize mechanisms of metal and metalloid salt-induced cell transformation.

II. MATERIALS AND METHODS

10T1/2 mouse embryo fibroblasts were cultured as per Reznikoff et al. (1973a), chemically induced cytotoxicity and morphological transformations were quantitated in them as per Reznikoff et al. (1973b), Landolph and Heidelberger (1979), and reviewed in Landolph (1985). Chemically induced mutation to ouabain resistance in 10T1/2 cells was quantitated as per Landolph and Heidelberger (1979) and Miura et al. (1989).

Diploid human fibroblasts were derived from circumcised human foreskins and cultured, and assays to detect metal-induced cytotoxicity, anchorage independence, and mutation to 6-thioguanine or ouabain resistance were performed as described in Biedermann and Landolph (1987, 1990).

III. RESULTS

A. INDUCTION OF MORPHOLOGICAL TRANSFORMATION OF 10T1/2 CELLS BY CARCINOGENIC ARSENIC, NICKEL, AND CHROMIUM COMPOUNDS

C3H/10T1/2 Cl 8 (10T1/2) cells are an aneuploid, immortal mouse cell line derived from embryos of C3H mice that is contact-inhibited, nontumorigenic, and has a low frequency of spontaneous transformation (Reznikoff et al., 1973a, 1973b). Treatment of 10T1/2 cells with chemical carcinogens or radiation induces a high frequency of morphological transformations in them (Reznikoff et al., 1973b; Landolph and Heidelberger, 1979; Patierno et al., 1988; Miura et al., 1989; reviewed in Landolph, 1985, 1990, 1994). 10T1/2 cells are very useful model cell culture systems for detecting carcinogens by measuring the ability of carcinogens to induce morphological transformations in these cells. 10T1/2 cells are also very useful to determine molecular and cellular mechanisms of chemically induced morphological and neoplastic transformation to gain insight into molecular mechanisms of chemical carcinogenesis (reviewed in Landolph, 1985a, 1985b, 1989, 1990, 1994).

Our laboratory has been studying the ability of carcinogenic metal salts to induce morphological transformation of 10T1/2 cells to gain insight into mechanisms of metal carcinogenesis. We have shown that carcinogenic arsenic, nickel, and chromium compounds induce morphological and neoplastic transformation of 10T1/2 cells (Table 1) (reviewed in Landolph, 1989, 1990, 1994a). Firstly, we found that 10T1/2 cells readily phagocytosed nickel subsulfide particles, which were readily seen in phagocytic vesicles (Miura et al., 1989). Phagocytosis of nickel subsulfide and crystalline nickel monosulfide by 10T1/2 cells occurred when low concentrations of these compounds were added to the medium (0 to 50 μM, Miura et al., 1989). We observed a concentration-dependent cytotoxicity in 10T1/2 cells treated with the insoluble nickel compounds, nickel subsulfide and nickel monosulfide, over concentrations of nickel compounds where phagocytosis occurred (Miura et al., 1989). With nickel oxide, we did not observe phagocytic vesicles, but we did observe cytotoxicity in 10T1/2 cells treated with up to 1 mM nickel oxide (Miura et al., 1989).

Next, we studied the ability of carcinogenic nickel compounds to induce morphological transformation in 10T1/2 cells (Miura et al., 1989). The insoluble carcinogenic nickel compounds, nickel subsulfide, nickel monosulfide, and nickel oxide (greenish preparation), all induced dose-dependent transformation in 10T1/2 cells (see Table 1) over the same concentration ranges at which these compounds were phagocytosed and induced cytotoxicity (0 to 50 μM virtual concentrations, calculated as if all insoluble nickel compounds dissolved (Miura et al., 1989). Therefore, phagocytosis of particulate nickel compounds is an initial step in the processes of cytotoxicity and cell transformation in 10T1/2 cells (Miura et al., 1989). Similar observations were made by Costa and co-workers in Syrian hamster cells (Costa and Mollenhauer, 1980; Heck and Costa, 1983). Nickel subsulfide induced primarily type II foci, and cell lines derived from these foci did not grow in soft agarose (see Table 1) (Miura et al., 1989). Nickel monosulfide induced type II and occasionally type III foci. Nickel monosulfide-induced type II and type III foci gave rise to cell lines that grew in soft agarose. Nickel oxide induced type II and occasionally type III foci, and one of these type II and one type III foci gave rise to cell lines that grew in soft agarose and formed fibrosarcomas in nude mice (see Table 1) (Miura et al., 1989).

In contrast, the soluble noncarcinogenic or substantially less carcinogenic compounds, nickel sulfate and nickel chloride, did not induce morphological transformation of 10T1/2 cells (see Table 1) (Miura et al., 1989; reviewed in Landolph, 1989, 1990, 1994a). The inability of soluble nickel compounds to

Table 1 Summary of *In Vitro* Morphological, Anchorage Independent, and Neoplastic Transformation in C3H/10T1/2 Cl 8 Mouse Embryo Cells by Carcinogenic Arsenic, Nickel, and Chromium Compounds[a]

Inducing compounds	Morphological transformation	Anchorage independent transformation	Neoplastic transformation
Nickel subsulfide	+	–	–
Nickel monosulfide	+	+	–
Nickel oxide	+	+	+
Nickel sulfate	–	–	–
Nickel chloride	–	–	–
Lead chromate	+	+	+
Calcium chromate	–	–	–
Potassium dichromate	–	–	–
Strontium chromate	–	–	–
Sodium arsenite	+	+	+
Sodium arsenate	–	–	–

[a] Data in this table are a summary of transformation results for nickel compounds from Miura et al., 1989; for chromium compounds, from Patierno et al., 1988; for arsenic compounds, from Landolph and Troesch, manuscript in preparation.

induce cell transformation is consistent with whole-animal carcinogenicity studies (Ottolenghi et al., 1974; Sunderman and Maenza, 1976; Sunderman, 1978, 1989). Therefore, our results for induction of morphological transformation in 10T1/2 cells with nickel compounds correlated with results of whole-animal carcinogenesis assays using nickel compounds. The morphological transformation we observed with insoluble nickel compounds are therefore specific responses.

Earlier, we developed an assay for detecting chemically induced (Landolph and Heidelberger, 1979), stable (Landolph et al., 1980a), and specific (Landolph et al., 1980b) base substitution mutations (Landolph and Jones, 1982) to ouabain resistance resulting from mutation in a gene encoding (Na,K)-ATPase activity on murine chromosome 3 (Landolph and Fournier, 1983) that confers a ouabain-resistant (Na,K)-ATPase activity (Shibuya et al., 1989) and hence a ouabain-resistant (Na,K)-ATPase-driven potassium transport (Landolph et al., 1980a) (reviewed in Landolph, 1985a; Shibuya et al., 1989). We then studied the ability of nickel compounds to induce mutations in this assay. Interestingly, concentrations of nickel subsulfide and nickel oxide that induced cytotoxicity and morphological transformation in 10T1/2 cells did not induce base substitution mutations to ouabain resistance in 10T1/2 cells (Miura et al., 1989). This indicated that nickel compounds likely do not induce morphological transformation by inducing the specific, restricted type of base substitution mutations that are detected in assays for mutation to ouabain resistance (Miura et al., 1989).

Current preliminary studies from our laboratory indicate that nickel monosulfide- and nickel oxide-transformed 10T1/2 cell lines express from higher steady-state levels of c-*myc* RNA (see Table 2). We hypothesize that transformation of 10T1/2 cells by nickel compounds results in small deletions or mutations in the 3′ or 5′ regulatory regions of specific protooncogenes. This could lead to activation of these protooncogenes via this mechanism, hence stabilization of the protooncogene RNAs due to a longer half-life of the RNAs, and then to higher steady-state levels of oncogene RNAs. Studies to test these hypotheses critically in our laboratory are in progress (Sakuramoto, T., Verma, A., Evans, D., Miura, T., and Landolph, J. R., manuscript in preparation).

In addition, we found that lead chromate induced a low, dose-dependent, and reproducible frequency of type III morphological transformation (Patierno et al., 1988), the strongest type of morphological transformation in 10T1/2 cells (Reznikoff et al., 1973b, reviewed in Landolph, 1985b). The transformed cells displayed a stable focus-forming phenotype, grew in soft agarose, and formed fibrosarcomas when injected into nude mice (Patierno et al., 1988). Lead chromate-treated cells had many vacuoles and extruded cytoplasm over the particles of lead chromate, likely in an attempt to phagocytose lead chromate particles. Calcium chromate, potassium dichromate, and strontium chromate did not induce morphological transformation in 10T1/2 cells. Strontium chromate was slightly soluble in culture medium and eventually dissolved, and calcium chromate did not induce transformation even when added to the cells as a particulate in acetone suspension. These results suggested that the unique physicochemical properties of insoluble lead chromate particles were responsible for its uptake, likely by phagocytosis, and its ability to induce cytotoxicity and morphological transformation (Patierno et al., 1988). These results are consistent with earlier observations that the slightly soluble hexavalent chromium compounds are carcinogenic in animals (reviewed in Norseth, 1977; Hernberg, 1977; Landolph, 1989, 1990, 1994a).

We showed that lead chromate induced morphological transformation in 10T1/2 cells but no mutation to ouabain resistance in 10T1/2 or CHO cells, nor mutation to 6-thioguanine resistance in CHO cells. In addition, calcium chromate did not induce morphological transformation in 10T1/2 cells or mutation to ouabain resistance in 10T1/2 or CHO cells, but did induce mutation to 6-thioguanine resistance in CHO cells (Patierno et al., 1988). Hence, we speculated that lead chromate caused morphological transformation by a mechanism not involving the specific types of base substitution or frameshift mutations detectable in assays for mutation to ouabain or 6-thioguanine resistance. Recently, we obtained preliminary evidence that in two lead chromate-transformed 10T1/2 cell lines there are four- to eightfold higher steady-state levels of c-*myc* RNA and up to twofold higher steady-state levels of c-*myc* protein, and the half-life of c-*myc* RNA is increased in these cell lines (Dews, M., Ozbun, L., Krishnan, K., and Landolph, J. R., manuscript in preparation; Table 2). Our current working hypothesis is that lead chromate induces mutations or small deletions in the 5' or 3' regulatory regions of the c-*myc* gene, leading to a c-*myc* RNA with an increased half-life. This may be due either to deletions or mutations generated by the reduction of Cr(VI), leading to either Cr(V) or oxygen radicals generated by intracellular reduction of Cr(VI).

Table 2 Summary of Induction of Anchorage Independence in Diploid Human Fibroblasts by Carcinogenic Arsenic, Nickel, and Chromium Compounds[a]

Inducing compounds	Induction of anchorage independent cell transformation
Nickel subsulfide	+
Nickel sulfate	+
Nickel acetate	+
Lead chromate	+
Potassium dichromate	+
Calcium chromate	+
Chromium trioxide	+
Sodium arsenite	+
Sodium arsenate	+

[a] Data in this table are a summary of transformation results for nickel, chromium, and arsenic compounds from Biedermann and Landolph, 1990.

Thirdly, we observed that sodium arsenite induced a low but reproducible yield of morphological transformation in 10T1/2 cells (Landolph, J. R. and Troesch, C., manuscript in preparation; see Table 1). Cloning of arsenite-transformed foci yielded cell lines that formed type II and type III foci, grew in soft agarose, and formed fibrosarcomas in nude mice. We also found that sodium arsenite promoted cell transformation initiated by 3-methylcholanthrene. Pentavalent sodium arsenate and potassium arsenate did not induce morphological transformation, indicating that this transformation was specific for trivalent arsenic (Landolph, J. R. and Troesch, C. T., manuscript in preparation; see Table 1). These data are consistent with epidemiological studies indicating that arsenic compounds are carcinogenic to humans (reviewed in IARC, 1980, 1982a, 1982b; Peters et al., 1986; Landolph, 1989, 1990, 1994a; Smith et al., 1992). A resolution to the apparent paradox of the noncarcinogenicity of arsenic in animal bioassays could be that sodium arsenite is a weak cell-transforming agent and a promoting or cocarcinogenic agent not easily detected in the relatively insensitive animal bioassays. A second possibility is that humans are more sensitive to arsenic-induced carcinogenesis than rats and mice. Further, these arsenic-induced transformed cell lines have elevated steady-state levels of c-*myc* RNA (Lillehaug, J. R., Troesch, C., and Landolph, J. R., manuscript in preparation).

Preliminary work from our laboratory has first ruled out the hypothesis that carcinogenic metal compounds induce morphological transformation in 10T1/2 cells by inducing amplification or gross rearrangement of known protooncogenes, such as c-*myc*, converting them into oncogenes. We did not observe gene amplification, gross rearrangements, or large deletions in protooncogenes in transformed cell lines induced by treating 10T1/2 cells with arsenic, nickel, or chromium compounds (manuscript in preparation). We are now testing a second hypothesis that carcinogenic metal compounds cause mutations or small deletions in the 5′ or 3′ regulatory regions of specific protooncogenes, converting them into activated oncogenes. We are also testing a third hypothesis that carcinogenic metal salts cause mutational inactivation or deletion of tumor suppressor genes, inactivating them. To date, we have found preliminary evidence for increased steady-state levels of c-*myc* RNA in transformed cell lines induced by treating 10T1/2 cells with lead chromate, nickel oxide, and sodium arsenite. We have not found amplification or rearrangements of the c-*myc* genes in these cell lines (Sakuramoto, T., Miura, T., Dews, M., Lillehaug, J. R., and Landolph, J. R., manuscripts in preparation; see Table 2). We are currently studying whether there are small deletions or mutations in the 3′ or 5′ regulatory regions of this gene in these transformed cell lines that might account for c-*myc* activation and higher steady-state levels of c-*myc* transcripts.

B. INDUCTION OF ANCHORAGE INDEPENDENCE IN DIPLOID HUMAN FIBROBLASTS

Recently, we began studying molecular mechanisms by which carcinogenic metal salts induce transformation of cultured diploid human fibroblasts. We employed circumcised human neonatal foreskins as a source of human fibroblasts. Previous studies showed that organic carcinogens induced anchorage independence in cultured human fibroblasts (Milo and DiPaolo, 1978; Greiner et al., 1981; Silinskas et al., 1981). We showed that nickel compounds such as nickel subsulfide, nickel acetate, and nickel subsulfide (Biedermann and Landolph, 1987); hexavalent chromium compounds such as lead chromate, potassium dichromate, calcium chromate, and chromium trioxide (Biedermann and Landolph, 1987, 1990); and arsenic compounds such as sodium arsenite and sodium arsenate (Biedermann and Landolph, 1990) induced dose-dependent anchorage independence (AI) in diploid human fibroblasts as summarized in Table 3. Metal-induced AI was a stable phenotype. To date, we have not observed other transformation phenotypes, such as morphological transformation. All the metal-induced AI cell strains had saturation densities comparable to those of normal human fibroblasts, and they all eventually senescenced (Biedermann and Landolph, 1987, 1990). Carcinogenic metal salt-induced AI is specific, because manganese chloride, mercuric acetate, and calcium chloride, which are not carcinogenic, did not induce AI (Biedermann and Landolph, 1987).

Table 3 Summary of Preliminary Data on the Expression and Structure of the c-*Myc* Gene and c-*Myc* Protein in Transformed C3H/10T1/2 Cell Lines Induced by Lead Chromate, Nickel Oxide, Nickel Monosulfide, and Sodium Arsenite[a]

In transformed C3H/10T1/2 cell lines induced by	Amplification or rearrangement of the c-*Myc* gene	Higher steady-state levels of c-*Myc* RNA	Increased half-life of c-*Myc* RNA	Higher steady-state levels of c-*Myc* protein
Nickel oxide	−	+	N.D.	N.D.
Nickel monosulfide	−	+	N.D.	N.D.
Lead chromate	−	+	+	+
Sodium arsenite	−	+	+	N.D.

[a] Data in this table are a summary of preliminary results from studies in progress from Sakuramoto, T., Verma, A., Miura, T., and Landolph, J. R., manuscript in preparation for nickel transformation studies; from Dews, M., Ozbun, L., Krishnan, K. and Landolph, J. R., manuscript in preparation for studies on lead chromate transformation; and from Lillehaug, J. R., Evans, D. and Landolph, J. R., manuscript in preparation, for studies of sodium arsenite transformation.

Nickel subsulfide did not induce mutation to ouabain or 6-thioguanine resistance (Biedermann and Landolph, 1987), and lead chromate did not induce mutation to 6-thioguanine resistance (Biedermann and Landolph, 1990) at concentrations that were cytotoxic and induced AI in diploid human fibroblasts. Hence, nickel subsulfide and lead chromate likely induced AI by mutations of the type not easily measured in assays for 6-thioguanine resistance. We speculate that metal-induced oxygen radical generation and consequent radical-induced mutations might be part of the mechanism of nickel subsulfide and lead chromate-induced AI in diploid human fibroblasts. Calcium chromate, potassium dichromate, and MNNG did induce mutation to 6-thioguanine resistance and AI over the same concentration ranges, indicating that base substitution or frameshift mutations might be a mechanism by which these compounds induced AI (Biedermann and Landolph, 1990).

Recently, we began testing the hypothesis that activation of arachidonic metabolism and consequent generation of oxygen radicals is part of the molecular mechanism by which metal compounds induce

AI in diploid human fibroblasts. We found that nickel subsulfide-induced AI in human fibroblasts was inhibited by dexamethasone, aspirin, and nordihydroguaieretic acid. Similarly, MNNG-induced AI was also inhibited by these three inhibitors of arachidonic acid release (dexamethasone) and metabolism (aspirin, which inhibits cyclooxygenase activity), and oxygen radical persistence (salicylic acid, which scavenges oxygen radicals) (Biedermann, K. A., Nwankwo, J. O., Weng, J., and Landolph, J. R., manuscript in preparation; review in Landolph, 1994b).

IV. CONCLUSIONS

The mechanisms of metal carcinogenesis are clearly complex and only beginning to become understood. Further, the mechanisms of carcinogenesis for each metal are specific to that metal (reviewed in Wetterhahn et al., 1992). However, a number of conclusions may be drawn and speculations made from work in our laboratory on mechanisms of metal-induced cell transformation. Insoluble carcinogenic nickel compounds such as nickel subsulfide are phagocytosed, and large amounts of these compounds are taken up into individual murine (Miura et al., 1989), hamster (Heck and Costa, 1983), and human (Biedermann and Landolph, 1987) fibroblasts. In addition, our recent preliminary data indicate that inhibitors of arachidonic release and its metabolism by cyclooxygenase inhibitors and an oxygen radical scavenger (salicylic acid) inhibit nickel subsulfide-induced AI in diploid human fibroblasts. We therefore speculate that nickel subsulfide induces membrane perturbations that activate the prostaglandin synthesis cascade, resulting in generation of oxygen radicals. This could explain our inability to measure mutation to ouabain resistance in 10T1/2 cells (Miura et al., 1989), or mutation to ouabain resistance or to 6-thioguanine resistance in diploid human fibroblasts (Biedermann and Landolph, 1987) following treatment of these cells with nickel subsulfide, since oxygen radicals do not induce the type of mutation that is easily measured in these assays. Further work needs to be done to determine the fraction of nickel-generated oxygen radicals derived from stimulation of arachidonic acid metabolism vs. the fraction that originates from protein-bound nickel ions that bind to DNA (Ciccarelli and Wetterhahn, 1982, 1984; Cohen et al., 1990) and generate oxygen radicals.

Secondly, we have preliminary evidence that there are increased steady-state levels of c-*myc* RNAs in nickel oxide- and nickel monosulfide-transformed 10T1/2 mouse embryo cells (Sakuramoto, T., Verma, A., Miura, T., Evans, D., and Landolph, J. R., manuscript in preparation; see Table 2). We speculate that this may result from nickel compound-generated oxygen radicals, which may induce mutations in the 3' or 5' regulatory regions of these protooncogenes, leading to enhanced stability of these RNAs and contributing to induction and maintenance of the transformed phenotype. We know that multiple activated oncogenes may cooperate in cell transformation (reviewed in Land et al., 1983; Weinberg, 1985; Bishop, 1987). Hence, we are now studying how many oncogenes are activated in nickel-transformed cell lines.

We also hypothesize that nickel ions generate oxygen radicals, and that these oxygen radicals are responsible for inactivating tumor suppressor genes (Klein, 1987) — both the known tumor suppressor genes such as the retinoblastoma gene Rb (Horwitz et al., 1988) and the p53 suppressor gene (Hollstein et al., 1991) as well as tumor suppressor genes that are just becoming identified and understood such as the suppressor gene on human chromosome 11 (Weissman et al., 1987) and novel suppressor genes just being discovered such as the senescence-mediating suppressor gene that is a target for nickel-induced cell transformation discovered by Costa's group (Klein et al., 1991). We speculate that nickel compounds could both inactivate genes such as the p53 gene and also activate them to dominantly acting negative oncogenes. Further work is in progress in our laboratory to test these hypotheses. Recently, Salnikow et al. (1994) showed that transcription of the thromboplastin gene was lost in nickel-transformed Chinese hamster embryo cells without deletion or rearrangement of this gene.

The active ionic species that induce chromium carcinogenesis have not been identified. Chromate induces DNA cross-links in rat liver and kidney (Tsapakos et al., 1981) and binds to chromatin and DNA (Cupo and Wetterhahn, 1985). Current speculation is that Cr(VI) is reduced to Cr(V), which is a proximate carcinogen, or to lower oxidation states, which generate oxygen radicals (reviewed in Wetterhahn et al., 1992). Our work with the strong carcinogen lead chromate (Furst et al., 1976) has given preliminary results that lead chromate-transformed 10T1/2 cell lines have higher steady-state levels of c-*myc* RNA, an increased half-life of c-*myc* RNA mediating the higher steady-state levels of c-*myc* RNA, and also higher steady-state levels of c-*myc* protein (Dews, M., Ozbun, L., Krishnan, K. and Landolph, J. R., manuscript in preparation; see Table 2). We are testing the hypothesis that these higher steady-state levels of c-*myc* RNA are due to mutations in 3' or 5' regulatory regions of the c-*myc* gene, leading to a longer half-life of c-*myc* RNA and contributing to maintenance of the transformed state. Our results also indicate that lead chromate does not induce mutation to ouabain resistance in 10T1/2 or CHO cells or mutation to 6-thioguanine resistance in CHO cells (Patierno et al., 1988). Hence, we speculate that lead chromate may either generate Cr(V) and/or generate oxygen radicals that induce mutations or small deletions in the 3′ or 5′ regulatory regions of the c-*myc* gene, activating it to an oncogene encoding a c-*myc* RNA with increased half-life. Further work is in progress in our laboratory to test this hypothesis.

Finally, we also found preliminary evidence for increased steady-state levels of c-*myc* expression in sodium arsenite-transformed 10T1/2 cell lines without amplification of the c-*myc* (Lillehaug, J. R., Evans, D., and Landolph, J. R., manuscript in preparation; see Table 2). This is interesting, because Lee et al. (1985) found that arsenite induced amplification of the dihydrofolate reductase gene in mouse 3T6 cells. Since we found that sodium arsenite did not induce mutation to ouabain resistance in 10T1/2 cells (Landolph, J. R. and Troesch, C., manuscript in preparation), we also postulate that arsenite induces mutations in the 3′ or 5′ regulatory regions of the c-*myc* gene in arsenite-transformed cell lines. Our laboratory is currently working to determine the type of mutations arsenite induces in the c-*myc* gene that contributes an increased half-life of c-*myc* RNA and to maintenance of the transformed phenotype (Lillehaug, J. R., Evans, D., and Landolph, J. R., manuscript in preparation).

In conclusion, we hypothesize that carcinogenic arsenic, nickel, and chromium compounds also cause mutation and/or deletions in tumor suppressor genes that inactivate these genes. We postulate that a combination of mutation or small deletion-induced activation of single or multiple protooncogenes into oncogenes, plus mutational or deletional inactivations of single or multiple tumor suppressor genes, is part of the mechanism of metal carcinogenesis.

REFERENCES

Biedermann, K. A. and Landolph, J. R., *Cancer Res.*, 47, 3815–3823, 1987.
Biedermann, K. A. and Landolph, J. R., *Cancer Res.*, 50, 7835–7842, 1990.
Bishop, J. M., *Science*, 235, 305–311, 1987.
Christie, N. T. and Katsifis, S. P., in *Biological Effects of Heavy Metals*, Vol. II, Foulkes, E. C., Ed., CRC Press, Boca Raton, FL, 95, 1990.
Ciccarelli, R. B. and Wetterhahn, K. E., *Cancer Res.*, 42, 3544–3549, 1982.
Ciccarelli, R. B. and Wetterhahn, K. E., *Cancer Res.*, 44, 3892–3897, 1984.
Cohen, M., Latta, D., Coga, T., and Costa, M., in *Biological Effects of Heavy Metals*, Vol. II, Foulkes, E. C., Ed., CRC Press, Boca Raton, FL, 19, 1990.
Costa, M., Nye, J. S., Sunderman, F. W., Allpass, P. R., and Gordos, B., *Cancer Res.*, 39, 3591–3597, 1979.
Costa, M. and Mollenhauer, H., *Science*, 209, 515–517, 1980.
Costa, M., Salnikow, K., Cosentino, S., Klein, C. B., Huang, X., and Zhuang, Z., *Environ. Health Perspect.*, 102 (Suppl. 3), 127–130, 1994.
Cupo, D. Y. and Wetterhahn, K. E., *Cancer Res.*, 45, 1146–1151, 1985.
DiPaolo, J. A. and Casto, B., *Cancer Res.*, 39, 1008–1013, 1979.
Furst, A., Schlander, M., and Sasmore, D. P., *Cancer Res.*, 36, 1779–1783, 1976.
Greiner, J. W., Evans, C. A., and DiPaolo, J. A., *Carcinogenesis*, 2, 359–362, 1981.
Heck, J. D. and Costa, M., *Biol. Trace Elem. Res.*, 4, 319–330, 1982.
Heck, J. D. and Costa, M., *Cancer Res.*, 43, 5652–5658, 1983.
Hernberg, S., in *Origins of Human Cancer, Book A: Incidence of Cancer in Humans*, Vol. 4, Hiatt, H. H., Watson, J. D., and Winsten, J. A., Eds., Cold Spring Harbor Press, Cold Spring Harbor, NY, 147, 1977.
Higinbotham, K. G., Rice, J. M., Diwan, B. A., Kasprzak, K. S., Reed, C. D., and Perantoni, A. O., *Cancer Res.*, 52, 4747–4751, 1992.
Hollstein, M., Sidransky, D., Vogelstein, B., and Harris, C. C., *Science*, 253, 49–53, 1991.
Horwitz, J. M., Yardell, D. W., Bark, S.-H., Canning, S., Whyte, P., Buchkoritch, K., Harlow, E., and Dryja, T. P., *Science*, 243, 937–945, 1988.
IARC Monographs on the Evaluation of the Carcinogenic Risk of Chemicals to Humans. Vol. 1–29, Suppl. 4, International Agency for Cancer Research, Lyon, France, 167, 1980.
IARC Monographs on the Evaluation of the Carcinogenic Risk of Chemicals to Humans, Chromium and Chromium Compounds, Vol. 23, International Agency for Cancer Research, Lyon, France, 205, 1982.
IARC Monographs on the Evaluation of the Carcinogenic Risk of Chemicals to Humans. Some Metallic Compounds, Arsenic and Arsenic Compounds, Vol. 23, International Agency for Cancer Research, Lyon, France, 105, 1983.
Kargacin, B., Klein, C. B., and Costa, M., *Mutat. Res.*, 300, 63–72, 1993.
Klein, C. B., Conway, K., Wany, X. W., Bhamra, R. K., Lin, X. H., Cohen, M. D., Annab, L., Barrett, J. C., and Costa, M., *Science*, 251, 796–799, 1991.
Klein, G., *Science*, 238, 1539–1546, 1987.
Land, H., Parada, L., and Weinberg, R. A., *Science*, 222, 771–776, 1983.

Landolph, J. R. and Heidelberger, C., *Proc. Natl. Acad. Sci. U.S.A.,* 76, 930–934, 1979.
Landolph, J. R., Telfer, N., and Heidelberger, C., *Mutat. Res.,* 72, 295–310, 1980a.
Landolph, J. R., Bhatt, R. S., Telfer, N., and Heidelberger, C., *Cancer Res.,* 40, 4581–4588, 1980b.
Landolph, J. R. and Jones, P. A., *Cancer Res.,* 42, 817–823, 1982.
Landolph, J. R. and Fournier, R. E. K., *Mutat. Res.,* 107, 447–463, 1983.
Landolph, J. R., in *The Role of Chemicals and Radiation in the Etiology of Neoplasia: Carcinogenesis,* Vol. 10, Huberman, E. and Barr, S. H., Eds., Raven Press, New York, 211, 1985a.
Landolph, J. R., Chemical transformation in C3H/10T1/2 Cl 8 mouse embryo fibroblasts: historical background, assessment of the transformation assay, and evolution and optimization of the transformation assay protocol, Transformation Assay of Established Cell Lines: Mechanisms and Application, Vol. 146, IARC Sci. Publ., Takunaga, T. and Yamasaki, H., Eds., International Agency for Cancer Research, Lyon, France, 185, 1985b.
Landolph, J. R., *Biol. Trace Elem. Res.,* 21, 459–467, 1989.
Landolph, J. R., in *Biological Effects of Heavy Metals,* Vol. II, Foulkes, E. C., Ed., CRC Press, Boca Raton, FL, 1, 1990.
Landolph, J. R., *Environ. Health Perspect.,* 102 (Suppl. 3), 119–125, 1994.
Landolph, J. R., in *Biological Oxidants and Antioxidants,* Packer, L. and Cadenas, E. C., Eds., Hippokrates Verlag, Stuttgart, Germany, 133, 1994.
Lavramendy, M. L., Popescu, N. C., and DiPaolo, J. A., *Environ. Mutagen.,* 3, 597–606, 1981.
Lee, T. C., Oshimura, M., and Barrett, J. C., *Carcinogenesis,* 6, 1421–1426, 1985.
Lee, T., Tanaki, N., Lamb, P. W., Gilmer, T. M., and Barrett, J. C., *Science,* 241, 79–84, 1988.
Lee, Y. W., Pons, C., Tummolo, D. M., Klein, C. B., Rossman, T. G., and Christie, N. T., *Environ. Mol. Mutagen.,* 21, 365–371, 1993.
Milo, G. E. and DiPaolo, J. A., *Nature (London),* 275, 130–132, 1978.
Miura, T., Patierno, S. R., Sakuramoto, T., and Landolph, J. R., *Environ. Mol. Mutagen.,* 14, 65–78, 1989.
Nishimura, M. and Umeda, M., *Mutat. Res.,* 68, 337–349, 1978.
Norseth, T., in *Origins of Human Cancer, Book A: Incidence of Cancer in Humans,* Vol. 4, Hiatt, H. H., Watson, J. D., and Winsten, J. A., Eds., Cold Springs Harbor Press Cold Springs Harbor, NY, 159, 1977.
Ottolenghi, A. D., Haseman, J. K., Pavne, W. W., Falk, H. L., and MacFarland, H. N., *J. Natl. Cancer Inst.,* 54, 1165–1172, 1974.
Patierno, S. R., Banh, D., and Landolph, J. R., *Cancer Res.,* 48, 5280–5288, 1988.
Peters, J. M., Thomas, D., Falk, H., Oberdorster, G., and Smith, T. J., *Environ. Health Perspect.,* 70, 71–83, 1986.
Reid, T. M., Feig, D. I., and Loeb, L. A., *Environ. Health Perspect.,* 102 (Suppl. 3), 57–62, 1994.
Reznikoff, C. A., Brankow, D. W., and Heidelberger, C., *Cancer Res.,* 33, 3231–3238, 1973a.
Reznikoff, C. A., Bertram, J. S., Brankow, D. W., and Heidelberger, C., *Cancer Res.,* 33, 3239–3249, 1973b.
Rivedal, E. and Sanner, T., *Cancer Res.,* 41, 2950–2953, 1981.
Salnikow, K., Cosentino, S., Klein, C., and Costa, M., *Cell. Mol. Biol.,* 14, 851–858, 1994.
Saxholm, H. J. K., Reith, A., and Brogger, A., *Cancer Res.,* 41, 4136–4139, 1981.
Shibuya, M. L., Miura, T., Lillehaug, J. R., Farley, R. A., and Landolph, J. R., *J. Mol. Toxicol.,* 2, 75–98, 1989.
Silinskas, K. C., Kateley, S. A., Tower, J. E., Maher, V. M., and McCormick, J., *Cancer Res.,* 41, 1620–1627, 1981.
Smith, A. H., Hopen-havn, R. C., Bates, M. N., Goeden, H. M., Hentz-Picciotto, I., Duggan, H. M. N., Wood, R., Kosrett, M. J., and Smith, M. T., *Environ. Health Perspect.,* 97, 259–267, 1992.
Sunderman, F. W. and Maenza, R. M., *Res. Commun. Chem. Pathol. Pharmacol.,* 14, 319–330, 1976.
Sunderman, F. W., *Fed. Proc.,* 37, 40–44, 1978.
Sunderman, F. W., Jr., *Scand. J. Work Environ. Health,* 15, 1–12, 1989.
Tsapakos, M. J., Hampton, T. H., and Jennette, R. W., *J. Biol. Chem.,* 256, 3623–3626, 1981.
Weinberg, R. A., *Science,* 230, 770–776, 1985.
Weissman, B. E., Saxon, P. J., Pasquale, S. R., Jones, G. R., Geiser, A. G., and Stanbridge, E. J., *Science,* 236, 175–180, 1987.
Wetterhahn, K. E., People, B., Kulesz-Martin, M., and Copeland, E. S., *Cancer Res.,* 52, 4058–4063, 1992.
Yu, M. C., Garabrant, D. A., Peters, J. M., and Mack, T. M., *Cancer Res.,* 48, 3843–3849, 1988.

Section IV
Clinical Aspects of Metal Toxicity

Overview

An Introduction to Clinical Aspects of Metal Toxicology

Kazuo T. Suzuki and Tsuguyoshi Suzuki

Metals are components of all biological systems. During evolutionary processes, the biological systems have utilized a large number of metals for their beneficial functions, while specialized mechanisms for either detoxification or elimination have been developed for others that are not desirable or toxic. Even the nutritionally essential metals produce a variety of undesirable health effects or tissue pathology at exposure or intake levels that are not too far from their desirable or beneficial levels. This section of the treatise deals with the clinical aspects of metals where health problems are manifest, either due to their deficiency or excess encountered in ordinary situations. An additional chapter has been devoted to chelation therapies that are employed in different metal exposures. The chapters have not been arranged in any particular sequence with regard to the toxicity of various metals or any other criteria that would imply their importance.

The metals that are considered essential nutrients include chromium (Cr), cobalt (Co), copper (Cu), iron (Fe), magnesium (Mg), manganese (Mn), molybdenum (Mo), selenium (Se), and zinc (Zn). Inadequate dietary intake of these metals results in a variety of deficiency syndromes. In many metals, such as Cu and Se, the range of desirable to toxic levels may be less than one order of magnitude. Nearly all metals produce toxic effects when their exposures are excessive, and a reduction of such exposures should be a prudent practice wherever possible, particularly when no obvious beneficial effects are known. Metals that are generally considered toxic because they are not required for normal functioning of living processes and no beneficial health effects have been known by their presence in the system are exemplified by aluminum (Al), antimony (Sb), arsenic (As), bismuth (Bi), cadmium (Cd), gold (Au), lead (Pb), lithium (Li), mercury (Hg), nickel (Ni), platinum (Pt), and thallium (Tl). In the past, and to a limited extent even at the present time, a few of the metal compounds are used for therapeutic purposes. Clinical aspects of most of these metals are described in the chapters that follow.

The detailed description of Minamata disease (Chapter 20) by R. Hamada and M. Osame (Hamada and Osame, 1995) is an elegant update on this topic. Organic Hg poisoning has been known for a long time, however, it is by no means an event of the distant past. Occupational and environmental exposures of people to both inorganic and organic Hg are still continuing, and an effective removal of Hg from the exposure media is far from being achieved. The presentation of various clinical parameters that are essential for identification of Hg poisoning, particularly the computerized topography (CT) scans, should be a valuable aid to the diagnostician. In addition to other inadvertent environmental exposures, exposure to Hg via the amalgam used in dental fillings presents a considerable health risk, and this aspect is discussed appropriately by J. A. Weiner and M. Nylander (Weiner and Nylander, 1996) in Chapter 30.

The role of Cd as an etiology of itai-itai disease has been well established. A detailed version of risk factors in the development of this malady, along with symptoms and tests required for its diagnosis, and specific biomarker, has been summarized by K. Nogawa and T. Kido (Nogawa and Kido, 1996) in Chapter 21. Of particular significance is a review of studies on the epidemiology of exposed populations. Recent information on cancer incidence and mortality from Cd-polluted areas has been presented.

Copper is one of the metals which is both essential and toxic. In particular, the deficiency of this metal because of the genetic disorders of its metabolism has been reported. The exact genetic basis of two of these problems, i.e., Wilson disease and Menkes disease, has been recently identified. H. Kodama has provided a very nice and balanced view of these disorders of Cu metabolism (Kodama, 1996) in Chapter 22. Both of these diseases, inherited by recessive genes which have similarities but are located on different chromosomes, are caused by an inadequate expression of Cu-transporting ATPases. Lesions of Wilson disease are produced largely by a lack of this enzyme in the hepatic tissue, which causes an overaccumulation of Cu in the liver, while producing a Cu deficiency in other organs; Cu load in the liver beyond the detoxifying capability of a specific metal-binding protein (metallothionein) leads to hepatic damage and subsequent jaundice. In the case of Menkes disease, an absence of Cu-transporting enzyme interferes with the appropriate absorption of Cu from the intestines and produces an overall deficiency of this metal, despite an adequate dietary supply. Animal models to investigate the pathogenesis of these diseases have recently been developed. Clinical symptoms and pathology, along with diagnosis and therapy and possible implications of these disorders in pregnancy have been discussed.

An interesting discussion of the role of Al in Alzheimer's disease has been presented by D. R. McLachlan and co-workers (McLachlan et al., 1996) in Chapter 23. Although largely circumstantial, taken together the observations presented lend credence to the hypothesis that Al may be an important risk factor in Alzheimer's disease, or possibly even play some role in Parkinson's disease. A number of neurotoxic properties of Al, including accelerated generation of a specific amyloid (AD amyloid) by Al, and neurofibrillary degeneration associated with the reduced DNA transcription caused by this metal, suggest the implication of Al in the pathogenesis of these disorders and that unnecessary exposure to this metal should be controlled. A further support to this hypothesis is provided by the incidence of dialysis encephalopathy with Al compounds and similar disorders due to the occupational exposures of this metal. Regardless of the exact role of Al in the etiology of Alzheimer's disease and Parkinson's disease, the neurotoxic properties of this metal warrant further considerations and avoidance of overexposures.

Public health problems associated with lead exposure have been recognized for a long time. Problems caused by the occupational and environmental exposures of various lead preparations have led to their phase-out in commercial use, i.e., as in gasoline additives and paints. Of particular importance is the sequel of environmental lead exposure in infants and young children. Subclinical symptoms like reduction in IQ, cognitive functions, and learning processes have long been considered ill effects of lead exposure in children; however, it has only been recently that a definitive clinical association and epidemiological evidence have emerged. The current status of childhood lead exposure at low doses has been presented by H. Needleman, a noted authority in the field (Needleman, 1996), in Chapter 24. It is because of these subtle effects of lead exposure that levels of blood lead considered at the threshold of toxicity have been gradually reduced. Concerns still persist about lead exposure during pregnancy; exact implications of any occupational or environmental lead exposure in this situation are still not fully understood.

A concise yet complete discussion of the neurological aspects of Mn exposure has been provided in Chapter 25 by N. Inoue and Y. Makita. (Inoue and Makita, 1996). This metal has been a known occupational toxin, particularly in Mn mining and ore processing, for a long time; still, much consideration was not accorded to it due to its relatively low acute toxicity. The delayed onset of nervous system disorders after occupational exposures has been well documented. In many cases, however, an exact diagnosis of the problem is very difficult due to lack of consistent symptomology and clinical indices of Mn exposure.

G. G. Garcia-Vargas and M. E. Cebrián in Chapter 26 have presented a comprehensive account of As toxicity (Garcia-Vargas and Cebrián, 1996). The syndromes associated with As exposures have been highly varied, depending on the chemical form of the compound, the route of exposure, and the status of exposure (i.e., peracute, acute, subacute, or chronic). The consequences of respiratory and dermal exposure of As have been appreciated for quite a while, and the evidence of this metal being a human carcinogen after pulmonary exposures in occupational situations has been established. A variety of arsenicals have been employed as therapeutic agents in the past; some of these may still be used in some folk medicines in certain parts of the world. Organic arsenicals are used as growth stimulants in poultry and to some extent in other livestock. Because of its carcinogenic potential, exposure to As via pulmonary and dermal routes warrants continued consideration.

Toxicology associated with a variety of miscellaneous metals, such as Li, Tl, Sb, Au, and Pt, is discussed in Chapter 27 by A. Slikkerveer and F.A. de Wolff (Slikkerveer and de Wolff, 1996) and in Chapter 28 by G. B. van der Voet and F. A. de Wolff, respectively. All of these metals possess considerable toxic properties (van der Voet and de Wolff, 1996), although the mechanisms of their action are still not fully understood. In spite of their inherent toxicity, compounds of Au and Pt (e.g., gold-thioglucose, *cis*-platinum) have therapeutic applications. Additional studies are needed to investigate the mechanisms of toxic action and establish biomonitoring indices for most of these metals.

The implications of the occupational exposures to Co and Ni, primarily with respect to the development of hard metal asthma, has been provided in Chapter 29 by Y. Kusaka. Irritation and sensitization of bronchial airways is caused by these metals (Kusaka, 1996). Cobalt, in addition to sensitization, induces interstitial pneumonitis and pulmonary fibrosis. The role of immunological responses in the induction of hard metal asthma has been discussed.

Finally, a complete discussion of chelation therapies for metal intoxications has been presented by C. Angle in Chapter 31. A variety of chelating agents have been in practice for a while, yet many of these have their inherent toxicity as well. A careful monitoring of the metal burden is highly important in chelation therapy. Some of the metals may be inaccessible due to their intracellular location or protein binding (e.g., Cd). Continued investigations in developing new and effective chelating agents, and the proper applications of the currently used chelation therapies, are imperative.

Chapter 20

Minamata Disease and Other Mercury Syndromes

Rikuzo Hamada and Mitsuhiro Osame

I. INTRODUCTION

Mercury has three chemical forms: elemental mercury (Hg^0), inorganic mercury salts (Hg^+, Hg^{2+}), and organic mercury (e.g., alkylmercury, phenylmercury, methoxyethylmercury). Although each form of mercury can have specific toxic effects in humans, each form undergoes conversion to one another. Elemental mercury vapor (Hg^0) is converted to Hg^{2+}, and Hg^{2+} is also reduced to Hg^0 in the natural environment and in mammals (WHO, 1990). Inorganic mercury is converted to some methylated form, and methylmercury is in turn converted to inorganic mercury in experimental animals and humans (WHO, 1990). Lability of the carbon-mercury bond is different in each organic mercury compound and the toxicity is also different. From a clinical point of view, methylmercury and mercury vapor are the most important forms with regard to the source for human exposure and toxic effects in humans. Methylmercury exhibits the most important and typical toxic effects of all organic mercury. Methylmercury is the only organic form of mercury known to be produced by natural processes (Clarkson, 1983). The main routes of methylmercury exposure are by ingestion (e.g., grain treated by an antifungal agent, fish, fish products, or other sea food) and by inhalation (occupational hazards, e.g., dust or vapor). A part of the inhaled dust is transported from the respiratory tract to the pharynx by ciliary movement, and swallowed to the alimentary tract. Ingested methylmercury is almost completely absorbed. Methylmercury is degraded to inorganic mercury in humans. After high oral intakes of methylmercury for 2 months, the percentages of inorganic mercury per total mercury in the tissues are 7% in whole blood, 22% in plasma, 39% in breast milk, 73% in urine, and 16–40% in liver (WHO, 1990). A considerable amount (above 80%) of mercury in the human brain is reported to be in the form of inorganic mercury in the cases with chronic Minamata disease (WHO, 1991). Although methylmercury is distributed to all the tissues within about four days, its toxic effects are selective to the nervous system.

II. MINAMATA DISEASE

Some sporadic episodes of organic mercury intoxication were reported before the epidemic in Minamata; however, this epidemic disaster in Minamata was different from previous episodic intoxications in scale, severity, and clinical variety. The sporadic episodes had occurred by accident or through occupational exposure only in high-risk groups and were limited to a small number of patients. Minamata disease was the first episode of epidemic intoxication occurring in an extraordinarily large area and affecting a large number of patients. It is known as Minamata disease because it first occurred in Minamata, Japan. The cause of the disease was methylmercury contained in waste water and discharged from a chemical plant (acetaldehyde plant) into the sea facing Minamata. The first patient was recorded in 1954, but industrial discharge of methylmercury was continued for more than 10 years. Methylmercury was discharged into Minamata Bay at first, and the pollution was limited to areas along the bay. Later, methylmercury was discharged outside the bay, thus spreading the pollution to extremely wide areas. Methylmercury was concentrated by food chains and accumulated in fish. Fish being the main source of animal protein for the inhabitants in those days, almost all of them ingested methylmercury-contaminated fish. Methylmercury exposure continued for a very long period of time. By the end of November 1993, 2256 patients had been officially recognized in the Minamata area (Table 1). These patients fall into three groups: patients definitely, probably, or possibly affected by the disease. Although almost all the inhabitants were exposed to methylmercury below or above the threshold of poisoning, fishermen and their families were obviously more affected since their consumption of fish is normally greater than for people of other occupations (Table 2).

Table 1 Number of Patients

	Minamata			Niigata	Iraq
	Kumamoto	Kagoshima	Total		
Male		282		389	3,144
Female		205		301	3,384
Total	1,769	487	2,256	690	6,528

Table 2 Occupations of the Patients in the Kagoshima Area

Fishermen	316
Fishmongers	23
Farmers	40
Others	108
Total	487

Minamata is situated at the southern end of Kumamoto Prefecture and borders on Kagoshima Prefecture. The authors' group studied the people living in the areas of Kagoshima Prefecture neighboring on Minamata. The total number of inhabitants in this area was approximately 74,000. We surveyed all of them from 1971 to 1974, and found 65 patients with Minamata disease (the first Kagoshima study). After this study, some additional patients were officially recognized sporadically along with changes in social conditions (the second Kagoshima study).

Unfortunately, almost the same phenomenon was to be repeated ten years later in Niigata. The only difference was that waste water was discharged into a river in Niigata instead of the sea in Minamata. Since river fish are consumed by fewer people compared with sea fish, methylmercury poisoning was restricted to a relatively small number of people in Niigata (Table 1). Heeding the lessons gained from the Minamata disease, high-risk pregnant women chose abortion, with the result that there was no congenital poisoning in Niigata, except for one case. In addition to the Japanese disaster, organic mercury poisoning has occurred on a large scale in Pakistan (Haq, 1963), Guatemala (Ordóñez et al., 1966), and twice in Iraq (Jalili and Abbasi, 1961; Damluji, 1962; Bakir et al., 1973). These episodes were all induced by the misuse of organomercury-treated seed grain.

A. CLINICAL MANIFESTATIONS OF MINAMATA DISEASE

A precise clinical report of methylmercury poisoning was first published by Hunter et al. (1940). They reported four cases of methylmercury poisoning as a result of industrial exposure and, after an interval of 15 years, an autopsy of one of the four cases was reported by Hunter and Russell (1954). Based on their reports, sensory impairment, cerebellar ataxia, dysarthria, concentric constriction of the visual field, and hearing impairment were regarded as the syndrome of methylmercury poisoning and was named the Hunter-Russell syndrome.

There is a latent period between methylmercury exposure and the initiation of the symptoms. This latent period was evident in the Iraqi epidemic. The first symptom appeared after consumption of the contaminated bread had stopped. The mean latent period ranged from 16 to 38 days. It is difficult to determine the beginning and the end of exposure together with the exact length of the latent period in the Japanese epidemic because contamination lasted for more than 10 years.

The clinical manifestations detected in Minamata, Niigata, and Iraq are fundamentally the same as those of the Hunter-Russell syndrome, though somewhat different from one another.

The rates of incidence of the signs and symptoms in methylmercury poisoning are shown in Table 3. These clinical features are established by the analysis of the patients who were clinically diagnosed as having been poisoned. However, such a traditional way implies a contradiction, that is, the patient whose clinical features were analyzed was diagnosed in accordance with the accepted clinical features of poisoning. For this reason, multi-variant analyses were introduced. Factor analysis and discriminant analysis were applied to the data of the first Kagoshima study. The results of factor analysis illustrate the clinical features of the Minamata disease while those of discriminant analysis provide relative weights to each sign and symptom in making such a diagnosis (Table 4) (for details, see Hamada et al., 1978a,b).

Table 3 Signs and Symptoms (in %)

	Minamata				Niigata[b]	Iraq[c]
	Kumamoto area[a]		Kagoshima area			
	1950s	1969	1971–73	1986–88		
Sensory impairments, superficial	100.0	82.6	93.0	98.0	93	32.6
Sensory impairments, deep	100.0	60.0	83.7	88.0	88	60.5
Tremors	75.5	65.2	44.2	60.0	35	
Dysarthria	88.2	73.9	41.9	40.0	37	71.7
Adiadochokinesis	93.5	87.0	62.8	62.0		90.0
Finger-nose test	80.6	73.9	74.4	46.0	65	82.5
Knee-heel test	100.0	68.2	55.8	38.0		87.8
Gait disturbance	82.4	56.5	72.1	76.0	30	92.3
Romberg's sign	42.9	57.1	16.3	10.0		
Visual field constriction	100.0	100.0	100.0	8.0	37	37.5
Abnormal EOM			34.9	26.0		
Hearing impairment	85.3	69.6	88.4	78.0	63	19.0

[a] Tokuomi and Okajima (1969)
[b] Tsubaki (1968)
[c] Rustam and Hamdi (1974)

1. Sensory Impairments

Sensory impairment is a critical symptom as well as the most basic one. It is agreed that the first exhibited symptom of poisoning is sensory impairment and that sensory impairment is exhibited even in the mildest cases (Bakir et al., 1973). Paresthesia or dysesthesia are the first symptoms. Generally, they are initiated from the most peripheral regions, i.e., the fingers or toes. Then they extend to proximal regions, showing a glove-and-stocking type of sensory disturbance. Occasionally, in more severe cases, the lips, perioral region, tip of the tongue, or midline of the anterior chest and abdomen may be involved. Some patients complain of generalized anesthesia, although this complaint can be due to a psychological overlay. Every modality of sensory perception like light touch, pain, thermal sense, vibration, and position sense may be disturbed. It is similar to the sensory polyneuropathy arising from other causes but, in general, decreased deep tendon reflexes, muscular atrophy, or distal weakness are less common. In an acute phase, deep tendon reflexes are increased in 38.2% and decreased in 8.8% of the patients (Tokuomi, 1968). Babinski's signs may be observed in acute severe cases. In Iraq, definite Babinski's signs were noted in 20 out of 53 patients. Hunter et al. (1940) reported that paresthesia is the initial symptom in all four of their cases, but superficial sensation, such as pin-click or light touch, was spared, whereas two-point discrimination, position sense, and stereognosis were affected.

Rustam and Hamdi (1974) suggested that the site of lesion may be the cerebral cortex, because available histopathological and electrophysiological information did not confirm the clinical concept of peripheral neuropathy caused by methylmercury poisoning in humans. Should sensory impairment be attributed to cortical lesion or peripheral nerve lesion? Although the glove-and-stocking type of sensory disorder is usually caused by polyneuropathy, this still remains an unsolved question.

The pathological study revealed that both the sensory cortex (postcentral gyrus) and peripheral sensory nerves were impaired. However, no particular correspondence between the lesion of the sensory cortex and the glove-and-stocking type distribution of sensory impairment was found.

Table 4 Signs and Symptoms of Minamata Disease in the Kagoshima Area

#	Signs and symptoms	Factor loading	Discriminant coefficient	Incidences (%) '71–'73 a)	'86–'88 b)	b) – a)
1	Mental deterioration	0.322	−0.648	48.8	62.0	13.2
2	Emotional change	0.521	4.053	53.5	26.0	−27.5
3	Hyposmia	0.105	0.301	30.2	50.0	19.8
4	Perioral hypesthesia	0.631	2.673	58.1	28.0	−30.1
5	Facial palsy	−0.028	−1.427	7.0	8.0	1.0
6	Dysphagia	−0.119	0.975	2.3	4.0	1.7
7	Dysarthria	0.469	−1.458	41.9	40.0	−1.9
8	Tongue movement	0.520	4.125	46.5	40.0	−6.5
9	Accesory nerve	0.021	−5.249	4.7	14.0	9.3
10	Lateralities of above	−0.018	0.967	11.6	6.0	−5.6
11	ROM of neck	0.146	−0.432	37.2	50.0	12.8
12	Nuchal stiffness	−0.037	1.604	0.0	0.0	0.0
13	Spurling's sign	0.105	0.033	30.2	20.0	−10.2
14	Involuntary movement	0.440	0.105	44.2	60.0	15.8
15	Muscular atrophy	0.005	−1.836	11.6	8.0	−3.6
16	Muscular weakness	0.012	2.250	32.6	52.0	19.4
17	Increased muscular tonus	0.097	0.002	7.0	20.0	13.0
18	Fine finger movement	0.363	−0.393	25.6	62.0	36.4
19	Dysdiadochokinesis	0.640	0.534	62.8	62.0	−0.8
20	Finger-nose test	0.778	5.023	74.4	46.0	−28.4
21	Lateralities of above	−0.022	−1.786	11.6	30.0	18.4
22	Trunk	0.099	0.247	4.7	8.0	3.3
23	Muscular atrophy	−0.049	2.221	4.7	8.0	3.3
24	Increased muscular tonus	0.007	−3.119	7.0	16.0	9.0
25	Muscular weakness	0.093	1.461	34.9	58.0	23.1
26	Knee-heel test	0.752	1.343	55.8	38.0	−17.8
27	Lateralities of above	0.004	−4.663	4.7	12.0	7.3
28	Bladder dysfunction	−0.205	−1.824	0.0	4.0	4.0
29	Superficial sensation	0.476	0.223	93.0	98.0	5.0
30	Deep sensation	0.582	0.614	83.7	88.0	4.3
31	Segmental distribution	−0.075	−1.315	9.3	14.0	4.7
32	Lateralities of above	−0.010	−1.017	14.0	14.0	0.0
33	Biceps increased	0.110	−1.741	23.3	26.0	2.7
34	decreased	−0.050	0.277	2.3	16.0	13.7
35	Triceps increased	0.131	0.786	25.6	18.0	−7.6
36	decreased	−0.014	−2.048	4.7	16.0	11.3
37	Radialis increased	−0.018	−0.778	16.3	22.0	5.7
38	decreased	−0.021	0.604	7.0	24.0	17.0
39	PTR increased	0.023	−1.082	39.5	46.0	6.5
40	decreased	0.041	−0.952	9.3	22.0	12.7
41	ATR increased	0.054	1.376	27.9	26.0	−1.9
42	decreased	0.079	0.609	30.2	42.0	11.8
43	Babinski's sign	−0.219	1.344	4.7	4.0	−0.7
44	Lateralities of above	−0.179	−1.107	4.7	22.0	17.3
45	Gait disturbance	0.466	0.666	72.1	76.0	3.9
46	Romberg's sign	0.263	−0.571	16.3	10.0	−6.3
47	Visual field constriction	0.399	1.246	100.0	8.0	−92.0
48	Laterality of visual field	−0.238	−2.020	30.2	0.0	−30.2
49	Abnormal EOM	0.360	2.532	34.9	26.0	−8.9
50	Nystagmus	−0.077	−2.655	2.3	2.0	−0.3
51	Hearing impairment	0.155	−0.166	88.4	78.0	−10.4
52	Laterality of hearing	−0.031	0.947	16.3	20.0	3.7
53	Abnormal dram	−0.097	0.016	27.9	32.0	4.1

Note: #3–10: Cranial nerves.
#14–21: Upper limbs.
#23–27: Lower limbs.
#29–32: Sensory disturbances.
#33–44: Deep tendon reflex.

Abnormalities in sural nerve action potentials were reported by Nagaki (1975). The potentials evoked by foot stimulation were split up into several waves and the duration was prolonged, although the conduction velocity in the fastest groups of fibers was nearly normal. A sural nerve biopsy also showed evidence of peripheral neuropathy (Igata, 1986; Takeuchi and Eto, 1975).

A nerve conduction study, a sural nerve biopsy, and a pathological study of autopsied cases (Takeuchi, 1968) all indicated that at least peripheral nerve lesion occurs in methylmercury poisoning, but at the same time these findings fail to explain the whole spectrum of clinical manifestations. It may be that the site of the lesion is in part the cerebral cortex, i.e., the postcentral gyrus, and in part the peripheral nerves.

2. Ataxia

Cerebellar ataxia presents no particular problem. Patients noticed disturbances in gait, speech, and hand movement. Cerebellar ataxia includes ataxic dysarthria (speech becomes slow, inarticulate, and sometimes explosive or coquettish), incoordination (e.g., clumsiness in writing), terminal tremors, clumsiness in finger-nose and finger-finger tests, adiadochokinesis, truncal ataxia, and ataxic gait.

Hunter et al. (1940) described cerebellar ataxia in all their cases. It included ataxic dysarthria, limb ataxia, and ataxic gait. These findings are almost the same as those in Minamata, Niigata, and Iraq.

In Minamata, ataxia was seen in 80.6 to 93.5% of the patients in the initial study; in 65% of the patients for limb ataxia and 37% for dysarthria in Niigata; and in 95% of the patients in an Iraqi study.

Nystagmus showed a very low incidence compared with cerebellar signs. No case was detected in the initial Kumamoto study, 2.0 to 2.3% in Kagoshima, and only one case in Iraq (Rustam and Hamdi, 1974).

3. Visual Symptoms

Symmetric concentric constriction of the visual field is the most specific finding in Minamata disease. Visual field constriction is due to a lesion in the visual cortex (calcarine). Occasionally, patients do not notice their peripheral visual loss, but most of the time they do. Along with visual field constriction, depression was also found in all isopters in Goldmann's perimetry. Fundamental deterioration of the visual field was not pure constriction, but rather a peripheral depression with normal central sensitivity (Iwata et al., 1975). In the initial studies of Kumamoto and Kagoshima, visual field constriction was seen in nearly all cases, but it was detected in only 37% of the cases in Niigata, and 37.7% of the cases in the Iraqi study (Rustam and Hamdi, 1974). Visual symptoms in the Japanese and Iraqi epidemics showed some differences. Visual acuity impairment was detected in 37.7% (20/53) in Iraq. It was the same frequency for visual field constriction. Out of 20 visual field constriction cases, 8 showed both visual field constriction and visual acuity impairment. Loss of visual acuity was also reported in the Iraqi study, though in some patients it was due to retrobulbar neuritis. Optic atrophy was observed in six patients (Rustam and Hamdi, 1974). In contrast with the Iraqi epidemic, pupils, retinas, or optic nerves were all normal, and visual acuity together with color vision were unaffected in the Japanese episodes. Color vision was also preserved in the adult cases in Iraq (Rustam and Hamd 1974). The reason for the disparity in the visual symptoms in Japan and Iraq is not clearly known. One factor may be the time difference in exposure. Methylmercury exposure was relatively acute for a short period of time in the Iraqi epidemic — the mean period of ingestion in Iraq was 41 to 63 days. However, it lasted more than 10 years in Japan, implying that the Iraqi epidemic involved a relatively more acute intoxication than in Japan.

4. Hearing Impairment

Hearing impairment is partially caused by both retrocochlear deafness and by inner ear deafness. Hearing was impaired in 85.3% of the initial Kumamoto cases and 88.4% of the cases in the first Kagoshima study. In Niigata, normal hearing was detected in 47.2% of the patients, mild hearing loss (10 to 30 dB) in 36.1%, and moderate hearing loss (30 to 60 dB) in 16.7%. Hearing impairment was detected by a pure tone audiogram and a speech discrimination test. High-frequency tones were more affected than low-frequency ones. The use of a tuning fork and watch in Iraq revealed that 11 (19%) out of 53 patients had hearing defects, but none of them were totally deaf.

Here is a typical patient's complaint: "I can hear the voice, but cannot ascertain the words." A similar finding was described by Hunter et al. (1940): "He hears a watch normally, but he cannot quickly comprehend the meaning of spoken speech." These findings suggest that retrocochlear deafness may share the hearing impairment, at least partially. The hypothesis of retrocochlear deafness is supported by the presence of pathological changes in the temporal cortical lesions, especially in the transverse temporal gyri.

The recruitment phenomenon in SISI (short increment sensitivity index) suggests that hearing impairment in the early or middle stages of intoxication may be caused by cochlear lesions, and at a later stage may arise from the temporal cortical lesions (Ino et al., 1975).

5. Other Manifestations

Ocular Movement — Abnormal ocular movement was detected in the first Kagoshima study (34.9%). The main finding is saccadic ocular movement. The patient cannot follow a slowly moving object smoothly. It was detected as a staircase pattern in an electroophthalmograph (EOG). Overshoot or other types of abnormal EOG seen in the cerebellar dysfunction are less common. They can be attributed to the occipital lobe lesion. The occipital visual association areas maintain a smooth pursuit movement, and a precentral motor area deals with the saccadic ocular movement. As a result of dysfunction of the smooth pursuit movement by the occipital lobe lesion, ocular movement was controlled only by the residual saccadic ocular movement, itself controlled by the precentral motor area.

Involuntary Movements — Resting finger tremors, with or without rigidity and akinesia, are sometimes associated with methylmercury poisoning (75% in the initial Kumamoto study, 44.2% in the first Kagoshima study, and 60% in the second Kagoshima study). Typical pill-rolling parkinsonian tremors are rare, though fine finger tremors, mostly bilaterally, can be observed. Though akinesia, slowness of movement, or mild rigidity may be associated with methylmercury poisoning, an oily face or a parkinsonian gait are not associated with it. Some parts of the extrapyramidal signs may be attributed to degraded inorganic mercury. In an acute phase of the disease, involuntary movements were seen in both Kumamoto and Iraq: 17 patients (30%) had involuntary movements, including myoclonus and choreo-athetosis in Iraq (Rustam and Hamdi, 1974). Ballism (14.7%), chorea (14.7%), and athetosis (8.8%) were detected in the initial Kumamoto study (Tokuomi, 1968).

Autonomic Nerve Dysfunctions — Excessive salivation and sweating were noted in 23.5% of the patients in the initial Kumamoto study (Tokuomi, 1968), and autonomic disorders in 19% of the patients in Iraq (Rustam and Hamdi, 1974). Autonomic nerve dysfunction was also detected by thermography (Hamada et al., 1982a).

B. PATHOLOGY

The toxic effects of methylmercury are selective to the nervous system. The central nervous system and peripheral nerves are both affected. The most characteristic feature is the distribution of lesions. In acute and subacute autopsied cases, the brain was swollen and there were no particular macroscopic lesions on sectioning. In chronic cases, atrophy of the calcarine cortex, precentral gyrus, postcentral gyrus, and temporal cortex was detected in the cerebrum. As vulnerable areas, the calcarine regions were first identified, followed by the precentral and postcentral gyri. The superior temporal gyri and visual association areas were noted thirdly. The normal cortical striation of Gennari is absent. The precentral motor cortex is more affected than the postcentral sensory cortex. Disintegration of nerve cells and nerve fibers, and nerve cell loss with astrocytosis and glial fiber proliferation were demonstrated microscopically (Takeuchi, 1968; Takeuchi and Eto, 1975). Disturbance of the nerve cells was much more pronounced in the anterior portions of the calcarine cortex. This prominent distribution of nerve cell losses corresponds to the concentric constriction of the visual field. There was no specific distribution of damage in the postcentral sensory cortex corresponding to the glove-and-stocking type distribution of sensory disturbances.

In the cerebellum, there was gross atrophy of the folia, both in the lateral lobes and in the vermis. These distributions correspond to clinically manifested incoordination, ataxic speech, and ataxic gait. The granular cells, most susceptible to disintegration, actually began to disintegrate beneath the Purkinje cell layer at the crests of the gyri, resulting in the "apical scar". Purkinje cells are more resistant than granular cells.

Changes in the posterior roots and sensory peripheral nerves were characterized by an intense involvement of sensory nerve fibers in which disappearance of nerve fibers with collagen increases, there was irregular regeneration in size and arrangement of nerve fibers with incomplete myelination, and proliferation of Schwan's cells were noticed (Takeuchi and Eto, 1975).

The results of sural nerve biopsy were characterized by the general loss of myelinated fibers with a relative increase in small-size myelinated fibers (Igata, 1986).

C. CLINICAL COURSE

Clinical manifestations usually show gradual improvement following exposure (Table 3). The degree of improvement differs depending on the symptoms.

In Iraq, mildly and moderately affected patients recovered remarkably, with or without specific treatment. Recovery was notable in the cerebellar functions, superficial sensations, and visual acuity (Rustam and Hamdi, 1974).

The differences in the number of incidences are shown in Table 3. After a 10-year interval, all the signs and symptoms were less marked. Apart from the rate of incidences, improvement was observed in 90% of the cases for the visual field, in 50% for dysarthria, in 45.5% for superficial sensory disturbance, in 55% for hearing, and in 54.5% for upper limb coordination in the Kumamoto study. On the other hand, deterioration was observed in 9.5% of the cases for the visual field, in 13.6% for superficial sensory disturbance, in 10% for hearing impairment, and in 4.5% for upper limb coordination (Tokuomi and Okajima, 1969; Okajima et al., 1976).

Comparing the initial and second studies in Kagoshima, it can be said that emotional changes, perioral hypesthesia, clumsiness in the finger-nose test, clumsiness in the knee-heel test, and visual field constriction apparently decreased. The signs and symptoms which were relatively increased were hyposmia, muscular weakness, disturbance of fine finger movements, laterality of findings, and changes in deep tendon reflexes (Table 4). Although these two groups are from different populations, these results indicate that the clinical features of the Minamata disease are becoming mild and atypical. In Iraq, over a relatively short period of observation, the patients who recovered the fastest from visual acuity disturbance were those affected with retrobulbar neuritis.

Mortalities attributed to methylmercury poisoning amounted to 459 hospital deaths out of 6530 cases admitted to hospitals in Iraq (Bakir et al., 1973)

The mortality rate in Japan was 36.9% (41/111) in the initial Kumamoto study (Tokuomi, 1968), and 16.7% (5/30) in Niigata (Tsubaki, 1968).

Delayed Onset — Although it is widely accepted that there is a latent period between exposure and the initial symptoms in methylmercury poisoning (delayed onset), some patients go through a very long latent period of more than several years. Other patients show worsening several years after cessation of exposure. It is possible that the latent period following cessation of exposure may extend to one year or thereabouts (WHO, 1990), but such a long delay is difficult to understand. This delayed onset might be accounted for: (1) by the effect of aging on latent poisoning, and/or (2) by the psychological stress connected with financial compensation (Igata, 1986).

D. DIAGNOSIS

Clinical suspicion is the first step in diagnosing poisoning. A history of exposure or signs of clinical resemblance will lead to suspicion of poisoning. Exposure can be confirmed by measuring mercury in blood, hair, or urine. Hair is rich in sulfhydryl groups, and the concentration of methylmercury in hair is 250 times the simultaneous concentration in blood. Human head hair grows at approximately 1 cm per month and accumulates mercury. Thus, analysis of consecutive segments of a hair specimen provides a recapitulation of the methylmercury levels in blood (WHO, 1990).

It is not difficult to make a diagnosis of methylmercury poisoning clinically when the patient shows the typical combination of signs and symptoms of the Hunter-Russell syndrome. In a mild or atypical case, it is difficult to make the diagnosis relying only on clinical features. Clinical manifestations are dose dependent. With the lowest dose above the threshold, only sensory impairment can be manifested. Sensory impairment is a nonspecific common symptom in neurological disorders. It is very rare to find someone who never eats fish or fish products. Since almost everyone consumes fish or fish products, it follows that all the inhabitants of the polluted areas have been subjected to methylmercury exposure to a lesser or greater degree. Thus, clinical diagnosis is very difficult in the cases with sensory impairment alone. Although concentric constriction of the visual field is relatively specific to the Minamata disease, a combination of the signs and symptoms is fundamental in reaching the right diagnosis.

It is important to know that the clinical manifestations may be modified by an array of complications, since poisoning may occur in any person, at any age, and with a great variety of complications.

As in pathological findings, computed tomography (Figure 1) may demonstrate specific cerebral and cerebellar cortical atrophy. In mild cases or in cases affecting aged people, diffuse cortical atrophy or other changes due to aging make it difficult to identify the specific distribution of the atrophy.

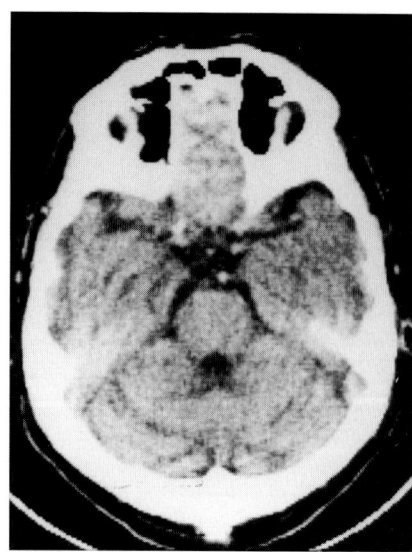

Figure 1 CT of Minamata disease in a 62-year-old male. Cerebellar atrophy is shown as the enlargement of cerebellar folia.

E. TREATMENTS
1. Removal of Mercury

It is urgent to find the route of methylmercury intake when methylmercury poisoning is suspected and to stop the intake immediately.

One should attempt to accelerate the clearance of methylmercury by means of chelating agents, dialysis, or perioral administration of thiol resin, since the biological half-time of methylmercury is too long. The mean half-time of clearance was 65 days and the range of individual values varied from 40 to 105 days in Iraq (Bakir et al., 1973). Biological half-time was assessed at 40 to 120 days in human experience by radioisotope.

The main purpose of excreting methylmercury is to prevent development of the expected symptoms. Once the symptoms have developed, satisfactory clinical improvement is not to be expected. Clarkson et al. (1981) reported a case whereby further deterioration was possibly prevented by a treatment leading to the lowering of the blood level of methylmercury. The patient, whose blood level at the start of the treatment was 3622 ng/g, was treated by thiol resin. According to the dose-response, dose-effect data by Bakir et al. (1973), she was expected to develop very severe signs and symptoms of poisoning, but no further deterioration was observed following the treatment.

Dialysis — Methylmercury exists mainly in the red blood cells of peripheral blood. Erythrocytapheresis may be a theoretically proper way to discharge methylmercury from the blood in an acute very severe case, although there has never been any report of its clinical application. Although hemodialysis is an attempt to excrete methylmercury from the blood in a severe case of an acute phase, effective removal is not expected by a conventional method since methylmercury is found almost exclusively in the red blood cells of peripheral blood and not in a diffusible form. Extracorporeal complexing hemodialysis was used clinically in an outbreak of poisoning in Iraq. It proved to be an effective procedure for removing methylmercury from the body. A small molecular weight complexing agent (L-cysteine) is infused into arterial blood entering the dialyzer in order to convert methylmercury into a diffusible form (Al-Abbasi et al., 1978). Within the hemodialyzer, both free cysteine and the methylmercury-cysteine complex, formed in the blood, diffuse across the membrane into the dialysate. Treatment of the patients by this procedure resulted in a reduction of the concentrations of mercury in the blood compartment and extravascular body compartments.

Chelating Agents — Dimercaprol (BAL) is contraindicated in methylmercury poisoning because it increases the concentration of mercury in the brain of experimental animals. D-penicillamine (PEN) and N-acetyl-DL-penicillamine (NAP) were administered in the Iraqi study. 2,3-Dimercaptopropane-1-sulfonate (DMPS) was also tested on humans exposed to methylmercury, specifically in an Iraqi patient, and proved more effective than PEN and NAP (Clarkson et al., 1981). Although they removed mercury, clinical improvement was not clearly related to the reduction of mercury (Bakir et al., 1973).

Thiol Resin — The most important pathway of methylmercury excretion is the bile, but a large fraction of the methylmercury secreted into the bile was subsequently reabsorbed from the intestinal tract. Fecal excretion of methylmercury is limited to only a small amount by this enterohepatic recirculation. Orally administered thiol resin, which is nonabsorbable, traps the methylmercury secreted in the bile, prevents its reabsorption, and thereby enhances fecal excretion (Clarkson et al., 1981). It does not result in redistribution of methylmercury to the nervous system, and any adverse effect of thiol resin itself is negligible because of its nonabsorbability. The effect of methylmercury clearance is almost equivalent to penicillamine. Thiol resin may block the physiologically necessary enterohepatic recirculation, but no adverse effect was observed.

2. Neurological Deficit

There is no satisfactorily effective therapy for the neurological deficit. Rehabilitation or physical therapy produces some results, especially in cerebellar ataxia. Physical therapy to prevent contracture is needed in a patient with paresis or severe extrapyramidal symptoms. Thyrotropin releasing hormone (TRH) exerts some positive effects on cerebellar symptoms.

F. CONGENITAL OR FETAL MINAMATA DISEASE

Methylmercury passes through the placenta into the fetus in pregnant women, thus causing fetal poisoning. It may also cause abortion when the intoxication is so severe as to terminate the pregnancy, although an unusually high incidence of abortions during the epidemic was not confirmed epidemiologically.

Fetal alkylmercury ("Panogen", a methylmercury compound) poisoning was first reported by Engleson and Herner (1952). They reported a girl with psychomotor retardation from an alkylmercury-intoxicated family. The girl was born to an asymptomatic mother.

Harada (1968) reported 22 fetally poisoned children in Minamata. An extremely high incidence of cerebral palsy was noticed in highly exposed areas in Minamata; 14 children with cerebral palsy were detected out of 190 newborn children (7.5%) in the 4 most exposed villages comprising a population of 2100 inhabitants from 1955 to 1958. The highest percentage of cerebral palsy in a village amounted to 12% of the number of births, and the average frequency was 5.8%. It was ten times as high as that of the common cerebral palsy. These 22 children were selected as confirmed cases of infantile poisoning. As for maternal abnormalities during pregnancy, minimal signs and symptoms of poisoning were observed in only five cases, and they soon disappeared except in one case. Forceps delivery was performed in one case, but the others were normal deliveries.

Clinically detectable cases of fetal methylmercury poisoning are also reported from the U.S. (Snyder, 1971) and Iraq (Amin-Zaki et al., 1974; Marsh et al., 1980).

Two of the children with fetal methylmercury poisoning in Minamata were autopsied, and neuropathological studies disclosed evidence of impaired brain development (Matsumoto et al., 1965). Generalized developmental defects such as sulcal and mild ventricular enlargement were seen in the central nervous system. The cerebral hemispheres were underdeveloped and convolutions were narrow. The cerebellum was symmetrically reduced in size and showed atrophy of its folia. Underdevelopment of the cerebral and cerebellar white matter was evident. Choi et al. (1978) reported two autopsied cases of fetal methylmercury poisoning. The main findings were abnormal migration of neurons to the cerebral and cerebellar cortices together with deranged cortical organization. A relatively short frontal lobe, an irregular pattern of gyri and sulci, or a simplified gyral pattern were also noticed in the macroscopic studies.

The clinical features of severe fetally poisoned children are the same as those for common cerebral palsy. Most of the patients were included either in the mixed form or the unclassified form. Retardation of physical and mental development was the predominant feature. Disturbance of motility and abnormal muscle tone, involuntary movements (72.7%), and excessive salivation (77.2%) were seen. Although visual acuity and hearing were relatively spared in the Japanese episodes (Harada, 1968), 4 out of 15 babies were totally blind and had severely impaired hearing in Iraq (Amin-Zaki et al., 1974).

The developing central nervous system is more sensitive to damage from methylmercury than the adult nervous system. Mothers who were slightly poisoned gave birth to infants with severe cerebral palsy. It was suspected that boys were more susceptible than girls (Marsh et al., 1987).

Pathologically, the effects of methylmercury are distributed over the whole brain, and do not show any specific distribution. This distribution contrasts well with the specific distribution in the adult type of methylmercury poisoning. Brainstem auditory evoked potentials show abnormality of the brainstem, although the brainstem itself was relatively spared in the pathological study (Hamada et al., 1982b). Computed tomography also shows diffuse cortical and subcortical atrophy (Figure 2).

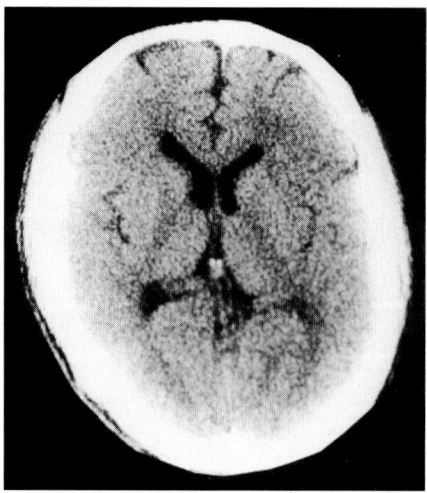

Figure 2 CT of fetal Minamata disease in a 26-year-old male. Diffuse cortical and subcortical atrophy is shown as the enlargement of cerebral fissures, cortical sulci, and ventricles.

G. RELATIONS BETWEEN DOSE– AND EFFECT–RESPONSE

Tsubaki (1968) reported that repeated analyses of hair samples from patients in Niigata showed that the lowest hair mercury concentration at the onset of symptoms was 52 µg/g. It was reanalyzed by atomic absorption and corrected to 82.6 µg/g.

In the outbreak in Iraq, dose-response relationship was clearly demonstrated. The lowest threshold body burden calculated from the linear regression line occurred when approximately 25 mg of mercury was present as methylmercury. This was associated with the onset of paresthesia. The threshold for ataxia occurred at 55 mg of mercury, for dysarthria at 90 mg, for deafness at 170 mg, and for death at 200 mg of mercury (Bakir et al., 1973). This analysis suggests the existence of a population threshold for the neurological effects of methylmercury. However, a population threshold may simply not exist.

Marsh et al. (1987) reported a dose-response relationship for fetal effects of methylmercury. For the maximum hair mercury concentration during gestation, single hair strands were measured by X-ray fluorescence spectrometry. According to the results, males were more susceptible than females. These results were statistically reanalyzed by Cox et al. (1989). Although the estimated threshold depends on the assumed background frequency, motor retardation ought to occur in children who are prenatally exposed to maternal hair concentrations of less than 50 µg/g and Hg may be expected in the range of 10 to 20 µg/g. Assuming the existence of a population threshold, the statistical best estimate places the value at about 10 µg/g, but the 95% range of uncertainty includes 0 to 13.6 µg/g.

III. INORGANIC MERCURY POISONING

A. INTRODUCTION

Inorganic mercury exists in three forms of oxidation states: (1) the metallic form (elemental mercury, Hg^0); (2) the mercurous form (monovalent cation, Hg^+, e.g., calomel); and (3) the mercuric form (divalent cation, Hg^{2+}, e.g., mercuric chloride). Elemental mercury vapor is highly diffusible and lipid soluble. Approximately 80% of the inhaled vapor is retained in humans (Clarkson, 1983). Mercury vapor following inhalation easily penetrates the alveolar membrane and rapidly distributes from the blood into the tissues, especially the nervous system, but ingested elemental mercury is poorly absorbed (less than 0.01% in rats) from the gastrointestinal tract. Elemental mercury is rapidly oxidized to the mercuric cation (Hg^{2+}) in the blood and tissues. The brain mercury levels were ten times higher than they were after equal doses of mercuric mercury were administered intravenously (WHO, 1991).

Gastrointestinal absorption of mercuric salts is less than 10% of the amount ingested, although absorption in young children may be considerably greater. After absorption, the highest concentration of Hg^{2+} is found in the kidney (50 to 90%) (WHO, 1991).

B. CHRONIC ELEMENTAL MERCURY (MERCURY VAPOR) POISONING

Occupational exposure has classically been emphasized as a hazard of chronic elemental mercury poisoning (miners, gold-smiths, mirror-makers, workers in thermometer factories or chlor-alkali plants, dentists, and scientists in laboratories). The potential hazards of dental amalgams (Clarkson, 1990) and smoking (Suzuki, 1976) are also given a warning.

Mercury contamination, caused by gold mining and refining, in the Amazon has been drawing attention recently. Mercury is used to amalgamate gold in the first phase. Part of this mercury inevitably escapes into the river. In the second phase, the gold is purified by heating the amalgam and evaporating the mercury, thus producing mercury vapor. Mercury levels in muscles and eggs from fish collected along the Amazon Basin ranged from 0.04 to 3.81 µg/g (Martinelli et al., 1988).

Gold miners, gold shop workers, and neighbors of gold shops in the Amazon have shown abnormal mercury levels in urine (total mercury, nondetectable level to 151 µg/l), blood (total mercury, 0.4 to 13 µg/dl), and symptoms of metallic mercury intoxication (dizziness, 51%; headache, 49%; palpitations, 47%; tremors, 42%). River-side residents, most of them with no previous contact with metallic mercury and its compounds, have shown high total mercury and methylmercury contamination in the hair. Although no case similar to classical Minamata disease was found, among the four persons with hair mercury levels higher than 50 µg/g three presented dysdiadochokinesia at physical examination (Branches et al., 1993).

Mercury vapor has more affinity to the nervous system than mercuric or mercurous mercury. Elemental mercury enters the nervous system before oxidation. Therefore, the toxic effects of elemental mercury are significant to the nervous system in chronic intoxications, although acute pneumonitis is the main result of acute mercury vapor poisoning. Usually the kidney is less susceptible to elemental mercury vapor than the central nervous system, but nephrotoxicity may be seen at lower exposure levels than those of central nervous system toxicity. The immunological process is suspected to play a role in nephrotoxicity. Nephrotoxicity is less severe than with the monovalent or divalent forms of chronic inorganic mercury poisoning. There is a linear relationship between plasma concentration and urinary excretion of mercury after exposure to vapor.

For the threshold of mercury vapor toxicity, Smith et al. (1970) surveyed the effects of low-dose mercury exposure in the manufacture of chlorine. Significant findings were observed in those workers exposed to time-weighted average exposures in excess of 0.1 mg/m^3. Most strongly related to excessive exposure were the symptoms of loss of appetite and weight loss, tremors in the fingers, eyelids and tongue, insomnia, shyness, and nervousness.

Subjects with urine mercury peak levels above 0.6 mg/l demonstrated significantly decreased strength, decreased coordination, increased tremors, decreased sensation, and increased prevalence of Babinski and snout reflexes when compared with the remaining subjects. Subjects with clinical polyneuropathy had significantly higher peak levels than normal subjects (Albers et al., 1988). The tremor changes were associated with those workers whose urinary mercury exceeded 0.5 mg/l in two or more months of the previous year (Langolf et al., 1981).

The initial symptoms or critical signs or symptoms are mental changes or tremors as mentioned above.

The common manifestations are mental changes, i.e., erethism (irritability, excitability, insomnia, and difficulty of concentration followed by hallucinations), excessive salivation or cessation of salivation, gingivitis, stomatitis, or contact dermatitis provided that mercury comes in direct contact with the skin. Tremor is one of the most common signs. Tremors vary from fine finger tremor to ataxic intention tremor or rotatory tremor. In addition, involuntary movements are detected in the tongue or eyelids, and choreic type movements are also observed. Thus, cerebellar ataxia (intention tremor, ataxic gait, titubation), and extrapyramidal signs (rigidity, resting tremor, other involuntary movements) are included.

Roels et al. (1985) proposed a biological threshold limit value of mercury concentration in urine of 50 µg/g creatinine for workers chronically exposed to mercury vapor. Several symptoms mainly related to the central nervous system (memory disturbances, depressive feelings, fatigue, irritability) were more prevalent in the mercury-exposed subjects. They were not, however, related to exposure parameters. Only slight renal tubular effects were detected in mercury-exposed males and females — that is, an increased urinary beta-galactosidase activity and an increased urinary excretion of retinol-binding protein.

Unlike the renal tubular effects, the preclinical signs of tremor were more related to the integrated exposure than to the current exposure.

Peripheral nerves are also affected by elemental mercury as sensorimotor neuropathy. Peripheral nerve conduction, both motor and sensory, was significantly affected in the group with urine mercury concentrations exceeding 0.5 mg/l. Low conduction velocity and low compound muscle action potential were reported (Levine et al., 1982). Levine et al. (1982) also reported against the threshold concept of toxicity of mercury — in other words, that peripheral nerves can be affected by elemental mercury even when average urine concentrations do not exceed 0.5 mg/l.

Kark et al. (1971) reported on a patient with chronic elemental mercury poisoning who had dysarthria, ataxia, hearing impairment, and constricted visual fields, i.e., the major signs of organic mercury poisoning, in addition to the cardinal signs of elemental mercury poisoning.

Mercurialentis is attributed to the absorption of mercury through the cornea and its subsequent deposition on the anterior surface of the lens (reviewed by Jeselow, 1972).

Treatment — Immediate cessation of exposure is the first step when poisoning is suspected. Chelation therapy with dimercaprol, D-penicillamine or N-acetyl-DL-penicillamine is required if exposure is confirmed by measurement of a biological sample. The newer derivatives of dimercaprol succimer (2,3-dimercaptosuccinic acid; DMSA) and DMPS appear promising for the treatment of inorganic mercury poisoning (reviewed by Klaassen, 1990).

C. CHRONIC INORGANIC MERCURY SALTS POISONING

Soap containing inorganic mercury salt was used for cosmetic needs. Mercurous chloride laxatives were also commonly used in former days. Though they are all prohibited now, they are still used.

Davis et al. (1974) reported two autopsied cases of inorganic mercury poisoning from mercurous chloride laxatives. The two patients developed dementia, erethism, colitis, and renal failure following chronic ingestion of a laxative containing calomel (240 mg of USP grade mercurous chloride daily for 6 to 25 years). The brains of both patients were small and showed loss of cerebellar granular cells. The configuration of the Purkinje cells were normal, although in some areas of the cerebellum there appeared to be a mild loss of Purkinje cells. These findings bear close similarities to what actually happens in methylmercury poisoning. Tissue mercury levels were highest in the colon (526 µg/g) and kidney (421.5 to 25.0 µg/g), and the levels from 21 areas within the brain ranged from 105.85 µg/g (inferior olive) to 0.16 µg/g (hippocampus). The highest levels were in the inferior olive, red nucleus, and choroid plexus (Wands et al., 1974).

Prenatal poisoning was also reported. The mother of the patient used soap containing mercury over a long period of time. The 3-month-old boy showed cataracta, renal tubular dysfunction, and anemia. Prenatal exposure and early postnatal exposure through lactation overlapped (Lauwerys et al., 1987).

Nephrotic syndrome may occur as a result of the exposure to inorganic mercury. Of 44 adult African women with nephrotic syndrome, 31 (70%) were using or had used skin-lightening cream which contains 10–15% aminomercuric chloride. By contrast only 11% of female general medical inpatients use these creams (Barr et al., 1972). This form of the nephrotic syndrome is usually independent of the features of systemic mercurialism, and an abnormal immune response is suspected to play a role in the pathogenesis. The findings of renal biopsies from the patients with nephrotic syndrome due to skin-lightening cream favor an immune-complex pathogenesis. Immunofluorescence showed a deposition of γ-globulin and C_3 along the glomerular basement membrane. Electron microscopic studies showed electron-dense deposits, presumably of γ-globulin (Kibukamusoke et al., 1974; Lindqvist et al., 1974).

D. ACUTE MERCURY VAPOR POISONING

Acute mercury vapor poisoning is caused by a very high concentration of elemental mercury vapor in inspired air. The main symptoms are due to acute pneumonitis. In general, the nervous system is spared, though McFarland and Reigel (1978) reported unusual cases with very acute intoxication. Six patients who had inhaled high concentrations of mercury vapor in a single brief exposure had symptoms of acute respiratory disorders with fever, chills, chest pain, and weakness. Three men had diffuse pulmonary infiltrates on chest X-ray, suggesting chemical pneumonitis. Two of the men excreted 1060 to 1160 µg/24 h of mercury in their urine. They also complained of nervousness, irritability, lack of ambition, and loss of sexual desire.

Levin et al. (1988) described four men who had symptoms of acute mercury poisoning (shortness of breath, dry cough, myalgias, wheezing and hemoptysis, fever, and laryngitis) following exposure to mercury vapor. They were attempting home gold ore purification by using a gold-mercury amalgam and sulfuric acid. A chest X-ray film showed a bilateral interstitial infiltrate pattern. Pulmonary function tests suggested a restrictive pattern, while no evidence of an obstructive disease was found. Three of the four patients required treatment with penicillamine. The 24-h urine mercury excretion was 520,169 µg/l.

In fatal cases, nephrotoxicity and neurotoxicity can be seen. A family of four was exposed to toxic levels of mercury vapor while attempting to extract silver from a mercury amalgam. Within 24 h of the incident, all occupants began having shortness of breath necessitating hospitalization. All underwent rapid deterioration with respiratory failure. Chest roentgenograms in all four cases were consistent with the adult respiratory distress syndrome. All the patients were treated with dimercaprol, but they all died within 9 to 23 days following exposure. There were no clinical signs of extrapulmonary manifestations despite toxic serum mercury levels. Although the serum mercury levels decreased in response to the mercury chelating agent dimercaprol, the serum levels remained in the toxic range and no clinical response was observed (Rowens et al., 1991). Autopsies were performed on all four patients. The lungs in all cases were heavy, firm, and airless. Histologic examination revealed severe diffuse alveolar damage, with variable amounts of fibrosis, conforming with acute lung injury in various stages of organization. Additional post-mortem findings included acute proximal renal tubular necrosis and vacuolar hepatoxicity. A spectrum of the alterations in the central nervous system is nonspecific including multifocal ischemic necrosis, gliosis, and vasculitis (Kanluen et al., 1991).

Treatment — In addition to the treatment for the pulmonary system, chelation therapy with dimercaprol or penicillamine is required to prevent nephrotoxicity or neurotoxicity. Pulmonary dysfunctions may persist unless prompt chelation therapy is performed (Levin et al., 1988).

E. ACUTE MERCURIC (INORGANIC MERCURY SALTS) POISONING

Suicide or accidental use of mercuric chloride are the most frequent causes of acute inorganic mercury poisoning. The main target organs are the kidney and the nervous system. The initial manifestation of acute mercuric salt (mercuric chloride) poisoning originates from the mucosa attached to the mercuric salt. Erosion or ulcers in the oral mucosa, esophagus, and gastrointestinal tract induced epigastric pain, vomiting, colic, severe bloody diarrhea followed by gingivitis, and excessive salivation or fetid odor (metallic smell). These are the first signs and symptoms. This gastrointestinal damage induces dehydration together with an imbalance of electrolytes. It may cause circulatory collapse. Inorganic mercury salts are readily absorbed from the gastrointestinal tract. Acute renal failure will follow. Oliguria or anuria will be observed as the results of dehydration and renal failure. Renal failure is directly due to the toxicity of the renal tubules. Fever and leukocytosis are also seen.

Irritability, delirium, hallucinations or weakness of the lower limbs may follow. If the patient survives, esophageal, gastric, or intestinal stenosis may occur.

Longitudinal analysis of hair revealed a peak in inorganic mercury corresponding to the time of mercury ingestion (Suzuki et al., 1992).

Treatment — Removal of ingested mercury by gastric lavage or emesis together with treatment for shock and renal failure are required as emergency therapy. Suitable replacement of fluid, electrolytes, or serum albumin is urged against circulatory collapse. Dimercaprol is used to remove the mercury for high-level exposure or symptomatic patients. The dimercaprol-mercury chelate is excreted into both bile and urine, whereas the penicillamine-mercury chelate is excreted only into urine. Thus, penicillamine should be used with extreme caution when the renal function is impaired (Klaassen, 1990). Hemodialysis is applied for renal failure and also to accelerate the discharge of the mercury-dimercaprol complex. A woman who ingested a lethal dose of sublimate survived and recovered in response to a combination of therapies, including dimercaprol, plasma exchange, hemodialysis, and peritoneal dialysis (Suzuki et al., 1992).

IV. ACRODYNIA

Acrodynia is a syndrome which consists of autonomic nerve dysfunction and disturbance of the central nervous system. The peripheral nervous system may also be involved. Manifestations arising from the central nervous system include personality changes, irritability, apathy, photophobia, and insomnia. A generalized rash and pink or red hands and feet are the predominant signs. The feet become cold, moist, painful, and later swollen (Kark, 1979). Renal failure may also occur.

Acrodynia was known as a common syndrome of chronic inorganic mercury poisoning in infants or children in the early twentieth century when mercury was widely used without fear of poisoning. After the withdrawal of mercurous chloride in teething powder by the leading United Kingdom manufacturers in 1953, there was a dramatic decline in the occurrence of acrodynia (WHO, 1991). Although the hazards of mercury poisoning are widely known today, cases with acrodynia are still reported.

Agocs et al. (1990) reported the case of a 4-year-old boy with acrodynia which occurred 10 days after the inside of his home was painted with interior latex paint containing phenylmercuric acetate. Leg cramps, a generalized rash, pruritus, sweating, tachycardia, an intermittent low-grade fever, marked personality changes, erythema, and desquamation of the hands, feet, and nose, weakness of the pelvic and pectoral girdles, and lower-extremity nerve dysfunction developed sequentially. A 24-h urine sample contained 324 nmol/l Hg. Potentially hazardous exposure to mercury had occurred among persons whose homes were painted with the same brand of latex paint. In Argentina, three infants with acrodynia were the clue to the massive exposure of infants to phenylmercury. Phenylmercury was absorbed through the skin from contaminated diapers, which had been treated with phenylmercury fungicide (Gotelli et al., 1985). Clarkson (1990) stressed that if one case is diagnosed, it is highly probable that many more people have been exposed. Mercury dispersed from broken fluorescent bulbs and long-term injection of gamma-globulin preserved with ethylmercurithiosalicylate have also been responsible for acrodynia (WHO, 1991).

Neither the occurrence of acrodynia nor its severity was dose related (WHO, 1991). Immunological effects are suspected to contribute to the etiology of acrodynia.

Treatment — Chelating agents should be used. Nifedipine, a Ca-channel blocker, was reported as useful for the treatment of acrodynia (Özsoylu et al., 1989). Following administration of nifedipine, the pain, restlessness, and pinkish color all disappeared. Ambudkar et al. (1988) suggested that $HgCl_2$-induced renal cell damage involves the influx of Ca^{2+} from the extracellular milieu which makes the progression of cellular injury possible. Thus, it is possible that the Ca-channel blocker acted not only as a vasodilator, but also as a cytoprotector.

REFERENCES

Agocs, M. M., Etzel, R. A., Parrish, R. G., Paschal, D. C., Campagna, P. R., Cohen, D. S., Kilbourne, E. M., and Hesse, J. L., *N. Engl. J. Med.*, 323, 1096–1101, 1990.
Al-Abbasi, A. H., Kostyniak, P. J., and Clarkson, T. W., *J. Pharmacol. Exp. Ther.*, 207, 249–254, 1978.
Albers, J. W., Kallenbach, L. R., Fine, L. J., Langolf, G. D., Wolfe, R. A., Donofrio, P. D., Alessi, A. G., Stolp-Smith, K. A., Bromberg, M. B., and the Mercury Workers Study Group, *Ann. Neurol.*, 24, 651–659, 1988.
Ambudkar, I. S., Smith, M. W., Phelps, P. C., Regec, A. L., and Trump, B. F., *Toxicol. Ind. Health*, 4, 107–123, 1988.
Amin-Zaki, L., Elhassani, S., Majeed, M. A., Clarkson, T. W., Doherty, R. A., and Greenwood, M., *Pediatrics*, 54, 587–595, 1974.
Bakir, F., Damluji, S. F., Amin-Zaki, L., Murtadha, M., Khalidi, A., Al-Rawi, N. Y., Tikriti, S., Dhahir, H. I., Clarkson, T. W., Smith, J. C., and Doherty, R. A., *Science*, 181, 230–241, 1973.
Barr, R. D., Rees, P. H., Cordy, P. E., Kungu, A., Woodger, B. A., and Cameron, H. M., *Br. Med. J.*, 2, 131–134, 1972.
Branches, F. J. P., Harada, M., Akagi, H., Malm, O., Kato, H., and Pfeiffer, W. C., Human mercury contamination as a consequence of goldmining activity in the Tapajos River Basin, Amazon, Brazil, in Proc. Int. Symp. "Assessment of Environmental Pollution and Health Effects From Methylmercury", WHO and National Institute for Minamata Disease, Kumamoto, Japan, 19, 1993.
Choi, B. H., Lapham, L. W., Amin-Zaki, L., and Saleem, T., *J. Neuropathol. Exp. Neurol.*, 37, 719–733, 1978.
Clarkson, T. W., Magos, L., Cox, C., Greenwood, M. R., Amin-Zaki, L., Majeed, M. A., and Al-Damluji, S. F., *J. Pharmacol. Exp. Ther.*, 218, 74–83, 1981.
Clarkson, T. W., *Ann. Rev. Public Health*, 4, 375–380, 1983.
Clarkson, T. W., *N. Engl. J. Med.*, 323, 1137–1139, 1990.
Cox, C., Clarkson, T. W., Marsh, D. O., Amin-Zaki, L., Tikriti, S., and Myers, G. G., *Environ. Res.*, 49, 318–332, 1989.
Damluji, S., *J. Fac. Med. Baghdad*, 4, 83–103, 1962.
Davis, L. E., Wands, J. R., Weiss, S. A., Price, D. L., and Girling, E. F., *Arch. Neurol.*, 30, 428–431, 1974.
Engleson, G. and Herner, T., *Acta Paediatr.*, 41, 289–294, 1952.
Gotelli, C., Astolfi, E., Cox, C., Cernichiari, E., and Clarkson, T. W., *Science*, 227, 638–640, 1985.
Hamada, R., Igata, A., and Yanai, H., *Saishin Igaku (Jpn.)*, 33, 98–99, 1978a.
Hamada, R., Igata, A., and Yanai, H., *Saishin Igaku (Jpn.)*, 33, 62–63, 1978b.
Hamada, R., Yoshida, Y., Kurosu, T., Nakashima, H., and Igata, A., *Autonomic Nerv. Sys. (Jpn.)*, 19, 283–286, 1982a.
Hamada, R., Yoshida, Y., and Kuwano, A., *Neurol. Med. (Jpn.)*, 16, 283–285, 1982b.
Haq, I. U., *Br. Med. J.*, 1, 1579–1582, 1963.
Harada, Y., in *Minamata Disease*, Kumamoto University, Kumamoto, Japan, 93, 1968.
Hunter, D., Bomford, R. R., and Russell, D. S., *Q. J. Med.*, 9, 193–213, 1940.
Hunter, D. and Russell, D. S., *J. Neurol. Neurosurg. Psychiatr.*, 17, 235–241, 1954.
Igata, A., in *Recent Advances in Minamata Disease Studies*, Tsubaki, T. and Takahashi, H., Eds., Kodansha, Tokyo, 41, 1986.

Ino, H., Kato, I., Ohno, Y., Ishikawa, K., and Mizukoshi, K., Otorhinolaryngological aspects of intoxication by organic mercury compounds, in Studies on the Health Effects of Alkylmercury in Japan, Environment Agency of Japan, Tokyo, 186, 1985.
Iwata, K., Nanba, K., Kojima, M., and Abe, H., Neuroophthalmological findings of organic mercury poisoning, "Minamata Disease" in Niigata Prefecture, in Studies on the Health Effects of Alkylmercury in Japan, Environment Agency of Japan, Tokyo, 202, 1975.
Jalili, M. A. and Abbasi, A. H., *Br. J. Ind. Med.*, 18, 303–308, 1961.
Joselow, M. M., Louria, D. B., and Browder, A. A., *Ann. Int. Med.*, 76, 119–130, 1972.
Kanluen, S. and Gottlieb, C. A., *Arch. Pathol. Lab. Med.*, 115, 56–60, 1991.
Kark, R. A. P., Poskanzer, D. C., Bullock, J. D., and Boylen, G., *N. Engl. J. Med.*, 285, 10–16, 1971.
Kark, R. A. P., in *Handbook of Clinical Neurology*, Vinken, P.J. and Bruyn, G.W., Eds., Elsevier/North-Holland, Amsterdam, 147, 1979.
Kibukamusoke, J. W., Davies, D. R., and Hutt, M. S. R., *Br. Med. J.*, 2, 646–647, 1974.
Klaassen, C. D., in *Goodman and Gilman's The Pharmacological Basis of Therapeutics*, Gilman, A.G., Rall, T.W., Nies, A.S., and Taylor, P., Eds., Pergamon Press, New York, 1592, 1990.
Langolf, G. D., Smith, P. J., Henderson, R., and Whittle, H. P., *Ann. Occup. Hyg.*, 24, 293–296, 1981.
Lauwerys, R., Bonnier, C., Evrard, P., Gennart, J. P., and Bernard, A., *Hum. Toxicol.*, 6, 253–256, 1987.
Levin, M., Jacobs, J., and Polos, P. G., *Chest*, 94, 554–556, 1988.
Levine, S. P., Cavender, G. D., Langolf, G. D., and Albers, J. W., *Br. J. Ind. Med.*, 39, 136–139, 1982.
Lindqvist, K. J., Makene, W. J., Shaba, J. K., and Nantulya, V., *East Afr. Med. J.*, 51, 168–169, 1974.
Marsh, D. O., Myers, G. J., Clarkson, T. W., Amin-Zaki, L., Tikriti, S., and Majeed, M. A., *Ann. Neurol.*, 7, 348–353, 1980.
Marsh, D. O., Clarkson, T. W., Cox, C., Myers, G. J., Amin-Zaki, L., and Al-Tikriti, S., *Arch. Neurol.*, 44, 1017–1022, 1987.
Martinelli, L. A., Ferreira, J. R., Forsberg, B. R., and Victoria, R. L., *Ambio*, 17, 252–254, 1988.
Matsumoto, H., Koya, G., and Takeuchi, T., *J. Neuropathol. Exp. Neurol.*, 24, 563–574, 1965.
McFarland, R. B. and Reigel, H., *J. Occup. Med.*, 20, 532–534, 1978.
Nagaki, J., Abnormalities in sural nerve action potentials recorded from methyl mercury poisoned patients, in Studies on the Health Effects of Alkylmercury in Japan, Environment Agency of Japan, Tokyo, 182, 1975.
Okajima, T., Mishima, I., and Tokuomi, H., *Int. J. Neurol.*, 11, 62–72, 1976.
Ordòñez, J. V., Carrillo, J. A., Miranda, M., and Gale, J. L., *Bol. Of. Sanit. Panam.*, 60, 510–519, 1966.
Özsoylu, S., Sarikayalar, F., and Aksoy, A., *Turk. J. Pediatr.*, 31, 159–161, 1989.
Roels, H., Gennart, J. P., Lauwerys, R., Buchet, J. P., Malchaire, J., and Bernard, A., *Am. J. Ind. Med.*, 7, 45–71, 1985.
Rowens, B., Guerrero-Betancourt, D., Gottlieb, C. A., Boyes, R. J., and Eichenhorn, M. S., *Chest*, 99, 185–90, 1991.
Rustam, H. and Hamdi, T., *Brain*, 97, 499–510, 1974.
Smith, R. G., Vorwald, A. J., Patil, L. S., and Mooney, T. F., Jr., *Am. Ind. Hyg. Assoc. J.*, 31, 687–700, 1970.
Snyder, R. D., *N. Engl. J. Med.*, 284, 1014–1016, 1971.
Suzuki, T., Shishido, S., and Urushiyama, K., *Tohoku J. Exp. Med.*, 119, 353–356, 1976.
Suzuki, T., Hongo, T., Matsuo, N., Imai, H., Nakazawa, M., Abe, T., Yamamura, Y., Yoshida, M., and Aoyama, H., *Hum. Exp. Toxicol.*, 11, 53–57, 1992.
Takeuchi, T., in *Minamata Disease*, Kumamoto University, Kumamoto, Japan, 141, 1968.
Takeuchi, T. and Eto, K., Minamata disease; chronic occurrence from pathological viewpoints. In Studies on the Health Effects of Alkylmercury in Japan, Environment Agency of Japan, Tokyo, 28, 1975.
Tokuomi, H., in *Minamata Disease*, Kumamoto University, Kumamoto, Japan, 37, 1968.
Tokuomi, H. and Okajima, T., *Adv. Neurol. Sci. (Jpn.)*, 13, 69–75, 1969.
Tsubaki, T., *Clin. Neurol. (Jpn.)*, 8, 511–520, 1968.
Wands, J. R., Weiss, S. W., Yardley, J. H., and Maddrey, W. C., *Am. J. Med.*, 57, 92–101, 1974.
WHO, Environmental Health Criteria 101: Methylmercury, World Health Organization, Geneva, 1990.
WHO, Environmental Health Criteria 118: Inorganic Mercury, World Health Organization, Geneva, 1991.

Chapter 21

Itai-Itai Disease and Health Effects of Cadmium

Koji Nogawa and Teruhiko Kido

I. INTRODUCTION

Itai-itai disease has been endemic among elderly women in the Jinzu River basin in Toyama Prefecture since World War II. In 1955, Dr. Noboru Hagino first reported his studies to an academic society and called this syndrome itai-itai disease. The Japanese word "itai" means "ouch" or "painful" in English. The pain results from unusual changes in bone, and the administration of large doses of vitamin D has been demonstrated to be effective in alleviating it. It was not until 1962 that extensive medical research into the itai-itai disease was performed in this area.

II. CLINICAL PICTURE OF ITAI-ITAI DISEASE

The clinical picture of itai-itai disease has been described by many researchers and research groups. The main symptoms and laboratory findings can be summarized as follows.

A. SYMPTOMS AND SIGNS

Most of the patients are postmenopausal women. Femoral pain and lumbago are frequently seen as the initial manifestations, after which painful sites gradually spread all over the body. A duck-like gait is characteristic of the disease. These conditions continue for several years, after which patients are finally confined to bed. The clinical condition deteriorates rapidly once the patient is bedridden. Bone fractures can be caused by the slightest external pressure, such as coughing, and skeletal deformation takes place. Because of the severe pain, patient can't sleep and even respiratory movement is restricted. The patient is undernourished and small, and in severe cases body height is reduced from the normal level by 30 cm. In 1978, Katoh et al. (1978) investigated the symptoms of 47 itai-itai patients; 44 patients (94%) always felt pain, 46 (98%) had difficulty in walking and 5 (11%) were confined to bed.

B. LABORATORY EXAMINATIONS
1. Blood Findings

Nakagawa (1960) described 30 patients from the Jinzu River basin who were studied in 1955 to 1958. Blood examinations in his study showed a decrease of inorganic phosphorus in serum in most patients, while an increase of alkaline phosphatase in serum was observed in all patients. The levels of calcium in serum were within the normal or low-normal range. The levels of total protein, nonprotein nitrogen, albumin, globulin, and albumin to globulin ratio in serum were also within the normal ranges. The carbon dioxide content in serum and the alkali reserve in blood were low. The levels of chloride in serum were slightly high. These findings demonstrated the presence of metabolic acidosis. Slight anemia was also observed in most patients.

In 1975, Nogawa et al. (1979e) analyzed blood specimens from 39 itai-itai disease patients, 53 suspected patients (persons with a history of residence in a cadmium-polluted area and with renal tubular dysfunction, but no significant findings of osteomalacia by X-ray examination), and 21 control subjects. Itai-itai disease patients showed low levels of plasma inorganic phosphorus, potassium, total protein, and uric acid. They also showed high levels of sodium, chloride, GOT, alkaline phosphatase, LDH, BUN, and creatinine. Erythrocyte counts and hemoglobin values were decreased in the itai-itai disease patients. No significant differences in the blood findings other than alkaline phosphatase and total protein were observed between the itai-itai disease patients and the suspected patients. In 1976 and 1977, a study group organized by the Japan Environment Agency performed clinical examinations in 53 itai-itai disease patients and 16 suspected patients who were admitted to hospital for a week (Shinoda and Yuri, 1978). The results of blood examinations were almost the same as those in 1975. Blood gas analysis showed decrease of pH, bicarbonate (HCO_3^-), and base excess in the itai-itai disease patients and in the suspected ones (Figure 1).

2. Urinary Findings and Renal Function Tests

All patients examined by Nakagawa (1960) in 1955 to 1958 had slight proteinuria and aminoaciduria, and some had glucosuria. Although urinary excretion of calcium was normal, excretion of phosphorus was decreased. On the PSP excretion test, seven of ten patients showed values between 10–22.5% and three 25% (the lower normal limit was considered to be 25% after 15 min). Clearance tests were performed in six patients. Urea clearance and glomerular filtration rates were within the normal ranges. PAH clearance showed slightly low values. Tubular reabsorptive mass (Tm) was examined in four patients, all of whom showed low values. Nakagawa concluded that renal functions, especially tubular function, were decreased in the patients. In 1968, Takeuchi et al. (1968) examined four patients. The PSP excretion was 6 to 12% after 15 min. The concentration test, dilution test, and phosphorus reabsorption test (% TRP) showed slight kidney damage in all four cases.

In 1975, Nogawa et al. (1979d) analyzed urine samples collected from 45 itai-itai disease patients, 71 suspected patients, and 81 control subjects. The itai-itai disease patients had high excretions of total protein, glucose, amino-N, proline, calcium, and cadmium (Cd) and low molecular weight proteins such as β_2-microglobulin (β_2-mg), retinol binding protein (RBP), and lysozyme. The Ca/P ratio in urine was highest in the patients with itai-itai disease. There were no significant differences in the urinary findings between the itai-itai disease patients and suspected patients.

In a 1976 and 1977 clinical study which was performed by the study groups organized by Japan Environment Agency, the results of urine analysis were almost the same as those in 1975 (Honda et al., 1978). As for renal function tests, mean values of creatinine clearance and % TRP in 53 itai-itai disease patients (46–87 years old) were 41.9 ml/min and 56.5%, respectively (Shinoda and Yuri, 1978). Both renal tubular and glomerular dysfunctions were observed in almost all patients.

Urine specimens were collected from 40 itai-itai disease patients, 17 suspected patients, and 10 controls, and analyzed for individual free amino acids (Nogawa et al., 1980). Generalized aminoaciduria was found in both the itai-itai disease patients and suspected patients. The amino acid pattern was very similar in the three groups. Endogenous renal transport of free amino acids was also determined in 6 suspected patients, 13 women with Cd-induced renal damage, and 3 control subjects (Kobayashi et al., 1981c). The clearance of most individual amino acids was much higher and the percentage of tubular reabsorption of most amino acids was characteristically lower in the Cd-exposed groups than in the control group.

Tohyama et al. (1982) determined metallothionein (MT) levels for the first time in the urine of itai-itai disease patients, suspected patients, and other Japanese women environmentally exposed to Cd. On a group basis, the urinary MT levels of the itai-itai disease patients and suspected patients were significantly higher than those of women living in a Cd-polluted area. Women in a Cd-nonpolluted area excreted significantly less MT than women living in Cd-polluted areas.

3. Radiographs of Bone

The characteristic radiological findings of bone are osteomalacia and marked decalcification. In the 30 cases examined by Nakagawa (1960), Looser's zones (a narrow radiolucency which transects one or both cortical margins of a bone and is a certain indication of osteomalacia) were found in all cases and

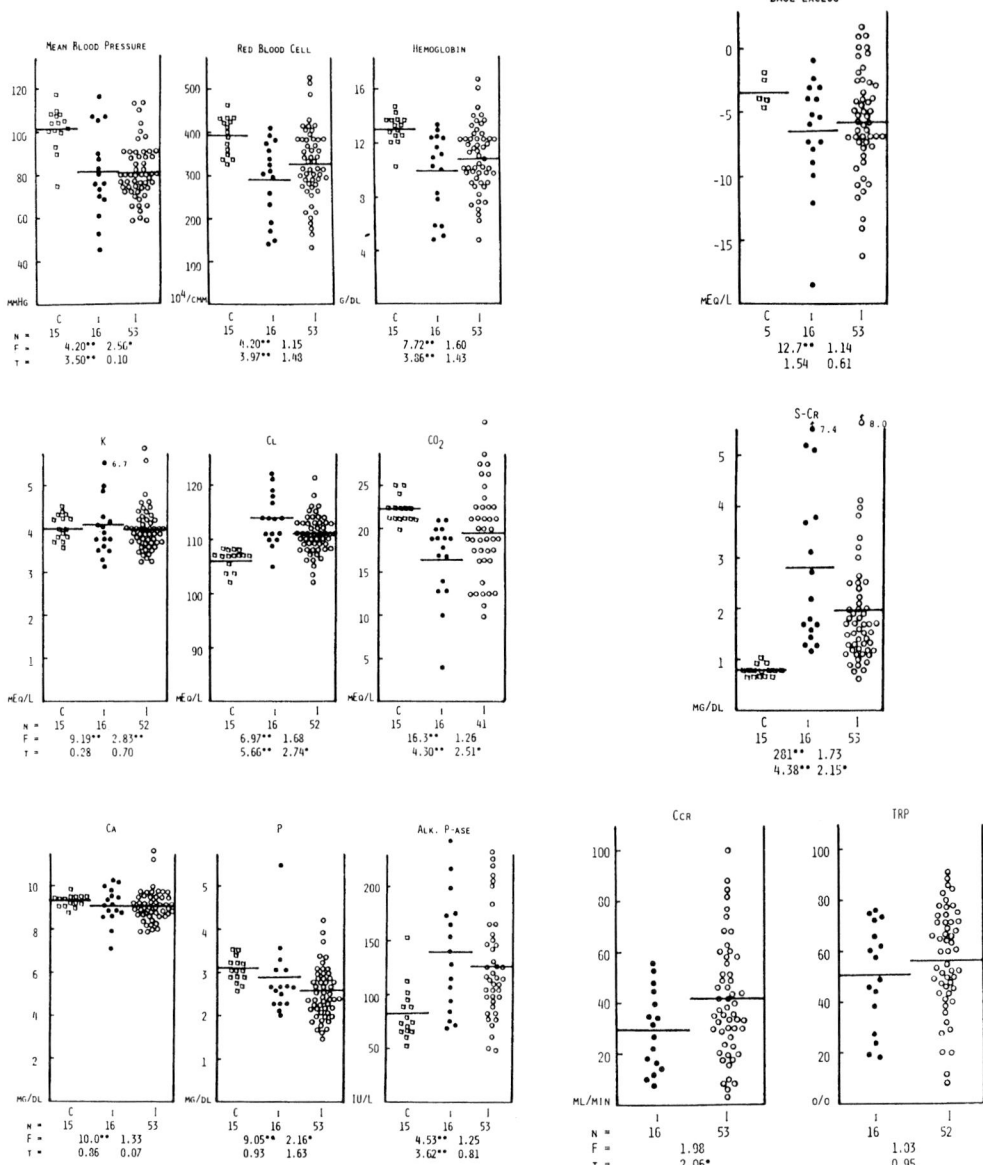

Figure 1 Laboratory findings on itai-itai disease patients (I), suspected patients (i), and controls (C). (From Shinoda, S. and Yuri, K., *Kanyo Hoken Rep.*, 44, 119, 1978. With permission.)

appeared in 342 sites. They were frequently noted at sites where pressure caused pain. Marked bone atrophy was observed throughout the skeleton. Deformities of the frame were frequently seen in pelvic bone, costae, and thoracic and lumbar vertebrae. In 1977, Nakagawa et al. (1977) performed roentgenological examinations of the bones of 53 itai-itai disease patients. They reported that most of the patients showed spine deformities; in particular, kyphosis and scoliosis were observed in 69.8 and 49.1% of the patients, respectively. Although Looser's zones were found in 32.1% of the patients, mostly in the femur, the zones were either in the process of healing or were very small. Pseudofracture was found in 9.4% of the patients. Osteoporotic changes indicated by trabecular pattern were observed in almost all cases. Tohda et al. (1978) also performed roentgenological examinations of bones in 16 suspected patients in 1978 and found Looser's zones in 5 cases (31%).

4. Pathological Findings

The principal pathological changes are observed in bones and kidneys. The findings in bones are consistent with those found in osteomalacia. Osteoporosis is also found in combination with osteomalacia. Nakagawa (1960) performed bone biopsies in seven cases and concluded that they had osteomalacia. Kajikawa et al. (1974) reported 11 autopsied cases from 1955 to 1971. Two cases had osteitis fibrosa and two had osteoporosis without osteomalacia. However, all other cases showed combined findings of osteomalacia and osteoporosis. Kitagawa and Yamasita (1993) found Looser's zones in 36 (47%) of 71 autopsied itai-itai disease patients from 1979 to 1991.

Although the kidneys were highly contracted, there were no obvious changes in the glomeruli. Tubules, however, showed so-called tubulopathy consisting of marked atrophy, degeneration, and dilatation. Takeuchi et al. (1968) found atrophic changes in the tubules without marked changes in the glomeruli in three biopsy specimens of itai-itai disease patients. Kajikawa et al. (1974) reported tubular changes in 5 of 11 cases up to 1971. In 73 autopsied cases (1979–1990) studied by Kitagawa, almost all cases showed tubular changes (Kitagawa and Yasuda, 1992). It is thought that both the clinical and pathological changes are in accordance with the features of an acquired Fanconi syndrome.

5. Treatment

The long-term administration of large doses of vitamin D is effective in the treatment of the bone symptoms. Nakagawa (1960) treated 30 itai-itai disease patients with vitamin D and followed them for 3 years. Outpatients were given vitamin D liquid (20,000 IU/day = calciferol 0.5 mg) and inpatients received i.m. injections of 100,000 IU every day for 2 to 3 months. After this, the patients received i.m. injections of 100,000 IU/day every second day plus 10,000 IU/day as a liquid. Pain decreased after about 3 weeks of treatment in the inpatients and after 2 months of treatment in the outpatients, but impaired mobility persisted for a long time. Looser's zones disappeared after 6 months of treatment in slight cases, but in severe cases persisted even after treatment for 3 years. Abnormal biochemical findings such as low inorganic phosphorus and high alkaline phosphatase activity slowly returned to normal values.

Hagino (1973) treated 83 itai-itai disease patients and 48 suspected patients with vitamin D from 1967 to 1971. The total doses of vitamin D administered were 2,600,000 to 138,800,000 IU to the itai-itai disease patients and 1,600,000 to 80,200,000 IU to the suspected patients. Both groups of patients showed improvement of subjective symptoms, radiographical findings of bone, and biochemical findings of blood. No significant changes were observed in urinary protein or glucose excretion by the administration of vitamin D. Ten patients complained of poor appetite, and two patients showed vomiting after the treatment of three to five months, but these symptoms soon resolved after discontinuation of the treatment.

Kasuya et al. (1983a) administered 1 α-hydroxycholecalciferol (1α-OH-D_3) to two itai-itai disease patients and seven suspected patients from 1980 to 1982 and investigated changes in subjective symptoms, bone, and blood findings. This treatment was demonstrated to be effective in seven cases, but the degree of improvement was almost equal to that obtained by vitamin D treatment.

6. Heavy Metal Concentrations in Blood, Urine, and Tissues

The earliest data on urinary Cd concentrations of itai-itai disease patients, suspected patients, and inhabitants in endemic areas were reported by Ishizaki et al. (1965). High urinary Cd concentrations were found in the patients and inhabitants of the endemic areas. In 1975, Nogawa et al. (1979d) measured the urinary Cd concentrations of 44 itai-itai disease patients, 66 suspected patients, and 62 control subjects and found mean urinary Cd concentrations of 24.6, 30.2, and 7.1 µg/g creatinine, respectively. All were females and mean ages were 69.6 years for the itai-itai disease patients, 67.9 years for the suspected patients, and 59.5 years for the control subjects. Median values of urinary Cd concentrations in 53 itai-itai disease patients and 16 suspected patients were 20.4 µg/g creatinine and 18.4 µg/g creatinine, respectively, in the 1976–1977 examination (Honda et al., 1978).

The blood Cd concentrations in nine itai-itai disease patients were measured from 1980 to 1982 at least once a month (Kasuya et al., 1983b). The arithmetic mean of each subject was in the range of 10.7–46.7 µg/l and clearly higher than the reference values (less than 10 µg/l). Geometric means of blood Cd concentrations in 37 suspected patients and control subjects examined in 1981 were reported to be 11.8 and 4.3 ng/g, respectively (Shinmura et al., 1984). Shinmura et al. (1984) also measured blood concentrations of iron (Fe), copper (Cu), zinc (Zn), and selenium (Se), and serum concentrations of Fe, Cu, and Zn in the same subjects. Blood concentrations of Fe and Se were significantly lower in the suspected patients than in the control subjects. However, significant differences were not found in the serum concentration of the metals between the suspected patients and the control subjects. Kobayashi (1981a; 1981b) determined Zn and Cu concentrations in serum and urine samples from 19 itai-itai disease patients, 21 suspected patients, and 31 control subjects. Cu concentrations in the urine of itai-itai disease patients and suspected patients were much higher than those of the control subjects. Zn concentrations in the urine of the patients, however, were not different from those of the control subjects. Zn levels in the serum of the itai-itai disease patients were somewhat lower than those of the control subjects. Serum Cu was almost equal in the patients and the control subjects.

Ishizaki et al. (1971) first reported Cd and Zn concentrations in organs obtained from five autopsied itai-itai disease patients. The liver Cd concentrations were in the range of 63–132 µg/g wet weight and five to ten times higher in the itai-itai disease patients than in similarly aged controls. The Cd concentrations in the kidneys of the patients, however, were lower than those of the controls. All of Cd values of the other organ were higher in the itai-itai disease patients. The authors contended that the low kidney values in the itai-itai disease patients could be explained by the advanced kidney damage. Nogawa et al. (1986a) determined the Cd concentrations in the kidneys and livers of 51 persons from Cd-polluted areas (18 itai-itai disease patients, 28 suspected patients, and five inhabitants of Cd-polluted areas) aged between 61 and 94 years, and 122 controls aged between 3 and 90 years. The mean Cd concentrations for the Cd-exposed group were 35.2 µg/g wet weight for kidney cortex and 66.7 µg/g wet weight for liver. Corresponding values for the control group aged over 60 were 90.1 and 10.7 µg/g wet weight, respectively. Cu and Zn were also determined in kidney and liver from the same subjects. In the liver, Zn levels were significantly higher in the Cd-exposed group, and Cu levels lower. On the other hand, the levels of these metals were significantly lower in the kidney cortex of the Cd-exposed group. In the kidney medulla, such a significant difference was not found. Kuzuhara et al. (1992) measured Cd, Cu, lead (Pb), and Zn concentrations in 9 organs from 74 cases of itai-itai disease and suspected ones, and 50 control subjects. Cd concentrations of all organs except for kidney, were significantly higher in the patient group (Table 1).

Se concentrations were determined in the kidney and liver of 21 Cd-exposed (8 itai-itai disease patients and 13 suspected patients) and 15 nonexposed autopsied subjects (Kido et al., 1988a). In the liver, the geometric means of Se concentrations in the Cd-exposed subjects were 0.35 µg/g wet weight and did not show any significant differences from those of the control subjects. In the kidney cortex and medulla, Se concentrations were 0.30 and 0.23 µg/g wet weight, respectively and were significantly lower than those of the controls.

Fe concentrations in liver in the itai-itai disease patients and suspected patients were not significantly different from those of the control subjects (Nogawa et al., 1984).

7. Prognosis

Although the administration of large doses of vitamin D improves subjective symptoms, especially pain and bone damage, a continuous deterioration of renal function was observed in long-term follow-up studies. Murata et al. (1972) administered serial laboratory examinations such as the PSP excretion test, urea clearance test, and blood urea-nitrogen to five itai-itai disease patients, and documented decreasing renal functions. They stated that the deterioration of renal function was caused by the administration of large doses of vitamin D. Serum creatinine was measured regularly in 67 suspected patients from 1972 to 1980 (Shiroishi et al., 1984). In 1972, the suspected patients with serum creatinine levels more than 2.0 mg/dl comprised 12.5% of 64 subjects. This percentage gradually increased year by year and reached 54.1% of 61 subjects in 1980. The maximum value of serum creatinine was 7.2 mg/dl. Nogawa et al. (1979f) analyzed urinary protein levels in 39 itai-itai disease patients and 26 suspected patients in 1967 and 1975, and demonstrated that the excretion of protein was significantly increased in 1975 as compared with 1967 in both groups. Similar studies in other Cd-polluted areas showed similar deterioration of renal function in the Cd-exposed inhabitants (Kido et al., 1990b; Harada et al., 1992). Therefore, it is thought that renal function progressively worsens following exposure to environmental Cd, with some patients eventually dying of uremia.

Nakagawa et al. (1990) conducted a follow-up study from 1967 to 1987 on itai-itai disease patients, suspected patients, and control subjects. Per category, 95 subjects were selected after matching for age, sex, and residence area. The cumulative survival rate of the itai-itai disease patients was significantly lower than that of the control group in every period after the first 3 years. The cumulative survival rate in the suspected patient group was virtually the same as that in the control group during the first 10 years, but after 12 years it tended to decrease, and was significantly lower after 18 years. The rate in the suspected subject group was generally higher than that in the itai-itai disease patient group (Figure 2).

Table 1 Cadmium, Copper, Lead, and Zinc Concentrations (μg/g Wet Weight) in Tissues of Cadmium-Exposed and Nonexposed People

	Cadmium (M ± SD)		Copper (M ± SD)		Lead (M ± SD)		Zinc (M ± SD)	
	Exposed	Nonexposed	Exposed	Nonexposed	Exposed	Nonexposed	Exposed	Nonexposed
Kidney cortex	40.3 ± 25.7 / 9.7 ~ 128[c]	84.4 ± 47.7[b] / 5.7 ~ 206	1.65 ± 0.94 / 0.70 ~ 7.87	2.76 ± 1.89[b] / 0.77 ~ 9.44	0.12 ± 0.09[b] / 0.03 ~ 0.54	0.09 ± 0.06 / 0.03 ~ 0.37	32.0 ± 10.7 / 18.0 ~ 81.6	58.4 ± 26.4[b] / 19.3 ~ 147
Kidney medulla	28.3 ± 14.7 / 9.8 ~ 80.1	38.5 ± 22.4[b] / 3.8 ~ 107	1.49 ± 0.73 / 0.59 ~ 5.44	2.02 ± 0.98[b] / 0.67 ~ 5.46	0.11 ± 0.05[b] / 0.04 ~ 0.32	0.08 ± 0.05 / 0.02 ~ 0.24	24.2 ± 6.6 / 9.8 ~ 44.3	36.3 ± 16.1[b] / 14.6 ~ 90.5
Liver	77.0 ± 54.3[b] / 6.5 ~ 301	11.7 ± 10.4 / 0.8 ~ 48.7	4.85 ± 3.52 / 0.85 ~ 22.2	6.87 ± 5.03[a] / 1.49 ~ 30.9	0.52 ± 0.67[b] / 0.04 ~ 3.80	0.15 ± 0.10 / 0.05 ~ 0.53	124 ± 45.4[b] / 49.3 ~ 261	80.9 ± 47.2 / 17.1 ~ 227
Pancreas	57.0 ± 27.1[b] / 17.2 ~ 147	10.8 ± 10.1 / 1.8 ~ 58.6	1.38 ± 0.59 / 0.30 ~ 3.08	1.39 ± 0.55 / 0.70 ~ 3.20	0.13 ± 0.11[a] / 0.03 ~ 0.68	0.09 ± 0.06 / 0.02 ~ 0.36	61.3 ± 15.1[b] / 30.4 ~ 132	44.0 ± 13.8 / 18.5 ~ 74.7
Thoracic muscle	13.0 ± 6.2[b] / 4.1 ~ 38.0	1.5 ± 1.4 / 0.2 ~ 7.7	0.81 ± 0.24 / 0.35 ~ 1.65	0.82 ± 0.24 / 0.45 ~ 1.55	0.05 ± 0.04 / 0.01 ~ 0.20	0.05 ± 0.04 / 0.01 ~ 0.23	61.7 ± 15.0 / 30.4 ~ 107	57.7 ± 12.4 / 36.9 ~ 89.2
Thyroid	62.5 ± 38.4[b] / 8.2 ~ 185	10.0 ± 7.5 / 1.7 ~ 40.8	1.07 ± 0.30 / 0.24 ~ 1.73	0.99 ± 0.33 / 0.41 ~ 2.07	0.16 ± 0.19 / 0.02 ~ 1.16	0.12 ± 0.30 / 0.02 ~ 2.11	72.0 ± 22.6[b] / 34.9 ~ 175	48.0 ± 14.0 / 15.1 ~ 88.1
Ovary	19.7 ± 9.3[b] / 4.5 ~ 52.8	2.9 ± 2.1 / 0.6 ~ 10.3	1.24 ± 0.59[b] / 0.25 ~ 3.14	0.91 ± 0.36 / 0.36 ~ 2.24	0.39 ± 0.87[b] / 0.01 ~ 5.49	0.11 ± 0.15 / 0.02 ~ 0.68	17.3 ± 6.2[b] / 9.3 ~ 39.7	11.2 ± 2.6 / 6.9 ~ 18.5
Aorta	3.5 ± 1.6[b] / 1.2 ~ 10.0	1.1 ± 0.8 / 0.2 ~ 4.6	0.88 ± 0.26 / 0.32 ~ 1.52	0.98 ± 0.36 / 0.49 ~ 2.56	0.57 ± 0.37 / 0.07 ~ 1.99	0.49 ± 0.38 / 0.01 ~ 1.97	17.9 ± 3.9 / 11.1 ~ 29.5	20.1 ± 4.9[a] / 12.1 ~ 40.2
Rib	2.0 ± 0.9[b] / 0.5 ~ 6.0	0.5 ± 0.3 / 0.1 ~ 1.5	0.41 ± 0.16 / 0.15 ~ 0.86	0.51 ± 0.35 / 0.21 ~ 2.64	2.11 ± 1.47 / 0.43 ~ 8.20	1.68 ± 1.26 / 0.20 ~ 8.50	63.5 ± 18.9[b] / 26.8 ~ 116	55.1 ± 14.1 / 30.4 ~ 117
Heart	1.4 ± 1.3[b] / 0.2 ~ 10.5	0.4 ± 0.3 / 0.1 ~ 1.5	2.84 ± 0.60 / 1.50 ~ 4.86	2.94 ± 0.96 / 1.90 ~ 6.72	0.09 ± 0.06 / 0.02 ~ 0.23	0.10 ± 0.24 / 0.02 ~ 1.73	24.9 ± 3.8 / 13.8 ~ 33.4	26.9 ± 4.4[b] / 14.8 ~ 39.1

Note: M: Mean, SD: Standard deviation, tissue samples are collected from 74 cadmium-exposed and 50 nonexposed people.

[a] $p < 0.05$
[b] $p < 0.01$
[c] Range

(Kuzuhara Y., Sumino, K., Hayashi, C., and Kitamura, S., *Kankyo Hoken Rep.*, 59, 154, 1992. With permission.)

Itai-Itai Disease and Health Effects of Cadmium

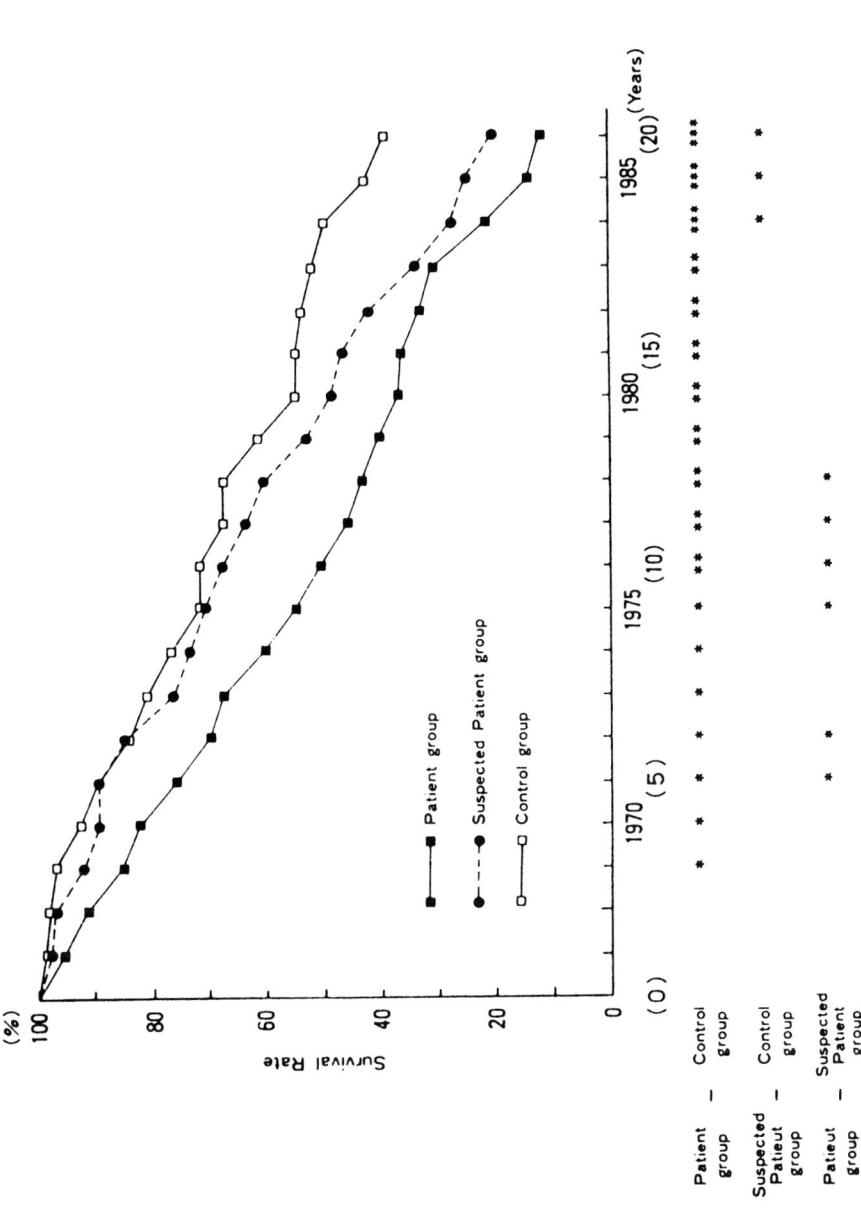

Figure 2 Survival rates for the patient, suspected subject, and control groups from the life expectancy table; $*p < .05$; $**p < .01$; $***p < .0.001$. (From Nakagawa, H., Tabata, M., Morikawa, Y., Senma, M., Kitagawa, Y., Kawano, S., and Kido, T., Arch. Environ. Health, 45, 283, 1990. With permission.)

These findings constitute strong evidence of the detrimental effects of itai-itai disease on life-span and of Cd-induced renal dysfunction on health.

III. EPIDEMIOLOGY OF ITAI-ITAI DISEASE

A. EPIDEMIOLOGICAL STUDY IN 1967

In 1967, an extensive epidemiological study was undertaken to identify endemic districts and to identify all patients with itai-itai disease. The target areas were Fuchu-machi, a part of Toyama City, Osawano-machi and Yatsuo-machi, through which the Jinzu River flows, and neighboring districts. The target population comprised all inhabitants over 30 years of age in these areas, the total number of whom was 6711. This study identified 50 itai-itai disease patients, 48 suspected patients, and 136 persons requiring observation (Fukushima et al., 1974, 1975). Although most of the patients were women, there was also one male itai-itai disease patient as well as four male suspected patients. All patients were 45 years of age or over, and only two observation cases were under 44 years. Until 1992, 160 cases of itai-itai disease and 275 suspected cases have been officially recognized. This study revealed that the patients resided only in the area around the Jinzu River, the water of which is used for irrigation of rice fields. No patients resided in areas where rice fields are irrigated by the water of other rivers. The 1967 study also revealed a high prevalence of proteinuria and glucosuria in the endemic area. The final examination of the 1967 study showed that high values of alkaline phosphatase and low values of inorganic phosphorus were more prevalent in inhabitants with typical bone signs than in those who had no obvious bone changes. Cd concentrations in urine were markedly higher in the endemic area. In the endemic area, patients with typical bone signs showed the highest urinary Cd concentrations, and persons without bone signs had lower values. Hereditary, dietary, and socioeconomic factors were examined in relation to the causes of itai-itai disease. However, the investigations demonstrated no significant associations between these factors and the occurrence of itai-itai disease.

B. ENVIRONMENTAL CD POLLUTION IN THE ENDEMIC AREA

Cd had been transported by the Jinzu River from the Kamioka mine to the endemic areas. The Kamioka mine, located just beside the Jinzu River about 55 km south of Toyama City was established about 400 years ago, and mining and ore processing have been performed on a large scale since the 1870s. The mine produces mainly Zn, as well as relatively small amounts of Pb and Cd.

In 1967, Fukushima et al. (1970) analyzed Cd, Pb, and Zn concentrations in paddy soils in 34 rice fields irrigated by Jinzu River water and 16 adjacent fields irrigated by other water systems. Cd, Pb, and Zn concentrations in paddy soils irrigated by the Jinzu River were clearly higher than those in rice fields irrigated by other river systems. In a patch of rice field watered from the Jinzu River, the concentrations of heavy metals in soils were highest near the water-inlet point of the field and upper layer of the rice field ground. The mean concentrations of 34 rice fields were 4.04 ppm (wet weight) at the entrance of the water, 2.42 ppm at the center, and 2.24 ppm at the outlet of the irrigation system. The highest concentration of Cd in the upper layer was 8.0 ppm. Cd concentrations in paddy fields irrigated by other water system were less than 1 ppm in almost all samples. From 1971 to 1975, the Toyama Prefecture Department of Health (1976) analyzed heavy metal concentrations in 1619 rice paddy soils of the Jinzu River basin, of which 746 showed concentrations of Cd over 1.0 ppm (wet weight); the maximum concentration was 6.88 ppm.

The greatest part of the ingested Cd in the polluted areas derives from rice. Since the endemic district cultivated mostly rice, with beans and vegetables relatively rare, total dietary Cd derives mostly from the rice grown in the area. Rice Cd concentrations are clearly higher in the endemic area than in nonendemic areas. Fukushima et al. (1973) measured Cd concentrations in rice on 77 farms that had been harvested in 1967 in the Jinzu River basin. The average Cd concentration in nonglutinous rice samples was 0.53 ppm (wet weight) in the highly endemic area, 0.38 ppm in the slightly endemic area, and 0.11 ppm in areas irrigated by other water systems. The Association to Consider Countermeasures Against Itai-Itai Disease (a group of residents who takes concerted action for itai-itai disease) reported Cd concentrations in 799 samples of unpolished nonglutinous rice from 49 hamlets harvested in 1972 in the endemic area. The arithmetic means and median Cd concentrations in these samples were 0.53 ppm (wet weight) and 0.50 ppm, respectively. The maximum concentration of Cd in rice was 2.71 ppm. The Toyama Prefecture Department of Health (1976) analyzed Cd concentrations in unpolished nonglutinous rice in the endemic district. The area was divided into about 2000 subareas, each measuring 25,000 m^2. A sample of mature rice on the stalk was taken from the center of each subarea. From 1971 to 1975, 2495 samples were analyzed of which 224 (9.0%) showed concentrations of Cd over 1.0 ppm (wet weight), the maximum being 5.20 ppm.

C. DOSE-RESPONSE RELATIONSHIP OF CD EXPOSURE IN THE ENDEMIC AREA

Fukushima et al. (1973) investigated the relationship between Cd concentrations in rice and abnormal urinary findings. Cd concentrations in glutinous and nonglutinous village rice were employed as indices of Cd exposure. As indices of the effects of Cd on health, the percentages of men and women over 50 years old with proteinuria and glucosuria in each village were used. Correlation coefficients between them were +0.68 for nonglutinous rice and +0.82 for glutinous rice. The rice samples used for this study were collected from 77 farm families and may not be adequate to reflect the level of Cd exposure in each village.

Nogawa and Ishizaki (1979a) also examined the relationship between the Cd concentration in rice and abnormal urinary findings in the inhabitants of the Jinzu River basin. Proteinuria and glucosuria were used as indices of the effects of Cd on health and average rice Cd concentration in hamlets as indices of Cd exposure. The analytical results from 2741 inhabitants in the Cd-polluted area, 2150 reference inhabitants, and 583 unpolished nonglutinous rice samples were used for this study. A close relationship between Cd exposure and health effects was found to exist when the inhabitants were grouped according to their village average rice-Cd concentrations. Increased prevalences of proteinuria and proteinuria with glucosuria in the Cd-exposed population over 50 years of age were observed at average Cd concentrations of 0.30–0.49 µg/g wet weight in unpolished rice, which corresponds to a daily intake of 201–287 µg of Cd.

Nogawa et al. (1979c) also investigated the relationship between urinary Cd concentration expressed as micrograms per gram of creatinine and renal effects in 542 inhabitants of the Jinzu River basin. The Cd concentration in urine was employed as the index of Cd exposure. Total protein with glucose, β_2-mg, RBP, and proline served as indices of renal effects. The prevalence of renal effects increased with increasing Cd concentration in urine, and probit regression lines could be calculated from the groups. The urinary Cd concentrations corresponding to a 1% prevalence rate of β_2-mg were 3.2 and 5.2 µg/g creatinine for men and women, respectively. Nogawa et al. (1979b) also found a clear relationship between Cd concentration in urine and the prevalences of hypocalcemia and hypophosphatemia in females, and increased activity of serum alkaline phosphatase in both sexes in the inhabitants of the Jinzu River basin.

The relationship between the prevalence of itai-itai disease in each hamlet and the hamlet's average Cd concentration in rice was investigated. In this study 598 women over 50 years of age in 32 Cd-polluted hamlets, 408 women in 16 nonpolluted hamlets, and 568 rice samples were examined. (Nogawa et al., 1983). It was found that the prevalence of itai-itai disease increased with increasing Cd concentration in rice. It can be said from these studies that a dose-response relationship between exposure to environmental Cd and health effects, especially renal effects, exists in the endemic areas.

IV. HEALTH EFFECTS OF CD IN THE CD-POLLUTED AREAS IN JAPAN

A. RENAL EFFECTS

Exposure to environmental Cd induces renal tubular dysfunctions indicated by urinary abnormalities such as proteinuria, glucosuria, and especially low molecular weight proteinuria. Epidemiological studies performed in the Cd-polluted areas in Japan showed higher prevalences of low-molecular weight proteinuria in the inhabitants of Cd-polluted areas than in those in nonpolluted areas. Kobayashi (1982) collected 596 urine samples from inhabitants aged over 5 years in 9 heavily Cd-polluted hamlets in the Jinzu River basin in 1976 and analyzed for β_2-mg and RBP. The prevalence of β_2-mg-uria was clearly higher in the Cd-polluted area than in the reference areas and increased with advancing age from 10.3% (age 20–29) to 90.5% (age >70 years) in men, and 6.8% (age 20–29) to 100% (age >70 years) in women. Prevalences of β_2-mg-uria in the reference areas were 0% (age <70 years) and 6.1% (age >70 years) in men, and 0% (age; <60 years) and 7.7% (age >70 years) in women.

In 1983 and 1984, Aoshima (1987) collected 187 urine samples from female inhabitants aged 55 to 65 years in 11 hamlets in the Cd-polluted Jinzu River basin and 46 urine samples from nonpolluted areas and analyzed for β_2-mg, α_1-microglobulin, amino nitrogen, glucose, Cd, calcium and phosphorus. Prevalence rates of renal tubular dysfunction, defined as the urinary β_2-mg level, exceeding 1000 µg/g creatinine and a urinary glucose level exceeding 100 mg/g creatinine were 38.3% in the Cd-polluted group and zero in the reference group.

In 1981 and 1982, urine samples collected from 3178 inhabitants of the Kakehashi River basin, an area polluted by Cd, and 294 nonexposed inhabitants aged over 50 years were analyzed for protein, glucose, amino acid, β_2-mg and Cd (Kido et al., 1987). The concurrent prevalences of proteinuria and glucosuria, as well as those of aminoaciduria and β_2-mg-uria were higher in the Cd-exposed subjects than in the nonexposed subjects. Among them, β_2-mg was considered to be the most sensitive indicator of the renal effects of Cd exposure. MT in urine was also determined in this study. The prevalence of MT-uria was calculated to be 4.6% in men and 8.4% in women from the Cd-polluted area, when the 97.5% upper limits of MT concentration in the nonexposed subjects were employed as the cut-off values of MT-uria (Shaikh et al., 1990).

From 1976 to 1984, a health survey was conducted on a total of 20,766 persons including 13,570 persons aged 50 years or over residing in Cd-polluted areas of 8 prefectures and 7196 residents of nonpolluted areas (Research Committee, 1989). This survey covered almost all of the major Cd-polluted areas in Japan, and the participation rates in the first screening were 92.5% and 89.4% in the polluted and nonpolluted areas, respectively. With the exception of one prefecture, the number of individuals who had or were suspected of having proximal renal tubular dysfunctions or related findings tended to be greater in the Cd-polluted areas than in the nonpolluted areas, and this was often significantly related to the degree of pollution. Of the 438 participants of the tertiary screenings (426 in the polluted areas and 12 in the nonpolluted areas), findings of "possible proximal renal tubular dysfunction" were noted in 334 persons (333 in the polluted areas and one in the nonpolluted area). Among these cases, 202 in the polluted areas were determined to have proximal renal tubular dysfunction, and 116 of them were considered to require medical supervision in view of the severity of the dysfunction. This study clearly demonstrated that renal tubular dysfunction is very specific to Cd exposure.

Studies have been conducted on the reversibility of β_2-mg-uria in Cd-exposed persons in the general environment. Urinary β_2-mg concentrations of 74 Cd-exposed inhabitants in the Kakehashi River basin were measured in 1981 and 1986 (Kido et al., 1988b) (Figure 3). The association between β_2-mg-uria in 1981 and 1986 clearly differs depending on whether or not the values of β_2-mg-uria in 1981 were higher than 1000 µg/g creatinine. When the value of β_2-mg-uria in 1981 was less than 1000 µg/g creatinine, 23 subjects had higher values of, and 19 subjects had lower values of β_2-mg-uria in 1986 than in 1981. On the other hand, when the values of β_2-mg-uria exceeded 1000 µg/g creatinine in 1981, 4 subjects had lower values and the remaining 28 subjects had higher values in 1986 than in 1981. Kasuya et al. (1991) and Iwata et al. (1993) performed similar studies in the Jinzu River basin and Tsushima Island, respectively, and showed that urinary β_2-mg levels of approximately more than 1000 µg/l or 1000 µg/g creatinine tended to be irreversible.

Harada et al. (1992) measured % TRP, serum HCO_3^-, and creatinine clearance in 25 cases with renal tubular dysfunction (urinary β_2-mg >1 mg/dl, % TRP <80%, HCO_3^- <23 mEq/1) once a year from 1976 to 1991, and showed that these abnormalities were persistent and gradually progressive in some cases. Kido et al. (1990b) determined serum creatinine and blood pH in 21 subjects with renal dysfunction in the Cd-polluted Kakehashi River basin annually for 9 to 14 years. Mean serum creatinine was significantly increased from 1.19 to 1.68 mg/100 ml. The most severe case had a serum creatinine value of 4.4 mg/100 ml, and manifested generalized edema, suggesting the presence of renal failure (Figure 4). The mean arterial blood pH values decreased significantly from 7.400 to 7.361 during this period. The 11 subjects also showed a significant decrease and progression of the dysfunction after Cd exposure ceased. These studies indicated that Cd-induced renal tubular dysfunction and decreased glomerular filtration are aggravated, even after cessation of environmental Cd exposure, and that in some cases the renal dysfunction may progress to renal failure.

B. BONE EFFECTS

Kido et al. (1989) performed a quantitative assessment of osteopenia using a microdensitometer in 28 women with itai-itai disease, 92 men and 114 women with Cd-induced renal dysfunctions, and 44 men and 66 women living in three different nonpolluted areas. The values of both indices corresponding to cortical width and bone mineral content were significantly lower in the itai-itai disease patients than in the Cd-exposed women with renal dysfunction and nonexposed subjects. The Cd-exposed women also showed a decrease in bone density compared with the nonexposed subjects. A significant decrease in bone density was also observed between the Cd-exposed men and nonexposed subjects. This study indicated that exposure to environmental Cd could cause marked osteopenia, particularly in women. Furthermore, Kido et al. (1990a) investigated the relationship between Cd-induced renal dysfunction and osteopenia using multivariate analysis. A significant association between indices of renal dysfunction

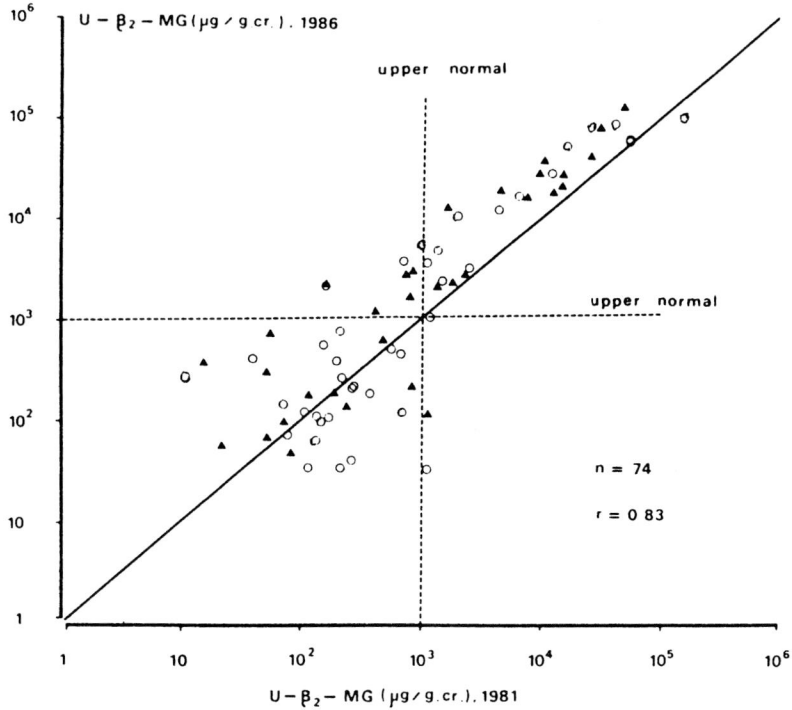

Figure 3 Urinary excretion of β_2-microglobulin, (β_2-MG) (μg/g creatine) in 74 Cd-exposed inhabitants examined in 1981 and 1986; ▲ = males, ○ = females. The r denotes the correlation coefficient; * significant r value at $p < .001$. (From Kido, T., Honda, R., Tsuritani, I., Yamaya, H., Ishizaki, M., Yamada, Y., and Nogawa, K., Arch. Environ. Health, 43, 213, 1988. With permission.)

and each index of the MD method was found to exist. Aoshima et al. (1988a) also demonstrated a significant association between renal dysfunction and osteopenia in Cd-exposed subjects in the Jinzu River basin. They divided the subjects into five groups on the basis of the level of fractional excretion of β_2-mg. MD indices demonstrated deterioration of osteopenia in proportion to the severity of renal dysfunction. These studies showed the existence of osteopenia in the Cd-exposed subjects and suggested a causal relationship between renal dysfunction and osteopenia.

The mechanism of bone damage is still not known. One hypothesis is that vitamin D metabolism in the kidney is disturbed by Cd exposure, which leads to internal vitamin D deficiency, causing bone damage. Nogawa et al. (1987) reported that serum $1\alpha,25$-dihydroxyvitamin D ($1\alpha,25(OH)_2D$) levels were lower and serum parathyroid hormone (PTH) levels higher in Cd-exposed subjects with renal damage than in nonexposed subjects. Decreases in serum $1\alpha,25(OH)_2D$ levels are closely related to serum concentrations of PTH, β_2-mg, and % TRP (Figure 5). They also found decreased serum concentrations of 24,25-dihydroxyvitamin D ($24,25(OH)_2D$) in the Cd-exposed subjects with normal serum $1\alpha,25(OH)_2D$) (Nogawa et al., 1990). Cd initially inhibits hydroxylation from $25(OH)D$ to $24,25(OH)_2D$ and then inhibits hydroxylation from $25(OH)D$ to $1\alpha,25(OH)_2D$ in the kidney. These findings suggest that Cd-induced bone damage is related to disturbances in vitamin D and PTH metabolisms.

C. OTHER HEALTH EFFECTS OF ENVIRONMENTAL Cd
1. Blood Pressure, Cerebrovascular Disease, and Heart Disease

In 1964 and 1965, Nogawa and Kawano (1969) measured blood pressures in 471 and 2308 inhabitants of itai-itai disease endemic districts and their reference areas, respectively. They found that both the systolic and diastolic blood pressures were lower in the inhabitants of the itai-itai disease endemic districts compared to those of the reference areas. Shinoda and Yuri (1978) measured the blood pressures of 53 itai-itai disease patients, 16 suspected patients, and 16 control subjects. The mean blood pressure was 80.9 mmHg for itai-itai disease patients, 81.4 for suspected patients, and 101.6 for control subjects, with a significant difference observed between the patients and controls. Aoshima and Kasuya (1988b) also reported that systolic and diastolic blood pressure levels were the lowest in Cd-exposed groups with

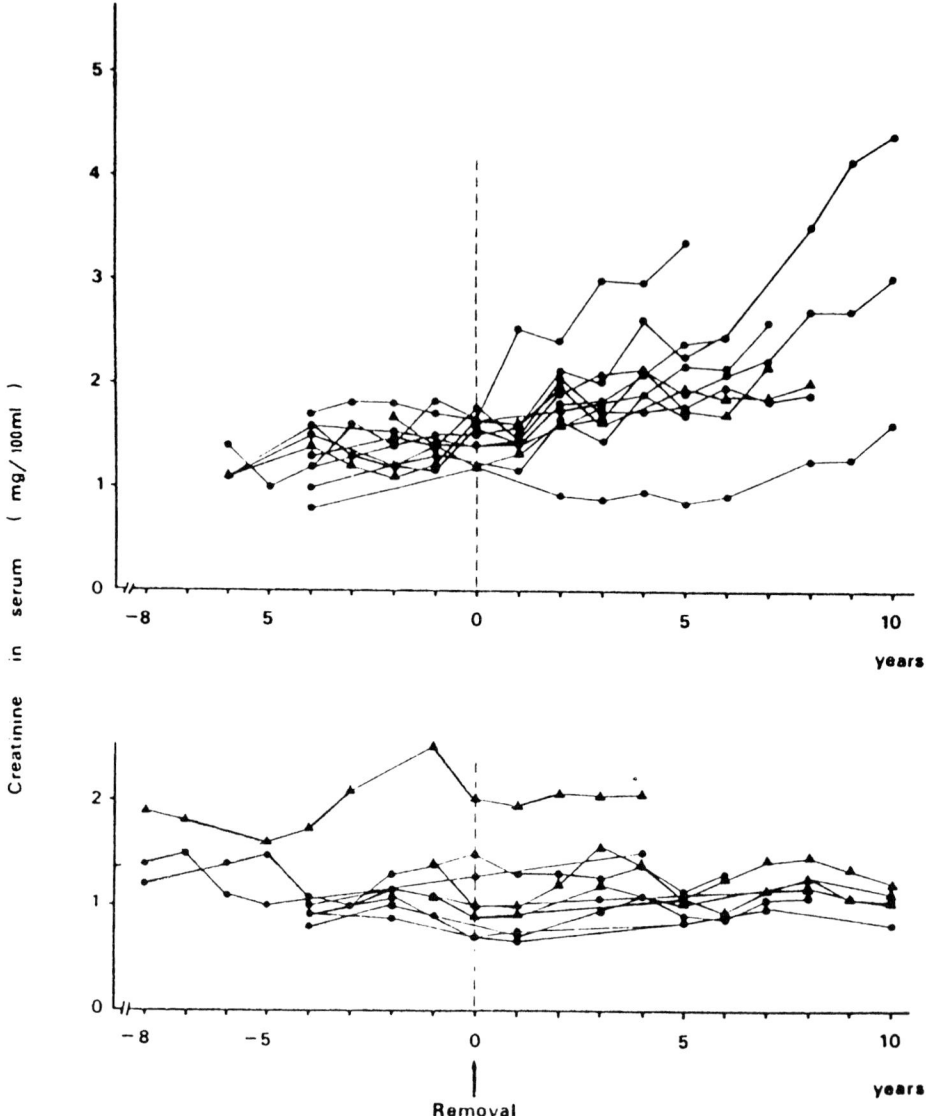

Figure 4 Annual changes of serum creatinine levels before and after cessation of Cd exposure. Top: subjects whose serum creatinine levels increased by >20% above baseline values at their most recent examination. Bottom: subjects whose levels increased <20% above baseline levels; ▲ – ▲ = males, • – • = females. (From Kido, T., Nogawa, K., Ishizaki, M., Honda, R., Tsuritani, I., Yamada, Y., Nakagawa, H., and Nishi, M., Arch. Environ. Health, 45, 35, 1990. With permission.)

the most severe renal tubular dysfunction. According to a health survey conducted from 1976 to 1984 in eight Cd-polluted areas by the Japan Environment Agency, the prevalence of hypertension was often lower in the polluted areas (Research Committee, 1989). These studies seem to indicate that exposure to environmental Cd does not induce hypertension.

A retrospective mortality study in the Cd-polluted areas of four prefectures revealed that the SMR from heart disease, hypertensive disease, and cerebrovascular disease among inhabitants of Cd-polluted areas are, in general, lower than in residents of nonpolluted areas (Shigematsu et al., 1980, 1982). In contrast, a mortality study in the Cd-polluted Kakehashi River basin revealed that the high mortality of inhabitants in the β_2-mg-positive group was caused by increases in cardiovascular disease and cerebrovascular disease (Nakagawa et al., 1994).

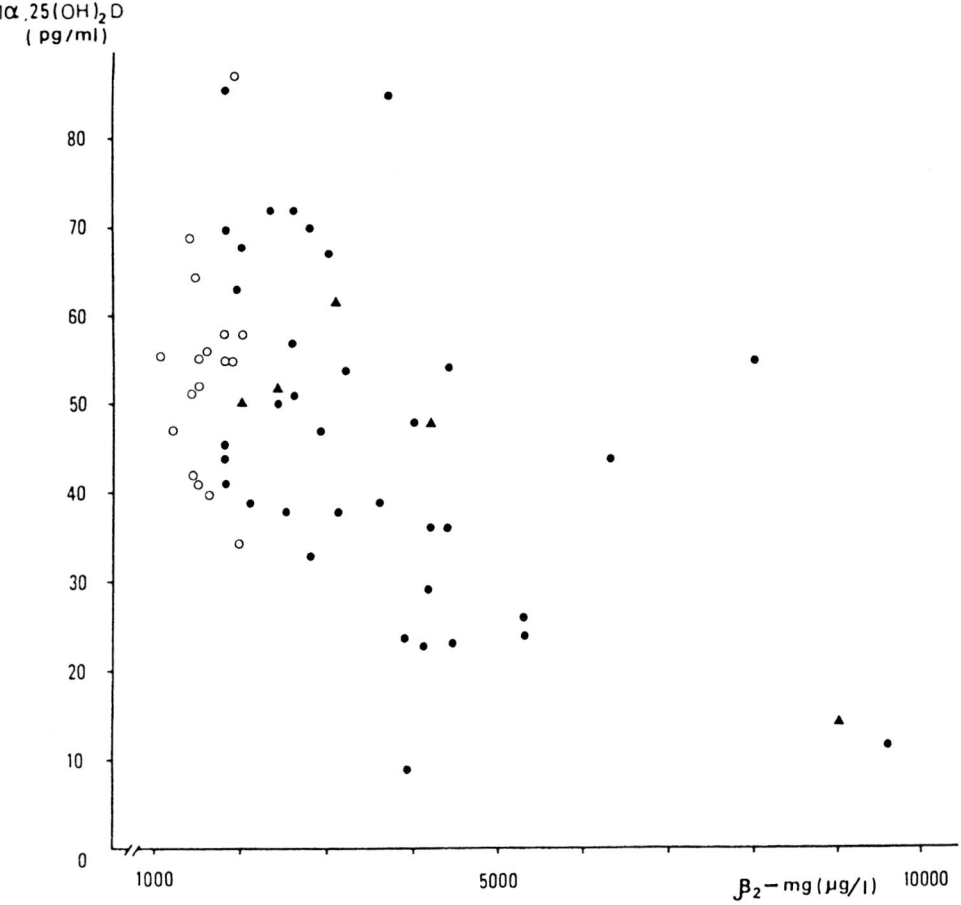

Figure 5 Relationship between 1α, 25(OH)$_2$D and β$_2$-microglobulin in serum of itai-itai disease patients (▲), Cd-exposed (•), and nonexposed subjects (○). (From Nogawa, K., Tsuritani, I., Kido, T., Honda, R., Yamada, Y., and Ishizaki, M., Int. Arch. Occup. Environ. Health, 59, 21, 1987. With permission.)

2. Cancer

Mortality studies in the Kakehashi River basin in Ishikawa Prefecture and Cd-polluted areas of Akita, Miyagi, Toyama, and Nagasaki Prefectures did not detect a significant increase in the mortality resulting from cancer for all or for selected sites (Nakagawa et al., 1994; Shigematsu et al., 1980, 1982).

3. Other Health Effects

Anemia can be seen in itai-itai disease patients and Cd-exposed subjects with renal dysfunction (Nakagawa, 1960; Shinoda and Yuri, 1978; Nogawa et al., 1979g). This anemia is not due to iron deficiency since serum and liver Fe concentrations were within the normal ranges (Nogawa et al., 1984).

Decreased thyroid function has been reported in itai-itai disease patients and residents in the Cd-polluted areas in Nagasaki and Ishikawa Prefecture (Saito et al., 1984; Nakagawa et al., 1993).

Most itai-itai disease patients suffer from peripheral neuropathy (Shinoda et al., 1977).

Shiraishi (1975) found an increased frequency of chromosomal aberrations in lymphocytes obtained from 12 itai-itai disease patients as compared to female controls. However, Bui et al. (1975) could not confirm this finding in a study of four itai-itai disease patients and four controls. Nogawa et al. (1986b) did not find evidence for increased sister chromatid exchange in residents exposed to environmental Cd.

V. MORTALITY

Shigematsu et al. (1980) investigated the mortality in Cd-polluted areas of Akita, Miyagi, Toyama, and Nagasaki Prefectures. The past mortality data on about 333,000 persons in both the Cd-polluted and the nonpolluted areas were collected retrospectively for the period of 6 to 30 years, based on vital statistics or death certificates. The results showed that the mortality from all causes and cardiovascular diseases such as cerebrovascular and hypertensive disease among the general population in the Cd-polluted areas was not different from or was even lower than in the nonpolluted areas. Mortality from cancer of all and selected sites did not differ significantly between Cd-polluted and nonpolluted areas. The mortality from other causes of death such as liver cirrhosis, congenital anomalies, and renal disease also did not show distinct differences between both areas.

Shigematsu et al. (1982) performed a more detailed analysis of the mortality data in the Jinzu River basin in Toyama Prefecture. The areas were divided into three categories: control, slightly polluted, and heavily polluted areas. The results showed that SMRs in all causes of death: heart disease, cerebrovascular disease, and suicide decreased with increasing Cd-pollution. SMRs in itai-itai disease and neuralgia increased with increasing Cd-pollution.

Nakagawa et al. (1993) performed a follow-up study of 3178 persons living in the Cd-polluted Kakehashi River basin from 1981 to 1991 (Figure 6). The SMRs of urinary β_2-mg-positive subjects (more than 1000 µg/g creatinine) of both sexes were higher than those of the general Japanese population, whereas the cumulative survival curves were lower than those of the urinary β_2-mg-negative group. A significant association was also found between urinary β_2-mg and mortality, using a Cox's proportional hazards model. Moreover, mortality rates increased in proportion to increases in the amount of urinary β_2-mg excreted. This study strongly indicates that the prognosis for Cd-exposed subjects with proximal tubular dysfunction is unfavorable. The high mortality rate of inhabitants in the β_2-mg-positive groups was caused by increases in cardiovascular disease and cerebrovascular disease in both sexes. Further statistical analysis between mortality and urinary findings of protein, glucose, β_2-mg, and amino acid in the same population revealed that urinary protein and urinary β_2-mg are important prognostic factors, with the latter, in particular, considered to be useful as an early index of premature mortality (Nishijo et al., 1994).

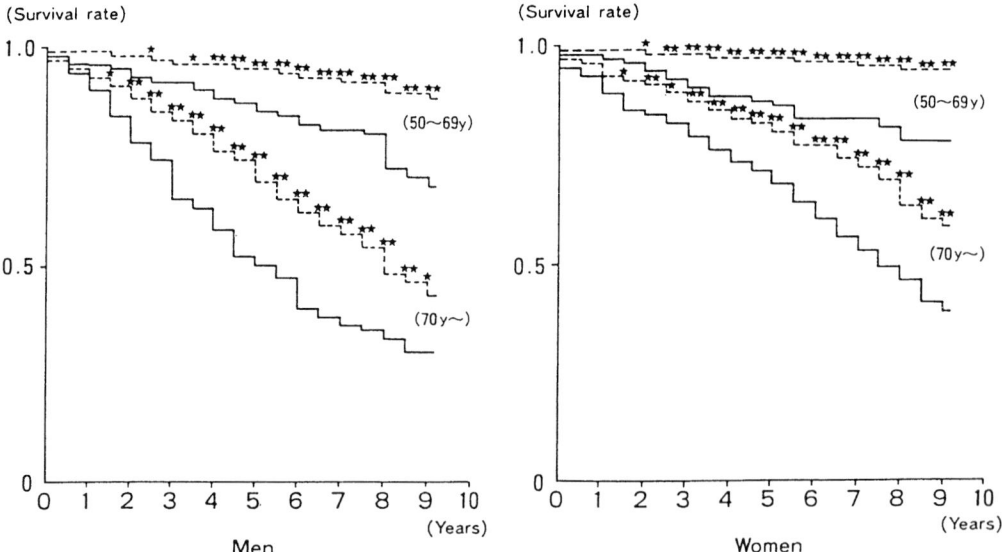

Figure 6 Survival rates, by urinary β_2-MG concentrations, for inhabitants from the life expectancy table. Accumulative survival rates for inhabitants with urinary β_2-MG at each period; ★ = significant difference ($p < .05$); ★★ = significant difference ($p < .01$); - - - - - = inhabitants without urinary β_2-MG; — = inhabitants with urinary β_2-MG; 50~ 69 y = 50 to 69-year-old subjects; 70 y~ = subjects more than 70 years of age. (From Nakagawa, H., Nishijo, M., Morikawa, Y., Tabata, M., Senma, M., Kitagawa, Y., Kawano, S., Ishizaki, M., Sugita, M., Nishi, M., Kido, T., and Nogawa, K., Arch. Environ. Health, 48, 428, 1993. With permission.)

Iwata et al. (1991) followed up 275 residents of Tsushima Island from 1982 until 1989 and found that the number of observed deaths was greater than that expected in both men and women with urinary β_2-mg concentrations greater than 1000 µg/g creatinine. Analysis using a Cox's proportional hazards model showed that, in both sexes, serum β_2-mg and creatinine as well as urinary protein and β_2-mg were significantly or marginally significantly related to mortality, independent of age.

VI. DOSE-RESPONSE RELATIONSHIP AND MAXIMUM ALLOWABLE INTAKE OF Cd IN THE GENERAL ENVIRONMENT

Because of the long biological half-life of Cd in the kidney, it is not easy to clarify the dose-response relationship and to determine the maximum allowable intake of Cd in human subjects. However, some reports have demonstrated its existence in Cd-polluted areas (Fukushima et al., 1973; Nogawa et al., 1978, 1979a, 1983; Saito et al., 1976). In the Kakehashi River basin, Nogawa et al. (1978) performed a statistical analysis on the data of the epidemiological study carried out in 1974 and 1975 and found a close dose-response relationship between village average Cd concentrations in rice or in urine and the prevalence of RBP-uria, and of renal tubular dysfunction when the inhabitants were stratified according to Cd concentrations in the village average rice or urinary Cd concentrations. The prevalence of tubular proteinuria increased continuously from the nonpolluted level of Cd in the inhabitants older than 70 years.

It is thought that the urinary Cd concentration mainly reflects the body burden of Cd (Nogawa et al., 1979f; Kido et al., 1992) and can be used as an index of Cd exposure. In the Kakehashi River basin, urinary Cd, β_2-mg, and MT were measured in 3119 inhabitants over 50 years of age in the Cd-polluted areas. The inhabitants were divided into 12 groups of men and 13 groups of women according to their urinary Cd concentrations. The prevalence rates of β_2-mg-uria and MT-uria in each group increased proportionally with increasing urinary Cd concentration, and probit linear regression lines could be calculated between them. The urinary Cd concentrations corresponding to the nonexposed subjects were 3.8–4.0 µg/g creatinine for men and 3.8–4.1 µg/g creatinine for women (Ishizaki et al., 1989). The urinary Cd concentrations corresponding to the prevalence of MT-uria in each control were 4.2 µg/g creatinine for men and 4.8 µg/g creatinine for women (Kido et al., 1991a).

The dose-response relationship between total Cd intake and β_2-mg-uria or MT-uria was investigated in 1850 Cd-exposed and 284 nonexposed inhabitants of the Kakehashi River basin (Nogawa et al., 1989; Kido et al., 1991b). Total Cd intake was found to affect the prevalence of β_2-mg-uria or MT-uria in a dose-related manner. The total Cd intake that induced an adverse effect on health was calculated as approximately 2000 mg for both men and women, using simple regression analysis (Figure 7). Logistic regression analysis was also performed for this dose-response relationship, and total Cd intake predicted by the logistic model was the same as that predicted by the simple linear regression model (Kido et al., 1993). According to Saito et al. (1976, 1978), on the basis of work done in Kosaka, Akita Prefecture, consumption of rice with approximately 0.3 µg Cd/g, could cause significant urinary excretion of β_2-mg, and daily intake of 177 µg/day of Cd for 15 years (total Cd intake 1 g) could result in tubular dysfunction.

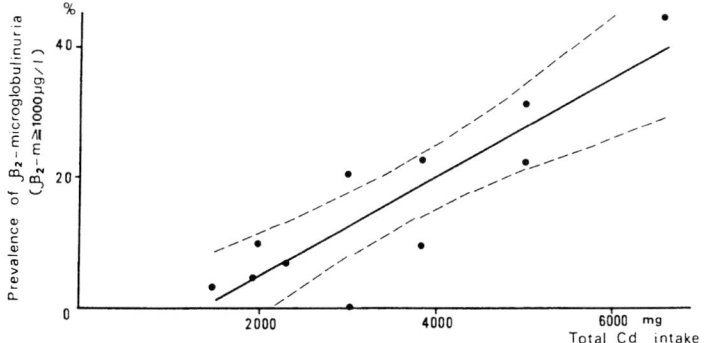

Figure 7 Correlation between total Cd intake and prevalence of β_2-m-uria (β_2-m $\geq$1000 µg/l) for the Cd-exposed male group. The solid line indicates the regression line $Y = 0.0076X - 10.33$ ($r = 0.88$; $p < .001$); dashed lines indicate the 95% confidence interval. (From Nogawa, K., Honda, R., Kido, T., Tsuritani, I., Yamada, Y., Ishizaki, M., and Yamaya, H., *Environ. Res.*, 48, 7, 1989. With permission.)

REFERENCES

Aoshima, K., *Tohoku J. Exp. Med.*, 152, 151–172, 1987.
Aoshima, K., Iwata, K., and Kasuya, M., *Jpn. J. Hyg.*, 43, 864–871, 1988a (in Japanese).
Aoshima, K. and Kasuya, M., *Jpn. J. Hyg.*, 43, 949–955, 1988b (in Japanese).
Bui, T. H., Lindsten, J., and Nordberg, G. F., *Environ. Res.*, 9, 187–195, 1975.
Fukushima, M., Ishizaki, A., Sakamoto, M., and Hayashi, E., *Jpn. J. Hyg.*, 24, 526–535, 1970 (in Japanese).
Fukushima, M., Ishizaki, A., Sakamoto, M., and Kobayashi, E., *Jpn. J. Hyg.*, 28, 406–415, 1973 (in Japanese).
Fukushima, M., Ishizaki, A., Nogawa, K., Sakamoto, M., and Kobayashi, E., I. *Jpn. J. Public Health*, 21, 65–73. II. *Jpn. J. Public Health*, 22, 217–224, 1974 (in Japanese).
Hagino, N., *Rohdoh Kagaku*, 28, 32–46, 1973 (in Japanese).
Harada, K., Hara, K., Shigeno, T., Shirai, G., and Ogata, T., *Kankyo Hoken Rep.*, 59, 264–270, 1992 (in Japanese).
Honda, R., Kobayashi, E., Nogawa, K., Ishizaki, A., Watanabe, M., Shiroishi, K., and Katoh, T., *Kankyo Hoken Rep.*, 44, 132–135, 1978 (in Japanese).
Ishizaki, A., Nomura, K., Tanabe, S., and Sakamoto, M., *Jpn. J. Hyg.*, 20, 261–267, 1965 (in Japanese).
Ishizaki, A., Fukushima, M., and Sakamoto, M., *Jpn. J. Hyg.*, 26, 268–273, 1971 (in Japanese).
Ishizaki, M., Kido, T., Honda, R., Tsuritani, I., Yamada, Y., Nakagawa, H., and Nogawa, K., *Toxicology*, 58, 121–131, 1989.
Iwata, K., Saito, H., and Nakano, A., *Tohoku J. Exp. Med.*, 164, 319–330, 1991.
Iwata, K., Saitoh, H., Moriyama, M., and Nakano, A., *Arch. Environ. Health*, 48, 157–163, 1993.
Japan Public Health Association Research Committee, *Kankyo Hoken Rep.*, 56, 53–345, 1989 (in Japanese).
Kajikawa, K., Kitagawa, M., Nakanishi, I., Ueshima, H., Katsuda, S., and Kuroda, K., *J. Juzen Med. Soc.*, 83, 309–347, 1974 (in Japanese).
Kasuya, M., Teranishi, H., Aoshima, K., Katoh, T., Kubota, Y., Kobashi, K., Kondoh, M., Tanaka, M., and Hagino, N., *Kankyo Hoken Rep.*, 49, 125–139, 1983a (in Japanese).
Kasuya, M., Teranishi, H., Aoshima, K., Katoh, T., Kubota, Y., Kobashi, K., Kondoh, M., and Hagino, N., *Kankyo Hoken Rep.*, 49, 156–159, 1983b (in Japanese).
Kasuya, M., Teranishi, H., Horiguchi, H., Katoh, T., Aoshima, K., Morikawa, Y., Nishijoh, M., and Kawai, M., *Kankyo Hoken Rep.*, 58, 120–122, 1991 (in Japanese).
Katoh, T., Yamamoto, S., Ohmura, T., Nakagawa, H., and Kanamori, C., *Kankyo Hoken Rep.*, 44, 116–118, 1978 (in Japanese).
Kido, T., Honda, R., Tsuritani, I., Yamaya, H., Ishizaki, M., Yamada, Y., and Nogawa, K., *Jpn. J. Hyg.*, 42, 964–972, 1987 (in Japanese).
Kido, T., Tsuritani, I., Honda, R., Yamaya, H., Ishizaki, M., Yamada, Y., and Nogawa, K., *J. Trace Elem. Electrolytes Health Dis.*, 2, 101–104, 1988a.
Kido, Tsuritani, I., Yamaya, H., Ishizaki, M., Yamada, Y., and Nogawa, K., *Arch. Environ. Health*, 43, 213–217, 1988b.
Kido, T., Honda, R., Nogawa, K., Yamada, Y., Honda, R., Tsuritani, I., Ishizaki, M., and Yamaya, H., *Int. Arch. Occup. Environ. Health*, 61, 271–276, 1989.
Kido, T., Nogawa, K., Honda, R., Tsuritani, I., Ishizaki, M., Yamada, Y., and Nakagawa, H., *Environ. Res.*, 51, 71–82, 1990a.
Kido, T., Nogawa, K., Ishizaki, M., Honda, R., Tsuritani, I., Yamada, Y., Nakagawa, H., and Nishi, M., *Arch. Environ. Health*, 45, 35–41, 1990b.
Kido, T., Shaikh, Z. A., Kito, H., Honda, R., and Nogawa, K., *Toxicology*, 65, 325–332, 1991a.
Kido, T., Shaikh, Z. A., Kito, H., Honda, R., and Nogawa, K., *Toxicology*, 66, 271–278, 1991b.
Kido, T., Nogawa, K., Ohmichi, M., Honda, R., Tsuritani, I., Ishizaki, M., and Yamada, Y., *Arch. Environ. Health*, 47, 196–202, 1992.
Kido, T., Shaikh, Z. A., Kito, H., Honda, R., and Nogawa, K., *Toxicology*, 80, 207–215, 1993.
Kitagawa, M. and Yasuda, M., *Kankyo Hoken Rep.*, 59, 141–153, 1992 (in Japanese).
Kitagawa, M. and Yamasita, H., *Kankyo Hoken Rep.*, 60, 172–174, 1993 (in Japanese).
Kobayashi, E., *J. Kanazawa Med. Univ.*, 6, 123–126, 1981a (in Japanese).
Kobayashi, E., *J. Kanazawa Med. Univ.*, 6, 119–122, 1981b (in Japanese).
Kobayashi, E., Honda, R., Nogawa, K., Kawano, S., and Sakamoto, M., *Jpn. J. Hyg.*, 36, 734–741, 1981c (in Japanese).
Kobayashi, E., *Jpn. J. Public Health*, 29, 123–133, 1982 (in Japanese).
Kuzuhara, Y., Sumino, K., Hayashi, C., and Kitamura, S., *Kankyo Hoken Rep.*, 59, 154–174, 1992 (in Japanese).
Murata, I., Nakagawa, S., and Hirono, T., *Kankyo Hoken Rep.*, 11, 132–138, 1972 (in Japanese).
Nakagawa, S., *J. Radiol. Phys. Ther. Univ. Kanazawa*, 56, 1–51, 1960 (in Japanese).
Nakagawa, S., Tohda, N., and Kobayashi, S., *Kankyo Hoken Rep.*, 41, 53–60, 1977 (in Japanese).
Nakagawa, H., Tabata, M., Morikawa, Y., Senma, M., Kitagawa, Y., Kawano, S., and Kido, T., *Arch. Environ. Health*, 45, 283–287, 1990.
Nakagawa, H., Nishijo, M., Morikawa, Y., Tabata, M., Senma, M., Yosida, K., Okumura, Y., Miura, K., Kawano, S., Tsuritani, I., Honda, R., Mizukoshi, K., Sugita, N., Nishi, M., and Kido, T., *Kankyo Hoken Rep.*, 60, 136–139, 1993 (in Japanese).
Nakagawa, H., Nishijo, M., Morikawa, Y., Tabata, M., Senma, M., Kitagawa, Y., Kawano, S., Ishizaki, M., Sugita, N., Nishi, M., Kido, T., and Nogawa, K., *Arch. Environ. Health*, 48, 428–435, 1993.
Nishijo, M., Nakagawa, H., Morikawa, Y., Tabata, M., Senma, M., Kitagawa, Y., Kawano, S., Ishizaki, M., Sugita, N., Nishi, M., Kido, T., and Nogawa, K., *Environ. Res.*, 64, 112–121, 1994.

Nogawa, K. and Kawano, S., *J. Juzen Med. Assoc. Kanazawa*, 3, 357–363, 1969 (in Japanese).
Nogawa, K., Ishizaki, A., and Kawano, S., *Environ. Res.*, 15, 185–198, 1978.
Nogawa, K. and Ishizaki, A., *Environ. Res.*, 18, 410–420, 1979a.
Nogawa, K., Ishizaki, A., and Kobayashi, E., *Environ. Res.*, 18, 397–409, 1979b.
Nogawa, K., Kobayashi, E., and Honda, R., *Environ. Health Perspect.*, 28, 161–168, 1979c.
Nogawa, K., Kobayashi, E., Honda, R., and Ishizaki, A., *Jpn. J. Hyg.*, 34, 407–414, 1979d (in Japanese).
Nogawa, K., Kobayashi, E., Honda, R., and Ishizaki, A., *Jpn. J. Hyg.*, 34, 415–419, 1979e (in Japanese).
Nogawa, K., Kobayashi, E., Honda, R., and Ishizaki, A., *Jpn. J. Public Health*, 26, 25–31, 1979f (in Japanese).
Nogawa, K., Kobayashi, E., Ishizaki, A., Katoh, T., and Kanamori, C., *Jpn. J. Hyg.*, 34, 574–579, 1979g (in Japanese).
Nogawa, K., Honda, R., Kobayashi, E., and Ishizaki, A., *Jpn. J. Hyg.*, 34, 723–732, 1980 (in Japanese).
Nogawa, K., Yamada, Y., Honda, R., Ishizaki, M., Tsuritani, I., Kawano, S., and Katoh, T., *Toxicol. Lett.*, 17, 263–266, 1983.
Nogawa, K., Honda, R., Yamada, Y., Kido, T., Tsuritani, I., and Ishizaki, M., *Toxicol. Lett.*, 21, 209–212, 1984.
Nogawa, K., Honda, R., Yamada, Y., Kido, T., Tsuritani, I., Ishizaki, M., and Yamaya, H., *Environ. Res.*, 40, 251–260, 1986a.
Nogawa, K., Tsuritani, I., Yamada, Y., Kido, T., Honda, R., Ishizaki, M., and Kurihara, T., *Toxicol. Lett.*, 32, 283–288, 1986b.
Nogawa, K., Tsuritani, I., Kido, T., Honda, R., Yamada, Y., and Ishizaki, M., *Int. Arch. Occup. Environ. Health*, 59, 21–30, 1987.
Nogawa, K., Honda, R., Kido, T., Tsuritani, I., Yamada, Y., Ishizaki, M., and Yamaya, H., *Environ. Res.*, 48, 7–16, 1989.
Nogawa, K., Tsuritani, I., Kido, T., Honda, R., Ishizaki, M., and Yamada, Y., *Int. Arch. Occup. Environ. Health*, 62, 189–193, 1990.
Saito, H., Nagai, K., Shioji, R., Furukawa, Y., Arikawa, T., Saito, T., and Furukawa, T., *Kankyo Hoken Rep.*, 38, 78–81, 1976 (in Japanese).
Saito, H., Kusakabe, K., Arikawa, T., Sudo, K., Furukawa, T., and Yoshinaga, K., *Kankyo Hoken Rep.*, 44, 175–177, 1978 (in Japanese).
Saito, H., Nakano, A., Tohyama, C., Mitane, Y., Sugihira, N., and Ishihara, Y., *Kankyo Hoken Rep.*, 50, 134–136, 1984 (in Japanese).
Shaikh, Z. A., Kido, T., Kito, H., Honda, R., and Nogawa, K., *Toxicology*, 64, 59–69, 1990.
Shigematsu, I., Takeuchi, J., Minowa, M., Nagai, M., Usui, T., and Fukushima, M., *Kankyo Hoken Rep.*, 46(2), 1–71, 1980 (in Japanese).
Shigematsu, I., Minowa, M., Nagai, M., Ohmura, T., and Takeuchi, K., *Kankyo Hoken Rep.*, 48, 118–136, 1982 (in Japanese).
Shinoda, S., Yuri, K., and Nakagawa, S., *Kankyo Hoken Rep.*, 41, 44–52, 1977 (in Japanese).
Shinoda, S. and Yuri, K., *Kankyo Hoken Rep.*, 44, 119–125, 1978 (in Japanese).
Shinmura, T., Shiroishi, K., Shimizu, R., and Uetake, H., *Kankyo Hoken Rep.*, 50, 186–190, 1984 (in Japanese).
Shiraishi, Y., *Humangenetik*, 27, 31–44, 1975.
Shiroishi, K., Shimizu, R., and Uetake, H., *Kankyo Hoken Rep.*, 50, 182–185, 1984 (in Japanese).
Takeuchi, J., Shinoda, S., Kobayashi, K., Nakamoto, Y., Takazawa, I., and Kurosaki, M., *Intern Med.*, 21, 876–884, 1968 (in Japanese).
Tohda, N., Kobayashi, S., and Nakagawa, S., *Kankyo Hoken Rep.*, 44, 126–130, 1978 (in Japanese).
Tohyama, C., Shaikh, Z. A., Nogawa, K., Kobayashi, E., and Honda, R., *Toxicology*, 20, 289–297, 1982.
Toyama Prefecture Department of Health, White Paper in Environmental Pollution, Toyama, 128–129, 1976 (in Japanese).

Chapter 22

Genetic Disorders of Copper Metabolism

Hiroko Kodama

ABSTRACT

Genetic disorders of copper metabolism reviewed are, mainly, Wilson's disease (WD) and Menkes disease (MNKD). WD shows autosomal recessive inheritance and the prevalence rate is high. WD is associated with the toxic effects of copper accumulated in many tissues such as the liver, brain, cornea, and kidney. The clinical symptoms mainly appear as liver diseases and neurological diseases. However, various other symptoms are also observed in patients. A serious problem of this disease is that these various symptoms sometimes make early diagnosis difficult. The characteristic findings are Kayser-Fleischer rings, low levels of serum ceruloplasmin and copper, an increase of the copper concentration in the liver, and an increase of urinary copper excretion after penicillamine challenge. All the patients should be treated with chelating agents or zinc. These treatments are beneficial. However, these treatments are ineffective in patients with fulminant hepatic failure. Moreover, some patients with neurological diseases show poor response to chelation therapy. Liver transplantation is accepted for these patients. In these cases also, the disturbances are prevented by early treatment with chelating agents. Thus, early diagnosis and treatment are most important for WD. Mass screening for WD of newborn babies or infants should be established for early diagnosis.

MNKD shows X-linked recessive inheritance. In this disease, orally administered copper accumulates in the intestine, resulting in the failure of copper absorption through the digestive tract. The primary metabolic defect which causes copper accumulation in the intestine is present in almost all extrahepatic tissues, such as the kidney and cultured skin fibroblasts. The blood, liver, and brain are in a state of copper deficiency which is due to defective intestinal copper absorption. The characteristic features of this disease, such as neurological degeneration, arterial degeneration, hair abnormalities, and hypothermia can be explained by the decrease of cuproenzyme activity. The treatment generally accepted so far is parenteral administration of copper. When the treatment is started in patients above the age of 2 months, however, it does not improve the neurological degeneration. When the treatment is initiated in fetuses *in utero* or newborn babies with this disease, the neurological degeneration may be prevented. In view of this, a reliable prenatal diagnostic method should be established.

Recently, the genes for both WD and MNKD have been identified. The gene analysis of both patients as well as biochemical and histochemical studies of the gene products will provide for more beneficial treatments, as well as simple and clear methods for diagnosis of these diseases.

I. INTRODUCTION

The genetic disorders of copper metabolism in humans are grossly classified into two groups: (1) copper toxicosis, namely Wilson's disease (WD), and (2) copper deficiency diseases, that is, Menkes disease (MNKD), milder forms of MNKD, and occipital horn syndrome (Ehlers Danlos syndrome type IX, X-linked cutis laxa). Neither the genes nor the primary metabolic defects of these diseases had been elucidated until 1993. In 1993, the MNKD gene was identified first (Vulpe et al., 1993; Chelly et al., 1993; Mercer et al., 1993a), followed by the identification of the WD gene. (Bull et al., 1993; Petrukhin et al., 1993; Tanzi et al., 1993). Both cDNA sequences were very similar, and both gene products were predicted to be P-type ATPases with six putative metal binding regions. However, the symptoms and signs of both diseases are completely different. WD is associated with the toxic effects of copper accumulation in many tissues such as the liver, brain, and kidney. Copper accumulation is also observed in many tissues of patients with MNKD, such as the intestine and kidney. However, the symptoms and signs of MNKD are associated with a decrease of cuproenzyme activities, which is probably caused by defective intestinal copper absorption and a disturbance of intracellular copper transport. The immunocytological, physiological, and biochemical studies of these two gene products should be necessary to clarify the mechanisms of copper transport in the cells and between the tissues, and the pathogenesis of these diseases.

II. WILSON'S DISEASE

Wilson's disease (WD) was first described by Kinner Wilson in 1912 as progressive lenticular degeneration: a familial nervous disease associated with cirrhosis of the liver (Wilson, 1912). The disease was reported to be related to the toxic effects of copper accumulation in the liver and brain in the 1940s. An effective treatment with penicillamine, a chelating agent, was introduced by Walshe in 1956 (Walshe, 1956).

A. PREVALENCE AND GENETIC ANALYSIS

The inheritance is autosomal recessive. Reilly et al. (1993) reported that the adjusted birth incidence rate was 17.0 per million live births, and that a gene carrier was 1 in 122 of the population in Ireland. The incidence rate in Japan (Saito, 1985), Israel (Bonné-Tamir et al., 1990) and Sardinia (Giagheddu et al., 1985) is generally higher than that found by the European studies. However, the high incidence rate in these countries probably relates to the high rate of consanguinity.

The WD gene is located on chromosome 13q14.3, closely linked to the locus of esterase D and retinoblastoma (Frydman et al., 1985; Bull and Cox, 1993). Recently, the WD gene has been identified. Tanzi et al. (1993) identified four disease-specific mutations by gene analysis of patients with WD. Bull et al. (1993) reported a seven-base deletion within the coding region of the WD gene in two patients with WD.

B. PATHOGENESIS

Gene analytical studies (Bull et al., 1993; Petrukhin et al., 1993; Tanzi et al., 1993) suggest that the primary metabolic defect of WD is a defect of copper-transporting ATPase. Although the serum ceruloplasmin level is significantly low in most patients with WD, the decreased copper is holo-ceruloplasmin (copper-bound ceruloplasmin). The serum level of apo-ceruloplasmin (copper-unbound ceruloplasmin), which is about 10% of the ceruloplasmin in the serum of normal adults, is normal in patients, indicating that WD is not caused by a defect of ceruloplasmin synthesis (Matsuda et al., 1974). The WD gene is located on chromosome 13, while the ceruloplasmin gene is found on chromosome 3. These findings also indicate that the primary metabolic defect of WD is not a defect of ceruloplasmin synthesis.

The copper accumulation in the liver of patients with WD is caused by a disturbance of biliary excretion of copper, probably as copper-ceruloplasmin (Iyengar et al., 1988), and a reduced incorporation rate of copper into ceruloplasmin. In the liver of patients, copper distributes diffusely as metallothionein-copper in the hepatocyte cytosol during the early stages of the disease. As the disease progresses, copper accumulates in the lysosomes (Goldfischer and Sternlieb, 1968; Nartey et al., 1987). Copper concentrated in lysosomes shows cytotoxicity (Lindquist, 1968). The excessive copper in the lysosomes is probably the cause of liver damage in WD.

Disturbances of copper incorporation into ceruloplasmin decrease the serum levels of ceruloplasmin-bound copper as well as ceruloplasmin. At the same time, an increase of the serum nonceruloplasmin-bound copper is observed, which has been considered to be the cause of the elevation of urinary copper excretion and copper deposits in various extrahepatic tissues, such as the kidney, brain, cornea, muscle, bone, and joint. However, gene expression studies of normal tissues (Bull et al., 1993; Tanzi et al., 1993) revealed that the WD gene was expressed predominantly in the liver, kidneys, and placenta. It was also expressed slightly in the heart, brain, lungs, muscles, and pancreas. These findings suggest that the WD gene product participates in the copper metabolism of these extrahepatic tissues as well as the liver. Studies on the WD gene product in these tissues would clarify the mechanism of copper toxicity in the extrahepatic tissues of patients with WD.

C. CLINICAL MANIFESTATIONS

The main clinical manifestations of WD are liver diseases and/or neurological diseases. Patients with neurological diseases potentially suffer from liver diseases, though this may not be clinically obvious. The hepatic and neurological symptoms appear at any age beyond 6 and 12 years, respectively. However, the hepatic symptoms appear most frequently between the age of 8 and 16 years, whereas the neurological symptoms appear later, in adolescence and the young adult. In rare cases, various symptoms such as Fanconi syndrome, degenerative arthritis, anemia, hypersalivation (Oder et al., 1991) and hematuria (Hoppe et al., 1993) appear initially. These nonspecific initial symptoms of WD make early diagnosis difficult. Even now, the median time interval between the initial symptoms and the diagnosis is 18 months (range 1–72 months) in neurological patients and 6 months (range 2–108 months) in hepatic patients (Oder et al., 1991).

1. Hepatic Manifestations

Hepatic symptoms of WD show a broad spectrum of chronic, chronic active, and acute liver diseases, cirrhosis, and fulminant hepatic failure (FHF). In most patients with hepatic diseases, the diseases progress chronically followed by cirrhosis (Nazer et al., 1986; Stremmel et al., 1991). Since the symptoms of patients with WD are nonspecific, as shown in Table 1, patients may be misdiagnosed as having either viral or drug-induced chronic hepatitis. Hemolysis, which may cause gallstones and cholecystitis occasionally occurs, resulting in abdominal pain (Rosenfield et al., 1978).

Table 1 Symptoms and Signs of Wilson's Disease

No. of patients	Nazer et al. (1986) (34 patients)(%)	Stremmel et al. (1991) (51 patients)(%)	Saito (1987) (276 patients)(%)
Lethargy, anorexia	70		10
Jaundice	56	14	28
Abdominal pain	48	41	5
Recurrent epistaxes	22		3
Hepatomegaly	81	49	9
Splenomegaly	70	49	5
Ascites		14	5
Gynecomastia		14	1
Diarrhea		8	1
Vomiting		4	6

Some patients suffer from FHF with severe jaundice, coagulopathy, ascites, encephalopathy, renal failure, and hemolysis. The diagnosis of patients with FHF may often be difficult, because their serum levels of copper and ceruloplasmin may sometimes be within normal ranges. The patients with FHF are associated with (1) Coomb's negative hemolytic anemia, (2) low serum levels of alkaline phosphatase and 5'-nucleotidase, and (3) disproportionately low serum alanine transaminase (ALT) compared with serum aspartate transaminase (AST). Therefore, these associations seem to be useful for diagnosis, though the mechanism is uncertain (Willson and Hartleb, 1988). FHF occurs frequently when patients do not continue a chelation therapy (Walshe and Dixon, 1986). In patients with FHF, medical treatment is generally ineffective, thus the mortality rate is very high. Liver transplantation is the only effective treatment.

Hepatocellular carcinoma is a well-recognized complication of other forms of cirrhosis, such as hemochromatosis. However, it is uncommon in patients with WD. Only 11 cases with hepatocellular carcinoma have been reported in patients with WD (Cheng et al., 1992).

Table 2 Neuropsychiatric Symptoms and Brain Lesions of Wilson's Disease*

Symptoms	Lesions
First group bradykinesia, rigidity, cognitive impairment organic mood syndrome (depression)	dilatation of the third ventricle
Second group ataxia, tremor, reduced functional capacity (adaptive functioning in work place competence, financial competence)	thalamic lesions
Third group dyskinesia, dysarthria, an organic personality syndrome (including affective instability, irritability, temper outbursts, impaired social judgment)	focal lesions in the putamen and the pallidum

Note: *Modified from the data of Order et al., (1993).

Histological observation of the liver specimens of patients with WD shows fat deposition, increased range of nuclear size, lipofuscin deposition, vacuolated nuclei, ballooning of hepatocytes, fibrosis, and cirrhosis. These are found in asymptomatic and neurological patients as well as hepatic patients (Stremmel et al., 1991).

2. Neuropsychiatric Manifestations

Neurological and psychiatric abnormalities are caused mainly by neuronal damage due to copper deposition in the brain. Sometimes, they are caused by encephalopathy due to hepatic dysfunction. Most frequent symptoms are dysarthria, dystonia, dysdiadochokinesia, rigidity, bradykinesia, and tremor (Starosta-Rubinstein et al., 1987; Oder et al., 1991). Oder et al. (1993) recommended to classify neurological abnormalities into three subgroups by the clinical findings and brain lesions (Table 2). Patients with loss of cerebral white matter have rarely been reported (Schulman and Barbeau, 1963; Kodama et al., 1988).

Cranial computed tomography (CT) and magnetic resonance imaging (MRI) reveal cortical atrophy, brainstem atrophy, and anatomic disruption of basal ganglia (Walshe, 1988a). MRI is thought to be sensitive to the changes of the basal ganglia, and thus is useful for documenting the effects of the treatment (Thuomas et al., 1993). Electromyographic responses evoked by transcranial magnetic brain stimulation (Meyer et al., 1991) and positron emission tomography (Volder et al., 1988) are also abnormal in patients with neurological symptoms. The copper level in the cerebrospinal fluid increases in patients with neurological symptoms (Weisner et al., 1987; Kodama et al., 1988). These laboratory tests seem to be useful markers for monitoring therapy.

3. Other Manifestations

Table 3 shows other manifestations of WD. Renal diseases are common in WD. Copper deposition is found in the tubular epithelium, causing tubular dysfunction, the most frequent manifestation being renal tubular acidosis. Renal stones were observed in 7 of 54 patients (16%) with WD (Wiebers et al., 1979). The reabsorptions of amino acids (Uzman, 1953) and urate (Wilson and Goldstein, 1973) are also disturbed. An adolescent patient with a 6-year history of hypercalciuria and nephrocalcinosis, who was finally diagnosed as WD, was reported recently (Hoppe et al., 1993). Bone and joint diseases are also found frequently. Most of these changes are asymptomatic, though these disorders rarely cause joint pain or stiffness (Golding and Walshe, 1977). Patients with WD rarely have cardiac symptoms, however, their cardiac function tests such as electrocardiography are frequently abnormal. These abnormalities are observed even in patients who are treated with chelating agents for a long period. Cardiac death rarely occurs (Kuan, 1987).

D. DIAGNOSIS

As described above, various symptoms are observed in patients with WD. Therefore, patients over the age of 6 years with liver disease, unclear neurologic and psychiatric diseases, renal stone, renal tubular dysfunction or hematologic abnormalities such as hemolysis, anemia, or thrombocytopenia should be tested for WD. Once a patient is diagnosed as having WD, all close relatives should be screened for WD. Treatment should be offered to them, if they are diagnosed by the biochemical lesions.

Table 3 Symptoms and Signs Found Occasionally in Patients With Wilson's Disease

Renal
  Renal tubular dysfunction, renal stones
Bone and joint
  Osteoporosis, osteomalacia, osteoarthritis, peri- and intraarticular calcification, arthralgia, joint hypermobility
Cardiac
  Congestive heart failure, cardiac hypertrophy, cardiac arrhysmia, cardiomyopathy, autonomic dysfunction
Endocrine
  Gynecomastia, amenorrhea and menstrual dysfunction,[a] parathyroid dysfunction,[b] glucose intolerance[c]
Skin
  Hyperpigmentation, acanthosis nigricans[d]
Ocular
  Kayser-Fleischer rings, sunflower cataracts
Spontaneous bacterial peritonitis[e]

[a] Kaushansky et al., 1987.
[b] Carpenter et al., 1983.
[c] Johansen et al., 1972.
[d] Leu et al., 1970.
[e] Person et al., 1987.

Diagnosis by DNA polymorphism analysis using closely linked markers has already been reported for heterozygotes, fetuses, and patients with WD (Cossu et al., 1992; Gaffney et al., 1992). Recently, the WD gene was identified. Thus, DNA-based diagnosis of WD should be established as a more precise diagnostic method.

Table 4 shows laboratory findings for the diagnosis of WD. Kayser-Fleischer rings are absent in some patients with hepatic WD and in asymptomatic patients, although they are present in most of the patients with neurologic WD (Oder et al., 1991; Stremmel et al., 1991). The serum levels of copper and ceruloplasmin decrease in 80–95% of patients with WD. However these levels are sometimes normal in patients with severe hepatic failure and hemolysis (Scott et al., 1978; McCullough et al., 1983; Stremmel et al., 1991). Urinary copper excretion after penicillamine challenge is a very useful diagnostic test for WD (Martins da Costa et al., 1992). High concentration of hepatic copper is observed even in asymptomatic patients and patients with neurologic WD (Stremmel et al., 1991). However, it should be noted that characteristic findings for WD, such as Kayser-Fleischer rings, the increase of urinary copper excretion after penicillamine challenge, and the increase of the liver copper content are sometimes observed in patients with other liver diseases, such as primary biliary cirrhosis, virus hepatitis, chronic active hepatitis, intrahepatic cholestasis, and Indian childhood cirrhosis (Yarze et al., 1992; Gregorio and Mieli-Vergani, 1993). These diseases can usually be differentiated from WD by clinical, biochemical, and histological analysis. However, correct diagnosis in these cases is sometimes difficult. ^{64}Cu administration test is proposed to be useful for the diagnosis of patients who cannot be identified by the examinations described above (Yarze et al., 1992). It is also very difficult to distinguish heterozygotes from asymptomatic patients by the biochemical examinations (Marecek and Nevsimalova, 1984; Walshe, 1988b).

Mass screening for WD in newborn babies should be established immediately. Generally, measuring serum ceruloplasmin is thought to be difficult, because the serum level of normal newborn babies is very low (Danks, 1989). Recently, Hiyamuta et al. (1993) have indicated that the detection of active ceruloplasmin (holo-cereloplasmin) by sandwich enzyme-linked immunosorbent assay is useful as an early diagnostic tool for WD.

E. TREATMENT

Patients with WD die either due to progressive liver failure or neurological disease, if they are not treated. All the patients with WD should be treated with chelating agents or zinc, and a copper-restricted diet. Food with a high copper content, such as shellfish, chocolate, mushrooms, nuts, and liver should be avoided. However, as a negative copper balance cannot be achieved by the diet alone, patients should also be treated with a chelator or zinc. These treatments should also be offered to asymptomatic or pregnant patients.

Table 4 Laboratory Data for the Diagnosis of Wilson's Disease[a]

	Normal	Wilson's disease
Serum ceruloplasmine (mg/dl)	25–35	<5
Serum copper (µg/dl)	70–150	<30
Nonceruloplasmin-bound plasma copper (µg/dl)	10–15	25–100
Urinary copper (µg/day)		
Untreated	<40	100–1000
On penicillamine	100–600	1500–3000
Liver copper (µg/g dry weight)	20–50	>250
Liver histology		See text
Incorporation ^{64}Cu into ceruloplasmin at 24 h		
Oral, ratio to initial peak	0.8–1.2	0.1–0.3
Intravenous, % of dose	>4	<1

[a] Modified from the data of Tankanow (1991). With permission.

1. Medicines

Table 5 lists medicines which have been used for the treatment of WD. All the medicines should be given on an empty stomach, for example, 2 h after meals or 60–30 min before meals. When patients receive a chelator, 24 h excretion of urinary copper should be monitored to know the therapeutic effect. A considerable amount of copper is excreted into the urine, usually rising to 2 to 5 mg/day during the early stage of treatment (initial treatment). Although the copper excretion rate decreases gradually during the therapy, about 500 µg/day is maintained after several months or years. Plasma nonceruloplasmin copper is also useful to determine the therapeutic effects. If the therapy is effective, the level is kept below 20 µg/dl during the treatment with zinc as well as with chelating agents (Brewer and Yuzbasiyan-Gurkan, 1989)

Table 5 Medicines Used For Treatment of Wilson's Disease

D-penicillamine	
Adults	1–2 g/day
Children	500 mg/day (5–11-year-old) or 20 mg/kg/day
Triethylene tetramine dihydrochloride	1.2–2.4 g/day[a]
Tetrathiomolybdate ammonium	100–200 mg/day[b]
Zinc	
Zinc sulfate	600–1200 mg/day (adults) 300–600 mg/day (children)[c]
Zinc acetate	150 mg/day (as zinc)[d]

[a] Walshe, 1982.
[b] Brewer et al., 1991.
[c] Hoogenraad et al., 1987.
[d] Brewer et al., 1993.

a. D-Penicillamine

D-Penicillamine is so far the drug of first choice for the treatment of WD. Orally administered D-penicillamine removes copper from the liver (Gibbs and Walshe, 1990). In addition, it seems to be able to detoxify hepatic copper (Scheinberg et al., 1987; Gibbs and Walshe, 1990). During the treatment with penicillamine, pyridoxine should be given orally 25 mg/day in adults and 12.5 mg/day in children, because of the antipyridoxine effect of penicillamine.

However, penicillamine induces adverse effects such as skin rash, fever, thrombocytopenia, leukopenia, aplastic anemia, nephrotic syndrome, and lupus erythematosus in about 20% of patients (Tankanow, 1991). If adverse effects occur, penicillamine should be withheld until the effect subsides. Subsequently, it can be reintroduced at a lower dosage. Some patients are unable to continue penicillamine therapy due to intolerable adverse effects.

b. Triethylene Tetramine Dihydrochloride (Trientine)

Trientine is an chelating agent that is generally used in penicillamine-intolerant patients (Walshe, 1982). About 10% of the administered trientine is absorbed through the digestive tract, and most of the absorbed trientine is metabolized to acetyltrientine (Gibbs and Walshe, 1986; Kodama et al., 1993a). The decoppering effect of trientine is less than that of penicillamine, however, no serious adverse effects have been reported for trientine. In rare cases, patients treated with trientine were reported to suffer from anorexia, gastrointestinal disturbances, iron deficiency anemia, mild elevations of serum amylase and lipase (Tankanow, 1991), and sideroblastic anemia (Condamine et al., 1993).

c. Zinc

Orally administered zinc inhibits copper absorption through the intestine (Friedman, 1993). Oral zinc therapy is probably a safe and effective maintenance treatment for patients who were initially treated with chelating agents, and seems to be an appropriate first-line therapy for presymptomatic and pregnant patients. Combination drug therapy of a chelating agent and zinc has no advantages when compared with zinc only, during maintenance therapy (Brewer et al., 1993).

2. Liver Transplantation

Liver transplantation is now accepted as the only effective treatment for FHF combined WD (Rela et al., 1993). The neurological symptoms have been reported to abate after liver transplantation. A patient was reported to have recovered from severe neurological symptoms following liver transplantation (Mason et al., 1993). Chelation therapy is not necessary after transplantation.

F. PROGNOSIS OF THE NEUROLOGICAL SYMPTOMS

Patients with neurological symptoms often become clinically worse within the first few weeks of therapy with penicillamine. One case is reported for a patient who developed new brain lesions during this period (Brewer et al., 1987). Pall et al. (1989) suggests that vitamin E prevents the deterioration of neurological symptoms. The deterioration is observed even in treatments with zinc (Lang et al., 1993). Asymptomatic patients rarely develop neurological signs after starting treatment with penicillamine (Walshe, 1988b). Brewer et al. (1991) reported that all the five patients with neurological symptoms, who were treated with ammonium tetrathiomolybdate, did not have a deterioration of neurological symptoms. Thus they recommended tetrathiomolybdate as the initial treatment of neurologically affected patients. Further studies are necessary regarding this treatment.

Walshe and Yealland (1993) treated 137 neurological patients with chelating agents, of whom 57 patients showed excellent, 36 showed good, and 24 showed poor response. In spite of the therapy, 11 patients died. Such poor response is a major problem in this treatment. Therefore, it is important to diagnose and treat patients before the neurological symptoms appear.

G. PREGNANCY

Successful pregnancy is very rare in patients with untreated WD. Oga et al. (1993) reported that a baby who was born from a mother with untreated WD showed hepatomegaly and liver damage, indicating that an untreated pregnancy causes fetal liver damage and copper accumulation in the placenta. Therefore, pregnant women with WD should be treated with penicillamine, trientine, or zinc (Tankanow, 1991). These treatments are considered to be safe for both the pregnant patients and the fetuses. Reversible cutis laxa, probably due to copper deficiency, was reported in a baby born from a patient with WD who was taking penicillamine 1.5 g/day (Linares et al., 1979). Thus, pregnant patients are recommended to continue the chelating agents therapy in a low dosage (Tankanow, 1991).

III. OTHER DISEASES IN WHICH COPPER ACCUMULATES IN THE LIVER

Copper is often retained in the liver in various chronic liver diseases with a cholestatic state. A few forms of liver diseases are more consistently associated with copper retention in the liver.

A. INDIAN CHILDHOOD CIRRHOSIS

Indian childhood cirrhosis, a rapidly progressive hepatic disease in children aged from 6 months to 3 years in India, is associated with extremely high concentrations of copper in the liver. The serum levels of copper and ceruloplasmin are normal (Adamson et al., 1992). These findings are different from those of WD. Such patients have rarely been reported in other countries (Adamson et al., 1992). D-penicillamine is beneficial (Bhusnurmath et al., 1991). A familial predilection is often observed (Sethi et al., 1993). The incidence of the disease, however, has significantly declined due to the improvement of nutrition for children and replacement of copper utensils with steel ones in the household (Bhave et al., 1992; Ramakrishna et al., 1993), showing that environmental factors contribute to the pathogenesis of the disease.

B. PRIMARY BILIARY CIRRHOSIS

This is an adult disease with no genetic inheritance which shows severe copper retention in the liver as in WD. The etiology has not been elucidated. However, the number of patients worldwide is increasing, especially in Western countries and Japan. Hepatocellular carcinoma has been reported to be found in 4.1% of patients with this disease (Nakanuma et al., 1990).

IV. ANIMAL MODELS FOR COPPER ACCUMULATION IN THE LIVER

Long-Evans rats with a cinnamon-like coat color (LEC rats), Bedlington terriers, and toxic milk mutation mice are known to have copper accumulation in the liver with an autosomal recessive inheritance. Among the above animals, the LEC rats seem to be most likely to contract WD according to various laboratory findings. Some differences, however, are observed between LEC rats and WD. In LEC rats, the serum ceruloplasmin level examined by a Western blotting method is almost normal; copper does not accumulate in the brain; hepatocellular carcinoma occurs frequently, while cirrhosis does not occur (Table 6). It is unclear whether these differences are due to the differences between the species (humans and rats) or that of the primary defect.

Table 6 Comparison of LEC Rats With Wilson's Disease

	Wilson's disease	LEC rats
Inheritance	Autosomal recessive	Autosomal recessive
Copper concentration		
Liver	Severely high	Severely high[d,e,f,g]
Biliary	Low	Low[e]
Serum	Low	Low[d,e,f,g]
Brain	High	Normal[d,e,g]
Kidney	High	High[d,e,g]
Urinary	High	High[d,g]
Serum ceruloplasmin		
Oxidase activity	Low	Low[d,e,f,g]
Amount of the protein[a]	Low[b]	Almost normal[f]
Serum AST, ALT levels	Increase	Increase[d,e,g]
Serum IgG level	High[c]	Low[h]
Hepatic pathology	See text	Similar to Wilson's disease[g]
Outcome	Liver cirrhosis	Hepatocellular carcinoma[g]
Therapy with chelators	Effective	Effective[g]

[a] Total protein amount measured by immunoassay.
[b] Some patients show normal level.
[c] Strickland and Leu, 1975.
[d] Okayasu et al., 1992.
[e] Sugawara et al., 1992.
[f] Yamada et al., 1993.
[g] Li et al., 1991.
[h] Matsumoto et al., 1989.

Bedlington terriers also have severe copper accumulation in the liver and reduced biliary excretion of copper (Brewer et al., 1992). However, the plasma ceruloplasmin of Bedlington terriers does not decrease. Moreover, the blood ^{64}Cu concentration 24 h after oral administration of ^{64}Cu has been reported to show no difference between affected and unaffected dogs (Brewer et al., 1992). These findings suggest that ceruloplasmin secretion into the blood from the liver is not disturbed in the dogs. Bedlington terriers

Genetic Disorders of Copper Metabolism

might be considered as a model for Indian childhood cirrhosis as indicated by the biochemical characteristics (Adamson et al., 1992).

V. MENKES DISEASE

Menkes disease (MNKD) is a neurodegenerative disorder described first in 1962 by Menkes and coworkers. Danks et al. (1972) demonstrated that MNKD is associated with copper deficiency due to defective intestinal copper absorption that results from copper accumulation in the intestine. This fact, together with the discovery of animal models, such as mottled mutant mice (Hunt, 1974) and macular mice (Nishimura, 1975) led to many studies on copper metabolism in MNKD. Recently, the MNKD gene has been identified in humans as well as in mice (Vulpe et al., 1993; Chelly et al., 1993; Mercer et al., 1993a,b; Levinson et al., 1993a). However, an effective treatment for MNKD has not yet been elucidated.

A. PREVALENCE AND GENETIC ANALYSIS

The prevalence rate seems to be 1 in 250,000–300,000 (Tønnesen et al., 1991a). The true rate, however, is unknown, because most patients die during early infancy. The inheritance is X-linked recessive. Thus, patients are male and their mothers are heterozygotes, that is, carriers of MNKD. However, a female patient with a cytologically balanced translocation, t(X;2)(q13;q32.3) has been reported (Kapur et al., 1987; Verga et al., 1991).

The MNKD gene is located on Xq13 (Tümer et al., 1992) and was recently identified. The gene deletions, which show variable sizes and localization of breakpoints, were detected in 16 out of 100 unrelated patients (Chelly et al., 1993). Northern blot analysis shows reduced or altered expression of transcript in some patients with MNKD (Mercer et al., 1993a). Two mottled (Mo) alleles, dappled (Mo^{dp}) and blotchy, have abnormalities in the murine MNKD mRNA and Mo^{dp} has a partial gene deletion, showing that this strain is a true animal model of MNKD (B. Levinson, personal communication).

B. PATHOGENESIS

Defective copper absorption results from copper accumulation in the intestine of patients with MNKD as well as that of mutant mice. The primary metabolic defect which causes copper accumulation in the intestine is present in almost all extrahepatic tissues, such as the kidney, muscle, and cultured fibroblasts (Danks, 1989). In the affected cells, the excessive copper is bound to metallothionein, the level of which is higher than that of control cells (Packman et al., 1987). However, the organelles in cells such as the mitochondria are in a state of copper deficiency, and then the activity of cuproenzymes is low (Kodama et al., 1989; Gacheru et al., 1993). Histochemical studies showed that copper concentrated in the cytosol (Kodama et al., 1993b). These findings suggest that the copper transport from the cytosol to the organelles is disturbed.

Defective copper absorption leads to copper deficiency especially in the serum, liver, and brain. The low copper concentration in the liver is improved by the parenteral administration of copper, suggesting that the decrease of copper in the liver results from the defective intestinal copper absorption (Grover and Scrutton, 1975). Vulpe et al. (1993) reported that the MNKD gene was expressed in the heart, brain, lungs, muscles, kidneys, and pancreas, but not in the liver of normal humans, indicating that copper metabolism in the liver is different from that in the extrahepatic tissues. Their findings are also consistent with the clinical observation that the liver is unaffected in patients with MNKD.

The copper concentration in the brain of affected fetuses increases (Heydorn et al., 1975). Moreover, ^{64}Cu retention is observed in the brain as well as in the intestine and kidneys of mutant mice after intracardiac injection of ^{64}Cu (Mann et al., 1979). These findings suggest that the brain is also an affected organ. However, the copper concentration in the brain of patients and mutant mice is lower than normal, showing that in MNKD the brain is in a state of copper deficiency due to the decrease of available copper in the brain resulting from postnatal intestinal malabsorption of copper (Danks, 1989). In patients, however, the cerebral degeneration due to the brain copper deficiency cannot be improved by the postnatal parenteral copper administration. In contrast, mutant mice survive and show almost normal growth when they are administered copper around postnatal day 7 (Nagara et al., 1981). The 7th postnatal day of mice is known to correspond with the mid-third trimester of humans. These findings suggest that the timing of parenterally administered copper to patients is important to prevent neurological degeneration. Thus, copper should be administered at the critical stage of brain development which seems to be associated with myelination and the development of the central nervous system, including the blood-brain barrier.

In mutant mice, copper accumulates in the astrocytes and the vascular endothelium, which are components of the blood-brain barrier (Kodama et al., 1991; Kodama, 1993). The blood-borne copper is transported to the neurons via the blood-brain barrier. Therefore, the blood-borne copper in patients as well as mutant mice is probably trapped in the blood vessels and the astrocytes, and thus is not transported to the neurons after the blood-brain barrier matures. This accords well with the fact that intravenously administered copper does not improve the neuronal degeneration in patients with MNKD.

The activities of cuproenzymes decrease in the affected tissues as well as in the liver and brain. Most features of MNKD can be explained by the decrease of the enzyme activities (Danks, 1993; Table 7). The decreased activities of dopamine-β-hydroxylase and cytochrome c oxidase in the brain result in abnormal levels of catecholamines and a high level of lactate in the cerebrospinal fluid. These laboratory tests seem to be useful markers for monitoring treatment (Kaler et al., 1993; Kreuder et al., 1993).

Table 7 Copper Enzymes in Humans

Common name	Functional role	Known or expected consequence of deficiency
Cytochrome oxidase	Electron transport chain	Neurological damage; muscle weakness; hypothermia
Superoxide dismutase	Free radical detoxification	Membrane peroxidation
Tyrosinase	Melanin production	Failure of pigmentation
Dopamine β-hydroxylase	Catecholamine production	Neurological effects; possibly hypothermia
Lysyl oxidase	Cross-linking of elastin and collagen	Arterial abnormalities; bladder diverticula, loose skin and joints
Ceruloplasmin	Copper transport, ferroxidase	Anemia; secondary copper deficiency
Enzyme not known	Cross-linking of keratin (disulfide bonds)	Pili torti
Coagulation factor V	Blood coagulation	Bleeding (not observed)

From Danks, D.M., *Connective Tissue and Its Inheritable Disorders*, Royce, P.M., and Steinman, B., Eds., Wiley-Liss, New York, 1993. With permission.

C. PATHOLOGY

Very striking pathological changes are found in the arteries and the brain (Danks, 1989, 1993). Angiography reveals elongation, aneurysmal dilatation, stenosis, and rupture of arteries in the brain and other organs. Microscopic and ultrastructural examinations of arteries show the fragmentation and disruption of internal elastic lamina with proliferation of bundles of poorly organized elastic fibers and intimal thickening. Skin elastic fibers have similar abnormalities.

Neuropathological changes are observed, especially in the cerebral cortex and cerebellum. The prominent changes are loss of many Purkinje cells as well as neuronal loss in the internal granule cell layer and molecular layer of the cerebellum. The remaining Purkinje cells show an unusual elaboration of dendritic sprouts from the cell body and a grotesque proliferation of the dendritic tree (torpedo-like), which have been considered to be characteristic abnormalities of MNKD. Ultramicroscopic studies show filamentous inclusion and abnormal mitochondria in the Purkinje cells and granule cells. Spinal cords are also affected, with severe demyelination in both ascending and descending tracts (Uno and Arya, 1987). Biochemically, 2′,3′-cyclic nucleotide 3′-phosphodiesterase (CNP) activity has been reported to decrease in the brain of mutant mice, suggesting the disturbance of myelination (Sasahara et al., 1988).

Ragged red fibers which are observed in the muscles of patients with mitochondrial myopathies are also seen in the muscles of patients with MNKD (Morgello et al., 1988). Suzuki and Nagara (1981) reported that increased numbers of mitoses, numerous vacuoles, and bizarre nuclei are observed in the proximal tubular epithelium of the kidney from mutant mice.

D. CLINICAL MANIFESTATIONS

Table 8 shows the symptoms and signs of patients with MNKD. The most serious symptoms are neuronal deterioration and arterial abnormalities. Premature delivery is very frequent. In the neonatal period, some patients have hypothermia, hyperbilirubinemia, and mild hair abnormalities (Danks, 1993).

However, patients are seldom diagnosed in the neonatal period, because these abnormalities sometimes appear in unaffected premature babies. Usually patients seem to be normal after the neonatal period until 2 to 3 months of age. After this period, however, various symptoms appear such as developmental delay, loss of early development skills, hypothermia, hair abnormality, and convulsion that is usually intractable (Figure 1).

Table 8 Symptoms and Signs of Menkes Disease

Premature delivery
Neurological deterioration: Intractable convulsions, developmental delay, failure to thrive
Hair abnormalities: Coarse, sparse, and tangled hair, sometimes no hair
Arterial abnormalities: Subdural hematoma, hemorrhage
Odd face: Pudgy cheeks, sagging jowls
Bone changes: Osteoporosis, wormian bones, flared metaphyses, fractures
Hypothermia
Eczema
Diarrhea
Bladder diverticula
Recurrent urinary infection
Retinal degeneration, iris cysts
Gastrointestinal polyps[a]

[a] Kaler, S.G., Goldstein, D.S., Holmes, C., Salerno, J.A., and Gahl, W.A., *Ann. Neurol.*, 33, 171, 1993.

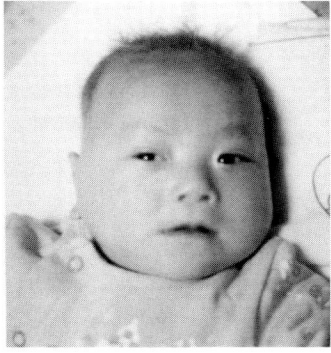

Figure 1 A boy with Menkes disease, at the age of four months.

Neurological deterioration becomes severe. Electroencephalograms (EEG) show moderate or severe abnormalities with multifocal spike and wave discharges. The degree of EEG abnormality seems to be related to the severity of neurodevelopmental impairment (White et al., 1993). Brain CT scan reveals brain atrophy, enlarged ventricles, or a low-density area in the cerebrum. MRI findings show severe cortical and cerebellar atrophy, deranged cerebral vasculature, and subdural fluid collection (Johnsen et al., 1991).

The hair abnormalities are characteristic to MNKD. These come into prominence at the age of 2 to 3 months. Bone changes are also present. Diverticula of the bladder and ureters are sometimes ruptured. An excessive amount of copper accumulates in the kidney of patients with MNKD. Copper accumulation is also observed in patients with WD, who frequently suffer from renal diseases such as renal tubular dysfunction, due to copper toxicity (see Section II.C). In patients with MNKD, however, renal diseases due to copper toxicity have seldom been reported (Garnica et al., 1977; Kreuder et al., 1993). The renal tubular function is not disturbed in patients with MNKD (Kodama et al., 1992). In both diseases, copper accumulates to the same degree in the kidneys (Kodama et al., 1992). The differences in the toxic effects of copper must be ascribed to the difference in the copper accumulation mechanism in the kidneys.

Most patients die from infection or neurological degeneration by the age of 3 years, though some patients with classical MNKD are reported to have survived until 14 years of age (Sander et al., 1988; Gerdes et al., 1988).

E. DIAGNOSIS

It is not difficult to diagnose patients with classical (typical) MNKD based on the clinical features of their symptoms: low levels of serum copper and ceruloplasmin, and the characteristic radiological changes. Orally administered copper does not improve the low levels of serum copper and ceruloplasmin. The copper concentration of the cultured fibroblasts and lymphoid cells is significantly high.

Heterozygote females can be diagnosed on Piti torti in their hairs, patchy skin depigmentation, and/or an increased copper level in cultured fibroblasts. However, the degree of these abnormalities is different among the heterozygotes. Sometimes they show no abnormality. This variety can be explained by random X chromosome inactivation. Therefore, heterozygotes cannot be excluded even if these abnormalities are not found.

Prenatal diagnosis is made by copper analysis or ^{64}Cu uptake examination in cultured amniotic cells or in chorionic villi. Tønnesen et al. (1989) reported that a combination of ^{64}Cu uptake with chase experiments on the amniotic cells can give a good distinction between affected and unaffected fetuses. However, there is no definitive biochemical method for the diagnosis of heterozygotes and fetuses. DNA-based diagnosis should be established as a more reliable diagnosis method of heterozygotes and fetuses with MNKD.

F. TREATMENT

Copper administered orally or intrarectally is too poorly absorbed to improve the serum copper level. Copper has been injected intravenously or subcutaneously in many different forms, such as copper-histidine, copper-acetate, and copper-EDTA. None of these postnatal treatments has yet been established as an adequately beneficial treatment for neurological deterioration, while they do improve the serum copper and ceruloplasmin levels (Grover and Scrutton, 1975; Sherwood et al., 1989; Kollros et al., 1991). Among the different forms of copper, copper-histidine has been reported to be taken up by the brain most efficiently, though the mechanism has not yet been clarified (Barnea and Katz, 1990). Thus, it has been used for the treatment of patients with MNKD. The dosage of copper as copper-histidine is recommended to be 50 to 150 μg/kg once a day (Sarker et al., 1993), or 100 μg/day in newborn babies and 1 mg/day in older children (Danks, 1993). In order to know that the dosage is adequate it is necessary to monitor the serum ceruloplasmin and copper levels, urinary copper excretion, or copper concentration in the liver (Danks, 1993; Kreuder et al., 1993).

It is important to begin treatment as early as possible. When treatment with copper-histidine is initiated to the fetuses *in utero* or newborn babies, the neurological deterioration may be prevented (Sarker et al., 1993; Danks, 1993; Kreuder et al., 1993). When treatment is initiated after 2 month of age, it does not improve the neurological deterioration. When copper is injected into the patient, almost all extrahepatic tissues accumulate a large amount of copper because of the primary metabolic defect. However, no serious copper toxicity has been observed in these cases (Kreuder et al., 1993). This is probably because administered copper becomes nontoxic by binding with metallothionein in these tissues. The treatment does not completely improve some of the nonneurological problems, such as connective tissue laxity (Sarker et al., 1993). This coincides with the fact that administered copper binds with metallothionein and is not available for the synthesis of cuproenzymes in these tissues. Parenteral copper administration in combination with either D-penicillamine (Nadal and Baerlocher, 1988) and vitamin E (Tada et al., 1988) has been reported. However, it is not clear whether these treatments are beneficial.

In contrast, animal models are known to survive and show almost normal growth when they are administered copper around postnatal day 7 (see Section V.B). During the months following treatment, biochemical data and morphological observations of the brain gradually improve (Meguro et al., 1991). Therefore, parenteral administration of copper seems to be necessary, especially during early infancy in patients with MNKD.

VI. MILD MENKES DISEASE, OCCIPITAL HORN SYNDROME, AND INHERITED COPPER DEFICIENCY

Patients with a milder form of MNKD have rarely been reported (Danks, 1988; Gerdes et al., 1988; Danks, 1993). These patients have some characteristic symptoms of MNKD, such as abnormal hair, skin and joint laxity, and facial appearance. However, neurological diseases are milder than in classical MNKD. The predominant problems in these cases are ataxia and dysarthria. Onset of the symptoms is later than that of classical MNKD. The serum levels of copper and ceruloplasmin are also low. Moreover, the ^{64}Cu uptake and retention values in cultured fibroblasts from these mild forms are indistinguishable from those of classical MNKD (Tønnesen et al., 1991b).

Occipital horn syndrome, first described by Lazoff et al. in 1975, is an X-linked recessive disorder characterized by skeletal dysplasia, soft easily bruisable skin, hyperextensible joints, diarrhea, obstructive uropathy, and occipital exostoses. The abnormalities observed in patients with occipital horn syndrome are very similar to those of patients with MNKD, such as the low levels of serum copper and ceruloplasmin, defective intestinal copper absorption, increased copper content of the cultured fibroblasts, and arterial changes. However, patients with occipital horn syndrome show only mild or minor neurological impairment. The activity of lysyl oxidase, a cuproenzyme, in the skin and cultured fibroblasts is significantly low. From the clinical and biochemical characteristics, occipital horn syndrome has been considered to be probably allelic to MNKD (Danks, 1993). Levinson et al. (1993b) demonstrated by Northern blot analysis that expression of the MNKD gene was markedly reduced in fibroblasts of two patients with occipital horn syndrome, indicating that this disease is allelic to MNKD.

Other copper deficiency diseases with X-linked or autosomal inheritance have been reported (Haas et al., 1981; Mehes and Petrovicz, 1982; Fujii et al., 1991; Iwakawa et al., 1993). Their clinical and biochemical findings differ from MNKD and occipital horn syndrome. The elucidation of these disease from various aspects, especially that of gene analysis, will provide a clue to solve the complex mechanism of copper metabolism in humans.

REFERENCES

Adamson, M., Reiner, B., Olson, J. L., Goodman, Z., Plotnick, L., Bernardini, I., and Gahl, W. A., *Gastroenterology,* 162, 1771–1777, 1992.
Barnea, A. and Katz, B. M., *J. Inorg. Biochem.,* 40, 81–93, 1990.
Bhave, S. A., Pandit, A. N., Singh, S., Walia, B. N. S., and Tanner, M. S., *Ann. Trop. Paediatr.,* 12, 23–30, 1992.
Bhusnurmath, S. R., Walia, B. N. S., Singh, S., Parkash, D., Radotra, B. D., and Nath, R., *Hum. Pathol.,* 22, 653–658, 1991.
Bonné-Tamir, B., Frydman, M., Agger, M. S., Bekeer, R., Bowcock, A. M., Hebert, J. M., Cavalli-Sforza, L. L., and Farrer, L. A., *Ann. Hum. Genet.,* 54, 155–168, 1990.
Brewer, G. J., Dick, R. D., Yuzbasiyan-Gurkin, V., Tankanow, R., Young, A. B., and Kluin, K. J., *Arch. Neurol.,* 48, 42–47, 1991.
Brewer, G. J., Schall, W., Dick, R., Yuzbasiyan-Gurkan, V., Thomas, M., and Padgett, G., *J. Vet. Intern. Med.,* 6, 41–43, 1992.
Brewer, G. J., Terry, C. A., Aisen, A. M., and Hill, G. M., *Arch. Neurol.,* 44, 490–493, 1987.
Brewer, G. J. and Yuzbasiyan-Gurkan, V., *Dig. Dis. Sci.,* 7, 178–193, 1989.
Brewer, G. J., Yuzbasiyam-Gurkan, V., Johnson, V., Dick, R. D., and Wang, Y., *J. Am. Coll. Nutr.,* 12, 26–30, 1993.
Bull, P. C. and Cox, D. W., *Genomics,* 16, 593–598, 1993.
Bull, P. C., Thomas, G. R., Rommens, J. M., Forbes, J. R., and Cox, D. W., *Nature Genet.,* 5, 327–337, 1993.
Carpenter, T. O., Carnes, D. L., and Anast, C. S., *N. Engl. J. Med.,* 309, 873–877, 1983.
Chelly, J., Tümer, Z., Tønnesen, T., Petterson, A., Ishikawa, Y., Tommerup, N., Horn, N., and Monaco, A. P., *Nature Genet.,* 3, 14–19, 1993.
Cheng, W. S. C., Govindarajan, S., and Redeker, A. G., *Liver,* 12, 42–45, 1992.
Condamine, L., Hermine, O., Alvin, P., Levine, M., Rey, C., and Courtecuisse, V., *Br. J. Haematol.,* 83, 166–168, 1993.
Cossu, P., Pirastu, M., and Nucaro, A., *N. Engl. J. Med.,* 327, 57, 1992.
Danks, D. M., *Am. J. Med. Genet.,* 30, 859–864, 1988.
Danks, D. M., in *The Metabolic Basis of Inherited Disease,* 6th ed., Scriver, C. R., Beaudet, A. I., Sly, W. S., and Walle, D., Eds., McGraw-Hill, New York, 1411, 1989.
Danks, D. M., in *Connective Tissue and Its Heritable Disorders,* Royce, P. M. and Steinmann, B., Eds., Wiley-Liss, New York, 487, 1993.
Danks, D. M., Campbell, P. E., Stevens, B. J., Mayne, V., and Cartwright, E., *Pediatrics,* 50, 188–201, 1972.
Friedman, L. S., *Gastroenterology,* 104, 1566–1568, 1993.
Frydman, M., Bonné-Tamir, B., Farrer, L. A., Conneally, P. M., Magazanik, A., Ashbel, S., and Goldwitch, Z., *Proc. Natl. Acad. Sci. U.S.A.,* 82, 1819–1821, 1985.
Fujii, T., Okuno, T., Ito, M., Kaji, M., Mutoh, K., Mikawa, H., and Saida, K., *Neurology,* 41, 1263–1266, 1991.
Gacheru, S., McGee, C., Uriu-Hare, J. Y., Kosonen, T., Packman, S., Tinker, D., Krawetz, S. A., Reiser, K., Keen, C. L., and Rucker, R. B., *Arch. Biochem. Biophys.,* 301, 325–329, 1993.
Gaffney, D., Walker, J. L., O'Donnell, J. G., Fell, G. S., O'Neill, K. F., Park, R. H., and Russell, R. I., *J. Inherit. Metab. Dis.,* 15, 161–170, 1992.
Garnica, A. D., Frias, J. L., and Rennert, O. M., *Clin. Genet.,* 11, 154–161, 1977.
Gerdes, A. M., Tønnesen, T., Pergament, E., Sander, C., Baerlocher, K. E., Wartha, R., Gutter, F., and Horn, N., *Eur. J. Pediatr.,* 148, 132–135, 1988.

Giagheddu, A., Demelia, L., Puggioni, G., Nurchi, A. M., Contu, L., Pirari, G., Deplano, A., and Rachele, M. G., *Acta Neurol. Scand.,* 72, 43–55, 1985.
Gibbs, K. and Walshe, J. M., *J. Gastroenterol. Hepatol.,* 5, 420–424, 1990.
Gibbs, K. R. and Walshe, J. M., in *Orphan Diseases and Orphan Drugs,* Scheinberg, I. H. and Walshe, J. M., Eds., Manchester University Press, London, 33, 1986.
Goldfischer, S. and Sternlieb, I., *Am. J. Pathol.,* 53, 883–901, 1968.
Golding, D. N. and Walshe, J. M., *Ann. Rheum. Dis.,* 36, 99–111, 1977.
Gregorio, G. V., and Mieli-Vergani, G., *Hepatology,* 18, 706–707, 1993.
Grover, W. D. and Scrutton, M. C., *J. Pediatr.,* 86, 216–220, 1975.
Haas, R. H., Robinson, A., Evans, K., Lascelles, P. T., and Dubowitz, V., *Neurology,* 31, 852–589, 1981.
Heydorn, K., Damsgaard, E., Horn, N., Mikkelsen, M., Tygstrup, I., and Vestermark, S., *Hum. Genet.,* 29, 171–175, 1975.
Hiyamuta, S., Shimizu, K., and Aoki, T., *Lancet,* 342, 56–57, 1993.
Hoogenraad, T. U., Van Hattum, J., and Van den Hamer, C. J. A., *J. Neurol. Sci.,* 77, 137–146, 1987.
Hoppe, B., Neuhaus, T., Superti-Furga, A., Forster, I., and Leumann, E., *Nephron,* 65, 460–462, 1993.
Hunt, D. M., *Nature,* 249, 852–853, 1974.
Iwakawa, Y., Shimohira, M., Kohyama, J., and Kodama, H., *Eur. J. Pediatr.,* 152, 368–371, 1993.
Iyengar, V., Brewer, G. J., Dick, R. D., and Owyang, C.,, *J. Lab. Clin. Med.,* 111, 267–274, 1988.
Johnsen, D. E., Coleman, L., and Poe, L., *Neuroradiology,* 33, 181–182, 1991.
Johansen, K., Gregersen, G., and Denmark, A., *Arch. Intern. Med.,* 129, 587–590, 1972.
Kaler, S. G., Westman, J. A., Bernes, S. M., Elsayed, A. M., Bowe, C. M., Freeman, K. L. B., Wu, C. D., and Wallach, M. T., *J. Pediatr.,* 122, 93–95, 1993.
Kaler, S. G., Goldstein, D. S., Holmes, C., Salerno, J. A., and Gahl, W. A., *Ann. Neurol.,* 33, 171–175, 1993.
Kapur, S., Higgins, J. V., Delp, K., and Rogers, B., *Am. J. Med. Genet.,* 26, 503–510, 1987.
Kaushansky, A., Frydman, M., Kaufman, H., and Homburg, R., *Fertil. Steril.,* 47, 270–273, 1987.
Kodama, H., *J. Inherit. Metab. Dis.,* 16, 791–799, 1993.
Kodama, H., Abe, T., Takama, M., Takahashi, I., Kadama, M., and Nishimura, M., *Histochem. Cytochem.,* 41, 1529–1535, 1993b.
Kodama, H., Meguro, Y., Abe, T., Rayner, M. H., Suzuki, K. T., Kobayashi, S., and Nishimura, M., *J. Inherit. Metab. Dis.,* 14, 896–901, 1991.
Kodama, H., Meguro, Y., Tsunakawa, A., Nakazato, Y., Abe, T., and Murakita, H., *Tohoku J. Exp. Med.,* 169, 59–66, 1993a.
Kodama, H., Okabe, I., Kihara, A., Mori, Y., and Okaniwa, M., *J. Inherit. Metab. Dis.,* 15, 157–158, 1992.
Kodama, H., Okabe, I., Yanagisawa, M., and Kodama, Y., *J. Inherit. Metab. Dis.,* 12, 386–389, 1989.
Kodama, H., Okabe, I., Yanagisawa, M., Nomiyama, H., Nomiyama, K., Nose, O., and Kamoshita, S., *Pediatr. Neurol.,* 4, 35–37, 1988.
Kollros, P. R., Dick, R. D., and Brewer, G. J., *Pediatr. Neurol.,* 7, 305–307, 1991.
Kreuder, J., Otten, A., Fuder, H., Tümer, Z., Tønnesen, T., Horn, N., and Dralle, D., *Eur. J. Pediatr.,* 152, 828–832, 1993.
Kuan, P., *Chest,* 91, 579–583, 1987.
Lang, C. J. G., Rabas-Kolominsky, P., and Engelhardt, A., *Arch. Neurol.,* 50, 1007–1008, 1993.
Lazoff , S. G., Ryback, J. J., Parker, B. R., and Luzzatti, L., *Birth Defects,* 11, 71–74, 1975.
Leu, M. L., Strickland, T., Wang, C. C., and Chen, T. S. N., *J. Am. Med. Assoc.,* 211, 1542–1543, 1970.
Levinson, B., Gitshier, J., Vulpe, C., Whitney, S., Yang, S., and Packman, S., *Nature Genet.,* 3, 6, 1993b.
Levinson, B., Vulpe, C., Das, S., Whitney, S., Martin, C., Elder, B., Verley, F., Packman, S., and Gitshier, J., *Am. J. Hum. Genet.,* 53 (Suppl. 920) (abstract), 1993a.
Li, Y., Togashi, Y., and Takeichi, N., in *The LEC Rat: A New Model for Hepatitis and Liver Cancer,* Mori, M., Yoshida, M. C., Takeichi, N., and Taniguchi, N., Eds., Springer-Verlag, Tokyo, 122, 1991.
Linares, A., Zarranz, J. J., Rodriguez-Alarcon, J., and Diaz-Perez, J. L., *Lancet,* 1, 43, 1979.
Lindquist, R. R., *Am. J. Pathol.,* 53, 903–927, 1968.
Mann, J. R., Camakaris, J., and Danks, D. M., *Biochem. J.,* 180, 613–619, 1979.
Marecek, Z. and Nevsimalova, S., *J. Inherit. Metab. Dis.,* 7, 41–45, 1984.
Martins da Costa, C., Baldwin, D., Portmann, B., Lolin, Y., Mowat, A. P., and Mieli-Vergani, G., *Hepatology,* 15, 609–615, 1992.
Mason, A. L., Marsh, W., and Alpers, D. H., *Dig. Dis. Sci.,* 38, 1746–1750, 1993.
Matsuda, I., Pearson, T., and Holtzman, N. A., *Pediatr. Res.,* 8, 821–824, 1974.
Matsumoto, K., Takeichi, N., Izumi, K., and Otsuka, H., *Transplant. Proc.,* 21, 3259, 1989.
McCullough, A. J., Fleming, C. R., Thistle, J. L., Baldus, W. P., Ludwig, J., McCall, J. T., and Dickson, E. R., *Gastroenterology,* 84, 161–167, 1983.
Meguro, Y., Kodama, H., Abe, T., Kobayashi, S., Kodama, Y., and Nishimura, M., *Brain Dev.,* 13, 184–186, 1991.
Mehes, K. and Petrovicz, E., *Arch. Dis. Child.,* 57, 716–718, 1982.
Menkes, J. H., Alter, M., Steigleder, G. K., Weakley, D. R., and Sung, J. H., *Pediatrics,* 29, 764–779, 1962.
Mercer, J. F. B., Grimes, A., Paynter, J. A., Lockhart, P., Ambrosini, L., Dierick, H., Danks, D. M., and Glover, T. W., *Am. J. Hum. Genet.,* 53 (Suppl. 711) (abstract), 1993b.
Mercer, J. F. B., Livingston, J., Hall, B., Paynter, J. A., Begy, C., Chandrasekharappa, S., Lockhart, P., Grimes, A., Bhave, M., Siemieniak, D., and Glover, T. W., *Nature Genet.,* 3, 20–25, 1993a.

Meyer, B.-U., Britton, T. C., Bischoff, C., Machetanz, J., Benecke, R., and Conrad, B., *Mov. Disord.*, 6, 320–323, 1991.
Morgello, S., Peterson, H. D. C., Kahn, L. J., and Laufer, H., *Dev. Med. Child. Neurol.*, 30, 812–816, 1988.
Nadal, D. and Baerlocher, K., *Eur. J. Pediatr.*, 147, 621–625, 1988.
Nagara, H., Yajima, K., and Suzuki, K., *J. Neuropathol. Exp. Neurol.*, 40, 428–446, 1981.
Nakanuma, Y., Terada, T., Doishita, K., and Miwa, A., *Hepatology*, 11, 1010–1016, 1990.
Nartey, N. O., Frei, J. V., and Cherian, M. G., *Lab. Invest.*, 57, 397–401, 1987.
Nazer, H., Ede, R. J., Mowat, A. P., and Williams, R., *Gut*, 27, 1377–1381, 1986.
Nishimura, M., *Exp. Anim. Tokyo*, 24, 185, 1975 (in Japanese).
Oder, W., Grimm, G., Kollegger, H., Ferenci, P., Schneider, B., and Deecke, L., *J. Neurol.*, 238, 281–287, 1991.
Oder, W., Prayer, L., Grimm, G., Spatt, J., Ferenci, P., Kollegger, H., Schneider, B., Gangl, A., and Deecke, L., *Neurology*, 43, 120–124, 1993.
Oga, M., Matsui, N., Anai, T., Yoshimatsu, J., Inoue, I., and Miyakawa, I., *Am. J. Obstet. Gynecol.*, 169, 196–198, 1993.
Okayasu, T., Tochimaru, H., Hyuga, T., Takahashi, T., Takekoshi, Y., Li, Y., Togashi, Y., Takeichi, N., Kasai, N., and Arashima, S., *Pediatr. Res.*, 31, 253–257, 1992.
Packman, S., Palmiter, R. D., Karin, M., and O'Toole, C., *J. Clin. Invest.*, 79, 1338–1342, 1987.
Pall, H. S., Williams, A. C., and Blake, D. R., *Arch. Neurol.*, 46, 359–361, 1989.
Person, J. L., Anderson, D. S., and Brower, R. A., *Am. J. Gastroenterol.*, 82, 66–68, 1987.
Petrukhin, K., Fischer, S. G., Pirastu, M., Tanzi, R. E., Chernov, I., Devoto, M., Brzustowicz, L. M., Cayanis, E., Vitale, E., Russo, J. J., Matseoane, D., Boukhgalter, B., Wasco, W., Figus, A. L., Loudianos, J., Cao, A., Sternlieb, I., Evgrafov, O., Parano, E., Pavone, L., Warburton, D., Ott, J., Penchaszadeh, G. K., Scheinberg, I. H., and Gilliam, T. C., *Nature Genet.*, 5, 338–343, 1993.
Ramakrishna, B., Date, A., Kirubakaran, C., and Raghupathy, P., *Ann. Trop. Paediatr.*, 13, 159–163, 1993.
Reilly, M., Daly, L., and Hutchinson, M., *J. Neurol. Neurosurg. Psychiatry*, 56, 298–300, 1993.
Rela, M., Heaton, N. D., Vougas, V., McEntee, G., Gane, E., Farhat, B., Chiyende, J., Mieli-Vergani, G., Mowat, A. P., Portmann, B., Williams, R., and Tan, K. C., *Br. J. Surg.*, 80, 909–911, 1993.
Rosenfield, N., Grand, R. J., Watkins, J. B., Ballantine, T. V. N., and Levey, R. H., *J. Pediatr.*, 92, 210–213, 1978.
Saito, T., *Jpn. J. Hum. Genet.*, 30, 249–253, 1985.
Saito, T., *Eur. J. Pediatr.*, 146, 261–265, 1987.
Sander, C., Niederhoff, H., and Horn, N., *Clin. Genet.*, 33, 228–233, 1988.
Sarker, B., Lingertat-Walsh, K., and Clarke, J. T. R., *J. Pediatr.*, 123, 828–830, 1993.
Sasahara, A., Yamasaki, S., Tachiiri, T., Kawasaki, H., Yamano, T., Ohya, N., and Shimada, M., *Brain Dev.*, 10, 54–56, 1988.
Scheinberg, I. H., Sternlieb, I., Schilsky, M., and Stockert, R. J., *Lancet*, 2, 95, 1987.
Schulman, S. and Barbeau, A., *J. Neuropathol. Exp. Neurol.*, 22, 105–119, 1963.
Scott, J., Gollan, J. L., Samourian, S., and Sherlock, S., *Gastroenterology*, 74, 645–651, 1978.
Sethi, S., Grover, S., and Khodaskar, M. B., *Ann. Trop. Paediatr.*, 13, 3–6, 1993.
Sherwood, G., Sarkar, B., and Kortsak, A. S., *J. Inherit. Metab. Dis.*, 12, 393–396, 1989.
Starosta-Rubinstein, S., Young, A. B., Kluin, K., Hill, G., Aisen, A. M., Gabrielsen, T., and Brewer, G. J., *Arch. Neurol.*, 44, 365–370, 1987.
Stremmel, W., Meyerrose, K.-W., Niederau, C., Hefter, H., Kreuzpaintner, G., and Strohmeyer, G., *Ann. Intern. Med.*, 115, 720–726, 1991.
Strickland, G. T. and Leu, M.-L., *Medicine*, 54, 113–137, 1975.
Sugawara, N., Sugawara, C., Sato, M., Takahashi, H., and Mori, M., *Pharmacol. Toxicol.*, 71, 321–324, 1992.
Suzuki, K. and Nagara, H., *Acta Neuropathol.*, 55, 251–255, 1981.
Tada, K., Tanaka, M., Inada, E., Koga, K., Morooka, K., Arimoto, K., and Matsuo, T., *No to Hattatsu*, 20, 514–516, 1988 (in Japanese).
Tankanow, R. M., *Clin. Pharmacol.*, 10, 839–849, 1991.
Tanzi, R. E., Petrukhin, K., Chernov, I., Pellequer, J. L., Wasco, W., Ross, B., Romano, D. M., Parano, E., Pavone, L., Brzustowicz, L. M., Devoto, M., Peppercorn, J., Bush, A. I., Sternlieb, I., Pirastu, M., Gusella, J. F., Evgrafov, O., Penchaszadeh, G. K., Honing, B., Edelman, I. S., Soares, M. B., Scheinberg, I. H., and Gilliam, T. C., *Nature Genet.*, 5, 344–350, 1993.
Thuomas, K. Å., Aquilonius, S. M., Bergstrom, K., and Westermark, K., *Neuroradiology*, 35, 134–141, 1993.
Tønnesen, T., Garret, C., and Gerdes, A. M., *J. Med. Genet.*, 28, 615–618, 1991b.
Tønnesen, T., Gerdes, A. M., Damsgaard, E., Miny, P., Holzgreve, W., Søndergaard, F., and Horn, N., *Prenat. Diagn.*, 9, 159–165, 1989.
Tønnesen, T., Kleijer, W. J., and Horn, N., *Hum. Genet.*, 86, 408–410, 1991a.
Tümer, Z., Tommerup, N., Tønnesen, T., Kreuder, J., Craig, I. W., and Horn, N., *Hum. Genet.*, 88, 668–672, 1992.
Uno, H. and Arya, S., *Am. J. Med. Genet.*, Suppl. 3, 367–377, 1987.
Uzman, L. L., *Am. J. Med. Sci.*, 226, 645–652, 1953.
Verga, V., Hall, B. K., Wang, S., Johson, S., Higgins, J. V., and Glover, T. W., *Am. J. Hum. Genet.*, 48, 1133–1138, 1991.
Volder, A. D., Sindic, C. J. M., and Goffinet, A. M., *J. Neurol. Neurosurg. Psychiatry*, 51, 947–949, 1988.
Vulpe, C., Levinson, B., Whitney, S., Packman, S., and Gitschier, J., *Nature Genet.*, 3, 7–13, 1993.
Walshe, J. M., *Am. J. Med.*, 21, 487–495, 1956.
Walshe, J. M., *Lancet*, 1, 643–647, 1982.

Walshe, J. M., *Mov. Disord.,* 3, 10–29, 1988a.
Walshe, J. M., *Lancet,* 2, 435–437, 1988b.
Walshe, J. M. and Dixon, A. K., *Lancet,* 1, 845–847, 1986.
Walshe, J. M. and Yealland, M., *Q. J. Med.,* 86, 197–204, 1993.
Weisner, B., Hartard, C., and Dieu, C., *J. Neurol. Sci.,* 79, 229–237, 1987.
White, S. R., Reese, K., Sato, S., and Kaler, S. G., *Electroencephalogr. Clin. Neurophysiol.,* 87, 57–61, 1993.
Wiebers, D. O., Wilson, D. M., McLeod, R. A., and Goldstein, N. P., *Am. J. Med.,* 67, 249–254, 1979.
Willson, R. A. and Hartleb, M., *Am. J. Gastroenterol.,* 83, 1309–1310, 1988.
Wilson, D. M. and Goldstein, N. P., *Kidney Int.,* 4, 331–336, 1973.
Wilson, S. A. K., *Brain,* 34, 295–509, 1912.
Yamada, T., Agui, T., Suzuki, Y., Sato, M., and Matsumoto, K., *J. Biol. Chem.,* 268, 8965–8971, 1993.
Yarze, J. C., Martin, P., and Muñoz, S. J., *Am. J. Med.,* 92, 643–654, 1992.

Chapter 23

Alzheimer's Disease and Other Aluminum-Associated Health Conditions

D.R. McLachlan, P.E. Fraser, E. Jaikaran, and W.J. Lukiw

I. INTRODUCTION

Aluminum has long been recognized as highly toxic to the nervous system. Aluminum is also essential to the welfare of advanced civilizations. Not only have aluminum compounds been widely used for water treatment, food preservation, and pharmaceuticals, but the metal is also increasingly important in numerous industrial applications. The long-accepted commercial use of aluminum and its compounds has led to considerable resistance to investigating the possible role of aluminum in human disease. Indeed, the complex chemistry of aluminum, the slow accumulation in nervous tissues, and the delayed toxic expression of this element have compounded the difficulty in assigning a precise role in human disease. However, increasing evidence, accumulated in the last five years, has raised important public health concerns regarding excess aluminum exposure in some segments of the population.

Excess nervous tissue aluminum concentrations, reaching the neurotoxic range for certain experimental animals, have been reported in a number of neurodegenerative diseases including Alzheimer's disease, renal failure with "dialysis dementia", high incidence foci of amyotrophic lateral sclerosis and parkinsonism/dementia complex in islands in the western Pacific Ocean, prolonged industrial exposure to aluminum fumes and powder in certain neurons of the substantia nigra in idiopathic parkinsonism, and in premature infants exposed to high concentrations of aluminum in parenteral fluids.

II. ALUMINUM AND ALZHEIMER'S DISEASE

The elderly are not only the most rapidly growing segment of the population in developed countries, but are also at considerable risk of developing a lethal, mind-destroying chronic illness, Alzheimer's disease (AD). AD is the most common cause of senile dementia, with a prevalence ranging between 5% (McDowell and Lindsay, 1994) and 9% of individuals over age 65 years (Katzman, 1976). There are three principal brain changes in AD: senile plaques (SP), neurons undergoing neurofibrillary degeneration (NFD), and neuronal loss. SP are composed of degenerating nerve terminals intermeshed with extracellular accumulations of a brain specific protein, Aβ, and reactive glial cells. NFD is an intraneuronal accumulation of large amounts of posttranslationally modified filamentous protein assembled into paired helical filament (PHF) arrays. Neuronal loss is also associated with loss of synaptic connections, the histopathological change which most strongly correlates with the clinically important event in AD, dementia (DeKosky, 1990; Masliah et al., 1992; Arriagada et al., 1992a). Similar, but much less abundant histopathological changes are found in nondemented elderly brains (Arriagada et al., 1992b). Identification of the molecular events which permit the transition from the normal low-density plaque and tangle state of normal aging to the high-density pathological state, are essential to the prevention of this lethal disease of the elderly. We submit that the evidence outlined below implicates environmental aluminum as a major factor in this transition.

A. RISK FACTORS FOR AD

Rarely, probably in less than 10% of all cases, AD appears to be inherited as an autosomal dominant disorder. Reported mutations in the amyloid protein precursor (APP) gene appear to account for AD in probably not more than 25 families worldwide (Goate et al., 1991; Murrell et al., 1991; Chartier-Harlin et al., 1991; Karlinsky et al., 1992; Mullan et al., 1992). A mutation in an unidentified gene on chromosome 14 (Schellenberg et al., 1992; St. George-Hyslop et al., 1992) appears to account for the largest group of early onset families with AD. However, the e4 allele of the apolipoprotein ApoE is a much more important risk factor for both late onset familial and sporadic AD (Pericak-Vance et al., 1991; Corder et al., 1993; Saunders et al., 1993). The e4 allele accounts for about 60% of all cases of AD, and there appears to be a dose effect resulting in earlier age of onset of the disease when homozygous for e4 (Table 1). An important conclusion is that AD is etiologically a heterogeneous condition in which several factors control expression of the disease.

We have recently reported that genotypes at the ApoE and APP locus may interact. In an extended pedigree segregating a Val → Ile missense mutation at codon 717 of the APP gene, two individuals exhibit the disease and have at least one copy of the e4 allele, whereas a sibling now two standard deviations above the mean age of onset of dementia in this family remains unaffected, and does not carry the e4 allele (St. George-Hyslop et al., 1994). Examination of Table 1 reveals that 50% of individuals heterozygous and 9% homozygous for the ApoE-e4 allele escape AD. Thus, complex genetic and nongenetic factors must be operative.

Table 1 ApoE Allele Risk For Alzheimer's Disease

Amount	% Affected	Age of onset	Hazard ratio
No e4	20%	84.3 yr	1.00
1 Dose e4	47%	75.5 yr	2.84
2 Doses e4	91%	64.4 yr	8.07

Note: n = 95 affected; 139 unaffected; Corder et al., 1993.

Among nongenetic environmental risk factors for AD, several have been investigated, including head injury (Breteler et al., 1992), level of education (Katzman, 1993), estrogen replacement (Paganini-Hill et al., 1993), antiinflammatory agents for the treatment of rheumatoid arthritis or leprosy (McGeer et al., 1990), and aluminum in the drinking water (Martyn et al., 1989; Flaten, 1990; Neri and Hewitt, 1991; Forbes and McAiney, 1992). Since human exposure to aluminum, particularly in the elderly population, could be greatly reduced, consideration in depth of this risk factor for AD is important. Indeed, the evidence appears sufficiently robust that public health action may be warranted to reduce the prevalence of AD in aging populations.

B. AD POPULATION-ATTRIBUTABLE RISK FROM ALUMINUM IN THE WATER SUPPLY

Epidemiological studies seek to assess the magnitude of the disease burden in a population in relation to a risk factor. Several epidemiological studies relating aluminum in drinking water to risk for AD support the idea that aluminum neurotoxicity becomes manifest early in the course of the illness. The studies of Martyn et al. (1989), Neri and Hewitt (1991), and Forbes and McAiney (1992) found a statistically significant association between the risk of first diagnosis of AD or early cognitive impairment in the elderly, and elevated concentrations of aluminum in the finished drinking water supply. These and other published studies demonstrating an association between aluminum in drinking water and deaths attributed to AD (Flaten, 1990; Frecker, 1991) do not differentiate between the following three possibilities: (1) excess aluminum causes AD; (2) excess aluminum at a preclinical stage of AD accelerates the progression of clinical dementia; (3) aluminum induces cognitive impairment unrelated to AD histopathology, but clinically indistinguishable from the dementia of AD.

The drinking water studies of Neri et al. (1992) focused on a large sample drawn from first admission hospital discharge data from 21 of the more populous cities in the province of Ontario, Canada. The number of first diagnosed AD cases demonstrated an increasing linear trend in the ratio of AD subjects to controls at higher concentrations of aluminum in drinking water. There was an approximate doubling of the risk for AD over the range of aluminum concentrations occurring in this survey (i.e., 0.004 to 0.203 mg Al/l). With no aluminum in the finished water, the expected case-control ratio would be only 0.595 or 40.5% below the general level of 1.00 for the province.

To evaluate clinical diagnostic accuracy in hospital practice, brain histopathological diagnoses of Ontario residents were drawn from the files of the Canadian Brain Tissue Bank (McLachlan et al., unpublished). At this facility, extensive histopathology was conducted by neuropathologists, and diagnoses were assigned in compliance with internationally recognized criteria, including the presence of widespread SP and NFD in neocortical and subcortical structures. Histopathological sections were routinely examined from 12 standard anatomical loci in all brains. From this series, 668 brains were available from geographical regions with known drinking water quality.

The overall clinical diagnostic accuracy was 87% correct for autopsy-verified AD, with 6% false negative clinical diagnoses. Thus the clinical diagnostic accuracy employed in the reports of Neri and Hewitt (1991) and Neri et al. (1992) drawn from the province of Ontario must be considered high, thus tending to validate their findings.

The Brain Tissue Bank sample included 376 autopsy-verified AD cases (mean age 78 years) and 292 autopsy-verified controls for which drinking water aluminum concentrations were available. Based upon a residential history which permitted a weighted annual average for aluminum exposure from municipal drinking water sources 0 to 10 years prior to death, comparison of the case to control ratio in communities with aluminum concentrations less than 0.1 mg/l, the relative risk in communities with drinking water aluminum concentrations greater than 0.1 mg/l was 2.3 (95% confidence interval: 1.8–3.2). As expected, the risk is somewhat higher than that reported by Martyne et al. (1989) and Neri and Hewett (1991) based on clinical criteria alone. Moreover, the evidence does not support the idea that aluminum affects cognitive function in the absence of AD, thereby excluding possibility (3) above.

In order to calculate the risk to a population resulting from aluminum in the drinking water, information regarding the proportion of the population exposed to elevated concentrations is required. Using the ratio of autopsy-verified controls coming from geographic areas of Ontario exposed to more than 0.1 mg/l aluminum in drinking water to those exposed to less than 0.1 mg/l, an estimate of 23% is obtained. Employing a relative risk for AD of 2.3, estimated from the weighted average annual exposure in the 10 years prior to death, the fraction of cases which might be prevented by reducing aluminum concentrations to below 0.1 mg/l is 23%. In the Province of Ontario the prevalence of AD is estimated at 90,200 cases over age 65 years (McDowell and Lindsay, 1994) and the estimated reduction in prevalence of AD could represent as many as 20,800 cases. If it were commercially feasible to reduce drinking water to the lowest level of aluminum encountered in the Neri et al. (1992) study, which resulted in a reduced odds ratio of 40.5% below the general level of 1 for that population, the number of presumed "preventable" cases would be 36,500 cases in that province alone.

Not all epidemiological studies report a relation between impaired cognitive function and aluminum exposure in the drinking water. Two studies, Wood et al. (1988) and Wettstein et al. (1991), failed to find a relation to aluminum in the drinking water supply. One explanation for the failure to detect a relation may represent geochemical differences in the drinking water supply. Exley et al. (1993) point out that epidemiological studies which take into account the influence of silicon in the water supply upon the bioavailability of aluminum have not yet been published. Since silicon will bind aluminum and both prevent toxic effects to fish (Birchall et al., 1989) and reduce intestinal absorption of aluminum in humans (Edwardson et al., 1993), the exact risk attributable to aluminum is probably not well estimated in studies which do not consider factors regulating absorption and bioavailability of aluminum. Similar conclusions were reached for the protective effect of the fluoride content in drinking water upon cognitive function in elderly males (Forbes et al., 1992).

Despite a large number of confounding factors, aluminum concentration in the drinking water supply, at least in some geographic regions, is a significant risk factor for AD. The population risk of AD from other sources of aluminum such as food additives, pharmaceuticals, adjuvants, cosmetics, deodorants, and respiratory dust (Rifat et al., 1990) is required before the precise risk for AD from environmental aluminum is fully evaluated.

C. BRAIN ALUMINUM CONCENTRATION IN AD

Although widely recognized as a potent neurotoxin, analytical difficulties surrounding the detection of low amounts of aluminum in a biological matrix have resulted in some skepticism about whether aluminum concentrations are elevated in AD brain, and could therefore be considered as a candidate

Table 2 Human Central and Peripheral Nervous System Effects of Aluminum Elevated in Neurological Disease

First author	Year	Technique	Site	Tissue source
Alzheimer's Disease				
Crapper et al.	1973	AA	Neocortex	Canada
Crapper et al.	1976	AA	Neocortex	Canada
Duckett and Galle	1976	XMA	Neuritic plaques	France
Trapp et al.	1978	AA	Neocortex	Central U.S.
Crapper et al.	1980	AA	Cortical nuclei	Canada
Perl and Brody	1980	XMA	Hippocampal nuclei	N.E. U.S.
Perl and Pendlebury	1984	XMA	Tangles	N.E. U.S.
Masters et al.	1985	AA	Tangle cores	Australia
Yoshimasu et al.	1985	INAA	Neocortex	Japan
Joshi et al.	1985	AA	Ferritin	S.E. U.S.
Candy et al.	1986	XMA	Neuritic plaques	England
Perl et al.	1986	LAMMA	Tangles	N.E. U.S.
Lukiw et al.	1992	EAA	Euchromatin	Canada
Good et al.	1992a	LAMMA	Tangles	N.E. U.S.
Yumoto et al.	1992	XMA	Neocortical nuclei	Japan
Xu et al.	1992	AA	Neocortex	E. U.S.
Corrigan et al.	1993	ICP-MS	Neocortex	England
Renal Disease/Dialysis Encephalopathy				
Alfrey et al.	1976	EAA	Neocortex	N.E. U.S.
Moreno et al.	1991	NA	Blood plasma	Europe
Bolla et al.	1992	NA	Blood plasma	N.E. U.S.
Candy et al.	1992	EAA/SIMS	Neocortex	England
Romanski et al.	1993	EAA	Blood plasma	Central U.S.
Downs Syndrome with Alzheimer's Disease				
Crapper et al.	1976	EAA	Neocortex	Canada
Guam and Kii Peninsula (Japan) Amyotrophic Lateral Sclerosis and Parkinson Dementia with Neurofibrillary Degeneration				
Yase et al.	1972	INAA	CNS tissues	Guam
Perl et al.	1982	INAA	CNS tissues	Guam
Garruto et al.	1984	XMA	Hippocampus	Guam
Yasui et al.	1991a	INAA	Neocortex	Japan
Yasui et al.	1991b	INAA	Neocortex	Japan
Industrial Exposure: Aluminum Workers/Cognitive Effects/Encephalopathy				
McLaughlin et al.	1962	NA	CNS tissues	England
Rifat et al.	1990	EAA	CNS tissues	Canada
Parkinson's Disease				
Hirsch et al.	1991	XMA	Substantia nigra	France
Good et al.	1992b	LAMMA	Neuromelanin	N.E. U.S.

Note: Analytical methods: AA = atomic absorption; EAA = electrothermal atomic absorption; INAA = instrumental neutron activation analysis; LAMMA = laser microprobe mass analysis; SIMS = secondary ion mass spectrometry; XMA = X-ray microanalysis; ICP-MS = inductively coupled plasma source mass spectroscopy; NA = data not available; "Tangles" refer to neurofibrillary tangles present in the cytoplasm of the diseased brain.

pathogenic factor in the disease. More than 11 laboratories in 6 countries employing 6 analytic techniques have reported, in peer-reviewed journals, elevated aluminum concentrations in AD and related neurodegenerative conditions compared to suitable control tissues (Table 2). Furthermore, brain destruction alone does not elevate brain aluminum concentration (Traub et al., 1981). In contrast to the weight of the evidence in Table 2, some laboratories have failed to detect elevated concentrations in AD brain tissue. The discrepancy in findings will be examined.

Landsberg et al. (1992a) applied nuclear microscopy to the analysis of *in situ* neuritic plaque cores in both stained and unstained AD tissue. A 3-MeV proton beam was focused and particle-induced X-ray emission and Rutherford backscattering spectrometry were employed for analysis. These researchers did not detect a signal in AD tissue that could be related to senile plaques. They concluded that the previously published observations that aluminum is involved in the formation of SP is a probable result of contamination of tissue by aluminosilicates present in laboratory reagents. Since nuclear microscopy, in the laboratory of these workers, is relatively insensitive (15 ppm; Landsberg et al., 1992b), compared to the many different techniques which have detected aluminum in fixed and unfixed, stained and unstained AD tissue and not in appropriate control tissue, Table 2 (e.g., for electrothermal AA, 2×10^{-11} g Al gives a reliable signal twice background), their conclusions are not valid. Indeed, in a subsequent publication, Landsberg et al. (1993) did report that at a sensitivity of 50 ppm or greater, aluminum and silicon were detected in 20% of senile plaques. These workers argue that until plaques can be unequivocally identified in untreated tissue, no conclusion can be reached on whether senile plaques contain aluminum and silicon. In contrast, application of a similar technique (PIXE) by Yumoto et al. (1993) detected aluminum in brain tissue and isolated nuclei from AD.

The low sensitivity of the energy dispersive X-ray microprobe as applied by Jacobs et al. (1989) may explain the failure to detect aluminum in AD neurofibrillary tangles. Employing a Cameca Camebax scanning electron microscope, with an electron microprobe in tandem with a Kevex 8000 energy detector at 5 KeV, 40 nA, 400 s, and a standard reference material containing very large concentrations of aluminum, 546 ppm, an aluminum signal was detected. However, 20 to 25 ppm sensitivity was reported when standard Al solutions were placed on the surface of a cortical "control" section. No estimate of signal loss in the 40-µm-thick section was reported nor was the attenuation of signal resulting from laying the section over the standard calibration material reported. Since aluminum in AD is within the tissue, it is unlikely that sufficient instrumental sensitivity was achieved to satisfactorily detect the very small absolute amounts of aluminum which occur within restricted volumes of subcellular particles such as a neurofibrillary tangles.

Lovell et al. (1993), using laser microprobe analysis, failed to find a statistically significant increase in aluminum in AD tangles, although they report a nine- to tenfold increase in the number of neurons whose cytoplasm and nuclei contained aluminum concentrations exceeding the mean plus three standard deviations from the mean obtained in control tissue. The instrument employed was similar to that used by Perl (1986), who reported the presence of high concentrations of aluminum associated specifically with neurofibrillary degeneration. With an advanced model of this instrument, Good et al. (1992a) detected aluminum signals in frozen, unfixed AD hippocampus which were subsequently demonstrated to arise from neurofibrillary tangles. Contamination during fixation and tissue preparation for analysis are improbable explanations for the different findings.

On balance, there is convincing evidence that aluminum is elevated in some compartments of AD-affected brain tissue compared to control. The elevation cannot be explained by spurious contamination or nonspecific brain damage. Analytical difficulties remain, and techniques with two orders of magnitude greater sensitivity than currently available are highly desirable.

D. ALUMINUM AND THE HISTOPATHOLOGY OF AD

An argument frequently employed against considering aluminum important in AD is that direct intracranial injection does not result in AD histopathology (Wisnieweski, 1991). Indeed, no manipulation in the laboratory of any putative "cause" of AD, including genetic, has reproduced the illness in animals. Current evidence supports the concept that AD results from a cascade of events ending with the characteristic histopathology and dementia. It is now apparent that aluminum has a number of neurotoxic properties which contribute to the pathogenesis of the disease.

E. AMYLOID

AD amyloid has an unique amino acid sequence of about 40 amino acids and is derived by proteolytic processing from Aβ precursor protein (APP). SP or neuritic plaques are complex brain tissue structures composed of extracellular amyloid fibrils in association with a large number of different proteins present in low mole ratios, such as serum amyloid P, basement membrane derivatives including proteoglycans, glycosaminoglycans, and α-1 antichymotrypsin, surrounded by abnormal neural processes and reactive glial cells (Rozemuller et al., 1989; Mann et al., 1992; Fraser et al., 1993a).

Large amounts of an amorphous form of amyloid accumulates in the brain extracellular space in healthy older individuals and in young Down's syndrome patients. This amorphous, presumed random, coil conformation is not associated with degenerating neurites or glial reaction. Some additional event appears to transform the random coil conformation of the peptide into a fibrillary, β-sheet conformation. Despite recent controversy, there appears little question now that organized, fibrillar β-sheet amyloid is toxic to nerve terminals (Fraser et al., 1994). Whether amyloid-induced neurite degeneration leads to neurofibrillary degeneration is uncertain.

Candy et al. (1986), employing an energy dispersive X-ray microanalytical system and high-resolution solid state nuclear magnetic resonance with magic-angle spinning, reported the localization of amorphous aluminosilicates in the core of senile plaques. An important question is whether aluminum facilitates a shift in amyloid conformation, perhaps in association with peptides, to aggregated fibrillary Aβ. Several lines of evidence indicate that aluminum, together with other factors, may contribute to amyloid formation:

1. Clauberg and Joshi (1993) have recently published *in vitro* evidence indicating that aluminum may accelerate proteolytic processing of APP by suppressing the inhibitory domain on proteolytic inhibitors, thus contributing to the accumulation of Aβ.

2. Al^{3+}, Fe^{3+}, and Zn^{2+} dramatically accelerate the process of aggregation, *in vitro*, of Aβ whereas Ca^{2+}, Co^{2+}, Hg^{2+}, Mn^{2+}, Mg^{2+}, Pb^{2+}, K^+, and Na^+ have no effect upon aggregation (Mantyh et al., 1993). Maggio et al. (1992) have demonstrated that once nucleating aggregates of Aβ have formed, even a low concentration of the peptide will support its continued growth. By promoting the initial aggregation of Aβ, certain metals including aluminum could contribute to the early step of amyloidogenesis. Exley et al. (1993) employed circular dichroism to demonstrate direct aluminum interactions at physiological concentrations with Aβ peptide which resulted in changes in the α- and β-helical content.

3. Aluminum salts, *in vitro*, greatly increase peroxidation of membrane lipids induced by Fe^{2+} at acid pH (Gutteridge et al., 1985). Aluminosilicates, *in vitro*, also stimulate free radical production by microglia (Evans et al., 1992). Furthermore, Al^{3+} appears to form a complex with O_2^- which is a stronger oxidant than is O_2^- itself (Kong et al., 1992). Aluminosilicates may contribute to the aggregation of Aβ in the extracellular space by oxidative damage to APP, partially degraded forms (Haass et al., 1992) or the cognate degradative enzymes. Support for this putative role of aluminum is the report of Dyrks et al. (1992) who induced highly insoluble amyloid aggregations by the addition of metal-catalyzed oxidation systems. These workers report inhibition of the amino acid oxidative process of protein cross-linking of APP by free radical scavengers. The induced conformational changes may prevent a proportion of the peptide from entering normal degradative pathways, resulting in slow accumulation of Aβ over many years. Even in those rare families with early onset AD resulting from mutations in the APP gene which may result in increased Aβ production, aluminum could contribute to the phenotypic expression of the disease.

4. Aluminum salts injected directly into rabbit cerebral spinal fluid or brain parenchyma are well known to be neurotoxic but the acutely induced neurofibrillary changes differ ultrastructurally from those found in AD. The 50% lethal intracranial dose, applied as aluminum maltol, raises bulk brain aluminum concentrations in rabbits less than fourfold (control = 1.5 µg/g dry weight, LD_{50} = 5.5 µg/g dry weight). The aluminum encephalopathy in rabbits is accompanied by a striking and long-lasting accumulation of APP in affected neurites as well as activated microglia and macrophages (Shigematsu and McGeer, 1992). While no extracellular amyloid fibrils were seen in this acute model, aluminum-induced neurotoxicity results in accumulation of APP in axons, which is a potential step in the formation of extracellular Aβ. Further, among several loci of accumulation in neurons, high aluminum concentrations are found in lysosomes (Galle et al., 1980). Evidence such as the colocalization of lysosomal proteases with amyloid plaques and inhibition of Aβ and Aβ-containing fragments by lysosomal inhibitors (e.g., choloroquin and $NHCl_4$) has implicated lysosomes in the amyloidogenic pathway (Cataldo and Nixon, 1990; Cataldo et al., 1991; Caporaso et al., 1992), a pathway with which aluminum may interfere.

5. Further support for the idea that aluminum may contribute to amyloid formation in humans comes from studies of brains exposed for approximately one decade to moderately elevated aluminum concentrations in blood secondary to renal failure. Candy et al. (1992) examined the clinical records, histopathology, and brain aluminum concentrations in 15 individuals who had undergone prolonged dialysis. Five, or 33%, exhibited Aβ-positive amorphous senile plaques in the cerebral cortex compared to 8% of age-matched controls (n = 12). The brains exhibiting Aβ-positive plaques were from subjects who were on average 61.6 years and who had suffered from uremia for an average of 13.2 years. Remarkably, the serum concentration of aluminum, on average, was 24.2 µg/l (control <10 µg/l) and

the average bulk neocortical gray matter aluminum concentration was 4 µg/g dry weight (control = 2.5 µg/g). These observations indicate that prolonged low-level elevations in serum aluminum concentration result in small increases in brain aluminum content, possibly via the lysomal pathway. Based on the findings in dialysis dementia (Alfrey et al., 1976), small increases of this magnitude in aluminum concentration in serum and brain are not usually considered neurotoxic. However, systematic application of modern immunohistochemical methods for amyloid indicate that low-level elevations in aluminum concentration are associated with a fourfold increase in risk for the premature accumulation of amyloid, a hallmark of brain aging and AD.

In conclusion, considerable circumstantial evidence indicates that aluminum could plausibly act as a cofactor in several stages of SP formation: (1) accelerated proteolytic processing of APP; (2) aggregation of amyloid filaments through electrostatic cross-links, seeding on aluminum silicates, and/or metal-catalyzed oxidative damage; (3) acute high dose effects resulting in altered axonal transport and accumulation of APP; and (4) prolonged low level brain exposure to aluminum, as in renal failure, resulting in increased risk for premature accumulation of extracellular amorphous Aβ accumulation.

F. NEUROFIBRILLARY DEGENERATION

AD neurofibrillary degeneration is an intraneuronal aggregate of filaments composed mainly of abnormally phosphorylated isoforms of the microtubule-associated protein, tau (Grundke-Iqbal et al., 1986). Khatoon et al. (1992) report AD brain contains about an eightfold increase in tau protein, mainly in a hyperphosphorylated form in the cytosol, possibly associated with ribosomes and assembled into characteristic paired helical filaments (PHFs). Mukaetova-Ladinska et al. (1993), employing an assay system reported to recognize tau from PHF cores, observed a PHF content 19-fold greater in AD than age-matched controls, but a 3-fold decrease in soluble tau protein. Despite changes in tau pool sizes, the size of the messenger RNA pool is not altered in AD, suggesting that the AD abnormality is not detected by the homeostatic mechanisms regulating tau transcription.

In AD brain, aluminum in bulk tissue correlated best with the density of NFD in unfixed neocortex (Crapper et al., 1976). The presence of aluminum in NFD has been repeatedly demonstrated in fixed, stained, and unstained AD neurofibrillary tangles by Perl and co-workers (Perl and Brody, 1980; Perl and Pendlebury, 1984; Good et al., 1992a), and independently by others (Kobayashi et al., 1987, Garruto et al., 1984) (Table 2). Another trivalent metal, iron, has also been localized to tangles in AD (Good et al., 1992a). Elevated concentrations of aluminum have been reported in the AD type neurofibrillary tangles of the Guam parkinsonism/amyotrophic lateral sclerosis syndromes (Table 2). Neurons with NFD may have intact membranes and normal appearing cytoplasmic and nuclear organelles. The significance of aluminum in association with the neurofibrillary tangle remains uncertain. One interpretation is that the presence of aluminum simply represents a nonspecific secondary event in which aluminum accumulates in an already damaged neuron and is trapped on the neurofibrillary tangle. Indeed, in purified PHF preparations, Sparkman (1993) failed to detect aluminum, although it is probable that instrumental sensitivity was insufficient to detect aluminum in these preparations.

There is considerable evidence that aluminum alone, or in combination with ligands, increases the phosphorylation of cytoskeletal proteins H and M neurofilament subunits and induces strong cross-linking between neurofilaments (Leterrier et al., 1992; Delamarche, 1993). Aluminum salts, *in vitro*, at concentrations ≥100 µM induce aggregation of tau that prevent entry into SDS-polyacrylamide gels (Abdel-Ghany et al., 1993). When incubated with ATP, GTP, or CTP, aluminum catalyzes a nonenzymatic covalent cross-linkage that results in incorporation of α and γ phosphates into the tau. Scott et al. (1993) reported that at 400 µM aluminum and 10 µM bovine or nonphosphorylated recombinant human tau, tau aggregated but without formation of fibrils.

To examine a possible role for aluminum in neurofibrillary degeneration, Fraser et al. (1993b) examined the effect of aluminum on the *in vitro* assembly of tau filaments prepared from AD neocortex. Soluble AD tau isoforms were extracted with 3% SDS under reducing conditions from a PHF/A68 preparation (Greenberg and Davies, 1990). Several major tau isoforms of about 55 to 68 kDa revealed phosphorylation-dependent epitopes. Following extensive dialysis against water, fine fibrillar aggregates consistent with normal tau could be visualized using rotary shadowing electronmicroscopy. Dialysis of these AD tau isoforms against 10 mM aluminum citrate, and to a much lesser extent iron chloride, at pH 7.2, promoted assembly of 10-nm filaments which aggregated laterally into PHFs indistinguishable from AD PHFs. In view of the observations of Scott et al. (1993) on nonphosphorylated tau, the observations of Fraser et al. (1993b) support a mechanism in which electrostatic cross-links between oxygens in hyperphosphorylated AD tau by aluminum produce tau-PHF fibers. Since iron metabolism is stringently regulated in brain and has not been reported in epidemiological studies to increase the risk for AD, these findings implicate environmental aluminum as an important toxic cofactor in the pathogenesis of neurofibrillary degeneration.

Aluminum exposure alone has not been observed to induce AD type PHFs in any cellular system, *in vivo*. However, human neuroblastoma cells in tissue culture exposed to aluminum exhibit epitopes found in AD neurofibrillary tangles. Mesco et al. (1991) reported aluminum induction of the well-known Alz50 epitope recognizing NFD, and Guy et al. (1991) reported the development of an epitope recognized by a polyclonal antibody staining for NFD and neuropil threads. Alz50 expression is also observed in experimental aluminum encephalopathy (Shigematsu and McGeer, 1992). These data indicate that aluminum may have several roles in the formation of PHFs. In addition to assembly of hyperphosphorylated tau into PHFs, *in vitro*, at neutral pH, aluminum induces at least some of the phosphorylation/conformational changes in tau which occur in AD. Neurofibrillary degeneration in AD may be a multistage sequence in which tau phosphorylation increases from a normal value of 3 mol to 8, 10, or more moles/mole of tau over a prolonged time interval. Fourier transformed infrared spectroscopy indicates that the hyperphosphorylation of tau in AD is associated with conformational changes (Fraser et al., unpublished) which, in the presence of aluminum, promotes assembly into PHFs.

Tau constructs (Wille et al., 1992), or isolated tau (Montejo de Garcini et al., 1986), lacking or with low phosphorylation, can be polymerized into fibrous structures. However, nonphysiological conditions are required including high ionic strength, low pH (5.0 to 5.5), and chemical cross-linking to produce dimers. The presence of aluminum under more physiological conditions may be necessary, but probably not sufficient for all steps in AD neurofibrillary degeneration.

G. ALUMINUM AND GENE EXPRESSION IN AD

Elevated concentrations of aluminum are found in the nuclei of mixed neurons and glia extracted from AD neocortex (n = 12), compared to neuropathologically healthy age-matched control brains (n = 17), brains from non-AD dementia-associated brain disorders (n = 5), and dialysis dementia (n = 5) (Crapper et al., 1980). In nuclear preparations from both control and AD, fractionation employing centrifugation revealed that highly condensed heterochromatin contained 17 times more aluminum per gram DNA than H1-depleted euchromatin. In these crude fractions, intact nuclei and heterochromatin fractions from AD contained about twice as much aluminum per gram DNA than controls ($p < 0.025$).

Employing micrococcal nuclease and collecting the dinucleosome fraction released by digestion of accessible linker regions, a 6.8-fold increase in aluminum concentration per gram DNA was found in AD superior temporal cortex compared to the age-matched control (superior temporal) cortex (Lukiw et al., 1992). Comparing the aluminum content of dinucleosomes collected randomly from all neocortical regions from healthy control brains, brains with occasional SP (<8/mm^2) and brains from non-AD dementia associated disorders (total n = 17 brains) to all AD (n = 21), a 4.3-fold increase in AD was found ($p < 0.0001$). Within AD brains, aluminum content was 28% higher in the temporal than the frontal neocortex, corresponding to the more profound histopathological changes usually found in the temporal cortex in this disease. Furthermore, a significant positive correlation was found between dinucleosome aluminum content and histone H1° content, a linker histone usually associated with repressed genes in terminal differentiated tissues. These observations strongly support the idea that aluminum acts as a high-affinity electrostatic cross-linker between brain specific, transcription repressing, linker histones and DNA. A putative aluminum coordination between oxygen of DNA phosphate and terminal oxygens on aspartic acid 98 and glutamic acid 99 of human histone H1° could plausibly contribute to altered chromatin conformation and ultimately to changes in the transcription of DNA as observed in AD (McLachlan et al., 1988; Lukiw et al., 1990).

Several studies indicate that aluminum, at concentrations found in the AD-affected human brain, reduces DNA transcription *in vitro* (Sarkander et al., 1982), in part through an effect upon chromatin conformation. This neurotoxic property of aluminum has been reproduced *in vivo*. A remarkably similar profile of messenger RNA pool sizes has been reported in the experimentally induced aluminum encephalopathy in rabbits and AD neocortex for the genes coding for critical cytoskeletal proteins, actin, α-tubulin, and the low molecular weight constituent of neurofilaments, NF-L (Muma et al., 1988; McLachlan et al., 1988). Down regulation in NF-L would further disrupt axoplasmic flow and the supply of essential metabolites to synaptic processes. Thus, the evidence in hand strongly supports a gene repression role for aluminum in AD.

H. ALUMINUM CHELATION AND THE CLINICAL COURSE OF AD

A further test of the role of aluminum in AD has been to examine the clinical course of AD following the daily use of a trivalent metal chelating agent, desferrioxamine (DFO). In three brains of AD patients who received 500 mg DFO intramuscular, twice each day, 5 days each week for a total of between 23.5 and 54 g in the weeks immediately preceding death unrelated to AD, the average brain aluminum concentration for 20 neocortical regions was 2.69 µg/g dry weight (Crapper McLachlan et al., 1993), or within 1 standard deviation of the average concentration in healthy brain (Crapper et al., 1976). Three brains of AD patients who had died without DFO treatment or who had received less than 3.5 g immediately prior to death had an average content of 4.09 µg/g dry weight ($p < 0.05$). In a two-year randomized, oral placebo, clinical trial which employed 125 mg DFO, the oral placebo group declined on tests of daily living skills and motor dyspraxia at twice the rate of the DFO-treated group (Crapper McLachlan et al., 1991; Crapper McLachlan et al., 1993). While no patients improved or arrested in progression, the results further support an important role for aluminum, and possibly iron, in the pathogenesis of the disorder.

I. CONCLUSION: ALUMINUM IN AD

With the rapid development of knowledge concerning the pathogenesis of AD, robust evidence indicates that aluminum is involved in several steps in the AD degenerative process. The critical question for the future is: how large a reduction in prevalence of AD would be achieved by stringent reduction in aluminum exposure in the elderly population?

III. ATYPICAL NEURODEGENERATIVE DISEASES OF THE WESTERN PACIFIC OCEAN

High incidence foci of atypical neurodegenerative diseases associated with amyotrophic lateral sclerosis (ALS) and/or parkinsonism with dementia have been described in the Kii peninsula of Japan, the Mariana Islands, and in southern Irian Jaya, Indonesia. An unique characteristic of these diseases is that affected brains contain high numbers of AD type neurofibrillary tangles. The etiology of these diseases is unknown; genetic and transmissible factors appear unlikely. Two hypotheses have been vigorously investigated: a neurotoxin from the palm *Cycas circinalis* (Spencer et al., 1987), and a chronic dietary deficiency of calcium and magnesium with excess intake of aluminum and manganese (Yase, 1972; Yangihara et al., 1983). Post-mortem nervous tissue has only been available from the Kii peninsula and Guam. Elevated concentrations of aluminum and calcium have been consistently reported in affected bulk tissue and on neurofibrillary tangles (Table 2). Is aluminum a plausible factor in the pathogenesis of these atypical Western Pacific foci?

Monkeys exposed to a calcium and magnesium deficient diet with excess aluminum, resembling dietary conditions reported in the high incidence foci, developed muscle atrophy, chromatolytic spinal neurons with eccentric nuclei, increased numbers of argyrophilic bodies and accumulations of 10-nm neurofilaments in swollen axons (Yano et al., 1989; Garruto et al., 1989). Furthermore, an aluminum-induced chronic myelopathy in rabbits has features of ALS (Strong et al., 1991). Aluminum may also play the same role in the formation of neurofibrillary tangles as in AD: by the assembly of paired helical filaments and by stimulating hyperphosphorylation and conformational changes in tau. As indicated in Section II.E, aluminum increases lipid peroxidation induced by Fe(II) at acid pH and aluminosilicates stimulate free radical production by microglia. Furthermore, aluminum appears to form a complex with O_2^- which is a stronger oxidant than O_2^- itself. Muntasser et al. (1994) have presented evidence that motor neurons in sporadic ALS of North America have a 42% increase in messenger RNA pool size for Cu/Zn superoxide dismutase (SOD). This suggests that motor neurons in sporadic ALS may be responding to oxidative stress by upregulation of SOD transcription. Taken together with the discovery that several genetic mutations in the SOD gene in familial ALS (Rosen et al., 1993) are linked to the disease, it is plausible that environmental agents, including trace elements such as aluminum, may contribute to free radical dysmetabolism and oxidative stress in the high incidence foci of the Western Pacific. Indeed, two cases from the Kii peninsula, reported by Yasui et al. (1991a), had remarkably elevated precental neocortical and spinal cord bulk aluminum concentrations ranging up to 23 fold higher than that found in control tissue. Taken together, considerable evidence implicates aluminum in the pathogenesis of the Western Pacific disease.

One probable environmental source of aluminum on Guam comes from the unusually high concentrations of bioavailable aluminum which are eluted from the soils by rain water at pH 7.0, particularly in the southwestern region of the island, (Crapper McLachlan et al., 1989). Aluminum from this source could gain access to the central nervous system via the respiratory tract from dusts and the gastrointestinal tract through aluminum concentrating plants such as taro skin and leaf, arrowroot, turmeric, and the drinking of river water (Zolan and Ellis-Neill, 1986).

Preliminary studies (McLachlan et al., unpublished) indicating a disorder in aluminum metabolism associated with ALS in the Irian Jaya focus has been obtained from the examination of four cases among the Ctak, Auyu, and Jakai peoples. The ALS cases had elevated fasting serum aluminum concentrations, on average, of 1074 nM/l, compared to 259 nM/l for 3 age- and sex-matched healthy relatives and 218 (105 SD) nM/l for 10 healthy Caucasian controls. One hour after an oral dose of aluminum citrate (2.9 g Al) the ALS patients had an average aluminum concentration in serum of 6102 nM/l, the controls 1000 nM/l, and the Caucasian controls 2121 (1588 SD) nM/l. The reasons for the elevated serum aluminum concentrations in the ALS patients are unclear, as are the mechanisms responsible for brain uptake. However, these preliminary observations further support a role for aluminum, and suggest that an intensive search for other free radical inducers and metabolic factors would be fruitful in the neurodegenerative diseases in the high incidence region of Irian Jaya.

IV. ALUMINUM IN PARKINSON'S DISEASE

Parkinson's disease (PD) is characterized by loss of dopaminergic neurons in the substantia nigra and globus pallidus. Studies using X-ray microanalysis and a laser microprobe with a time of flight mass spectrometer capable of measuring trace amounts of elements in subcellular compartments have demonstrated increased aluminum and iron concentrations in Lewy bodies, which are a histopathological hallmark of the idiopathic form of the disease (Hirsch et al., 1991) and neuromelanin granules of neurons in the substantia nigra of patients with PD (Goode et al., 1992b), respectively.

Neuromelanin binds metals and may be responsible for the site-specific accumulation of iron and reduction to Fe^{2+}. In the presence of aluminum and peroxides generated from the metabolism of dopamine, the Fe^{2+} could trigger a cascade of reactions resulting in the formation of cytoxic radicals and neuronal degeneration (Olanow, 1992). Neurons of the substantia nigra in PD also exhibit a failure to increase ferritin levels despite increased iron concentration. The observed high concentrations of aluminum could result in displacement of iron and an increased pool of Fe^{2+} ligated to low molecular weight substances capable of promoting the formation of free radicals by mechanisms listed under Amyloid (Section II.E) above. Thus, aluminum, in conjunction with different cellular derangements for each disease, may act in the final common pathway of neuronal destruction in AD, PD, and the Guam PD/ALS syndromes. Further support for this mechanism requires evidence of oxidative damage to proteins in the target neurons.

V. RENAL DISEASE AND DIALYSIS ENCEPHALOPATHY ASSOCIATED WITH ALUMINUM

In the 1960s, aluminum hydroxide was widely introduced as a phosphate binding agent used in the hemodialyzing fluid of patients undergoing treatment for chronic renal failure. By the early 1970s, the first outbreaks of encephalopathy in dialysis wards were reported by Alfrey and colleagues, and over the next several years there were widespread reports throughout the scientific literature of similar epidemics associated with the aluminum hydroxide dialysis protocol (Alfrey et al., 1972; Table 1). The features of this hemodialysis-associated encephalopathy dementia included dysphagia, speech dyspraxia, myoclonic and epileptic episodes, and a progressive dementia. Over a 6- to 9-month period following the onset of symptoms, the patients became mute, were unable to perform any purposeful movements, and death rapidly ensued (Alfrey et al., 1972). The cause of the dialysis dementia was attributed to a number of possible etiologies including hypophosphatemia, hypertension, and conventional or slow virus infections of the brain. Strong positive correlations between the duration of the dialysis treatment and the index of the encephalopathy were apparent. Alfrey and colleagues gathered evidence which identified aluminum in the dialysate as the cause of this hitherto undescribed dialysis dementia. The source of the aluminum was either from the actual aluminum hydroxide buffers employed in the dialysis preparation or from the high aluminum content of the tap waters used to prepare the dialysate itself (Alfrey et al., 1976, 1992). It was demonstrated that the aluminum was transferred from the dialysate into the patient's blood during the dialysis treatment, however, as a result of plasma protein binding of the aluminum, both the transferred aluminum was retained in the circulation and the gradient from the dialysate to the blood was maintained to further enhance aluminum uptake (Kaehny et al., 1977). Kidney dysfunction also prevented the normal elimination of aluminum from the body during the interdialytic period.

These accidental aluminum-induced dementias in humans were associated with elevated aluminum in blood serum of up to 200 µg/l, although it has more recently been shown that serum aluminum concentrations of 60 µg/l are associated with impaired cognitive function (Romanski et al., 1993). Notably, concentrations of aluminum in the central nervous system (CNS), and in particular, the neocortex of the brain are markedly elevated (Candy et al., 1992). Cytologically, the accumulation of aluminum in the lysosomes of dialysis dementia patients is a consistent feature, suggesting that in this syndrome aluminum is preferentially routed and sequestered into specific cellular compartments. Removal of aluminum from patients by including desferrioxamine during the dialysis procedure has been shown to be effective in removing CNS aluminum, causing substantial reductions in plasma aluminum concentrations and general improvements in the intellectual functioning and short-term memory of these patients (Chang and Barre, 1983). Renal transplantation has been shown to benefit patients with early features of dialysis encephalopathy. Very substantial evidence indicates that aluminum in high concentrations is toxic to an otherwise healthy human brain.

Dialysis dementia is not associated with specific histopathology. While aluminum accumulation is an accepted etiological agent in this condition, the aluminum exposure is superimposed upon an otherwise healthy human brain. This fact about dialysis dementia is often used as an argument against a role of aluminum in other neurodegenerative diseases associated with neurofibrillary degeneration. However, in these latter conditions, the bulk of the evidence indicates that the neurotoxic effects of aluminum are operating upon a much altered metabolic state.

VI. ALUMINUM IN THE WORKPLACE

While there is strong evidence that aluminum may induce an encephalopathy in the absence of renal function, there is also increasing evidence that excess aluminum exposure in the presence of normal renal function presents a risk of nervous tissue damage. Examination of 16 clinical reports and studies (Table 3) from 8 industrialized nations, involving aluminum exposure in the form of dusts, fumes, and/or skin contact in the workplace, strongly support the conclusion that among the neurological manifestations of aluminum, neurotoxicity in otherwise healthy humans is focal damage to the neuro-substrates of cognitive function, motor control and peripheral nerves. Indeed, a syndrome characterized, in part, by impairment in some cognitive functions including memory and concentration, disorders in motor control, and peripheral neuropathy appear to be related to the inhalation of aluminum dusts or fumes. The constellation of clinical signs seen in these relatively acute exposures are more closely related to dialysis encephalopathy than to the other neurodegenerative diseases discussed above. However, unanswered is the delay risk for the dementia/parkinsonism syndromes following long-term chronic exposure.

Table 3 Examination of 16 Clinical Reports and Study Involving Aluminum Exposure

Author (1st), year	Sample size	Clinical
1. Spofforth, 1921	1	Progressive encephalopathy
2. McLaughlin, 1962	1	Progressive encephalopathy
3. Langauer-Lewowicka, 1983	444	Cognitive/motor control
4. Vida, 1983	331	Paresthesia, pain
5. Longstreth, 1985	3	Cognitive and motor control impairment
6. Alessio, 1987	22	Neuroendocrine alteration
7. Kobayashi, 1987	1	Dementia with Balint's syndrome
8. Popovic, 1989	47	Cognitive impairment
9. Banicevic, 1989a	123	Cognitive impairment
10. Banicevic, 1989b	24	Peripheral neuropathy
11. Hosovski, 1989	141	Cognitive and motor control impairment
12. Ljunggren, 1991	13	Aluminum body retention
13. Sjogren, 1990	65	Neuropsychiatric difficulty
14. Hosovski, 1990	87	Cognitive impairment
15. Rifat, 1990	262	Cognitive impairment
16. White, 1992	25	Cognitive and motor control, neuropathy

A. LITERATURE REPORTS

In several of the following reports aluminum exposure occurred in the presence of ligands or solvents which may also be neurotoxic. However, the constellation of clinical features which emerge from these reports are submitted as evidence that the major toxin is aluminum.

1. Spofforth (1921) reported that a 46-year-old British worker exhibited loss of memory, tremor, jerking movements, impaired coordination and persistent vomiting. He had been dipping red-hot metal articles, contained in an aluminum holder, into concentrated nitric acid. Urine analysis revealed large amounts of aluminum. The author concluded that "aluminum produces a rather slow intoxication".
2. McLaughlin et al. (1962) reported the clinical findings for a 49-year-old British aluminum ball-mill worker, who was exposed for 13.5 years to stearic acid coated aluminum dust and who developed a lethal progressive encephalopathy of 10 months duration. He was first noted to have forgetfulness and speech difficulty. He exhibited clonic jerking movements of limbs. The disease progressed and the patient developed hemiparesis, nominal dysphasia, disorientation, severe memory loss, dysarthria, complete aphasia, general convulsions, coma, and death. Brain tissue examination revealed no specific histopathology but contained 17 times the aluminum concentration found in healthy age-matched control brains. After excluding all known causes of an encephalopathy of the type reported, the authors concluded that it seemed "possible that the encephalopathy was due to aluminum intoxication".
3. Langauer-Lewowicka and Braszczynska (1983) reported the results of a health survey of 444 Polish aluminum electrolysis workers. The workers' mean age was 44 years and the mean duration of exposure was 17 years — 20% complained of memory dysfunction, equilibrium disturbances, headaches, dizziness, sleep disturbance, and general weakness, while 6.5% of the workers were found on examination to exhibit motor control impairment considered to result from exposure in the workplace after other medical conditions had been excluded. The neurological signs involved the pyramidal and/or the cerebellar equilibrium system; 14% of a sample of 71 workers who had electroencephalograms were classified as distinctly pathological and a further 14% were considered borderline. Tests of cognitive function were not reported. The authors, *a priori,* assumed that aluminum does not damage the nervous system and attributed the findings to other chemical exposures including fluoride, polycyclic hydrocarbons, carbon dioxide, and electromagnetic radiation.
4. Vida and Vido (1983) reported on complaints of Polish metallurgists involved in the electrolytic production of aluminum. In a group of 331 workers exposed for more than 10 years, a high incidence (29%) complained of paresthesia and intermittent pains in the extremities that could not be attributed to fluorosis.
5. Longstreth et al. (1985) reported the neurological and cognitive status of 3 workers who had worked for more than 12 years in an aluminum smelting plant in the U.S. on potlines which were without effective hoods for about the first 6 years of their employment. The workers complained of dizziness, severe headaches, joint pains, severe lack of energy and strength, frequent loss of balance, tremor, and two complained of trouble sleeping, memory difficulty, pins and needles sensations, and had to retire from the workforce. On examination, formal neuropsychological testing revealed significant impairment of short-term memory, visual motor speed tasks in all of the workers, and impaired Wechsler Memory Quotients, and problem-solving in two of the three workers. All three demonstrated motor incoordination, intention tremor, and ataxia of gait. One demonstrated bilateral delays in visual evoked and somatosensory evoked potentials. In the absence of other neurological diagnoses and/or probable explanation, and firm evidence for aluminum as the etiological agent, the authors concluded "this report serves to alert practioners to a potential association between an occupational exposure in the potroom of an aluminum plant and a neurological disorder that begins with incoordination".
6. Alessio et al. (1987) examined 22 Italian workers prior to and 18 months after exposure to aluminum dust or welding fumes (n = 8) at concentrations of less than 5 mg/m^3. They reported that two pituitary hormones, prolactin and thyroid stimulating hormone, were depressed within 3 months of exposure and remained depressed for 18 months by about 33%. The authors concluded "that the affect observed is the consequence of a direct mechanism on the hypothalamus-hypophysis axis" and "exposures to aluminum concentrations considered 'safe' can produce in a short time, effects upon the adenohypophysis".
7. Kobayashi et al. (1987) reported the history and pathological findings for a 65-year-old male who had worked as an aluminum refiner for 30 years. At age 55 he developed Balint's syndrome and was found to have Alzheimer's disease at post-mortem. Microanalysis revealed focal accumulations of aluminum in the brain nuclei and on neurofibrillary tangles. The instrument employed for microanalysis did not detect aluminum accumulation in tangle-bearing neurons in a case of senile dementia who was not

exposed to aluminum in the workplace. The authors concluded "that aluminum deposition may be a concomitant phenomenon due to the exposure to environmental aluminum". Thus, elevated aluminum concentrations were found in two brain tissue compartments in an aluminum worker.

8. Popovic and Banicevic (1989) reported the results of cognitive assessment employing Coblentz's dementia rating scale, an Alzheimer disease scale, and Hamilton's depression scale on 47 Yugoslavian workers, mean age 40 years and mean duration of exposure to aluminum in the workplace 11.5 years. A control group without professional aluminum exposure was composed of 32 workers of average age 42 years. A group of 17 tests of cognitive function were delivered which examined memory, speech, praxis, and 10 noncognitive tests of mood and behavior. Overall scores on 11 tests from the Alzheimer disease scale revealed that the exposed group scored significantly lower, $p < 0.01$, than the control group. On five tests of memory, speech, and praxis, the exposed group performed significantly more poorly, $p < 0.001$, and on three tests of mood and behavior the exposed group performed less well, $p < 0.05$, than the unexposed control group. On six tests of word remembering, the exposed group was significantly more impaired, $p < 0.001$, but there was no difference in word recognition between the two groups. The authors concluded "their tests indicated a significant influence of occupational exposure to aluminum on amnestic functions and total cognitive behavior of tested workers".

9. Banicevic et al. (1989a) reported the results of neuropsychological testing employing Coblentz's dementia scale and Hamilton's depression scale on 123 workers exposed to aluminum in Valjaonica Aluminijuma Sevojno, Yugoslavia and 33 unexposed workers. As a group, the unexposed workers exhibited twice the tendency toward depressive moods compared to the exposed group. However, the exposed group achieved lower overall scores on the Coblentz dementia scale, $p < 0.005$, with particularly robust differences on tests of attention and memory, $p < 0.002$ for each test compared to the control group. When exposure duration was analyzed "poor scoring is directly affected by aluminum exposure over 15 years". When scores of alcohol abstainers were compared, the risk of cognitive deficit was threefold higher in the exposed group. No difference was observed between workers considered alcoholic.

10. Banicevic et al. (1989b) performed nerve conduction studies on 24 workers exposed to aluminum in Valjaonica Aluminijuma Sevojno and 52 workers not exposed to aluminum or any other known neurotoxin at the Institute of Occupational Medicine, Belgrade. Electromyoneurographic diagnostic studies were carried out on the peroneal and sural nerves by modern techniques. The mean age of the exposed workers was 42.6 years and mean exposure 11.6 years; the average age of the control group was 43 years. For the exposed group, 37% complained of paresthesia in the hands and 62% in the legs; pains in the hands 25% and legs 27%; and weakness in hands 29% and in legs 29%. Neurological examination revealed hyporeflexia in legs in 8.3%, reduced sensation in 8.4%, and ataxia in 16.7%. Conduction velocity was significantly impaired in the sural ($p < 0.001$) but not the peroneal nerve in 28% of the exposed workers, and nerve conduction velocity reduction was about 9.3 m/s per tested worker in relation to the control group. Based on electroneurographic findings in workers occupationally exposed to aluminum, neuropathy was noted in 46%. "Among workers with subclinical signs of neuropathy, most frequent are sensomotoric disturbances (21%), then sensoric (17%), and finally motoric (8%)." Alcohol abuse was excluded as a significant factor in the etiopathology of the neuropathies. The authors concluded that "negative testing results of effects of other noxae which might have contributed to occurrence of discrepancies noted, (alcohol, avitaminosis, drugs, metabolic disturbances), suggest that aluminum could be the cause of findings noted".

11. Hosovski [edited by Selena (1989)] summarized the results of long-term studies on 141 aluminum-exposed workers involved in a resmelting operation in Yugoslavia. The subjects' mean age of 40.7 years and 18.9 years of exposure were compared with 60 control workers, mean age 41.9 years and an average of 18.2 years of employment. A total of 71 workers were admitted to a hospital and administered a desferrioxamine challenge test: 0.5 g/12 h intramuscular for 3 days. The investigative team included 35 professionals who delivered about 600 biological and psychological tests. No specific effects upon liver or kidney function were reported. However, in addition to extending evidence for patchy deficits in cognitive function, peripheral neuropathy, ataxia, and motor-incoordination, this study revealed diminished activity in the serum enzymes lactate dehydrogenase, hydroxybutyric dehydrogenase, glucose-6-phosphate dehydrogenase, glutamate dehydrogenase, gamma-glutamic transpeptidase, alkaline phosphatase, 5-nucleotidase, and leucine-aryl aminopeptidase. Furthermore, this study reports changes in the electrocardiogram indicative of conduction changes including prolonged QRS-complexes and T-wave reduction. The summary concludes "workers in whom aluminum intoxication has been proven should be immediately removed from exposure to this metal". Finally, "the occurrence of clinical

signs of dementia, (initially manifested by loss of memory and creative abilities, changed behavior-stupor and agitation, loss of balance, joint pains, dizziness, sleep disturbances, tremor, ataxia, etc.) indicates complete working disability of exposed workers. In cases of progressive symptoms, they usually require help and care from third persons. Therefore, aluminum intoxicated workers may become a serious social problem, especially if adequate health protection is not undertaken".

12. Ljunggren et al. (1991) reported on the blood and urine concentration of 13 Swedish aluminum flake powder workers who were investigated before and after 4 to 5 weeks of vacation. The mean age was 32.6 years and mean exposure time 5.2 years. Their data were compared to five unexposed healthy control subjects. Ten retired workers, mean age 67.9 years were examined, 8 of whom were exposed to aluminum powder. Workers exposed to aluminum flake powders had urinary excretions of the metal 80 to 90 times higher than controls and the calculated half-life was 4 to 5 weeks. Among retired workers, the half-life of excretion varied up to 8 years. The authors concluded that "these results indicate that aluminum is retained and stored in several compartments of the body and eliminated from these compartments at different rates".

13. Sjogren et al. (1990) reported that Swedish welders exposed to aluminum, lead, or manganese for a prolonged period of time had significantly more neuropsychiatric symptoms than welders not exposed to these metals. A group of 65 aluminum welders were compared to 217 iron track welders. Aluminum welders exposed for 20,000 h corresponding to 13 years of full-time exposure were twice as likely to complain of neuropsychiatric difficulty: e.g., difficulty in concentrating (odds ratio = 2.34, 95% CI = 1.15–4.76) compared to iron welders with similar exposure. The authors concluded that "as aluminum is a neurotoxic metal, it seems biologically plausible that high and long term occupational exposure to welding fume-containing aluminum may have neurotoxic effects".

14. Hosovski et al. (1990) reported the results of psychometric tests performed on 87 aluminum foundry workers in Yugoslavia. The workers were exposed from 6 to 18 years to aluminum fumes and dust at concentrations ranging between 4.6 and 11.5 mg/m^3 and of particle size distribution: 1 μm, 65.6%; 1–5 μm 26.6%; >5 μm, 7.6%. A desferrioxamine challenge test induced a significant elevation in both blood (mean increase 24%), and urine (mean increase 129%), compared to unexposed workers, n = 60. Wechsler's tests of intelligence and Bender's test for cerebral damage were delivered. While global deterioration in cognitive function was not observed, exposed workers had significant impairment in tests of memory, picture completion, object assembly, coding, attention, learning ability, concept formation, and visual motor coordination. The authors concluded that "the observed changes in psychomotor and intellectual abilities could be a consequence of the long-lasting toxic effects of aluminum."

15. Rifat et al. (1990) measured cognitive function in Northern Ontario miners, exposed to McIntyre powder as a prophylaxis for silicosis. Originally the powder was analyzed as 15% aluminum and 85% aluminum oxide. Modern analysis revealed particles of bayerite, gibbsite, norstrandite (polymorphs of aluminum trihydroxyde), triangular plates of aluminum, and traces of haematite. From a sample of 262 exposed miners, those exposed for 20 or more years had a 4.5-fold greater risk of impaired scores on tests of cognitive function (CI = 1.6–11.1) and those exposed for 10 to 19.9 years had a relative risk of 3.1 (CI = 1.4–6.8) compared to nickel/copper miners with comparable underground time. Criteria for impairment on the cognitive tests were stringent and this study did not have the confounding factors of cyclic hydrocarbons and other possible neurotoxic agents accompanying aluminum smelting.

16. White et al. (1992) reported on 25 U.S. workers exposed on the same potline as #14 above. The mean age was 47.0 years, and the mean duration of exposure was 18.7 years. The constellation of symptoms in this group of workers included frequent loss of balance: 88%; memory loss: 84%; joint pain: 84%; dizziness: 80%; numbness: 80%; severe weakness: 80%; concentration difficulties: 76%; sleep disturbances: 76%; paresthesia: 72%; tremor: 68%; severe headache: 68%. Detailed cognitive testing revealed mild to moderate impairment (>2 SD) on immediate stories recall: 70%; 30-min delayed stories recall: 50%; immediate visual reproduction: 75%; 30-min delayed visual reproduction: 70%. Incoordination was documented in 84%. The authors conclude: "Aluminum exposure in the potroom seems the most likely cause".

Epidemiological studies addressing what, if any, risk of developing dementia, parkinsonism, or other neurological conditions after long-term chronic aluminum exposure would now be justified.

VII. CONCLUSION

Evidence has emerged over the past five years to indicate that aluminum neurotoxicity is of considerable importance in the pathogenesis of several human neurodegenerative diseases. There is no convincing evidence that aluminum initiates the neurodegenerative process. At least two mechanisms of toxicity appear to occur: electrostatic cross-links which alter the function of certain biological molecules, and an indirect effect upon free radical production. Although many questions remain unanswered, the public health implications for the prophylaxis of certain neurodegenerative diseases, particularly as a risk factor for AD, require systematic investigation. Geochemical factors may alter the risk among geographic areas. However, in contrast to AD genetic risk factors for which no preventative measures can be instituted in the foreseeable future, human exposure to aluminum, especially the aging population, can be substantially reduced. Such steps hold potential for a substantial reduction in prevalence, or slowing, of the progress of AD.

ACKNOWLEDGMENTS

Supported by the Ontario Mental Health Foundation, Medical Research Council of Canada, Scottish Rite Charitable Foundation, and the Azlheimer Association of Ontario. Special thanks to Dr. D. Hewitt for critical discussion.

REFERENCES

Abdel-Ghany, M., El-Sebae, A. K., Shalloway, D., *J. Biol. Chem.,* 268(16), 11976–11981, 1993.

Alessio, L., Mussi, I., Di Sipio, I., Catenacci, G., Trace elements in human health and disease: Extended abstracts. Second Nordic Symp., August 17–24, Odense, Denmark, 1987.

Alfrey, A. C., LeGendre, G. R., and Kheany, W. D., *N. Engl. J. Med.,* 294, 184–188, 1976.

Alfrey, A. C., Mishell, J. M., and Burks, J., *Trans. ASAIO,* 18, 257–261, 1972.

Alfrey, A. C., In Proc. 2nd Int. Conf. Aluminum and Health, Tampa, FL, Feb. 2–6, 1992, 5, 1992.

Arriagada, P. V., Growdon, J. H., Hedley-Whyte, E. T., and Hyman, B. T., *Neurology,* 42(3pt1), 631–639, 1992a.

Arriagada, P. V., Marzloff, K., and Hyman, B. T., *Neurology,* 42(9), 1681–1688, 1992b.

Banicevic, R., Popovic, R., Trpeski, L. J., and Bulat, P., Proc. Symp. Copper and Aluminium Toxicology, Uzice, No. 51:10–21, 1989a.

Banicevic, R., Popovic, R., Vidakovic, Z., Rakic, G., Trpeski, L. J., and Bulat, P., Proc. Symp. Copper and Aluminium Toxicology, Uzice, 52:22–32, 1989b.

Birchall, J. D., Exley, C., Chappell, J. S., and Phillips, M. J., *Nature,* 338, 146–148, 1989.

Bolla, K., Briefel, G., Spector, D., Schwartz, B., Weiler, L., Herron, J., and Gimenez, L., *Arch. Neurol.,* 49, 1021–1026, 1992.

Candy, J. M., McArthur, F. K., Oakley, A. E., Taylor, G. A., Chen, C. P., Mountfort, S. A., Thompson, J. E., Chalker, P. R., Bishop, H. E., Beyreuther, K., Perry, G., Ward, M. K., Martyn, C. N., and Edwardson, J. A., *J. Neurol. Sci.,* 107, 210–218, 1992.

Candy, J. M., Klinowski, R. H., Perry, E. K., Fairbairn, A., Oakley, A. E., Carpenter, T. A., Atack, J. R., Blessed, G., and Edwardson, J. A., *Lancet,* 1, 354–357, 1986.

Caporaso, G. L., Gandy, S. E., Buxbaum, J. D., and Greengard, P., *Proc. Natl. Acad. Sci. U.S.A.,* 89, 2252–6, 1992.

Cataldo, A. M. and Nixon, R. A., *Proc. Natl. Acad. Sci. U.S.A.,* 87, 3861–3865, 1990.

Cataldo, A. M., Paskevich, P. A., Kominami, E., and Nixon, R. A., *Proc. Natl. Acad. Sci. U.S.A.,* 88, 10998–1002, 1991.

Chang, T. M. and Barre, P., *Lancet,* 2(8358), 1051–1053, 1983.

Chartier-Harlin, M. C., Crawford, F., Houlden, H., Warren, A., Hughes, D., Fidani, L., Goate, A., Rossor, M., Roques, P., Hardy, J., et al., *Nature,* 353(6347), 844–846, 1991.

Clauberg, M. and Joshi, J. G., *Proc. Natl. Acad. Sci. U.S.A.,* 90(3), 1009–1012, 1993.

Corder, E. H., Saunders, A. M., Strittmatter, W. J., Schmechel, D. E., Gaskell, P. C., Small, G. W., Roses, A. D., Haines, J. L., and Pericak-Vance, M. A., *Science,* 261(5123), 921–923, 1993.

Corrigan, F. M., Reynolds, G. P., and Ward, N. I., *Biometals,* 6:3, 149–154, 1993.

Crapper McLachlan, D. R., Smith, W. L., and Kruck, T. P., *Ther. Drug Monit.,* 15, 602–607, 1993.

Crapper McLachlan, D. R., Dalton, A. J., Kruck, T. P. A., Bell, M. Y., Smith, W. L., Kalow, W., and Andrews, D. R., *Lancet,* 337, 1304–1308, 1991.

Crapper McLachlan, D. R., McLachlan, C. D., Krishnan, B., Krishnan, S. S., Dalton, A. J., and Steele, J. C., *Environ. Geochem. Health,* 11(2), 45–53, 1989.

Crapper, D. R., Quittkat, S., and Krishnan, S. S., *Acta Neuropathol. (Berlin),* 50, 19–24, 1980.

Crapper, D. R., Krishnan, S. S., and Quittkat, S., *Brain,* 99, 67–69, 1976.

Crapper, D. R., Krishnan, S. S., and Dalton, A. J., *Science,* 180, 511–513, 1973.
DeKosky, S. T. and Scheff, S. W., *Ann. Neurol.,* 27(5), 457–464, 1990.
Delamarche, C., *J. Neurochem.,* 60(1), 384, 1993.
Duckett, F. and Galle, J., *C. R. Acad. Sci. (Paris),* 282, 393–396, 1976.
Dyrks, T., Dyrks, E., Hartmann, T., Masters, C., and Beyreuther, K., *J. Biol. Chem.,* 267(25), 18210–18217, 1992.
Evans, P. H., Yano, E., Klinowski, J., and Peterhans, E., *Oxidative Damage in Alzheimer's Dementia and the Potential Etiopathogenic Role of Aluminosilicates, Microglia and Micronutrient Interactions,* Emerit, I. and Chance, B., Eds., Birhauser Verlag, Switzerland, 178, 1992.
Edwardson, J. A., Moore, P. B., Ferrier, I. N., Lilley, J. S., Newton, G. W. A., Barker, J., Templar, J., and Day, J. P., *Lancet,* 342, 211–212, 1993.
Exley, C., Price, N. C., Kelly, S. M., and Birchall, J. D., *FEBS Lett.,* 324(3), 293–295, 1993.
Flaten, T. P., *Environ. Geochem. Health,* 12(1/2), 152–167, 1990.
Forbes, W. F., Hayward, L. M., and Agwani, N., *Can. J. Aging,* 11(3), 269–280, 1992.
Forbes, W. F. and McAiney, C. A., *Lancet,* 340, 668–669, 1992.
Fraser, P. E., Levesque, L., and McLachlan, D. R., *J. Neurochem.,* 1994.
Fraser, P. E., Levesque, L., and McLachlan, D. R., *Clin. Biochem.,* 26, 339–349, 1993a.
Fraser, P. E., Tam, C., and McLachlan, D. R., Aluminum promotes in vitro assembly of Alzheimer tau into paired helical filaments, Soc. Neurochemistry 24th Annu. Meet. Richmond, VA, March, 1993b.
Frecker, M. F., *J. Epidemiol. Commun. Health,* 45, 307–311, 1991.
Galle, P., Berry, J. P., and Duckett, S., *Acta Neuropathol.,* 4, 245–247, 1980.
Garruto, R. M., Shankar, S. K., Yanagihara, R., Salazar, A. M., Amyx, H. L., and Gajdusek, D. C., *Acta Neuropathol.,* 78, 210–219, 1989.
Garruto, R. M., Fukatsu, R., Yanagihara, R., Gajdusek, D. C., Hook, G., and Fiori, C. E., *Proc. Natl. Acad. Sci. U.S.A.,* 81, 1875–1879, 1984.
Greenberg, S. G. and Davies, P., *Proc. Natl. Acad. Sci. U.S.A.,* 87. 5927–5831, 1990.
Goate, A., Chartier-Harlin, M. C., Mullan, M., Brown, J., Crawford, F., Fidani, L., Giuffra, L., Haynes, A., Irving, N., James, L., et al., *Nature,* 349(6311), 704–706, 1991.
Good, P. F., Perl, D. P., Bierer, L. M., and Schmeidler, J., *Ann. Neurol.,* 31, 286–292, 1992a.
Good, P. F., Olanow, C. W., and Perl, D. P., *Brain Res.,* 593, 343–346, 1992b.
Grundke-Iqbal, I., Iqbal, K., Tung, Y.-C., Quinlan, M., Wisniewski, H. M., and Binder, L. I., *Proc. Natl. Acad. Sci. U.S.A.,* 83, 4913–4917, 1986.
Gutteridge, J. M. C., Quinlan, G. J., Clark, I., and Halliwell, B., *Biochim. Biophys. Acta,* 835, 441–447, 1985.
Guy, S., Jones, D., Mann, D., and Itzhaki, R., *Neurosci. Lett.,* 121, 166–168, 1991.
Haass, C., Schlossmacher, M. G., Hung, A. Y., Vigo-Pelfrey, C., Mellon, A., Ostaszweski, B. L., Lieberburg, I., Koo, E. H., Schenk, D., Teplow, D. B., and Selkoe, D. J., *Nature,* 359, 322–325, 1992.
Hirsch, E. C., Brandel, J. P., Galle, P., Javoy-Agid, F., and Agid, Y., *J. Neurochem.,* 56(2), 446–451, 1991.
Hosovski, E. (Quoted in Selena, Branko (Ed.)), *Summary of Symposium on Toxicology of Copper and Aluminum.* In: Rev. Pap. Vol. 29, Spec. Issue 1989, Izdavac: Niro Zastita Rada, 11000, Beograd, Jelene Cetkovic 3, 1989.
Hosovski, E., Mastelica, Z., Sunderic, D., and Radulovic, D., *Med. Lav.,* 81(2), 119–123, 1990.
Jacobs, R. W., Duong, T., Jones, R. E., Trapp, G. A., and Scheibel, A. B., *Can. J. Neurol. Sci.,* 16 (Suppl. 4), 498–503, 1989.
Joshi, J. G., Fleming, J., and Zimmerman, A., *J. Neurol.,* (Suppl. 232) 61, 1985.
Kaehny, W. D., Alfrey, A. C., and Holman, R. E., *Kidney Int.,* 12, 361–369, 1977.
Karlinsky, H., Vaula, G., Haines, J. L., Ridgley, J., Bergeron, C., Mortilla, M., Tupler, R. G., Percy, M. E., Robitaille, Y., Noldy, N. E., et al., *Neurology,* 42(8), 1445–1453, 1992.
Katzman, R., *Neurology,* 43, 13–20, 1993.
Katzman, R., *Arch. Neurol.,* 33(4), 217–218, 1976.
Khatoon, S., Grundke-Iqbal, I., Iqbal, K., *J. Neurochem.,* 59(2), 750–753, 1992.
Kobayashi, S., Hirota, N., Saito, K., and Utsuyama, M., *Acta Neuropathol (Berlin),* 74, 47–52, 1987.
Kong, S., Liochev, S., and Fridovich, I., *Free Rad. Biol. Med.,* 13, 79–81, 1992.
Landsberg, J., McDonald, B., Grime, G., and Watt, F., *J. Geriatr. Psychiatr. Neurol.,* 6, 97–104, 1993.
Landsberg, J. P., McDonald, B., and Watt, F., *Nature,* 360, 65–68, 1992a.
Landsberg, J. P., McDonald, B., and Watt, F., Identification and Analysis of Neuritic Plaques. Proc. Second Int. Conf. Aluminum and Health, Tampa, FL, Feb. 2–6, 121, 1992b.
Langauer-Lewowicka, H. and Braszczynska, Z., *Neurol. Neurosurg., Poland,* XV11 (XXX111) No. 1, 1983.
Leterrier, J. D., Langui, D., Probst, A., and Ulrich, J., *J. Neurochem.,* 60(1), 385–387, 1993.
Leterrier, J. F., Langui, D., Probst, A., and Ulrich, J., *J. Neurochem.,* 58(6), 2060–2070, 1992.
Ljunggren, K. G., Lidums, V., and Sjogren, B., *Br. J. Ind. Med.,* 48, 106–109, 1991.
Longstreth, W. T., Rosenstock, L., and Heyer, N. J., *Arch. Intern. Med.,* 145, 1972–1975, 1985.
Lovell, M. A., Ehmann, W. D., and Markesbery, W. R., *Ann. Neurol.,* 33, 36–42, 1993.
Lukiw, W. J., Bergeron, C., Wong, L., Kruck, T. P. A., Krishnan, B., and Crapper McLachlan, D. R., *Neurobiol. Aging,* 13, 115–121, 1992.
Lukiw, W. J., and Crapper McLachlan, D. R., *Mol. Brain Res.,* 7, 227, 233, 1990.

Maggio, J. E., Stimson, E. R., Ghilardi, J. R., Allen, C. J., Dahl, C. E., Whitcomb, D. C., Vigna, S. R., Vinters, H. V., Labenski, M. E., and Mantyh, P. W., *Proc. Natl. Acad. Sci. U.S.A.*, 89, 5462–5466, 1992.
Mann, D. M. A., Younis, N., Jones, D., and Stoddart, R. W., *Neurodegeneration*, 1, 201–215, 1992.
Mantyh, P. W., Ghilardi, J. R., Rogers, S., DeMaster, E., Allen, C. J., Stimson, E. R., and Maggio, J. E., *J. Neurochem.*, 1993.
Martyn, C. N., Barker, D. J., Osmond, C., Harris, E. C., Edwardson, J. A., and Lacey, R. F., *Lancet*, 1 (8629), 59–62, 1989.
Masliah, E., Ellisman, M., Carragher, B., Mallory, M., Young, S., Hansen, L., DeTeresa, R., and Terry, R. D., *J. Neuropathol. Exp. Neurol.*, 51(4), 404–414, 1992.
Masters, C., Multhaup, G., Simms, G., Pottgiesser, J., Martins, R. N., and Beyreuther, K., *EMBO J.*, 4, 2757–2763, 1985.
McDowell, I. and Lindsay, J., *Can. Med. Assoc. J.*, 1994.
McGeer, P. L., McGeer, E., Rogers, J., and Sibley, J., *Lancet*, 335(8696), 1037, 1990.
McLachlan, D. R. C., Lukiw, W. J., Wong, L., Bergeron, C., and Bech-Hansen, N. T., *Mol. Brain Res.*, 3, 255–262, 1988.
McLaughlin, A. I. G., Kazantzis, G., King, E., Teare, D., Porter, R. J., and Owen, R., *Br. J. Ind. Med.*, 19, 253–263, 1962.
Mesco, E. R., Kachen, C., and Timiras, P. S., *Mol. Chem. Neuropathol.*, 14, 199–212, 1991.
Montejo de Garcini, E., Serrano, L., and Avila, J., *Biochem. Biophys. Res. Commun.*, 141, 790–796, 1986.
Moreno, A., Dominguez, P., Dominguez, C., and Ballabriga, A., *Eur. J. Pediatr.*, 150, 513–514, 1991.
Mukaetova-Ladinska, E. B., Harrington, C. R., Roth, M., and Wischik, C. M., *Am. J. Pathol.*, 143(2), 565–578, 1993.
Mullan, M., Crawford, F., Axelman, K., Houlden, H., Lilius, L., Winblad, B., and Lannfelt, L., *Nature Genet.*, 1(5), 345–347, 1992.
Muma, N., Troncoso, J., Hoffman, P., Koo, E., and Price, D., *Mol. Brain Res.*, 3, 115–122, 1988.
Muntasser, S., Percy, M. E., Somerville, M. J., Weyer, L., and Bergeron, C., *J. Neurochem.*, 1994.
Murrell, J., Farlow, M., Ghetti, B., and Benson, M. D., *Science*, 254(5028), 97–99, 1991.
Neri, L. C. and Hewitt, D., *Lancet*, 338, 390, 1991.
Neri, L. C., Hewitt, D., and Rifat, S. L., Aluminium concentrations in drinking water and population risk for diagnoses of Alzheimer's disease. Paper presented at the Third Int. Conf. Alzheimer's Disease and Related Disorders, Padova, Italy, July, 1992.
Olanow, C. W., *Ann. Neurol.*, 32, S2–S9, 1992.
Paganini-Hill, A., Buckwalter, J. G., Logan, C. G., and Henderson, V. W., *Soc. Neurosci. Abstr.*, 19 (Abstr.), 425.12, 1046, 1993.
Pericak-Vance, M. A., Bebout, J. L., Gaskell, P. C., Jr., Yamaoka, L. H., Hung, W. Y., Alberts, M. J., Walker, A. P., Bartlett, R. J., Haynes, C. A., Welsh, K. A., et al., *Am. J. Hum. Genet.*, 48(6), 1034–1050, 1991.
Perl, D. P. and Brody, A. R., *Science*, 208, 297–299, 1991.
Perl, D. P. and Pendlebury, W. W., *J. Neuropathol. Exp. Neurol.*, 43, 349–359, 1984.
Perl, D., Gajdusek, C., Garruto, R., Yanagihara, R., and Gibbs, C., *Science*, 217, 1053–1055, 1982.
Perl, D. P., Munoz-Garcia, D., Good, P., et al., in *Alzheimer's Disease and Parkinson's Diseases*, Fisher, A., Hanin, I., and Lachman, C., Eds., Plenum Press, New York, 241, 1986.
Popovic, R. and Banicevic, R., The application of Alzheimer's disease scale for workers professionally exposed to aluminium. Proc. Symp. Copper Aluminium Toxicology, Uzice, No. 50: 1–9, 1989.
Rifat, S. L., Eastwood, M. R., McLachlan, D. R. C., and Corey, P. N., *Lancet*, 336, 1162–1165, 1990.
Rifat, S. L., Eastwood, M. R., McLachlan, D. R., and Corey, P. N., *Lancet*, 336, 1162–1165, 1990.
Romanski, S. A., McCarthy, J. T., Kluge, K., Fitzpatrick, L. A., *Mayo Clin. Proc.*, 68, 419–426, 1993.
Rosen, D. R., et al., *Nature*, 362, 59–63, 1993.
Rozemuller, J. M., Eikelenboom, P., Stam, F. C., Beyreuther, K., and Masters, C. L., *J. Neuropathol. Exp. Neurol.*, 48, 674–691, 1989.
Sarkander, H. I., Lux, R., and Cervos-Navarro, J., *Exp. Brain Res.*, 5, 45–50, 1982.
Saunders, A. M., Strittmatter, W. J., Schmechel, D., George-Hyslop, P. H., Pericak-Vance, M. A., Joo, S. H., Rosi, B. L., Gusella, J. F., Crapper-McLachlan, D. R., Alberts, M. J., et al., *Neurology*, 43(8), 1467–1472, 1993.
Schellenberg, G. D., Bird, T. D., Wijsman, E. M., Orr, H. T., Anderson, L., Nemens, E., White, J. A., Bonnycastle, L., Weber, J. L., Alonso, M. E., et al., *Science*, 258(5082), 668–671, 1992.
Scott, C. W., Fieles, A., Sygowski, L. A., and Caputo, C. B., *Brain Res.*, 628, 77–84, 1993.
Shigematsu, K. and McGeer, P. L., *Brain Res.*, 593, 117–123, 1992.
Sjogren, B., Gustavsson, P., and Hogstedt, C., *Br. J. Ind. Med.*, 47, 704–707, 1990.
Sparkman, D. R., *Neurosci. Lett.*, 151, 153–157, 1993.
Spencer, P. S., Nunn, P. B., Hugon, J., et al., *Science*, 237, 517–522, 1987.
Spofforth, J., Edin, L. R. C., and Eng, M. R. C. S., *Lancet*, 30, 1921.
St. George-Hyslop, P., Crapper McLachlan, D., Tuda, T., and Rogaeve, E., *Science*, 263, 537, 1994.
St. George-Hyslop, P., Haines, J., Rogaev, E., et al., *Nature Genet.*, 2, 330–334, 1992.
Strong, M. J., Wolff, A. V., Wakayama, I., and Garruto, R. M., *Neurotoxicology*, 12, 9–22, 1991.
Trapp, G. A., Miner, G. D., Zimmerman, R. L., Mastri, A. R., and Heston, L. L., *Biol. Psychiatr.*, 13(6), 709–718, 1978.
Traub, R. D., Rains, T. C., Garruto, R. M., Gajdusek, D. C., and Gibbs, C. J., Brain destruction alone does not elevate brain aluminum, 1981.
Vida, A. and Vido, M., *Prac. Lek.*, 35(6), 254–258, 1983.
Wettstein, A., Aeppli, J., Gautschi, K., and Peters, M., *Int. Arch. Occup. Environ. Health*, 63, 97–103, 1991.

White, D. M., Longstreth, W. T., Rosenstock, L., Claypoole, K. H. J., Brodkin, C. A., and Townes, B. D., *Arch. Intern. Med.,* 152, 1443–1448, 1992.

Wille, H., Drewes, G., Biernat, J., Mandelkow, E.-M., and Mandelkow, E., *J. Cell Biol.,* 118, 573–584, 1992.

Wisniewski, H. M., in *Alzheimer's Disease and the Environment,* Lord Walter of Detchant, Ed., Royal Society of Medicine Services, Round Table Ser. 26, London, 35, 1991.

Wood, D. J., Cooper, C., Stevens, J., and Edwardson, J., *Age Ageing,* 17, 415–419, 1988.

Xu, N., Majidi, V., Markesbery, W. R., and Ehmann, W. D., *Neurotoxicology,* 13, 735–744, 1992.

Yanagihara, R., Garruto, R. M., and Gajdusek, D. C., *Ann. Neurol.,* 13, 79–86, 1983.

Yano, I., Yoshida, S., Uebayashi, Y., Yoshimasu, F., and Yase, Y., *Biomed. Res.,* 10(1), 33–41, 1989.

Yase, Y., *Lancet,* 2, 292–296, 1972.

Yasui, M., Yase, Y., Ota, K., Mukoyama, M., and Adachi, K., *Neurotoxicology,* 12, 277–283, 1991a.

Yasui, M., Yase, Y., Ota, K., and Garruto, R. M., *Neurotoxicology,* 12, 615–620, 1991b.

Yoshimasu, F., Yasui, M., Yoshida, H., Yoshida, S., Labayashi, Y., Yase, Y., Gadjusek, D. C., and Chen, K., Aluminum in Alzheimer's disease in Japan and Parkinsonism-dementia in Guam. XII World Congr. Neurology, Hamburg, Abstr. 15-07-02, 1985.

Yumoto, S., Ohashi, H., Nagai, H., Kakimi, S., Ogawa, Y., Iwata, Y., and Ishii, K., *Int. J. PIXE,* 2, 493–504, 1992.

Yumoto, S., Kakimi, S., Ohashi, H., Nagai, H., Ogawa, Y., Mizutani, T., and Ishii, K., *J. Neurochem.,* Suppl. 61, 1993.

Zolan, W. J. and Ellis-Neill, L., *Water and Energy Research Institute of the Western Pacific,* Tech. Rep. No. 64, University of Guam, Phillipines, 1986.

Chapter 24

Current Status of Childhood Lead Exposure At Low Dose

Herbert L. Needleman

I. INTRODUCTION

Knowledge of the effects of lead on human health has advanced dramatically in the past decade. As recently as 1940 it was believed that if a child did not die of acute disease, he or she was left with no long-term effects. Later it was believed that only individuals who displayed symptoms of encephalopathy would have long-term CNS deficits. New tools in the experimental laboratory and new methods in epidemiologic design have brought increased precision to bear on the measurement of lead effects, and this has resulted in lowering the threshold at which toxicity has been observed. With this lowering of the toxic threshold, the numbers of affected individuals have increased, and populations previously thought to be free of effect are now recognized as at risk. Governmental policies, recognizing the implications of the new data, have shifted prevailing strategies. Public health activities, once designed to detect and treat exposed individuals, are changing their emphasis from case finding to primary prevention of exposure (CDC, 1991).

Lead has been recognized as a potent hazard since antiquity. Pliny, in the second century B.C., warned of breathing lead fumes and described its risks to both animals and humans (Major, 1931). Childhood lead poisoning was first described in Australia in 1894 (Turner, 1897). The source, paint from flaking powdering porch rails painted with white lead, was identified 10 years later (Gibson, 1904). The disease was first described in the U.S. in 1914 (McKhann and Vogt, 1926; Blackfan, 1917). The prevailing pediatric wisdom at that time was that if a child did not die of the acute illness, he or she recovered without any residua. This perception was challenged in 1943 when Randolph Byers followed up 20 children who had recovered from acute toxicity, and who were assumed to be untouched by their past illness. In 19 of the 20 he found school failures, behavior disorders, or learning problems. Byers suggested that a portion of the neurocognitive difficulties in American children could be a result of undiagnosed lead toxicity. This report opened the modern period of lead toxicology (Byers, 1943).

In the 1960s the defined toxic dose of lead was 60 µg/dl. When screening studies in eastern U.S. cities showed that substantial numbers of children bore blood levels over 40 µg/dl, the question of unrecognized toxicity came under scrutiny.

Early studies of lead at doses below the symptomatic level tended to use small samples, exert limited control of covariates, and employ relatively unsophisticated statistical tools. These design problems resulted in widely differing conclusions in the first studies. De la Burde and Choate followed up subjects enrolled in the large federal Perinatal Collaborative Study, a longitudinal investigation of children identified during gestation (de la Burde and Choate, 1975). Children with elevated lead levels (>30 µg/dl) and pica had fine motor dysfunction, impaired concept formation, and impaired behavior compared to subjects without pica. When followed into their seventh year of life, continued CNS difficulties were found, and the rate of grade retention or referral for counseling was seven times higher in the high lead group. Kotok (1972) found no differences in cognition and sensory function between children with blood lead levels <40 µg/dl and those above 60 µg/dl. The sample size in this study was 64 subjects.

Perino and Ernhart (1974), in perhaps the most rigorous study of this period, compared children with blood levels >40 µg/dl to those with blood lead levels <30 µg/dl. Significant differences on the McCarthy Scales of Child Development were found after adjustment for age and parental education. The number of subjects was 80. The authors concluded: "While the effects of subclinical lead intoxication may not be noted in individual cases ... analysis of group data indicate quite clearly that performance on an intelligence test is impaired."

Seven years later Ernhart et al. (1981) followed up 63 of these subjects. When IQ was measured in association with the contemporary blood lead level, three of seven outcomes showed significant lead effects after covariate adjustment. The relationship between preschool blood lead levels and IQ was inverse and approached, but did not reach, statistical significance at the .05 level. Statistical power was substantially reduced by subject loss; consequently the power to find an effect was quite small. Despite this limitation, and the positive findings, Ernhart chose to interpret this as evidence for either a trivial lead effect, or no effect at all.

Most studies of this period used blood lead as the exposure marker. Lead has a relatively short residence time in blood, and may be normal in exposed subjects after contact with the toxin has ended. In an attempt to circumvent this problem, Needleman et al. (1979) measured lead in shed deciduous teeth. They compared IQ, speech, language performance, and attentional function in high and low tooth lead subjects, adjusting for maternal IQ, education, age at birth, fathers' education and occupation, and family size. High lead children had significantly lower IQ scores, poorer attention, and diminished speech and language function. Teachers' ratings on 11 classroom items were measured in relation to dentine lead level for over 2000 subjects. A dose-dependent increase in maladaptive behavior was reported by teachers for every item (Figure 1).

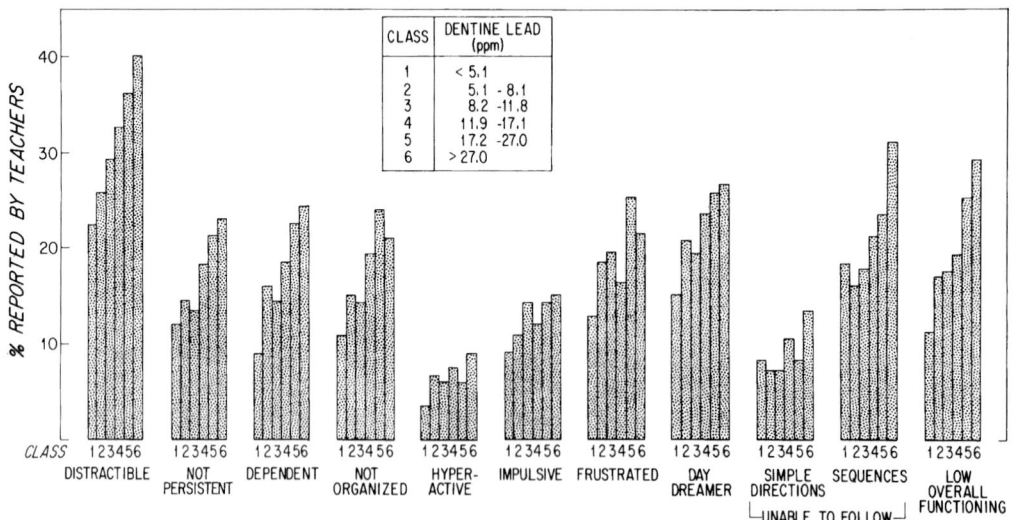

Figure 1 The teachers' ratings of 2146 students on 11 behaviors. Students are grouped according to dentine lead level. Teachers knew the students at least two months, but were blind to their lead measurements. A dose-related increase in the rate of nonadaptive behavior is seen on all 11 items.

Schroeder et al. 1986 evaluated 104 lower SES North Carolina children and adjusting for maternal IQ, rearing, and SES, reported that lead was significantly related to IQ. After five years, the relationship between lead and outcome was still negative, but it no longer was statistically significant. The sample size in the follow-up was substantially reduced, however (N = 50). A second sample of 75 subjects studied by the same group (Hawk et al., 1986) showed a significant lead effect after covariate adjustment.

Studies of lead in asymptomatic children began to be published from not only the U.S. but also from Europe, Scandinavia, Australia, and New Zealand in the 1980s. Yule and colleagues (Yule et al., 1981) studied 166 English school children aged 6 to 12. After control for age and socioeconomic status, lead was significantly related to IQ scores, reading, and spelling. These investigators also examined the relationship between lead levels and behavior in the same sample (Lansdown et al., 1983). When subjects were classified into quartiles by blood lead level, a dose-related increase in poor behavior (i.e., restlessness, conduct disorder, oppositional behavior, etc.) was observed. Subjects were also rated on the Rutter B2 Scale, a widely used measure of behavior. On 20 of 26 items, children above the median for lead scored more deviant than those below the median. Lead level was also related to three of four scales of the Connors factors (conduct problems, inattentive-passive, and hyperactivity). Lansdown et al. (1983) examined blood lead levels and IQ in 194 children, almost evenly divided between offspring of manually and nonmanually employed parents. There was no relationship between lead and IQ in the nonmanual group. The children of manual workers whose blood levels were above 10 µg/dl had consistently lower psychometric scores than those with higher blood lead levels.

Smith et al. (1983) studied 403 London children classified by tooth lead level. Before covariate adjustment, there was a strong relationship between lead and IQ. This was attenuated when covariates were introduced into the model, and no longer was significant at the .05 level. When concentration was plotted against adjusted IQ, a clear dose-related decline in IQ was found. This was only violated in the highest tooth lead group. The investigators also found that high lead subjects showed more behavior problems, although this did not achieve statistical significance. In a later reanalysis of the same data set, a significant lead effect was observed in male subjects (Pocock et al., 1987).

Winneke et al. (1982) studied children living in the industrial town of Duisberg, Germany. When children with mean tooth lead levels of 2.4 ppm (N = 26) were compared to those with a mean tooth lead of 7 ppm (N = 16) the high lead subjects had a mean decrease in IQ of 5 to 7 points ($p < .1$). Scores on a visual motor task were significantly decreased ($p < .05$). In a second study (N = 115), children's psychometric IQ, reaction time, and behavioral rating were evaluated in relation to tooth lead by multiple regression. Lead was significantly related to perceptual motor integration, reaction time, and behavior after adjustment for covariates. The IQ relationship was not statistically significant. A third study (Winneke et al., 1983) found no relationship between IQ and lead, but a clear effect on reaction time.

Fulton and colleagues (1987) measured the relationship between the blood lead level and scores on the British Ability Scales in 501 middle class Scottish children. After adjustment for a large group of covariates, a significant dose-dependent relationship between lead and IQ was found. When the same children were examined for attentional performance and behavior, lead significantly impaired the performances (Thomson et al., 1989; Raab et al., 1990).

Hansen et al. (1989) collected teeth from 2412 Danish schoolchildren. The upper 8%, with dentine lead levels >18.7 ppm, were designated as the "high lead" group, and were matched by sex and SES with a "low lead" group whose dentine lead levels were <5 ppm. Children from this group with defined neurologic syndromes and well-defined risk factors were excluded from the study. The final sample size was 162. Two strategies were used to control for confounding. In the first, the two groups were matched on sex and SES, and the data analyzed by t-test for matched pairs. In the second method, the association of a large number of factors with lead was evaluated by logistic regression. Those related at $p < .2$ were considered confounders, and entered into a multiple regression model. By both methods, lead was significantly related to verbal and full-scale IQ, and performance on the Bender Gestalt test of visual motor integration. The contemporary blood lead level was significantly related to motor speed and performance on a test of auditory memory and attention. Both blood and dentine lead were related to measured on-task performance.

Hatzakis et al. (1987) examined IQ performance and classroom ratings of 509 students in Lavrion, Greece at the site of an ancient lead and silver mine. There were 17 covariates entered into the multiple regression model, including parental IQ and education, family size, birth weight, and medical history. Blood lead was closely related to verbal and performance IQ after covariate control ($p = .0004$ and $p = .002$, respectively). Lead was also related to reaction time. The difference in intelligence between the extreme lead groups (45 µg/dl and 15 µg/dl, was 9 points.) This group also showed a dose-related effect between blood lead levels and teachers ratings of classroom behavior, using the same scale as Needleman et al. The finding persisted after control of age, sex, and socioeconomic status.

In New Zealand, Silva et al. (1988) evaluated 579 children 11 years of age, enrolled in the Dundedin forward study of child development. Blood lead was used as the exposure index, and the outcomes measured were IQ, reading, and behavior using the Rutter Teacher Behavior Questionnaire. Lead was not related to IQ. After control for social and environmental factors, including maternal intelligence, lead was significantly related to both teachers' and parents' ratings of behavior, primarily inattention and hyperactivity.

Fergusson et al. (1988), in another forward follow-up of a larger cohort of New Zealand children, this group from Christchurch, collected teeth from 724 children aged 8 to 9 years, and examined the relationship between lead and IQ, reading, and teachers' ratings of school performance. After adjustment for a large battery of covariates, lead was significantly related to reading scores and the teachers' rating of reading, written expression, spelling, mathematics, and handwriting. In a companion study, lead was significantly related to inattention and restlessness. A recent follow-up of these subjects into their 13th year, discussed below, found more severe delays in reading and academic achievement (Fergusson et al., 1993).

II. FORWARD STUDIES

Critics of the lead-IQ hypothesis have suggested that the causal arrow operates in the opposite direction: the findings could be explained by cognitively impaired children having more oral behavior and ingesting more lead-containing substances. To test this, and to evaluate the progression of lead effects, forward studies of children identified at birth or earlier were designed.

The first of these studies measured umbilical blood lead levels in a two-year cohort of children born at the Boston Hospital for Women between 1979 and 1981. The number of subjects was slightly less than 12,000 births. From this group, the relationship between umbilical cord blood and congenital anomalies was studied. An increased blood lead level was associated with an increase in the rate of minor congenital anomalies (Needleman et al., 1984). Major anomalies were not related to lead level. A subsample of 247 of these subjects was followed forward at 1, 6, 12, 24, 57, and 120 months of age. Blood lead levels and cognitive development was measured at each epoch. At 24 months of age, intrauterine exposure, as reflected in the umbilical cord blood lead level, was significantly related to IQ. At 57 and 120 months, the effects of intrauterine exposure were no longer significant (Bellinger et al., 1987). The effect of blood lead concentration at 24 months was significantly related to outcome at both 57 and 120 months, after covariate adjustment (Bellinger et al., 1992).

Dietrich et al. (1993) followed a cohort of 245 children forward from birth and measured blood lead levels and neurobehavioral development. After adjustment for covariates, the neonatal blood lead level was significantly associated with poorer performance of upper limb speed and dexterity and fine motor function. Postnatal scores were associated with diminished visual motor coordination speed and fine motor function.

Baghurst et al. (1992) studied a cohort of 494 Australian children living in proximity to a lead mine and mill. Umbilical cord blood lead was measured and the determination repeated at regular intervals thereafter. Between the subjects' seventh and eighth years, IQ and a battery of covariates were measured. The lifetime average blood lead level was significantly related to verbal and full-scale IQ score after covariate adjustment.

In Cleveland, Ernhart et al. (1993) studied an unusual sample of pregnant women: half were self reported abusers of alcohol. The maximum number of subjects was 153. Neither alcohol exposure nor lead was found to be significantly related to IQ scores. Although the authors assert this is evidence for either a small or no lead effect, the power to find an effect with a sample of this size (N = 153) is about .25, clearly inadequate to base a null assertion, even if the outcome was not clouded by alcoholic intake. Recently, this group reported on the relationship between tooth lead and outcome in this sample. Lead exposure was discovered to be significantly related to verbal IQ ($p = .011$). The relationship between lead and performance IQ was smaller. As a result, the lead full-scale IQ effect was of borderline significance ($p = .063$). Many others, including the first dentine lead IQ study reported 14 years earlier, have noted that lead has a stronger effect on verbal IQ.

III. LONG-TERM EFFECTS OF LEAD

Enough time has past for follow-up studies of these subjects to begin to appear. Bellinger et al. (1984) followed into the fifth-grade subjects from the 1979 Boston study. High lead subjects had lower intelligence scores on a school-based test, more need for remedial services, and more grade retention. When followed into their 18th year, the findings of the initial report seemed to have increased in severity. High lead subjects were seven times more likely to fail out of high school than the lowest lead group, and six times more likely to have a reading disability. In addition, they ranked lower in class placement in their final year of high school, had more absenteeism, and poorer vocabulary, reasoning, and fine motor function (Needleman et al., 1990). These data suggest that the effects of earlier exposure are permanent, and affect general life success (Figure 2 and Figure 3).

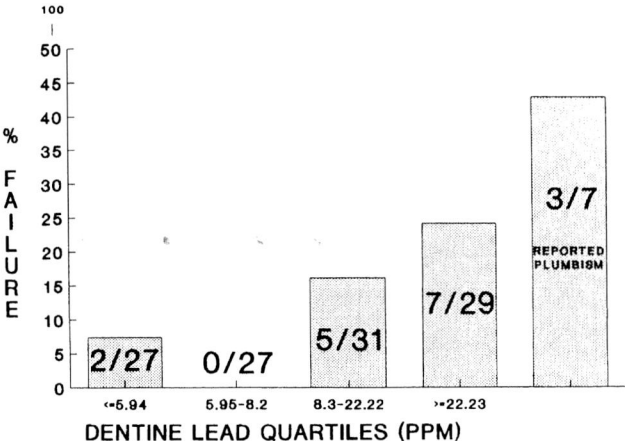

Figure 2 The relationship between early dentine lead levels and academic failure. Subjects are divided into quartiles by dentine lead concentration. Ten subjects were reported to have had clinical lead toxicity, and were inadvertently admitted into the study. Three of those were still in school, three had failed out.

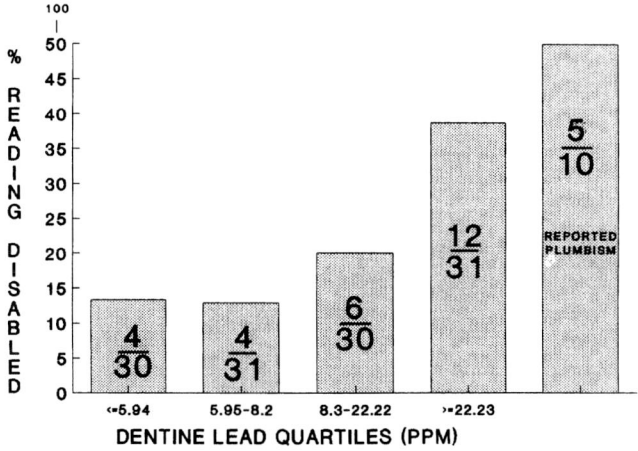

Figure 3 The relationship between early dentine lead levels and reading disability. Disability is defined as reading two years below age level. Of the ten lead-intoxicated subjects inadvertently studied, five had reading disability. (Reprinted with the permission of the *New England Journal of Medicine*. 322:83–88, 1990).

Most recently, Fergusson et al. (1993) have reported on the performance of their sample followed forward to age 13 years. Outcome measures were word recognition, reading comprehension, the Test of Scholastic Abilities, teachers' ratings of performance, and items of the Rutter and Connors Scales measuring inattention and attentiveness. The number of subjects followed was 690 to 891. After covariate adjustment, past dentine lead levels were significantly related to all outcomes. It should be remembered that IQ deficits at the initial testing period did not achieve statistical significance. This careful study suggests that IQ may not be the most sensitive target for lead toxicity, and that neurocognitive effects increase over time. This is consistent with the report of Needleman et al. (1990) in their 12-year follow up.

IV. LIMITS OF EPIDEMIOLOGY

The discipline of epidemiology confronts certain intractable limits when studying a toxin such as lead. It cannot administer lead at precise doses to subjects matched on important covariates, but must measure lead burden and effect as they occur in nature. Lead exposure is differentially distributed in populations that have other risk factors. Lead levels generally are higher in individuals of minority status, of lesser education, lower incomes, and many other factors. These factors also affect development. A covariate that is correlated with both the exposure variable and outcome is a confounder. Confounders, if not controlled, can bias the observed relationship either away from the null or towards it.

Multivariate space is infinite; there are a limitless number of factors that could affect outcome, many of which are not identified. Epidemiologic studies must study a finite number of subjects, and given the state of research budgets, the sample size is generally small. A study that has one measure of lead divided into three groups and measures ten outcomes and 25 covariates, each with five classes, will have 3750 cells. The maximum number of subjects has never exceeded 891. This results in an unsaturated causal model. Unsaturated causal models have the property of fitting to many different regression lines. This is virtually the same as saying that an epidemiologic investigation cannot prove a causal relationship, and cannot rule out the presence of an unidentified confounder.

Some who are skeptical of the low lead-CNS impairment hypothesis employ this fixed constraint as a challenge to the positive conclusions that lead is damaging at low dose. They suggest that there may be unidentified factors that could confound lead's effect. This is always a possibility, but given the bulk of knowledge about the important influences on child development, it is unlikely that an important factor correlated with lead and influential in psychometric intelligence, attention, reaction time, language function, and classroom behavior has gone unrecognized.

There are two important responses to this dilemma: quantitative synthesis of all suitable studies, or metaanalysis and studies in the experimental animal. In the latter, precise control of dose and of potential confounders is readily achievable, social factors are irrelevant, and the direction of the causal arrow is unequivocal.

V. METAANALYSIS OF HUMAN LEAD STUDIES

Metaanalysis is a relatively new form of synthesis of studies of a given research question. It attempts to reduce the subjectivity and bias that adheres to narrative reviews of the literature. In this method, all suitable studies are included, an attempt is made to find those that are unpublished, a common metric such as p value or effect size is chosen for effect, and each study is treated as a subject in a study of studies. A joint measure of probability or effect size is estimated and used to synthesize the import of the studies.

This is not a perfect method. Bias can enter in choosing studies for inclusion, weighting of studies can be arbitrary, and studies of widely varied merit are often treated as identical. Still, metaanalysis represents a huge step forward from narrative syntheses, which are often replete with personal and idiosyncratic bias, and which often rely on simple tallies of the positive and negative reports to draw a conclusion. Counting votes severely degrades the information contained in a group of investigations, and generally overstates the null point of view.

A single example illuminates this point. Consider a series of studies of the efficacy of drug X on tuberculosis. Of the twenty studies reported, ten find no effect; ten find a beneficial effect. A narrative synthesis might conclude that the evidence for an effect is weak or absent. But if the null hypothesis is true, and there is no effect of the drug, one would expect one study in 20 to show an effect by chance at $\alpha = .05$. Metaanalysis, by mathematically summarizing the information in each study, reaches a more accurate synthesis of the joint import of the studies.

There have been four metaanalyses of the question of lead and IQ at low dose. Schwartz and colleagues (1985) used Fisher's aggregation technique to synthesize the joint probability of six cross-sectional studies published between 1979 and 1983, three of which showed a statistically significant effect ($p < .05$) and three did not. The joint p value for the studies under the null hypothesis was $p < 0.005$.

Needleman and Bellinger (1989) metaanalyzed 14 studies published between 1974 and 1986, 9 of which were significant at the conventional level. The joint p value of this analysis was $p = 3 \times 10^{-12}$. In a later metaanalysis, Needleman and Gatsonis (1990) identified 24 studies from 1972 to 1987, 12 of which used multiple regression. The joint p value for the seven studies that used blood lead as the exposure measure was $p < .0001$. No single study was influential. The joint p value for the five studies that used tooth lead was $p < .0005$. No single study was influential. The the partial r for the blood studies was $-.152$ $CL_{95} = -.2$ to $-.1$. For the tooth studies the partial r was $-.13$, $CL_{95} = -.13$ to $-.03$.

Schwartz (1993) metaanalyzed seven recent studies that used multiple regression to evaluate blood lead and full-scale IQ. He linearized the relationship between 5 and 15 µg/dl and computed the mean regression coefficient. He reported that the $\beta = 0.245 \pm 0.039$, $p < .0001$. The results of these four metaanalyses provide strong support for the lead-IQ association.

VI. EXPERIMENTAL STUDIES OF LEAD AND THE CNS

Experimental studies of lead offer an opportunity to deepen our understanding of possible mechanisms through which the metal damages organs and disturbs behavior. In the case of lead, there is no shortage of demonstrated vulnerable sites. It had been thought that the most critical targets were the heme pathway and the integrity of the capillary wall. Research over the past decade has shown effects in almost all organ systems, at levels far below those previously thought to be toxic.

The current problem is not finding candidate mechanisms; it is sorting out the many demonstrated lead-induced disturbances, and relating them to disturbances in human function. This is a forbidding task. In the past 3 years there are 4000 titles listed in *Medline* under the keywords "Lead" and "Experimental."

Most of lead's toxicity is ascribed to its action on proteins, where it binds to sulfhydryl groups. As a result, many enzymes have reduced activity in the presence of lead. New work has emphasized lead's actions at other sites than protein structure and enzyme activity. Brown et al. (1983) measured lead binding to yeast tRNA. At pH 7.4, they observed site-specific cleavage of the ribophosphate backbone. This took place between sites D17 and G18, and was catalytic rather than stoichiometric. This may account for lead's ability to rapidly depolymerize tRNA, and may provide a toxic mechanism of at least as much generalized importance as sulfhydryl binding, with no apparent threshold. It is also a toxic effect which would be expected to have a long latency, and the effect could indeed increase over time.

Marcovac and Goldstein (1988) studied phosphokinase c, a calcium and phospholipid enzyme involved in growth, differentiation, and other cellular functions. Lead alone, of all metals tested, activated phosphokinase c at picomolar concentrations. The authors suggest that this may be a fundamental route for lead toxicity in CNS and other organs. Goldstein (1992) incorporated this finding into a new pathogenetic hypothesis. He notes that an essential element in brain development is the pruning back of neuronal fibers to fewer, more specific connections, and that this process is determined by both genetic mechanisms and experiential input to synapses. If lead increases neuronal responsivity by increased phosphokinase c activity, then a given stimulus in a lead-bearing brain would result in greater excitation of more neurones than in brains with less lead. A stimulated synapse has a higher probability of persistence during development. This could account for lead-induced changes in ultrastructural anatomy and cerebral connectivity which could underlie behaviors such as distractibility, impulsivity, and motor restlessness. This model also predicts that early exposure could produce effects that would be visible years later.

Another new area of study is opioid receptor development and function. Winder et al. (1984) found that administration of low doses of lead to nursing rodent dams produced decreased development of proencephalin, an endogenous opioid precursor in offspring pups. This observation spawned a number of related studies. Kitchen et al. (1984), from the same lab, found that the antinociceptive action of morphine was strongly diminished by low levels of lead in early life. They inferred that altered development of the µ opioid receptor was responsible for the observed effect. Similar blunting of the analgesic action of keto-cyclazocine, a delta opioid receptor agonist, was reported. Bailey and Kitchen (1985) then showed that the four putative peptide products of proencephalin were markedly depressed by small doses of lead at 10 days of age.

Whether these intriguing findings in the opioid system have relevance for human behavior remains to be determined. Studies of rodent consumption of alcohol or other drugs in relation to lead exposure suggest that they may. Nation et al. (1986) have shown that naive rats find 15% solutions of alcohol aversive, but when given lead in their diet sufficient to raise their blood leads to 61 µg/dl, the rats increase their alcohol intake in both forced choice and free choice paradigms. Lead-fed rodents were also more responsive in the avoidance training period. The authors infer that lead increases emotionality, and that the rodents seek alcohol for its anxiolytic properties. Scholars of delinquency have suggested that disturbances in the reward system are among the aberrations underlying criminal behavior.

VII. THE LEAD "CONTROVERSY"

At the time this chapter is being written there are about 30 published studies of lead at low dose in children. Most of the recent (since 1980) studies of large sample size and rigorous design show an effect of lead on IQ or behavior after control of covariates. This is supported to an unusual degree by data from carefully conducted studies in experimental animals. Yet many consider the subject still a controversy. It is reasonable to ask: why, in the face of such an extensive data base, do some parties consider the effects of lead to be surrounded by mystery?

Part of the reason derives from the basically conservative position of science when it confronts new questions. Kuhn (1962) demonstrated this constricting power of scientific world views, which he found enveloping enough to be termed "paradigms." A dominant paradigm dictates the questions to be asked, the tools to be used, and the interpretation of data. Findings that are at variance with the dominant paradigm are initially ignored, then contested, and when the mass of conflicting data becomes indigestible, the regnant paradigm is overthrown. The long-held belief that, short of death, there were no long-term effects of lead exposure, followed by the belief that without clinical symptoms no central nervous system damage of consequence occurred, are difficult to dislodge.

Another expression of the conservatism derived from the dominance of "normal science", is the preferential weighting of Type I error adversiveness among scientists, and the limited attention given to risk of Type II errors. Avoiding Type I errors, (the accepting of spurious relationships as true) is considered good scientific behavior. Considerably less attention and value is given to avoiding Type II errors (dismissing true relationships as spurious). Most studies accept an α error (Type I) of 1 in 20, while ignoring the β (Type II) error, or accepting a much larger risk (1 in 5 or 6). As a result, many studies of inadequate power (defined as $1 - \beta$) are conducted. Many of these, when they find no effect, report it as evidence of no relationship in nature, even though their study had insufficient power to find an effect were it there. The Environmental Protection Agency examined this issue and determined that an N of more than 400 subjects was required to detect a small effect size with a power of .8 at an α of .05 (U.S. EPA, 1992).

Not only cognitive reasons are found for the controversy. The discourse has been clouded by commercial interests who have insisted that the evidence for toxicity at low dose is tenuous if not false. The arguments to this point have utilized those addressed above: human studies are imperfect because they cannot achieve complete covariate control, and animal data are inapplicable because of species differences. Were this syllogism applied to all toxicants, none would be identified as hazardous at low dose.

REFERENCES

Baghurst, P. A., McMichael, A. J., Wigg, W., Vimpani, G. V., Robertson, E. F., Roberts, R. J., and Tong, S. L., *N. Engl. J. Med.*, 327, 1280–1284, 1992.
Bailey, C. and Kitchen, I., *Dev. Brain. Res.*, 22, 75–79, 1985.
Bellinger, D., Needleman, H. L., Bromfield, R., and Mintz, M., *Biol. Trace Elem. Res.*, 6, 207–223, 1984.
Bellinger, D., Leviton, A., Waternaux, C., Needleman, H. L., and Rabinowitz, M., *N. Engl. J. Med.*, 316, 1037–1043, 1987.
Bellinger, D. C., Stiles, K. M., and Needleman, H. L., *Pediatrics*, 90(6), 855–861, 1992.
Blackfan, K. D., *Am. J. Med. Sci.*, 53, 877–887, 1917.
Brown, R. S., Hingerty, B. E., Dewan, J. C., and Klug, A., *Nature*, 303, 543–546, 1983.
Centers for Disease Control, Strategic Plan for the Elimination of Childhood Lead Poisoning, Department of Health and Human Services, Atlanta, GA, 1991.
de la Burde, B. and Choate, M. S., *J. Pediatr.*, 87, 638–642, 1975.
Dietrich, K. N., Berger, O. G., and Succop, P. A., *Pediatrics*, 91, 301–307, 1993.
Ernhart, C. B., Landa, B., and Schell, N. B., *Pediatrics*, 67, 911–919, 1981.
Fergusson, D. M., Fergusson, J. E., Horwood, L. J., and Kinzett, N. G., *J. Child Psychol. Psychiatr.*, 29, 793–809, 1988.
Fergusson, D. M., Horwood, J., and Lynskey, M. T., *J. Child Psychol. Psychiatr.*, 34, 215–227, 1993.
Fulton, M., Raab, G., Thomson, G., Laxen, D., Hunter, R., and Hepburn, W., *Lancet*, 1, 1221–1226, 1987.
Gibson, J. L., *Aust. Med. Gaz.*, 23, 149–153, 1904.
Goldstein, G., in *Human Lead Exposure*, Needleman, H. L., Ed., CRC Press, Boca Raton, FL, 1992.
Greene, T. and Ernhart, C., *J. Clin. Epidemiol.*, 46(4), 323–339, 1993.
Hansen, O. N., Trillingsgaard, A., Beese, I., Lyngbye, T., Grandjean, P., *Neurotoxicol. Teratol.*, 11, 205–213, 1989.
Hatzakis, A., Kokkevi, A., Katsouyanni, K. et al., in *Heavy Metals in the Environment*, CEP Consultants, Edinburgh, 204, 1987.
Hawk, B., Schroeder, S., and Robinson, G., *Am. J. Ment. Def.*, 91, 178–183, 1986.

Kitchen, I., McDowell, J., Winder, C., and Wilson, J. M., *Toxicol. Lett.*, 22, 119–123, 1984.
Kotok, D., *J. Pediatr.*, 80, 57–61, 1972.
Kuhn, T., *The Structure of Scientific Revolutions*, University of Chicago Press, Chicago, IL, 1962.
Lansdown, R., Yule, W., Urbanowicz, M., and Millar, I., in *Lead Versus Health*, Rutter, M. and Jones, R. R., Eds., John Wiley & Sons, New York, 1983.
Major, R. H., *Ann. Med. Hist.*, 3, 218–227, 1931.
Marcovac, J. and Goldstein, G. W., *Nature*, 334, 71–72, 1988.
McKhann, C. F. and Vogt, E. C., *Am. J. Dis. Child.*, 32, 386–392, 1926.
Nation, J. R., Baker, D. M., Taylor, B., and Clark, D. E., *Behav. Neurosci.*, 100, 525–530, 1986.
Needleman, H. L., Schell, A., Bellinger, D., Leviton, A., and Allred, E. N., *N. Engl. J. Med.*, 322, 83–88, 1990.
Needleman, H. L. and Gatsonis, C., *J. Am. Med. Assoc.*, 263(5), 673–678, 1990.
Needleman, H. L., Rabinowitz, M., Leviton, A., Linn, S., and Schoenbaum, S., *J. Am. Med. Assoc.*, 25, 2956–2959, 1984.
Needleman, H. L., Gunnoe, C., Leviton, A., Peresie, H., Maher, C., and Barret, P., *N. Engl. J. Med.*, 300, 689–695, 1979.
Needleman, H. L. and Bellinger, D. C., in *Lead Exposure and Child Development: An International Assessment*, Smith, M.A., Grant, L.D., and Sors, A.I., Eds., Kluwer Academic, Boston, 1989.
Perino, J. and Ernhart, C. B., *J. Learn. Disabil.*, 7, 26–30, 1974.
Pocock, S. J., Ashby, D., and Smith, M. A., *Int. J. Epidemiol.*, 16, 57–67, 1987.
Raab, G. M., Thomson, G. O. B., Boyd, L., Fulton, M., and Laxen, D. P. H., *Br. J. Dev. Psychol.*, 8, 101–118, 1990.
Schroeder, S., Hawk, R., Otto, D., Mushak, P., and Hicks, R. E., *Environ. Res.*, 91, 178–183, 1986.
Schwartz, J., Pitcher, H., Levin, R., Ostro, B., and Nichols, A. L., Costs and Benefits of Reducing Lead in Gasoline: Final Regulatory Analysis, USEPA/OPA, U.S. Environmental Protection Agency, Washington, D.C., 1985.
Schwartz, J., *Neurotoxicology* 14, 237–246, 1993.
Silva, P. A., Hughes, P., Williams, S., and Faid, J. M., *J. Child Psychol. Psychiatr.*, 29, 43–52, 1988.
Smith, M., Delves, T., Lansdown, R., Clayton, B., and Graham, P., *Dev. Med. Child Neurol.*, 25 (Suppl. 47), 1–54, 1983.
Thomson, G. O. B., Raab, G. M., Hepburn, W. S., Hunter, R., Fulton, M., and Laxen, D. P. H., *J. Child Psychol. Psychiatr.*, 30, 515–528, 1989.
Turner, A. J., *Austr. Med. Gaz.*, 16, 475–479, 1897.
U.S. EPA, An SA Report: Review of the Uptake Biokinetic (UBK) Model for Lead, Science Advisory Board, EPA rep. no. EPA-SA-IAQC-92-016, U.S. Environmental Protection Agency, Washington, D.C., 1992.
Winder, C., Kitchen, I., Clayton, L. B., et al., *Toxicol. Appl. Pharmacol.*, 73, 30–34, 1984.
Winneke, G., Hrdina, K. G., and Brockhaus, A., *Int. Arch. Occup. Environ. Health*, 51, 169–183, 1982.
Winneke, G., Kramer, U., Brockhaus, A., Ewers, U., Kujanek, G., Lechner, H., and Janke, W., *Int. Arch. Occup. Environ. Health*, 51, 231–252, 1983.
Yule, W., Lansdown, R., Millar, I. B., and Urbanowicz, M. A., *Dev. Med. Child. Neurol.*, 23, 567–576, 1981.

Chapter 25

Neurological Aspects in Human Exposures to Manganese

Naohide Inoue and Yuji Makita

I. HISTORY

Manganese was first discovered by Sheele in Sweden in 1774. It was in 1785 that Guyton de Morveau proposed the name "manganese" to distinguish a metal called "magnesium". It was isolated in a pure form by Gahn in 1807.

Industrial manganese poisoning was first described by Couper in 1837 in five workers who made bleaching powders using manganese dioxide in France. He found that long-term exposure to manganese dusts causes a peculiar extrapyramidal syndrome. However, his observations were forgotten for a long time. This was followed by reports of similar cases by Embden and von Jaksch in Germany in the early 1900s. Von Jaksch, from 1901 to 1907, reported 15 cases of the extrapyramidal syndrome among manganese workers. It was not until 1919 that a definite relationship between the epidemiologic, clinical, and pathological effects of manganese poisoning on the central nervous system was established by Edsall, Wilbur, and Drinker (1919) in the U.S. Their review drew attention to the serious effects on the central nervous system of this occupational disease. Reports from many countries identified the source of manganese poisoning in manganese ore mills and foundries. Up to 1934, 70 cases of manganese poisoning had been reported in the literature. Fairhall (1945) found records of 353 cases of manganese poisoning since the first report in 1837. Rodier in 1955 described 150 cases among miners in Morocco.

With the onset of World War II, the accelerated growth of the steel industry led to a similar growth of manganese mining. Pneumatic drilling improved the efficacy of mining, but also generated inordinate amounts of dust from the manganese ores (Mena, 1979). The incidence of manganese poisoning in the mines climbed to very high levels, crippling as many as 25% of their working population, as was reported by Schuler et al. (1957), who described in extensive detail the extrapyramidal syndrome of this condition in 15 patients. Rodier in Morocco (1958) reported 223 cases of manganese poisoning, while Penalver (1957) reported on 120 cases in Cuba. Abd El Naby and Hassanein (1963) commented on 32 cases with neuropsychiatric manifestations secondary to exposure to manganese ore in the mines in Egypt.

In the U.S. Flinn et al. (1941) reported 11 cases in workers from a manganese ore crushing plant. Fluctuations in the sanitary control of that factory has led to reports of new cases by Tanaka and Lieben (1969) and by Cook et al. (1974). In 1967, Mena reported in detail the neurological characteristics of 18 patients suffering from manganese poisoning in Chile and correlated these findings with metabolic disturbances of manganese metabolism.

In 1989, Wang et al. reported an outbreak of manganese poisoning due to an unrepaired ventilation control system in a ferromanganese smelter in Taiwan.

In this decade, manganese poisoning has become rare, even in the developing countries.

II. INDUSTRIAL USES AND EXPOSURES

Manganese has three principal uses: (1) in the production of steel as a reagent to reduce oxygen and sulfur, and as an alloying agent for special steels, aluminum, and copper; and (2) in the manufacture of dry-cell batteries, and for the production of potassium permanganate and other manganese chemicals as well as an oxidizing agent in the chemical industry (Saric, 1991). Manganese is also used for electrode coating in welding rods. Several salts are used as driers for linseed oil, glass and textile bleaching, and in fertilizers (Saric, 1991).

The most hazardous manganese exposures occur in mining and smelting of ore (Hamilton and Hardy, 1974). In plants manufacturing alloys of manganese and steel, harmful levels may exist. Less dangerous industrial uses of manganese occur in dry-battery manufacture, electric arc welding, as well as to some extent in production of paint, vanish, enamel, linoleum, fireworks, and fertilizer. Of these, only battery manufacture has produced manganese poisoning.

III. ABSORPTION, DISTRIBUTION, AND EXCRETION

In occupational exposure, manganese is absorbed mainly by inhalation. As manganese dioxide and other manganese compounds used or produced are practically insoluble in water, only particles small enough to reach the alveoli are eventually absorbed into the blood. Large inhaled particles may be cleared from the respiratory tract and swallowed (Saric, 1991). Manganese may also enter the gastrointestinal tract with contaminated food and water. Absorption of manganese through the skin is negligible.

The average adult contains about 12 mg of manganese (Gilmore and Bronstein, 1992). The skeletal system contains about 43%, with the rest in soft tissues including the liver, pancreas, kidneys, and central nervous system.

After inhalation, or after parenteral and oral exposure, the absorbed manganese is rapidly eliminated from the blood and distributed mainly to the liver. Manganese preferentially accumulates in tissues rich in mitochondria. It also penetrates the blood-brain barrier. The biological half-time for manganese is between 36 and 41 days, but for manganese in the brain, it is considerably longer than for the whole body. In the blood, manganese is bound to the erythrocyte porphyrin complex.

Bile flow is the main route of excretion of manganese. Consequently, it is eliminated almost entirely with feces, and only 0.1 to 1.3% of daily intake through urine.

IV. SIGNS AND SYMPTOMS

This disease afflicts workers employed in industries which deal with manganese dioxide among many other manganese compounds.

In general, there is a delay of onset of some years after exposure ceases. In the cases reported by Schuler et al. (1957), the average time of exposure was 8 years and a maximum of 16 years. The neurological signs and symptoms usually become apparent 1 to 2 years after exposure. The neurological findings are almost uniform.

Manganese poisoning is clinically characterized by the central nervous system involvement including psychiatric symptoms, extrapyramidal signs, and other neurological manifestations.

The onset of symptoms is usually insidious and progressive (Cook et al., 1974). The symptoms progress only for a relatively short time, but over the span of years they are nonprogressive. The initial manifestations are usually vague complaints of asthenia, anorexia, apathy, insomnia or drowsiness, and a slowing down in performing motor acts. In miners, the initial phase may be one of psychomotor excitement. This manganese psychosis (Rodier, 1955) or manganese madness (Mena et al., 1967) lasts for about one month whether the miner is removed from the mine or not. Other frequent symptoms of the early and established phases of the poisoning are malaise, somnolence, imbalance while walking or on arising, slurred speech, difficulty with fine movements (handwriting), limb stiffness, diminished libido or impotence. Sometimes, mental languor and lack of energy are prominent symptoms at the onset. Tremor, paresthesia, muscle cramps, memory loss, swallowing difficulty, urinary urgency or incontinence, lumbosacral pain, metallic taste, anorexia, and nervousness are less frequent (Cook et al., 1974).

Psychiatric symptoms are well described in the manganese miners and include sleep disturbance, disorientation, emotional lability, compulsive acts, hallucinations, illusions, and delusions. A marked somnolence is also observed, most often to be replaced later by stubborn insomnia. Most of the cases show emotional incontinence, particularly forced laughing (Rosenstock et al., 1971). The patients may abruptly burst into laughter or (more rarely) into tears without any apparent reason. Frequent irritability and nervousness resulted in arguments and friction among the miners, occasionally approaching violence (Mena, 1979). Psychomotor irritability frequently has been observed, and which leads to impulsive acts such as a strong desire to walk. In Mena's cases (1979), one patient chased passing cars until exhausted. Others run during the night or could sleep only in the open air, and others displayed compulsive behavior and were unable to control it. Some patients heard hallucinatory voices calling them by name, or experienced visions in which their tools appeared to be either gigantic or microscopic. These same patients had clear visions of their favorite foods. All knew that they were experiencing hallucinations. In most of the cases, memory and intelligence are unimpaired.

Other characteristic neurological manifestations usually began shortly after the appearance of those psychiatric symptoms. Extrapyramidal manifestations appear 1 to 2 months after the first symptoms (Barbeau et al. 1976). The later neurologic signs include progressive bradykinesia, dystonia, and disturbance of gait (Rosenstock et al., 1971). The facial expression is somewhat fixed and the patients do not talk a great deal. Speech difficulty is frequently observed. The patients develop slurring and stuttering speech with diminished volume. The voice becomes monotonous and sinks to a whisper and speech is slow and irregular, perhaps with a stammer. Sialorrhea is present to varying degrees and is usually most marked during periods of psychomotor excitation. Spasmodic torticollis is not so rare in severe cases. Clumsiness in movements and increased tone of limb musculature are frequently observed (Barbeau et al., 1976).

The dystonic posture of the limb is often accompanied by painful cramps. This attitudinal hypertonia had a tendency to decrease or disappear in the supine position and to increase in orthostation. Thus, dystonic postural abnormality is one of the characteristic findings in manganese poisoning (Figure 1). Fine finger movements are usually slow and limited. Bradykinesia is one of the most important findings. There is a remarkable slowing of both active and passive movements of the extremities (Schuler et al., 1957). Micrographia is frequently observed and a characteristic finding. The patients may show some tremor, which usually is not so marked. Tremor observed in manganese poisoning is symmetrical and is seen with hyperextension of the arms and hands, which become exaggerated on movement. The tremor varies from a fine twitching of the hands to gross rhythmical movements of the limbs, trunk, or head. The tremor is quite different from that seen in Parkinson's disease. It has much more of an attitudinal or flapping quality, resembling that seen in Wilson's disease.

Figure 1 Typical dystonic posture in walking.

Cog-wheel rigidity is also elicited on passive movement of all extremities. The most typical symptom reported is the gait disturbance (McNally, 1935; Schuler et al., 1957). The patients stand on a wide base and are very unsteady and uncertain of themselves. They need help to walk and move with a propulsive gait. If asked to step back, they take a few steps and fall into the examiner's arms. In the typical severe case, cock-gait as described by von Jaksch (1907) and Seelert (1913) has been observed. The patient uses small steps, but has a tendency to elevate the heels and to rotate them outward. He progresses without pressing on the flat of his feet, but only upon the metatarsophalangeal articulations, mainly of the fourth and fifth toes (Figure 2). This posture may be absent upon initiation of walking, but eventually appeared after a certain distance in every patient, unless he frequently stopped for a rest (Barbeau et al., 1976). No involvement of the cranial nerves is noted. The peripheral nerves are intact. There is no definite sensory disturbance. In general, the snout and glabella reflexes are absent. The jaw jerk was present, but not increased. Deep reflexes are usually normal or hyperactive. Hyperreflexia is not so marked. The pathological reflexes are usually negative. Sphincter disturbances have not been reported.

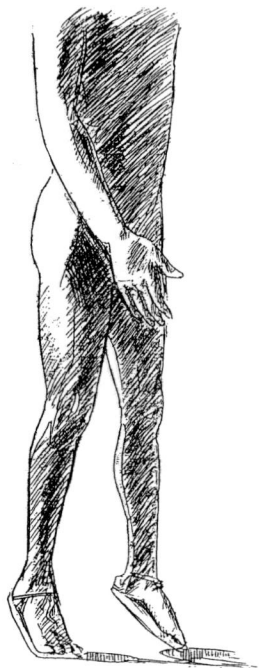

Figure 2 Cock gait.

V. COURSE AND PROGNOSIS

Manganese poisoning ultimately becomes chronic. However, if the disease is diagnosed while still at the early stages and the patient is removed from exposure, the course may be reversed (Marti-Feced, 1991). Once well established, it becomes progressive and irreversible, even when exposure is terminated. The neurological signs show no tendency to regress and may be followed by deformation of the joints. Gait remains permanently affected.

Although seriously poisoned individuals are lifelong cripples, the condition is not lethal. The psychiatric symptoms observed in the early stages are usually transient. However, it is widely accepted that the extrapyramidal symptom and signs tend to persist. In any group of workmen showing such extrapyramidal symptoms and signs, it is unusual for as many as 10% to recover sufficiently to resume work. In the remainder, the extrapyramidal sign of bradykinesia with gait disturbance render impossible any return to the former employment.

VI. PATHOLOGY

The first description of the pathology of manganese poisoning is due to Casamajor in 1913, but Ashizawa (1927) was the first to properly describe the brain changes. He particularly emphasized the pallidal degeneration. Further studies were carried out by Flinzer (1931), Canavan et al. (1934), Trendtel (1936), Voss (1939), Parnitzke and Peiffer (1954), Bernheimer et al. (1973), and more recently by Yamada et al. (1986). Most of the authors found the damage mainly within the striatum in the central nervous system.

Yamada et al. (1986) reported neuropathological findings of a case with manganese poisoning. In his case, the pallidum bilaterally presented a loss of nerve cells, which was marked in the medial segments and moderate in the lateral ones. A marked decrease of myelinated fibers was also found in the pallidum. In the putamen and caudate nucleus, the large nerve cells showed a moderate decrease in number, with shrinkage of those remaining. The small nerve cells often appeared shrunken. A microscopic focus of gliosis was observed in the head of the right caudate nucleus. The nerve cells of the subthalamic nucleus and thalamus were found to be shrunken, but well preserved in number. Shrinkage of the cells was also detected scattered in the mammillary body and in the third and fifth layer of the cerebral cortex.

The cerebellum showed a mild degree of granular cells in the cortex. The pigment cells of the substantia nigra were intact. There was no remarkable change in the spinal cord and peripheral nerves. All authors found many brain lesions evident in the basal ganglia, pallidum, caudate nucleus, and putamen. The lesions are always bilateral and more or less symmetrical. Thus, a review of the accumulated neuropathological evidence indicates that the pallidum-subthalamic nucleus system may be preferentially damaged in manganese encephalopathy. Caudate nucleus and putamen are also constantly and severely involved. Only Bernheimer et al. (1973) reported depigmentation and cell damage in the substantia nigra. Such involvement of the substantia nigra in manganese poisoning is unusual.

In 1986, Yamada et al. analyzed distribution and concentration of manganese in the brain in a typical case of manganese poisoning. They showed no elevation in average concentration of manganese. This study might be suggestive of no relationship between extrapyramidal manifestations and manganese concentration in the brain.

VII. LABORATORY EXAMINATIONS

The blood counts are usually normal. Any specific hematological change has not been recorded. Urinalysis is also intact. Glycosuria and proteinuria have not been observed. Blood chemistry including liver function, renal function, serum protein, and electrolytes is normal. Isolated reports of liver dysfunction have been recorded, but there is no confirmatory evidence for chronic liver toxicity. Roentgenogram of the skull, pneumoencephalogram and computed tomography of the brain revealed no specific abnormalities. Spinal fluid is normal in every respect. Neurophysiological evaluation including electromyography, nerve conduction velocity studies, electroencephalography, and somatosensory-evoked potentials are all within normal limits. The mean whole-blood manganese concentration is 9 µg/l (range 3.9 to 15.0), while mean serum manganese concentration is 1.8 µg/l (range 0.9 to 2.9), (Gilmore and Bronstein, 1992). Blood manganese concentration may increase to some extent, while urinary concentration remains within normal limits.

VIII. DIAGNOSIS

There is no specific diagnostic test for manganese poisoning, and the diagnosis depends chiefly upon a combination of the characteristic neurological features and an occupational history of exposure (Cook et al., 1974).

Diagnosis has been made based on the following points.

1. Symptoms and signs appear after exposure to dusts or fumes of manganese for more than at least three months.
2. Symptoms and signs appear gradually and then become progressive to some extent.
3. The initial symptoms are various and vague. They are asthenia, anorexia, apathy, insomnia and daytime-drowsiness, and a slowing down in performing motor acts.
4. Main manifestations are psychiatric and neurological. Psychiatric symptoms are followed by neurological signs and symptoms.

5. Psychiatric symptoms are disturbance of sleep rhythm, emotional incontinence (forced laughing), and alteration of character. Hallucination, delusion, and intellectual impairment rarely appear.
6. Extrapyramidal signs and symptoms are the main manifestations; parkinsonism with dystonia is characteristic. Other signs include a mask-like face, speech disturbance (low and monotonous voice), limited fine finger movement, micrographia, tremor, difficulty in walking (cock-gait), and so on. In the severest cases, deep reflexes may be hyperactive.

Neither blood nor urinary manganese levels correlate with any neurological manifestations. Therefore, the determinations of blood and urinary concentrations of manganese are of no value in the diagnosis of the poisoning.

Parkinson's disease, vascular parkinsonism, postencephalitic parkinsonism, toxic parkinsonism, and Wilson's disease should be differentiated in any case. The symptoms developed in manganese poisoning are frequently very much like those of Parkinson's disease. However, the detailed neurological evaluation will disclose the different symptomatology in manganese poisoning. However, there may be some difficulty in distinguishing between manganese poisoning and the above-mentioned diseases. Also, it must be differentiated from malingering among manganese workers.

IX. TREATMENT

For many decades, it has been known that the best way to treat manganese poisoning has been the removal from exposure to dusts or fumes containing manganese. This has also been valuable for prevention of further development of the neurological manifestations.

The successful treatment of Wilson's disease with chelating agents was expected to provide a precedent for treating manganese poisoning. Chelating agents, particularly calcium EDTA, seem to produce some improvement in manganese poisoning if applied in its early stage, when there presumably is neuronal degeneration in the basal ganglia. However, the efficacy of calcium EDTA in the acute stage is not well defined. Since manganese poisoning exhibits some neurological similarities with Parkinson's disease, several therapeutic effects have been reported with the oral administration of a dopamine precursor, levodopa, with or without dopa-decarboxylase inhibitor. In 1969, Cotzias, who first treated the neurological symptoms of manganese poisoning with levodopa and observed a definite improvement of most of the symptoms, concluded that the neurological damage was related to the overwhelming manganese levels in the brain which caused selective damage to dopaminergic neurons. Rosenstock et al. (1971) reported their clinical experiences with a therapeutic trial of levodopa, and documented improvement in speech disturbance, dysdiadochokinesis, and bradykinesia. This treatment of chronic manganese poisoning has been based on a better understanding of the pathophysiology of the condition. Namely, it was shown experimentally that alterations in the brain levels of dopamine play a significant role in the pathogenesis of the disease. Administration of a dopamine precursor increases the striatal concentration of dopamine. However, recent accumulating results may be pessimistic. Greenhouse (1971) observed therapeutic failure in his four patients. Thereafter, Cook et al. (1974) also tried levodopa in their three patients and no definite improvement was obtained. More recently, Huang et al. (1989) reported some favorable responses with levodopa-carbidopa in six patients. Therefore, it seems that the beneficial therapeutic effect of levodopa may result only if the dopaminergic fibers are not completely degenerated. In any case of manganese poisoning, levodopa-carbidopa should be tried as a first choice. The other dopaminergic drugs, bromocriptine and amantadine, may be effective to some extent.

REFERENCES

Abd El Naby, S. and Hassanein, M., *J. Neurol. Neurosurg. Psychiatr.*, 28, 282–288, 1965.
Ashizawa, R., *Int. Med. Pediatr. Psychiatry*, 1, 173–191, 1927.
Barbeau, A., Inoue, N., and Cloutier, T., in *Advances in Neurology*, Vol. 14, Eldridge, R. and Fahn, S., Eds., Raven Press, New York, 339, 1976.
Bernheimer, H., Birkmayer, W., Hornykiewicz, O., Jellinger, K., and Seitelberger, F., *J. Neurol. Sci.*, 20, 415–425, 1973.
Canavan, M. M., Cobbs, S., and Drinker, C. K., *Arch. Neurol. Psychiatr.*, 32, 501–512, 1934.
Casamajor, L., *J. Am. Med. Assoc.*, 60, 646–647, 1913.
Cook, D. J., Fahn, S., and Brait, K. A., *Arch. Neurol.*, 30, 59–64, 1974.
Cotzias, G. C., *J. Am. Med. Assoc.*, 210, 1255–1262, 1969.
Couper, J., *Br. Ann. Med. Pharmacol.*, 1, 41–42, 1837.
Edsall, D. L., Wilbur, F. P., and Drinker, C. K., *J. Ind. Hyg.*, 1, 183–193, 1919.

Embden, H., *Dtsch. Med. Worchenschr.,* 27, 795–796, 1901.
Fairhall, L. T., *Physiol. Rev.,* 25, 182–202, 1945.
Flinn, R. H., Neal, P. A., and Fulton, W. B., *J. Ind. Hyg. Toxicol.,* 23, 374–387, 1941.
Flinzer, F., *Arch. Psychiatr.,* 93, 84–115, 1931.
Gilmore, D. A. and Bronstein, A. C., in *Clinical Principles of Environmental Health,* Sullivan, J. B. and Krieger, J. R., Eds., Williams & Wilkins, Baltimore, 896, 1992.
Greenhouse, A. H., *Trans. Am. Neurol. Assoc.,* 96, 248–249, 1971.
Hamilton, A. and Hardy, H. L., *Manganese, Industrial Toxicology,* 3rd ed., Publishing Sciences Group, Acton, 127, 1974.
Huang, C.-C., Chu, N.-S., Lu, C.-S., Wang, J.-D., Tsai, J.-L., Tzeng, J.-L., Wolters, E. C., and Calne, D. B., *Arch. Neurol.,* 46, 1104–1107, 1989.
Marti-Feced, C., in *Encyclopaedia of Occupational Health and Safety,* 3rd ed., Vol. 2, Parmeggiani, L., Ed., International Labor Office, Geneva, 1281, 1991.
McNally, W. D., *Ind. Med.,* 4, 349–350, 1935.
Mena, I., *Handbook of Clinical Neurology,* Vinken, P.J. and Bruyn, G.W., Eds., North-Holland, Amsterdam. 1979
Mena, I., Marin, O., Fuenzarida, S., and Cotzias, G. C., *Neurology,* 17, 128–136, 1967.
Parnitzke, K. H. and Peiffer, J., *Arch. Psychiatr. Z. Neurol.,* 192, 405–429, 1954.
Penalver, R., *Arch. Ind. Health,* 16, 64–66, 1957.
Rodier, J., *Maroc Med.,* 37, 429–454, 1958.
Rodier, J., *Br. J Ind. Med.,* 12, 21–35, 1955.
Rosenstock, H. A., Simons, D. G., and Meyer, J. S., *J. Am. Med. Assoc.,* 217, 1354, 1971.
Saric, M., in *Encyclopaedia of Occupational Health and Safety,* 3rd ed., Vol. 2, Parmeggiani, L., Ed., International Labor Office, Geneva, 1279, 1991.
Schuler, P., Oyangren, H., Maturana, V., Valenzuela, A., Cruz, E., Plaza, V., Schmidt, E., and Haddad, R., *Ind. Med. Surg.,* 26, 167–173, 1957.
Seelert, H., *Monatschr. Psychol. Neurol.,* 24, 82–92, 1913.
Tanaka, S. and Lieben, J., *Arch. Environ. Health,* 19, 674–684, 1969.
Trendtel, I., *Monatsschr. Unfallheilk.,* 43, 69–84, 1936.
von Jaksch, R., *Munch. Med. Wochenschrift.,* 20, 969–972, 1907.
von Jaksch, R., *Wien. Klin. Rundschau,* 15, 729–733, 1901.
Voss, H., *Arch. Gewerbepathol. Gewerbehyg.,* 9, 464–476, 1939.
Wang, J.-D., Huang, C.-C., Hwang, Y.-H., Chiang, J.-R., Lin, J.-M., and Chen, J.-S., *Br. J. Ind. Med.,* 46, 856–859, 1989.
Yamada, M., Ohno, S., Okayasu, I., Okeda, R., Hatakeyama, S., Watanabe, H., Ushio, K., and Tsukagoshi, H., *Acta Neuropathol.,* 70, 273–278, 1986.

Chapter 26

Health Effects of Arsenic

Gonzalo G. García-Vargas and Mariano E. Cebrián

I. INTRODUCTION

The primary purpose of this chapter is to provide interested readers with an overall perspective of the adverse health effects produced by arsenic and the exposure levels associated to them. The analysis of the toxic effects of arsenic is complicated by the fact that arsenic is present in many different inorganic and organic compounds. The acute and chronic toxicity largely depend on the chemical form and physical state of the compound involved. Inorganic trivalent arsenic is generally regarded as being more acutely toxic than inorganic pentavalent arsenic, which in turn is more toxic than the methylated forms (EPA, 1984). Several organic arsenicals, mainly arsenobetaine and arsenocholine, are found to accumulate in fish and shellfish and are generally regarded as nontoxic (Yamauchi et al., 1986; Brown et al., 1990). Elemental arsenic (the metalloid) is nontoxic even if eaten in substantial amounts (Winship, 1984). Thus, it is important to identify the arsenic compound involved when assessing the relationships between effects and exposure. The most common inorganic arsenical in air is arsenic trioxide, whereas a variety of inorganic arsenates or arsenites occur in water, soil, or food. A number of studies have noted differences in the relative toxicity of these compounds, with trivalent arsenites being more toxic than pentavalent arsenates (Byron et al., 1967; Cebrián et al., 1988; Gaines, 1960; Maitani et al., 1987; Willhite, 1981). However, it is difficult to use this information in human situations since: (1) differences in relative potencies depend on the end point studied; (2) different forms of arsenic may be interconverted, both in the environment and/or the body; and (3) in many cases, the chemical species involved is not known. Gallium arsenide (GaAs) is another inorganic arsenic compound of potential human health concern, due to its widespread use in the microelectronics industry. Available toxicokinetic data suggest that although gallium arsenide is poorly soluble, it undergoes slow dissolution and oxidation to form gallium trioxide and arsenite (Webb et al., 1984, 1986). Therefore, its toxic effects are attributable to the arsenite plus the additional effects of gallium. Organic arsenicals are usually considered less toxic than inorganics; however, several methyl and phenyl derivatives widely used in agriculture are potentially toxic for humans. Chief among these are monomethyl arsonic acid (MMA) and its salts (monosodium methane arsonate [MSMA] and disodium methane arsonate [DSMA]), dimethyl arsinic acid (DMA, also known as cacodylic acid) and its sodium salt (sodium dimethyl arsinite, or sodium cacodylate), and roxarsone (3-nitro-4-hydroxyphenylarsonic acid). As with inorganic compounds, there are toxicological differences between these organic derivatives, but data are rarely adequate to allow rigorous quantitative comparisons (Kaise et al., 1985).

An additional complexity to the analysis of arsenic toxicity is that most laboratory animals appear to be substantially less susceptible to inorganic arsenic than humans, but it is not known if this also applies to organic arsenicals (Schaumburg, 1980). Lethal doses in animals are higher than the estimated lethal dose in humans. For example, acute LD_{50} values for arsenate and arsenite in rats and mice range from 15 to 110 mg of As per kilogram (Gaines, 1960; Harrisson et al., 1958). Human data indicate that

As doses of 1 to 3 mg/kg are usually fatal (Vallee et al., 1960). Chronic oral exposure of humans to inorganic arsenic at doses of 0.05 to 0.1 mg/kg/day is frequently associated with neurological or hematological signs of toxicity, but none of these signs was detected in monkeys, dogs, or rats chronically exposed to arsenate or arsenite at As doses of 0.7 to 2.8 mg/kg/day (Byron et al., 1967; Heywood and Sortwell, 1979). Furthermore, whereas good evidence exists that arsenic is a human carcinogen by both oral and inhalation routes, evidence of arsenic-induced cancer in animals is mostly negative (IARC, 1987). For these reasons, animal studies will be considered only when human data are insufficient to suggest an effect. Recent reviews on the health effects of arsenic include ATSDR (1992), Bates et al. (1992), Fielder et al. (1986), EPA (1984).

II. ACUTE EFFECTS

A. INHALATION EXPOSURE

Arsine is considered the most toxic arsenic compound, levels as low as 3 to 10 ppm produce severe toxicity in humans. Its extreme toxicity is attributed to its potent hemolytic activity, causing severe hemolytic anemia followed by acute renal failure (Fowler and Weissburg, 1974). Although there are many studies on workers exposed to high exposure levels (1 to 100 mg/m^3) of arsenic trioxide in air, no cases of lethality directly attributable to short-term exposure have been reported (Enterline and Marsh, 1982; Jarup et al., 1989; Lee-Feldstein, 1986). No studies are available regarding death or severe poisoning in humans after inhalation of organic arsenicals. The high LC$_{50}$ values needed to produce death to rats by inhalation exposure to DMA (2100 mg/m^3 As) suggest a low acute lethal potential (Stevens et al., 1979).

B. ORAL EXPOSURE

There are numerous reports on the acute effects of ingested arsenic following accidental, suicidal, or homicidal incidents; consumption of contaminated food or drinking water also account for a high proportion of cases. However, little information is available on actual doses and type of compounds ingested. Most data are available on arsenic trioxide; doses of 70 to 180 mg (about 1 to 3 mg/kg) are usually fatal and death occurs 12 to 48 h after ingestion (Vallee et al., 1960). The most immediate effects are hemorrhagic gastritis and gastroenteritis, vomiting, diarrhea, convulsions and hypotension; death may ensue from fluid loss and circulatory collapse (Levin-Scherz et al., 1987; Saady et al., 1989). Acute high-dose As exposures (1 mg/kg/day or above) often lead to encephalopathy, with signs and symptoms such as headache, lethargy, mental confusion, hallucination, seizures, and coma (Armstrong et al., 1984; Fincher and Koerker, 1987). Other toxic effects include muscular cramps, facial edema, hematological abnormalities (anemia, leukopenia, and especially granulocytopenia), renal insufficiency and, in moderate to severe cases, pulmonary edema and hemorrhagic bronchitis (Campbell and Alvarez, 1989; Espinoza, 1963). Electrocardiographic alterations, most commonly a prolonged QT interval and T wave abnormalities, have also been described (Fennell and Stacy, 1981). There are no studies regarding death in humans after oral exposure to organic arsenicals, but the acute lethality of MMA, DMA, and roxarsone have been investigated in several animal studies and most As values range from about 15 to 70 mg/kg (NTP, 1989).

III. CHRONIC EFFECTS

A. CARDIOVASCULAR EFFECTS
1. Inhalation Exposure

Several studies on smelter workers suggest that chronic inhalation exposure to arsenic trioxide increases the risk of dying from cardiovascular and cerebrovascular diseases. Pinto and Bennett (1963) reported an increased risk of dying from cardiovascular diseases especially at younger ages. Welch et al. (1982) found in smoking workers higher standard mortality ratios (SMRs) of ischemic heart diseases, but not of cerebrovascular diseases. Elevated SMRs for ischemic heart disease and other heart and cerebrovascular diseases have also been reported (Wall, 1980; Axelson et al., 1978). However, these effects have not been observed in other studies (Rencher et al., 1977; Enterline and Marsh, 1982). Quantitative estimates of the exposure levels leading to these effects are not available and other risk factors besides arsenic, such as lead and smoking, may also have contributed. An increased incidence of Raynaud's disease and vasospastic reactivity at As exposure levels about 0.05 to 0.5 mg/m^3 has been reported (Lagerkvist et al., 1988). These findings suggest that long-term inhalation of arsenic trioxide injures blood vessels.

2. Oral Exposure

There is evidence indicating that arsenic ingestion also affects the cardiovascular system. Characteristic effects on the heart from both acute and long-term exposure include altered myocardial depolarization (prolonged Q-T interval, nonspecific S-T segment changes) and cardiac arrhythmias (Glazener et al., 1968; Goldsmith and From, 1986; Little et al., 1990; Mizuta et al., 1956). Long-term low-level exposures may also damage the vascular system. Blackfoot disease is characterized by a progressive loss of circulation in hands and feet, leading ultimately to necrosis and gangrene. Studies in an endemic area of Taiwan, where arsenic drinking water levels ranged from 0.17 to 0.80 ppm, have shown that this condition had an overall mortality rate of 50% (Chen et al., 1988; Chi and Blackwell, 1968; Tseng 1977), corresponding to As doses of about 0.014 to 0.065 mg/kg/day. Although most studies lack quantitative data on the duration of exposure that caused these effects, Chen et al. (1988) have reported a minimum of five years for these effects to become evident. Studies in Chile indicate that exposure to 0.6 to 0.8 mg/l of As in drinking water increase the incidence of Raynaud's disease and cyanosis of fingers and toes (Borgoño and Greiber, 1972; Zaldivar and Guillier, 1977). Autopsy of five children with cutaneous signs of arsenicism showed a marked thickening of small- and medium-sized arteries in tissues throughout the body, especially in coronaries, cerebral and mesenteric arteries, with myocardial infarction in two cases (Rosenberg, 1974). Peripheral vascular alterations have also been described in Región Lagunera, México where exposure to 0.41 mg/l of As resulted in a 4% prevalence of peripheral vascular alterations in several stages of progress and a 0.7% prevalence of Blackfoot disease which led to amputations (Cebrián, 1987). Garcia-Salcedo et al. (1984) studied five villages in the same area having As concentrations ranging from 0.27 to 0.51 mg/l and reported a prevalence of 0.75% (1.18% in men and 0.27% in women).

A high prevalence of peripheral vascular alterations, ranging from abnormal temperature and acrocyanosis in toes and fingers to gangrene, has also been reported in German vine-growers exposed to arsenical pesticides (Butzengeiger, 1940; Roth, 1957; Grobe, 1976).

It has been suggested that the combination of arsenic and humic substances found in the well water of Blackfoot disease endemic areas in Taiwan shortens prothrombin time of human plasma, and that these substances may play a role in the etiology of this disease (Lu, 1990). Studies on cultured human umbilical vein endothelial cells suggest that arsenic plays an important role in the pathogenesis of Blackfoot disease by damaging the endothelial cells (Chen et al., 1990). Thus, the clear association between the occurrence of Blackfoot disease and the intake of elevated arsenic levels indicates that arsenic is an important contributing factor. Furthermore, the effects of arsenic on the vascular system have also been reported in other exposure situations (e.g., arsenical pesticides), where the presence of humic substances is unlikely.

Chen et al. (1992b) examined the association between ingested inorganic arsenic and prevalence of hypertension in the Blackfoot disease area of Taiwan. Age and sex-adjusted odds ratios were 4.1 and 5.3 for residents who had a cumulative arsenic exposure of 0.1 to 15.0 and more than 15.0 ppm per year, respectively, as compared to those without arsenic exposure. The association between cumulative arsenic exposure and hypertension remained significant after other risk factors were considered. There is no information available on cardiovascular effects in humans after human oral exposure to organic arsenicals.

B. NEUROLOGICAL EFFECTS
1. Oral Exposure

Peripheral neuropathy and encephalopathy have frequently been reported in individuals surviving acute arsenic poisoning. Neurological involvement usually starts with paresthesias, hyperesthesias, and neuralgias, later developing into a painful sensation accompanied by considerable muscle weakness, progressing from distal to proximal nerves and muscles (Heyman et al., 1956). In general terms, these clinical signs are indicative of progressive peripheral polyneuropathy involving both sensory and motor nerves and most intensively affecting long-axon neurons with histopathological evidence of Wallerian degeneration of myelinated nerve fibers (Le Quesne and McLeod, 1977; Goebel et al., 1990; Hindmarsh and McCurdy, 1986). Clinical and electromyographic features presenting a Guillain-Barré-like syndrome have also been described (Donofrio et al., 1987). Subchronic and chronic exposures to low arsenic levels (0.019 to 0.5 mg/kg/day) also produce symmetrical peripheral neuropathy (Chhuttani et al., 1967;

Hindmarsh et al., 1977; Mizuta et al., 1956). A slow recovery, usually incomplete, may occur following cessation of exposure (Fincher and Koerker, 1987; Murphy et al., 1981).

Southwick et al. (1981) conducted a study on individuals aged 47 years and younger exposed to arsenic in drinking water (0.18 to 0.27 mg/l) in Utah. Neurological examinations revealed that nerve velocity conductions did not vary significantly with respect to age and community; however, a slightly greater proportion of individuals exhibiting below normal velocities was seen in the exposed group, which had at least five years of exposure. No neurological effects could be detected in populations chronically exposed to As doses of 0.01 mg/kg/day or less (Harrington et al., 1978; Hindmarsh et al., 1977; Southwick et al., 1981; Valentine et al., 1985). Most studies lack quantitative data on the duration of exposure that caused these effects. Several studies suggest that children are more susceptible to arsenic-induced central nervous system (CNS) damage. For example, severe CNS deficits have been observed in children exposed for several months as babies to arsenic-contaminated powdered milk formulas in Morinaga, Japan (Hamamoto, 1955; Okamura et al., 1956). Follow-up studies have revealed a higher incidence (18%) of severe hearing loss (<30 dB) and of abnormal electroencephalographic brain wave patterns, and other indications of severe brain damage such as mental retardation and epilepsy (14%). Neurological effects have not been reported in dogs or monkeys chronically exposed to arsenate or arsenite by the oral route (Byron et al., 1967; Heywood and Sortwell, 1979).

There is little information regarding neurological effects in humans after oral exposure to organic arsenicals. Gherardi et al. (1990) reported a case of Guillain-Barre-like syndrome due to organic arsenic (melarsoprol) treatment. However, studies in pigs indicate that roxarsone (0.87 to 5.8 mg/kg/day of As for 1 month) produce muscle tremors, partial paralysis, and seizures (Edmonds and Baker, 1986; Rice et al., 1985). Histological examinations of the spinal cord revealed a time-dependent degeneration of myelin and axons (Kennedy et al., 1986). Such prominent signs of neurological effects were not detected in rodents, although hyperexcitability, ataxia, and trembling were noted at the highest As dose (11.4 mg/kg/day) (NTP, 1989). These data suggest that organic arsenicals are neurotoxic at high doses.

2. Inhalation Exposure

Clinical evidence of peripheral neuropathy, including both sensory and motor neurons, was found in copper smelter workers, with the frequency of sensory neuropathies predominating. Apart from a slight reduction in peritoneal motor nerve velocity conduction, no impairment was seen in electrophysiological measurements (Feldman et al., 1979; Landau et al., 1977). Frank encephalopathy (hallucinations, agitation, emotional lability, memory loss) have also been described (Beckett et al., 1986; Morton and Caron, 1989). The effects tend to diminish after exposure ceases (Bolla-Wilson and Bleecker, 1987), but some effects may persist (Beckett et al., 1986). Bencko and Syman (1977) reported both air and bone conduction hearing losses in children exposed to arsenic derived from emissions of a power plant burning arsenic-rich coal. However, Milham (1977) did not find analogous arsenic-induced damage to the inner ear in children living in the vicinity of a copper smelter in the U.S. There is an obvious need to characterize the dose-response relationships for arsenic-induced central and peripheric functional deficits.

The pattern of development of peripheral neurophatic effects is closely associated with the type of exposure. Acute exposure to single high doses can produce a rapid onset of both sensorial and motor dysfunctions. Under more chronic occupational exposure to lower levels, the neuropathy develops in a gradual and insidious way, such as unilateral polyneuropathies without motor involvement (Ishinishi et al., 1973). More subtle peripheral neurotoxic effects arising from nonoccupational groups are more difficult to establish, particularly as assessed by abnormal electromyographic or nerve conduction velocity findings.

C. RESPIRATORY EFFECTS
1. Inhalation Exposure

Symptoms of irritant effects on the respiratory mucosa (rhinitis, laryngitis, bronchitis) have frequently been reported in workers exposed to arsenic dusts in copper smelters. Very high exposures can cause perforation of the nasal septum (Pinto and McGill, 1953). A high incidence of these effects and pronounced hyperplasia and atrophy throughout the upper respiratory tract of workers have been reported where As exposure has reached levels as high as 7 mg/m^3; however, concurrent exposure to sulfur dioxide was a confounding factor (Lundgren, 1954). It appears that these effects are rare or absent at exposure levels of about 0.1 to 1.0 mg/m^3 (Ide and Bullough, 1988).

There are no studies reporting respiratory effects in humans exposed to organic arsenicals. However, short-term exposure of rodents to high concentrations (2170 to 2760 mg/m^3) of MMA or DMA caused respiratory distress (Stevens et al., 1979).

D. GASTROINTESTINAL EFFECTS
1. Inhalation Exposure

Several studies have reported nausea, vomiting, and diarrhea in workers exposed to arsenic dusts or fumes (Beckett et al., 1986; Morton and Caron, 1989). Quantitative data on the exposure duration or exposure level that cause these effects are not available. These effects usually disappear if exposure ceases and are rarely reported in workers exposed to As levels lower than 0.11 mg/m^3 (Ide and Bullough, 1988).

2. Oral Exposure

Clinical signs of gastrointestinal irritation, including nausea, vomiting, diarrhea, and abdominal pain have been reported in most cases of acute high-dose exposure to inorganic arsenic (Armstrong et al., 1984; Franzblau and Lilis, 1989; Mizuta et al., 1956). Less severe symptoms are also frequently observed in individuals with longer-term lower-dose exposures via drinking water (Albores et al., 1979; Huang et al., 1985; Mazumder et al., 1988). These effects are usually not detectable at medium-term As exposure levels between 0.050 and 0.393 mg/l (Harrington et al. 1978; Valentine et al., 1985) and generally decline within a short time after exposure ceases. Gastrointestinal effects in humans have not been studied after oral exposure to organic arsenicals.

E. HEPATIC EFFECTS
1. Oral Exposure

Several studies in humans exposed orally to inorganic arsenic have noted signs or symptoms of hepatic injury. Clinical examination often reveals a swollen and tender liver in individuals treated with arsenicals (Silver and Wainman, 1952; Wade and Frazer, 1953) or exposed to contaminated food (Mizuta et al., 1956). Analysis of blood sometimes shows elevated levels of hepatic enzymes (Armstrong et al., 1984; Franzblau and Lilis, 1989). Liver enlargement was reported in 35.5% of individuals exposed to an average As concentration of 0.64 mg/l in drinking water, whereas people exposed to 0.21 mg/l were unaffected (Chakraborty and Saha, 1987). Histological examination of livers of individuals chronically exposed to similar levels has revealed a consistent finding of portal tract fibrosis (Mazumder et al., 1988; Piontek et al., 1989) leading, in some cases, to portal hypertension and bleeding from esophageal varices (Szuler et al., 1979). This effect may also occur after acute exposures to higher doses (Armstrong et al., 1984; Morris et al., 1974). Nevens et al. (1990) have also reported noncirrhotic portal hypertension secondary to Fowler's solution treatment. It has been suggested that these hepatic effects are secondary to damage to the hepatic blood vessels (Morris et al., 1974; Rosenberg, 1974).

No studies on hepatic effects after human oral exposure to organic arsenicals were found. However, histological examination of liver from rabbits given repeated oral doses of MMA showed diffuse inflammation and mild hepatocellular degeneration (Jaghabir et al., 1989). No effects were observed in rats exposed to DMA (Siewicki, 1981).

2. Inhalation Exposure

There is little evidence to suggest that inhalation exposure to arsenic is associated with liver damage. Kodama et al. (1976) reported that workers involved in the production of arsenic trioxide at a copper smelter showed a slight increase in the mean values of serum aspartate aminotransferase and lactic dehydrogenase levels, as compared to office workers; however, all values were within normal limits. No evidence of hepatic dysfunction was detected by clinical examination of workers exposed to arsenic dusts (Bolla-Wilson and Bleecker, 1987; Ide and Bullough, 1988). Further studies are needed to draw firm conclusions on the hepatotoxic potential of arsenic after inhalation exposure. There are no animal studies regarding hepatic effects of exposure to inorganic or organic arsenicals.

F. RENAL EFFECTS
1. Oral Exposure

Sublethal arsenic poisoning has resulted in renal necrosis and insufficiency (Gerhardt et al., 1978). In the contaminated powdered milk incident in Morinaga, renal injury was diagnosed by the presence of hematuria, leukocyturia, and glycosuria (Hamamoto, 1955). Proteinuria has also been described secondary to ingestion of ground water contaminated by an arsenic plant (Terada, 1960). In some cases of acute or subchronic exposure, elevated serum levels of creatinine and mild proteinuria have been described (Glazener et al., 1968; Tay and Seah, 1975; Armstrong et al., 1984) and renal failure has been reported in rare cases (Fincher and Koerker, 1987). However, most studies do not report clinical signs of significant renal injury, even when other systems are severely impaired (Franzblau and Lilis, 1989; Jenkins, 1966; Mizuta et al., 1956). There is little information available on renal damage in chronically exposed populations. Similar findings have also been reported in animal studies, suggesting that the kidney is relatively less sensitive to inorganic arsenic than most other organ systems.

There is not enough information to assess nephrotoxicity in humans after oral exposure to organic arsenicals. However, animal studies have shown that organic arsenicals can lead to significant renal injury, in contrast to findings in other organs and systems. Roxarsone was reported to cause tubular degeneration and necrosis (Abdo et al., 1989; NTP, 1989), whereas MMA caused interstitial nephritis and tubular nephrosis (Jaghabir et al., 1989).

2. Inhalation Exposure

Renal functional impairment is mainly seen in acute poisoning by arsine, with oliguria and anuria progressing to renal failure and sequelae including chronic renal insufficiency and hypertension (Fowler, 1977). Routine clinical urinalysis of workers exposed to arsenic dusts has not revealed evidence of kidney damage (Ide and Bullough, 1988; Morton and Caron, 1989). Foa et al. (1987) reported no increases in urinary levels of several proteins indicative of glomerular damage or tubular cell exfoliation. No studies were located regarding renal effects in animals after inhalation exposure to inorganic arsenicals.

G. HEMATOLOGICAL EFFECTS
1. Oral Exposure

Anemia and leukopenia are common effects of arsenic poisoning in humans and have been reported following acute (Armstrong et al., 1984), subchronic (Franzblau and Lilis, 1989; Mizuta et al., 1956; Westhoff et al., 1975), and chronic oral exposures (Kyle and Pease, 1965; Tay and Seah, 1975). Several explanations have been put forward for these effects — a direct cytotoxicity on bone marrow cells, an hemolytic effect and/or suppression of erythropoiesis (Fincher and Koerker, 1987; Goldsmith and From, 1986; Lerman et al., 1980). Arsenic intoxication also causes myelodysplastic syndrome involving dysmyelopoietic changes in all three marrow cell lines (Rezuke et al., 1991). Hematological effects are usually not observed in humans exposed to As levels of 0.07 mg/kg/day or less (Harrington et al., 1978; Huang et al., 1985; Southwick et al., 1981), although subacute exposure for 2 to 3 weeks to 0.05 mg/kg/day resulted in mild anemia (Mizuta et al., 1956). There is not enough information available on hematological effects after oral exposure to organic arsenicals, and studies in rodents have not detected significant effects from repeated exposure to MMA (Prukop and Savage, 1986), DMA (Siewicki, 1981), or roxarsone (NTP, 1989).

2. Inhalation Exposure

Although anemia is often noted in humans exposed to arsenic by the oral route, red and white blood cell counts are usually normal in workers exposed to inorganic arsenicals by inhalation (Beckett et al., 1986; Bolla-Wilson and Bleecker, 1987; Ide and Bullough, 1988; Morton and Caron, 1989). Negative findings were also reported by Watrous and McCaughey (1945) in workers exposed to organic arsenicals (arsanilic acid) dusts (0.13 mg/m^3). The reason for this apparent route specificity is not clear, but might simply be related to dose.

H. ENDOCRINOLOGICAL EFFECTS
1. Oral Exposure

There is limited information relating endocrinological effects and chronic exposure. Chang et al. (1991) reported a higher prevalence of goiter in school children living in the Blackfoot disease endemic area of Taiwan, as compared to children living outside the area (3.44 vs. 2.08%; $p < .01$). Although differences were statistically significant, the presence in this area of other water contaminants, particularly humic substances, which have been reported to have goitrogenic activity, makes it difficult to draw firm conclusions. Chen et al. (1992a) examined the association between ingested inorganic arsenic and prevalence of diabetes mellitus in the Blackfoot disease area of Taiwan, using the oral glucose tolerance test as an indicator. Age and sex-adjusted odds ratios were 5.6 and 8.1, respectively, for residents who had a cumulative arsenic exposure of 0.1 to 15.0 and more than 15.0 ppm per year, as compared to those without high arsenic exposure, thus suggesting that arsenic may induce diabetes mellitus. This finding is consistent with the hyperglycemia and glucose intolerance reported in animal studies (Ghafghazy et al., 1980). There is no available information on endocrine effects after inhalation exposure, nor after exposure to organic arsenicals.

I. DERMAL EFFECTS
1. Oral Exposure

A common and characteristic effect of arsenic ingestion is a pattern of skin changes that include generalized areas of hyperpigmentation interspersed with small areas of hypopigmentation ("raindrop"-like appearance) on the neck, chest, and back. Palmoplantar keratosis characterized by small corn-like elevations accompanied by diffuse keratosis is also characteristic. These effects have been noted in most human studies involving chronic oral exposure via drinking water at As levels around 0.400 mg/l (Borgoño and Greiber, 1972; Cebrián et al., 1983; Huang et al., 1985; Mazumder et al., 1988; Saha and Poddar, 1986; Tseng et al., 1968) and after arsenic exposure for therapeutic purposes (Tay and Seah, 1975). These skin lesions have been considered appropriate for deriving maximum concentration limits in drinking water (ATSDR, 1992). However, other effects (hepatic injury, vascular disease, neurological effects) also appear at similar levels of exposure. Studies on small populations (20 to 200 people) exposed to arsenic through drinking water have detected no dermal or other effects at chronic As exposures of 0.007 mg/l (Cebrián et al., 1983; Harrington et al., 1978; Mazumder et al., 1988; Southwick et al., 1981; Valentine et al., 1985), and one very large study based on 17,000 people reported no effects at an average total As daily intake (from water plus food) of 0.0008 mg/kg/day (Tseng et al., 1968). Dermal effects appear to be highly dependent on time of exposure; the shortest time of exposure after which lesions were detected was 8 year for hypopigmentation, 12 years for hyperpigmentation and palmoplantar keratosis, 25 years for papular keratosis, and 38 years for ulcerative lesions (Cebrián et al., 1983). Some malignant lesions may evolve from hyperkeratotic areas, whereas altered pigmentation areas are not considered to be precancerous (EPA, 1988); however, it is noteworthy that some patients considered that papular keratosis arose from depigmented spots (Cebrián et al., 1983). Dermatological effects in humans have not been studied after oral exposure to organic arsenicals. Another prominent dermal effect associated with chronic ingestion of inorganic arsenic is skin cancer, which will be dealt with separately. Dermal alterations similar to those observed in humans have not been noted in oral exposure studies in laboratory animals (Byron et al., 1967; Kroes et al., 1974; Schroeder et al., 1968).

2. Inhalation Exposure

Hyperkeratoses and hyperpigmentation are very common in people exposed to inorganic arsenic by the oral route, but similar effects are not as frequently mentioned in workers exposed primarily by inhalation. However, Perry et al. (1945) reported a 90% (28/31) prevalence of gross skin pigmentation changes and hyperkeratosis and 29% of warts in workers at an arsenical pesticides factory. Their mean duration of exposure was 36 years and they were exposed to As concentrations ranging 384 to 1034 µg/m^3. The basis for this apparent route distinction is not clear, but could simply be related to dose and duration of exposure. Workers exposed to low levels of arsanilic acid to an average concentration of 0.13 mg/m^3) for several years did not complain to doctors about dermal or ocular effects (Watrous and McCaughey, 1945).

3. Dermal Exposure

A high incidence of eczematous and follicular dermatitis has been described in smelter workers exposed to inorganic arsenic; however, the dermal contact rates causing these effects was not quantified (Holmqvist, 1951; Pinto and McGill, 1953). A similar direct skin irritation has been noted in mice exposed to 2.5 mg/kg as sodium arsenite (Boutwell, 1963). In contrast, no significant dermal irritation was noted in guinea pigs exposed to aqueous As solutions containing 4000 mg/l as arsenate or 580 mg/l as arsenite (Wahlberg and Boman, 1986). These studies suggest that direct contact at low levels of exposure are unlikely to cause significant irritation. Dermal application of MMA resulted in mild dermal irritation in rabbits (Jaghabir et al., 1989).

J. DEVELOPMENTAL EFFECTS
1. Inhalation Exposure

Babies born to women exposed to arsenic dusts in a copper smelter during pregnancy had a slightly below average birth weight and a higher than expected incidence of congenital malformations (Nordstrom et al., 1978, 1979b). The incidence of spontaneous abortion in women who lived near the smelter decreased as a function of distance from the smelter (Nordstrom et al., 1979a). These data are consistent with a possible effect of arsenic on development, but the presence of lead, cadmium, sulfur dioxide, and other chemicals makes it difficult to judge the role of arsenic. Animal studies have shown that high levels of arsenic in air cause visceral and skeletal malformations (Nagymajtenyi et al., 1985).

2. Oral Exposure

Developmental effects of arsenic oral exposure in humans has not been extensively investigated. Zierler et al. (1988) reported no overall association between arsenic in drinking water and congenital heart defects in a case-control study in Boston, although association with coarctation of the aorta was noted. Aschengrau et al. (1989) reported a very weak association between arsenic levels in drinking water ranging from 0.0008 to 0.0019 mg/l and the occurrence of spontaneous abortion (odds ratios ranging from 1.2 to 1.7). However, a similar association was noted for mercury, potassium, silica, and water hardness. A decreased incidence of abortion was associated with sulfate, nitrate, and alkalinity. Thus, further studies are needed to ascertain whether arsenic ingestion at these levels causes developmental toxicity in humans.

Experimental studies have shown teratogenic and embryotoxic effects following prenatal treatment with high doses (2 to 20 mg/kg) of inorganic arsenic in several animal species. Among the alterations reported are increased fetal resorption; decreased fetal weights and malformations such as exencephaly, agnathia, anophthalmia, and hydroencephaly; and renal and gonadal agenesis (Hood et al., 1977). There is little information available on the effects of oral exposure to organic arsenicals in human development. However, animal studies indicate that methylated arsenicals are less fetotoxic and teratogenic than inorganics on an equivalent dose basis, but the nature of the effects is comparable (Hood et al., 1982).

K. REPRODUCTIVE EFFECTS

Reproductive effects in humans have not been studied after oral exposure to inorganic or organic arsenicals. However, a three-generation study in mice given sodium arsenite in drinking water at an average dose of 1 mg/kg/day, reported no significant effects on a number of reproductive parameters, although a trend towards a decreased number of pups per litter and a slightly altered male:female sex ratio were observed (Schroeder and Mitchener, 1971). Cebrián et al. (1984) studied the reproductive and teratogenic effects of comparatively low doses of arsenite in drinking water — 0, 5, 50, or 100 mg/l — in female rats prior to mating to nonexposed males and during pregnancy. No malformations were found, but a significant increase in the number of pre- and post-implantation losses, accompanied by significant decreases in fetal weights in the medium and high dose groups was observed. Male and female mice dosed with MMA (55 mg/kg/day) prior to mating and during pregnancy, produced fewer litters than normal — an effect which was attributed mainly to decreased male fertility (Prukop and Savage, 1986). This observation suggests that spermatogenesis and/or sperm function might be impaired by organic arsenicals.

L. HEME METABOLISM EFFECTS
1. Oral Exposure

Garcia-Vargas et al. (1991) conducted a pilot study in Mexico aimed to investigate if the porphyrinuria produced by arsenic in rodents was present in humans chronically exposed to 0.40 mg/l via drinking water. No significant increases in total porphyrin excretion were found; however, an inversion of the coproporphyrin/uroporphyrin (COPRO/URO) ratio caused both by a decrease in coproporphyrin excretion and increases in that of uroporphyrin, was observed. Garcia-Vargas et al. (1994), in a follow-up study, reported that the major abnormalities in the urinary porphyrin excretion pattern of arsenic-exposed individuals were (1) significant reductions in coproporphyrin III excretion, resulting in decreases in the COPRO III/COPRO I ratio, and (2) significant increases in uroporphyrin excretion. Both alterations were responsible for the decrease in the COPRO/URO ratio. The urinary excretion of porphyrins did not greatly change at urinary As concentrations below 1000 µg/g of creatinine; however, as arsenic exceeded this value, the excretion of porphyrins (except coproporphyrin III) increased proportionally. The prevalence of clinical signs of arsenicism also showed a direct relationship both to As concentration in urine and time-weighted exposure, whereas alterations in the urinary porphyrin excretion ratios were only related to time-weighted exposure. The authors suggest that the alterations found are compatible with a lower uroporphyrinogen decarboxylase activity in arsenic-exposed populations; however, the similarity in the urinary porphyrin excretion pattern between As-exposed individuals and Dubin-Johnson patients also suggests that an impairment in coproporphyrin isomers excretion may contribute to the pattern observed. Notably, the alterations in the porphyrin ratios seem to occur earlier than the appearance of cutaneous signs of arsenicism. Therefore, the authors proposed that the profile of urinary porphyrins could be used as early biomarkers for arsenic toxicity in humans chronically exposed via drinking water.

2. Inhalation Exposure

A recent study on smelter workers reported that arsenic trioxide exposure was associated with coproporphyrinuria, but no increases in uroporphyrin excretion nor alterations on the COPRO/URO or COPRO III/COPRO I ratios were found (Telolahy et al., 1993). Further studies are needed to ascertain the effects of different routes and patterns of exposure to distinct arsenic species on the profile of urinary porphyrins.

M. IMMUNOLOGICAL EFFECTS
1. Oral Exposure

There are few studies regarding immunological effects in humans after oral exposure to inorganic arsenicals. Ostrosky-Wegman et al. (1991) conducted a pilot study in individuals exposed to arsenic in drinking water (0.390 mg/l) in Región Lagunera, Mexico. The main finding was related to a slower lymphocyte proliferation kinetics in arsenic-exposed individuals. A higher percentage of first divisions was found in cultures from exposed subjects, who also displayed a low proportion of third divisions. The average generation time was approximately 28 h in the exposed group, as compared to 19 h in both laboratory control individuals and local controls. Gonsebatt et al. (1992) exposed *in vitro* human lymphocytes from control individuals to inorganic arsenic at concentrations similar to those found in blood of exposed subjects (10^{-9} to 10^{-7} M), and reported a dose-related depressed response to phytohemagglutinin stimulation and a delayed cell cycle progression. The authors suggested that arsenic impairs the cellular immune response, an effect which could play a role in the increased incidence of cancer observed in the exposed individuals.

The effects of arsenic on the mitogenic response of mononuclear cells derived from Taiwanese patients with arsenical skin cancers were evaluated by Yu et al. (1992a). Several groups of patients were studied, including arsenical skin cancers, nonarsenical skin cancers, and healthy controls from endemic and nonendemic areas. Phytohemagglutinin stimulated [^{3}H]-thymidine incorporation in mononuclear cells in all groups, except in the arsenical skin cancer group. However, when a low concentration of arsenic trioxide (2.5×10^{-7} M) was added to phytohemagglutinin-stimulated cells, a significant amplification in [^{3}H]-thymidine uptake was noticed in cells from patients with arsenical skin cancers. This phenomenon did not occur in cancers not related to arsenic. The authors suggest that arsenic plays a role as a costimulant of phytohemagglutinin, similar to interleukin 1. Yu et al. (1992b) have also reported peritumoral mononuclear infiltration in Bowen's disease and squamous cell carcinoma in patients from the endemic arsenicism area of Taiwan, suggesting a selective recruitment of helper T cells to tumor infiltrates in these arsenical skin cancers. Immunological effects in humans have not been studied after oral exposure to organic arsenicals.

2. Inhalation Exposure

No abnormalities in serum levels of immunoglobins were detected in workers exposed to arsenic in a coal-burning power plant (Bencko et al., 1988). In animals, single exposures of mice to arsenic trioxide (0.94 mg/m^3 As) led to increased susceptibility to respiratory bacterial pathogens, apparently as a result of injury to alveolar macrophages (Aranyi et al., 1985). A decreased humoral response to antigens and decreases in several complement proteins were noted in mice given an intratracheal As dose of 5.7 mg/kg as sodium arsenite (Sikorski et al., 1989). Recent animal studies have shown that intratracheal instillation of gallium arsenide selectively inhibits T cell proliferation, possibly by interfering with primary and secondary signals in both mitogenic and antigen-driven responses (Burns and Munson, 1993). Previously, this research group showed that the arsenic component of gallium arsenide is the major contributor to the gallium arsenide-induced immunosuppression (Burns et al., 1991). The existing evidence indicates that inorganic arsenic interferes, in both animals and humans, with the cellular immune response producing selective alterations in the proliferation of T cell subpopulations and their response to mitogenic signals. Overall, animal studies suggest that inhalation of inorganic arsenicals can interfere with the immune system function.

3. Dermal Exposure

Holmqvist (1951) reported that repeated dermal contact with arsenic trioxide dusts in a copper smelter leads to dermal sensitization. A positive patch test was found in 80% of the exposed workers, as compared to 30% in the control population. However, a much lower response rate (0.5%) was noted in a more recent study (Wahlberg and Boman, 1986). Studies in guinea pigs did not yield evidence of a sensitization reaction (Wahlberg and Boman, 1986); further studies are needed to ascertain this effect.

N. GENOTOXIC EFFECTS
1. Oral Exposure

Research on genotoxic effects of ingested arsenic have yielded mixed results. Burgdorf et al. (1977) reported an increased frequency of sister chromatid exchanges without changes in the frequency of chromosomal aberrations in lymphocytes from patients under Fowler's solution, whereas increased chromosomal aberrations, but no increase in sister chromatid exchanges, was reported in other study (Nordenson et al., 1979). Lymphocytes from psoriasis patients or vintners showed a ninefold elevation in chromosomal aberrations, as compared to control individuals (Petres et al., 1977). These parameters were studied by Vig et al. (1984) in individuals exposed to As via drinking water to more than 0.05 mg/l for at least five years; neither endpoint was significantly increased in the exposed group. Wen et al. (1981) reported that the frequency of sister chromatid exchanges was significantly higher in Taiwanese Blackfoot disease patients, as compared to control individuals. Ostrosky-Wegman et al. (1991) conducted a study on individuals chronically exposed to 0.390 mg/l of As in drinking water in Mexico. No significant increases in the frequencies of sister chromatid exchanges and chromosomal aberrations were found. However, the frequency of complex chromosomal aberrations (dicentrics, rings, and translocations) in the exposed group was 0.73%, whereas in the control group and laboratory controls they were 0.16 and 0.30%, respectively. On average, the frequency of lymphocytes resistant to thioguanine (HGPRT locus assay) in the exposed individuals was twice as high as in the control individuals; however, no significance was attached to these results by the authors. Studies on genetic damage induced by urine samples in the *Bacillus subtilis Rec* assay were negative for all samples. Significant increases in the frequency of chromosomal aberrations and sister chromatid exchanges has been reported in studies *in vitro* using several human cell lines, including lymphocytes, leukocytes, and fibroblasts (Nordenson and Beckman, 1991; Tsutsui et al., 1991).

An increased incidence of chromosomal abnormalities was detected in rats given oral doses of sodium arsenate (4 mg/kg/day As) for 2 to 3 weeks (Datta et al., 1986), but no consistent increase in chromosomal aberrations was detected in bone marrow cells or spermatogonia from mice given sodium arsenite (about 50 mg/kg/day As) for up to 8 weeks (Poma et al., 1987). Recently, Tinwell et al. (1991) reported that sodium arsenite, potassium arsenite, and Fowler's solution were equally active in the mouse bone marrow micronucleus assay. These studies suggest that ingested arsenic cause chromosomal effects and genetic damage in rodents. Genotoxic effects in humans have not been studied after oral exposure to organic arsenicals. An increased number of DNA strand breaks were detected in lung and other tissues of mice given oral doses of DMA (Yamanaka et al., 1989) and in human alveolar type II cells (L-132) (Tezuka et al., 1993), an effect which appeared to be related to the formation of some active oxygen species producing a putative dimethylarsenic peroxyl radical (Yamanaka et al., 1990). These breaks were largely repaired within 24 h, so their relevance in respect to genetic risks is uncertain. More recently, the same group has reported that DMA causes cross-link formation between DNA and nuclear proteins in human (L-132) cells *in vitro,* and in lung isolated nuclei from mice orally treated with a dose of 1500 mg of DMA per kilogram of body weight (Yamanaka et al., 1993).

2. Inhalation Exposure

It has been reported that arsenic trioxide increases the frequency of chromosomal aberrations in peripheral lymphocytes of smelter workers (Beckman et al., 1977; Nordenson et al., 1978) and in the liver of fetuses from mice exposed to 22 mg/m^3 As on days 9 to 12 of gestation (Nagymajtenyi et al., 1985). These data indicate that arsenic is clastogenic, but do not indicate whether it is mutagenic.

3. Systemic Effects Derived From Dermal Exposure

There is no available information regarding respiratory, cardiovascular, gastrointestinal, hematological, musculoskeletal, neurological, developmental, reproductive, genotoxic, hepatic, or renal effects in humans or animals after dermal exposure to inorganic or organic arsenicals.

O. CANCER
1. Inhalation Exposure

A large number of epidemiological studies have shown that inorganic arsenic increases the risk of lung cancer. Most studies have involved workers exposed primarily to arsenic trioxide in air at copper smelters (Axelson et al., 1978; Jarup et al., 1989; Lee-Feldstein, 1986; Pinto et al., 1978; Welch et al., 1982). Copper ore frequently contains small amounts of arsenic which are driven off as arsenic trioxide when the ore is roasted during the early stages of the smelting process. Interpretation of these studies is complicated by exposure to numerous other chemicals (including irritant gases, principally sulfur dioxide, and other heavy metals such as copper, lead, nickel, antimony, and tin) and the little information available on the characteristics of exposure. However, the demonstration of a dose-response relationship with arsenic exposure, and not to other substances, strongly suggest an association.

Based on studies providing adequate exposure data, EPA (1984) derived risk estimates of lung cancer associated to lifetime exposures. In general, the data indicate an approximately linear increase in relative risk of lung cancer as a function of increasing cumulative exposure. Enterline et al. (1987) reexamined the dose-response relationship, using historical records of airborne arsenic levels in the smelters, complemented with records of urinary arsenic levels in exposed workers. It was concluded that arsenic is a more potent lung carcinogen than previously believed, with a dose-response relationship that becomes steeper at exposure levels below cumulative doses of 0.01 mg/yr/m^3. An increased incidence of lung cancer has also been observed at chemical plants where exposure was primarily to arsenates (Mabuchi et al., 1979; Sobel et al., 1988). Ott et al. (1974) studied worker mortality in a plant formulating and packaging arsenic-based insecticides (arsenates of lead, calcium, and magnesium, and copper acetoarsenite) and reported, besides lung cancer, significant increases in lymphatic and hematopoietic cancers, except for leukemia. Limited proportional mortality studies in vine growers using arsenate-containing pesticides also suggested an increased incidence of deaths due to lung cancer (Luchtrath, 1983).

In addition, several studies suggest that residents living near smelters or arsenical chemical plants may also have increased risk of lung cancer. The lung cancer mortality in 71 counties with smelters and refineries was studied; comparisons were made with the remaining 2985 counties in 48 U.S. states. Lung cancer mortality was significantly higher in both males and females in 36 counties having smelters processing copper, lead, or zinc ores; however, occupational cancers were included in the analyses and contributed to some of the excess risk (Blot and Fraumeni, 1975). Pershagen et al. (1987) performed a similar study in the area surrounding the Ronnskar smelter in Sweden, but when the occupationally exposed cases were excluded, significant increases in respiratory cancer were no longer detected. Recent studies have suggested that residents living near smelters or chemical plants may also have an increased risk of lung cancer (Brown et al., 1984; Cordier et al., 1983; Matanoski et al., 1981; Pershagen, 1985). In some cases the increases are small and not clearly detectable (Frost et al., 1987). It appears that arsenic does not specifically increase the incidence of one particular type of lung cancer (Axelson et al., 1978; Pershagen et al., 1987). No studies have been reported regarding respiratory effects in animals after inhalation exposure to inorganic arsenicals, although intratracheal instillation of arsenic trioxide (13 mg/kg As) or gallium arsenide (1.5 to 52 mg/kg) cause marked irritation and hyperplasia in the lung of rats and hamsters (Ohyama et al., 1988; Webb et al., 1987). However, this response is not specific to arsenic.

2. Oral Exposure

There is convincing evidence from a large number of epidemiological studies and case reports that chronic ingestion of high quantities of inorganic arsenic increases the risk of developing skin cancer either via drinking water (Tseng et al., 1968; Zaldivar et al., 1981; Cebrián et al., 1983; Wu et al., 1989), for medicinal purposes (Sommers and McManus, 1953; Fierz, 1965), or in wine growers (Luchtrath, 1983). The most common lesions are multiple epidermoid carcinomas, basal cell carcinomas, intraepidermal carcinomas, and Bowen's disease (Yeh et al., 1968). In most cases, skin cancer develops only after prolonged exposure; no cases were found in individuals exposed for less than 20 years (Tseng et al., 1968; Cebrián et al., 1983); however, isolated case-studies have reported the appearance of skin cancer in individuals exposed for 12 years at higher doses (Wagner et al., 1979). Metastases and invasion of arsenic-induced squamous cell carcinomas and basal cell carcinomas may occur, but their progression is rarely life threatening (Shannon and Strayer, 1989).

Based on the study of Tseng et al. (1968), in which the prevalence of skin cancer was measured as a function of exposure level in over 40,000 people in Taiwan, EPA (1988) developed a model to estimate cancer risks which was tested with information from studies in Mexico (Cebrián et al., 1983) and

Germany (Fierz, 1965). The precision of the dose-response relationship calculated with this model has been questioned; however, this does not alter the conclusion that chronic arsenic ingestion is associated with an increased risk of skin cancer. Grantham and Jones (1977) reported no significant relationships between exposure to arsenic concentrations below 0.100 mg/l and moderate signs of chronic arsenicism; however, such relationship was found above those levels. In contrast, several studies performed in the U.S. have not detected an increased frequency of skin cancer in individuals consuming water containing 0.1 mg/l As (Goldsmith et al., 1972; Harrington et al., 1978; Morton et al., 1976; Southwick et al., 1981). These studies were performed in small populations and lacked sufficient statistical power to detect the small increases in skin cancer that might have occurred at these low doses (Andelman and Barnett, 1983). Besides, no information on duration of exposure was given.

There is increasing evidence indicating that arsenic ingestion may increase the risks of internal cancers. Many case studies have noted the occurrence of internal tumors in patients with arsenic-induced skin cancer. Liver angiosarcoma has been reported after exposure to arsenical pesticides (Roth, 1957; Luchtrath, 1983), treatment with Fowler's solution (Falk et al., 1981; Kasper et al., 1984), and arsenic in drinking water (Zaldivar et al., 1981). Cancers in the kidney, bladder, and other urinary organs have been reported after arsenic exposure in workers and nearby residents of a mining town (Tsuda et al., 1990) and treatment with Fowler's solution (Cuzick et al., 1992; Sommers and McManus, 1953; Tay and Seah, 1975). Lung cancer has been reported after treatment with Fowler's solution (Robson and Jelliffe, 1963; Kasper et al., 1984), exposure to arsenical pesticides (Luchtrath, 1983), and arsenic in drinking water (Biagini et al., 1978). These studies are supported by the remarkable large-scale epidemiological studies in Taiwan, conducted by Chen's group, where clear associations and/or dose-response trends between arsenic exposure via drinking water and tumors of the bladder, kidney, liver, lung, and colon have been described (Chen et al., 1985, 1986, 1988; Chiang et al., 1988; Chen and Wang, 1990; Chen et al., 1992c). Associations were also described with prostate cancer, but not with cancer of nasopharynx, esophagus, stomach, colon, uterine cervix, or leukemia (Wu et al., 1989). A recent review on internal cancers was published by Bates et al. (1992).

IARC (1987) considers that arsenic is one of the few human carcinogens for which animal demonstration of such activity has been relatively unsuccessful, since most experimental studies have not detected any clear evidence (Byron et al. 1967; Kroes et al., 1974; Schroeder et al., 1968). Animal studies have suggested that arsenic may distinctly affect the different types of neoplastic cells, perhaps acting mainly as a tumor promoter (Schrauzer and Ishmael, 1974; Shirachi et al., 1987).

Cancer in humans has not been studied after oral exposure to organic arsenicals. Studies on several animal species yielded no evidence of roxarsone carcinogenicity (Prier et al., 1963) except for a slight increase in pancreatic tumors in male rats (NTP, 1989). The incidence of basophilic foci, considered a precancerous lesion, in liver of rats initiated with diethylnitrosamine was increased by subsequent exposure to DMA, suggesting that DMA acts as a cancer promoter (Johansen et al., 1984).

IV. NEEDS FOR FURTHER RESEARCH

A major constraint in establishing dose-response relationships for the various health effects associated with arsenic exposure is the limited availability of adequate data to assess exposure. Data on this type and oxidation states of arsenic compounds present in different exposure situations would be of great value.

As most laboratory animals appear to be substantially less susceptible to inorganic arsenic than humans, animal studies are less useful for regulatory purposes. Thus, studies on human beings chronically exposed to low doses of inorganic arsenic should be encouraged. In addition, much less is known on the chronic effects of organic arsenicals in humans.

Finally, the understanding of the mechanisms involved in the genesis of health effects associated to arsenic exposure would provide the rationale for biomarkers of toxicity and be of great value in the process of risk assessment. In this regard, due to the increasing reports on immunological and neurological damage produced by arsenic, there is a great need to develop biomarkers for these effects. Recent advances in the knowledge of arsenic detoxification pathways have raised the possibility of identifying useful markers of individual susceptibility to arsenic effects.

REFERENCES

Abdo, K. M., Elwell, M. R., Montgomery, C. A., et al., *Toxicol. Lett.*, 45, 55–56, 1989.
Albores, A., Cebrián, M. E., Tellez, I., and Valdéz, B., *Bol. Of. Sanit. Panam.*, 86, 196–205, 1979.

Andelman, J. B. and Barnett, M., Feasibility study to resolve questions on the relationship of arsenic in drinking water to skin cancer. Final rep. U.S. Environmental Protection Agency, Center for Environmental Epidemiology, University of Pittsburgh, Pittsburgh, PA, 1983.
Aranyi, C., Bradof, J. N., O'Shea, W. J., et al., *J. Toxicol. Environ. Health,* 15, 163–172, 1985.
Armstrong, C. W., Stroube, R. B., Rubio, T., et al., *Arch. Environ. Health,* 39, 276–279, 1984.
Aschengrau, A., Zierler, S., and Cohen, A., *Arch. Environ. Health,* 44, 283–290, 1989.
ATSDR, Toxicological Profile for Arsenic, Agency for Toxic Substances and Disease Registry, Atlanta, GA, 1992.
Axelson, O., Dahlgren, E., Jansson, C. D., et al., *Br. J. Ind. Med.,* 35, 8–15, 1978.
Bates, M. N., Smith, A. H., and Hopenhayn-Rich, C., *Am. J. Epidemiol.,* 135, 462–476, 1992.
Beckett, W. S., Moore, J. L., Keogh, J. P., et al., *Br. J. Ind. Med.,* 43, 66–67, 1986.
Beckman, G., Beckman, L., and Nordenson, I. *Environ. Health Perspect.,* 19, 145–146, 1977.
Bencko, U. and Symon, K., *Environ. Health Perspect.,* 19, 95–101, 1977.
Bencko, V., Wagner, V., Wagnerova, M., et al., *J. Hyg. Epidemiol. Microbiol. Immunol.,* 32, 137–146, 1988.
Biagini, R., Rivero, M., Salvador, M., and Cordoba, S., *Arch. Argent. Dermatol.,* 28, 151–158, 1978.
Blot, W. J. and Fraumeni, J. F., Jr., *Lancet,* 2, 142–144, 1975.
Bolla-Wilson, K. and Bleecker, M. L., *J. Occup. Med.,* 29, 500–503, 1987.
Borgoño, J. M. and Greiber, R., in *Trace Substances Environmental Health,* V, Hemphill, D.D., Ed., University of Missouri, Columbia, 13, 1972.
Boutwell, R. K., *Agric. Food Chem.,* 11, 381–385, 1963.
Brown, L. M., Pottern, L. M., and Blot, W. J., *Environ. Res.,* 34, 250–261, 1984.
Brown, R. M., Newton, D., Pickford, C. J., et al., *Hum. Exp. Toxicol.,* 9, 41–46, 1990.
Burgdorf, W., Kurvink, K., and Cervenka, J., *Hum. Genet.,* 36, 69–72, 1977.
Burns, L. A. and Munson, A. E., *J. Pharmacol. Exp. Ther.,* 265, 178–265, 1993.
Burns, L. A., Sikorski, E. E., Saady, J. J., et al., *Toxicol. Appl. Pharmacol.,* 110, 157–169, 1991.
Butzengeiger, C. H., *Klin. Wochenschr.,* 19, 523–527, 1940.
Byron, W. R., Bierbower, G. W., Brouwer, J. B., et al., *Toxicol. Appl. Pharmacol.,* 10, 132–147, 1967.
Campbell, J. P. and Alvarez, J. A., *Am. Fam. Physician,* 40, 93–97, 1989.
Cebrián, M. E., Some potential problems in assessing the effects of chronic arsenic exposure in north Mexico. American Chemical Society, 194th Natl. Meet. New Orleans, (Abstr.), 1987.
Cebrián, M. E., Albores, A., Aguilar, M., et al. *Hum. Toxicol.,* 2, 121–133, 1983.
Cebrián, M. E., Chamorro, G., and Chávez, C., Absence of teratogenic effects of low doses of arsenic in the rat, Proc. 9th Int. Congr. Pharmacology, London, 1984.
Cebrián, M. E., Albores, A., Connelly, J. C., et al., *J. Biochem. Toxicol.,* 3, 77–86, 1988.
Chakraborty, A. K. and Saha, K. C., *Indian J. Med. Res.,* 85, 326–334, 1987.
Chang, T. C., Hong, M. C., and Chen, C. J., *Hsueh Hui Tsa Chih,* 90, 941–946, 1991.
Chen, C. J., Chuang, Y. C., Lin, T. M., et al., *Cancer Res.,* 45, 5895–5899, 1985.
Chen, C. J., Chuang, Y. C., You, S.-S., et al., *Br. J. Cancer,* 53, 399–405, 1986.
Chen, C. J., Wu, M.-M., Lee, S.-S., et al., *Arteriosclerosis,* 8, 452–460, 1988.
Chen, C. J. and Wang, C. J., *Cancer Res.,* 50, 5470–5474, 1990.
Chen, C. J., Lai, M., Hsu, M., et al., Dose-response relationship between ingested inorganic arsenic and diabetes mellitus, 4th Annu. Meet. Int. Soc. Environ. Epidemiology, Cuernavaca, Morelos, (Abstr.), 1992a.
Chen, C. J., Hsu, M., Chen, S. and Wu, M., Increased prevalence of hypertension among residents in hyperendemic villages of chronic arsenicism in Taiwan. 4th Annu. Meet. Int. Soc. Environ. Epidemiology, Cuernavaca, Mexico, (Abstr.), 1992b.
Chen, C. J., Cen, C. W., Wu, M. M., and Kuo, T. L., *Br. J. Cancer,* 66, 882–892, 1992c.
Chen, G. S., Asai, T., Suzuki, Y., et al., *J. Dermatol.,* 17, 599–608, 1990.
Chi, I.-C. and Blackwell, R. Q., *Am. J. Epidemiol.,* 88, 7–24, 1968.
Chiang, H. S., Hong, C. L., Guo, H. R., et al., *J. Formosan Med. Assoc.,* 87, 1074–1080, 1988.
Chhuttani, P. N., Chawla, L. S., and Sharma, T. D., *Neurology,* 17, 269–274, 1967.
Cordier, S., Theriault, G., and Iturra, H., *Environ. Res.,* 31, 311–322, 1983.
Cuzick, J., Sasieni, P., and Evans, S., *Am. J. Epidemiol.,* 136, 417–421, 1992.
Datta, S., Talukder, G., and Sharma, A., *Sci. Cult.,* 52, 196–198, 1986.
Donofrio, P. D., Wilboum, A. J., Albers, J. W., et al., *Muscle Nerve,* 10, 114–120, 1987.
Edmonds, M. S. and Baker, D. H., *J. Anim. Sci.,* 63, 533–537, 1986.
Enterline, P. E. and Marsh, G. M., *Am. J. Epidemiol.,* 116, 895–911, 1982.
Enterline, P. E., Henderson, V. L., and Marsh, G. M., *Am. J. Epidemiol.,* 125, 929–938, 1987.
EPA, Health assessment document for inorganic arsenic. Final report. U.S. Environmental Protection Agency, Environmental Criteria and Assessment Office, 2–1-3–22, 9–1-9–4, EPA 600/8–83–021F, Research Triangle Park, NC, 1984.
EPA, Special report on ingested inorganic arsenic. Skin Cancer; Nutritional Essentiality, U.S. Environmental Protection Agency, Office of Pesticide Programs, NTIS No. PB89–102842, Washington, D.C., 1988.
Espinoza, E., *Bol. Epidemiol. (México, D.F.),* 4, 213–220, 1963.
Falk, H., Herbert, J. T., Edmonds, L., et al., *Cancer,* 47, 382–391, 1981.
Feldman, R. G., Niles, C. A., Kelly-Hayes, M., et al., *Neurology,* 29, 939–944, 1979.

Fennell, J. S. and Stacy, W. K., *Ir. J. Med. Sci.,* 150, 338–339, 1981.
Fielder, R. J., Dale, E. A., and Williams, D., Inorganic Arsenic Compounds. Toxicity Review 16. Health and Safety Executive, Her Majesty's Stationery Office, London, 1986.
Fierz, U., *Dermatologica,* 131, 41–48, 1965.
Fincher, R. M. and Koerker, R. M., *Am. J. Med.,* 82, 549–552, 1987.
Foa, V., Colombi, A., Maroni, M., et al., in *Occupational and Environmental Chemical Hazards,* De Rosa, E., Bartolucci, G. B., and Foa, V., Eds., Halsted Press, Horwood, NY, 362, 1987.
Fowler, B. A., in *Toxicology of Trace Elements,* Goyer, R.A. and Mehlman, M.A., Eds., eds. Halstead Press, Horwood, NY, 79, 1977.
Fowler, A. B. and Weissburg, J. B., *N. Engl. J. Med.,* 91, 1171–1174, 1974.
Franzblau, A. and Lilis, R., *Arch. Environ. Health,* 44, 385–390, 1989.
Frost, F., Harter, L., Milham, S., et al., *Arch. Environ. Health,* 42, 148–152, 1987.
Gaines, T. B., *Toxicol. Appl. Pharmacol.,* 2, 88–99, 1960.
García-Salcedo, J. J., Portales, A., Blakely, E., and Díaz, R., *Rev. Fac. Med. (Torreón),* 1, 12–16, 1984.
García-Vargas, G. G., García-Rangel, A., Aguilar-Romo, M., et al., *Hum. Exp. Toxicol.,* 10, 189–193, 1991.
García-Vargas, G. G., Del Razo, L. M., Cebrián, M. E., et al., *Hum. Exp. Toxicol.,* 13, 1994 (in press).
Gerhardt, R. E., Hudson, M., Rao, R. N., et al., *Arch. Intern. Med.,* 138, 1267–1269, 1978.
Ghafghazy, T., Ridlington, J. W., and Fowler, B., *Toxicol., Appl. Pharmacol.,* 55, 126–130, 1980.
Gherardi, R. K., Chariot, P., Vanderstigel, M., et al., *Muscle Nerve,* 13, 637–645, 1990.
Glazener, F. S., Ellis, J. G., and Johnson, P. K., *Calif. Med.,* 109, 158–162, 1968.
Goebel, H. H., Schmidt, P. F., Bohl, J., et al., *J. Neuropathol. Exp. Neurol.,* 49, 137–149, 1990.
Goldsmith, S. and From, A. H., *N. Engl. J. Med.,* 303, 1096–1097, 1986.
Goldsmith, J. R., Deane, M., Thom, J., et al., *Water Res.,* 6, 1133–1136, 1972.
Gonsebatt, M. E., Vega, L., and Herrera, L. A., *Mutat. Res.,* 283, 91–95, 1992.
Grantham, S. A. and Jones, J. F., *J. Am. Water Works Assoc.,* 69, 653–657, 1977.
Grobe, J. W., *Berufsdermatosen,* 24, 78–84, 1976.
Hamamoto, E., *Jpn. Med. J.,* 1649, 3–12, 1955 (in Japanese).
Harrington, J. M., Middaugh, J. P., Morse, D. L., et al., *Am. J. Epidemiol.,* 108, 377–385, 1978.
Harrisson, J. W., Packman, E. W., and Abbott, D. D., *Arch. Ind. Health,* 17, 118–123, 1958.
Heyman, A., Pfeiffer, J. B., Willett, R. W., et al., *N. Engl. J. Med.,* 254, 401–409, 1956.
Heywood, R. and Sortwell, R. J., *Toxicol. Lett.,* 3, 137–144, 1979.
Hindmarsh, J. T. and McCurdy, R. F., *Crit. Rev. Clin. Lab. Sci.,* 23, 315–347, 1986.
Hindmarsh, J. T., McLetchie, O. R., Heffeman, L. P., et al., *J. Anal. Toxicol.,* 1, 270–276, 1977.
Hood, R. D., Thacker, G. T., and Patterson, B. L., *Environ. Health Perspect.,* 19, 219–222, 1977.
Hood, R. D., Harrison, W. P., and Vedel, J. C., *Bull. Environ. Contam. Toxicol.,* 29, 679–687, 1982.
Holmqvist, I., *Acta Derm. Venereol.,* 31, 26–29, 44–45, 110–112, 195–204, 1951.
Huang, Y. Z., Qian, X. C., Wang, G. Q., et al., *Chin. Med. J. (Eng.)* 98, 219–222, 1985.
IARC, IARC monographs on the evaluation of carcinogenic risk of chemicals to humans. Suppl 7. Overall evaluations of carcinogenicity: updating of IARC monogr. Vol. 1–42, International Agency for Research on Cancer, Lyon, France, 29, 1987.
Ide, C. W. and Bullough, G. R., *J. Soc. Occup. Med.,* 38, 85–88, 1988.
Ishinishi, N., Kodama, Y., Kunitake, E., et al., *Nihon Rinsho,* 31, 1991–1999, 1973.
Jaghabir, M. T., Abdeighani, A. A., and Anderson, A. C., *Bull. Environ. Contam. Toxicol.,* 42, 289–293, 1989.
Jarup, L., Pershagen, G., and Wall, S., *Am. J. Ind. Med.,* 15, 31–41, 1989.
Jenkins, R. B., *Brain,* 89, 479–498, 1966.
Johansen, M. G., McGowan, J. P., Tu, S. H., et al., *Proc. West. Pharmacol. Soc.,* 27, 289–291, 1984.
Kaise, T., Watanabe, S., and Itoh, K., *Chemosphere,* 14, 1327–1332, 1985.
Kasper, M. L., Schoenfield, L., Strom, R. L., et al., *J. Am. Med. Assoc.,* 252, 3407–3408, 1984.
Kennedy, S., Rice, D. A., and Cush, P. F., *Vet. Pathol.,* 23, 454–461, 1986.
Kodama, Y., Ishinishi, N., Kunitake, E., Inamasu, T., and Nobutomo, K., in *Effects and Dose Response Relationship of Toxic Metals,* Nordberg, G.F., Ed., Elsevier, Amsterdam, 464, 1976.
Kroes, R., van Logten, M. J., Berkvens, J. M., et al., *Toxicology,* 12, 671–679, 1974.
Kyle, R. A. and Pease, G. L., *N. Engl. J. Med.,* 273, 18–23, 1965.
Lagerkvist, B., Linderholm, H., and Nordberg, G. F., *Int. Arch. Occup. Environ. Health,* 60, 361–364, 1988.
Landau, E. D., Thompson, D. J., Feldman, R. G., et al., Selected noncarcinogenic effects of industrial exposure to inorganic arsenic, U.S. Environmental Protection Agency, EPA 560/6–77–018, Washington, D.C., 1977.
Le Quesne, P. M. and McLeod, J. G., *J. Neurol. Sci.,* 32, 437–451, 1977.
Lee-Feldstein, A., *J. Occup. Med.,* 28, 296–302, 1986.
Lerman, B. B., Ali, N., and Green, D., *Ann. Clin. Lab. Sci.,* 10, 515–517, 1980.
Levin-Scherz, J. K., Patrick, J. D., Weber, F. H., et al., *Ann. Emerg. Med.,* 16, 702–704, 1987.
Little, R. E., Kay, G. N., Cavender, J. B., et al., *PACE,* 13, 164–170, 1990.
Lu, F. J., *Thromb. Res.,* 58, 537–541, 1990.
Luchtrath, H., *J. Cancer Res. Clin. Oncol.,* 105, 173–182, 1983.

Lundgren, K. D., *Nord. Hyg. Tidskr.,* 3, 66–82, 1954.
Mabuchi, K., Lilienfeld, A. M., and Snell, L. M., *Arch. Environ. Health,* 34, 312–320, 1979.
Maitani, T., Saito, N., Abe, M., et al., *Toxicol. Lett.,* 39, 63–70, 1987.
Matanoski, G., Landau, E., Tonascia, J., et al., *Environ. Res.,* 25, 8–28, 1981.
Mazumder, D. N., Chakraborty, A. K., Ghose, A., et al., *Bull. WHO,* 66, 499–506, 1988.
Milham, S., Jr., *Environ. Health Perspect.,* 19, 131–132, 1977.
Mizuta, N., Mizuta, M., Ito, F., et al., *Bull. Yamaguchi Med. Sch.,* 4, 131–149, 1956.
Morris, J. S., Schmid, M., Newman, S., et al., *Gastroenterology,* 64, 86–94, 1974.
Morton, W. E. and Caron, G. A., *Am. J. Ind. Med.,* 15, 1–5, 1989.
Morton, W., Starr, G., Pohl, D., et al., *Cancer,* 37, 2523–2532, 1976.
Murphy, A. U., Lyon, L. W., and Taylor, J. W., *J. Neurol. Neurosurg. Psychiatry,* 44, 896–900, 1981.
Nagymajtenyi, L., Selypes, A., and Berencsi, G., *J. Appl. Toxicol.,* 5, 61–63, 1985.
Nevens, F., Fevery, J., Van Steenbergen, W., et al., *J. Hepatol.,* 11, 80–85, 1990.
Nordenson, I. and Beckman, L., *Hum. Hered.,* 41, 171–173, 1991.
Nordenson, I., Beckman, G., Beckman, L., et al., *Hereditas,* 88, 47–50, 1978.
Nordenson, I., Salmonsson, S., Brun, E., et al., *Hum. Genet.,* 48, 1–6, 1979.
Nordstrom, S., Beckman, L., and Nordenson, I., *Hereditas,* 88, 43–46, 1978.
Nordstrom, S., Beckman, L., and Nordenson, I., *Hereditas,* 90, 291–296, 1979a.
Nordstrom, S., Beckman, L., and Nordenson, I., *Hereditas,* 90, 297–302, 1979b.
NTP, National Toxicology Program — technical report ser. no. 345. Toxicology and Carcinogenesis Studies of Roxarsone (CAS No. 121–19–7) in F344/N rats and B6C3F, mice (feed studies), U.S. Department of Health and Human Services, Research Triangle Park, NC, 1989.
Ohyama, S., Ishinishi, N., Hisanaga, A., et al., *Appl. Organomet. Chem.,* 2, 333–337, 1988.
Okamura, K., Ota, T., Horiuchi, K., et al., *Diagn. Ther. (Shinryo),* 9, 240–249, 1956.
Ostrosky-Wegman, P., Gonsebatt, M. E., Montero, R., et al., *Mutat. Res.* 250, 477–482, 1991.
Ott, M. G., Holder, B. B., and Gordon, H. L., *Arch. Environ. Health,* 29, 250–255, 1974.
Perry, K., Bowler, R. G., Buckell, H. M., et al., *Br. J. Ind. Med.,* 5, 6–15, 1945.
Pershagen, G., *Am. J. Epidemiol.,* 122, 684–694, 1985.
Pershagen, G., Elinder, C. G., and Bolander, A. M., *Environ. Health Perspect.,* 19, 133–137, 1977.
Pershagen, G., Bergman, F., Klominek, J., et al., *Br. J. Ind. Med.,* 44, 454–458, 1987.
Petres, J., Baron, D., and Hagedorn, M., *Environ. Health Perspect.,* 19, 223–227, 1977.
Pinto, S. S. and McGill, C. M., *Ind. Med. Surg.,* 22, 281–287, 1953.
Pinto, S. S. and Bennett, B. M., *Arch. Environ. Health,* 7, 583–591, 1963.
Pinto, S. S., Henderson, V., and Enterline, P. E., *Arch. Environ. Health,* 33, 325–331, 1978.
Piontek, M., Hengels, K. J., Borchard, S., et al., *Dtsch. Med. Wochenschr.,* 114, 1653–1657, 1989 (in German).
Poma, K., Degraeve, N., and Susanne, C., *Cytologia,* 52, 445–449, 1987.
Prier, R. F., Nees, P. O., and Derse, P. H., *Toxicol. Appl. Pharmacol.,* 5, 526–542, 1963.
Prukop, J. A. and Savage, N. L., *Bull. Environ. Contam. Toxicol.,* 36, 337–341, 1986.
Rencher, A. C., Carter, M. W., and McKee, D. W., *J. Occup. Med.,* 19, 754–758, 1977.
Rezuke, W. N., Anderson, C., Pastuszak, W. T., Conway, S. R., and Firshein, S. I., *Am. J. Hematol.,* 36, 291–293, 1991.
Rice, D. A., Kennedy, S., McMurray, C. H., et al., *Res. Vet. Sci.,* 39, 47–51, 1985.
Robson, A. O. and Jelliffe, A. M., *Br. Med. J.,* 5351, 207–209, 1963.
Rosenberg, H. G., *Arch. Pathol.,* 97, 360–365, 1974.
Roth, F., *Ger. Med. Mon.,* 2, 172–175, 1957.
Saady, J. J., Blanke, R. V., and Poklis, A., *J. Anal. Toxicol.,* 13, 310–312, 1989.
Saha, K. C. and Poddar, D., *Indian J. Dermatol.,* 31, 29–33, 1986.
Schaumburg, H. A., Failure to produce arsenic neurotoxicity in the rat. An experimental study, Office of Toxic Substances, No. EPA 560/11–80–022, NTIS no. PB80–209505, U.S. Environmental Protection Agency, Washington, D.C., 1980.
Schrauzer, G. N. and Ishmael, D., *Ann. Clin. Lab. Sci.,* 4, 441–447, 1974.
Schroeder, H. A. and Mitchener, M., *Arch. Environ. Health,* 23, 102–106, 1971.
Schroeder, H. A., Kanisawa, M., Frost, D. V., et al., *J. Nutr.,* 96, 37–45, 1968.
Shannon, R. L. and Strayer, D. S., *Hum. Toxicol.,* 8, 99–104, 1989.
Shirachi, D. Y., Tu, S. H., and McGowan, J. P., Carcinogenic effects of arsenic compounds in drinking water, Health Effects Research Laboratory, NTIS PB87–232542, EPA 600/1–87–007, U.S. Environmental Protection Agency, Cincinnati, OH, 1987.
Siewicki, T. C., *J. Nutr.,* 111, 602–609, 1981.
Sikorski, E. E., McCay, J. A., White, K. L., Jr., et al., *Fundam. Appl. Toxicol.,* 13, 843–858, 1989.
Silver, A. S. and Wainman, P. L., *J. Am. Med. Assoc.,* 150, 584–585, 1952.
Sobel, W., Bond, G. G., Valdwin, C. L., et al., *Am. J. Ind. Med.,* 13, 263–270, 1988.
Sommers, S. C. and McManus, R. G., *Cancer,* 6, 347–359, 1953.
Southwick, J. W., Western, A. E., Beck, M. M., et al., Community health associated with arsenic in drinking water in Millard County, Utah, Health Effects Research Laboratory, EPA-600/1–81–064, NTIS no. PB82–108374, U.S. Environmental Protection Agency, Cincinnati, OH, 1981.

Stevens, J. T., DiPasquale, L. C., and Farmer, J. D., *Bull Environ. Contam. Toxicol.,* 21, 304–311, 1979.
Szuler, I. M., Williams, C. N., Hindmarsh, J. T., et al., *Can. Med. Assoc. J.,* 120, 168–171, 1979.
Tay, C.-H. and Seah, C.-S., *Med. J. Aust.,* 2, 424–428, 1975.
Telolahy, P., Javelaud, B., Cluet, J., de Ceaurriz, J., and Boudene, C., *Toxicol. Lett.,* 66, 89–95, 1993.
Terada, H., *Nihon Rinsho,* 18, 118–127, 1960.
Tezuka, M., Hanioka, K., Yamanaka, K., Okada, S., *Biochem. Biophys. Res. Commun.,* 191, 1178–1183, 1993.
Tinwell, H., Stephens, S. C., and Ashby, J., *Environ. Health Perspect.,* 95, 205–210, 1991.
Tseng, W. P., *Environ. Health Perspect.,* 19, 109–199, 1977.
Tseng, W. P., Chu, H. M., How, S. W., et al., *J. Natl. Cancer Inst.,* 40, 453–463, 1968.
Tsuda, T., Nagira, T., Yamamoto, M., and Kume, Y., *Ind. Health,* 28, 53–62, 1990.
Tsutsui, T., Kawamoto, Y., Suzuki, N., et al., *Toxicol. In Vitro,* 5, 353–361, 1991.
Valentine, J. L., Reisbord, L. S., Kang, H. K., et al., Arsenic effects on population health histories. In Mills, C.F., Bremner, I., Chesters, J.K., Eds., Trace Elements in Man and Animals — TEMA 5. Proc. Fifth Int. Symp. Trace Elements in Man and Animals, Commonwealth Agricultural Bureau, Slogh, U.K., 189, 1985.
Vallee, B. L., Ulmer, D. D., and Wacker, W. E., *Arch. Ind. Health,* 21, 132–151, 1960.
Vig, B. K., Figueroa, M. L., Cornforth, M. N., et al., *Am. J. Ind. Med.,* 6, 325–338, 1984.
Wade, H. J. and Frazer, E. S., *Lancet,* 1, 269–271, 1953.
Wagner, S. L., Maliner, J. S., Morton, W. E., et al., *Arch. Dermatol.,* 115, 1205–1207, 1979.
Wahlberg, J. E. and Boman, A., *Derm. Beruf. Umwelt.,* 34, 10–12, 1986.
Wall, S., *Int. J. Epidemiol.,* 9, 73–87, 1980.
Watrous, R. M. and McCaughey, M. B., *Ind. Med.,* 14, 639–646, 1945.
Webb, D. R., Sipes, I. G., and Carter, D. E., *Toxicol. Appl. Pharmacol.,* 76, 96–104, 1984.
Webb, D. R., Wilson, S. E., and Carter, D. E., *Toxicol. Appl. Pharmacol.,* 82, 405–416, 1986.
Webb, D. R., Wilson, S. E., and Carter, D. E., *Am. Ind. Hyg. Assoc. J.,* 48, 660–667, 1987.
Welch, K., Higgins, I. Oh, M., et al., *Arch. Environ. Health,* 37, 325–335, 1982.
Wen, W., Lieu, T.-L., Chang, H.-J., Wuu, S. W., Yau, M.-L., and Jan, K. Y., *Hum. Genet.,* 59, 201–201, 1981.
Westhoff, D. D., Samaha, R. J., and Barnes, A., *Blood,* 45, 241–246, 1975.
Willhite, C. C., *Exp. Mol. Pathol.,* 34, 145–158, 1981.
Winship, K. A., *Adverse Drug React. Acute Poisoning Rev.,* 3, 129–160, 1984.
Wu, M.-M., Kuo, T.-L., Hwang, Y.-H., et al., *Am. J. Epidemiol.,* 130, 1123–1132, 1989.
Yamanaka, K., Hasegawa, A., Sawamura, R., et al., *Biochem. Biophys. Res. Commun.,* 165, 43–50, 1989.
Yamanaka, K., Hoshino, M., Okamoto, M., et al., *Biochem. Biophys. Res. Commun.,* 168, 58–64, 1990.
Yamanaka, K., Tezuka, M., Kato, K., et al., *Biochem. Biophys. Res. Commun.,* 191, 1184–1191, 1993.
Yamauchi, H, Kaise, T., and Yamamura, Y., *Bull. Environ. Contam. Toxicol.,* 36, 350–355, 1986.
Yeh, S., How, S. W., and Lin, C. S., *Cancer,* 21, 312–339, 1968.
Yu, H. S., Chang, K. L., Wang, C. M., and Yu, C. L., *J. Dermatol. (Japan),* 19, 710–714, 1992a.
Yu, H. S., Chen, G. S., Sheu, H. M., et al., *Proc. Natl. Sci. Counc. Rep. China,* 16, 17–22, 1992b.
Zaldivar, R. and Guillier, A., *Zentralbl. Bakteriol. Hyg.,* 165, 226–234, 1977.
Zaldivar, R., Prunes, L., and Ghai, G., *Arch. Toxicol.,* 47, 145–154, 1981.
Zierler, S., Theodore, M., and Cohen, A., *Int. J. Epidemiol.,* 17, 589–594, 1988.

Chapter 27

Toxicity of Bismuth and Its Compounds

Anja Slikkerveer and Frederik A. de Wolff

I. INTRODUCTION

Human exposure to bismuth (Bi) is limited, except for its use in medicine. The significance of occupational and environmental exposure is unknown, although Bi is increasingly used as a replacement for lead in many of its technical and chemical applications. Replacement of lead shot by Bi shot and the increasing amounts of Bi entering the environment after human use (Mueller, 1989) may draw attention to the question of the environmental safety of the metal. In medicine, Bi compounds have been prescribed for over a century. The intake of Bi has been related to nephrotoxicity and a reversible encephalopathy, the latter mainly in France. Because of its epidemiological characteristics and limited geographic distribution, 20 years after its occurrence, this encephalopathy is still an intriguing phenomenon in metal toxicology.

Even though large amounts of Bi-containing medications are consumed worldwide in often uncontrolled situations, the risk for Bi-related toxicity in the population seems to be very low. Nephrotoxicity is related to very high doses of Bi or to the intake of organic Bi compounds which have long since been withdrawn from the market. The mechanism of the neurotoxicity is unknown and it is impossible to predict which individuals might be at risk. Awareness among clinicians of the rare possibility that Bi may be involved in neurological disease will help to prevent the development of serious Bi-related disease.

II. MEDICAL USES OF Bi COMPOUNDS

The use of Bi-containing medications has a long history and indications have been many. Organic Bi preparations, such as the tartrate and the salicylate, were used more or less successfully to treat syphilis until the 1940s (Kolmer et al., 1939). Bi-triglycollamate was used as an oral or intramuscular therapy for warts, stomatitis, and infections of the upper respiratory tract, causing more side effects than benefits (Urizar and Vernier, 1966). These applications are now outdated, but inorganic Bi compounds such as Bi subnitrate (BSN), subcarbonate (BSC), and subgallate (BSG) are still being used in the treatment of a variety of gastrointestinal complaints including diarrhea, flatulence, constipation, and cramps. BSG is also used on the skin as an antiseptic and as a hemostatic agent after adenotonsillectomy. In most countries, Bi-containing preparations are available as over-the-counter products. Colloidal Bi subcitrate (CBS) is available on prescription and is used in 4- to 8-week courses in the treatment of peptic ulcer (Wagstaff et al., 1988) and *Helicobacter pylori*-associated gastritis (Marshall et al., 1987; Rauws et al., 1988). Bi-subsalicylate (BSS) is used in the U.S. for the treatment and prevention of traveler's diarrhea (Bierer, 1990) and in the treatment of *Helicobacter pylori* infections of the gastrointestinal tract (Drumm et al., 1988; Avunduk et al., 1991). New developments in the application of Bi-containing medications are the probable causal association between the development of gastric cancer and the presence of *H. pylori* infection of the stomach (Nomura et al., 1991; Parsonnet et al., 1991), the development of ranitidine Bi citrate for use in combined peptic ulcer and *H. pylori* infection (Ciociola et al., 1992; Stables et al., 1993), the successful pilot studies in the treatment of ulcerative proctitis and colitis with CBS and BSS enemas (Ryder et al., 1990; Srivastava et al., 1990), and the beneficial effect of pretreatment with BSN on the nephrotoxicity of *cis*-diammine-dichloroplatin (cisplatin) without compromising antitumor activity (Naganuma et al., 1985, 1987; Kondo et al., 1992).

III. CHEMISTRY AND ANALYSIS

Bi is located in group V, below nitrogen, phosphorus, arsenic, and antimony in the periodic system of the elements. It is a heavy metal with an atomic number of 83 and an atomic weight of 208.9 Da. The only stable valency is 3+. The composition of its salts is very complex and elucidation of its behavior in aqueous solutions is still far away (Billon et al., 1976; Lazarini, 1981; Asato et al., 1991). The solubility of most Bi salts in water is very low, but can be affected by acidity and the presence of sulfhydryl- or hydroxy-group-containing compounds. CBS has been shown to form stable complexes with dithiols and unstable complexes with monothiols such as glutathione (Beil et al., 1993). Careful descriptions of the preparation of experimental Bi formulations should be given, because small differences in preparation may change the properties of the experimental formulation and consequently the Bi species present (Funck et al., 1991).

At present the method of choice for the measurement of Bi in biological matrices is flame or electrothermal atomic absorption spectrophotometry (EAAS) (Froomes et al., 1988; Slikkerveer et al., 1991, 1993; Minagawa et al., 1993).

IV. BIOKINETICS

A. ABSORPTION

The bioavailability of Bi from the gastrointestinal tract is very low ($\ll 1\%$). The amount of Bi actually entering the body and the factors affecting this process, however, are relevant for the development of Bi toxicity. Although the composition and the structure of the Bi-containing molecule — the Bi species — that is absorbed from the intestine is unknown, the Bi-containing molecule absorbed is usually not the compound that was ingested. To complicate matters, differences in physiological and physicochemical conditions such as dissolution of tablets, stomach pH, and gastric emptying (Conso et al., 1975; Lee, 1981; Hespe et al., 1988) add to the large interindividual and intraindividual variation in absorption. Absorption of Bi from the gastrointestinal tract was dose dependent, although nonlinear, in an *in vivo* perfusion system of rat small intestine for CBS and $BiCl_3$ in citrate buffer (pH = 6.3) (Slikkerveer et al., 1995). Absorption as estimated by the total amount of Bi excreted in urine for ranitidine Bi citrate in human volunteers was also dose dependent; at an oral dose of 192 mg twice daily the Bi elimination was 36.4 µg and at a dose of 782 mg twice daily, 68.7 µg (median) (Prewett et al., 1991). The use of urinary Bi excretion for determination of the absorption underestimates the absorption, because some of the metal will at least temporarily be stored in body compartments.

Using whole-body retention and accumulated urinary excretion, the bioavailability of ^{205}Bi from various pharmaceutical Bi preparations was studied in rats. The bioavailability for the citrate-containing compounds (CBS, soluble Bi-citrate, and basic Bi-citrate) was significantly higher (0.26–0.33% of the dose) compared with basic Bi-nitrate, salicylate, gallate and Bi-aluminate (0.04–0.11% of the dose) (Dresow et al., 1991). Similar differences between Bi compounds were found after intestinal perfusion of rat small intestine (Slikkerveer et al., 1995). Perfusion with Bi-citrate and CBS resulted in higher Bi concentrations in blood. After administration of ^{205}Bi-labeled compounds to human volunteers the bioavailability proved to be even lower than in rats: $0.043 \pm 0.008\%$ for CBS, $0.039 \pm 0.001\%$ for basic BSG, and significantly less (<0.005%) for basic Bi-nitrate, basic Bi-subsalicylate, and Bi-aluminate (Dresow et al., 1992).

After administration of a single dose of 108 mg Bi as CBS to fasted volunteers, a maximum concentration in blood between 4.7 and 21 µg/l was found after 15 to 60 min (Hespe et al., 1988). The intake of 216 mg Bi as CBS resulted in maximum plasma concentrations (C_{max}) ranging between 25 and 300 µg/l (Nwokolo et al., 1989) after a mean time of 30 min (range 15 to 105 min). After intake of 432 mg of Bi as CBS, a C_{max} of 24.7 µg/l (range 5.5 to 57.5 µg/l) was observed after 30 to 60 min

(Froomes et al., 1989). With multiple doses, however, C_{max} was highly variable in the same person. Similar experiments with single doses of BSN and BSS showed lower maximum concentrations (Nwokolo et al., 1990a,b; Raedsch et al., 1990; Brosius et al., 1990). All single-dose experiments were performed after an overnight fast. Intake of Bi compounds after a standard breakfast resulted in lower maximum concentrations in plasma (Madaus et al., 1992). The C_{max} after a single dose of CBS decreased from 64.5 ± 15.3 (mean ± SE) to 10.9 ± 6.3 µg/l. The contribution of the relatively high maximum concentrations shortly after intake of CBS to the steady-state Bi concentration during a course of CBS has been calculated to be less than 4% (Benet, 1991).

After multiple dosing with CBS, steady state was reached after 3–4 weeks (Lee, 1981; McLean et al., 1988) with no difference between healthy volunteers and gastritis patients (Froomes et al., 1989). The Bi concentration in blood ranged from 5 to 20 µg/l for both liquid and coated-tablet formulations (Lee, 1981; Dekker and Reisma, 1979; Dekker et al., 1986); median steady-state concentrations of 12 to 17 µg/l were found in plasma after 3 to 6 weeks (Gavey et al., 1989). After 30 to 40 days, steady-state plasma concentrations of 38.3 ± 3.4 µg/l (mean ± SE; n = 12) were observed (Froomes et al., 1989). From urinary excretion levels at the end of a 6-week course of CBS it was calculated that 0.2% of the ingested Bi had been absorbed (Gavey et al., 1989). This is four times higher than calculated from reliable radiolabel studies (Dresow et al., 1992).

1. Enhancement of Absorption

Citrate was shown to increase the intestinal absorption of Bi from BSN, BSS, Bi-citrate, $BiCl_3$, and CBS during *in vivo* perfusion of rat small intestine (Slikkerveer et al., 1992a). Simultaneous intake of citrate and BSN by human volunteers significantly increased the Bi absorption and showed that the absorption profiles for Bi in blood and citrate in serum were parallel. This suggests the formation of a Bi-citrate complex (Slikkerveer et al., 1992b). A significant rise in the Bi concentration in blood was seen after oral administration of BSN and sulfhydryl-group-containing compounds to rats. The increase was largest for 3-mercaptopropionic acid, penicillamine, cysteine, and homocysteine; no effect was found for methionine, serine, and alanine (Chaleil et al., 1981). Simultaneous intraperitoneal administration of BSN and cysteine also increased Bi absorption (Chaleil et al., 1979a). In rats the absorption of Bi from BSS was enhanced by cysteine; this effect was absent when cysteine was given simultaneously with the Bi salt in a cysteine-rich diet (D'Souza and Francis, 1987).

Absorption-promoting effects have also been suggested for sorbitol (Lechat et al., 1964), lactic acid (Hartemann and Maugras, 1976), and several other hydroxy-group-containing compounds, but experimental proof is lacking. Circulating complexing substances may promote the absorption of Bi from the gastrointestinal tract, too, as dimercaprol administered intramuscularly to a beagle dog increased the absorption of Bi from orally administered BSG (Thomas et al., 1977).

Predosing with ranitidine or omeprazole before the intake of CBS increased the absorption of Bi three- to fourfold. For ranitidine, the median integrated 8-h plasma concentration of Bi (AUC_{0-8}) increased from 61 µg·h/l to 147 µg·h/l. This increase was paralleled by an increase in the urinary elimination of Bi over the first 8 h from 213 to 686 µg Bi (Nwokolo et al., 1991). For omeprazole the AUC increased from 46 ± 33 to 172 ± 158 µg·h/l (Treiber et al., 1994). The absorption of Bi from BSN and BSS was not affected by ranitidine (Nwokolo et al., 1991). Measurement of gastric pH showed that in the omeprazole study the increase in absorption was related to the increase in gastric pH (Treiber et al., 1994), supporting the suggestion that increased gastric pH may prevent rapid precipitation of Bi as Bi_xCit_y and BiOCl (Williams, 1977).

2. Route(s) of Absorption

Bi is thought to share absorption routes with other substances, because it is not an essential element. In the intestine several unidentified Bi species may form, in which the counter-ion could be responsible for absorption. The actual sites of absorption are unknown. It has been suggested that absorption of inorganic Bi compounds is dependent on the solubility in the gastrointestinal tract (Billon et al., 1976; Thomas et al., 1977; Thomas et al., 1983; Lechat and Kisch, 1986). The bioavailability of radiolabeled Bi compounds corresponded with the solubility of the compound in artificial duodenal juice, but not with solubility in gastric juice (Dresow et al., 1991). Study of the citrate-enhanced absorption of Bi showed that both active and passive transport blockers could reduce the concentration of Bi in blood (Slikkerveer et al., 1992a). This suggests that both transcellular and paracellular routes may be involved and that Bi absorption is a mixed process. Absorption of Bi particles may take place, but the relative importance of this route still has to be assessed. CBS particles were seen by electron microscopy in the epithelium of the esophagus, and in the enterocytes of the duodenum and of the proximal small bowel after oral intake of CBS in humans. No passage of particles through the gastric mucous membrane was observed (Coghill et al., 1983). Nwokolo et al. (1992) demonstrated the presence of Bi-containing particles with a diameter of 4.5 nm between and in the epithelial cells of the antrum of the stomach of volunteers within 30 to 60 min after oral intake of CBS. The particles were also seen in the lamina propria. After intake of BSS, no penetration through the gastric mucosa was present. Endocytosis of Bi particles was suggested as a mechanism for absorption into the enterocyte (Volkheimer, 1977; Coghill et al., 1983). Since the presence of particles in blood after an oral dose with CBS has not been demonstrated, persorption as a major factor in the uptake of CBS remains improbable.

B. DISTRIBUTION

The chemical form in which Bi is present in blood is still unknown. No data are available on the carriers of Bi in blood after *oral* intake of Bi. Russ et al. (1975) showed that *in vitro* 17% of the radioactive Bi citrate was associated with erythrocytes. The remainder was bound to serum proteins. Twenty days after a single oral dose of CBS (n = 4), relatively large amounts of Bi were associated with red blood cells ($8.7 \pm 3.9 \times 10^{-5}$% of dose) (Dresow et al., 1991). Within 2 h after the addition of BSN to whole blood, 82.3 ± 5.0% was found in red blood cells at 25 µg/l and 97.7 ± 0.7% at 2500 µg/l (Rao and Feldman, 1990a). Below 200 to 300 µg/l the overall binding to blood components seemed to be stronger than above that level. Higher concentrations were easily removed from the blood by hemodialysis (Allain et al., 1980a; Olive, 1980; Chaleil et al., 1981). Gel filtration of human blood after incubation with BSG showed an association of Bi with the high molecular fraction ($\geq$200,000 Da), consisting of α_2-macroglobulin, IgM, β-lipoprotein, and haptoglobulin (Thomas et al., 1983).

The distribution of Bi over tissues is largely independent of the Bi compound administered or the route of administration. The highest concentration per gram wet weight was always found in the kidney (Sollman et al., 1938; Sollman and Seifter, 1942; Van Der Werff, 1965; Hursh and Brown, 1969; Hall and Farber, 1972; Russ et al., 1975; Lechat et al., 1976; Badinand and Quincy, 1980; Pieri and Wegmann, 1981; Vienet et al., 1983; Zidenberg-Cherr et al., 1987). However, more detailed studies suggest that exceptions do exist. When Bi was adsorbed onto charcoal or administered as a colloid, distribution after intravenous administration is depended on particle size. The Bi-containing particles could be found in the lung and the reticuloendothelial system (Coenegracht and Dorleyn, 1961; Erdani et al., 1964; Russ et al., 1975). To study tissue distribution BSN, BSS, CBS, Bi-citrate, and BSN with added citrate buffer were fed to rats for 14 days so that comparable Bi concentrations in blood were reached (Slikkerveer et al., 1992c). At concentrations of 66 ± 14.2 and 552 ± 71 µg/l Bi in blood, rats fed with BSS had a lower Bi concentration in the kidney and a higher amount in the liver compared to all other Bi compounds. BSS intake led to the lowest Bi content in the kidney, but it was still the only compound which demonstrated transient nephrotoxicity at the highest blood concentration (Slikkerveer et al., 1992c). After oral administration of trimethyl-Bi to dogs, the concentration of the metal was higher in the liver than in the kidney, probably due to the organic character of the molecule (Sollman and Seifter, 1939).

The retention time of Bi in the kidney is longer than in any other organ. While in other organs no, or hardly any, Bi could be demonstrated 144 h after intravenous injection of ^{206}Bi-citrate, 12% of the injected dose remained in the kidneys and 0.9% in bone (Russ et al., 1975). After administration of CBS to rats for 14 months, Bi concentrations in the organs ranked from high to low in kidney, lung, spleen, liver, brain, and muscle, respectively, with 13.9 µg/g wet weight in kidney and 0.13 µg/g in muscle (Lee et al., 1980). In the studies in which the concentration in bone was measured the level was usually 10 to 20 times lower than in the kidney (Jeanrot et al., 1982; Gregus and Klaassen, 1986; Zidenberg-Cherr et al., 1987). Analysis in humans not recently exposed to Bi suggests that the metal is not detectable in their tissues (Slikkerveer et al., 1993).

In the guinea pig, placental transfer of Bi was shown to be less than 1%, 1 h after intracardiac injection with ^{207}BiCl$_3$ in 0.1 *N* HCl. The majority of the radioactivity in the fetoplacental unit was found in the placenta and the amniotic sac (Richardson and Wills, 1992).

C. METABOLISM

No metabolism of Bi is known. The metal affects its own binding profile to proteins in the kidney by inducing the *de novo* synthesis of Bi-metal-binding protein, a kind of metallothionein (Szymanska et al., 1977; Szymanska and Zelazowski, 1979; Szymanska and Piotrowski, 1980; Zidenberg-Cherr et al., 1989). Simultaneous administration of selenium inhibits the binding of Bi to this Bi-metal-binding protein in the kidney (Szymanska et al., 1978).

After subcutaneous administration of high doses of BSN to rats, swelling and distortion of the inner mitochondrial membrane have been demonstrated in liver and proximal renal tubules. In liver, the activity of δ-aminolevulinic acid (ALA)-synthetase, hemesynthetase, and ALA-dehydrogenase was reduced by 35 to 50%. Similarly, ALA-synthetase and ALA-dehydrogenase activity was reduced in the kidney (Woods and Fowler, 1987). Anemia, however, has never been associated with Bi ingestion.

D. ELIMINATION

Bi is cleared from the body through urine and feces. The exact contribution of the urinary and fecal route to the elimination of Bi is still subject to discussion and may be dependent on the compound and on the dose. After intravenous injection of ^{206}Bi-nitrate in rats, 38% of the metal was excreted in 4 days: 21% in urine and 17% in feces (Gregus and Klaassen, 1986). In another study with intravenous radiolabeled Bi, 55% was excreted after 2 h, 40% in the urine and 15% in the feces (Zidenberg-Cherr et al., 1987). After intramuscular administration of Bi-butylthiolaureate to rats, equal amounts were excreted in urine and feces and after 90 days, 10% of the Bi dose remained in the body (Chaleil et al., 1979b). In humans, 99.9% of an oral dose of several radiolabeled Bi compounds was excreted in the feces within 7 days. On day 12 after intake, the ratio between feces (bile) and urine was estimated to be 1:1. The elimination of different Bi compounds showed no differences (Dresow et al., 1992).

In the presence of biliary drainage, the metal could still be found in the intestine after intravenous administration (Vienet et al., 1983). This suggests that Bi is eliminated into the intestine, partly via bile and partly via intestinal secretion. In rats, the cecum is suggested as a possible location for intestinal secretion, because after intravenous injection of Bi, a relatively high amount of the metal was found there (Chaleil et al., 1979b; Pieri and Wegmann, 1981; Wieriks et al., 1982).

In rat, 2 h after intravenous administration 5 to 8% of the Bi was detected in bile. Bi in bile was 10 times more concentrated than in plasma. This biliary excretion mechanism was not saturated at 10 mg/kg (Gregus and Klaassen, 1986). After intravenous administration of 0.75 mg/kg Bi (n = 5) as BSN, the elimination of Bi in bile of the rat was 0.020 ± 0.006 of the dose. In this experiment the process showed saturation (Rao and Feldman, 1990b). The transport of Bi into bile, after intravenous injection of 150 nmol/kg of the soluble Bi ammonium citrate (±30 µg/kg Bi), was dependent on glutathione (GSH) (Gyurasics et al., 1992). One molecule of Bi was cotransported with about three molecules of GSH. In 2 h 3% of the i.v. dose was excreted in bile.

The essential role of the kidneys in Bi elimination is illustrated by the inverse relationship between the creatinine clearance and the steady-state concentration of Bi in plasma (Treiber et al., 1991). A higher steady-state level of Bi in blood was reached after four weeks of treatment with CBS in patients with severely impaired renal function and normal intestinal absorption. The authors suggested to adapt the oral dosage of CBS when the creatinine clearance is less than 20 ml/min. Four to eight weeks after cessation of CBS therapy the total Bi elimination in urine was still above the baseline (Gavey et al., 1989; Treiber et al., 1991; Nwokolo et al., 1994). Clearance of Bi from the body, absorbed from CBS, was estimated to be between 50 and 95 ml/min (Benet, 1991). In another study in healthy volunteers and gastritis patients treated with CBS, renal clearance of Bi was estimated to be 22.0 ± 2.0 ml/min (mean ± SE, n = 12) (Froomes et al., 1989).

After intravenous administration of ^{206}Bi to rats half-life times in blood have been calculated of 1 and 27.7 h, <0.5, 13, and 122 h, and 0.2, 2.4, and 122 h, respectively (Russ et al., 1975; Pieri and Wegmann, 1981; Vienet et al., 1983). These studies suggested a two- or three-compartment model to describe the elimination kinetics of Bi, with the kidneys as an important compartment. In a study with intravenous ^{206}Bi in two human volunteers and three patients, recovered from Bi encephalopathy, half-lives of 3.5 min and 0.25 and 3.2 h were established (De Bray et al., 1977). After administration of a single oral dose of ^{205}Bi as CBS to rats the whole-body kinetics, the urinary and fecal elimination could be described by a three-compartment model. The biological half-life was estimated to be 10, 36, and 295 h, respectively (Dresow et al., 1991). The elimination half-life was in human plasma after multiple dosing was estimated at 20.7 ± 3.9 days (mean ± SE, n = 6) (Froomes et al., 1989). In general, there seems to be little consistency and considerable variation in these data, possibly caused by the use of different, sometimes even unnamed, Bi compounds.

V. TOXICITY

The main target organs for Bi toxicity are bone, kidney, and brain. "Insoluble" inorganic Bi compounds are associated with neurotoxicity, while the compounds used in the past for treatment of syphilis caused kidney and bone disease. Hepatotoxicity has been suggested, but this suspicion is only based on two reports. The first is about a 6-year-old boy, who received Bi-thioglycollate and developed jaundice and anuria (Karelitz and Freedman, 1951). The second described 121 inmates of an American prison with liver damage, who had received Bi and organic arsenic compounds in the treatment of syphilis. As arsenic (As) compounds are potentially hepatotoxic, it is impossible to distinguish between the effects of As and Bi (Kulchar and Reynolds, 1942). In both reports the relation with Bi was not convincing and no Bi concentrations were measured. No support can be found in animal experiments either, so it is most unlikely that Bi is hepatotoxic in humans.

As rare complications after treatment with sodium Bi-tartrate and an unspecified Bi compound, stomatitis and gingivitis have been described (Dowds, 1939; McClendon, 1941). A Bi line (blackish-blue) has been observed on the gingiva of some patients (Goodman, 1948; Winship, 1983). Intravenous injection of sodium Bi-tartrate (sobita), 10 g in aqueous solution, which was three times the usual dose for treatment of yaws, caused the death of three African women. Two of them died within 2 min and the third died of renal failure after 10 days. At autopsy, the latter showed gingivitis, blue discoloration of the gums, hemorrhagic colitis, and destruction of the epithelium of the renal tubuli (Goodman, 1948). Erythroderma with edema and vesiculae was seen after treatment of syphilis with Bi and the side effects were treated successfully with dimercaprol (Dérot et al., 1947). Metal-gray pigmentation of the skin and the mucous membranes observed in a 70-year-old woman was shown to be caused by deposits of Bi in the papillary dermis and the basal membranes of sweat glands. More than 25 years earlier, the patient had been treated with injections with an unknown Bi compound for over 10 years (Zala et al., 1993). A paste containing BSG, which is used to lower the incidence of hemorrhage after adenotonsillectomy, has been shown to cause acute pneumonia followed by a histiocytic foreign body response after intratracheal administration to rats (Cozzi et al., 1992).

An *in vivo* mutagenicity study for chromosomal aberrations in mouse bone marrow after oral intake of an aqueous suspension of Bi_2O_3 for up to 21 days, showed a dose-dependent increase (top dose 1000 mg/kg) in chromosomal aberrations with gaps (Gurnani et al., 1993). No abnormal sperm heads were detected in the same experiment. Because absorption from Bi_2O_3 is considered to be truly negligible, it is unfortunate that the results were not supported by the measurement of Bi concentrations in blood, ensuring that the metal had indeed reached the bone marrow. An old life-time study with 20 rats fed 2% BSC in the diet did not show a decrease of survival nor an increase in tumors caused by the material (Von Osswald, 1968).

A. NEPHROTOXICITY

The use of Bi-tri- or thioglycollamate and Bi-diallylacetate has been known to cause acute renal reactions. Both compounds are not used in medicine anymore. Functional impairment and sometimes necrosis of the proximal tubuli of the kidney, developing soon after administration of Bi, was associated with acute but reversible renal failure. Still, several children died because hemodialysis had not yet been introduced. The nephrotoxicity was generally reversible (McClendon, 1941; Boyette and Ahoskie, 1946; O'Brien, 1959; Gryboski and Gotoff, 1961; Czerwinski and Ginn, 1964; Urizar and Vernier, 1966; James, 1968; Randall et al., 1972). Sharply defined round structures with a diameter of 5 μm were present in the cell nuclei, in the cytoplasm, and possibly also in the lysosomes of 86% of the kidneys of syphilis patients (Beaver and Burr, 1963a). These inclusions contained Bi, protein, carbohydrates, lipids, and sulfur (Beaver and Burr, 1963b; Müller and Ramin, 1963; Burr et al., 1965; Fowler and Goyer, 1975; Ghadially, 1979). Their implication for renal function remains unclear. More recently, delayed reversible renal failure was seen after intake of large single doses of CBS (39 to 100 tablets) (Hudson and Mowat, 1989; Taylor and Klenerman, 1990; Huwez et al., 1992; Schweiberer and Vogelsang, 1993). One patient died within four days from perforation of a preexistent duodenal ulcer. The Bi concentration in two kidney samples from this patient were 11 and 16 mg/g wet weight (Taylor and Klenerman, 1990). These high Bi concentrations were caused by autointoxication with 80 tablets after a preceding intake of four weeks of CBS in standard doses. In a nephrotoxicity study with CBS in female Wistar rats, the damage to the kidney was shown to be dose dependent, with the lowest adverse effect dose being 1500 μmol Bi/kg (±200 tablets of CBS). The damage was seen as a band of tubular epithelial necrosis in the corticomedullary transition zone, starting with vacuolation and ending in total destruction of the tubuli (Slikkerveer et al., 1992d). Distribution of Bi over the liver and kidney suggested that the liver may play a role in the development of Bi nephrotoxicity.

In a follow-up study it could be demonstrated that this renal damage was rapidly and completely reversible. Damage had disappeared four days after a large single dose (Slikkerveer et al., 1995).

B. OSTEOARTHROPATHY

After Bi treatment for syphilis generalized osteoporosis was observed, sometimes in combination with osteomalacia. Lesions were localized in the pelvis, the head of the femur, and the vertebrae. It has been suggested that the presence of Bi in bone may have aggravated a preexisting tendency towards osteoporosis (Revault et al., 1958; De Sèze et al., 1958; Gaucher et al., 1980). The bone complications which were seen in some patients with a Bi encephalopathy had a different pattern. Fractures of the thoracic vertebrae were observed together with demineralization and necrosis of the humeral head, osteolysis, and fractures of the proximal humerus. Disturbances of bone mineralization and microfractures were thought to be caused by trauma from the combination of serious myoclonia and the restriction of the restless patients (Buge et al., 1975; Lhermitte et al., 1978; Sany et al., 1978; Murray, 1979; Emile et al., 1979; Garcia et al., 1980; Mabille et al., 1980).

C. NEUROTOXICITY
1. Clinical Presentation

Five recent representative cases of Bi encephalopathy are followed by a more general description of the clinical picture of Bi neurotoxicity. A 69-year-old German patient developed depression, marked insomnia, loss of concentration, nervousness, and panic attacks at night after intake of Bi-nitrate, 8 g/day for 3 weeks, followed by 4 g/day for 16 months. After 12 months of Bi intake he was diagnosed as suffering from a major depressive disorder and treated with tricyclic antidepressants and benzodiazepines, which improved his symptoms. Gradually he developed problems with short- and long-term memory, difficulty in reading and writing, tremor, vertigo, and dysarthria. After 15 months of Bi intake the syndrome had developed into a full myoclonic Bi encephalopathy. After discontinuation of his medication the patient was treated with steroids and recovered within 4 months (Von Bose and Zaudig, 1991).

The second patient was treated with large doses of BSS (5.2 to 9.4 g/day) for diarrhea as a complication of AIDS. He developed lethargy, dysarthria, and myoclonic jerking of his facial and axial muscles. The Bi concentration in blood was 20 µg/l and in urine 2.96 mg/l. Bi encephalopathy was diagnosed and the patient was treated with D-penicillamine. He died within 24 h after the start of chelation therapy (Mendelowitz et al., 1990).

The third patient had chronic renal function impairment (creatinine clearance 15 ml/min) and took twice the recommended dosage of CBS for 2 months (864 mg Bi/day), besides several other medications. His symptoms were general cerebral dysfunction, urinary incontinence, ataxia, visual hallucinations, and bilateral grasp reflexes (Playford et al., 1990). The Bi concentration in blood was 880 µg/l and in urine 230 µg/l. He recovered after discontinuation of the Bi intake and 10 days of treatment with dimercaptopropane sulfonic acid.

The fourth patient was an American who suffered from paranoid ideas, memory impairment, and defective gustatory and olfactory function while using BSG as a colostomy deodorant. The Bi concentration in blood was 67 µg/l, and 24-h urinary Bi excretion was 889 µg (Friedland et al., 1993).

The fifth case history describes a German woman who developed progressive dementia with abnormal coordination, occasional tremors, and dysarthria. She had been using a BSG-containing stomach powder for more than 15 years. Six days after discontinuation of BSG, the Bi concentration in blood was 70 µg/l. The patient was discharged from hospital after $4^{1}/_{2}$ months in good condition with restored intellectual functions, indicating the reversible character of Bi encephalopathy (Kendel et al., 1993).

Before encephalopathy became evident, generally a prodromal period with vague symptoms had been present. Symptoms during the prodromal period could be many: impairment of walking, standing, or writing, deterioration of memory, changes in behavior, insomnia, and muscle cramps. Psychiatric symptoms such as anxiety, depression, excitation, or hallucinations could also be present and sometimes resulted in psychiatric treatment. The characteristic symptoms of the encephalopathy are myoclonia, changes in awareness, astasia and/or abasia, and dysarthria. From 698 patients 80% showed myoclonia, 79% confusion, 60% loss of concentration capacity, 59% trembling, 56% memory disturbances, and 53% writing disturbances (Loiseau et al., 1976; Reinicke, 1982). The changes in awareness varied from disorientation in time and place to coma. Myoclonias were seen most frequently in the distal parts of the arms, but also in the legs, trunk, face, and tongue. They were at first only present at night and could be provoked by passive motion of the limb or by touch, light, sound, or emotion. Electromyographic recordings in the acute phase showed characteristic bouts of myoclonia with simultaneous activity of agonists and antagonists (Gastaut et al., 1975; Lagueny et al., 1976). The pathogenesis of the walking, standing, and speaking impairment is unclear, but may have been caused by the involuntary muscle movements. The combination of impairment of consciousness and frequent myoclonias was also held responsible for the occurrence of urine incontinence in these patients. When consciousness was unimpaired, psychic disturbances such as deceleration of thought (58%), excitation (35%), depression (32%), anxiety (27%), hallucinations (17%), and delirium (16%) (Mattle et al., 1982), formed an important part of the clinical picture, as did neuropsychological changes such as impairment of intellect and memory (50%), loss of interest, and a general mental deterioration (Loiseau et al., 1976; Mol et al., 1979; Reinicke, 1982).

After discontinuation of Bi-containing medication the encephalopathy disappeared. Recovery was generally complete in two to six weeks, the symptoms disappearing in reverse order of appearance. Most patients recovered completely, but it was suggested that residual phenomena such as intellectual impairment, affective disorders, headache, tremor, insomnia, and osteoarthropathy sometimes remained when a patient had experienced severe, prolonged, or complicated forms of the disease (Buge et al., 1977). This suggestion is difficult to prove because retrospectively it is impossible to assess the neuropsychological and psychiatric condition of a patient in the period before the disease. The neuropsychological recovery of the patients appeared to be related to the disappearance of the changes in EEG and on CT scan, when present, and to the decrease of the Bi concentrations in CSF, rather than to the fall in the Bi concentration in blood (Bès et al., 1976; Buge et al., 1979).

2. History of Bi Encephalopathy

The first patients with Bi encephalopathy were described in Australia (Morrow, 1973). All 28 patients had a colostomy or ileostomy and took BSG (about 700 mg/day) to improve the consistency of their stools. No Bi concentrations were measured in these patients, but retrospectively the clinical picture was strongly suggestive for Bi encephalopathy. The syndrome regressed when BSG was discontinued and reoccurred on rechallenge (Burns et al., 1974; Lowe, 1974; Robertson, 1974; Winship, 1983). Withdrawal of BSG for oral administration in 1974 caused this problem to disappear.

In France, where a wide range of Bi compounds were used for a large number of gastrointestinal complaints, an outbreak of encephalopathy possibly related to Bi intake drew public attention in 1974 (Buge et al., 1974). All Bi-containing preparations on the French market at the time were associated with Bi encephalopathy. BSN was involved most frequently, possibly because it was used by the majority of the patients (Martin-Bouyer, 1976). The disease presented as an "epidemic", which is the more remarkable since Bi had already been used on a large scale in France for years. The decline in Bi sales (and indirectly the decline in the number of patients with Bi encephalopathy) was achieved by imposing legal restrictions on the sale of all Bi-containing preparations in France (Winship, 1983). In 1980, a total of 942 cases were reported in France, 72 of which ended in death (Martin-Bouyer et al., 1981a). It has never been proven that these deaths were directly related to Bi intake. Bi intoxication was rare in countries other than France, where 682 tons of Bi in drugs were consumed in 1973. Even though large amounts of Bi were sold in the U.S., the U.K., and especially the Netherlands, no cases of poisoning were reported in these countries (Buge et al., 1978). Intoxications were seen only in the countries neighboring France: 18 in Belgium (Monseu et al., 1976; Indekeu and Laterre, 1978; Loseke et al., 1978; Liessens et al., 1978; Collignon et al., 1979; de Mol et al., 1979; Lotstra et al., 1983), 9 in Spain (Zarranz et al., 1977; Relat et al., 1979; Cervello et al., 1982; Nogué et al., 1985; Molina et al., 1989), 4 in Germany (Cramer et al., 1978), and 5 in Switzerland (Grandjean et al., 1976). One case was reported in Morocco concerning an employee of the French embassy (Lagier, 1980). Separate mention is to be made of two patients with encephalopathy and elevated Bi levels in cerebral venous blood after long-term use of a skin cream containing Bi and mercury (Krüger et al., 1976). Concentrations of the latter were not measured, so that the etiology of this encephalopathy can not be attributed exclusively to Bi. Since 1980, sporadic cases of Bi encephalopathy have been reported in France (Stahl et al., 1982; Ferrer et al., 1984; Vidialhet et al., 1987).

3. Etiology

It is generally accepted that Bi neurotoxicity has been caused by more than Bi intake alone. Only recently some light has been shed on how Bi could have caused reversible neurotoxicity. After a large intraperitoneal dose of BSN (2500 mg/kg), which acts as a depot, autometallography showed that in the brains of mice Bi was located adjacent to fenestrated blood vessels. This suggests diffusion to circumventricular organs as the mechanism for the entrance of Bi in brain (Ross et al., 1994). This localization of the metal is consistent with the prodromal signs of Bi-related neurotoxicity in humans. The mice all showed hydrocephalus, but no signs of microscopic pathology or (neuro)toxicity. Based on significant treatment-related changes in the choroid plexus, the authors suggested that the choroid plexus may be the key structure in the pathogenesis of Bi neurotoxicity (Ross et al., 1994). The factor which turns Bi into a neurotoxicant is, however, still unknown. Increased absorption is relevant in the development of toxicity, but we suspect that the formation of a Bi complex or aggregate with specific entry into the brain or a specific toxicity for neural tissue may play an important role. Recent studies have demonstrated that in patients treated for eight weeks with CBS or BSN no signs of overt or subclinical neurotoxicity could be found. In these studies magnetic resonance imaging (MRI), visual evoked potentials, spectral electroencephalography, nerve conduction studies, and neuropsychological tests were used (Nwokolo et al., 1994; Noach et al., 1995).

Most hypotheses have proposed that an unknown factor enhances Bi absorption and increases Bi concentrations in plasma or blood. In the literature the presence of absorption-enhancing factors, such as SH-group-containing compounds, in the intestine, pharmaceutical characteristics of the salts (Moindrot et al., 1977; Buge et al., 1978), or prolongation of intestinal transit time (Martin-Bouyer et al., 1978, 1981a; Carruzzo, 1980; Lechat and Kisch, 1986) have been suggested. Comparison of patients with encephalopathy and a healthy control group of Bi consumers showed no difference in age profile, sex, quantity of drug consumed, duration of intake, concurrent use of other medicaments (especially psychiatric drugs), or diet. The only significant parameter in which the patients who developed encephalopathy differed from Bi users without neurological symptoms was the indication for prescription of Bi compounds. Constipation was more frequently the indication for Bi-containing medication in those users who developed encephalopathy (Martin-Bouyer et al., 1978, 1980, 1981a).

A microbiological explanation for the individual reaction to Bi could not be proven. The hypothesis was that microorganisms present in the intestine of encephalopathic patients changed Bi salts into better absorbed or toxic compounds (Martin-Bouyer et al., 1978, 1981a,b). Two chemical forms have been suggested: trimethyl-Bi (because its inhalation resulted in encephalopathy in dogs [Sollman and Seifter, 1942; Martin-Bouyer et al., 1980]) and a Bi-lactate complex (Hartemann and Maugras, 1976). Comparable microorganisms were, however, identified in the intestinal tract of encephalopathic patients and healthy Bi users (Buge et al., 1978; Lechat and Kisch, 1986) while inoculation of germ-free rats with methanogenic bacteria from man did not increase the Bi concentrations in blood and tissues after repeated oral intake of BSN (Chaleil et al., 1988).

Even though large amounts of Bi-containing medications are consumed worldwide, often in uncontrolled situations, the risk for Bi-related toxicity in the population seems to be very low. It still is impossible to predict which individuals might be at risk. Increased risk for patients on Bi-containing medication, however, may be present when renal function is impaired (creatinine clearance <20 ml/min) or absorption is strongly enhanced, e.g., by extreme diets with very large daily amounts of citrate-containing foods. Furthermore some individuals may be predisposed to form a toxic Bi complex or compound under specific, but yet unidentified conditions. The diet, comedication, or endogenous factors may play a role.

4. Pathology

The death of only one patient can be directly related to Bi neurotoxicity. This patient's brain showed necrosis of the nucleus dentatus and the putamen, and of the cerebral and cerebellar cortex (Buge et al., 1979). Other deaths of encephalopathy patients are not well documented. Out of 12 patients, 3 died from decubitus and 1 from cardial disease (Buge et al., 1979). It is remarkable that in encephalopathy patients no kidney function disturbances have been described.

Bismuth encephalopathy has no morphological basis. Anoxia was held responsible for the loss of cerebellar Purkinje cells and cerebral cortex neurons (Burns et al., 1974; Liessens et al., 1978; Ribadeau-Dumas et al., 1978). Recently it has been suggested that changes in the choroid plexus and hydrocephalus may have been related to the prodromal phase in humans (Ross et al., 1994), but in humans hydrocephalus has only rarely been described.

In humans and animals receiving Bi, the concentration in brain tissue was higher than in controls (Hamilton et al. 1972; Lagueny et al., 1976; Lee et al., 1980; Warren et al., 1983), showing that Bi is able to pass the blood-brain barrier, although in very small amounts. Patients who died with Bi encephalopathy had Bi concentrations in the gray matter (1.2 to 29 $\mu g/g$) that were generally twice as high as that in white matter (1.3 to 15.9 $\mu g/g$). The highest concentrations could be found in the thalamus (10.0 ± 8.6 $\mu g/g$), frontal (11.4 ± 8.5 $\mu g/g$), and cerebellar (11.1 ± 5.6 $\mu g/g$) cortex (Escourolle et al., 1977). In another study the highest concentrations were found in the brain stem (9.1 ± 1.9 $\mu g/g$) and the pons (12.3 ± 3.7 $\mu g/g$)(Ribadeau Dumas et al., 1978).

In rats, 2 h after intravenous administration of radiolabeled Bi citrate, the highest activity was found in the spinal cord, medulla oblongata, hypothalamus, and pons (Pieri and Wegmann, 1981). The presence of Bi in the nucleus ruber, nucleus dentatus, and cerebellar cortex may explain the presence of myoclonia. After eight weeks of treatment with CBS in a clinical study, aimed at screening for neurotoxicity with sensitive methods, one patient was found with a high-intensity signal in the globus pallidus on the MRI scan, without signs of neurological abnormalities (Nwokolo et al., 1994).

Astrocytes were shown to be much more sensitive to Bi sodium tartrate than nerve cells in *in vitro* cultures of brain, meninges, and neuronal retina cells from chicks (Bruinink et al., 1992). Damage to the astrocytes may be responsible for the early "vague" symptoms of Bi neurotoxicity, while nerve cells are only damaged after prolonged exposure. Prolonged exposure was also necessary to demonstrate neuronal cell degradation in rat hippocampal slices, while acute exposure showed no effect on the bioelectric activity of pyramidal cells (Bruinink et al., 1992).

After intraperitoneal administration of 2500 mg/kg BSN to mice, 20% of the mice showed signs of neurotoxicity. All had Bi concentrations in the brain exceeding 6 µg/g (8.4 ± 2.0 µg/g). The blood concentrations of Bi did not predict the development of neurotoxicity, but relatively high concentrations in blood during several weeks seemed to be necessary for the development of brain effects (Ross et al., 1988). The highest Bi concentrations were found in the olfactory bulb (the ventral side showing higher Bi concentrations than the dorsal side), hypothalamus, septum, and brainstem (Ross et al., 1994). High concentrations of Bi were found adjacent to fenestrated blood vessels and were associated with damage to the neuropil. All treated mice, symptomatic and asymptomatic, showed hydrocephalus and a limited amount of kidney damage. Electronmicroscopic examination showed expansion of the extracellular space between the epithelial cells of the choroid plexus. Vacuolation and membranous debris were found in the dendrites in the neuropil of the hypothalamus and septum. The presence of high Bi concentrations and pathology in the choroid plexus suggests a role for the choroid plexus in the development of hydrocephalus and possibly also in the development of Bi neurotoxicity.

Furthermore, in symptomatic mice only, fiber tract degeneration was observed in the anterior and lateral funiculi of the cervical and lumbosacral spinal cord. No lesions of the spinal cord have been described in humans, but these effects were possibly not looked for. This interesting finding may also be an explanation for the myoclonia seen during Bi encephalopathy (Ross et al., 1988).

VI. DIAGNOSIS OF Bi TOXICITY

A. Bi CONCENTRATIONS IN BIOLOGICAL MATRICES

Bi nephrotoxicity can be diagnosed by the combination of acute renal failure with a history of Bi intake or high Bi concentrations in blood. Bi encephalopathy can only be diagnosed when involvement of the metal can be proven by elevated concentrations of Bi in blood, plasma, serum, or CSF. Blood is the preferred matrix since the relation between the Bi concentration in serum and blood has still not been elucidated (Rao and Feldman, 1990b; Dresow et al. 1991). High concentrations of Bi in urine also support the diagnosis of Bi encephalopathy, but are more difficult to relate to the clinical condition. No Bi should be detectable in blood or serum from persons not recently exposed to Bi therapy (Froomes et al., 1988; Slikkerveer et al., 1991).

The severity of the encephalopathy can not be predicted by the concentration of Bi in blood (Martin-Bouyer et al., 1978; Ross et al., 1988). The decline of the Bi concentration in blood and the recovery of the patient were parallel (Boiteau et al., 1976), but the concentration in CSF may probably reflect the clinical condition of the patient better (Bès et al., 1976; Indekeu and Laterre, 1978; Liessens et al., 1978; Buge et al., 1979). The Bi concentration in CSF was 50 to 60 µg/l in patients with Bi encephalopathy, but variation was large. Allain et al. (1980b) estimated the elimination half-life from CSF to be roughly 16 days. No relationship could be found between the Bi concentration in blood and the concentration in urine or in CSF.

Hillemand et al. proposed highly provisional limits for the French clinicians (Hillemand and Cottet, 1976; Hillemand et al., 1977): Bi concentrations in blood below 50 µg/l can be considered as safe, between 50 and 100 µg/l the patient should be checked regularly, and above 100 µg/l treatment with Bi should be stopped. Their tentative rule of thumb, however, was never intended to be (mis)used as international safety levels by drug regulatory authorities unaware of its original limitations. Reading of the original publications is strongly recommended.

B. ADDITIONAL INFORMATION

Sensory evoked potentials measured in a patient with encephalopathy after intake of BSN showed giant potentials after cortical stimulation (Von Bose and Zaudig, 1991). On CT scans, made of a few encephalopathy patients only, an increase in density was seen in the cortex of the cerebral hemispheres and in the basal ganglia. One patient also showed increased density in the cerebellar cortex. It was suggested that this increase was caused directly by the presence of Bi (Gardeur et al., 1978; Monseu et al., 1976). In some other patients, however, the CT scan was normal (Lotstra et al., 1983; Mendelowitz et al., 1990; Playford et al., 1990). The presence of Bi in the intestine is visible as a radiopaque substance on a plain abdominal X-ray (Heller and Waitsmann, 1970; Emile et al., 1974; Guard et al., 1977).

Analysis of CSF revealed no significant changes. A slight increase of cells or of total protein content was seen in several patients, but in most cases the values were within normal limits (Emile et al., 1974; Bès et al., 1976; Guard et al., 1977; Von Bose and Zaudig, 1991). Probenecid tests showed a decrease in the accumulation of homovanillic acid (HVA) (but in some patients an increase) and a decrease of cyclic guanosine monophosphate (cGMP) in CSF (Cramer et al., 1978; Passouant et al., 1979; Molina et al., 1989).

1. Electroencephalography (EEG)

A characteristic EEG showed monomorphic 4–6 cycles/s activity bilaterally in temperofrontorolandic regions that was unaffected by eye opening and photostimulation in 31 of 45 intoxicated patients. Over the occipital regions α-rhythm was absent, sometimes a β-rhythm was observed. Bursts of δ- and θ-activity were seen over frontal regions in comatose patients. Although convulsions were described in many patients, no paroxysmal activity has been recorded on the EEG during these seizures. The cause of the convulsions was most probably generalized myoclonia (Supino-Viterbo et al., 1977a; Buge et al., 1981). The EEG abnormalities were remarkably similar despite the variations in clinical condition and in concentrations of Bi in blood (between 150 and 1500 µg/l) (Gastaut et al., 1975; Hazemann et al., 1975; Mabin et al., 1976; Supino-Viterbo et al., 1977b; Supino-Viterbo et al., 1977a).

An almost complete absence of sleep (average 4 min/night) on polygraphic recordings explained complaints of insomnia and somnolence in three patients. On recovery, there was a gradual return of the successive stages of nonrapid eye movement (NREM) sleep parallel to reestablishment of REM sleep. The return to a normal sleep pattern lagged behind clinical recovery (Billiard et al., 1977).

VII. TREATMENT OF Bi INTOXICATION

The most elemental therapeutic measure is to discontinue Bi intake. Evacuation of Bi from the intestine should be achieved by gastrointestinal lavage followed by high-dose charcoal and laxation. The absorption of Bi is very low ($\ll 1\%$), but the continuing presence of Bi in the intestine may increase or sustain the Bi body load. Intestinal lavage and forced diuresis have been used in France without any documented positive effect on the condition of the patient or on the Bi concentration in blood (Buge et al., 1974). In case of encephalopathy, general supportive measures were sometimes taken in the form of artificial respiration and benzodiazepines or other anticonvulsants for the suppression of myoclonia. The positive effect of this medication, however, has not been demonstrated.

In cases of nephrotoxicity the damage is usually reversible, but hemodialysis may be necessary as a temporary measure. Hemodialysis in a patient with renal failure after an overdose of CBS did not convincingly decrease the Bi body burden. The decrease in the Bi concentration in blood was 101 µg/l on the first day and 5, 40, and 15 µg/l on the following three days, with a slight increase between two sessions (Hudson and Mowat, 1989).

On the basis of an acute overdosage treatment study in mice, penicillamine has been suggested as the antidote of choice in the treatment of Bi intoxication (Basinger et al., 1983). However, D-penicillamine was shown *not* to increase the Bi elimination in patients who had recently received a course of CBS (Nwokolo and Pounder, 1990b). It was also used unsuccessfully in a patient with Bi encephalopathy (Mendelowitz et al., 1990). We feel that D-penicillamine treatment is contraindicated for Bi poisoning, since it is nephrotoxic and better alternatives are present.

The effect of chelators on the Bi concentration in the tissues was studied in rats loaded with Bi (Slikkerveer et al., 1992e). Dimercaprol (BAL), *meso*-2,3-dimercaptosuccinic acid (DMSA), and DL-2,3-dimercapto-propane-1-sulfonic acid (DMPS) reduced the Bi concentration in most organs (especially in kidney and liver) and increased elimination of Bi in urine. BAL was the only chelator effective in lowering brain Bi concentrations. In this model, treatment with EDTA resulted in *increased* brain Bi levels. BAL has been tried in five patients with Bi encephalopathy with varying effects. No significant rise in elimination of Bi was seen in one of these patients (Goule et al., 1975) after 48 h of BAL 720 mg/day. In another patient a large increase in urinary excretion and a slight rise in the Bi concentration in blood was seen after BAL 800 mg/day (Liessens et al., 1978).

In the third patient BAL was administered for 8 days in reducing doses, resulting in a 60-fold increase in the renal clearance of Bi; Bi concentrations in blood were raised during the first 4 days, but declined later. The clinical recovery was claimed to be faster than without therapy (Nogué et al., 1985). The last 2 patients received BAL in reducing doses for 10 to 13 days, starting with 300 mg/4 h, and marked improvement of the clinical condition was claimed after 2 days of therapy (Molina et al., 1989). Treatment of a patient with Bi encephalopathy with DMPS (100 mg twice daily) for 10 days resulted in a 10-fold increase in the renal clearance of Bi (Playford et al., 1990).

For nephrotoxicity the extent of renal damage as measured in serum creatinine and urea, and for neurotoxicity the clinical picture plus the concentration of Bi in blood are the best parameters to judge the severity of an intoxication. Chelator therapy is indicated on overdosage with Bi-containing medications and when high concentrations of Bi in blood have been found. The use of DMSA (30 mg/kg/day) or DMPS in the same dosage, is advised to be given orally to symptomatic patients or patients with high Bi concentrations in blood until the symptoms have disappeared. Treatment with BAL should be reserved for patients with life-threatening nephrotoxicity or severe Bi encephalopathy because of its own strong toxic potential and its painful intramuscular administration (Aposhian, 1983). The use of EDTA is contraindicated because of the risk of redistribution of the body load of Bi to the brain.

REFERENCES

Allain, P., Alquier, P., and Dumont, A.-M., *Therapie*, 35, 703–706, 1980a.
Allain, P., Chaleil, D., and Emile, J., *Therapie*, 35, 303–304, 1980b.
Aposhian, H. V., *Annu. Rev. Pharmacol. Toxicol.*, 23, 193–215, 1983.
Asato, E., Driessen, W. L., de Graaff, R. A. G., Hulsbergen, F. B., and Reedijk, J., *Inorg. Chem.*, 30, 4210–4218, 1991.
Avunduk, C., Suliman, M., Gang, D., Polakowski, N., and Eastwood, G. L., *Dig. Dis. Sci.*, 36, 431–434, 1991.
Badinand, A. and Quincy, Cl., *Therapie*, 35, 303–304, 1980.
Basinger, M. A., Jones, M. M., and McCroskey, S. A., *J. Toxicol. Clin. Toxicol.*, 20, 159–165, 1983.
Beaver, D. L. and Burr, R. E., *Arch. Pathol.*, 76, 89–94, 1963a.
Beaver, D. L. and Burr, R. E., *Am. J. Pathol.*, 42, 609–613, 1963b.
Beil, W., Bierbaum, S., and Sewing, K.-F., *Pharmacology*, 47, 135–140, 1993.
Benet, L. Z., *Scand. J. Gastroenterol.*, 26 (Suppl 185), 29–35, 1991.
Bierer, D. W., *Rev. Infect. Dis.*, 12 (Suppl. 1), S3–S8, 1990.
Billiard, M., Besset, A., Renaud, R., Baldy-Moulinier, M., and Passouant, P., *Rev. Electr. Neurophys. Clin.*, 7, 147–152, 1977.
Billon, J. P., Gernez, G., Gourdin, J. C., Martin, Ch., and Pallière, M., *Ann. Pharm. Fr.*, 34, 161–171, 1976.
Boiteau, H.-L., Cler, J.-M., Mathé, J.-F., Delobel, R., and Fève, J.-R., *Eur. J. Toxicol.*, 9, 233–239, 1976.
Boyette, D. P. and Ahoskie, A. C., *J. Pediatr.*, 28, 493–497, 1946.
Brosius, B., Blessing, J., Hopert, R., Gregor, M., and Menge, H., *Rev. Esp. Enferm. Apar. Dig.*, 108, P234, 1990.
Bruinink, A., Reiser, P., Müller, M., Gähwiler, B. H., and Zbinden, G., *Toxicol. In Vitro*, 6, 285–293, 1992.
Buge, A., Rancurel, G., Poisson, M., Gazengel, J., Dechy, H., Fressinaud, L., and Emile, J., *Ann. Med. Intern.*, 125, 877–888, 1974.
Buge, A., Hubault, A., and Rancurel, G., *Rev. Rhum. Mal. Osteo-Articulaires*, 42, 721–729, 1975.
Buge, A., Rancurel, G., and Dechy, H., *Rev. Neurol.*, 133, 401–415, 1977.
Buge, A., Rancurel, G., and Dechy, H., *Nouv. Presse Med.*, 7, 3531–3534, 1978.
Buge, A., Supino-Viterbo, V., Rancurel, G., Metzber, J., Dechy, H., and Gardeur, D., *Sem. Hop. Paris*, 55, 1466–1472, 1979.
Buge, A., Supino-Viterbo, V., Rancurel, G., and Pontes, C., *J. Neurol. Neurosurg. Ps.*, 44, 62–67, 1981.
Burns, R., Thomas, D. W., and Barron, V. J., *Br. Med. J.*, 1, 220–223, 1974.
Burr, R. E., Gotto, A. M., and Beaver, D. L., *Toxicol. Appl. Pharmacol.*, 7, 588–591, 1965.
Bès, A., Caussanel, J. P., Géraud, G., Jauzac, Ph., and Géraud, J., *Rev. Med. Toul.*, 12, 801–813, 1976.
Carruzzo, F., *Rev. Med.*, 23, 1171–1177, 1980.
Cervello, M. A., Alfaro, A., Antolin, M. A., and Penarrocha, M., *Med. Clin. Barcelona*, 79, 137–140, 1982.
Chaleil, D., Hugot, P., and Allain, P., *Therapie*, 34, 397–399, 1979a.
Chaleil, D., Allain, P., and Emile, J., *Pathol. Biol.*, 27, 417–420, 1979b.
Chaleil, D., Lefevre, F., Allain, P., and Martin, G. J., *J. Inorg. Biochem.*, 15, 213–221, 1981.
Chaleil, D., Regnault, J. P., Allain, P., Motta, R., Raynaud, G., et al., *Ann. Pharm. Fr.*, 46, 133–137, 1988.
Ciociola, A. A., Koch, K. M., McSorley, D., Smith, J. T. L., Dwyer, M. A., and Webb, D. D., *Gastroenterology*, 102, A50, 1992.
Coenegracht, J. M. and Dorleyn, M., *J. Belg. Radiol.*, 44, 485–504, 1961.

Coghill, S. B., Hopwood, D., McPherson, S., and Hislop, S., *J. Pathol.*, 139, 105–114, 1983.
Collignon, R., Bruyer, R., Rectem, D., Indekeu, P., and Laterre, E. C., *Acta Neurol. Belg.*, 79, 73–91, 1979.
Conso, F., Bourdon, R., and Gaultier, M., *Eur. J. Toxicol.*, 8, 137–141, 1975.
Cozzi, L. M., Megerian, C. A., Dugue, C., Barcello, M., Abdul-Karim, F. W., Arnold, J. E., and Maniglia, A. J., *Laryngoscope*, 102, 597–599, 1992.
Cramer, H., Renaud, B., Billiard, M., Mouret, J., and Hammers, R., *Arch. Psychiatr. Nervenkr.*, 226, 173–181, 1978.
Czerwinski, A. W. and Ginn, H. E., *Am. J. Med.*, 37, 969–975, 1964.
D'Souza, R. W. and Francis, W. R., *Pharm. Res.*, 4, S115, 1987.
De Bray, J.-M., Emile, J., Moreau, R., Daver, A., Cheguillaume, J., and Allain, P., *Nouv. Presse Med.*, 6, 1394, 1977.
De Sèze, S., Hioco, D., Mazabraud, A., and Bordier, Ph., *Rev. Rhum. Mal. Osteo-Articulaires*, 25, 623–634, 1958.
Dekker, W., Dal Monte, P. R., Bianchi Porro, G., van Bentum, N., Boekhorst, J. C., Crowe, J. P., Robinson, T. J., Thijs, O., and van Driel, A., *Scand. J. Gastroenterol.*, 21 (Suppl 122), 46–50, 1986.
Dekker, W. and Reisma, K., *Ann. Clin. Res.*, 11, 94–97, 1979.
Dérot, M., Tanret, P., and Grivaut, M., *Bul. Mem. Soc. Med. H. Paris*, 1090–1093, 1947.
Dowds, J. H., *Lancet*, ii, 1039–1040, 1939.
Dresow, B., Nielsen, P., Fischer, R., Wendel, J., Gabbe, E. E., and Heinrich, H. C., *Arch. Toxicol.*, 65, 646–650, 1991.
Dresow, B., Fischer, R., Gabbe, E. E., Wendel, J., and Heinrich, H. C., *Scand. J. Gastroenterol.*, 27, 333–336, 1992.
Drumm, D., Sherman, P., Chiasson, D., Karmali, M., and Cutz, E., *J. Pediatr.*, 113, 908–912, 1988.
Emile, J., Feve, J., Bastard, J., Clerc, P., De Bray, J.-M., Mathe, J.-F., Truelle, J.-L., Derouet, E. M., and Fressinaud, L., *Arch. Med. L'Ouest*, 6, 699–722, 1974.
Emile, J., De Bray, J.-M., Bernat, M., Morer, T., and Allain, P., *Ann. Med. Intern.*, 130, 75–80, 1979.
Erdani, S., Balzarini, M., Taglioretti, D., Romussi, M., and Valentini, R., *Br. J. Radiol.*, 37, 311–314, 1964.
Escourolle, R., Bourdon, R., Galli, A., Galle, P., Jaudon, M. C., Hauw, J. J., and Gray, F., *Rev. Neurol.*, 133, 153–163, 1977.
Ferrer, X., Raymond, J. M., Begaud, B., Raptopoulo, F., Languency, A., and Julien, J., *Presse Med.*, 13, 1749–1750, 1984.
Fowler, B. A. and Goyer, R. A., *J. Histochem. Cytochem.*, 23, 722–726, 1975.
Friedland, R. P., Lerner, A. J., Hedera, P., and Brass, E. P., *Clin. Neuropharmacol.*, 16, 173–176, 1993.
Froomes, P. R. A., Wan, A. T., Harrison, P. M., and McClean, A. J., *Clin. Chem.*, 34, 382–384, 1988.
Froomes, P. R. A., Wan, A. T., Keech, A. C., McNeil, J. J., and McLean, A. J., *Eur. J. Clin. Pharmacol.*, 37, 533–536, 1989.
Funck, J. A., Schnaare, R. L., Schwartz, J. B., and Sugita, E. T., *Drug Dev. Ind. Pharmacol.*, 17, 1957–1970, 1991.
Garcia, J. L., Penin, F., Weber, M., Regent, M. Ch., and Cuny, G., *Rev. Rhum. Mal. Osteo-Articulaires*, 47, 275–280, 1980.
Gardeur, D., Buge, A., Rancurel, G., Dechy, H., and Metzger, J., *J. Comput. Assist. Tomogr.*, 2, 436–438, 1978.
Gastaut, J. L., Tassarini, C. A., Terzano, G., and Picornell, I., *Rev. Electr. Neurophys. Clin.*, 5, 295–302, 1975.
Gaucher, A., Netter, P., Faure, G., Pourel, J., Hutin, M. F., and Burnel, D., *Rev. Rhum. Mal. Osteo-Articulaires*, 47, 31–35, 1980.
Gavey, C. J., Szeto, M.-L., Nwokolo, C. U., Sercombe, J., and Pounder, R. E., *Aliment. Pharmacol. Ther.*, 3, 21–28, 1989.
Ghadially, F. N., *Crit. Rev. Toxicol.*, 6, 303–350, 1979.
Goodman, L., *Br. Med. J.*, 1, 978–979, 1948.
Goule, J. P., Husson, A., Fondimare, A., Rapoport, F., Lebreton, M., and Lajarige, V., *Nouv. Presse Med.*, 4, 1366, 1975.
Grandjean, E. M., Ducommun, E., Gauthier, G., and Courvoisier, B., *Schweiz. Med. Wochenschr.*, 106, 1006–1011, 1976.
Gregus, Z. and Klaassen, C. D., *Toxicol. Appl. Pharmacol.*, 85, 24–38, 1986.
Gryboski, J. D. and Gotoff, S. P., *N. Engl. J. Med.*, 265, 1289–1291, 1961.
Guard, O., Soichot, P., Dumas, R., Besancenot, J.-B., and Gaudet, M., *J. Med. Lyon*, 58, 403–416, 1977.
Gurnani, N., Sharma, A., and Talukder, G., *Biol. Trace Elem. Res.*, 37, 281–292, 1993.
Gyurasics, A., Koszorús, L., Varga, F., and Gregus, Z., *Biochem. Pharmacol.*, 44, 1275–1281, 1992.
Hall, R. J. and Farber, T., *J.A.O.A.C.*, 55, 639–642, 1972.
Hamilton, E. I., Minski, M. J., and Cleary, J. J., *Sci. Total Environ.*, 1, 341–374, 1972.
Hartemann, P. and Maugras, M., *Nouv. Presse Med.*, 5, 1001, 1976.
Hazemann, P., Rebelo, F., and Landau, J., *Rev. Electr. Neurophys. Clin.*, 5, 291–294, 1975.
Heller, R. H. and Waitsmann, E. S., *J. Pediatr.*, 76, 637–638, 1970.
Hespe, W., Staal, H. J. M., and Hall, D. W. R., *Lancet*, 2, 1258, 1988.
Hillemand, P. and Cottet, J., *Bull. Acad. Nat. Med. Paris*, 160, 274–278, 1976.
Hillemand, P., Pallière, M., Laquais, B., and Bouvet, P., *Sem. Hop. Paris*, 53, 1663–1669, 1977.
Hudson, M. and Mowat, N. A. G., *Br. Med. J.*, 299, 159, 1989.
Hursh, J. B. and Brown, C., *Proc. Soc. Exp. Biol. Med.*, 131, 116–120, 1969.
Huwez, F., Pall, A., Lyons, D., and Stewart, M. J., *Lancet*, 340, 1298, 1992.
Indekeu, P. and Laterre, Ch., *Acta Clin. Belg.*, 33, 350–362, 1978.
James, J. J., *Calif. Med.*, 109, 317–319, 1968.
Jeanrot, R., Chaleil, D., Allain, P., Papillon, C., Raynaud, G., and Bouvet, P., *Toxicol. Eur. Res.*, 4, 181–185, 1982.
Karelitz, S. and Freedman, A. D., *Pediatrics*, 8, 772–777, 1951.
Kendel, K., Kabiri, R., and Schäffer, S., *Dtsch. Med. Wochenschr.* 118, 221–224, 1993.
Kolmer, J. A., Brown, H., and Rule, A. M., *Am. J. Syph. Gonorrhea Vener. Dis.*, 23, 7–40, 1939.
Kondo, Y., Satoh, M., Imura, N., and Akimoto, M., *Anticancer Res.*, 12, 2303–2308, 1992.
Krüger, G., Thomas, D. J., Weinhardt, F., and Hoyer, S., *Lancet*, 2, 485–487, 1976.

Kulchar, G. V. and Reynolds, W. J., *J. Am. Med. Assoc.*, 120, 343–348, 1942.
Lagier, G., *Therapie*, 35, 315–317, 1980.
Lagueny, A., Vallat, J.-M., and Julien, J., *Nouv. Presse Med.*, 5, 2252–2253, 1976.
Lazarini, F., *Bull. Bism. Inst.*, 32, 3–8, 1981.
Lechat, P. and Kisch, R., *Gastroenterol. Clin. Biol.*, 10, 562–569, 1986.
Lechat, P., Majoie, B., Levillain, R., Cluzan, R., and Deleau, D., *Therapie*, 19, 551–556, 1964.
Lechat, P., Pallière, M., Gernez, G., Dechy, H., and Letteron, N., *Ann. Pharm. Fr.*, 34, 179–182, 1976.
Lee, S. P., Lim, T. H., Pybus, J., and Clarke, A. C., *Clin. Exp. Pharmacol P.*, 7, 319–324, 1980.
Lee, S. P., *Res. Commun. Chem. Pathol.*, 34, 359–364, 1981.
Lhermitte, F., Mallecourt, J., Chedru, F., Kuperwasser, B., and Glaser, P., *Nouv. Presse Med.*, 7, 2170–2171, 1978.
Liessens, J. L., Monstrey, J., Eeckhout, vanden, E., Djudzman, R., and Martin, J. J., *Acta Neurol. Belg.*, 78, 301–309, 1978.
Loiseau, P., Henry, P., Jallon, P., and Legroux, M., *J. Neurol. Sci.*, 27, 133–143, 1976.
Loseke, N., Retif, J., Coune, A., Soebert, A., and Molle, L., *Acta Gastroenterol. Belg.*, 41, 72–80, 1978.
Lotstra, F., Soebert, A., Linkowski, P., and Mendlewicz, J., *Acta Neurol. Belg.*, 83, 135–141, 1983.
Lowe, D. J., *Med. J. Aust.*, 2, 664–666, 1974.
Mabille, J. P., Gaudet, M., and Charpin, J. F., *Ann. Radiol.*, 23, 515–517, 1980.
Mabin, D., Goas, J.-Y., Bedou, G., Besson, G., and Tuset, M.-C., *Sem. Hop. Paris*, 52, 109–116, 1976.
Madaus, S., Schulte-Frolinde, E., Scherer, C., Kämmereit, A., Schusdziarra, V., and Classen, M., *Aliment. Pharmacol. Ther.*, 6, 241–249, 1992.
Marshall, B. J., Armstrong, J. A., Francis, G. J., Nokes, N. T., and Wee, S. H., *Digestion*, 37 (Suppl. 2), 16–30, 1987.
Martin-Bouyer, G., *Therapie*, 31, 683–702, 1976.
Martin-Bouyer, G., Barin, C., Beugnet, A., Cordier, J., and Guerbois, H., *Gastroenterol. Clin. Biol.*, 2, 349–356, 1978.
Martin-Bouyer, G., Foulon, B., Guerbois, H., and Barin, C., *Therapie*, 35, 307–313, 1980.
Martin-Bouyer, G., Nevot, P., Singlas, E., Foulon, B., Philippon, A., Duret, M., Frevet, J. H., Clavel, J. P., and Simon, P., *Therapie*, 36, 483–488, 1981a.
Martin-Bouyer, G., Foulon, B., Guerbois, H., and Barin, C., *Clin. Toxicol.*, 18, 1277–1283, 1981b.
Mattle, H., Henn, V., and Baumgartner, G., *Schweiz. Med. Wochenschr.*, 112, 1308–1311, 1982.
McClendon, S. J., *Am. J. Dis. Child.*, 61, 339–341, 1941.
McLean, A. J., Froomes, P., McNeil, J. J., Wan, A. T., and Harrison, P. M., *Clin. Pharmacol. Ther.*, 43, 186, 1988.
Mendelowitz, P. C., Hoffman, R. S., and Weber, S., *Ann. Intern. Med.*, 112, 140–141, 1990.
Minagawa, A., Sawada, K., and Suzuki, T., *Anal. Chim. Acta*, 278, 287–292, 1993.
Moindrot, J., Evreux, J. C., Savet, J. F., and Aimard, G., *Lyon Med.*, 238, 537–541, 1977.
Mol, de, J., Loseke, N., and Leleux, C., *Acta Psychiatr. Belg.*, 79, 185–197, 1979.
Molina, J. A., Calandre, L., and Bermejo, F., *Acta Neurol. Scand.*, 79, 200–203, 1989.
Monseu, G., Struelens, M., and Roland, M., *Acta Neurol. Belg.*, 76, 301–308, 1976.
Morrow, A. W., *Med. J. Aust.*, 1, 192, 1973.
Mueller, R. I., *Zbl. Hyg.*, 189, 117–124, 1989.
Murray, J. R., *Med. J. Aust.*, 1, 522, 1979.
Müller, H.-A. and Ramin, von, D., *Beitr. Pathol. Anat. Allg. Pathol.*, 128, 445–467, 1963.
Naganuma, A., Satoh, M., Koyama, Y., and Imura, N., *Toxicol. Lett.*, 24, 203–207, 1985.
Naganuma, A., Satoh, M., and Imura, N., *Cancer Res.*, 47, 983–987, 1987.
Noach, L. A., Eekhof, J. L. A., Bour, L. J., Posthumus Meyjes, F. E., Tytgat, G. N. J., and Ongerboer de Visser, B. W., *Hum. Exp. Toxicol.*, 14, 349–355, 1995.
Nogué, S., Mas, A., Parés, A., Nadal, P., Bertrán, A., Torra, M., Bachs, M., and Blesa, R., *Med. Clin. Barcelona*, 84, 530–532, 1985.
Nomura, A., Stemmermann, G. N., Chyou, P. H., Kato, I., Perez-Perez, G. I., and Blaser, M. J., *N. Engl. J. Med.*, 325, 1132–1136, 1991.
Nwokolo, C. U., Gavey, C. J., Smith, J. T. L., and Pounder, R. E., *Aliment. Pharmacol. Ther.*, 3, 29–39, 1989.
Nwokolo, C. U., Mistry, P., and Pounder, R. E., *Aliment. Pharmacol. Ther.*, 4, 163–169, 1990a.
Nwokolo, C. U., Prewett, E. J., Sawyerr, Af. M., and Pounder, R. E., *Eur. J. Gastroenterol. Hepatol.*, 2, 433–435, 1990b.
Nwokolo, C. U. and Pounder, R. E., *Br. J. Pharmacol.*, 30, 648–650, 1990c.
Nwokolo, C. U., Prewett, E. J., Sawyerr, Af. M., Hudson, M., and Pounder, R. E., *Gastroenterology*, 101, 889–894, 1991.
Nwokolo, C. U., Lewin, J. F., Hudson, M., and Pounder, R. E., *Gastroenterology*, 102, 163–167, 1992.
Nwokolo, C. U., Fitzpatrick, J. D., Paul, R., Dyal, R., Smits, B. J., and Loft, D. E., *Aliment. Pharmacol. Ther.*, 8, 45–53, 1994.
O'Brien, D., *Am. J. Dis. Child.*, 97, 384–386, 1959.
Olive, G., *Therapie*, 35, 305–306, 1980.
Parsonnet, J., Friedman, G. D., Vandersteen, D. P., Chang, Y., Vogelman, J. H., Orentreich, N., and Sibley, R. K., *N. Engl. J. Med.*, 325, 1127–1131, 1991.
Passouant, P., Billiard, M., Besset, A., and Renaud, B., *Int. J. Neurol.*, 12, 199–204, 1979.
Pieri, F. and Wegmann, R., *Cell Mol. Biol.*, 27, 57–60, 1981.
Playford, R. J., Matthews, C. H., Campbell, M. J., Delves, H. T., Hla, K. K., Hodgson, H. J. F., and Calam, J., *Gut*, 31, 359–360, 1990.

Prewett, E. J., Nwokolo, C. U., Hudson, M., Sawyerr, A. M., Fraser, A., and Pounder, R. E., *Aliment. Pharmacol. Ther.*, 5, 481–490, 1991.
Raedsch, R., Walter-Sack, I., Weber, E., and Blessing, J., *Klin. Wochenschr.*, 68, 488, 1990.
Randall, R. E., Osterhoff, R. J., Bakerman, S., and Setter, J. G., *Ann. Intern. Med.*, 77, 481–482, 1972.
Rao, N. and Feldman, S., *Pharm. Res.*, 7, 188–191, 1990a.
Rao, N. and Feldman, S., *Pharm. Res.*, 7, 237–241, 1990b.
Rauws, E. A. J., Langenberg, W., Houthoff, H. J., Zanen, H. C., and Tytgat, G. N. J., *Gastroenterology*, 94, 33–40, 1988.
Reinicke, C., in *Dukes Side Effects of Drugs*, Annu. 6, Dukes, Ed., Excerpta Medica, Amsterdam, 217–221, 1982.
Relat, C.-J. F., Barbera, A. M., Matos, M. J. A., Pousa, L. S., and Vilalta, M. J. L., *Rev. Neurol.*, 7, 289–296, 1979.
Revault, P., Lejeune, E., and Pellet, M., *Rev. Lyon Med.*, 7, 827–837, 1958.
Ribadeau-Dumas, J.-L., Lechevalier, B., Bretau, M., and Allain, Y., *Nouv. Presse Med.*, 7, 4021–4025, 1978.
Richardson, R. B. and Wills, H. H., *Radiat. Prot. Doosimet.*, 41, 169–172, 1992.
Robertson, J. F., *Med. J. Aust.*, 1, 887–888, 1974.
Ross, J. F., Sahenk, Z., Hyser, C., Mendell, J. R., and Alden, C. L., *Neurotoxicology*, 9, 581–586, 1988.
Ross, J. F., Broadwell, R. D., Poston, M. R., and Lawhorn, G. T., *Toxicol. Appl. Pharmacol.*, 124, 191–200, 1994.
Russ, G. A., Bigler, R. E., Tilbury, R. S., Woodard, H. Q., and Laughlin, J. S., *Radiat. Res.*, 63, 443–454, 1975.
Ryder, S. D., Walker, R. J., Jones, H., and Rhodes, J. M., *Aliment. Pharmacol. Ther.*, 4, 333–338, 1990.
Sany, J., Bataille, R., Rosenberg, F., Antoine, L., Serre, H., Monstrey, J., and Labauge, R., *Rev. Rhum. Mal. Osteo-Articulaires*, 45, 729–732, 1978.
Schweiberer, G. and Vogelsang, W., *Nieren und Hochdruckkrankheiten*, 22, 381–385, 1993.
Slikkerveer, A., Helmich, R. B., Edelbroek, P. M., van der Voet, G. B., and de Wolff, F. A., *Clin. Chim. Acta*, 201, 17–26, 1991.
Slikkerveer, A., Helmich, R. B., van der Voet, G. B., and De Wolff, F. A., *J. Pharm. Sci.*, 84, 512–515, 1995.
Slikkerveer, A., Helmich, R. B., van der Voet, G. B., and De Wolff, F. A., Effect of citrate on the intestinal absorption of bismuth compounds. In Bismuth: Biokinetics, Toxicity and Experimental Therapy of Overdosage. Thesis pp. 89–102, 1992a.
Slikkerveer, A., Terpstra, I. J., and De Wolff, F. A., Citrate enhances intestinal absorption of bismuth in man. In Bismuth: Biokinetics, Toxicity and Experimental Therapy of Overdosage. Thesis pp. 103–116, 1992b.
Slikkerveer, A., Helmich, R. B., and De Wolff, F. A., Comparative tissue distribution of bismuth: effects of compound and dose. In Bismuth: Biokinetics, Toxicity and Experimental Therapy of Overdosage. Thesis pp. 117–132, 1992c.
Slikkerveer, A., Schroijen, V. J. H. M., De Wolff, F. A., and Bruijn, J. A., Characterization of renal damage caused by bismuth overdosage. In Bismuth: Biokinetics Toxicity and Experimental Treatment of Overdosage. Thesis pp. 133–148, 1992d.
Slikkerveer, A., Jong, H. B., Helmich, R. B., and De Wolff, F. A., *J. Lab. Clin. Med.*, 119, 529–537, 1992e.
Slikkerveer, A., Helmich, R. B., and De Wolff, F. A., *Clin. Chem.*, 39, 800–803, 1993.
Slikkerveer, A., Engelbrecht, M. R. W., de Heer, E., De Wolff, F. A., and Bruijn, J. A., *Exp. Nephrol.*, submitted.
Sollman, T., Cole, H. N., and Henderson, K., *Am. J. Syph. Gonorrhea Vener. Dis.*, 22, 555–583, 1938.
Sollman, T. and Seifter, J., *J. Pharmacol. Exp. Ther.*, 67, 17–49, 1939.
Sollman, T. and Seifter, J., *J. Pharmacol. Exp. Ther.*, 74, 134–154, 1942.
Srivastava, E. D., Swift, G. L., Wilkinson, S., Williams, G. T., Evans, B. K., and Rhodes, J. R., *Aliment. Pharmacol. Ther.*, 4, 577–581, 1990.
Stables, R., Campbell, C. J., Clayton, N. M., Clitherow, J. W., Grinham, C. J., McColm, A. A., McLaren, A., and Trevethick, M. A., *Aliment. Pharmacol. Ther.*, 7, 237–246, 1993.
Stahl, J. P., Gaillat, J., Leverve, X., Carpentier, F., Guignier, M., and Micoud, M., *Nouv. Presse Med.*, 11, 3856, 1982.
Supino-Viterbo, V., Sicard, C., Risvegliato, M., Rancurel, G., and Buge, A., *J. Neurol. Neurosurg. Ps.*, 40, 748–752, 1977a.
Supino-Viterbo, V., Sicard, C., and Cathala, H.-P., *Rev. Electr. Neurophys. Clin.*, 7, 139–146, 1977b.
Szymanska, J. A., Mogilnicka, E. M., and Kaszper, B. W., *Biochem. Pharmacol.*, 26, 257–258, 1977.
Szymanska, J. A. and Piotrowski, J. K., *Biochem. Pharmacol.*, 29, 2913–2918, 1980.
Szymanska, J. A., Zychowics, M., Zelazowski, A. J., and Piotrowski, J. K., *Arch. Toxicol.*, 40, 131–141, 1978.
Szymanska, J. A. and Zelazowski, A. J., *Biochem. Pharmacol.*, 26, 139–146, 1979.
Taylor, E. G. and Klenerman, P., *Lancet*, 335, 670–671, 1990.
Thomas, D. W., Hartley, T. F., Coyle, P., and Sobecki, S., in *Clinical Chemistry and Clinical Toxicology of Metals*, Brown, S.S. and Savory, J., Eds., Elsevier/North Holland, Biomedical Press, Amsterdam, 293–296, 1977.
Thomas, D. W., Sobecki, S., Hartley, T. F., Coyle, P., and Alp, M. H., in *Chemical Toxicology and Clinical Chemistry of Metals*, Brown, S.S. and Savory, J., Eds., Academic Press, London, 391, 1983.
Treiber, G., Gladziwa, U., Ittel, T. H., Walker, S., Schweinsberg, F., and Klotz, U., *Aliment. Pharmacol. Ther.*, 5, 491–502, 1991.
Treiber, G., Walker, S., and Klotz, U., *Clin. Pharmacol. Ther.*, 55, 486–491, 1994.
Urizar, R. and Vernier, R. L., *J. Am. Med. Assoc.*, 198, 207–209, 1966.
Van Der Werff, J. Th., *Acta Radiol.*, Suppl. 243, 2–49, 1965.
Vidailhet, M., Le Thi Huong Du, Wechsler, B., and Godeau, P., *Presse Med.*, 16, 1054, 1987.
Vienet, R., Bouvet, P., and Istin, M., *Int. J. Appl. Radiat. Is.*, 34, 747–753, 1983.
Volkheimer, G., *Adv. Pharmacol. Chemother.*, 14, 163–187, 1977.
Von Bose, M. J. and Zaudig, M., *Br. J. Psychiatr.*, 158, 278–280, 1991.

Von Osswald, H., *Arzneimittelforschung*, 18, 1064–1065, 1968.
Wagstaff, A. J., Benfield, P., and Monk, J. P., *Drugs*, 36, 132–157, 1988.
Warren, H. V., Horsky, S. J., and Gould, C. E., *Sci. Total Environ.*, 29, 163–169, 1983.
Wieriks, J., Hespe, W., Jaitly, K. D., Koekkoek, P. H., and Lavy, U., *Scand. J. Gastroenterol.*, 17 (Suppl. 80), 11–16, 1982.
Williams, D. R., *J. Inorg. Nucl. Chem.*, 39, 711–714, 1977.
Winship, K. A., *Adv. Drug React. Acute Poison Rev.*, 2, 103–121, 1983.
Woods, J. S. and Fowler, B. A., *Toxicol. Appl. Pharmacol.*, 90, 274–283, 1987.
Zala, L., Hunziker, T., and Braathen, L. R., *Dermatology*, 187, 288–289, 1993.
Zarranz, J. J., Forcadas, I., Larracoechea, J., and Caldera, A., *Med. Clin. Barcelona*, 68, 78–80, 1977.
Zidenberg-Cherr, S., Parks, N. J., and Keen, C. L., *Radiat. Res.*, 111, 119–129, 1987.
Zidenberg-Cherr, S., Clegg, M. S., Parko, N. J., and Keen, C. L., *Biol. Trace Elem. Res.*, 19, 185–194, 1989.

Chapter 28

Human Exposure to Lithium, Thallium, Antimony, Gold, and Platinum

Gijsbert B. van der Voet and Frederik A. de Wolff

I. LITHIUM

A. CHEMISTRY, USE, AND EXPOSURE

Lithium (Li) is found in group 1A of the periodic table. Its atomic number is 3, and its relative molecular mass is 6.94 Da. It is the lightest metallic element, sharing its group with sodium and potassium. It is less reactive than the other two, and it is widely distributed in nature. No physiological function has been reported for Li. Industrially, it is used in the space industry, in batteries, and with photo materials. In the past, Li was used in the treatment of podagra. As yet, Li succinate is medically used in dermatology, but the main medical use of Li salts is in psychiatry to counteract mood changes; Li carbonate and Li citrate are now widely accepted as the treatment of choice for acute mania and for the prophylactic treatment of recurrent bipolar affective disorders. Intoxications related to Li exposure are mainly reported for this group of patients.

B. POISONING

Acute intoxication can occur in the initial phase in a course of therapy, but also at any point of time during long-lasting treatment or after an acute overdose. At plasma levels between 1.5 and 2.5 mmol/l, signs of toxicity include anorexia, dry mouth, nausea, vomiting, diarrhea, tremor of the hands, faintness of musculature, thirst, leucocytosis, and concentration and memory disturbances (especially with older people). These phenomena are often seen in the initial phase of a course of treatment and usually disappear when treatment continues, except with the tremor of the hands. In elderly people, reversible delirious conditions can occur with confusion, restlessness, and ataxia.

At plasma levels above 2.5 mmol/l, serious toxic symptoms occur; fasciculations, muscle contractions, hyperreflexia and hypertonia, drowsiness, confusion, sometimes epileptiform insults, hypotension, coma, collapse. Independent of the plasma level, changes can occur in the ECG and in the EEC, with symptoms such as polyuria and polydipsia, seldom nephrogenic diabetes insipidus, ulcers of the leg, enhancement of acne and psoriasis, transient hyperglycemia, pruritus, and a metal taste. In about 5% of the cases, a (usually reversible) hypothyroidia develops.

C. BIOKINETICS AND MECHANISM

Nearly complete Li absorption occurs from the gastrointestinal tract, with peak therapeutic levels appearing 30 min. to 2 to 3 h postingestion. However, clinically significant delayed absorption may develop in an overdose situation up to 72 h postingestion. Li is not bound to plasma proteins; it is associated with red blood cells, and the volume of distribution of Li is equivalent to the total body water (0.7 L/kg). The plasma half-life (in healthy volunteers) shows a considerable variability: from 5 to 40 h, with most values between 15 and 30 h, it depends on the duration of treatment as well as on kidney function and age. Li is not metabolized in the liver or anywhere else. Glomerular filtration eliminates the entire dose. The proximal tubule, however, reabsorbs 70 to 80% of the filtered Li together with sodium. This reabsorption of Li is affected by the sodium balance. Hyponatremia reduces Li clearance, while an alkaline urine increases Li clearance.

Mechanistically, the Li effect results from its substitution for body cations, e.g., sodium and potassium, resulting in multisystemic actions. A partial substitution for normal cations causes changes in ion exchange and transfer in cellular processes. Incorporation of Li into membrane structures may alter responses to hormones and the coupling of energy processes.

Many interactions of Li with other drugs exist; in fact, drugs reducing the renal clearance of Li (diuretics) can produce the most serious adverse effects. When used in combination with psychiatric drugs affecting mood, the therapeutic potential of Li may be enhanced. Among the factors that may modify Li toxicity and kinetics are the type of the poisoning, the presence of the underlying disease, and renal impairment.

D. DIAGNOSIS AND TREATMENT

Li is a drug with a very narrow therapeutic range. Formerly, a 0.6 to 1.5 mmol/l plasma level was listed as the general therapeutic range for prophylactic Li use. Later, however, 0.7 to 1.2 mmol/l was used as the adequate maintenance blood level. At present, the following interpretations of serum Li levels are considered appropriate: levels between 1.2 to 1.5 mmol/l require care to avoid toxicity; 1.5–2.5 mmol/l are levels at which mild toxicity is usually seen; over 3.5 mmol/l, severe symptoms such as seizures and coma are more common. Clearly, Li has a narrow margin of safety which is much less effective at half the optimal plasma level and toxic at twice that level. Decontamination measures (emesis or lavage) may be effective more than several hours postingestion, due to possible delays in absorption of overdose or sustained release tablets. No specific antidotes exist. Hemodialysis is indicated above 3.5 mmol/l, which significantly increases Li clearance, with Li extraction higher from serum than from whole blood or red blood cells. No general and rigid indication for hemodialysis can be set, but the need for hemodialysis should be based on clinical and kinetic data determined during the 12 h following admission. Supportive care is required.

II. THALLIUM

A. CHEMISTRY, USE, AND EXPOSURE

The element thallium (Tl) is found in group IIIA of the periodic table below aluminum and gallium and between mercury (group IIb) and lead (group IVa). Its atomic number is 81, and its relative atomic mass is 204.39. As an element, it has a crystalline (tetragonal) structure and a blue-white color. It occurs as mono- and trivalent compounds such as Tl sulfate, -nitrate, -acetate, Tl(I) oxide, Tl(I) carbonate, and Tl(I) sulfide. In nature, Tl occurs in various minerals, in magmatic and sedimentary rock, and, consequently, in soil, water, and in the air. Industrially, Tl is used in lenses and prisms for the transmission of long wavelength radiation, as an alloy with mercury in low temperature thermometers, and in the preparation of high density liquids. Historically, Tl compounds were used medically in the treatment of venereal disease, ringworm, gout, dysenteria, and tuberculosis. Low dosage Tl was used earlier to remove hair for cosmetic purposes. Tl(I) sulfate has been used on a large scale as a rodenticide and as a pesticide. Exposure to Tl as a rodenticide still dominates the human experience as compared to exposure to Tl compounds processed and used for other purposes.

B. POISONING

Tl compounds are extremely toxic. Because they are tasteless, there are many accounts of their criminal use. Acute poisoning is the type prominently reported in the literature. The fatal dose of Tl is approximately 0.2 to 1 g of absorbed Tl. Large doses cause acute gastrointestinal symptoms (diarrhea, nausea and vomiting, abdominal pain) with neurological effects hours to days later, including burning thirst, insomnia, psychologic change, neuralgic pains especially in the soles, combined motor and sensory neuropathy, and neuritis retrobulbaris tachycardia with hypertension. Lethargia, aphasia, tremors, choreatic movements, convulsions, fear, confusion, drowsiness to deep coma may also occur. After about

2 weeks, late gastrointestinal symptoms such as obstipation may be caused by reduced motor activity of the intestinal tract. Further cardiovascular symptoms are hypertension, tachycardia, and dysrhythmia.

Alopecia occurring about 2 weeks post-ingestion, is the most characteristic sign of Tl poisoning. It should be realized that in the first days after exposure in the roots of the hairs a black pigment can already be observed as an indicator of Tl poisoning. Atrophic changes of the skin can occur. White, transversal "lunula" bands on the nails of fingers and toes can be seen. Mental disturbances can occur in a rather late phase (2 to 4 weeks) including psychosis, paranoia, and hallucinations. Less frequent are kidney and liver damage. As to chronic poisoning, the integumental changes, alopecia and atrophic changes in the skin, as well as gastrointestinal changes, are also common. Less often, renal damage and functional changes of the endocrine system can occur.

C. BIOKINETICS AND MECHANISM

Tl compounds are rapidly absorbed following ingestion, inhalation, or skin contact and may be complete after ingestion. Tl is widely distributed in the tissues, mainly intracellularly, the highest concentrations of which are found in the kidneys, (followed by heart, brain, skin, liver, bones, and muscles). A high volume of distribution is reported (e.g., 3.6 L/kg). Tl is mainly excreted by the kidneys and the intestines, perhaps because of enterohepatic recirculation, as well as in small part via hair and into milk. It can cross the placental barrier. Tl is soluble at physiological pH and does not form complexes in bone, unlike arsenic and lead. Elimination of Tl after overdose occurs mainly via the feces and, less frequently, in urine.

The mechanism of toxicity is based on (1) substitution for potassium with a higher affinity for the sodium/potassium ATPase pump, causing membrane depolymerization and (2) binding to sulfhydryl-groups interfering with sulfhydryl containing enzymes, especially in the mitochondrial respiratory chain.

D. DIAGNOSIS AND THERAPY

Exposure is usually monitored by measuring Tl in urine using a method for atomic absorption spectrometry. In severe poisoning episodes, though, blood, serum, or plasma may also be taken for analysis. Chelating agents such as dimercaprol (BAL) and edetic acid (EDTA) cause severe redistribution phenomena and are not effective for treatment. The recent use of the oral chelator dimercaptosuccinic acid (DMSA) may offer some advantages.

Treatment includes gastrointestinal decontamination. Activated charcoal may be effective (even in cases of delayed presentation). Slow intravenous potassium chloride infusions enhance Tl diuresis. The use of hemoperfusion and hemodialysis is disputed with combinations of these methods with forced diuresis currently being explored. Oral Prussian blue (potassium ferric cyanoferrate II) prevents enterohepatic recirculation of Tl from the gut and increases fecal elimination.

III. ANTIMONY

A. CHEMISTRY, USE, AND EXPOSURE

The element antimony (Sb, stibium) is found in group VA of the periodic table directly below arsenic, with which it shares many characteristics. Its atomic number is 51, and its relative molecular mass is 121.76. As an element, it is a crystalline (hexagonal) silver-white metal. It occurs in tri- and pentavalent compounds. In nature, Sb is associated with sulfur as stibnite, and it often occurs in ores associated with arsenic. Industrially, Sb is a common constituent of metal alloys, for example, with lead and copper. Stibine gas (i.e., arsine gas) appears when alloys are treated with acid. Sb is used in the manufacture of paints, ceramics, glass, solders, typemetal, explosives, batteries, bearing metals, and semiconductors. Sb compounds are also used for flame proofing and as abrasives. Historically, Sb compounds were known as expectorants and emetics (the Romans called their winebeakers manufactured from Sb alloys "calices vomitorum"). More recently, Sb compounds are used medically as antihelminthic (schistosomiasis) and antiprotozoic (leishmaniasis) drugs. The use of trivalent Sb compounds is greatly banned because they are more toxic than pentavalent compounds. Exposure to Sb compounds can take place in the mining and extraction industries. Toxicity is often reported in relation to the medical use of Sb compounds.

B. POISONING

The fatal dose for Sb compounds after ingestion is 100 to 200 mg. Fatalities from Sb poisoning are rare. Acute systemic exposure to Sb compounds causes loss of hair, dry scaly skin, and weight loss. Damage to the heart, liver, and kidneys can occur, and death from myocardial failure may follow. Chronic exposure during the industrial processing of antimony compounds may cause local toxicity, affecting the skin, mucus membranes, and lungs. Pruritic papules progressing to pustular skin eruptions, "antimony spots", are sometimes seen in persons working with Sb and Sb salts. These eruptions are transient and mainly affect skin areas exposed to heat and those areas where sweating occurs. Pneumoconiosis is regularly reported from the Sb industry and is commonly regarded as a relatively benign condition; nevertheless, chronic respiratory effects have been reported in a number of studies.

The systemic toxicity of trivalent compounds Sb(III) sulfide and Sb(III) oxide is significantly less than the pentavalent compounds Sb(V) sulfide and Sb(V) oxide. This is particularly important for the medically used Sb compounds. The use of trivalent compounds, e.g., Sb(III) potassiumtartrate, in the treatment of helminthic infections is abandoned due to toxicity. Cardiovascular effects as a result of parasitic disease are reported.

Stibine shows an analogous but essentially less fulminant toxicity than arsine; symptoms include nausea, profuse vomiting, abdominal and low back pain, headache, hemolytic anemia, myoglobinuria, hematuria, and renal failure. Stibine is mentioned in sudden infant death syndrome (SIDS).

C. BIOKINETICS AND MECHANISM

The intestinal absorption of trivalent Sb is lower than the pentavalent compound. Trivalent Sb is mainly bound to erythrocytes; therefore, the plasma level is low.

After acute or chronic oral or parenteral exposure to Sb, the highest concentrations are found in the thyroid, adrenals, and kidneys. Unlike arsenic, inorganic trivalent antimony is not methylated *in vivo*. It is excreted in the bile after conjugation with gluthathione as well as in urine. A significant proportion of Sb excreted in bile undergoes enterohepatic circulation. Renal excretion is slow while the excretion of pentavalent Sb, which is not bound to erythrocytes, is much faster.

Mechanistically, Sb has properties and biological activities similar to arsenic, although it is considerably less toxic. It shares its affinity for sulfhydrylgroups on many enzymes.

D. DIAGNOSIS AND THERAPY

Occupational exposure is monitored by measuring Sb in urine in suspected toxicity or investigation of antimony therapeutics; blood concentrations may also be determined using a method for atomic absorption spectrometry. In overdoses, decontamination measures include gastric lavage or emesis. As an antidote, dimercaprol may be used; as to stibine intoxication, the role of dimercaprol is not clear. Treatment may be modeled analogous to arsenic treatment. Dysrythmia should be monitored.

IV. GOLD

A. CHEMISTRY, USE, AND EXPOSURE

Gold (Au, aureum) is found in group 1B of the periodic table below copper and silver. Its atomic number is 79, and its relative atomic mass is 197.0. It occurs in free form in nature, sometimes in larger pieces, i.e., "nuggets", but more often as fine granules distributed in sandstone or rock. Au only dissolves in a mixture of hydrochloric acid and nitric acid, forming complex gold-chloride ions. Au can be extracted from its matrix using mercury, forming an amalgam. Since the pure metal is too soft, alloys are made for the fabrication of jewelry, utensils, and coins. Au has a long history in science and culture. Historically, Au has been used for centuries for treatment of many diseases and later on rheumatic diseases and tumor therapy as well as for gold fillings in the dental medicine. Au compounds show antibacterial activity in the treatment of tuberculosis and syphilis. Au compounds are given orally (auranofine) or intramuscularly (aurothiomalate acid, aurothioglucose) to rheumatoid patients to delay the disease processes. The therapeutic use of Au in the management of selected patients with rheumatoid arthritis has made gold one of the more common causes of metal toxicity.

B. POISONING

The therapeutic use of Au in rheumatoid arthritis must be carefully monitored. In every type of goldtherapy, a regular bloodcell- and urine-control should be performed, and the skin should be checked. Gastrointestinal symptoms, i.e., diarrhea, nausea, and vomiting may occur with oral Au preparations. Mucocutaneous effects are relatively frequent, including allergic reactions, dermatitis, stomatitis, and chrysiasis. Further, nephrotic syndrome with light proteïnuria and hematological abnormalities (megacaryocytopoiesis), seldom bone marrow suppression (leukopenia, thrombocytopenea and anemia) may also occur. Liver toxicity may include cholestatic jaundice and hepatitis.

C. BIOKINETICS AND MECHANISM

Using an oral preparation, such as auranofine, only 25% of the Au is absorbed; using aurothiomalate intramuscularly, though, an optimal absorption can be achieved. Most of the Au is bound to plasma proteins. Initially, the plasma half-life is about 7 days. With continuing therapy, this period can be extended to weeks or months due to accumulation. The highest concentrations are found in the kidneys. Au does not pass the blood-brain barrier, although it can pass the placental barrier. Excretion occurs for the main part in urine, as well as in feces. It can also be excreted in breast milk.

The mechanism of action for Au is not known.

D. DIAGNOSTICS AND THERAPY

Therapy should be withdrawn at symptoms of toxicity. No clear relationship between the plasma or urine levels or the onset of toxicity exists. Therefore, analysis could be performed for diagnostic reasons and may have some value in the assessment of patients being treated for toxicity by chelation therapy. Therapy includes chelation with dimercaprol, administration of steroids, and supportive care.

V. PLATINUM

A. CHEMISTRY, USE, AND EXPOSURE

Platinum (Pt) is found in group VII of the periodic table below nickel and palladium. Its atomic number is 78, and its relative atomic mass is 193.09. As an element, it occurs in free form in nature. Only as a mixture of hydrochloric acid and nitric acid can Pt be dissolved, forming a complex chloride ion. Industrially, Pt is used as catalyst as well as in laboratory equipment, electrodes, and jewelry. Pt containing drugs such as cisplatin and the more recently introduced carboplatin are used as therapeutic agents in a range of neoplastic diseases. The drugs work by interrupting normal DNA replication by forming inter- and intra-strand crosslinks. Toxicity is mainly reported in relation to treatment of cancer patients.

B. POISONING

Pure Pt is relatively nontoxic. Complex salts containing Pt, however, give rise to allergic symptoms, including irritation and hypersensitivity of the skin, dermatitis, "platinosis", asthma ("platinum asthma"), and rhinorrhea. In the treatment of cancer patients, Pt compounds can have severe toxic effects, including cumulative nephrotoxicity, nausea and vomiting, myelosuppression, ototoxicity, and sometimes neurotoxicity. Excess renal magnesium loss with hypomagnesemia may also be a feature. The effects are less severe with carboplatin than with cisplatin.

C. BIOKINETICS AND MECHANISM

After systemic administration, Pt is rapidly distributed to most tissues of the body, with a large proportion of the dose being excreted within a few hours. This initial high plasma concentration falls quickly, and the great majority of the remaining Pt is protein-bound. Pt is mainly excreted in the urine in a biphasic profile with a half-life of about 1 and 60, respectively, for cisplatin.

As to the mechanism, nephrotoxicity is mainly reported for the proximal tubule. Recently, pretreatment of experimental animals and patients with bismuth — apparently leading to induction of metallothionein — was shown to significantly reduce nephrotoxicity of cisplatin. Combined with aminoglycosides, Pt enhances nephrotoxicity and ototoxicity.

D. DIAGNOSIS AND THERAPY

In occupational settings, Pt exposure should be stopped. Treatment should be supportive. Pt analysis in plasma is not useful. In cancer patients, renal function should be monitored. In cancer patients who have developed renal failure, but in whom continued Pt treatment is appropriate, measurements of plasma concentration may be useful.

REFERENCES

Bailly, R., Lauwerijs, R., Buchet, J. P., Mahieu, P. and Konings J., *Br. J. Ind. Med.* 48, 93–97, 1991.

De Wolff, F. A., *Br. J. Med.*, 310, 1216–1217, 1995.

Dreisbach, R. H. and Robertson, W. O. Eds., *Handbook of Poisoning* (12th ed.), Appleton & Lange, Norwalk, CT, 1990.

Ellenhorn, M. J. and Barceloux, D. G. Eds., *Medical Toxicology: Diagnosis and Treatment of Human Poisoning,* Elsevier Science, New York, 1988.

Fernandez Casares, M., Casas, H. A., and Leczycki, H., *Medicina* 51, 59–61, 1991.

Gregg, R. W., Molepo, J. M., Monpetit, V. J. A., Mikael, N. Z., Redmond, D., Gadia, D., and Stuart, D. J., *J. Clin. Oncol.* 10, 795–803, 1992.

Haddad, L. M. and Winchester, J. F. Eds., *Clinical Management of Poisoning and Drug Overdose,* 2nd ed., W.B. Saunders, Philadelphia, 1990.

Hansen, R. M., Varma, R. R. and Hanson, G. A., *J. Rheumatol.,* 18, 1251–1253, 1991.

Harvey, N. S. and Merriman, S., *Drug Safety* 10, 455–463, 1994.

Hurlbut, K. M., Dart, R. C., Sullivan, J. B., and Campbell, D. S., *Vet. Hum. Toxicol* 32, 363, 1990.

Jaeger, A., Sauder, P., Kopferschmitt, J., Tritch, L., and Flesch, F., *J. Toxicol. Clin. Toxic.* 3, 429–447, 1993.

Kondo, Y., Satoh, M., Imura, N., and Akimoto, M., *Anticancer Res.* 12, 2303–2308, 1992.

Moeschlin, S., Klinik und Therapie der Vergiftungen 6. Auflage Georg Thieme Verlag, Stuttgart, 1980.

Mulkey, J. P. and Oehme, F. W., *Vet. Hum. Toxicol.* 35, 445–453, 1993.

Naganuma, N., Satoh, M., Kondo, Y. and Imura, N., Abstract of *The 5th Nordic Symposium on Trace Elements in Human Health and Disease,* June 19–22, Loen, Norway, No. O-21, 1994.

Sing, G., Fries, J. F., Williams, C. A., Zatarain, E., Spitz, B., and Bloch, D. A., *J. Rheumatol.,* 18, 188–194, 1991.

Sisam, D. A. and Sheehan, J. P., *J. Am. Osteop. Assoc.,* 90, 83–86, 1990.

Toxicological profile for antimony. Govt Reports Announcements & Index (GRA&I), 19. Agency for Toxic Substances and Disease Registry, Atlanta, GA, Environmental Protection Agency, Washington D.C., 1993.

Vicellio, P., Ed., *Handbook of Medical Toxicology,* 1st Ed., Little, Brown and Company, Boston, 1993.

Villanueva, E., Hernandez-Cueto, C., Lachica, E., Rodrigo, M. D., and Ramiros, V., *Drug Safety* 5, 384–389, 1993.

Van Kempen, G. M. J., in *Handbook of Clinical Neurology,* Vol. 20 (64), *Intoxications of the Nervous System, Part I,* F.A. de Wolff, Ed., Elsevier Science BV, Amsterdam, 1994.

Chapter 29

Cobalt and Nickel Induced Hard Metal Asthma

Yukinori Kusaka

I. INTRODUCTION

The metals affecting the respiratory system in humans may be classified according to the entity of respiratory diseases induced. Although animal experimental studies have shown various effects of metals such as bismuth (Bi), gallium (Ga), germanium (Ge), indium (In), molybdenum (Mo), niobium (Nb), strontium (Sr), tantalum (Ta), and zirconium (Zr) on the respiratory system, these metals are not known to pose health hazards to the human respiratory system. The metals hazardous to human respiratory system are grouped and listed in Table 1.

Cobalt belongs to the same trace element group as nickel and iron and has been known to cause both irritation and sensitization of the bronchial airway. Cobalt can also induce interstitial pneumonitis and pulmonary fibrosis, especially among workers exposed to hard metal (HM). This chapter will focus on the induction of hard metal asthma (HMA) caused by exposure to cobalt and nickel.

II. OVERVIEW

Exposure to HM has been known to cause two distinct respiratory diseases: interstitial lung disease and bronchial asthma (Morgan and Seaton, 1984). Population-based studies have revealed that HMA was prevalent as 0.6% (Coates et al., 1973) to 5.6% (Kusaka et al., 1986a). Case studies (Coates et al., 1973; Scherrer and Maillard, 1982; Davison et al., 1983; Cirla, 1985; Pisati et al., 1986; Kusaka et al., 1986a; Shirakawa et al., 1988) have shown that cobalt, a matrix of hard metal, is capable of provoking asthma attacks in the bronchial provocation test (BPT). The BPT with HM dust or cobalt (metallic cobalt dust or cobalt salt) generates an immediate asthmatic response (IAR) (Scherrer and Maillard, 1982; Kusaka et al., 1986a; Shirakawa et al., 1988; Knape, 1990), late asthmatic response (LAR) (Coates et al., 1973; Sjögren et al., 1980; Scherrer and Maillard, 1982; Ebihara, 1983; Cirla, 1985; Pisati et al., 1986; Kusaka et al., 1986a; Shirakawa et al., 1988), and dual asthmatic response (DAR) (Hartmann et al., 1982; Davison et al., 1983; Cirla, 1985; Pisati et al., 1986; Kusaka et al., 1986a; Shirakawa et al., 1988). Prevalence of work-related wheezing and chest tightness or asthma, which may not be compatible with occupational asthma, have also been reported among HM workers with reference to job type and cobalt exposure (Alexandersson and Swensson, 1979; Sprince et al., 1988; Meyer-Bisch et al., 1989). Our own studies (Kusaka et al., 1986a; Shirakawa et al., 1988) were among the first, to our knowledge, to describe HMA from an immunoallergological viewpoint. Positive skin tests (patch or scratch test) (Sjögren et al., 1980; Hartmann et al., 1982; Ebihara, 1983) or the presence of specific IgG to cobalt (Cirla, 1985) have also been reported.

Table 1 Metals Inducing the Diseases of the Respiratory System in Human

Disease	Metals
Rhinitis	Arsenic, chromium, lithium, nickel, platinum
Acute tracheobronchitis	Mercury, platinum,
Bronchial asthma	Aluminum, chromium, cobalt, manganese, nickel, platinum, osmium, vanadium,
Pulmonary emphysema	Cadmium
Chronic bronchitis	Gold, vanadium
Metal fume fever	Cadmium, copper, magnesium, zinc
Acute chemical pneumonitis	Beryllium, cadmium, manganese, nickel, selenium, lithium
Interstitial pneumonitis	Cobalt, nickel, tungsten
Pulmonary fibrosis	Aluminum, beryllium
Pneumoconiosis	Barium, iron, silver, tin, antimony, bismuth, boron, cerium, zirconium, yttrium

Nickel belongs to the same element group (VIII) as cobalt, and exposure to nickel has been known to cause occupational asthma in metal-plating workers (McConnell et al., 1973; Malo et al., 1982; Block and Yeung, 1982; Novey et al., 1983; Cirla et al., 1985; Malo et al., 1985). Metallic nickel has been used as an alternative HM matrix and was therefore suggested to result possibly in HM-related respiratory disease (Coates and Watson, 1971; Kusaka et al., 1986a; Cugell et al., 1990). Excess amounts of nickel have been identified in the lungs and blood of a patient with HM pneumonitis (Rizzato et al., 1986). Contact dermatitis from sensitization to both cobalt and nickel was also prevalent in the Swedish HM industry (Fischer and Rystedt, 1983; Rystedt and Fischer, 1983; Fischer and Rystedt, 1985). The significance of hypersensitivity to nickel in HMA has also been demonstrated by the authors (Shirakawa et al., 1990; Kusaka et al., 1991a).

Since our cross-sectional study on HM workers in 1981 (Kusaka et al., 1982), we have conducted surveys including medical examinations, as well as environmental and biological monitoring, at the plant (Ichikawa et al., 1985; Kusaka et al., 1986b) . In 1983, HMA was seen among 5.6% (18/319) of workers (Kusaka et al., 1986a), and new cases of occupational asthma have occurred since then (Kusaka et al., 1991b). Our immunoallergological studies on some of these patients with HMA have revealed humoral and cell-mediated immunity to both cobalt and nickel (Shirakawa et al., 1988; Kusaka et al., 1989; Shirakawa et al., 1990; Kusaka et al., 1991a), and studies have revealed both cobalt and nickel in biological samples, including urine, blood, and hair from the workers as well as in the atmosphere at the plant (Ichikawa et al., 1992; Kusaka et al., 1992).

We have described evidence for sensitization with cobalt or nickel separately (Shirakawa et al., 1988; Kusaka et al., 1989; Shirakawa et al., 1990; Kusaka et al., 1991a), but the data reported were derived from the same patients having HMA. The results from biological and environmental monitoring of exposure to cobalt and nickel for HM workers at the plant have been reported by us previously without special reference to respiratory disorders of the patients. In the following chapter, we will present the results of a comprehensive study, highlighting what we consider to be an important new contribution to the understanding of HMA pathophysiology.

III. SUBJECT SELECTION FOR HMA STUDY

Since our longitudinal study began at an HM plant in 1981, we have found nearly 30 individuals with HMA (Kusaka et al., 1991b), including 18 cases reported in the previous study (Kusaka et al., 1986a). While some of them are completely asymptomatic, others have still been suffering from asthma attacks. In order to determine causative agents in the latter, they were recommended by the author, who was a contracted Industrial Health Consultant physician for the company, to take extensive medical examinations including the BPT with metals. A total of 13 workers agreed to hospital admission (National Kinki Chuo Hospital for Chest Diseases, Takatsuki Red Cross Hospital). Interstitial lung diseases were ruled out in these patients, and asthma was confirmed as described previously (Kusaka et al., 1986a; Shirakawa et al., 1988). The patients were also dermatologically examined. The present study concerns

8 of the 13 who undertook immunoallergological tests with both cobalt and nickel. Appropriate controls were taken for each test as described elsewhere (Shirakawa et al., 1988; Kusaka et al., 1989; Shirakawa et al., 1990; Kusaka et al., 1991a).

IV. METHODS FOR THE STUDY

HM is made using a process of powder metallurgy, which involves several manufacturing steps: powder mixing, pressing, sintering, shaping and forming, and grinding. Workers were exposed to both HM dust and coolants used in the grinding of HM tools. Patients were interviewed and data on the working history of the patients were collected. The latent period was defined as the duration between the start of the dusty work and the development of asthma. For the patients who remained at dusty worksites at the time of the current study, exposure duration was considered as the duration between the beginning of exposure and the biological monitoring done for the metals.

Personal samplings were done for individuals who had not changed their work since the onset of the asthma. It was also understood that the working environments and conditions had not changed since then. Air samples including total dust were collected on more than 3 workdays within 1 year of the time when the patients were diagnosed. Cobalt and nickel concentrations were determined by using atomic absorption spectrophotometry (AAS), and the aerodynamic size distribution of the airborne particles in a forming room and a grinding room was estimated with an Andersen sampler as described elsewhere (Kusaka et al., 1992).

Urine and blood were collected from the subjects toward the end of working day, and cobalt in blood and urine were determined by using the Zeeman-effect electrothermal AAS (Ichikawa et al., 1992). Hair was taken and analyzed for cobalt and nickel by neutron activation analysis as recommended by International Atomic Energy Agency (Ohmori et al., 1981). Neither urine nor blood was assessed for nickel at the time of the current study because of our lack of knowledge concerning nickel involvement.

Immunoallergological tests were carried out as follows. The methods and criteria for the tests are described here only briefly with references (Shirakawa et al., 1988; Kusaka et al., 1989; Shirakawa et al., 1990; Kusaka et al., 1991a).

Total and differential white blood cell counting was done, and frequency of eosinophils in white blood cells of greater than 5% was defined as hypereosinophilia. Total serum IgE titer was determined using radioimmunosorbent assay with a commercial kit (Shionogi, Japan), and the upper limit of the normal range in this reference was 400 IU/ml. Specific IgE antibodies to common inhalant allergens were measured using a Phadebas commercial kit (radioallergosorbent test, RAST), and a RAST score of two or more was considered positive. A case showing more than three positive RASTs was defined as atopic. Specific IgE against cobalt-human serum albumin conjugate (Co-HSA) was measured using a modification of the method by Cromwell et al. (1979), and a RAST score of two or more was considered positive. Specific IgE against nickel-HSA conjugate was determined by a modification of the method by Malo et al. (1982, 1985), and a RAST score of two or more was defined as positive. The lymphocyte transformation test (LTT) with cobalt or nickel was accomplished according to a modification of the method by Al-Tawil et al. (1981, 1984), and a stimulation index of more than two was considered positive.

Intradermal skin test (IDST) with cobalt or nickel salt solution was achieved and a wheal with a diameter of 10 mm or more induced by 0.1% solution of cobalt chloride or a 2% solution of nickel sulfate, respectively, was defined as positive. A patch test (PT) with metal salts was done using the European Standard Series, and a reading was done according to the criteria recommended by the ICDRG (Fregert and Bandmann, 1975). Since the patients engaged in grinding HM tools were also exposed to coolants, patch tests with four kinds of coolant dilutions in the water were done (Calnan, 1967).

A methacholine challenge test was performed according to a modification of the method described by Hargreave et al. (1981), after which PC_{20} was calculated. A PC_{20} of less than 1000 µg/ml was defined as bronchial hyperreactivity (BHR). A specific bronchial provocation test (BPT) with cobalt chloride and nickel sulfate was performed in a fashion similar to the one done by the Asthma and Allergic Disease Center (Chai et al., 1975). A fall in $PEV_{1.0}$ of more than 20% following the relevant metal solution of up to 2% or less was considered positive. The reaction patterns were divided into three groups (Pepys and Hutchcroft, 1975): (1) IAR, onset within 20 min; (2) LAR, onset more than 30 min.; (3) DAR, combination of the first two groups.

V. RESULTS OF THE STUDY

Results from studies are summarized in the Table 2, Table 3, and Table 4. For a comparison, the initials of HMA patients in the tables are the same as those appearing in previous papers on LTT with cobalt and nickel (Kusaka et al., 1989; Kusaka et al., 1991a) except for one patient, (N) who was not included in these studies.

Table 2 Characteristics of Patients with Hard Metal Asthma

Patient[x,y]	Age (at onset)	Smoking habit (at onset)	Work (at onset)	Latent Period (in months)	Exposure (duration in years)
A[7,8]	42	S[a]	P[c]	6	
B[2,5]	28	E[b]	P	2	
D[10,7]	31	S	F[d]	12	
E[5,4]	41	S	P	24	
F[15,6]	36	E	G[e]	48	11
G[N,3]	48	S	G	240	19
H[9,2]	44	S	S[f]	120	17
N[11,1]	47	S	F	3	4

Note: [x], indicates a case number designated in the reference (Kusaka et al., 1986a) and [y], indicates a subject number designated in the reference (Shirakawa et al., 1988). [N], means that the case was not subjected in the above reference by Kusaka et al. (1986a). The initials of patients are the same as those appearing in the author's previous papers on LTT with cobalt or nickel (Kusaka et al., 1989; Kusaka et al., 1991a) except for one patient (N), who was not included in these studies.

[a] S, smoker; [b]E, ex-smoker; [c]P, powder mixing; [d]F, forming or shaping; [e]G, grinding; [f]S, sintering.

Table 3 Results for Hard Metal Asthma Patients from Occupational Hygiene Study and Biological Monitoring

	Cobalt						Nickel	
	Air							
Patient[x,y]	Range (No.[a])	Ma (SD$_a$[b]) (μg/m³)	Mg (SD$_g$[c])	Urine (μg/l)	Blood (μg/dl)	Hair (ppm)	Air (μg/m³)	Hair (ppm)
A[7,8]			ND[d]				ND	
B[2,5]			ND				ND	
D[10,7]			ND				ND	
E[6,4]			ND				ND	
F[15,6]	21–40 (3)	31 (8)	30 (1)	29	0.40	9.11	18	44.3
G[N,3]	2–12 (10)	6 (3)	5 (2)	9	0.32	3.42	<1	58.1
H[9,2]	7–51 (18)	25 (13)	22 (2)	2	0.28		<1	
N[11,1]	9–436 (9)	141 (129)	81 (3)	1	0.42	25.8		64.8
Reference value**	Mean (SD[#])			2 (1)	0.19 (0.11)	0.042 (2.5)		2.2 1.6

Note: [x], indicates a case number designated in the reference (Kusaka et al., 1986a) and [y], indicates a subject number designated in the reference (Shirakawa et al., 1988). [N], means that the case was not subjected in the above reference by Kusaka et al. (1986a). The initials of patients are the same as those appearing in the author's previous papers on LTT with cobalt or nickel (Kusaka et al., 1989; Kusaka et al., 1991a) except for one patient (N) who was not included in these studies. ** Reference value, the values are derived from the study by Ichikawa et al. (1985) and by Ohmori et al. (1981) for cobalt levels in urine and blood, and for cobalt and nickel concentrations in hair, respectively; [#]SD represent arithmetic standard deviation and geometric standard deviation for cobalt level in urine and blood, and cobalt concentration in hair, respectively; [a]No., number of air samples collected by personal monitoring; [b]$M_a \pm SD_a$, arithmetic mean ± arithmetic standard deviation; [c]$M_g \pm SD_g$, geometric mean ± geometric standard deviation; [d]ND, not determined.

Table 4 Results from Immuno-Allergological Studies of Patients

Patient	Atopy	Dermatitis	High IgE	Eosinophilia	BHR[a]	Cobalt BPT[b]	Cobalt IDST[c]	Cobalt IgE[d]	Cobalt PT[e]	Cobalt LTT[f]	Nickel BPT	Nickel IDST	Nickel IgE	Nickel PT	Nickel LTT
A		Y[g]		Y	Y	IAR[h]	Y	Y	Y		IAR	Y	Y	Y	Y
B	Y		Y	Y	Y	DAR[i]	Y	Y	Y		IAR	Y	Y	Y	
D	Y			Y	Y	LAR[j]	Y				IAR	Y			
E						IAR					LAR				
F	Y		Y	Y	Y	LAR	Y	Y		Y	LAR	Y	Y		Y
G	Y		Y		Y	DAR	Y								
H					Y	LAR					IAR				
N					Y	LAR	Y	Y		ND[k]	LAR	Y	Y		ND

Note: [a]BHR, bronchial hyperreactivity; [b]BPT, bronchial provocation test; [c]IDST, intradermal skin test; [d]IgE, specific IgE antibody against metal-human serum albumin conjugate; [e]PT, patch test; [f]LTT, lymphocyte transformation test; [g]Y, yes means a positive finding, otherwise a negative result is indicated; [h]IAR, immediate asthmatic reaction; [i]DAR, dual asthmatic reaction; [j]LAR, late asthmatic reaction; [k]ND, not done.

In the forming room and the grinding room, the respirable fraction (less than 7.0 μm) formed 75 and 66% of the total dust, respectively, and the cobalt and nickel contents in the respirable fraction were 70 and 85% of total metal content in the total dust, respectively. As shown in Table 2, five of the eight workers developed asthma within 2 years of exposure to HM. Cobalt exposure levels under which individuals contracted asthma could be determined in four patients (F, G, H, and N). As summarized in Table 3, two of these patients had cobalt concentrations on average below the current Swedish, American, and Japanese exposure limits for cobalt (50 μg/m^3). The monitoring of air samples for patient F clearly showed that he inhaled both cobalt and nickel at the workplace. The levels of cobalt and nickel in biological samples of some patients clearly increased in comparison with reference values determined by the authors.

As shown in Table 4, four (B, D, F, and G) of the eight patients (i.e., 50%) were atopic. Contact dermatitis was observed in one patient (A). PT showed a simultaneous positive result to cobalt and nickel for patient A and B, while the coolants did not generate positive PT in any patients.

BHR was observed in seven of the eight patients. It is notable that all cases reacted to inhalation of cobalt chloride of 1% in the BPT. All three patterns of response were seen, with IAR being present in two patients (A and E), LAR in four patients (D, F, H, and N), and DAR in two patients (B and G). Of these eight patients, four (A, B, F, and N) showed specific IgE antibody to Co-HSA, while IDST with cobalt chloride was positive in seven patients (including the four). A positive LTT with cobalt were found in one (F) of the four patients with concomitant IgE antibody to Co-HSA. Positive reaction to nickel in BPT was observed in seven of the eight patients who underwent BPT with nickel sulfate of 1 or 2%. IAR was provoked in four patients (A, B, D, and H), and LAR was provoked in three patients (E, F, and N), while DAR was not seen. Four (A, B, F, and N) of the seven patients showed positive IgE to Ni-HSA, while five, including these four, reacted to nickel sulfate in IDST. A positive LTT (patients A and F) was seen among the four patients with specific IgE to Ni-HSA.

No clear relationships were found among the types of asthma attack induced on BPT, the status of humoral or cell-mediated immunity to the metals, or the reactions in the skin tests. In the controls taken for each test, positive BPTs, IDSTs, PTs, nor LTTs were not seen with cobalt or nickel.

VI. DISCUSSION

It is notable that all the cases in the current study correlated to cobalt under the BPT. In the controls (n = 8) who under went the BPT in the same way, 1% cobalt chloride solution did not provoke any asthma attack (Shirakawa et al., 1988). In addition, 2 ml of 2% cobalt chloride solution failed to provoke an asthma attack in one atopic HM worker whose asthma did not show a clear time relation to hard metal exposure (Kusaka et al., 1986a). Therefore, inhalation of 1% cobalt chloride solution in the BPT can distinguish between specific sensitization to cobalt and nonspecific irritation. This finding is in general agreement with the report by Roto (1980), in which 2 ml of 1% cobalt chloride solution was applied to the respiratory challenge test among cobalt production workers, resulting in positive reactions in individuals with work-related asthma.

McConnell et al. (1973) and Malo et al. (1982, 1985) reported that inhalation of 1% nickel sulfate solution could provoke asthma attacks in individuals who were hypersensitive to nickel. In the present study, positive reactions were caused in some patients by inhalation of 1% nickel sulfate and in others by 2% nickel sulfate. Eight controls inhaling 2% nickel sulfate solution showed negative responses (Shirakawa et al., 1990). Thus, taking these results into consideration, the concentration of nickel sulfate solution of 1 to 2% seems to be appropriate for BPT.

Among the eight patients in the current study, three reaction patterns of IAR, LAR, and DAR following inhalation of cobalt appeared in almost equal proportions. As stated in the introduction, 13 workers from the industry underwent BPT with cobalt. Among the 13, IAR was observed in 5, LAR in 4, and DAR in 4 (Kusaka et al., 1991b). Therefore, it seems that incidence of the three reaction patterns in almost equal proportions in individuals is one of features of HMA. It has been claimed that isolated late asthmatic reactions are characteristic of occupational asthma caused by exposure to low molecular weight industrial substances (Chan-Yeung and Lam, 1986b).

Although the patients in the present study were not randomly selected, it can be concluded that humoral immunity (HI) to cobalt-haptenated protein, *via* particularly type I hypersensitivity (suggested by circulating specific IgE antibody to Co-HSA), plays a role in HMA; cobalt specific IgE antibody, as shown by Cirla (1985), may also be involved.

For cobalt, a positive LTT indicating cell-mediated immunity (CMI) accounted for in one patient (F) was not seen in individuals with a positive PT, indicative of delayed type hypersensitivity (DTH) in the skin in two patients (A and B). As shown in the results, these three patients contracted HI to cobalt. With respect to nickel, CMI in patients (A and F) and/or DTH in patients (A and B) were again associated with HI.

In a LTT study done at the same time by the author, two female HM workers with contact dermatitis due to both cobalt and nickel, diagnosed on the basis of positive PT, also reacted to the two metals. Hence, it can be said that the LTT applied in the present study is sensitive enough for detection of CMI. CMI to common inhalant allergens has been reported in atopic asthma (Lanzavecchia et al., 1983; Rawle et al., 1984), and *in vitro* evidence of cell-mediated immunity (CMI) in occupational asthma due to small molecular weight chemicals has been reported (Gallagher et al., 1981) although its immunological relevance has not been fully clarified. The significance of T lymphocyte activation reported in acute severe asthma (Corrigan et al., 1988) may also be involved in the pathogenesis of HMA. The relationship between HI, CMI, and DTH is of importance in understanding the pathophysiology of HMA.

Sjögren et al. (1980) reported that individuals with hard metal disease (allergic alveolitis or asthma) developed eczema associated with positive patch test to cobalt prior to onset of the respiratory disorders. Thus, they argued for a significance of skin sensitization. Our results, in which only two patients (A and B) showed positive patch test with cobalt, are partly consistent with the above observation. Regarding the other subjects, however, this finding was not confirmed, which is in concordance with reports by Calnan (1967) and Hartmann et al. (1982). A relationship between skin sensitization and respiratory sensitization due to airborne agents of small molecular weight warrants further study.

All of the patients with type I allergies to nickel (A, B, F, and N) also had type I allergies to cobalt. Sensitization to nickel alone was not observed. In the case of PT, there were two patients (A and B) who reacted to both cobalt and nickel. This raised the concern as to whether concomitant reaction between cobalt and nickel is cross-reactivity or simultaneous sensitization (Menne, 1980; Joost and Everdingen, 1981; Fischer and Rystedt, 1983; Rystedt and Fischer, 1983; Fischer and Rystedt, 1985). We can only speculate on this from our own data, but the fact that not all individuals with sensitization to cobalt were sensitized to nickel does not support a concept of cross-sensitivity. Nevertheless, simultaneous exposure to both cobalt and nickel, which could cause an interaction in immune response as shown in the metal-sensitized guinea pig model (Lammintausta et al., 1985), is a problem to be tackled.

In our preliminary study at the HM plant, atopy was observed in 24% (12 out of 51) of the asymptomatic HM-exposed workers (Kusaka et al., 1991b), and in a cohort of Japanese schoolchildren, atopy was reported to be about 14%, which decreased with age (Kagamimori et al., 1982). In contrast, atopy was seen as 50% (4 out of 8) of the patients in the present study and as 48% (15 out of 31) in individuals with HMA observed during the follow-up study (Kusaka et al., 1991b). Hence, it is apparent that atopy is a strong predisposing factor in the development of HMA. This finding also seems to be a common finding in other types of occupational asthma (Chan-Yeung and Lam, 1986). Of more concern is the fact that many individuals developed HMA within a few years of work, and some developed work-related asthma with cobalt exposure levels less than 50 $\mu g/m^3$. Thus, both individual and environmental factors relating to HMA should be further investigated.

Not all of the patients who responded to cobalt in BPT showed positive IDST with cobalt, and not all the patients with positive IDST had specific IgE against Co-HSA. As described previously for LTT in patients with nickel contact dermatitis (Blomberg-van der Flier et al., 1987), the RAST or LTT used, as well as in the present study, may lack sensitivity because of problems such as the relevancy of the cobalt-haptenated antigen. In addition, skin reactivity to nickel is not always reported to correspond to specific RAST scores (Malo et al., 1985). More valid immunoallergological tests for metallic haptens need to be developed.

Both cobalt and nickel were detected in biological samples, including hair, as reported by Rizzato et al. (1986). As far as the subjects were concerned, we did not discover any toxicological abnormalities related to the metals in addition to respiratory problems. By contrast, changes in serum immunoglobulins and inflammatory proteins have been described among workers exposed to cobalt-containing dust (Bencko et al., 1986) and among patients having cobalt contact eczema (Janeckova et al. 1989). The implications of the accumulated metals need to be studied with respect to the possible effects on immune response.

The mechanisms underlying HMA still remain to be clarified, especially for patients without any evidence of metal sensitization. The strong toxicity of cobalt to the bronchial epithelium and the lung, as shown in animal experimental studies (Delahant, 1955; Schepers, 1955; Kaplun and Mezencewa, 1960), the potential of cobalt to form the very toxic hydroxyl radical from H_2O_2 (Moorhouse et al., 1985), or the enhanced cytotoxicity of cobalt to pulmonary macrophages owing to tungsten as an adjuvant (Lison and Lauwerys, 1990) might all be implicated in HMA. Nonspecific toxicity-mediated bronchial obstruction and bronchial irritation arising from these effects of cobalt among HM workers (Alexandersson and Swensson, 1979; Kusaka et al., 1986b; Meyer-Bisch et al., 1989) might precede or aid HMA breakthrough.

ACKNOWLEDGMENT

The author would like to thank Dr. S. Ohmori, Professor, Department of Environmental and Health Science, School of Social Information Studies, Otsuma Women's University, for determining cobalt and nickel in hair.

REFERENCES

Alexandersson, R. and Swensson, A., *Arh. Hig. Rada. Toksikol.*, 30 (Suppl), 355–361, 1979.
Al-Tawil, N. G., Marcusson, J. A., and Moller, E., *Acta Derm. Venereol.*, 61, 511–515, 1981.
Al-Tawil, N. G., Marcusson, J. A., and Moller, E., *Acta Derm. Venereol.*, 64, 203–208, 1984.
Bencko, V., Wagner, V., Wagnerová, M., and Zavázal, V., *Environ. Res.*, 40, 399–410, 1986.
Block, G. T. and Yeung, M., *JAMA*, 247, 1600–1602, 1982.
Blomberg-van der Flier, M., van der Burg, C. K. H., Pos, O., van de Plassche-Boers, E. M., Bruynzeel, D. P., Garotta, G., and Scheper, R. J., *J. Invest. Dermatol.*, 88, 362–368, 1987.
Calnan, C. D., *Br. J. Dermatol.*, 79, 60–61, 1967.
Chai, H., Farr, R. S., Froerich, L. A., Mathison, D. A., McLean, J. A., Rosenthal, R. R., Sheffer, A . L., Spector, S. L., and Townley, R. G., *J. Allergy Clin. Immunol.*, 56, 323–327, 1975.
Chan-Yeung, M. and Lam, S., *Am. Rev. Respir. Dis.*, 133, 686–703, 1986.
Cirla, A. M., *Folia Allergol. Immunol. Clin.*, 32, 21–28, 1985.
Cirla, A. M., Bernabeo, F., Ottoboni, F., and Ratti, R., in *Progress in Nickel Toxicology*, S.S. Brown and F.W. Sunderman, Eds., Blackwell Scientific, Oxford, 165–168, 1985
Coates, E. O. and Watson, J. H. L., *Ann. Intern Med.*, 75, 709–716, 1971.
Coates, E. O., Sawyer, H. J., Rebuck, J. W., Kvale, P. A., and Sweat, L. W., *Chest.* 64, 390, 1973.
Corrigan, C. J., Hartneil, A., and Kay, A. B., *Lancet*, i, 1129–1132, 1988.
Cromwell, O., Pepys, J., Parish, W. E., and Hughes, E. G., *Clin. Allergy.* 9, 109–117, 1979.
Cugell, D. W., Morgan, W. K. C., Perkins, D. G., and Rubin, A., *Arch. Intern. Med.*, 150, 177–183, 1990.
Davison, A. G., Haslam, P. L., Corrin, B., Coutts, I. I., Dewar, A., Riding, W. D., Studdy, P. R., and Newman-Taylor, A. J., *Thorax,* 38, 119–128, 1983.
Delahant, A. B., *Arch. Ind. Health,* 12, 116–120, 1955.
Ebihara, I., *J. Sci. Labour,* 59, 321–325, 1983.
Fischer, T. and Rystedt, I., *Contact Dermatitis,* 9, 115–121, 1983.
Fischer, T. and Rystedt, I., *Am. J. Ind. Med.*, 8, 381–394, 1985.
Fregert, S. and Bandmann, H.-J., *Patch Test.,* Springer-Verlag, Berlin, 24, 1975.
Gallagher, J. S., Tse, C. S. T., Brooks, S. M., and Bernstein, I. L., *J. Occup. Med.*, 23, 610–616, 1981.

Hargreave, F. E., Ryan, G., Thomson, N. C., O'Byrne, P. M., Latimer, K., Juniper, E. F., and Dolovich, J., *J. Allergy Clin. Immunol.,* 67, 347–355, 1981.
Hartmann, A., Wüthrich, B., and Bolognini, G., *Schweiz. Med. Woohenschr.,* 112, 1137–1141, 1982.
Ichikawa, Y., Kusaka, Y., and Goto, S., *Int. Arch. Occup. Environ. Health,* 55, 269–276, 1985.
Ichikawa, E. Y., Kusaka, Y., Ogawa, Y., and Goto, S., *J. Sci. Labour,* 68, 11–22, 1992.
Janečkova, V., Znojemská, S., Korčákova, L., Wagnerová, M., Kalenský, J., and Svobodová, J., *J. Hyg. Epidemiol. Microbiol. Immunol.,* 33, 121–127, 1989.
Joost, T. and Everdingen, J. J. E., *Acta Derm. Venereol.,* 62, 525–529, 1981.
Kagamimori, S., Naruse, Y., Watanabe, S., Nohara, S., and Okada, A., *Clin. Allergy,* 12, 561–568, 1982.
Kaplun, Z. S. and Mezencewa, N. W., *J. Hyg. Epidemiol. Microbiol. Immunol.,* 4, 390–399, 1960.
Knape, H., *Pneumologie,* 44, 862–865, 1990.
Kusaka, Y., Sugimoto, K., Seki, Y., Goto, S., Yokoyama, K., Yamamoto, S., Sera, Y., Kyono, H., and Kohyama N., *Jpn. J. Ind. Health,* 24, 636–648, 1982.
Kusaka, Y., Yokoyama, K., Sera, Y., Yamamoto, S., Sone, S., Kyono, H., Shirakawa, T., and Goto, S., *Br. J. Ind. Med.,* 43, 474–485, 1986a.
Kusaka, Y., Ichikawa, Y., Shirakawa, T., and Goto, S., *Br. J. Ind. Med.,* 43, 486–489, 1986b.
Kusaka, Y., Nakano, Y., Shirakawa, T., and Morimoto, K., *Ind. Health.,* 27, 155–163, 1989.
Kusaka, Y., Nakano, Y., Shirakawa, T., Fujimura, N., Kato, M., and Heki, S., *Ind. Health,* 29, 153–160, 1991a.
Kusaka, Y., Fujimura, N., and Morimoto, K., in *Advances in Asthmology 1990,* J.A. Ballanti and S. Kobayashi, Eds., Elsevier Science, Amsterdam, 271–276, 1991b.
Kusaka, Y., Kumagai, S., Kyono, H., and Shirakawa, T., *Ann. Occup. Hyg.,* 36, 497–507, 1992.
Lammintausta, K., Pitkanen, O.-P., Kalimo, K., and Jansen, C. T., *Contact Dermatitis,* 13, 148–152, 1985.
Lanzavecchia, A., Santini, P., Maggi, E., Delprete, G. F., Falagian, P., and Romagnani, S., *Clin. Exp. Immunol.,* 52, 21–28, 1983.
Lison, D. and Lauwerys, R., *Environ. Res.,* 52, 187–198, 1990.
Malo, J.-L., Cartier, A., Doepner, M., Nieboer, E., Evans, S., and Dolovich, J., *J. Allergy Clin. Immunol.,* 69, 55–59, 1982.
Malo, J. L., Cartier, A., Gagnon, G., Evans, S., and Dolovich, J., *Clin. Allergy,* 15, 95–99, 1985.
McConnell, L. H., Fink, J. F., Schlueter, D. P., and Schmidt, M. G., *Ann. Intern. Med.,* 78, 888–890, 1973.
Menne, T., *Contact Dermatitis,* 6, 337–340, 1980.
Meyer-Bisch, C., Pham, Q. T., Mur, J.-M., Massin, N., Moulin, J.-J., Teculescu, D., Carton, B., Pierre, F. and Baruthio, F., *Br. J. Ind. Med.,* 46, 302–309, 1989.
Moorhouse, C. P., Halliwell, B., Grootveld, M., and Gutteridge, J. M. C., *Biochem. Biophys. Acta,* 843, 261–268, 1985.
Morgan, W. K. C. and Seaton, A., *Occupational Lung Diseases,* 2nd ed., W. B. Saunders, Philadelphia, 1986, 486–489.
Novey, H. S., Habib, M., and Wells, I. D., *J. Allergy Clin. Immunol.,* 72, 407–412, 1983.
Ohmori, S., Tsuji, H., Kusaka, Y., Takeuchi, T., Hayashi, T., Takada, J., Koyama, M., Kozuka, H., Shinogi, M., Aoki, A., Katayama, K., and Tomiyama, T., *J. Radioanal. Chem.,* 63, 269–282, 1981.
Pepys, J. and Hutchcroft, B. J., *Am. Rev. Respir. Dis.,* 112, 829–859, 1975.
Pisati, G., Bernabeo, F., and Cirla, A. M., *Med. Lav.,* 77, 538–546, 1986.
Rawle, F. C., Mitchell, E. B., and Platts-mills, A. E., *J. Immunol.,* 133, 195–201, 1984.
Rizzato, G., Cicero, S. Lo., Barberis, M., Torre, M., Pietra, R., and Sabbioni, E., *Chest,* 89, 101–106, 1986.
Roto, P., *Scand. J. Work Environ. Health,* 6 (Suppl 1), 1–49, 1980.
Rystedt, I. and Fischer, T., *Contact Dermatitis,* 9, 195–200, 1983.
Schepers, G. W. H., *Arch. Ind. Health,* 12, 121–146, 1955.
Scherrer, M. and Maillard, J-M., *Schweiz. Med. Wochenschr.,* 112, 198–207, 1982.
Shirakawa, T., Kusaka, Y., Fujimura, N., Goto, S., and Morimoto, K., *Clin. Allergy,* 18, 451–460, 1988.
Shirakawa, T., Kusaka, Y., Fujimura, N., Kato, M., and Heki, S., *Thorax,* 45, 267–271, 1990.
Sjögren, I., Hillerdal, G., Andersson, A., and Zetterstrom, O., *Thorax,* 35, 653–659, 1980.
Sprince, N. L., Oliver, L. C., Eisen, E. A., Greene, R. E., and Chamberlin, R. I., *Am. Rev. Respir. Dis.,* 138, 1220–1226, 1988.

Chapter 30

Aspects on Health Risks of Mercury From Dental Amalgams

Jan A. Weiner and Magnus Nylander

I. GENERAL INTRODUCTION

Mercury (Hg) is a nonessential heavy metal. It exists in three different chemical states: elemental (Hg^0), mercurous (Hg^+), and mercuric (Hg^{++}) mercury. It forms both inorganic and organic compounds, which differ in metabolism and toxicity (WHO, 1990; WHO, 1991). Mercury compounds have an affinity for sulfhydryl groups. Binding of mercury to proteins of enzymes and membranes affects enzyme activity and membrane function (Berlin, 1986). A toxic effect on membranes may also be due to oxidative damage induced by mercury (Lund et al., 1993; Miller and Woods, 1993). The central nervous system is considered the critical organ after exposure to (organic) methyl-Hg and elemental mercury vapor. The kidneys are considered the critical organ after exposure to inorganic mercuric mercury (WHO, 1990; WHO, 1991).

II. DENTAL AMALGAM

Dental amalgam has been the dominant restorative material in cavities of molars and premolars during this century. In the beginning of the 1980s, it was estimated that dental amalgam accounted for 75 to 80% of all single tooth fillings (Bauer and First, 1982; Wollf et al., 1983; SoS, 1987).

Conventional dental amalgam or "silver amalgam" is an alloy in which metallic Hg (about 50% by weight) is mixed with a powder containing mainly silver (about 35%) and tin (about 10%), but also smaller amounts of copper (1 to 6%) and zinc (0 to 2%). The composition of conventional amalgam has been about the same since the beginning of this century. Another type of amalgam in use is the non-gamma-two amalgam, where part of the silver is replaced by copper. This type of amalgam with up to about 25% copper is claimed to be more resistant to corrosion (SoS, 1987).

After mixing metallic Hg and the powder of other metals to a plastic amalgam mass, it is inserted under manual pressure, so-called "condensation," into the prepared dental cavity. Excess Hg, less than 5%, is removed before or at the condensation of the plastic amalgam mass in the prepared tooth cavity. The amalgam filling hardens within minutes and must be anatomically and functionally modeled, preferably without any contact with saliva, during this time. The amalgam filling continues to set and strengthen over the next 24 h and is then ready to be finished by polishing with rotating instruments. The hardening of the amalgam filling will then continue over several months (ADA, 1985).

Yet another type of amalgam is called "copper amalgam". This type of amalgam was previously the predominant type used in pediatric dentistry (treatment of children). It contains 60 to 70% Hg and 30 to 40% copper. This amalgam is easier to insert in dental cavities, but it has to be prepared in the dental surgery by heating, which cause a considerable exposure to Hg vapor. Copper amalgam disintegrates quickly in the mouth. This was actually considered an advantage in that the release of Hg and copper ions were thought to have a bactericidal effect, thus inhibiting bacterial growth and secondary caries. This type of copper amalgam was more frequently used a few decades ago and earlier (SoS, 1987).

Amalgam for dental restorations was introduced in Europe and the U.S. in the mid- to late 19th century. Intense debates regarding the possible risk for systemic health effects due to its content and release of Hg were already roused at the time of introduction and have flared several times since then (Stock, 1939; Frykholm, 1957).

In the late 1970s and during the 1980s studies were reported that verified Stock's reports from the 1920s and 1930s of a continuous emission of considerable amounts of Hg vapor from dental amalgam fillings and an increased urinary excretion of mercury as a result of amalgam fillings (Stock and Cucuel, 1934; Stock, 1939; Gay et al., 1979; Svare et al., 1981; Nilsson and Nilsson, 1986a; Olstad et al., 1987). Other studies showed a significant retention in human tissues of Hg released from amalgam fillings (Nylander, 1986; Eggleston and Nylander, 1987; Nylander et al., 1987; Schiele, 1988; Nylander et al., 1989).

As a result of this information there are continuing intense debates in several countries on possible systemic health hazards emanating from dental amalgam fillings. Those who claim that Hg from amalgam fillings is a health hazard refer to case reports of such alleged cases of adverse reactions and information on the release and uptake of Hg from amalgam fillings. Those who deny a causal relationship between dental amalgams and health problems claim that amalgam has been used for more than 100 years on millions of people with no proven negative effects on general health and/or that the release of Hg from the fillings is too low to result in systemic health effects (Ziff, 1984; Enwonwu, 1987; Langan et al., 1987; SoS, 1987; Knolle, 1988; Weiner et al., 1990; Horstedt-Bindslev et al., 1991; Swedish MRC, 1992; U.S. D.H.H.S., 1993).

The present work will deal only with the issue of possible systemic effects due to the uptake and distribution in the human body of mercury released from amalgam fillings. Any local reactions due to amalgam fillings will not be considered and the possibility of a toxic effect due to the release of other metals than Hg will not be dealt with either. The text will focus on the release, subsequent absorption, and distribution of Hg from fillings in the human body. This is followed by a discussion on potential risks due to the exposure to Hg from amalgam fillings. The possibility of an effect on the fetus will be briefly discussed.

III. EXPOSURE AND METABOLISM OF MERCURY FROM DENTAL AMALGAM FILLINGS

A. RELEASE AND UPTAKE OF MERCURY FROM AMALGAM FILLINGS

Hg is released from amalgam fillings in different forms and may be absorbed by several routes. Hg released in the elemental form is volatile and evaporates; part of this Hg vapor will be inhaled. Elemental Hg may also be dissolved in the saliva and swallowed with it (Stock, 1939; Gay et al., 1979).

Uptake of Hg vapor in the lungs is approximately 80%. Hg that is absorbed in the elemental form, e.g., as elemental vapor in the lungs, is enzymatically oxidized to Hg^{2+} in the blood and different tissues. In blood this oxidation takes a few minutes and during this time the elemental mercury readily passes biological barriers, e.g., the blood-brain barrier and the placental barrier, and thus penetrates into all tissues (Clarkson et al., 1988a).

Electrochemical dissolution of fillings may result in the release of mercury in the divalent, Hg^{2+}, form. Divalent Hg ions that are dissolved in the saliva and swallowed will be partly absorbed in the gastrointestinal tract; such absorption is about 15% (Clarkson et al., 1988a). It has been indicated, though, that it is mainly the other metals of amalgam that undergo electrochemical dissolution and that Hg is released mainly in the elemental form (Gross and Harrison, 1989).

Possible routes of uptake of mercury from fillings also include a direct absorption of mercury in the mucosa of the oral cavity and a migration of mercury through the dentin to the dental pulp and the adjacent bones (Schiele et al., 1987; SoS, 1987; Hahn et al., 1989; Willerhausen-Zönnchen et al., 1992).

The magnitude of the uptake of mercury from amalgam fillings may be determined based on measurements at the source, i.e., measurements of the release of mercury from fillings and the subsequent estimate of the fraction absorbed. Alternatively, it can be based on measurements of concentrations of mercury in a biological medium, e.g., brain tissue, kidneys or urine, and the subsequent calculation of the uptake that would result in the measured amounts.

B. PULMONARY UPTAKE OF MERCURY VAPOR FROM AMALGAM FILLINGS

It is generally believed that the dominant route of uptake of Hg from amalgam fillings is through pulmonary absorption of inhaled Hg vapor (Clarkson, 1992; Swedish MRC, 1992; U.S. D.H.H.S, 1993). A number of studies show that amalgam fillings continuously emit Hg vapor and that the emission increases as a result of chewing, toothbrushing, or with intake of hot beverages (Gay et al., 1979; Svare et al., 1981; Abraham et al., 1984; Patterson et al., 1985; Vimy and Lorscheider, 1985a; Björkman and Lind, 1992). The return to baseline rates of emission after chewing occurs gradually, during a period that may extend to one or more hours (Vimy and Lorscheider, 1985b; Berglund, 1990). Despite many investigations, there is still a great uncertainty regarding the pulmonary uptake of mercury vapor (see Table 1).

Table 1 Estimates of Average Uptake of Mercury from Amalgam Fillings in Different Studies*

Study	Estimated average uptake of mercury from amalgam fillings (μg/day)	Amount of amalgam in studied group. Number of tooth surfaces restored with amalgam
Patterson et al. (1985)	≥27	Upper tenth percentile of study group, with 1–75 amalgam restored surfaces[a]
Vimy and Lorscheider (1985b)	20	1–16 occlusal surfaces, average 8.6
Mackert (1987)[b]	1.2	1–16 occlusal surfaces, average 8.6
Clarkson et al. (1988b)	3–18	Data from subjects in four different studies[c]
Aronsson et al. (1989)	7–10	24–63 amalgam surfaces
Berglund (1990)	1.7	13–48 amalgam surfaces, average 27
WHO (1991)	3–17	
Snapp et al. (1989)	≥1.3	Average of 14 amalgam restored surfaces
Jokstad et al. (1992)	10–12	≥36 amalgam restored surfaces

*Note:** The first 7 estimates are based on measurements of concentrations of mercury vapor intraorally or in expired breath. Snapp et al. (1989) is based on changes in blood concentration of mercury before and after amalgam removal. Jokstad et al. (1992) is based on urinary concentrations of mercury. Remaining data taken from Weiner, J. A. and Nylander, M., *Sci. Total Environ.*, 168, 255–265, 1995.

[a] Amalgam status examined for 94 out of 172 subjects.
[b] Estimate based on recalculation of the data presented by Vimy & Lorscheider (1985b).
[c] Estimate based on data on mercury vapor emission in four different studies. The amalgam load in these studies were Svare et al. (1981): 0–21 amalgam fillings; Abraham et al. (1984): occlusal area of posterior amalgam fillings 0.2–4.2 cm^2; Patterson et al. (1985): 1–75 amalgam restored surfaces; Vimy and Lorscheider (1985b): 1–16 occlusal surfaces restored with amalgam, average 8.2 occlusal surfaces. The different data give different estimates of average uptake of mercury vapor in the lungs, thus the range given. Average of the four different estimates of uptake is 8 μg/day.

Clarkson et al. (1988b) estimated the average daily uptake of Hg in the lungs by using experimental data on Hg vapor emission from four different studies (Svare et al., 1981; Abraham et al., 1984; Patterson et al., 1985; Vimy and Lorscheider, 1985b). Depending on which set of data that was used the model employed gave highly different result. An average daily uptake of between 3 and 18 μg was thus estimated. Of the studies utilized, the one by Abraham et al. (1984) is the one that reports the most carefully designed and standardized sampling methodology. Data from this study generated an estimated average uptake approximately in the middle of the presented interval, i.e., an uptake of 8 μg Hg daily. The average amount of amalgam of the 47 subjects of the latter study is low in comparison with average amounts in the Swedish studies that will be discussed below (Berglund, 1990; Molin et al., 1990; Åkesson et al., 1991).

Based on a kinetic model, the authors also estimated the resulting steady-state concentrations in blood, urine, brain, and kidneys. A pulmonary uptake of 8 μg Hg vapor daily was thus estimated to correspond to 0.77 μg of Hg per l in blood, 2.4 μg Hg/l in urine, 12 μg Hg/kg in the brain, and 714 μg Hg/kg in the kidneys.

It should be emphasized that the range given by Clarkson et al. (1988b), i.e., 3 to 18 µg Hg/day, concerns the uncertainty, depending on which set of data that was used, in the estimated **average** pulmonary uptake of Hg vapor from fillings. Available data on Hg vapor emission show very large interindividual variation. Using the model of calculation of pulmonary uptake from Clarkson et al. (1988b) and employing it for maximum values of mercury vapor emission from Abraham et al. (1984) results in an estimated uptake of approximately 80 µg Hg/day. This should only be considered as a tentative figure of possible maximum levels of exposure to Hg vapor from fillings, but it is fillings indicative of very large individual variation in the emission of Hg vapor from amalgam.

There are a number of further estimates of average pulmonary uptake of Hg vapor from amalgam fillings. As can be seen from Table 1, presented estimates vary an order of magnitude, from slightly above 1 µg/day to approximately 20 µg/day (Patterson et al., 1985; Vimy and Lorscheider, 1985b; Mackert, 1987; Clarkson et al., 1988b; Aronsson et al., 1989; Berglund, 1990). Part of this variation might obviously be explained by variation in the average amount of amalgam fillings in the different groups studied, but this factor can explain only a smaller part of the variation. The reason for the large differences in these estimates is variations in the measured rates of emission and in the assumptions made in estimating the concentrations or amounts of mercury reaching the lungs.

The figure on average pulmonary uptake of mercury vapor from fillings reported by Berglund (1990), i.e., 1.7 µg Hg/day, has had a large impact on major assessments of the potential risks associated with dental amalgam fillings; see for example that of the Swedish Medical Research Council (Swedish MRC, 1992; Bergman, 1992), the U.S. Department of Health and Human Services (1993), or the British Dental Association (Eley and Cox, 1993). However, the calculations of pulmonary uptake require a number of assumptions some of which may be questioned. The uncertainties in the necessary assumptions and the available data were emphasized by Clarkson et al. (1988b).

The basic assumption made by Berglund (1990) and in the majority of the other estimates of pulmonary uptake of Hg vapor is that emission of Hg vapor from fillings is independent of air flow through the oral cavity. The rate of emission of mercury vapor thus reported by Berglund (1990) correspond to an average of 18 µg Hg emitted/day. This contrasts to the estimated uptake, 1.7 µg Hg/day. The fate of the approximately 90% of Hg vapor that is estimated to be released but not absorbed is not accounted for. Obviously part of it is exhaled during expiration through the mouth, but the discrepancy between the emitted amount and the estimated uptake is largely dependent on the assumption of no pulmonary uptake during nasal breathing. There is no discussion of this assumption or on what happens to any mercury emitted in this situation. With a basic assumption that emission of Hg is independent of air flow through the oral cavity, i.e., independent of mouse/nose breathing, a detailed discussion on this would have been desirable.

In figures given later from the same research group on the fate of the emitted Hg that is not taken up in the lungs, it is assumed that 15% is exhaled and that the major part, 75%, is dissolved in the saliva and swallowed (see Olsson and Bergman, 1992). These assumptions are not substantiated. Regarding the absorption of elemental mercury dissolved in saliva, it was furthermore stated that the gastrointestinal uptake of such Hg is practically nil, i.e., 0.01%. The referred figure (0.01%) (Bornmann et al., 1970) is, however, most likely not relevant. It applies to absorption of ingested metallic, i.e., liquid, mercury. The very low absorption of metallic mercury is due to its poor solubility in water. There appears to be no experimental data on the gastrointestinal absorption of elemental Hg dissolved in saliva, but elemental Hg, which is highly lipid soluble and readily passes biological membranes, is likely to have a very high degree of absorption. If it is oxidized to Hg^{2+} in the gastrointestinal tract before absorption this would decrease absorption markedly, but it would still be significant, i.e., about 15% (Clarkson et al., 1988a). The same figure concerning the absorption of elemental Hg dissolved in the saliva is repeated in the risk evaluation initiated by the U.S. Department of Health and Human Services (1993).

C. GASTROINTESTINAL ABSORPTION OF MERCURY RELEASED FROM AMALGAM FILLINGS

In addition to the release of Hg in the elemental form, there may also be a release to the saliva of ionic mercury (Hg^{2+}) and of small particles due to the wear of amalgam fillings (Brune et al., 1983; Brune and Evje, 1985). A Swedish study of the amount of Hg in feces showed an average of approximately 60 µg/day (range 27 to 190) in nine amalgam bearers (Skare and Engqvist, 1994). It is presently not possible to determine which part of this Hg that has just passed through the gastrointestinal tract. There are no detailed estimates available of the amount of Hg that is absorbed via the gastrointestinal route. The uncertainty concerning the magnitude of this absorption of Hg is thus even greater than the uncertainty in the estimates of pulmonary absorption of Hg vapor.

D. RELEASE OF MERCURY FROM COPPER AMALGAMS

A special consideration must be given the so-called "copper amalgams", which are especially prone to degradation. In an evaluation of a possible worst case situation regarding the release of mercury form such fillings NIOM estimated that the release might reach 6.3 mg/day in a patient with big copper amalgams in all the molars of the incidious teeth. The form of Hg that would be released was not specified (NIOM, 1981). Assuming the major part being released as ionized Hg to the saliva, then the gastrointestinal absorption would be about 15% (Clarkson et al., 1988a), i.e., a worst case estimate of the absorption is approximately 900 μg Hg/day. It might be added that it is clear from the same report by NIOM that copper amalgam may contain significant amounts of cadmium. Likewise, a worst case scenario of a release of up to 50 μg cadmium/day was calculated. It is unclear to which extent these copper amalgams are used presently; however, usage does occur (Statens Helsetilsyn, 1994).

E. URINARY EXCRETION OF MERCURY FROM FILLINGS AND ESTIMATES OF UPTAKE OF MERCURY FROM AMALGAM FILLING BASED ON CONCENTRATIONS IN URINE

Inorganic mercury is eliminated mainly in the urine and feces (Rahola et al., 1973; Cherian et al., 1978). Urinary elimination of mercury after intake of methyl-mercury is only about 10%. With an average intake of methyl-mercury with food of a few micrograms per day in the general population, this form of Hg should contribute relatively little to the urinary excretion of Hg. Uptake of inorganic Hg from sources other than amalgam fillings is of minor importance (Clarkson et al., 1988a; WHO, 1991). Urinary data from Swedish subjects confirm a limited effect from non-amalgam sources of mercury on urinary excretion of mercury (Berglund, 1990; Åkesson et al., 1991; Svensson et al., 1992).

A number of studies have shown a significant correlation between measures of the amount of amalgam fillings and urinary excretion of Hg (Nilsson and Nilsson, 1986a; Olstad et al., 1987; Berglund, 1990; Skare et al., 1990; Åkesson et al., 1991; Aposhian et al., 1992; Herrman and Schweinsberg, 1993).

Data on urinary excretion from different studies of Swedish subjects without occupational exposure are highly consistent and show average urinary concentrations of 1.9 to 2.3 μg Hg/g creatinine in morning spot samples (Berglund, 1990; Åkesson et al., 1991; Langworth et al., 1991) or in average approximately 3 μg Hg in 24 h-urine (Aronsson, 1988; Skare et al., 1990). The average number of amalgam surfaces in the subjects of these studies range from 32 to 42 surfaces. Urinary data from German subjects indicate somewhat lower average excretion, i.e., 1.7 μg/24 h, largely reflecting a somewhat lower amount of amalgam fillings, in average approximately 23 tooth surfaces restored with amalgam (with totally filled occlusal surfaces counted as two surfaces) (Herrman and Schweinsberg, 1993).

Weiner and Nylander (1995) estimated based on data on urinary excretion of mercury from six Swedish studies (Aronsson, 1988; Aronsson et al., 1989; Berglund, 1990; Molin et al., 1990; Skare et al., 1990; Åkesson et al., 1991; Langworth et al., 1991) the average uptake of mercury from amalgam fillings in Swedish subjects. The assumptions for this estimate were discussed in detail. The major ones were that a steady state for the turnover of mercury absorbed from amalgam fillings was reached, i.e., that uptake equaled elimination, and that nonurinary elimination of Hg absorbed from amalgam fillings is 2.5 times the urinary excretion, but with an uncertainty in this figure of ±1.5 times, i.e., a nonurinary elimination from equal to up to 4 times the urinary excretion. The average daily urinary excretion of Hg in the utilized studies was 3.1 μg, and the average contribution from non-amalgam sources was estimated to be 0.4 μg/day. Subtracting the latter from the former and adding the estimated nonurinary elimination results in an estimated total elimination, i.e., equal to uptake, of approximately 9 μg Hg/day. With respect to the uncertainty in the assumptions made and the data used the uncertainty in this estimate was found to range from 4 to 19 μg Hg/day. The large uncertainty in the estimate of nonurinary excretion is the major reason for the comparatively wide range. The overall average number of tooth surfaces restored with amalgam fillings for the subjects of the studies utilized was 33 surfaces.

Notwithstanding the wide range in this estimate, it excludes figures on the lower end of those presented for the pulmonary uptake of Hg vapor (see Table 1). It should be emphasized that none of those lower estimates contain any quantification of the uncertainties in the data used or the assumptions made.

Jokstad et al. (1992) have also presented an estimate of uptake of Hg from amalgam fillings based on urinary excretion of Hg. For a group with more than 36 amalgam surfaces, it was estimated that the daily uptake was 10 to 12 μg Hg/day. The report is not explicit on all assumptions made, but the elimination of Hg from fillings through the feces was assumed to be equal to the urinary elimination.

A result similar to Weiner and Nylander (1995) and Jokstad et al. (1992) was obtained by Clarkson et al. (1988b), who based on a entirely different model predicted a urinary concentration of 2.4 µg Hg/l in individuals with a daily pulmonary uptake of 8 µg Hg, i.e., a urinary concentration in agreement with empirical data from Swedish subjects, see above.

An estimate on the uptake of mercury from amalgam fillings may also be obtained by noticing that the average contribution to urinary Hg from amalgam fillings in Swedish subjects is approximately equal to the average contribution to urinary Hg from occupational exposure in dental personnel in Sweden (Nilsson and Nilsson, 1986a; Skare et al., 1990; Åkesson et al., 1991). It may therefore be assumed that the uptake of Hg from amalgam fillings is approximately equal to the uptake from the occupational environment in Swedish dental personnel. Based on measurements of the concentration of Hg vapor in Swedish dental offices (Nilsson and Nilsson, 1986b; Sällsten et al., 1992), the average pulmonary uptake of Hg from the occupational environment can be estimated to approximately 12 µg/day, (see also Weiner and Nylander, 1995).

The discrepancy between the estimates of uptake of Hg vapor based on urinary excretion of Hg and the lower estimates of pulmonary uptake of Hg vapor (see Table 1) is exemplified by the study by Berglund (1990) where data on urinary excretion of Hg is in agreement with the other Swedish studies, i.e., indicating a higher total uptake of Hg than the presented estimate of pulmonary uptake. It appears either that the pulmonary uptake is underestimated or that there is a significant contribution from other routes of uptake.

Measurements of concentrations of Hg in the urine indicate, in agreement with the reports of emission of mercury vapor from fillings, that the exposure to mercury may be highly varying even for individuals with comparable amounts of amalgam fillings (Aronsson et al., 1989; Åkesson et al., 1991; Langworth et al., 1991; Herrman and Schweinsberg, 1993).

There have been very few reports aiming at obtaining data on maximum values of Hg uptake from amalgam fillings. Sällsten and Barregård (1992), however, reported one case in which, based on urinary excretion, the estimated uptake of mercury from amalgam fillings was 50 to 100 µg/day. Other sources of exposure were carefully sought, but could not be found. Recently, two further cases were reported with a daily excretion of Hg in urine of 52 and 60 µg and an excretion of Hg in feces of 140 and 150 µg, respectively. No exposure to Hg at work or otherwise except amalgam fillings could be traced. Habitual chewing of gum and bruxism i.e., grinding of teeth were suggested as factors that could have been responsible for a high emission of mercury from amalgam fillings in these cases (Barregård and Sällsten, 1993). Epidemiologic studies show signs of bruxism in 20% or more in the adult Swedish population and that about 2% suffer from severe bruxism with extensive wear of teeth and restorative materials (Ekfeldt, 1989). Assuming a steady state in the turnover of Hg from fillings and that the part of the Hg that is eliminated in feces after absorption (as distinguished from the part that has just passed through) is at least as large as the amount that is eliminated in the urine (see also Weiner and Nylander, 1995) would indicate a daily uptake of more than 100 µg Hg from amalgam fillings in these cases.

Large amounts of Hg vapor and particulate matter are released during grinding of amalgam fillings, e.g., in connection with removal of old fillings. This is especially so if water-cooling is not employed (Buchwald, 1972; Reinhardt et al., 1983; Richards and Warren, 1985). Based on measurements of exhaled Hg vapor, Reinhardt et al. (1979) estimated an uptake of 169 µg as a result of removal of one amalgam filling covering two tooth surfaces. Water-cooling was not used.

Taskinen et al. (1989) reported on a patient who had been subjected to dental treatment involving extensive grinding of old amalgam fillings, both in order to prepare for bridge work and to replace broken amalgam fillings. The dentist used a high-speed drill, water-cooling, and aspiration during the treatment. The patient was treated during twelve visits over a period of 2.5 mo. No mercury analyses were carried out during the treatment, but 3 months after the cessation of the treatment the urinary Hg was about 20 µg/l. Subsequent measurements of Hg in the urine over 9 months showed concentrations decreasing to approximately 1 µg/l. The measured concentrations followed closely a curve with an elimination half-life of approximately 60 days. No details on the total amount of amalgam fillings before and after the treatment were given.

Based on the data presented, an assumed half-life of 60 days for the mercury retained during the treatment, which would indicate a urinary concentration of 55 µg/l at the cessation of the treatment, and an assumed urine volume of 1.5 l daily, it is possible to calculate the amount of Hg that was excreted in the urine as the result of the treatment. With these assumptions, approximately 7000 µg Hg attributable to the treatments was excreted in the urine after cessation of the treatment.* During the treatment period of 2.5 mo, the authors assumed that the urinary concentration had been 50 to 80 µg Hg/l. Using the lower figure as an average yields approximately 6000 μg Hg excreted in the urine during the treatment period. Adding these figures and assuming a nonurinary elimination of Hg that is as large as the urinary excretion (see above) indicates an average uptake during the twelve treatments of approximately 2000 μg Hg. This is obviously a rough estimate, but it is likely to at least show the order of magnitude of the absorption that may occur as a result of extensive grinding of old fillings.

G. MERCURY IN TISSUES OF AMALGAM BEARERS

A Swedish study based on a limited number of autopsy cases reported that cadavers with amalgam fillings had markedly higher levels of Hg in the brain cortex and pituitary glands than those with complete dentures, i.e., without amalgam fillings (Nylander, 1986).

In continued Swedish autopsy studies cases with no amalgam fillings showed a mean Hg concentration of 6.7 ng/g (2.4 to 12.2) in the occipital cortex, whereas cases with a moderate amount of amalgam fillings, i.e., an average of 21 tooth surfaces restored with amalgam, showed a mean concentration of 12.3 ng/g (4.8 to 28.7). There was a significant correlation between Hg concentration and number of amalgam surfaces (Nylander et al., 1987). Based on the reported regression coefficient, the predicted contribution to mercury levels in occipital cortex from 30 tooth surfaces restored with amalgam is approximately 7 ng Hg per gram of tissue.

In a study of American autopsy cases (Eggleston and Nylander, 1987), those with 0 to 1 occlusal surface restored with amalgam showed an average mercury level of 6.7 ng/g (1.9 to 22.1) and 3.8 ng/g (1.4 to 7.1) in occipital cortex and occipital medulla, respectively. In specimens from cases with 5 to 15 occlusal tooth surfaces restored with amalgam, the mean levels were 15.2 ng of Hg per gram (3.0 to 121.4) and 11.2 ng per gram (1.78 to 110.1) for occipital cortex and medulla, respectively.

In a similar German study, no amalgam-free group was included, but there was a significant correlation between number of amalgam surfaces and concentrations of Hg in brain tissue. The highest values were about 40 ng Hg per gram in three individuals with 25 to 30 amalgam surfaces (Schiele, 1988). Another German study showed a Hg concentration of on average approximately 20 ng/g tissue in amalgam bearers, 5 ng/g of which was organic Hg, i.e., indicating approximately 15 ng/g from amalgam fillings (Schupp, 1994).

In the Swedish autopsy studies, average Hg concentration in the renal cortex was 49 ng/g (range 21 to 105) in amalgam-free subjects, whereas subjects with moderate amounts of amalgam fillings showed a mean concentration of 433 ng/g Hg (range 48 to 810). The latter group had on the average 18 tooth surfaces restored with amalgam (Nylander et al., 1987).

In the German study by Schiele (1988), Hg concentrations in renal tissue were significantly correlated to the number of amalgam surfaces. Two of these cases had Hg levels of about 1500 ng/g. The reported regression coefficient predicted a contribution to mercury levels in renal tissue of approximately 800 ng Hg per gram as a result of 30 tooth surfaces restored with amalgam.

Another German study (Drasch et al., 1989) showed that inorganic Hg in renal cortex with a range from 2 to 1563 ng Hg/g was significantly correlated to the number of teeth with amalgam fillings or the number of tooth surfaces restored with amalgam. The regression coefficient was 21.3 (ng Hg/g tissue and surface), i.e., with 30 amalgam surfaces the regression equation predicted a contribution to the Hg concentration of approximately 640 ng Hg/g tissue. The concentration of organic mercury was low, i.e., range <0.5 to 22 ng Hg/g, and did not correlate to the number of amalgam surfaces.

These empirical data on concentrations of Hg in the brain and the kidneys are in agreement with the predicted contributions from a pulmonary uptake of 8 μg Hg vapor per day according to Clarkson et al. (1988b).

The Swedish studies also showed that Hg in pituitary glands (average 25.0 ng/g; range 6.3 to 77) was significantly correlated to number of amalgam surfaces (Weiner and Nylander, 1993). High concentrations of mercury have furthermore been shown in the dental pulp of teeth restored with amalgam fillings (Schiele et al., 1987).

Further analyses on the relationship between Hg concentrations in tissues and possible predictor variables have shown that Hg concentrations in the brain were more accurately predicted by number of amalgam surfaces multiplied with age than by just amalgam surfaces (Schiele, 1988; Weiner and Nylander, 1993). It was noted that the number of amalgam surfaces times age at the time of death for the Swedish subjects might be seen as a rough estimate of the life-time cumulated exposure. Thus, one hypothesis for this finding was that there is a significant compartment with very long biological half-

---

*$\int_0^\infty 1.5 * C_0 \exp(-\ln2/t_{1/2} * t) \, dt = 1.5 * 55 * [-t_{1/2}/\ln2 * \exp(-\ln/t_{1/2} * t)]_0^\infty = 1.5 * 55 * 60/0.69 > 7000.$

life of inorganic mercury in the brain. Another hypothesis that was put forward to explain the finding was that the capacity of elimination of inorganic mercury decreases with advancing age, thus resulting in increasing concentrations in the brain with age. In the Swedish material, this latter hypothesis corresponded with decreasing concentration of Hg in renal tissue with age, which might indicate a decreasing capacity for urinary excretion with age (Weiner and Nylander, 1993).

Experimental studies in sheep with radioactively labeled Hg in amalgam fillings placed in the teeth of the animals have confirmed the human data. Mercury from amalgam fillings was widely distributed and could be visualized in the CNS, pituitary gland, kidneys, liver, and gastrointestinal tract; interestingly, a very high activity was also shown in the jaw bone (Hahn et al., 1989). This might indicate a migration of Hg through the dentin into the pulp and surrounding tissues.

With regard to the question on the form and routes of the uptake of Hg from fillings, it might be noted that the distribution of Hg from amalgam fillings to the central nervous system clearly indicates a significant uptake in the form of elemental Hg. Uptake of Hg in the form of Hg^{2+} is less likely to explain the accumulation in the brain, as Hg^{2+} has a limited penetration of the blood-brain barrier (Berlin et al., 1966).

H. SUMMARY ON RELEASE, UPTAKE, AND DISTRIBUTION OF MERCURY FROM AMALGAM FILLINGS

Hg is released in substantial quantities from amalgam fillings. Such Hg is inhaled as elemental vapor and absorbed in the lungs. There is also probably an absorption of Hg from fillings in the gastrointestinal tract. The amount of Hg that is absorbed is not well known. The most reliable estimates of the total uptake on Hg from amalgam fillings are obtained by using data on urinary concentrations of Hg. Different models of calculating the uptake indicate a daily absorption of in average around 10 µg Hg in a group with an average of slightly more than 30 amalgam surfaces. Due to the uncertainty inherent in the models and the data used for such estimates, the uncertainty in this figure is quite large.

It should be stressed that the individual variation in the release and uptake of Hg from fillings is very large. There is a paucity of data on maximum exposure to Hg from fillings, but available data on excretion of Hg in urine and feces indicate that there are individuals that have an uptake that is in the magnitude of 100 µg/day. The occurrence of such high exposures may be related to bruxism (grinding of teeth) or habitual gum chewing.

Significant amounts of Hg from amalgam fillings are retained in different tissues, e.g., brain tissue, pituitary gland and renal tissue. An ordinary amount of amalgam fillings, i.e., of 25 to 30 tooth surfaces restored with amalgam, increase the brain Hg concentration by an average of approximately 10 µg/kg. The corresponding contribution to kidney concentration is about 600 to 700 µg Hg/kg. Details on the distribution to different possible target tissues or structures are not well known.

In addition to the continuous release of Hg from amalgam fillings, dental treatment involving grinding of old amalgam fillings results in an added exposure and uptake . Available data are insufficient for any detailed estimates of the magnitude of such uptake, but it might extend to the order of a thousand micrograms.

IV. EVALUATION OF POSSIBLE HEALTH RISKS ASSOCIATED WITH EXPOSURE TO MERCURY FROM AMALGAM FILLINGS

As shown above, restorative treatment with amalgam fillings results in a substantial exposure to a highly toxic substance, i.e., inorganic Hg. This suggests that amalgam in dental restorations should be looked upon as a pharmaceutical drug used to prevent illness (Berlin et al., 1992). It is therefore natural to compare with the situation for pharmaceuticals regarding documentation of safety and other regulations.

Assessment of the safety of pharmacological drugs, before registration and marketing, include animal studies and clinical trials. During their use in clinical practice, the safety is followed-up through reporting, registration, and investigation of suspected cases of adverse effects. In Sweden, restorative materials for teeth have not been included in the legislation concerning the safety of pharmaceuticals. From July 1993, materials for restoration of teeth are regulated by the Medical Device Act, which includes a demand of safety of the product. Before this date, there was no legislation specifically dealing with these products and their safety. However, it is still the manufacturer who decides if the requirements are fulfilled, and there is no demand on presentation of documentation of the safety and subsequent registration before marketing of the products.

In contrast to the situation for pharmaceuticals, in Sweden there is presently no legislation demanding that the manufacturers of dental restorative material declare the contents or quality of their products. It appears that the legal status is similar in other countries. It should be noticed that the Swedish legislation regarding dental restorative materials as well as that of other European countries are under development as a result of the legislative process in the European community.

A. ANIMAL STUDIES ON THE EFFECT OF AMALGAM FILLINGS

A study in sheep reported that implantation of amalgam fillings in the teeth of these animals resulted in effects on kidney function, e.g., a reduced glomerular filtration rate (GFR) as indicated by the inulin clearence (Boyd et al., 1991). This study was discussed at the evaluation by the Swedish MRC (Vimy and Lorscheider, 1992), but questions were raised as to the reliability of inulin clearande as a measure of GFR, and the reported result was not considered conclusive (Swedish MRC, 1992). Studies on primates have suggested that amalgam fillings might provoke an increase in both mercury- and antibiotic-resistant bacteria in the oral and intestinal flora (Summers et al., 1993). The latter effect although somewhat indirect might be of considerable importance if it can be verified. Both these studies need to be followed up before any conclusions may be drawn.

B. CASE REPORTS

A major source of information regarding possible side-effects of pharmaceuticals is reports of suspected cases. In Sweden, the reporting of suspected cases of adverse side-effects of drug therapy is mandatory for physicians, minor and well-known side-effects excluded. Such reports are registered and if necessary further investigated. Similar systems for surveillance and registration of possible side-effects of drug therapy are generally maintained in industrialized countries.

For restorative treatment with amalgam fillings there are in the literature a number of cases where alleged systemic side-effects have been reported, hardly any of which have been thoroughly enough investigated to allow any conclusions. A thorough evaluation, possibly including further investigation, of alleged cases and reports of suspected adverse reaction of systemic nature from dental amalgam restorative treatment would be highly desirable. The absence of such an evaluation is a major limitation in the, from several other aspects, meritable work by the committee that was appointed by the Swedish National Board of Health and Welfare, which reported their work in 1987 (SoS, 1987). Neither is such an evaluation included in other major and more recent evaluations initiated by governmental institutions (Swedish MRC, 1992; U.S. D.H.H.S, 1993). Such an evaluation is not within the limits of this work; however, a brief outline of reports available in the literature will be given.

Patients with suspected reaction of systemic nature due to Hg from amalgam fillings commonly report symptoms related to the central nervous system, e.g., fatigue, headaches, vertigo, visual disturbances, impaired short-term memory, insomnia, anxiety, difficulties in concentrating, and depression. Other symptoms oftenly reported are muscle and articular pains, rashes, gastrointestinal disturbances (including weight loss), metallic taste in the mouth, heart problems, and increased susceptibility to infections. In reported cases of suspected systemic side-effects of amalgam fillings, there is often a combination of several of these symptoms in addition to the presence of amalgam fillings, absence of any other plausible explanation and a longstanding improvement after the removal of these fillings. It is also reported that improvement in such cases often occurs gradually over a long period of time, from months to years (Roussy, 1891; Tuthill, 1898; Stock, 1926; Fleischmann, 1928; Stock, 1928; Stock, 1939; Strassburg and Schubel, 1967; Pleva, 1983; Hansson, 1986; Klock et al., 1989; Jaakkola and Grans, 1994). It has been speculated that improvement in health after the removal of amalgam fillings may be due to a placebo effect of the dental treatment (SoS, 1987; Klock et al., 1989).

C. SIDE-EFFECTS ASSOCIATED WITH EXTENSIVE GRINDING OF OLD AMALGAM FILLINGS

Patients with symptoms that are suspected to arise from amalgam filling often report acutely worsened symptoms in connection with drilling in old amalgam fillings. Such changes for the worse have been reported to occur after a couple of days, continuing for a day up to a few weeks (Stock, 1928; Hansson, 1986; SoS, 1987). In the 1920s, Alfred Stock already claimed that measures to minimize exposure as a result of grinding during the removal of amalgam fillings were of importance in avoiding worsened symptoms in cases of illness allegedly due to amalgam fillings (Stock, 1928). It might be of interest to note that cases of intoxication due to occupational exposure have been reported to develop increased susceptibility to the effects of mercury (Baader and Holstein, 1933).

Cases of adverse effects after extensive grinding of amalgam fillings have also been reported in individuals without any previous suspicion of amalgam related disease. Taskinen et al. (1989) reported one such case where the patient had been subjected to extensive amalgam work during 12 visits to her dentist over a period of 2.5 mo. One week after the beginning of the treatment, the patient began to develop symptoms. These included a sore throat, stomatitis, a bad taste in her mouth, a loss of the sense of smell, headaches, and dizziness. Later, she developed pains in the thorax, fever, an elevated sedimentation rate, a weakened sense of touch in her left hand, cold fingers, and a weakened hand grip. The patient lost 9 kg of weight and became labile and depressed. Analyses of Hg in urine showed that the patient had been subjected to a large amounts of Hg during the dental treatment (see also above). The authors concluded that the presented case indicated a potential risk to the patient, and the dentist, from extensive grinding of amalgam fillings.

In order to study the effect of amalgam removal on certain biological parameters Molin et al. (1990) on one occasion removed all amalgam fillings in ten healthy volunteers. The fillings were removed with ordinary methods used in general dentistry, i.e., by means of water-spray cutting and a vacuum evacuator. Of these subjects, two displayed symptoms following the removal. A male subject experienced severe dizziness about 18 h after the removal of amalgam fillings. The symptoms persisted for approximately 4 h and then slowly disappeared during the day. A female subject had vigorous attacks of vomiting about 8 h after the removal. Her vomiting disappeared after a few hours. These subjects were those with the largest number of amalgam fillings among those studied, i.e., amalgam fillings covering 42 and 36 tooth surfaces, respectively, were removed.

D. RISK ESTIMATION OF MERCURY FROM AMALGAM FILLINGS

In the absence of relevant animal studies, clinical trials, or thoroughly evaluated case reports that would allow any certain conclusions, one is forced to try to assess the possibility of a health risk associated with amalgam therapy from more indirect data. Available information that may be used is mainly data on the toxicity of mercury derived from animal studies or occupational studies and epidemiological studies of the effect of amalgam fillings.

E. COMPARISON WITH DATA ON EFFECTS OF OCCUPATIONAL EXPOSURES TO MERCURY

The major source of knowledge of the effects on humans of exposure to Hg vapor comes from studies of occupationally exposed individuals. Thus, data on the uptake of Hg from amalgam fillings might be compared with dose-response and dose-effect relationships obtained in occupational studies. A further advantage, apart from those already mentioned, in using urinary mercury as an index of exposure to Hg from fillings is that quantification of exposure in occupational studies is usually based on urinary concentrations of Hg. Therefore, comparison between exposure to mercury from fillings and in occupational environments is naturally made based on urinary mercury. There are, however, several difficulties involved in this approach, which may lead to an underestimation of the risk of exposure from amalgam fillings. Some reasons for this will be discussed below.

- **Limited sensitivity:** occupational studies often only comprise a moderate number of subjects. In the studies referred to below, there is typically around 50 subjects. Furthermore, most occupational studies only deal with a limited number of more or less well known effects on the critical organs, i.e., the brain and the kidneys. It is therefore not unlikely that there may be effects of exposure that are not detected in available studies, either because the effect is fairly uncommon (see also below) or because they are not included in the category of effects usually studied.
- **Selection bias:** generalizing the results from studies of the effect of a noxious agent in occupational groups to the general population may (due to selection of subjects under study) give rise to a negative bias. The selection operating may be divided into two parts. First, there is a **selection in** of individuals that are healthier than the general population into industrial cohorts, i.e., individuals with diseases or health problem that might affect the sensitivity to a noxious agent are less likely than healthy individuals to be entering the work force. This may to a large extent be due to self-selection. A special case of this selection concerns the exclusion of children and aging individuals, who might be more sensitive to noxious agents than adults. Second, there may be a **selection out** from the work force of individuals that are particularly sensitive to a noxious agent or other strain imposed by the work environment

(Rothman, 1986). The latter type of selection is particularly important to bear in mind as the major part of occupational data on the effect of exposure to Hg vapor come from of cross-sectional studies.

Furthermore, selection may also occur when subjects are identified for a study of the effect of Hg exposure; individuals with neurological, renal or other disorders possibly affecting the parameters commonly studied are usually excluded from the studied group. This is generally done without any investigation as to whether the disorder could be related to exposure (Roels et al., 1985; Barregård et al., 1988; Piikivi, 1989; Piikivi and Hänninen, 1989; Piikivi and Tolonen, 1989; Cárdenas et al., 1993).

- **Duration of exposure;** if the duration of exposure is of importance for the development of an effect, which cannot be excluded, then data from occupational studies may underestimate a possible risk of exposure from fillings. The typical duration of exposure in studies of occupational exposure to mercury is a median exposure of approximately 10 years with exposure durations of more than 20 years being less common. Obviously, the duration of exposure to mercury from fillings in many instances is much longer.

These factors may give rise to an underestimation of the risk of exposure to Hg if dose-response or dose-effects relationships obtained in occupational studies are extrapolated to the general population, e.g., applied also to amalgam exposure. In the absence of more conclusive data and bearing these facts in mind, it is still relevant to make such a comparison.

Several studies have shown effects on the group level of an exposure to Hg vapor that correspond to an average urinary excretion of Hg in the interval 20 to 50 µg Hg/l or g creatinine in morning urine samples. The effects shown in this exposure interval include effects on the central nervous system as indicated by influence on spontaneous tremor and psychomotor function, subjective symptoms such as fatigue and irritability, and effects on the kidneys as indicated by an increased urinary excretion of a number of endogenous substances, e.g., increased urinary excretion of lyzosomal enzymes (Foa et al., 1976; Fawer et al., 1983; Piikivi et al., 1984; Roels et al., 1985; Barregård et al., 1988; Piikivi, 1989; Piikivi and Hänninen, 1989; Piikivi and Tolonen, 1989; Soleo et al., 1990; Langworth et al., 1992; Cárdenas et al., 1993). Assuming a urinary excretion of somewhat less than 2 g creatinine/24 h (Ganong, 1975) shows that individuals with a high level of exposure from their fillings have an exposure that is of the same magnitude. It should be noticed, though, that no clinical significance has been established in connection with these effects.

The typical case of a chronic ***intoxication*** due to exposure to mercury vapor, mercurialism, is considered to include a triad of

1. mental symptoms
2. tremor
3. stomatitis and gingivitis

The mental symptoms include erethism, i.e., the victim becomes excitable and easy to disturb (Friberg and Vostal, 1972). However, a large number of effects may be seen as a result of exposure to high levels of Hg vapor. Other symptoms of toxicity from the nervous system are fatigue, insomnia, poor short-term memory, and difficulties in concentrating. Signs and symptoms from other organs include metallic taste in the mouth, weight loss, gastrointestinal disturbances, pain in the joints and limbs, renal disturbances, and increased susceptibility to infections (Baader and Holstein, 1933; Baldi et al., 1953; Granati and Scavo, 1961; Smith et al., 1970; Joselow et al., 1972).

Presently, clinical cases of Hg intoxication are rare, but they were not uncommon earlier. Such cases generally occurred at exposure levels that were higher than those reported above, but there are reported cases with comparatively low excretion of Hg in the urine, i.e., about 100 µg/24 h or even less (Agate and Buckel, 1949; Bidstrup et al., 1951; Friberg, 1951; Ladd et al., 1966). A problem in interpreting these data is that the analytical quality is not well known. Furthermore, a low excretion of Hg in connection with intoxication might be due to kidney damage.

A comparatively recent study of workers in chlor-alkali plants indicated an increase in symptoms of toxicity at exposures exceeding a time-weighted average of 100 µgHg/m^3. With exposures exceeding 200 µgHg/m^3, the prevalence of individuals showing signs of toxicity rose dramatically (Smith et al., 1970). The former level would correspond to a urinary excretion of approximately 100 µg Hg/g creatinine (Roels et al., 1987). Two recently reported cases of classical mercury poisoning had an urinary excretion of 160 and 340 µg/g creatinine, respectively (Skerfving, 1991). With a maximum uptake of Hg from fillings corresponding to a urinary excretion of approximately 50 µg/day, as indicated by available data, the margin of safety to the level where significant toxic effects may occur is very small.

The worst case scenario regarding exposure from copper amalgam as reported by NIOM would indicate the possibility of an exposure from such amalgams that is higher than level of exposure where significant toxic effects have been seen.

F. IMMUNOLOGICAL EFFECTS OF MERCURY

In animal studies, it has been shown that $HgCl_2$ given at doses not otherwise toxic may affect the the immune system. Depending upon the species and the strain tested, both autoimmunity and immunosuppression are observed. Susceptibility to autoimmunity is genetically controlled; four genes localized within and outside the major histocompatibility complex are involved. A number of cases of autoimmune glomerulonephritis in Hg-exposed individuals also suggest an association between this exposure and autoimmunity in humans (Druet, 1991). WHO (1991) concluded, based upon the evaluation in animals, that the most sensitive adverse effect for inorganic mercury risk assessment is the formation of mercuric-mercury-induced autoimmune glomerulonephritis. They also concluded that as a consequence of an immunological etiology there may well be a fraction of the population that is particularly sensitive to the effects of mercury and that it is presently not scientifically possible to set a level of exposure below which Hg-related symptoms will not occur in individual cases.

G. EPIDEMIOLOGICAL STUDIES

There are a limited number of epidemiological studies available. This is probably partly due to the difficulties in answering the question of possible health risks from amalgam fillings through such studies. Major difficulties are due to power problems and problems with confounding. A few studies have however received much attention in this respect, see for example the conference arranged by the Swedish Medical Research Council (Swedish MRC, 1992; Bengtsson, 1992) or the British Dental Association (Eley and Cox, 1993). In the absence of human data in the form of clinical trials, epidemiological studies are potentially a major source of knowledge of the effects of amalgam fillings. Therefore this matter will be discussed rather thoroughly.

It is well known that when the proportion of a population that may be negatively affected by a certain exposure is small, then a very large population is required to detect such an effect in a cohort or cross-sectional study. This is especially so in the case of possible adverse effects of amalgam fillings, since the symptoms that have been claimed to be due to amalgam fillings are common in many populations. For example, assuming that the symptom investigated occurs with a frequence of 1% due to a certain exposure and with a frequency of 10% among individuals not exposed, then, in the simplest case where one group of exposed and one group of non-exposed are compared, more than 14,000 individuals in each group will be needed to have a reasonably good chance (80% power) to obtain a statistically significant result. A less common occurrence among nonexposed and a more common incidence due to exposure implies that a smaller number is needed and vice versa (Armitage, 1971).

It also appears that there may be bias that is difficult to control for in estimates of the effect on health of amalgam filling obtained in epidemiological studies, since dental health may be related to other health parameters. Kampe et al. (1986) studied personality traits of Swedish adolescents. The personality patterns of 29 subjects aged 15 years with intact dentitions were studied by means of a personality inventory and compared with that of 41 subjects of the same age with repaired dentitions. The subjects with repaired dentitions showed significantly higher scores in the somatic anxiety and muscular tension scales; inferior, though nonsignificantly, results were also demonstrated for several other scales. The difference in personality patterns between the groups may be an effect of the fillings per se or may have other explanations, e.g., that being prone to anxiety may influence or be associated with factors that contribute to development of caries.

The result is in agreement with that of Havland and Larsson (1976), who studied 800 Swedish conscripts and reported that individuals with lower performance in the psychological tests displayed a higher caries index; this negative correlation was statistically significant.

Lavstedt and Sundberg (1989) investigated possible associations between dental fillings and ten different symptoms in a group of 1204 individuals. The data were collected in 1970 without any intent of studying this association and before the debate on dental amalgam had flared. The dentulous individuals were divided into five groups depending on the number of restored tooth surfaces (with amalgam or other restorative materials). A comparison was made between each of these exposure groups and the group with the lowest number of restored surfaces. A test for a possible trend in prevalence of symptoms among the exposure groups was also made. Dentulous individuals were furthermore compared to the edentulous. Standardization was made for a number of possible confounding factors, i.e., age group, socioeconomic status, gender, smoking habits, and number of remaining teeth.

There was a general tendency toward negative associations between amount of amalgam and symptoms. In 8 out 14 symptoms studied, the group with largest amount of fillings had a lower prevalence of symptoms than the least-exposed group. However, edentulous individuals, i.e., those without amalgam, had a lower prevalence of symptoms than dentulous individuals in 10 out of the 14 studied symptoms. For gastrointestinal disturbances, this difference was statistically significant ($p < 0.05$). Pointing in the same direction, there was also a significantly increased incidence of gastrointestinal disturbances among the group with the largest number of restored tooth surfaces($p < 0.05$). For depression, there was a decreased incidence ($p = 0.09$) among edentulous individuals as compared to the dentulous. For this symptom, there was also a borderline significant ($p = 0.10$) positive trend in the prevalence among the dentulous individuals. Thus, this study suggests a possible effect of amalgam fillings on the prevalence of gastrointestinal disturbances and depression. The findings might be due to chance, since a comparatively large number of associations were studied, or to confounding that was not possible to control for. A further problem relates to the poor exposure assessement.

Ahlqwist et al. (1988) studied the prevalence of 30 symptoms among 1024 dentolous Swedish women aged 38 to 72 years. Most of the symptoms studied occurred with a prevalence of 10 to 50%. Number of amalgam surfaces was estimated from a panoramic X-ray of the teeth. Age-adjusted risk ratios comparing women with more than 20 tooth surfaces restored with amalgam to women with 0 to 4 amalgam surfaces were calculated. For 14 symptoms, these were statistically significant. In all of these cases the associations were negative, i.e., less symptoms with more amalgam. It should be noticed that there was a strong correlation between number of amalgam surfaces and the number of remaining teeth, i.e., in this material a small number of amalgam surfaces was generally due to a significant loss of teeth. Thus, a low number of amalgam fillings, generally indicating poor dental health, was associated with poor general health. After adjusting calculation of correlations between number of amalgam surfaces and symptoms for number of remaining teeth three significantly negative correlations remained. After adjusting the correlations for socioeconomic group instead of number of remaining teeth, five significantly negative correlations remained.

Later, the same group reported similar results on the association between amalgam fillings and some "hard endpoints", i.e., cardiovascular disease, diabetes, and death, indicating a general tendency toward better health with more amalgam fillings (Ahlqwist et al., 1993).

In this context, it should also be added that the measure of exposure used, i.e., current exposure at the time of inclusion in the study, might be discussed. Based on studies on concentrations of Hg in postmortem material, it might be hypothesized that brain concentrations of Hg are related to accumulated exposure rather than to current exposure (Schiele, 1988; Weiner and Nylander, 1993; see also above).

Studies of patients with symptoms claimed to be related to dental amalgams have not shown higher mercury levels in plasma as compared with control groups (Björkman et al., 1993; Molin et. al., 1987). Assuming that amalgam fillings do not have a positive effect on health per se, then some of the associations between dental status and other health parameters reported by Ahlqwist et al. (1988) and Ahlqwist et al. (1993) may be due to a positive association between dental health and good health. Such an association between dental health status and general health is also a possible explanation for the associations reported by Kampe et al. (1986) and Havland and Larsson (1975). The findings of Lavstedt and Sundberg (1989) may differ since for some health parameters (depression and gastrointestinal disturbances) individuals with more fillings (comparatively poor dental status) displayed more symptoms than individuals with less fillings and at the same time, individuals without permanent teeth (the poorest dental status) showed fewer symptoms than individuals with their own teeth. These findings are difficult to explain with the type of confoundings outlined above.

H. SUMMARY ON RISK ESTIMATION

There are no clinical trials or animal studies that allow conclusions on the possibility of health risks associated with amalgam fillings. There are a considerable number of individuals who claim that they have become seriously ill from their amalgam fillings. Available reports are difficult to interpret, but it is problematic that several evaluations of the possibility of health risks related to amalgam fillings dismisses available case reports without any detailed motivation or any further investigation.

Comparison of the continuous exposure to Hg from amalgam fillings with dose-effect and dose-response relationships as obtained in occupational studies show no definite safety margin to the level of exposure where effects, e.g., on the nervous system and the kidneys, have been shown. It should be noticed though that no clinical significance has been established in connection with these effects. Still, the absence of a substantial safety margin is clearly worrisome, especially as dose-response relationships obtained from occupationally exposed groups are likely to underestimate the risk in the whole population. The level of exposure to Hg vapor where significant toxic effect may occur is not well established, but it appears that the safety margin for cases of maximum exposure to mercury from amalgam fillings is not satisfactory.

Special consideration must be given to the possibility that an effect on the immune system, resulting in autoimmunity or immunosuppression, might occur in susceptible individuals at a level of exposure where other effects are not seen. Available epidemiological studies suffer from lack of statistical power to detect reasonably uncommon effects of amalgam exposure, a likely bias in the risk estimates, and poor exposure assessments. Therefore they do not allow any conclusions on the possibility of systemic side-effects from amalgam therapy. Hence further studies are needed.

In addition to continuous exposure there also might be a potential risk of an unacceptable acute exposure of mercury from extensive work involving grinding of old amalgam fillings.

Copper amalgams, which disintegrate very fast, must no doubt be considered a potential health hazard.

V. EFFECTS ON THE FETUS

A special case concerns the possibility of effects on the unborn. Hg in the form of methyl-Hg is a well known neuroteratogen (WHO, 1990). Studies in squirrel monkeys have shown histopathological changes in the fetal brain after exposure to Hg vapor similar to those that appear after exposure to methyl-Hg during pregnancy (Berlin et al., 1992). There are no studies on the possible effects of fetal exposure to Hg from amalgam fillings, but it has been recommended that extensive dental work involving amalgams should be avoided during pregnancy (SoS, 1987). This recommendation has, however, later been stated to be scientifically unfounded. (Swedish MRC, 1992).

A considerable transfer of Hg vapor to the fetus has been shown in studies of exposure to Hg vapor in pregnant guinea pigs. Of particular importance seems to be the fact that considerable redistribution of Hg took place from the liver to the central nervous system and particularly to the kidneys, after delivery (Yoshida et al., 1986; Yoshida et al., 1990). Studies in guinea pigs also indicate that inorganic Hg may be transferred to the newborn via breast milk (Yoshida et al., 1992). Vimy et al. (1990) confirmed a transfer to the fetus of Hg from radioactively labeled amalgam fillings implanted in pregnant sheep. Hg was demonstrated in several organs of the fetuses, e.g., CNS, pituitary gland, kidneys, and liver. A possible transfer of Hg via milk was also indicated.

The magnitude of fetal exposure to inorganic Hg resulting from exposure to Hg vapor in humans is not well known. Eggleston and Nylander (1987) reported concentrations of Hg in autopsy specimens from a pregnant woman with a relatively small amount of amalgam fillings as well as from the 7 months fetus. Analysis of the mother's occipital cortex revealed 9.9 ng Hg/g with 6.7 ng/g also in the occipital medulla. Analysis of the fetal brain revealed somewhat lower levels or 6.7 ng/g in the cortex with 2.8 ng/g also in the medulla. Drasch et al. (1994) showed a significant association between the number of teeth with amalgam fillings in the mother and concentrations of Hg in the liver and the kidneys of autopsied fetuses. Hg concentrations in the kidneys and the brain of infants 11 to 50 weeks of life were also associated with number of teeth with amalgam fillings in the mother.

Human data on the effects on spontaneous abortions from exposure to Hg vapor derived from occupational exposure are available, but they are limited and inconclusive (WHO, 1991).

Sikorski et al. (1987) reported 6 congenital malformations 5 of which were spina bifida out of 117 pregnancies in a group of dental personnel. This is an extremely high rate of spina bifida. A Swedish epidemiological register study did not show a higher incidence of spina bifida in children of dental workers. (Ericson and Källén, 1989).

Animal studies have shown an effect on the anatomical and functional development of the nervous system as a result of prenatal or neonatal exposure to Hg vapor (Berlin et al., 1992; Fredriksson et al., 1992; Danielsson et al., 1993). The concentrations of Hg vapor and the duration of exposure used were such that for a human they would correspond to a daily uptake of between 10,000 µg/day and 25 µg/day. The assumptions in these calculations were a pulmonary ventilation of 10 l/min and 80% absorption in the lungs.

Human data on a possible association between exposure to inorganic mercury and the functional development of organs are sparse, but a recent case-referent study reported an association between mental retardation and occupational exposure to mercury in the late pregnancy (Roelevald et al., 1993). It should be noticed that there was no *a priori* hypothesis regarding Hg and a large number of associations were investigated; thus the result may be a chance finding, and further studies are needed.

A. SUMMARY ON EFFECTS ON THE FETUS

Available data indicate that Hg from amalgam fillings is transported to the fetus. Animal data show an effect of exposure to Hg vapor on the anatomical and functional development of the nervous system. It is difficult to extrapolate the exposure levels, but the very high exposure that may result from dental treatment involving extensive grinding of old amalgam fillings must be considered a potential risk to the fetus and should thus be avoided. The data also give rise to a concern for a potential risk of an effect on the functional development of organs as a result of Hg that is continuously transported from the mother to the fetus during pregnancy and breastfeeding.

VI. CONCLUSION

Mercury is continuously released from dental amalgam fillings. In addition to this continuous release, there may be a large added exposure as a result of dental treatments involving drilling in old amalgam fillings. The Hg released from amalgam fillings is taken up in the form of vapor in the lungs and most likely also in the gastrointestinal tract. It is distributed to various tissues, including the brain and the kidneys. The distribution to different possible target tissues is not well known though.

There are a considerable number of individuals who claim that they have become ill from their amalgam fillings. It is problematic that so little effort has been made to evaluate such cases thoroughly. In the absence of a thorough evaluation of such cases, they cannot just be dismissed.

Comparison of data on the release and uptake of Hg from amalgam fillings with available human and animal data on the toxicity of inorganic Hg indicates that the level of exposure to Hg from amalgam fillings should be a matter of serious concern. A special consideration must be given to the potential risk of effects on the fetus. There is little doubt that the recommendation put forward in 1987 that pregnant women should avoid extensive dental treatments involving amalgam is warranted.

It follows that the practice of using amalgams as tooth restorative materials cannot be defended by claims that they are safe, for there is no basis for such a statement. It can only be motivated based on a judgement that, as is the case for pharmaceuticals, the benefits outweigh the risks. However, presently available data may not allow for any conclusions. In this context, the question also arises as to whether the potential risk with alternative materials are the same, greater, or less. It appears, though, that data for the toxicological evaluation of possible side-effects of many alternative material are also insufficient.

In our view, the potential dangers with these materials in the past have been grossly neglected. An illustrative example is the usage of the copper amalgam in child dentistry. This restorative material must be considered a health risk due to its large release of toxic metals. Previously, this release (instead of being viewed as a potential danger) was actually considered an advantage from the odontological treatment point of view, in that it inhibited bacterial growth and thus reduced the risk of secondary caries.

It is disconcerting that large proportions of the population in developed countries is exposed to toxic substances from a category of products that are so poorly evaluated from a toxicological point of view. It appears that the legal status for many materials used in dentistry do not even include a demand on documentation of content and quality of components of the products used. The difference compared with the demands on pharmaceuticals is most striking, and yet many dental restorative materials have a far wider usage than even the most commonly used pharmaceuticals. Negligence in the past with respect to the possible health hazards cannot justify a continuing lack of concern in the potential risks with products used in dentistry.

REFERENCES

Abraham, J. E., Svare, C. W., and Frank, C. W., *J. Dent. Res.*, 63(1), 71–73, 1984.
ADA, *Dental Amalgam*, American Dental Association, 1985.
Agate, J. N. and Buckel, M., *Lancet*, ii, 451–454, 1949.

Ahlqwist, M., Bengtsson, C., Furunes, B., Hollender, L., and Lapidus, L., *Community Dent. Oral. Epidemiol.*, 16, 227–231, 1988.
Ahlqwist, M., Bengtsson, C., and Lapidus, L., *Community Dent. Oral Epidemiol.*, 21, 40–44, 1993.
Åkesson, I., Schutz, A., Attewell, R., Skerfving, S., and Glantz, P. O., *Arch. Environ. Health,* 46, 102–109, 1991.
Aposhian, H. V., Bruce, D. C., Alter, W., Dart, R. C., Hurlbut, K. M., and Aposhian, M. M., *FASEB J.,* 6, 2472–2476, 1992.
Armitage, P., Blackwell, London, 1971.
Aronsson, A.-M., Release of mercury vapour from amalgam fillings, Department of Hygiene, Karolinska Institute, Solna, Sweden, 1988 (In Swedish).
Aronsson, A.-M., Lind, B., Nylander, M. and Nordberg, M., *Biol. Metals,* 2, 25–30, 1989.
Baader, E. W. and Holstein, E., *Das Quecksilber seine gewinnung, technische Verwendung und Giftwirkung mit eingehender Darstellung der Quecksilbervergiftung nebst Therapie und Prophylaxe,* Veröffentlichungen aus dem Gebiete der Medizinalverwaltung, Verlagsbuchhandlung von Richard Schoetz, Berlin, 1933.
Baldi, G., Vigliani, E. C., and Zurlo, N., *Med. Lav.,* 44, 182–198, 1953.
Barregård, L., Hultberg, B., Schütz, A., and Sällsten, G., *Int. Arch. Occup. Environ. Health,* 61, 65–69, 1988.
Barregård, L. and Sällsten, G., *Hygiea,* 102(3), 118–118, 1993 (In Swedish).
Bauer, J. G. and First, H. A., *Calif. Dent. J.,* 47–61, 1982.
Bengtsson, C., in *Potential Biological Consequences of Mercury Released from Dental Amalgams, Proceedings from a Conference,* B. Bergman, H. Boström, K.S. Larsson, H. Löe, Eds., Swedish Medical Research Council, Stockholm, 33–42, 1992.
Bergman, M., in *Potential Biological Consequences of Mercury Released from Dental Amalgams, Proceedings from a Conference,* B. Bergman, H. Boström, K.S. Larsson, H. Löe, Eds., Swedish Medical Research Council, Stockholm, 43–58, 1992.
Berglund, A., *J. Dent. Res.,* 69(10), 1646–1651, 1990.
Berlin, M., in *Handbook on the Toxicology of Metals, 2nd edition, Vol II: Specific Metals,* L. Friberg, G. F. Nordberg, and V. B. Vouk, Eds., pp. 387–445. Elsevier, Amsterdam, 1986.
Berlin, M., Hua, J., Lögdberg, B., and Warfvinge, K., *Fundam. Appl. Toxicol.,* 19, 324–326, 1992.
Berlin, M., Jeksell, L. G., and von Ubish, H., *Arch. Environ. Health,* 12, 33–42, 1966.
Bidstrup, P. L., Bonnel, J. A., Harvey, D. G., and Locket, S., *Lancet,* ii, 856–861, 1951.
Björkman, L. and Lind, B., *Scand. J. Dent. Res.,* 100, 354–360, 1992.
Björkman, L., Langworth, S., Lind, B., Elinder, C., Nordberg, M. J., *Trace Elem. Electrolytes Health Dis.*, 7, 157–164, 1993.
Bornmann, G., Henke, G., Alfes, H., and Möllman, H., *Arch. Toxikol.,* 26, 203–209, 1970.
Boyd, N. D., Benediktsson, H., Vimy, M. J., Hooper, D. E., and Lorscheider, F. L., *Am. J. Physiol.,* 261, R1010–R1014, 1991.
Brune, D. and Evje, D. M., *Sci. Total Environ.,* 44, 51–63, 1985.
Brune, D., Gjerdet, N., and Paulsen, G., *Scand. J. Dent. Res.,* 91, 66–71, 1983.
Buchwald, H., *Am. Ind. Hyg. Assoc. J.,* 33, 492–502, 1972.
Cárdenas, A., Roels, H., et al., *Br. J. Ind. Med.,* 50, 17–27, 1993.
Cherian, G., Hursch, J. B., Clarkson, T. W., and Allen, J., *Arch. Environ. Health,* 33, 109–114, 1978.
Clarkson, T., *Fundam. Appl. Toxicol.,* 19, 320–321, 1992.
Clarkson, T. W., Friberg, L., Nordberg, G. F., and Sager, P. R., in *Biological Monitoring of Toxic Metals,* T.W. Clarkson, L. Friberg, G.F. Nordberg, and P.R. Sager, Eds., Plenum Press, New York, 199–246, 1988a.
Clarkson, T. W., Friberg, L., Hursh, J. B., and Nylander, M., in *Biological Monitoring of Toxic Metals,* T.W. Clarkson, L. Friberg, G.F. Nordberg and P.R. Sager, Eds., Plenum Press, New York, 247–264, 1988b.
Danielsson, B. R. G., Fredriksson, A., Dahlgren, L., Teiling Gårdlund, A., Olsson, L., Dencker, L., and Archer, T., *Neurotoxicol. Toxicol.,* 15, 391–396, 1993.
Drasch, G., Schupp, I., and Günther, G., in *Proc. of the 6th International Trace Element Symposium* in Leipzig, M. Anke, W. Baumann, H. Bräunlich, C. Brückner, B. Groppel, and M. Grün, Eds., Der Friedrich Schiller Universität, Jena, 1653–1659, 1989.
Drasch, G., Schupp, I., Höfl, H., Reinke, R., and Roider, G., *Eur. J. Pediatr.,* 153(8), 607–610, 1994.
Druet, P., in *Advances in Mercury Toxicology,* T. Suzuki, Ed., Plenum Press, New York, 315–409, 1991.
Eggleston, D. W. and Nylander, M., *J. Prosthet. Dent.,* 58, 704–707, 1987.
Ekfeldt, A., Incisal and occlusal tooth wear and wear of prosthodontics materials. An epidemilogical and clinical study, Academic Thesis, *Swed. Dent. J.* (Suppl 65), 1989.
Eley, B. M. and Cox, S. W., *Br. Dent. J.,* 175, 161–168, 1993.
Enwonwu, C. O., *Environ. Res.* 42, 257–274, 1987.
Ericson, A. and Källén, B., *Int. Arch. Occup. Environ. Health,* 61, 329–333, 1989.
Fawer, R. F., Ribaupierre, Y., Guillemin, M. P., Berode, M., and Lob, M., *Br. J. Ind. Med.,* 40, 204–208, 1983.
Fleischmann, P., *Dtsch. Med. Wochenschr.,* 54, 304–307, 1928.
Foa, V., Caimi, L., Amante, L., Antonini, C., Gattinoni, A., Tettamanti, G., Lombardo, A., and Guiliani, A., *Int. Arch. Occup. Environ. Health,* 37, 115–124, 1976.
Fredriksson, A., Dahlgren, L., Danielsson, B., Eriksson, P., Dencker, L., and Archer, T., *Toxicology,* 74, 151–160, 1992.
Friberg, L., *Nordisk Hygienisk Tidskrift,* 32, 240–249, 1951.
Friberg, L. and Vostal, J., *Mercury in the Environment,* CRC Press, Boca Raton, Florida, 1972.

Frykholm, K. O., Mercury from dental amalgam its toxic and allergic effects and some comments on occupational hygiene, *Acta Odontol. Scand.,* 15, 1957 (Suppl. 22), 1–108.
Gay, D. D., Cox, R. D., and Reinhardt, J. W., *Lancet,* i, 985–986, 1979.
Granati, A. and Scavo, D., *Folia Med.,* 44, 529–545, 1961.
Gross, M. J. and Harrison, J. A., *J. Appl. Electrochem.,* 19, 301–310, 1989.
Hahn, L. J., Kloiber, R., Vimy, M. J., Takahashi, Y., and Lorscheider, F. L., *FASEB J.,* 3, 2641–2646, 1989.
Hansson, M., *TF-bladet/Tidskrift för tandhälsa,* 7(1), 3–11, 1986 (In Swedish).
Havland, A. and Larsson, P.-G., *Tandläkartidningen,* 68, 6–11, 1976 (In Swedish).
Herrman, M. and Scweinsberg, F., *Zbl. Hyg.,* 194, 271–291, 1993.
Horstedt-Bindslev, P., Magos, L., Holmstrup, P., and Arenholdt-Bindslev, D., Eds., *Dental Amalgam — A Health Hazard?,* Munksgaard, Copenhagen, 1991.
Jaakkola, K. and Grans, L., *Amalgaamisairaudet ja antioksidanttihoito,* Mividata OY, Tampere, 1994 (In Finnish).
Jokstad, A., Thomassen, Y., Bye, E., Clench-Aas, J., and Aaseth, J., *Pharmacol. Toxicol.,* 70, 308–313, 1992.
Joselow, M. M., Louria, D. B., and Browder, A. A., *Ann. Intern. Med.,* 76, 119–130, 1972.
Kampe, T., Edman, G., and Molin, C., *Acta Odontol. Scand.,* 44, 95–102, 1986.
Klock, B., Blomgren, J., Ripa, U., and Andrup, B., *Tandläkartidningen,* 81(23), 1297–1302, 1989 (In Swedish).
Knolle, G., Ed., *Amalgam — Pro und Contra,* Deutscher Ärzte-Verlag, Köln, 1988.
Ladd, A. C., Zuskin, E., Valic, F., Almonte, J. B., and Gonzales, T. V., *J. Occup. Med.,* 8(3), 127–131, 1966.
Langan, D. C., Fan, P. L., and Hoos, A. A., *J. Am. Dent. Assoc.,* 115, 867–880, 1987.
Langworth, S., Almkvist, O., Söderman, E. and Wikström, B.-O., *Br. J. Ind. Med.,* 49, 545–555, 1992.
Langworth, S., Elinder, C-G., Göthe, C-J., and Vesterberg, O., *Int. Arch. Occup. Environ. Health,* 63, 161–167, 1991.
Lavstedt, S. and Sundberg, H., *Tandläkartidningen,* 81(3), 81–88, 1989 (In Swedish).
Lund, B. O., Miller, D. M., and Woods, J. S., *Biochem. Pharmacol.,* 45(10), 2017–2024, 1993.
Mackert, J. R., *J. Dent. Res.,* 66, 1775–1780, 1987.
Miller, D. M. and Woods, J. S., *Chem. Biol. Interact.,* 88(1), 23–35, 1993.
Molin, M., Bergman, B., Marklund, S. L., Schütz, A., and Skerfving, S., *Acta Odontol. Scand.,* 48(3), 189–202, 1990.
Molin, M., Marklund, S.., Bergman, B., Bergman, M., Stenman, E., *Scand. J. Dent. Res.,* 95, 328–334, 1987.
Nilsson, B. and Nilsson, B., *Swed. Dent. J.,* 10, 221–232, 1986a.
Nilsson, B. and Nilsson, B., *Swed. Dent. J.,* 10, 1–14, 1986b.
NIOM, Cadmium in copper amalgams, Scandinavian Institute of Dental Materials, no. 51/81, 1981 (In Norwegian).
Nylander, M., *Lancet,* i, 442, 1986.
Nylander, M., Friberg, L., Eggleston, D., and Björkman, L., *Swed. Dent. J.,* 13, 235–243, 1989.
Nylander, M., Friberg, L., and Lind, B., *Swed. Dent. J.,* 11, 179–187, 1987.
Olsson, S. and Bergman, M., *J. Dent. Res.,* 71(2), 414–423, 1992.
Olstad, M. L., Holland, R. I., Wandel, N., and Pettersen, A. H., *J. Dent. Res.,* 66, 1179–1182, 1987.
Patterson, J. E., Weissberg, B. G., and Dennison, P. J., *Bull. Environ. Contam. Toxicol.,* 34, 459–468, 1985.
Piikivi, L., *Int. Arch. Occup. Environ. Health,* 61, 391–395, 1989.
Piikivi, L. and Hänninen, H., *Scand. J. Work Environ. Health,* 15, 69–74, 1989.
Piikivi, L., Hänninen, H., Martelin, T., and Mantere, P., *Scand. J. Work Environ. Health,* 10, 35–41, 1984.
Piikivi, L. and Tolonen, U., *Br. J. Ind. Med.,* 46, 370–375, 1989.
Pleva, J., *J. Orthomol. Psychiatr.,* 12, 184–193, 1983.
Rahola, T., Hattula, T., Korolainen, A., and Miettinen, J. K., *Ann. Clin. Res.,* 5, 214–219, 1973.
Reinhardt, J. W., Boyer, D. B., Gay, D. D., Cox, R., Frank, C. W., and Svare, C. W., *J. Dent. Res.,* 58(10), 2005, 1979.
Reinhardt, J. W., Chan, K. C. and Schulein, T. M., *J. Prosthet. Dent.,* 50, 62–64, 1983.
Richards, J. M. and Warren, P. J., *Br. Dent. J.,* 155, 231–232, 1985.
Roelevald, N., Zielhuis, G. A., and Gabreëls, F., *Br. J. Ind. Med.* 50, 945–954, 1993.
Roels, H., Abdeladim, S., Ceulemans, E., and Lauwerys, R., *Ann. Occup. Hyg.,* 31, 135–145, 1987.
Roels, H., Gennart, J.-P., Lauwerys, R., Buchet, J.-P., Malchaire, J., and Bernard, A., *Am. J. Ind. Med.,* 7, 45–71, 1985.
Rothman, K. J., *Modern Epidemiology,* Little, Brown, Boston, 1986.
Roussy, M. L., *Schweiz. Vierteljahrschr. Zahnheilk.,* 1, 97–103, 1891.
Sällsten, G., Barregård, L., Langworth, S., and Vesterberg, O., *Appl. Occup. Environ. Hyg.,* 7, 434–440, 1992.
Schiele, R., in *Amalgam — Pro und Contra Statements,* Knolle, G., Ed., Deutsche Ärzte-Verlag, Köln, 123–131, 1988.
Schiele, R., Hilbert, M., Schaller, K.-H., Weltle, D., Valentin, H., and Kröncke, A., *Dtsch. Zahnärztl. Z.,* 42, 885–889, 1987.
Schupp, I. B., Untersuchungen an menschlichen organen zur frage der Quecksilberbelastung durch Zahnamalgam und weitere faktoren, Universität München, Thesis, 1994.
Sikorski, R., Juszkiewicz, T., Paszkowski, T. and Szprengier-Juskiewicz, T., *Int. Arch. Occup. Environ. Health,* 59, 551–557, 1987.
Skare, I. and Engqvist, A., *Arch. Environ. Health,* 49, 384–394, 1994.
Skare, S., Bergström, T., Engqvist, A. and Weiner, J. A., *Scand. J. Work Environ. Health,* 16, 340–347, 1990.
Skerfving, S., in *Advances in Mercury Toxicology,* T. Suzuki, Eds., Plenum Press, New York, 411–425, 1991.
Smith, R. G., Vorwald, A. J., Patil, L. S., and Mooney, T. F., *Am. Ind. Hyg. Assoc. J.,* 31, 687–700, 1970.
Snapp, K. R., Boyer, D. B., Peterson, L. C., and Svare, C. W., *J. Dent. Res.,* 68, 780–785, 1989.
Soleo, L., L, U. M., Petrera, V., and Ambrosi, L., *Br. J. Ind. Med.,* 47, 105–109, 1990.

SoS, *Mercury/Amalgam, Health Risks,* Socialstyrelsen Redovisar 1987:10, National Board of Health and Welfare, Stockholm, 1987. (summary of each chapter in English).
Statens Helsetilsyn, Dentale biomateriale: NIOMS lister over sertifiierte 1994/95 — Fremtidig godkjenningsordning etter EU-direktiv, registrering av biverkninger och journalföring, no. IK-11/94, Statens Helsentilsyn, Oslo, 1994.
Stock, A., *Z. Angew. Chemie,* 39, 984–989, 1926.
Stock, A., *Z. Angew. Chemie,* 41, 663–686, 1928.
Stock, A., *Zahnärtzl. Rundsch.,* 48, 403–407, 1939.
Stock, A. and Cucuel, F., *Z. Angew. Chemie,* 47, 641–647, 1934.
Strassburg, M. and Schubel, F., *Dtsch. Zahnärtzl.,* Z. 22, 3–9, 1967.
Summers, A. O., Wireman, J., et al., *Antimicrob. Agents Chemother.,* 37(4), 825–834, 1993.
Svare, C. W., Peterson, L. C., Reinhardt, J. W., Boyer, D. B., Frank, C. W., Gay, D. D., and Cox, R. D., *J. Dent. Res.,* 60(9), 1668–1671, 1981.
Svensson, B-G., Schütz, A., Nilsson, A., Åkesson, I., Åkesson, B., and Skerfving, S., *Sci. Total Environ.,* 126, 61–74, 1992.
Swedish MRC, *Potential Biological Consequences of Mercury Released from Dental Amalgam,* A State of the Art Document, Swedish Medical Research Council, Stockholm, 1992.
Taskinen, H., Kinnunen, E., and Riihimäki, V., *Scand. J. Work Environ. Health,* 15, 302–304, 1989.
Tuthill, J. Y., *Brooklyn Med. J.,* 12(12), 725–742, 1898.
U.S. D.H.H.S., *Dental Amalgam: A Scientific Review and Recommended Public Health Service Strategy for Research, Education and Regulation,* U.S. Department of Health and Human Services, Washington, D.C., 1993.
Vimy, M. J. and Lorscheider, F. L., *J. Dent. Res.,* 64(8), 1069–1071, 1985a.
Vimy, M. J. and Lorscheider, F. L., *J. Dent. Res.* 64(8), 1072–1075, 1985b.
Vimy, M. J. and Lorscheider, F. L., in *Potential Biological Consequences of Mercury Released from Dental Amalgam,* B. Bergman, H. Boström, K.S. Larsson, and H. Löe Eds., Swedish Medical Research Council, Stockholm, 191–200, 1992.
Vimy, M. J., Takahashi, Y., and Lorscheider, F. J., *Am. J. Physiol.,* 258, R939–R945, 1990.
Weiner, J. A. and Nylander, M., *Sci. Total Environ.,* 138, 101–115, 1993.
Weiner, J. A. and Nylander, M., *Sci. Total Environ.,* 168, 255–265, 1995.
Weiner, J. A., Nylander, M., and Berglund, F., *Sci. Total Environ.,* 99, 1–22, 1990.
WHO, *Environmental Health Criteria 86: Mercury — Environmental Aspects,* World Health Organization, Geneva, 1990.
WHO, *Environmental Health Criteria 118: Inorganic Mercury,* World Health Organization, Geneva, 1991.
Willerhausen-Zönnchen, B., Zimmerman, M., Defregger, A., Schramel, P., and Hamm, G., *Dtsch. Med. Wochenschr.,* 117, 1743–1747, 1992.
Wollf, M., Osborne, J. W., and Hansson, A. L., *Neurotoxicology,* 4(3), 201–204, 1983.
Yoshida, M., Satoh, H., Kishimoto, T., and Yamamura, Y., *J. Toxicol. Environ. Health,* 35, 135–139, 1992.
Yoshida, M., Satoh, H., Kojima, S., and Yamamura, Y., *J. Trace Elements Exp. Med.,* 3, 91–109, 1990.
Yoshida, M., Yamamura, Y., and Satoh, H., *Arch. Toxicol.,* 58, 225–228, 1986.
Ziff, S., *Silver Dental Fillings — The Toxic Time Bomb. Can the Mercury in Your Dental Fillings Poison You?,* Aurora Press, New York, 1984.

Chapter 31

Chelation Therapies for Metal Intoxication

Carol R. Angle

The relative infrequency of metal poisoning in humans delays the information necessary to approve new agents and new uses of established pharmaceuticals. Controlled clinical trials to document safety and efficacy are often lacking. This review is an attempt to present consensus opinions of current therapy and the rationale for careful consideration of what, in most cases, is still experimental treatment.

The structures of the twelve chelating agents discussed in this chapter are presented in Table 1 along with their FDA-approved and non-approved indications. The usual therapeutic dose of each chelator is given in the table with exceptions noted in the text. The molecular weights are supplied to facilitate the calculation of equimolar doses.

The initial selection of these chelating agents is derived from their specific affinity for and the stability of their complexes with target metals. The octanol to water partition predicts oral absorption of the water soluble and the relative intracellular distribution of the lipophilic chelates. The majority of chelates have a predominantly extracellular distribution and are most effective during the distribution phase of a metal and in the decorporation of metals characterized by relatively rapid exchanges between intracellular and extracellular fluid. Toxicity increases with intracellular distribution. As shown in Figure 1, the mouse LD_{50} i.p. of dimercaprol (BAL), *D*-penicillamine (PCN) and desferrioxamine (DFO), all of which have a significant intracellular distribution, approaches or exceeds the LD_{50} oral of many metal compounds. Intracellular distribution also explains the relatively greater toxicity of 2,3 dimercapto-1-propanesulfonic acid (DMPS) compared with *meso* 2,3 dimercaptosuccinic acid (DMSA) and of calcium trisodium diethylenetetramine pentetic acid (DTPA) compared with calcium disodium ethylene diaminetetraacetic acid ($CaNa_2EDTA$) (Jones, 1991). The metal toxicities reviewed are those for which there are human data defining the safety and efficacy of chelation therapy: aluminum, arsenic, copper, iron, lead, manganese, mercury, thallium, and briefly, the radionuclides of the lanthanides and transuranics.

I. ALUMINUM

A. MICROCYTIC ANEMIA AND ALUMINUM TOXICITY

Anemia due to aluminum toxicity continues to occur in chronic renal failure patients despite the elimination of aluminum from dialysate, parenteral and alimentary solutions, oral phosphate binders and medications (Min and D'Elia, 1992; Shoskes et al., 1992).

Aluminum anemia is characterized as a microcytic, hypochromic, or normochromic anemia with increased erythrocyte protoporphyrins (EP) and inhibition of red cell delta-amino levulinic acid dehydratase (δ ALA D) activity. It is poorly responsive to the usual 3 months of erythropoietin treatment of renal failure patients with anemia, defined as a hemoglobin ≤7 g/dl. The anemia may be complicated

Table 1 Therapeutic Chelators

Name	Common Name	Indications (non approved in italics)	Dosage
Calcium trisodium pentetate	DTPA Na₃Ca DTPA Pentetate Calcium trisodium	*Lanthanide and transuranic radionuclides*	15 mg/kg/d IV, IM, or SC in 4–6 doses
	Ditripentate® (Heyl, Berlin)* MW 497.4	*Reversal DFO toxicity*	0.5–1 g SC × 5 d (Wonke et al., 1989)
Desferrioxamine Mesylate Deferoxamine Mesylate	DFO Desferrioxamine Desferal Mesylate MW 656.8	Iron intoxication *Aluminum intoxication*	Acute: 1 g IV at 15 mg/kg/h; 0.5 g every 4–12 h; maximum 6 g/day; do not continue at 15 mg/kg/h beyond 24 h (Tenenbein, 1992) Chronic: 0.5–1 g/d IV, IM, SC; 5 mg/kg IV over 1 h
Deferiprone 1,2 dimethyl-3-hydroxypyrid-4-one	L1 MW 139.2	*Transfusion hemosiderosis (investigational)*	75–100 mg/kg/d in 4–6 oral doses/d
Dimercaprol	BAL in Oil MW 124.2	Arsenic, gold or mercury poisoning Acute lead encephalopathy	Arsenic or Gold Intoxication Mild: 2.5 mg/kg IM 4x/d × 2d; 2x/d × 1d; 1x/d × 10d Severe: 3.0 mg/kg IM 6x/d × 2d; 4x/d × 1d; 2x/d × 10d Mercury 5 mg/kg IM × 1; 2.5 mg/kg 1–2x/d × 10d Acute Lead Encephalopathy 4 mg/kg IM; 4 mg/kg 6x/d × 2–7d with CaNa₂ EDTA

Structures shown:

Calcium trisodium pentetate (DTPA):
$$HOOCCH_2\text{-}N(CH_2COOH)\text{-}CH_2\text{-}CH_2\text{-}N(CH_2COOH)\text{-}CH_2\text{-}CH_2\text{-}N(CH_2COOH)_2$$

Desferrioxamine Mesylate:
$$H_2N(CH_2)_5NC(=O)(CH_2)_2C(=O)NH(CH_2)_5NC(=O)(CH_2)_2C(=O)NH(CH_2)_5NCCH_3 \cdot CH_3SO_3H$$
(with OH groups on the hydroxamate nitrogens)

Deferiprone: 1,2-dimethyl-3-hydroxypyrid-4-one (pyridinone ring with OH, two CH₃ groups, and carbonyl)

Dimercaprol: $CH_2(SH)\text{-}CH(SH)\text{-}CH_2OH$

Chelation Therapies for Metal Intoxication

Compound	Name/MW	Indication	Dosage
2,3-Dimercapto-1-propanesulfonic acid, sodium salt $$CH_2-CH-CH_2SO_3Na$$ $$||$$ $$SHSH$$	DMPS Unithiol Dimaval® (Heyl, Berlin) MW 210.3	Lead, mercury, arsenic poisoning	Adult: 100 mg 3x/d × 10–21 d (Reynolds: Martindale, 1994) Child: 50–100 mg/m² 4x/d × 5d (Chisholm and Thomas, 1985) 200 mg 2x/d (Walshe, 1985)
Meso-2,3-Dimercapto-Succinic acid $$HOOC-CH-CH-COOH$$ $$||$$ $$SHSH$$	DMSA Succimer Chemet® MW 182.2	Wilson's disease Childhood blood lead > 45 μg/dL Aluminum intoxication Lead, arsenic, mercury poisoning Blood lead > 20 μg/dL	10 mg/kg or 350 mg/m² 3x/d × 5–7 d; 2x/d × 14 d
D-Penicillamine $$CH_3O$$ $$|\|$$ $$CH_3-C-CH_2-C$$ $$||OH$$ $$SHNH_2$$	PCN Cuprimine® Depen® MW 149.2	Wilson's disease; biliary cirrhosis; Lead, gold or mercury poisoning Cystinuria; rheumatoid arthritis	Adult: Metals: 0.5–1.0 g 4x/d × 1–2 m Arthritis: 0.125–0.25 g 1–2x/d × 1–3 m Cystinuria: 7.5 mg/kg 4x/d Wilson's: 0.5–1 g 4x/d Child: Metals: 10–13 mg/kg 3x/d × 1–6 m Wilson's: 62.5–250 mg 4x/d
Edetate Calcium Disodium $$HOOCCH_2CH_2COOH$$ $$\backslash/$$ $$N-CH_2-CH_2-N$$ $$/\backslash$$ $$HOOCCH_2CH_2COOH$$	CaNa₂ EDTA Versene Ca Calcium EDTA Calcium Versenate MW 374.28	Lead poisoning	Adult: 1 g IV in 250–500 mL 0.9% NaCl over 1 h 2x/d × 3–5 d 35 mg/kg IM in 0.5% procaine HCl 2x/d × 5 d Child: 35 mg/kg or 850 mg/m² IV in isotonic fluid 2x/d × 5 d or IM in 0.5% procaine hydrochloride
N-Acetyl-*L*-Cysteine $$HSCH_2CHCOOH$$ $$|$$ $$NHCOCH_3$$	NAC Acetylcysteine *L*-cysteine, *N*-acetyl Mucomyst® MW 185.2	Pulmonary disorders Mucolytic (by inhalation) Acetaminophen poisoning *Inorganic mercury poisoning* *Cobalt poisoning*	140 mg/kg × 1; 70 mg/kg 6x/d × 3 d Into arterial line of dialyzer at 10 mm/L blood (Lund et al., 1984) Similar to treatment acetaminophen poisoning (Martin et al., 1990)

Table 1 (continued) Therapeutic Chelators

Name	Common Name	Indications (non approved in italics)	Dosage
N-Acetyl-DL-Penicillamine $\begin{array}{c} CH_3 \\ | \\ CH_3C{-}CHCOOH \\ | \quad | \\ SH \quad NHCOCH_3 \end{array}$	NAP reagent grade (Sigma Chemical) MW 149.2	*Inorganic mercury poisoning*	Adult: 250–500 mg 4x/d .25–.5 g 4x/d Child: 7.5 mg/kg 4x/d (maximum 1 g) × 10 d (Florentine and Sanfilippo, 1991; Bluhm et al., 1992)
Prussian Blue Ferric Ferrocyanide $Fe_4[Fe(CN)_6]_3$	Radiogardase-Cs® (Heyl, Berlin) Prussian Blue, insoluble (Sigma) MW 859.2	[137]Cesium exposure *Thallium poisoning*	10 g or 125 mg/kg 2x/d by duodenal tube × 1–6 m 62.5 mg/kg in 100 mL 15% mannitol 4x/d by duodenal tube until urine thallium < 0.5 mg/d (Moore et al., 1993)
Prussian Blue Potassium Ferricyanide $K_3Fe[Fe(CN)_6]$	Prussian Blue photographic or histologic grade (Sigma Chemical) MW 306.9	*Same*	Same
Trientine dihydrochloride $H_2N{-}CH_2CH_2{-}NHCH_2CH_2NH{-}CH_2CH_2NH_2$	Trientine Trien Syprine® Cuprid® MW 219.2	Wilson's disease	Adult: 250–500 mg 2–4x/d to maximum 2 g/d Child: 250 mg 1–3x/d to maximum 1.5 g/d

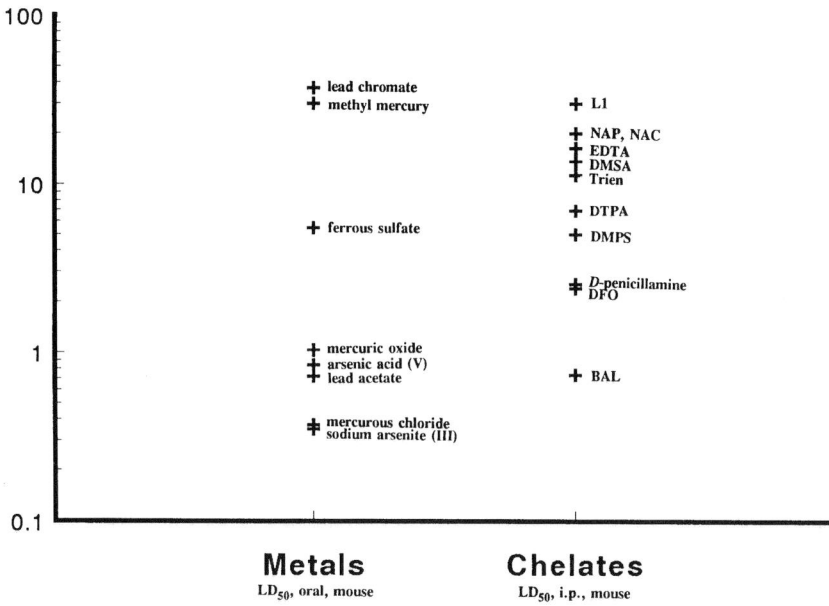

Figure 1 Comparison of the lowest LD_{50} mmol/kg for mice, oral, of representative metal salts from data cited by Sax and Lewis (1989) and the lowest LD_{50} mmol/kg for mice, ip, as cited by Jones (1990) plus data for L1 (Kontoghiorges 1990 and NAC Goldenthal 1971).

by iron deficiency (serum ferritin <100 µg/l) or by iron overload (serum ferritin >500 µg/l). Bone biopsy shows the local accumulation of aluminum. The diagnosis is confirmed by a predialysis serum aluminum >2 µM/l (60 µg/l) increasing to >5 µM/l (150 µg/l) after DFO 5 mg/kg i.v. administered over the last hour of hemodialysis. In renal failure, the serum aluminum peaks and stabilizes at this new level at 12 to 24 h and can be sampled at the start of the next dialysis (Ackrill and Day, 1993; TEDTA, 1993; Yaqoob et al., 1993). A positive DFO challenge test predicts a beneficial response to DFO therapy.

Current protocols for the diagnosis and treatment of aluminum overload, developed in 1992 by the European Dialysis and Transplant Association (TEDTA, 1993), are reproduced in Figures 2 and 3. Serum aluminum is 95% bound to transferrin and poorly dialyzable, and DFO does not affect serum aluminum or red cell aluminum, which declines only as red cells are replaced. DFO does mobilize tissue aluminum and iron to the serum as the ultrafilterable and potentially toxic aluminoxamine and ferrioxamine (Ackrill and Day, 1993). DFO also enters the liver cells and chelates lysosomal aluminum and iron resulting in significant fecal excretion of iron and aluminum.

As with other intracellular chelators, the toxicity of DFO may exceed that of the metal overload. DFO infusions as low as 2 g may be associated with severe adverse effects, including acute cochlear and ocular toxicity, the latter characterized by a loss of visual acuity and color vision. Mucormycosis, a frequently fatal infection, is a risk attributed to the availability of ferrioxamine as a growth promotor for *Rhizopus sp.* (Ackrill and Day, 1993). Treatment with DFO has been reduced accordingly to 5 mg/kg infused over 1 h/week × 12 weeks during a simultaneous high flux dialysis capable of removing ferrioxamine and aluminoxamine. Two positive challenge tests with DFO, both establishing serum aluminum >2.5 µM/l (75 µg/l) 1 month after completion of 3 months of therapy, are recommended before repeat therapy (TEDTA, 1993).

B. ALUMINUM AND OSTEODYSTROPHY

A primary criterion for aluminum overload as the cause of erythropoietin-unresponsive anemia in chronic renal failure patients is biopsy evidence of low turnover osteodystrophy with increased bone aluminum. Low turnover osteodystrophy, found in about half of chronic dialysis patients, is primarily due to decreased production of and decreased bone response to 1,25 dihydroxy vitamin D_3 ($1,25(OH)_2D_3$) in the absence of hypocalcemia and secondary hyperparathyroidism. Conversely, Hercz et al. (1993) found evidence of aluminum toxicity in less than one-third of patients with low turnover osteodystrophy. This aplastic osteodystrophy responds to treatment with $1,25(OH)_2D_3$ 0.25 to 0.5 µg/d and 24,25

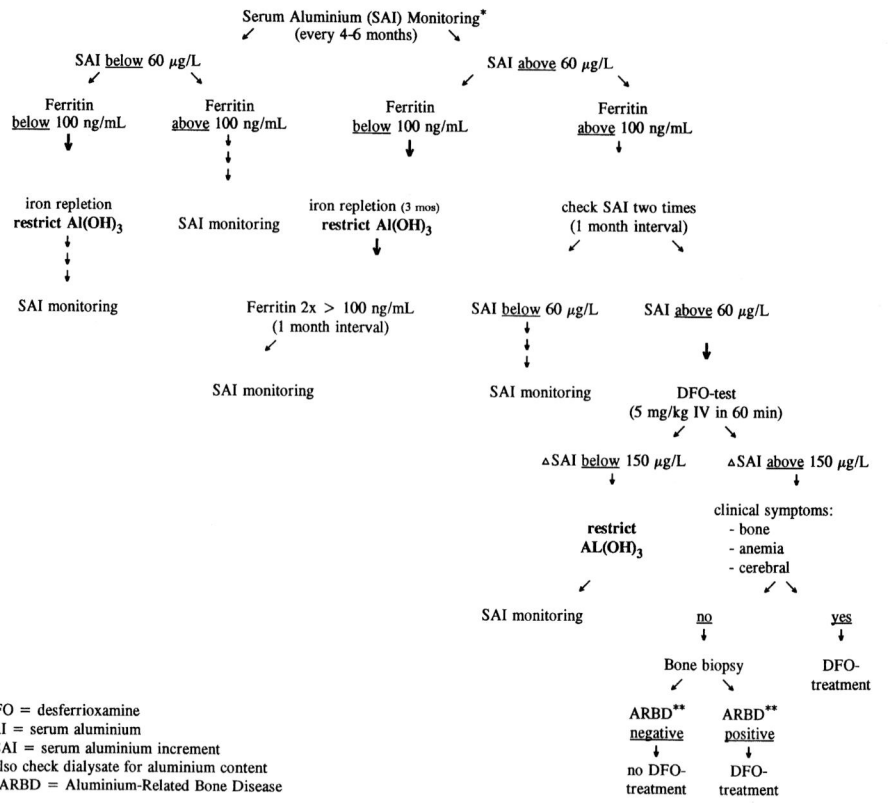

Figure 2 Reprinted with permission from European Dialysis and Transplant Association European Renal Association, *Nephrol. Dial. Transplant. Suppl.* (Suppl. 1) 1-4, by permission of Oxford University Press, 1993.

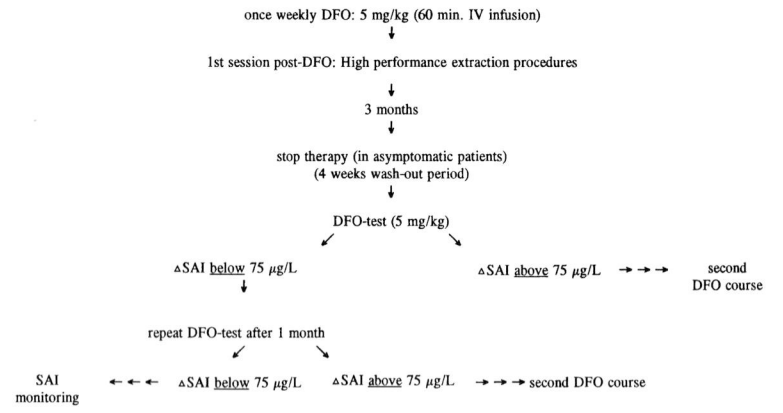

Figure 3 Reprinted with permission from European Dialysis and Transplant Association European Renal Association, *Nephrol. Dial. Transplant. Suppl.* (Suppl. 1) 1-4, by permission of Oxford University Press, 1993.

dihydroxy vitamin D_3 ($24,25(OH)_2D_3$), which also suppress the parathyroid hyperactivity responsible for osteitis fibrosa (Mortensen et al., 1993). Aluminum deposition may be only secondary and coincidental. DFO is recommended as a supplement to $1,25(OH)_2D_3$ therapy if the patient has symptomatic aplastic osteodystrophy with an aluminum overload established by both the DFO challenge test and a bone biopsy demonstrating the histochemical localization and biochemical quantitation of excessive aluminum on the osteoid front (TEDTA, 1993).

C. ALUMINUM ENCEPHALOPATHY

The high-aluminum dialysis dementia syndrome of altered mental status, seizures, and myoclonus is reversed by aluminum-free dialysate (Russo et al., 1992) and by DFO. To avoid DFO toxicity such as the fatalities of four patients with advanced dialysis dementia treated with DFO 41.7 ± 11 mg/kg/week (McCarthy et al., 1990), the current recommendations are a low dose of DFO, 5 mg/kg/week, (see Figures 1 and 2) in combination with high flux dialysis.

II. ARSENIC

A. ARSINE

Arsine gas (AsH_3) causes a fulminant intravascular hemolysis with hemoglobinemia, hemoglobinuria, renal failure, and skin bronzing. Treatment includes circulatory support, transfusions, and hemodialysis. At lesser exposures, the primary toxic effect of AsH_3 is a macrocytic hyperchromic anemia with bone marrow hyperplasia, reticulocytosis, and increased δ ALA D activity (Blair et al., 1990 a,b). Oxidative injury of the erythrocytes is evident in the decreased red cell glutathione and the refractile inclusions of denatured hemoglobin (Heinz bodies).

Chemical warfare with AsH_3 or dichloro (2-chlorovinyl) arsine (lewisite) was the impetus for the first therapeutic chelator, 2,3-dimercaptopropanol (dimercaprol; British AntiLewisite; BAL). Over 50 years later, there is still no safe and effective antidote for lewisite which is ten times as toxic as sodium arsenite. Inns et al. (1988) report the lewisite LD_{50} i.v. in rabbits as 8.6 μmol/kg (1.8 mg/kg) or one tenth that of sodium arsenite 86 μmol/kg (7.6 mg/kg). Protection against lewisite at 2 LD_{50} was afforded by repetitive doses of 160 μmol/kg DMSA, DMPS, or BAL, but BAL was toxic at 40 μmol/kg (Inns and Rice, 1993). The best that can be said is that nonfatal doses of BAL do not exacerbate anemia, but can cause a reversible granulocytopenia.

In most clinical reports of arsenic poisoning not due to arsine, the hematologic effects are less severe. Hemolysis and anemia are treated supportively, with attention to the provision of glutathione and methyl donors.

Evaluations of poisoning by lewisite in rabbits (Inns et al., 1990; Inns and Rice, 1993) and of arsenic trioxide in mice (Kreppel et al., 1990, 1993) and guinea pigs (Reichl et al., 1991), all favor treatment with DMSA and DMPS over BAL. Reichl et al. (1992) reported that the biliary excretion of arsenic from perfused guinea pig livers increased from 0.1% with BAL to 12.3% with DMPS.

The significantly lower toxicity, the ease of oral administration, and the enhanced biliary clearance of arsenic all contribute to the clinical consensus that DMSA and DMPS, not BAL, are first choice therapy for arsenic poisoning (Kelafant et al., 1993; Kew et al., 1993; Marcovigi et al., 1993). Organoarsenates are the least toxic of the arsenic compounds, and BAL therapy usually causes more symptoms than the poison. Shum and Whitehead (1994) report treatment of an adult who ingested 80 g methanearsenate. DMSA 30 mg/kg/d × 5 d × 4 weeks over 1 month reduced serum arsenic from 2871 μg/l to 6 μg/l.

Mathieu et al. (1992) suggest that BAL may still be advantageous in patients requiring hemodialysis since dialyzer membranes are permeable to arsenic-BAL complexes, but not to arsenic-DMSA (Sheabar et al., 1989). This is inconsistent with the report by Kostyniak et al. (1990) of more dialysis removal of DMSA-mercury than of BAL-mercury on extracorporeal chelation.

B. ARSENIC NEUROPATHY

Arsenic is a major cause of a chronic polyneuropathy characterized by Wallerian degeneration of myelinated fibers (Goebel et al., 1990). It is controversial as to whether chelation with BAL prevents neuropathy (Tallis, 1989). Reversibility decreases with the duration of exposure. Kew et al. (1993) found no improvement in peripheral neuropathy of 4 months duration after DMPS 300 mg/d × 3 weeks and DMSA 1.2 g/d × 2 weeks. In contrast, Bansal et al. (1991) reported significant improvement in neuropathy of 6 wks duration after 2 to 4 weeks PCN 750 mg/d, even though PCN is experimentally ineffective in relief of the systemic symptoms of acute arsenic poisoning (Kreppel et al., 1989). Brouwer et al. (1992) report severe paraparesis after arsenic exposure in a girl with congenital deficiency of 5,10-methylenetetrahydrofolate reductase, suggesting the possible benefits of an adequate supply of methyl donors. This is also inferred from the report by Martin et al. (1990) of the better response of acute symptoms to i.v. NAC than to IM BAL.

Doses of oral DMSA and DMPS equivalent to intramuscular BAL, assuming 50% absorption, are outlined in Table 2. DMPS, which has some intracellular distribution, is approximately twice as toxic as DMSA.

Table 2 Dimercaprol Analogues

	BAL	DMSA	DMPS
MW	124.2	182.2	210.3
Equimolar Dose	1 mg IM	2.9 mg oral*	3.4 mg oral*
Arsenic Poisoning			
Mild	2.5 mg/kg[†]	10 mg/kg[‡]	100 mg[‡]
	4x/d × 2d	3x/d × 5–7d	3x/d × 10–21d
	2x/d × 1d	2x/d × 10–14d	
	1x/d × 10d		
Severe	3.0 mg/kg[†]	18 mg/kg*	
	6x/d × 2d	3x/d × 5–7d	
	4x/d × 1d	2x/d × 10–14d	
	2x/d × 10d		
Mercury Poisoning			
	5 mg/kg × 1 dose[†]	10 mg/kg	
	2.5 mg/kg	3x/d × 10–21d[§,¶]	100 mg[‡]
	1–2x/d × 10d		3x/d × 10–21d
Lead Poisoning			
PbB < 70 µg/dL		10 mg/kg[†,‡]	100 mg[‡]
		3x/d × 5–7d	3x/d × 10–21d
		2x/d × 10–14d	
PbB ≥ 70 µg/dL; no	3 mg/kg[†]	18 mg/kg*	
CNS symptoms	6x/d × 2–7d	3x/d × 7d	
CNS Symptoms	4 mg/kg[†]	18 mg/kg*	
	6x/d × 2–7d	3x/d × 7d	

* assumes 50% absorption of oral preparation; dose calculated as equimolar to BAL in the absence of clinical data; [†] PDR, 1994; [‡] Reynolds (Martindale) 1994; [§] Mortensen and Valenzuela, 1990; [¶] Fournier et al., 1988.

BAL evokes hypertension, but any cardiovascular effects of DMSA and DMPS in humans have not been defined. In both lead-treated and control rats, three courses of DMSA between 6 and 12 months were associated with an increased glomerular filtration rate and a decrease in blood pressure, protracted for at least 2 months (Khalil-Manesh et al., 1992). On the basis of the acute hypotensive effects of DMSA and DMPS in beagles, Klimmek et al. (1993) recommend a maximal human dose i.v. of 12 mg/kg DMSA, 15 mg/kg DMPS. This may be particularly relevant in the treatment of arsenical cardiomyopathy.

III. BISMUTH

Bismuth encephalopathy, often misdiagnosed as neurosis, psychosis, or Alzheimer's or Creutzfeldt-Jakob's disease is reported to be completely reversible 4 to 12 months after termination of bismuth nitrate (Von Bose and Zaudig, 1991), bismuth subgallate (Kendel et al., 1993), and other bismuth salts. The spontaneous and complete recovery relates to the absence of specific pathologic findings in brain tissue and is consistent with the primary therapeutic recommendations: cessation of oral bismuth; catharsis or intestinal lavage; respiratory support and the treatment of seizures. After evacuation of the residual intestinal bismuth, chelation may speed recovery, but this is not established. In rats, DMPS, DMSA, and BAL were effective in reducing tissue bismuth, but only BAL reduced brain bismuth (Slikkerveer et al., 1992). Efficacy was confirmed in human subjects treated with dimercaprol 800 to 1200 µg/d (Slikkerveer and de Wolff, 1989) and in a patient with severe renal insufficiency given DMPS (Playford et al., 1990). In renal failure, hemodialysis with concurrent chelation may be indicated (Hudson and Mowat, 1989). PCN was reported as effective in mice (Basinger et al., 1983), but it did not increase the urine bismuth of five humans with chronic bismuth exposure (Nwokolo and Powder, 1990). Controlled trials of DMSA seem appropriate to evaluate the effect on symptoms, blood and CSF bismuth, and the renal clearance of bismuth.

IV. COPPER

A. WILSON'S DISEASE

Wilson's disease is an inherited, chromosome 13 disorder of hepatobiliary copper excretion manifested as progressive hepatic and neurologic toxicity (Yarze et al., 1992). Traditional chelation therapy, summarized in Table 3, usually begins with PCN. This mobilization of tissue copper stabilizes or improves the clinical status despite increased cytosolic copper, continued inability to remove copper from metallothionein, no reversal of hepatomegaly and cirrhosis, and an initial neurologic worsening (Stremmel et al., 1991; McQuaid et al., 1992). Adverse reactions to PCN are common, and even a few days of interruption in treatment can evoke a hypersensitivity reaction. When PCN is no longer tolerated, trientine is available as an approved (orphan drug status) alternate chelator. Oral zinc acetate at doses of 25 to 50 mg 3 ×/d is also useful in stable patients since it induces intestinal metallothionein, thus decreasing copper absorption (Brewer and Yuzbasiyan-Gurkan, 1992). There has also been considerable clinical and experimental investigation of ammonium tetrathiomolybdate, which forms inactive complexes with copper, decreasing copper uptake by hepatocytes and removing copper from liver cells at oral doses of 25 mg 3 ×/d (Tankanow, 1991). Walshe (1985) has used DMPS 200 mg 2 ×/d. In 137 cases of Wilson's disease, Walshe and Yealland (1993) report the outcome of chelation therapy, all types, as 42% symptom free, 23% improved, and 35% poor. A poor response may be an indication for orthoptic liver transplantation. With graft survival, copper metabolism returns to normal, and the short term outcome, despite the hazards of immunosuppression, is superior to chelation (Isai et al., 1992; Mason et al., 1993; Rela et al., 1993).

Table 3 Wilson's Disease — Treatment

D-penicillamine	500–2000 mg/d
child	250–1000 mg/d
Trientine	750–2000 mg/d
child	500–1500 mg/d
Zinc acetate	75–150 mg/d
Ammonium tetrathiomolybdate	25 mg 3x/d

V. IRON

DFO continues to be the only available and approved iron chelator. The oral iron chelator 1,2-dimethyl-3-hydroxypyrid-4-one (L-1; deferiprone) has been of major interest in the treatment of the iron overload of chronic transfusion therapy in beta thalassemia, sickle cell disease, and other hemolytic disorders. In sickle cell patients, Collins et al. (1994) found a comparable urine iron excretion with L-1 75 mg/kg/d oral and DFO 50 mg/kg/d i.m. or i.v., but the high dose of DFO evoked a significantly greater stool iron. Continued treatment with oral L-1 75–100 mg/kg/d normalizes the iron load and upregulates the erythropoietin response (Vreugdenhil et al., 1993).

Long-term use of L-1 has been complicated by arthropathies, usually reversible (Berkovitch et al., 1992), and relatively short term therapy by agranulocytosis (al-Refaie et al., 1992). This has resulted in discontinuation of major clinical trials (Kontoghiorghes et al., 1993). Berkovitch et al. (1992) suggest that the adverse effects may relate to the formation of 1:1 and 1:2 iron to deferiprone complexes capable of generating free radicals and that lower doses may be less toxic. An active search continues for a safe oral alternate to parenteral DFO.

Parabactin (Bergeron et al., 1993), like DFO, is also a bacterial siderophore. It has an iron formation constant 10^{52} mol/l vs. 10^{30} mol/l for DFO. When given parenterally, parabactin evokes five to six times the biliary clearance of iron from iron overloaded *Cebus* monkeys as DFO and may be a promising alternate.

In the chronic iron overload of patients with beta thalassemia major on transfusion therapy, DFO, infused subcutaneously or intravenously overnight, mobilizes 10 to 20 mg iron. Since a 500 ml unit of blood delivers 200 to 250 mg of iron and every increase of serum ferritin of 1 µg/l reflects an increase in iron stores of 65 mg, DFO must be given consistently, usually 4 nights a week, and toxicity may occur. Visual and auditory impairment may reverse with interruption of DFO (Marciani et al., 1991). Wonke et al. (1989) report reversal of deafness due to DFO by calcium trisodium pentetate (Na_3Ca DTPA) with zinc supplements.

Acute iron ingestion with fulminant hepatic failure continues to be a major cause of fatality from childhood poisons (MMWR, 1993). Serum iron >500 µg/dl is associated with coma, intestinal radiopacities, leukocytosis, an elevated anion gap, and a high risk of hepatic failure (Chyka and Butler, 1993; Cheney et al., 1994). Whole bowel irrigation (8 l or more polyethylene glycol intestinal lavage fluid administered intragastrically over 2 to 4 h) with gastrotomy if needed for impacted iron may remove sufficient quantities of ingested iron to prevent hepatic failure (Tenenbein et al., 1991). Kaczorowski and Wax (1994) report the use of 44 l lavage fluid over 5 d to clear retained ferrous sulfate tablets in a 15-kg child. Tenenbein et al. (1992) advise limitation of DFO to 15 mg/kg/h for a maximum of 24 h. More intensive therapy is considered to increase the risk of respiratory distress syndrome due to pulmonary free radicals generated by iron in the presence of DFO (Adamson et al., 1993).

VI. LEAD

Disruption of porphyrin metabolism is an early and sensitive index of lead poisoning. Chelation with DMSA (Graziano et al., 1992), $CaNa_2EDTA$ (Alessio, 1988), or PCN (Zuniga-Charles et al., 1981) results in a prompt restoration of red cell δ ALA D activity and a decrease in the urinary δ ALA that parallel the decrease in blood lead. The more rapid decrease in blood lead by DMSA than by $CaNa_2EDTA$ (Graziano et al., 1992) may evoke a more rapid restoration of heme synthesis. Chelation therapy of subacute or acute lead poisoning normalizes the red cell zinc erythrocyte protoporphyrin (EP) within 3 months, the time required for regeneration of the circulating red cells. DMSA, for example, decreases blood lead 2 to 4 times as effectively as penicillamine, but the effect on EP at 1 month is equivalent (Shannon, 1994). In chronic lead poisoning, elevation of the zinc EP is persistent and is more closely related to the chelatable lead burden than to the blood lead (Graziano et al., 1985), suggesting that the bone marrow lead regulates the EP.

The hemolytic component of lead-induced anemia is characterized by a normochromic, spherocytic anemia with reticulocytosis and basophilic stippling, the latter representing aggregated ribosomes. There is an accumulation of red cell cytidine triphosphate (CTP) and a deficiency of pyrimidine 5′ nucleotidase (P5N) resembling congenital deficiency of P5N (Paglia et al., 1975). Aly et al. (1993) confirm that the erythrocyte P5N is immediately and inversely related to the blood lead on treatment with DMSA. The clearance of red cell CTP, however, requires the usual 3 months for maturation of a new generation of red cells (Angle et al., 1980; Tomokuni et al., 1989).

An associated iron deficiency is common. This is characterized by hypochromia, microcytosis, decreased MCV and serum iron, and the absence of the marrow sideroblasts considered diagnostic of impaired heme synthesis in an iron-replete bone marrow.

Concurrent oral iron therapy is well tolerated during DMSA therapy. Haust et al. (1989) reported that i.m. iron therapy over the last 2 of 3 weeks of DMSA elicited a rapid and safe hematopoietic response. This is in contrast to the BAL iron complex which dissociates in an acid urine posing a risk of renal failure. An additional reason to withhold iron therapy until completion of $CaNa_2EDTA$ therapy is that major losses of zinc may prevent effective hematopoiesis.

The current CDC recommendations for the treatment of lead poisoning are summarized in Table 4.

In the treatment of lead poisoning with nephropathy, DMSA, which has not been associated with significant nephrotoxicity, would seem preferable to $CaNa_2EDTA$ in the treatment of lead poisoning with nephropathy (Grandjean et al., 1991).

Table 4 Treatment of Lead Poisoning

Blood Lead µg/dL (µmol/L)	Treatment
10–24 (0.5–1.19)	Identification; remediation or removal; nutritional counselling (all children with blood lead ≥ 0.5 µmol/l)
25*–39 (1.2–1.89)	Optional DMSA 350 mg/m² 3x/d × 5–7d, then 350 mg/m² 2x/d × 21d
40*–69 (1.9–3.3)	DMSA or $CaNa^2$ EDTA 1000 mg/m²/d × 5d BAL + $CaNa^2$ EDTA 1500 mg/m²/d × 5
≥70 (3.38)	? DMSA + $CaNa^2$ EDTA

* Alternate recommendations are that chelation therapy is optional at blood leads as low as 20 µg/dl and up to 45 µg/dl (CDC, 1991; Mortensen 1994).

In established lead nephropathy, the response to chelation may depend on the stage and potential reversibility of the renal pathology; creatinine clearances reduced to 60 to 80% by lead poisoning usually stabilize or improve spontaneously on termination of exposure. Wedeen et al. (1979) treated eight adults with mild lead nephropathy with $CaNa_2EDTA$ 1 g 3 times per week until the chelatable lead response was normal; in half of the subjects the glomerular filtration rate increased 20% or more over 6 to 50 mo.

Chelation is of uncertain value in therapy of saturnine gout, defined as an increased body burden of lead and a decreased fractional excretion of uric acid (Ball and Sorenson, 1969). Our clinic has noted exacerbations of gout in chronic lead poisoning cases treated intensively with $CaNa_2EDTA$ for the relief of other symptoms and considered the possibility that additional retention of uric acid might be evoked by the renal tubular toxicity of edetate or by the increased renal load of mobilized lead.

Clinical trials of $CaNa_2EDTA$ in the 1950s established it as lifesaving in lead poisoning despite its deleterious effect on CSF lead and the symptoms of lead encephalopathy. These resulted in the long standing recommendations for combined therapy of lead encephalopathy or of any child with a blood lead ≥ 3.3 µmol/l (70 µg/dl): BAL 75 mg/m^2 IM every 4 h × 2–7 d with the first dose of BAL at least 4 h before beginning $CaNa_2EDTA$ 1500 mg/m^2/d i.v. or i.m. × 5 d (CDC, 1991).

Since $CaNa_2EDTA$ increases brain lead, a current question is its safety for treatment at low levels of blood lead associated with modest changes in cognitive function and behavior. Ruff et al. (1993) found cognitive improvement at 6 mo but not at 7 weeks in 154 children treated with $CaNa_2EDTA$ 100 mg/m^2/d i.v. or i.m. × 5 d for blood lead levels 1.2 to 2.7 µmol/l (25 to 55 µg/dl), but there was no control group, and the results were confounded by the absence of benefit in children with iron deficiency. In rats with low level lead poisoning, Cory-Slechta and Weiss (1989) found no effect of $CaNa_2EDTA$ on the fixed-interval response performance.

In contrast to $CaNa_2EDTA$, DMSA does not increase brain lead (Smith and Flegal, 1992; Maiorino et al., 1993). In rats, Flora et al. (1994) found greater reduction in brain lead, early and late, from DMSA plus $CaNa_2EDTA$ than DMSA alone. $CaNa_2EDTA$ alone increased brain lead above that of untreated animals. DMSA is reported safe and effective in case reports of adults with relatively mild encephalopathy (Haust et al., 1989; Grandjean et al., 1991; Linz et al., 1992; Frumkin and Gerr, 1993). In evaluation of a series of monoalkyl esters of DMSA, Walker et al. (1992) found that brain lead in mice was reduced about 35% by DMSA × 5 d and 50 to 75% by the monoesters. The ability to reduce brain lead and toxicity both related to intracellular distribution.

In humans, but not rats, 5 to 30% of urine DMSA is excreted as the 1:1 and 1:2 DMSA-cysteine metabolite with significant individual differences as well as greater red cell distribution and decreased renal clearance of DMSA in children with lead poisoning (Dart et al., 1994). As emphasized by Mortensen (1994), considerably more data are needed before the indiscriminate DMSA treatment of all children with blood lead above 1 µmol/l (20 µg/dl) and before clinical trials of DMSA in the treatment of lead encephalopathy.

The effect of DMSA treatment on cognitive function is currently under investigation in the Treatment of Lead Poisoned Children cooperative study sponsored by the National Institute of Environmental Health Sciences. Approximately 1000 preschool children with blood lead levels of 1.0 to 2.1 µmol/l (20 to 44 µg/dl) are to have behavioral and cognitive evaluation at 6 months, 1 year, and 3 years after treatment with 1 to 3 courses of placebo or DMSA 300 or 600 mg/d × 5 d and 200 or 400 mg/d × 21 d.

Organic lead encephalopathy is diagnosed almost exclusively in chronic, intentional inhalation of leaded gasoline, in which it may be difficult to distinguish the toxicity of organic lead from that of the hydrocarbons. Although subjects with the highest blood leads have the most severe encephalopathy, chelation therapy has little effect on the irreversible encephalopathy (Goodheart and Dunne, 1994).

VII. MANGANESE

There is little evidence for effective treatment of manganese encephalopathy beyond control of Parkinsonian symptoms. The diagnosis is rarely made during the early manifestations of cognitive dysfunction and disorders of mood and affect. The blood and urine manganese correlate poorly with neurologic symptoms. The late findings of dementia, postural instability, and bradykinesia are permanent as are the losses of the neurons and myelinated fibers of the putamen and caudate nuclei (Yamada et al., 1986), despite the clearance of manganese from the mid brain (Newland and Weiss, 1989). Patients with significant Parkinsonian symptoms are said to respond to L-dopa (Wang et al., 1969; Rosenstock et al., 1971; Hine and Pasi, 1975; Huang et al., 1989), but as in idiopathic Parkinsonism, the response is variable or transient (Greenhouse, 1971; Cook et al., 1974).

Manganese excretion is predominantly biliary. In healthy subjects, $CaNa_2EDTA$ 1 g i.v. increased urine manganese from 2.9 to 16.5 µg/d (Allain et al., 1991). Challenge tests with $CaNa_2EDTA$ 2 g i.v. in five men with occupational manganism (Smyth et al., 1973) increased 24 h urine manganese from <10 µg/l to 30 to 40 µg/l, although Whitlock et al. (1966) noted a peak urine manganese of 1000 µg/24 h in response to $CaNa_2EDTA$ 2 g i.v. Although $CaNa_2EDTA$ therapy has been associated with gradual clinical improvement of manganese encephalopathy (Kosai and Boyle, 1956; Whitlock et al., 1966; Neff et al., 1969; Tanaka and Leiben, 1969), little benefit is reported by others (Penalver, 1954; Hunter, 1969; Cook et al., 1974). BAL is also considered ineffective.

There is a provocative but solitary report by Ky et al. (1992) of improvement in the chronic advanced Parkinsonism and dementia of two subjects treated with paraminosalicylic acid-sodium (PAS-Na) 6 g/d i.v. 4 d/week for 3 1/2 mo despite little increase in urine manganese. PAS-Na increases the fecal excretion of manganese in rabbits (Tandon, 1978; Tandon and Khandelwal, 1982), but it is not known if its putative clinical benefits relate to chelation or to antioxidant activity (Allgayer et al., 1992).

VIII. MERCURY

After acute exposures to inorganic mercury, the primary therapeutic problem is acute tubular necrosis. In acute renal failure due to mercuric chloride, hemodialysis and hemoperfusion remove negligible amounts of mercury (Koppel et al., 1982; Pellinen et al., 1983).

Dimercaprol (BAL) has not been remarkably effective in the prevention or reversal of acute renal failure, although therapy is usually initiated after the onset of anuria (Murray and Hedgepeth, 1988).

In three mercury overloaded patients without anemia and with normal blood mercury, Fournier et al. (1988) reported that oral DMSA 30 mg/kg/d increased urinary mercury 1.5 to 8.4 times. Mortensen and Valenzuela (1990) reported up to 3 × urine mercury on DMSA 30 mg/kg/d. In chronic occupational exposure to mercury vapor, a single oral dose of DMSA 2 g significantly increased the 24 h urine mercury (Roels et al., 1991). Stremski et al. (1994) recovered 1078 µg Hg for 7 d urine in an adult burned with mercurous perchlorate and treated with DMSA 3 g/d × 5 d and 2 g/d × 21 d. DMPS 300 to 400 mg/d reversed the hematuria in chronic occupational mercury poisoning (Campbell et al., 1986); BAL has reversed proteinuria (Damrau, 1990). The experimentally protective effects of captopril, a sulfhydryl chelator (Chavez et al., 1991), have not been tested in humans with subacute or acute mercury exposure, but may be associated with increased urine mercury.

In the treatment of renal failure in a 10 year old who had ingested 1.425 g of mercuric chloride, Kostyniak et al. (1990) found extracorporeal complexing with DMSA in the arterial line to result in significantly more hemodialysate mercury than BAL. In a case of mercuric chloride ingestion, Nadig et al. (1985) found the hemodialysis clearance of mercury by DMSA to be 3.5 to 5.0 ml/min vs. 0.6 ml/min with BAL. In rats, equimolar DMPS increases urine mercury more than DMSA (Gale et al., 1993) and is an effective rescue agent for up to 24 h after mercuric chloride nephrotoxicity (Zalups et al., 1991).

Treatment of encephalopathy and peripheral neuropathy due to elemental mercury, inorganic mercury salts, and unstable organic mercurials such as phenyl mercuric acetate, is more effective than that of organic mercury poisoning.

Acrodynia ("painful extremities"), a manifestation of metallic, inorganic, and unstable organic mercury poisoning of young children, includes fever, irritability, and red, swollen hands and feet that subsequently desquamate. Treatment with NAP, DMSA, and DMPS have all been effective.

Florentine and Sanfilippo (1991) found almost complete recovery of two children with neuropathic acrodynia due to inhaled mercury after treatment with NAP 30 mg/kg/d for five courses of 10 d each. A similarly good response was reported for three other children after DMPS 60 mg/kg/d (Muhlendahl, 1990) and in adults with neuropathy treated with DMPS 300 to 400 mg/d for 8 weeks (Campbell et al., 1986).

Bluhm et al. (1992) evaluated 53 construction workers acutely exposed to elemental mercury. At 5 to 7 weeks post exposure, 26 of the 53 were treated with either DMSA 90 mg/kg/d or NAP 100 mg/kg/d. Urine mercury increased threefold with DMSA and doubled with NAP, but the doses were not equimolar and there were more adverse reactions in the NAP group. At 10 d after the start of chelation, cognitive performance scores on Trails A and B were significantly improved, although the scores on the Grooved Pegboard worsened. The ten workers reevaluated at 570 d continued to have abnormal self-evaluation scores in nine areas of psychological performance.

Although BAL is the only drug in the U.S. pharmacopeia approved for the treatment of mercury poisoning, DMPS (Unithiol; Dimaval, Heyl, Berlin®) at oral doses of 300 to 400 mg/d is the current treatment of choice outside the U.S. (Hruby and Donner, 1987; Aposhian et al., 1992) with DMSA preferred over BAL in the U.S.

Alternate chelating agents under investigation include the monoalkyl esters of DMSA, particularly monoisoamyl DMSA, which are associated with more effective reduction of brain mercury in mice than DMPS (Kostial et al., 1993). Benov et al. (1990) propose that antioxidant activity may be a better index of renal protection than mobilization of the metal and that chelators in combination with glutathione or NAC as a glutathione source may be particularly effective. Agents enhancing the hepatic excretion of mercury mobilized from the kidney such as the alpha-mercapto-beta-aryl acrylic acids may also be promising (Kachru et al., 1989).

A. METHYL MERCURY

As reviewed by Clarkson (1993) and by Atchison and Hare (1994), methyl mercury causes irreversible destruction of neurons. Treatment centers on the termination of methyl mercury exposure, intestinal evacuation, and inactivation of the ingested mercury compound. In acute events, lavage, charcoal, and cathartics prevent absorption, and enterohepatic recycling can be interrupted by polythiol resin.

The distribution of absorbed methyl mercury to the red cells and plasma is complete in about 30 h with distribution to the brain complete in about 3 d. Early use of sulfhydryl compounds is essential for removal from the vascular compartment and late treatment is ineffective. DMPS and DMSA are favored, although PCN and N-acetyl-D, L-penicillamine (NAP) (250 mg reagent grade in a gelatin capsule every 4 h) have been used in epidemics.

In a 20 year-old with methyl mercury ingestion, Lund et al. (1984) reported the urinary clearance of organic mercury to be significantly increased after hemodialysis with N-acetylcysteine (NAC); oral DMPS 800 mg/d resulted in a urine mercury excretion five to six times that associated with oral PCN 2 g/d; the combination of NAC and DMPS was considered beneficial.

IX. THALLIUM

The pathognomonic clinical triad of thallium poisoning involves gastroenteritis, alopecia, and central or peripheral neuropathy, the last often requiring prolonged respiratory support.

The absorption of thallium in the first 4 hours after ingestion may be reduced by gastric lavage and the gastrointestinal instillation of charcoal, ferric ferrocyanide ($Fe_4[Fe(CN)_6]_3$, prussian blue, Radiogardase-Cs® Heyl, Berlin) or of potassium ferric hexacyanoferrate ($K_3Fe(CN)_6$, prussian blue, histologic or photographic grade (Sigma St. Louis, MO; Aldrich, Milwaukee, WI). Although not approved as pharmaceuticals except in Germany, oral prussian blue salts are not absorbed. The calculated release of cyanide ions from 20 g prussian blue is 2.9 mg CN^- from potassium ferric hexacyanoferrate and 1.6 mg CN^- from ferric ferrocyanide; both are safe if blood CN <0.5 μg/mL (Verzijl et al., 1993).

Treatment other than prussian blue during the initial 48 h phase of vascular and central nervous system distribution of thallium is uncertain. The kinetics of thallium resemble those of potassium, but there is no clinical evidence that potassium salts can effectively block tissue uptake. Experimentally, the early administration of oral potassium salts increases the renal and cardiac uptake of thallium (Careaga-Olivares and Morales-Aguilera, 1990). An adverse effect on the central nervous system uptake of thallium and the severity of the encephalopathy is evoked by parenteral potassium salts and by furosemide (Leloux et al., 1990). Dithiocarbamate (Kamerbeek et al., 1970; Moore et al., 1993) and PCN (Rios and Monroy-Noyola, 1992) are contraindicated because of similar deleterious effects of increasing neurotoxicity. Dimercaprol and $CaNa_2EDTA$ are ineffective.

When tissue distribution is complete (at about 48 h) and the excretory phase begins, potassium supplements may increase removal of thallium from the central nervous system (Papp et al., 1969). The most effective enhancement of excretion is by intestinal trapping of enterically secreted thallium by $K_3Fe(CN)_6$. The recommended dose of prussian blue is 250 mg/kg/d in aqueous solution in four divided doses instilled by nasogastric tube. The addition of mannitol 5 to 15% to the prussian blue solution prevents or treats a paralytic ileus secondary to large doses of prussian blue (Moore et al., 1993; Mulkey and Oehme, 1993). Prussian blue reduces the elimination half life of thallium, 1.7 to 30 d, by about two thirds (Papp et al., 1969). It has been administered for prolonged periods until urine thallium is negligible. Its efficacy is not enhanced by the addition of NAC (Meggs et al., 1994). A patient described by Chandler et al. (1990) required 95 d of assisted ventilation but continued to improve over almost 2 years of follow-up, by which time urine thallium was below 1 nmol/l. Dumitru and Kalantri (1990) report a similarly protracted partial recovery of distal axonopathy over more than 2 years and stress the importance of long-term neurorehabilitative therapy.

X. RADIONUCLIDES

Exposure to the radioactive isotopes of metals poses the risk of acute radiation syndrome and of chronic contamination via wound, pulmonary, and intestinal absorption. Treatment of ingested radioactive metals may include emesis or lavage, cathartics or intestinal lavage, dilution and water diuresis, and administration of a specific chelating or blocking agent. Advice and assistance regarding patient care are available at all times from the Radiation Emergency Assistance Center/Training Site (REAC/TS), Oak Ridge, TN (423–576–1004).

To increase intestinal trapping of strontium, Gaviscon, a sodium alginate antacid, is recommended (Ivannikov et al., 1993). For cesium, prussian blue, if available, is preferred over BioRex 40®, an oral cation exchange resin (Lipsztein et al., 1991).

DTPA is available from REAC/TS for the decorporation of most of the radioactive transuranics: plutonium, americum, curium, and californicum; the lanthanides: cesium, lanthanum, samarium, lutetium; and also for scandium, copper, zinc, yttrium, and niobium (Guilmette et al., 1992). In rats, 3,4,3-LIHOPO, a siderophobe analog, was at least 10 times as effective as DTPA in removal of inhaled plutonium (Poncy et al., 1993) and of simultaneously injected americum and plutonium (Volf et al., 1993).

Chelation with DTPA does not effectively mobilize strontium (Llobet et al., 1992), uranium, and polonium bound to tissue and bone. None of the eleven chelating agents reduced bone uranium when given more than 24 h after the administration of uranium to rats (Domingo et al., 1992). Rencova et al. (1993) present preliminary data for the possible efficacy of dithiol and dithiocarbamate in mobilizing polonium. Reduction of the body burden of uranium, polonium, and other radionuclides continues to be a major challenge.

XI. SUMMARY

Decorporation is most successful for metals with an active equilibrium between extracellular and intracellular sites. Metals such as cadmium are essentially inaccessible to chelation by virtue of their intracellular protein binding. The pulmonary sequestration of inhaled metals is particularly resistant to chelation as are the lipophilic complexes of organic mercury and lead in the brain.

A list of metals for which there are adequate human clinical data supporting the efficacy of current chelating agents is given in Table 5. Metals for which there is limited information such as zinc are not included.

Table 5 Current Chelation Therapy

Metal	Chelate
Al	DFO
As	DMSA, DMPS, BAL, NAC
Au	BAL, PCN, DMSA
Bi	BAL, DMPS, DMSA
Co	EDTA (?)
Cs	Prussian Blue
Cu	PCN, Trien, DMSA, DMPS
Fe	DFO, L1
Hg	DMPS, DMSA, BAL, PCN
Lanthanides	DTPA
Pb	DMSA, EDTA, BAL, PCN
Tl	Prussian Blue
Transuranics	DTPA

The clinical availability of DMSA, despite the limitation of approved indications, has resulted in a consensus that it should supplant BAL as an equally effective but much less toxic thiol chelator. Since humans (but not rats) metabolize DMSA to the active metabolite DMSA-cysteine (1:2) (Maiorino et al., 1993), additional human data are needed to verify all experimental data on DMSA, particularly its effects on tissue distribution. In the interim, clinical trials of DMSA, conducted with informed consent, appear justified by the relatively low toxicity and the rapidly accumulating data on efficacy.

REFERENCES

Ackrill, P. and Day, J. P., *Contrib. Nephrol.,* 102, 125–134, 1993.
Adamson, I. Y., Sienko, A., and Tenenbein, M., *Toxicol. Appl. Pharmacol.,* 120, 13–19, 1993.
Alessio, L., *Sci. Total Environ.,* 71, 293–299, 1988.
Al Refaie, F. N., Wonke, B., Hoffbrand, A. V., Wickens, D. G., Nortey, P., and Kontoghiorghes, G. J., *Blood,* 80, 593–599, 1992.
Allain, P., Mauras, Y., Premel-Cabic, A., Islam, S., Herve, J. P., and Cledes, J., *Br. J. Clin. Pharmacol.,* 31, 347–349, 1991.
Allgayer, H., Gugler, R., Bohne, P., Schmidt, M., and Kruis, W., *Gastroenterology,* 103, 1991–1992, 1992.
Aly, M. H., Kim, H. C., Renner, S. W., Boyarsky, A., Kosmin, M., and Paglia, D. E., *Am. J. Hematol.,* 44, 280–283, 1993.
Angle, C. R., Stohs, S. J., McIntire, M. S., Swanson, M. S., and Rovang, K. S., *Toxicol. Appl. Pharmacol.,* 54, 161–167, 1980.
Aposhian, H. V., Maiorino, R. M., Rivera, M., Bruce, D. C., Dart, R. C., Hurlbut, K. M., Levine, D. J., Zheng, W., Fernando, Q., Carter, D., et al., *J. Toxicol. Clin. Toxicol.,* 30, 505–528, 1992.
Atchison, W. D. and Hare, M. F., *FASEB J.,* 8, 622–629, 1994.
Ball, G. V. and Sorensen, L. B., *N. Engl. J. Med.,* 280, 1199–1202, 1969.
Bansal, S. K., Haldar, N., Dhand, U. K., and Chopra, J. S., *Chest,* 100, 878–880, 1991.
Basinger, M. A., Jones, M. M., and McCroskey, S. A., *J. Toxicol. Clin. Toxicol.,* 20, 159–165, 1983.
Benov, L. C., Benchev, I. C., and Monovich, O. H., *Chem. Biol. Interact.,* 76, 321–332, 1990.
Bergeron, R., Streiff, R. R., King, W., Daniels, R. D., and Wiegand, J., *Blood,* 82, 2552–2557, 1993.
Berkovitch, M., Laxer, R. M., Matsui, D., et al., *Blood,* 80 (Suppl. 1), 7a, 1992.
Blair, P. C., Thompson, M. B., Morrissey, R. E., Moorman, M. P., Sloane, R. A., and Fowler, B. A., *Fund. Appl. Toxicol.,* 14, 776–787, 1990a.
Blair, P. C., Thompson, M. B., Bechtold, M., Wilson, R. E., Moorman, M. P., and Fowler, B. A., *Toxicology,* 63, 25–34, 1990b.
Bluhm, R. E., Bobbitt, R. G., Welch, L. W., Wood, A. J., Bonfiglio, J. F., Sarzen, C., Heath, A. J., and Branch, R. A., *Hum. Exp. Toxicol.,* 11, 201–210, 1992.
Brewer, G. J. and Yuzbasiyan-Gurkan, V., *Medicine (Baltimore),* 71, 139–164, 1992.
Brouwer, O. F., Onkenhout, W., Edelbroek, P. M., De Kom, J. F., De Wolff, F. A., and Peters, A. C., *Clin. Neurol. Neurosurg.,* 94, 307–310, 1992.
Campbell, J. R., Clarkson, T. W., and Omar, M. D., *JAMA,* 256, 3127–3130, 1986.
Careaga-Olivares, J. and Morales-Aguilera A., *Arch. Invest. Med. Mex.,* 21, 273–277, 1990.
Centers for Disease Control, Preventing Lead Poisoning in Young Children: A Statement by the Centers for Disease Control, Atlanta, GA: U.S. Department of Health and Human Services, 1991.
Chandler, H. A., Archbold, G. P., Gibson, J. M., O'Callaghan, P., Marks, J. N., and Pethybridge, R. J., *Clin. Chem.* 36, 1506–1509, 1990.
Chavez, E., Zazueta, C., Osornio, A., Holguin, J. A., and Miranda, M. E., *J. Pharmacol. Exp. Ther.,* 256, 385–390, 1991.
Cheney, K. K., Gumbiner, C., Benson, B. E., and Tenenbein, M., *J. Toxicol. Clin. Toxicol.,* 32, 33, 61–66, 1995.
Chisolm, J. J. and Thomas, D. J., *J. Pharmacol. Exp. Ther.,* 235, 665–669, 1985.
Chyka, P. A. and Butler, A. Y., *Am. J. Emerg. Med.,* 11, 99–103, 1993.
Clarkson, T. W., *Environ. Health Perspect.,* 100, 31–38, 1993.
Collins, A. F., Fassos, F. F., Stobie, S., Lewis, N., Shaw, D., Fry, M., Templeton, D. M., McClelland, R. A., Koren, G., and Olivieri, N. F., *Blood,* 83, 2329–2333, 1994.
Cook, D. G., Fahn, S., and Brait, K. A., *Arch. Neurol.,* 30, 59–64, 1974.
Cory-Slechta, D. A. and Weiss, B., *Neurotoxicology,* 10, 685–697, 1989.
Damrau, J., *Z. Gesamte. Inn. Med.,* 45, 89–92, 1990.
Dart, R. C., Hurlbut, K. M., Mairoino, R. M., Mayersohn, M., Aposhian, H. V., and Hassen, L. V. B., *J. Pediatr.,* 125, 309–316, 1994.
Domingo, J. L., Colomina, M. T., Llobet, J. M., Jones, M. M., Singh, P. K., and Campbell, R. A., *Fund. Appl. Toxicol.,* 19, 350–357, 1992.
Dumitru, D. and Kalantri, A., *Muscle Nerve,* 13, 433–437, 1990.
The European Dialysis and Transplant Association — European Renal Association, Consensus Conference 1992, *Nephrol. Dial. Transplant.,* (Suppl 1) 1–4, 1993.
Flora, S. J. S., Battacharya, R., and Vijayaraghvan, R., *Fund. Appl. Tox.,* 25, 233–240, 1995.
Florentine, M. J. and Sanfilippo, D. J., *Clin. Pharm.,* 10, 213–221, 1991.
Fournier, L., Thomas, G., Garnier, R., Buisine, A., Houze, P., Pradier, F., and Dally, S., *Med. Toxicol. Adverse Drug Exp.,* 3, 499–504, 1988.
Frumkin, H. and Gerr, F., *Am. J. Ind. Med.,* 24, 701–706, 1993.
Gale, G. R., Smith, A. B., Jones, M. M., and Singh, P. K., *Toxicology,* 81, 49–56, 1993.
Goebel, H. H., Schmidt, P. F., Bohl, J., Tettenborn, B., Kramer, G., and Gutmann, L., *J. Neuropathol. Exp. Neurol.,* 49, 137–149, 1990.

Goldenthal, E. I., *Tox. Appl. Pharmacol.,* 18, 185–207, 1971.
Goodheart, R. S. and Dunne, J. W., *Med. J. Aust.,* 160, 178–181, 1994.
Grandjean, P., Jacobsen, I. A., and Jorgensen, P. J., *Pharmacol. Toxicol.,* 68, 266–269, 1991.
Graziano, J. H., Siris, E. S., Lolacono, N., Silverberg, S. J., and Turgeon, L., *Clin. Pharmacol. Ther.,* 37, 431–438, 1985.
Graziano, J. H., Lolacono, N. J., Moulton, T., Mitchell, M. E., Slavkovich, V., and Zarate, C., *J. Pediatr.,* 120, 133–139, 1992.
Greenhouse, A. H., *Trans. Am. Neurol. Assoc.,* 69, 248–249, 1971.
Guilmette, R. A. and Muggenburg, B. A., *Health Phys.,* 62, 311–318, 1992.
Haust, H. L., Inwood, M., Spence, J. D., Poon, H. C., and Peter, F., *Clin. Biochem.,* 22, 189–196, 1989.
Hercz, G., Pei, Y., Greenwood, C., Manuel A., Saiphoo, C., Goodman, W. G., Segre, G. V., Fenton, S., and Sherrard, D. J., *Kidney Int.,* 44, 860–866, 1993.
Hine, C. H. and Pasi, A., *West. J. Med.,* 123, 101–107, 1975.
Hruby, K. and Donner, A., *Med. Toxicol. Adverse Drug Exp.,* 2, 317–323, 1987.
Huang, C. C., Chu, N. S., Lu, C. S., Wang, J. D., Tsai, J. L., Tzeng, J. L., Wolters, E. C., and Calne, D. B., *Arch. Neurol.,* 46, 1104–1106, 1989.
Hudson, M. and Mowat, N. A. G., *Br. Med. J.,* 299, 159, 1989.
Hunter, D., in *The Diseases of Occupations,* The English Universities Press Ltd., London, 459–465, 1969.
Inns, R. H., Bright, J. E., and Marrs, T. C., *Toxicology,* 51, 213–222, 1988.
Inns, R. H., Rice, P., Bright, J. E., and Marrs, T. C., *Hum. Exp. Toxicol.,* 9, 215–220, 1990.
Inns, R. H. and Rice, P., *Hum. Exp. Toxicol.,* 12, 241–246, 1993.
Isai, H., Sheil, A. G., McCaughan, G. W., Thompson, J. F., Dorney, S. F., Dolan, P. M., Shun, A., Strasser, S., and Uchino, J., *Transplant. Proc.,* 24, 1475–1476, 1992.
Ivannikov, A. T., Altukhova, G. A., Zhorova, E. S., and Parfenova, I. M., *Radiobiologiia,* 33, 297–301, 1993.
Jones, M. M., *Crit. Rev. Toxicol.,* 21, 209–233, 1991.
Jones, M. M. and Cherian, M. G., *Toxicology,* 62, 1–25, 1990.
Kachru, D. N., Khandelwal, S., Sharma, B. L., and Tandon, S. K., *Pharmacol. Toxicol.,* 64, 182–184, 1989.
Kaczorowski, J. and Wax, P., *Vet. Human. Toxicol.,* 36, 340, 1994.
Kamerbeek, H. H., Van Heijst, A. N., Rauws, A. G., and Ten Ham, M., *Ned. Tijdschr. Geneeskd.,* 114, 457–460, 1970.
Kelafant, G. A., Kasarskis, E. J., Horstman, S. W., Cohen, C., and Frank, A. L., *Am. J. Ind. Med.,* 24, 723–726, 1993.
Kendel, K., Kabiri, R., and Schaffer, S., *Dtsch. Med. Wochenschr.,* 118, 221–224, 1993.
Kew, J., Morris, C., Aihie, A., Fysh, R., Jones, S., and Brooks, D., *Br. Med. J.,* 306, 506–507, 1993.
Khalil-Manesh, F., Gonick, H. C., Cohen, A., Bergamaschi, E., and Mutti, A., *Environ. Res.,* 58, 35–54, 1992.
Klimmek, R., Krettek, C., and Werner, H. W., *Arch. Toxicol.,* 67, 428–434, 1993.
Kontoghiorghes, G. J., Agarwal, M. B., Tondury, P., Kersten, M. J., Jaeger, M., Vreugdenhil, G., Vania, A., and Rahman, Y. E., *Lancet,* 341, 1479–1480, 1993.
Koppel, C., Baudisch, H., and Keller, F., , *J. Toxicol. Clin. Toxicol.,* 19, 391–400, 1982.
Kosai, M. F. and Boyle, A. J., *Ind. Med. Surg.,* 25, 1, 1956.
Kostial, K., Blanusa, M., Simonovic, I., Jones, M. M., and Singh, P. K., *J. Appl. Toxicol.,* 13, 321–325, 1993.
Kostyniak, P. J., Greizerstein, H. B., Goldstein, J., Lachaal, M., Reddy, P., Clarkson, T. W., Walshe, J., and Cunningham, E., *Hum. Exp. Toxicol.,* 9, 137–141, 1990.
Kreppel, H., Reichl, F. X., Forth, W., and Fichtl, B., *Vet. Hum. Toxicol.,* 31, 1–5, 1989.
Kreppel, H., Reichl, F. X., Szinicz, L., Fichtl, B., and Forth, W., *Arch. Toxicol.,* 64, 387–392, 1990.
Kreppel, H., Paepcke, U., Thiermann, H., Szinicz, L., Reichl, F. X., Singh, P. K., and Jones, M. M., *Arch. Toxicol.,* 67, 580–585, 1993.
Ky, S. Q., Deng, H. S., Xie, P. Y., and Hu, W., *Br. J. Ind. Med.,* 49, 66–69, 1992.
Leloux, M. S., Nguyen, P. L., and Claude, J. R., *J. Toxicol. Clin. Exp.,* 10, 147–156, 1990.
Linz, D. H., Barrett, E. T., Pflaumer, J. E., and Keith, R. E., *J. Occup. Med.,* 34, 638–641, 1992.
Lipsztein, J. L., Bertelli, L., Oliveira, C. A., and Dantas, B. M., *Health Phys.,* 60, 57–61, 1990.
Llobet, J. M., Colomina, M. T., Domingo, J. L., and Corbella, J., *Vet. Hum. Toxicol.,* 34, 7–9, 1992.
Lund, M. E., Banner, W., Clarkson, T. W., and Berlin, M., *J. Toxicol. Clin. Toxicol.,* 22, 31–49, 1984.
Maiorino, R. M., Aposhian M. M., Xu, A.-F., Li, Y., Polt, R. L., and Aphosian H. V., *J. Pharmacol. Exp. Therap.,* 267, 1221–1226, 1993.
Marciani, M. G., Cianciulli, P., Stefani, N., Stefanini, F., Peroni, L., Sabbadini, M., Maschio, M., Trua, G., and Papa, G., *Haematologica,* 76, 131–134, 1991.
Marcovigi, P., Calbi, G., Valtancoli, E., and Calbi, P., *Minerva Anestesiol.,* 59, 339–341, 1993.
Martin, D. S., Willis, S. E., and Cline, D. M., *J. Am. Board Fam. Pract.,* 3, 293–296, 1990.
Mason, A. L., Marsh, W., and Alpers, D. H., *Dig. Dis. Sci.,* 38, 1746–1750, 1993.
Mathieu, D., Mathieu-Nolf, M., Germain-Alonso, M., Neviere, R., Furon, D., and Wattel, F., *Intensive Care Med.,* 18, 47–50, 1992.
McCarthy, J. T., Milliner, D. S., and Johnson, W. J., *Q. J. Med.,* 74, 257–276, 1990.
McQuaid, A., Lamand, M., and Mason, J., *J. Lab. Clin. Med.,* 119, 744–750, 1992.
Meggs, W. J., Morasco, R., Shih, R. D., et al.,*Vet. Hum. Toxicol.,* 36, 364, 1994.
Min, D. I. and D'Elia, J. A., *Clin. Pharm.,* 11, 636–639, 1992.
Moore, D., House, I., and Dixon, A., *Br. Med. J.,* 1527–1529, 1993.

Morbidity and Mortality Weekly Report, *MMWR,* 42, 111–113, 1993.
Mortensen, M. E., *J. Pediatr.,* 125, 233–234, 1994.
Mortensen, B. M., Aarseth, H. P., Ganss, R., Haug, E., Gautvik, K. M., and Gordeladze, J. O., *Bone,* 14, 125–131, 1993.
Mortensen, M. E. and Valenzuela, M. C., *Vet. Hum. Toxicol.,* 32, 362, 1990.
Muhlendahl, K. E., *Lancet,* 336, 1578, 1990.
Mulkey, J. P. and Oehme, F. W., *Vet. Hum. Toxicol.,* 35, 445–453, 1993.
Murray, K. M. and Hedgepeth, J. C., *Drug Intell. Clin. Pharm.,* 22, 972–975, 1988.
Nadig, J., Knutti, R., and Hany, A., *Schweiz. Med. Wochenschr.,* 115, 507–511, 1985.
Neff, N. H., Barrett, R. E., and Costa, E., *Experientia,* 25, 1140–1141, 1969.
Newland, M. C. and Weiss, B., *Toxicol. Appl. Pharmacol.,* 113, 87–97, 1989.
Nwokolo, C. U. and Powder, R. E., *Br. J. Clin. Pharmacol.,* 30, 648–650, 1990.
Paglia, D. E., Valentine, W. N., and Dahlgren, J. G., *J. Clin. Invest.,* 56, 1164–1169, 1975.
Papp, J. P., Gay, P. C., Dodson, V. N., and Pollard, H. M., *Ann. Intern. Med.,* 71, 119–123, 1969.
Pellinen, T. J., Karjalainen, K., and Haapanen, E. J., *J. Toxicol. Clin. Toxicol.,* 20, 187–189, 1983.
Penalver, R., *Ind. Med. Surg.,* 1–7, 1954.
Physicians Drug Reference, 48th Edition, Medical Economics Data Production Co., Montvale, NJ, 1994.
Playford, R. J., Matthews, C. H., Campbell, J. J., Delves, H. T., Hla, K. K., Hodgson, H. J., and Calam, J., *Gut,* 31, 359–360, 1990.
Poncy, J. L., Rateau, G., Burgada, R., Bailly, T., Leroux, Y., Raymond, K. N., Durbin, P. W., and Masse, R., *Int. J. Radiat. Biol.,* 64, 431–436, 1993.
Reichl, F. X., Kreppel, H., and Forth, W., *Arch. Toxicol.,* 65, 235–238, 1991.
Reichl, F. X., Muckter, H., Kreppel, H., and Forth, W., *Pharmacol. Toxicol.,* 70, 352–356, 1992.
Rela, M., Heaton, N. D., Vougas, V., McEntee, G., Gane, E., Farhat, B., Chiyende, J., Mieli-Vergani, G., Mowat, A. P., Portmann, B., et al., *Br. J. Surg.,* 80, 909–911, 1993.
Rencová, J., Volf, V., Jones, M. M., and Singh, P. K., *Int. J. Radiat. Biol.,* 63, 223, 1993.
Reynolds, J. E. F., ed., *Martindale: The Extra Pharmacopeia* (electronic version), Micromedex Inc., Denver, CO, 1994.
Rios, C., and Monroy-Noyola, A., *Toxicology,* 74, 69–76, 1992.
Roels, H. A., Boeckx, M., Ceulemans, E., and Lauwerys, R. R., *Br. J. Ind. Med.,* 48, 247–253, 1991.
Rosenstock, H. A., Simons, D. G., and Meyer, J. S., *JAMA,* 217, 1354–1358, 1971.
Ruff, H. A., Bijur, P. E., Markowitz, M., Ma, Y. C., and Rosen, J. F., *JAMA,* 269, 1641–1646, 1993.
Russo, L. S., Beal, G., Sandroni, S., and Ballinger, W. E., *J. Neurol. Neurosurg. Psychiatry,* 55, 697–700, 1992.
Sax, N. I., and Lewis, R. J. Sr., Eds., *Dangerous Properties of Industrial Materials,* 7th Ed., Van Nostrand Reinhold, New York, 1989.
Shannon, M. W., *Vet. Hum. Toxicol.,* 36, 339, 1994.
Sheabar, F. Z., Yannai, S., and Taitelman, V., *Pharmacol. Toxicol.,* 64, 329–333, 1989.
Shoskes, D. A., Radzinski, C. A., Struthers, N. W., and Honey, R. J., *J. Urol.,* 147, 697–699, 1992.
Shum, S. and Whitehead, J., *Vet. Hum. Toxicol.,* 36, 341, 1994.
Slikkerveer, A. and de Wolff, F. A., *Med. Toxicol. Adverse Drug Exp.,* 4, 303–323, 1989.
Slikkerveer, A., Jong, H. B., Helmich, R. B., and de Wolff, F. A., *J. Lab. Clin. Med.,* 119, 529–537, 1992.
Smith, D. R. and Flegal, A. R., *Toxicol. Appl. Pharmacol.,* 116, 85–91, 1992.
Smyth, L. T., Ruhf, R. C., Whitman, N. E., and Dugan, T., *J. Occup. Med.,* 15, 101–109, 1973.
Stremmel, W., Meyerrose, K. W., Niederau, C., Hefter, H., Kreuzpaintner, G., and Strohmeyer, G., *Ann. Intern. Med.,* 115, 720–726, 1991.
Stremski, E., Yousif, J., and Furbee, B., *Vet. Hum. Toxicol.,* 36, 341, 1994.
Tallis, G. A., *Aust. N. Z. J. Med.,* 19, 730–732, 1989.
Tanaka, S., and Leiben, J., *Arch. Environ. Health,* 19, 674–684, 1969.
Tandon, S. K., *Toxicology,* 9, 379–385, 1978.
Tandon, S. K., and Khandelwal, S., *Arch. Toxicol.,* 50, 19–25, 1982.
Tankanow, R. M., *Clin. Pharm.,* 10, 839–849, 1991.
Tennebein, M., Wiseman, N., and Yatscoff, R. W., *Pediatr. Emerg. Care,* 7, 286–288, 1991.
Tenenbein, M., Kowalski, S., Sienko, A., Bowden, D. H., and Adamson, I. Y., *Lancet,* 339, 699–701, 1992.
Tomokuni, K., Ichiba, M., and Hirai, Y., *Arch. Toxicol.,* 63, 23–28, 1989.
Verzijl, J. M., Joore, H. C. A., van Dijk, A., Wierckx, F. C. J., Savelkoul, J. F., and Glerum, J. H., *J. Toxicol. Clin. Toxicol.,* 31, 553–562, 1993.
Volf, V., Burgada, R., Raymond, K. N., and Durbin, P. W., *Int. J. Radiat. Biol.,* 63, 785–793, 1993.
Von Bose, M. J., and Zaudig, M., *Br. J. Psychia.* 158, 278–280, 1991.
Vreugdenhil, G., Smeets, M., Feelders, R. A., and van Eijk, H. G., *Acta. Haematol.* 89, 57–60, 1993.
Walker, E. M., Jr., Stone, A., Milligan, L. B., Gale, G. R., Atkins, L. M., Smith, A. B., Jones, M. M., Singh, P. K., and Basinger, M. A., *Toxicology,* 76, 79–87, 1992.
Walshe, J. M., *Br. Med. J. Clin. Res. Ed.,* 290, 673–674, 1985.
Walshe, J. M., and Yealland, M., *Q. J. Med.,* 86, 197–204, 1993.
Wang, J.-D., Huang, C.-C., Hwang, Y.-H., Chaing, J.-R., Lin, J.-M., and Chen, J.-S., *Br. J. Ind. Med.,* 46, 856–859, 1969.
Wedeen, R. P., Mallik, D. K., and Batuman, V., *Arch. Intern. Med.,* 139, 53–57, 1979.

Whitlock, C. M., Amuso, S. J., and Bittenbender, J. B., *Am. Ind. Hyg. Assoc. J.,* 454–459, 1966.
Wonke, B., Hoffbrand, A. V., Aldouri, M., Wickens, D., Flynn, D., Stearns, M., and Warner, P., *Arch. Dis. Child.,* 64, 77–82, 1989.
Yamada, M., Ohno, S., Okayasu, I., Okeda, R., Hatakeyama, S., Watanabe, H., Ushio, K., and Tsukagoshi, H., *Acta. Neuropathol. (Berlin),* 70, 273–278, 1986.
Yaqoob, M., Ahmad, R., McClelland, P., Shivakumar, K. A., Sallomi, D. F., Fahal, I. H., Roberts, N. B., and Helliwell, T., *Postgrad. Med. J.,* 69, 124–128, 1993.
Yarze, J. C., Martin, P., Munoz, S. J., and Friedman, L. S., *Am. J. Med.,* 92, 643–654, 1992.
Zalups, R. K., Gelein, R. M., and Cernichiari, E., *J. Pharmacol. Exp. Ther.,* 256, 1–10, 1991.
Zuniga-Charles, M. A., Gonzalez-Ramirez, J. D., and Molina-Ballesteros, G., *Arch. Environ. Health,* 36, 40–43, 1981.

PART 2
Target Organ Toxicology

Section V
Neurotoxicology of Metals

Overview

An Introduction to Neurotoxicology of Metals

Louis W. Chang

There is little argument that the nervous system is one of the prime target sites of metal toxicity. Indeed, characteristic neurological dysfunctions, behavioral changes, and neuropathology have been described in both humans and in animals as a result of exposure to mercury, lead, cadmium, managanese, aluminum, arsenic, alkyltins, etc. The clinical aspects of such metals have been presented in Part 1 of this text. Furthermore, in a recent publication by this author (Chang and Dyer, 1995), the specific neurotoxicology of the major neurotoxic metals was extensively and individually reviewed.

In the present volume, we plan to review "metals" as a unified category of neurotoxicants rather than as individual neurotoxic chemicals. In this fashion, we hope to provide the reader a common ground to compare, correlate, and cross-reference between metals by their toxic endpoints, effects, and mechanisms of action. Therefore, the chapters in this section are issue or conceptually oriented. We believe that this approach will help readers and students to gain a much more global concept on "metal neurotoxicology" as a whole.

Neurotoxicology has been generally defined as the adverse *structural* or *functional* changes in the nervous system produced by exposure to chemical or physical agents. It becomes obvious, therefore, that neuropathology and neurobehavior are the two primary endpoints of neurotoxicity. The first two chapters provide comprehensive reviews on the neuropathology (Chang) and neurobehavioral changes (Cory-Slechta) induced by heavy metals. When one wants to investigate the toxic actions of a chemical, it is important to identify and to examine the specific tissue or cellular components which are closely related to the toxic consequences of that chemical. Current concepts in metal neurotoxicology indicate that at least five tissue or cellular components play important roles in metal neurotoxicity. These components are the blood-brain barrier, astrocytes, the choroid plexus, neurotransmitters, and neurocytoskeletons. Effects on and roles of these components in metal neurotoxicity are being extensively reviewed and discussed by Romero et al. (blood-brain barrier), Aschner and Kimelberg (astrocytes), Zheng (choroid plexus), Mailman et al. (neurotransmitters), and Graff and Reuhl (neurocytoskeletons). The investigations and studies of metal toxicity are always plagued by difficulties or lack of clinical correlates with experimental findings. This concern is particularly serious in terms of lead poisoning. A special chapter (Rice and Silbergeld) is devoted to addressing this issue, with an attempt to "bridge" the clinical observations with experimental findings.

As one may expect, the mechanisms of action for metals are complex and frequently multifaceted. Despite the seemingly complex involvements of metal neurotoxicity, several "common-ground" biomolecular mechanisms seem to underlie most, if not all, metal neurotoxicity. These mechanisms of action are ion channel disruptions, reactive oxygen species (free radical formations), and disruption of protein synthesis and metabolism. The biomolecular bases of these toxic mechanisms are discussed in chapters by Narahashi, Bondy, and Verity, respectively.

In sum, in this section, through these eleven chapters by the most respected experts in the field of metal neurotoxicology, we have addressed the three most important issues of metal neurotoxicology: primary adverse effects and endpoints, tissue and cellular components involved, and the biomolecular bases in metal neurotoxicity. We hope that our present "issues and conceptual" approach will compliment and supplement our previous publication (Chang and Dyer, 1995), where the toxic effects and mechanisms of toxic metals were addressed individually.

REFERENCE

Chang, L. W. and Dyer, R. S. (1995), Handbook of Neurotoxicology, Marcel Dekker, New York, 1995.

Chapter 32

Toxico-Neurology and Neuropathology Induced by Metals

Louis W. Chang

I. INTRODUCTION

More than 40 elements in nature can be classified as metals. While some of these, such as copper, iron, and zinc are important to life, many, for example, mercury, lead, and cadmium, have deleterious effects on the biological system. These metals are known as toxic metals. Among the toxic metals, several have a primary target action on the nervous system. These groups of metals are referred to as neurotoxic metals. The most noticeable neurotoxic metals are mercury, lead, arsenic, cadmium, aluminum, and manganese. These, together with several organometals, such as methylmercury, alkyl leads, triethyltin, and trimethyltin which are known potent neurotoxicants, will be included in this chapter. Other metals such as lithium, thallium, copper, etc., which are also known to have neurotoxic potentials, have been presented in other chapters in Part 1 of this volume and will not be further discussed here.

This chapter will focus on the neurological alterations and clinicopathology induced by metals. Other effects of metals such as those on neurotransmitter, neurobehavior, cytoskeletons, and protein metabolism will be discussed separately in other chapters in this volume. The neurotoxic effects and mechanisms of mercury, lead, cadmium, aluminum, manganese, organoleads and organotins have been extensively reviewed by Chang and Verity (1995), Cory-Slechta and Pounds (1995), Hastings (1995), Lukiw and McLachlan (1995), Chu et al. (1995), and Chang (1995), respectively, in a recent publication, Handbook of Neurotoxicology (Chang and Dyer, 1995). The present chapter represents an excerpt and synoptic summary from these reviews. Readers are encouraged to refer to these excellent reviews if more detailed information is desired.

II. MERCURY

Mercury is perhaps one of the most insidious neurotoxic chemicals known to man. Mercury exists in different forms: elemental (metallic) mercury, mercury vapor, inorganic mercury salts (mercurous or mercuric), and organic mercury compounds (aryl- or alkylmercury). Among the various forms of mercury, mercury vapor and alkylmercuric compounds are considered to be most neurotoxic. A detailed account on the biometabolism and cytotoxicity of mercurials is presented by Dr. Massaro (Chapter 66) and therefore will not be reviewed in this chapter.

A. MERCURY VAPOR

Metallic mercury is rather volatile and vaporizes readily even at room temperature. Elemental mercury ("quicksilver"), when ingested, is poorly absorbed from the gastrointestinal tract and poses little toxic consequences. Mercury vapor, when inhaled, is efficiently absorbed through the alveolar membrane (Berlin et al., 1969) and has high affinity for the central nervous system. It was found that after exposure to mercury vapor, most mercury is distributed to the gray matter of the cerebral and cerebellar cortices and to various nuclei in the brainstem (Berlin et al., 1969; Nordberg and Serenius, 1969; Takahata et al., 1970; Berlin et al., 1975). The average biological half-time of inhaled mercury in the body is found to be about 60 days (Cherian et al., 1978).

Mercury vapor poisoning is characterized by unspecific symptoms including fatigue, gingivitis, ptyalism, disturbance of gastrointestinal functions, general weakness, and erethism (insomnia, shyness, increased excitability, loss of memory, personality changes, and depression). These early syndromes are sometimes referred to as "micromercurialism" (Trachtenberg, 1969; Friberg and Vostal, 1972). In more severe conditions, a fine intentional tremor involving fingers, tongue, eyelids, and lips usually follows. In some cases, the tremors may develop into a generalized body tremor with spasms of the extremities (Stopford, 1979). Constriction of visual field (Rosen, 1950) and ALS-like symptoms (Vroom and Greer, 1972) have also been reported.

B. INORGANIC MERCURY SALTS
1. Mercurous Salt

Human episodes of mercurous mercury poisoning have been reported in children using calomel as teething powder in the early 20th century (Swift, 1914; Warkang and Hubbard, 1953). Patients exhibited redness of the hands and feet, thus the term, "pink disease." This condition is accompanied by painful extremities (acrodynia), which is believed to be due to stimulation of the sympathetic nervous system by mercury (Cheek, 1980). The patients also experience photophobia, profuse sweating, anorexia, and insomnia.

Two cases of adult poisoning involving ingestion of laxative containing mercurous chloride have been reported. At autopsy, atrophy of the brain and loss of cerebellar granule cells were found (Davis et al., 1974).

2. Mercuric Salt

Micromercuralism, erethism, tremor, and incoordination ("mad hatter" syndrome) similar to that observed in mercury vapor poisoning can also be induced by chronic exposures to inorganic mercuric salts such as mercuric oxide and mercuric nitrate (WHO, 1976; Stopford, 1979). Experimental investigations also revealed neuronal changes in the cerebellum and in the dorsal root ganglia of rats after exposure to mercuric chloride (Chang and Hartmann, 1972).

C. ORGANOMERCURY COMPOUNDS
1. Arylmercury and Alkoxyalkylmercury

This category of mercury is best exemplified by phenylmercury (arylmercury) and by methoxyethylmercury (alkoxyalkylmercury). These compounds are biologically unstable and rapidly degraded, mainly in the liver, into inorganic mercury (Hg^{2+}) (Daniel et al., 1971, 1972; Gage, 1975; Beliles, 1975), and thus exert toxic actions similar to those of inorganic mercuric salts.

There have been some suggestions as to the induction of ALS-like or motor neuron disease-like syndromes by inorganic mercury and by phenylmercuric compounds (Vroom and Greer, 1972; Brown, 1954; Kantarjian, 1964; Adams et al., 1983). A recent observation by Arvidson (1992) also demonstrated an accumulation of mercury (Hg^{2+}) in spinal and brainstem motor neurons following intramuscular injection, lending some support to this hypothesis. Controversial findings and observations also exist. A distributional study by Gage and Swain (1961) failed to demonstrate significant mercury in the CNS following systemic exposure to arylmercury. Other investigators also reported no pathological lesions in the CNS in animals and humans exposed to phenylmercury (Goldwater, 1963; Ladd et al., 1964; Currier and Haerer, 1968; WHO, 1976; Stopford, 1979; Roberts et al., 1979; Conradi et al., 1982; Spencer and Schaumburg, 1982; Yanagihara, 1982).

2. Alkylmercury

The most neurotoxic examples of alkylmercury are methylmercury and ethylmercury; both of these are short-chain organomercuric compounds. The best known of these is methylmercury, due to its

Table 1 Frequency of Clinical Signs and Symptoms in Minamata Disease

Symptom or Sign	Frequency (%)
Constriction of visual fields	100
Sensory disturbance	100
Ataxia	100
Impairment of speech	88
Impairment of hearing	85
Impairment of gain	82
Tremor	76
Mental disturbance	71
Exaggerated tendon reflexes	38
Hypersalivation	24
Hyperhydrosis	24
Muscular rigidity	21
Ballism	15
Chorea	15
Pathologic reflexes	12
Athetosis	9
Contractures	9

After Takeuchi et al. (1968).

association with the massive outbreak of poisonings in Japan in the 1950s and 1960s ("Minamata disease") (Takeuchi, 1968, 1977) and in Iraq in the 1970s (Bakir et al., 1973; Amin-Zaki et al., 1974, 1976, 1978).

The clinical symptoms in alkylmercury poisoning may vary with the age and sex of the patients. The overall clinical signs and symptoms in methylmercury poisoning as seen in Minamata disease is summarized in Table 1. The major clinical symptoms and signs are visual disturbance (constriction of visual field), sensory disturbance, and cerebellar ataxia.

In both human autopsy and experimental animals with methylmercury poisoning, the most consistent pathological lesions were found in calcarine cortices (visual cortices), dorsal root ganglia, and cerebellum (Takeuchi, 1977; Chang, 1979, 1980). This topographical distribution of lesions correlates well with the neurological signs and symptoms (constriction of visual field, sensory disturbance, and cerebellar ataxia) observed in patients with Minamata disease.

Primary sensory neuropathy is probably one of the most sensitive parameters in methylmercury poisoning. Chang and co-workers first demonstrated the extensive damage of dorsal root ganglion neurons (Figure 1A and B) and fibers in rats after exposure to methylmercury (Chang and Hartmann, 1972). These observations were later confirmed by other investigators (Hermann et al., 1973; Jacob et al., 1977).

Histopathological changes in the cerebellum may serve as a characteristic diagnostic criterion for methylmercury poisoning. Cerebellar granule cell loss acquires a characteristic pattern, involving severe cell losses at the depth of the sulci (Figure 2) with proliferation of Bergmann's glial fibers. Widespread destruction of the granule cells throughout the cerebellum eventually occurs in prolonged intoxication. Most of the Purkinje neurons, however, are spared.

D. MECHANISM OF ACTIONS FOR MERCURY NEUROTOXICITY

The mechanistic bases for mercury neurotoxicity are complex and multifaceted. Major thoughts include: (1) disturbances of macromolecular synthesis and metabolism, such as those of protein and nucleic acids; (2) disturbance of Ca^{2+} homeostasis; (3) oxidative injury; and (4) aberrant protein phosphorylation. The role of glial cells in neuronal injury in metal intoxication has also been suggested. Detailed discussions on the various aspects of the mechanism of mercury neurotoxicity have been recently presented in a chapter by Chang and Verity (1995) and will not be further elaborated in the present chapter. The interrelationships of these various mechanism of actions is summarized in Diagram 1.

III. LEAD

The neurotoxic effects of lead are less foci oriented than those of mercury. The most prominent neuropathological feature of lead-induced encephalopathy is generalized interstitial edema. This change

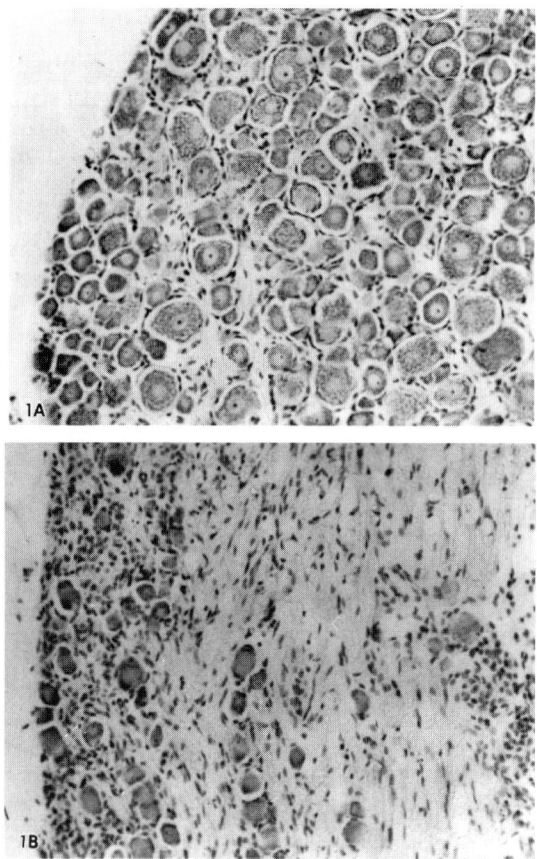

Figure 1. (A) Dorsal root ganglion, rat. Note the abundance of neurons in the ganglion. (Original magnification ×250.) (B) Dorsal root ganglion, rat, McHg-treated. Note the significant loss of neurons as compared with (A). (Original magnification ×250.) (From Chang, L. W., Desnoyers, P. A., and Hartmann, H. A. (1972), *J. Neuropathol. Exp. Neurol.*, 31, 389–395. With permission.)

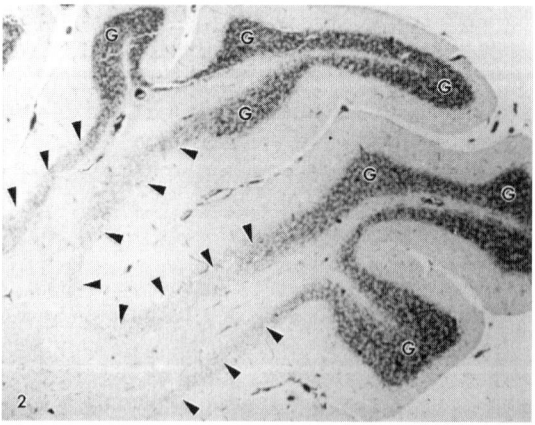

Figure 2. Cerebellum, rat, McHg-treated. Note the extensive loss of granule cells (G) at the depth of the sulci (→). This is a characteristic pathological lesion in methyl mercury poisoning. (Original magnification ×100.)

is believed to be the result of changes in the microvasculatures and the barrier properties of the capillary endothelium (Winder and Lewis, 1985; Bressler and Goldstein, 1991). Astrocytes, an integral part of the blood-brain barrier (BBB), are also affected by lead, possibly through alterations of calcium homeostasis or by activation of protein kinase C (Gebhart and Goldstein, 1988; Bressler and Goldstein, 1991). Changes in astrocytes will also compromise the integrity and function of the BBB.

Toxico-Neurology and Neuropathology Induced by Metals

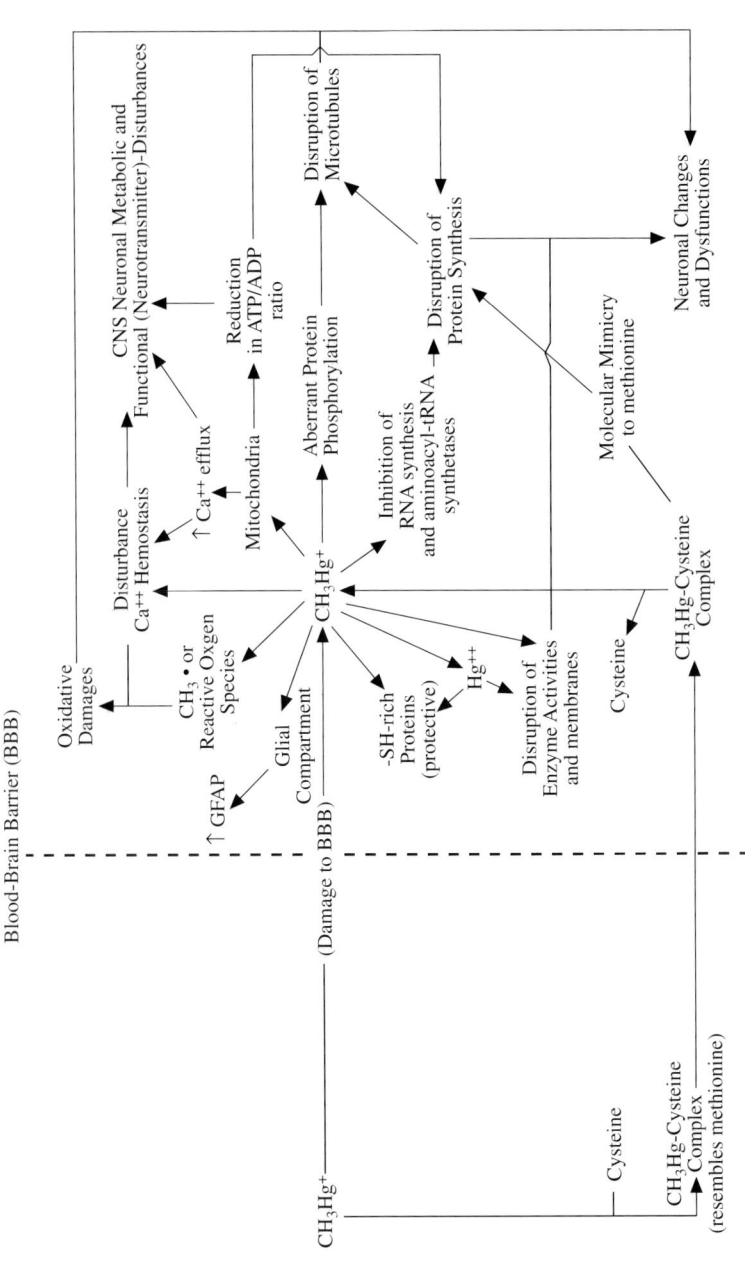

Diagram 1. Neurotoxic mechanism of actions of methylmercury. (From Chang, L. W., in *Handbook of Neurotoxicology*, Chang, L. W. and Dyer, R. S., Eds., Marcel Dekker, New York, 1995, 143–170. With permission.)

Acute encephalopathy is one of the most serious consequences of acute lead poisoning and is most common in young children. The most characteristic signs of lead encephalopathy include sporadic vomiting, ataxia, apathy, coma, and convulsions. The most common morphological changes in the brain include diffused cerebral edema, proliferation and swelling of endothelial cells, and focal necrosis of degeneration (Waldron and Stofen, 1974; Nordberg, 1975; Winder and Lewis, 1985). A diffuse astrocytic proliferation in gray and white matter is also frequently observed (Nordberg, 1975; Winder and Lewis, 1985).

Studies concerned with delineating neuropathological correlates of lead exposure have focused primarily on three brain regions: cerebellum, cerebral cortex, and hippocampus.

A. CEREBELLUM

In rats exposed to lead during development, histopathological changes are seen mostly in the cerebellum. Edema and focal and perivascular hemorrhages in the gray matter were observed. Regional atrophy and hypoplasia, cavitation, and degeneration of the white matter also occurred. Gitter cells together with degeneration and loss of Purkinje cells were frequently found (Clasen et al., 1974; Hirano and Kochen, 1979).

B. CEREBRAL CORTEX

Edema was mainly in the white matter of the corpus striatum, with proliferation of capillaries. It is noteworthy that these neuropathological changes were observed only in very young animals (<3 weeks of age for rats) and subsequently dissipated when the exposure was discontinued (Pentschew and Garro, 1966; Goldstein et al., 1974). McCauley and colleagues (McCauley and Bull, 1978; McCauley et al., 1979; McCauley et al., 1982) reported delays in cerebral cortical development and synaptogenesis in pre- and postnatally lead-exposed rats, which they attributed to alterations in cerebral energy metabolism. These experiments, however, were performed with relatively high levels of lead. In support of a lead-induced CNS developmental delay, various investigators (Cookman et al., 1987; Hasan et al., 1989; Regan, 1989; Regan et al., 1989) reported lead-induced inhibition of neural cell acquisition, particularly of the postnatal structuring of the central nervous system and an impaired desialylation of the D2-CAM/N-CAM protein. This inhibition of normal desialylation was attributed to improper guidance of neuronal cells and their fibers. Such findings could be related to the reduced synaptic elaboration as well as to subsequent altered neuronal structuring contributing to a reduction in fine motor skills and other manifestations of toxicity.

C. HIPPOCAMPUS

Studies suggested that developmental exposure to lead also has significant effects on hippocampal weight and size (Petit et al., 1983; Louis-Ferdinand et al., 1978; Campbell et al., 1983). Since these changes were found in the absence of lead-induced alterations in body or brain weight, these results were not attributed to nonspecific or nutritional effects of lead on general brain development. Morphometric studies of hippocampus indicate that lead exposure decreases the size of the mossy fiber zone, the numerical density of mossy fiber boutons, and also granule cell layer size. Pyramidal and granule cell dendrites are found to exhibit spine loss and a decrease in the extent and length of branching (Alfano et al., 1982; Campbell et al., 1983). Measurements of the dendritic pattern of the dentate granule cells and the mossy fiber pathways also revealed a delay in the refinement of the dendritic tree of dentate granule cells and a significant reduction in the overall length of mossy fiber pathway and terminals in lead-treated animals (Petit et al., 1983; Campbell et al., 1983).

A recent study by Slomianka et al. (1989) demonstrated that lead could induce structural changes in the hippocampus of developing animals even at low blood lead levels. Significant changes were noted in the size of the mossy fiber zone, the granule cell layer and the commissural-associational zone of the dentate molecular layer when evaluated at 28 days of age following lactational lead exposure. These structural changes, together with changes in neurotransmitters (Lai et al., 1985; Cory-Slechta, 1995; Cory-Slechta and Pounds, 1995; Zarriello, 1996), may be responsible for the various behavioral changes in animals exposed to lead.

D. NEUROPATHOLOGICAL EFFECTS ON THE PERIPHERAL NERVOUS SYSTEM

It is well known that lead induces neuropathy in the peripheral nervous system, and the ensuing peripheral nervous system changes affecting predominantly the large myelinated nerve fibers. Pathological changes in peripheral nerves can include marked edema of the nerves, segmental demyelination and axonal degeneration (Dyck et al., 1980; Winder and Lewis, 1985). Changes in myelination have been suggested to arise from injury to the blood-nerve barrier, which subsequently permits entry of lead-containing fluids into the endoneurium (Lampert and Schochet, 1968; Ohnishi and Dyck, 1981).

E. MECHANISTIC CONSIDERATIONS FOR LEAD NEUROTOXICITY

Like mercury, the mechanism of actions for lead neurotoxicity are multifaceted and complex. The influences of lead on neurotransmitters and on ion channels are presented in great detail in other chapters in this volume (see Chapters 37 and 40). Biomolecular studies also revealed the impact of lead on gene expression, signal transduction, and calcium messenger system. Detailed discussions on the mechanistic actions of lead have been recently presented in an excellent review by Cory-Slechta and Pounds (1995) and Zarriello (1996). Interested readers are encouraged to seek further information from this review. The mechanistic considerations for lead neurotoxicity are schematically summarized in Diagram 2.

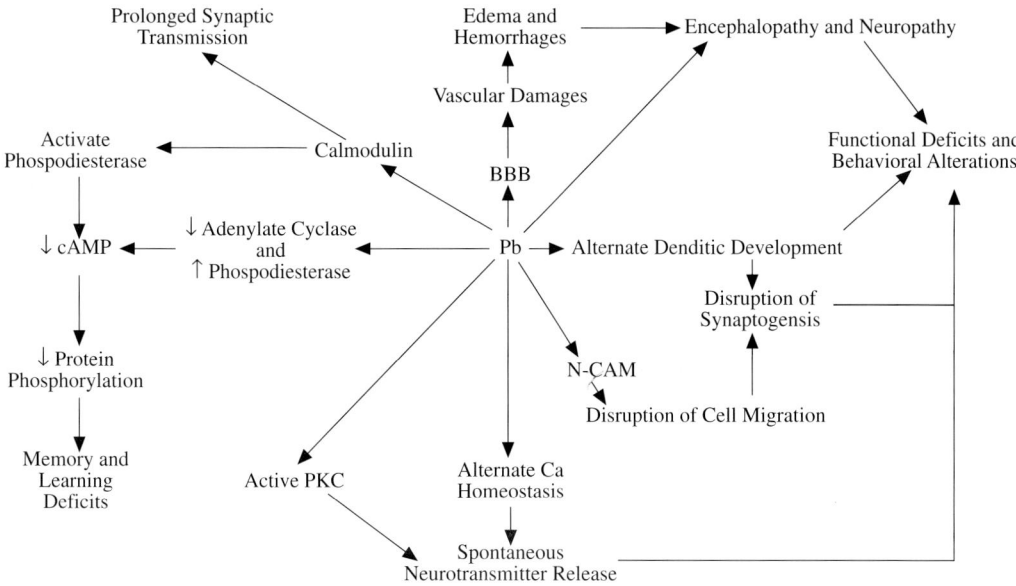

Diagram 2. Mechanistic considerations of Lead neurotoxicity.

IV. ARSENIC

During World War II, arsenical compounds were used as war gases (chlorophenyl-dichlorarsine or lewisite). There seems to be little qualitative difference between poisoning by inorganic and organic arsenical compounds. In acute poisoning, involvement of the nervous system is indicated by headache, vomiting, confusion, epileptic convulsions, and coma.

In the CNS, petechial hemorrhages in the corpus callosum, in the deep cerebral white matter near the inferior and posterior horns of the lateral ventricle, in the internal capsule, in the cerebral peduncles, and in the peripheral parts of pons and medulla are the main pathological features. Foci of perivascular demyelineation and necrosis in cerebral cortex and white matter may also develop (Meyer, 1963; Smith, 1976).

In chronic exposures, peripheral polyneuropathy is the predominant compliction. This condition is most frequently encountered from occupational exposure in the smelting industry. A syndrome known as Ronnskär disease is developed (the term was derived from smelting workers in Ronnskär, Sweden) (Politis et al., 1980). Three phases of the disease may be identified: an initial phase of weaknesses, anorexia, and vomiting, followed by irritations of the respiratory tract, cutaneous pigmentation, and hyperkeratosis. The final stage is represented by peripheral neuropathy characterized by paresthesias of lower and upper extremities, extensor palsy, distal weakness, and wasting of forearms and legs (especially smaller muscles of hands and feet). Myelin and axonal degeneration resemble those in distal axonopathy.

Cranial nerve involvement may also develop. Sensory fibers appear to be more affected than motor fibers. Segmental demyelination may occur. Involvement of posterior root ganglia and anterior horn neurons has also been reported (Smith, 1976).

Arsenic may act on the tricarboxylic acid cycle at the level of pyruvate metabolism by inactivating the SH-rich cofactor thioic acid. Arsenic may also react with the SH-groups in the enzyme succinate dehydrogenase, as succinate is sometimes found in urine of patients with arsenic poisoning.

V. CADMIUM

The primary source of cadmium (Cd^{2+}) exposure, in the general population, is through food consumption. Inhalation constitutes the second major source of Cd^{2+} exposure for the general population. However, for occupational situations, inhalation is the primary route of exposure. Cd^{2+} is only poorly absorbed in the gastrointestinal tract of adults (ATSDR, 1989), but is fairly high, up to 55%, in neonates (Clarkson et al., 1985). Ca^{2+} absorption via inhalation is much higher, ranging from 30–90% (Lee and Oberdorster, 1985). After absorption, the excretion of cadmium, primarily via urine, is very slow. The biological half-life for Cd^{2+} is estimated to be between 25 and 30 years in humans (Friberg et al., 1986). Thus, even with very low levels of exposure for a prolonged period of time, significant elevated body burden of Cd^{2+} can result.

A. CADMIUM-INDUCED NEUROPATHOLOGY

Gabbiani et al. (1967b) first documented the neurotoxic effects of Cd^{2+} in adult rats. Acute hemorrhagic lesions in the trigeminal and sensory spinal ganglia were reported. These lesions were characterized by hemorrhagic suffusions around ganglion cells and neurons showing nuclear pyknosis with lysis of the cytoplasm. However, when neonates were exposed to cadmium, hemorrhagic lesions in the cerebrum and cerebellum and not in the sensory ganglia were observed (Gabbiani et al., 1967a; Webster and Valois, 1981). Wong and Klaassen (1982) also found lesions in the cerebellum as well as in the caudate-putamen and corpus callosum with neonatal exposure. The Cd^{2+}-induced hemorrhage in the CNS was characterized by vasculature changes including vacuolization of the capillary wall, thinning of the basement membrane, widening of intercellular junctions, and denudation of the endothelial lining. Damage to the neural elements may therefore be secondary to vascular damages (Nolan and Shaikh, 1986). It was also found that cadmium exposure produced lesions in the central nervous system only up to postnatal day (PND) 20. Exposure after PND 30 produced lesions only in the sensory ganglia of the peripheral nervous system (PNS), reflecting that the CNS could be protected from Cd^{2+} by the maturation of the blood-brain barrier.

Unlike cerebral capillaries which have tight interendothelial junctions and the absence of fenestrations, capillaries of the sensory ganglia have fenestrations which permit Cd^{2+} to reach cells of the ganglia via the vasculature. The damage seen in the sensory ganglia involved both capillary and venular endothelial cells with a direct or primary effect of Cd^{2+} on the endothelial cells (Schlaepfer, 1971; Gabbiani et al., 1974). Arvidson (1983) found lesions appeared much earlier in the trigeminal ganglion (by PND 12) than in the dorsal root ganglion (not until PND 22). He also noted that damage to nerve cells and axons was restricted to sensory, but not sympathetic, ganglia (Arvidson, 1980). These differences may be due to variations in embryonic development of the tissues. Lesions in the CNS could also be induced in weanling (PND 21) rats after chronic exposure to cadmium via the drinking water (Murthy et al., 1987). Degenerative changes were observed in the Purkinje cells of the cerebellum only and not in the capillary endothelium.

In summary, CNS lesions can be induced by exposure of neonates to Cd^{2+}. Lesions in adults, on the other hand, are confined to the peripheral nervous system. The primary toxic effect of the Cd^{2+} in adults therefore appeared to primarily involve the vascular system, with secondary damage to the neural components at a later stage.

Although Ca^{2+} does not cross the BBB in sufficient amounts to cause morphological damage to the neurons, it exerts noticeable effects on neurotransmitters (Lai et al., 1985). This effect is most likely responsible for the behavioral alterations in adult animals exposed to cadmium (see Chapter 33). An excellent review on the effects and mechanisms of cadmium neurotoxicity has been recently published by Hastings (1995). Readers are encouraged to seek more complete information there.

VI. MANGANESE

Under normal circumstances, manganese (Mn) in blood is preferentially bound to transferrin (Gibbons et al., 1976) and only a small amount is bound to albumin (Scheuhammer and Cherian, 1985). Transport of Mn across the BBB likely occurs via transferrin receptor-mediated endocytosis. Within the brain, manganese primarily accumulates in the globus pallidus (GP) and substantia nigra pars reticularis (SNr). Accumulation also occurs in the striatum, pineal gland, olfactory bulb, and, to a lesser degree, in the substantia nigra pars compacta (SNc) (Larsen et al., 1979; Yamada et al., 1986; Newland et al., 1989). With cells, manganese preferentially binds to the outer membrane of mitochondria and to nuclear structures (Maynard and Cotzias, 1955).

A. HUMAN MANGANISM

Edsall et al. (1919) first established the relation between occupational exposure to manganese and the clinical syndrome and pathological effects of manganese neurotoxicity. Ashizawa (1927) emphasized the vulnerability of the globus pallidus and, particularly, its medial segment to the toxic actions of manganese. Since then, hundreds of cases of manganese poisoning have been reported in miners as well as in industrial and agricultural workers (Fahn, 1977; Mena, 1979; Ferraz et al., 1988). Chronic manganism causes an extrapyramidal syndrome with features resembling those seen in Parkinson's disease, Wilson's disease, and postencephalitic parkinsonian (Schwab and England, 1968; Barbeau et al., 1976; Fahn, 1977; Mena, 1979). The clinical course of manganism can be divided into three phases: an initial phase of subjective symptoms with or without psychotic episodes, an intermediate phase of evolving neurological symptoms and signs, and finally an established phase with persisting neurological deficits (Rodier, 1955; Fahn, 1977; Mena, 1979).

The initial symptoms are usually subjective and nonspecific. These may include fatigue, anorexia, headache, poor memory, reduced concentration, apathy, insomnia, diminished libido, and generalized slowing of movements (Mena, 1979). The initial phase of manganese intoxication is usually followed within 1–2 months by neurologic symptoms related to speech, writing, dexterity, movement, facial expression, posture, and gait. The face becomes expressionless with a dazed appearance (masque manganigue), and may be interrupted by spasmodic laughing or a dystonic grimace. Handwriting can become tremulous, micrographic, and cramped. Movements are generally slow, clumsy, and uncertain, with difficulty rising from a supine or sitting position. Gait is often impaired with anteropulsion and retropulsion. Turns tend to be "en bloc". Walking backwards may be particularly difficult and tends to be one of the earliest and most prominent features of manganese intoxication (Huang et al., 1989).

In the last phase of the illness, there is aggravation of neurologic dysfunction and disorders of walking become more pronounced. Gait becomes slow with small steps and shuffling or high-stepping and swinging. Inability to walk backwards due to severe retropulsion is generally the most more striking feature. Dystonic posturing of the foot causes some patients to have a characteristic gait described by von Jaksch (1907) as a "cockwalk" or "coq au pied".

B. NEUROPATHOLOGICAL FEATURES

Degeneration of the basal ganglia, principally confined to the medial segment of the GP and the SNr, constitute the hallmark of neuropathology by manganese. The putamen and the caudate nucleus are often affected and, to a lesser degree, the SNc may may also be involved. Other areas of the brain that may also be affected include the cerebral cortex, thalamus, subthalamus, hypothalamus, and red nucleus. Manganese neurotoxicity may be differentiated from Parkinson's disease (PD) both pathologically and clinically. Manganese specifically affects the striatum, GP, and SNr, whereas PD preferentially affects dopaminergic neurons of the SNc.

Recent studies indicate that clinical features of PD are most likely to be associated with an asymmetric presentation, resting tremor, and a good response to levodopa (Hughes et al., 1992). In contrast, manganese, which primarily affects the GP and striatum, presents with a clinical syndrome that more closely resembles "atypical parkinsonism". These features include speech disturbance, gait impairment, relative absence of tremor and little or no response to levodopa (Huang et al., 1989).

Observations on the effect of manganese on dopaminergic neurons in rodent models and in humans appear to conflict. On one hand, there is evidence that dopamine neurons in rodent models may be associated with a depletion of striatal dopamine (Brouillet et al., 1992; Neff et al., 1969). On the other hand, striatal fluorodopa uptake on PET scan in patients with indisputable basal ganglia dysfunction due to manganese is consistently normal, suggesting that the nigro-striatal pathway is relatively preserved (Wolters et al., 1989). These paradoxal findings may be explained by the sites where manganese accumulates in the CNS of rodents and of man. Direct injection of manganese into the SN or striatum in rodent models causes toxic effects on nigro-striatal dopaminergic neurons. However, in man and in nonhuman primates, manganese intoxication results in damage primarily in the GP and striatum, with relative sparing of dopaminergic neurons (Chu et al., 1995). Therefore, rodents may not serve as the ideal animal model for human manganism.

Pathological involvement of the cerebellum and spinal cord have also been demonstrated. Indeed, patients with ALS-Parkinson dementia of Guam showed 5- to10- folds of manganese elevation in their spinal cord tissues as compared with control cases (Donaldson and Barbeau, 1985). The relationship of manganese and motor neuron diseases cannot be totally ignored.

C. MECHANISTIC CONSIDERATIONS

Manganese (Mn) is a transition metal that can exist in different valence states and has the capacity to promote redox reactions with the formation of cytotoxic free radicals. The redox potential of manganese varies with its particular valence state. Donaldson and colleagues (Donaldson et al., 1981; Donaldson and Barbeau, 1985) have proposed that manganese can enhance the autooxidation of dopamine with the formation of reactive oxicant species. These investigators suggest that this mechanism could account for, at least in part, the cell damage associated with manganese neurotoxicity. It has been postulated that melanized neurons in the SNc might be particularly vulnerable to neurointoxication by manganese (Graham, 1984). Inhibition of neurotransmitter uptakes, such as dopamine, choline, GABA, and glutamate, by Mn has also been described (Lai et al., 1985).

It was found that, when injected directly into the rodent striatum, manganese impairs oxidative metabolism and decreased ATP synthesis (Brouillet et al., 1992). This will also lead to a rise in cytosolic free calcium. A rise in cytosolic calcium can result in activation of calcium-dependent protease, endonuclease, and lipase enzymes with consequent cell degeneration. An increased cytosolic calcium can also activate calpain and nitric oxide synthase (NOS) enzymes which may produce superoxide (O_2^-) and nitric oxide (NO·) radicals and cellular damages.

Manganese is known to have a high affinity for and accumulates in mitochondria (Maynard and Cotzias, 1955). Brouillet et al. (1992) have recently postulated that manganese is a primary mitochondrial toxin. Indeed, pathological damage associated with manganese intoxication is primarily confined to the GP, a distribution consistent to that seen with other mitochondrial toxins such as cyanide or carbon monoxide (Beal, 1992). It was proposed that manganese accumulates by way of the calcium uniporter and promotes an increase in mitochondrial calcium (Gavin et al., 1990) thereby potentiating the development of mitochondrial damages and oxidative stress.

Excitotoxic lesions are associated with selective sparing of NADPH-diaphorase positive neurons (somatostatin and neuropeptide Y) and the selective loss of GABA and substance P neurons (Beal et al., 1986). These neurochemical changes are also observed in the striatum with manganese-induced lesions, suggesting that excitotoxins may play a role. This hypothesis is further supported by the observation that manganese-induced basal ganglia damages can be blocked by prior decortication, with removal of the cortical glutamatergic input, or by treatment with the NMDA receptor antagonist, MK801. Thus, manganese neurotoxicity might be mediated through excitotoxic activity consequent to a primary mitochondrial lesion with disrupted oxidative metabolism.

VII. ALUMINUM

Neurological and neuropathological changes in the central nervous system of animal models are termed "experimental aluminum encephalopathy" (EAE). Aluminum, at concentrations found in EAE, is also found in a number of human neurological disorders (Crapper-McLachlan and Farnell, 1985), most notably Parkinson's disease (PD) (Hirsch et al., 1991), amyotrophic lateral sclerosis (ALS) and motoneuron disease (Kobayashi et al., 1990; Yasni et al., 1991a,b), the ALS-parkinsonian dementia of Guam (ALS-PD/G) (Perl et al., 1982; Garruto, 1991), dialysis encephalopathy (Allfrey et al., 1976; Crapper et al., 1980; Bolla et al., 1992), and Alzheimer's disease (AD) (Crapper et al., 1983; 1976; Perl and Brody, 1980; Perl and Pendleburg, 1984; Lukiw et al., 1992a; Spink, 1992; Good et al., 1992). The actual sites of aluminum accumulation may vary. Detailed discussions on these various conditions have been recently presented in an excellent review (Lukiw and McLachlan, 1995) as well as in a chapter by these authors in this volume (Chapter 23). The present chapter is confined to the discussion on EAE.

A. EXPERIMENTAL ALUMINUM ENCEPHALOPATHY (EAE)

Experimental aluminum encephalopathy (EAE) can be induced in aluminum-susceptible animals by intracranial (Crapper, 1973; Muma et al., 1988), subcutaneous (De Boni et al., 1974; Crapper, 1974), or intravenous (Wen and Wisniewski, 1985) injection. Aluminum is found to accumulate primarily in the chromatin of both neurons and glias (Crapper et al., 1976; 1980; Wen and Wisniewski, 1985). Neurological dysfunctions were similar to those of Alzheimer's disease, including progressive decline in higher cortical functions, an impairment of short-term memory, and motor disturbance (Crapper et al., 1980; Crapper and De Boni, 1980). Inhibitions of synaptosomal ATPase and various neurotransmitter uptake by Al have also been reported (Lai et al., 1985). High concentrations of aluminum were found in the glial cells (De Boni et al., 1980; Crapper-McLachlan et al., 1991; Young, 1992) and neuronal lysosomes (Steckhoven et al., 1990). Significant binding of aluminum was also found specifically on the neocortical neuronal chromatin (De Boni et al., 1974; Crapper et al., 1980), which probably represents the target site of aluminum neurotoxicity (Wen and Wisniewski, 1985; Lukiw et al., 1987; Crapper-McLachlan and Farrell, 1986; Crapper-McLachlan, 1986; Lukiw et al., 1992). Such nuclear binding of aluminum is found in both EAE and in human AD patients.

It has been suggested that the genes which are important for cytoskeletal proteins are the specific target sites for aluminum neurotoxicity (Muma et al., 1988; Muma et al., 1990; Lukiw, 1991; Roberson et al., 1992). Indeed, alterations in cytoskeletons in the nerve cells are the primary pathological features in both EAE and AD (Vemura and Ireland, 1984; Kosik et al., 1985; Muma et al., 1988; Katsetos et al., 1990; Troncoso et al., 1990). In EAE, CNS of aluminum-treated animals showed various neuropathological changes, including neuritic shrinkage, axonal atrophy, and accumulations of abnormal in neurofilaments in the neuronal perikarya (Figure 3) and axons (Troncosco et al., 1986; Muma et al., 1988; Bizzi and Gambetti, 1986; Johnson and Jope, 1988). Neurobehavioral changes induced by aluminum are presented in Chapter 33.

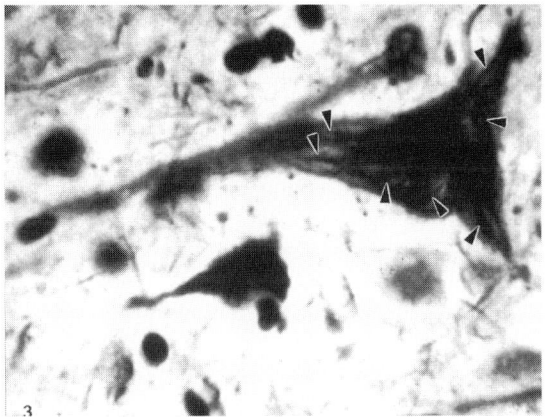

Figure 3. Hippocampus, rat, aluminum intoxication, Bielchowsky's stain. Accumulation of bundles of argentophilic filaments in the neuronal cytoplasm is demonstrated (→). (Original magnification ×650.) (From Chang, L. W. (1994), in *Principles of Neurotoxicology*, Chang, L. W., Ed., Marcel Dekker, New York, 3–34. With permission.)

VIII. ALKYLTINS

While there is no known neurotoxicity of inorganic tin, organic tin compounds, especially triethyl- and trimethyltin, are potent neurotoxicants capable of inducing characteristic neural lesions in the mammalian central nervous system.

A. TRIETHYLTIN (TET)

The most important human episode of human exposure to TET occurred in France in the 1950s. A medication, Stalinon, was inadvertently contaminated with 10% triethyltin (TET). Over 100 patients died. Patients showed various neurological problems, including persistent headache, vertigo, visual disturbances, abdominal pain, psychic disturbances, muscular weakness, EEG changes, increased cerebral spinal fluid (CSF) pressure, and convulsion (Alajouanine et al., 1958). In more severe cases, patients developed a flaccid type paraplegia, sensory loss, absence of reflexes, severe psychiatric disturbances, convulsion, coma, and death. Autopsies revealed severe edema in the white matter of the brain and spinal cord (Alajouanine et al., 1958; Barnes and Stoner, 1959; Cossa et al., 1958; Stoner et al., 1955).

Animal studies confirmed that massive cerebral edema, confined to the white matter of the CNS, is the primary and characteristic lesion induced by TET (Figure 4) (Magee et al., 1957; Torak et al., 1960, 1970; Wenger et al., 1986; McMillan et al., 1986; Chang, 1987). Electron microscopic examination revealed that the edema is intramyelinic. The accumulation of fluid splits the myelin sheath at the interperiod line to form fluid-filled vacuoles (Aleu et al., 1963; Hirano et al., 1968; Graham and Gontas, 1973; Jacobs et al., 1977). The edematous effect apparently is quite specific in the CNS, with little or no effect on the peripheral nerves (Graham and Gontas, 1973).

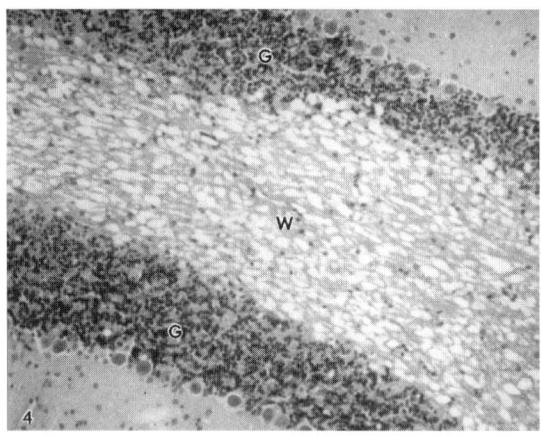

Figure 4. Cerebellum, rat, TET-treated. The foamy or bubbly appearance in the white matter (W) indicates myelinic edema of the fibers. G, granule cells. (Original magnification ×250.)

The precise mechanism of actions of TET on the central myelin is still obscure. Various investigations suggest that these actions would include: (1) free radical formation which promotes lipid peroxidation on the myelin membranes (Prough et al., 1981; Wiebkin et al., 1982); (2) effect on mitochondrial phosphorylation and ATP production (Stockdale et al., 1970; Rose and Aldridge, 1972; Kirschner and Sapirstein, 1982); and (3) inhibition of ATPase, 5-nucleotidase, and phosphodiesterase in various brain regions (Wassenaar and Kroon, 1973), which would contribute to myelinic edema. A more detailed discussion on these aspects have been recently presented in a review by this author (Chang, 1995). The proposed mechanism of action of TET on the nervous system is summarized in Diagram 3.

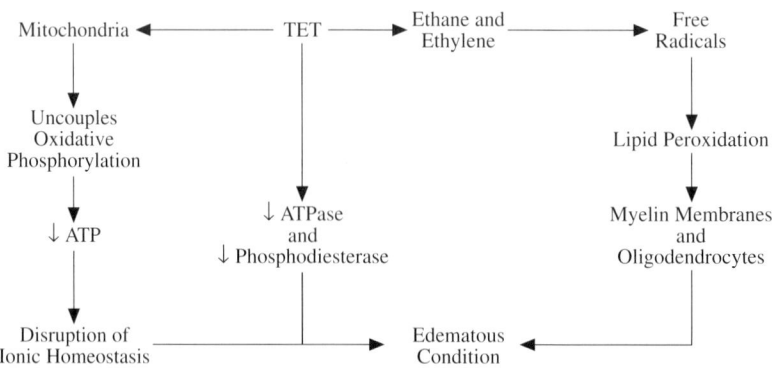

Diagram 3. Mechanistic considerations for TET neurotoxicity. (From Chang, L. W., in *Handbook of Neurotoxicity*, Chang, L. W. and Dyer, R. S., Eds., Marcel Dekker, New York, 1995, 143–170. With permission.)

B. TRIMETHYLTIN (TMT)

In 1978, Fortemps et al. described two cases of accidental human exposure to trimethyltin (TMT) (Fortemps et al., 1978). The patients suffered mental confusion, headaches, seizures, and psychic disturbances. Both patients seemed to have recovered from these toxic effects. In the early 1980s, a number of German industrial workers also suffered from exposure to TMT (Ross et al., 1981; Rey et al., 1984). These patients displayed a wide range of psychomotor symptoms including personality changes, irritability, memory deficits, insomnia, aggressiveness, headaches, tremors, convulsion, and changes of libido. No histopathological information on the CNS in these patients was available.

TMT also shows prominent and characteristic neurotoxic effects on animal models. The behavioral changes in rats exposed to TMT include aggression, hyperirritability, tremor, spontaneous seizures, hyperreactivity, and changes in schedule-controlled behavior (Brown et al., 1979; Wenger et al., 1982, 1984a,b; Dyer et al., 1982b). These changes in behavior have been referred to as the "trimethyltin syndrome" (Dyer et al., 1982b). Rats showed selective sensitivity to TMT in areas of the limbic system, including the entorhinal cortex and the hippocampus (Chang and Dyer, 1983; Chang et al., 1983c).

Mice were found to be more sensitive to TMT toxicity than rats. Rapid neurological changes were induced within 24 hours by a single low-dose exposure (Chang et al., 1984). The primary CNS lesions induced in the mice were found in the hippocampus, brainstem (Figure 5), and spinal cord (Chang et al., 1982a,b,c; Chang et al., 1983a,b,c; Chang et al., 1984).

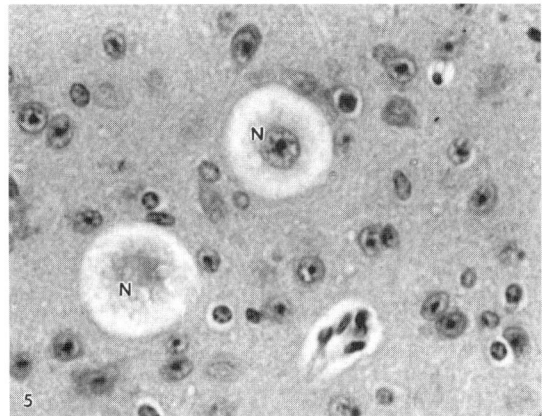

Figure 5. Brainstem, mouse, TMT. Two edematous and degenerating neurons (N) appeared to be highly distended with foamy cytoplasm. (Original magnification ×400.) (From Chang, L. W. (1994), in *Principles of Neurotoxicology,* Chang, L. W., Ed., Marcel Dekker, New York, 3–34. With permission.)

Within the limbic system, mice showed lesion involvement primarily in the fascia dentate (granule cells) (Figures 6A, B) with little involvement of the hippocampal Ammon's horn (pyramidal neurons) and the entorhinal cortex. Rats, on the other hand, showed more prominent involvement in the pyramidal neurons of the Ammon's horn with less involvement in the fascia dentate (Figures 7A, B). Considerable pathology at the entorhinal cortex was also observed in rats. These comparative pathological lesions in the limbic system between mice and rats are summarized in Table 2.

Subsequent investigations with neonatal rats (Chang, 1984a,b) further demonstrated that the vulnerability of the Ammon's horn to TMT was closely associated with and heavily dependent upon the functional maturity and integrity of the neurons and the neuronal circuitry in the hippocampal formation (Table 3).

Chang (1986) proposed that the TMT-induced unique pattern of neuronal damage in the limbic system is related to hyperexcitation of the neuron groups along the neural circuitry of the limbic system (Diagram 3). A surge of hyperexcitation of electrical impulses may produce damages on nerve cells along the path of this electrical surge.

Biochemical investigations in TMT-poisoning revealed a reduction in glutamate and GABA uptake and synthesis (Doctor et al., 1982a,b,c; De Haven et al., 1984; Mailman et al., 1983; Naalsund et al., 1985; Patel et al., 1990) with an increased synaptic release of glutamate in the hippocampus. This release of glutamate, together with a depletion of hippocampal zinc (Chang and Dyer, 1984) and an inhibition and damage of dentate basket cells by TMT (Dyer et al., 1982a; Chang and Dyer, 1985), will also

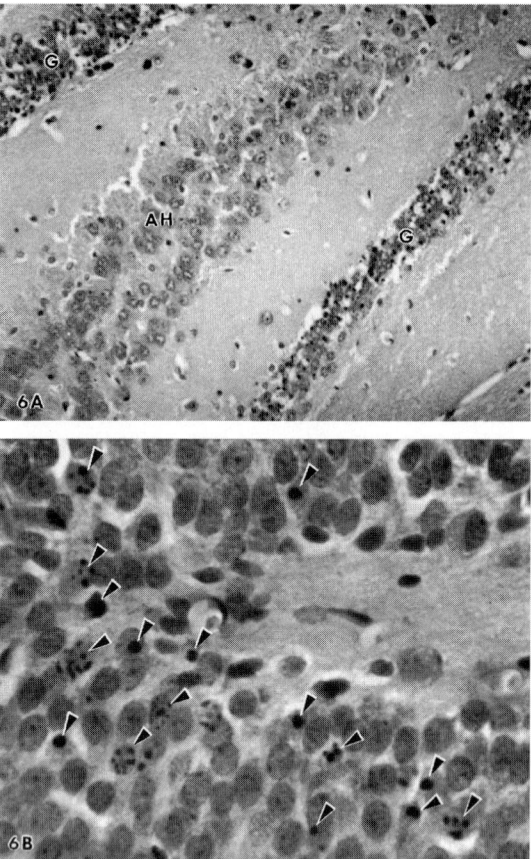

Figure 6. (A) Hippocampus, mouse, TMT-treated. Note the extensive cellular destruction in the granule layer (G) of the fascia dentata with sparing of the pyramidal cells of the Ammon's horn (AH). (Original magnification ×250.) (B) Fascia dentata, mouse, TMT-treated. Large number of necrotic neurons (→) with pyknotic nuclei (densely stained) are noted. (Original magnification ×450.) (From Chang, L. W. (1994), in *Principles of Neurotoxicology*, Chang, L. W., Ed., Marcel Dekker, New York, 3–34. With permission.)

Table 2 Comparison of Lesion Development in the Limbic System Between Mice and Rats

	Mice	Rats
Entorhinal cortex	±	++
Fascia dentate granule cells	+++	+
Ammon's horn neurons	–	+++

From Chang, L. W. (1995), in *Handbook of Neurotoxicology*, Chang, L. W. and Dyer, R. S., Eds., Marcel Dekker, New York, 143–170. With permission.

promote neuronal hyperexcitation. Aldridge and co-workers (Aldridge and Street, 1971; Aldridge, 1976) described the effects of TMT on mitochondrial respiration leading to a "hypoxic" state of the neurons. One of the consequences of the hypoxic condition of the nervous system is the release of glutamate and neuronal excitation. This neuronal excitation may be initiated at the entorhinal cortex, and the cascade of excitation will progress along the limbic circuitry as outlined in Diagram 4. The overall mechanistic base of TMT neurotoxicity on the hippocampal formation has been recently reviewed by the same author (Chang, 1990; 1995) and is represented in Diagram 5.

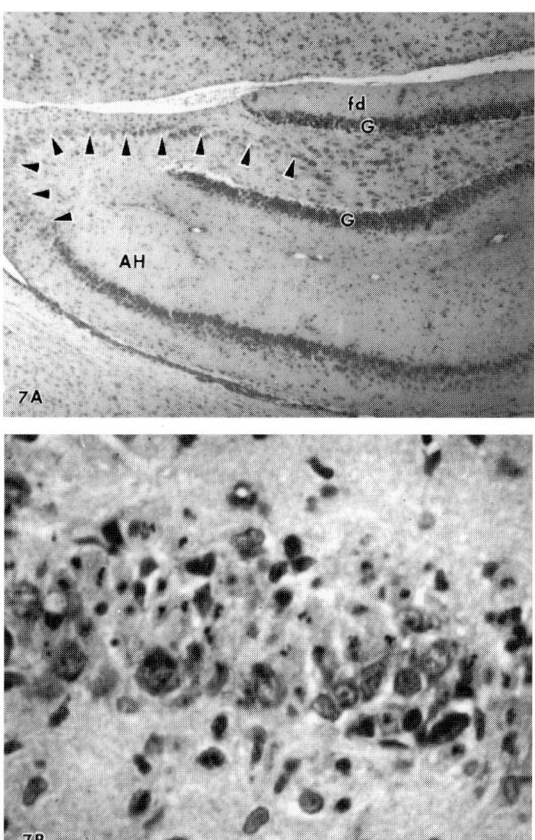

Figure 7. (A) Hippocampus, rat, TMT-treated. Significant thinning of CA_3 sector of Ammon's horn (AH) (→) due to loss of pyramidal neurons is demonstrated. Note the fascia dentata (fd) remains relatively intact, with little damage to the granule cells (G). (Original magnification ×150.) (B) Ammon's horn, rat, TMT-treated. Significant cell death and cell loss are demonstrated. (Original magnification ×450.)

Table 3 Correlation Between Hippocampal Development and TMT-Induced Lesions in Neonatal Rats

	PND 1–4	PND 5–6	PND 7	PND 8–10	PND 11–12	PND 13–15
Mossy fiber and synaptic development[a]	+ (CA_{3b})	++ (CA_{3b})	++ ($CA_{3a,b}$)	+++ ($CA_{3a,b,c}$)	+++	++++
Functional efficiency[a] (electrical stimulation response)	None	Weak	Stronger response	Responsive but inconsistent	More mature and responsive	Strong and consistent
Damages in Ammon's horn as a result of TMT exposure	None	+ (CA_{3b} only)	++ ($CA_{3a,b}$)	++ ($CA_{3a,b,c}$)	+++ (CA_2, CA_3)	++++ ($CA_{1,2,3}$)

[a] Data interpreted from: Bliss et al., 1974; Stirling and Bliss, 1978; Cowan et al., 1980.

From Chang, L. W. (1984a), *Neurotoxicology*, 5(2), 205–216. With permission.

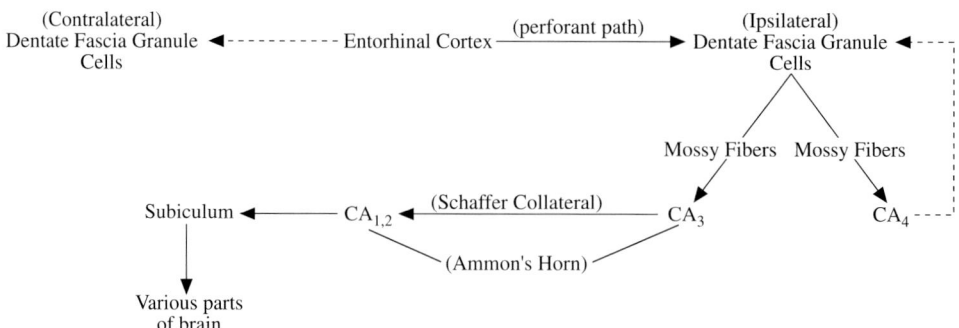

1. Elimination or destruction of entorhinal cortex (rats, 0.8 mg/kg/day, 14 days), spares damage to the entire hippocampal formation (dentate fasica and Ammon's horn).

2. Elimination or destruction of dentate fascia granule cells (mice, 3.0 mg/kg; rats, ventral hippocampus, 6.0 mg/kg; rats, dorsal hippocampus, 12.5 mg/kg), spares damage to the Ammon's horn neurons.

3. Elimination or destruction of Ammon's horn CA_3 neurons at the septal portion of the hippocampus (rats, 6.0 mg/kg), spares damage to the $CA_{1,2}$ neurons.

   Sparing of the Ammon's horn CA_3 neurons at the temporal portion of the hippocampus (rats, 6.0 mg/kg), results in damages to the $CA_{1,2}$ neurons.

Diagram 4. Limbic path vs. TMT-induced lesion development. (From Chang, L. W., *J. Toxicol. Sci.*, 15 (suppl. 4), 125–151, 1990. With permission.)

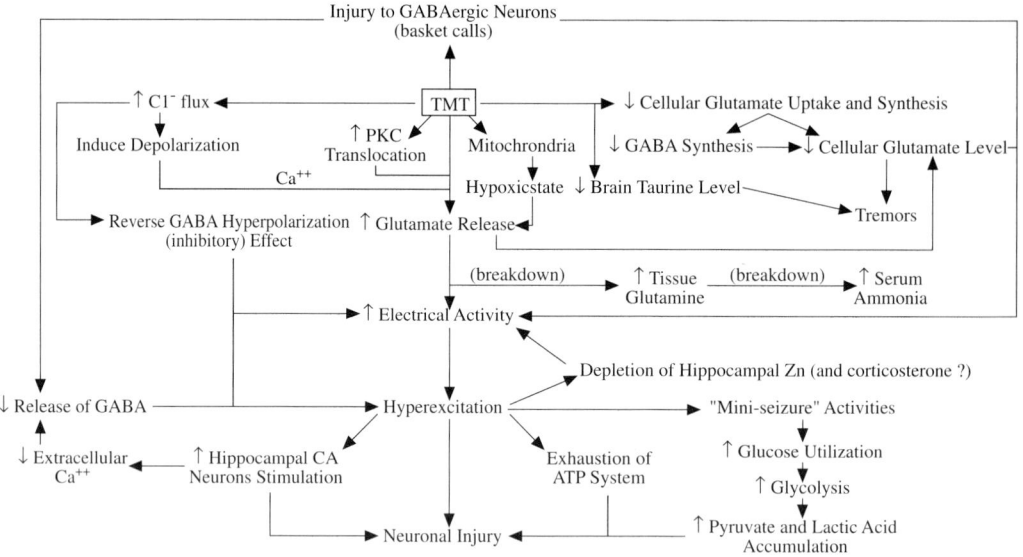

Diagram 5. Proposed neurotoxic mechanism of TMT on rat hippocampus. (From Chang, L. W., in *Handbook of Neurotoxicology*, Chang, L. W. and Dyer, R. S., Eds., Marcel Dekker, New York, 1995, 143–170. With permission.)

IX. ALKYLLEAD

Alkylleads are highly lipophilic and can penetrate skin readily. Inhalation of vapors of alkyllead compounds also results in rapid absorption of alkyllead through the lungs. Although tetraalkylleads (Me_4Pb or Et_4Pb) are not very toxic, these compounds are degraded in the liver to highly toxic trialkylleads, with the central nervous system being the target organ (Creamer, 1965; Springman et al., 1963; Nikowitz, 1974).

The general clinical signs and symptoms of R_3Pb are presented in Table 4. The first comprehensive studies in the neuropathological effects of alkyllead compounds were reported by Davis et al. (1963)

and Schepers (1964). The overall neuropathological involvement (Table 5) is also very similar to that observed in TMT poisoning. Aside from the pathological changes in limbic neurons and neurons in the brain stem and spinal cord, scattered neuronal degeneration was also observed in the neocortex, cerebellum, thalamus, and basal nuclei (Davis et al., 1963; Schepers, 1964).

Table 4 Clinical Signs and Symptoms of Et_4Pb or Et_3Pb Exposures

Phase I	Lethargy
Phase II	Inappetance, tremor, hypermotility, hyperexcitability, aggression
Phase III	Hypothermia, convulsion, incoordination, ataxia, paralysis
Phase IV	Death

N.B., symptoms and signs resemble those observed in trimethyltin, TMT, intoxication.

Chang, L. W. (1990), *J. Toxicol. Sci.*, 15 (Suppl. 4), 125–151. With permission.

Table 5 Neuropathology Involvement in Acute Exposure to Et_4Pb

The neuropathological changes are quite similar to those observed in trimethyltin (TMT) poisoning:
Neuronal necrosis and pycnosis primarily in pyriform/entorhinal cortex, hippocampal formation: fascia dentata and Ammon's horn, amygdaloid nuclei, neocortex
Neuronal chromatolysis, swelling, and necrosis in the brainstem and midbrain nuclei, pontine nuclei, basal nuclei, anterior cervical spinal cord
In certain species, involvement of cerebellar Purkinji cells is also observed

Chang, L. W. (1990), *J. Toxicol. Sci.*, 15 (Suppl. 4), 125–151. With permission.

With electron microscopy, nuclear condensations, hypertrophy of the Golgi saccules, swelling of the mitochondria, dilatation of the endoplasmic reticulum and dispersion of polyribosomes, proliferation of neurofilaments, disruption of microtubules, and accumulation of dense, multilaminar bodies were observed in neurons affected by trialkyllead (Niklowitz, 1974, 1975; Menthos et al., 1980; Seawright et al., 1984; Roderer and Doenges, 1983; Bondy and Hall, 1986).

Recent investigations by Chang et al. (1984a) and Walsh et al. (1986) examined further the neurotoxic effects of both triethyllead (TEL) and trimethyllead (TML) in rats. TEL-induced sensory disturbances and prominent mitochondrial changes in the dorsal root ganglion neurons. TML, on the other hand, induced extensive chromatolytic and degenerative changes in large brain stem neurons (Figure 8) and anterior horn motoneurons of the spinal cord (Figure 9). Neuronal swelling and isolated neuronal necrosis were also observed in the hippocampus after either TEL or TML exposure.

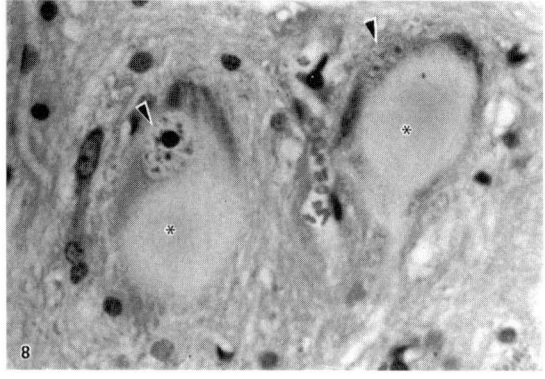

Figure 8. Brainstem, rat, TML-treated. Two large neurons displayed central chromatolysis (*) with eccentric nuclei (→). (Original magnification ×400.) (From Chang, L. W., Ed., (1994), in *Principles of Neurotoxicology*, Marcel Dekker, New York, 3–34. With permission.)

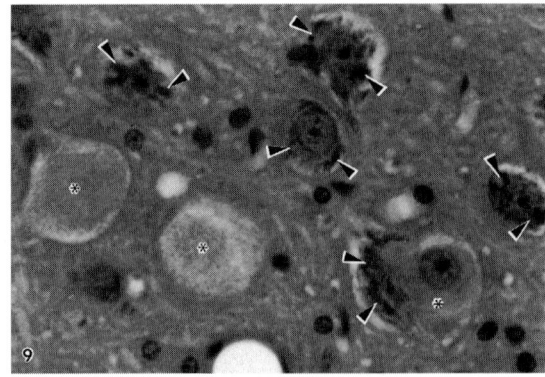

Figure 9. Spinal cord, rat, TML. Chromatolytic changes of the anterior horn motoneurons (*). Note some of the motoneurons still retain their Nissl pattern (→). (Original magnification ×400.) (From Chang, L. W. (1995), in *Handbook of Neurotoxicology,* Chang, L. W. and Dyer, R. S., Eds., Marcel Dekker, New York, 143–170. With permission.)

The mechanisms of action for alkyllead neurotoxicity are multifaceted and complex. They probably involve (1) disturbance of the Cl⁻ ionic influx and transport system in the neurons and neuronal hyperexcitation, (2) induction of Cl⁻/OH⁻ exchange across the biological membranes, (3) suppression of mitochondrial function, (4) inhibition of ATP synthesis, and (5) disruption of calcium homeostasis and stimulation of synaptosomal Ca^{2+} influx and neurotransmitter release. Detailed discussion on these aspects of the effects and mechanisms of alkyllead neurotoxicity has been recently reviewed by this author (Chang, 1995) and will not be further elaborated here. An overall mechanistic scheme for alkyllead neurotoxicity is summarized in Diagram 6.

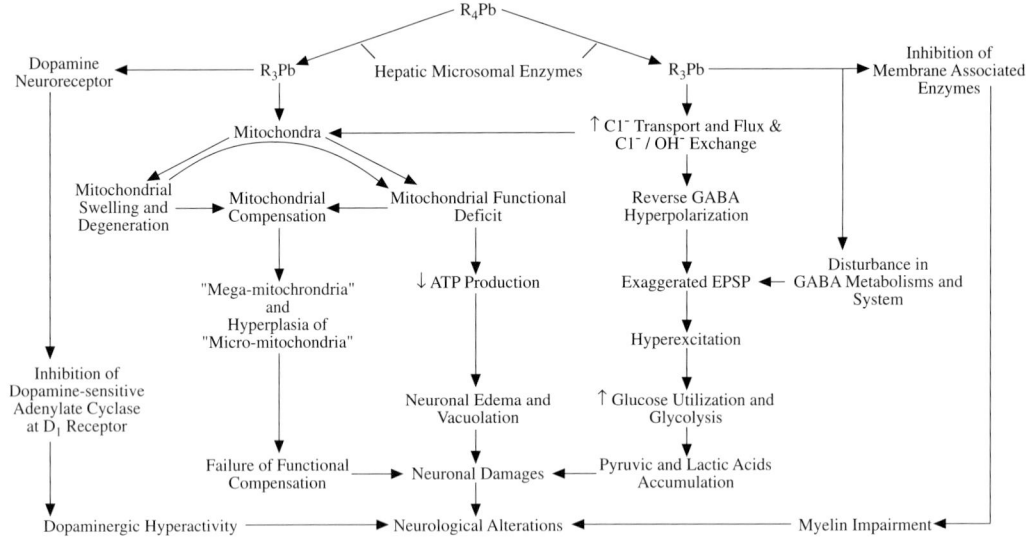

Diagram 6. Neurotoxic mechanisms of alkyllead compounds. (From Chang, L. W., *J. Toxicol. Sci.,* 15 (suppl. 4), 125–151, 1990. With permission.)

X. THALLIUM

The clinical toxicity of thallium has been covered in Chapter 28 of this volume (see chapter by Prof. van der Voet and Prof. de Wolff). Only the essential information on thallium neurotoxicity will be summarized and presented in this chapter.

The three major routes of exposure to thallium are inhalation, ingestion, and dermal contact. Thallium is rapidly absorbed into the body, especially via inhalation and ingestion.

The most prominent clinical signs and symptoms for thallium toxicity are alopecia, painful extremities ("burning feet"), and peripheral neuropathies. Preservation of reflexes early in the disease is also a characteristic feature. Other neurological phenomena include choreathetosis, lethargy, ataxia, tremor, cranial palsies, psychosis, convulsion, and coma. Autonomic dysfunctions manifested as tachycardia, hypertension, increased salivation, and fever have also been reported in patients exposed to thallium.

Pathological studies in human patients suffering from thallium poisoning are very limited. Autopsy cases from patients who died from acute thallium poisoning revealed localized areas of edema in the cerebral hemispheres and brainstem; disseminated and focal neuronal atrophy in the cortex; pyramidal degeneration in the pons', chromatolytic changes in the neurons in the motor cortex, substantia nigra, and globus pallidus; and glial proliferation in the dentate nucleus and in the inferior olive. Peripheral nerves also displayed axonal degeneration and demyelination resembling that of dying-back axonopathy. The sensory nerves and the posterior roots are most prominently involved. Degeneration is also observed in the posterior column of the spinal cord and in the anterior horn motoneurons.

Experimental models for thallium neurotoxicity have not met with much success. In rabbits, thallium produced neuronal changes in the hypothalamic nuclei, corpus mammillare, and the nucleus supraopticus. Axonal degenerations have been reproduced in rats after high doses of thallium. Mitochondrial swelling appeared to be the most characteristic lesion induced by thallium. This mitochondrial change represents an early pathological change in the cells and axons. Such changes were most conspicuous in the nerve fibers and their terminals.

The mechanism of action for thallium neurotoxicity is still obscure. It has been pointed out that the characteristic "triad" of thallium poisoning (alopecia, polyneuropathy, and "burning feet") is consistent with vitamin B deficiency. Indeed, clinical symptoms related to changes in the spinal roots, vertebral ganglia, and peripheral nerves are all analogous to those in states of deficiencies of thiamine, nicotinic acid, and panthothenic acid. Thus, it is not unreasonable to postulate that thallium toxicity may be related to disturbance of riboflavin metabolism. Indeed, thallium is known to complex and form an insoluble salt with riboflavin.

For more detailed information, readers are encouraged to refer to Osetowska (1971), Bank (1980), and Chang and Reuhl (1996).

XI. CONCLUDING REMARKS

Many metals are found to be neurotoxic. In the present chapter, only the most prominent neurotoxic metals are discussed. It is of interest to note that while some metals, such as lead and inorganic mercury, induce rather nonspecific or generalized neurological alterations, other metals, such as manganese, alkylmercury, and trimethyltin, induce very specific neurological syndromes and lesions. A comparative summary of this metal-induced toxiconeurology and neuropathology is presented in Table 6 and Table 7, respectively.

Table 6 Comparative Toxico-neurology Induced by Metals

Metal	Primary Toxico Neurology
Mercury vapor	"Mad-hatter's" syndrome
	Asthenic-vegetative syndrome, erethism, and micromercurialism (fatigue, ptyalism, gingivitis, insomnia, personality changes, memory deficits, depression, and general weakness)
	Intentional tremors
Inorganic mercury	
mercurous salt	Pink's disease
	Acrodynia
Mercuric salt	Resembles those seen in mercury vapor poisoning
Organic mercury	
Aryl- and alkoxyalkyl-mercury	ALS-like and motor neuron disease-like syndromes (rare and controversial)
Alkylmercury	Minamata disease (sensory disturbance, constriction of visual fields, and cerebellar ataxia)

Table 6 (continued)

Metal	Primary Toxico Neurology
Inorganic lead	In adults: epileptiform convulsions, delirium, papilloedema, dementia, peripheral neuropathies (esp. motor nerves and muscles)
	Associated with "lead-line" of the gums, basophilic stippling of the red cells, and evidence of renal impairment
	In children: mental retardation, cognitive impairments, mental deterioration, encephalopathy, behavioral disorders
Alkyllead	Delirium, convulsive seizures, generalized encephalopathy, tremor, hyperexcitability, incoordination, ataxia, coma
Arsenic	In acute exposure: epileptic convulsion, raised intracranial pressure, confusion, coma
	In chronic conditions: "Ronnskär disease," peripheral neuropathies, paresthesia, extensor palsy, wasting of small muscles in forearm and legs, general weakness
Cadmium	Headaches, vertigo, insomnia, tremor, dermographia, sweating, changes in sensory, optical, dermal, and motoric chronaxia
Manganese	Extrapyramidal syndrome resembling those in Parkinson's disease
	Initial symptoms: fatique, headaches, poor memory and concentration, apathy, insomnia
	Expressionless face (masque manganique), spasmodic laughing, dystonic grimaces, tremors, slowing of movements, difficulties in rising and walking backwards, impaired gaits, "en bloc" turns
	Final stage: "coq au pied" gaits, inability to walk backwards, aggravation of neurologic dysfunctions
Aluminum	Alzheimer's disease-like syndrome
	Progressive decline in higher cortical functions
	Short-term memory impairment
	Motor disturbance
Alkyltin	
Triethyltin (TET)	Persistent headaches, vertigo, visual disturbance, psychic disturbances, EEG changes, increased CSF pressure, convulsions
	Flaccid paraplegia, sensory loss, absence of reflexes, convulsions and coma
Trimethyltin (TMT)	Mental confusion, headaches, seizures, psychic disturbances
	Personality changes, irritability, memory deficit, aggressiveness, tremor, changes of libido, convulsions, coma
Thallium	Most prominent "triad": alopecia, painful extremities ("burning feet"), and peripheral neuropathy
	Choreathetosis, ataxia, tremor, cranial palsies, psychosis, convulsions, autonomic dysfunctions

Table 7 Comparative Neuropathology Induced by Metals

Metal	Primary Neuropathology
Mercury vapor	Generalized neuronal degeneration and necrosis in the cerebral gray cerebellar cortex, and brainstem nuclei
Inorganic Mercury	
Mercurous salt	Isolated neuronal degeneration and necrosis in the cerebral and cerebellar cortices
Mercuric salt	Resemble those of mercury vapor intoxication but of lesser extent
Organic mercury	
Aryl- and alkoxyalkyl-mercury	Resemble those of inorganic mercuric salt intoxication
	Some reports on lesions in the anterior horns of the spinal cord and in the motor nuclei
Alkylmercury	Selected neuronal degeneration and necrosis in the cerebellar granule layer (esp. those cells at the depth of the sulci), the calcarine cortex, and the dorsal root ganglia
Inorganic Lead	Generalized encephalopathy of the CNS
	Edematous changes of the meninges, corpus callosum, and brain matter
	Vascular and glial proliferations
	Focal necrosis in the pons and laminar necrosis of the cerebral cortex
	Thinning of the hippocampal mossy fiber zone and granule cell layer
	Segmental degeneration of the peripheral nerve with or without axonal degeneration

Table 7 (continued)

Metal	Primary Neuropathology
Alkyllead	Neuronal changes in the cerebral and cerebellar cortices, thalamus, basal nuclei
	Some lesions resemble those in TMT poisoning with neuronal changes and necrosis in the limbic system, brainstem, spinal cord, and dorsal root ganglia
Arsenic	Petechial hemorrhages in the corpus callosum and various brain structures
	Foci of perivascular demyelination and cortical necrosis
	Peripheral neuropathy with degeneration of myelin and axons (sensory motor)
	Changes in posterior root ganglia and anterior horn neurons
Cadmium	In adults: hemorrhagic ganglionic lesions with no specific CNS involvement
	In neonates: hemorrhagic lesions with neuronal degeneration in the CNS, including cerebral and cerebellar cortices, caudate-putamen, and corpus callosum
Manganese	Primary changes in the medial segment of globus pallidus, substantia nigra pars reticularis
	Lesions also occur in putamen caudate nucleus, and sustantia nigra pars compacta
Aluminum	Affects both cerebrum and cerebellum
	Neurofilabrillary accumulations in various nerve cells in the CNS
	Neuritic shrinkage and axonal atrophy
Alkyltin	
Triethyltin (TET)	Myelinic edema of the white matter in the CNS
Trimethyltin (TMT)	Selective neuronal necrosis in the limbic system (entorhinal cortex, hippocampus dentate gyri), brainstem nuclei, and spinal cord
	Also involves sensory neurons such as inner ear hair cells, retinal ganglia, olfactory and pyriform cortex, and dorsal root ganglia
Thallium	Edema in cerebral cortex and brainstem, neuronal changes in the pons, motor cortex, substantia nigra, and globus pallidus
	Glia proliferations in the inferior olive
	Degeneration in posterior column and anterior horn of spinal cord
	Peripheral neuropathy resembles that of dying-back axonopathy, especially in sensory nerves and posterior roots

REFERENCES

Adams, C. R., Ziegler, D. K., and Lin, J. T. (1983), *JAMA*, 250, 642–643.
Aldridge, W. N. (1976), *Adv. Chem. Ser.*, 157, 186–192.
Aldridge, W. N. and Street, B. W. (1971), *Biochem. J.*, 124, 221–234.
Alfano, D. P., LeBoutillier, J. C., and Petit, T. L. (1982), *Exp. Neurol.*, 75, 308–319.
Alfrey, A. C., LeGendre, G. R., and Kheany, W. D. (1976), *N. Engl. J. Med.*, 294, 184–188.
Amin-Zaki, L., Elhassani, S., Majeed, M. A., Clarkson, T. W., Doherty, R. A., and Greenwood, M. (1974), *Pediatrics*, 54, 587–595.
Amin-Zaki, L., Elhassani, S., Majeed, M. A., Clarkson, T. W., Doherty, B. A., Greenwood, M., and Giovanoli-Jakubczak, T. (1976), *Am. J. Dis. Child.*, 130, 1070–1076.
Amin-Zaki, L., Majeed, M. A., Clarkson, T. W., and Greenwood, M. R. (1978), *Br. J. Med.*, 1, 613–616.
ATSDR (Agency for Toxic Substances and Disease Registry) (1989), *Toxicological Profile for Cadmium*, U.S. Public Health Service.
Arvidson, B. (1980), *Acta Neuropathol. (Berl.)*, 49, 213–224.
Arvidson, B. (1983), *Environ. Res.*, 32, 240–246.
Arvidson, B. (1992), *Muscle Nerve*, 15, 1089–1094.
Ashizawa, R. (1927), *Jpn. J. Med. Sci. Trans. Intern. Med. Pediat. Psychiat.*, 1, 173–191.
Bakir, F., Damluji, S. F., Amin-Zaki, L., Murthadha, M., Khaldidi, A., Al-Rawi, N. Y., Tikriti, S., Dhahir, H. I., Clarkson, T. W., Smith, J. C., and Doherty, R. A. (1973), *Science*, 181, 230–241.
Bank, W. J. (1980), in *Experimental and Clinical Neurotoxicology*, Spencer, P. S. and Schaumburg, H. H., Eds., Williams & Wilkins, Baltimore, 570–577.
Barbeau, A., Inoue, N., and Cloutier, T. (1976), *Adv. Neurol.*, 14, 339–352.
Beal, M. F. (1992), *Ann. Neurol.*, 31, 119–130.
Beal, M. F., Kowall, N. W., and Ellison, D. W. (1986), *Nature*, 321, 168–171.
Beliles, R. P. (1975), in *Toxicology — The Basic Sciences of Poisons*, Casarett, L. T. and Doull, Eds., Macmillan, New York, 454–502.
Berlin, M., Carlsen, J., and Norseth, T. (1975), *Arch. Environ. Health*, 30, 307–313.
Berlin, M., Nordberg, G., and Serenius, F. (1969), *Arch. Environ. Health*, 18, 42–50.

Bizzi, A., Crane, R. C., Autilio-Gambetti, and Gambetti, P. (1986), *J. Neurosci.*, 4, 722–731.
Bock, J. L. and Ash, A. E. (1980), *J. Inorg. Biochem.*, 13, 105–110.
Bolla, K., Briefel, G.,Spector, D., Schwartz, B., Weiler, L., Herron, J., and Gimenez, L. (1992), *Arch. Neurol.*, 49, 1021–1026.
Bressler, J. P. and Goldstein, G. W. (1991), *Biochem. Pharmacol.*, 41, 479–487.
Brouillet, E. P., Shinobu, L., McGarvey, U., Hochberg, F., and Beal, M. F. (1992), *Exp. Neurol.*, 119, 86–98.
Brown, I. (1954), *A.M.A. Arch. Neurol. Psychiatry*, 674–681.
Brown, A. W., Aldridge, W. N., Street, B. W., and Verschoyle, R. D. (1979), *Am. J. Pathol.*, 97, 59–82.
Campbell, J. B., Wooley, D. E., Vijayan, V. K., and Overmann, S. R. (1983), *Dev. Brain Res.*, 3, 595–612.
Chang, L. W. (1979), in *Biogeochemistry of Mercury*, Nriagu, J. O., Ed., Elsevier, New York, 519–580.
Chang, L. W. (1980), in *Experimental and Clinical Neurotoxicology*, Spencer, P. S. and Schaumberg, H. H., Eds., Williams & Wilkins, Baltimore, 508–526.
Chang, L. W. (1984a), *Neurotoxicology*, 5(2), 205–216.
Chang, L. W. (1984b), *Bull. Environ. Contam. Toxicol.*, 33, 295–301.
Chang, L. W. (1986), *Toxicol. Appl. Pharmacol.*, 6, 217–232.
Chang, L. W. (1990), *J. Toxicol. Sci.*, 15 (Suppl. 4), 125–151.
Chang, L. W. (1994), in *Principles of Neurotoxicology*, Chang, L. W., Ed., Marcel Dekker, New York, 3–34.
Chang, L. W. (1995), in *Handbook of Neurotoxicology*, Chang, L. W. and Dyer, R. S., Eds., Marcel Dekker, New York, 143–170.
Chang, L. W., Desnoyers, P. A., and Hartmann, H. A. (1972), *J. Neuropathol. Exp. Neurol.*, 31, 389–395.
Chang, L. W. and Dyer, R. S. (1983), *Neurobehav. Toxicol. Teratol.*, 5, 443–459.
Chang, L. W. and Dyer, R. S. (1984), *Neurobiology of Zinc*, Frederickson, C. and Howell, G., Eds., Alan R. Liss, New York, 275–190.
Chang, L. W. and Dyer, R. S. (1985), *J. Toxicol. Environ. Health*, 16, 641–653.
Chang, L. W. and Dyer, R. S., Eds. (1985), *Handbook of Neurotoxicology*, Marcel Dekker, New York.
Chang, L. W. and Hartmann, H. A. (1972), *Acta Neuropathol.*, 20, 316–334.
Chang, L. W., McMillan, D. E., Wenger, G. R., and Dyer, R. S. (1983a), *Toxicologist*, 3(1):76 (Abstr.).
Chang, L. W. and Reuhl, K. (1996), in *Comprehensive Toxicology, Vol. 8, Neurotoxicology*, Lowndes, H. and Reuhl, K. Eds., Raven Press, in press.
Chang, L. W., Tiemeyer, T. M., Wenger, G. R., and McMillan, D. E. (1982a), *Neurobehav. Toxicol. Teratol.*, 4, 149–156.
Chang, L. W., Tiemeyer, T. M., Wenger, G. R., McMillan, D. E., and Reuhl, K. R. (1982b), *Environ. Res.*, 29, 435–444.
Chang, L. W., Tiemeyer, T. M., Wenger, G. R., McMillan, D. E., and Reuhl, K. R. (1982c), *Environ. Res.*, 29, 445–458.
Chang, L. W., Tiemeyer, T. M., Wenger, G. R., and McMillan, D. E. (1983b), *Environ. Res.*, 30, 399–411.
Chang, L. W., Wenger, G. R., and McMillan, D. E. (1984), *Environ. Res.*, 34, 123–134.
Chang, L. W., Wenger, G. R., McMillan, D. E., and Dyer, R. S. (1983c), *Neurobehav. Toxicol. Teratol.*, 5, 337–350.
Chang, L. W. and Verity, M. A. (1995), in *Handbook of Neurotoxicology*, Chang, L. W. and Dyer, R. S., Eds., Marcel Dekker, New York, 31–60.
Cheek, D. (1980), *Brennemann's Practice of Pediatrics*, Harper & Row, Hagerstown, N. Y., 110–124.
Cherian, M. G., Hursh, J. B., Clarkson, T. W., and Allen, J. (1978), *Arch. Environ. Health*, 33, 109–114.
Chu, N. S., Hochberg, F. H., Calne, D. B., and Olanow, C. W. (1995), in *Handbook of Neurotoxicology*, Chang, L. W. and Dyer, R. S., Eds., Marcel Dekker, New York, 91–104.
Clarkson, T. W., Nordberg, G. F., Sager, P. R., Berlin, M., Friberg, L., Mattison, D. R., Miller, R. K., Mottet, N. K., Nelson, N., Parizek, J., Rodier, P. M., and Sandstead, H. (1985), in *Reproductive and Developmental Toxicity of Metals*, Clarkson T. W., Nordberg, G. F., and Sager, P. F., Eds., Plenum Press, New York, 1–26.
Clasen, R. A., Hartmann, J. F., Starr, A. J., Coogan, P. S., Pandolfi, S., Laing, I., Becker, R., and Hass, G. M. (1974), *Am. J. Pathol.*, 74, 215–224.
Conradi, S., Ronnevi, D., and Norris, F. (1982), in *Human Motor Neuron Diseases*, Rowland, L. P., Ed., Raven Press, New York, 201–231.
Cookman, G. R., King, W. B., and Regan, C. M. (1987), *J. Neurochem.*, 49, 399–403.
Cory-Slechta, D. A. (1995), in *Neurotoxicology: Approaches and Methods*, Chang, L. W. and Slikker, W., Eds., Academic Press, San Francisco, 385–398.
Cory-Slechta, D. A. and Pounds, J. (1995), in *Handbook of Neurotoxicology*, Chang, L. W. and Dyer, R. S., Eds., Marcel Dekker, New York, 61–90.
Crapper, D. R. (1973), *Electroenceph. Clin Neurophysiol.*, 35, 575–588.
Crapper, D. R. (1974), *Frontiers in Neurology and Neuroscience Research*, Parkinson Foundation, Symp. No. 1 of the Neuroscience Institute of the University of Toronto, Seeman, P. and Brown, G. M., Eds., University of Toronto Press, Toronto, 97–111.
Crapper, D. R., Krishnan, S. S., and Quittkat, S. (1976), *Brain*, 99, 67–79.
Crapper-McLachlan, D. R. (1986), *Neurobiol. Aging*, 7, 525–532.
Crapper-McLachlan, D. R. and de Boni, U. (1980), *Neurotoxicology*, 1, 3–16.
Crapper-McLachlan, D. R. and Farnell, B. J. (1985), in *Metal Ions in Neurology and Psychiatry*, Gabay, S., Harris, J., and Ho, B. T., Eds., Alan R. Liss, New York, 69–87.
Crapper-McLachlan, D. R. and Farnell, B. J. (1986), *Ann. 1st. Super. Sanita.*, 2, 697–702.

Crapper-McLachlan, D. R., Farnell, B., Galin, H., Karlik, S., Eichhorn, G., and de Boni, U. (1983), in *Biological Aspects of Metals and Metal Related Diseases,* Sarkar, B., Ed., Raven Press, New York, 209–218.
Crapper-McLachlan, D. R., Kruck, T. P. A., Lukiw, W. J., and Krishnan, S. S. (1991), *Can. Med. Assoc. J.,* 145, 793–804.
Crapper, D. R., Quittkat, S., and Krishnan, S. S. (1980), *Acta Neuropath. (Berlin),* 50, 19–24.
Currier, R. D. and Haerer, A. F. (1968), *Arch. Environ. Health,* 17, 712–719.
Daniel, J. W., Gage, J. C., and Lefevre, P. A. (1971), *Biochem. J.,* 121, 411–415.
Daniel, J. W., Gage, J. C., and Lefevre, P. A. (1972), *Biochem. J.,* 129, 962–967.
Davis, L. E., Wands, J. R., Weiss, S. A., Price, D. L., and Girling, E. F. (1974), *Arch. Neurol.,* 30, 428–431.
De Boni, U., Scott, J. W., and Crapper, D. R. (1974), *Histochemistry,* 40, 31–37.
De Boni, U., Seger, M., and Crapper-McLachlan, D. R. (1980), *Neurotoxicology,* 1, 65–81.
De Haven, D. L., Walsh, T. J., and Mailman, R. B. (1984), *Toxicol. Appl. Pharmacol.,* 74, 182–189.
Doctor, S. V., Costa, L. G., Kendall, D. A., Enna, S. J., and Murphy, S. D. (1982a), *Toxicologist,* 2, 86 (abstr.).
Doctor, S. V., Costa, L. G., Kendall, D. A., and Murphy, S. D. (1982b), *Toxicology,* 25, 213–223.
Doctor, S. V., Costa, L. G., and Murphy, S. D. (1982c), *Toxicol. Lett.,* 13, 217–223.
Donaldson, J. and Barbeau, A. (1985), in *Metal Ions in Neurology and Psychiatry,* Gabay, S., Harris, J., and Ho, B. T., Eds., Alan R. Liss, New York, 259–285.
Donaldson, J., Labella, F. S., and Gesser, D. (1981), *Nerotoxicology,* 2, 53–64.
Dyck, P. J., Windebank, A. J., Low, P. A., and Barrmann, W. J. (1980), *J. Neuropathol. Exp. Neurol.,* 39, 700–712.
Dyer, R. S., Wonderlin, W. F., Walsh, T. J., and Boyes, (1982a), *Soc. Neurosci.,* Abst. 8(No. 23.7):82.
Dyer, R. S., Walsh, T. J., Wonderlin, W. F., and Bercegeay, M. (1982b), *Neurobehav. Toxicol. Teratol.,* 4, 141–147.
Edsall, D. L., Wilbur, F. P., and Drinker, C. K. (1919), *J. Ind. Hyg.,* 1, 183–193.
Fahn, S. (1977), in *Scientific Approaches to Clinical Neurology,* Vol. 2, Goldensohn, E. S. and Appel, S. H. Eds., 71:1159–1189, Lea and Febiger, Philadelphia.
Ferraz, H. B., Bertolucci, P. H. F., Pereira, J. S., Lima, J. G. C., and Andrade, L. A. F. (1988), *Neurology,* 38, 550–553.
Fortemps, E., Amand, G., Bombois, A., Lauwerys, R., and Laterre, E. G. (1978), *Int. Arch. Occup. Environ. Health,* 41, 1–6.
Friberg, L., Elinder, C. G., Kjellstrom, T., and Nordberg, G. F., Eds. (1986), *Cadmium and Health. A Toxicological and Epidemiological Appraisal,* Vol. 2, *Effects and Response,* CRC Press, Boca Raton, FL.
Friberg, L. and Vostal, J., Eds. (1972), *Mercury in the Environment,* CRC Press, Boca Raton, FL.
Gabbiani, G., Badonnel, M.-C., Mathewson, S. M., Ryan, G. B. (1974), *Lab. Invest.,* 30, 686–695.
Gabbiani, G., Baic, D., and Deziel, C. (1967a), *Exp. Neurol.,* 18, 154–160.
Gabbiani, G., Gregory, A., and Baic, D. (1967b), *J. Neuropathol. Exp. Neurol.,* 26, 498–506.
Gage, J. C. (1975), *Toxicol. Appl. Pharmacol.,* 32, 225–238.
Gage, J. C. and Swain, A. A. B. (1961), *Biochem. Pharmacol.,* 8, 77 (Abst. No. 250).
Garruto, R. M. (1991), *Neurotoxicology,* 12, 347–378.
Gavin, C. E., Gunter, K. K., and Gunter, T. E. (1990), *Biochem. J.,* 266, 329–334.
Gebhart, A. M. and Goldstein, G. W. (1988), *Toxicol. Appl. Pharmacol.,* 94, 191–206.
Gibbons, R., Dixon, S., and Hallis, K. (1976), *Biochem. Biophys. Acta,* 444, 1–10.
Goldstein, G. W., Asbury, A. K., and Diamond I. (1974), *Arch. Neurol.,* 13, 281–391.
Goldwater, L. J. (1963), in *Mercury, Mercurials, and Mercaptans,* Charles C Thomas, Springfield, IL, 56–67.
Good, P. F., Perl, D. P., Bierer, L. M., and Schmeidler, J. (1992), *Ann. Neurol.,* 31, 286–292.
Graham, D. G. (1984), *Neurotoxicology,* 5, 83–96.
Hasan, F., Cookman, G. R., Keane, G. J., Bannigan, J. G., King, W. B., and Regan, C. M. (1989), *Neurotoxicol. Teratol.,* 11, 433–440.
Hastings, L. (1995), in *Handbook of Neurotoxiciology,* Chang, L. W., and Dyer, R. S., Eds., Marcel Dekker, New York, 171–212.
Herman, S. P., Klein, R., Talley, F. A., and Kirgman, M. R. (1973), *Lab. Invest.,* 28, 104–118.
Hirano, A. and Kochen, J. A. (1979), *Progress in Neuropathology,* Vol. 3, Zimmerman, H. M., Ed., Grune & Stratton, New York, Chap. 2.
Hirsch, E. C., Brandel, J. P., Galle, P., Javoy-Agid, F., and Agid, Y. (1991), *J. Neurochem.,* 56, 446–451.
Huang, C. C., Chu, N. S., Lu, C. S., Wang, J. D., Tsai, J. L., Tseng, J. L., Wolters, E. C., and Calne, D. B. (1984), *Arch. Neurol.,* 46, 1104–1106.
Hughes, A. J., Daniel, S. E., Kilford, L., and Lees, A. J. (1992), *J. Neurol. Neurosurg. Psychiatry,* 54, 388–396.
Jacobs, J. M., Carmichael, N., and Cavanagh, J. B. (1977), *Toxicol. Appl. Pharmacol.,* 39, 249–261.
Johnson, G. V. and Jope, R. S. (1988), *Brain Res.,* 456, 95–103.
Kantarjian, A. (1964), *Neurology,* 15, 639–644.
Katsetos, C. C., Savory, J., Herman, M. M., Carpenter, R. M., Frankfurter, A., Hewitt, C. D., and Wills, M. R. (1990), *Neuropathol. App. Neurobiol.,* 16, 511–518.
Kobayashi, K., Yumoto, S., Nagai, H., Hosoyama, Y., Imamura, M., Masuzawa, S., Koizumi, Y., and Yamashita, H. (1990), *Proc. Jpn. Acad.,* 66, 189–192.
Kosik, K. S., McCluskey, A. H., Walsh, F. X., and Selkoe, D. J. (1985), *Neurochem. Pathol.,* 3, 99–108.
Ladd, A. C., Goldwater, L. J., and Jacobs, M. B. (1964), *Arch. Environ. Health,* 9, 43–52.
Lai, J. C. K., Leung, T. K. C., and Lim, L. (1985), in *Metal Ions in Neurology and Psychiatry,* Gabay, S., Harris, J., and Ho, B. T., Eds., Alan R. Liss, New York, 177–197.

Lampert, P. W. and Schochet, S. S., Jr. (1968), *J. Neuropathol. Exp. Neurol.*, 27, 527–545.
Larsen, N., Pakkenberg, H., Damsgaard, E., and Heydorn, K. (1979), *J. Neurol. Sci.*, 19, 407–416.
Lee, H. Y. and Oberdorster, G. (1985), *Toxicologist*, 5, 178.
Louis-Ferdinand, R. T., Brown, D. R., Fiddler, S. F., Daughtry, W. C., and Klein, A. W. (1978), *Toxicol. Appl. Pharmacol.*, 43, 351–360.
Lukiw, W. J. (1991), Chromatin Structure, Gene Expression and Nuclear Aluminum in Alzheimer's Disease, Ph.D. thesis, National Library of Canada, Ottawa, Canada.
Lukiw, W. J., Bergeron, C., Wong, L., Kruck, T. P. A., Krishnan, B., and Crapper-McLachlan, D. R. (1992), *Neurobiol. Aging*, 13, 115–121.
Lukiw, W. J. and McLachlan, D. R. (1995), in *Handbook of Neurotoxicology*, Chang, L. W. and Dyer, R. S., Eds., Marcel Dekker, New York, 105–143.
Mailman, R. B., Krigman, M. R., Frye, G. D., and Hannin, Z. (1983), *J. Neurochem.*, 40, 1423–1429.
Maynard, L. and Cotzias, G. (1955), *J. Biol. Chem.*, 214, 489–495.
McCauley, P. T. and Bull, R. J. (1978), *Fed. Proc. Fed. Am. Soc. Exp. Biol.*, 37, 40.
McCauley, P. T., Bull, R. J., and Lutkenhoff, S. D. (1979), *Neuropharmacology*, 18, 93–101.
McCauley, P. T., Bull, R. J., Tonti, A. P., Lutkenhoff, S. D., Meister, M. V., Doerger, J. U., and Stober, J. A. (1982), *J. Toxicol. Environ. Health*, 10, 639–651.
Meana, L. (1979), Manganese poisoning, in *Handbook of Clinical Neurology*, Vinken, P. J. and Bruyn, G. W. Eds, North Holland, Amsterdam, 217–237.
Meyer, A. (1963), in *Neuropathology*, Blackwood, W., McMenemey, W. H., Meyer, A., Norman, R. M., and Russell, D. S., Eds., Arnold, London, 261–262.
Muma, N. A., Troncoso, J. C., Hoffman, P., Koo, E. H., and Price, D. (1988), *Mol. Brain Res.*, 3, 115–122.
Muma, N., Hoffman, P., Slunt, H., Applegate, M., Lieberburg, I., and Price, D. (1990), *Exp. Neurol.*, 107, 230–235.
Murthy, R. C., Saxena, D. K., Sunderaraman, V., and Chandra, S. V. (1987), *Ind. Health*, 25, 159–162.
Naalsund, L. V., Suen, C. N., and Fonnum, F. (1985), *Neurotoxicology*, 6, 145–158.
Neff, N. H., Barrett, R. E., and Costa, E. (1964), *Experientia*, 25, 1140–41.
Newland, M., Cox, C., Hamada, R. (1987), *Fund. App. Toxicol.*, 9, 314–328.
Newland, M., Ceckler, T., Kordower, J., and Weiss, B. (1989), *Exp. Neurol.*, 106, 251–258.
Nolan, C. V. and Shaikh, Z. A. (1986), *Life Sci.*, 39, 1403–1409.
Nordberg, G. (1975), *Effects and Dise-Response Relationships of Toxic Metals*, Elsevier, Armsterdam, 1–111.
Nordberg, G. and Serenius, F. (1969), *Acta Pharmacol.*, 27, 269–283.
Ohnishi, A and Dyck, P. J. (1981), *Ann. Neurol.*, 10, 469–477.
Osetowska, E. (1971), in *Pathology of the Nervous System, Vol. 2*, Minckler, J. Ed., McGraw-Hill, New York, 1644–1651.
Pentschew, A. and Garro, F. (1966), *Acta Neuoropathol.*, (Berlin), 6, 266–278.
Perl, D. P. and Brody, A. R. (1980), *Science*, 208, 297–299.
Perl, D., Gajdusek, C., Garruto, R., Yanagihara, R., and Gibbs, C. (1982), *Science*, 217, 1053–1055.
Perl, D. P. and Pendlebury, W. W. (1984), *J. Neuropathol. Exp. Neurol.*, 43, 349–359.
Petit, T. L., Aflano, D. P., and LeBoutillier, J. C. (1983), *Neurotoxicology*, 41(1), 79–84.
Politis, M. J., Schaumburg, H. H., and Spencer, P. S. (1980), in *Experimental and Clinical Neurotoxicology*, Spencer, P. S. and Schaumbug, H. H., Eds., Williams & Wilkins, Baltimore, 613–630.
Regan, C. M. (1989), *Neurotoxicol. Teratol.*, 11, 533–537.
Regan, C. M., Cookman, G. R., Keane, G. J., King, W., and Hemmens, S. E. (1989), in *Lead Exposure and Child Development*, Smith, M. A., Grant, L. D., and Sors, A. I., Eds., Kluwer Academic Publishers, Dordrecht, 440–452.
Rey, C. H., Reinecke, H. J., and Besser, R. (1984), *Vet. Human Toxicol.*, 26, 121–122.
Roberts, M. C., Seawright, A. A., and Ng, J. C. (1979), *Vet. Human Toxicol.*, 25, 321–327.
Roberson, M. D., Toews, A. D., Goodrum, J. F., and Morell, P. (1992), *J. Neurosci. Res.*, 33, 156–162.
Rodier, J. (1955), *Br. J. Ind. Med.*, 12, 21–35.
Ross, W. D., Emmett, E. A., Steiner, J., and Tureen, R. (1981), *Am. J. Psychiatry*, 138, 1092–1095.
Scheuhammer, A. and Cherian, M. (1985), *Biochem. Biophys. Acta.*, 840, 163–169.
Schlaepfer, W. W. (1971), *Lab. Invest.*, 25:556.
Schwab, R. S. and England, A. C. (1968), in *Handbook of Clinical Neurology*, Vinken, P. J., and Bruyn, G. W., Eds., North Holland, Amsterdam, 227–247.
Slomianka, L., Rungby, J., West, M. J., Danscher, G., and Andersen, A. H. (1989), *Neurotoxicology*, 10, 1771–190.
Smith, T. (1976), in *Greenfield's Neuropathology*, Blackwood, W. and Corsellis, J. A., Eds., Year Book Medical Publishers, Chicago, IL, 152–153.
Spencer, P. S. and Schaumberg, H. H. (1982), in *Human Motor Neuron Diseases*, Rowland, L. P., Ed., Raven Press, New York, 249–266.
Spink, D. (1992), *Can. Med. Assoc. J.*, 146, 431–432.
Steckhoven, J., Renkawek, K., Otte-Holler, I., and Stols, A. (1990), *Neurosci. Let.*, 119, 71–74.
Stopford, W. (1979), in *The Biogeochemistry of Mercury in the Environment*, Nriagu, J. O., Ed., Elsevier/North Holland, Amsterdam, 367–397.
Swift, H. (1914), *Australian Medical Congress*, Trans. 10th Session, Auckland, New Zealand.
Takahata, N., Hayashi, H., Watanabe, B., and Anso, T. (1970), *Folia Psychiatr. Neurol. Jpn.*, 24, 59–69.

Takeuchi, T. (1968), in *Minamata Disease,* Kutsuma, M., Ed., Study Group of Minamata Disease, Kumamoto University, Japan, 141–228.
Takeuchi, T. (1977), in *Neurotoxicology,* Vol. 1, Roizin, L., Shiraki, H., and Grcevic, N., Eds., Raven Press, New York, 235–246.
Trachtenberg, I. M. (1969), *Zdorv'ja, Kiev,* pp. 292–301 (in Russian; translation available through EPA).
Troncoso, J., Sternberger, N., Sternberger, P., Hoffman, P., and Price, D. (1986), *Brain Res.,* 364, 295–300.
Troncoso, J., March, J. L., Haner, M., and Aebi, U. (1990), *J. Struct. Biol.,* 103, 2–12.
Uemura, E. and Ireland, W. P. (1984), *Exp. Neurol.,* 85, 1–9.
von Jaksch, R. (1907). *Münch. Med. Wochenschr.,* 54, 969–972.
Vroom, F. Q. and Greer, M. (1972), *Brain,* 95, 305–318.
Waldron, H. A. and Stofen, D. (1974), *Sub-clinical Lead Poisoning,* Academic Press, New York.
Warkany, J. and Hubbard, D. M. (1953), *J. Pediatrics,* 42, 365–369.
Webster, W. S. and Valois, A. A. (1981), *J. Neuropathol. Exp. Neurol.,* 40, 247–257.
Wen, G. and Wisniewski, H. M., (1985), *Acta Neuropath. (Berl).,* 68, 175–184.
Wenger, G. R., McMillan, D. E., and Chang, L. W. (1982), *Neurobehav. Toxicol. Teratol.,* 4, 157–161.
Wenger, G. R., McMillan, D. E., and Chang, L. W. (1984a), *Toxicol. Appl. Pharmacol.,* 73, 78–88.
Wenger, G. R., McMillan, D. E., and Chang, L. W. (1984b), *Toxicol. Appl. Pharmacol.,* 73, 89–96.
Winder, C. and Lewis, P. D. (1985), in *Metal Ions in Neurology and Psychiatry,* Gabay, S., Harris, J., and Ho, B. T., Eds., Alan R. Liss, New York, 231–245.
Wolters, E. C., Huang, C. C., Clark, C., Peppard, R. F., Okada, J., Chu, N. S., Adam, M. J., Ruth, J. J., Li, D., and Calne, D. B. (1989), *Ann. Neurol.,* 26, 647–651.
Wong, K.-L. and Klaassen, C. D. (1982), *Toxicol. Appl. Pharmacol.,* 63, 330–337.
WHO (1976), *Environmental Health Criteria 1. Mercury,* World Health Organization, Geneva.
Yamada, M., Ohno, S., Okayasu, I., Okeda, R., Hatakeyama, S., Watanabe, H., Ushio, K., and Tsukagoshi, H. (1986), *Acta Neuropathol. (Berl.)* 70, 273–278.
Yanagihara, R. (1982), in *Human Motor Neuron Diseases,* Rowland, L. P., Ed., Raven Press, New York, 233–247.
Yasni, M., Yase, Y., Ota, K., Mukoyama, M., and Adachi, K. (1991a), *Neurotoxicology,* 12, 277–183.
Yasni, M., Yase, Y., Ota, K., and Garruto, R. M. (1991b), *Neurotoxicology,* 12, 615–620.
Young, J. K. (1992), *Med. Hypoth.,* 38, 1–4.
Zarriello, J. J. (1996), in *Toxicology and Risk Assessment,* Fan, A. M. and Chang, L.W., Eds., Marcel Dekker, Inc., New York, 573–600.

Chapter 33

Comparative Neurobehavioral Toxicology of Heavy Metals

Deborah A. Cory-Slechta

I. INTRODUCTION

It is clear that metals play a critical role in functioning of the central nervous system. Numerous studies attest to the requirements for adequate levels of zinc, copper, and iron, for example, in normal brain performance. It should be no surprise, then, that deficiencies or excesses of those same metals can have adverse effects. Likewise, although levels of these essential metals in brain appear to be tightly regulated, nonessential metals may also gain access to the CNS, by virtue of the fact that they share properties such as ionic charge or other characteristics with essential metals. Lead, aluminum, and mercury are examples of such nonessential metals, and ones for which brain appears to be a primary target organ. Thus, all three are associated with characteristic patterns of neurotoxicity that include behavioral impairments. This chapter is not intended to be a compendium of the behavioral effects associated with each of these metals, but rather to highlight those aspects of behavioral deficits that appear to be the primary manifestations of each, and to point out deficiencies in our current understanding of their behavioral toxicity.

Understanding the behavioral toxicology of metals is critical for several reasons. It is only with such studies, for example, that it is possible to precisely define directions for efforts aimed at establishing underlying neurobiological mechanisms of behavioral manifestations. Moreover, studies of behavioral toxicity permit a determination of the behavioral mechanisms of effect as well, information that is critical to defining pharmacological or behavioral therapeutics and to precisely defining risk.

In general, studies on the behavioral toxicity of lead, aluminum, and mercury have followed analogous pathways. In all three cases, studies in human populations have primarily pursued questions related to general domains of functional impairment and exposure levels with which they may be associated. Corresponding efforts in experimental animal studies have been critical to confirming such effects and to explicitly defining the nature of these behavioral impairments. Currently, our understanding of the behavioral toxicity of lead is significantly more advanced than is our corresponding comprehension of either aluminum or mercury, a fact that reflects the extent of experimental effort that has been directed at lead over the past 20 years as a significant public health problem. However, growing concerns over environmental exposures to mercury and aluminum and their associations with detrimental behavioral consequences portend a parallel evolution.

II. LEAD

Realization that exposure to lead could give rise to an array of toxic manifestations related to behavior and nervous system function came even with the first uses of this metal by mankind. As early as the second century B.C., some of the most common signs of overexposure were described by the Greek physician and poet, Nicander, in the Alexipharmaca. The Romans made even greater use of this metal than did the Greeks, and likewise knew the toxicity associated with lead. Pliny, for example, described attempts of workers to avoid lead dusts by placing their faces in loose bags. Activities associated with the Industrial Revolution added further to knowledge about the extent of lead's toxicity.

The most extensive problems with lead, however, have occurred in more modern history, as a result of its use in paints and gasoline, and, to a lesser extent in products such as batteries. As an antiknock additive in gasoline, lead has been widely dispersed into the environment. While some countries have since banned or restricted the use of unleaded gasoline, it remains the primary fuel in many places in the world even today. The addition of lead to paint for its drying properties and as a pigment have provided another source of exposure, particularly for pediatric populations. The ingestion of chips or flakes of lead-based paint in poorly maintained homes continues to provide a major exposure source to millions of children worldwide. Moreover, the accumulation of lead into dusts and soils from weathering, flaking, or even intentional removal of lead-based paints contributes to atmospheric exposure sources. Together, the atmospheric dispersal of lead from automobile exhaust and from paint has resulted in a ubiquitous incorporation of this toxicant into dusts and soils, as well as food and water supplies, resulting in universal exposure of the general population.

Much of the focus of the literature addressing the behavioral toxicity of lead has been on cognitive function, no doubt a result of early reports that exposure to high levels of lead, i.e., blood lead levels of 80 μ/dl and above, could result in permanent mental retardation and other neurological sequelae in children. However, lead is by no means a selective behavioral toxicant, and induces apparent sensory and motor deficits as well, disturbances which may adversely impact cognitive function. Nor does it necessarily spare adults, as similar changes may ensue, albeit at higher exposure levels, in response to occupational lead exposures.

A. EFFECTS ON COGNITIVE FUNCTION
1. Studies in Human Populations
a. Pediatric Lead Studies

To date, assessments of cognitive deficits as a result of lead exposure have been aimed primarily at pediatric populations. Children clearly exhibit greater vulnerability to the neurotoxic properties of lead than do adults, largely owing to the immaturity of the blood brain barrier, the ongoing development of the nervous system, and their greater relative absorption of lead.

The recognition that lead poisoning even without associated acute encephalopathy could engender permanent behavioral sequelae (Byers and Lord, 1943) was followed by numerous clinical and smelter studies in children identified as having elevated lead burdens (U.S. Environmental Protection Agency, 1986). These studies were important for determining areas of function related to lead exposure. Collectively, however, they suffered from methodological problems that included insufficient sample sizes, undocumented exposure histories and thus possible improper assignment of subjects to exposure groups, and inadequate control of potential confounding factors for measures of intelligence test scores.

Cross-sectional epidemiological studies relying on blood lead or, in some instances, tooth lead levels, to assign exposure classifications to subjects constituted the next major approach pursued to relate lead exposure to cognitive impairment (U.S. Environmental Protection Agency, 1986). These cross-sectional studies exhibited significant methodological differences, including the range of potential confounders of neuropsychological performance measured (e.g., parental intelligence, socioeconomic status), and the criteria for inclusion of potential confounders in statistical analyses. The assignment of subjects to exposure groups on the basis of indices such as blood lead, moreover, engendered additional problems, since blood lead reflects only relatively recent lead exposure and is not indicative of total body burden. The frequent reliance on a single blood lead measurement was also problematic since it provided no information on the period during which exposure actually occurred, the pattern of exposure over time, etc. Nevertheless, many of these investigations found a negative association between measures of lead exposure and intelligence test scores, even when the data were controlled for potential covariates of IQ (intelligence quotient), such as parental intelligence and social demographic measures. Furthermore, these effects were noted at far lower levels of lead in blood, i.e., ~ 30 μg/dl, than had previously been regarded as detrimental. Additional strength was added to these conclusions in the outcomes of meta-analyses carried out across studies confirming that this pattern of effects was unlikely to be due to chance (Needleman and Gatsonis, 1990; Schwartz and Otto, 1987).

The inherent limitations of cross-sectional study designs and the controversial nature of their findings served as the impetus for the implementation of better controlled and designed prospective longitudinal epidemiological studies. These prospective longitudinal studies, several of which remain ongoing, were initiated in cohorts both in the U.S. and abroad, and shared many common elements of design, including pre- or perinatal subject recruitment, longitudinal assessment of blood lead beginning antenatally or at birth, the use of similar well-standardized, validated instruments for determination of cognitive function, and assessments of such functions in infancy, late preschool age, and, where possible, during the school-age years. There have also been differences between the studies, however, such as the degree to which other correlates of the outcome measures exist, (particularly with respect to socioeconomic status), differences in the extent of lead exposure, in sample sizes, and in the manner in which the data are reported.

These pediatric longitudinal studies have focused largely on three issues: (1) the verification of cognitive impairments in human pediatric populations, as based on alterations in group IQ scores or other psychometric and neurodevelopmental indices; (2) on differentiating the contribution of lead from other environmental, sociological, and genetic contributions to any observed deficits; and (3) to defining the nature of the associated dose-effect function.

While the pattern of effects noted in measurements made during infancy or early preschool in these prospective endeavors exhibited some inconsistencies, a more uniform set of findings has emerged with respect to later preschool and school age assessments, where most studies find significant inverse correlations between indices of blood lead and IQ measures (World Health Organization, 1994). The greater uniformity of this inverse relationship in school-aged children probably reflects the ability to obtain more precise measurement of behavioral function at this age. Alternatively, or concomitantly, it may indicate a preferential impact of lead on behavioral functions such as higher order cognitive processes which cannot be readily evaluated during infancy.

The relationship between blood lead level and intelligence test scores obtained from a prospective study cohort based in Cincinnati as obtained by Dietrich et al. (1993) is exhibited in Figure 1. Its authors reported a 7 point decrease in the Performance IQ scores of the Weschler Intelligence Scale for Children-Revised (WISC-R) at lifetime blood lead concentrations in excess of 20 µg/dl at approximately 6.5 years of age. In fact, the specific threshold for lead-induced alterations in intelligence test scores is not yet fully defined, but in a cohort from Boston studied by Bellinger et al. (1991), the mean blood lead level for the group was only 7.0 µg/dl, and the decrement in intelligence test score on the WISC-R relevant to a blood lead range of approximately 4 to 14 µg/dl, thus even further decreasing the blood lead levels associated with cognitive impairments.

Another facet of the lead–IQ association revealed by the prospective epidemiological studies is its apparent interaction with socioeconomic status. This is best exemplified in the studies of Bellinger and colleagues. Figure 2 depicts the greater vulnerability of less advantaged children to lead. As it indicates, when plotted against 6-month blood lead concentrations, the decline in the Mental Development Index scores of the Bayley Scales at 18 and 24 months occurred only in children of lower socioeconomic status, and these effects were particularly evident at the higher lead exposure levels.

The longevity and/or reversibility of lead-induced changes in behavioral function as measured in the human pediatric studies is still unknown. A study by Needleman et al. (1990) following up the cohort from the original 1979 Needleman et al. study (1979) indicated that of those subjects located and re-examined, increased dentin (tooth) lead levels were associated with higher risks of dropping out of high school, having a reading disability, lower class standing in high school, increased absenteeism, lower vocabulary and grammatical-reasoning scores, poorer eye–hand coordination, longer reaction times and slower finger tapping. These findings suggest a permanency to the effect of lead. An alternative interpretation to a permanent direct effect of lead per se that must be considered, however, is that early academic problems induced by lead exposure are magnified as students progress through the school years with their increasing demands and requirements. Clearly, greater attention needs to be directed to this issue, but of course either scenario portends deleterious consequences for future success.

The human pediatric studies, particularly the prospective longitudinal studies, have been instrumental in confirming the existence of a relationship between lead exposure and cognitive function in human populations and in beginning to define the associated dose-effect functions. What has not been provided by these studies, however, is any precise characterization of the nature of the behavioral deficit(s)

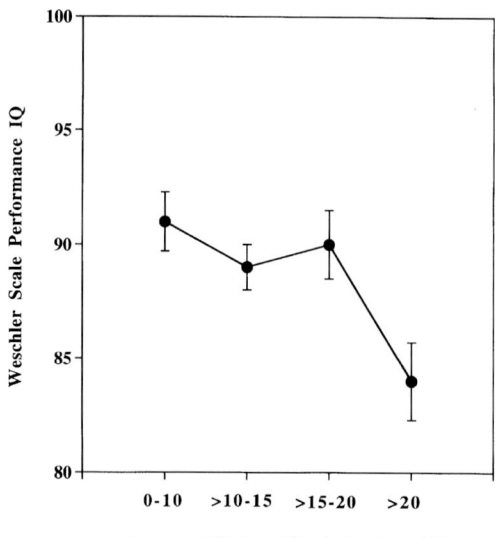

Figure 1. Adjusted dose-effect relationships between average lifetime blood lead (PbB) concentrations and Weschler Performance Scale IQ. Mean (M) lifetime PbB concentrations and sample sizes (n) within each group were: 0–10: n = 68, M = 7.7 ± 1.4 µg/dl; >10–15: n = 89, M = 12.3 ± 1.4 µg/dl; >15–20: n = 53, M = 17.1 ± 1.2 µg/dl; >20: n = 41, M = 26.3 ± 5.0 µg/dl. (Adapted from Dietrich et al., 1993.)

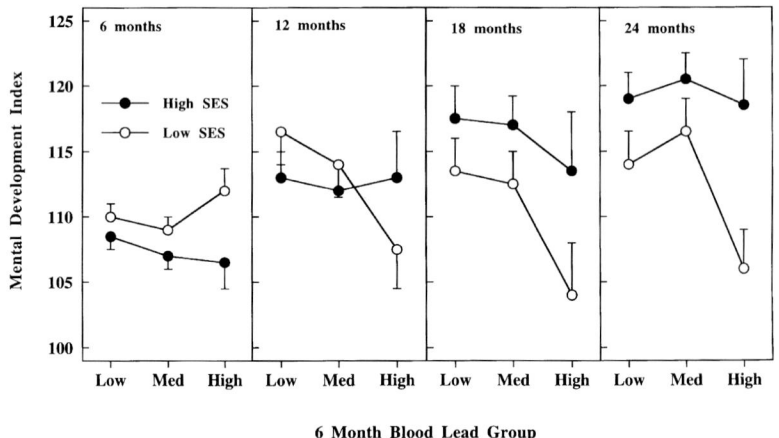

Figure 2. Least-squares mean Mental Development Index scores ± standard error of infants stratitifed by blood lead group as measured at 6 months of age and social class. The performance of infants in the "upper" social class (•) is contrasted with the performance of infants in the "lower" social class (o). The number of infants in the six groups defined by 6-month blood lead group and social class group varied between 11 and 53. (Adapted from Bellinger et al., 1991.)

contributing to or leading to changes in IQ scores (Bellinger, 1985; Cory-Slechta, 1995). Some studies have noted effects expressed primarily in nonverbal or motor skills, others in verbal skills, and still others note roughly equal effects in the two domains. The reason for such discrepancies remain unclear, but may well involve interactions with the differing characteristics of the cohort populations (World Health Organization, 1994).

b. Occupationally-Exposed Populations

Cognitive changes related to lead exposure are also described in studies of occupationally lead-exposed individuals, demonstrating that such effects, even while they may be manifest at higher exposure levels, are not restricted to exposures occurring during the earliest stages of development. This point is important because it emphasizes the need to pursue neurobiological mechanisms that are not solely based on developmental aberrations, as these would obviously not be adequate to explain cognitive manifestations that arise from lead exposures occurring long after most developmental processes have ceased.

Like the pediatric studies, many of the occupational lead exposure studies have focused on the potential for cognitive changes, utilizing standardized intelligence tests as their primary outcome measure. Most such studies have compared the performance of a group of workers exposed to lead at some defined level to a matched control group with minimal lead exposure. Under these experimental conditions, several, though not all, studies have noted changes in various tests designed to measure intelligence and memory (Hanninen et al., 1979; Hogstedt et al., 1983; Parkinson et al., 1986; Araki et al., 1986; Yokoyama et al., 1988; Stollery et al., 1989; Stollery et al., 1991). In an attempt to more precisely delineate the nature of the behavioral functions contributing to changes in cognitive function, Stollery et al. (1989) used a computerized neurobehavioral test battery comprised of tests of various behavioral functions. Their results demonstrated that exposure levels producing blood lead concentrations of 40 µg/dl or above resulted in a general slowing of sensory-motor reaction time and mild impairment of attention, verbal memory, and linguistic processing. A subsequent study by these authors (Stollery et al., 1991) assessed behavioral functions longitudinally, i. e., three times over an 8-month period, demonstrating that in the high lead group (mean blood lead of 51.8 µg/dl), the slowing of sensory-motor reaction time persisted, was not affected by practice, and was most evident when the cognitive demands of the task were low. The second study also revealed difficulties in the recall of incidental information. Such results challenged previously held conceptions of 80 µg/dl and above as the cutoff for nervous system manifestations in adults.

2. Studies in Experimental Animals

Studies carried out in experimental animals provide firm support for the existence of lead-associated impairment of cognitive function. Reported effects include changes in various learning paradigms, including discrimination learning and reversal learning, concept learning, and repeated learning paradigms, as well as deficits in the acquisition of characteristic behavior maintained by a fixed interval (FI) schedule of reinforcement. Less clear at the present time is the exact contribution of developmental period of exposure to such effects, their reversibility or permanency, and the behavioral and neurobiological mechanisms by which they occur.

Lead-induced impairments in the speed and/or accuracy with which animals learn to discriminate between relevant environmental stimuli has been described both in rodents and nonhuman primates (Bushnell and Bowman, 1979a; Bushnell and Bowman, 1979b; Munoz et al., 1986; Rice, 1990). This has included experimental paradigms in which the correct (associated with reward) and incorrect stimuli are presented either simultaneously or in succession. The collective studies also suggest that stimulus dimension may not be a particularly important determinant of these effects, since lead-related performance impairments have been found in brightness, shape, form, pattern, and color discrimination (Cory-Slechta, 1984). In addition, behavioral deficits have been found in spatial discrimination problems (Cory-Slechta, 1984), but only when lead was administered orally. The lack of effectiveness of intraperitoneal lead exposure paradigms to produce spatial discrimination impairments probably reflects the poor and erratic absorption of lead into the bloodstream with this exposure route (Jugo and Kello, 1975). That reported impairments in discrimination learning are not due to nonspecific behavioral changes such as inadequate motivational level was addressed by findings such as those of Bushnell and Bowman (1979a) that deficits in discrimination performance were still evident even with a change from a food reward to a non-food-based reward.

In many of these discrimination studies, the most notable effects of lead occurred in the context of discrimination reversal. That is, after acquisition of the original discrimination problem, the stimuli designated as correct and incorrect were repeatedly reversed each time some criterion accuracy level had been achieved. This scheme mandates behavioral transitions, i.e., requires the subject to change its behavioral performance as the environmental stimuli change their association with reinforcement (reward). An example of the ensuing difficulties for a group of high-dose lead-exposed monkeys relative to controls in a horizontal shape discrimination problem is evident in Figure 3 (Bushnell and Bowman, 1979b), depicting the increase in the mean number of trials to criterion across reversals. Experimental paradigms mandating behavioral transitions appear to be particularly vulnerable to disruption by lead as a function of the tendency of lead to enhance repetitive (i.e., perseverative) responding, i.e., reiteration of the previously correct response. Impairments of this nature have been observed at blood lead concentrations as low as 11 to 15 µg/dl in nonhuman primate studies (Rice, 1985).

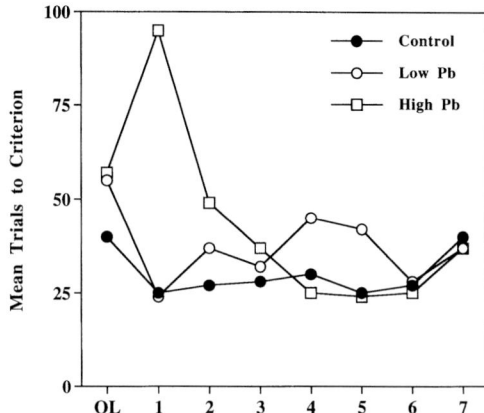

Figure 3. Mean trials to criterion across reversal learning sets from groups of monkeys exposed to lead throughout the first year of life. Stimuli were colored planometric patterns. OL indicates performance during the original learning; OT indicates an overtraining period that was imposed prior to the first reversal. (Adapted from Bushnell and Bowman, 1979b.)

Additional evidence supportive of learning deficits produced by lead exposure derives from studies involving learning set formation and repeated acquisition of response chains. Lilienthal et al. (1986) utilized a learning set paradigm in which successive discrimination problems are typically solved with increasing speed and efficiency. These investigators reported that monkeys exposed to lead both pre- and postnatally were slower to acquire a learning set than nonexposed controls, and that these effects were not due to increased activity levels of the monkeys. When lead-treated rats were required to learn a new sequence of three responses during each successive experimental session (repeated acquisition), lead-induced deficits were noted at blood lead concentrations of 20 to 25 µg/dl (Cohn et al., 1993). That this represented a selective learning deficit was indicated by the fact that no deficits were evident when lead-exposed rats were simply required to perform a sequence of three responses that had already been learned.

While changes in memory function have been described in studies of occupationally lead-exposed individuals, they are not reliably borne out in experimental animal studies using paradigms involving recall procedures such as delayed alternation. In this the paradigm, reinforcement is contingent upon the alternation of responses between manipulanda each time a delay period has ended. Choosing the appropriate response option depends upon remembering which of the options was correct before the delay period began and selecting the alternative option. In delayed alternation paradigms, however, rodents and nonhuman primates exposed to lead have been reported to show not only impairments of performance, but in other contexts, to show improved performance (Rice and Karpinski, 1988; Levin and Bowman, 1986; Levin and Bowman, 1989; Rice and Gilbert, 1990b; Cory-Slechta et al., 1991; Rice, 1992c).

3. Mechanisms of Learning Deficits

Much evidence points to response perseveration as a basis of lead-induced learning deficits. That is, lead-exposed subjects continue to emit previously reinforced responses, rather than changing behavior as demanded by the environmental contingencies. This has been shown, for example, by analyses of error patterns in experimental animal studies (Davis et al., 1990; Cory-Slechta et al., 1991; Cohn et al., 1993; Rice, 1993). Some congruent evidence from human studies has also been described. In an effort to determine a neuropsychological basis for lead-associated changes in cognitive abilities in children, Stiles and Bellinger (1993) measured perseverative behavior on both the California Verbal Learning Test for Children and the Wisconsin Card Sorting Test in a subset of the population of children in the Boston cohort prospective study mentioned above. In conjunction with the reports from the experimental animal studies, a significant association of higher concurrent or recent blood lead levels and perseverative responding was observed, an effect deemed surprising by the authors in light of the very low and restricted range of lead levels at 10 years of age and the relatively small numbers of subjects on which these data were available.

A slightly different, although potentially overlapping, mechanism is suggested by the recent report of Newland et al. (1994). These authors used a behavioral baseline in which food reinforcement was available for responding on either of two response levers, but one lever was sometimes "richer", i.e., had a greater density of reinforcement than the other, and the reinforcement density associated with one lever relative to the other was changed intermittently over the course of the experiment. Control squirrel monkeys were sensitive to changes in reinforcement density, and preferential responding followed the "richer" lever. In contrast, monkeys postnatally exposed to lead and achieving blood lead levels above 40 μg/dl changed response preferences either very slowly, not at all, or even in the wrong direction (Figure 4). These authors postulated that insensitivity to changes in reinforcement contingencies may serve as a mechanism for lead-related learning impairments.

Deficits in attention are also frequently cited as a basis of lead-associated cognitive impairments, as described in numerous human studies (Needleman et al., 1979; Hunter et al., 1985; Hatzakis et al., 1989; Raab et al., 1990). While some experimental animal studies using irrelevant stimuli in discrimination acquisition and reversal studies have likewise postulated attentional deficits, in fact, behavioral paradigms specifically designed to evaluate attention, such as complex vigilance performance, have not been utilized to date in experimental animal or in human studies. Moreover, attention may have numerous meanings. For example, distractibility, as cited in some human studies, may suggest little attention to the relevant environmental stimuli. Response perseveration could also be construed as an attentional deficit, in that the subject is paying attention to stimuli other than the relevant ones. Thus, both distractibility and perseveration may be defined as attentional deficits, even though they obviously constitute quite different patterns of behavior. Little consideration has yet been afforded to the various potential operational definitions of this term and how these might differentially contribute to lead-induced cognitive deficits.

B. CHANGES IN FIXED-INTERVAL SCHEDULE CONTROLLED BEHAVIOR

That lead exposure affects cognitive and other behavior functions is further supported by studies of schedule-controlled operant behavior, in particular, fixed-interval (FI) schedule-controlled behavior. The FI schedule of reinforcement stipulates that the first response occurring after a specified interval of time has elapsed will result in reinforcement delivery; responses occurring during the fixed interval itself have no specified consequences. This schedule typically generates a characteristic scallop type of responding, in which rate of responding is low early in the interval, when reinforcement availability is not imminent, and increases later in the interval as the time for reinforcement delivery approaches.

Studies both in rodents (Cory-Slechta and Thompson, 1979; Cory-Slechta et al., 1983; Cory-Slechta, 1984; Cory-Slechta et al., 1985; Cory-Slechta and Pokora, 1991) and nonhuman primates (Rice et al., 1979; Rice, 1988a; Rice, 1992a) consistently show changes in the rate of acquisition of the characteristic scalloped pattern of responding on this schedule. As can be seen in Figure 5 (Cory-Slechta, 1994), low levels of exposure to lead reliably increase rate of responding measured across experimental sessions on this baseline, while relatively higher exposure levels initially produce lower response rates than are observed in nonexposed controls. A similar function emerges when blood lead, rather than dose, is used as the index of exposure (Cory-Slechta, 1984). Microanalyses of performance under such conditions reveal that lead exposure generally does not appear to change the temporal patterning of responding during the interval, only the rate at which responding occurs once it has begun, with low level lead increasing these local rates of responding, and higher exposures decreasing them.

These findings contrast with the absence of lead effects on reinforcement schedules based on response rather than time requirements. Little, or often transient effects of lead have been noted on fixed ratio (FR) schedules, in which a specified number of responses are required for reinforcement delivery (Padich and Zenick, 1977; Angell and Weiss, 1982; Rice, 1988a; Rice, 1992a; Cory-Slechta, 1986, 1994). The basis for the preferential sensitivity of the FI schedule to lead is not yet known, but may be related to the differential contingencies of the FI and FR. Specifically, on the FI schedule, response rate can vary widely without affecting the density of reinforcement (number of reinforcers earned per unit time), as long as the single response requirement is met after the interval has elapsed. On the FR schedule, however, significant changes in response rate will alter the density of reinforcement, since reinforcement delivery is based upon completion of the required number of responses.

Significance of lead-induced changes in schedule-controlled responding should not be overlooked. The characteristic scalloped pattern of responding on the FI schedule has been exhibited by widely

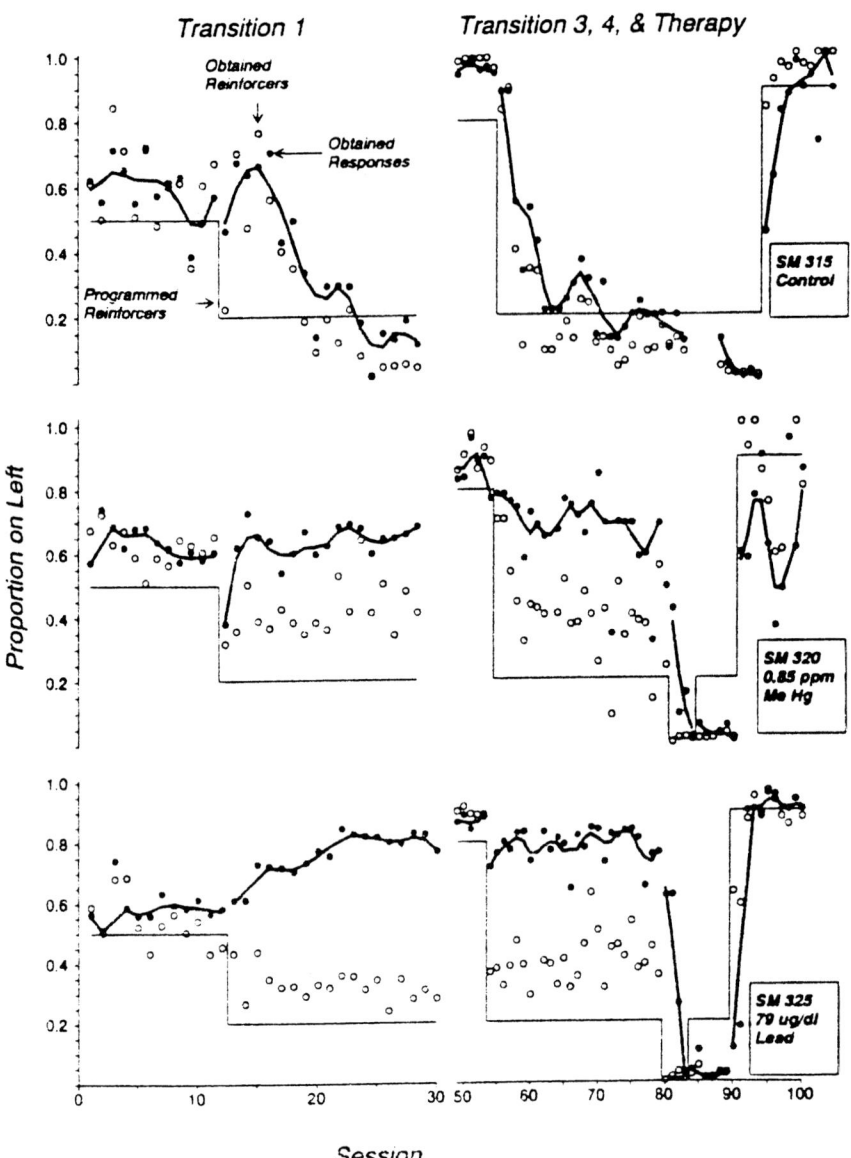

Figure 4. Representative transition performance showing behavioral changes subsequent to a change in the reinforcement density on the two levers. The ordinate represents relative response rates on the left lever. The thin line designates programmed relative reinforcement rates. Open circles indicate the obtained relative reinforcement rates on the left lever (or left lever reinforcers divided by all reinforcers). Filled circles represent obtained relative response rates on the left lever, with the thick line being a smoothed version of these response rates. The first transition is shown in the left column for a representative control monkey (top), methylmercury exposed monkey (middle) and lead-exposed monkey (bottom). The third and fourth transitions are shown in the right column. It also shows the behavioral therapeutic intervention imposed on the two treated monkeys to finally force a behavioral transition. Exposure levels for the treated monkeys are indicated in the box to the right of the right column. (From Newland, M. C., Yezhou, S., Logdberg, B., and Berlin, M., *Toxicol. Appl. Pharmacol.*, 126, 6–15, 1994. With permission.)

divergent species in which this performance has been examined, ranging from the housefly to the human. A delay in the rate at which it is acquired suggests deficits in learning or memory processes. Moreover, a sustained change in response rates over time, as compared to normal or control performance, is consistent with the type of perseverative behavior noted above and may have important consequences for other behavioral functions as well (Cory-Slechta, 1994).

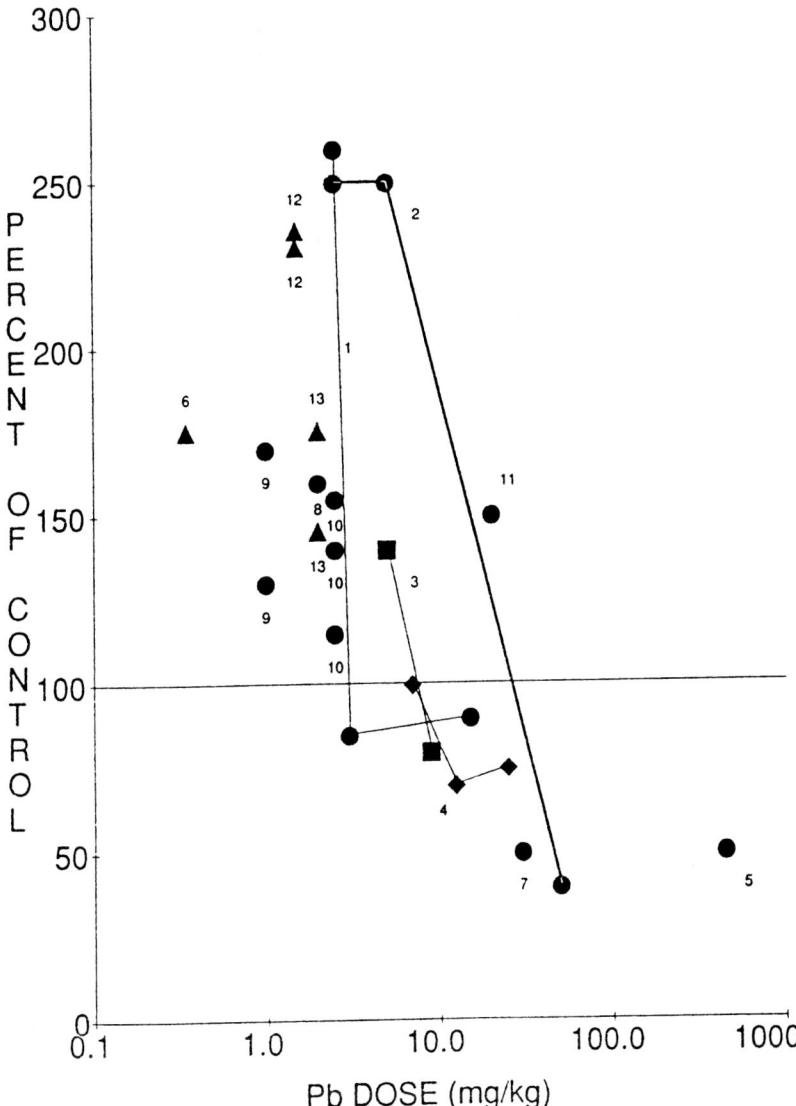

Figure 5. Dose-effect function for lead-induced changes in fixed-interval performance. The lead effect (response rate, interresponse time, or percentage of reinforcement) was plotted as a percentage of the control group value for sessions in which peak effects were observed. Different symbols represent different experimental species: circles, rats; triangles, monkeys; squares, sheep; diamonds, pigeons. Numbers next to curves or selected points represent data from the following studies: 1. (Cory-Slechta et al., 1983); 2. (Cory-Slechta and Thompson, 1979), 3. (Van Gleder et al., 1973), 4. (Barthalmus et al., 1977), 5. (Zenick et al., 1979), 6. (Rice et al., 1979), 7. (Angell and Weiss, 1982), 8. (Cory-Slechta and Pokora, 1991), 9. (Cory-Slechta et al., 1985), 10. (Cory-Slechta and Weiss, 1989), 11. (Nation et al., 1989), 12. (Rice, 1992a), 13. (Rice, 1988a). (From Cory-Slechta, D. A., in *Principles of Neurotoxicology*, Chang, L. W., Ed., Marcel Dekker, New York, 313–344. With permission.)

C. CHANGES IN SENSORY-MOTOR FUNCTION

Lead exposure can affect both visual and auditory function, effects that have been detected in human as well as experimental animal investigations (Fox, 1992; Otto and Fox, 1993). Changes in auditory thresholds have been reported at blood lead levels of 70 µg/dl in occupationally exposed workers. More recently, elevations of hearing thresholds with increasing lead levels were reported in children at blood lead values as low as 10 µg/dl (Schwartz and Otto, 1987; Schwartz and Otto, 1991). In conjunction with these findings, brainstem auditory evoked potentials, a measure used to evaluate auditory nerve dysfunction, have been found to be impaired in lead-exposed monkeys, children, and occupationally-exposed workers (Fox, 1992; Otto and Fox, 1993).

Lead-induced impairment of visual function is known on the basis of studies describing scotopic visual deficits in lead-exposed nonhuman primates (Bushnell et al., 1977) and rats (Fox et al., 1982), and occupationally exposed workers (Caverelli et al., 1982). Acute lead poisoning has long been known to impair visual function as well as to induce visual system pathology. Alterations in visual function are likewise indicated by changes in flash-evoked potentials and in pattern-reversal visual evoked potentials. Further efforts to define the extent of functional visual and auditory loss resulting from lead exposure and their contribution to lead-induced alterations in other behavior processes are warranted.

Alterations in motor function are another consequence of lead exposure. High levels of lead exposure are well known to produce peripheral neuropathy in adults. Case reports of manifestations of lead poisoning in children include reports of impaired motor function at elevated blood lead levels (>45 µg/dl; see Davis et al., 1990). The studies of Bhattacharya et al. (1988, 1990) provide evidence for effects at even lower levels of exposure. These investigators have measured postural sway, a reflex involving motor adjustments to proprioceptive, visual, vestibular, and cutaneous sensory stimuli, in relationship to lead exposure in children. In their studies, the area of postural sway related positively to blood lead concentration in a population of children with a mean age of 5.7 years and a blood lead level of only 20.7 µg/dl. Recently, similar deficits were described in workers exposed to lead, when postural sway was measured in an eyes-closed position (Chia et al., 1994). Further, several of the cross-sectional pediatric studies have reported changes in reaction time in lead-exposed children, which could involve motor deficits (Needleman et al., 1979; Yule et al., 1981). In addition, developmental motor delays have been reported as a function of exposure, including prolonged time to walk, and to sit up (Schwartz and Otto, 1987). Few corresponding efforts from experimental animal studies are available, although changes in reaction time were not found in nonhuman primates exposed to lead (Rice, 1988b). Given the relatively sophisticated technologies that have evolved for measuring various aspects of motor behavior, further attention to this topic would certainly be useful.

III. METHYLMERCURY

Like lead, utilization of mercury, as well as the recognition of its toxicity, extends back to the time of antiquity, when Pliny warned of the hazards of this metal. Nevertheless, mercury has been used continuously over the years for various medicinal purposes as well as for numerous industrial processes, most notably in the production of mirrors, felt hats, and paints, as described by Maurissen (1981).

The neurotoxicity accruing from exposures to high levels of methylmercury became painfully evident from episodes of poisonings such as those at Minamata in the 1950s and in Iraq in the 1970s. Methylmercury causes adverse central nervous system effects in adults as well as children, although children clearly appear to be more vulnerable to such effects. Acute high level exposure during development resulted in a range of neurotoxic effects that included cerebral palsy and mental deficiency, as well as motor retardation, and sensory deficits such as blindness and deafness (Harada, 1977; Amin-Zaki et al., 1980; Marsh, 1987). The disaster in Iraq occurred as a result of the ingestion of seed that had been treated with methylmercury fungicide, while that in Minamata was caused by the consumption of shellfish that had accumulated mercury dumped into Minamata Bay as an industrial effluent.

Concerns over the potential for adverse effects of chronic low level exposures or sustained exposure to methylmercury have recently arisen, much as they did 20 years ago with lead. These apprehensions are based on the growing recognition of the accumulation of mercury into the food chain, particularly in larger species of fish that serve as the primary food source for many populations of the world. While studies aimed at subclinical effects of methylmercury document both sensory and motor changes, questions involving alterations in cognitive function, associated exposure levels, and developmental periods of vulnerability remain to be resolved.

A. EFFECTS ON SENSORY FUNCTION

Changes in sensory function are a well-known consequence of methylmercury exposure. Such effects have significant detrimental effects for many aspects of normal function, including the potential to adversely undermine cognitive processes, since sensory input is critical to learning and memory functions.

1. Visual System Function

As noted above, blindness and visual system impairments were frequent accompaniments of human methylmercury intoxication. The nature and magnitude of deficits in visual function depend upon the extent of exposure, as well as the developmental period of exposure, and have included visual field constriction and deficits in both temporal and spatial visual function, as well as in flicker fusion sensitivity (Sabelaish and Himli, 1976; Iwata, 1974; Iwata, 1980; Ishikawa et al., 1979). Similar impairments have been described in nonhuman adult primates exposed to methylmercury. Merigan (Merigan, 1980; Merigan et al., 1983), for example, described visual field constriction in methylmercury poisoned monkeys. These same authors also reported deficits in flicker fusion (Merigan, 1980) in monkeys that had been chronically exposed to relatively high levels of methylmercury, even though exposure had ceased 3 years previously. Similar effects on flicker fusion were reported by Berlin et al. (1975) in squirrel monkeys exposed to methylmercury.

As is the case with lead, the developing organism is considered to be more vulnerable to the consequences of methylmercury. Rice and Gilbert (Rice and Gilbert, 1982, 1990a) provided a detailed characterization of visual function in two groups of macaque monkeys exposed developmentally to methylmercury, one from birth onward to 50 µg/kg/d, the other exposed in utero via dosing of the mother at levels of 10, 25, or 50 µg/kg/d with offspring treated postnatally with the same dose. Blood mercury levels in the group exposed from birth peaked at 1.2 to 1.4 ppm, declining after weaning to levels of 0.6 to 0.9 ppm. In the group exposed in utero and postnatally, mothers' blood mercury levels averaged 0.37, 0.75, and 1.42 ppm, respectively, in the 10, 25, and 50 µg/kg/d groups, with infants' blood mercury levels approximately 1.7 times higher. These levels subsequently declined to values of 0.21, 0.35, and 0.65 ppm.

Spatial visual function was affected by both exposure protocols, with individual differences in susceptibility between treated monkeys evident. Some monkeys exposed from birth onward demonstrated impairments at high frequencies under high luminance conditions; another showed impairments at middle and high frequencies, and all monkeys were impaired at low luminances. In utero exposure was associated with deficits in both high and low luminance spatial vision, with the high luminance deficits occurring at middle and high spatial frequencies. Temporal visual function was primarily affected by in utero exposure, with deficits in middle to low frequencies under high luminance conditions. No visual field constriction was observed under these conditions. An improvement in low-luminance temporal vision noted in both the group exposed from birth onward and that exposed in utero and postnatally was suggested by the authors to possibly result from remodeling during development of some areas of visual function as others were damaged.

In conjunction with the results reported in human and nonhuman primates, Dyer et al. (1978) found alterations in visual evoked potentials in rats given a single 5 mg/kg dose of methylmercury on the 7th day of gestation. Furthermore, continuous exposure of rats to 2.5 mg/kg/d via the dam produced abnormal visual evoked potentials in offspring, even if they had been cross-fostered to nonexposed dams (Zenick, 1976).

2. Changes in Auditory Function

Exposure to high levels of methylmercury can produce deafness in humans. In fact, the percentages of hearing deficits in adults exposed to methylmercury have been quite high, ranging up to 85% in some cases (Harada, 1977). Hearing impairments have also been reported following in utero exposure to methylmercury. These percentages may be even more alarming when it is considered that, in most cases, the methods used for auditory assessments in human populations often rely on relatively crude technologies and thus may reflect only the more severely affected cases.

More detailed information on auditory system dysfunction as obtained from experimental animal studies is quite limited and has focused almost solely on high level exposure of adult animals. Such studies, nevertheless, tend to confirm not only the neuropathology of methylmercury, but also its functional correspondences. For example, damage to temporal cortex (Garman et al., 1975) and hair cells of the organ of Corti are described (Anniko and Sarkady, 1978), as are changes in brain stem auditory evoked potentials in mice, effects that were noted across all auditory frequencies (Wassick and Yonovich, 1985).

The impact of postnatal methylmercury exposure of primates on auditory function was examined by Rice and Gilbert (1992) in a cohort exposed from birth to 7 years of age to 50 µg/kg/d of mercury and exhibiting steady state blood mercury levels of 0.6 to 0.9 ppm. Pure tone detection thresholds across various frequencies were determined by operant psychophysical procedures when these monkeys were 14 years of age, i.e., 7 years past the cessation of methylmercury exposure. As with visual function, individual differences in susceptibility to auditory system dysfunction were noted within the exposed group. One of five treated monkeys, for example, showed no evidence of dysfunction across a range of frequencies from 125 to 31,500 Hz. Three treated monkeys showed elevated thresholds at the second highest frequency tested (25,000 Hz) and were not tested at 31,500 Hz. A final monkey displayed elevated thresholds at intermediate frequencies as well, i.e., at 10,000 to 12,500 Hz values. Taken together, these data suggest a uniform and permanent impairment in high-frequency hearing in response to developmental methylmercury exposure. As the authors point out, it is unclear whether this effect occurred during the period of dosing, and, moreover, that although the exposure levels used here were higher than those that would be encountered environmentally, the threshold for such effects has yet to be determined.

B. EFFECTS ON MOTOR DEVELOPMENT

Alterations in motor performance as a consequence of exposure to methylmercury have been noted across species, with the magnitude of the effects increasing with the degree of exposure. In humans poisoned by methylmercury in Minamata, Japan, for example, ataxia and severely delayed motor development were noted, with some children failing to crawl or stand before the age of 3, or, in some cases, to walk before the age of 7 (Harada, 1977). Weakness, increased tone, abnormal plantar reflexes, and delayed motor development were among the characteristics of a milder syndrome described on the basis of exposures in Iraq (Marsh, 1987). An analysis of the relationship between methylmercury exposure in maternal hair in Iraq and the impact on retarded walking as recently reported by Cox et al. (1989) is shown in Figure 6.

Similar deficits are observed in experimental animals treated with methylmercury. Cerebral palsy, spasticity, and seizures have been described in experimental animal studies, as shown in Table 1 (Burbacher et al., 1990). Swimming behavior is mice in altered following methylmercury exposure (Spyker et al., 1972). The Collaborative Behavior Teratology study (Buelke-Sam et al., 1985) utilized rats exposed to either 2 or 6 mg/kg methylmercury on days 6 and 9 of gestation. Consistent findings across the participating laboratories included changes in an auditory startle habituation response which could reflect motor and/or sensory damage. An extended version of this study (Vorhees, 1985) indicated delays in surface righting and in swimming ontogeny. Likewise Geyer et al. (1985) noted effects on negative geotaxis and pivoting. Schalock et al. (1981) described hypoactivity in rats prenatally and postnatally exposed to methylmercury. Similarly, decreases in exploratory behavior and spontaneous locomotor activity were found in mice exposed to methylmercury prenatally at doses of 8 mg/kg and above (Su and Okita, 1976).

Apparent aging-related effects in the form of delayed development of motor impairments were reported by Rice (1989) in a cohort of monkeys dosed from birth to ~7 years of age with 50 μg/kg/d and sustaining blood mercury levels during that period ranging from 0.6 to 0.9 ppm. When these monkeys were approximately 13 years of age, clumsiness was noted during routine exercise periods. Subsequent quantification of fine motor performance as well as clinical neurological assessments were carried out. Treated monkeys required more time to retrieve raisins from a food well than did controls and were also apparently insensitive to touch and pin prick. As the author notes, these data indicate that overt signs of toxicity may not be manifest until long after exposure has ceased.

Evidence for the persistence of motor impairments can also be found in the human literature. Kishi et al. (1993) compared results of neurobehavioral tests of 76 male ex-mercury mine workers 18 years after the cessation of mercury exposure to controls matched for age, sex, and years of education. The most consistent differences between the groups were in the domain of motor function, with deficits in grip strength, reaction time, tapping, pegboard dexterity and hand-eye coordination tests in the ex-mine workers.

C. CHANGES IN COGNITIVE FUNCTIONS

Delays in, as well as abnormalities of, intellectual functioning were reported in cases from both the Japanese and Iraq cohorts (Harada, 1977; Marsh, 1987) poisoned with methylmercury. Questions about the extent to which changes in cognitive functions occur in response to even lower levels of methylmercury exposure, particularly chronic exposures, and the levels with which they may be associated, constitute the focus of current endeavors. A pilot component of a well-controlled pediatric prospective longitudinal study carried out in a fish-eating population in the Seychelles Islands revealed an increase in questionable outcomes on the Denver Developmental Screening Tests when children were evaluated at 6 months of age (Cox et al., 1994). The 25th, 50th, and 75th percentiles of the distribution of maternal hair mercury concentrations with which these effects were associated were 4.19, 6.62, and 10.27 ppm, respectively. However, in the subsequently enrolled cohort of the main study (maternal hair mercury levels were 3.29, 5.86, and 9.26 ppm for the 25th, 50th, and 75th percentiles of this distribution), there was no significant increase in abnormal outcomes on the Denver or as revealed by neurological exams

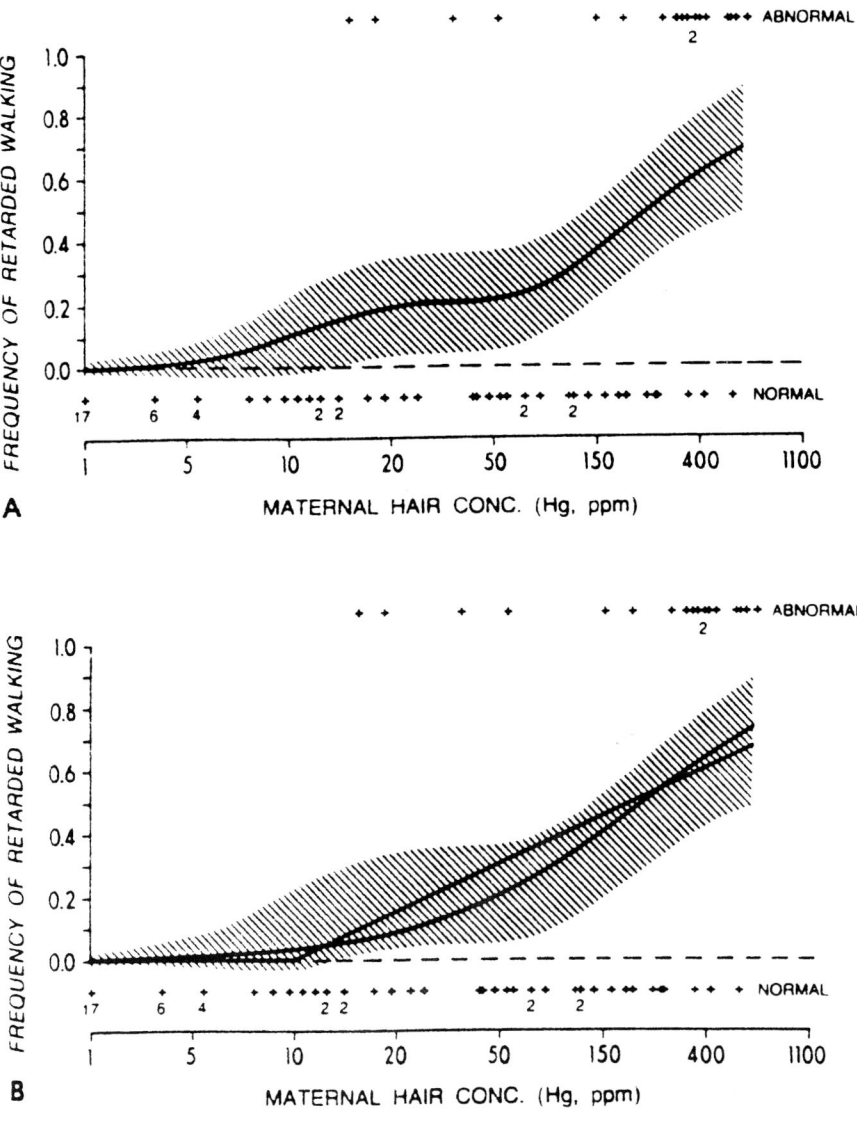

Figure 6. (A) Nonparametric kernel-smoothing analysis of the relationship between concentrations of mercury in maternal hair and retarded walking in offspring. The exposure value represents the maximum level obtained during gestation. The shaded area denotes the nonsimultaneous 95% confidence limits for individual points on the smoothed curve. Maternal hair concentrations for normal and abnormal infants are plotted, respectively, below and above the graph. (B) Plots of the logit and hockey-stick dose-response analysis (solid lines) of the relationship between maternal hair concentrations of mercury during gestation and retarded walking. The shaded area denotes the 95% confidence limits from kernel smoothing. (From Cox C., Clarkson, T. W., Marsh, D. O., and Amin-Zaki, L., Environ. Res., 49, 318–332, 1989. With permission.)

at 6 months. Nor were differences related to methylmercury evident on the Bayley Scales of Infant Development measured at either 19 months or at 29 months of age (Davidson et al., 1994). In contrast, Kjellstrom et al. (1986) reported a significant relationship between maternal hair mercury levels and outcome on the Denver Developmental Screening Test that suggested increased risk at hair mercury levels above 6 ppm (fetal brain mercury of 0.3 ppm) (Burbacher et al., 1990). In many respects, these inconsistencies in outcomes at early developmental stages are reminiscent of the discrepancies in the outcomes of the prospective longitudinal studies with lead in pediatric populations (see above) and may be a reflection of the more subjective nature of the tests utilized in children of younger ages and the greater difficulty in assessing cognitive outcomes at this stage.

Table 1 Neurobehavioral Effects of Developmental Methylmercury Exposure

Humans	Nonhuman Primates	Small Mammals
	High Brain Doses (12–20 ppm)	
Blindness, deafness, cerebral palsy, spasticity, mental deficiency, seizures	Blindness, cerebral palsy, spasticity, seizures	Blindness, cerebral palsy, spasticity, seizures
	Moderate Brain Doses (3–11 ppm)	
Mental deficiency, abnormal reflexes and muscle tone, retarded motor development	Retarded development of object permanence, visual recognition memory and social behavior, visual disturbances, reduced weight at puberty (males)	Abnormal on water maze, auditory startle, visual evoked potentials, escape and avoidance, operant task (DRH), activity, response to drug challenge
	Low Brain Doses (<3ppm)	
Delayed psychomotor development	No data	Response to drug challenge, active-avoidance, operant DRH

Modified from Burbacher et al., 1990.

Experimental animal studies have yet to resolve these questions. Two nonhuman primate cohorts exposed to methylmercury have been tested on a number of relevant behavioral endpoints. One such cohort was exposed to methylmercury throughout pregnancy at levels of 0, 50, 70, or 90 μg/kg/d, resulting in blood mercury levels at birth ranging from 1.04 to 2.46 ppm. Deficits in visual recognition memory and retardation of object permanence development, considered a spatial memory task for infants, were observed in these monkeys during infancy (Burbacher et al., 1986; Gunderson et al, 1986; Gunderson et al., 1988), indicative of cognitive deficits.

However, when tested as adults, i.e., at approximately 7 to 9 years of age, there was little evidence for any residual memory deficits. Gilbert et al. (1993) examined the performance of these same monkeys on a spatial delayed alternation task using delay values ranging from 0.5 to 15 seconds. Monkeys were first tested on a series of fixed delay values, with values increasing across sessions. This was followed by a series of sessions incorporating variable delay values. Treated monkeys actually exhibited a greater number of correct responses and made fewer errors, perseverative responses, and delay responses than controls in the fixed delay procedure, while performances of the treated and control groups did not significantly differ in the variable delay procedure.

Another cohort of nonhuman primates was exposed in utero and postnatally to 0, 10, 25, or 50 μg/kg/d mercury as methylmercury chloride, as described above. Behavior on a nonspatial discrimination reversal task was assessed during infancy and when subjects were juveniles (Rice, 1992b). No substantive treatment-related deficits in performance were noted at either time point. After completion of discrimination testing in infancy, a chained fixed ratio 1- fixed interval 2-minute schedule of reinforcement was imposed. On this schedule, a single response (fixed ratio 1) produced a fixed interval component in which the first response occurring after 2 min had elapsed resulted in reinforcer (formula) delivery. The change in schedule from the fixed ratio 1- to the fixed interval 2-min schedule was signaled to the monkeys by a change in the color of light on the response button from yellow to blue. A total of 30 sessions of 16 or 21 h in length were carried out. Treated monkeys actually earned more reinforcers than did control monkeys on this schedule. However, some evidence of possible deficiencies in temporal discrimination capabilities was suggested by the fact that median pause times and quarter life values (time taken to emit the first 25% of the responses in the fixed interval) were shorter in the treated monkeys relative to controls. Such deficits may be related to cognitive deficits in that characteristic performance under these conditions was apparently not fully acquired by methylmercury-exposed monkeys in the 30 sessions carried out.

A recent report by Newland et al. (1994) does support changes in complex cognitive function as a result of methylmercury exposure. In that study, squirrel monkey mothers were exposed to levels of methylmercury that produced blood mercury levels of 0.7 to 0.9 ppm during the last half to two thirds of gestation. Offspring were trained on the concurrent random interval-random interval schedule of reinforcement described above at about 5 to 6 years of age. Like lead-exposed monkeys, the ability of methylmercury-treated monkeys to behaviorally track transitions in reinforcement density was impaired, as shown in Figure 4.

Studies examining the effects of methylmercury on complex behavior in rodents provide some additional support for changes in complex function, although for the most part, these effects have been observed at relatively high levels of exposure. For example, Hughes and Annau (1976) found that mice exposed to methylmercury hydroxide at doses of 3 mg/kg or above on day 8 of gestation required a greater number of trials to learn an active avoidance response when tested beginning at 56 days of age. Similar changes were reported in adult rats prenatally exposed to methylmercury by Eccles and Annau (1982). Schalock et al. (1981) exposed rats to 10 mg/kg methylmercury chloride during gestation and postnatally until day 21. Behavioral effects, as characterized at 110 to 140 days of age, included deficits in the acquisition of operant avoidance and escape behavior. This same group of rats was also tested on an appetitive learning task which consisted of continuous reinforcement (fixed ratio 1 schedule; phase I), extinction (phase II), and reacquisition of continuous reinforcement (phase III). Methylmercury-treated rats evidenced altered performance in all three phases, consisting of the emission of fewer responses during both the acquisition and reacquisition phases and an increased number of responses during extinction. As described by the authors, these effects reflected slower learning and a tendency towards response perseveration, effects, interestingly, that are consistent with those described by Newland et al. (1994, see above) in squirrel monkeys exposed prenatally to methylmercury.

A study by Olson and Boush (1975) also relates to the possibility of methylmercury-induced learning deficits. These investigators exposed rat dams to either a marlin/mouse chow (group 1) or tuna/mouse chow diet (group 2) laced with methylmercury hydroxide, producing a total mercury concentration of 2 ppm. A motivational assessment was first carried out in a maze with no barriers to the food reward (45-mg pellet) and the locomotion speed to obtain the food pellet on each of 12 trials per day for 8 days revealed no differences between control and treated rats. Subsequently, using a symmetrical maze in which the path to reinforcement could be changed, i.e., a type of repeated learning paradigm, group 1 exhibited deficits in the form of a greater number of errors in learning the four maze problems presented. While provocative, these results are especially difficult to interpret because brain mercury levels of group 2, who showed no deficits in the learning paradigm, were reported to exhibit an increase of 7.3 times that of group 1.

Similarly intriguing results at extremely low brain doses of methylmercury (see Table 1) were described by Müsch et al. (1978) and Bornhausen et al. (1980). In those studies, rats were exposed prenatally to methylmercury in doses ranging from 0.005 to 2.0 mg/kg/d on days 6 and 9 of gestation. Behavior of offspring was tested at 3 to 4 months of age on a differential reinforcement of high rates (DRH) schedule of reinforcement which required the completion of a specified number of responses within a specified period of time for reinforcement delivery. The number of responses per unit time was increased from two responses in 1 second, to four responses in 2 s, to eight responses in 4 s. As shown in Figure 7, doses of only 0.01 mg/kg methylmercury administered on days 6 and 9 to the dam resulted in a significantly lower success rate (percentage of earned reinforcers). In contrast, the ability of rats to discriminate light on (a period associated with reinforcement availability) from light off (nonavailability of reinforcement) was only impacted at the highest dose of methylmercury and only in males, suggesting a specificity to the DRH effects of methylmercury. Moreover, motor performance was said not to be altered by these exposures, although no explicit details of how this was measured were provided. A particularly striking aspect of these data is the low levels at which they occur; the dose of 4×0.01 mg/kg (0.04 mg/kg) is 100 times lower than the lowest effective doses for impairment of avoidance behavior, 5 mg/kg (Eccles and Annau, 1982).

Thus, to date, there are certainly indications both from human pediatric and experimental animal studies that lower level exposure to methylmercury or sustained exposures to methylmercury can deleteriously impact cognitive functions. However, the range of behavioral conditions under which such putative effects have been examined is limited, as is information on effective exposure levels, behavioral mechanisms, differences in species susceptibility, and whether such effects might be indirect reflections of alterations in sensory and/or motor function. It is clear that additional efforts will be required to fully resolve questions regarding the impact of methylmercury exposures on cognitive function.

IV. ALUMINUM

Interest in the neurotoxic properties of aluminum is engendered by the accumulating evidence linking this metal to neurodegenerative diseases such as Alzheimer's disease, amyotrophic lateral sclerosis (ALS)

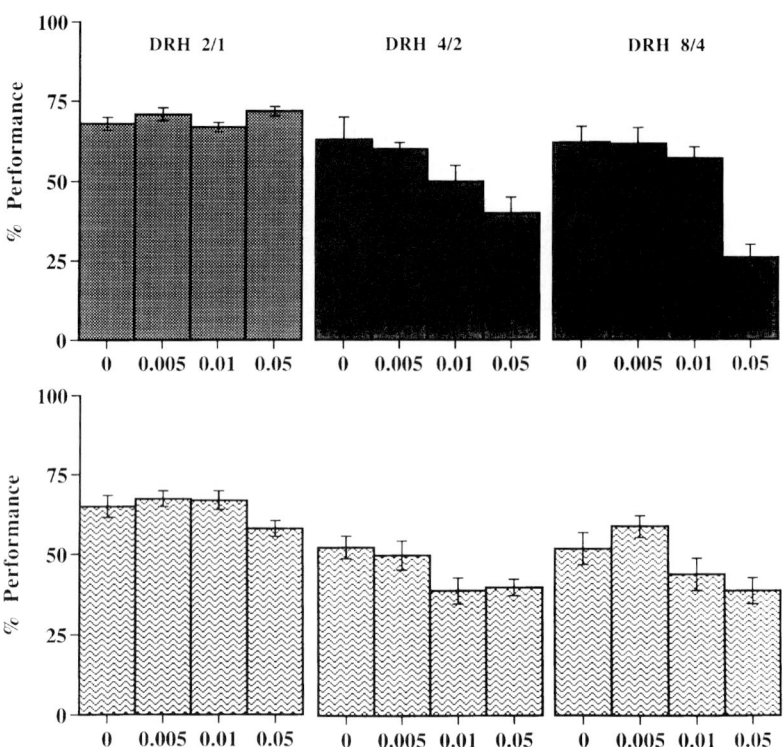

Figure 7. Methylmercury-induced deficits in differential reinforcement of high rate schedule performance of prenatally treated males (top) and females (bottom). Bars represent group mean ± S.E. of the percentages of successful operant performance in test sessions in which two responses in 1 second (left column), four responses in 2 s (middle column) or eight responses in 4 s (right column) were required for reinforcement. (Adapted from Bornhausen et al., 1980.)

and Parkinson-dementia (PD). Like lead and mercury, aluminum is also a nonessential metal. It is widely used as a building material and industrially as an abrasive and catalyst. In addition, salts of aluminum have been used as astringents, styptics, and antiseptics. Insoluble salts are used as antacids and antidiarrheal agents.

In contrast to lead and mercury, however, aluminum is widely prevalent as a common element of the earth's crust, where it is usually found in combination with either oxygen or silicon. While lead and mercury are estimated to constitute 15 and 0.5 ppm of the earth's crust, respectively, the corresponding figure for aluminum is 81,300 ppm and human daily intake is estimated at approximately 36 mg/d per 70 kg-man. Studies linking aluminum to the high rates of ALS and PD in Guam and West New Guinea reveal high prevalence rates of both diseases in areas where aluminum content of soil or water is high, while calcium and magnesium levels are low. Combined with other evidence, this has resulted in the hypothesis that defects in mineral metabolism provoked by these geochemical conditions induce a form of secondary hyperparathyroidism, resulting in enhanced uptake of aluminum and the deposition of calcium, aluminum, and silicon as hydroxyappetites or aluminosilicates in neurons, presumably serving as the ultimate basis of the dementias associated with ALS and PD (Garruto, 1991).

Interestingly, the neurotoxicity of aluminum was described over 100 years ago. Döllken (1898) cites the studies of Siem in 1886 in which aluminum was shown to elicit nerve and muscle paralysis in frogs, cats, dogs, and rabbits. A surge of interest in aluminum neurotoxicity was provided in more recent times, however, by reports such as those of Klatzo et al. (1965) that aluminum given to rabbits either intracerebrally or intracisternally resulted in the development of neurofibrillary tangles that were similar, although not identical, to those associated with Alzheimer's disease. Furthermore, experimental and clinical evidence continues to accumulate linking environmental aluminum with the high focal rates of ALD and PD in Guam and West New Guinea (Garruto, 1991).

The association of elevated aluminum levels with neurodegenerative diseases such as Alzheimer's, PD, and ALS has led to a focus of research efforts primarily on two components of behavior, namely, motor performance and cognitive function. Surprisingly, the breadth of approaches to evaluating changes in cognitive function has been relatively restricted when the possibilities are considered. As with the earlier studies of methylmercury and lead, most experimental investigations of aluminum to date have utilized either acute, subacute or, at best, subchronic regimens combined with relatively high dose exposures. More recently, the importance of examining the effects of chronic lower level exposures has been advanced as a strategy more relevant to human exposure conditions.

A. EFFECTS ON MOTOR FUNCTION

Impaired motor performance appears to be a hallmark of aluminum exposures, particularly of acute high level exposures, and has been observed in a wide range of experimental species, including mice, rats, cats, rabbits and humans. Golub and colleagues, have studied the effects of exposures of mice to aluminum at various points in the life cycle. Exposures of mice to aluminum during early periods of development have been somewhat conflicting. Donald et al. (1989), for example, demonstrated that exposures of female mice to either 25 (control), 500, or 1000 µg Al per gram diet from conception through weaning resulted in motor deficits in offspring at day 25 that included increases in landing foot spread and inability to climb a screen. In contrast, however, enhanced forelimb and hindlimb grip strength was detected, and negative geotaxis and startle were not affected by these aluminum exposures. In a subsequent study (Golub et al., 1992b), however, again using 1000 µg Al per gram diet exposures, hindlimb grip strength was improved, but forelimb grip strength was impaired, and time to negative geotaxis extended when assessed at 21 days of age. The differences between these studies include the day of testing (21 vs. 25 days of age), inclusion or absence of fostering procedures, inclusion or absence of a 4-d period of recovery from the dietary exposure, and use of additional behavioral tests. Additional evidence for motor deficits following prenatal aluminum exposure, however, comes for the study of Bernuzzi et al. (1989), who reported deficits in the righting reflex and negative geotaxis, as well as impairments in locomotor coordination in rats exposed throughout gestation at levels of 300 mg kg/d of aluminum chloride.

Adult mice likewise exhibit motor deficits in response to aluminum administration. Six-week-old female mice, administered 1000 µg aluminum per gram diet for a period of 5 or 7 weeks showed lowered grip strength and greater startle responsiveness than control mice (Oteiza et al., 1993). Moreover, the 1000 µg Al per gram diet exposure was associated with decrements in motor activity measures (Golub et al., 1989, 1992a). Deriving effective brain levels of aluminum for these effects, however, is complicated by the fact that brain aluminum levels differed across these studies, even under relatively similar exposure conditions.

Although it was originally reported that rats did not develop progressive encephalopathy or chronic learning deficits in response to aluminum exposure (King et al., 1975), more recent studies have provided evidence of motoric difficulties in this species. An aluminum encephalopathic model was reported in rats by Lipman et al. (1988) following intracerrebroventricular administration of 1.9 mg aluminum tartrate, with locomotor disturbances and discoordination observed approximately 19 d after the exposure. A study by Bowdler et al. (1979) exposed rats to doses of 0, 550, 1100, or 1650 mg aluminum chloride per kg body weight per day by intubation for 21 d and reported no changes in behavioral measures that included assessment of motor function. Because of the high mortality in the first study, however, a second study was carried out in which rats were exposed to 0, 200, 400, or 600 mg aluminum chloride per kg body weight for 28 d. Of the various measures of behavioral function used, a decrement in performance was detected on the rotarod and in the number of meters traveled in an open field.

Forrester and Yokel (1985) reported a similar pattern of neurotoxicity in rabbits given 20 to 28 subcutaneous injections of 400 µmol/kg of aluminum lactate over 1 month or a single 2.5 to 5 µmol injection of aluminum into each cerebral ventricle. Under these conditions, rabbits displayed an encephalopathic syndrome that included hindlimb weakness, splayed limbs, and postural changes as assessed by a standard test developed for rabbits. Motor impairment in aluminum-treated rabbits was also described by Strong et al. (1991) following intracisternal injections of 100 µg aluminum chloride once a month for eight months. Treated rabbits were reported to gradually develop hindlimb followed by forelimb hyperreflexia, with focal hindlimb deficits, subsequent hypertonia, impaired righting reflexes, limb splaying, gait abnormalities, and paraspinal hypotonia.

Crapper and Dalton (1973b) exposed cats to 6 to 8 µmoles of aluminum (AlCl$_3$; 1540 µg/g) administered either bilaterally into hippocampus, internal capsule, or the cisterna cerebellomedularis. After 8 to 21 days, cats began to exhibit motor impairments which consisted of difficulty in maintaining balance after jumping. A progressive deterioration in movement was subsequently noted, with initial manifestations consisting of motor dyspraxia, followed by truncal ataxia, head tremor, and difficulty in maintaining balance. In addition, an unusual high frequency low amplitude short duration proximal limb movement appeared that was superimposed on purposeful movement of the animals. The syndrome then progressed to retarded motor activity, myoclonic jerks, and general motor seizures.

The study of Bowdler et al. (1979) cited above included a study correlating performance levels of elderly humans with serum aluminum levels. Male and female volunteers (out of a total population of 93) between the ages 56 and 90 years of age were divided into two groups of 25 individuals with the highest serum aluminum levels (504 ng/ml) and the 25 with the lowest aluminum levels (387 ng/ml), but not differing in comparisons of age, race, sex, urban vs. rural background, education level, disease, or medications (other than antacids) taken. Although not a direct demonstration of aluminum-induced motor deficits in humans, it was nonetheless interesting to note that two of the three significant correlations of serum aluminum with performance were noted with tests of visuo-motor function and coordination, i.e., the Trail Making Test and the Digit Symbol Test.

B. EFFECTS ON COGNITIVE FUNCTIONS

Since direct experimental exposures of humans to aluminum is obviously not feasible, most information on the role of this metal in learning and/or memory studies derives from investigations using experimental animals, or assessments of behavior in humans as related to serum aluminum levels or to changes as a result of aluminum chelation with compounds such as desferroxamine.

Surprisingly, the range of approaches to addressing the question of aluminum-induced cognitive deficits has been a relatively narrow one. Most experimental animal studies have relied on conditioned avoidance responding, some using passive avoidance and others active avoidance paradigms. This singular strategy no doubt derives from the emphasis that has been placed on issues such as effective exposure protocols for aluminum, species differences in response, and the relation between learning and memory deficits and neurofibrillary tangles, rather than on questions such as the generality of cognitive changes and their behavioral mechanisms. Using conditioned avoidance paradigms, effects of aluminum on both acquisition (learning) and retention (memory) have been assessed.

Studies based on conditioned avoidance have, in general, uniformly demonstrated adverse effects of aluminum exposures both on acquisition as well as on subsequent retention. Some investigators, moreover, have concluded that these cognitive impairments derive from the neuropathological changes that accompany aluminum exposure. This contention was primarily based on the findings of Crapper and Dalton (1973b) and King et al. (1975), demonstrating that aluminum exposure produces both conditioned avoidance deficits as well as neurofibrillary degeneration in cats, but produces neither in rats. More recent studies, however, do report deficits in avoidance conditioning and other behavioral manifestations in rats (Thorne et al., 1986; Lipman et al., 1988), although it is not clear whether these deficits are due to neurofibrillary degeneration, since brains were apparently not evaluated for these lesions in the latter rat studies. Thus, neurofibrillary degeneration as a basis of aluminum-related learning and memory deficits remains an unresolved issue.

Crapper and Dalton (1973b) first reported that acquisition of a relatively simple one-way active avoidance response was impaired by exposure to aluminum. Cats that had received either hippocampal or internal capsule injections of 6 to 9 µmoles of aluminum showed a significant increase in the number of trials required to reach the specified training criteria, as shown in Figure 8, when tested 9 d after aluminum treatment. Alternative interpretations, such as the possibility that treated cats were motorically or sensorily impaired, or evidenced altered pain sensitivity, seemed negligible, moreover, since these cats showed response speeds comparable to controls during the first five escape trials, the first five avoidance trials, and the last five avoidance trials on day 3 of training. Moreover, both treated and control animals exhibited the same pattern of behavior, as visually observed in the apparatus, seeming to rule out lack of attention or vigilance as the basis of these differences in performance. As pointed out by the authors, these behavioral manifestations occurred several days after the onset of neurofibrillary degeneration in the brain, but well before the terminal stages of encephalopathy with accompanying focal neurological signs (Crapper and Dalton, 1973a). Similarly, Petit et al. (1980) described deficits in the acquisition and retention of an active avoidance task in rabbits 10 d after exposure to 5 µM aluminum tartrate in the lateral ventricles.

A subsequent series of studies by Crapper and colleagues (King et al., 1975) reported only a transient deficit in acquisition of avoidance behavior following aluminum administration in rats, as assessed in either hooded rats from two different sources or in Wistar rats given doses ranging from 0.125 to 0.25 M bilaterally into the ventral hippocampus. Deficits in avoidance acquisition were observed only in the

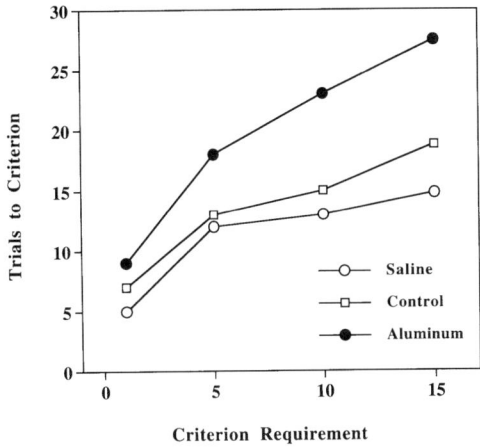

Figure 8. Mean number of training trials required for groups of cats exposed to saline (○), nontreated (□), or exposed to aluminum (●) to achieve criteria of 1, 5, 10, and 15 consecutive avoidance responses. (Adapted from Crapper and Dalton, 1973b.)

first day of a 3-day testing protocol and did not correlate with brain aluminum concentrations, even though these were reported to be 5 to 6 times higher than those in cats and ranged from 2.48 to >45 µg/g. Nor did rats exhibit any signs of neurofibrillary degeneration or progressive encephalopathy over a subsequent year-long period of observation.

In contrast to the report of King and colleagues (1975), however, Lipman et al. (1988) reported changes in both active and passive shuttlebox avoidance learning in rats 7 to 8 d after aluminum tartrate administration and in the absence of other concomitant behavioral aberrations. Male Sprague-Dawley rats were treated with 1.9 mg of the salt intracerebroventricularly, a dose chosen to reproduce tissue levels found in human brain in dialysis dementia. Three training sessions were utilized, with the deficit in aluminum-treated rats detected in the third session, as shown in Figure 9. Interestingly, in the passive avoidance component of the experiment, aluminum-treated rats displayed a notably higher level of intertrial crossings despite receiving shock for doing so, leading the authors to suggest a hyperkinetic preservative performance resistant to conditioning. Brain aluminum concentrations arising from these exposures were not reported in this study. Commissaris et al. (1982) also noted an impaired acquisition of shuttlebox avoidance in Sprague Dawley rats following a chronic oral high dose aluminum chloride administration, again without other behavioral impairments characteristic of aluminum encephalopathy. Similarly, Thorne et al. (1986) reported that oral administration of aluminum at doses of 1500 to 3500 mg/kg resulted in an adverse relationship between neocortical or hippocampal aluminum level and performance on a passive avoidance paradigm.

The findings of Connor et al. (1988), like those earlier reported by Crapper and Dalton (1973b) in cats, were actually suggestive of selective effects of aluminum on passive avoidance. In their study, rats were exposed to 0.3% aluminum sulfate octadecahydrate ad libitum for a 1-month period. Subsequently, acquisition and retention of a passive avoidance task, an active avoidance task, and a radial arm maze were evaluated. Only behavior on the passive avoidance task was impaired following aluminum administration, with slower acquisition and a more rapid extinction of the passive avoidance response. These deficits were only detected, moreover, under specific experimental parameters, such as attenuating the saliency of environmental cues in the maze by eliminating a prehabituation phase. The authors were also able to rule out deficits in motor function, sensitivity to foot shock, or decreased fluid consumption in aluminum-treated rats as contributing to these deficits. In a subsequent study, Connor et al. (1989) demonstrated that a partial reversal of the reduction in number of days to reach extinction in the passive avoidance paradigm could be achieved by termination of aluminum exposure for 2 weeks prior to testing or, in a dose-dependent fashion, by the administration of the aluminum chelator, deferoxamine (Figure 10).

In contrast to the findings of the rodent studies cited above, Bowdler et al. (1979) failed to detect effects of aluminum chloride in rats intubated with doses ranging up to 1650 mg aluminum chloride per kilogram body weight for 21 to 28 days. However, in this study, only a single shuttlebox avoidance training session was carried out, which included a 45-s habituation prior to the first trial. The findings

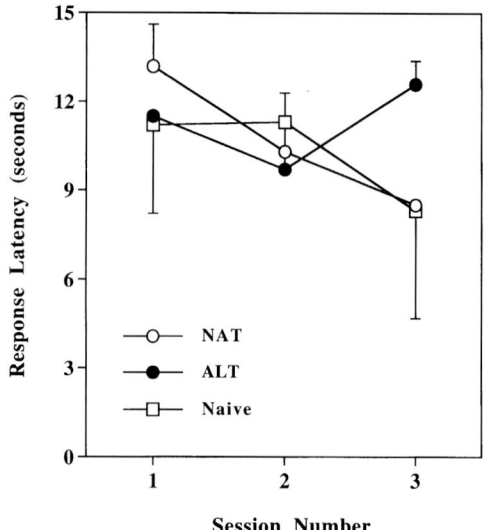

Figure 9. Mean latency to avoid or escape (in seconds) in a shuttlebox avoidance test across sessions in groups of rats injected with sodium tartrate (NAT), aluminum tartrate (ALT), or receiving no injection (Naive). (Adapted from Lipman et al., 1988.)

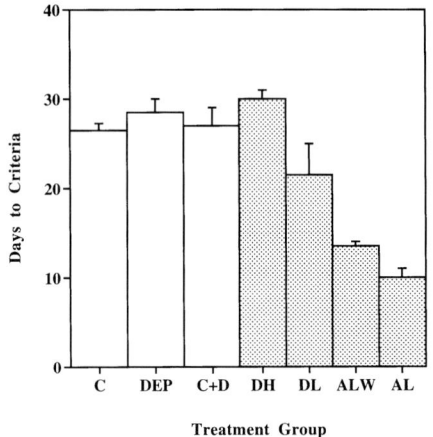

Figure 10. Reversal of the effects of aluminum sulfate administration on extinction of a learned passive avoidance response following various manipulations. Extinction criterion was achieved when a rat entered the darkened side of the chamber in less than 1 min on 3 out of 4 consecutive days, and the number of days (sessions) to achieve this criterion are shown. Indicated groups are as follows: C = control, DEP = water-deprived controls, C + D = controls receiving deferoxamine, DH = aluminum-treated rats receiving the high dose of deferoxamine, DL = aluminum-treated rats receiving the low dose of deferoxamine, ALW = aluminum-treated rats whose exposure was terminated 2 weeks before testing, AL = aluminum-treated rats maintained on aluminum throughout the study. (Modified from Connor et al., 1989.)

of Connor et al. (1988), cited above, suggest that such conditions would not have been sufficiently sensitive to detect aluminum-induced changes.

A recent study based on postnatal aluminum lactate exposure of rats (Cherroret et al., 1992) also failed to observe altered learning in an avoidance paradigm based on the opportunity to respond to turn off or avoid the onset of a bright light in rats tested at either 50 or 100 days of age. Whether the absence of aluminum effects on avoidance responding in this study reflected the differential period of exposure utilized here, differential exposures of brain to aluminum, or the differences in the behavioral paradigm is not clear.

Respondent conditioning paradigms, specifically, acquisition and retention of the nictitating membrane response in rabbits also reveal cognitive impairments arising from aluminum exposure, at least in rabbits. Pendlebury et al. (1987), for example, described deficits in both the acquisition and retention of the conditioned response in rabbits that were treated with 1% aluminum chloride intraventricularly. In their experiments, some groups of animals received aluminum treatment prior to the acquisition of the conditioned nictitating membrane response, whereas other groups were treated after attaining a specified level of performance. Solomon et al. (1988) also noted that retention of the classically conditioned nictitating membrane response in rabbits was disrupted by 100 µl of 1% aluminum chloride administered intraventricularly, with rabbits tested 10 d after surgery. A significant negative relationship, moreover, was found between conditioned response retention and the rating of the severity of neurofibrillary tangles.

Failure to observe aluminum-related alterations in other behavioral baselines which measure learning and or memory, however, raises questions about the generalized nature of these putative effects. Both Cherroret et al. (1992) and Connor et al. (1988) examined the effects of aluminum treatment on performance in a radial arm maze. In neither study was a significant effect of aluminum observed. Moreover, no influence of aluminum administration on discrimination and/or reversal learning paradigms were noted by either Crapper and Dalton (1973b) in a study using cats, or by Cherroret et al. (1992) in a study of rats exposed to aluminum postnatally. In contrast, Thorne et al. (1986) did report that whole brain aluminum level was significantly related to error scores on the initial discrimination task, and aluminum content in "rest-of-brain" samples to error scores in the first reversal of a black/white visual discrimination paradigm after a 30-d treatment with rat chow supplemented with levels of aluminum up to 3500 mg/kg.

Reports of performance impairments and aluminum exposure in humans are indirect in nature. In a study cited above, Bowdler et al. (1979) described an association in elderly humans between serum aluminum levels and long-term memory impairment as indicated by poorer performance of that component of the population with the highest serum aluminum levels on the Serial Sevens Test when compared to that component of the population with the lowest serum aluminum levels. A more recent study by Crapper-McLachlan et al. (1993) used a video home behavior assessment technology to compare the performance of a group of probable Alzheimer's patients receiving the aluminum chelator, desferrioxamine, to a group receiving placebo. When the rate of behavioral decline over time of the groups was compared, the rate exhibited by the untreated group was shown to be twice as rapid as that of the group treated with desferrioxamine, suggesting an involvement of aluminum in this disease state.

Rifat et al. (1990) examined the potential neurotoxicity of aluminum that had been administered from 1944 to 1979 as a finely ground powder (McIntyre Powder) prophylactically against silicotic lung disease in gold and uranium miners. They reported a higher proportion of men with scores in the impaired range in the exposed group as well as lower cognitive test scores based on the Mini-Mental State Exam, Ravens Colored Progressive Matrices, and the Symbol Digit Modalities Test.

V. SUMMARY AND CONCLUSIONS

As yet, a complete understanding of the behavioral toxicity of any of the three metals covered in this chapter, i.e., lead, methylmercury, and aluminum, has yet to emerge. The behavioral deficits arising from lead exposure have been the most extensively studied and best characterized. In addition to both sensory and motor system problems, impairments in cognitive function, primarily in learning processes, are indicated by human study outcomes and supported by corresponding efforts in experimental animal studies. Evidence for memory impairments per se as noted in occupational exposure studies in humans have not been systematically borne out in animal studies. Unclear at the present time are the behavioral mechanisms underlying lead-induced cognitive impairments. Whether they derive from a loss of stimulus control over behavior, as, for example, in attentional deficits, or failure to detect changes in reinforcement density, or some combination of these, remains undetermined.

While it is quite apparent that exposure to methylmercury has deleterious impacts on sensory as well as motor function, its other behavioral consequences are less apparent. Intriguing results regarding potential changes in complex cognitive function have been suggested by some studies, both in humans and rodents. Yet, other investigations, based on human as well as nonhuman primate models, have failed to support this potential connection. It is important to note, though, that the conditions under which these questions have been explored are extremely limited, not only with respect to various methylmercury dosing parameters and developmental periods, but also in regard to the behavioral paradigms utilized.

There seems to be little doubt that exposure to aluminum has deleterious consequences for motor performance, as such deficits have been demonstrated across a range of species. The contention that aluminum overexposure underlies cognitive deficits, however, such as those associated with Alzheimer's disease and Parkinson-dementia is still a more tenuous one. Particularly when dialysis dementia is considered (not reviewed here), human studies are clearly suggestive of a link between aluminum and impaired cognitive function. Moreover, such conclusions are supported by some experimental animal studies reporting deficits in acquisition and/or retention of conditioned avoidance responding. Detracting from the persuasiveness of such studies, however, is the apparent lack of aluminum-related effects on other behavioral baselines measuring cognitive processes, such as the radial arm maze and some types of discrimination and reversal paradigms. Moreover, like methylmercury, few if any studies of aluminum treatment using more advanced behavioral technologies for measuring complex cognitive function have as yet been undertaken.

Resolving the questions posed here is of particular significance for several reasons. First, an explicit understanding of the behavioral consequences of each metal provides the most precise and direct guidance to underlying neurobiological mechanisms of effect. Secondly, an understanding of the specific behavioral toxicity permits the development and/or implementation of possible "behavioral therapeutic" approaches to resolving the behavioral deficits produced by heavy metals. Finally, the more explicitly defined the nature of the behavioral consequences, the more precisely the risks of exposure can be gauged.

REFERENCES

Amin-Zaki, L., Elshassani, S. B., Majeed, M. A., Clarkson, T. W., Doherty, R. A., and Greenwood, M. R. (1980), in *Mechanisms of Toxicity and Hazard Evaluation,* Holmstedt, B., Lauwerys, R., Mercier, M., and Roberfroid, M., Eds., Elsevier, Amsterdam, 75–78.
Angell, N. F. and Weiss, B. (1982), *Toxicol. Appl. Pharmacol.,* 63, 62–71.
Anniko, M. and Sarkady, L. (1978), *Acta Otolaryngol.,* 85, 213–224.
Araki, S., Yokoyama, K., Aono, H., and Murata, K. (1986), *Am. J. Ind. Med.,* 9, 535–542.
Barthalmus, G. T., Leander, J. D., McMillan, D. E., Mushak, P., and Krigman, M. R. (1977), *Toxicol. Appl. Pharmacol.,* 42, 271–284.
Bellinger, D., Sloman, J., Leviton, A., Rabinowitz, M., Needleman, H. L., and Waternaux, C. (1991), *Pediatrics,* 87, 219–227.
Bellinger, D. C. (1995), *Neurotoxicol. Teratol.* (in Press).
Berlin, M., Grant, C., Hellberg, J., Hellstrom, J., and Schutz, A. (1975), *Arch. Environ. Health,* 30, 340–348.
Bernuzzi, V., Desor, D., and Lehr, P. R. (1989), *Teratology,* 40, 21–27.
Bhattacharya, A., Shukla, R., Bornschein, R. L., Dietrich, K. N., and Kopke, J. E. (1988), *Neurotoxicology,* 9, 327–340.
Bhattacharya, A., Shukla, R., Bornschein, R. L., Dietrich, K. N., and Keith, R. (1990), *Environ. Health Perspect.,* 89, 35–42.
Bornhausen, M., Müsch, H. R., and Greim, H. (1980), *Toxicol. Appl. Pharmacol.,* 56, 305–310.
Bowdler, N. C., Beasley, D. S., Fritze, E. C., Goulette, A. M., Hatton, J. D., Hession, J., Ostman, D. L., Rugg, D. J., and Schmittdiel, C. J. (1979), *Pharmacol. Biochem. Behav.,* 10, 505–512.
Buelke-Sam, J., Kimmel, C. A., Adams, J., Nelson, C. J., Vorhees, C. V., Wright, D. C., St. Omer, V., Korol, B. A., Butcher, R. E., Geyer, M. A., Holson, J. F., Kutscher, C. L., and Wayner, M. J. (1985), *Neurotoxicol. Teratol.,* 7, 591–624.
Burbacher, T. M., Grant, K. S., and Mottet, N. K. (1986), *Dev. Psychobiol.,* 22, 1–7.
Burbacher, T. M., Rodier, P. M., and Weiss, B. (1990), *Neurotoxicol. Teratol.,* 12, 191–202.
Bushnell, P. J., Bowman, R. E., Allen, J. R., and Marlar, R. J. (1977), *Science,* 196, 333–335.
Bushnell, P. J. and Bowman, R. E. (1979a), *Pharmacol. Biochem. Behav.,* 10, 733–742.
Bushnell, P. J. and Bowman, R. E. (1979b), *J. Toxicol. Environ. Health,* 5, 1015–1023.
Byers, R. and Lord, E. (1943), *Am. J. Dis. Child,* 66, 471–494.
Cavelleri, A., Trimarchi, F., Gelmi, C., Baruffini, A., Minoia, C., Biscaldi, G., and Gallo, G. (1982), *Scand. J. Work Environ. Health,* 8 (Suppl. 1), 148–152.
Cherroret, G., Bernuzzi, V., Desor, D., Hutin, M.-F., Burnel, D., and Lehr, P. R. (1992), *Neurotoxicol. Teratol.,* 14, 259–264.
Chia, S. E., Chua, L. H., Ng, T. P., Foo, S. C., and Jeyaratnam, J. (1994), *Occup. Environ. Med.,* 51, 768–771.
Cohn, J., Cox, C., and Cory-Slechta, D. A. (1993), *Neurotoxicology,* 14, 329–346.
Commissaris, R. L., Cordon, J. J., Sprague, S., Keiser, J., Mayor, G. H., and Rech, R. H. (1982), *Neurotoxicol. Teratol.,* 4, 403–410.
Connor, D. J., Jope, R. S., and Harrell, L. E. (1988), *Pharmacol. Biochem. Behav.,* 31, 467–474.
Connor, D. J., Harrell, L. E., and Jope, R. S. (1989), *Behav. Neurosci.,* 103, 779–783.
Cory-Slechta, D. A. and Thompson, T. (1979), *Toxicol. Appl. Pharmacol.,* 47, 151–159.
Cory-Slechta, D. A., Weiss, B., and Cox, C. (1983), *Toxicol. Appl. Pharmacol.,* 71, 342–352.
Cory-Slechta, D. A. (1984), in *Advances in Behavioral Pharmacology,* Vol. 4, Thompson, T., Dews, P. B., and Barrett, J. E., Eds., Academic Press, New York, 211–255.
Cory-Slechta, D. A., Cox, C., and Weiss, B. (1985), *Toxicol. Appl. Pharmacol.,* 78, 291–299.
Cory-Slechta, D. A. (1986), *Neurotoxicol. Teratol.,* 8, 237–244.
Cory-Slechta, D. A. and Weiss, B. (1989), *Neurotoxicology,* 10, 685–698.

Cory-Slechta, D. A. and Pokora, M. J. (1991), *Neurotoxicology,* 12, 745–760.
Cory-Slechta, D. A., Pokora, M. J., and Widzowski, D. V. (1991), *Neurotoxicology,* 12, 761–776.
Cory-Slechta, D. A. (1994), in *Principles of Neurotoxicology,* Chang, L. W., Ed., Marcel Dekker, New York, 313–344.
Cory-Slechta, D. A. (1995), *Neurotoxicol. Teratol.*
Cox, C., Clarkson, T. W., Marsh, D. O., and Amin-Zaki, L. (1989), *Environ. Res.,* 49, 318–332.
Cox, C., Myers, G., Davidson, P., Marsh, D. O., Shamlaye, C., Choi, A., Berlin, M., and Clarkson, T. W., *Neurotoxicology,* 15, (in press).
Crapper-McLachlan, D. R., Smith, W. L., and Kruck, T. P. (1993), *Therap. Drug. Monitor.,* 15, 602–607.
Crapper, D. R. and Dalton, A. J. (1973a), *Physiol. Behav.,* 10, 935–945.
Crapper, D. R. and Dalton, A. J. (1973b), *Physiol. Behav.,* 10, 925–933.
Davidson, P. W., Myers, G., Cox, C., Sloane-Reeves, J., Marsh, D. O., Shamlaye, C., and Clarkson, T. W., *Neurotoxicology,* (in press).
Davis, J. M., Otto, D. A., Weil, D. E., and Grant, L. D. (1990), *Neurotoxicol. Teratol.,* 12, 215–229.
Dietrich, K. N., Berger, O. G., Succop, P. A., Hammond, P. B., and Bornschein, R. L. (1993), *Neurotoxicol. Teratol.,* 15, 37–44.
Döllken, A. (1898). *Arch. Exper. Pathol. Pharmacol.,* 40, 98–120.
Donald, J. M., Golub, M. S., Gershwin, M. E., and Keen, C. L. (1989), *Neurotoxicol, Teratol.,* 11, 345–351.
Dyer, R. S., Eccles, C. U., and Annau, Z. (1978), *Pharmacol. Biochem. Behav.,* 8, 137–141.
Eccles, C. U. and Annau, Z. (1982), *Neurotoxicol. Teratol.,* 4, 377–382.
Forrester, T. M. and Yokel, R. A. (1985), *Neurotoxicology,* 6, 71–80.
Fox, D. A., Wright, A. A., and Costa, L. G. (1982), *Neurotoxicol. Teratol.,* 4, 689–693.
Fox, D. A. (1992), in *Human Lead Exposure,* Needleman, H. L., Ed., CRC Press, Boca Raton, FL, 105–123.
Garman, R. H., Weiss, B., and Evans, H. L. (1975), *Acta Neuropath.,* 32, 61–74.
Garruto, R. M. (1991). *Neurotoxicology,* 12, 347–378.
Geyer, M. A, Butcher, R. E., and Fite, K. (1985), *Neurotoxicol. Teratol.,* 7, 759–765.
Gilbert, S. G., Burbacher, T. M., and Rice, D. C. (1993), *Toxicol. Appl. Pharmacol.,* 123, 130–136.
Golub, M. S., Donald, J. M., Gershwin, M. E., and Keen, C. L. (1989), *Neurotoxicol. Teratol.,* 11, 231–235.
Golub, M. S., Han, B., Keen, C. L., and Gershwin, M. E. (1992a), *Toxicol. Appl. Pharmacol.,* 112, 154–160.
Golub, M. S., Keen, C. L., and Gershwin, M. E. (1992b), *Neurotoxicol. Teratol.,* 14, 177–182.
Gunderson, V., Grant, K. S., Burbacher, T. M., Fagan, J., and Mottet, N. K. (1986), *Child Dev.,* 57, 1076–1083.
Gunderson, V. M., Grant-Webster, K. S., Burbacher, T. M., and Mottet, N. K. (1988), *Neurotoxicol. Teratol.,* 10, 373–379.
Hanninen, H., Mantere, P., Hernberg, S., Seppalainen, A. M., and Kock, B. (1979), *Neurotoxicology,* 1, 333–347.
Harada, Y. (1977), in *Minamata Disease: Methyl Mercury Poisoning in Minamata and Niigata, Japan,* Tsubak, R. and Irukayama, K., Eds., Kodansha, Tokyo, 209–239.
Hatzakis, A., Kokkevi, A., Maravelias, C., Katsouyanni, K., Salaminios, F., Kalandidi, A., Koutselinis, A., Stefanis, C., and Trichopoulos, D. (1989), in *Lead Exposure and Child Development. An International Assessment,* Smith, M., Grant, L. D., and Sors, A. I., Eds., Kluwer, Dordrecht, 260–270.
Hogstedt, C., Hane, M., Agrell, A., and Bodin, L. (1983), *Br. J. Ind. Med.,* 40, 99–105.
Hughes, J. A. and Annau, Z. (1976), *Pharmacol. Biochem. Behav.,* 4, 385–391.
Hunter, J., Urbanowicz, M. A., Yule, W., and Lansdown, R. (1985), *Int. Arch. Occup. Environ. Health,* 57, 27–34.
Ishikawa, S., Okamura, R., and Mukuno, K. (1979), *Nippon Ganka Gakkai Zasshi,* 83, 336–343.
Iwata, K. (1974), *Acta Soc. Opthamol. Jpn.,* 77, 1788–1834.
Iwata, K. (1980), in *Neurotoxicity of the Visual System,* Merigan, W. H. and Weiss, B., Eds., Raven Press, New York, 165–185.
Jugo, S. and Kello, D. (1975), *Health Phys.,* 28, 617–620.
King, G. A., DeBoni, U., and Crapper, D. R. (1975), *Pharmacol. Biochem. Behav.,* 3, 1003–1009.
Kishi, R., Doi, R., Fukuchi, Y., Satoh, H., Satoh, T., Ono, A., Moriwaka, F., Tashior, K., Takahata, N., Sasatani, H., Shirakashi, H., Kamada, T., and Nakagawa, L. (1993), *Environ. Res.,* 62, 289–302.
Kjellstron, T., Kennedy, P., Wallis, S., and Mantell, C. (1986), Physical and Mental Development of Children with Prenatal Exposure to Mercury from Fish. Stage 1: Preliminary Tests at Age 4, National Swedish Environmental Protection Board, Solna.
Klatzo, I., Wisniewski, H., and Streicher, E. (1965), *J. Neuropathol. Exp. Neurol.,* 24, 187–199.
Levin, E. D. and Bowman, R. E. (1986), *Neurotoxicol. Teratol.,* 8, 219–224.
Levin, E. D. and Bowman, R. E. (1989), *Neurotoxicol. Teratol.,* 10, 505–510.
Lilienthal, H., Winneke, G., Brockhaus, A., and Molik, B. (1986), *Neurotoxicol. Teratol.,* 8, 265–272.
Lipman, J. J., Colowick, S. P., Lawrence, P. L., and Abumrad, N. N. (1988), *Life Sci.,* 42, 863–875.
Marsh, D. O. (1987), in *The Toxicity of Methylmercury,* Eccles, C. U. and Annau, Z., Eds., Johns Hopkins Press, Baltimore, 45–53.
Maurissen, J. P. J. (1981), *N. Y. St. J. Med.,* 81, 1902–1909.
Merigan, W. H. (1980), in *Neurotoxicity of the Visual System,* Merigan, W. H. and Weiss, B., Eds., Raven Press, New York, 149–163.
Merigan, W. H., Maurissen, J. P. J., Weiss, B., Eskin, T., and Lapham, L. W. (1983), *Neurotoxicol. Teratol.,* 5, 649–658.
Munoz, C., Garbe, K., Lilienthal, H., and Winneke, G. (1986), *Neurotoxicology,* 7, 569–580.
Müsch, H. R., Bornhausen, M., Kriegel, H., and Greim, H. (1978), *Arch. Toxicol.* 40, 103–108.

Nation, J. R., Fry, G. D., Von Stultz, J., and Bratton, G. R. (1989), *Behav. Neurosci.,* 5, 1108–1114.
Needleman, H. L., Gunnoe, C., Leviton, A., Reed, R., Peresie, H., Maher, C., and Barrett, P. (1979), *N. Eng. J. Med.,* 300, 689–695.
Needleman, H. L., Schell, A., Bellinger, D., Leviton, A., and Allred, B. (1990), *N. Engl. J. Med.,* 322, 83–88.
Needleman, H. L. and Gatsonis, C. A. (1990), *J. Am. Med. Assoc.,* 263, 673–678.
Newland, M. C., Yezhou, S., Logdberg, B., and Berlin, M. (1994), *Toxicol. Appl. Pharmacol.,* 126, 6–15.
Olson, K. and Boush, G. M. (1975), *Bull. Environ. Contam. Toxicol.,* 13, 73–79.
Oteiza, P. I., Keen, C. L., Han, B., and Golub, M. S. (1993), *Metabolism,* 42, 1296–1300.
Otto, D. A., and Fox, D. A. (1993), *Neurotoxicology,* 14, 191–210.
Padich, R. and Zenick, H. (1977), *Pharmacol. Biochem. Behav.,* 6, 371–375.
Parkinson, D. K., Ryan, C., Bromet, E. J., and Connell, M. M. (1986), *Am. J. Epidemiology,* 123, 261–269.
Pendlebury, W. W., Beal, M. F., Kowall, N. W., and Solomon, P. R. (1987), *J. Neural Transm. (Suppl.),* 24, 213–217.
Petit, T. L., Biederman, G. B., and McMullen, P. A. (1980), *Exp. Neurol.,* 67, 152–162.
Raab, G., Thomson, G., Boyd, L., Fulton, M., and Laxen, D. (1990), *Br. J. Dev. Psychol.,* 8, 101–118.
Rice, D. C., Gilbert, S. G., and Willes, R. F. (1979), *Toxicol. Appl. Pharmacol.,* 51, 503–513.
Rice, D. C. and Gilbert, S. G. (1982), *Science,* 216, 759–761.
Rice, D. C. (1985), *Toxicol. Appl. Pharmacol.,* 75, 201–210.
Rice, D. C. (1988a), *Neurotoxicology,* 9, 75–88.
Rice, D. C. (1988b), *Neurotoxicology,* 9, 105–108.
Rice, D. C. and Karpinski, K. F. (1988), *Neurotoxicol. Teratol.,* 10, 207–214.
Rice, D. C. (1989), *Neurotoxicology,* 10, 645–650.
Rice, D. C. (1990), *Toxicol. Appl. Pharmacol.,* 106, 327–333.
Rice, D. C. and Gilbert, S. G. (1990a), *Toxicol. Appl. Pharmacol.,* 102, 151–163.
Rice, D. C. and Gilbert, S. G. (1990b), *Toxicol. Appl. Pharmacol.,* 103, 364–373.
Rice, D. C. (1992a), *Neurotoxicology,* 13, 757–770.
Rice, D. C. (1992b), *Neurotoxicology,* 13, 443–452.
Rice, D. C. (1992c), *Neurotoxicol. Teratol.,* 14, 235–245.
Rice, D. C. and Gilbert, S. G. (1992), *Toxicol. Appl. Pharmacol.,* 115, 6–10.
Rice, D. C. (1993). *Neurotoxicology,* 14, 167–178.
Rifat, S. L., Eastwood, M. R., Crapper-McLachlan, D. R., and Corey, P. N. (1990), *Lancet,* 336, 1162–1165.
Sabelaish, S. and Himli, G. (1976), *Bull. WHO (Suppl.),* 53, 83–86.
Schwartz, J. and Otto, D. A. (1987), *Arch. Environ. Health,* 42, 153–160.
Schwartz, J. and Otto, D. A. (1991), *Arch. Environ. Health,* 46, 300–305.
Schalock, R. L., Brown, W. J., Kark, R. A., and Menon, N. K. (1981), *Dev. Psychobiol.,* 14, 213–219.
Solomon, P. R., Pingree, T. M., Baldwin, D., Koota, D., Perl, D. P., and Pendlebury, W. W. (1988), *Neurotoxicology,* 9, 209–222.
Spyker, J. M., Sparber, S. B., and Goldberg, A. M. (1972), *Science,* 177, 621–623.
Stiles, K. M. and Bellinger, D. C. (1993), *Neurotoxicol. Teratol.,* 15, 27–35.
Stollery, B. T., Banks, H. A., Broadbent, D. E., and Lee, W. R. (1989), *Br. J. Ind. Med.,* 46, 698–707.
Stollery, B. T., Broadbent, D. E., Banks, H. A., and Lee, W. R. (1991), *Br. J. Ind. Med.,* 48, 739–749.
Strong, M. J., Wolff, A. V., Wakayama, I., and Garruto, R. M. (1991), *Neurotoxicology,* 12, 9–22.
Su, M. and Okita, G. T. (1976), *Toxicol. Appl. Pharmacol.,* 38, 195–205.
Thorne, B. M., Donohoe, T., Ker-neng, L., Lyon, S., Medeiros, D. M., and Weaver, M. L. (1986), *Physiol. Behav.,* 36, 63–67.
U.S. Environmental Protection Agency (1986), Air Quality Criteria for Lead. Rep. No. EPA-600/8–83/028dF, Environmental Criteria and Assessment Office, U.S. Environmental Protection Agency, Research Triangle Park, NC.
Van Gelder, G. A., Carson, T., Smith, R. M., and Buck, W. B. (1973), *Clin. Toxicol.,* 6, 405–418.
Vorhees, C. V. (1985), *Neurotoxicol. Teratol.,* 7, 717–725.
Wassick, K. H. and Yonovich, A. (1985), *Acta Otolaryngol.,* 99, 35–45.
World Health Organization (in press), *Environmental Health Criteria on Inorganic Lead,* World Health Organization, Geneva.
Yokoyama, K., Araki, S., and Aono, H. (1988), *Neurotoxicology,* 9, 405–410.
Yule, W. R., Landsdown, R., Millar, I. B., and Urbanowicz, M. A. (1981), *Dev. Med. Child Neurol.,* 23, 567–576.
Zenick, H. (1976), *Pharmacol. Biochem. Behav.,* 5, 253–255.
Zenick, H., Rodriguez, W., Ward, J., and Elkington, B. (1979), *Dev. Psychobiol.,* 12, 509–514.

Chapter 34

The Blood-Brain Barrier in Normal CNS and in Metal-Induced Neurotoxicity

Ignacio A. Romero, N. Joan Abbott, and Michael W. B. Bradbury

I. GENERAL INTRODUCTION

In this chapter, we review the interaction of metals with the blood-brain barrier (BBB), stressing interactions that may underlie the neurotoxic effects of these metals. It will be necessary to review the ways in which the metals cross the blood-brain barrier (by simple diffusion or by specific transport mechanisms), as well as ways in which a secondary neurotoxicity can be produced by a primary toxic action of the metals on the BBB. To set the context for the discussion, we begin with a review of current understanding of the structure and general transport functions of the BBB.

II. NORMAL ANATOMY AND PHYSIOLOGY

A. STRUCTURE AND FUNCTION OF THE BLOOD-BRAIN BARRIER
1. Barrier Layers and Brain Fluids

The blood and the brain come into close contact at two important sites: the blood vessels that course through the parenchyma or run on the pial surface below the arachnoid membrane, and the blood vessels that supply the choroid plexuses, responsible for secretion of the cerebrospinal fluid (CSF) (reviewed in Davson et al., 1987). The CSF flows through the ventricles and subarachnoid space, before draining out through the arachnoid granulations and the sheaths surrounding the cranial nerve roots (Cserr and Patlak, 1992). In the majority of tissues, the capillary wall is relatively permeable to small solutes, so that the bulk of blood-tissue exchange occurs through the interendothelial clefts. However, in the brain, there is a significant restraint on diffusion between blood and tissue: the BBB is formed by the endothelial cells lining parenchymal and pial vessels, and the blood-CSF barrier is formed by the specialized ependyma that forms the choroid plexus epithelium (reviewed in Davson et al., 1987). Together with the tight layer of the arachnoid membrane, these barriers effectively isolate the brain environment from the vascular compartment. This isolation, combined with a flowing interstitial fluid (ISF) and CSF, allows for greater control over the fluid microenvironment of the brain than of most other tissues. There may have been several evolutionary pressures for development of a barrier, but a major factor was probably the need to provide good ionic homeostasis around synaptic zones processing patterned information (Abbott et al., 1986).

As the barrier tightened, it became necessary to develop specific transport mechanisms in the endothelial cells to sustain the entry of essential nutrients. The same argument applies to metals: metals essential for brain metabolism and function require mechanisms guaranteeing their entry into brain.

Conversely, the BBB and blood-CSF barrier together with brain fluid drainage contribute to the exclusion of nonessential hydrophilic substances from the brain. A secondary consequence of the BBB is therefore protection from a range of potentially toxic compounds (including metals) that may reach high concentrations in plasma. The advantage of this system of isolation is that it provides protection not only against substances that are components of the earth's natural environment, but also from some of the additional pollutants that have arisen as a result of human industrial activity. However, the system is not perfect, and toxic effects do occur. In order to understand how native metals and certain organic metal compounds interact with the BBB and exert toxic actions, we will first describe the anatomy, physiology, and ontogeny of the BBB, and the mechanisms responsible for inducing, maintaining, and modulating the barrier properties.

2. Anatomy of the Adult Mammalian BBB and Blood-CSF Barrier

Figure 1 shows a 3-dimensional diagram of the cell types associated with a brain parenchymal microvessel. The lumen is lined by a continuous layer of endothelial cells, which are connected to each other by "zonular" tight junctions (*zonulae occludentes*), i.e., junctions that run circumferentially around the whole cell margin. Vesicular profiles, that may mediate non-specific fluid-phase transcytosis in other vessels, are relatively sparse in brain endothelium (Brightman and Reese, 1969). Electron-dense tracer studies with horseradish peroxidase (HRP, 5 nm diameter) and ionic lanthanum (0.42 nm diameter) show that both tracers are effectively blocked by the occluding tight junctions of the brain endothelium, while lanthanum but not HRP is able to penetrate the choroid plexus tight junctions (reviewed by Davson et al., 1987). Freeze-fracture examination shows that the tight junctions of cerebral microvessels consist of a much more complete anastomosing network of intramembranous particle strands than those of other tissues, the difference being particularly striking in the venular end of the vascular tree (Nagy et al., 1984). While the endothelial tight junctions of microvessels such as those of the heart can be seen to have a gap of around 4 nm between the outer leaflets of adjacent cell membranes, no such gap can be detected in mature brain microvessels (Schulze and Firth, 1992). The tighter microvascular endothelium of brain compared to nonbrain tissues is confirmed in studies using microelectrodes to measure the transendothelial resistance, the resistance being ~2000 Ω cm^2 or more in brain, and 3 to 20 Ω cm^2 in nonbrain vessels (reviewed in Olesen, 1989; Butt et al., 1990). All these studies lead to the conclusion that the tight junctions of the brain endothelium can be as much as 100 times tighter than those of microvessels elsewhere.

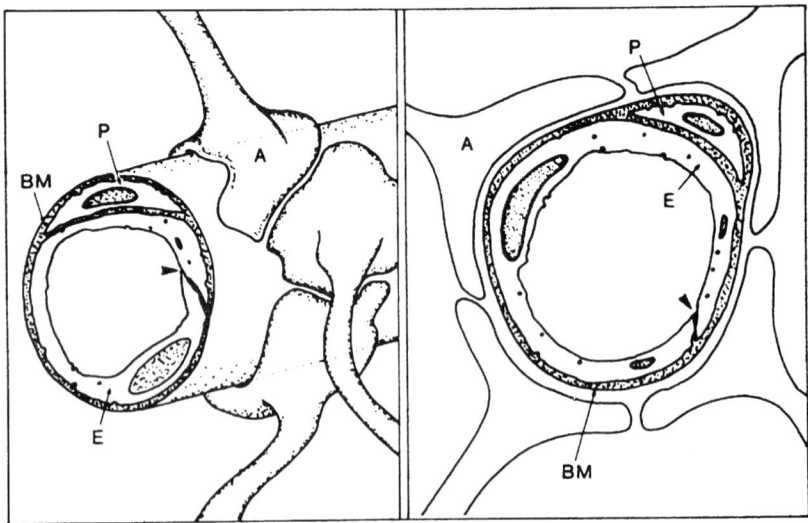

Figure 1. The cell types forming the blood-brain interface of a capillary in mammalian brain parenchyma: left, three-dimensional view; right, transverse section. E, endothelium; BM, basement membrane; P, pericyte; A, astrocytic glial process. Arrowheads indicate the interendothelial clefts closed by tight junctions. In larger arterial and venous vessels, smooth muscle occupies the zone between the endothelium and glia. (From Abbott, N.J., Bundgaard, M., and Hughes, C.C.W., *Progr. Appl. Microcirc.*, 16, 1–19, 1990. With permission.)

The greater permeability of the choroid plexus barrier appears to be an adaptation ensuring the high fluid flow required for CSF production; in general the choroid plexus is not a major route for entry of molecules into the brain interstitium, as a result of the much greater surface area of brain endothelium compared with choroid plexus epithelium, and the continuous flow of CSF that tends to make the ventricles a sink rather than a source for exchange with the brain (see Davson et al., 1987). There is evidence that the ISF is also a flowing fluid system, probably produced as a secretion across the brain endothelium, so that both ISF and CSF can act as routes for clearance of potentially toxic agents (Cserr and Patlak, 1992).

B. DEVELOPMENT OF THE BBB AND BLOOD-CSF BARRIERS: ONTOGENY

A major issue concerning the neurotoxic actions of xenobiotics is whether particular features of brain development render the fetus more vulnerable than the adult. Other chapters in this volume deal with neuronal development and the vulnerability of the fetus to neurotoxins because of the limited regeneration capacity of neurons after birth. There has been controversy concerning the ontogeny of the BBB and blood-CSF barriers. However, it is now clear that the BBB to large proteins is established relatively early in ontogenetic development (Saunders, 1992). Quantitative studies to derive permeability coefficients (P, cm · s^{-1}) for the fetal BBB are harder to interpret because of the changing effects of CSF:brain geometry, and the developing dynamics of CSF secretion and sink (Habgood, 1992, 1993). However, there is some evidence from microelectrode studies that the final tightening of the barrier to small solutes such as ions may not occur until around the time of birth, at least in the rat (Butt et al., 1990). It is also worth mentioning that the fetus may have routes for penetration into brain no longer available in the adult. Thus, while brain entry of albumin is relatively low in the adult, a measurable uptake mediated by a specific transport process can be detected in the fetus (Dziegielewska et al., 1991). This pathway could serve as a route by which metal ions bound to albumin could reach the fetal brain.

C. SPECIFIC TRANSPORT AT THE BRAIN ENDOTHELIUM AND CHOROID PLEXUS EPITHELIUM

The brain endothelium contains specific systems for glucose, amino acids (several different transporters with some overlap in specificity), some peptides, some vitamins, ions (especially the Na$^+$, K$^+$ ATPase), and nucleosides (Figure 2, reviewed in Bradbury, 1992; Davson et al., 1993). The carriers include examples of facilitated diffusion (glucose, L-system amino acid transporter), active transport requiring breakdown of ATP (Na$^+$, K$^+$, ATPase) and secondary active transport where an actively maintained gradient of one substance (e.g., Na$^+$) drives the uphill transport of another molecule (e.g., Na$^+$-dependent A system amino acid transporter). Some of the transporters show a markedly polarized localization (e.g., predominantly on the abluminal membrane: Na$^+$, K$^+$, ATPase, A system amino acid transporter), while others are more symmetrical (L-system transporter). The choroid plexus has some transporters similar to those of brain endothelium (L and A system carriers), but with some differences (prominent ASC transporter, more intense GABA uptake), and for some vitamins such as folates, the choroid plexus may be a major route into the brain (see Pratt, 1992; Lefauconnier, 1992).

D. INDUCTION OF BARRIER PROPERTIES BY BRAIN TISSUE

Early grafting studies established that certain BBB properties were induced in microvessels growing into brain tissue, but not into nonbrain (reviewed in Abbott et al., 1992). Janzer and Raff (1987) demonstrated that rat astrocytic cultures implanted into sites containing permeable vessels showed relatively little leak of albumin, suggesting that astrocytes might be a source of at least some of the factors responsible for barrier tightening. Tout et al. (1993) found a similar effect with implanted retinal Müller (glial) cells, validated at the electron microscopic level with HRP. However, such results have not been universally accepted, and other interpretations have been advanced (Small et al., 1993; Holash et al., 1993). *In vitro,* the complexity of tight junctions between brain endothelial cells and the transendothelial resistance can be increased by coculture with astrocytes (Tao-Cheng et al., 1987; Dehouck et al., 1990), or exposure to astrocyte-conditioned medium together with elevation of cAMP (Rubin et al., 1991). The latter is further evidence for the production of inductive factors by glia. BBB enzymes have also been shown to be subject to induction, but in studies by Tonsch and Bauer (1991), an increased expression of gamma-glutamyl transpeptidase and Na$^+$, K$^+$, ATPase was induced by neuronal as well as glial sources, and membrane fractions were effective, while conditioned medium was not. In the mouse, the induction of barrier properties in the choroid plexus occurs earlier than in the brain endothelium,

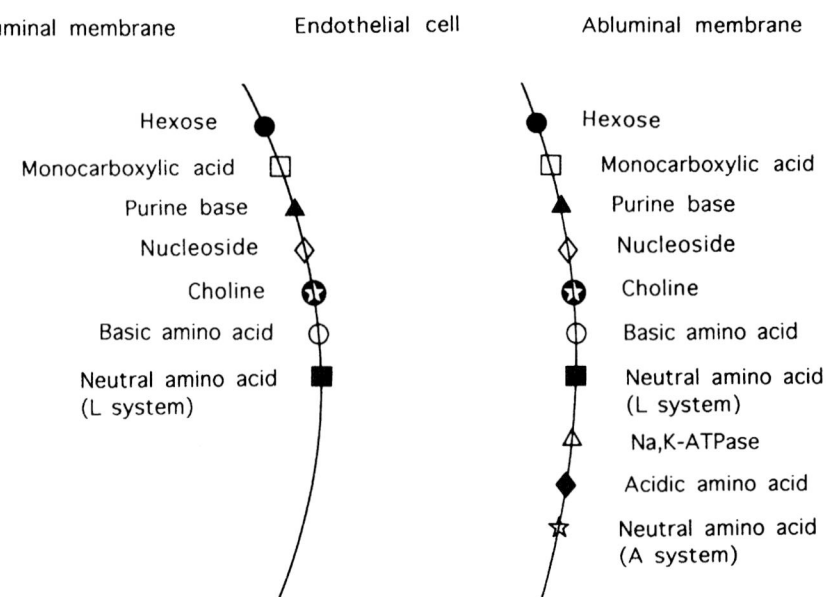

Figure 2. Polarized properties of the brain endothelium forming the blood-brain barrier. The transport systems essential for nutrients and substrates are present on both luminal and abluminal membranes. Certain amino acid transporters and the Na+, K+ ATPase are predominantly abluminal, enabling them to exert control over the composition of the brain interstitial fluid. Additional transport systems are present for some vitamins and peptides (see Davson et al., 1993), but as less is known about their membrane location, they have not been included in the diagram. (Based on Cornford, 1985; Davson et al., 1993.)

and corresponds to the period of neuronal proliferation, before glia are present in large numbers (Risau et al., 1986). Taken together, these studies suggest that the barrier phenotype in brain endothelium and choroid plexus epithelium is induced by a combination of neuronal and glial factors. The extent to which the factors must be continuously present to maintain the barrier properties is not yet clear. In a toxicological context, it is possible that toxic effects on the neurons or glial of the nervous system could cause a decline in the production of inductive factors, leading to a down-regulation of barrier properties, increased permeability of the BBB, and exacerbation of the neurotoxic effects. An indirect action on the barrier may indeed be involved in some types of metal-induced toxicity (see below).

E. MODULATION OF THE BBB

It is well known that a range of inflammatory mediators can increase the permeability of the BBB, acting predominantly at the level of tight junctions (reviewed in Olesen, 1989; Greenwood, 1992). Recently, Fraser and co-workers have identified two phases of permeability increase in rat pial vessels, a first phase measured as a 10- to 25-fold increase in permeability to sucrose, readily reversed, and a second more severe increase in permeability (>25-fold) that may be difficult to reverse under experimental conditions (Easton and Fraser, 1992, 1994). Pharmacological studies show that the first phase can be mimicked by histamine, acting via H2 receptors coupled to a rise in intracellular [Ca^{2+}], while the second phase can be mimicked by application of arachidonic acid and involves release of free radicals. Agents such as bradykinin and ATP may also cause phase 1 opening since they induce a rise in [Ca^{2+}] (Revest et al., 1991; Nobles et al., 1995). Several agents able to open the BBB are released in the vicinity of the endothelium, either from the blood side (platelets) or from the brain side (nerve terminals, damaged tissue), and since cytokines from macrophages/activated microglia may exacerbate BBB opening (Quagliariello et al., 1991; Kim et al., 1992), a variety of mechanisms may contribute to BBB opening under inflammatory conditions. The observations also raise the intriguing possibility that a certain degree of modulatory activity occurs in normal physiology, the BBB being a dynamic and not a static entity. It is then to be expected that certain toxic actions of metals will be mediated by interactions with the modulatory process, certainly in pathological conditions, and possibly in normal function. The evidence is best for the production of free radicals (Section IV.E.1 below), but as we learn more about the cellular and neural basis of BBB modulation, a role for other toxic metals, acting at other levels of the modulatory mechanisms, may emerge.

III. TRANSPORT OF METALS AT THE BLOOD-BRAIN BARRIER

A. INTRODUCTION

If a toxic metal is to influence the function or development of cells within the nervous system, it must obviously penetrate any barriers between blood and these cells. In the case of brain, there must be transport across the BBB, the most important component of which is the tight endothelium of the cerebral microvessels. Since this barrier has a very low permeability to hydrophilic solutes, there are likely to be specialized mechanisms involved. Several essential trace metals have their own specific systems. A toxic metal may utilize such a mechanism or an organic complex with it may use a transporter which normally carries a metabolic precursor.

B. CHEMICAL SPECIES OF METALS IN PLASMA

Knowledge of the chemical species of a metal in blood plasma will indicate which species of the metal are available for transport. Most metals in plasma can occur in up to four general chemical forms. These include (1) the free ion, (2) simple inorganic or organic complexes of low molecular weight, (3) exchangeable complexes with albumin, (4) a tightly bound complex with a specific binding protein, or such a complex with the normal binding protein of a similar essential trace metal (Table 1). The chemical distributions of metals in blood plasma have generally been computed from the known formation constants of putative metal–ligand complexes in relation to the known concentrations of the ligands in normal plasma (May et al., 1977). Aluminum, a widely distributed nonessential and potentially neurotoxic trace metal, illustrates the proposition of four general species in plasma very well. The total aluminum concentration is about 5 µg/l or 0.2 µM. The concentration of free Al^{3+} possible in the presence of transferrin with 50 µM of its metal-binding sites free is <10^{-15} M (Martin, 1986). While at pH 7.4 in water the main Al complex of low molecular weight is $Al(OH)_4^-$ (Martin, 1986), the presence of 0.1 mM citrate in plasma leads to Al-citrate being the main such complex. It comprises about 5% of the total plasma aluminum (Harris and Sheldon, 1990). Al binds with albumin in the absence of transferrin and such a complex may become significant when Al in plasma is raised to the levels found in dialysis dementia, i.e., ~5 µM (Fatemi et al., 1991). The high formation constants of Al-transferrins, i.e., K_1 12.9–13.5 and K_2 12.3–12.5 (Martin et al., 1987; Harris and Sheldon, 1990), ensure that at equilibrium most Al in plasma must be bound to this specific iron-binding protein.

Table 1 Chemical State and Associated Ligands of Some Trace Metals in Blood Plasma

Dynamic Equilibrium			High Affinity, Slow
Free	Low MW	High MW	Or Nonexchangeable
Zn^{2+}	Histidine	Albumin	α_2-Macroglobulin
Pb^{2+}	Cysteine	Albumin	α_1-Intertrypsin inhibitor
Fe^{3+}	(Citrate)	(Albumin)	Transferrin
Al^{3+}	Citrate	(Albumin)	Transferrin

Note: Ligands in parenthesis only form complexes when saturation of transferrin is high.

For a particular metal, the free ion or one or more of the complexes in plasma may be important in transport across the BBB. The following sections review observations in the literature which indicate mechanisms involved in transport of individual trace metals — essential and neurotoxic. Information on essential metals, e.g., zinc, manganese, copper, and iron, has been included, first, because a neurotoxic metal may utilize the normal mechanism for an essential metal and, second, because most essential metals may become toxic in excess or in other particular conditions. Pb, Hg, and Al have been included as metals of no known function which are widely distributed in the environment and are neurotoxic.

C. LEAD

It has been supposed that lead might be restricted in its entry into brain and hence that its influence on the CNS might be due to a primary effect on transport or its regulation at the BBB. Since there is little evidence that lead levels of 20 to 80 µg/dl in blood cause appreciable change in transport of neurotransmitter precursors or metabolic substrates into brain in young rats (Section IV.B.4 below), and lead appears to be rapidly transported into brain in both young and adult rats, a direct effect on neurones or glia is quite feasible.

Bradbury and Deane (1986) maintained a near constant level of ^{203}Pb in the blood of rats with little increase in the concentration of total lead in plasma. The radiotracer entered different brain regions at a linear rate, the concentration in cerebellum reaching about 35% of that in blood plasma at 4 hours. This rapid rate of entry suggested that ^{203}Pb uptake into rat brain might be studied during cerebrovascular perfusion of one cerebral hemisphere of the rat with a buffered salt solution (Deane and Bradbury, 1990a). Uptake was again linear with time, in this case over 90 s. In frontal cortex it reached 10% of the activity in the perfusion fluid at 1 min. The flux into brain was also linearly related to total lead concentration between 0.1 and 4 µM, but was rendered insignificantly small by the presence 5% albumin, 200 µM L-cysteine or 1 mM EDTA (Figure 3). The results led to the conclusion that lead enters brain from blood either as the free lead ion Pb^{2+} or as a simple inorganic complex closely related to it. The effect of pH on transport mirrored that of the predicted concentration of $PbOH^+$, suggesting that this is the transported species. In postnatal rats, transport of ^{203}Pb was considerably greater, being 205% and 242% of the adult values at 16 to 17 and 26 days, respectively (R. Deane, J. Adu, and M. W. B. Bradbury, unpublished observations).

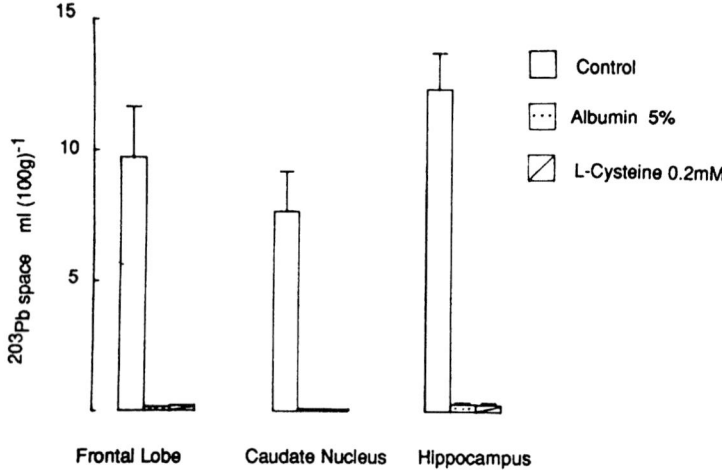

Figure 3. Effect of albumin and of cysteine on ^{203}Pb uptake into three regions of brain after 1 min of carotid perfusion of ^{203}Pb (1 µM total lead) in saline buffered to pH 7.4 with 20 mM HEPES. (From Deane and Bradbury (1990)).

If rat plasma is comparable to that of the human, ~60% of lead is bound to thiol compounds, including cysteine, and the remaining 40% is largely bound to protein, particularly albumin. Free lead has been computed to be about $10^{-12}M$ (Al-Modhefer et al., 1991). At pH 7.4, the concentration of $PbOH^+$ is nearly three times that of Pb^{2+} (Simons, 1986). Since albumin does not appreciably penetrate the blood-brain barrier and since it and cysteine inhibit ^{203}Pb transport presumably by binding ionic lead, it seems likely that the lead is transported into brain as a simple ionic complex, i.e., as its monohydroxide.

D. ZINC

After intravenous administration, ^{65}Zn entered brain at a rather similar rate in relation to its serum activity to that of ^{203}Pb (Adu et al., 1990; Pullen et al., 1991). The flux into brain is very much higher in the case of zinc because of the very much greater serum concentration of total zinc than of total lead, perhaps 5000 times greater. When L- or D-histidine were infused intravenously into rats with injection of ^{65}Zn, transport of zinc into brain was enhanced equally by each stereoisomer (Aiken et al., 1992a). Mechanisms of transport into brain have, as in the rat, been studied by cerebrovascular perfusion of one hemisphere (Buxani-Rice et al., 1994). Since constants for high affinity binding of zinc to albumin and for formation of the two zinc histine complexes are available, it was possible to buffer $[Zn^{2+}]$ to levels approaching its physiological concentration in serum, i.e., 10^{-10} to $10^{-9}M$ (Giroux and Henkin, 1974; Magneson et al., 1987). When free zinc concentration, $[Zn^{2+}]$, was buffered with albumin, ^{65}Zn flux into brain followed Michaelis-Menten kinetics. Half saturation of transport, K_m, occurred at a $[Zn^{2+}]$ of

16 nM and the flux at saturation, V_{max}, into brain was 44 nmol/kg/min. Specific saturable transport of Zn^{2+} itself has been observed in other preparations (Tacnet et al., 1990; Gachot et al., 1991; Raffaniello et al., 1992; Taylor and Simons, 1994), as well as in endothelial cells (Bobilya et al., 1992).

When zinc in the perfusion fluid was buffered with histidine, flux into brain at a constant $[Zn^{2+}]$ was enhanced by increasing histidine concentrations up to 1 mM. This effect was equal for L- and D-histidine, as in the experiments *in vivo* of Aiken et al. (1992). Transport of zinc in the presence of 100 μM L-histidine was unaffected by either 500 μM L-phenylalanine or 500 μM L-arginine. Hence, in contrast to zinc transport in the rat red cell *in vitro* (Aiken et al., 1992b) or that of methylmercury cysteine across rat cerebral microvessels under various conditions (see below), an amino acid transporter, such as that for large neutral amino acids (L) or for basic amino acids (y^+) does not seem to be involved. The kinetics can best be modeled by a relation of transport to the complex of Zn^{2+} with one histidine, i.e., $ZnHis^+$. It has been argued that this small positively charged complex allows diffusion of appreciable amounts of zinc through unstirred layers of a macromolecular matrix through which the very low concentration of free zinc only permits a limited diffusion flux of this ion and to which the zinc-albumin complex is not accessible (Buxani-Rice et al., 1994).

The normal concentration of zinc in both human and rat plasma is about 15 μM (Foote and Delves, 1984; Buxani-Rice, 1992). About 15 to 20% is bound to α_2-macroglobulin (Foote and Delves, 1984) and is not accessible to exchange with ^{65}Zn *in vitro*. Of the remainder, about 88% is bound exchangeably to albumin, 11% to transferrin, and ~1% as low molecular weight complexes with histidine and cysteine (Harris and Keen, 1989). The Zn-α_2-macroglobulin-complex is not known to be involved in transport; and, even if receptor-mediated endocytosis of zinc-transferrin occurs as fast as that of iron-transferrin, it is not likely to contribute much to zinc fluxes into tissues. It does seem that one or more of the low molecular weight complexes of zinc with amino acids may be important in delivering Zn^{2+} to specific sites for transport.

E. MANGANESE

Few experiments have been made on manganese entry into brain from blood. After introduction into the body, manganese in blood plasma appears to end up as largely Mn^{3+}-transferrin (Davidsson et al., 1989). There is good evidence that radiolabeled Mn-transferrin is bound by and taken up into certain cells *in vitro*, e.g., cultured neuroblastoma cells, via receptor-mediated endocytosis (Morris et al., 1987), and that this is inhibited by the presence of cold Fe-transferrin (Suarez and Eriksson, 1993). In the latter experiments, interestingly, uptake of ^{54}Mn-transferrin reached saturation at 2 hours, whereas that of ^{59}Fe from transferrin continued linearly for up to 25 h.

When ^{54}Mn as manganous chloride was infused intravenously, uptake of the radiotracer into brain was very rapid (Murphy et al., 1991; Michison and Parma, 1992). This transport into brain was saturable with increasing concentration, the K_m being ~1 μM. In contrast, movement into CSF was linear, with plasma concentration up to 80 μM (Murphy et al., 1991).

Key observations concerning ^{54}Mn uptake from blood into rat brain were made by Pullen and Franklin (1992). $^{54}MnCl_2$ was kept for 48 h in three solutions — (1) in Hank's salt solution, (2) in Hank's with 2 mg/ml apotransferrin, and (3) in fresh rat plasma. Each solution was given as a single intravenous injection, and uptake into rat brain was studied by multiple time-point analysis (Patlak et al., 1983). Uptake from the salt solution was very fast, the brain concentration being 8% of that in plasma at 10 min. In the presence of transferrin, the rate of uptake was 1/250 and in the presence of plasma 1/90 of that in the absence of protein. A rapid entry of ^{54}Mn-Cl_2 into brain has also been observed during short vascular perfusion of one cerebral hemisphere with either a physiological saline or blood (Rabin et al., 1993). Uptake from saline was 3 to 4 times greater than from blood, a fact which could be attributed to binding of Mn to albumin, α_2-macroglobulin or transferrin. It seems clear that the Mn^{2+} ion or an immediately formed complex penetrates the BBB very rapidly, whereas Mn^{3+}-transferrin once formed enters brain much more slowly. The mechanism involved in the fast transport is unclear, but might be a calcium channel or transporter.

The precise normal distribution of manganese in blood plasma is not known. The main putative ligands are bicarbonate, citrate, albumin, and transferrin (May et al., 1977). After intravenous or oral administration of $^{54}MnCl_2$ into the rat, most of the tracer in plasma was found by a combination of anion exchange and gel filtration chromatography to be bound to transferrin (Davidsson et al., 1989). Similarly, if $^{54}MnCl_2$ is incubated with plasma or with transferrin plus a buffer *in vitro*, the ^{54}Mn associated with the transferrin peak increases over several hours, the process in the latter case being hastened by addition of the oxidase ceruloplasmin (Critchfield and Keen, 1992; Aschner and Aschner, 1990a). The findings suggest slow oxidation of Mn^{2+} to Mn^{3+} before it associates with transferrin. The normal concentration of the free ion Mn^{2+} is certainly less than $10^{-8} M$ and probably closer to $10^{-12} M$ (May et al., 1977). Normal transport across the blood-brain barrier is probably related to receptor-mediated endocytosis of Mn-transferrin at the luminal surface of the capillary endothelium (see transport of Fe and of Al below), but if appreciable Mn^{2+} is present, uptake into brain is much faster by another mechanism.

F. MERCURY

More than 99% of mercuric ion in blood serum is bound to protein (Berlin and Gibson, 1963; Jakubowski et al., 1970), and the remaining <1% which is ultrafilterable probably represents mercuric complexes with sulfhydryl compounds of low molecular weight. Aihara and Sharma (1986) compared the uptake of mercury into brain and other tissues of mice after a single intraperitoneal injection of either mercuric chloride or methylmercuric chloride. While either form of mercury entered kidney and liver much faster than brain, brain uptake of CH_3Hg was more than an order of magnitude faster than that of Hg^{2+}, the rate constants being 0.6×10^{-3} and 4.8×10^{-3} h^{-1}, respectively. Glutathione depletion by diethylmaleate may have increased mercury penetration into brain after CH_3Hg administration, whereas L-cysteine had no effect on Hg^{2+} entry into brain. Farris et al. (1993), using a model for mercury pharmacokinetics after administration of CH_3Hg, computed transfer constants for penetration of CH_3Hg and of Hg^{2+} into brain of 2.1×10^{-3} and 1.9×10^{-3} ml blood per gram tissue per hour, respectively. They suggested that Hg^{2+} might cross the blood-brain barrier by simple diffusion of a low molecular weight complex. Exchange of Hg^{2+} across available sulfhydryl binding sites might also be part of the process.

Methylmercury, however, appears to use the transporter for large neutral amino acids. Methylmercury combines with L-cysteine to form a complex which is structurally very similar to L-methionine (Aschner and Clarkson, 1988). These authors used the intracarotid injection method of Oldendorf to show that the L-cysteine complex was rapidly taken up into brain and was preferred to the D-complex. L-Methionine caused marked inhibition, but the acidic amino acid L-aspartic did not. Such studies were extended by Kerper et al. (1992), who established saturation kinetics for uptake for the methyl-Hg–L-cysteine complex and much slower linear kinetics for the D-cysteine complex. Both L-methionine and a specific substrate for the L-transporter, BCH, inhibited transport of the L-complex which itself inhibited transport of L-methionine. Thus the role of the transporter for large neutral amino acids in Hg transport is quite secure.

Exposure of rats to mercury vapor (Hg^0) may lead to rapid accumulation of mercury in brain. Since Hg^0 is only slowly oxidized to Hg^{2+}, the blood leaving the lungs has been calculated to lose only 3% of its Hg^0 on its way to the cerebral capillaries. Entry into brain is then favored by the high octanol-water distribution coefficient of Hg^0. Once in the brain, Hg^0 is oxidized to Hg^{2+}; (Hursh et al., 1988). In rats exposed to Hg vapor 1 mg/m^3 in their air for 5 weeks, mean blood Hg was 0.25 µg/g, whereas that in brain was no less than 5 µg/g. The distribution pattern of inorganic Hg within brain cells corresponded to that seen after administration of CH_3Hg (Warfvinger et al., 1992).

G. COPPER

Copper transport into brain is of particular interest, because of the degeneration especially in the basal ganglia due to copper toxicity in Wilson's disease. Copper levels in the brains of those dying of the disease are generally 10 times normal (Brewer and Yuzbasiyan-Gurkan, 1992). This is due to the association of low ceruloplasmin and raised concentrations of nonceruloplasmin copper in the blood plasma. The latter is largely bound to albumin with a small low molecular weight fraction, complexed with amino acids.

Saturable transport of copper, presented as cupric ion, has been demonstrated in a number of *in vitro* preparations, the hepatocyte being the most commonly used. Harris (1991) has listed 11 K_ms for such transport, ranging from 3 to 40 µM in different studies. The significance of such transport is uncertain when it is considered that the concentration of free copper in plasma is of the order of 10^{-18} M (May et al., 1977). Transport of copper into tissues is thus probably dependent on the ion being delivered to the cells, either in association with ceruloplasmin or with an amino acid, of which L-histidine is probably the most important. A number of cell types, including hepatic endothelium (Kataoka and Tavassoli, 1985), have been reported to have high affinity receptors for ceruloplasmin, there being no cross-reaction of the receptors with other carrier proteins such as transferrin, although excess copper, as a complex with nitrolo-triacetic acid, will displace ceruloplasmin from its receptor. It seems that the copper-ceruloplasmin complex is not endocytosed, as in the case of iron-transferrin, but the copper ions are released in contact with the membrane to enter the cell. The latter process is potentiated by ascorbate and hence may involve reduction of Cu^{2+} to Cu^+. Entry of cuprous ions into the cells may involve sulfhydryl groups, since it is inhibited both by N-ethylmaleimide and by iodoacetamide (Harris, 1991). It is not known whether such a process is involved in transport of copper across the endothelium of cerebral microvessels.

The abnormal accumulation of copper in Wilson's disease is most likely to depend on transport of amino acid-bound copper across the blood-brain barrier. The presence of an amino acid bound fraction of copper in plasma was demonstrated by Neumann and Sass-Kortsak (1967) and a role for this fraction in copper uptake into liver slices was shown by Harris and Sass-Kortsak (1967). Copper can form a tridentate complex with histidine and albumin (Lau and Sarkar, 1971), and this may allow transport of the albumin-Cu-histidine complex in blood to tissues, where Cu-histidine may be made available. In hepatocytes, histidine potentiates uptake of copper in the presence of albumin. The kinetics and interactions are best explained by $Cu(His)_2$ interacting with the hypothetical transport protein to release Cu^{2+} for transport into the cell (Darwish et al., 1984). In hypothalamic slices, there appears to be specific uptake of $Cu(His)_2$ itself on an amino acid transporter (Hartter and Barnea, 1988; Katz and Barnea, 1990). The mechanism by which amino acid-complexed copper crosses the BBB is unknown. The normal availability of copper in serum seems to be 70 to 90% as ceruloplasmin, <1% as amino acid complexes, and the remainder largely albumin-bound.

H. IRON

Although iron is an essential metal, under certain conditions it may cause cellular damage or degeneration in brain cells (see below, Section IV. E.1). Hence its transport is considered here. Iron enters most cells by receptor-mediated endocystosis of iron-transferrin. In the case of brain, it has initially to cross the endothelium of cerebral microvessels. Transferrin receptors are present on the luminal surface of this endothelium (Jefferies, et al., 1984), and it has been supposed that intact iron-transferrin crosses this endothelium by transcytosis, to be released as the same complex into cerebral interstitial fluid (Fishman et al., 1987). However, there is now strong evidence that, as in other cells, iron is released from transferrin within the endothelium. It must then be reabsorbed from the vesicles, move to the basal pole of the cell, and pass across the abluminal plasma membrane into the cerebral interstitial fluid where it will again combine with apotransferrin. The apotransferrin formed within the endothelium recycles back into blood. The basis of this model is that when ^{59}Fe-^{125}I-transferrin is introduced into the blood, the ^{125}I (representing intact transferrin) rapidly, i.e., within 15 min, attains a low distribution volume in brain and then barely increases, whereas ^{59}Fe goes on accumulating in brain at a much higher rate (Taylor and Morgan, 1990; Morris et al., 1992; Strahan et al., 1992). A similar separation of uptake of ^{59}Fe from that of ^{125}I-transferrin has been observed in cultured brain endothelium (Figure 4) (Taylor, 1993). The form in which iron, released by acidification in vesicles in the endothelium, is taken up and transferred across the abluminal plasma membrane is unknown, but there is circumstantial evidence that it may be as Fe^{2+} (Nunez et al., 1990). Certainly, during cerebrovascular perfusion of one cerebral hemisphere in the rat, movement of Fe^{2+} into brain occurs from a buffered solution not containing transferrin (Bradbury, 1994). Similarly, ^{59}Fe-ferrous chloride infused intravenously into hypo-transferrinemic mice (*hpx/hpx*), almost lacking transferrin, enters brain some 100 times faster than in control mice (Ueda et al., 1993).

I. ALUMINUM

The cells of mammals are normally protected from the toxic effects of aluminum by the dual features of limited absorption from the gut and ready excretion by the kidney. However, when these restraints are overwhelmed, as in dialysis of those with renal failure, aluminum may reach high levels in serum and cause an acute dementia, dialysis encephalopathy. Neurotoxicity of aluminum plays a less certain role in Alzheimer's disease and in the chronic neurodegenerations of the Southwest Pacific. Nevertheless, Al does enter brain and hence there must be a mechanism for this transport. It has been supposed that since aluminum is largely combined with transferrin in blood (see Section III.B above), it may also enter brain by the mechanism which handles iron-transferrin. There is a relation between regional aluminum uptake into brain and the regional density of transferrin receptors in chronic dialysis patients (Morris et al., 1989). Similarly in the experimental rat, ^{67}Ga, used as a marker for Al, entered brain regions at rates which varied with the distribution of these receptors (Pullen et al., 1990).

Few measurements of the rate of Al entry into brain have been made. Pullen et al. (1990), who used single intravenous injection of ^{67}Ga in the rat followed by multiple time point analysis, found a very slow entry rate, K_{in}, of 1.5×10^{-4} ml/g/h. More recently, Radunovič et al. (1994) have compared the brain uptake of ^{67}Ga with net uptake of Al, near constant levels of each metal being maintained by prolonged intravenous infusion in the rat. Gallium and aluminum were given as both the chloride and

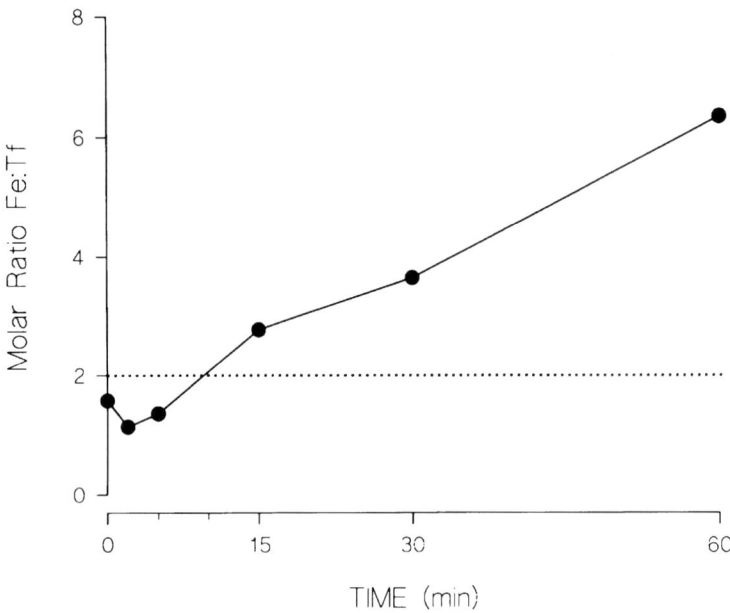

Figure 4. The molar ratio of ^{59}Fe to ^{125}I (i.e., transferrin) uptake into brain endothelial cells from ^{59}Fe-^{125}I-transferrin. An increase in the ratio, to a level greater than 2, indicates that iron accumulates in concentrations above that present in the transferrin. (From Taylor, E.M., Transfer of Iron across Cellular Barriers, Ph.D. thesis, University of London, London. With permission.)

citrate. $K_{in}s$ for ^{67}Ga were in the region 0.9–1.4×10^{-3} ml/g/h and were not influenced by the chemical state in which Ga was administered. Net aluminum uptakes, measured over longer time periods, occurred at very similar rates, namely, 0.9–1.6×10^{-3} ml/g/h. Additionally, a careful comparison was made at 3.5 h between ^{67}Ga entry into brain, when it was given either as the purified transferrin complex or with enough citrate to raise plasma citrate levels to about 1 mM. No significant difference in brain uptake occurred between the two conditions. These results are compatible, with most infused ^{67}Ga or aluminum being converted to the transferrin complex and being initially taken into the endothelial cells of brain microvessels in this chemical form by receptor-mediated endocytosis.

Uptake of ^{59}Fe into rat brain, measured by similar methods in the same laboratory, was 6 to 8 times faster than that of ^{67}Ga or of Al (Ueda et al., 1993). A similar greater uptake of ^{59}Fe-transferrin compared with ^{67}Ga was seen when the results of Morris et al. (1992) were related to those of their colleagues, Pullen et al. (1990). However, Murphy and Rapoport (1992) found no great difference between ^{55}Fe and ^{67}Ga uptakes into rat brain. If, as seems likely, aluminum and ^{67}Ga bound to transferrin enter brain more slowly than ^{59}Fe complexed with transferrin, the difference may be attributed to processes occurring after transferrin-receptor-mediated endocytosis into the endothelium. Since Fe^{3+} can be reduced to Fe^{2+}, subsequent transport of this ion within the cell and across the abluminal membrane may be much faster than that of Al^{3+} and Ga^{3+}, which cannot be so reduced but readily form poorly soluble hydroxides.

IV. TOXICITY OF METALS TO BLOOD-BRAIN BARRIER

A. INTRODUCTION

Toxic compounds may induce changes in the permeability of the BBB by three main mechanisms:

1. A direct action on BBB integrity which results in free passage of plasma into the brain parenchyma, leading to formation of brain edema and secondary neurological dysfunction. Gross and generalized changes in BBB permeability invariably lead to brain edema. Cerebral edema can be defined as an increase in the brain content of water, sodium, and plasma constituents. Cellular mechanisms which may promote the passage of plasma solutes into the brain parenchyma include opening of the tight junctions, transendothelial fenestration, increased vesicular transport, and lysis of endothelial plasmalemma. Concerning the pathogenesis, Klatzo (1967) defined two types of edema: **vasogenic,** in which

the walls of cerebral blood vessels lose their integrity and allow free passage of water and plasma constituents, and **cytotoxic,** in which a damaging agent directly affects the structural elements of the brain parenchyma, resulting in intracellular swelling. An increase in the intracellular sodium content may occur mainly through failure of the sodium pump or through activation of sodium channels (Betz, et al., 1989). When the BBB is damaged, hydrostatic pressure becomes an additional driving force favoring the movement of solutes and water into the brain.

2. An alteration of specific transport systems which could lead to inhibition in the influx of substrates necessary for energy metabolism, synthesis of proteins and neurotransmitters, and neuronal function. Specific changes in brain uptake of metabolic substrates may be classified according to whether they are irreversible (denaturation of the carrier system) or reversible (Pardridge et al. 1975). Reversible mechanisms may represent alterations in the affinity of the transporter for the solute (by carrier conformational changes or competitive inhibition) or in the maximal rate of transport (by alterations in the concentration of the carrier or in its rate of movement within the membrane). These changes could lead to alteration in cerebral metabolism as a result of inadequate supply of key metabolic substrates, and could have toxicological consequences.
3. A direct action on neuronal or glial function leading to an alteration of the normal CNS metabolism, which may in turn result in secondary changes in influx rates of metabolites across the BBB.

The following sections will deal with the five metals for which BBB toxicity information is available, namely, lead, mercury, aluminum, iron, and cadmium.

B. LEAD
1. Introduction

Toxic effects of lead in the CNS are more common in children than in adults and may involve symptoms of acute encephalopathy such as ataxia, headache, convulsions, and coma at blood levels higher than 100 µg/dl (Blackman, 1937). Lesions in brains of children occur in gray and white matter, particularly in the cerebellar molecular layer, and consist of dilated and narrowed capillaries, necrotic vessels, capillary thrombi, hemorrhages, vacuolation, foci of necrosis, reactive gliosis, and edema (Blackman, 1937; Bressler and Goldstein, 1991). Low blood lead levels, 13 to 35 µg/dl, are associated with lesser deficits in children, including learning disorders and hyperactive behavior (Needleman et al., 1990).

2. Experimental Approaches

An experimental model of lead encephalopathy was first developed by Pentschew and Garro (1966) in suckling rats. Maternal diet contained 4% lead carbonate and was fed to rats from parturition, achieving concentrations of lead in the milk high enough to produce neurological damage in the offspring by the third postnatal week. Subsequent studies have used this model or direct administration of lead to neonatal rats (reviewed in Winder et al., 1983). Lead toxicity to the rat nervous system is a dose-related phenomenon and can be divided into high and low dose effects. The doses required to induce neurotoxic effects are higher than in humans.

3. High Dose Studies

High doses of lead, resulting in blood levels around 200 to 500 µg/dl and above, induce morphological changes in the rat brain very similar to those observed in humans with lower doses, and include scattered hemorrhages, vacuolation, necrosis, and edema. The cerebellum seems to be particularly sensitive to damage, although the hippocampus also shows structural changes (Press, 1977; Campbell et al., 1982). Microscopically, vascular damage is manifested in endothelial cells by swelling of cell bodies, vacuolation of mitochondria and endoplasmic reticulum, and increase in pinocytotic vesicles. Astrocytes show reactive proliferation, and foot processes occasionally appear vacuolated or absent. Neuronal changes include a decrease in synaptic contact, dendritic branching, and cell number (Krigman et al., 1974).

The primary target in lead poisoning at these high doses seems to be the cerebral blood vessels, since swelling of endothelial cells occurs early, without associated edema or neuronal necrosis (Goldstein et al., 1974). Changes in glial cells and the onset of gliosis also appear to be secondary. In addition, cerebral blood vessels have a higher concentration of lead than whole brain (Toews et al., 1978) and lead accumulates within endothelial cells, as demonstrated autoradiographically (Stumpf et al., 1980). This may be associated with the selective changes observed in the cerebral vascular bed (summarized in Table 2).

Table 2 Summary of the Effects of Lead on BBB Integrity and Function

Model	Concentration (μM)	BBB Disruption	Transport Systems	Ref.
In vitro				
Rat, rabbit, calf	10–25	Yes (hemorrhages, ↑ HRP, ↓ Evans blue-albumin)	↓ Tryptophan ↓ Glucose	Lorenzo and Gewirtz, 1977 Pelling et al., 1989 Kolber et al., 1980 Ahrens, 1993
			Glucose, phenylalanine, lysine, proline, pyruvate or uridine not affected	Lefauconnier et al., 1980
	<5	No	↑ Thiamine, ↑ lysine, ↑ histidine (adenine, choline, pyruvate, 3-hydroxybutyrate, glucose or mannitol not affected)	Moorhouse et al., 1988
			Choline or tyrosine not affected	Michaelson and Bradbury, 1982
In vitro				
Rat brain capillaries	0.1		↓ 3-*o*-Methylglucose	Kolber et al., 1980
	10		↓ Na$^+$, K$^+$ ATPase	Caspers et al., 1990
Mouse brain endothelial cells	10		↓ Glucose	Maxwell et al., 1986
Rat astrocytes	10		↑ Rb$^+$ efflux	Aschner et al., 1991

a. Effects on BBB Permeability

Disruption of the BBB is manifested by the extravasation of molecules such as albumin and horseradish peroxidase into the extracellular space. In some instances, this precedes the occurrence of brain hemorrhages (Lampert et al., 1967; Clasen et al., 1974; Lefauconnier et al., 1983; Press, 1985; Sundstrom et al., 1985). Although occasionally little endothelial injury is observed accompanying these permeability changes, localized degenerating or necrotic endothelium has been identified in immature blood vessels of lead-poisoned experimental animals (Pentschew, 1965; Press, 1985) and children (Blackman, 1937). This observation supports the idea that intravascular tracers probably reach the extracellular space through many small focal disruption points in endothelial cells and arrive by diffusion on the abluminal surface of vessels with morphologically intact junctions. However, an alteration in the permeability of endothelial tight junctions has been proposed by other groups (Lampert et al., 1967; Hirano and Kochen, 1973; Vistica and Ahrens, 1977), and a combination of the two mechanisms of BBB opening cannot be ruled out.

b. Effects on Transport Systems Across the BBB

It has been proposed that the apparent primary effect on the cerebral vasculature may result secondarily in an impairment of cerebral function leading to the neuronal changes observed in experimental lead intoxication. In support of this hypothesis, Lorenzo and Gewirtz (1977) observed a reduced uptake of tryptophan into brain of lead-treated neonatal rabbits. Glucose uptake was reduced in all brain regions after acute high lead level exposure (Pelling et al., 1989), and microvessels isolated from lead-treated neonatal rat brains also showed a dose-dependent decrease in facilitated transport of monosaccharides, especially in younger animals (Kolber et al., 1980). In contrast, Lefauconnier et al. (1980) found no impairment in the transport of D-glucose, L-phenylalanine, L-lysine, proline, pyruvate, or uridine across the BBB in neonatal rats with blood levels around 350 μg Pb/dl. The differing results of these studies may have explanation in the effects of growth retardation, a common consequence of the malnutrition observed in high dose lead intoxication. Growth retardation may alter brain development and cerebral function, making it difficult to distinguish between specific lead-induced effects and those of undernutrition (Bedi et al., 1980). Indeed, poor nutritional status induced by a low-protein diet has been shown to exacerbate the severity of hemorrhagic lead encephalopathy in suckling rats (Sundström et al., 1984). However, a later study has shown that toxic doses of lead which do not induce growth retardation can still affect the permeability of the BBB to albumin (Sundström et al., 1985).

4. Low Dose Studies

Lower levels of lead in blood (< 100 µg/dl) do not produce vascular damage. However, a reduction in cortical synaptic density and hippocampal cell number and alterations in dopamine and opioid systems have been observed (for reviews see Deane and Bradbury, 1990b; Cory-Slechta et al., 1993: Kitchen, 1993). These disturbances have been associated with the impairment of learning performance also observed at these blood levels in experimental animals (Winneke, 1986). Although undernutrition has been shown to affect neuronal branching and synaptic formation (Bedi et al., 1980), growth retardation at these low lead levels is negligible, suggesting a specific effect of lead on neuronal function. In addition, the transport of nutrients across the BBB does not seem to be affected; there is no evidence of increased cerebral vascular transport of tyrosine and choline in young rats (Michaelson and Bradbury, 1982). Another study with suckling rats with blood levels of 20 to 80 µg/dl revealed an increased entry of thiamine, L-lysine, and L-histidine in some regions of the cerebral hemispheres, but not of adenine, choline, pyruvate, 3-hydroxybutyrate, D-glucose, or mannitol (Moorhouse et al., 1988). Clearly, the psychological changes observed at these low lead doses both in children (Needleman et al., 1990) and in rats (Winneke, 1986) may be due to causes other than impairment of transport of essential substrates into brain.

5. *In Vitro* Studies

In contrast to *in vivo* studies, lead has been shown to alter transport systems in cultured brain endothelial cells and isolated microvessels from rat and bovine brain, such as glucose uptake and Na^+, K^+ ATPase (see Table 2; Kolber et al., 1980; Maxwell et al., 1986; Caspers et al., 1990; Ahrens 1993). However, lead salts are only sparingly soluble in aqueous solutions at physiological pH (Matthews et al., 1993), and lead may interact with proteins and other metals, making the interpretation of *in vitro* results difficult.

6. Mechanism of Action

Lead may exert its toxic effects on the CNS through different mechanisms, including alterations in brain neurotransmitter release and in calcium and energy metabolism (for reviews, see Holtzman et al., 1984; Bressler and Goldstein, 1991). However, two main mechanisms may be considered as underlying acute lead toxicity in the brain microvasculature:

1. Intracellular Second Messenger Activity
   X-ray analysis of capillary endothelial cells isolated from rat cortex and exposed to lead *in vitro* indicated that lead accumulated in the same intramitochondrial areas as calcium (Silbergeld et al., 1980). An increase in the steady-state calcium uptake by these cells (Goldstein et al., 1977) also suggests that calcium efflux from cells may have been impaired and that alterations in calcium-mediated cellular function might contribute at least in part to lead-induced toxicity. In addition, lead has been shown to enter cells through voltage-dependent calcium channels at a higher rate than calcium (Simons and Pocock, 1987). Therefore, it is not surprising that lead can substitute for calcium as an intracellular second messenger. Interactions between lead and two second messenger mediators of calcium signals, calmodulin, and protein kinase C, have been studied extensively (reviwed in Goldstein, 1993). Calmodulin exhibits a higher affinity for lead than it does for calcium, leading to an up-regulation of the enzyme phosphodiesterase (Habermann et al., 1983; Richardson et al., 1986). Concentrations of lead as low as 10^{-14} M induce a dose-dependent stimulation of protein kinase C in isolated cerebral capillaries and brain homogenates (Figure 5; Markovac and Goldstein, 1988a,b). The activation of protein kinase C has been associated with the translocation of the enzyme from the cytosol to a membrane-bound form (Kraft and Anderson, 1983; Hirota et al., 1985). Interestingly, the activity of protein kinase C is present in the cytosol in immature capillaries, while it is membrane bound in microvessels isolated from adult rat brains (Markovac and Goldstein, 1988b). It is possible that, during development, the premature activation of protein kinase C may then disrupt the sequence of proliferation and differentiation of brain endothelial cells, explaining the sensitivity of the immature cerebral vasculature to lead.
2. Loss of Induction of BBB Properties
   Astrocyte-endothelial cell interactions are crucial for the maintenance of a tight BBB, which in turn will influence the sensitivity of the BBB to different xenobiotics (see Section II.D) Damage can then result from the breakdown of inductive action caused by astrocytic damage and retraction of glial end feet. In fact, this mechanism of action has been proposed for lead toxicity to the brain microvasculature

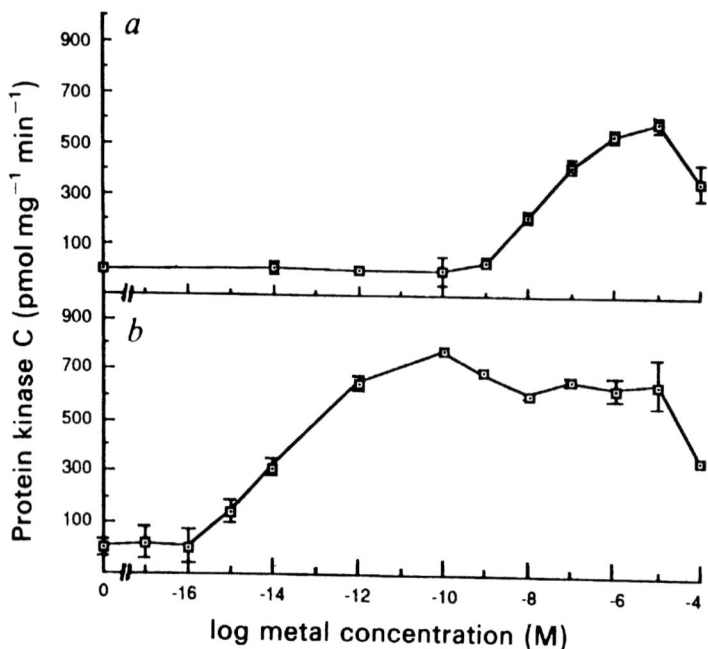

Figure 5. Dose-dependent stimulation of protein kinase C in brain homogenates by calcium (a) and lead (b). (From Markovac, J. and Goldstein, G.W., *Toxicol. Appl. Pharmacol.,* 96, 14–23, 1988. With permission.)

(Bressler and Goldstein, 1991; Rönnback and Hansson, 1992). Astrocytes appear to be more sensitive to the toxic effects of lead compared with cerebellar granule neurons and brain capillary endothelial cells in culture (Holtzman et al., 1987; Gebhart and Goldstein, 1988). The ability of cultured astrocytes to maintain a transmembrane K^+ gradient is compromised by lead at concentrations between 10 and 500 μM (Aschner et al., 1991). In addition, *in vitro* studies have shown that lead can inhibit astroglial-induced tube formation by brain endothelial cells (Laterra et al., 1992). Injury to astrocytes, therefore, may play a role in the loss of barrier properties of brain endothelium. Interestingly, astrocytes have been shown to sequester lead in mature animals, which is associated with the development of resistance (Holtzman et al., 1984).

C. MERCURY
1. Introduction

As with other heavy metals, it is important to consider the different chemical forms of mercury in terms of potential hazard to man. Human exposure to mercury may derive from two major sources, (1) occupational exposure to the vapor of metallic mercury (Hg^0) or (2) to methylmercury via the food chain, mainly fish and its products or seed grains treated with a mercurial fungicide. While mercury ions predominantly damage the kidney, due to the hydrophilic nature of Hg^{2+}, with moderate and reversible involvement of the CNS, alkylmercurials such as methylmercury and ethylmercury are potent neurotoxicants in humans, as proved by the outbreaks of methylmercury poisoning in Japan (Takeuchi et al., 1962) and Iraq (Bakir et al., 1973). In humans, there is a latent period between exposure to alkylmercury compounds and the onset of neurological signs and cerebellar lesions, referred to as the "silent phase" (see previous chapters).

2. Experimental Approaches

Cerebellar lesions similar to those observed in humans have been described after methylmercury experimental intoxication in rats and rabbits, also following a "silent phase" after exposure to the toxin (Cavanagh and Chen, 1971). In the context of the BBB, Choi et al. (1981) showed vacuolation of the cytoplasm of vascular endothelium and swelling of perivascular glia in the cerebellum of neonatal mice exposed to methylmercury. Although the pathology in the developing nervous system is more widespread than in adults, alteration of BBB function seems a possible general mechanism of toxicity of mercury compounds (Steinwall and Klatzo, 1966). However, although all mercury compounds cause the same type of pathology to the brain parenchyma, the extent of damage will depend on the dose and the solubility of the chemical form and thus its capacity to enter nervous tissue and remain in cells for prolonged periods. Consequently, the effects of mercury compounds on the BBB will also depend on these factors and mercuric ions and alkylmercurials should be considered separately.

a. Inorganic Mercury

Early studies using direct cerebral intravascular perfusion of inorganic mercury compounds at doses between 30 and 1000 μM showed extravasation of intravascular tracers into the brain parenchyma of the rat and rabbit (Steinwall and Klatzo, 1966; Pardridge, 1976). The process appears to be molecular size related, since small tracers such as inulin and sucrose entered the brain more readily than albumin and gammaglobulin (Steinwall and Klatzo, 1966). Neutral amino acid carrier transport into the brain was inhibited by almost 100% by mercuric chloride as proved by the low brain uptake indices obtained for cycloleucine and tryptophan compared with control animals (Figure 6; Pardridge, 1976). The uptake of 3-o-methylglucose and pyruvate into brain is also inhibited, but to a lesser extent, while uptake of salicylic acid is increased, suggesting an impairment in the active acid efflux system operating at the BBB level. These selective effects on transport systems seem to be more pronounced at doses that do not induce generalized changes in BBB permeability (Steinwall and Klatzo, 1966; Pardridge, 1976). More recently, Szumanska et al., (1993) demonstrated histochemically a reversible inhibition of the Na$^+$, K$^+$, ATPase in the rat cerebral cortical microvessels following a single intraperitoneal injection of 6 mg/kg mercuric chloride. This inhibition coincided with morphological alterations in endothelial cells and surrounding perivascular astrocytes. This effect is not unique to mercury, and other metals have also been shown to induce inhibition of the Na$^+$, K$^+$, ATPase (see rest of this chapter).

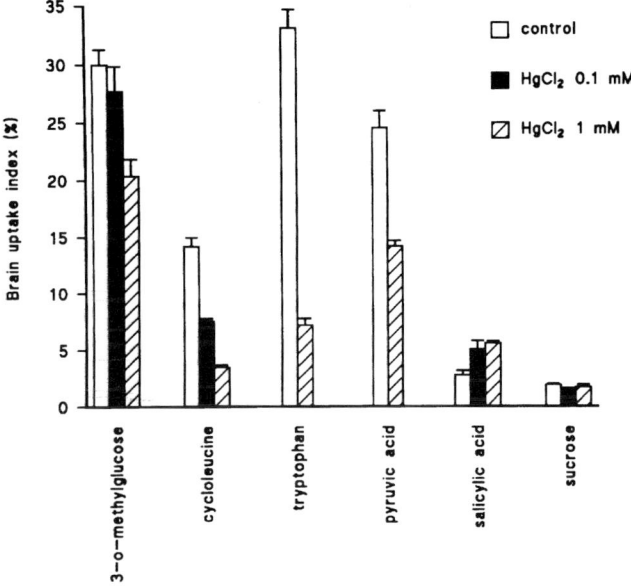

Figure 6. Selective effects of HgCl$_2$ on blood-brain barrier transport systems. Tracer concentrations of the respective [^{14}C] metabolite in the carotid injection are: 3-o-methylglucose, 0.02 mM; cycloleucine, 0.04 mM; tryptophan, 0.02 mM; pyruvic acid, 0.03 mM; salycilic acid, 0.07 mM; sucrose, 0.29 mM. (From Pardridge, W.M., J. Neurochem., 27, 333–335, 1976. With permission.)

b. Organic Mercury

Continuing the early work by Steinwall and Olsson (1969), Ware et al., (1974) found direct evidence of pathological changes in the cerebral endothelium underlying alkylmercury CNS toxicity, in a study with adult rats given methylmercuric chloride in a single intraperitoneal dose of 10 mg/kg. Permeation of horseradish peroxidase into the cerebellum and calcarine cortex was observed as early as 4 to 6 h following methylmercury administration and was accompanied by morphological alterations in capillary endothelial cells, including mitochondrial damage, increased pinocytotic activity, and, in some instances, widening of lateral intercellular spaces without disruption of tight junctions. This was followed by swelling of perivascular astrocytic end-feet. Intracarotid perfusion of the small lipid-soluble p-chloromercuribenzoic acid gave rise to effects on transport of neutral amino acids and glucose similar to those observed with mercuric chloride, but associated with extravasation of intravascular tracers (Hervonen and Steinwall, 1984). Contrary to these results, the transport mechanisms for glucose and aminoacids did not appear to be altered when a single dose (10 mg/kg) of methylmercury chloride was given to rats (Cremer, 1981). Similar studies on rabbits have not shown any disturbances in vascular permeability (Jacobs et al., 1977).

Although the evidence for alteration of the BBB by acute methylmercury intoxication appears conflicting, it would seem that organic mercury compounds cross the BBB rapidly (see above), while mercuric ions may be retained in blood for longer, allowing interaction with transport mechanisms. It would appear, therefore, that chronic or subacute administration of methylmercury should have a more pronounced toxic effect on the BBB. However, the permeability to mannitol remained unaltered during both phases of intoxication following 6-day exposure to 8 mg/kg of methylmercuric chloride in the rat (Eley et al., 1983). Although unidirectional glucose influx seemed to be increased in several brain regions, including the cerebellum and cerebral cortex, this occurred during the silent phase only. There was no evidence for changes in glucose phosphorylation or supply to the CNS, whereas the regional rates of cerebral blood flow were significantly reduced during this phase (Hargreaves et al., 1988). The linear relationship across brain regions between blood flow and metabolic rates of glucose utilization was maintained in methylmercury-treated rats, but the operational level of metabolically regulated blood flow appeared to have been reset to a lower level. The contrast between the effects observed in the early acute studies and those of the later chronic studies may reflect differences in effective concentrations of mercuric ions both in plasma and/or within endothelial cells. Whether the neurotoxic action of alkyl-mercurials is a function of the alkylmercury radical or of inorganic mercury released from it still remains to be determined.

3. Mechanism of Action

The earliest general metabolic change described after mercury poisoning is a disruption of the rough endoplasmic reticulum and loss of ribosomes, leading to an inability of the target tissue to synthesize proteins (Cremer, 1981). Other factors that may be involved in the general biological action of mercury are the high affinity of mercury compounds for sulfhydryl groups and their involvement in redox reactions (Aschner and Aschner, 1990b). Two major sites of action at the BBB have been described: the endothelial cell membrane, the first barrier met by a neurotoxic agent, and the astrocytes, involved in maintaining cerebral homeostasis.

a. Membrane Toxicant

The effects of mercury compounds on passive ion permeability have been studied extensively in erythrocytes (Kinter and Pritchard, 1977). Although the anionic aqueous channel appears to be unaffected by mercury, concentrations of mercury as low as 5 μM increase the cationic permeability to Na^+ and K^+ (Kinter and Pritchard, 1977). An increase in the uptake and efflux of Rb^+ in astrocytes after incubation with 10 to 100 μM mercury has also been demonstrated (Aschner et al., 1991), indicating an alteration in the ability of astrocytes to maintain a transmembrane K^+ gradient. Methylmercury has also been shown to induce changes in the distribution of anionic groups on the surface membrane of mouse astrocytes (Peckham and Choi, 1986). In common with other metals such as lead and aluminum, mercury has a general inhibitory effect on the enzyme Na^+, K^+ ATPase in many tissues (Aschner and Aschner, 1990b). As mentioned earlier, this effect has also been demonstrated *in vivo* in cerebral capillaries (Szumanska et al., 1993).

b. Involvement of Astroglia

An involvement of astroglia in methylmercury intoxication has been demonstrated histochemically in rats (Hargreaves et al., 1985). Organic mercury accumulated initially in the brain during the "silent phase", mainly in glial cells, while at the onset of neurological signs, the same staining was also observed in neurons. These observations may suggest a process involving cellular translocation of the metal from astrocytes to the more susceptible neurons.

Mercury may also have a direct deleterious effect on astrocytes, leading to a loss of cerebral homeostasis and secondary neuronal death. Methylmercury has been shown to induce irreversible inhibition of DNA and protein synthesis in astrocytes, while the effect of mercuric chloride was reversible (Choi et al., 1980; Choi and Kim, 1984; Brookes and Kristt, 1989). The effects of methylmercury on astrocytic K^+ flux may also compromise extracellular K^+ homeostasis either by spatial buffering or active uptake, resulting in cellular swelling (Aschner et al., 1990, 1991). More recently, alterations in the transport of excitatory amino acids in astrocytes have been shown to play an important role in mercury neurotoxicity. Mercuric chloride has been shown to lower the threshold for glutamate neurotoxicity in neurons in organotypic cultures (Matyja and Albrecht, 1993). Submicromolar concentrations of mercuric chloride inhibit L-glutamate uptake selectively and reversibly in astrocytes (Brookes, 1988), while inhibition by methylmercury appears to be a nonselective consequence of a decrease in protein synthesis (Brookes and Kristt, 1989). The uptake of alpha-aminoisobutyric acid by the A or L amino acid transporter and of gamma-aminobutyric acid appears to be inhibited by mercuric chloride, albeit to a lesser extent (Brookes and Kristt, 1989). Efflux of both L-glutamate and D-aspartate from preloaded astrocytes is also increased by methylmercury and mercuric chloride (Albrecht et al., 1993; Aschner et al., 1993). These effects of mercurials on amino acid transport appear to be mediated by the interactions of the metal with -SH groups located intracellularly in astrocytes (Albrecht et al., 1993).

D. ALUMINUM
1. Introduction

In the mid-1970s, Alfrey et al. (1976) demonstrated a direct link between the accumulation of large quantities of aluminum and the development of dialysis encephalopathy in patients affected by renal failure. Since then, the ability of aluminum to act as a neurotoxin has become well established. Its toxicity in humans is characterized by speech difficulties, EEG changes, dementia, myoclonus, and seizures (King et al., 1981). Aluminum accumulation may also occur in several other types of dementia, including Alzheimer's disease, Down syndrome, and the Parkinson-dementia syndrome of Guam, although evidence for a direct etiological link is still highly controversial. Deloncle and Guillard (1990) suggested that in these conditions an alteration in selective transfer through biological barriers in general, and BBB in particular, resulted in an imbalance in mineral metabolism. The immediate consequence of this is a replacement of calcium and magnesium by metals (mainly aluminum) present *in vivo*, yielding complexes more stable with amino acids such as aspartic and glutamic acid. As these authors suggest, it is possible that these complexes cross the BBB and deposit in neuronal structures where they exert their toxic effects.

2. Experimental Approaches

The systemic, central or oral administration of aluminum to certain experimental animals gives rise to severe neurological signs such as impairment of learning and retention, and to cellular abnormalities in the brain, including brainstem demyelination, disruption of neurofilaments, and neurofibrillary degeneration (Klatzo et al., 1965; Crapper and Dalton, 1973; Ebina et al., 1984). In the rat brain, aluminum has been shown to induce an increase in local cerebral glucose utilization, which is particularly pronounced in the structures of the limbic system, and a generalized inhibition of energy metabolism (Stefanovich and Joó, 1990).

a. Effects on BBB Permeability

There is a considerable literature advancing the view that high parenteral doses of aluminum in experimental animals open the BBB either to specific solutes or in a more general way. Aluminum has been most commonly given as aluminum chloride in a dose of 100 mg/kg body weight of elemental Al, intraperitoneally or intravenously (Banks and colleagues, several references below; Kim et al., 1986). If this dose were to completely dissolve in body fluids, it would represent 6.2 mmol $AlCl_3$ per liter body water (60 ml/100 g) or ~24.7 mOsm. In fact, $AlCl_3$ in the peritoneum or in blood plasma will hydrolyze to Al hydroxides, releasing a maximum of 18.5 mmol of hydrochloric acid per liter body water.

The evaluation of experiments using high doses of $AlCl_3$ must take account of both severe metabolic acidosis and possible hyperosmolality. In experiments in which a similar dose of either aluminum citrate or lactate is injected, some hyperosmolality without acidosis is to be anticipated.

Cerebral blood flow and blood volume, cerebral hemodynamics, and brain blood hematocrit do not seem to be affected by treatment with aluminum (Banks and Kastin, 1989). However, the occurrence of aluminum deposits in the microvessels forming the BBB observed in patients affected with dialysis dementia (Perl et al., 1980) and in rats injected with aluminum systemically (Wen and Wisniewski, 1985) may suggest an alteration in membrane structure and function at the BBB level. The permeability of the blood-brain barrier to molecules which cross the BBB to a small degree such as neuropeptides (delta-sleep-inducing peptide and β-endorphin), sucrose, and other nonpeptides was altered in several studies where aluminum was administered systemically to rats and mice at doses ranging from 2 to 100 mg/kg of essential Al (Banks and Kastin, 1983, 1985a,b; Kim et al., 1986; Favarato et al., 1992). In these studies, the concentration of aluminum in blood was similar to that found in patients with dialysis dementia (Sideman and Manor, 1982). Acute and chronic administration of aluminum were also found to produce edema in the rat brain, as reflected by an increase in sodium and water content (Stefanovich and Joó, 1990).

The observed increases in the rate of entry of peptides and nonpeptides into the brain were reversible within 24 h, although the effect was highly dependent on the type of aluminum compound used (Kim et al., 1986; Stefanovich and Joó, 1990; Favarato et al., 1992). Although compounds with a marked effect on BBB permeability such as aluminum-gluconate give rise to higher levels of aluminum in blood than the lactate, chloride, or hydroxide forms (Leblondel and Allain, 1980), the physicochemical properties of the metal coordination sphere of the compound and the pH of the injected solution should also be taken into account, the most effective agent at modifying BBB permeability being the most lipophilic one. Aluminum maltolate injected at physiological pH has thus been shown to produce irreversible changes in BBB permeability to sucrose (Favarato et al., 1992).

The mechanism by which intravascular tracers may reach the brain extracellular space after aluminum administration is probably by enhanced transmembrane diffusion across the endothelium, and the increase in permeability induced by aluminum may be predicted by the lipid solubilities of the different aluminum compounds (Banks and Kastin, 1989). An increase in the permeability of the BBB induced by aluminum has also been confirmed *in vitro* using brain microvessel endothelial cell monolayers and sodium fluorescein as a tracer (Audus et al., 1988). Neither pinocytotic activity nor alterations in the membrane lipid order seem to contribute to the aluminum-induced alterations on BBB permeability.

There are contradictory reports on the effects of aluminum on the integrity of the blood-brain barrier, as assessed by extravasation of high molecular weight tracers. Banks and colleagues (1988, 1989) found no evidence for increased exudation of radiolabeled albumin after treatment with aluminum, while Stefanovich and Joó (1990) observed extravasation of Evans blue (albumin) in the whole brain. More recently, Vorbrodt et al. (1994) studied the effect of chronic exposure of mice to drinking water containing 0.06 M aluminum lactate on BBB function. Focal leakage of HRP and albumin was observed only in large vessels of some areas such as the basal ganglia and amygdaloid nuclei, while capillaries remained unaffected. The localization of the plasmalemma-bound enzymatic activities of alkaline phosphatase and Ca^{2+}-activated ATPase and the distribution of anionic sites on the luminal surface of the cerebral endothelium were also essentially unchanged. In view of these results, whether alteration of BBB function plays an important role in aluminum-induced neurotoxicity still remains to be determined.

In the study by Stefanovich and Joó (1990), aluminum-induced albumin leakage and the development of brain edema could be prevented by the simultaneous systemic administration of propentofylline, a neuroprotective xanthine derivative which acts as an adenosine uptake blocker. This compound also seems to reverse the inhibition of energy metabolism but not the stimulation of glucose utilization induced by aluminum. The protective mechanism of action of propentofylline is not known, but the authors suggested a modulation of the metabolism of cerebral endothelial cells by either increasing cerebral blood flow or attenuating the excitatory effect of aluminum. The administration of vitamin D analogues prevents the decrease in brain gangliosides observed in aluminum intoxication and induces perivascular astrocytes to sequester the metal from the extracellular space, which is associated with the development of resistance to aluminum neurotoxicity (Vukicevic et al., 1992).

b. Effect on Transport Systems Across the BBB

In addition to an increase in the rate of nonsaturable free diffusion across endothelial plasma membranes, aluminum may selectively alter saturable transport systems of substances such as 2-deoxyglucose, *N*-tyrosinated peptides, enkephalins, and thyroxine (Banks and Kastin, 1985a, 1987; Banks et al., 1988; Stefanovich and Joó, 1990). This may contribute, at least in part, to the mechanism of aluminum-induced encephalopathy.

3. Mechanism of Action

Although the basis for aluminum toxicity is unknown, it has been suggested that many of its effects on the CNS as well as on peripheral tissues can be explained by its actions as a membrane toxin (Banks and Kastin, 1989). In *in vitro* studies with artificial membranes, aluminum has been found to bind to both positively and negatively charged sites (Deleers, 1985; Tracey and Boivin, 1983) and to alter membrane structure by increasing rigidification and fusion of membranes (Deleers et al., 1986). However, evidence for this happening at the BBB is not conclusive (Vorbrodt et al., 1994).

On the other hand, the toxic effects of aluminum may also involve the displacement of essential metals with important functions for the normal metabolism of the CNS. Aluminum binds to transferrin and ferritin with possible consequences in iron metabolism (see previous pages), and also displaces calcium from artificial membranes (Deleers, 1985) and from calmodulin and acetylcholinesterase (Siegel et al., 1983; Marquis, 1983). In addition, aluminum may displace zinc (Sugawara et al., 1987), leading to alterations in the activity of zinc enzymes such as glutamate dehydrogenase, superoxide dismutase, carbonic anhydrase, lactate dehydrogenase, and those involved in the metabolism of DNA, neurotransmitters, neuropeptides, etc.

E. IRON
1. Introduction

Iron in the ferrous state has the capacity for catalyzing the formation of free radicals by the so-called Fenton reaction and hence of inducing or potentiating brain damage (Aust et al., 1993):

$$Fe^{2+} + H_2O_2 \rightarrow \text{intermediate complex} \rightarrow Fe^{3+} + OH^{\cdot} + OH^{-}$$

Normally iron in blood plasma and in extracellular fluid is very strongly complexed as Fe^{3+} with its binding protein transferrin, which is normally only about one-third saturated. Hence the concentration of low molecular weight iron in these fluids is negligible. Similarly, stored iron within cells, e.g., those of liver and spleen, is tightly bound within ferritin. An important function of transferrin and of ferritin is probably to keep iron in an innocuous form suitable for transport in the circulation and storage within cells, respectively. CSF contains little iron-binding capacity, together with low molecular weight iron, high levels of ascorbate, and a lack of the ferroxidase ceruloplasmin. Hence, excess iron in CSF may well be present as Fe^{2+} (Gutteridge, 1992). Since the composition of cerebral interstitial fluid is similar to that of CSF, the brain is at risk from generation of free radicals due to the Fenton reaction. Formation of OH· may secondarily increase cerebral damage after tissue injury, primarily due to trauma or ischemia (Halliwell and Gutteridge, 1989).

Alternatively, catalytically active iron may contribute to damage by other heavy metals such as lead by favoring the production of these highly reactive hydroxyl radicals (Braughler and Hall, 1989). Hemoglobin released from extravasated erythrocytes may form a good source of iron. There is ample evidence that superoxide radicals may be produced by brain vessels in different pathological states, such as acute hypertension, cerebral ischemia, seizures, and brain injury (Kontos, 1989). This has been confirmed separately in isolated cultured endothelial cells, vascular smooth muscle, and astrocytes (Chan et al., 1988; Steele et al., 1991; Terada et al., 1991).

2. Pathology

The toxic ion species, superoxide, hydrogen peroxide, and hydroxyl radicals, generated via the hypoxanthine/xanthine oxidase system, have been shown to relax cerebral vascular smooth muscle and induce pial and cerebral dilatation (Rosenblum, 1983; Kontos et al., 1984; Wei et al., 1985). When the application is at low concentration and short-lived, the relaxation is reversible. Application of higher doses for longer times, however, results in sustained changes characterized by relaxation which outlasts the application of the radical-generating agents, and is associated with morphological evidence of vascular damage. In transmission electron microscopy, the endothelial lesions consist of localized destructive damage to the luminal membrane or focal endothelial swelling. With scanning electron microscopy, the lesions appear as protrusions into the lumen or as craters. The density of the lesions is however very low, 0.6 to 0.8 lesion per 100 µm² of tissue section (Wei et al., 1985), and therefore difficult to detect. Histochemical evidence shows that these lesions are sites of production of superoxide (Povlishock et al., 1988). The lesions of the vascular smooth muscle are also focal and affect usually less than 5% of the smooth muscle cells in the vessel wall; they involve vacuolation, inclusion bodies, and, more rarely, necrosis.

a. Effects on BBB Permeability

There is varied susceptibility to the action of radicals on permeability, but penetrating arterioles seem to be particularly vulnerable (Wei et al., 1986). However, Del Maestro et al. (1981) showed marked extravasation of fluorescein-labeled dextran predominantly in postcapillary venules, accompanied by delayed petechial hemorrhages, in the hamster cheek pouch after application of the hypoxanthine/xanthine oxidase system. In contrast, Unterberg et al. (1988) did not favor the concept that oxygen-derived free radicals have a marked effect on normal cerebral vessels, since only moderate arterial dilatation and BBB dysfunction were obtained during superfusion with the hypoxanthine/xanthine oxidase system. They suggested that the findings reported by other authors were due to the hypertonicity of undialyzed superfusates. However, topical application of the hypoxanthine/xanthine oxidase system to the pial surface of the frog brain causes a reversible decrease in electrical resistance of the endothelium (Crone and Olesen, 1982), which was counteracted by the addition of allopurinol and superoxide dismutase + catalase (Olesen, 1987). Fraser and colleagues have recently reported similar effects of free radicals on pial venules of rats (Easton and Fraser, 1994).

A similar sequence of events has been shown to take place in cerebral vessels on application of arachidonic acid, which is responsible for the release of superoxide via the action of cyclooxygenase (Kontos et al., 1984, Wei et al., 1986). It was found that superoxide dismutase and catalase eliminated the arterial responses induced by topical application of arachidonic acid. This further supports the hypothesis of a direct action of free radicals on the cerebral vasculature.

b. Effects on Vascular Function

Cerebral arterioles which have undergone vascular injury from oxygen radicals display altered vasodilator and vasoconstrictor responses (Kontos, 1989). These include reduced vasodilator responses to hypercapnia and vasoconstrictor responses to hypocapnia, and disturbed autoregulatory responses to changes in blood pressure (Wei et al., 1985; Kontos, 1989). Dilatation of cerebral arterioles initiated by acetylcholine and nitric oxide can be reversed by free radicals generated by application of arachidonic acid or percussion brain injury (Marshall et al., 1988; Kontos and Wei, 1992). The elimination of endothelium-dependent dilatation is either due to inactivation of EDRF (nitric oxide or a nitric oxide-containing compound) or due to direct damage to the endothelium. Both effects are probably by mechanisms mediated by free radicals, involving the superoxide anion radical (Gryglewski et al., 1986) and/or the hydroxyl radical (Kontos and Wei, 1992). These abnormalities of vascular function are likely to lead to severe alteration in the ability of the brain to regulate and maintain an adequate blood flow and may lead to severe dysfunction.

F. CADMIUM

Cadmium has become of toxicological importance due to increasing levels in the environment and workplace as a result of industrial practices. Studies on adult rat brains have shown that the BBB is able to exclude cadmium from the CNS, resulting in low concentrations in the brain (Klaasen and Wong, 1982). Chronic exposure to 100 ppm Cd^{2+} in drinking water leads to degenerative changes in the choroid plexus of adult mice, characterized by loss of microvilli, rupturing of the apical surface, increased cytoplasmic vacuolation and cellular debris (Valois and Webster, 1989). However, capillary changes and hemorrhages have been described in spinal and cranial ganglia (Schlaepfer, 1971; Gabbiani et al., 1974). The primary sensory neurons of the olfactory mucosa have been identified as target cells in experimental intoxication by cadmium via inhalation and may represent an alternative route of entry of cadmium and other airborne toxicants into the brain (Hastings and Evans, 1991; Evans and Hastings, 1992).

Greater concentrations of cadmium are found in brains of newborn rats and mice than in adults, indicating that the barrier to cadmium is not fully developed in neonatal rodents as to other heavy metals (Klaasen and Wong, 1982; Valois and Webster, 1987). While the immature sensory ganglia remain unaffected after exposure to cadmium (Arvidson, 1983), pathological changes in the brain of newborn animals include hemorrhagic lesions, edema and cellular pycnosis in cerebrum and cerebellum (Gabbiani et al., 1967; Webster and Valois, 1981). Petechial hemorrhages are accompanied by thinning and vacuolation of the capillary walls and widening of interendothelial junctions, especially in partially differentiated capillaries, and precede degenerative changes in the brain cells (Webster and Valois, 1981). Prenatal exposure to cadmium has also been shown to induce changes in the cerebral vasculature of the fetus (Rohrer et al., 1978). Not all endothelial cells are equally susceptible to cadmium, and necrotized subpopulations of susceptible endothelial cells are replaced by proliferative resistant ones (Nolan and Shaikh, 1986).

The reasons for the variation of susceptibility of the cerebral vascular endothelium and for the differences between stages of development are unclear, but modulation of sensitivity or resistance to cadmium by steroid hormones has been suggested as a probable mechanism (Nolan and Shaikh, 1986). Cadmium also modulates the synthesis of proteins in type 1 but not type 2 endothelial cells cultured from bovine aorta (Hannan and McAuslan, 1982). Metallothionein may also be involved in the detoxification of cadmium by binding the metal ions, and differences in the content of the enzyme between different subpopulations of endothelial cells may reflect differences in sensitivity to the metal. Cadmium administration to adult rats induced metallothionein expression in endothelial cells and surrounding glia along with other specific cell types with high initial content of metallothionein such as ependymal cells and choroid plexus epithelium (Nishimura et al., 1992).

V. SUMMARY

The normal structure, function, and development of the blood-brain barrier are described. Possible interactions between this system and toxic metals are discussed.

There are a variety of mechanisms by which metal ions may cross the blood-brain barrier. Access to such mechanisms depends on the chemical state of the metal in blood plasma. The mechanisms themselves may include transport of the free ion, transport of a low molecular weight complex (often a metal–amino acid complex) or a process involving the metal and its specific binding protein in blood plasma.

The possible toxic effects of metals on components of the BBB are comprehensively reviewed. While lead may open the BBB at high doses in young children, as may also mercury and aluminum at such levels in experimental animals, numerous more subtle influences are possible on types of specific transport at the barrier and on the neural and glial mechanisms beyond.

REFERENCES

Abbott, N. J., Bundgaard, M., and Cserr, H. F. (1986), in *The Blood-Brain Barrier in Health and Disease,* Suckling, A. J., Rumsby, M. G., and Bradbury, M. W. B., Eds., Ellis Horwood, Chichester, 52–72.
Abbott, N. J., Bundgaard, M., and Hughes, C. C. W. (1990), *Progr. Appl. Microcirc.,* 16, 1–19.
Abbott, N. J. and Revest, P. A. (1991), *Cerebrovasc. Brain Metab. Revs.,* 3, 1–34.
Abbott, N. J., Revest, P. A., and Romero, I. A. (1992), *Neuropath. Appl. Neurobiol.,* 18, 424–433.
Adu, J., Bradbury, M. W. B., and Buxani, S. (1990), *J. Physiol. (Lond.)* 423, 40P.
Ahrens, F. A. (1993), *Am. J. Vet. Res.,* 54, 808–812.
Aihara, M. and Sharma, R. P. (1986), *Arch. Environ. Contam. Toxicol.,* 15, 629–636.
Aiken, S. P., Horn, N. M., and Saunders, N. R. (1992a), *Biometals,* 5, 235–243.
Aiken, S. P., Horn, N. M., and Saunders, N. R. (1992b), *J. Physiol. (Lond.),* 445, 69–80.
Al-Modhefer, A. J. A., Bradbury, M. W. B., and Simons, T. J. B. (1991), *Clin. Sci.,* 81, 823–829.
Albrecht, J., Talbot, M., Kimelberg, H. K., and Aschner, M. (1993), *Brain. Res.,* 607, 249–254.
Alfrey, A. C., Le Gendre, G. R., and Kaehny, W. D. (1976), *N. Engl. J. Med.,* 294, 184–188.
Arvidson, B. (1983), *Environ. Res.,* 32, 240–246.
Aschner, M. and Aschner, J. L. (1990a), *Brain Res. Bull.,* 24, 857–860.
Aschner, M. and Aschner, J. L (1990b), *Neurosci. Biobehav. Rev.,* 14, 169–176.
Aschner, M. and Clarkson, T. W. (1988), *Brain Res.,* 462, 31–39.
Aschner, M., Chen, R., and Kimelberg, H. K. (1991), *Brain. Res. Bull.,* 26, 639–642.
Aschner, M., Du Y. L., Gannon, M., and Kimelberg, H. K. (1993), *Brain. Res.,* 602, 181–186.
Aschner, M., Eberle, N. B., Miller, K., and Kimelberg, H. K. (1990), *Brain. Res.,* 530, 245–250.
Audus, K. L., Shinogle, J. A., Guillot, F. L., and Holthaus, S. R. (1988), *Int. J. Pharm.,* 45, 249–257.
Aust, S. D., Chignell, C. F., Bray, T. M., Kalyanaraman, B., and Mason, R. P. (1993), *Toxicol. Appl. Pharmacol.,* 120, 168–178.
Bakir, F., Damluji, S. F., Amin Zaki, L., Murtadha, M., Khalidi, A., Al Rawi, N. Y., Tikriti, S., Dhaahir, H. I., Clarkson, T. W., Smith, J. C. and Doherty, R. A. (1973), *Science,* 181, 230–241.
Banks, W. A. and Kastin, A. J. (1983), *Lancet,* ii, 1227–1229.
Banks, W. A. and Kastin, A. J. (1985a), *Neuropharmacology,* 24, 407–412.
Banks, W. A. and Kastin, A. J. (1985b), *Psychopharmacology,* 86, 84–89.
Banks, W. A. and Kastin, A. J. (1987), *Life Sci.,* 41, 1319–1338.
Banks, W. A. and Kastin, A. J. (1989), *Neurosci. Biobehav. Rev.,* 13, 47–53.
Banks, W. A., Kastin, A. J., and Fasold, M. B. (1988), *J. Pharmacol. Exp. Ther.,* 244, 579–585.
Bedi, K. S., Thomas, Y. M., Davies, C. A., and Dobbing, J. (1980), *J. Comp. Neurol.,* 193, 49–56.
Berlin, M. and Gibson, S. (1963), *Arch. Environ. Health,* 6, 617–625.
Betz, A. L., Iannotti, F., and Hoff, J. T. (1989), *Cerebrovasc. Metab. Rev.,* 1, 133–154.
Blackman, S. S. (1937), *Bull. Johns Hopkins Hosp.,* 61, 1–61.
Bobilya, D. J., Briske-Anderson, M., and Reeves, P. G. (1992), *J. Cell Physiol.,* 151, 1–7.
Bradbury, M. W. B., Ed., (1992), *Physiology and Pharmacology of the Blood-Brain Barrier,* Springer Verlag, Berlin.
Bradbury, M. W. B. *J. Physiol.,* 479.P, 37p.
Bradbury, M. W. B. and Deane, R. (1986), *Ann. N.Y. Acad. Sci.,* 481, 142–160.

Bradbury, M. W. B. and Deane, R. (1993), *Neurotoxicology,* 14, 131–136.
Braughler, J. M. and Hall, E. D. (1989), *Free Radic. Biol. Med.,* 6, 289–301.
Bressler, J. P. and Goldstein, G. W. (1991), *Biochem. Pharmacol.,* 41(4), 479–484.
Brewer, G. J. and Yuzbasiyan-Gurkan, V. (1992), *Medicine,* 71, 139–164.
Brightman, M. W. B. and Reese, T. S. (1969), *J. Cell Biol.,* 40, 648–677.
Brookes, N. (1988), *J. Neurochem.,* 50, 1117–1122.
Brookes, N. and Kristt, D. A. (1989), *J. Neurochem.,* 53, 1228–1237.
Butt, A. M., Jones, H. C., and Abbott, N. J. (1990), *J. Physiol.,* 429, 47–62.
Buxani-Rice, S. (1992), Characterization of ^{65}Zn Flux from Circulation into Brain in the Anaesthetized rat, Ph.D. thesis, University of London, London.
Buxani-Rice, S., Ueda, F., and Bradbury, M. W. B. (1994), *J Neurochem.,* 62, 665–672.
Campbell, J. B., Woolley, D. E., Vijayan, V. K., and Overmann, S. R. (1982), *Dev. Brain Res.,* 3, 595–612.
Caspers, M. L., Kwaiser, T. M., and Grammas, P. (1990), *Biochem. Pharmacol.,* 39, 1891–1895.
Cavanagh, J. B. and Chen, F. C. (1971), *Acta Neuropathol. (Berl.),* 19, 216–224.
Chan, P. H., Chen, S. F., and Yu, A. C. H. (1988), *J. Neurochem.,* 50, 1185–1193.
Choi, B. H., Cho, K. H., and Lapham, L. W. (1980), *Brain. Res.,* 202, 238–242.
Choi, B. H. and Kim, R. C. (1984). *Exp. Mol. Pathol.,* 41, 371–376.
Choi, B. H., Kudo, M., and Lapham, L. W. (1981), *Acta Neuropathol. (Berl.),* 54, 233–237.
Clasen, R. A., Hartman, J. F., Starr, A. J., Coogan, P. S., Pandolfi, S., Laing, I., Becker, R., and Hass, G. M. (1974), *Am. J. Pathol.,* 74, 215–240.
Cornford, E. M. (1985), *Mol. Physiol.,* 7, 219–260.
Cory-Slechta, D. A., Widzowski, D. V., and Pokora, M. J. (1993), *Neurotoxicology,* 14, 105–114.
Crapper, D. C. and Dalton, A. J. (1973), *Physiol. Behav.,* 10, 935–945.
Cremer, J. E. (1981), in *The Molecular Basis of Neuropathology,* Davison, A. N. and Thompson, R. H. S. Eds., Edward Arnold, London, 234–264.
Critchfield, J. W. and Keen, C. L. (1992), *Metabolism,* 41, 1087–1092.
Crone, C. and Olesen, S. P. (1982), *Brain Res.,* 241, 49–55.
Cserr, H. F. and Patlak, C. S. (1992), in *Physiology and Pharmacology of the Blood-Brain Barrier,* Bradbury, M. W. B., Ed., Springer Verlag, Berlin, 245–261.
Darwish, H. M., Cheney, J. C., Schmitt, R. C., and Ettinger, M. J. (1984), *Am. J. Physiol.,* 246, G72–G79.
Davidsson, L., Lonnerdal, B., Sandstrom, B., Kunz, C., and Keen, C. L. (1989), *J. Nutr.,* 119, 1461–1464.
Davson, H., Welch, K., and Segal, M. B. (1987), *The Physiology and Pathophysiology of the Cerebrospinal Fluid,* Churchill Livingston, Edinburgh.
Davson, H., Zlokovic, B., Rakic, L., and Segal, M. B. (1993), *An Introduction to the Blood-Brain Barrier,* Macmillan, London.
Deane, R. and Bradbury, M. W. B. (1990a), *J. Neurochem.,* 54, 905–914.
Deane, R. and Bradbury, M. W. B. (1990b), in *Pathophysiology of the Blood-Brain Barrier,* Johansson, B. B., Owman, C., and Widner, H. Eds., Elsevier, Amsterdam, 279–290.
Dehouck, M.-P., Méresse, S., Delorme, P., Fruchart, J.-C., and Cecchelli, R. (1990), *J. Neurochem.,* 54, 1798–1801.
Del Maestro, R. F., Björk, J., and Arfors, K. E. (1981), *Microvasc. Res.,* 22, 239–254.
Deleers, M. (1985), *Res. Commun. Chem. Pathol. Pharmacol.,* 49, 277–294.
Deleers, M., Servais, J. P., and Wulfert, E. (1986), *Biochim. Biophys. Acta,* 855, 271–276.
Deloncle, R. and Guillard, O. (1990), *Neurochem. Res.,* 15, 1239–1245.
Dziegielewska, K. M., Habgood, M. D., Møllgård, K., Staagard, M., and Saunders, N. R. (1991), *J. Physiol.,* 439, 215–237.
Easton, A. S. and Fraser, P. A. (1992), *J. Physiol.,* 446, 503P.
Easton, A. S. and Fraser, P. A. (1994), *J. Physiol.,* 475, 71P.
Ebina, Y., Okada, S., Hamazaki, S., and Midorikawa, O. (1984), *Toxicol. Appl. Pharmcol.,* 75, 211–218.
Eley, B. P., Gangolli, S. D., Hargreaves, R. J., Moorhouse, S. R., and Pelling, D. (1983), *J. Physiol.,* 342, 71P–72P.
Evans, J. and Hastings, L. (1992), *Fundam. Appl. Toxicol.,* 19, 275–278.
Farris, F. F., Dedrick, R. L., Allen, P. V., and Smith, J. C. (1993), *Toxicol. Appl. Pharmacol.,* 119, 74–90.
Fatemi, S. J. A., Kadir, F. H. A., and Moore, G. R. (1991), *Biochem. J.,* 280, 527–532.
Favarato, M., Zatta, P., Perazzolo, M., Fontana, L., and Nicolini, M. (1992), *Brain, Res.,* 569, 330–335.
Fishman, J. B., Rubin, J. B., Handrahan, J. V., Connor, J. R., and Fine, R. E. (1987), *J. Neurosci. Res.,* 18, 299–304.
Foote, J. W. and Delves, H. T. (1984), *J. Clin. Pathol.,* 37, 1050–1054.
Gabbiani, G., Bodonnel, M. C., Matthewson, S. M., and Ryan, G. B. (1974), *Lab. Invest.,* 30, 686–693.
Gabbiani, G., Gregory, A., and Baric, D. (1967), *J. Neuropathol. Exp. Neurol.,* 26, 498–501.
Gachot, B., Tauc, M., Morat, L. and Poujeol, P. (1991), *Eur. J. Physiol.,* 419, 583–587.
Gebhart, A. M. and Goldstein, G. W. (1988), *Toxicol. Appl. Pharmacol.,* 94, 191–206.
Giroux, E. L. and Henkin, R. I. (1974), *Biochim. Biophys. Acta,* 273, 64–72.
Goldstein, G. W. (1993), *Neurotoxicology,* 14, 97–101.
Goldstein, G. W., Asbury, A. K., and Diamond, I. (1974), *Arch. Neurol.,* 31, 382–389.
Goldstein, G. W., Wolinsky, J. S., and Gzejtey, J. (1977), *Ann. Neurol.,* 1, 235–239.

Greenwood, J. (1992), in *Physiology and Pharmacology of the Blood-Brain Barrier,* Bradbury, M. W. B., Ed., Springer Verlag, Berlin, 460–486.
Gryglewski, R. J., Palmer, R. M. J., and Moncada, S. (1986), *Nature,* 320, 454–456.
Gutteridge, J. M. C. (1992), *Clin. Sci.,* 82, 315–320.
Habermann, E., Crowell, K., and Janicki, P. (1983), *Arch. Toxicol.,* 54, 61–70.
Habgood, M. D., Knott, G. W., Dziegielewska, K. M., and Saunders, N. R. (1993), *J. Physiol.,* 468, 73–83.
Habgood, M. D., Sedgwick, J. E. C., Dziegielewska, K. M., and Saunders, N. R. (1992), *J. Physiol.,* 456, 181–192.
Halliwell, B. and Gutteridge, J. M. C. (1989), *Free Radicals in Biology and Medicine, 2nd ed.,* Clarendon Press, Oxford.
Hannan, G. N. and McAuslan, B. R. (1982), *J. Cell Physiol.,* 111, 207–212.
Hargreaves, R. J., Eley, B. P., Moorhouse, S. R., and Pelling, D. (1988), *J. Neurochem.,* 51, 1350–1355.
Hargreaves, R. J., Foster, J. R., Pelling, D., Moorhouse, S. R., Gangoli, S. D., and Rowland, I. R. (1985), *Neuropathol. Appl. Neurobiol.,* 11, 383–401.
Harris, D. I. M. and Sass-Kortsak, A. (1967), *J. Clin. Invest.,* 46, 659–677.
Harris, E. D. (1991), *Proc. Soc. Exp. Biol. Med.,* 196, 130–140.
Harris, W. H. and Sheldon, J. (1990), *Inorg. Chem.,* 29, 119–124.
Harris, W. R. and Keen, C. (1989), *J. Nutr.,* 119, 1677–1682.
Hartter, D. E. and Barnea, A. (1988), *J. Biol. Chem.,* 263, 799–805.
Hastings, L. and Evans, J. E. (1991), *Neurotoxicology,* 12, 707–714.
Hervonen, H. and Steinwall, O. (1984), *Acta Physiol. Scand.,* 121, 343–351.
Hirano, A. and Kochen, J. A. (1973), *Lab. Invest.,* 29, 659–668.
Hirota, K., Hirota, T., Aguilera, G., and Catt, K. J. (1985), *J. Biol. Chem.,* 260, 3243–3246.
Holash, J. A., Noden, D. M., and Stewart, P. A. (1993), *Dev. Dyn.,* 197, 14–25.
Holtzman, D., De Vries, C., Nguyen, H., Olson, J., and Bensch, K. (1984), *Neurotoxicology,* 5, 97–124.
Holtzman, D., Olson, J. E., Devries, C., and Bensch, K. (1987), *Toxicol. Appl. Pharmacol.,* 89, 211–225.
Hursh, J. B., Sichak, S. P., and Clarkson, T. W. (1988), *Pharmacol. Toxicol.,* 63, 266–273.
Jacobs, J. M., Carmichael, N., and Cavanagh, J. B. (1977), *Toxicol. Appl. Pharmacol.,* 39, 249–261.
Jakubowski, M., Piotrowski, J., and Trojanowska, B. (1970), *Toxicol. Appl. Pharmacol.,* 16, 743–753.
Janzer, R. C. and Raff, M. C. (1987), *Nature,* 325, 253–257.
Jefferies, W. A., Brandon, M. R., Hunt, S. V., Williams, A. F., Gatter, K. C., and Mason, D. Y. (1984), *Nature,* 312, 162–163.
Kataoka, M. and Tavassoli, M. (1985), *J. Ultrastruct. Res.,* 90, 194–202.
Katz, B. M. and Barnea, A. (1990), *J. Biol. Chem.,* 265, 2017–2021.
Kerper, L. E., Ballatori, N., and Clarkson, T. W. (1992), *Am. J. Physiol.,* 262, R761–R765.
Kim, K. S., Wass, C. A., Cross, A. S., and Opal, S. M. (1992), *Lymphokine Cytokine Res.,* 11, 293–298.
Kim, Y. S., Lee, M. H., and Wisniewski, H. M. (1986), *Brain. Res.,* 377, 286–291.
King, S. W., Savory, J., and Wills, M. R. (1981), *Crit. Rev. Clin. Lab. Sci.,* 14, 1–20.
Kinter, W. B. and Pritchard, J. B. (1977), in *Handbook of Physiology — Reactions to Environmental Agents,* Sect. 9, Lee, D.H.K., Ed., American Physiology Society, Baltimore, MD, 563–576.
Kitchen, I. (1993), *Neurotoxicology,* 14, 115–124.
Klaasen, C. D. and Wong, K. L. (1982), *Can. J. Physiol Pharmacol.,* 60, 1027–1036.
Klatzo, I. (1967), *J. Neuropathol. Exp. Neurol.,* 26, 1–14.
Klatzo, L., Wisniewski, H., and Streicher, E. (1965), *J. Neuropathol. Exp. Neurol.,* 24, 187–197.
Kolber, A. R., Krigman, M. R., and Morell, P. (1980), *Brain. Res.,* 192, 513–521.
Kontos, H. A. (1989), *Chem.-Biol. Interact.,* 72, 229–255.
Kontos, H. A. and Wei, E. P. (1992), *J. Neurotrauma,* 9, 349–354.
Kontos, H. A., Wei, E. P., Povlishock, J. T., and Christman, C. W. (1984), *Circ. Res.* 55, 295–303.
Kraft, A. S. and Anderson, W. B. (1983), *Nature,* 301, 621–623.
Krigman, M. R., Druse, M. J., Traylor, T. D., Wilson, M. H., Newell, L. R., and Hogan, E. L. (1974), *J. Neuropath. Exp. Neurol.,* 33, 671–686.
Lampert, P., Garro, F., and Pentschew, A. (1967), in *Brain Edema,* Klatzo, I. and Seitelberger, F. Eds., Springer-Verlag, Berlin, 207–222.
Laterra, J., Bressler, J. P., Indurti, R. R., Belloni-Olivi, L., and Goldstein, G. W. (1992), *Proc. Natl. Acad. Sci. U.S.A.,* 89, 10748–10752.
Lau, S.-J. and Sarkar, B. (1971), *J. Biol. Chem.,* 246, 5938–5943.
Leblondel, G. and Allain, P. (1980), *Res. Commun. Chem. Pathol. Pharmacol.,* 227, 579–586.
Lefauconnier, J.-M. (1992), in *Physiology and Pharmacology of the Blood-Brain Barrier,* Bradbury, M.W.B., Ed., Springer Verlag, Berlin, 117–150.
Lefauconnier, J. M., Hauw, J. J., and Bernard, G. (1983), *J. Neuropathol. Exp. Neurol.,* 42, 177–190.
Lefauconnier, J. M., Lavielle, E., Terrien, N., Bernard, G., and Fournier, E. (1980), *Toxicol. Appl. Pharmacol.,* 55, 467–476.
Lorenzo, A. Y. and Gewirtz, M. (1977), *Brain. Res.,* 132, 386–392.
Magneson, G. R., Puvathingal, J. M., and Ray, W. J. (1987), *J. Biol. Chem.* 262, 11140–11148.
Markovac, J. and Goldstein, G. W. (1988a), *Nature,* 334, 71–73.
Markovac, J. and Goldstein, G. W. (1988b), *Toxicol. Appl. Pharmacol.,* 96, 14–23.
Marquis, J. K. (1983), *Bull. Environ. Contam. Toxicol.,* 31, 164–169.

Marshall, J. J., Wei, E. P., and Kontos, H. A. (1988), *Am. J. Physiol.,* 255, H847–H854.
Martin, R. B. (1986), *Clin. Chem.,* 32, 1791–1806.
Martin, R. B., Savory, J., Brown, S., Bertholf, R. L., and Wills, M. R. (1987), *Clin. Chem.,* 33, 405–407.
Matthews, M. R., Parsons, P. J., and Carpenter, D. O. (1993), *Neurotoxicology,* 14, 283–290.
Matyja, E. and Albrecht, J. (1993), *Neurosci. Lett.,* 158, 155–158.
Maxwell, K., Vinters, H. V., Berliner, J. A., Bready, J. V., and Cancilla, P. A. (1986), *Toxicol. Appl. Pharmacol.,* 84, 389–399.
May, P. M., Linder, P. W., and Williams, D. R. (1977), *J. Chem. Soc. Dalton,* 588–595.
Michaelson, I. A. and Bradbury, M. W. B. (1982), *Biochem. Pharmacol.,* 31, 1881–1885.
Michison, J. and Palma, T. S. (1992), quoted in Bradbury, M.W.B. (1992), in *Physiology and Pharmacology of the Blood-Brain Barrier,* Bradbury M.W.B., Ed., Springer-Verlag, Berlin, Heidelberg, 269.
Moorhouse, S. R., Carden, S., Drewitt, P. N., Eley, B. P., Hargreaves, R. J., and Pelling, D. (1988), *Biochem. Pharmacol.,* 37, 4539–4547.
Morris, C. M., Candy, J. M., Court, J. A., Whitford, C. A., and Edwardson, J. A. (1987), *Biochem. Soc. Trans.,* 15, 498.
Morris, C. M., Candy, J. M., Oakley, A. E., Taylor, G. A., Mountfort, S., Bishop, H., Ward, M. K., Bloxham, C. A., and Edwardson, J. A. (1989), *J. Neurol. Sci.,* 94, 295–306.
Morris, C. M., Keith, A. B., Edwardson, J. A., and Pullen, R. G. L. (1992), *J. Neurochem.,* 59, 300–306.
Murphy, V. A. and Rapoport, S. I. (1992), *J. Neurochem.,* 58, 898–902.
Murphy, V. A., Wadhwani, K. C., Smith, Q. R., and Rapoport, S. I. (1991), *J. Neurochem.,* 57, 948–954.
Nagy, Z., Peters, H., and Hüttner, I. (1984), *Lab. Invest.,* 50, 313–322.
Needleman, H. L., Schell, A., Bellinger, D., Leviton, A., and Allred, E. N. (1990), *N. Eng. J. Med.,* 322, 83–88.
Neumann, P. Z. and Sass-Kortsak, A. (1967), *J. Clin. Invest.,* 46, 646–658.
Nishimura, N., Nishimura, H., Ghaffar, A., and Tohyama, C. (1992), *J. Histochem. Cytochem.,* 40, 309–315.
Nobles, M., Revest, P. A., Couraud, P.-O., and Abbott, N. J. (1995), *Brit. J. Pharmacol.,* 115, 1245–1252.
Nolan, C. V. and Shaikh, Z. A. (1986), *Life Sci.,* 39, 1403–1409.
Nunez, M.-T., Gaete, V., Watkins, J. A., and Glass, J. (1990), *J. Biol. Chem.,* 265, 6688–6692.
Olesen, S. P. (1987), *Acta Physiol. Scand.,* 129, 181–187.
Olesen, S. P. (1989), *Acta Physiol. Scand.,* 136 (suppl. 579) 1–28.
Pardridge, W. M. (1976), *J. Neurochem.,* 27, 333–335.
Pardridge, W. M., Connor, J. D., and Crawford, I. L. (1975), *Crit. Rev. Toxicol.,* 3, 159–199.
Patlak, C. S., Blasberg, R. G., and Fenstermacher, J. D. (1983), *J. Cerebr. Blood Flow Metab.,* 3, 1–7.
Peckham, N. H. and Choi, B. H. (1986), *Exp. Mol. Pathol.,* 44, 230–234.
Pelling, D., Hargreaves, R. J., and Moorhouse, S. R. (1989), in *Lead Exposure and Child Development. An International Assessment,* Smith, M. A., Grand, L. D., and Sors, A. I., Eds., Kluwer Academic, London.
Pentschew, A. (1965), *Acta Neuropathol.,* 5, 133–160.
Pentschew, A. and Garro, F. (1966), *Acta Neuropathol. (Berl.),* 6, 266–278.
Perl, D. P., Gajdusek, D. C., Garruto, R. M., Yanagihara, R. T., and Gibbs, C. J. (1980), *Acta Neuropathol. (Berl.),* 50, 19–24.
Povlishock, J. T., Wei, E. P., and Kontos, H. A. (1988), *FASEB J.,* 2, A–835.
Pratt, O. E. (1992), in *Physiology and Pharmacology of the Blood-Brain Barrier,* Bradbury, M. W. B., Ed., Springer Verlag, Berlin, 205–220.
Press, M. (1977), *Acta Neuropathol. (Berl.),* 40, 259–268.
Press, M. (1985), *Acta Neuropathol. (Berl.),* 67, 86–95.
Pullen, R. G. L., Candy, J. M., Morris, C. M., Taylor, G., Keith, A. B., and Edwardson, J. A. (1990), *J. Neurochem.,* 55, 251–259.
Pullen, R. G. L. and Franklin, P. A. (1992), *J. Physiol.,* 446, 507P.
Pullen, R. G. L., Franklin, P. A., and Hall, G. H. (1991), *J. Neurochem.,* 56, 485–489.
Quagliarello, V. J., Wispelwey, B., Long, W. J., and Scheld, W. M. (1991), *J. Clin. Invest.,* 87, 1360–1366.
Rabin, O., Hegedus, L., Bourre, J.-M., and Smith, Q. R. (1993), *J. Neurochem.,* 61, 509–517.
Radunovič, A., Transport of aluminium compared with 67-6a and 59-Fe into brain and other tissues of the young rat, Ph.D. Thesis, University of London, London.
Raffaniello, R. D., Lee, S. Y., Teichberg, S., and Wapnir, R. A. (1992), *J. Cell Physiol.,* 152, 356–361.
Revest, P. A., Abbott, N. J., and Gillespie, J. I. (1991), *Brain Res.,* 549, 159–161.
Richardt, G., Federolf, G., and Habermann, E. (1986), *Biochem. Pharmacol.,* 35, 1331–1335.
Risau, W., Hallmann, R., Albrecht, U., and Henke-Fahle, S. (1986), *Dev. Biol.,* 117, 537–545.
Rohrer, S. R., Shaw, S. M., and Lamar, C. H. (1978), *Acta Neuropathol.,* 44, 147–149.
Ronnbäck, L. and Hansson, E. (1992), *Br. J. Ind. Med.,* 49, 233–240.
Rosenblum, W. I. (1983), *Am. J. Physiol.,* 245, H139–H142.
Rubin, L. L. (1992), *Curr. Opinion Cell Biol.,* 4, 830–833.
Rubin, L. L., Hall, D. E., Porter, S., Barbu, K., Cannon, C., Horner, H. C., Janatpour, M., Liaw, C. W., Manning, K., Morales, J., Tanner, L. I., Tomaselli, K. J., and Bard, F. (1991), *J. Cell Biol.,* 115, 1725–1735.
Rutten, M. J., Hoover, R. L., and Karnovsky, M. J. (1987), *Brain Res.,* 425, 301–310.
Saunders, N. R. (1992), in *Physiology and Pharmacology of the Blood-Brain Barrier,* Bradbury, M. W. B., Ed., Springer Verlag, Berlin, 327–369.
Schlaepher, W. W. (1971), *Lab. Invest.,* 25, 556–564.

Schulze, C. and Firth, J. A. (1992), *Dev. Brain Res.,* 69, 85–95.
Sideman, S. and Manor, D. (1982), *Nephron,* 31, 1–10.
Siegel, N., Coughlin, R., and Haug, A. (1983), *Biochem. Biophys. Res. Commun.,* 115, 512–517.
Silbergeld, E. K., Wolinsky, J. S., and Goldstein, G. W. (1980), *Brain. Res.,* 189, 369–376.
Simons, T. J. B. (1986), *J. Physiol. (Lond.),* 378, 267–286.
Simons, T. J. B. and Pocock, G. (1987), *J. Neurochem.,* 48, 383–389.
Small, R. K., Watkins, B. A., Munro, R. M., and Liu, D. (1993), *Glia,* 7, 158–169.
Steele, J. A., Stockbridge, N., Maljkovic, G., and Weir, B. (1991), *Circ. Res.,* 68, 416–423.
Stefanovich, V. and Joó, F. (1990), *Metab. Brain Dis.,* 5, 7–17.
Steinwall, O. and Klatzo, I. (1966), *J. Neuropathol. Exp. Neurol.,* 25, 542–549.
Steinwall, O. and Olsson, Y. (1969), *Acta Neurol. Scand.,* 45, 351–361.
Strahan, M. E., Crowe, A., and Morgan, E. H. (1992), *Am. J. Physiol.,* 263, R924–R929.
Stumpf, W. E., Sar, M., and Grant, L. D. (1980), *Neurotoxicology,* 1, 593–606.
Suarez, N. and Eriksson, H. (1993), *J. Neurochem.,* 61, 127–131.
Sugawara, C., Sugawara, N., Ikeda, N., Okawa, H., Okazaki, T., Otaki, J., Taguchi, K., Yokokawa, K., and Miyake, H. (1987), *Drug Chem. Toxicol.,* 10, 195–207.
Sundström, R., Conradi, N. G., and Sourander, P. (1984), *Acta Neuropathol. (Berl.),* 62, 276–283.
Sundström, R., Müntzing, K., Kalimo, H., and Sourander, P. (1985), *Acta Neuropathol. (Berl.),* 68, 1–9.
Szumanska, G., Gadamski, R., and Albrecht, J. (1993), *Acta Neuropathol. (Berl.),* 86, 65–70.
Tacnet, F., Watkins, D. W., and Ripoche, P. (1990), *Biochim. Biophys. Acta,* 1024, 323–330.
Takeuchi, T., Morikawa, N., Matsumoto, H., and Shiraishi, Y. 1962, *Acta Neuropathol. (Berl.),* 2, 40–57.
Tao-Cheng, J.-H., Nagy, Z., and Brightman, M. W. (1987), *J. Neurosci.,* 7, 3293–3299.
Taylor, E. M. (1993), Transfer of Iron across Cellular Barriers, Ph.D. thesis, University of London, London.
Taylor, E. M. and Morgan, E. H. (1990), *Dev. Brain Res.,* 55, 35–42.
Taylor, J. A., and Simons, T. J. B. (1994), *J. Physiol. (Lond.),* 474, 55–64.
Terada, L. S., Willingham, I. R., Rosandich, M. E., Leff, J. A., Kindt, G. W., and Repine, J. E. (1991), *Cell. Physiol.,* 148, 191–196.
Toews, A. D., Kolber, A., Hayward, J., Krigman, M. R., and Morrell, P. (1978), *Brain Res.,* 147, 131–138.
Tonsch, U. and Bauer, H. C. (1991), *Brain. Res.,* 539, 247–253.
Tout, S., Chan-Ling, T., Hollander, H., and Stone, J. (1993), *Neuroscience,* 55, 291–301.
Tracey, A. S. and Boivin, T. L. (1983), *J. Am. Chem. Soc.,* 105, 4901–4905.
Ueda, F., Raja, K. B., Simpson, R. J., Trowbridge, I. S., and Bradbury, M. W. B. (1993), *J. Neurochem.,* 60, 106–113.
Unterberg, A., Wahl, M., and Baethmann, A. (1988), *Acta Neuropathol.,* 76, 238–244.
Valois, A. A. and Webster, W. S. (1987), *Toxicology,* 46, 43–55.
Valois, A. A. and Webster, W. S. (1989), *Toxicology,* 55, 193–205.
Vistica, D. T. and Ahrens, F. A. (1977), *Exp. Mol. Pathol.,* 26, 139–154.
Vorbrodt, A. W., Dobrogowska, D. H., and Lossinsky, A. S. (1994), *J. Histochem. Cytochem.,* 42, 203–212.
Vukicevic, S., Kracun, I., Vukelic, Z., Krempien, B., Rosner, H., and Cosovic, C. (1992), *Neurochem. Int.,* 20, 391–399.
Ware, R. A., Chang, L. W., and Burkholder, P. M. (1974), *Acta Neuropathol. (Berl.),* 30, 211–224.
Warfvinger, K., Hua, J., and Berlin, M. (1992), *Toxicol. Appl. Pharmacol.,* 117, 46–52.
Webster, W. S. and Valois, A. A. (1981), *J. Neuropathol. Exp. Neurol.,* 40, 247–257.
Wei, E. P., Christman, C. W., Kontos, H. A., and Povlishock, J. T. (1985), *Am. J. Physiol.,* 248, H157–H162.
Wei, E. P., Ellison, M. D., Kontos, H. A., and Povlishock, J. T. (1986), *Am. J. Physiol.,* 251, H693–H699.
Wen, G. Y. and Wisniewski, H. M. (1985), *Acta Neuropathol. (Berl.),* 68, 175–184.
Winder, C., Garten, L. L., and Lewis, P. D. (1983), *Neuropath. Appl. Neurobiol.,* 9, 87–108.
Winneke, G. (1986), in, *The Lead Debate: The Environment, Toxicology and Child Death,* Lansdown, R. and Yule, W., Eds., Croom Helm, London, 217–234.

Chapter 

Astrocytes: Potential Modulators of Heavy Metal-Induced Neurotoxicity

Michael Aschner and Harold K. Kimelberg

I. INTRODUCTION

Metals are inextricably bound into many facets of modern human existence. While some are biologically essential (zinc, copper, vanadium, manganese), others are not and are actually toxic (lead, cadmium, mercury). For the environmental health sciences, positioned at the juncture of toxicology and social health regulation, the ultimate question is the estimation of risk. This is often based on extrapolation from the dose-response curve. However, with the exception of catastrophic accidents or unique circumstances of occupational exposure, it is chronic exposure at the lower end of the dose-response curve that poses the greatest threat to mankind and the greatest challenge to toxicologists. Progress in neurotoxicology, like any other science, is based on the pillars of new developments in experimental methodologies; these have been critical to the rapid advances in the understanding of nervous system physiology and pathology. In the field of neurotoxicology one significant example of such advances is the development of methodologies for the *in vitro* culturing of a variety of CNS-derived cells, including astrocytes.

A concept unique to the nervous system is that its functions are overwhelmingly due to the properties of its electrically excitable cells, the neurons. However, nonexcitable cells comprise the majority of cells in the nervous system. The major class of nonexcitable, nonneuronal cells are collectively referred to as the neuroglia, and are comprised of astrocytes, oligodendrocytes, and microglia. While all subserve crucial functions, this review will focus only on the astrocytes, addressing persistent issues of selected heavy metal toxicity from the vantage point of the biochemical and cellular mechanisms that may potentially contribute to astrocyte-mediated neurotoxic effects.

This chapter starts with a brief review of recent research on the function of astrocytes (summarized in Table 1) in homeostatic mechanisms in the CNS, such as the maintenance of normal extracellular ion concentrations, the uptake of K^+, and the control of extracellular pH. Astrocytes in primary culture have been found to contain a number of receptors and also some of the uptake systems for CNS transmitters, properties which were formerly thought to be exclusively neuronal. The uptake of transmitters, usually by Na^+-dependent mechanisms, and their subsequent inactivation by metabolism are then briefly described, and possible functions for the receptors are surveyed (summarized in Table 2). Important roles of astrocytes during early brain development, especially in neuronal migration and the production of neurotrophic factors important for neuronal division and differentiation, as well as the potential role of astrocytes in inflammatory or immunological responses in the brain are briefly discussed. The role of astrocytes in brain pathology emphasizing the gliotic and swelling responses is also described. The information currently available on how heavy metals, specifically lead (Pb), mercury (Hg), and

Table 1 Current Views of Properties and Roles of Astrocytes

1. Development
   a. Neuronal and axonal guidance and migration in development, especially associated with radial glia which then develop into astrocytes
   b. Influence synaptogenesis and neuronal development and survial
   c. Induction of tight junctions between endothelial cells (blood-brain barrier)
2. Ion and pH homeostasis involving voltage and ligand gated ion channels and HCO_3^- and $Na^+ - H^+$ carrier systems
3. Neurotransmitter uptake and homeostasis
4. Receptors for neurotansmitters
5. Release of neurotransmitters and synthesis of neuropeptides
6. Immune responses — Release of immune system signals, response to interleukins, and possible antigen presentation
7. Phagocytosis associated with a number of lysosomal hydrolyses
8. Compartmentation of neurons, nonmyelinated axons, dendrites, and synapses (glomeruli)
9. Glycogen storage and metabolic interactions, including supply of substrates to neurons and other CNS cells (endothelial, microglial, oligodendrocytes, etc.)
10. Pathology
    a. Swelling of astrocytes associated with trauma, ischemia, and hepatic encephalopathy
    b. Structural cellular injury leads to astrogliosis or glial scars formed by reactive astrocytes
    c. Environmentally induced neuropathies, e.g., Parkinsonism via an MPTP, via effects on radial glia (Hg, Pb), or direct effects on mature astrocytes
    d. Role in growth or transplantation of neural or nonneural tissue to the brain and regeneration of damaged neurons
    e. Epilepsy and psychiatric disorders because of involvement with transmitter functions and neuronal excitability
    f. Protective or promotive roles in the aging brain and degenerative diseases and multiple sclerosis

manganese (Mn) interfere with the above-mentioned astrocytic functions in the development and maintenance of brain function are then reviewed.

II. HOMEOSTATIC FUNCTIONS OF ASTROCYTES

A. ION CHANNELS

The control of extracellular K^+ was one of the earliest physiological functions attributed to astrocytes, and studies associated with it led to considerable information about the K^+ transporting systems in these cells. These include different K^+ channels, carrier systems for K^+, and the Na^+/K^+ pump. Astrocytic $[K^+]_o$ buffering plays a vital role in maintaining normal nerve function, but exactly how astrocytes buffer $[K^+]_o$ is unknown. Three models have been proposed: (1) spatial buffering of K^+ which involves diffusion of K^+ in an electrically coupled glial syncytium driven by differences in the electrical potential between different regions; (2) the related K^+ siphoning, specifically referring to it in retinal Müller cells where K^+ is siphoned into the vitreous humor; and (3) K^+ accumulation associated with passive influx of K^+, Cl^- and water (reviewed by Barres and Chun, 1990; Bevan et al., 1985; Walz, 1988). Although it is beyond the scope of this review to discuss the details of all these models, common basic principles to all three models include: reduction of locally increased $[K^+]_o$ by uptake of K^+ and subsequent spatial redistribution of K^+ ions. How sequestered K^+ is returned to the neuron is unknown.

K^+ channels are the most diverse ionic channel type in astrocytes. [This topic has been recently reviewed by Barres and Chun (1990), Duffy and MacVicar (1993), and Kimelberg et al. (1993).] A wide variety of K^+ channels have been found in astrocytes. These include an inward rectifying K^+ channel (K_{in}), a Ca^{2+}- dependent K^+ channel (K^+_{Ca}), delayed rectifying channels (K_d), and an inactivating potassium channel (K_a). K^+ channels sensitive to ATP have also been found in astrocytes. Some of these channels may be related to the K^+ spatial buffering attributed to astrocytes (as described above).

Voltage-dependent Na^+ channels, similar in their properties to those found in neurons where they are responsible for electrical excitability, are surprisingly also found in cultured primary astrocytes (Bevan et al. 1985; Bowman et al. 1984). Like their neuronal counterparts, some astrocytic Na^+ channels have been found to be sensitive to tetrodotoxin (TTX). However, there are both TTX-sensitive and relatively TTX-insensitive Na^+ channels which have different characteristics in terms of the depolarization required to activate them (Sontheimer 1991). Recent work (Sontheimer et al. 1992) has also identified that astrocytes from certain CNS regions, such as the spinal cord, have a very high density of Na^+ channels which have open probability at the resting membrane potential of these cells (approximately –70 mV). It was hypothesized that the Na^+ channels function in regulating Na^+ entry into the astrocytes to activate the Na^+/K^+ pump when active uptake of K^+ is required, such as when $[K^+]_o$ rises from its normal level

Table 2 Receptors in Cultured Astrocytes and *In Situ*, and Their Proposed Effects

Receptor Type		Response/Functional Effects
1. Adrenergic and Other Aminergic Receptors in Cultured Astrocytes		
α1	(i)	Receptor activation leads to increased intracellular Ca^{2+} levels and inositol polyphosphate (IP) turnover
	(ii)	Inhibits increases in cAMP
	(iii)	Functions to regulate Ca^{2+} concentration and Ca^{2+} waves within the astroglial syncytium
β2	(i)	Receptor activation leads to increased intracellular cAMP
	(ii)	Functions to regulate: glycogen metabolism, taurine and NGF release, membrane potential, early response genes, and astrocytic morphology
5-HT	(i)	Receptor activation leads to increased second messenger levels (increased Ca^{2+} and increased K^+ conductance)
	(ii)	Functions to increase glycogen metabolism, regulate serotonergic neuronal growth, and astrocytic gene expression and/or stability
2. Amino acid and Peptide Receptors		
$GABA_A$	(i)	Membrane depolarization
	(ii)	Receptor activation leads to efflux of Cl^-, and HCO_3^-
	(iii)	Functions to regulate ion concentrations and pH in the vicinity of active neurons
$GABA_B$	(i)	Membrane hyperpolarization
	(ii)	Ionic basis of response unclear, possibly increased K^+ conductance
	(iii)	Inhibits agonist-evoked Ca^{2+} fluxes, IP metabolism, and eicosanoid release
Glutamate — ionotropic	(i)	Membrane depolarization in response to glutamate, alpha-amino-3-hydroxy-5-methyl-isoxazole (AMPA), and kainate (KA), but not *n*-methyl-D-aspartate (NMDA)
	(ii)	Receptor activation leads to influx of Na^+ and Ca^{2+} and efflux of K^+, Cl^- and HCO_3^-
	(iii)	Functions to regulate ion concentrations at nodal regions of axons, or reversal of GABA transport system
Glutamate — metabotropic	(i)	Functions to regulate extracellular ion concentrations
	(ii)	Astrocyte-derived release factor (ADRF) release
	(iii)	Inhibits proliferation and induces filopodia formation
3. Peptide Receptors		
Natriuretic	(i)	Receptor activation leads to increased cGMP and guanylate cyclase but functions unclear
Angiotensin II	(i)	Functions to release plasminogen activator inhibitor (PAI)
Endothelins		
Bradykinins	(ii)	Functions in the release of prostaglandins
Substance P	(iii)	Potentiates NE-induced increases in cAMP
VIP	(i)	Functions to stimulate glycogenolysis, mitogenesis, and increases neuronal survival (via secretion of neurotrophic factors)
	(ii)	Functions to increase intracellular cAMP
Opioid [delta subtype]	(i)	Functions in glycogen metabolism and suppresses astrocytic DNA synthesis
Substance P	(i)	Receptor activation leads to increased IP turnover
	(ii)	Functions in the release of prostaglandins, and potentiates norepinephrine-induced increases in cAMP
Bradykynin	(i)	Functions in the release of prostaglandins
	(ii)	Receptor activation leads to increased IP turnover

of 3 m*M* to 5 to 10 m*M* during periods of sustained neuronal activity. It is postulated that these Na^+ channels represent an autoregulating mechanism for active K^+ clearance by astrocytes that does not require any special properties of the Na^+/K^+ pump (Sontheimer, 1991; Sontheimer et al. 1992).

Astrocytes, at least those grown in primary cultures, also contain voltage-gated L-type Ca^{2+} channels (McVicar, 1984). This finding was surprising because the prevailing thought at the time was that such channels were exclusive to neurons and were responsible for such properties as Ca^{2+} action potential and the depolarization-induced Ca^{2+} influx at nerve terminals required for transmitter exocytosis. Although a topic of wide speculation, the function of these Ca^{2+} channels in astrocytes has not yet been fully explained; the occurrence of large changes in $[Ca^{2+}]_i$ levels in astrocytes upon stimulation by receptor agonists or neurotransmitters (see below), such as glutamate, as well as mechanical stimulation and swelling, suggest that such channels are needed for Ca^{2+} to enter the cells under such conditions to raise $[Ca^{2+}]_i$ or replenish intracellular Ca^{2+} (Duffy and MacVicar, 1993). In addition, as in many other cell types, an array of regulatory processes in astrocytes requires the release of Ca^{2+} from intracellular stores.

B. ANION CHANNELS

A number of anion channels have been identified in astrocytes, including small conductance chloride channels (Cl_s), and a high conductance chloride channel (Cl_H) (Jalonen 1993; Kimelberg et al., 1993). These channels transport both Cl^- and HCO_3^- and are likely to be involved in the uptake of HCO_3^- or Cl^- when $[K^+]_o$ rises, or release of KCl and/or amino acids during volume regulation (see below). They can obviously contribute to the resting membrane potential since both Cl^- and HCO_3^- can be out of equilibrium with the membrane potential. However, they are often closed at resting membrane potential. They can be both chemically and voltage-gated, so they will contribute to current flow across the membrane and change the membrane potential.

1. Ion Carriers

Ion carriers are distinct from channels in that a synchronous movement of more than one ion always occurs, rather than the independent diffusional movement of a single ion down its electrochemical gradient characteristics of channels. An important ion carrier is the $Na^+/K^+/2Cl^-$ uptake system, utilized by cells for active uptake of Cl^- driven by the inward Na^+ gradient and involved in volume regulation and active absorption of Cl^-. Astrocytes, like many other cell types, are known to express this carrier in primary culture (Kimelberg et al., 1993). Intracellular Cl^- concentrations in astrocyte cultures have been found to be several-fold higher than expected from the electrochemical equilibrium. This high Cl^- concentration may serve, when required, to maintain extracellular Cl^- levels, or for the efflux of KCl during volume regulatory processes (Kimelberg and Frangakis, 1985).

C. NA+/K+ PUMP

Like all mammalian cells, astrocytes contain an active Na^+/K^+ pump. This pump functions to accumulate K^+ and extrude intracellular Na^+. A number of isoforms ($\alpha 1$, $\alpha 2$, and $\alpha 3$) comprise this pump. Whereas neurons exhibit all three isoforms, astrocytes express $\alpha 1$ or $\alpha 2$ or both, but not $\alpha 3$ (Sweadner, 1991). In terms of the kinetics of the different isoforms, there is evidence both for and against a specialized role of astrocytic Na^+/K^+ ATPase in uptake of K^+ (Sweadner, 1991). As with other Na^+ pumps, the astrocytic Na^+/K^+ seems likely to be driven mainly by intracellular Na^+. It has a high affinity for K^+ on the outside, and a midactivation level for Na^+ of about 10 mM on the inside.

D. pH CARRIER

Other carrier systems for Cl^- or Na^+ involve cotransport or exchange transport with pH equivalents, such as H^+, HCO_3^-, or OH^-. These carrier systems in astrocytes include the Na^+/H^+ and Cl^-/HCO_3^-, or OH^- exchangers, as well as a variety of electrogenic or nonelectrogenic cotransport systems of Na^+ plus $nHCO_3^-$ (where n can be one to three; Newman, 1991). Since astrocytes *in situ* can undergo large pH changes (such as in ischemia), often in the opposite direction to the extracellular pH, it has been postulated that the existence of these carriers on astrocytes renders them critically important in pH homeostasis in the CNS (Chesler and Chen, 1991). The operation of such pH transporting systems may also be important in astrocytic volume changes. For example, the simultaneous operation of the Na^+/H^+ and Cl^-/HCO_3^- exchangers, driven by intracellular hydration of CO_2 to H^+ and HCO_3^-, could lead to a net uptake of Na^+ and Cl^- with concomitant astrocytic swelling.

E. NEUROTRANSMITTER UPTAKE SYSTEMS

Uptake systems for a number of amino acid neurotransmitters, such as glutamate, glycine, taurine, and γ-aminobutyric acid (GABA) have been identified on the astrocytic membrane. These systems are Na^+-dependent and can also be electrogenic. There is considerable evidence for extremely active electrogenic uptake of glutamate in both cultured and acutely isolated astrocytes. Uptake *in situ* has been clearly shown by means of autoradiography and immunocytochemistry. Three members of a family of glutamate transporters have been recently identified, two of which are exclusively located on astrocytes (Kanner, 1993).

Uptake of a number of monamine transmitters has been reported in cultured astrocytes (Kimelberg et al., 1993). These uptake systems resemble their counterparts in nerve terminals in being both Na^+-dependent and inhibitable by specific inhibitors such as antidepressants (fluoxetine). Uptake systems for adenosine (Matz and Hertz, 1990), taurine (Holopainen et al., 1988; Lee et al., 1992; and Shain and Martin, 1990) and histamine (Huszti, 1990) have also been described. The relevance of these uptake systems to both astrocytic function and brain homeostasis has not yet been clearly established.

F. RECEPTORS FOR NEUROTRANSMITTERS

A number of neurotransmitter receptors have been localized on cultured primary astrocytes (see Table 2). β-Adrenergic and the ionotropic kainic acid/alpha-amino-3-hydroxy-5-methyl-isoxazole (AMPA) receptors have also been identified on astrocytes *in vivo*. Thus, it appears likely that astrocytes respond to the very same transmitters that at one time were thought to be exclusively located on post- or presynaptic neuronal membranes. The perisynaptic location of many astrocyte processes which form glial nets around neurons (Bruckner et al., 1993) puts both transmitter receptors and uptake systems in direct apposition to their sites of release (namely, presynaptic boutons on the neuronal soma or dendrites), and thus optimizes their potential for neurotransmission and neuromodulation. Indeed, electron microscopy studies combined with immunocytochemistry (Aoki, 1992) show that β_2 receptors are located on astrocyte processes close to the synaptic cleft, and Derouiche et al. (1993) have described glutamine-positive astrocyte processes directly next to, or even penetrating to a limited extent into, glutamatergic synapses.

The functional implications of the above mentioned receptors on the astrocytic membrane are still subject to speculation. These, of course, initially include the activation of second messenger systems, in turn leading to a variety of functional effects. What is known thus far for these receptors is summarized in Table 2. For example, the activation of the KA/AMPA glutamate receptor has been shown to lead to membrane potential depolarization and Na^+ and K^+ inward currents (Bowman and Kimelberg, 1987; Sontheimer et al., 1988). New studies have implicated this receptor as a glial-specific type of AMPA receptor, which can also transport Ca^{2+} (Muller et al., 1992).

While the minute-to-minute Ca^{2+} buffering in the cell at submicromolar basal Ca^{2+} levels is carried out by endoplasmic reticulum (Becker et al., 1980), local spikes of cytosolic Ca^{2+} (Cobbold and Rink, 1987) may reach μM levels, triggering mitochondrial influx of Ca^{2+} (Unitt et al., 1989). Several neurotransmitters (5-HT and glutamate) have been shown to induce Ca^{2+} spikes in astrocytes (van den Pol et al., 1992; Cornell-Bell and Finkbeiner, 1991; Charles et al., 1992; Charles et al., 1993). The Ca^{2+} spikes or intracellular Ca^{2+} oscillations are both spatially uniform, or show intracellular and intercellular spreading. The Ca^{2+} rise can begin in one part and spread as a wave through the rest of the cell. With high enough agonist concentrations, an astrocyte syncytium supports intercellular waves which propagate from cell to cell over relatively long distances, and it has been suggested that networks of astrocytes may constitute a long-range signaling system (Cornell-Bell and Finkbeiner, 1991; Charles et al., 1992; Charles et al., 1993). Endothelins, a relatively recently described family of vasoactive peptides, also have profound effects on $[Ca^{2+}]_i$ in cultured astrocytes (Goldman et al., 1991), inducing Ca^{2+} spiking within seconds, and maintaining increased $[Ca^{2+}]_i$ for minutes. Recent findings are also consistent with inositol triphosphate (IP_3) mediating the propagation of Ca^{2+} waves in astrocytes (Charles et al., 1993). This, at a minimum, will allow astrocytes to signal changes over a wide region of the brain much as a neuronal network might function. Because of this there have been suggestions that these Ca^{2+} waves are involved in some way in information processing (Van den Pol et al., 1992).

III. DEVELOPMENT AND IMMUNOLOGY

A. DEVELOPMENT AND NEUROTROPHIC FACTORS

Interactions between astrocytes and developing neurons have been found to be of the utmost importance in CNS development. In the immature CNS, neuronal cell body migration and axonal outgrowth occur in radial glia which later on lose their longitudinal orientation and are thought to mature into astrocytes (Rakic, 1990; Rakic and Siedman, 1973; Reichenbach, 1989). Neuronal migration is the basis of CNS pattern formation and is characterized by astrocyte-guided migration of nerve cells from the subventricular zone to their final destination. This process is perhaps best exemplified by the "outside-in" migration of cerebellar granule cells along radial fibers of the Bergmann glial cells seen both *in vivo* and *in vitro* (Hatten et al., 1990; Rakic, 1990). Astrocyte-neuron interactions were a subject of a detailed review, recently published by LoPachin and Aschner (1993). Some important features of these interactions are summarized below.

Astrocyte-neuron interactions are now recognized as being due to a number of adhesive and recognition molecules that are expressed by both cell types. Astrocyte-neuron cell-to-cell contact mediated by adhesion molecules is well exemplified by the migration of cerebellar granule cells along Bergmann fibers. Translocation of neuronal cell bodies requires the coordinated temporal and spatial expression of different adhesive molecules, e.g., N-CAM, astrotactin, and L1 (Chuong, 1990; Rakic, 1990; Stitt and Hatten, 1990). Similarly, neurite outgrowth along astrocytic cell surfaces and the extracellular matrix is characterized by a specific spatiotemporal elaboration of a number of adhesive molecules including L1, N-CAM, N-cadherin, and integrin-class extracellular matrix receptors (Smith et al., 1990). Moreover, adhesion molecule binding mediates a number of additional processes that are also critical for development and regeneration; nerve fiber fasciculation (e.g., fasciclins, cadherins), pathway cues for guidance and target connectivity (e.g., L2/HNK-1 carbohydrate epitope), demarcation of topographic boundaries between laminar neuronal assemblies (e.g., J1 antiadhesion molecules), astroglial differentiation (e.g., L1), regulation of intra- and extracellular ion composition (i.e., adhesion molecule on glia, AMOG), and nerve-target adhesion (e.g., N-CAM) (Gloor et al., 1990; Kruse et al., 1985; Schachner, 1991). Adhesion molecule interactions promote growth cone motility along astrocytic surfaces, and provide the neurite with directional cues and other pertinent information regarding the surrounding microenvironment. Processing of this information clearly requires complex signal transduction and intracellular integration. How this occurs is not currently known, although it has been shown that the cytoplasmic domains of certain adhesion molecules (e.g., integrins, N-CAM 180, L1) are linked to various cytoskeletal elements (Chamak et al., 1987; Pollerberg et al., 1987). In addition, specific binding-induced changes in these membrane-cytoskeleton linkage complexes might be responsible for alterations in cytoskeletal components that are the basis of cell-cell stabilization and growth cone extension (Chuong, 1990). *In vitro* evidence suggests that adhesion molecule binding influences second messenger turnover (i.e., inositol phosphates, Ca^{2+}), which might mediate appropriate changes in neurite metabolism and membrane ion channel function (Acheson and Thoenen, 1983; Acheson and Rutishauser, 1988; Schuch et al., 1989).

It is now clear that central neurons can modulate astrocyte differentiation and proliferation (Hatten, 1985; Gasser and Hatten, 1990), as well as ion channel and neurotransmitter receptor phenotypes (Sontheimer et al., 1992; Thio et al., 1993). Astrocytic-neuronal interactions are now also viewed as essential in determining the characteristic morphology of the neurons. For example, morphological features of mesencephalic neurons (e.g., branching and varicosities) are dependent upon whether co-cultured astroglia are prepared from mesencephalic tissue (homotopic origin) or from a different brain region (heterotopic origin) (Denis-Donini et al., 1984). Soluble factors from astrocyte cultures have also been shown to promote neurite extension by neurons (Miller et al., 1990). On the other hand, adult astroglial scar tissue inhibits neurite extension (Rudge and Silver, 1990)(see also Reactive Gliosis).

Astrocytes are also implicated in lineage promotion of oligodendrocytes, the myelin-producing cells of the CNS. When grown in fetal bovine serum (FBS), a role for astroglia in the initiation or maintenance of a myelinogenic state is predicted by comparing the differentiation of enriched oligodendroglia in subculture with their counterparts remaining on an astrocyte underlayer. Once separated, the oligodendrocyte lineage exhibits signs of impaired cytoskeletal progression and plasticity, exemplified by the prolonged retention of vimentin (indicating immaturity) and induced expression of glial-fibrillary acidic protein (GFAP) (Ingraham and McCarthy, 1989; Meyer et al., 1989). With respect to myelinogenic products, newly separated cells within the oligodendrocyte lineage fail to initiate or sustain production of the oligodendrocyte-specific differentiation marker, galactocerebroside (GalC) (Saneto and De Vellis, 1985; Ingraham and McCarthy, 1989). Coculture experiments suggest that GalC expression and the loss of vimentin, two events that normally coincide with terminal oligodendrocyte differentiation *in vivo* (Raff et al., 1984), require direct cell-cell contact with the astrocytes and the presence of fetal bovine serum (FBS) (Keilhauer et al., 1985; Aloisi et al., 1988).

Expression of myelin basic protein (MBP) by oligodendrocyte in culture requires the presence of exogenous insulin-like growth factor-I (IGF-1) or insulin at concentrations sufficient to cross-react with IGF-I receptors (McMorris et al., 1986). IGF-I is produced at maximal concentrations in the premyelinating CNS and in short-term cultures of embryonic astrocytes (Rotwein et al., 1988; Ballotti et al., 1987). Could a similar relationship between astrocytes and oligodendrocyte precursors be instrumental in triggering the early phases of remyelination? Perhaps, because after cuprizone-mediated demyelination in the adult mouse CNS (Komoly et al., 1991), the induction of IGF-1 messenger RNA levels and peptide production in astrocytes was localized to lesion foci, coinciding with increased expression of IGF-I receptors on oligodendrocyte precursors.

Other *in vitro* studies indicate that astrocytes can exert a major mitogenic influence on oligodendrocyte development. When O-2A cells are dissociated from optic nerve and placed into culture, they stop dividing and some cells prematurely differentiate into oligodendrocytes, unless provided with a sufficient number of astrocytes (Raff et al., 1985). It is postulated that astrocytes provide to the oligodendrocyte lineage the necessary driving mitogen for proliferation. Evidence suggests that this is platelet-derived growth factor (PDGF) (Noble et al., 1988).

Recent work *in vitro* has now indicated that oligodendrocyte survival may depend on astrocyte-derived factors. O4 antigen-bearing oligodendrocytes from dissociated cerebellum of postnatal rats are unable to survive in culture unless they are cocultured with astrocytes (Meier and Schachner, 1982). It has also been found that defined medium, conditioned by astrocytes, prolonged the survival of subcultured O-2A cells (Hunter and Bottenstein, 1990), suggesting release of a diffusible survival factor from the astrocytes into the environment. Likewise, long-term survival of cortical oligodendrocytes progenitors isolated directly from postnatal germinal cortex also requires a soluble factor, one that is supplied specifically in culture medium conditioned by type 1 astrocytes (Gard and Pfeiffer, 1991). Two growth factors selected by cultured astrocytes, IGF-I and PDGF, promote the short-term survival of O-2A progenitor cells (Barres et al., 1992). These data lead to speculation that oligodendrocyte development (and possibly maintenance) requires trophic support from astrocytes in a relationship subject to disruption under some conditions of dysmyelination. This may already be evident in the case of congenital, hypomyelinating mouse mutation, jimpy mice. Although mapped to a defective structural gene encoding the integral myelin proteolipid protein, the mutation is also characterized by accelerated death of premyelinating oligodendrocytes. Preliminary evidence obtained in culture suggests that other glial cells, perhaps astrocytes, in the jimpy mouse are defective in providing oligodendrocytes with a gliotrophic signal (Bartlett et al., 1988).

IV. BLOOD-BRAIN BARRIER

Astrocytic perivascular "end-feet" surround the brain capillaries. Lack of fenestrations between the endothelial cells distinguishes the brain capillaries (with the exception of those in the circumventricular organs) from those in peripheral tissues and form the basis of the passive impermeability of the blood-brain barrier (BBB). For the better part of this century it was believed that astrocytic foot-processes actually formed the BBB, since this was the most obvious distinguishing feature between the brain capillaries and those in the periphery. However, electron microscope studies in the 1950s, using electron-dense markers, showed that the barrier to the diffusion of these markers resided in tight occluding junctions (*Zonula occludens*) between the endothelial cells, but that there was free passage of such markers between the astrocytic "end feet" (Goldstein and Betz, 1986).

Recent work has now indicated that astrocytic "end-feet" processes may play an important role in the induction of the BBB, which occurs in late gestation or early postnatal life in most mammals, such as in humans and rats, respectively. Transplantation experiments showed that the formation of the BBB depended largely on the CNS environment since it formed in systemic capillaries growing into CNS tissue (Stewart and Wiley, 1981). Janzer and Raff (1987) showed that injection of primary astrocyte cultures into the rat anterior eye chamber or the chick chorioallantoic membrane induces a permeability barrier in the endothelial cells of the capillaries of these tissues that would otherwise lack such a barrier. Additional work has used cocultures of endothelial cells and astrocytes (Dehouck et al., 1990) to study whether astrocytes are responsible for the induction of tight junctional complexes between the endothelial cells. Some properties that are lost in cultured endothelial cells, such as the expression of gamma-glutamyl transpeptidase (γ-GT), a specific marker for endothelium, can be reinduced in the co-culturing system. The BBB is characterized by a high electrical resistance of around 2000 ohm $\times$ cm^2, indicative of a low conductance to even small ions (Rissau and Wolfburg, 1990). In the cocultures, a trans-filter resistance of 661 ohm $\times$ cm^2 was found. However, Dehouck et al. (1990) also reported that the endothelial cells, when grown alone, had a resistance of 416 $\pm$ 58 ohm $\times$ cm^2. The resistance of the astrocyte cultures alone was not reported by these authors. It therefore remains to be established whether the reconstructed barrier merely restricts solute permeability by virtue of the addition of an astrocyte culture to the other side of the filter (and thus additive resistance), or whether it truly represents an induced "tight" barrier in the endothelial barrier.

Another interesting feature of the BBB is the very high density of intermembranous particle assemblies in astrocytic membranes facing the blood capillaries (Landis and Reese, 1982). The exact nature of these systems is unknown, but it is thought that they might be K$^+$ channels, because a high K$^+$ conductance has been specifically found on these membranes. If they are K$^+$ channels, they should be involved in the spatial buffering concept or siphoning of K$^+$ from the neuropil into the blood (Newman, 1986). Since astrocytes are so often observed in contact with both axons and blood vessels, it has been proposed that a neuron to astrocyte to endothelial cell signaling mechanism initiates compensatory changes in brain microcirculation (Newman, 1986). The identity of the signal mediating these changes remains to be determined; nitric oxide is a possible candidate since astrocytes are now known to contain an inducible form of nitric oxide synthetase (Murphy, 1993).

V. IMMUNE AND INFLAMMATORY RESPONSES

Cells of the CNS constitutively express very low levels of antigens encoded for by major histocompatibility complex (MHC) genes whose products play a fundamental role in the induction and regulation of immune responses in the body. However, the long-standing view that the brain is insulated from the effects of the immune system is now being challenged (Fierz et al., 1985; Fontana et al., 1986; Schnyder et al., 1986).

A prominent feature of inflammatory and degenerative diseases of the CNS is the accumulation of macrophages, recruited from circulating blood monocytes or from the resident CNS macrophages (microglia). The signals leading to this accumulation are, as yet, poorly understood (reviewed by Benveniste, 1993). Nevertheless, astrocytes have been implicated as active participants in this process, in view of their ability to secrete an interleukin-3 (IL-3)-like factor which induces growth of cultured mouse peritoneal exudate cells (PEC) and brain tissue macrophages (Frei et al., 1987). Astrocytes also secrete granulocyte-macrophage colony-stimulating factor (GM-CSF), as shown by induction of colony formation in bone marrow cells and growth of FDC-P1 cells (Malpiero et al., 1990). GM-CSF is a cytokine necessary for growth and differentiation lesions. GM-CSF enhances a number of functional activities of mature macrophages, such as their phagocytic, cytotoxic, and microbicidal activities. GM-CSF produced locally by astrocytes may therefore provide an essential element for the recruitment and activation of macrophages. Malpiero et al. (1990) have also demonstrated the presence of the mRNA for GM-CSF in cultured astrocytes. It would appear, therefore, that after an initial penetration of T cells into the CNS, astrocytes can further support the intracerebral T cell activation process.

Astrocytes have also been proposed to function as antigen-presenting cells (APCs), i.e., those cells with the ability to present antigens to lymphocytes (Erb et al., 1986; Fontana et al., 1987). Study of the capacity of astrocytes to function as APCs has been shown using a myelin basic protein (MBP)-specific T lymphocyte line derived from mice or rats immunized with MBP in complete Freund's adjuvant. Astrocytes from Lewis rats cocultured with a syngenic, MBP-specific, Ia-restricted T cell line of Lewis rat origin, stimulated proliferation of these T cells. This process is antigen specific and restricted to the major histocompatibility complex (MHC)(Fontana et al., 1984). The astrocytes in such cocultures are induced by the preactivated T cells to express MHC type II molecules (also termed-Ia antigens) (Fontana et al., 1984). It has also been shown that interferon-γ (IFN-γ)-containing supernatants of lectin-stimulated spleen cells can induce murine astrocytes in culture to express Ia antigens (Hirsch et al., 1983), underscoring the dependence of astrocytes as APCs on the presence of Ia-inducing signals, such as IFN-γ. However, the validity of the studies depends on the absolute astrocyte purity of the cultures and the absence of microglia, which are very active APCs (Giulian and Baker, 1986).

VI. PATHOLOGICAL REACTIONS OF ASTROCYTES

A. REACTIVE GLIOSIS

Gliosis, also known as reactive gliosis or reactive astrocytosis, represents the major response of astrocytes to brain injury (Lindsay, 1986; Reier, 1986; Eng, 1987; Eng, 1988 Norton et al., 1992). Its hallmark is the accumulation of the intermediate filament protein — GFAP (glial fibrillary acidic protein)(Norton et al., 1992; Eng, 1988). By definition, therefore, gliosis is accompanied by increased expression of GFAP. Gliosis occurs in response to a number of etiologies, including, but not limited to, physical trauma, chemical, and vascular damage (O'Callaghan, 1993). It is a generic response triggered not only by damaged neurons but also by other cell types. For example, damaged oligodendroglial myelin sheaths are known to elicit reactive gliosis in the absence of accompanying neuronal damage (Smith et al., 1983). When damaged, astrocytes themselves may elicit gliosis, directly activating neighboring astrocytes (Takada et al., 1990). Gliosis occurs in astrocytes throughout the CNS, although in general, it is more extensive in the white matter than in the gray matter (reviewed by Kimelberg and Norenberg, 1993).

Reactive gliosis occurs as an early feature of brain injury, and is normally detected within the first 24 hours. In rats, a peak response is reached within 3 to 4 days (Amaducci et al., 1981), subsiding over the succeeding 14 to 21 days. The degree as well as the reversibility of the response are closely correlated with the intensity of the injury. Thus, in mild injuries, scar tissue is localized predominately in the vicinity of the injury and resolves within weeks (Streit and Kreutzberg, 1988; Petito et al., 1990), whereas in severe injuries, gliosis is permanent and widespread (Carbonell and Boya, 1988; Petito et al., 1990). It has been generally assumed that reactive gliosis is a property unique to mature and fully differentiated astrocytes (Barrett et al., 1984) and astrocytes in the developing animal were generally believed to be nonresponsive. In fact, gliosis was believed to be absent during the early stages of development (Osterberg and Wattenberg, 1963; Bignami and Dahl, 1976). It is now evident, however, that prenatal damage to the CNS can result in enhanced expression of GFAP in a variety of experimental conditions, followed by GFAP abatement to normal levels with time (reviewed by O'Callaghan, 1993).

In injuries where the BBB remains intact, the enhanced expression of GFAP in astrocytosis seems to be due to astrocytic hypertrophy and not hyperplasia. Most recent studies with [^{3}H]-thymidine autoradiography combined with GFAP immunocytochemistry indicate that a very small number of astrocytes, perhaps as few as 1%, divide in the process of reactive gliosis (reviewed by O'Callaghan, 1993). However, hyperplastic responses (Cavanagh, 1970; Latov et al., 1979; Aldskogius, 1982) often occur where the integrity of the BBB is compromised, allowing for the entry of blood-borne factors, mitogens, and other agents (e.g., PDGF, thrombin, fibronectin) into the CNS. In keeping with enhanced proliferative activity are the findings of the presence of the early response genes, c-fos (Arenander et al., 1989; Dragunow and Robertson, 1989; Dragunow and Robertson, 1990) and ras p21 (Charman et al., 1988) in reactive gliosis, with breaching of the BBB.

Characteristics of reactive gliosis include increases in astrocytic nuclear diameter, elevated DNA levels, accumulation of intermediate filaments, elevated oxidoreductive enzyme activity, and increased GFAP and vimentin. Increased glycogenolysis, increased numbers of mitochondria, endoplasmic reticulum, lysosomes, microtubules, dense bodies, and lipofuscin pigments are also seen (Maxwell and Gruger, 1965; Vaughn and Pease, 1970; Nathaniel and Nathaniel, 1981; Petito and Babiak, 1982). Additionally, an increase in gap-junctions has been documented (Nathaniel and Nathaniel, 1981; Lafarga et al., 1991), perhaps facilitating the spread of signals or substrates through an astrocytic network.

As noted above, one of the most striking findings in reactive gliosis is an increase in astrocytic GFAP (reviewed by Eng, 1987; Eng, 1988; Kimelberg and Norenberg, 1993). Increased GFAP is noted as early as 30 min after an injury (Amaducci et al., 1981), and GFAP mRNA expression is elevated within 6 h of injury, peaking at 1 to 3 days (Cavicchioli et al., 1988; Rataboul et al., 1988; Condorelli et al., 1990). Increments in mRNA are not confined solely to the lesion site, and can be encountered quite some distance from the site of injury.

Levels of a second intermediate filament, vimentin, are also increased in reactive gliosis (Pixley and de Vellis, 1984). Vimentin is normally found in cells of mesenchymal origin (Dahl et al., 1981). Although widely expressed in developing astrocytes, it disappears upon cell maturation (Pixley and de Vellis, 1984). Upon injury, astrocytes regain the capacity to synthesize vimentin, providing a sensitive marker for reactive gliosis (Schiffer et al, 1986; Petito et al., 1990).

Gliosis is also associated with elevated oxidoreductive enzyme activity, presumably reflective of an enhanced metabolic activity of reactive astrocytes. While a controversial issue has been whether glutamine synthetase is uniformly increased in gliosis (Norenberg, 1983, Sandberg et al., 1985; Condorelli et al., 1988; Politis, 1989; Petito et al., 1992) it is generally agreed that the activities of Ca^{2+}-ATPase (Kawai et al., 1989), glutathione-S-transferase (Cammer et al., 1990), LDH (Politis, 1989), and lysosomal enzymes (Vijayan and Cotman, 1983) are increased in reactive astrocytes. Increased levels of glycogen are also seen (Shimizu and Hamuro, 1958).

Additional changes commonly encountered in reactive gliosis include increased levels of M1 antigen (Schachner, 1982), membrane assemblies (Anders and Brightman, 1979; Landis and Reese, 1981), S-100 protein (Griffin et al., 1989), laminin (Liesi et al., 1984), G_{D3} ganglioside (Seyfried and Yu, 1985), epidermal growth factor (EGF) receptors (Nieto-Sampedro et al., 1988), beta-amyloid precursor protein (Siman et al., 1989), prion protein (Gonzales et al., 1988), lipocortin-1 (Johnson et al., 1989), and peripheral-type benzodiazepine receptors (Schoemaker et al., 1982).

The molecular mechanisms involved in reactive astrogliosis remain elusive. Figure 1 represents a schematic of possible pathways leading to reactive gliosis. Triggers may originate from damaged cellular components of the CNS (e.g., astrocytes, neurons, oligodendrocytes, endothelial cells), invasion of the CNS by blood-borne factors following the disruption of the blood-brain barrier, microglial activation (Murabe et al., 1981), inflammation (Goldmuntz et al., 1986), or the production of cytokines.

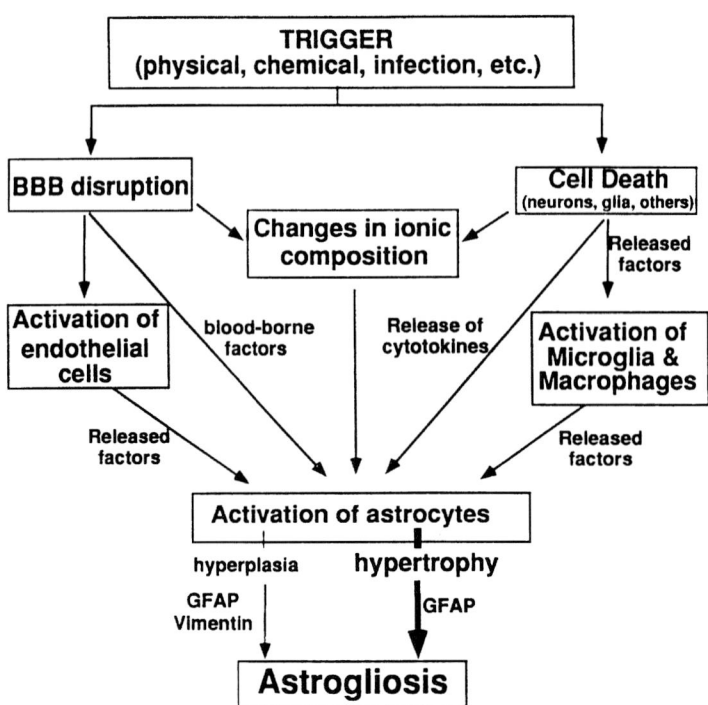

Figure 1. Schematic of possible pathways leading to reactive gliosis. The molecular mechanisms involved in reactive astrogliosis remain elusive. For further discussion please refer to the text.

Epidermal growth factor (EGF), has also been implicated as a trigger for gliosis (Leutz and Schachner, 1981; Simpson et al., 1982). Astrocytes express EGF receptors, the density of which is markedly increased in reactive astrocytes. Tyrosine kinase serves as a second messenger upon receptor activation, a common finding in proliferating cells (Hunter and Cooper, 1985). Growth factors, such as platelet-derived growth factor (PDGF) (Heldin et al., 1977) and fibroblast growth factor (FGF) (Pettmann et al., 1985; Pettmann et al., 1986; Perraud et al., 1988), may also play a role in glitoic responses, as well as pituitary fibroblast growth factor (Brockes et al., 1980; Pruss et al., 1982), prostaglandin F-2α, glia maturation factor (Lim et al., 1985), lymphocyte-derived glial cell-stimulating factor (Fontana et al., 1980), and an insulin-like growth factor (Han et al., 1987). Other postulated triggers include myelin basic protein (Bologa et al., 1985), extracellular ATP (Norenberg et al., 1990; Neary and Norenberg, 1992), cyclic AMP (Federoff et al., 1984), prion protein (Gonzales et al., 1988), fibronectin (Goetschy et al., 1987), and plasminogen activator (Kalderon and William, 1986).

The significance of astrogliosis remains unknown. Unlike connective tissue scars, astrocytic scars do not seem to provide tensile support (Kimelberg and Norenberg, in press). Nevertheless, astrocytic scars may delineate the damaged tissue, walling-off the injured area from intact tissue and providing a barrier against entry of noneuronal components into the CNS (Cavanagh, 1970; Blakemore, 1976). In view of the many functions subserved by astrocytes in both the developing and mature brain, the view that gliosis simply impedes neuronal regeneration merits further scrutiny. Astrocytes contribute to the removal of excitotoxins and neurotransmitters (Rosenberg and Azienman, 1989; Kimelberg et al., 1990; Aschner et al., 1990), produce neuronotrophic factors (Lindsay, 1979; Banker, 1980; Noble et al., 1984; Johnson et al., 1989), and regulate ion composition in the microenvironment (Kimelberg and Ransom, 1986; Vernadakis, 1991). Furthermore, they elaborate growth factors (Furukawa et al., 1986; Azimitia et al., 1990), extracellular matrix proteins (Liesi et al., 1984; Price and Hynes, 1985; Liesi and Silver, 1988; Matthiessen et al., 1989; Yamamoto and Kawana, 1990), adhesion molecules (Neugebauer et al., 1988), surface molecules (Gasser and Hatten, 1990), and protease inhibitors (Rosenblatt et al., 1987; Nakanishi and Guroff, 1988; Shea, 1991). They possess phagocytic properties and therefore may assist in the removal of synaptic endings and other debris (Hoff et al., 1982), and they provide axonal guidance to newly developing neurons (Raisman and Field, 1973). All of these functions are of importance for the restoration of homeostatic function to a site of injury and for provision of migratory guidance to neurons in the regenerative process.

B. ASTROCYTIC SWELLING

Astrocytic swelling also occurs uniformly in response to a wide variety of pathological states, such as trauma and ischemia. Unlike gliosis it occurs rapidly, within 1 hour of injury (Kimelberg and Norenberg, 1993), and may reverse slowly with time. Like many of the other functions addressed in this review, the consequences and the mechanisms of astrocytic swelling, are as yet unclear but are beginning to yield to experimental analysis *in vitro,* using primary astrocyte cultures as a model system. It appears, at least in its exaggerated form, that astrocytic swelling is deleterious or a true pathology and can be viewed as a pathological extension of more limited and controlled volume changes which are otherwise part of the normal homeostatic function of astrocytes.

When cell suspensions or cultured cells are exposed to hypotonic medium they reshrink toward normal size, a process called regulatory volume decrease (RVD). In many cell types, including astrocytes, the mechanisms include swelling-induced activation of conductive K^+ and Cl^- channels, allowing escape of KCl and osmotically obligated water (Eveloff and Warnock, 1987; Hoffman, 1985; Hoffman and Simonson, 1989), and also increased efflux of a number of amino acids, including aspartate, glutamate, and taurine (Pasantes-Morales et al., 1990; Kimelberg et al., 1990). RVD can be blocked by a number of agents that inhibit volume activated K^+ or Cl^- pathways (Hofman and Simonson, 1989; Kimelberg et al., 1990).

Astrocytes are commonly and specifically swollen in a number of pathological states (Bakay and Lee, 1968; Barron et al., 1988; Castejon, 1980; Garcia et al., 1977; Siesjo, 1981; Norenberg, 1981; Yu et al., 1972; Jenkins et al., 1984), all of which are associated with elevated extracellular glutamate levels. Astrocytic swelling and volume regulatory processes *in vitro* are associated with loss of intracellular amino acids, such as taurine, glutamate, and aspartate (Pasantes-Morales and Schousboe, 1988; Pasantes-Morales et al., 1990; Kimelberg et al., 1990; Kimelberg and Goderie, 1991). A variety of mechanisms for such swelling have been proposed (reviewed by Kimelberg et al., 1993). In brief, these include released glutamate or kainic acid stimulating neurons, increasing production of the metabolic products, CO_2 or H^+, and thus leading to astrocytic swelling by coupling. Alternatively, astrocytic swelling may be induced directly by glutamate uptake. It is well established that astrocytes take up glutamate by a Na^+-and Cl^--dependent mechanism (Hertz, 1979; Waniewski and Martin, 1986). Increased K^+ levels up to 80 to 100 mM are encountered in stroke and trauma (Kimelberg, 1992). Such increased K^+ leads to net uptake of KCl with resultant swelling. Such uptake is commonly driven by Donnan forces, driving K^+ and Cl^- through separate K^+ and Cl^- channels (reviewed by Kimelberg et al., 1993). Inhibition of astrocytic swelling *in situ* is associated with improved outcome in an animal head injury and focal ischemia model (Kimelberg, 1992; Kohut et al., 1992). [The topic of astrocytic swelling was extensively reviewed by Kimelberg et al., 1993.]

VII. ASTROCYTES: TARGETS FOR HEAVY METALS

There is compelling evidence that astrocytes are involved in the etiology of heavy metal neurotoxicity. As early as 1966, Oyake et al. showed that, in humans, methylmercury (MeHg) preferentially accumulates within astrocytes. This was further confirmed by Garman et al. (1975). Lead (Pb) has also been reported to concentrate in astrocytes (Holtzman et al., 1984; Holtzman et al., 1987), leading to the "sink hypothesis" and the implication that the resistance of the mature CNS to the development of lead encephalopathy is dependent upon the capacity of astroglia to take up and sequester lead into nontoxic "dump" sites. Although our understanding of the involvement of astrocytes in heavy metal-induced pathology has been enormously extended over the last decade, many questions remain. The remainder of this chapter addresses pertinent information regarding these processes, focusing on lead (Pb), mercury (Hg), and manganese (Mn).

A. LEAD (Pb)
1. Observations on the Role of Astrocytes in Pb-Mediated Neurotoxicity

In the CNS, Pb concentrates in lysosomes, the cytoplasm, mitochondria, as well as the other organelles of all cells, but it appears to preferentially accumulate within GFAP-positive astrocytes (Lach et al., 1977; Young, 1988; Holtzman et al., 1982; Holtzman et al., 1984; Holtzman et al., 1987; Tiffany-Castiglioni et al., 1986; Tiffany-Castiglioni et al., 1987; Young, 1988). *In vitro* studies reveal that a number of conditions, such as culture age, the duration of Pb exposure, Pb level, and the recovery period modify the degree of intracellular Pb accumulation. This ability of astrocytes to preferentially accumulate Pb brought about the "Pb sink" theory (Holtzman et al., 1984), implying that astrocytes have more Pb uptake systems or specific sequestration mechanisms which, as postulated by these authors, render them more resistant to Pb-induced toxicity compared with neurons. Recent studies (Shemancik et al., 1992; Shemancik and Finkelstein, 1993) suggest that low-level Pb exposure influences gene transcription in astrocytes, increasing the synthesis of a 23-kDa protein which persists throughout a 14-day *in vitro* exposure time, providing the cell with enhanced capacity to tolerate exposure to Pb. However, it is also known that intracellular accumulation of Pb per se leads to aberrant astrocytic functions, compromising their ability to maintain optimal extracellular levels of K^+ and glutamate, and other functions crucial in homeostatic control of the brain microenvironment (see below).

The astrocytic contribution to the development of Pb neurotoxicity has been the subject of much recent research (Tiffany-Castiglioni et al., 1987). While astrocytes sequester Pb up to a 55-fold greater concentration than the surrounding extracellular fluid, thereby perhaps limiting neural exposure to Pb (Tiffany-Castiglioni et al., 1986), astrocytic vulnerability to the toxic effects of Pb has also been suggested as a key etiologic factor in Pb-induced neurotoxicity (Gebhart and Goldstein, 1988; Sierra and Tiffany-Castiglioni, 1991). Altered homeostatic functions in Pb-exposed astrocytes include the inhibition of aerobic energy metabolism (Holtzman et al., 1987), increased efflux of K^+ (Aschner et al., 1991), and altered intracellular iron and copper concentrations (Tiffany-Castiglioni et al., 1987). Pb is also known to inhibit a large number of enzymes, probably by binding to sulfhydryl groups, and these are presumably also inhibited in astrocytes (Vallee and Ulmer, 1972).

A prominent feature of prenatal Pb poisoning is interference with neuronal migration (Stowe et al., 1973; Krigman et al., 1974). Since neurons migrate along radial glial processes (Rakic, 1972), damage to surrounding astrocytes may be one of the reasons for the documented effects of Pb on CNS development. Effects of Pb on neuronal migration in culture reveal cessation of cell movement at 10 μM Pb. Lead acetate also inhibits mitotic activity in primary cultures of cell suspensions of fetal cerebrum. This is in agreement with *in vivo* studies demonstrating abnormalities of neuronal migration in Pb-affected monkey brains (Lögdberg, et al., 1987; Lögdberg et al., 1988). These are consistent, but direct evidence of damage to radial glia is still needed.

Astrocytic functions crucial in glutamate metabolism are also altered by low-level Pb exposure. *In vitro* studies by Sierra and Tiffany-Castiglioni (1991) reveal that Pb directly reduces astrocytic glutamine synthetase (GS) activity in a dose-dependent manner. Engle and Volpe (1990) report a 70% inhibition in astrocytic GS activity at Pb concentrations as low as 2.5 μM. Although extrapolation from enzymatic analyses *in vitro* is fraught with uncertainties, it has been suggested that such Pb concentrations might be achieved with chronic exposure to an extracellular concentration of 0.05 μM Pb or approximately 1.0 μg/dl (Engle and Volpe, 1990). This value is one order of magnitude lower than the blood Pb levels presently designated by the Centers for Disease Control as acceptable (10 μg Pb/dl).

Pb is also known to induce astrocytic swelling *in vivo* (Thomas et al., 1972). The findings of persistent astrocytic swelling *in vitro* (unpublished observations from our laboratory, Vitarella et al.) also suggest that Pb interferes with regulatory volume decrease (RVD) processes, perhaps by interfering with Ca^{2+} signaling mechanisms. The release of K^+ in RVD is Ca^{2+} dependent, as evident by the inhibitory effect of Ca^{2+} uptake blockers and quenching of intracellular Ca^{2+} (Vitarella et al., submitted). The consequences of the inhibitory effects of Pb^{2+} on astrocytic RVD are yet to be determined.

2. Pb Accumulation in Astrocytes

There is a lack of information on the mechanism/s of Pb transport into astrocytes. Studies by Simons (1984, 1986) on the transport of Pb into ghost erythrocytes postulate two pathways for its transport. The first pathway is dependent upon HCO_3^-, since it is blocked by inhibitors of the anion transport exchange system, such as 4'-isothiocyanostilbene-2,2'-disulfonic acid (SITS), while the second pathway is independent of HCO_3^- concentration and is SITS-insensitive. Elevated concentrations of HCO_3^- in the media favor the anion-exchange pathway, while in the presence of perchlorate (ClO_4^-) (and in the absence of HCO_3^-) this transport pathway is minimized (Simons, 1986). Pb uptake depends on the presence of a second anion, and in the presence of HCO_3^- the rate is stimulated as follows: $ClO_4^- < NO_3^- < CH_3CO_2^-$, $F^- < Cl^- < Br^- < I^-$ (Simons, 1986). Transport of other cations by the anion-exchange transport system is well established (i.e., Li^+ and Na^+). In both cases, the uptake is inhibited by SITS (Cabantchik et al., 1978). The transport kinetics of Pb in ghost erythrocytes are consistent with the hypothesis that before transport of Pb can occur, a $PbCO_3$ complex must form, because the rate of Pb uptake is directly proportional to the concentrations of both Pb^{2+} and HCO_3^-, and inversely proportional to the concentration of H^+. Furthermore, the temperature dependence of Pb uptake into ghost erythrocytes is similar to that of the HCO_3^-/Cl^- exchange. Accordingly, it was postulated that the transport of Pb may occur either via a nonelectroneural exchange of $PbCO_3$ with an anion, or via exchange of an anion-ternary complex of $PbCO_3$ with another anion (Simons, 1986).

A second mechanism in adrenal medullary cells (Simons and Pocock, 1987; Tomsig and Suszkiw, 1990; Tomsig and Suszkiw, 1991) shows that Pb uptake occurs via Ca^{2+} channels. K^+ and veratridine stimulate the uptake of both Ca^{2+} and Pb^{2+}. Ca^{2+} acts as a competitive inhibitor for Pb^{2+} uptake with the K_i for the inhibitory effect of Ca^{2+} on Pb^{2+} uptake exhibiting a similar value to the K_m for Ca^{2+} uptake. The entry of Pb into chromaffin cells consists of both voltage-independent and voltage-dependent (K^+ stimulated) components. The Ca^{2+} channel blockers, D-600, and nifedipine block Pb^{2+} uptake, while Bay K 8644, a Ca^{2+} channel agonist, stimulates the uptake of Pb, suggesting the involvement of the L-type Ca^{2+} channels (Simons and Pocock, 1987; Tomsig and Suszkiw, 1990; Tomsig and Suszkiw, 1991). In contrast to ghost erythrocytes, SITS is ineffective in blocking Pb^{2+} and Ca^{2+} uptake in chromaffin medullary cells (Simons and Pocock, 1987). Astrocytes express both voltage-dependent and -independent Ca^{2+} channels, as well as the HCO_3^-/Cl^- anion exchange transporter (reviewed by Kimelberg et al., 1993). Thus there is the possibility that one or both of these mechanisms are operative in the uptake of Pb^{2+} by astrocytes.

B. MANGANESE (Mn)

Mn, like Pb is an abundant transition metal. However, unlike Pb, it is a biologically essential metal, present in many enzymes (see below). Mn is especially critical during development. At higher levels of exposure, Mn can also be toxic, causing an irreversible brain disease characterized by both psychological and neurological disturbances. The neurological signs have received close attention because they resemble several clinical disorders collectively described as "extrapyramidal motor system dysfunction," and in particular, Parkinson's disease and dystonia (Cotzias et al., 1968; Barbeaue et al., 1976; Barbeau, 1984; Donaldson, 1987). Since several countries, including the U.S., have now replaced, or are in the process of replacing lead (Pb) in gasoline with a Mn antiknock compound (Wedler, 1993), methylcylopentadienyl manganese tricarbonyl (MMT), the biological hazards of Mn need to be studied. MMT has indeed been reported to cause increased health problems in heavily air-polluted areas (Cooper, 1984; Fishman et al., 1987; Hakkinen et al., 1983). A new source of Mn exposure is the street drug called "Bazooka," which is cocaine contaminated with manganese-carbonate from free-base preparation methods (Ensing, 1985).

I. Effects of Mn on Astrocytic Functions

In vivo, Mn^{2+} has been shown to accumulate in mitochondria of brain areas associated with neurological symptoms of manganism (Liccione and Maines, 1988; 1989). Studies with cultured glia reveal that these cells can accumulate Mn to concentrations 50-fold higher than those present in the culturing medium (Wedler et al., 1989; Aschner et al., 1992). The distribution of Mn in cultured chick glia and rat astrocytes has been found to be roughly 40% cytoplasmic and 60% mitochondrial. A significant function of Mn normally present within the CNS is typified by the mitochondrial enzyme superoxide-dismutase (SOD) that catalyzes the disproportionate conversion of superoxide (O_2^-), the univalent reduction product of dioxygen, to H_2O_2. SOD functions in mitochondria to protect against the toxicity of oxygen radicals, catalyzing the formation of H_2O_2 from reactive oxygen species. Other primary Mn-activated or Mn-containing enzymes in mitochondria include phosphoenolpyruvate carboxykinase, pyruvate carboxylase, and malic enzyme (reviewed by Prohaska, 1987; Wedler, 1993), key enzymes of glucose storage, mobilization, and catabolic and anabolic metabolism.

Glutamine synthetase (GS) is an abundant manganoprotein present in the cytoplasm of astrocytes (Wedler and Toms, 1986). Glutamate is an excitotoxic amino acid (Olney, 1979), and as described earlier, GS, an astrocyte-specific enzyme, catalyzes the conversion of glutamate to glutamine (Norenberg, 1979). GS, a Mn-rich enzyme, contains 8 Mn ions per octamer (Wedler and Toms, 1986; Fressinaud et al., 1991). In fact, astrocytes contain 80% of the total Mn in the brain (Wedler and Denman, 1984).

$[Ca^{2+}]_i$ serves an important role in regulating the activity of the ATP-dependent GS (Benjamin, 1987). GS has a K_d for Mn near 1 μM, and responds to changes in Mn ion concentration at this level, even in the presence of 1 to 3 mM Mg (Wedler et al., 1993). The free concentration of Mn in chick glia and rat astrocytic cytoplasm is also near 1 μM (Wedler et al., 1986; Aschner et al., 1992), thus allowing changes in free Mn to regulate GS activity. It also follows that frequent Ca^{2+} spikes may, in addition to elevating $[Ca^{2+}]_i$ increase the mitochondrial uptake of Mn^{2+} at the expense of free cytosolic Mn^{2+}. Consequently, GS activity may decrease in direct proportion to the decrease in free cytoplasmic Mn^{2+} concentration, providing a rapid, enzyme-level mechanism for altering GS activity. The overall effect of reduced GS activity would lead to imbalances in glutamate homeostasis. Alternatively, as postulated by Wedler et al. (1993), should cytoplasmic Ca^{2+} stimulate the release of mitrochondrial Mn, it would serve to increase the cytoplasmic free levels of Mn^{2+}. The transient "burst" of Mn^{2+} from the mitochondria would catalyze intra-astrocytic glutamate to glutamine. Since glutamate receptors are themselves regulated by Ca^{2+}, this mechanism can modulate the effects of extracellular glutamate.

Intra-astrocytic Ca^{2+} spikes, although short lived, may have prolonged effects on Ca^{2+} homeostasis (Gavin et al., 1990). In neuronal mitochondria, considerable evidence has accumulated to support the existence of separate sites on the uniporter for transport and for activation (Vinogradov and Scarpa, 1973). Once bound to the activation site, Ca^{2+} remains bound for up to 1 min (Kroner, 1988). Therefore, activation of the mitochondrial Ca^{2+} transporter may continue even after a Ca^{2+} spike has passed. Notably, Mn^{2+} also stimulates incorporation of myoinositol into phosphatidylinositol (PI)(Schoepp, 1985). Rapid PI turnover causes increased $[Ca^{2+}]_i$ by mobilizing Ca^{2+} from endoplasmic reticulum, leading to further increases in the uptake of Ca^{2+} (and Mn^{2+}) into the mitochondria. In several pathological states associated with excess cytosolic Ca^{2+}, mitochondria are known to take up large quantities of Ca^{2+}. In one such example, excessive Ca^{2+} uptake by mitochondria of cultured hepatocytes is known to cause oxidative stress, accompanied by a decrease in mitochondrial glutathione (GSH) levels (Olafsdotirr et al., 1988; Crompton, 1990).

2. Mn Uptake in Astrocytes

In view of the preferential accumulation of brain Mn in astrocytes, recent studies were directed to determine the transport mechanisms of Mn in these cells. The time-dependent uptake of ^{54}Mn into cultured astrocytes was recently studied by Aschner et al. (1992). The initial rate (1 min) of Mn uptake by astrocytes is concentration dependent. The apparent kinetic constants are $K_m = 0.30 \pm 0.03$ μM and $V_{max} = 0.30 \pm 0.02$ nmol Mn $\times$ min^{-1} $\times$ mg protein^{-1}. It was found that both Ca^{2+} and Mn^{2+} compete for the same transport systems in astrocyte plasma membrane which may include Ca^{2+} channels and/or Na^+/Ca^{2+} exchange. Co^{2+}, Zn^{2+}, and Pb^{2+} do not have any effect on the initial rate of ^{54}Mn uptake, suggesting that only Ca^{2+} and Mn^{2+} share the same transport system. Transport of ^{54}Mn into astrocytes is also independent of the presence of $[Na^+]_o$.

The kinetic data showing saturation kinetics, competition by related substrates (i.e., Ca^{2+}), and a probable counter transport mechanism exhibit the accepted criteria of a transport process facilitated by a specific membrane transport protein. The relatively high capacity (V_{max}) of 0.30 nmol $\times$ mg protein^{-1} $\times$ min^{-1} and the apparent K_m are consistent with an efficient uptake of Mn^{2+} by astrocytes. Metal exchange is not obligatory for Mn efflux. However, increasing $[Mn^{2+}]_o$ has a marked effect on the initial rate of release of Mn. The simplest interpretation of these data is that extracellular Mn *trans*-stimulates the efflux of intracellular Mn, most consistent with uptake and efflux occurring on the same transport system or a separate exchange system. The biphasic kinetic profiles of Mn efflux suggest several types of intracellular Mn pools or different transport systems (Aschner et al., 1992). The fast phase of efflux (up to 10 min) must occur from a pool that is in rapid equilibrium with the extracellular medium. This pool can be assumed to contain all intracellular complexes whose dissociation rate constants are higher than the rate constant for efflux. A second pool must be in slow equilibrium with the rapid equilibrium pool to account for the slower phase of Mn release between 10 to 120 min. Finally, a third pool of Mn (~25 to 30%) is unavailable for efflux, representing irreversibly bound Mn. This pool may account for the principal concentrative mechanism for Mn in the astrocytes and presumably represents Mn uptake into mitochondria.

C. MERCURY (Hg)

Methylation of inorganic mercury species to methylmercury (MeHg) by microorganisms is known to take place in waterways (Wood et al., 1968; Jensen and Jernelov, 1969), resulting in its accumulation in the food chain. Human poisoning outbreaks as a result of food-borne MeHg consumption have led to its recognition as a ubiquitous environmental contaminant, capable of inflicting characteristic toxic effects in both humans and animals. MeHg is a particular threat to the CNS in humans, as evidenced by the tragic epidemics of MeHg poisoning in Japan (Takeuchi et al., 1962), and Iraq (Bakir et al., 1973).

1. Effects of MeHg on Astrocytic Homeostasis

A prominent feature of prenatal MeHg poisoning is a reduction in CNS mitotic activity (Rodier et al., 1984), and interference with neuron migration (Choi et al., 1981; Choi, 1983). Effects of MeHg on neuronal migration in culture reveal cessation of cell movement at 10 μM MeHg. Abnormalities of neuronal migration are also prominent pathological features of MeHg-affected human brains (review, Reuhl and Chang, 1979). Since neurons migrate along radial glial processes (Rakic, 1972), it is possible that a major effect of MeHg is damage to the astrocytes. Indeed, effects on astrocytes have been shown. The initial site of MeHg injury appears to be the astrocytic plasma membrane, with a marked shift in the distribution of anionic groups and loss of filopodial activity (Peckham and Choi, 1986). Electron spin resonance studies on cultured astrocyte membranes exposed to MeHg indicate a significant reduction in membrane fluidity following exposure to this organometal (Choi, 1988).

It has long been noted that MeHg compounds can influence ion, water, and nonelectrolyte transport in a variety of cells and tissues (Kinter and Pritchard, 1977; Nechay, 1975; Webb, 1966). Recently Aschner et al. (1990) have shown that K^+ uptake and efflux (as measured by $^{86}Rb^+$) in cultured astrocytes are altered by exposure to MeHg concentrations as low as 10^{-5} M, extending previous observations on the sensitivity of Na^+/K^+-ATPase activity and ion permeability in other cell types to MeHg (Webb, 1966). Altered membrane transport systems would also be expected to compromise the ability of astrocytes to exclude Na^+ and would be expected to result in astrocytic swelling. Indeed, MeHg induces astrocytic swelling in a dose-dependent fashion (Aschner et al., 1990).

MeHg leads to a dose-dependent decrease in the uptake of both $[^3H]$-L-glutamate and $[^3H]$-D-aspartate in primary astrocyte cultures. Also, exposure of astrocytes to MeHg leads to a dose- and time-dependent increase in the release of L-glutamate and D-aspartate from astrocytes. Anion transport inhibitors block the MeHg-induced release of these amino acids (Aschner et al., 1993). However, the rank order of the effectiveness of these blockers differs from the rank order of the blockers for reversal of swelling-induced amino acid release.

Two sulfhydryl (-SH) protecting agents, a cell membrane nonpenetrating compound, reduced glutathione (GSH), and the membrane permeable dithiothreitol (DTT), were recently found to inhibit the stimulatory action of MeHg on the efflux of radiolabeled D-aspartate as well as ^{86}Rb (Mullaney et al., in press). MeHg-induced $[^3H]$-D-aspartate and ^{86}Rb release were completely inhibited by the addition DTT or GSH during the 5 min exposure period with MeHg (10 μM). However, when added after MeHgCl treatment, the membrane impermeable GSH was only partially effective in reversing, while the membrane permeable DTT fully reversed the MeHgCl-induced release of D-aspartate. Thus, the stimulatory effect exerted by MeHgCl on astrocytic D-aspartate release may be associated with vulnerable -SH groups located within, but not on the surface of the cell membrane. Omission of Na^+ from the perfusion solution did not accelerate MeHgCl-induced D-aspartate release, suggesting that reversal of the D-aspartate carrier is unlikely to explain MeHgCl-induced D-aspartate release. Omission of Ca^{2+} from the perfusion solution increased the time-dependent MeHgCl-induced D-aspartate release.

Given the complex interrelationships between proteins and lipids and the ability of MeHgCl to interact with both, its effect on the cell membrane is extremely complex, reflecting upon the varied roles played by sensitive -SH membrane components (Kinter and Pritchard, 1977). While the role of amino acid release in volume regulation is still debatable, released L-aspartate and L-glutamate may lead to excitotoxic effects. MeHg-damaged astrocytes can also serve as a source for the release of excitatory amino acids, further increasing their extracellular concentrations. Abnormally high levels of excitatory amino acids causing exaggerated stimulation of excitatory amino acid receptors on the surface of adjacent neurons can trigger a destructive cascade of events that can damage neurons *en masse* (Choi, 1988; Olney, 1979).

2. MeHg Transport in Astrocytes

MeHg in the environment is produced by methylation of inorganic mercury by microorganisms. An aerobic pathway involves methylation of homocysteine-bound inorganic Hg^{2+} by those processes in the cell normally responsible for the formation of methionine (Wood et al., 1968; Jensen and Jernelov, 1969). In other words, the Hg-cysteine complex is methylated by "mistake". Aschner and Clarkson (1988; 1989) drew attention to the close structural similarity between MeHg-L-cysteine complexes and methionine and demonstrated cysteine-facilitated transport of MeHg across the BBB. This transport is inhibited by coadministration of neutral amino acids, both those that do and do not contain -SH groups. The structural similarity of the MeHg-L-cysteine complex to methionine provides the theoretical basis for these observations. Structurally, the MeHg-L-cysteine conjugate differs from methionine by the presence of the Hg atom between the -S- and CH_3 groups; both possess linear bonds at an angle of 180° (Carty and Malone, 1979).

Work by Aschner et al. (1990; 1991) suggests that the uptake of $[^{203}Hg]$-MeHg by astrocytes when added to the media as the L-cysteine conjugate exhibits the kinetic criteria of a specific transport system.

Saturation kinetics, substrate specificity and inhibition, and trans stimulation were demonstrated in the presence of this -SH-containing amino acid. Cysteine-mediated uptake of MeHg was inhibited by the co-administration of L-methionine, and 2-aminobicyclo-[2, 2, 1]-heptane-2-carboxylic acid (BCH), a specific substrate for the neural amino acid transport system L. 2-Methylaminoisobutyric acid (MeAIB) was ineffective in inhibiting the uptake of the MeHg-cysteine conjugate. Pre-loading of the astrocytes with glutamate was moderately effective in trans-stimulating the uptake of MeHg-cysteine conjugates, while in the absence of cysteine, uptake of [^{203}Hg]-MeHg was unchanged. These results indicate the presence in astrocytes of a neutral amino acid carrier transport system L, capable of selectively mediating cysteine-MeHg uptake.

Additional studies indicate that the putative MeHg transport system in astrocytes can also mediate MeHg transport in the net efflux direction (Aschner et al., 1991). Metal exchange is not required for efflux, since efflux of [^{203}Hg]-MeHgCl occurs in MeHg-free buffer. The apparent stimulatory effects of extracellular L-cysteine-MeHg conjugates on the rate and amount of ^{203}Hg efflux are clearly shown to reflect displacement of [^{203}Hg]-MeHgCl from intracellular pools via a specific effect on the transport system itself. The simplest interpretation is that the efflux observed represents transport by the same neutral amino acid system-L that facilitates MeHg uptake (Aschner et al., 1990).

IV. CONCLUSIONS

This brief review cannot do justice to the rapid advances in experimental data and concepts now being made within the "astrocytic" field (see Section I, Introduction for reviews), and this is just as applicable to the interactions between these cells and the heavy metals.

There is a wealth of data on the effects of acute, prenatal and prolonged exposure to Pb, Mn, and Hg on the development and properties of neurons. This is understandable because of the variety of CNS dysfunctions that occur as a result of prenatal and adult exposure to these metals. Recently there has been growing evidence of the important role of glia, and especially astroglia (astrocytes), in brain function. At present, these roles principally relate to brain development, homeostatic uptake functions, receptor interactions, and pathological responses (Kimelberg, 1983; Kimelberg and Aschner, 1993; LoPachin and Aschner, 1993; Hatten and Mason, 1986; Fedoroff, 1986a; 1986b; 1986c; Althaus and Seifert, 1987; Abbott, 1991). In the past 15 years this work has produced a growing appreciation of the many diverse functions that astrocytes perform in the developing and mature brain (Kimelberg and Norenberg, 1989; Murphy, 1993). It is quite likely that many of the effects of Pb, Mn, and Hg on development and other CNS functions will not be unraveled without considering the role of astrocytes. This view is supported by the relatively few studies that have been done so far on heavy metal uptake into these cells and their ability to interfere with astrocytic functions.

In addition to a purely emphasized support role of astrocytes for developing neurons, new information now provides evidence that astrocytic contact with neurons significantly affects the morphological and functional differentiation of the latter. This reciprocity between neurons and astrocytes suggests that the morphological and physiological attributes of neurons are a product of this cell-cell interaction and vice versa. In addition, such reciprocity appears to exist with other CNS cell-types such as oligodendrocytes, microglia, and endothelial cells. The wide diversity of astrocytic functions both in maintaining homeostasis and their potential in modulating damage and repair is reflected in this review. The number of functions assumed by these cells is rapidly growing as novel biochemical and molecular tools are developed. We hope that the diversity of their functions described herein provides the reader with an appreciation for their complexity and functional integration, within both the normal and abnormal CNS. Their potential in modulating damage to the CNS by heavy metals and their capacity to be damaged when their own defenses are overwhelmed is also reflected in this review. New methodologies have begun to alert us to the important roles astrocytes play in heavy metal neurotoxicity. Expanded investigation of the astrocytic involvement in heavy metal neurotoxicity is clearly warranted to improve our understanding of these processes.

ACKNOWLEDGMENTS

Preparation of this review was supported in part by NIEHS Grant 05223 and USEPA 819210 awarded to MA, and NIH grants NS 23750, NS 19492, and NS 30303 awarded to HKK.

REFERENCES

Abbott, N. J., Ed. (1991), in *Glial-Neuronal Interactions, Ann. N.Y. Acad. Sci.,* Vol. 633, New York
Acheson, A. and Thoenen, H. (1983), *J. Cell Biol.,* 97, 925–928.
Acheson, A. and Rutishauser, U. (1988), *J. Cell Biol.,* 106, 479–486.
Aldskogius, H. (1982), *Neuropathol. Appl. Neurobiol.,* 8, 341–349.
Aloisi, F., Agresti, C., D'Urso, D., and Levi, G. (1988), *Proc. Natl. Acad. Sci. U.S.A.,* 85, 6167–6171.
Althaus, H. H. and Seifert, W., Eds. (1987), in *Glial-Neuronal Communication in Development and Regeneration,* Springer Verlag, Berlin.
Amaducci, L., Forno, K. I., and Eng, L. F. (1981), *Neurosci. Lett.,* 21, 27–32.
Anders, J. J. and Brightman, M. W. (1979), *J. Neurocytol.,* 8, 777–795.
Aoki, C. (1992), *J. Neurosci.,* 12, 781–792.
Arai, K., Lee, F., Miyajima, A., Miyatake, S., Arai, N., and Yokota, T. (1990), *Annu. Rev. Biochem.,* 59, 783–836.
Arenander, A. T., Lim, R. W., Varnum, B. C., Cole, R., de Vellis, J., and Herschman, H. R. (1989), *J. Neurosc. Res.,* 23, 247–256.
Aschner, M. and Clarkson, T. W. (1988), *Brain Res.,* 462, 31–39.
Aschner, M. and Clarkson, T. W. (1989), *Pharmacol. and Toxicol.,* 64, 293–297.
Aschner, M., Eberle, N. B., Goderie, S., and Kimelberg, H. K. (1990), *Brain Res.,* 524, 221–228.
Aschner, M., Eberle, N. B., Miller, K., and Kimelberg, H. K. (1990), *Brain Res.,* 530:245–250.
Aschner, M., Chen, R., and Kimelberg, H. K. (1991), *Brain Res. Bull.,* 26, 639–642.
Aschner, M., Eberle, N., and Kimelberg, H. K. (1991), *Brain Res.,* 554, 10–14.
Aschner, M., Gannon M., and Kimelberg, H. K. (1992), *J. Neurochem.,* 58, 730–735.
Aschner, M., Du, Y.-L., Gannon, M., and Kimelberg, H. K. (1993), *Brain Res.,* 602, 181–186.
Azmitia, E. C., Dolan, K., and Whitaker-Azmitia, P. M. (1990), *Brain Res.,* 516, 354–356.
Bakay, L. and Lee, J. C.(1968), *Brain,* 91, 697–706.
Bakir, F., Damluji, S. F., Amin-Zaki, L., Murthada, M., Khalidi, A., Al-Rawi, N. Y., Tikriti, S., Dhahrir, H. I., Clarkson, T. W., Smith, J. C., and Doherty, R. A. (1973), *Science,* 181, 230–242.
Balloti, R., Nielsen, F. C., Pringle, N., Kowalski, A., Richardson, W. D., Van Obberghen, E., and Gammeltoft, S. (1987), *EMBO J,* 6, 3633–3639.
Banker, G. A. (1980), *Science,* 209, 809–810.
Barbeau, A., Inoué, N., and Cloutier, T. (1976), *Adv Neurol.,* 14, 339–352.
Barbeau, A. (1984), *Neurotoxicology,* 5, 13–16.
Barres, B. A., Chun, L. Y., and Corey, D. P. (1989), *J. Neurosci.,* 9, 3169–3175.
Barres, B. A. and Chun, L. L. Y. (1990), *Annu. Rev. Neurosci.,* 13, 441–474.
Barres, B. A., Hart, I. K., Coles, H. S. R., Burne, J. F., Voyvodic, J. T., Richardson, W. D., and Raff, M. C. (1992), *Cell,* 70, 31–46.
Barrett, C. P., Donati, E. J., and Guth, L. (1984), *Exp. Neurol.,* 84, 374–385.
Barron, K. D., Dentinger, M. P., Kimelberg, H. K., Nelson, L. R., Bourke, R. S., Keegan, S., Mankes, R. F., and Cragoe, E. J., Jr. (1988), *Acta Neuropathol. (Berl.),* 75, 295–307.
Bartlett, W. P., Knapp, P. E., and Sokoloff, R. P. (1988), *Glia,* 1, 253–259.
Becker, G. L., Fiskum, G., and Lehninger, A. L. (1980), *Biochim. Biophys. Acta,* 591, 234–239.
Benjamin, A. M. (1987), *J. Neurochem.,* 48, 1157–1164.
Benveniste, E. N. (1993), in *Astrocytes: Pharmacology and Function,* Murphy, S. Ed., Academic Press, San Diego, CA, 355–395.
Bevan, S., Chiu, S. Y., Gray, P. T. A., and Ritchie, J., M., (1985), *Phil. Trans. Roy. Soc. Biol.,* 225, 299–313.
Bignami, A and Dahl, D. (1976), *Neuropathol. Appl. Neurobiol.,* 2, 99–110.
Blakemore, W. F. (1976), *J. Neuropathol. Appl. Neurobiol.,* 2, 21–39.
Bologa, L., Deugnier, M., Joubert, R., and Bisconte, J. (1985), *Brain Res.,* 346, 199–203.
Bowman, C. L., Kimelberg, H. K., Frangakis, M. V., Berwald-Netter, Y., and Edwards, C. (1984), *J. Neurosci.,* 4, 1527–1534.
Bowman, C. L. and Kimelberg, H. K. (1987), *Brain Res.,* 423, 403–407.
Brockes, J. P., Lemke, G. E., and Balzer, D. R. (1980), *J. Biol. Chem.,* 255, 8374–8377.
Bruckner, G., Brauer, K., Hartig, W., Wolff, J. R., Rickmann, M. J., Derouiche, A., Delpech, B., Girard, N., Oertel, W. H., and Reichenbach, A. (1993), *Glia,* 8, 183–200.
Cabantchik, Z. I., Knauf, P. A., and Rothstein, A. (1978), *Biochim. Biophys. Acta,* 515, 239–302.
Cammer, W., Tansey, F. A., and Brosnan, C. F. (1990), *J. Neuroimmunol.,* 27, 111–120.
Carbonell, A. L. and Boya, J., (1988). *Brain Res.,* 439, 337–341.
Carty, A. J. and Malone, S. F. (1979). in *The Biogeochemistry of Mercury in the Environment,* Nrigau, J. O., Ed., Elsevier/North Holland Biomedical Press, Amsterdam, 433–479.
Castejon, O. J. (1980), *J. Neuropathol. Exp. Neurol.,* 29, 296–327.
Cavanagh, J. B. (1970), *J. Anat.,* 106, 471–487.
Cavicchioli, L., Dickson, G., Prentice, H. et al. (1988), *Pharmacol. Res. Commun.,* 20, 609–610.
Chamak, B., Fellous, A., Glowinski, J., and Prochiantz, A. (1987), *J. Neurosci.,* 7, 3163–3170.
Charles, A. C., Naus, C. C., Zhu, D., Kidder, G. M., Dirksen, E. R., and Sanderson, M. J. (1992), *J. Cell Biol.,* 118, 195–201.

Charles, A. C., Dirksen, E. R., Merrill, J. E., and Sanderson, M. J. (1993), *Glia,* 7, 134–145.
Charman, H., Gonzales, M., Renaut, P., and DeArmond, S. J. (1988), *J. Neuropathol. Exp. Neurol.,* 47, 345.
Chesler, M. and Chen, J. C. T. (1991), *Can. J. Physiol. Pharmacol.,* 70, S286–S292.
Choi, B. H., Cho, K. H., and Lapham, L. W. (1981), *Environ. Res.* 24, 61–74.
Choi, B. H. (1983). in *Reproductive and Developmental Toxicity of Metals,* Clarkson, T. W., Nordberg, G. F., and Sager, P. R. Eds., Plenum Press, New York, 473–495.
Choi, B. H. (1988), in *The Biochemical Pathology of Astrocytes,* Alan R. Liss, New York, 219–230.
Choi, D. W. (1988), *Neuron,* 1, 623–634.
Chuong, C.-M. (1990), *Experientia,* 46, 892–899.
Cobbold, P. H. and Rink, T. J. (1987), *Biochem. J.,* 248, 313–328.
Condorelli, D. F., Dell Albani, P., Kaczmarek, L., Messina, L., Spampinato, G., Avola, R., Messina, A., and Giuffrida-Stella, A-M. (1990), *J. Neurosci. Res.,* 26, 251–257.
Cooper, W. C. (1984), *J. Toxicol. Environ. Health,* 14, 23–46.
Cornell-Bell, A. H. and Finkbeiner, S. M. (1991), *Cell-Calcium,* 12, 185–204.
Cotzias, G. C., Horiuchi, K., Fuenzalida, S., and Mena, I. (1968), *Neurology,* 18, 376–382.
Crompton, M. (1990), in *Calcium and the Heart,* Langer, G. A., Ed., Raven Press, New York, 167–197.
Dahl, D., Bignami, A., Weber, K., and Osborn, M. (1981), *Exp. Neurol.,* 73:496–506.
Dehouck, M.-P., Meresse, S., Delorme, P., Fruchart, J-C., and Cecchelli, R. (1990), *J. Neurochem.,* 54, 1798–1801.
Denis-Donini, S., Glowinski, J., and Prochiantz, A. (1984), *Nature,* 307, 641–643.
Derouiche, A., Heimrich, B., and Frotscher, M. (1993), *Eur. J. Neurosci.,* 5, 122–127.
Donaldson, J. (1987), *Neurotoxicology,* 8, 451–462.
Dragunow, M. and Robertson, H. A. (1989), *Brain Res.,* 455, 295–299.
Dragunow, M., Goulding, M., Faull, R. L., Ralph, R., Mee, E., and Frith, R. (1990), *Exp. Neurol.,* 107, 236–248
Duffy, S. and MacVicar, B. A. (1993), in *Astrocytes, Pharmacology and Function,* Murphy, S., Ed., Academic Press, San Diego, 137.
Eng, L. F. (1987). in *Glial-Neuronal Communication in Development and Regeneration,* Althaus, H. H. and Seifert, W., Eds., Springer-Verlag, Heidelberg, 27–40.
Eng, L. F. (1988), in *Biochemical Pathology of Astrocytes,* Norenberg, M. D., Hertz, L., and Schousboe, A., Eds., A. R. Liss, New York, 79–90.
Engle, M. J. and Volpe, J. J. (1990), *Dev. Brain Res.,* 55, 283–287.
Ensing, J. G. (1985), *J. Anal. Toxicol.,* 9, 45–46.
Erb, P., Kennedy, M., Hagmann, I., Wassmer, P., Huegli, G., Fierz, W., and Fontana, A. (1986), in *Regulation of Immune Gene Expression,* Feldmann, M. and McMichael, A., Eds., Humana Press, Clifton, NJ, 187.
Eveloff, J. L. and Warnock, D. G., (1987), *Am. J. Physiol.,* 252, F1–F10.
Fedoroff, S., McAuley, W. A. J., Houle, J. D., and Devon, R. M. (1984), *J. Neurosci. Res.,* 12, 15–27.
Fedoroff, S. (1986a), in *Astrocytes: Development, Morphology, and Regional Specialization,* Fedoroff, S. and Vernadakis, A., Eds., Vol. 1, Academic Press, New York.
Fedoroff, S. (1986b), in *Astrocytes: Biochemistry, Physiology and Pharmacology of Astrocytes,* Fedoroff, S. and Vernadakis, A., Eds., Vol. 2, Academic Press, New York.
Fedoroff, S. (1986c), in *Astrocytes: Cell Biology and Pathology of Astrocytes,* Fedoroff, S. and Vernadakis, A., Eds., Vol. 3, Academic Press, New York.
Fierz, W., Endler, B., Reske, K., Wekerle, H., and Fontana, A. (1985), *J. Immunol.,* 134, 3785–3793.
Fishman, B. E., McGinley, P. A., and Gianutsos, G. (1987), *Toxicology,* 45, 193–201.
Fontana, A., Grieder, A., Arrenbrecht, S. T., and Grob, P. (1980), *J. Neurol. Sci.,* 46, 55–62.
Fontana, A., Fierz, W., Wekerle, H. (1984), *Nature,* 307, 273–276.
Fontana, A., Erb, P., Pircher, H., Zinkernagel, R., Weber, E., and Fierz, W. (1986), *J. Neuroimmunol.,* 12, 15–28.
Fontana, A., Frei, K., Bodmer, S., Hofer, E. (1987), *Immunol. Rev.,* 100, 185–201.
Frei, K., Siepl, C., Groscurth, P., Bodmer, S., Schwerdel, C., and Fontana, A. (1987), *Eur. J. Immunol.,* 17, 1271–1278.
Fressinaud, C., Weinrauder, H., Delaunoy, J. P., Tholey, G., Labourdette, G., and Sarlieve, L. L. (1991), *J. Cell. Physiol.,* 149, 459–468.
Furukawa, S., Furukawa, Y., Satoyoshi, E., and Hayashi, K. (1986), *Biochem. Biophys. Res. Commun.,* 136, 57–63.
Garcia, J. H., Kalimo, H., Kamijyo, Y., and Trump, B. F. (1977), *Virchows Arch. B Cell. Path.,* 15, 191–206.
Gard, A. L. and Pfeiffer, S. E. (1989), *Development,* 106, 119–132.
Garman, R. H., Weiss, B., and Evans, H. L. (1975), *Acta Neuropathol.,* 32, 61–74.
Gavin, C. E., Gunter, K. K., and Gunter, T. E. (1990), *Biochem. J.,* 266, 329–334.
Gasser, U. E. and Hatten, M. E. (1990), *J. Neurosci.,* 10, 1276–1285.
Giulian, D. and Baker, T. J. (1986), *Neuroscience,* 6, 2163–2178.
Gloor, S., Antonicek, H., Sweadner, K. J., Pagliusi, S., Frank, R., Moos, M., and Schachner, M. (1990), *J. Cell Biol.,* 110, 165–174.
Goetschy, J.-F., Ulrich, G., Aunis, D., and Ciesielski-Treska, J. (1987), *Int. J. Dev. Neurosci.,* 5, 63–70.
Goldman, R. S., Finkbeiner, S. M., and Smith, S. J. (1991), *Neurosci. Lett.* 123, 4–8.
Goldmuntz, E. A., Brosnan, C. F., Chiu, F.-C., and Norton, W. T. (1986), *Brain Res.,* 397, 16–26.
Goldstein, G. W. and Betz, A. L. (1986), *Sci. Am.,* 255, 74–83.

Gonzales, M. F., Renaut, P., Stowring, L., Charman, H. P., Prusiner, S. B., and DeArmond, S. J. (1988), *J. Neuropathol. Exp. Neurol.,* 47, 345.
Griffin, W. S. T., Stanley, L. C., Ling, C. et al. (1989), *Proc. Natl. Acad. Sci. U.S.A.,* 86, 7611–7615.
Hakkinen, P. J., Morse, C. C., Martin, F. M., Dalbey, W. E., Haschek, W. M., and Witschi, H. R. (1983), *Toxicol. Appl. Pharmacol.,* 67, 55–69.
Han, V. K. M., Lauder, J. M., and D'Ercoli, A. J. (1987), *J. Neurosci.,* 7, 501–511.
Hatten, M. E. (1985), *J. Cell Biol.,* 100, 384–396.
Hatten, M. E., Fishell, G., Stitt, T. N., and Mason, C. A. (1990), *Sem. Neurosci.,* 2, 455–465.
Hatten, M. E. and Mason, C. A. (1986), *Trends Neurosci.,* 9, 168–174.
Heldin, C., Wasteson, A., and Westermark, B. (1977), *Exp. Cell. Res.,* 109, 429–437.
Hertz, L. (1979), in *Progress in Neurobiology 13,* Pergamon, Oxford, 277–323.
Hirsch, M.-R., Wietzerbin, J., Pierres, M., and Goridis, C. (1983), *Neurosci. Lett.,* 41, 199–204.
Hoff, S. F., Scheff, S. W., and Cotman, C. W. (1982), *J. Comp. Neurol.,* 205, 253–259.
Hoffman, E. K. (1985), *Fed. Proc.,* 44, 2513–2519.
Hoffman, E. K. and Simonson, L. O. (1989), *Physiol. Rev.,* 69, 315–382.
Hoffmann, E. K., Simonsen, L. O., and Lambert, I. H. (1990), *J. Membr. Biol.,* 91, 227–244.
Holapainen, I. (1988), *Neurochem. Res.,* 13, 853–858.
Holtzman, D., DeVries, C., Nguyen, H., Jameson, N., Olson, J. E., Carrithers, M., and Bensch, K. (1982), *J. Neuropath. Exp. Neurol.,* 41, 652–663.
Holtzman, D., DeVries, C., Nguyen, H., Olson, J., and Bensch, K. (1984), *Neurotoxicology,* 5, 97–124.
Holtzman, D., Olson, J. E., DeVries, C., and Bensch, K. (1987), *Toxicol. Appl. Pharmacol.,* 89, 211–225.
Hunter, T. and Cooper, J. A. (1985), *Annu. Rev. Biochem.,* 54, 897–930.
Hunter, S. F. and Bottenstein, J. E. (1990), *Dev. Brain Res.,* 54, 235–248.
Huszti, Z., Rimanoczy, A., Juhasz, A., and Magyar, K. (1990), *Glia,* 3, 159–168.
Ingraham, C. A. and McCarthy, K. D. (1989), *J. Neurosci.,* 9, 63–69.
Jalonen, T. (1993), *Glia,* 9, 227–237.
Janzer, R. C. and Raff, M. C. (1987), *Nature,* 325, 253–257.
Jenkins, L. W., Becker, D. P., and Coburn, T. H. (1984), in *Recent Progress in the Study and Therapy of Brain Edema,* Go, K. G. and Baethmann, A., Eds., Plenum Press, New York, 523–537.
Jensen, S. and Jernelov, A. (1969), *Nature,* 223, 753–754.
Johnson, M. I., Higgins, D., and Ard, D. (1989), *Dev. Brain Res.,* 47, 289–292.
Johnson, M. D., Kamso-Pratt, J. M., Whetsell, W. O., Jr., and Pepinsky, R. B. (1989), *Am. J. Clin. Pathol.,* 92, 424–429.
Kalderon, N. and William, C. A. (1986), *Brain Res.,* 390, 1–9.
Kanner, B. I. (1993), *FEBS Lett.,* 325, 95–99.
Kawai, K., Takahashi, H., Wakabayashi, K., and Ikuta, F. (1989), *Acta Neuropathol. (Berl.),* 78, 449–454.
Keilhauer, G., Meier, D. H., Kuhlmann-Krieg, S., Nieke, J., and Schachner, M. (1985), *EMBO J.,* 4, 2499–2504.
Kimelberg, H. K. (1983), *Cell Mol. Neurobiol.,* 3, 1–16.
Kimelberg, H. K. and Frangakis, M. V. (1985), *Brain Res.,* 361, 125–134.
Kimelberg, H. K. and Ransom, B. R. (1986), in *Astrocytes,* Vol. 3, Fedoroff, S. and Vernadakis, A., Eds. Academic Press, Orlando, FL, 129–166.
Kimelberg, H. K. and Norenberg, M. D. (1989), *Sci. Am.,* 260, 66–76.
Kimelberg, H. K., Goderie, S. K., Higman, S., Pang, S., and Waniewski, R. A. (1990), *J. Neurosci.,* 10, 1583–1591.
Kimelberg, H. K. and Goderie, S. (1991), in *Glial-Neuronal Interactions, Ann. NY Acad. Sci.,* Vol. 633, Abbott, N. J., Ed., 619–622.
Kimelberg, H. K. (1992), *J. Neurotrauma Suppl.,* 9, S71–S81.
Kimelberg, H. K., O'Connor, E., Goderie, S. K., Higman, S., and Jalonen, T. (1993), in *Astrocytes, Pharmacology and Function,* Murphy, S. Ed., Academic Press, New York, 193–228.
Kimelberg, H. K. and Aschner, M., in *Proc. Neuroscience Satellite Meeting on Alcohol,* Washington, DC, November, 1993.
Kimelberg, H. K. and Norenberg, M. D. (in press), Astrocytic responses to central nervous system trauma.
Kinter, W. B. and Pritchard, J. B. (1977), in *Handbook of Physiology-Reactions to Environmental Agents,* Lee, D. H. K., Ed., American Physiological Society, Baltimore, MD, 563–576.
Kohut, J. J., Bednar, M. M., Kimelberg, H. K., and Gross, C. E. (1992), *Stroke,* 23, 93–97, 1992.
Komoly, S., Hudson, L. D., Webster, H., and Bondy, C. A. (1992), *Proc. Natl. Acad. Sci. U.S.A.,* 89, 1894–1898.
Krigman, M. R., Druse, M. J., Traylor, M. J., Wilson, M. H., Newell, L. R., and Hogan, E. L. (1974), *J. Neuropathol. Exp. Neurol.,* 33, 671–686.
Kroner, H. (1988), *Biol. Chem. Hoppe-Seyler,* 369, 149–155.
Kruse, J., Keilhauer, G., Faissner, A., Timpl, R., and Schachner, M. (1985), *Nature,* 316, 146–148.
Lach, H., Dziubek, K., Krawczyk, S., and Szaroma, W. (1977), *Acta Biol. Acad. Sci. Hung.,* 28, 367–373.
Lafarga, M., Berciano, M. T., Suarez, I., Viadero, C. F., Andres, M. A., and Berciano, J. (1991), *Neuroscience,* 40, 337–352.
Landis, D. M. D. and Reese, T. S. (1981), *J. Exp. Biol.,* 95, 35–48.
Landis, D. M. D. and Reese, T. S. (1982), *Neuroscience,* 7, 937–950.
Latov, N., Nilaver, G., Zimmerman, E. A., Johnson, W. G., Silverman, A., Defendindi, R., and Cote, L. (1979), *Dev. Biol.,* 72, 381–384.

Lee, I. S., Renno, W. M., and Beitz, A. J. (1992), *J. Comp. Neurol.,* 321, 65–82.
Leutz, A. and Schachner, M. (1981), *Cell Tissue Res.,* 220, 393–404.
Liccione, J. J. and Maines, D. M. (1988), *J. Pharmacol. Exp. Ther.,* 247, 151–161.
Liccione, J. J. and Maines, D. M. (1989), *J. Pharmacol. Exp. Ther.,* 248, 222–228.
Liesi, P., Kaakkola, S., Dahl, D., and Vaheri, A. (1984), *EMBO J.,* 3, 683–686.
Liesi, P. and Silver, J. (1988), *Dev. Biol.,* 130, 774–785.
Lim, R., Miller, J. F., Hicklin, D. J., Holm, A. C., and Ginsberg, B. H. (1985), *Exp. Cell Res.,* 159, 335–343.
Lindsay, R. M. (1979), *Nature,* 282, 80–82.
Lindsay, R. M. (1986), in *Astrocytes: Cell Biology and Pathology of Astrocytes,* Vol. 3, Fedoroff, S. and Vernadakis, A., Eds., Academic Press, Orlando, FL, 231–262.
Lögdberg, B., Berlin, M., and Schütz, A. (1987), *Scand. J. Work Environ. Health,* 13, 135–145.
Lögdberg, B., Brun, A., Berlin, M., and Schütz, A. (1988), *Acta Neuropathol. (Berl.),* 77, 120–127.
LoPachin, R. and Aschner, M. (1993), *Toxicol. Appl. Pharmacol.,* 118, 141–158.
MacVicar, B. A. (1984), *Science,* 2265, 1345–1247.
Malpiero, U. V., Frei, K., and Fontana, A. (1990), *J. Immunol.,* 144, 3816–3821.
Matthiessen, H. P., Schmalenbach, C., and Muller, H. W. (1989), *Glia,* 2, 177–188.
Matz, H. and Hertz, L. (1990), *Brain Res.,* 515, 168–172.
Maxwell, D. S. and Kruger, L. (1965), *J. Cell. Biol.,* 25, 141–157.
McMorris, F. A., Smith, T. M., DeSalvo, S., and Furlanetto, R. W. (1986), *Proc. Natl. Acad. Sci. U.S.A.,* 83, 822–826.
Meier, D. H. and Schachner, M. (1982), *J. Neurosci. Res.,* 7, 135–145.
Meyer, S. A., Ingraham, C. A., and McCarthy, K. D. (1989), *J. Neurosci. Res.,* 24, 251–259.
Miller, R. H., Smith, G. M., Rutishauser, U., and Silver, J. (1990), *Dev. Biol.,* 138, 377–390.
Muller, T., Moller, T., Berger, T., Schnitzer, J., and Kettenmann, H. (1992), *Science,* 256, 1563–1566.
Murabe, Y., Ibata, Y., and Sano, Y. (1981), *Cell Tissue Res.,* 216, 569–580.
Murphy, V. A., Wadhwani, K. C., Smith, Q. R., and Rapoport, S. I. (1991), *J. Neurochem.,* 57, 948–954.
Nakanishi, N. and Guroff, G. (1988), in *Neuronal and Glial Proteins,* Marangos, P. J., Campell, L. C., and Cohen, R. M., Eds., Academic Press, New York, 169–208.
Nathaniel, E. J. H. and Nathaniel, D. R. (1981), *Adv. Cell Neurobiol.,* 2, 249–301.
Neary, J. T. and Norenberg, M. D. (1992), in *Neuronal-Astrocytic Interactions: Pathological Implications,* Yu, A. C. H., Hertz, L., Norenberg, M. D., Sykova, E., and Waxman, S. G., Eds., Elsevier, New York.
Nechay, B. R., (1975), in *Mercury, Mercurials and Mercaptens,* Miller, M. W. and Clarkson, T. W., Eds., Charles C Thomas, Springfield, IL, 111–123.
Neugebauer, K. M., Tomaselli, K. J., Lilien, J., and Reichardt, L. F. (1988), *J. Cell. Biol.,* 107, 1177–1187.
Newman, E. A. (1986), *Science,* 233, 453–454.
Newman, E. A. (1991), *J. Neurosci.,* 11, 3972–3983.
Nieto-Sampedro, M. (1988), *Science,* 240, 1784–1786.
Noble, M., Fok-Seang, J., and Cohen, J. (1984), *J. Neurosci.,* 4, 1892–1903.
Noble, M., Murray, K., Stroobant, P., Waterfield, M. D., and Riddle, P. (1988), *Nature,* 333, 560–562.
Norenberg, M. D. (1979), *J. Histochem. Cytochem.,* 27, 469–475.
Norenberg, M. D. (1981), in *Advances in Cellular Neurology,* Vol. 2, Fedoroff, S. and Hertz, L., Eds., Academic Press, New York, 304–338.
Norenberg, M. D. (1983), in *Glutamine Glutamate and GABA in the Central Nervous System,* Hertz, L., Kvamme, E., McGeer, E. G., and Schousboe, A., Eds., Alan R. Liss, New York, 95–111.
Norenberg, M. D., Neary, J. T., Baker, L., Blicharska, J., and Norenberg, L. O. B. (1990), *Soc. Neurosci. Abstr.,* 16, 667.
Norton, W. T., Aouino, D. A., Hozumi, I., Chiu, F.-C., and Brosnan, C. F. (1992), *Neurochem. Res.,* 17, 877–885.
O'Callaghan, J. P. (1993), in *Markers of Neuronal Injury And Degeneration,* Johannessen, J. N., Ed., *Ann. NY Acad. Sci.,* 679, 188–210.
Olafsdottir, K., Pascoe, G. A., and Reed, D. J. (1988), *Arch. Biochem. Biophys.,* 263, 226–235.
Olney, J. W. (1979), in *Advances in Neurology,* Vol. 23, Chase, T. N., Wexler, N. S., and Barbeau, A., Eds., Raven Press, New York, 609–624.
Osterberg, K. and Watterberg, L. (1963), *Proc. Soc. Exp. Biol. Med.,* 113, 145–147.
Oyake, Y., Tanaka, M., Kubo, H., and Cichibu, H. (1966), *Adv. Neurol. Sci.,* 10, 744–750.
Pasantes-Morales, H. and Schousboe, A. (1988), *J. Neurosci. Res.,* 20, 505–509.
Pasantes-Morales, H., Moran, J., and Schousboe, A. (1990), *Glia,* 3, 427–432.
Peckham, N. H. and Choi, B. H. (1986), *Exp. Mol. Pathol.,* 44, 230–234.
Perraud, F., Besnard, F., Pettmann, B., Sensenbrenner, M., and Labourdette, G. (1988), *Glia,* 1, 124–131.
Petito, C. K. and Babiak, T. (1982), *Ann. Neurol.,* 11, 510–518.
Petito, C. K., Morgello, S., Felix, I. C., and Lesser, M. L. (1990), *J. Cereb. Blood Flow Metab.,* 10, 850–859.
Petito, C. K., Chung, M., Verkhovsky, L. M., and Cooper, A. J. L. (1992), *Brain Res.,* 569, 275–280.
Pettmann, B., Weibel, N., Sensenbrenner, M., and Labourdette, G. (1985), *FEBS Lett.,* 189, 102–108.
Pettmann, B., Labourdette, G., Weibel, M., and Sensenbrenner, M. (1986), *Neurosci Lett.,* 68, 175–180.
Pixley, S. K. R. and de Vellis, J. (1984), *Dev. Brain Res.,* 15, 201–210.
Politis, M. J. (1989), *J. Neurol. Sci.,* 92, 71–79.

Pollerberg, G. E., Burridge, K., Krebs, K. E., Goodman, S. R., and Schachner, M. (1987), *Cell Tiss. Res.*, 250, 227–236.
Price, J. and Hynes, R. O. (1985), *J. Neurosci.*, 5, 2205–2211.
Prohaska, J. R. (1987), *Physiol. Rev.*, 67, 858–901.
Pruss, R. M., Bartlett, P. F., Gavrilovic, J., Lisak, R. P., and Rattray, S. (1982), *Dev. Brain Res.*, 2, 19–35.
Raff, M. C., Williams, B. P., and Miller, R. H. (1984), *EMBO J.*, 3, 1857–1864.
Raff, M. C., Abney, E. R., and Fok-Seang, J. (1985), *Cell*, 42, 61–69.
Raisman, G. and Field, P. M. (1973), *Brain Res.*, 50, 241–264.
Rakic, P. (1972), *J. Comp. Neurol.*, 145, 61–84.
Rakic, P. (1990), *Experientia*, 46, 882–891.
Rakic, P. and Siedman, R. L. (1973), *Proc. Natl. Acad. Sci. U.S.A.*, 70, 240–244.
Rataboul, P., Faucon-Biguet, N. F., Vernier, P., De-Vitry, F., Boularand, S., Privat, A., Mallet, J. (1988), *J. Neurosci. Res.*, 20, 165–175.
Reichenbach, A. (1989), *Glia*, 2, 71–77.
Reier, P. J. (1986), in *Astrocytes: Cell Biology and Pathology of Astrocytes*, Fedoroff, S. and Vemadakis, A., Eds., Academic Press, Orlando, FL 263–324.
Reuhl, K. R. and Chang, L. W. (1979), *Neurotoxicology*, 1, 21–55.
Risau, W. and Wolfburg, H. (1990), *Trends Neurosci.*, 13, 174–178.
Rodier, P. M., Aschner, M., and Sager, P. R. (1984), *Neurobehav. Toxicol. Teratol.*, 6, 379–385.
Rosenberg, P. A. and Aizenman, E. (1989), *Neurosci. Lett.*, 103, 162–168.
Rosenblatt, D. E., Cotman, C. W., Nieto-Sampedro, M., Rowe, J. W., and Knauer, D. J. (1987), *Brain Res.*, 415, 40–48.
Rotwein, P., Burgess, S. K., Milbrandt, J. D., and Krause, J. E. (1988), *Proc. Natl. Acad. Sci. U.S.A.*, 85, 265–269.
Rudge, J. S. and Silver, J. (1990), *J. Neurosci.*, 10, 3594–3603.
Sandberg, M., Ward, H. K., and Bradford, H. F. (1985), *J. Neurochem.*, 44, 42–47.
Saneto, R. P. and De Velis, J. (1985), *Proc. Natl. Acad. Sci. U.S.A.*, 82, 3509–3513.
Schachner, M. (1982), *Trends Neurosci.*, 5, 225–228.
Schachner, M. (1991), in *Glial-Neuronal Interaction*, Abbot, N. J., Ed., New York Academy of Science, New York, 105–112.
Schiffer, D., Giordana, M. T., Migueli, A., Giaccone, G., Pezzotta, S., and Mauro, A. (1986), *Brain Res.*, 374, 110–118.
Schnyder, B., Weber, E., Fierz, W., and Fontana, A. (1986), *J. Neuroimmunol.*, 10, 209–218.
Schoemaker, H., Morelli, M., Deshmukh, P., and Yamamura, H., (1982), *Brain Res.*, 248, 396–401.
Schoepp, D. D. (1985), *J. Neurochem.*, 45, 1481–1486.
Schuch, U., Lohse, M. J., and Schachner, M. (1989), *Neuron*, 3, 13–20.
Seyfried, T. and Yu, R. K. (1985), *Mol. Cell. Biochem.*, 68, 3–10.
Shain, W. and Martin, D. L. (1990), in *Taurine, Functional Neurochemistry, Physiology and Cardiology,* Pasantes-Morales, H., Martin, D. L., Shain, W., Martin, and del Rio, R., Eds., Wiley-Liss, Inc, New York, 243–252.
Shea, T. B. (1991), *Cell. Biol. Int. Rep.*, 15, 437–443.
Shemancik, L., Cory-Slechta, D., and Finkelstein, J. N. (1992), *Neurosci. Abstr.*, 18, 1461.
Shemancik, L. and Finkelstein, J. N. (1993), *Toxicologist*, 13, 167.
Shimizu, N. and Hamuro, Y. (1958), *Nature*, 181, 781–782.
Sierra, E. M. and Tiffany-Castiglioni, E. (1991), *Toxicology*, 65, 295–304.
Siesjo, B. K. (1981), *J. Cereb. Blood Flow Metab.*, 1, 155–185.
Siman, R., Card, I. P., Nelson, R. B., and Davis, L. G. (1989), *Neuron*, 3, 275–285.
Simons, T. J. B. (1984), *FEBS Lett.*, 172, 250–264.
Simons, T. J. B. (1986), *J. Physiol.*, 378, 287–312.
Simons, T. J. B. and Pocock, G. (1987), *J. Neurochem.*, 48, 383–389.
Simpson, D. L., Morrison, R., de Vellis, J., and Herschman, H. R. (1982), *J. Neurosci. Res.*, 8, 453–462.
Smith, G. M., Rutishauser, U., Silver, J., and Miller, R. H. (1990), *Dev. Biol.*, 138, 377–390.
Smith, M. E., Somera, F. P., and Eng, L. F. (1983), *Brain Res.*, 264, 241–253.
Sontheimer, H., Kettenmann, H., Backus, K. H., and Schachner, M. (1988), *Glia*, 1, 328–336.
Sontheimer, H. (1991), *Can. J. Physiol. Pharmacol.*, 70, S223–S238.
Sontheimer, H., Black, J. A., Ransom, B. R., and Waxman, S. G. (1992), *J. Neurophysiol.*, 68, 985–1000.
Stewart, P. A. and Wiley, M. J. (1981), *Dev. Biol.*, 84, 183–193.
Stitt, T. N. and Hatten, M. E. (1990), *Neuron*, 5, 639–649.
Stowe, H. D., Goyer, R. A, Krigman, M. R., Wilson, M. A., and Cates, M. (1973), *Arch. Pathol.*, 95, 106–116.
Streit, W. J. and Kreutzberg, G. W. (1988), *J. Comp. Neurol.*, 268, 248–263.
Sweadner, K. J. (1991), *Can. J. Physiol. Pharmacol.*, 70, S255–S259.
Takada, M., Li, Z. K., and Hattori, T. (1990), *Brain Res.*, 509, 55–61.
Takeuchi, T., Morikawa, N., Matsumoto, H., and Shiraishi, A. (1962), *Acta Neuropathol.*, 2, 40–57.
Thio, C. L., Waxman, S. G., and Sontheimer, H. (1993), *J. Neuropathol.*, 69, 819–831.
Thomas, J. A., Dallenbach, F. D., and Thomas, M. (1972), *J. Pathol.*, 109, 45–50.
Tiffany-Castiglioni, E., Zmudzki, J., Wu, J.-N., and Bratton, G. R. (1986), *Toxicology*, 42, 303–315.
Tiffany-Castiglioni, E., Sierra, E., Wu, J. N., and Rowles, T. (1989), *Neurotoxicol.*, 10, 417–444.
Tomsig, J. L. and Suszkiw, J. B. (1990), *Am. J. Physiol.*, 259, C762–C768.
Tomsig, J. L. and Suszkiw, J. B (1991), *Biochim. Biophys. Acta*, 1069, 197–200.

Unitt, J. F., McCormack, J. G., Reid, D., MacLachlan, L. K., and England, P. J. (1989), *Biochem. J.*, 262, 293–301.
van den Pol, A. N., Finkbeiner, S. M., and Cornell-Bell, A. H. (1992), *J. Neurosci.*, 12, 2648–2664.
Vaughn, I. E. and Pease, D. C. (1970), *J. Comp. Neurol.*, 140, 207–226.
Vernadakis, A. (1991), in *Encyclopedia of Human Biology,* Academic Press, Orlando, FL, 433–446.
Vijayan, V. K. and Cotman, C. W. (1983), *Neurobiol. Aging,* 4, 13–23.
Vinogradov, A. and Scarpa, A. (1973), *J. Biol. Chem.,* 248, 5527–5531.
Vitarella, D., DiRisio, D., Kimelberg, H. K., and Aschner, M. *J. Neurochem.*
Walz, W. (1988), in *Glial Cell Receptors,* Kimelberg, H. K., Ed., Raven Press, New York, 121–130.
Waniewski, R. A. and Martin, D. L. (1986), *J. Neurochem.*, 47, 304–313.
Webb, J. L. (1966), in *Enzyme and Metabolic Inhibitors,* Vol. 2, Webb, J. L., Ed., Academic Press, New York, 729–1070.
Wedler, F. C. and Denman, R. B. (1984), *Curr. Top. Cell. Regul.,* 24, 153–169.
Wedler, F. C., Ley, B. W., and Grippo. A. A. (1989), *Biocem. Res.,* 14, 1129–1135.
Wedler, F. C. and Toms, R. (1986), in *Manganese in Metabolism and Enzyme Function,* Schramm V. L. and Wedler F. C., Eds., Academic Press, New York, 221–228.
Wedler, F. C. (1993), in *Progress in Medicinal Chemistry,* Vol. 30, Ellis, G. P. and Luscombe, D. K., Eds., Elsevier Science Publishers, Berlin, 89–133.
Wood, J. M., Kennedy, F. S., and Rosen, C. E. (1968), *Nature,* 220, 173–174.
Yamamoto, C. and Kawana, E. (1990), *Okajimas Folia Anat. Jpn.,* 67, 21–29.
Young, J. K. (1988), *Brain Res. Bull.* 20, 97–104.
Yu, M. C., Bakay, L., and Lee, J. C. (1972), *Acta Neuropathol. (Berl.),* 22, 235–244.

Chapter 36

Choroid Plexus and Metal Toxicity

Wei Zheng

I. INTRODUCTION

The choroid plexus constitutes a barrier between the blood and cerebrospinal fluid (CSF). To maintain the homeostasis of the CSF, the choroid plexus excludes most water-soluble materials from the blood and allows only selective exchange between the two body fluids. Thus, the integrity of the barrier, both structurally and functionally, is critical to its role in maintaining the chemical stability of the internal milieu of the central nervous system (CNS). Increasing evidence indicates that heavy metals and metalloids accumulate in the choroid plexus at concentrations much greater than those found in the CSF and elsewhere in brain tissues. Such a high deposition of toxic metals in the choroid plexus has suggested that this tissue may serve as a target for toxicities associated with environmental exposure to heavy metals.

From the toxicological point of view, however, the overall significance and importance of the choroid plexus has received surprisingly little attention. This is due, in part, to our past ignorance of its role relative to whole brain function. Indeed, until recently, some had viewed the choroid plexus as physiologically insignificant tissue because of its small tissue mass.

The present chapter deals with current progress in unraveling the toxicological aspect of the choroid plexus, particularly in metal-induced toxicities. The structure and normal function of the choroid plexus will first be introduced. Inasmuch as our understanding of the toxic effects of heavy metals on the choroid plexus have arisen primarily through observations of metal accumulation in that tissue, examples of these phenomena are presented in detail. Finally, future research needs concerning the toxicology of the choroid plexus are discussed.

II. STRUCTURE AND FUNCTION OF THE CHOROID PLEXUS

A. STRUCTURE

The choroid plexus is developed primarily from spongioblasts. In humans, the choroid plexus first appears in roof of the fourth ventricle at the 9th week of ontogeny; it subsequently appears in the lateral and third ventricles at approximately 9 to 10 weeks (Nishmimura, 1983). During the 10th week, the choroid plexus becomes granulated and begins to perform its secretory functions, although the entire structure is not matured or fully developed until the 6th gestational month (Milhorat, 1976; Nishimura, 1983). The adult choroid plexus extends along the floor of the lateral ventricle, hangs down from the roof of the third ventricle, and overlies the roof of the fourth ventricle. The size of the mature choroid plexus relative to brain varies widely among vertebrates, but is typically less than 1% of brain weight (Cserr et al., 1980). In normal human subjects, the choroid plexus weighs about 2 to 3 grams.

The blood supply of the choroid plexus is derived from two posterior choroidal arteries, which branch from the internal carotid arteries. On the basis of the data from experimental animals, the blood flow rate to the choroid plexus is about 4 to 6 ml/min/g tissue (Maktabi et al., 1990; Mayhan et al., 1989; Page et al., 1980; Schalk et al., 1989).

Histologically, the choroid plexus consists of three cellular layers: apical epithelial cells, underlying supporting connective tissue, and an inner layer of endothelial cells (Figure 1). The apical layer consists of numerous closely packed, cuboidal or columnar epithelial cells. On the cell surface are many primary microvilli projecting into the cerebral ventricles. The choroidal epithelial cells possess tight junctions (zonulae occludentes) near their apical surface, which seal one epithelial cell to another, providing a structural basis for the blood-CSF barrier. The cytoplasm of the epithelia is rich in mitochondria, rough endoplasmic reticulum, and has well-developed Golgi apparatus. These features suggest an active secretory process for the production of CSF. The choroidal epithelial intracellular pH of 7.0 is slightly more acidic than that of blood (pH = 7.4) or CSF (pH = 7.3) (Johanson, 1978).

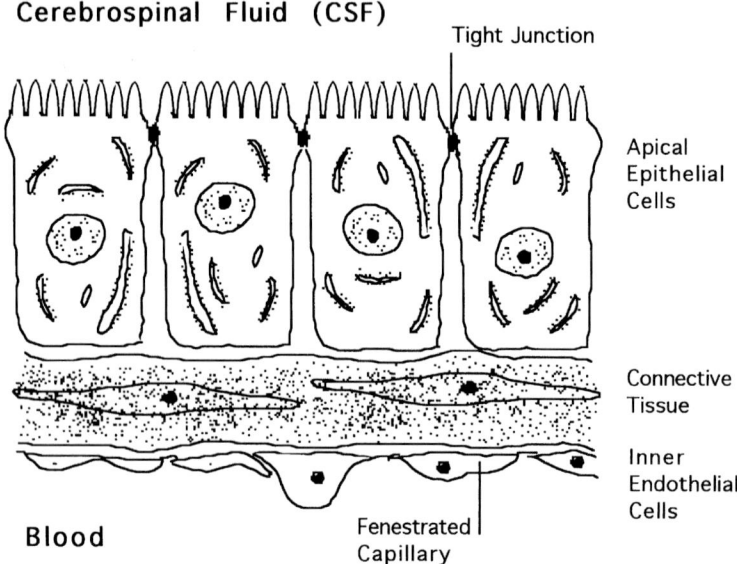

Figure 1. The choroid plexus consists of three layers. The apical columnar epithelial cells contact directly with CSF. The tight junctions between epithelial cells provide the structural basis for the blood-CSF barrier. Underneath the epithelial cells is a thin connective tissue. The inner endothelial cells lining the choroidal capillaries are in direct contact with blood. The endothelial cells are extremely fenestrated.

The basal side of the epithelial cells rests on a thin stromal connective tissue that contain a few many macrophages (Milhorat, 1976; Matyszak et al., 1992). Next to the connective tissue are the endothelial cells lining the choroidal capillaries. The endothelial cells are extremely fenestrated and are considered to be functionally "leaky". Large molecules such as proteins can pass from the blood through the fenestrated capillary and into the connective tissue. Yet, most of these materials are prevented from entering the CSF by the tight junctions between the epithelial cells. One must keep in mind that in the blood-brain barrier, tight junctions are located between the endothelial cells, whereas in the blood-CSF barrier, the tight junctions exist between the epithelial cells. Studies among a variety of species reveal that there are no appreciable species differences in the ultrastructure of the choroid plexus (Milhorat, 1976).

In comparison to its small tissue mass, the choroid plexus has a very large surface area. Besides the primary microvilli, the entire choroid plexus is so pleated that it creates secondary multiform macrovilli, which greatly increase the total surface area of the tissue. As estimated from the tissues of 1-month old rats, the total apical surface area of the choroidal epithelium approximates 75 cm^2, about one half that of the blood-brain barrier (155 cm^2) (Keep and Jones, 1990). The broad surface area together with the fast blood flow of the blood-CSF barrier ensures an efficient exchange of materials between the CSF and blood.

In summary, the tight junctions between the epithelial cells, the fenestrated endothelial cells lining the blood vessels, and a large surface area of the choroidal epithelia represent the most important anatomical features of the blood-CSF barrier.

B. BIOCHEMISTRY AND PHARMACOLOGY

Early metabolic study suggested that the general metabolic activities of the choroid plexus were about one third to one half the values of the kidney (Fisher and Copenhaver, 1957). The estimate, however, depended on a limited experimental approach whereby only the activities of anaerobic glycolysis, succinic dehydrogenase, and cytochrome oxidase were compared between two tissues. In fact, many enzymes have been identified in the choroid plexus by using more recent enzymatic histochemical techniques (Kim et al., 1990; Masuzawa and Sato, 1983; Riachi and Harik, 1992). The choroidal epithelium contains high activities of Na^+/K^+-ATPase and carbonic anhydrase, enzymes also found to be actively involved in fluid production in the renal proximal tubule (Harik et al., 1985; Masuzawa and Sato, 1983; Masuzawa et al., 1984). Inhibition of both enzymes by selective inhibitors, ouabain and acetazolamide, respectively, resulted in a significant reduction in the production of CSF (Faraci et al., 1990; Johanson and Murphy, 1990; Pollay et al., 1985; Vogh, 1980).

The choroid plexus also possesses many enzymes that are of particular toxicologic interest, such as protein kinase C (PKC) (Zheng et al., 1995), cytochrome P450 enzymes (Volk et al., 1991), glutathione (GSH) -s-transferase (GST) (Carder et al., 1990; Senjo et al., 1986), γ-glutamyltranspeptidase (Prusiner and Prusiner, 1978a,b), cytochrome or monoamine oxidase (Kim et al., 1990; Riachi and Harik, 1992), superoxide dismutase (SOD), GSH peroxidase and reductase, and catalase (Tayarani et al., 1989). Among these enzymes, GST catalyzes the conjugation reaction between GSH and a wide range of electrophiles. GSH reductase and catalase function to protect tissues against oxidative damage. It is likely that these abundant enzymes in the choroid plexus may play a role in maintaining the integrity of the blood-CSF barrier.

With regard to drug-metabolizing enzymes, the choroid plexus contains phenytoin (a widely used antielepy drug)-inducible P450IIB1 isozyme. By using polyclonal antibodies against phenytoin-induced mouse cerebral microsomal P450, Volk et al. (1991) have reported that the choroid plexus is strongly P450IIB1 immunoactive. The pharmacological implication of this finding remains unclear. One can speculate, however, that this enzyme might contribute, to a certain extent, to the metabolism of phenytoin in brain.

Physiologically, the choroid plexus is innervated by sympathetic, parasympathetic, and peptidergic nerve endings (Lindvall and Owman, 1981; Nilsson et al., 1990). Many pharmacologically important receptors commonly found on blood vessels have been identified in the vasculature of the choroid plexus, including α and β-receptors (Lindvall and Owman, 1981), muscarinic receptors (Rotter et al., 1979), and dopamine receptors (Nicklaus et al., 1988; Townsend et al., 1984). In addition, choroidal epithelial cells possess receptors for neurotransmitters and polypeptide hormones. For example, the epithelial cells contain rich serotonin receptors (Conn et al., 1986; Yagaloff and Hartig, 1985). Activation of serotonin receptors produced an increase in phosphatidyl-inositol turnover in the choroidal epithelia, a reaction that is 10-fold more potent in the choroid plexus than in the brain cortex (Conn et al., 1986; Hoyer et al., 1989). Other receptors found in the epithelial cells include prolectin receptors (Di Carlo et al., 1992; Lai et al., 1992), $GABA_A$ receptors (Amenta et al., 1989), atrial natriuretic peptide receptor (Konrad et al., 1990), nerve growth factor receptors (Vega et al., 1992), and insulin receptors (Nilsson et al., 1992).

Binding of various ligands to their receptors in the choroid plexus regulates the production and constitution of the CSF. The regulation can occur at the inner endothelial layer so as to adjust the blood flow to the choroid plexus, or at the apical epithelial layer to govern the enzyme activities necessary for iron transport. A recent investigation has found that interactions between the ligands and receptors can alter the permeability of the blood-CSF barrier (Nag, 1991). Binding of atrial natriuretic factor (ANF) to its receptor in the blood-CSF barrier caused an accelerated passage of a protein marker, horseradish peroxidase (HRP), and an altered permeability of the ion, lanthanum, from blood into the choroid plexus. It is highly possible that drugs and chemicals may act upon these receptors and thereby alter the production and content of CSF.

C. FUNCTIONS OF THE CHOROID PLEXUS
1. The Blood-CSF Barrier

Figure 2 illustrates the four fluid compartments of the mammalian brain and their interrelationships. As the brain must function in a chemically stable intramural environment, the brain relies heavily on protective mechanisms that rigorously restrict access of substances from the systemic circulation into the other three compartments. The blood-brain barrier regulates the exchange between blood and interstitial fluid (ISF) that surrounds neuronal cells, whereas the choroid plexus constitutes another defense line between the blood and CSF. Since there is no barrier evident between the ISF and the CSF, substances in these two fluid compartments can exchange freely. In principle, substances that enter the CSF can ultimately come into contact with neurons.

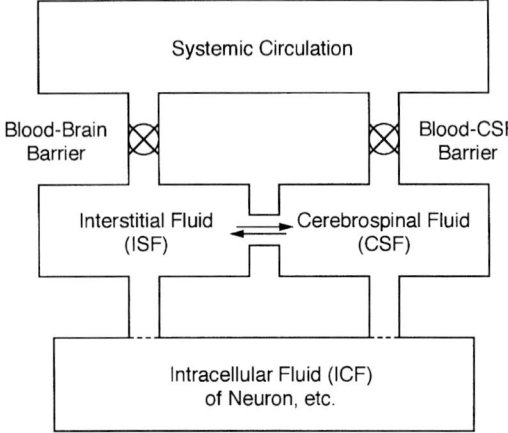

Figure 2. The blood-brain barrier and the blood-CSF barrier restrict access of substances from the systemic circulation into ISF and CSF. The ISF provides an intracerebral transport system that allows movement of nutrients, toxins, and products of metabolism. Direct continuity of the CSF and ISF allows free exchange of substances within these two compartments. Thus, substances entering the CSF can come into contact with neurons.

As mentioned above, tight junctions between the choroidal epithelial cells constitute the blood-CSF barrier. These tight junctions impede the diffusion of water soluble molecules or ions from the blood. In early classical studies using dye-staining or silver tracer techniques, the dyes or tracers, when given to animals, caused a substantial stain in the choroid plexus, while surrounding brain parenchymal tissues were spared (van Deurs and Amtorp, 1978). The tight junctions also hinder the movement of proteins and other macromolecules across the blood-CSF barrier. For example, HRP, cytochrome c, and microperoxidase can pervade plexus connective tissue and trespass the epithelial basement membrane, yet fail to reach the CSF plexus due to the presence of tight junctions (van Deurs, 1980).

It is necessary to point out that the tight junctions between the epithelial cells in blood-CSF barrier seem less effective, or somewhat more "leaky", than those between endothelial cells of the blood-brain barrier (Davson et al., 1987). This is particularly true in infants and children of very young age whose plexus structures are not fully developed. Hence, the incomplete or impaired barrier structure due to the immaturity or some disease states may increase the potential risk of exposure of the CNS to environmental substances. This latter notion, however, lacks clear-cut experimental proof.

2. Secretory Function

A major function of the choroid plexus is to secrete CSF. The human choroid plexus produces approximately 80 to 90% of CSF at a rate of 0.3 to 0.4 ml per min (Davson et al., 1987; Milhorat, 1976). The remaining 10 to 20% of CSF are secreted by the ventricular ependyma. The secretion of CSF by the choroid plexus is an active process including transport of Na^+, Cl^- and HCO_3^- from capillaries to the ventricles (Johanson et al., 1979; Johanson, 1995). Movement of these ions results in an osmotic pressure across the cell surface of the epithelium. The plasma water component is then driven by this osmotic force into the ventricles.

Of many ion pumps present in choroidal epithelium, the Na^+/K^+ exchange pump plays the most fundamental role in secretion of CSF. Na^+/K^+ exchange occurs on the apical side of the epithelium (Ernst et al., 1986, Masuzawa et al., 1981). Wright (1979) estimated that one single choroidal epithelial cell may have about 10^6 pumps. Each pump operates in such a way that it transports 3 Na^+ ions from the epithelium to the CSF with 2 K^+ ions being transported in the opposite direction. The efficiency of Na^+/K^+ exchange essentially depends on the activity of Na^+/K^+-ATPase. Inhibition of Na^+/K^+-ATPase activity abolished Na^+/K^+ exchange, resulting in a significant reduction in the rate of CSF formation (Lindvall-Axelsson et al., 1989; Pollay et al., 1985). Therefore, Na^+/K^+-ATPase activity may reflect the functional integrity of the choroid plexus. The toxicologic significance of the enzyme will be discussed later.

In addition to manufacture of the CSF, the choroid plexus secretes various proteins and peptide hormones into the CSF. The barrier structure guarantees much lower protein concentrations in the CSF than those in plasma. Nevertheless, the choroid plexus itself synthesizes proteins and releases them into the CSF. Transthyretin, transferrin, and ceruloplasmin, for example, are synthesized in the choroidal epithelial cells where the mRNAs for those molecules are expressed at levels higher in the other parts of the brain (Schreiber and Aldred, 1993). Transthyretin is responsible for transport of thyroid hormone (thyroxine) from the blood to the CSF, while transferrin and ceruloplasmin may participate in transport of iron and copper into the CSF, respectively.

3. Transport Function

Unlike a simple barrier, the choroid plexus is actively engaged in regulating the composition of the CSF. Bidirectional transport, rather than unidirectional influx or outflow, regulates CSF homeostasis. The choroid plexus, by extracting useful materials from the blood, controls the influx of substances into the CSF. Essential metal ions, for instance, are transported by ionic pumps or transport systems, such as the Na^+-H^+ exchanger (Mayer and Sanders-Bush, 1993; Murphy and Johanson, 1989) and the Na^+-Cl^- co-transporter (Johanson et al., 1990), Na^+/K^+-ATPase (Parmelee and Johanson, 1989; Parmelee et al., 1991; and Bairamian et al., 1991). The choroid plexus extracts or obtains micronutrients from the systemic circulation for support of the brain's function. Some of these molecules, such as ascorbate, nucleotides, vitamin B12, and folate, are presumably taken into the epithelial cells at the basolateral surface by active transport, followed by their release into the CSF (at the apical surface) by facilitated diffusion (Spector, 1989). Many other examples include certain amino acids (glycine, L-alanine) (Preston and Segal, 1992; Stoll et al., 1993), hormones (thyroid hormones, melatonin, growth hormone) (Dratman, et al., 1991; Lai et al., 1991; Schreiber et al., 1990), and peptides (atriopeptin, vasopressin) (Nilsson et al., 1992).

The choroid plexus also transports materials in the opposite direction, i.e., from the CSF to blood. Since there is no lymphatic system in the CNS, the constant removal of wastes by the choroid plexus, in addition to the cerebral circulation via the arachnoid villi, may provide a useful mechanism to prevent a build-up of the brain's metabolites. An active transport system that has been shown to remove cephalosporin antibiotics from the CSF plays a major role in determining the pharmacokinetics of β-lactam antibiotics in the CSF (Nohjoh et al., 1989). Cimetidine, an H_2-receptor antagonist, is eliminated from the CSF by a saturable, carrier-mediated active transport process in the choroid plexus (Suzuki et al., 1986; Whittico et al., 1990). Ca^{2+} pumps located in the CSF-facing membrane (Borke et al., 1989) ensure a low concentration of Ca^{2+} in the CSF even under extreme conditions, such as those in experiments by Murphy et al., (1986, 1989), where the plasma Ca^{2+} was either decreased by a low Ca diet or increased by a high Ca diet. The uptake of amino acids from the CSF by choroid plexus ensures a low level of amino acids in the CSF (Davson et al., 1982). The choroidal epithelium also transports protein, by vesicular and micropinocytic mechanisms, from the apical surface to the intercellular spaces (van Deurs and Amtorp, 1978). Other examples include inorganic anions such as I^- and SCN^- (Davson and Pollay, 1963; Pollay and Kaplan, 1972); organic anions such as benzylpenicillin (Suzuki et al., 1987), bilirubin (Jakobson, 1991), and anionic pesticides (Kim et al., 1983; Pritchard, 1980); organic cations such as quaternary ammonium compounds (Gosling and Lu, 1968; Lanman and Schanker, 1980; Miller and Ross, 1975); prostaglandins (DiBenedetto and Bito, 1986); some peptides (Huang, 1982), and many drugs and chemicals (Table 1).

Table 1 Chemicals Transported Out of Brain by the Choroid Plexus

Chemicals	Species/Modal	Refs.
Anionic pesticides	Rabbit/*in vivo*	Kim et al., 1983
	Rabbit/*in vitro*	Pritchard, 1980
Antipyrine and barbital	Rat/*in vivo*	Johanson and Woodbury, 1977
Atropine, methylatropine	Rabbit/*in vitro*	Winbladh, 1972
	Rabbit/*in vitro*	Eriksson and Winbladh, 1971
Cefodizime	Rat/*in vitro*	Nohjoh, et al, 1989
Cimetidine	Rat/*in vitro*	Suzuki et al., 1986
	Cow/*in vitro*	Whittico et al., 1990
Digoxin	Human/post-mortem	Bertler et al., 1973
	Human/post-mortem	Andersson et al., 1975
	Human/post-mortem	Krakauer and Steiness, 1978
Diphenhydramine	Rabbit/*in vivo*	Goldberg et al., 1987
Gentamicin	Rabbit/*in vitro*	Spector, 1975
Lignocaine (lidocaine)	Rabbit/*in vitro*	Spector, 1980
Methadone	Rabbit/*in vitro*	Huang and Takemori, 1975
Methotrexate	Rabbit/*in vitro*	Rubin et al., 1968
Morphine	Rabbit/*in vitro*	Takemori and Stenwick, 1966
	Rabbit/*in vitro*	Craig et al., 1971
	Rat/*in vitro*	Huang et al., 1980
Morphine-3-glucuronide	Rabbit/*in vitro*	Muraki, 1971
Narcotic analgesics	Rabbit, dog/*in vitro*	Hug 1967
Penicillin	Rabbit/*in vivo, in vitro*	Spector and Lorenzo, 1974
	Dog/*in vivo*	Dixon et al., 1969
Proline	Rabbit/*in vitro*	Agnew and Yuen, 1975
Salicylic acid	Rabbit, cat/*in vitro*	Lorenzo and Spector, 1973
	Rabbit, cat/*in vivo*	Spector and Lorenzo, 1973
Tetrahydrocannabinol (Marihuana)	Rabbit/*in vitro*	Agnew et al., 1976

The identity of the transporters in the choroidal epithelial membrane has been further investigated at the molecular level. An anion exchanger has been cloned from rat choroid plexus. Functional expression of its cDNA indicates that the gene encodes a Cl/HCO_3 exchanger (Julius et al., 1988; Lindsey et al., 1990). By examining the expression of mRNA in brain tissues using a cDNA probe, Smith's group in NIH (Stoll et al., 1993) demonstrated that the cationic amino acid transporter gene is present in both the choroid plexus and cerebral microvessels.

Thus, over the past 25 years, extensive experimental evidence indicates that the choroid plexus regulates bidirectional transport and provides a highly selective pathway for materials that communicate between the periphery and the brain. For a more comprehensive discussion, the reader is referred to review articles by a number of leading investigators in this filed (Bradbury, 1992; Johanson, 1989; Smith, 1991; Spector, 1990).

4. Immunological Function

The immunological function of the choroid plexus is not completely understood at present, although the tissue has been found to have an abundant macrophage population (Matyszak et al., 1992). The macrophages in the plexus, also called Kolmer or epiplexus cells, are anchored to the epithelial cell surface (Carpenter et al., 1970). These apical macrophages may function to remove debris from the CSF. Recently, Matyszak et al. (1992) demonstrated that macrophages were also present between the fenestrated capillaries and the basal surface of the epithelium. These cells seemed to be resident cells rather than cells that had migrated from elsewhere, because they express a wide range of antigens that were common to the macrophages from other tissues. Thus, the macrophages may play a role in precluding the entry of foreign antigens into the CNS, or in preventing CNS antigens from being exposed to the peripheral immune system. Nathanson and Chun (1989) have suggested a possible role of the choroid plexus in immunological communication between the CNS and periphery.

III. THE CHOROID PLEXUS AS A TARGET TISSUE FOR METAL TOXICITY

Because of its affluent blood supply and leaky structure, the choroid plexus is a target tissue for toxicities caused by a variety of toxicants. For example, an oral dose of cyclophosphamide (alkylating agent) induced severe choroid plexitis in rats. The damage was located primarily in the endothelial vessel (Levine and Sowinski, 1974). In contrast, oral administration of certain tertiary amines produced direct ultrastructural disruptions in the epithelial cells (Wenk et al., 1979). Damage caused by other toxic compounds, such as octanoic acid (Kim et al., 1990), 2,4,5-trichlorophenoxy-acetic acid (Kim et al., 1987), bis(4-amino-3-methylcyclohexyl)methane (Shibata et al., 1990), and piperamide (Benitz and Kramer, 1968), has also been documented. In addition to the morphological alterations, mitochondrial functions and cellular enzyme activities were also reduced by some of these agents.

The choroid plexus is also a target tissue for metal-induced toxicities. Like toxic chemicals, heavy metals may concentrate in epithelial cells via either the blood or CSF sides. Metals present in the choroid plexus most likely, however, come from the systemic circulation. The leaky endothelial lining of the choroidal blood capillaries allows ready access of epithelial cells to toxic metals in blood. The accumulation of metals resulting from the passage of metals from the blood to the CSF literally exposes the choroidal epithelial cells to high concentrations of toxic metals. Similarly, metals can be reabsorbed from the CSF into the blood. This process can also lead to a high accumulation of metals in the choroid plexus. The reabsorption of heavy metals from the CSF is possible because metals such as Pb and Hg are capable of passing across the blood-brain barrier, via proposed transport mechanisms, to enter the ISF and thereby the CSF.

A. NORMAL VALUES OF METAL IONS IN THE CHOROID PLEXUS, CSF, AND BLOOD

Appropriate concentrations of some essential and nonessential ions and trace elements in the choroid plexus, CSF and blood of the normal human subjects are listed in Table 2. It is noteworthy that the essential ions, normalized as to per gram of choroid plexus tissue, are generally much higher than those reported in CSF and blood. For example, while calcium concentration of the CSF is about half the value of the blood, the calcium of the choroid plexus is about a hundred times greater than either that of the CSF or that of the blood. As discussed above, the choroid plexus actively transports calcium across the blood-CSF barrier. The high concentrations of calcium in the choroid plexus may be somehow relevant to its regulatory function on calcium homeostasis in the CSF.

Table 2 Concentrations of Metals in the Choroid Plexus, CSF and Blood of Normal Human Subjects[a]

	Choroid Plexus (μg/g)	CSF (μg/ml)	Blood (μg/ml)
Potassium (K)	1,450	109	137
Sodium (Na)	2,520	3,173	3,173
Calcium (Ca)	18,700	84	192
Magnesium (Mg)	450	56	41
Chloride (Cl)	—	4,219	3,616
Phosphorus (P)	9,300	34[b]	47[b]
Barium (Ba)	0.90	0.04[c]	—
Copper (Cu)	0.70	—	1.06[d]
Iron (Fe)	60.00	—	1.2
Lead (Pb)	3.00	—	1–2
Lithium (Li)	—	0.01[c]	—
Rubidium (Rb)	—	0.06[e]	—
Silicon (Si)	15.00	—	—
Zinc (Zn)	39.00	—	0.92[d]

[a] Unless otherwise indicated, the values for the choroid plexus are primarily derived from Hershey et al. (1987); for the CSF and blood, mainly from Crill (1989).
[b] Davson et al., 1987, page 24.
[c] Palavinskas and Schulten (1984).
[d] Meret and Henkin (1971).
[e] El-Yazigi et al. (1992).

B. SEQUESTRATION OF HEAVY METALS BY THE CHOROID PLEXUS

As early as 1963, Berlin and Ullberg (1963a,b) reported depositions of Cd and Hg in a brain area that corresponded to the choroid plexus. Accumulation of heavy metals and metalloids by the mammalian choroid plexus has been further verified by results of studies in both humans and animals. Zheng and Aposhian observed that animals exposed to doses of Pb, Hg, Cd, and As approaching 25 to 33% of their LD_{50} retained much higher concentrations of those metal ions in the choroid plexus than in the brain cortex or CSF (Zheng et al., 1991). The concentration of Pb in the choroid plexus, for example, was about 57 times greater than that in the brain cortex; Hg was 12 times greater, Cd, 33 times greater; and As, 13 times greater. CSF concentrations of Pb, Hg, and As were about 70, 95, and 40 times less, respectively, than those found in the choroid plexus. This finding, along with many others (see below), indicates that the choroid plexus has the capacity to sequester toxic heavy metals. Some typical examples are discussed in the following.

1. Lead (Pb)

Of the heavy metals and metalloids causing neurotoxicity, Pb captures particular interest. Environmental exposure to Pb in children has been associated with cognitive deficits in children. Moreover, there appears to be no threshold with regard to the association between blood Pb concentration and intelligence (Wasserman et al., 1994). The mechanism whereby Pb adversely affects childhood brain development is not known.

Friedheim et al. (1983) indicates for the first time in the literature that Pb in the choroid plexus increased significantly with age, while Pb in the brain did not. Their conclusion was based on autopsy data from 51 human subjects who had lived in New York City, and who had died from causes other than Pb-induced encephalopathy. This conclusion was further confirmed by Manton et al. (1984), who reported a 100-fold increase of Pb in human choroid plexus compared with that in the brain cortex. A remarkable aspect of these findings is that an age-related accumulation of Pb in a particular tissue inside the brain was first emphasized, a phenomenon possibly associated with environmental exposure.

A dose-time study in rats indicates that accumulation of Pb in the choroid plexus is both dose-dependent and time-related (Zheng et al., 1991). The concentrations of Pb in the choroid plexus increased proportionally with the increase in dose, while Pb concentrations in the brain cortex and CSF were not significantly changed. Following acute administration of Pb acetate (50 mg/kg, i.p.), the choroid plexus Pb continued to increase. The accumulation level did not reach a plateau even at 24 h (Zheng et al., 1991). O'Tuama et al. (1976) studied the mechanism whereby Pb accumulates in the choroid plexus. Pretreatment of animals with ouabain, which inhibits Na^+, K^+-ATPase, significantly reduced Pb uptake by the choroid plexus. These results led to the conclusion that in the rat, an active transport is involved in Pb uptake by the choroid plexus. Recent studies on Pb transport at the blood-brain barrier suggest that the blood-brain barrier is highly permeable to Pb. The net influx of Pb into the brain seems dependent upon two transport mechanisms that operate in opposite directions. A rapid influx of Pb at the brain endothelium may occur by a passive transport mechanism involving a $PbOH^+$ species, whereas the outflow in the opposite direction that transport Pb back into the capillary lumen may be via an active Ca^{2+}-ATP-dependent mechanism (Bradbury and Deane, 1988; Deane and Bradbury, 1990).

2. Mercury (Hg)

Hg is another important metal pollutant in the environment (Clarkson, 1987; Swain et al., 1992). The release of organic mercurial compounds into the environment has led to human epidemic tragedies in Japan and Iraq (Clarkson, 1987). Some mercury compounds, particularly alkyl derivatives, are severely neurotoxic. On their way to enter the brain, mercury compounds were highly concentrated in the choroid plexus. In one clinical case, a 63-year-old lady took inorganic mercury-containing laxative for 25 years for the treatment of chronic constipation. Pathological examination showed that many small punctuate granules, representing the deposits of mercury, accumulated in the choroid plexus. Further analysis of the tissue samples from different brain regions revealed that the choroid plexus contained high amounts of Hg (26 µg/g), about 5 and 45 times higher than those seen in brain cortex and in the CSF, respectively (Davis et al., 1974).

In animal studies, the choroid plexus accumulates Hg following administration of either inorganic Hg (Berlin and Ullberg, 1963a; Steinwall and Olsson, 1969; Suda et al., 1989; Zheng et al., 1991) or organic Hg (Moller-Madsen, 1990, 1991; Suda et al., 1989). Using photo-emulsion histochemical techniques, Suda et al. (1989) identified mercury granules in choroidal epithelial cells of rats treated with both inorganic and organic Hg. Further, they found that these granules represented inorganic mercury other than organic derivatives. Subcellularly, mercury and its derivatives rested primarily within the lysosomes of the choroid plexus, which was similar to the deposition or mercury in other brain tissues (Moller-Madsen, 1990, 1991).

An active transport process at the blood-brain barrier may drive the organic mercurial compounds into the brain (Aschner and Clarkson, 1988, 1989; Kerper et al., 1992; Hirayama, 1980). Clarkson's group proposed that the complex formed between methylmercury and cysteine, which has a structural similarity to methionine, may be transported by a carrier-mediated transporter in brain capillaries. In addition, they proposed that plasma MeHg-glutathione complex may serve as a source of MeHg-cysteine (Kerper et al., 1992). It will be interesting to find out whether the same transport mechanism(s) operate in the blood-CSF barrier.

3. Cadmium (Cd)

The brain is not a primary target for Cd toxicity, although some behavioral changes in animal models have indeed been observed following administration of Cd (Nation et al., 1983; Rastogi, et al., 1977). The effectiveness of the barrier systems that impede passage of Cd ions into the CNS may play a key role in protecting the brain from Cd toxicity. Evidence indicates that Cd does accumulate in the choroid plexus. In both chronic (22 weeks) and acute (1 to 24 days) exposure models, the level of Cd in the choroid plexus was found to be high, while Cd in the CSF fell below the detection limit (Arvidson and Tjalve, 1986; Valois and Webster, 1987, 1989a; Zheng et al., 1991). A postmortem human study revealed that Cd concentration in the choroid plexus was about 2 to 3 times higher than that found in the brain cortex (Manton et al., 1984).

In addition to entry from the blood side, the uptake of Cd by the choroid plexus may take place at the apical side of the epithelium. When the choroid plexus was pretreated with ouabain *in vitro*, the amount of Cd taken up by the choroid plexus was reduced to 43% of control. The results suggest that an active transport process might be responsible for Cd movement across the blood-CSF barrier (Zheng et al., 1991).

4. Manganese (Mn)

Excessive exposure to Mn produces an irreversible encephalopathy similar to Parkinson's disease (Donaldson et al., 1984). In a series of carefully designed experiments using an *in situ* brain perfusion technique, Smith's group found that Mn ions were rapidly transported from the blood to the choroid plexus and brain parenchymal tissues in a saturable manner (Murphy et al., 1991; Rabin et al., 1993). The influx constant of Mn to the choroid plexus was about 150 and 1000 times greater than that of cerebral cortex and CSF, respectively, when plasma concentration of Mn was maintained at a constant infusion rate of 4.4 nmol/g/s (Murphy et al., 1991). This result concurs with other studies where accumulation of Mn in the choroid plexus was determined by either autoradiography (Valois and Webster, 1989b) or atomic absorption spectrophotometry (Ingersoll and Aposhian, 1995).

The entry of Mn into the CNS occurs mainly at the blood-brain barrier. However, Mn may enter the brain via the choroid plexus if there is a substantial increase in plasma concentration of Mn (Murphy et al., 1991; Rabin et al., 1993). The form in which Mn enters the brain appears to be free ion. The exact role of the choroid plexus in regulating CNS Mn levels remains uncertain.

5. Silver (Ag), Gold (Au), Tellurium (Te), and Arsenic (As)

Silver-induced neurotoxicity is rare. In rats, exposure to Ag during pregnancy led to Ag accumulation in the choroid plexus of the offspring (Rungby and Danscher, 1983). In humans, one clinical case showed that a 72-year-old patient who had taken nose drops containing a silver preparation for 2 to 5 years developed argyria, symptoms representing a diffuse deposition of silver in tissues after prolonged exposure to silver compounds. The necropsy data revealed that the choroid plexus displayed a distinct deposition of silver in the epithelial basal lamina. In contrast, the brain parenchymal tissues were generally free of silver (Goebel and Muller, 1973).

The results of an experiment in which radioactive ^{198}Au was introduced into the lateral ventricle of experimental dogs showed that the choroid plexus had a level of radioactivity about 550 times higher than that found in the brain cortex. Thus, the choroidal epithelium appears to take up Au from the CSF

(Rish and Meacham, 1967). Based on the experiments with dogs and rats, the choroid plexus appears to retain arsenic compounds, such as melaminvelthioarsenite and sodium arsenate (Friedheim et al., 1983; Zheng et al., 1991).

In addition, Te has been found to accumulate in the choroid plexus. The mechanism for this accumulation may be due to intracellular binding rather than active transport (Agnew, 1972; Agnew et al., 1974).

6. Iron (Fe) and Zinc (Zn)

Aside from the toxic heavy metals, many essential trace elements are transported by or sequestered in the choroid plexus. Fe, for instance, showed a selective uptake by the choroid plexus following intravenous injection to rats (Morris et al., 1992). In human subjects with a calcified choroid plexus, Fe concentration in the choroid plexus was about 5 times greater than the values obtained from other brain tissues (Michotte et al., 1977). One human study also suggested that the choroid plexus may help protect the brain against iron overload (Divork, 1995). Since the choroid plexus secretes and regulates CSF transferrin (a glycoprotein for iron transport), it is possible that the choroid plexus may indirectly regulate the iron concentration of the CSF. Again, the actual role of the choroid plexus in this respect remains unclear.

Zn also accumulates in the choroid plexus as determined by autoradiography (Franklin et al., 1992). The permeability of choroid plexus to Zn was about 12 times higher than that of the cerebral capillaries, although the overall contribution of the choroid plexus to Zn influx into the brain is insignificant in comparison with that of the blood-brain barrier (Franklin et al., 1992).

C. STRUCTURAL AND FUNCTIONAL DAMAGE

Accumulation of heavy metals in the choroid plexus can cause the structural damage, which, in turn, induces the functional alteration. In some cases, metals may be secluded in the tissue as intracellular inclusion bodies.

1. Structural Damage

Toxic metals in the choroid plexus can produce a deterioration of the plexus structure (Arvidson and Tjalve, 1986; Berlin and Ullberg, 1963a,b; Valois and Webster, 1987, 1989a,b). When mice received Cd in a chronic-exposure model, the choroid plexus showed the loss of microvilli, a rupture of the apical surface, and an increased number of blebs (Valois and Webster, 1987). Cellular debris present in the ventricular lumen may have resulted from the rupture of the apical membrane. Subcellularly, the epithelial cells displayed an abnormally high number of cytoplasmic vacuoles and lysosomes with condensed or irregular nuclei (Valois and Webster, 1987). Using fluorescent Evans blue-protein complex as a permeability indicator, Steinwall and Olsson (1969) demonstrated that treatment with mercury produced a permeability to the complex, resulting in heavy fluorescence in choroidal stroma and epithelia. The result indicated an impaired integrity of choroidal epithelial cells.

Metal exposure often produces condensed granules in the choroid plexus. The ultradense particles or granules have been identified in the basement membrane beneath the choroidal epithelial cells in patients or experimental animals exposed to both inorganic and organic mercury compounds (Davis et al., 1974; Suda et al., 1989). Similar solid granules have been seen in the choroid plexus of patients with silver poisoning (Goebel and Muller, 1973). It is unclear, however, if those metal-induced granules are similar to the cellular inclusion bodies as identified by other investigators.

Metals can reside in cytoplasmic or intranuclear inclusion bodies in other tissues as well. A protein present in intranuclear inclusion bodies in kidneys of lead-intoxicated animals has been suggested to bind Pb and minimize Pb toxicity (Klaan and Shelton, 1989; Shelton and Egle, 1982). The brain also has this low-abundance 32-kDa nuclear protein, which can be identified in many parts of the brain (Shelton et al., 1990). Fowler's group demonstrated the presence of high affinity Pb-binding protein in kidney and brain cytosol (DuVal and Fowler, 1989; Fowler and DuVal, 1991)

It is noteworthy that cytoplasmic inclusions are frequently reported in human autopsy choroid plexus specimens. The frequency is reportedly about 25% out of 197 autopsied patients (Ohama et al., 1988). No direct evidence, however, indicates that these inclusions are associated with heavy metal exposure.

2. Functional Alteration

There have been few studies of functional alterations of the choroid plexus as the result of metal exposure. Of the limited studies which have been done, those examining interactions between Pb and amino acid transport in the choroid plexus are worthy of discussion. The choroid plexus transports amino acid in a bidirectional manner. In an *in vitro* system where transport is considered to be primarily via the apical epithelium, Pb significantly inhibits the uptake of L-tyrosine by the choroid plexus (Kim and O'Tuama, 1978). As tyrosine is a known precursor of several neurotransmitters, the authors raised an interesting hypothesis that Pb poisoning may be accompanied by abnormal amino acid concentrations in the brain extracellular compartment. However, no further attempt has been made to determine experimentally the compositions of amino acids in the CSF as influenced by metal exposure.

IV. PERSPECTIVES

A. IS THERE A POTENTIAL RISK ASSOCIATED WITH ENVIRONMENTAL EXPOSURE TO HEAVY METALS?

It appears plausible, even likely, that the choroid plexus may be a target for metal toxicity. This may add it to a growing list of tissues whose impaired function results from environmental exposure to toxic metals. Though it is still too early to draw a firm conclusion, this hypothesis is supported by a number of observations.

As discussed earlier, the structural characteristics of the choroid plexus render the tissue highly susceptible to the xenobiotics present in blood. Along with the toxic chemicals and metals discussed in the previous section, clinical evidence indicates that some pathogens and systemic diseases can injure the choroid plexus. In cases of liver cirrhosis, 83% of patients had a similar type of infection in the choroid plexus (Pittella and Bambirra, 1991). Recently, a number of reports indicate that the choroid plexus of AIDS patients can be infected with human immunodeficiency virus (Falangola and Petito, 1993; Harouse et al., 1989; Lackner et al., 1991). Further, the metastasis of tumors to the choroid plexus has been seen in patients with lung tumors (Tanimoto et al., 1991). These examples further strengthen the view that the choroid plexus is not an isolated, well-protected tissue. Rather, it is the tissue that can be easily attacked by toxicants, pathogens, and viruses present in the blood. Therefore, there is a sound reason to postulate that environmentally derived metals present in blood can gain access to the choroid plexus. Our recent study found that Pb prompted the translocation of PKC from the cytosol to membrane in cultured not choroidal epithelial cells *in vitro*. However, no significant alteration in plexus PKC activity was observed in a long term, low dose Pb exposure model *in vivo* (Zheng, 1996).

Second, if the metals in the choroid plexus are the results of life-time environmental exposure, one might expect to see an age-related accumulation. Such a relationship has been established in humans with Pb exposure (Friedheim et al., 1983). Animal studies also provide a positive dose-response relationship of Pb in the choroid plexus (Zheng et al., 1991). It is important to note that a functionally intact choroid plexus barrier is crucial for excluding or eliminating hazardous molecules from the brain. In one clinic case, an 11-year-old child diagnosed with evident encephalopathy of lead poisoning displayed an abnormally high level of CSF Pb (Manton et al., 1984). This, as the authors stated, could have resulted from the failure of the choroid plexus to regulate the composition of CSF. Thus, the neurological consequences of childhood Pb exposure might be due, at least in part, to the inadequate development and/or function of the choroid plexus.

Finally, is sequestration of toxic heavy metals by the choroid plexus toxicologically or pathologically important? At present, we do not have a clear answer to this question so far as metals are concerned. However, it is known that the clinical impairment of the blood-CSF barrier can lead to encephalopathy. Of particular interest are findings that schizophrenia and certain forms of idiopathic mental retardation may result from dysfunction of the choroid plexus (Rudin, 1979, 1980, 1981). According to Rudin's hypothesis, damage by immune complexes allows easy access of exogenous psychopeptides into the CSF and surrounding limbic brain tissue. This, in turn, induces abnormal behavior. Other CNS disorders, possibly associated with the choroid plexus dysfunction, include Reye's syndrome (Levine, 1987), endogenous depression (Jorgensen, 1988), and African sleeping sickness (Ormerod and Venkatesan, 1970). These findings suggest a major role of the choroid plexus in certain CNS diseases.

On the other hand, many CNS disorders could be due to an environmental exposure to heavy metals, such as lead in the retardation of child's brain development, manganese in symptoms analogous to

Parkinson's disease, and mercury compounds in degeneration and necrosis of neurons in specific areas of the brain (Clarkson, 1987). However, many questions remain. For example, how do these metals get into the brain? Are the metals removed from the CNS by the choroid plexus, or by other mechanisms? Does the disease status become significant only after the barrier loses its protective effect? Hence, a clear challenge for the future is to demonstrate how the blood-CSF barrier safeguards normal function of the brain, and how the metal toxicities due to environmental exposure may affect such a role of the choroid plexus.

B. CONSEQUENCES OF METAL SEQUESTRATION IN THE CHOROID PLEXUS

The bulk of evidence strongly supports the view that the choroid plexus is a vital physiological compartment which readily accumulates toxic metals. However, our knowledge of the consequences of metal sequestration in the choroid plexus, particularly in the functional aspect, is strikingly incomplete. Keeping in mind the role of the choroid plexus in production and regulation of the CSF, we should be able to anticipate possible negative outcomes due to the exposure of the tissue to toxic metals.

For example, it appears reasonable to hypothesize that the accumulation of heavy metals in the choroid plexus might interfere with the activities of a variety of enzymes that are important to CSF production. As mentioned in Section II, the choroid plexus contains a high activity of Na^+/K^+-ATPase. Experimental results from other tissues demonstrate that Pb and Hg are potent inhibitors of Na^+/K^+-ATPase activity (Anner et al., 1992; Fox et al., 1991; Imesch et al., 1992; Rajanna et al., 1991; Tiffany-Castiglioni et al., 1987; Vig et al., 1989). Recent works indicate that Pb inhibits Na^+/K^+-ATPase by interfering with phosphorylation of enzyme molecules and dephosphorylation of the enzyme-phosphoryl complex (Rajanna et al., 1991). Hg inactivates Na^+/K^+-ATPase by binding to a metal-binding domain in this enzyme (Ahammadsahib et al., 1987; Anner et al., 1992). Because of the high activity of this enzyme in the choroid plexus, and because of the excessive accumulation of toxic metals in this tissue, it is possible that metals may exert an effect on Na^+/K^+-ATPase.

Pb also stimulates the activity of protein kinase C (PKC). The interaction of Pb and Ca has long been recognized (Bressler and Goldstein, 1991; Vig et al., 1989). The modulation of PKC activity by Pb was found in the brain extracts (Murakami et al., 1993), blood-brain barrier (Laterra et al., 1992; Markovac and Goldstein, 1988), peripheral blood vessel (Chai and Webb, 1988), and osteoblastic cells (Long and Rosen, 1992). At the blood-brain barrier, Pb appears to increase the PKC activity, resulting in an inhibition of CNS endothelial differentiation (Laterra et al., 1992). These findings suggest that modulation of PKC activity by Pb at the subcellular level may be one of the mechanisms that leads to Pb cytotoxicity in the CNS.

Accumulation of heavy metals in the choroid plexus may impair the transport functions of the choroid plexus. Since many essential ions and nutrients are transported into or out of the CSF by the choroid plexus, impairment of transport systems by toxic metals may alter the homeostasis of these molecules in the CNS, thereby influencing the brain's function. The possible effects of heavy metals on choroid plexus transport systems have not yet been studied.

Metals in the choroid plexus may also alter the cell's metabolic activity. Pb, for example, is well known to interrupt heme synthesis. Loss of heme-containing enzymes in turn affects mitochondrial function. Mn also alters heme metabolism, resulting in changes in cytochrome P450-dependent mixed-function oxidase activities in rat brain (Qato and Maines, 1985). The choroidal epithelium requires abundant mitochondria for its secretory functions. It will be interesting to see if Pb or Mn affects mitochondria function in the choroid plexus. Moreover, interruption of heme synthesis by Pb or Mn could conceivably alter the biotransformation of xenobiotics in the blood-CSF barrier.

Finally, accumulation of heavy metals in the choroid plexus may be related to the aging of the choroid plexus. It is known that the choroid plexus deteriorates with age (Huang, 1984; Shuangshorti and Netsky, 1970; Tayarani et al., 1989). One outcome of this deterioration may be cerebral malnutrition with resulting impaired cerebral functioning in the elderly.

C. BIOMARKERS OF CHOROID PLEXUS INTEGRITY

One of the most pressing needs for study of the overall functions of the choroid plexus is the means by which one can estimate the functional integrity of the choroid plexus. Obviously, a well-defined biomarker would be of enormous help in assessing and predicting the function of the choroid plexus.

Transthyretin, also known as prealbumin, is a plasma protein which is primarily derived from the liver, where it is synthesized and secreted (Navab et al., 1977). Transthyretin plays an important role in the transport of vitamin A and thyroid hormones by plasma. Herbert et al. (1986) demonstrated that transthyretin in the mammalian CNS is exclusively synthesized in choroidal epithelial cells. In terms of concentration per gram of tissue, the choroid plexus contains at least 100 times more transthyretin mRNA than the liver (Dickson et al., 1985). Concentrations of transthyretin in the ventricular CSF range between 2 to 4 mg/dl, making up 10 to 25% of the total ventricular protein (Herbert et al., 1986). Because the choroid plexus is the only site for transthyretin production in the CNS, it is tempting to speculate that the level of transthyretin in the CSF may reflect the functional integrity of the choroid plexus. In our recent experiment, rats were exposed to Pb in drinking water at doses of 0 to 250 μg Pb/mL for 1 to 3 months. Pb treatment caused about 32 to 42% reduction in CSF TTR. The percent reduction of CSF TTR was directly associated with Pb concentration in the choroid plexus (Zheng et al., 1996). A number of investigators have suggested the use of CSF transthyretin concentration as a functional marker for study of the neurohumoral regulation of the choroid plexus (Weisner and Roethig, 1983; Herbert et al., 1990).

D. METAL BINDING LIGAND(S)

The choroid plexus retains metals, suggesting some special ligands may be present that can chelate heavy metals. Thiol-containing molecules are the most common ligands to which a variety of heavy metals are bound. Glutathione (GSH), a cysteine-containing tripeptide, has been suggested to protect against cadmium toxicity (Kang, 1992; Singhal et al., 1987). The choroid plexus manufactures, secretes, and regulates GSH in the CSF (Anderson et al., 1989). Although the choroid plexus contains less GSH than the brain cortex (Zheng et al., 1991), a number of GSH-related enzymes, such as GSH-*s*-transferase (Senjo et al., 1986; Carder et al., 1990), γ-glutamyltranspeptidase (Prusiner and Prusiner, 1978a,b), GSH peroxidase and reductase (Tayarani et al., 1989), are found high in concentrations in the choroid plexus. Metallothioneins (MT) are a group of low molecular weight proteins with a high cysteine content. MT binds essential metals such as Zn and Cu and toxic metals such as Cd and Hg (Kaji and Nordberg, 1979). In a recent study using a sensitive immunoblotting technique, Nishimura et al., (1992) identified the presence of MT in the choroid plexus. However, the exact role of GSH and MT in metal sequestration and detoxification in the choroid plexus is unknown. The identification of the specific ligand(s) to which the metals bind would be of obvious interest.

E. THE CHOROID PLEXUS: A "KIDNEY" TO BRAIN?

The choroid plexus has much in common with the renal proximal tubule, both structurally and functionally. Table 3 compares some of the physiological parameters of these two organs. Similar to the choroidal epithelium, the renal proximal tubule is composed of columnar epithelial cells that possess the characteristic tight junctions. The renal epithelial cells also have abundant mitochondria. Although the mass of the choroid plexus is small, the percentage of the tissue weight to total brain weight is very similar to that of kidneys to the whole body weight. Maktabi et al. (1990) found that the high blood flow of the choroid plexus is comparable with that of the kidney in rabbit.

Table 3 Comparison of Some Physiological Parameters Between the Choroid Plexus (CP) and Kidney

	Choroid Plexus	Kidney[a]
Weight ratio	3 g/1.3 kg (CP/brain)	300 g/70 kg (kidney/body)
% of weight	0.2–0.3% of total brain weight	0.4% of total body weight
Blood flow	12–18 ml/min	1200 ml/min
	4–6 ml/min/g[b]	4 ml/min/g
Fluid production	0.35 ml/min[c]	130 ml/min
	0.18 ml/min/g	0.43 ml/min/g
Epithelia	Effective tight junctions	Less effective tight junctions

[a] Stirling (1989), 929, 1061–1097.
[b] Data are derived from sheep (Page et al., 1980; Mayhan et al., 1989; Schalk et al., 1989). In same animal species, the blood flow to the choroid plexus is highly comparable to that of kidney (Maktabi et al., 1990). Data from humans are not available.
[c] Davson et al. (1987), 200.

Functionally, the choroid plexus acts even more like a "kidney" to brain (Spector and Johanson, 1989). As the kidneys function to maintain the homeostasis of body's extracellular fluid, the choroid plexus maintains the homeostasis of brain extracellular fluid. Kidneys reabsorb many useful materials from the urine filtrate into the blood, while the choroid plexus transports useful materials from the blood to the CSF. Further, the kidneys transport and eliminate wastes from the blood to urine. The choroid plexus operates in the opposite direction, moving wastes from the CSF to the blood.

V. SUMMARY

"The chemical changes in the brain underlie all thinking, learning, and behavior" (Koshland, 1993). Thus, the function of the choroid plexus in maintaining CSF chemical stability must be crucial to brain function. The bulk of evidence clearly indicates that the choroid plexus is a compartment where heavy metals accumulate. As a consequence, the metals accumulated in the choroid plexus may ultimately alter the structure and function of the blood-CSF barrier. Our knowledge on this aspect is glaringly incomplete. The mechanism whereby the choroid plexus sequesters toxic metals remains largely unknown. Indeed, a better understanding of the relationship between metals in the choroid plexus and their possible effects on its vital functions may lead to a better mechanistic understanding of certain metal-induced neurotoxicities. This may lead to the consequent protection through the development of preventive measures. Finally, the concept of the choroid plexus as a "kidney" to the brain seems plausible, but much remains to be learned from future studies.

ACKNOWLEDGMENT

The author is indebted to Drs. Joseph H. Graziano at Columbia, Jack B. Bishop at NIEHS, and Conrad E. Johanson at Brown University for their helpful reviews and suggestions with this manuscript. The author is supported in part by Grant P20-ES-06831–01 and by the Division of Environmental Health Sciences, Columbia University.

REFERENCES

Agnew, W. F. (1972), *Teratology,* 6, 331–338.
Agnew, W. F., Snyder, D. A., and Cheng, J. T. (1974), *Microvas. Res.,* 8, 156–163.
Agnew, W. F. and Yuen, T. G. H. (1975), *Brain Res.,* 93, 343–348.
Agnew, W. F., Rumbaugh, C. L., and Cheng, J. T. (1976), *Brain Res.,* 109, 355–66.
Ahammadsahib, K. I., Ramamurthi, R., and Dusaiah, D. (1987), *J. Biochem. Toxicol.,* 2, 169–180.
Amenta, F., Cavallotti, C., Collier, W. L., Ferrante, F., and Napoleone, P. (1989), *Pharmacol. Res.,* 21, 369–73.
Andersson, K. E., Bertler, A., and Wettrell, G. (1975), *Acta Paediatr. Scand.,* 64, 497–504.
Anderson, M. E., Underwood, M., Bridge, R. J., and Meister, A. (1989), *FASEB J.,* 3, 2527–2531.
Anner, B. M., Moosmayer, M., and Imesch, F. (1992), *Am. J. Physiol.,* 262(5 Pt. 2), F830–836.
Arvidson, B. and Tjalve, H. (1986), *Acta Neuropathol.,* 69, 111–116.
Aschner, M. and Clarkson, T. W. (1988), *Brain Res.,* 462, 31–39.
Aschner, M. and Clarkson, T. W. (1989), *Pharmacol. Toxicol.,* 64, 293–299.
Bairamian, D., Johanson, C. E., Parmelee, J. T., and Epstein, M. H. (1991), *J. Neurochem.,* 56, 1623–1629.
Benitz, K. F. and Kramer, A. W. (1968), *Food Cosmet Toxicol.,* 6, 125.
Berlin, M. and Ullberg, S. (1963a), *Arch. Environ. Health,* 6, 589–601.
Berlin, M. and Ullberg, S. (1963b), *Arch. Environ. Health,* 7, 686–693.
Bertler, A., Andersson, K. E., and Wettrell, G. (1973), *Lancet.,* 2(843), 1453–1454.
Borke, J. L., Caride, A. J., Yaksh, T. L., Penniston, J. T., and Kumar, R. (1989), *Brain Res.,* 489:355–360.
Bradbury, M. W. (1992), *Prog. Brain Res.,* 91, 133–8.
Bradbury, M. W. B. and Deane, R. (1988), *Ann. N. Y. Acad. Sci.,* 529, 1–8.
Bressler, J. P. and Goldstein, G. W. (1991), *Biochem. Pharmacol.,* 41, 479–484.
Carder, P. J., Hume, R., Fryer, A. A., Strange, R. C., Lauder, J., and Bell, J. E. (1990), *Neuropath. Appl. Neurobil.,* 16, 293–303.
Carpenter, S. J., McCarthy, L. E., and Borison, H. L. (1970), *Z. Zellforsch.,* 110, 471–486.
Chai, S. S. and Webb, R. C. (1988), *Environ. Health Perspect.,* 78, 85–89.
Christensen, O., Simon, M., and Randlev, T. (1989), *Pflügers Arch.,* 415, 37–46.
Clarkson, T. W. (1987), *Environ. Health Perspect.,* 75, 59–64.
Conn, P. J., Sanders-Bush, E., Hoffman, B. J., and Hartig, P. R., (1986), *Proc. Natl. Acad. Sci. U. S. A.,* 83, 4086–4088.
Craig, A. L., O'Dea, R. F., and Takemori, A. E. (1971), *Neuropharmacology,* 10, 709–714.
Crill, W. E. (1989), in *Textbook of Physiology,* Vol. 1, Patton, H. D., Fuchs, A. F., Hille, B., Scher, A. M., and Steiner, R., Eds., 2nd ed., Saunders, Philadelphia, 759–769.

Cserr, H. F., Bundgaard, M., Ashby, J. K., and Murray, M. (1980), *Am. J. Physiol.,* 238, R76–81. (25)
Davis, L. E., Wands, J. R., Weiss, S. A., Price, D. L., and Girling, E. F. (1974), *Arch. Neurol.,* 30, 428–31.
Davson, H. and Pollay, M. (1963), *J. Physiol.,* 167, 239–246.
Davson, H., Hollingsworth, J. G., Carey, M. B., and Fenstermacher, J. D. (1982), *J. Neurobiol.,* 13, 293–318.
Davson, H., Welch, K., and Segal, M. B. (1987), *Physiology and Pathophysiology of the Cerebrospinal Fluid,* Churchill Livingstone, NY, pp. 189–220; p.p. 375–451.
Deane, R. and Bradbury, M. W. B. (1990), *J. Neurochem.,* 54, 905–914.
Di Carlo, R., Muccioli, G., Papotti, M., and Bussolati, G. (1992), *Brain Res.,* 570, 341–346.
DiBenedetto, F. E. and Bito, L. Z. (1986), *J. Neurochem.,* 46, 1725–1731.
Dickson, P. W., Aldred, A. R., Marley, P. D., Guo-Fen, T., Howlett, G. J., and Schreiber, G. (1985), *Biochem. Biophys. Res. Commun.,* 127, 890–895.
Dixon, R. L., Owens, E. S., and Rall, D. P. (1969), *J. Pharmaceut. Sci.,* 58, 1106–1109.
Donaldson, J., Cranmer, J. M., and Dryk, G. G. (1984), *Neurotoxicology of Manganese,* Intox Press, Little Rock, AR.
Dratman, M. B., Crutchfield, F. L., and Schoenhoff, M. B. (1991), *Brain Res.,* 554, 229–236.
DuVal, G. and Fowler, B. A. (1989), *Biochem. Biophys. Res. Commun.,* 159, 177–184.
Dwork, A. J. (1995), *J. Neurol. Sci.,* 135 (Suppl), 45–51.
El-Yazigi, A., Kanaan, I., and Raines D. A. (1992), *Trace Elem. Med.,* 9, 183–189.
Eriksson, K. H. and Winbladh, B. (1971), *Acta Physiol. Scand.,* 83, 300–308.
Ernst, S. A., Palacios, J. R., and Siegel, G. J. (1986), *J. Histochem. Cytochem.,* 34, 189–195.
Falangola, M. F. and Petito, C. K. (1993), *Neurology,* 43, 2035–2040.
Faraci, F. M., Mayhan, W. G., and Heistad, D. D. (1990), *J. Pharmacol. Exp. Ther.,* 254, 23–27.
Fisher, R. G. and Copenhaver, J. H. (1957), *J. Neurosurg.,* 16, 167–176.
Fowler, B. A. and DuVal, G. (1991), *Environ. Health Perspect.,* 91, 77–80.
Fox, D. A., Katz, L. M., and Farber, D B. (1991), *Neurotoxicology,* 12, 641–654.
Franklin, P. A., Pullen, R. G. L., and Hall, G. H. (1992), *Neurochem. Res.,* 17, 767–771.
Friedheim, E., Corvi, C., Graziano, J., Donnelli, T., and Breslin, D. (1983), *Lancet,* i(8331), 981–982.
Goebel, H. H. and Muller, J. (1973), *Acta Neuropath. (Berl.),* 26(3), 247–51.
Goldberg, M. J., Spector, R., and Chiang, C. K. (1987), *J. Pharmacol. Exp. Ther.,* 240, 717–722.
Gosling, J. A. and Lu, T. C. (1968), *J. Pharmacol. Exp. Ther.,* 167, 56–62.
Harik, S. I., Doull, G. H., and Dick, A. P. (1985), *J. Cereb. Blood Flow Metab.,* 5, 156–160.
Harouse, J. M., Wroblewska, Z., Laughlin, M. A., Hickey, W. F., Schonwetter, B. S., and Gonzalez-Scarano, F. (1989), *Ann. Neurol.,* 25, 406–11.
Herbert, J., Wilcox, J. N., Pham, K. C., Fremeau, R. T., Zeviani, M., Dwork, A., Soprano, D. R., Makover, A., Goodman, D. S., Zimmerman, E. A., Roberts, J. L., and Schon, E. A. (1986), *Neurology,* 36, 900–911.
Herbert, J., Cavallaro, T., and Dwork, A. J. (1990), *Am. J. Pathol.,* 136, 1317–1325.
Hershey, C. O., Hershey, L. A., Varnes, A. W., Wongmongkolrit, T., and Breslau, D. (1987), *Trace Elem. Med.,* 4, 21–24.
Hirayama, K. (1980), *Toxicol. Appl. Pharmacol.,* 55, 318–323.
Hoyer, D., Waeber, C., Schoeffer, P., Palacios, J. M., and Dravid, A. (1989), *Naunyn-Schmiedeberg's Arch. Pharmacol.,* 339, 252–258.
Huang, J. T. (1980), *Res. Commun. Chem. Pathol. Pharmacol.,* 28, 567–570.
Huang, J. T. (1982), *Neurochem. Res.,* 7, 1541–1548.
Huang, J. T. (1984), *Age,* 7, 63–65.
Huang, J. T. and Takemori, A. E. (1975), *Neuropharmacology,* 14, 241–246.
Hug, C. C., Jr. (1967), *Biochem. Pharmacol.,* 16, 345–359.
Imesch, E., Moosmayer, M., and Anner, B. M. (1992), *Am. J. Physiol.,* 262(5 Pt 2), F837–842.
Ingersoll, R., Montgomery, E. B., and Aposhian, H. V. (1995), *Fundam. Appl. Toxicol.,* 27, 106–113.
Jakobson, A. M. (1991), *Biol. Neonate,* 60, 221–229.
Johanson, C. E. (1978), *Life Sci.,* 23, 861–868.
Johanson, C. E. (1989), in *Implications of the blood-brain barrier and its manipulation.* Vol. 1, Neuwelt, E. A., Ed., Plenum, New York, 223–260.
Johanson, C. E. (1995), in *Neuroscience in Medicine,* Conn, P. M., Ed., Lippincott, Philadelphia, PA, 171–196.
Johanson, C. E., Reed, D. J., and Woodbury, D. M. (1974), *J. Physiol.,* 241, 359–372.
Johanson, C. E. and Woodbury, D. M. (1977), *Exp. Brain Res.,* 30, 65–74.
Johanson, C. E., Sweeney, S. M., Parmelee, J. T., and Epstein, M. H. (1990), *Am. J. Physiol.,* 258, C211–216.
Johanson, C. E. and Murphy, V. A. (1990), *Am. J. Physiol.,* 258, F1538–1546.
Jorgensen, O. S. (1988), *Acta Psychiatr. Scand. (Suppl.),* 345, 29–37.
Julius, D., MacDermott, A. B., Axel, R., and Jessell, T. M. (1988), *Science,* 241, 558–564.
Kaji, J. H. R. and Nordberg, M. (1979), *Metallothionein,* 1st ed., Birkhauser Verlag, Boston
Kang, Y. J. (1992), *Drug Metab. Disp.,* 20, 714–718.
Keep, R. F. and Jones, H. C. (1990), *Dev. Brain Res.,* 56, 47–53.
Kerper, L. E., Ballatori, N., and Clarkson, T. W. (1992), *Am. J. Physiol.,* 262, R761–765.
Kim, C. S. and O'Tuama, L. A. (1978), *Toxicol. Appl. Pharmacol.,* 45, 213–217.

Kim, C. S., O'Tuama, L. A., Mann, J. D., and Roe C. R. (1983), *J. Pharmacol. Exp. Ther.,* 225, 699–704.
Kim, C. S., Keizer, R. F., Ambrose, W. W., and Breese, G. R. (1987), *Toxicol. Appl. Pharmacol.,* 90, 436–44.
Kim, C. S., Roe, C. R., and Ambrose, W. W. (1990), *Brain Res.,* 536, 335–338.
Klaan, E. and Shelton, K. R. (1989), *J. Biol. Chem.,* 264, 16969–16972.
Konrad, E. M., Bianchi, C., Thibault, G., Garcia, R., Pelletier, S., Genest, J., and Cantin, M. (1990), *Neuroendocrinology,* 51, 304–314.
Koshland, D. E., Jr. (1993), *Science,* 262, 635.
Krakauer, R. and Steiness, E. (1978), *Clin. Pharmacol. Ther.,* 24, 454–458.
Lackner, A. A., Smith, M. O., Munn, R. J., Martfeld, D. J., Gardner, M. B., Marx, P. A., and Dandekar, S. (1991), *Am. J. Pathol.,* 139, 609–21.
Lai, Z. N., Emtner, M., Roos, P., and Nyberg, F. (1991), *Brain Res.,* 546, 222–226.
Lai, Z., Roos, P., Olsson, Y., Larsson, C., and Nyberg, F. (1992), *Neuroendocrinology,* 56, 225–233.
Lanman, R. C. and Schanker, L. S. (1980), *J. Pharmacol. Exp. Ther.,* 215, 563–568.
Laterra, J., Bressler, J. P., Indurti, R. R., Belloni-Olivi, L., and Goldstein, G. W. (1992), *Proc. Natl. Acad. Sci. U.S.A.,* 89, 10748–10752.
Levine, S. (1987), *Lab. Invest.,* 56, 231–3.
Levine, R. and Sowinski, R. (1974), *Arch. Pathol.,* 98, 177–182.
Lindsey, A. E., Schneider, K., Simmons, D. M., Baron, R., Lee, B. S., and Kopito, R. R. (1990), *Proc. Natl. Acad. Sci. U.S.A.,* 87, 5278–5282.
Lindvall, M. and Owman, C. (1981), *J. Cereb. Blood Flow Metab.,* 1, 245–266.
Lindvall-Axelsson, M., Hedner, P., and Owman, C. (1989), *Exp. Brain Res.,* 77, 605–610.
Long, G. J. and Rosen, J. F. (1992), *Toxicol. Appl. Pharmacol.,* 114, 63–70.
Lorenzo, A. V. and Spector, R. (1973), *J. Pharmacol. Exp. Ther.,* 184, 465–471.
Maktabi, M. A., Heistad, D. D., and Faraci, F. M. (1990), *Am. J. Physiol.,* 258, H414–H418.
Manton, W. I., Kirkpatrick, J. B., and Cook, J. D. (1984), *Lancet,* ii(8398), 351.
Markovac, J. and Goldstein, G. W. (1988), *Nature,* 334, 71–73.
Masuzawa, T., Saito, T., and Sato, F. (1981), *Brain Res.,* 222, 309–322.
Masuzawa, T. and Sato, F. (1983), *Brain,* 106, 55–99.
Masuzawa, T., Hasegawa, T., Nakahara, N., Iida, K., and Sato, F. (1984), *Ann. N.Y. Acad. Sci.,* 429, 405–407.
Matyszak, M. K., Lawson, L. J., Perry, V. H., and Gordon, S. (1992), *J. Neurochem.,* 40, 173–182.
Mayer, S. E. and Sanders-Bush, E. (1993), *J. Neurochem.,* 60, 1308–1316.
Mayhan, W. C., Faraci, F. M., Spector, R., and Heistad, D. D. (1989), *Am. J. Physiol.,* 257, H834–838.
Meret, S. and Henkin, R. I. (1971), *Clin. Chem.,* 17, 369–373.
Michotte, Y., Massart, D. L., Lowenthal, A., Knaepen, L., Pelsmaekers, J., and Collard, M. (1977), *J. Neurol.,* 216, 127–133.
Milhorat, T. H. (1976), *Int. Rev. Cytol.,* 47, 225–288.
Miller, T. B. and Ross, C. R. (1975), *J. Pharmacol. Exp. Ther.,* 196, 771–777.
Moller-Madsen, B. (1990), *Toxicol. Appl. Pharmacol.,* 103, 303–323.
Moller-Madsen, B. (1991), *Fundam. Appl. Toxicol.,* 16, 172–187.
Morris, C. M., Keith, A. B., Edwardson, J. A., and Pullen, R. G. (1992), *J. Neurochem.,* 59, 300–6.
Murakami, K., Feng, G., and Chen, S. G. (1993), *J. Pharmacol. Exp. Ther.,* 264, 757–761.
Muraki, T. (1971), *Eur. J. Pharmacol.,* 15, 393–395.
Murphy, V. A., Smith, Q. R., and Rapoport, S. I. (1986), *J. Neurochem.,* 47, 1735–1741.
Murphy, V. A., Smith, Q. R., and Rapoport, S. I. (1989), *Brain Res.,* 484: 65–70.
Murphy, V. A. and Johanson, C. E. (1989), *Biochem. Biophys. Acta,* 979, 187–92.
Murphy, V. A., Wadhwani, K. C., Smith, Q. R., and Rapoport, S. I. (1991), *J. Neurochem.,* 57, 948–954.
Nag, S. (1991), *Acta Neuropathol.,* 82, 274–279.
Nathanson, J. A. and Chun, L. L. Y. (1989), *Proc. Natl. Acad. Sci. U.S.A.,* 86, 1684–1688.
Nation, J. R., Clark, D. E., Bourgeois, A. J., and Baker, D. M. (1983), *Neurobehav. Toxicol. Teratol.,* 5, 275.
Navab, M., Mallia, A. K., Kanda, Y., and Goodman, D. S. (1977), *J. Biol. Chem.,* 252, 5100–5106.
Nicklaus, K. J., McGonigle, P., and Molinoff, P. B. (1988), *J. Pharmacol. Exp. Ther.,* 247, 343–348.
Nilsson, C., Ekman, R., Lindvall-Axelsson, M., and Owman, C. (1990), *Regul. Pept.,* 27, 11–26.
Nilsson, C., Lindvall-Axelsson, M., and Owman, C. (1992a), *Brain Res. Rev.,* 17, 109–138.
Nilsson, C., Blay, P., Nielsen, F. C., and Gammeltoft, S. (1992), *J. Neurochem.,* 58, 923–930.
Nishimura, H. (1983), *Atlas of Human Prenatal Histology,* 1st ed., Igaku-Shon, New York, 20–28.
Nishimura, N., Nishimura, H., Ghaffar, A., and Tohyama, C. T. I. (1992), *J. Histochem. Cytochem.,* 40, 309–315.
Nohjoh, T., Suzuki, H., Sawada, Y., Sugiyama, Y., Iga, T., and Hanano, M. (1989), *J. Pharmacol. Exp. Ther.,* 250, 324–328.
Ohama, E., Takeda, S., and Ikuta, F. (1988), *Acta Neuropathol.,* 76, 11–16.
Ormerod, W. E. and Venkatesan, S. (1970), *Lancet,* 2(676), 777.
O'Tuama, L. A., Kim, C. S., Gatzy, J. T., Krigman, M. R., and Mushak, P. (1976), *Toxicol. Appl. Pharmacol.,* 36, 1–9.
Page, R. B., Funsch, D. J., Brennan, R. W., and Hernandez, M. J. (1980), *Brain Res.,* 197, 532–537.
Palavinskas, R. and Schulten, H. R. (1984), *Trace Elem. Med.,* 1, 29–34.
Parmelee, J. T. and Johanson, C. E. (1989), *Am. J. Physiol.,* 256, R786–791.

Parmelee, J. T., Bairamian, D., and Johanson, C. E. (1991), *Dev. Brain Res.,* 60, 229–233.
Pitella, J. E. and Bambirra, E. A. (1991), *Arch. Pathol. Lab. Med.,* 115, 220–2.
Pollay, M. and Kaplan, R. (1972), *J. Neurobiol.,* 3, 339–346.
Pollay, M., Hisey, B., Reynolds, E., Tomkins, P., Stevens, A., and Smith, R. (1985), *Neurosurgery,* 17, 768–772.
Preston, J. E. and Segal, M. B. (1992), *Brain Res.,* 581, 351–355.
Pritchard, J. B. (1980), *J. Pharmacol. Exp. Ther.,* 212, 354–359.
Prusiner, P. E. and Prusiner, S. B. (1978a), *J. Neurochem.,* 30, 1253–1259.
Prusiner, P. E. and Prusiner, S. B. (1978b), *J. Neurochem.,* 30, 1261–1267.
Qato, M. K. and Maines, M. D. (1985), *Biochem. Biophys. Res. Commun.,* 128, 18–24.
Rabin, O., Hegedus, L., Bourre, J. M., and Smith, Q. R. (1993), *J. Neurochem.,* 61, 509–517.
Rajanna, B., Chetty, C. S., Stewart, T. C., and Rajanna, S. (1991), *Biomed. Environ. Sci.,* 4, 441–451.
Rastogi, R. B., Merali, Z., and Singhal, R. L. (1977), *J. Neurochem.,* 28, 789.
Riachi, N. J. and Harik, S. I. (1992), *Exp. Neurol.,* 115, 212–217.
Rish, B. L. and Meacham, W. F. (1967), *J. Neurosurg.,* 27, 15–20.
Rotter, A., Birdsall, N. J. M., Burgen, A. S. V., Field, P. M., Hulme, E. C., and Raisman, G. (1979), *Brain Res. Rev.,* 1, 141–165.
Rubin, R., Owens, E., and Rall, D. (1968), *Cancer Res.,* 28, 689–694.
Rudin, D. O. (1979), *Schizophr. Bull.,* 5, 623–6.
Rudin, D. O. (1980), *Biol. Psychiatry,* 15, 517–39.
Rudin, D. O. (1981), *Biol. Psychiatry,* 16, 373–97.
Rungby, J. and Danscher, G. (1983), *Acta Neuropathol.,* 61, 258–262.
Schalk, K. A., Williams, J. L., and Heistad, D. D. (1989), *Am. J. Physiol.,* 257, R1365–1369.
Schreiber, G., Aldred, A. R., Jaworowski, A., Nilsson, C., Achen, M. G., and Segal, M. B. (1990), *Am. J. Physiol.,* 258, R338–345.
Schreiber, G. and Aldred, A. R. (1993), in *The Blood-Brain Barrier: Cellular and Molecular Biology,* Paardridge, W. M., Ed., Raven Press, New York, 441–459.
Senjo, M., Ishibashi, T., Terashima, T., and Inoue, Y. (1986), *Neurosci. Lett.,* 66, 131–134.
Shelton, K. R. and Egle, P. M. (1982), *J. Biol. Chem.,* 257, 11802–11807.
Shelton, K. R., Cunningham, J. G., Klann, E., Merchant, R. E., Egle, P. M., and Bigbee, J. W. (1990), *J. Neurosci. Res.,* 25, 287–294.
Shibata, T., Ohshima, S., Shimizu, Y., Suzuki, M., Ishizuka, M., Sasaki, N., and Nakayama, E. (1990), *Virchows Archiv. Pathol. Anat.,* 417, 203–212.
Shuangshoti, S. and Netsky, M. (1970), *Am. J. Anat.,* 128, 73–96.
Singhal, M. E., Anderson, M. E., and Meister, A. (1987), *FASEB J.,* 1, 220–223.
Smith, Q. R. (1991), *Adv. Exp. Med. Biol.,* 291, 55–71.
Spector, R. (1975), *J. Pharmacol. Exp. Ther.,* 194, 82–88.
Spector, R. (1980), *Clin. Sci.,* 58, 107–109.
Spector, R. (1989), *J. Neurochem.,* 53, 1667–1674.
Spector, R. (1990), *Pharmacology,* 41, 113–8.
Spector, R. and Johanson, C. E. (1989), *Sci. Amer.,* 261, 68–74.
Spector, R. and Lorenzo, A. V. (1973), *J. Pharmacol. Exp. Ther.,* 185, 276–286.
Spector, R. and Lorenzo, A. V. (1974), *J. Clin. Invest.,* 54, 316–325.
Steinwall, O. and Olsson, Y. (1969), *Acta Neurol. Scandinav.,* 45, 351–361.
Stirling, C. E. (1989), *Textbook of Physiology,* 2nd ed., Vol. 2, Patton, H. D., Fuchs, A. F., Hille, B., Scher, A. M. and Steiner, R., Eds., Saunders, Philadelphia, pp1061–1097; p929.
Stoll, J., Wadhwani, K. C., and Smith, Q. R. (1993), *J. Neurochem.,* 60, 1956–1959.
Suda, I., Eto, K., Tokunaga, H., Furusawa, R., Suetomi, K., and Takahashi, H. (1989), *Neurotoxicology,* 10, 113–126.
Suzuki, H., Sawada, Y., Sugiyama, Y., Iga, T., and Hanano, M. (1986), *J. Pharmacol. Exp. Ther.,* 239, 927–935.
Suzuki, H., Sawada, Y., Sugiyama, Y., Iga, T., and Hanano, M. (1987), *J. Pharmacol. Exp. Ther.,* 243, 1145–1152.
Swain, E. B., Engstrom, D. R., Brigham, M. E., Henning, T. A., and Brezonik, P. L., (1992), *Science,* 257, 784–787.
Takemori, A. E. and Stenwick, M. W. (1966), *J. Pharmacol. Exp. Ther.,* 154, 586–594.
Tanimoto, M., Tatsumi, S., Tominaga, S., Kamikawa, S., Nagao, T., Tamaki, N., and Matsumoto, S. (1991), *Neurol. Med. Chir. (Tokyo),* 31, 152–5.
Tayarani, I., Cloez, I., Clement, M., and Bourre, J. M. (1989), *J. Neurochem.,* 53, 817–824.
Tiffany-Castiglioni, E., Zmudzki, J., Wu, J. N., and Bratton, G. R. (1987), *Metab. Brain Dis.,* 2, 61–79.
Townsend, J. B., Ziedonis, D. M., Bryan, R. M., Brennan, R. W., and Page, R. B. (1984), *Brain Res.,* 290, 165–169.
Valois, A. A. and Webster, W. S. (1987), *Toxicology,* 46, 43–55.
Valois, A. A. and Webster, W. S. (1989a), *Toxicology,* 55, 193–205.
Valois, A. A. and Webster, W. S. (1989b), *Toxicology,* 57, 315–328.
van Deurs (1980), *Int. Rev. Cytol.,* 65, 117–191.
van Deurs, B. and Amtorp, O. (1978), *Cell Tissue Res.,* 187, 215–234.
Vega, J. A., Del Valle, M. E., Calzada, B., Bengoechea, M. E., and Peres-Casas, A. (1992), *Cell. Mol. Biol.,* 38, 145–149.

Vig, P. J., Nath, R., and Desaiah, D. (1989), *J. Appl. Toxicol.,* 9, 313–316.
Vogh, B. P. (1980), *J. Pharmacol. Exp. Ther.,* 213, 321–331.
Volk, B., Hettmannsperger, U., Papp, T., Amelizad, Z., Oesch, F., and Knoth, R. (1991), *Neuroscience,* 42, 215–235.
Walsh, R. J., Mangurian, L. P., and Posner, B. I. (1990), *J. Anat.,* 168, 137–141.
Wasserman, G., Graziano, J. H., Factor-Litvak, P., Popovac, D., Morina, N., Musabegovic, A., Vrenezi, N., Capuni-Paracka, S., Lekic, V., Preteni-Redgepi, E., Hadjialjevic, S., Slavkovich, V., Kline, J., Shrout, P., and Stein, Z. (1994), *Neurotoxicol. Neurotetratol.,* in press.
Weisner, B. and Roethig, H. J. (1983), *Eur. Neurol.,* 22, 96–105.
Wenk, E. J., Levine, S., and Hoenig, E. M. (1979), *J. Neuropathol. Exp. Neurol.,* 38, 1–9.
Whittico, M. T., Gang, Y. A., and Giacomini, K. M. (1990), *J. Pharmacol. Exp. Ther.,* 255, 615–623.
Winbladh, B. (1972), *Acta Physiol. Scand.,* 84, 109–114.
Wright, E. M. (1979), *Trends Neurosci.,* 2, 13–15.
Yagaloff, K. A. and Hartig, P. R. (1985), *J. Neurosci.,* 5, 3178–3183.
Zheng, W., Slakovich, V., and Luo, J. (1995), *Toxicologist,* 15, 9.
Zheng, W., Ren, X., Shen, H., Blanner, W. S., and Graziano, J. H. (1996), *Fundam. Appl. Toxicol.,* 30 (Suppl), 90.
Zheng, Q., Steinberg, S., and Zheng, W. (1996), *Fundam. Appl. Toxicol.,* 30 (Suppl), 295.
Zheng, W., Perry, D. F., Nelson, D. L., and Aposhian, H. V. (1991), *FASEB J.,* 5, 2188–2193.

Chapter 37

Effects of Toxic Metals on Neurotransmitters

Richard B. Mailman, Mechelle Mayleben, and Cindy P. Lawler

I. INTRODUCTION

The present chapter deals with the effects of toxic metals on neurotransmitter function. *A priori*, there are several reasons for wanting to investigate why, how, or if some aspect of neurotransmitter function has been affected by a metal or metal-containing compound. In some cases, the primary signs of intoxication may suggest that selected aspects of chemical neurotransmission have been affected. In other cases, such studies may be designed to determine if such changes have occurred, and if they can be used as a toxic endpoint. Finally, a metal (whether in organic or inorganic form) may in some cases provide a useful tool to study CNS mechanisms of general interest. This chapter will attempt to provide a broad review of how such studies can be approached, in the process highlighting specific approaches, and illustrating these with examples from the literature. It is obvious, however, that the explosion of the neurosciences during the last decade has made the notion of summarizing this field in a few dozen pages impossible. The fact that a review limited only to neurochemical effects of organic forms of toxic metals was a daunting task (Morell and Mailman, 1987) should make it clear that an exhaustive review or critique of the available literature is not possible in one chapter. What this chapter is designed to do is provide an overview of the neurochemical approaches often used for studies of metal and other toxicants, their strengths and limitations, and examples of data with various toxic metals.

The central nervous system (CNS), the brain and the spinal cord, provides the source and control of emotions, thoughts, control of movement, and the regulation of cardiovascular, neuroendocrine, and even immune function. Toxic metals may affect the CNS in dramatic fashion with gross physical signs (e.g., frank encephalopathy or death), but they also may cause more subtle lesions; it is the detection of the latter that forms the basis for this chapter. Although there are many more neuroglia (supporting cells) in brain than there are neurons (nerve cells), it is the specialized physiology of neurons that permits the exquisitely controlled process of neurotransmission to occur, and it is on neurons that this chapter will be focused. During the past decade, however, it has been widely recognized that glia are not passive entities but themselves play an active role in the function of the CNS. Nonetheless, for the purposes of this chapter, the focus will be on the transfer of chemical information between neurons and how toxic metals may affect this process. The study of neurotoxicity is generally thought to be approached best by integrated techniques, with the neurochemical fashion discussed in this chapter one important component of this armamentarium. The failure to discuss other elegant approaches (e.g., physiological ones) should not be construed as undervaluing them, but rather reflects the focus of this chapter

II. NEUROTRANSMISSION

The primary mechanism by which the billions of CNS neurons communicate with each other is chemical in nature. Historically, the term "neurotransmitter" was reserved for compounds that met several criteria. The compound had to be synthesized and stored in neurons, and released upon stimulation of the neurons. There had to be receptors present to recognize the compound with high affinity. Application of exogenous compounds had to have the same effect at the synapse as did electrical stimulation of the presynaptic pathway. Finally, high capacity inactivation mechanisms had to be present to terminate the action of the compound rapidly. In 1970, only about a dozen compounds were thought to meet these criteria; these compounds included acetylcholine, serotonin, the catecholamines (norepinephrine, dopamine, and epinephrine), glutamate, and γ-aminobutryic acid (GABA). Since then, advances in the neurosciences have increased greatly the complexity of this picture. The first advance was the finding that almost all peripheral hormones also functioned as chemical transmitters in the CNS, having functions distinct from their endocrine roles. The use of molecular methods has led to the discovery of new molecules believed to subserve the function of information transfer in the nervous system. The number of such molecules is now greater than 100, and increases continually. The study of these newer chemical messengers has also resulted in conceptual changes, discussed below, in what constitutes a neurotransmitter.

Many neurotransmitters come from groups of cells restricted to a few areas of the brain. For example, most dopamine neurons (i.e., those that utilize dopamine as a neurotransmitter) are located in, or are adjacent to, a midbrain region called the substantia nigra. Yet these dopamine neurons send axons to many parts of the brain, thus providing dopamine innervation to many areas of the cortex, limbic system, and basal ganglia. On the other hand, some neurotransmitters (e.g., glutamate or GABA) are ubiquitous, with both neuronal cell bodies and processes occurring throughout the brain. Such regional differences in neurochemical architecture of the nervous system are often an important issue in neurotoxicological studies.

In addition to the term neurotransmitter, one often sees the term "neuromodulator" used. This concept arose when it was found that the newer neurotransmitters that were being discovered did not meet the formal criteria listed above. For example, a mechanism for rapid inactivation often was not found since these compounds may work more slowly and over greater distances. In this chapter, the use of the term neurotransmitter or chemical messenger is meant to include both the classical neurotransmitters and the newer neuromodulators.

Another important concept that has come to light in recent years is the phenomenon in which neurons release more than one neurotransmitter. One of these is often a classical transmitter (e.g., dopamine), while the second may be a neuropeptide (e.g., substance P). There have been several elegant studies that have demonstrated that such **cotransmission** provides an exquisite mechanism for regulation of cellular function. While the **mechanisms** of cotransmission have been studied in only a few defined *in vitro* systems, colocalization (i.e., two or more neurotransmitters in one cell) is now a common observation in all areas of the nervous system. This mechanism is likely to be an important site for action of toxicants like toxic metals, but has yet received little attention.

A. THE SYNAPSE

Although neurons vary widely in size and shape, they generally are composed of a cell body (soma or perikaryon), a long process extending from the soma (axon), and shorter processes (dendrites) that extend from the cell body and receive local information. Chemical transmission is usually thought to involve a specialized structure called the synapse from which neurotransmitters are released. The synapse acts as a rectifier to ensure unidirectional flow of information under physiological conditions. It is the events that occur at the synapse that are often important as mechanisms or markers of neurotoxicity. We have previously reviewed the general approaches that can be used to assess changes in synaptic function, as well as the effects of specific toxicants (DeHaven and Mailman, 1983; Mailman and DeHaven, 1984; Morell and Mailman, 1987). The discussion that follows will show how toxic metals and other toxicants may affect neurochemical indices of neurotransmission, and also will describe some newer methods that have become available. As background, it is useful to review the biochemistry of the synapse, using as an example a synapse that uses the neurotransmitter dopamine (Figure 1).

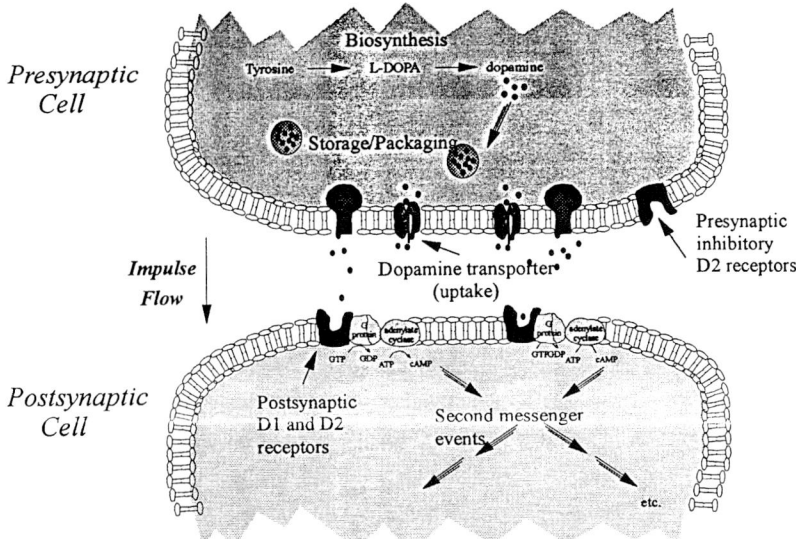

Figure 1. Model of synapse using dopamine as a neurotransmitter. This figure details many of the biochemical loci whereby toxic metals can alter neurochemical function, including the enzymes responsible for synthesis, the release mechanisms, regulatory presynaptic receptors, postsynaptic receptors, second messenger events, etc.

B. BIOCHEMICAL RESPONSES TO RECEPTOR OCCUPATION: SECOND MESSENGERS AND OTHER MECHANISMS

When the neurotransmitter binds to the receptor, it causes perturbations resulting in either direct or allosteric interactions that, ultimately, alter enzyme activity that generates or inhibits further electrical impulses. Several mechanisms may mediate these changes. One type of mechanism is the direct effect of receptor occupation on an ion channel, resulting in modulation of ion fluxes that ultimately change the intracellular concentration of ions that affect cellular metabolism (e.g., Ca^{2+}) or alter the electrical potential of cells (e.g., Na^+ or Cl^-). Study of such processes usually entails electrophysiological approaches that are beyond the scope of a text in biochemical toxicology.

Of more direct relevance to this text are the biochemical synaptic events that can be studied. One of the most important of these is the generation of cyclic nucleotides (cAMP or cGMP) as a consequence of receptor-transmitter interactions. For example, certain receptors have been demonstrated to be associated with the enzyme adenylate cyclase. A sequence of neural events results in the binding of a neurotransmitter to a receptor with subsequent activation of adenylate cyclase. This increases the synthesis of cAMP from ATP, resulting in alterations in intracellular cAMP levels that change the activity of certain enzymes — enzymes that ultimately mediate many of the changes caused by the neurotransmitter. For example, there are protein kinases in the brain whose activity is dependent upon these cyclic nucleotides; the presence or absence of cAMP alters the rate at which these kinases phosphorylate other proteins (using ATP as substrate). One example of this type of reaction is the transport of cations (e.g., Na^+, K^+) by the enzyme adenosine triphosphatase (APTase). These processes are all possible loci for biochemical attack by toxicants — events that can markedly alter neural transmission and result in toxicity in the CNS.

Many unanswered questions still exist regarding the definitive roles of cAMP and cGMP in neuronal functioning, but it is assumed that cyclic nucleotide systems probably play a major role in direct impulse transmission as well as in feedback control in the CNS. The evidence for this includes the following observations: alterations in nucleotide levels correlate temporally with physiological events; enzymatic machinery is present in sufficient concentration for rapid synthesis and degradation of the nucleotides; and there are enzymes present whose activities in ion transport, protein phosphorylation, or related processes are markedly altered by cyclic nucleotides.

It has been suggested that some parameters related to neurotransmission may serve as a uniquely sensitive screen for neurotoxicity (Bondy, 1982), although this approach has been criticized (DeHaven and Mailman, 1983; DeHaven and Mailman, 1986; Mailman, 1987). Nonetheless, in the following section, we have attempted to summarize some illustrative phenomenological data of how some of these compounds may affect CNS neurotransmission.

III. SITES OF TOXIC METAL NEUROCHEMICAL ACTION

A. TOXICANT-INDUCED CHANGES IN NEUROTRANSMITTER METABOLISM

The intermediary metabolism of a neurotransmitter involves its synthesis and metabolism, mechanisms that are tightly coupled to neurotransmitter release. As such, these biochemical steps may be affected, directly or indirectly, by toxic metals. For example, a toxicant may cause profound damage to the nervous system, such that there is actual loss of neuronal elements or the nerve cells themselves. In such cases, a measurable decrease in the total concentration of neurotransmitter can be detected. Measurement of steady-state concentrations, however, reveal little of how a toxicant perturbs the CNS, and such measurements are relatively insensitive to important toxicant-induced events that do not alter total transmitter content.

The first widespread availability of high performance liquid chromatography (HPLC) around 1980 greatly facilitated the task of accurately measuring neurotransmitters, their precursors, and their metabolites in discrete brain regions and nuclei. HPLC with electrochemical detection was first applied in neurochemistry to the measurement of dopamine (DA) and norepinephrine (NE) (Keller et al., 1976) and, later, to metabolites and precursors of these compounds (Crombeen et al., 1978; Kilts et al., 1981; Scratchley et al., 1979; Van Valkenburg et al., 1982) and related monoamines such as serotonin (Kilts et al., 1981; Lyness, 1982; Mefford, 1981). Methods were soon developed to measure neuroactive α-amino acids and GABA (Caudill et al., 1982; Lindroth and Mopper, 1979; Smith et al., 1975), acetylcholine (Potter et al., 1983), and various neuropeptides (e.g., Desiderio et al., 1981; Meek, 1980). These early techniques have been improved and modified during the past decade, and coupled with improved analytical hardware, have allowed ever more refined hypotheses to be tested.

B. TOXICANT EFFECTS ON CYCLIC NUCLEOTIDES

It is obvious that toxicants that interfere with the levels of cyclic nucleotides via either synthetic or degradative enzymes can markedly alter CNS function. Several alkaloids known to have these actions include the methylxanthines, caffeine and theophylline, which are found in coffee and tea. While the principal mode of action of these compounds is via purine receptors, at higher concentrations they also affect cyclic nucleotide systems by inhibiting the phosphodiesterase enzymes that degrade cAMP or cGMP. Chronic administration of high concentrations of caffeine to rats during development is known to cause a condition similar to the Lesch-Nyhan syndrome seen in humans. The mechanisms for this action are unknown but may involve actions on both cyclic nucleotide systems and adenosine receptors in the brain.

There are three well-studied toxins that affect cAMP metabolism directly. These agents have actions primarily in the periphery (e.g., because of blood-brain barrier transport), but are interesting mechanistically. Pertussis toxin is produced by the Gram-negative bacterium *Bordetella pertussis*. Primarily responsible for the pathogenesis of whooping cough, the toxin can act on a number of cell types. A part of the toxin binds to a cell surface receptor, facilitating internalization of the active form of the toxin. Once inside the cell, the toxin catalyzes the transfer of an ADP-ribose moiety from NAD^+ to an inhibitory G-protein. This ribosylation prevents the normal inhibitory action of the G-protein, increasing cellular cAMP. In a similar way, cholera toxin (secreted by the Gram-negative bacterium *Vibria cholerae*) also enters the cell by having one part of the toxin bind to a specific site on the cell surface (in this case, a GM1 ganglioside). Like pertussis toxin, part of the toxin catalyzes the transfer of an ADP-ribose unit from NAD^+ to the G-protein, in this case preventing normal stimulatory events from being turned off. Finally, forskolin (an alkaloid of the plant *Coleus forskolin*) directly affects the catalytic adenylate cyclase unit, perpetually turning on the enzyme.

Although cyclic nucleotides were the first and most extensively studied second messengers, other compounds are now known to serve similar roles. One area of great current interest involves certain products of phospholipid metabolism. For example, phospholipids of the phosphoinositide class are hydrolyzed by enzymes that are sensitive to many cellular signals, including receptor-mediated events. The products of this hydrolysis are the phosphoinositols and diacylglycerol, all of which play important roles in further intracellular changes. Other phospholipids free arachidonic acid when hydrolyzed, and this results in the synthesis of a series of eicosanoid messengers of the prostaglandin, prostacyclin,

C. TOXICANT-INDUCED CHANGES IN ENZYME ACTIVITY

Changes in the characteristics of an enzyme that is critical for neurotransmitter activity, synthesis, or degradation may reflect neurotoxicant-induced perturbations in neuronal activity. The enzymes of most interest are usually those at specific regulatory points (e.g., rate-limiting enzymes). With the monoamine transmitters, methods exist for the assessment of tyrosine hydroxylase (McMillen, 1982), dopamine-β-hydroxylase (Wallace et al., 1973), DOPA decarboxylase, 5-HTP decarboxylase (Melamed et al., 1980), tryptophan hydroxylase (Boadle-Biber, 1982) and monoamine oxidase (Garrick and Murphy, 1980). The best studies usually determine toxicant effects on kinetic parameters of the enzyme (e.g., changes in K_m, V_{max}), or when a toxicant is an enzyme inhibitor, the rate of reaction or determination of K_I is performed.

One recent example of such a study was the report by Leung et al. (1992) that tested the hypothesis that neurotoxic metals can exert their toxicity through the direct inhibition of monoamine oxidases, the enzymes that are responsible for scavenging catecholamines, serotonin, and other related neurotransmitters. They investigated how several neurotoxic metal ions affected type A (MAO-A) and type B (MAO-B) monoamine oxidases in both forebrain and liver mitochondria. MAO-A was slightly more sensitive than MAO-B to inhibition by Cu^{2+} or Cd^{2+}, but effects were generally seen at high concentrations (> 10 μM). Conversely, Al^{3+} only inhibited either MAO isoform at concentrations greater than 1 mM. While the authors concluded that their data supported the hypothesis that Cu^{2+} or Cd^{2+} exerted their toxicity through the direct inhibition of isoforms of MAO, they did not evaluate the relative effects of these ions on other neuronal processes. Assumptions about the pathophysiological importance of such changes in enzyme activity should be part of more global investigations that make such comparisons.

D. BRAIN PERFUSION AND DIALYSIS

Recently, technical breakthroughs in two areas have resulted in wider use of methods to assess neurochemical changes in freely moving animals. Although similar in principle to the push-pull perfusion techniques described by Tilson and Sparber (1970), one important new technique is cerebral microdialysis, in which a narrow piece of tubing with a semipermeable wall (dialysis tubing) is stereotaxically placed in brain (Ungerstedt et al., 1982). The temporal effects of exogenous compounds (e.g., toxic metals) on neurotransmitter release (overflow) in specific brain structures or nuclei can be examined in the awake animal, often in conjunction with behavioral experiments (e.g., Stahle, 1992; Young, 1993; Zetterstrom et al., 1984). The second widely used method involves implanting a chemically pretreated microelectrode into the brain, thus making it possible to measure the extracellular availability of neurotransmitters in real time (called *in vivo* voltammetry). For example, the catecholamines (e.g., dopamine) and the indoleamine serotonin can be selectively detected. Some of the initial problems encountered with this method, such as interference from other endogenous compounds that oxidize at similar potentials and overlapping of peaks that oxidize at the same potential, have been resolved by technical advances (see Brazell and Marsden, 1982; Ewing et al., 1982; Wiedeman et al., 1991). The advantages offered by this method are numerous. As with brain dialysis, one can monitor ongoing processes in the awake animal, but unlike these latter techniques, the electrode is the only exogenous element introduced into the brain region under study, thus eliminating the need for perfusion fluid or radioactive tracers. Changes seen by *in vivo* voltammetry can be verified by HPLC analysis, and may be correlated with behavioral changes.

E. EFFECTS ON SYNTHESIS OR RELEASE
1. Synthesis and Turnover

Toxic insult to metabolic processes at the synapse can occur by several distinct mechanisms (e.g., the inhibition of transmitter inactivation or disruption of the storage and/or release of transmitters). To avoid some of the pitfalls previously cited, a cogent strategy to pinpoint the effects of a toxicant on CNS function is to examine its effects on the rates of formation and degradation of the neurotransmitter in question (see Weiner, 1974 for review). This can be accomplished by using several approaches. For example, a radioactive precursor can be injected and the rate of formation of its metabolites measured. Similarly, neuronal pools can be prelabeled with the radioactive neurotransmitter itself and the decline in specific activity (or rate of decay) can be determined. Although these methods have the theoretical advantage of not perturbing the system being measured, the values derived from the use of these techniques are influenced by many factors. These include diffusion from the site of injection, blood-brain permeability, equilibration with, or alteration in the size of, endogenous pools of amine, metabolism to compounds other than those being studied, uptake into nonneuronal elements, and different rates of turnover for the same neurotransmitter in different brain regions.

Non-steady-state methods involve the use of drugs as enzyme or transport inhibitors, with concurrent measurement of the rate of accumulation or decline of the amine and its metabolites. Several factors may confound these types of studies, including diffusion out of the brain, the possibility of incomplete inhibition of enzyme activity or metabolite efflux, and secondary effects of enzyme blockade on elevation of precursor concentrations and feedback inhibition. Because each of these methods may have a particular disadvantage, sound design necessitates the use of two or more of these approaches in a given study. If similar changes are seen with several techniques, resulting conclusions are likely to be due to actual effects on synthesis or turnover.

In addition to these important direct effects of toxicants, neurochemical changes can also occur as indirect sequelae to intoxication. In either case, neurotoxicologists can use a variety of neurochemical techniques to assess whether the activity of neurons has been perturbed by toxicants. Such approaches are based on basic biochemical principles (e.g., the intermediary metabolism of the transmitter being studied and the rate-limiting steps in the synthesis and/or degradation of the transmitter). As stated in the previous paragraph, several approaches (each with different strengths and limitations) are often combined to provide an estimate of toxicant-induced changes in neural activity.

In general, all approaches require quantification of the concentrations of neurotransmitters and their metabolites in appropriate regions of the CNS. In cases where the metabolites of the neurotransmitter are relatively stable, the ratio of metabolite to parent transmitter may be an estimate of increased utilization. Radiometric techniques have also been used commonly. One can measure the rate of formation of a transmitter and its intermediates after injection of radiolabeled precursor. Biochemical inhibitors are also useful. In the presence of a synthesis inhibitor, toxicant-induced increases in transmitter "turnover" (utilization) would be seen as an increased rate of loss of the neurotransmitter.

2. Uptake and Release

Toxic agents may also affect transmitter uptake and release. Many procedures currently utilized are designed to determine if toxicants alter the release of neurotransmitters (e.g., from synaptic vesicles or brain slices), or alter transmitter uptake, a major mechanism of inactivation or recycling. Toxicants can potentially affect packaging or storage of neurotransmitters in synaptic vesicles, transport of vesicles to the synaptic cleft, or uptake mechanisms involved in recycling of the transmitter. In release studies, either membrane or tissue slice preparations are used, and an uptake inhibitor is often incorporated (Sanders-Bush and Martin, 1982; Schoepp and Azzaro, 1982; Stoof et al., 1979). Release can be induced by either electrical stimulation or by exposure to buffers with high concentrations of potassium. Electrical stimulation has an advantage in that the frequency of stimulation can be regulated, and under appropriate conditions release seems to solely represent that of vesicular neurotransmitter. In contrast, potassium-induced stimulation may cause release of both vesicular and nonvesicular transmitter, and the amount of release cannot be adequately controlled (Vargas et al., 1977). Studies with synaptosomal preparations are technically easier to execute, but have the disadvantages of using a homogenate of disrupted nerve cells and of being unstable following prolonged incubation. Tissue slices, however, provide an intact preparation of nerve terminals with which one can study release of a given neurotransmitter, as well as the interactions of two or more transmitter systems coexisting in the same brain region (e.g., DA and ACh in the striatum). Release can also be studied *in vivo* using push-pull perfusion or voltammetry (*vide infra*).

Uptake experiments generally use membrane preparations to determine the rate of accumulation of labeled amine (Baldessarini and Vogt, 1971; Komiskey et al., 1978; Ross, 1982). Uptake is dependent upon transport across the neuronal membrane, vesicular storage, rate of release, and presynaptic MAO activity. The main confounding factor is that the neurotransmitter can be taken up and bound nonspecifically in other cellular compartments. However, with appropriate assay conditions and pharmacological controls, many of these problems can be overcome.

F. TOXICANT EFFECTS ON RECEPTORS

Of the major receptor superfamilies, the two that are of greatest importance to neurotoxicologists are the G-protein-coupled receptors and ligand-gated ion channels (see Mailman and Lawler [1993] for discussion). Metals may affect receptors in two ways: they may affect the normal recognition of ligands by competitive or noncompetitive mechanisms, or they may induce changes in the receptor as a secondary consequence (e.g., homeostatic, compensatory, or sequelae of injury). The most commonly used technique to explore this mechanism is the radioreceptor assay, one of the most utilized techniques in neurobiology (cf. DeHaven and Mailman, 1983). This method can be used to assess the effects of prior toxicant exposure on receptor characteristics (e.g., density and affinity), as well as the ability of the toxicant to interact directly with a receptor population and alter the binding of specific ligands (such as the endogenous transmitter). The technique is simple and reliable as long as specific criteria (e.g., saturability, specificity, and reversibility) are met. The availability of molecular biological techniques has resulted in a dramatic acceleration in the cloning of CNS receptors. One unexpected but consistent finding has been a multiplicity of cloned receptor subtypes, often greater than that predicted on the basis of available pharmacological data. Because the physiological significance of such subtype multiplicity is often unclear, it complicates studies of toxic metals.

There are two ways that neurotoxicologists have explored the involvement of receptors in neurotoxicity. Without exception, cations can affect the binding of ligands to every neuroreceptor that has been studied. Because physiological ions (such as Na^+, K^+, Mg^{2+}, Mn^{2+}, Ca^{2+}, etc.) can cause such effects, one might hypothesize that toxic metals might affect receptors directly. In fact, such effects can be observed. For example, Treherne et al. (1991) have reported that the binding of radioligands to the histamine H_1-receptor was inhibited by a series of mono- and divalent cations (order of potency: Hg^{2+} > Cd^{2+} > Zn^{2+} > Ni^{2+} > Co^{2+} > Mn^{2+} > Ca^{2+} > Mg^{2+} > Li^+ = Na^+ > K^+ > Cs^+). The binding of [^{3}H]mepyramine, a tertiary amine, was inhibited by the divalent cations to a similar extent as the binding of [^{3}H]-QMDP. In fact, Hg^{2+}, presumably by its ability to bind to critical sulfhydryl groups, has been reported to affect the binding of many receptors. Thus, methylmercury inhibits the binding of cholinergic ligands to purified *Torpedo* acetylcholine receptors, or crude brain membrane fractions (Shamoo et al., 1976; Eldefrawi et al., 1977). Yet Von Burg et al. (1980) evaluated how *in vivo* and *in vitro* exposure to methylmercury affected the binding of the muscarinic ligand [^{3}H]-QNB. They suggested that since mercuric chloride was about two orders of magnitude more potent than methylmercury, a conversion of even small amounts of methylmercury to Hg^{2+} *in situ* might contribute to the toxic effects of methylmercury. These examples, in a fashion similar to that discussed earlier for enzyme activity, indicate the difficulty of ascertaining whether such effects are relevant to the neurotoxicity of toxic metals.

The other major use of receptor studies is related to how toxic metals cause sequelae that may affect receptor recognition. For example, although available evidence suggests a role for dopaminergic systems in the neurotoxicity of lead, the exact nature of this dysfunction remains unknown. Using [^{3}H]-sulpiride to label D_2 dopamine receptors in the brains of rats exposed to lead, Lucchi et al. (1981) found an increase in the number of D_2 receptors in striatum and a decrease in nucleus accumbens. This suggested to these investigators that lead may preferentially affect D_2 receptors. In most cases, however, investigators have failed to detect permanent changes that persist after lead exposure has ceased. In addition, we have not found significant changes in the number of striatal DA receptors using [^{3}H]-spiperone as the radioligand (DeHaven et al., 1984). Similarly, cadmium exposure via the drinking water results in decreased binding of the muscarinic cholinergic ligand [^{3}H]-QNB in striatum and cerebral cortex (Hedlund et al., 1979), suggesting changes in muscarinic acetylcholine receptors; this was also reported in rat pups whose mothers were exposed to cadmium (Donaldson et al., 1981). Cadmium also has been reported to inhibit [^{3}H]-imipramine binding in hypothalamus both *in vitro* and *in vivo* (Peterson and Bartfai, 1983). From the above data, it is evident that exposure to inorganic heavy metals causes a variety of neurochemical alterations that include, but are not limited to, dopaminergic systems. Not surprisingly, these changes are not similar for all toxic metals.

IV. THE STUDY OF TOXIC METAL-INDUCED CHANGES IN NEUROTRANSMITTER METABOLISM: SOME EXAMPLES

As noted in the Introduction, even a cursory review of the literature transcends the space allotted to this chapter. On the other hand, we believe that the application of, and issues with, neurochemical techniques can be illustrated best by specific examples.

A. MANGANESE EFFECTS ON NEUROTRANSMISSION

Among the toxic metals, manganese is unique in that it appears to be selective (at least relative to other toxic metals) in the types of damage it causes. The finding that symptoms of manganese poisoning resemble Parkinson's disease (Cotzias et al., 1974) suggested a role for dopaminergic systems in manganese neurotoxicity. Administration of manganese resulted in decreased levels of DA in brain (Cotzias et al., 1974), that occurred selectively in striatum and were accompanied by a decrease in turnover in this region (Autissier et al., 1982). In rats exposed neonatally to manganese, decreases in DA concentrations and tyrosine hydroxylase activity and increases in MAO activity occurred in the hypothalamus (Deskin et al., 1980). It has been suggested that manganese toxicity in dopaminergic neurons may result in part from its ability to oxidize DA (Donaldson et al., 1980), although it clearly affects the function of other neurotransmitter systems as well.

As might be expected from a toxicant that targets, at least somewhat preferentially, specific neurotransmitters, manganese intoxication can alter dopamine receptors. Treatment of adult rats with 10 mg/kg manganese chloride for 15 days results in increased striatal [^{3}H]-spiperone binding (Seth et al., 1981), a marker for the dopamine D_2-like receptors. At 15 mg/kg, decreases were seen in striatal [^{3}H]-QNB (muscarinic receptors) binding, frontal cortical [^{3}H]-5-HT binding, and cerebellar [^{3}H]-muscimol (GABA receptors) binding. Exposure to manganese via the drinking water also increased [^{3}H]-spiperone binding in striatum and decreased [^{3}H]-muscimol binding in hippocampus (Gerhart and Tilson, 1982). It appears that manganese can cause widespread changes, reflecting the underlying damage caused by this metal.

More recently, aspects of manganese neurotoxicity have been explored using modern imaging techniques. Eriksson et al. (1992a) performed positron emission tomography (PET) scans on monkeys receiving manganese by subcutaneous injection over a period of 16 months. As markers for the integrity of the presynaptic terminal, they used [^{11}C]-nomifensine (a label for the dopamine transporter) and [^{11}C]-L-DOPA (reflecting dopamine turnover). The expected degeneration of dopaminergic nerve endings during manganese intoxication was supported by a decrease (up to 60%) in [^{11}C]-nomifensine binding. Conversely, [^{11}C]-L-DOPA decarboxylation was not affected, suggesting a compensatory increase in the activity of the remaining nerve terminals. Later, postmortem studies of these animals after 26 months of exposure was performed using quantitative receptor autoradiography. Consistent with the PET studies, the binding of [^{3}H]-mazindol to the dopamine uptake sites was reduced by 75% in areas of the brain known to be involved in the motor deficits induced by manganese, but not elsewhere. Interestingly, while neither muscarinic cholinergic nor $GABA_A$ receptors were altered, D_1 dopamine receptors were significantly decreased. Since the D_1 receptors are believed to occur only on target cells, this suggests that manganese intoxication may also affect nondopaminergic neurons containing the D_1 receptors.

B. EFFECTS OF METHYLMERCURY AND MERCURIC ION ON NEUROTRANSMISSION

The well-documented instances of methylmercury poisoning have provided the impetus to understand the mechanisms and sequelae of intoxication by this agent. A brief review of the manifold effects of this agent underscores the complex effects that toxic metals may have on neurotransmission. Taylor and DiStefano (1976) found that postpartum administration of methylmercury to rats decreased whole brain concentrations of several monoamine neurotransmitters when assayed at day 8, whereas increases were found at day 15. Neonatal exposure to methylmercury for 21 days results in increased dopamine turnover, followed by persistent decreases in synaptosomal [^{3}H]-dopamine uptake after 20 days of age (Bartolome et al., 1982). Alterations in uptake and release of dopamine, serotonin, and norepinephrine have also been observed following methylmercury exposure (Komulainen and Tuomisto, 1981, 1982). Prasad et al. (1979) reported that methylmercury inhibited the uptake of glutamate and glycine in C-6 glioma cells, but markedly increased uptake of these transmitters. It is unclear how these latter observations may be related to the changes found in mammalian brain (Bondy et al., 1979).

Many investigators have attempted to derive a unifying hypothesis to explain the actions of mercury. O'Kusky and McGeer (1985) found that the activity of glutamic acid decarboxylase, a marker for GABA neurons, but not its substrate affinity, was decreased by postnatal methylmercury treatment in proportion to the severity of intoxication. Conversely, the activity of another marker enzyme, choline-acetyltransferase, was not significantly affected. They suggested that such alterations in GABA neurons may be responsible, in part, for both the visual and motor disturbances associated with methylmercury intoxication. More recently, Arakawa et al. (1991) have studied the effects of mercuric chloride and methylmercury on primary cultures of rat dorsal root ganglion neurons using the whole-cell patch clamp technique. GABA-induced chloride currents were augmented by mercuric chloride, yet decreased by methylmercury. Since GABA systems are known to modulate monoaminergic neurotransmission, it may be that the monoamine changes noted earlier are related to direct effects on GABA systems.

On the other hand, Brookes (1992) has suggested that glutamate is key in the neurotoxicity of mercuric ion. Hg^{2+} was shown to inhibit glutamate uptake in astrocyte and spinal cord cultures, as well as decreasing glutamine content and export in astrocyte cultures without affecting neuronal viability in spinal cord cultures in the absence of excitotoxic accumulations of glutamate. These data were felt to support the hypothesis that mercuric ion in the brain can cause neurotoxicity by selectively inhibiting the uptake of synaptically released glutamate, with consequent elevation of glutamate levels in the extracellular space. Yet Levesque et al. (1992) reported that methylmercury also reduces choline uptake into nerve terminals, as well as increasing the release of acetylcholine via mechanisms that involve both extracellular Ca^{2+} and mitochondrial actions that induced the release of bound intraterminal Ca^{2+} stores. Thus, it was suggested that methylmercury can interfere with cholinergic neurotransmission in several ways.

While these previous studies have focused on specific neurotransmitter systems, other studies have suggested that mechanisms common to all neuronal systems may play a major role. Anner et al. (1992) have suggested that mercury is a potent inhibitor of Na-K-ATPase activity, although Chetty et al. (1993) indicated that such an effect might be common to any toxic metal that could chelate with protein sulfhydryl groups. Similarly, Saijoh et al. (1993) have posited that protein kinases A and C are inhibited by methylmercury, and that such effects might impair intracellular signal transduction and play a role in methylmercury toxicity. Sarafian and Verity (1990) have reported that methylmercury stimulates protein [^{32}P]-phospholabeling in cerebellar granule cell culture, an effect opposite from that predicted by the report of methylmercury on protein kinases (Saijoh et al., 1993). Yet the fact that Hg^{2+} can react with the free -SH groups of proteins prompted Vignes et al. (1993) to investigate the effects of Hg^{2+} and the sulfhydryl reagent p-chloromercuric benzosulfonic acid (PCMBS) on accumulation of the second messenger inositol phosphate (IP). Hg^{2+} and PCMBS, depending on their concentration, had two distinct effects on IP accumulation: at low concentrations, Hg^{2+} (from 1 to 10 µM) stimulated PI breakdown alone, yet interacted in different ways with IP accumulation induced by glutamate or cholinergic receptors. At higher concentrations, Hg^{2+} (from 0.01 to 1 mM) inhibited both basal and transmitter-stimulated IP accumulation. The actions of Hg^{2+} were similar, but not identical, to those of sulfhydryl reagent p-chloromercuric benzosulfonic acid.

On the other hand, Palkiewicz et al. (1994) reported that ADP-ribosylation of brain neuronal proteins is altered by *in vitro* and *in vivo* exposure to inorganic mercury. They examined the effect of Hg^{2+} on the ADP-ribosylation of tubulin and actin, cytoskeletal proteins found in neurons, and B-50/43-kDa growth-associated protein (B-50/GAP-43), a neuronal tissue-specific phosphoprotein. Both *in vitro* and *in vivo* studies indicated that $HgCl_2$ markedly inhibited the ADP-ribosylation of tubulin and actin, suggesting that Hg^{2+} may affect specific neurochemical reactions involved in the development and maintenance of brain structure.

Such data underscore the dilemma faced by neurotoxicologists when dealing with the toxic metals in general, and mercury in particular. These examples make clear that neurochemical perturbations of known neurotoxic metals can be found in many systems. Many (most?) parameters that are assessed may be affected, suggesting either that the neurotoxicity is nonselective and often unpredictable, or that most of the phenomena that are reported are epiphenomena unrelated to the neurotoxicity. These data also indicate the difficulty of knowing which of the changes that have been found are actually relevant to the initial intoxication and subsequent sequelae. It may well be that the nonselective nature of mercury and methylmercury make all such changes of equal interest. It also may be that these are epiphenomena that do not play a critical role in the toxicity of these metals.

V. SUMMARY: THE ISSUES OF TOXIC METALS AND NEUROTRANSMISSION

Although the examples listed previously (and summarized in Table 1) illustrate that one can obtain much useful information by application of these neurochemical techniques to the study of metal intoxication, there are often significant problems that are necessary to resolve. For example, the problem of separating proximal mechanism from epiphenomenon (relevant or otherwise) is often daunting. Such epiphenomena may be useful as markers of the consequences of neurointoxication, even when such changes represent events quite distal from the primary insult. On the other hand, unlike toxicants with relatively selective modes of action, the toxic metals are often "nonselective" and may perturb simultaneously a number of metabolic processes. Neurochemical investigations made soon after intoxication may reveal changes closely related to the acute toxic effects of the compound. These may be very different from those observed in chronic neurotoxicity, and gross neurochemical perturbations may not be seen until many days or weeks after exposure. Yet these delayed alterations, while only distantly related to the initial insult, may offer clues as to the behavioral or physiological alterations associated

Table 1 Some Strategies Used to Assess Neurotransmitter Function in the Central Nervous System

Synaptic Mechanism	Technique Used to Assess
Assessment of "turnover"	Brain dialysis
	Injection of labeled precursor or amine
	Use of enzyme inhibitors
	Metabolite measurements
Neurotransmitter release	In vitro (slice or synaptosome studies)
	Cerebral microdialysis
	In vivo voltammetry
Neurotransmitter inactivation	
Uptake	Enzyme activity (e.g., transporter activity)
	Binding or density of transporters (using radioaligno methods)
Degradation	Enzyme activity (e.g., acetyl cholinesterase or MAO)
Receptor properties	Membrane radioreceptor studies
	Quantitiative receptor autoradiography
Second messenger function	Activity of metabolic enzymes (e.g., adenylate cyclases)
	Changes in concentration of intermediates (e.g., cAMP)
	Protein phosphorylation

with intoxication. The above discussion emphasizes that the results of biochemical studies late in the sequence of neurotoxic events need to be interpreted not so much in terms of the mechanism initiating the cascade leading to neurotoxicity, but rather as an assay for terminal and/or compensatory events.

As this general discussion makes clear, many studies examining toxic metal-induced functional changes in neurotransmission do not have a well-defined hypothesis. With such "nonselective" toxicants, there may be no example of discovery of a mechanism that explains preferential toxicity, or the pattern of effects, for the nervous system. It should be noted that this problem is related more to the nature of these toxicants than to the ability of the scientists studying such problems. The key issue is often when a specific perturbation is the primary event, or when widespread neurological dysfunction is secondary, even to peripheral events such as kidney or liver damage. In the present chapter, we have highlighted some of the mechanisms that have been implicated in toxic metal action, and their relationship to mechanisms of, or sequelae to, toxicity.

As was noted in the Introduction, the goal of determining mechanisms of neurotoxicity is a formidable task that is not often achieved unless the toxicant of interest is one with high specificity. The toxic metals, in both their organic and inorganic forms, seldom have such specificity. Thus, one is faced with ascertaining when a measured change is of relevance to the toxicity under investigation, or is just one of many secondary alterations. Often, however, one can relate neurochemical changes to observed behavioral, morphological, pharmacological, and/or physiological perturbations. Even if such neurochemical alterations are clearly dissociated from the mechanism(s) of toxicity, the consequences of injury (and the resulting compensatory responses found because of the considerable plasticity of the nervous system) can often provide scientifically interesting problems.

This chapter has discussed the wide variety of neurochemical techniques available for the assessment of neurotoxicity and how these techniques have been employed in the study of several toxic metals. A multifaceted strategy (e.g., combining appropriate neurochemical techniques with tests assessing those behaviors affected by the neurochemical changes) can often provide important information in evaluating the neurotoxicity of a particular compound, as well as clues for more detailed biochemical and neurochemical studies. It should be emphasized that almost any neurotoxicant regimen can be carried out to the point where a given neurochemical parameter may be perturbed, yet the change is not of relevance to the primary actions of the toxicant. It is imperative that consideration be given to differentiating such cases from those toxicant-induced compensatory effects (e.g., neuronal sprouting, changes in neurotransmission, receptor supersensitivity, etc.) that may be of great importance. Such issues continue to make the effects of toxic metals on neurotransmission a challenging area of study.

ACKNOWLEDGMENTS

This work was supported, in part, by training grants ES-07126, DA-07244, center grants HD03110 and MH33127, and program grant ES-01104.

REFERENCES

Anner, B. M., Moosmayer, M., and Imesch, E. (1992), *Am. J. Physiol.*, 262, F830–6.
Arakawa, O., Nakahiro, M., and Narahashi, T. (1991), *Brain Res,* 551, 58–63.
Autissier, N., Rochette, L., Dumas, P., Beley, A., Loireau, A., and Bralat, J. (1982), *Toxicology,* 24, 175–182.
Baldessarini, R. J. and Vogt, M. (1971), *J. Neurochem.,* 18, 2519–2533.
Bartolome, J., Trepanier, P., Chait, E. A., Seidler, F. J., Deskin, R., and Slotkin, T. A. (1982), *Toxicol. Appl. Pharmacol,* 65:92–99.
Boadle-Biber, M. C. (1982), in *Biology of Serotonergic Transmission,* Osborne, N. N., Ed., John Wiley & Sons, New York, 63–94.
Bondy, S. C., Anderson, C. L., Harrington, M. E., and Prasad, K. N. (1979), *Environ. Res.,* 19, 102–111.
Bondy, S. C. (1982), in *Mechanisms of Actions of Neurotoxic Substances,* Prasad, K. N. and Vernadakis, A. Eds., Raven Press, New York, 25–50.
Brazell, M. P. and Marsden, C. A. (1982), *Br. J. Pharmacol.,* 75, 539–547.
Brookes, N. (1992), *Toxicology,* 76:245–256.
Caudill, W. L., Houck, G. P., and Wightman, R. M. (1982). *J. Chromatog.,* 227, 331–339.
Chetty, C. S., Stewart, T. C., Cooper, A., Rajanna, B., and Rajanna, S. (1993), *Drug Chem. Toxicol.,* 16, 101–110.
Cotzias, G. C., Papavasiliou, P. S., Mena, I., Tang, L. D., and Miller, S. T. (1974). *Adv. Neurol.,* 5:235–243.
Crombeen, J. P., Kraak, J. C., and Poppe, H. (1978). *J. Chromatog.,* 167, 219–230.
DeHaven, D. L. and Mailman, R. B. (1983). *Rev. Biochem. Toxicol.,* 5, 193–238.
DeHaven, D. L., Krigman, M. R., Gaynor, J. J., and Mailman, R. B. (1984), *Brain Res.,* 297, 297–304.
DeHaven, D. L. and Mailman, R. B. (1986), in *Neurobehavioral Toxicology,* Annau, Z., Ed., Johns Hopkins University Press, Baltimore, 214–243.
Desiderio, D. M., Yamada, S., Tanzer, F. S., Horton, J., and Trimble, J. (1981). *J. Chromatog,* 217, 437–452.
Deskin, R., Bursian, S. J., and Edens, F. W. (1980), *Neurotoxicology,* 2, 65–73.
Donaldson, J., LaBella, F. S., and Gesser, D. (1980). *Neurotoxicology,* 2, 53–64.
Donaldson, J., McGregor, D., Gesser, H., and LaBella, F. S. (1981), *Toxicologist,* 1, 61–62.
Eldefrawi, M. E., Mansour, N. A., and Eldefrawi, A. T. (1977). *Adv. Exp. Med. Biol.,* 84, 449–463.
Eriksson, H., Tedroff, J., Thuomas, K. A., Aquilonius, S. M., Hartvig, P., Fasth, K. J., Bjurling, P., Langstrom, B., Hedstrom, K. G., and Heilbronn, E. (1992a), *Arch. Toxicol.,* 66, 403–407.
Eriksson, H., Gillberg, P. G., Aquilonius, S. M., Hedstrom, K. G., and Heilbronn, E. (1992b), *Arch Toxicol.,* 66, 359–364.
Ewing, A. G., Wightman, R. M., and Dayton, M. A. (1982). *Brain Res,* 249, 361–370.
Garrick, N. A. and Murphy, D. L. (1980), *Psychopharmacology,* 72, 27–33.
Gerhart, J. M. and Tilson, H. A. (1982), *Toxicologist,* 2, 87.
Hedlund, B., Gamarra, M., and Bartfai, T. (1979), *Brain Res.,* 168, 216–218.
Keller, R., Oke, A., Mefford, I. and Adams, R. N. (1976), *Life Sci.,* 19, 995–1004.
Kilts, C. D., Breese, G. R., and Mailman, R. B. (1981), *J. Chromatog.,* 225, 347–357.
Komiskey, H. L., Hsu, F. L., Bossart, F. J., Fowble, J. W., Miller, D. D., and Patil, P. N. (1978), *Eur. J. Pharmacol.,* 52, 37–45.
Komulainen, H. and Tuomisto, J. (1981), *Acta Pharmacol. Toxicol.,* 48, 214–222.
Komulainen, H. and Tuomisto, J. (1982), *Neurobehav. Toxicol. Teratol.,* 4, 647–649.
Leung, T. K., Lim, L., and Lai, J. C. (1992), *Metab. Brain Dis.,* 7:139–146.
Levesque, P. C., Hare, M. F., and Atchison, W. D. (1992), *Toxicol. Appl. Pharmacol.,* 115, 11–20.
Lindroth, P. and Mopper, K. (1979), *Anal. Chem.,* 51, 1667–1674.
Lucchi, L., Memo, M., Airaghi, M. L., Spano, P. F., and Trabucchi, M. (1981), *Brain Res.,* 213, 397–404.
Lyness, W. H. (1982), *Life Sci.,* 31, 1435–1443.
Mailman, R. B. and DeHaven, D. L. (1984), *Cellular and Molecular Neurotoxicology,* Narahashi, T., Ed., Raven Press, New York, 207–224.
Mailman, R. B. (1987), *Neurotoxicol. Teratol.,* 9, 417–426.
Mailman, R. B. and Lawler, C. P. (1993). , in *Introduction to Biochemical Toxicology,* Hodgson, E. and Levi, P. Eds., Elsevier, Amsterdam.
McMillen, B. A. (1982), *Biochem. Pharmacol.,* 31, 2643–2647.
Meek, J. L. (1980), *Proc. Natl. Acad. Sci. U.S.A.,* 77, 1632–1636.
Mefford, I. N. (1981), *J. Neurosci. Methods,* 3, 207–224.
Melamed, E., Hefti, F., and Wurtman, R. J. (1980), *J. Neurochem.,* 34, 1753–1756.
Morell, P. and Mailman, R. B. (1987), in *Neurotoxicants and Neurobiological Function: Effects of Organoheavy Metals,* Tilson, H. A. and Sparber, S., Eds., 201–229.
O'Kusky, J. R. and McGeer, E. G. (1985), *Brain Res.,* 353, 299–306.
Palkiewicz, P., Zwiers, H., and Lorscheider, F. L., (1994), *J. Neurochem.,* 62, 2049–2052.
Peterson, L. L. and Bartfai, T. (1983), *Eur. J. Pharmacol.,* 90, 289–292.
Potter, P. E., Meek, J. L., and Neff, N. H. (1983), *J. Neurochem.,* 41, 188–194.
Prassad, K. N., Nobles, E., and Ramanujam, M. (1979), *Environ. Res.,* 19, 189–201.
Ross, S. B. (1982), in *Biology of Serotonergic Transmission,* Osborne, N. N., Ed., John Wiley & Sons, New York, 159–195.
Saijoh, K., Fukanaga, T., Katsuyama, H., Lee, M. J., and Sumino, K. (1993), *Environ Res.,* 63, 264–273.

Sanders-Bush, E. and Martin, L. L. (1982),in *Biology of Serotonergic Transmission,* Osborne, N. N. Ed., John Wiley & Sons, New York, 95–118.
Sarafian, T. and Verity, M. A. (1990), *J. Neurochem.,* 55, 913–921.
Schoepp, D. D. and Azzaro, A. J. (1982), *Biochem. Pharmacol.,* 31, 2961–2968.
Scratchley, G. A., Masoud, A. N., Stohs, S. J., and Wingard, D. W. (1979), *J. Chromatog.,* 169, 313–319.
Seth, P. K., Hong, J. S., Kilts, C. D., and Bondy, S. C. (1981), *Toxicol. Lett.,* 9, 247–254.
Shamoo, A. E., Maclennan, D. H., and Elderfrawi, M. E. (1976), *Chem. Biol. Interact.,* 12, 41–52.
Smith, J. E., Lane, J. D., Shea, P. A., McBride, W. J., and Aprison, M. H. (1975), *Anal. Biochem.,* 64, 149–169.
Stahle, L. (1992), *Psychopharmacology,* 106, 1–13.
Stoof, J. C., Den Breejen, E. J. S., and Mulder, A. H., (1979), *Eur. J. Pharmacol.,* 57, 35–42.
Taylor, L. L. and DiStefano, V. (1976), *Toxicol. Appl. Pharmacol.,* 38, 489–497.
Tilson, H. A. and Sparber, S. B. (1970), *Behav. Res. Methods Instrum.,* 2, 131–134.
Treherne, J. M., Stern, J. S., Flack, W. J., and Young, J. M. (1991), *Agents Actions Suppl.,* 33, 271–276.
Ungerstedt, U., Herrera-Marschitz, M., Jungnelius, U., Stahle, L., Tossman, U., and Zetterstrom, T. (1982), *Adv. Biosci.,* 37, 219–231.
Van Valkenburg, C., Tjaden, U., Van der Krogt, J., and Van der Leden, B. (1982), *J. Neurochem.,* 39, 990–997.
Vargas, O., deLorenzo, M. C. D., Saldate, M. C., and Orrego, F. (1977), *J. Neurochem.,* 28, 165–170.
Vignes, M., Giuramand, J., Sassetti, I., and Recasens, M. (1993), *Eur. J. Neurosci.,* 5, 327–334.
Von Burg, R., Northington, F. K., and Shamoo, A. (1980), *Toxicol. Appl. Pharmacol.,* 53:285–292.
Wallace, E. F., Krantz, M. J., and Lovenberg, W. (1973), *Proc. Natl. Acad. Sci. U.S.A.,* 70, 2253–2255.
Weiner, N. 1974, in *Neuropsychopharmacology of Monoamines and Their Regulatory Enzymes,* Usdin, E. Ed., Raven Press, New York 143–159.
Wiedemann, D. J., Kawagoe, K. T., Kennedy, R. T., Ciolkowski, E. L., Wightman, R. M. (1991), *Annal. Chem.,* 63, 2965–2970.
Young, A. M. (1993), *Rev. Neurosci.,* 4, 373–395.
Zetterstrom, T., Sharp, T., and Ungerstedt, U. (1984). *Eur. J. Pharmacol.,* 106, 27–37.

Chapter 38

Cytoskeletal Toxicity of Heavy Metals

Ronald D. Graff and Kenneth R. Reuhl

I. INTRODUCTION

It has been over a century since investigators first recognized that exposure to toxic metals results in alterations of neural cell structure. These "degenerative" changes, identified by the accumulation of fibrils which stained with silver reagents, were the subject of many classical investigations of neuronal injury. From these studies and the work of early cytologists evolved the concept that the fibrils formed part of a cell framework or "cytoskeleton" which provides the physical basis of cell shape. However, it was not until the advent of electron microscopy that various components of the cytoskeleton were clearly identified and their functions began to be elucidated.

Initially envisioned as being merely a scaffold in support of cell structure, the cytoskeleton is now recognized to consist of diverse families of dynamic and integrated proteins which provide a physicochemical continuum, bridging extracellular matrix with the nucleus. The principle of "dynamic reciprocity" holds that influences from the extracellular milieu may be transduced by transmembrane molecules to affect the cytoskeleton, and hence to affect cytoskeleton-associated organelles, chromatin, and the nuclear matrix. Virtually all cellular functions are directly or indirectly influenced by cytoskeleton. Accordingly, disruption of cytoskeletal organization by toxicants is reflected in a spectrum of altered cell function, with consequences to the cell ranging from the innocuous to the lethal (for review, see Abou-Donia and Gupta, 1994). Conversely, the cytoskeleton will respond to disturbances of other cell systems by rearrangement of its architecture. Regardless of the source of the eliciting event, changes in cytoskeletal organization and integrity can have significant adverse effects on cell functions and responses to toxicants.

II. ELEMENTS OF THE CYTOSKELETON

The fibrillary proteins of the cytoskeleton are divided into three families according to size. Largest are the microtubules (MT), rod-shaped structures approximately 25 nm in diameter. The second class, intermediate filaments (IF), is a complex family of polypeptides classified into five groups differentially expressed in diverse tissues, and includes neurofilaments and astrocytic glial filaments. Intermediate filaments are small, approximately 10 nm in diameter. Smaller still are the microfilaments, filamentous actin protein polymers 3 to 5 nm in diameter.

Associated with each of these cytoskeletal classes are families of proteins which modulate stability and participate in linkage of cytoskeleton with other cell structures. In recent years, a bewildering number of these cytoskeleton-associated proteins have been described and at least partially characterized. Some, such as the microtubule-associated proteins (MAPs), have been studied in considerable detail (for reviews, see Matus, 1988; Matus and Riederer, 1986; Olmsted, 1986). Others, including the intermediate filament associated proteins (IFAPs) and actin-linking proteins, are less well understood (Wang, 1985) but appear to serve critical roles as linker molecules between the various components of the cytoskeleton, as well as linking cytoskeletal elements with cellular elements, including cytoplasmic organelles, membrane receptor systems, and extracellular matrix (Traub and Shoeman, 1994). Since little is known regarding the involvement of these poorly characterized cytoskeletal structures in metal toxicology, they receive only brief consideration in this chapter. However, it is probable that as their biochemistry becomes better understood, they too will be identified as targets for toxicants.

The numerous components of cytoskeleton may respond to injury singly or in a cooperative fashion, depending upon the nature and severity of insult. The general cytoskeletal response to a stimulus is to alter its organization, accomplished by changing either the physical orientation of its components or by differentially increasing assembly or disassembly of specific cytoskeletal proteins. In this manner the cytoskeleton may act as both a signal transduction and a signal effector system. Perturbation of one cytoskeletal element most often results in rearrangement of one or more additional elements. For example, disassembly of microtubules by methylmercury is associated with secondary collapse of intermediate filaments to a perinuclear position (Cadrin et al., 1988). However, such secondary responses are not inevitable and highly selective changes may be restricted to one cytoskeletal element. Redistribution of vimentin filaments by acrylamide, for example, does not appear to secondarily alter normal microtubule architecture.

A. POTENTIAL SITES OF METAL ACTION ON CYTOSKELETON

The susceptibility of the neural cytoskeleton to toxic metals has not been systematically examined and it is often necessary to make assumptions about neural vulnerability based on evidence derived from nonneural cells or *in vitro* preparations. The degree to which such extrapolations are valid is not known. Nevertheless, it is unnecessary conservatism and poor toxicology to discard out of hand available data from nonneural sources. Evidence to date suggests that the response of neural cytoskeleton can be described in the same general terms as cytoskeleton in other organs, although specific differences can be demonstrated.

Metals may affect cytoskeleton by binding **directly** to the cytoskeletal element or its associated proteins, or **indirectly** by changing the intracellular environment necessary for the cytoskeleton to function normally. Given the reactivity of metals, both direct and indirect influences may be present simultaneously.

Direct actions of metals usually influence assembly/disassembly kinetics of cytoskeleton, either by blocking addition of subunits during assembly or by binding to cytoskeleton and placing steric strain on the protein. The net result of either process is eventual disassembly of the cytoskeletal polymers. Metals may also affect posttranslational modifications of the cytoskeleton required for normal function; these include changes to the filament itself (such as phosphorylation) or the addition of cytoskeletal-associated proteins which link elements of the cytoskeleton together. Rarely, perturbation of posttranslational modification may increase rather than decrease stability of the cytoskeletal element.

Metals may indirectly affect cytoskeletal integrity by altering the intracellular environment. Microtubules and microfilaments in particular are extremely sensitive to changes in ionic homeostasis. For example, elevations in intracellular Ca^{2+} concentration will induce disassembly of microtubules and microfilaments, and can activate proteases involved in neurofilament degradation. Ca^{2+} influx may also initiate proteolysis or phosphoryalation of microtubule-associated-proteins (Fischer et al., 1991; Yamamoto et al., 1983). Damage to biosynthetic organelles such as endoplasmic reticulum or Golgi apparatus will affect the availability of cytoskeletal precursor molecules for assembly. Metals can also activate (or inactivate) enzymes, such as proteases, involved in cytoskeleton homeostasis, resulting in premature or delayed degradation of the proteins.

B. INTERPRETATION OF CYTOSKELETAL CHANGES FOLLOWING INJURY — A CAUTIONARY NOTE

The central role played by cytoskeleton in cell architecture and physiology, as well as the dynamic nature of cytoskeletal assemblies, make this system an attractive target for toxic action. A wide variety of chemicals, including toxic metals, have been shown to alter one or more elements in the cytoskeletal network *in vivo* and *in vitro*. Nevertheless, changes in cytoskeleton following toxicant exposure must be carefully interpreted and mechanistic extrapolations applied with caution. Most studies of toxicant effects on cytoskeleton have employed *in vitro* or cell culture models. These systems have inherent limitations, particularly in their inability to fully mimic the chemical and architectural complexity of intact tissue. Many specialized cytoskeletal processes (e.g., posttranslational modifications) or responses to hormones or circulating factors active *in vivo* may not be fully operative in cultures. Because of the inherent artificiality of *in vitro* systems, disturbances of cellular microenvironment are more likely to occur in cultured cell systems than *in vivo*. Alterations in calcium homeostasis, oxidative stress, or membrane permeability, frequently encountered in culture, are deleterious to cytoskeletal integrity and may be misinterpreted as direct rather than secondary toxic actions.

Evaluation of cytoskeleton *in vivo* is similarly complicated, primarily by our technical limitations. Cytoskeleton, particularly the microtubule, is difficult to preserve unless extreme care is taken during sacrifice and tissue preparation. Reorganization of many cytoskeletal structures occurs rapidly following death and in response to fixation. Furthermore, the small size of cytoskeletal elements makes them difficult to resolve and study in tissues. Finally, there is great heterogeneity of response to toxicants in different brain areas. Distinguishing a primary effect of a toxic agent from secondary cellular responses is difficult both *in vivo* and *in vitro*, and must be approached with caution.

III. MICROTUBULES

A. STRUCTURE AND FUNCTION OF MICROTUBULES

Microtubules are filamentous proteins about 25 nm in diameter and are present in all eukaryotic cells. They are composed of heterodimers of two isotypes of tubulin, α- and β-, which polymerize in a helical, end-to-end fashion, to form cylindrical polymers of variable length. A third tubulin subtype, gamma tubulin, is not a major component of cellular microtubules but is believed to be responsible for microtubule nucleation in the centrosome (Joshi et al., 1992; Stearns and Kirschner, 1994). Microtubule polymers form complex patterns in cells, ranging from a radial pattern originating from the centrosome in interphase cells to the mitotic spindle of dividing cells and linear staggered arrays found in neurites (Cassimeris, 1993). These complex microtubular arrays, along with a variety of microtubule-associated proteins are involved in virtually every cellular activity, including cell division, motility, intracellular transport, establishment of cell morphology and polarity, and positioning of organelles (for reviews, see Dustin, 1984; Gelfand and Bershadsky, 1991; Avila, 1990).

Assembly of microtubules occurs by the preferential addition of subunit monomers to one end of the polymer, termed the plus (+) end, while the opposite or minus (–) end shows little or no net addition of monomer, thus establishing an intrinsic structural polarity believed to be important in establishing cell shape. For example, most microtubules of interphase cells elongate with their plus ends directed toward the cell periphery; in neurons, however, microtubules in axons and dendrites are longitudinally oriented along the neurite axis. In addition, axonal microtubules are all oriented with the plus ends directed away from the perikaryon, while dendritic microtubules may be oriented in either direction (Heidemann, 1981; Baas et al., 1988). This difference in microtubule polarity may be important for directing MAP-based axonal transport (Vale, 1987; Vallee et al., 1989) and might also explain the localization of neuronal proteins in different subcellular compartments (Garner et al., 1988; Baas and Black, 1989).

Tubulin heterodimers possess two guanine nucleotide binding sites. The nonexchangeable site (N-site) binds irreversibly either GTP or GDP and may be located on α-tubulin. The exchangeable site or E-site, identified on β-tubulin, binds and readily exchanges GTP or GDP (Geahlen and Haley, 1977). Tubulin polymerization requires binding of a guanine nucleotide and is coupled to the hydrolysis of GTP at the E-site (Schilstra et al., 1987; Stewart et al., 1990).

The steady-state behavior of microtubules and the relationship between elongation and shortening phases was termed "dynamic instability" by Mitchison and Kirschner (1984). In this model, individual microtubules persist for some time in phases of rapid elongation or shortening and can abruptly shift between phases (Cassimeris, 1993). Elongating microtubules are presumed to be inherently unstable but may be stabilized by a "cap" of multiple subunits at the growing end which contain unhydrolyzed GTP (Hill and Carlier, 1983; Mitchison and Kirschner, 1984; Bayley, 1990; Wilson et al., 1990). A GTP cap arises due to the apparent lag between assembly of tubulin subunits and the hydrolysis of GTP on the exchangeable site of β-tubulin; should this cap be lost, depolymerization will continue until such time as the polymer is completely disassembled or the cap is restored and disassembly is arrested (Walker et al., 1988).

B. MICROTUBULE ISOTYPES

The vertebrate tubulin gene family contains at least seven distinct β-tubulin and six α-tubulin genes, though the exact number varies somewhat between species. The polypeptide products of β-tubulin genes can be grouped into six major isotype classes denoted I–VI, which are highly conserved among species and have distinct regional and temporal patterns of expression (Sullivan, 1988; Oblinger and Kost, 1994; see also Figure 1). The six α-tubulin genes are less conserved but there is evidence of tissue-specific patterns of expression (Sullivan, 1988).

There are ample data to suggest that neurons distinguish between β-tubulin isotypes in the formation of microtubules. Neurite outgrowth in neuroblastoma cells is associated with phosphorylation of only one of the six β-tubulin isotypes (Gard and Kirschner, 1985; Luduena et al., 1988; Diaz-Nido et al., 1990), and during PC12 cell neurite extension two isotypes of β-tubulin are more efficiently incorporated into assembled microtubules, regardless of the relative composition of soluble tubulin pools (Joshi and Cleveland, 1989). In embryonal carcinoma cells, differentiating neurons are capable of sorting β-tubulin isotypes into colchicine-stable and -labile microtubules (Falconer et al., 1992). The function of β-tubulin isotypes may lie partly in their different assembly properties. While it is unclear if different tubulin isotypes represent distinct targets for neurotoxicants *in vivo,* microtubule proteins are developmentally regulated (see Figure 1) and β-tubulin isotypes bind colchicine at different rates (Banerjee and Luduena, 1991, 1992). Preferential sorting of tubulin dimers into more stable polymers may help to explain the apparent decrease in microtubule vulnerability to toxicants during later stages of development (Graff and Reuhl, 1995).

C. MARKERS OF MICROTUBULE STABILIZATION — POSTTRANSLATIONAL MODIFICATIONS

Purified microtubules and microtubules of interphase cells are inherently unstable, and possess half-lives of but a few seconds or minutes. However, neurons also possess subpopulations of microtubules which may persist for several hours. These subpopulations are termed "stable" and are not readily susceptible to disassembly by colchicine or other compounds which bind to the ends of growing microtubules. Although the mechanism(s) of microtubule stabilization are as yet unknown, accumulation of posttranslationally modified tubulin proteins play a key role (Gundersen and Bulinski, 1986). The best characterized posttranslational modifications include reversible tyrosination-detyrosination of the carboxy terminus (Raybin and Flavin, 1977; Thompson, 1980; Kumar and Flavin, 1981) and acetylation of lysyl residues (L'Hernault and Rosenbaum, 1985). More recently, the list of tubulin modifications has been expanded to include polyglutamylation and polyglycylation of both α- and β-tubulins (Edde et al., 1990, 1992; Mary et al., 1994; Redeker et al., 1994) as well as phosphorylation of several isotypes (Gard and Kirschner, 1985).

D. SIGNIFICANCE OF POSTTRANSLATIONAL MODIFICATIONS IN NEUROTOXICOLOGY

Posttranslational modifications of α-tubulin appear to be indicators of microtubule stability rather than stabilizing processes in themselves. Stable microtubules are resistant to microtubule inhibitors such as colchicine and nocodazole, which affect only growing polymers, and should be similarly resistant to metals which target microtubule assembly. However, posttranslationally modified microtubules are not cold-stable (Khawaja et al., 1988), indicating that the presence of modified α-tubulin is not an infallible index of microtubule resistance to disruption.

Certain isotypes of β-tubulin are better able to associate with MAPs than others, and there is some suggestion that posttranslational modifications of β-tubulins can mediate interactions with MAPs (Lee et al., 1990). Thus, it is possible that through these interactions, posttranslational modifications of β-tubulin could indirectly affect microtubule vulnerability to toxicant exposure. There have been no attempts to correlate the degree of β-tubulin modification with sensitivity of cellular microtubules to toxic metals.

E. MICROTUBULE-ASSOCIATED PROTEINS (MAPS)

A critical element in the generation of stable neuronal microtubular arrays is the decoration of microtubules with MAPs which promote microtubule assembly (Nunez, 1986; Matus, 1988). In the absence of MAPs, microtubule assembly requires substantially higher tubulin concentrations (Fellous et al., 1977). At least four different MAP classes are known, each differing in molecular weight, function, and differential expression within and between cell types during development. The best studied MAPs are the high molecular weight MAP1 (in its various forms) and MAP2, and the low molecular weight

Cytoskeletal Toxicity of Heavy Metals

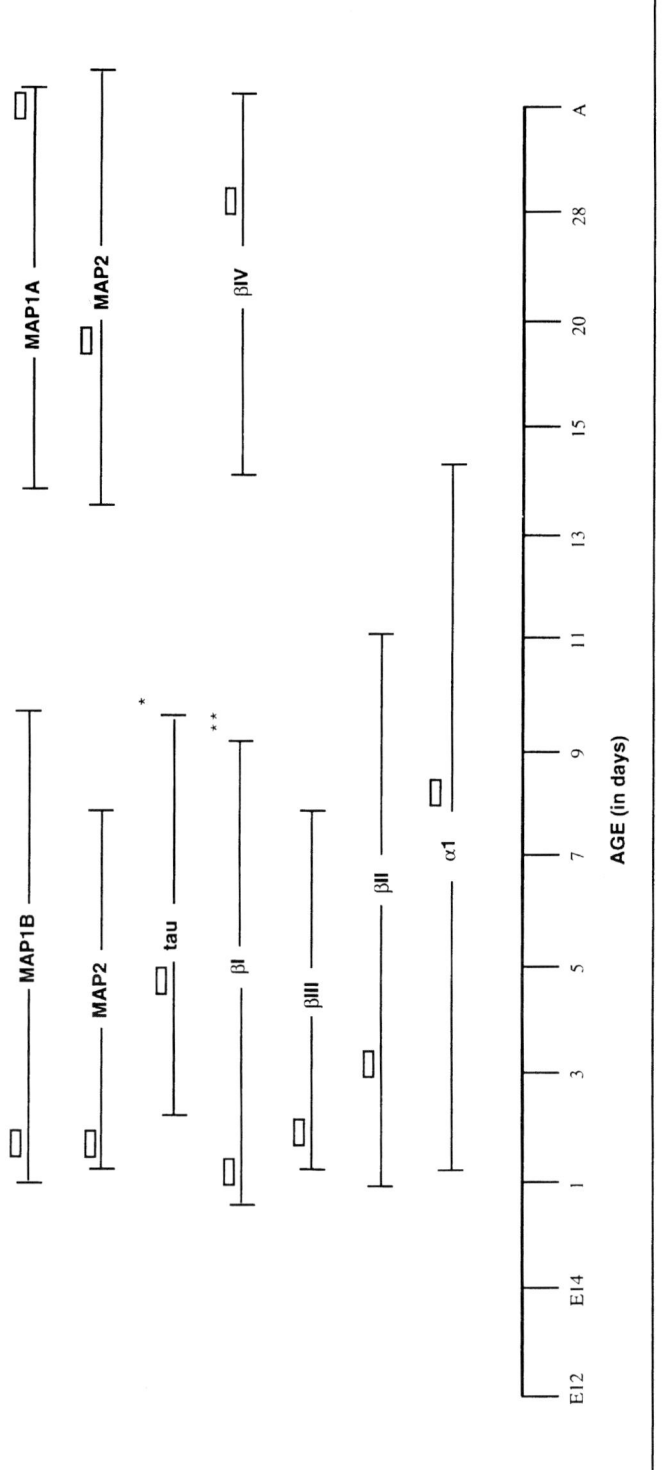

Figure 1. Developmental changes in microtubule expression. (Adapted from Oblinger and Kost, 1994).

MAPs, including the tau family and MAP2c. Two other microtubule-associated proteins, MAP3 and MAP4, are also high molecular weights, but much less is known of their structure or function within the cell (Matus, 1988).

MAP1 designates a pair of structurally related proteins, MAP1A and MAP1B, of which MAP1A is the largest at 350 kDa. MAP1A is a heat-labile polypeptide found primarily in dendrites (Huber and Matus, 1984), where it colocalizes with high molecular weight (HMW) MAP2 to form cross-bridges between adjacent microtubules (Shiomura and Hirokawa, 1987). MAP1B (also called MAP1x or MAP5) is also a component of microtubule cross-bridges but its expression in rodents is strongly down-regulated after postnatal day 10 (Matus, 1988). In the rodent brain MAP1B is evenly distributed in axons, dendrites, and cell bodies including glia (Bloom et al., 1985). MAP1c is unrelated to either MAP1A or MAP1B and functions as cytoplasmic dynein, involved in axonal transport (Vallee et al., 1989).

High molecular weight (HMW) MAP2 (280 kDa) is actually a pair of polypeptides, MAP2a and MAP2b. MAP2b is expressed throughout development and adulthood, while MAP2a is present only after the third postnatal week, replacing the juvenile form, MAP2c (see below). The HMW MAP2 peptides contain both a microtubule-binding domain and a long spacer arm which is believed to contribute to the stabilization of the neuronal cytoskeleton. The spacer arm also possesses binding sites for neurofilaments as well as organelles such as Golgi apparatus and mitochondria (Matus, 1988). In contrast to MAP2a/b, the low molecular weight MAP2c lacks a spacer arm and contains little more than the microtubule-binding domain of the parent transcript.

Similar in size to MAP2c is a family of proteins referred to collectively as tau, which were the first MAPs identified to promote microtubule assembly (Weingarten et al., 1975). Present mostly in axons, tau proteins range in size from about 50 kDa to 65 kDa, with different forms resulting from developmentally regulated splicing of the transcribed mRNA and from varying degrees of phosphorylation. The shift in tau expression from its "juvenile" to "mature" forms occurs at approximately the same time as similar changes in MAP2 proteins. The microtubule-binding domain of tau is also very similar to that of MAP2.

The best described posttranslational modification of MAPs is phosphorylation, which alters microtubule binding and the assembly-promoting properties of MAPs. The phosphorylated form of MAP1B has been identified in growing axons and in cross-bridges between microtubules (Sato-Yoshitake et al., 1989), suggesting a role in neural development. This possibility is supported by the observation of increased expression, and phosphorylation of MAP1B is associated with neurite outgrowth in PC12 cells (Aletta et al., 1988; Brugg and Matus, 1988).

Phosphorylation appears to modulate the nature of interactions between MAP2 and many other cytoskeletal proteins. Phosphorylation decreases the affinity of MAP2 for tubulin (Brugg and Matus, 1991), lessening its microtubule assembly-promoting properties, and increasing its affinity for neurofilaments and actin.

F. DEVELOPMENTAL REGULATION OF MICROTUBULE PROTEINS

Development of the nervous system involves a complex and carefully orchestrated series of morphological changes which include modulation of the neuronal cytoskeleton at every step. Microtubules play a central role not only in the stability, but also in the growth and plasticity of neurites and are themselves highly regulated during crucial periods of neuronal development.

Because of the sizeable number of tubulin isotypes in brain, each possessing its own biochemical properties, it has been held that different isotypes of tubulin yield microtubules of different function (Fulton and Simpson, 1976; Cleveland, 1987). A corollary of this hypothesis might suggest that the different functions required of the cytoskeleton during periods of morphogenesis would be subserved by microtubules consisting of different tubulin isotypes. It is known that isotypes of α- and β-tubulin are differentially expressed during development (Lewis et al., 1985; Bond et al., 1986). Differential expression of several tubulin isotypes and MAPs during the normal development of the hamster forebrain was recently determined (Oblinger and Kost, 1994). This pattern of expression, considered generally representative for rodents, is illustrated in Figure 1.

High levels of mRNA for βI, βII, βIII and α1 tubulin are detected at early postnatal times; however, these levels drop substantially after the first postnatal week (Figure 1). In contrast, expression of βIV tubulin becomes significant only after the third week. Expression of microtubule-associated proteins is also temporally regulated (Lewis et al., 1986; Charrière-Bertrand et al., 1991). While MAP1B expression is highest at early postnatal periods, MAP1A mRNA accumulates only at much later periods. In terms of mRNA levels, MAP1B expression is coincident to that of the βII and βIII isotypes of neuronal tubulin

(Oblinger and Kost, 1994). Because of its late expression, the level MAP1A is temporally similar to that of βIV tubulin. High levels of tau proteins are expressed early, but are less abundant later in development, similar to the patterns observed for α1 and βII tubulins.

Although little is known about the normal turnover of MAPs, it too appears to be developmental stage-specific, given the sequential appearance of juvenile and adult MAP isoforms. This is best seen with MAP2, where both translation and transcription appear to be under developmental control (Charrière-Bertrand et al., 1991; Charrière-Bertrand and Nunez, 1992). In addition, MAPs have differing sensitivities to degradation by proteases, and the Ca^{2+}-activated neutral protease calpain appears to be of particular importance in their breakdown (Fischer et al., 1991). Consequently, conditions which alter Ca^{2+} homeostasis and raise intracellular calcium may enhance MAP degradation.

The significance of such expression patterns is not yet fully apparent. Certainly the sequential appearance of juvenile and adult MAPs has been previously documented (reviewed in Matus, 1988). The studies of Oblinger and Kost (1994) illustrate coordinately regulated expression of both tubulin (and therefore microtubules) and microtubule-associated proteins, and are strongly suggestive of different tubulin isotypes, by way of their slightly different carboxy-terminal domains, preferentially binding with subsets of MAPs. This may provide a means by which MAPs differentially promote microtubule assembly and neurite outgrowth during development, but promote microtubule stability in adulthood (Baas et al., 1994). Such binding may also be involved in differential susceptibility of microtubules to toxicants at different developmental stages.

IV. DISRUPTION OF MICROTUBULES BY METALS
A. ALTERATION OF MICROTUBULE DYNAMICS

Studies on the assembly of pure tubulin into microtubules led to the model of dynamic instability (Mitchison and Kirschner, 1984; Kirschner and Mitchison, 1986). The dynamic instability model has also been identified as an important pathway of microtubule turnover in cultured cells (Sammak and Borisy, 1988; Tanaka and Kirschner, 1991; Sheldon and Wadsworth, 1993), where the bulk of interphase microtubules are dynamic and considered labile. Nondynamic microtubules do not elongate, but they do eventually depolymerize (Schultz et al., 1987; Webster et al., 1987; Webster and Borisy, 1989), making it difficult to distinguish between mechanisms of microtubule disruption and inhibition of microtubule assembly in cultured cells, especially in instances where exposure to a toxicant is prolonged. Several well-known microtubule inhibitors such as colchicine and nocodazole bind substoichiometrically to tubulin at the growing ends of polymers and inhibit further elongation (Olmsted and Borisy, 1973). Similar alteration of dynamic instability is also observed in microtubules exposed *in vitro* to some metals. Table 1 lists metals which have been reported to affect the cytoskeleton through interactions with microtubules. Of these, mercury most notably inhibits microtubule assembly at substoichiometric levels *in vitro*.

B. INTERACTION OF METALS WITH SPECIFIC SITES ON TUBULIN
1. Nucleotide Binding Sites

As previously mentioned, tubulin heterodimers possess two nucleotide binding sites, one exchangeable and one nonexchangeable. In addition to guanine nucleotide binding, association of Mg^{2+} at these sites is also necessary for optimal microtubule polymerization (Olmsted and Borisy, 1973). Tubulin can interact with other divalent cations at these sites, and some metals, for example beryllium, not only substitute for magnesium in the support of microtubule polymerization, but actually stabilize the polymer by inhibiting hydrolysis of GTP (Hamel et al., 1992) as a consequence of tight binding to the exchangeable site. Most other metals tested, specifically aluminum, cadmium, cobalt, manganese, tin, and zinc, while able to support MAP-dependent microtubule assembly, do not affect subsequent GTP hydrolysis (Hamel et al., 1992). Duhr et al. (1993) examined metal ions for interaction with GTP binding sites. Of the several metals tested, only cadmium ion inhibited photoincorporation of GTP analogs at the E-site. While cadmium effects were abolished by the addition of EDTA, the effects of mercuric ion were vastly increased, suggesting that chelation had improved affinity of the mercuric ion for binding sites on tubulin (Duhr et al., 1993).

2. Tubulin Sulfhydryl Groups

Kuriyama and Sakai (1974) documented the role of tubulin sulfhydryl groups in microtubule assembly. Of the 15 free sulfhydryls on the tubulin dimer, oxidation or binding of only two is sufficient to inhibit

Table 1 Metals and Metal Compounds Which Perturb Microtubules

Metal	Ref.	Comments
Aluminum	MacDonald et al., 1987	Supports microtubule polymerization *in vitro*.
	Johnson et al., 1992	Decreases MAP2 *in vivo*
		Alters neurofilaments
Arsenic	Li and Chou, 1992	Affects MTs only at very high concentrations
	Chou, 1989	
Cadmium	Miura et al., 1984	Some interaction with MTs; microfilaments a more likely
	Perrino and Chou, 1989	target. May complex with calmodulin and MAPs
	Chou, 1989	
	Li et al., 1993	
	Mills and Ferm, 1989	
Chromium	Li and Chou, 1992	Affects MTs at high levels only.
	Miura et al., 1984	
	Chou, 1989	
Cobalt	Chou, 1989	High levels only. Doubtful significance.
Copper	Miura et al., 1984	Not a likely MT inhibitor
Lead	Zimmerman et al., 1985	Inhibits assembly, disassembles polymers
	Roderer and Doenges, 1983	
Mercury	Duhr et al., 1993	Inhibits assembly and disrupts formed MTs.
	Keates and Yott, 1984	Binds sulfhydryl bases; enhanced by chelation.
	Wallin et al., 1977	
	Abe et al., 1975	
Methylmercury	Brown et al., 1988	Inhibits MTs *in vitro;* disassembles MTs in cultured cells.
	Wasteneys et al., 1988	Binds –SH groups on tubulin.
	Cadrin et al., 1988	
	Graff and Reuhl, 1990	
	Graff et al., 1993	
	Vogel et al., 1985	
	Sager et al., 1983	
	Sager and Syversen, 1984	
	Miura and Imura, 1989	
	Miura et al., 1984	
Nickel	Chou, 1989	Disruption at high levels.
	Lin and Chou, 1990	
Platinum	Boeckelheide et al., 1992	Decreases dynamics.
	Kopf-Maiert and Mulhausen, 1992	Alters polymerization.
	Peyrot et al., 1986	Cross-linking possible.
Zinc	Eagle et al., 1983	Abnormal polymers.
	Mills and Ferm, 1992	Most likely binds actin rather than MTs

assembly (Kuriyama and Sakai, 1974). Wallin and co-workers (1977) suggested that mercuric ion (Hg^{2+}) inhibits *in vitro* polymerization of tubulin by binding to sulfhydryl groups. Many other metal cations such as cadmium and chromium also have high affinities for sulfhydryl bases, and binding to tubulin sulfhydryl groups has been proposed as a general metal effect on microtubule kinetics (Chou, 1989).

Because of the high affinity of some metals for tissue sulfhydryls, it was proposed that the tripeptide glutathione is important as a regulator of tubulin sulfhydryls and for binding and sequestering reactive metals away from the cytoskeleton (Li and Chou, 1992). In the case of methylmercury, depletion of intracellular glutathione has little direct effect on the sensitivity of microtubules to disruption (Graff et al., 1993). Duhr et al. (1993) suggested that multiple cysteine residues in the primary structure of tubulin form a sulfhydryl pair or dithiol in a region close to the exchangeable site of β-tubulin, and that the binding affinity of this dithiol for chelated mercuric ion is very much higher than that of a single sulfhydryl, including glutathione. This represents a novel mechanism for tubulin disruption in chelation by EDTA, but it is uncertain whether such high affinity dithiols could interact with other complexes of mercury or with other cations (cadmium chelates were shown to be ineffective in inhibiting photoincorporation of GTP analogs).

V. METAL EFFECTS ON MICROTUBULES

A. MERCURY

Mercury compounds occur in several organic and inorganic states and are highly toxic. Even elemental mercury (Hg^0), once considered relatively inert, has a high vapor pressure and is extremely toxic. Mercury vapor and the alkylmercurials are the most neurotoxic mercurials, due to their relative ease of entry into the brain. Inorganic mercury salts are less neurotoxic due to their exclusion by the blood-brain barrier. Several comprehensive reviews of mercury compounds and their toxicity are available (Choi, 1983; Clarkson and Marsh, 1982; Chang and Reuhl, 1983; Clarkson, 1987; Chang and Verity, 1994).

B. INORGANIC MERCURY

Mercuric salts inhibit *in vitro* polymerization of microtubules from assembly-competent tubulin preparations (Wallin et al., 1977; Keates and Yott, 1984; Miura et al., 1984) and depolymerize assembled microtubules (Abe et al., 1975; Miura and Imura, 1989). Microtubule disruption also occurs in cultured cells (Miura et al., 1984). Wallin et al. (1977) suggested that Hg^{2+} inhibits polymerization by binding to or interfering with free sulfhydryls on the tubulin dimer necessary for microtubule assembly (Kuriyama and Sakai, 1974). Since these reports, no alternative mechanism has been presented to explain microtubule disruption caused by any mercurials.

Inhibition of microtubule assembly by Hg^{2+} *in vitro* was shown not to be hindered by the addition of metal chelators (Miura et al., 1984). Addition of EDTA or EGTA actually enhances microtubule disruption by reducing the mercury concentration necessary for complete inhibition or disruption of microtubules (Keates and Yott, 1984). Recently, Duhr et al. (1993) reported that the Hg-EDTA complex has a higher affinity for tubulin than Hg^{2+}, and in contrast to the ion alone, the Hg-EDTA complex abolishes photoinsertion of GTP analogs into the exchangeable binding site of β-tubulin. These results suggest different sites of interaction for complexed and uncomplexed Hg^{2+} and support the observation that the specificity of the mercuric ion for the highly reactive site on β-tubulin is increased by chelation (Duhr et al., 1993). An interesting clinical aspect of these results is the implication that widespread use of EDTA in medicines and foodstuffs, coupled with the prevalence of mercury in the environment, may lead to formation of mercury complexes which could pose possible human health risks (Duhr et al., 1993).

C. METHYLMERCURY

Most mechanistic studies concerning mercury compounds have been performed using organomercury compounds. Methylmercury gained notoriety following the description by Takeuchi et al. (1962) of Minamata disease, the clinical expression of methylmercury poisoning. The most consistent clinical observations are constriction of the visual fields, sensory disturbances, and cerebellar ataxia (Takeuchi et al., 1968). Infants exposed to methylmercury *in utero* are often profoundly retarded or suffer high mortality (Amin-Zaki et al., 1974) and, at the time of autopsy, have small brains with abnormal cortical gyri and permanent fetal architecture characteristic of migrational disturbances (Takeuchi, 1968; Choi et al., 1978). Serious congenital poisoning may occur in the absence of signs in the mother, suggesting greater vulnerability of the developing nervous system to methylmercury. Developmental effects of MeHg observed in animal studies includes mitotic inhibition (Rodier et al., 1984) and disturbances of cell migration and differentiation (Choi et al., 1978; Reuhl and Chang, 1979), all of which may be at least partially explained by perturbation of cellular microtubules (Choi, 1991).

Methylmercury inhibits the polymerization of purified brain tubulin and causes collapse of assembled microtubules *in vitro* (Sager et al., 1983; Miura et al., 1984; Sager and Syversen, 1984; Vogel et al., 1985). The requirement for free sulfhydryl groups for the polymerization of microtubules (Kuriyama and Sakai, 1974; Mellon and Rebhun, 1976) and the affinity of methylmercury for sulfhydryl bases strongly supports sulfhydryl binding as the mechanism of methylmercury-induced microtubule disruption. Indeed, Vogel et al. (1989) measured 15 sulfhydryl groups exposed on the surface and ends of tubulin dimers, each of them able to bind methylmercury. Further, it was determined that inhibition of assembly required binding of methylmercury to as few as 2 of the 15 sites.

Microtubules of cultured cells are highly sensitive to disruption by methylmercury (Sager et al., 1983; Miura et al., 1984; Sager and Syversen, 1984; Sager, 1988; Brown et al., 1988; Wasteneys et al., 1988; Cadrin et al., 1988; Graff and Reuhl, 1990; Graff et al., 1993, 1995). Examination of cytoskeletal proteins following methylmercury exposure reveals a profound loss of polymerized microtubules and the consequent perinuclear rearrangement of intermediate filaments and microfilaments (Wasteneys et al., 1988). Microtubules of mitotic spindles are particularly sensitive to methylmercury, and inhibition of mitosis at prometaphase and metaphase is commonly observed (Wasteneys et al., 1988). In differentiated neurons, however, there are subsets of microtubules which are significantly less sensitive to the effects of methylmercury (Cadrin et al., 1988). During neuronal differentiation, microtubules become increasingly resistant to disruption by methylmercury, coincident with accumulation of posttranslationally modified α-tubulin (Figure 2). This is most clearly seen in neurites which contain large numbers of acetylated microtubules and are less affected by methylmercury than are the primarily tyrosinated perikaryal microtubules (Reuhl et al., 1994; Graff and Reuhl, 1995).

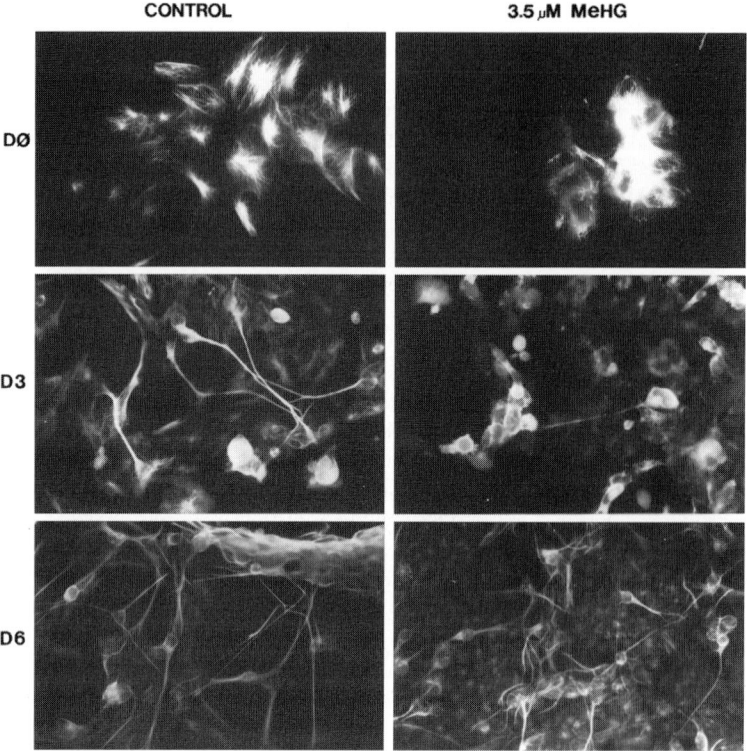

Figure 2. Methylmercury effects on microtubules of differentiating neurons in culture. Shown are representative cultures of control and methylmercury-treated embryonal carcinoma-derived neurons labeled with anti-tubulin antibody. Undifferentiated cells (upper left) have a radial pattern of labile microtubules originating from the centrosome. Microtubules in these cells are severely disrupted following methylmercury treatment (3.5 µM, 2 hours; upper right). Three days after induction with retinoic acid, cells have extended short but prominent neurites with dense parallel arrays of microtubules (middle left). Most of these microtubules remain highly sensitive to methylmercury, and both perikaryal and neuritic microtubules are extensively disrupted (middle, right). Note that the microtubules are extensively disrupted but that the neurite remains extended as long as a few MTs are intact. After 6 days of differentiation, neurites are much longer and often branching or fasciculated. Neuritic microtubules are compact and brightly labeled (lower left). These microtubules have become relatively stable to methylmercury at concentrations which caused near complete disassembly at earlier periods of differentiation (lower right). Note, however, that perikaryal microtubules remain labile and sensitive to disruption by methylmercury.

While posttranslational modifications of microtubules correlate with increasing stability of the polymer to methylmercury, it may well be that such modifications are simply markers for microtubule stabilization and do not cause stabilization, since even tyrosinated microtubules begin to demonstrate a degree of stability with time in culture (Baas and Black, 1992; Graff et al., 1995). Nevertheless, *in vitro* experiments with methylmercury serve to illustrate the differing vulnerability of microtubule isoforms to toxicant-induced disruption, a factor which is often overlooked when examining total microtubule populations.

D. CADMIUM

Like methylmercury, cadmium ions have a high affinity for sulfhydryl bases and it is therefore not surprising that cadmium also interacts with microtubule proteins. Cadmium has been reported to inhibit microtubule assembly from purified brain tubulin (Wallin et al., 1977). This effect may be due to inhibition of GTP incorporation by uncomplexed cadmium ion, an effect abolished by chelation with EDTA (Duhr et al., 1993).

Miura et al. (1984) showed cadmium ion to be much less cytotoxic to cultured glioma cells than divalent mercuric ion; even high concentrations of cadmium did not appreciably disassemble cellular microtubules. Similarly, Manin-Darby canine kidney cells show little microtubule disruption after cadmium exposure and illustrate that cadmium interacts more with actin filaments than with tubulin (Mills and Ferm, 1989; Mills et al., 1992).

In contrast, studies by Chou and co-workers have documented disassembly of fibroblast microtubules following exposure to cadmium and other metal ions (Chou, 1989). In these experiments cellular microtubules are disrupted, resulting in rearrangement of microfilaments (Li et al., 1994). Disruption of microtubules is suggested to involve a cadmium-calmodulin complex which interacts with microtubule-associated proteins to inhibit polymerization and disassemble polymerized microtubules (Perrino and Chou, 1986). In this way, cadmium evokes an effect similar to that of calcium complexed with calmodulin (Lee and Wolf, 1982, 1984). These results have yet to be confirmed with neuronal microtubules.

E. CHROMIUM

Microtubule depolymerization and inhibited tubulin assembly have been described for chromium, but it is doubtful that microtubules represent a significant target for chromium neurotoxicity. Inhibition of polymerization *in vitro* requires high concentrations (200 μM) of Cr^{3+} (Miura et al., 1984); there was no effect on microtubules of glioma cells after 1 hour, and in every assay chromium was less cytotoxic to glioma cells than were mercuric ions. In other studies, high chromium concentrations (100 μM) caused "thinning" of microtubules and perinuclear rearrangement of the surviving microtubules (Chou, 1989). Loss of peripheral microtubules and alterations in microfilament distribution were also seen following hexavalent chromium exposure (Li et al., 1992).

F. ALUMINUM

Interaction of aluminum with microtubules has been reported *in vitro*, but rather than inhibiting the assembly of microtubules from brain tubulin preparations, Al^{3+} ions compete with magnesium for support of tubular polymerization (MacDonald et al., 1987). The morphology of aluminum-promoted microtubules is normal, but some changes in the rates of GTP hydrolysis are readily observed.

Considerable interest surrounds the interaction of aluminum with tau proteins and neurofilaments in the pathogenesis of Alzheimer's dementia. The involvement of these cytoskeletal proteins in neurodegenerative disease states has been reviewed elsewhere (Doering, 1993). Some studies have examined the levels of high molecular weight MAP2 in brains of rats given elevated dietary aluminum. Johnson et al. (1992) documented reduction of MAP2 levels in the hippocampus of developing and adult rats with high dietary aluminum. The mechanism for loss of MAP2 is as yet uncertain.

G. NICKEL

The effects of nickel on microtubule assembly and the stability of cellular microtubules has been examined by Lin and Chou (1990). Prolonged exposure of fibroblasts to nickel chloride (2 mM, 20 h) results in reversible perinuclear bundling and aggregation of microtubules. *In vitro* assembly measurements indicate significant promotion of the rate and extent of polymerization, suggesting that the effects of nickel on cultured cells include altered microtubule kinetics (Lin and Chou, 1990). The neurotoxicological relevance of these observations is unclear.

H. LEAD

Inhibition of microtubule assembly by organic and inorganic lead compounds has been demonstrated (Zimmerman et al., 1985; Rodierer and Doenger, 1993). At concentrations of 50 μM, triethyllead inhibits microtubule polymerization and causes the disassembly of preformed polymers. At concentrations greater than 50 μM, microtubules assemble in aberrant patterns, forming wide fibers and amorphous aggregates.

In cultured cells, triethyllead depolymerizes cytoplasmic and spindle microtubules at concentrations of 1 to 50 μM, levels similar to those measured in brains of acutely poisoned humans (Zimmerman et al., 1985). Vimentin rearrangement subsequent to microtubule loss was also seen, but no changes were observed in microfilaments. Both microtubule effects and changes in vimentin were ameliorated upon removal of triethyllead from culture media, suggesting reversible binding of the toxicant to microtubule proteins, possibly the free sulfhydryl groups on tubulin dimers (Zimmerman et al., 1985).

VI. OTHER METALS

In addition to those described above, some less-studied metals have been reported to cause alterations in microtubules. Chou and co-workers have described the cytoskeletal alterations observed following exposure of fibroblast cells to arsenite and cobalt (Chou, 1989). Toxicant concentrations at which these alterations were seen were very high, and thus the specificity of the interaction of these metals with microtubules is suspect. However, such cytoskeletal changes may be used as an index of cellular toxicity and have been used to study cellular responses to toxic insult, such as alterations in glutathione levels (Li and Chou, 1992; Li et al, 1994).

Platinum compounds, namely, cis-dichlorodiammine platinum (II) (cisplatin), also have an affinity for microtubules (Kopf-Maiert and Mulhausen, 1992; Peyrot et al, 1986). This chemotherapeutic agent inhibits *in vitro* microtubule assembly, but also increases resistance of microtubules to cold-induced microtubule disassembly and decreases binding of other drugs such as colchicine or vinca alkaloids, implying a direct conformational change to tubulin such as cross-linking (Peyrot et al, 1986; Boekelheide et al, 1992).

Some evidence suggests that microtubules might be a target of zinc (Eagle et al., 1983). While zinc caused abnormal microtubule polymerization *in vitro,* disruption was not observed in cultured cells below 50 μM. Significant microfilament changes were seen at 10 μM.

VII. INTERMEDIATE FILAMENTS

A. STRUCTURE AND FUNCTION OF INTERMEDIATE FILAMENTS

Intermediate filaments (IFs) represent a complex family of fibrillar proteins. There are at least five major classes of IFs within this family; of these, class III (which includes glial filaments) and class IV (the neurofilament triplet proteins) are of primary interest here. Amino acid sequences of the various family members suggest that they all evolved from a single ancestral gene early in evolution. More than 30 IF subunit proteins have been identified (Shaw, 1991), each showing specific patterns of tissue and cell distribution.

Three of these IF subunit proteins have been particularly well characterized in neurons. These proteins, referred to as the neurofilament (NF) triplet proteins, are classified according to their apparent molecular weight as 200-kDa (NF-H), 145-kDa (NF-M), and 68-kDa (NF-L) NF proteins. These molecular weights have since been recalibrated downward (115 kDa, 100 kDa, and 60 kDa, respectively) but the earlier weights are still established in the literature. Several other proteins contribute to the structure of the NF, with the triplet proteins representing the major components.

Since NFs can be separated only under extreme denaturing conditions, the characteristics of the NF have only recently been elucidated. Only a small percentage of total NF protein is in a soluble, unpolymerized state (Morris and Lasek, 1982), unlike the more dynamic microtubules and microfilaments. Each NF consists of an α-helical rod domain of 40 kDa; the remainder of the molecule is composed of a head piece and long carboxy-terminal tail. The backbone of the NF is assembled from the NF-L protein, with NF-H and NF-M added to this core, while charged tail pieces appear to form interconnections between NFs and microtubules (Heins and Aebi, 1994). Studies indicate that prior to axonal transport NF proteins self-assemble in the cell body from four helically wound protofibrils, each in turn composed of two helically twisted protofilaments (Shoeman and Traub, 1993). Neurofilament proteins, particularly NF-H and NF-M, are highly phosphorylated at the carboxy-terminal domain; the process of phosphorylation occurs progressively as the filament is transported down the axon by slow axonal transport (Julien and Mushynski, 1982; Komiya et al., 1986; Nixon and Lewis, 1986). At the nerve terminal, the NF is disassembled by proteases and subunits or fragments of subunits are returned by retrograde transport to the neuronal cell body, where they may contribute to regulation of NF synthesis.

The functions of NF in the neuron are by no means clear. Conservative interpretation of the data attributes a relatively minor role to NF in neuronal processes such as axonal transport, neurite outgrowth, or intracellular compartmentalization, but significant roles may be identified in the future. A role for NF in early neural development is not apparent, as NF first appear in neurons only after the last mitotic division; prior to this time vimentin represents the major IF present (Nixon and Shea, 1992). Further, a number of studies have suggested that the presence of a NF network may not be essential for cellular viability (Albers and Fuchs, 1994). The primary role of NF may be to occupy space and provide the cell with relatively stable spatial architecture, a function that would require the NF to possess considerable resistance to depolymerization. In this capacity, NF would represent the "cytoskeleton" as envisioned by early cytologists. Hoffman and co-workers have provided evidence that NF in axons act as primary determinants of axon diameter (Hoffman et al., 1984, 1987), perhaps by regulating distance between NF and the coaxially oriented microtubules (Shaw, 1991). Over-expression of NF-H in transgenic mice results in abnormal filamentous aggregates and progressive neurological deficits (Cote et al., 1993).

An alternative view provided by Traub and co-workers (for review, see Traub and Shoeman, 1994) argues that intermediate filaments represent a system by which signals are transduced from the extracellular matrix through the cytoplasm to the nucleus, where they participate at the transcriptional or translational level in gene regulatory activities. Evidence linking NF to other elements of the cytoskeleton provides tantalizing support for this proposal, but data are still inconclusive.

B. INTERMEDIATE FILAMENTS IN METAL NEUROTOXICITY

NF are affected by metals *in vivo* and *in vitro*. As with other cytoskeletal components, both primary and secondary effects are implicated in alterations of NF networks. Many secondary actions appear to be consequences of alterations in calcium homeostasis which activate neutral proteases responsible for NF degradation. Theoretically, elevated calcium concentrations and protease activation could occur regionally within a neuron, leading to highly localized effects on NF. In addition, loss of other cytoskeletal elements may result in secondary reorganization of NFs, as has been reported in cultured neurons exposed to methylmercury (Cadrin et al., 1988).

Direct effects of metals on NF have been well characterized. Several metal cations have been shown to bind to NF proteins with varying affinity (Pierson and Evenson, 1988). Exposure of native NF to multivalent cations (chloride salts of aluminum, zinc, magnesium, manganese, and gadolinium) results in neurofilamentous aggregations (Troncoso et al., 1990); however, the concentrations of metals required to induce NF aggregation are extremely high (ranging from 0.75 mM for Al to 30 mM for Mg) and are of questionable biological relevance. Specific mechanistic data regarding metal effects on NF are available only for aluminum and lead.

C. ALUMINUM

The effects of aluminum on NF have attracted enormous attention, due primarily to speculation regarding the role of this element in neurodegenerative conditions like dialysis dementia, Alzheimer's disease, and amyotrophic lateral sclerosis/Parkinsonian dementia of Guam (McLachlan and De Boni, 1980; Yasui et al., 1991; Candy et al., 1992). While the role of aluminum in the etiology of these diseases is complex and still hotly debated, the topic has stimulated a substantial body of research. Studies have shown that aluminum indeed affects NF homeostasis both *in vivo* and *in vitro,* inducing the formation of neurofibrillary tangles in experimental animals (particularly rabbits). Production of the tangles usually requires that the aluminum be administered directly into the brain or cerebral spinal fluid, but supplementation of a calcium-deficient diet with aluminum will also induce the NF lesion in nonhuman primates (Garruto et al., 1989). These NF tangles are morphologically distinct from the paired neurofibrillary tangles observed in Alzheimer's disease (De Boni et al., 1976; Wisniewski et al., 1980; Bertholf et al., 1989; Katsetos et al., 1990). The tangles are often seen in both axons and dendrites as well as the perikaryon, where they may appear as small, basophilic inclusion bodies or larger, "hyaline" bodies (Strong, 1994). The NF aggregates stain intensely with SMI 31, a monoclonal antibody for phosphorylated NF-H and NF-M neurofilaments (Hewitt et al., 1991). Cortical layers III and V, ventral hippocampus, basal forebrain, dorsal raphe nucleus, brainstem nuclei, and lower motor neurons appear particularly vulnerable (Triarhou et al., 1985; Pendlebury et al., 1988). The neurofibrillary tangles are thought to arise as a result of aluminum-induced defects of axonal transport, with sequestration of NF in the cell body (Bizzi et al., 1984; Troncoso et al., 1985).

Other pathogenetic mechanisms such as disturbance of phosphorylation-dephosphorylation systems (Kihira et al., 1994) or Ca^{2+}-calmodulin regulation (Triarhou et al., 1985) cannot be excluded in aluminum neurotoxicity. Since NFs are generally phosphorylated only after entering the axon, the presence of heavily phosphorylated NF aggregations in the perikaryon following aluminum exposure suggests that the metal may induce premature phosphorylation (Bizzi and Gambetti, 1986). Recent studies, however, have not identified increased phosphorylation of NF following aluminum toxicity (Savory et al., 1993;

Strong and Jakowec, 1994). Accumulation of the NF in perikaryon and neurites is associated with either reduction (Muma et al., 1988) or normal levels (Savory et al., 1993) of the mRNAs for the 68- and 200-kDa NF proteins, suggesting that increased production of NF protein is not occurring. Some studies have suggested that aluminum inhibits calpain-mediated proteolysis (Nixon et al., 1990; Shea et al., 1992), while others have found no evidence of defective proteolytic degradation of NF (Ginkel et al., 1993). The discrepancies in these data may result from differences in experimental design or the chemical form of aluminum employed.

D. LEAD

Both inorganic and organic lead compounds disrupt NF networks *in vivo* and *in vitro*. Using Neuro-2a neuroblastoma cells, Zimmerman and colleagues (1987) investigated the effects of triethyllead on preformed and *in vitro*-assembled NF. Triethyllead caused unraveling of the NF into protofilaments and inhibited the *in vitro* assembly of NF. Intact cells treated with triethyllead exhibited the collapse of NF into a perinuclear coil. These rearrangements were reversible following removal of the toxicant. The authors speculated that triethyllead may weaken the hydrophobic interaction between the protofilaments or bind with cysteine residues on the protofilament subunits, destabilizing the assembled NF and permitting disassembly without requiring proteolysis.

In contrast to disassembly of NF by organolead compounds reported *in vitro* by Zimmerman et al. (1987) and Niklowitz (1974, 1975) reported neurofibrillary accumulation in hippocampal pyramidal neurons of rabbits following administration of tri- and tetraethyllead. Inorganic lead has been associated with NF accumulation in a human case report (Niklowitz and Mandybur, 1975), but similar changes have not been reported in a wide range of experimental animals exposed to lead salts. These conflicting data underscore the difficulties of direct comparison of *in vivo* and *in vitro* studies.

E. INTERMEDIATE FILAMENTS IN ASTROCYTES

Astrocytes of the central nervous system, ependymal cells, and immature oligodendroglia and peripheral nervous system Schwann cells all express class III intermediate filaments, usually termed glial fibrillary acidic protein (GFAP). The function of this 51-kDa protein is largely unknown. It has been speculated that GFAP is crucial for astrocytic functions, such as astrocytic guidance of migrating neurons, nutritional and structural support of postmigratory neurons, and local regulation of concentrations of ions and neurotransmitters (Kimelberg et al., 1993). Astrocytes demonstrate a remarkable capacity to respond to neural injury (LoPachin and Aschner, 1993; Norenberg, 1994). One element of this response is hypertrophy of the astrocyte and the production of long, thick processes filled with large amounts of GFAP (Figure 3). The induction of both message and protein for GFAP following stimulation is extremely rapid. O'Callaghan and co-workers have demonstrated that GFAP may peak as early as 1 day following insult and return to baseline within 2 weeks (O'Callaghan et al., 1990), although the duration of the GFAP response will clearly depend upon the nature of the eliciting event and may persist for weeks or months.

Despite the intensity of the GFAP response and the variety of conditions which elicit it (O'Callaghan, 1992; O'Callaghan et al., 1995), the ability of astrocytes to make GFAP does not appear to be essential to brain development or astrocytic response to injury. Using gene targeting techniques, Gomi et al. (1995) produced mutant mice which in their homozygous form were devoid of GFAP. These mice developed normally and had no detectable neurological deficits. Absence of GFAP did not result in compensatory expression of other IFs, such as vimentin. Moreover, astrocytes of transgenic animals responded like those of controls when the brain was infected with scrapie prions, suggesting that the astrocytic reaction to injury was not dependent on GFAP expression.

F. GFAP IN METAL NEUROTOXICOLOGY

While GFAP expression may not be crucial to the astrocyte reaction, it is observed as a consequence of exposure to numerous toxic metals. This has been particularly well described for trimethyltin (TMT), an index metal used to validate neurotoxicity protocols. A single administration of TMT to rats resulted in a dose-dependent increase of hippocampal GFAP by as much as 600% of control. When the most affected region of the hippocampus was microdissected and analyzed, GFAP levels were 6000% of control (O'Callaghan et al., 1993). Elevated levels of GFAP were seen in areas other than those classically reported to be damaged by TMT. However, when these areas were subsequently examined by silver stains for degeneration, the GFAP staining colocalized to areas of subtle neuronal damage (Balaban et al., 1988).

Figure 3. Astrocyte GFAP staining in rat hippocampal CA-3 region 21 days following administration of trimethyltin (8.0 mg/kg). Abundant expression of GFAP in reactive astrocytes is demonstrated by immunocytochemistry. (Micrograph courtesy of Dr. James O'Callaghan).

GFAP is heavily expressed in brain regions damaged by a variety of heavy metals, including inorganic and methylmercury, inorganic and organic lead, trimethyl-, triethyl-, and tributyltin, and cadmium (O'Callaghan, 1993; Peters et al., 1994). The sensitivity of the GFAP response to toxic metals has led to its inclusion in neurotoxicity screening batteries. However, despite the robustness of the GFAP response to toxic metals, the role it plays in astrocytic response to neurotoxicity is unclear.

VIII. MICROFILAMENTS

A. STRUCTURE AND FUNCTION OF ACTIN MICROFILAMENTS

Actin microfilaments (MF) are the thinnest of the three fibrillary proteins comprising the cytoskeleton. Actin is the product of a gene family which includes at least three isoforms (α, β, γ). While all actin isoforms are highly conserved, neural actin is composed of polymers of the β and gamma isoforms. MF are assembled from globular (G-actin) monomers to form filaments which organize into tight helical pairs of actin polymers (F-actin). A dynamic equilibrium exists between polymerized filaments and the unpolymerized actin, and exact ratios of the two forms differ according to cell type and physiological state. Polymerized actin filaments have polarity, with one end described as "barbed" and the other end "pointed." Elongation of the filament occurs preferentially at the barbed end, although addition at either end is possible. Regulation of actin MF assembly and disassembly is not well understood, but a variety of extracellular and cellular signals influence the formation and distribution of the molecule (Hall, 1994).

While MF are widely distributed within neural cells, many are localized at the plasma membrane, to which they are linked by actin-associated molecules such as ankyrin, α-actinin, fodrin/spectrin, talin, and vinculin (Pollard and Cooper, 1986; Bamburg and Bernstein, 1991; Hitt and Luna, 1994). These linking molecules participate in signal transduction and the positional stabilization of ion channels, receptors, and other membrane-spanning molecules within the plasma membrane. MFs are also linked to extracellular matrix through transmembrane proteins, the integrins (Juliano and Haskill, 1993).

Actin MF are involved in critical neural functions. While ubiquitous in neural cells throughout life, the distribution of MF in the axonal growth cone is of particular importance during brain development. As the axon elongates toward its synaptic partners, it extends multiple MF-filled filopodial extensions from its growing end. MF assembly and disassembly are rapid and this lability plays a critical role in supporting the exploratory movement of the growth cone. Disruption of MF polymerization, as occurs following treatment with the fungal toxin cytochalasin B, causes prompt cessation of growth cone motion. Guidance of the growth cone is complex and poorly understood, but appears to also involve interactions between MF, adhesion molecules, second messenger systems (Beggs et al., 1994; Ignelzi et al., 1994), and transmembrane adhesion molecule signals (Sobue, 1993).

In late development and adulthood, MF play a critical role in the formation and maintenance of synapses, particularly the dendritic spine. A heavy compliment of MF located at the dendrite stabilizes the postsynaptic component of the synaptic complex, in part by anchoring transmembrane adhesion molecules such as the neural cell adhesion molecule, NCAM180 (Pollerberg et al., 1985, 1987). Postsynaptic MF are also important in synaptic remodeling during learning and memory, and in synaptic plasticity.

B. DISRUPTION OF MICROFILAMENTS BY METALS

Little is known regarding the effects of specific toxic metals on the MF system. While a priori considerations of the dynamic nature of MF and their dependency upon stable ionic microenvironment would suggest that they are prime targets for metals, they are difficult to resolve in neural tissues *in vivo*. Moreover, their tendency to reorganize in response to alterations of other cytoskeletal elements has discouraged investigators from systematic studies of metal-MF interactions.

Despite the paucity of data, MF changes have been implicated in several major events observed in neurons and glia both in culture and *in vivo* following metal exposure. Cytoplasmic blebbing, a common cytologic response of injured cells, involves a focal loss of submembranous actin MFs, permitting herniation of small amounts of cytoplasm. It is probable that the MF loss is the secondary result of a regional change in calcium microenvironment leading to actin disassembly. Similarly, loss of synaptic integrity following exposure to toxic metals such as trimethyltin has been correlated with the appearance of breakdown products of MFs and their associated proteins, such as spectrin (Dey et al., 1994). These changes reflect complex reorganization of the synaptic complex, including disassembly of MF. Again, this change is usually a secondary consequence of disturbance of the ionic microenvironment of the synapse.

Several toxic metals have shown direct effects on MF, but there are insufficient data upon which to formulate mechanistic explanations. Zinc ion is required for normal MF assembly and has been shown to be involved in phorbol ester-induced MF reorganization (Hedberg et al., 1991). Zinc exposure of cultured kidney cells causes retraction of submembranous actin MFs into the cytoplasm, loss of junctional complexes, and consequent changes in cell shape (Mills et al., 1992). These changes occur at concentrations (10 μM) which have no apparent effect on cellular microtubules. Similar findings are observed in neurons and astrocytes derived from embryonal carcinoma cells (Reuhl and Brown, unpublished observations).

Cadmium also appears to alter MFs. Cadmium treatment of cultured cells results in loss of cytoplasmic F-actin bundles and MFs from the submembranous region of the plasma membrane (Mills et al., 1992). Cadmium is known to disrupt cell junctions, perhaps by affecting calcium-dependent adhesion processes. Loss of tight junctions in blood vessels, particularly in developing nervous system, results in the edema and hemorrhage characteristic of cadmium toxicity (Prozialeck and Niewenhuis, 1975). At this stage, changes in regional brain microenvironment result in changes of all elements of the cytoskeleton.

ACKNOWLEDGMENTS

The authors wish to thank Dr. Herbert Lowndes, Dr. Martin Philbert, Dr. Sally Pyle, and P. Markus Dey for their discussions and critical review, and Ms. Patricia Pailing and Ms. Jaimie Graff for their assistance in the preparation of this manuscript. Supported by NIEHS Grants ES 04976 and ES 05022.

REFERENCES

Abe, T., Haga, T., and Hirokawa, M. (1975), *Brain Res.*, 86, 504–508.
Abou-Donia, M. B. and Gupta, R. P. (1994), in *Principles of Neurotoxicology*, Chang, L. W., Ed., Marcel Dekker, New York, 153–210.
Albers, K. and Fuchs, E. (1992), *Int. Rev. Cytol.*, 134, 243–279.
Aletta, J. M., Lewis, S. A., Cowan, N. J., and Greene, L. A. (1988), *J. Cell. Biol.*, 106, 1573–1581.

Amin-Zaki, L., Elhassani, S., Majeed, M. A., Clarkson, T. W., Doherty, R. A., and Greenwood, M. (1974), *Pediatrics*, 54, 587–595.
Avila, J. (1990), *FASEB*, 4, 3284–3290.
Baas, P. W., Deitch, J. S., Black, M. M., and Banker, G. A. (1988), *Proc. Natl. Acad. Sci. U.S.A.*, 85, 8335–8339.
Baas, P. W. and Black, M. M. (1989), *Neuroscience*, 30, 795–803.
Baas, P. W., Pienkowski, T., Cimbalnik, K. A., Toyama, K., Bakalis, S., and Ahmad, F. J. (1994), *J. Cell Sci.*, 107: 135–143.
Balaban, C. D., O'Callaghan, J. P., and Billingsley, M. L. (1988), *Neuroscience*, 26, 337–361.
Bamburg, J. R. and Bernstein, B. W. (1991), in *The Neuronal Cytoskeleton*, Wiley-Liss, New York, 121–160.
Banerjee, A. and Luduena, R. F. (1991), *J. Biol. Chem.*, 266, 1689–1691.
Banerjee, A. and Luduena, R. F. (1992), *J. Biol. Chem.*, 267, 13335–13339.
Bayley, P. M. (1990), *J. Cell Sci.*, 95, 329–334.
Beggs, H. E., Soriano, P., and Maness, P. F. (1994), *J. Cell Biol.*, 127, 825–833.
Bertholf, R. L., Herman, M. M., Savory, J., Carpenter, R. M., Sturgill, B. C., Katsetos, C. D., Vandenberg, S. R., and Wills, M. R. (1989), *Toxicol. Appl. Pharmacol.*, 98, 58–74.
Bizzi, A., Clark-Crane, R. C., Autilio-Gambetti, L., and Gambetti, P. (1984), *J. Neurosci.*, 4, 722–731.
Bizzi, A. and Gambetti, P. (1986), *Acta Neuropathol. (Berl.)*, 71, 154–158.
Bloom, G. S., Luca, F. C., and Vallee, R. B. (1985), *Proc. Natl. Acad. Sci. U.S.A.*, 82, 5404–5408.
Boekelheide, K., Arcila, M. E., and Eveleth, J. (1992), *Toxicol. Appl. Pharmacol.*, 116, 146–151.
Bond, J. F., Fridovich-Keil, J. L., Pillus, L., Mulliqan, R. C., and Soloman, F. (1986), *Cell*, 44, 461–468.
Brown, D. L., Reuhl, K. R., Bormann, S., and Little, J. E. (1988), *Toxicol. Appl. Pharmacol.*, 94, 66–75.
Brugg, B. and Matus, A. (1988), *J. Cell. Biol.*, 107, 643–650.
Brugg, B. and Matus, A. (1991), *J. Cell Biol.*, 114, 735–748.
Cadrin, M., Wasteneys, G. O., Jones-Villeneuve, E., Brown, D. L., and Reuhl, K. R. (1988), *Toxicol. Cell Biol.*, 4, 61–80.
Candy, J. M., McArthur, F. K., Oakley, A. E., Taylor, G. A., Chem, C. P., Mountfort, S. A., Thompson, J. E., Chalker, P. R., Bishop, H. E., Beyreuther, K., Perry, G., Ward, M. K., Martyn, C. N., and Edwardson, J. A. (1992), *J. Neurol. Sci.*, 107, 210–218.
Cassimeris, L. (1993), *Cell Motil. Cytoskeleton*, 26, 275–281.
Chang, L. W. and Reuhl, K. R. (1983), in *Trace Elements in Health*, Butterworths, London, 132–149.
Chang, L. W. and Verity, A. (1994), in *Handbook of Neurotoxicology*, Chang, L. W. and Dyer, R. S., Eds., Marcel Dekker, New York, 31–59.
Charrière-Bertrand, C., Garner, C., Tardy, M., and Nunez, J. (1991), *J. Neurochem.*, 56, 385–391.
Charrière-Bertrand, C. and Nunez, J. (1992), *Neurochem. Int.*, 21, 249–267.
Choi, B. H., Lapham, L. W., Amin-Zaki, L., and Saleem, T. (1978), *J. Neuropath. Exp. Neurol.*, 37, 719–733.
Choi, B. H. (1983), in *Neurobiology of the Trace Elements*, vol. 1, Dreosti, I. and Smith, R., Eds., The Humana Press, Englewood Cliffs, NJ, 197–235.
Choi, B. H. (1991), *Acta Neuropath.*, 81, 359–365.
Chou, I. N. (1989), *Biomed. Environ. Sci.*, 2, 358–365.
Clarkson, T. W. and Marsh, D. O. (1982), in *Clinical, Biochemical, and Nutritional Aspects of Trace Elements*, Pasad, A. S., Ed., Alan R. Liss, New York, 549–568.
Clarkson, T. W. (1987), *Environ. Health Perspect.*, 75, 59–64.
Cleveland, D. W. (1987), *J. Cell Biol.*, 104, 381–383.
Cote, F., Collard, J. F., and Julien, J. P. (1993), *Cell*, 73, 35–46.
De Boni, U., Orvos, A., Scott, J. W., and Crapper, D. R. (1976), *Acta Neuropathol. (Berl.)*, 35, 285–294.
Dey, P. M., Graff, R. D., Lagunowich, L. A., and Reuhl, K. R. (1994), *Toxicol. Appl. Pharmacol.*, 126, 69–74.
Diaz-Nido, J., Hernandez, M. A., and Avila, J. (1990), in *Microtubule Proteins*, Avila, J., Ed., CRC Press, Raton Boca, FL, 193–257.
Doering, L. C. (1993), *Mol. Neurobiol.*, 7, 265–291.
Duhr, E. F., Pendergrass, J. C., Slevin, J. T., and Haley, B. E. (1993), *Toxicol. Appl. Pharmacol.*, 122, 273–280.
Dustin, P. (1984), *Microtubules*, 2nd ed., Springer-Verlag, New York.
Edde, B., Rossier, J., Le Caer, J. P., Desbruyeres, E., Gros, F., and Denoulet, P. (1990), *Science*, 247, 83–85.
Edde, B., Rossier, J., Le Caer, J. P., Prome, J. C., Desbruyeres, E., Gros, F., and Denoulet, P. (1992), *Biochemistry*, 31, 403–410.
Falconer, M. M., Echeverri, C. J., and Brown, D. L. (1992), *Cell Motil. Cytoskeleton*, 21, 313–325.
Fellous, A., Francon, J., Lennon, A. M., and Nunez, J. (1977), *Eur. J. Biochem.*, 78, 167–174.
Fischer, I., Romano-Clarke, G., and Grynspan, F. (1991), *Neurosci. Res.*, 16, 891–898.
Fuchs, E. and Weber, K. (1994), *Annu. Rev. Biochem.*, 63, 345–382.
Fulton, C. and Simpson, P. A. (1976), in *Cell Motility*, Goldman, R., Pollard, T., and Rosenbaum, J., Eds., Cold Spring Harbor Press, Cold Spring Harbor, NY, 987–1005.
Gard, D. L. and Kirschner, M. W. (1985), *J. Cell Biol.*, 100, 764–774.
Garner, C. C., Tucker, R. P., and Matus, A. (1988), *Nature*, 336, 674–677.
Geahlen, R. L. and Haley, B. E. (1977), *Proc. Natl. Acad. Sci. U.S.A.*, 74, 4375–4377.
Gelfand, V. I. and Bershadsky, A. D. (1991), *Annu. Rev. Cell Biol.*, 7, 93–116.
Ginkel, M. F., Heijink, E., Dekker, P., Miralem, T., Voet, G. B., and Wolff, F. A. (1993), *Neurotoxicology*, 14, 13–18.

Gomi, H., Yokoyama, T., Fujimoto, K., Ikeda, T., Katoh, A., Itoh, T., and Itohara, S. (1995), *Neuron,* 14, 29–41.
Graff, R. D. and Reuhl, K. R. (1990), *The Toxicologist,* (Abstr.) 10, 138.
Graff, R. D., Philbert, M. A., Lowndes, H. E., and Reuhl, K. R. (1993), *Toxicol. Appl. Pharmacol.,* 120, 20–28.
Graff, R. D., Dey, P. M., and Reuhl, K. R. (1995), *Neurotoxicology,* (Abstr.), 15, 966.
Gundersen, G. G. and Bulinski, J. C. (1986), *Eur. J. Cell Biol.,* 42, 288–294.
Hall, A. (1994), *Annu. Rev. Cell Biol.,* 10, 31–54.
Hamel, E., Lin, C. M., Kenney, S., Skehan, P., and Vaughns, J. (1992), *Arch. Biochem. Biophys.,* 295: 327–339.
Hedberg, K. K., Birrell, G. B., and Griffith, O. H. (1991), *Cell Regulation,* 2, 1067–1079.
Heidemann, S. R., Landers, J. M., and Hamborg, M. A. (1981), *J. Cell Biol.,* 91, 661–665.
Heins, S. and Aebi, U. (1994), *Curr. Opinion Cell Biol.,* 6, 25–33.
Hewitt, C. D., Herman, M. M., Lopes, M. B. S., Savory, J., and Wills, M. R. (1991), *Neuropathol. Appl. Neurobiol.,* 17, 47–60.
Hill, T. and Carlier, M. F. (1983), *Proc. Natl. Acad. Sci. U.S.A.,* 80, 7234–7238.
Hitt, A. L. and Luna, E. J. (1994), *Curr. Opinion Cell Biol.,* 6, 120–130.
Hoffman, P. N., Griffin, J. W., and Price, D. L. (1984), *J. Cell Biol.,* 99, 705–714.
Hoffman, P. N., Cleveland, D. W., Griffin, J. W., Landes, P. W., Cowan, N. J., and Price, D. L. (1987), *Proc. Natl. Acad. Sci. U.S.A.,* 84, 3472–3476.
Ignelzi, M. A., Miller, D. R., Soriano, P., and Maness, P. F. (1994), *Neuron,* 12, 873–884.
Johnson, G. V., Watson, A., Lartius, R., Uemura, E., and Jope, R. S. (1992), *Neurotoxicology,* 13, 463–474.
Joshi, H. C. and Cleveland, D. W. (1989), *J. Cell Biol.,* 109, 663–673.
Joshi, H. C., Monica, J. P., McNamara, L., and Cleveland, D. W. (1992), *Nature,* 356, 80–83.
Juliano, R. L. and Haskill, S. (1993), *J. Cell Biol.,* 120, 577–585.
Julien, J.-P. and Mushynski, W. E. (1982), *J. Biol. Chem.* 257, 10467–10470.
Katsetos, C. D., Savory, J., Herman, M. M., Carpenter, R. M., Frankfurter, A., Hewitt, C. D., and Wills, M. R. (1990), *Neuropathol. Appl. Neurobiol.,* 16, 511–528.
Keates, R. A. and Yott, B. (1984), *Can. J. Biochem. Cell Biol.,* 62, 814–818.
Khawaja, S., Gundersen, G. G., and Bulinski, J. C. (1988), *J. Cell Biol.,* 106, 141–150.
Kihira, T., Yoshida, S., Uebayashi, Y., Wakayama, I., and Yase, Y. (1994), *Biomed. Res.,* 15, 27–36.
Kimelberg, H. K., Jalonen, T., and Walz, W. (1993), in *Astrocytes. Pharmacology and Function,* Murphy, S. Ed., Academic Press, New York, 193–228.
Kirschner, M. and Mitchison, T. (1986), *Cell,* 45, 329–342.
Komiya, Y., Tashiro, T., and Kurokawa, M. (1986), *Biomed. Res.,* 7, 345–348.
Kopf-Maier, P. and Muhlhausen, S. K. (1992), *Chem.-Biol. Interact.,* 82, 295–316.
Kumar, N. and Flavin, M. (1981), *J. Biol. Chem.,* 256, 7678–7686.
Kuriyama, R. and Sakai, H. (1974), *J. Biochem.,* 76, 651–654.
Lee, M. K., Tuttle, J. B., Rebhun, L. I., Cleveland, D. W., and Frankfurter, A. (1990), *Cell Motil. Cytoskeleton,* 17, 118–132.
Lee, Y. C. and Wolff, J. (1982), *J. Biol. Chem.,* 257, 6306–6310.
Lee, Y. C. and Wolff, J. (1984), *J. Biol. Chem.,* 259, 1226–1230.
Lewis, S. A., Lee, M. G., and Cowan, N. J. (1985), *J. Cell Biol.,* 101, 852–861.
Lewis, S. A., Sherline, P., Cowan, N. J. (1986), *J. Cell Biol.,* 102, 2106–2114.
L'Hernault, S. W. and Rosenbaum, J. L. (1985), *Biochemistry,* 24, 473–478.
Li, W. and Chou, I. N. (1992), *Toxicol. Appl. Pharmacol.,* 114, 132–139.
Li, W., Zhao, Y., and Chou, I.-N. (1992), *Toxicol. In Vitro.,* 49, 122–167.
Li, W., Zhao, Y., and Chou, I. (1993), *Toxicology,* 77, 65–79.
Li, W., Kagan, H. M., and Chou, I. (1994), *Toxicol. Appl. Pharmacol.,* 126, 114–123.
Lin, K. C. and Chou, I. N. (1990), *Toxicol. Appl. Pharmacol.,* 106, 209–221.
LoPachin, R. M. and Aschner, M. (1993), *Toxicol. Appl. Pharmacol.,* 118, 141–158.
Luduena, R. F., Zimmermann, H.-P., and Little, M. (1988), *FEBS Lett.,* 230, 142–146.
MacDonald, T. L., Humphreys, W. G., and Martin, R. B. (1987), *Science,* 236, 183–186.
Mary, J., Redeker, V., Le Caer, J. P., Prome, J. C., and Rossier, J. (1994), *FEBS Lett.,* 353, 89–94.
Matus, A. and Riederer, B. (1986), *Ann. N. Y. Acad. Sci.,* 466, 167–179.
Matus, A. (1988), *Annu. Rev. Neurosci.,* 11, 29–44.
Matus, A. (1988), *Trends Neurosci.,* 11, 291–292.
Matus, A. (1994), *Trends Neurosci.,* 17, 19–22.
McLachlan, D. R. and De Boni, U. (1980), *Neurotoxicology,* 2(Suppl.), 3–16.
Mellon, M. G. and Rebhun, C. I. (1976), *J. Cell Biol.,* 70, 226–238.
Mills, J. W. and Ferm, V. H. (1989), *Toxicol. Appl. Pharmacol.,* 101, 245–254.
Mills, J. W., Zhou, J. H., Cardoza, L., and Ferm, V. H. (1992), *Toxicol. Appl. Pharmacol.,* 116, 92–100.
Mitchison, T. and Kirschner, M. (1984), *Nature (Lond.),* 312, 237–242.
Miura, K., Inokawa, M., and Imura, N. (1984), *Toxicol. Appl. Pharmacol.,* 73, 218–231.
Miura, K. and Imura, N. (1989), *Biol. Trace Elem. Res.,* 21, 313–316.
Morris, J. R. and Lasek, R. J. (1982), *J. Cell Biol.,* 92, 192–198.
Muma, N. A., Troncoso, J. C., Hoffman, P. N., Koo, E. H., and Price, D. L. (1988), *Mol. Brain Res.,* 3, 115–122.
Niklowitz, W. J. (1974), *Environ. Res.,* 8, 17–36.

Niklowitz, W. J. (1975), *Neurology,* 25, 927–934.
Niklowitz, W. J. and Mandybur, T. I. (1975), *J. Neuropathol. Exp. Neurol.,* 34, 445–455.
Nixon, R. A. and Lewis, S. E. (1986), *J. Biol. Chem.,* 261, 16298–16301.
Nixon, R. A., Clarke, J. F., Logvinenko, K. B., Tan, M. K., Hoult, M., and Grynspan, F. (1990), *J. Neurochem.,* 55, 1950–1959.
Nixon, R. A. and Shea, T. B. (1992), *Cell Motil. Cytoskeleton,* 22, 81–91.
Norenberg, M. D. (1994), *J. Neuropathol. Exp. Neurol.,* 53, 213–220.
Nunez, J. (1986), *Dev. Neurosci.,* 8, 125–141.
Oblinger, M. M. and Kost, S. A. (1994), *Dev. Brain Res.,* 77, 45–54.
O'Callaghan, J. P., Miller, D. B., and Reinhard, J. F., Jr. (1990), *Brain Res.,* 521, 73–80.
O'Callaghan, J. P. (1992), in *Neurotoxicology,* Abou-Donia, M. B., Ed., CRC Press, Boca Raton, FL, 61–78.
O'Callaghan, J. P. (1993), *Ann. N.Y. Acad. Sci.,* 679, 195–210.
O'Callaghan, J. P., Jensen, K. F., and Miller, D. B. (1995), *Neurochem. Int.,* 26, 115–124.
Olmsted, J. B. and Borisy, G. G. (1973), *Biochemistry,* 12, 4282–4289.
Olmsted, J. B. (1986), *Annu. Rev. Cell Biol.,* 2, 421–457.
Pendlebury, W. W., Beal, M. F., Kowall, N. W., and Solomon, P. R. (1988), *Neurotoxicology,* 9, 503–510.
Perrino, B. A. and Chou, I. N. (1986), *Cell Biol. Int. Rep.,* 10, 565–573.
Peters, B., Stoltenburg, G., Hummel, M., Herbst, H., Altmann, L., and Wiegand, H. (1994), *Neurotoxicology,* 15, 685–694.
Peyrot, V., Briand, C., Momburg, R., and Sari, J. C. (1986), *Biochem. Pharmacol.,* 35, 371–375.
Pierson, K. B. and Evenson, M. A. (1988), *Biochem. Biophys. Res. Commun.,* 152, 598–604.
Pollard, T. D. and Cooper, J. A. (1986), *Annu. Rev. Biochem.,* 55, 987–1035.
Pollerberg, G. E., Sadoul, R., Goridis, C., and Schachner, M. (1985), *J. Cell Biol.,* 101, 1921–1929.
Pollerberg, G. E., Burridge, K., Krebs, K. E., Goodman, S. R., and Schachner, M. (1987), *Cell Tissue Res.,* 250, 227–236.
Prozialeck, W. C. and Niewenhuis, R. J. (1991), *Toxicol. Appl. Pharmacol.,* 107, 81–97.
Raybin, D. and Flavin, M. (1977), *Biochemistry,* 16, 2189–2194.
Redeker, V., Levilliers, N., Schmitter, J. M., Le Caer, J. P., Rossier, J., Adoutte, A., and Bre, M. H. (1994), *Science,* 266, 1688–1691.
Reuhl, K. R. and Chang, L. W. (1979), *Neurotoxicology,* 1, 21–55.
Reuhl, K. R., Lagunowich, L. A., and Brown, D. L. (1994), *Neurotoxicology,* 15, 133–144.
Roderer, G. and Doenges, K. H. (1983), *Neurotoxicology,* 4, 171–180.
Rodier, P. M., Aschner, M., and Sager, P. R. (1984), *Neurobehav. Toxicol. Teratol.,* 6, 379–385.
Sager, P. R., Doherty, R. A., and Olmsted, J. B. (1983), *Exp. Cell Res.,* 146, 127–137.
Sager, P. R. and Syversen, T. L. (1984), *Exp. Neurol.,* 85, 371–382.
Sager, P. R. (1988), *Toxicol. Appl. Pharmacol.,* 94, 473–86.
Sammak, P. J. and Borisy, G. G. (1988), *Cell Motil. Cytoskeleton,* 10, 237–245.
Sato-Yoshitake, R., Shiomura, Y., Miyasaka, H., and Hirokawa, N. (1989), *Neuron,* 3, 229–238.
Savory, J., Herman, M. M., Hundley, J. C., Seward, R. L., Griggs, C. M., Katsetos, C. D., and Wills, M. R. (1993), *Neurotoxicology,* 14, 9–12.
Schilstra, M. J., Mardin, S. R., and Bayley, P. M. (1987), *Biochem. Biophys. Res. Commun.,* 147, 585–595.
Schulze, E., Asai, D. J., Bulinski, J. C., and Kirschner, M. (1987), *J. Cell Biol.,* 105, 2167–2177.
Shaw, G. (1991), in *The Neuronal Cytoskeleton,* Wiley-Liss, New York, 185–214.
Shea, T. B., Balikian, P., and Beermann, M. L. (1992), *FASEB Lett.,* 307, 195–198.
Sheldon, E. and Wadsworth, P. (1993), *J. Cell Biol.,* 120, 935–945.
Shiomura, Y. and Hirokawa, N. (1987), *J. Cell Biol.,* 104, 1575–1578.
Shoeman, R. L. and Traub, P. (1993), *BioEssays,* 15, 605–611.
Sobue, K. (1993), *Neurosci. Res.,* 18, 91–102.
Stearns, M. and Kirschner, M. (1994), *Cell,* 76, 623–637.
Stewart, R. J., Farrell, K. W., and Wilson, L. (1990), *Biochemistry,* 27, 6489–6498.
Strong, M. J. (1994), *J. Neurol. Sci.,* 124, 20–26.
Strong, M. J. and Jakowec, D. M. (1994), *Neurotoxicology,* 15, 799–808.
Sullivan, K. F. (1988), *Annu. Rev. Cell Biol.,* 4, 687–716.
Takeuchi, T., Morikawa, N., Matsumoto, H., and Shiraishi, Y. (1962), *Acta Neuropathol.,* 2, 40–57.
Takeuchi, T. (1968), in *Minamata Disease (Organic Mercury Poisoning),* Study Group of Minamata disease, Kumamoto University, Kumamoto, Japan, 141–252.
Tanaka, E. M. and Kirschner, M. W. (1991), *J. Cell Biol.,* 115, 345–363.
Thompson, W. C. (1980), *Methods Cell Biol.,* 24, 235–255.
Traub, P. and Shoeman, R. L. (1994), *Int. Rev. Cytol.,* 154, 1–103.
Triarhou, L. C., Norton, J., Bugiani, O., and Ghetti, B. (1985), *Neuropathol. Appl. Neurobiol.,* 11, 407–330.
Troncoso, J. C., Hoffman, P. N., Griffin, J. W., Hess-Kozlow, K. M., and Price, D. L. (1985), *Brain Res.,* 342, 172–175.
Troncoso, J. C., March, J. L., Haner, M., and Aebi, U. (1990), *J. Struct. Biol.,* 103, 2–12.
Vale, R. D. (1987), *Annu. Rev. Cell Biol.,* 3, 347–378.
Vallee, R. B., Shpetner, H. S., and Paschal, B. M. (1989), *Trends Neurosci.,* 12, 66–70.
Vogel, D. G., Margolis, R. L., and Mottet, N. K. (1985), *Toxicol. Appl. Pharmacol.,* 80, 473–486.

Vogel, D. G., Margolis, R. L., and Mottet, N. K., (1989), *Pharmacol. Toxicol.,* 64, 196–201.
Walker, R. A., O'Brien, E. T., Pryer, N. K., Soboeiro, M. F., Voter, W. A., et al. (1988), *J. Cell Biol.,* 107, 1437–48.
Wallin, M., Larsson, H., and Edstrom, A. (1977), *Exp. Cell Res.,* 107, 219–225.
Wang, E. (1985), *Ann. N.Y. Acad. Sci.,* 455, 32–56.
Wasteneys, G. O., Cadrin, M., Jones-Villeneuve, E. M., Reuhl, K. R., and Brown, D. L. (1988), *Toxicol. Cell Biol.,* 4, 41–60.
Webster, D. R., Gundersen, G. G., Bulinski, J. C., and Borisy, G. G. (1987), *J. Cell Biol.,* 105, 265–276.
Webster, D. R. and Borisy, G. G. (1989), *J. Cell Sci.,* 92, 57–65.
Weingarten, M., Lockwood, A., Hwo, S., and Kirschner, M. (1975), *Proc. Natl. Acad. Sci. U.S.A.,* 66, 436–439.
Wilson, L., Miller, H. P., and Farrell, K. W. (1990), *J. Cell Biol.,* 111, 28a.
Wisniewski, H. M., Sturman, J. A., and Shek, J. W. (1980), *Ann. Neurol.,* 8, 479–490.
Yamamoto, H., Fukunaga, K., Tanaka, E., and Miyamoto, E. (1983), *J. Neurochem.,* 41, 1119–1125.
Yasui, M., Yase, Y., Ota, K., Mukoyama, M., and Adachi, K. (1991), *Neurotoxicology,* 12, 277–284.
Zimmerman, U. and Schlaepfer, W. W. (1984), *Prog. Neurobiol.,* 23, 63–78.
Zimmerman, H., Doenges, K. H., and Roderer, G. (1985), *Exp. Cell Res.,* 156, 140–152.
Zimmermann, H. P., Plagens, U., Vorgias, C. E., and Traub, P. (1986), *Exp. Cell Res.,* 167, 360–368.
Zimmermann, H. P., Plagens, U., and Traub, P. (1987), *Neurotoxicology,* 8, 569–578.

Chapter 39

Lead Neurotoxicity: Concordance of Human and Animal Research

Deborah Rice and Ellen Silbergeld

I. INTRODUCTION

Lead poisoning is generally recognized as the most significant preventable disease of environmental origin affecting young children in industrialized and industrializing countries (CDC, 1991; ATSDR, 1988; Feldman and White, 1992). In addition, in many countries, occupational and perioccupational exposures to lead are associated with intoxications of hundreds of thousands of workers (Silbergeld, et al, 1991; MacDiarmid and Weaver, 1993). Over the course of the 20th Century, lead poisoning has come to be characterized primarily as a disease of chronic exposure whose earliest and most significant effects are persistent neurotoxicity involving both central and peripheral targets. Based upon epidemiological studies conducted over the past decade, many public health authorities have concluded that there may be no "acceptable" level of lead exposure, at least in postindustrial societies (CDC, 1991).

These conclusions have been supported by a strong and highly productive interaction between clinical observation and experimental research, which has provided extensive information on consequences of exposure to lead. This chapter will not review that history. Our purpose is to provide a comparative analysis of critical findings from both clinical and experimental research, as these findings have illuminated critical issues: the nature of the behavioral effects produced by lead neurotoxicity; dose:response relationships for lead neurotoxicity; the existence of specially susceptible groups, and periods of heightened vulnerability; and issues related to reversibility or persistence.

The history of research on lead neurotoxicity offers important lessons to both clinical and experimental toxicology. In several instances, experimental research has provided insights unavailable from clinical studies; in other instances, experimental models have been developed to replicate findings first observed in human populations under conditions of greater control over dose and potential confounders. Experimental toxicology provides the investigator with certain advantages over clinical/observational study: confounders can be eliminated; exposures can be controlled in amount, route, and timing; mechanisms can be studied at the organ and cellular level. While there are limits on inference from experimental to clinical toxicology, many of these do not hold in the case of lead. Knowledge of lead neurotoxicity in humans antedates research in animal models (see Needleman [1992] for a recent history). Carefully conducted epidemiological studies over the past 2 decades have provided extraordinarily robust information on human dose:response. In fact, as discussed in this chapter, the concordance of experimental with human findings assists us in evaluating the potential significance of confounding variables.

II. LESSONS FROM CLINICAL RESEARCH

The neurotoxic effects of lead exposure were among the first recorded clinical findings of this disease in exposed workers. That lead could affect both the central and peripheral nervous system was known to both Greek and Roman medicine (Maino, 1973). In the 1850s, Tanquerel des Planches provided a complete description of lead-induced peripheral neuropathy in lead workers. However, understanding of the mechanism remained unclear, as discussed by Hamilton in 1925 (Hamilton, 1925). As late as 1947, Reznikoff and Aub suggested that lead affected muscle contraction, rather than neuronal function, to produce the symptoms of fatigue, weakness, and a characteristic wrist or foot "drop" (Reznikoff and Aub, 1947). In 1957, Kostial and Vouk demonstrated that lead could block release of acetylcholine from superior cervical ganglionic cells. Silbergeld et al. (1974) and Manalis and Cooper (1973) reported similar effects at the neuromuscular junction; from that time, lead-induced peripheral neuropathy has been recognized as a prejunctional dysfunction of reduced transmitter release. Over prolonged exposure, demyelination and axonopathy are also observed (Windebank and Dyck, 1984; Weerasuriya et al, 1990).

At the end of the 19th century, clinical attention turned to the risks of lead to central nervous system manifestations in young children. The first observations concerned exposures of young children from lead transported home from the workplace by their parents (Oliver, 1911; Hamilton, 1925). This phenomenon, termed "fouling the nest" by Chisolm, continues to this day, as noted by MacDiarmid and Weaver (1993) in a recent review of perioccupational exposures. In the same period, the first reports of nonoccupational exposures to lead in young children were published from Australia (Fee, 1990). In these reports, the hazards of lead-based paint in residences were first recognized; the earliest symptoms of these relatively high dose episodes of pediatric lead poisoning were neurologic and renal in expression. The course of lead intoxication in young children was recognized to be rapid and particularly devastating (Reiss and Needleman, 1992). However, not until the mid-1940s was it recognized that subencephalopathic intoxications could also induce persistent, significant neurologic damage. Byers and Lord (1943) reported upon 35 children who had been diagnosed as lead exposed without florid neurotoxic signs (seizures, coma). As adolescents they were significantly impaired in learning ability. This report created an enormous reaction, particularly among the lead industry whose members feared government regulation of lead used in gasoline, plumbing, and paints (Lin-Fu, 1985; Hays, 1993).

Several types of studies have now been conducted to investigate the associations between lead exposure and neurotoxicity in children: retrospective, case:control, cross-sectional, and prospective. Some of the earliest studies were cross-sectional. Among these, the pioneering studies by Perlstein and Attala (1966), de la Burdé and Choate (1972) and Perino and Ernhart (1975) were among the first to suggest that even lower exposures could induce measurable neurocognitive dysfunction in young children. Because these first studies were all cross-sectional in nature, children were identified at school age and classified as to lead exposure on the basis of blood lead determinations. This approach was somewhat unsatisfactory, since it was suspected that the critical period for lead exposure and toxicity was probably during the first 6 years of life. David et al. (1972) utilized a case:control approach in a study of lead and hyperactivity; they identified a group of children diagnosed with attention deficit disorder/hyperactivity, and then attempted to establish **prior** lead exposure through the use of chelation challenge. This study suggested that body burdens of lead and, by inference, prior lead exposures were higher in children with no known cause of hyperactivity, such as head trauma or difficult birth as compared to nonhyperactive, matched controls. Case:control studies examining lead exposures in mentally retarded children with no known etiology also found a higher than expected incidence of elevated blood lead levels and contaminated drinking water (Beattie et al, 1975).

These early studies were limited in several respects. First, as indicated by Bellinger et al. (1989), cross-sectional studies cannot definitely identify the direction of causality: it was possible that mentally impaired children were more likely to be exposed to lead (particularly given the assumption of the time, that much of childhood lead exposure to lead-based paint involved actual eating of paint chips, a practice close to if not identical to pica); thus, rather than lead causing mental impairment, mental impairment could cause lead exposure. Another aspect of the debate over causality involved assertions that poor parenting could cause lead poisoning, based upon assumptions that neglected children were more likely to have pica behavior. Second, the classification of children as lead exposed was dependent upon blood lead measurement; however, blood lead provides only a limited picture of lead exposures, given its half life of approximately 35 days (Rabinowitz et al., 1989). Thus, children could have been exposed to lead years before the time of study, and the damage being detected might, as in Byers and Lord's earlier study, have been associated with that earlier exposure rather than with exposures at the time of testing, as indicated by blood lead determinations. The case:control design eliminated some of these problems, but could not adequately account for all potential intercurrent variables, particularly for complex outcomes such as learning and social behaviors. Third, it was recognized by some observers that increased risks of lead exposure coincided with other sociocultural conditions of disadvantage — race, income, inner city residence (Bellinger et al., 1989). Because these factors also contribute to determining neurocognitive development and school success in the U.S., it was recognized that sorting out the specific contribution of lead would be difficult. Considerable controversy has attended this issue, involving the debate over the role of genetics in intelligence (Ernhart, 1992).

Despite these uncertainties, it was generally assumed by the mid 1970s that the young child was the most susceptible to lead neurotoxicity (NAS, 1972). While relatively low levels of lead exposure were associated with signs of peripheral neurotoxicity in adults chronically exposed (Seppäläinen and Landrigan, 1988), the more subtle impairments in neurocognitive performance reported in children were thought to occur at lower doses than neurotoxic signs in adults. Moreover, several reports in the occupational literature indicated that lead-induced neurotoxicity in adults might be reversible with reductions in exposure (Matte et al., 1992); in children, there is little evidence to support the hypothesis of reversibility (see below).

By the end of the 1970s, Needleman and co-workers had developed an innovative study design to deal with the concerns raised above. With respect to understanding causality, they made great efforts to collect a great deal of information on other factors related to childrearing and other influences on children's neurocognitive development (including home environment, maternal IQ, education, and mothering skills). They added to the criteria for determining lead exposure a marker of chronic exposure, concentrations of lead in circumpulpal dentine of shed teeth (de la Burdé and Shapiro, 1975). This marker allowed them to study children aged 8 to 12 years old, when relatively complex neurocognitive measures could be applied and school performance assessed, with information on past exposures for these children. They also attended to the issues of covariates not only by collecting extensive additional information within a study design of sufficient power to support multivariate analysis but also by selecting a population for study with relatively less risk factors for adverse neurocognitive development. This landmark study (Needleman et al., 1979), since replicated in several countries by many other investigators (Davis and Svendsgaard, 1989; Needleman and Gatsonis, 1990), clearly demonstrated an association between elevated body burdens of lead (past lead exposure) and decrements in measured IQ, reaction time, and teachers' evaluations of school performance and social behavior.

The success of this study stimulated major **prospective** studies in at least four other centers: Cincinnati, Cleveland, London, Port Pirie, and Christchurch, NZ. Such prospective designs provide the strongest controls on causality and undetected covariates, through extensive data collection and periodic reassessment. Although the major prospective studies share considerable features in design, there are differences among them, related to the types of populations sampled, cohort size, timing of measurements, and outcomes assessed (Bellinger and Stiles, 1993). Although some (Volpe et al., 1992; Ernhart, 1992) have emphasized the differences in reported results among these studies as proving lack of effect because of lack of replicability, they are sufficiently similar to support careful meta-analysis (Table 1) (Needleman and Gatsonis, 1990; Schwartz, 1994).

Table 1

Study	Effect[a]	SE	SES Status[b]	N	Mean Blood Lead in Exposed Group
Hawk et al. (1986)	2.55	1.5	Low	75	21 μg/dL
Hatzakis et al. (1987)	2.66	0.7	Average	509	23
Fulton et al. (1987)	2.56	0.91	Average	501	12
Yule et al. (1981)	5.6	3.2	Low	166	13
Bellinger et al. (1992)	5.8	2.1	High	147	6.5
Dietrich et al. (1991)	1.3	0.9	Low	231	15
Baghurst et al. (1992)	3.33	1.46	Average	494	20
Silva et al. (1988)	1.51	—	Average	579	11

[a] Estimated loss in IQ for an increase from 10 to 20 μg/dl in blood lead.
[b] Low, disadvantaged; average, normal; high, advantaged.

Adapted from Schwartz (1994).

The prospective studies have provided some insight as to potential windows of susceptibility during perinatal development. In several of these studies, the subjects were enrolled prenatally through the recruitment of pregnant women. Prenatal lead exposures were inferred by measurement of mothers' blood lead during the last trimester of pregnancy or of cord blood at birth, because data indicate that fetal and maternal blood lead levels are almost identical (Goyer, 1990). These levels were correlated with measures of early infant neurodevelopment; the associations between infant status and cord/prenatal lead measures were compared to those between infant status and lead exposure at the time of assessment. While the data are not completely consistent, the largest study (Bellinger et al, 1991, 1992) suggests that prenatal exposures may influence early infant development through the first 4 years of life; after that time children's neurocognitive performance is more highly correlated with lead exposures measured at 18 to 24 months of age. Observed effects of prenatal lead exposure on early infant physical growth and development are also reported to be attenuated by age 4 in prospective studies (Mushak et al., 1989; Shukla et al., 1986), although a cross-sectional analysis of the NHANES II data suggested more stable associations (Schwartz et al., 1986).

Clinical studies have provided only limited information on the potential reversibility of lead neurotoxicity. Reversibility of lead neurotoxicity is of considerable concern in public health. It is difficult to control lead exposures in human populations, so that one cannot exclude the possibility that exposures have continued when apparently persistent effects are observed and associated with earlier exposures. As noted above, some long-term followup studies indicate that exposures of children before 6 years of age are still associated with school performance and sociobehavioral outcome in adolescents (Needleman et al., 1990; Fergusson et al., 1988; 1993). In a retrospective study, White et al. (1993) reported measurable decrements in neuropsychometric performance among persons with positive history of childhood lead poisoning. However, these studies may be confounded by age-related changes in lead storage within the body, such that apparent persistence of neurotoxicity associated with childhood exposures may actually reflect mobilization of lead stored in bone during aging (Silbergeld, 1992; White et al., 1993).

The results of clinical research have been interpreted by one research group as providing "little evidence that lead exposure has a distinctive 'behavioral signature'." (Bellinger et al., 1992) because both verbal and performance IQ have been reported to be affected by lead. In contrast, experimental researchers (including these authors) assume that the neurotoxic effects of lead are likely to be specific in site and mechanism (Silbergeld, 1992; Winder and Kitchen, 1984). Some of this apparent lack of specificity of clinical lead neurotoxicity may result from differences among exposures as to timing, dose rate, and other toxicokinetic factors that cannot be accurately determined in clinical studies; it may also result from lack of precision of the IQ tests and educational and behavioral assessments; and it may also reflect real differences in response among children due to variability in host factors, such as genetics.

The behavioral endpoints chosen for assessment obviously greatly influenced the results and interpretation of the epidemiological studies just described, yet this critical issue has received almost no discussion in the epidemiological literature relating to lead. Most studies have included standardized tests of intelligence as the major dependant variable. Infants and very young children are typically assessed using the Bayley Scales of Infant Development, while older children are assessed using the McCarthy Scales of Child Development or Wechsler Intelligence Scale for Children — Revised. The advantage of these tests is that they are extensively used in psychological research and are standardized for the population. It is well recognized, however, that early measures of intelligence such as the Bayley Scale have little power in predicting later intelligence in individual children, although low scores on the Bayley may be predictive for poor academic performance and intelligence scores later (Rubin and Barlow, 1979). Nonetheless, prospective studies have revealed deficits as a function of lead burden on both early tests such as the Bayley and tests that are predictive of results of later IQ tests such as the WISC-R. All of these intelligence tests assess global processes, which may not be the most sensitive indicators of behavioral impairment produced by lead. Assessment of specific behavioral processes affected by lead exposure, if these were known, would undoubtedly result in increased sensitivity of the behavioral endpoints in detecting an effect.

There are in fact some indications from the human literature concerning behavioral processes responsible for lead-induced impairment. For example, in Needleman's 1979 study, assessment by classroom teachers on a rating scale revealed a dose-dependent increase in distractibility and inability to follow complex sequences of directions as a function of increased body burden of lead (Needlemen et al., 1979), which was subsequently replicated by others (Yule et al., 1984). These results are suggestive of impairment of attentional processes in lead-exposed children, although other deficits may also explain these findings. In a direct examination of attentional processes, Winneke et al. (1989) examined 223 children, aged 72 to 119 months, recruited from two German cities with differing intensities of environmental lead exposures. They utilized a device measuring serial choice reaction time with both visual and auditory signals. There was a greater effect on the number of response errors (making a response inappropriately) than on failure to make correct responses, which indicates inability to inhibit inappropriate responding. Yule et al. (1984) also reported that high-lead children exhibited more deviant performance in tests of conduct problem, inattentive-passive, and hyperactive scale.

In a follow-up of the Boston prospective study (Stiles and Bellinger, 1993), children were tested at age 10 on the Wisconsin Card Sorting Test, which assesses abstract thinking, sustained attention, and ability to change response strategy according to changing environmental requirements. Recent blood lead levels were associated with perseverative behavior (i.e., staying with a formerly effective strategy despite a change in the rules of the task). The consequences of early poor performance as a result of lead exposure in terms of grade retention or need for special education have also been investigated. In a follow-up of the children from the Needleman et al. 1979 study, tooth lead levels as 5- and 6-year-olds predicted an increase in grade retention and a 2-fold increase in the need for academic aid in teenagers (Bellinger et al., 1984) as well as an increased incidence of dropping out of high school and deficits on various performance measures (Needleman et al., 1990). These results are not surprising in view of the effects of lead on classroom behavior cited above. It is known that early attentional deficits and their associated behaviors place children at risk for academic failure (Horn and Packard, 1985).

There are additional issues related to interpretation of results of neuropsychological tests in children. Since many of the tests call upon both central and peripheral elements of somatosensory and visuomotor function, performance decrements may result from effects at any point in this functional complex. The WISC has been criticized for many reasons (Smith, 1985), including cultural bias and insensitivity, and, although it has been normed for age, it remains an issue of some controversy as to whether further covariate adjustment must be done when comparing groups that differ in age or other factors (Ernhart, 1992). On the other hand, covariate adjustment can overcontrol and thus obscure real effects of lead. Overcontrol for social class in some English studies by Yule and co-workers, and control for school assignment in the studies by Winneke and co-workers (1982, 1983) may have resulted in type II errors (Winneke and Kraemer, 1984; Mushak et al., 1989).

Relatively little investigation has followed effects of chronic lead exposure on peripheral nervous system function in children (Seppäläinen and Landrigan, 1988). An analysis of data collected on young children exposed to lead emissions from a smelter suggested that significant decrements in nerve conduction velocity are not observed until blood lead levels exceed 30 mcg/dl (Schwartz et al., 1988). This might indicate that the PNS is less sensitive to lead; however, there are differences in sensitivity of outcome measurement between neuropsychological tests and neurophysiological measurement of compound nerve conduction velocity. As noted by Buchtal and Behse (1979), significant neuropathology of the peripheral nerve must be induced before decreases in nerve conduction velocity can be detected.

III. CONTRIBUTION FROM THE ANIMAL LITERATURE

The neurotoxic effects of lead in experimental (animal) models has been extensively investigated over the last 2 decades, and it has been clear for at least a decade that lead exposure reliability produces behavioral impairment (see Cory-Slechta, 1984, for review). The use of animal models allows direct investigation of issues such as sensitive period, reversibility, dose-response relationships, and underlying behavioral processes responsible for lead-induced behavioral toxicity. Of particular relevance to the issue of lead toxicity is that in experimental studies individuals are assigned randomly to treatment groups, and potential confounders such as socio-economic status, nutrition, and parental IQ are eliminated. The present section is not a review of the animal literature, but rather focuses on relatively subtle impairment produced by reasonably low-level developmental exposure to lead in monkeys, with selected relevant studies in the rodent also included.

A substantial literature exists on the behavioral consequences of developmental lead exposure in the monkeys; most of the research has been performed at our laboratory at the Health Protection Branch in Ottawa and at the University of Wisconsin primate laboratory (see Rice, 1992a, for review). These laboratories utilized two different species of macaque monkeys, different dosing regimens, and different types of equipment for behavioral testing. While performance on some of the same tasks was assessed in both laboratories, details of the task requirement differed between laboratories. Both laboratories have tested a number of different cohorts of monkeys, born over a number of years, with the inevitable concomitant changes in general rearing procedures. Despite these procedural differences, however, the

Table 2 Summary of Effects of Developmental Lead Exposure in Monkeys (Rat)

Task	Effects of Lead Exposure
Nonspatial discrimination reversal	More errors across reversals
	Deficit in acquisition when changing stimulus dimension (higher doses)
	More attention to irrelevant cues
Spatial discrimination reversal	More errors across reversals, especially in the presence of irrelevant cues
	More attention to irrelevant cues
Learning set	Retarded learning within problems
	Retarded learning across problems
Concurrent discrimination	Retarded learning
	Perseveration for position
Spatial delayed alternation	Deficits in acquisition, more marked at higher doses
	More errors at longer but not shorter delays
	Marked perseveration for position
Delayed matching to sample (nonspatial and spatial)	No effect on acquisition
	More errors at longer but not shorter delays
	Perseveration for position on spatial task
Repeated acquisition (rat)	More errors on the repeated acquisition baseline
	Systematic perseveration for position
Fixed interval (monkey and rat)	Increased rate of response
DRL	More nonreinforced responses
	Fewer reinforced responses
	Inability to inhibit inappropriate responding

types of behavioral deficits observed are gratifyingly similar between laboratories, between experiments within laboratories, and across time. Moreover, careful studies in the rat performed at the University of Rochester are congruent with effects in monkeys in terms of the presumed behavioral processes responsible for the observed behavioral deficits. The body of data described in the present section suggests that perseveration, increased distractibility, inability to inhibit responding, and inability to adapt to complex behavioral requirements are at least partly responsible for the deficits in learning and memory exhibited by lead-treated animals (Table 2). The results of these studies also suggest that lead-induced behavioral deficits are not reversible, and that there is no specific "sensitive period" for lead-induced impairment as a result of developmental exposure in the monkey, or in the rat if careful behavioral analyses are performed. Theoretically, lack of reversibility has been suggested to be due to lack of repair in the CNS (Goldstein, 1992; Silbergeld, 1992) It does not appear to depend upon long-term persistence of elevated lead levels in brain, after reductions in external exposure sources. Cory Slechta et al. (1992) reported that lead-induced supersensitivity of dopamine receptors (determined by drug-induced behavioral response) was observed in 60-day old rats, exposed only through lactation (day 21 postnatally), although at the time both blood and brain levels were no different between prior exposed and control animals. Certain neurochemical parameters were also persistently altered at this point (Widzowski and Cory Slechta, 1991). Other investigators have reported apparent recovery of lead-induced changes in brain morphology (Campbell et al., 1982). However, as noted by Slomianka et al. (1989) it is important to distinguish between recovery, when defined as a return to control size or brain regions or volume of cell layers within regions, and a different or progressive type of toxic response involving hypertrophy.

For the past 50 years it has been assumed that the young child is most susceptible to the neurotoxic effects of lead for reasons related to exposure, absorption, toxicokinetics, and sensitivity of target sites within the nervous system (Reiss and Needleman, 1992). Within human populations, it is known that the first three variables generally place the young child at increased risk (CDC, 1991).

The latter aspect has been investigated in experimental studies. The precise definition of this window of sensitivity is somewhat controversial. Brown (1975) had suggested that the early preweaning period in the rat was the most sensitive period for neurobehavioral effects. Structural alterations in neurodevelopment are reported to differ with timing of exposure. Early postnatal exposures in rats and primates results in decreased dendritic arborization in cortex, cerebellum, and hippocampus (Alfano and Petit, 1982: Bull et al., 1983; Holtzman et al., 1984; Lorton and Anderson, 1986a,b; Reuhl et al., 1989). However, no effects on dendrites were observed in guinea pigs exposed pre- or postnatally (Legare et al., 1993). Olson et al. (1984) found age-related differences in both neurophysiologic and histomorphologic effects of lead, which differed among various brain regions.

But an important limit on interpreting these studies is that very few of them include toxicokinetic evaluation which is important since the observed differences in sensitivity may be due to different concentrations of lead in brain attained at different periods of development owing to maturation of the blood:brain barrier, among other factors (Regan, 1989; Cory Slechta, 1990; Bradbury and Deane, 1993). There is also some evidence for a bimodal effect of lead on hippocampal neuromorphology (Slomianka et al., 1989), which points to the importance of determining target organ dose.

The overall greater sensitivity of the young organism is thought to relate to specific aspects of neurodevelopment. A number of investigators have proposed that susceptibility to lead neurotoxicity is highest during perinatal stages of neurodevelopment on the basis of proposed mechanisms of action (Silbergeld, 1992; Goldstein, 1992; Olson et al., 1984; Cory Slechta et al., 1992). Bressler and Goldstein (1991) for example, suggest that lead can affect perinatal synaptogenesis by inhibiting processes of pruning and synaptic strengthening through calcium-dependent mechanisms. Two different sets of investigators have independently proposed that developmental lead neurotoxicity depends upon specific interactions between lead and glutamate receptors that change with age (Ujihara and Albuquerque, 1992; and Guilarte and Miceli, 1992). In the *in vivo* studies, binding of H-MK-801, a receptor ligand for the NMDA-responsive ionophore in glutaminoceptive neurons, was inhibited in 14-day old, but not 56-day old, rats exposed to lead from conception onwards (Guilarte and Miceli, 1992). *In vitro*, lead was a more potent inhibitor of ligand binding to synaptic membranes prepared from 14-day old rats (Guilarte and Miceli, 1992); other studies using hippocampal cell cultures demonstrated loss of lead effect as cells remained in culture to 4 weeks (Ujihara and Albuquerque, 1992). Another mechanistic hypothesis for age-related changes in sensitivity is proposed by work on neural cell adhesion molecules by Regan and co-workers. They propose that lead affects the conversion of NCAMs from the embryonic to adult form (Regan, 1993). However, the interactions between lead and age may be very complex: Breen and Regan (1988) also report that from days 4 through 16 lead *in vitro* inhibits Golgi-associated sialytransferase, whereas in adult rats lead *in vitro* stimulates enzyme activity.

This review discusses the results of animal studies of lead neurobehavioral toxicity by type of test instruments applied.

A. TESTS OF COMPLEX LEARNING

A test that has proven sensitive to the effects of developmental lead exposure is the discrimination reversal task. In this paradigm, the monkey is presented with two or more stimuli which vary in one or more ways (i.e., form, shape, color, position). The monkey must respond to a specified stimulus (e.g., always choose the red rather than the green, irrespective of position or shape) in order to be rewarded, usually with a preferred food or juice. When the monkey learns the task to some predetermined criterion, the "rule" is changed (reversed) so that the previously incorrect stimulus becomes the correct one (e.g., green rather than red). Typically a number of such reversals are instituted. A normal monkey will learn each successive reversal more quickly, displaying a "learning curve". In addition, the task may change in terms of the relevant stimulus dimension; for example, "attend to the color and ignore the shape" may change to "attend to the shape and ignore the color". Both of these manipulations tax intellectual capabilities different from those required by the simple acquisition of a discrimination, in that they require changing established response strategies.

In 1979, researchers from the Primate Center at the University of Wisconsin reported impaired reversal learning performance in rhesus monkeys (*Macaca mulatta*) exposed postnatally to lead (Bushnell and Bowman, 1979a,b). Blood lead levels of approximately 50 or 90 µg/dl were associated with impairment early in life on a series discrimination reversal tasks. A subset of these monkeys was found to be impaired on a series of spatial reversal tasks with irrelevant color cues at the age of 4 years, despite the fact that lead exposure ceased at 1 year of age, and blood lead levels at the time of testing were at control levels.

In the same year, deficits were reported on a simple nonspatial form discrimination reversal task in 2-to-3-year-old cynomolgus monkeys (*Macaca fascicularis*) exposed continuously to lead from birth (Rice and Willes, 1979). Blood levels of these monkeys peaked at approximately 50 µg/dl, and decreased after infancy to stable levels of about 30 µg/dl. This apparently robust effect of lead on discrimination reversal performance was pursued with a group of monkeys, consisting of a control and two dosed groups, exposed to lower levels of lead continuously from birth. Blood levels peaked during infancy at 25 or 15 µg/dl for the higher and lower dose groups, respectively, then decreased to steady-state levels of 13 or 11 µg/dl. These monkeys were tested as juveniles (3-year-olds) on a series of nonspatial discrimination reversal problems, including problems with irrelevant form or color cues (Rice, 1985a). This afforded the opportunity to change two sets of "rules": the positive and negative stimulus within a pair of relevant stimuli (such as cross vs. square), as well as the relevant stimulus dimension (such as form vs. color). The inclusion of irrelevant cues provided the opportunity to study distractibility in these monkeys. Lead-treated monkeys were impaired over the set of reversals on the first discrimination, their introduction to a discrimination reversal task, and on second problem, a color discrimination with irrelevant cues, their introduction to irrelevant cues. Analysis of the kinds of errors made by treated monkeys revealed that they were attending to irrelevant cues in systematic ways, either responding on or avoiding a particular position or stimulus. This suggests that lead-treated monkeys were being distracted by these irrelevant cues to a greater degree than controls, which may have been responsible at least in part for their poorer performance.

When these monkeys were mature adults (9 to 10 years old), they were tested on a series of three spatial discrimination reversal tasks, the last two of which included irrelevant cues (Gilbert and Rice, 1987). In this task, the monkey was required to respond to a particular position irrespective of what stimuli appeared on the response buttons. Treated monkeys were impaired relative to controls in the presence but not in the absence of irrelevant stimuli. Moreover, the lower dose group was impaired only during the first task after the introduction of irrelevant cues, but not on the second task with irrelevant cues, when irrelevant stimuli were familiar. As in the nonspatial discrimination reversal task, there was evidence that lead-exposed monkeys were attending to the irrelevant stimuli in systematic ways, suggesting that this behavior was responsible for, or at least contributing to, the impairment in performance. This is also suggested by the fact that lead-treated monkeys were impaired in the presence but not the absence of irrelevant stimuli.

In a subsequent study on possible sensitive periods for deleterious effects produced by lead, monkeys were exposed to lead either continuously from birth, during infancy only, or beginning after infancy. Lead levels were about 30 to 35 µg/dl when monkeys were exposed to lead and given access to infant formula, and 19 to 22 µg/dl when dosed with lead after withdrawal of infant formula (Rice and Gilbert, 1990a). When these monkeys were juveniles, they were tested on the same nonspatial discrimination reversal tasks as described above (Rice and Gilbert, 1990a), and were tested on the spatial discrimination reversal task as adults (Rice, 1990). On the nonspatial series of tasks, both the group dosed continuously from birth and the group dosed beginning after infancy were impaired over the course of the reversals in a way similar to that observed in the study discussed above. The higher exposure levels in this study were reflected in impairment on all three tasks, whereas in the previous study lead-treated monkeys were impaired on only the first two tasks. The group exposed only during infancy was unimpaired on these tasks. On the spatial version of the task, all three treated groups were impaired to an approximately equal degree. In the spatial discrimination reversal study, treated monkeys were the most impaired over the series of reversals on the first task after the introduction of irrelevant cues, although all three dose groups were impaired on all three tasks. These data suggest that spatial and nonspatial tasks may be affected differentially depending on the developmental period of lead exposure.

There are several generalizations that may be drawn from these studies. It may be stated with some degree of confidence that lead impairs performance on both spatial and nonspatial discrimination reversal performance in monkeys exposed developmentally (postnatally) to lead. It is clear that the discrimination reversal paradigm is more sensitive to lead-induced behavioral impairment than is simply the acquisition of a discrimination task. In the studies just described, impairment in task acquisition was observed only at high blood lead levels, or when the "rules" were changed (i.e., when changing between tasks requiring attention to different stimulus dimensions). It also appears that lead-exposed monkeys exhibit a greater degree of impairment across a set of reversals in the presence of distracting irrelevant stimuli, although performance was sometimes impaired in the absence of irrelevant stimuli, especially at higher doses. This constellation of effects points to decreased adaptability and/or increased distractibility in lead-treated monkeys, although other types of impairment may also be present.

Another example of the types of intellectual deficits produced in monkeys as a result of developmental lead exposure was performed in Winneke's laboratory in Germany (Lilienthal et al., 1986). Rhesus monkeys were exposed to lead *in utero* and continuing during infancy at doses sufficient to produce blood lead values up to 50 µ/dl in the lower dose group and 110 µ/dl in the high dose group. These monkeys were tested on a learning set formation ("learning to learn") task as juveniles. The task is comprised of a sequential series of visual discrimination problems; when the monkey learns one task to a pre-set criterion, another is presented. Different stimulus sets are used for each discrimination. Normal monkeys will learn successive discriminations more quickly as a result of exposure to the learning situation. Lead-exposed monkeys were impaired both in terms of improvement in performance across trials on any given problem, as well as inability to learn successive problems more quickly as the experiment progressed. Such a deficit represents impairment in the ability to take advantage of previous exposure to a particular set of "rules". This deficit is reminiscent of failure of lead-treated monkeys to improve as quickly as controls over a series of discrimination reversals.

A task similar in some ways to the learning set task just described is the concurrent discrimination task. Instead of learning a series of visual discrimination problems sequentially, subjects are required to learn a number of such problems at the same time: i.e., a number of different stimulus pairs are presented across trials within the same session. Performance on such a task was assessed in our laboratory in the group of monkeys in which the contribution of developmental period of exposure to the behavioral toxicity of lead was explored, described above, in which monkeys were exposed continuously from birth, during infancy only, or beginning after infancy (Rice 1992d). Monkeys were required to learn two sets of concurrent discrimination problems. All three treated groups learned more slowly than controls, although monkeys dosed during infancy only were less impaired than the other two groups. In addition, all three treated groups exhibited perseverative behavior, responding incorrectly more often than controls at the same position that had been responded on in the previous trial. These results reinforce results from other studies in several respects. First, treated monkeys were most impaired on the first task, upon introduction of a new set of contingencies. This pattern is consistent with results from the discrimination reversal tasks in all groups of monkeys. Second, analysis of error pattern revealed perseveration on a previous position (response button) to be in part responsible for the increased errors by treated monkeys. Third, while all three treated groups were impaired, the group dosed during infancy only was the least affected. This is consistent with results from other experiments in this group of monkeys.

B. ASSESSMENT OF SPATIAL MEMORY

A relatively simple task that has proved sensitive to disruption by lead, in both the Ottawa and the Wisconsin laboratory, is spatial delayed alternation. This task assesses attentional processes as well as short-term spatial memory. The monkey alternates responding between two stimuli, with each correct alternation being reinforced. The stimuli are identical, with no cue to indicate the correct position on any trial. Delays of various lengths can be instituted between opportunities to respond, in order to assess short-term memory. A study at the University of Wisconsin revealed that rhesus monkeys exposed to lead from birth to 1 year of age, with peak blood levels as high as 300 µg/dl and levels of 90 µg/dl for the remainder of the first year of life, were markedly impaired on this task as adults (Levin and Bowman, 1986).

In a study in our laboratory, monkeys with steady-state blood lead levels of 11 or 13 µg/dl, described above, were tested on a spatial delayed alternation task at 7 to 8 years of age (Rice and Karpinski, 1988). Both treated groups were impaired relative to controls, during the initial acquisition of the task as well as at the longer but not shorter delay values (Figure 1). The deficit in acquisition in lead-treated monkeys was the result of "pounding" on both buttons indiscriminately, which represents a failure to inhibit inappropriate responding. At the longer delays, some monkeys in both treated groups displayed marked perseverative behavior, responding in some cases on the same incorrect response button for hours at a time. The severe deficit observed in these monkeys was comparable to that observed following extensive brain lesions in certain parts of the cerebral cortex, and were certainly unexpected in monkeys with a history of such moderate lead exposure.

Performance on this same task was assessed in the group of monkeys described above, exposed to lead during different developmental periods (Rice and Gilbert, 1990b). The results were very similar to those just described. Lead-exposed monkeys were impaired on the acquisition of the task because of indiscriminate responding on both buttons. As in the previous study, treated monkeys were unimpaired at short delay values and increasingly more impaired as the delay period was lengthened. Treated monkeys also made more perseverative responses (i.e., continuing to respond repeatedly on the wrong button). These effects are indicative of perseveration and lack of ability to inhibit inappropriate responding, which have proven hallmarks of behavioral impairment produced by lead in the monkey. In this study, all three lead-exposed groups were impaired to an approximately equal degree, as was the case on the spatial version of the discrimination reversal task. These findings provide further evidence of a lack of sensitive period for lead-induced impairment on spatial tasks.

In a study in rats, improved performance on delayed alternation was observed in young (exposed postweaning) and old animals, but not rats exposed as adults (Cory-Slechta et al., 1991). The training procedure consisted of many sessions of a cued alternation procedure; i.e., the rat had only to respond on the lever associated with a cue light as it alternated between positions from trial to trial. The authors interpreted the improved performance of the lead-treated groups to be the result of perseveration of

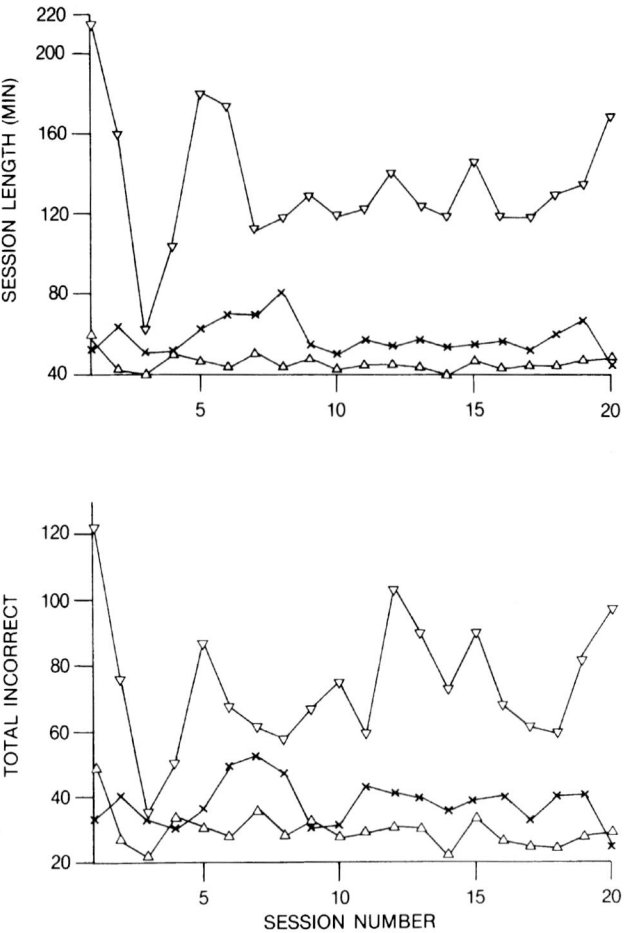

Figure 1. Session length and total number of incorrect responses over the course of the session on the longest (15 s) delay value of a spatial delayed alternation task. Session length was increased by each error, since a session consisted of 100 correct trials (responses). △ = control, × = 11 µg/dl blood level group ▽ = 13 µg/dl blood level group. Both treated groups were impaired relative to controls. (From Rice, D. C. and Karpinski, K. F., *Neurotoxicol. Teratol.*, 10, 207–214, 1988. With permission.)

alternation behavior as a result of the extensive training procedure. This explanation is consistent with the interpretation of the results of the monkey studies.

A delayed matching to sample paradigm was used to assess learning and short-term memory in the group of monkeys, discussed above, with preweaning blood lead levels of 50 µg/dl and postweaning steady-state blood lead levels of 30 µg/dl (Rice, 1984). These monkeys began testing at 3 to 4 years of age. They were tested on both a spatial and nonspatial version of the task. In the nonspatial task, a sample color was displayed on a push button on which the monkey was required to respond. Following a delay period which varied from a few seconds to several minutes, three different buttons were lit with three different colors, and the monkey was required to respond on the button displaying the color shown previously to obtain a juice reward. The color of the sample button varied from trial to trial, and the positions on which the colors appeared on the test buttons also varied between trials. For the spatial task, the monkey was required to remember in which position a sample stimulus appeared, and respond at that same position after the interposition of a delay. Lead-exposed monkeys were impaired on both the spatial and nonspatial versions of this task. They were not impaired in their ability to learn the matching task per se, but were increasingly impaired as the delay between exposure to the sample stimulus and the set of stimuli to be matched was increased (Figure 2). Investigation of the types of errors revealed that for the nonspatial matching task, lead-exposed monkeys were responding incorrectly on the position that had been correct on the previous trial. This type of behavior may be considered to represent perseverative behavior and is reminiscent of the perseverative errors observed in other groups

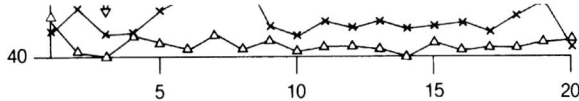
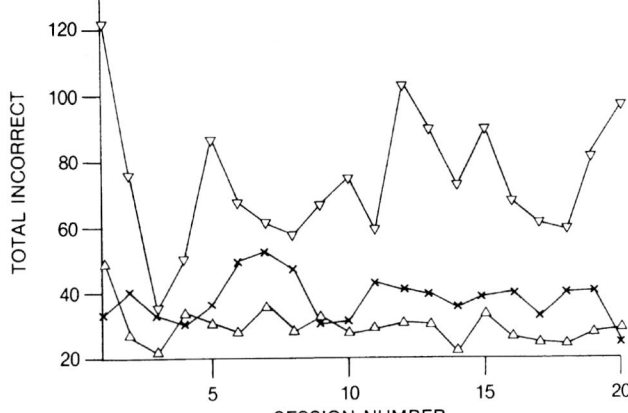

Figure 2. Delay values at which monkeys reached chance performance on nonspatial and spatial matching to sample tasks. The nonspatial task was tested twice under different conditions. Each symbol represents an individual monkey. C = control, T = monkeys with steady-state blood levels at 30 µg/dl. For the nonspatial task first time, three control monkeys were performing above chance at 120 s, but were not tested further since all treated monkeys were performing at chance. Treated monkeys reached chance performance at shorter delays than controls on both tasks. (From Rice, D. C., *Toxicol. Appl. Pharmacol.*, 75, 337–345, 1984. With permission.)

on delayed alternation. On the other hand, it may be considered to be the result of increased distractibility by irrelevant cues by lead-treated monkeys, similar to the increased attention to irrelevant cues displayed in the discrimination reversal tasks. (These interpretations are not mutually exclusive.) This behavior is at least partly responsible for the apparent deficit in short-term memory observed in lead-treated monkeys on the nonspatial matching to sample task, although other mechanisms may also play a part. The lack of interference from previous trials on the spatial version of the task, however, may indicate a pure deficit in spatial short-term memory on that task.

The Wisconsin group has utilized the Hamilton Search Task to study attention and spatial memory. In this task, a row of boxes was baited with food, and then closed. The monkey lifted the lids to obtain the food. The most efficient performance requires that each box be opened only once, necessitating that the monkey remember which boxes have already been opened. Monkeys exposed to doses of lead sufficient to produce blood lead levels of approximately 45 or 90 µg/dl during the first year of life or 50 µg/dl *in utero* were impaired in their ability to perform this task at 4 to 5 years of age (Levin and Bowman, 1986). These results were replicated in another group exposed postnatally to higher lead levels, and tested at 5 to 6 years of age (Levin and Bowman, 1983). The effects of lead on the Hamilton Search Task were in general less robust than effects on delayed spatial alternation tested in the same monkeys, despite the fact that both tests presumably assess attention and spatial memory. The greater deficit observed on the delayed alternation task may have been due to the requirement for alternation or adaptation of response pattern, an ability which seems to be globally impaired in lead-exposed monkeys.

A recent assessment of spatial learning and memory in the rat revealed an interesting pattern of errors responsible for the overall poorer performance of lead-treated subjects (Cohn et al., 1993). Rats were exposed to lead in drinking water beginning at weaning, and tested beginning at 55 days of age on a task with two components. One component of the schedule (repeated acquisition component) required the rat to learn a new sequence of lever presses every day. In the other component, the rat was required to perform the same sequence of lever presses every session (performance component). Significant impairment of performance was observed on the repeated acquisition component but not on the performance component in lead-exposed rats compared to control. Analyses of error patterns revealed that the decrease in the percent of correctly completed sequences in lead-treated rats on the repeated acquisition component derived from two sources. The first consisted of a perseveration of performance-like sequence responding (left center right) even during the repeated acquisition component (Figure 3). Second, lead exposure increased perseverative responding on a single lever even though the schedule itself never directly reinforced such repetitive responding. This very systematic and specific perseverative behavior provides evidence that perseveration is a consistently observed effect of developmental lead exposure across species.

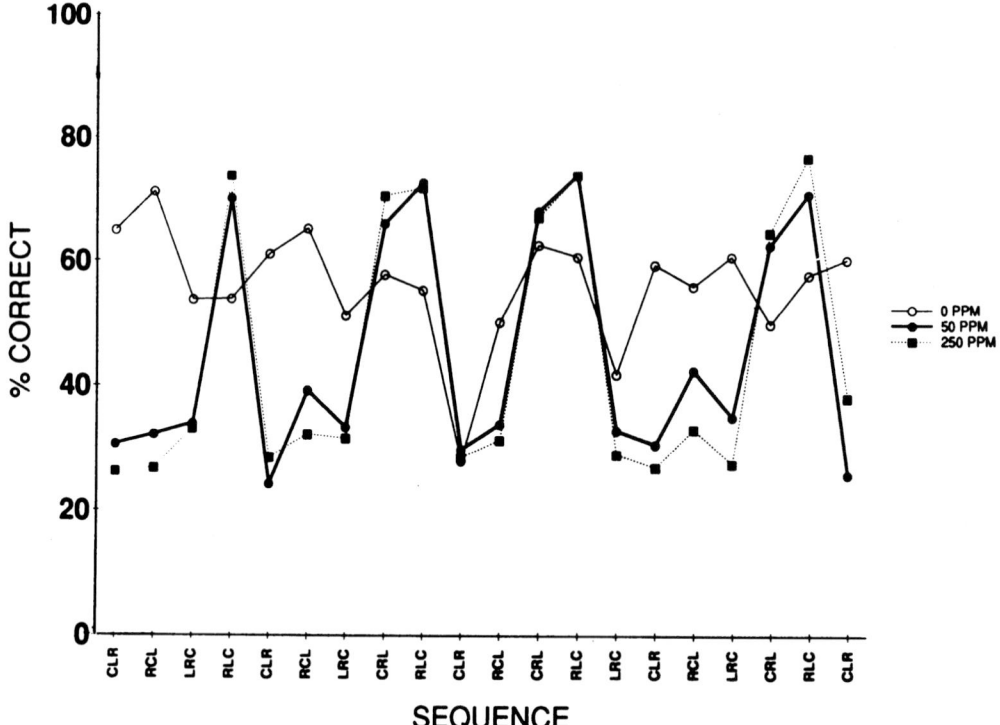

Figure 3. Percent of correct response sequences on a repeated acquisition task for control and two groups of rats exposed to lead beginning at weaning. Treated rats completed fewer correct response sequences for sequences that were unlike those of the performance sequence (LCR). This error pattern is indicative of perseverative behavior in lead-treated but not control rats. L = left, C = center, R = right. (From Cohn, J., Cox, C., and Cory-Slechta, D. A., *Neurotoxicology,* 14, 329–346, 1993. With permission.)

C. INTERMITTENT SCHEDULES OF REINFORCEMENT — LEAD AND ACTIVITY

Intermittent schedules of reinforcement have been studied extensively by experimental psychologists, and performance thus generated is orderly, predictable, and reproducible. (See Rice, 1988 for review of the use of intermittent schedules in behavioral toxicology.) The intermittent schedule most studied with lead is the fixed interval (FI) schedule of reinforcement. In this schedule, the subject is rewarded for the first response (typically a lever press) after a specified time has elapsed. Responding before the specified time has no scheduled consequences, even though there is no requirement for more than a single response at the end of the interval. The rate of response is in some sense a measure of the activity of the animal.

In 1979, it was reported that monkeys exposed to lead exhibited an increased rate of response on an FI schedule of reinforcement (Rice et al., 1979). This effect was observed in the group already described which had steady-state blood lead concentrations of approximately 30 µg/dl. The rate-increasing effect of lead on FI performance was later replicated in monkeys exposed to lower doses of lead (steady-state blood lead concentrations of 11 or 13 µg/dl) (Rice, 1985b). The three dose groups exhibited an orderly dose-related increase in rate of responding as a consequence of lead exposure. This same dose-dependent increase in response rate on FI performance has also been observed in rats exposed to lead beginning postweaning, with blood levels comparable to those in monkeys (Cory-Slechta and Thompson, 1979; Cory-Slechta et al., 1983, 1985). Performance on the FI schedule was also assessed in the monkeys exposed continuously from birth, during infancy only, or beginning after infancy, described above, with steady-state blood lead levels during dosing of 19 to 26 µg/dl (Rice, 1992b). Increased rates of response were observed in all three treated groups. These results indicate that exposure to lead during infancy is not necessary for FI performance to be affected, and that exposure only during infancy is sufficient to produce effects.

Since lead-exposed monkeys exhibited higher rates of responding on the FI schedule, which does not specify any particular response rate, it was of interest to determine whether these monkeys would inhibit responding if required to do so. This issue was examined by assessing performance on a DRL (differential reinforcement of low rate) schedule of reinforcement. The DRL schedule required that the monkey inhibit responding for at least 30 seconds after the preceding response in order to receive a reward. Responding before the 30 seconds elapsed reset the clock, and another 30-s wait was in effect. Thus the DRL schedule "punishes" all but very low rates of responding by postponing the opportunity for reinforcement. There is no external cue to signify the end of the wait period; the monkey must resort to internal timing mechanisms. Performance on this schedule was initially examined in the groups of monkeys having steady-state blood lead levels of 11 or 13 µg/dl (Rice and Gilbert, 1985). Lead-treated monkeys were able to perform the DRL task in a way that was indistinguishable from controls. However, they learned the task at a slower rate, as measured by the increment in reinforced responses and decrement of nonreinforced responses over the course of the early sessions. This schedule was also examined in monkeys having peak blood lead levels early in life of 100 µg/dl, with steady state levels of 40 µg/dl. These monkeys exhibited a higher number of nonreinforced responses, a lower number of reinforced responses, and shorter average time between responses over the course of the experiment than did control monkeys (Rice, 1992c).

The increases in rate of response on the FI may be considered in some sense to represent failure to inhibit inappropriate responding, since treated monkeys made more responses than controls without increasing reinforcement density: i.e., their behavior was less efficient. The increase in button-pushing behavior may also be considered to represent perseveration. The higher rate of response of treated monkeys on the DRL schedule, which actually resulted in fewer reinforcements, clearly represents a failure to inhibit inappropriate responding, and may reflect the fact that lead-treated monkeys are less able to utilize internal cues for timing.

D. SOCIAL BEHAVIOR AND ENVIRONMENTAL REACTIVITY

The effects of developmental lead exposure have been assessed on other behaviors, including mother-infant and peer social behavior and reaction to environmental manipulation, by the group at the University of Wisconsin. Infants exposed to lead either prenatally or postnatally engaged in more clinging to the mother than did control monkeys (Schantz et al., 1986); this was observed in mothers not exposed to lead but taking care of lead-exposed infants as well as in mothers exposed to lead. Monkeys treated with lead postnatally and raised with peers exhibited suppressed play behavior and increased social clinging (Bushnell and Bowman, 1979c). Moreover, these abnormal patterns were exacerbated when these infant monkeys were introduced to a novel environment. Monkeys exposed to lead during the first year of life were also more reluctant than controls to enter a novel environment at 4 years of age (Ferguson and Bowman, 1990). Infant monkeys exposed to lead also exhibited decreased muscle tonus and increased arousal or agitation on a neonatal developmental battery analogous to the Brazelton Neonatal Behavioral Assessment Battery developed for human infants (Levin et al, 1988). These monkeys also displayed decreased looking at novel stimuli in a test of visual exploration. A test similar to this has proved to be highly predictive of future intelligence in human infants (Rose and Wallace, 1985). It appears, then, that lead has important effects on social behavior and reactivity to environmental circumstances, which may be manifested early in life. The latter effect in particular may be an important contributing factor to the deficits on intellectual tasks observed as a result of lead exposure.

E. SUMMARY OF EXPERIMENTAL DATA

It is well established that monkeys exhibit behavioral impairment as a result of developmental exposure to lead. This has been demonstrated in two different species of monkey, in different laboratories, in a number of different groups of monkeys, and on a variety of behavioral tasks. Moreover, behavioral impairment has been observed in every group tested, with no evidence of a threshold for effect. Specifically, research has consistently demonstrated impairment in a group of monkeys with steady-state blood lead levels of 11 µg/dl and a peak early in life of 15 µg/dl, compared to a "control" group with blood lead levels of 3 to 5 µg/dl. Moreover, deficits persisted into adulthood. This is the lowest dose group in the colony; there is therefore not an opportunity to determine a no-observable-effect level. A blood lead level of 10 µg/dl is considered safe by regulatory agencies in the U.S.; in most countries blood lead levels believed to represent a risk to children are considerably higher. The data from monkeys suggest that 10 µg/dl in blood does not represent a body burden that protects from behavioral deficits.

Inferences may also be drawn from this body of data concerning the issue of sensitive period for lead-induced behavioral impairment. Research at the University of Wisconsin revealed permanent behavioral sequelae as a result of *in utero* exposure, or exposure during the first year of life in a number of cohorts of monkeys. This was replicated in Ottawa in one cohort dosed for the first 400 days of life, revealing deficits on a number of tasks. On the other hand, exposure beginning at 300 days of age produced impairment roughly equivalent to ongoing exposure beginning at birth. Data from rats from the University of Rochester also suggest that very early (prenatal or pre-weaning) exposure is not necessary to produce lead-induced behavioral impairment. It thus appears that (1) the effects of early exposure to lead are permanent, and (2) early exposure (during infancy or before) is not necessary to produce robust behavioral deficits.

Another issue illustrated by the examples presented here is the possibility that the types of behavioral impairment observed may depend on the developmental period of exposure. In studies in the Ottawa laboratory, monkeys exposed to lead beginning at birth or after the period of infancy were impaired on both spatial and nonspatial tasks. Monkeys exposed to lead only during infancy, however, were impaired on spatial but not nonspatial tasks. This serves as a caution for interpretation of results of epidemiological studies; it may be misleading to assume that there is a single constellation or pattern of effects in children as a result of lead exposure.

The body of behavioral data suggests that the poorer performance of lead-exposed animals on a variety of behavioral tasks is at least in part the result of the constellation of perseveration/distractibility/inability to inhibit inappropriate responding (Rice, 1993). These impairments are remarkably similar to those observed in children. Such congruence between children and animals in the apparent behavioral mechanisms responsible for the global deficits in cognitive ability as measured by IQ (children) or various learning tasks (animals) is both satisfying and alarming. These robust, reproducible effects in an animal model, under conditions not subject to the potential confounds of epidemiological studies, strongly support the interpretation that the positive findings in children at low blood levels are in fact the result of lead exposure.

IV. CONCLUSIONS

1. Data from both the animal and human literature suggest that a concurrent blood lead level of 10 µg/dl does not protect against lead-induced behavioral deficits, that is, there are effects observable in both animals and children at lead exposures at or below current pediatric guidelines.
2. There is considerable evidence that lead-induced cognitive deficits are not reversible. Ongoing deficits have been observed in monkeys exposed to moderate doses of lead (blood lead levels of 30 µg/dl) for the first year of life 9 years after cessation of exposure, when blood lead levels were "normal". Permanent behavioral deficits have also been documented during adulthood following *in utero* exposure in the monkey. Increased behavioral problems have been observed in teenagers, linked to both lead levels as 6- and 7-year olds. In addition, prospective studies have revealed that behavioral deficits may be better correlated to blood lead levels in past years than to concurrent blood lead levels. The extent to which this effect represents an ongoing elevated lead body burden (e.g., brain or bone stores) or the failure to repair earlier damage is unknown.
3. There are suggestions from the animal data that exposure early in infancy (or prenatally) is not necessary for lead-induced behavioral impairment. In the monkey, exposure beginning at about a year of age (comparable to approximately 3 years in humans) results in robust behavioral deficits. In the rat, reliable behavioral impairment is produced by exposure beginning at weaning. There are no comparable studies in which assessment was performed in children with blood levels that were low at birth and increased following infancy. However, the results from the Boston study linking long-term impairment with blood lead levels at 2 years rather than earlier in development are suggestive that the child continues to be vulnerable after birth and the very early neonatal period.
4. Data from monkeys suggest that the pattern of behavioral impairment may depend on the period (or pattern) of lead exposure. It has been suggested that the results from different studies in children are "inconsistent", based on the fact that effects are observed on different functional domains (different subtests of the intelligence tests). It is likely that differences in pattern of exposure during development, as well as other variables, are responsible for these differences.

5. Exploration of underlying behavioral processes responsible for lead-induced behavioral impairment has revealed congruence between effects in animals and children. A constellation of perseveration, increased distractibility, inability to inhibit inappropriate responding, and inability to adapt to changes in behavioral requirements has been demonstrated in several cohorts of monkeys over a number of behavioral tasks. Perseveration and increased responding has also been demonstrated in the rat. Tests of specific functions in children have revealed perseveration, increased distractibility, inability to inhibit inappropriate responding, and decreased ability to follow sequences of directions in addition to global decrements in IQ. Determination of the underlying behavioral mechanisms responsible for impairment produced by lead represents an important future challenge. It may be that such understanding would suggest methods for teaching children with lead-induced behavioral problems that would attenuate the damage and ameliorate potential long-term sequalae such as school drop-out and increased antisocial behavior.

REFERENCES

Agency for Toxic Substances and Disease Registry (ATSDR). (1988), *The Nature and Extent of Lead Exposures of Children in the U.S.*, Public Health Service, U.S. Dept. of Health and Human Services, Atlanta.
Alfano, D. P. and Petit, T. L. (1982), *Exp. Neurol.*, 75, 275–288.
Beattie, A. D., Moore, M. R., and Goldberg, A. (1975), *Lancet*, 7907, 499–592.
Bellinger, D. C. and Stiles, K. M. (1993), *Neurotoxicology*, 14(2–3), 151–160.
Bellinger, D., Leviton, A., Waternaux, C. et al. (1989), *Int. J. Epidemiol.*, 18, 180–185.
Bellinger, D., Needleman, H. L., Bromfield, R., and Mintz, M. A. (1984), *Biol. Trace Elem. Res.*, 6, 207–223.
Bellinger, D., Sloman, J., Leviton, A. et al. (1991), *Pediatrics*, 87, 219–227.
Bellinger, D. C., Stiles, K. M., and Needleman, H. L. (1992), *Pediatrics*, 90, 855–861.
Bradbury, M. W. B. and Deane, R. (1993), *Neurotoxicology*, 14(2–3), 131–136.
Breen, K. C. and Regan, C. M. (1988), *Toxicology*, 49, 71–76.
Bressler, J. and Goldstein, G. W. (1991), *Biochem. Pharmacol.*, 41, 479–484.
Brown, D. R. (1975), *Toxicol. Appl. Pharmacol.*, 32, 1–10.
Buchtal, F. and Behse, F. (1979), *Br. J. Ind. Med.*, 36, 135–147.
Bull, R. J., McCauley, P. T., Taylor, D. H. et al. (1983), *Neurotoxicology*, 4(1), 1–18.
Bushnell, P. J. and Bowman, R. E. (1979a), *Pharmacol. Biochem. Behav.*, 10, 733–742.
Bushnell, P. J. and Bowman, R. E. (1979b), *J. Toxicol. Environ. Health*, 5, 1015–1023.
Bushnell, P. J. and Bowman, R. E. (1979c), *Neurobehav. Toxicol.*, 1, 207–219.
Byers, R. K. and Lord, E. E. (1943), *Am. J. Dis. Child.*, 66, 471.
Campbell, J. B., Woolley, D. E., Vijayan, V. K. et al. (1982), *Dev. Brain Res.*, 3, 595–612.
Centers for Disease Control (CDC). (1991), *Preventing Lead Poisoning in Young Children*, U.S. Public Health Service, U.S. Dept. of Health and Human Services, Atlanta.
Cohn, J., Cox, C., and Cory-Slechta, D. A. (1993), *Neurotoxicology*, 14, 329–346.
Cory-Slechta, D. A. (1984), in *Advances in Behavioral Pharmacology*, Vol. 4, Thompson, T. and Dews, P., Eds., Academic Press, New York, 211–255.
Cory-Slechta, D. A., Pokora, M. J., and Widzowski, D. V. (1992), *Brain Res.*, 598, 162–172.
Cory-Slechta, D. A., Pokora, M. J., and Widzowski, D. V. (1991), *Neurotoxicology*, 12, 761–776.
Cory-Slechta, D. A. and Thompson, T. (1979), *Toxicol. Appl. Pharmacol.*, 47, 151–159.
Cory-Slechta, D. A., Weiss, B., and Cox, C. (1983), *Toxicol. Appl. Pharmacol.*, 71, 342–352.
Cory-Slechta, D. A., Weiss, B., and Cox, C. (1985), *Toxicol. Appl. Pharmacol.*, 78, 291–299.
David, O., Clark, J., and Voeller, K. (1972), *Lancet*, ii, 900–902.
Davis, J. M. and Svendsgaard, D. J. (1987), *Nature (London)*, 329, 297–300.
de la Burdé, B. and Choate, M. S., Jr. (1972), *J. Pediatr.*, 81(6), 1088–1091.
de la Burdé, B. and Shapiro, I. M. (1975), *Arch. Environ. Health*, 30, 281–284.
Dietrich, K., Succop, P., Berger, O. et al. (1991), *Neurotoxicol. Teratol.*, 13, 203–211.
Ernhart, C. B. (1992), *Reprod. Toxicol.*, 6, 21–40.
Fee, E. (1990), *J. Hist. Med. Allied Sci.*, 45, 570–606.
Feldman, R. G. and White, R. F. (1992), *J. Child Neurol.*, 7, 354–359.
Ferguson, S. A. and Bowman, R. E. (1990), *Neurotoxicol. Teratol.*, 91–97.
Fergusson, D. M., Fergusson, J. E., Horwood, L. J. et al. (1988), *J. Child Psychol. Psychiatr.*, 29(6), 811–824.
Fergusson, D. M., Horwood, L. J., and Lynskey, M. T. (1993), *J. Child Psychol. Psychiat.*, 34, 215–227.
Gilbert, S. G. and Rice, D. C. (1987), *Toxicol. Appl. Pharmacol.*, 91, 484–490.
Goldstein, G. (1992), in *Human Lead Exposure*, Needleman, H. L., Ed., CRC Press, Boca Raton, FL.
Goyer, R. A. (1990), *Environ. Health Perspect.*, 89, 101–105.
Guilarte, T. R. and Miceli, R. C. (1992), *Neurosci. Lett.*, 148, 27–30.
Hamilton, A. (1925), *Industrial Poisons in the United States*, Macmillan, New York.

Hays, S. (1992), in *Human Lead Exposure,* Needleman, H. L., Ed., CRC Press, Boca Raton, FL, pp. 1992.
Holtzman, D., DeVries, C., Nguyen, H. et al. (1984), *Neurotoxicology,* 5, 97–124.
Horn, W. F. and Packard, T. (1985), *J. Educat. Psychol.,* 77, 597–607.
Kostial, K. and Vouk, V. B. (1957), *Br. J. Pharmacol.,* 12, 219–222.
Legare, M. E., Castiglioni, A. J., and Rowles, T. K. (1993), *Neurotoxicology,* 14(1), 77–80.
Levin, E. D. and Bowman, R. E. (1983), *Neurobehav. Toxicol. Teratol.,* 5, 391–394.
Levin, E. D. and Bowman, R. E. (1986), *Neurobehav. Toxicol. Teratol.,* 8, 219–224.
Levin, E. D., Schneider, M. L., Ferguson, S. A., Schantz, S. L., and Bowman, R. E. (1988), *Dev. Psychobiol.,* 21, 371–382.
Lilienthal, H., Winneke, G., Brockhaus, A., and Molik, B. (1986), *Neurobehav. Toxicol. Teratol.,* 8, 265–272.
Lin-Fu, J. S. (1985), in *Dietary and Environmental Lead: Human Health Effects,* Mahaffey, K., Ed., Elsevier Science, New York.
Lorton, D. and Anderson, W. J. (1986a), *Neurobehav. Toxicol. Teratol.,* 8, 45–50.
Lorton, D. and Anderson, W. J. (1986b), *Neurobehav. Toxicol. Teratol.,* 8, 51–59.
MacDiarmid, M. A. and Weaver, V. (1993), *Am. J. Industr. Med.,* 24, 1–9.
Maino, G. (1993), *The Healing Hand.,* Harvard University Press, Cambridge.
Manalis, R. S. and Cooper, G. P. (1973), *Nature,* 143, 354–356.
Matte, T., Landrigan, P. J., and Baker, E. L. (1992), in *Human Lead Exposure,* Needleman, H., Ed., CRC Press, Boca Raton, FL, 155–168.
Mushak, P., Davis, J. M., Crocetti, A. F. et al. (1989), *Environ. Res.,* 50, 11–36.
National Academy of Sciences. (1972), *Airborne Lead in Perspective,* Committee on Medical and Biological Effects of Atmospheric Pollutants, Washington, DC.
Needleman, H. L., Ed. (1992), *Human Lead Exposure,* CRC Press, Boca Raton, FL.
Needleman, H. L. (1993), *Neurotoxicology,* 14(2–3), 161–166.
Needleman, H. L. and Gatsonis, C. A. (1990), *JAMA,* 263(5), 673–678.
Needleman, H. L., Gunnoe, C., Leviton, A., Reed, R., Peresie, H., Maher, C., and Barrett, P. (1979), *N. Engl. J. Med.,* 300, 689–695.
Needleman, H. L. Schell, A., Bellinger, P., Lenton, A., and Alfred, E. N. (1990), *N. Engl. J. Med.,* 322, 83–88.
Oliver, T. (1911), *Br. Med. J.,* 1, 1096–1098.
Olson, L., Björklund, H., Henschen, A. et al. (1984), *Acta Neurol. Scand.,* 70(Suppl. 100), 77–87.
Perino, J. and Ernhart, C. B. (1974), *J. Learn. Dis.,* 7, 616–620.
Perlstein, M. A. and Attala, R. (1966), *Clin. Pediatr.,* 5, 292–298.
Rabinowitz, M. B., Leviton, A., and Bellinger, D. C. (1989), *Bull. Environ. Contam. Toxicol.,* 43, 485–492.
Regan, C. M. (1993), *Neurotoxicology,* 14, 69–76.
Reiss, J. A., and Needleman, H. L. (1992), in *The Vulnerable Brain and Environmental Risk,* Vol. 2, Isaacson, R. L. and Jensen, K. F., Eds., Plenum Press, New York, 111–126.
Reuhl, K. R., Rice, D. C., Gilbert, S. G. et al. (1989), *Toxicol. Appl. Pharmacol.,* 99, 501–509.
Reznikoff, P. and Aub, J. (1947), *J. Arch. Neurol. Psych.,* 17, 444–546.
Rice, D. C. (1984), *Toxicol. Appl. Pharmacol.,* 75, 337–345.
Rice, D. C. (1985a), *Toxicol. Appl. Pharmacol.,* 77, 201–210.
Rice, D. C. (1985b), in *Behavioral Pharmacology: The Current Status,* Seiden, L. S. and Balster, R. L., Eds., Alan R. Liss, New York, 473–486.
Rice, D. C. (1988), *Toxicol. Lett.,* 43, 361–379.
Rice, D. C. (1990), *Toxicol. Appl. Pharmacol.,* 106, 327–333.
Rice, D. C. (1992a), in *Human Lead Exposure,* Needleman, H. L., Ed., CRC Press, Boca Raton, FL, 137–152.
Rice, D. C. (1992b), *Neurotoxicology,* 13, 757–770.
Rice, D. C. (1992c), *Neurobehav. Toxicol., Teratol.,* 14, 235–245.
Rice, D. C. (1992d), *Neurotoxicology,* 13, 583–592.
Rice, D. C. (1993), *Neurotoxicology,* 4, 167–178.
Rice, D. C. and Gilbert, S. G. (1985), *Toxicol. Appl. Pharmacol.,* 80, 421–426.
Rice, D. C. and Gilbert, S. G. (1990a), *Toxicol. Appl. Pharmacol.,* 103, 364–373.
Rice, D. C. and Gilbert, S. G. (1990b), *Toxicol. Appl. Pharmacol.,* 102, 101–109.
Rice, D. C. and Karpinski, K. F. (1988), *Neurotoxicol. Teratol.,* 10, 207–214.
Rice, D. C. and Willes, R. F. (1979), *J. Environ. Pathol. Toxicol.,* 2, 1195–1203.
Rice, D. C., Gilbert, S. G., and Willes, R. F. (1979), *Toxicol. Appl. Pharmacol.,* 51, 503–513.
Rose, S. A. and Wallace, I. F. (1985), *Child Develop.,* 56, 843–852.
Rubin, R. A. and Barlow, B. (1979), *Dev. Psychol.,* 15, 225–227.
Schantz, S. L., Laughlin, N. K., Van Valkenberg, H. C., and Bowman, R. E. (1986), *Neurotoxicology,* 7, 641–654.
Schwartz, J., Landrigan, P. J., Feldman, R. G. et al. (1988), *J. Pediatr.* 112(1), 12–17.
Schwartz, J. (1994), *Environ. Res.,* 65, 42–55.
Schwartz, J. Angle, C., and Pitcher, H. (1986), *Pediatrics,* 77, 281–288.
Seppäläinen, A. N. and Landrigan, P. J. (1988), *Crit. Rev. Toxicol.,* 18(4), 245–298.
Shukla, R., Bornschein, R. L., Dietrick, K. N. et al. (1989), *Pediatrics,* 84, 604–612.
Silbergeld, E. K. (1992), *FASEB J.,* 6, 3201–3206.

Silbergeld, E. K., Fales, J. T., and Goldberg, A. M. (1974), *Nature,* 247, 49–50.
Silbergeld, E. K., Landrigan, P. J., Froines, J. R. et al. (1991), *New Solutions,* 3, 18–28.
Slomianka, L., Rungby, J., West, M. J. et al. (1989), *Neurotoxicology,* 10, 177–190.
Smith, M. (1985), *J. Am. Acad. Child Psychiatry,* 24, 24–32.
Stiles, K. M. and Bellinger, D. C. (1993), *Neurotoxicol. Teratol.,* 15, 27–35.
Ujihara, H. and Albuquerque, E. X. (1992), *J. Pharmacol Exp. Ther.,* 263(2), 868, 875.
Volpe, R. A., Cole, J. F., and Borelko, C. J. (1992) *Environ. Geochem. Health,* 14(4), 133–140.
Weerasuriya, A., Curran, G. L., and Poduslo, J. F. (1990), *Brain Res.,* 517, 1–16.
White, R. F. Diamond, R., Practor, S., Morey, C., and Hu, H. (1993), *Br. J. Ind. Med.,* 50, 613–622.
Widzowski, D. V. and Cory-Slechta, D. A. (1991), *Soc. Neurosci. Abstr.,* 17, 1350.
Windebank, A. J. and Dyck, P. J. (1984), in *Peripheral Neuropathy,* Dyck, P. J., Thomas, P. K., Lambert, E. H., and Bunge, R., Eds., W. B. Saunders, Philadelphia.
Winder, C. and Kitchen, I. (1984), *Prog. Neurobiol.,* 22, 59–87.
Winneke, G. and Kraemer U. (1984), *Neuropsychology,* 11, 195–202.
Winneke, G., Brockhaus, A., Collet, W. et al. (1989), *Neurotoxicol. Teratol.,* 11, 587–592.
Winneke, G., Hrdina, K.-G. and Brockhaus, A. (1982), *Int. Arch. Occup. Environ. Health,* 51, 169–183.
Winneke, G., Kraemer, U., Brockhaus, A. et al. (1983), *Int. Arch. Occup. Environ. Health,* 51, 231–252.
Yule, W., Urbanowicz, M.-A., Lansdown, R. et al. (1984), *Br. J. Dev. Psychol.,* 2, 295–305.

Chapter 40

Effects of Metals on Ion Channels

Toshio Narahashi

I. INTRODUCTION

In order to elucidate the mechanism of action of heavy metals on the nervous system, four major approaches have been used, i.e., behavioral, structural, biochemical/molecular, and functional studies. Behavioral responses of animals to toxicants are sensitive parameters, and have been used to assess the overall toxic effects. However, behavioral studies will not provide us with mechanistic information. Structural changes in various regions of the nervous system have been observed especially in chronic intoxication with certain heavy metals. However, unless combined with biochemical or physiological studies, this approach alone will not provide us with the information about the mechanism of action. Biochemical/molecular approaches are very useful in the sense that the nature of biochemical and molecular lesions caused by toxicants can be identified. However, this line of study alone will not give the answer to the question as to how functional changes in the nervous system take place as a result of intoxication. Functional studies, most of which utilize electrophysiological techniques, will provide information as to how toxicants alter excitability of the nervous system, thereby causing behavioral changes. When combined with the information obtained by biochemical/molecular studies, electrophysiological experiments, especially those utilizing advanced technologies, such as voltage clamp and patch clamp, will constitute a powerful approach to the cellular and molecular mechanisms of action of neurotoxic heavy metals. One of the reasons why progress in the field of neurotoxicology of heavy metals has not occurred as fast as one would hope is that such powerful advanced electrophysiological techniques have not been effectively utilized. It is important to use mammalian neuron preparations which are more closely related to human neurons than invertebrate neurons.

Another important aspect of heavy metal neurotoxicology is the question of acute vs. chronic toxicity. It is well known that significant toxic effects of heavy metals, especially mercury and lead, manifest themselves with a long delay after exposure. It goes without saying that the mechanism of chronic toxicity will be the eventual goal in heavy metal neurotoxicology. However, aside from gross changes such as those in synaptic transmission and conduction velocity, it is extremely difficult to assess changes in receptor/channel functions in the CNS in chronically intoxicated animals, because large fluctuations of data, even in control animals, as obtained by patch clamp techniques from individual neurons will make it very difficult to compare test and control animals in a statistically meaningful manner. Furthermore, without sufficient knowledge of the acute toxic mechanism, it is almost impossible to study the mechanism underlying chronic toxicity.

If we are to pursue the mechanism underlying acute toxicity, then a question arises as to whether the observed acute changes in receptor/channel function are related to changes in chronic conditions. There is obviously no clear-cut answer to this question at present, as we know little about either acute or chronic toxicity. However, there are some data in the literature suggesting the same type of effect or the same target site for both acute and chronic intoxication with heavy metals. For instance, $HgCl_2$ inhibited binding of quinuclidinyl benzilate (QNB) to brain muscarinic acetylcholine (ACh) receptors in acute and chronic intoxication (Abdallah and Shamoo, 1984). Large increases in the frequency of miniature postsynaptic potentials in the guinea pig superior cervical ganglia were observed in both acute and chronic intoxication with methylmercury (Juang and Yonemura, 1975). Interesting results were obtained with snail neurons in which lead suppressed the calcium channel current in acute intoxication and augmented it in chronic intoxication (Audesirk, 1987). In this case, the same target channel is affected, but the direction of modulation changes with time. [^{3}H]Nitrendipine binding to rat brain was increased by lead after both acute and chronic intoxication (Rius et al., 1986). Lead suppressed the activity of Na,K-ATPase in both acute *in vitro* treatment and chronic exposure, and in both types of treatment the retinal enzyme was much more sensitive to lead than the renal enzyme (Fox et al., 1991). These data indicate that the same target site could be affected by a heavy metal in both acute and chronic intoxication.

II. CHEMISTRY OF METALS IN AQUEOUS SOLUTIONS

When dissolved in aqueous solutions containing chloride, many metals including mercury, lead, zinc, and cadmium are not simply ionized but form complexes with chloride and hydroxy groups (Webb, 1966; Hahne and Kroontje, 1973; Rickard and Nriagu, 1978; Gutknecht, 1981; Matthews et al., 1993). For example, mercuric chloride, when dissolved in 100 mM Cl$^-$ solution at pH 8.5, is not ionized into Hg^{2+} and Cl$^-$, but exists largely in four forms, $HgCl_2$, $HgCl_3^-$, $HgCl_4^{2-}$, and $Hg(OH)_2$ in roughly the same proportion. The relative fraction of each form changes with pH and Cl$^-$ concentration. The situation is equally complex for lead. At pH 8.5 and 100 mM Cl$^-$, lead exists largely (98%) as $PbOH^+$, with small fractions of $PbCl^+$ and $PbCl_2$ (1% each). Again the relative fraction changes with pH and Cl$^-$ concentration. Cadmium under this condition exists as $CdCl_2$ (40%), $CdCl^+$ (40%), Cd^{2+} (15%), and $CdCl_3^-$ (5%). Zinc exists mostly (~99%) as $Zn(OH)_2$, and other forms such as Zn^{2+}, $ZnCl_2$, $ZnCl^+$, $ZnCl_3^-$, and $ZnCl_4^{2-}$ are almost negligible. This situation has not received much attention or has even been ignored in many studies dealing with the actions of heavy metals on membranes and channels.

The impact of heavy metals existing in various forms on neurotoxicology is potentially very significant. For example, it is traditionally assumed that methylmercury is toxic to the brain because it is capable of penetrating the blood-brain barrier and that mercuric chloride is primarily kidney poison, as it cannot penetrate the blood-brain barrier, as Hg^{2+} ions. However, since in the normal physiological condition, mercuric chloride is not ionized to form Hg^{2+} and Cl$^-$ but exists in the forms of $HgCl_2$, $HgCl_3^-$, $HgCl_4^{2-}$, and $Hg(OH)_2$, it is likely to be able to cross the blood-brain barrier as $HgCl_2$. From the mechanistic point of view, the situation may require modification, to some extent at least, of the interpretation of data on the interactions of metals with ion channel function. The effects of various divalent and trivalent cations on channel gating have traditionally been interpreted on the basis of the assumption that these cations are totally ionized. This may not be the case, and the concentration of ionized cations may be much less than assumed. Furthermore, there are no data to document that only the ionized polyvalent cation forms are active.

III. MERCURY

A. MECHANISMS OF ACTION OF MERCURY ON NEURONAL RECEPTORS/CHANNELS

A wealth of data has been accumulated regarding the effects of mercury on neuromuscular and peripheral synaptic transmission. The popularity of these preparations was due to the fact that intracellular microelectrode techniques were easily applicable to them before patch clamp techniques were developed. Methylmercury and mercuric chloride increased the frequency of miniature end-plate potentials (EPPs) (Juang, 1976). The same effects were obtained with the guinea pig ganglia (Juang and Yonemura, 1975). Mercuric chloride was also found to increase the amplitude of EPP prior to its depression (Cooper et al., 1984). A series of experiments with rat neuromuscular preparations have led to the conclusions that methylmercury irreversibly alters presynaptic function (Atchison and Narahashi, 1982); that the decrease in EPP amplitude is due to block of calcium entry to the nerve terminals (Atchison et al., 1986; Traxinger and Atchison, 1987); and that the increase in MEPP frequency is due to intracellular release of Ca^{2+} from stores such as mitochondria (Atchison, 1986, 1987; Traxinger and Atchison, 1987; Levesque and Atchison, 1987, 1988; Hare et al., 1993). Mercuric chloride gains access to intracellular site via sodium and calcium channels, whereas methylmercury does so via the nerve membrane lipid phase (Miyamoto, 1983).

In contrast to the extensive studies performed with mercury-intoxicated neuromuscular preparations, only a limited amount of data is available for the action of mercury on ion channels. Sodium and potassium currents were suppressed by methylmercury in squid axons and neuroblastoma cells (Quandt et al., 1982; Shrivastav et al., 1976). $^{45}Ca^{2+}$ influx into synaptosome and PC12 cells was inhibited by methylmercury (Shafer et al., 1990). N-type and L-type calcium channels of PC12 cells were blocked by low concentrations of methylmercury, 10 μM methylmercury causing 77% and 70% block of N- and L-type, respectively (Shafer and Atchison, 1991). Recent patch clamp experiments have shown that mercuric chloride blocks the L/N-type calcium channels of rat DRG neurons with an IC_{50} of 1.1 μM, the T-type channels of rat DRG neurons in the same concentration range of 0.5 to 2 μM, and the calcium channels of Aplysia neurons at 5 to 50 μM (Pekel et al., 1993). Thus the calcium channels of mammalian neurons are highly sensitive to the blocking action of mercuric chloride and methylmercury and appear to be one of the important target sites. Kainate-activated currents were suppressed by low concentrations of Hg^{2+} (K_i = 70 nM) (Kiskin et al., 1986; Umbach and Gundersen, 1989). Leakage current was increased by mercuric chloride and methylmercury in various preparations, leading to a membrane depolarization (Quandt et al., 1982; Shrivastav et al., 1976). A drastic increase in GABA-activated chloride currents by low concentrations of mercuric chloride has recently been observed (Arakawa et al., 1991), as will be described later.

Binding experiments have shown some interactions of mercury with receptors. QNB or pilocarpine binding to the muscarinic ACh receptors was inhibited by mercuric chloride, but methylmercury was much less potent (Abdallah and Shamoo, 1984; Von Burg et al., 1980). Nicotine or ACh binding to the nicotinic ACh receptors was also inhibited by methylmercury (Eldefrawi et al., 1977). In keeping with these binding data, the responses of a nicotinic ACh receptor and two kinds of muscarinic ACh receptors were suppressed by methylmercury (Quandt et al., 1982).

Release of neurotransmitters other than ACh has also been shown to be affected by mercury. Spontaneous release of dopamine was augmented by mercuric chloride and methylmercury, whereas K-evoked dopamine release was increased or decreased, depending on the kind of mercury and the Ca^{2+} concentration (McKay et al., 1986). Spontaneous release of norepinephrine, 5-hydroxytryptamine, and dopamine was stimulated by methylmercury (Komulainen and Tuomisto, 1981).

Methylmercury is known to be converted to inorganic mercury in animals through biotransformation (Evans et al., 1977; Neville and Berlin, 1974; Norseth and Clarkson, 1970a, b). Although some studies indicated low levels (1 to 6%) of inorganic mercury relative to the total mercury in the brain after administration of methylmercury (Omata et al., 1980; Komulainen, 1988), recent studies with humans (the Minamata case) and monkeys have clearly indicated that the percentages of inorganic mercury out of the total mercury in the brain are very high, ranging from 72% to 88% (Friberg and Mottet, 1989). Daily oral administration of subtoxic methylmercury in doses of 50 and 90 µg Hg/kg b.w./day to monkeys for 0.5 to 1.5 years resulted in the accumulation of inorganic mercury in the brain in an amount as much as 10 to 33% and 90% of the total mercury, respectively (Lind et al., 1988). This is partly because inorganic mercury formed in the brain as a result of biotransformation from methylmercury has a very long half-life extending several years. It has also been shown that mercuric chloride is highly permeant to planar lipid bilayer membranes due to its unchanged form, $HgCl_2$, at pH 7.0 and chloride concentrations of 10 to 1000 mM (Gutknecht, 1981). These aspects are important, as mercuric chloride is more potent than methylmercury on various receptors and channels (Abd-Elfattah and Shamoo, 1981; Arakawa et al., 1991; Von Burg et al., 1980).

Experiments reported in the literature were performed in many cases using relatively high concentrations (10 to 100 μM) of mercury compounds. Mercury concentration in the brain of patients chronically intoxicated with methylmercury in the Minamata case was estimated to be 1.7 to 26 μM (Friberg and Mottet, 1989). In the monkeys chronically intoxicated with methylmercury, clinical signs comparable to humans appeared when the blood mercury level exceeded 14 μM. As will be described later, we have found that $GABA_A$-mediated chloride currents are greatly augmented by low concentrations (~1 μM) of mercuric chloride. Methylmercury (100 μM) has an opposite effect, decreasing the current. Since methylmercury is known to be effectively biotransformed to inorganic mercury in the brain as described above, the modulation of the GABA system by mercuric chloride appears to play an important role in intoxication with methylmercury. Methylmercury was found in our previous study (Quandt et al., 1982) to suppress the muscarinic and nicotinic ACh receptor-channel currents, but the action was not potent.

The EAA receptor-channel complex is another potential target site of mercury, because mercuric chloride is known to block the kainate-induced currents with a K_i of 70 nM (Umbach and Gundersen, 1989).

B. MERCURY MODULATION OF GABA-ACTIVATED CHLORIDE CHANNELS AND NONSPECIFIC CATION CHANNELS

We have found that the GABA-activated chloride channel is modulated by mercuric chloride in a highly potent and efficacious manner (Arakawa et al., 1991). It was also found that mercuric chloride and methylmercury generated a slow inward current at high concentrations. Experiments were performed with neurons dissociated from the dorsal root ganglion (DRG) of newborn rats (1 to 5 days postnatal) and maintained in culture. The neurons cultured for 1 to 5 days were used for experiments. Ionic currents were recorded by a whole-cell variation of patch clamp techniques (Hamill et al., 1981).

Bath application of GABA generated an inward Cl⁻ current in a dose-dependent manner with an EC_{50} value of about 60 μM. At a concentration of 10 μM or higher, the current showed an initial peak which decayed to a smaller steady-state level due to desensitization. The effect of $HgCl_2$ on the nondesensitized peak current induced by GABA was examined. In the presence or absence of $HgCl_2$, 30 μM GABA was applied for a brief period of time (~5 s) at an interval of 2 to 3 min, which was long enough for the GABA system to recover from desensitization. An example of such an experiment is illustrated in Figure 1 in which 10 μM $HgCl_2$ generated a slow inward current and greatly enhanced GABA-induced current. Both effects were irreversible after washing with $HgCl_2$-free solution for 6 to 8 min. Slow inward current was evoked by $HgCl_2$ at a concentration of 1 μM or higher, but its amplitude varied greatly in different cells. $HgCl_2$ was potent in enhancing GABA-induced currents in a dose-dependent manner (0.1 to 100 μM, Figure 2).

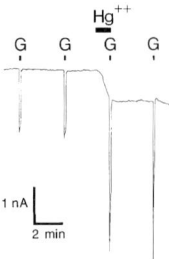

Figure 1. $HgCl_2$ (10 μM) generates a slow inward current and enhances the peak current induced by 30 μM GABA (G) in a rat DRG neuron. Both effects are irreversible after washing with $HgCl_2$-free solution. GABA and $HgCl_2$ were applied as indicated by bars above the record. (From Arakawa, O., Nakahiro, M., and Narahashi, T., *Brain Res.*, 551, 58–63, 1991. With permission.)

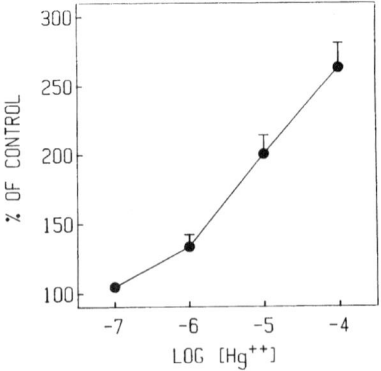

Figure 2. Dose-response relationship for $HgCl_2$-induced enhancement of the peak current evoked by 30 μM GABA in rat DRG neurons. Data are given as mean ± S.D. (n = 4–5). At 0.1 μM, S.D. is smaller than the size of the symbol. (From Arakawa, O., Nakahiro, M., and Narahashi, T., *Brain Res.*, 551, 58–63, 1991. With permission.)

Methylmercury (100 µM) also generated a slow current similar to that produced by HgCl$_2$, but suppressed GABA-induced current (Figure 3). The amplitude was decreased to 82.4 ± 6.1% of control (n = 4).

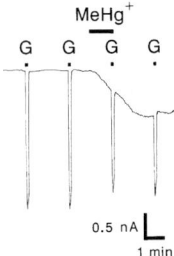

Figure 3. Methylmercury (100 µM) generates a slow inward current and suppresses the peak current induced by 30 µM GABA in a rat DRG neuron. GABA (G) and methylmercury were applied as indicated by bars above the record. (From Arakawa, O., Nakahiro, M., and Narahashi, T., *Brain Res.*, 551, 58–63, 1991. With permission.)

Pharmacological characteristics of slow inward current were examined using several drugs known to exert specific effects on channels. HgCl$_2$ (10 µM) generated slow inward current in the presence of 30 µM bicuculline (BIC), which inhibited GABA-induced current almost completely both in the presence and absence of HgCl$_2$. Therefore, generation of slow inward current is not due to the interaction of HgCl$_2$ with the GABA binding site. Tetrodotoxin (TTX, 1 µM) also failed to abolish slow inward current. Thus, no voltage-activated sodium channels are involved. Like bicuculline, picrotoxin (PTX) did not prevent HgCl$_2$ from generating slow inward current and abolished GABA-induced current.

Figure 4 illustrates the effect of La^{3+}. In this experiment, depolarizing 0.5 s pulses to 0 mV were applied every 2 min to monitor the voltage-activated Ca^{2+} current (arrows). HgCl$_2$ (10 µM) again generated slow inward current in the presence of 10 µM La^{3+}. At this concentration, La^{3+} markedly inhibited inward Ca^{2+} currents induced by the depolarizing pulses, and slightly enhanced GABA-induced current, the amplitude being increased to 113.6 ± 1.2% of control (n = 3). This La^{3+} augmentation accounts for the slight decrease in HgCl$_2$-augmented GABA-induced current following washout of La^{3+}. The voltage-activated calcium channels are not the site of slow inward current generation caused by HgCl$_2$.

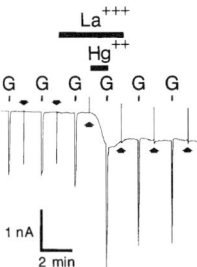

Figure 4. La^{3+} (10 µM) fails to inhibit the slow inward current induced by 10 µM HgCl$_2$ in a rat DRG neuron. La^{3+} was applied as indicated by a bar above the records. Depolarizing 0.5-s pulses to 0 mV were applied at arrows to monitor the voltage-activated Ca^{2+} currents which are distorted in shape and direction due to their fast time course. (From Arakawa, O., Nakahiro, M., and Narahashi, T., *Brain Res.*, 551, 58–63, 1991. With permission.)

In order to determine the ion species contributing to slow inward current, the reversal potential for slow inward current was measured with different internal Cl$^-$ concentrations, one containing 142 mM (normal) and the other containing 20 mM Cl$^-$. In the latter, 122 mM CsCl was replaced by the equimolar amount of cesium glutamate. Current-voltage relationships for slow inward current at two different internal Cl$^-$ are shown in Figure 5. Slow inward current became zero at 9.1 mV with 142 mM internal Cl$^-$ (●), and at 1.5 mV with 20 mM internal Cl$^-$ (◆). The average reversal potentials for slow inward current with normal and low internal Cl$^-$ were 10.6 ± 1.2 mV (n = 4) and 2.2 ± 1.5 mV (n = 4), respectively. Thus the reversal potential was shifted in the negative direction when internal Cl$^-$ was

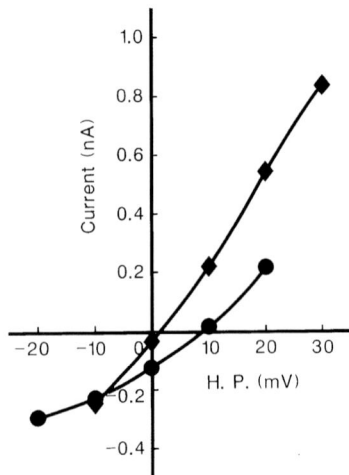

Figure 5. Current-voltage relationship for the slow inward current induced by 10 μM HgCl$_2$ at two different internal Cl$^-$ concentrations, one at 142 mM (●) and the other at 20 mM (♦) in rat DRG neurons. (From Arakawa, O., Nakahiro, M., and Narahashi, T., *Brain Res.*, 551, 58–63, 1991. With permission.)

reduced. However, the magnitude of the shift (8 mV) was much less than that predicted by the Nernst potential for Cl$^-$ (50 mV), indicating that contribution of Cl$^-$ to slow inward current is small, and that an ion or ions other than Cl$^-$ contribute to slow inward current.

The slow current is not mediated by the GABA-activated chloride channels, nor by voltage-activated sodium or calcium channels. The possible contribution of potassium channels can also be excluded because neither external nor internal solution contained K$^+$. Polyvalent cations such as Pb^{2+}, Cd^{2+}, Ca^{2+}, and Al^{3+} have been shown to generate slow inward currents via nonspecific cation channels (Oortgiesen et al., 1990a,b; Weinreich and Wonderlin, 1987). It appears that the mercury-induced slow inward current is responsible for the membrane depolarization and the decrease in membrane resistance observed in the nerve and muscle membranes exposed to mercury compounds (Juang, 1976; Quandt et al., 1982; Shrivastav et al., 1976).

IV. LEAD

A. MECHANISMS OF ACTION OF LEAD ON NEURONAL RECEPTORS/CHANNELS

There are some similarities between mercury and lead on their effects on neuromuscular transmission. Lead also increased the frequency of MEPPs and decreased the amplitude of EPP in frogs and rats (Atchison and Narahashi, 1984; Cooper et al., 1984; Kolton and Yaari, 1982; Suszkiw et al., 1984). The stimulation of MEPP was ascribed to an increase in intracellular Ca^{2+} concentration in the presynaptic terminals as a result of disruption of the Ca^{2+} sequestering activity of mitochondria and/or intraterminal organelles (Cooper et al., 1984; Minnema et al., 1988; Suszkiw et al., 1984). The EPP suppression appears to be due to inhibition of calcium entry to the nerve terminals as a result of Pb^{2+}–Ca^{2+} antagonism (Atchison and Narahashi, 1984; Cooper and Manalis, 1984). The end-plate current suppressed by 1 μM Pb^{2+} recovered after continuous exposure for 30 min (Oortgiesen et al., 1990c).

Calcium channels have indeed been shown to be blocked reversibly by lead at micromolar concentrations in recent patch clamp experiments with neuroblastoma cells (Audesirk and Audesirk, 1993; Oortgiesen et al., 1990a, 1993; Reuveny and Narahashi, 1991). In contrast, certain types of neurons in snails were blocked irreversibly by lead even at nanomolar concentrations (Audesirk, 1987; Audesirk and Audesirk, 1989; Büsselberg et al., 1991).

In rat DRG neurons, IC$_{50}$ values for lead block of the L-type, L/N-type, and T-type calcium channel currents are estimated to be 1.03 μM, 0.64 μM, and 6 μM, respectively (Büsselberg et al., 1993). Sodium and potassium channel currents are not affected by lead at 1 μM or 200 μM. Since lead exists in various forms in aqueous solutions, it was emphasized that free Pb^{2+} concentrations are important in interpreting the data (Audesirk, 1993). When measured with a calibrated Pb^{2+}-selective electrode, the IC$_{50}$ values of free Pb^{2+} to block calcium channel currents are: in N1E-115 neuroblastoma cells, ~700 nM and ~1300 nM for the L-type and T-type calcium channels, respectively; and in E18 rat hippocampal neurons, ~30 nM (in 10 mM Ba^{2+}) and ~55 nM (in 50 mM Ba^{2+}) for the L-type calcium channels, and ~80 nM (in 10 mM Ba^{2+}) and ~200 nM (in 50 mM Ba^{2+}) for the N-type calcium channels (Audesirk and Audesirk, 1993). Although the acceptable blood level of lead is now set at 10 μg/dl (~0.5 μM), most of the lead in whole blood is in the blood cells. In fact, total lead in plasma of humans not exposed to lead was estimated to be 30 nM, and total lead in cerebrospinal fluid was 25 nM (Cavallin et al., 1984). Most of the total lead in the plasma is bound to proteins and anions, and Al-Modhefer et al. (1991) measured free Pb^{2+} in human serum and found it very low indeed, in the order of picomolar. However, these arguments are based on the assumption that lead acts on channels in its divalent cation form. Since lead exists mostly in bound forms other than Pb^{2+} in aqueous solutions mimicking serum or physiological saline solutions, the assumption is unlikely to be justified.

Calcium-activated potassium channels were also suppressed by lead (Oortgiesen et al., 1993). Low conductance channels were more sensitive to lead than high conductance channels, with IC$_{50}$ values of < 1 μM and > 1 μM, respectively. 5-Hydroxytryptamine (5-HT)-activated currents in neuroblastoma cells were also blocked by lead but in less potent manner, with an IC$_{50}$ of 49 μM (Oortgiesen et al., 1990a). ACh-activated currents (nicotinic) of neuroblastoma cells have been found very sensitive to lead with an IC$_{50}$ of 19 nM (Oortgiesen et al., 1990a). However, the block was relieved with increasing concentration of lead with an EC$_{50}$ of 21 μM resulting in a bell-shaped dose-dependent curve (Oortgiesen et al., 1990a). This experiment points to the importance of testing chemicals over a wide range of concentrations. Contrary to the neuronal nicotinic ACh receptors/channels, those at the frog end-plate were not affected by 10 μM Pb^{2+}, with some decrease only at 100 μM (Manalis and Cooper, 1973). Thus, one cannot necessarily extrapolate the data on the muscle ACh receptors/channels to the CNS. Voltage-activated sodium and potassium channels of neuroblastoma cells were not affected by lead at 10 to 100 μM (Oortgiesen et al., 1990a).

Excitatory amino acid (EAA)-activated channels of hippocampal neurons have been found to be blocked by lead (Alkondon et al., 1990). The currents evoked by N-methyl-d-aspartate (NMDA) were suppressed reversibly by lead with an IC$_{50}$ of 10 μM, whereas those evoked by quisqualate or kainate were only slightly suppressed by 50 μM Pb^{2+}. The frequency of openings of NMDA-activated channels was decreased by lead, and the block was not voltage dependent and not antagonized by glycine, indicating that the glycine site was not involved. Binding of [^{3}H]MK-801 was decreased by lead with an IC$_{50}$ of 7 μM. Lead appears to bind to the MK-801 binding site in a manner similar to Zn^{2+}, but with a higher potency. Our preliminary patch clamp studies of EAA-activated channels have shown that lead at 10 μM suppressed the NMDA-induced currents by approximately 35% in a reversible manner, and at 1 μM decreased the currents only slightly (~5%).

Binding of QNB to the muscarinic ACh receptors of rat brain was inhibited by lead at ~50 μM (Aronstam and Eldefrawi, 1979). In keeping with the electrophysiological data, K-stimulated ^{45}Ca^{2+} influx to rat brain synaptosomes was inhibited by lead with a K$_i$ of 1.1 μM (Cooper et al., 1984; Suszkiw et al., 1984). As might be expected from these calcium data, release of ACh was suppressed by lead with a K$_i$ at 16 μM (Suszkiw et al., 1984).

Like studies of mercury, many experiments in the literature were performed with high concentrations (10 to 100 μM) of lead. However, hyperactivity of chronically intoxicated children and experimental animals was observed at the low blood levels of lead ranging from 1.9 μM to 3.8 μM (David, 1974; Silbergeld and Goldberg, 1974). The blood levels of lead in children exhibiting neuropsychological disorders were estimated to be 1.5 to 2.5 μM (Needleman et al., 1979).

The modulation of calcium channels, GABA-activated channels and EAA-activated channels by mercury and lead has a direct bearing on acute and chronic symptoms of poisoning in humans and experimental animals. These complex behavioral changes appear to be the result of a series of cellular and subcellular alterations. For example, block of voltage-activated calcium channels by lead or mercury will affect the intracellular level of calcium, which in turn changes many calcium-dependent processes such as second messengers, transmitter and hormone release, and the activity of certain enzymes. These secondary and tertiary changes will influence the activity of other receptor/channel systems (Pounds, 1984; Pounds and Rosen, 1988). Inasmuch as lead and mercury block the calcium channels with K$_i$ values of 1 μM (Reuveny and Narahashi, 1991) and <10 μM (Shafer and Atchison, 1991), respectively, the toxicological significance speaks for itself. Similar arguments apply to the modulation of GABA- and EAA-activated channels by mercury and lead.

Lead has been found to suppress the chloride current evoked by GABA in the rat DRG neurons. However, the potency was not very high; Pb at 10 μM, 100 μM, and 1 mM blocked the current by 4, 19, and 65%, respectively (Ma and Narahashi, 1993a). The effect was reversible after washing with drug-free media and is comparable with that of some other divalent cations as described below. Lead is also known to block the serotonin (5-HT)-induced current in mouse neuroblastoma cells with a K_i of 49 μM (Oortgiesen et al., 1990a). We have been able to confirm this effect in our preliminary experiments: 1, 10, and 100 μM Pb^{2+} suppressed the 5-HT-induced current by 0, 3, and 66%, respectively.

B. BLOCKING ACTION OF LEAD ON VOLTAGE-ACTIVATED CALCIUM CHANNELS IN HUMAN NEUROBLASTOMA CELLS

We have found that two distinct types of voltage-activated calcium channels in human neuroblastoma cells are blocked by lead in a potent and reversible manner (Reuveny and Narahashi, 1991). SH-SY5Y cells (Ross et al., 1983) were kindly provided by Dr. J. A. Biedler. Cells were grown at 37°C with 95% air/5% CO_2 in monolayer. Growth medium consisted of a 1:1 mixture of modified Eagle medium and nutrient supplement F12 with 10% fetal calf serum containing 40 μg/ml gentamicin. Dibutyryl cyclic adenosine monophosphate (dBcAMP) (1 mM) was used to induce morphological and physiological differentiation of confluent SH-SY5Y cells.

Voltage-activated calcium channel currents carried by barium ions were elicited by a step depolarization from a holding potential of –80 mV to +40 mV for 500 ms. The current trace consisted of two components, a transient current which decayed with a time constant of ~100 ms (N-type) and a long-lasting current which decayed with a time constant of ~1000 ms (L-type) (Reuveny and Narahashi, 1993). The peak amplitude of the current consisted of the N- and L-type components, and the current amplitude at the end of depolarizing pulse (~480 ms) mainly consisted of the L-type component. After application of lead acetate (1 to 30 μM), a rapid decrease in current amplitude was observed (Figure 6). Concentration-response relationships for lead block of two types of calcium channel currents are shown in Figure 7. The inhibition of calcium channel currents by lead was concentration dependent. Both types of calcium channels were almost equally affected by lead. The K_i values for both types were estimated to be ~1 μM.

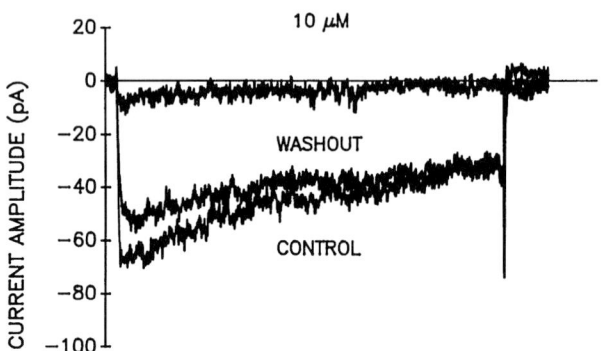

Figure 6. Calcium channel currents recorded from human neuroblastoma SH-SY5Y cells before (control) and 5 min after application of 10 μM lead, and 5 min after washing with lead-free solution. (From Reuveny E., and Narahashi, T., *Brain Res.*, 545, 312–314, 1991. With permission.)

Sodium currents were elicited by a step depolarization from a holding potential of –100 mV to +10 mV for 10 ms (Figure 8A). Potassium currents were evoked by a step depolarization from a holding potential of –80 mV to +50 mV (Figure 8B). Lead (10 μM) had no effect on the voltage-activated sodium channel currents in two cells tested. The minute reduction seen in Figure 8A was not due to block of sodium channels by lead, but due to rundown of the preparation. Moreover, lead did not shift the voltage dependence of activation. The effect of lead on the steady-state amplitude of potassium channel currents was negligible. In the experiment shown in Figure 8B, a small decrease in current amplitude was observed, but in two other cells, lead was without effect on the amplitude of the steady-state current. However, the rising phase of potassium current tended to be slowed slightly in the presence of lead in all three cases.

Lead was more potent on the N- and L-types of calcium channels of human neuroblastoma cells (K_i ~1 μM) than on the T-type of calcium channels of mouse neuroblastoma cells (K_i = 4.8 μM) (Oortgiesen et al., 1990a). Our results are in contrast to previous observations on the effect of lead on molluscan neurons in which the block by lead was neither concentration dependent nor reversible (Audesirk, 1987).

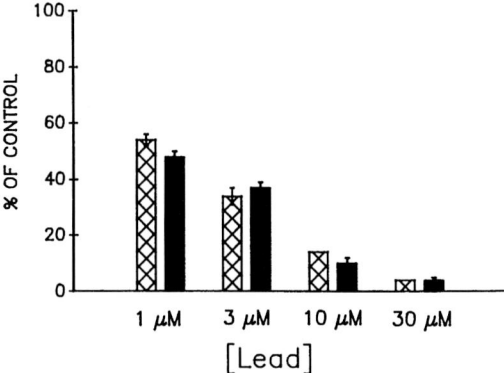

Figure 7. Concentration-response relationships for lead-induced block of two types of calcium channels in SH-SY5Y cells. Current amplitudes at the peak of current (hatched bars) and at the end of depolarization pulse (filled bars) are expressed as percent of control (mean ± S.E.M., n = 4). (From Reuveny, E. and Narahashi, T., *Brain Res.*, 545, 312–314, 1991. With permission.)

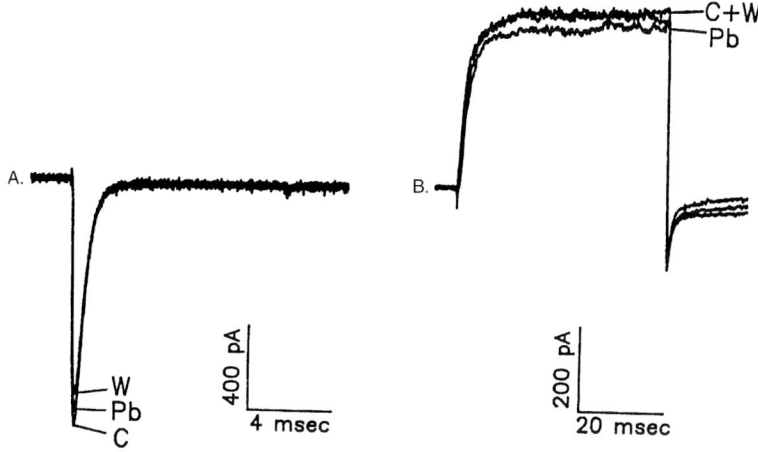

Figure 8. The effect of lead on voltage-activated sodium (**A**) and delayed rectifier potassium channels (**B**) in SH-SY5Y cells. Sodium and potassium channel currents were recorded in control solution (C), after exposure to 10 μM lead (Pb), and after washout with lead-free solution (W). (From Reuveny, E. and Narahashi, T., *Brain Res.*, 545, 312–314, 1991. With permission.)

Lead is more potent in blocking calcium channels in SH-SY5Y cells than in invertebrate neurons (K_i ~90 μM) (Büsselberg et al., 1990) and is almost equipotent to the inhibition of ^{45}Ca uptake into rat synaptosomes (50%, <1 μM) (Nachshen, 1984). The lack of 10 μM lead to significantly reduce sodium and potassium currents indicates that lead has a specific affinity for calcium channels. The concentrations of lead used in this study are in the range of blood levels in children exhibiting neuropsychological disorders (30 to 50 μg/100 ml or 1.5 to 2.5 μM) (Needleman et al., 1979). Block of the two types of calcium channels by lead is expected to lead to disturbances of synaptic transmission in the central nervous system.

V. EFFECTS OF VARIOUS METALS ON GABA RECEPTORS/CHANNELS

A variety of metals other than mercury and lead have been found to affect the GABA-activated chloride channel current. Experiments were performed with rat DRG neurons using the whole-cell patch clamp technique (Ma and Narahashi, 1993a). Most metals were tested at a relatively high concentration of 1 mM. Several divalent metals suppressed the chloride current in the following order of potency, with the percentage of suppression by 1 mM test metals given in parenthesis: (Zn (97%) > Cu (94%) > Cd (83%) > Ni (80%) > Co (70%) > Pb (65%) > Ag (50%) > Al (23%) > Mn (5%). Two polyvalent metals, mercury and lanthanum, augmented the chloride current. The mechanism of action of La^{3+} to augment the GABA-induced current will be described later.

VI. MECHANISM OF ACTION OF LANTHANIDES, ZINC AND COPPER ON GABA RECEPTORS/CHANNELS

Lanthanides, or rare earth metals, comprise a series of 15 metals starting with lanthanum (atomic number 57) and ending with lutetium (atomic number 71). Lanthanides have been used extensively in industry, ophthalmic and camera lenses, petroleum cracking, nuclear reactors, television tubes, and mirrors, to mention a few applications. Some lanthanides were used in the treatment of tuberculosis, as anticoagulant agents, and as antinausea agents, and their potential uses in other areas of medicine are being explored, including inflammation, arthritis, tumor scanning, diagnosis of bone disease, and atherosclerosis.

Symptoms of acute lanthanide intoxication include defecation, writhing, ataxia, sedation, and labored respiration. Also included are pulmonary edema, hyperemia, liver edema and necrosis, pleural effusion, and granulomatous peritonitis (Luckey and Venugopal, 1977). The mechanisms of action of lanthanides at the cellular and molecular level have not been studied extensively until recently. We have recently demonstrated that lanthanum augments GABA-induced current (Arakawa et al., 1991). This unique action was pursued to elucidate the site and mechanism of action (Ma and Narahashi, 1993a,b; Ma et al., 1994).

Lanthanum reversibly potentiated the GABA-induced current in rat DRG neurons in primary culture (Figure 9A; Ma and Narahashi, 1993a). When co-applied with 30 μM GABA, La^{3+} potentiated the current with an EC_{50} of 231 μM and a Hill coefficient of 0.94 (Figure 9B). Dose-response analyses of GABA induction of current indicated that 100 μM La^{3+} shifted the GABA dose-response curve in the direction of lower concentrations of GABA (Figure 9C). La^{3+} potentiation of current was almost voltage independent, indicating that it bound to a receptor-channel site near the external surface.

The $GABA_A$ receptor-chloride channel complex is known to be endowed with three allosteric binding sites, i.e., the benzodiazepine site, the barbiturate site, and the picrotoxin (PTX) site. La^{3+} has been demonstrated not to bind any of the allosteric sites through a series of competition experiments (Ma and Narahashi, 1993a). One such example of an experiment is shown in Figure 10. Pentobarbital at 100 μM increased the GABA-induced current to about 750% of control. La^{3+} at 100 μM augmented the GABA-induced current to 130% of control, and the degree of augmentation was not changed by the presence of 30 and 100 μM pentobarbital. Therefore, La^{3+} does not seem to compete with pentobarbital for the barbiturate site. Similar results were obtained with chlordiazepoxide and picrotoxin.

Zinc and copper have been shown to suppress the GABA-induced chloride current (Ma and Narahashi, 1993a; Narahashi et al., 1994). An example of such an experiment with Cu^{2+} is shown in Figure 11A. Cu^{2+} at 15 μM suppressed the current to approximately 50% of control. The dose-response analysis indicated an EC_{50} of 16 μM and a Hill coefficient of 0.89 (Figure 11B). The Cu^{2+} block was voltage independent. Zn^{2+} exerted almost the same effect as Cu^{2+} on the GABA-induced current with an EC_{50} of 19 μM and a Hill coefficient of 0.82.

Cu^{2+} competes with Zn^{2+} but not with La^{3+} for binding sites (Ma and Narahashi, 1993a). Zn^{2+} at 10 μM suppressed the GABA-induced current to 67% of control. In the presence of increasing concentrations of Cu^{2+} from 10 to 300 μM, the degree of Zn^{2+}-induced suppression was reduced (Figure 12A). However, the situation was different for La^{3+}. La^{3+} block was not significantly influenced by the presence of 10 μM Cu^{2+} (Figure 12B) or 15 μM Zn^{2+} (Figure 12C). It was concluded that Cu^{2+} and Zn^{2+} bind to a negative modulation site, while La^{3+} binds to a positive modulation site, both of which are located near the external surface of the GABA receptor-channel complex.

Lanthanum and six other lanthanides tested enhanced the GABA responses and produced inward currents by themselves (Ma and Narahashi, 1993b). The rank order of efficacies of lanthanides to potentiate the GABA responses and to generate currents correlated inversely with the hydrated ionic radii of these ions (Figure 13). Lanthanides at 1 mM reversibly potentiated the GABA responses (times of control): lutetium (Lu^{3+}) 12.4 ± 0.1 > erbium (Er^{3+}) 11.4 ± 0.2 > terbium (Tb^{3+}) 9.3 ± 0.5 > europium (Eu^{3+}) 5.4 ± 0.4 > neodymium (Nd^{3+}) 4.1 ± 0.2 > cerium (Ce^{3+}) 3.3 ± 0.2 > lanthanum (La^{3+}) 2.4 ± 0.2 (n = 10). The enhancing effect of Tb^{3+} was dose dependent and weakly voltage dependent, increasing with hyperpolarization. The amplitudes of currents induced by various lanthanides (1 mM) were (percentage of 10 μM GABA-induced current) Lu^{3+} 101 ± 8 > Er^{3+} 57 ± 8 > Tb^3 + 33 ± 3 > Eu^{3+} 20 ± 2 > Nd^{3+} 14 ± 2 ≈ Ce^{3+} 13 ± 0.6 > La^{3+} 10 ± 0.2 (n = 4) (Figure 14). The Tb^{3+}-induced currents, which were not voltage dependent, were reversed in polarity at the chloride equilibrium potential. Both pentobarbital

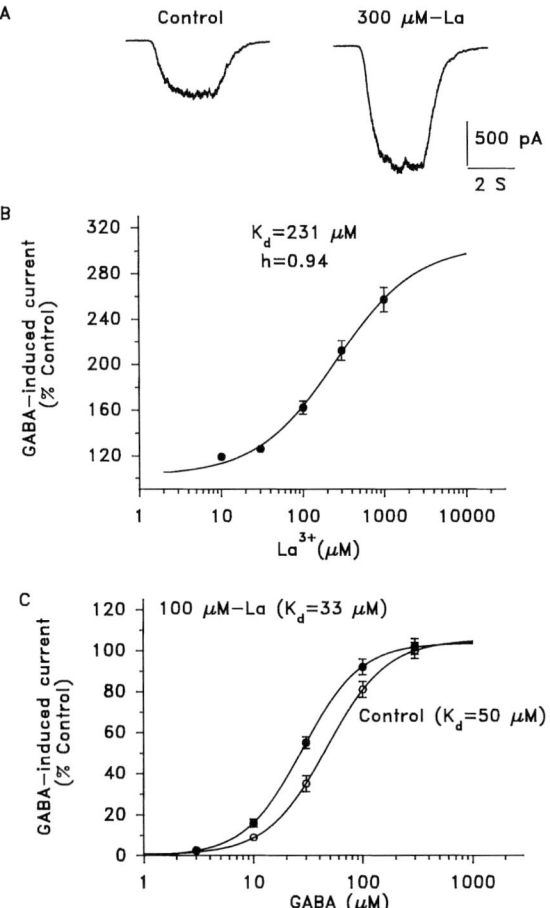

Figure 9. La^{3+} potentiation of GABA-induced currents in cultured DRG neurons. (**A**) enhancement by 300 μM La^{3+} of the current induced by 10 μM GABA at holding potential of −60 mV. (**B**) the dose-response relationship for La^{3+} potentiation of the current induced by 30 μM GABA (n = 4). The current amplitude is given as the percentage of control recorded in the absence of La^{3+}. (**C**) shift of the dose-response curve for GABA-induced current by 100 μM La^{3+} in the direction of low concentrations without changing the maximum response (n = 4). The current amplitude is given as the percentage of current induced by 300 μM GABA in the absence of La^{3+}. (From Ma, J.Y. and Narahashi, T., *Brain Res.*, 607, 222–232, 1993. With permission.)

(100 μM) and chlordiazepoxide (50 μM) potentiated both GABA- and Tb^{3+}-induced currents equally. The GABA receptor-channel antagonists bicuculline (10 μM), picrotoxin (10 μM), penicillin (500 μM), and Zn^{2+} (20 and 100 μM) all suppressed the Tb^{3+}-induced current. It is suggested that there is a distinct binding site on the GABA receptor-channel complex for the lanthanides and, at high concentrations, lanthanides may act on the GABA site or some other site to open the GABA-gated chloride channels.

The above study has been extended to the single-channel level (Ma et al., 1994). In outside-out membrane patches held at a holding potential of −60 mV, channel openings were rarely observed in the absence of GABA. Following application of 10 μM GABA, inward single-channel currents occurred singly and in the form of bursts (Figure 15A). With increasing time resolution, it is clearly seen that the channel openings induced by GABA are brief and occur in isolation and in groups separated by brief closures (Figure 15A). Tb^{3+} at 100 μM increased the single-channel activity induced by 10 μM GABA (Figure 15B). At least two different current amplitudes were commonly recorded, and the histogram of current amplitude (0.05 pA bin width) was best fitted by two Gaussian functions (Figure 16A). The larger current amplitudes were recorded much more frequently (about 70 to 80% of events) than the smaller ones (about 20 to 30% of events). The current amplitudes and the relative proportions of the main- and subconductance-state currents with respect to the number of events were unchanged by Tb^{3+} (Figure 16).

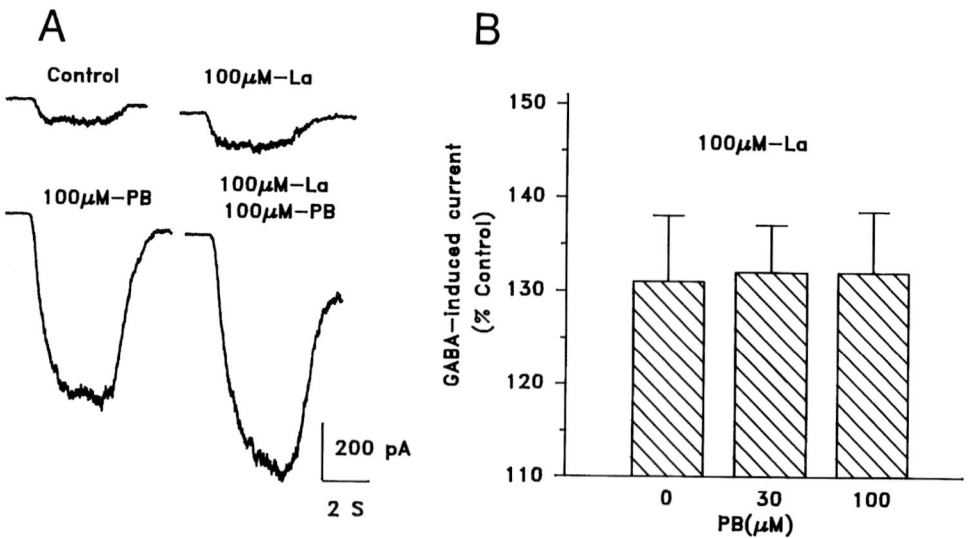

Figure 10. La^{3+} does not compete with pentobarbital (PB) for binding sites in rat DRG neurons. (**A**) augmentation of 10 μM GABA-induced current by 100 μM La^{3+} without and with 100 μM PB. (**B**) summary of data (n = 4) obtained by the experiments shown in A. The ordinate represents the amplitude of 10 μM GABA-induced current as the percentage of respective control. The current in 100 μM La^{3+} and 0 μM PB is expressed as the percentage of current in 0 μM La^{3+} and 0 μM PB. The currents in 100 μM La^{3+} and 30/100 μM PB are expressed as the percentages of current in 0 μM La^{3+} and 30/100 μM PB, indicating that the presence of PB does not affect the potentiating effect of La^{3+}. (From Ma, J.Y. and Narahashi, T., *Brain Res.*, 607, 222–232, 1993. With permission.)

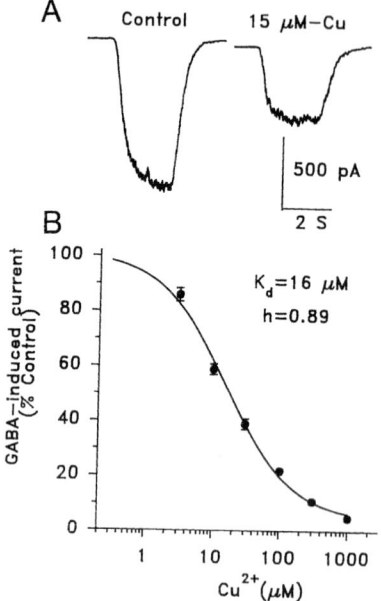

Figure 11. Cu^{2+} suppresses the GABA-induced current in DRG neurons. (**A**) suppression by 15 μM Cu^{2+} of the current induced by 10 μM GABA at a holding potential of –60 mV. (**B**) the dose-response relationship for Cu^{2+} suppression of the current induced by 30 μM GABA (n = 5). (From Ma, J.Y. and Narahashi, T., *Brain Res.*, 607, 222–232, 1993. With permission.)

Tb^{3+} increased the overall mean open time. In the presence of GABA alone, the channels opened 7.16% of the time with the mean open time of 3.00 ms. In the presence of GABA plus Tb^{3+}, the channels opened 12.2% of the time with the mean open time of 4.76 ms. Therefore, the percentage of open time in the presence of Tb^{3+} was almost doubled, in good agreement with the whole-cell data. To determine

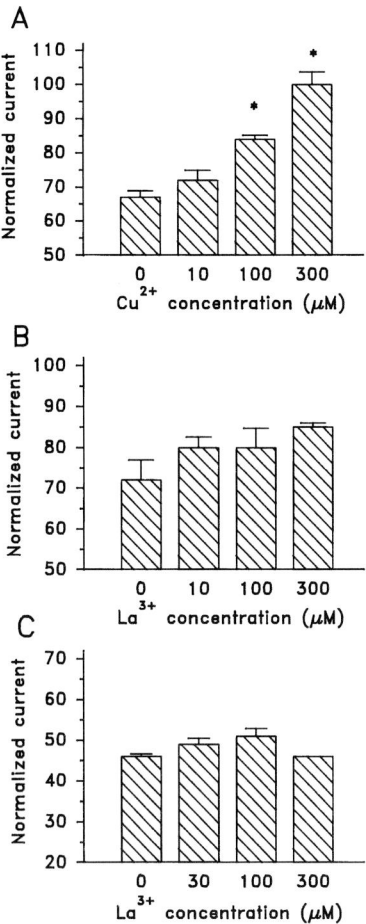

Figure 12. Cu^{2+} competes with Zn^{2+} but not with La^{3+} for binding sites in rat DRG neurons. (**A**) the ordinate represents the amplitude of 10 μM GABA-induced current as the percentage of respective control. The current in 10 μM Zn^{2+} and 0 μM Cu^{2+} is expressed as the percentage of current in 0 μM Zn^{2+} and 0 μM Cu^{2+}. The currents in 10 μM Zn^{2+} and 10/100/300 μM Cu^{2+} are expressed as the percentages of current in 0 μM Zn^{2+} and 10/100/300 μM Cu^{2+}, indicating that Cu^{2+} antagonizes the blocking action of Zn^{2+} (n = 5). Asterisks denote significant differences with respect to the control group ($P < 0.05$). (**B**) the ordinate represents the amplitude of GABA-induced current as the percentage of respective control. The current in 10 μM Cu^{2+} and 0 μM La^{3+} is expressed as the percentage of current in 0 μM Cu^{2+} and 0 μM La^{3+}. The currents in 10 μM Cu^{2+} and 10/100/300 μM La^{3+} are expressed as the percentages of current in 0 μM Cu^{2+} and 10/100/300 μM La^{3+}, indicating that the presence of La^{3+} has little or no effect on the blocking action of Cu^{2+} (n = 5). (**C**) experiments similar to those in B, but with 15 μM Zn^{2+}. The presence of La^{3+} does not affect the blocking action of Zn^{2+} (n = 5). Data in B and C with La^{3+} are not statistically different from those without La^{3+} ($P > 0.05$). Error bars represent S.E.M. (From Ma, J.Y. and Narahashi, T., *Brain Res.*, 607, 222–232, 1993. With permission.)

the basis for the Tb^{3+} increase in mean open time, the durations of openings of main-conductance channels were collated into frequency histograms. The frequency histogram of GABA receptor channel openings was best fitted to a sum of three exponential functions in a range of 0.4 to 80 ms with the time constants of 0.53 ± 0.04, 2.94 ± 0.42, and 9.46 ± 1.3 ms (Figure 17A). In the presence of Tb^{3+}, the frequency histogram was also best fitted to a sum of three exponential functions (Figure 17B). The time constants were 0.48 ± 0.05, 2.86 ± 0.37, and 11.82 ± 1.1 ms and are not significantly different from the corresponding time constants in the presence of GABA alone. Thus, Tb^{3+} did not increase the open time by changing the individual time constants. The relative area of each time constant in the histogram is a measure of the relative frequency of openings contributed by each channel component. In the presence of GABA alone, the relative areas of these times were estimated to be $49 \pm 0\%$, $38 \pm 0\%$, and $13 \pm 0\%$. In the presence of GABA and Tb^{3+}, the relative areas of these time constants were $45 \pm 0\%$, $32 \pm 0\%$, and $23 \pm 0\%$. Therefore, the Tb^{3+} increase in the GABA receptor channel open time is attributed in part

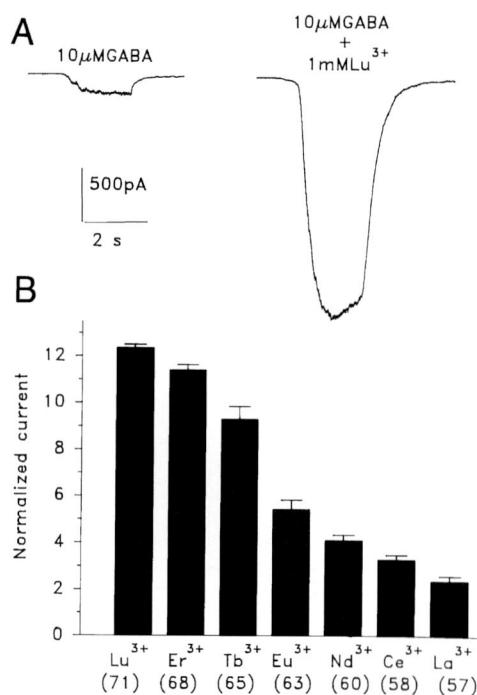

Figure 13. Enhancement of GABA-induced currents by lanthanides in cultured rat DRG neurons. (**A**) GABA (10 μM)-induced inward chloride current was greatly enhanced (12.6 times of control) by adding 1 mM Lu^{3+}. Holding potential, –60 mV. The effect of Lu^{3+} was completely reversible. (**B**) the chloride currents induced by 10 μM GABA were potentiated by different lanthanides to different degrees. The ordinate represents the current amplitude with S.E.M. (n = 10) normalized to 10 μM GABA-induced current. The numbers in parentheses are atomic numbers of lanthanides. The rank order of efficacies of lanthanide potentiation correlates with the atomic numbers of these ions. (From Ma, J.Y. and Narahashi, T., *J. Neurosci.*, 13, 4872–4879, 1993. With permission.)

to an increase in the relative frequency of occurrence of the longest time constant while the relative frequencies of occurrence of the shortest and medium time constants remain largely unchanged.

The mean closed time was decreased by Tb^{3+} from the control value of 31.1 ms in GABA alone to 24.5 ms in GABA plus Tb^{3+}. Although each of the three time constants for the mean closed time was not affected by Tb^{3+}, the portion of the longest time constant was decreased. Therefore, Tb^{3+} decreased the relative frequency of occurrence of the longest closed time constant.

The mean duration of bursts was increased from 10 ms in the presence of GABA alone to 16.6 ms in the presence of GABA and Tb^{3+}. Consistent with this, the percentage of time for the channel to spend in a burst increased from 8.75% in GABA alone to 13.3% in GABA plus Tb^{3+}. Although each of the three time constants for the burst duration was not changed by Tb^{3+}, the relative frequency of occurrence of the longest burst component was increased by Tb^{3+}.

In summary, although Tb^{3+} does not change any of the three time constants of each of channel open, closed, and burst durations, it increases the relative proportions of the longest open and burst duration time constants and decreases the relative proportion of the longest closed time constant, thereby increasing the amplitude of whole-cell GABA-induced current.

VII. EFFECTS OF VARIOUS METALS ON THE GLUTAMATE RECEPTOR-CHANNEL COMPLEX

A variety of divalent cations have been shown to block the NMDA receptor channel. Some of them, including Mg^{2+} and Zn^{2+}, have been studied extensively, and they can serve as useful tools for elucidating the mechanism of action of various drugs on the NMDA receptor channel. Mg^{2+} causes open channel block in a voltage-dependent manner, block being intensified with hyperpolarization (Mayer et al., 1984; Nowak et al., 1984). Based on Woodhull's equation (Woodhull, 1973), the Mg^{2+} binding site is located at halfway in the channel pore. However, Mg^{2+} is also effective in blocking the NMDA receptor channels

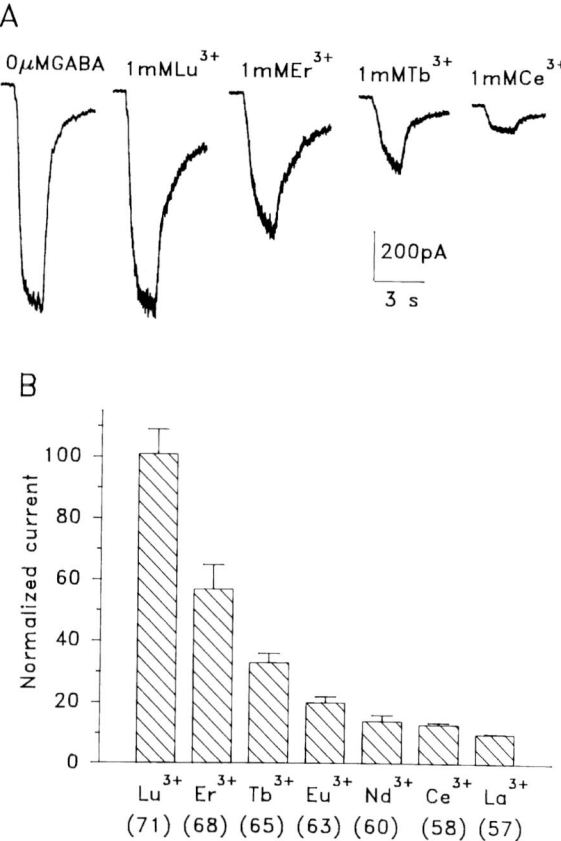

Figure 14. Lanthanides themselves generate inward currents in rat DRG neurons. (**A**) the responses to 1 mM Lu^{3+}-, Er^{3+}-, Tb^{3+}-, and Ce^{3+}-induced currents at a holding potential of –60 mV. All records were obtained from the same neuron. The current induced by 10 μM GABA is shown for comparison. (**B**) summary of the data obtained in A. The ordinate represents the current amplitude with S.E.M. (n = 4) normalized to 10 μM GABA-induced current. The numbers in parentheses are atomic numbers of lanthanides. The rank order correlates with the atomic numbers of lanthanides as well as the rank order of the efficacies for potentiation of GABA-induced current. (From Ma, J.Y. and Narahashi, T., *J. Neurosci.*, 13, 4872–4879, 1993. With permission.)

when applied to the internal membrane surface, and the binding site is different from that calculated from the data on the external application of Mg^{2+} (Johnson and Ascher, 1990). Thus reservations must be made for applying Woodhull's calculation to determine the geographical binding site. At large hyperpolarized membrane potentials (–150 mV to –200 mV), Mg^{2+} block is relieved, resulting in an N-shaped current-voltage (I-V) curve (Mayer and Westbrook, 1987). This relief may be due to Mg^{2+} being driven out of the inner orifice of the channel by hyperpolarizing inward current. In single-channel experiments, Mg^{2+} was seen to cause fast block (flickering block) (MacDonald and Nowak, 1990). However, the simple sequential block model proposed by Neher and Steinbach (1978) for local anesthetic block of acetylcholine-activated channels does not apply here, because the characteristics of Mg^{2+} block fail to meet all the criteria of the sequential block. More complex models are required.

Zn^{2+} also blocks the NMDA receptor channels. Unlike Mg^{2+} block, Zn^{2+} block is voltage independent, suggesting its binding site is located on or near the external orifice of the channel (Westbrook and Mayer, 1987). The action of Zn^{2+} is not limited to the NMDA receptor channels and is more complex, because it augments the currents induced by kainate, quisqualate or 1-glutamate in glycine-free conditions (Mayer et al., 1989). Furthermore, depending on the presence or absence of Mg^{2+}, Zn^{2+} either suppresses or increases the currents induced by 1-glutamate, AMPA, or kainate (Rassendren et al., 1990). At the single-channel level, Zn^{2+} (1 to 10 μM), decreases the probability of channel openings in a voltage-independent manner, and at higher concentrations (10 to 100 μM) Zn^{2+} decreases the single-channel current amplitude, suggesting fast block (Christine and Choi, 1990).

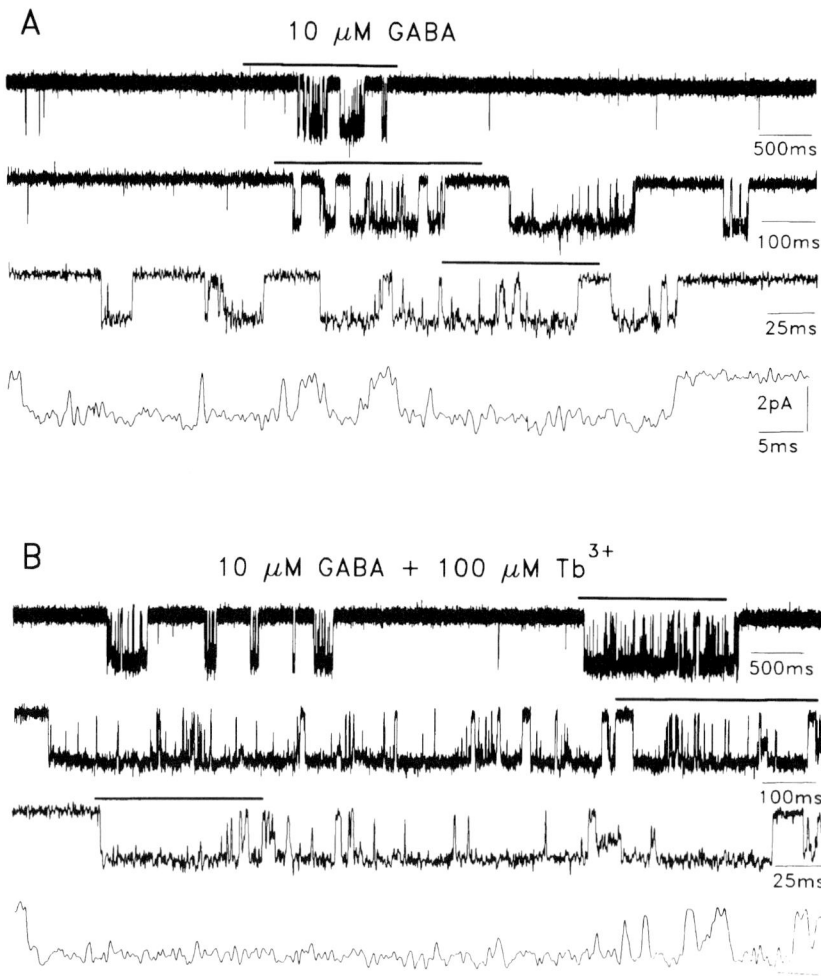

Figure 15. Alteration of GABA-induced single-channel currents by Tb^{3+} in an outside-out membrane patch isolated from a rat DRG neuron. Currents were recorded at a holding potential of −60 mV. (**A**) GABA (10 μM)-induced bursting single-channel currents. (**B**) same as A, but prolonged by 100 μM Tb^{3+}. Inward currents are shown on various time scales. The portion of current record under the horizontal line in each current trace is shown on an expanded time-scale in the trace below. (From Ma, J.Y. Reuveny, E., and Narahashi, T., *J. Neurosci.*, 14, 3835–3841, 1994. With permission.)

Other divalent cations have also been found to block the NMDA receptor channels with the potency sequence of $Ni^{2+} > Co^{2+} > Mg^{2+} > Mn^{2+} > Ca^{2+}$. Ions that give up their water slowly are considered blockers (e.g., Ni^{2+}, Co^{2+}, Mg^{2+}), whereas ions that dehydrate more quickly are considered to be permeant (e.g., Ca^{2+}, Ba^{2+}, Cd^{2+}, Sr^{2+}) (Collingridge and Lester, 1989). However, this classification is rather arbitrary, because many of these divalent cations permeate and block (e.g., Ca^{2+}, Mg^{2+}, Cd^{2+}).

VIII. EFFECTS OF VARIOUS METALS ON VOLTAGE-ACTIVATED CALCIUM CHANNELS

As described before, mercury and lead exert potent blocking action on the voltage-activated calcium channels. Some other metals are also known to block the calcium channels. Aluminum is the third most abundant element in nature and is chemically active, forming a variety of complexes in aqueous solutions (Martin, 1988). In rat DRG neurons, aluminum has been shown to block both transient and sustained calcium channel currents with an IC_{50} of 83 μM in a manner dependent on stimulation (use dependence)

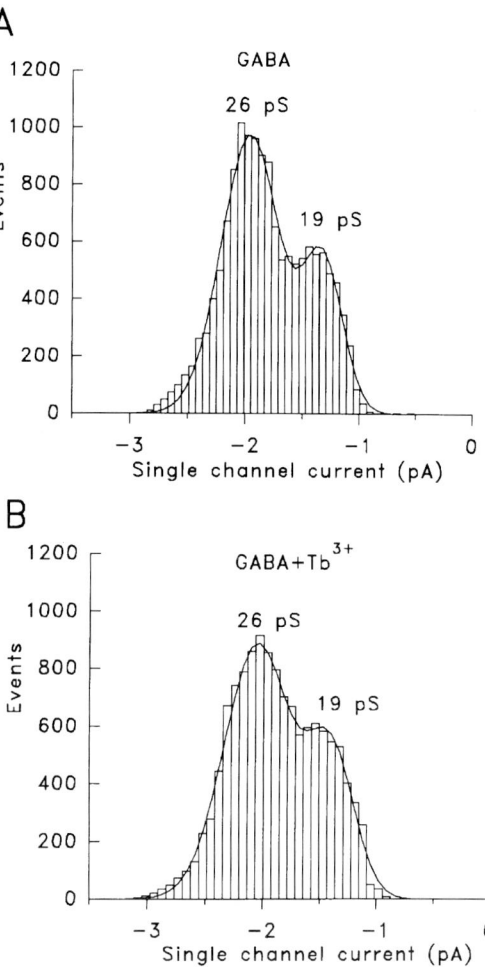

Figure 16. Tb^{3+} does not change the relative proportion of the main- and subconductance currents in rat DRG neurons. (**A**) amplitude histogram of single-channel currents in the presence of 10 μM GABA alone. (**B**) amplitude histogram in the presence of GABA plus 100 μM Tb^{3+}. (From Ma, J.Y. Reuveny, E., and Narahashi, T., *J. Neurosci.*, 14, 3835–3841, 1994. With permission.)

and pH (Büsselberg et al., 1993; Platt et al., 1993). A decrease in pH augmented the aluminum block, and this pH dependence appears to be due to changes in the form of aluminum.

Lanthanum is a well-known blocker of calcium channels (Hagiwara and Byerly, 1981; Lansman, 1990). It blocks the channels with an IC$_{50}$ of 163 nM in rat dorsal horn neurons (Reichling and MacDermott, 1991). Lanthanum is more potent than divalent cations such as Cd^{2+}, Ni^{2+}, and Co^{2+} in blocking the calcium channels, and the sequence of blocking potencies of these polyvalent cations differs significantly among calcium channels of different tissues (Hagiwara and Byerly, 1981). In mouse neuroblastoma N1E-115 line cells, the sequence of blocking potencies is (IC$_{50}$ in μM): La^{3+} (1.5) >> Ni^{2+} (47) > Cd^{2+} (160) = Co^{2+} (160) for type I (T-type) calcium channels; and La^{3+} (0.9) > Cd^{2+} (7.0) > > Ni^{2+} (280) > Co^{2+} (560) for type II (L-type) calcium channels (Narahashi et al., 1987). Figure 18 illustrates the dose-response curves for polyvalent cation block of two types of calcium channels. Particularly notable is the potency difference between the two types of calcium channels for Cd^{2+} which amounts to more than 20 times. A similar differential blocking action of Cd^{2+} was found for two kinds of calcium channels in sensory neurons (Nowycky et al., 1985). Thus, Cd^{2+} has often been used to distinguish the two types of calcium channels. High potencies of Cd^{2+} and La^{3+} to block calcium channels were also shown by measuring ^{45}Ca^{2+} influx in rat brain synaptosomes as induced by K$^+$: Cd^{2+}, La^{3+} ≥ Mn^{2+}, Co^{2+}, Ni^{2+} >> Sr^{2+}, Ba^{2+} Mg^{2+} (Nachshen, 1984). However, ^{45}Ca^{2+} flux measurements do not allow us to distinguish between two (or three) types of calcium channels. The calcium channel-blocking action

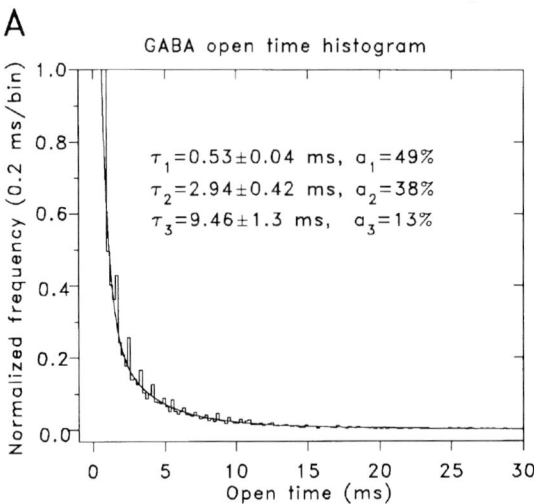

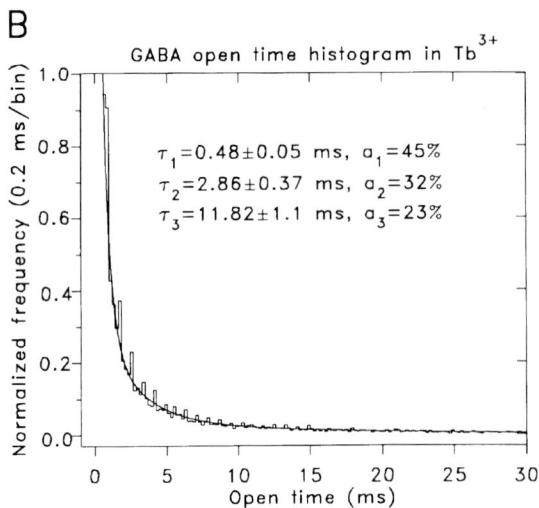

Figure 17. Channel open states are increased by Tb^{3+} in rat DRG neurons. (**A**) the histogram of 10 μM GABA-induced channel openings. (**B**) the histogram of 10 μM GABA-induced channel openings in the presence of 100 μM Tb^{3+}. The open time was binned into 0.2 ms and fitted with three exponential functions in the range of 0.4 to 80 ms. For display, the distribution was normalized and overlaid in a range of 0.4 to 30 ms. τ_1, τ_2, and τ_3 are time constants, and α_1, α_2, and α_3 are the relative areas under the respective time constant. The relative area under the longest time constant τ_3, α_3, is increased by Tb^{3+}, although the three open time constants are not changed significantly. (From Ma, J.Y. Reuveny, E., and Narahashi, T., *J. Neurosci.,* 14, 3835–3841, 1994. With permission.)

of Cd^{2+} is responsible for the suppression of end-plate potential observed in the presence of Cd^{2+} (Cooper and Manalis, 1984; Cooper et al., 1984).

The mechanisms underlying calcium channel block caused by polyvalent cations were analyzed in detail by single-channel experiments (Lansman et al., 1986; Lansman, 1990; Winegar and Lansman, 1990; Winegar et al., 1991). These cations exhibit permeability and blocking activity in vastly different degrees. The different permeability and blocking potency do not arise from ionic radii, as Cd^{2+}, Ca^{2+}, and Na^+ have almost the same radii yet show different behavior. It was proposed that differences in the affinity of cations for the intrachannel binding sites account for different behavior: very high-affinity binding results in potent block but slow dissociation makes permeation extremely small (e.g., La^{3+}, Cd^{2+}); very low-affinity binding causes large flux and no block (e.g., Na^+, Li^+, K^+, Cs^+); and intermediate-

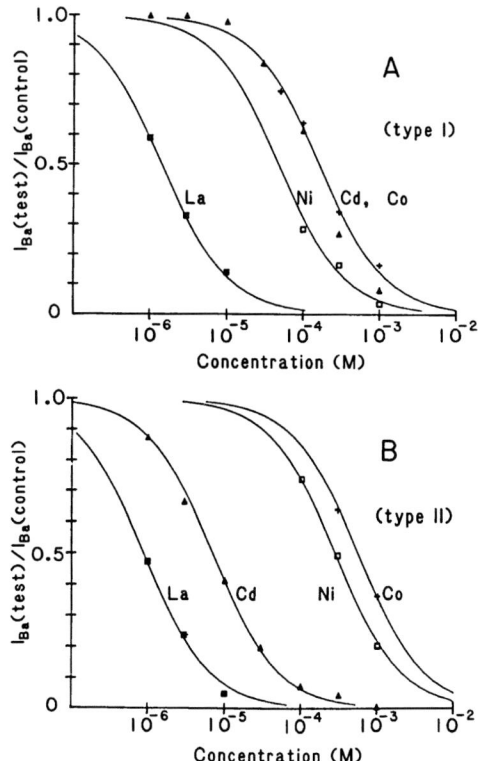

Figure 18. Dose-response relations for various polyvalent metal cations in blocking the two components of Ba^{2+} current through calcium channels in N1E-115 neuroblastoma cells. Relative amplitudes of the transient inward currents (**A**) and long-lasting inward currents (**B**) recorded at –20 and +10 mV, respectively, are plotted against the concentration of the blocking cations. The amplitudes of the currents are normalized to their maximum in the control. Continuous curves are drawn based on the one-to-one binding stoichiometry. I_{Ba} (test) and I_{Ba} (control) refer to Ba^{2+} currents in test polyvalent cations and in control, respectively. (From Narahashi, T., Tsunoo, A., and Yoshii, M., *J. Physiol.*, 383, 231–249, 1987. With permission.)

affinity binding causes block and permeation (e.g., Ba^{2+}, Sr^{2+}, Ca^{2+}, Mn^{2+}) (Lansman et al., 1986). Mg^{2+} does not measurably permeate and does not block, presumably because it is slowly dehydrated.

IX. EFFECTS OF VARIOUS METALS ON VOLTAGE-ACTIVATED SODIUM CHANNELS

Most metals do not have a potent effect on the voltage-activated sodium channels. As an endogenous component, calcium modulates the gating kinetics of the sodium channel in a manner explicable by surface charge effect and binding effect (McLaughlin et al., 1970; Ohmori and Yoshii, 1977). The conductance-voltage curve and the steady-state inactivation curve are both shifted in the depolarizing direction when the calcium concentration is increased, the effect that can be explained by the ability of divalent Ca^{2+} ions to neutralize negative surface charges on the external membrane surface (Frankenhaeuser and Hodgkin, 1957; Vogel, 1974). High concentrations of Sr^{2+}, Mg^{2+}, and Ba^{2+} also cause a shift of the steady-state sodium inactivation curve in the depolarizing direction (Narahashi, 1966).

Lanthanum exerts a potent action on the sodium channel. It shifts the voltage dependence of activation gating in the depolarizing direction in a manner similar to that of high calcium, but with a much higher potency (Arhem, 1980; Brismar, 1980, Takata et al., 1966; Vogel 1974; Hille et al., 1975). This effect was usually attributed to a surface charge effect. However, recent analysis has indicated that a positive shift of the channel opening rate and a negative shift of the channel closing rate caused by 10 μM La^{3+} cannot be explained by surface charge theory (Armstrong and Cota, 1990). An additional factor must be taken into consideration to explain the opposing effect.

Dorsal root ganglion neurons of rats are endowed with two types of sodium channel, one sensitive to TTX (TTX-S channel) and the other resistant to TTX (TTX-R channel) (Elliott and Elliott, 1993;

Kostyuk et al., 1981; Ogata and Tatebayashi, 1993; Roy and Narahashi, 1992). Lead and cadmium exhibited differential blocking actions on the two types of sodium channel, being more potent on the TTX-R channel than the TTX-S channel (Roy and Narahashi, 1992). The blocking action appears to be due to direct binding of the metals to the external surface of the channels. The channel activation curve was shifted in the depolarizing direction by both metals due to surface charge effect.

X. EFFECTS OF VARIOUS METALS ON THE ATP-ACTIVATED CATION CHANNEL

ATP is not only an energy source but has a multiple action on various ion channels. There is a cation channel that is opened by external ATP (Friel, 1988; Nakazawa and Matsuki, 1987); a high potassium-selective channel is activated by external ATP (Bean and Friel, 1990); and a potassium channel which is normally closed at physiological intracellular ATP concentrations opens when the intracellular ATP is diminished (Ashcroft et al., 1984; Cook and Hales, 1984; Noma, 1983; Noma and Shibasaki, 1988; De Weille and Lazdunski, 1990).

Zinc has been found to potentiate the ATP-activated cation channel current in rat nodose ganglion neurons, celiac and superior cervical ganglion neurons, and spinal cord neurons (Li et al., 1993). In nodose ganglion neurons, the EC_{50} value for current potentiation was estimated to be 11 μM and the maximum response reached 585% of control. As expected from these results, Zn^{2+} increased membrane depolarization and action potential firing elicited by ATP.

In contrast to potentiation by Zn^{2+}, several other divalent cations have been shown to block the ATP-activated cation channels (Nakazawa and Hess, 1993). The order of blocking potencies was $Cd^{2+} > Mn^{2+} > Mg^{2+} \sim Ca^{2+} > Ba^{2+}$.

XI. EFFECTS OF GADOLINIUM ON STRETCH-MODULATED CATION CHANNELS

Skeletal muscle myotubes from mdx mice are used as an animal model for human Duchenne muscular dystrophy (Bulfield et al., 1984). The myotubes have two types of mechano-transducing cation channels: one is open at resting potential and closes upon stretching the muscle (Franco and Lansman, 1990a); and the other opens upon stretch (Franco and Lansman, 1990b; Guhary and Sachs, 1984). Gadolinium (Gd) is a member of lanthanides with an ionic radius of 0.105 nm, close to that of Ca^{2+} (0.106 nm). Gadolinium has been found to block stretch-activated cation channels (Franco and Lansman, 1990b; Yang and Sachs, 1989) as well as stretch-inactivated cation channels (Franco et al., 1991). In the latter channels, Gd^{3+} block was voltage dependent, hyperpolarization increasing the rate of unblocking. This suggests that the Gd^{3+} ions can exit the channel by hyperpolarizing voltage field. The Gd^{3+} block is similar to the block of calcium channels by Ca^{2+} and La^{3+} cations (Lansman et al., 1986; Lansman, 1990).

ACKNOWLEDGMENT

I wish to thank Vicky James-Houff for her unfailing secretarial assistance.

REFERENCES

Abd-Elfattah, A.-S. A. and Shamoo, A. E. (1981), *Mol. Pharmacol.,* 20, 492–497.
Abdallah, E. A. M. and Shamoo, A. E. (1984), *Pesticide Biochem. Physiol.,* 21, 385–393.
Alkondon, M., Costa, A. C. S., Radhakrishnan, V., Aronstam, R. S., and Albuquerque, E. X. (1990), *FEBS,* 261, 124–130.
Al-Modhefer, A. J. A., Bradbury, M. W. B., and Simons, T. J. B. (1991), *Clin. Sci.,* 81, 823–829.
Arakawa, O., Nakahiro, M., and Narahashi, T. (1991), *Brain Res.,* 551, 58–63.
Arhem, P. (1980), *Acta Physiol. Scand.,* 108, 7–16.
Armstrong, C. M. and Cota, G. (1990), *J. Gen. Physiol.,* 96, 1129–1140.
Aronstam, R. S. and Eldefrawi, M. E. (1979), *Toxicol. Appl. Pharmacol.* 48, 489–496.
Ashcroft, F. M., Harrison, D. E., and Ashcroft, S. J. H. (1984), *Nature,* 312, 446–447.
Atchison, W. D. (1986), *J. Pharmacol. Exp. Ther.,* 237, 672–680.
Atchison, W. D. (1987), *J. Pharmacol. Exp. Ther.,* 241, 131–139.
Atchison, W. D. and Narahashi, T. (1982), *Neurotoxicology,* 3, 37–50.
Atchison, W. D. and Narahashi, T. (1984), *Neurotoxicology,* 5(3), 267–282.
Atchison, W. D., Joshi, U., and Thornburg, J. E. (1986), *J. Pharmacol. Exp. Ther.,* 238, 618–624.
Audesirk, G. (1987), *Neurotoxicology,* 8, 579–592.
Audesirk, G. (1993), *Neurotoxicology,* 14(2–3), 137–148.
Audesirk, G. and Audesirk, T. (1989), *Neurotoxicology,* 10, 659–670.

Audesirk, G. and Audesirk, T. (1993), *Neurotoxicology,* 14(2–3), 259–266.
Bean, B. P. and Friel, D. D. (1990), in *Ion Channels,* Vol. 2, Narahashi, T., Ed., Plenum, New York, 169–203.
Brismar, T. (1980), *Acta Physiol. Scand.,* 108, 23–29.
Bulfield, G., Siller, W. G., Wight, P. A., and Moore, K. J. (1984), *Proc. Natl. Acad. Sci. U.S.A.,* 81, 1189–1192.
Büsselberg, D., Evans, M. L., Rahmann, H., and Carpenter, D. O. (1990), *Toxicol. Lett.,* 51, 51–57.
Büsselberg, D., Evans, M. L., Rahmann, H., and Carpenter, D. O. (1991), *J. Neurophysiol.,* 65, 786–795.
Büsselberg, D., Platt, B., Haas, H. L., and Carpenter, D. O. (1993), *Brain Res.,* 622, 163–168.
Cavalleri, A., Minoia, C., Ceroni, M., and Poloni, M. (1984), *J. Appl. Toxicol.,* 4, 63–65.
Christine, C. W. and Choi, D. W. (1990), *J. Neurosci.,* 10, 108–116.
Collingridge, G. L. and Lester, R. A. J. (1989), *Pharmacol. Rev.,* 40, 143–210.
Cook, D. L. and Hales, N. (1984), *Nature,* 311, 271–273.
Cooper, G. P. and Manalis, R. S. (1984), *Toxicol. Appl. Pharmacol.,* 74, 411–416.
Cooper, G. P., Suszkiw, J. B., and Manalis, R. S. (1984), *NeuroToxicology,* 5(3), 247–266.
David, O. J. (1974), *Environ. Health Perspect.,* 7, 17–25.
De Weille, J. R. and Lazdunski, M. (1990) in *Ion Channels,* Vol. 2, Narahashi, T., Ed., Plenum, New York, 205–222.
Eldefrawi, M. E., Mansour, N. A., and Eldefrawi, A. T. (1977), in *Membrane Toxicology,* Miller, M.W. and Shamoo, A.E., Eds., Plenum, New York, 449–463.
Elliott, A. A. and Elliott, J. R., (1993), *J. Physiol.,* 463, 39–56.
Evans, H. L., Garman, R. H., and Weiss, B. (1977), *Toxicol. Appl. Pharmacol.,* 41, 15–33.
Fox, D. A., Rubinstein, S. D., and Hsu, P. (1991), *Toxicol. Appl. Pharmacol.,* 109, 482–493.
Franco, A., Jr. and Lansman, J. B. (1990a), *Nature (London),* 344, 670–673.
Franco, A., Jr. and Lansman, J. B. (1990b), *J. Physiol.,* 427, 361–380.
Franco, A., Jr., Winegar, B. D., and Lansman, J. B. (1991), *Biophys. J.,* 59, 1164–1170.
Frankenhauser, B. and Hodgkin, A. L. (1957), *J. Physiol.,* 137, 218–244.
Friberg, L. and Mottet, N. K. (1989), *Biol. Trace Element Res.,* 21, 201–206.
Friel, D. D. (1988), *J. Physiol.,* 401, 361–380.
Guhary, F. and Sachs, F. (1984), *J. Physiol.,* 352, 685–701.
Gutknecht, J. (1981), *J. Membrane Biol.,* 61, 61–66.
Hagiwara, S. and Byerly, L. (1981), *Annu. Rev. Neurosci.,* 4, 69–125.
Hahne, H. C. H. and Kroontje, W. (1973), *J. Environ. Qual.,* 2, 444–450.
Hamill, O. P., Marty, A., Neher, E., Sakmann, B., and Sigworth, F. J. (1981), *Pflügers Arch.,* 391, 85–100.
Hare, M. F., McGinnis, K. M., and Atchison, W. D. (1993), *J. Pharmacol. Exp. Ther.,* 266, 1626–1635.
Hille, B., Woodhull, A. M., and Shapiro, B. I. (1975), *Philos. Trans. R. Soc. London,* B270, 301–318.
Johnson, J. W. and Ascher, P. (1990), *Biophys. J.,* 57, 1085–1090.
Juang, M. S. (1976), *Toxicol. Appl. Pharmacol.,* 37, 339–348.
Juang, M. S. and Yonemura, K. (1975), *Nature (London),* 256, 211–213.
Kiskin, N. A., Krishtal, O. A., Tsyndrenko, A.-Y., and Akaike, N. (1986), *Neurosci. Lett.,* 66, 305–310.
Kolton, L. and Yaari, Y. (1982), *Isr. J. Med. Sci.,* 18, 165–170.
Komulainen, H. (1988), in *Metal Neurotoxicity,* Bondy, S. C. and Prasad, K. N., Eds., CRC Press, Boca Raton, FL, 167–182.
Komulainen, H. and Tuomisto, J. (1981), *Acta Pharmacol. Toxicol.,* 48, 214–222.
Kostyuk, P. G., Veselovsky, N. S., and Tsyndrenko, Y. (1981), *Neuroscience,* 6, 2423–2430.
Lansman, J. B. (1990), *J. Gen. Physiol.,* 95, 679–696.
Lansman, J. B., Hess, P., and Tsien, R. W. (1986), *J. Gen. Physiol.,* 88, 321–347.
Levesque, P. C. and Atchison, W. D. (1987), *Toxicol. Appl. Pharmacol.,* 87, 315–324.
Levesque, P. C. and Atchison, W. D. (1988), *Toxicol. Appl. Pharmacol.,* 94, 55–65.
Li, C., Peoples, R. W., Li, Z., and Weight, F. F. (1993), *Proc. Natl. Acad. Sci. U.S.A.,* 90, 8264–8267.
Lind, B., Friberg, L., and Nylander, M. (1988), *J. Trace Elements Exp. Med.,* 1, 49–56.
Luckey, T. D. and Venugopal, B. (1977), *Metal Toxicity in Mammals,* Vol. 1., Plenum, New York, 238 pp.
Ma, J. Y. and Narahashi, T. (1993a), *Brain Res.,* 607, 222–232.
Ma, J. Y. and Narahashi, T. (1993b), *J. Neurosci.,* 13, 4872–4879.
Ma, J. Y., Reuveny, E. and Narahashi, T. (1994), *J. Neurosci,* 14, 3835–3841.
MacDonald, J. F. and Nowak, L. M. (1990), *Trends Pharmacol. Sci.,* 11, 167–172.
Manalis, R. S. and Cooper G. P. (1973), *Nature (London),* 243, 354–356.
Martin, B. R. (1988), in *Metal Ions in Biological Systems. Aluminum and Its Role in Biology,* Vol. 24, Siegel, H. and Sigl, A., Eds., Dekker, New York, 1–57.
Matthews, M. R., Parsons, P. J., and Carpenter, D. O. (1993), *Neurotoxicology,* 14(2–3), 283–290.
Mayer, M. L. and Westbrook, G. L. (1987), *J. Physiol.,* 394, 501–527.
Mayer, M. L., Westbrook, G. L. and Guthrie, P. B. (1984), *Nature (London),* 309, 261–263.
Mayer, M. L., Vyklicky, L. Jr., and Westbrook, G. L. (1989), *J. Physiol.,* 415, 329–350.
McKay, S. J., Reynolds, J. N., and Racz, W. J. (1986), *Can. J. Physiol. Pharmacol.,* 64, 1507–1514.
McLaughlin, S. G. A., Szabo, G., Eisenman, G., and Ciani, S. M. (1970), *Proc. Natl. Acad. Sci. U.S.A.,* 67, 1268–1275.
Minnema, D. J., Michaelson, I. A., and Cooper, G. P. (1988), *Toxicol. Appl. Pharmacol.,* 92, 351–357.
Miyamoto, M. D. (1983), *Brain Res.,* 267, 375–379.

Nachshen, D. A. (1984), *J. Gen. Physiol.*, 83, 941–967.
Nakazawa, K. and Hess, P. (1993), *J. Gen. Physiol.*, 101, 377–392.
Nakazawa, K. and Matsuki N. (1987), *Pflügers Arch.*, 409, 644–646.
Narahashi, T. (1966), *Comp. Biochem. Physiol.*, 19, 759–774.
Narahashi, T., Ma, J. Y., Arakawa, O., Reuveny, E., and Nakahiro, M. (1994), *Cell Mol. Neurosci.*, 14, 599–621.
Narahashi, T., Tsunoo, A., and Yoshii, M. (1987), *J. Physiol.*, 383, 231–249.
Needleman, H. L., Gunnoe, C., Leviton, A., Reed, R., Peresie, H., Maher, C., and Barrett, P. (1979), *N. Engl. J. Med.*, 300, 689–695.
Neher, E. and Steinbach, J. H. (1978), *J. Physiol.*, 277, 153–176.
Neville, G. A. and Berlin, M. (1974), *Environ. Res.*, 7, 75–82.
Noma, A. (1983), *Nature*, 305, 147–148.
Noma, A. and Shibasaki T. (1988), in *Ion Channels*, Vol. 1, Narahashi T., Ed., Plenum, New York, 183–212.
Norseth, T. and Clarkson, T. W. (1970a), *Arch. Environ. Health*, 21, 717–727.
Norseth, T. and Clarkson, T. W. (1970b), *Biochem. Pharmacol.*, 19, 2775–2783.
Nowak, L., Bregestovski, P., Ascher, P., Herbet A., and Prochiantz, A. (1984), *Nature*, 307, 462–465.
Nowycky, M. C., Fox, A. P., and Tsien, R. W. (1985), *Nature (London)*, 316, 440–443.
Ohmori, H. and Yoshii, M. (1977), *J. Physiol.*, 267, 429–463.
Ogata, N. and Tatebayashi, H. (1993), *J. Physiol.*, 466, 9–37.
Omata, S., Sato, M., Sakimura, K., and Sugano, H. (1980), *Arch. Toxicol.*, 44, 231–241.
Oortgiesen, M., van Kleef, R. G. D. M., Bajnath, R. B., and Vijverberg, H. P. M. (1990a), *Toxicol. Appl. Pharmacol.*, 103, 165–174.
Oortgiesen, M., van Kleef, R. G. D. M., and Vijverberg, H. P. M. (1990b), *J. Membr. Biol.*, 113, 261–268.
Oortgiesen, M., Lewis, B. K., Bierkamper, G. G., and Vijverberg, H. P. M. (1990c), *Neurotoxicology*, 11, 87–92.
Oortgiesen, M., Leinders, M. T., van Kleef, R. G. D. M., and Vijverberg, H. P. M. (1993), *Neurotoxicology*, 14(2–3), 87–96.
Pekel, M., Platt, B., and Büsselberg, D. (1993), *Brain Res.*, 632, 121–126.
Platt, B., Haas, H., and Büsselberg, D. (1993), *NeuroReport*, 4, 1251–1254.
Pounds, J. G. (1984), *NeuroToxicology*, 5(3), 295–332.
Pounds, J. G. and Rosen, J. F. (1988), *Toxicol. Appl. Pharmacol.*, 94, 331–341.
Quandt, F. N., Kato, E., and Narahashi, T. (1982), *Neurotoxicology*, 3, 205–220.
Rassendren, F.-A., Lory, P., Pin, J.-P., and Nargeot, J. (1990), *Neuron*, 4, 733–740.
Reichling, D. B. and MacDermott, A. B. (1991), *J. Physiol.*, 441, 199–218.
Reuveny, E. and Narahashi, T. (1991), *Brain Res.*, 545, 312–314.
Reuveny, E. and Narahashi, T. (1993), *Brain Res.*, 630, 64–73.
Rickard, D. T. and Nriagu, J. O. (1978), in *The Biogeochemistry of Lead in the Environment*, Nriagu, J. O., Ed., Elsevier, Amsterdam, 219–284.
Rius, R. A., Govoni, S., and Tsabucchi, M. (1986), *Toxicology*, 40, 191–197.
Ross, R. A., Spengler, B. A., and Biedler, J. A. (1983), *J. Nat. Cancer Inst.*, 71, 741–747.
Roy, M. L. and Narahashi T. (1992), *J. Neurosci.*, 12, 2104–2111.
Shafer, T. J. and Atchison, W. D. (1991), *J. Pharmacol. Exp. Ther.*, 258, 149–157.
Shafer, T. J., Contreras, M. L., and Atchison, W. D. (1990), *Mol. Pharmacol.*, 38, 102–113.
Shrivastav, B. B., Brodowick, M. S., and Narahashi, T. (1976), *Life Sci.*, 18, 1077–1082.
Silbergeld, E. K. and Goldberg, A. M. (1974), *Exp. Neurol.*, 42, 146–157.
Suszkiw, J., Toth, G., Murawsky, M., and Cooper, G. P. (1984), *Brain Res.*, 323, 31–46.
Takata, M., Pickard, W. F., Lettvin, J. Y., and Moore, J. W. (1966), *J. Gen. Physiol.*, 50, 461–471.
Traxinger, D. L. and Atchison, W. D. (1987), *J. Pharmacol. Exp. Ther.*, 240, 451–459.
Umbach, J. A. and Gundersen, C. B. (1989), *Mol. Pharmacol.*, 36, 582–588.
Vogel, W. (1974), *Pflügers Arch.*, 350, 25–39.
Von Burg, R., Northington, F. K., and Shamoo, A. (1980), *Toxicol. Appl. Pharmacol.*, 53, 285–292.
Webb, J. L. (1966), *Enzyme and Metabolic Inhibitors*, Vol. 2. chap. 7, Mercurials, Academic Press, New York.
Weinreich, D. and Wonderlin, W. F. (1987), *J. Physiol.*, 394, 429–443.
Westbrook, G. L. and Mayer, M. L. (1987), *Nature*, 328, 640–643.
Winegar, B. D. and Lansman J. B. (1990), *J. Physiol.*, 425, 563–578.
Winegar, B. D., Kelly, R., and Lansman, J. B. (1991), *J. Gen. Physiol.*, 97, 351–367.
Woodhull, A. N. (1973), *J. Gen. Physiol.*, 61, 687–708.
Yang, X. and Sachs, F. (1989), *Science*, 243, 1068–1071.

Chapter 41

Oxygen Generation as a Basis for Neurotoxicity by Metals

Stephen C. Bondy

I. INTRODUCTION

There are several means by which metal ions can catalyze the intracellular production of reactive oxygen species. Perhaps the most significant of these are plurivalent metals, which are able to alter their electronic configurations under biological conditions. The promotion of redox cycling by such transitions forms the basis of neurological damage effected by iron, copper, and manganese. These metals are not only essential but often owe their effectiveness in their normal biological role to those very same transition features that, when acting in an uncontrolled manner, promote adverse oxidative events. The fact that many metals are essential cofactors for several enzyme complexes ensures their continuing presence within neural tissue and may set the stage for their aberrant distribution or metabolic handling. Oxygen species can be stabilized and rendered innocuous by the metal-containing region of many metalloenzymes concerned with oxidative and reductive processes (Czapiski and Goldstein, 1986).

No metals are exclusively neurotoxic. While a wide range of tissues may be adversely affected by the abnormal presence of a metal, some organs are clearly more vulnerable to induced oxidative damage. The nervous system and the kidney are, for different reasons, among the most susceptible to such injury. Both of these organs receive a disproportionate amount of arterial blood and both have very high oxidative metabolic rates. In the case of the kidney, the pH changes associated with glomerular filtration and tubular production of urine can lead to concentration and precipitation of metals. The brain is a target because of its relatively low levels of enzymes protecting against oxidative stress (Savolainen, 1978), and its high myelin-associated lipid content, and consequent susceptibility to propagation of peroxiodative events (Bondy, 1992).

Neurotoxicity induced by, and consequent to, the presence of excessive levels of metals, may be associated with several conditions including:

  a. Genetic defects involving failure of normal metabolism, and thus accumulation of a metal.
  b. Xenobiotic exposure from environmental or dietary sources.
  c. Abnormal presence of a metal within tissue due to physiological disruption, e.g., extravasation of hemoglobin and, thus, iron into cerebral tissue.
  d. Increased excitability of neural tissue with consequent elevation of cytosolic calcium levels can be a stimulus for generation of excess prooxidant events.

While the blood-brain barrier is effective in limiting access of many metal cations to the brain, this protection is incomplete and can be circumvented by several means:

a. Metals in an organic form can readily enter the brain by virtue of their lipophilic properties. Amphiphilic compounds are among the most deleterious forms of metals, perhaps because, possessing both a lipophilic and a water soluble component, they will concentrate in and align themselves within intracellular membranes. Following therapeutic intervention with chelators, metals may also enter the CNS in the form of sequestered complexes. Thus, while acute toxicity is prevented in this way, the possibility of subsequent neurological damage remains. This may constitute a significant hazard, especially since loosely chelated metals can be the most potent inducers of oxidative events.
b. Xenobiotic ions can also gain such access to neural cells by entry through ion channels which are not completely specific for normal biologically occurring cations. This is due to the resemblance of several toxic metals to a biologically essential counterpart.
c. The brain has regions, especially in the hypothalamus, that have an incompletely formed blood-brain barrier, probably to permit monitoring of the endocrine composition of circulating blood. This can enable localized entry of ionic species that would otherwise be excluded. The immature brain also has no effective means of excluding charged compounds and thus the fetus may be particularly vulnerable to penetration by undesirable metal ions.

The damage effected by most neurotoxic metals is unlikely to be solely attributable to generation of excess reactive oxygen species (ROS). However, all such metals appear to possess a free radical-inducing component relating to their toxicity. The toxicity of each metal is distinctive and involves a characteristic range of morphological and biochemical abnormalities. Superimposed on this may be an increased rate of ROS formation, which can enhance the more selectively damaging effects of the agent. Thus, catalysis of generation of ROS may represent a final common path provoked by disparate metal-containing chemicals. This overview will describe salient features of several neurotoxic metals insofar as this may involve their capacity to induce ROS and consequently provoke oxidative stress. Metals have been grouped by classes believed to involve similar mechanisms underlying their prooxidant potential (Figure 1).

Fluctuating valency	**Sulfhydryl affinity**
Iron	Mercury
Copper	Arsenic
Manganese	Thallium
Vanadium	Cadmium
Excitation by mimicking calcium	**Unknown**
	Aluminum
Lead	
Tin	

Figure 1. Metals grouped by potential atomic basis for inducing oxidative events.

II. METALS WHERE VALENCE FLUX MAY SUBSERVE ENHANCED ROS PRODUCTION (Fe, Cu, Mn)

Transition metals may induce oxidative stress by cycling between two valence states, thereby alternately donating and receiving electrons. Copper, iron, and manganese are in this class. This property is related to the presence of an unfilled inner electronic shell. The first metal after this series, zinc, has completely filled inner electronic orbitals and thus has no valence ambiguity and no capacity to induce ROS. Other metals with plurivalent potential that may also induce oxidative stress, such as mercury, tin, thallium, and lead, do not readily undergo such valence changes under physiological conditions, and so are active in ROS stimulation primarily by other means.

All three of the transition metals mentioned above are essential elements and constitute part of the active sites of many enzymes and transport proteins. These enzymes include those such as superoxide dismutase and catalase, which are important in removal of free radicals or their precursors. In addition, many enzymes involved in bringing about regulated oxidations are metalloenzymes. The importance of keeping the free concentration of ionic salts of these metals within biological tissues very low is implied by the existence of proteins with high affinity for a specific metal, which are thus able to sequester that metal as an inactive high molecular weight complex. Ceruloplasmin is the main copper-chelating plasma protein, while iron is served by several proteins including ferritin. It has been proposed that most biological molecules are not readily autooxidized and that low molecular weight complexes of a transition metal, especially iron, are needed to catalyze all such oxidations, whether beneficial or harmful (Aust and Miller, 1991).

A. IRON AND COPPER

While superoxide anion and hydrogen peroxide are not in themselves very active oxidants, they are able to interact in the presence of trace amounts of iron, and by the Haber-Weiss reaction, give rise to the highly reactive, short-lived hydroxyl radical.

A wide range of enzymic and nonenzymic reactive free radical-generating processes can thus be activated by Fe^{3+} (or Cu^{2+} to a lesser degree) in the presence of an electron donor. Such activation can lead to lipid peroxidative events and also to the oxidative degradation of proteins, and thence to enzyme inactivation and membrane damage. Metal ion-catalyzed modification of proteins can be site-specific and may occur near an iron-binding lysine residue (Stadtman, 1990). The susceptibility to oxidation of amino acid residues within proteins differs from that of free amino acids, since the proximity of amino acids in the peptide chain to metal binding sites largely governs their vulnerability (Stadtman, 1993).

The situations under which copper and iron can be neurotoxic do not generally involve excessive ingestion. Iron penetrance into the brain can follow traumatic brain injury or hemorrhagic stroke when low molecular weight iron is liberated during degradation of free hemoglobin (Sadrzadeh et al., 1987). The consequent elevation of ROS may in part account for the delayed appearance of seizure activity bleeding into brain tissue (Willmore and Triggs, 1991). Excessive levels of unsequestered iron have also been associated with several other neurological disorders, including various lipofuscinoses such as Batten's disease (Gutteridge et al., 1983), Alzheimer's disease (Grundke-Iqbal et al., 1990; Connor et al., 1992), Parkinson's disease (Youdim et al., 1989; Dexter et al., 1990), seizure disorders (Shoham, et al., 1992), stroke (Prat and Turrens, 1992), ischemia (Rosenthal, 1992), Hallervorden-Spatz Disease (Perry et al., 1985), and edema (Ikeda et al., 1989). Evidence exists that ethanol can lead to elevated levels of lipid peroxidation in liver and brain. This has been attributed, at least in part, to liberation of protein-bound iron by ethanol (Cederbaum, 1989; Shaw, 1989; Rouach et al., 1990; Bondy and Pearson, 1993).

Chelation therapy using deferoxamine, a potent iron chelator, is being used experimentally in the treatment of Alzheimer's disease (Crapper-McLachlan, et al., 1992) and multiple sclerosis (Le Vine 1992). However, deferoxamine, like all chelators, is not completely specific for a single metal. In addition to metal sequestration, its mode of action as an antioxidant may also involve direct scavenging of free radicals (Halliwell, 1989). Chelation therapy as a means of decorporating toxic metals has, however, many adverse side effects including the potential for *enhancement* of ROS formation (Klebanoff et al., 1989; Wahba et al., 1990; Kruck et al., 1990). Some of these undesirable side effects are probably due to removal of essential metals (Thomas and Chisholm, 1986).

A caveat concerning iron-related neurotoxicity is that association does not always imply causation. Thus, while hyperexcitation can elevate intracellular iron levels, this is not necessarily the cause of accompanying neuronal damage (Shoham et al., 1992). The molecular state of cytosolic iron is critical. Under normal homeostatic conditions, iron is bound to intracellular proteins. However, oxidative events can lead to decompartmentation of this metal and consequent appearance of its pro-oxidant properties. Liberation of iron by nitric oxide has been proposed as occurring in Parkinson's disease (Rief and Simmonds, 1990; Youdim et al., 1993).

Sometimes pharmacological therapy has the potential for inadvertently stimulating ROS. For example, l-DOPA, used in the therapy of Parkinsonism, is capable of reducing ferric iron, the form in which iron forms a stable complex ferritin. This may effect the liberation of ferrous iron as a low molecular weight species with significant ROS-enhancing potential (Ogawa et al., 1993). Basal iron levels are especially high in the substantia nigra and globus pallidus (Wurtman and Wurtman, 1990). Ferritin levels have been reported as being both elevated and depressed in these regions in Parkinson's disease (Dexter et al., 1990; Jellinger et al., 1993), implying a deranged iron metabolism.

In the case of copper, which is present in bodily tissues to a much lesser extent than iron, the only clear appearance of neurotoxicity is due to a genetic defect in copper metabolism, Wilson's disease. This disorder is characterized by the virtual absence of ceruloplasmin, leading to greatly elevated levels of low molecular weight diffusible copper. The psychiatric and neurological consequences of this disease can be greatly alleviated by chelation with drugs such as penicillamine. Copper is known to enhance catecholamine autooxidation, and catalysis of ROS together with distortion of the normal role of this metal is likely to underlie this many-faceted disease (Scheinberg, 1988).

Oxidases are enzymes utilizing molecular oxygen. The scission of the O_2 molecule is an intrinsically hazardous process wherein directed oxidation is required. The control of O_2 utilization is regulated at the catalytic center of these enzymes which generally contains a metal with plurivalent potential (Fe, Cu, or Mn). However, all oxidases "leak" nonspecific ROS to a certain extent. There is especially efficient control in the mitochondrial respiratory chain where the uncoupled leakage rate is less than 2% (Boveris and Chance, 1973; Turrens and Boveris, 1980). The presence of both copper and iron within cytochrome oxidase may allow this tight regulation, and enzymes containing iron alone, such as the mixed function oxidases and monoamine oxidases, may be less effective in this respect (Babcock and Wikstrom, 1992). Inhibition of monoamine oxidase B has proved effective in slowing the rate of progression of Parkinson's disease (Birkmayer et al., 1985), and since antioxidant therapy has a similar effect (Shoulson, 1989), the mechanism of the protective effect of MAO-B blockers may be by way of inhibition of free radical generation.

B. MANGANESE

Manganese is another member of this group of metals with the potential for several valence states. Neurological disease is found following excessive exposure to this mineral in the mining, handling, or processing of manganese ore. Manganism presents in several distinct phases, including an initial hyperactive, manic state ("manganese madness"), an intermediate Parkinsonian-like phase involving tremor and incoordination, and a terminal spastic, rigid phase (Seth and Chandra, 1988). The resemblance of the latter two stages to Parkinson's disease (PD) is striking, especially because of the involvement of the basal ganglia. There is increasing evidence for the role of oxidative stress in PD, and trials involving antioxidant therapy have met with some success (Parkinson's Study Group, 1989). In view of the readiness of dopamine and norepinephrine to spontaneously oxidize, catecholaminergic pathways seem to be especially susceptible to metal-catalyzed oxidative damage. This is especially true of dopamine autooxidation, and such a mechanism may underlie manganism (Donaldson et al., 1981). Manganese is a transition metal that can exist in at least four oxidative states (with a valence of 2, 3, 4 or 5), and like iron, has been shown to be a potent ROS enhancer in isolated systems. There is also evidence for a role of iron in PD (see previous section).

The situation is complicated by reports describing significant free radical scavenging properties of many manganese complexes. Hydrogen peroxide and superoxide may be quenched by such chelates (Cheton and Archibald, 1988). In fact, manganese has been stated to "fulfill the requirements of a physiologically relevant antioxidant" (Coasin et al., 1992). These apparent contradictions can be reconciled by recognition of the essential role of the ionic status of metal ions and the readiness of their interconvertibility in enabling or retarding oxidative events (Minotti and Aust, 1989).

Although not shown to be neurotoxic in man, another transition metal, vanadium, has also been found to induce lipid peroxidation in the brains of experimental animals (Haider and El-Fakhri, 1991).

III. METALS WITH A HIGH AFFINITY FOR SULFHYDRYL GROUPS AND FOR SELENIUM (Hg, Pb, Tl, Cd, AND As)

The capacity of several metals to form covalent linkages to sulfhydryl groups of peptides can lead to the formation of nonionic complexes. A key low molecular weight peptide is glutathione, which is present at millimolar concentrations within the cytosol. Glutathione (GSH) plays three vital roles within the cell; it is a source of reducing power and thus constitutes an important water-soluble defense against excess ROS. Glutathione is also able to detoxify xenobiotic agents by direct conjugation of xenobiotic agents using glutathione transferase. Finally, glutathione can destroy peroxides in the presence of glutathione peroxidase. Under normal circumstances, the oxidized glutathione formed (GSSG) is reconverted to GSH by glutathione reductase and once again its reducing power becomes available. However, excessive ROS levels may overwhelm such regenerative capacity. GSSG usually constitutes a very minor fraction of total intracellular glutathione, but if its level is elevated, this non-ionic lipophilic molecule can readily become lost to the cell by diffusion across the limiting membrane.

The heavy metals, mercury, lead, thallium, and cadmium all have electron-sharing tendencies that can lead to formation of covalent attachments. Metals of this family have an avidity for sulfhydryl groups and can deplete cellular glutathione levels (Naganuma et al., 1990), increase levels of lipid peroxidation

(Hasan and Ali, 1983, Yonaha et al., 1983) and accelerate lipofuscinogenesis. This latter effect is enhanced in the presence of hyperbaric oxygen (Marzabadi and Jones, 1992).

α-Tocopherol has frequently been found to be protective against neurotoxicity induced by this class of metals (Chang et al., 1978; Shukla et al., 1988). This may reflect a predominantly lipophilic site of the -SH groups reacting with these metals. In addition, these metals may displace selenium from glutathione peroxidase (Reddy and Massaro, 1983). The consequent inactivation of this enzyme removes a key antioxidant sequence, and this may account for the ability of selenium to protect against the toxicity of methylmercury (Chang and Suber, 1982). Methylmercury can induce generation of ROS within the CNS (LeBel et al., 1990), and this property can be blocked by deferoxamine. Since deferoxamine does not chelate methylmercury, iron-catalyzed pro-oxidant reactions may play a role in methylmercury neurotoxicity (LeBel et al., 1992).

The proportion of the overall toxicity of salts of these metals that is expressed as neurotoxicity is related to their ability to cross the blood-brain barrier. Thus, the charged ionic mercuric chloride form of this element is predominantly harmful to nonneural tissues such as the kidney, and inorganic lead has a large range of adverse systemic effects. On the other hand, the more amphiphilic methylmercuric halides and triethyllead salts are primarily neurotoxic.

Cadmium salts are also largely nephrotoxic, but can enter the CNS by the olfactory route where the blood-brain barrier is attenuated. Anosmia is characteristic of cadmium intoxication. Interestingly, the complex that cadmium forms with diethyldithiocarbamate increases the access of this metal to the brain but reduces its neurotoxicity (O'Callaghan and Miller, 1986). However, the extended presence of a relatively inert metal chelate within the CNS may have unforseeable consequences.

Although peroxidative events are unlikely to constitute the primary mechanism of lead toxicity, there is evidence that both inorganic and organic lead derivatives can enhance the generation of ROS (Sifri and Hoekstra, 1978; Ramstoek et al., 1980; Munter et al., 1989). This enhancement has also been specifically demonstrated within the CNS (Gelman et al., 1979; Rehman, 1984; Ali and Bondy, 1989). Since 5-aminolevulinic acid, a heme precursor, accumulates during lead poisoning and is capable of inducing ROS, the mechanism by which lead can generate free radicals may be indirect (Hermes-Lima et al., 1991).

The next element of the sulfur-containing family in the periodic table is selenium, and metals binding to sulfur also have an affinity for selenium. Selenium has been shown to attenuate the toxicity of cadmium, arsenic, mercury, thallium, and copper. The only known mammalian enzyme in which selenium is an essential cofactor is glutathione peroxidase. This enzyme uses GSH to reduce organic hydroperoxides and is an important defense against oxidant damage (Hoekstra, 1975). Thus the protective properties of selenium imply oxidative events as a basis for the toxicity of the above elements.

Cadmium and mercury, but not lead, are able to induce metallothionein in several organs (Kojima and Kagi, 1978). This cysteine-rich protein, which is capable of sequestering sulfhydryl-binding metals, is not present to a significant extent in neural tissues.

Complex interactions with both synergistic and antagonistic potential exist between cadmium, mercury, and lead (Shubert et al., 1978). Such synergism implies action at different sites, while antagonism may be related to competition of a less toxic metal with a more toxic metal, for a common -SH target.

IV. METALS WITH THE CAPACITY TO INCREASE NEURAL EXCITABILITY Ca, Sn (ORGANIC), Pb (ORGANIC)

The normal physiological route of elevation of rates of neuronal firing involves influx of calcium into the presynaptic area, thereby eliciting neurotransmitter release. Cytosolic calcium levels are transiently but greatly elevated, before homeostasis is restored by rapid sequestration of excess calcium into endoplasmic reticulum and mitochondria, followed by eventual extrusion of calcium into the extracellular space using ion pumping or exchange processes. There is evidence that chronic excitation may result in failure to restore resting levels of calcium and that this can initiate excess ROS generation by several means. Mechanisms may include activation of phospholipases and the arachidonic acid cascade (Pazdernik et al., 1992). Calcium can also activate superoxide production by polymorphonuclear lymphocytes which can be present in postischemic neural tissues (Zimmerman et al., 1989). By these means, calcium may act as a mediator of oxidative stress within neurons in a variety of disease states involving persistent excitation, including ischemia, epilepsy, and chemically induced hyperactivity (White et al., 1984).

Both exogenous antioxidants and calcium chelators can inhibit lipid peroxidation and generation of superoxide anion following postischemic reperfusion (Vanella et al., 1992). The degree to which cytosolic calcium elevation and pro-oxidant events occur independently or act in concert remains to be unraveled.

Several metals can indirectly evoke ROS by way of disruption of normal calcium homeostasis enabling release of excess neurotransmitter. Lead is known to interfere with calcium metabolism, and lead compounds have been reported to enhance rates of lipid peroxidation within the brain (Rehman, 1984). Organic lead compounds, which are unlikely to closely mimic calcium, can also induce oxidant conditions in cerebral tissues (Ali and Bondy, 1989). In this case, depolarization-induced excitotoxicity may underlie such elevations in ROS generation.

Aliphatic organic tin compounds, such as trimethyl- and triethyltin, are potent neurotoxic agents. Trimethyltin is an excitatory agent with a rather high degree of selectivity toward the hippocampus. While this agent has no ability to induce ROS in isolated preparations, it can specifically elevate hippocampal ROS in treated rats (LeBel et al., 1990). This region-specific stimulation is likely to be due to the susceptibility of hippocampal circuitry to excitatory events, rather than to any distinctive biochemical features of this area. Calcium may here once again be the primary effector of excess production of ROS.

V. ALUMINUM AND ALZHEIMER'S DISEASE

Aluminum does not fit clearly into any of the classes of metals described here. However, this element may be capable of inducing ROS within neural tissues, and this property may play a role in the etiology of Alzheimer's disease (AD). While such a role is controversial, aluminum has been recognized as a significant factor in dialysis dementia for some time. However, it is uncertain whether an excessive intraneuronal accumulation of aluminum is found in AD, and whether this has a causal relation to the disease. Recent studies have reported the following:

1. Aluminosilicate complexes are capable of stimulating ROS generation in isolated glial cultures (Evans et al., 1992).
2. Levels of superoxide dismutase and catalase are elevated in AD brains, implying a response to oxidative stress (Papparella et al., 1992).
3. Chelation therapy with deferoxamine, in order to reduce the body burden of aluminum, may retard the rate loss of intellectual performance in AD (Crapper-McLachlan et al., 1991). Since iron is also sequestered by this chelator, this report cannot be considered conclusive evidence for a role of aluminum in AD. However, aluminum salts have been reported to strongly enhance the pro-oxidant potential of iron salts in an isolated system (Gutteridge et al., 1985; Fraga et al., 1990; Oteiza et al., 1993).
4. Glutamine synthetase, an enzyme that is very susceptible to oxidative degradation, is depressed in brains from AD patients in relation to an aged control group (Smith et al., 1992). This depression is confined to the frontal cortex, in parallel with the neuropathologic involvement of this region.
5. The iron-binding protein, transferrin, is depressed in AD (Connor et al., 1992), and levels of lipid peroxidation are elevated (Subbarao et al., 1990).

While this evidence is circumstantial, and great advances have been made in understanding of the role of amyloid protein in AD, there is a good possibility that pro-oxidant events and derangements of metal ion balance contribute to AD. Aluminum and iron have recently been reported to promote the aggregation of β-amyloid peptide (Mantyh et al., 1993).

The mechanism by which aluminum can promote oxidative stress is unclear. Subtle modification of membrane strucures allowing increased availability of peroxidizable fatty acids has been proposed (Gutteridge et al., 1985, Oteiza et al., 1993), and another possibility is that formation of aluminium/iron-containing mineral particulates can promote oxidative injury by stimulation of macrophage-like microglial events (Evans et al., 1991). The issue of whether ionic or complexed aluminum is the active agent in these events is currently controversial (Oteiza et al., 1993, Garrel et al., 1993) and its resolution would clarify processes underlying aluminum-catalyzed oxidative events.

REFERENCES

Ali, S. F. and Bondy, S. C. (1989), *J. Toxicol. Environ. Health*, 26, 235–242.

Aust, S. D. and Miller, D. M. (1991), in *New Horizons in Molecular Toxicology*, Lilly Research Laboratories Symp., pp. 29–34.

Babcock, G. M. and Wikstrom, M. (1993), *Nature,* 356, 301–309.
Birkmayer, W., Knoll, J., and Reiderer, P. (1985), *J. Neur. Trans.,* 64, 113–117.
Bondy, S. C. (1992), *Neurotoxicology,* 13, 87–100.
Bondy, S. C. and Pearson, K. R. (1993), *Alc. Clin. Exp. Res.,* 17, 651–654.
Boveris, A. and Chance, B. (1973), *Biochem. J.,* 134, 707–716.
Cederbaum, A. I. (1989), *Free Radicals Med. Biol.,* 7, 559–567.
Chang, L. W., Gilbert, M., and Sprecher, J. (1978), *Environ. Res.,* 17, 356–366.
Chang, L. W. and Suber, R. (1982), *Bull. Environ. Contam. Toxicol.,* 29, 285–289.
Cheton, P. L. B. and Archibald, F. S. (1988), *Free Radicals Biol. Med.,* 5, 325–333.
Coassin, M., Ursini, F., and Bindoli, A. (1992), *Arch. Biochem. Biophys.,* 299, 330–333.
Connor, J. R., Snyder, B. S., Beard, J. L., Fine, R. E., and Mufson, E. J. (1992), *J. Neurosci. Res.,* 31, 327–335.
Crapper-McLachlan, T. P. A., Dalton, A. J., Kruck, T. P., Bell, M. Y., Smith, W. L., Kalow, W., and Andrews, D. F. (1991), *Lancet,* 337, 1304–1308.
Czapiski, G. and Goldstein, S. (1986), *Free Radical Res. Commun.,* 3, 157–161.
Dexter, D. T., Carayon, A., Vidailhet, M., Ruberg, M., Agid, F., Agid, Y., Lees, A. J., Wells, F. R., Jenner, F. R., and Marsden, C. D. (1990), *J. Neurochem.,* 55, 16–20.
Donaldson, J., Labella, F. S., and Gesser, D. (1981), *Neurotoxicology,* 2, 53–64.
Evans, P. H., Klinowski, J., and Vano, E. (1991), *Med. Hypoth.,* 35, 209–219.
Evans, P. H., Peterhans, E., Burge, T., and Klinowski, J. (1992), *Dementia,* 3, 1–6.
Fraga, C. G., Oteiza, P. I., Golub, M. S., Gershwin, M. E., and Keen, C. L. (1990), *Toxicol. Lett.,* 51, 213–219.
Garrel, C., Lafond, J. L., Faure, P., and Favier, A. (1993), *Proc. 2nd. Int. Symp. Reactive Oxygen Species,* II. 18.
Gelman, B. B., Michealson, I. A., and Bornschein, R. L. (1979), *J. Toxicol. Environ. Health,* 5, 683–698.
Grundke-Iqbal, I., Fleming, J., Tung, Y. C., Lassmann, H., Iqbal, K., and Joshi, J. G. (1990), *Acta Neuropathol.,* 81, 105–110.
Gutteridge, J. M. C., Westermark, T., and Santavuori, P. (1983), *Acta Neurol. Scand.,* 68, 365–370.
Gutteridge, J. M. C., Quinlan, G. J., Clark, I., and Halliwell, B. (1985), *Biophys. Biochim. Acta,* 835, 441–447.
Haider, S. S. and El-Fakhri, M. (1991), *Neurotoxicology,* 12, 79–85.
Halliwell, B. (1989), *Free Radicals Biol. Med.,* 7, 645–651.
Hasan, M. and Ali, S. F. (1981), *Toxicol. Appl. Pharmacol.,* 57, 8–13.
Hermes-Lima, M., Periera, B., and Bechara, E. J. H. (1991), *Xenobiotica,* 21, 1085–1090.
Hoekstra, W. G. (1975), *Fed. Proc.,* 34, 2083–2089.
Ikeda, Y., Ikeda, K., and Long, D. M. (1989), *J. Neurosurg.,* 71, 233–238.
Jellinger, K. A., Kienzl, E., Rumpelmaier, G., Paulus, W., Reiderer, P., Stachelberger, H., Youdim, M. B. H., and Ben-Shacker, D. (1993), *Adv. Neurol.,* 60, 267–272.
Klebanoff, S. J., Waltersdorph, A. M., Michel, B. R., and Rosen, H. (1989), *J. Biol. Chem.,* 264, 19765–19771.
Kojima, Y. and Kagi, J. H. R. (1978), *Trends Biochem. Sci.,* 3, 90–92.
Kruck, T. P. A., Fisher, E. A., and McLachlan, D. R. C. (1990), *Clin. Pharm. Therap.,* 48, 439–446.
LeBel, C. P., Ali, S. F., McKee, M., and Bondy, S. C. (1990), *Toxicol. Appl. Pharmacol.,* 104, 17–24.
LeBel, C. P., Ali, S. F., and Bondy, S. C. (1992), *Toxicol. Appl. Pharmacol.,* 112, 161–165.
Le Vine, S. M. (1992), *Med. Hypoth.,* 39, 271–274.
Mantyh, P. W., Ghilardi, J. R., Rogers, S., Demasters, E., Allen, C. J., Stimson, E. R., and Maggio, J. E. (1993), *J. Neurochem.,* 61, 1171–1174.
Marzabadi, M. R. and Jones, C. B. (1992), *Mech. Ageing Dev.,* 66, 159–171.
Minotti, G. and Aust, S. D. (1989), *Chem. Biol. Interact.,* 71, 1–19.
Munter, K., Athanasiou, M., and Stounaras, C. (1989), *Biochem. Pharmacol.,* 38, 3941–3945.
Naganuma, A., Anderson, M. E., and Meister, A. (1990), *Biochem. Pharmacol.,* 40, 693–697.
O'Callaghan, J. P. and Miller, D. B. (1986), *Brain Res.,* 370, 354–358.
Ogawa, N., Edamatsu, R., Mizukawa, K., Asanuma, M., Kohno, M., and Mori, A. (1993), *Adv. Neurol.,* 60, 242–250.
Oteiza, P., Fraga, C. G., and Keen, C. L. (1993), *Arch. Biochem. Biophys.,* 300, 517–521.
Papparella, M. A., Omar, R. A., Kim, K. S., and Robakis, N. K. (1992), *Am. J. Pathol.,* 104, 621–628.
Parkinson's Study Group (1989), *Arch. Neurol.,* 46, 1052–1060.
Pazdernik, T. L., Layton, M., Nelson, S. R., and Samson, F. E. (1992), *Neurochem. Res.,* 17, 11–21.
Perry, T. L., Norman, M. G., Yong, W. V., Whiting, S., Crichton, J. U., Hansen, S., and Kish, S. J. (1985), *Ann. Neurol.,* 18, 482–489.
Prat, A. G., and Turrens, J. F. (1992), *Free Radicals Biol. Med.,* 8, 319–325.
Ramstoek, E. R., Hoekstra, W. G., and Ganther, H. E. (1980), *Toxicol. Appl. Pharmacol.,* 54, 251–257.
Reddy, C. C. and Massaro, E. J. (1983), *Fundam. Appl. Toxicol.,* 3, 431–436.
Rehman, S. U. (1984), *Toxicol. Lett.,* 21, 333–337.
Reif, D. W. and Simmonds, R. D. (1990), *Arch. Biochem. Biophys.,* 283, 537–541.
Rosenthal, R. E., Chanderbahn, R., Marshall, G., and Fiskum, G. (1992), *Free Radicals Biol. Med.,* 12, 29–33.
Rouach, H., Houze, P., Orfanelli, M. T., Gentil, M., Bourdon, R., and Nordmann, R. (1990), *Biochem. Pharmacol.,* 39, 1095–1100.
Sadrzadeh, S. M., Anderson, D. K., Panter, S. S., Hallaway, P. E., and Eaton, J. W. (1987), *J. Clin. Invest.,* 79, 662–664.
Savoliainen, K. (1978), *Res. Commun. Chem. Pathol. Pharmacol.,* 21, 173–175.

Scheinberg, I. H. (1988), in *Metal Neurotoxicity,* Bondy, S. C. and Prasad, K. N., Eds., CRC Press, Boca Raton, FL, 56–60.
Schubert, J., Riley, E. J., and Tyler, S. A. (1978), *J. Toxicol. Environ. Health,* 4, 763–776.
Seth, P. K. and Chandra, S. V. (1988), in *Metal Neurotoxicity,* Bondy, S. E. and Prasad, K. N., Eds., CRC Press, Boca Raton, FL, 19–33.
Shaw, S. (1989), *Free Radicals Biol. Med.,* 7, 541–547.
Shoham, S., Wertman, E., and Ebstein, R. P. (1992), *Exp. Neurol.,* 118, 227–242.
Shoulson, I. (1989), *Acta Neurol. Scand.,* 80 (Suppl. 126), 171–175.
Shukla, G. S., Srivastava, R. S., and Chandra, S. V. (1988), *J. Appl. Toxicol.,* 8, 355–358.
Sifri, E. M. and Hoekstra, W. G. (1978), *Fed. Proc.,* 37, 757.
Smith, C. D., Carney, J. M., Tatsumo, T., Stadtman, E. R., Floyd, R. A., and Markesbery, W. R. (1992), *Ann. N. Y. Acad. Sci.,* 636, 110–119.
Stadtman, E. R. (1990), *Free Radicals Biol. Med.,* 9, 315–325.
Stadtman, E. R. (1993), *Annu. Rev. Biochem.,* 62, 797–821.
Subbarao, K. V., Richardson, J. S., and Ang, L. C. (1990), *J. Neurochem.,* 55, 342–355.
Thomas, D. J. and Chisholm, J. J. (1986), *J. Pharmacol. Exp. Ther.,* 239, 829–833.
Turrens, J. F. and Boveris, A. (1980), *Biochem. J.,* 191, 421–427.
Vanella, A., Sorrentini, V., Castorina, C., Campisi, A., Di Giacomo, C., Russo, A., and Perez-Polo, J. R. (1992), *Int. J. Dev. Neurosci.,* 10, 75–80.
Wahba, Z. Z., Murray, W. J., and Stohs, S. J. (1990), *J. Appl. Toxicol.,* 10, 119–124.
White, B. C., Aust, S. D., Arfors, K. E., and Aronson, L. D. (1984), *Ann. Emerg. Med.,* 13, 862–867.
Willmore, L. J. and Triggs, W. J. (1991), *Int. J. Dev. Neurosci.,* 9, 175–180.
Wurtman, R. J. and Wurtman, J. J. (1990), in *Nutrition and the Brain,* Wurtman, R. J. and Wurtman, J. J. Eds., Raven Press, New York, 59–74.
Yonaha, M., Saito, M., and Sagan, M. (1983), *Life Sci.,* 32, 1507–1514.
Youdim, M. B. H., Ben-Shachar, D., and Reiderer, P. (1989), *Acta Neurol. Scand.,* 126, 47–66.
Youdim, M. B. H., Ben-Shacher, D., Eshel, G., Finberg, J. P. M., and Riederer, P. (1993), *Adv. Neurol.,* 60, 259–266.
Zimmerman, J. J., Zuk, S. M., and Millard, (1989), *Biochem. Pharmacol.,* 38, 3601–3610.

Chapter 42

Disruption of Protein Synthesis as Mechanistic Basis of Metal Neurotoxicity

M. Anthony Verity

The purpose of this review is to identify events in the overall process of protein synthesis which may be modified by metals. Specifically, attention will be directed to identified steps in translation within neurosystem-derived tissue known to be influenced by heavy metals or their alkyl derivatives. In order to identify and understand potential or identified steps in translation, a short discussion of translation will be presented identifying portions of the overall synthetic pathway which may be particularly sensitive to such interruption. Appropriate references for this description include Schimmel (1987); Mitra et al. (1982); Kaziro (1978); Proud (1986); Moldave (1985); Jimenez (1976); Hershey (1991). An analysis of selected metal or organometal species will follow with final reference to modification of posttranslational states.

I. GENERAL DESCRIPTION OF PROTEIN SYNTHESIS

Proteins are synthesized in a process called translation on ribosomes by the stepwise addition of amino acids to the carboxyl end of a peptide chain. The activated precursor amino acids are presented in the form of aminoacyl-tRNA formed by the linking of an amino acid to its specific tRNA by an aminoacyl-tRNA synthetase. Synthesis takes place in three stages: initiation, elongation, and termination. **Initiation** represents the binding of initiator tRNA to the start codon of mRNA whereby the initiator tRNA occupies the peptidyl site on the ribosome. **Elongation** is the addition of activated amino acids to a growing polypeptide sequentially dictated by messenger RNA. Finally, **termination** occurs when a stop signal on the messenger RNA is recognized by a specific protein-releasing factor leading to separation of the completed polypeptide chain from the ribosome.

A. PROCESS OF INITIATION

The first event in the initiation pathway is the binding of initiator tRNA to the 40S ribosome subunit. The complex thus formed (often called the ternary complex) needs the controlled interaction of GTP and initiation factor-2 (IF-2). The subsequent binding to the 40S ribosomal subunit is modulated by the presence of IF-3 which prevents the 40S subunit from associating with the 60S subunit, thereby forming the complete 40S-met-tRNA-IF-GTP complex. It is of interest that while GTP is needed, the reaction also proceeds in the presence of nonhydrolyzable analogs of GTP. Once the ternary complex-40S subunit is formed, the assembled complex is now competent to bind messenger RNA. This event requires additional factors, collectively ATP and IF-4 (in this event nonhydrolyzable analogs of ATP will not substitute). The 60S subunit probably joins the 40S-mRNA-met-tRNA complex when the anticodon has engaged the initiator AUG codon. Initiation factor IF-5 is needed as well as GTP hydrolysis, with ultimate binding to the P-site of the 60S subunit from which they can form a peptide bond with incoming amino acid or with puromycin. Such 60S ribosome binding cannot occur until IF-3 has left, since this factor inhibits subunit association. Moreover, release of all the initiation factors absolutely requires GTP hydrolysis. The 80S initiation complex now occupies the P-site on the ribosome and is poised for the elongation phase of protein synthesis. The A-site is empty.

B. PROCESS OF ELONGATION

The elongation cycle in protein synthesis begins with the insertion of an aminoacyl-tRNA into the empty A-site on the ribosome. The particular kind inserted depends on the mRNA codon that is positioned at the A-site. Amino acids are activated and linked to specific transfer RNA by specific synthetases. The first step is the formation of an aminoacyl-adenylate from amino acid and ATP, followed by the transfer of the aminoacyl group to a specific tRNA molecule to form the activated intermediate, aminoacyl-tRNA. Two high-energy phosphate bonds are utilized. At least one aminoacyl-tRNA synthetase exists for each amino acid, differing markedly in molecular size, and such synthetases are highly selective in their ability to recognize both tRNA and amino acid to be activated. The complementary aminoacyl-tRNA is delivered to the A-site by a protein, elongation factor 1 (EF-1) homologous to elongation factor-Tu (EF-Tu) in prokaryotes. EF-1 contains a guanyl nucleotide which cycles between GTP and GDP. Once the factor has positioned the aminoacyl-tRNA in the A-site, then GTP is hydrolyzed and the GDP form of EF-1 dissociates from the ribosome. The accuracy of protein synthesis depends on the correct insertion of aminoacyl-tRNA in the A-site.

Peptide-bond formation is catalyzed by the peptidyl transferase site of the ribosome. This is an enzymatic site on the 60S subunit and allows the elongating chain to be transferred to the amino group of the incoming aminoacyl-tRNA. At this stage, the P-site is occupied by an uncharged tRNA, whereas a dipeptidyl-tRNA occupies the A-site. The next phase of elongation is translocation whereby the peptidyl-tRNA moves from the A-site to the P-site and the mRNA moves a distance of three nucleotides representing positioning of the next codon for reading the incoming aminoacyl-tRNA. Again, translocation requires a further elongation factor, EF-2, a translocase which cycles between a GTP and GDP form.

Diphtheria toxin blocks protein synthesis by inhibiting EF-2 mediated translocation. Inhibition occurs by the toxin-mediated transfer of the ADP-ribose unit of NAD to a modified histidine group in EF-2. Such ADP-ribosylation of EF-2 blocks subsequent translocation of the growing polypeptide chain with ultimate cessation of protein synthesis and cellular toxicity.

C. PROCESS OF TERMINATION

Aminoacyl-tRNA does not bind to the A-site of a ribosome if the codon in place is UAA, UGA, or UAG. If these codons are in place, a release factor (eRF) is recognized that binds with high specificity and activates peptidyl transferase which hydrolyzes the bond between polypeptide and tRNA in the P-site. As the polypeptide chain leaves the ribosome, the ribosome dissociates into 40- and 60S subunits prior to the beginning of initiation again.

II. CONTROL OF PROTEIN SYNTHESIS

A. MODULATION BY ENERGY CHARGE

Macromolecular synthetic reactions, especially protein synthesis *in vivo* or *in vitro*, are responsive to small changes in the energy state of the system. Minor increases in either (ADP) or (AMP) have been shown to regulate the overall rate of protein biosynthesis acting at the level of initiation (Young, 1969; Hucul et al., 1985). Using cell-free protein-synthesizing systems, evidence that both initiation and elongation may be influenced by changes in energy charge has been obtained (Henshaw and Panniers, 1983). Further, an inhibitory effect of AMP on the aminoacylation reaction has also been reported (Marshall and Zamecnik, 1970), while Mosca et al., (1983) found that AMP specifically inhibited peptide chain elongation not coupled to the aminoacylation of tRNA, but reversed by enzymatic removal of AMP. While the modulation of *in vivo*/intact cell protein synthesis may be directly coupled to (ATP) and/or energy charge, some secondary mechanisms may be influenced by (ATP), e.g., with activity of Na+K-ATPase. Also, amino acid activation and linkage to specific transfer RNA by specific synthetase requires ATP and the utilization of two ~P bonds in the synthesis of an aminoacyl-tRNA. Notably, a molecule of AMP and pyrophosphate is produced, the latter shown to be inhibitory.

The cycling of GTP:GDP is relevant at numerous steps. Walton and Gill (1976) have demonstrated the potential of the GTP:GDP ratio to determine the extent of formation of the ternary initiation complex through competition for the nucleotide binding site on IF-2. Moreover, the guanine nucleotide ratio is responsive to the adenylate energy charge via nucleoside diphosphate kinase (Hucul et al., 1985). Similarly, a well-controlled reaction cycle for GTP/GDP in the activity of EF-1 and activity of the translocase (EF-2) are well documented, details of which are not significant here, but may be found (Proud, 1986; Kaziro, 1978).

Protein synthesis in reticulocyte extracts ceases in the absence of heme due to the formation of a heme-controlled inhibitor discovered to be a cyclic AMP independent kinase that phosphorylates the alpha subunit of IF-2, the initiation factor binding met-tRNA to the 40S ribosomal subunit. Phosphorylated IF-2 molecules cannot initiate protein synthesis because they continually bind the guanyl nucleotide exchange factor providing an irreversible complex inhibiting normal GDP:GTP exchange. Consequently, insufficient GTP is available to pick up IF-2 to initiate a further round of synthesis. It is possible that the "energy charge" of the cell (see above) may also regulate initiation as GDP has a 100-fold higher affinity for IF-2 than GTP and does not allow the binding of initiator met-tRNA.

B. MODULATION OF TRANSLATION BY PHOSPHORYLATION AND PROTEIN KINASE ACTIVITY

From the studies of heme regulation of translation in reticulocytes, and the identification of a hemin-regulated protein kinase, kinase-dependent activation has revealed the phosphorylation of the alpha subunit of eIF-2. Phosphorylation of IF-2 blocks the recycling of the factor because the phosphorylated form has such a high affinity for the guanyl nucleotide exchange factor (GEF), thereby stoichiometrically binding available GEF and irreversibly complexing the molecule. Hence, phosphorylated IF-2 is unable to initiate protein synthesis. The phosphorylated IF-2 becomes functional again following removal of the phosphate group by a specific phosphatase. Although well described in the reticulocyte, it appears likely that IF-2 phosphorylation by different kinases may control translation in other cells. For instance, translational repression by transition metal compounds occurs by different pathways, one of which is associated with alterations in protein phosphorylation (Duncan and Hershey, 1987).

C. ROLE OF GLUTATHIONE AND SULFHYDRYL GROUP OXIDATION-REDUCTION

The activity of IF-2 may be regulated by processes unrelated to the degree of phosphorylation, or phosphatase activity, or IF-2 alpha kinase activity. Changes in the oxidation-reduction state of -SH groups of proteins affect the *in vitro* and *in vivo* characteristics. Protein sulfhydryl groups may be regulated by disulfide exchange or by oxidative reduction, for instance glutathione reductase. Jagus and Safer (1981) have demonstrated that IF-2 in lysates is modified by the oxidation-reduction state of its -SH groups. Particularly, they found that glucose was required for the generation of NADPH which, in the presence of catalytic amounts of glutathione, maintained the sulfhydryl groups of IF-2 in an active form. Hence, oxidation of such groups will interfere with the capacity of IF-2 to interact with GTP.

Earlier, Ernst et al. (1978) had revealed that oxidized glutathione (GSSG) inhibited protein synthesis in hemin-supplemented lysates by the activation of the inhibitory protein kinase that phosphorylates IF-2. In this respect GSSG inhibition mirrors the inhibition produced by heme deficiency (*vide supra*). The mechanism by which GSSG activates the kinase is not understood, but some data reveal that glucose-6-phosphate can reverse the inhibition by GSSG, revealing a direct role for the phosphorylated sugar in the regulation of initiation, possibly by NADPH regeneration.

III. PROTEIN METABOLISM DISRUPTION AS A MECHANISTIC COMPONENT OF METAL NEUROTOXICITY

In this section, selected inorganic and organic metal compounds will be analyzed for their role in producing defects in translation, *in vivo* or *in vitro*, and/or changes in posttranslational protein modification. Limited data are available for most metals, but key studies in some instances have identified either a sensitive defect in translation both *in vivo* or *in vitro*, e.g., alkylmercury or a primary posttranslational modification of a critical neuronal protein, e.g., neurofilaments of aluminum neurotoxicity. In many instances, the data may prove causal to the ultimate disturbance in neurofunction. The general interactions of metals on the protein synthesis path are schematically represented in Figure 1.

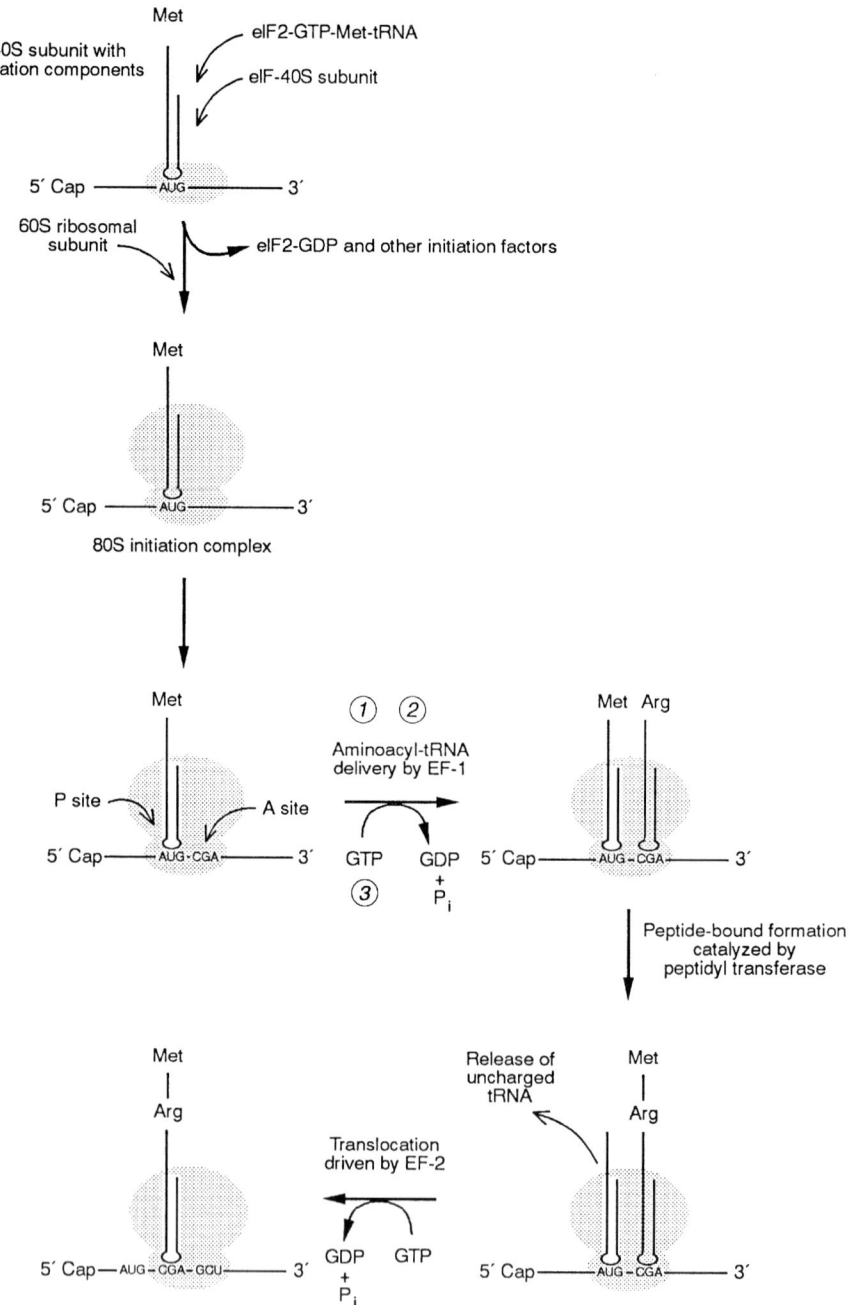

Figure 1. Schematic for protein synthesis (translation) pathway referred to in text. Protein synthesis occurs in three phases: initiation, elongation, and termination. The initiator tRNA is methionine, which is bound to the 40S ribosome subunit by the GTP form of IF-2. The larger 60S subunit binds to the smaller subunit to form the 80S initiation complex under the influence of ATP and GTP hydrolysis. The elongation cycle of protein synthesis consists of the binding of an aminoacyl-tRNA to the A-site; peptide-bond formation; and translocation. Aminoacylation of tRNA is carried out by selective synthetases and activated (ATP) aminoacids. Translocation consists of removal of uncharged tRNA from the P-site, followed by movement of the peptidyl-tRNA from the A- to the P-site under the influence of a translocase, coupled to GTP-GDP cycling. Numbers on figure indicate sites of metal-induced modification. (1) Inhibition of aminoacyl-adenylation secondary to change in ATP/ADP ratio, e.g, methylmercury *in vivo;* alkyltin *in vivo.* (2) Selective aminoacyl-tRNA synthetase inhibition, e.g., methylmercury. (3) Change in ratio/availability GTP secondary to shift in ATP/ADP or energy charge, e.g., methylmercury, trialkyllead, or trialkyltin.

A. ALUMINUM

Aluminum has been implicated in numerous neurological disorders, including Alzheimer's disease, amyotrophic lateral sclerosis/dementia complex of Guam, and dialysis encephalopathy. Even so, the causal role of aluminum in the genesis of Alzheimer's disease remains very controversial. Acute and chronic experimental models of aluminum neurotoxicity have proven valuable, especially in revealing the genesis of disturbed neurofilament aggregation in aluminum-treated cells in culture or *in vivo*. This review will concentrate on the experimental model, paying little attention to Alzheimer's disease per se. The acute intracerebral injection of aluminum in rabbits (DeBoni, 1974; Crapper et al., 1980) revealed aluminum to be almost exclusively associated with nuclear chromatin. Selective accumulation appeared rapidly, and suggested an early effect on transcription. However, no major effect on certain messenger RNA levels has been found, e.g., neurofilament mRNA and no major difference in protein synthesis by free or membrane-bound polyribosomes in a cell-free system was observed in preparations from aluminum-treated rabbits (Czosnek et al., 1978). It is therefore likely that the major protein biosynthetic abnormalities reside in posttranslational processing of cytoskeletal proteins. Table 1 identifies some events coupled to defects in posttranslational processing. Two major abnormalities have been studied in detail, (1) an increase in neurofilament phosphorylation (Bizzi and Gambetti, 1986; Johnson and Jope, 1988; Johnson et al., 1989); and (2) induction of neurofilament aggregation and formation of intraneuronal neurofilament tangles (Miller and Levine 1974; Hewitt et al., 1991). Based on these observations, it is likely that the appearance of Al^{3+}-induced neurofilament aggregates lead to a continuous change in their phosphorylation state under the dynamic influence of protein kinase and phosphatase modulation. Shea et al. (1988; 1989) have revealed the formation of phosphorylated neurofilaments in neuroblastoma cell culture associated with posttranslational modification and incorporation into the cytoskeleton during differentiation. These studies provide a model of a differentiating neuronal cell culture for elucidating the role of Al^{3+}-neurofilament interaction. Such reasoning is supported by the fact that Al^{3+} induces high molecular weight neurofilament protein complexes *in vitro* (Nixon et al., 1990) that are protease resistant. Moreover, phosphorylated neurofilament protein epitopes are found associated with a selective impairment in the axonal transport of such neurofilaments (Troncoso et al., 1986). Other biochemical studies have shown that the phosphorylated state of the neurofilament proteins *in vitro* alters their electrophoretic property (Shea et al., 1992).

Table 1 Aluminum-induced Changes in Protein Expression and Posttranslation Modification

	Ref.
Al^{3+} binds to nuclear chromatin	Karlik et al., 1980
Suppression of gene transcription	Krekoski et al., 1988
	Muma et al., 1988
Defective axonal transport of neurofilaments	Bizzi et al., 1984
	Troncoso et al., 1985
	Kosik et al., 1985
Impaired proteolysis of neurofilament proteins	Nixon et al., 1990
Neuronal neurofilament aggregation,	Miller and Levine, 1934
in vivo and *in vitro*	Selkoe et al., 1979
	Shea et al., 1989
	Hewitt et al., 1991
Perikaryal accumulation of phosphorylated	Bizzi and Gambetti, 1986
NF epitopes	Shea et al., 1989
	Johnson and Jope, 1988

B. TIN

The well-documented neurotoxicity produced by organotin stands in marked contrast to the limited studies on inorganic tin neurotoxicity. Hence, the most neurotoxic organotin compounds, triethyltin (TET) and trimethyltin (TMT) have been used to delineate the mechanisms of neurochemical and behavioral toxicity, recently reviewed by Aschner and Aschner (1992). Triethyltin (TET) has been shown to produce generalized edema in the CNS, a characteristic myelinopathy, and a disturbance of cerebral energy metabolism likely associated with the mitochondriopathy induced by disturbed ion flux across the mitochondrial membrane. In contrast, triethyltin (TET) produces specific neuronal degeneration, especially in primary sensory neurons, the hippocampus and brain stem often accompanied by alterations in regional dopamine and GABA.

Although not necessarily mechanistically coupled, a close association has been found between TET-induced brain edema, vacuolar myelinopathy, and the activity of aminoacyl-tRNA synthetases (Wender et al., 1975, 1974). Following a single intraperitoneal injection of TET (4 mg/kg body weight) a selective inhibitory effect on the activity of several RNA-aminoacyl-synthetases was observed in the white matter. The data, however, may only partially explain previous observations of a decrease in amino acid incorporation. For instance, the leucyl-specific enzymes revealed increased activity at a time when this amino acid was incorporated at a decreased rate. O'Callaghan et al. (1983) showed that acute postnatal exposure to TET in the rat produced defects in specific protein composition of subcellular factions. For instance, administration of TET at postnatal day 5 produced a dose-related decrease in the concentration of myelin basic protein. Myelin basic protein may play a role in stabilizing the compaction of myelin-lamellae (Smith, 1977) allowing for speculation on the mechanism of TET-induced vacuolar myelinopathy. Tributyltin differentially inhibited macromolecular synthesis in T cells, associated with a concomitant change of ATP (Snoeij et al., 1988). Such studies draw attention to a further mechanism for organotin inhibition of protein synthesis, coupled to a mediated Cl–OH exchange across mitochondrial membranes, ATPase activation, and inhibition of oxidative phosphorylation (Selwyn et al., 1970; Aldridge et al., 1977; Kauppinen et al., 1988).

Trimethyltin (TMT) has a cytotoxic effect limited to neurons of the brain, retina, and peripheral nervous system. The neuronal injury begins with alterations in the Golgi apparatus and endoplasmic reticulum, suggesting that these may be target organelles for the neurotoxicity (Bouldin et al., 1984; Brown et al., 1984). Such target organelles are involved in macromolecular synthesis and posttranslational processing. In a study of protein and glycoprotein synthesis in the retina following TMT by gastric intubation (4 mg/kg body weight), Toews et al. (1986) revealed an increased synthesis of membrane macromolecules as an early response of retinal neurons to TMT intoxication. Rare focal receptor cell necrosis was observed, but the changes previously described in the more acute model (Bouldin et al., 1984) were found rarely.

C. MERCURY

The majority of studies associating mercury with neurotoxicity have concentrated on the alkyl mercurials. Early morphological studies revealed dissolution and ribosomal disaggregation in the rough endoplasmic reticulum of brain cells (Brown and Yoshida, 1965; Chang and Hartmann, 1972; Jacobs et al., 1975). Not surprisingly, the studies of Yoshino et al. (1966) demonstrated an inhibition of protein synthesis in brain slices during the latent phase of organic mercury intoxication in rats, an observation confirmed *in vivo* and *in vitro* by subsequent studies (Cavanagh and Chen, 1971; Verity et al., 1977; Syversen 1977; Omata et al., 1978; Cheung and Verity, 1983; Sarafian et al., 1984). Since those studies, the underlying mechanistic basis for the inhibition of protein synthesis in neural derived systems has been sought. Major mechanisms which have been well documented to modulate translation include first, a decrease in ATP or in ADP/ATP ratio as a manifestation of disturbed energy charge; second, a primary inhibition of aminoacyl-adenylation reflecting a disturbance in ADP/ATP ratio; and/or third, the selective inhibition by the alkylmercury of certain aminoacyl-tRNA synthetases.

1. Modification of Energy Charge

We have previously discussed the control of translation by the energy charge of the system *in vivo* or *in vitro*. Methylmercury has been found to inhibit oxidative phosphorylation, uncouple mitochondrial respiration, and inhibit ATP production in numerous systems (Verity et al., 1975; Sone et al., 1977; Cheung and Verity, 1981; Kauppinen et al., 1989). Other studies have revealed a proximate effect of the alkylmercury compounds on mitochondrial function, including Ca^{2+} flux (Levesque and Atchinson, 1991; Hare and Atchinson, 1992). Grundt and Bakken (1986) demonstrated alterations in the intracellular adenine nucleotide pool reflecting an aberrant energy charge state, documented in numerous studies, some of which simultaneously measured the coincident inhibition of protein synthesis (Cheung and Verity, 1981; Sarafian et al., 1984; Sarafian and Verity, 1985; Kuznetsov et al., 1986).

a. Studies on Mercurial Inhibition of Aminoacyl-Adenylation and Selective Aminoacyl-tRNA Synthetase Activity

Aminoacyl-tRNA synthesis involves two interconnective steps: the first representing an activated amino acid or aminoacyl-adenylate followed by the synthetase-mediated coupling to the appropriate aminoacyl-tRNA. Although the amino acid activation step and subsequent linkage to tRNA are considered separate events, they are catalyzed by the same aminoacyl-tRNA synthetase. The aminoacyl-AMP intermediate is tightly bound to the active site of the enzyme, and while normally a transient intermediate it is quite stable. As mentioned previously, two high-energy phosphate bonds are consumed in the formation of an aminoacyl-tRNA, and a decrease of ATP or alteration in energy charge may be expected to influence the formation of aminoacyl-adenylates.

In a comprehensive study, Cheung and Verity (1985) using a brain-derived cell-free translation system, observed both *in vivo* and *in vitro* methylmercury inhibition of polyuridylic acid-directed incorporation of [^{3}H] phenylalanine, pinpointing an early and sensitive defect in elongation. While no abnormality was found associated with the steps of initiation, ribosomal function or in the formation of peptidyl [^{3}H] puromycin, significant MeHg inhibition of phenylalanyl-tRNA synthetase (Phe-tRNA) was observed in the pH 5 enzyme fraction derived from the brain postmitochondrial supernatant of mercurial poisoned rats. Kuznetsov et al. (1987) also observed suppression of aminoacyladenylate synthesis by methylmercury *in vivo* and *in vitro*.

It is useful to note that aminoacyl-tRNA synthetases exist as multienzyme complexes (Dang and Dang, 1986). Numerous studies have identified the specific subcellular topology for individual synthetases indicating possible involvement of lipids in maintaining the structural integrity of the complex (Sihag and Deutscher, 1983) variation in size and subunit structure. Such observations may account for the considerable variability in the activity of aminoacyl-tRNA synthetases in rat brain following *in vivo* or *in vitro* methylmercury (Cheung et al., 1985; Hashegawa et al., 1988; Omata et al., 1991). While the studies support an inhibition of brain protein synthesis following *in vivo* methylmercury as a result of limiting amounts of aminoacyl-tRNA molecules, differences in the individual synthetase activities were found.

2. Mercurial Inhibition of Translation due to Altered GSH/GSSG Ratio and/or Oxidant-Induced Lipid Peroxidation

An activation of oxidative mechanisms has been shown to be associated with methylmercury neurotoxicity (Yonaha et al., 1983; Sarafian and Verity, 1991). A stimulation of lipoperoxide formation and reduction of cellular GSH was found in cerebellar granule cell suspensions or culture. Studies have revealed a relationship between diminished protein synthesis and lipid peroxidation (Gravela and Dianzani, 1970; Fraga et al., 1989). While the mechanism is unclear and is likely multifactorial, involving diminished energy charge, ribosomal dissociation, and increased membrane permeability, the oxidation of intracellular glutathione and/or drainage of NADPH may well represent a powerful signal for the inhibition of protein synthesis initiation. Intracellular elevation of oxidized glutathione (GSSG) activates a translation inhibitor (I-GSSG) which is accompanied by increased cyclic AMP-independent protein kinase activity capable of phosphorylating the alpha subunit of IF-2. GSH loss, binding, and increased GSSG occurs with methylmercury. Such phosphorylation prevents the binding of met-tRNA to the 40S ribosomal subunit. While direct evidence for the activation of this system in methylmercury inhibition of translation has not been provided, the key events are in place. (See also Section IV on stress protein induction).

3. Protein Phosphorylation in Neuron Culture after Exposure to Methylmercury

Exposure of cultured cerebellar granule neurons to methylmercury produces a dose-dependent stimulation of protein phosphorylation (Sarafian and Verity, 1990). The neuron-specific stimulation was not observed in cultured glial cells, not coupled to intracellular (^{32}P) ATP-specific activity or accounted for by the increased (^{32}P) incorporation into TCA-insoluble lipid. 2-D polyacrylamide gel electrophoresis revealed a nonuniform phosphorylation among specific proteins. Many proteins revealed no change or a slight decrease in labeling, while others displayed increases three- to fourfold (Kasama et al., 1989; Sarafian and Verity, 1990; Omata et al., 1991). We could not account for such variability in terms of activation or inhibition of specific protein kinases. Changes in ^{45}Ca uptake and inositol phosphate production imply an activation of phospholipase C by methylmercury which would be coupled to protein kinase C stimulation via diacylglycerol (Sarafian, 1993). Similar observations on select protein phosphorylation in the cytosol or particular proteins of methylmercury-treated C-6 glioma cells have been observed (Ramanujam and Prasad, 1979).

The vulnerability and high sensitivity of the developing brain to toxic effects of methylmercury are well established. An experimental study on the effect of methylmercury on neuroepithelial germinal cells in mice revealed a decrease in mitotic activity, abnormalities in neuroblast migration, and apoptotic degeneration (Choi, 1991). These findings lend further credence to the hypothesis linking methylmercury-induced abnormalities in microtubule polymerization/depolymerization to the genesis of neuronal dysplasia. Methylmercury has been shown to disassemble microtubules at the mitotic spindle (Miura et al., 1978); inhibit microtubule polymerization *in vitro* and promote disassembly (Imura et al., 1980; Sager et al., 1983; Miura et al., 1984; Sager, 1988), demonstrate microtubule loss *in vivo* (Choi 1991) and block lymphocyte response to the mitogen concanavalin (Brown et al., 1980). Miura and Imura (1991) have demonstrated microtubule disruption by methylmercury in cultured glioma cells prior to other morphological abnormality but associated with inhibition of cell proliferation. The increase in the cellular pool of tubulin subunits following methylmercury resulted in an inhibition of tubulin biosynthesis coupled to a specific decline of mRNA β-tubulin.

Cadmium, chromium, and copper also inhibited tubulin polymerization *in vitro*, but when present at growth inhibitory concentrations in glioma or neuroblastoma culture, had no effect on the microtubular network (Miura et al., 1984).

IV. METALS AND STRESS PROTEIN INDUCTION

Exposure of eukaryotic cells, including neurons and glia, to elevated temperature induces the synthesis of four species of "stress" proteins with molecular weights of 110, 90, 70 and approximately 30 kDa. A similar response is promoted by exposure to thiol-binding metal species including cadmium, zinc, mercury, and arsenite (Levinson et al., 1980; Caltabiano et al., 1986). Shelton et al. (1986) provided evidence that glutathione participated in the induction of the "stress" proteins, especially a 30- to 32-kDa protein. This concept was supported by Freeman and Meredith (1989). Such studies suggested that both glutathione oxidation and glutathione conjugation acted as possible signals for the induction of 32-kDa protein. Evidence for a similar induction of the 32-kDa protein was obtained following methylmercury exposure of cerebellar granule cell culture (Sarafian and Verity, 1990). It is likely, however, that such "stress" protein induction occurred principally in the small number of astrocytes in culture, as Marini et al. (1990) demonstrated a 70-kDa heat shock protein induction in cerebellar astrocytes and not cerebellar granule cells when both astrocytes and neurons were cocultured *in vitro*. However, an intense immunoreactivity for the 70-kDa protein was observed in the granule cell layer following hyperthermic stress *in vivo*, a distribution suggesting that it cannot be accounted for solely by the astrocytes present in this region. Nishimura et al. (1991) revealed that neurons synthesize much less 68 kDa protein than similarly heat-shocked astrocytes and concluded that cultured cerebellar cortical neurons were more susceptible to injury than the heat-resistant astrocytes. Keyse and Tyrrell (1989) have confirmed that heme oxygenase is the major 32-kDa stress protein induced by arsenite, cadmium, H_2O_2, and UV radiation. Hence, the induction of heme oxygenase may be a general response to oxidant or methylmercury-induced stress, known to be present in various isoforms in rat brain (Sun et al., 1990).

The heavy metal–stress protein induction response is associated with other transcriptional or translational alterations. Such alterations include changes in mRNA transcription and processing (Yost and Lindquist, 1986), alterations in protein phosphorylation (Welch, 1985), and most commonly an accompanying severe inhibition in the translation of normal cellular mRNA (McKenzie and Meselson, 1977; Duncan and Hershey, 1984). However, translational repression by chemical or metal inducers of stress protein induction appears to occur by different pathways (Duncan and Hershey, 1987) and is not linked uniquely to the enhanced phosphorylation of IF-2, thought originally to account for the inhibition of initiation associated with stress protein induction (Proud, 1986).

REFERENCES

Aldridge, W. N., Street, B. W., and Skilleter, D. N. (1977), *Biochem J.*, 168, 353–364.
Aschner, M. and Aschner, J. L. (1992), *Neurosci. and Biobehav. Rev.*, 16, 427–435.
Bizzi, A. and Gambetti, B. (1986), *Acta Neuropathol.*, 71, 154–158.
Bouldin, T. W., Goines, N. D., and Krigman, M. R. (1984), *J. Neuropathol. Exp. Neurol.*, 43, 162–174.
Brown, A. W., Cavanagh, J. V., Berschoyle, R. D., Gysbers, M. F., Jones, H. B., and Aldridge, W. N. (1984), *Neuropathol. Appl. Neurobiol.*, 10, 267–283.
Brown, W. J. and Yoshida, N. (1965), *Adv. Neurol. Sci. (Tokyo)* 9, 34–42.
Caltabiano, M. M., Koestler, T. P., Poste, G., and Greig, R. G. (1986), *J. Biol. Chem.*, 261, 13381–13386.
Cavanagh, J. V. and Chen, F. C. K. (1971), *Acta Neuropathol.*, 19, 216–224.
Chang, L. W. and Hartman, N. (1972), *Acta Neuropathol.*, 20, 122–138.
Cheung, M. and Verity, M. A. (1983), *Exp. Mol. Pathol.*, 38, 230–242.

Cheung, M. K. and Verity, M. A. (1981), *Environ. Res.,* 24, 286–298.
Cheung, M. K. and Verity, M. A. (1985), *J. Neurochem.,* 44, 1799–1803.
Crapper, D. R., Quittkat, S., Krishnan, S. S., Dalton, A. J., and DeBoni, U. (1980), *Acta Neuropathol.,* 50, 19–27.
Dang, D. (1986), *Mol. Cell Biochem.,* 71, 107–120.
DeBoni, U., Scott, J. W., and Crapper, D. R. (1974), *Histochemistry,* 40, 31–37.
Duncan, R. F. and Hershey, J. W. V. (1987), *Arch. Biochim. Biophys.,* 256, 651–661.
Ernst, V., Levin, D. H., and London, I. M. (1978), *Proc. Natl. Acad. Sci. U.S.A.,* 75, 4110–4114.
Freeman, M. L. and Meredith, M. J. (1989), *Biochem. Pharmacol.,* 38, 299–304.
Gravela, E. and Dianzani, M. U. (1970), *FEBS Lett.,* 9, 93–96.
Grundt, I. K. and Bakken, A. M. (1986), *Acta Pharmacol. Toxicol.,* 59, 11–16.
Hare, M. F. and Atchison, W. D. (1992), *J. Pharmacol. Exp. Ther.,* 261, 166–172.
Hashegawa, K., Omata, S., and Sugano, H. (1988), *Arch. Toxicol.,* 62, 470–472.
Henshaw, C. and Panniers, R. (1983), *Methods Enzymol.,* 101, 616–629.
Hewitt, C. D., Herman, M. M., Lopes, M. B. S., Savory, J., and Wills, M. R. (1991), *Neuropathol. Appl. Neurobiol.,* 17, 47–60.
Hucul, J. A., Henshaw, E. Z., and Young, D. A. (1985), *J. Biol. Chem.,* 260, 15585–15591.
Jacobs, J. M., Carmichael, N., and Cavanagh, J. B. (1975), *Neuropathol. Appl. Neurobiol.,* 11–19.
Jagus, R. and Safer, B. (1981), *J. Biol. Chem.,* 256, 1324–1329.
Jimenez, A. (1976), *Trans. Biochem. Sci.,* 1, 28–29.
Johnson, G. V. W. and Jope, R. S. (1988), *Brain Res.,* 456, 95–103.
Johnson, G. V. W., Xiahua, L., and Jope, R. S. (1989), *J. Neurochem.,* 53, 258–263.
Karlik, S. J., Eichhorn, J. O., Lewis, P. N., and Crapper, D. R. (1980), *Biochemistry,* 19, 5991–5998.
Kasama, H., Itoh, K., Omata, S., and Sugano, H. (1989), *Arch. Toxicol.,* 63, 226–230.
Kauppinen, R. A., Komulainen, H., and Taipale, H. T. (1988), *J. Neurochem,* 51, 1617–1625.
Kauppinen, R. A., Komulainen, H., and Taipale, H. (1989), *J. Pharmacol. Exp. Ther.,* 248, 1248–1254.
Kaziro, Y. (1978), *Biochim. Biophys. Acta,* 505, 95–127.
Keyse, S. M. and Tyrrell, R. M. (1989), *Proc. Natl. Acad. Sci. U. S. A.,* 86, 99–103.
Krekoski, C. A., Matthew, A., and Parhad, I. M. (1988), *J. Cell Biol.,* 9, 123–138.
Kuznetsov, D. A., Zavijalov, N. V., Guborkov, A. V., and Ivanov-Snaryad, A. A. (1986), *Toxicol. Lett.,* 30, 267–271.
Kuznetsov, D. A., Zavijalov, N. V., Guborkov, A. V., and Richter, V. (1987), *Toxicol. Lett.,* 36, 161–165.
Levesque, P. C. and Atchison, W. D. (1991), *J. Pharm. Exp. Ther.,* 256, 236–256.
Levinson, W., Oppermann, H., and Jackson, J. (1980), *Biochim. Biophys. Acta,* 606, 170–180.
Marini, A. M., Kozuka, M., Lipsky, R. H., and Nowak, T. S. (1990), *J. Neurochem.,* 54, 1509–1516.
Marshall, R. D. and Zamecnik, P. C. (1970), *Biochim. Biophys. Acta,* 198, 376–385.
Miller, C. A., and Levine, E. M. (1974), *J. Neurochem.,* 22, 751–758.
Mitra, U., Stringer, E. A., and Chaudhuri, A. (1982), *Annu. Rev. Biochem.,* 51, 869–900.
Moldave, K. (1985), *Annu. Rev. Biochem.,* 54, 1109–1149.
Mosca, J. D., Wu, J. M., and Suhadolnik, R. J. (1983), *Biochemistry,* 22, 346–354.
Muma, N. A., Troncoso, J. C., and Hoffman, P. N. (1988), *Mol. Brain Res.,* 3, 115–122.
Nishimura, R. N., Dwyer, B. E., Vinters, H. V., DeVellis, J., and Cole, R. (1991), *Neuropathol. Appl. Neurobiol.,* 17(2), 139–47.
Nixon, R. A., Clarke, J. F., Logvinenko, K. B., Tan, M. K. H., Hoult, M., and Grynspan, F. (1990), *J. Neurochem.,* 55, 1950–1959.
O'Callaghan, J. P., Miller, D. B., and Reiter, L. W. (1983), *J. Pharmacol. Exp. Ther.,* 224, 466–472.
Omata, S., Sakimura, K., Tsubaki, H. and Sugano, H. (1978), *Appl. Pharmacol.,* 44, 367–378.
Omata, S., Terui, Y., Kasama, H., Ichimura, T., Horigome, T., and Sugano, H. (1991), in *Advances in Mercury Toxicology,* Suzuki, T., Imura, N., and Clarkson, T. W., Eds. Plenum Press, New York, 223–240.
Proud, C. G. (1986), *Trans. Biochem. Sci.,* 11, 73–77.
Ramanujam, M. and Prasad, K. N. (1979), *Biochem. Pharmacol.,* 28, 2979–2984.
Sarafian, T., Cheung, M. K., and Verity, M. A. (1984), *Neuropathol. Appl. Neurobiol.,* 10, 85–100.
Sarafian, T. and Verity, M. A. (1985), *Neurochem. Pathol.,* 3, 27–39.
Sarafian, T. and Verity, M. A. (1990), *J. Neurochem.,* 55, 913–921.
Sarafian, T. and Verity, M. A. (1991), in *J. Dev. Neurosci.,* 9, 147–153.
Sarafian, T. A. (1993), *J. Neurochem.,* 61, 648–657.
Schimmel, P. (1987), *Annu. Rev. Biochem.,* 56, 125–158.
Selwyn, M. J., Dawson, A. P., Stockdale, M., and Gains, N. (1970), *Eur. J. Biochem.,* 14, 120–126.
Shea, T. B., Bermann, M. L., and Nixon, R. A. (1992), *J. Neurochem.,* 58, 542–547.
Shea, T. B., Clarke, J. F., Wheelock, T. R., Paskevich, V., and Nixon, R. A. (1989), *Brain Res.,* 492, 53–64.
Shea, T. B., Sihag, R., and Nixon, R. A. (1988), *Dev. Brain Res.,* 41, 97–109.
Shelton, K. R., Egle, P. M., and Todd, J. M. (1986), *Biochem. Biophys. Res. Commun.,* 134, 492–498.
Sihag, R. K. and Deutscher, M. P. (1983), *J. Biol. Chem.,* 258, 11846–11850.
Smith, R. (1977), *Biochim. Biophys. Acta,* 470, 170–184.
Snoeij, N. J., Bul-Schoenmakers, M., Penninks, A. H., and Seinen, W. (1988), *Int. J. Immunopharmacol.,* 10, 29–37.
Sone, E., Larrstuvold, M. K., and Kagawa, Y. (1977), *J. Biochem.,* 82, 859–868.

Sun, Y., Rotenberg, M. O., and Maines, M. D. (1990), *J. Biol. Chem.*, 265, 8212–8217.
Syversen, T. L. M. (1977), *Neuropathol. Appl. Neurobiol.*, 3, 225–236.
Toews, A. V., Ray, B., Goines, N. D., and Bouldin, T. W. (1986), *Brain Res.*, 398, 298–304.
Troncoso, J. C., Sternberger, N. H., Sternberger, L. A., Hoffman, P. N., and Price, D. L. (1986), *Brain Res.*, 364, 295–300.
Verity, M. A., Brown, W. J., and Cheung, M. (1975), *J. Neurochem.*, 25, 759–766.
Verity, M. A., Brown, W. J., Cheung, M., and Czer, G. (1977), *J. Neurochem.*, 29, 673–679.
Walton, G. M. and Gill, G. N. (1976), *Biochim. Biophys Acta.*, 418, 195–203.
Wender, M., Zgorzalewizc, V., and Peizhowski, A. (1975), *Neuropatol. Pol.*, 23, 415–421.
Wender, M., Zgorzalewizc, V., and Peizhowski, A. (1974), *Acta Neurol. Scand.*, 50, 103–108.
Yonaha, M., Satio, M., and Sagai, M. (1983), *Life Sci.*, 32, 1507–1514.
Yoshino, Y., Mozai, T., and Nakao, K. (1966), *J. Neurochem.*, 13, 1223–1230.
Young, D. A. (1969), *J. Biol. Chem.*, 244, 2210–2217.
Zosnek, H., Soifer, D., and Wisniewski, H. (1978), *J. Neuropathol. Exp. Neurol.*, 37, 604–612.

Section VI
Renal Toxicology of Metals

Overview

An Introduction to Metal-Induced Nephrotoxicity

Bruce A. Fowler

There are a number of metals which exert marked toxic effects on the kidney following acute or chronic exposure. The kidney is a common target organ for toxic metals, since it forms a major excretory pathway for metals from the body and has a high metabolic activity with a number of sensitive metabolic processes. Renal proximal tubule cells which reabsorb a number of substances including metals are a primary target cell population, but other essential components such as the glomeruli and interstitial capillaries are also frequently affected. The chapters in this section are focused on examining mechanisms of metal-induced nephrotoxicity from a variety of perspectives. The biochemical, morphological, and physiological mechanisms by which metals produce renal damage will be examined in relation to analytical aspects of renal metal concentrations, cellular mechanisms of metal uptake, and intracellular binding patterns.

It is hoped that this integrative approach will provide the reader with a more comprehensive view of the interrelationships which must exist between the accumulation of metals in the kidney and mechanisms of toxicity. The chapters will examine the nephrotoxic effects of the more well-studied metals such as lead, mercury, and cadmium, as well as those of metals/metalloids like gallium, indium, arsenic, bismuth, and thallium, which are used in high technology processes such as the manufacture of semiconductors and high-temperature superconductors. It is important to consider that future health problems may arise from these metals, which also accumulate in the kidney and are known to produce nephrotoxic effects. Exposure of the general population to these less-studied elements, due to their expanded use in modern electronic devices, may occur following incineration, a common disposal mechanism. The issue of interactions between metals in the kidney will be discussed in relation to biomarkers of nephrotoxicity and the use of these tests for detecting early manifestations of renal damage. There is a pressing need for biomarkers capable of detecting early signs of metal-induced nephrotoxicity resulting from exposure to one metal or several metals in combination. The effects of biological factors such as age, sex, and species on biomarker responses will also be considered in relation to alterations in renal gene expression patterns following *in vivo* or *in vitro* metal exposure.

Differences in renal tubule cell responses following *in vivo* or *in vitro* exposure to metals will also be discussed in relation to understanding mechanisms of toxicity, since there appear to be a number of marked differences in cellular responsiveness depending on how renal tubule cells are exposed.

Chapter 43

The Nephropathology of Metals

Bruce A. Fowler

I. INTRODUCTION

The pathological manifestations of metals in the kidney vary with chemical form of the metal, the dose, and whether the exposure is acute or chronic in nature. In general, the proximal tubule frequently shows marked manifestations of toxicity for most metal species following acute or chronic exposure, but interstitial fibrosis of the peritubular capillaries is also a common finding following chronic exposure to metals such as lead, cadmium, and mercury. The interactive nature of metal-induced damage to the proximal tubules and the peritubular capillary system has not been investigated in detail but represents a needed area of research. The role of metals such as lead in the development of renal adenocarcinoma will also be examined since both experimental animal studies and, more recently, epidemiological studies in humans have indicated that lead is a renal carcinogen. The possible role(s) of renal lead-binding proteins in mediating the carcinogenic response in both animals and humans will be discussed in relation to the formation of pathognomonic lead inclusion bodies in renal proximal tubule cells.

The following discussion will examine the pathological manifestations of metals in the kidney on both an individual and multielement basis, since it is clear that interactions between metals do occur in the kidney and that these interactions will influence the characteristic pathological manifestations of a single metal. The molecular processes underlying these morphological effects will be discussed in relation to metal–metal interactions at the level of molecular binding patterns and the bioavailability of toxic metal species to disrupt essential biochemical processes localized in specific organelle compartments. Finally, the chapter will consider the effects of metals on cell signaling and the induction of stress proteins in renal proximal tubule cells as important molecular mechanisms for mediating metal toxicity and protection, respectively.

II. LEAD

Lead-induced nephropathy in humans has been known for many years (Goyer and Rhyne, 1973) and is morphologically characterized by the presence of pathognomonic lead inclusion bodies in renal proximal tubule cells (Figure 1) and mitochondrial swelling (Goyer, 1968; Goyer and Krall, 1969; Fowler, et al., 1980). In addition to the renal tubular effects, interstitial fibrosis of the peritubular capillaries lead to an increased incidence of renal failure in lead-exposed workers (Landrigan et al., 1984; Steenland et al., 1990, 1992). A high incidence of renal adenocarcinoma has been observed in rodents exposed to lead in drinking water for prolonged time periods (Goyer and Rhyne, 1973), and an increased incidence of renal adenocarcinoma has been recently reported (Steenland et al., 1992) in lead-exposed workers. Overall, the data from these studies indicate that the pathological effects of lead in the kidney of both animal models and humans involve tubular toxicity, interstitial disease processes, and renal adenocarcinoma formation. The molecular mechanisms producing this spectrum of effects is complex, but it appears that at low lead concentrations, renal lead-binding proteins play a major role (Fowler, 1992; Fowler et al., 1994) in the intracellular handling of lead. The known and hypothesized mechanisms of renal lead toxicity are discussed below on an organelle system basis.

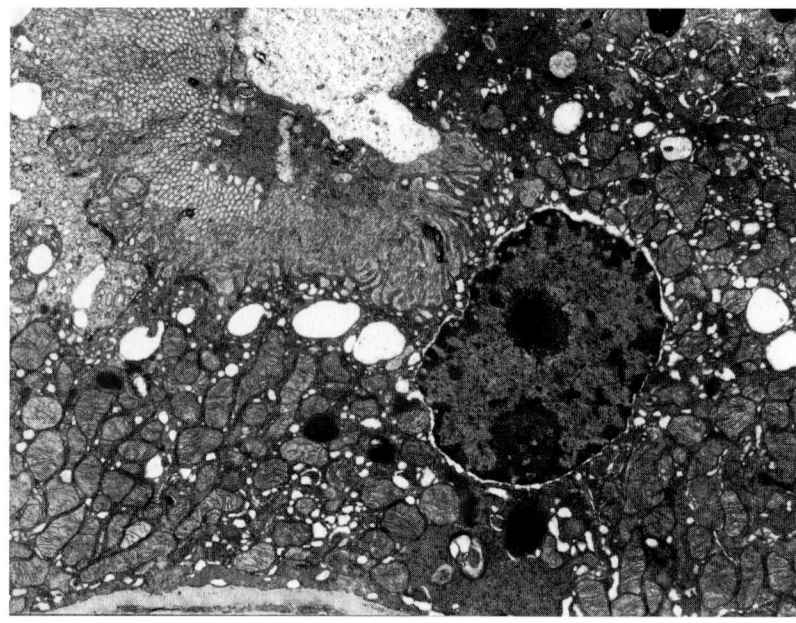

Figure 1. Electron micrograph of a renal proximal tubule cell from a rat exposed to lead showing the pathognomonic lead intranuclear inclusion body and marked mitochondrial swelling. (Original magnification × 12,000.) (From Fowler, B. A., *Univ. Ill. Bull.*, 71, 65–76, 1974.)

A. RENAL UPTAKE OF LEAD

Lead in the blood is primarily transported bound to the red cells, but the biologically active or tissue-diffusible fraction (Goyer and Mushak, 1977) represents only a relatively small component. Based on *in vitro* studies of ^{203}Pb uptake by renal brush border vesicles (Victery et al., 1985), it appears that uptake of Pb^{2+} is not an active process, but that there are extensive amounts of nonspecific membrane binding. Alternatively, it is more likely that lead is taken up by renal proximal tubule cells bound to filterable proteins in the blood. The renal lead-binding proteins with molecular masses of 7,000 to 18,000 kDa appear to be synthesized in the liver and possibly other organ systems and released into the circulation, where they may be taken up into proximal tubule cells by protein reabsorption mechanisms. In the rat, the lead-binding protein is a cleavage product of alpha-2-µglobulin, which appears to be resistant to proteolysis and to be capable of translocating lead into the nucleus (Mistry et al., 1985, 1986) and binding to chromatin. These molecules may, hence, play a role in the formation of the pathognomonic lead intranuclear inclusion bodies described below.

B. INTRACELLULAR COMPARTMENTALIZATION OF LEAD

At low dose exposures, lead is initially bound to acidic low molecular weight cytosolic proteins in rats (Oskarsson et al., 1982), monkeys (Fowler et al., 1992), and humans (Kahng, et al., 1992; Smith et al., 1993, 1994). These proteins are not identical across species, but share common chemical characteristics which include: low molecular weight, a highly anionic nature, and a high content of glutamic and aspartic amino acids. The lead-binding proteins appear to act as the initial intracellular binding sites for lead in the kidney prior to formation of the pathognomonic lead-containing cytosolic or intranuclear inclusion bodies (Figure 1) which also contain acidic proteins high in aspartic and glutamic amino acids (Moore and Goyer, 1973; Moore et al., 1973). It has been hypothesized (Fowler and DuVal, 1991) that the cytosolic proteins may play a role in the formation of the intranuclear inclusions through a process of lead-induced aggregation.

C. MITOCHONDRIA

The mitochondria are a major target organelle for lead in the kidney (Goyer, 1968; Goyer and Krall; 1969, Goyer et al., 1971). They show high amplitude swelling (Figure 1) following both *in vivo* and *in vitro* exposure to lead (Goyer et al., 1971; Fowler et al., 1980), with loss of biochemical functionality and membrane structural integrity (Oskarsson and Fowler, 1985a; Fowler et al., 1987). The loss of mitochondrial functionality could play a central role in processes of cell injury in proximal tubule cells, since many of the membrane transport processes localized in the renal tubule cells are dependent upon ATP.

D. RENAL HEME BIOSYNTHESIS

The effects of lead on heme biosynthesis have been extensively studied (Fowler et al., 1980; Oskarsson and Fowler, 1985a; Goering and Fowler, 1984, 1985) and found to be mediated, at least in part, by the renal lead-binding proteins described above. The cytosolic enzyme aminolevulinic acid dehydratase (ALAD) in the kidney has been found to be relatively resistant to Pb^{2+}, inhibition while the mitochondrial enzymes aminolevulinic acid synthetase (ALAS) and ferrochelatase have been shown to be readily inhibited by Pb^{2+} exposure following both *in vitro* and *in vivo* exposure. It is hypothesized that the renal lead-binding proteins may play several roles in this process by chelating lead, donating zinc to ALAD (Goering and Fowler, 1984, 1985, 1987), and by permitting the movement of lead across mitochondrial membranes (Oskarsson and Fowler, 1985b; Fowler et al., 1987) such that it could interact with mitochondrial enzymes such as ALAS and ferrochelatase. Further studies are needed to evaluate the molecular mechanisms of this process.

E. RENAL GENE EXPRESSION

The chronic or acute administration of lead to rodents has been known for many years to produce both increased mitosis (Choie and Richter, 1972, 1973, 1974a,b) and an increased incidence of renal adenocarcinoma (IARC, 1980). Studies of alterations in renal gene expression (Mistry et al., 1985) have demonstrated both up- and down-regulation of a number gene products, suggesting that prolonged exposure to lead may alter normal renal gene expression patterns and that this may play a role in the carcinogenesis process. The hypothesized roles of the renal lead-binding proteins in mediating the observed alterations in renal gene expression via their capacity to bind to chromatin (Mistry et al., 1985) have been recently discussed in detail (Fowler et al., 1994). Further research is needed into the molecular biology of the interactions of the lead-binding proteins and specific 5' flanking region regulatory/promotion sites on those genes showing altered expression patterns (Fowler, 1992, Fowler et al., 1994).

III. CADMIUM

The nephropathological effects of cadmium are largely related to the protein metallothionein, which acts as the main transport vehicle for cadmium in the circulation following its release from the liver (Garvey and Chang, 1981) and for the intracellular binding of cadmium ions in kidney tubule cells following degradation of the reabsorbed cadmium metallothionein (CdMT) complex (Squibb et al., 1979). A diagram of the hypothesized pathway for cadmium handling by the body is presented in Figure 2. The pathological effects of this process appear to be manifested in the renal tubule cells once the capacity of the renal tubule cells to produce metallothionein is exceeded. Following acute parenteral administration of CdMT there is a characteristic vesiculation (Figure 3) which can be attenuated by prior administration of zinc, which will increase the intracellular metallothionein pool (Squibb et al., 1984). Recent *in vitro* studies (Liu et al., 1995) using Cd^{2+} have also demonstrated the induction of the 70-kDa stress protein family by prior *in vivo* zinc administration, indicating that the mechanisms of the observed protective effects involve a number of cellular protective mechanisms in addition to metallothionein induction. Under more chronic exposure conditions, interstitial fibrosis is also observed (Fowler et al., 1975).

IV. MERCURY

The nephropathological effects of mercury have been known for many years and are largely dependent upon the chemical form of mercury exposure. Inorganic mercurials such as Hg^+ and Hg^{2+} preferentially exert toxic effects on the third segment of the renal proximal tubule while organomercurials such as

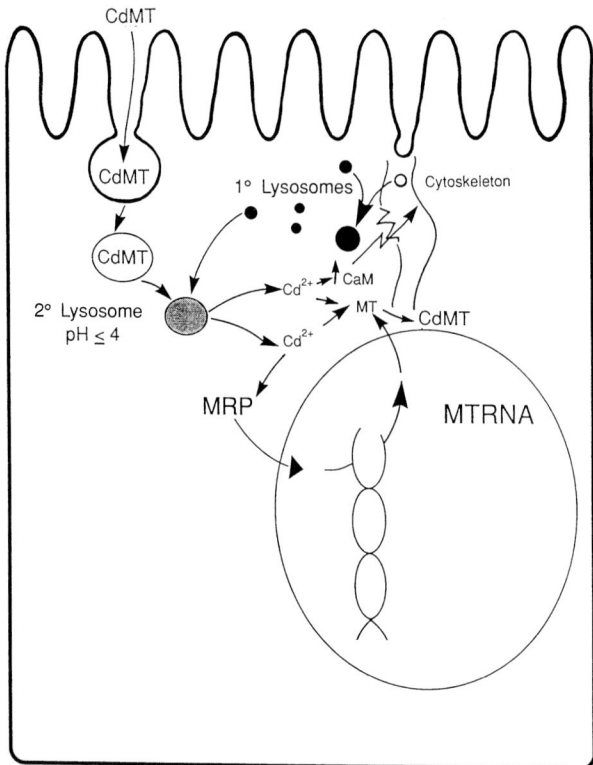

Figure 2. Diagram of a hypothesized pathway by which circulating CdMT is reabsorbed from the glomerular filtrate by renal proximal tubule cells and degraded by secondary lysosomes leading to the intracellular release of Cd^{+2}. (From Fowler, B. A. et al., in *Metallothionein in Biology and Medicine,* Klassen, C. D. and Suzuki, K. T., Eds., CRC Press, Boca Raton, FL, 322–331. With permission.)

phenylmercury and methylmercury exert toxic effects more extensively in all segments of the proximal tubules. The following discussion will hence separate the morphological effects of these mercurials on the basis of chemical form.

V. INORGANIC MERCURY

The nephrotoxic effects of the inorganic mercurials Hg^+ and Hg^{2+} on the third segment of the proximal renal tubule are characterized by both damage to the cellular membrane, and mitochondrial damage and loss of mitochondrial functionality (Gritzka and Trump, 1968; Ganote et al., 1975). Morphologically, at the light microscope level, the damage is characterized by cloudy swelling and the development of necrosis. Acute renal failure is the physiological consequence of this process. Exposure of rat kidney proximal tubule cells to 50 μ*M* $HgCl_2$ results in marked ultrastructural changes in mitochondria and the endoplasmic reticulum that are indicative of irreversible cell injury (Figure 4).

VI. ORGANOMERCURIALS

Nephropathy is also observed following acute (Klein et al., 1973) or chronic (Fowler, 1972a,b) exposure to methylmercury, and marked differences in responsiveness between sexes of rodents has been observed (Fowler, 1972b; Yasutake et al., 1990). The primary morphological effects (Figure 5) are characterized by exocytosis of cytoplasmic blebs frequently containing endoplasmic reticulum and showing mitochondrial damage (Fowler, 1972a,b; Fowler et al., 1975; Fowler and Woods, 1977) with an accumulation of lysosomes (Fowler, et al., 1975).

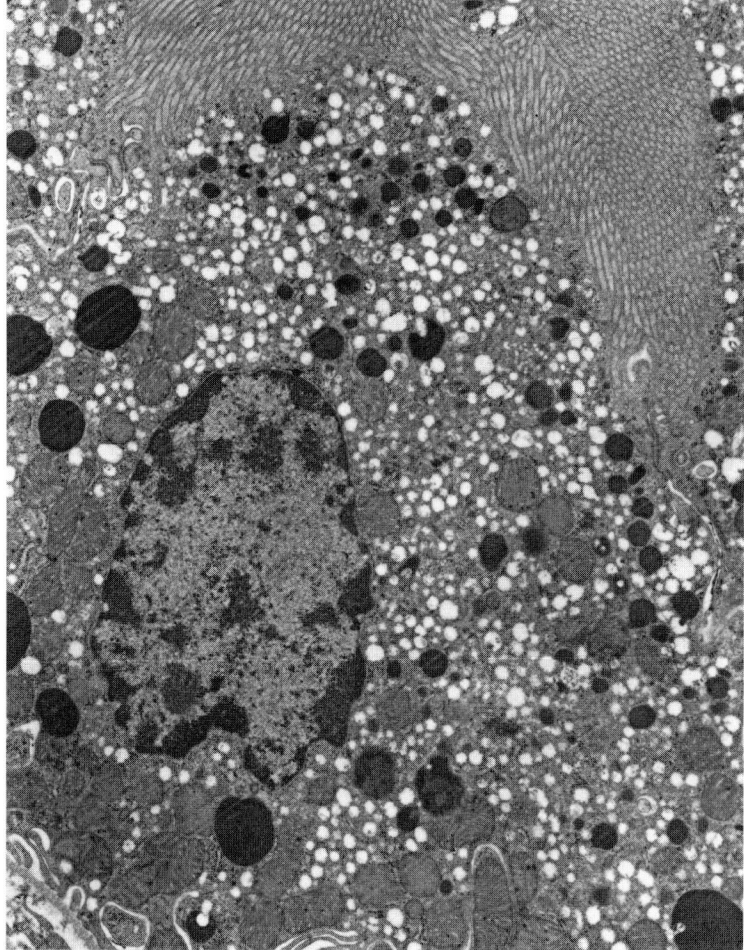

Figure 3. Electron micrograph of a renal proximal tubule cell from a rat given a single intraperitoneal injection of CdMT, showing the characteristic apical vesiculation and formation of primary lysosomes. (Original magnification × 16,375.)

VII. ARSENIC

The morphological effects of arsenic have received less attention than those of other metals, but studies by Brown et al. (1975) demonstrated structural and functional damage to the mitochondria of the renal proximal tubules and an increased presence of autophagic lysosomes following prolonged oral exposure to sodium arsenate in drinking water (Figure 6). Similar morphological findings were reported in proximal tubule cells by Frejaville et al. (1972) in four clinical cases of human poisonings with arsenic.

VIII. GALLIUM ARSENIDE

Goering et al. (1988) reported that a single intratracheal instillation of GaAs at 100 mg/kg produced mitochondrial swelling in renal proximal tubule cells similar to that observed with arsenical exposure.

IX. INDIUM

Studies by Castronovo and Wagner (1971) demonstrated the nephrotoxicity of indium chloride by histopathology, while ultrastructural/biochemical studies by Woods and Fowler (1982) reported that a single intraperitoneal injection of indium chloride produced an increase in smooth endoplasmic reticulum

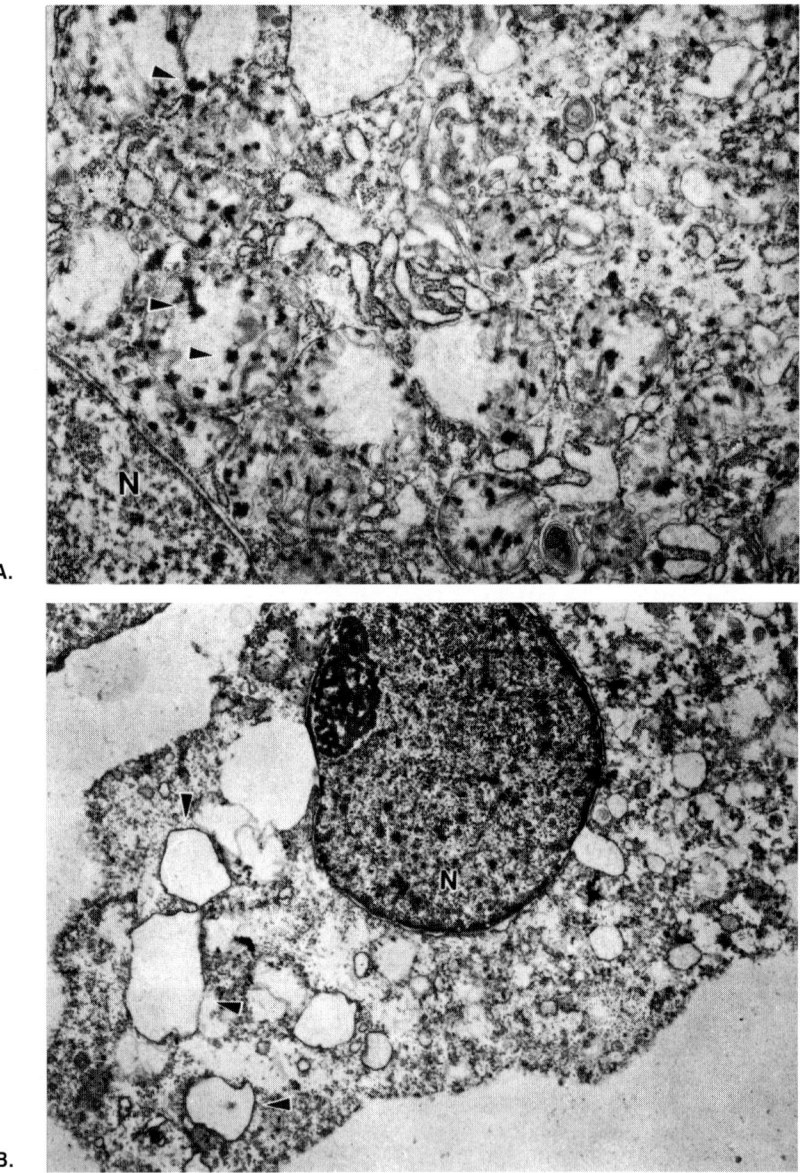

Figure 4. Transmission electron micrographs of rat kidney proximal tubule cells (PTE) following exposure to 50 µM $HgCl_2$. (A) PTE treated for 10 min showing advanced ultrastructural alterations indicative of advanced injury. Mitochondria are swollen with pale matrices and contain flocculent densities (arrowheads). Note also the dilatation of the endoplasmic reticulum. Nucleus (N). (Original magnification × 5,500.) (B) PTE treated for 15 min showing more advanced injury with disruption of the plasma membrane and marked dilatation of the endoplasmic reticulum (arrowheads). The nuclear membrane is focally dilated and there is margination and clumping of the chromatin in the nucleus (N). (Original magnification × 5,500). (From Elliget et al., 1995. With permission.)

in renal proximal tubule cells which was associated with an increase in the specific activity of renal heme oxygenase.

X. URANIUM

Clinical studies have reported (Berlin and Rudell, 1986) human exposure to uranium produces necrosis of the renal proximal tubule cells and development of renal failure. Experimental studies by Stone et al. (1961) also reported hyaline cast formation and necrosis of renal tubular cells following acute injection

The Nephropathology of Metals

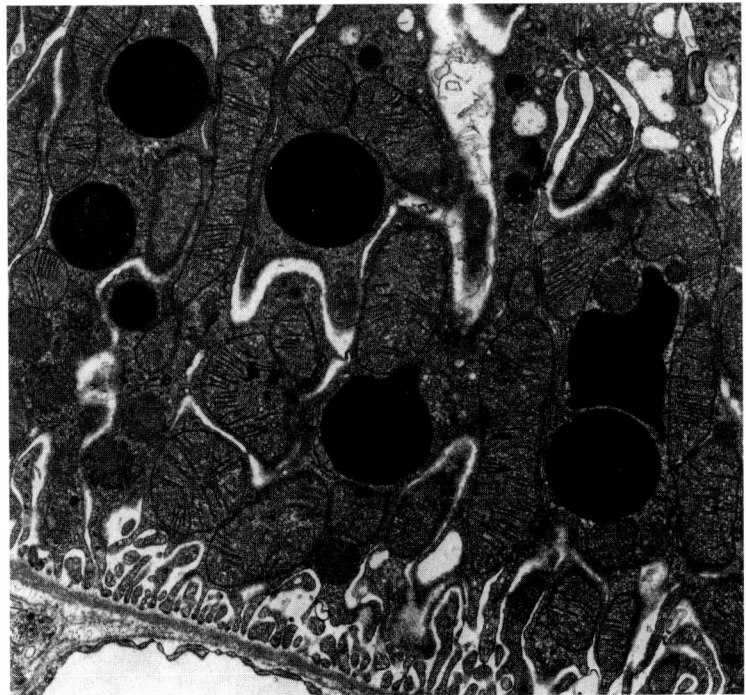

Figure 5. Electron micrograph of a renal proximal tubule cell from a rat exposed to methyl mercury in drinking water showing exocytosis of endoplasmic reticulum in cytoplasmic casts. (From Fowler, B. A., *Univ. Ill. Bull.*, 71, 65–76, 1974.)

of uranium. Other studies (Braunlich and Fleck, 1981) demonstrated increased excretion of alkaline phosphatase in animals injected with uranyl nitrate.

XI. CHROMATE

Human exposure to chromate has been shown to produce acute renal tubular damage (Langard and Norseth, 1986). Experimental studies by Evan and Dail (1974) reported a similar destruction of the renal proximal tubular cell membrane followed by mitochondrial damage after a single injection of chromate. These morphological effects were associated with alterations in renal tubular transport functions (Evan and Dail, 1974; Berndt, 1976; Franchini et al., 1978; Kumar and Rana, 1984).

XII. LEAD, CADMIUM, ARSENIC INTERACTIONS

It is also important to note that human exposure to metals usually involves a number of elements in combination, and that this may alter the usual morphological effects of a single metal. An example of this are the studies reported by Mahaffey and Fowler (1977), Fowler and Mahaffey (1978), and Mahaffey et al. (1981) which showed that concomitant exposure to lead and cadmium attenuated formation of the pathognomonic lead intranuclear inclusion bodies. This altered effect was associated with decreased renal lead burdens, but an increase in urinary porphyrins, indicating an increased bioavailability of metals to sensitive sites in the heme pathway.

XIII. MECHANISMS OF METAL-INDUCED RENAL PATHOGENESIS

It should be clear from the above discussion that metals may preferentially attack a number of different organelle systems within the kidney. The renal proximal tubular cell membrane is an apparent early target for metals with strong oxidization potential such as Hg^{2+}, U^{3+}, and Cr^{6+}. Other metals, such as those bound to metallothionein or organo-metal compounds which readily traverse the cellular membrane, may

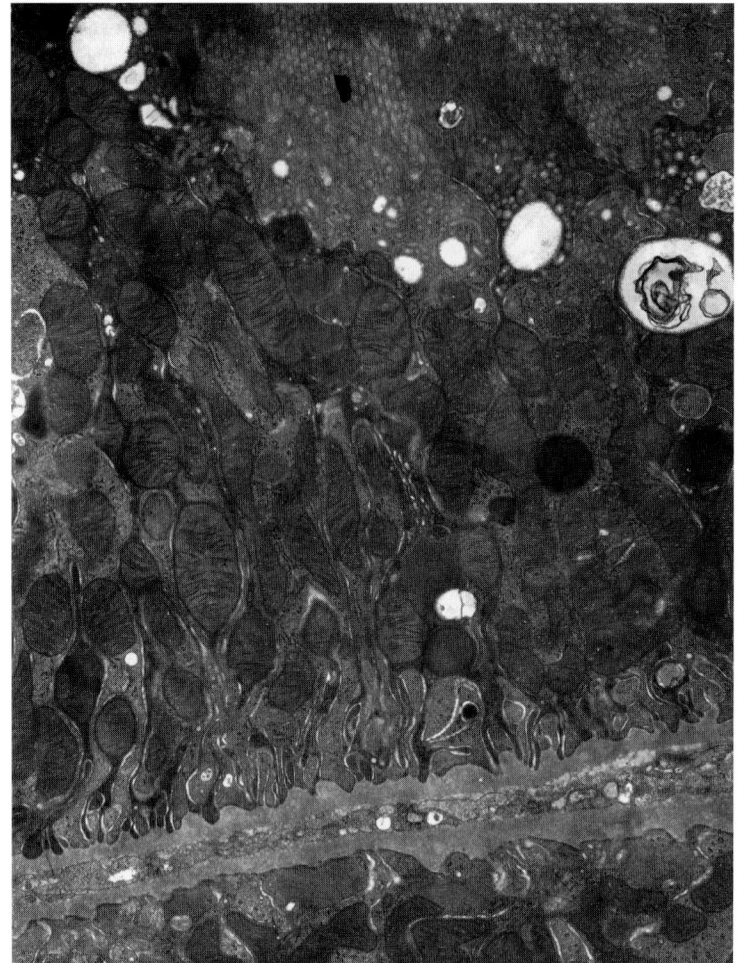

Figure 6. Electron micrograph of renal proximal tubule cell from a rat exposed to sodium arsenate in drinking water showing marked mitochondrial swelling. (From Brown, M. M. et al., *J. Toxicol. Environ. Health*, 1, 507–516.)

act against organelle systems such as the mitochondria, lysosomal apparatus, or endoplasmic reticulum. There are hence a number of molecular target sites within the proximal tubule cell that are differentially sensitive to various toxic elements, either alone or in combination. The primary mechanisms of organelle toxicity are therefore determined in large part by the specific chemical species of the element involved.

REFERENCES

Aoki, Y., Lipsky, M. M., and Fowler, B. A. (1990), *Toxicol. Appl. Pharmacol.*, 106, 462–468.
Baker, E. L., Jr., Goyer, R. A., Fowler, B. A., Khettry, U., Barnard, D., deVere White, R., Babayan, R., and Feldman, R. G. (1980), *Am. J. Ind. Med.*, 1, 139–148.
Berlin, M. and Rudell, B. (1986), in *Handbook on the Toxicology of Metals*, Vol. 2, 2nd ed., L. Friberg, G. G. Nordberg, and V. B. Vouk, Eds., Elsevier Science, New York,
Berndt, W. O. (1976), *J. Toxicol. Environ. Health*, 1, 449–459.
Braunlich, H. and Fleck, C. (1981), *Exp. Pathol.*, 20, 182.
Brown, M. M., Rhyne, B. C., Goyer, R. A., and Fowler, B. A. (1976), *J. Toxicol. Environ. Health*, 1, 507–516.
Castronovo, F. P. and Wagner, H. N. (1971), *Br. J. Exp. Pathol.*, 52, 543–559.
Choie, D. D. and Richter, G. W. (1972), *Am. J. Pathol.*, 66, 265–276.
Choie, D. D. and Richter, G. W. (1974a), *Lab. Invest.*, 30, 647–651.
Choie, D. D. and Richter, G. W. (1974b), *Lab. Invest.*, 30, 652–656.
Choie, D. D., Richter, G. W., and Young, L. B. (1975), *Beitr. Pathol.*, 155, 197–203.
Evan, A. P. and Dail, W. G. (1974), *Lab. Invest.*, 30, 704–715.

Fowler, B. A. (1974), *Univ. Ill. Bull.,* 71, 65–76.
Fowler, B. A. (1972a), *Science,* 175, 780–781.
Fowler, B. A. (1972b), *Am. J. Pathol.,* 69, 163–174.
Fowler, B. A. (1992), *Environ. Health. Perspect.,* 100, 57–63.
Fowler, B. A. and Akkerman, M. (1992), in *Int. Agency Res. Cancer,* G. F. Nordberg, R. F. M. Herber, and L. Allessio, Eds., 271–277.
Fowler, B. A. and Woods, J. S. (1977), *Exp. Mol. Pathol.,* 27, 402–412.
Fowler, B. A. and Mahaffey, K. R. (1978), *Environ. Health. Perspect.,* 25, 87–97.
Fowler, B. A. and DuVal, G. E. (1991), *Environ. Health. Perspect.,* 91, 77–80.
Fowler, B. A., Jones, H. S., Brown, H. W., and Haseman, J. K. (1975), *Toxicol. Appl. Pharmacol.,* 34, 233–252.
Fowler, B. A., Brown, H. W., Lucier, G. W., and Krigman, M. R. (1975), *Lab. Invest.,* 32, 313–322.
Fowler, B. A., Kimmel, C. A., Woods, J. S., McConnell, E. E., and Grant, L. D. (1980), *Toxicol. Appl. Pharmacol.,* 56, 59–77.
Fowler, B. A., Oskarsson, A., and Woods, J. S. (1987), *Ann. N.Y. Acad. Sci.,* 514, 172–182.
Fowler, B. A., Gandley, R. E., Akkerman, M., Lipsky, M. M., and Smith, M. (1991), in *Metallothionein in Biology and Medicine,* Klassen, C. D. and Suzuki, K. T., Eds., CRC Press, Boca Raton, FL, 322–321.
Fowler, B. A., Kahng, M. W., Smith, D. R., Conner, E. A., and Laughlin, N. K. (1992), *J. Exp. Anal. Environ. Epidemiol.,* 3, 441–448.
Fowler, B. A., Kahng, M. W, and Smith, D. R. (1994), *Environ. Health. Perspect.,* 102 (Suppl. 3), 115–116.
Frejaville, J. P., Bescol, J., Leclerc, J. P., Gulliam, L., Crabie, P., Consco, F., Gervias, P., and Faultier, M. (1972), *Ann. Med. Intern.,* 123, 713–722.
Franchini, I., Mutti, A., and Cavatorata, A. (1978), *Contrib. Nephrol.,* 10, 98–119.
Ganote, C. E., Reimer, K. A., and Jennings, R. B. (1975), *Lab. Invest.,* 31, 633.
Garvey, J. S. and Chang, C. C. (1981), *Science,* 214, 805–807.
Goering, P. S. and Fowler, B. A. (1984), *J. Pharmacol. Exp. Ther.,* 231, 66–71.
Goering, P. S. and Fowler, B. A. (1985), *J. Pharmacol. Exp. Ther.,* 234, 365–371.
Goering, P. L. and Fowler, B. A. (1987), *Ann. N. Y. Acad. Sci.,* 514, 235–247.
Goering, P. L., Maronpot, R. R., and Fowler, B. A. (1988), *Toxicol. Appl. Pharmacol.,* 92, 179–193.
Goyer, R. A. (1968), *Lab. Invest.,* 19, 71–77.
Goyer, R. A. and Krall, A. R. (1969), *J. Cell Biol.,* 41, 393–400.
Goyer, R. A. and Mushak, P. (1977), in *Toxicology of Trace Elements,* R. A. Goyer, and M. A. Mehlman, Eds., John Wiley & Sons, New York, 41–78.
Goyer, R. A. and Rhyne, B. C. (1973), *Int. Rev. Exp. Pathol.,* 12, 1–77.
Goyer, R. A., Leonard, D. L., Moore, J. F., Ryne, B., and Krigman, M. R. (1971), *Arch. Environ. Health.,* 20, 704–711.
Gritzka, T. L. and Trump, B. F. (1968), *Am. J. Pathol.,* 52, 1225.
IARC (International Agency for Research on Cancer) (1980), in *IARC Monographs on the Evaluation of the Carcinogenic Risk of Chemicals to Humans: Some Metals and Metallic Compounds,* Vol. 23, October, 1979, Lyon, France, 325–416.
Kahng, M. W., Conner, E. A., and Fowler, B. A. (1992), *Toxicologist,* 12, 214.
Klein, R., Herman, S. P., Bullock, B. C., and Talley, F. A. (1973), *Arch. Pathol.,* 96, 83.
Kumar, A. and Rana, S. W. S. (1984), *Int. J. Tissue React.,* 6, 135.
Landrigan, P. J., Goyer, R. A., Clarkson, T. W., Sandler, D. P., Smith, H., Thun, M. J., and Wedeen, R. P. (1984), *Arch. Environ. Health,* 39, 225–230.
Langard, S. and Norseth, T. (1986), *Handbook on the Toxicology of Metals,* 2nd ed., L. Friberg, G. G. Nordberg, and V. B. Vouk, Eds., Elsevier Science New York.
Liu, J., Squibb, K. S., Akkerman, M., and Fowler, B. A. (1995), *Toxicologist,* 15, 299.
Mahaffey, K. R. and Fowler, B. A. (1977), *Environ. Health. Perspect.,* 19, 165–171.
Mahaffey, K. R., Capar, S. G., Gladen, B. C., and Fowler, B. A. (1981), *J. Lab. Clin. Med.,* 98, 463–481.
Mistry, P., Lucier, G. W., and Fowler, B. A. (1985), *J. Pharmacol. Exp. Ther.,* 232, 462–469.
Mistry, P., Mastri, C., and Fowler, B. A. (1986), *Biochem. Pharmacol.,* 35, 711–713.
Moore, J. F. and Goyer, R. A. (1973), *Environ. Health. Perspect.,* 7, 121–127.
Moore, J. F., Goyer, R. A., and Wilson, M. (1973), *Lab. Invest.,* 29, 488–494.
Oskarsson, A. and Fowler, B. A. (1985a), *Exp. Mol. Pathol.,* 43, 409–417.
Oskarsson, A. and Fowler, B. A. (1985b), *Exp. Mol. Pathol.,* 43, 397–408.
Oskarsson, A., Squibb, K. S., and Fowler, B. A. (1982), *Biochem. Biophys. Res. Commun.,* 104, 290–298.
Smith, D. R., Kahng, M. W., Conner, E. A., and Fowler, B. A. (1993), *Toxicologist,* 13, 443.
Smith, D. R., Kahng, M. W., Quintanilla, B., and Fowler, B. A. (1994), *Toxicologist,* 14, 84.
Squibb, K. S., Ridlington, J. W., Carmichael, N. G., and Fowler, B. A. (1979), *Environ. Health. Perspect.,* 28, 287–296.
Squibb, K. S., Pritchard, J. B., and Fowler, B. A. (1984), *J. Pharmacol. Exp. Ther.,* 229, 311–321.
Steenland, N. K., Thun, M. J., and Fergusson, C. W. (1990), *Am. J. Public Health,* 80, 153.
Steenland, N. K., Selevan, S., and Landrigan, P. (1992), *Am. J. Public Health,* 82, 1641–1644.
Stone, R. W., Benxcome, S. A., Latta, H., and Madden, S. C. (1961), *Arch. Pathol.,* 71, 160.
Victery, W. W., Miller, C. R., and Fowler, B. A. (1984), *J. Pharmacol. Exp. Ther.,* 231, 589–596.
Woods, J. S. and Fowler, B. A. (1982), *Exp. Mol. Pathol.,* 36, 306–315.
Yasutake, A., Hirayama, K., and Inouye, M. (1990), *Renal Failure,* 12, 233–240.

Chapter 44

Roles of Metal-Binding Proteins in Mechanisms of Nephrotoxicity of Metals

Katherine S. Squibb

I. INTRODUCTION

Since the discovery of metallothionein by Margoshes and Vallee in 1957 (Margoshes and Vallee, 1957), it has become increasingly apparent that metal-binding proteins play an important role in determining the target organ toxicity of metals Studies over the past 30 years have shown that all aspects of metal ion metabolism and toxicity, from intestinal uptake to DNA binding, are controlled in some way by extracellular and intracellular proteins. Some of these proteins serve as transport or storage forms for the metals; others are enzymes or regulatory proteins. The balance between the accumulation of metals in cells in nontoxic forms vs. the binding of metals to proteins that alter normal cell metabolic pathways is a key to metal ion toxicity.

The purpose of this chapter is to review the role of metal-binding proteins in the nephrotoxicity of metals. Two of the most well-studied metals in this regard are cadmium (Cd) and lead (Pb). Both produce proximal tubule cell damage following chronic exposure, but they differ in their mechanisms of toxicity and they accumulate in proximal tubule cells in much different forms. This accumulation and its significance with respect to the cellular toxic effects of these metals will be discussed.

II. CADMIUM

Acute exposure to Cd results primarily in liver damage in experimental animals (Dudley et al., 1982), while chronic exposure causes renal proximal tubule cell necrosis (Dudley et al., 1985; Kobota et al., 1970; Goyer et al., 1989). These effects in rats are consistent with the nephrotoxic effects of chronic Cd exposure observed in humans, which are characterized by a low molecular weight proteinuria and glucosuria that are indicative of renal tubular cell damage (Adams et al., 1969; Kjellstrom, 1986; Nordberg et al., 1986).

The key to the development of renal damage vs. hepatic damage following chronic exposure to Cd appears to be the ability of cells to detoxify Cd when it enters cells slowly, and a difference in the form in which Cd is transported to the kidney under chronic exposure conditions. It is well known that Cd accumulates in cells bound to a low molecular weight, high affinity metal-binding protein called metallothionein (MT). This well-characterized 6000-kDa protein contains 20 cysteine residues per polypeptide chain and binds 7 g-atoms of metal per mole of protein (Kagi, 1993). It serves normally as a cellular storage form for the essential metals, zinc (Zn) and copper (Cu) (Cherian and Chan, 1993; Bremner, 1993). However, due to the higher binding affinity of Cd for MT and the high kinetic lability of

MT-bound metals, Cd can displace Zn from the protein molecule when it enters the cell and generally remains in this protein-bound form (as CdMT) (Vallee and Maret, 1993; Robbins et al., 1991; Kagi and Schaffer, 1988). Because Cd ions bound to MT are unavailable for interactions with other sensitive cellular ligands, MT acts as a metal ion-buffering system for Cd, effectively decreasing its intracellular toxicity (Goering and Klassen, 1984; Kagi and Schaffer, 1988).

A. HEPATIC VS. RENAL TOXICITY

Hepatic toxicity can occur when Cd is administered in high, acute doses to an animal, however, because Cd ions are taken up by the liver at such a high rate that the capacity of cells to synthesize sufficient MT is overwhelmed. Renal damage does not occur under these conditions, however, because much lower concentrations of Cd are taken up by the kidney compared to the liver, and the Cd is relatively evenly distributed between the different cell types along the renal tubules (Dorian et al., 1992). Thus, the kidney MT detoxification system is able to cope with the Cd entering the kidney cells, and renal damage does not occur under acute exposure conditions.

Under chronic exposure conditions, however, Cd accumulates in hepatic cells in the form of CdMT without causing damage, but renal damage then occurs, with time, when hepatic CdMT is released into the blood during normal cell turnover or due to hepatic cell damage (Dudley et al., 1985; Goyer et al., 1989; Cain and Griffith, 1980). Under these circumstances, circulating CdMT is filtered through the glomerulus due to its small size and is reabsorbed by the normal protein reabsorption system present on the luminal side of the cells of the S1 and S2 segments of the proximal tubule. By this mechanism, all of the Cd leaving the liver is deposited directly within a small number of cells in the renal cortex, leading to a high dose of Cd to these cells (Figure 1).

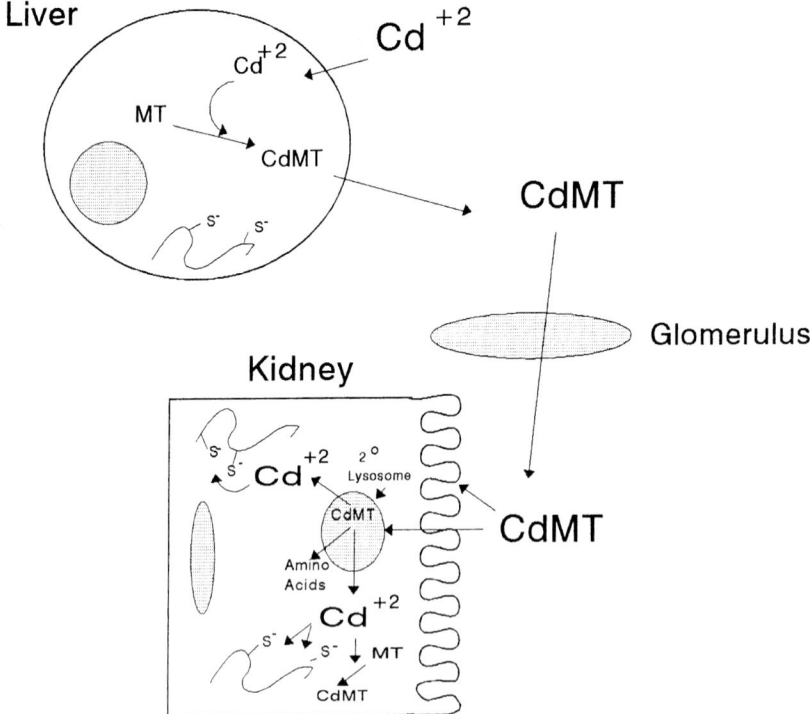

Figure 1. Diagram illustrating the movement of Cd as CdMT from the liver to the kidney following chronic exposure to Cd. Reabsorption of circulating CdMT by renal proximal tubule cells results in the intracellular release of Cd ions when the CdMT is degraded by cellular lysosomes. (Concept adapted from Squibb, K. S., Pritchard, J. B., and Fowler, B. A., *J. Pharmacol. Exp. Ther.,* 229, 311–321, 1984.)

Evidence for this hypothesized role of CdMT in Cd-induced nephrotoxicity takes various forms. Injection of CdMT *in vivo* produces a renal toxicity very similar to that observed following chronic exposure to the Cd ion (Cherian et al., 1976; Nordberg et al., 1975; Squibb et al., 1984; Wang et al.,

1993). LD_{50} values for CdMT are a tenth of those reported for Cd^{2+} (Webb and Etienne, 1977), and Cd is deposited almost exclusively in the kidney when it is injected in the form of CdMT (Nordberg et al., 1975). Dorian and co-workers (1992) have shown that injected CdMT is specifically reabsorbed by cells in the S1 and S2 segments of the renal tubule. Other investigators have shown by radioimmunoassay that MT is present in plasma samples taken from animals and humans chronically exposed to Cd (Chang et al., 1980; Garvey and Chang, 1981; Shaikh and Hirayama, 1979; Tohyama and Shaikh, 1981), thus providing evidence that Cd reaches the kidney bound to MT under chronic exposure conditions.

There are currently two theories as to the mechanism by which filtered CdMT damages the S1 and S2 segments. Cherian and co-workers (1976) have suggested that CdMT damages the cell membrane when it binds to the luminal side during the reabsorption process. Others (Squibb et al., 1979; 1982, and 1984; Sato and Nagai, 1986) have proposed that cellular toxicity occurs when Cd ions are released directly within the tubule cells after reabsorption and lysosomal degradation of the CdMT molecule (Figure 1). By this process, Cd is released within the cell at high enough concentrations to overwhelm the kidney cell MT detoxification system and produce cell injury or cell death.

III. CIS-PLATINUM

Recent studies have shown that metallothionein is also important in regulating the renal toxicity of the anticancer agent, cis-platinum (cis-DDP) (Kondo et al., 1993; Naganuma et al., 1987; Saijo et al., 1993; Satoh et al., 1988). Co-treatment of cancer patients with bismuth subnitrate increases kidney MT concentrations and decreases the nephrotoxic side-effects of cis-DDP treatment (Kondo et al., 1993; Saijo et al., 1993). This finding is consistent with numerous studies indicating that cultured cells with elevated levels of MT are resistant to cis-platinum and other alkylating agents (Andrews et al., 1990; Lazo et al., 1993).

There appear to be multiple mechanisms by which MT decreases the cellular toxicity of cis-DDP. Some studies suggest that MT increases DNA repair (Farnworth et al., 1990; Fujiwara et al., 1990; Hospers et al., 1988). Others indicate that direct scavenging of cis-DDP by MT occurs. Petering and co-workers (Pattanaik et al., 1992; Petering et al., 1993) have shown that incubation of cis-DDP with MT *in vitro* results in displacement of Zn and stoichiometric binding of platinum to the protein's sulfhydryl groups. Thus, direct sequestration of cis-DDP may occur in tubule cells with elevated MT. Additional work is clearly needed to establish the various functions of MT in protecting against cellular damaging agents.

IV. LEAD

High affinity, low molecular weight lead-binding proteins (PbBPs) present in specific tissues of mammalian organisms also appear to regulate the intracellular distribution and toxicity of this heavy metal. PbBPs with dissociation constants (K_d) of about 10^{-8} M and molecular weights ranging from 6,000 to 20,000 daltons have been isolated from rat, monkey, and human tissues (Fowler and DuVal, 1991). Because they are present in highest concentrations in the two primary target organs for Pb, the kidney and the brain, it has been proposed that they play a role in mediating Pb toxicity (Fowler, 1989).

A. RAT LEAD-BINDING PROTEINS

The major renal PbBP in rat proximal tubule cells has been shown to be a cleavage product of α_2-microglobulin, a member of the retinol-binding protein superfamily (Fowler and DuVal, 1991). With a molecular weight of approximately 18,600 daltons, α_2-microglobulin is synthesized primarily in the livers of male rats, under positive androgen control (Swenberg et al., 1989). In the kidney, both the cleaved and noncleaved forms of α_2-microglobulin are present. The primary difference between these forms is the absence of the first nine N-terminal amino acid residues. These low molecular weight protein chains are rich in aspartic and glutamic acids, but low in cysteine, a characteristic that sets them apart from the metallothionein family of metal-binding proteins (Fowler and DuVal, 1991).

Immunohistochemical studies have shown that α_2-microglobulin in rat kidney is present primarily in the lysosomes of cells in the descending segments (S_2) of proximal tubules, with limited amounts also present in the cytoplasm of these cells and some cells in the third (S_3) segments (Fowler and DuVal, 1991). This is consistent with the hypothesis that the protein is released from the liver and reabsorbed in the proximal tubules. The cellular specificity for uptake may be due to the presence of specific membrane receptors for the protein in cells in the S_2 segments of the tubules.

Within tubule cells, it has been hypothesized that the cleavage form of α_2-microglobulin plays a role in the formation of lead inclusion bodies (Fowler, 1989) (Figure 2). *In vitro* studies have shown that the PbBP isolated from rat kidneys aggregates in the presence of Zn and Pb (DuVal et al., 1989; DuVal and Fowler, manuscript in preparation) forming tetramers with two metal ions bound per tetramer. During high dose or chronic Pb exposure, Pb accumulates in kidney cells in the form of cytoplasmic and nuclear inclusion bodies (Fowler et al., 1980; Goyer et al., 1970). The formation of these inclusion bodies is reversible and may be initiated by the aggregation of α_2-microglobulin in the presence of Pb (Mistry et al., 1985). Mistry and co-workers (1985) have shown in *in vitro* studies that Cd is able to displace Pb from the cleavage form of α_2-microglobulin (Fowler and DuVal, 1991). Consistent with this is the finding from *in vivo* studies that kidney Pb concentrations are much lower in rats exposed to both Cd and Pb (Mahaffey et al., 1981), and no nuclear inclusion bodies were present in Cd/Pb-treated animals (Mahaffey and Fowler, 1977). These results suggest that Cd alters Pb uptake and accumulation in the kidney by altering the intracellular binding of the Pb ion and the formation of inclusion bodies.

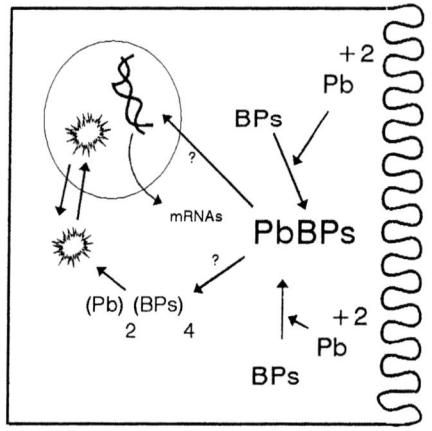

Figure 2 Proposed role of PbBPs in the formation of inclusion bodies and the translocation of Pb into the nucleus in renal proximal tubule cells. (From Fowler, B. A. and DuVal, G., *Environ. Health Perspect.*, 91, 77–80, 1991. With permission.)

In vitro nuclear uptake studies have also shown that the low molecular weight, high affinity PbBP of the rat cytosol is capable of translocating Pb into the nucleus of the cell (Mistry et al., 1985, 1986). These results have led to the hypothesis that cytosolic PbBPs may act as specific receptors for Pb in cells, facilitating the movement of Pb into the nucleus and perhaps mediating the nuclear effects of Pb (Fowler and DuVal, 1991). Early studies have shown that Pb exposure causes cell proliferation and increased total DNA, RNA, and protein synthesis in rodent kidneys (Choie and Richter, 1972, 1974a, 1974b). More recently, Fowler and co-workers (Fowler et al., 1985; Mistry et al., 1987) have reported that Pb exposure leads to alterations in the expression of specific genes in renal proximal tubule cells. Further studies in this area are needed to determine the mechanisms by which these effects occur.

B. HUMAN LEAD-BINDING PROTEINS

The ability to extrapolate the results discussed above from rats to humans is dependent upon the identification of similar low molecular weight PbPBs in human tissues. Small proteins with Pb binding constants similar to that of α_2-microglobulin have been identified in nonhuman primate and human renal tissue; however these proteins are antigenically different from the rat protein (Quintanilla-Vega et al., 1994; Smith et al., 1994). The two primary low molecular weight human kidney PbBPs have been identified as the 9000-dalton diazepam binding inhibitor (DBI, also known as acyl-CoA binding protein) and thymosin beta-4 (Smith et al., 1994). The interaction of Pb with these regulatory proteins suggest that they may play a role in effects of Pb on cellular metabolism, homeostasis, and cytoskeletal functions.

Studies are in progress to determine whether these PbBPs also play a role in the translocation of Pb into the nucleus in humans as α_2-microglobulin appears to do in rats. Humans exposed to Pb develop nuclear inclusion bodies similar to those seen in rats (Baker et al., 1980), and epidemiological data have shown that humans chronically exposed to Pb develop renal tubular damage and have a higher incidence of renal cancer (Baker et al., 1980; Lilis, 1984; Selevan et al., 1984). Clearly, further studies are needed to delineate the role of these low molecular weight PbBPs in the development of renal disease and cancer in humans.

V. SUMMARY

It is becoming increasingly clear that high affinity metal-binding proteins play a key role in determining and regulating the target organ toxicity of metals. In general, metal storage proteins such as metallothionein decrease the intracellular toxicity of metals such as cadmium by preventing the interaction of these ions with ligands within the cell that are critical to normal cellular metabolic processes. Due to the protein reabsorption function of the kidney, however, metals bound to circulating low molecular weight proteins will preferentially deposit in specific cells within renal proximal tubules, increasing the accumulation of the metals in these cells. Thus, chronic effects of Cd on the kidney appear to be due to the fact that it is transported to the kidney from other tissues in the form of CdMT.

Recent studies have shown that low molecular weight, high affinity lead-binding proteins (PbBPs) are present in rat, nonhuman primate, and human kidney tissues. Results suggest that the rat PbBP, identified as α_2-microglobulin, is involved in the formation of Pb inclusion bodies and perhaps in the nuclear translocation and nuclear effects of Pb in renal proximal tubule cells. The importance of PbBPs in determining the nephrotoxic effects of Pb in humans is currently under investigation.

REFERENCES

Adams, R. G., Harrison, J. F., and Scott, P. (1969), *Q. J. Med.*, 38, 425–436.
Andrews, P. A. and Howell, S. B. (1990), *Cancer Cells*, 2, 35–43.
Baker, E. L., Goyer, R. A., and Fowler, B. A. (1980), *Am. J. Ind. Med.*, 1, 139–148.
Bremner, I. (1993), in *Metallothionein III*, K. T. Suzuki, N. Imura, and M. Kimura, Eds., Birkhauser, Basel, Switzerland, 111–124.
Cain, K. and Griffiths, B. (1980), *Biochem. Pharmacol.*, 29, 1852–1855.
Chang, C. C., Van der Mallie, R. J., and Garvey, J. S. (1980), *Toxicol. Appl. Pharmacol.*, 55, 94–102.
Cherian, M. G. and Chan, H. M. (1993), in *Metallothionein III*, K. T. Suzuki, N. Imura, and M. Kimura, Eds., Birkhauser, Basel, Switzerland, 87–109.
Cherian, M. G., Goyer, R. A., and Delaquerriere-Richardson, L. (1976), *Toxicol. Appl. Pharmacol.*, 38, 399–408.
Choie, D. D. and Richter, G. W. (1972), *Am. J. Pathol.*, 68, 359–370.
Choie, D. D. and Richter, G. W. (1974a), *Lab. Invest.*, 30, 647–651.
Choie, D. D. and Richter, G. W. (1974b), *Lab. Invest.*, 30, 652–656.
Dorian, C., Gattone, V. H., II, and Klaassen, C. D. (1992), *Toxicol. Appl. Pharmacol.*, 114, 173–181.
Dudley, R. E., Svoboda, D. J., and Klaassen, C. D. (1982), *Toxicol. Appl. Pharmacol.*, 65: 302–313.
Dudley, R. E., Gammal, L. M., and Klaassen, C. D. (1985), *Toxicol. Appl. Pharmacol.*, 77, 414–426.
DuVal, G. E., Jett, D. A., and Fowler, B. A. (1989), Lead-induced aggregation of α_2-microglobulin *in vitro*, *Toxicologist*, 9, 98.
Farnworth, P., Hillcoat, B., and Roos, I. (1990), *Cancer Chemother. Pharmacol.*, 25, 411–417.
Fowler, B. A. (1989), *Comments Toxicology*, 3, 27–46.
Fowler, B. A. and DuVal, G. (1991), *Environ. Health Perspect.*, 91, 77–80.
Fowler, B. A., Kimmel, C. A., Woods, J. S., McConnell, E. E., and Grant, L. (1980), *Toxicol. Appl. Pharmacol.*, 56, 59–77.
Fowler, B. A., Mistry, P., and Victery, W. W. (1985), *Toxicologist*, 5, 53.
Fujiwara, Y., Sugimoto, T., Kasahara, K., Bungo, M., Yamakido, M., Tew, K. S., and Saijo, N. (1990), *Jpn. J. Cancer Res.*, 81, 527–535.
Garvey, J. S. and Chang, C. C. (1981), *Science*, 214, 805–807.
Goering, P. L. and Klassen, C. D. (1984), *Toxicol. Appl. Pharmacol.*, 74, 308–313.
Goyer, R. A., Leonard, D. L., Moore, J. F., Rhyne, B., and Krigman, M. R. (1970), *Arch. Environ. Health*, 20, 704–711.
Goyer, R. A., Miller, C. R., Shi-Ya, Z., and Victery, W. (1989), *Toxicol. Appl. Pharmacol.*, 101, 232–244.
Hospers, G. A. P., Mulder, N. H., de Jong, B., de Ley, L., Uges, D. R., Fichtinger Schepman, A. M. J., Sleper, R. J., and de Vries, E. G. E. (1988), *Cancer Res.*, 48, 6803–6807.
Kagi, J. H. R. (1993), in *Metallothionein III*, Suzuki, K. T., Imura, N., and Kimura, M., Eds., Birkhauser, Basel, Switzerland, 29–55.
Kagi, J. H. R. and Schaffer, A. (1988), *Biochemistry*, 27, 8508–8515.
Kjellstrom, T. (1986), in *Cadmium and Health: A Toxicological and Epidemiological Appraisal*, Vol. 2, Friberg, L., Elinder, C.-G., and Kjellstrom, T., Eds., CRC Press, Boca Raton, FL, 47–86.
Kobota, K., Matsui, R., Fukuyama, Y., Ishimoto, M., and Takayanagi, N. (1970), *Igaku To Seibutsugaku*, 80, 315–320.
Kondo, Y., Yamagata, K., Satoh, M., Naganuma, A., Imura, N., and Akimoto, M. (1993), in *Metallothionein III*, Suzuki, K. T., Imura, N., and Kimura, M., Eds., Birkhauser, Basel, Switzerland, 269–278.

Lazo, J. S., Yang, Y.-Y., Woo, E., Kuo, S.-M., and Saijo, N. (1993), in *Metallothionein III,* Suzuki, K. T., Imura, N., and Kimura, M., Eds., Birkhauser, Basel, Switzerland, 293–302.

Lilis, R. (1981), *Am. J. Ind. Med.,* 2, 293–297.

Mahaffey, K. R. and Fowler, B. A. (1977), *Environ. Health Perspect.,* 19, 165–171.

Mahaffey, K. R., Capar, S. G., Gladen, B. C., and Fowler, B. A. (1981), *J. Lab. Clin. Med.,* 98, 463–481.

Margoshes, M. and Vallee, B. L. (1957), *J. Am. Chem. Soc.,* 79, 4813–4814.

Mistry, P., Lucier, G. W., and Fowler, B. A. (1985), *J. Pharmacol. Exp. Ther.,* 232, 462–469.

Mistry, P., Mastri, C., and Fowler, B. A. (1986), *Biochem. Pharmacol.,* 35, 711–713.

Mistry, P., Mastri, C., and Fowler, B. A. (1987), *Toxicologist,* 7, 78.

Naganuma, A., Satoh, M., and Imura, N. (1987), *Cancer Res.,* 47, 983–987.

Nordberg, G. F., Goyer, R. A., and Nordberg, M. (1975), *Arch. Pathol.,* 99, 192–197.

Nordberg, G. F., Kjellstrom, T., and Nordberg, M. (1986), in *Cadmium and Health: A Toxicological and Epidemiological Appraisal,* Vol. 2, Friberg, L., Elinder, C.-G., and Kjellstrom, T., Eds., Boca Raton, FL, CRC Press.

Pattanaik, A., Bachowski, G., Laib, J., Lemkuil, D., Shaw, C. F., III, Petering, D. H., Hitchcock, A., and Saryan, L. (1992), *J. Biol. Chem.,* 267, 16121–16128.

Petering, D. H., Quesada, A., Dughish, M., Krull, S., Gan, T., Lemkuil, D., Pattanaik, A., Byrnes, R. W., Savas, M., Shelan, H., and Shaw, C. F., III (1993), in *Metallothionein III,* Suzuki, K. T., Imura, N., and Kimura, M., Eds., Birkhauser, Basel, Switzerland, 329–346.

Quintanilla-Vega, B., Kahng, M. W., Smith, D. R., and Fowler, B. A. (1994), *Toxicologist,* 14, 255.

Robbins, A. H., McRee, D. E., Williamson, M., Collett, S. A., Xuong, N. H., Furey, W. F., Wang, B. C., and Stout, C. D. (1991), *J. Mol. Biol.,* 221, 1269–1293.

Saijo, N., Miura, K., and Kasahara, K. (1993), in *Metallothionein III,* Suzuki, K. T., Imura, N., and Kimura, M., Eds., Birkhauser, Basel, Switzerland, 279–291.

Sato, M. and Nagai, Y. (1986), *J. Toxicol. Sci.,* 11, 29–39.

Satoh, M., Naganumaa, A., and Imura, N. (1988), *Toxicology,* 53, 231–237.

Shaikh, Z. A. and Hirayama, K. (1979), *Environ. Health Perspect.,* 28, 267–271.

Selevan, S. G., Landrigan, P. J., Stern, F. B., and Jones, J. H. (1984), *Am. J. Epidemiol.,* 122, 673–683.

Smith, D. R., Kahng, M. W., Quintanilla, B., and Fowler, B. A. (1994), *Toxicologist,* 14, 254.

Squibb, K. S., Ridlington, J. W., Carmichael, N. G., and Fowler, B. A. (1979), *Environ. Health Perspect.,* 28, in *Biological Roles of Metallothionein,* Foulkes, E. C., Ed., Elsevier North Holland, Amsterdam, 181–192.

Squibb, K. S., Pritchard, J. B., and Fowler, B. A. (1984), *J. Pharmacol. Exp. Ther.,* 229, 311–321.

Swenberg, J. A., Short, B., Borghoff, S., Strasser, J., and Charboneau, M. (1989), *Toxicol. Appl. Pharmacol.,* 97, 35–46.

Tohyama, C. and Shaikh, Z. A. (1981), *Fundam. Appl. Toxicol.,* 1, 1–7.

Vallee, B. L. and Maret, W. (1993), in *Metallothionein III,* Suzuki, K. T., Imura, N., and Kimura, M., Eds., Birkhauser, Basel, Switzerland, 1–27.

Wang, X.-P., Chan, H. M., Goyer, R. A., and Cherian, M. G. (1993), *Toxicol. Appl. Pharmacol.,* 119, 11–16.

Webb, M. and Etienne, A. T. (1977), *Biochem. Pharmacol.,* 26, 25–30.

Chapter 45

In Vivo Measurement and Speciation of Nephrotoxic Metals

Donald R. Smith and Fiona McNeill

I. INTRODUCTION

The measurement of toxic metals in the kidney and, specifically, their biomolecular speciation are of primary importance in evaluating the risk of individuals for metal-induced renal injury. As discussed elsewhere in this volume, the kidney is an important and sensitive target organ for the toxic effects of numerous metals, including cadmium (Cd), lead (Pb), mercury (Hg), and arsenic (As), among others (e.g., Fowler, 1992; Nolan and Shaikh, 1992; Conner and Fowler, 1993; Squibb, this volume). There are multiple important factors controlling the levels and toxicity of metals in the kidney, including the magnitude and duration of exposure, the kinetic parameters (e.g., residence time) of the metals within the renal compartment, and the co-occurrence of other metals, as well as the intraorgan and intracellular distribution and biomolecular speciation of the metals (McLellan et al., 1975; Woods and Fowler, 1977; Fowler et al., 1980; Mahaffey et al., 1981; Svartengren et al., 1986; Goering and Fowler, 1987; Nordberg, 1989; Foulkes, 1990; Fowler and DuVal, 1991). Moreover, these and other studies have substantiated the importance of measuring renal metal levels and metal speciation, as opposed to measuring only metal levels in other physiologic compartments such as blood and urine, in order to derive more meaningful concentration-effect relationships within the kidney.

Although the measurement of the overall concentration of toxic metals within the kidney has been of importance in elucidating the relationship between metal concentration and the clinical expression of disease (Barham and Tarara, 1982; Verbueken et al., 1985), it is now well recognized that the biomolecular speciation and distribution of metals are the primary factors controlling toxicity (Axelsson and Piscator, 1966; Goering and Fowler, 1984; Singhal et al., 1987; Woods et al., 1990a,b; Fowler and DuVal, 1991; Cornelis, 1992; Fowler, 1992; Van Loon and Barefoot, 1992; Fowler et al., 1994; Lauwerys et al., 1994; Squibb, this volume). For example, Woods et al. (1990a,b) investigated the etiology of Hg-induced porphyrinuria by testing a hypothesis that mercuric ions (Hg^{2+}) promote free radical-mediated oxidation of reduced porphyrins by compromising the antioxidant potential of endogenous thiols. Further, Fowler and associates have shown *in vitro* that the heme pathway enzyme delta-aminolevulinic acid dehydratase (ALAD) from rat kidney can be protected from inhibition by Pb with the presence of a specific Pb-binding cytosolic protein (Goering and Fowler, 1984; Fowler and DuVal, 1991).

The detection, quantification, and localization of trace elements in biological tissues *in vivo* and *in situ* continue to present major challenges to investigators. The "*in vivo*" measurement of kidney metal burdens strictly implies measurements that are made noninvasively within a living organism. Several radiative techniques have been investigated regarding their feasibility for *in vivo* measurements of target organ metal burdens, with two currently available techniques in particular, neutron activation analysis (NAA) and X-ray fluorescence (XRF), being most successful in clinical investigations of metal toxicity. An important underlying assumption surrounding both *in vivo* and *in vitro* measurements of renal metal levels is that the region (*in vivo* analyses) or sample (*in vitro* biopsy analyses) being measured is representative of the sample or organ as a whole, or that there is an established relationship between the measured region and the whole organ.

This chapter will review the methodologies available for the *in vivo* measurement and speciation of nephrotoxic metals, including some of the advantages and limitations of those techniques. This is not intended to be an exhaustive review, but rather a general perspective on this important toxicological issue. Several reviews specific to metal speciation in both biological and environmental matrices have been published recently, as well as several review articles on the use of radiative techniques for the *in vivo* determination of body trace metal burdens (e.g., Harrison and Rapsomanikis, 1989; Barnes, 1991; Skerfving and Nilsson, 1992; Cornelis and DeKimpe, 1994), and we refer the reader to those discussions for a broader presentation of the topics. Here we will focus largely on studies relating to the measurement of nephrotoxic metals in the kidney and, in some cases, to studies and techniques that have shown promise for application to the measurement and speciation of nephrotoxic metals.

II. *IN VIVO* MEASUREMENTS

A. NEUTRON ACTIVATION ANALYSIS

Toxic elements, when irradiated with the appropriate energy of neutrons, can often be converted into radioactive atoms or atoms in high energy states. These radioactinides can emit gamma-ray or X-ray spectra which are characteristic of and specific to that particular actinide. The gamma- or X-rays can be detected and counted, and when compared against the signals from appropriate calibration standards, allow a quantitative estimate of the elemental content of the sample. This principle can be applied both *in vitro* and *in vivo* and is known as neutron activation analysis (NAA).

Generally, *in vivo* measurements have been made of the essential elements such as nitrogen, calcium, sodium, and iron (Chettle and Fremlin, 1984) and have been used in studies of body composition and nutritional status in both disease and health. This is in part due to the fact that the human body contains larger masses of these elements, making the measurement, in principle, slightly easier than with trace elements. However, it is not just the sheer number of atoms in the sample which make an element suitable for measurement by NAA. Importantly, trace elements must have an adequate cross section for the reaction, i.e., the probability of the element reacting with the neutrons and converting to another radioisotope must be high. The actinide must emit gamma-rays or X-rays which can be detected, and there should preferably be no interfering reactions. Given these considerations, the most successful application of the technique in the areas of occupational health and toxicology has been in studies of Cd exposure (Franklin, 1986; Armstrong, 1989; Fedorowicz et al., 1993; Ralston et al., 1994), although other nephrotoxic elements such as mercury (Smith et al., 1982) have also been studied.

Cadmium has a large cross section for thermal (low energy) neutrons and promptly emits a series of gamma-rays which are readily detectable, so detection and activation must take place simultaneously. Systems have been built to measure liver and kidney Cd using both Pu-Be and Am neutron sources (Franklin, 1986; Armstrong, 1989; Fedorowicz et al., 1993; Ralston et al., 1994) housed in a collimator/shielding system which is portable. The neutron beam enters the body, where the neutrons are thermalized. Cadmium atoms in the kidney are activated and the detector, placed at 90° to the beam direction, detects the emitted spectrum. Subjects are measured and their signals compared with calibration measurements made in a water phantom to provide a quantitative estimate of the Cd content of the kidney. Such systems have been successfully used in several "in field" surveys because of their ready transportability.

Using this technique, Mason et al. (1988) determined the levels at which occupational Cd exposure affected renal function in a survey of 75 workers employed in the manufacture of CuCd alloy. Many indicators of tubular and glomerular function were correlated with both an index of cumulative air exposure and liver Cd burden (used as a biomarker of long term exposure). The study identified a threshold level above which changes in renal function occur, and it also developed a two-phase inflection model which fit many of the biochemical variables reflecting exposure effects. Such studies were successful enough so as to effect change in the British and European legislation regarding air Cd levels.

B. X-RAY FLUORESCENCE ANALYSIS

The observation that elements, when irradiated with X-rays, electrons, or protons are stimulated to emit their own series of characteristic secondary fluorescence X-rays, was originally made early this century (Barkla and Sadler, 1908). This basic principle of irradiating a sample and detecting and quantifying the characteristic X-rays underlies the methodology of XRF as well as those of proton-induced X-ray emission (PIXE), synchrotron radiation-induced X-ray emission (SRIXE), and electron microprobes. These latter techniques are discussed elsewhere in this chapter. XRF as discussed here will be in the context of macroscopic volumes on the order of several cubic centimeters, analyzed either *in vitro* or *in vivo*, while discussion of the other techniques (PIXE, SRIXE, microprobes) will be in the context of analyzing microscopic distributions *in vitro* by using narrowly focused particle beams.

The first reported use of XRF as a method of elemental analysis was in studies of mineral composition performed in the 1920s using X-ray diffraction systems (Hadding, 1922). However, it was not until the 1960s, with the advent of readily available semiconductor technologies, that *in vivo* measurements became possible (Hoffer et al., 1968; Tinney, 1968). The most widespread use of the technique of *in vivo* XRF has been in bone Pb measurement systems (Ahlgren et al., 1976; Chettle et al., 1991) to develop estimates of long-term Pb exposure. XRF kidney studies have also been pursued.

While the *in vivo* measurement of only a few elements will be discussed here, possibilities also exist for measuring other toxic elements *in vivo* by XRF, especially the heavy metals. The characteristic X-rays produced by elements are dependent on their atomic number; the higher the number, the higher the X-ray energy. Since organs such as the kidney and liver often have several centimeters of tissue overlay, X-rays emitted from the organ must penetrate through the tissue to be recorded and therefore must be of higher energy. In addition, there are other factors, including the fluorescence yield (ratio of fluorescence X-rays to Auger electrons), that favor measurement of the higher atomic number metals.

Although the basic technique of XRF is theoretically very simple, there are practical considerations that make the method somewhat more complicated. The choice of fluorescing source is therefore not always obvious nor are the detector-source geometries. Such considerations have led to what may seem from the literature to be a sometimes confusing multiplicity of systems with respect to the type of source and source-detector geometry for measuring different elements. However, XRF studies can be roughly divided into two groups, those investigating clinical uses of nephrotoxic elements, and those concerned with occupational exposure. The former group represents possibly the most successful use of XRF in the study of nephrotoxicity and has focused on patients undergoing treatment with the widely used chemotherapeutic drug cisplatin (Jonson et al., 1991; Borjesson et al., 1993; Todd et al., 1993), and of subjects receiving gold (Au) chrysotherapy for the treatment of rheumatoid arthritis (Borjesson et al., 1993; Shakeshaft et al., 1993).

Cisplatin is nephrotoxic (Cerny, 1991), although the exact mechanisms of nephrotoxicity are not completely understood. *In vivo* measurement studies of platinum (Pt) have been conducted to help determine whether Pt retention in the kidney was a factor for nephrotoxicity. XRF clinical studies were designed to determine the metabolism of Pt within the body and to elucidate the uptake of the drug within tumors and its retention within the kidney over the span of hours to days. The first reported system used polarized X-rays produced by an X-ray generator as the fluorescing source, and studies using this system showed that Pt concentrations fluctuated rapidly over a short period of time (Borjesson, 1991; Jonson et al., 1991; Borjesson et al., 1993). In a subsequent study, El-Sharkawi et al. (1986) used a system with a ^{57}Co radioisotope as the fluorescing source in a 90° geometry with respect to the detector. The results of this study were broadly consistent with the previous data, and they also showed Pt accumulated in the kidney after cessation of treatment. This, in combination with the observations that Pt concentrations declined quickly over the first few hours after treatment, provided further clues regarding cisplatin's nephrotoxicity.

These latter researchers also noted a high level of Pb appearing in the kidney in some patients during the course of cisplatin therapy (El-Sharkawi et al., 1986), and it was suggested that cisplatin treatment caused a mobilization of skeletal Pb stores. To investigate this, a XRF system was constructed to simultaneously measure Pb and Pt in the kidney using a ^{99m}Tc fluorescing source in a backscatter (180°) geometry (Todd et al., 1993). Although little evidence supporting a mobilization of Pb under cisplatin chemotherapy was found, this work illustrated the feasibility of XRF systems for the simultaneous *in vivo* measurement of multiple elements.

In other studies, there is concern that gold chrysotherapy for rheumatoid arthritis may lead to Au accumulation in the kidney and the possible development of transient proteinuria or nephritis (Vernon-

Roberts, 1976; Tozman and Gottlieb, 1987). Animal studies have determined that Au is a strong nephrotoxin, causing acute extensive necrosis of the proximal tubules (Kaizu et al., 1986). To investigate this concern, the polarized X-ray system used in the Pt studies described above was adapted and used to study 27 rheumatoid arthritis patients who had been treated with Au salt (Borjesson, 1991). While levels of kidney Au were found to be high in some cases (ranging from 9 to 97 µg Au/g tissue), there was no correlation between the administered dose of Au and the levels accumulated in the kidney. These observations were also corroborated, using a XRF system with a ^{153}Gd source in a 90° geometry, by Shakeshaft et al. (1993), who further observed that kidney Au levels decreased over time after the cessation of therapy, which may explain the transient nature of some of the health effects.

The second major group of XRF studies are those conducted in occupational settings, where XRF was used to measure kidney burdens of toxic elements, primarily Cd (Christoffersson, 1986). Measurements were originally performed on the kidney cortex using an ^{241}Am source, although significant improvements in detection limits have since been made by switching to an X-ray generator (Christoffersson and Mattsson, 1983). The system was used in studies of occupationally exposed workers (Christofferson et al., 1986), who exhibited Cd concentrations in the kidney cortex ranging from 47 to 317 µg/g tissue, illustrating the usefulness of the system in studies of both occupational and environmental Cd exposure. The use of an X-ray generator may, however, limit the system's transportability.

C. OTHER RADIATIVE TECHNIQUES

There are other radiative techniques which have been investigated for use in *in vivo* body composition studies, and some of these could possibly be adapted to the analysis of trace metals in the kidney. As an example, techniques such as nuclear resonant scattering and neutron inelastic scattering could possibly be applied to measuring nephrotoxic metals. Further research into other possible methods is certainly warranted.

III. *IN SITU* MEASUREMENT METHODOLOGIES

A number of techniques are available that are capable of determining the cellular and subcellular distribution of metals *in situ*, within tissue or cell preparations. While these techniques generally require varying levels of sample preparation (e.g., tissue fixation, cell suspensions, etc.), they provide powerful tools for probing different cellular compartments and processes involving nephrotoxic metals. Thus, the site-specific distribution of metals can be used to evaluate local toxicity, which can be correlated with pathological alterations of tissues.

A. LAMMA

Lasar microprobe mass analyses (LAMMA) is an analytical methodology capable of simultaneous multi-element analysis of metals and their distribution on a cellular or subcellular scale, with a measurement sensitivity of 10^{-17} to 10^{-20} g for most metals. For example, the detection sensitivity for Cd and Pb is 30 and 5 µg/g, respectively, with a limit of detection of approximately 3×10^{-17} and 2×10^{-18} g, respectively (Schmidt and Barckhaus, 1991). LAMMA has received somewhat widespread use in the analysis of essential and toxic metals in tissues and cell preparations (Drüeke, 1980; Schmidt et al., 1980; Verbueken et al., 1984; Visser et al., 1984; Vandeputte et al., 1985). In a study to investigate Cd accumulations in different structural elements of the kidney cortex of rats, Schmidt and Barckhaus (1991) observed a nonhomogenous distribution of Cd within the proximal tubule cells. The cytoplasm of cells of the proximal renal tubule near the glomeruli exhibited the highest concentration of Cd, whereas the glomeruli, the distal tubules, and the collecting tubules showed relatively lower Cd content. Further, the Cd concentration was found to be highly variable, depending on the site of the specific structural element (Schmidt et al., 1986). This is consistent with what is known about the transport mechanisms of Cd involving metallothionein (MT), whereby the Cd-MT complex is released from the liver into the blood and then filtered through the glomeruli and reabsorbed and accumulated within renal tubule cells (Nordberg et al., 1975).

Other studies using LAMMA have demonstrated its usefulness as a method to investigate the cellular distribution of metals within nervous and vascular tissues, as well as within the kidney. Goebel et al. (1990) used LAMMA to locate As within myelinated nerve fibers in biopsied sural nerve specimens prior to chelation treatment, while Schmidt et al. (1985) studied the morphological distribution of Pb within substructure microsamples (1 to 3 mm) of the vascular wall. The topochemical distribution of Pb in the human arterial wall has also been determined using LAMMA (Linton et al., 1985; Schmidt et al., 1985), as has the localization of Pt in renal proximal tubule cells of dogs receiving cisplatin (Verbueken et al., 1984).

However, LAMMA is a sample-destructive technique, and quantification of the local ion concentrations is still under investigation. Moreover, it is often assumed that the inherent metal binding/distribution of the elements under study is not altered by processing and analyses, although few studies have systematically investigated the loss or addition (i.e., contamination) of elements during tissue processing for microanalyses (Morgan et al., 1975; Lane and Martin, 1982; Vandeputte et al., 1985, 1990; Blaineau et al., 1988). Vandeputte et al. (1990) observed that Pb-induced inclusion bodies in rat kidney tissue fixed with glutaraldehyde and OsO_4 contained large amounts of Ca but no Pb, an observation also reported by Zhong et al. (1987) in a study using electron microprobe. It was determined that postfixation of glutaraldehyde-fixed tissue with OsO_4 caused a notable decrease in the LAMMA-detected Pb signal, suggesting some modification of the protein binding of Pb within the inclusions. Based upon these observations, sample preparation using cryotechniques (snap-freezing and vacuum drying), rather than fixative procedures, may be preferable to minimize the possible loss or contamination of metals within the sample.

B. MICROPROBES (ELECTRON, PROTON, PHOTON)

As discussed in the *in vivo* X-ray fluorescence measurement section, most microprobe systems are based on the basic principle that the elements within the sample can be made to emit characteristic fluorescence X-rays. For electron- and proton-based microprobes, characteristic fluorescence X-rays are generated from elements in the sample by an electron or proton beam that is focused by magnetic fields. The beam can be scanned over a sample and the X-ray spectrum from specific locations recorded by a detector, thereby generating a two-dimensional image of the sample's elemental composition. Photon-based microprobes, using either X-ray or gamma-ray beams, are used in a similar fashion, but the photon beam is generally focused by means of collimating and shielding a wider signal, since photons do not carry a charge. The focusing of charged beams (electrons, protons) is much easier, and so typically they have a better spatial resolution than photon sources. The elements which can be detected by these methods depend on the energy of the proton, electron, or X-ray beam, but in general all of these techniques allow multiple elements to be detected simultaneously. A brief discussion of several of these techniques follows.

1. Electron X-ray Microprobe

X-ray microprobe measurements have been used to identify regions of relatively high metal concentrations at the cellular and intracellular level. Notable studies relevant to toxicology include the identification of Pb-rich cytosolic and nuclear inclusion bodies in rat renal tubule cells following elevated exposures to Pb (Fowler et al., 1980; Oskarsson and Johnasson, 1987).

2. Pixe

Proton-induced X-ray emission (PIXE), a relatively nondestructive technique with detection limits in the parts per million range (Torok and Van Grieken, 1994), has been useful in studies of the microdistribution of toxic metals in thin-sliced sections of tissues (Lindh et al., 1978; Schidlovsky et al., 1990; Lowe et al., 1993). The simultaneous quantification of a large number of elements in a sample makes PIXE an important analytical tool for use in diverse toxicological studies (Johansson and Campbell, 1988). However, a possible limitation of the technique is the shallow sampling depth (tens of microns) of the proton beam; thus only surface distributions can be studied. The usefulness of PIXE in animal toxicity studies has recently been discussed in a study on the elemental analysis of renal slices (Lowe et al., 1993).

3. Srixe

Synchrotron radiation-induced X-ray emission (SRIXE) utilizes a synchrotron radiation X-ray source to induce the emission of characteristic fluorescence X-rays. Synchrotrons are capable of producing extremely high intensity pulsed beams of X-rays which allow the detection of low levels of trace elements, but because of the limited number of such instruments, there currently exist only a few laboratories in the U.S. that can make SRIXE measurements using X-rays in the keV energy range. Spatial resolution is typically in the tens of millimeters, though it may be improved to about 1 mm (Jones and Gordon, 1989). An apparent advantage of SRIXE over other microprobe sources is that the damage to the analyzed sample due to the photon beam is estimated to be at least an order of magnitude less than the damage caused by the charged particles of proton and electron beams (Slatkin et al., 1984). The lower sample damage from radiation in SRIXE analyses may be important for the analyses of living tissue/cells, where the preservation of tissue/cell integrity and metal distribution is a primary concern.

4. Exafs

EXAFS (extended X-ray absorption fine structure) and XANES (X-ray absorption near edge structure) are types of X-ray spectroscopy that utilize synchrotron radiation to investigate structural and speciation information in an atomic environment within a prepared sample target (Gordon and Jones, 1991). These techniques, while not widely available, may complement other analytical techniques (e.g., X-ray microprobe) to provide important additional information on the chemical structure of metal complexes at a spatial resolution of 1 mm^2 and detection limits in the 5 ppm range. Applications of this technique in studies to quantitate speciation, however, generally require a comparison of the sample absorption spectrum with those of model compounds that contain the absorbing atoms of interest, including the range of its oxidation states, types of ligands, etc. (Gordon and Jones, 1991). The potential application of EXAFS and XANES for *in situ* speciation studies of toxic metals in renal tissue is evident, based on numerous studies of other metals in biological systems (i.e., Fe, Cu, Pt, Ga) (Hitchcock et al., 1982; Cramer, 1988; Bockman et al., 1990a,b).

C. NUCLEAR MAGNETIC RESONANCE

A principal benefit of *in vivo* nuclear magnetic resonance (NMR) spectroscopy is its usefulness in determining the biomolecular speciation of metals in a preselected region of tissue (or cells), providing the metal species is NMR active. Although there have been a limited number of studies investigating metal toxicity in renal tissue, work done with other tissues has demonstrated the applicability of *in vivo* NMR for the study of nephrotoxic metals (e.g., Pb, Cd, Hg), including the characteristics of ligand interactions and mechanisms of toxicity (Spencer et al., 1985; Vasak et al., 1985; Arkowitz et al., 1987; Jones et al., 1988; Pesek and Schneider, 1988; Reid and Podanyi, 1988; Gartland et al., 1989; Schanne et al., 1989a,b; Long et al., 1994). These studies have further substantiated the view that particular biomolecular species of metals elicit specific effects which may manifest into overt cellular or organ toxicity.

For example, NMR has been used by Rosen and colleagues to investigate the toxic effects of Pb in bone cells. These studies used ^{19}F-NMR and the divalent cation indicator 1,2-bis(2-amino-5-fluorophenoxy)ethane-N,N,N',N'-tetraacetic acid to investigate the levels of biologically labile Pb within cultured bone cells, as well as the effects of Pb on other calcium-mediated cellular processes (Schanne et al., 1989a,b; Dowd and Gupta, 1991; Long et al., 1994). In other studies, the interactions of Cd with red blood cell 2,3-bisphosphoglycerate (DPG) was studied with ^{31}P NMR, which provided evidence of Cd-DPG complexation in aqueous solution (Arkowitz et al., 1987), while ^{113}Cd-NMR was used in studies of the mechanism of metal ion activation of yeast enolase (Spencer et al., 1985). ^{113}Cd-NMR was also used to study the structure flexibility of Cd-metallothionein (MT) (Vasak et al., 1985). Studies on the binding characteristics of Pb and Hg to hen egg-white lysozyme demonstrated site-specific binding of Pb and Hg, which proved to be different than the binding of methylmercury (Pesek and Schneider, 1988). Collectively, these studies evidence the applicability of *in vivo* NMR to investigate free and complexed metal levels within the cell, as well as their effects on normal cellular processes.

IV. FRACTIONATION AND SEPARATION OF CELLULAR COMPONENTS AND METAL COMPLEXES

The toxicological importance of elucidating the subcellular distribution of metals is based upon the numerous studies demonstrating a relationship between specific metal distributions and observable toxic effects (e.g., Goyer and Krall, 1969; Fowler et al., 1980; Squibb and Fowler, 1983; Oskarsson and Fowler, 1985a,b). In addition to the *in situ* measurement techniques discussed above, the subcellular distribution of nephrotoxic metals can also be investigated using density gradients and differential centrifugation to fractionate the cellular components into subcellular fractions (e.g., Storrie and Madden, 1990; Cornelis, 1992), which can then be analyzed for their metal concentrations using a number of methods.

Centrifugation fractionation studies are generally simplest when a radioisotopic tracer is used (assuming isotopic equilibrium), opposed to using inherent metal levels, due to the general ease of radiotracer analyses. For example, the radiotracer ^{203}Pb and organellar fractionation techniques were used to investigate the effects of renal Pb inclusion bodies on the subcellular distribution of Pb (Oskarsson and Fowler,

1985a). That study showed that nuclear, mitochondrial, and cytosolic Pb behavior was different in control vs. Pb-treated rats, suggesting that Pb influences its own subcellular distribution in the kidney. A more recent study of the metabolism of tracer amounts of radiolead in rats found that within the kidney, the majority of radiolead was associated with the kidney cell cytosol after 1 to 24 hours post exposure, and with the nucleus at 144 hours post exposure (Keen et al., 1990). Bose et al. (1993) investigated the distribution of Hg (as mercuric nitrate) in the subcellular fractions of rat and fish liver. That study, which utilized the radiotracer ^{203}Hg and the separation of cellular components into subcellular fractions, determined that Hg follows a definite distribution pattern in the subcellular fractions of liver of both animal species, and that the cytosol was the major site of Hg accumulation.

V. MEASUREMENT OF BIOMOLECULAR SPECIATION

The speciation of an element is defined here to include the individual physicochemical forms of the element in a sample (Florence, 1982), though more specifically it may be considered to be the biologically active compounds to which the metal is bound in a biological matrix (Cornelis, 1990). Specific information on the biomolecular speciation of toxic metals, together with their distribution within the kidney has provided valuable insights into the factors mediating metal-induced nephrotoxicity. This has been well demonstrated with Pb-protein species (Oskarsson and Fowler, 1985a; Fowler and DuVal, 1991; Fowler, 1992; Squibb, this volume), as well as with Cd-metallothionein, which plays an important role in Cd-induced renal injury (Squibb et al., 1984; Squibb and Fowler, 1984; Squibb, this volume).

There are many methodologies that have been applied to elucidate metal speciation, each possessing inherent strengths and limitations. In general, methodologies have utilized a fractionation technique, such as ultrafiltration, ultracentrifugation, size exclusion, ion-exchange, and reversed-phase liquid chromatography, as well as polyarcylamide gel-electrophoresis (PAGE) to separate biomolecular components, and a detection technique such as measurement of a radiotracer, or atomic absorption or mass spectrometry. In particular, we will draw upon the studies of Fowler and co-workers (Oskarsson et al., 1982; Goering and Fowler, 1984; Mistry et al., 1985, 1986; Oskarsson and Fowler, 1985a; Fowler and DuVal, 1991; Fowler, 1992), as well as studies by several others to illustrate the importance of metal biomolecular speciation in efforts to understand the roles of toxic metals as mediators of cell toxicity and injury.

As discussed below, there are difficulties associated with conducting successful investigations of metal biomolecular speciation, in addition to those presented by limitations in the methodologies. Chief among these is the difficulty in demonstrating that a particular biomolecule and toxic metal that are detected in the same isolate are actually associated with one another, and/or that the metal complex has not been altered during its isolation and analysis (Cornelis and DeKimpe, 1994). Several of the factors which can confound determining the *in situ* relationship between a co-isolated metal and biomolecule include the complexity of the mixture of major and trace biocompounds in the isolate, the contamination of the preparation/isolate with the metal of interest or biologically competitive metals, and of course the alteration of the original metal-biomolecule binding.

A. CHROMATOGRAPHY

There are numerous methodologies and separation media available for the separation of (metal-bound) biomolecules that are based upon the principles of chromatography. Both gas and liquid chromatography are commonly utilized as separation techniques in speciation studies. However, separations by gas chromatography may be somewhat limited in application, because they require the metal species to be volatile and thermally stable under the temperature regimes used for analyses. This may often require derivitization of the analyte prior to analyses (e.g., Ebdon and co-workers, 1986, 1988). These requirements, and their possible effects on the distribution and stability of the inherent metal-biomolecular species, has limited the uses of gas chromatography in studies of metal-biomolecular speciation.

Liquid chromatography is better suited for the separation of nonvolatile, high relative molecular mass compounds, provided that a suitable separation medium and eluants compatible with sample components are available (Barnes, 1991). Separation of biomolecules can be achieved with one or a combination of several chromatographic modes, including size exclusion, normal phase, reversed-phase, paired ion reversed-phase, ion-exchange (anionic and cationic), and supercritical fluid chromatography. Size exclusion chromatography (SEC) may be particularly flexible, since the pH and ionic strength of the buffer can be selected to correspond with that of the sample. However, there may be associated difficulties with liquid chromatography which must be considered. These include poor separation resolution of species, trace metal contamination of the column and buffer, sample dilution on the column which may lead to a shift in the chemical equilibrium or require sensitive detection, and irreversible interaction of some species with the separation medium resulting in the loss of species from the sample (Wallaeys et al., 1987; Borguet et al., 1990; Barnes, 1991; Cornelis and DeKimpe, 1994).

As will be apparent from the discussion of Pb speciation below, simple SEC of complex mixtures, such as renal cortex cytosol preparations, is seldom adequate to separate and resolve individual biocompounds. Rather, SEC fractions will often contain a complex mixtures of semi-purified proteins, which must be further isolated using additional separation modes (e.g., ion-exchange and reversed-phase chromatography, PAGE). Moreover, the studies of Fowler and co-workers, as well as others, have demonstrated the necessity to minimize the time required for species separation, purification, and analyses to minimize alteration of the metal-biomolecular complex (Borguet et al., 1990; DuVal et al., 1996). Thus, the many opportunities for incurring spurious results often requires the use of several independent separation/analyses schemes to correlate and verify observations.

B. RADIOISOTOPES

Radioisotopes have been commonly used as tracers of metal distribution and biomolecular speciation via subcellular fractionation of tissues using differential centrifugation and other biomolecular fractionation techniques, as mentioned above (e.g., Sarkar, 1989; Jayawickreme and Chatt, 1990; Fowler and DuVal, 1991; Cornelis, 1992; DuVal et al., 1996). Radioisotopes exist for many of the metals known to be nephrotoxic (e.g., Pb, Cd, As, Hg, Se, Cr, In), although not all of them are ideally suited as tracers in biological systems due to limitations in the radioactive properties of the isotope, including the half-life and the type and energy of the emitted decay particle. For example, the short half-life of the gamma-emitting isotope ^{203}Pb ($t_{1/2}$ = 52 hours) limits its experimental usefulness to days or weeks, depending on the starting specific activity. In general, gamma emitters are the most suitable for tracer work because the self-absorption of the radiation is negligible and the samples are easy to prepare (Cornelis, 1992). Further, carrier-free radioisotopes are preferred because less is needed to achieve sufficient labeling of the metal pool, thereby minimizing the perturbation of the concentration of the metal in the physiologic compartment.

Much of the available information on the binding of Pb to low molecular weight cytosolic proteins in renal tissue has been obtained using the radioisotope ^{203}Pb, in combination with the subcellular and biomolecular fractionation techniques mentioned above (Goering and Fowler, 1984, 1987; Mistry et al., 1985, 1986; Fowler et al., 1993; DuVal et al., 1996). For example, in studies reported by Fowler et al. (1993), monkey kidney tissue cytosol preparations obtained using ultracentrifugation were incubated with tracer amounts of ^{203}Pb followed by protein separation using a series of chromatographic modes. The ^{203}Pb-binding peak in the low molecular weight protein fractions separated by sephadex G-75 SEC (arrow, Figure 1) was collected and further purified by sephadex DEAE ion-exchange chromatography (arrow, Figure 2) to locate the major Pb-protein complexes. Semipurified protein mixtures were further purified using reversed-phase HPLC and PAGE, so that the isolated proteins could be evaluated for their metal-binding properties. Data from these and other studies have been invaluable in the development and testing of hypotheses on the role of Pb-binding proteins in mediating cell toxicity, including alterations in renal proximal tubule cell gene expression (Fowler and DuVal, 1991; Fowler, 1992; Fowler et al., 1993; Fowler et al., 1994; Squibb, this volume).

Radiotracers have also been effectively used to study the biomolecular speciation of selenium (Se) in tissues (Behne et al., 1990a,b) and chromium (Cr) in blood serum (Borguet et al., 1990). In the former studies, SDS-PAGE and autoradiography were used to identify ^{75}Se-labeled proteins in tissues of rats administered ^{75}Se, revealing a series of 13 new selenoproteins in addition to the Se-containing enzyme glutathion peroxidase. The latter study used ^{51}Cr to investigate the biomolecular speciation of Cr in plasma of rats as a model of plasma Cr distribution in humans. As in many studies utilizing radioisotopes, the use here of the gamma-emitter ^{51}Cr, which was easily measured with relatively good sensitivity, partly alleviated the limitation of poor analytical sensitivity in samples where the Cr-protein complex was diluted by up to 100-fold during the separation (SEC) process.

However, the use of radiotracers to study the distribution and speciation of metals in a living system can entail many artifacts and limitations (Cornelis, 1992). These include factors affecting the reaching of true isotopic equilibrium between the inherent (stable) and added (radioactive) metal, as well as the physiological relevance of the tracer dose. The amount of isotope added should be kept within the context of the amount of metal inherent to the system (tissue preparation, etc.), so that biologically relevant binding to biomolecules is observed, rather than ubiquitous nonspecific interactions, as may be the case when the radiotracer is added in large excess. A basic assumption in tracer methodologies is that the

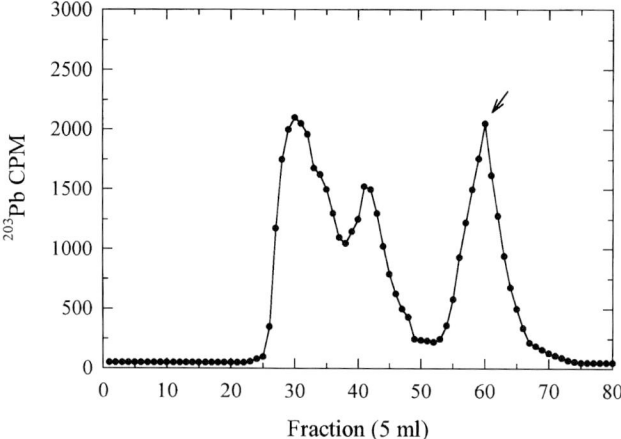

Figure 1. Sephadex G-75 column chromatography of kidney tissue cytosol from a low-lead-exposed monkey. Cytosol was incubated with ^{203}Pb prior to chromatographic separation. The low molecular weight peak fractions (arrow) were collected for further purification (see Figure 2). (From Fowler, B. et al., *J. Expos. Anal. Environ. Epidemiol.*, 3, 441–448, 1993. With permission.)

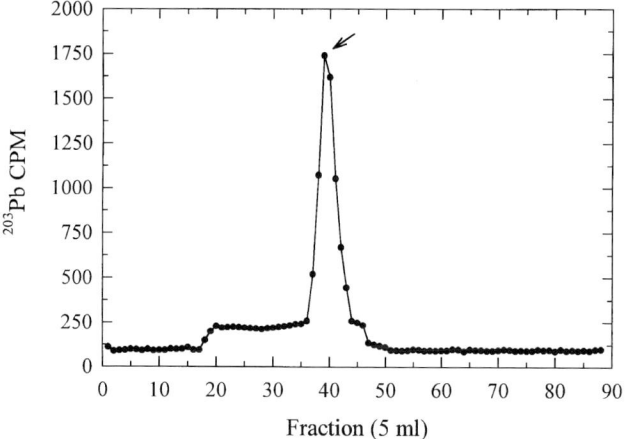

Figure 2. DEAE anion exchange chromatography of specified sephadex G-75 fractions (arrow, Figure 1). ^{203}Pb-binding peak fractions (arrow) were collected and further purified using reversed-phase HPLC and PAGE (data not shown). (From Fowler, B., et al., *J. Expos. Anal. Environ. Epidemiol.*, 3, 441–448, 1993. With permission.)

tracer behaves in the same way as the inherent metal that it is being used to characterize. But this may not be the case, as shown by Borguet et al. (1990), who observed differences in the biomolecular speciation of Cr in plasma depending upon whether the Cr was added *in vivo* (becoming associated mainly with transferrin) or *in vitro* (associated with albumin).

C. HYBRID TECHNIQUES

The above concerns over the use of exogenous (i.e., noninherent) tracers of metal metabolism and speciation, such as radioisotopes, are substantiated by the difficulty in determining whether the tracer is taking part in all the reactions contributing to the distribution of the metal without altering those reactions. Moreover, relatively lengthy separation/analysis schemes may contribute to an alteration of the inherent binding of the metal-biomolecule complex (e.g., Borguet et al., 1990; DuVal et al., 1996). The development and refinement of coupled or "hybrid" methodologies for the on-line separation and analyses of inherent metal speciation in biological materials may circumvent many of these concerns (Crews et al., 1989: Ebdon and Hill, 1989; Barnes, 1991).

Coupled-instrumentation techniques are analytical methodologies involving the interfacing of two or more established analytical techniques, such as a subcellular or biomolecular separation technique with an analyte detector. A major advantage of these coupled techniques is that they generally offer excellent analytical sensitivity and the capability of on-line analyses of inherent or endogenously bound metal species, thereby facilitating relatively rapid analyses (Barnes, 1991). However, interfacing techniques such as liquid chromatography with atomic spectroscopy have presented a number of challenges, particularly with HPLC applications, which have led to the development of several innovative designs for the separation and analyses of metal-biomolecular complexes (Brinkman et al., 1977; Ebdon et al., 1986, 1987a; Gustavsson, 1987; Gustavsson and Nygren, 1987; Ebdon and Hill, 1989; Uden, 1989).

HPLC is the most suitable chromatographic method for the separation of biomolecules in complex samples, as discussed above. There are several available HPLC separation methods (SEC, ion-exchange, normal and reversed-phase) which are all suitable for use coupled with atomic and/or mass spectrometry (Ebdon et al., 1987b; Uden, 1989; Barnes, 1991; Gercken and Barnes, 1991; Owen et al., 1992; Cleland et al., 1994; Tomlinson et al., 1994). Gas chromatography has also been coupled with atomic spectrometry for trace metal speciation (Chana and Smith, 1987; Heitkemper et al., 1989; Bulska et al., 1992; Liang et al., 1994; Vahter et al., 1994), although it is limited to volatile and thermally stable organometallic species or metal-chelates (Ebdon et al., 1986). Analyte detectors in coupled instrument configurations may be single (e.g., atomic absorption spectrometry, AAS) or multi-element (e.g., inductively coupled plasma–mass spectrometry, ICP-MS) in nature (Ebdon et al., 1986, 1987b; Uden, 1989; Barnes, 1991).

ICP-MS has shown particular promise for a number of applications in metal speciation studies (Bushee, 1988; Crews et al., 1989; Hietkemper et al., 1989; Bushee et al., 1990; Barnes, 1991; Brown et al., 1994). The most important advantages of ICP-MS when coupled with a suitable chromatographic methodology are (i) its high elemental sensitivity and selectivity, providing detection limits comparable to graphite furnace AAS (i.e., sub-ppb), (ii) metals can be monitored directly in the eluent of the chromatographic system, (iii) many metals can be measured simultaneously, and (iv) its stable isotope measurement capabilities for stable isotopic tracer applications (Dean et al., 1987; Delves and Campbell, 1988; Paschal, 1990; Barnes, 1991; Gercken and Barnes, 1991). Moreover, plasma spectrometric detectors, in contrast to flame or furnace AAS systems, exhibit lower matrix interferences and longer linear dynamic concentration ranges (Gardiner, 1988). Studies using SEC coupled with ICP-MS detection have investigated several protein-bound metals, including Pb and Cd (Dean et al., 1987; Crews et al., 1989; Gercken and Barnes, 1991; Owen et al., 1992; Brown et al., 1994). Mercury speciation detection by ICP-MS has also been reported by some workers (Bushee, 1988; Bushee et al., 1990).

A potentially valuable, though relatively unexplored application of chromatography coupled with ICP-MS, is its use in stable isotope tracer studies. Stable isotope tracer methodologies have proven to be powerful techniques for investigating the metabolism of nephrotoxic metals at low dose exposures (e.g., Smith et al., 1992; Smith and Flegal, 1992; Smith et al., 1994). The sensitivity and mass resolution of quadrapole mass spectrometers, and particularly the new generation of magnetic sector ICP-mass spectrometers, is sufficient for quantitative isotope abundance analyses of many metals. Several major benefits of methodologies utilizing stable isotopic tracers include (i) stable isotopes do not possess any radiation hazard and they are not limited by their nuclear decay properties (i.e., half-life, energy of decay), (ii) the high sensitivity of ICP-MS allows for the use of low tracer doses sufficient to alter the inherent stable isotopic abundances within the biologic system without measurably changing the total metal concentration, and (iii) stable isotope methodologies may be directly applied in human studies. In studies of metals possessing multiple measurable stable isotopes, such as Pb, the different stable isotopes may be used to evaluate several factors simultaneously, such as the temporal effects of exposure on metal speciation, differences in intrinsic vs. extrinsic tracer labeling, etc.

Other methodologies which have proven useful in speciation analyses of volatile nephrotoxic metals such as Hg and As include HPLC separation coupled with hydride generation and cryogenic trapping with GC-FAAS analyses (Chana and Smith, 1987; Heitkemper et al., 1989), and combined gas chromatography/cold vapor atomic fluorescence spectrometry and GC/microwave-induced plasma atomic emission spectrometry (Bulska et al., 1992; Liang et al., 1994; Vahter et al., 1994).

VI. SUMMARY AND FUTURE NEEDS

In vivo measurement techniques have proven very useful in studies of occupational exposure to metals, particularly Cd, although these techniques have not yet received widespread use for this purpose. The feasibility of *in vivo* measurement by XRF of metals other than those discussed here should be pursued, as should the continued pursuit of more detailed studies of measurements of Cd and Pb. Continued advances in detector technologies (e.g., larger detector volumes) will improve *in vivo* XRF measurement capabilities. These improvements may be particularly useful for Pb studies, since they may facilitate applications in some clinical studies of Pb disposition in humans (e.g., chelation-mediated changes in organ disposition) when used in conjunction with other biomarkers of effect. In addition, recent advances in detector and electronic technologies also make elemental analysis by NAA more feasible, and continued application in Cd exposure studies is warranted, since Cd is still widely used and occupational exposures continue to occur. Finally, other toxic metals, such as those used in the semiconductor industries should be investigated using *in vivo* measurement techniques to address the occurrence and extent of occupational exposures to semiconductor metals.

The collection, treatment, and preservation of samples for qualitative and quantitative analyses of nephrotoxic metal distribution and speciation requires careful consideration and planning. The central concern in all methodologies evaluating the biomolecular speciation of metals should be the preservation of the integrity of the species in the sample, and their representativeness of the speciation within the cell/organ/region of interest. Sample and species alteration may occur at any stage of sample collection, processing, and analyses, and it is important that any factors associated with processing and analyses that are liable to influence the distribution of trace element species in the system be well understood. In particular, the susceptibility and effects of metal contamination and/or metal loss on speciation should be considered independently and systematically evaluated at each stage of the investigation, especially for ubiquitous contaminants like Pb and Cr that may be present in high levels in buffers and other reagents. Our experience with Pb-binding proteins in environmentally exposed individuals, as well as the experience of others, has demonstrated the necessity of adhering to trace metal-clean techniques throughout sample processing and analyses, including the use of purified buffers and reagents to avoid contamination and invalidation of the results (i.e., Flegal and Smith, 1992; Cornelis and DeKimpe, 1994; Smith et al., 1994). While these methodological concerns represent unique challenges in studies of metal biomolecular speciation, their proper attention will contribute substantially to the accuracy and toxicological relevance of the investigations.

REFERENCES

Ahlgren, L., Liden, K., Mattsson, S., and Tejning, S. (1976), *Scand. J. Work. Environ. Health*, 2, 82–286.
Arkowitz, R., Hoehn-Berlage, M., and Gersonde, K. (1987), *FEBS. Lett.*, 217(1), 21–24.
Armstrong, R. (1989). Ph.D. thesis, University of Birmingham, Birmingham, England.
Axelsson, B. and Piscator, M. (1966), *Arch. Environ. Health*, 12(3), 360–373.
Barham, S. and Tarara, J. E. (1982), in *40th Annu. Proc. Electron Microsc. Soc. Am.*, Bailey, G. W., Ed., Washington, D.C., 266.
Barkla, C. G. and Sadler C. A. (1908), *Phil. Mag.*, 16, 550–584.
Barnes, R. M. (1991), in *Biological Trace Element Research: Multidisciplinary Perspectives*, Subramanian, K. S., Iyengar, G. V., and Okamoto, K., Eds., ACS Series 445, Washington, D. C., 158–180.
Behne, D., Scheid, S., Hilmert, H., Gessner, H., Gawlik, D., and Kyriakopoulos, A. (1990a), *Biol. Trace Elem. Res.*, 26–27, 439–447.
Behne, D., Scheid, S., Kyriakopoulos, A., and Hilmert, H. (1990b), *Biochim. Biophys. Acta*, 1033(3), 219–225.
Blaineau, S., Amsellem, J., and Nicaise, G. (1988), *Stain Technol.*, 63(6), 339–349.
Bockman, R. S., Repo, M. A., Warrell, R. P., Jr., Pounds, J. G., Schidlovsky, G., Gordon, B. M., and Jones, K. W. (1990a), *Proc. Natl. Acad. Sci. U.S.A.*, 87(11), 4149–4153.
Bockman, R. S., Warrell, R. P., Jr., Levine, B., Pounds, J. G., Schidlovsky, G., and Jones, K. W. (1990b), *Basic Life Sci.*, 55, 293–296.
Borguet, F., Cornelis, R., and Lameire, N. (1990), *Biol. Trace Elem. Res.*, 26–27, 449–460.
Borjesson, J. (1991), Ph.D. thesis, University of Gothenburg, Gothenburg, Sweden.
Borjesson, J., Alpsten, M., Huang, S., and Jonson, R. (1993), *Basic Life Sci.*, 60, 275–280.
Bose, S., Ghosh, P., Ghosh, S., Chaudhury, S., and Bhattacharya, S. (1993), *Biomed. Environ. Sci.*, 6(2), 195–206.
Brinckman, F., Blair, W., Jewett, K., and Inverson, W. P. J. (1977), *J. Chromatogr. Sci.*, 15, 493–501.
Brown, A., Ebdon, L., and Hill, S. (1994), *Anal. Chim. Acta*, 286, 391–399.
Bulska, E., Emteborg, H., Baxter, D. C., Frech, W., Ellingsen, D., and Thomassen, Y. (1992), *Analyst*, 117(3), 657–663.
Bushee, D.S. (1988), *Analyst*, 113, 1167–1170.
Bushee, D. S., Moody, J. R., and May, J. C. (1990), *J. Anal. At. Spectrom.*, 4, 773–775.
Cerny, T. (1991), *Topics on Supportive Care in Oncology Trimestrial*, 2, 8–10.
Chana, B. S. and Smith, N. J. (1987), *Anal. Chim. Acta*, 197, 177–186.
Chettle, D. R. and Fremlin, J. H. (1984), *Phys. Med. Biol.*, 29, 1011–1043.
Chettle, D. R., Scott, M. C., and Somervaille, L. J. (1991), *Environ. Health Perspect.*, 91, 49–55.
Christoffersson, J.-O. (1986), Ph.D. thesis, Lund University, Lund Sweden.

Christoffersson, J.-O. and Mattsson, S. (1983), *Phys. Med. Biol.*, 28, 1135–1144.
Christoffersson, J.-O., Schutz, A., Ahlgren, L., Haeger-Aronsen, B., Mattsson, S., and Skerfving, S. (1986), *Am. J. Ind. Med.*, 6, 447–457.
Cleland, S. L., Olson, L. K., and Caruso, J. A. (1994), *J. Anal. At. Spectrom.*, 9, 975–980.
Conner, E. A. and Fowler, B. A. (1993), in *Toxicology of the Kidney,* 2nd ed., Hook, J. and Goldstein, R., Eds., Raven Press, New York, 437–457.
Cornelis, R. (1990), in *Metal Speciation in the Environment,* Broekaert, J. A. C., Gücer, S., and Adams, F., Eds., NATO ASI Series G, Ecological Sciences, Vol. 23, Springer-Verlag, Berlin, 169–194.
Cornelis, R. (1992), *Analyst,* 117(3), 583–588.
Cornelis, R. and DeKimpe, J. (1994), *J. Anal. At. Spectrom.*, 9, 945–950.
Cramer, S. P. (1988), in *X-Ray Absorption,* Koningsberger, D. C., and Prins, R., Eds., Wiley & Sons, New York, 257–320.
Crews, H. M., Dean, J. R., Ebdon, L., and Massey, R. C. (1989), *Analyst,* 114(8), 895–899.
Dean, J., Ebdon, L., and Massey, R. (1987), *J. Anal. At. Spectrom.*, 2, 369–374.
Delves, H. and Campbell, M. (1988), *J. Anal. At. Spectrom.*, 3, 343–348.
Dowd, T. L. and Gupta, R. K. (1991), *Biochim. Biophys. Acta,* 1092(3), 341–346.
Drüeke, T. (1980), *Nephron.*, 26, 207–210.
DuVal, G., Gilg, D., Garvey, J., and Fowler, B. (1995), Demonstration of the high affinity renal lead-binding protein (PbBP) to be a member of the $\alpha 2\mu$-globulin family, manuscript submitted.
Ebdon, L. and Hill, S. (1989), in *Environmental Analysis Using Chromatography Interfaced with Atomic Spectroscopy,* Harrison, R. and Rapsomanikis, S., Eds., Ellis Horwood, Chichester, 165–187.
Ebdon, L., Hill, S., and Ward, R. (1986), *Analyst,* 111, 1113–1138.
Ebdon, L., Hill, S., and Jones, P. (1987a), *Analyst,* 112, 437–440.
Ebdon, L., Hill, S., and Ward, W. (1987b), *Analyst,* 112, 1–16.
Ebdon, L., Hill, S., Walton, A., and Ward, W. (1988), *Analyst,* 113, 1159–1165.
El-Sharkawi, A. M., Morgan, W. D., Cobbold, S., Jaib, M. B. M., Evans, C. J., Somervaille, L. J., Chettle, D. R., and Scott, M. C. (1986), *Lancet,* 2, 249–250.
Fedorowicz, R. P., Chettle, D. R., Kennett, T. J., Prestwich, W. V., and Webber, C. E. (1993), *Basic Life Sci.*, 60, 323–324.
Flegal, A. R. and Smith, D. R. (1992), *Environ. Res.*, 58, 125–133.
Foulkes, E. C. (1990), *Crit. Rev. Toxicol.*, 20(5), 327–339.
Fowler, B. A. (1992), *Environ. Health Perspect.*, 100, 57–63.
Fowler, B. A. and DuVal, G. (1991), *Environ Health Perspect.*, 91, 77–80.
Fowler, B. A., Kahng, M. W., and Smith, D. R. (1994), *Environ. Health Perspect.*, 102 (Suppl. 3), 115–116.
Fowler, B., Kahng, M., Smith, D., Conner, E., and Laughlin, N. (1993), *J. Expos. Anal. Environ. Epidemiol.*, 3, 441–448.
Fowler, B. A., Kimmel, C., Woods, J., McConnell, E., and Grant, L. (1980), *Toxicol. Appl. Pharmacol.*, 56, 59–77.
Franklin, D. M. (1986), Ph.D. thesis. University of Birmingham, Birmingham, England.
Gardiner, P. E. (1988), *J. Anal. At. Spectrom.*, 3, 163–168.
Gartland, K. P., Bonner, F. W., and Nicholson, J. K. (1989), *Mol. Pharmacol.*, 35(2), 242–250.
Gercken, B. and Barnes, R. M. (1991), *Anal. Chem.*, 63(3), 283–287.
Goebel, H. H., Schmidt, F., Bohl, J., Tettenborn, B., Kramer, G., and Gutmann, L. (1990), *J. Neuropath. Exp. Neurol.*, 49(2), 137–149.
Goering, P. L. and Fowler, B. A. (1984), *J. Pharmacol. Exp. Ther.*, 231(1), 66–71.
Goering, P. L. and Fowler, B. A. (1987), *Arch. Biochem. Biophys.*, 253(1), 48–55.
Gordon, B. and Jones, K. (1991), in *Biological Trace Element Research: Multidisciplinary Perspectives,* Subramanian, K. S., Iyengar, G. V., and Okamoto, K., Eds., ACS Series 445, Washington D.C., 290–305.
Goyer, R. A. and Krall, R. (1969), *J. Cell Biol.*, 41(2), 393–400.
Gustavsson, A. (1987), *Spectrochim. Acta Part B,* 42, 111–118.
Gustavsson, A. and Nygren, O. (1987), *Spectrochim. Acta Part B,* 42, 883–888.
Hadding, A. (1922), *Methode. Zeit. Anorg. Allg. Chem.*, 122, 192–200.
Harrison, R. and Rapsomanikis, S., Eds. (1989), *Environmental Analysis Using Chromatography Interfaced with Atomic Spectroscopy,* Ellis Horwood, Chichester, 370 pp.
Heitkemper, D., Creed, J., and Caruso, J. (1989), *J. Anal. At. Spectrom.*, 4, 279–284.
Hitchcock, A., Lock, C., and Pratt, W. (1982), *Inorg. Chim. Acta,* 66, L45–L47.
Hoffer, P. B., Jones, W. B., Crawford, R. B., Beck, R., and Gottschalk, A. (1968), *Radiology,* 90, 342–344.
Jayawickreme, C. K. and Chatt, A. (1990), *Biol. Trace Elem. Res.*, 26–27, 503–512.
Johansson, S. and Campbell, J. (1988), *PIXE: A Novel Technique for Elemental Analysis,* John Wiley & Sons, New York, 134–140.
Jones, M., Basinger, M., Topping, R., Gale, G., Jones, S., and Holscher, M. (1988), *Arch. Toxicol.*, 62, 29–36.
Jones, K. and Gordon, B. (1989), *Anal. Chem.*, 61, 349A–358A.
Jonson, R., Borjesson, J., Mattsson, S., Unsgaard, B., and Wallgren, A. (1991), *Acta. Oncol.*, 30, 315–319.
Kaizu, K., Matsuno, K., Kodoma, Y., and Etoh, S. (1986), Nephrotoxic effect of gold sodium thiomalate in rats — ultrastructural observations using electron microscopy and X-ray energy dispersive analysis, *Sangyo Ika Daigaku Zasshi,* 8, 27–33.

Keen, C. L., Zidenberg-Cherr, S., and Parks, N. J. (1990), in *Final Annual Report,* Laboratory for Energy-Related Health Research, University of California, Davis. Prepared for the Department of Energy, 110–114.
Lane, D. S. and Martin, E. S. (1982), *Z. Pflanzenphysiol.,* 107, 33–41.
Lauwerys, R., Bernard, A., Roels, H., and Buchet, J. (1994), *Clin. Chem.,* 40, 1391–1394.
Liang, L., Bloom, N. S., and Horvat, M. (1994), *Clin. Chem.,* 40(4), 602–607.
Lindh, U., Brune, D., and Nordberg, G. (1978), *Sci. Total Environ.,* 10(1), 31–37.
Linton, A. L., Richmond, J. M., Clark, W. F., Lindsay, R. M., Driedger, A. A., and Lamki, L. M. (1985), *Clin. Nephrol.,* 24(2), 84–87.
Long, G. J., Rosen, J. F., and Schanne, F. A. (1994), *J. Biol. Chem.,* 269(2), 834–837.
Lowe, T., Chen, Q., Fernando, Q., Keith, R., and Gandolfi, A. J. (1993), *Environ. Health Perspect.,* 101(4), 302–308.
Mahaffey, K., Capar, S., Gladen, B., and Fowler, B. (1981), *J. Lab. Clin. Med.,* 98, 463–481.
Mason, H. J., Davison, A. G., Wright, A. L., Guthrie, C. J., Fayers, P., Venablesm, K., Smith, N. J., Chettle, D. R., Franklin, D. M., Scott, M. C., et al. (1988), *Br. J. Ind. Med.,* 45, 793–802.
McLellan, J. S., Thomas, B. J., Fremlin, J. H., and Harvey, T. C. (1975), *Phys. Med. Biol.,* 20(1), 88–95.
Mistry, P., Lucier, G. W., and Fowler, B. A. (1985), *J. Pharmacol. Exp. Ther.,* 232(2), 462–469.
Mistry, P., Mastri, C., and Fowler, B. A. (1986), *Biochem. Pharmacol.,* 35(4), 711–713.
Morgan, A., Davies, T., and Erasmus, D. (1975), *Micron,* 6, 11–15.
Nolan, C. V. and Shaikh, Z. A. (1992), *Toxicology,* 73(2), 127–146.
Nordberg, G. F. (1989), *Biol. Trace Elem. Res.,* 21, 131–135.
Nordberg, G. F., Goyer, R., and Nordberg, M. (1975), *Arch. Pathol.,* 99(4), 192–197.
Oskarsson, A. and Fowler, B. A. (1985a), *Exp. Mol. Pathol.,* 43(3), 397–408.
Oskarsson, A. and Fowler, B. A. (1985b), *Exp. Mol. Pathol.,* 43(3), 409–417.
Oskarsson, A. and Johansson, A. (1987), *Toxicology,* 44(1), 61–72.
Oskarsson, A., Squibb, K. S., and Fowler, B. A. (1982), *Biochem. Biophys. Res. Commun.,* 104(1), 290–298.
Owen, L. M., Crews, H. M., Hutton, R. C., and Walsh, A. (1992), *Analyst,* 117(3), 649–655.
Paschal, D. (1990), *Spectrochim. Acta,* 44B, 1229–1236.
Pesek, J. J. and Schneider, J. F. (1988), *J. Inorgan. Biochem.,* 32(4), 233–238.
Ralston, A., Utteridge, T., Paix, D., and Beddoe, A. (1994), *Aust. Phys. Eng. Sci. Med.,* 17, 34–42.
Reid, R. S. and Podanyi, B. (1988), *J. Inorg. Biochem.,* 32, 183–195.
Sarkar, B. (1989), *Biol. Trace Elem. Res.,* 21, 137–144.
Schanne, F. A., Moskal, J. R., and Gupta, R. K. (1989a), *Brain Res.,* 503(2), 308, 311.
Schanne, F. A., Dowd, T. L., Gupta, R. K., and Rosen, J. F. (1989b), *Proc. Natl. Acad. Sci. U.S.A.,* 86(13), 5133–5135.
Schidlovsky, G., Jones, K. W., Burger, D. E., Milder, F. L., and Hu, H. (1990), *Basic Life Sci.,* 55, 275–280.
Schmidt, P. F. and Barckhaus, R. H. (1991), *Prog. Histochem. Cytochem.,* 23(1–4), 342–354.
Schmidt, P. F., Barckhaus, R. H., and Kleimeier, W. (1986), *Trace Elem. Med.,* 3, 19–24.
Schmidt, P. F., Fromme, H. G., and Pfefferkorn, G. (1980), *Scan. Electron. Microsc.,* Pt. 2, 623–634.
Schmidt, P. F., Lehmann, R. R., Ilsemann, K., and Wilhelm, A. H. (1985), *Artery,* 12(5), 277–285.
Shakeshaft, J., Clarke A. K., Evans, M. J., and Lillicrap, S. C. (1993), *Basic Life Sci.,* 60, 307–310.
Singhal, R. K., Anderson, M. E., and Meister, A. (1987), *FASEB J.,* 1(3), 220–223.
Skerfving, S. and Nilsson, U. (1992), *Toxicol. Lett.,* 64–65, 17–24.
Slatkin, D., Hanson, A., Jones, K., Kraner, H., Warren, J., and Finkel, G. (1984), *Nucl. Instrum. Methods,* 227, 378–384.
Smith, J. R., Athwal, S. S., Chettle, D. R., and Scott, M. C. (1982), *Int. J. Appl. Radiat. Isot.,* 33, 557–561.
Smith, D. R. and Flegal, A. R. (1992), *Toxicol. Appl. Pharmacol.,* 116(1), 85–91.
Smith, D., Markowitz, M., Crick, J., Rosen, J., and Flegal, A. R. (1994), *Environ. Res.,* 67, 39–57.
Smith, D. R., Osterloh, J. D., Niemeyer, S., and Flegal, A. R. (1992), *Environ. Res.,* 57(2), 190–207.
Spencer, S. G., Brewer, J. M., and Ellis, P. D. (1985), *J. Inorg. Biochem.,* 24(1), 47–57.
Squibb, K. S. (1996), in *Toxicology of Metals,* Vol. 2: *Target Organ Toxicology,* Chang, L. W., Ed., CRC Press, Boca Raton, FL.
Squibb, K. S. and Fowler, B. A. (1983), in *Biological and Environmental Effects of Arsenic,* Fowler, B. A., Ed., Elsevier/North-Holland, Amsterdam, 233–270.
Squibb, K. S. and Fowler, B. A. (1984), *Environ. Health Perspect.,* 54, 31–35.
Squibb, K. S., Pritchard, J. B., and Fowler, B. A. (1984), *J. Pharmacol. Exp. Ther.,* 229(1), 311–321.
Storrie, B., and Madden, E. (1990), *Guide to Protein Purification,* Deutscher, M., Ed., Academic Press, San Diego, 203–224.
Svartengren, M., Elinder, C. G., Friberg, L., and Lind B. (1986), *Environ. Res.,* 39(1), 1–7.
Tinney, J. F. (1968), Ph.D. thesis, University of Oklahoma, Norman.
Todd, A. C., Chettle, D. R., Scott, M. C., and Somervaille, L. J. (1993), *Nucl. Med. Biol.,* 20, 589–595.
Tomlinson, M., Wang, J., and Caruso, J. (1994), *J. Anal. At. Spectrom.,* 9, 957–964.
Torok, S. B. and Van Grieken, R. E. (1994), *Anal. Chem.,* 66(12), 186R–206R.
Tozman, E. C. S. and Gottlieb, N. L. (1987), *Med. Toxicol.,* 2, 177–189.
Uden, P. (1989), in *Environmental Analysis Using Chromatography Interfaced with Atomic Spectroscopy,* Harrison, R. and Rapsomanikis, S., Eds., Ellis Horwood, Chichester, 96–126.
Vahter, M., Mottet, N. K., Friberg, L., Lind, B., Shen, D. D., and Burbacher, T. (1994), *Toxicol. Appl. Pharmacol.,* 124(2), 221–229.

Vandeputte, D., Jocob, W., and Van Grieken, R. (1990), *J. Histochem. Cytochem.,* 38, 331–337.
Vandeputte, D., Verbueken, A., Jacob, W., and Van Grieken, R. (1985), *Acta Pharmacol. Toxicol.,* 59 (Suppl 7), 617–629.
Vasak, M., Hawkes, G. E., Nicholson, J. K., and Sadler, P. J. (1985), *Biochemistry,* 24(3), 740–747.
Van Loon, J. C. and Barefoot, R. R. (1992), *Analyst,* 117(3), 563–570.
Verbueken, A. H., Van Grieken, R. E., Paulus, G. J., Verpoten, G. A., and De Broe, M. E. (1984), *Biomed. Mass Spectrom.,* 11(4), 159–163.
Verbueken, A. H., Bruynseels, F. J., and Van Grieken, R. E. (1985), *Biomed. Mass Spectrom.,* 12(9), 438–463.
Vernon-Roberts, B. (1976), *Ann. Rheum. Dis.,* 35, 477–486.
Visser, W., Van de Vyver, F., Verbueken, A., Lentferink, M., Van Grieken, R., and De Broe, M. (1984), *Calcif. Tissue Int.* 36, S22.
Wallaeys, B., Cornelis, R., Lameire, N., and Belpaire, F. (1987), in *Toxicology of Metals: Clinical and Experimental Research,* Brown, S. and Kodoma, Y., Eds., Ellis Horwood, Chichester, 367–368.
Woods, J. S., Calas, C. A., Aicher, L. D., Robinson, B. H., and Mailer, C. (1990a), *Mol. Pharmacol.,* 38(2), 253–260.
Woods, J. S., Calas, C. A., Aicher, L. D. (1990b), *Mol. Pharmacol.,* 38(2), 261–266.
Woods, J. S. and Fowler, B. A. (1977), *J. Lab. Clin. Med.,* 90(2), 266–272.
Zhong, C., Ling, Y., Wu, Z., Zhu, S., and Hu, R. (1987), *J. Electron Microsc. Tech.,* 7, 91–99.

Chapter 46

Metal-Induced Alterations in Renal Gene Expression

Hiroshi Yamauchi and Bruce A. Fowler

I. INTRODUCTION

Acute and chronic nephropathy caused by metals exposure has been reported in many epidemiological and animal studies. Cadmium, mercury, lead, and inorganic arsenic are known, typical metals that generate nephropathy (IPCS, 1991). Depending upon the type of metal and the length of exposure, injury can occur at many places along the renal nephron, including the glomerulus, portions of the renal tubule, or both. Functional alterations occurring as a result of metal exposure have been well characterized, but our understanding of the precise mechanisms by which metals cause injury at the cellular level is still limited. The development of molecular biology techniques in recent years, however, has greatly enhanced our ability to study mechanisms of toxicity and has begun to show that early, important effects of metals on cells involve alterations in gene expression. An important area of research at this time is studies of the relationship between cellular death and the stress response.

The elevated expression of stress proteins has been observed in a broad distribution of kidney diseases and other instances of cell and tissue damage. Studies of the relationship between renal disease and stress protein induction have recently been reported (Dodd et al., 1993). Numerous studies have shown that metals and metalloids can induce stress proteins (heat shock proteins; HSP) in a variety of cell types and organisms. This chapter will review recent studies on the relationship of metals/metalloids-induced alterations in renal gene expression and their renal toxicity.

II. ARSENIC

The chemical form and oxidation state of the arsenical of concern are of great importance for understanding mechanisms of cell injury. Inorganic pentavalent arsenic (As^{5+}) and trivalent arsenic (As^{3+}) vary markedly in their acute toxicity and mechanisms of biological action. Methylation of these inorganic forms to methylarsonic acid and dimethylarsinic acid in the liver further complicates an understanding of *in vivo* renal toxicity, because they are excreted by the kidney in the urine (Yamauchi and Fowler, 1994). Furthermore, early studies by Ginsburg and co-workers (Ginsburg and Lotspeich, 1963; Ginsburg, 1965) showed arsenate (As^{5+}) is actively transported by the kidney tubules and that a small fraction of this form is reduced to As^{3+} which is the more acutely toxic chemical form. These types of data greatly complicate any understanding of the mechanisms of arsenical-induced nephrotoxicity because metabolic interconversion of these arsenicals by the renal proximal tubule cells would yield both species present. At the organelle level of biological organization, the mitochondrion is a major target site of action for all inorganic arsenicals (Squibb and Fowler, 1983). Combined ultrastructural/biochemical studies (Brown et al., 1976) conducted on kidneys of rats exposed to arsenate (As^{5+}) in drinking water for prolonged periods of time showed *in situ* swelling associated with decreased respiratory function.

More recent studies (Brown and Rush, 1984) have shown that exposure of animals to arsenic causes induction of several stress proteins in the kidney, indicating that the genetic machinery in the nuclei is also being affected by arsenical exposure. That arsenic (mainly arsenite) induces stress proteins has been shown by *in vitro* studies with many various kinds of cells, but studies of stress protein induction specifically in kidney proximal tubule epithelial cells are few. Studies in our laboratory have shown that induction of stress proteins occurs after exposure to arsenical semiconductor compounds in *in vitro* studies with human and hamster male and female renal proximal tubule epithelial cells, and also in *in vivo* studies with male hamsters. In *in vivo* studies, alterations in male hamster proximal tubule cell gene expression were monitored based upon similar tissue concentrations of indium or arsenic at various time-points, following a single subcutaneous injection of indium, arsenic, or indium arsenide at 10 and 30 days. Two-dimensional gel electrophoresis patterns showed that exposure to arsenic and InAs stimulated the synthesis of a number of proteins in the < 10-, 28-, 32-, 60-, 70-, and 90-kDa gene families at 10 days (Conner et al., 1993). In contrast, at 30 days after InAs exposure, the overall synthesis of the proteins was decreased. These results are consistent with the specific accumulation of indium that occurs in the kidney over time (Yamauchi et al., 1992, 1993).

On the other hand, in *in vitro* studies (Fowler et al., 1995), primary cultures of hamster and human kidney epithelial cells were exposed to 100 µM arsenic, gallium, or indium, or 50 µM and 100 µM combinations of arsenic + indium or arsenic + gallium. There were similarities between the cells from the two animal species with regard to induction of the 60-, 70-, and 90-kDa stress protein gene families. The combination of arsenic + gallium showed marked inhibition of the stress protein response compared to arsenic alone, indicating that the combination of these elements alters normal gene responsiveness to an individual element. The induction of stress proteins in the 60-kDa gene families was observed after exposure to arsenic, gallium, and indium in the male and female hamster kidney epithelial cells, but induction of this gene family was not observed in male and female human kidney epithelial cells. Results of these studies indicate both similarities and differences between hamster and human kidney epithelial cells with regard to induction of the stress protein response by these semiconductor elements. Experiments in rat kidney proximal tubule epithelial cells indicate that similar effects on gene expression occur in this mammalian species (Aoki et al., 1990).

Of particular interest with respect to arsenic is the induction of the 32-kDa stress protein, which has been shown to be heme oxygenase (Shibahara et al., 1987; Taketani et al., 1989). This enzyme, which is thought to be the rate-limiting enzyme in heme degradation, has been shown to be induced by arsenic in numerous systems. In the studies discussed above, induction of a 32-kDa stress protein was observed in the *in vivo* studies conducted with the hamster kidney epithelial cells (Conner et al., 1993), but was not observed in the *in vitro* studies. It is possible that the induction of the 32-kDa stress protein was inhibited by high concentrations of arsenic in the cells.

III. INDIUM

It is clear from animal experiments that indium compounds are nephrotoxic (Castronovo and Wagner, 1971, 1973; Fowler et al., 1983; Yamauchi et al., 1993), although no cases have been reported of renal effects in humans exposed to indium (Fowler, 1986). When ionic indium compounds $InCl_3$ are administered to animals, indium specifically accumulates in the kidney. It is thought that this accumulation is related to its manifestation of toxicity. *In vitro* experiments with human and hamster kidney proximal tubule epithelial cells exposed to indium ($InCl_3$, 100 µM) (Fowler et al., 1995) have shown that induction of the 60-, 70-, and 90-kDa stress protein gene families occurs in both species. On the other hand, in *in vivo* studies in hamsters, the induction of stress proteins was observed after exposure to $InCl_3$ at 10 days; similar patterns were observed after exposure to arsenic and indium arsenide (Conner et al., 1993). However, at 30 days after $InCl_3$ exposure, the overall synthesis of the stress proteins was decreased. Indium accumulated in the kidneys in this experiment in a manner consistent with the observed effects (Yamauchi et al., 1992; 1993).

IV. GALLIUM

No cases have been reported of systemic effects in human kidneys exposed to gallium compounds. Gallium is not recognized as an element that specifically accumulates in renal tissue in animal experiments (Webb et al., 1984; Yamauchi et al., 1993). However, Newman et al. (1979) reported that diuresis reduced the severity of gallium-induced renal lithiasis and subsequent renal accumulation of gallium by diluting the urinary concentration of gallium and calcium, thereby lowering the incidence of interaction of these two elements within the kidney tubule.

Patterns of protein synthesis in primary cultures of hamster and human kidney proximal tubule epithelial cells were examined following exposure to gallium chloride (100 µM) *in vitro* (Fowler et al., 1995). As expected, these results showed a similar pattern of induction of the 60-, 70-, and 90-kDa stress protein gene families in the two species. Sex differences, however, were apparent in the stress protein induction response, with females responding to a greater extent than males.

V. CADMIUM

Nephropathy is a well-recognized result of cadmium poisoning, as evidenced by both epidemiological studies and animal experiments (Friberg et al., 1986; IPCS, 1992). The metal-binding protein metallothionein (MT) has been shown by workers in a number of laboratories to play several central roles in the mechanism of cadmium-induced renal toxicity. There have been a number recent reviews on this protein with regard to its chemistry and biochemistry (Petering and Fowler, 1986; Dunnick and Fowler, 1986). Briefly, cadmium-metallothionein (CdMT), synthesized in the liver in response to cadmium exposure, is released into circulation and transported to the kidney where it is reabsorbed with great efficiency from the tubular lumen (Cherian and Shaikh, 1975; Squibb et al., 1979) by proximal tubule cells. Once intracellular, the CdMT is rapidly degraded (Cherian and Shaikh, 1975, Squibb et al., 1984), with the release of Cd^{2+} ions that stimulate the synthesis of MT within the tubule cells of the S1 and S2 segments (Squibb et al., 1979, 1982, 1984; Squibb and Fowler, 1984). This process continues until the finite capacity of the cells to sequester Cd as CdMT is exceeded, either as a function of chronic exposure or dose. Studies of the mechanisms of Cd-induced nephrotoxic effects have shown that Cd injected as CdMT deposits in the S1 and S2 segments of the proximal tubule (Dorian et al., 1992) and causes a nephrotoxicity similar to that observed following long-term Cd exposure (Wang et al., 1993). It has been proposed that Cd-induced nephropathy occurs either through an alteration of membrane integrity and function by exogenous CdMT present in the tubular lumen (Cherian et al., 1976; Cherian, 1978) or through interactions of Cd^{2+} with sensitive intracellular ligands that are critical in cellular metabolic processes. Most important, it appears that it is the chemical species of Cd present in the kidney and not just the total concentration of Cd in the kidney that determines its toxicity (Wang et al., 1993).

Studies from this laboratory have shown that it is the Cd^{2+} associated with the non-MT fraction that is temporally associated with cytotoxicity (Squibb and Fowler, 1984; Squibb et al., 1984). The ultrastructural appearance of cells exposed to parenteral doses of CdMT shows a characteristic pattern of vesiculation and an increased number of electron-dense lysosomes which is temporally associated with decreases in lysosomal protease activity (Squibb et al., 1984), low molecular weight proteinuria (Squibb and Fowler, 1984; Squibb et al., 1984), and calcuria (Fowler et al., 1987a; Jin et al., 1987; Leffler et al., 1990; Liu et al., 1992). These alterations are similar to those observed in persons exposed to cadmium for prolonged time periods (Kanzantzis, 1979). The interpretation of these findings is that non-MT-bound Cd^{2+} is capable of interfering with the normal process of lysosomal biogenesis, which results in decreased reabsorption and degradation of low molecular weight proteins from the urinary filtrate, thus resulting in a tubular proteinuria. Some data (Fowler et al., 1987) indicate that most of the calcium in the urine is non-ionized and probably protein bound, suggesting that it is secondary to the proteinuria, whereas more recent studies using shorter collection times (Liu et al., 1992) suggest that the increased excretion of calcium in the urine occurs before the main increase in protein excretion.

To elucidate the mechanisms of cadmium-induced renal cell injury, recent preliminary *in vitro* studies (Fowler et al., 1991; Fowler and Akkerman, 1992) using CdMT have focused on relationships between morphological alterations in proximal tubule cells in culture, changes in intracellular Ca^{2+} concentrations, and induction of stress proteins. Results indicate that induction of stress proteins followed by the cellular vesiculation phenomenon occur well before measurable alterations in intracellular Ca^{2+}, which occur only as the cells begin to die. Such data are important because they suggest that the observed toxic phenomena are not secondary to altered calcium-induced mechanisms but rather are a function of Cd^{2+} binding to effector molecules early in the toxic process. A primary candidate for this role is calmodulin, because *in vitro* studies from a number of laboratories (Chao et al., 1984; Mills and Johnson, 1985; Suzuki et al., 1985) have shown that Cd^{2+} is capable of activating this protein. Activation of calmodulin by the Cd^{2+} ions not bound to MT could damage the cytoskeleton, which is thought to play an important role in the process of lysosomal biogenesis (Wall and Maack, 1985). Studies are currently in progress to evaluate this hypothesis with regard to competition between MT and calmodulin for the Cd^{2+} ions and to elucidate the role(s) of stress proteins in mediating Cd^{2+}-induced cell injury. Studies have shown that calmodulin inhibitors, such as trifluoperazine, N-(6-aminohexyl)-5-chloro-1-naphthalene sulfonamide, calmidazolium, and chlorpromazine, as well as the calcium ionophore A23187, have a marked stimulatory effect on renal MT gene expression, and that calcium regulatory pathways may play an important role in MT induction (Shiraishi and Waalkes, 1994).

VI. LEAD

Lead is the most ubiquitous of the nephrotoxic metals, and humans are exposed to this agent in air, food, and water. Clinical studies in workers (Landrigan et al., 1984; Vacca et al., 1986; Verschoor et al., 1986; Wedeen et al., 1986; Craswell, 1987) have shown the development of renal insufficiency after exposure to lead, and there have been several case reports of renal cancer (Baker et al., 1980; Lilis, 1981; Selevan et al., 1985). Irreversible chronic interstitial nephropathy resulting from occupational lead exposure is well known (Baker et al., 1980), as is renal tubular cell injury in acute lead poisoning in children (Klaassen, 1990). In these instances, Fanconi's syndrome is observed clinically, accompanied by proteinuria, erythruria, and urinary casts. Depositions of immunoglobulin exist in glomeruli, and histological alterations are evident in renal tubules (Klaassen, 1990). Effects of lead on heme synthesis are recognized in the hematopoietic system and kidney, which are evidenced by the development of chemical porphyria (Fowler et al., 1987b). In recent years, studies of renal lead binding proteins have progressed, and the presence of two lead binding proteins of 11.5 and 63 kDa in the kidney has been confirmed (Mistry et al., 1983). It has been proposed that these lead binding proteins are involved in the accumulation of lead in the kidney and alter its cellular toxicity (Egle and Shelton, 1986; Fowler, 1989).

The mechanisms underlying these phenomena are not fully understood, but animal studies involving chronic exposures have demonstrated renal tubular damage characterized by development of pathognomonic lead intranuclear inclusion bodies and renal cancer in rodents after high-dose exposure (Goyer and Rhyne, 1973). The toxic effects of lead on the kidney appear to be primarily localized in the proximal tubule (Goyer, 1968; Goyer et al., 1970; Choie and Richter, 1972; Moore et al., 1973; Moore and Goyer, 1973; Richter, 1976; Fowler et al., 1980; Murakami et al., 1983; Stiller and Friedrich, 1983; Oskarsson and Fowler, 1985a,b). Physiological studies of lead transport in the kidney (Vander et al., 1977, 1979) have shown that this metal is taken up by proximal tubule cells by a process that is inhibited by tin and several other metabolic inhibitors. Brush-border membrane vesicle transport studies (Victery et al., 1984) have shown that lead is taken up by extensive membrane binding and possibly by a passive transport mechanism.

The intracellular distribution and binding of lead appears to be mediated at low doses by soluble lead binding proteins (Oskarsson et al., 1982), which mediate the bioavailability of Pb^{2+} to sensitive enzymes such as d-aminolevulinic acid dehydratase (ALAD) (Murakami et al., 1983, Oskarsson and Fowler, 1985a, Goering and Fowler, 1984, 1985; Goering et al., 1986; Mistry et al., 1985) and the intranuclear transport and chromatin binding of lead (Mistry et al., 1985; Goering et al., 1986) with attendant changes in renal gene expression (Fowler et al., 1985). A cleavage product of alpha-2-microglobulin (Swenberg et al., 1989; Fowler and DuVal, 1991) is the principal lead binding protein (PbBP) in renal proximal tubule cells of rats. In humans and monkeys, the low molecular weight PbBPs are not fully characterized; however, they appear to have similar ion exchange characteristics to the rat PbBP and are chemically similar, with high percentages of glutamate and aspartate amino acids (Kahng et al., 1992; Fowler, unpublished data).

The importance of these data is derived from the fact that the PbBPs appear to bind lead at low doses and mediate the metal ion's biological activity within target cell populations. It has been previously hypothesized (Fowler and DuVal, 1991; Fowler, 1989) that the well-known lead-induced alterations in renal gene expression (Mistry et al., 1983; Fowler, 1989; Shelton and Egle, 1986) are mediated by the binding of these PbBPs to the 5′ flanking regions of genes showing altered expression patterns. In addition, if the reported (DuVal et al., 1986) polymerization of alpha-2-μglobulin plays a role in the formation of the cytoplasmic lead-containing inclusion bodies (Murakami et al., 1983; Stiller and Friedrich, 1983; Oskarsson and Fowler, 1985b), then the observed temporal relationship (Fowler et al., 1985) between formation of these inclusions and coincident alterations in renal gene expression becomes even more closely linked. Further research is needed to complete ongoing molecular biology studies concerning the relationships between Pb^{2+} binding to the PbBP in rats, monkeys, and humans and altered gene expression in proximal tubule cells.

A. ORGANELLE SYSTEM EFFECTS OF LEAD
1. Mitochondria

Renal proximal tubule cell mitochondria have long been known for their sensitivity to lead (Goyer, 1968; Goyer and Rhyne, 1973; Fowler et al., 1980; Oskarsson and Fowler, 1985a,b), with both morphological and biochemical alterations demonstrated in structure/functional relationships. In particular, decreased respiratory function, which has been linked to decreased morphological transformational capability (Goyer and Krall, 1969), is of clear importance with regard to cell injury from this metal. Decreases in the specific activities of mitochondrial-based heme enzymes have also been reported (Fowler et al., 1980; Oskarsson and Fowler, 1985a). Overall, it is clear that this organelle system is a highly sensitive, early target for lead in the kidney.

2. Nuclei

Lead-induced alterations in renal gene expression associated with formation of intranuclear inclusion bodies and tubular mitosis have been known for many years (Goyer, 1968; Goyer et al., 1970; Choie and Richter, 1972; Goyer and Rhyne, 1973; Moore et al., 1973; Moore and Goyer, 1973; Richter, 1976; Fowler et al., 1980; Murakami et al., 1983; Stiller and Friedrich, 1983; Oskarsson and Fowler, 1985b). Studies from a number of laboratories (Choie and Richter, 1974; Fowler et al., 1985, Hitzfeld et al., 1989) have shown that these changes in renal gene expression and mitosis are not secondary to a cell death and replacement phenomenon. Such findings are consistent with the hypotheses discussed above regarding the receptor-like nature of the PbBPs in the kidney in mediating these effects of lead at doses below those that precede overt cell death.

VII. SUMMARY

As molecular biology techniques improve, our ability to study metal-induced alterations in gene expression will increase enormously. Sufficient data exist at this time to suggest that metal ions can dramatically alter gene expression both directly through interactions with DNA regulatory proteins and indirectly through effects on cell functions that lead to decreased energy production, loss of membrane integrity, altered Ca homeostasis, or oxidative damage. The induction of stress proteins is an important response to metals in kidney proximal tubule cells. Further research is needed to understand the relationship between the stress protein response and cell injury/cell death processes that occur following metal ion exposure.

REFERENCES

Aoki, Y., Lipsky, M. M., and Fowler, B. A. (1990), *Toxicol. Appl. Pharmacol.*, 106, 462–468.
Baker, E. L., Jr., Goyer, R. A., Fowler, B. A., Khettry, U., Barnard, D., deVere White, R., Babayan, R., and Feldman, R. G. (1980), *Am. J. Ind. Med.*, 1, 139–148.
Brown, M. M., Rhyne, B. C., Goyer, R. A., and Fowler, B. A. (1976), *J. Toxic. Environ. Health*, 1, 507–516.
Brown, I. R. and Rush, S. J. (1984), *Biochem. Biophys. Res. Commun.*, 120, 150–155.
Castronovo, F. P. and Wagner, H. N. (1971), *Br. J. Exp. Pathol.*, 52, 543–559.
Castronovo, F. P. and Wagner, H. N. (1973), *J. Nucl. Med.*, 14, 677–682.
Chao, S. H., Suzuki, Y., Zysk, J. R., and Cheung, W. Y. (1984), *Mol. Pharmacol.*, 26, 75–82.
Cherian, M. G. (1978), *Biochem. Pharmacol.*, 27, 1163–1166.
Cherian, M. G. and Shaikh, Z. A. (1975), *Biochem. Biophy. Res. Commun.*, 65, 863–869.
Cherian, M. G., Goyer, R. A., and Richardson, L. D. (1976), *Toxicol. Appl. Pharmacol.*, 38, 399–408.
Choie, D. D. and Richter, G. W. (1972), *Am. J. Pathol.*, 66, 265–276.
Choie, D. D. and Richter, G. W. (1974), *Lab. Invest.*, 30, 652–656.
Conner, E. A., Yamauchi, H., Fowler, B. A., and Akkerman, M. (1993), *J. Expo. Anal. Environ. Epi.*, 3, 431–440.
Craswell, P. W. (1987), *Annu. Rev. Med.*, 38, 169–173.
Dodd, S. M., Martin, J. E., Swash, M., and Mather, K. (1993), *Clin. Nephrol.*, 39, 239–244.
Dorian, C., Gattone, V. H., and Klaassen, C. D. (1992), *Toxicol. Appl. Pharmacol.*, 114, 173–181.
Dunnick, J. and Fowler, B. A. (1986), in *Handbook on the Toxicity of Inorganic Compounds* Seiler, H. G. and Sigel, H., Eds., Marcel Dekker, New York, 155–174.
DuVal, G. E., Jett, D. A., and Fowler, B. A. (1986), *Toxicologist*, 9, 98.

Egle, P. M. and Shelton, K. P. (1986), *J. Biol. Chem.*, 261, 2294–2298.
Fowler, B. A. (1986), in *Handbook on the Toxicology of Metals,* 2nd ed., L. Friberg, G. F. Nordberg, and V. B. Vouk, Eds., Elsevier, Amsterdam, 267–275.
Fowler, B. A. (1989), *Comments Toxicol.*, 3, 27–46.
Fowler, B. A. and DuVal, G. E. (1991), *Environ. Health Perspect.*, 91, 77–89.
Fowler, B. A. and Akkerman, M. (1992), in *Cadmium in the Human Environment: Toxicity and Carcinogenicity,* Nordberg, G. F., Alessio, L., and Herber, R. F. M., Eds., IARC, Lyon, 1–7.
Fowler, B. A., Kimmer, C. A., Woods, J. S., McConnell, E. E., and Grant, L. D. (1980), *Toxicol. Appl. Pharmacol.*, 56, 59–77.
Fowler, B. A., Kardish, R., and Woods, J. S. (1983), *Lab. Invest.*, 48, 471–478
Fowler, B. A., Mistry, P., and Victery, W. W. (1985), *Toxicologist*, 5, 53.
Fowler, B. A., Goering, P. L., and Squibb, K. S. (1987a), in *Metallothionein, Second International Meeting on Metallothionein and Other Low Molecular Weight-Binding Proteins,* Kagi, J. H. R., Ed., Birkhauser Verlag, Basel, 661–668.
Fowler, B. A., Oskarsson, A., and Woods, J. S. (1987b), in *Mechanisms of Chemical-Induced Porphyrinopathies,* Silbergeld, E. K. and B. A. Fowler, Eds., *Ann. N.Y. Acad. Sci.*, 514, 172–182.
Fowler, B. A., Gandley, R. E., Akkerman, M., and Lipsky, M. M. (1991), in *Metallothionein in Biology and Medicine,* Klaassen, C. D. and Suzuki, K. T., Eds., CRC Press, Boca Raton, FL, 322–331.
Fowler, B. A., Yamauchi, H., Akkerman, M., Conner, E. A., and Squibb, K. S. (1995), *Toxicologist*, 15, 299.
Friberg, L., Kjellstrom, T., and Nordberg, G. F. (1986), in *Handbook on the Toxicology of Metals,* 2nd ed., L. Friberg, G. F. Nordberg, and V. Vouk, Eds., Elsevier, Amsterdam, 152–175.
Fullmer, C. S., Edelstein, S., and Wasserman, R. H. (1985), *J. Biol. Chem.*, 260, 6816–6819.
Ginsberg, J. M. and Lotspeich, W. W. (1963), *Am. J. Physiol.*, 205, 707.
Ginsberg, J. M. (1965), *Am. J. Physiol.*, 208, 832.
Goering, P. L. and Fowler, B. A. (1984), *J. Pharmacol. Exp. Ther.*, 231, 66–71.
Goering, P. L. and Fowler, B. A. (1985), *J. Pharmacol. Exp. Ther.*, 234, 365–371.
Goering, P. L., Mistry, P., and Fowler, B. A. (1986), *J. Pharmacol. Exp. Ther.*, 237, 220–225.
Goyer, R. A. (1968), *Lab. Invest.*, 19, 71–77.
Goyer, R. A. and Krall, A. R. (1969), *J. Cell Biol.*, 41, 393–400.
Goyer, R. A. and Rhyne, B. C. (1973), *Int. Rev. Exp. Pathol.*, 12, 1–77.
Goyer, R. A., Leonard, D. L., Moore, J. F., Ryne, B., and Krigman, M. R. (1970), *Arch. Environ. Health*, 20, 704–711.
Hitzfeld, P., Planas, F., and Taylor, D. (1989), *Bio. Trace Elem. Res.*, 21, 87–95.
IPCS (1991), in *Environmental Health Criteria 119,* World Health Organization, Geneva, 134–147.
IPCS (1992), in *Environmental Health Criteria 134,* World Health Organization, Geneva, 97–195.
Jin, T., Leffler, P., and Nordberg, G. F. (1987), *Toxicology*, 45, 307–317.
Kahng, M. W., Conner, E. A., and Fowler, B. A. (1992), *Toxicologist*, 12, 214.
Kanzantzis, G. (1979), *Environ. Health Perspect.*, 28, 155–159.
Klaassen, C. D. (1990), in *The Pharmacological Basis of Therapeutics,* A., Goodman Gilman, T. W., Rall, A. S., Nies, and P. Taylor, Eds., Pergamon Press, New York, 1592–1598.
Landrigan, P. J., Gover, A., Clarkson, T. W., Sandler, D. P., Smith, H., Thun, M. J., and Wedeen, R. P. (1984), *Arch. Environ. Health*, 39, 225–230.
Leffler, P., Jin, T., and Nordberg, G. F. (1990), *Toxicol. Appl. Pharmacol.* 103, 180–184.
Lilis, R. (1981), *Am. J. Ind. Med.*, 2, 293–297.
Liu, X. Y., Jin, T. Y., Nordberg, G. F., Rannar, M., Sjostrom, M., and Zhou, Y. (1992), *Toxicol. Appl. Pharmacol.*, 114, 239–245.
Mills, J. S. and Johnson, J. D. (1985), *J. Biol. Chem.*, 260, 15100–15105.
Mistry, P., Mastri, C., and Fowler, B. A. (1983), *Biochem. Pharmacol.*, 35, 711–713.
Mistry, P., Lucier, G. W., and Fowler, B. A. (1985), *J. Pharmacol. Exp. Ther.*, 232, 462–469.
Moore, J. F. and Goyer, R. A. (1973), *Environ. Health Perspect.*, 7, 121–127.
Moore, J. F., Goyer, R. A., and Wilson, M. (1973), *Lab. Invest.*, 29, 488–494.
Murakami, M., Kawamura, R., Nishi, S., and Karsunuua, H. (1983), *Br. J. Exp. Pathol.*, 64, 144–155.
Newman, R. A., Brody, A. R., and Krakoff, I. H. (1979), *Cancer,* 44, 1728–1979.
Oskarsson, A. and Fowler, B. A. (1985a), *Exp. Mol. Pathol.*, 43, 397–408.
Oskarsson, A. and Fowler, B. A. (1985b), *Exp. Mol. Pathol.*, 43, 409–417.
Oskarsson, A., Squibb, K. S., and Fowler, B. A. (1982), *Biochem. Biophys. Commun.*, 104, 290–298.
Petering, D. H. and Fowler, B. A. (1986), *Environ. Health Perspect.*, 65, 217–224.
Richter, G. W. (1976), *Am. J. Pathol.*, 83, 135–149.
Selevan, S. G., Landrigan, P. J., Stern, F. B., and Jones, J. M. (1985), *Am. J. Epidemiol.*, 122, 673–683.
Shelton, K. R. and Egle, P. M. (1986), *J. Biol. Chem.*, 257, 11802–11807.
Shelton, K. R., Todd, J. M., and Egle, P. M. (1986), *J. Biol. Chem.*, 261, 1935–1940.
Shibahara, S., Muller, R., and Taguchi, H. (1987), *J. Biol. Chem.*, 262, 12889–12892.
Shiraishi, N. and Waalkes, M. P. (1994), *Toxicol. Appl. Pharmacol.*, 125, 97–103.
Squibb, K. S. and Fowler, B. A. (1983), in *Biological and Environmental Effects of Arsenic,* Fowler, B. A. Ed., Elsevier/North-Holland, Amsterdam, 233–270.
Squibb, K. S. and Fowler, B. A. (1984), *Environ. Health Perspect.*, 54, 31–35.

Squibb, K. S., Ridlington, J. W., Carmichael, N. G., and Fowler, B. A. (1979), *Environ. Health Perspect.,* 28, 287–296.
Squibb, K. S., Pritchard, J. B., and Fowler, B. A. (1982), in *Biological Roles of Metallothionein Proceedings of a USA-Japan Workshop on Metallothionein,* Foulkes, E. C., Ed., Elsevier/North-Holland, Amsterdam, 181–192.
Squibb, K. S., Pritchard, J. B., and Fowler, B. A. (1984), *J. Pharmacol. Exp. Ther.,* 229, 311–321.
Stiller, D. and Friedrich, H. J. (1983), *Exp. Pathol.,* 24, 133–141.
Suzuki, Y., Charo, S. H., Zysk, J. R., and Cheung, W. Y. (1985), *Arch. Toxicol.,* 57, 205–211.
Swenberg, J. A., Short, B., Borghoff, S., Strasser, J., and Charboneau, M. (1989), *Toxicol. Appl. Pharmacol.,* 97, 35–46.
Taketani, S., Kohno, H., Yoshinaga, T., and Tokunaga, R. (1989), *FEBS Lett.,* 245, 173–176.
Vacca, C. V., Hines, J. D., and Hall, P. W. (1986), *Environ. Res.,* 41, 440–449.
Vander, A. J., Taylor, D. L., Kalitis, K., Mouw, D. R., and Victery, W. (1977), *Am. J. Physiol.,* 2, 532.
Vander, A. J., Mouw, D. R., Cox, J., and Johnson, B. (1979), *Am. J. Physiol.,* 236, 373.
Verschoor, M. A., Wibowo, A. A. E., Van Hennen, J. J., Herber, R. F. M., and Zielhuis, R. L. (1986), *Acta Pharmacol. Toxicol.,* 59 (Suppl. 7), 80–82.
Victery, W. W., Miller, C. R., and Fowler, B. A. (1984), *J. Pharmacol. Exp. Ther.,* 231, 589–596.
Wall, D. A. and Maack, T. (1985), *Am. J. Physiol.,* 248, C12–C20.
Wang, X. P., Chan, H. M., Goyer, R. A., and Cherian, M. G. (1993), *Toxicol. Appl. Pharmacol.,* 119, 11–16.
Webb, D., Sipes, I. G., and Carter, D. E. (1984), *Toxicol. Appl. Pharmacol.,* 76, 96–104.
Wedeen, R. P., D'Hasese, P., Van de Vyver, F. L. Verpooten, G. A., and DeBroe, M. E. (1986), *Am. J. Ind. Kidney Dis.,* 8, 380–383.
Yamauchi, H., Takahashi, K., Yamamura, Y., and Fowler, B. A. (1992), *Toxicol. Appl. Pharmacol.,* 116, 66–70.
Yamauchi, H., Takahashi, K., Conner, E. A., Akkerman, M., Yamamura, Y., and Fowler, B. A. (1993), in *Hazard Assessment and Control Technology in Semiconductor Manufacturing,* Vol. II, American Conference of Governmental Industrial Hygienists, Cincinnati, OH, 109–120.
Yamauchi, H. and Fowler, B. A. (1994), in *Arsenic in the Environment, Part II: Human Health and Ecosystem Effects. Advances in Environmental Science and Technology,* Nriagu, J. O., Ed., Wiley, New York, 35–54.

Chapter 47

Biomarkers of Metal-Induced Nephrotoxicity

Bruce A. Fowler and Monica Nordberg

I. INTRODUCTION

Biomarkers or biological indicators of metal-induced renal toxicity have been employed for several decades in monitoring the early clinical effects of nephrotoxic metals such as lead (Pb), cadmium (Cd), and mercury (Hg). These indicators, which are early biochemical tests for assessing the biochemical responses of the kidney to these metals prior to overt clinical disease, have proven useful for indicating not only metal exposure, but that a sufficient quantity of a given metal is biologically available to produce a cellular response. In order to be really useful, a good biomarker must be highly sensitive, easily measurable, relatively chemical-specific, and interpretable as to whether it is an index of exposure or toxic effect. A major research need is to provide correlative morphological/biochemical data so that the prognostic value to a given biomarker may be determined (Fowler, 1983).

This is particularly an issue for the kidney, where the existence of extensive reserve capacity produces an apparently fine line between a kidney undergoing chronic metal-induced damage and overt renal failure such that putative biomarkers are sometimes difficult to delineate. For this reason, an understanding of the biochemical mechanisms underlying a given biomarker is imperative for more precise interpretation.

The types of renal biomarkers discussed below fall into both the physiological and biochemical categories. For each category of biomarker, an up-to-date assessment of the current state of knowledge regarding underlying mechanisms will be included. It is hoped that this discussion will provide the reader with a perspective on how the various types of biomarkers may be interpreted as measures for monitoring cell injury, cell death, and prospects for the future overall functionality of the kidney. A discussion of future research needs for each category will also be included, since as with any area of research, there is always a need to improve understanding of current biomarkers and develop new ones with even greater sensitivity, specificity, and interpretability. It is hoped that this chapter will provide the reader with insights into how biomarkers of metal-induced renal toxicity may be used for clinical studies and for understanding how metals damage the kidney.

II. PROTEINURIAS

Increased excretion of proteins into the urine may be an early indicator of metal-induced renal damage. The nature and size of the proteins may be a relatively good marker of which cellular populations are being damaged and whether the lesions are glomerular (basement membrane proteins) or tubular (low molecular weight proteins/brush border proteins) in nature. The application of two-dimensional gel electrophoresis coupled with computerized image analysis to urine samples has proven to be a powerful tool in the evaluation of specific proteinuria patterns. Use of this technology (Conner et al., 1993) for evaluation of metal-induced nephrotoxicity from III–V semiconductors has given new insights into chemical-specific alterations that may be of future value for monitoring worker exposure to these agents.

Increased urinary concentrations of retinol-binding protein, beta-2-microglobulin, alpha-1-microglobulin, metallothionein, and protein 1,N-acetyl glucosamidase (NAG), calcium, and amino acids are all potentially useful markers for metal-induced alterations in renal function.

Previous studies (Cardenas et al., 1993; Holmquist et al., 1993) have demonstrated proteinuria patterns in both workers and persons in the general population exposed to a variety of chemical agents such as cadmium. The IUPAC (Herber et al., 1994) has published a standardized method for the estimation of beta-2-microglobulin, retinol-binding protein, and albumin in urine, since the increased urinary excretion of a number of these proteins is known to occur following exposure of workers to agents such as cadmium and lead.

III. URINARY EXCRETION OF ENZYMES

The urinary excretion of enzymes such as NAG has been used as a marker for metal-induced nephrotoxicity in a number of both experimental (Suzuki and Cherian, 1987) and clinical studies (Gutherie et al., 1994) involving metal-induced nephrotoxicity. There are a number of problems related to the sensitivity of this enzyme assay which limit its utility for subclinical investigations (NAS, 1995).

IV. CALCURIA

The increased excretion of calcium in the urine has been previously observed in both clinical studies (Bernard, 1992) and in experimental animal model systems (Goering et al., 1986; Fowler et al., 1987) following exposure to cadmium.

The increased excretion of this element into the urine has been hypothesized to play a role in the development of osteomalacia in workers and multiparous women in Japan (Fowler et al., 1987). The increased excretion of this element into the urine is hence a potential biomarker of both cadmium-induced renal and skeletal toxicity.

V. STANDARDIZATION OF URINARY BIOMARKERS

There is an ongoing discussion of how to best standardize urinary biomarkers against creatinine or 24-hour total urinary excretion (NAS, 1995). At present, it appears that both approaches provide useful information. The practical utility of biomarker/creatinine ratios for spot urine samples must be weighed against errors in assuming that creatinine excretion may be used as a general marker substitute for glomerular filtration and that a given nephrotoxic agent is not simultaneously exerting both glomerular and tubular toxicities.

VI. THE HEME BIOSYNTHETIC PATHWAY AND RENAL PORPHYRINURIAS

The inhibitory effects of metals such as lead on the heme biosynthetic pathway in the hematopoietic system have been known for many years. The effects of lead on the renal heme biosynthetic pathway have received less attention, but metal effects in the kidney may in part be related to the action of these agents on this highly sensitive pathway. In examining the effects of metals on enzymes in this essential pathway, it is important to consider that the kidney becomes a target for metal toxicity in part because it accumulates metals to a high degree.

Of the total amount of a given metal present in the kidney, only a relatively small fraction is probably biologically available at any one time due to complexation with metal-binding proteins such as metallothionein (Ridlington et al., 1981; Goering and Fowler, 1987) PbBPs (Oskarsson et al., 1982; Goering and Fowler, 1984, 1985), and precipitates such as cytoplasmic or intranuclear inclusion bodies (Baker et al., 1980; Fowler et al., 1980; Oskarsson and Fowler 1985a,b; Carmichael and Fowler, 1979). This means that the interpretation of metal-induced effects on the renal heme biosynthetic pathway must factor into account this biologically based limitation on metal intracellular bioavailability and the prospect that this relationship may change as a function of time and dose.

The renal heme biosynthetic pathway has been shown to be measurably sensitive to a number of toxic metals such as lead, mercury, gallium, indium, and bismuth. Disturbance of this pathway by several metals has been linked to the excretion of heme precursors in the urine, suggesting the use of renal porphyrinurias as biomarkers for metal-induced nephrotoxicity. The discussion below is intended to review the known effects of metals, the renal heme pathway, and in particular those studies that have monitored concomitant urinary excretion heme precursors in the urine as possible biomarkers. Specific attention will also be given to those studies with other correlative ultrastructural/biochemical data that are linked to the porphyrinuria.

A. LEAD

The effects of lead on renal heme biosynthesis have been studied by a number of investigators (Fowler et al., 1980; Oskarsson and Fowler, 1985b; Goering and Fowler, 1984, 1985; Oskarsson and Fowler, 1987). The results of these studies have generally indicated that chronic or long-term *in vivo* exposure to lead produces inhibitory effects on the mitochondrial heme biosynthetic pathway enzymes, delta aminolevulinic acid synthetase and ferrochelatase, but that the cytosolic enzyme delta aminolevulinic acid dehydratase is remarkably resistant to inhibition by lead. Studies by Goering and Fowler (1984, 1985) showed that this enzyme activity was approximately seven times more resistant to lead inhibition than that found in liver following *in vitro* incubation. This effect appears to be the result of the presence of lead-binding proteins which are capable of both chelating lead and donating zinc to this zinc-dependent enzyme. The sensitivity of the mitochondrial enzymes to lead appears to be related to the extensive uptake of lead by this organelle following *in vivo* (Oskarsson and Fowler, 1985) or *in vitro* (Fowler and Oskarsson, 1987) exposure. The relative contribution of lead inhibition of renal heme biosynthesis to the urinary excretion of specific porphyrin isomers has not been delineated, but it is likely that, given the extensive heme biosynthetic capacity of this organ, the contribution of the kidney to the increased urinary excretion of uroporphyrin and coproporphyrin isomers following lead exposure is considerable. It is also worth noting that interactions between lead, cadmium, and arsenic (Mahaffey et al., 1981) have been documented with respect to formation of lead intranuclear inclusions and alterations in urinary excretion of total uroporphyrins and coproporphyrin.

B. MERCURY

Studies by Woods and Fowler (1996) in rats demonstrated increased excretion of coproporphyrin with lesser amounts of uroporphyrin in urine with primary alterations in major enzymes of the renal heme biosynthetic pathway indicating a renal porphyrinuria.

These changes were associated with a number of other ultrastructural/biochemical indices of toxicity to renal proximal tubule cells suggesting the use of the porphyrinuria as a potential biomarker for mercury-induced cellular toxicity. Subsequent studies (Ove-Lund et al., 1991) indicated that mercurial inhibition of mitochondrial respiration with formation of H_2O_2 and generation of reactive oxygen species played a major role in mediating inhibition of the renal heme biosynthesis. The concept of a mercury-specific porphyrinuria pattern linked to a thorough mechanistic understanding of mechanisms of toxicity increases the potential value of this alteration in the renal heme biosynthetic pathway as a biomarker for early detection of cell injury before the onset of clinical renal disease.

C. GALLIUM

Gallium is a metal from group III of the Periodic Table of elements, which is used for anticancer chemotherapy and in the production of III–V semiconductors such as GaAs. Studies by Goering et al. (1987) demonstrated that the enzyme ALA-dehydratase (ALAD) is highly sensitive to this metal, with an LC_{50} of approximately 10^{-6} M. *In vivo* studies indicated that renal ALAD showed a marked inhibition following intratracheal instillation of GaAs particles.

D. INDIUM

Studies by Woods and Fowler (1983) showed that the group III element indium (In) also exerted marked inhibitory actions on the enzyme ALAD in kidney following acute *in vivo* exposure to indium chloride. Subsequent studies (Conner et al., 1995), using particles of the III–V semiconductor InAs, showed similar results following subcutaneous instillation. *In vitro* studies also showed that In possessed an LC_{50} for ALAD of 10^{-6} M. In addition, an increased excretion of several specific porphyrins in the urine was also noted, indicating that several other enzymes in the heme pathway are also being altered in a highly specific manner.

E. BISMUTH

Acute injection studies with bismuth (Woods and Fowler, 1985) demonstrated marked inhibition of the renal heme biosynthetic pathway including ALAD.

VII. FACTORS INFLUENCING THE BIOAVAILABILITY OF METALS TO THE HEME BIOSYNTHETIC PATHWAY AND OTHER SENSITIVE MOLECULAR TARGETS

The ability of metals to interact with enzymes in the heme biosynthetic pathway and other sensitive molecular targets within the kidney is, in part, determined by how they are complexed in renal epithelial cells. As noted above, metal-binding proteins such as metallothionein for cadmium, mercury ion, and bismuth may greatly influence bioavailability to a number of other sensitive processes, while the Pb binding proteins have been shown to influence lead bioavailability to ALAD and a number of other sensitive biochemical processes such as gene expression Mistry et al., 1985, 1986) and tubular transport mechanisms (Zalups and Barfuss, 1993; Zalups et al., 1991). Formation of intranuclear inclusions following elevated exposures to lead and bismuth is another important intracellular depot for these elements which will influence their bioavailability to sensitive biochemical processes such as the heme biosynthetic pathway and intracellular gene regulation in renal proximal tubule cells.

REFERENCES

Aoki, Y., Lipsky, M. M., and Fowler, B. A. (1990), *Toxicol. Appl. Pharmacol.*, 106, 462–468.
Baker, E. L., Jr., Goyer, R. A., Fowler, B. A., Khettry, U., Barnard, D., DeVere-White, R., Babayan, R., and Feldman, R. G. (1980), *Am. J. Ind. Med.*, 1, 139–148.
Bernard, A., Roels, H., Thielemans, N., VanLierde, M., and Lauwerys, R. (1992), in *Cadmium in the Human Environment: Toxicity and Carcinogenicity*, Nordberg, G. F., Herber, R. F. M., and Alessio, L., Eds., IARC Scientific Publications No. 118, IARC, Lyon, France, 341–346.
Cardenas, A., Roels, H., Bernard, A. M., Barbon, R., Buchet, J. P., Lauwerys, R., Rosello, J., Ramis, I., Mutti, A., Franchini, I., Fels, L. M., Stolte, H., Debroe, M. E., Nuyts, G. D. Taylor, S., and Price, R. G. (1993), *Br. J. Ind. Med.*, 50, 28–36.
Carmichael, N. G. and Fowler, B. A. (1979), *J. Env. Pathol. Toxicol.*, 3, 399–412.
Conner, E. A., Yamauchi, H., and Fowler, B. A. (1995), Alterations in the heme biosynthetic pathway from III–V semiconductor metal, indium arsenide (InAs), 96, 273–285.
Conner, E. A., Yamauchi, H., Fowler, B. A., and Akkerman, M. (1993), *J. Expos. Anal. Environ. Epidemiol.*, 3, 431–440.
Fowler, B. A. (1983), *Fed. Proc.*, 42, 2957–2964.
Fowler, B. A., Brown. H. W., Lucier, G. W., and Krigman, M. R. (1974), *Lab. Invest.*, 32, 313–322.
Fowler, B. A., Goering, P. L., and Squibb, K. S. (1987), in *Metallothionein, 2nd Int. Meeting on Metallothionein and Other Low Molecular Weight Metal-Binding Proteins*, Kagi, J. H. R., and Kojima, Y., Eds., Birkhauser Verlag, Basel, 613–616.
Fowler, B. A., Kimmel, C. A., Woods, J. S., McConnell, E. E., and Grant, L. D. (1980), *Toxicol. Appl. Pharmacol.*, 56, 59–77.
Fowler, B. A. and Nordberg, G. F. (1978), *Toxicol. Appl. Pharmacol.*, 46, 609–624.
Fowler, B. A. and Woods, J. S. (1977), *Exp. Mol. Pathol.*, 27, 403–412.
Goering, P. L. and Fowler, B. A. (1984), *J. Pharmacol. Exp. Ther.*, 231, 66–71.
Goering, P. L. and Fowler, B. A. (1985), *J. Pharmacol. Exp. Ther.*, 234, 365–371.
Goering, P. L. and Fowler, B. A. (1987), *Arch. Biochem. Biophys.*, 253, 48–55.
Goering, P. L., Maronpot, R. R., and Fowler, B. A. (1988), *Toxicol Appl. Pharmacol.*, 92, 179–193.
Goering, P. L., Squibb, K. S., and Fowler, B. A. (1986), in *Trace Substances in Environmental Health*, XIX. Hemphill, D. D., Ed., University of Missouri Press, Columbia, 22–36.
Gutherie, C. J. G., Chettle, D. R., Franklin, D. M., Scott, M. C., Mason, H. J., Wright, A. L., Gompertz, D. R., Davidson, A. G., Fayers, P. M., and Newman-Taylor, A. J. (1994), *Environ. Res.*, 65, 22–41.
Herber, R. F. M., Bernard, A., and Schaller, K.-H. (1994), *Pure Appl. Chem.*, 66, 915–930.
Holmquist, L., Vesterberg, O., Persson, B. (1993), *Int. Arch. Occup. Environ. Health*, 64, 469–472.
Mahaffey, K. R., Capar, S. G., Gladen, B. C., and Fowler, B. A. (1981), *J. Lab. Clin. Med.*, 98, 463–481.
Mistry, P., Lucier, G. W., and Fowler, B. A. (1985), *J. Pharmacol. Exp. Ther.*, 232, 462–469.
Mistry, P., Mastri, C., and Fowler, B. A. (1986), *Biochem. Pharmacol.*, 35, 711–713.
NAS (National Academy of Sciences/National Research Council) (1995) *Report of the Committee on Biological Markers of Urinary Toxicology.* NAS/NRC Press, Washington, D.C.
Oskarsson, A., Squibb, K. S., and Fowler, B. A. (1982), *Biochem. Biophys. Res. Commun.*, 104, 290–298.
Oskarsson, A. and Fowler, B. A. (1985b), *Exp. Mol. Pathol.*, 43, 409–417.
Oskarsson, A. and Fowler, B. A. (1985a), *Exp. Mol. Pathol.*, 43, 397–408.
Ove-Lund, B., Miller, D. M., and Woods, J. S. (1991), *Biochem. Pharmacol.*, 42, S181–S187.
Ridlington, J. W., Winge, D. R., and Fowler, B. A. (1981), *Biochim. Biophys. Acta*, 673, 177–183.

Squibb, K. S., Pritchard, J. B., and Fowler, B. A. (1984), *J. Pharmacol. Exp. Ther.,* 231, 311–321.
Suzuki, C. A. M. and Cherian, M. G. (1987), *J. Pharmacol. Exp. Ther.,* 240, 314–319.
Woods, J. S. and Fowler, B. A. (1977), *J. Lab. Cin. Med.,* 90, 266–272.
Woods, J. S. and Fowler, B. A. (1982), *Exp. Mol. Pathol.,* 36, 306–315.
Woods, J. S. and Fowler, B. A. (1987), *Toxicol. Appl. Pharmacol.,* 90, 274–283.
Zalups, R. and Barfuss, D. W. (1993), *Toxicol. Appl. Pharmacol.,* 121, 176–185.
Zalups, R., Robinson, M. K., and Barfuss, D. W. (1991), *J. Am. Soc. Nephrol.,* 2, 866–878.

Chapter 48

Study on the Transport and Toxicity of Metals Along the Nephron by Means of the Isolated Perfused Tubule Technique

Rudolfs K. Zalups and Delon W. Barfuss

I. INTRODUCTION

Prior to the advent of renal tubular microtechniques, the kidney was viewed as a black box when it came to the study of the renal handling and toxicology of metals. Measurements were restricted, for the most part, to what entered into the kidney via the renal artery and what came out of the kidney in the venous blood and urine. This approach has provided the scientific literature with a fair amount of information regarding the renal handling of metals and the effects of metals on the various functions of the kidney. However, it has provided little insight into the specific tubular sites involved in the renal tubular handling of metals and the mechanisms involved in the transport and toxicity of metals along the numerous discrete segments making up the nephron and collecting duct.

Much of our current understanding of how the nephron and collecting duct function is based on data collected from studies implementing microtechniques, such as the micropuncture, microperfusion, and isolated perfused tubule techniques. In recent years, investigators have begun to apply these techniques to the study of the renal tubular transport and toxicity of metals. These techniques hold much promise for discovering and characterizing mechanisms involved in the uptake and transport of metals along the nephron and collecting duct.

In this chapter, we will describe and compare some of the details of these microtechniques as they relate to the study of the renal tubular transport of metals. We will also discuss some of the advantages and limitations of each of the techniques. Focus, however, will be placed on the advantages of using the isolated perfused tubule technique to study and characterize mechanisms involved in the transport and toxicity of metals along the neprhon and collecting duct. As part of this focus, we will describe, in some detail, various methodological and technical aspects of the isolated perfused tubule in relation to the study of the renal tubular transport and toxicity of metals.

II. MICROTECHNIQUES USED TO STUDY RENAL TUBULAR TRANSPORT OF HEAVY METALS

Cikrt and Heller (1980) were apparently the first investigators to use a microtechnique to study the renal tubular handling of a heavy metal. They studied the fate of radiolabeled inorganic mercury ($^{203}Hg^{2+}$) after it was microinjected into the superficial proximal tubules of the dog. In the late 1980s, Felley-Bosco and Diezi (1987, 1989) also used microtechniques to study the renal tubular transport of another heavy metal. They examined the transport of radiolabeled inorganic cadmium ($^{109}Cd^{2+}$) and $^{109}Cd^{2+}$ bound to metallothionein in superficial nephrons using both the microinjection (microperfusion) and micropuncture techniques. Besides the work of the investigators mentioned above, there do not appear to be any other studies where either the micropuncture or microperfusion techniques have been used to study the renal tubular handling of a heavy metal.

In both the micropuncture and microperfusion techniques, glass pipettes are pulled to a fine point and are polished until a sharp beveled opening of about 3 to 5 µm is achieved. These pipettes are used to collect samples of fluid (10 to 100 nl) from the lumen of the segment of the nephron studied (micropuncture) and/or are used to perfuse fluid into the lumen of a particular segment of the nephron (microperfusion). Both of these techniques require that the animal be maintained under anesthesia and that one of the kidneys be exposed or exteriorized, with the renal vessels remaining attached and intact. This is accomplished by carefully placing the kidney in a specialized cup that permits the decapsulated kidney to be immersed in mineral oil. The oil prevents the kidney from desiccating during the course of the experiment, which can last for several hours. The same style pipettes can also be used to collect blood from, or perfuse fluid into, the peritubular capillaries, which can be used to evaluate or control the composition of the peritubular environment of the nephron segment being studied. Data collected from studies where micropuncture or microperfusion techniques have been used have provided much of our understanding of the mechanisms involved in the transport and handling of normal solutes along selected segments of the nephron. However, the techniques have been used to a limited degree to study the transport and toxicity of metals along specific segments of the nephron. Despite their usefulness, these techniques do have several limitations.

A. LIMITATIONS OF THE MICROPUNCTURE AND MICROPERFUSION TECHNIQUES

One of the first limitations of the micropuncture and microperfusion techniques is that they can be applied to studying directly only a few segments of the nephron and collecting duct. Any segment that does not reach the surface of the kidney or does not reach the outer surface of the renal papilla cannot be accessed for study. Thus, only S1 segments of the proximal tubule, distal convoluted tubules (of which there are few), and cortical collecting ducts that reach the outer subcapsular surface of the kidney and the terminal segments of the inner medullary collecting ducts and the hairpin loops of the thin limbs of Henle that are present at the tip of the papilla can be studied.

A second limitation is that transport is assessed primarily by the disappearance of a substance from the luminal fluid (absorption) or by the appearance of a substance in the lumen (secretion). Although the measurement of the rate of appearance of a substance that is transepithelially transported from the luminal fluid into the blood is not impossible, it is very difficult, and the results can be very difficult to interpret.

Another limitation in the use of these techniques in studying renal epithelial transport is that the cellular levels of transported substances cannot be determined.

A fourth limitation is that the environment of the tubular segment being studied is not controlled. The tubular segment being studied is under all the normal external control mechanisms that regulate renal function (hormonal, neural, hemodynamic, etc.), which are probably different in each experimental animal. This places in question whether each experiment is controlled. Most of these limitations, however, can be circumvented with the isolated perfused tubule technique.

III. THE ISOLATED PERFUSED TUBULE TECHNIQUE

A. HISTORICAL DEVELOPMENT AND APPLICATION TO RENAL TOXICOLOGY

The isolated perfused renal tubule technique for studying transport and metabolism of solutes in individual segments of the nephron was developed in the mid 1960s at the National Institutes of Health under the direction of Maurice Burg (Burg et al., 1966). After initially working out the mechanics for perfusing tubular segments, the investigators found that all the dissected tubular segments they obtained contained holes. The holes were found to be caused by collagenase digestion, which was used as an aid in the dissection of individual segments of the nephron. After several months of futile attempts to obtain suitable specimens for perfusing, Jarad Gratham (who was collaborating with Maurice Burg at the time) attempted to dissect segments of the nephron from the rabbit without the aid of collagenase. His attempts proved to be successful. However, another problem arose subsequently. It was discovered, during the initial experiments with the isolated segments of the nephron from rabbits, that the collection pipettes did not seal tightly around the tubular segments during the perfusing process. This lack of a tight seal allowed bathing fluid to leak into the collection pipette. Gratham solved this problem by constricting the collection pipette and coating the inside of the pipette with Sylgard, which forms a tight seal between the perfused segment and the collection pipette. Without these early modifications, it would not have possible for Burg and his collaborators to perform successfully the first isolated perfused tubule experiment.

A diagram illustrating the salient features of the technique is presented in Figure 1. As shown, an isolated segment of the nephron or collecting duct is connected to the holding and collection pipette, while fluid is perfused though the lumen of the tubular segment via the perfusion pipette, which is inserted into the lumen of the tubular segment. Fluid collected at the opposite end of the tubular segment by the collection pipette and is subsequently taken up for analysis in a constant volume or volumetric pipette.

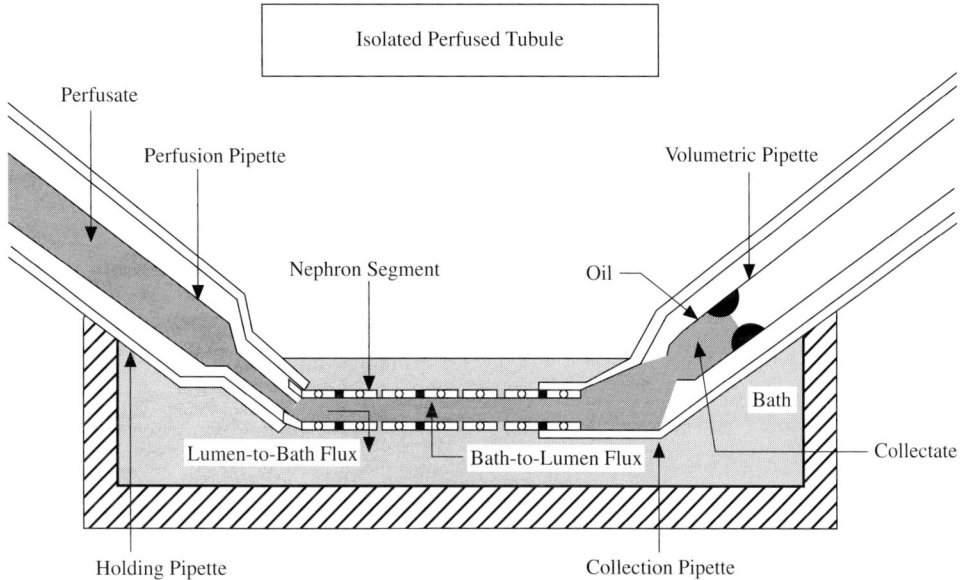

Figure 1. A schematic drawing illustrating the important features of the isolated perfused tubule technique. A segment of the nephron is suspended between a holding pipette (on the left) and a collection pipette (on the right) while it is immersed in a bathing solution in the perfusion chamber. Generally the base of the perfusion (bathing) chamber is made from a cover slip for a microscope slide and is positioned over the objectives of an inverted biological microscope. This allows the investigator to visualize all aspects of the experiment. After the tubular segment is suspended between the holding and collection pipettes, a perfusion pipette is guided down inside the middle of the holding pipette until the tip of the perfusion pipette is in the lumen of the isolated segment of the nephron. Once the perfusion pipette is in place, fluid that was preloaded in the pipette is perfused into the lumen of the tubule. Fluid that is not absorbed, or that is secreted into the lumen, is collected at the opposite end of the isolated perfused segment by the collection pipette. This fluid can be collected and analyzed for solutes after it is taken up by a constant-volume (volumetric) pipette, which is placed down the middle of the collection pipette. This pipette is generally calibrated to hold 20 to 50 nl. With the isolated perfused tubule technique, one can measure both the lumen-to-bath transport or bath-to-lumen transport of solutes, including heavy metals.

The isolated perfused tubule technique has been used extensively to study and characterize the renal tubular handling of numerous plasma solutes. However, until recently, the technique had never been applied to study the renal tubular transport, accumulation, metabolism, or toxicity of nephrotoxic metals.

Beginning in the early 1990s, we were the first investigators to study the toxicity, transport, accumulation, and metabolism of heavy metals along isolated segments of the nephron using the isolated perfused tubule technique. One of our first studies involved examining the toxicity and transport of radiolabelled inorganic mercury ($^{203}Hg^{2+}$) along isolated S1, S2, and S3 segments of the proximal tubule of the rabbit (Barfuss and Zalups, 1990). One example of a toxicological observation made in this study is presented in Figure 2, which shows some cellular swelling in the proximal portion of an S1 (initial convoluted) segment of a proximal tubule that was perfused through the lumen with 18.4 μM $^{203}Hg^{2+}$ for 7 min.

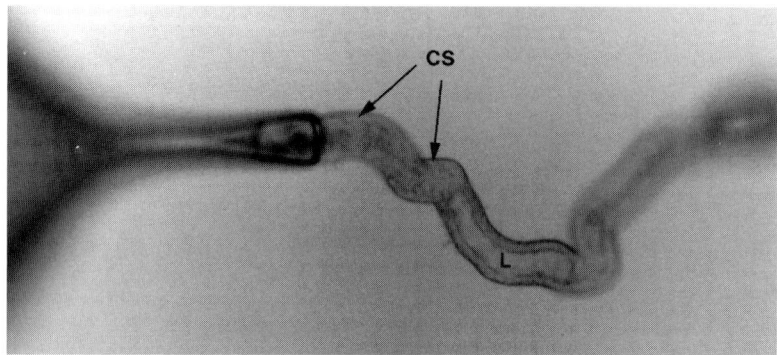

Figure 2. A photomicrograph of an S1 segment of a proximal tubule isolated from the kidney of a rabbit after it was perfused *in vitro* through the lumen with 18.4 μM inorganic mercury ($^{203}Hg^{2+}$) for 7 min. Cellular swelling (CS) occurred in the proximal one third portion of this perfused segment, which is attached to the holding pipette on the left. Necrosis was not a feature of the pathological changes at this time, since the swollen epithelial cells did take up the vital dye FD&C green, which was present in the perfusate. The swelling was severe enough that the lumen of the tubule could not be observed under microscopic observation. Note, however, that the lumen (L) of the tubule can be clearly seen beyond the portion of the perfused segment that is displaying signs of cellular swelling. Both the perfusate and the bathing solution were composed of essential electrolytes. With the exception of FD&C green dye and glucose in the perfusate, and glutamine and glucose in bathing solution, no other organic molecules were present in either the perfusate or bathing solution.

We have subsequently studied the transport and toxicity of radiolabeled inorganic mercury when it is bound to compounds containing free sulfhydryl groups, such as cysteine and glutathione (Zalups et al., 1991). We have also studied the toxicity, transport, accumulation, and metabolism of cadmium (Robinson et al., 1993) and methylmercury (Zalups and Barfuss, 1993) in isolated perfused segments of the proximal tubule.

B. ADVANTAGES OF THE TECHNIQUE
1. Direct Study of Any Segment of the Nephron

One of the greatest advantages for using the isolated perfused tubule technique is that virtually all the different segments of the nephron and collecting duct can be studied. Segments from both superficial and juxtamedullary nephrons can be studied. Some of these segments include the S1, S2, and S3 segments of the proximal tubule, the thin descending and thin ascending limbs of Henle's loop, the medullary and cortical ascending thick limbs of Henle's loop, the macula densa, the distal convoluted tubule, the connecting tubule, the cortical collecting duct, the outer medulla collecting duct, and the papillary collecting duct. The axial heterogeneity of the nephron and collecting duct was originally established on the basis of histological studies, but since the advent of the isolated perfused tubule technique, functional features have been correlated with histological characterization in most segments of the nephron and collecting duct.

2. Ability to Measure Lumen-to-Bath Transepithelial Fluxes of Metals

Another advantage of the isolated perfused tubule technique is that transepithelial transport processes can be studied in much greater detail than in any other *in vitro* or *in vivo* system. Rates of flux of a particular metal can be measured by its rate of disappearance from the lumen while simultaneous measurements can be made on the rate of appearance of the metal in the bathing solution. With some solutes that are transepithelially transported intact (not metabolized), the rate at which the solute disappears from the lumen equals the rate of appearance of the solute in the bath. This is the case with the transport of some organic substances like glucose (Barfuss and Schafer, 1981) and organic metals like methylmercury (Zalups and Barfuss, 1993) in segments of the proximal tubule. In contrast, when solutes are metabolized or sequestered by renal tubular epithelial cells, the appearance flux of that solute in the bath is substantially less than the disappearance flux from the lumen. This is the case for some organic solutes like glycylsarcosine and adenosine (Barfuss et al., 1988, 1992) and inorganic forms of the heavy metals mercury and cadmium (Barfuss et al., 1990; Robinson et al., 1993).

3. Ability to Measure Bath-to-Lumen Transepithelial Fluxes of Metals

One can also measure the bath-to-lumen transepithelial flux of a metal using the isolated perfused tubule technique. It is possible that some metals may be secreted *in vivo* by certain segments of the nephron from the peritubular blood into the luminal fluid. If this is the case, then the rate of secretion of a metal in a specific segment of the nephron can be determined by exposing the basolateral membrane of the segment to a bathing solution containing the metal, and by subsequently measuring the rate of appearance and concentration of the metal in the luminal fluid. If the concentration of the metal in the collected fluid is greater than that in the bathing fluid, active transport of the metal would be indicated. Location of the site of active transport can be determined by measuring cellular concentrations of the metal (see below). If the intracellular concentration of the metal is greater than the concentration of the metal in the luminal fluid, then the site of active transport of the metal is probably at the basolateral membrane. However, if the intracellular concentration of the metal is less than the concentration of the metal in the luminal fluid, then the site of active secretion of the metal is probably occuring at the luminal membrane.

4. Ability to Measure Cellular Concentrations of Metals

In conjunction with measuring rates of flux of a given metal, one can also measure cellular concentrations of that metal in isolated perfused segments of the nephron. This is accomplished by rapidly pulling the perfused tubular segment free from the holding and collections pipettes with fine forceps and placing it in 10 µl of a 3% solution (w/v) of tricholoracetic acid (TCA). This process takes less than a second to complete and is rapid enough that very little cellular content of metal is lost to the bathing solution after perfusion is interrupted. The TCA is used to separate the protein fraction from the cytosolic fraction of the tubular epithelial cells.

Knowing the cellular concentration and form of transported metals along with the appearance and disappearance rates of flux helps the investigator to determine the location of the transport mechanism(s) and types of transport processes that are involved in the transepithelial transport of that particular metal. For example, during the transport of inorganic mercury in S3 segments of the proximal tubule, we have found that the concentration of inorganic mercury in a TCA-soluble fraction of the tubular epithelial cells is 10 times greater than the concentration of inorganic mercury in the luminal fluid and is infinitely greater than that in the bathing fluid (about 0 mM). Assuming that the inorganic mercury in this TCA extract is transportable, then our findings tend to indicate that the site for initial transport of inorganic mercury (in this segment of the nephron) is at the luminal membrane, with an exit step for the transport process occurring at the basolateral membrane. The rate of exit of inorganic mercury across the basolateral membrane appears to be limited, as indicated by the high concentration of inorganic mercury in the epithelial cells. This approach allows the investigator to assess the relative importance of the luminal and basolateral membranes in the transepithelial transport of a metal.

5. Control of Tubule Environment

The extracellular environment surrounding a perfused tubular segment can be controlled to a much greater extent than in other approaches used to study the function of that same tubular segment. One can very carefully regulate the composition of both luminal and bathing fluids by adding or deleting regulatory influences such as hormones, certain electrolytes, growth factors, and toxicants.

The exact composition of the extracellular fluid that comes in contact with the basolateral membrane of tubular segments, *in vivo,* has never been determined precisely. The various segments of the nephron are probably bathed, in part, by fluid that has been absorbed from the lumen. Since some of the absorptive processes involve translocation of selective solutes, the absorbed fluid is distinctly different in composition from the luminal fluid and plasma in the peritubular capillaries. Examination of absorbed fluid collected from the basolateral surface of tubules perfused *in vitro* with an ultrafiltrate of plasma, while immersed in light mineral oil, has revealed that the fluid has a very high concentration in solutes, such as glucose and bicarbonate. As the concentration of preferentially absorbed solutes increases in the absorbed fluid, the osmolality in this fluid increases to as much as 18 mOsmol greater than that in the luminal fluid or plasma (Barfuss and Schafer, 1984a,b). The use of this same type of experimental preparation (Figure 3) could be very useful in determining the concentration of a heavy metal that is absorbed across the basolateral membrane in segments of the proximal tubule.

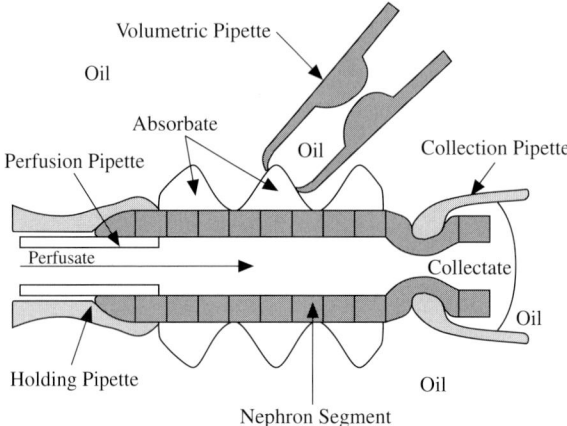

Figure 3. Schematic drawing of a modification of the isolated perfused tubule technique used to evaluate the composition of fluid and solutes emanating from the basolateral membrane of an isolated perfused segment of the nephron. In this technique, the isolated segment of the nephron is suspended and perfused between the holding and the collection pipettes in the normal manner. However, after the tubular segment is suspended between these pipettes, the bathing solution is removed in a retrograde manner and is replaced with a light weight mineral oil. When the tubular segment is completely surrounded with oil, torus-shaped droplets form on the extracellular surface of the basolateral membrane of the epithelial cells lining the perfused segment. These droplets can be taken up in timed collections with a constant-volume (volumetric) pipette and the rate of fluid formation and the concentration of solutes in the fluid can be determined subsequently. This technique may prove to be very useful in analyzing the transport and leak of metals across the basolateral membrane after they have left the luminal fluid. By knowing the composition of the fluid environment bathing the basolateral membranes of tubular segments, much can be gained in understanding the factors and mechanisms involved in the renal tubular handling and toxicity of metals.

6. Intact Epithelium

Isolated segments of the nephron used in the isolated perfused tubule technique represent samples of intact renal tubular epithelia, which have an intact basal lamina. Unlike other *in vitro* preparations, all of the structural and corresponding functional features of the various segments of the nephron are maintained after the tubular segment is removed from the kidney. The epithelial cells in an isolated perfused tubular segment maintain the same polarity as found *in vivo,* that is luminal vs. basolateral orientation. The cells are also joined by intercellular junctions and they are attached to a tubular basal lamina. As a whole, the unique characteristics of the epithelium from the various segments of the nephron are maintained *in vitro* as they would be *in vivo.*

C. LIMITATIONS

One of the major limitations of the isolated perfused tubule technique is that collected samples are very small, and analytical assays can be quite difficult. This can be a major obstacle to overcome. However, in many cases it can be overcome by using radiolabeled compounds, especially ones with high specific activites.

Another disadvantage of the isolated perfused tubule technique is that rabbit is about the only small mammal from which various segments of the nephron can be readily obtained without the aid of collagenase. There has been some work done on isolated cortical collecting ducts from the kidneys of rats and segments of the proximal tubule from the kidneys of rats and mice. However, most of the published work has been done on segments of the nephron from rabbits. Unfortunately, there is very little whole animal data that has been collected from the rabbit for correlative purposes. The reason that other mammals are not chosen for studies employing the isolated perfused tubule is that their kidneys generally contain too much connective tissue, which in most cases makes it impossible to dissect freely segments of the nephron that are suitable for perfusion.

It should be pointed out that renal tubular segments can be obtained for perfusion from nonmammalian species such as the snake, frog, salamander, and some fish. Using segments of the nephron from these species could be useful in comparative studies directed at evaluating the renal tubular handling and toxicity of metals.

IV. PROCEDURES INVOLVED IN PERFUSING ISOLATED SEGMENTS OF THE NEPHRON

A typical setup used to perfuse isolated segments of the nephron is displayed in Figures 4A and 4B. These figures show some of the necessary components needed to perfuse renal tubules

To master the perfusion of isolated renal tubules one must learn to hand craft glass pipettes; harvest the kidney from the rabbit; identify, dissect, and isolate individual segments of the nephron; transfer the segments to the perfusion chamber; and attach the segments to the holding and collection pipettes (Figures 1 and 3).

A. HANDCRAFTING PIPETTES

Handcrafting the pipettes used to perfuse individual segments of the nephron is the rate-limiting procedure on the learning curve in mastering the isolated perfused tubule technique. The investigator must learn several microforging techniques. The following section describes briefly how to craft each of the four pipettes used in the perfusion process. Figure 5 shows some of the characteristics and dimensions of these four pipettes.

The holding and collection pipettes are both made from 6-inches-long borosilicate R-6 glass tubing having an outer diameter (OD) of 0.064 inches and inner diameter (ID) of 0.084 inches. Both of these pipettes are made by first pulling the glass tubing to a needle with a heavy-duty vertical pipette-puller. The pipettes are subsequently crafted to the appropriate size and shape (Figure 5) on a commercially available microforge.

The perfusion pipette is made from 6-inches-long borosilicate R-6 glass tubing having an OD of 0.047 inches and an ID of 0.040 inches. The glass tubing is first pulled on the vertical pipette-puller to form the tip. The tip portion of the pipette is further crafted on the microforge. It is crafted so that the shaft of the tip is 2 to 3 mm in length and 10 µm in diameter, with the very end of the pipette being drawn to a size of 3 to 5 µm in diameter (Figure 5).

The constant volume or volumetric pipette is made from 6-inches-long borosilicate R-6 glass tubing having an OD of 0.020 inches and an ID of 0.014 inches. This pipette is crafted entirely on the microforge. A volume-chamber that is about 2 mm long and about 300 µm in diameter is formed at the tip of the pipette (Figure 5). Approximately 50 nl of fluid can be contained in a chamber having these dimensions. This pipette is calibrated for volume by measuring samples of a stock solution with known amount of radioactivity per volume.

B. HARVESTING THE KIDNEY

Harvesting the kidney from the rabbit is quite simple. The rabbit (the younger the better) is anesthetized with an intermuscular injection of a cocktail containing ketamine and xylazine (70 and 30 mg/kg, respectively). Once the rabbit is anesthetized, one or both kidney(s) is/are removed through a midline abdominal incision. After the kidney(s) is/are removed from the animal, 1-mm thick coronal sections are obtained with a single-edge razor blade. The slices are placed in a cold (on ice) phosphate/sucrose buffer solution (Pirie and Potts, 1986). Each kidney will provide at least 8 to 10 slices. These slices can be used for up to 8 hours for dissection of tubules if kept on ice.

The kidneys can also be perfused *in situ* (through the renal artery) with 60 ml of the phosphate/sucrose buffer before harvesting the kidney (Pirie and Potts, 1986). In our laboratories, this more extensive procedure has not proven necessary. The only advantage it provides is that the kidney is cleared of blood, which can make the dissection of tubular segments easier since clotting factors are removed. In the absence of clotting factors, segments of the nephron are less "sticky", which makes it less difficult to isolate them.

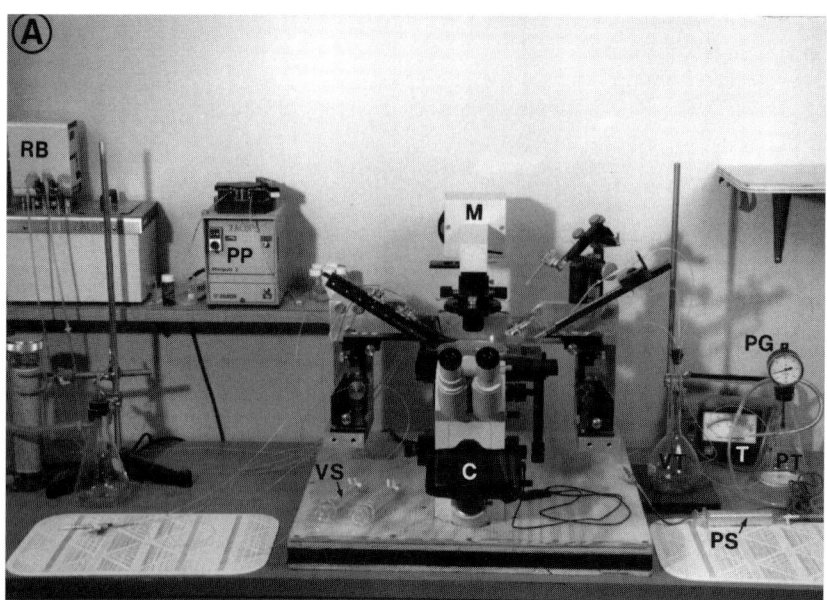

Figure 4A. Photograph showing a typical setup used in the perfusion and study of isolated segments of the nephron. In the center of the field is one of the central pieces of equipment needed to perfuse isolated renal tubules. This is an inverted biological microscope (M) with phase contrast optics. The microscope has a mechanical stage to which an aluminum support system for micromanipulators is attached. In the center of the stage is a thin cylindrical plate of Plexiglass, in the center of which is the chamber used for the perfusion of isolated segments of the nephron. The microscope also has a camera (C) attached for photomicrography. An X, Y, and Z micromanipulator is attached on the base portion of both the right and left sides of the aluminum support bar. Just to the left of the base of the microscope, two vacuum syringes (VS) can be seen. These syringes are used to help mount isolated tubular segments on the holding and collection pipettes and to help draw up collected fluid during perfusion. On the right side of the photograph, below the platform for the microscope, a pressure syringe (PS) and pressure trap (PT) can be seen. The pressure syringe is used to generate positive air pressure in the pressure trap, which is used as reservoir for pressure needed to drive the perfusate out of the perfusion pipette into and through the perfused tubular segment. The pressure trap is equipped with a pressure gauge (PG) to monitor the perfusion pressure. To the left of the pressure trap is a telethermometer (T), which is used to monitor the temperature of the bathing solution in the perfusion chamber via a temperature probe. Next to the telethermometer is a volumetric flask, which is used as a vacuum trap (VT). Polyethylene lines go into the vacuum trap to a vacuum system and reservoir (far left of photograph). The vacuum trap is connected to a pipette on the micromanipulator set on the top of the aluminum support bar. This pipette is used to maintain a constant volume of bathing fluid in the perfusion chamber. On the left side of the photograph, a recirculating bath (RB) and peristaltic pump (PP) can be seen. The recirculating bath is used to heat the perfusion chamber and bathing fluid to the desired temperature. The peristaltic pump is used to deliver the solution bathing the isolated tubular segment in the perfusion chamber.

C. SEGMENT IDENTIFICATION

Identification, dissection, and isolation of segments of the nephron and collecting duct is not difficult in the rabbit. These steps can usually be mastered in less than a week with substantial practice. Generally, the best approach to dissect and isolate a segment of the nephron is to first pull a wedge-shaped strip of renal tissue from one of the 1-mm thick slices of kidney while viewing the slice of tissue with a dissecting microscope. This strip should be taken with a fine pair of forceps (Dumont #5) and should include tubular segments from the tip of the papilla to the cortex. The wedge is then pulled apart along its long axis, which is the natural cleavage plane running parallel to the direction of most straight segments of the nephron. This results in the remaining wedge becoming progressively smaller and smaller until there is only a small bundle of tubules remaining. During the process of dissection, one must keep in mind, at all times, where the representative zones of the kidney are located in the wedge of tissue. This will aid in identifying specific segments of the nephron. When the wedge of tissue has become small enough, an individual tubular segment can be identified for isolation. The remaining tubules near the prospective tubular segment will be stripped away, avoiding pulling on the prospective tubule, until it is attached only to one or two other segments. The end of the prospective tubule is grabbed and gently

Study on the Transport and Toxicity of Metals 773

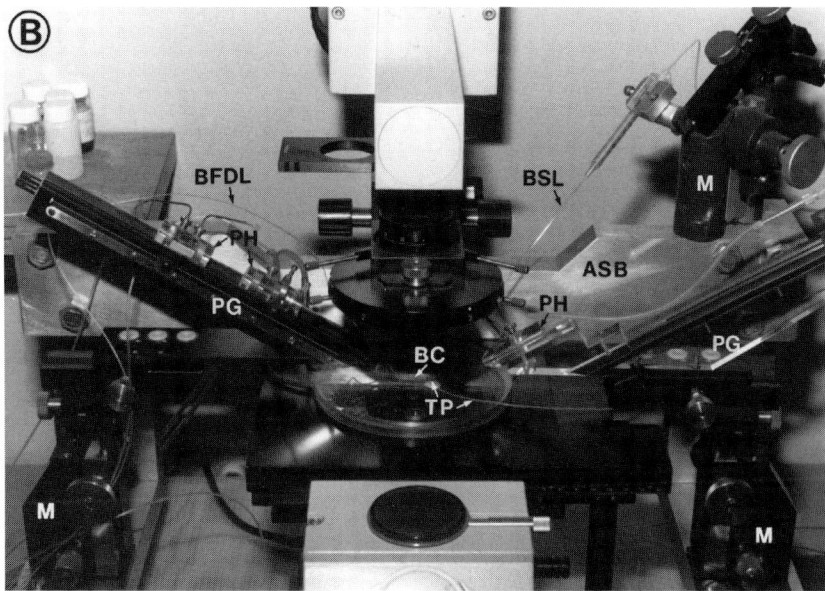

Figure 4B. This photograph shows a higher power view of the apparatus surrounding the bathing (perfusion) chamber (BC). The micromanipulators (M) mounted on the lower left and right platforms of the aluminum support bar (ASB) have pipette guides (PG) mounted on them. Pipette holders (PH), which are used to move and secure in place the pipettes used in perfusing a tubular segment, are mounted on the guides. The pipette guide and holders on the left side of the stage are used to hold and move the holding and perfusion pipettes, while the pipette guide and holder on the right side of the stage are used to hold and move the collection and constant-volume pipettes. A temperature probe (TP) can be seen at the side of the bathing chamber. The bathing fluid delivery line (BFDL) coming from the peristaltic pump and bath siphon line (BSL) going to the vacuum trap can also be seen clearly. It should be pointed out that the setup seen in this photograph is only one example of a setup used to perfuse isolated segments of the nephron.

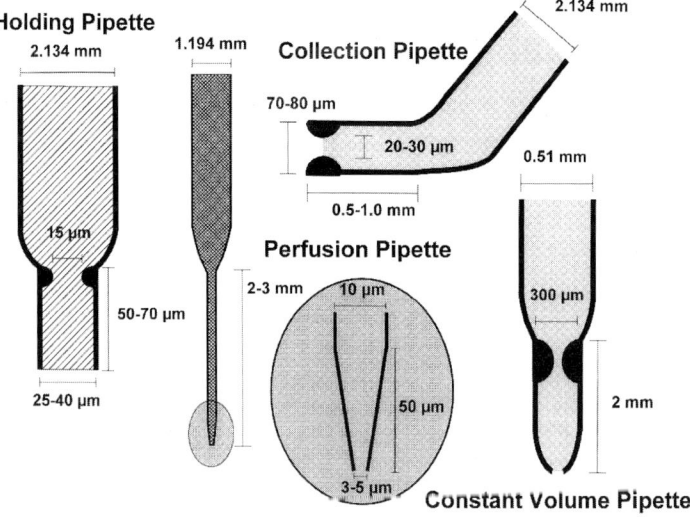

Figure 5. Schematic drawing showing some of the characteristics of the four pipettes used to perfuse and study isolated segments of the nephron. The shapes of the various pipettes are achieved by pulling glass tubing on a pipette puller and forging the pulled glass on a microforge. See text for more details.

stripped away from the remaining tubules. It must come unattached very easily or it can be stretched to the point of disrupting or damaging the junctional complexes between the tubular epithelial cells and/or tubular basal lamina.

Once the tubular segment is free, it can be visually inspected for disrupted or damages areas that will appear as opaque (milky) spots. If the segment passes initial inspection, then the ends of the segment are trimmed. This is usually necessary because the ends have been damaged by the forceps. Trimming is accomplished by using the sharp edges of two 20-gauge hypodermic needles as scissors. One needle is placed across the tubule at the location to be cut with the needle point held on the bottom of the dissecting dish. The second needle is then drawn across the tubule in such a manner that the two sharp edges of the needles slice past each other like the blades of a pair of scissors. This trimming approach prevents crushing the ends of the tubule closed, which renders the tubule nonperfusable. After trimming, the tubular segment is ready to be transferred to the perfusion chamber.

Identification of the various segments of the nephron is accomplished primarily by anatomical or zonal location and with a familiarity of the structural characteristics unique to the segments.

The following are some characteristics used to identify the three segments of the proximal tubule, which are generally the main segments used currently in the study of the toxicity and transport of heavy metals in the kidney: the S1 segments of the proximal tubule are dissected from the subcapsular surface of the renal slices and are identified by their larger outside diameter and very convoluted shape. Occasionally a renal corpuscle is found to be attached to the isolated S1 segment, which serves as positive identification for this segment. The S2 segments are identified as tubules that span the entire length of the cortex and have a very straight shape. The S3 segments are identified as the last 1 mm of the proximal tubule that are attached to the descending thin limb of Henle's loop.

D. TRANSFERRING THE TUBULE TO PERFUSION CHAMBER

Transferring an isolated segment of the nephron to the perfusion chamber can be somewhat difficult. It is accomplished by using a 1-ml plastic syringe attached to a 10-cm glass tube (0.047 inches OD by 0.040 inches ID) by a 50 cm-long piece of polyethylene tubing. The glass tubing is bent at the end to a 30° angle and the end is fire-polished. While viewing through a dissecting microscope, the tip of the transfer pipette is brought very near to, but not touching, the tubular segment, which is lying on the bottom of a dissection dish. With the free hand, the plunger of the 1-ml syringe is drawn out slowly. The plunger is drawn out until the tubule and a little of the dissecting fluid is sucked up about 1 to 2 mm into the transfer pipette. If not too much fluid is drawn into the transfer pipette, the fluid and the tubule will remain stationary in the tip while the transfer pipette is relocated to the perfusion chamber. While viewing the tubular segment in the tip of the transfer pipette with the microscope (to which the perfusion apparatus is attached), the plunger on the syringe is gently pushed downward to force the fluid and tubule in the transfer pipette out into the bathing fluid in perfusion chamber. Caution should be taken to avoid pushing too hard or quickly on the plunger or forming air bubbles. The tubular segment will eventually settle on the bottom of the perfusion chamber, which consists of a cover slip. If the transfer pipette is new and has not been used, there is a strong possibility that a tubular segment, that is drawn up into the pipette, will stick to the inside wall of the pipette. This problem can be minimized by repeatedly using the same transfer pipette and by not cleaning the transfer pipette too thoroughly after use. It is best if the pipette is cleaned with a quick rinse of distilled water and then is dried with acetone and air suction.

E. PERFUSION CHAMBER

The perfusion chamber is a cavity milled out in a piece of Lucite. The cavity is generally milled to dimensions of about 1.5 cm in length and 0.3 cm in depth and width. Approximately 0.3 ml of fluid can be contained in this size of a chamber. On the underside of the piece of Lucite, a trench is milled out around the chamber to contain a copper water jacket to heat the fluid bathing the perfused tubular segment. The Lucite wall of the perfusion chamber should be as thin as possible to allow efficient exchange of heat from the copper tubing to the bathing fluid. Warmed water is delivered from a recirculating bath to the copper water jacket fitted around the perfusion chamber. Generally, the water in the copper jacket must be kept at about 54°C to maintain a temperature of 37 to 38°C in the bathing solution, which is continuously exchanged at a rate of 0.3 ml/min by a peristaltic pump (Figure 4A). In addition, the bathing fluid that is being pumped to the perfusion chamber must be preheated with a heat-exchanger or the bathing solution will be too cool. The heat-exchanger is made by simply wrapping the polyethylene tubing (carrying the bathing solution) around the portion of the copper tubing just before, or just after, the tubing feeds around the perfusion chamber.

In experiments where the bath-to-lumen flux of a metal is being studied, the bathing solution is typically not exchanged. To prevent evaporation of bathing fluid in these experiments, a light weight mineral oil is placed over the surface of the bathing solution in the prefusion chamber. Under these conditions (where the bathing solution is not exchanged), the temperature of the water in the heating jacket must be kept at a lower 38°C. In this state of maintaining a constant volume of bathing solution, unstirred layers are prevented from occurring by connecting a reciprocating pump to the bathing through the oil layer by polyethylene tubing. The reciprocating action of the pump draws about 50 μl of bathing solution in and out of the chamber, resulting in a complete mixing of the bathing solution. It is important to place a small hole (needle puncture) in the polyethylene tubing. This allows for equilibration of air pressure between the ambient air and air confined inside the polyethylene tubing as the room temperature and pressure changes during the course of an experiment.

F. ATTACHMENT OF DISSECTED TUBULAR SEGMENTS TO PIPETTES

Connecting an isolated segment of the nephron to the holding and collection pipettes is accomplished by first placing the opening of the holding pipette very near, but not touching, one end of the dissected tubule while it is lying on the bottom of the perfusion chamber, and then pulling the tubule into the holding pipette by suction applied by a hand-controlled syringe (preferably a 50- to 60-cm^3 glass syringe). The pipette and tubule are then raised off the bottom of the chamber and are positioned in the middle of the perfusion chamber with the aid of a micromanipulator assembly that is used to hold and guide the pipettes (Figure 4A). Next, the perfusion pipette is advanced down the middle of the holding pipette until the tip of the perfusion pipette enters into the lumen of the tubule.

Air pressure (10 to 30 mmHg) is applied to the perfusion fluid from an air reservoir. This forces the fluid in the perfusion pipette to enter into the lumen of the tubule. Continued pressure causes the lumen of the tubule to open along the entire length of the segment in a progressive manner until fluid is perfusing freely through the tubule. In some instances, when the vital dye FD&C green is added to the perfusion solution (0.25 mM), the dye can be seen exiting the free end of the tubule.

After the holding and perfusion pipettes are in place, the collection pipette is attached. The collection pipette is brought into place next to the free end of the perfusing tubule using another micromanipulator assembly that is attached to the mechanical stage of the inverted microscope. The tip or end of the collection pipette is plugged with liquid Sylgard (Dow Corning Sylgard 184) to prevent the bathing fluid from being drawn into the collection pipette by capillary action. A pulse of suction is applied to the collection pipette and about 100 μm of the tubular segment is pulled into the collection pipette through the Sylgard. If this maneuver is done well, less than 5 nl of bathing fluid is drawn into the collection pipette with the tubule.

G. PERFUSION OF DISSECTED TUBULAR SEGMENTS

After the tubular segment is attached to all the pipettes, it is allowed to perfuse and warm to 37 to 38°C. Figure 6 shows an example of an S1 segment of a proximal tubule that was being perfused through the lumen with a balanced electrolyte solution. One can see in this figure that the ends of the tubule are mounted to the holding and collection pipettes and that the tip of the perfusion pipette is present in the lumen of the tubule.

During the "equilibration" or "warm-up" time, Sylgard flows down the inside of the collection pipette and lodges between the tubule and the wall of the glass pipette. This makes a very tight mechanical and electrical seal. While the tubule is equilibrating, the researcher will place light mineral oil in the collection pipette in contact with the surface of the perfused fluid to prevent evaporation of water from this collectate.

During the course of an experiment where the lumen-to-bath flux of a substance is being studied, the bathing solution will be continuously flowing through the perfusion chamber at the rate of 0.3 ml/min. This exchanges the bath once every minute during the experiment. As stated above, flow of bathing fluid through the perfusion chamber is accomplished by pumping the bathing fluid to the perfusion chamber with a peristaltic pump. Fluid is removed from the top of the chamber by aspiration, which is achieved by vacuum. The aspirated samples of bathing fluid can be collected into a scintillation vial for analysis or can be emptied into a vacuum trap.

The rate at which fluid is perfusing through the tubule is controlled by positive pressure and is measured by the time required to fill a constant-volume pipette (30 to 100 nl), that has been positioned in the collection pipette. The perfused tubular segment is generally not subjected to elevated transmural hydrostatic pressures because the pressure delivered to the perfusion pipette (15 to 30 mmHg) is dissipated as the perfusing fluid flows through the terminal portion of the pipette, which has an inside diameter of 10 to 12 μm for most of its 2 to 3 mm of length (Figure 5). Normally, the tubular segment is subjected to less than 10 mmHg transmural hydrostatic pressure if the segment is properly sealed in

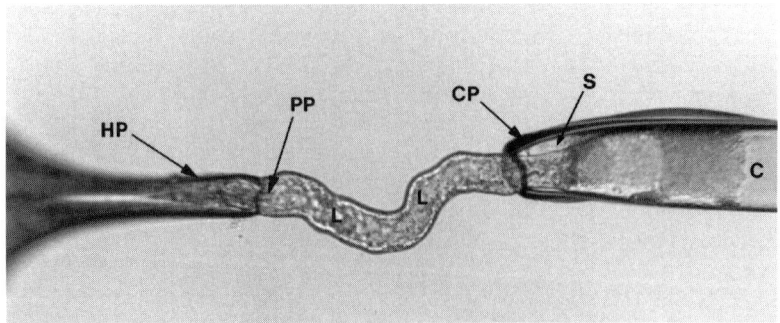

Figure 6. A photomicrograph of an S1 segment of the proximal tubule from the kidney of a rabbit perfused through the lumen *in vitro* with a balanced electrolyte solution. The tubule is attached on the left to the holding pipette (HP) and on the right to the collection pipette (CP). The tip of the perfusion pipette (PP) can be seen in the lumen on the left side of the perfused segment. When this S1 segment was warmed to 37°C, Sylgard (S), which was placed in the tip of the collection pipette, flowed around the portion of the tubular segment inside the pipette to form a water-resistant seal. On the far right side of the photomicrograph, collectate (C) can be seen inside the collection pipette.

the collection pipette. If the seal is too tight around the tubule, that is, the opening of the collection pipette is less than 70% of the outside diameter of the tubule, the transmural pressure across the epithelium of the tubule will approach that of the pressure at the pressure reservoir. If this occurs, the epithelial cells will appear very flat, instead of being cuboidal in shape. Continued elevated pressure will cause injury to the tubular epithelial cells.

H. PERFUSION OF TUBULAR SEGMENTS UNDER MINERAL OIL

For perfusing isolated segments of the nephron under oil, the tubular segment is initially set up in the normal manner as described above (Figure 3). A 3-mm layer of light mineral oil is then placed on the surface of the bathing fluid. Subsequently, the bathing fluid is sucked back out of the bathing chamber into a syringe. While all of the bathing fluid is being removed, the oil that covers the surface of the bathing solution comes down into the perfusion chamber and eventually surrounds the perfused tubular segment. S1 segments of the proximal tubule usually slide into mineral oil very easily, while S2 and S3 segments resist leaving the aqueous environment. Consequently, suction must be applied to the holding and collection pipettes to keep the S2 or S3 segments from being pulled off the pipettes. Once a tubular segment is completely immersed in oil, it will usually stay in place. However, it is difficult to move the tubule through the water-oil interface. It requires greater air pressure (20 to 60 mmHg) to force perfusing fluid through a tubular segment under oil than in an aqueous bathing solution. However, this increased perfusing pressure does not alter significantly the transmural pressure in the perfused segment, which is reflected by the tubular epithelial cells maintaining their normal cuboidal shape.

After the tubular segment is warmed to temperature, absorbed fluid (absorbate) will begin to appear on the basolateral surface of the perfused tubule as torus-shaped droplets evenly distributed along the tubule (Figure 3). These droplets can be moved closer together manually with a glass needle. The volume of the absorbed fluid and the rate of the absorptive process can be determined by timed collections into a 5- to 10-nl constant volume pipette. The collected absorbate can be subsequently analyzed to determine the concentration and rate of transport of the solute studied, which can include a radiolabeled heavy metal.

I. SOLUTIONS USED IN PERFUSION EXPERIMENTS

Isolated segments of the nephron are generally perfused and bathed with buffered electrolyte solutions. As an example, the bathing and perfusing solutions used most in our experiments generally contain the following: Na^+, 145 mM; Cl^-, 140 mM; K^+, 5 mM; Ca^{2+}, 2.5 mM; Mg^{2+}, 1.2 mM; SO_4^-, 1.2 mM; $HPO_4^-/H_2PO_4^-$, 2 mM; D-glucose, 1 mM; with 0.5 mM glutamine in the bathing solution only as a substrate for metabolism.

When segments of the proximal convoluted tubule are studied, it is generally necessary to add an essential amino acid as a substrate for metabolism to prevent rapid epithelial degeneration. L-Alanine, L-glutamate, L-glycine, and L-glutamine can be used, but we have found that L-glutamine is the best for preserving tubular integrity.

In some cases, it may be desirable to perfuse a tubular segment through the lumen with an ultrafiltrate of rabbit plasma. The ultrafiltrate is obtained by centrifuging rabbit plasma through a filter that excludes compounds having a molecular weight greater than 50,000 daltons. Serum of rabbit blood can be used to bath the perfused segments of the nephron. By perfusing with an ultrafiltrate and using serum in the bath, one can emulate very closely the luminal and basolateral environment found *in vivo*. This may be useful as part of the study of the toxicity and transport of heavy metals if one wants to emulate the conditions involved in the complex metal–ligand exchange interactions that occur during the renal tubular uptake of a heavy metal, such as mercury, cadmium, or lead.

When studying the toxicity and transport of heavy metals along isolated perfused segments of the nephron, the unique binding and solubility characteristics of each metal, and the solutions in which the metal (radioactive) is delivered, must be taken into account when mixing it in the bathing or perfusing solutions. For example, before we add radioactive inorganic mercury ($^{203}Hg^{2+}$) to the perfusate, we remove the desired volume (generally 1 to 2 µl) of solution (which is usually in 1.0 N HCl) containing the radioactive mercury from the manufacturer's vessel and place it in a well of a bacteriological slide. The volatile components of the solution are allowed to evaporate. Subsequently, the desiccated sample of radioactive inorganic mercury is put back into solution in the well by placing the desired volume (generally 50 µl) of the normal buffered perfusing solution over the desiccated sample. This procedure prevents a significant change in pH of the perfusing solution after the addition of the radioactive inorganic mercury. If the acidic radioactive solution from the manufacturer were mixed directly with the perfusate, the pH of the perfusate would drop well below physiological levels (7.4), making the perfusate useless for experiments.

J. CALCULATIONS USED TO DETERMINE RATES OF FLUX OF METALS
1. Measuring Lumen-to-Bath Flux

When one is perfusing a metal through the lumen of a tubular segment and wants to evaluate the lumen-to-bath transport of that metal, one has to determine the rate at which the metal is disappearing from the luminal fluid and the rate at which the metal is entering into the bathing fluid.

The rate of disappearance (J_D, fmol min^{-1} mm^{-1}) of a metal from the luminal fluid can be measured by the equation:

$$J_D = \left([M]_P \times \left([VM]_C/[VM]_P\right) \times V_C\right) - \left([M]_C \times V_C\right)$$

where $[M]_P$ and $[M]_C$ are the concentrations (fmol nl^{-1}) of the metal in the perfusate and collectate, and V_C (nl min^{-1}) is the volume collection rate. Collection rates are calculated from the time required to fill the constant volume pipette (30 to 100 nl). $[VM]_C$ and $[VM]_P$ are the concentrations (activities of the radioisotope per unit volume, dpm nl^{-1}) of the selected volume marker (usually ^{3}H-L-glucose, ^{3}H-inulin, ^{14}C-polyethyleneglycol) in the collectate and perfusate, respectively. $[M]_P$ and $[M]_C$ are determined from the specific activity (SA_M, dpm fmol^{-1}) of the radioisotope of the metal (e.g., $^{203}Hg^{2+}$ or $^{109}Cd^{2+}$).

The rate of appearance (J_A, fmol min^{-1} mm^{-1}) of the metal in the bathing solution is calculated by the equation:

$$J_A = DPM/(SA_M \times T)$$

where DPM is the activity of the radioisotope (dpm) that appears in the bathing solution in T (minutes) time.

2. Measuring Bath-to-Lumen Flux

In experiments where the bath-to-lumen transport of a metal is to be evaluated, only the rate of appearance of the metal in the luminal fluid from the bathing solution can be determined. This rate (J_{BL}, fmol min^{-1} mm^{-1}) is computed by the following equation:

$$J_{B-L} = DPM/(SA_M \times T)$$

where DPM is the activity of the radioisotope (dpm) appearing in the luminal fluid, SA_M (dpm fmol^{-1}) is the specific activity of the radioactive metal, and T (min) is the time required to collect the sample. All rates of flux are normalized to tubular length (I) in millimeters.

3. Measuring the Cellular Concentration of Metal

When one desires to compute the concentration of metal in a perfused tubular segment, the cellular volume of water needs to be determined first.

For segments of the proximal tubule, cellular volume of water (V_{Cell}) is calculated in picoliters (pl) from the equation:

$$V_{Cell} = \pi(OR^2 - IR^2) \times (1 \times 0.7)$$

where OR and IR are the average outer and inner tubular radii (µm), respectively, I is the tubular length (mm), and 0.7 accounts for the fraction of cellular volume that is water (Tune and Burg, 1971).

Once the cellular volume of water has been computed, the cellular concentration of metal ($[M]_{Cell}$, mM) can be calculated by the equation:

$$[M]_{Cell} = (DPM/SA_M)/V_{Cell}$$

where DPM is the activity of radioisotope (dpm) in the TCA soluble fraction of the perfused segment, SA_M is the specific activity of the radioactive metal taken from standards (dpm fmol^{-1}), and V_{Cell} is cellular volume. Values for DPM of the metal in the TCA-soluble fraction of the tubule are corrected for any radioactivity that may have remained in the lumen of the tubule. This is calculated by the amount of volume marker extracted with the tubule. This correction is always very small (about 1%).

4. Determining the Rate of Leak

When considering any aspect of the transport of metals by isolated segments of the nephron, leak is an important factor that needs to be evaluated. If a metal such as mercury causes degeneration or necrosis along a portion of the perfused segment of the nephron, there is an increased probability that the metal in the lumen will enter the bathing fluid by means of leaking between tubular epithelial cells or actually going through necrotic epithelial cells when cellular necrosis has been induced. It should also be pointed out that segments of the proximal tubule are inherently leaky, especially the S1 segments (Barfuss and Schafer, 1981). Thus, the magnitude of leak in any experimental situation needs to be assessed carefully before conclusions regarding the transport of a metal can be made.

The leak of the metal from the tubular lumen can be measured by the appearance (leak) of the volume marker into the bathing solution (L_V, nl min^{-1}) during experiments where the lumen-to-bath flux of a metal is being studied. The leak of the volume marker can be calculated by the equation:

$$L_V = DPM/(SA_{VM} \times T)$$

where DPM is the activity (dpm) of the radiolabeled volume marker appearing in the bathing solution in T time (min), and SA_M is the specific volumetric activity of the radiolabeled volume marker (dpm nl^{-1}). The expected leak of a metal (L_M, fmol min^{-1} mm^{-1}) can be computed by the equation:

$$L_M = (L_V \times [M]_{ML})/l$$

where L_V is the volume leak (nL min^{-1}), $[M]_{ML}$ is the mean luminal concentration (fmol nl^{-1}) of the metal, and I is the length (mm) of the tubule.

Leak of a metal into the lumen during the bath-to-lumen flux measurements is calculated with the same equations used to compute the lumen-to-bath leak, except that the respective concentrations in the bathing solution are substituted for appropriate concentrations in the luminal fluid.

K. EVALUATION OF THE TOXICITY OF METALS

There are generally two ways in which toxicity of a metal in perfused segments of the nephron can be evaluated. One method is by simply observing and documenting, with photomicrography and in writing, the sequence and time of occurrence of pathological changes in a segment of the nephron exposed to a metal at the luminal membrane and/or basolateral membrane. Videomicrography also proves to be very useful. When the vital dye FD&C green is added to the perfusate (0.25 mM), cellular necrosis can be demonstrated. In addition, the progression, extent, and pattern of cellular necrosis can be documented throughout the entire perfusion process.

Another useful way in which to document epithelial injury involves correlating the leak of the volume marker with the progression of visually demonstrable changes along isolated perfused segments of the nephron. As stated above, when injury progress along the tubular epithelium, there is an increased probability of intercellular and even transcellular leak. We have found, when using the volume marker ^{3}H-L-glucose, that there is a correlation between the leak of this volume marker and the severity of injury and necrosis along segments of the proximal tubule.

V. SUMMARY

This chapter has provided a brief summary of tubular microtechniques and their use in the study of the transport and toxicity of metals along the nephron. The majority of this chapter has focused, however, on the utility and methods used to study the transport and toxicity of metals along isolated segments of the nephron using the isolated perfused tubule technique. Although this technique is complicated, and at times cumbersome, it provides one of the most useful tools to evaluate and characterize mechanisms involved in the transport, accumulation, metabolism, and toxicity of metals in intact renal tubular epithelia.

REFERENCES

Barfuss, D. W., Ganapathy, V., and Leibach, F. H. (1988), *Am. J. Physiol.*, 255, F177–F181.
Barfuss, D. W., McCann, W. P., and Katholi, R. E. (1992), *Kidney Int.*, 41, 1143–1149.
Barfuss, D. W., Robinson, M. K., and Zalups, R. K. (1990), *J. Am. Soc. Nephrol.*, 1, 910–917.
Barfuss, D. W. and Schafer, J. (1981), *Am. J. Physiol.*, 240, F322–F332.
Barfuss, D. W. and Schafer, J. A. (1984a), *Am. J. Physiol.*, 247, F117–F129.
Barfuss, D. W. and Schafer, J. A. (1984b), *Am. J. Physiol.*, 247, F130–F139.
Burg, M., Grantham, J., Abramov, M., and Orloff, J. (1966), *Am. J. Physiol.*, 210, 1293–1298.
Cikrt, M. and Heller, J. (1980), *Environ. Res.*, 21, 308–331.
Felley-Bosco, E. and Diezi, J. (1987), *Toxicol. Appl. Pharmacol.*, 91, 204–211.
Felley-Bosco, E. and Diezi, J. (1989), *Toxicol. Appl. Pharmacol.*, 98, 243–251.
Pirie, S. C. and Potts, D. J. (1986), *Clin. Sci.*, 70, 443–452.
Robinson, M. K., Barfuss, D. W., and Zalups, R. K. (1993), *Toxicol. Appl. Pharmacol.*, 14, 221–225.
Tune, B. M. and Burg, M. B. (1971), *Am. J. Physiol.*, 221, 580–585.
Zalups, R. K. and Barfuss, D. W. (1993), *Toxicol. Appl. Pharmacol.*, 121, 176–185.
Zalups, R. K. and Barfuss, D. W. (1993), *Toxicol. Appl. Pharmacol.*, 121, 176–185.
Zalups, R. K., Robinson, M. K., and Barfuss, D. W. (1991), *J. Am. Soc. Nephrol.*, 2, 866–878.

Section VII
Immunomodulation by Metals

Overview

An Introduction to Immunomodulation By Metals

Donald Gardner and Judith T. Zelikoff

Considerable progress has been made in improving our understanding of the sensitivity of various target organ systems to a wide range of toxic substances. Although the discipline of immunotoxicology is relatively young, considerable progress has been made in demonstrating that xenobiotic substances can modulate immunity, and in some cases the immune system can be the primary target. For example, lead, polychlorinated biphenyls, and some pesticides have been shown to affect the immune system at exposure levels lower than those affecting other organ systems. Because of their widespread use and persistence in biological systems, attention has been given to the possible immunomodulations by metals. The immune system has been shown to be vulnerable to such exposures, producing a wide variety of undesirable effects. While a normal functioning immune system is highly effective in recognizing and handling foreign substances, many chemicals are capable of altering this delicate balance, compromising the well-being of the host, and resulting in an increased risk of disease. Metal exposure can result in three undesirable immunological effects: (1) the breakdown of recognition of self, leading to immune dysregulation and to autoimmune disease; (2) the compromise or suppression of normal immune defense mechanisms, leading to an increased risk of infectious disease and the development of neoplasia; and (3) augmented immune responses that can lead to hypersensitivity.

Immunotoxicology is an important and rapidly developing discipline, and this chapter reviews the clinical and experimental animal evidence demonstrating that both inorganic and organic metals can result in a variety of immunological alterations. Since it is widely accepted that the immune systems of humans and animals are comparable, the data obtained from toxicological studies can often be extrapolated to humans. However, while the principles of immunology are similar and comparable, it must be recognized that the response which may occur in humans may be different from those observed in animal models and at different toxicant concentrations. Although we have assumed that readers have some understanding of immunology, this chapter should be of interest to any individual interested in metal toxicology.

The topics of the following chapters were selected to illustrate the action of metals on the immune system and to explore the biological mechanisms associated with these responses. Often, exposure to the same chemical can result in multiple host responses, such as increased risk of infectious diseases, development of neoplasia, autoimmune disorders, and/or allergies. Chapters are included which describe adverse consequences of metals on the immune system that may result in dysfunction leading to either immunostimulation or immunosuppression.

Chapter 49

Effects of Metals on Cell-Mediated Immunity and Biological Response Modulators

Raghubir P. Sharma and Raviprakash R. Dugyala

I. INTRODUCTION

Heavy metals were some of the first chemicals ever tested for their immunotoxic potential. In as early as 1966, Selye et al. (1966) reported on the effect of lead on susceptibility of rats to endotoxins of Gram-negative bacteria. Administration of lead acetate increased sensitivity of rats to endotoxins nearly 100,000 times compared to untreated animals. These studies were later confirmed by lead-induced suppression of mouse resistance to *Salmonella typhimurium* (Hemphill et al., 1971). Since that time interest in metal immunotoxicity has grown considerably. In has only been a recent trend that emphasis is placed on investigating the mechanisms of such action. The importance of studying mechanisms over merely screening the effects has been emphasized (Sharma and Zeeman, 1980).

Exposure to metals is almost unavoidable, both in environmental and occupational situations. Attempts have been made to minimize the intake of metals to avoid any health problems. However, metals are ubiquitous in the environment, and exposure to at least some metals in unavoidable. Episodes of epidemiologic proportions after ingestion of mercury and cadmium via diet have been recorded; clinically unacceptable levels of lead in blood and tissues of children and its effect on development are still a matter of public health concern. Most heavy metals have multiple types of toxicities; immunotoxic effects are only one of these. However, since the immune system is vulnerable to toxic insults over a long period of time, possible effects of metal exposure should be carefully evaluated on this system.

Metals are virtually nondestructible and, therefore, highly persistent in the biological systems. Like other toxic effects, immunotoxicity is also determined by the chemical form of the metal to which an organism is exposed; however, effects are often related to the metal and its valence state, a chemical transformation may affect the bioavailability and translocation in the system. Several organometallic compounds are biologically active and their immunotoxic effects are often more potent than inorganic ions. Metals also have a large spectrum of effects, influencing various enzyme activities, transport mechanisms, or even gene expression. All these processes may have an indirect influence on the immune responses which depend a great deal on homeostatic mechanisms of the organism. Most toxic heavy metals displace some of the essential metals in biologic functions and can have profound effects on cellular metabolism, including those of lymphocytes. This chapter summarizes effects of metals on cell-mediated immunity, in particular on T lymphocytes and selected biological response modifiers. The information available on several metals is, however, sketchy and even contradictory in terms of their specific effects on cell-mediated immune functions.

Cell-mediated immune functions are not independent of other immunological mechanisms. Several interleukins produced by macrophages influence the T cell-dependent immune functions and are considered elsewhere in this monograph. At the same time T cell functions can sometimes be very subtle to be observed and are often expressed in animals as an alteration of clonal proliferation and subsequent T cell-dependent antibody production. It is possible that some of the discussion covered here may, therefore, have some overlap with other chapters in this book. Many of the immune functions in immunotoxicity evaluation are measured simultaneously and, therefore, T cell effects are often only one of the many parameters observed. Studies on isolated or specific cell populations for immunotoxicity of metals are relatively new in the literature.

II. ARSENIC

There are few immunotoxicological studies on arsenic (As) compounds compared with those on other metals such as lead, cadmium, and mercury. The immunotoxic effects of arsenicals may be due to their capacity to bind sulfhydryl groups in T cell subsets, thereby affecting the signal transduction.

A. EFFECTS OF *IN VIVO* EXPOSURE

Treatment of 4-week old Swiss BALB/c and CD-1 mice with sodium arsenite ($NaAsO_2$, 3.13 to 6.25 mg/kg, subcutaneously, or 0.001 to 0.002 M in drinking water), arsenic trioxide (5 mg/kg, subcutaneously), or 4-hydroxy-3-nitrobenzenearsonic acid (0.002 M in drinking water) produced splenomegaly and increased viral titers in spleen and plasma and increased mortality when infected with pseudorabies, encephalomyocarditis, and St. Louis encephalitis viruses (Gainer and Pry, 1972).

B. *IN VITRO* EFFECTS OF ARSENIC

Effects of $NaAsO_2$ (10^{-5} to 2×10^{-5} M) and sodium arsenate ($NaHAsO_4$, 10^{-6} to 6×10^{-4} M) on human and bovine lymphocytes *in vitro* were studied (McCabe et al., 1983). The authors reported an augmentation of PHA-stimulation at low concentrations and inhibition at higher concentrations. In another study, Yoshida et al. (1987) reported a 30% drop in viability of T cells exposed to 50 ng/ml of $NaAsO_2$ for 4 h. Their studies suggested that enhanced plaque-forming cell response was due to the inhibition of precursor and suppresser T cells.

C. BIOLOGICAL RESPONSE MODIFIERS

Interferon (IFN) production and action were inhibited *in vivo* and *in vitro* in the mouse by arsenicals (Gainer, 1972). In this study, $NaAsO_2$ (3.13 mg/kg) decreased the protective immunity in mice to encephalomyocarditis virus by treatment with various interferons. *In vitro* treatment with $NaAsO_2$ ($10^{-4.5}$ M) or $NaHAsO_4$ ($10^{-3.7}$ M) decreased the action of interferons on primary Swiss mouse embryonic (ME) cells.

Recently a trivalent arsenical compound (phenylarsine oxide) was found to be a potent protein tyrosine phosphatase (PTPase) inhibitor (Liao et al., 1991). Phenylarsine oxide inhibited anti-CD3-stimulated tyrosine phosphorylation of phospholipase Caγ1, production of inositol 1, 4, 5-triphosphate, and cellular Ca^{2+}, but itself increased intracellular free calcium in T cells (Fletcher et al., 1993). Further studies are needed to determine the sensitivity of different subsets to T lymphocytes towards the arsenicals as cell number of free sulfhydryl groups in T cells varies (Lawrence, 1981b).

III. CADMIUM

A. EFFECTS OF *IN VIVO* TREATMENT

Effects of cadmium on various immunological functions are ambiguous. Early studies suggested that overall immunity is compromised by cadmium, resulting in increased fatality after challenge to sublethal doses by injection (Koller and Vos, 1981). Gaworski and Sharma (1978) reported that treatment of mice with 0.18 and 1.42 mM cadmium in water for 30 days decreased the lymphocytic activation in the presence of PHA. Humoral response to a T-dependent antigen, SRBC, in mice produced paradoxical results (Koller et al., 1976). Antibody response was increased after a single intraperitoneal injection of 0.15 mg in each mouse, whereas after a similar oral dose these responses were decreased.

A continuous oral administration of 25 ppm cadmium acetate in water for periods of up to 77 d failed to produce any effect on rat blood lymphocytes when stimulated with concanavalin A (Con A) (Ramsten-Wasenberg and Wasenberg, 1983). In another report, mice treated with cadmium chloride exhibited an insignificant increase in DNA synthesis when their splenic lymphocytes were cultured in the presence of Con A (Blakley, 1985). Blakley and Tomar (1986) reported that oral treatment of mice with cadmium suppressed antibody responses to T cell-dependent antigen (SRBC), but not to T cell-independent antigen lipopolysaccharide (LPS) or dinitrophenol-aminoethylcarbamylmethyl-Ficoll (DNP-Ficoll). This observation strongly suggests involvement of T cell-mediated processes in cadmium-induced immunomodulation.

A strain-related difference to thymotoxicity of cadmium was reported in rats (Morselt et al., 1988). These authors reported that brown Norway rat showed a decrease in thymocytes in the S-phase and an increase in G-2 phase and mitosis, whereas Lewis rats do not show such an effect after a subcutaneous cadmium injection. This effect is not related to either cadmium uptake by the thymus or differential ability of the two strains to induce thymic metallothionein, a specific cadmium-binding protein. The basal metallothionein levels were higher in the Lewis rat compared to the other strain.

B. *IN VITRO* EFFECTS OF CADMIUM

Incubation of lymphocytes in the presence of T cell mitogen has been affected by simultaneous exposure to cadmium. Presence of 6 to 25 µM cadmium in mouse splenic cultures reduced the PHA-induced proliferation of cells (Gaworski and Sharma, 1978). *In vitro* addition of cadmium at 1 µM or higher concentrations inhibited the mixed lymphocyte culture response, whereas suppresser cell activity was not influenced by cadmium (Bozelka and Burkholder, 1982). Presence of cadmium at 10 to 200 µM caused a concentration-dependent decrease in cell proliferation of mouse splenic cultures in the presence of Con A (Ciavarra and Simeone, 1990). Cadmium complexed with metallothionein caused a mitogen-like response in mouse spleens (Lynes et al., 1990); however, similar effects were also observed with the apoprotein alone.

C. BIOLOGICAL RESPONSE MODIFIERS

No effects of cadmium on T cell-associated biological response modifiers have been reported. Gainer (1977a) reported that presence of cadmium at toxic doses reduced the activity of interferon in protecting mouse embryonic cultures from vesicular stomatitis virus.

IV. LEAD

A. *IN VIVO* EXPOSURE AND T CELL FUNCTIONS

Male Swiss-Webster mice exposed to relatively large oral doses of lead acetate, 1.2 and 9.7 mM in drinking water for 15 or 30 d, reduced [^{3}H]thymidine uptake by splenocytes exposed to a T cell mitogen, phytohemagglutinin (PHA) (Gaworski and Sharma, 1978). The effect showed a statistical significance, however, only at the high dose exposure for 30 d, suggesting that relatively large doses of lead are immunotoxic. This observation compared well with early experiments where the metal salt was injected intraperitoneally (Selye et al., 1966). Similar observations were noted in later studies; up to 10 mM lead acetate given to mice for 10 weeks did not alter response to a T cell-dependent antigen, sheep erythrocytes (SRBC); mixed lymphocyte responses were also not altered (Lawrence, 1981a). In these studies the absorption of lead was not considered, and since oral lead is poorly absorbed when animals are dosed with high amounts, it is difficult to ascertain the influence of lead on these immune functions.

Blakley and Archer (1981) used a similar approach, but measured blood lead levels. At their highest dose level, blood lead concentrations were 0.8 µg/ml, the levels associated with clinical lead toxicity. In these experiments, although responses to T cell-dependent antigens were suppressed, the authors demonstrated that the effects were indirect, perhaps due to the response of macrophages to lead. Effects of lead exposure in cells exposed *in vitro* are indeed different than those of *in vivo* exposure.

Similar treatment of lead to mice in drinking water (827 ppm/lead) for 8 weeks that resulted in 0.4 µg/ml blood lead decreased antibody-dependent cell cytotoxicity (ADCC), but produced no effect on the activity of natural killer (NK) cells (Neilan et al., 1983). This report confirmed the observation from a different group that oral lead exposure did not alter T cell-mediated cytotoxicity, although it augmented the cytotoxic responses in spleen or lymph nodes of Moloney sarcoma virus–tumor-bearing mice (Kerkvliet and Baecher-Steppan, 1982). Intraperitoneal injections of various lead salts, however, decreased the delayed type hypersensitivity (DTH) response to SRBC in mice (Descotes et al., 1984).

B. *IN VITRO* EXPOSURE

When cells are exposed to lead *in vitro,* their response is different than the *in vivo* exposures to this metal. A synergistic effect of lead with PHA was observed when splenocytes were cultured for 3 d (Gaworski and Sharma, 1978). The effects were directly related to lead concentration in the range of 0.1 to 1 mM. At lower concentrations of lead, the effects of PHA were not evident (Lawrence, 1981a). Later studies by McCabe and Lawrence (1991) suggested that *in vitro* exposure to 0.1 mM lead enhanced T cell activity as well as factor-mediated T cell help. In this study, the antigen specificity was evaluated using cell co-cultures, and the final measurements involved antibody production. The authors demonstrated that lead somehow differentially interferes with antigen presentation to T_{H1} and T_{H2} cell clones; the former were inhibited and the latter were enhanced. They suggested that these effects of lead may involve interleukin receptor density on various lymphocyte populations; the ultimate effect of lead may be a dysfunction of immune responses characteristic of autoimmune disease induction. Overall, the conclusion of various studies suggests a mitogen-like effect of lead on selective cell populations, perhaps via its effects on cell membranes.

C. BIOLOGICAL RESPONSE MODIFIERS

The only T cell-associated biological response modifier that has been evaluated after *in vivo* lead treatment is viral-induced interferon (Blakley et al., 1982). Lead did not alter the production of interferon when T lymphocytes from lead-treated mice were activated by staphylococcal enterotoxin A, Con A, or PHA. To evaluate the *in vivo* production of immune interferon, mice were orally treated with lead for 3 weeks, and 18 h before sampling tilorone was given orally. Again no effect on the production of interferon was observed.

Inhalation exposure of rabbits to lead oxide (30 µg/m^3, 3 h/d for 4 d) altered the biological activity of tumor necrosis factor α (TNF α) produced by pulmonary macrophages (Zelikoff et al., 1993). At 24 hours the LPS-induced TNFα was decreased; however, a significant enhancement was noted after 72 h.

The effects of lead on various immune functions are somewhat paradoxical. Effects were observed only when blood levels clearly approached levels toxic to other body functions. Immune functions were somewhat decreased, but this is probably not related to T cell function. The *in vivo* effects produced on macrophage function apparently are not mediated by interference with interferon production by T cells. *In vitro* effects of lead mimic mitogen responses and are probably due to membrane effects on various immuno-competent cells.

V. MERCURY

A. *IN VIVO* EXPOSURE

Both methylmercury and inorganic mercury salts (e.g., $HgCl_2$) are widespread in the environment and have been demonstrated to exert immunotoxic properties. Male New Zealand white rabbits given $HgCl_2$ (10 ppm) in feed for 70 d had a decrease in resistance to pseudorabies virus (Koller, 1973). In a similar study, Gainer (1977b) demonstrated that as low as 0.01 ppm $HgCl_2$ intensified the mortality of male CD-1 mice in response to encephalomyocarditis virus. A decrease in [^{3}H]thymidine uptake by splenocytes from male B6C3F1 mice treated with 3 to 75 ppm $HgCl_2$ for 7 weeks was observed when the splenocytes were cultured with T cell mitogens (PHA and Con A) (Dieter et al., 1983). There was also a decrease in the mixed leukocyte response (MLR) in the same study, suggesting that T cells are more sensitive to the effects of $HgCl_2$ exposure. The effect of 12 weeks of exposure to methylmercury (3.9 µg/g diet) on the immune functions was studied in female BALB/c mice (Illback, 1991). Increased activity of T and B lymphocytes in response to Con A and LPS and a 44% reduction of NK cell activity were observed. In an earlier study, Gaworski and Sharma (1978) reported a decrease in PHA stimulation of splenocytes in male Swiss mice treated with $HgCl_2$ (0.10 and 0.5 mM in drinking water) and a decrease in pokeweed mitogen (PWM) stimulation observed only at the higher dose (0.5 mM), indicating that T cells are more susceptible to mercury.

B. *IN VITRO* EFFECTS

In vitro exposure to $HgCl_2$ (5 to 20 mM) decreased the response of murine splenocytes to PHA more than the PWM-induced response, suggesting that T cells are the more sensitive cell population (Gaworski and Sharma, 1978). A decreased T cell response was also observed when 0.1 to 10 mM $HgCl_2$ was added to splenocytes for 2 d and activated by Con A (Nakatsuru et al., 1985). A decrease in MLR at low concentrations of mercury was also reported in the same study.

In another study (Warner and Lawrence, 1986), peak proliferation of human peripheral blood lymphocytes (PBL) occurred on day 5 or 6 in response to 30 μM $HgCl_2$. Similar proliferation of splenic and lymph node lymphocytes from BALB/c mice was observed when cultures were treated with 10 μM $HgCl_2$ for 144 h (Reardon and Lucas, 1987). Using purified T cells, the investigators also found that $HgCl_2$ was mitogenic for T cells.

From the above studies, it appears that mercury compounds mainly affect T cells. Very little is known about the kind of T cell types affected and the mechanisms of immunotoxicity. Pelletier et al. (1988) reported that W3/25+ T cells from brown Norway rats exposed to $HgCl_2$ develop autoimmune responses. Their report suggested that activation of autoreactive T cells and suppression of T suppresser cells are responsible for autoimmunity.

It is well known that metals react with thiol groups (Valee and Ulmer, 1972). Oxidation or reduction of thiol or disulfide groups has been shown to modulate the proliferation of murine T cells (Noelle and Lawrence, 1981). Mercury may dimerize the thiol-group bearing proteins on the cell membrane and induce tyrosine phosphorylation and cell death (Rahman et al., 1993). Preliminary studies from these same authors reported that one tyrosine kinase, namely, p56^{lck}, is involved. In another study using methylmercury (0.02 to 2 mM) and $HgCl_2$ (0.01 to 1 mM), free Ca^{2+} levels in lymphocytes from rat spleens were affected in a concentration-dependent manner (Tan et al., 1993). These studies indicated that methylmercury caused a rapid increase in intracellular cell Ca^{2+}, both by increasing its extracellular influx and by increasing the mobilization from intracellular stores, whereas $HgCl_2$ increased Ca^{2+} influx from the extracellular source. Further studies are needed to determine the exact relationship between Ca^{2+} influx and the induction of tyrosine phosphorylation, as well as the kinds of proteins involved in order to elucidate the mechanisms of mercury immunotoxicity.

VI. NICKEL

Nickel is a toxic metal that causes significant pathological conditions in man and animals. Metallic nickel, nickel subsulfide, nickel oxide, and nickel carbonyl cause immunomodulation and hypersensitivity reactions (Sunderman, 1977; Romaguera et al., 1988).

A. *IN VIVO* EXPOSURE

Experimental inhalation exposure of nickel chloride (499.2 μg Ni/m^3) in rats or mice for 2 h reduced the clearance of streptococcal infection and increased the mortality (Adkins et al., 1979). Inhalation of nickel chloride in rabbits (0.3 mg/m^3, 5 d/week, 6 h/d for 1 month) reduced the activity of macrophages against *S. aureus* (Wiernik et al., 1983).

A single dose of nickel chloride (18.3 mg/kg) in male CBA/J mice decreased the number of theta-positive T lymphocytes (Smialowicz et al., 1984a). A significant reduction in the lymphoproliferation response of T lymphocytes against PHA and Con A, but not to the B lymphocyte mitogen, LPS, indicates that T lymphocytes are more susceptible to nickel. Suppression of primary antibody response to the T cell-dependent antigens and NK cell activity were also reported in this study. The NK cell suppression was not due to the increased production of suppresser T cells (Smialowicz et al., 1985) suggesting that it affects other phases of cell-mediated immunity.

Subchronic inhalation exposure (5 d/week for 65 d) of three nickel compounds (Ni_3S_2, $NiSO_4$ and NiO) in B6C3F1 female mice was evaluated for immunotoxic effects (Haley et al., 1990). Among the three compounds, nickel subsulfide (Ni_3S_2, 1.8 mg/m^3) seems to be more immunotoxic than others. Ni_3S_2 suppressed T-dependent antibody production, MLR and NK cell activity, whereas all compounds decreased the phagocytic activity of macrophages. This difference may be due to a slower clearance of Ni_3S_2 in the body than the other compounds. NK cell activity in Fischer 344 rats injected with 10 to 20 mg/kg $NiCl_2$ was also suppressed (Smialowicz et al., 1987b), suggesting that NK cell activity is affected more than other cell types. Similar results were observed in nickel subsulfate-treated cynomologus monkeys (Haley et al., 1987). Nickel-induced suppression of NK-cell activity was antagonized by manganese (Smialowicz et al., 1987a). The mechanism of such antagonism may be induction of α and β interferons by manganese, which enhances NK cell activity or by competition between Mn and Ni for the common surface receptors on NK cells.

B. *IN VITRO* EFFECTS

Pretreatment of lymphocyte cultures with nickel severely inhibited interferon production, while chromium or cadmium had no effect on IFN production (Treapen and Furst, 1970; Pribyl and Treapen,

1977; Hahon et al., 1980). Crystalline NiS (0.1 µg to 20 µg/ml) inhibited the production of α/β interferon induced by Poly I:C in C3H/HI mouse embryo fibroblast cultures. In another study (Jaramillo and Sonnenfeld, 1992), crystalline nickel sulfide (NiS) induced IL-1 and IL-2 production when mice splenocytes were cultured with it.

VII. TIN COMPOUNDS

Because of the industrial application of a variety of alkyltin compounds and the fact that some of these have shown thymotoxic effects in early toxicity studies, several of these compounds have thoroughly been tested for their effect on several rat immune functions. In particular, di-*n*-octyl-tindichloride (DOTC) and di-*n*-butyl tin dichloride (DBTC) have specific toxic effects on the lymphoid organs (Seinen, 1981). Various organotin compounds are extremely cytotoxic; however, after *in vivo* treatment of animals with DOTC and DBTC, the cell-mediated responses are specifically influenced.

A. *IN VIVO* EFFECTS

The organotin compounds DOTC and DBTC did not affect the antibody production to T-independent antigens (LPS), whereas antibody response to T-dependent antigen (SRBC) was greatly reduced (Seinen et al., 1977). A dose-related suppression of tuberculin DTH-response was also reported. Organotin compounds cause inhibition of proliferation in thymus, spleen, or blood lymphocyte cultures in the presence of PHA and Con A, but not in response of LPS (a B cell mitogen). Skin graft rejection and graft- vs. -host reactivity were similarly affected in rats (Seinen et al., 1979).

Inorganic tin compounds, e.g., stannous chloride, also causes depression of immunity, but the effects are nonspecific and also observed in other species like mice. Hayashi et al. (1984) reported on the effects of tin chloride on humoral immune responses in mice. Effects of inorganic tin compounds on T cell functions have not been evaluated.

It is indeed interesting to note that the T cell effects of DOTC and DBTC are limited to rats and perhaps can also be predicted in humans. Neonatal rats gavaged with 5 mg/kg of DBTC 3 times a week from day 2 after birth showed growth stunting and mortality without other pathologic lesions except for severe thymic atrophy (Seinen, 1981). Similar effects are noted in pups born to mothers fed 150 ppm of DOTC in their diet. The effects are similar to those produced by neonatal thymectomy in other species.

B. EFFECTS *IN VITRO*

Of particular interest is the finding that lymphotoxic and immunotoxic effects of organotin compounds are observed in rats, but not in mice, guinea pig, or Japanese quail (Seinen et al., 1977). It is also noteworthy that human thymocytes exhibit a time- and dose-related inhibition of SRBC rosette formation when cultured in the presence of DOTC or DBTC. A similar selective effect of triphenyltin hydroxide on T cell response of rats was also observed (Vos et al., 1984).

VIII. ESSENTIAL TRACE METALS

A. COPPER

Copper is required for oxidative enzymes like catalase, peroxidase, cytochrome oxidase, etc. and is transported in serum mainly via α-ceruloplasmin (Goyer, 1991). Massie et al. (1993) reported that lymphocytes containing excessive serum copper (0.223 to 1.715 ppm) in old mice showed reduced mitogenic responses compared to young mice. Copper deficiency results in decreased humoral and cell-mediated as well as decreased nonspecific immunity by phagocytic cells such as macrophages and neutrophils (Prohaska and Lukasewycz, 1981).

1. *In Vivo* Effects

Phenotypic analysis of lymphocytes from male rats fed a copper-deficient (0.6 µg/g) diet showed a decrease in relative and absolute numbers of T cells and the CD4+ and CD8+ cells, compared to animals fed a copper-adequate diet (6 µg/g) (Bala et al., 1991). These rats also showed reduced (75 to 80%) mitogenic responses to PHA and Con A, indicating that T cells are the most sensitive to copper-deficiency. Excessive oral administration of copper (50 to 300 ppm) in mice had inhibitory effects on Con A-induced lymphocyte proliferation (100 and 200 ppm/8 weeks), and on DTH (at 300 ppm for 5 to 10 weeks) (Pocino et al., 1991).

2. In Vitro Effects

In *in vitro* spleen cultures (Flynn, 1984), mitogenic response to Con A and PHA was depressed by copper depletion of the media, whereas B lymphocyte response to LPS was unaffected. However, in T lymphocyte-enriched culture, proliferation in response to PHA and Con A was unaffected by copper deprivation. The disparity in the responses of the two systems may be due to modulation of lymphokine production by monocytes in the spleen cultures caused by copper deprivation.

3. Biological Response Modifiers

Interleukin-1 levels were unaffected when splenocytes from C57Bl/6 mice were cultured with alloantigen in the absence of copper (Flynn et al., 1984), although the IL-2 levels were relatively low in copper-deficient cultures. But in copper-deficient (0.6 µg/g) rats, IL-2R levels were increased when compared to rats with adequate (6 µg/g) copper diet. Despite normal expression of IL-2R, affinity of IL-2 to its receptor may be decreased in Cu-deficient animals (Bala et al., 1991).

Deficiency of copper may affect copper-dependent enzymes, which have antioxidant activity, thereby increasing the production of free radicals, which may then damage membranes or signal transport during lymphocyte activation (Korte and Prohaska, 1987).

B. IRON

Iron is an essential trace metal in the body that is involved in electron transport and energy production, as well as in cell growth and division. Iron may regulate the relative relationship of CD4+ and CD8+ lymphocytes, and iron overload causes decrease of CD4+ cells in individuals with thalassemia intermedia or major (Pardalos et al., 1987). Abnormally high CD4+ to CD8+ ratios have been shown in subgroups of patients with idiopathic hemochromatosis (Reimao et al., 1989).

1. In Vivo Effects

Cell-mediated cytotoxicity, immunity, and resistance to infection are decreased due to iron deficiency (Ronulund et al., 1982; Dhur et al., 1989). Four weeks of iron overloading (animals fed 2.5% w/w carbonyl iron) in rats stimulated the mitogenic response of lymphocytes to Con A, increased lipid peroxidation, and prostaglandin formation (Wu et al., 1990), but produced no change in response with PHA.

There was no influence of iron overloading (2.5% w/w carbonyl iron) for 4 or 11 weeks on IL-2 production in rats, suggesting that iron toxicity does not affect lymphocytes *in vivo* (Wu et al., 1990). These findings suggest that lymphocytes have both high priority access to iron when the supply is low and high level protection against iron-related toxicity when the supply is in excess (Kemp, 1993).

2. In Vitro Effects

In vitro exposure of human lymphocytes to iron (0.003 to 0.3 mM) enhanced their response to PWM, but suppressed their responsiveness to PHA and Con A at the highest metal concentration. Increasing concentrations (0.005 to 100 mM) of iron citrate caused a moderate inhibition of lymphocyte DNA synthesis and CD8+ cell number in human lymphocytes (Carvalho and Bravo, 1993).

C. SELENIUM

Selenium is an important micronutrient in humans and livestock (Gissel-Nielson et al., 1984) as well as an established component in glutathione peroxidase. Acting as an antioxidant along with vitamin E, it forms part of the cell defense against reactive metabolites of oxygen (Neve, 1989). Selenium affects the functions of almost all components of the immune system, i.e., nonspecific, humoral, and cell-mediated responses (Kiremidjian-Schumacher and Stotzky, 1987; Turner and Finch, 1991). It can augment and/or restore effector mechanisms or mediators of host defenses depending on its concentrations in the body.

1. In Vivo Effects

Guinea pigs supplemented with 1 or 3 ppm of selenium showed significantly more sensitivity to dinitrochlorobenzene (DNCB) than controls, indicating that selenium supplementation augments DTH reaction (Martin and Spallholz, 1977). Selenium-supplemented rats (0.5 to 5 ppm in diet) however, showed an inhibition of the DTH response to bovine serum albumin (Koller et al., 1986) suggesting that immunomodulation by selenium is species-dependent. In a study of Kiremidjian-Schumacher et al. (1990), 6-week-old male (C57Bl/6J) mice were fed a selenium-deficient (0.02 ppm), normal (0.2 ppm), or supplemented (2 ppm) diet for 8 weeks. Mice maintained on the supplemental diet showed an increase in MLR against mitomycin C-treated DBA/2 spleen cells and an increase in proliferation when stimulated with PHA compared to selenium-deficient mice. In a similar study (Roy et al., 1990), selenium-supplemented mice showed an increased CTL activity; however, in another study (Petrie et al., 1989b), mice fed a basal (0.1 ppm) diet for 3 to 4 weeks showed activation of CTL and NK cells when compared to mice fed the supplemented diet (4 ppm).

Many studies have shown that vitamin E can modulate the immunotoxic effects of selenium (Turner and Finch, 1991). Four groups of calves were given Se-supplemented (Se+), Se-deficient (Se–), vitamin E-supplemented (E+), or vitamin E-deficient (E–) diets (Pollock et al., 1994). The respective groups were: I group: E–/Se–; 4.9 nmol g^{-1} DL-α-tocopherol; II group: E+/Se– 0.46 µmol g^{-1} DL-α-tocopherol; III group: E–/Se+, 2.53 nmol Se, as sodium selenite g^{-1}; and the IV group: E+/Se+. The animals were studied for their cellular immune response towards PWM and keyhole limpet-hemocyanin (KLH). The lymphoproliferative response to PWM in the presence of autologous and pooled serum was more in E+/Se– and E+/Se+ groups, whereas the E–/Se+ group responded more to KLH antigen than any other group when cultured in the presence of autologous serum. The response was maximum when cells from Se-supplemented animals were cultured in the presence of serum from calves supplemented with vitamin E, suggesting that interaction between vitamin E and selenium is required for cellular response to KLH. It was inferred that selenium is required on a long-term basis and is stored in the lymphocytes, while vitamin E is more critical in the immediate microenvironment. In another study (Meeker et al., 1985), a marked depression of NK cell and CTL activity was observed in mice fed vitamin E/selenium deficient diet for 8 weeks.

2. *In Vitro* Effects

Human mononuclear peripheral blood lymphocytes (PBL) cultured with 1 to 3 × 10^{-6} M sodium selenite (Na$_2$SeO$_3$) showed a concentration-dependent proliferation with alloantigen; no difference in the cytotoxic activity was observed when compared to the controls (Watson et al., 1986). However, a high concentration (10^{-4} M) of sodium selenite, sodium selenate, and selenium dioxide strongly inhibited NK cell activity, and Con A and PWM induced lymphocyte proliferation at 10^{-5} and 10^{-4} M concentrations. Suppressor cell activity was decreased with increasing selenium, suggesting that immuno-modulating effects of selenium are due to the inhibition of suppressor cell activity (Petrie et al., 1989a).

3. Biological Response Modifiers

Selenium also modifies lymphokine production. Interferon production was enhanced at low levels of selenium dioxide (10^{-9} to 10^{-6} M). Higher concentrations (10^{-5} to 10^{-4} M) inhibited its production in human PBL (Watson et al., 1986). However, levels of IL-1 and IL-2 did not differ in selenium-deficient (0.02 ppm), normal (0.2 ppm), or supplemented (2 ppm) mice, whereas proliferation of lymphocytes with PHA was increased in supplemented animals (Kiremidjian-Schumacher et al., 1990). This suggests that lymphocyte proliferation in selenium-supplemented animals is independent of the levels of IL-2 and IL-1. Supplementation with selenium results in a significant increase in the number of high affinity IL-2-binding sites (kDa 10^{-11} M) on the surface of Con A-stimulated lymphocytes from mice (Roy et al., 1990), suggesting that selenium in the diet alters the kinetics of IL-2R expression. Selenium-supplemented mice (2 ppm in diet) showed enhanced lymphotoxin production (121% increase).

From all of the above studies it appears that selenium augments cell-mediated immune responses, while its deficiency is immunosuppressive. In another study (Sordillo et al., 1993), selenium-supplemented bovine PBL had no difference in IL-2 and IL-2R expression compared to cells from animals fed deficient diets.

D. ZINC

Zinc is an essential metal and its deficiency results in severe nutritionally related health problems. Zinc toxicity is uncommon as it is eliminated effectively from the body. In both humans and animals, zinc deficiency causes rapid and severe depression of immune functions (Fraker et al., 1986). Large numbers of enzymes associated with DNA and RNA synthesis are dependent on zinc for function. Furthermore, zinc is essential for adequate membrane function, protein synthesis, and association of nuclear proteins with DNA (Prasad, 1984). Thus, zinc may affect the proliferation of lymphocytes and the production of cytokines.

1. *In Vivo* Effects

Immunosuppression and increased susceptibility to infection are important features of Zn deficiency. Lambs fed marginal (basal diet + 5 mg zinc/kg) diet for 58 d showed a decreased number of lymphocytes and increased susceptibility to *Pasteurella hemolytica* (Droke et al., 1993). Zinc-deficient CD-1 mice were not protected against encephalomyocarditis virus (Gainer, 1977b).

Zinc-deficient animals showed thymic atrophy and lymphophenia, but no alteration in T lymphocytes functions. When A/J mice were provided Zn-deficient or Zn-adequate diets for 30 d (Cook-Mills and Fraker, 1993), splenocytes from deficient animals showed a higher degree of proliferation (150%) compared to mice provided a Zn-adequate diet. However, lymphocyte number was diminished in Zn-deficient animals. In another study, Zn-deficient C57Bl/KS female mouse spleen cells showed reduced cytotoxic lymphocytes (CTL), NK cell and ADCC activity, and a decrease in the proportion of T lymphocytes (Fernandes et al., 1979).

2. *In Vitro* Effects

In vitro Zn (25 to 200 μM) increased [^{3}H]thymidine incorporation into lymph node lymphocytes, as well as their cytotoxicity (Reardon and Lucas, 1987). Enriched T cells from Nu/Nu mice showed a significant increase in the mitogenic response to zinc. Adherent cells were also important for activation of Zn-activated cytotoxic cells. Zinc (2×10^{-5} to 2×10^{-4} M) caused a suppression of the proliferative response of lymphocytes against allogeneic HeLa cells, whereas stimulation was observed with PHA, Con A, and 12-*O*-tetradecanoyl phorbol-acetate (TPA). Peak stimulation of human peripheral lymphocytes was observed with 100 μM ZnCl$_2$, and only T cells were responsive (Warner and Lawrence, 1986). In another study (Rao et al., 1979), zinc (50 to 200 μM) decreased proliferation of human lymphocytes activated by PHA or Con A. A decrease in mouse NK cell activity was reported when 0.5 to 1 mM of zinc was added (Ferry and Donner, 1984).

3. Biological Response Modifiers

In a study by Reardon and Lucas (1987), splenic lymphocytes from BALB/c mice stimulated with 10 to 200 mM zinc did not induce IFN production. Zinc also did not inhibit the Con A-induced production of IFN. In elderly people, zinc deficiency produced a decrease in IL-1 production from lymphocytes, which was restored when the deficiency was corrected (Prasad et al., 1993). Increase in the production of IL-2 was observed in Zn-deficient mice (Cook-Mills and Fraker, 1993) and rats (Dowd et al., 1986) when compared to animals fed a Zn-adequate diet.

IX. MISCELLANEOUS METALS

Dietary aluminum (Al), 1000 μg/g diet from conception to 6 months of age in Swiss Webster mice, depressed CD4+ cells, IL-2, IFN-γ and TNF levels in splenic lymphocytes (Golub et al., 1993).

Chromium (Cr) is one of the transition metals causing toxicity, cancer, and delayed-type hypersensitivity at higher metal concentrations (Cohen et al., 1993; Ischii et al., 1993). Hexavalent chromium had stimulating effects at low concentrations (10^{-8} to 10^{-6} M) and inhibitory effects at higher concentrations (10^{-6} to 10^{-4} M) *in vitro* when human lymphocytes were activated with PHA (Borella et al., 1990). However, in other *in vitro* studies with human lymphocytes, T lymphocytes were not affected with Cr^{VI+} or Cr^{III+} (Bravo et al., 1990). In another study with cobalt-chromium alloy or titanium-aluminum-vanadium particles which are used in prosthetics, immune regulation in rats was evaluated (Haynes et al., 1993). Exposure to titanium-aluminum-vanadium increased prostaglandin E2, IL-1, TNF, and IL-6, whereas cobalt-chromium particles decreased PGE$_2$ and IL-6.

Cobalt (CO^{2+}) inhibited the E-rosette reaction and CD2+ lymphocyte cells *in vitro* in human lymphocytes (Bravo et al., 1990).

Manganese (Mn), a ubiquitous trace metal and a product of numerous mining and industrial processes, causes various types of toxicities. A single injection of 40 to 50 μg/MnCl$_2$/g body weight in mice produced enhanced NK cell activity and IFN levels (Rogers et al., 1983; Smialowicz et al., 1984b). The addition of MnCl$_2$ (10 to 20 μg per culture) to spleen cultures in mice produced enhanced NK cell activity and interferon α and β levels, indicating that manganese may augment cell-mediated immunity (Smialowicz et al., 1986). Manganese was shown to augment the conjugation of IL-2-activated lymphocytes to tumor cells through activation of the leukocyte function-associated antigen-1 (LFA-1) which is needed for proper adhesion and T cell activation (Jameson et al., 1993).

X. CONCLUSIONS

Based on the above review it appears that many heavy metals affect cell-mediated immunity, and that T cells are relatively more sensitive to their effects. In the case of essential trace metals, both a deficiency and an excess can result in modification of the immune responses. Effects of metals may be associated with membrane alterations, modifications in signal transduction, displacement of essential metals, or in some cases, with interactions with cellular proteins or enzymes. Few studies have investigated the molecular mechanisms involved in metal-induced immunotoxicity. Some heavy metals like nickel cause hypersensitivity. In most cases of metal exposure, effects on the immune system are observed at exposure levels that also cause other systemic effects. The mechanisms of heavy metal effects on T cell functions and biological response modifiers are not fully investigated. Because of the environmental importance of heavy metals and their widespread biological effects, the mechanistic studies have great importance and should be persued.

REFERENCES

Adkins, B., Richards, J. H., and Gardner, D. E. (1979), *Environ. Res.*, 20, 30–42.
Bala, S., Lunney, J. K., and Failla, M. L. (1991), *J. Nutr.*, 121, 745–753.
Blakley, B. R. (1985), *Can. J. Comp. Med.*, 49, 104–108.
Blakley, B. R. and Archer, D. L. (1981), *Toxicol. Appl. Pharmacol.*, 61, 18–26.
Blakley, B. R. and Tomar, R. S. (1986), *Int. J. Immunopharmacol.*, 8, 1009–1015.
Blakley, B. R., Archer, D. L., and Osberne, L. (1982), *Can. J. Comp. Med.*, 46, 43–46.
Borella, P., Manni, S., and Giardiuo, A. (1990), *J. Trace Elem. Electrolytes Health Dis.*, 4, 87–95.
Bozelka, B. E. and Burkholder, P. M. (1982), *Environ. Res.*, 27, 421–432.
Bravo, I., Carvalho, G. S., Barbosa, M. A., and Desousa, M. (1990), *J. Biomed. Mater. Res.*, 24, 1059–1068.
Bryan, C. F. and Leecho, S. H. (1983), *Cell. Immunol.*, 75, 71–79.
Carvalho, G. S. and Bravo, I. (1993), *J. Mater. Sci.-Mater. Med.*, 4, 366–311.
Ciavarra, R. P. and Simeone, A. (1990), *Cell. Immunol.*, 131, 11–26.
Cohen, M. D., Kargdelin, B., Kleus, C. B., and Costa, M. (1993), *Crit. Rev. Toxicol.*, 23, 255–281.
Cook-Mills, J. M. and Fraker, P. J. (1993), *Br. J. Nutr.*, 69, 835–848.
Descotes, J., Evreux, J. C., Laschi-Locquerie, A., and Tachon, P. (1984), *J. Appl. Toxicol.*, 4, 265–266.
Dhur, A., Gralan, P., and Hercberg, S. (1989), *Comp. Biochem. Physiol.*, 94, 11–19.
Dieter, M. P., Luster, M. I., Boorman, G. A., Jameson, C. W., Dean, J. H., and Cox, J. W. (1983), *Toxicol. Appl. Pharmacol.*, 68, 218–227.
Dowd, P. S., Kelleher, J., and Giuillore, P. J. (1986), *Br. J. Nutr.*, 55, 59–69.
Droke, E. A., Spears, J. W., Brown, T. T., and Qureshi, M. A. (1993), *Nutr. Res.*, 13, 1213–1226.
Fernandes, G., Nair, M., Ohoe, K, Tanaka, T., Floyd, R., and Good, R. A. (1979), *Proc. Natl. Acad. Sci. U.S.A.*, 76, 457–461.
Ferry, F. and Donner, M. (1984), *Scand. J. Immunol.*, 192, 435–445.
Fletcher, M. C., Samelson, L. E., and June, C. H. (1993), *J. Biol. Chem.*, 268, 23697–23703.
Flynn, A. (1984), *J. Nutr.*, 114, 2034–2042.
Flynn, A., Margaret, A. L., and Finke, J. H. (1984), *Nutr. Res.*, 4, 673–679.
Fraker, P. J., Gershwin, M. E., Good, R. A., and Prasad, A. (1986), *Fed. Proc.*, 45, 1474–1479.
Gainer, J. H. (1972), *Am. J. Vet. Res.*, 33, 2579–2586.
Gainer, J. H. (1977a), *Am. J. Vet. Res.*, 38, 863–867.
Gainer, J. H. (1977b), *Am. J. Vet. Res.*, 38, 869–872.
Gainer, J. H. and Pry, T. W. (1972), *Am. J. Vet. Res.*, 33, 2299–2307.
Gaworski, C. L. and Sharma, R. P. (1978), *Toxicol. Appl. Pharmacol.*, 46, 305–313.
Gissel-Nielson, G., Gupta, U. C., Lamand, M., and Westermark, T. (1984), *Adv. Agron.*, 37, 397–460.
Golub, M. S., Takeuch, P. T., Gershwin, M. E., and Yoshida, S. H. (1993), *Immunopharmacol. Immunotoxicol.*, 15, 605–619.
Goyer, R. A. (1991), *Toxic Effects of Metals in Toxicology* Amdur, M. O., Doull, J., and Klaassen, C. D., Eds., Pergamon Press, New York, 653–655.
Hahon, N., Booth, J. A., and Pearson, D. J. (1980), *The In Vitro Effects of Mineral Dusts,* Academic Press, New York, 219–228.
Haley, P. J., Rice, D. E., Muggenburg, B. A., Hahn, F. F., and Benjamin, S. A. (1987), *Toxicol. Appl. Pharmcol.*, 88, 1–12.
Haley, P. J., Shopp, G. M., Benson, J. M., Chang, Y. S., Bice, D. E., Luster, M. I., Dunnick, J. K., and Hobbs, C. H. (1990), *Fundam. Appl. Toxicol.*, 15, 476–487.
Hayashi, O., Chiba, M., and Kikuchi, M. (1984), *Toxicol. Lett.*, 21, 279–285.
Haynes, D. R., Rogers, S. D., Hayes, S., Pearcy, M. J., and Howe, D. W. (1993), *J. Bone Joint Surg. Am.*, 75, 825–834.
Hemphill, F. E., Kaeberle, M. L., and Buck, W. B. (1971), *Salmonella typhimurium, Science,* 172, 1031–1032.
Illback, N. A. (1991), *Toxicology,* 67, 117–124.

Ischii, N., Takahashi, K., Kawaguichi, H., Nakajima, H., Tanaka, S., and Aoki, I. (1993), *Int. Arch. Allerg. Immunol.*, 100, 333–337.
Jameson, A. M., Alexandroff, A. B., Lappin, M. B., Esuvaranathainin, K., James, K., and Chisholm, G. D. (1993), *Immunology,* 81, 120–126.
Jaramillo, A. and Sonnerfeld, G. (1992), *Oncology,* 49, 396–406.
Kemp, J. D. (1933), *J. Clin. Immunol.*, 13, 84–92.
Kerkvliet, N. I. and Baecher-Steppan, L. (1982), *Immunopharmocology,* 4, 213–224.
Kiremidjian-Schumacher, L. and Stotzky, G. (1987), *Environ. Res.,* 42, 277–303.
Kiremidjian-Schumacher, L., Roy, M., Wishe, H. I., Cohen, M. W., and Stotzky, G. (1990), *Proc. Soc. Exp. Biol. Med.,* 193, 136–142.
Koller, L. D. (1973), *Am. J. Vet. Res.,* 34, 1457–1458.
Koller, L. D. and Vos, J. G. (1981), in *Immunologic Considerations in Toxicology,* Vol. 1, Sharma, R. P., Ed., CRC Press, Boca Raton, 67–78.
Koller, L. D., Exon, J. H., and Roan, J. G. (1976), *Proc. Soc. Exp. Biol. Med.,* 151, 339–342.
Koller, L. D., Exon, J. H., Talkott, P., Osborne, C. A., and Henningren, G. M. (1986), *Clin. Exp. Immunol.,* 63, 570–576.
Korte, J. and Prohaska, J. (1987), *J. Nutr.,* 117, 1076–1084.
Lawrence, D. A. (1981a), *Infect. Immun.*, 31, 136–143.
Lawrence, D. A. (1981b), in *The Handbook of Cancer Immunology,* Vol. 8, *Tumor Antigens: Structure and Functions,* (Watey, H., Ed.), Garland STPM Press, New York, 295–303.
Liao, K., Hoffman, R. D., and Lane, M. D. (1991), *J. Biol. Chem.,* 266, 6544–6553.
Lynes, M. A., Garvey, J. S., and Lawrence, D. A. (1990), *Mol. Immunol.,* 27, 211–219.
Martin, J. L., and Spallholz, J. E. (1977), Selenium in the immune response, in Proc. Symp. Se-Te in the Environment, Industrial Health Foundation, Pittsburgh, PA, 204–225.
Massie, H. R., Obosuappiah, W., and Aiello, V. R. (1933), *Gerontology,* 39, 136–145.
McCabe, M. J., Jr. and Lawrence, D. A. (1991), *Toxicol. Appl. Pharmacol.,* 111, 13–23.
McCabe, M., Maguire, D., and Nowak, M. (1983), *Environ. Res.,* 31, 323–331.
Meeker, H. C., Eskew, M. L., Scheuchenzuber, W., Scholz, R. W., and Zarkower, A. (1985), *J. Leukoc. Biol.* 38, 451–458.
Morselt, A. F. W., Leene, W., De Groot, C., Kipp, J. B. A., Evers, M., Roelofsen, A. M., and Bosch, K. S. (1988), *Toxicology,* 48, 127–139.
Nakatsuru, S., Oohashi, J., Nozaki, H., Nakada, S., and Imura, N. (1985), *Toxicology,* 36, 297–305.
Neilan, B. A., O'Neill, K., and Handwerger, B. S. (1983), *Toxicol. Appl. Pharmacol.,* 69, 272–275.
Neve, J. (1989), in *Selenium in Medicine and Biology,* Neve, J. and Favier, A. Eds., Walter de Gruyter, New York, 97–111.
Noelle, R. J. and Lawrence, D. A. (1981), *Cell. Immunol.,* 60, 453–469.
Pardalos, G., Kanakoudi-Tsakalidis, F., Malaka-Zafiriu, M., Tsantali, H., Athanasiou-Metaxa, M., Kallinikos, G., and Papaevangelou, G. (1987), *Clin. Exp. Immunol.,* 68, 138–145.
Pelletier, L., Pasquier, R., Rossert, J., Vial, M. C., Mandet, C., and Druet, P. (1988), *J. Immunol.,* 140, 750–754.
Petrie, H. T., Klassen, L. W., and Kay, H. D. (1989a), *J. Leukoc. Biol.,* 45, 207–214.
Petrie, H. T., Klassen, L. W., Klassen, P. S., Odell, J. R., and Kay, H. D. (1989b), *J. Leukoc. Biol.,* 45, 215–220.
Pocino, M., Baute, L., and Malone, I. (1991), *Fundam. Appl. Toxicol.,* 16, 249–256.
Pollock, J. M., McNair, J., Kennedy, S. Kennedy, D. G., Walsh, D. M., Goodall, E. A., Mackie, D. P., and Crockard, A. D. (1994), *Res. Vet. Sci.,* 56, 100–107.
Prasad, A. (1984), *Fed. Proc.,* 43, 2829–2834.
Prasad, A. S., Fitzgerald, J. T., Hess, J. W., Kaplan, J., Pelen, F., and Dardenne, M. (1993), *Nutrition,* 9, 218–224.
Pribyl, D. and Treapen, L. (1977), *Acta Virol.,* 21, 507.
Prohaska, J. R. and Lukasewycz, O. A. (1981), *Science,* 213, 559–561.
Rahman, S. M. J., Du, M.-Y., Hamaguchi, M., Iwamito, T., Isobe, K.-I., and Nakashima, I. (1993), *FEBS Lett.,* 317, 35–38.
Ramsten-Wesenberg, G. B. and Wesenberg, F. (1983), *Environ. Res.,* 31, 413–419.
Rao, K. M. K., Schwartz, S. A., and Good, R. A. (1979), *Cell Immunol.,* 42, 270–278.
Reardon, C. L. and Lucas, D. O. (1987), *Immunobiology,* 175, 455–469.
Reimao, R., Proto, G. and Desousa, M. (1989), *C.R. Acad. Sci. Paris,* 313, 481–488.
Rogers, R. R., Garner, R. J., Riddle, M. M., Luebke, R. W., and Smialowicz, R. J. (1983), *Toxicol. Appl. Pharmacol.,* 70, 7–17.
Romaguera, C., Grimalt, F. and Vilaplana, J. (1988), *Contact Dermatol.,* 19, 52–57.
Ronulund, R. D., Rohner, S. D., Sliga, S., and Suskind, R. M. (1982), *Fed. Proc.,* 1982, Abstr. 422.
Roy, M., Kiremidjian-Schumacher, L., Wishe, H. I., Cohen, M. W., and Stotzky, G. (1990), *Proc. Soc. Exp. Biol. Med.,* 193, 143–148.
Seinen, W. (1981), in *Immunological Considerations in Toxicology,* Vol. 1, R. P., Sharma, Ed., CRC Press, Boca Raton, FL, 103–119.
Seinen, W., Vos, J. G., Brands, R., and Hooykaas, H. (1979), *Immunopharmacology,* 1, 343–355.
Seinen, W., Vos, J. G., Spanje, I., Snoek, M., Brands, R., and Hooykaas, H. (1977), *Toxicol. Appl. Pharmacol.,* 42, 197–212.
Selye, H. B., Tuchweber, B., and Bertok, L. (1966), *J. Bacteriol.,* 91, 884–890.
Sharma, R. P. and Zeeman, M. G. (1980), *J. Immunopharmacol.,* 2, 285–307.
Smialowicz, R. J., Rogers, R. R., Riddle, M. M., and Stott, G. A. (1984a), *Environ. Res.,* 33, 413–427.

Smialowicz, R. J., Rogers, R. R., Riddle, M. M., Garner, R. J., Rowe, D. G., and Luebke, R. W. (1985), *Environ. Res.,* 36, 56–66.
Smialowicz, R. J., Rogers, R. R., Riddle, M. M., Leubke, R. W., Rowe, D. G., and Garner, R. J. (1984b), *J. Immunopharmacol.,* 6, 1–23.
Smialowicz, R. J., Rogers, R. R., Riddle, M. M., Luebke, R. W., and Fogelson, L. D. (1987a), *J. Toxicol. Environ. Health,* 20, 67–80.
Smialowicz, R. J., Rogers, R. R., Riddle, M. M., Rowe, D. O., and Luebke, R. W. (1986), *J. Toxicol. Environ. Health,* 19, 243–254.
Smialowicz, R. J., Rogers, R. R., Rowe, D. G., Riddle, M. M., and Luebke, R. W. (1987b), *Toxicology,* 44, 271–281.
Sordillo, L. M., Hicres, C. R., Wilson, R., and Maddox, J. (1993), *J. Vet. Med. Ser. A.,* 40, 615–623.
Sunderman, F. W., Jr. (1977), *Ann. Clin. Lab. Sci.,* 7, 377–398
Tan, X-X., Tang, C., Castoldi, A. F., Manzo, L., and Costa, L. G. (1993), *J. Toxicol. Environ. Health,* 38, 159–170.
Treapen, L. and Furst, A. (1970), *Res. Commun. Chem. Pathol. Pharmacol.,* 1, 395–402.
Turner, R. and Finch, J. M. (1991), *Proc. Nutr. Soc.,* 50, 275–285.
Vallee, B. L. and Ulmer, D. D. (1972), *Annu. Rev. Biochem.,* 41, 91–128.
Vos, J. G., Van Logten, M. J., Kreeftenberg, J. G., and Kruizinga, W. (1984), *Toxicology,* 29, 325–336.
Warner, G. L. and Lawrence, D. A. (1986), *Fed. Proc.,* 46, 1322.
Watson, R. R., Moriguchi, S., McRae, B., Tobis, L., Mayberry, J. C., and Lucas, D. (1986), *J. Leukoc. Biol.,* 39, 447–457.
Wiernik, A., Johnson, A., Jarstrand, C., and Camner, P. (1983), *Environ. Res.,* 30, 129–141.
Wu, W-H., Meydani, M., Meydani, S. N., Burklund, P. M., Blumberg, J. B., and Munro, H. N. (1990), *J. Nutr.,* 120, 280–289.
Yoshida, T., Shimamura, T., and Shigata, S. (1987), *Int. J. Immunopharmacol.,* 9, 411–415.
Zelikoff, J. T., Parsons, E., and Schlesinger, R. B. (1993), *Environ. Res.,* 62, 207–222.

Chapter 50

Effects of Metals on the Humoral Immune Response

Jerry H. Exon, Elizabeth H. South, and Kathleen Hendrix

I. INTRODUCTION

The effects of metals on the humoral arm of the immune system are reviewed in this chapter. Most of the information presented will focus on studies conducted during the last 10 years, since data previous to that time have been the subject of a number of excellent reviews (Koller, 1982, 1985; Exon and Koller, 1986; Kiremidjian-Schumacher and Stotzky, 1987; Gleichmann et al., 1989; Dhur et al., 1990; Keen and Gershwin, 1990). The effects of metals on several parameters of humoral immunity will be presented. These include effects on antibody-forming B cells, antibody titers, antibody-dependent cell-mediated cytotoxic (ADCC) responses, B cell proliferation/differentiation, and production/secretion of humoral factors (cytokines) which may regulate humoral or cell-mediated immune responses. Several of the metals discussed in this chapter are essential trace minerals, such as zinc, copper, and selenium. These essential metals have deficiency syndromes associated with them that involve in all instances serious immune dysfunction. The specific effects of metal deficiencies and reconstitution studies on immunity will not be discussed in any detail here, but there are extensive reviews on this subject (Chandra, 1985; Chandra and Chandra, 1986; Kiremidjian-Schumacher and Stotzky, 1987; Fletcher et al., 1988; Spallholz and Stewart, 1989; Sherman, 1992). Discussions here will be limited to supplementation studies in animals with presumed adequate levels of essential trace metals present.

II. ARSENIC (As)

The majority of the literature on the effects of As on immune function indicate that this metal is immunosuppressive. A suppressive effect of As on primary IgM and secondary IgG plaque-forming cell (PFC) responses was reported by Blakley et al. (1980). This effect was seen in male Swiss mice exposed to 0.5, 2.0, and 10.0 ppm sodium arsenite in the drinking water for 3 weeks. A clear dose response was not evident in that study. Sikorski et al. (1989) compared the effects of different As compounds on immune function in mice. They observed that B6C3F1 female mice exposed to the industrial chemical, gallium arsenide (GaAs), intratracheally at single doses ranging from 50 to 200 mg/kg also had reduced primary (IgM) and secondary (IgG) PFC responses 14 days after treatment. The IgM PFC response was reduced to the T-dependent antigen sheep red blood cells (SRBC), but not the T-independent antigen, dinitrophenol-Ficoll (DNP-Ficoll). The IgG PFC response to GaAs was also significantly decreased at all doses. In the same study, intratracheal treatment of mice with 10 mg/kg sodium arsenite decreased the IgM PFC by 24%, but had no effect on the IgG PFC response. Lipopolysaccharide (LPS)-induced mitogen responses were increased in GaAs-exposed mice, but not in a dose-related manner. There were also dose-related decreases in splenic B cell numbers and dose-related increases in C_3 complement levels at the higher doses. In contrast, sodium arsenite treated mice had decreased C_3 and C_5 serum levels. It appears that the form of As may be important in producing certain effects on immune function. In a subsequent study, Sikorski et al. (1991) reported that splenocytes of B6C3F1 mice which had been treated with a single intratracheal dose of GaAs *in vivo* had reduced *in vitro*-generated IgM PFC responses to the T-dependent antigen, SRBC, and the T-independent antigen, trinitrophenol-Ficoll (TNP-Ficoll). They also reported that PFC responses were decreased in mice treated i.p. with GaAs. Other effects noted were decreases in spleen cellularity and splenocyte B cell populations. Mitogen responses in these GaAs treated mice were not affected. The authors also studied the effects of different combinations of reconstituted cell populations from GaAs- or vehicle-treated mice. They concluded that the suppression of the IgM PFC response was due to adherent cells that were not suppressor macrophages. The suppression was also not due to release of prostaglandins.

Burns et al. (1991) also studied the effects of GaAs on the *in vitro* IgM antibody response in splenocytes of B6C3F1 mice. They found that GaAs at doses of 6.5, 13, 26, and 78 μM inhibited the *in vitro* primary IgM PFC response to SRBC in a dose-dependent manner. Sodium arsenite at a dose of 1.0 μM, but not lower, also suppressed the PFC response. The timing of exposure to GaAs was important. Suppression of the PFC response occurred when GaAs was added within 24 h after SRBC immunization *in vitro*. The same doses of GaAs given 48, 72, or 96 h after SRBC had no effect on the PFC response. The effects of GaAs on the PFC response were blocked by an arsenic chelator but not by a gallium chelator. The authors concluded that arsenic was the major immunosuppressive agent in GaAs, and the compound induced its effects within the first 36 h of a 5-day culture period for the *in vitro* IgM response. Burns et al. (1993) reported that *in vivo* exposure to GaAs significantly reduced splenic IgM PFC responses in B6C3F1 mice. The GaAs was administered intratracheally at a dose of 200 mg/kg. The IgM PFC response in these mice was reduced 89% compared to controls. Treatment of splenocytes of GaAs exposed mice *in vitro* with the metal chelator; meso, 2–3-dimercaptosuccinic acid (DMSA) failed to reverse the suppression of IgM PFC responses. However, a synthetic form of DMSA, L-cysteine DMSA, was able to reverse the *in vivo* or *in vitro* GaAs-induced PFC suppression in a dose-dependent manner. This was further evidence that As and not Ga was the immunosuppressive agent.

In a subsequent study, Burns and Munson (1993) reported that suppression of antibody responses after *in vivo* exposure to GaAs could be reversed by addition of supernatants from splenocytes of vehicle-treated mice and that this effect was dependent on proteins of 5K to 50K in the cell cultures. They further demonstrated that lymphokine (IL-2, -4, -5, and -6) concentrations were altered in splenocyte cultures of GaAs-treated mice. These alterations differed according to GaAs exposure. The GaAs-induced suppression of the IgM PFC *in vitro* could be reversed in a dose-dependent manner by the addition of IL-2, and this effect was both time and dose dependent. The addition of IL-4 to GaAs-suppressed cultures failed to reverse the effect. Both IL-5 and IL-6 were able to reverse the suppressive effects of GaAs, but not in a dose-related manner. The authors conclude that altered antibody production in GaAs-exposed mice may be due to changes in lymphokine production, and IL-2 may be a primary target.

III. CADMIUM (Cd)

The effects of Cd on immune function (including humoral) have been reviewed by Exon and Koller (1986). Depending on the study, Cd exposures have been reported to stimulate, suppress, or have no effect on humoral immune responses. Studies performed more recently have not clarified the previous disparities in regard to humoral immunity. There is a trend, however, which suggests that low doses of Cd in young adult animals are stimulatory to the PFC response, while high doses in older animals are more often related to suppression. Some data have been more difficult to interpret, however, since actual intake of Cd was not documented.

Thomas et al. (1985) reported that primary IgM PFC responses to SRBCs were unaffected in B6C3F1 female mice exposed to 10, 50, or 250 ppm $CdCl_2$ in the drinking water for 90 days, even in the presence of increasing Cd levels in the spleen. Conversely, mitogenic responses to LPS of splenocytes from Cd-treated mice were significantly impaired but not in a dose-related manner.

Chowdhury et al. (1987) studied the effect of $CdCl_2$ and $ZnCl_2$ on primary and secondary PFC response to SRBC in male C57BL/6 mice. The mice were exposed to 50 ppm $CdCl_2$ or 500 ppm $ZnCl_2$ via the drinking water, either separately or concomitantly for 3 weeks. The PFC responses were assessed at 0, 3, and 6 weeks after treatment. Mice exposed to $CdCl_2$ for 3 weeks had elevated IgM and IgG PFC responses. Mice treated with the combination of $CdCl_2$ and $ZnCl_2$ had PFC responses similar to controls. No significant effects were observed at 3 or 6 weeks after treatment. An enhancement of B cell function was also observed in Balb/c mice treated with 0.11, 0.33, or 1 mg Cd as $CdCl_2$/kg BW daily by s.c. injection for 5 weeks (Hurtenbach et al., 1988). These mice had augmented IgM and IgG PFC responses to a T-dependent antigen, Protein A.

Borgman et al. (1986) treated CD1 male mice with 50 ppm Cd ($CdCl_2$) in the drinking water for 3 weeks. The mice were sacrificed at 0, 3, or 6 weeks after completion of Cd exposure, and selected immune functions were assessed. The primary IgM PFC response was suppressed in Cd mice at 5 but not 7 days after Cd exposure. No effects were reported at any other time period. Serum IgG levels were not affected. Similar effects were observed by Krzystyniak et al. (1987) after short-term inhalation exposure to $CdCl_2$ and longer-term oral treatment in female C57BL/6 mice. Mice treated by nose only inhalation exposure with 0.88 mg Cd/m^3 for 60 min had significantly reduced IgM PFC responses to the T-dependent antigen SRBC for up to 18 d following exposure. The effect seemed to be correlated to reduced splenocyte viability in the Cd-treated mice. Mice treated for 12 weeks with 3 ppm $CdCl_2$ in the drinking water also had reduced PFC responses, but splenocyte viability was no different than controls. The LPS-induced lymphoproliferative response was also reduced in the inhalation-exposed mice. Blakley (1985) also reported that splenic PFC responses were suppressed following *in vivo* exposure of BDF_1 male mice to $CdCl_2$ in the drinking water for 3 weeks. Mice treated with 5, 10, or 50 ppm $CdCl_2$ had dose-dependent decreases in the primary IgM PFC response to SRBC. On the other hand, lymphocyte blastogenesis induced by LPS was significantly enhanced in these mice in a dose-dependent manner. Blakley and Tomar (1986) further investigated the effect of $CdCl_2$ on the humoral arm of the immune system in mice by comparing responses to different antigens. Female CD1 mice exposed to 5, 10, or 50 ppm $CdCl_2$ in the drinking water had enhanced splenic PFC response to DNP-Ficoll, a T cell-independent/macrophage (Mφ)-dependent antigen. A similar enhancement was seen in PFC responses to the T-independent/Mφ-independent antigen, LPS, but only at the two higher doses of $CdCl_2$. Conversely, the PFC response to the T cell-dependent/Mφ-dependent antigen SRBC was suppressed in $CdCl_2$ treated mice. There was no clear dose response observed in any of these studies. Further studies examined the *in vitro* SRBC PFC response in BDF_1 mice splenocytes. Using selective depletion/repletion studies with Cd-treated or untreated populations of cells, the authors showed that the Cd-induced suppression of the SRBC PFC response was indirectly due to effects on T lymphocytes.

Burchiel et al. (1987) reported a time-dependent suppression of the *in vitro* primary (IgM) antibody response to SRBC, TNP-LPS, or TNP-Ficoll in splenocytes of B6C3F1 mice treated with a single injection of 0.9 mg/kg Cd acetate. The mitogenic response to LPS was also suppressed in a time-dependent manner in splenocytes of Cd-treated mice. In another study, it was noted that the *in vitro* antibody response to the T-independent antigen DNP-Ficoll was significantly enhanced at low doses of $CdCl_2$ (8 μM) but suppressed at higher doses (20 to 40 μM) in splenocyte cultures from male Balb/c mice (Fujimaki, 1985b). They also found the primary IgM antibody response to SRBC was suppressed in splenocytes of Balb/c mice given a single i.p. injection of 1.8 mg $CdCl_2$/kg BW (Fujimaki, 1985a). *In vitro* reconstitution studies showed the major effect was on the B cell population as opposed to Mφ or T cells. In a subsequent study, Fujimaki (1987) compared the effects of Cd *in vitro* on IgM PFC responses of young (3 week) and adult (5 month) male Balb/c mice. He reported that exposure of splenocytes to 4 μM $CdCl_2$ stimulated the primary PFC response in young and adult animals. Conversely, treatment of splenocytes from young or old mice with 40 μM $CdCl_2$ resulted in significant suppression of the PFC response. A reduction in B cell viability was also reported for splenocyte cultures exposed to 40 μM $CdCl_2$. Blakley (1988) reported that the primary PFC response to SRBC in older (12 month) BDF_1 mice was not suppressed by treatment with 10 or 50 μg/l $CdCl_2$ in the drinking water for 26 d. It was surmised that there was already an age-related suppression of humoral immunity in these animals which masked the suppressive effect of $CdCl_2$ treatment seen in younger animals.

Differences have been observed on the effects of *in vivo* and *in vitro* treatment with Cd on mitogenic responses in different strains of mice (C3H/He, Balb/c, DBA/2) (Ohsawa et al., 1986). Splenocytes from C3H mice treated *in vivo* with Cd (0.5 or 1.0 mg/kg/d) for 5 d had depressed responses to LPS stimulation. The other two strains of mice were resistant to Cd-induced changes following *in vivo* treatment. Following *in vitro* treatment of splenocytes of different mouse strains with Cd, the DBA mice were the most sensitive to suppression. Unlike *in vivo* exposure results, C3H mice splenocytes treated with Cd *in vitro* were the most resistant of the strains to effects on mitogenic stimulation. Daum et al. (1993) also studied the *in vitro* effects of $CdCl_2$ on B lymphocyte activation of DBA/J2 mice. They reported B cell RNA and DNA synthesis was inhibited in splenocytes treated with $CdCl_2$ in cultures at levels above 3 μM.

This effect occurred within the first 2 hours of metal exposure. LPS stimulated production of Ig isotypes was also inhibited in Cd-exposed cells; this inhibitory effect varied markedly between isotypes. Expression of B cell surface antigens, class I and II MHC, as stimulated with LPS, was also decreased by Cd treatment. The authors conclude that Cd exerts early inhibitory effects on B cell activation. At least one study has implicated Cd exposure in autoimmune activity. Ohsawa et al. (1988) reported that serum antinuclear antibodies (ANA) were increased in a dose-dependent manner in ICR mice treated with 3, 30, or 300 ppm $CdCl_2$ in drinking water for 10 weeks. Splenocytes of these mice also had enhanced nonspecific direct PFC responses to SRBC. Mice that were primed with SRBC and treated with Cd had suppressed PFC responses at the 300 ppm dose. The authors concluded that lower doses of Cd exposure could cause the induction of ANA and nonspecific stimulation of antibody production.

At least two investigators have studied the effects of Cd treatment on antibody-dependent cell-mediated cytotoxicity (ADCC). Chopra et al. (1984) reported that male Wistar rats treated with Cd (5 mg/kg BW/d) by gavage for 7 weeks had enhanced splenocyte ADCC responses to challenge with chicken RBC. Conversely, Stacey (1986) reported that human PBL exposed to 25 to 250 μM $CdCl_2$ *in vitro* had reduced ADCC responses. Cotreatment with $ZnSO_4$ at doses of 100 to 250 μM had an additional additive suppressive effect on ADCC reactions.

In fish, Robohm (1986) found that the effects of $CdCl_2$ on antibody responses were species dependent. Exposure of cunners to 12 µg Cd/ml in water caused a significant inhibition of serum antibody titers to *Bacillus cereus*. Conversely, antibody titers in striped bass similarly exposed were increased up to 6-fold. The differences in responses could not be explained by tissue Cd levels or general toxicity with reference to LC_{50}.

IV. COPPER (Cu)

Most studies on the effects of Cu on immune function have been focused on reductions in immune competency secondary to Cu deficiency. It is apparent that adequate Cu is necessary for normal immune function. Cu-deficient animals have decreased humoral, cell-mediated, and nonspecific immune response. These effects appear to be mostly reversible with Cu repletion. It appears that B cell-related dysfunction in Cu deficiency is mostly due to T-dependent antigens. (For reviews see Spallholz and Stewart, 1989; Fletcher et al., 1988; Chandra and Chandra, 1986.) Fewer studies have examined the effects of excess Cu on immune functions. As with Cu deficiency, immunotoxicity of excess exposure is affected by nutritional status of other metals such as Zn.

Several studies indicate that exposure to Cu in excess of normal requirements is detrimental to immune function inclusive of humoral immunity. Yamauchi and Yamamoto (1990) reported that splenic IgM PFC responses of female Balb/c mice were suppressed following 4-d exposure to 2 μM $CuSO_4$ *in vitro*. Mitogen responses to LPS were also decreased in a dose-response manner in splenocytes exposed to 0.1 to 8 μM $CuSO_4$ *in vitro*. The addition of 500 U/ml catalase to the cultures almost completely reversed the effects on PFC and mitogen responses. C57BL/6 male and female mice exposed to 100 or 200 ppm $CuSO_4$ in drinking water for 3 or 8 weeks had significantly reduced splenic IgM PFC responses to SRBC (Pocino et al., 1991). No effects on PFC responses were seen, however, in mice given 50 ppm $CuSO_4$ in the water. In this same study, LPS-induced mitogenesis was suppressed in animals exposed to 200 ppm $CuSO_4$ for 3 or 8 weeks. Conversely, LPS mitogenesis was enhanced at lower doses of $CuSO_4$ (50 and 100 ppm) at 3 weeks exposure. The authors also reported that the development of anti-bromelain mouse RBC spontaneous PFC responses were significantly increased in mice given 50 or 200 ppm $CuSO_4$ in their water for 3 or 8 weeks. Pocino et al. (1990) also studied the effects of Zn administration on altered immune responses of C57BL/6 male and female mice given Cu. Administration of 100 or 200 ppm $CuSO_4$ in the drinking water for 8 weeks resulted in reduced splenocyte responses to LPS, decreased IgM PFC responses to SRBC, and increased occurrence of autoantibodies (200 ppm group only). The mitogen responses and IgM PFC responses were completely restored in mice also treated with weekly i.p. injections of 1.14 mg/kg/BW of $ZnSO_4$. The increase in autoantibodies in Cu treated mice were not affected by Zn exposure.

Scuderi (1990) studied the effects of excess Cu on cytokine production by human peripheral blood leukocytes (PBL) *in vitro*. They found that $CuSO_4$ at concentrations ranging from 0.12 to 0.5 mM stimulated the production of TNF but suppressed the production of IL1β and IL6. No effects were seen on the secretion of IL1α and IFNγ. These Cu-treated cells also had substantially higher levels of TNF mRNA. The addition of substimulatory doses of LPS to Cu-treated cultures synergized the stimulation of TNF secretion. This effect was not seen with other cytokines. Treatment with Cu also resulted in a capacity for sustained secretion of TNF compared to controls. This study suggests that Cu may play a role in regulating the production of immunoregulatory cytokines.

Female C57BL/6 mice pretreated with a single injection of 8 mg/kg BW s.c. of a synthetic copper complex (copper II) (3,5-disopropylsalicylate) (Cu-DIPS) recovered immune function much faster following immunosuppressive irradiation given 3 h after treatment (Soderberg et al., 1987). The Cu-DIPS-treated mice recovered 55% of normal antibody producing capacity by 42 d, compared to only 20% of the nontreated irradiated mice. These mice also had greater ability to respond to B cell mitogens (LPS) 24 d after irradiation than mice which did not receive Cu-DIPS treatment. It appears from this study that pretreatment with the synthetic copper compound accelerated recovery of B and T cell activity following irradiation.

Kornegay et al. (1989) found no consistent effects of prolonged $CuSO_4$ supplementation on immune responses of weanling pigs. These feeding trials lasted for 5 weeks at $CuSO_4$ levels of 200 to 400 ppm. Humoral responses were assessed by hemagglutination titers to SRBC and specific antibody induction by the protein lysozyme.

V. LEAD (Pb)

The immunotoxicologic properties of Pb were the subject of a review by Koller (1985). Numerous reports cited in that review implicate lead as modulator of humoral (and other) immune responses. Lead exposure had been reported to suppress primary and secondary antibody titers in several species. Effects on mitogenesis (LPS) and B cell parameters have been mixed. Exposure to Pb also renders animals more susceptible to infectious agents. Others have shown that splenocytes exposed to Pb *in vitro* have enhanced B cell activity.

Talcott and Koller (1983) reported that serum antibodies to the T-dependent antigen bovine serum albumin were significantly suppressed in 8-week-old offspring of Swiss Webster (SW) mice exposed to 1000 ppm Pb acetate in the drinking water throughout gestation and lactation. Secondary antibody responses to the T-dependent antigen keyhole limpet hemocyanin (KLH) were also suppressed in male Sprague Dawley (SD) rats exposed to 10 or 1000 ppm Pb acetate in the drinking water for 10 weeks (Exon et al., 1985). These same authors also found that Con A-induced splenocyte IL-2 synthesis was depressed following exposure to Pb either *in vivo* or *in vitro*. Burchiel et al. (1987) reported that B6C3F1 mice treated with a single injection of 12 mg/kg Pb acetate had decreased primary antibody response to T-dependent (SRBC) and T-independent (TNP-Ficoll, TNP-LPS) antigens *in vitro*. Mitogenic responses to LPS were also suppressed in splenocytes of Pb-treated mice. Blakley et al. (1980) treated Swiss mice with 0.5, 1.0, 2.2, and 10.0 ppm tetraethyl lead in the drinking water for 3 weeks. The primary and secondary splenic PFC responses to SRBC were significantly suppressed at all doses. Hemagglutination titers to SRBC and LPS were also significantly lower than controls as were specific antibody concentrations to LPS measured by radial immunodiffusion. Numerous other reports have shown that *in vivo* treatment with Pb suppresses both the primary and secondary antibody-forming cell response in animals (see Koller, 1985 review).

Functional impairments of lymphocytes from workers occupationally exposed to Pb were reported by Fischbein et al. (1993). The mitogenic response to PWM was significantly reduced in Pb-exposed workers, but the response to the T-independent B cell blastogen *Staphyloccocus aureus* was not affected. Also, $CD3^+$ and $CD4^+$ cell numbers were significantly reduced.

Lawrence (1981a,b) reported that *in vitro* exposure of murine splenocytes to Pb produced activation of B cells. McCabe and Lawrence (1990) found that mouse (CBA/J and C3H/HeJ) splenocytes exposed to 10 µM $PbCl_2$ *in vitro* had augmented T cell-independent B cell responses as measured by LPS-induced Ig production. Exposure to $PbCl_2$ significantly enhanced IgM synthesis from LPS-stimulated splenocytes or semipurified B cell populations. The proliferation of B cells was not altered in the $PbCl_2$-treated cultures. Lead exposure in this study was also shown to increase the density of B cell MHC class II (Ia) surface markers but not class I (H-2K) markers. Other cell surface B cell markers were also altered in Pb-treated cultures. Surface IgD was lowered while FcεR densities were increased. The authors concluded that Pb may be acting as an activator of B cells, which could have implications in autoimmune diseases. Subsequent studies by McCabe and Lawrence (1991) indicate that the effects of Pb augmentation on B cell differentiation occurs indirectly via effects on B–T cell interaction. The authors report that Pb augments the antibody-forming cell response by both enhancing the production of T helper factors and enhancing the response of B cells to these factors. They also report that direct B–T cell contact is not absolutely necessary for increased B cell activity in the presence of Pb, but the response was optimal when this interaction occurred.

Warner and Lawrence (1988) studied the *in vitro* effect of $PbCl_2$ on IL-2 synthesis and responsiveness by CBA/J female mouse splenocytes. Metal concentrations used were 20 to 200 μM. $PbCl_2$ treatment had little effect on Con A-stimulated proliferation of HT-2 cell line in the presence of exogenous IL-2. Therefore, it appears that Pb does not affect the binding of IL-2 to its receptor. Conversely, splenocyte cultures were stimulated to produce IL-2 in the presence of $PbCl_2$ after 6 to 8 days of treatment, and there were elevated numbers of cells. The flow cytometry data suggest that the number of cells bearing IL-2 receptors increased but the density per cell decreased. Treatment of cultures with anti-IFN-γ, but not anti-IL-2 receptor, prevented Pb-induced lymphoproliferation. The authors suggest that IFN-γ is involved in the Pb-induced lymphoproliferative response, maybe to a greater extent than IL-2.

In an avian species, the red-tailed hawk, Redig et al. (1991) studied the short- and longer-term effects of Pb acetate on antibody titer to SRBC. No effects on antibody titers were seen in hawks given 0.82 mg Pb/kg BW daily by gavage for 24 d or in hawks given increasingly higher doses 1.64 to 6.55 mg/kg BW/d for up to 75 d.

VI. MERCURY (Hg)

Systemic autoimmune diseases can be induced experimentally in rodents by several chemicals, including $HgCl_2$. Rats, mice, and rabbits of certain strains develop autoimmune responses following low dose repeated exposure to Hg. Several common features are induction of ANA, immune glomerulonephritis, MHC class II hyperexpression on B cells, hyper IgE, increased IL-4 activity, and impaired IL-2 production (reviewed by Goldman, 1991). Strain-specific rat models produce high levels of autoantibodies to lamin after Hg treatment. The BN rat strain produce autoantibodies to type IV collagen, heparin, and intact thyroglobulin. The PVG rat strain produces autoantibodies to nuclear antigens. Some mouse strains produce antinuclear or antinucleolar antibodies, although no autoantibodies to kidney, thyroid, or skin have been observed in Hg-treated mice (review by Bigazzi, 1992). Other investigators have reported suppressed humoral immune function following Hg exposure.

Several investigators have reported increased blastogenesis following exposure of human and animal lymphocytes *in vitro* to mercury compounds (Caron et al., 1970; Hutchison et al., 1976; Nordlind and Henze, 1984). This effect appears, however, to be dose dependent, as high levels of exposure will inhibit DNA synthesis (Caron et al., 1970; Pauly et al., 1969; Reardon and Lucas, 1987). Dieter et al. (1983) observed enhanced LPS-induced mitogenesis in B6C3F1 mice treated with $HgCl_2$, and llback (1991) also reported female Balb/c mice treated with 1.2 μg/g methylmercury (MeHg) in the diet had increased B cell responses to mitogens. Reardon and Lucas (1987) also found that splenic or lymph node cells from Balb/c female mice had increased lymphoproliferative responses following *in vitro* treatment with $HgCl_2$ alone. The optimal mitogenic concentration of $HgCl_2$ was 200 μM/l. No effect of $HgCl_2$ was noted on LPS-induced mitogenesis. Hultman et al. (1991) studied mouse strains known to be sensitive or resistant to induction of autoimmunity by mercury and demonstrated there were differences in responses to mitogen exposure following either *in vitro* or *in vivo* treatment with $HgCl_2$. SJL/N mice (autoimmune sensitive) had two-fold increases in LPS-induced splenic cell mitogenesis following 2 weeks of exposure to 5 ppm $HgCl_2$ in the drinking water. No effect on LPS mitogenesis was seen in splenocytes of DBA mice (autoimmune resistant) given the same treatment. *In vitro* treatment of SJL/N splenocytes with 1×10^{-7} or 10^{-8} M $HgCl_2$ also resulted in a dose-dependent increase in thymidine incorporation. No effect or a slight decrease was seen in splenocytes from DBA mice receiving the same or higher doses of $HgCl_2$ *in vitro*.

SJL mice injected s.c. with 1.6 mg $HgCl_2$ per kilogram BW every 3rd day for 2, 4, 8, or 12 weeks developed ANA (Hultman and Enerstrom, 1988). This reaction may be involved in autoimmune renal disease associated with Hg exposure. They also reported polyclonal activation of splenic B cells in SJL/N and C57BL/6J(B6) female mice treated *in vivo* with 1.6 mg/kg $HgCl_2$ s.c. every 3rd day for 4 weeks (Hultman and Enestrom, 1989b). Mice treated with $HgCl_2$ had a 2- to 3-fold increase in the number of splenic cells spontaneously secreting IgG. The nonspecific IgM-producing cells were also elevated in number but only transiently and only in SLJ/N mice. The secretion of anti-TNP antibodies was also elevated, as were serum total Ig concentrations and ANA (in C57 mice). The induction of ANA following exposure to subtoxic doses of inorganic Hg has been reported to occur in several strains of mice (Mirtcheva et al., 1989; Hultman et al., 1989a,b).

The effects of HgCl$_2$ in vivo on antibody production in several strains of mice were reported by Pietsch et al. (1989). Strains tested were C57BL/6J, DBA/2J and A.SW. The mice were injected s.c. 3 times weekly for 5 weeks with 0.5 mg/kg HgCl$_2$, and serum was collected for Ig analyses by a sandwich ELISA. IgE and IgG levels were significantly elevated up to 30-fold in the A.SW mice. C57BL/6J mice also had elevated IgE levels following Hg treatment, but this effect was transitory and returned to normal by week 3. IgM levels were not affected in any strain. These same strain specific responses have been reported with regard to the autoimmunizing effect inducible by HgCl$_2$.

Nordlind and Henze (1984) studied the mitogenic activity of HgCl$_2$ using human thymocytes and PBL in vitro. Exposure to 5.5 or 1.4 × 10^{-6} HgCl$_2$ resulted in increased DNA synthesis in thymocytes and PBL. Although significant elevation in proliferative activity was seen in both types of cultures treated with Hg, they were fewer than seen with classical polyclonal mitogens such as PWM. Nordlind and Liden (1993) compared the effects of several metal cations on in vitro lymphocyte responses using PBL of patients with oral mucosal changes following amalgam restorations and a group of normal controls. The HgCl$_2$-treated (5.5 mol/l) lymphocyte cultures of patients had significantly increased IFN levels compared to controls, but no effects were seen on IL-2 receptor expression. Exposure to HgCl$_2$ did stimulate lymphoproliferation; however, the magnitude was similar between patients and controls. Cultures exposed to Al, Ni, Zn, Cu, Sn, or Cd were similar to controls. Exposure to 5.5 or 1.4 × 10^{-6} M HgCl$_2$ resulted in increased DNA synthesis in thymocytes and PBL. Although significant elevation in proliferative activity was seen in both types of cultures treated with Hg, it was less than that seen with classical polyclonal mitogens such as PWM. Reardon and Lucas (1987) also reported elevated levels of IFN in Balb/c mice splenocytes treated with 10 µM HgCl$_2$. No effect was evident, however, on Con A-induced IFN production.

Prouvost-Danon et al. (1981) reported that serum IgE was elevated in BN (autoimmune sensitive) but not Lewis strain (autoimmune resistant) rats following treatment with HgCl$_2$. The treatment schedule was s.c. injections at doses of 0.05 or 0.2 mg HgCl$_2$ per 100 g BW 3 times weekly for 1 or 2 months. These rats also had increased IgE responses to immunization with ovalbumin. Murdoch and Pepys (1986) found that serum IgE levels were significantly enhanced in hooded Lister rats treated 3 times a week for 3 weeks with 50 µg/100 g BW HgCl$_2$. The IgE levels increased up to day 8, then plateaued and returned to normal by day 50 (treatment terminated on day 25). Pelletier et al. (1988) also investigated the effects of in vivo HgCl$_2$ on Ig synthesis in BN rats. The serum IgE levels were dramatically increased in rats injected with 100 µg HgCl$_2$ per 100 g BW 3 times a week for 6 weeks. Serum levels of IgM, IgG$_1$, IgG$_{2b}$, IgG$_{2c}$, and IgA were also increased compared to controls. Levels of IgG$_{2a}$ were not different from controls. The peak levels of the Ig were reached between 2 to 3 weeks, and then all values except IgE progressively declined to normal by the last day (day 42). The IgE remained significantly higher than controls throughout the study. Mercury treatment also resulted in lymphoproliferative responses of lymph node cells and splenocytes and an increase in the number of cells with surface Ig. A subsequent study showed the treatment of A. SW mice with anti-IL-4 antibody completely prevents the IgE increase and partially blocks the increase in IgG$_1$ but not in IgG$_2$ (Ochel et al. 1991).

Other studies have shown Hg decreases B cell activity. Nakatsuru et al. (1985) found that in vitro treatment of splenocytes from Balb/c and C57BL/6 male mice with MeHg or HgCl$_2$ strongly inhibited polyclonal B cell activation by LPS. Cultures of splenic lymphocytes were treated with the mercury compounds at doses of 1 × 10^{-6} to 10^{-7} M for 2 d. MeHg was 20 times more potent as an inhibitor than HgCl$_2$. It was noted that the mercuric compounds seemed to have their effects early in the proliferation process, since addition during the final few hours of incubation had no effect. Daum et al. (1993) also found that in vitro treatment of murine splenocytes with HgCl$_2$ resulted in early inhibitory effects on B cell activation. Semipurified B cells from adult DBA/2J female mice had decreased RNA and DNA synthesis and impaired differentiation to Ig secretion following in vitro exposure to HgCl$_2$ at doses of 0.01 to 100 µM. These effects were present after only 2 h of exposure.

Several investigators have reported reduced antibody-forming cell responses in Hg-treated animals. Blakley et al. (1980) exposed mice to 0.5, 2.0, or 10.0 ppm MeHg in the drinking water for 3 weeks. The primary (IgM) and secondary (IgG) PFC response and hemagglutination titers to SRBCs were significantly suppressed at all doses in these animals. Specific antibodies as measured by radial immunodiffusion to LPS were also suppressed at all MeHg doses. Koller and Roan (1980) also reported that secondary antibody responses (IgG) to SRBC were suppressed in SW mice treated with 10 ppm MeHg in the feed for 10 weeks. Dieter et al. (1983) found similar suppression of T-dependent and T-independent PFC responses in B6C3F1 mice exposed to HgCl$_2$.

VII. NICKEL (Ni)

Nickel is known to be a potent allergen in both occupational and nonoccupational exposures. The effects of nickel on humoral immune function is varied, ranging from enhancement to suppression to no effect.

Smialowicz et al. (1984) reported that the primary IgM antibody response to the T-dependent antigen SRBC but not the T-independent antigen polyvinylpyrrolidone was reduced in CBA/J mice given a single im injection of 18.3 mg $NiCl_2$ per kilogram. Mitogenic response to LPS were not affected. They concluded T cells were a main target population for Ni. In a different study, $NiCl_2$ was delivered by minipumps to pregnant C57BL/J6 females from days 5 to 19 gestation at doses ranging from 9.1 to 73.2 µg/g in two separate experiments (Smialowicz et al., 1986). Lymphoproliferation induced in splenocytes of 8- to 10-week-old offspring by the B cell mitogen LPS was significantly suppressed at all Ni exposures in one experiment but not in the other. The IgM PFC response to SRBC was enhanced at the higher doses of Ni in one experiment but not the other. When the PFC response was expressed as total spleen cell number there were no significant changes seen in either study. The authors stated their results were inconclusive for effects of Ni on the murine immune system and may be of no significant biological relevance since their susceptibility to tumor challenge was not altered. These same investigators also did not observe consistent effects of Ni treatment on immune function of rats (Smialowicz et al., 1987). No changes were seen in IgM PFC responses to SRBC or PWM, or *Salmonella typhimurium* mitogenesis in rats treated with one i.v. injection of 10, 15, or 20 mg/kg $NiCl_2$. The most consistent result of this study indicates that Ni causes a transient decrease in NK cell activity. It also appears the Ni does not affect T cell function in the rat as it does in the mouse.

The effect of *in vivo* treatment of SJL female mice with $NiSO_4$ on antigen- (KLH) and mitogen- (LPS) induced blastogenesis and antibody synthesis was examined by Schiffer et al. (1991). Mice treated with 2.7 g/kg $NiSO_4$ in the diet for 4 weeks had depressed IgG and IgM serum antibody responses to KLH. These animals also had decreased proliferative responses to LPS and KLH *in vitro* which persisted for up to 5 days.

Dieter et al. (1988) studied the effects of prolonged exposure (180 d) of B6C3F1 mice to $NiSO_4$ in the drinking water at doses 1, 5, and 10 g/l. The authors reported dose-related decreases in LPS-induced mitogenesis and an increased IgM PFC response to SRBC at the highest dose. However, when the PFC response was expressed as per total spleen cells there was no change from control. Effects on the immune system of mice in this study are difficult to interpret due to other toxic effects such as histopathologic changes in the liver and kidney. Thus, any effect on immune function may be secondary.

Lawrence (1981a,b) investigated the effects of a number of heavy metals on primary humoral immune response *in vitro* using splenocytes from CBA/J female mice. Metals were added to splenocyte cultures at doses ranging from 1×10^{-4} to 10^{-7} M. Pb and Ni enhanced the primary antibody response *in vitro*. Haley et al. (1990) reported the effects of prolonged whole body exposure inhalation (6 h/d, 5 d/week, 65 d) in B6C3F1 mice to the Ni compounds, nickel subsulfide (Ni_3S_2), nickel oxide (NiO), and nickel sulfate hexahydrate ($NiSO_4 \cdot 6H_2O$). The IgM PFC cell response to SRBC in lung-associated lymphoid (LALN) cells was significantly augmented in animals exposed to 2.0 and 8.0 mg Ni/m³ NiO or 1.8 mg Ni/m³ Ni_3S_2. There was also a dose-related increase in size of LALNs in mice exposed to each of the Ni compounds. The IgM PFC response to splenocytes from mice exposed to 0.47 and 2.0 mg Ni/m³ NiO was significantly decreased. This effect was also seen in mice exposed to a high dose of Ni_3S_2 (1.8 mg Ni/m³). These results indicate varying effects of Ni inhalation on the humoral immune system, depending on the form of Ni and the source of cells used for assay.

The role of $NiSO_4$ in secretion of humoral factors was reported by Silvennoinen-Kassinen et al. (1991). The T cell clones used in the study were isolated from a Ni-sensitive male human subject. The 12 T cell clones were found to secrete IL-1β (4/12), IL-2R (all), IL-4 (6/12), and IL-6 (7/12). When supernatants of these clones were added to human PBL, all clones augmented spontaneous IgA synthesis, all but one augmented IgM production, and all but two helped IgG synthesis. The authors surmised that humoral factors may be involved in nickel-induced allergies.

Jaramillo and Sonnenfeld (1992a) reported that splenocytes from CD rats exposed to nickel sulfide (NiS) *in vitro* had altered blastogenic responses and cytokine production. Exposure of splenocyte to NiS ranging from 5 to 40 µg/ml increased proliferative activity which appeared to be enhanced by increased cell density. Splenocytes treated with 20 µg/ml of NiS had increased levels of IL-1 but not IL-2. Splenocytes treated with NiS and LPS had proliferative responses much greater than would be expected from an additive effect. The author postulated that blastogenic properties of Ni may be important in the carcinogenic action of this metal. Jaramillo and Sonnenfeld (1992b) also reported that 10 µg/ml NiS inhibited the production/secretion of IFN-α/β from L929 cells *in vitro*. This effect was greatest when the cells were pre-treated with NiS for 24 h prior to stimulation with PolyI/C. The duration of suppression of IFN-α/β secretion/synthesis was up to 120 h, the longest time period measured. The NiS, however, did not have any effect on the antiviral activity of IFN-α/β. Other investigators have reported the effects of Ni treatment *in vitro* on humoral factors such as IL-2 and IFN-γ produced by CBA/J mouse splenocytes (Blakley et al., 1980). A dose of $NiCl_2$ (100 µM) that stimulated T cell proliferation had no effect on IL-2 (exogenous)-induced proliferation of HT-2 cells. IL-2 secretion and IL-2 receptor expression were significantly enhanced in $NiCl_2$-treated splenocyte cultures. The metal-induced proliferation of T cells could be blocked by pretreatment of the cultures with either anti-IFN-γ antibodies or anti-IL-2 receptor antibodies. Daniels et al. (1987) reported that IFN production was not affected in female C_3H/HeJ or CD1 mice treated with $NiCl_2$ by im injection or inhalation. Others have reported inhibition of IFN synthesis following *in vitro* treatment with $NiCl_2$ (Treagan and Furst, 1970). They found that *in vitro* treatment of mouse L929 cultures with 30 µg $NiCl_2$/ml inhibited both induction and activity of IFN induced by Newcastle's disease virus. This effect persisted for several days after the single exposure.

Bencko et al. (1986) assessed serum protein levels including Ig in Ni exposed workers. They found elevated levels of IgG, IgA, and IgM compared to age-matched controls. Levels of IgE were significantly decreased in the Ni-exposed workers. The authors concluded that occupational Ni exposure leads to stimulation of synthesis of serum proteins (except IgE) that may be involved in allergic reactions. Unfortunately, Ni concentrations in the air did not correlate well with altered Ig synthesis.

Thymocytes and PBL from human donors had increased mitogenic activity when exposed to $NiCl_2$ *in vitro* at doses of 3.3 to 1.4×10^{-6} M (Nordlind and Henze, 1984). The blastogenic activity was not as pronounced as that induced by PWM, PHA, or Con A. The authors also noted an age-related decrease in lectin or $NiCl_2$-induced mitogenesis. In a later study, Nordlind and Liden (1993) reported 3.8×10^{-5} mol/l $NiCl_2$ stimulated proliferation of human PBL but had no effect on expression of IL-2 receptors or IFN synthesis/release.

VIII. SELENIUM (Se)

The effects of Se on immune function, including humoral immunity, have been reviewed (Kiremidjian-Schumacher and Stotzky, 1987; Dhur et al., 1990). In general, it has been shown that Se has effects on all major arms of the immune system, cell-mediated, humoral, and nonspecific. Deficiency of Se most commonly results in immunosuppression, while low doses of Se usually result in immunopotentiation. Higher doses given for prolonged periods have been shown to be immunosuppressive.

Spallholz et al. (1973a) reported that Swiss mice fed 0.7 to 2.8 ppm sodium selenite for 5 weeks had enhanced IgM PFC responses to SRBC. These mice also had enhanced serum antibody titers to SRBC. Mice fed diets containing 14 or 42 ppm Se had reduced PFC responses and serum antibody titers. It appears, therefore, that Se supplementation can enhance the IgM antibody response at low doses, but is suppressive at more toxic doses. In a related study, these investigators showed that IgG as well as IgM PFC responses were enhanced in Swiss mice fed 1 to 3 ppm sodium selenite for 6 weeks (Spallholz et al., 1973b). The authors also reported that Se (selenite) injected i.p. one time at doses of 3 or 5 µg resulted in enhanced IgG and IgM antibody titers to SRBC in Swiss mice (Spallholz et al., 1975). The highest titers were obtained when Se was given before or simultaneously with the SRBC injection.

Koller et al. (1986) reported that the IgG antibody response was suppressed in female S/D rats exposed to 5.0 ppm Se (as sodium selenite) in the drinking water for 10 weeks. No effects on humoral immunity were seen at doses of 0.5 or 2.0 ppm Se. Koller et al. (1979) had previously reported that antibody synthesis was enhanced in Se-treated mice.

A dose-related enhancement/suppression of antibody synthesis in human PBL exposed to Se compounds *in vitro* was reported by Reinhold et al. (1989). Sodium selenite and selenomethionine at doses of 1×10^{-3} to 10^{-5} M significantly reduced IgG and IgM antibody titers to PWM. Lower levels of these Se compounds (1×10^{-6} to 10^{-8} M) tended to enhance antibody formation. Treatment of the cultures with DL-selenocystine appeared to be mostly suppressive to antibody synthesis.

Stabel et al. (1991) studied the *in vitro* effects of sodium selenite and two organic selenium compounds on IgM antibody synthesis by bovine PBL. They reported that IgM synthesis was elevated in PWM-stimulated cultures treated with 100 mg DL-selnomethionine or DL-selenocystine, but not sodium selenite. The effect of Se supplementation on antibody responses of calves at various times during a 70-day feeding trial was studied by Swecker et al. (1989). Animals consuming feed supplemented with mineral mixes containing 80 or 100 ppm Se as sodium selenite had elevated primary and secondary antibody responses to challenge with egg lysozyme. Interestingly, calves given 160 or 200 ppm Se in the feed had antibody responses similar to the control group. Also, the effect of Se on antibody responses did not correlate with blood Se levels. It appears that Se supplementation can enhance the humoral immune system, but there is also an exposure limit to how much is beneficial in calves. A similar effect on antibody synthesis was reported by Larsen et al. (1988) in lambs following Se supplementation in the feed. Se-deficient (0.1 mg Se/kg) animals had reduced antibody responses and serum Ig levels, while those fed intermediate levels (0.5 mg/kg) had enhanced responses, and those fed the highest levels (1.0 mg/kg) had responses similar to controls.

Selenium treatment has also been shown to alter cytokine production. Kiremidjian-Schumacher et al. (1990) reported that IL-1 and IL-2 production/secretion was not affected in male C57BL/6J mice treated with 2 ppm Se in the diet for 8 weeks. These investigators did show in a subsequent study that Se supplementation at the same level, although not affecting IL-1 or IL-2 production, did significantly enhance IL-2R expression by splenocytes (Kiremidjian-Schumacher et al., 1992; Roy et al., 1992). In related experiments, Roy et al. (1990) reported that lymphotoxin release was enhanced in this same strain at the same dietary exposure level. In a different study using human PBL, it was reported that IL-2 production/secretion was suppressed following 3 days in culture with doses of 2 or 3 μM sodium selenite *in vitro* (Petrie et al., 1989). No effects were noted, however, at Se doses of 1, 4, and 5 μM or by any dose groups on days 1, 2, or 4 of exposure.

IX. TIN (Sn)

The major effects of exposure to compounds containing organic tin are on the cell-mediated immune responses (Penninks et al., 1990). However, other investigators have reported effects of tin on humoral immunity. Miller et al. (1986) reported that female Balb/c mice exposed orally to 20, 100, or 500 mg/kg di(octyl) tin dichloride for 8 weeks had decreased PFC responses to rat erythrocytes in the highest dose group from week 1 through 8. Also, autoantibodies to erythrocytes were suppressed compared to controls at the 500 mg/kg dose. Reduced PFC responses were also reported following *in vivo* exposure to the organotin compounds, di-*n*-butyltindichloride (DBTC) and di-*n*-actyltindichloride (DOTC), in inbred Wistar rats (Seinen et al., 1977). The primary IgM antibody responses to SRBC was significantly suppressed in female rats treated with 50 or 150 ppm DBTC in the feed for 6 weeks. These animals also had reduced hemolysin and hemagglutination titers. Rats treated with the same schedule of DOTC had reduced primary antibody responses at the 150 ppm dose. Secondary IgG responses were not altered in any exposure groups, nor was the primary response to the T-independent antigen LPS. Also, no effects on humoral immune responses were seen in Swiss mice exposed to either organotin compound in these studies.

Levine et al. (1987) reported that a 20% solution of metallic tin in saline given i.p. or i.v. to male and female Lewis rats resulted in elevated IgM PFC responses to SRBC measured 4 d postinjection. These PFC responses were elevated 3- and 17-fold, respectively, in the spleen and mediastinal lymph nodes. The authors also measured plasma cell number in mediastinal lymph nodes in rats injected in the footpad with 100 mg of powdered metallic tin 14 d previous. The relative number of IgM-secreting cells was decreased and the ratio of IgG/IgM increased. IgA-staining cells were not affected. The authors' overall conclusion was that metallic tin acts as an adjuvant to the primary PFC response in rats and is responsible for polyclonal B cell activation.

Earlier, Levine and Sowinski (1982) reported that i.p. injections of 200 mg metallic tin powder into Lewis rats resulted in plasma cell hyperplasia and elevated serum IgG levels; the effect was transient. In a subsequent study (Levine et al., 1983) male Lewis rats injected twice i.v. with 200 mg metallic tin powder had enlarged spleens and marked hyperplasia of plasma cells. Rats exposed to a single i.v. or i.p. injection of tin failed to respond similarly. The authors assumed some prior sensitization was needed to elicit the response.

X. ZINC (Zn)

Studies in experimental animals have revealed that multiple immune responses are altered in Zn-deficient or Zn-restricted animals and humans (reviewed by Fraker et al., 1986; Keen and Gershwin, 1990). Studies concerning the administration of Zn to normal subjects have been less frequent. Duchateau et al. (1981) reported Zn supplementation of elderly humans (>70 years old) could improve immune function. Subjects given 220 mg $ZnSO_4$ twice daily for 1 month had enhanced IgG antibody response to tetanus vaccine and other reactions mediated by T cells. There was no effect on response to mitogens including PWM. Similar results were reported by Wagner et al. (1983), who found daily oral doses of 55 mg Zn reversed depressed immune function in elderly human volunteers. These studies imply that increasing Zn intake above normal can play a role in restoring reduced immune responses in the aged. Similar results have been reported from animal studies, but the mechanism of the effect is not clear. Winchurch et al. (1984) studied the *in vitro* effects of $ZnCl_2$ on the splenic T-dependent PFC response of young adult (6 to 8 week) and aged (not specified) C57/B16 mice. PFC responses to SRBCs were stimulated within a narrow dose response range which peaked at 1×10^{-4} M $ZnCl_2$. Higher levels of exposure were inhibitory to the PFC response. Responses to Zn treatment between different age groups appeared to be similar, although baseline PFC responses were much lower in splenocytes of aged mice. In some cases, however, PFC responses in aged mice could be restored to levels similar to young adult mice with Zn treatment. Winchurch et al. (1987) also found that depressed *in vitro* T-dependent antibody-forming responses of splenocytes of aged C57BL/J mice could be restored with additional of 10 to 15 $\times 10^{-5}$ M $ZnCl_2$. They also found that IL-1, but not IL-2, levels in Zn-supplemented cultures were increased 300% over nonsupplemented controls, and these supernatants also supported *in vitro* antibody function in splenocytes of aged mice not treated with Zn. In a later study, it was observed that addition of $ZnCl_2$ to cultures of splenocytes from young adult and aged C57/B16 mice inhibited production of IL-2 (Winchurch et al., 1988). The addition of Zn to PFC cultures of aged mice dramatically increased the primary response to SRBCs. Cultures of isolated B lymphocytes had good PFC responses when provided with IL-1 and Zn. Addition of IL-2 to isolated B cells did not restore the PFC response in aged mice.

Other studies indicate cytokine levels are altered in cultures of human lymphocytes treated with Zn. Tanaka et al. (1989) investigated the effect of Zn on the *in vitro* antibody synthesis of human PBL. Concentrations of $ZnCl_2$ at ranges of 1×10^{-4} to 10^{-6} mol/l inhibited PWM-induced generation of Ig secreting cells. This inhibition may have been through the stimulation of CD8 Leu8 helper-inducer T cell subsets. Exposure to $ZnCl_2$ *in vitro* in this study did not enhance the proliferation of B cells as it did T cells. The authors surmise that humoral factors produced by T cells are involved. In a later study, it was reported that human peripheral blood mononuclear cells exposed to 1×10^{-4} M $ZnCl_2$ are stimulated *in vitro* to produce higher levels of IL-2 than controls (Tanaka et al., 1990). IL-2-induced T cell proliferation was also enhanced in the presence of Zn but was blocked by anti-IL-2 receptor antibodies. The proliferation of B cells was not enhanced at this dose of Zn. Zinc became toxic to cultured cells at doses of 1×10^{-3} M or greater.

Scuderi (1990) reported the addition of $ZnCl_2$ to cultures of human PBL resulted in stimulation of TNF and IL-1β but had no effect on IL-1α, IFN-γ, or IL-6 levels. Doses of $ZnSO_4$ in cultures ranged from 0.12 to 2.0 mM. Addition of substimulatory doses of LPS to Zn-treated cell cultures resulted in a synergistic secretion of TNF but not other cytokines. Also, the ability to sustain secretion of TNF was much greater in Zn-treated cells. The study suggests that Zn may play an important role in regulating secretion of TNF and IL-1β. Weide et al. (1991) reported that exogenous Zn given as zinc glucuronate at a dose of 20 mg 3 times daily for 12 treatments resulted in significant elevation of IFN in sera of human cancer patients. Also, Salas and Kirchner (1987) determined that IFN-γ levels were significantly elevated in supernates of cultures of human PBL treated with 5×10^{-4} to 5×10^{-5} $ZnCl_2$. Different results were obtained in studies using animal splenocytes treated with Zn. Balb/c mouse splenocytes and lymph node cells had increased proliferative responses following *in vitro* exposure to 50 to 400 μM ZnCl (Reardon and Lucas, 1987). The LPS-induced mitogenesis was not altered by Zn treatment. The production/secretion of IFN was induced in splenocytes exposed to 25 μM $ZnCl_2$ *in vitro*, but Con A-stimulated IFN production was not altered. Nordlind and Liden (1993) found no effect of $ZnCl_2$ on IFN synthesis by human PBL *in vitro;* exposure of these lymphocytes to 9×10^{-5} mol/l $ZnCl_2$ did stimulate lymphoproliferation.

Singh et al. (1992) reported that the humoral immune response was suppressed while T cell-mediated or macrophage functions were enhanced in Swiss albino mice treated i.p. with a single injection of 3 mg of Zn acetate per kilogram body weight. Humoral immunity was assessed by the IgM PFC response to SRBC on day 5. Interestingly, mice in this study given 2 i.p. injections of Zn on consecutive days had splenic PFC responses similar to controls. Mice which received the two injection treatments with Zn were also protected from endotoxin-induced shock.

Mulhern et al. (1985) studied the effects of Zn supplementation during different stages of development on C57BL/J mice. The authors found primary (IgM) PFC responses to SRBC were reduced in mice exposed to 50/2000/2000 or 2000/2000/2000 ppm Zn during gestation/lactation/postweaning development, respectively. It was concluded from this study that the PFC response was most sensitive to dietary Zn supplementation during development rather than postdevelopment of immune responsiveness.

Elemental Zn supplementation of 3.2 mg/l or 2 mg/kg in formula of malnourished infants has been shown to increase serum or salivary Ig levels (especially IgA) in two separate studies (Castillo-Duren et al., 1987; Schlesinger et al., 1992). These children may have been marginally Zn deficient at the beginning of the study.

Cunningham-Rundels et al. (1980) studied the effects of $ZnCl_2$ in vitro on several different sources of lymphocytes from human donors. These included PBL from healthy donors, regional lymph nodes removed during mastectomies, and splenocytes from patients with Hodgkins disease. Cells from all three sources were incubated for 6 d in the presence of 6 to 12×10^{-6} mmol Zn/ml. All cultures were found to contain activated lymphocytes capable of lysing SRBCs by the PFC assay. The highest PFC responses were produced in splenocytes, followed by lymph node cells and then PBL. The effect of Zn treatment was just as effective as treatment of the cultures with known lymphocyte stimulators such as PWM, purified protein derivative (PPD), or diphtheria toxin. Co-treatment of the cultures with $ZnCl_2$ and PPD resulted in a synergistic stimulus of lymphocyte activation. This study shows that Zn activates B cells as had been previously reported for T cells. Increases in B cells are a constant finding in Zn-deficiency diseases.

Others have failed to observe effects of Zn supplementation on humoral immunity in animal studies. Chowdhury et al. (1987) reported no effect on primary or secondary antibody PFC of male C57BL/6 mice exposed to $ZnCl_2$ in the drinking water for 3 weeks. They did report, however, that treatment with 500 ppm $ZnCl_2$ reversed an enhanced effect of $CdCl_2$ treatment on primary and secondary PFC response. Schiffer et al. (1991) reported that SJL mice exposed to $ZnSO_4$ in the diet for 4 weeks at levels of 0.63 g Zn/kg had normal antibody responses to KLH antigen and lymphoproliferative responses to LPS or KLH. Chopra et al. (1984) reported no effects of administration of 20 mg Zn/kg BW/d for 7 weeks by gavage on splenic ADCC responses in male Wistar rats. Conversely, Stacey (1986) reported the ADCC responses were significantly decreased in a dose-response manner in human PBL exposed to 100 to 250 μM $ZnSO_4$ in vitro.

Pimentel et al. (1991) studied the effects of Zn dietary supplementation on various breeds of chickens. Zinc concentrations of 10 to 125 ppm in diet for 4 or 7 weeks resulted in a variable weight loss of the bursa of Fabricius, the organ for B cell maturation. However, no effects were noted on primary or secondary antibody titers to SRBC. Stahl et al. (1989) also reported no effect on antibody production in chickens fed Zn-supplemented diets.

REFERENCES

Bencko, V., Wagner, V., Wagnerova, M., and Zavazal, V. (1986), *Environ. Res.*, 40, 399–410.
Bigazzi, P. (1992), *Clin. Immunol. Immunopathol.*, 62, 81–84.
Blakley, B. R. (1985), *Can. J. Comp. Med.*, 49, 104–108.
Blakley, B. R. (1988), *Can. J. Vet. Res.*, 52, 291–292.
Blakley, B. R. and Tomar, R. S. (1986), *Int. J. Immunopharmacol.*, 8, 1009–1015.
Blakley, B. R., Sisodia, C. S., and Mukkur, T. K. (1980), *Toxicol. Appl. Pharmacol.*, 52, 245–254.
Borgman, R. F., Au, B., and Chandra, R. K. (1986), *Int. J. Immunopharmacol.*, 8, 813–817.
Burchiel, S. W., Hadley, W. M., Cameron, C. L., Fincher, R. H., Lim, T.-W., Elias, L., and Stewart, C. C. (1987), *Int. J. Immunopharmacol.*, 9, 597–610.
Burns, L. A. and Munson, A. E. (1993), *J. Pharmacol. Exp. Ther.*, 256, 150–158.
Burns, L. A., Butterworth, L. F., and Munson, A. E. (1993), *J. Pharmacol. Exp. Ther.*, 264, 695–700.
Burns, L. A., Sikorski, E. E., Saady, J. J., and Munson, A. E. (1991), *Toxicol. Appl. Pharmacol.*, 110, 157–169.
Caron, G. A., Poutala, S., and Provost, T. T. (1970), *Int. Arch. Allergy Appl. Immunol.*, 37, 76.
Castillo-Duran, C., Heresi, G., Fisberg, M., and Uauy, R. (1987), *Am. J. Clin. Nutr.*, 45, 602–608.
Chandra, R. K. (1985), *J. Am. Coll. Nutri.*, 4, 5–16.
Chandra, S. and Chandra, R. K. (1986), *Prog. Food Nutri. Sci.*, 10, 1–65.
Chopra, R. K., Sehgal, S., and Nath, R. (1984), *Toxicology*, 33, 303–310.
Chowdhury, B. A., Friel, J. K., and Chandra, R. K. (1987), *J. Nutr.*, 117, 1788–1794.
Cunningham-Rundles, S., Cunningham-Rundles, C., Dupont, B., and Good, R. A. (1980), *Clin. Immunol. Immunopathol.*, 16, 115–122.
Daniels, M. J., Menache, M. G., Burleson, G. R., Graham, J. A., and Selgrade, M. K. (1987), *Fundam. Appl. Toxicol.*, 8, 433–453.
Daum, J. R., Shepherd, D. M., and Noelle, R. J. (1993), *Int. J. Immunopharmacol.*, 15, 383–394.

Dhur, A., Galan, P., and Hercberg, S. (1990), *Comp. Biochem. Physiol.,* 96C, 271–280.
Dieter, M. P., Luster, M. I., Boorman, G. A., Jameson, C. W., Dean, J. H., and Cox, J. W. (1983), *Toxicol. Appl. Pharmacol.,* 68, 218–228.
Dieter, M. P., Jameson, C. W., Tucker, A. N., Luster, M. I., French, J. E., Hong, H. L., and Boorman, G. A. (1988), *J. Toxicol. Environ. Health,* 24, 357–372.
Duchateau, J., Delepesse, G., Vrijens, R., and Collet, H. (1981), *Am. J. Med.,* 70, 1001–1004.
Exon, J. H. and Koller, L. D. (1986), in *Handbook of Experimental Pharmacology,* Foulkes, E.C., Ed., 80, Springer Verlag, Heidelberg, 338–350.
Exon, J. H., Talcott, P. A., and Koller, L. D. (1985), *Fundam. Appl. Toxicol.,* 5, 158–164.
Fischbein, A., Tsang, P., Luo, J.-C. J., Roboz, J. P., Jiang, J. D., and Bekesi, J. G. (1993), *Clin. Immunol. Immunopathol.,* 66, 163–168.
Fletcher, M. P., Gershwin, M. E., Keen, C. L., and Hurley, L. (1988), in *Contemporary Issues In Clinical Nutrition,* Vol. 11, Chandra, R. K., Ed., Alan R. Liss, New York, 215–239.
Fraker, P. J., Gershwin, M. E., Good, R. A., and Prasad, A. (1986), *Fed. Proc. Soc. Exp. Biol.,* 45, 1474–1479.
Fujimaki, H. (1985a), *Toxicol. Lett.,* 25, 69–74.
Fujimaki, H. (1985b), *Toxicol. Lett.,* 24, 21–24.
Fujimaki, H. (1987), *J.E.P.T.O.,* 7, 39–46.
Gleichmann, W., Kimber, I., and Purchase, I. F. H. (1989), *Arch. Toxicol.,* 63, 257–273.
Goldman, M., Druet, P., and Gleichmann, E. (1991), *Immunol. Today,* 12, 223–227.
Haley, P. J., Shopp, G. M., Benson, J. M., Cheng, Y.-S., Bice, D. E., Luster, M. I., Dunnick, J. K., and Hobbs, C. H. (1990), *Fundam. Appl. Toxicol.,* 15, 476–487.
Hultman, P. and Enestrom, S. (1988), *Clin. Exp. Immunol.,* 71, 269–274.
Hultman, P. and Enestrom, S. (1989b), *J. Clin. Lab. Immunol.,* 28, 143–150.
Hultman, P., Enestrom, S., Pollard, K. M., and Tan, E. M. (1989a), *Clin. Exp. Immunol.,* 78, 470–472.
Hultman, P. and Johansson, U. (1991), *Fed. Chem. Toxicol.,* 29, 633–638.
Hurtenbach, U., Oberbarnscheidt, J., and Gleichmann, E. (1988), *Arch. Toxicol.,* 62, 22–28.
Hutchinson, F., MacLeod, T. M., and Raffle, E. J. (1976), *Clin. Exp. Immunol.,* 26, 531–533.
Ilback, N.-G. (1991), *Toxicology,* 67, 117–124.
Jaramillo, A. and Sonnenfeld, G. (1992a), *Oncology,* 49, 396–406.
Jaramillo, A. and Sonnenfeld, G. (1992b), *Environ. Res.,* 57, 88–95.
Keen, C. L. and Gershwin, M. E. (1990), *Annu. Rev. Nutr.,* 10, 415–431.
Kiremidjian-Schumacher, L. and Stotzky, G. (1987), *Environ. Res.,* 42, 277–303.
Kiremidjian-Schumacher, L., Roy, M., Wishe, H. I., Cohen, M. W., and Stotzky, G. (1990), *Proc. Soc. Exp. Biol. Med.,* 193, 136–142.
Kiremidjian-Schumacher, L., Roy, M., Wishe, H. I., Cohen, M. W., and Stotzky, G. (1992), *Biol. Trace Elem. Res.,* 33, 23–35.
Koller, L. D. (1982), *Environ. Health Perspect.,* 43, 37–39.
Koller, L. D. (1985), in *Dietary and Environmental Lead: Human Health Effects* (Mahaffey, Ed.), Elsevier, New York, 339–353.
Koller, L. D., Exon, J. H., Talcott, P. A., Osborne, C. A., and Henningsen, G. M. (1986), *Clin. Exp. Immunol.,* 63, 570–576.
Koller, L. D. and Roan, J. G. (1980), *J. Environ. Pathol. Toxicol.,* 4, 47–52.
Koller, L. D., Kerkvliet, N. J., and Exon, J. H. (1979), *Arch. Environ. Health,* 34, 248–251.
Kornegay, E. T., Van Heugten, P. H. G., Lindemann, M. D., and Blodgett, D. J. (1989), *J. Anim. Sci.,* 67, 1471–1477.
Krzystyniak, K., Fournier, M., Trottier, B., Nadeau, D., and Chevalier, G. (1987), *Toxicol. Lett.,* 38, 1–12.
Larsen, H. J., Moksnes, K., and Overnes, G. (1988), *Res. Vet. Sci.,* 45, 4–10.
Lawrence, D. A. (1981a), *Toxicol. Appl. Pharmacol.,* 57, 439–451.
Lawrence, D. A. (1981b), *Infect. Immun.,* 31, 136–143.
Levine, S. and Sowinski, R. (1982), *Exp. Mol. Pathol.,* 36, 86–98.
Levine, S., Sowinski, R., and Koulish, S. (1983), *Exp. Mol. Pathol.,* 39, 364–376.
Levine, S., Saad, A., and Rappaport, I. (1987), *Immunol. Invest.,* 16, 201–212.
McCabe, J. M. and Lawrence, D. A. (1990), *J. Immunol.,* 145, 671–677.
McCabe, M. J. and Lawrence, D. A. (1991), *Toxicol. Appl. Pharmacol.,* 111, 13–23.
Miller, K., Maisey, J., and Nicklin, S. (1986), *Environ. Res.,* 39, 434–441.
Mirtcheva, J., Pfeiffer, C., De Bruijn, J. A., Jacquesmart, F., and Gleichmann, E. (1989), *Eur. J. Immunol.,* 19, 2257–2261.
Mulhern, S. A., Vessey, A. R., Taylor, G. L., and Magruder, L. E. (1985), *Proc. Soc. Exp. Biol. Med.,* 180, 453–461.
Murdoch, R. D. and Pepys, J. (1986), *Int. Arch. Allergy Appl. Immunol.,* 80, 405–411.
Nakatsuru, S., Oohashi, J., Nozaki, H., Nakada, S., and Imura, N. (1985), *Toxicology.,* 36, 297–305.
Nordlind, K. and Henze, A. (1984), *Int. Arch. Allergy Appl. Immunol.,* 73, 162–165.
Nordlind, K. and Liden, S. (1993), *Br. J. Dermatol.,* 128, 38–41.
Ochel, M., Vohr, H-W., Pfeiffer, C., and Gleichmann, E. (1991), *J. Immunol.,* 146, 3006–3011.
Ohsawa, M., Masuko-Sato, K., Takahashi, K., and Otsuka, F. (1986), *Toxicol. Appl. Pharmacol.,* 84, 379–388.
Ohsawa, M., Takahashi, K., and Otsuka, F. (1988), *Clin. Exp. Immunol.,* 73, 98–102.
Pauly, J. L., Caron, G. A., and Suskind, R. R. (1969), *J. Cell Biol.,* 40, 847.

Pelletier, L., Pasquier, R., Guettier, C., Vial, M.-C., Mandet, C., Nochy, D., Bazin, H., and Druet, P. (1988), *Clin. Exp. Immunol.,* 71, 336–342.
Penninks, A. (1990), *Immunotoxicology of Metals and Immunotoxicology,* Dayan, A. D. et al., Eds., Plenum, New York, 191–207.
Petrie, H. T., Klassen, L. W., and Kay, H. D. (1989), *J. Leukocyte Biol.,* 45, 207–214.
Pietsch, P., Vohr, H.-W., Degitz, K., and Gleichmann, E. (1989), *Int. Arch. Allergy Appl. Immunol.,* 90, 47–53.
Pimentel, J. L., Cook, M. E., and Greger, J. L. (1991), *Poultry Sci.,* 70, 947–954.
Pocino, M., Baute, L., and Malave, I. (1991), *Fundam. Appl. Toxicol.,* 16, 249–256.
Pocino, M., Malave, I., and Baute, L. (1990), *Immunopharmacol. Immunotoxicol.,* 12, 697–713.
Prouvost-Danon, A., Abadie, A., Sapin, C., Bazin, H., and Druet, P. (1981), *J. Immunol.,* 126, 699–702.
Reardon, C. L. and Lucas, D. O. (1987), *Immunobiology,* 175, 455–469.
Redig, P. T., Lawler, E. M., Schwartz, S., Dunnette, J. L., Stephenson, B., and Duke, G. E. (1991), *Arch. Environ. Contam. Toxicol.,* 21, 72–77.
Reinhold, U., Pawelec, G., Enczmann, J., and Wernet, P. (1989), *Biol. Trace Elem. Res.,* 20, 45–58.
Robohm, R. A. (1986), *Vet. Immunol. Immunopathol.,* 12, 251–262.
Roy, M., Kiremidjian-Schumacher, L., Wishe, H. I., Cohen, M. W., and Stotzky, G. (1990), *Proc. Soc. Exp. Biol. Med.,* 193, 143–148.
Roy, M., Kiremidjian-Schumacher, L., Wishe, H. I., Cohen, M. W., and Stotzky, G. (1992), *Proc. Soc. Exp. Biol. Med.,* 200, 36–43.
Salas, M. and Kirchner, H. (1987), *Immunol. Immunopathol.,* 45, 139–142.
Schiffer, R. B., Sunderman, F. W., Jr., Baggs, R. B., and Moynihan, J. A. (1991), *J. Neuroimmunol.,* 34, 229–239.
Schlesinger, L., Arevalo, M., Arredondo, S., Diaz, M., Lonnerdal, B., and Stekel, A. (1992), *Am. J. Clin. Nutr.,* 56, 491–498.
Scuderi, P. (1990), *Cell. Immunol.,* 126, 391–405.
Seinen, W., Vos, J. G., Van Krieken, R., Penninks, A., Brands, R., and Hooykaas, H. (1977), *Toxicol. Appl. Pharmacol.,* 42, 213–224.
Sherman, A. R. (1992), *J. Nutr.,* 122, 604–609.
Sikorski, E. E., McCay, J. A., White, K. L., Jr., Bradley, S. G., and Munson, A. E. (1989), *Fundam. Appl. Toxicol.,* 13, 843–858.
Sikorski, E. E., Burns, L. A., Stern, M. L., Luster, M. I., and Munson, A. E. (1991), *Toxicol. Appl. Pharmacol.,* 110, 129–142.
Silvennoinen-Kassinen, S., Poikonen, K., and Ikaheimo, I. (1991), *Scand. J. Immunol.,* 33, 429–434.
Singh, K. P., Zaidi, S. I. A., Raisuddin, S., Saxena, A. K., Murthy, R. C., and Ray, P. K. (1992), *Immunopharmacol. Immunotoxicol.,* 14, 813–840.
Smialowicz, R. J., Rogers, R. R., Riddle, M. M., Rowe, D. G., and Luebke, R. W. (1986), *Toxicology,* 38, 293–303.
Smialowicz, R. J., Rogers, R. R., Riddle, M. M., and Stott, G. A. (1984), *Environ. Res.,* 33, 413–427.
Smialowicz, R. J., Rogers, R. R., Rowe, D. G., Riddle, M. M., and Luebke, R. W. (1987), *Toxicology,* 44, 271–281.
Soderberg, L. S. F., Barnett, J. B., Baker, M. L., Salari, H., and Sorenson, R. J. (1987), *Scand. J. Immunol.,* 26, 495–502.
Spallholz, J. E. and Stewart, J. R. (1989), *Biol. Trace Elem. Res.,* 19, 129–151.
Spallholz, J. E., Martin, J. L., Gerlach, M. L., and Heinzerling, R. H. (1973a), *Proc. Soc. Exp. Biol. Med.,* 143, 685–689.
Spallholz, J. E., Martin, J. L., Gerlach, M. L., and Heinzerling, R. H. (1973b), *Infect. Immun.,* 8, 841–842.
Spallholz, J. E., Martin, J. L., Gerlach, M. L., and Heinzerling, R. H. (1975), *Proc. Soc. Exp. Biol. Med.,* 148, 37–40.
Stabel, J. R., Reinhardt, T. A. and Nonnecke, B. J. (1991), *J. Dairy Sci.,* 74, 2501–2506.
Stacey, N. H. (1986), *J. Toxicol. Environ. Health,* 18, 293–300.
Stahl, J. L., Cook, M. E., Sunde, M. J., and Greger, J. L. (1989), *Appl. Agric. Res.,* 4, 86–89.
Swecker, Jr., W. S., Eversole, D. E., Thatcher, C. D., Blodgett, D. J., Schurig, G. C., and Meldrum, J. B. (1989), *Am. J. Vet. Res.,* 50, 1760–1763.
Talcott, P. A. and Koller, L. D. (1983), *J. Toxicol. Environ. Health,* 12, 337–352.
Tanaka, Y., Shiozawa, S., Morimoto, I., and Fujita, T. (1989), *Int. J. Immunopharmacol.,* 11, 673–679.
Tanaka, Y., Shiozawa, S., Morimoto, I., and Fujita, T. (1990), *Scand. J. Immunol.,* 31, 547–552.
Thomas, P. T., Ratajczak, H. V., Aranyi, C., Gibbons, R., and Fenters, J. D. (1985), *Toxicol. Appl. Pharmacol.,* 80, 446–456.
Treagan, L. and Furst, A. (1970), *Res. Commun. Chem. Pathol. Pharmacol.,* 1, 395–402.
Wagner, P. A., Jemigan, J. A., Bailey, L. B., Nickens, C., and Brazzi, G. A. (1983), *Int. J. Vitam. Nutr. Res.,* 53, 94–101.
Warner, G. L. and Lawrence, D. A. (1988), *Int. J. Immunopharmacol.,* 10, 629–637.
Weide, M., Zhaoming, D., Baoliang, L., and Huibi, X. (1991), *Biol. Trace Elem. Res.,* 28, 11–19.
Winchurch, R. A., Togo, J., and Adler, W. H. (1987), *Eur. J. Immunol.,* 17, 127–132.
Winchurch, R. A., Togo, J., and Adler, W. H. (1988), *Clin. Immunol. Immunopathol.,* 49, 215–222.
Winchurch, R. A., Thomas, D. J., Adler, W. H., and Lindsay, T. J. (1984), *J. Immunol.,* 133, 569–571.
Yamauchi, T. and Yamamoto, I. (1990), *Jpn. J. Pharmacol.,* 54, 455–460.

Chapter 51

Metal-Induced Alterations in Innate Immunity*

Judith T. Zelikoff and R.J. Smialowicz

I. INTRODUCTION

Within the last few decades, experimental data have shown that low level exposure to certain metals induces subtle changes within a host, including altered immunological competence. Environmental stressors like metals may act directly to kill the exposed organisms or indirectly to exacerbate disease states by lowering resistance and allowing the invasion of infectious pathogens.

This chapter reviews results from numerous immunotoxicological studies and demonstrates that certain inorganic and organic metal compounds are capable of altering macrophage and natural killer (NK) cell activity following acute and repeated exposure both *in vivo* and *in vitro*. It outlines the reported effects of six different environmentally/occupationally relevant metals of varying immunotoxic potential on two immune cell types critical for host resistance against infectious agents, viruses, and/or developing neoplasms. The metals discussed in this chapter include: arsenic (As), cadmium (Cd), lead (Pb), mercury (Hg), nickel (Ni), and tin (Sn).

Immunotoxicological studies were selected for discussion which demonstrated the immunosuppressive effects (both at the clinical and mechanistic levels, where possible) of the individual metals. The reviewed studies demonstrate metal-induced alterations that could potentially offset the balance necessary for immunoregulation and, thus, produce a cascade of detrimental secondary events, including compromised host resistance, hypersensitivity, or local tissue damage.

This chapter concludes with a brief discussion of the needs and directions for future research in the general area of metal immunotoxicology.

II. MACROPHAGES

A. GENERAL OVERVIEW

Bone marrow-derived stem cells give rise to the mononuclear phagocyte system in which the tissue macrophages form a network. These cells arise from proliferating monoblasts that differentiate into promonocytes and, finally, into mature circulating monocytes which are thought to be a replacement pool for tissue-resident macrophages.

Macrophages are involved at all stages of the immune response. They act as a rapid protective mechanism which can respond before T cell-mediated amplification has taken place. Then they take part

---

* This report has been reviewed by the Environmental Protection Agency's Office of Research and Development, and approved for publication. Approval does not signify that the contents necessarily reflect the views and policies of the Agency nor does mention of trade names or commercial products constitute endorsement or recommendation for use.

in the initiation of T cell activation by processing and presenting antigen. Finally, macrophage are important in the effector phase of the cell-mediated response following T cell-mediated activation, as inflammatory, secretory, tumoricidal, and microbicidal cells (Nathan et al., 1980; Nathan, 1987).

As secretory cells, macrophages synthesize and secrete a number of diverse products, including prostaglandins, interleukin-1, tumor necrosis factor-alpha, complement components, reactive oxygen and nitrogen intermediates, and chemical products important in macrophage and lymphoid cell regulation and differentiation. Many of these products, as well as certain macrophage functional and biochemical properties, appear to be essential *in vivo* for maintaining host defense against infectious agents and developing neoplasms (Marino and Adams, 1980).

Clearly, the macrophage and its functional, biochemical, and secretory activities are crucial for proper immune surveillance and, thus, for adequate host defense. As such, metal-induced effects on any of the aforementioned macrophage functions could bring about alterations in host resistance to disease and, possibly, cancer.

B. METAL-INDUCED EFFECTS
1. Arsenic (As)

Arsenic, a major constituent of the earth's crust (~2 to 5 mg As/kg), is present in various concentrations and in different oxidation states in the atmosphere, soil, and water (World Health Organization, 1980; Bencko, 1985). Airborne As, the result of both natural and anthropogenic sources, is released into the air at an estimated total of 73,000 metric tons/year. Arsenic also occurs "naturally" in the soil, with concentrations ranging from 1 to 40 ppm As (Bencko, 1985). Levels may be substantially higher, however, if As-containing pesticides have been used continuously in the area.

In unpolluted, fresh waters, the As content is typically less than a few micrograms per liter (U.S. EPA, 1985; Science Applications International Corporation, 1987; Marcus and Rispin, 1988), while in salt water, As concentrations are much greater. As a result, unusually high concentrations of As are found in many types of ingested seafood (Malachowski, 1990).

In general, inorganic As compounds are more toxic than their organic counterparts of similar valence states (Malachowski, 1990). Inorganic As exists in several different oxidation states which result in markedly different biological behaviors and overall toxicities of its compounds. For example, arsine gas (arsenic hydride, AsH_3), containing the trivalent As anion, represents the most toxic form of As, most likely because of its hemolytic effects on red blood cells (Fowler and Weissberg, 1974). Trivalent As compounds or arsenites appear to be more toxic to most mammalian organisms than pentavalent arsenate compounds. This is probably due to the ability of arsenites to react with, and bind to, sulfhydryl groups, leading to the inactivation of essential enzymes (Vahter, 1988).

Several aspects of the medical side of the As story suggest that it acts as an immunosuppressant (National Academy of Sciences, 1977). Yet, despite the historical epidemiological evidence suggesting the effects of As on host immunocompetence (Buchanan, 1962), few laboratory studies have been performed to "sort out" these implications and/or to help elucidate possible underlying mechanisms. Although a number of investigations have implicated the macrophage as the primary immune cell responsible for As-induced alterations in host resistance and/or immunocompetence, direct effects of As on macrophages have not been well studied.

Inhalation of arsenic trioxide (As_2O_3) has been shown to reduce pulmonary resistance to pathogenic bacterial infections most likely by depressing alveolar macrophage bactericidal activity (Aranyi et al., 1985). In a previous study by the same investigators, *in vitro* exposure to As-containing copper smelter dust (13% soluble As) was found to be cytotoxic to rabbit alveolar macrophages (Aranyi, 1981). Arsenic has also been shown to depress interferon (IFN)-γ production/activity (Gainer, 1972). Since this cytokine plays an important role in the activation of T lymphocytes and macrophages through immune "networks", the inhibitory effects of trivalent As observed on macrophage activity may have been due to changes in IFN-γ metabolism.

Additional *in vitro* experiments have shown that exposure of bovine pulmonary macrophages to sodium arsenite ($NaAsO_2$) at concentrations above 0.5 µg As^{3+}/ml depressed phagocytic activity in a dose-dependent manner. In this same study, lower concentrations of arsenite stimulated phagocytic uptake (Fisher et al., 1986). Prenatal exposure to low, nonteratogenic doses of $NaAsO_2$ (dam exposure doses = 1 and 2 mg/kg) produced the same stimulatory effects on macrophage phagocytosis in mice exposed *in utero* and examined 4 weeks after exposure ceased; effects on macrophage function in the offspring were gender-dependent and occurred in the absence of effects on the directly exposed dams (Zelikoff et al., 1989). In this same study, peritoneal macrophage number was also increased in the NaAsO$_2$-exposed offspring.

Less well known and well studied of the As-containing compounds is gallium arsenide (GaAs), currently of great interest because of its use in semiconductor devices necessary for supercomputers and telecommunications. Gallium arsenide, like most other arsenic compounds, also modulates immune responses in mammalian species (Sikorski et al., 1989; Sikorski et al., 1991a,b; Burns et al., 1991). For example, 14 days after intratracheal instillation of GaAs at 50, 100, or 200 mg/kg, peritoneal exudate cells (PEC) were dose-dependently decreased in number and the adherent PEC population demonstrated decreased phagocytosis of latex particles and increased phagocytosis of chicken erythrocytes (Sikorski et al., 1989). Inasmuch as GaAs has been shown to dissociate, in certain mammalian species, to gallium and trivalent As (Sikorski et al., 1991a), it is likely that its immunotoxicity is directly related to the As species (Burns et al., 1991).

2. Cadmium (Cd)

Cadmium, a commonly occurring environmental contaminant of food, water, and air (Vallee and Ulmer, 1972), is recovered as a by-product during the roasting, smelting, or other processing of zinc and lead ores. There are also several inadvertent emission sources of Cd that are significant and widespread, including: municipal incineration of waste materials, burning of fossil fuels, use of phosphate fertilizers, and the disposal of sewage sludge. For the nonindustrially exposed individual, food represents the major source of Cd exposure, while intake via drinking water is negligible. Cadmium is also present in tobacco and, thus, smoking can also be a major contributor to nonindustrial exposures.

The toxicology of Cd is complex because of the large number of biological systems that can be affected and the many different toxic effects that may be elicited. Cadmium has a long biological half-life, is present in most tissues, and accumulates with age. Once absorbed from the lungs or gut, Cd is transported by the blood to a number of different targets. Deposition occurs principally in the liver and kidneys, with the latter generally containing about one third of the total body burden. Smaller amounts are distributed to the bone, testes, spleen, various endocrine organs, brain, and muscle tissue (Lee and White, 1980).

In addition to its well-documented toxic effects on the kidney, Cd is a potent immunotoxicant for a variety of mammalian and nonmammalian animal species (Greenspan and Morrow, 1984; Zelikoff et al., 1995; Zelikoff, 1994; Elasser et al., 1986; Koller and Vos, 1981). Studies have shown that both *in vitro* and *in vivo* exposure to Cd effects alveolar macrophage morphological and functional parameters in most cases by suppressing immunological responsiveness. *In vitro* studies have also shown that the toxic effects of Cd on mammalian macrophage viability, morphology, and/or functionality can be reduced by coexposure with nontoxic concentrations of sodium selenite, zinc, or copper (Huisingh et al., 1977; Murthy and Holovack, 1991).

Early studies by Waters and Gardner (1975) have demonstrated that *in vitro* exposure (20 h) of rabbit alveolar macrophages to Cd at concentrations between 0.0089 and 0.89 mM altered macrophage surface characteristics in a manner that correlated with increasing exposure concentrations and cell viabilities. In the same study, exposure to the Cd LC$_{50}$ concentration (0.1 mM) decreased specific activity of the lysosomal indicator enzyme, acid phosphatase. In other *in vitro* studies, Loose et al. (1977, 1978) have shown a significant depression in the phagocytic capacity and respiratory burst of mouse elicited peritoneal and pulmonary alveolar macrophages following short-term incubation with either cadmium chloride (CdCl$_2$) or cadmium acetate [Cd(C$_2$H$_3$O$_2$)$_2$] at concentrations between 16 and 1600 µM; microbicidal activity of the Cd-exposed pulmonary macrophages was also depressed. The authors suggested that the observed depression in phagocytosis and in the resulting respiratory burst may have been due to the inhibitory action of Cd on ATPase activity.

Studies using nonmammalian models have demonstrated similar suppressive effects of Cd on macrophage phagocytosis. For example, *in vitro* exposure of rainbow trout peritoneal macrophages for 5 h to CdCl$_2$ at concentrations between 30 and 250 µM reduced the phagocytic uptake of polystyrene latex particles by bacterially elicited macrophages. However, unlike the effects observed for rodents (Loose et al., 1977), exposure of trout macrophages to Cd increased, rather than decreased, hydrogen peroxide production. In this same study using fish, Cd-induced changes in macrophage ultrastructure correlated with increasing metal concentrations and cell viability (Zelikoff, 1994). The author suggested the potential usefulness of altered surface morphology as a biomarker for predicting the effects of environmental Cd exposure. Cd-induced changes in macrophage ultrastructure have also been reported following inhalation exposure of rabbits for 1 month to CdCl$_2$ at 0.4 to 0.6 mg/m^3 (Johansson et al., 1983); inhalation of Cd aerosols also increased bactericidal activity, as well as oxidative metabolic activity after stimulation with *Escherichia coli* bacteria.

In vitro exposure to $CdCl_2$ at concentrations between 1 and 75 µM has been shown to increase both the attachment and adherence of mouse peritoneal macrophages. In this study, effects were greatest on those cells recovered from male animals (Hernandez et al., 1991). Observed gender differences were thought to be due to differences in steroid levels between the sexes (steroid levels in rodents are higher in males than in female animals). In another *in vitro* study investigating the effects of relatively low, nontoxic Cd concentrations on macrophage activity, Hadley et al. (1977) demonstrated that exposure for 30 min to 20 µM Cd inhibited the IgG-mediated rosette formation by alveolar macrophages. In a more recent study (Greenspan and Morrow, 1984), *in vitro* exposure of rat pulmonary macrophages for 1 h to soluble Cd at concentrations between 20 and 200 µM produced a dose-dependent decrease in cell viability, percentage of cells taking up latex particles, and the total number of particles phagocytosed. Similar effects on phagocytic activity were observed following a single inhalation exposure of rats to $CdCl_2$ at 5.0 mg/m^3; reduced phagocytic activity was observed for up to 8 days post exposure (Greenspan and Morrow, 1984). In contrast, however, inhalation of Cd at a lower concentration (1.5 mg/m^3) significantly increased phagocytic activity immediately and 24 h following exposure. Similar stimulatory effects on macrophage phagocytic activity have also been observed in other rodent studies (Thomas et al., 1985), as well as in studies using nonmammalian animal models (Zelikoff et al., 1994). For example, Thomas et al. (1985) demonstrated that macrophage recovered from adult female mice that received water containing 10, 50, or 250 ppm of $CdCl_2$ for 90 d had significantly increased phagocytic activity in the absence of effects on humoral immunity. Additionally, peritoneal macrophages recovered from rainbow trout exposed to waterborne Cd for 8 d demonstrated almost a 2-fold increase in phagocytic activity (Zelikoff et al., 1994).

3. Lead (Pb)

Lead is a naturally occurring element that may be found in the Earth's crust and in all components of the biosphere. Although efforts have been undertaken to reduce the environmental sources of Pb (i.e., through a ban on Pb in interior paints and reconfiguring car engines to utilize unleaded gasoline), Pb is still readily encountered in soil, water, and air; as well as in the plumbing of buildings, and on the walls of millions of older homes (U.S. EPA, 1986a,b,c).

The primary route of exposure to environmental Pb is via the diet. Food crops and livestock contain Pb in varying proportions from direct and indirect atmospheric sources (U.S. EPA, 1986b,c; leonard et al., 1988; Marcus, 1990). Another major source of human exposure to Pb is contaminated water. In drinking water, Pb is "normally" found at concentrations between 6 to 8 µg Pb/l (U.S. EPA, 1986c; Marcus, 1990), but water softeners passed through leaden pipes may infuse the water with several milligrams Pb/l.

Of particular importance for human exposure are emissions of Pb to the atmosphere (U.S. EPA, 1986c). The greatest risk for exposure to airborne Pb, however, is not the ambient environment, but, rather, from occupational settings (Froines et al., 1990). Because of the adverse health effects associated with exposure to high Pb concentrations (U.S. EPA, 1986c; Marcus, 1990; Leonard et al., 1988; Goyer, 1991), and the increased risk of exposure in "high-Pb" industrial settings, an occupational air Pb standard of 50 µg Pb/m^3 has been promulgated by the Occupational Safety and Health Administration (OSHA) (Froines et al., 1990).

Pb is inherently toxic to most animals, including humans (Goyer et al., 1986). The biological basis of Pb toxicity is thought to be due to its ability to bind ligands on biomolecular substances crucial for various physiological functions. Thus, Pb can act as a toxicant by interfering with the functions of biomolecules. This may occur as a result of Pb competing with native essential metals for binding sites, inhibiting enzyme activities, or altering essential ion transport.

In addition to its other toxic effects (Goyer, 1986; Beck, 1992), Pb, at levels below those associated with overt toxicity, alters immune defense mechanisms important for maintaining host resistance against infectious agents and, possibly, cancer (Nielan et al., 1980; Lawrence, 1981; Nielan et al., 1983; Lawrence, 1985; Koller, 1990). In general, most immunotoxicological studies indicate that Pb acts to suppress normal immune responses.

Macrophages, like lymphocytes, are sensitive targets for the immunotoxic effects of Pb. Both soluble and particulate Pb compounds can alter macrophage activity (Graham et al., 1975; Kaminski et al., 1977; DeVries et al., 1983; Castranova et al., 1983; Hilbertz et al., 1986; Kowolenko et al., 1988; Mauel et al., 1989; Buchmuller-Rouiller et al., 1989; Gulyas et al., 1990; Zelikoff et al., 1993; Cohen et al., 1994).

While *in vitro* exposure to soluble Pb ions had no effect on phagocytosis or interleukin-1 (IL-1) production, exposure to these same Pb compounds altered the ability of macrophages to present antigen during an autologous mixed-lymphocyte reaction (AMLR). It was hypothesized in this study that Pb may have interfered with antigen-specific interactions between macrophages and T cells (Kowolenko et al., 1988). Further studies have demonstrated that *in vitro* exposure of macrophages to soluble Pb reduced both the intracellular killing of parasites and the extracellular cytolysis of target cells (Mauel et al., 1989). This suppression was most likely due to the ability of Pb to inhibit the oxidative metabolism (respiratory burst) of the cells, an event critical to the microbicidal/tumoricidal activity of macrophages (Hilbertz et al., 1986; Buchmuller-Rouiller et al., 1989; Gulyas et al., 1990). In this same study, Pb inhibited the acquisition of the activated state and reduced receptor turnover and degradation of surface-bound IFN-γ by the Pb-treated macrophages (Maüel et al., 1989).

Exposure to moderately insoluble Pb compounds can both inhibit and enhance macrophage activity (Zelikoff et al., 1993; Cohen et al., 1994). For example, inhalation of Pb oxide (PbO) at a concentration below the acceptable limit for occupational exposure (i.e., 30 $\mu g/m^3$), reduced the phagocytosis of inert particles and recoverable tumor necrosis factor activity (at specific postexposure time points) by rabbit alveolar macrophage, while stimulating the production of reactive oxygen intermediates (ROI) (i.e., superoxide anion and hydrogen peroxide) in a time-dependent manner (Zelikoff et al., 1993). The enhanced ROI production was thought to be due to effects of Pb ions on the cell membrane which, in turn, promote activation of specific surface receptors involved in triggering the respiratory burst (Buchmuller-Rouiller et al., 1989).

4. Mercury (Hg)

Since the outbreak of Minamata disease (methylmercury poisoning) in Japan in the early 1950s (Clarkson, 1972), Hg, particularly organic methylmercury (MeHg), has been recognized as an extremely hazardous environmental pollutant. Mercury is ubiquitous in the environment and its levels are reported to be increasing in water and air as a consequence of discharges from various industries, as well as from medical and scientific waste (Sunderman, 1988; Franchi et al., 1994). Chronic exposure to inorganic Hg may also occur from dental "silver" amalgams (Hahn et al., 1989).

Mercury, like As, exists in a number of chemical forms which are all neurotoxic (Yang et al., 1994). While exposure to inorganic Hg produces tremor and a neuropsychiatric syndrome, organic Hg, often found in contaminated water and in the inhabiting species (Franchi et al., 1994), produces encephalopathy with persistent neurological disabilities. Elemental Hg, found as a contaminant in a variety of industries, as well as in the manufacture of chlorine, thermometers, and themonuclear weapons, is readily absorbed through the lungs into the blood stream and then distributed to different parts of the body.

In addition to the well-known toxic effects of Hg on the kidneys and nervous system (Clarkson, 1972), studies have demonstrated its effects on host immunocompetence. Adverse consequences of inorganic Hg on the immune system range from immunosuppression to autoimmune disease (Contrino et al., 1992; Kubicka-Muranyi et al., 1993; Kosuda et al., 1994). In humans, the best-known immunologic alteration induced by inorganic Hg exposure is contact dermatitis (Friberg and Enestrom, 1990). In genetically susceptible rodents, a systemic autoimmune disease can be experimentally induced by repeated injections of mercuric chloride ($HgCl_2$) (Bigazzi, 1988).

However, effects of Hg on immune aspects other than hypersensitivity reactions and autoimmunity are not as well defined. In fact, despite its ability to augment immunological responses, only a handful of investigations have evaluated the effects of Hg on macrophage-mediated immunity (Castranova et al., 1980; Lison et al., 1988; Contrino et al., 1992; Boroskova et al., 1993). Studies by Contrino et al. (1992) have demonstrated that *in vitro* exposure to nontoxic concentrations of $HgCl_2$ at 0.1 pM, 10 pM, 0.1 μM, and 1 μM stimulated hydrogen peroxide (H_2O_2) release by elicited peritoneal macrophages collected from Lewis (LEW), but not from brown Norway (BN) rats; exposure of both elicited and resident macrophages from LEW rats to $HgCl_2$ at 0.1 nM to 0.01 pM inhibited H_2O_2 production. Immunosuppression was thought to be due to the inhibitory effects of other immune cells present in the samples examined, since purified cell cultures consisting of ≥98% resident peritoneal macrophages were stimulated to release significantly greater amounts of H_2O_2 following exposure to $HgCl_2$. The authors suggested that the genetic background of the host influenced Hg-induced effects on macrophage activity and that this factor may contribute to the difficulties in fully understanding the hazards associated with environmental Hg.

In other *in vitro* studies, Lison et al. (1988) demonstrated that at concentrations of inorganic Hg that impair cell viability (5 and 10 μM), mouse peritoneal macrophages produced less superoxide than nonexposed cells, due probably to an effect upon cellular NADPH. However, the authors concluded that this effect was probably of little human significance because of the high Hg concentration required to produce such an effect. Lower concentrations of Hg (1 and 2 μM) stimulated (twofold) the effect of phorbol 12,-myristate 13-acetate (PMA) on the activity of macrophage plasminogen activator (PA). The authors hypothesized that this effect was not due to direct effects of Hg on protein kinase C activity (Lison et al., 1990), but rather to the enhanced synthesis of PA synthesis, its translocation to the cell surface, or its binding to the membrane receptor. Studies by Guillard and Lauwerys (1989) demonstrated that *in vitro* exposure of human whole blood to inorganic Hg also inhibited phagocyte production of free radicals as measured by chemiluminescence, but the same effect was not observed in the blood of workers exposed to Hg vapor. A similar depressive effect of Hg on phagocyte metabolic activity was observed in guinea pigs following injection of heavy metals emission from a Hg-producing plant (Boroskova et al., 1993). In the same study, administration of the emission for 5 days also suppressed peritoneal macrophage phagocytic activity during the migration phase of the parasite *Ascaris suum*. The complexity of the emission, however, made interpretation difficult.

Despite the potent toxicity of MeHg, particularly as a neurotoxicant, and its ability to accumulate in the food chain, not much is known concerning its impact on the immune system. Methylmercury has been shown to affect the immunocompetence of laboratory animals by altering specific aspects of humoral immunity (Ohi et al., 1976; Koller et al., 1977; Koller et al., 1980), and by reducing host resistance to infectious virus (Koller, 1975). However, very little is known concerning the effects of MeHg on macrophage-mediated immunity. In one reported study, Koller et al. (1980) demonstrated that at concentrations as high as 10 ppm (for 10 weeks), *in vivo* exposure of mice to MeHg failed to alter peritoneal macrophage phagocytic activity or their numbers of Fc receptors.

5. Nickel (Ni)

Nickel is widely distributed in the ambient environment, and human exposure is ubiquitous from anthropogenic sources including tobacco. Iatrogenic exposures to Ni derive from: (a) Ni-containing prostheses (joint replacements, intraosseous pins, cardiac valve replacements, cardiac pacemaker wires, and dental prostheses); (b) intravenous fluids and medications that are contaminated with Ni; and (c) extracorporeal hemodialysis, owing to Ni contamination of the dialysate fluid (Sunderman, 1983).

The toxicity of Ni is dependent not only upon the route of metal exposure, but also upon the solubility of the specific compound (DiPaolo and Casto, 1979; Sen and Costa, 1985). Nickel toxicity is thought to occur because of its ability to interact with a variety of essential metal cations (i.e., magnesium, zinc, calcium, and manganese) which may then act to: (a) block essential biological functional groups of biomolecules; (b) displace essential metal ions in critical biomolecules; or (c) modify the active conformation of biomolecules (Coogan et al., 1989).

Exposure to Ni has been shown to alter host immunocompetence. The effects of Ni exposure on the immune system can be divided into two subcategories. Nickel can either augment an immune response, resulting in allergic contact dermatitis or asthma, or can be immunosuppressive, altering the activity of specific cell types involved in the normal immune response.

Since the primary route of Ni exposure in humans is via inhalation, a number of studies have been performed to examine the effects of inhaled or instilled soluble and/or insoluble Ni compounds on alveolar macrophage functions. In terms of morphology/ultrastruture, macrophages recovered from rats exposed to insoluble nickel oxide (NiO) demonstrated distinct changes in cell size as well as alterations in cytoplasmic inclusion bodies, increased numbers of rough and smooth endoplasmic reticuli, pigment aggregations, and eccentrically placed nuclei with peripheral chromatinization (Camner et al., 1978; Migally et al., 1982; Murthy and Niklowitz, 1983). Exposure to higher-doses and/or longer exposure durations of NiO: increased macrophage size, increased the number of multinucleated macrophages, decreased lavageable macrophage numbers, and at the highest Ni concentration, drastically reduced macrophage viability (Bingham et al., 1972; Spiegelberg et al., 1984). Nickel-induced changes in macrophage ultrastructure are similar to those observed following *in vitro* exposure of rabbit alveolar cells to soluble Ni salts (Waters and Gardner, 1975).

Inhaled Ni also affects pulmonary macrophage functionality. In one study, effects on phagocytic activity were biphasic and depended upon metal dose; phagocytosis was enhanced at a midrange concentration, but depressed by exposure to a 2-fold higher Ni concentration (Spiegelberg et al., 1984). The authors suggested that effects observed at the highest dose were most likely due to metal cytotoxicity. In a study by Gardner (1980), both insoluble NiO and nickel subsulfide, at occupationally relevant concentrations, depressed alveolar macrophage phagocytic activity; no effects on the phagocytic activity of peritoneal macrophages were observed. In contrast, macrophages recovered from rabbits exposed for 6 months to metal particles from a Ni refinery dump demonstrated increased phagocytosis and increased lysosomal enzyme activities, as well as a reduction in antibody-mediated rosette formation which appeared to be due to a dose-dependent Ni-induced inhibition of F_c receptor activity (Reichrtova et al., 1986).

Similar to the effects produced in the lungs by insoluble Ni, inhalation of soluble Ni compounds increased macrophage cell diameter, as well as their capacity to reduce nitroblue tetrazolium at rest and after stimulation with *E. coli* (Wiernik et al., 1983). In addition, inhaled nickel chloride ($NiCl_2$) also reduced pulmonary macrophage bactericidal capacity; effects on bactericidal activity were thought to have been due to reduced lysozyme activity in the exposed macrophage.

In vitro results support those findings observed *in vivo*. Rabbit alveolar macrophages exposed *in vitro* to soluble Ni salts also reduced lysosomal enzymes (Waters and Gardner, 1975), as well as ROI production and oxygen consumption (Castranova et al., 1980). Alterations in alveolar macrophage functions similar to those observed following Ni inhalation/instillation have also been observed following parenteral exposure (Sunderman et al., 1989). For example, parenteral injections of nickel chloride ($NiCl_2$) resulted in reduced alveolar macrophage phagocytic capacity, hypertrophy, and altered macrophage ultrastructure; injections of $NiCl_2$ also increased the macrophage activation state.

In addition to the influence of Ni solubility on toxicity (DiPaolo and Casto, 1979; Sen and Costa, 1985), interspecies differences in sensitivity to Ni compounds have also been shown to be important in assessing the potential risks from Ni exposure (Benson et al., 1986; Benson et al., 1988). Dog alveolar macrophages were at least 10-times more sensitive to the immunotoxic effects of certain Ni compounds than were rat cells (Benson et al., 1986), with the differences appearing to correlate with macrophage phagocytic ability (Benson et al., 1988). Macrophages from fish have also been shown to be sensitive to the immunomodulating effects of soluble Ni (Bowser et al., 1994).

6. Tin (Sn)

In general, environmental contamination by inorganic Sn compounds is only slight. While food represents the main source of Sn exposure for both humans and animals (Kent and McCance, 1941; Tipton et al., 1966), most ambient pollution arises from the gases and fumes, as well as waste slags, derived during Sn processing. Based upon the predicted low levels of Sn present in cooking/drinking water, average daily intakes of Sn by humans are estimated to be <30 µg Sn/d.

The predominant acute toxicity from the ingestion of inorganic Sn is the result of its oxidizing/irritant properties. In humans, the lungs, rather than clearing deposited Sn as occurs in the liver and kidneys, accumulates the metal with increasing host age. Several studies have recently demonstrated that inorganic Sn, administered as a powder, in a needle form, or as a liquid, displays a marked tendency for concentrating within the lymphoid system (Hamilton et al., 1973; Cardarelli, 1986).

Relatively few studies have described specific immunotoxicologic effects of inorganic Sn compounds and of those, most have focused upon lymphocyte-associated activities. This results, in part, from the observations that exposures to metallic Sn gives rise to increased lymph node size, with a greater incidence of granulomas and a proliferation of plasma cells and Russell body cells within these draining lymph nodes (Levine and Sowinski, 1982, 1983).

The time-dependent dissolution of metallic Sn and its impact on lymphocytes has also been observed with respect to macrophage functions. In a study of the effects of silver-Sn alloys (AG_3Sn, the g-phase of dental amalgam), prolonged *in situ* or *in vitro* exposures of macrophages to the alloy particles resulted in macrophage dysfunction and cellular abnormalities (Syrjanen et al., 1986). In peritoneal macrophages obtained from rodent hosts injected i.p. with a single dose of alloy, the numbers of alloy particles per recovered cell were significantly reduced as compared with the numbers of particles recovered in macrophages obtained from hosts injected with nonopsonized latex particles (of the same mass median diameter). In a second study, when macrophages obtained from alloy treated hosts were later presented with latex beads, phagocytic activity was again reduced compared to cells from alloy-free hosts. This subsequent response resulted in the suggestion by the authors that macrophage phagocytic function was nonspecifically impaired by alloy exposure. With increasing postexposure time, the incidence of macrophage autolysis was also increased. This effect was evidenced by both a reduction in the numbers of recoverable macrophages within peritoneal washes and by direct analysis of cultured macrophages incubated with the alloy particles. Solubilization of the ingested particles was thought to be adequate to cause intracellular toxicity in the macrophages, with the subsequent release of Sn ions into the surrounding milieu causing damage to previously unaffected macrophages. This mechanism is similar to that which has been observed within alveolar macrophages after either *in vivo* or *in vitro* exposures to water-insoluble Pb oxide particles (Zelikoff et al., 1993; Cohen et al., 1994).

Although a number of studies have documented the effects of organotin compounds on natural killer cell activity (as described later in this chapter), as well as on specific aspects of cell and humoral immune functions (Smialowicz et al., 1988; Nishida et al., 1990; Penninks et al., 1991), relatively little is known concerning the effects of organic Sn compounds on macrophage-mediated immunity. Tributyltin oxide (TBTO) is thought to impair the mononuclear phagocyte system based on host infectivity studies performed with *Listeria monocytogenes* (Vos et al., 1984, 1985). Lending support to this hypothesis was the observation that TBTO fed rats had decreased numbers of splenic and peritoneal macrophages, as well as decreased phagocytosis and intracellular killing of bacterial target cells. In contrast to TBTO, triphenyltin hydroxide failed to alter splenic clearance of *L. monocytogenes* (Vos et al., 1984), and in another *in vivo* study, oral administration of diocytltin dichloride (DOTC) did not impair the phagocytic capacity of macrophages as determined by carbon clearance (Opacka and Sparrow, 1985). In addition, rats fed 20 or 80 ppm TBTO in the diet for 6 weeks had suppressed natural cytotoxic activity of recovered peritoneal macrophages; exposure to the higher TBTO concentration also suppressed natural killer cell activity. As noted by Vos et al. (1984), prolonged TBTO suppression of the immune functions of cytotoxic macrophages and natural killer cells could have led to increased incidences of neoplastic diseases as well as infections.

III. NATURAL KILLER CELLS

A. GENERAL OVERVIEW

Natural killer (NK) cells are $CD3^-TCR^-$lymphocytes which are spontaneously cytotoxic for a variety of tumor and virally infected cells (Moretta et al., 1992). The cytotoxic activity of NK cells occurs in the absence of immunologic priming, and unlike T cells NK cells do not display target-specific memory. Natural killer cells are considered to play a primary role in resistance to certain tumors, viruses, and infectious agents, as well as participating in the regulation of lymphoid and other hematopoietic cell populations (Kiessling and Wigzell, 1979; Herberman and Ortaldo, 1981; Hanna and Burton, 1981; Abruzzo and Rowley, 1983).

The precise lineage of NK cells is not known for certain. Natural killer cells can be generated from a specific phenotype population of thymocytes; however, this may be a result of the fact that the thymus is one of several homing organs for NK precursors as opposed to being a specific maturation pathway. Also, nonlytic murine NK cell precursors from the bone marrow have been shown to develop into lytic NK cells (Karre et al., 1991).

Natural killer cells are closely associated with a morphological subpopulation of cells called large granular lymphocytes (LGL), which has enabled their isolation and characterization (Herberman et al., 1986). However, not all NK cells are LGLs, and not all LGLs display NK activity. It appears that LGL morphology is more a characteristic of activated cytotoxic lymphocytes rather than a unique property of NK cells (Karre et al., 1991).

A variety of cytokines are capable of augmenting NK cell activity, including interferon (IFN) and IFN inducers (Djeu et al., 1979; Platsoucas et al., 1986). Pharmacologic doses of IL-2 can also activate and induce the proliferation of NK cells. However, it is not known if this is a direct induction and activation by IL-2 alone or an indirect effect mediated by IFN and/or other cytokines (Karre et al., 1991). IL-12, which was originally called NK-stimulatory factor (NKSF), is produced by B lymphocytes, monocytes/macrophages, and possibly other accessory cells. IL-12/NKSF can induce NK to produce IFN-γ, IFN-α, granulocyte-macrophage colony-stimulating factor (GM-CSF) and IL-3 (Cuturi et al., 1989; Kobayashi et al., 1989). NK cells can also be down-regulated by a number of cytokines. For example, transforming growth factor β (TGF-β), IL-10, prostaglandins, and IL-3 have all been reported to significantly down-regulate NK cell activity, proliferation, and/or induction (Brunda et al., 1980; Kalland, 1986; Su et al., 1991; D'Andrea et al., 1993).

Direct and indirect experimental evidence suggests that NK cells play a role in protection against certain tumors and infectious agents (Herberman and Ortaldo, 1981). For example, the *in vivo* growth of implanted tumor cell lines correlated with the levels of NK activity in different strains of inbred mice, such that strains with high NK activity have fewer tumors than those with low NK activity (Kiessling et al., 1975; Hanna, 1986). The growth of transplanted tumors in homozygous C57BL/6 (bg/bg) beige mice, which exhibit deficient NK activity, was more rapid and progressive compared with control heterozygous C57BL/6 (bg/+) littermates (Karre et al., 1980). Tumor metastasis and the clearance of injected radiolabeled tumor cells have also been found to correlate with NK activity in mice (Gorelik et al., 1982).

Natural killer cells also play a role in natural immunity against infectious agents, as demonstrated by the work of several investigators. For example, the injection of mice 4 to 6 h prior to viral infection with anti-asialo GM1 serum, which decreases NK activity but does not alter the cytotoxic activity of macrophages, resulted in higher viral titers in the liver and spleen compared to controls (Bukowski et al., 1983). Resistance to the lethal effects of murine cytomegalovirus (MCMV) has been reported to correlated with the degree of virus-augmented NK activity in several strains of mice with different susceptibility to MCMV (Bancroft et al., 1981; Shellman et al., 1985). Resistance of mice to the protozoan parasite *Babesia microti* has also been reported to be associated with NK cell activity (Ruebush and Burgess, 1982). These cells have also been suggested to play a role in natural resistance against infection of mice with the fungus *Cryptococcus neoformans* (Murphy, 1982).

Natural killer cells have been shown to serve an early and important role as nonspecific effectors in a number of experimental animal host resistance models. Therefore, alterations in the integrity of this lymphocyte population by toxic chemicals, including metals, have the potential to result in increased host susceptibility and severity of disease.

B. METAL-INDUCED EFFECTS
1. Arsenic

There have been limited reports that various arsenic compounds have the ability to alter immune function, viral resistance, and/or tumor growth (Gainer and Pry, 1972; Blakely et al., 1980; Kerkvliet et al., 1980; Sikorski et al., 1989; McCabe et al., 1983). Moderately high concentrations of $NaAsO_2$ have also been shown to inhibit interferon production (Gainer, 1972). Alterations in viral resistance and tumor growth by As compounds suggests that NK cells may be affected by this metal.

In a recent study, a single intratracheal instillation of $NaAsO_2$ in mice failed to alter splenic NK cell activity, as measured *in vitro* by ^{51}Cr release from YAC-1 target cells, or to modify the number of B16F10 melanoma tumor nodules in the lungs of exposed mice (Sikorski et al., 1989). Intratracheal instillation of GaAs, however, caused an enhancement of splenic NK activity, as well as an increase in the number of B16F10 tumor colonies in the lungs of exposed mice. The observed increase in tumor burden was unrelated to a possible direct injury of the lungs by GaAs, with a resultant inability of the lungs to remove tumor cells, since clearance of radiolabled B16F10 cells from the lungs of GaAs-exposed mice was not reduced (Sikorski et al., 1989). The apparent discrepancy in the enhanced splenic activity and increased lung tumor burden in these mice is difficult to reconcile. However, the cytotoxic activity of lung-associated NK cells was not determined in this study. This is an important consideration, given that suppression of regional (i.e., lung-associated), but not systemic (i.e., splenic) NK cell activity has been reported to be related to the inhalation dose of phosgene in rats (Burleson and Keyes, 1989).

The differential effects of $NaAsO_2$ and GaAs on NK cell activity (Sikorski et al., 1989) makes it difficult to implicate As itself as altering NK cell activity. Any conclusions about the potential immunotoxic effects of As on NK cells, based solely on the results of this single study, is premature.

2. Cadmium

There are conflicting reports of the effects that Cd has on immune function in experimental animals (Descotes, 1992). For example, exposure to Cd has been reported to suppress (Koller and Roan, 1975), enhance (Malave an De Ruffino, 1984), or not affect (Thomas et al., 1985) humoral immunity. These discrepancies have been attributed to differences in the doses and chemical form of Cd employed, the route of administration, the duration of exposure, and genetic susceptibility of the animals studied (Cifone et al., 1989).

Variable effects on NK activity have also been reported following Cd exposure. Exposure of mice for 90 days to $CdCl_2$ via their drinking water, at a concentration as high as 250 ppm, enhanced, albeit not significantly, splenic NK cell activity as determined by the *in vitro* ^{51}Cr release assay (Thomas et al., 1985). In contrast to this earlier study in mice, Chowdhury and Chandra (1989) reported that mice given Cd in their drinking water at 50 ppm for 3 weeks had lower NK cell activity than controls in a 12-h *in vitro* ^{51}Cr release assay. These investigators also reported that the concurrent administration of zinc with Cd prevented Cd-induced suppression.

In a subchronic study, rats were exposed to Cd via their drinking water for 170 days at concentrations of 200 or 400 ppm and evaluated for NK activity at various time points (Cifone et al., 1989). At 10, 20, and 30 days into the Cd exposure, splenic and peripheral blood lymphocyte (PBL) NK cell activity was reduced. However, by 40 days into exposure and until the end of the 170-d exposure period, splenic and PBL NK activity was enhanced compared with controls. A pattern of initial suppression during the first month and subsequent enhancement of NK activity in the percentage of LGL in peripheral blood was correlated with a change in LGL percentage and NK activity in Cd-treated rats. In addition, the percentages of asialo GM1+ and OX8+ peripheral blood cells were also initially decreased and then increased after the first month of exposure compared with controls. Examination of rats 2 months after cessation of Cd exposure for 170 days indicated that all endpoints returned to control values (Cifone et al., 1989).

Acute parenteral administration of Cd in rodents has been found to suppress NK cell activity. The intraperitoneal injection of rats with 4 mg/kg $CdCl_2$, but not 0.25 or 1 mg/kg, resulted in suppression of splenic NK activity (Stacey et al., 1988). Suppression of NK activity at 4 mg Cd/kg was associated with decreased spleen weight and cellularity. However, the authors suggest that the lack of an effect on the mitogen-stimulated response of splenocytes from these same animals indicated that the observed decrease in NK activity was more specific than a general inhibition of splenocyte function (Stacey et al., 1988). Acute intramuscular, but not inhalation, exposure of mice to 6.25 mg/kg $CdCl_2$ also resulted in suppression NK activity by Cd, however, did not correlate with increased susceptibility to infection with MCMV.

In vitro studies using human PBLs indicate that Cd inhibits NK cell activity. Cadmium chloride inhibited human NK activity against K526 target cells in a dose-dependent manner at concentrations ranging from 25 to 250 μM, with significant inhibition observed at 100 and 250 μM (Stacey, 1986). In this same study, Stacey (1986) reported that zinc also inhibited NK activity and that in combination with Cd, the effects on NK activity were generally additive. Cifone et al. (1990) also demonstrated that Cd inhibits *in vitro* human PBL NK activity in a dose- and time-dependent manner. Maximal inhibition occurred when Cd was present during the early period of the NK assay. Preincubation of either NK effector cells or K562 target cells with Cd did not result in any inhibition Furthermore, the results of Cifone et al. (1990) indicated that Cd appears to interfere with the hydrolysis of phosphoinositides, as demonstrated by Cd-induced decreases in PBL inositol triphosphate.

The evidence presented above indicates that Cd is capable of modulating NK cell activity both *in vivo* and *in vitro*, as determined primarily by the ^{51}Cr release assay. However, the biological significance of Cd-induced NK modulation is open to question because of the variable responses reported. Furthermore, in the only paper that actually compared NK activity with a host resistance model, which is dependent on intact NK activity, no correlation was observed between activity and host resistance to MCMV infection (Daniels et al., 1987).

3. Lead

Lead has been shown to alter humoral and cell-mediated immunity and increase susceptibility of rodents to bacterial and viral infections (Koller and Vos, 1974; Koller, 1980). In contrast to the effects of Pb on humoral and cell-mediated immunity, *in vivo* exposure of rodents to Pb has not been shown to alter innate immunity mediated by NK cells. For example, exposure of mice to 1300 ppm lead acetate (827 ppm Pb) in their drinking water for 8 weeks did not alter spontaneous or poly I:C-induced splenic NK cell activity (Neilan et al., 1983). Similarly, splenic NK activity was unaffected in rats exposed for 10 weeks to 10 or 1000 ppm lead acetate in their drinking water (Talcott et al., 1985; Exon et al., 1985). In another study, native and IFN-activated splenic NK activity was not different from controls in rats exposed to 100 or 1000 ppm lead acetate in drinking water for 8 weeks (Kimber et al., 1986). On the other hand, *in vitro* exposure of rat splenocytes to lead acetate at concentrations of 0.4 and 20 $\mu g/ml$ resulted in suppression of NK activity (Talcott et al., 1985) and IL-2 production (Exon et al., 1985) in the absence of a direct cytotoxic effect of Pb.

Lead has been reported to alter host defenses to certain viral infections (Gainer, 1974; Thind and Singh, 1977). Since NK cells have been implicated as playing an important role in host defense against viral infections, it is not unreasonable to expect that Pb might alter NK activity. However, from the studies cited above there is insufficient evidence to implicate Pb as a modulator of NK activity.

4. Mercury

Mercury compounds have been found to cause autoimmunity or immunosuppression in experimental animals. Mercury-induced autoimmunity is characterized by a T cell-dependent polyclonal B cell activation with lymphadenopathy, increased IgE and IgG serum levels, and the appearance of IgG autoantibodies (Hirsch et al., 1986; Ochel et al., 1991; Hultman and Enestrom, 1992). On the other hand, mecuric chloride (HgCl$_2$) has been reported to suppress humoral and cell-mediated immune function under certain conditions (Koller, 1973; Lawrence, 1981; Dieter et al., 1983; Pelletier et al., 1987). Organic mercury (i.e., MeHg) has also been reported to suppress antibody responses (Ohi et al., 1976; Koller et al., 1977) and to result in increased susceptibility to viral infections (Koller, 1975).

The effect that mercury has on NK activity has not been well studied. In fact, we are aware of only two such studies. Mice exposed to MeHg for 12 weeks via their diets had reduced splenic and peripheral blood NK cell activity as determined by the *in vitro* ^{51}Cr release assay (Ilback, 1991). Splenic NK cell activity was also suppressed in 15-day-old rats exposed perinatally to MeHg *in utero* and during lactation (Ilback et al., 1991). While these findings are interesting, neither study provided dose-response data for Hg-induced NK suppression. Further work is necessary to corroborate these findings.

5. Nickel

Nickel has been found to adversely affect the immune system of experimental animals. For example, nickel compounds depress interferon production *in vivo* and *in vitro* (Treagan and Furst, 1970; Gainer, 1977) and suppress T lymphocyte-mediated reactions on NK cell activity (Smialowicz et al., 1984a; 1985b; 1986a; 1987a; 1987b).

A single intramuscular (i.m.) injection of nickel chloride (NiCl$_2$) in mice resulted in a dose-related suppression of splenic NK activity, reduced *in vivo* clearance of radiolabeled tumor cells, and increased lung tumor burdens following challenge with NK-sensitive B16F10 melanoma cells (Smialowicz et al., 1984a; 1985a; 1987b). Young adult (i.e., 8 to 10 weeks old) mice which were exposed to NiCl$_2$ *in utero* via subcutaneously implanted osmotic minipumps, from gestation day 5 to birth, had reduced NK activity, but did not display increased susceptibility to B16F10 tumors (Smialowicz et al., 1986a). However, mice infected with MCMV and subsequently given a single im injection of NiCl$_2$ had reduced virus-augmented NK activity and increased mortality to MCMV infection (Daniels et al., 1987).

In contrast to Ni-induced suppression of NK activity, a single i.m. injection of manganese chloride (MnCl$_2$) resulted in enhanced splenic NK activity, increased *in vivo* clearance of radiolabeled tumor cells, and decreased B16F10 lung tumor burdens (Rogers et al., 1983; Smialowicz et al., 1984b; 1988a). Manganese also enhanced macrophage phagocytic function as well as tumor cytostatic and cytolytic activity in mice (Smialowicz et al., 1985b). Manganese-induced enhancement of NK activity was found to be mediated via the induction of IFN both *in vivo* (Smialowicz et al., 1984b, 1988a) and *in vitro* (Smialowicz et al., 1986b). Finally, the co-administration of NiCl$_2$ and MnCl$_2$ in mice resulted in the reversal of suppressed NK activity in mice injected with NiCl$_2$ alone (Smialowicz et al., 1987b).

Rat NK activity is also suppressed by Ni exposure. For example, rats given a single i.m. injection of NiCl$_2$ had reduced splenic NK activity and decreased resistance to challenge with NK-sensitive MADB106 rat mammary adenocarcinoma cells (Smialowicz et al., 1987a). In another study, rats given a single i.m. injection of either Ni or nickel subsulfide (Ni$_3$S$_2$) had reduced PBL NK activity, and those rats which had persistent depressed NK activity subsequently developed Ni-induced rhabdomyosarcomas (Judde et al., 1987). In this same study, the inclusion of Mn with Ni compounds injected i.m. in rats inhibited the development of tumors and also prevented the depression of NK cell activity produced by Ni alone (Judde et al. 1987).

While a consistent suppression of NK activity has been observed following parenteral administration of Ni, inhalation or lung deposition of Ni compounds has been reported to result in variable changes in NK activity. Splenic NK activity of mice was not altered following exposure to aerosols of Ni$_3$S$_2$ for 6 h/d for 12 d (Benson et al., 1987b; 1988). However, more prolonged exposure of mice to Ni$_3$S$_2$ or nickel sulfate (NiSO$_4$) (i.e., 6 h/d, 5 d per week for 65 d) resulted in either decreased splenic NK activity or increased lung tumor burdens following challenge with NK-sensitive B16F10 melanoma cells, respectively (Haley et al., 1990). Unfortunately, alveolar NK activity was not determined in any of these three inhalation studies. On the other hand, lung-associated NK activity was evaluated in cynomolgus monkeys which had been previously immunized and repeatedly challenged with SRBC in specific lung lobes followed by instillation of Ni$_3$S$_2$ (Haley et al., 1987). In this study, NK cell activity was enhanced in all lung lobes examined regardless of SRBC or nickel exposure in all nickel-exposed monkeys.

While Ni is undoubtedly capable of modulating NK cell activity both *in vivo* and *in vitro*, the biological relevance of Ni-induced NK modulation is debatable given that *in vivo* suppression occurs following parenteral, but not inhalation, exposure. Future work should focus on determining if long-term inhalation exposure to Ni alters lung-associated NK activity and host resistance models which are dependent upon intact lung NK function.

6. Tin

The immunomodulatory effects of inorganic tin compounds have not been extensively studied; however there are several studies which have documented effects of organotin compounds on NK cell activity. These effects are apparently related to the organotin compound used and to the duration and route of exposure. Prenatal (day 10 to 20 of gestation), pre- and postnatal (day 11 to 20 of gestation and 2 to 11 days of age), early postnatal (2 to 13 days of age), or adult (12 weeks old) oral exposure of rats to diocyltin dichloride (DOTC), at a dosage as high as 50 mg/kg/d, did not alter splenic NK activity (Smialowicz et al., 1988b). Adult rats given a single intraperitoneal (i.p.) injection of 10 mg/kg trimethyltin (TMT) displayed NK activity similar to that of controls (Hioe and Jones, 1984). Similarly, early postnatal (3 to 24 days of age) or adult (9 weeks old) oral exposure of rats to tributyltin oxide (TBTO), at a dosage as high as 10 mg/kg/d, failed to alter NK activity compared with controls (Smialowicz et al., 1989). In contrast, weaned rats maintained on a diet containing TBTO for 6 weeks at 80, but not 20 mg/kg TBTO had suppressed splenic NK activity (Vos et al., 1984). Maintenance on diets containing either 20 or 80 mg/kg TBTO for 6 weeks, however, resulted in suppressed NK activity in the lungs of rats (Van Loveren et al., 1990). Long-term exposure of weaned rats maintained on diets containing 0.5, 5, or 50 mg/kg TBTO had suppressed splenic NK activity after 16 months, but not after 4.5 months of exposure (Vos et al., 1990). Finally, mice dosed for one week with 10 or 1000 ppm tributyltin chloride (TBTC), but not inorganic tin, had reduced splenic LGL numbers and NK activity (Ghoneum et al., 1990). In this study, the preincubation of mouse splenocytes with 0.01 or 0.05 ppm TBT, but not inorganic tin, also suppressed NK activity.

Taken together, organotins but not inorganic tin are capable of altering NK activity. However, the organotin used and the route and duration of exposure appear to be critical elements for subsequent suppression of NK activity.

IV. CONCLUSIONS

Metals clearly act to alter normal macrophage and natural killer cell activity in a variety of animal species, including humans. In some cases, metal-induced suppression of innate immunity has also been correlated with increased susceptibility to bacterial, viral, or tumor challenges. Consistency of results among various studies with a given metal, however, have not been obtained very frequently. These conflicting results may be partially explained by diferences in the metal compound employed, in the dose and route of exposure, and/or in species and strain susceptibility.

Many of the metals discussed in this chapter appear to augment host innate immunity at low levels, yet are immunosuppressive at higher concentrations. While enhancements of immune responses may, on initial inspection, appear advantageous, any disruption of the delicately balanced immune "network" could result in adverse consequences for the exposed host, such as hypersensitivity reactions and autoimmune disorders. Conversely, no advantage whatsoever could be gained by even a minimal reduction in host immunoresponsiveness. The increased incidences of viral and bacterial diseases, as well as cancers, are most commonly the ultimate consequences of this type of metal-induced immunomodulation.

It appears clear that future experimental work should focus on studies designed to determine the effects and underlying mechanisms of long-term, environmentally relevant levels of metals on macrophage and NK cell activity and on host resistance to infectivity models. Such studies would provide important information to the database for immunotoxicology risk assessment of metals.

REFERENCES

Abruzzo, L. V. and Rowley, D. A. (1983), *Science,* 222, 581–585.

Aranyi, C. (1981), Effect of industrial particulate samples on alveolar macrophages, U.S. EPA Rep. EPA-600/S1–81–003.

Aranyi, C., Bradof, J. N., O'Shea, J., Graham, J. A., and Miller, F. J. (1985), *J. Toxicol. Environ. Health,* 15, 163–170.

Bancroft, G. J., Shellam, G. R., and Chalmer, J. E. (1981), *J. Immunol.,* 126, 988–994.

Beck, B. D. (1992), *Fundam. Appl. Toxicol.,* 18, 1–10.

Bencko, V. (1985), in *Genotoxic and Carcinogenic Metals: Environmental and Occupational Occurrence and Exposure,* Fishbein, L., Furst, A., and Mehlman, M. A., Eds., Princeton Scientific, Princeton, NJ, 1–25.

Benson, J. M., Henderson, R. F., and McClellan, R. O. (1986), *J. Toxicol. Environ. Health,* 19, 105–113.

Benson, J. M., Henderson, R. F., and Pickrell, J. A. (1988), *J. Toxicol. Environ. Health,* 24, 373–382.

Benson, J. M., Burt, D. G., Carpenter, R. L., Eidson, A. F., Hahn, F. F., Haley, P. J., Hanson, R. L., Hobbs, C. H., Pickrell, J. A., and Dunnick, J. K. (1987a), *Fundam. Appl. Toxicol.,* 10, 164–178.

Benson, J. M., Carpenter, R. L., Hahn, F. F., Haley, P. J., Hanson, R. L., Hobbs, C. H., Pickrell, J. A., and Dunnick, J. K. (1987b), *Fundam. Appl. Toxicol.,* 9, 251–265.
Bigazzi, P. E. (1988), *J. Toxicol. Clin. Toxicol.,* 26, 125–156.
Bingham, E., Barkley, W., Zerwas, M., Stemmer, K., and Taylor, P. (1972), *Arch. Environ. Health,* 25, 406–414.
Blakely, B. R., Sisodia, C. S., and Mukkur, T. K. (1980), *Toxicol. Appl. Pharmacol.,* 52, 245–254.
Boroskova, Z., Benkova, M., Soltys, J., Krupicer, I., and Simo, K. (1993), *Vet. Parasitol.,* 47, 245–254.
Bowser, D. H., Frenkel, K., and Zelikoff, J. T. (1994), *Bull. Environ. Contam. Toxicol.,* 52, 367–373.
Boyer, I. J. (1989), *Toxicology,* 55, 253–259.
Brunda, M. J., Herberman, R. B., and Holden, H. T. (1980), *J. Immunol.,* 124, 2682–2687.
Buchanan, W. D. (1962), in *Elsevier Monographs on Toxic Agents,* Browning, E., Ed., Elsevier Publishing Co., New York, 1–50.
Buchmuller-Rouiller, Y., Ransijn, A., and Auel, J. (1989), *Biochem J.,* 260, 325–333.
Bukowski, J. F., Woda, R. A., Habu, S., Okumura, K., and Welsh, R. M. (1983), *J. Immunol.,* 131, 1531–1538.
Burleson, G. R. and Keyes, L. L. (1989), *Immunopharmacol. Immunotoxicol.,* 11, 421–443.
Burns, L. A., Sikorski, E. E., Saady, J. J., and Munson, A. E. (1991), *Toxicol. Appl. Pharmacol.,* 110, 157–164.
Camner, P., Johansson, A., and Lundborg, M. (1978), *Environ. Res.,* 16, 226–234.
Cardarelli, N. F. (1986), in *Tin as a Vital Nutrient,* Cardarelli, N. F., Ed., CRC Press, Boca Raton, FL, 41–60.
Castranova, V., Bowman, L., Reasor, M. J., and Miles, P. R. (1980), *Toxicol. Appl. Pharmacol.,* 53, 14–22.
Chowdhury, B. A. and Chandra, R. K. (1989), *Immunol. Lett.,* 22, 287–292.
Cifone, M. G., Alesse, E., Di Eugenio, R., Napolitano, T., Morrone, S., Paolini, R., Santoni, G., and Santoni, A. (1989), *Immunopharmacology,* 18, 149–156.
Clarkson, T. W. (1972), *Annu. Rev. Pharmacol.,* 12, 375–406.
Cohen, M. D., Yang, Z., and Zelikoff, J. T. (1994), *J. Toxicol. Environ. Health,* 42, 377–392.
Contrino, J., Kosuda, L. L., Marucha, P., Kreutzer, D. L., and Bigazzi, P. E. (1992), *Int. J. Immunopharmacol.,* 14, 1051–1059.
Coogan, T. P., Latta, D. M., Snow, E. T., and Costa, M. (1989), *Crit. Rev. Toxicol.,* 20, 135–152.
Cuturi, M. C., Anegon, I., Sherman, F., Loudon, R., Clark, S. C., Perussia, B., and Trinchieri, G. (1989), *J. Exp. Med.,* 169, 569–583.
D'Andrea, A., Aste-Amezaga, M., Valiante, N. M., Ma, X., Kubin, M., and Trinchieri, G. (1993), *J. Exp. Med.,* 178, 1041–1048.
Daniels, M. J., Menache, M. G., Burleson, G. R., Graham, J. A., and Selgrade, M. J. K. (1987), *Fundam. Appl. Toxicol.,* 8, 443–453.
Descote, J. (1992), *IARC Sci. Publ.,* 118, 385–390.
DeVries, C. R., Ingram, P., Walker, S. R., Linton, R. W., Gutknecht, W. F., and Shelburne, J. D. (1983), *Clin. Invest.,* 48, 35–45.
Dieter, M. J., Luster, M. I., Boorman, G. A., Jameson, C. E., Dean, J. H., and Cox, J. W. (1983), *Toxicol. Appl. Pharmacol.,* 68, 218–228.
DiPaolo, J. A. and Casto, B. C. (1979), *Cancer Res.,* 39, 1008–1010.
Djeu, J. Y., Heinbaugh, J. A., Holden, H. T., and Herberman, R. B. (1979), *J. Immunol.,* 122, 175–181.
Elasser, M. S., Roberson, B. S., and Hetrick, F. M. (1986), *Vet. Immunol. Immunopathol.,* 12, 243–250.
Fisher, G. L., McNeill, K. L., and Democko, C. J. (1986), *Environ. Res.,* 39, 164–171.
Fowler, B. A. and Weissberg, J. B. (1974), *N. Engl. J. Med.,* 291, 1171–1175.
Franchi, E., Loprieno, G., Ballardin, M., Petrozzi, L., and Migliore, L. (1994), *Mutat. Res.,* 320, 23–29.
Friberg, L. and Enestrom, S. (1990), in *Immunotoxicity of Metals and Immunotoxicology,* Proc. Int. Workshop, Dayan E. Ed., Plenum Press, London, 139–152.
Froines, J. R., Baron, S., Wegman, D. H., and O'Rourke, S. (1990), *Am. J. Indust. Med.,* 18, 1–8.
Gainer, J. H. (1972), *Am. J. Vet. Res.,* 33, 2579–2585.
Gainer, J. H. (1974), *Environ. Health Perspect.,* 7, 113–119.
Gainer, J. H. (1977), *Am. J. Vet. Res.,* 38, 869–872.
Gardner, D. E. (1980), in *Nickel Toxicology,* Brown, S. S. and Sunderman F. W., Jr., Eds. Acadmic Press, London, 121–140.
Ghoneum, M., Hussein, A. E., Gill, G., and Alfred, L. J. (1990), *Environ. Res.,* 52, 178–186.
Gorelik, E., Wiltrout, R., Okumura, K., Habu, S., and Herberman, R. B. (1982), in *NK Cells and Other Natural Effector Cells,* Herberman, R. B., Academic Press, New York, 1331–1337.
Goyer, R. A. (1986), in *Casarett and Doull's Toxicology: The Basic Science of Poisons,* Klaassen, C. D., Amdur, M. O., and Doull, J., Eds. Macmillan, New York, 582–605.
Goyer, R. A. (1991), in *Toxicology, the Basic Science of Poisons,* Amdur, M. O. Doull, J., and Klaassen, C. D., Eds., McGraw-Hill, New York, chap. 19.
Graham, J. A., Gardner, D. E., Waters, M. D., and Coffin, D. L. (1975), *Infect. Immunol.,* 11, 1278–1288.
Greenspan, B. J. and Morrow, P. E. (1984), *Fundam. Appl. Toxicol.,* 4, 48–57.
Guillard, O. and Lauwerys, R. (1989), *Biochem. Pharmacol.,* 38, 2819–2823.
Gulyas, H., Labedzka, M., and Gercken, G. (1990), *Environ. Res.,* 51, 218–230.
Hadley, J. G., Gardner, D. E., Coffin, D. L., and Menzel, D. B. (1977), *J. Reticuloendothel. Soc.,* 22, 417–425.
Hahn, L. J., Kloiber, R., Vimy, M. J., Takahasi, Y., and Lorscheider, F. L. (1989), *FASEB J.,* 3, 2641–2646.
Haley, P. J., Bice, D. E., Muggenburg, B. A., Hahn, F. F., and Benjamin, S. (1987), *Toxicol. Appl. Pharmacol.,* 88, 1–12.

Haley, P. J., Shopp, G. M., Benson, J. M., Cheng, Y.-S., Bice, D. E., Luster, M. I., Dunnick, J. K., and Hobbs, C. H. (1990), *Fundam. Appl. Toxicol.,* 15, 476–487.

Hamilton, E. I., Minski, M. J., and Cleary, J. J. (1973), *Sci. Total Environ.,* 1, 341–349.

Hanna, N. (1986), in *Immunobiology of Natural Killer Cells,* Lotzova, E. and Herberman, R. B., Eds., Vol. 2 CRC Press, Boca Raton, FL, 1–10.

Hanna, N. and Burton, R. C. (1981), *J. Immunol.,* 127, 1754–1758.

Herberman, R. B. and Ortaldo, J. R. (1981), *Science,* 214, 24–30.

Herberman, R. B., Reynolds, C. W., and Ortaldo, J. R. (1986), *Annu. Rev. Immunol.,* 4, 651–680.

Hernandez, M., Macia, M., Conde, J. L., and De La Fuente, M. (1991), *Int. J. Biochem.,* 23, 541–544.

Hioe, K. M. and Jones, J. M. (1984), *Toxicol. Lett.,* 20, 317–323.

Hirsch, F., Kuhn, J., Ventura, M., Vial, M.-C., Fournie, G., and Druet, P. (1986), *J. Immunol.,* 136, 3272–3281.

Huisingh, J. L., Campbell, J. A., and Waters, M. D. (1977), in *Pulmonary Macrophage and Epithelial Cells,* Sanders, C. L., Schneider, R. P., Dagle, G. E., and Ragan, H. A., Eds., Conf. 760972, ERDA Symp. Ser. 43, Tech. Inf. Center, Energy Research and Development Administration, Washington, D.C., 346–357.

Ilback, N.-G. (1991), *Toxicology,* 67, 117–124.

Ilback, N.-G., Sundberg, J., and Oskarsson, A. (1991), *Toxicol. Lett.,* 58, 149–158.

Johansson, A., Camner, P., Jarstrand, C., and Wiernik, A. (1983), *Environ. Res.,* 31, 340–354.

Judde, J. G., Breillout, F., Cleenceau, C., Poupon, M. F., and Jasmin, C. (1987), *J.N.C.I.,* 78, 1185–1190.

Kalland, T. (1986), *J. Immunol.,* 137, 2268–2271.

Kaminski, E. J., Fischer, C. A., Kennedy, G. L., and Calandra, J. C. (1977), *Br. J. Exp. Pathol.,* 58, 9–17.

Karre, K., Hansson, M., and Keissling, R. (1991), *Immunol. Today,* 12, 343–345.

Karre, K., Klein, G. O., Kiessling, R., Klein, G., and Roder, J. C. (1980), *Nature,* 284, 624–626.

Kent, N. L. and McCance, V. R. A. (1941), *Biochem. J.,* 35, 877–880.

Kerkvliet, N. I., Steppan, L. B., Koller, L. D., and Exon, J. H. (1980), *J. Environ. Pathol. Toxicol.,* 4, 65–79.

Kiessling, R. and Wigzell, H. (1979), *Immunol. Rev.,* 44, 165–208.

Kiessling, R., Petranyi, G. Klein, G., and Wigzell, H. (1975), *Int. J. Cancer,* 15, 933–940.

Kimber I., Jackson, J. A., and Stonard, M. D. (1986), *Toxicol. Lett.,* 31, 211–218.

Kobayashi, M., Fitz, L., Ryan, M., Hewick, R. M., Clark, S. C., Chan, S., Loudon, R., Sherman, F., Perussia, B., and Trinchieri, G. (1989), *J. Exp. Med.,* 170, 827–845.

Koller, L. D. (1973), *Am. J. Vet. Res.,* 34, 1457–1458.

Koller, L. D. (1975), *Am. J. Vet. Res.,* 36, 1501–1504.

Koller, L. D. (1980), *Int. J. Immunopharmacol.,* 2, 269–279.

Koller, L. D. (1990), *Ann. N. Y. Acad. Sci.,* 230, 162–175.

Koller, L. D. and Vos, J. G. (1974), in *Immunologic Considerations in Toxicology,* Vol. 1 Sharma, R. P., Ed., CRC Press, Boca Raton, FL, 67–78.

Koller, L. D. and Roan, J. G. (1975), *Arch. Envir. Health,* 30, 15–19.

Koller, L. D. and Vos. J. G. (1981), in *Immunologic Considerations in Toxicology,* Sharma R. P., Ed., CRC Press, Boca Raton, FL, 67–78.

Koller, L. D., Exon, J. H., and Brauner, J. A. (1977), *Proc. Soc. Exp. Biol. Med.,* 155, 602–604.

Koller, L. D., Roan, J. G., and Brauner, J. A. (1980), *J. Environ. Pathol. Toxicol.,* 3, 407–411.

Kosuda, L. L., Greiner, D. L., and Bigazzi, P. L. (1994), *Cell. Immunol.,* 155, 77–94.

Kowolenko, M., Tracy, L., Mudzinski, S., and Lawrence, D. A. (1988), *J. Leukocyte Biol.,* 43, 357–363.

Kubicka-Muranyi, M., Behmer, O., Uhrberg, M., Klonowski, H., Bister, J., Gleichmann, E. (1993), *Int. J. Immunopharmacol.,* 15, 151–161.

Lawrence, D. A. (1981a), *Infect. Immun.,* 31, 136–141.

Lawrence, D. A. (1981b), *Toxicol. Appl. Pharmacol.,* 57, 238–239.

Lawrence, D. A. (1985), in *Immunotoxicology and Immunopharmacology,* Dean, J., Luster, M. I., Munson, A. E., and Amos, H., Eds. Raven Press, New York, 341–360.

Lee, J. S. and White, K. L. (1980), *Am. J. Industrial Med.,* 1, 307–317.

Leonard, A., Gerber, G. B., Jacquet, P., and Lauwerys, R. R. (1988), in *Mutagenicity, Carcinogenicity, and Teratogenicity of Industrial Pollutants,* Kirsch-Volders, M., Ed., Plenum Press, New York, 28–40.

Levine, S. and Sowinski, R. (1982), *Exp. Mol. Pathol.,* 36, 86–93.

Levine, S. and Sowinski, R. (1983), *Toxicol. Appl. Pharmacol.,* 68, 110–123.

Lison, D., Dubois, P., and Lauwerys, R. (1988), *Toxicol. Lett.,* 40, 29–36.

Loose, L. D., Silkworth, J. B., and Simpson, D. W. (1978), *Infect. Immunol.,* 22, 378–381.

Loose, L. D., Silkworth, J. B., and Warrington, D. (1977), *Biochem. Biophys. Res. Commun.,* 79, 326–332.

Malachowski, M. E. (1990), *Clin. Toxicol.,* 10, 459–470.

Malave, I. and DeRuffino, D. T. (1984), *Toxicol. Appl. Pharmacol.,* 4, 46–56.

Marcus, W. L. (1990), in *Advances in Modern Environment Toxicology, Environmental and Occupational Cancer: Scientific Update,* Mehlman, M. A., Ed., Princeton Scientific, Princeton, New Jersey.

Marcus, W. L. and Rispin, A. S. (1988), in *Risk Assessment and Risk Management of Industrial and Environmental Chemicals,* Vol. 15, Cothern, C. R., Mehlman, M. A., and Marcus, W. L., Eds., pp. 10–35. Princeton Scientific, Princeton, New Jersey.

Marino, P. A. and Adams, D. O. (1980), *Cell Immunol.,* 54, 11–25.
Mauel, J., Ransijn, A., and Buchmuller-Rouiller, Y. (1989), *J. Leukocyte Biol.,* 45, 401–411.
McCabe, M., Maguire, D., and Nowak, M. (1983), *Environ. Res.,* 31, 323–331.
Migally, N., Murthy, R. C., Doye, A., and Zambernard, J. (1982), *J. Submicrosc. Cytol.,* 14, 62–70.
Moretta, L., Ciccone, E., Moretta, A., Hoglund, P., Ohlen, C., and Karre, K. (1992), *Immunol. Today,* 13, 300–305.
Murthy, R. C. and Holovack, M. J. (1991), *J. Submicrosc. Cytol. Pathol.,* 23, 289–293.
Murthy, R. C. and Niklowitz, W. J. (1983), *J. Submicrosc. Cytol.,* 15, 655–673.
Nathan, C. F. (1987), *J. Clin. Invest.,* 79, 319–326.
Nathan, C. F., Murray, H. W., and Cohn, Z. A. (1980), *N. Eng. J. Med.,* 303, 622–626.
National Academy of Sciences (1977), in *Arsenic,* National Academy of Sciences, Washington, D.C., 189–199.
Nielan, B. A., O'Neill, K., and Handwerger, B. S. (1983), *Toxicol. Appl. Pharmacol.,* 69, 272–283.
Nielan, B. A., Taddeini, L., McJilton, C. E., and Handwerger, B. S. (1980), *Clin. Exp. Immunol.,* 39, 746–753.
Nishida, H., Matsui, H., Sugiura, H., Kitagaki, K., Fuchigami, M., Inagaki, N., Nagai, H., and Koda, A. (1990), *J. Pharmacobio-dynamics,* 13, 543–548.
Ochel, M., Vohr, H.-W., Pfeiffer, C., and Gleichmann, E. (1991), *J. Immunol.,* 146, 3006–3011.
Ohi, G., Fukuda, M., Seto, H., and Yagyu, H. (1976), *Bull. Environ. Contam. Toxicol.,* 15, 175–180.
Opacka, J. and Sparrow, S. (1985), *Toxicol. Lett.,* 27, 97–103.
Pelletier, L., Pasquier, R., Rossert, J., and Druet, P. (1987), *Eur. J. Immunol.,* 17, 49–54.
Penninks, A. H., Snoeij, N. J., Pieters, R. H. H., and Seinen, W. (1991), in *Immunotoxicity of Metals and Immunotoxicology,* IPCS Joint Symposium, 15, Dayan, A. Hertel, R. F. and Heseltine, E., Eds., Plenum Press, New York, 197–202.
Platsoucas, C. D., Fernandes, G., Good, R. A., and Gupta, S. (1986), *Int. Arch. Allergy Appl. Immunol.,* 79, 1–7.
Reichrtova, E., Takac, L., and Kovacikova, Z. (1986), *Environ. Pollut.,* 40, 101–112.
Rogers, R. R., Garner, R. J., Riddle, M. M., Luebke, R. W., and Smialowicz, R. J. (1983), *Toxicol. Appl. Pharmacol.,* 70, 7–17.
Science Applications International Corporation (1987), Estimated National Occurrence and Exposure to Arsenic in Public Drinking Water Supplies (revised draft), U.S. Environmental Protection Agency, Washington, D. C. (EPA contract no. 68–01–7166).
Sen, P. and Costa, M. (1985), *Cancer Res.,* 45, 2320–2323.
Shellam, G. R., Flexman, J. P., Farrell, H. E., and Papadimitriou, J. M. (1985), in *Genetic Control of Host Resistance to Infection and Malignancy,* Skamene, E. Ed., Alan R. Liss, New York, 175–180.
Sikorski, E. E., Burns, L. A., McCoy, K. L., Stern, M., and Munson, A. E. (1991a), *Toxicol. Appl. Pharmacol.,* 110, 143–150.
Sikorski, E. E., Burns, L. A., Stern, M. L., Luster, M. I., and Munson, A. E. (1991b), *Toxicol. Appl. Pharmacol.,* 110, 129–141.
Sikorski, E. E., McCay, J. A., White, K. L., Bradley, S. G., and Munson, A. E. (1989), *Fundam. Appl. Toxicol.,* 13, 843–850.
Smialowicz, R. J., Rogers, R. R., Riddle, M. M., Garner, R. J., Rowe, D. G., and Luebke, R. W. (1985b), *Environ Res.,* 36, 56–66.
Smialowicz, R. J., Luebke, R. W., Rogers, R. R., Riddle, M. M., and Rowe, D. G. (1985a), *Immunopharmacology,* 9, 1–11.
Smialowicz, R. J., Riddle, M. M., Rogers, R. R., Luebke, R. W., and Copeland, C. B. (1989), *Toxicology,* 57, 97–111.
Smialowicz, R. J., Riddle, M. M., Rogers, R. R., Luebke, R. W., and Burleson, G. R. (1988a), *Immunopharmacol. Immunotoxicol.,* 10, 93–107.
Smialowicz, R. J., Riddle, M. M., Rogers, R. R., Rowe D. G., Luebke, R. W., Fogelson, L. D., and Copeland, C. B. (1988b), *J. Toxicol. Environ. Health,* 25, 403–422.
Smialowicz, R. J., Rogers, R. R., Riddle, M. M., and Stott, G. A. (1984a), *Environ. Res.,* 33, 413–427.
Smialowicz, R. J., Rogers, R. R., Riddle M. M., Luebke, R. W., Rowe, D. G., and Garner, R. J. (1984b), *J. Immunopharmacol.,* 6, 1–23.
Smialowicz, R. J., Rogers, R. R., Riddle, M. M., Rowe, D. G., and Luebke, R. W. (1986a), *Toxicology,* 38, 293–303.
Smialowicz, R. J., Rogers, R. R., Riddle, M. M., Rowe, D. G., and Luebke, R. W. (1986b), *J. Toxicol. Environ. Health,* 19, 243–254.
Smialowicz, R. J., Rogers, R. R., Rowe, D. G., Riddle, M. M., and Luebke, R. W. (1987a), *Toxicology,* 44, 271–281.
Smialowicz, R. J., Rogers, R. R., Riddle, M. M., Luebke, R. W., Fogelson, L. D., and Rowe, D. G. (1987b), *J. Toxicol. Environ. Health,* 20, 67–80.
Spiegelberg, T., Kordel, W., and Hochrainer, D. (1984), *Ecotoxicol. Environ. Safety,* 8, 516–520.
Stacey, N. H. (1986), *J. Toxicol. Environ. Health,* 18, 293–300.
Stacey, N. H., Craig, G., and Muller, L. (1988), *Environ. Res.,* 45, 71–77.
Su, H. C., Leite-Morris, K. A., Braun, L., and Biron, C. A. (1991), *J. Immunol.,* 147, 2717–2727.
Sunderman, F. W. (1988), *Ann. Clin. Lab. Sci.,* 18, 89–101.
Sunderman, F. W., Jr. (1983), *Ann. Clin. Lab. Sci.,* 13, 1–9.
Sunderman, F. W., Jr., Hopfer, S. M., Lin, S. M., Plowman, M. C., Stojanovic, T., Wong, S. H. Y., Zaharia, O., and Ziebka, L. (1989), *Toxicol. Appl. Pharmacol.,* 100, 107–116.
Syrjanen, S., Nilner, K., and Hensten-Pettersen, A. (1986), *J. Biomed. Mater. Res.,* 20, 1111–1115.
Thind, I. S. and Singh, N. P. (1977), *Acta Virol.,* 21, 317–325.
Thomas, P. T., Ratajczak, H. V., Aranyi, C., Gibbons, R., and Fenters, J. D. (1985), *Toxicol. Appl. Pharmacol.,* 80, 446–456.
Treagan, L. and Furst, A. (1970), *Res. Commun. Chem. Pathol. Pharmacol.,* 1, 395–402.
U.S. Environmental Protection Agency (1985), Health Assessment Document for Inorganic Arsenic, Environmental Criteria and Assessment Office, Research Triangle Park, NC.

U.S. Environmental Protection Agency (1986a), Air Quality Criteria for Lead, Vol. 1, Office of Research and Development, Environmental Criteria and Assessment Office, EPA-600/883/028aF, Research Triangle Park, NC.
U.S. Environmental Protection Agency (1986b), Air Quality Criteria for Lead, Vol. 2, Office of Research and Development, Environmental Criteria and Assessment Office. EPA-600/883/028bF, Research Triangle Park, NC.
U.S. Environmental Protection Agency (1986c), Air Quality Criteria for Lead, Vol. 4, Office of Research and development, Environmental Criteria and Assessment Office, EPA-600/883/028bF, Research Triangle Park, NC.
Vahter, M. E. (1988), in *Biological Monitoring of Toxic Metals,* Clarkson, T. W., Friberg, L., Nordberg, G. F., and Sager, P. R., Eds. Plenum Press, New York, 303–340.
Vallee, B. L. and Ulmer, D. D. (1972), *Annu. Rev. Biochem.,* 41, 78–128.
Van Loveren, H., Kranjnc, E. I., Rombout, P. J. A., Blommaert, F. A., and Vos, J. G. (1990), *Toxicol. Appl. Pharmacol.,* 102, 21–33.
Vos, J. G., de Klerk, E. I., Krajnc, W., Kruizinga, B., van Ommen, B., and Rozing, J. (1984), *Toxicol. Appl. Pharmacol.,* 75, 387–394.
Vos, J. G., De Klerk, A., Kranjnc, E. I., Van Loveren, H., and Rozing, J. (1990), *Toxicol. Appl. Pharmacol.,* 105, 144–155.
Vos, J. G., Krajnc, E. I., and Wester, P. W. (1985), in *Immunotoxicology and Immunopharmacology,* Dean, J. H., Luster, M. I., Munson, A. E., and Amos, H., Eds., Raven Press, New York, 327–350.
Waters, M. D. and Gardner, D. E. (1975), *Environ. Res.,* 9, 32–37.
Wiernik, A., Johansson, A., Jarstrand, C., and Camner, P. (1983), *Environ. Res.,* 30, 129–134.
World Health Organization, (1980), *Report of a Study Group: Recommended Health-Based Limits in Occupational Exposure to Heavy Metals,* Tech. Rep. Ser. No. 647, World Health Organization, Geneva.
Yang, Y.-J., Huang, C.-C., Shih, T.-S., and Yang, S.-S. (1994), *Occup. Environ. Med.,* 51, 267–270.
Zelikoff, J. T. (1994), in *Modulators of Fish Immune Responses,* Stolen, J. S., and Fletcher, T. C., Ed., SOS Publications, Fair Haven, NJ, 101–110.
Zelikoff, J. T., Bowser, D., Squibb, K. S., and Frenkel, K. (1995), *J. Toxicol. Environ. Health,*
Zelikoff, J. T., Parsons, E., and Schlesinger, R. B. (1993), *Environ. Res.,* 62, 207–214.
Zelikoff, J. T., Reynolds, C., Bowser, D., Valle, C., and Snyder, C. A. (1989), *Proc. Eastern Regional Symp. Mech. Immunotoxicol.,* 5, 102.

Chapter 52

Contact Hypersensitivity to Metals

Ian Kimber and David A. Basketter

I. INTRODUCTION

Allergic contact dermatitis (ACD) to metals is fairly common, due partly to the relatively widespread occurrence of and opportunities for skin contact with those metals which can sensitize. Metal allergy has been reviewed by Fowler (1990), who identifies 15 metals as potential causes of ACD. The majority of these are, at most, weak allergens. However, nickel chrome, cobalt, and mercury are relatively common allergens for man, and palladium and gold also may be giving rise to more episodes of ACD than have been appreciated previously. In this chapter, it is these more important metal contact allergens which will be discussed. With respect to man, most information is of course available only for those metals included in standard patch test batteries (nickel sulfate, 2.5%, potassium dichromate 0.25%, cobalt chloride, 1%). The other metals discussed are only patch-tested when ACD is suspected (gold sodium thiosulfate, 0.5 or 1%, palladium chloride, 1%, ammoniated mercury, 1%).

As a note of caution, it should be remembered that the reported incidences of ACD almost always refer to patient populations attending dermatology clinics, and thus are not necessarily representative of the general population. Furthermore, the prevalence of metal allergy may vary with the population tested, geographic location, and the patch test techniques, vehicles, and concentrations employed.

II. ALLERGIC CONTACT DERMATITIS TO METALS IN MAN

A. NICKEL

Nickel and its salts are among the commonest causes of metal allergic contact dermatitis (ACD) and indeed are generally regarded as the most common of all causes of human skin sensitization. As a result, the immunology and toxicology of nickel formed the sole topic of a book edited by Maibach and Menné (1989). Since then, data from patch test clinics have continued to demonstrate that a high percentage of those tested (up to 40%) may be patch test positive to nickel. In many cases, the majority of those who react to nickel are women — see, for example, the review of Fowler (1990). The clinical evidence supporting the predominance of nickel allergy in women and demonstrating its increase in recent decades has been summarized recently by Fowler (1990) and by Basketter et al. (1993). At present, the prevalence of nickel allergy among the general female population appears to be as high as 10% in some countries (Menné et al., 1989).

The sources of nickel which give rise to this high rate of ACD are widespread and occur in both occupational and domestic settings. Of primary importance is metallic nickel in close contact with skin; thus jewelry, wrist watches with metal backs, and jeans buttons are common causes of nickel ACD (Romaguera et al., 1988; Grandjean et al., 1989; Santucci et al., 1989). However, the most important primary cause of allergic sensitization to nickel appears to be the jewelry associated with ear piercing, where the combination of prolonged contact between allergen and damaged skin is a major predisposing factor (Emmett et al., 1988; Santucci et al., 1989).

Once sensitized to nickel, it can be difficult for an individual to avoid the further contact which may then precipitate episodes of ACD. The problem lies in the widespread distribution and use of nickel, for example, as nickel plate on taps, coins, scissors, zippers, buttons, clips, hairpins, metal spectacle frames, costume jewelry, wrist watches, etc. Furthermore, both silver and gold objects may contain (and release) significant quantities of nickel (Grandjean et al., 1989). Consequently, it is unfortunate that sensitive individuals can respond to quite low levels of nickel (down to 0.5 ppm), when applied to damaged skin under occlusive conditions (Allenby and Basketter, 1993). The elicitation threshold may be much higher, however, on undamaged skin and in the absence of occlusion (Menné and Calvin, 1993).

In recognition of the problems of nickel ACD, the Danish government have introduced a nickel release limit of 0.5 $\mu g/cm^2$/week for metal objects which may be worn in prolonged close contact with the skin (Menné and Rasmussen, 1990). A similar regulation is now being adopted by the EEC, and consumer products such as detergents, household cleaning agents, and personal products should be (and usually are) formulated to contain only low levels of nickel (Basketter et al., 1993).

B. CHROMIUM

In contrast to nickel, chromium most commonly is a cause of occupational ACD. Chromium metal is insoluble and its sensitization potential cannot be expressed. However, corrosive action, such as that induced by sweat, can cause solubilization of chromium and thus realize its potential. The sensitizing capacity of chromium salts depend upon concentration, valency, solubility, pH, and presence of organic matter. Hexavalent salts, which are more soluble than trivalent salts, penetrate more easily through the skin where they are reduced to Cr^{3+}, which is considered to be the sensitizing agent (Polak, 1983).

Chromate sensitivity in cement workers, first described by Bonnevie (1939) and Stauffer (1939), was thought to be due to chromium compounds in leather gloves. Pirilä and Kilpio (1949) and Jaeger and Pelloni (1950) demonstrated the relationship between cement dermatitis and chromium allergy. Subsequently cement was recognized by Cronin (1971, 1980) as the most common cause of primary sensitization to chromate.

In certain countries, bleaches and liquid detergents containing chromates were reported to be a cause of chrome dermatitis (Garcia-Perez et al., 1973; Dooms-Goossens et al., 1980; Lachapelle et al., 1980). In other countries, bleaches had only trace levels of chromate, which were no different from other consumer products (Hostynek and Maibach, 1988). Cessation of deliberate addition of chromate salts to bleach resolved the outbreak of ACD (Burrows, 1983).

Historically the incidence of positive patch test reactions to chromate and nickel has remained remarkably constant (9% in 1937, 12% in 1970; Baer et al., 1973). However, during recent years, in contrast to the increasing trend observed for nickel ACD, a decreasing incidence (to about half of earlier values) was found for chromate (Gailhofer and Ludvan, 1987; Kiec-Swierzcynska, 1988; Storrs et al., 1989). Reasons for the decrease with chromium may be better workplace protective habits which reduce contact with construction materials, supported by data from Finland (Estlander, 1990) and the addition of ferrous sulfate to cement to complex the chromium (Burrows, 1983).

Chromates are used to cover the protective zinc coating on stainless steel and thus prevent rapid corrosion (Wass and Wahlberg, 1991). ACD due to the release of hexavalent chromium from chromated surfaces has been described (Fregert et al., 1970). Repeated handling of chromated objects can result in the release of hexavalent chromium due to the action of human sweat. A recent study with chromate-sensitive individuals showed that discs releasing 0.6 $\mu g/cm^2$ of Cr^{6+} during a 20-min period or more elicited a positive response in all panelists. The authors proposed that mean release of Cr^{6+} from chromated parts should not exceed 0.5 $\mu g/cm^2$ (Wass and Wahlberg, 1991).

Contact with chromium salts can be intimate and exaggerated especially in the building industry, for example the hand-mixing of cement. Other occupational causes of chromate ACD include primer paints, galvanizing, antirust agents in coolants, welding fumes, leather goods, pigments, and printing. Nonoccupational exposure to chromium salts occurs through contact with leather products, rubber, stainless steel utensils, chrome alloys, and chrome-plated objects (Burrows, 1983).

It does appear that prolonged and/or repeated contact with low (ppm) levels of chromate in cement does give rise to relatively chronic ACD (Polak, 1983). However, although similar low levels can elicit reactions under 48-h occlusive patch test conditions (Allenby and Goodwin, 1983), in the great majority of circumstances, even substantially higher levels are likely to present minimal risk of sensitization (Nethercott et al., 1994).

Because of the risks associated with chromium and its salts, various regulations are in place or are proposed. Thus, in the U.S., chromium in cosmetic coloring agents is restricted to 300 mg/kg (IRPTC, 1987). In European Economic Community countries, the marketing of cosmetic products containing chromium is prohibited (ECC, 1976). Last, as mentioned above, Wass and Wahlberg (1991) have proposed a limit of release of 0.5 µg/cm^2/h for chromated sheet.

C. COBALT

Cobalt also causes allergic reactions in man, and many metal alloys contain cobalt together with nickel (Cronin, 1980; Dooms-Goossens et al., 1980; Fisher, 1995; Fowler, 1990; Shehade et al., 1991). The latter authors reported on 4721 subjects, of whom 5.7% were patch test positive to cobalt. Simultaneous allergy to nickel and cobalt is frequent, and cobalt has been considered of significance in persistent hand eczema in patients with positive patch tests to nickel and cobalt (Menné 1980).

Most comparative studies demonstrate that the incidence of positive patch tests to cobalt has increased in the last 20 years, but not with such a clear effect as nickel (Storrs et al., 1989). In Poland, an increase in positive patch tests to cobalt was found in women, but with the opposite tendency in men (Kiec-Swierzcynska, 1988).

Cobalt is widely distributed naturally, always in association with nickel (Domingo, 1989). Thus, simultaneous exposure to nickel and cobalt is frequent. Cobalt has been used for the coloring of pottery and glass since pre-Christian times. Today, more than 75% of the world's production of cobalt is used in the manufacture of alloys. Occupational exposure to cobalt in airborne dust occurs mainly in the tungsten and cemented carbide industries. Cobalt compounds are used as dyeing agents in lacquers, varnishes, paints, inks, pigments, and enamels. Cobalt is also an important catalyst for the petroleum industry (Domingo, 1989).

Because nickel and cobalt are present together in most alloys, exposure to nickel, such as in jewelry, may result frequently in concomitant sensitization to cobalt (Menné, 1980; Fischer, 1989). There are reports of ACD due to the cobalt content of metal frames for glasses and in wrist watches (Grimm, 1971).

Patch test dose response studies have shown that cobalt-sensitized individuals can react to low (ppm) levels of cobalt when it is applied to surfactant-damaged skin (Allenby and Basketter, 1989).

In the U.S., the quantity of cobalt allowed in coloring agents for cosmetics is limited to 200 mg/kg (IRPTC, 1987).

D. MERCURY

Although mercury and its salts are toxic by both ingestion and dermal adsorption, mercury salts have found considerable use, particularly as preservatives and antiseptics. In this setting, ACD can be quite common even today, despite the fact that the problems associated with such compounds as thimerosal, ammoniated mercury, and other organic and especially inorganic mercury salts were recognized many years ago (Cronin, 1980; Fisher, 1986). In particular, thimerosal is a well-recognized cause of ACD through its use as an antiseptic and preservative in medicaments (Fisher, 1981).

E. PALLADIUM

ACD to palladium has given rise to steady stream of reports in the clinical literature over the last 25 years. Munro-Ashman et al. (1969) described a case of contact dermatitis to palladium in a research worker investigating precious metals. Many of the cases since then have been discussed by Fowler (1990) and by Aberer et al. (1993). The primary causes of palladium ACD appear to be contact with jewelry and from dental alloys. The incidence of patch test positives reported by some has reached 8.3% (Aberer et al., 1993), although the relevance of some of these reactions is unclear.

F. GOLD

The story of ACD to gold has some similarities to chromium in that the metal has little potential to sensitize, while salts such as gold chloride produced by the action of sweat are relatively effective sensitizers (Fisher, 1986). However, the issue is complicated by the fact that gold often contains other metals such as palladium and nickel. The main source of skin contact with gold is via jewelry and this is the primary cause of ACD (Fisher, 1995; Fowler, 1988).

III. CONTACT SENSITIZATION OF ANIMALS TO METAL SALTS

In many communities nickel represents the most frequent cause of ACD. Contact sensitization to the majority of significant human skin allergens can, in most instances, be induced readily in rodents. By comparison, it has proven less easy to induce in experimental animals skin sensitization to nickel salts. A review of the results obtained with salts of nickel in guinea pig and mouse tests for the predictive evaluation of skin sensitizing potential illustrates this point. Original studies with the occluded patch test of Buehler failed to identify as positive nickel salts (Buehler, 1965). In other guinea pig methods, responses to nickel have proven variable. In a comparison of three guinea pig assays performed by Goodwin et al. (1981), it was found that nickel sulfate induced sensitization in only 10% of the test animals using the guinea pig maximization test developed by Magnusson and Kligman (1970) and failed completely to sensitize using a modified Draize procedure (Draize, 1959; Sharp, 1978). In the same study, nickel sulfate induced contact sensitization in 20% of animals using a single injection adjuvant test, but employing the same assay nickel chloride was without activity (Goodwin et al., 1981). In a second comparative investigation, Lammintausta et al. (1985) found open epicutaneous exposure provided the most effective method for sensitization to nickel sulfate, with 50% of the guinea pigs responding. Intradermal injection with Freund's complete adjuvant resulted in a reduced number of responders, and using the guinea pig maximization test only 23% of the animals were sensitized (Lammintausta et al., 1985). Although it has been reported that nickel sulfate is able to provoke responses in all exposed animals using the optimization test (Maurer, 1985), it is apparent that, under normal circumstances, sensitization of guinea pigs to nickel salts is at best variable. Contact sensitization of mice to nickel is, if anything, more difficult. Contact hypersensitivity is measured usually in mice as a function of challenge-induced ear swelling in previously sensitized animals. Using this method, Moller (1984) examined the sensitization of mice to nickel sulfate using various exposure regimes. All short-term procedures failed to induce measurable levels of sensitization. Weak responses were achieved only by use of a concentrated (20%) nickel sulfate solution applied by repeated epicutaneous exposure over a 3-week period. Treatment of mice with either cyclophosphamide or Freund's complete adjuvant, procedures considered to favor the initiation of skin sensitization, failed to induce or promote contact hypersensitivity to nickel (Moller, 1984). A number of attempts have been made to develop predictive test methods in the mouse based upon changes in ear thickness following challenge. The most thoroughly examined of these is the mouse ear swelling test (MEST) reported by Gad et al. (1986). In the original description of the test, nickel sulfate was found to induce contact sensitization in 38% of the test animals and to cause a mean increase in ear thickness of 18%. Using a similar method, the mouse ear sensitization assay, Descotes (1988) also reported a modest ear swelling response following sensitization and challenge with nickel chloride. In subsequent studies of the MEST, however, nickel sulfate was found not to induce sensitization (Cornacoff et al., 1988; Dunn et al., 1990), although in one of the two investigations a variable response was observed when hypersensitivity was measured instead by the infiltration of radiolabeled cells into the challenge site (Cornacoff et al., 1988). Variable responses to nickel sulfate applied epicutaneously to mice under occlusion were recorded also by Kimber et al., (1990). Contact sensitization, measured again as a function of changes in ear thickness resulting from challenge, was found to be weak and to vary between experiments and between mice. Nevertheless, evidence for significant levels of sensitization to nickel was found in four of seven independent experiments (Kimber et al., 1990).

An alternative approach to the predictive assessment of skin-sensitizing potential is the murine local lymph node assay in which contact allergens are identified on the basis of lymphocyte proliferative responses induced in lymph nodes draining the site of exposure (Kimber and Weisenberger, 1989; Kimber et al., 1989; Kimber and Basketter, 1992). A number of studies have demonstrated that topical exposure of mice to nickel sulfate, dissolved in either dimethylsulfoxide or ethanol, provokes a modest proliferative response by draining lymph node cells (Kimber et al., 1990; Ikarashi et al., 1992a,b). Experiments in which proliferative activity was measured *in vitro* revealed a greater than 2-fold increase in ^{3}H-thymidine incorporation by draining lymph node cells from mice exposed to nickel sulfate compared with cells prepared from vehicle-treated controls (Kimber et al., 1990). Similar responses were reported by Ikarashi et al., (1992 a,b) using the same methods. It has been proposed that coadministration with skin irritants may serve to increase lymph node cell proliferative responses to metal salts (Ikarashi et al., 1993). Nevertheless, it is apparent that despite the induction of modest proliferative responses, nickel sulfate does not exhibit activity in the local lymph node assay characteristic of chemicals known to cause allergic contact dermatitis in man. In some, but not all instances, nickel salts fail to register a formal positive response in the assay, a current criterion for which is the stimulation of a 3-fold or greater increase in proliferative activity compared with concurrent vehicle controls (Basketter and Scholes, 1992; Gerberick et al., 1992; Kimber and Basketter, 1992).

The general inability to induce robust contact sensitization to nickel in guinea pigs and mice is intriguing. One could speculate that the data reflect a fundamental difference between the immune systems of man and experimental animals with respect to the recognition of and response to nickel, perhaps secondary to variations in the T lymphocyte repertoire. It must be emphasized, however, that while nickel is a frequent cause of allergic contact dermatitis in man, it is not necessarily a potent skin sensitizer. Moreover, there is evidence that nickel is immunogenic in mice. Robinson and Sneller (1990) demonstrated that draining lymph node cells isolated from mice which had received repeated epicutaneous exposure to nickel sulfate were able to mount vigorous proliferative responses *in vitro* when cultured with syngeneic epidermal cells modified with the same salt. Despite evidence for the stimulation of nickel-specific immune responses, mice sensitized in this way failed to exhibit significant changes in ear thickness following challenge with nickel sulfate. It was found, however, that if the ears of sensitized animals were abraded gently immediately prior to challenge, a substantial ear swelling response was obtained (Robinson and Sneller, 1990). Taken together, the available data suggest that while nickel is not a potent skin allergen in experimental animals, topical exposure is able to cause sensitization. It is probable that an important factor which serves to limit effective induction and elicitation of contact hypersensitivity in mice and guinea pigs is the comparatively inefficient penetration of the allergen into the epidermis. This may not be the whole story, however.

Van Hoogstraten et al., (1992, 1993) have demonstrated that immunological tolerance to nickel can be induced in both mice and guinea pigs by oral administration prior to cutaneous sensitization, and that in mice tolerance to nickel could be demonstrated in animals reared in cages which released nickel. These data are apparently consistent with a clinical epidemiologic study which provided some evidence for tolerance to nickel in humans. In a retrospective study of more than 2000 patients attending European patch test clinics, it was found that individuals who had experienced nickel at an early age, through the use of orthodontic braces, but only if prior to ear piercing (which strongly favors the development of nickel allergy), exhibited a reduced frequency of hypersensitivity to nickel. Importantly, the incidence of other contact hypersensitivities was unaffected by the early use of nickel-releasing dental braces (Van Hoogstraten et al., 1991).

It is possible, therefore, that both the relatively poor penetration of nickel salts into the epidermis and the inadvertent generation of oral tolerance may, in some circumstances, combine to impair or prevent sensitization of experimental animals to this allergen.

By comparison with nickel salts, it has been found that both guinea pigs and mice can be sensitized readily with potassium dichromate. In all well-established guinea pig predictive tests, other than the occluded patch test of Buehler, potassium dichromate stimulates vigorous responses (Botham et al., 1991). The same salt induces contact sensitization in mice (Mor et al., 1988; Kimber et al., 1990; Vreeburg et al., 1991), is positive in the MEST (Gad et al., 1986), and provokes robust proliferative responses by draining lymph node cells (Kimber et al., 1990, 1991; Ikarashi et al., 1992a,b). Also able to stimulate local lymph node assay responses is cobalt chloride (Ikarashi et al., 1992a,b; Basketter and Scholes, 1992), a metal salt classified as an extreme sensitizer in the guinea pig maximization test (Basketter and Scholes, 1992). It has been possible also to achieve contact sensitization to mercury in mice (Vreeburg et al., 1991) and to palladium in guinea pigs (Wahlberg and Boman, 1992). Finally, two independent studies have found copper chloride to provoke local lymph node assay responses (Ikarashi et al., 1992b; Basketter and Scholes, 1992). The significance of this is presently unclear as the available evidence indicates that copper is only a rare cause of contact sensitization in man (Karlberg et al., 1983).

IV. MOLECULAR MECHANISMS OF METAL HYPERSENSITIVITY

Allergic contact dermatitis is a form of delayed-type hypersensitivity reaction dependent upon the activity of T lymphocytes. Classically, T lymphocytes recognize and respond to immunogenic fragments of foreign proteins presented in association with gene products of the major histocompatibility complex (MHC class I and class II molecules). There is no doubt that in susceptible animals and humans appropriate exposure to nickel results in nickel-specific T lymphocytes (Sinigaglia et al., 1985). An interesting question is raised regarding the nature of the stimulating immunogenic complex presented to responsive T lymphocytes. Two possibilities can be considered, either that nickel associates directly with the MHC molecule or, alternatively, that nickel interacts with peptide located within the "groove" of MHC determinants. Studies by Romagnoli et al., (1991) suggest that the latter is in fact the case, and that T lymphocytes recognize nickel associated with MHC-bound peptides. It is likely that nickel interacts with the MHC class II-peptide complex at the cell surface and that, although certain residues may be necessary for binding, effective association may occur with many peptide sequences (Sinigaglia, 1994). It should be borne in mind, however, that other metals capable of inducing contact sensitization may be recognized by responsive T lymphocytes in different contexts (Sinigaglia, 1994).

V. CONCLUSIONS

Metals are an important cause of allergic contact dermatitis. Investigative studies indicate that the effective induction of sensitization to nickel, and possibly other metal allergens, may be influenced by epidermal penetration and by immunoregulatory mechanisms resulting possibly from previous exposure. The cellular and molecular mechanisms which result in the stimulation of specific T lymphocyte responses to metals may provide a paradigm for sensitization to chemical allergens in general.

REFERENCES

Aberer, W., Holub, H., Strohal, R., and Slavicek, R. (1993), *Contact Derm.,* 28, 163–165.
Allenby, C. F. and Basketter, D. A. (1989), *Contact Derm.,* 20, 186–190.
Allenby, C. F. and Goodwin, B. F. J. (1983), *Contact Derm.,* 9, 491–499.
Baer, R. L., Ramsey, D. L., and Biondi, E. (1973), *Arch. Dermatol.,* 108, 74–78.
Basketter, D. A., Briatico-Vangosa, G., Kaestner, W., Lally, C., and Bontinck, W. (1993), *Contact Derm.,* 28, 15–25.
Basketter, D. A. and Scholes, E. W. (1992), *Food Chem. Toxicol.,* 30, 65–69.
Bonnevie, P. (1939), *Aetiologie und Pathogenese der Ekzemkrankheiten,* Busck, Copenhagen-Leipzig.
Botham, P. A., Basketter, D. A., Maurer, T., Mueller, D., Potokar, M., and Bontinck, W. J. (1991), *Food Chem. Toxicol.,* 29, 275–286.
Buehler, E. V. (1965), *Arch. Dermatol.,* 91, 171–177.
Burrows, D. (1983), in *Chromium: Metabolism and Toxicity,* Burrows, D., Ed., CRC Press, Boca Raton, FL, 137–163.
Cornacoff, J. B., House, R. V., and Dean, J. H. (1988), *Fundam. Appl. Toxicol.,* 10, 40–44.
Cronin, E. (1971), *Br. J. Dermatol.,* 85, 95–96.
Cronin, E. (1980), *Contact Dermatitis,* Churchill Livingstone, London.
Descotes, J. (1988), *J. Toxicol — Cut. Ocular Toxicol.,* 7, 263–272.
Domingo, J. L. (1989), *Rev. Environ. Contam. Toxicol.,* 108, 105–112.
Dooms-Goossens, A., Ceuterick, A., Vanmaele, N., and Degreef, H. (1980), *Dermatologica,* 160, 249–255.
Draize, J. H. (1995), *Dermal Toxicity,* Association of Food and Drug Officials of the United States, Texas State Department of Health, Austin, 46.
Dunn, B. J., Rusch, G. M., Siglin, J. C., and Blaszcak, D. L. (1990), *Fundam. Appl. Toxicol.,* 15, 242–248.
EEC (1976), Council Directive for Cosmetics 76/768/EEC, *Off. J.,* L 262 of 27 September 1976.
Emmet, A. E., Risby, T. H., Jiang, L., Ng, S. K., and Feinman, S. (1988), *J. Am. Acad. Dermatol.,* 19, 314–322.
Estlander, T. (1990), *Acta Derm. Venereol.,* 1990, Suppl. 155.
Fischer, T. (1989), in *Nickel and the Skin. Immunology and Toxicology.,* T., Maibach, H. I. and Menné, T., Eds., CRC Press, Boca Raton, FL, 117–132.
Fisher, A. A. (1981), *Cutis,* 27, 580–585.
Fisher, A. A. (1986), *Contact Dermatitis,* Lea & Febiger, Philadelphia.
Fowler, J. F. (1988), *Arch. of Derm.,* 124, 181–182.
Fowler, J. F. (1990), *Am. J. Contact Derm.,* 1, 212–223.
Fregert, S., Gruvberger, B., and Heijer, A. (1970), *Berufdermatosen,* 18, 254–260.
Gad, S. C., Dunn, B. J., Dobbs, D. W., Reilly, C., and Walsh, R. D. (1986), *Toxicol. Appl. Pharmacol.,* 84, 93–114.
Gailhofer, G. and Ludvan, M. (1987), *Dermatosen,* 35, 12–16.
Garcia-Perez, A., Martin-Pascual, A., and Sanchez-Misiego, A. (1973), *Acta Dermatol.,* 53, 353–358.
Gerberick, G. F., House, R. V., Fletcher, E. R., and Ryan, C. A. (1992), *Fundam. Appl. Toxicol.,* 19, 438–445.
Goodwin, B. F. J., Crevel, R. W. R., and Johnson, A. W. (1981), *Contact Derm.,* 7, 248–258.
Grandjean, P., Nielsen, G. D., and Andersen, O. (1989), *Nickel and the Skin. Immunology and Toxicology,* Maibach, H. I. And Menné, T., Eds. CRC Press, Boca Raton, FL, 9–34.
Grimm, I. (1971), *Berufsdermatosen,* 19, 39–42.
Hostynek, J. J. and Maibach, H. I. (1988), *Contact Derm.,* 18, 206–209.
Ikarashi, Y., Ohno, K., Tsuchiya, T., and Nakamura, A. (1992a), *Toxicology,* 76, 283–292.
Ikarashi, Y., Tsuchiya, T., and Nakamura, A. (1992b), *Toxicol. Lett.,* 62, 53–61.
Ikarashi, Y., Tsukamuto, Y., Tsuchiya, T., and Nakamura, A. (1993), *Contact Derm.,* 29, 128–132.
IRPTC (1987), International Register of Potentially Toxic Chemicals, Legal File 1986, United Nations Environment

Jaeger, H. and Pelloni, E. (1950), *Dermatologica,* 100, 207.
Karlberg, A. T., Boman, A, and Wahlberg, J. E. (1983), *Contact Derm.,* 9, 134–139.
Kiec-Swierzcynska, M. (1990), *Contact Derm.,* 22, 229–231.
Kimber, I. and Basketter, D. A. (1992), *Food Chem. Toxicol.,* 30, 165–169.
Kimber, I., Bentley, A. N., and Hilton, J. (1990), *Contact Derm.,* 23, 325–330.
Kimber, I., Hilton, J., Botham, P. A., Basketter, D. A., Scholes, E. W., Miller, K., Robbins, M. C., Harrison, P. T. C., Gray, T. J. B., and Waite, S. J. (1991), *Toxicol. Lett.,* 55, 203–213.
Kimber, I., Hilton, J., and Weisenberger, C. (1989), *Contact Derm.,* 21, 215–220.
Kimber, I. and Weisenberger, C. (1989), *Arch. Toxicol.,* 63, 274–282.
Lachapelle, J. M., Lauwerys, R., Tennstedt, D., Andanson, J., Benezra, C., Chabeau, G., Ducombs, G., Foussereau, J., Lacroix, M., and Martin, P. (1980), *Contact Derm.,* 6, 107–110.
Lammintausta, K., Kalimo, K., and Jansen, C. T. (1985), *Contact Derm.,* 12, 258–262.
Magnusson, B. and Kligman, A. M. (1970), *Identification of Contact Allergens,* Charles C Thomas, Springfield, IL.
Maibach, H. I. and Menné, T. (1989), *Nickel and Skin: Immunology and Toxicology,* CRC Press, Boca Raton, FL.
Maurer, T. (1985), *Contact Allergy Predictive Tests in Guinea Pigs. Current Problems in Dermatology,* Vol. 14, Andersen, K. E. and Maibach, H. I., Eds., Karger, Basel, 114–151.
Menné, T. (1980), *Contact Derm.,* 6, 337–340.
Menné, T. and Calvin, G. (1993), *Contact Derm.,* 29, 180–184.
Menné, T., Christopherson, J., and Green, A. (1989), in *Nickel and the Skin. Immunology and Toxicology,* Maibach, H. I. and Menné, T., Eds. CRC Press, Boca Raton, FL, 109–116.
Menné, T. and Rasmussen, K. (1990), *Contact Derm.,* 23, 57–59.
Moller, H. (1984), *Contact Derm.,* 10, 65–68.
Mor, S., Ben-Efraim, S., Leibovici, J., and Ben-David, A. (1988), *Int. Arch. Allergy Appl. Immunol.,* 85, 452–457.
Munro-Ashman, D., Munro, D. D., and Hughes, T. H. (1969), *Trans. St. John's Hosp. Derm. Soc.,* 55, 196–197.
Nethercott, J., Paustenbach, D., Adams, R., Fowler, J., Marks, J., and Morton, C. (1994), *Occup. Environ. Med.,* 51, 371–380.
Pirilä, V. and Kilpio, O. (1949), *Acta Derm. Venereol.,* 29, 550–552.
Polak, L. (1983), in *Chromium: Metabolism and Toxicity,* Burrows, D., Ed. CRC Press, Boca Raton, FL, 51–136.
Robinson, M. K. and Sneller, D. L. (1990), *Toxicol. Appl. Pharmacol.,* 104, 106–116.
Romagnoli, P., Labhardt, A. M., and Sinigaglia, F. (1991), *EMBO J.,* 10, 1103–1109.
Romaguera, C., Grimalt, F., and Vilaplana, J. (1988), *Contact Derm.,* 19, 52–57.
Santucci, B., Ferrari, P. V., Cristaudo, A., Cannistraci, C., and Picardo, M. (1989), *Contact Derm.,* 21, 245–248.
Sharp, D. W. (1978), *Toxicology,* 9, 261–271.
Shehade, S. A., Beck, M. H., and Hillier, V. F. (1991), *Contact Derm.,* 24, 119–122.
Sinigaglia, F. (1994), *J. Invest. Dermatol.,* 102, 398–401.
Sinigaglia, F., Scheidegger, D., Garotta, G., Scheper, R., Pletscher, M., and Lanzavecchia, A. (1985), *J. Immunol.,* 135, 3929–3931.
Stauffer, H. (1939), *Arch. Dermatol. Syphilol.,* 162, 517–522.
Storrs, F. J., Rosenthal, L. E., Adams, R. M., Clendening, W., Emmet, E. A., Fisher, A. A., Larsen, W. G., Maibach, H. I., Reitschel, R. L., Schorr, W. F., and Taylor, J. S. (1989), *J. Am. Acad. Dermatol.,* 20, 1038–1044.
Van Hoogstraten, I. M. W., Anderson, K. E., Von Blomberg, B. M. E., Boden, D., Bruynzeel, D. P., Burrows, D., Camarasa, J. G., Dooms-Goossens, A., Kraal, G., Lahti, A., Menné, T., Rycroft, R. J. G., Shaw, S., Todd, D., Vreeburg, K. J. J., Wilkinson, J. D., and Scheper, R. J. (1991), *Clin. Exp. Immunol.,* 85, 441–445.
Van Hoogstraten, I. M. W., Boden, D., Von Blomberg, B. M. E., Kraal, G., and Scheper, R. J. (1992), *J. Invest. Dermatol.,* 99, 608–616.
Van Hoogstraten, I. M. W., Boos, C., Boden, D., Von Blomberg B. M. E., Scheper, R. J., and Kraal, G. (1993), *J. Invest. Dermatol.,* 101, 26–31.
Vreeburg, K. J. J., De Groot, K., Van Hoogstraten, I. M. W., Von Blomberg, B. M. E., and Scheper, R. J. (1991), *Int. Arch. Allergy Appl. Immunol.,* 96, 179–183.
Wahlberg, J. E. and Boman, A. S. (1992), *Acta Derm. Venereol.,* 72, 95–97.
Wass, U. and Wahlberg, J. E. (1991), *Contact Derm.,* 24, 114–118.

Chapter 53

Autoimmunity Induced by Metals

Pierluigi E. Bigazzi

I. INTRODUCTION

The term "autoimmunity" comprises both autoimmune response and autoimmune disease. Immune responses directed against normal components of the body (antigens of "self" or "autoantigens") are termed "autoimmune". They may be humoral, characterized by the production of antibodies directed against an autoantigen ("autoantibodies"), and cellular, mediated by effector T cells. Autoimmune responses may produce structural and/or functional damage, resulting in pathologic conditions that are defined as "autoimmune diseases" (Bigazzi, 1991). The distinction between autoimmune response and disease is critically important, since the former does not necessarily result in the latter. For example, with the availability of increasingly sensitive methods of detection one can demonstrate autoimmune responses to an increasing number of autoantigens. Some of these responses are found in apparently healthy, "normal" individuals. Some are associated with disease of unknown or uncertain etiology, which we still hesitate to call "autoimmune". Some are clearly associated with autoimmune disease.

No organ or tissue is spared by autoimmune-mediated damage; thus, at least 30 human diseases are autoimmune or have autoimmune manifestations. Traditionally, these disorders are divided into *systemic* (nonorgan-specific, generalized) and *organ-* or *tissue-specific*. In the former, the targets of autoimmune responses are autoantigens common to various organs and tissues, e.g., nuclear antigens and immunoglobulins. Examples of systemic autoimmune diseases are systemic lupus erythematosus (SLE), rheumatoid arthritis, and systemic sclerosis. In the organ- or tissue-specific disorders, the targets of autoimmune responses are autoantigens present in a single organ or tissue, e.g., thyroid, pancreas, striated muscle, etc. Examples of organ- or tissue-specific autoimmune diseases are Hashimoto's thyroiditis, Graves' disease, insulin-dependent diabetes mellitus, and myasthenia gravis. In addition, there are disorders that are both organ- and nonorgan-specific, i.e., they are characterized by autoimmune responses to various autoantigens, some of which are organ-specific, others systemic. An example of this type of autoimmune disease is provided by Sjögren's syndrome.

Despite years of research, the causes of autoimmunity are still unknown, but numerous studies have involved both infectious microorganisms and environmental chemicals as causative agents. Autoimmune responses to a variety of self-antigens are indeed caused by xenobiotics. Depending on circumstances still unknown, these responses may be relatively harmless or may result in autoimmune disease. Thus, autoimmunity induced by xenobiotics is increasingly recognized as an environmental hazard that may affect genetically predisposed individuals, chronically exposed to certain chemicals (reviewed in Bigazzi, 1988; Kammuller et al., 1989; Goldman et al., 1991; Pelletier et al., 1992; Kosuda and Bigazzi, 1996).

Studies of autoimmunity induced by xenobiotics in humans are usually quite difficult. The absorption of low doses of environmental chemicals (e.g., through the gastrointestinal or respiratory systems) may occur over a long period of time and escape notice. Once the disease develops, a retrospective study may not reveal the etiologic agent involved or not establish a causal relationship. In addition, the patient may have been exposed to a combination of xenobiotics, further complicating an etiologic investigation. Therefore, studies of drug-induced autoimmunity and investigations of experimental animal models have provided some of the best evidence in this area of immunotoxicology. It has long been recognized that certain drugs can induce autoimmune responses and (less often) autoimmune disease. In these cases, the etiologic relationship between administration of the drug and its autoimmune consequences is quite clear. The disappearance of the problem after withdrawal of the drug usually confirms its role. Therefore, these iatrogenic disorders can be used as models for autoimmunity induced by environmental chemicals. Another opportunity for mechanistic investigations is provided by autoimmune responses and diseases caused in experimental animals by exposure to xenobiotics. Chemical-induced autoimmune disease of laboratory animals seldom corresponds exactly to its human counterpart, as is the case for other experimental models of autoimmunity. For example, drug-induced lupus and myasthenia gravis have been quite difficult to reproduce and are often different from the pathologic conditions observed in humans. There are no experimental models of penicillamine-induced pemphigus and just a few of drug-induced hepatitis. On the other hand, renal disease induced by chemicals shows striking similarities in animals and humans. Finally, studies in experimental animals have convincingly demonstrated the role of iodine in autoimmune thyroiditis. In conclusion, these animal models are useful to ascertain the relationship between administration of the chemical, kinetics of autoimmune responses, and eventual development of disease. In addition, they can provide illustrative examples of the role of genetic factors as well as an understanding of the immunologic, biochemical, and molecular aspects of chemical-induced autoimmunity.

Various metals have well-known toxic and/or immunotoxic effects (Descotes, 1986; Goyer, 1991). For example, aluminum, arsenic, germanium, iron, magnesium, manganese, molybdenum, nickel, platinum, selenium, tin salts, and vanadium can affect cellular and/or humoral immune responses. However, in a recent literature search we have found no evidence associating autoimmunity with exposure to these metals. Therefore, we will not discuss them further in this chapter and will review only the metals that have been associated with autoimmune or putative autoimmune disease (Table 1). Because of space limitations, we will not review the autoimmune effects of silicon and iodine, that are not usually considered to be metals: their association with autoimmunity is reviewed elsewhere (Kosuda and Bigazzi, 1996). With a few exceptions, we will also limit most of our references to reviews and recent papers. Some reports have claimed that environmental metals such as beryllium, lead, zinc, and cadmium induce autoimmune responses. The evidence is far from solid, and for this reason we will review those metals first. We will then examine cobalt, chromium, and lithium as inducers of autoimmunity, a possibility that is suggested by some investigators. Finally, we will discuss in detail the abundant clinical and experimental evidence demonstrating that exposure to gold and mercury may result in both autoimmune responses and disease. For each metal, we will first summarize the clinical evidence and then will present the data obtained from experimental animal studies. As a conclusion, we will provide a summary outline of metal-induced autoimmunity, i.e., the general aspects of humoral and cellular autoimmunity, immunopathology, pathogenesis, and etiology.

II. BERYLLIUM

It is well known that protracted exposure to airborne dusts or fumes of beryllium (or its oxides, salts, and alloys) can result in chronic berylliosis, a disease characterized by focal noncaseating granulomas scattered throughout the lungs and other sites (Cullen et al., 1987). Recent studies of T cells obtained from peripheral blood and by bronchoalveolar lavage have identified CD4+ T cells capable of proliferating in response to beryllium *in vitro* (Saltini et al., 1989). Lines and clones of cells developed from T cells from the patients' lungs showed dose-dependent proliferation in response to beryllium, but did not respond to other metals. This evidence suggests that berylliosis is a hypersensitivity disease in which beryllium is the specific antigen (Saltini et al., 1989; Deodhar and Barna, 1991). However, it does not exclude the possibility that autoimmune responses may also be involved, as was hypothesized in older studies (Descotes, 1986).

Table 1 A List of Metals That Have Been Associated With Autoimmune (or Putative Autoimmune) Diseases

Metal[a]	Human Disease (Autoimmune or Putative Autoimmune)	Evidence of Autoimmune Effects
Beryllium	Berylliosis	Very scarce, old, and inconclusive
Cadmium	Autoimmune kidney disease	Scarce, old, and inconclusive
Chromium	SLE-like syndrome, pemphigus	Recent reports of individual cases
Cobalt	Goodpasture's syndrome	Recent report of individual case
Gold	Autoimmune kidney disease, autoimmune hemolytic anemia, SLE-like syndrome pemphigus	Solid, with numerous reports of autoimmune effects
Lead	Autoimmune kidney disease	Very scarce, old, and inconclusive
Lithium	Autoimmune thyroid disease, autoimmune kidney disease, SLE-like syndrome, insulin-dependent diabetes mellitus	Adequate, with various reports of autoimmune effects
Mercury	Autoimmune kidney disease	Solid, with various reports of autoimmune effects
Zinc	Multiple sclerosis	Very scarce and inconclusive

[a] Listed in alphabetical order.

III. LEAD

Lead is a very diffuse environmental contaminant. There is an abundance of literature showing the immunotoxicologic effects of this metal (Descotes, 1986). Lead induces T cell proliferation *in vitro*, enhances differentiation and MHC class II expression of B lymphocytes (McCabe and Lawrence, 1990), and alters the ability of macrophages to present antigen (Kowolenko et al., 1988). Therefore, it is not surprising that some investigators have speculated that "immune abnormalities leading to self-reactivity may be related to accumulation of environmental lead" (McCabe and Lawrence, 1990). However, this hypothesis is not supported by clinical studies of subjects with kidney pathology caused by exposure to lead.

Chronic lead nephropathy does not have specific or unique glomerular lesions. The main abnormalities are patchy tubular atrophy and interstitial fibrosis, occasionally with small numbers of mononuclear cells infiltrating the interstitium (Heptinstall, 1992; Wedeen, 1992). Most glomeruli are normal or show mesangial increase. Focal glomerular sclerosis may be present. Reports of immunohistopathology findings are scarce, with immunoglobulins and complement observed without any regularity (Heptinstall, 1992). Several years ago, kidney biopsies from twelve patients with occupational lead nephropathy were examined by light and fluorescence microscopy (Wedeen et al., 1979). Light microscopy showed normal kidneys in six of these cases. In the other six, there was focal tubular atrophy and interstitial disease. In two biopsy specimens, tubular interstitial disease was associated with occasional glomerular sclerosis. When eight of these biopsies were examined by direct immunofluorescence, finely granular deposits of IgG were found in five specimens and deposits of complement in two. In addition, six of these eight biopsies had linear immunoglobulins and, less often, complement deposits in the basement membrane of most proximal tubules. Indirect immunofluorescence did not detect antibodies against basement membranes in any of the sera from these patients. In conclusion, the immunofluorescence findings in the glomeruli suggested deposition of immune complexes of the granular type, but no binding of anti-GBM antibodies, whereas the tubular pattern suggested binding of anti-TBM antibodies. Histopathology did not show any active inflammatory processes in the glomeruli but only interstitial nephritis at the tubular level. A report on the association of Goodpasture's syndrome with environmental chemicals noted that one of their two patients had worked in a ceramic factory and had been exposed to both silica dust and lead derivatives (Perez Garcia et al., 1980). This patient had diffuse pulmonary silicosis and necrotizing glomerulonephritis, with linear deposits of IgG at the level of the glomerular basement membranes. An association between silica and autoimmunity has been suggested by other reports (Kosuda and Bigazzi, in press). On the other hand, unusual exposure to hydrocarbons (lead-containing or not) was once associated with the development of Goodpasture's syndrome, but this possibility is difficult to verify, because nowadays exposure to organic fumes and dusts is universal (Rees and Lockwood, 1988). Deposition of immunoglobulins, with or without complement, in a damaged kidney does not per se demonstrate the presence of autoimmune disease. Therefore, more recent studies have focused on the detection of autoimmune responses in subjects exposed to lead. The prevalence of circulating antilaminin antibodies was not significantly increased in workers that were employed in a lead smelter and in a ceramic factory (Bernard et al., 1987). Similarly, a Scandinavian study did not detect clinical signs of renal impairment in a cohort comprising both active and retired lead smelter workers (Gerhardsson et al., 1992). Interestingly, in a recent review of the pathologic effects of heavy metals on the kidney, Wedeen does not mention the possibility of autoimmune-mediated renal pathology, but suggests that the "primary renal injury from lead is in the microvascular endothelium" (Wedeen, 1992).

IV. COPPER

Very little information is currently available on the immunotoxic effects of copper. Mice and rabbits treated with copper show deficiencies in either humoral immunity or non-specific host defenses (Descotes, 1986). Thus, CBA/J mice exposed to various levels of copper had decreases in their T-dependent antibody responses to sheep red blood cells. Similarly, rabbits treated with cuprous chloride had decreased chemotaxis of polymorphonuclear leukocytes. In a recent review of the literature on xenobiotics we have found no convincing evidence of an association between copper and human autoimmune disease. On the other hand, it is interesting to note that mice exposed to copper in their drinking water show an increased production of autoantibodies to red cells, an effect that was not modified by an increment in their zinc supply (Pocino et al., 1990). In conclusion, there is evidence that copper can induce autoimmune responses in experimental animals, but to date no demonstration of similar effects in human beings.

V. COBALT

There are numerous reports of hypersensitivity to cobalt, either alone or as a component of so-called hard metal, that is a mixture of tungsten carbide and cobalt with small amounts of other metals. Hard metal is widely used for industrial purposes, e.g., cutting tools, jet engine exhaust ports, and oil well drilling bits. The hard metal diseases usually result from exposure to cobalt, either in the production of hard metal, machining hard metal parts, or other sources (Cugell, 1992). Lung disease includes asthma, hypersensitivity pneumonitis, interstitial fibrosis, and giant cell interstitial pneumonia. The pathogenesis of these conditions is still unclear, but may include both immediate and delayed hypersensitivity reactions to cobalt. However, autoimmune responses induced by this metal cannot be excluded, especially after a recent report of Goodpasture's syndrome in a 26-year-old man with occupational exposure to hard metal dust (Lechleitner et al., 1993). This patient developed a life-threatening interstitial lung disease that 2 months later was followed by a rapidly progressive glomerulonephritis. A kidney biopsy revealed linear binding of IgG and C3 along the renal glomerular basement membrane (GBM). His serum contained IgG antibodies against the GBM, confirming the diagnosis of Goodpasture's syndrome. Therefore, it appears that cobalt may be capable of inducing autoimmune responses and disease in genetically susceptible individuals.

VI. CHROMIUM

It is well known that chromium is a very common skin sensitizer and ranks only second to nickel as a cause of contact hypersensitivity. It also has effects on humoral and cellular immune responses as well as nonspecific host defenses (Descotes, 1986). However, in a recent review of the literature we have found that there are very few reports of its association with autoimmunity. Long-term low-dose exposure to inorganic chromium (and other chemicals) in contaminated well water has been reported to result in an SLE-like syndrome (Kilburn and Warshaw, 1992). Similarly, occupational contact with a compound containing basic chromium sulfate preceded the appearance and development of a vesiculobullous eruption that on the basis of clinical symptoms as well as histological and immunological findings was diagnosed as induced pemphigus (Tsankov et al., 1990). This is an interesting observation because to date only a few drugs (e.g., penicillamine) have been associated with the onset of this skin disorder.

VII. ZINC

The role of zinc in the maintenance of adequate immune responses is well established. Deficiencies of zinc result in decreased humoral and cellular immunity (Descotes, 1986). As far as autoimmunity is concerned, zinc deprivation was found to exert beneficial effects on autoimmune disorders in murine lupus (Descotes, 1986). In humans this metal has been implicated in the etiology of multiple sclerosis (MS), a disease that many investigators believe to be autoimmune (McFarland and Dhib-Jalbut, 1989; Raine, 1991).

The etiology of MS is unknown, even though infectious agents have been postulated as a possible cause. To date, the search for a viral cause of this disease has not been successful. Among other risk factors for MS, metals have recently received some attention (Stein et al., 1987; Schiffer et al., 1988, 1990, 1991). Exposure to lead or zinc has been associated with MS, even though studies of twins do not seem to support a causal association between exposure to metals and MS (Juntunen et al., 1989). However, concentrations of soluble zinc (and various other metals) in the soil are significantly higher in areas with a high MS prevalence rate as compared to control areas with a lower MS prevalence rate. Zinc levels are significantly elevated in erythrocytes from patients with MS and vary with disease activity (Ho et al., 1986). Finally, a zinc cluster of MS has been reported (Stein et al., 1987). Eleven cases of MS occurred within a 10-year period in a zinc-related manufacturing plant, an incidence greater than expected from population data. All employees of the plant had an increased body burden of zinc, as demonstrated by the finding that both MS patients and controls working in the plant had higher serum zinc levels than subjects (MS and controls) not working there.

To date, no additional reports of zinc clusters of MS have been published. However, experimental studies in laboratory animals do not appear to confirm the autoimmune effects of zinc. Experimental allergic encephalomyelitis (EAE) is an autoimmune disorder induced in animals of various species by immunization with CNS antigens. EAE is thought by various investigators to be an animal model of MS (Swanborg, 1990). A recent study has demonstrated that dietary supplements of nickel and zinc can affect the incidence and severity of EAE induced in SJL/J mice by immunization with syngeneic spinal cord homogenate (Schiffer et al., 1988, 1990, 1991). The total incidence of EAE was 17% in mice on a control diet, 25% in mice on a high zinc diet, and 5% in mice on a high nickel diet. Severity of inflammation and demyelination were also higher in zinc-treated mice than in the other groups. However, when incidence, severity, and time course of EAE were analyzed, the differences observed fell short of statistical significance. Preliminary investigations from our laboratory have also shown that Lewis rats injected with high doses of zinc chloride do not experience autoimmune responses to CNS antigens and have no EAE (Bigazzi et al., unpublished). In conclusion, studies in experimental animals have provided some suggestive clues, but not a convincing demonstration that zinc can induce or modulate EAE.

VIII. LITHIUM

Lithium, a light metal used in the treatment of manic-depressive patients, has been associated with the appearance of ANA, an SLE-like syndrome, or an occasional nephrotic syndrome with minimal glomerular changes (Hart 1991). Its administration has also been implicated in the exacerbation of myasthenia gravis but not of MS, even though lithium has been used to treat affective disorders in MS patients (Hart, 1991a; Hart 1991a,b).

More solid is the evidence of lithium's adverse effects on thyroid function (Hassman and McGregor, 1988; Kushner and Wartofsky, 1988). Numerous reports have associated this chemical with both thyroid disease and formation of thyroid autoantibodies. Goiters have been observed in patients treated with lithium for periods up to 2 years with a reported frequency that varies from 4% to 60%, depending on the study, duration of treatment, etc. Hypothyroidism, subclinical or overt, has been reported in 2 to 15% of patients after lithium treatment. A higher incidence of autoantibodies to thyroglobulin and/or the thyroid microsomal antigen (now identified as thyroid peroxidase, TPO) has been detected in patients with lithium. This metal is also associated with the production of antigastric (antiparietal cell) antibodies, that are often observed in patients treated with thyroid autoimmunity. It is uncertain whether lithium actually induces autoimmune thyroid disease, stimulates an existing predisposition, or causes hypothyroidism by direct stimulation of autoimmune responses to thyroid antigens. Thyroid abnormalities might be expected to occur in patients with psychiatric disturbances because of the possible association between thyroid dysfunction and psychiatric problems. In this case, thyroid autoantibodies would occur independently of lithium treatment. However, there is also an increased frequency of other, nonthyroid-related autoantibodies (ANA, antiparietal cell antibodies), suggesting that the occurrence of autoimmune disease may be the consequence of lithium treatment. Other studies of manic-depressive patients tested before and after lithium therapy have shown that no patient developed higher levels of thyroid autoantibodies who did not have them before therapy. For example, three patients, all with detectable titers of anti-TPO antibodies before therapy, showed an increase in antibody titers and hypothyroidism following therapy. These results have been confirmed by various groups. Individuals who had thyroglobulin antibodies (but not anti-TPO) before treatment, developed anti-TPO antibodies following lithium treatment. All this evidence is interpreted in favor of the concept that either autoimmune thyroid disease or a predisposition to it is present in patients subsequently developing evidence of thyroid failure and autoantibodies after lithium treatment (Hassman and McGregor, 1988). All the antibody-positive patients showed increases in antibody titers and several developed overt disease, whereas age-matched nonlithium-treated controls did not. There is an apparent dissociation between hypothyroidism and thyroid antibodies, that are not detected in all cases of lithium-induced hypothyroidism. In a smaller number of cases, lithium is associated with hyperthyroidism (Kushner and Wartofsky, 1988). Increased thyroid activity and clinical thyrotoxicosis associated with lithium therapy are relatively uncommon. Some cases have occurred upon discontinuation of lithium, while others were observed during lithium use. Exophthalmos has also been noted after treatment with lithium. No investigations of immune mechanisms (anti-receptor antibodies?) have been performed in these cases. It is not clear how lithium alters thyroid physiology (Kushner and Wartofsky, 1988). Lithium has direct effects (inhibition of hormone release, inhibition of hormone synthesis, and peripheral thyroxine degradation) on the thyroid, independently of any immune consequence. It may also have a variety of immunological effects, similar to those postulated or demonstrated for many other chemicals (reviewed in Hart, 1990, 1991a,b). This metal has effects on PMN, NK cells, mast cells, macrophages, and lymphocytes. Lithium has been reported to inhibit the activity of suppressor T lymphocytes by altering a plaque-forming assay believed to be inhibited by a suppressor cell population (Gallicchio, 1991; Messino, 1991). Lithium increases the production of various lymphokines (IL-2, colony-stimulating factor CSF, gamma interferon).

The effects of lithium on experimental animals are controversial (reviewed in Hassman and McGregor, 1988). Early studies demonstrated that treatment with lithium significantly suppressed Arthus and DTH reactions to thyroglobulin and to a lesser extent production of antithyroglobulin antibodies and thyroiditis in an adjuvant-induced rat model of autoimmune thyroiditis. These findings were not confirmed by another study that investigated the effects of lithium treatment at pharmacological doses (0.4 to 1.1 mmol/l) on the development of thyroiditis in female rats of the August (AUG) strain. Rats were immunized with rat thyroglobulin in CFA at 6 weeks of age, boosted 1 week later, and then exposed to lithium treatment (2 mmol/kg i.p.) at different stages of the development of autoimmune thyroiditis. Lithium increased the production of autoantibodies to thyroglobulin during the initial stages of EAT, but decreased such antibodies when administered at later stages, when the disease was spontaneously resolving. It had no effects on the degree of lymphocytic infiltration of the thyroid or on serum TSH. Lithium did not induce antithyroglobulin production in a group of normal unimmunized rats, suggesting that while the drug may have an immunomodulatory effect on existing autoimmune disease, it cannot induce autoimmune thyroid disease *de novo.*

IX. CADMIUM

Cadmium has well-demonstrated immunotoxic properties. Suppression of antibody responses has been observed both *in vitro* and *in vivo;* however, immune potentiation has also been observed (Descotes, 1986). Similarly, cellular immunity can be decreased or increased. These variations may depend on levels and route of exposure, animal species, and inbred strain. On the other hand, there is scarce and not very convincing evidence of cadmium-induced autoimmune effects in humans. Workers exposed to cadmium may develop proteinuria and tubulo-interstitial nephritis (Wedeen, 1992). Some investigators have suggested that glomerular damage may also occur (Nomiyama, 1981; Bernard and Lauwerys, 1986). These lesions might be immunologically mediated, but there is no evidence of renal immune deposits. In addition, one would want the demonstration of circulating autoantibodies against nuclear or other self antigens. Serum autoantibodies against laminin have been detected in approximately 17% of workers currently exposed to cadmium and 19% of workers that had a previous exposure to this metal (Bernard et al., 1987). The prevalences were 2 to 4 times higher than those observed in age-matched controls, but did not reach the level of statistical significance. However, antilaminin antibodies were found in a significantly higher percentage of cadmium-exposed workers whose urinary cadmium levels exceeded 20 µg/g creatinine. Autoantibodies to laminin were found in cadmium workers with normal renal function as well as those with increased proteinuria. No detailed immunopathological studies of kidneys from these patients were published, even though on the basis of previous studies it was suggested that immunoglobulins might be present in the GBM of many of these subjects (Bernard et al., 1987).

Some studies in rats and mice have suggested that exposure to cadmium may result in autoimmune effects, but the available evidence is rather inconclusive. The oral administration of cadmium to Sprague Dawley rats induced a diffuse membranous glomerulonephritis after 30 weeks of exposure (Joshi, 1981). Electron microscopy revealed irregular thickening of the glomerular basement membrane, as well as electron-opaque deposits within mesangial cells. Granular IgG deposits were observed within most glomeruli. In a more recent investigation (Bernard et al., 1984), Sprague Dawley and Brown Norway (BN) rats were chronically exposed to cadmium administered either in their drinking water or by intraperitoneal injection. Antilaminin antibodies were transiently detected in the sera of Sprague-Dawley, but not BN rats. Serum antibodies to type IV procollagen and linear deposits of immunoglobulins in the kidneys were not present in either group of rats. After 13 months of exposure to cadmium, granular renal deposits of immunoglobulins were observed in 25% of animals, i.e., a prevalence that was not significantly different between control and cadmium-treated rats. Finally, when C57Bl mice were exposed to cadmium in their drinking water, their kidneys did not contain detectable immune deposits (Chowdhury et al., 1987). On the other hand, antinuclear antibodies (ANA) were detected in the sera of 50, 89, and 90% of ICR mice that received, respectively, 3, 30, and 300 ppm cadmium in their drinking water for a period of 10 weeks (Ohsawa et al., 1988). Cadmium-treated inbred BALB/c mice were less susceptible than ICR mice to the induction of ANA, that was observed only in mice receiving 300 ppm. Thus, studies in rats and mice suggest that the autoimmune effects of cadmium may vary and are possibly conditioned by genetic (immunogenetic and pharmacogenetic) factors.

X. GOLD

Gold salts, e.g., sodium aurothiomalate and aurothioglucose, are widely used in the treatment of rheumatoid arthritis. However, relatively little is known about the immunotoxicologic properties of gold (Descotes, 1986). In contrast, the autoimmune effects of this metal are well demonstrated. Autoimmune thrombocytopenia (ATP) has been associated with the administration of gold salts (reviewed in Chong, 1991). ATP occurs in 1 to 3% of patients receiving either oral or parenteral forms of gold (Kosty et al., 1989). ATP usually develops after the patient has been taking the drug for several weeks or months. Patients with gold-induced ATP show petechiae, ecchymosis, and mucosal bleeding. Platelet counts usually return to normal within 5 to 7 days after cessation of therapy. However, there are reports of ATP persisting for longer than 30 days. Patients' sera contain drug-dependent antibodies to platelets, i.e., antibodies that bind to platelets in the presence of the drug or metabolites of the drug. This phenomenon can be detected by incubating normal platelets with patients' serum plus the drug and then measuring the binding of immunoglobulins. Thus, an immunologic mechanism may be involved in this condition, as also suggested by a reported association with HLA-DR3.

Exposure to gold can induce a nephrotic syndrome and immune complex-mediated glomerulonephritis, with an incidence that varies, but may be more than 17% of all patients treated (Hoitsma et al., 1991). Its severity is not correlated with the dose of gold compound or the level of gold present in blood or urine, and the nephrotic syndrome usually disappears after gold therapy is discontinued (Hall, 1988, 1989). Mild proteinuria is observed in approximately 10% of patients with rheumatoid arthritis after treatment with gold salts. Massive proteinuria is observed in 1% of these subjects. Kidney biopsies show diffuse granular deposits of immunoglobulins and complement in the glomeruli. Gold is not present in the glomerular deposits, but may be found in proximal tubules and mesangial cells (Katz and Little, 1973; Watanabe et al., 1976). The histopathology varies and may show membranous glomerulonephritis, mesangial glomerulonephritis, or minimal change nephropathy. An immunogenetic influence is suggested by the finding that the relative risk of proteinuria during gold treatment of rheumatoid arthritis is increased 32 times in patients who are HLA-DR3 positive. There are close similarities between gold and penicillamine nephropathy, suggesting that similar autoimmune effects may be caused by different xenobiotics (Hall, 1989).

Experimental animal studies have demonstrated without question the autoimmune effects of gold. Guinea pigs and rabbits exposed to this metal develop autoimmune tubulointerstitial nephritis and/or immune complex nephropathy (reviewed in Bigazzi, 1988). Similarly, BN rats and mice that bear the H-2^s haplotype develop autoimmune responses after treatment with gold (Pietsch et al., 1989; Tournade et al., 1991a,b). In addition, a glomerulonephritis mediated by antibodies to renal glomerular basement membrane and, occasionally, a membranous glomerulopathy have been demonstrated in BN rats (Tournade et al., 1991a,b). The autoimmune effects of different gold compounds, including some thiol-containing salts, have been examined in both rats and mice (Schuhmann et al., 1990; Tournade et al., 1991). These investigations have shown that sulfur-containing groups may potentiate the autoimmune effects of gold, but by themselves do not induce autoimmune responses.

XI. MERCURY

Both inorganic and organic forms of mercury have immunotoxic effects (Descotes, 1986). Suppression or potentiation of cellular immunity and antibody responses have been observed both *in vitro* and *in vivo*, with variations that may depend on levels and route of exposure, animal species and strain. There is also solid evidence that mercury can induce autoimmunity.

Once upon a time, mercury was widely used in medical practice in laxatives, teething powders, diuretics, and ointments. Thus, there are abundant reports of renal complications (e.g., proteinuria and nephrotic syndrome) resulting from therapeutic administration of this metal (Wedeen, 1992). Occupational exposure may also occur, even though currently this is a very rare phenomenon. Individuals exposed to mercury may develop a diffuse membranous nephropathy. Kidney biopsies from these patients usually show no major abnormalities by light microscopy, but direct immunofluorescence reveals granular deposits of immunoglobulins and complement at the level of the glomerular basement membrane (Charpentier et al., 1981). Subepithelial glomerular deposits may be observed by electron microscopy. On the other hand, immunohistopathology findings may be more complex. Kidney biopsies from eight African patients with a nephrotic syndrome associated with the use of skin-lightening creams containing mercury compounds showed linear GBM staining in four cases, granular in one, and mixed granular and linear in three (Lindqvist et al., 1974).

As previously mentioned when discussing lead, deposits of immunoglobulins and complement in kidneys may be suggestive of, but not proof of autoimmunity. It is still questionable whether workers occupationally exposed to mercury actually develop immune responses to self-antigens. Belgian investigators have reported that circulating antilaminin antibodies were present in eight workers exposed to mercury vapors but not in controls (Lauwerys et al., 1983). However, a later study by the same group did not detect an increased prevalence of antilaminin antibodies in a larger number of workers from a chloralkali plant (Bernard et al., 1987). More recent studies have not detected significant autoimmune responses to renal autoantigens in two different cohorts of workers exposed to mercury. In the first, a Scandinavian cohort did not show any evidence of glomerular damage or tubular reabsorption defect (Langworth et al., 1992). In addition, the subjects had no changes in serum immunoglobulin concentrations, no increased levels of anti-GBM or anti-laminin antibodies. At the same time, an international collaborative study has been carried out within the European Community by Belgian, Spanish, Italian, German, and British investigators on a second cohort of 50 male Belgian workers exposed to mercury vapor in a chloralkali plant for at least one year (Cárdenas et al., 1993). These workers had significant changes in urinary excretion of prostaglandins and tubular antigens. The mean level of anti-GBM autoantibodies was 21 in controls and 19.6 in exposed workers (a difference that was not significant). The prevalence was 4.1% in controls and 11.4% in exposed workers (again, not a significant change). Means and prevalence of total IgE and rheumatoid factor were also not significant. On the other hand, the mean levels of anti-DNA were 2.2 in controls and 3.1 in exposed workers ($p < 0.01$), with a prevalence of 4.1% in controls vs. 18.2% in exposed workers ($p < 0.05$). Therefore, this study showed a slight increase in antibodies against DNA (Cárdenas et al., 1993).

The repeated administration of $HgCl_2$ to rats, mice and rabbits results in both autoimmune responses and disease. We wish to stress at this point that studies of autoimmunity induced by treatment with $HgCl_2$ differ from the older, classic investigations of nephrotoxicity because of the much lower dosages employed. Levels of mercury that do not cause major alterations of renal structures are still capable of inducing autoimmune effects (Michaelson et al., 1985). These animal models show striking similarities as well as differences, that may be related to animal species and/or inbred strain (Bigazzi, 1992). Humoral autoimmune responses to a variety of autoantigens have been observed in all models. However, a few autoantigens (that could be identified as "major") seem to be the target of the strongest, possibly pathogenic, autoimmune responses. For example, mercury-treated BN, MAXX, and Dorus Zadel black

(DZB) rats produce high levels of autoantibodies to laminin (Fukatsu et al., 1987; Bigazzi et al., 1989; Aten et al., 1992). Autoantibodies to other autoantigens (e.g., type IV collagen, heparan sulfate proteoglycan, entactin, thyroglobulin, etc.) have also been detected in BN rats after exposure to mercury, but they are present both in lower concentrations and for shorter periods of time (Bowman et al., 1987; Goldman et al., 1991). In addition, they may not be correlated with disease. Rats of the PVG strain produce autoantibodies to nuclear antigens after mercury treatment, but no antibodies to laminin or other autoantibodies have been detected in these animals (Weening et al., 1981). BN rats treated with mercury may also develop a syndrome with some features in common with graft-vs.-host disease, including skin lesions (Mathieson et al., 1992). The kinetics of humoral autoimmune responses also differ with animal species and/or strain. Mercury-treated BN, MAXX, and DZB rats show a self-limited production of autoantibodies to renal GBM/laminin: these autoantibodies are first detectable in the circulation after 6 to 7 days, reach peak titers and incidence by day 14 to 15, and then decrease within 20 to 30 days (Bowman et al., 1984; Michaelson et al., 1985; Fukatsu et al., 1987; Henry et al., 1988; Bigazzi et al., 1989). Other effects of mercury include splenomegaly, lymph node hyperplasia, and thymic atrophy in BN rats. Changes in peripheral lymphocyte subpopulations of mercury-treated rats have also been reported by various groups (Bowman et al., 1987; Aten et al., 1988; Pelletier et al., 1988; Kosuda et al., 1991, 1993, 1994a, b). In our laboratory, we have found that spleens from mercury-treated BN rats contained 21% RT6+ cells on day (D) 10 of treatment, 13% on D17, 16% on D24, and 20% on D30. Lymph nodes from the same rats had 36% RT6+ cells on D10, 23% on D17, 29% on D24, and 28% on D30. The decrease in RT6+ cells preceded the development of autoimmune responses to GBM, which peaked on days 17 to 24. It also correlated inversely with such responses that declined by day 30. Both immunopathology and histopathology of the mercury-induced models have received detailed attention. Renal immune deposits observed after mercury treatment in BN and MAXX rats as well as outbred rabbits are quite similar in appearance and are characterized by two-stage kinetics. The first stage shows linear deposits of IgG at the level of both glomerular basement membrane (GBM) and tubular basement membrane (TBM). Eluates from these deposits contain autoantibodies to laminin (Fukatsu et al., 1987; Bigazzi et al., 1989; Aten et al., 1992). The second stage shows granular deposits of IgG both in GBM and TBM. These deposits also contain autoantibodies that react most strongly with laminin (Fukatsu et al., 1987). A granular pattern (*not* preceded by the linear staining) is observed in GBM of mercury-treated PVG rats (Weening et al., 1981). Histopathologically, all animal models of mercury autoimmunity show a membranous glomerulopathy (MGP), but lesions in other tissues are uncommon (Aten et al., 1992; Hultman et al., 1992). The immunogenetics of autoimmunity induced by mercury in rats have been recently reviewed (Goldman et al., 1991). Rats of the RT-1^n haplotype are susceptible to anti-GBM/laminin autoimmunity, whereas those of the RT-1^l haplotype are resistant, and rats with other haplotypes (RT-$1^{c,a,k,f,b}$) show intermediate susceptibility. It has been suggested that susceptibility to the first phase (anti-GBM antibodies) depends on several genes, one of which is RT-1 linked, whereas the second phase (immune complex-type glomerulonephritis) depends on one major RT-1-linked gene or cluster of genes, with a role for other non-RT-1-linked genes controlling the magnitude of the response. Interestingly, rats that are congenic at the MHC between the "responder" BN strain and the "resistant" LEW strain, i.e., BN.1L and LEW.1N, do not show any autoimmune response to the administration of mercury (Druet et al., 1977; Sapin et al., 1982; Aten et al., 1991; Kosuda et al., 1994).

Finally, mercuric chloride administered to mice of the H-2^s phenotype (A.SW, B10.S, SJL/J, etc.) induces the production of both antinucleolar and antinuclear antibodies (reviewed in Goldman et al., 1991). The former react with a 34-kDa nucleolar protein (U3 RNP protein, also defined as "fibrillarin") and occasionally with other nucleolar proteins of 60 to 70 and 10 to 15 kDa. The latter (antichromatin and/or antihistone antibodies) may occur in a small percentage of H-2^s and H2^k mice after mercury treatment. However, autoantibodies against nucleolar antigens are usually present in higher concentrations. No autoantibodies to kidney, thyroid, skin, etc., have been observed in mercury-treated mice (Hultman et al., 1992). Murine autoantibodies to fibrillarin seem to recognize the same epitopes as the autoantibodies from certain patients with scleroderma (Reimer, 1990). Thus, exposure to mercury results in a murine model that can allow the study of environmental factors in scleroderma and other autoimmune diseases. Antinucleolar autoimmune responses of H2^s mice persist for up to 10 to 12 weeks or longer; thus they are longer-lasting than autoimmune responses observed in mercury-treated BN rats (Hultman et al., 1989). Granular deposits of IgG, IgM, and C3 are detected in kidneys of mice after exposure to mercury, and eluates from the immune deposits of SJL mice have been found to contain antinucleolar antibodies (Hultman and Eneström, 1988; Hultman et al., 1992). Immune complex deposits can also be observed in the vessels of other tissues (e.g., spleen, heart) (Hultman and Eneström, 1988; Hultman et al., 1992). Histopathologically, mice with mercury autoimmunity show a membranous glomerulopathy (MGP), but lesions in other tissues are uncommon. Other effects of mercury include lymph node hyperplasia in H-2^s mice. Studies of the immunogenetics of mercury-induced autoimmunity in mice have shown that both MHC class II loci and unknown non-MHC loci govern susceptibility. Susceptibility to the development of antifibrillarin antibodies can be mapped to the H2A region (Hultman et al., 1992).

The pathogenesis of mercury-induced autoimmunity in animal models is not completely clear. The major target organ in all models is the kidney, which exhibits MGP, whereas other organs and tissues show minor effects or seem to be spared. Circulating autoantibodies against GBM/laminin, present in BN (as well as MAXX and DZB) rats and outbred NZW rabbits, may have a pathogenetic role as demonstrated by their correlation with proteinuria. Activation of the complement cascade (especially in rabbits, where C3 has been noted in the renal immune deposits) or direct autoantibody effects (especially in rats, where scarce or no binding of complement components occurs, both *in vivo* and *in vitro*) may cause the proteinuria in animals with autoantibodies to laminin. Rats and mice with circulating autoantibodies to nuclear and nucleolar antigens likely experience proteinuria through the renal deposition of immune complexes and complement activation. Lymphocytes and macrophages, often present in the renal interstitium, may be an additional damaging factor, but direct proof of their pathogenetic involvement is lacking. The etiology of mercury-induced animal models of autoimmunity is obvious, i.e., they are caused by treatment with mercury in its various chemical forms, using different routes of administration (subcutaneous injection, oral ingestion, inhalation). Extremely low doses of mercury have been found effective in both BN rats and SJL/J mice. Less obvious are the mechanisms by which mercury stimulates the cells of the immune system to cause autoimmunity. T lymphocytes play a central role, as shown by the absence of mercury-induced effects in T cell-deprived animals: indeed, mercury may sequentially activate T helper and regulatory lymphocytes (Goldman et al., 1991). Various subpopulations of T lymphocytes are affected by mercury. The RT6+ subset of T cells (that have a regulatory role in various rat models of autoimmune disease) decreases in mercury-treated BN rats (Kosuda et al., 1991, 1993, 1994a,b). Conversely, CD8+ T cells were reported to be increased in LEW rats, that are resistant to the autoimmune effects of this metal (Rossert et al., 1991). B lymphocytes may be activated by mercury directly, or indirectly through IL-4 production by T cells (Goldman et al., 1991). Another effect of mercury is an increased expression of MHC class II molecules on B cells, detected as early as 3 days after the first injection of the metal.

Changes in cytokine production have been observed in animal models of mercury-induced autoimmunity. A deficiency in IL-2 production was noted in BN rats after mercury treatment (Baran et al., 1988). Druet and his co-workers have reported that treatment with mAb against the IL-2 receptor supports a role for Th1-like cells during the regulation phase of $HgCl_2$-induced autoimmunity (Dubey et al., 1993). Weening's group has shown that $HgCl_2$ down-regulates IFN-γ production in BN but not LEW rats (van der Meide et al., 1993). The presence of hyper-IgE and hyper-IgG1 have suggested that IL-4 could be an important mediator of T cell-dependent B cell activation. Increased levels of IL-4 mRNA have been detected within CD4+ T cells of mercury-treated H-2^s mice, and inhibition of IL-4 has dramatic effects. *In vivo* treatment of mercury-injected SJL/J mice with anti-IL-4 mAb completely abrogated the increase in serum IgE and reduced titers of IgG1 anti-nucleolar antibodies (Ochel et al., 1991). On the other hand, it also resulted in increased levels of IgG2a, IgG2b, and IgG3 antinucleolar antibodies, possibly as a result of compensatory activity of Th1-like T lymphocytes (Ochel et al., 1991). Therefore, at least in H-2^s mice, Th2 cells secreting IL-4 may be responsible for B cell stimulation, and Th2 hyperactivity may be associated with decreased Th1 functions (e.g., impaired ability to produce IL-2 *in vitro*). Highly susceptible strains such as A.SW, SJL, and B10.S (H-2^s haplotype) do not develop contact dermatitis in an ear swelling test (a form of delayed-type hypersensitivity reaction). On the other hand, H-2^d mice, resistant or low responders to mercury, develop contact dermatitis. In conclusion, both Gleichmann and Druet have proposed the hypothesis that mercury activates Th2-like T cells in susceptible and resistant strains of rats and mice (Dubey et al., 1991; Goldman et al., 1991). In susceptible strains, a defect of the Th1 subset would lead to autoimmunity, whereas in the resistant strains the early and efficient activation of the Th1 subset would result in DTH responses and resistance to humoral autoimmunity (Dubey et al., 1991). The suggested role of T helper lymphocyte subsets in mercury-treated rats and mice still lacks a detailed and complete analysis of cytokine profiles (Goldman et al., 1991). In addition, a role of suppressor cells, the idiotype–anti-idiotype network and regulatory cytokines has been postulated, at least for those models where mercury-induced autoimmunity is spontaneously down-regulated (Goldman et al., 1991). As yet, there is no explanation for the lack of a similar down-regulation in rats and mice with antinuclear and antinucleolar autoimmune responses. There is also no explanation for the difference in the autoantigens involved in the various models: e.g., why is laminin the "major" autoantigen in BN rats and fibrillarin the "major" autoantigen in SJL/J mice? Similarly, there are only a few studies on the role of other immune cells, e.g., antigen-presenting and/or cytokine-producing cells, in animal models of autoimmunity caused by mercury. Both macrophages and neutrophils from inbred rats are affected by mercury *in vitro,* and their functions may be stimulated or inhibited depending on mercury concentration and rat strain (Contrino et al., 1992). Finally, the biochemical pathway(s) utilized by mercury to directly activate the cells of the immune system are still unknown.

XII. THE LESSONS OF METAL-INDUCED AUTOIMMUNITY

In the preceding sections of this chapter we have examined the autoimmune responses and diseases resulting from exposure to various metals. From the data available it is apparent that autoimmunity induced by metals is a reality, at least for gold and mercury. We will now examine the lessons that can be drawn from an analysis of this phenomenon.

The variety of compounds involved in metal-induced autoimmunity makes it very difficult to establish structure-activity relationships. A possible role of thiol groups was initially suggested by studies with gold compounds. The formation of disulfide bonds between –SH groups and thiols of self-proteins might change the structure of autoantigens. It would render them foreign or expose hidden epitopes. Alternatively, disulfide bonds might be formed with surface structures of immune cells, causing their stimulation. However, aurothiomalate and aurothioglucose contain a sulfur atom, not as sulfhydryl or disulfide, and still have similar autoimmune effects. Studies of autoimmunity induced by other xenobiotics that do not contain thiol groups confirm that these structures are not involved in all cases of autoimmunity caused by chemical exposure (Kosuda and Bigazzi, 1996).

Another aspect to be considered is possible variations in susceptibility according to species. Human beings are quite prone to xenobiotic-induced autoimmune disease, but their susceptibility is likely controlled by endogenous, genetically determined factors. This may explain why not all individuals exposed to metals develop autoimmunity and why it is often difficult to reproduce in experimental animals the various syndromes observed in humans. However, once the appropriate species and/or inbred strain are selected, often by serendipity, it is possible to obtain parallel (albeit not always analogous) models of human autoimmune disease. Studies of experimental animals have provided a clear-cut causal correlation between exposure to gold and mercury and onset of autoimmune responses. Each model has unique characteristics, again dependent on the metal as well as the animal species and the strain involved. For example, mercury-treated BN, MAXX, and DZB rats produce high levels of autoantibodies to laminin (reviewed in Bigazzi, 1992). Rats of the PVG strain produce autoantibodies to nuclear antigens after mercury treatment, but no antibodies to laminin or other autoantibodies have been reported in these animals. Finally, mice of the $H-2^s$ phenotype (A.SW, B10.S, SJL/J, etc.) produce both antinucleolar and antinuclear antibodies. The former react with a 34-kDa nucleolar protein (U3 RNP protein, also defined as "fibrillarin") and occasionally with other nucleolar proteins of 60 to 70 and 10 to 15 kDa. The latter (antichromatin and/or antihistone antibodies) may occur in a small percentage of $H-2^s$ and $H2^k$ mice after mercury treatment. However, autoantibodies against nucleolar antigens are usually present in higher concentrations. No autoantibodies to kidney, thyroid, skin, etc. have been observed in mercury-treated mice. To date, we have no explanation for the difference in the autoantigens involved in the various models. However, the animal models of mercury-induced autoimmunity seem to share two fundamental properties: they may be induced by relatively low doses of the metal, and the genetic makeup of the animal is extremely important for the induction of autoimmunity. Thus, the role of genetic factors observed in humans is confirmed by experimental animal studies.

Autoimmune disease associated with exposure to gold or mercury is characterized by an immunologic and pathologic profile that may be similar to that of "naturally occurring" autoimmune disease. Individuals exposed to these metals experience autoimmune responses against various autoantigens, as demonstrated by the production of autoantibodies to tissue-specific as well as not-tissue-specific antigens. There have been no studies of cellular immunity in patients with autoimmunity induced by metals, but T lymphocytes and their subsets must obviously be involved. Macrophages and other antigen-presenting cells are also affected by metals and may participate both in the induction and effector phases. As for "naturally occurring" autoimmunity, the pathogenesis of metal-induced autoimmunity may involve numerous immunologic mechanisms. Cytotoxic autoantibodies, a group that comprises the "classical" tissue-specific (anti-red cell, antiplatelets, antilaminin) autoantibodies, are observed in autoimmunity induced by gold and mercury. Autoantibodies to erythrocytes and platelets may destroy these cells.

Chemically induced anti-red cell antibodies may cause hemolysis through the activity of the reticuloendothelial system or, less frequently, through complement activation. Autoantibodies against laminin may cause damage by an *in situ* immune complex formation, with activation of the complement cascade and inflammation. They may also act without activating complement. Cytotoxic T lymphocytes (CTL) and T lymphocytes producing cytokines may also have a pathogenetic role in metal-induced autoimmunity. However, to date we have no evidence of their role in human autoimmunity induced by metals. Studies of cytokines in autoimmunity induced by mercury in experimental animals have been initiated by a few groups of investigators, and their outcome promises to be quite intriguing (Goldman et al., 1991).

In contrast to "naturally occurring" autoimmune disease, where the etiologic agents are unknown, the cause of metal-induced autoimmune disease seems obvious, i.e., the autoimmune process is initiated by exposure to an environmental metal. This is simple to determine in an experimental situation, where we know the initiation of the exposure to mercury or gold. We also know dosage and duration of treatment. The correlation is less easy to ascertain in humans, especially when the exposure to the metal did not occur in the workplace or if multiple xenobiotics were involved. In any case, metals may induce autoimmune responses and disease through a variety of effects at the cellular, biochemical, and molecular level (Bigazzi, 1988; Kosuda and Bigazzi, 1996). They may be indirectly responsible for autoimmunity by altering the structural integrity and/or functions of organs and tissues and therefore, releasing autoantigens that normally would be present in low concentrations, if at all. Alternatively, or in addition, metals may cause autoimmunity by their direct immunotoxic effects. They may act on lymphocytes, inhibiting T suppressor lymphocytes, stimulating T helper cells or B lymphocytes. Some metals (lead, mercury) may affect macrophage and neutrophil functions (Contrino et al., 1988; Kowolenko et al., 1988; Contrino et al., 1992).

The indirect effects of metals are related to their well-known toxicity on various tissues and organs of the body. In this case, autoimmune responses and disease may result from modifications of the structure of autoantigens, so that they are no longer recognized as self. Conformational changes may induce new epitopes or expose hidden antigenic determinants. Metals may also cause the release of cellular or other tissue components which are normally present at very low levels (or completely absent) in the circulation. This phenomenon, that occurs after exposure to gold and other metals, results in higher levels of autoantigens that may stimulate the immune system. Molecular mimicry, i.e., the sharing of epitopes with autoantigens, is usually observed with microorganisms; however, metals may also possess determinants that are found in cells and tissues. Finally, copper, mercury and zinc can induce heat shock proteins (also called stress proteins) in various types of cells (Levinson, 1979; Levinson et al., 1980). These substances may have a role in autoimmunity (Cohen, 1992).

Metals may also cause aberrant MHC class II expression on target organs. This effect might result directly from the interaction of metals with cell surfaces or be a consequence of the increased production of cytokines (e.g., interferon-γ). Increased percentages of lymphocytes positive for MHC class II have been observed in BN rats treated with mercury; similarly, other cells including epithelial cells may have increased expression of MHC class II (Goldman et al., 1991). Lead has been reported to increase murine B cell MHC class II expression. Treatment of mouse spleen cell populations (containing macrophages) with anti-Ia antibodies plus complement completely ablates proliferative responses induced by lead, nickel, and zinc, suggesting that metals may act, directly or indirectly, to enhance the expression of MHC class II antigens (Warner and Lawrence, 1986a,b). Other investigators have postulated that aberrant expression of MHC class II antigens by epithelial cells is a key event in the initiation and maintenance of organ-specific autoimmunity (Bottazzo et al., 1986). The presence of class II antigens would convert the epithelial cells to functional antigen-presenting cells, enabling them to present their autoantigens to T helper cells, by-passing the need for "conventional" antigen-presenting cells (macrophages and dendritic cells) and resulting in autoimmune reactions specifically targeted on the class II positive epithelial cells. This hypothesis originated from the observation of the aberrant expression of MHC class II molecules on thyroid epithelial cells, which normally are class II negative. Abnormal expression of class II molecules by specific cells of target organs has later been reported in other human conditions (insulin-dependent diabetes mellitus, Sjögren's syndrome, inflammatory bowel diseases) and experimental autoimmune diseases (glomerulonephritis, experimental allergic encephalomyelitis). Thyroid cells expressing class II MHC are capable of activating cloned T cells, a phenomenon that can be blocked by monoclonal anti-class II antibodies. A variety of factors have been found to modulate class II MHC expression in epithelial cells, in particular gamma interferon (IFN-γ). Interestingly, MHC class II expression occurs naturally in other organs that are only occasional targets of autoimmune responses, e.g., the adrenals. Normal adrenals express class II antigens, particularly in the zona reticularis. In addition to the negative results of this "experiment of nature" at the adrenal level, investigations of transgenic mice with inappropriate expression of MHC class II antigens on beta cells of their pancreas have not revealed any autoimmune response to those cells. Finally, studies of thyroids from Hashimoto's thyroiditis patients support the view that MHC class II expression is actually secondary to lymphocytic infiltration of those organs. Thus, it may have a role in the perpetuation of autoimmune damage, but is unlikely to be the initiating etiologic event.

Metals may act as immunogens or haptens. Monoclonal antibodies that react with soluble mercuric ions have been produced in mice (Wylie et al., 1992). Mercury-specific murine T helper cells reportedly react to mercury or a mercury-protein complex stored in macrophages (Kubicka-Muranyi et al., 1993). Beryllium-specific T cell clones have been derived from T lymphocytes obtained by bronchial lavage of patients with berylliosis (Saltini et al., 1989). Similarly, gold-specific T lymphocyte clones were isolated from a patient with rheumatoid arthritis who developed delayed type hypersensitivity reactions to gold (Romagnoli et al., 1992). These observations lead to the obvious conclusion that metals can specifically stimulate cells of the immune system. Mercury and other metals may activate T helper lymphocytes. They may stimulate B cells, both directly or indirectly through their effects on T helper cells. Macrophages and other antigen-presenting cells may also be stimulated by metals to become more active in epitope presentation and production of cytokines (IL-1, etc.). Mercury has effects on neutrophils, macrophages, and leukocyte migration (Contrino et al., 1988; Nordlind and Lidén, 1990; Contrino et al., 1992). The result of these interactions between metals and cells of the immune system may actually be changes in the production of cytokines by any of the various components of the immune system, thus altering the cytokine network and breaking tolerance to autoantigens. Possible mechanisms of xenobiotic-induced activation of immune cells have been recently reviewed (Pelletier et al., 1992). They comprise ligand mimicry, modification of the membrane lipid bilayer, modification of ligand-receptor interaction, modification of activation via TCR-class II peptide interaction, activation via accessory molecules (CD4, CD2), modulation of transducers, modification of the metabolism of second messengers, modulation of enzymatic functions (kinases, phosphodiesterases), modification of the oxidative metabolism of the cell or intracellular ionic activities, and, finally, interactions with nucleoproteins.

Metals may inhibit T suppressor cells. Inhibition of T suppressor lymphocytes as a possible mechanism of autoimmunity was particularly favored during the 1970s, when autoimmune diseases were considered the result of immunoregulatory dysfunction due to a loss of suppressor or regulatory T cells (reviewed in Bigazzi, 1991). However, the actual existence of suppressor T cells has later been questioned by some investigators. This is due primarily to our fragmentary knowledge of their function, lack of unique markers, scarcity (or lack) of suppressor T cell clones and no (or variable) T cell receptor gene rearrangements in T cell hybridomas that produce antigen-specific suppressor factors. Further complications include the lack of identified genes for antigen-specific T cell suppressor factors as well as a gene for the I-J marker. There is no doubt that suppression remains as important as ever to cell biologists and clinical immunologists, but progress in identifying unique cells and molecules has been very disappointing. As far as metal-induced autoimmunity is concerned, some investigators have suggested that mercury activates suppressor cells in certain strains of rats and inhibits them in others (Goldman et al., 1991). However, even though alterations of T lymphocyte subsets have been observed in experimental animals with mercury-induced autoimmunity, a detailed functional analysis of various T cell subsets is still lacking.

Metals may cause alterations of the idiotype–anti-idiotype network. Various investigators have suggested that certain autoimmune diseases, and in particular those with fluctuating clinical symptoms (e.g., systemic lupus erythematosus and myasthenia gravis), may result from a failure of the idiotype–anti-idiotype (id-anti-id) network to regulate immune responses against self-components (Bigazzi, 1986; Bigazzi, 1991). Id-anti-Id interactions *in vivo* may also contribute to the continuous stimulation of autoreactive clones and be responsible for the initiation and perpetuation of autoimmune disease. Indeed, auto-anti-Id immunity is an essential component of the Id-anti-Id network, and its involvement in autoimmune disease is a necessary corollary. Similarly, metal-induced autoimmunity may result from a derangement of the idiotype–anti-idiotype network. Alterations of this network may occur through an immune response against the metal, followed by an anti-idiotypic response characterized by anti-idiotypic antibodies that are the internal image of one or more autoantigens, therefore, generating anti-anti-idiotypic responses that are actually autoimmune responses to tissue antigens and eventually result in autoimmune disease. To date, there is no evidence that such a mechanism has an initiating role in metal-induced autoimmunity. On the other hand, alterations of the network may enhance and/or perpetuate autoimmune disease, as we have observed in BN rats immunized with idiotype (Bigazzi et al., 1989; Bigazzi, 1991).

Metal-induced autoimmunity is conditioned by genetic factors. The mechanisms we have just described, acting alone or in combination, may induce autoimmunity only in the presence of the appropriate genetic terrain. Experimental animal studies indicate that a genetically controlled predisposition is indispensable for the expression of autoimmune disease induced by metals. This phenomenon is quite complex, since one must consider both immunogenetic and pharmacogenetic influences (Bigazzi, 1988; Kosuda and Bigazzi, 1996). The study of inherited variations of enzymes resulting in abnormal and untoward drug reactions has been called "pharmacogenetics" (reviewed in Vogel and Motulsky, 1979; Shear and Bhimji, 1989; Nebert and Weber, 1990). Genetic polymorphisms of drug-metabolizing enzymes (e.g., acetylating enzymes, P450 monooxygenase) give rise to distinct subpopulations that differ in their ability to perform certain biotransformation reactions of drugs (Meyer et al., 1990). The concept of "ecogenetics" has historically evolved from pharmacogenetics and extends to other chemicals the central idea of genetically determined variable drug responses (Vogel and Motulsky, 1979; Nebert and Weber, 1990). The environment contains numerous potentially toxic agents that may damage a genetically predisposed fraction of the population. In this view, different individuals exhibit profound variations in their response to xenobiotics because of genetic variations in enzymes or proteins that play a role in the metabolism of chemicals. Susceptibility to the autoimmune effects of xenobiotics may be caused by the acetylator, sulfoxidizer, aromatic hydrocarbon (Ah) receptor, P450, and metallothionein phenotypes. In the case of autoimmunity induced by metals it is important to consider an interesting group of polypeptides, the metallothioneins, characterized by multiple isoforms (Kägi and Schäffer, 1988; Bremner, 1991; Templeton and Cherian, 1991). They may be involved in the autoimmune effects of various metals, but to date we know nothing about this function or of the possible role of additional ecogenetic phenotypes. As far as immunogenetics are concerned, investigations of "naturally" occurring autoimmunity have shown that the major histocompatibility complex (MHC), the T cell receptor (TCR), certain complement deficiencies, and particular immunoglobulin genes may all have important effects. As far as MHC class II associations are concerned, HLA-DR3 seems to be strongly connected with predisposition to SLE and other autoimmune disorders (reviewed in Liszewski and Atkinson, 1991; Reveille and Arnett, 1991). On the other hand, a study of four polymorphic class II genes has shown that the presence or absence of an aspartic acid at position 57 of the DQ_β chain is correlated with susceptibility to or protection against IDDM. When aspartic acid is present at position 57, no IDDM is found, whereas the disorder occurs when alanine, valine, or serine are present at position 57. DQ_β polymorphisms, particularly at position 57, might determine the specificity and extent of the autoimmune response against islet cell antigens through T cell help and/or suppression. It is still unknown whether similar DQ_β polymorphisms occur in other autoimmune disorders (reviewed in Bigazzi, 1993). Finally, recent studies of rheumatoid arthritis have demonstrated that a specific structural component of HLA-DR molecules may be the critical disease-associated determinant in this condition. Sequence comparisons of the third hypervariable regions of the DR beta-1 chains expressed by the various HLA-DR4 subtypes have shown that the aminoacid sequence associated with rheumatoid arthritis is glutamine (Q), lysine or arginine (K or R), alanine (A), and alanine (A). Because the determinant is found on more than one HLA-DR molecule, it has been named the "shared epitope" (Albani et al., 1992; Nepom, 1992; Nepom and Nepom, 1992; Winchester et al., 1992). To date, we do not know its role in autoimmune disease other than rheumatoid arthritis or its possible relevance in xenobiotic-induced autoimmunity. Specific T cell receptor gene combinations may also influence the development of autoimmune disease (Brostoff and Howell, 1992; Marguerie et al., 1992; Moss et al., 1992). Possible restrictions of TCR V gene usage are currently being explored by molecular biology techniques in a variety of autoimmune disorders. At present, studies of "naturally occurring" rheumatoid arthritis, multiple sclerosis, Graves' disease, and Sjögren's syndrome have given conflicting evidence, either in favor of or against a restricted TCR repertoire. Finally, malfunctions of the complement system and in particular C4 deficiencies are associated with autoimmune syndromes, possibly affecting the clearance of immune complexes (Liszewski and Atkinson, 1991). These effects are likely operative only in those conditions characterized by the formation of antigen-antibody complexes, e.g., SLE. The abundant information on the immunogenetics of "naturally occurring" autoimmune diseases contrasts with the preliminary and incomplete evidence obtained from patients with autoimmunity induced by xenobiotics. Certain MHC haplotypes (e.g., HLA-DR3) may be more susceptible to the autoimmune effects of some metals. However, this does not seem to be a general phenomenon, and the "shared epitope" theory may explain the observed variations. Complement C4 null haplotypes are associated with autoimmune disease induced by gold (Clarkson et al., 1992). To our knowledge, there is no information on TCR usage in metal-induced autoimmunity.

Considering the complexity of the immune system, it is likely that a combination of factors rather than a single mechanism is responsible for the induction of autoimmune responses and disease by metals. The inductive event is the encounter of the metal with an individual that is genetically predisposed because of pharmacogenetic and immunogenetic factors. The other most likely requirement is chronic exposure, which may take place over a period of several months if not years. The metal may act by itself or in association with other environmental factors, such as other chemicals or infectious microorganisms. Based on the analysis of several animal models of autoimmune disease, we suggest that interactions between microorganisms and chemicals may actually result in potentiation of their autoimmune effects. The next likely step is a change in the cytokine network. This might be initiated through the stimulation of macrophages by the chemical, resulting in the production of interleukin 1 (IL-1) and oxygen radicals. Until recently only a few structurally defined substances (including muramyldipeptide, poly(I)-poly(C), PHA, C5a, and lipopeptide) had been shown to be IL-1 inducers. However, it has now been demonstrated that IL-1 production is induced *in vitro* by fibronectin, urate crystals, and kaolin. Various drugs may also affect IL-1 synthesis: for example, rats treated with cyclophosphamide or avridine have significantly increased IL-1 production, whereas rats treated with dexamethasone produce significantly less IL-1. The release of IL-1 activates the cytokine network and its various effects on T and B lymphocytes. At the same time, the oxygen radicals produced by macrophages may have deleterious effects on surrounding cells, releasing a variety of self-epitopes. It should be noted that macrophages are not necessarily the first cells to be involved by metals. IL-1 is a cytokine that initially was strictly defined on the basis of its production by mononuclear phagocytes and its stimulation of T lymphocytes. However, more recent studies have shown that IL-1-like molecules are produced by cells other than mononuclear phagocytes (e.g., epithelial cells, astrocytes, renal mesangial cells, endothelial cells, etc.). In addition, it has been demonstrated that IL-1 participates in the stimulation of B-lymphocytes and can also affect a variety of nonlymphoid cells, such as fibroblasts, hepatocytes or hypothalamic cells. A number of agents stimulate IL-1 production: among the stimulatory agents are microorganisms and their products (including endotoxin), antigen-antibody complexes, inflammatory substances, lectins and lymphokines. Similarly, the production of other cytokines by a variety of cells may be stimulated by metals. At the same time, autoantigens are released and/or changed, new tissue epitopes are expressed, and self peptides processed by antigen-presenting cells. Some metals, e.g., lead and mercury, increase MHC expression, which may help in the presentation of autoantigens. Another possibility is metal-induced activation of Th2-like or Th1-like T cells that would lead to autoimmune disease mediated by either autoantibodies or DTH responses (Dubey et al., 1991; Romagnani, 1994). In any case, IL-1 stimulation or activation of T helper subsets (or both) unleashes the autoimmune response, generating effector CD4+ and CD8+ T lymphocytes as well as proliferating B lymphocytes. Damage of target tissues and disease eventually occur. The rather complex scenario that we have presented is quite difficult to dissect, which explains our relative lack of information.

To date, what may be the most interesting aspect of metal-induced autoimmunity has not been investigated, i.e., why some metals are capable of inducing autoimmune responses and others are not. For example, lead has well-known immunotoxic properties, however, there are no confirmed observations of its autoimmune effects. In a recent review of the literature, we have found a few reports of chromium-induced autoimmunity, but no published cases of autoimmune disease induced by aluminum, a metal with good adjuvant capabilities. In addition, one would not be surprised if chronic ingestion or inhalation of various other metals that affect lymphocytes and macrophages resulted in autoimmune responses. Perhaps the correlation with these metals has not been made because of inadvertent exposure to extremely low environmental levels. Thus, it is important to ascertain the actual incidence of autoimmune disease induced by environmental chemicals and identify the substances involved. We have suggested the establishment of a *dedicated national registry* of both confirmed and suspected cases of autoimmune disease associated with xenobiotics, as the first step to obtain a national data base in this area (Kosuda and Bigazzi, 1996). A similar registry has recently been established in France (personal communication from Prof. Descotes). In our opinion, it would be extremely beneficial if a registry of xenobiotic-induced autoimmunity were more than an impersonal computerized listing of cases published in medical journals or reported to health authorities. Instead, it should be closely associated with an established diagnostic immunology laboratory and an academic group of experts in autoimmune disease. This would allow the verification of autoimmune responses through studies of sera and immune cells obtained from individuals allegedly affected by chemically induced autoimmunity. Ideally, there ought to be a section within the

National Institute of Environmental Health Sciences of the NIH devoted to intramural research on xenobiotic-induced autoimmunity. This section should be in close contact with the registry and initiate laboratory studies of specific areas of xenobiotic-induced autoimmunity suggested by case reports. Epidemiological and mechanistic investigations should then be performed to confirm possible findings and explain the autoimmune effects of the chemicals involved. Finally, perusal of the data base obtained by a research-oriented registry of xenobiotic-induced autoimmune disease might reveal that various metals are incapable of causing autoimmunity, a phenomenon that by itself would deserve a thorough investigation.

In conclusion, there are numerous advantages that can result from the increasing interest in metal-induced autoimmunity. The etiology of many autoimmune diseases is still unknown, making their prevention difficult or impossible. If future investigations will reveal that environmental metals and other xenobiotics (either alone or in association with other factors) play a role in those disorders, we will be able to establish better preventive measures than currently available and possibly reduce the incidence of autoimmune disease.

ACKNOWLEDGMENTS

Our studies of xenobiotic-induced autoimmunity are supported by USPHS Grant ES03230 and RG-2396 from the National Multiple Sclerosis Society.

REFERENCES

Albani, S., Carson, D. A., and Roudier, J. (1992), *Rheum. Dis. Clin. N. Am.,* 18, 729–740.
Aten, J., Bosman, C. B., Rozing, J., Stijnen, T., Hoedemaeker, P. J., and Weening, J. J. (1988), *Am. J. Pathol.,* 133, 127–138.
Aten, J., Stet, R. J. M., Wagenaar-Hilbers, J. P. A., Weening, J. J., Fleuren, G. J., and Nieuwenhuis, P. (1992a), *Scand. J. Immunol.,* 35, 93–105.
Aten, J., Veninga, A., Bruijn, J. A., Prins, F. A., de Heer, E., and Weening, J. J. (1992b), *Clin. Immunol. Immunopathol.,* 63, 89–102.
Aten, J., Veninga, A., de Heer, E., Rozing, J., Nieuwenhuis, P., Hoedemaeker, P. J., and Weening, J. J. (1991), *Eur. J. Immunol.,* 21, 611–616.
Baran, D., Lantz, O., Dosquet, P., Sfaksi, A., and Druet, P. (1988), *Clin. Exp. Immunol.,* 73, 401–405.
Bernard, A. and Lauwerys, R. (1986), in *Cadmium,* Foulkes, E. C., Ed. Springer-Verlag, Berlin, 135–177.
Bernard, A., Lauwerys, R., Gengoux, P., Mahieu, P., Foidart, J. M., Druet, P., and Weening, J. J. (1984), *Toxicology,* 31, 307–313.
Bernard, A. M., Roels, H. R., Foidart, J. M., and Lauwerys, R. L. (1987), *Int. Arch. Occup. Environ. Health,* 59, 303–309.
Bigazzi, P. E. (1986), *Ann. N.Y. Acad. Sci.,* 475, 66–80.
Bigazzi, P. E. (1988), *J. Toxicol. Clin. Toxicol.,* 26, 125–156.
Bigazzi, P. E. (1991), in *Systemic Autoimmunity,* Bigazzi, P. E. and Reichlin, M., Ed., Marcel Dekker, New York, 39–64.
Bigazzi, P. E. (1992), *Clin. Immunol. Immunopathol.,* 65, 81–84.
Bigazzi, P. E. (1993), in *The Molecular Pathology of Autoimmune Disease* Bona, C. A., Siminovitch, K. Theofilopoulos, A. N., and Zanetti, M., Eds., Harwood Academic Publishers, Chur, Switzerland, 493–510.
Bigazzi, P. E., Michaelson, J. H., and Potter, N. T. (1989), *Autoimmunity,* 5, 3–16.
Bottazzo, G. F., Todd, I., Mirakian, R., Belfiore, A., and Pujol-Borrell, R. (1986), *Immunological Rev.,* 94, 137–169.
Bowman, C., Green, C., Borysiewicz, L., and Lockwood, C. M. (1987), *Immunology,* 61, 515–520.
Bowman, C., Mason, D. W., Pusey, C. D., and Lockwood, C. M. (1984), *Eur. J. Immunol.,* 14, 464–470.
Bremner, I. (1991), *Methods Enzymol.,* 205, 25–35.
Brostoff, S. W. and Howell, M. D. (1992), *Clin. Immunol. Immunopathol.,* 62, 1–7.
Cárdenas, A., Roels, H., Bernard, A. M., Barbon, R., Buchet, J. P., Lauwerys, R. R., Roselló, J., Hotter, G., Mutti, A., Franchini, I., Fels, L. M., Stolte, H., De Broe, M. E., Nuyts, G. D., Taylor, S. A., and Price, R. G. (1993), *Br. J. Ind. Med.,* 50, 17–27.
Charpentier, B., Moullot, P., Faux, N., Manigand, G., and Fries, D. (1981), *Néphrologie,* 2, 153–157.
Chong, B. H. (1991), *Platelets,* 2, 173–181.
Chowdhury, B. A., Friel, J. K., and Chandra, R. K. (1987), *J. Nutr.,* 117, 1788–1794.
Clarkson, R. W., Sanders, P. A., and Grennan, D. M. (1992), *Br. J. Rheumatol.,* 31, 53–54.
Cohen, I. R. (1992), *Adv. Int. Med.,* 37, 295–311.
Contrino, J., Kosuda, L. L., Marucha, P., Kreutzer, D. L., and Bigazzi, P. E. (1992), *Int. J. Immunopharmacol.,* 14, 1051–1059.
Contrino, J., Marucha, P., Ribaudo, R., Ference, R., Bigazzi, P. E., and Kreutzer, D. L. (1988), *Am. J. Pathol.,* 132, 110–118.
Cugell, D. W. (1992), *Clin. Chest Med.,* 13, 269–279.
Cullen, M. R., Kominsky, J. R., Rossman, M. D., Cherniak, M. G., Rankin, J. A., Balmes, J. R., Kern, J. A., Daniele, R. P., Palmer, L., Naegel, G. P., McManus, K., and Cruz, R. (1987), *Am. Rev. Resp. Dis.,* 135, 201–208.

Deodhar, S. D. and Barna, B. P. (1991), *Cleve. Clin. J. Med.,* 58, 157–160.
Descotes, J. (1986), *Immunotoxicology of Drugs and Chemicals,* Elsevier, New York.
Druet, E., Sapin, C., Günther, E., Feingold, N., and Druet, P. (1977), *Eur. J. Immunol.,* 7, 348–351.
Dubey, C., Bellon, B., and Druet, P. (1991), *Eur. Cytokine Net.,* 2, 147–152.
Dubey, D., Kuhn, J., Vial, M. C., Druet, P., and Bellon, B. (1993), *Scand. J. Immunol,* 37, 406–412.
Fukatsu, A., Brentjens, J. R., Killen, P. D., Kleinman, H. K., Martin, G. R., and Andres, G. A. (1987), *Clin. Immunol. Immunopathol.,* 45, 35–47.
Gallicchio, V. S. (1991), *Ther. Monog.,* 4, 1–17.
Gerhardsson, L., Chettle, D. R., Englyst, V., Nordberg, G. F., Nyhlin, H., Scott, M. C., Todd, A. C., and Vesterberg, O. (1992), *Br. J. Ind. Med.,* 49, 186–192.
Goldman, M., Druet, P., and Gleichmann, E. (1991), *Immunol. Today,* 12, 223–227.
Goyer, R. A. (1991), in *Casarett and Doull's Toxicology,* Amdur, M. O., Doull, J., and Klaassen, C. D., Eds., Pergamon Press, New York, 623–680.
Hall, C. L. (1988), *Nephron,* 50, 265–272.
Hall, C. L. (1989), *Adv. Exp. Med. Biol.,* 252, 247–256.
Hart, D. A. (1990), in *Lithium and Cell Physiology,* Bach, R. O. and Gallicchio, V. S., Eds., Springer-Verlag, New York, 58–81.
Hart, D. A. (1991a), *Lithium Ther. Monogr.,* 4, 46–67.
Hart, D. A. (1991b), *Lithium Ther. Monogr.,* 4, 68–78.
Hassman, R. A. and McGregor, A. M. (1988), *Lithium Ther. Monogr.,* 2, 134–146.
Henry, G. A., Jarnot, B. M., Steinhoff, M. M., and Bigazzi, P. E. (1988), *Clin. Immunol. Immunopathol.,* 49, 187–203.
Heptinstall, R. H. (1992), in *Pathology of the Kidney,* Heptinstall, R. H., Ed., Little, Brown, Boston, 2085–2111.
Ho, S.-Y., Catalanotto, F. A., Lisak, R. P., and Dore-Duffy, P. (1986), *Ann. Neurol.,* 20, 712–715.
Hoitsma, A. J., Wetzels, J. F. M., and Koene, R. A. P. (1991), *Drug Safety,* 6, 131–147.
Hultman, P. and Eneström, S. (1988), *Clin. Exp. Immunol.,* 71, 269–274.
Hultman, P., Bell, L. J., Eneström, S., and Pollard, K. M. (1992), *Clin. Immunol. Immunopathol.,* 65, 98–109.
Hultman, P., Eneström, S., Pollard, K. M., and Tan, E. M. (1989), *Clin. Exp. Immunol.,* 78, 470–477.
Joshi, B. C. (1981), *J. Comp. Pathol.,* 91, 11–15.
Juntunen, J., Kinnunen, E., Antti-Poika, M., and Koskenvuo, M. (1989), *Br. J. Ind. Med.,* 46, 417–419.
Kägi, J. H. R. and Schäffer, A. (1988), *Biochemistry,* 27, 8509–8515.
Kammuller, M. E., Bloksma, N., and Seinen, W. (1989), in *Autoimmunity and Toxicology. Immune Dysregulation Induced by Drugs and Chemicals,* Kammuller, M. E., Bloksma, N., and Seinen, W., Eds., Elsevier, Amsterdam, 3–34.
Katz, A. and Little, A. H. (1973), *Arch. Pathol.,* 96, 133–136.
Kilburn, K. H. and Warshaw, R. H. (1992), *Environ. Res.,* 57, 1–9.
Kosty, M. P., Hench, P. K., Tani, P., and McMillan, R. (1989), *Am. J. Hematol.,* 30, 236–239.
Kosuda, L. L. and Bigazzi, P. E. (1996), in *Experimental Immunotoxicology,* Smialowicz, R. J. and Holsapple, M. P., Eds., CRC Press, Boca Raton, FL.
Kosuda, L. L., Greiner, D. L., and Bigazzi, P. E. (1993), *Environ. Health Perspect.,* 101, 178–185.
Kosuda, L. L., Greiner, D. L., and Bigazzi, P. E. (1994a), *Cell. Immunol.,* 155, 77–94.
Kosuda, L. L., Hosseinzadeh, H., Greiner, D. L., and Bigazzi, P. E. (1994b), *J. Toxicol. Environ. Health,* 42, 303–321.
Kosuda, L. L., Wayne, A., Nahounou, M., Greiner, D. L., and Bigazzi, P. E. (1991), *Cell. Immunol.,* 135, 154–167.
Kowolenko, M., Tracy, L., Mudzinski, S., and Lawrence, D. A. (1988), *J. Leuk. Biol.,* 43, 357–364.
Kubicka-Muranyi, M., Behmer, O., Uhrberg, M., Klonowski, H., Bister, J., and Gleichmann, E. (1993), *Int. J. Immunopharmacol.*, 15, 151–161.
Kushner, J. P. and Wartofsky, L. (1988), Lithium-thyroid interactions. An overview, *Lithium Ther. Monogr.,* 2, 74–98.
Langworth, S., Elinder, C. G., Sundquist, K. G., and Vesterberg, O. (1992), *Br. J. Ind. Med.,* 49, 394–401.
Lauwerys, R., Bernard, A., Roels, H., Buchet, J. P., Gennart, J. P., Mahieu, P., and Foidart, J. M. (1983), *Toxicol. Lett.,* 17, 113–116.
Lechleitner, P., Defregger, M., Lhotta, K., Totsch, M., and Fend, F. (1993), *Chest,* 103, 956–957.
Levinson, W. (1979), *Biol. Trace Element Res.,* 1, 15–23.
Levinson, W., Oppermann, H., and Jackson, J. (1980), *Biochim. Biophys. Acta,* 606, 170–180.
Lindqvist, K. J., Makene, W. J., Shaba, J. K., and Nantulya, V. (1974), *E. Afr Med. J.,* 51, 168–169.
Liszewski, M. K. and Atkinson, J. P. (1991), in *Systemic Autoimmunity,* Bigazzi, P. E. and Reichlin, M., Eds., Marcel Dekker, New York, 13–37.
Marguerie, C., Lunardi, C., and So, A. (1992), *Immunol. Today,* 13, 336–338.
Mathieson, P. W., Thiru, S., and Oliveira, D. B. G. (1992), *Lab. Invest.,* 67, 121–129.
McCabe, M. J. and Lawrence, D. A. (1990), *J. Immunol.,* 145, 671–677.
McFarland, H. F. and Dhib-Jalbut, S. (1989), *Clin. Immunol. Immunopathol.,* 50, S96–105.
Messino, M. J. (1991), *Lithium Ther. Monogr.,* 4, 18–29.
Meyer, U. A., Zanger, U. M., Skoda, R. C., Grant, D., and Blum, M. (1990), *Progr. Liver Dis.,* 9, 307–323.
Michaelson, J. H., McCoy, J. P. J., Hirszel, P., and Bigazzi, P. E. (1985), *Surv. Synth. Pathol. Res.,* 4, 401–411.
Moss, P. A. H., Rosenberg, W. M. C., and Bell, J. I. (1992), *Annu. Rev. Immunol.,* 10, 71–96.

Nebert, D. W. and Weber, W. W. (1990), in *Principles of Drug Action. The Basis of Pharmacology,* Pratt, W. B. and Taylor, P., Eds., Churchill Livingstone, New York, 469–531.
Nepom, G. T. (1992), *Rheum. Dis. Clin. N. Am.,* 18, 719–727.
Nepom, G. T. and Nepom, B. S. (1992), *Rheum. Dis. Clin. N. Am.,* 18, 785–792.
Nomiyama, K. (1981), in *Cadmium in the Environment,* Nriagu, J. O. Ed., Wiley, New York, 643–689.
Nordlind, K. and Lidén, S. (1990), *Immunopharm. Immunotoxicol.,* 12, 715–721.
Ochel, M., Vohr, H. W., Pfeiffer, C., and Gleichmann, E. (1991), *J. Immunol.,* 146, 3006–3011.
Ohsawa, M., Takahashi, K., and Otsuka, F. (1988), *Clin. Exper. Immunol.,* 73, 98–102.
Pelletier, L., Bellon, B., Tournade, H., Dubey, C., Guery, J. C., Saoudi, A., Hirsch, F., and Druet, P. (1992), in *Molecular Immunobiology of Self-Reactivity,* Bona, C. A. and Kaushik, A. K., Eds., Marcel Dekker, New York, 315–353.
Pelletier, L., Pasquier, R., Guettier, C., Vial, M. C., Mandet, C., Nochy, D., Bazin, H., and Druet, P. (1988), *Clin. Exp. Immunol.,* 71, 336–342.
Perez Garcia, A., Panadero Sandoval, J., Martin Abad, L., Garcia Martinez, J., and Cruz Rodriguez, J. M. (1980), *Rev. Clin. Españ.,* 156, 203–206.
Pietsch, P., Vohr, H.-W., Degitz, K., and Gleichmann, E. (1989), *Int. Arch. Allergy Appl. Immunol.,* 90, 47–53.
Pocino, M., Malavé, I., and Baute, L. (1990), *Immunopharm. Immunotoxicol.,* 12, 697–713.
Raine, C. S. (1991), in *Textbook of Neuropathology,* Davis, R. L. and Robertson, D. M., Eds., Williams & Wilkins, Baltimore, 535–620.
Rees, A. J. and Lockwood, C. M. (1988), in *Diseases of the Kidney,* Schrier, R. W. and Gottschalk. C. W., Eds., Little, Brown, Boston, 2091–2126.
Reimer, G. (1990), *Rheum. Dis. Clin. N. Am.,* 16, 169–183.
Reveille, J. D. and Arnett, F. C. (1991), in *Systemic Autoimmunity,* Bigazzi, P. E. and Reichlin, M. Eds., Marcel Dekker, New York, 97–140.
Romagnani, S. (1994), *Annu. Rev. Immunol.,* 12, 227–257.
Romagnoli, P., Spinas, G. A., and Sinigaglia, F. (1992), *J. Clin. Invest.,* 89, 254–258.
Rossert, J., Pelletier, L., Pasquier, R., Villaroya, H., Oriol, R., and Druet, P. (1991), *Cell. Immunol.,* 137, 367–378.
Saltini, C., Winestock, K., Kirby, M., Pinkston, P., and Crystal, R. G. (1989), *N. Engl. J. Med.,* 320, 1103–1109.
Sapin, C., Mandet, C., Druet, E., Gunther, E., and Druet, P. (1982), *Clin. Exp. Immunol.,* 48, 700–704.
Schiffer, R. B., Herndon, R. M., and Eskin, T. (1990), *Neurotoxicology,* 11, 443–450.
Schiffer, R. B., Herndon, R. M. and Stabrowski, A. (1988), *Ann. Neurol.,* 24, 141.
Schiffer, R. B., Sunderman, F. W. J., Baggs, R. B., and Moynihan, J. A. (1991), *J. Neuroimmunol.,* 34, 229–239.
Schuhmann, D., Kubicka-Muranyi, M., Mirtschewa, J., Günther, J., Kind, P., and Gleichmann, E. (1990), *J. Immunol.,* 145, 2132–2139.
Shear, N. H. and Bhimji, S. (1989), *Semin. Dermatol.,* 8, 219–226.
Stein, E. C., Schiffer, R. B., Hall, W. J., and Young, N. (1987), *Neurology,* 37, 1672–1677.
Swanborg, R. H. (1990). in *Organ-Specific Autoimmunity,* Bigazzi, P. E., Wick, G., and Wicher, K., Eds., Marcel Dekker, New York, 155–167.
Templeton, D. M. and Cherian, M. G. (1991), *Methods Enzymol.,* 205, 11–24.
Tournade, H., Guery, J.-C., Pasquier, R., Nochy, D., Hinglais, N., Guilbert, B., Druet, P., and Pelletier, L. (1991a), *Nephrol. Dial. Transplant.,* 6, 621–630.
Tournade, H., Guery, J. C., Pasquier, R., Vial, M. C., Mandet, C., Druet, E., Dansette, P. M., Druet, P., and Pelletier, L. (1991b), *Arthritis Rheum.,* 34, 1594–1599.
Tsankov, N., Stransky, L., Kostowa, M., Mitrowa, T., and Obreschkowa, E. (1990). *Derm. Beruf Umwelt — Occup. Environ. Dermat.,* 38, 91–93.
van der Meide, P. H., de Labie, M. C. D. C., Botman, C. A. D., van Bennekom, W. P., Olsson, T., Aten, J., and Weening, J. J. (1993), *Eur. J. Immunol.,* 23, 675–681.
Vogel, F. and Motulsky, A. G. (1979), *Human Genetics,* Springer-Verlag, Berlin.
Warner, G. L. and Lawrence, D. A. (1986a), *Eur. J. Immunol.,* 16, 1337–1342.
Warner, G. L. and Lawrence, D. A. (1986b), *Cell. Immunol.,* 101, 425–439.
Watanabe, I., Whittier, F. C., Moore, J., and Cuppage, F. E. (1976), *Arch. Pathol. Lab. Med.,* 100, 632–635.
Wedeen, R. P. (1992), in *Oxford Textbook of Clinical Nephrology,* Cameron, S., Davison, A. M., Grunfeld, J.-P., Kerr, D., and Ritz, E., Eds., Oxford University Press, Oxford, 837–848.
Wedeen, R. P., Mallik, D. K., and Batuman, V. (1979), *Arch Intern. Med.,* 139, 53–57.
Weening, J. J., Hoedemaeker, P. J., and Bakker, W. W. (1981), *Clin. Exp. Immunol.,* 45, 64–71.
Winchester, R., Dwyer, E., and Rose, S. (1992), *Rheum. Dis. Clin. N. Am.,* 18, 761–783.
Wylie, D. E., Lu, D., Carlson, L. D., Carlson, R., Babacan, K. F., Schuster, S. M., and Wagner, F. W. (1992), *Proc. Natl. Acad. Sci. U.S.A.,* 89, 4104–4108.

Chapter 54

Altered Host Defenses and Resistance to Respiratory Infections Following Exposure to Airborne Metals*

Mary Jane K. Selgrade and Donald E. Gardner

I. INTRODUCTION

The respiratory system has a number of complex, highly regulated, cooperative host defense mechanisms which represent the first line of defense against inhaled microorganisms and also may have a role in controlling pulmonary tumors. Despite great differences in body size and respiratory tract morphology, most mammals appear to have functionally similar pulmonary defense systems (Green, 1969; Green, 1984). When these local defenses are compromised, the potential risk of microbial infections in the lung is significantly increased (Gardner, 1988; Reynolds, 1991). Since contact between a toxicant and cells of the pulmonary defense system is more direct following inhalation as compared to other types of exposure, the impact that inhaled metals may have on these cells and on susceptibility to respiratory infection may not be reflected in more conventional immunotoxicity studies of parenterally administered compounds (reviewed in other chapters). This chapter will focus on studies of the effects of airborne metals on pulmonary host defenses with special emphasis on susceptibility to respiratory infections.

II. EXPOSURE TO AIRBORNE METALS

Metals are widely distributed in the environment, and the potential for airborne exposure, either in ambient air or the workplace, is significant. The emissions of metal aerosols is heavily influenced by primary and secondary metal production. Sources of atmospheric pollutants includes those emitted from refineries, chemical plants, cement manufacturers, power plants, smelters, trash burning and tobacco smoke (Duffus, 1980; Friberg et al., 1986; Gee et al., 1984; Harte et al., 1991). Vehicles emit metals, particularly older automobiles using leaded gasolines. However newer vehicles, as a result of certain fuel additives, may also contribute to the production of certain airborne metals, e.g., chromium, copper,

---

* Disclaimer: The research chapter has been reviewed by the Health Effects Research Laboratory, U.S. Environmental Protection Agency and approved for publication. Approval does not signify that the contents necessarily reflects the views and policies of the agency, nor does mention of trade names or commercial products constitute endorsement or recommendation for use.

nickel, and manganese (Nriagu and Davidson, 1986; Watson et al., 1988). Metals are also present in working environments where welding, grinding, soldering, painting, and certain types of manufacturing take place.

Metals such as cadmium, lead, and mercury are classified as heavy metals because their densities are at least five times greater than water. These metals are also frequently referred to as trace metals, because they tend to be present in very low concentrations. However, even at low concentrations, the inhalation of these metal particles may have a significant effect on the respiratory system, increasing an individual's risk of disease. Table 1 provides examples of airborne metals, their sources, and reported effects associated with inhaling these aerosols. Table 2 gives examples of typical concentrations of a few airborne metals in different environmental areas and threshold limit values (time-weighted averages) for a normal 8-hour work day and a 40-hour work week.

Table 1 Airborne Metals

Chemical	Source/Description	Associated Effects
Aluminum	Coal combustion, cigarette smoke, electrical conductors	Pneumoconiosis
Arsenic	Smelting, used in oils, pesticides, wood preservatives, metallurgy	Perforated nasal septum, cancer
Beryllium	Oil and coal combustion, mining, metal production, cement plants, ceramics, rocket propellants	Inflammation of mucous membranes, cancer, chronic berylliosis
Cadmium	Fossil fuel combustion, fertilizers, batteries, electroplating, pesticides, sewage sludge	Respiratory irritation, edema, pneumonitis, emphysema, kidney disease
Chromium	Chrome plating, paints, leather tanning, wood preservation	Bronchitis, lung cancer, liver and kidney damage, perforation of nasal septum
Copper	Fertilizers, fungicides, coal combustion, electrical conductors, alloys	Metal fume fever, changes in pulmonary cell types, pulmonary granulomas, irritation
Lead	Production of lead batteries, gas additives, smelters and refineries	Altered macrophage function, increase in respiratory disease
Manganese	Ore processing, battery production, fuel additive	Neurotoxicity, male reproductive dysfunction, pulmonary function changes, respiratory infections
Mercury	Ore and fuel contaminants, fungicides, electrolysis process, paints	Autoimmune effects, neurotoxicity
Nickel	Metal products, metal refineries	Lung and nasal cancer, fibrosis, decreased macrophage function
Selenium	Combustion of coal and oil, sewage sludge and municipal waste, copper smelters and refineries, glass production	Pulmonary irritation, edema

III. DEPOSITION AND RETENTION OF INHALED METALS

The lung is the organ with the greatest direct continuous contact between the external environment and the internal components of the body. For example, in the course of a year, an adult breathes approximately 7 million liters of air. Industrial workers breathe about 20 l/min, and runners can breathe up to 80 l/min (Gardner, 1994). As a primary route of entry into the body, the respiratory system is a vulnerable target for toxic substances. It has nearly 4 times the total surface area interfacing with the environment as does the total combination of the skin and the gastrointestinal tract. When one compares the adult daily intake of food (1.5 kg) and water (2.0 kg) and air (15 kg) (Gardner and Kennedy, 1993), the potential exposure through inhalation appears to be very significant.

Airborne particles exist in the atmosphere in many sizes. Metals occur in both the fine (<2.5 µm) and coarse (>2.5 µm) mode. Usually, particles from anthropogenic sources are in the fine mode, while those from the earth's crust are in the coarse mode. Also, many metals are contained on and in fly ash which is typically in the fine mode. The host response to the disposition of inhaled metals in the respiratory system can generally be expected to occur at the site of deposition, which could be in either the conducting airways or the lung's gas exchange region or both. The specific site of deposition is dependent on the aerodynamic size of the particle as well as the morphology of the lung and ventilatory characteristics of the exposed subject (Schlesinger, 1989; Muir, 1991). Particles larger than 2.5 µm in

Table 2 Typical Concentrations of Metals in Air and TLV for the Workplace

	Concentration in Air (ng/m³)	Environment Area	TLV Standards Metal (mg/m³)
Cadmium	5.0	Urban/industrial	0.01 (dust)
	0.003–0.62	Rural	
Arsenic	3.0	Urban	0.2
	30.0	Near smelters	
Nickel	10–60	Urban	1.0 (metal)
	1–20	Rural	
	>100	Heavily industrial	
	2000	Near facility	
Aluminum	100–5,000	Urban	10 (dust)
	50–500	Rural	
Beryllium	0.001–0.28	Ambient air	0.002
Lead	100–10,000	Urban	0.15 (dust)
	8–10	Rural	
	100–75,000	Near smelters	
	8,200–18,000	Near freeway	
Manganese	30–70	Urban	5 (dust)
	10–30	Rural	
	200–500	Near foundry	

diameter are predominantly deposited in the upper respiratory tract. Particles smaller than 2.5 µm in diameter remain in suspension in the airflow, pass through the airways, and are deposited predominantly in the gas exchange (alveoli) region of the lung. Also, the route of breathing (oral, nasal, or oronasal) influences the efficiency with which the respiratory tract filters out inhaled particles (including metals) and thus impacts the dose delivered to the lower respiratory tract.

Once particles are deposited within the respiratory system, their pulmonary retention can vary widely. Soluble particles are rapidly cleared and translocated to other tissues or excreted. While insoluble particles are also cleared rather rapidly (within hours or days) from the upper airways (nasopharynx and tracheobronchial region) by mechanical processes, they are removed more slowly (within months) from the pulmonary gas exchange region. For insoluble particles the pulmonary burden may increase with exposure concentration and duration. If the deposited metal is cytotoxic, then the retention time could be increased significantly. Removal of such particles involves the alveolar macrophage coupled with mucocillary activity that transports the engulfed particles out of this region. Inhaled metals such as nickel, cadmium, and manganese have been shown to significantly impair the function of these pulmonary clearance mechanisms (Gardner, 1979) The deposition, clearance, and retention of inhaled particles have been extensively reviewed (Schlesinger, 1989; Gardner, 1994).

IV. PULMONARY HOST DEFENSES

A number of reviews of pulmonary host defenses have been published elsewhere (Danielle, 1988; NRC, 1989; Schlesinger, 1990; Reynolds, 1991, Koren and Becker, 1992). Briefly, the anatomical structure of the lung with its many bifurcations serves as a filter to prevent the penetration of particles to the alveolar (gas exchange) region. The upper airways are lined with ciliated epithelium overlayed by mucus. This mucociliary escalator traps particles (including some microorganisms and metals) and propels them up and out of the respiratory tract.

The alveolar macrophage (AM) is the principal phagocytic cell in the respiratory tract and as such plays an important role in removing particles from the lung. AM are known to be the critical host defense in the inactivation of some Gram-positive bacteria (Green and Kass, 1964). In addition to AM, a number of extracellular factors in both the mucous layer of the upper airways and the surfactant-containing fraction of the alveolar lining material have bactericidal activity (Nugent and Pesanti, 1982; Konstan et al., 1982; Conrad, 1987). While usually not present in normal bronchoalveolar lavage fluid (BAL), polymorphonuclear leukocytes (PMNs) can be recruited into the air spaces in response to exposure to microbial agents or other inhaled materials. Migration of PMNs into the lung appears to be stimulated by cytokines secreted by AM and epithelial cells. This inflammatory response provides a secondary line of phagocytic defenses, which is particularly important in the inactivation of pneumococci and Gram-negative bacteria in the lung (Toews, 1986). PMNs also produce defensins which inactivate viruses and other microorganisms (Daher et al., 1986).

While few lymphocytes are found in BAL, natural killer (NK) cells, which respond nonspecifically to microorganisms, and T and B lymphocytes, responsible for antigen-specific responses, are found in the lung interstitium, lung-associated lymphoid tissue, and draining lymph nodes. Augmentation of pulmonary NK activity in response to viral infection and virus-specific cytotoxic T cell activity have been demonstrated in whole lung homogenates (Stein-Streilein et al., 1983; Ehrlich et al., 1989), and antibody-forming cells in lung-associated lymph nodes have been found following instillation of antigen (Bice and Shopp, 1988). IgG and IgA are the major antibody subtypes found in BAL and may represent both local production and serum exudate. Antibodies are known to aid in phagocytosis of bacteria and neutralization of virus. The time required to generate these specific responses suggests they may play a greater role in preventing recurring infection and in the later stages of primary infection, whereas nonspecific responses are more likely to be important in the early stages of primary infections.

Inhaled metals have been shown to alter AM function (reviewed in Chapter III.7), mucociliary function, NK cell activity, and antibody-forming cell numbers in local lymph nodes (as well as spleen). These effects and the consequences in terms of altered susceptibility to respiratory infection are reviewed below. Inhaled metals may also gain access to the circulation and cause the same types of systemic immune suppression (detailed in other chapters) seen following parenteral exposure. These effects may impact the outcome of respiratory infections (particularly if dissemination of the microbe to other organs is a possibility).

V. SUSCEPTIBILITY TO BACTERIAL INFECTION

A. NICKEL

One of the most well-characterized pulmonary host resistance models is the *Streptococcus zooepidemicus** mouse infectivity model. In this system, mice are exposed to a pollutant or filtered air and then challenged with an aerosol of bacteria resulting in deposition in the lung of between 400 and 4000 microorganisms. Mortality and bacterial clearance from the lung are subsequently assessed. Enhanced mortality and decreased clearance of bacteria from the lung as a result of pollutant exposure correspond to suppression of AM phagocytic function (Gardner, 1980; Selgrade and Gilmour, in press). Mice exposed for 2 h to $NiCl_2$ (500 µg Ni/m^3) or $NiSO_4$ (455 µg Ni/m^3) and infected 24 h after exposure with *S. zooepidemicus* showed significantly enhanced mortality and decreased mean survival times compared to air controls (Adkins et al., 1979). In these same studies, clearance of bacteria from the lungs of mice similarly exposed to $NiCl_2$ was impaired, and the capacity of AM obtained from these mice to ingest latex beads *in vitro* was diminished. When mice were infected immediately as opposed to 24 h after exposure, susceptibility to *S. zooepidemicus* was unaffected.

In other studies, phagocytosis of latex spheres by rabbit AMs was impaired following *in vitro* exposure to $NiCl_2$ at concentrations that had very little effect on viability (ranging from 30 to 65 µg/ml) (Graham et al., 1975), and *in vitro* bactericidal activity for *Staphylococcus aureus* was depressed in AM harvested from rabbits exposed for 1 month, 5 d/week, 6 h/d to an aerosol of 0.3 mg/m^3 $NiCl_2$ (Wiernik et al., 1983). In addition to effects on AM function, $NiCl_2$ has also been shown to suppress ciliary activity following *in vitro* exposure of hamster tracheal ring cultures to concentrations as low as 0.011 nM Ni and following *in vivo* exposure of hamsters to concentrations ranging from 100 to 275 µg Ni/m^3 (Adalis et al., 1978). (It should be noted however, that in the aerosolized *S. zooepidemicus* infectivity model, most of the bacteria are deposited in the lung below the mucociliary escalator.) Also, lysozyme levels in lavage fluid of rabbits exposed for 4 to 6 weeks, 5 d/week, 6 h/d to 0.3 mg/m^3 $NiCl_2$ were lower than in air controls, suggesting that alterations in extracellular bactericidal components might also contribute to enhanced susceptibility to bacteria (Lundborg and Camner, 1982). Finally, a 2-h exposure to 250 to 500 µg Ni/m^3 caused significant suppression of the splenic antibody-forming cell response in mice immunized with sheep erythrocytes on the day of metal exposure (Graham et al., 1978). Systemic

---

* *S. zooepidemicus* was isolated from a pneumonic guinea pig lung and originally described as *S. pyogenes* or group C *Streptococcus* sp., until reclassification according to its ability to ferment trehalose. The same organism was used in all the studies cited here, although earlier studies may refer to it by a different name.

suppression of antibody responses could contribute to dissemination of bacteria from the lung, which occurs in the later stages of streptococcal infection.

Other nickel-containing compounds in addition to $NiCl_2$ have been shown to affect macrophage function. AM phagocytosis of opsonized erythrocytes was suppressed following exposure of mice 6 h/d, 5 d/week, for 65 d to concentrations as low as 0.47 mg Ni/m^3 for NiO and 0.45 mg Ni/m^3 for Ni_3S_2, but phagocytosis was not affected in mice exposed to concentrations as high as 0.45 mg Ni/m^3 as $NiSO_4$ (Haley et al., 1990). Suppression of AM phagocytosis of unopsonized erthrocytes was also demonstrated in cynomologus monkeys instilled with Ni_3S_2 at a dose of 0.06 μM/g lung (Haley et al., 1987). In contrast, exposure of rabbits to 2 mg/m^3 metallic nickel dust, for 4 to 5 wks, 5 d/week, 6 h/d, appeared to activate AM and enhanced phagocytosis (Jarstrand et al., 1978). However, lavage fluid lysozyme levels were significantly lower in rabbits exposed to 0.1 mg/m^3 metallic nickel dust, for 6 or 8 months, 5 d/week, 6 h/d (Lundborg and Camner, 1982).

B. CADMIUM

Enhanced susceptibility to *S. zooepidemicus* was observed in mice that were infected following a 2-h exposure to 100 to 1600 µg Cd/m^3 as $CdCl_2$ (Gardner et al., 1977). Unlike $NiCl_2$, enhanced mortality was more pronounced in mice infected immediately after exposure, and higher exposure concentrations were required to enhance mortality and decrease mean survival time in mice infected 24-h postexposure. Enhanced mortality following $CdCl_2$ exposure was accompanied by impaired clearance of bacteria from the lung and decreased numbers of AM recovered in BAL immediately after exposure. AM numbers had returned to normal by 24 h postexposure, and an influx of PMNs was observed. Suppression of AM phagocytosis was demonstrated following *in vitro* exposure to 2.5 µg/ml Cd (Graham et al., 1975). As with nickel, *in vitro* exposure of hamster tracheal ring cultures to concentrations of $CdCl_2$ as low as 0.006 mM caused significant reductions in ciliary activity, and suppression of ciliary beat frequency was also demonstrated in hamsters exposed *in vivo* for 2 h to concentrations of $CdCl_2$ ranging from 50 to 1420 µg/m^3 (Adalis et al., 1977). Again, the data suggest that Cd has the potential to affect mucociliary clearance. Also like nickel, exposure to $CdCl_2$ (190 µg Cd/m^3) caused a significant reduction in the systemic antibody response to sheep erythrocytes as indicated by reduced splenic antibody-forming cells in mice immunized on the day of metal exposure (Graham et al., 1978).

In addition to enhanced susceptibility to infection with Gram-positive bacteria (*Streptococcus*), effects of Cd on susceptibility to Gram-negative bacteria have also been reported. Following cadmium oxide exposure (10 mg Cd/m^3) for 15 min, enhanced mortality and decreased clearance of bacteria from the lung was demonstrated in rats infected by inhalation to *Salmonella enteritidis* and in mice infected with *Pasteurella multocida* (Bouley et al., 1977). These studies also demonstrated an initial decrease and then an increase in AM numbers and an influx of PMNs following Cd exposure.

C. ARSENIC AND MANGANESE

Enhanced mortality due to *S. zooepidemicus* was observed following a single 3-h exposure to 270, 500, and 940 µg arsenic trioxide per cubic meter (Aranyi et al., 1985). Increased susceptibility to infection was also observed following 5 or 20 multiple exposures (3 h/d, 5 d/week) to 500 µg As/m^3. *In vivo* bactericidal activity in the lungs of mice exposed simultaneously to the As and ^{35}S-labeled *Klebsiella pneumoniae* was also suppressed.

Intratracheal instillation of 200 mg/kg gallium arsenide (GaAs) caused suppression of both IgM and IgG antibody-forming cell responses in the spleens of mice immunized 14 days after exposure (Sikorski et al., 1989). Burns et al. (1991) demonstrated that this GaAs-induced immune suppression was due primarily to arsenic as a result of dissociation of gallium and arsenic from GaAs in the lung. Hence, as with nickel and cadmium, exposure of the respiratory tract to arsenic can result in suppression of systemic antibody responses. However, it should be noted that GaAs treatment 24 h prior to challenge increased resistance to two systemic bacterial infections, and evidence was presented suggesting that this effect was due to the chemotherapeutic properties of arsenic (inhibition of bacterial growth) present in the serum (Burns et al., 1993).

Single and 3 to 4 daily, 3-h exposures to MnO_2 enhanced susceptibility to airborne *K. pneumoniae* when bacterial challenge occurred within 1 h of the last metal exposure (Maigetter et al., 1976). Also, a single 2-h exposure to Mn_3O_4 (0.5 mg/m^3 or greater) enhanced mortality following challenge with *S. zooepidemicus*, delayed clearance of this microorganism from the lung, and subsequently resulted in enhanced growth of the microbe in the lung and increased incidence of septicemia (Adkins et al., 1980a). These effects were attributed to suppression of AM function, since *in vitro* exposure to MnO_2 impaired phagocytic activity (Graham et al., 1975) and 2-h inhalation exposure to Mn_3O_4 (879 µg Mn/m^3) reduced the total number, viability, and phagocytic function of AM (Adkins et al., 1980b). It should be noted, however, that in another study both splenic and peritoneal macrophage activity was enhanced by intramuscular injection of $MnCl_2$ (Smialowicz et al., 1985). In this case, spleen cell antibody-dependent cell-mediated cytotoxicity (a macrophage function) and peritoneal macrophage phagocytosis of chicken erythrocytes (with or without opsonizing antibody) were enhanced. These effects were attributed to increased interferon levels in mice injected with $MnCl_2$, i.e., it appears that $MnCl_2$ administered intramuscularly induces interferon (Smialowicz et al., 1985). Hence, effects of manganese on macrophage function may depend on the route of exposure, the form of the metal, the source of the macrophages, or other unidentified variables.

VI. SUSCEPTIBILITY TO VIRAL INFECTION

While there is a large body of literature that suggest that exposure to airborne metals enhances susceptibility to bacterial infections in the lung, far fewer studies have been reported on the effects of metal exposures on viral respiratory infections, and both protection against and enhancement of infection have been demonstrated, depending on the study.

Increased frequency of colds (most probably due to rhinoviruses) and influenza infection was demonstrated in lead workers with blood levels of 21 to 85 µg Pb/dl (Ewers et al., 1982), and increased incidence of colds was also demonstrated in lead workers with blood levels greater than 40 µg/100g (Horiguchi et al., 1992). Maigetter et al. (1976) also demonstrated increased mortality rates, reduced survival times and increased pulmonary lesions in mice infected with influenza 24 or 48 h before a 3-h exposure to 109 mg/m^3 MnO_2. These are the only studies which suggest that resistance to viral infection is impaired by exposure to airborne metals.

In contrast, mice exposed for 15 min to cadmium microparticles (CdO, 9 mg Cd/m^3, mass median diameter 1.79 µm) and infected 48 h later with influenza virus showed significantly lower death rates than air controls, and protection was also observed in mice similarly exposed once a day, 5 d/week for 4 weeks and infected midway through this chronic exposure regimen (Chaumard et al. 1983). In a later study, exposure to CdO particles did not affect either the antibody response or the interferon response to infection. The protective effect of Cd against influenza infection was attributed to an increase in inflammatory cells in the lung (Chaumard et al., 1991); however, it might be anticipated that such an increase in inflammatory cells could enhance lung lesions associated with the viral infection.

Similarly, resistance to viral infection was not impaired in mice exposed to concentrations as high as 2000 µg Cd/m^3 ($CdCl_2$) or to concentrations as high as 1000 µg Ni/m^3 ($NiCl_2$) for 2 h/d for 4 days and infected with murine cytomegalovirus on the first day of exposure (Daniels et al., 1987). It should be noted that both $NiCl_2$ and $CdCl_2$ enhanced susceptibility to this infection and suppressed virus-augmented splenic natural killer cell activity when administered parenterally. The fact that mice could tolerate a much higher dose of metal administered by the intramuscular as compared to the inhalation route may explain this discrepancy. In conclusion, the evidence that aerosol metal exposures impair host defenses to viral respiratory infections is equivocal. There may, however, be other interactions that result from the combination of viral infection and metal exposures which impact health. For example, *in vitro* exposure to nickel sulfate caused increased proliferation of Epstein-Barr virus (EBV) positive lymphoblastoid cell lines and increased early antigen expression (Wu et al., 1986). In certain "high risk" areas (China), EBV is associated with nasopharyngeal carcinoma. Since nickel is also found in high levels in the environment of high risk areas, the authors proposed that nickel could contribute to the development of EBV-associated nasopharyngeal carcinoma. Also, it should be noted that the distribution of both nickel and cadmium (administered intravenously to mice) was altered by coxsackievirus B3 infection. For nickel, the result was a greatly increased accumulation in the pancreas and wall of the ventricular myocardium (Ilback et al., 1992), and for cadmium an increased accumulation in the renal and adrenal cortices and (in some cases) the spleen (Ilback et al., 1992). Hence, the effects that the combination of metal exposure and viral infection may have on the toxicity of the compound need to be considered, in addition to the effects that may occur as a result of impaired host defenses against the infectious agent.

VII. HOST DEFENSES AGAINST LUNG TUMORS

Intratracheal exposure to 200 mg/kg of gallium arsenide resulted in a 7-fold increase in the lung tumor burden of mice challenged intravenously with B16F10 melanoma tumor cells (Sikorski et al.,

1989). Natural killer cells (both bloodborne and lung-associated) appear to have a significant role in limiting B16 tumor cell growth (Wiltrout et al., 1985), although other immune mechanisms may also be involved. Intratracheal GaAs exposure at concentrations ranging from 50 to 200 mg/kg suppressed splenic NK cell activity (Sikorski et al., 1989). Haley et al. (1990) demonstrated suppression of splenic NK cell activity following subchronic exposure of mice to 1.8 mg Ni/m^3 (Ni$_3$S$_2$); however, this compound, when instilled in cynomologus monkeys at a final dose of 0.06 µmol/g lung, enhanced pulmonary NK cell activity. Additional studies are needed before conclusions can be drawn concerning the effects of metals on antitumor defenses.

VIII. SUMMARY

A number of studies have shown that a variety of airborne metal exposure regimens cause increased susceptibility to bacterial infection in the lung (summarized in Table 3). These metals also suppress AM function, ciliary beat frequency, and splenic antibody-forming cell responses. All of these may contribute to enhanced mortality due to respiratory bacterial infections. While the evidence that metal exposures may increase the risk of bacterial infections in the lung is strong, the role that metals may have in compromising host defenses against respiratory viral infection or pulmonary tumors is unclear and requires further research.

Table 3 Metal Exposures Causing Enhanced Susceptibility to Bacteria in Mouse Infectivity Models

Metal Compound	Lowest Effective Exposure Concentraton (µg metal/m^3)	Time of Infection Relative to Metal Exposure	Bacteria[a]	Ref.
NiCl$_2$	500 (2 h)	24 h post	S. zooepidemicus	Adkins et al., 1979
NiSO$_4$	455 (2 h)	24 h post	S. zooepidemicus	Adkins et al., 1979
CdCl$_2$	100 (2 h)	0 h post	S. zooepidemicus	Gardner et al., 1976
CdCl$_2$	550 (2 h)	24 h post	S. zooepidemicus	Gardner et al., 1976
CdO	10,000 (15 min)	48 h post	P. multocida	Bouley et al., 1977
As$_2$O$_3$	270 (3 h)	0 h post	S. zooepidemicus	Aranyi et al., 1985
MnO$_3$	109 (3h/d for 3 days)	1 or 5 h post	K. pneumoniae	Maigetter et al., 1976
Mn$_3$O$_4$	≥0.5 mg/m^3 (2 h)	0 h post	S. zooepidemicus	Adkins et al., 1980a

[a] S. zooepidemicus was isolated from a pneumonic guinea pig lung and originally described as S. pyogenes or group C Streptococcus sp. until reclassification according to its ability to ferment trehalose. The same organism was used in all the studies cited here, although earlier studies may refer to it by a different name.

REFERENCES

Adalis, D., Gardner, D. E., Miller, F. J., and Coffin, D. L. (1977), *Environ. Res.*, 13, 111–120.
Adalis, D., Gardner, D. E., and Miller, F. J. (1978), *Am. Rev. Respir. Dis.*, 118, 347–354.
Adkins, B., Richards, J. H., and Gardner, D. E. (1979), *Environ. Res.*, 20, 33–42.
Adkins, B., Luginbuhl, G. H., Miller, F. J., and Gardner, D. E. (1980a), *Environ. Res.*, 23, 110–120.
Adkins, B., Luginbuhl, G. H., and Gardner, D. E. (1980b), *J. Toxicol. Environ. Health*, 6, 445–454.
Aranyi, C., Bradof, J. N., O'Shea, W. J., Graham, J. A., and Miller F. J. (1985), *J. Toxicol. Environ. Health*, 15, 163–172.
Bice, D. E. and Shopp, G. M. (1988), *Exp. Lung Res.*, 14, 133–155.
Bouley, G., Dubreuil, A., Despaux, N., and Boudene, C. (1977), *Scand. J. Work Environ. Health*, 3, 116–121.
Burns, L. A., Sikorski, E. E., Saady, J. J., and Munson, A. E. (1991), *Toxicol. Appl. Pharmacol.*, 110, 157–169.
Burns, L. A., McCay, J. A., Brown, R., and Munson, A. E. (1993), *J. Pharmacol. Exp. Ther.*, 265, 795–800.
Chaumard, C., Quero, A. M., Bouley, G., Girard, F., Boudene, C., and German, A. (1983), *Environ. Res.*, 31, 428–439.
Chaumard, C., Forestier, F., and Quero, A. M. (1991), *Arch. Environ. Health*, 46, 50–56.
Coonrad, J. D. (1987), *Eur. J. Respir. Dis.*, 71/Suppl. 153, 209–214.
Daher, K. A., Salsted, M. E., and Lehrer, R. I. (1986), *J. Virol.*, 60, 1068–1074.
Daniele, R. P. (1988), *Immunology and Immunological Diseases of the Lung*, Blackwell Scientific, Boston.
Daniels, M. J., Menache, M. G., Burleson, G. R., Graham, J. A., and Selgrade, M. J. K. (1987), *Fundam. Appl. Toxicol.*, 8, 443–454.
Duffus, J. H., Ed. (1980), *Environmental Toxicology*, John Wiley & Sons, New York.
Ehrlich, J. P., Gunnison, A. F., and Burleson, G. R. (1989), *Inhalation Toxicol.*, 1, 129–138.
Ewers, U., Stiller-Winkler, R., and Idel, H. (1982), *Environ. Res.*, 29, 351–357.

Fribero, L., Nordberg, G. F., and Vouk, V. B. (1986), *Handbook on the Toxicology of Metals,* Elsevier, New York.
Gardner, D. E., Miller, F. J., Illing, J. W., and Kirtz, J. M. (1977), *Bull. Europ. Physiopathol. Resp.,* 13, 157–174.
Gardner, D. E. (1979), in *Aerosols in Science, Medicine and Technology. The Biochemical Influence of the Aerosol,* Stöber, W. and Jaenicke, R., Eds., Gesellschaft für Aerosolforschung, Mainz, Germany.
Gardner, D. E. (1980), in *Nickel Toxicology,* Brown, S. S. and Sunderman, F. W., Jr., Eds., Academic Press, 121–124.
Gardner, D. E. (1988), *J. Appl. Toxicol.,* 6, 385–388.
Gardner, D. E. and Kennedy, G. L., Jr., (1993), *Methodologies and Technology for Animal Inhalation Toxicology Studies,* in Gardner, D. E., Crapo, J. D., and McClellan, R. O., Eds., Raven Press, New York, 1–30.
Gardner, D. E. (1994), Direct and indirect injury to the respiratory tract, CRC Press, Boca Raton, FL, 19–47.
Gee, J. B. L., Morgan, W. K. C., and Brooks, S. M. Eds. (1984), *Occupational Lung Diseases,* Raven Press, New York.
Graham, J. A., Gardner, D. E., Waters, M. D., and Coffin, D. L. (1975), *Infect. Immun.,* 11, 1278–1283.
Graham, J. A., Miller, F. J., Daniels, M. J., Payne, E. A., and Gardner, D. E. (1978), *Environ. Res.,* 16, 77–87.
Green, G. M. and Kass, E. H. (1964), *J. Exp. Med.,* 119, 167–176.
Green, G. M. (1969), *Arch. Environ. Health,* 18, 548–550.
Green, G. M. (1984), in *Fundamentals of Extrapolation Modeling of Inhaled Toxicants,* Miller, F. J. and Menzel, D. B., Eds., Hemisphere Press, Washington, D.C., 291–298.
Haley, P. J., Bice, D. E., Muggenburg, B. A., Hahn, F. F., and Benjamin, S. A. (1987), *Toxicol. Appl. Pharmacol.,* 88, 1–12.
Haley, P. J., Shopp, G. M., Benson, J. M., Cheng, Y. S., Bice, D. E., Luster, M. I., Dunnick, J. K., and Hobbs, C. H. (1990), *Fundam. Appl. Toxicol.,* 15, 476–487.
Harte, J., Holdren, C., Schneider, R., Shirley, C., eds. (1991), *Toxics A to Z,* University of California Press, Berkley.
Horiguchi, S., Endo, G., Kiyota, I., Teramoto, K., Shinagawa, K., Wakitani, F., Tanaka, H., Konishi, Y., Kiyota, A., Ota, A., and Fukui, M. (1992), *Osaka City Med. J.,* 38, 79–81.
Ilback, N. G., Folman, J., and Friman, G. (1992a), *Toxicol. Appl. Pharmacol.,* 114, 166–170.
Ilback, N. G., Folman, J., and Friman, G. (1992b), *Toxicology,* 71, 193–202.
Jarstrand, C., Lundborg, M., Wiernik, A., and Camner, P. (1978), *Toxicology,* 11, 353–358.
Konstan, M. W., Cheng, P. W., Sherman, J. M., et al. (1981), *Am. Rev. Respir. Dis.,* 123, 120–124.
Koren, H. S. and Becker, S. (1992), in *A Comprehensive Treatise on Pulmonary Toxicology: Comparative Pulmonary Biology of the Normal Lung,* Plopper, C., Costa, D., Raub, J., Boelkel, N. F., Schlesinger, R., and Kelly, C. J., 747–769.
Lundborg, M. and Camner, P. (1982), *Toxicology,* 22, 353–358.
Maigetter, R. Z., Ehrlich, R., Fenters, J. D., and Gardner, D. E. (1976), *Environ. Res.,* 11, 386–391.
Miller, F. J., Graham, J. A., Raub, J. A., House, D. E., and Gardner, D. E. (1987), *J. Toxicol. Environ. Health,* 21, 99–112.
Muir, D. C. F. (1991), in *The Lung,* Crystal, R. G. and West, J. B., Eds., Raven Press, New York, 1839–1843.
NRC (1989), in *Biologic Markers in Pulmonary Toxicology,* National Academy, Washington, D. C., 91–103.
Nriagu, J. O. and Davidson, C. I., Eds. (1986), *Toxic Metals in the Atmosphere,* John Wiley & Son, New York.
Nugent, K. M. and Pesanti, E. L. (1982), *Infect. Immunity,* 36, 1185–1191.
Reynolds, H. Y. (1991), in *The Lung Scientific Foundations,* Crystal, R. G., West, J. B., Barnes, P. J., Cherniack, N. S., and Weibel, E. R. Vol. 2, 1899–1911.
Schlesinger, R. B. (1989), in *Concepts in Inhalation Toxicology,* McClellan, R. O. and Henderson, R. F., Eds., Hemisphere Publishing, Washington, D.C., 163–192.
Schlesinger, R. B. (1990), *Crit. Rev. Toxicol.,* 20, 257–286.
Selgrade, M. J. K. and Gilmour, M. I. (in press), in *Immunotoxicity and Immunopharmacology,* 2nd ed., Dean, J. H., Luster, M. I., Munson, A. E., and Kimber, I., Raven Press, New York.
Sikorski, E. E., McCay, J. A., White, K. L., Bradley, S. G., and Munson, A. E. (1989), *Fundam. Appl. Toxicol.,* 13, 843–858.
Smialowicz, R. J., Luebke, R. W., Rogers, R. R., Riddle, M. M., and Rowe, D. G. (1985), *Immunopharmacology,* 9:1–11.
Stein-Streilein, J., Bennett, M., Mann, D., and Kumar, V. (1983), *J. Immunol.,* 131:2699–2704.
Toews, G. B. (1986), *Semin. Respir. Infect.,* 1:68.
Watson, A. Y., Bates, R. R., and Kennedy, D., Eds. (1988), *Air Pollution: The Automobile and Public Health,* National Academy Press, Washington, D.C.
Wiernik, A., Johansson, A. Jarstand, C., and Camner, P. (1983), *Environ. Res.,* 30, 129–141
Wiltrout, R. H., Herberman, R. B., Zhang, S. R., Chirigos, M. A., Ortaldo, J. R., Green, K. M., and Talmadge, J. E. (1985), *J. Immunol.,* 134, 4267–4275.
Wu, Y., Luo, H. and Johnson, D. R. (1986), *Cancer Lett.,* 32, 171–179.

Chapter 55

Concepts of Immunological Biomarkers of Metal Toxicity

Hassan A. N. El-Fawal

I. INTRODUCTION

In recent years, a strong emphasis has been placed on the development of biological markers for toxic exposures and/or subclinical effects in hope of preventing some of the debilitating effects associated with these exposures (Fowle and Sexton, 1992; Silbergeld, 1993). The distinction between markers of **exposure** and markers of **effect** are often blurred. For example, while blood lead levels are used to indicate exposure to lead, these values do not provide information on the biological effects of such exposures. From the outset, it is important to distinguish our discussion on concepts of immunological biomarkers of metal toxicity from the immunotoxic effects of metals. The latter subject is dealt with in other chapters in this section. The concept of utilizing immunological indicators to reflect metal-induced damage to other systems derives from the definition of the immune response as a functional system responsible for surveillance of the host's organs. The immune system provides a unique opportunity for experimental toxicologists to screen thousands of potential toxicants and to reduce the number of experimental animals, as well as for the clinical toxicologist, to assess the potential pathology associated with subclinical exposures. This is facilitated by the relatively simple collection of sera and immune cells. It also draws on the very nature of the immune system as a defense system with sensitivity, in terms of threshold of xenobiotic required to elicit an immune response and specificity in terms of humoral and cellular responses to a specific xenobiotic. Furthermore, for some systems, such as the nervous system, some similarities in function and response may allow the accessible immune system to act as a surrogate for the inaccessible nervous system. Along these lines, Snyder (1991) has remarked on the use of immune cells as surrogates to assess the neurotoxic potential of chemical exposure. For example, human lymphocytes can be used to assess the potential for development of delayed neuropathy following exposure to organophosphorus compounds (Lotti, 1987). However, despite the advantages provided by the immune system, to date, few toxicological studies have capitalized on the immune system's accessibility in documenting exposure or the biological effects associated with toxic metal exposures.

The following discussion will deal with two potential areas where the use of immunological biomarkers show great promise. These include their use as markers of exposure to heavy matals and as markers of toxic effects. Immunological markers of effect will be elaborated by using the nervous system as an example to illustrate how the immune system may be exploited to detect subclinical toxic effects.

II. IMMUNOLOGICAL MARKERS OF EXPOSURE TO METALS

The use of immune functional assays as biological markers of metal exposure has been demonstrated by Snyder and Valle (1991a,b). These authors measured proliferative responses of T and B lymphocytes when stimulated *in vitro* by concanavalin A (Con A) and lipopolysaccharide, respectively. They found that when splenocytes isolated from rats exposed *in vivo* to potassium chromate or cadmium chloride for 6 weeks were cocultured *in vitro* with exogenous chromate or cadmium, lymphoproliferative response to mitogen stimulation was altered. Exogenously added cadmium induced a greater depression of thymidine uptake by splenocytes from chromate-exposed rats during Con A stimulation than did exogenous chromate. On the other hand, exogenously added chromate induced a greater depression in the uptake of thymidine by cells isolated from cadmium-exposed rats than did exogenous cadmium. The authors suggested that, based on these preliminary findings, it may be possible to screen for an individual exposure to a specific metal by performing lymphocyte proliferation assays in the presence of a battery of exogenously added metals.

The best established immunological marker of metal exposure, and one that is used clinically, is that of beryllium. Occupational exposure to beryllium results in chronic beryllium disease (CBD), where sensitized individuals accumulate pathologic granulomas surrounding beryllium particles in the alveolar wall (Newman, 1993). Specific diagnosis of CBD is based on the lymphocyte proliferation tests (BeLT), similar to those mentioned above, where blood lymphocytes from exposed individuals proliferate *in vitro* when cultured in the presence of beryllium salts (Kreiss et al., 1989). Another immunological approach for the detection of metal exposure and possible susceptibility to CBD is the detection of antibodies to beryllium using an enzyme-linked immunosorbent assay (ELISA) (Clarke, 1991). Recently, Richeldi et al. (1993) described a genetic marker in workers with CBD, the amino acid glutamate present in a critical location of an allele coding for the major histocompatibility complex (MHC), which participates in antigen recognition and presentation. Glutamic acid at residue 69 of this allele (HLA-DPB* 0201) is also associated with susceptibility to autoimmune disorders as insulin-dependent diabetes and rheumatoid arthritis (Todd et al., 1987). The case of beryllium provides an example of how immunological markers may be used to indicate exposure, effect, and possibly risk to metal-induced toxicity.

III. IMMUNOLOGICAL MARKERS OF BIOLOGICAL EFFECTS

It should be recognized that in evaluating any biological marker of toxicity, stringent criteria must be satisfied. An appropriate biomarker of toxic effects should possess the following attributes: (1) specificity for the target system; (2) sensitivity to low level subacute or chronic exposures; (3) be informative as to the extent and type of pathology; (4) be objective and quantitative; (5) be readily accessible for sample collection; and (6) be relatively inexpensive to perform.

An area where the use of immunological markers of metal toxicity shows greatest promise is organ-specific autoimmunity. The hypotheses delineating how autoimmune responses may be precipitated have been reviewed in several immunology texts (Sell, 1987; Roitt et al., 1993). Briefly, immune responses to self, in terms of autoantibody production, may reflect an immune activation against previously sequestered proteins with which the immune system is not familiar (e.g., the case of the nervous system and blood-barriers), or proteins to which the immune system has not established tolerance because of barriers and their late development in the life of the organism (e.g., antisperm antibodies; Hendry, 1992). Autoimmune responses may also reflect an environmentally induced disarray of self-tolerance and immunoregulatory clone deletion (Varela et al., 1991). The nervous system, long considered "immuno-privileged" and whose cytoskeletal structures are sequestered prior to the development of immune tolerance, will be discussed in detail to illustrate the applicability of immune markers to assess the biological effects of metal toxicity.

A. IMMUNOLOGICAL MARKERS MEET THE CHALLENGE

The inaccessibility of the nervous system has long posed a challenge to neuroscientists and, in particular, to neurotoxicologists. This inaccessibility has impeded the evaluation of cellular and molecular changes which represent the initial changes during neurotoxicity, prior to overt clinical symptoms. Current methods of assessing the development of (and recovery from) neurotoxic insult include behavioral, electrophysiological, and brain-imaging techniques (Ehle, 1986; Anger, 1992). However, these evaluations require highly trained personnel, are costly, and do not identify specific cellular substrates or mechanisms. This imposes limitations to their utility in clinical diagnostics, risk assessment, and their applicability to populations in the exposure arena (e.g., agriculture, industry). Furthermore, the detection of deficits often does not occur in the absence of overt manifestations.

O'Callaghan (1988, 1992) has proposed that neurotypic and gliotypic proteins can be used to detect and characterize the cellular response to toxicant-induced injury. Alterations in nervous system-specific proteins, for example, glial fibrillary acidic protein (GFAP), the astrocytic intermediate filament, may be expressed as a result of low level exposure to neurotoxic metals (O'Callaghan, 1988; Evans et al., 1992; El-Fawal et al., 1992; Evans and El-Fawal, 1994) and other nervous system insults. This "reactive gliosis" occurs secondary to neuronal insult and degeneration. Although direct measurement of brain proteins, in experimental animals, provides insights into the cellular targets of neurotoxicity in animal studies (O'Callaghan, 1988; Evans et al., 1993), it are not applicable to human populations. However, this evidence is invaluable in evaluating potential biomarkers of metal neurotoxicity.

Our laboratory has proposed that exposure to neurotoxic agents, particularly metals, and the attendant neurodegeneration with the liberation of neural proteins, induce an immune response which can be easily measured in the blood. This response is reflected in the production of serum antibody against nervous system proteins. The magnitude of this response correlates with the extent of exposure and nervous system damage.

IV. RATIONALE

A. IMMUNOPRIVILEGE AND NEURO-IMMUNE INTERACTIONS

The nervous systems, both central (CNS) and peripheral (PNS), are relatively "immunoprivileged", as provided by the blood-brain barrier (BBB) and blood-nerve barrier (BNB). The BBB consists of specialized endothelial cells, pericytes, and bone marrow-derived perivascular elements. These are enclosed within a basal lamina and astrocyte foot processes (Jacobs, 1994). On the other hand, the BNB is formed by the permeable epineurium, the tight junctions of the perineurium, and the impermeable endoneurium (Jacobs, 1994). These barriers control the selective entry of essential biomolecules and exclude potentially harmful elements, including immune and inflammatory cells.

It has become evident in recent years that glial cells (astrocytes, Schwann cells, and microglia) play more than a structural role in both the CNS (Prochiantz and Mallat, 1988) and PNS (Hanson et al., 1989). In the case of the astrocyte, this includes the formation of the blood-brain barrier interface, synthesis and release of interleukins, and antigen presentation (Prochiantz and Mallat, 1988). Furthermore, astrogliosis, common in nervous system trauma of varying etiologies is stimulated by gamma interferon (IFNγ; Yong et al., 1991). Similarly, in the PNS, Schwann cells have been shown to respond, *in vivo* and *in vitro*, to IFN-γ and tumor necrosis factor, and to synthesize and express major histocompatibility complex (MHC) I and II (Armati et al., 1990; Kingston and Bergsteindottir, 1989). This ability of myelinating Schwann cells to act as antigen-presenting cells has been documented following peripheral nerve crush and in peripheral nerve disease (Bergsteinsdottir et al., 1992). In addition, GFAP (Tanaka et al., 1989) and myelin basic protein (MBP; Sheffield and Kim, 1977) have been shown to be mitogenic to cultured lymphocytes, *in vitro*. Myelin basic protein is also mitogenic to cultured glial cells and lymphocytes (Sheffield and Kim, 1977). This evidence suggests that nervous system proteins may activate immune mechanisms following neuronal damage and demyelination (Mucke and Eddlesteon, 1993; Fabry et al., 1994; Madden and Felten, 1995).

Consistent with the possible role of glial cells as antigen-presenting cells are reports of lymphocyte infiltration across the BBB following neurotrauma (Hughes et al., 1988; Kaijiwara et al., 1990; Raine et al., 1990; Irani and Griffin, 1991). The presence of T and B lymphocytes in nervous system tissues and cerebral spinal fluid (CSF) has been documented in humans (e.g., multiple sclerosis) and experimental models of nervous system disease (Chan et al., 1989; Cash et al., 1992). Furthermore, experimental evidence indicates that activated T lymphocytes are capable of entry into the CNS without a need for antigen-specific activation (Hickey et al., 1991). This potential interaction between cells of the nervous and immune systems has implications for the use of antibody generation to assess the neurotoxicity of metals. For example, under certain conditions, lead has been shown to augment immune responses. Lead enhances B lymphocyte differentiation *in vitro* and enhances the activity of B lymphocytes towards T cell-dependent antigens *in vivo* (Lawrence, 1981a,b,c; Koller et al., 1976). Lead also enhances the production of antibodies and directly activates B cells (McCabe et al., 1990), as well as potentiates the production and release of interleukin-2 (IL-2) from T lymphocytes (Warner and Lawrence, 1988). These observations have resulted in speculation that exposure to lead may result in an autoimmune response (McCabe and Lawrence, 1991).

B. AUTOIMMUNE RESPONSES IN THE NERVOUS SYSTEM

Autoimmune mechanisms have been recognized in neurological diseases such as myasthenia gravis (Drachman et al., 1988), Guillain-Barré syndrome (Hartung et al., 1988), and multiple sclerosis (Tabira,

1988). As noted earlier, the nervous system itself may play a role in the activation of immunological mechanisms in these diseases (e.g., antigen presentation and cytokine release). How nervous system system antigens encounter and interact with cells of the systemic circulation and whether this occurs *in situ* in the nervous system or in the periphery has not been fully delineated, since assays for these proteins in peripheral blood as markers of nervous system insult have not met with success. On the other hand, autoantibodies against neurofilament triplet proteins (NF) as well as MBP and GFAP have been detected in sera and CSF of subjects suffering from neurological disorders, including Alzheimer's disease, Parkinson dementia, amyotrophic lateral sclerosis (ALS), Creutzfeldt-Jakob disease, and kuru (1986; Toh et al., 1985; Tanaka et al., 1988, 1989; Matsiota et al., 1988; Gorny et al., 1990; Mitrova and Maye, 1989; Braxton et al., 1989; Vedeler et al., 1988), as well as in animal models of allergic encephalomyelitis and electroconvulsive shock (Walls et al., 1988; Vlajkovic and Jankovic, 1991). Thus, the immune system can provide evidence for the occurrence of damage in the nervous system.

C. AUTOANTIBODIES FOR DETECTING SYSTEM-SPECIFIC TOXIC EFFECTS: HYPOTHESIS

Metal-induced degeneration of target organs results in cell death and degeneration. This exposes and liberates intracellular antigens, often antigens which the immune system has not encountered previously. For example, in the case of the nervous system and neurotoxicity, where axonal degeneration, demyelination, and glial degeneration are commonly encountered following exposure to lead and mercury, proteins specific to these structures, being perceived as foreign due to the "immunoprotected" status of the nervous system, are presented as autoantigens and antibodies are raised against them. The generation of these autoantibodies would likely occur particularly with metal intoxication in light of their ability to compromise blood-barriers (Chang, 1980; Krigman et al., 1980; Bressler and Goldstein, 1991). In this way, the immume system provides a means whereby cellular damage in the nervous system can be documented and measured in serum, eliminating the problems associated with the inaccessibility of the nervous system. A summary of the hypothesis is presented in Figure 1.

V. AUTOANTIBODIES TO NERVOUS SYSTEM PROTEINS IN OCCUPATIONALLY EXPOSED POPULATIONS

A. ANTIBODIES AND EXPOSURE

Field testing of autoantibody assays was performed by our laboratory in male workers occupationally exposed to either lead, at a battery factory, or mercury vapor, at a fluorescent light factory. In addition, a reference group, from a frozen food packing plant with no prior work history of exposure to either metal, was also recruited (Shamy and El-Fawal, 1993; Abdel-Moneim et al., 1994; El-Gazzar et al., 1994). All participants in the study were matched on the basis of demographics and socioeconomic status, as well as in years of metal exposure. Ambient mercury (as vapor) and lead (as dust or fumes), 0.05 and 0.09 mg/m^3, respectively, are below or at the TLV-TWA adopted by the American Conference of Governmental Industrial Hygienists (1992).

Blood lead (PbB) and urinary mercury (HgU) were significantly elevated in the lead- and mercury-exposed population, respectively, compared to the reference population. Titers of autoantibodies (IgM and IgG isotypes) to neurofilament triplet (NF) proteins: NF-68, NF-160 and NF-200, GFAP, and MBP were determined in sera of exposed and reference populations by an ELISA developed by El-Fawal against these proteins. A summary of the percent of each population with detectable immunoglobulins (M and G isotypes) against these five neural antigens is given in Figure 2. Autoantibodies against neuroproteins predominated in metal-exposed populations compared to the reference population. The detection of autoantibodies in a small percentage of the reference population probably reflects natural autoantibodies, usually IgM, which are found in some individuals (Stephenson et al., 1985; Matsiota et al., 1988; Vrethem et al., 1991). In exposed populations, antibody titers which significantly correlated with exposure and sensorimotor function were predominantly IgG. Immunoglobulin G is the isotype most commonly associated with secondary antigen challenge or antigen persistsnce and pathology (Cohen and Cooke, 1986). This is further illustrated in Figure 3, where individual titers of anti-NF-200 are shown for the reference and mercury-exposed groups. Immunoglobulin G was absent in the reference population altogether, but was the dominant isotype in the mercury-exposed population. Anti-NF-200 (IgG) for this population significantly correlated with HgU and sensorimotor deficits. In contrast, anti-NF-68 (IgG) was the best correlated with PbB and sensorimotor deficits. It should also be noted that profiles of autoantibodies to nervous system antigens differed for the two metal-exposed groups (Figure

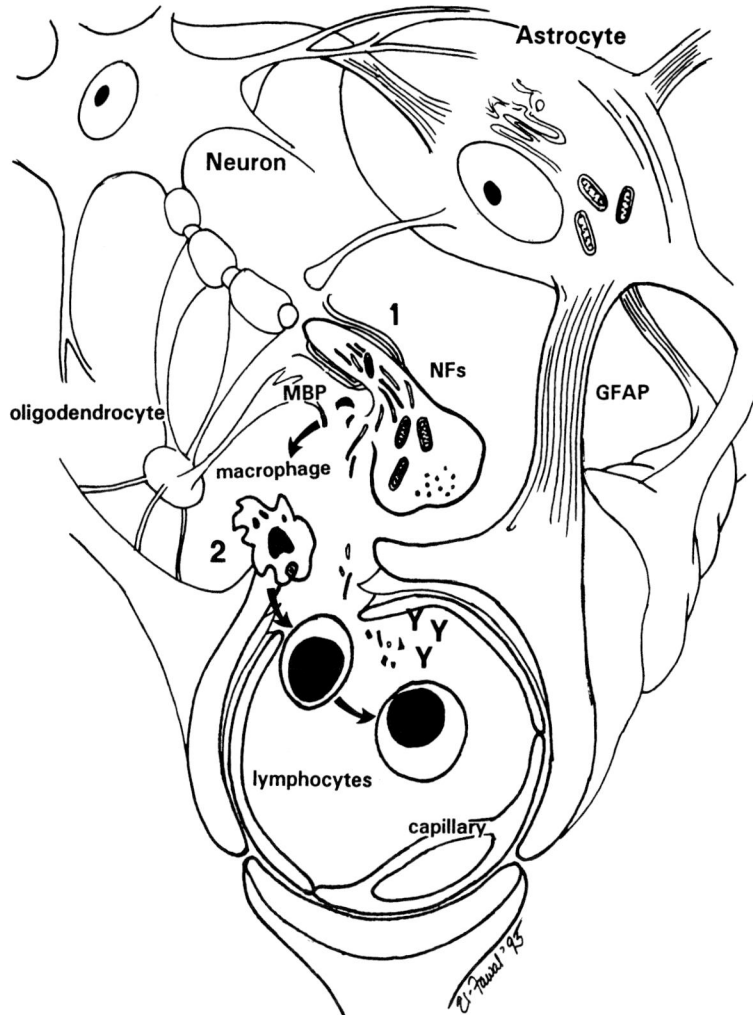

Figure 1. Hypothesis for the induction of autoantibodies to nervous system proteins during metal exposure. Following the initial insult (1) by neurotoxic metals to cells of the nervous system, autoantigens [which may include synaptic proteins, axonal neurofilaments (NFs), myelin sheath proteins (MBP), and astrocytic proteins (GFAP)] are released and (2) undergo antigen processing and presentation by microglia, astrocytes, or macrophages. This results in T lymphocyte activation and stimulation of B lymphocytes to produce autoantibodies to these autoantigens. The perception that these antigens are foreign is facilitated by the presence of blood-barriers (BBB).

2). Since the two populations were matched in all aspects except exposure, these profile differences likely reflect the differences in neurotoxicity at the levels studied.

A significant dose-response relationship between the total number of detectable autoantibodies of the G isotype (IgG score) to the five antigen proteins and the total number of upper and lower limb sensorimotor deficits (clinical score) was observed (Figure 4). This suggests an association between the appearance of antibodies in the sera and the biological effects of metal exposure.

The generation of autoantibodies to these antigens during metal exposure has been confirmed in our laboratory using animal models (El-Fawal et al., 1993; El-Fawal et al., 1994). In studies with lead (El-Fawal et al., 1993), male Fisher 344 rats (>42 days of age) were exposed to 50 or 450 ppm lead acetate in the drinking water. No overt signs of toxicity or changes in home-cage behavior (Evans, 1989) were evident in any of the rats, including those monitored for 42 days of exposure. In these same groups, PbB levels reached 15 and 50 µg/dl in rats exposed to 50 and 450 ppm, respectively. While control rats had no detectable titers to these particular antigens, titers of autoantibodies, particularly IgM, were

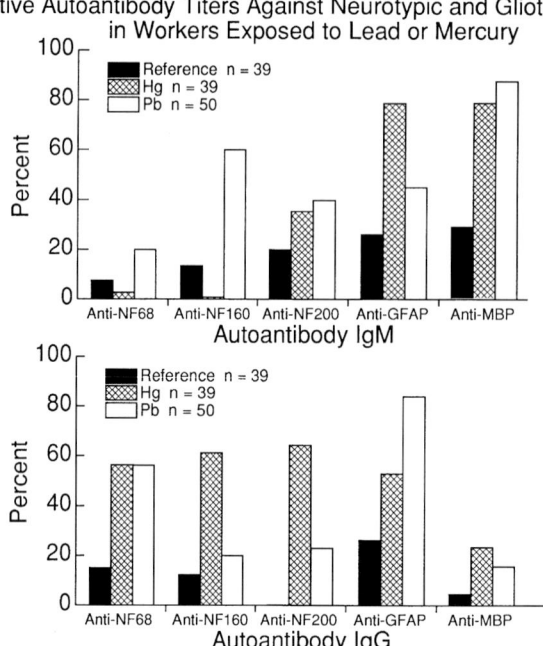

Figure 2. Percent of three cohorts: reference, Hg-exposed, or Pb-exposed, displaying detectable IgM and IgG titers against nervous system-specific proteins. Autoantibody detection predominated in exposed populations, compared to the reference group.

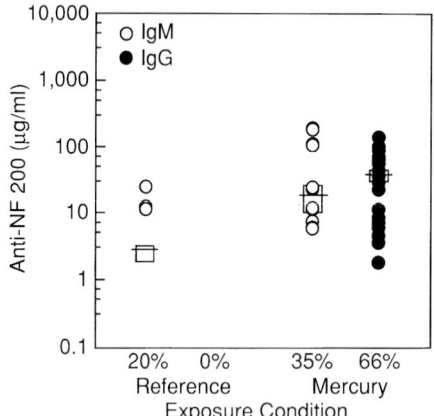

Figure 3. Individual anti-NF 200 titers in reference and Hg-exposed cohorts. Bars indicate means ± S.E.M. Note the absence of detectable IgG in the reference group and the predominance of IgG in the Hg-exposed group. IgG is the isotype prevalent in the pathological states (see text).

detected and quantified as early as 4 d after the initiation of Pb exposure (PbB as low as 5 μg/dl). Consistent with a primary antigen challenge (in this case autoantigens) there was a significant elevation of IgM titers, followed by isotype switching to IgG at later durations (42 d). As in the human studies, anti-NF-68 (IgG) titers correlated significantly with PbB.

B. ANTIBODIES AND CELLULAR TARGETS

The neuropathology associated with chronic exposures to mercury or lead primarily involves the neuro-axon with secondary demyelination (Chang, 1977; Krigman et al., 1980; Windebank et al., 1984). In the human studies, anti-NFs, IgG isotype, were the most frequently detected antibodies. These were

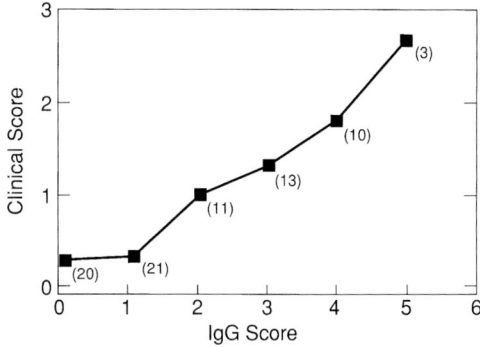

Figure 4. Dose-response relationship between clinical scores and the number of detectable IgG titers against nervous system proteins in mercury-exposed and reference populations, combined (n = 78). These data indicate that as more autoantibodies against nervous system proteins are detected, the greater the likelihood of clinical deficits (numbers in parenthesis represent the number of individuals at each score).

also the antibodies and the isotype best correlated with HgU and clinical scores of sensorimotor deficits. This is consistent with the type of neuropathy in humans which manifests primarily neuro-axonal damage (Chang, 1980). Furthermore, the detection of anti-GFAP titers in these studies supports the targeting of the CNS and astrocytes by heavy metals (Garman et al., 1975; Evans et al., 1982; Rowles et al., 1989), since astrocytes are exclusively found in the CNS. In this manner, autoantibodies may provide a means whereby subcellular targets and the progression of neuropathy may be documented.

In this context, the temporal appearance of autoantibodies in serum provides information on the extent of target cell involvement and the progression of insult. For example, it is likely that anti-NF titers would be the first to appear with primary neuronal involvement (e.g., mercury), while anti-MBP would precede all others in the case of metals inducing primary demyelination (e.g., triethyltin).

C. METALS AND ALTERED ANTIGEN IMMUNOGENICITY

The role of environmental factors in the production or augmentation of autoimmune responses has been suggested for some time. The interaction of drugs or environmental chemicals, including metals (e.g., gold, mercury), with endogenous protein constituents may alter their immunogenicity and result in the induction of immune responses reflected in the generation of autoantibodies. Having detected antibodies to nervous system antigens in humans and experimental animals, it was speculated that further alteration of these antigens may occur with chronic exposures to metals. This may provide a basis for differential antibody profiles between metals and possibly metal specificity, as a result of altered immunogenicity. Indeed, this may be the case. Waterman et al. (1994) have demonstrated that *in vitro* treatment of the neural antigens MBP and GFAP with lead, followed by inoculation of CBA/J mice, resulted in antibody titers that were significantly higher than those produced by the antigens alone, thus providing evidence that lead can alter the immunogenicity of antigen and enhance the magnitude of the autoimmune response.

VI. CONCLUSIONS

The studies discussed in this chapter illustrate the usefulness of the immune system's functional status for developing markers of metal exposure and their biological effect. The lymphoproliferation assays, particularly in the case of beryllium, and the autoantibody assays, in the case of lead and mercury, provide promising examples of these applications. In the case of the autoantibody assays, these studies suggest a strong and promising association between the appearance of autoantibody titers against nervous system proteins and exposure to subclinical levels of known neurotoxicants. This approach can be expanded to include other organ systems with characterized organ-specific antigens (e.g., reproductive, cardiovascular, and pulmonary systems). The use of these immunological indicators to assess organ system integrity following metal exposure provides a relatively simple means of documenting pathogenesis in human populations and animal studies.

REFERENCES

Abdel-Moneim, I., Shamy, M. Y., El-Gazzar, R. M., and El-Fawal, H. A. N. (1994), *Toxicologist,* 14, 291.
Anger, W. K. (1992), in *Neurotoxicology,* Tilson, H. A. and Mitchell, C. L., Eds., Raven Press, New York, 363–386.
Armati, P. J., Pollard, J. D., and Gatenby, P. (1990), *Muscle Nerve,* 13, 106–116.
Bergsteinsdottir, K., Kingston, A., and Jessen, K. R. (1992), *J. Neurocytol.,* 21, 382–390.
Braxton, D. B., Williams, M., Kamali, D., Chin, S., Liem, R., and Latov, N. (1989), *J. Neuroimmunol.,* 21, 193–203.
Bressler, J. P. and Goldstein, G. W. (1991), *Biochem. Pharmacol.,* 41, 479–484.
Cash, E., Weert, S., Voltz, R., and Kornhuber, M. (1992), *Scand. J. Immunol.,* 35, 695–701.
Chan, W. L., Javanovic, T., and Lukic, M. L. (1989), *J. Neuroimmunol.,* 232, 195–201.
Chang, L. W. (1977), *Environ. Res.,* 14, 329–373.
Chang, L. W. (1980), in *Experimental and Clinical Neurotoxicology,* Spencer, P. S. and Schaumburg, H. H. Eds., Williams & Wilkins, Baltimore, 508–526.
Clarke, S. M. (1991), *J. Immunol. Methods,* 137, 65–72.
Cohen, I, R. and Cooke, A. (1986), *Immunol. Today,* 7, 363–364.
Drachman, D. B., McIntosh, K. R., De Silva, S., Kuncl, R. W., and Kahn, C. (1988), *Ann. N. Y. Acad. Sci.,* 540, 176–186.
Ehle, A. L. (1986), *Neurotoxicology,* 7, 203–216.
El-Fawal, H. A. N., Little, A. R., Gong, Z., and Evans, H. L. (1992), *Toxicologist,* 12, 312.
El-Fawal, H. A. N., Little, A. R., Gong, Z. L., and Evans, H. L. (1993), *Neurosci. Abstr.,* 19, 1483.
El-Fawal, H. A. N., Gong, Z. L., Little, A. R., and Evans, H. L. (1994), *Neurotoxicology,* 15, 969.
El-Gazzar, R. M., Shamy, M. Y., Abdel-Moneim, I., and El-Fawal, H. A. N. (1994), *Toxicologist,* 14, 291.
Evans, H. L. (1989), *J. Am. Coll. Toxicol.,* 8, 35–52.
Evans, H. L. and El-Fawal, H. A. N. (1994), *NeuroToxicology,* 15, 969.
Evans, H. L., Garman, R. H., and Laties, V. G. (1982), *Neurotoxicology,* 3, 21–36.
Evans, H. L., Jortner, B. S., and El-Fawal, H. A. N. (1992), *Toxicologist,* 12, 317.
Evans, H. L., Little, A., Gong, Z. L., Duffy, J. S., and El-Fawal, H. A. N. (1993), *Ann. N.Y. Acad. Sci.,* 642, 402–406.
Fabry, Z., Raine, C. S., and Hart, M. N. (1994), *Immunol. Today,* 15, 218–221.
Federsppiel, B. S., Karcher, D., and Lowenthal, A. (1987), *Gerentology,* 33, 193–196.
Fowle, J. R., 3rd and Sexton, K. (1992), *Environ. Health Perspect.,* 98, 235–241.
Garman, R. H., Weiss, B., and Evans, H. L. (1975), *Acta Neuropathol.,* 32, 61–74.
Gorny, M., Losy, J., and Wender, M. (1990), *Neurologia I Neurochirurgia Polska,* 24, 17–22.
Hanson, S. H., Stagaard, M., and Mollgard, K. (1989), *J. Neurocytol.,* 18, 427–436.
Harrington M. G. and Merril, C. R. (1988), *J. Chromatography,* 429, 345–358.
Hartung, H. P., Heininger, K., Schafer, B., Fierz, W., and Toyka, K. V. (1988), *Ann. N.Y. Acad. Sci.,* 540, 122–161.
Hendry, W. F. (1992), *Clin. Endocrin.,* 36, 219–221.
Hickey, W. F., Hsu, B. L., and Kimura, H. (1991), *J. Neurosci. Res.,* 28, 254–260.
Hughes, C. C., Male, D. K., and Lantos, P. L. (1988), *Immunology,* 64, 677–681.
Irani, D. N. and Griffith, D. E. (1991), *J. Immunol. Methods,* 139, 223–231.
Jacobs, J. M. (1994), in *Principles of Neurotoxicology,* Chang, L. W., Ed., Marcel Dekker, New York, 35–68.
Kajiwara, K., Ito, H., and Fukumoto, T. (1990), . *Neuroimmunology,* 27, 133–140.
Kingston, A. E. and Bergsteindottir, K. (1989), *Eur. J. Immunol.,* 19, 177–183.
Koller, L. D., Exon, J. H., and Roan, J. G. (1976), *Proc. Soc. Exp. Biol. Med.,* 151, 339–342.
Kreiss, K., Newman, L. S., Mroz, M. M., and Campbell, P. A. (1989), *J. Occup. Med.,* 31, 603–608.
Krigman, M. R., Bouldin, T. W., and Mushak, P. (1980), in *Experimental and Clinical Neurotoxicology,* Spencer P. S. and Schaumburg, H. H., Eds., Williams & Wilkins, Baltimore, 490–507.
Lawrence, D. A. (1981a), *Infect. Immunol.,* 31, 136–143.
Lawrence, D. A. (1981b), *Toxicol. Appl. Pharmacol.,* 57, 439–451.
Lawrence, D. A. (1981c), *Int. J. Immunopharmacol.,* 3, 153–161.
Lotti, M. (1987), *Trends Pharmacol. Sci.,* 8, 176–177.
Male, D., Pyrce, G., Hughes, C., and Lantos, P. (1990), *Cell. Immunol.,* 127, 1–11.
Matsiota, P., Blancher, A., Doyon, B., Guilbert, B., Clanet, M., Kouvelas, E. D., and Avrameas, S. (1988), *Ann. Inst. Pasteur,* 139, 99–108.
McCabe, M. J. and Lawrence, D. A. (1991), *Toxicol. Appl. Pharmacol.,* 111, 13–23.
Mitrova, E. and Maye, V. (1989), *Acta Virolo.,* 33, 371–374.
Mucke, L. and Eddleston, M. (1993), *FASEB J.,* 7, 1226–1232.
Nakamura, K., Takeda, M., Tanaka, T., Tanaka, J., Kato, Y., Nishinuma, K., and Nishimura, T. (1992), *Methods Find. Exp. Clin. Pharmacol.,* 14, 141–149.
Newman, L. S. (1993), *Science,* 262, 197–198.
O'Callaghan, J. P. (1988), *Neurotoxicol. Teratol.,* 10, 445–452.
O'Callaghan, J. P. (1992), in *Neurotoxicology,* Tilson, H. and Mitchell, C., Eds., Raven Press, New York, 83–100.
Prochiantz, A. and Mallat, M. (1988), *Ann. N.Y. Acad. Sci.,* 540, 52–63.
Raine, C. S., Cannella, B., Duijvestijn, A. M., and Cross, A. H. (1990), *Lab. Invest.,* 63, 476–489.
Richeldi, L., Sorrentino, R., and Saltini, C. (1993), *Science,* 262, 242–244.

Roitt, I. (1993), in *Immunology,* I., Roitt, J., Brostoff, and D., Male, Eds. C.V. Mosby, St Louis, 24.2–24.11.
Rowles, T. K., Womac, C., Bratton, G. R., and Tiffany-Castiglioni, E. (1989), *Metabolic Brain Dis.,* 4, 187–201.
Sell, S. (1987), *Immunology, Immunopathology and Immunity,* Elsevier, New York, 235–260.
Shamy, M. Y. and El-Fawal, H. A. N. (1993), *Toxicologist,* 13, 124.
Sheffield, W. D. and Kim, S. U. (1977), *Brain Res.,* 132, 580–584.
Silbergeld, E. K. (1993), *Environ. Res.,* 63, 274–286.
Snyder, C. A. (1989), *Environ. Health Perspect.,* 81, 165–166.
Snyder, C. A. and Valle, C. (1991a), *J. Toxicol. Environ. Health,* 34, 127–139.
Snyder, C. A. and Valle, C. (1991b), *Environ. Health Perspect.,* 92, 83–86.
Stephenson, K., Marton, L. S., Dieperink, M. E. et al. (1985), *Science,* 228, 1117–1119.
Tanaka, J., Murakoshi, K., Takeda, M., Kato, Y., Tada, K., Hariguchi, S., and Nishimura, T. (1988), *Biomed. Res.,* 9, 209–216.
Tanaka, J., Nakamura, K., Takeda, M., Tada, K., Suzuki, H., Morita, H., Okada, T., Hariguchi, S., and Nishimura, T. (1989), *Acta Neurol. Scand.,* 80, 554–560.
Tabira, T. (1988), *Ann. N.Y. Acad. Sci.,* 540, 187–201.
Todd, J. A., Acha-Orbea, H., Bell, J. I., Chao, N., Fronek, Z., Jacob, C. O., McDermott, M., Sinha, A. A., Timmerman, L., Steinman, L. et al. (1988), *Science,* 240, 1003–1009.
Toh, B. H., Gibbs, J. H., Gajdusek, D. C., Tuthill, D. D., and Dahl, D. (1985), *Proc. Natl. Acad. Sci. U.S.A.,* 82, 3485–3489.
Varela, F., Andersson, A., Dietrich, G., Sundblad, A., Holmberg, D., Kazatchkine, M., and Coutinho, A. (1991), *Proc. Natl. Acad. Sci. U.S.A.,* 88, 5917–5921.
Vedeler, C. A., Matre, R., and Nyland, H. (1988), *Acta Neurol. Scand.,* 78, 401–407.
Vermuyten, K. (1989), *Acta Neurol. Belg.,* 89, 318.
Vlajkovic, S. and Jankovic, B. D. (1991), *Int. J. Neurosci.,* 59, 205–211.
Walls, A. E., Suckling, A. J., and Rumsby, M. G. (1988), *Acta Neurol. Scand.,* 78, 422–428.
Warner, G. L. and Lawrence, D. A. (1988), *Int. J. Immunopharmacol.,* 10, 629–637.
Waterman, S. J., El-Fawal, H. A. N., and Snyder, C. A. (1994), *Environ. Health Perspect.,* 102, 1052–1056.
Wells, M. R., Racis, S. P., and Vaidya, U. (1992), *J. Neuroimmunol.,* 39, 261–268.
Windebank, A. J., McCall, J. T., and Dyck, P. J. (1984), in *Peripheral Neuropathy,* Vol. 2, Dyk, P. J., Thomas, P. K., Lambert, E. H., and Bunge, R. Eds., W. B. Saunders, Philadelphia, 2133–2161.
Yong, V. W., Moumdjian, R., Young, F. P., Ruijs, T. C., Freedman, M. S., Cashman, N., and Antel, J. P. (1991), *Proc. Natl. Acad. Sci. U.S.A.,* 88, 7016–7020.

Chapter 56

Modulation of Immune Functions by Metallic Compounds

Lee S. F. Soderberg, John R. J. Sorenson, and Louis W. Chang

I. INTRODUCTION

Copper, iron, manganese, and zinc are essential metalloelements like sodium, potassium, calcium, magnesium, chromium, cobalt, and vanadium. Just like essential amino acids, essential fatty acids, and essential cofactors (vitamins), these metalloelements are required by all cells for normal metabolic processes but cannot be synthesized *de novo,* and dietary intake and absorption are required to obtain them. Amounts of Cu, Fe, Mn, and Zn found in adult body tissues and fluids (Tipton and Cook, 1963; Iyengar et al., 1978) correlate with the number and kind of metabolic processes requiring them. However, there may be an aging-related decrease in tissues due to inadequate dietary intake and absorption.

Ionic forms of these metalloelements have particularly high affinities for organic ligands found in biological systems and rapidly undergo bonding interactions to form complexes or chelates in biological systems. Calculated amounts of ionic Cu (10^{-18} M), Fe (10^{-23} M), Mn (10^{-12} M), and Zn (10^{-9} M) in plasma are extremely small and not measurable with any existing equipment (Sorenson, 1989) and they are likely to be still smaller in solid tissues which contain many more bonding sites. Consequently, measured tissue Cu, Fe, Mn, and Zn contents reflect tissue contents of chelates which are primarily metalloelement-dependent enzymes (Table 1); proteins such as hemoglobin, ferritin, metallothioneines, β_2-macroglobulins, transferrin, transmanganin, and transcuprein; RNA and DNA chelates; and small molecular mass amino acid, carboxylic acid, phosphate, amine, diamine, and thiol chelates, as well as various small molecular mass peptide chelates (May and Williams, 1977). To study effects of essential metalloelements in a biological system it is not appropriate to use ionic forms, but it is appropriate to use relevant chelates, which have biological activity.

II. ESSENTIAL METALLOELEMENTS AND IMMUNITY

Ingested foods and beverages contain these essential metalloelements in chelated forms which may yield other chelates as a result of ligand exchange in the digest or following absorption (Sorenson, 1989). However, some of the absorbed chelates may exist in the digest and in blood as the original chelate or a more stable ternary chelate. Absorbed metalloelement chelates undergo systemic circulation to all tissues and utilization by all cells following ligand exchange with small molecular mass ligands, apoproteins, and apoenzymes to form metalloproteins and metalloenzymes in *de novo* syntheses. These metalloelements are stored in the appropriate tissues as metallothioneines (MTs) or Fe-ferritin, or they are excreted in the event tissue needs have been met and stores replenished (Sorenson, 1989). There are

Table 1 Copper-, Iron-, Manganese-, and Zinc-Dependent Mammalian Enzymes

Metal-dependent enzymes	Function
Copper-dependent enzymes	
Cytochrome c oxidase	Reduction of oxygen
Cu-Zn-superoxide dismutases (Cu_2Zn_2SOD)	Superoxide disproportionation
Tyrosinase	Synthesis of dihydroxyphenylalanine
Dopamine-β-monooxygenase	Synthesis of norepinephrine and epinephrine
Neurocuprein	Synthesis of norepinephrine and epinephrine
Amine oxidases	Metabolism of primary amines
Factor V	Blood clotting
Ceruloplasmin	Cu transport, mobilization of Fe, angiogenesis
α-Amidating Monooxygenases	Synthesis of neuroendocrine hormones
Procollagen proelastin, peptidyllysyl oxidases	Collagen and elastin cross-linking
Other possible copper-dependent enzymes	
Adenylate cyclase	Synthesis of c-AMP
Guanylate cyclase	Synthesis of c-GMP
Lipolytic protein	Lipolysis
ACE1 and CUP2	Metallothioneine gene regulatory proteins
Iron-dependent enzymes	
Electron transport cytochromes	Oxidation-reduction
Catalase	Disproportionation of hydrogen peroxide
P450	Activation of oxygen
Lipoxygenase	Conversion of arachadonic acid
Cyclo-oxygenase	Conversion of aracadonic acid to PGG_2
Xanthine oxidase	Purine metabolism
Peptidylprolyl and peptidyllysyl oxidases	Hydroxylation of procollagen and proelastin
Manganese-dependent enzymes	
Mn-Dependent superioxide dismutase	Disproportionation of superoxide
Arginase	Syntheses of urea and ornithine
Pyruvate carboxylase	Synthesis of oxaloacetate
Pseudocatalase	Disproportionation of H_2O_2
Calmodulin-dependent protein phosphatase	Phosphate ester hydrolysis
α-Isopropylmalate synthetase	Synthesis of α-isopropylmalate
A brain adenylate cyclase	Synthesis of c-AMP
Catechol-O-methyltransferase	Synthesis of catechol methyl ester
Farnesyl pyrophosphate synthetase	Synthesis of farnesyl pyrophosphate
Glycosyl synthetase	Synthesis of glycosyl-phosphonucleotides
Extradiol-cleaving dioxygenase	Insertion of dioxygen
A ganglioside galactosyl-transferase	Galactose transfer
Saccharide polymerases	Synthesis of polymeric carbohydrates
Glycosyltransferases	synthesis of glycosides
Galactosylhydroxylysl glucosyltransferase	Glycosaminoglycan and glycoprotein syntheses
Zinc-dependent enzymes	
Cu_2Zn_2SOD	Disproportionation of superoxide
Aminopeptidase	Protein hydrolysis
Aldehyde hydrase	Aldehyde hydration
Esterase	Ester hydrolysis
Methylmalonyl-oxaloacetate transcarboxylase	Transcarboxylation
Carboxypeptidases A and B	Protein hydrolysis
NAD-dependent dehydrogenases	Oxidations
Carbonic anhydrase	Dehydration of carbonic acid
α-Hydroxyacid dehydrogenase	Oxidation of α-hydroxy acids
Alkaline phosphatase	Phosphorylation
Purine and pyrimidine, nucleoside kinases	Phosphorylation of nucleosides
DNA polymerase, and gyrase	DNA synthesis
"Zinc-finger" proteins	Transcription-regulating proteins

no inducible excessive storage diseases known in normal individuals. However, there must be upper limits dictated by physiological requirements.

Stored essential metalloelements are released as chelates via ligand exchange to meet normal metabolic needs. This homeostatic release of relatively small amounts of essential metalloelement chelates meets normal physiologic requirements. Release of larger quantities in a pronounced mobilization of these metalloelements is a feature of interleukin-1 (IL-1)-mediated acute and chronic responses to many disease states (Sorenson, 1989 and cited references).

It is likely that the magnitude of these responses is adequate in normal well-nourished individuals. However, it is now generally recognized that dietary intakes of Cu, Fe, Zn (Klevay, 1990, Wright et al., 1991), and most likely Mn are less than the recommended daily intakes for the U.S. population and may be no better for individuals living in other developed countries. Rats maintained on a copper-deficient diet had reduced levels of serum ceruloplasmin (Davis et al., 1987; Kramer et al., 1988), a copper-containing acute phase reactant involved in scavenging oxygen radicals from activated neutrophils (Broadley and Hoover, 1989). Ceruloplasmin is induced by IL-1, but in copper-deficient rats, ceruloplasmin was defective in that it lacked oxidase activity (Barber and Cousins, 1988). In addition, copper deficiency has been associated with depressed T cell mitogenic responsiveness (Davis et al., 1987; Kramer et al., 1988) and T-dependent antibody induction (Blakley and Hamilton, 1987). T cell reactivity was shown to be compromised by a reduced capacity to synthesize IL-2, since added exogenous IL-2 restored responsiveness (Bala and Failla, 1992). Also reported to be impaired in copper-deficient animals is natural killer (NK) cell function (Koller et al., 1987). Reductions in NK cell activity, which is important in tumor immunity, may be a reflection of T cell function, since they are activated by the T cell cytokines, IL-2 or γ-interferon. Thymus mass is also reportedly reduced in copper-deficient rats (Koller et al., 1987), suggesting that T cells may be affected at all levels of differentiation.

While at high concentrations various metalloelements are known to be immunotoxic (Carpentieri et al., 1988), at lower concentrations or in chelated forms metalloelements can enhance immune responses, even in the absence of any deficiency. Copper chelates were reported to increase NK cell antitumor activity (Elo, 1987). Copper and iron at subtoxic concentrations synergistically increased human peripheral blood mononuclear cell responses to mitogens (Carpentieri, et al., 1988) and, in the presence of endotoxin, copper and zinc increased monocyte production of the cytokine, tumor necrosis factor-α (TNF-α) (Scuderi, 1990). Studying the copper chelate, $Cu_2(II)$ 3,5-diisopropylsalicylate$_4$ [$Cu(II)_2$(3,5-DIPS)$_4$], we have found that 80 mg/kg injected subcutaneously into normal, well-nourished mice enhanced T-dependent antibody induction (Figure 1). Specific antibody induction was increased by 80% 7 days after treatment with $Cu(II)_2$(3,5-DIPS)$_4$. The enhancement subsequently decreased, but was still elevated 24 days after treatment. T cell responses to the mitogen, concanavalin A (Con A), were not affected by treatment with the copper compound. B cell responses, on the other hand, were increased by the copper compound by up to 60%.

III. HEMATOPOIETIC STIMULATION BY ESSENTIAL METALLOELEMENTS

Much more dramatically than its effects on lymphocyte function, $Cu(II)_2$(3,5-DIPS)$_4$ stimulated hematopoietic activity. Mice injected subcutaneously with this copper compound had increases of greater than 10-fold in granulocyte-macrophage progenitor cells (GM-CFU) in the spleen 24 days after treatment (Figure 2). Multipotent myeloid progenitor cells (multi-CFU) were also elevated by 5-fold in the spleen, reaching peak levels 7 d after treatment. Bone marrow hematopoietic activity was not stimulated by treatment with $Cu(II)_2$(3,5-DIPS)$_4$. Other investigators (Wieczorek et al., 1983a) have reported that treatment with copper complexes increased the number of pluripotent stem cells, CFU-S. Increased accessory cell production through hematopoietic stimulation could elevate immune responses.

IV. RADIATION PROTECTION AND RECOVERY WITH COPPER CHELATES

The modulation of immune and hematopoietic cell functions become dramatically important when some other influence, such as exposure to radiation, challenges those functions. Exposure of mice to 8.0 Gy radiation destroys 98 to 99% of lymphocyte and myeloid progenitor cell functions (Soderberg et al., 1987; Soderberg et al., 1988; Soderberg et al., 1990), and these mice die of hematopoietic failure. $Cu(II)_2$(3,5-DIPS)$_4$ treatment of these mice did not prevent the hematopoietic and immunologic damage produced by the radiation. However, $Cu(II)_2$(3,5-DIPS)$_4$ did accelerate the repopulating of bone marrow and spleen following irradiation. This was demonstrated as increased cellularity and the earlier recovery of myeloid progenitor cells, both GM-CFU and Multi-CFU (Soderberg et al., 1988; Soderberg et al.,

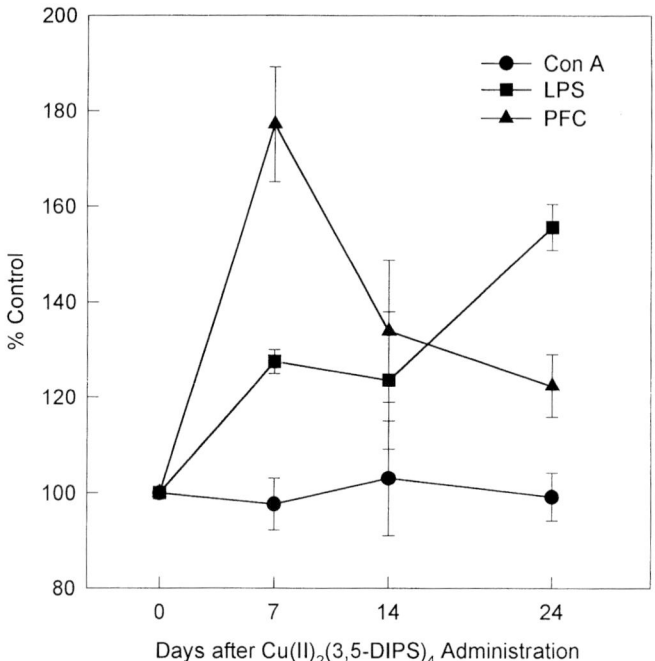

Figure 1. Groups of 5 normal, female C57BL/6 mice were injected with 0 or 80 μmol/kg Cu(II)$_2$(3,5-DIPS)$_4$. After 7, 14, or 24 days, spleen cell suspensions were prepared and assayed for proliferative responses to Con A or LPS. For antibody induction, mice were injected intraperitoneally with sheep red blood cells 5 d prior to a standard plaque-forming assay (PFC). Results are reported as the mean percent of control responses ± SE.

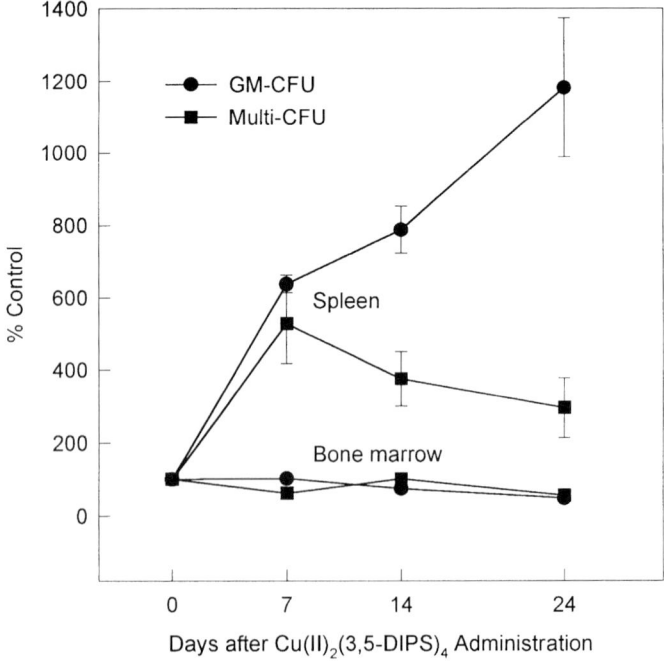

Figure 2. Spleen and bone marrow cell suspension prepared from mice treated as in Figure 1 were cultured for 7 days in the presence of GM-CSF or IL-3. Colonies of 50 or more cells were counted using an inverted microscope.

1990). By day 24 after irradiation, mice treated with vehicle had recovered 25% of normal levels of splenic Multi-CFU and 10% of GM-CFU. Mice treated with $Cu(II)_2(3,5\text{-DIPS})_4$ had much higher levels of splenic myeloid progenitor cells, 150% of normal levels of Multi-CFU, and nearly 300% of normal levels of GM-CFU. By 42 days after irradiation, mice treated with the copper complex had recovered to 100% of normal Multi-CFU and 70% GM-CFU, while vehicle-treated mice had recovered only about 10% of each progenitor cell type. The recovery of hematopoietic activity was accelerated whether mice were treated with the copper compound 3 h before irradiation or 3 h after irradiation. Mice treated with $Cu(II)_2(3,5\text{-DIPS})_4$ also had a 6-fold increased level of pluripotent stem cells, endogenous CFU-S, which can generate both myeloid and lymphoid progenitor cells. Consistent with this, mice treated with this compound recovered responses to T and B cell mitogens earlier than control irradiated mice and generated uniformly better antibody responses to a T-dependent antigen (Soderberg et al., 1987). By 24 days after irradiation, mice treated with the copper compound had recovered approximately 30% of normal T cell responses to the mitogen, Con A, and 30% of B cell responses to LPS, while vehicle-treated mice could only manage 2% and 6% of normal T and B cell responses, respectively. Also at 24 days, T-dependent antibody responses of mice treated with $Cu(II)_2(3,5\text{-DIPS})_4$ had recovered to 35% of normal, unirradiated levels, while vehicle-treated mice could produce only 5% of control antibody responses.

In addition to hematopoietic recovery, mice treated with the copper complex recovered normal histological characteristics more rapidly. Histopathological studies of spleen, bone marrow, thymus, and small intestine were conducted in parallel with immunological studies of animals exposed to $LD_{50/30}$ irradiation (8.0 Gy, 1.55 Gy/min) alone, 40 μmol/kg $Cu(II)_2(3,5\text{-DIPS})_4$ alone, irradiation and a single dose of vehicle, or $Cu(II)_2(3,5\text{-DIPS})_4$ treatment. The protective effects of $Cu(II)_2(3,5\text{-DIPS})_4$ treatment against irradiation-induced histopathology were apparent in these tissues. These effects can be best exemplified by changes that occurred in the spleen. After 24 h following irradiation, total disintegration of splenic follicles (white pulp) and total depletion of lymphocytes were observed in animals unprotected by $Cu(II)_2(3,5\text{-DIPS})_4$ (Figure 3A and 3B). Severe atrophy and fibrosis of this organ was observed on the 7th (Figure 4) and 14th (Figure 5) day, respectively, after irradiation. Both reduction of tissue damage and enhancement of tissue recovery were evident in animals treated with $Cu(II)_2(3,5\text{-DIPS})_4$. While there were still signs of splenic follicle disintegration 24 h after irradiation, many lymphocytes in these follicles were preserved (Figure 6). Signs of recovery were also demonstrable by the 7th day after irradiation in $Cu(II)_2(3,5\text{-DIPS})_4$-treated mice. A definitive recovery in splenic size and a gradual increase in splenic cellularity was evident by the 14th day after irridiation (Figure 7).

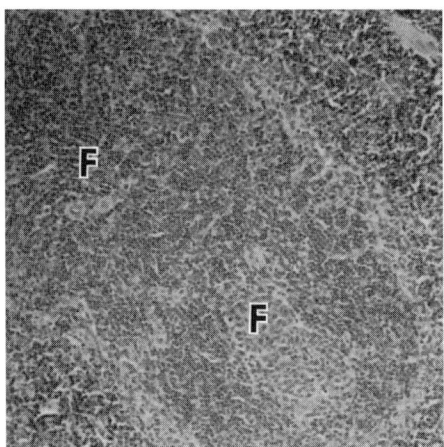 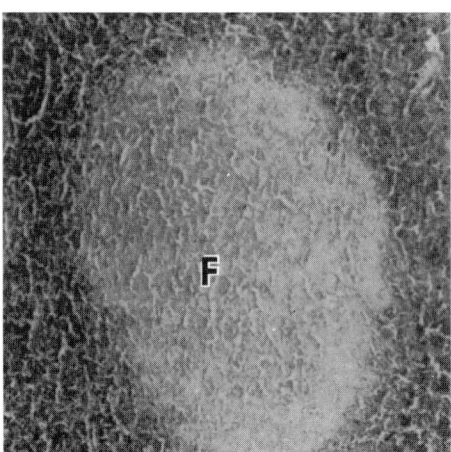

Figure 3. (Left) Spleen mouse, normal. Note the abundance of cellularity (lymphocytes) in the follicles (F), original magnification ×400. (Right) Spleen, mouse, 24 h after irradiation. Note total depletion of cells (lymphocytes) in the follicle (F). Original magnification ×400. (From Sorenson, J.R.J., Soderberg, L.S.F., and Chang, L.W., Proc. Soc. Exp. Biol. Med., in press. With permission.)

Similar reductions in tissue damage and enhancement of tissue repair were observed in bone marrow, thymus, and intestine from irradiated animals treated with $Cu(II)_2(3,5\text{-DIPS})_4$. These observations lead to the general conclusion that although $Cu(II)_2(3,5\text{-DIPS})_4$ does not prevent tissue injury when it is given

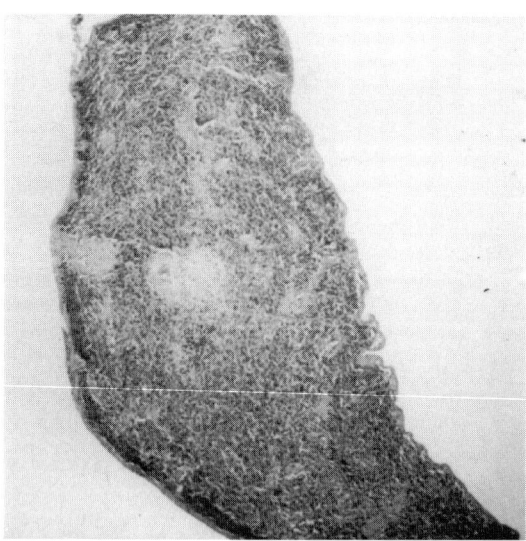

Figure 4. Spleen, mouse, 7 days after irradiation. Severe atrophy (reduction in size) of the entire spleen was noted. Original magnification ×400. (From Sorenson, J.R.J., Soderberg, L.S.F., and Chang, L.W., *Proc. Soc. Exp. Biol. Med.*, in press. With permission.)

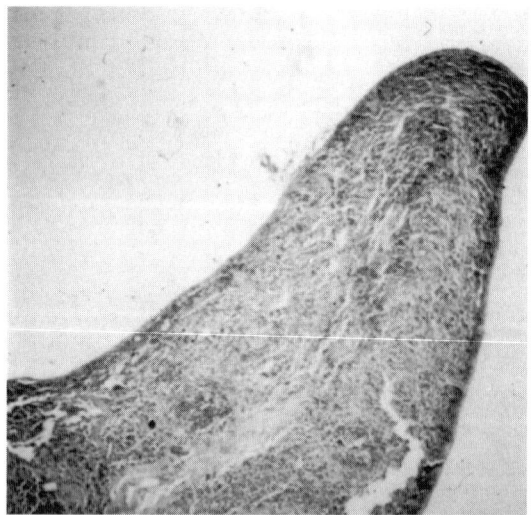

Figure 5. Spleen, mouse, 14 days after irradiation. The atrophied spleen became fibrotic with little cellularity. Original magnification ×400. (From Sorenson, J.R.J., Soderberg, L.S.F., and Chang, L.W., *Proc. Soc. Exp. Biol. Med.*, in press. With permission.)

before irradiation, it reduces tissue damage as well as facilitates rapid tissue recovery from radiation-induced injuries. These recoveries from radiation-induced histopathology offer a biological marker for the reestablishment of the integrity of those tissues which are known to be most vulnerable to irradiation. It also serves as a basis for the increase in survival and recovery of immunological function following $Cu(II)_2(3,5\text{-DIPS})_4$ treatment.

Enhanced recovery of hematopoietic and immune activities is likely responsible for the increased survival of irradiated mice treated with $Cu(II)_2(3,5\text{-DIPS})_4$. This SOD-mimetic and lipophilic complex was found to produce 58% survival in $LD_{100/30}$-irradiated (10 Gy, 0.4 Gy/min) mice (Sorenson, 1984), when $Cu(II)_2(3,5\text{-DIPS})_4$ was given subcutaneously at a dose of 80 µmol/kg of body mass 24 h before irradiation (Sorenson et al., 1987). These results have been extended by Steel et al. (1988) with the report that 25 or 50 µmol/kg $Cu(II)_2(3,5\text{-DIPS})_4$ given subcutaneously is effective in increasing survival of $LD_{100/30}$ irradiated mice.

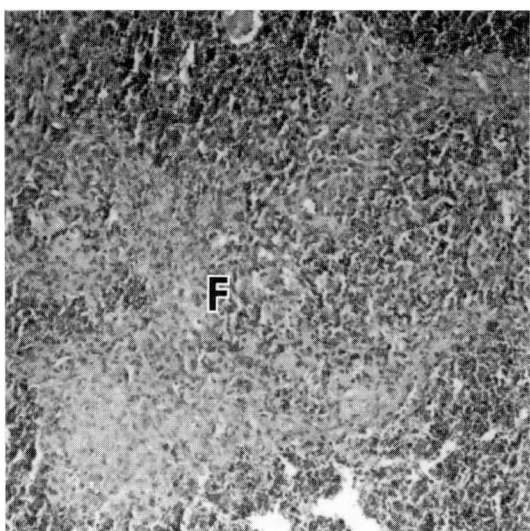

Figure 6. Spleen, Cu(II)$_2$(3,5-DIPS)$_4$-treated mouse, 24 h after irradiation. While a significant loss of lymphocytes in the follicle (F) was seen, it was also evident that many lymphocytes were still preserved (compare with Figure 3B). Original magnification ×400. (From Sorenson, J.R.J., Soderberg, L.S.F., and Chang, L.W., *Proc. Soc. Exp. Biol. Med.,* in press. With permission.)

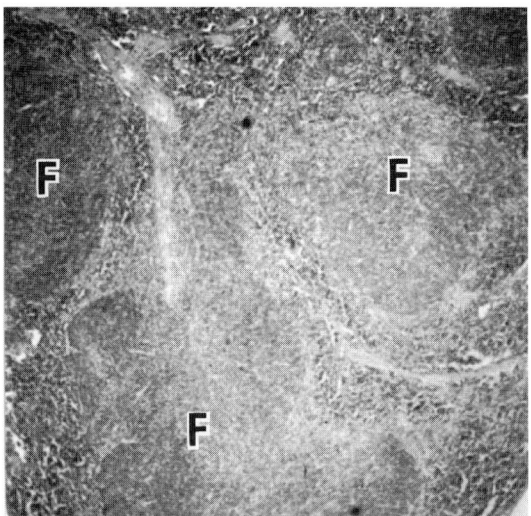

Figure 7. Spleen, Cu(II)$_2$(3,5-DIPS)$_4$-treated mouse, 14 days after irradiation. The spleen returned to almost full size with formation of many new follicles (F). Original magnification ×400 (From Sorenson, J.R.J., Soderberg, L.S.F., and Chang, L.W., *Proc. Soc. Exp. Biol. Med.,* in press. With permission.)

It was subsequently determined that a 20 μmol/kg dose of Cu(II)$_2$(3,5-DIPS)$_4$ was as effective as 80 μmol/kg when it was given subcutaneously 3 h before an LD$_{50/30}$-irradiation (8 Gy, 1.55 Gy/min) producing 92% or 80% survival in male or female mice, respectively (Sorenson et al., 1990). Subcutaneous treatment with much smaller doses of 2.5 or 5.0 μmol/kg 3 h after irradiation increased survival from 48% for vehicle-treated mice to 88% or 92%, respectively. These results have been confirmed and extended by Suriyachan et al. (1989), with the report that Cu(II)$_2$(3,5-DIPS)$_4$ given intraperitoneally at a dose of 10 μmol/kg either before or after irradiation (8 Gy, 0.51 Gy/min) produced 60% or 75% survival, respectively, in male rats vs. 40% for vehicle-treated rats. Male mice treated orally with 25 or 50 μmol/kg either 24 or 4 h before irradiation had increased survivals of 75% or 95%, respectively, compared to 50% survival for the vehicle-treated group (Sorenson et al., 1990; Sorenson et al., 1993).

All doses of $Cu(II)_2(3,5\text{-DIPS})_4$ used in these studies were nontoxic and ranged from 1/5 to 1/1000 of acutely toxic dose for male or female mice (Sorenson et al., 1990).

Steel et al. (1988) found that 24 to 48 µmol/kg doses of $Cu(II)_2(3,5\text{-DIPS})_4$ and 49 to 96 µmol/kg doses of $Cu(II)_2(3,5\text{-DIPS})_4$ given subcutaneously 24 h prior to irradiation were radioprotective in mice irradiated with 10 Gy (0.4 Gy/min); however, doses of 133 to 901 µmol/kg 3,5-DIPS acid produced no radiation protection, which has since been confirmed (Sorenson et al., 1993). However, treatment with 10 or 20 µmol/kg $Cu(II)_2(3,5\text{-DIPS})_4$ after $LD_{50/30}$ irradiation revealed that $Cu(II)_2(3,5\text{-DIPS})_4$ was more effective than $Cu(II)Cl_2$. The two doses of $Cu(II)Cl_2$ only increased survival 20% or 29% above vehicle-treated control female mice (56% survival), while the two copper equivalent doses of $Cu(II)_2(3,5\text{-DIPS})_4$ increased survival 75% or 88% above vehicle-treated control mice (32% survival) (Irving et al., in press). It is most likely that Cu chelates formed *in vivo* following the administration of $Cu(II)Cl_2$ ultimately account for its radioprotectant effects. However, use of chelates avoids the irritation caused by the injection of $Cu(II)Cl_2$ or $Cu(II)SO_4$, which are strong Lewis acids. Copper chelates are less acutely toxic than these inorganic forms of Cu, and the use of chelates also enables tissue targeting with regard to essential metalloelement distribution.

Immunological properties of the 3-morpholine-2-hydroxypropyl ether of dextran (Wieczorek et al., 1983b) led to the discovery of superior immunostimulant properties of the Cu chelate of the 3-mercapto-2-hydroxypropyl ether of dextran (Wieczorek et al., 1983c), which afforded nearly 100% survival of mice infected with *Shigella sonnei* dysentery bacilli (Wieczorek et al., 1983d), attenuated tumor growth and increased survival of sarcoma 180-implanted mice, suppressed spontaneous leukemia and prolonged survival of AKR-mice (Wieczorek et al., 1983e), and has some radioprotectant activity in 6 Gy-irradiated 129 Ao/Boy female mice (Wieczorek et al., 1983f).

$Cu(II)_2(3,5\text{-DIPS})_4$ was originally examined for radioprotectant activity based upon reports that Cu_2Zn_2SOD had radioprotectant activity (Petkau, 1978). However, two recent reports suggest that superoxide (O_2^-) disproportionation by Cu_2Zn_2SOD does not account for its radioprotectant activity. Inactivated polyethylene-glycol derivatized Cu_2Zn_2SOD was as effective as the enzymatically active derivative (Westland and Marklund, 1987). This result and the possibility that these proteins are in fact antigenic (Vaille et al., 1990), although reduced antigenicity is always claimed, may lead to phagocyte elaborated IL-1 and/or some other cytokine that initiates immune and repair responses that account for increased survival. Another report by Scott et al. (1989) demonstrated that increased radiation sensitivity of three *E. coli* clones was related to increased SOD activity. However, radiation sensitivity in these clones may actually be due to O_2^- produced and allowed to accumulate in these clones following radiation-induced loss of Fe from this Fe-dependent enzyme and production of inactive apoenzyme. If differing concentrations of FeSOD in these clones are due to relative need to remove O_2^- produced under normal metabolic conditions, then more O_2^- is produced by the clone having the highest concentration of FeSOD, and this clone is most labile to radiation-induced death with radiolysis of the coordinate-covalent bonded Fe in this enzyme.

The hypothesis that there is radiolytic loss of essential metalloelement cofactor has merit and accounts for the 20% loss of both Cu-dependent and Mn-dependent SODs in intestinal smooth muscle of rats immediately (1 h) following irradiation (15 Gy) (Summers et al., 1989). Activities of both enzymes increased over a 20-h period to a level 45% higher than normal with a further increase in Mn SOD by 72 h accounting for the further increase in total SOD activity, a likely result of *de novo* synthesis in response to injury.

Another possible mechanism accounting for the radioprotectant/radiorecovery activity of Cu chelates involves Cu-dependent nuclear metabolism (Agarwal et al., 1989). It is known that 20% of cellular Cu is found in the nucleus. Copper has the highest bonding affinity for nucleic acids of all of the essential metalloelements. Nucleic acid phosphate and base donor sites yield Cu chelates which are much more stable than Cu chelates of most amino acids and near the stability of protein chelates. This stability has been used to support suggestions of intrastrand and interstrand DNA bridging. Higher physiologic concentrations account for inhibition of ribonuclease syntheses: initiation, nucleotide condensations, and release of newly formed mRNA. However, the original observation that injection of $Cu(II)Cl_2$ into rats induced the synthesis of mRNA coding for the synthesis of MT has been explained as due to the formation of a Cu-dependent DNA-binding protein ACE1 or CUP2, the protein products of *S. cerevisiae* (*ACE1*) and *E. coli* (*CUP2*) genes, which recognize upstream activation sequences required for active transcription of *CUP1*, a MT gene (Ecker et al., 1989; O'Halloran, 1989; Buchman et al., 1989; Buchman et al., 1990). Since radiation injury and injection of Cu chelates induce MT synthesis, and increased MT synthesis correlates with increased survival, Cu-dependent transcription and translation are important nuclear responses in overcoming radiation injury.

Spassky and Sigman (1985) recognized nuclease-mimetic activity for the Cu(I)(o-phenanthroline)$_2$ chelate as a valuable footprinting reagent useful in revealing details of RNA polymerase bonding to the *lac* promoter. Predictable chromatin DNA cleavage by this chelate due to preferential bonding of the chelate to DNA spacer segments was also suggested to be a useful characterizing feature (Skalka and Cejkova, 1985). Interpretations that these and other DNA strand scissions resulting from Cu(I) chelates are damaging (Swauger et al., 1991; Reed and Douglas, 1991) without recognizing that these scissions may be due to expected Cu chelate (Modak et al., 1991) or other essential metalloelement chelate (Basile and Barton, 1989) catalyzed diphosphate ester hydrolyses which may be viewed as endonuclease- (or exonuclease-) mimetic activity merit reconsideration.

Two genes, *SoxR* and *SoxS,* are involved in activating the *soxR* regulon gene (Demple, 1990, Greenberg et al., 1991, Ramotar et al., 1991, Amobile-Cuevas and Demple, 1991). The predicted SoxR and SoxS proteins may constitute a novel two-compartment regulatory system in which these two proteins act sequentially to activate transcription of various regulon genes including *soxR* in response to O_2^- stress. Two soxR regulon proteins, endonuclease IV (*E. coli*) and Apn1 (*S. cerevisiae*), have recently been suggested to be metalloelement-dependent endonucleases (Levin et al., 1991). Both proteins were found to contain various essential metalloelements including: Cu, Fe, Mn, and Zn. The order of abundance in endonuclease IV is Zn > Mn > Cu > Fe, and in Apn1 it is Zn > Fe > Mn > Cu. Both proteins are active in removal of base- or sugar-damaged nucleotides, the penultimate steps in repair of damaged DNA.

V. RADIATION PROTECTION AND RECOVERY WITH OTHER ESSENTIAL METALLOELEMENT CHELATES

Other essential metalloelement chelates proved to have similar radioprotective activity. The iron chelate, Fe(III)-(3,5-DIPS)$_3$, increased the survival of irradiated mice from 50% for control LD$_{50/30}$ irradiated mice up to 84%. Mn(II)Cl$_2$ given 24 h prior to 6.3 Gy irradiation produced 94% survival (Matsubara et al., 1987). This dose also increased liver MT synthesis 6-fold above nontreated and nonirradiated control values demonstrating radiation-induced MT synthesis in association with host radioresistance, consistent with an earlier report by Shiraishi et al. (1983) of radiation-induced MT synthesis in rat liver. Manganese(II)Cl$_2$ treatment was also reported to *prevent* the 90% decrease in leukocytes observed for control mice with 6.3 Gy irradiation; however the decrease in leukocyte count following 8 Gy was not prevented. Manganese(II)Cl$_2$-treated 8 Gy-irradiated mice had leukocyte counts which were 19% of those found for Mn-treated nonirradiated mice whereas nontreated irradiated controls had leukocyte counts of only 10% of normal. Leukocyte counts also recovered more rapidly in Mn-treated mice than in nontreated irradiated controls. The Mn chelate, Mn(III)$_2$(II)(O)(3,5-DIPS)$_6$ when given at a dose of 80 µmol/kg 3 h before irradiation increased the survival of irratiated mice from 50% up to 100% (Sorenson et al., 1990; Sorenson et al., 1993). It was also found that Zn(II)$_2$(acetate)$_4$ given 24 h before an LD$_{50/30}$ dose of radiation produced 81% survival, respectively, vs. 25% for control mice given saline (Matsubara et al., 1987). Subcutaneous injection of 37 µmol/kg of ^{65}Zn(II)Cl$_2$ 1 and 3 days before LD$_{30/30}$ irradiation produced 100% survival (Matsubara et al., 1986). Floersheim and Floersheim (1986) reported that Zn(II)(D,L-aspartate)$_2$ given 10, 30, or 120 min before a 9 Gy irradiation produced 88%, 80%, or 90% survival, respectively, while treatment before the 12 Gy irradiation produced 3%, 57%, or 35% survival. Treatment with Zn(II)Cl$_2$, Zn(II)SO$_4$(H$_2$O)$_7$, Cu(II)D,L-aspartate, or Na aspartate was either less effective or ineffective. Treatment with Cu$_2$Zn$_2$ SOD intraperitoneally 10 min before irradiation failed to increase survival. Treatment with 60 µM Zn(II)$_2$(3,5-DIPS)$_4$/kg produced 95% survival in LD$_{50/30}$ irradiated male mice (Steele et al., 1988; Suriyachan et al., 1989).

Recognizing that loss of essential metalloelement-dependent enzyme activity may at least partially account for lethality of ionizing radiation and that Cu-, Fe-, Mn-, and Zn-dependent enzymes have roles in protecting against accumulation of O_2^- as well as facilitating repair (Sorenson, 1978) may explain the radiation protection and radiation recovery activities of Cu, Fe, Mn, and Zn compounds. It is suggested that the IL-1 mediated redistribution of essential metalloelements have a general role in responding to radiation-injury. These redistributions may also account for subsequent *de novo* synthesis of metalloelement-dependent enzymes required for biochemical repair and replacement of cellular and extracellular components needed for recovery from radiolytic damage.

De novo syntheses of metalloelement-dependent enzymes required for utilization of oxygen and prevention of O_2^- accumulation as well as tissue repair processes including metalloelement-dependent

DNA and RNA repair are key to the hypothesis that essential metalloelement chelates decrease and/or facilitate recovery from radiation-induced pathology. A widely held understanding is that O_2^- accumulation due to the lack of normal concentrations of CuSODs, reduction of oxygen leading to relatively large steady-state concentrations of O_2^-, or inappropriate release of O_2^- from oxygen-activating centers lead to the formation of much more reactive oxygen radicals. These more reactive oxygen radicals include: singlet oxygen and hydrogen peroxide produced as a result of acid-dependent self disproportionation or water-catalyzed disproportionation of O_2^-, hydroxyl radical, and hydroperoxyl radical (Fee and Valentine, 1977; Corey et al., 1987). Facilitated syntheses of CuSODs and catalase and prevention of the formation of these oxygen radicals may account for some of the recovery from pathological changes associated with irradiation.

VI. CONCLUSIONS

While high concentrations of essential metalloelement salts may be toxic for the mammalian immune system, lower concentrations or complexes have been shown to stimulate immune responses. These complexes, notably complexes of copper, iron, manganese, and zinc have radioprotectant and radiorecovery activities. Increased survival following lethal irradiation generally represents improved hematopoietic recovery, regenerating both myeloid and lymphoid lineages of cells. The increased survival produced by these compounds may be due to hematopoietic stimulation and functional regeneration of metalloelement-dependent enzymes, including those required for DNA repair mechanisms. Additional work will be required to determine which mechanisms are, in fact, most important in immune reactivity.

REFERENCES

Agarwal, K., Sharma, A., and Talukder, G. (1989), *Chem.-Biol. Interact.,* 69, 1–16.
Amobile-Cuevas, C. F. and Demple, B. (1991), *Nucleic Acids Res.,* 19, 4479–4484.
Bala, S. and Failla, M. L. (1992), *Proc. Natl. Acad. Sci. U.S.A.,* 89, 6794–6797.
Barber, E. F. and Cousins, R. J. (1988), *J. Nutr.,* 118, 375–381.
Basile, L. A. and Barton, J. K. (1989), Metallonucleases: real and artificial, in Sigel, H. and Sigel, A., Eds., *Metal Ions in Biological Systems,* Marcel Dekker, New York, 31.
Blakley, B. R. and Hamilton, D. L. (1987), *Drug-Nutr. Interact.,* 5, 103–111.
Broadley, C. and Hoover, R. L. (1989), *Am. J. Pathol.,* 135, 647–655.
Buchman, C., Skroch, P., Dixon, W., Tullius, T. D., and Karin, M. (1990), *Mol. Cell Biol.,* 10, 4778–4787.
Buchman, C., Skroch, P., Welch, J., Fogel, S., and Karin, M. (1989), *Mol. Cell. Biol.,* 9, 4091–4095.
Carpertieri, U., Myers, J., Deaschner, C. W. I., and Haggard, M. E. (1988), in *Biological Trace Element Research,* Humana Press, Clifton, NJ, 165–176.
Corey, E. J., Mehrota, M., and Khan, A. U. (1987), *Biochem. Biophys. Res. Commun.,* 145, 842–846.
Davis, M. A., Johnson, W. T., and Briske-Anderson, M. (1987), *Nutr. Res.,* 7, 211–222.
Demple, B. (1990), in *UCLA Symp. Mol. Cell. Biol. — New Series Ionizing Radiation Damage to DNA: Molecular Aspects,* Vol. 136, New York, Wiley-Liss, 297.
Ecker, D. J., Butt, T. R., and Crooke, S. T. (1989), in *Metal Ions in Biological Systems,* Sigel, H. and Sigel, A., Eds., Marcel Dekker, New York, 147.
Elo, H. (1987), *Cancer Lett.,* 36, 333–339.
Fee, J. A. and Valentine, J. S. (1977), in *Superoxide and Superoxide Dismutases.* Michelson, A. M., McCord, J. M., and Fridovich, I., Eds., Academic Press, New York, 19.
Floersheim, G. L. and Floersheim, P. (1986), *Br. J. Radiobiol.,* 59, 597–602.
Greenberg, J. T., Chow, J. H., Monach, P. A., and Demple, B. (1991), *J. Bacteriol.,* 173, 4433–4439.
Irving, H. J., Henderson, T. D., Henderson, R. D., Williams, E. L., Willingham, E. L., and Sorenson, J. R. J. (1995), *Inflammopharmacology,* 3 in press.
Iyengar, G. V., Kollmer, W. E., and Bowen, J. M. (1978), *The Elemental Composition of Human Tissues and Body Fluids,* Springer, New York.
Klevay, L. M. (1990), in *Role of Copper in Lipid Metabolism,* Lei, K. Y. and Carr, T. P., Eds, CRC Press, Boca Raton, FL, 233.
Koller, L. D., Mulhern, S. A., Frankel, N. C., Steven, M. G., and Williams, J. R. (1987), *Am. J. Clin. Nutr.,* 45, 997–1006.
Kramer, T. R., Johnson, W. T., and Briske-Anderson, M. (1988), *J. Nutr.,* 118, 214–221.
Levin, J., Shapiro, R., and Demple, B. (1991), *J. Biol. Chem.,* 266, 2893–2898.
Matsubara, J., Shida, T., Ishioka, K., Egawa, S., Inada, T., and Machida, K. (1986), *Environ. Res.,* 41, 558–567.
Matsubara, J., Tajima, Y., and Karasawa, M. (1987), *Environ. Res.,* 43, 66–74.
May, P. M. and Williams, D. R. (1977), *FEBS Lett.,* 78, 134–138.
Modak, A. S., Gard, J. K., Merriman, M. C., Winkler, K. A., Bashkin, J. K., and Stern, M. K. (1991), *J. Am. Chem. Soc.,* 113, 283–291.

O'Halloran, T. V. (1989), in *Metal Ions in Biological Systems,* Sigel, H. and Sigel, A., Eds., Marcel Dekker, New York, 105. p 355.
Petkau, A. (1978), *Photochem. Photobiol.,* 28, 765–774.
Ramotar, D., Popoff, S. C., Gralla, E. B., and Demple, B. (1991), *Mol. Cell. Biol.,* 11, 4537–4544.
Reed, C. J. and Douglas, K. T. (1991), *Biochem. J.,* 275, 601–608.
Scott, M. D., Meschnick, S. R., and Eaton, J. W. (1989), *J. Biol. Chem.,* 264, 2498–2501.
Scuderi, P. (1990), *Cell. Immunol.,* 126, 391–405.
Skalka, M. and Cejkova, M. (1985), *Mol. Biol. Rep.,* 10, 163–168.
Soderberg, L. S. F., Barnett, J. B., Baker, M. L., Salari, H., and Sorenson, J. R. J. (1987), *Scand. J. Immunol.,* 26, 495–502.
Soderberg, L. S. F., Barnett, J. B., Baker, M. L., Salari, H., and Sorenson, J. R. J. (1988), *Exp. Hematol.,* 16, 577–580.
Soderberg, L. S. F., Barnett, J. B., Baker, M. L., Salari, H., and Sorenson, J. R. J. (1990), *Exp. Hematol.,* 18, 801–805.
Sorenson, J. R. J. (1978), *Inorg. Perspect. Biol. Med.,* 2, 1–26.
Sorenson, J. R. J. (1984), *J. Med. Chem.,* 27, 1747–1749.
Sorenson, J. R. J. (1989), *Prog. Med. Chem.,* 26, 437–568.
Sorenson, J. R. J., Soderberg, L. S. F., Baker, M. L., Barnett, J. B., Chang, L. W., Salari, H., and Willingham, W. M. (1990), in *Antioxidants in Therapy and Preventive Medicine,* Emerit, I., Auclair, C., and Packer, L., Eds., Plenum Press, New York, Vol 264: 69.
Sorenson, J. R. J., Soderberg, L. S. F., Barnett, J. B., Baker, M. L., Salari, H., and Bond, K. (1987), *Rec. Trav. Chim.,* 106, 391.
Sorenson, J. R. J., Soderberg, L. S. F., Chang, L. W., Willingham, W. M., Baker, M. L., Barnett, J. B., Salari, H., and Bond, K. (1993), *Eur. J. Med. Chem.,* 28, 221–229.
Sorenson, J. R. J., Soderberg, L. S. F., and Chang, L. W. (1995), *Proc. Soc. Exp. Biol. Med.,* in press.
Spassky, A. and Sigman, D. S. (1985), *Biochemistry,* 24, 8050–8056.
Steel, L., Seneviratne, S., and Jackson, W. E., III (1988), in *Anticarcinogenesis and Radiation Protection,* Cerutti, P., Nygaard, O. F., and Simic, M. G., Eds., Plenum Press, New York.
Summers, R. W., Maves, B. V., Reeves, R. D., Arjes, L. J., and Oberley, L. W. (1989), *Free Rad. Biol. Med.,* 6, 261–270.
Suriyachan, D., Patarakitvanit, S., and Ratananakorn, P. (1989), *Royal Thai Army Med. J.,* 42, 145–148.
Swauger, J. E., Dolan, P. M., Zeier, J. L., Kuppusamy, P., and Kensler, T. W. (1991), *Chem. Res. Toxicol.,* 4, 223–228.
Tipton, I. H. and Cook. M. J. (1963), *Health Phys.,* 1, 103–145.
Vaille, A., Jadot, G., and Elizagaray, A. (1990), *Biochem. Pharmacol.,* 39, 247–255.
Westman, N. G. and Marklund, S. L. (1987), *Acta. Oncol.,* 26, 483–487.
Wieczorek, Z., Gieldanowski, J., Hraba, T., Zimecki, M., and Mioduszewski, J. Z. (1983a), *Arch. Immunol. Ther. Exp.,* 31, 707–713.
Wieczorek, Z., Gieldanowski, J., Zimecki, M., Mioduszewski, J. Z., Szymaniec, S., and Daczynska, R. (1983b), *Arch. Immunol. Ther. Exp.,* 31, 715–729.
Wieczorek, Z., Gieldanowski, J., Zimecki, M., Mioduszewski, J. Z., Szymaniec, S., and Daczynska, R. (1983c), *Arch. Immunol. Ther. Exp.,* 31, 715–729.
Wieczorek, Z., Gieldanowski, J., Zimecki, M., Skibinski, G., and Mioduszewski, J. Z. (1983d) *Arch. Immunol. Ther. Exp.,* 31, 677–689.
Wieczorek, Z., Gieldanowski, J., Zimecki, M., Skibinski, G., and Mioduszewski, J. Z. (1983e), *Arch. Immunol. Ther. Exp.,* 31, 691–700.
Wieczorek, Z., Kowalewska, D., Gieldanowski, J, Zimecki, M, and Mioduszewski, J. Z. (1983f), *Arch. Immunol. Ther. Exp.,* 31, 701–705.
Wright, H. S., Guthrie, H. A., Wong, M.-Q., and Bernardo, V. (1990), *Nutr. Today,* 26, 21–27.

Section VIII
Effects of Metals on Other Organ Systems

Overview

An Introduction to Metal Effects on Other Organ Systems

Louis W. Chang

The predominant clinical syndromes in human populations suffering from heavy metal intoxications, such as those by mercury, lead, manganese, cadmium, etc., are usually related to either neural or renal dysfunctions. Because of this predominance of clinical syndromes, there is frequently a misconception that most metals are either "neurotoxic" or "renal toxic." Some may even refer to certain metals as "neurotoxic metals" or "renal toxic metals." The facts cannot be further from the truth. It must be emphasized that metals should be considered "systemic" toxicants. They have the potential to affect many other organs besides the brain and the kidney.

Because the predominance of effects (and therefore investigations) of metals is on the nervous system and on the kidneys, two full sections have been devoted to covering the toxicology of metals on these two organ systems. In this section, an effort is made to present, through the various chapters, the toxic effects of metals on other organ systems such as the hepatic, gastrointestinal, cardiovascular, respiratory, hematopoietic, and skeletal systems. Each of these chapters (systems) is presented by some of the most distinguished scientists in these special fields of metal toxicology: Nieminen and Lemasters (hepatic), Keogh and Siegers (gastrointestinal), Klevay (cardiovascular), Benson and Zelikoff (respiratory), Woods (hematopoietic), and Bhattacharyya et al. (skeletal).

Each chapter is devoted to providing the reader with information on specific organ toxicity of heavy metals in terms of metabolism, toxicity, pathology, and mechanistic elucidation. I am confident that our readers will find this section stimulating and informative.

Chapter 57

Hepatic Injury by Metal Accumulation*

Anna-Liisa Nieminen and John J. Lemasters

I. INTRODUCTION

The liver is the largest gland and, with the exception of the skin, it is the largest organ in the body. The adult human liver weighs about 1500 g, and a major function of the liver is to metabolize, transform, and store or excrete a wide variety of substances, especially those arising from the digestive tract. About 75% of blood comes to the liver from the intestinal tract via the portal vein. The remainder, 25%, is supplied by the hepatic artery. The somewhat deoxygenated blood from the intestinal tract and the well-oxygenated blood from hepatic artery mix together in the hepatic sinusoids.

Histologically, the liver consists of ill-defined lobules that lack separating connective tissue septae (Figure 1). Each lobule is formed of anastamosing plates of hepatocytes, one cell in thickness. These plates course from the periphery of the lobule to its center, forming a spongelike structure. The spaces between the hepatocyte plates are occupied by specialized capillaries, the liver sinusoids. Fenestrated endothelial cells line the sinusoids and permit free movement of solutes between the sinusoidal lumen and the subendothelial space of Disse. This exchange is physiologically important since many blood proteins, such as ceruloplasmin and transferrin, are synthesized by hepatocytes and released directly into the blood. Other macromolecules are taken up by hepatocytes, such as asialglycoprotein, low density lipoprotein, transferrin, and polymeric IgA. The sinusoidal lining also contains macrophages, called Kupffer cells. A third type of nonparenchymal cell is the fat-storing cell, also called a lipocyte or Ito cell. Fat-storing cells reside in the space of Disse between the sinusoidal lining and the hepatocytes. A main function of fat-storing cells is storage of vitamin A. In pathological states, such as in chronic iron and copper overload, fat-storing cells can produce collagen, leading to fibrosis and cirrhosis.

Another important function of hepatocytes is production of bile. Bile is secreted into bile canaliculi that are formed between adjacent hepatocytes. Individual bile canaliculi are connected in a complex network resembling chicken wire in appearance. Ultimately, the bile canaliculi drain into bile ducts located in portal spaces at the periphery of the lobules. There bile ducts coalesce to form a single bile duct leaving the liver and terminating in the duodenum. Portal spaces also contain branches of the portal vein and hepatic artery that bring blood into the liver. This blood moves through the sinusoids and is drained to the central vein at the center of each lobule. Central veins coalesce to form the hepatic veins that empty into the vena cava.

As blood moves through the lobules, hepatic metabolism causes oxygen extraction, uptake of hormones and nutrients, and release of urea, CO_2 and a variety of metabolic products into the sinusoidal blood (Figure 1). As a consequence, large gradients of oxygen and other metabolites are established

---

* This work was supported, in part, by Grants AG13318, AG07218, and DK37034 from the National Institutes of Health.

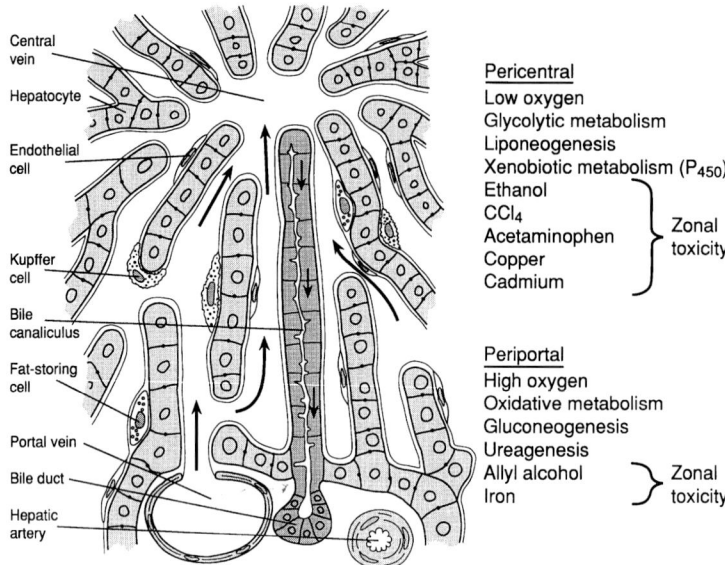

Figure 1. Diagram of the liver lobule. Blood empties from branches of the portal vein and hepatic artery and flows in sinusoids between plates of hepatocytes to the central vein. Sinusoids are lined by fenestrated endothelial cells and macrophages, called Kupffer cells. In the subendothelial space (of Disse) between the endothelial cells and the hepatocytes are scattered fat-storing (Ito) cells, mesenchymal cells capable of generating collagen fibrils. Bile forms inside canaliculi between individual hepatocytes, which empty into bile ducts. As blood flows through the liver lobule, gradients of oxygen and other metabolites are created as a consequence of tissue metabolism. Additionally, there are gradients in the distribution of enzymes within the lobule. Together, these gradients may account for the zonal specificity of hepatotoxins, including heavy metals.

between periportal regions, where blood first enters the lobules, and pericentral regions, where it is drained. In addition, gradients of hepatic enzymes exist across the liver lobule, notably in enzymes for drug metabolism, glycolysis, and lipogenesis, which are concentrated in pericentral regions, and enzymes for oxidative phosphorylation, gluconeogenesis, and ureagenesis, which are concentrated in periportal regions. Flow-induced metabolite gradients together with regional differences in enzyme content together account for the striking regional specificity of many hepatotoxins, including heavy metals. The heavy metals of toxicological interest to the liver are cadmium, mercury, copper, and iron.

II. CADMIUM

Cadmium is a toxic heavy metal used in galvanizing, as a pigment in paints and plastics, and as a cathode material in batteries. Airborne cadmium in urban environments is about 0.02 μg/m^3 (Kneip et al., 1970). Cadmium is also present in aquatic ecosystems. Shellfish, such as mussels, scallops, and oysters, and fish are a major source of dietary cadmium (Nriagu and Pacyna, 1988). Humans are also exposed to cadmium through cigarette smoking.

In animals, a single intravenous dose of 3 mg/kg cadmium causes hepatic failure. Plasma enzyme activities, such as aspartate aminotransferase and sorbitol dehydrogenase, rise quickly, and histology reveals dose-dependent cell swelling, cytoplasmic eosinophilia, pyknosis, karyorrhexis, and necrosis advancing from pericentral to periportal regions of the liver lobule (Goering and Klaassen, 1984b; Theocharis et al., 1991, 1994).

A. CADMIUM UPTAKE BY HEPATOCYTES

The main route of cadmium entry into the body is via respiration and gastrointestinal absorption, depending on the type of exposure. In the gastrointestinal tract, the primary site for cadmium absorption is the proximal part of the duodenum (Sorensen et al., 1993). In the body, cadmium accumulates in kidney and liver. The half-life of cadmium in man is long, as much as 30 years. The kinetics of hepatic cadmium uptake has been studied in isolated perfused livers and hepatocytes. Uptake is rapid and biphasic and does not require energy, since inhibition of respiration or uncoupling of mitochondria does not change uptake (Stacey and Klaassen, 1980). Cadmium uptake by the perfused liver occurs via carrier-mediated diffusion (Kingsley and Frazier, 1979). Calcium, copper, and zinc compete with cadmium uptake, suggesting a common uptake pathway (Blazka and Shaikh, 1992). Cadmium uptake is inhibited by SH-reactive agents, such as *N*-ethylmaleimide and parachloromercuribenzenesulfonate (Gerson and Shaikh, 1984). Thiol-dependent cadmium uptake appears to occur partly through a receptor-operated calcium-channel (Blazka and Shaikh, 1991).

B. INTRACELLULAR DISTRIBUTION OF CADMIUM IN HEPATOCYTES

Cadmium taken up by hepatocytes binds predominantly to metallothionein, a low molecular weight protein of 6.5 to 7.0 kDa (Figure 2). Metallothionein contains 61 amino acids of which 20 are cysteine residues and none are aromatic. Six isoforms of metallothionein have been isolated from human hepatic cytosol (Hunziker and Kagi, 1985). The structure of metallothionein is characterized by two domains. An α domain contains 11 cysteine residues and binds 4 atoms of zinc or cadmium or 5 or 6 atoms of copper. A β domain contains 9 cysteine residues and binds 3 atoms of zinc or cadmium or 6 atoms of copper. Due to the high affinity of metallothionein for cadmium, most intracellular cadmium is localized in cytosol in a bound form. Cadmium, copper, and zinc increase liver metallothionein content by inducing gene induction and synthesis of new protein, and induction of metallothionein synthesis before the acute exposure to cadmium markedly reduces liver necrosis in rats (Goering and Klaassen, 1984a,b; Hiratsuka et al., 1993). This tolerance is associated with a metallothionein-induced change in the hepatic subcellular distribution of cadmium. Increased binding of cadmium to metallothionein results in a smaller portion of cadmium binding to organelles, such as mitochondria, endoplasmic reticulum and nuclei (Figure 2) (Goering and Klaassen, 1983; Coogan et al., 1992). Thus, metallothionein plays an important role in protecting against cadmium-induced hepatic toxicity.

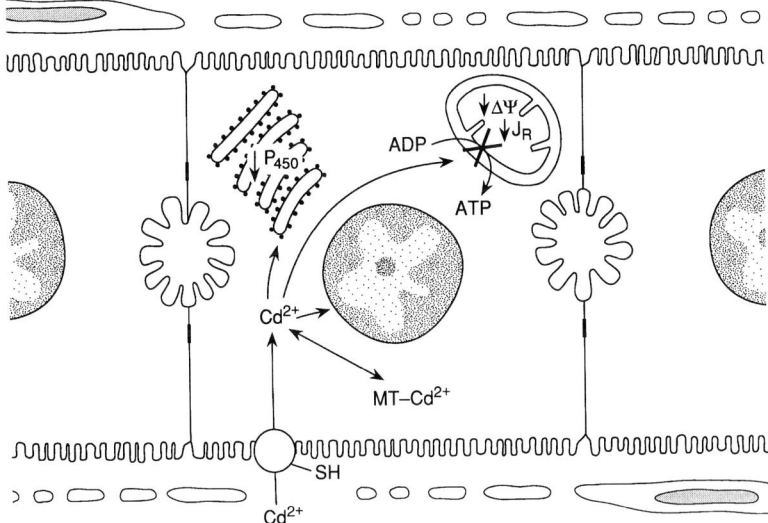

Figure 2. Uptake and intracellular distribution of cadmium in hepatocytes. Cadmium (Cd^{2+}) enters hepatocytes via a thiol-dependent transporter. Inside the cells, cadmium binds to metallothionein (MT). Cadmium also binds to mitochondria, leading to inhibition of respiration, collapse of the mitochondrial membrane potential ($\Delta\psi$), and ATP depletion. Cadmium also binds to endoplasmic reticulum, decreasing cytochrome P_{450} in microsomes, and to the nucleus, resulting in DNA degradation.

C. EFFECT OF CADMIUM ON ENERGY METABOLISM IN HEPATOCYTES

In hepatocytes, 10 to 15% of total cellular cadmium accumulates into mitochondria, binding to the mitochondrial inner membrane (Goering and Klaassen, 1983; Muller, 1986; Martel et al., 1990). This suggests that mitochondria may be a primary target for cadmium toxicity in hepatocytes. In support of this concept, Martel et al. (1990) showed that cadmium causes collapse of the mitochondrial membrane potential, which is followed by plasma membrane depolarization and cell death. The mechanism causing collapse of the mitochondrial membrane potential is unclear. However, inhibition of sulfhydryl-containing enzymes in the citric acid cycle and in the electron transport chain may be involved (Prasada Rao et al.,

1983; Cameron et al., 1986). Cadmium has recently been shown to be a potent inhibitor of uncoupler-stimulated oxidation of various NAD-linked substrates and succinate (Miccadei and Floridi, 1993). An alternative mechanism proposed by Cameron et al. (1986) is that cadmium perturbs mitochondrial calcium homeostasis by binding to the functional sulfhydryl group of the mitochondrial calcium uniporter and inhibiting calcium uptake. This leads to inhibition of calcium-dependent intramitochondrial dehydrogenases that are responsible for controlling the overall rate of mitochondrial oxidative metabolism (Denton and McCormack, 1980), although this alone would not account for mitochondrial depolarization. Overall, the effect of cadmium seems to be inhibition of oxidative phosphorylation, resulting in ATP depletion and bioenergetic death (Muller, 1986).

Cadmium also depletes an important antioxidant, α-tocopherol, and causes lipid peroxidation in isolated hepatocytes (Fariss, 1991). However, maintaining cellular stores of α-tocopherol during cadmium exposure does not protect against the onset of cell death. This indicates that loss of cellular α-tocopherol is a *consequence* rather than a *cause* of cadmium toxicity. Interestingly, α-tocopheryl succinate protects completely against cadmium toxicity. Cytoprotection correlates not with an increase in cellular α-tocopherol content, but rather with cellular accumulation of the intact tocopheryl succinate molecule. The mechanism for this protection is not known. Fariss (1991) suggests that tocopheryl succinate acts as a lipophilic carrier for the succinate moiety, facilitating its cellular uptake. Succinate is a substrate for complex II of the mitochondrial electon transport. Since cadmium inhibits complex I more effectively than complex II of the electron transport chain (Kisling et al., 1987), externally added succinate may help maintain mitochondrial respiration and energy production. Cadmium also inhibits succinate-supported mitochondrial respiration, but 2 to 3 times higher concentrations are required to achieve the same inhibition (Miccadei and Floridi, 1993).

D. EFFECT OF CADMIUM ON THE FUNCTION OF ENDOPLASMIC RETICULUM AND NUCLEUS

After a single lethal dose of cadmium, about 20% of total cellular cadmium is localized in the endoplasmic reticulum of hepatocytes (Goering and Klaassen, 1983). Electron microscopy reveals heavy proliferation and vesicular dilatation of smooth endoplasmic reticulum after acute exposure (Meiss et al., 1982). A single dose of cadmium causing hepatic necrosis decreases hepatic cytochrome P_{450} in microsomes by 30 to 40% and inhibits metabolism of several xenobiotics (Hadley et al., 1974; Gregus et al., 1982; Goering and Klaassen, 1984b; Iscan et al., 1993).

About 25% of the subcellular cadmium is distributed in the liver nuclei (Goering and Klaassen, 1983). Cadmium induces single-strand DNA damage in a liver-derived cell line, TRL-1215. Induction of metallothionein gene expression by low doses of cadmium pretreatments provides protection against cadmium-induced DNA damage (Coogan et al., 1994). The mechanism of this protective role of metallothionein in unclear. Since metallothionein has been detected in hepatocyte nuclei (Leyshon-Sorland and Stang, 1993) metallothionein may act as a scavenger of cadmium-induced active oxygen species and thus protect DNA from cadmium-induced damage (Ochi et al., 1983; Thornalley and Vasak, 1985; Abel and de Ruiter, 1989). Cadmium is also a strong inhibitor of liver O^6-alkylguanine DNA alkyltransferase (Scicchitano and Pegg, 1987). This enzyme removes methyl groups from DNA and is important for DNA repair after exposure to DNA-methylating agents, like methylnitrosourea.

III. MERCURY

Mercury exists as three forms: elemental, inorganic, and organic. The major source of mercury is natural degassing of the earth's crust, including land areas, rivers, and the ocean. Organic and inorganic forms of mercury undergo environmental transformation. Metallic mercury may be oxidized to inorganic divalent mercury. Divalent inorganic mercury may, in turn, be reduced to metallic mercury in the presence of reducing equivalents. Divalent mercury can also be methylated to methylmercury salt, CH_3HgCl, and dimethylmercury by anaerobic bacteria. These forms of methylmercury can diffuse into the atmosphere and redistribute in land and water during rainfall. If taken up by fish in the food chain, methylmercury and dimethylmercury may eventually cycle through humans. Compared to cadmium, the half-life of mercury compounds is short. The biologic half-life for methylmercury is about 120 days, whereas the half-life for salts of inorganic mercury is about 70 days.

The primary site of mercury accumulation is the kidney. Other organs and cells where mercury accumulates are liver, mucous membranes of the intestinal tract, epithelium of the skin, spleen, interstitial cells of the testicles, and some parts of the brain. In the liver, the highest concentrations of mercury are found in periportal regions of the liver lobule (Berlin and Ullberg, 1963). Acute mercury overload due to accidental or suicidal ingestion causes abdominal pain, diarrhea, and necrosis of the intestinal mucosa that may lead to circulatory collapse and death. However, if patients survive the gastrointestinal damage, renal failure due to necrosis of the proximal tubular epithelium occurs, leading to anuria and uremia.

A. MERCURY UPTAKE BY HEPATOCYTES

Like uptake of cadmium, uptake of Hg^{2+} by hepatocytes is rapid and biphasic. However, the uptake of these two metals occurs by different processes. Hg^{2+} is taken up in a temperature-independent pathway, whereas cadmium uptake is temperature dependent (Blazka and Shaikh, 1992). Unlike cadmium, Hg^{2+} does not seem to involve a SH-dependent transport process (Figure 3) (Blazka and Shaikh, 1991, 1992). Transporters are not needed for uptake of hydrophobic organometallic compounds, such as methylmercury, which easily cross lipid membranes. This may contribute to their more potent toxicity.

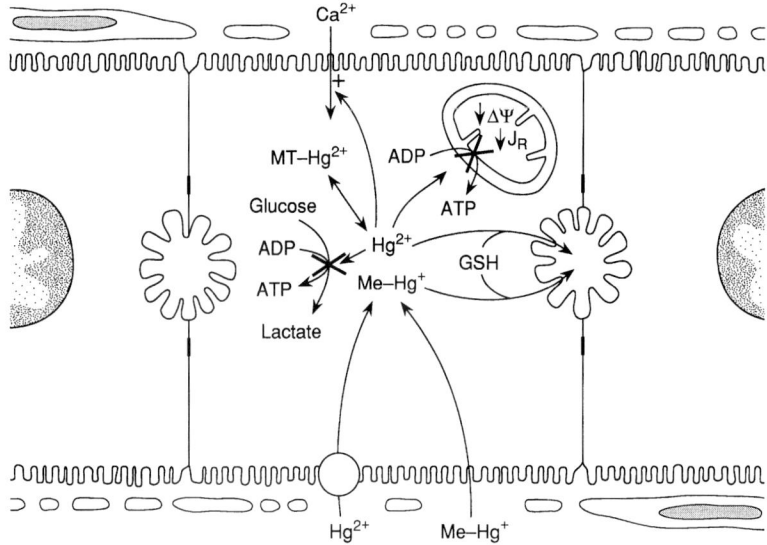

Figure 3. Uptake and intracellular distribution of mercury in hepatocytes. Hg^{2+} and organic mercury compounds, such as methylmercury (Me-Hg$^+$), enter hepatocytes via a transporter or by simple diffusion. Inside the cells, mercury binds to metallothionein (MT) or reacts with glutathione (GSH) to form conjugates that are excreted into the bile. Hg^{2+} is extremely reactive, with thiols and inactivates key enzymes in glycolysis and oxidative phosphorylation, leading to inhibition of respiration, depolarization of mitochondria ($\Delta\psi$), and depletion of ATP. Hg^{2+} also promotes Ca^{2+} entry into hepatocytes.

B. INTRACELLULAR DISTRIBUTION OF MERCURY IN HEPATOCYTES

Intracellular Hg^{2+} binds to metallothionein located in the cytosol (Figure 3) (Liu et al., 1991). Zinc alters the subcellular distribution of mercury by increasing the synthesis of metallothionein. This causes more mercury binding in the cytosol and less to intracellular organelles, such as mitochondria and endoplasmic reticulum, an effect associated with increased resistance to mercury toxicity (Liu et al., 1991). Mercury has very high affinity to thiols. Gagne et al. (1990) showed that mercury in rainbow trout hepatocytes forms an adduct with a low molecular weight cytosolic component that is probably glutathione. A methylmercury-glutathione complex has been identified in rat liver (Omata et al., 1978), and glutathione seems to play an important role in the biliary excretion of methylmercury (Refsvik, 1978).

C. CELLULAR MECHANISMS OF MERCURY TOXICITY IN HEPATOCYTES

The degree of mercury-dependent hepatotoxicity depends on the chemical form of the heavy metal. Studies by Stacey and Klaassen (1981) and Ashour et al. (1993) showed that methylmercury is far more toxic to rat hepatocytes than mercuric chloride, causing both pathological and biochemical changes in the organ. In the study by Ashour et al. (1993), toxicity of different concentrations of methylmercury was evaluated by aspartate and alanine transaminase leakage. Methylmercury caused increases of serum alanine transaminase activity at concentrations that did not increase aspartate transaminase. Since aspartate transaminase is localized mainly to mitochondria (Clampitt, 1978), the authors concluded that the major site of methylmercury toxicity was cytosolic (Ashour et al., 1993). However, injured hepatocytes may simply release cytosolic alanine transaminase more readily than aspartate transaminase held inside mitochondria.

The liver contains the highest concentration of glutathione (about 5 mM) of any major organ. Glutathione acts as a reducing and conjugating agent to detoxify many potential toxicants. In hepatocytes, glutathione forms a complex with methylmercury that is rapidly excreted into the bile to reduce the hepatic content of methylmercury (Figure 3) (Refsvik and Norseth, 1975; Hirata and Takahashi, 1981; Alexander and Aseth, 1982). Thus, glutathione protects hepatocytes from overload with methylmercury. Approximately 90% of cellular glutathione is found in the cytosol with the remainder sequestered in mitochondria. Hepatocytes can maintain viability in the face of cytosolic glutathione depletion. However, even small alterations of mitochondrial glutathione promote irreversible cell injury (Meredith and Reed, 1982; Comporti, 1987). Mercury compounds appear to preferentially target mitochondria and may cause rapid mitochondrial glutathione depletion leading to mitochondrial dysfunction and ultimately cell death (Weinberg et al., 1982a,b; Nieminen et al., 1990a,b).

Although less toxic than methylmercury in hepatocytes, mercuric chloride is also quite reactive in liver cells. In rat hepatocytes, mercuric chloride causes rapid oxidation of pyridine nucleotides and glutathione, followed closely by mitochondrial depolarization, as indicated by the release of the cationic fluorophore, rhodamine 123 (Figure 4) (Lemasters et al., 1987; Nieminen et al., 1990a,b). Uncoupling and depolarization caused by mercury chloride have also been shown in isolated mitochondria from liver, heart, and kidney (Scott and Gamble, 1961; Southard et al., 1974; Weinberg et al., 1982a,b).

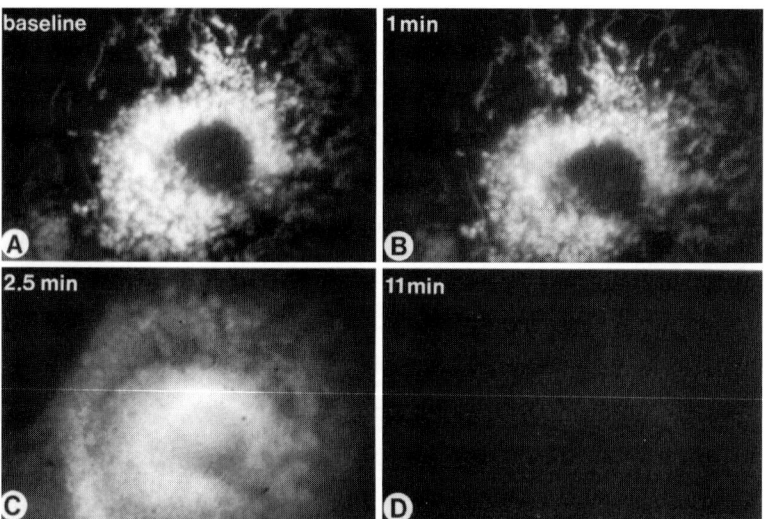

Figure 4. Effect of $HgCl_2$ on mitochondrial membrane potential in hepatocytes. A cultured hepatocyte was loaded with the mitochondrial membrane potential-indicating fluorophore, rhodamine 123. Before addition of $HgCl_2$, mitochondria stained brightly with rhodamine 123 (A). After 1 min of exposure to 50 μM $HgCl_2$, fluorescence was unchanged (B). After 2.5 min, rhodamine 123 had leaked from mitochondria to diffusely stain the cytosol (C). After 11 min, most dye was lost from the cell (D). (From Nieminen A.-L. et al., *J. Biol. Chem.*, 265, 2399–2408. With permission.)

ATP depletion ensues soon after mitochondrial depolarization in mercuric chloride-treated hepatocytes (Nieminen et al., 1990a). Fructose is an effective ATP-generating glycolytic substrate in liver that prevents cell death from several toxic chemicals. Fructose does not, however, protect against mercuric chloride-induced toxicity (Nieminen, et al., 1990c). Lack of protection is related to the fact that mercuric chloride inhibits the glycolytic enzyme, glyceraldehyde 3-phosphate dehydrogenase, most likely by binding to its catalytic sulfhydryl group.

Inorganic and organic mercury compounds induce lipid peroxidation in liver (Stacey and Kappus, 1982; Yonaha et al., 1983; Andersen and Andersen, 1993). High levels of dietary α-tocopherol protect against methylmercury-induced hepatic lipid peroxidation, whereas an α-tocopherol-deficient diet enhances methylmercury-induced hepatic lipid peroxidation (Andersen and Andersen, 1993). α-Tocopherol is stored in liver in association with membrane lipids, particularly those in endoplasmic reticulum and mitochondria. One can speculate that high levels of dietary α-tocopherol specifically protect against methylmercury-induced dysfunction and endoplasmic reticulum and mitochondria. High levels of dietary β-carotene, on the other hand, promote methylmercury-induced lipid peroxidation (Andersen and Andersen, 1993). This may be due to a prooxidant effect of β-carotene, since at high concentrations, β-carotene can undergo autooxidation (Burton and Ingold, 1984). An alternative explanation is that high concentrations of β-carotene reduce the concentration of α-tocopherol in liver (Blakley et al., 1990).

Mercuric chloride also perturbs intracellular calcium homeostasis in hepatocytes. Even extremely low concentrations of mercuric chloride (0.1 μM) that do not cause toxicity, increase total intracellular calcium in rainbow trout hepatocytes (Gagne et al., 1990). In another study, an toxic dose of 50 μM mercuric chloride rapidly increased cytosolic free Ca^{2+} in single intact rat hepatocytes (Nieminen et al., 1990a). This increase preceded the onset of cell death. Removal of Ca^{2+} from the extracellular medium prevented the increase of cytosolic free Ca^{2+}, indicating that mercuric chloride induces net influx of Ca^{2+} across the plasma membrane. Although low extracellular Ca^{2+} prevented cytosolic Ca^{2+} from rising after $HgCl_2$, cell killing was not retarded. Thus, increased free Ca^{2+} does not appear to contribute directly to the onset of cell death. Rather, mitochondria are likely the most important target for mercuric chloride toxicity.

IV. COPPER

Unlike cadmium and mercury, copper is an essential metal for several metabolic pathways. Copper is a prosthetic group of numerous enzymes and can serve as a cofactor for enzyme activation. Catalase, peroxidase, cytochrome c oxidase, superoxide dismutase, alcohol dehydrogenase, and alkaline phosphatase all require copper for activity. The normal diet usually provides a sufficient amount of copper so that a nutritional deficiency seldom occurs. Acute poisoning, however, due to ingestion of excessive amounts of oral copper salts causes vomiting, hypotension, and coma leading eventually to death. Histology reveals pericentral hepatic necrosis (Chuttani et al., 1965). Excess copper stored in the liver causes hepatitis, leading to hepatic failure, cirrhosis, and ultimately death. the two well-known copper storage diseases are Wilson's disease and Indian childhood cirrhosis.

Wilson's disease is a genetic metabolic defect that leads to accumulation of copper first in the liver, and later in the central nervous system, eyes, kidneys, and other organs. The primary genetic defect that is responsible for the copper accumulation in Wilson's disease remains unclear. It is known that biliary excretion of copper is decreased during the disease. Synthesis of ceruloplasmin is also decreased, leading to defective copper secretion as ceruloplasmin.

In Wilson's disease, symptoms of liver injury seldom occur before 4 to 5 years of age. After this age, various clinical signs and symptoms of hepatic failure develop, including anorexia, fatigue, malaise, various gastrointestinal complaints, and episodes of jaundice. This initial acute hepatitis may rapidly progress to severe hepatic and renal failure. At this stage of disease, serum copper levels are elevated, due to release of hepatic copper into the circulation. Subsequent redistribution of copper to the central nervous system causes neurological symptoms, such as movement disorder and dystonia. Copper may also accumulate in the ocular cornea leading to the characteristic Kayser-Fleischer ring, a greenish brown ring at the periphery of the cornea on its posterior surface. If patients do not respond to the copper chelating agent, D-penicillamine, liver transplantation becomes necessary. Chronic active hepatitis is a more common clinical feature among young adults than acute hepatitis. Patients at this stage of disease respond well to the copper chelation therapy. Cirrhosis is associated with the patients who also have central nervous system symptoms.

Indian childhood cirrhosis is a fatal disease that affects children under 3 years of age in India. The disease occurs in babies who have been fed boiled animal milk stored in brass utensils. It is likely that the disease is due to increased dietary intake of copper, although hereditary factors have not been totally ruled out.

A. INTRACELLULAR DISTRIBUTION OF COPPER IN HEPATOCYTES

The liver has a central role in regulating copper homeostasis. About 50% of the dietary copper is absorbed from the stomach and small intestine into the blood. In blood, copper is bound to serum albumin to form an albumin-copper complex (Bearn and Kunkel, 1954). The detailed mechanism of copper transport into hepatocytes remains unclear. Waldrop et al. (1990) suggest that hepatocytes do not take up copper as an intact albumin-copper complex. Instead, hepatocytes may have one or more cell-specific high affinity copper-binding carriers that facilitate copper uptake from the plasma. This transport mechanism does not seem to be temperature dependent (Schmitt et al., 1983).

About 80% of intracellular copper is distributed in the cytosol where it is bound predominantly to cytosolic metallothionein, cytosolic cuprophyllin, lysosomal metallothionein and, to a lesser extent, to specific copper proteins, such as superoxide dismutase (Figure 5) (Fridovich, 1974). The remaining 20% of copper is associated with the membranous organelles, particularly cytochrome c oxidase of mitochondria.

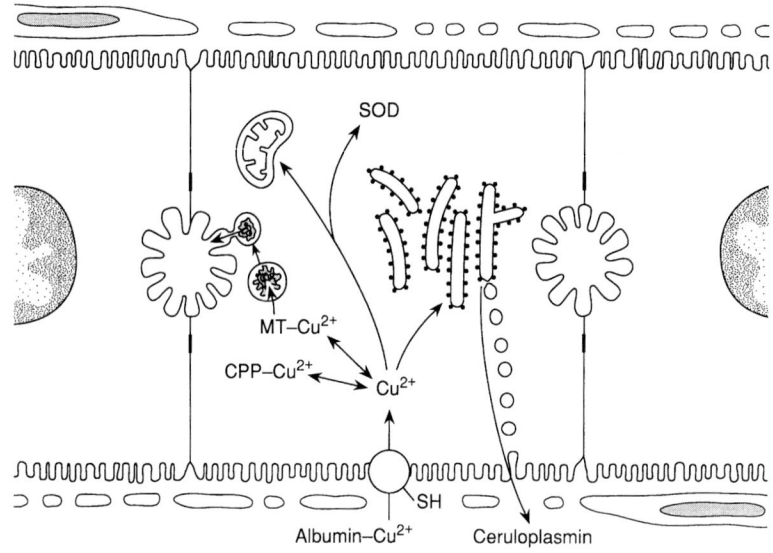

Figure 5. Copper uptake and distribution in hepatocytes. The majority of copper in the plasma is bound to ceruloplasmin and albumin. Uptake is by a thiol-dependent transporter. Inside the hepatocyte, copper binds to cuprophyllin (CPP) and metallothionein (MT). Metallothionein-Cu^{2+} complexes gain access to lysosomes and are excreted into the bile. Cu^{2+} also binds to ceruloplasmin, the major copper-carrying protein in the plasma, which is synthesized and secreted by hepatocytes. (After Sokol, 1992).

The liver has two distinct pathways to discharge excess copper. The first involves ceruloplasmin, an important transport protein for copper in hepatocytes. Apoceruloplasmin is first synthesized in liver as a single polypeptide chain and is then secreted from hepatocytes into plasma as a holoceruloplasmin containing 6 atoms of copper (Takahashi et al., 1984). About 90% of copper in plasma is bound to ceruloplasmin. The remainder is bound to albumin and amino acids. A second and more important route for hepatic copper excretion is the bile. More than 80% of absorbed copper is excreted into bile. It is generally believed that most biliary copper originates from lysosomes in hepatocytes (DeDuve and Wattiaux, 1966; Gross et al., 1989). This hypothesis is supported by the observation that in the liver from the copper-overloaded rat, increased biliary copper excretion is associated with an increased excretion of lysosomal enzymes into the bile (Gross et al., 1989). The precise mechanism for transport of copper from lysosomes to bile remains unclear. Several carriers have been implicated, including small peptides (Evans and Cornatzer, 1971; Terao and Owen, 1983; Martin et al., 1986), amino acids (Evans and Cornatzer, 1971; Martin et al., 1986), bile acids (Lewis, 1973), metallothionein (Sato and Bremner, 1984), and macromolecular complexes (Gollan and Deller, 1973; Terao and Owen, 1983; Kressner et al., 1984). In addition, the study by Harada et al. (1992) suggests that some of the intracellular copper is excreted into bile in a microtubule-dependent fashion, possibly via vesicular transport.

B. MECHANISMS OF TOXICITY FROM COPPER OVERLOAD

Copper overload affects a wide variety of functions by membranes, cytosolic proteins, and subcellular organelles. Copper initiates lipid peroxidation by participating in the Haber-Weiss reaction (Figure 6) (Lindquist, 1968; Aust et al., 1985; Halliwell, 1989). Activation of protein kinase C is also suggested to be involved in copper-induced lipid peroxidation and cell death (Mudassar et al., 1992). Protein kinase C activation by copper may be mediated through reactive oxygen species generated in response to copper.

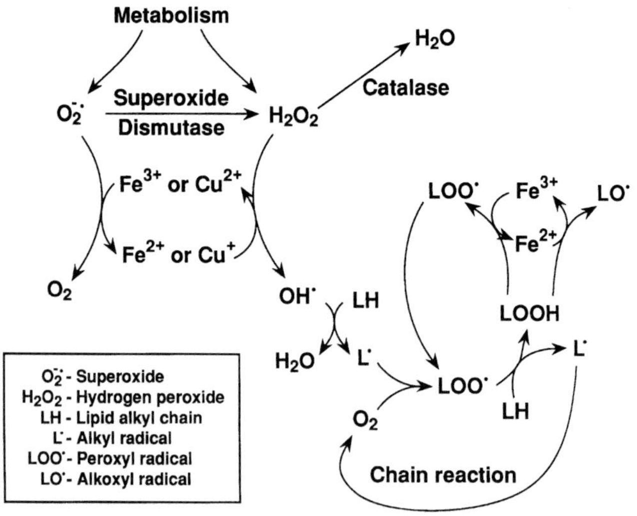

Figure 6. Transition metal-catalyzed free radical formation. The transition metals, iron and copper, catalyze hydroxyl radical (OH·) formation by the Haber-Weiss reaction. OH· reacts with lipids to generate alkyl radicals (L·) that participate in an oxygen-dependent chain reaction generating lipid peroxides (LOOH). LOOH can again react with free iron or copper to generate peroxyl (LOO·) and alkoxyl (LO·) radicals.

Another possibility is that protein kinace C is directly activated by the metal, as has been shown with zinc (Csermely et al., 1988). Since the main route of excretion of intracellular copper is through the bile, any impairment of bile excretion can result in excessive lysosomal copper accumulation. This leads to decreased lysosomal membrane fluidity, increased lysosomal pH, breakdown of membranes and leakage of lysosomal enzymes, such as acid phosphatases, into the cytosol (Lindquist, 1967; McNatt et al., 1971; Myers et al., 1993). Acute copper poisoning also causes oxidation of hepatic mitochondria *in vivo* (Nakatani et al., 1994).

Cytosolic components containing sulfhydryl groups are sensitive to excess copper. Glutathione depletion precedes cell death after exposure of hepatocytes to copper due to formation of copper-glutathione complexes (Khan et al., 1992). Copper also inhibits the polymerization of tubulin leading to degradation of microtubules. This affects spindle formation during mitosis and export of proteins and triglycerides (Wallin et al., 1977).

V. IRON

Since iron is essential in a wide variety of biochemical reactions but is toxic in excess, body iron levels are tightly regulated. Acute iron overload occurs frequently among children and often leads to death. The most common reason is ingestion of maternal iron supplements. In adults, acute iron overload usually occurs after suicide attempts with iron supplements. The early symptoms of acute iron overload are vomiting, diarrhea, abdominal pain, and gastrointestinal hemorrhage. These are followed by circulatory shock, periportal hepatic necrosis, and liver failure. Parenteral iron chelation therapy with deferoxamine is used for acute iron poisoning, but is limited by the hypotensive effect of the drug. Deferoxamine conjugated to dextran or hydroxyethylstarch is less toxic and shows therapeutic promise (Hallaway and Hedlund, 1992). Chronic iron overload results from increased absorption of iron in the diet or excess parenteral iron administration, such as from repeated blood transfusions. The most common liver disease associated with chronic iron overload is a hereditary hemochromatosis, which leads to portal fibrosis and ultimately cirrhosis.

A. IRON DISTRIBUTION IN HEPATOCYTES

Iron is absorbed from the intestinal lumen through mucosal epithelial cells into the plasma where it binds to transferrin, an 83-kDa glycoprotein. Transferrin, the major transport protein for iron, is synthesized by hepatocytes, which take up iron as an iron-transferrin complex. The complex binds to a transferrin receptor on the plasma membrane and is internalized by receptor-mediated endocytosis (Figure 7). Low pH in the endosomal vesicles causes dissociation of iron from the transferrin-receptor complex.

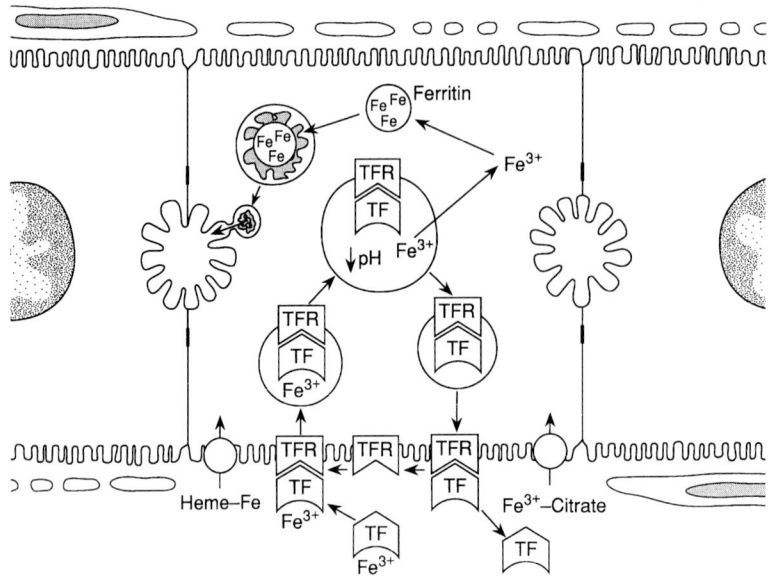

Figure 7. Uptake and intracellular distribution of iron in hepatocytes. The majority of iron (Fe^{3+}) enters hepatocytes as an iron-transferrin complex (TF-Fe^{3+}) that binds to a transferrin receptor (TFR) on the plasma membrane and is internalized by receptor-mediated endocytosis. Low pH in the endosomal vesicles causes dissociation of iron from the transferrin-receptor complex. The transferrin-receptor complex recycles to the plasma membrane, where transferrin (TF) is released back into the circulation. Iron can also enter hepatocytes as FE^{3+} citrate complexes or Fe^{3+}-heme complexes through special carriers. Intracellular iron is stored in the form of ferritin, some of which is degraded to hemosiderin and accumulates in lysosomes.

This free iron can enter other cellular compartments. The transferrin-receptor complex recycles to the plasma membrane, where transferrin is released back into the circulation. Normally more than 99% of the iron in plasma is bound to transferrin. The remaining 1% consists of so called nontransferrin-bound iron (Brissot et al., 1985). This pool of iron is taken up by hepatocytes as low molecular weight chelates, such as Fe^{3+}-citrate complexes, through a special carrier mechanism that is driven by the membrane potential (Wright et al., 1988).

Iron is also taken up by hepatocytes as heme. At least three different heme complexes can enter: hemoglobin-haptoglobin, heme-hemopexin, and heme-albumin (Smith and Morgan, 1979; Kino et al., 1980; Sinclair et al., 1988). Once inside the hepatocyte, the heme is broken down. Iron is released and enters the rapidly exchangeable iron pool.

Intracellular iron is stored mainly in the form of ferritin. Increased intracellular iron stimulates the synthesis of apoferritin by increasing the translation of preexisting apoferritin mRNA. Mature ferritin is composed of 24 subunits with a molecular weight of 450 kDa and is capable of storing 4500 atoms of iron as hydrated ferric oxide-phosphate (Seligman et al., 1987). When needed, ferric iron is released from ferritin by reducing agents, such as cysteine, glutathione, ascorbic acid, and various flavin-containing compounds. Ferritin has a half-life of 24 h and is degraded to hemosiderin, which accumulates in the lysosomes of iron-overloaded cells (Figure 7) (Masuda et al., 1993). The biliary excretion of endogenous ferritin during iron overload is a chloroquine-sensitive, microfilament-dependent process (Ramm et al., 1994). Iron binding by transferrin and ferritin serves not only a transport and storage function, but also acts to prevent iron from catalyzing toxic free radical reactions.

B. MECHANISMS OF TOXICITY FROM IRON OVERLOAD

The most common form of iron overload is hereditary hemochromatosis. In hemochromatosis, increased iron absorption from gastrointestinal tract leads to cellular accumulation of iron, especially in hepatocytes. Like copper, iron is a transition metal that catalyzes free radical formation by the Fenton reaction (Figure 6). Free iron also catalyzes alkoxyl and peroxyl radical formation from lipid peroxides (Halliwell, 1989). Thus, an important component of iron toxicity in hepatocytes may be free radical injury.

In suspended hepatocytes exposed to free iron, lipid peroxidation precedes cell death (Hogberg et al., 1975). The antioxidant, α-tocopherol, and the iron chelators, deferoxamine and apotransferrin, significantly reduce lipid peroxidation and improve cell viability (Poli et al., 1985; Bacon and Britton, 1989). Several studies have attempted to elucidate which cellular membranes specifically are injured during iron overload. The study by LeSage et al. (1986) showed that in rats fed with carbonyl iron-supplemented diets, latent liver lysosomal enzyme activities were reduced, indicating the breakdown of lysosomes (Peters et al., 1985). Myers et al. (1991) also showed that iron overload increased lysosomal lipid peroxidation, decreased lysosomal membrane fluidity, and alkalinized lysosomal pH.

Microsomal dysfunction is also associated with iron overload. Microsomal conjugated diene formation, an indication of lipid peroxidation *in vivo*, is increased and cytochrome P_{450} enzymes are decreased in iron overload to rats (Bacon et al., 1986; Britton et al., 1991). However, it is unclear whether lipid peroxidation actually causes inhibition of the enzymes. In human liver microsomes, iron also promotes hydroxyl radical formation and lipid peroxidation (Rashba-Step and Cederbaum, 1994). Iron overload also decreases microsomal calcium sequestration that may perturb intracellular calcium homeostasis and contribute to cell death (Britton et al., 1991).

Several studies have investigated the effects of iron overload on mitochondrial function. Mitochondrial swelling, inhibition of mitochondrial oxidative phosphorylation, peroxidation of mitochondrial membrane lipids, and alterations in membrane proteins follow iron overload (Ganote and Nahara, 1973; Hanstein et al., 1981; Castilho et al., 1994). Mitochondrial membrane potential also decreases, even at doses of iron that do not cause respiratory inhibition (Masini et al., 1984). Thus, mitochondrial membrane potential may be a more sensitive indicator of iron-induced mitochondrial injury than rates of respiration and oxidative phosphorylation. Studies by Bacon et al. (1993) further characterized the location in the electron transport chain where iron-induced damage occurs during iron overload. Chronic iron overload decreased hepatic mitochondrial cytochrome *c* oxidase activity. Cytochrome *c* oxidase activity depends on phospholipids, especially cardiolipin (Tzagoloff and MacLennan, 1965; Fry and Green, 1980). Cardiolipin contains polyunsaturated fatty acids that are highly sensitive to peroxidation (Mitchell et al., 1980). Thus, peroxidation of these polyunsaturated fatty acids may be responsible for the inactivation of cytochrome *c* oxidase and subsequent disruption of mitochondrial oxidative phosphorylation.

VI. CONCLUSION

Cadmium, mercury, copper, and iron are potently toxic to liver cells. Cadmium and mercury both interfere with the energy metabolism of hepatocytes, which suggests that mitochondria are the primary target of their toxicity. Metallothionein plays an important role in protecting against cadmium- and mercury-induced liver toxicity, possibly by preventing the binding of these metals to intracellular organelles, especially mitochondria. Copper and iron as essential metals are important for several biochemical reactions, but cause toxicity in excess. For these transition metals, one likely mechanism of toxicity is peroxidation of membrane lipids initiated by the Haber-Weiss reaction, causing lysosomal, microsomal, and mitochondrial dysfunction. Prevention remains the most effective means of avoiding acute heavy metal hepatotoxicity. But for inherited metal-storage diseases, increased understanding of the mechanisms of heavy metal hepatotoxicity is needed to develop more effective means of therapy.

REFERENCES

Abel, J. and De Ruiter, N. (1989), *Toxicol. Lett.*, 47, 191–196.
Alexander, J. and Aaseth, J. (1982), *Biochem. Pharmacol.*, 31, 685–690.
Andersen, H. R. and Andersen, O. (1993), *Pharmacol. Toxicol.*, 73, 192–201.
Ashour, H., Abdel-Rahman, M., and Khodair, A. (1993), *Toxicol. Lett.*, 69, 87–96.
Aust, S. D., Morehouse, L. A., and Thomas, C. E. (1985), *Free Radicals Biol. Med.*, 1, 3–25.
Bacon, B. R. and Britton, R. S. (1989), *Chem. Biol. Interact.*, 70, 183–226.
Bacon, B. R., Healey, J. F., Brittenham, G. M., Park, C. H., Nunnari, J., Tavill, A. S., and Bonkovsky, H. L. (1986), *Gastroenterology*, 90, 1844–1853.
Bacon, B. R., O'Neill, R., and Britton, R. S. (1993), *Gastroenterology*, 105, 1134–1140.
Bearn, A. G. and Kunkel, H. G. (1954), *Proc. Soc. Exp. Biol. Med.*, 85, 44–48.
Berlin, M. and Ullberg, S. (1963), *Arch. Environ. Health*, 6, 586–616.
Blakley, S. R., Grundel, E., Jenkins, M. Y., and Michell, G. V. (1990), *Nutr. Res.*, 10, 1035–1044.
Blazka, M. E. and Shaikh, Z. A. (1991), *Toxicol. Appl. Pharmacol.*, 110, 355–363.
Blazka, M. E. and Shaikh, Z. A. (1992), *Toxicol. Appl. Pharmacol.*, 113, 118–125.
Brissot, P., Wright, T. L., Ma. W.-L., and Weisiger, R. A. (1985), *J. Clin. Invest.*, 76, 1463–1470.
Britton, R. S., O'Neill, R., and Bacon, B. R. (1991), *Gastroenterology*, 101, 806–811.

Burton, G. W. and Ingold, K. U. (1984), *Science,* 224, 569–573.
Cameron, I., McNamee, P. M., Markham, A., Morgan, R. M., and Wood, M. (1986), *J. Appl. Toxicol.,* 6, 325–330.
Castilho, R. F., Meinicke, A. R., Almeida, A. M., Hermes-Lima, M., and Vercesi, A. E. (1994), *Arch. Biochem. Biophys.,* 308, 158–163.
Chuttani, H. K., Gupta, P. S., Gulati, S., and Gupta, D. N. (1965), *Am. J. Med.,* 39, 849–854.
Clampitt, R. B. (1978), *Arch. Toxicol. Suppl.,* 1, 1–13.
Comporti, M. (1987), *Chem. Phys. Lipids,* 45, 143–169.
Coogan, T. P., Bare, R. M., and Waalkes, M. P. (1992), *Toxicol. Appl. Pharmacol.,* 113, 227–233.
Coogan, T. P., Bare, R. M., Bjornson, E. J., and Waalkes, M. P. (1994), *J. Toxicol. Environ. Health,* 41, 233–245.
Csermely, P., Szamel, M., Resch, K., and Somogyi, J. (1988), *J. Biol. Chem.,* 263, 6487–6490.
DeDuve, C. and Wattiaux, R. (1966), *Annu. Rev. Physiol.,* 28, 435–492.
Denton, R. M. and McCormack, J. G. (1980), *FEBS Lett.,* 119, 1–8.
Evans, G. W. and Cornatzer, W. E. (1971), *Proc. Soc. Exp. Biol. Med.,* 136, 719–721.
Fariss, M. W. (1991), *Toxicology,* 69, 63–77.
Fridovich, I. (1974), *Adv. Enzymol.,* 41, 35–97.
Fry, M. and Green, D. E. (1980), *Biochem. Biophys. Res. Commun.,* 93, 1238–1246.
Gagne, F., Marion, M., and Denizeau, F. (1990), *Toxicol. Lett.,* 51, 99–107.
Ganote, C. E. and Nahara, G. (1973), *Lab. Invest.,* 28, 426–436.
Gerson, R. J. and Shaikh, Z. A. (1984), *Biochem. Pharmacol.,* 33, 199–203.
Goering, P. L. and Klaassen, C. D. (1983), *Toxicol. Appl. Pharmacol.,* 70, 195–203.
Goering, P. L. and Klaassen, C. D. (1984a), *Toxicol. Appl. Pharmacol.,* 74, 299–307.
Goering, P. L. and Klaassen, C. D. (1984b), *Toxicol. Appl. Pharmacol.,* 74, 308–313.
Gollan, J. L. and Deller, D. J. (1973), *Clin. Sci.,* 44, 9–15.
Gregus, Z., Watkins, J. B., Thompson, T. N., and Klaassen, C. D. (1982), *J. Pharmacol. Exp. Ther.,* 222, 471–479.
Gross, J. B., Myers, B. M., Kost, L. J., Kuntz, S. M., and LaRusso, N. F. (1989), *J. Clin. Invest.,* 83, 30–39.
Hadley, W. M., Miya, T. S., and Bousquet, W. F. (1974), *Toxicol. Appl. Pharmacol.,* 28, 284–291.
Hallaway, P. E. and Hedlund, B. E. (1992), in *Iron and Human Disease,* Lauffer, R. B., Ed., CRC Press, Boca Raton, FL, 477–508.
Halliwell, B. (1989), *Br. J. Exp. Path.,* 70, 737–757.
Hanstein, W. G., Heitmann, T. D., Sandy, A., Biesterfeldt, H. L., Liem, H. H., and Muller-Eberhard, U. (1981), *Biochim. Biophys. Acta,* 678, 293–299.
Harada, M., Sakisaka, S., Yoshitake, M., Shakadoh, S., Gondoh, K., Sata, M., and Tanikawa, K. (1992), *Hepatology,* 17, 111–117.
Hirata, E. and Takahashi, H. (1981), *Toxicol. Appl. Pharmacol.,* 58, 483–491.
Hiratsuka, H., Katsuta, O., Iwata, H., Matsumoto, J., and Umemura, T. (1993), *J. Toxicol. Sci.,* 18, 197–201.
Hogberg, J., Orrenius, S., and O'Brien, P. J. (1975), *Eur. J. Biochem.,* 59, 449–455.
Hunziker, P. E. and Kagi, J. H. R. (1985), *Biochem. J.,* 231, 375–382.
Iscan, M., Coban, T., Eke, B. C., and Iscan, M. (1993), *Biol. Trace Elem. Res.,* 38, 129–137.
Khan, M. F., Ohno, Y., and Takanaka, A. (1992), *Arch. Toxicol.,* 66, 587–591.
Kingsley, B. S. and Frazier, J. M. (1979), *Am. J. Physiol.,* 236, C139–C143.
Kino, K., Tsunoo, H., Higa, Y., Takami, M., Hamaguchi, H., and Nakajima, H. (1980), *J. Biol. Chem.,* 255, 9616–9620.
Kisling, G. M., Kopp, S. J., Paulson, D. J., Hawley, P. L., and Tow, J. P. (1987), *Toxicol. Appl. Pharmacol.,* 89, 295–304.
Kneip, T. J., Eisenbud, M., Strehlow, C. D., and Freudenthal, P. C. (1970), *J. Air Pollut. Control Assoc.,* 20, 144–149.
Kressner, M. S., Stockert, R. J., Morell, A. G., and Sternlieb, I. (1984), *Hepatology,* 4, 867–870.
Lemasters, J. J., DiGuiseppi, J., Nieminen, A.-L., and Herman, B. (1987), *Nature,* 325, 78–81.
LeSage, G. D., Kost, L. J., Barham, S. S., and LaRusso, N. F. (1986), *J. Clin. Invest.,* 77, 90–97.
Lewis, K. O. (1973), *Gut,* 14, 221–232.
Leyshon-Sorland, K. and Stang, E. (1993), *Histochem. J.,* 25, 857–864.
Lindquist, R. R. (1967), *Am. J. Pathol.,* 51, 471–481.
Lindquist, R. R. (1968), *Am. J. Pathol.,* 53, 903–927.
Liu, J., Kershaw, W. C., and Klaassen, C. D. (1991), *Toxicol. Appl. Pharmacol.,* 107, 27–34.
Martel, J., Marion, M., and Denizeau, F. (1990), *Toxicology,* 60, 161–172.
Martin, M. T., Jacobs, F. A., and Brushmiller, J. G. (1986), *Proc. Soc. Exp. Biol. Med.,* 181, 249–255.
Masini, A., Trenti, T., Ventura, E., Ceccarelli-Stanzani, D., and Muscatello, U. (1984), *Biochem. Biophys. Res. Commun.,* 124, 462–469.
Masuda, T., Kasai, T., and Satodate, R. (1993), *Anal. Quant. Cytol. Histol.,* 15, 379–382.
McNatt, E. N., Campbell, W. G., and Callahan, B. C. (1971), *Am. J. Pathol.,* 64, 123–144.
Meiss, R., Robenek, H., Rassat, J., Themann, H., and Reichert, B. (1982), *Arch. Environ. Contam. Toxicol.,* 11, 283–289.
Meredith, M. J. and Reed, D. J. (1982), *J. Biol. Chem.,* 257, 3747–3753.
Miccadei, S. and Floridi, A. (1993), *Chem. Biol. Interact.,* 89, 159–167.
Mitchell, J. R., Smith, C. V., Hughes, H., Lauterburg, B. H., and Horning, M. G. (1980), *Semin. Liver Dis.,* 1, 143–150.
Mudassar, S., Andrabi, K. I., Khullar, M., Ganguly, N. K., and Walia, B. N. S. (1992), *J. Pharm. Pharmacol.,* 44, 609–611.
Muller, L. (1986), *Toxicology,* 40, 285–295.

Myers, B. M., Prendergast, F. G., Holman, R., Kuntz, S. M., and LaRusso, N. F. (1991), *J. Clin. Invest.*, 88, 1207–1215.
Myers, B. M., Prendergast, F. G., Holman, R., Kuntz, S. M., and LaRusso, N. F. (1993), *Gastroenterology*, 105, 1814–1823.
Nakatani, T., Spolter, L., and Kobayashi, K. (1994), *Life Sci.*, 54, 967–974.
Nieminen, A.-L., Gores, G. J., Dawson, T. L., Herman, B., and Lemasters, J. J. (1990a), *J. Biol. Chem.*, 265, 2399–2408.
Nieminen, A.-L., Gores, G. J., Dawson, T. L., Herman, B., and Lemasters, J. J. (1990b), in *Optical Microscopy for Biology*, Herman, B. and Jacobson, K., Eds., Alan R Liss, New York, 323–335.
Nieminen, A.-L., Dawson, T. L., Gores, G. J., Kawanishi, T., Herman, B., and Lemasters, J. J. (1990c), *Biochem. Biophys. Res. Commun.*, 167, 600–606.
Nriagu, J. O. and Pacyna, J. M. (1988), *Nature*, 333, 134–139.
Ochi, T., Ishiguro, T., and Ohsawa, M. (1983), *Mutat. Res.*, 122, 169–175.
Omata, S., Sakimura, K., Ishii, T., and Sugano, H. (1978), *Biochem. Pharmacol.*, 27, 1700–1701.
Peters, T. J., O'Connel, M. J., and Ward, R. J. (1985), in *Free Radicals in Liver Injury*, Poli, G., Cheeseman, K. H., Dianzani, M. U., and Slater, T. F., Eds., IRL Press, Oxford, 107–115.
Poli, G., Albano, E., Biasi, F., Cecchini, G., Carini, R., Bellomo, G., and Dianzani, M. U. (1985), in *Free Radicals in Liver Injury*, Poli, G., Cheeseman, K. H., Dianzani, M. U., and Slater, T. F., Eds. IRL Press, Oxford, 207–215.
Prasada Rao, P. V. V., Sridhar, M. K. C., and Desalu, A. B. O. (1983), *Arch. Environ. Contam. Toxicol.*, 12, 293–297.
Ramm, G. A., Powell, L. W., and Halliday, J. W. (1994), *Hepatology*, 19, 504–513.
Rashba-Step, J. and Cederbaum, A. I. (1994), *Mol. Pharmacol.*, 45, 150–157.
Refsvik, T. (1978), *Acta Pharmacol. Toxicol.*, 42, 135–141.
Refsvik, T. and Norseth, T. (1975), *Acta Pharmacol. Toxicol.*, 36, 67–78.
Sato, M. and Bremner, I. (1984), *Biochem. J.*, 223, 475–479.
Schmitt, R. C., Darwish, H. M., Cheney, J. C., and Ettinger, M. J. (1983), *Am. J. Physiol.*, 244, G183–G191.
Scicchitano, D. A. and Pegg, A. E. (1987), *Mutat. Res.*, 192, 207–210.
Scott, R. L. and Gamble, J. L. (1961), *J. Biol. Chem.*, 236, 570–573.
Sinclair, P. R., Bement, W. J., Gorman, N., Liem, H. H., Wolkoff, A. W., and Muller-Eberhard, U. (1988), *Biochem. J.*, 256, 159–165.
Seligman, P. A., Klausner, R. D., and Huebers, H. A. (1987), in *The Molecular Basis of Blood Diseases*, Stamatoyannopoulos, G., Nienhuis, A. W., Leder, P., and Majerus, P. W., Eds. W. B. Saunders, Philadelphia, 219–244.
Smith, A. and Morgan, W. T. (1979), *Biochem. J.*, 182, 47–54.
Sokol, R. J. (1992), in *Liver and Biliary Diseases*, Kaplowitz, N., Ed. Williams & Wilkins, Baltimore, 322–333.
Sorensen, J. A., Nielsen, J. B., and Andersen, O. (1993), *Pharmacol. Toxicol.*, 73, 169–173.
Southard, J., Nitisewojo, P., and Green, D. E. (1974), *Fed. Proc.*, 33, 2147–2153.
Stacey, N. H. and Kappus, H. (1982), *Toxicol. Appl. Pharmacol.*, 63, 29–35.
Stacey, N. H. and Klaassen, C. D. (1980), *Toxicol. Appl. Pharmacol.*, 55, 448–455.
Stacey, N. H. and Klaassen, C. D. (1981), *J. Toxicol. Environ. Health*, 7, 19–147.
Takahashi, N., Ortel, T. L., and Putnam, F. W. (1984), *Proc. Natl. Acad. Sci. U.S.A.*, 81, 390–394.
Terao, T. and Owen, C. A. (1973), *Am. J. Physiol.*, 224, 682–686.
Theocharis, S., Margeli, A., Fasitsas, C., Loizidou, M., and Deliconstantinos, G. (1991), *Comp. Biochem. Physiol.*, 99C, 127–130.
Theocharis, S. E., Margeli, A. P., Alevizou, V., and Varonos, D. (1994), *Toxicol. Lett.*, 71, 1–7.
Thornalley, P. J. and Vasak, M. (1985), *Biochim. Biophys. Acta*, 827, 36–44.
Tzagoloff, A. and MacLennan, D. H. (1965), *Biochim. Biophys. Acta*, 99, 476–485.
Waldrop, G. L., Palida, F. A., Hadi, M., Lonergan, P. A., and Ettinger, M. J. (1990), *Am. J. Physiol.*, 259, G219–G225.
Wallin, M., Larsson, H., and Edstrom, A. (1977), *Exp. Cell Res.*, 107, 219–225.
Weinberg, J. M., Harding, P. G., and Humes, H. D. (1982a), *J. Biol. Chem.*, 257, 60–67.
Weinberg, J. M., Harding, P. G., and Humes, H. D. (1982b), *J. Biol. Chem.*, 257, 68–74.
Wright, T. L., Fitz, J. G., and Weisiger, R. A. (1988), *J. Biol. Chem.*, 263, 1842–1847.
Yonaha, M., Saito, M., and Sagai, M. (1983), *Life Sci.*, 32, 1507–1514.

Chapter 58

Influences on the Gastrointestinal System by Essential and Toxic Metals

Julian P. Keogh and Claus-Peter Siegers

I. INTRODUCTION

The gastrointestinal (GI) tract is the first organ susceptible to attack by ingested xenobiotics. Consequently, local concentrations encountered by this tissue may often be many times higher than those endured elsewhere in the body. Despite the fact that for metals, ingestion is the major route of exposure for the general population, comprehensive data describing GI toxic effects of metals are lacking, probably because most cases of intoxication arise from either intentional or accidental overexposure, which is rare in comparison to chronic intoxications, which may induce pathologies elsewhere in the body.

Metal toxicity often arises after an accumulation of nonspecific effects. Their small size and charged nature allow them to act as substrates for many ionic binding sites on both structural and enzymatic biomolecules with differing consequence. They may also interfere competitively with functions of physiologically important cations such as sodium, potassium, calcium, and other essential elements. Metal toxicity often displays a threshold over which toxic effects rapidly develop, presumably because the body has a high capacity to bind many metals and prevent them from interacting with target molecules whose binding would predispose the development of pathology or even lethality.

Although metal intoxication is usually perceived as arising from unintentional intake of heavy metals of largely anthropogenic origin, such a view does not comprehensively describe the nature of this problem. Metals of toxicological significance may be characterized physiologically as either essential or nonessential, or chemically as either elemental, cationic, anionic (where they are associated with nonmetallic elements which confer their anionic character), or as organometallic compounds. Metal toxicity will be discussed in relation to what is already known about the GI system concerning xenobiotic-induced pathology. Several noteworthy reviews and general texts which partially describe the GI toxicity of various metals have been used to help provide what we hope is a fresh perspective on this interesting subject, (Friberg et al., 1979; Mykännen 1986, Merian, 1991; Sullivan and Kruger, 1992; and van Riel, 1988). This review will not attempt to make any special distinction between essential and nonessential metals, since more often than not, GI toxicity arising form essential metals tends not to be related to their normal physiological function.

II. SOURCES OF METAL EXPOSURE

The three major routes by which xenobiotics enter the body are through ingestion, inhalation, and dermal contact. Parenteral injection may be considered a fourth route of exposure for pharmaceuticals.

Although many metals can be volatile, respiratory exposure is usually only significant in industrial occupational environments where metal-containing dusts may become airborne. Dermal exposure also tends to be higher in industrial environments; however, toxicity tends to be limited to the skin, since inorganic solutes are not generally lipophilic, and the skin is not otherwise specialized as an organ of absorption. GI metal toxicity tends to be uncommon, since it is primarily observed after ingestion, and uptake by this route is usually voluntary. Involuntary exposure to toxic doses through environmental means is also uncommon, although the potential for accidental mass intoxications of this manner always exists (Ferner, 1993). Intentional exposure through homicide or suicide attempts is even less frequent. High exposures of metals are usually required to exert overt toxicity to the GI tract, probably because this organ tends to be well suited to cope with high concentrations of potentially damaging xenobiotics. In some cases, however, significant GI toxicity can arise even after parenteral adminstration of metal compounds. For example, the anticancer compound cisplatin causes intestinal degeneration after injection (Choie et al. 1981), whereas gold compounds, used as antirheumatics, can induce enterocolitis (Fortson and Tedesco, 1984). Otherwise, potassium chromate can induce intestinal bleeding after dermal exposure (Schlatter and Kissling 1973). Nausea and vomiting may often be induced by central activation of the vomiting reflex, and not from direct effects on the GI system, as can be shown after experimental injection of copper sulfate (Mitchelson, 1992).

A. MEDICINAL PREPARATIONS

Overdosing of metal-containing drugs, or mineral prophylactic supplements, is perhaps the most frequent cause of severe metal-induced GI toxicity. Often these preparations are available over the counter, and better health education presents one way of reducing this kind of exposure. Children are at particular risk to accidental overexposure, since they tend to explore their environment orally, and may be particularly attracted to colored mineral salt preparations. Zinc and iron salts in particular have had a history of presenting GI tract problems, since they are often recommended to patients suffering, respectively, from zinc deficiency or anemia (Mykännen, 1986). Typically they exert strong corrosive action to GI mucous membranes.

Metal use in medicine has been on the decrease since the advent of organic pharmaceuticals, but its use remains significant, and there is now even renewed interest in the effectiveness of metal-containing medications for the treatment of various disorders. Many of these agents present problems not only when taken inappropriately, but during regular therapeutic courses.

A major source of metal medicinal exposure remains the use of aluminum- and bismuth-containing antacids and antiulcer compounds (van Riel, 1988). These preparations may be considered unusual in that they confer beneficial effects towards the functioning of the GI tract in patients suffering GI complications, and bismuth compounds in particular have been recommended for treatment of reflux esophagitis, gastritis, ulceration, diarrhea, and indigestion (Thomas, 1991). They are generally safe, with prolonged administration rarely resulting in GI toxicity. Colloidal bismuth subcitrate in particular displays improved efficacy over antacids and other, more effective antiulcer agents, through its ability to act as an antibacterial against *Helicobacter pylori*, implicated in the recurrence of chronic peptic ulcer disease (Gorbach, 1990). Certain zinc and copper preparations (organic salts and complexes) also prevent the development of experimental ulcers in animal models (Frechilla et al., 1991; Goel et al., 1992; Bulbena et al. 1993), although a common feature of many agents under investigation as potential antiulcer agents is that protection is only attributable to a mild irritant activity where higher doses may, in fact, be injurous (Silen, 1988).

Adverse GI effects from metal medicinals have been reported in patients taking gold compounds for rheumatoid arthritis (Fortson and Tedesco, 1984) and lithium carbonate for manic depression (Cannon, 1982). However, platinum compounds possess probably the most frequent and problematic GI toxicity of any metal medication currently in use, although since these compounds are invaluable for tumor therapy, the benefits of therapy outweigh its disadvantages (Mitchelson, 1992). Indeed, major research efforts in the pharmaceuticals sector have been undertaken aimed solely at controlling these GI effects. In certain situations, medications may exert GI toxicity as a desirable effect, such as with the use of copper sulfate or saline emetics to alleviate the effects of more serious intoxications. Barium sulfate is used clinically in roentgenographic diagnostic procedures because of its high X-ray density, high tolerability, and poor absorption (due to low solubility) from the GI tract. It is usually safe, although through its widespread use occasional cases involving colon perforation have been reported (Han and Tishler, 1982). Recently, oral magnetic particles (ferric iron complexes) have come under scrutiny as contrast media for diagnostic magnetic resonance imaging of the GI tract. Although they are reasonably well tolerated, they do induce nausea and vomiting (Rinck et al., 1992).

B. ENVIRONMENTAL, OCCUPATIONAL, AND HOUSEHOLD SOURCES

The environment and households also represent significant sources of exposure to GI active metals. GI metal toxicity results almost exclusively from contamination of food and water supplies. Although environmental exposure does not often result in damage as severe as that observed with medicinal exposure, a greater proportion of the population is affected. Furthermore, the risks of intense exposure arising from industrial accidents are not likely to decrease, particularly in industrialized societies where economic austerity may hinder enactment of rational environmental protective measures. The most significant environmental metal toxicants affecting the GI tract appear to be cadmium, mercury, lead, and arsenic. Thallium is also growing in significance through its use as a rodenticide, particularly in underdeveloped nations, where more specific compounds may be less economically viable.

Lead is clearly an important environmental pollutant, and through its use as an antiknock agent in gasoline, its levels tend to be highest around heavily populated areas. It is also found in piping, paint pigments, alloys, plastics, and battery oxides, all of which may be common in households (Keogh, 1992). Its use in colored printed material and paint presumably predisposes children to a greater risk of oral exposure. Lead invariably contaminates food, but its levels are not usually sufficient to evoke GI toxicity. The GI tract is nevertheless sensitive to lead, which induces a specific pathology characterized by spastic pain (colic), and alternate diarrhea and constipation.

Contamination of food and water supplies arises generally from industrial pollution. However, even in heavily industrialized areas, one rarely finds enough metal contaminant to induce GI disturbances. An exception to this rule, however, may be considered for the role of cadmium in itai-itai disease, whereby ingestion of cadmium-contaminated rice foodstuffs (in Japan between about 1890 and 1950) resulted in disruption of calcium uptake in the intestine, and thereafter impaired bone function (Tsuchiya, 1978). Hence, metal effects on the GI system can potentially predispose more serious pathologies elsewhere in the body. However, these patients did also show signs of duodenal enteropathy, characterized by shortening of the villi (Murita et al., 1969).

Mercury, unlike cadmium, has always presented itself as an environmental contaminant, since a large proportion arises from natural (e.g., volcanic) rather than anthropogenic sources. The GI toxicity of the divalent form is well known, and is one of the most significantly toxic of all metals to the GI system, causing ulceration, bleeding, and ultimately death through these effects (Aaranson and Spiro, 1973; Campbell et al., 1992). Ingestion is the major route of exposure for inorganic mercurials. However, although they continue to be used in a number of occupational settings and are still used in household goods such as batteries, environmental sources represent an insignificant fraction of inorganic mercurial intoxications. Environmental contamination by organic mercury compounds still remains a significant hazard, especially through their use in agrochemical products, and their ability thereafter to accumulate in the food-chain. However, although these compounds enter primarily through ingestion, they are efficiently absorbed and exert few if any significant GI toxic effects.

Arsenic often finds its way into food and water supplies through its use in pesticides. It is also significant by-product of a number of heavy industries (Léonard, 1991; Dart, 1992). Occasionally, high concentrations of arsenic are found in natural sources, such as artesian well water (Chen and Wang, 1990). Most cases of acute intoxication arise from ingestion, and GI effects thereafter include garlic metallic taste, mucosal burning, nausea and vomiting, diarrhea, hematemesis, and melena. Stomatitis and diarrhea can also result from chronic intoxications. The potential for large-scale exposure to arsenic remains significant and serious, as demonstrated by the mass poisoning of over 12,000 Japanese children by arsenic-tainted dry milk, an incident which caused 130 deaths (Yamashiti et al., 1972). Of all metals, arsenic perhaps is that which is most commonly employed for inducing death by suicide or homicide.

Thallium is extensively used as a rodenticide and pesticide, and because of warfarin resistance, its use has not decreased, particularly in developing countries (Sullivan, 1992). Furthermore, contamination of drinking water has been found in areas surrounding the metal industry. Thallium exposure has resulted from both suicidal or accidental exposure to monovalent thallium salts, the latter of which occurs most often from ingestion of pesticide-contaminated fruit. Acute exposure to thallium results initially in nausea and vomiting, which may subside. Much later, however (7 to 14 days), further GI symptoms may develop, such as pain, constipation, a bloating sensation, and bleeding.

III. FACTORS AFFECTING METAL POISONING IN THE GUT

Metals have diverse physical and chemical properties which naturally influence the degree to which they exert GI toxicity. Among the most important factors affecting toxicity are solubility, speciation, molecular form, efficiency of absorption, and metal–nutrient interactions. The major barrier to the movement of toxic agents from the gut lumen to the mucosa and beyond is the mucosal epithelial layer. In order to exert toxicity, substances must be able for the most part to penetrate this layer. The mucus layer adherent to most parts of the GI tract does not provide a significant barrier function to many, especially small molecules, and exists primarily as a lubricant and in the upper GI tract as a mixing barrier in which backdiffusing luminal acid can be neutralized by a countercurrent epithelial alkaline secretion (Allen et al., 1988). Recent studies also suggest that mucus even facilitates transport of metal cations to the mucosa (Conrad et al., 1991). In the stomach, the epithelium is tight, a property afforded by a junctional complex, the so called zonula occludens, which forms a continuous barrier around the cells. Although this restricts diffusion of larger molecules, water passes through pores at the apical intercellular junctions which also show selectivity for transporting metal cations by way of the negatively charged groups which line their inner walls (Powell, 1983). Organic lipophilic molecules readily partition in the cellular membranes, and the absorption of organic acids is even facilitated by the presence of the large gastric lumen to blood pH gradient. Such mechanisms are useful for facilitating entry of orally administered pharmaceuticals, but also serve to exacerbate toxicity induced by various gastric toxicants such as nonsteroidal anti-inflammatory drugs (Aungst and Shen, 1986). The small intestine is an organ specialized for absorption, and consequently has a higher surface area and a relatively leaky epithelium. Absorption of inorganic charged solutes (including metals) therefore tends to be more effective here, particularly in the upper (duodenal) and mid (jejunal) small intestinal regions. Furthermore, metal ions can compete for specific transport mechanisms present in small intestine, such as those which already exist for essential metals such as calcium and iron (Foulkes, 1984). The large intestine has in turn a tighter epithelium and is generally specialized for reabsorption of water from the intestinal contents before defecation.

The importance of speciation in determining the GI toxicities of ingested metals is well illustrated in the case of mercury. Most elemental metals presented to the GI system are unreactive, of very poor solubility, and tend to be in large particulate form. Very often the most significant problems presented are merely physical. Only in the stomach are significant amounts of metal ion likely to be released as a result of acid action, a process which in any case is greatly limited by rapid gastric emptying and neutralization. Consequently, ingested elemental metals tend to be well tolerated in comparison with their ionic counterparts. Being insoluble in both aqueous and nonaqueous media, they can not enter the GI tract wall through the mechanisms described above. Elemental mercury, which systemically is rapidly converted to the GI toxic mercuric form, is hardly absorbed from the intestine when ingested, and reports have shown that individuals have ingested as much as 3 kg without significant GI toxicity (Lin and Lim, 1993).

Marked differences in toxicity are often observed depending on the ionic form of a metal. Toxicity tends not to be dependent on differences in solubility, but rather on differences in the way ions affect organ biochemistry. Many biochemical processes are flexible with regard to the elemental form of an ion which they may require, but are highly dependent on their charge and ionic radius. Cobalt, for example, can readily substitute other essential divalent cations in many enzymes *in vitro* without apparent effect on their biological activity (Lindskog, 1970). With regard to GI toxicity, mercury again illustrates the importance of speciation. Mercuric chloride is highly corrosive and toxic to the GI tract, and in rats exhibits a lower oral LD_{50} (37 mg/kg) than does mercurous chloride (210 mg/kg), which, although as soluble, only exerts mild GI toxicity (NIOSH, 1976; Aaronson and Spiro, 1973). Indeed, for this reason mercurous chloride has been used historically as a purgative (Wands et al., 1974). The GI toxicities of both arsenic and chromium are also very dependent on speciation, with trivalent arsenic being more toxic than the pentavalent form (Léonard, 1991), and hexavalent chromium more corrosive than the trivalent form (Cohen et al., 1993). Sometimes the toxicity of a less toxic form of an ion is dependent on the degree to which it can be converted to the more toxic form.

Inorganic compounds are unlike many organic compounds in that when not water soluble, they are not particularly lipid soluble, and are therefore poorly absorbed by the GI tract. Insoluble salts of toxic inorganic ions therefore tend to be less toxic than more soluble salts. Organic metal compounds, however, are often lipophilic when water insoluble, and are especially well absorbed if dispersed as colloids. Despite the high permeability of the small intestine to inorganic charged solutes, overall intestinal uptake of organic metal species is much higher than their inorganic counterparts, even when the inorganic ions are fully soluble. This is reflected by the higher oral (as distinct from GI) toxicity of methylmercury, which is almost completely absorbed, compared to mercuric chloride, of which only 10% is absorbed (NIOSH 1976; Campbell et al., 1992). Furthermore, the target organ for methylmercury is the brain, whereas that for mercuric chloride is both the GI tract and the kidneys. Most metals which produce GI effects are even less efficiently absorbed than mercuric chloride, and those which are efficiently absorbed tend to produce pathologies elsewhere in the body without causing much GI toxicity. Efficient absorption therefore seems to present less GI problems than inefficient absorption and high latency in the GI tract, possibly because once absorbed, metals are efficiently transported into the blood (Mykännen, 1986). Another important factor may be that a poorly absorbed substance is more likely to induce GI damage towards the distal end of the GI tract where protective and regenerative mechanisms may not be as well developed as those in the upper GI tract. However, some metals, such as thallium, copper, zinc, and arsenic do exert significant GI toxicity despite their efficient transport and uptake into the blood. One can therefore not generalize the GI toxicity of metals, taking into account only differences in their respective toxicokinetics.

Nutrient–metal interactions are important in influencing both GI and systemic toxicity in several ways. For example, zinc, copper, and iron deficiencies enhance the adverse systemic effects of cadmium poisoning (Bremner, 1978), whereas malnutrition increases tissue retention and toxicity of lead (Foulkes, 1984). Selenium is an essential element in some enzyme systems, and its ability to antagonize the effects of arsenic mutually suggest that part of its toxicity may arise from its ability to substitute for selenium (Levander, 1977). Selenium also protects against toxicity induced by cadmium, mercury, copper, thallium, and cisplatin, possibly by enhancing selenium-dependent GSH-peroxidase activity and other antioxidant mechanisms which might antagonize the lipid peroxidation-stimulating effects of various metals (Fishbein, 1991). Another type of detoxification arising from metal-nutrient interactions occurs as a result of chelation by dietary constituents such as histidine or cysteine in proteins, citrate, ascorbate and various food additives (Mykännen 1986).

IV. PATHOPHYSIOLOGY OF METAL POISONING IN THE GUT

Metals induce a diverse and often nonspecific pathology to the GI system. Indeed, many of the common symptoms and signs of intestinal poisoning can be induced by different metals to varying degrees. In pathological doses, most metals induce signs of general discomfort shortly after ingestion, such as abdominal pain, nausea, and vomiting. These serve both as early warning signals and crude mechanisms for the removal of noxious substances from the gut. Other, usually later-developing symptoms or signs may indicate disruption of intestinal homeostasis (diarrhea, constipation), or frank tissue cytotoxicity (GI ulceration and bleeding and GI infection, which can follow extensive damage). Carcinogenesis and tumor promotion are other important aspects of intestinal metal toxicology to consider. Unfortunately, since cases of metal poisoning in the GI tract are rare, and comprehensive clinical or experimental studies are limited, information concerning GI metal toxicity is necessarily qualitative.

A. NAUSEA AND VOMITING

Although nausea and vomiting are often interpreted as detrimental to one's well-being, they represent necessary phenomena important in removing ingested toxic agents. In addition to the involvement of nausea and vomiting as responses to acute poisoning, they are also important features of various diseases, motion sickness, pregnancy, and cancer chemotherapy, and in all these cases confer little or no advantage for those individuals who must endure it.

With regard to metals, of which most induce this reaction to some degree, knowledge about the specific mechanisms involved in inducing nausea and emesis is limited. However, their induction does not necessarily imply that metals (and indeed other substances) exert their primary toxicity by influencing the GI tract. Early studies using copper sulfate as an emetic indicated that vomiting was induced by activation of sensory neurons within the GI tract (Borison and Wang, 1953). Messages from the GI tract are conveyed to the nervous system via both sympathetic and parasympathetic afferents. However, even after nerve ligation, vomiting could still be induced after absorption of high doses by direct activation of the chemoreceptor trigger zone (CTZ), located in the area postrema on the upper surface of the medulla, and from which neurons pass to the vomiting centre. The CTZ has a fenestrated epithelium surrounded by a double basement membrane which forms a perivascular space into which noxious substances in the blood can enter freely without having to cross the blood-brain barrier. This allows access to small components, such as metal ions, which through as yet unknown mechanisms set the CTZ into action. Motor neurons arising from the vomiting center are responsible for the final execution of the vomiting reflex, which involves contraction of the respiratory diaphragm rather than alteration in the contractile status of the gastric musculature. Nausea is often considered to result from low level stimulation of the vomiting reflex, although vomiting can occur without nausea.

The degree to which specific orally absorbed metals induce nausea and emesis through either central (CTZ) or GI-mediated mechanisms is generally unknown; however, much information has been gained about one particular metal complex, cisplatin, which is unusual in that it is a particularly powerful emetic, and that circumstances have demanded that its mechanism of action be elucidated (Mitchelson, 1992). Nausea and emesis are among the most significant problems associated with cancer chemotherapy. Significantly, cisplatin, used for ovarian cancer treatment, is one of the most effective of these agents in inducing nausea, which occurs in a delayed manner in over 90% of patients receiving injections (Plezia and Alberts, 1985). With this agent, an intact CTZ is required, whereas with other cytotoxic drugs this is not always the case (Mitchelson, 1992). Indeed, animal studies suggest that not all anti-cancer drugs induce emesis by the same common pathway, and cisplatin-induced emesis has proven the hardest to control pharmacologically (Sanger, 1990). Activated serotinergic mechanisms are believed responsible for mediating the emetic properties of cisplatin. The importance of these pathways has been best illustrated by the proven clinical effectiveness of the specific 5-HT$_3$ receptor antagonist ondansetron, in controlling cisplatin-induced emesis (Marty et al., 1990). Cisplatin induces serotonin release indirectly from intestinal enterochromaffin cells by activation of excitatory interneurons (Schwörer et al., 1991). This study also suggested that cisplatin may activate 5-HT$_3$ receptor, both in the gut on the autonomic afferents, and in the area postrema where these messages are relayed to the vomiting center. Another feature of cisplatin's intestinal toxicity is that it causes a histological necrosis, shortening of crypts, and decrease in cell number in ileum and jejunum, where the intestinal stem cells appear to be the primary target (Choie et al., 1981). Recent studies have suggested that cisplatin exerts toxicity to rapidly proliferating cells by activating programmed cell death, or apoptosis (Barry et al., 1990; Evans and Dive, 1993). It remains to be seen whether this mechanism underlies cisplatin's ability to induce enteropathy. Furthermore, as other anticancer agents are also known to induce enteropathy (Ijiri and Potten, 1987), the link between this effect and their ability to induce nausea and emesis also warrants investigation.

Gallium nitrate is another compound which has shown potential as an antitumor agent, since it is cytotoxic and accumulates in certain types of tumor (van Riel, 1988). It is also effective in treating malignancy-associated hypercalcemia, since it inhibits bone resorption (Hughes and Hanson, 1992). Phase II studies have revealed that like other antitumor agents it induces vomiting and nausea, although only in about 33% of patients receiving injections.

B. ULCERATION AND BLEEDING

While the occurrence of nausea and vomiting does not specifically indicate that events which lead up to them result from intestinal cytotoxicity, the same can not be said for ulceration and bleeding. Furthermore, in serious cases the latter pathologies may be life threatening. The pathophysiology of metal-induced ulceration and bleeding has not been extensively investigated; however, important advances in the understanding of xenobiotic-induced ulceration have come from studies using other toxicants, such as corrosive agents, alcohol, and aspirin. These studies concentrated on gastric and duodenal mucosal effects, as part of an ongoing effort to understand the pathophysiology of acute and chronic ulcer diseases.

Maintenance of normal intestinal mucosal function results from a balance between aggressive luminal forces (provided by endogenous acid, enzymes, bile salts, and exogenous toxicants) and protective mechanisms within the gut wall (Allen et al., 1988). Gastric studies have shown that the gut wall can accommodate limited damage without lasting pathologically significant effects. Graded increases in alcohol concentration, for example, produce progressively deeper damage, which even after exposure to relatively high concentrations can be reversible (Silen and Ito, 1985). Ethanol readily permeates through the mucus to the stomach's tight surface epithelium and causes its destruction by desquamation into the adherent mucus layer. Mucous neck cells in the gastric pits remain relatively protected. The resulting increase in permeability of the epithelium allows blood components (notably fibrin) to diffuse from the perivascular space into the mucus and form an impermeable structure which prevents further diffusion of both toxicant and acid into the mucosa (Figure 1). This then allows epithelial regeneration by migration of mucous cells from the gastric pits over the exposed lamina propria (Sellers et al., 1987). Acid damage

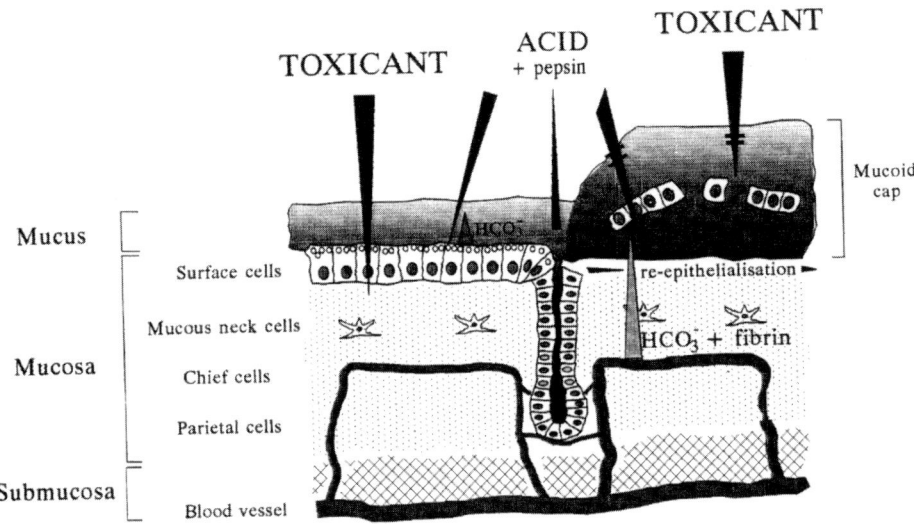

Figure 1. Mechanisms involved in protecting GI mucosa against moderate toxic challenges. This figure illustrates epithelial restitution such as is caused in the stomach by moderate challenges with various toxic agents (Silen and Ito, 1985; Sellers et al., 1987). Mechanisms with similar characteristics probably occur in other parts of the intestine such as the small intestine (Matovelo et al., 1990) and colon (Böhme et al., 1992a).

is also restricted because of a plasma-derived alkaline tide produced in areas of damage which far exceeds the bicarbonate secretion normally produced by intact gastric mucosa (Takeuchi and Okabe, 1983).

To above-mentioned phenomenon of "epithelial restitution" occurs in similar forms throughout the GI tract (Matovelo et al., 1990; Böhme et al., 1992a,b), and allows protection of the gut wall against xenobiotic challenges which might otherwise cause more extensive damage. After severe toxic challenges, however, damage may extend deeper into the mucosa, and hyperemia and hemorrhage may result. Restitution of the epithelium may also be prevented by destruction of the pit/crypt cells as well as the lamina propria. When healing occurs from this situation it takes several days, involving inflammation and other recovery processes (Silen and Ito, 1985).

Significantly, so long as all the components of mucosal protection remain intact after a moderate toxic challenge, factors such as luminal acid will not complicate the recovery process. However, xenobiotic-induced disruption of epithelial recovery may render the mucosa susceptible to acid/pepsin in the upper GI tract, or bile in lower intestinal regions. This may occur if damage is extensive, or if particular protective mechanisms are specifically disrupted. E-series prostaglandins (PGs), for example, are very important in maintaining gut wall function, especially when the mucosa undergoes deep damage (Szabo, 1987), and certain metals may interfere with their production (Fujimoto et al., 1991). Their presence may determine whether an agent will produce hemorrhage as well as extensive mucosal and lamina propria damage, which they accomplish by maintaining mucosal capillary function and preventing capillary congestion. PGs are generally important in maintaining or activating GI protective mechanisms, and different regions of the GI tract rely on these mechanisms to varying degrees. The duodenum, for example, possesses a leaky epithelium and relies heavily on a PG-sensitive bicarbonate secretion for neutralizing the luminal acid (Wilkes et al., 1988) which might complicate the effects of duodenum active toxicants. In the lower GI tract, PGs are involved in mediating diarrhea, which is an important mechanism for removing GI toxicants from the body (see next section). Figure 2 summarizes the mechanisms by which agents induce erosive damage to the GI tract and other problems developing from these effects.

Among the metals producing the most significant toxicity resulting in necrosis and bleeding are iron, chromium, zinc, mercuric, arsenic, and to a lesser extent, cadmium salts. Isolated case reports also exist for other metals inducing this kind of toxicity. GI toxicities of ferrous salts are well known from case reports of overconsumption of iron sulfate tablets, particularly in children, where of all known medications only aspirin causes more intoxications (Friberg et al., 1979; Hammond and Beliles, 1980). These and experimental studies have revealed an intense corrosive action in both stomach and duodenum, characterized by extensive subepithelial erythema and hemorrhage and a copious mucus secretion

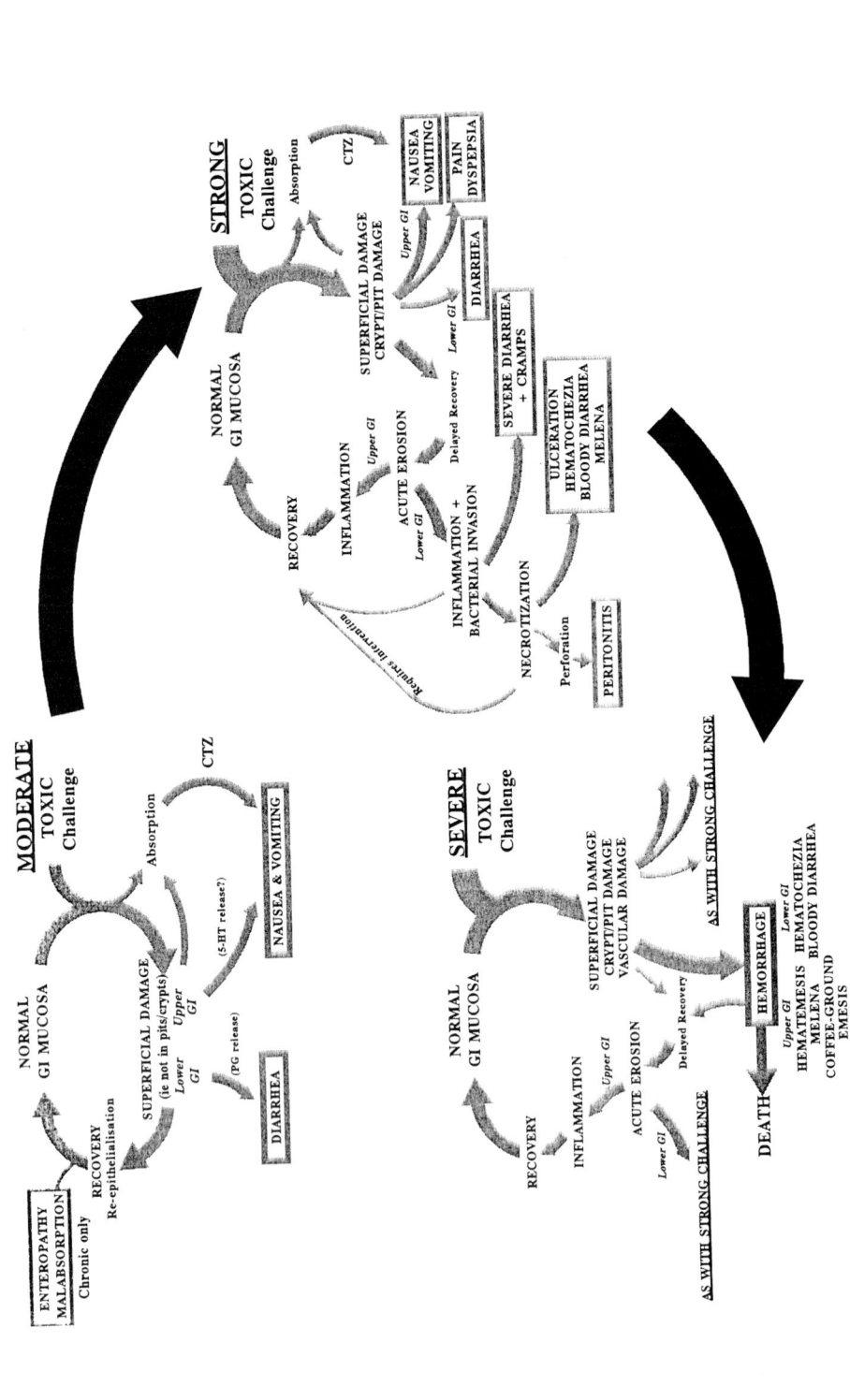

Figure 2. Symptoms associated with various degrees of toxicant damage. This figure summarizes the symptoms and pathologies usually involved with different grades of toxic damage induced by various nonspecific agents. Many metals, such as zinc, mercury, and iron approximately follow the scheme indicated here.

(Benoni et al., 1993; Mann et al, 1989). Stricture and infection have also been reported during the weeks following intoxication. The unique physical chemistry of iron probably plays a significant role in determining its upper GI toxicity. Both ferrous and ferric ions act as strong catalysts and participants of reactive oxygen metabolite (ROM) production, which can occur extracellularly in the gut lumen. Iron preferentially stimulates hydroxyl radical production which can lead to membrane damage via lipid peroxidation (Fodor and Marx, 1988). The upper GI tract is particularly sensitive to oxidative stress which results in production of both superoxide and hydroxyl radicals. This can be demonstrated by the experimental production of ulcers in intestine and stomach by ischemia and reperfusion (Parks et al, 1982; Itoh and Guth, 1985). Presumably under these conditions lipid peroxidation could increase the permeability of gastric mucosa to acid back diffusion. Iron also stimulates small intestinal granulocyte infiltration caused by ischemia-reperfusion (Zimmerman et al., 1990), and blocks desferrioxamine (an iron antagonist)-stimulated PG synthesis in rabbit gastric mucosa *in vitro* (Fujimoto et al., 1991), suggesting other mechanisms by which iron may be able to propagate its own toxic effects.

The mechanism of chromium toxicity also involves ROMs. Ingestion of 0.5 to 1 g of potassium chromate is fatal to man, causing diarrhea and upper GI hemorrhage (Schlatter and Kissling, 1973). Hexavalent chromium is normally rapidly absorbed by cells and then undergoes a rapid stepwise intracellular reduction to the trivalent form, a process accompanied by excessive ROM production (Cohen et al., 1993). Significantly, although trivalent chromium is comparatively nontoxic (with an oral LD_{50} about 100 times higher [NIOSH, 1976]), it is better absorbed from the GI tract than the hexavalent form (DeFlora et al., 1987).

Zinc, when administered therapeutically in tablets as zinc sulfate, causes gastric erosions when high doses are accidentally ingested. This is not a specific effect of zinc, but arises because in the stomach it is converted to the highly caustic combination of zinc chloride and sulfuric acid. Hence the course of its toxicity is similar to that of other caustic agents such as strong acids or bases, which produce progressively deeper damage with increasing dose, which can eventually culminate in hemorrhage. The lowest toxic dose in man estimated for zinc sulfate is 106 mg/kg (Ohnesorge and Wilhelm, 1991). Lower doses of zinc tend to cause cramps and diarrhea and are emetic, a property which usually prevents more damaging manifestations. A major factor determining GI toxicity with metal medications arises from physical factors, such as tablet formulation, which prevent transit and allow them to concentrate their effects at specific parts of the GI tract. Consequently clinical observations may not necessarily agree with *in vivo* experimental toxicity data, where for example, LD_{50} (oral) values of between 1 to 2.5 g/kg for most zinc salts are usually reported (NIOSH, 1976).

With regard to the GI tract, the mercuric cation in concentration terms is one of the most damaging of all metals. It is extremely corrosive to all mucous membranes, and the GI pathology often determines this substance's lethality. LD_{50}s in rats for parenteral and oral administration are similar (10 to 60 mg/kg; NIOSH, 1976) even though orally it is poorly absorbed (Section IV.D). As a result of this poor absorption, toxicity can be exerted along the entire length of the GI tract. The mercuric ion has a particularly strong affinity for sulfydryl groups on proteins and other biological molecules such as glutathione (Clarkson, 1972). It therefore has a strong capacity to precipitate cellular proteins and cause an overall tissue necrosis (Aaronson and Spiro, 1973). Mercury sulfide may also precipitate in vascular tissue, which can predispose capillary congestion culminating in hemorrhage. Less severe intoxications involving tissue necrosis may also lead to GI infection and subsequent complications (see Section IV.E). Specific changes induced by cytotoxic concentrations of $HgCl_2$ have been investigated recently in rat colonic mucosa *in vitro* (Böhme et al., 1992). Exposure of 30 min to 10 μM $HgCl_2$ is sufficient to cause loosening of tight junctions primarily between surface epithelial, and to a lesser extent, crypt cells. An increase in permeability results, thus allowing diffusion of Hg^{2+} into the mucosa to be further facilitated. Furthermore, the surface, but not the crypt cells, are severely damaged and exfoliate from the lamina propria. Paradoxically, prostaglandin production appears to be stimulated by this treatment, although the authors did point out that this damage was fully reversible, since crypt cells remained unaffected, and remaining cells flattened and developed lateral processes to cover denuded ares of lamina propria. In this manner, electrical integrity was restored within two and a half hours of the toxic challenge. Clearly what the authors had confirmed using $HgCl_2$ as a model agent, was the colonic equivalent of gastric epithelial restitution (as illustrated in Figure 1). The diarrhea which often accompanies mercuric (comparatively insevere) intoxications can be explained by the ability of the stimulated PGs to activate purgative mechanisms (see Section IV.C). Moreover, many GI toxicants share an ability at low doses to activate PG production (Miller, 1983), which serves to control further damage by activating mucosal defense mechanisms. Clearly the above study does not model deep mucosal injury which culminates in hemorrhage, although it does confirm the similarity of mechanisms, both with regard to tissue and toxicant type, which operate at lower levels of damage.

Acute cases of trivalent arsenic poisoning are associated with symptoms of severe erosion, primarily in the upper GI tract (melena, hematemesis), probably because most is absorbed before it reaches the colon (Dart, 1992). Although its lethal toxicity is higher than that of mercury (human lethal dose 70 to 180 mg), this arises from systemic effects which develop subsequent to absorption (Léonard, 1991). Arsenic affects tissues rich in oxidative enzymes (such as the stomach) and acts both by inhibiting ATP synthesis and binding thiol anions necessary for enzymatic reactions (Léonard, 1991).

Cadmium at high doses sometimes leads to GI toxicity similar to that induced by mercury; however, these symptoms are usually overshadowed by earlier developing systemic pathologies (Waalkes et al., 1992). Cadmium and mercury are chemically related and show similarities in absorption and sulfydryl binding characteristics (Foulkes, 1993). Intestinal metallothionein also plays a significant role in determining the experimental intestinal toxicity of cadmium, with intracellular metallothionein protecting and extracellular (extruded) metallothionein exacerbating such toxicity (Valberg et al., 1977). Significantly, 100 μM $CdCl_2$ exerts similar changes in mouse intestine (Valberg et al., 1977) to those observed after exposure of rat colonic mucosa to 10 μM $HgCl_2$ *in vitro* (Böhme et al., 1992), although the latter study showed no effect at 50 μM $CdCl_2$. These figures are in line with our own studies using epithelial jejunal cells, where cytotoxicity was determined *in vitro* (see Section V). Enteropathy was one of the main clinical features observed after chronic cadmium poisoning in patients with itai-itai disease (Murata et al., 1969). This may have resulted from a gradual attrition of the mucosal regenerative mechanisms (epithelial restitution) which may have been overactive as a result of repeated superficial damage to the intestinal mucosa.

C. CONSTIPATION AND DIARRHEA

Although constipation and diarrhea are often viewed as opposite ends of the same problem, their pathophysiology appears in many cases to be quite different. Constipation usually arises primarily from altered gut neural motor function, and rarely from changes in gut secretory or absorption capacity (Venho, 1986). The motor activity of the GI tract is controlled by the autonomic nervous system, the enteric nervous system, and locally acting hormones. Parasympathetic activity enhances, whereas sympathetic activity decreases, propulsive motor activity. Not surprisingly, adynamic constipation can be caused by anticholinergics which diminish propulsive activity by decreasing motility in a seemingly coordinated fashion. Spastic constipation, however, arises from dysfunction of gut neural motor mechanisms. Hence agents may even increase some aspects of gut motor activity, but through a lack of coordination, efficient intestinal transit is not achieved. Agents which alter specific neural motor components involved in transit may be particularly effective in this respect. Hence opiates, which act directly on the intestinal smooth muscle, cause spastic constipation (George, 1980).

With regard to metals, aluminum and calcium antacids cause adynamic constipation (van Riel, 1988). Although this could be related to their ability to decrease luminal pH, a factor known to inhibit intestinal motility, this can not satisfactorily explain why magnesium-containing antacids should cause diarrhea. Iron salts induce constipation in about 10% of poisoned patients, and diarrhea in another 10% (van Riel, 1988). Experimental studies in dogs show that at low doses, ferrous sulfate enhances activity by increasing the length of the migrating motor complex. At higher doses, however, the activity of this cycling is disrupted, while tonic contraction in the bowel ensues (De Ponti et al., 1991). A spastic constipation may therefore arise from a dysfunction of gut motor activity. Differential effects on components involved in maintaining gut motility may explain why agents under certain conditions may induce diarrhea, and under others constipation.

The mechanisms by which lead induces pain (colic) and constipation/diarrhea are largely unknown. Lead is not particularly cytotoxic and is believed to exert its effects by interfering with the activity of the enteric nervous system (Keogh, 1992). Studies in rats show that lead decreases the amplitude of contraction waves, thereby decreasing the efficiency of GI transit (Karmakar and Anand, 1989).

Although diarrhea can result from hypermotility alone, it more often results primarily from changes in gut absorption and secretion mechanisms (Binder and Sandle, 1987). Since the enteric nervous system is highly integrated, rarely does one see changes in secretory status which occur without some change in motility. Diarrhea will result either if small intestinal damage results in large increases in fluid presented to the colon, or colonic damage/irritation prevents normal water resorption (Venho, 1986). Diarrhea is a common reaction to the ingestion of many toxic substances, and like vomiting, is an important mechanism allowing their removal from the body. Recent evidence has even suggested that serotonin may be involved in activating purgative mechanisms in a manner analogous with that of its involvement in emesis (Ooe et al., 1993; von der Ohe et al., 1993). In many cases, local irritation by toxic agents may be sufficient to trigger diarrhea, since mediators such as prostaglandins (and possibly serotonin) may be produced which increase small intestinal fluid output and inhibit colonic water resorption.

Metals may induce diarrhea simply by bulk osmotic action. Magnesium sulfate is not soluble in the gut, and consequently can exert activity via this mechanism (Gullikson and Bass, 1984). Activation of motility, however, is also associated with osmotic fluid retention. Magnesium depresses smooth muscle activity, but in the small intestine contraction spike activity is elevated, whereafter the intestinal contents are swept into the colon and the absorption capacity is exceeded. A common side effect of gold therapy is diarrhea, which occurs in about a third of patients receiving the oral formulation (van Riel, 1988).

Many metals induce diarrhea as a result of their irritant activity. Corrosive salts, such as those of iron or mercury, can also cause bleeding which can manifest itself as melena if damage occurs in the upper GI tract, or bloody diarrhea if it occurs in lower GI regions (Sullivan, 1992). Studies on rat colonic mucosa *in vitro* confirm that mercuric chloride stimulates prostaglandin production, which results in increased chloride secretion via intramucosal secretomotor neuronal pathways, and thereafter increased fluid retention in the bowel (Böhme, 1992b). These studies also showed the importance of mercury's ability to break the mucosal barrier to exert this effect. Significantly, concentrations of cadmium which did not break the mucosal barrier function were only able to affect these mechanisms when administered to the serosal side of the mucosa. Hence, for diarrhea to be stimulated by this mechanism, it appears that some degree of intestinal cytotoxicity must first be induced.

D. MALABSORPTION

Malabsorption as a pathological response to intestinal metal intoxication is unusual in that the primary symptoms occurring as a result are systemic. Since the major function of the GI tract is nutrient absorption, metals can cause a generalized malabsorption if they induce cytotoxic damage. However, malabsorption may also be induced more specifically after chronic and lower exposures to these metals. Such intoxications tend to be more common than those causing acute pathology, and usually involve mechanisms involved in essential metal transport.

The major systems involved in essential metal transport include (1) the transferrin system for both ferric and ferrous iron uptake, (2) Vitamin D-facilitated calcium/magnesium transport, and (3) homeostatically regulated zinc and (4) copper transport. Intrinsic factor-mediated uptake of hydroxycobalamin, and sodium/potassium transport (by paracellular shunts, Na^+/K^+-ATPase, Na^+/Cl^- symport, and Na^+/H^+ exchange) will not be considered further here, since metals rarely interact with these systems to cause malabsorption. Importantly through, many metals do enter the body by competing with these mechanisms, including lithium and other alkali metals (paracellular shunts)(Schultz, 1979) and thallium (through the Na/K-ATPase and paracellular shunts) (Gehring and Hammond, 1967).

Iron, although it can be absorbed throughout the intestine, is absorbed mainly in the duodenum and upper jejunum (Huebers and Rummel, 1984). At low concentrations, iron absorption is carrier-mediated, and at higher levels it is determined by diffusion. Iron attached at the luminal surface of the enterocyte is transported by transferrin to the basolateral side where it is released to plasma transferrin. Another cellular protein, ferritin, binds but does not transport iron, and thus prevents its transfer to blood. Excess iron bound in this manner can then be excreted after epithelial desquamation.

Calcium/magnesium transport may occur by trans or paracellular pathways, and may be absorptive or secretory in direction (Forte, 1984). 1,25-Dihydroxycholecalciferol at physiological concentrations stimulates uptake in jejunum and ileum, and at supraphysiological concentrations stimulates duodenal and colonic uptake. Luminal calcium enters cells via a Ca^{2+} channel, after which it binds a calcium-binding protein which transports it to the serosal membrane to be secreted by a Ca^{2+} ATPase. Vitamin D regulates these components both directly via its receptor, and indirectly via gene induction.

Mechanisms involved in zinc and copper uptake are not well characterized (Foulkes, 1984). Generally, once absorbed to the intestinal wall, most zinc and copper is transported eventually to the bloodstream, and these transfers are subject to homeostatic control. Uptake of low concentrations is saturable, suggesting the presence of specific carriers, but uptake at higher concentrations is not, showing that they can also enter by simple diffusion. Zinc absorption is controlled by metallothionein, a protein with an analogous function to that of transferrin. A separate mechanism is suspected to be responsible for copper absorption, since in Menke's disease patients show a hereditary defect in copper, but not zinc absorption. However, a metallothionein protein has also been suggested to be involved in copper binding and transport (Felix et al., 1990).

Although it is attractive to look for single mechanisms by which heavy metals induce essential metal malabsorption, the situation in reality is complicated since transport mechanisms are inherently nonspecific, and interactions between metals may be complex. Lead for example, is mostly absorbed in duodenum, since it competes with Vitamin D-dependent calcium uptake (Fullmer, 1990). Vitamin D enhances lead absorption 4-fold in vitamin D-deficient rats, and calcium deficiency facilitates intestinal lead uptake. However, even at high doses, lead does not inhibit calcium absorption. On the other hand, iron deficiency increases body lead retention, and lead induces anemia, suggesting a role for transferrin in mediating lead uptake (Huebers and Rummel, 1984). Zinc and copper also inhibit calcium absorption, and zinc alone even promotes colonic calcium secretion, showing the level of competition that exists between essential elements. As further examples to illustrate the confusion of interactions among metals, copper binds transferrin, and iron inhibits zinc absorption (Foulkes, 1984).

Once cadmium is absorbed by the intestine, it is not efficiently transported to the blood. It inhibits calcium uptake directly or indirectly by altering hydroxylation of cholecalciferol in the kidneys (through renal damage)(Mykännen, 1986). This is particularly significant as it is the mechanism proven to underly the development of itai-itai disease, a disorder characterized by severe osteomalacia (bone decay; Murata et al., 1969). Cadmium also competitively inhibits zinc absorption, and binds and induces metallothionein (Foulkes, 1984). It has been suggested that the primary mechanism underlying cadmium's toxicity is its ability to compete with zinc for binding to various proteins. However, cadmium also interferes with iron and copper absorption, the former of which may explain the reversible fall in blood hemoglobin content seen after cadmium exposure (Schaefer et al., 1988).

Divalent mercury has particularly high avidity for intestinal tissue protein, and like lead and cadmium, is transferred only inefficiently to the blood (Foulkes, 1984). Thus, although a large percentage may initially bind to intestinal cells, only 2% is transported to the bloodstream. Although it is related chemically to zinc and cadmium, its transport characteristics are quite different. Indeed, studies on everted jejunal sacs suggest that mercury first binds as an anion, $HgCl_3^-$, to cationic sites before it is absorbed and transported (Foulkes and Bergman, 1993). This may explain the lack of information concerning specific effects of mercury on absorption of other ions. Although once in the cell mercury binds glutathione very strongly, glutathione does not seem to play a transferrin type role in binding of excess metal before excretion by desquamation (Böhme et al., 1992).

In addition to their effects on mineral malabsorption, cadmium, mercury, and copper have all been found to inhibit uptake of sugar and amino acids in rat jejunum, supposedly by binding to charges on protein functionally related to the sodium substrate cotransport systems (Rodriguez et al., 1989).

E. GASTROINTESTINAL INFECTION AND INFLAMMATION

Gastrointestinal infection is often a problem associated with extensive damage to a weakened, necrotically damaged, GI tract. Even organisms such as *E. coli* may invade necrotic tissue to exacerbate damage and cause localized inflammation (Simon and Gorbach, 1987). The lower GI tract is particularly susceptible to this kind of damage, where the most common long-lasting symptom is diarrhea. Significantly, however, iron appears to be especially effective in promoting intestinal infection following oral iron overload, possibly because as an essential nutrient for bacterial growth, it preferentially promotes the growth of various pathogenic organisms (Mykännen, 1986). In rats, chronic doses of iron too low to produce histological damage increase fecal output of *Clostridium perfringens* toxin (Benoni et al., 1993), an organism which in humans is implicated in the development of necrotizing enteritis.

Acute mercury intoxications which cause necrotic damage present good examples of how damage may arise from compounding GI infection (Aaronson and Spiro, 1973). In this case, bacteria invade necrotically damaged tissue, after which problems developing thereon may eventually prove even fatal. This is especially true where damage occurs in the colon, since in the upper GI tract luminal factors such as acid limit the development of infection. Delayed large intestinal symptoms include diarrhea, tenesmus, ulceration, and hemorrhage.

Generally, inflammation of the GI wall is a common consequence of necrotic damage. Chronic laxative abuse with mercurous chloride, for example, has also been found to result in colitis so severe that it requires surgery (Wands et al., 1974). The toxicity in this case arose from its conversion to the mercuric form, which deposited as mercuric sulfide crystals in the colon wall.

Gold, when administered either orally or injected, induces colitis, which, although rare, is a severe condition with a poor prognosis (mortality 40% among the 20 or so cases reported)(Fortson and Tedesco, 1984). The condition is characterized by severe bloody diarrhea and abdominal pain, and although it initially appears restricted to the colon, postmortem examinations reveal diffuse damage throughout the

GI tract. The colonic mucosa is often ulcerative, erythematous, and edematous, and histological examination reveals invasion of the lamina propria by lymphocytes and other plasma cells. The condition is difficult to distinguish from ulcerative colitis, but significantly, one patient responded well to cromolyn sodium therapy, suggesting an allergic etiology for this condition. As with other conditions where inflammation and necrosis are involved, a major problem is that of bacterial sepsis, which presumably contributes to the high mortality of this condition.

F. GASTROINTESTINAL CARCINOGENESIS

Arsenic, chromium, nickel, cadmium, and beryllium are the metals which are positively confirmed in humans as those which initiate and/or promote cancer disease (Snow, 1992). Some of these metals, because of their nonspecific activity, are considered complete carcinogens, since they can act both as promoters or initiators. Their roles in precipitating GI carcinogenesis, however, are questionable.

The role of arsenic in bringing about cancer disease is unequivocal, and it appears only to act as a tumor promoter. Occupational exposure is primarily by inhalation or dermal contact, as a result of which it can cause lung or skin cancer, respectively (Snow, 1992). With regard to oral exposure, however, a detailed study on Taiwanese who had ingested artesian well water contaminated with arsenic revealed an increased incidence of tumors in the skin, kidney, lung, liver, prostate, and the bladder (Chen and Wang, 1990). Tumor incidence in the GI tract, however, was not particularly high, suggesting that the organ was somewhat resistant to arsenic's effects. Arsenic is well absorbed by the GI tract, particularly in the duodenum, where it competes with phosphate transport mechanisms; hence little reaches the colon which is a major site at which primary cancers can develop in the GI tract (Léonard, 1991). Furthermore, when pentavalent arsenate is ingested, it must first be converted to trivalent arsenite before its tumor-promoting activity is revealed, which may occur more efficiently after absorption. The ability of arsenite to inhibit DNA ligase II (responsible for excision repair) may account for both its co-mutagenic and clastogenic activity. In summary, although arsenic is a tumor promoter over the oral route, its role in specifically promoting GI cancer remains dubious.

Chromium is both an environmental pollutant and an essential metal. Although it is a confirmed carcinogen among chromate workers, its site of action is restricted to its absorption site (i.e., primarily in the lungs, and less so in the skin)(Cohen et al., 1993). Although GI carcinomas have also been reported in these workers, controversy remains as to whether these were caused by chromium. Carcinogenicity arises from the ability of hexavalent chromium to enter the cells and then undergo stepwise reduction, a process resulting in various forms of DNA damage and free radical production. Chromium therefore can act both as promoter and initiator.

Nickel, although a powerful carcinogen, appears specific for the lung even after parenteral injection, and is not carcinogenic following ingestion (Snow, 1992). Beryllium, like nickel, is also a lung carcinogen without confirmed GI activity (Snow, 1992). The risks of exposure are again occupational.

Evidence for cadmium-induced carcinogenesis in man is weak although conclusive evidence has been obtained from animal studies (Stoeppler, 1991). The primary target organs in man are considered to be the prostate, testes, and lung. Recently it was argued that cadmium should not be considered a carcinogen by the oral route since (a) no evidence was found of GI tract cancers in occupationally exposed workers who developed prostatic cancer; (b) most experimental studies showed that high levels of orally administered cadmium were not carcinogenic to rats; and (c) the bulk of the evidence reviewed in that study only showed cadmium to be genotoxic *in vitro,* and not *in vivo* (Collins et al., 1992). This study did not rule out the possibility that cadmium was acting as a promoter. Cadmium induces DNA damage and lipid peroxidation and stimulates the oxidative burst response in macrophages, processes which are suggested as contributing towards both tumor initiating and promoting activity (Snow, 1992).

A variety of other metals have also been found to induce mutagenesis, DNA/chromosome damage or transformation in various *in vitro* models (Gebhart and Rossman, 1991). Tumors may also be induced *in vivo* at injection sites by zinc, cobalt, lead, and iron salts. Generally, little epidemiological evidence confirms the notion that these metals are carcinogenic to humans by normal routes of exposure. Significantly, in lead workers, increases in incidence of both GI and pulmonary cancers was noticed, although these studies did not rule out the possibility that confounding factors were responsible for the increased incidence (Cooper, 1985). There are clearly considerable difficulties involved in confirming agents to be carcinogenic in humans, and the true number of carcinogenic metals is probably higher than those listed. Furthermore, although many studies have investigated whether metals initiate carcinogenesis, there are only few studies investigating their abilities to act as tumor promoters. In our own group we found that dietary iron promotes colon carcinogenesis after induction by dimethylhydrazine in mice, an effect which was found to be dependent on stimulation of lipid peroxidation, and independent of the time or presence of the initial mutagenic challenge (Siegers et al., 1992). The significance of tumor promotion in giving rise to malignant disease is worthy of deeper investigation, especially since many mutagenic/clastogenic substances are not by themselves carcinogenic.

As well as being involved in tumor initiation and promotion, some metals are implicated in reducing the risks of developing active cancer. Some experimental studies have shown an antimutagenic activity for cobalt and germanium (Gebhart and Rossman 1991), but selenium is the metal which has most widely been confirmed for reducing cancer, and most particularly GI cancer, risk (Fishbein, 1991). Sodium selenite itself has been found to be a weak mutagen in several *in vitro* models, but only at high concentrations unlikely to be encountered in real life situations. More significantly, selenium inhibits the development of experimental carcinogenesis induced by many compounds, including metals such as cisplatin, which is carcinogenic to rodents (Rosner and Hertel, 1986). Most of these studies discussed by Gebhart and Rossman used amounts of dietary inorganic selenium (selenate or selenite) between 20 and 60 times higher than the usual daily requirement. However, strong negative associations exist between environmental selenium levels and the incidence of GI cancers (Clark, 1985), and blood selenium levels are lower in GI cancer patients than in the general population (Willett et al., 1983), suggesting that amounts adequate to prevent cancer disease may indeed be ingested by the general population. Among the mechanisms involved in the activity of selenium include its participation as a co-factor for selenium-dependent glutathione peroxidase (Siegers, 1989), and its ability to interact with the anti-oxidant vitamin E (Ip, 1985). Hence long-term protection is afforded by reducing the effects of free radical production and lipid peroxidation which would normally act to promote carcinogenesis. The fact that selenium reduces GI cancer risk through a mechanism involving prevention of tumor promotion illustrates again the importance of environmental tumor promoters in contributing to the development of active GI cancer disease.

V. EXPERIMENTAL INVESTIGATION OF GASTROINTESTINAL METAL TOXICOLOGY

An extensive understanding of the nature of GI metal toxicity naturally requires both quantitative and qualitative experimental *in vivo* and *in vitro* studies. In the absence of a large amount of data, some information can be found by comparing lethal toxicity observed after oral exposure with that observed after parenteral exposure. Large databases for this information do exist for upwards of 50 different types of metals, such as the NIOSH Registry of Toxic substances (NIOSH, 1976). In this case, one can assume that if parenteral toxicity is higher than oral toxicity, absorption from the gut is poor. With virtual equivalent toxicity, one can assume either that absorption is efficient, or that the toxic lethal effects arise from the GI symptoms caused by a poorly absorbed substance. A higher oral toxicity would provide a strong suggestion that lethal toxicity arises from GI effects irrespective of how efficiently the substance was absorbed. Analysis of NIOSH data, however, reveals that this situation is very rarely encountered, and with metals this is seen in only some cases of mercuric and thallic toxicity. More commonly, corrosive GI metal toxicants induce marginally lower toxicities when applied by oral as opposed to parenteral routes. Organic metal compounds, on the other hand, also exhibit high oral toxicity and lower ratios of parenteral/oral toxicity than their mother inorganic compounds, because they are efficiently absorbed from the gut. Most inorganic metallic compounds appear to be well tolerated when administered orally, although some of their number may be very toxic when administered parenterally. The most toxic of the orally administered salts are those of mercuric, cadmium, thallium, beryllium, and arsenates, as well as the cis-platinum compounds discussed before (less than 100 mg/kg). Iron, copper, cobalt, barium, lithium, manganese, mercurous, nickel, rhodium, and molybdate salts exert toxicity between 100 and 1000 mg/kg. Salts which appear to be better tolerated include among their number the more common alkali or alkali earth essential metals (Na^+, K^{2+}, Ca^{2+}, Mg^{2+}), other alkali metals (Rh^+, Cs^+, but not low atomic weight Li^+), and lead. Among other metals, lanthanides, actinides, and some other high atomic weight cations are also in this class, although when administered parenterally they can be highly toxic. Presumably this is because these salts are inefficiently absorbed, not being able to compete with the absorption mechanisms which exist for other essential metal ions.

Recently much interest has been focused on developing *in vitro* toxicity tests and evaluating their usefulness in predicting toxicities observed *in vivo*. Assuming a simple model for the distribution of substances *in vivo* throughout the total body water volume, good agreements can be observed between *in vivo* nephrotoxicity and *in vitro* toxicity in renal tubular cells, as well as *in vivo* hepatotoxicity and *in vitro* toxicity to hepatoma cells, for a variety of agents including heavy metals (Siegers and Samblebe

1992; Samblebe and Siegers 1992). Using a widely used and accepted procedure for determining *in vitro* toxicity, neutral red uptake (Borenfreund and Puerner, 1984), we determined toxicity induced in embryojejunal epithelial (I-407) cells by a range of toxicologically significant heavy metals. We aimed to determine whether these cells were especially resistant to those metals, and also the role of endogenous sulfhydryls, as factors known to be important in mucosal protection, in determining metal toxicity (see also Keogh et al., in press).

Generally our results showed that I-407 cells showed similar sensitivity to metals compared to hepatic or renal cells in culture (Siegers and Samblebe, 1992; Samblebe and Siegers, 1992), and did not seem to possess specific protection mechanisms against these toxicants. The results are summarized in Figure 3, in terms of the LC_{50} values obtained for each salt. Although mercuric chloride *in vitro* is one of the most toxic of the agents tested to the GI system, several agents were found to be more toxic *in vitro*. The higher toxicity of methylmercury was not surprising, since *in vivo* oral lethal toxicity of methylmercury (1 mg/kg in rat) is normally much higher than that of mercury chloride (10 to 40 mg/kg in rat) (NIOSH, 1976, Campbell et al., 1992). Being lipophilic, it can presumably gain access to sensitive thiols in various intracellular compartments more easily than inorganic mercury. However, gastrointestinal toxicity *in vivo* is limited because of rapid absorption into the blood and distribution to the brain, which is its target organ for toxicity. Similarly, trivalent arsenic was more toxic than mercuric chloride. It, too, is more toxic than inorganic mercury *in vivo*, is rapidly absorbed, and despite prominent GI effects, lethal toxicity again arises from systemic effects. Significantly, pentavalent arsenic (arsenate) was much less toxic than the trivalent form, consistent with the overall pattern of arsenic's cellular toxicity (Léonard, 1991). Presumably, biotransformation of arsenate to arsenite over the time of the experiment was insignificant. Zinc chloride was also found to be extremely toxic in this system. This was surprising, given the absence of the effects of luminal acid in this *in vitro* system, which would normally mediate the corrosive effects of zinc. Although sodium selenite was also found to be highly toxic, it should be stressed that the daily requirements for this essential metal are extremely low, as is normal dietary intake, which doe not often exceed 300 µg/d (Fishbein, 1991). Indeed, daily intake of only 0.5 ppm in water is considered to be dangerous to man. One study has suggested that selenite (though not selenate) at high concentrations interacts with cellular thiols, including glutathione, to produce ROMs and peroxides (Yan and Spallholz, 1993). Since they also found that seleno-cystine, but not seleno-methionine, could also act in this manner, our results might suggest that toxicity was being induced in I-407 cells by the same mechanisms. However, this needs to be further investigated using glutathione depletion experiments. Cadmium was only marginally less toxic than copper, although *in vivo* oral toxicities of these substances are all markedly different. This probably emphasizes the importance of mechanisms other than cytotoxicity in contributing towards the development of gastrointestinal hemorrhage. Surprisingly, thallium exerted weak toxicity, although it has an *in vivo* toxicity equivalent with that of mercury. Thallium is normally rapidly absorbed by the GI tract, and exerts delayed toxic symptoms to the whole body including the GI tract (Sullivan, 1992). The reasons for this delay in activity remain unclear, and it remains to be seen whether thallium induces such delayed effects in our *in vitro* system. Of all the metals tested, lead appeared to exert the least cytotoxicity. Although lead typically induces a marked specific pathology in the GI tract, it appears to be unrelated to cytotoxicity, and it is assumed to be due to specific effects on enteric neural pathways (Keogh, 1992). Oral LD_{50}s of lead salts are typically very high (1 to 10 g/kg in rats)(NIOSH, 1976).

Depletion of cellular glutathione by the gamma-glutamyl synthetase inhibitor buthionine sulfoximine caused a sensitization of the cells to the effects of mercury and copper, but not lead, thallium, or cadmium (Keogh et al., in press). With mercury and copper, toxicities induced by these agents were increased by 5.7-fold and 1.44-fold, respectively. Both divalent mercury and, to a lesser extent, copper sequester sulfydryl groups of glutathione and other thiols avidly (Vallee and Ulmer, 1972). With cadmium, however, it appears that no protection was afforded by the presence of intracellular glutathione, despite the fact that cadmium is also known to bind glutathione (Flinder and Nordberg, 1985). Foulkes (1993) discussed the relative distribution of metals among cellular thiols in rat jejunum, explaining that tissue cadmium binding appears much more specific than that of either mercury or copper, and preferentially associated with metallothionein, and not glutathione. Hence cadmium's toxicity may have reflected its binding to specific high affinity thiol groups, and not glutathione, whose binding led to lethality. In another experiment, we found that cadmium and mercury administered together at their respective LC_{50}s did not exert higher toxicity than either of the agents applied alone. This suggested that some of the mechanisms involved in bringing about toxicity may have been shared for the two metals. Hence, mercury by way of its nonspecific thiol binding activity was also competing with the cadmium-sensitive thiol binding

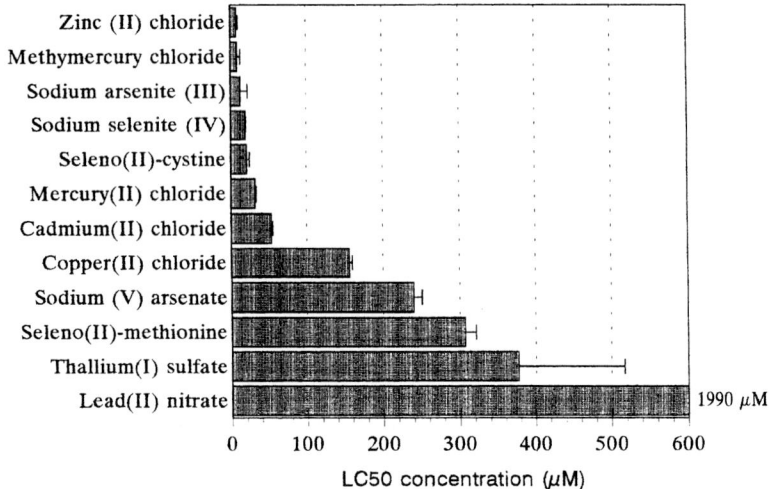

Figure 3. Comparative toxicities of different metal preparations on embryojejunal I-407 epithelial cells *in vitro*.

groups, but with a lower affinity than it possessed for glutathione. The two metals may have thus shared this common pathway for exerting toxicity. Considering that the mechanisms involved in jejunal cadmium and mercury absorption are different and nonspecific (Foulkes and Bergman, 1993), it seems unlikely that competition arose at the level of cellular uptake.

In another experiment, we found that although exogenous glutathione did not affect mercury toxicity, N-acetylcysteine reduced it, suggesting that mercury bound to glutathione was fully accessible to the cells, and that the mechanism by which complexed mercury entered cells was specific for glutathione. The fact that extracellular glutathione, but not N-acetylcysteine, exacerbated cadmium's toxicity suggested that this metal's cellular uptake may also have been specifically facilitated by glutathione. The role of sulfhydryls is clearly complex as shown by the marked differences in their effects on the toxicities of the two metals. Glutathione stimulates cellular extrusion of mercury, but not cadmium, from intestinal mucosa which may also in part account for these differences (Foulkes, 1993). Glutathione is involved in normal mucosal defense mechanisms. Its levels are high in the GI system (Mutoh et al., 1991; Siegers et al., 1989), and BSO treatment *in vivo* damages the GI tract (Martensson et al., 1990; Hirota et al., 1989), and sensitizes it towards the development of hemorrhage (Stein et al., 1990). The effectiveness of mercury as a GI toxicant might be related to its ability to sequester glutathione, but this would not satisfactorily explain why copper salts, which also appear to bind glutathione physiologically *in vitro*, are much less toxic orally (LD_{50}s between 200 to 1000 mg/kg; NIOSH, 1976).

Comparisons of *in vitro* with *in vivo* toxicity data are complicated by the complex intestinal symptomology of metals, and quantitative data with regard to specific symptoms or clinical signs are lacking (Mykännen, 1986). Acute lethal oral toxicities of $HgCl_2$ in rodents (between 10 and 40 mg/kg; NIOSH, 1976) arise from its ability to induce intestinal necrosis and hemorrhage. Even allowing generous assumptions on the topical concentrations likely to be encountered in those studies, our calculated *in vitro* toxicity (corresponding to 8.9 mg/ml) was about 2 orders of magnitude (ca. 100×) higher. Generally, this was also found to be true for the other metals we tested (NIOSH, 1976). Hence, using this kind of *in vitro* data to predict at least severe damage leading to hemorrhage is of limited value. One reason for these wide dicrepancies may be that before lethal hemorrhagic damage is induced, mechanisms involved in epithelial repair become activated (see Section IV.B). Severe necrosis and hemorrhage only develop if damage is so deep that regenerative mechanisms are destroyed and vascular damage ensues. Toxicant concentrations required to induce hemorrhage are thus likely to be much higher than those cytotoxic to the superficial cells. Indeed, it is likely that the only symptoms likely to develop from superficial cellular cytotoxicity are those of nausea and diarrhea, as discussed previously in Section IV.B.

VI. SUMMARY AND FUTURE DIRECTIONS

Although the lack of clinical/experimental data prevents a comprehensive analysis of the risks and course of GI metal intoxication, an overall view appears to indicate that acute toxicity rarely results from specific biochemical modifications, and in many cases follows a pathology similar to that induced by other corrosive, nonspecific gastrointestinal toxicants. Signs of low grade damage include emesis and diarrhea, with more intense damage causing local necrosis and ulceration, complicated in the lower GI tract by the effects of bacterial invasion of the damaged tissue. Hemorrhage represents a potentially fatal response seen only after severe damage.

Whereas acute toxicity usually occurs only in exceptional individual cases, chronic intoxications affect wider segments of the population, usually as a result of pollution. Pathologies which develop after ingestion are usually systemic and may be secondary to malabsorption syndromes initiated in the gut. Enteropathy may also be a feature, possibly because epithelial regenerative mechanisms become exhausted after long-term exposure to low, but cytotoxic challenges.

Because the GI tract is particularly sensitive to cancer, and certain metals in occupational environments are known to induce absorption site-specific carcinogenesis, interest has focused on whether metals are carcinogenic to the GI tract, or elsewhere, after ingestion. The bulk of the evidence reviewed has failed to confirm that this is the case under normal, even occupational, circumstances of exposure. Whether metals are involved in GI tumor promotion, however, has not been comprehensively investigated. Generally the question needs to remain open, especially since it is not easy to confirm from epidemiological studies alone whether agents are damaging in this respect.

The GI tract in general has a functional resistance to xenobiotic-induced intoxication, presumably because the concentrations it must endure are often many times higher than those endured elsewhere in the body. There are many mechanisms involved in gastrointestinal mucosal defense, of which some may be preferentially affected by certain metals so as to predispose the mucosa to serious forms of mucosal damage. This may be one reason why for some metals we found only poor correlation between lethal oral metal toxicities and intestinal cell toxicity *in vitro*. If substances act purely as corrosives, one might expect some correlation between superficial cellular damage and damage resulting in hemorrhage and death *in vivo*, although their sensitivities may be quite different. We partly investigated the role of one such mechanism, glutathione, in determining the toxicity of several metals in intestinal cells *in vitro*. Our results appeared to indicate that it was important in protecting the mucosa against mercury in particular. Hence, exhaustion of mucosal glutathione may represent one mechanism by which the tissue could become predisposed towards the development of more serious effects of mercury. However, other thiols, including metallothionein may also be important in protecting the GI mucosa against the effects of many metals including mercury, and this is one area of further investigation. Given the multiplicity of effects induced by metals, both with regard to tissue damage and carcinogenesis, another important field to investigate is that of combination effects of metal toxins on these processes. Since individual metal toxicants in the environment rarely exist without the presence of other environmentally significant toxicants, including metals, this also represents an important future field of investigation.

REFERENCES

Aaronson, R. M. and Spiro, H. M. (1973), *Dig. Dis. Sci.*, 18, 583–594.
Allen, A., Garner, A., Hunter, A. C., and Keogh, J. P. (1988), in *Advances in Eicosanoid Research: Eicosanoids and the Gastrointestinal Tract*, Hillier, K., Ed. MTP Press, Lancaster, England, 195–215.
Aungst, B. and Shen, D. D. (1986), in *Gastrointestinal Toxicology*, Rozman, K., and Hänninen, O. , Eds., Elsevier, Amsterdam, Holland, 29–56.
Balbena, D., Escolar, G., Navarro, G., Bravo, L., and Pfeiffer, C. J. (1993), *Dig. Dis. Sci.*, 38, 730–739.
Barry, M. A., Behnke, C. A., and Eastman, A. (1990), *Biochem. Pharmacol.*, 40, 2353–2362.
Benoni, G., Cuzzolini, L., Zambreri, D., Donin, M., Del-Soldato, P., and Caramazzi, I. (1993), *Pharmacol. Res.*, 27, 73–80.
Böhme, M., Diener, M., Mestres, P., and Rummel, W. (1992a), *Toxicol. Appl. Pharmacol.*, 114, 285–294.
Böhme, M., Diener, M., and Rummer, W. (1992b), *Toxicol. Appl. Pharmacol.*, 114, 295–301.
Borenfreund, E. and Puerner, J. A. (1984), *Cell. Biol. Toxicol.*, 1: 55.
Borison, H. L. and Wang, S. C. (1953), *Pharmacol. Rev.*, 5, 193–230.
Bremner, I. (1978), *World Rev. Nutr. Diet*, 32, 165–197.
Campbell, D., Gonzales, M., and Sullivan J. B. (1992), in *Hazardous Material Toxicology*, Sullivan, J. B., and Kruger, G. R., Eds., Williams & Wilkins, Baltimore, MD, 814–823.
Cannon, S. R. (1982), *Postgrad. Med. J.*, 58, 445–447.
Chen, C. J. and Wang, C. J. (1990), *Cancer Res.*, 50, 5470–5474.
Choie, D. D., Longnecker, D. S., and Copley, M. D. (1981), *Toxicol. Appl. Pharmacol.*, 60, 354–359.
Clark, L. C. (1985), *Fed. Proc.*, 44, 2584–2589.
Clarkson, T. W. (1972), *Annu. Rev. Pharmacol.*, 12, 375–406.

Cohen, M. S., Kargacin, B., Klein, C. B., and Costa, M. (1993), *Crit. Rev. Toxicol.*, 23, 255–281.
Collins, J. F., Brown, J. P., Painter, R. R., Ja,all, I. S., Zeise, L. A., Alexeef, G. V., Wade, M. J., Siegel, D. M., and Wong, J. J. (1992), *Regul. Toxicol. Pharmacol.*, 16, 57–72.
Conrad, M. E., Umbreit, J. N., and Moore, E. G. (1991), *Gastroenterology*, 100, 129–136.
Cooper, W. C. (1985), *Scand. J. Work. Environ. Health*, 11, 331–345.
Dart, R. C. (1992), in *Hazardous Material Toxicology*, Sullivan, J. B., and Kruger, G. R., Eds., Williams & Wilkins, Baltimore, MD, 818–823.
DeFlora, S., Badolati, G. S., and Serra, D. (1987), *Mutat. Res.*, 192, 169–174.
De Ponti, F., D'Angelo, L., Forster, R., Einaudi, A., and Crema, A. (1991), *Digestion*, 50, 72–81.
Elinder, C.-G. and Nordberg, M. (1985), in *Cadmium and Health*, Friberg, F. L., et al., Eds., CRC Press, Boca Raton, FL, 65–79.
Evans, D. L. and Dive, C. (1993), *Cancer Res.*, 53, 2133–2139.
Felix, K., Nagel, W., Hartmann, H. J., and Weser, U. (1990), *Biol. Met.*, 3, 141–145.
Ferner, R. E. (1993), *Pharmacol. Ther.*, 58, 157–172.
Fishbein, L. (1991), in *Metals and Their Compounds in the Environment: Occurrence, Analysis and Biological Relevance*, Merian, E., Ed., VCH Verlag, Weinheim, Germany, 1153–1190.
Fodor, I. and Marx, J. J. (1988), *Biochim. Biophys. Acta*, 961, 96–102.
Forte, L. R. (1984), *Handb. Exp. Pharmacol.*, 70(I), 465–511.
Fortson, W. C. and Tedesco, F. J. (1984), *Am. J. Gastroenterol.*, 79, 878–883.
Foulkes, E. C. (1984), *Handb. Exp. Pharmacol.*, 70(I), 543–566.
Foulkes, E. C. (1993), *Life Sci.*, 52, 1617–1620.
Foulkes, E. C. and Bergman, D. (1993), *Toxicol. Appl. Pharmacol.*, 120, 89–95.
Frechilla, D., Lasteras, B., Ucelay, M., Pavondo, E., Gracienescu, G., and Cenarruzabeitia, E. (1991), *Arzneimittelforschung*, 41, 247–249.
Friberg, L., Nordberg, G. F., and Vouk, V. B. (1979), *Handbook of the Toxicology of Metals*, Elsevier, Amsterdam, Holland.
Fujimoto, Y., Kondo, Y., Nakajima, M., Takai, S., Sakuma, S., and Fujita, T. (1991), *Biochem. Int.*, 24, 33–2.
Fullmer, C. S. (1990), *Proc. Soc. Exp. Biol. Med.*, 194, 258–264.
Gebhart, E. and Rossman, T. G. (1991), in *Metals and Their Compounds in the Environment: Occurrence, Analysis and Biological Relevance*, Merian, E., Ed., VCH Verlag, Weinheim, Germany, 617–640.
Gehring, P. J. and Hammond, P. B. (1967), *J. Pharmacol. Exp. Ther.*, 155, 187–201.
George, C. F. (1980), *J. Soc. Med.*, 73, 200–204.
Goel, R. K., Tavares, I. A., and Bennett, A. (1992), *J. Pharm. Pharmacol.*, 44, 862–864.
Gorbach, S. L. (1990), *Gastroenterology*, 99, 863–875.
Gullikson, G. W. and Bass, P. (1984), *Handb. Exp. Pharmacol.*, 70(II), 419–459.
Hammond, P. B. and Beliles, R. P. (1980), in *The Basic Science of Poisons*, 2nd ed; Doull, J. Klaasen, C. D., Amdur, M. O., Eds., Macmillan, New York, 409–467.
Han, S. Y. and Tishler, J. M. (1982), *Radiology*, 144, 253.
Hirota, M., Inoue, M., Ando, Y., Hirayama, K., Morino, Y., Sakamoto, K., Mori, K., and Akagi, M. (1989), *Gastroenterology*, 97: 853–859.
Huebers, H. and Rummel, W. (1984), *Handb. Exp. Pharmacol.*, 70(I), 465–512.
Hughes, T. E. and Hansen, L. A. (1992), *Ann. Pharmacother.*, 26, 354–362.
Ijiri, K. and Potten, C. S. (1987), *Br. J. Cancer*, 55, 113.
Ip, C. (1985), *Fed. Proc.*, 44, 2573–2578.
Itoh, M. and Guth, P. H. (1985), *Gastroenterology*, 88, 1162–1167.
Karmakar, N. and Anand, S. (1989), *Clin. Exp. Pharamacol. Physiol.*, 16, 745–750.
Keogh, James P. (1992), in *Hazardous Material Toxicology*, Sullivan, J. B. and Kruger, G. R., Eds., Williams & Wilkins, Baltimore, MD, 834–844.
Keogh, Julian P., Steffen, B., and Siegers, C.-P., *J. Tox. Environ. Health.*, in press.
Kusterer, K., Pihan, G., and Szabo, S. (1987), *Am. J. Physiol.*, 252, G811–G816.
Léonard, A. (1991), in *Metals and Their Compounds in the Environment: Occurrence, Analysis and Biological Relevance*, Merian, E. Ed., VCH Verlag, Weinheim, Germany, 751–774.
Levander, O. A. (1977), *Environ. Health Perspect.*, 19, 159–164.
Lin, J. L. and Lim, P. S. (1993), *J. Toxicol. Clin. Toxicol.*, 31, 487–492.
Lindskog, S. (1970), *Struct. Bond.*, 8, 153–196.
Martensson, J., Jain, A., and Meister, A. (1990), *Proc. Natl. Acad. Sci. U.S.A.*, 87, 1715–1719.
Marty, M., Pouillart, P., Scholl, S., Droz, J. P., and Azab, M. (1990), *N. Engl. J. Med.*, 322, 816–821.
Matovelo, J. A., Landsverk, T., and Sund, R. B. (1990), *A.P.M.I.S.*, 98, 887–895.
Merian, E., Ed., (1991), *Metals and Their Compounds in the Environment: Occurrence, Analysis and Biological Relevance*, VCH Verlag, Weinheim, Germany.
Miller, T. A. (1983), *Am. J. Physiol.*, 245, G601–G623.
Mitchelson, F. (1992), *Drugs*, 43, 295–315.
Murata, I., Hirono, T., Saeki, Y., and Nakashimi, S. (1969), *Bull. Soc. Int. Chir.*, 29, 34.
Mutoh, H., Ota, S., Hiraishi, H., Ivey, K., Terano, A., and Sugimoto, S. (1991), *Am. J. Physiol.*, 261, G65–G70.

Mykännen, H. M. (1986), in *Gastrointestinal Toxicology,* Rozman, K., Hänninen, O., Eds., Elsevier, Amsterdam, Holland, 416–435.
NIOSH (1976), *Registry of Toxic Effects of Chemical Substances,* National Institutes for Occupational Safety and Health, U.S. Dept. of Health, Education and Welfare, Cincinnati, OH.
von der Ohe, M. R., Camilleri, M., Kvols, L. K., and Thomforde, G. M. (1993), *N. Engl. J. Med.,* 329, 1073–1078.
Ohnesorge, F.-K. and Wilhelm, M. (1991), in *Metals and Their Compounds in the Environment: Occurrence, Analysis and Biological Relevance,* Merian, E., Ed., VCH Verlag, Weinheim, Germany, 1309–1342.
Ooe, M., Asano, K., Haga, K., and Setoguchi, M. (1993), *Nippon Yakurigaku Zasshi,* 101, 299–307.
Parks, D., Burkley, G., Granger, D., Hamilton, S., and McCord, J. (1982), *Gastroenterology,* 82, 9–15.
Plezia, P. M. and Alberts, D. S. (1985), *Clin. Oncol. (Eastbourne),* 4, 357–386.
Powell, D. W. (1981), *Am. J. Physiol.,* 241, G275–G278.
van Riel, P. L. C. M. (1988), in *Meyler's Side Effects of Drugs,* 11th ed. Dukes, M. N. G. Ed., Elsevier, Amsterdam, 437–460.
Rinck, P. A., Myhr, G., Smevik, O., and Borseth, A. (1992), *Roto. Fortschr. Geb. Rontgenstr. Neuen. Bildgeb. Verfahr.,* 157, 533–538.
Rodriguez, M. J., Lugea, A., Barber, A., Lluch M., and Ponz, F. (1989), *Rev. Esp. Fisiol.,* 45S, 207–214.
Rosner, G. and Hertel, R. F. (1986), *Staub Reinhalt. Luft,* 46, 281–285.
Samblebe, M. and Siegers, C.-P. (1992), *Toxicologist,* 12, 84.
Sanger, G. J. (1990), *Can. J. Physiol. Pharmacol.,* 68, 314–324.
Schaeffer, S. G., Elsenhans, B., and Forth, W. (1988), in *Environmental Toxins,* Hutzinger, O. and Safe, S. H., Eds., Vol. 2, *Cadmium,* Stoeppler, M., Piscator, M., Vol. Eds., Springer, Berlin, 27–31.
Schlatter, C. and Kissling, U. (1973), *Beitr. Gerichtl. Med.,* 30, 382–388.
Schultz, S. G. (1979), in *Membrane Transport in Biology,* Vol. IV B. Transport Organs, Giebisch, G. Ed., Springer, New York, 749–780.
Schwörer, H., Racke, K., and Kilbinger, H. (1991), *Archiv. Pharmacol.,* 344, 143–149.
Sellers, L. A., Allen, A., and Bennett, M. K. (1987), *Gut,* 28, 835–843.
Siegers, C.-P. (1989), in *Progress in Pharmacology and Clinical Pharmacology,* Vol. 7/2, Gustav Fischer Verlag, Stuttgart, Germany, 171–180.
Siegers, C.-P., Bartels, L., and Riemann, D. (1989), *Pharmacology,* 38, 121–128.
Siegers, C.-P., Bumann, D., Trepkau, H.-D., Schadwinkel, B., and Baretton, G. (1992), *Cancer Lett.,* 65, 245–249.
Siegers, C.-P. and Samblebe, M. (1992), *Toxicologist,* 12, 243.
Silen, W. (1988), *Am. J. Physiol.,* 255, G395–G402.
Silen, W. and Ito, S. (1985), *Annu. Rev. Physiol.,* 47, 217–229.
Simon, G. L. and Gorbach, S. L. (1987), in *Physiology of the Gastrointestinal Tract,* 2nd ed., Johnson, L. R., Ed., Raven Press, New York, 1729–1747.
Snow, E. T. (1992), *Pharmacol. Ther.,* 53, 31–65.
Stein, H. J., Hinder, R. A., and Oosthuizen, M. M. (1990), *Surgery,* 108, 467–473.
Stoeppler, M. (1991), in *Metals and Their Compounds in the Environment: Occurrence, Analysis and Biological Relevance,* Merian, E. Ed., VCH Verlag, Weinheim, Germany, 803–851.
Sullivan, J. B. (1992), in *Hazardous Material Toxicology,* Sullivan, J. B., and Kruger, G. R., Eds., Williams & Wilkins, Baltimore, MD, 908–911.
Sullivan, J. B. and Kruger, G. R., Eds. (1992), *Hazardous Material Toxicology,* Williams & Wilkins, Baltimore, MD.
Szabo, S. (1987), *Scand. J. Gastroenterol.,* S127, 21–28.
Takeuchi, K. and Okabe, S. (1983), *Dig. Dis. Sci.,* 28, 993–1000.
Thomas, D. W. (1991), in *Metals and Their Compounds in the Environment: Occurrence, Analysis and Biological Relevance,* Merian, E., Ed., VCH Verlag, Weinheim, Germany, 789–801.
Tsuchiya, K. (1978), *Cadmium Studies in Japan: a Review,* Kodansha, Elsevier, Tokyo, Japan.
Valberg, L. S., Haist, J., Cherian, M. G., Richardson, L. D., and Goyer, R. A. (1977), *J. Toxicol. Environ. Health,* 2, 963–975.
Vallee, B. L. and Ulmer, D. D. (1972), in *Annu. Review Biochem.,* 80, 91–128.
Venho, V. M. K. (1986), in *Gastrointestinal Toxicology,* Rozman, K. and Hänninen, O., Eds., Elsevier, Amsterdam, Holland, 363–396.
Waalkes, M. P., Wahba, Z. Z., and Rodriguez, R. E. (1992), in *Hazardous Material Toxicology,* Sullivan, J. B. and Kruger, G. R., Eds., Williams & Wilkins, Baltimore, MD, 845–852.
Wands, J. R., Whelan-Weiss, S., Yardley, J. H., and Maddrey, W. C. (1974), *Am. J. Med.,* 57, 92.
Wilkes, J. M., Garner, A., and Peters, T. J. (1988), *Dig. Dis. Sci.,* 33, 361–367.
Willett, W. C., Morris, J. S., Pressel, S., and Taylor, J. D. (1983), *Lancet,* 1983(II), 130–134.
Yamashiti, N., Doi, M., and Nishio, M. (1972), *Jpn. J. Hyg.,* 27, 364.
Yan, L. and Spallholz, J. E. (1993), *Biochem. Pharmacol.,* 45, 429–437.
Zimmerman, B. J., Grisham, M. B., and Granger, D. N. (1990), *Am. J. Physiol.,* 258, G185–G190.

Chapter 59

Copper and Other Chemical Elements that Affect the Cardiovascular System

Leslie M. Klevay

I. INTRODUCTION

Atherosclerosis is the leading cause of acute myocardial infarction and cerebral infarction in the industrialized world. Other cardiovascular pathology associated with metallic elements pales into insignificance in comparison to these manifestations of atherosclerosis. Atherosclerosis is incremental and relentless; this process begins early in life and often remains unnoticed until middle age. In contrast, infarction often is rapid and deadly.

The amount and location of atherosclerosis are difficult to measure in individuals before death, although methods are improving. Late lesions associated with illness are easier to evaluate than early, subclinical lesions. For these reasons, much of what is known about the etiology of atherosclerosis is inferred from the epidemiology of ischemic heart disease and stroke. Although there are certain discrepancies between the epidemiology of cerebral and myocardial infarction, it generally is assumed that both arise from similar arterial pathology.

During the first five decades of life, normal coronary and cerebral arteries, circular in cross section and lined with a smooth layer of endothelial cells, gradually change because of disordered metabolism of connective tissue, glucose, lipids, and uric acid. Chemical irritants, growth factors, hormones, and vasoactive amines act on foam cells, macrophages, monocytes, platelets, and smooth muscle cells; the arterial endothelium becomes damaged and more permeable. The encrustation/imbibition progresses gradually until the fifth decade, when the cross section of the arterial lumen may be reduced to a small crescent. Then, under the influence of arterial spasm or other hemodynamic factors, with or without ulceration or a superimposed blood clot, blood supply becomes insufficient and organ damage may be fatal. Because of its association with pathology of the coronary arteries, ischemic heart disease often is called coronary heart disease.[1]

The search for associations between chemical elements and human illness is complicated by the fact that not all of the measurements that characterize the illness necessarily involve measurement of elements. Also, interpretation of some of the measurements may become puzzling because of ill-defined biological influences. For example, Vallee[2] suggested that if it had been hypothesized — prior to the discovery of vitamin B_{12} — that cobalt is a significant factor in pernicious anemia, pharmacologic experiments with cobalt would have been negative. Measurement of cobalt in organs would have been similarly uninformative.

Some of this biological complexity occurs because essential trace elements often are constituents of enzymes. Mildvan[3] has estimated than more than 27% of known enzymes "have metals built into their structures, require metals for activity, or are further activated by metal ions." Green[4] suggests that "enzyme catalysis is the only rational explanation of how a trace of some substance can produce profound biological effects."

Although Green identified **one** way in which trace substances can have biological effects, surely his way is not the **only** way. Hormones are present in low concentrations in biological fluids, yet have important biological effects which are regulated by biorhythms, feedback relationships, and numbers of receptors.[5] Chemical elements, too, may modify hormone activity. For example, chromium may produce some of its effects by potentiating insulin.[6] Numerous small peptides with biological activity have been discovered during recent decades. Activity of some of these may be modified by chemical elements; e.g., mild zinc deficiency decreased serum thymulin and interleukin 2 of men and women.[7] Similarly, prostaglandin metabolism can be modified by dietary copper.[8,9]

Thus, chemical elements have "profound biological effects" via, *inter alia,* enzymes, hormones and various messenger molecules. Sometimes it may be more important to measure these materials than to measure the elements. If exposure to toxic elements or deficiency of essential elements consistently produces characteristic anatomical or physiological pathology, the findings can be extremely useful in guiding analysts toward measurement of appropriate enzymes, hormones or messenger molecules in the search for chemical pathology. Elemental measurement may be usefully restricted to specific cells, tissues or organs.

Some chemical and physiological pathology has been closely associated with the epidemiology of ischemic heart disease. It frequently is inferred that these characteristics also are closely associated with specific anatomical pathology — atherosclerosis. Several of these characteristics are more important than the others because they "are closely associated with risk, ... can be used to predict risk to a reasonable degree and ... seem to involve a potentially plausible mechanism of producing disease."[10] Insull summarized these: hypercholesterolemia, glucose intolerance, abnormalities of the electrocardiogram, male predisposition, hypertension, and smoking of cigarettes.[11] These characteristics are called risk factors. Consideration of these characteristics permits the identification of the 20% of the presymptomatic population with 40% of the risk.[12]

My early attempts to understand the etiology of ischemic heart disease in the early 1960s were influenced by the inverse relationship between risk of ischemic heart disease and the hardness of available drinking water first noted by Kobayashi.[13-15] This relationship led to a collection of 29 chemical elements related to the epidemiology of ischemic heart disease or to the metabolism of cholesterol or other lipids.[16] References that prompted inclusion of elements in the table have been cited in earlier publications[10,16-18] and will be omitted here for the sake of brevity. Figure 1 displays the 34 elements found to date. The concept guiding the development of this table continues to favor inclusion over exclusion, with a growing emphasis on atherosclerosis and related cardiovascular effects.

II. COPPER DEFICIENCY AND ATHEROSCLEROSIS

In the latest version of the table,[18] relationships between some of the elements and copper were documented. Some of the biological effects of the elements with darker backgrounds are mediated by enhancement or inhibition of copper. Understanding these enhancements, inhibitions, and relationships is impossible without knowledge of some of the biochemical and physiological roles of copper. Copper was found to be essential for health in 1928.[19] Hundreds of articles on the biology of copper have been published since; Owen,[20] Mertz and Davis,[21] and O'Dell[22] have reviewed some of these. Some effects of copper deficiency are understood incompletely; others are mediated via enzymes.

Table 1 contains enzymes, activity of which can be affected by copper nutriture. Some of these enzymes are known to contain copper. Guanylate cyclase (EC 4.6.1.2) is an enzyme containing copper[34] that has not been assayed in deficient animals.

Superoxide dismutase contributes to the primary defense against metabolically generated free radicals.[22] Lysyl oxidase catalyzes the cross-linking of collagen and elastin.[22] Copper deficiency impairs the activity of both enzymes.[22] Some of the damage to arteries in the atherosclerotic process and to hearts and brains after infarction may be caused by free radicals.[35,36] Insufficient lysyl oxidase leads directly to vascular disease.[22]

Lecithin:cholesterol acyltransferase (LCAT) and lipoprotein lipase are not known to be metalloenzymes. Decreased activity of these enzymes increases plasma cholesterol. Substantially purified LCAT can be stimulated *in vitro* by copper.[37] Although lipoprotein lipase can be activated by divalent cations, copper has not been studied in this context.[38] An increase in hydroxymethylglutaryl-coenzyme and a

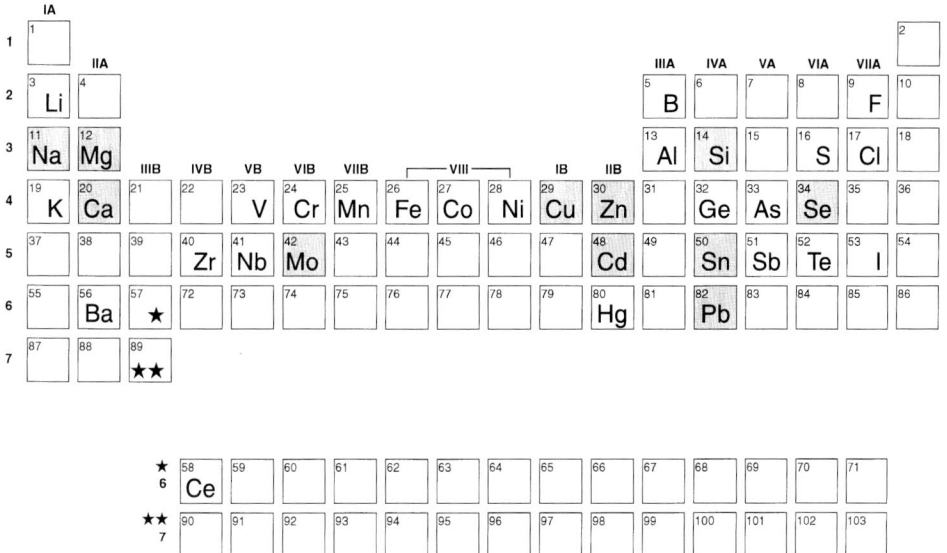

Figure 1. Elements of atherosclerosis. Klevay, *Perspect. Biol. Med.* 20:186, 1977; *Adv. Nutr. Res.* 1:227, 1977; CRC Press, 1984; Proc. Portugese Atherosclerosis Soc., in press.

Table 1 Enzyme Activity Altered by Copper Deficiency

Activity	Enzyme	Ref.
Decreased activity	Superoxide dismutase (EC 1.15.1.1)[a]	23
	Lysyl oxidase (EC 1.4.3.6)[a]	24
	Lecithin:cholesterol acytransferase (EC 2.3.1.43)	25, 26
	Lipoprotein lipase (EC 3.1.1.34)	27–29
	Choline phosphotransferase (EC 2.7.8.2)	30
Increased activity	Hydroxymethylglutaryl-coenzyme A reductase (EC 1.1.1.34)	28, 31, 32

[a] Cuproenzymes according to Prohaska.[33]

reductase activity secondary to faulty regulation of glutathione concentrations will increase the concentration of cholesterol in plasma or serum.[39,40] Thus, at least three enzymes contribute to the hypercholesterolemia of copper deficiency. Copper deficiency also substantially decreases the α_2 isoform of Na$^+$/K$^+$ ATPase in rat heart.[41]

After hypercholesterolemia, cardiac arrhythmia is the risk factor studied most intensively in copper deficiency. Numerous arrhythmias have been found[42] since the first description of the phenomenon.[43] Medeiros et al.[44,45] have been prominent in this research. The decreased Na$^+$/K$^+$ ATPase of copper deficiency[41] may contribute to both the increased cardiac sodium[46] and the frequency of cardiac arrhythmia. Glucose intolerance[47] and hypertension[48] have received less attention, although early data have been confirmed.[49–53]

Physiological amounts of copper lessen the pressor response of perfused mesenteric arteries of rats to norepinephrine and angiotensin.[54] Contractibility of aortic strips in response to norepinephrine *in vitro* was increased by copper deficiency.[55] Impaired vascular relaxation in copper deficiency[56,57] may be explained by the inactivation of endothelium-derived relaxing factor.[58]

Although hyperuricemia is not usually considered a risk factor for ischemic heart disease,[11] Frohlich[59] considers it as such, perhaps as an index of poor renal function. In any case, some of the multivariate analysis from Framingham are corrected for uric acid,[60] and risk in this population is related to uric acid value.[12,61]

Newland hypothesizes that uric acid is involved in the pathogenesis of coronary heart disease by promoting thrombosis and hypertension.[62] The hyperuricemia of copper deficiency[63] thus can augment the effects on thrombosis of altered prostaglandin metabolism. Some of these concepts have been reviewed.[10,64–66] The thrombogenic effect of a diet low in copper[67] can be explained by an increased ratio of thromboxane A$_2$ to prostacyclin I$_2$[68,69] and by impaired fibrinolytic activity.[70]

Other phenomena associated with ischemic heart disease have been found in animals deficient in copper. These phenomena are not understood as well as those relating to risk factors. "Falling disease" in cattle was the first cardiovascular phenomenon to be associated with copper deficiency, which is the leading deficiency of livestock world-wide.[71] Death of cattle was sometimes so sudden that they fell on the milker.[72,73] People with ischemic heart disease often die suddenly; 30 to 80% die within 1 hour of the onset of symptoms.[74] It generally is assumed that these sudden deaths are caused by cardiac arrhythmias. Perhaps the decreased Na+/K+ ATPase of copper deficiency is contributory because the α_2 isoform of this enzyme is one of the predominant isoforms of the conducting system of rat heart.[75]

The solubility of cholesterol in aqueous media such as blood plasma is aided by the incorporation of cholesterol into complex molecules called lipoproteins. These molecules generally are characterized by electrophoresis or centrifugation and are named for their centrifugal densities. Low density lipoprotein (LDL) carries cholesterol throughout the arterial tree, where it can become deposited via a complex series of events involving cellular receptors and enzymes.

Copper deficiency increases arterial lipid peroxides[76] and increases the susceptibility of lipoproteins to oxidation.[77] Oxidatively damaged LDL has been hypothesized to contribute to atherogenesis,[78–80] probably by disrupting the regulation of cholesterol deposition in arterial cells.[78]

Foam cells and proliferation of smooth muscle cells, which are thought to be the earliest signs of atherosclerosis in humans,[66] have been found in pigs[81,82] and rats[83] deficient in copper. All three classes of connective tissue (collagen, elastin, and ground substance) are adversely affected by copper deficiency[81,84–89] in several species of animals, probably secondary to decreased activity of lysyl oxidase. Copper-deficient arteries contain abnormal amounts of connective tissues with abnormal chemical and physical properties. Copper deficiency can produce anatomical change early in the atherosclerotic process.

Thus, when dietary copper is insufficient, there are many important, harmful effects on the cardiovascular system. Altered arterial[83] and myocardial[90] morphology and impaired cardiac function[91] in copper deficiency have been found in animals that would be considered normal by usual methods of clinical and nutritional assessment.

III. CUPROTROPIC ELEMENTS

Table 2 classifies the elements from Figure 1 that have some biological effects mediated by enhancement or inhibition of copper. These elements resemble the cuprotropic and cholesterotropic organic chemicals[92–95] which alter plasma cholesterol and copper nutriture in a reciprocal manner.[95] A few examples are useful to illustrate modification of copper utilization by other elements.

Table 2 Elements Affecting Cardiovascular Health Via Copper[a]

Enhancers	Inhibitors
Magnesium	Sodium
Silicon	Zinc
Calcium	Molybdenum
Selenium	Cadmium
	Tin
	Lead

[a] In order of atomic number. Experimental data have been reviewed.[18]

Zinc, an essential nutrient, is well-known to inhibit copper utilization.[64,96–98] Excess zinc can disrupt the utilization of copper and the metabolism of cholesterol.[90,96] That copper deficiency produces hypercholesterolemia generally is accepted.[33] Toxicological doses of cadmium can induce a hypercholesterolemia that can be relieved by pharmacological doses of copper.[99] Victery et al.[100] found that rats given a drinking solution containing large amounts of lead excreted extra copper and other elements in urine; data are in harmony with the belief of Klauder and Petering[101] that lead can antagonize copper in hematopoiesis. Pekelharing et al.[102] found that amounts of tin similar to those found in human diets can decrease copper nutriture, in confirmation of earlier data.[103] Cadmium, lead, and tin are without known nutritional benefit; they may have harmful nutritional effects, however.

Both cadmium and lead have been suggested as contributing to the origin of essential hypertension, that is, hypertension of unknown origin. Chronic, low level feeding of cadmium to animals increases blood pressure.[115,116] The dose-response relationship is not linear; hypertension is more likely at environmental than industrial levels.[117] Some of the early epidemiology has been reviewed.[118,119] Recent epidemiology confirms higher cadmium concentrations in people with hypertension[120] and the possibility that some of this cadmium is acquired from cigarettes.[121,122]

Schroeder and Balassa[123] reviewed environmental sources of lead; Selye[124] reviewed atherosclerosis, but did not mention hypertension. Goyer[125] reviewed other aspects of lead toxicology, as well. Recent evidence supports the hypothesis that hypertension is more closely associated with industrial exposure to lead[126,127] than to lead in the wider environment.[121,128–130] Thus, there is an interesting contrast between the epidemiology of hypertensive effects of cadmium and lead.

IV. COPPER DEPLETION IN HUMANS

There probably are hundreds of experiments on copper deficiency in animals. In contrast, there are only a few copper deficiency experiments in humans. These experiments reveal that people respond to copper deficiency similarly to animals.

More than 31 men and women have responded to diets low in copper with potentially harmful changes in lipids,[104,105] glucose tolerance,[106] blood pressure,[107] and electrocardiograms.[104,108] One of these experiments was terminated prematurely in the interest of safety; a myocardial infarction, a heart block, and two tachycardias were found in four men.[108] In addition, more than 76 men and women have responded to high doses of zinc with potentially harmful changes in lipid metabolism.[109–112] These dyslipidemias reflect the interference of zinc with copper utilization noted above.[90,96–98] Thus, although the number of human experiments is comparatively few, more than 100 men and women demonstrate biochemistry and physiology similar to animals.

Copper in the human depletion experiments[104–108] ranged from 0.65 to 1.02 mg/day, amounts readily accessible to the general population. Chemical analysis of 849 composite diets collected in Belgium, Canada, the U.K., and the U.S.[113] reveals that 32% of daily diets contain less than 1 mg of copper, and 61% contain less than 1.5 mg, the lower limit of the estimated safe and adequate intake in the U.S.[114]

V. POSSIBLE MECHANISMS

Copper utilization and coronary (ischemic) heart disease were linked more than 20 years ago by observations on populations, observations on human individuals, and experiments with animals.[96] The link was largely epidemiologic, and the only credible mechanism was based on hypercholesterolemia. Although much needs to be done to identify the boundaries of this link, much progress has been made. Specific, related hypotheses have been tested successfully by experiments involving animals and human volunteers. Extensive reading and searching have revealed appropriate experiments published since 1928[19] that need not be repeated and are useful in consideration of mechanism.

Early in life, diets low in copper induce smooth muscle cell proliferation and migration and the production of foam cells in arterial walls. Mucopolysaccharides increase. The smooth muscle cells and elastic tissues degenerate.

As blood cholesterol begins to rise, along with blood pressure and average concentrations of glucose and uric acid, elastic membranes in arteries become fragmented and arterial walls become more permeable. The deposition of lipids is facilitated by impaired defense against free radicals and the accumulation of peroxides. The increase in blood pressure is mediated by impaired vascular relaxation and abnormally high sensitivity to pressor hormones. High concentrations of glucose in blood may accentuate abnormalities of lipid metabolism via glycosylation of proteins. Gradually, the size of the lumen decreases as the arterial wall increases in thickness. Intramural hemorrhages occur and further increase the thickness of the arterial walls. Thrombus dissolution is impaired. As these various processes continue over long periods of time, the injuries to the arteries accumulate and atherosclerosis is the result.

VI. SUMMARY

In summary, the Western diet often is low in copper. Copper deficiency produces arteries with abnormal connective tissue. These arteries are subjected to abnormally high blood pressure and are bathed continuously with abnormally high concentrations of cholesterol, glucose, and uric acid. Abnormal lipids are deposited in arterial walls and structural proteins are glycosylated. Dissolution of blood clots is impaired. Gradually over the decades, these events produce atherosclerosis and, ultimately, the acute infarctions that often are fatal.

REFERENCES

1. Klevay, L. M. (1990), in *Copper Bioavailability and Metabolism*, (Adv. Exp. Med. Biol., Vol. 258,) Kies, C., Ed., Plenum Press, New York, 197–208.
2. Vallee, B. L. (1959), *J. Chron. Dis.*, 9, 74–79.
3. Mildvan, A. S. (1970), in *The Enzymes*, Vol. 2, Boyer, P. D., Ed., Academic Press, New York, 445–536.
4. Green, D. E. (1941), *Adv. Enzymol.*, 1, 177–198.
5. Wilson, J. D. and Foster, D. W. (1985), in *Williams Textbook of Endocrinology*, Wilson, J. D. and Foster, D. W., Eds., W. B. Saunders, Philadelphia, 1–8.
6. Pi-Sunyer, F. X. and Offenbacher, E. G. (1984), in *Present Knowledge in Nutrition*, Olson, R. E., Broquist, H. P., Chichester, C. O., Darby, W. J., Kolbye, A. C., Jr., and Stalvey, R. M., Eds., 5th ed, Nutrition Foundation, Washington, D.C., 571–586.
7. Prasad, A. S., Meftah, S., Abdallah, J., Kaplan, J., Brewer, G. J., Bach, J. F., and Dardenne, M. (1988), *J. Clin. Invest.*, 82, 1202–1210.
8. Mitchell, L. L., Allen, K. G. D., and Mathias, M. M. (1988), *Prostaglandins*, 35, 977.
9. Allen, K. G. D., Lampi, K. J., Bostwick, P. J., and Mathias, M. M. (1991), *Nutr. Res.*, 11, 61–70.
10. Klevay, L. M. (1984), in *Metabolism of Trace Metals in Man*, Vol. 1, Rennert, O. M. and Chan, W.-Y., Eds., CRC Press, Boca Raton, FL, 129–157.
11. Insull, W., Jr. (1973), *Coronary Risk Handbook*, American Heart Association, New York, 3.
12. Kannel, W. B. (1976), *Am. J. Cardiol.*, 37, 269–282.
13. Kobayashi, J. (1957), *Ber. Ohara Inst.*, 11, 12–21.
14. Klevay, L. M. (1980), *J. Environ. Pathol. Toxicol.*, 4, 281–287.
15. Klevay, L. M. (1987), *Clin. Med.*, 15, 20.
16. Klevay, L. M. (1977), *Perspect. Biol. Med.*, 20, 186–192.
17. Klevay, L. M. (1977), in *Advances in Nutritional Research*, Vol. 1, Draper, H. H., Ed., Plenum Publishing, New York, 227–252.
18. Klevay, L. M. (in press), in *Trace Elements in Medicine, Health and Cardiology*, Reis, M. F., Miguel, A. A. S. C., Machado, M., Abdulla, M., Eds., Smith-Gordon, London.
19. Hart, E. B., Steenbock, H., Waddell, J., and Elvehjem, C. A. (1928), *J. Biol. Chem.*, 77, 797–812.
20. Owen, C. A., Jr. (1981), *Copper Deficiency and Toxicity*, Noyes Publications, Park Ridge, NJ.
21. Mertz, W. and Davis, G. K. (1987), in *Trace Elements in Human and Animal Nutrition, 5th ed.*, Vol. 1, Mertz, W., Ed., Academic Press, San Diego, 301–364.
22. O'Dell, B. L. (1990), in *Present Knowledge in Nutrition*, Brown, M. L. Ed., 6th ed., International Life Sciences Institute–Nutrition Foundation, Washington, D.C., 261–267.
23. Mann, T. and Keilin, D. (1939), *Proc. Roy. Soc. Lond. B.*, 126, 303–315.
24. Kim, C. S. and Hill, C. H. (1966), *Biochem. Biophys. Res. Commun.*, 24, 395–400.
25. Lau, B. W. and Klevay, L. M. (1981), *J. Nutr.*, 111, 1698–1703.
26. Harvey, P. W. and Allen, K. G. (1981), *J. Nutr.*, 111, 1855–1858.
27. Lau, B. W. and Klevay, L. M. (1982), *J. Nutr.*, 112, 928–933.
28. Valsala, P. and Kurup, P. A. (1987), *J. Biosci.*, 12, 137–142.
29. Koo, S. I., Lee, C. C., and Norvell, J. E. (1988), *Proc. Soc. Exp. Biol. Med.*, 188, 410–419.
30. Cornatzer, W. E., Haning, J. A., and Klevay, L. M. (1986), *Int. J. Biochem.*, 18, 1083–1087.
31. Yount, N. Y., McNamara, D. J., Al-Othman, A. A., and Lei, K. Y. (1990), *J. Nutr. Biochem.*, 1, 21–27.
32. Kim, S., Chao, P. Y., and Allen, K. G. D. (1992), *FASEB J.*, 6, 2467–2471.
33. Prohaska, J. R. (1990), *J. Nutr. Biochem.*, 1, 452–461.
34. Gerzer, R., Böhme, E., Hofmann, F., and Schultz, G. (1981), *FEBS Lett.*, 132, 71–74.
35. McCord, J. M. (1985), *N. Engl. J. Med.*, 312, 159–163.
36. Southorn, P. A. and Powis, G. (1988), *Mayo Clin. Proc.*, 63, 390–408.
37. Suzue, G., Vezina, C., and Marcel, Y. L. (1980), *Can. J. Biochem.*, 58, 539–541.
38. Srinivasan, S. R., Radhakrishnamurthy, B., and Berenson, G. S. (1975), *Arch. Biochem. Biophys.*, 170, 334–340.
39. Bunce, G. E. (1993), *Nutr. Rev.*, 51, 305–307.
40. Allen, K. G. D. and Klevay, L. M. (1994), *Curr. Opinion Lipidol.*, 5, 22–28.
41. Huang, W.-H., Wang, Y., Askari, A., Klevay, L. M., Askari, A., and Chui, T. H. (1993), *FASEB J.*, 7, A103.
42. Klevay, L. M. (in press), in *Handbook on Metal-Ligand Interactions in Biological Fluids*, Berthon, G., Ed., Marcel Dekker, New York.
43. Klevay, L. M. and Viestenz, K. E. (1981), *Am. J. Physiol.*, 240, H185–H189.
44. Davidson, J., Medeiros, D. M., and Hamlin, R. L. (1992), *J. Nutr.*, 122, 1566–1575.
45. Medeiros, D. M., Liao, Z., and Hamlin, R. L. (1992), *Proc. Soc. Exp. Biol. Med.*, 200, 78–84.

46. Klevay, L. M. and Halas, E. S. (1991), *Physiol. Behav.,* 49, 309–314.
47. Keil, H. L. and Nelson, V. E. (1934), *J. Biol. Chem.,* 106, 343–349.
48. Klevay, L. M. (1987), *Nutr. Rep. Int.,* 35, 999–1005.
49. Klevay, L. M. (1982), *Nutr. Rep. Int.,* 26, 329–334.
50. Cohen, A. M., Teitelbaum, A., Miller, E., Ben-Tor, V., Hirt, R., and Fields, M. (1982), *Isr. J. Med. Sci.,* 18, 840–844.
51. Hassel, C. A., Marchello, J. A., and Lei, K. Y. (1983), *J. Nutr.,* 113, 1081–1083.
52. Fields, M., Ferretti, R. J., Smith, J. C., Jr., and Reiser, S. (1983), *J. Nutr.,* 113, 1335–1345.
53. Medeiros, D. M. (1987), *Nutr. Res.,* 7, 231–235.
54. Cunnane, S. C., Zinner, H., Horrobin, D. F., Manku, M. S., Morgan, R. O., Karmali, R. A., Ally, A. I., Karmazyn, M., Barnette, W. E. and Nicolaou, K. C. (1979), *Can. J. Physiol. Pharmacol.,* 57, 35–40.
55. Kitano, S. (1980), *Circ. Res.,* 46, 681–689.
56. Schuschke, D. A., Reed, M. W., Saari, J. T., and Miller, F. N. (1992), *J. Nutr.,* 122, 1547–1552.
57. Saari, J. T. (1992), *Proc. Soc. Exp. Biol. Med.,* 200, 19–24.
58. Furchgott, R. F. and Vanhoutte, P. M. (1989), *FASEB J.,* 3, 2007–2018.
59. Frohlich, E. D. (1993), *JAMA,* 270, 378–379.
60. Kannel, W. B., Castelli, W. P., and Gordon, T. (1979), *Ann. Intern. Med.,* 90, 85–91.
61. Brand, F. N., McGee, D. L., Kannel, W. B., Stokes, J., and Castelli, W. P. (1985), *Am. J. Epidemiol.,* 121, 11–18.
62. Newland, H. (1975), *Med. Hypotheses,* 1, 152–155.
63. Klevay, L. M. (1980), *Nutr. Rep. Int.,* 22, 617–621.
64. Klevay, L. M. (1980), *Ann. N. Y. Acad. Sci.,* 355, 140–151.
65. Klevay, L. M. (1983), *Biol. Trace Elem. Res.,* 5, 245–255.
66. Klevay, L. M. (1990), in *Role of Copper in Lipid Metabolism,* Lei, K. Y., and Carr, T. P., Eds., CRC Press, Boca Raton, FL, 233–267.
67. Klevay, L. M. (1985), *Atherosclerosis,* 54, 213–224.
68. Nelson, S. K., Huang, C. J., Mathias, M. M., and Allen, K. G. (1992), *J. Nutr.,* 122, 2101–2108.
69. Morin, C. L., Allen, K. G., and Mathias, M. M. (1993), *Proc. Soc. Exp. Biol. Med.,* 202, 167–173.
70. Lynch, S. M. and Klevay, L. M. (1993), *Nutr. Res.,* 13, 913–922.
71. Mills, C. F. (1987), in *Trace Elements in Man and Animals — TEMA 5,* Mills, C. F., Bremner, I., and Chesters, J. K., Eds., Commonwealth Agricultural Bureaux, Farnham Royal, U.K., 1–10.
72. Bennetts, H. W. and Hall, H. T. B. (1939), *Aust. Vet. J.,* 15, 152–159.
73. Bennetts, H. W., Harley, R., and Evans, S. T. (1942), *Aust. Vet. J.,* 18, 50–63.
74. Goldstein, S. (1974), *Sudden Death and Coronary Heart Disease,* Futura Publishing, Mt. Kisco, NY, 24–32.
75. Zahler, R., Brines, M., Kashgarian, M., Benz, E. J., Jr., and Gilmore-Hebert, M. (1992), *Proc. Natl. Acad. Sci. U.S.A.,* 89, 99–103.
76. Anon. (1993), *Nutr. Rev.,* 51, 88–89.
77. Rayssiguier, Y., Gueux, E., Bussiere, L., and Mazur, A. (1993), *J. Nutr.,* 123, 1343–1348.
78. Steinberg, D. and Witztum, J. L. (1990), *JAMA,* 264, 3047–3052.
79. Steinberg, D., Parthasarathy, S., Carew, T. E., Khoo, J. C., and Witztum, J. L. (1989), *N. Engl. J. Med.,* 320, 915–924.
80. Steinberg, D. (1993), *N. Engl. J. Med.,* 328, 1487–1489.
81. Carnes, W. H., Coulson, W. F., and Albino, A. M. (1965), *Ann. N. Y. Acad. Sci.,* 127, 800–810.
82. Hill, K. E. and Davidson, J. M. (1986), *Arteriosclerosis,* 6, 98–104.
83. Hunsaker, H. A., Morita, M., and Allen, K. G. (1984), *Atherosclerosis,* 51, 1–19.
84. Shields, G. S., Coulson, W. F., Kimball, D. A., Carnes, W. H., Cartwright, G. E., and Wintrobe, M. M. (1962), *Am. J. Pathol.,* 41, 603–621.
85. Carnes, W. H. (1968), *Int. Rev. Connect. Tissue Res.,* 4, 197–232.
86. Waisman, J., Cancilla, P. A., and Coulson, W. F. (1969), *Lab. Invest.,* 21, 548–554.
87. Prockop, D. J., Kivirikko, K. I., Tuderman, L., and Guzman, N. A. (1979), *N. Engl. J. Med.,* 301, 13–23.
88. Opsahl, W., Zeronian, H., Ellison, M., Lewis, D., Rucker, R. B., and Riggins, R. S. (1982), *J. Nutr.,* 112, 708–716.
89. Radhakrishnamurthy, B., Ruiz, H., Dalferes, E. R., Jr., Klevay, L. M., and Berenson, G. S. (1989), *Proc. Soc. Exp. Biol. Med.,* 190, 98–104.
90. Klevay, L. M., Pond, W. G., and Medeiros, D. M. (in press), *Nutr. Res.*
91. Klevay, L. M., Milne, D. B., and Wallwork, J. C. (1985), *Nutr. Rep. Int.,* 31, 963–971.
92. Klevay, L. M. (1983), *Drug Nutr. Interact.,* 2, 131–137.
93. Klevay, L. M. (1985), in *Trace Elements in Man and Animals — TEMA 5,* (Mills, C. F., Bremner, I., and Chesters, J. K., Eds.), Commonwealth Agricultural Bureaux, Farnham Royal, U. K., 180–183.
94. Klevay, L. M. (1986), *Nutr. Res.,* 6, 1281–1292.
95. Klevay, L. M. (1987), *Med. Hypotheses,* 24, 111–119.
96. Klevay, L. M. (1973), *Am. J. Clin. Nutr.,* 26, 1060–1068.
97. Underwood, E. J. (1977), *Trace Elements in Human and Animal Nutrition,* 4th ed., Academic Press, New York, 61, 226, 231.
98. Klevay, L. M. (1975), *Am. J. Clin. Nutr.,* 28, 764–774.
99. Bordas, E. and Gabor, S. (1982), *Rev. Roumaine Biochem.,* 19, 1, 3–7.
100. Victery, W., Miller, C. R., and Goyer, R. A. (1986), *J. Lab. Clin. Med.,* 107, 129–135.

101. Klauder, D. S. and Petering, H. G. (1977), *J. Nutr.,* 107, 1779–1785.
102. Pekelharing, H. L. M., Lemmens, A. G., and Beynen, A. C. (1994), *Br. J. Nutr.,* 71, 103–109.
103. Reicks, M. and Rader, J. I. (1990), *Proc. Soc. Exp. Biol. Med.,* 195, 123–128.
104. Klevay, L. M., Inman, L., Johnson, L. K. et al. (1984), *Metabolism,* 33, 1112–1118.
105. Reiser, S., Powell, A., Yang, C.-Y., and Canary, J. J. (1987), *Nutr. Rep. Int.,* 36, 641–649.
106. Klevay, L. M., Canfield, W. K., Gallagher, S. K. et al. (1986), *Nutr. Rep. Int.,* 33, 371–382.
107. Lukaski, H. C., Klevay, L. M., and Milne, D. B. (1988), *Eur. J. Appl. Physiol.,* 58, 74–80.
108. Reiser, S., Smith, J. C., Jr., Mertz, W. et al. (1985), *Am. J. Clin. Nutr.,* 42, 242–251.
109. Hooper, P. L., Visconti, L., Garry, P. J., and Johnson, G. E. (1980), *JAMA,* 244, 1960–1961.
110. Chandra, R. K. (1984), *JAMA,* 252, 1443–1446.
111. Goodwin, J. S., Hunt, W. C., Hooper, P., and Garry, P. J. (1985), *Metabolism,* 34, 519–523.
112. Black, M. R., Medeiros, D. M., Brunett, E., and Welke, R. (1988), *Am. J. Clin. Nutr.,* 47, 970–975.
113. Klevay, L. M., Buchet, J. P., Bunker, V. W., Clayton, B. E., Gibson, R. S., Medeiros, D. M., Moser-Veillon, P. B., Patterson, K. Y., Taper, L. J., and Wolf, W. R. (1993), in *Trace Elements in Man and Animals — TEMA 8,* Anke, M., Meissner, D., and Mills, C. F. Eds., Friedrich-Schiller-Universität, Jena, 207–210.
114. *Recommended Dietary Allowances* (1989), 10th ed., National Academy Press, Washington, D.C., 7, 227, 284.
115. Schroeder, H. A., Nason, A. P., and Mitchener, M. (1968), *Am. J. Physiol.,* 214, 796–800.
116. Perry, H. M., Jr., Erlanger, M., and Perry, E. F. (1977), *Proc. Soc. Exp. Biol. Med.,* 156, 173–176.
117. Kopp, S. J., Glonek, T., Perry, H. M., Jr., Erlanger, M., and Perry, E. F. (1982), *Science,* 217, 837–839.
118. Perry, H. M., Jr. (1971), *Geol. Soc. Am. Mem.,* Cannon, H. L., and Hopps, H. C., Eds., 123, 179.
119. Sandstead, H. H. and Klevay, L. M. (1975), *Geol. Soc. Am. Mem., Spec. Pap.,* 155, 73–81.
120. Thind, G. S. and Fischer, G. M. (1976), *Clin. Sci. Mol. Med.,* 51, 483–486.
121. Kromhout, D., Wibowo, A., Herber, R. F., Dalderup, L. M., Heerdink, H., de-Lezenne-Coulander, C., and Zielhuis, R. L. (1985), *Am. J. Epidemiol.,* 122, 378–385.
122. Mussalo-Rauhamaa, H., Leppanen, A., Salmela, S., and Pyysalo, H. (1986), *Arch. Environ. Health,* 41, 49–55.
123. Schroeder, H. A. and Balassa, J. (1961), *J. Chron. Dis.,* 14, 408–425.
124. Selye, H. (1970), *Experimental Cardiovascular Diseases,* Springer-Verlag, New York, 295, 307, 308, 595.
125. Goyer, R. A. (1993), *Environ. Health Perspect.,* 100, 177–187.
126. Egeland, G. M., Burkhart, G. A., Schnorr, T. M., Hornung, R. W., Fajen, J. M., and Lee, S. T. (1992), *Br. J. Ind. Med.,* 49, 287–293.
127. Maheswaran, R., Gill, J. S., and Beevers, D. G. (1993), *Am. J. Epidemiol.,* 137, 645–653.
128. Staessen, J., Sartor, F., Roels, H., Bulpitt, C. J., Claeys, F., Ducoffre, G., Fagard, R., Lauwerijs, R., Lijnen, P., and Rondia, D. (1991), *J. Hum. Hypertens.,* 5, 485–494.
129. Kromhout, D. (1988), *Environ. Health Perspect.,* 78, 43–46.
130. Dolenc, P., Staessen, J. A., Lauwerys, R., and Amery, A. (1993), *J. Hypertens.,* 11, 589–593.

Chapter 60

Respiratory Toxicology of Metals

Janet M. Benson and Judith T. Zelikoff

I. INTRODUCTION

Inhalation is a primary route for occupational and, to a lesser extent, environmental exposure of people to metals, and metals have high potential for producing toxic effects in the respiratory tract. The toxicity of metals to the human respiratory tract has been recently reviewed by Nemery (1990). The primary risk to people is associated with occupational exposures to metals during mining, refining, smelting, and end-use operations such as electroplating, machining, and welding. Epidemiological studies have indicated that occupational exposure to many metals results in a variety of acute and chronic lung diseases, including chemical pneumonitis, chronic obstructive lung disease, immune-mediated diseases (i.e., chronic beryllium disease [CBD]), fume fever, and cancer. The contribution of metals to respiratory tract cancer has been presented by Peters et al. (1986).

The purpose of this chapter is to review the neoplastic and nonneoplastic changes induced in the respiratory tract, resulting from metal exposure. Results of epidemiological and laboratory animal studies are included. The chapter focuses on metals with significant impact on the respiratory tract and is divided into sections discussing metals that primarily affect the respiratory tract and metals that affect the respiratory tract as well as other organ systems.

II. METALS THAT PRIMARILY AFFECT THE RESPIRATORY TRACT

A. BERYLLIUM
1. Humans

Exposure of people to relatively high concentrations of beryllium (Be) (>100 µg/m^3; Ridenour and Preus, 1991) causes acute beryllium disease, characterized by chemical pneumonitis. This disease is often fatal in cases of massive exposure, resulting in lung burdens ranging from 4 to 1800 µg/kg tissue (Freiman and Hardy, 1970). Acute Be disease occurred predominantly in the earlier days of the Be industry when industrial hygiene practices and engineering controls for minimizing worker exposure were not in place.

Some people inhaling low concentrations of Be develop chronic beryllium disease (CBD), a granulomatous lung disease characterized by dyspnea, cough, reduced pulmonary function, and a variety of other symptoms, including weight loss. Pathologically, the disease is a granulomatous interstitial pneumonitis characterized by lymphocytic infiltrates in the interstitium, noncaseating granulomas, and pulmonary fibrosis (Kriebel et al., 1988). Lymphocytes obtained by bronchopulmonary lavage from patients with CBD often exhibit strong proliferative responses to Be challenge *in vitro* (Rossman et al., 1988) and may exhibit a delayed hypersensitivity response when challenged dermally with Be. CBD is difficult to distinguish from sarcoidosis, and positive diagnosis requires histological evidence of pulmonary granulomas and a positive proliferative response of bronchoalveolar cells to Be or documentation of previous exposure to Be (Rossman, 1991). The lack of a dose-response relationship between the extent of exposure and development of the disease, long latency period between exposure and onset, and the low incidence among Be-exposed individuals suggests that the disease is immune mediated. Recently, Richeldi and co-workers (1993) described a modification of the major histocompatibility complex (MHC allele HLA-DPb-1-glu) of individuals with CBD which may serve as a useful marker for human susceptibility to CBD. Occupational risk associated with exposure to Be-containing alloys has been documented for individuals exposed to Be-Cu and Be-Ni alloys (Hooper, 1981; Lockey et al., 1983).

Beryllium is a suspected human carcinogen (ACGIH, 1986), based on results of animal data only. Epidemiologic evidence relating Be exposure to cancer in humans is inadequate to demonstrate or refute that Be is carcinogenic in humans, (US EPA, 1987), and the International Agency for Research on Cancer (IARC) lists the evidence for Be-induced carcinogenicity in humans as "limited".

2. Laboratory Animals

The pulmonary effects of inhaled Be have also been evaluated in a variety of laboratory animal species. Many of these studies have been reviewed in the EPA's Health Assessment Document for Beryllium (1987). Briefly, monkeys exposed to relatively high concentrations of Be compounds (200 µg/m^3 as beryllium fluoride (BeF$_2$), beryllium sulfate (BeSO$_4$), and beryllium phosphate (BePO$_4$) developed symptoms and histopathological findings consistent with acute Be disease (Schepers, 1964). Exposure of monkeys to lower concentrations of BeSO$_4$ (35 µg/m^3) led to lesions typical of CBD (Vorwald, 1966). More recently, Haley and co-workers (1994) have found that intratracheal instillation of Be metal, but not BeO (calcined at 500°C), into lungs of cynomolgus monkeys produces pulmonary immune granulomas. These investigators have demonstrated that a Be-induced lung disease with features characteristic of human CBD can also develop in BeO-exposed beagle dogs exposed by inhalation to BeO (Haley et al., 1989, 1992).

Granulomatous lung disease has also been produced in guinea pigs exposed to Be compounds by inhalation or by intratracheal instillation (Policard, 1950; Chiappino et al., 1969; Barna, 1981). Studies with guinea pigs showing strain differences in sensitivity and development of hypersensitivity responses furthered the argument that CBD is genetically controlled (Table 1).

Repeated inhalation of Be-containing materials, including soluble Be compounds and Be-containing ores, by various strains of laboratory rats has resulted in development of inflammatory and proliferative changes, granulomatous lung changes, and the development of lung tumors (Reeves et al., 1967; Wagner et al., 1969; Sanders et al., 1975; Groth, 1980; Finch et al., 1994). Although Be-exposed rats have developed various degrees of granulomatous lung disease, none have developed immunopathological responses in lung or Be hypersensitivity (Reeves, 1978).

More recent studies have shown that several strains of mice inhaling Be develop pulmonary lesions with features consistent with CBD (Huang et al., 1992; Finch et al., 1992; Nikula et al., 1992). Lung lesions consisted of infiltration of lymphocytes into the lung interstitium, development of microgranulomas consisting of T lymphocytes and macrophages, and the presence of some pulmonary fibrosis. Under certain exposure conditions, increased numbers of lymphocytes were recovered in bronchoalveolar lavage fluid from exposed animals. These results strongly suggest that A/J and C3H mouse strains may be relevant models for studying the inhalation toxicity of a variety of Be-containing materials, including Be-containing alloys.

B. CHROMIUM
1. Humans

The first case of chromium (Cr)-induced respiratory tract cancer was reported more than 100 years ago. While this report concerned the development of an adenocarcinoma in the nasal turbinate, results of numerous subsequent epidemiologic studies indicate inhalation of Cr compounds is also associated with an increased risk of developing lung cancer (see reviews by Hayes, 1988; Langard, 1990; Lees, 1991). Cigarette smoking and nonsmoking individuals involved in chromate production, electroplating, and use of Cr-containing pigments have developed lung tumors. Because data are lacking on the concentrations and specific types of Cr compounds encountered by workers in the various cohorts studied, it has been impossible to identify the risks attributable to specific Cr compounds and to establish exposure-risk relationships for individual compounds (Lees, 1991). However, an association exists between exposure to the more water-soluble Cr(VI) compounds (chromates, including zinc chromate)

Table 1 Metals that Primarily Affect the Respiratory Tract: Comparison of Effects in Man and Laboratory Animals

Metal	Humans	Laboratory Animals

Beryllium	Acute Be disease/ chemical pneumonitis after exposure to high Be concentrations
	Chronic Be disease/ granulomatous lung disease after exposure to low Be concentrations
	Limited evidence for cancer
		Monkeys
		  At high exposure concentrations have histological findings consistent with acute Be disease in humans
		  At low Be exposure concentrations have lesions consistent with Be-induced granulomatous lung disease in humans
		Dogs
		  Pulmonary lesions with features consistent with Be-induced granulomatous lung disease
		Rats
		  Inflammation, proliferative lung lesions, granulomas, lung tumors, little or no involvement of lymphocytes in the pulmonary response
		Guinea pigs
		  Granulomatous lung disease; strain differences in sensitivity
		Mice
		  Pulmonary lesions have features consistent with of chronic Be-induced lung disease

Chromium	Association between exposure to Cr(VI) compounds and lung cancers; some evidence for nasal cancers	Mice
		  No lung tumors after Cr_2O_3 exposure
		  Lifespan inhalation $CaCrO_4$ produced pulmonary adenomas, epithelial necrosis, alveolar bronchiolization, alveolar proteinosis, emphysema-like changes
		  No lung tumors following chronic inhalation of Cr roast material.
		Rats
		  CrO_2 inhalation produced cystic keratinizing squamous cell carcinoma in females; type II cell hyperplasia, fibrosis, cholesterol granulomas, alveolar proteinosis, fibrotic pleurisy, alveolar bronchiolarization found in males and females
		Guinea pigs
		  Chronic exposure to chromium roast material produced low incidence of lung tumors, bronchopneumonia, alveolar and interstitial inflammation, alveolar hyperplasia, and interstitial fibrosis

Nickel	Nasal and lung tumors associated with occupational exposure to oxidic and sulfidic forms of Ni; Ni inhalation not related to development of pneumoconiosis or chronic respiratory disease; asthma associated with exposure to hard metal	Rats
		  Lung tumors in rats chronically inhaling Ni_3S_2
		  Nonneoplastic lesions in rats inhaling Ni include inflammation, emphysema, atrophy of nasal olfactory epithelium
		Mice
		  Chronic inhalation of Ni produces pulmonary inflammation, fibrosis, atrophy of the olfactory epithelium
		Syrian golden hamster
		  NiO inhalation produced fibrosis, epithelial cell proliferation

and the incidence of lung cancer (Langard, 1990). Human exposure to Cr(III) compounds does not appear to be associated with an increased risk of developing lung cancer.

Exposure to high levels of chromate has been associated with an excess risk of developing nasal cancer (Davies et al., 1991), but excess risk for nasal cancers was not reported for lower-level exposures resulting from welding and spray painting with Cr-containing paints (Luce et al., 1993). Inhalation of corrosive hexavalent Cr compounds results in ulceration and perforation of the nasal septum (Kim et al., 1989; Goyer 1991).

2. Laboratory Animals

Many studies have been conducted to determine whether Cr compounds are pulmonary carcinogens in laboratory animals (see review by Langard, 1988). In a series of studies evaluating the inhalation toxicity of air pollutants, Nettesheim and co-workers (1969, 1970) exposed male and female C57BL/6 mice for their lifespans to chromic oxide (Cr_2O_3; 25 mg/m^3). There were no lung tumors among the Cr-exposed females, and the incidence of lung tumors in males was not considered to be significantly different from controls. In additon, no fibrosis was observed among the Cr-exposed mice. In contrast, C57BL/6 mice inhaling calcium chromate ($CaCrO_4$; approximately 13 mg/m^3) for their lifespans had a 4-fold increase in the incidence of pulmonary adenomas. Nonneoplastic alterations in the respiratory tract included epithelial necrosis, regenerative hyperplasia in the conducting airways, alveolar bronchiolization, alveolar proteinosis, and emphysema-like changes (Nettesheim et al., 1971). In studies conducted by Lee and co-workers (1989), male and female Sprague Dawley rats were exposed by inhalation to 0.5 or 25 mg/m^3 chromium dioxide (CrO_2); minimal lung lesions were present in rats exposed to 0.5 mg CrO_2/m^3. Nonneoplastic lesions present in the lungs of rats exposed to 25 mg/m^3 included type II cell hyperplasia of alveolar walls, and slightly collagenized fibrosis, cholesterol granulomas, alveolar proteinosis, minute fibrotic pleurisy, and alveolar bronchiolarization. Females, but not males, developed cystic keratinizing squamous cell carcinoma.

The incidence of lung tumors did not increase significantly in rats and mice exposed by inhalation to finely ground mixed chromium roast material (containing <5 mg/m^3 chromium trioxide) for up to 58 weeks (Steffee and Baetjer, 1965). Steffee and Baetjer (1965) also exposed guinea pigs and rabbits to the mixed chromium roast material to which finely ground potassium dichromate and a mist containing potassium dichromate were added (1 day/week), to sodium chromate mist (1 d/week), and to a dust from which the sodium chromate had been leached (1 d/week) for a total of 4 to 5 h/d, 4 d/week for the animals' lifespan. The average concentration of CrO_3 in the exposure atmosphere was 3 to 4 mg/m^3. No increase in lung tumors was observed among rabbits, but the incidence of pulmonary carcinomas in guinea pigs was 3/50. Interestingly, several nonneoplastic histopathological changes were observed among exposed as well as control animals. This was especially notable among the guinea pigs. Nonneoplastic changes in guinea pigs which could be attributable to Cr exposure included bronchopneumonia, alveolar and interstitial inflammation, alveolar hyperplasia, and interstitial fibrosis.

C. NICKEL
1. Humans

Excess risk of nasal and lung cancers has been reported among individuals employed in the nickel (Ni)-refining industry during the earlier part of this century (U.S. Environmental Protection Agency, 1985; Report of the International Committee on Nickel Carcinogenesis in Man, 1990). Epidemiological data indicated that excess risks of lung and nasal cancers were associated with exposure to sulfidic and oxidic forms of Ni produced during roasting, scintering, and calcining operations, but it was not clear whether the oxidic, sulfidic, or both forms of Ni were responsible for the lung and nasal cancers. Recently, extensive efforts have been made to revisit the existing epidemiology data to better define the human health risk associated with Ni inhalation. Specifically, 10 worker cohorts were examined to correlate the extent (airborne concentrations of Ni compounds and years of worker exposure), and type of Ni compound exposure with tumor incidence (Report of the International Committee on Nickel Carcinogenesis in Man, 1990). The results of this extensive investigation indicated that exposure of individuals to high concentrations of both sulfidic and oxidic forms of Ni or with high concentrations of oxidic Ni alone was associated with an increase in nasal and lung cancers. Exposure to soluble Ni increased the risk associated with exposure to the less soluble (oxidic and sulfidic) forms. No increased risk of lung and nasal cancer was associated with exposure to Ni metal.

Inhalation of Ni by metal plating workers has resulted in rare occurrence of asthma (McConnell et al., 1973; Block and Yeung, 1982; Malo et al., 1982, Cirla et al., 1985). Metallic nickel inhaled as a component of "hard metal" may contribute to development of "hard metal"-related respiratory disease, including asthma (Coates and Watson, 1971; Kussaka et al., 1986; Cugell et al., 1990). It does not appear that inhalation of Ni-containing compounds and dusts leads to specific pneumoconiosis or to chronic respiratory disease (Mastromatteo, 1986).

Nasal irritation, damage to the nasal mucosa, perforation of the nasal septum and loss of smell have also been reported among workers inhaling Ni-containing aerosols (Mastromatteo, 1986).

2. Laboratory Animals

Exposure of F344/N rats to > 15 mg $NiSO_4 \cdot 6H_2O/m^3$ and of $B6C3F_1$ mice to >7 mg $NiSO_4 \cdot 6H_2O/m^3$ or greater for 6 h/d, 5 d/week for 2 weeks has resulted in mortality (Dunnick et al., 1988; Benson et al., 1988). The lethal concentration of Ni_3S_2 in rats and mice exposed under similar conditions was 10 mg Ni_3S_2/m^3 (Benson et al., 1987; Dunnick et al., 1988).

Inhalation of various soluble or moderately soluble forms of Ni ($NiSO_4 \cdot 6H_2O$, Ni_3S_2) and a poorly soluble form (NiO) by laboratory animals has resulted in pulmonary inflammation (Wehner et al., 1975; Dunnick et al., 1988, 1989; Tanaka et al., 1988; Benson et al., 1990). The ability of Ni compounds to produce inflammation is related to their solubility, with $NiCl_2$ and $NiSO_4 \cdot 6H_2O$ being significantly more potent inflammagens than the less soluble nickel oxides (Benson et al., 1986, 1989; Dunnick et al., 1989). Emphysema has been reported among F344/N rats inhaling 5 or 10 mg Ni_3S_2 6 h/d, 5 d/week for 2 weeks (Benson et al., 1987) and in Wistar rats inhaling 1 mg NiO (green NiO) for 6 months (Haratake et al., 1992). Fibrosis has been reported among $B6C3F_1$ mice inhaling 5 mg Ni_3S_2 5 d/week for 2 weeks or inhaling 1.2 and 2.5 mg Ni_3S_2 5 d/week for 13 weeks (Benson et al., 1987, 1990). Wehner and co-workers (1975) reported fibrosis, epithelial cell proliferation from the bronchioles, and bronchiolization of alveoli among Syrian golden hamsters chronically exposed to 53 mg NiO/m^3 7 h/d, 5 d/week for their lifespans.

An early bioassay study investigating the carcinogenicity of chronically inhaled Ni_3S_2 in rats indicated this form of nickel to be carcinogenic. F344 rats were exposed to approximately 1 mg Ni_3S_2/m^3 6 h/d, 5 d/week for 78 weeks. The overall incidence of lung tumors in the Ni_3S_2-exposed rats was 14% compared with 1% in controls (Ottolenghi et al., 1974). In a more recent bioassay study, the relative toxicity and carcinogenicity of $NiSO_4 \cdot 6H_2O$, Ni_3S_2, and NiO (calcined at 1200°C) were studied in F344/N rats and $B6C3F_1$ mice (Dunnick et al., 1995). Animals were exposed 6 h/d, 5 d/week for 2 years. Nickel subsulfide (0.15 and 1 mg/m^3) and nickel oxide (1.25 and 2.5 mg/m^3) caused an exposure-related increase in alveolar/bronchiolar neoplasms in male and female rats. Nickel oxide caused equivocal exposure-related increases in alveolar/bronchiolar neoplasms in female mice. No exposure-related neoplastic responses occurred in rats or mice exposed to $NiSO_4 \cdot 6H_2O$ or in mice exposed to Ni_3S_2.

Inhalation of >3.5 mg $NiSO_4 \cdot 6H_2O/m^3$ or 1.2 mg Ni_3S_2 6 h/d, 5 d/week results in atrophy of the nasal olfactory epithelium (Dunnick et al., 1988). Subchronic exposure of rats to at least 0.5 mg $NiSO_4 \cdot 6H_2O/m^3$ or to 0.3 mg Ni_3S_2/m^3 6 h/d, 5 d/week for 13 weeks also results in atrophy of the nasal olfactory epithelium (Dunnick et al., 1989, Benson et al., 1990). Inhalation of NiO does not appear to affect nasal tissue.

III. OTHER METALS THAT AFFECT THE RESPIRATORY TRACT

A. ALUMINUM
1. Humans

Occupational exposure to aluminum (Al) is not associated with an excess risk of respiratory tract cancer, and insufficient epidemiological evidence is available on whether inhalation of aluminum can result in development of chronic obstructive lung disease (Abramson et al., 1989). Pulmonary fibrosis and interstitial pneumonia have been reported in Al arc welders (Vallyathan et al., 1982; Herbert et al., 1982) and in an Al polisher (De Vuyst et al., 1986). Occupational exposure to Al has also been associated with an asthma-like syndrome ("potroom asthma"), but it is unclear whether the aluminum initiates asthma or merely precipitates symptoms of asthma in predisposed individuals (Abramson et al., 1989).

2. Laboratory Animals

Inhaled Al powder appears to have poor solubility in the lung (Thompson et al., 1986). Acute inhalation exposure of F344 rats to 50 mg Al/m^3 resulted in increased numbers of polymorphonuclear lymphocytes recovered in the lavage fluid. Inhalation of 200 and 1000 mg Al/m^3 (resulting in estimated particle lung burdens of 2 to 6 mg Al/g lung) resulted in development of microgranulomas. Essentially no information is available on the inhalation toxicity associated with repeated inhalation of occupationally relevant concentrations of Al metal (Table 2).

B. ARSENIC
1. Humans

Occupational exposure to inorganic arsenic (As), especially in mining and copper smelting, has consistently been associated with an increased risk of cancer (IARC, 1987; Lubin et al., 1981; Rencher

Table 2 Other Metals that Affect the Respiratory Tract; Comparison of Effects in Man and Laboratory Animals

Metal	Humans	Laboratory Animals
Aluminum	No evidence Al exposure results in development of respiratory tract cancers; inhalation may result in pulmonary fibrosis and interstitial pneumonia; exposure associated with development of asthma-like syndrome	Rats — Inhalation of high concentrations of insoluble Al powder produced microgranulomas
Arsenic	Strong association between As inhalation and development of respiratory tract cancers	Syrian golden hamsters — Pulmonary adenomas produced by instillation of arsenic trioxide; no tumors with instillation of As_2O_3 or $As_2Ca_3O_8$
Cadmium	Some evidence that Cd is pulmonary carcinogen; CdO inhalation produces chronic bronchitis, pneumonitis, fibrosis, emphysema, and chronic obstructive lung disease	Rats — Inhalation of $CdCl_2$, CdO, $CdSO_4$, and CdS produces lung tumors
Copper/zinc	Inhalation of various forms of Cu and Zn results in metal fume fever Cu inhalation results in upper respiratory tract lesions	Rats — Pulmonary inflammation following ZnO or $CuSO_4$ inhalation Guinea pig — ZnO inhalation alters pulmonary function, produces pulmonary inflammation
Lead	Some evidence for increased risk of respiratory tract cancers.	Rabbits — PbO inhalation produces cytotoxic but not inflammatory responses in lung

et al., 1977). An almost 10-fold increase in the incidence of lung cancer was found in workers most heavily exposed to As (IARC, 1987). In a study carried out by Lee-Feldstein (1986), a direct relationship between cumulative exposure to arsenic and the incidence of respiratory cancers among employees in copper smelters was noted. Histological types of cancers reported include adenocarcinomas and oat-cell cancers (Wicks et al., 1981).

2. Laboratory Animals

Few studies have been conducted on the pulmonary carcinogenicity of arsenicals. Ishinishi and co-workers (1983) reported a 10 to 33% incidence of lung adenomas in Syrian golden hamsters administered a total dose of 5.2 or 3.7 mg As (as arsenic trioxide; As_2O_3) by intratracheal instillation. The tumorigenicity of As_2S_3 and $As_2Ca_3O_8$ administered by intratracheal instillation was also investigated in Syrian golden hamsters (Pershagen and Bjorklund, 1985). The hamsters were administered weekly doses of 3 mg As/kg body weight for 15 consecutive weeks. The incidence of adenomas was 1/28 for the arsenic trisulfide group and 4/35 for the calcium arsenate group. No adenomas occurred among the 26 control hamsters. Alveolar proteinosis was also present in the lungs of hamsters treated with arsenic trisulfide and calcium arsenate.

C. CADMIUM
1. Humans

The IARC lists the evidence for carcinogenicity to humans from Cd exposure as "limited" (IARC, 1987). However, even in the presence of potentially confounding factors such as coexposure to other carcinogenic metals and the presence of smoking individuals among the cohorts studied, the evidence implicating Cd as a pulmonary carcinogen in humans remains (IARC, 1987). Results of a recent evaluation of risk from lung cancer in occupationally exposed individuals showed a statistically significant dose-response relationship. The lifetime excess lung cancer risk for individuals exposed to Cd fumes at a level of 100 µg/m^3 was 50 to 111 lung cancer deaths per 1000 workers exposed to Cd for 45 years (Stayner et al., 1992). Occupational exposure to cadmium oxide (CdO) aerosol has been known to cause chronic bronchitis, pneumonitis, progressive fibrosis of the lower airways, and emphysema, leading eventually to the development of chronic obstructive lung disease (Goyer, 1991).

2. Laboratory Animals

Cadmium chloride, $CdSO_4$, CdO, and CdS pigments produce lung tumors when inhaled by rats (Takenaka et al., 1983; Heinrich, 1989; Heinrich et al., 1992). Inhalation of $CdCl_2$ produced a dose-dependent increase in adenocarcinomas, epidermoid (squamous cell) carcinomas, combined epidermoid and adenocarcinomas, and mucoepidermoid carcinomas. The incidence of lung carcinomas was 71.4%, 52.6%, and 15.4% in the groups exposed to 50, 25, and 12.5 µg Cd/m^3, respectively. CdSO$_4$, CdO, and cadmium sulfide pigments had similar carcinogenic potencies in rats. No increase in tumor rate was observed in mice or hamsters inhaling Cd compounds (Heinrich et al., 1989). Histopathological examination of the nasal tissue from CdCl$_2$-exposed rats revealed no lesions. Nonneoplastic lung lesions have been described by Asvadi and Hayes (1978). Ultrastructural changes in lungs of Cd-exposed rats and hamsters have been described by Thiedemann and co-workers (1989).

D. COPPER AND ZINC
1. Humans

Occupational inhalation exposure to Cu and Zn in the forms of CuO, Cu(CH$_3$CO$_2$)$_2$, Cu fumes, ZnO, or brass alloy has commonly been associated with the development of "metal fume fever" (Cohen, 1974; NIOSH, 1975). The symptoms associated with this syndrome include fever, chills, dyspnea, muscle soreness, nausea, and fatigue. The occurrence of this syndrome has been the basis for the current threshold limit values for these metals. The mechanisms underlying these effects remain unknown (Gordon et al., 1992). Occupational exposure to copper has also resulted in upper respiratory tract effects including congestion of the nasal mucous membranes and pharynx and perforation of the nasal septum (Cohen, 1974).

2. Laboratory Animals

Short-term inhalation exposure of guinea pigs to ZnO has resulted in altered pulmonary function parameters and inflammation in the proximal portion of the alveolar ducts and adjacent alveoli (Lam et al., 1985). Short-term (3 h) inhalation of ZnO (2.5 and 5.0 mg/m^3) also produces cytotoxic and inflammatory responses in lungs of rats and rabbits as determined by biochemical changes measured in bronchoalveolar lavage fluid (Gordon et al., 1992). CuSO$_4$ produced inflammatory responses in rats when administered by intratracheal instillation (Hirano et al., 1990). A dose of 5 µg Cu per rat was sufficient to elicit an inflammatory response, which peaked by the 3rd day after Cu instillation and was resolved by the 7th day after instillation.

E. LEAD
1. Humans

Little information is available on the respiratory tract toxicity of lead compound in humans or laboratory animals. One epidemiological study has evaluated mortality among cohorts working for at least 1 year in lead production facilities and battery plants during the period 1946 to 1970 (Cooper and Gaffey, 1975). Although original analysis of the data suggested an increased risk of respiratory tract cancers associated with lead exposure, subsequent statistical evaluation indicates statistically significant increase in cancers of the respiratory tract for both production and battery plant workers (Kang and Infante, 1980).

2. Laboratory Animals

Limited data are available on the respiratory tract toxicity of inhaled lead compounds. Zelikoff and co-workers (1993) have studied the effects of inhalation exposure on rabbits to 30 µg/m^3 PbO 3 h/d, for 4 days. The PbO was cytotoxic in lung, as indicated by increases in lactate dehydrogenase in bronchoalveolar lavage fluid, but did not produce an inflammatory response. PbO exposure also decreased alveolar macrophage phagocytic ability but increased macrophage production of H$_2$O$_2$ and reactive oxygen species.

IV. SUMMARY

Individuals with the greatest potential for inhalation exposure to toxic metals are those employed in metal refining, production, processing, and reclamation industries. The information provided in this chapter indicates that metals produce a wide spectrum of effects in the human respiratory tract; no two metals produce exactly the same pattern of responses. Further, responses to inhalation of a given metal often depend on its chemical form/solubility. Interestingly, lesions in the respiratory tracts of laboratory animals exposed to metals often do not parallel human respiratory tract lesions. Therefore identification of appropriate animal models for studying cancer and other respiratory tract diseases can be difficult.

Epidemiological and experimental research on the occurrence and mechanisms underlying respiratory tract toxicity of metals have tended to focus on known or suspected human carcinogens. The occurrence and mechanisms underlying Be-induced granulomatous lung disease, Zn and Cu-induced "metal fume fever", and Al-induced "potroom asthma" remain less well studied. Another fruitful area of research may include investigation of the respiratory tract effects of selected metals inhaled in combination. Such investigations could have direct applicability to evaluations of risk associated with inhalation of multiple metals as may occur in the metal refining and smelting industries, and in industries producing and using metal alloys.

REFERENCES

Abramson, M. J., Wlodarczyk, J. H., Saunders, N. A., and Hensley, M. J. (1989), *Am. Rev. Respir. Dis.,* 139, 1042–1057.
ACGIH (1986), *Threshold Limit Values for 1992–1993,* American Conference of Government Industrial Hygenists, Washington, D.C.
Asvadi, S. and Hayes, J. A. (1978), *Am. J. Pathol.,* 90, 89–98.
Barna, P., Chiang, T., Pillarisetti, S. G., Deodhar, S. D. (1981), *Clin. Immunol. Immunopathol.,* 20, 402–411.
Benson, J. M., Henderson, R. F., McClellan, R. O., Hanson, R. L., and Rebar, A. H. (1986), *Fundam. Appl. Toxicol.,* 7, 340–347.
Benson, J. M., Carpenter, R. L., Hahn, F. F., Haley, P. J., Hanson, R. L., Hobbs, C. H., Pickrell, J. A., and Dunnick, J. K. (1987), *Fundam. Appl. Toxicol.,* 9, 251–265.
Benson, J. M., Burt, D. G., Carpenter, R. L., Eidson, A. F., Hahn, F. F., Haley, P. J., Hanson, R. L., Hobbs, C. H., Pickrell, J. A., and Dunnick, J. K. (1988), *Fundam. Appl. Toxicol.,* 10, 164–178.
Benson, J. M., Burt, D. G., Cheng, Y. S., Hahn, F. F., Haley, P. J., Henderson, R. F., Hobbs, C. H., Pickrell, J. A., and Dunnick, J. K. (1989), *Toxicology,* 57, 225–266.
Benson, J. M., Burt, D. G., Cheng, Y. S., and Eidson, A. F. (1990), *Inhal. Toxicol.,* 9, 219–222.
Block, G. T. K. and Yeung, M. (1982), *J. Am. Med. Assoc.,* 247, 1600–1602.
Chiappino, G., Cirla, A., and Vigliani, E. C. (1969), *Arch. Pathol.,* 87, 131–140.
Cirla, A. M., Bernabeo, F., Ottoboni, F. K., and Ratti, R. (1985), in *Progress in Nickel Toxicity,* Brown, S. S. and Sunderman, F. W. Jr., Eds., Blackwell, Oxford, 165–168.
Coates, E. O. and Watson, J. H. L. (1971), *Ann. Int. Med.,* 75, 709–716.
Cohen, S. R. (1974), *J. Occup. Med.,* 16, 621–624.
Cooper, W. C. and Gaffey, W. R. (1975), *J. Occup. Med.,* 17, 100–107.
Cugell, D. W., Morgan, W. K. C., Perkins, D. G., and Rubin, A. (1990), *Arch. Intern. Med.,* 150, 177–183.
Davies, J. M., Easton, D. F., and Bidstrup, P. L. (1991), *Br. J. Ind. Med.,* 48, 299–313.
De Vuyst, P., Dumortier, P., Rickaert, F., Van De Weyer, R., Lenclud, C., and Yernault, J. C. (1986), *Eur. J. Respir. Dis.,* 68, 131–140.
Dunnick, J. K., Benson, J. M., Hobbs, C. H., Hahn, F. F., Cheng, Y. S., and Eidson, A. F. (1988), *Toxicology,* 50, 145–156.
Dunnick, J. K., Elwell, M., R., Radovsky, A. E., Benson, J. M., Hahn, F. F., Nikula, K. J., Barr, E. B., and Hobbs, C. H. (1995), *Cancer Res.,* 55, 5251–5256.
Dunnick, J. K., Elwell, M., R., Benson, J. M., Hobbs, C. H., Hahn, F. F., Haley, P. J., Cheng, Y. S., and Eidson, A. F. (1989), *Fundam. Appl. Toxicol.,* 12, 584–594.
Finch, G. L., Hoover, M. D., and Nikula, K. J. (1992), in *Inhalation Toxicology Research Institute Annual Report 1991–1992,* LMF-138, Finch, G. L., Nikula, K. J., and Bradley, P. L., Eds., National Technical Information Service, U. S. Department of Commerce, Springfield VA, 169–170.
Finch, G. L., Haley, P. J., Hoover, M. D., Snipes, M. B., and Cuddihy, R. G. (1994), *Inhal. Toxicol.,* 6, 205–224.
Freiman, D. G. and Hardy, H. L. (1970), *Hum. Pathol.,* 1, 25–44.
Gordon, T. K., Chen, L. C., Fine, J. M., Schlesinger, R. B., Su, W. Y., Kimmel, T. A., and Amdur, M. O. (1992), *Am. Ind. Hyg. Assoc. J.,* 53, 503–509.
Goyer, R. A. (1991), *Toxic effects of metals,* in *Toxicology, the Basic Science of Poisons,* Amdur, M. O., Doull, J., and Klassen, C. D., Eds., Pergamon Press, New York, 623–680.
Groth, D. H., Kommineni, C., and Mackay, G. R. (1980), *Environ. Res.,* 21, 63–84.
Haley, P. J., Finch, G. L., Mewhinney, J. A., Harmsen, A. G., Hahn, F. F., Hoover, M. D., Muggenburg, B. A., and Bice, D. E. (1989), *Lab Invest.,* 61, 219–227.
Haley, P. J., Pavia, K. F., Swafford, D. S., Davila, D. R., Hoover, M. D., and Finch, G. L. (1994), *Immunopharmacol. Immunotoxicol.,* 16, 627–644.
Haratake, J., Horie, A., Kodama, Y., and Tanaka, I. (1992), *Inhal. Toxicol.,* 4, 67–79.
Hayes, R. B. (1988), *Sci. Total. Environ.,* 71, 331–339.
Heinrich, U., Peters, L., Ernst, H., Rittinghausen, S., Dasenbrock, C., and Konig, H. (1989), *Exp. Pathol.,* 37, 253–258.
Heinrich, U. (1992), *IARC Sci. Publ.,* 118, 405–413.
Herbert, A., Sterling, G., Abraham, J., and Corrin, B. (1982), *Hum. Pathol.,* 13, 694–699.
Hirano, S., Sakai, S., Ebihara, H., Kodamma, N., and Suzuki, K. T. (1990), *Toxicology,* 64, 223–233.

Hooper, W. F (1981), *N. C. Med. J.,* 42, 551–553.
Huang, H., Meyer, K. C., Kubaii, L., and Auerbach, R. (1992), *Lab. Invest.,* 67, 138–146.
International Agency for Research on Cancer (1987), Overall Evaluation of Carcinogenicity: An Updating of IARC Monographs Vol. 1–42, Supplement 7, World Health Organization, Lyon.
Ishinishi, N., Yamamoto, A., Hisanaga, A., and Inamasu, T. (1983), *Cancer Lett.,* 21, 141–147.
Kang, H. K. and Infante, P. F. (1980), *Science,* 207, 180–181.
Kim, M. J., Lee, J. D., Choi, H. S., Kim, D. I., Chung, T. S., Suh, J. H., and Roh, J. H. (1989), *Yonsei Med. J.,* 30, 305–309.
Kriebel, D., Brain, J. D., Sprince, N. L., and Kazemi, H. (1988), *Am. Rev. Respir. Dis.,* 137, 464–473.
Kusaka, Y., Yokoyama, K., Sera, Y., Yamamoto, S., Sone, S., Kyono, H., Shirakawwa, T., and Goto, S. (1986), *Br. J. Ind. Med.,* 43, 474–485.
Lam, H. F., Conner, M. W., Rogers, A. E., Fitzgerald, S. K., and Amdur, M. O. (1985), *Toxicol. Appl. Pharmacol.,* 78, 29–38.
Langard, S. (1988), *Sci. Total Environ.,* 71, 341–350.
Langard, S. (1990), *Am. J. Ind. Med.,* 17, 189–215.
Lee, K. P., Ulrich, C. E., Geil, R. G., and Trochimowicz, H. J. (1989), *Sci. Total Environ.,* 86, 83–108.
Lee-Feldstein, A. (1986), *J. Occup. Med.,* 28, 296–302.
Lees, P. S. J. (1991), *Environ. Health Perspect.,* 24, 93–104.
Lockey, J. E., Rom., W. N., White, K. L., Lee, J. S., and Abraham, J. L. (1983), *Am. Rev. Respir. Dis.,* 127, 183.
Lubin, J. H., Pottern, L. M., Blot, W. J., Tokudome, S., Stone, B. J., and Fraumeni, J. F. (1981), *J. Occup. Med.,* 23, 779–784.
Luce, D., Gerin, M., Leclerc, A., Morcet, J.-F., Brugere, J., and Goldberg, M. (1993), *Int. J. Cancer,* 53, 224–231.
Malo, J.-L., Cartier, A., Doepner, N., Niebohr, E., Evans, S. K., and Dolovich, J. (1982), *J. Allergy Clin. Immunol.,* 69, 55–59.
Mastromatteo, E. (1986), *J. Am. Ind. Hyg. Assoc.,* 47, 589–601.
McConnell, L. H., Fink, J. N., Schlueter, D. P., and Schmidt, M. G. (1973), *Ann. Intern. Med.,* 78, 888–890.
National Institute for Occupational Safety and Health (1975), Occupational Exposure to Zinc Oxide (DHEW/NIOSH Pub. No. 77–104), Government Printing Office, Washington, D.C.
Nemery, B. (1990), *Eur. Resp. J.,* 3, 202–219.
Nettesheim, P., Hanna, M. G., Jr., Doherty, D. G., Newell, R. F., and Hellman, A. (1969), in *Inhalation Carcinogenesis,* Hanna, M. G., Jr., Nettesheim, P., and Gilbert, J. R., Eds., Division of Technical Information, U. S. Atomic Energy Commission, Symp. Ser. 18, 305–320.
Nettesheim, P., Hanna, M. G., Jr., Doherty, D. G., Newell, R. F., and Hellman, A. (1970), in *Morphology of Experimental Respiratory Carcinogenesis,* Nettesheim, P., Hanna, M. G., Jr., and Deatherage, J. W. Jr., Eds., Division of Technical Information, U. S. Atomic Energy Commission, Symp. Ser. 21, 437–448.
Nettesheim, P., Hanna, M. G., Doherty, D. G., Newell, R. F., and Hellman, A. (1971), *J. Natl. Cancer Inst.,* 47, 1129–1144.
Nikula, K. J., Tohulka, M. D., Swafford, D. S., Hoover, M. D., and Finch, G. L. (1992), in *Inhalation Toxicology Research Institute Annual Report 1991–1992,* LMF 138, Finch, G. L., Nikula, K. J., and Bradley, P. L., Eds., National Technical Information Service, U. S. Department of Commerce, Springfield, VA, 171–172.
Ottolenghi, A. D., Haseman, J. K., Payne, W. W., Falk, H. C., and MacFarland, H. N. (1974), *J. Natl. Cancer Inst.,* 54, 1165–1172.
Pershagen, G. and Bjorklund, N.-E. (1985), *Cancer Lett.,* 27, 99–104.
Peters, J. M., Thomas, D., Falk, H., Oberdorster, G., and Smith, T. J. (1986), *Environ. Health Perspect.,* 70, 71–83.
Policard, A. (1950), *Br. J. Ind. Med.,* 7, 117–121.
Reeves, A. L., Deitch, D., and Vorwald, A. J. (1967), *Cancer Res.,* 27, 439–445.
Reeves, A. L., (1978), in *Inorganic and Nutritional Aspects of Cancer,* Schrauzer, G. N., Ed., Plenum Publishing, New York, 13–27.
Rencher, A. C., Carter, M. W., and McKee, D. W. (1977), *J. Occup. Med.,* 19, 754–758.
Report of the International Committee on Nickel Carcinogenesis in Man (1990), *Scand. J. Work Environ. Health,* 16, 1–82.
Richeldi, L., Sorrentino, R., and Saltini, C. (1993), *Science,* 262, 242–244.
Ridenour, P. K. and Preuss, O. P. (1991), in *Beryllium, Biomedical and Environmental Aspects,* Rossman, M. D., Preuss, O. P., and Powers, M. B., Eds., Williams & Wilkins, Baltimore, 103–112.
Rossman, M. D., Kern, J. A., Elias, J. A., Cullen, M. R., Epstein, P. E., Preuss, O. P., Markham, T. N., and Daniele, P. (1988), *Ann. Intern. Med.,* 108, 687–693.
Rossman, M. D. (1991), in *Beryllium, Biomedical and Environmental Aspects,* Rossman, M. D., Preuss, O. P., and Powers, M. B., Eds., Williams & Wilkins, Baltimore, 167–177.
Sanders, C. L., Cannon, W. C., Powers, G. J., Adee, R. R., and Meier, D. M. (1975), *Arch. Environ. Health,* 30, 546–551
Schepers, G. W. H. (1964), *Ind Med. Surg.,* 33, 1–16.
Stayner, L., Smith, R., Thun, M., Schnorr, T., and Lemen, R. (1992), *Ann. Epidemiol.,* 2, 177–194.
Steffee, C. H. and Baetjer, A. M. (1965), *Arch. Environ. Health,* 11, 66–75.
Tanaka, I., Horie, A., Haratake, J., Kodama, Y., and Tsuchiya, K. (1988), *Biol. Trace Elem. Res.,* 16, 19–26.
Takenaka, S. K., Oldiges, H., Konig, H., Hochrainer, D., and Oberdorster, G. (1983), *JNCI,* 70, 367–371.
Thiedemann, K. U., Luthe, N., Paulini, I., Kreft, A., Heinrich, U., and Glaser, U. (1989), *Exp. Pathol.,* 37, 264–268.
Thompson, S. M., Burnett, D. C., Bergmann, J. D., and Hixson, C. J. (1986), *J. Appl. Toxicol.,* 6, 197–209.
U.S. Environmental Protection Agency (1985), Health Assessment Document for Nickel EPA/600/8–83/012F, National Technical Information Service, U.S. Department of Commerce, Springfield, VA.

U. S. Environmental Protection Agency (1987), Health Assessment Document for Beryllium, EPA/600/8–84/026F, National Technical Information Service, U. S. Department of Commerce, Springfield, VA.

Vallyathan, V., Bergeron, W. N., Robichaux, P. A., and Craighead, J. E. (1982), *Chest,* 81, 372–374.

Vorwald, A. J., Reeves, A. L., and Urban, E. C. J. (1966), *Experimental beryllium toxicology,* in: Beryllium, its industrial hygiene aspects, Stokinger H.E., Academic Press, New York, 201–234.

Wagner, W. D., Groth, D. H., Holtz, J. L., Madden, G. E., Stokinger, H. E. (1969), *Toxicol. Appl. Pharmacol.,* 15, 10–29.

Wehner, A. P., Busch, R. H., Olson, R. J. K., and Craig, D. K. (1975), *Am. Ind. Hyg. Assoc. J.,* 36, 801–810.

Wicks, M., Archer, V., Auerbach, O., and Kuschner, M. (1981), *Am. J. Ind. Med.,* 2, 25–31.

Zelikoff, J. T., Parsons, E., and Schlesinger, R. B. (1993), *Environ. Res.,* 62, 207–222.

Chapter 61

Effects of Metals on the Hematopoietic System and Heme Metabolism

James S. Woods

I. THE HEMATOPOIETIC SYSTEM

The hematopoietic system is a complex organ comprising the bone marrow, spleen, lymph nodes, and reticuloendothelial tissues, as well as the three major groups of formed elements of blood consisting of erythrocytes, leukocytes (granulocytes, monocytes, and lymphocytes), and platelets. "Hematopoiesis" refers generally to formation of all the formed blood elements. In humans, the liver is the principal site of hematopoiesis throughout most of gestation. Beginning in the 4th month of development, hematopoiesis begins in bone marrow and, by the time of birth, the marrow becomes virtually the sole source of blood cells. Up to the age of puberty, all the marrow throughout the skeleton is hematopoietically active, while in adults only about 50% of the marrow space is active in blood cell formation (Robbins et al., 1984). Metals exert principal toxic effects on hematopoiesis by altering blood cell formation in the bone marrow or by direct effects on the formed elements, particularly red blood cells, following their entrance into the circulation. Each of these principal sites of metal action on hematopoiesis is considered in this section.

A. EFFECTS OF METALS ON BONE MARROW FUNCTION

Inadequate blood cell production results largely from bone marrow failure. Chemicals toxic to bone marrow can impair or cause cessation of hematopoiesis, resulting in anemia or a general decrease in the circulating numbers of the three major groups of formed elements, a condition known as pancytopenia. While pancytopenia is most commonly associated with exposure to ionizing radiation or benzene and related chemicals, several metals are regularly associated with this condition.

Gold preparations employed therapeutically as antipyretics or for the treatment of inflammatory diseases such as arthritis have produced pancytopenia, characterized by thrombocytopenia, leukopenia, and aplastic anemia through suppression of bone marrow function (Flower et al., 1985). When pancytopenia results from gold therapy, the urinary concentrations of coproporphyrin and δ-aminolevulinic acid (ALA) may increase, similar to the situation in lead poisoning (Sassa, 1978). Whether this effect represents an action of gold on specific heme pathway enzymes or other heme pathway constituents in erythropoietic cells is not known.

Inorganic arsenic compounds, e.g., arsenates (As^{5+}) and arsenites (As^{3+}), can cause pancytopenia through impairment of bone marrow function. Kyle and Pease (1965) studied hematologic changes in six cases of arsenic intoxication and noted anemia, leukopenia, prominent basophilic stippling, and depression of bone marrow erythropoiesis. The anemia is secondary to hemolysis and bone marrow depression. Depression of erythropoiesis may be due to arsenic induced-inhibition of iron incorporation into red cells, since the 18-hour incorporation of ^{59}Fe into red blood cells of mice was reduced 50% when initiated 30 h after injection of normal mice with arsenate (0.24 nmol/kg) or arsenite (0.10 nmol/kg) (Morse et al., 1980). Inorganic arsenicals may also inhibit erythropoietin-induced erythroid cell differentiation, contributing further to decreased red cell formation.

A number of specific effects on the hematopoietic system associated with mercury exposure have been reported. Verschuuren et al. (1976) studied the hematologic effects of short-term organic mercury exposure in male and female weanling rats given methylmercury chloride (MMC) at 0, 0.1, 0.5, 2.5, and 25 ppm in the diet for 12 weeks. No hematological changes were observed at the 0.1 to 12.5 ppm levels. However, among animals given 25 ppm MMC, hematologic examination revealed significant decreases in hemoglobin concentration, packed cell volume, and erythrocyte count in males but not in females. Both sexes, however, displayed increased neutrophil and lowered lymphocyte counts. In humans, hematologic effects characterized by moderate to high level leukocytosis with neutrophilia have been reported following acute and chronic inhalation of metallic mercury vapor (Campbell, 1948; Matthes et al., 1958). These effects are assumed to represent depression by mercury of bone marrow function. There are no reports of oral or chronic exposure effects of mercury vapor in humans which would permit determination as to whether hematologic effects are dependent on route or duration of exposure, nor as to the mechanism(s) involved in these effects.

Aluminum accumulates in bone and has been reported to cause anemia in chronic kidney disease patients receiving hemodialysis (Wills and Savory, 1983). The mechanisms involved in alteration of bone marrow function by aluminum have not been defined. Aluminum salts have been reported to inhibit erythropoiesis *in vitro* (Mladenovic, 1988) and to inhibit iron incorporation into heme in rat bone marrow cell cultures (Zaman et al., 1990). Abreo et al., (1990) have reported that aluminum utilizes the transferrin uptake pathway for entry into erythroid precursor cells and have suggested that aluminum interferes with erythropoiesis at sites distal to that of iron uptake, possibly those associated with heme or globin synthesis. *In vivo*, aluminum as $AlCl_3$ or $Al(NO_3)_3$ has also been reported to produce a decrease in erythrocyte count, hemoglobin, and hematocrit as well as an increase in reticulocytes and polychromatophilic erythrocytes in rats (Zaman et al., 1993). Other investigators (Fulton and Jeffery, 1990), however, have not observed changes in hematocrit or other hematological indices following prolonged treatment of rabbits with $AlCl_3$ at doses from 100 to 500 mg/l in drinking water. The mechanisms associated with aluminum alteration of hematopoiesis in bone marrow cells, therefore, remain to be elucidated.

Cobalt can affect bone marrow function resulting in polycythemia, a disease in which there is an overproduction of red cells. Polycythemia was regularly observed in episodes of "beer drinker's cardiomyopathy" in the 1960s, which occurred coincidentally with the introduction of small amounts of cobalt into some brands of beer to reduce foaming (Gosselin et al., 1984). The basis of this effect of cobalt is thought to lie in the reciprocal relationship which exists between red cell production and oxidative metabolism in tissues, which is inhibited by cobalt (Thorling and Erslev, 1972). Thus, ingestion of cobalt results in tissue hypoxia, presumably reflecting impaired oxygen utilization because of inhibited cellular oxidative metabolism. Tissue hypoxia, in turn, results in increased rates of red cell production. The mechanism which links intracellular aerobic metabolism and red cell production in bone marrow is still not fully understood, but is known to be mediated by the renal erythropoietic hormone, erythropoietin (Erslev et al., 1980), of which cobalt is a potent inducer.

Changes in hematologic function have been observed in humans and animals after intermediate or chronic exposure to zinc or zinc-containing compounds. Long-term administration of zinc supplements has caused anemia in humans (Broun et al., 1990; Hale et al., 1988; Prasad et al, 1978) at doses as low as 2 mg/kg/d for 10 months (Hoffman et al., 1988). In animals, decreased hemoglobin, hematocrit, and erythrocyte levels have been reported in rats and mice (Maita et al., 1981), rabbits (Bentley and Grubb, 1991), dogs (Drinker et al., 1927), and ferrets (Straube et al., 1980) during zinc exposure. In contrast, zinc also appears to play a protective role against the toxic effects of other heavy metals on erythropoiesis. Lutton et al. (1984) studied the effects of gold, lead, and cadmium on rat bone marrow erythroid colony (CFU_E) formation *in vitro*. All three metals were found to be significantly toxic to CFU_E growth at concentrations from 10^{-9} to 10^{-7} M. In the presence of zinc (10^{-5} M $ZnCl_2$), the inhibitory effect of cadmium on erythropoiesis was reduced, with CFU_E growth reaching 53 and 86% of control levels at concentrations of 10^{-5} and 10^{-6} M and Cd^{2+}, as compared with 28 and 49% of control in the absence of zinc. Thus, zinc was able to overcome the inhibitory effect of Cd on CFU_E formation *in vitro*. The mechanism involved in the protective role of zinc on erythropoiesis is not known. However, since heme (as hemin) was also found to be protective with respect to the toxic action of heavy metals on CFU_E

formation *in vitro*, it is possible that zinc may act by impairing heme degradation with a resultant maintenance of cellular heme levels needed for erythropoiesis.

Lead interferes with hematopoiesis by impairing hemoglobin production. Both globin and heme moieties are affected. With respect to globin, lead impairs synthesis of both alpha and beta globin chains by interfering with amino acid incorporation into nascent polypeptides on polyribosomes (Traketeller et al., 1965). Also, when reticulocytes are incubated with lead, polyribosomes are disaggregated leading to a marked depression of globin formation. The inhibitory effect of lead is prevented by addition of hemin *in vitro* (Waxman and Rabinovitz, 1966). In reticulocytes or immature erythroid cells, heme stimulates the initiation of translation of globin chains by inhibiting a protein kinase (heme-regulated eukaryotic initiation factor [eIF-α] kinase) (HRI), which is responsible for inactivation of the eukaryotic initiation factor, eIF-2 (Chen et al., 1990). Matts et al. (1991) have demonstrated that Pb^{2+} (as well as Hg^{2+} and Cd^{2+}) impairs heme-dependent protein (globin) synthesis by inhibiting cellular reductases which are required for maintaining the protein kinase, HRI, in an inactive state. Lead also interferes with hemoglobin production by impairing synthesis of the heme moiety at several sites of the heme biosynthetic pathway (see below).

B. METAL EFFECTS ON ERYTHROCYTES

Erythrocytes perform the essential function of transporting oxygen, bound to hemoglobin, from lungs to tissues, and CO_2, from tissues to the lungs for excretion. Acute damage to red cells or its hemoglobin content can result in impaired oxygen transport with consequent hypoxia and, secondarily, damage to the CNS and the heart, the organs most sensitive to oxygen deficiency. Metals may directly impair red cell function either by inhibiting erythropoiesis in bone marrow or by decreasing red cell survival in the circulation leading to anemia. Hemolytic anemia can result if the rate of red cell destruction in peripheral blood exceeds the normal rate of production in the bone marrow.

Several metals are known to have acute and direct hemolytic effects. Most notorious is this respect is arsine (arsenous hydride). The rapid and fulminant hemolysis of red blood cells is a unique and characteristic feature of arsine poisoning. Blair et al. (1990) exposed male and female rats, mice, and hamsters to arsine at atmospheric concentrations of 0 to 5 ppm for 6-hour periods over intervals of 14 to 90 days. Exposure-related decreases in packed red cell volumes occurred in blood samples collected from both sexes of all species after 14, 28, and 90 d of exposure. The mechanisms of arsine-induced hemolysis may entail several mechanisms (Fowler and Weissberg, 1974). Hemoglobin fixes arsine in a nonvolatile form within the red cell, after which cell lysis occurs. Hemolysis takes place only in the presence of oxygen, and it has been suggested that elementary arsenic produced by the action of oxygen on arsine is the active hemolytic agent. Arsenic dihydride, an intermediate in the oxidation of arsine to arsenous oxide, has also been postulated to be the destructive agent. The inhibition of catalase by arsenic or arsenic compounds with the formation of hydrogen peroxide during oxidation of arsenic is also considered a possible mechanism of arsenic-mediated red cell hemolysis.

Lead can induce two types of anemia through effects on red cell production or survival. Acute high-level lead poisoning has been associated with hemolytic anemia by diminishing red blood cell survival associated with increased fragility of the red cell membrane. The biochemical basis of this effect is not known, but is accompanied by inhibition of sodium- and potassium-dependent ATPases (Hernberg and Nikkanen, 1970), an activity known to be intimately associated with active membrane transport. Moreover, erythrocytes from lead-exposed individuals lose K^+ during incubation in much larger amounts and augmented rates than are normally observed. The mechanistic relationship of these findings to erythrocyte destruction requires further investigation. In chronic lead poisoning, lead induces anemia both by interfering with erythropoiesis and by diminishing red blood cell survival. The effects of lead on erythropoiesis probably result from interference with hemoglobin production as described above, as well as with heme biosynthesis. The basophilic stippling of erythrocytes often observed in subjects with lead poisoning most likely results from accumulation of basophilic granules which are pyrimidine products that arise from the impaired degradation of reticulocyte ribosomal RNA that is caused by inhibition of pyrimidine 5'-nucleotidase activity by lead (Paglia et al., 1975).

Red cell hemolysis has been reported to result from ingestion of copper as copper sulfate as well as from accumulation of toxic amounts from copper-contaminated hemodialysis fluid (Klein et al., 1972; Manzler and Schreiner, 1970). The pathogenesis of this effect is not known but may be related to copper-induced oxidation of intracellular glutathione (GSH), hemoglobin, and NADPH as well as inhibition of glucose-6-phosphate dehydrogenase (G-6-PDase) by copper (Fairbanks, 1967).

The principal effects of metals on hematopoiesis are summarized in Table 1.

Table 1 Principal Effects of Metals on Hematopoiesis

Metal	Effect	Ref.
Aluminum (Al^{3+})	Anemia	Wills and Savory, 1983
	Inhibits erythropoiesis	Mladenovic, 1988
Arsenic		
As^{3+}, As^{5+}	Leukopenia, anemia	Kyle and Pease, 1965
As^{3+}, As^{5+}	Depression of erythropoiesis	Morse et al., 1980
Arsine	Red cell hemolysis, hemolytic anemia	Fowler and Weissberg, 1974
Copper (Cu^{2+})	Red cell hemolysis	Klein et al., 1972; Manzler and Schreiner, 1970
Cobalt (Co^{2+})	Polycythemia	Gosselin et al., 1985
Gold (Au^{1+}, Au^{3+})	Thrombocytopenia, leukopenia, aplastic anemia	Flower et al., 1985
Lead (Pb^{2+})	Inhibits heme-dependent globin synthesis	Trakeletter et al., 1965; Matts et al., 1991
	Hemolytic anemia by diminishing red cell survival	Sassa, 1978
Mercury		
CH_3Hg^+	Decreased hemoglobin and erythrocyte count	Verschuuren et al., 1976
Hg·	Leukocytosis, neutrophilia	Campbell, 1948; Matthes et al., 1958
Zinc (Zn^{2+})	Anemia	Broun et al., 1990; Hoffman et al., 1988
	Decreased hemoglobin, hematocrit	Maita et al., 1981; Bentley and Grubb, 1991
	Protects against depression of erythropoiesis by Pb^{2+}, Au^{3+}, and Cd^{2+}	Lutton et al., 1984

II. HEME METABOLISM

A. OVERVIEW OF THE HEME BIOSYNTHETIC PATHWAY

Distinct from hematopoiesis, heme biosynthesis functions is virtually all eukaryotic cells to provide heme, chlorophyll, and related structures. In animal tissues, the principal product of this pathway is heme (protoheme, iron protoporphyrin IX), an essential component of oxygen transport systems, mixed function oxidation reactions, and other oxidative metabolic processes.

The heme biosynthetic pathway (Figure 1) utilizes glycine and succinyl coenzyme A as initial substrates to form the 5-carbon aminoketone, δ-aminolevulinic acid (ALA). This step is catalyzed by ALA synthetase (EC 2.3.1.37), a mitochondrial matrix enzyme which is generally considered rate-limiting in this process in normal adult tissues (Granick and Sassa, 1971). ALA then passes into the cytoplasm of the cell, where two molecules are condensed by ALA dehydratase (porphobilinogen synthase) (EC 4.2.1.24) to form the pyrrole, porphobilinogen (PBG). Two cytoplasmic proteins, PBG deaminase (uroporphyrinogen I synthetase) (EC 4.3.1.8) and uroporphyrinogen III cosynthetase, then act in concert to convert four molecules of PBG to 8-carboxyl porphyrinogen (uroporphyrinogen). The physiologically relevant isomeric form of 8-carboxyl porphyrinogen is uroporphyrinogen III, in which the propionic and acetic acid side chains of the "D" ring are reversed in comparison with the arrangement of these groups on the other three pyrrole rings of the molecule (Figure 1). This form serves as substrate for another cytoplasmic enzyme, uroporphyrinogen decarboxylase (EC 4.1.1.37), which converts the four acetic acid side chains of uroporphyrinogen III to methyl groups, yielding the 4-carboxyl porphyrinogen, coproporphyrinogen III (Figure 1). The decarboxylation of uroporphyrinogen to coproporphyrinogen proceeds through the 7-, 6-, and 5-carboxyl porphyrinogen intermediates in clockwise sequence (Jackson et al., 1975). The rate of decarboxylation of the various side chains varies with the enzyme from different tissues. In adult rat liver (Kardish and Woods, 1980a) and spleen (Romeo and Levin, 1971) the conversion of 8- to 4-carboxyl porphyrinogen proceeds essentially stoichiometrically. In rat kidney (Woods et al., 1984) and chicken erythrocytes (Garcia et al., 1973) the decarboxylation of the first acetic acid side chain proceeds more rapidly than that of the next three, and 7-carboxyl porphyrinogen accumulates.

The remaining steps of the heme biosynthetic pathway, by which coproporphyrinogen is converted to heme, are mediated by several mitochondrial enzymes. Coproporphyrinogen oxidase (EC 1.3.3.3), which is loosely bound to the mitochondrial inner membrane (Elder and Evans, 1978), decarboxylates the two propionic acid side chains of rings "A" and "B" of coproporphyrinogen to vinyl groups, yielding protoporphyrinogen "IX." Protoporphyrinogen oxidase (EC 1.3.3.4), also associated with the mitochondrial fraction (Poulson, 1976), then oxidatively cleaves six hydrogen atoms from protoporphyrinogen IX to yield protoporphyrin IX. In the final step, Fe^{2+} is incorporated into the protoporphyrin ring by the

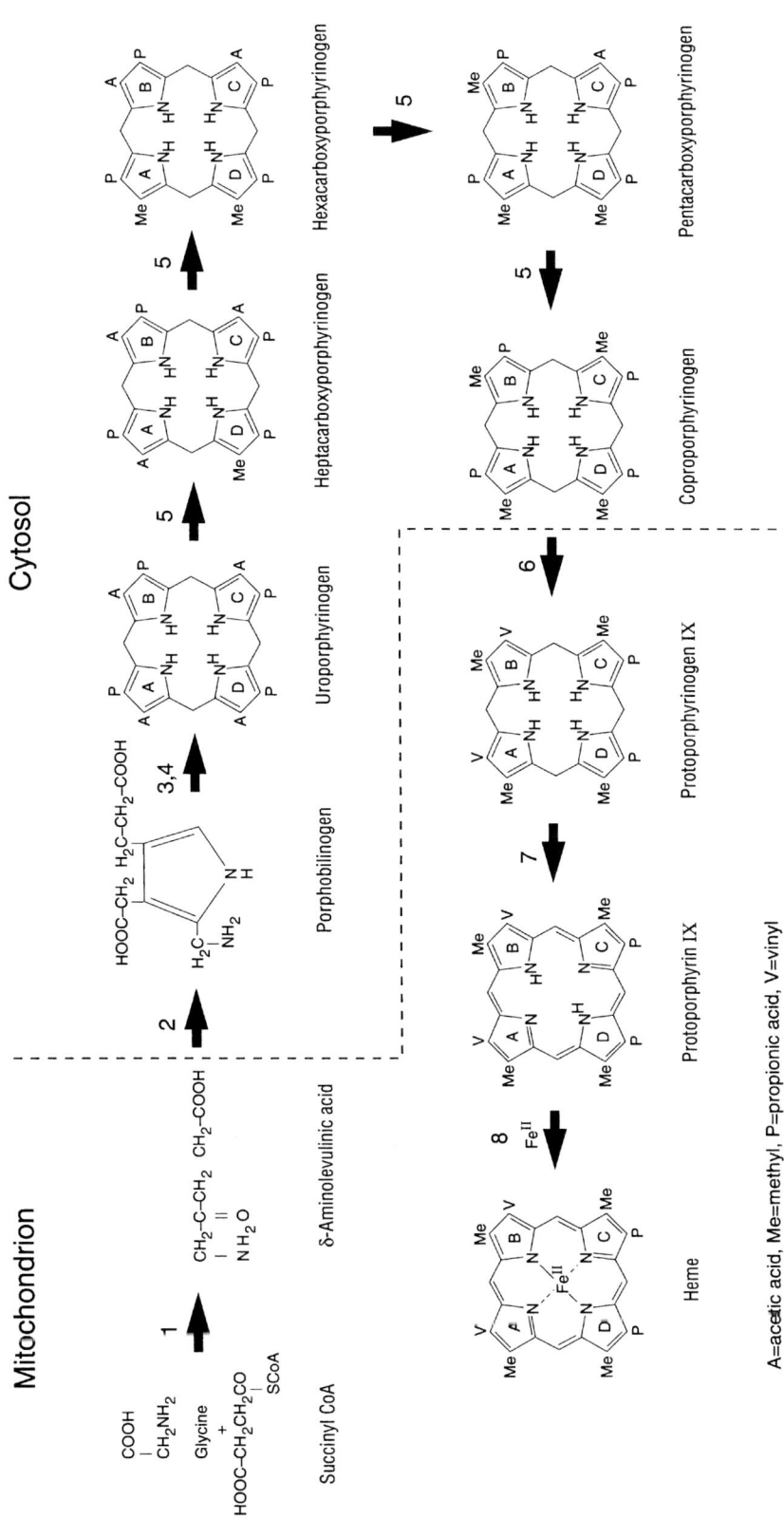

Figure 1. Heme biosynthetic pathway. Steps are catalyzed by (1) δ-Aminolevulinic acid (ALA) synthetase, (2) ALA dehydratase, (3) Uroporphyrinogen I synthetase (PBG deaminase), (4) Uroporphyrinogen III cosynthetase, (5) Uroporphyrinogen decarboxylase, (6) Coproporphyrinogen oxidase, (7) Protoporphyrinogen oxidase, and (8) Ferrochelatase (heme synthetase).

inner mitochondrial membrane enzyme, ferrochelatase (heme synthetase) (EC 4.99.1.1) (Jones and Jones, 1969).

In erythropoietic tissues (spleen, bone marrow, fetal liver) heme is utilized principally for hemoglobin synthesis. In nonerythropoietic tissues heme participates to a large extent in a variety of oxidative processes which are essential to cellular growth and function. From its site of synthesis in the mitochondria, heme is incorporated into cytochrome oxidase, in which it plays an essential role in the electron transport of intermediary metabolism. Outside the mitochondria, heme is incorporated into microsomal cytochromes P_{450} and b_5, in which it mediates the oxidative metabolism of various endogenous substances and xenobiotics. Heme is also incorporated into proteins, such as tryptophan pyrrolase, in the cytosol and catalase in peroxisomes. Bound heme in liver cells is considered to be in equilibrium with a "regulatory free heme pool," the concentration of which is estimated to be on the order of 10^{-7} to 10^{-8} M (Kappas et al., 1982). Functions performed by heme in hepatic cells are observed in other nonerythropoietic tissues such as kidney (Schwartz et al., 1976; Woods and Fowler, 1977), heart (Briggs et al., 1976; Sedman et al., 1982), testis (Tofilin and Piper, 1982), and adrenal gland (Condie et al., 1976). Since heme-mediated processes are essential to the viability of all biological tissues, compromise of heme biosynthetic capability by metals or other chemicals may lead to cell injury or death resulting from impairment of such processes.

B. MECHANISMS OF METAL ALTERATION OF HEME BIOSYNTHESIS

Metals are known to alter heme biosynthesis in biological tissues through several processes. Metals which are capable of forming porphyrin complexes, such as cobalt, may alter the regulation of heme synthesis through mechanisms which mimic the regulatory role of heme on ALA synthetase (Igaraski et al, 1978). Metals may also alter heme biosynthetic function directly as a result of metal-mercaptide bond formation with sulfhydryl groups of any of the enzymes of this pathway, or by induction of heme oxygenase. Additionally, metals may alter heme biosynthesis through mechanisms which do not involve a direct interaction with heme biosynthetic or degradative pathway enzymes per se. Principal among these are processes which involve perturbation of the mitochondrial membranes with which specific enzymes of the heme biosynthetic pathway are functionally associated, and promotion of free radical formation leading to oxidation of porphyrinogen intermediates *in vivo*. Each of these mechanisms is considered further below.

1. Effects of Metals on the Regulation of ALA Synthetase

The rate-limiting enzyme in the regulation of heme biosynthesis in normal adult mammalian tissues is considered to be ALA synthetase (Granick and Sassa, 1971). Current evidence supports the view that both transcriptional and posttranscriptional regulatory mechanisms involving heme and its iron moiety contribute to the modulation of ALA synthetase levels in various tissues. In liver, the tissue most studied, the regulation of ALA synthetase is considered to be mediated principally by the end-product, heme. Regulation by heme involves inhibition of the synthesis of ALA synthetase *de novo* at the genetic level (Granick, 1966) and suppression of the translocation of newly synthesized enzyme (or its precursor protein) from cytosol into mitochondria (Hayashi et al., 1972). Experiments in rat liver have suggested that heme may act to decrease the rate of transcription of the ALA synthetase gene. Srivastava et al. (1988) measured ALA synthetase gene transcription rates in nuclei from livers of rats treated with the ALA synthetase inducers, phenobarbital or allylisoproplyacetamide, and reported that concomitant treatment with heme significantly decreased ALA synthetase gene transcription. Similar findings were reported by Yamamoto et al. (1988). In contrast, Hamilton et al. (1991), using chick embryo hepatocytes, found that 1 μM heme, a physiologically more relevant concentration than that employed by other investigators, had no effect on the transcription rate of the ALA synthetase gene, but significantly decreased the half-life of ALA synthetase mRNA in this system. Thus, both transcriptional and post-transcriptional mechanisms have been demonstrated in the genetic regulation of hepatic ALA synthetase by heme. There is evidence that the gene for ALA synthetase expressed in liver is also expressed in other tissues, including kidney, heart, brain, testes, and spleen (Srivastava et al., 1988). Whether the regulation of ALA synthetase in different tissues is subject to modification at the genetic level by heme as shown to occur in the liver remains to be determined.

In addition to inhibiting ALA synthetase synthesis at the genetic level, heme also prevents transfer of newly formed ALA synthetase (or its precursor protein) from its site of synthesis in the cytosol to its final site of action in the mitochondrial matrix (Hayashi et al., 1972, 1983; May 1986). The mechanism by which heme inhibits transport of ALA synthetase is not known. May et al. (1990) have suggested the possibility that heme may bind to highly conserved regions of the precursor ALA synthetase protein in such a way as to prevent its transfer across the mitochondrial membrane. This possibility, however, requires further confirmation.

The regulation of ALA synthetase in liver by direct feed-back inhibition by heme was suggested by Scholnick et al. (1972). However, current evidence does not support the view that heme directly alters the activity or stability of ALA synthetase in liver mitochondria (May et al., 1986). Whether heme regulates ALA synthetase by direct feed-back inhibition in other nonerythropoietic tissues remains to be determined.

Unlike ALA synthetase in nonerythropoietic tissues, erythroid ALA synthetase does not appear to be regulated at either the transcriptional or posttranscriptional levels by heme per se. Woods (1974) first demonstrated that, in fetal rat liver mitochondria, heme neither repressed the activity of ALA synthetase nor impaired the uptake of newly synthesized enzyme into mitochondria, as is observed in adult liver tissue. More recently, Elferink et al. (1988) reported that transcription of the ALA synthetase gene in mouse erythroleukemia cells was not affected by heme or by inhibition of heme synthesis using succinylacetone, indicating that transcription of the erythroid gene is not subject to end-product repression by heme. Of interest is the finding that the 5'-untranslated region of the human erythroid ALA synthetase mRNA contains a distinctive domain similar to the iron-responsive elements (IREs) found in the 5'-untranslated region of the mRNA for the iron-storage protein, ferritin, and also in the 3'-untranslated region of the mRNA for the transferrin receptor involved in cellular iron uptake (May et al., 1990). The expression of ferritin and transferrin receptor is dependent on cellular iron availability, and the IREs in the mRNAs are essential to the iron-dependent posttranscriptional regulation of these proteins (Casey et al., 1988; Theil, 1990). The presence of an IRE-like domain in the 5'-untranslated region of the erythroid ALA synthetase gene strongly suggests that translation of this mRNA is controlled by cellular iron. Heme may affect this process indirectly, inasmuch as heme inhibits iron release from transferrin in reticulocytes (Ponka et al., 1988). Thus, although heme does not appear to directly regulate ALA synthetase in erythropoietic tissues, heme may indirectly modulate erythroid ALA synthetase synthesis through a translational mechanism which is primarily iron dependent (May et al., 1990).

2. Effects of Metals on Other Heme Biosynthetic Pathway Enzymes

All eight steps of the heme biosynthetic pathway are catalyzed by enzymes which require functional sulfhydryl (-SH) groups for optimal catalytic activity, either as part of the active site configuration or to maintain their structural integrity. Since most metals have a strong affinity for nucleophilic ligands, especially thiols, each step of the heme biosynthetic pathway is theoretically susceptible to direct inhibition as a result of metal-mercaptide bond formation with functional SH groups. Metals which have been shown to impair specific enzymes or other processes involved in heme biosynthesis in various tissues either *in vitro* or *in vivo* are summarized in Table 2. The following discussion considers the effects of metals as direct inhibitors of several of these enzymes.

a. ALA Dehydratase

ALA dehydratase (ALAD) catalyzes the second step in heme biosynthesis, in which two molecules of ALA are transformed into the pyrrole, porphobilinogen (PBG). ALA dehydratase is susceptible to inhibition by metals which readily complex with functional thiol groups, lead being particularly notable in this respect. Inhibition of ALA dehydratase in erythrocytes by low lead concentrations is regarded as a highly sensitive indicator of lead exposure (see Woods, 1995). Inhibition of ALA dehydratase in liver and erythrocytes *in vitro* has also been observed with silver, iron, manganese, copper, and zinc ions (Gibson et al., 1955), and *in vivo* with lead (Sassa, 1978) and cobalt (Nakemura et al., 1975). Interestingly, renal ALA dehydratase is somewhat resistant to inhibition *in vivo* by lead or other metals which substantially inhibit this enzyme in liver and erythrocytes. Chronic exposure of rats to lead acetate for up to 9 months failed to produce inhibition of ALA dehydratase in the kidney at dosage levels resulting in blood lead concentrations as high as 97 ppm (Fowler et al., 1980). This lack of effect of lead in the kidney may be attributable to the sequestration of lead within intranuclear inclusion bodies which form rapidly in renal proximal tubule cells following onset of lead exposure (Goyer, 1971). Renal ALA dehydratase is also not substantially inhibited *in vivo* following treatment with mercury compounds (Woods and Fowler, 1977), which also form inclusion bodies in the kidney. In contrast, the group III metal, indium, is a potent inhibitor of renal ALA dehydratase both *in vivo* and *in vitro*, owing largely to the preferential accumulation of indium in the cytoplasmic fraction of the renal proximal tubule cell without inclusion body formation (Woods and Fowler, 1982). Although lead is not an effective inhibitor

Table 2 Principal Effects of Metals on Heme Biosynthetic Pathway Enzymes and Processes

Metal	Enzyme or Process Affected	Tissue	Ref.
Arsenic			
As^{3+}, As^{5+}	Inhibits UD in vitro	Liver	Woods and Fowler, 1981
As^{5+}	Decreases FC in vivo	Liver	Woods and Fowler, 1978
As^{3+}	Inhibits CO in vitro	Liver	Batlle et al., 1965
As^{5+}	Decreases CO in vivo	Kidney	Woods and Southern, 1989
As^{5+}	Inhibits CO in vitro	Kidney	Woods and Southern, 1989
Bismuth (Bi^{3+})	Inhibits ALAS, ALAD, and FC in vivo and in vitro	Liver, kidney	Woods and Fowler, 1987
Cadmium (Cd^{2+})	Inhibits CO in vitro	Liver	Batlle et al., 1965
	Inhibits UD in vitro	Liver	Woods et al., 1981
	Inhibits FC in vitro	Liver	Labbe and Hubbard, 1961
Cobalt			
Co^{2+}	Inhibits ALAD in vivo	Erythrocytes	Nakemura et al., 1975
Co^{2+}-protoporphyrin	Inhibits ALAS	Liver	Igaraski et al., 1978
Co^{2+}	Inhibits UD	Liver	Woods et al., 1981
Indium (In^+)	Inhibits ALAD in vitro and in vivo	Kidney	Woods and Fowler, 1982
Iron			
Fe^{2+}	Regulates translation of ALAS mRNA	Erythroid tissue	
Fe^{2+}	Inhibits UD	Liver	Straka and Kushner, 1983; Mukerji and Pimstone, 1986
	Inhibits UD	Erythrocytes	
Fe^{2+}-protoporphyrin	Decreases rate of transcription of ALAS gene	Liver	Srivastava et al., 1988; Yamamoto et al., 1988
	Decreases half-life of ALAS mRNA	Liver	Hamilton et al., 1991
	Inhibits incorporation of ALAS precursor into mitochondria		Hayashi et al., 1972; May, 1986
Fe^{2+}	Promotes reactive oxygen ($OH^·$, O_2^-) mediated porphyrinogen oxidation		De Matteis et al., 1988
Lead (Pb^{2+})	Inhibits ALAD in vivo	Erythrocytes	Sassa, 1978
	Inhibits PBGD in vitro	Erythrocytes	Anderson and Desnick, 1980
	Decreases PBGD in vivo	Liver	Piper and van Lier, 1977
	Inhibits CO in vitro	Liver	Kardish et al., 1980; Batlle et al., 1965
	Inhibits CO in vitro	Kidney	Henderson and Toothill, 1983
	Inhibits FC in vitro	Liver	Labbe and Hubbard, 1961; Dailey and Fleming, 1983
	Decreases FC in vivo	Kidney	Fowler et al., 1980
	Decreases FC in vivo	Reticulocytes	Piomelli et al., 1987
	Impairs coproporphyrinogen incorporation by mitochondria	Reticulocytes	Rossi et al., 1993
	Impairs Fe^{2+} availability to FC	Liver	Taketani et al., 1985
	Impairs structural association of CO or FC with mitochondrial membranes		Woods and Fowler, 1987
Mercury			
CH_3Hg^+	Decreases PBGD in vivo	Kidney	Woods and Fowler, 1977
CH_3Hg^+	Decreases CO in vivo	Kidney	Woods and Southern, 1989
CH_3Hg^+	Inhibits FC in vitro	Kidney	Woods and Fowler, 1977
Hg^{2+}	Inhibits CO in vitro	Liver, kidney	Woods and Southern, 1989
Hg^{2+}	Inhibits UD in vitro	Kidney	Woods et al., 1984
Hg^{2+}	Inhibits FC in vitro	Liver	Daily and Fleming, 1983
Hg^{2+}	Promotes free radical porphyrinogen oxidation in vitro and in vivo	Kidney	Woods et al., 1990; Miller and Woods, 1993

Table 2 (Continued)

Metal	Enzyme or Process Affected	Tissue	Ref.
Platinum (Pt^{2+})	Decreases PBGD *in vivo*	Kidney	Maines and Kappas, 1977
Thallium (Tl^{3+})	Inhibits FC *in vitro*	Liver	Woods and Fowler, 1986
Zinc (Zn^{2+})	Activates ALAD	All tissues	Gurba et al., 1972; Anderson and Desnick, 1979
	Reactivates ALAD after Pb inhibition	Erythrocytes	Abdulla and Haeger-Aronson, 1971; Finelli et al., 1975
	Reactivates ALAD after In inhibition	Kidney	Woods et al., 1979
	Inhibits FC	Liver	Labbe and Hubbard, 1961

of renal ALA dehydratase *in vivo*, a synergistic response between lead acetate and indium chloride in the inhibition of renal ALA dehydratase in rats is observed when both metals are administered simultaneously (Woods et al., 1979). These observations suggest that specific metals may produce effects on ALA dehydratase activity through mechanisms which alter the enzyme at different loci. Thus, synergistic metal interactions might be observed when one metal compromises catalytic activity by binding at the active site, while a second metal might do so through binding at noncatalytic sites in such a way as to distort the physical structure of the enzyme. Alternatively, it is conceivable that two metals could function differently at the same catalytic site of the enzyme. By whatever mechanism the observed metal interactions on renal ALA dehydratase are eventually explained, these observations have important implications with respect to the effect of multiple metal exposures on heme biosynthetic capability in kidney cells.

ALA dehydratase from various tissues has been reported to have an absolute requirement for zinc. The purified enzyme from bovine liver has been shown to contain zinc in a stoichiometric ratio of 1 mole per mole of enzyme (Gurba et al., 1972), although up to 8 zinc atoms per molecule of native enzyme from various other preparations have been reported (Anderson and Desnick, 1979). ALA dehydratase from human erythrocytes is activated by zinc at concentrations as low as 0.02 mM *in vitro* (Anderson and Desnick, 1979). Zn has been shown to reactivate ALA dehydratase from red cells of rats previously exposed to lead (Finelli et al., 1975) in erythrocytes of lead-exposed humans (Abdulla and Haeger-Aronson, 1971) and in kidneys of indium-treated rats (Woods et al., 1979). The precise role of zinc in the activity of ALA dehydratase has not been determined. One suggestion is that zinc functions as a Lewis acid (i.e., a positively charged ion or proton donor), serving to polarize the carbonyl group of one molecule of ALA and rendering it more susceptible to nucleophilic attack by a lysine residue in the formation of a Schiff base at the enzyme's active site (Kappas et al., 1982). Another suggestion is that zinc enhances the nucleophilicity of the resulting carbanion, which facilitates aldol condensation with the second molecule of ALA leading to PBG formation (Kappas et al., 1982).

Genetically linked deficiencies of ALA dehydratase in human tissues have been described by a number of investigators (Bird et al., 1979; Beckmann et al., 1983; Doss et al., 1979; Wetmur et al., 1991), and it has been suggested that marked deficiencies of this enzyme in liver or other tissues might exacerbate the toxic effects of lead or other metals which inhibit this enzyme (Astrin et al, 1987; Doss and Muller, 1982; Wetmur et al., 1991). ALA deficiency has an autosomal recessive inheritance and in homozygous subjects, in whom ALA dehydratase concentrations may be less than 5% of normal, the clinical manifestations resemble those of acute intermittent porphyria, including acute abdominal pain, neuropathy, and anemia. Heterozygous subjects usually have sufficient ALA dehydratase activity to prevent symptoms, but it is known that they are vulnerable to the development of lead intoxication as a result of enzyme inhibition when there is only a moderate increase in blood lead levels (Batlle et al., 1987; Doss et al., 1984). The identification of polymorphisms in ALA dehydratase gene expression in blood cells or other tissues might serve as a useful biomarker of susceptibility to lead toxicity in human subjects.

b. Porphobilinogen Deaminase and Uroporphyrinogen III Cosynthetase

The third step of the heme biosynthetic pathway involves condensation of four molecules of PBG to form one molecule of the cyclic tetrapyrrole, uroporphyrinogen (Figure 1). Although uroporphyrinogen can conceivably occur in any of four isomeric forms, defined by the relative positions of the acetic and propionic acid side chains, only the I and III isomers and their derivatives are found in biological systems (Tait, 1978). Only the III isomer, formed through the joint action of PBG deaminase PGBD and uroporphyrinogen III cosynthetase, is an intermediate in the biosynthesis of protoporphyrinogen IX, because coproporphyrinogen oxidase utilizes only coproporphyrinogen III as a substrate. In mammalian tissues, the activity of PBG deaminase is generally quite low, second only to that of ALA synthetase in terms of its potential to become rate-limiting in heme biosynthesis. Uroporphyrinogen III cosynthetase, in contrast, is present in considerable excess in most tissues (Tait, 1978), but has no catalytic activity in the absence of PBG deaminase.

PBG deaminase has been purified from human erythrocytes (Anderson and Desnick, 1980; Miyagi et al, 1979; Stevens et al, 1968) as well as from other mammalian tissues (Levin and Coleman, 1967; Sancovich, et al., 1969), and has been shown to be readily inhibited by SH-binding agents including numerous trace metals. Lead is known to be a particularly effective inhibitor (Anderson and Desnick, 1980). Mercury, cadmium, copper, iron, magnesium, and calcium also impair the purified enzyme, although to a lesser degree than lead (Anderson and Desnick, 1980). Studies *in vivo* have determined that platinum significantly decreases renal PBG deaminase activity in kidney, whereas nickel only slightly impairs the renal enzyme (Maines and Kappas, 1977). Neither of the group III metals, indium (Woods et al., 1979) or thallium (Woods and Fowler, 1986), nor the group V metal, bismuth (Woods and Fowler, 1987a), has been observed to alter PBG deaminase in rat liver or kidney *in vivo*. Of interest, however, is the observation that PBG deaminase activity was increased by 135% as compared with control levels in livers of mice fed sodium arsenate in the drinking water at 85 ppm for 6 weeks (Woods and Fowler, 1978). Arsenate treatment of rats, however, failed to produce any changes in hepatic PBG deaminase activity (Woods and Fowler, 1978). In contrast, treatment of rats with methylmercury hydroxide in drinking water at dosage levels up to 10 ppm for 6 weeks produced a 50% decrease in renal PBG deaminase levels but did not alter hepatic deaminase activity (Woods and Fowler, 1977).

Of considerable interest with respect to metal effects on uroporphyrinogen formation in mammalian tissues is the observation by Piper and colleagues (Piper and van Lier, 1977) that PBG deaminase may be regulated by certain pteroylglutamate derivatives which also protect the enzyme from inhibition by lead and other inhibitors. Moreover, pteroylglutamates appear to be capable of activating PBG deaminase from rat liver *in vitro* (Piper et al., 1983). Further studies by this group have demonstrated that uroporphyrinogen III cosynthetase is also inhibited by various metals (K^{1+}, Na^{1+}, Mg^{2+}, Ca^{2+}, Zn^{2+}, Cd^{2+}, Cu^{2+}) (Piper et al., 1983; Clement et al., 1982), and that the cosynthetase is a folate-binding protein that supplies a reduced pteroylopolyglutamate to PBG deaminase, thereby increasing catalytic activity. These findings have important implications with respect to understanding the mechanisms of porphyrin formation in mammalian cells and the effects of metals on the regulation of heme synthesis at the level of uroporphyrinogen formation.

c. Uroporphyrinogen Decarboxylase

The cytosolic enzyme, uroporphyrinogen decarboxylase (UD), catalyzes the conversion of uroporphyrinogen to coproporphyrinogen in a step-wise process involving the sequential decarboxylation of each of the 4 acetic acid side chains of 8-carboxyl porphyrinogen to the 2-carboxyl porphyrinogen intermediate (Figure 1). Uroporphyrinogen (urogen) decarboxylase from all tissues thus far studied has an absolute requirement for free SH groups for both catalytic and immunoreactive activity (Edler et al., 1983; Elder and Urquhart, 1984). Five free SH groups per mole of native enzyme were detected on titration of the purified human red cell decarboxylase using 5,5′-dithiobis(2-nitrobenzoic acid) (DTNB) (De Verneuil et al., 1983), and at least one free SH group at the active site or sites appears to be required for catalytic activity. In light of these findings, it is not surprising that uroporphyrinogen decarboxylase is readily inhibited by numerous trace metals. Among the most potent of the metals evaluated is Hg^{2+}, which, in kidney cytosol *in vitro,* significantly inhibits enzyme activity at concentrations as low as 10 μM (Woods et al., 1984). Interestingly, uroporphyrinogen decarboxylase from rat kidney appears to be much more readily inhibited by Hg^{2+} than the enzyme from liver, by a factor of at least 10 when measured in crude enzyme preparations from rat tissues (Woods et al., 1984). Reasons underlying the increased sensitivity of renal urogen decarboxylase by Hg^{2+} are not clear, although the existence of slightly different isozymic forms of the enzyme in liver and kidney could account for the differences in responsiveness observed. Of particular relevance to this suggestion is the observation that renal urogen decarboxylase is exquisitely sensitive to the concentration of reduced glutathione (GSH) in reaction mixtures *in vitro* when the activity of the enzyme is assayed using either uroporphyrinogen I or III as substrate (Woods et al., 1984). The hepatic enzyme, in contrast, is not modulated by the GSH concentration. The principal effect of GSH in modulating the activity of renal urogen decarboxylase is to accelerate decarboxylation of 7- to 6-carboxyl porphyrinogens, the slowest step of the enzyme reaction. These observations suggest the possible requirement or participation of GSH in mediating renal urogen decarboxylase activity *in vivo*. Since Hg^{2+} is a potent SH binding agent (first order binding constant $>10^{20}$), it is possible that the increased sensitivity of renal urogen decarboxylase to Hg^{2+} inhibition could be due to the action of Hg^{2+} to deplete GSH required for optimal enzyme activity, in addition to the direct binding of mercuric ions to SH groups of the enzyme.

Urogen decarboxylase from several tissues has been reported to be inhibited by arsenic. Studies *in vitro* have demonstrated that both As^{3+} and As^{5+} inhibit urogen decarboxylase from rat liver (Woods et al., 1981) and kidney (Woods, unpublished results), and that the primary focus of this effect appears to be at the enzyme site which catalyzes 8- to lesser carboxylated porphyrinogens. This unusual effect is consistent with the urinary porphyrin pattern observed during prolonged exposure of rats and mice to arsenic compounds, which is characterized by substantially elevated uroporphyrin concentrations (Woods, in press; Woods and Fowler, 1978). However, inhibition of urogen decarboxylase by arsenic is observed in tissues only at considerably higher arsenic concentrations (e.g., As^{5+} = 10 to 50 mM) than are observed *in vivo* during prolonged arsenic treatment regimens (e.g., hepatic or renal total arsenic concentration measured after exposure of rats to sodium arsenate for 6 weeks = 150 to 200 µM). It is possible, therefore, that additional factors, such as the direct oxidation of porphyrinogens by reactive oxidants which could be produced in cells during the interaction of arsenic ions with mitochondrial or other electron transfer processes (Fowler et al., 1979; Mitchell et al., 1971; Yamanaka et al., 1990) might be involved in the etiology of arsenic induced-porphyrinuria.

Ferrous iron (Fe^{2+}) has long been postulated to directly inhibit urogen decarboxylase, and this effect has been considered to exacerbate the clinical manifestations of human porphyria cutanea tarda (PCT) (Kushner et al., 1975), a disorder characterized by an inherited or acquired deficiency of urogen decarboxylase in erythrocytes and/or liver. Studies of the direct action of Fe^{2+} on purified preparations of urogen decarboxylase from various sources, however, have provided conflicting observations regarding its effects on enzyme activity. Ferrous iron was ineffective in inhibiting urogen decarboxylase purified from human erythrocytes (De Verneuil et al., 1983) and also had no effect on preparations of the enzymes from rat kidney (Woods et al., 1984) or rat liver (Woods et al., 1981) at concentrations as high as 1 mM *in vitro*. In contrast, Straka and Kushner (1983) found significant inhibition of purified bovine liver urogen decarboxylase by Fe^{2+} in reaction mixtures containing buffer systems (Tris-HEPES) which do not complex ferrous ions. Fifty percent inhibition of urogen decarboxylase by 2 mM Fe^{2+} was observed in the presence of 100 mM Tris-HEPES buffer, whereas 10 mM Fe^{2+} was required to produce comparable inhibition when 100 mM phosphate buffer was used. In other studies, Mukerji and Pimstone (1986) reported inhibition of rat liver urogen decarboxylase by Fe^{2+} and described the direct interaction of ferrous ions with cysteinyl residues at the active site(s) of the enzyme. All steps of urogen decarboxylation were inhibited approximately 50% by 0.45 mM ferrous ions when urogen I was used as substrate. Additionally, cysteine (6.7 mM) was shown to protect urogen decarboxylase from inhibition by Fe^{2+} under anaerobic conditions but to enhance inhibition by the Fe^{2+} under aerobic conditions. The latter effect was reversed by antioxidants. These results support the view that iron may act in two ways to inhibit urogen decarboxylase: (1) through direct interaction with SH groups at the active site(s) of the enzyme; and (2) by facilitating the generation of free radicals in the presence of oxygen and cysteine, which oxidize porphyrinogen substrates or which oxidatively deactivate the enzyme. Further knowledge of the concentrations of ferrous iron in normal and PCT cells, and of the mechanisms by which iron is mobilized from ferritin or other storage sites in biological tissues, is required to confirm the mechanistic role of iron in perturbing urogen decarboxylase activity.

In contrast to Hg, As, and Fe, lead (Pb^{2+}) does not effectively inhibit urogen decarboxylase *in vivo* or *in vitro* except at relatively high concentrations. Thus, lead did not inhibit hepatic urogen decarboxylase when added to reaction mixtures as lead acetate at concentrations as high as 300 mM (Woods et al., 1981). Cadmium and cobalt salts are moderate inhibitors of hepatic urogen decarboxylase activity; 300 mM $CoCl_2$ and $CdCl_2$ inhibited hepatic urogen decarboxylase *in vitro* to 67 and 30% of control levels, respectively (Woods et al., 1981). Neither tin (De Verneuil et al., 1986) nor magnesium (Mauzerall and Granick, 1958) had an effect on decarboxylase activity *in vitro* when studied on preparations of the enzyme isolated from human erythrocytes.

d. Coproporphyrinogen Oxidase

Coproporphyrinogen oxidase (CO) catalyzes the oxidative decarboxylation of propionic acid groups of the A and B rings of coproporphyrinogen III to vinyl groups at these positions. The resultant molecule is protoporphyrinogen IX. Coproporphyrinogen (coprogen) oxidase is unusual as an oxidase in that metals do not appear to be involved in its oxidative function, nor is its activity affected by treatment with metal chelators (Yoshinaga and Sano, 1980a). Studies on preparations of coprogen oxidase from various sources have characterized it as an SH-containing protein (Yoshinaga and Sano, 1980b), although the extent to which functional SH groups are required for oxidative activity is not clear. A preparation of coprogen oxidase from bovine liver (Camadro et al., 1986) was reported to contain three SH groups in the intact enzyme and seven more when the protein was unfolded with detergent. A purified preparation of the oxidase from yeast has been described as being partially inhibited by thiol-directed agents, suggesting that at least one SH group is required for catalytic activity (Camadro et al., 1986).

A number of reports have described the effects of specific trace metals on coprogen oxidase activity (Batlle et al., 1965; Woods, 1988, 1989; Woods and Fowler, 1987b; Yoshihaga and Sano, 1980a). Batlle et al. (1965) reported inhibition of a purified preparation of coprogen oxidase from rat liver by As^{3+}, Cd^{2+}, and Pb^{2+}. Coprogen oxidase activity was not altered by monovalent cations in those studies. Similar findings were reported by Kardish and Woods (1980b) with respect to inhibition of coprogen oxidase by metal ions in enzyme preparations from rat liver. More recently, Woods and Southern (1989) studied coprogen oxidase from rat liver and kidney and found that the enzyme from either tissue was significantly inhibited by Hg^{2+} at concentrations as low as 0.1 mM in vitro. Renal coprogen oxidase activity was also decreased to 21% of control levels when measured in mitochondrial preparations 24 hours following acute treatment with $HgCl_2$ (2 mg/kg, i.p.). Also of interest was the finding that the specific activity of renal coprogen oxidase from untreated rats is approximately one third that of the liver enzyme. Hepatic and renal coprogen oxidase activities were 676 ± 62 and 247 ± 27 pmoles protoporphyrinogen produced per milligram of protein per hour, respectively, when assayed under identical conditions (Woods and Southern, 1989). These findings are of interest with respect to the overall regulation of heme metabolism in the kidney, inasmuch as reduction of the activity of coprogen oxidase by as much as two thirds could render this step rate-limiting in heme biosynthesis in kidney cells (Woods, 1988).

In studies in vivo (Woods and Fowler, 1977; Woods and Southern, 1989; Woods et al., 1991) prolonged treatment of male rats with MMH at 5 or 10 ppm in drinking water produced a pronounced porphyrinuria characterized by up to a 20-fold increase in the urinary penta- and coproporphyrin concentrations. Studies on the mechanism of this effect suggest that excess porphyrin excretion during MMH treatment is largely of renal etiology (Woods and Southern, 1989; Woods et al., 1991). Thus, renal coprogen oxidase activity was depressed by as much as 56% of control levels 3 weeks after initiation of MMH treatment, whereas the activity of the liver enzyme was not changed. Depression of renal coprogen oxidase activity correlated closely with increased mercury levels in the kidney, suggesting that the interaction of Hg^{2+} either with the enzyme itself or with structural components of the mitochondrial membrane upon which coprogen oxidase may be dependent for optimal activity was responsible for the decrease in activity observed.

The excretion of large amounts of coproporphyrin in the urine is commonly associated with lead exposure in both animals and humans, and it has long been assumed that lead inhibition of coproporphyrinogen oxidase in tissue cells underlies this effect. However, evidence for this effect in vivo is inconclusive. Batlle et al., (1965) showed that purified rat liver coprogen oxidase was inhibited by lead in vitro, but only at lead concentrations as high as 20 mM. Kardish et al. (1980b) also showed that the activity of coprogen oxidase measured in rat liver mitochondria was reduced only in the presence of lead at concentrations as high as 10 mM. More recently, Rossi et al. (1993) reported the lack of inhibition of coprogen oxidase from human lymphocytes by lead at 50 µM in reaction mixtures. In contrast, in studies in vivo, Henderson and Toothill (1983) reported a decrease in coprogen oxidase to 30% of control values when measured in mitochondria from kidneys of rabbits administered lead acetate in drinking water sufficient to achieve blood lead levels of 93 µg/dl and renal lead concentration of 41 µg/g protein. One explanation for this apparent inconsistency could be that lead indirectly affects coprogen oxidase in cells by compromising the structural integrity of membrane structures with which the enzyme is associated in tissue mitochondria. Woods and Fowler (1987b) have provided evidence that disruption of membrane integrity characterized by mitochondrial swelling and fragmentation of mitochondrial inner membrane segments occurs at tissue lead concentrations as low as 10 µM, considerably below those which have been shown to inhibit coprogen oxidase activity in vitro. It is possible, therefore, that disruption of the structural association between coprogen oxidase and the mitochondrial membrane could account for the effect of lead on enzyme activity observed in vivo.

Other studies have suggested that lead may also impair the transport of coproporphyrinogen into the mitochondria subsequent to its synthesis by uroporphyrinogen decarboxylase in the cytoplasmic portion of the cell (Rossi et al., 1992, 1993) (Figure 1). The processes by which coproporphyrinogen is transported into the mitochondria or through the outer to inner membrane spaces are not defined. If lead were to interfere with these processes, the accumulation and subsequent oxidation of coproporphyrinogen in the cytosol could constitute the source of excess coproporphyrin excreted in the urine during lead exposure.

e. Protoporphyrinogen Oxidase

Protoporphyrinogen oxidase catalyzes the removal of six hydrogen atoms from the nucleus of protoporphyrinogen IX to form protoporphyrin IX. As a mitochondrial inner membrane enzyme (Ferreira et al., 1988), protoporphyrinogen oxidase has an essential requirement for oxygen, but no metal or other co-factor has been demonstrated. The enzyme from rat liver possesses monothiol groups, but it is not known whether these are directly involved in the binding of substrate to its active site (Poulson, 1976). Several metals have been shown to inhibit protoporphyrinogen oxidase activity *in vitro*. Mercury as *p*-hydroxymercuribenzoate (0.5 mM) reduced enzyme to 6% of control levels, and this effect was largely reversed by the presence of GSH (1 mM) in reaction mixtures. Moderate concentrations of arsenite (e.g., 10 mM) did not affect enzyme activity, but inhibition was observed in the presence of 75 mM arsenite, suggesting the presence of monothiol rather than vicinal dithiol groups in the enzyme structure (Poulson, 1976). Dailey (1987) studied the effects of Mg^{2+}, Mn^{2+}, Ni^{2+}, Hg^{2+}, Cu^{2+}, Pb^{2+}, and Cd^{2+} (10 μM) on purified mouse liver protoporphyrinogen oxidase and found that none of these metals affected enzyme activity at the concentration employed. Little is known of the effects of metals on protoporphyrinogen oxidase *in vivo* or of the consequences of long-term metal exposures on enzyme function.

f. Ferrochelatase

The final step in heme biosynthesis, i.e., the insertion of ferrous iron into the protoporphyrinogen ring to form heme (protoheme), is catalyzed by ferrochelatase (FC) (heme synthetase) (Figure 1). An integral part of the mitochondrial inner membrane (Harbin and Dailey, 1985), ferrochelatase utilizes ferrous iron as well as the divalent cations of zinc and cobalt as substrates. The effects of metals on ferrochelatase activity in mammalian tissues have been examined by various investigators. Labbe and Hubbard (1961) first reported depression of rat liver ferrochelatase in crude preparations *in vitro* by Zn^{2+}, Mn^{2+}, Pb^{2+}, Cd^{2+}, Mg^{2+} and Ca^{2+}, but not by Sn^{2+}. In further studies, Tephly et al. (1971) suggested that metal ions at low concentrations inhibit ferrochelatase directly by metal-mercaptide bond formation with SH groups of the enzyme. These findings were confirmed in studies by Dailey (1984), which demonstrated that modification of a single sulfhydryl group is sufficient to inactivate the enzyme and that ferrous iron, but not the porphyrin substrate, protects ferrochelatase from inactivation. Arsenate at a concentration of 1 mM was found to inhibit the purified enzyme by over 90%. Inhibition of the purified bovine liver ferrochelatase by Mn^{2+}, Hg^{2+}, and Pb^{2+} has also been described (Dailey and Fleming, 1983). A purified preparation of ferrochelatase from rat liver (Taketani and Tokunaga, 1981) has also been shown to be readily inactivated by various divalent metal cations.

Studies *in vivo* have demonstrated widely varying and tissue-specific effects of trace metals on ferrochelatase activity. Chronic exposure of rats for up to 9 months to lead acetate in the drinking water at concentrations up to 85 ppm did not result in impaired ferrochelatase activity in the liver but caused a statistically significant decrease in renal ferrochelatase at blood lead levels as low as 11 μg/dl (Fowler et al., 1980). Similarly, exposure of rats to MMH for 6 weeks at drinking water concentrations up to 10 ppm inhibited ferrochelatase in the kidney by up to 50% of control levels, but had no measurable effect on the liver enzyme (Woods and Fowler, 1977). In contrast, hepatic ferrochelatase was inhibited by as much as 40% of control levels in both rats and mice exposed to sodium arsenate at 85 ppm in drinking water for 6 weeks (Woods and Fowler, 1978). In acute studies with the group III metal, indium, no significant changes in the activity of hepatic (Woods et al., 1979) or renal (Woods and Fowler, 1982) ferrochelatase were observed following treatment with indium chloride in doses up to 40 mg/kg 16 hours prior to sacrifice. Interestingly, another group III metal, thallium, did not inhibit, but significantly increased, the activity of ferrochelatase in rat liver 16 hours following treatment with thallium chloride at doses of 50 to 200 mg/kg (Woods and Fowler, 1986). Since thallium was found to be a potent inhibitor of ferrochelatase *in vitro*, these observations again suggest that the action of the metal on the enzyme *in vivo* is mediated through effects which involve alteration of the mitochondrial structures with which the enzyme is functionally associated in the intact cell, rather than via a direct effect on the enzyme per se.

Inhibition by lead of ferrochelatase in reticulocytes has been considered for many years as the principal cause of increased erythrocyte zinc protoporphyrin (ZPP) levels observed during lead exposure. However, evidence that Pb^{2+} directly inhibits ferrochelatase *in vivo* is not conclusive. Henderson and Toothill (1983) reported no decrease in ferrochelatase activity in kidneys of rabbits administered lead acetate at doses sufficient to produce mean blood lead levels of 93 μg/dl. More recently, Piomelli et al. (1987) reported that lead competitively inhibits ferrochelatase in human reticulocytes *in vitro* by binding to the enzyme in such as way as to modify the conformation of the active site. However, the effect of lead was shown to occur only when iron was not already bound to the active site of the enzyme. These findings suggest that, while lead may directly inhibit ferrochelatase *in vivo*, this effect is expressed or pronounced only in the case of iron deficiency.

An alternate mechanism to explain the etiology of elevated erythrocyte ZPP during lead exposure has its basis in the relative affinities of iron and zinc as substrates for ferrochelatase. Ferrous iron is the preferential substrate of ferrochelatase and is also an effective inhibitor of zinc utilization by this enzyme. However, when the concentration of iron as Fe^{2+} decreases to suboptimal levels, as in iron deficiency, zinc is utilized by ferrochelatase as a substrate. Studies have suggested that lead decreases the availability of Fe^{2+} as a substrate for ferrochelatase by inhibiting the enzymatic reduction of Fe^{3+} to Fe^{2+} within mitochondria (Taketani et al., 1985), as required for use by ferrochelatase. The result of this effect is the increased utilization of zinc as substrate by ferrochelatase, resulting in increased ZPP production. The effects of lead on iron availability and on ferrochelatase may, in fact, be inextricably linked, inasmuch as studies with bovine heart mitochondria have demonstrated that reduction of iron occurs in the same portion of the mitochondrial inner membrane (complex I of the electron transport chain) where ferrochelatase is localized (Taketani et al., 1986). Thus, reduction in the rate of iron use by ferrochelatase could be the result of inhibition of either step by lead. The implication of these findings is that lead mediates elevated erythrocyte ZPP levels primarily by reducing iron availability as a substrate for ferrochelatase, with direct compromise on ferrochelatase activity *in vivo* occurring subsequent to iron depletion.

g. Heme Oxygenase

Heme catabolism in mammalian tissues is mediated by the microsomal enzyme, heme oxygenase (EC 1.14.99.3). This enzyme catalyzes the breakdown of heme to the open chain tetrapyrrole, biliverdin, which is subsequently reduced to bilirubin by biliverdin reductase in the cytosol (Figure 2). The discovery that a wide variety of metal ions induce the synthesis of heme oxygenase in mammalian liver (Tenhunen et al., 1968) has served to focus attention on the property of metals to produce changes in cellular heme metabolism as an early subtoxic indicator of metal exposure. An extensive literature has developed regarding the biological basis of metal action on heme oxygenase in various tissues and the implications of this effect with respect to heme-mediated processes. The reader is referred to any of several comprehensive review articles (e.g., Maines, 1984, 1990; Sunderman, 1987) for a more detailed discussion of this topic.

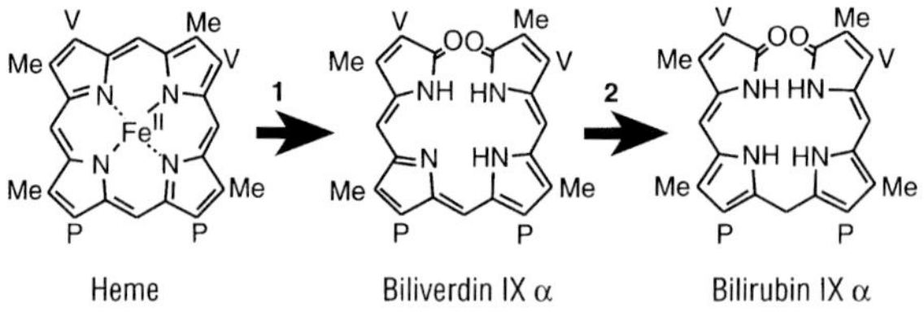

Figure 2. Heme catabolic pathway. Steps are catalyzed by (1) heme oxygenase and (2) biliverdin reductase.

3. Indirect Effects of Metals on Heme Metabolism
a. Perturbation of the Structural Integrity of Subcellular Organelles

Perturbation of the structural integrity of subcellular organelles, such as mitochondria and endoplasmic reticulum, may be an important mechanism by which metals indirectly alter heme metabolism in the intact cell. This action may occur independently of effects associated with direct binding of metals to regulatory sites of heme biosynthetic or degradative enzymes per se, and may be manifested as either impaired or enhanced catalytic activity (Woods and Fowler, 1986; Fowler et al., 1983) of membrane-associated enzymes, e.g., ALA synthetase, coproporphyrinogen oxidase, protoporphyrinogen oxidase, and ferrochelatase. For example, treatment of rats with subtoxic doses of $TlCl_3$ 16 hours prior to sacrifice produces a dose-related increase in the surface density of hepatic mitochondrial inner membranes, accompanied by a pronounced increase in the activity of ferrochelatase. This action is clearly associated with the disruptive effect of thallium on the integrity of the inner membrane structure, since no such effects are observed with respect to non-membrane-bound enzymes nor to enzymes associated with organellar membranes which are not perturbed by thallium treatment. Moreover, thallium is a potent inhibitor of ferrochelatase activity either when given at high dose levels *in vivo* or in low concentrations *in vitro* (Woods and Fowler, 1986). Similar observations have been made with respect to other metals including arsenic (Woods and Fowler, 1978; Fowler and Woods, 1979; Fowler et al, 1979), mercury (Fowler and Woods, 1977; Fowler et al., 1987), indium (Fowler et al., 1983; Woods and Fowler, 1982; Woods et al., 1979), lead (Fowler et al., 1980; 1987), and bismuth (Woods and Fowler, 1987).

Of particular relevance to metal effects on heme regulation is the observation that perturbation of mitochondrial membranes by metals may produce substantial alteration of mitochondrial-associated enzymes of the heme biosynthetic pathway, even though these enzymes are only loosely associated with mitochondrial membranal structures (Elder and Evans, 1978). Thus, coproporphyrinogen oxidase, which is considered to be associated with membrane surfaces surrounding the mitochondrial inner membrane, and therefore not an integral part of the membrane itself, is nonetheless significantly compromised by lead in rabbit kidney (Henderson and Toothill, 1983). This effect could result from damage to the mitochondrial structure leading to impaired transport of coproporphyrinogen across mitochondrial membranes from the cytosol, or from alteration of the juxtaposition of the enzyme within the intermembranal compartment in such a way that access to substrates is compromised. Similar explanations could account for changes in ALA synthetase activity observed during low level metal exposures under conditions when cellular heme levels are not substantially altered (Woods and Fowler, 1977; Fowler et al., 1980; Woods and Fowler, 1978; Fowler et al., 1987). From these observations, it is apparent that physical perturbation of subcellular structures may be an important mechanism by which metals alter heme regulation in mammalian cells, independently of direct effects on heme pathway enzymes themselves.

b. Porphyrinogen Oxidation

The distinction between reduced and oxidized forms of porphyrins is shown in Figure 3, using uroporphyrin(ogen) as example. With the exception of protoporphyrin, the substrate of ferrochelatase (Figure 1), only the reduced forms of porphyrins (porphyrinogens, hexahydroporphyrins) are utilized as substrates by heme biosynthetic pathway enzymes. In the oxidized form, porphyrins are known to serve no biological function and are excreted in urine or feces. Porphyrinogens are readily oxidized to the corresponding porphyrin by reactive oxygen species (e.g., O_2^-, OH·) and other free radicals *in vitro* and also endogenously by reactive oxidants which are produced by cellular oxidative processes (De Matteis, 1988; Francis and Smith, 1988; Jacobs et al., 1989; Woods and Calas, 1989). Numerous metals are known to promote free radical formation in tissue cells by catalyzing the one-electron reduction of oxygen, by impairing membrane ion or electron transport, by thiol depletion, or through other mechanisms (Amoruso et al., 1982; Cantoni et al., 1982; Christie and Costa, 1984; Gstraunthaler et al., 1983; Halliwell, 1987; Lund et al., 1991, 1993; Miller et al., 1991; Miller and Woods, 1993; Woods et al., 1990a,b). Transition metals (Fe, Co, Ni), as well as others which are capable of one-electron changes in redox status under physiological conditions (e.g., Cr, Cu, V), have been shown to promote reactive oxidant formation via oxygen reduction reactions *in vivo,* and this property has been associated with metal-mediated oxidation of cellular lipids, proteins, and other biomolecules, including reduced porphyrins (De Matteis, 1988; Francis and Smith, 1988; Woods and Sommer, 1990). Iron, the most extensively studied in this regard, has been shown to catalyze the NADPH-dependent oxidation of reduced prophyrins by liver microsomes (De Matteis, 1988; Elder et al., 1986; Sinclair et al., 1987) and also to stimulate free radical-mediated porphyrinogen oxidation by rat liver and kidney mitochondria (Woods and Calas, 1989; Woods and Sommer, 1990). These actions may contribute to the porphyrinogenic potential of metals and other chemicals which directly impair porphyrinogen metabolism via effects on specific enzymes of this process. These pro-oxidant actions of metals may also exacerbate the genetic defects in heme pathway enzymes which underlie specific forms of inherited porphyria (e.g., porphyria cutanea tarda), contributing to the highly increased rates of excretion of (oxidized) porphyrins observed in these conditions.

Nontransition metals which cause oxidative tissue injury may also promote porphyrinogen oxidation via free radical formation. Recent studies (Woods et al., 1990a,b) have demonstrated the property of Hg^{2+} ions to promote the oxidation of porphyrinogens *in vitro* and *in vivo* through mechanisms involving the increased production of reactive oxygen species (O_2^-, H_2O_2) by the mitochondrial electron transport chain. In the presence of iron (Fe^{2+}), which is abundant in mitochondria, increased H_2O_2 formed in the presence of Hg^{2+} is reduced to OH·, which readily oxidizes reduced porphyrins and other biomolecules (Lund et al., 1991, 1993). Coupled with impaired coprogen oxidase activity, this pro-oxidant effect of mercury predisposes to substantially increased oxidation and excretion of coproporphyrin as well as

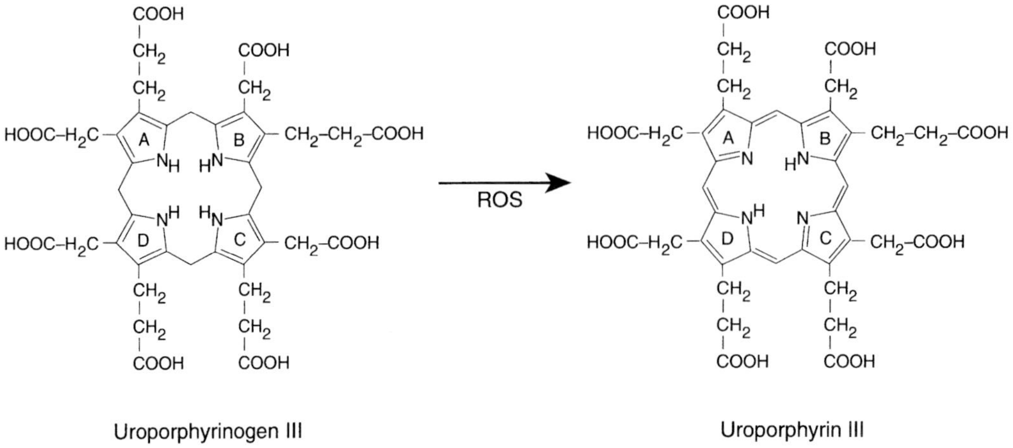

Figure 3. Structures of uroporphyrinogen III and uroporphyrin III. ROS = reactive oxygen species.

other immediate upstream porphyrinogens, contributing to the selective excretion of 5- and 4-carboxyl porphyrins observed during mercury exposure (Bowers et al., 1992; Woods et al., 1991). This effect of mercury has been demonstrated to underlie both porphyrinogen oxidation and lipid peroxidation by Hg in rat kidney *in vivo,* and thus establishes a mechanistic link between the pronounced and distinctive porphyrinuria and oxidative tissue damage associated with mercury exposure (Lund et al., 1993; Miller and Woods, 1993). Arsenic, lead, and other porphyrinogenic metals (Ribarov and Benov, 1981; Somashekaraiah et al., 1992; Sunderman, 1986; Yamanaka et al., 1990) also possess pro-oxidant properties. The oxidation of reduced porphyrins which accumulate in target tissue cells subsequent to their impaired use in heme biosynthesis as result of metal inhibition of specific enzymes of this process may contribute to the porphyrinurias observed during exposure to these metals (Woods, 1995).

c. Relationship to Cell Injury

It is apparent from the above discussion that trace metals of both physiologic and toxicologic importance may significantly effect heme or porphyrin metabolism through various mechanisms in mammalian tissues. The principal processes by which these effects are mediated include (1) alteration of the regulation of heme biosynthesis and degradation through effects of metals either as ions or porphyrin-chelates on the rate-limiting enzymes of these processes; (2) direct compromise of catalytic activity of heme biosynthetic pathway enzymes through formation of metal-mercaptide bonds with sulfhydryl groups at regulatory or catalytic sites; and (3) indirect effects on heme and porphyrin biosynthesis resulting from physical perturbation of the membranes of subcellular organelles with which specific enzymes of the heme biosynthetic pathway are structurally and functionally associated, or by oxidation of prophyrinogen substrates.

The relationship of these processes to metal toxicity and overall cell injury is apparent from several perspectives. First, heme-dependent metabolic functions, particularly those mediated by cytochrome P_{450} and cytochrome oxidase, are dependent upon a sufficient supply of cellular heme; thus, compromise of heme biosynthetic capacity by metals could lead to cellular toxicity as a result of impaired heme-dependent oxidative or metabolic processes. The prospects for cell toxicity resulting from heme depletion would most likely be of greatest concern in the occupational or environmental exposure context, inasmuch as compromise of heme-dependent processes occurs most probably as a function of impaired heme biosynthetic capability resulting from prolonged, rather than acute, metal exposures. Of additional concern is the possibility of impaired heme biosynthesis and associated porphyria-like symptoms resulting from metal-inhibition of heme pathway enzymes which may be present in tissues at significantly less than normal levels, as may occur, for example, with lead exposure in the case of an inherited ALA dehydratase deficiency.

Second, it is apparent from studies described above that altered heme metabolism by metals may occur secondarily as a result of ultrastructural damage to membranes of subcellular organelles with which specific enzymes of the heme biosynthetic pathway and other metabolic processes are functionally associated. While these subcellular effects of metals predispose to altered heme biosynthesis or porphyrin metabolism as a result of impaired heme biosynthetic enzyme activities, they may also induce cell injury or toxicity owing to compromise of other membrane-affiliated metabolic processes. Hence, the relationship of metal-induced alterations in heme metabolism to cell injury must be viewed from the perspective of the associated compromise of cell functions which accompany or occur as a result of altered heme biosynthetic capacity.

Finally, the observation that the porphyrinogenic response to metals involves the oxidation of endogenous reduced porphyrins by metal-induced reactive oxidants and other free radicals provides a biochemical and mechanistic explanation for the oxidative tissue damage often associated with exposure to porphyrinogenic metals. As depicted in Figure 4, numerous metals are capable of promoting reactive oxidant formation either through impairment of cellular electron transport processes or other mechanisms. These events, coupled with depletion of endogenous antioxidants, particularly thiols such as GSH, and functional impairment of heme pathway and other enzymes, promote increased oxidation of cell constituents, porphyrinogens being among the most highly susceptible to this effect. Since oxidized porphyrins are rapidly excreted in excess amounts prior to manifestations of metal toxicity and in patterns which are specific to metal effects on heme biosynthetic enzymes in target tissues, these circumstances establish the biochemical and diagnostic significance of metal-induced porphyrinurias as a pretoxic manifestation of metal exposure and toxicity in tissue cells. A review of the utility of porphyrins and other heme pathway parameters in the diagnosis and monitoring of metal exposures in human subjects has recently been published (Woods, 1995).

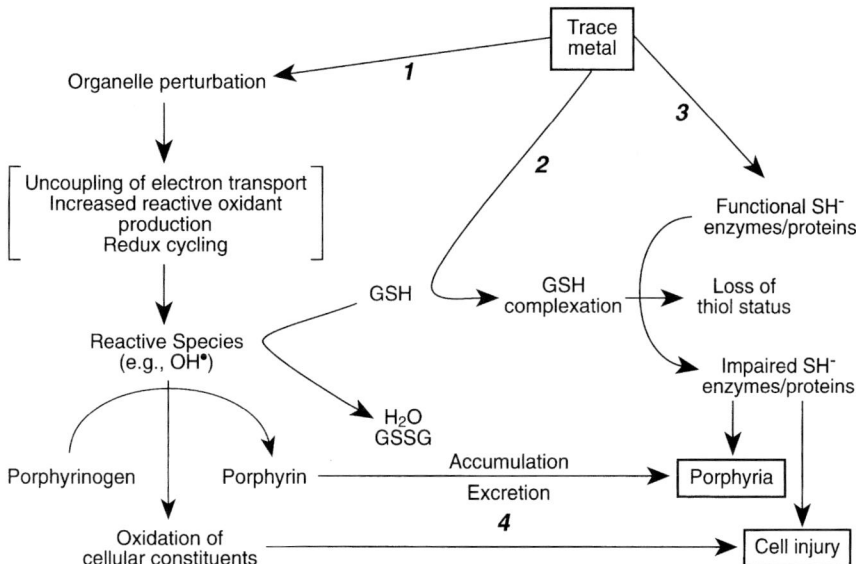

Figure 4. Mechanisms of trace metal-induced porphyria and cell injury. 1. Metals promote increased reactive oxidant formation. 2. Metals complex with GSH, compromising antioxidant and thiol status. 3. Metals impair-SH-dependent enzymes and other proteins via mercaptide bond formation and/or exchange reactions. 4. Metal-induced oxidant stress causes oxidation of reduced porphyrins (porphyrinuria) and other biomolecules (cell injury). (Reproduced with permission from Springer-Verlag (Berlin); Woods, 1995).

ACKNOWLEDGMENT

This work was supported by grants ES03628 and ES04696 from the National Institute of Environmental Health Sciences.

REFERENCES

Abdulla, M. and Haeger-Aronsen, B. (1971), *Enzyme,* 12, 708–710.
Abreo, K., Glass, J., and Sella, M. (1990), *Kidney Int.,* 37, 677–681.
Amoruso, M. A., Witz, G., and Goldstein, B. D. (1982), *Toxicol. Lett.,* 10, 133–138.
Anderson, P. M. and Desnick, R. J. (1979), *J. Biol. Chem.,* 254, 6924–6930.
Anderson, P. M. and Desnick, R. J. (1980), *J. Biol. Chem.,* 255, 1993–1999.

Astrin, K. H., Bishop, D. F., Wetmur, J. G., Kaul, B., Davidow, B., and Desnick, R. J. (1987), *Ann. N. Y. Acad. Sci.,* 514, 23–29.
Batlle, A. M. del C., Benson, A., and Rimington, C. (1965), *Biochem. J.,* 97, 731–740.
Batlle, A. M. del C., Fukuda, H., Parera, V. E., Wider, E., and Stella, A. M. (1987), *Int. J. Biochem.,* 19, 717–720.
Beckmann, H-G., Bogdanski, P., and Goedda, H. W. (1983), *Hum. Hered.,* 33, 61–64.
Bentley, P. J. and Grubb, B. R. (1991), *Trace Elem. Med.,* 8, 202–207.
Bird, T. D., Hamernyik, P., Nutter, J. Y., and Labbe, R. F. (1979), *Am. J. Hum. Genet.,* 31, 662–668.
Blair, P. C., Thompson, M. B., Morrissey, R. E., Moorman, M. P., Sloane, R. A., and Fowler, B. A. (1990), *Fundam. Appl. Toxicol.,* 14, 776–787.
Bowers, M. A., Aicher, L. D., Davis, H. A., and Woods, J. S. (1992), *J. Lab. Clin. Med.,* 120, 272–281.
Briggs, D. W., Condie, L. W., Sedman, R. M., and Tephly, T. R. (1976), *J. Biol. Chem.,* 251, 4996–5001.
Broun, E. R., Greist, A. Tricot, G., and Hoffman, R. (1990), *J. Am. Med. Assoc.,* 264, 1441–1443.
Camadro, J. M., Chambon, H., Jolles, J., and Labbe, P. (1986), *Eur. J. Biochem.,* 156, 579–587.
Campbell, J. (1948), *Can. Med. Assoc. J.,* 58, 72–75.
Cantoni, O., Evans, R. M., and Costa, M. (1982), *Biochem. Biophys. Res. Commun.,* 108, 614–619.
Casey, J. L., Hentze, M. W., Koeller, D. M., Caughman, S. W., Rouault, T. A., Klausner, T. A., and Harford, J. B. (1988), *Science,* 240, 924–928.
Chen, J.-J., Yang, J. M., Petryshyn, R., Kosower, N., and London, I. M. (1989), Disulfide bond formation in the regulation of eIF-2α kinase by heme.
Christie, N. T. and Costa, M. (1984), *Biol. Trace Elem. Res.,* 6, 139–158.
Clement, R. P., Kohashi, K., and Piper, W. N. (1982), *Arch. Biochem. Biophys.,* 214, 657–667.
Condie, L. W., Tephly, T. R., and Baron, J. (1976), *Ann. Clin. Res.,* 8 (Suppl. 17) 83.
Dailey, H. A. (1984), *J. Biol. Chem.,* 259, 2711–2715.
Dailey, H. A. (1987), *Ann. N. Y. Acad. Sci.,* 514, 81–86.
Dailey, H. A. and Fleming, J. E. (1983), *J. Biol. Chem.,* 258, 11453–11459.
De Matteis, F. (1988), *Mol. Pharmacol.,* 33, 463–469.
De Verneuil H., Sassa, S., and Kappas, A. (1983), *J. Biol. Chem.,* 258, 2454–2460.
Doss, M., Laubenthal, F., and Stoeppler, M. (1984), *Int. Arch. Occup. Environ. Health,* 54, 55–63.
Doss, M. and Muller, W. A. (1982), *Blut,* 45, 131–139.
Doss, M., Von Tiepermann, R., Schneider, R., and Schmid, H. (1979), *Klin. Wochenschr.,* 57, 1123–1127.
Drinker, K. R., Thompson, P. K., and Marsh, M. (1927), *Am. J. Physiol.,* 80, 31–64.
Elder, G. H. and Evans, J. O. (1978), *Biochem. J.,* 172, 345–351.
Elder, G. H., Roberts, A. G., and Urquhart, A. J. (1986), in *Porphyrins and Porphyrias,* Nordmann, Y., Ed., *Colloq. INSERM,* 134, 147–152.
Elder, G. H., Tovey, J. A., and Sheppard D. M. (1983), *Biochem. J.,* 215, 45–55.
Elder, G. H. and Urquhart A. J. (1984), *Biochem. Soc. Trans.,* 12, 663–664.
Elferink, C. J., Srivastava, G., Maguire, D. J., Borthwick, I. A., May, B. K., and Elliot, W. H. (1987), *J. Biol. Chem.,* 262, 3988–3992.
Erslev A. J., Caro, J., Miller, O., and Silver, R. (1980), *Ann. Clin. Lab. Sci.,* 10, 250–257.
Fairbanks, V. F. (1967), *Arch. Intern. Med.,* 120, 428–432.
Ferreira, G. C., Andrew, T. L., Karr, S. W., and Dailey, H. A. (1988), *J. Biol. Chem.,* 263, 3935–3839.
Finelli, V. N., Klauder, D. S., Karaffa, M. A., and Petering, H. G. (1975), *Biochem. Biophys. Res. Commun.,* 65, 303–311.
Flower, R. J., Moncada, S., and Vane, J. R. (1985), in *Goodman and Gilman's The Pharmacologic Basis of Therapeutics,* 7th ed. Gilman, A.G., Goodman, L. S., Rall, T. W., and Murad, F., Eds., Macmillan Publishing, New York, 674–715.
Fowler, B. A., Kardish, R. M., and Woods, J. S. (1983), *Lab. Invest.,* 48, 471–478.
Fowler, B. A., Kimmel, C. A., Woods, J. S., McConnell, E. E., and Grant, L. D. (1980), *Toxicol. Appl. Pharmacol.,* 56, 59–77.
Fowler, B. A., Oskarsson, A., and Woods, J. S. (1987), *Ann. N. Y. Acad. Sci.,* 514, 172–182.
Fowler, B. A. and Weissberg, J. B. (1974), *N. Engl. J. Med.,* 291, 1171–1174.
Fowler, B. A. and Woods, J. S. (1977), *Exp. Mol. Pathol.,* 27, 403–412.
Fowler, B. A. and Woods, J. S. (1979), *Toxicol. Appl. Pharmacol.,* 50, 177–187.
Fowler, B. A., Woods, J. S., and Schiller, C. M. (1979), *Lab. Invest.,* 41, 313–320.
Francis, J. E. and Smith, A. G. (1988), *FEBS Lett.,* 233, 311–314.
Fulton, B. and Jeffrey, E. H. (1990), *Fundam. Appl. Toxicol.,* 14, 788–796.
Garcia, R. C., San Martin de Viale, L. C., Tomio, J. M., and Grinstein, M. (1973), *Biochem. Biophys. Acta.,* 309, 203–210.
Gibson, R. D., Neuberger A., and Scott, J. J. (1955), *Biochem. J.,* 61, 618–629.
Gosselin, R. E., Smith, R. P., and Hodge, H. C. (1984), *Clinical Toxicology of Commercial Products,* 5th ed., Williams & Wilkins, Baltimore.
Goyer, R. A. (1971), *Am. J. Pathol.,* 64, 167–182.
Granick, S. (1966), *J. Biol. Chem.,* 241, 1359–1375.
Granick, S. and Sassa, S. (1971), *Metabol. Pathways,* V, 71–141.
Gstraunthaler, G., Pfaller, W., and Kotanko, P. (1983), *Biochem. Pharmacol.,* 32, 2969–2972.
Gurba, P. E., Sennet, R. E., and Kobes, R. D. (1972), *Arch. Biochem.,* 150, 130–136.
Hale, W. E., May, F. E., Thomas, R. G. et al. (1988), *J. Nutr. Elder.,* 8(2), 49–57.
Halliwell, B. (1987), *FASEB J.,* 1, 358–364.

Hamilton, J. W., Bement, W. J., Sinclair, P. R., Sinclair, J. F., Alcedo, J. A., and Wetterhahn, K. E. (1991), *Arch. Biochem. Biophys.*, 289, 387–392.
Harbin, B. M. and Dailey, H. A. (1985), *Biochemistry,* 24, 366–370.
Hayashi, N., Kurashima, Y., and Kikuchi, G. (1972), *Arch. Biochem. Biophys.*, 148, 10–21.
Hayashi, N., Watanabe, N., and Kikuchi, G. (1983), *Biochem. Biophys. Res. Commun.*, 115, 700–706.
Henderson, M. J. and Toothill, C. (1983), *Clin. Sci.*, 65, 527–532.
Hernberg, S. and Nikkanen, J. (1970), *Lancet,* 1, 63–65.
Hoffman, H. N., II., Phyliky, R. L., and Fleming, C. R. (1988), *Gastroenterology,* 94, 508–512.
Igaraski, J., Hayashi, N., and Kikuchi, G. (1978), *J. Biochem.*, 84, 997–1004.
Jackson, A. H., Sancovich, H. A., Ferramola, A. M., Evans, N., Games, D. E., Matlin, S. A., Elder, G. H., and Smith, S. G. (1975), *Philos. Trans. R. Soc. London,* B273, 119–134.
Jacobs, J. M., Sinclair, P. R., Lambrecht, R. W., Sinclair, J. F. (1989), *FEBS Lett.*, 250, 349–352.
Jones, M. S. and Jones, O. T. G. (1969), *Biochem. J.*, 113, 507–514.
Kappas, A., Sassa, S., and Anderson, K. E. (1982), in *Metabolic Basis of Inherited Disease,* (Stanbury, et al. Eds.), 5th ed., 1301–1384.
Kardish, R. M. and Woods, J. S. (1980a), *J. Appl. Biochem.*, 2, 159–167.
Kardish, R. M. and Woods, J. S. (1980b), *J. Appl. Biochem.*, 2, 168–174.
Klein, W. J., Jr., Metz, E. N., and Price, A. R. (1972), *Arch. Intern. Med.*, 129, 578–582.
Kushner, J. P., Steinmuller, D. P., and Lee, G. R. (1975), *J. Clin. Invest.*, 56, 661–667.
Kyle, P. A. and Pease, G. L. (1965), *N. Engl. J. Med.*, 272, 18–23.
Labbe, R. F. and Hubbard, N. (1961), *Biochem, Biophys. Acta,* 52, 131–135.
Levin, E. Y. and Coleman, D. L. (1967), *J. Biol. Chem.*, 242, 4248–4253.
Lund, B., Miller, D. M., and Woods, J. S. (1991), *Biochem. Pharmacol.*, 42, S181–S187.
Lund, B. O., Miller, D. M., and Woods, J. S. (1993), *Biochem. Pharmacol.*, 45, 2017–2024.
Lutton, J. D., Ibraham, N. G., Friedland, M., and Levere, R. D. (1984), *Environ. Res.*, 35, 97–103.
Maines, M. D. (1984), *Crit. Rev. Toxicol.*, 12, 241–314.
Maines, M. D. (1990), *Crit. Rev. Toxicol.*, 21, 1–20.
Maines, M. D. and Kappas, A. (1977), *J. Exp. Med.*, 146, 1286–1293.
Maita, K., Hirano, M., Mitsumori, K. et al. (1981), *J. Pest. Sci.*, 6, 327–336.
Manzler, A. D. and Schreiner, A. W. (1970), *Ann. Intern. Med.*, 73, 409–412.
Matthes, F., Kirschner, R., Yow, M. et al. (1958), *Pediatrics,* 22, 675–688.
Matts, R. L., Schatz, J. R., Hurst, R., and Kagen, R. (1991), *J. Biol. Chem.*, 266, 12695–12702.
Mauzerall, D. and Granick, S. (1958), *J. Biol. Chem.*, 232, 1141–1162.
May, B. K. and Bawden, M. J. (1989), *Semin. Hematol.*, 26, 150–156.
May, B. K., Bhasker, C. R., Bawden, M. J., and Cox, T. C. (1990), *Mol. Biol. Med.*, 7, 405–421.
Miller, D. M., Lund, B. O., and Woods, J. S. (1991), *J. Biochem. Toxicol.*, 6, 293–298.
Miller, D. M. and Woods, J. S. (1993), *Biochem. Pharmacol.*, 46, 2235–2241.
Mitchell, R. A., Chang, B. F., Huang, C. A., and DeMaster, E. G. (1971), *Biochemistry,* 10, 204–210.
Miyagi, K., Kaneshima, M., Kawakami, J, Nokada, F., Petrayka, Z. J., and Watson, C. J. (1979), *Proc. Natl. Acad. Sci. U.S.A.*, 76, 6172–6176.
Mladenovic, J. (1988), *J. Clin. Invest.*, 81, 1661–1665.
Morse, B. S., Conlan, M., Guiliani, D. G., and Nussbaum, M. (1980), *Am. J. Hematol.*, 8, 272–280.
Mukerji, S. K. and Pimstone, N. R. (1986), *Arch. Biochem. Biophys.*, 244, 619–629.
Nakemura, M., Yasuhochi, Y., and Minokami, S. (1975), *J. Biochem.*, 78, 373–380.
Paglia, D. E., Valentine, W. N., and Dahlgren, J. G. (1975), *J. Clin. Invest.*, 56, 1164–1169.
Piomelli, S. and Davidow, B. (1972), *Pediatr. Res.*, 6, 366.
Piper, W. N., Tse, J., Clement, R. P., and Kohashi, M. (1983), in *Chemistry and Biology of Pteridines,* Blair I.A., Ed., Walter de Gruyter, Berlin, 415–419.
Piper, W. N. and van Lier, R. B. L. (1977), *Mol. Pharmacol.*, 13, 1126–1135.
Ponka, P., Schulman, H. M., and Martinez-Medellin, J. (1988), *Biochem. J.*, 251, 105–109.
Poulson, R. (1976), *J. Bio. Chem.*, 251, 3730–3733.
Prasad, A. S., Brewer, G. J., Schoomaker, E. B. et al. (1978), *J. Am. Med. Assoc.*, 240, 2166–2168.
Ribarov, B. R. and Benov, L. C. (1981), *Biochem. Biophys. Acta,* 640, 722–726.
Robbins, S. L., Cotran, R. S., and Kumar, V. (1984), in *Pathologic Basis of Disease,* 3rd ed. W.B. Saunders, Philadelphia, 610–652.
Romeo, G. and Levin, E. Y. (1971), *Biochem. Biophys. Acta,* 230, 330–341.
Rossi, E., Attwood, P. V., and Garcia-Webb, P. (1992), *Biochem. Biophys. Acta,* 1135, 262–268.
Rossi, E., Taketani, S., and Garcia-Webb, P. (1993), *Biomed. Chromatog.*, 7, 1–6.
Sancovich, H. A., Batlle, A. M. C., and Grinstein, M. (1969), *Biochem. Biophys. Acta.*, 191, 130–143.
Sassa, S. (1978), in *Heme and Hemoproteins,* DeMatteis, F. and Aldridge, W.N., Eds., Springer-Verlag, Berlin, 333–371.
Scholnick, P. L., Hammaker, L. E., and Marver, H. S. (1972), *J. Biol. Chem.*, 247, 4126–4131.
Schwartz, S., Stephenson, B., Sarkar, D., Freyholtz, H., and Runge, W. (1976), in *Porphyrins in Human Diseases,* Doss, M., Ed., S. Karger, Basel, 370–379.

Sedman, R., Ingall, G., Rois, G., and Tephly, T. R. (1982), *Biochem. Pharmac.,* 31, 761–766.
Sinclair, P., Lambrecht, R. W., and Sinclair, J. (1987), *Biochem. Biophys. Res. Commun.,* 146, 1324–1329.
Somashekaraiah, B. V., Padmaja, K., and Prasad, A. R. K. (1992), *Free Radicals Biol. Med.,* 13, 107–114.
Srivastava, G., Borthwick, I. A., Maguire, D. J., Elferink, C. J., Bawden, M. J., Mercer, J. F. B., and May, B. K. (1988), *J. Biol. Chem.,* 263, 5202–5209.
Stevens, E., Frydman, R. B., and Frydman, B. (1968), *Biochem. Biophys. Acta,* 158, 496–498.
Straka, J. G. and Kushner, J. P. (1983), *Biochemistry,* 22, 4664–4672.
Straube, E. F., Schuster, N. H., and Sinclair, A. J. (1980), *J. Comp. Pathol.,* 90, 355–361.
Sunderman, W. F., Jr. (1986), *Acta Pharmacol. Toxicol.,* 59 (Suppl. 7), 248–255.
Sunderman, W. F., Jr. (1987), *Ann. N.Y. Acad. Sci.,* 514, 65–80.
Tait, G. H. (1978), in *Heme and Hemoproteins,* DeMatteis, F. and Aldridge, W.N., Eds., Springer-Verlag, Berlin (1978), 1–48.
Taketani, S., Tanaka, A., and Tokunaga, R. (1985), *Arch. Biochem. Biophys.,* 242, 291–296.
Taketani, S., Tanaka-Yoshioka, A., Masaki, R., Tashiro, Y., and Tokunaga, R. (1986), *Biochem. Biophys. Acta,* 883, 227–283.
Taketani, S. and Tokunaga, R. (1981), *J. Biol. Chem.,* 256, 12748–12753.
Tenhunen, R., Marver, H. S., and Schmid, R. (1968), *Proc. Natl. Acad. U.S.A.,* 61, 748–755.
Tephly, T. R., Hasegawa, D., and Baron, J. (1971), *Metabolism,* 20, 200–210.
Theil, E. C. (1990), *J. Biol. Chem.,* 265, 4771–4774.
Thorling, E. B. and Erslev, A. J. (1972), *Br. J. Haematol.,* 23, 483–490.
Tofilon, P. J. and Piper, W. N. (1982), *Arch. Biochem. Biophys.,* 201, 104–109.
Traketeller, A. C., Hemle, E. W., Montjar, M., Axelrod, A. E., and Jensen, W. N. (1965), *Arch. Biochem.,* 112, 89–97.
Verschuuren, H. G., Kroes, R., Dem Tonkelaar, E. M., Berknnvens, J. M., Hellman, P. W., Rauws, A. G., Schuller, P. L., and Van Esch, G. J. (1976), *Toxicology,* 6, 85–96.
Waxman, H. S. and Rabinovitz, M. (1966), *Biochem. Biophys. Acta,* 129, 369–379.
Wetmur, J. G., Lehnert, G., and Desnick, R. J. (1991), *Environ. Res.,* 56, 109–119.
Wills, M. R. and Savory, J. (1983), *Lancet,* 2, 29–34.
Woods, J. S. (1974), *Mol. Pharmacol.,* 10, 384–397.
Woods, J. S. (1988), *Semin. Hematol.,* 25, 336–348.
Woods, J. S. (1989), *Commun. Toxicol.,* 3, 3–25.
Woods, J. S. (1995), in *Handbook of Experimental Pharmacology, Toxicology of Metals — Biochemical Aspects,* Goyer, R. A. and Cherian, M. G., Eds., Springer-Verlag, Berlin, 19–52.
Woods, J. S., Bowers, M. A., and Davis, H. A. (1991), *Toxicol. Appl. Pharmacol.,* 110, 464–476.
Woods, J. S. and Calas, C. A. (1989), *Biochem. Biophys. Res. Commun.,* 160, 101–108.
Woods, J. S., Calas, C. A., and Aicher, L. D. (1990b), *Mol. Pharmacol.,* 38, 261–266.
Woods, J. S., Calas, C. A., Aicher, L. D., Robinson, B. H., and Mailer, C. (1990a), *Mol. Pharmacol.,* 38, 253–260.
Woods, J. S., Carver, G. T., and Fowler, B. A. (1979), *Tox. Appl. Pharmacol.,* 49, 455–461.
Woods, J. S., Eaton, D. L., and Lukens, C. B. (1984), *Mol. Pharmacol.,* 26, 336–341.
Woods, J. S. and Fowler, B. A. (1977), *J. Lab. Clin. Med.,* 90, 266–273.
Woods, J. S. and Fowler, B. A. (1978), *Toxicol. Appl. Pharmacol.,* 43, 361–371.
Woods, J. S. and Fowler, B. A. (1982), *Exp. Mol. Pathol.,* 36, 306–315.
Woods, J. S. and Fowler, B. A. (1986), *Toxicol. Appl. Pharmacol.,* 83, 218–229.
Woods, J. S. and Fowler, B. A. (1987a), *Toxicol. Appl. Pharmacol.,* 90, 274–283.
Woods, J. S. and Fowler, B. A. (1987b), *Ann. N. Y. Acad. Sci.,* 514, 55–64.
Woods, J. S., Kardish, R. M., and Fowler, B. A. (1981), *Biochem. Biophys. Res. Commun.,* 103, 264–271.
Woods, J. S. and Sommer, K. M., (1990), *Adv. Exp. Med. Biol.,* 238, 857–862.
Woods, J. S. and Southern, M. R. (1989), *Toxicol. Appl. Pharmacol.,* 97, 183–190.
Yamamoto, M., Kure, S., Engel, J. D., and Hiraga, K. (1988), *J. Biol. Chem.,* 263, 15973–15979.
Yamanaka, K., Hoshino, M., Okamoto, M., Sawamura, R., Hasegawa, A., and Okada, S. (1990), *Biochem. Biophys. Res. Commun.,* 168, 58–64.
Yoshinaga, T. and Sano, S. (1980a), *J. Biol. Chem.,* 255, 4722–4726.
Yoshinaga, T. and Sano, S. (1980b), *J. Biol. Chem.,* 255, 4727–4731.
Zaman, K., Dabrowski, Z., and Miszta, H. (1990), *Folia Hematol.,* 117, 375–383.
Zaman, K., Zaman, A., and Batcabe, J. (1993), *Comp. Biochem. Physiol.,* 106C, 285–293.

Chapter 62

Bone Metabolism: Effects of Essential and Toxic Trace Metals

Maryka H. Bhattacharyya, Elizabeth Jeffery, and Ellen K. Silbergeld

I. INTRODUCTION

Osteoporosis is an insidious disease prevalent mainly in the elderly, particularly postmenopausal women. While the detailed etiology remains to be determined, the underlying mechanism appears to be an imbalance between osteoblast-dependent bone matrix formation, which is coupled to mineralization, and osteoclast-dependent bone resorption, resulting in a net bone loss. The literature on Ca dynamics is legion, but leads inevitably to the conclusion that manipulation of Ca levels alone does not optimally counteract the progress of the disease (Bronner, 1994). While Ca is the major mineral of bone matrix, other essential and xenobiotic metals not only deposit in the matrix, but have specific, dose-dependent effects on bone cell function that may play major roles in the balance between osteoblast and osteoclast activities. This chapter focuses on some essential and xenobiotic metals that have been found to have just such a direct effect on bone cell function.

The cells responsible for bone growth and maintenance are osteoblasts and osteoclasts. Osteoblasts are the bone-forming cells that produce and secrete type I collagen to form the organic matrix of the bone that is subsequently mineralized. Osteoblasts also synthesize noncollagenous bone matrix proteins required for normal bone growth and development, such as osteocalcin and osteonectin, and enzymes including alkaline phosphatase (ALP), a marker of osteoblast differentiation that participates in matrix mineralization (Noda, 1993). Osteoclasts are the larger, multinucleated cells specialized for resorption, i.e., the dissolution of both mineral and matrix (Sauk and Somerman, 1991). Osteoblasts are of mesenchymal origin, whereas osteoclasts are formed from the fusion of precursor cells from the granulocyte-macrophage lineage (Aubin et al., 1993; Baron et al., 1993). Precursors for both types of cells are found in the bone marrow. Under normal conditions, the activities of osteoblasts and osteoclasts are closely coupled so that the amount of bone removed by osteoclasts is precisely replaced by osteoblasts. The tight coupling between these cells is probably due to cell-to-cell signalling by autocrine, paracrine, and systemic factors including granulocyte-macrophage colony stimulating factor, tumor necrosis factor-alpha, interleukins 1, 4, and 6, transforming growth factor-β, estradiol, parathyroid hormone, vitamin D, and calcium (Hruska et al., 1993; Zaidi, 1993). Although most of these factors are soluble, some are found trapped in the mineralized matrix and are only released by the resorbing activity of osteoclasts. Osteoblasts and osteoclasts are associated with the bone surfaces and are found in far greater preponderance in association with spongy or trabecular bone than cortical, because of the greater trabecular surface area (Jee, 1991). Thus 60% of their activity is associated with the 20% of bone mass that makes up trabecular bone (Williams and Frolik, 199). Results of disruption of the normal balance between osteoblast and osteoclast activities are seen far sooner in trabecular than cortical bone, giving rise to the typical picture of hip and spine fractures in osteoporosis.

The purpose of this chapter is to evaluate the biochemical mechanisms by which the essential trace metals (Mg, Zn, Cu, Mn) participate in the normal growth and remodeling of bone and the mechanisms by which several important osteotoxic trace metals (Cd, Pb, Al) disturb these processes.

II. ESSENTIAL TRACE METALS: Ca, Mg, Zn, Cu, AND Mn

A. OVERVIEW

Calcium (Ca) is the major cation present in hydroxyapatite $[Ca_{10}(PO_4)_6(OH)_2]$, the compound that makes up the mineral portion of bone. As such, calcium plays a critical and direct role in the structural integrity of the skeleton. However, calcium and other trace elements have also been shown to play critical roles in bone cell function, for example, as cofactors for key enzymes involved in bone-specific pathways or as members of important cellular second messenger systems. The purpose of Section II is to review the role of essential trace metals other than calcium — metals whose deficiencies lead to early defects of bone growth or structural integrity.

Table 1 provides an overview of these metals with respect to concentrations in the circulation and in the mineral portions of bone for humans exposed only through environmental routes (food, water, air). After calcium, magnesium (Mg) and zinc (Zn) are the most abundant essential trace metals in bone mineral (column 3), followed by copper (Cu) and manganese (Mn) at much lower levels. For both Mg and Mn, which have very high bone-to-plasma concentration ratios, bone mineral accounts for about half of the body store of the element. In fact, bone mineral is 0.5 to 1.0% Mg, making Mg potentially important with respect to the structural properties of hydroxyapatite.

Table 1. Concentrations of Essential and Toxic Metals in Blood, Plasma, and Bone of Humans in the General Population[a]

Metal	Whole Blood (µg/ml)	Plasma (µg/ml)	Bone (µg/g ash)	Bone:Plasma Ratio	% Total Body in Bone
Essential Metals					
Calcium	60	97	330,000	3,400	99.9
Magnesium	40	23	3,600 (4,800[b])	210	58
Zinc	6.5	1.9	158 (194[c])	100	21
Copper	1.1	1.2	2.4 (26[c])	22	10
Manganese	0.0087[d]	0.0010[d]	1.7 (14[c])	1,700	43
Toxic Metals					
Aluminum	0.0086[e]	0.0053[e]	6.9	16	34
Lead	0.05[f]	0.005[f]	36	3,600	92
Cadmium	0.00024[g]	0.00002[g]	—	—	—

[a] Except where indicated, values are from ICRP 23 (1975). Comparison values are given for some metal concentrations in bone.
[b] Results from Cohen et al. for human ileum as reported in Wallach (1990). Another reference confirms these values, with Mg in human rib at 3900 µg/g ash (Tsuboi et al., 1994).
[c] Nusbaum et al., 1965. Values are for human rib from analysis of 173 cases.
[d] Friedman et al., 1987
[e] Abreo et al., 1989.
[f] Brody et al., 1994
[g] Bhattacharyya et al., 1992

B. MAGNESIUM

Magnesium is a metal critical to normal osteoblast function. By 2 days after the start of a Mg-deficient diet, weanling rats had serum concentrations of osteocalcin, a noncollagenous bone matrix protein indicative of osteoblast activity, that were 80% of control values; by 8 days, serum osteocalcin concentrations had decreased to 50% of controls, with corresponding decreases in calvarial osteocalcin mRNA concentrations (Carpenter et al., 1992). Most striking was the chaotic bone formation in the femoral trabeculae by day 8 of Mg deprivation — formation so chaotic that the fluorescent calcein label used to determine bone formation rates could not be scored. Within 17 days after the start of a Mg-deficient diet, young growing rats were nearly the same as their pair-fed Mg-sufficient controls in body weight (90% of controls), but had serum Mg concentrations that were 3-fold lower (Boskey et al., 1992).

The Mg-deficient diet caused no changes in serum Ca or Zn concentrations or in the total bone mineral content, Ca content, or Ca:P molar ratios of the femur metaphysis and diaphysis, indicating that short-term Mg deficiency caused no gross defects in animal growth or calcium metabolism. By 17 days, however, Mg deficiency did cause statistically significant decreases in the femur metaphysis Mg concentration, trabecular bone volume, and maximum strength of the femur midshaft (values for Mg-deficient animals were 50%, 66%, and 75% of control values, respectively) (Boskey et al., 1992). In addition, the hydroxyapatite crystallite was significantly larger in the femur metaphysis of the Mg-deficient vs. control animals.

The above studies demonstrate the importance of Mg to normal bone growth and development. An excellent review of effects of magnesium on skeletal metabolism (Wallach, 1990) documents many additional studies showing that long-term Mg depletion ultimately has adverse effects on all phases of skeletal metabolism, including a cessation of bone growth, with a decrease in both osteoblast and osteoclast activity, decreased bone formation, altered responsiveness to PTH and vitamin D, osteopenia, increased fragility, and development of a form of "aplastic bone disease."

Two important questions follow: (1) Is Mg deficiency a significant cause of bone disease in humans? and (2) By what biochemical mechanism(s) does Mg participate in normal bone growth and development?

Numerous studies have attempted to answer the first question (Cohen, 1988; Abraham and Grewal, 1990; Wallach, 1991; Carpenter et al., 1992; Stendig-Lindberg et al., 1993). These studies establish magnesium deficiency as a nutritional problem in the U.S. A significant fraction of the US population have inadequate dietary intake of Mg (Morgan et al., 1985; USDA, 1985), with 85% of adolescent females in the U.S. consuming diets with Mg levels below the recommended daily allowance (Morgan et al., 1985) and a high prevalence of hypomagnesemia in a randomly sampled population (Whang, 1987). Bone Mg concentrations have been shown to be low in persons with postmenopausal osteoporosis (PMO) (Cohen, 1988), and Mg supplementation is important in the treatment of PMO (Abraham and Grewal, 1990; Tranquilli et al., 1994). Diabetic patients and alcoholic patients have bone disease that may be associated with low concentrations of Mg in bone (Cohen et al., 1985a; Wallach, 1988). Collectively, these results indicate that Mg inadequacy may have a significant impact on bone growth and development in members of the U.S. population, decreasing peak bone mass and increasing the probability of osteoporosis in later life.

Regarding biochemical mechanism, Carpenter et al. (1992) have clearly shown that Mg acts to maintain normal osteoblast-dependent bone formation and osteocalcin synthesis. The active form of bone alkaline phosphatase (ALP) is $MgZn_2ALP$ (Ciancaglini et al., 1989), and therefore Mg participates as an enzyme cofactor in normal chondrocyte and osteoblast function. Mg has also been shown to be involved in osteoclast-matrix interactions, participating in cell-matrix adhesion (Makgoba and Datta, 1992). Mg is required for normal parathyroid hormone (PTH) function (Wallach, 1990), and animals severely depleted of Mg ultimately develop a resistance to PTH action. Mg appears to play a central role in hydroxyapatite crystallite growth, with larger crystallites formed in Mg-deficient animals (Boskey et al., 1992) and smaller crystallites formed under conditions of Mg excess (Burnell et al., 1986). During the early stages of calcification, the inorganic phase of bone is hypothesized to be laid down inside the gap regions of the collagen structure as smaller crystallites relatively rich in Mg; on increasing calcification, apatite crystallites with reduced Mg content are deposited (Bigi et al., 1992). Which of these biochemical roles of Mg is most critical is not clear at this time, but the fact that Mg is an absolute requirement for normal skeletal growth and development is indisputable.

C. ZINC

Zinc is essential for bone growth (Burch et al., 1975). Zinc deficient humans exhibit dwarfism, and growing rats given a Zn-deficient diet (1 ppm) for 3 weeks show a significant decrease in femur weight gain and in serum ALP that is not reversed by growth hormone (Oner et al., 1984). Furthermore, Zn-adequate rats given an oral Zn supplement (10 mg Zn/kg) exhibit increases in ALP, bone growth, and mineralization (Yamaguchi and Inamoto, 1986).

The essentiality of Zn is thought to rest in its role in a number of enzymes, including ALP (Kirchgessner and Roth, 1983). There are at least three isozymes of ALP, all of which require Zn at the active center. Unlike intestinal and placental ALP, a tissue nonspecific ALP, which is the isoenzyme found in bone, also requires Mg for maximum activity (Coleman and Gettins, 1983). A Zn-free diet will produce signs of deficiency in a matter of days, with certain enzyme activities being lost far more rapidly than others. This is due in part to differing affinities for zinc among the enzymes, and in part because in some enzymes, Zn confers stability on the protein. Loss of the Zn then results in rapid degradation. Zinc appears necessary for both activity and stability of skeletal alkaline phosphatase (Ciancaglini et al., 1989; Genge et al., 1988), and rats on a Zn-deficient diet (1 ppm Zn) lose 50% of serum ALP within 4 days. Except under special conditions such as pregnancy, the majority of serum ALP is osteoblast derived, and a direct estimate of bone ALP shows loss of a similar order of magnitude, making lack of osteoblast ALP a possible cause in the retarded bone growth of Zn deficiency.

A number of *in vitro* studies have evaluated the effects of Zn on osteoblast activities. Zinc (10^{-6} to 10^{-3} M) added to the culture medium stimulated synthesis of collagen and caused a significant increase in ALP in calvaria of weanling rats (Yamaguchi et al., 1987). This increase in ALP activity was not due to a direct effect of Zn on the holoenzyme, since 10^{-7} to 10^{-6} M Zn had no effect on the semipurified enzyme, while 10^{-5} M or greater actually inhibited activity. This increase occurred over 2 to 3 days, and was blocked by inhibition of protein synthesis, causing the authors to propose that Zn stimulated the synthesis of new ALP enzyme (Yamaguchi et al., 1987). However, DNA synthesis, measured as incorporation of ^{3}H-thymidine, was not increased, suggesting no increase in osteoblast cell proliferation, even though this has been reported from *in vivo* studies (Yamaguchi and Inamoto, 1986).

Zinc not only has a direct effect on osteoblasts, but at least in *in vitro* systems, affects the response of these cells to other agents. The stimulatory effect of Zn and vitamin D together on osteoblast activity was twice that expected from the sum of their individual effects (Yamaguchi and Inamoto, 1986). Zinc (0.5 to 20 μM) prevents Cd accumulation in osteoblasts and prevents Cd-dependent stimulation of PGE_2 and subsequent increase in osteoclast-dependent resorption. Cadmium is considered to stimulate bone resorption in calvaria via a PGE_2-mediated mechanism, possibly by an increase in cyclooxygenase activity (Suzuki et al., 1989). Zn supplementation prevented Cd-induced inhibition of collagen synthesis, inhibited Cd-dependent ^{45}Ca loss from prelabeled calvaria, and inhibited Cd-dependent release of PGE_2 into the medium in a dose-dependent manner. However, the presence of Zn also greatly decreased Cd accumulation in osteoblasts, and therefore these effects of zinc may have more to do with altering Cd availability than with a direct effect of Zn on bone cell physiology. Indeed, all the *in vitro* work needs to be confirmed by whole animal studies. Since plasma Zn levels are in the range of 10^{-8} M (Hempe et al., 1991), direct extrapolation from the *in vitro* work suggests that if Zn levels rose above normal, then bone formation would be stimulated, which is in agreement with whole animal studies.

D. COPPER

Ceruloplasmin is the major Cu-binding protein in plasma. The Cu-ceruloplasmin complex has been shown to play a key role in iron mobilization from the liver for delivery of the ferrous ion to transferrin, with the result that anemia is an early sign of Cu deficiency (Evans and Abraham, 1973; Cohen et al., 1985b; Van Den Berg et al., 1992). In addition to ceruloplasmin, Cu-dependent enzymes include lysyl oxidase, cytochrome oxidase, superoxide dismutase, dopamine β-hydroxylase (dopamine→noradrenaline), amine oxidase, and tyrosinase (tyrosine→melanin). Erythrocyte superoxide dismutase, serum Cu, and serum ceruloplasmin are considered good indicators of Cu status in humans.

Bone defects in copper deficiency are normally attributed to decreases in bone lysyl oxidase and osteoblast function (Dollwet and Sorenson, 1988). Lysyl oxidase catalyzes the molecular cross-linking of lysine residues, which give structural integrity to both elastin and collagen. Two studies show that the tensile strength of bone matrix significantly decreased, with no accompanying decreases in either bone Ca or ash content, in rats fed a Cu-deficient diet (Jonas et al., 1993a,b). In addition, a large body of literature in farm animals (cows, pigs, lambs, and chicks) and humans (milk-fed infants and infants fed by total parenteral nutrition) link bone defects to Cu deficiency — defects of growth and susceptibility to fracture that show rapid improvements upon Cu supplementation (Dollwet and Sorenson, 1988).

According to guidelines by the Food and Nutrition Board of the National Research Council of the National Academy of Sciences, adults should consume 2 to 3 mg of Cu daily (Dollwet and Sorenson, 1988). Most adults in the U.S. seldom consume 2 mg/d and are borderline in Cu intake. One manifestation of marginal Cu intake may be the fact that treatment of postmenopausal osteoporosis is significantly more effective when supplements contain Ca + Cu, Mn, Zn instead of Ca alone (Strause et al., 1994) and that fracture healing can proceed faster with Cu supplements (Dollwet and Sorenson, 1988).

E. MANGANESE

One of the highest concentrations of Mn in the body is in bone (Wolinsky et al., 1994). Manganese is a cofactor required for the glycosyltransferase enzymes involved in proteoglycan synthesis (Leach,

1971, 1986). Proteoglycans are synthesized by osteoblasts, act as a cement substance for collagen fibrils, and may be involved in the induction of calcification (Choi et al., 1983; Boskey, 1990). A proteoglycan is a core protein with sulfated carbohydrates called glycosaminoglycans (GAGs) attached (Ruoslahti, 1988). The GAGs are attached to the protein backbone one at a time by the Mn-dependent glycosyltranferases. Manganese deficiency leads to skeletal defects associated with reductions in tissue proteoglycans.

For example, Leach and Muenster (1962) showed that the mucopolysaccaride concentration of chick bones had decreased nearly 3-fold by 4 weeks of Mn deficiency, and that this difference persisted even when controls were pair-fed such that there were no differences in body weight between the control and Mn-deficient chicks. In day-old chicks, exposure to a Mn-deficient diet caused no decrease in chick growth in the first 10 days, but did cause a dramatic, 5-fold decrease in bone Mn concentration and a 2-fold decrease in sulfation of GAGs (the final step in proteoglycan synthesis) that returned to normal by 24 h of Mn repletion (Bolze et al., 1985). Gylcosyltransferase activity, measured *in vitro* in the presence of Mn, increased nearly 2-fold by 10 d of Mn deficiency, indicating an increase in carbohydrate acceptor sites on the proteoglycan chains from Mn-deficient bones.

Mn deficiency during gestation in rats resulted in abnormal development of the otoliths, calcified structures of the inner ear required for body balance reflexes, a defect that was completely prevented by Mn supplements on or before day 14 of gestation (Hurley et al., 1958; Hurley, 1980). These results again demonstrate the importance of Mn in the development of calcified structures.

Strause et al. (1986) fed a diet to weanling rats that was marginally deficient in Mn in that it caused no decrease in animal growth and a 3-fold decrease in bone Mn concentrations over a 1-year period. These rats showed significant increases in both serum Ca and serum P and striking decreases in femur Ca concentration (66% of control values) in the Mn-deficient animals. Friedman et al. (1987) showed that young men, aged 19 to 22, developed significant increases in serum Ca and P concentrations after 10 d of Mn depletion, corresponding to the results of Strause et al. in young, growing rats. These results indicate that, in order to achieve peak bone mass, humans need to ingest diets adequate in Mn, because significant decreases in bone calcification occurred in the absence of any effects on growth when sufficient Mn was not provided (Strause et al., 1986). Results showing that supplements of Ca + Cu, Mn, Zn produced better results in treatment of postmenopausal osteoporosis than did Ca supplementation alone is another indication of the importance of Mn to optimal bone growth and development in humans (Strause, 1994).

F. SUMMARY: ESSENTIAL TRACE METALS

Table 2 summarizes the key steps in bone growth and remodeling that require essential trace metals. Deficiencies in each of these trace metals has been shown to result in significant, early defects in bone development. Careful studies demonstrate that a large fraction of the U.S. population consumes diets that are marginally deficient in each of these trace metals. Consequences of these marginal deficiencies that can be predicted from the results presented here are:

1. Less than optimal bone strength,
2. Less than optimal peak bone mass,
3. Less than optimal fracture healing,
4. Less than optimal prevention of bone loss after menopause, and
5. Susceptibility to trace element deficiencies upon consumption of a single trace element supplement.

Table 2 Key Steps in Bone Growth and Maintenance that Require Essential Trace Metals

Trace Metal	Trace Element-Dependent Enzyme	Early Biochemical Defect in Trace Element Deficiency	Ref.
Magnesium	Osteocalcin	Chaotic bone formation	Carpenter et al., 1992
	Alkaline phosphatase	Increased hydroxyapatite crystal growth	Boskey et al., 1992
Zinc	Alkaline phosphatase	Decreased osteoblast activity	Oner et al., 1984
		Decreased skeletal growth	Oner et al., 1984
Copper	Lysyl oxidase	Defective collagen crosslinks	Jonas et al., 1993a,b
Manganese	Glycosyltransferase	Decreased bone proteoglycan synthesis	Leach and Muenster, 1962
			Bolze et al., 1985

Prevention and treatment of osteoporosis are critical to maintaining an acceptable quality of life among the increasing number of elderly members of our population. It is clear from the data presented here that *multielement* trace metal nutrition is key to the development and maintenance of a healthy skeleton.

III. TOXIC TRACE METALS: Cd, Pb, Al, Mg, Zn, Cu

A. OVERVIEW

The role of the skeleton in metal toxicology is threefold; it is

1. A store for xenobiotic metals and thus an integrated dosimeter,
2. A source of these metals to the circulation, and
3. A site of toxic action.

Because many metals accumulate in the extracellular bone matrix, their adverse effects on bone physiology have frequently been considered to be a physical interference at the osteoid surface. More recently, however, research has focused on the effects of metals on bone cells directly. Intestines and kidneys can also play a role in the osteotoxic effects of xenobiotic metals, with toxic metals interfering with the intestinal absorption of essential metals or with renal calcium excretion or bioactivation of vitamin D. Partly because of its ability to accumulate metals, bone can also serve as an internal source of toxic metal exposure, particularly during periods of increased bone resorption, such as pregnancy, lactation, and after menopause in women. In Section III, we focus on the osteotoxicity of three chemically disparate metals: cadmium, lead, and aluminum.

In humans not occupationally exposed to metals, plasma concentrations of lead and aluminum are similar and are 5-fold higher than for cadmium (Table 1, column 2). In bone, lead and aluminum are both at concentrations higher than those of the required trace element, manganese.

B. CADMIUM

The possibility that cadmium has adverse effects on bone physiology was first recognized during the latter part of World War II in Japan. An endemic disease that included both renal dysfunction and osteodystrophy was identified in postmenopausal multiparous women living downstream from an old Pb-Zn mine that had contaminated their water supply with cadmium (Takeuchi, 1978; Lauwerys, 1979; Nogawa, 1981; Kjellstrom, 1986, 1992; Friberg et al., 1974, 1986). Due to the bone pain associated with the disease, it became known as "itai itai" or "ouch ouch" disease. On the basis of evidence provided by many careful studies, the Japanese Ministry of Health in May 1968 declared cadmium to be a causal factor in the etiopathology of itai-itai disease (Nogawa, 1981). Even though industrial exposure to cadmium in France and Sweden also led to bone disease (Kjellstrom, 1986; Nicaud et al., 1942; Valetas, 1946), investigators have continued to debate the causal role of cadmium in itai-itai disease. The lack of consensus with respect to cadmium's role probably stems from the many confounding factors associated with the disease, such as malnutrition, multiple childbirths, and the postmenopausal status of the afflicted individuals, which set them apart from the many others sharing their water supply but not their disease. Current hypotheses concerning the etiology of itai-itai disease indicate that the bone changes associated with pregnancy, lactation, and menopausal hormone depletion may have exacerbated cadmium osteotoxicity, either by enhancing dietary cadmium uptake or by enhancing cadmium-dependent bone loss (Bhattacharyya et al., 1995). In addition to environmental and industrial exposure, smoking (>20 cigarettes per day) can raise blood cadmium levels 10-fold to 2 µg/l (Elinder et al., 1983) and may be involved in the etiology of postmenopausal osteoporosis in women who smoke (Daniell, 1976; Aloia et al., 1985; Slemenda, 1989).

A prominent early hypothesis holds that cadmium acts on bone only secondary to its effects on the kidney, with inadequate renal tubular reabsorption of bone salts and disturbances in vitamin D metabolism leading to severe mineral imbalance and subsequent osteomalacia (Nomiyama et al., 1973, 1975; Kajikawa, 1981).

While a connection between the kidney and bone responses to cadmium is clearly important, during the 1970s and early 1980s, a number of investigators demonstrated a bone response to cadmium that occurs prior to the renal dysfunction that typically results from prolonged dietary cadmium exposure (Kimura et al., 1974; Yoshiki et al., 1975; Anderson and Danylchuk, 1978, 1979; Ando et al., 1978; Bonner et al., 1981). In addition, Furata (1978) showed a disruption of bone cell architecture shortly after an acute, high subcutaneous injection of cadmium. The concept developed that cadmium might act on bone prior to and independent of its action on the kidney.

Four studies most clearly demonstrate a prerenal bone response to cadmium. Three show that dietary cadmium at 25 to 50 ppm causes an increased release of ^{45}Ca from the mouse or rat skeleton within 3 to 10 days of the start of exposure (Ando et al., 1978; Bhattacharyya, 1988a; Wang and Bhattacharyya, 1993), well before the typical time for onset of renal tubular dysfunction characterized by proteinuria, aminoaciduria, and glucosuria — starting at 3 to 4 months after dietary Cd at 300 ppm (Nomiyama et al., 1975) and 5 to 8 months after 100 to 200 ppm (Bernard et al., 1981; Cardenas, 1991). This early release of bone calcium was specific in that it did not occur in response to lead exposure (Wang and Bhattacharyya, 1993). The fourth study (Ogoshi et al., 1989) demonstrates a significant decrease in compression strength of the femur metaphysis in young rats exposed to dietary cadmium at 10 ppm for only 4 weeks, while at the same dietary level, moderate ultrastructural changes in the kidney at the electron microscope level in smooth membranes of renal proximal tubules just begin at 6 months (Nishizumi, 1972).

In summary, for animals with a normal but not excessive calcium intake, cadmium causes a prolonged increase in the release of calcium from bone starting during the first week to 10 days of dietary exposure (Ando et al., 1978; Bhattacharyya et al., 1988a; Wang and Bhattacharyya, 1993). Cadmium also causes a decrease in bone formation (Anderson and Danylchuk, 1978, 1979), with the combined effects on resorption and formation eventually leading to osteoporotic bone loss and increased bone fragility (Ogoshi et al., 1989, 1990, 1992). These changes are important because they occur at very low bone Cd concentrations (0.1 µg/g dry weight) (Ogoshi et al., 1992) and have been shown to begin at blood cadmium concentrations of 2 to 5 µg/l (Sacco-Gibson et al., 1992) — concentrations below the current OSHA standard for increased biologic monitoring among industrial workers (5 µg/l blood) (*CFR* 1993). With time, enough cadmium accumulates in the kidneys to overwhelm renal defense mechanisms, and renal tubular damage occurs (Kajikawa et al., 1981) accompanied by impairment of vitamin D metabolism (Nogawa et al., 1990; Kido et al., 1991; Tsuritani et al., 1992). This renal damage ultimately exacerbates the early bone loss response to cadmium, transitioning from an osteoporotic to an osteomalacic condition.

With respect to biochemical mechanisms, cadmium has been shown to stimulate bone resorption and inhibit bone formation in both *in vivo* and *in vitro* systems. With respect to bone resorption, studies of fetal rat limb bones in culture demonstrated that as little as 10 nM Cd causes significant demineralization (Bhattacharyya et al., 1988a), while in neonatal mouse calvarial cultures, exposure to 250 nM was necessary to effect similar responses (Suzuki et al., 1989a; Miyahara et al., 1992). In the calvaria system and osteoblast-like cell line, MC3T3-E1 (Suzuki et al., 1989b), the increased calcium loss is dependent upon a prostaglandin-cAMP signaling system, being blocked by indomethacin (an inhibitor of cyclooxygenase) and stimulated by isobutylmethylxanthine (an inhibitor of phosphodiesterase) and involving induction of cyclooxygenase, the enzyme that converts arachidonic acid to PGE$_2$. A working hypothesis for the intramembranous bone systems (calvaria and parietal cultures) is that cadmium interacts with the osteoblast to decrease bone formation and stimulate PGE$_2$ formation, which in turn stimulates osteoclast-dependent bone resorption. Another mechanism for cadmium-induced bone loss may be the formation of new osteoclasts from precursor cells in the bone marrow; cadmium at 10 to 90 nM has been shown to stimulate the formation of osteoclastlike cells in bone marrow cultures (Konz et al., 1989; Miyahara et al., 1991; Bhattacharyya, 1991).

With respect to bone formation, cadmium at 2 to 20 µM causes a striking inhibition of bone formation in cultures of chick embryo long bones (tibita, femur) (Miyahara et al., 1978, 1983; Kaji et al., 1988a,b, 1990). These results in organ culture are strengthened by analogous results with MC3T3-Ea cells, an osteoblastlike cell line derived from neonatal mouse calvaria (Miyahara et al., 1988; Iwami and Moriyama, 1993). Other results in rats indicate that cadmium may compromise bone formation by inhibiting the Cu-dependent enzyme, lysyl oxidase (Iguchi et al., 1990). The key, early biochemical interactions of cadmium with bone cells that lead to decreased bone formation and stimulation of osteoclast-mediated bone resorption have yet to be determined.

C. LEAD

Lead is one of the most studied toxic agents, known to cause a plethora of ills, including anemia and neurological and reproductive disorders. Although the bone stores greater than 95% of the body burden of lead (Barry, 1975), there is little information on the adverse effects of lead on bone. Early in the century, children very heavily exposed to lead were reported to exhibit a rickets-like skeletal toxicity

(Nye, 1929). X-rays of heavily intoxicated children showed a diagnostic "lead line," a radiopaque area at the epiphyseal margins of the growing long bones. Because of reductions in lead exposure, other toxicities became the focus of lead studies, and the effect of lead on bone physiology was not further considered until very recently. Now it appears that lead has a number of adverse effects on bone during development, and it remains to be seen whether lead adversely affects the bone at any other time in life. A case:control study of Paget's disease has suggested that some individuals suffering from this disease have been exposed occupationally to high lead levels, leading Spencer to query the possibility of lead playing a role in the etiology of the disease (Spencer et al., 1992).

An examination of the NHANES II (National Health and Nutrition Examination Surveys) data shows an inverse correlation between skeletal length and blood lead in children (Schwartz, 1986). A number of other studies have confirmed that lead present during fetal growth and early postnatal development causes retarded bone growth (Schwartz, 1986; Shukla et al., 1989, 1991; Angle and Kuntzelman, 1989; Frisancho and Ryan, 1991). This finding has now been reproduced in experimental animals exposed to lead *in utero* and during development (Hamilton et al., 1994). With hindsight, one can look back at the histopathology of the "lead line" in bone, which is formed by an area of calcified cartilage that has not undergone regular remodeling (Eisenstein and Kawanoue, 1975).

Classically, identification of an organ toxicity depends upon determining a dose-response. Yet lead is heterogeneously sequestered in bone in such a manner that neither the plasma nor the bone cells are in equilibrium with the entire bone store. Even were measurement of bone lead easily accomplished, which it is not, no methods are available to calculate the dose to which bone cells are exposed. Lead has a relatively short half-life in blood (Rabinowitz, 1991), and therefore blood lead only reflects recent exposure, and not the accumulated burden stored in mineralized tissue. Chelation has been used to determine the body lead load. However, chelatable lead is a measure of exchangeable lead pools, including soft tissues and actively remodeling bone surfaces (Hammond et al., 1967), but does not accurately reflect deep bone stores (Tell et al., 1992). X-ray fluorescence, still predominantly a research tool, gives a reliable noninvasive measure of total bone lead for heavily exposed individuals (Ahlgren et al., 1976; Todd et al., 1992a,b; Somervaille et al., 1987, 1988; Wielopolski et al., 1986; Tell et al., 1992). However, because it is not sensitive enough to quantitate bone lead concentrations in persons with low-level lead exposure, the utility of this method is limited (Hu et al., 1990).

Because of its ability to accumulate lead, bone has been considered protective, decreasing the circulating lead concentrations. However, recent focus has been placed on bone as a possible internal source of exposure, particularly during periods of bone resorption. The significance of this exposure source depends upon both the amount of lead in bone and the rate at which it is mobilized. For example, a woman who had been exposed to high lead levels during childhood mobilized sufficient bone lead during the last trimester of her pregnancy that she intoxicated herself and her infant (Thompson et al., 1985). Because significant bone resorption can also occur during lactation and aging (particularly after menopause in women), these are periods in which individuals may additionally be at risk for increased exposure due to mobilization of lead stored in bone (Silbergeld et al., 1993). In an analysis of the NHANES II cross-sectional dataset, conducted in the U.S. from 1976 to 1980, blood lead levels in postmenopausal women were found to be higher than those in premenopausal women (Silbergeld et al., 1988). Several more recent population-based studies, all cross sectional, have found similar results (Silbergeld and Watson, in press). In all cases, women whose ages indicate that they are likely to be past the menopause (>50 years) have higher mean blood lead levels as compared to younger women — a pattern distinct from the decline in blood lead concentrations seen in older men (Schultz et al., 1987). These effects were observed in studies conducted at different times (the first U.S. study was conducted during the period in which lead was still extensively used in gasoline); independent of the actual lead levels, these relative differences can still be observed. Until now, potential toxicities associated with release of bone lead stores has received considerable attention but little research.

With respect to biochemical mechanisms, lead is thought to adversely affect the coordinate interaction of osteoblasts and osteoclasts (Puzas et al., 1992), slowing bone formation and enhancing resorption. Lead may impair osteoblast function by two general mechanisms. Lead affects osteoblasts indirectly through an inhibitory effect on hormones, cytokines, or other growth factors, for example by inhibiting the renal bioactivation of vitamin D (Sauk and Somerman, 1991). Second, lead may have a direct effect on the biochemistry of the osteoblast, increasing intracellular calcium levels, and thus perturbing intracellular calcium signal transduction (Long et al., 1992). A number of *in vitro* studies using the osteoblast-like ROS 17/2.8 cell have shown that while general protein synthesis appears unaffected, lead specifically inhibits synthesis and secretion of a number of proteins, including alkaline phosphatase (Klein and Wiren, 1993) and the matrix proteins osteonectin (Sauk et al., 1992) and osteocalcin (Long et al., 1990).

Lead has been shown to stimulate bone resorption *in vitro* (Miyahara et al., 1994). Osteoclasts from lead-intoxicated pigs contain intranuclear inclusion bodies, suggesting that lead released from the bone during resorption is taken up by the osteoclast (Hsu et al., 1973). When bone marrow cells were cultured in the presence of lead (2 to 10 μM), formation of osteoclast-like multinuclear cells was greatly enhanced (Miyahara et al., 1994). Lead stimulated PGE_2 synthesis in marrow cells, and the stimulation of formation of osteoclasts was inhibited by inhibition of PGE_2 synthesis, suggesting that the effect of lead on osteoclast formation was through stimulation of PGE_2 synthesis. At high lead levels (>40 μM), both formation and resorption were inhibited (Miyahara et al., 1994).

D. ALUMINUM

Many individuals with chronic renal failure exhibit an osteodystrophy reversible by vitamin D therapy. A few develop an osteodystrophy that does not respond to vitamin D, and appears to be associated with an accumulation of Al in bone (Pieridis, 1980). Aluminum administration to experimental animals produces, among other toxicoses, osteomalacia (Goodman, 1986). While Al is ubiquitous in the environment, being the third most prevalent element in the earth's crust, healthy people do not normally accumulate Al. Although a considerable fraction of ingested Al is absorbed, it is also rapidly excreted by the kidneys (Weberg and Berstad, 1986). However, there are a number of specific clinical conditions under which patients accumulate Al and subsequently exhibit Al toxicity, including osteodystrophy. These include children on total parenteral nutrition and adults on renal dialysis. Recognition of the problem has led to careful monitoring for aluminum in parenteral and dialysis fluids and the partial replacement of Al-binding gels with calcium hydroxide gels. Yet a small number of patients with chronic renal failure persist in accumulating aluminum to osteotoxic levels. Low plasma Al is associated with osteitis fibrosa, a state of enhanced bone remodeling (Chazan et al., 1991). Higher plasma Al levels are associated with the more severe osteomalacia and a substantial slowing of remodeling. The etiology remains to be determined.

Histopathological evaluation of bone from Al-intoxicated individuals has shown Al localized at the mineralization front (Cournot-Witmer et al., 1981), leading to the hypothesis that Al forms a physical barrier to mineralization. However Al inhibits calcification in a model system where no prior Al deposition has occurred at the bone surface (Severson et al., 1992). Also, normal bone is seen to deposit over Al following kidney transplant (Nordal et al., 1992). More recently, direct effects of Al on the function of bone cells have been identified. In keeping with the stimulation in remodeling seen at low plasma aluminum levels and the decrease in remodeling at higher levels, low doses of aluminum added to tissue cultures enhance osteoblast synthesis and activity as well as osteoclast activity, while higher levels are inhibitory to both cell types (Lieberherr et al., 1987). It appears that Al has far more effect on bone metabolism than merely forming a physical barrier to bone growth.

Minimally Al-intoxicated patients exhibiting osteitis fibrosa also exhibit hyperparathyroidism, with significantly increased plasma PTH (Chazan et al., 1991). Patients with higher plasma Al levels and osteomalacia have abnormally low serum PTH levels. PTH is considered to affect osteoblast-directed osteoclast activity, probably through insulin-like growth factors, and thus enhance remodeling. Experimental work has shown that Al is toxic to the parathyroid and disrupts PTH secretion, leading to the proposal that changes in bone remodeling are secondary to changes in plasma PTH levels (Morrisey et al., 1983). While the presence of PTH was necessary to see the mitogenic action of low levels of Al in beagle dogs (Quarles et al., 1989), the effect of Al on bone cell physiology does not appear to be limited to an effect on PTH levels. For example, in fetal limb bone cultures, high Al levels were seen to inhibit basal and PTH-dependent osteoclast activity (Lieberherr et al., 1987). Furthermore, kidney replacement in vitamin D-resistant renal failure patients reverses the bone disorder without immediate reversal of the low serum PTH levels (Nordal et al., 1992). Together these data suggest that, while changes in PTH may aggravate Al-dependent bone disease, they are not causal in the disease. It has been proposed that Al may interact with osteoblast physiology by altering the message sent to direct the osteoclast, for example through an effect on G protein function (Jeffery, 1994) or an effect on insulin-like growth factors (Lau et al., 1993). Very recently, a receptor has been identified on cultured MC3T3-E1 osteoblast-like cells that appears similar to the calcium-sensing membrane receptor present on parathyroid cells, and appears to respond to small, mitogenic doses of Al by stimulating DNA synthesis and enhancing GTP binding to G proteins (Quarles et al., 1994). Thus, it is possible that Al affects the parathyroid and the osteoblast through a shared mechanism.

E. TRACE ELEMENT EXCESSES AND INTERACTIONS

In addition to toxic trace metals, essential trace metals when taken in excess can lead to osteotoxicity. For example, one mechanism by which Mg appears to function in bone growth is to slow the growth of hydroxyapatite crystals. Correspondingly, rats given a high Mg diet for 6 weeks developed abnormal bones with a small crystal size in the mineral phase (Burnell et al., 1986). Using an *in vitro* system for culture of 9-d-old chick embryo femurs, Kaji and coworkers showed that Cu at ≥ 2.5 μM decreased both mineral and matrix accumulation, resulting in an osteoporotic-like condition, while high concentrations of Zn (100 μM) accumulated at the mineral-matrix interface and specifically decreased mineral accumulation, leaving an excess of uncalcified matrix similar to osteomalacia (Kaji et al., 1988a,b). These results imply that excesses of Cu or Zn could have related osteotoxic effects in the whole animal.

In ruminant animals grazing on pastures in the southwest of England and in New Zealand, excess molybdenum in the soil along with sulfur in the drinking water caused a Cu deficiency that was especially apparent in ruminant offspring, producing neurological and skeletal defects that were prevented by Cu supplementation (Howell and Gawthorne, 1987; Dollwet and Sorenson, 1988). The sulfur was reduced to sulfide in the ruminant digestive tract and formed a complex with molybdenum that bound tightly to Cu, preventing its intestinal absorption and causing a debilitating Cu deficiency. Excessive Zn intake has also been demonstrated to produce Cu deficiency in animals on a marginal Cu diet (Howell and Gawthorne, 1987).

The above results illustrate the importance of consuming a diet that achieves a balance of trace minerals in order to grow and maintain a healthy skeleton. Because persons in the U.S. currently consume diets that are marginal in a number of essential trace metals, including Mg, Cu, and Mn, those who take supplements of one particular trace metal, such as zinc, may create a condition of suboptimal nutrition with respect to another trace metal, such as Cu. Reflecting the importance of this balance, treatment protocols for osteoporosis that include trace metals in addition to calcium have shown clear advantages in terms of increasing bone mass (Abraham and Grewal, 1990; Tranquilli et al., 1994; Strause, 1994).

F. SUMMARY: TOXIC TRACE METALS

REFERENCES

Abraham, G. E. and Grewal, H. (1990), *J. Reprod. Med.,* 35, 503–507.
Abreao, K., Brown, S. T., Sella, M., and Trapp, G. (1989), *Lab. Clin. Med.,* 113, 50–57.
Ahlgren, L., Liden, K., Mattsson, S., and Tejning, S. (1976), *Scand. J. Work Environ. Health,* 2(2), 82–86.
Aloia, J. F., Cohn, S. H., Baswani, A., Yeh, J. K., Yven, K., and Ellis, K. (1985), *Am. J. Med.,* 78, 95–100.
Anderson, C. and Danylchuk, K. D. (1978), *Calcif. Tissue Res.,* 26, 143–148.
Anderson, C. and Danylchuck, K. D. (1979), *Calcif. Tissue Int.,* 27, 121–126.
Ando, M., Sayato, Y., and Osawa, T. (1978), *Toxicol. Appl. Pharmacol.,* 46, 625–632.
Angle, C. R. and Kuntzelman, D. R. (1989), *J. Toxicol. Environ. Health,* 26, 149–156.
Aubin, J. E., Turksen, K., and Heersche, J. N. M. (1993), in *Cellular and Molecular Biology of Bone,* Noda, M., Ed., Academic Press, San Diego, CA, 1–45.
Baron, R., Ravesloot, J.-H., Neff, L., Chakraborty, M., Chatterjee, D., Lomri, A., and Horne, W. (1993), in *Cellular and Molecular Biology of Bone,* Noda, M., Ed., Academic Press, San Diego, CA, 445–495.
Barry, P. S. I. (1975), *Br. J. Ind. Med.,* 32, 119–139.
Bernard, A., Lauwerys, R., and Gengoux, P. (1981), *Toxicology,* 20, 345–357.
Bhattacharyya, M. H., Whelton, B. D., and Peterson, D. P. (1981), *Toxicol. Appl. Pharmacol.,* 61, 335–342.
Bhattacharyya, M. H., Whelton, B. D., Stern P. H., and Peterson, D. P. (1988a), *Proc. Natl. Acad. Sci. U.S.A.,* 85, 8761–8765.
Bhattacharyya, M. H., Whelton, B. D., Peterson, D. P., Carnes, B. A., Moretti, E. S., Toomey, J. M., and Williams, L. L. (1988b), *Toxicology,* 50, 193–204.
Bhattacharyya, M. H. (1991), *Water Air Soil Pollut.,* 57/58, 665–674.
Bhattacharyya, M. H., Peterson, D. P., and Sacco-Gibson, N. (1992), in Edited Proceedings *7th Int. Cadmium Conf.* New Orleans, Cook, M. E., Hiscock, S. A., Morrow, H., and Volpe, R., Eds., Cadmium Association, London; Cadmium Council, Virginia; International Lead Zinc Research Organization, Raleigh-Durham, NC, 113–114.
Bhattacharyya, M. H., Wilson, A. K., Silbergeld, E. K., Watson, L., and Jeffery, E. (1995), in *Metal Toxicology,* Goyer, R., Waalkes, M., and Klaassen, C., Eds., Academic Press, San Diego, CA, 465–509.
Bonner, F. W., King, L. J., and Parke, D. V. (1981), *J. Inorg. Biochem.,* 14, 107–114.
Bigi, A., Foresti, E., Gregorini, R., Ripamonti, A., Roveri, N., and Shah, J. S. (1992), *Calcif. Tissue Int.,* 50, 439–444.
Bolze, M. S., Reeves, R. D., Lindbeck, F. E., Kemp, S. F., and Elders, M. J. (1985), *J. Nutr.,* 115, 352–358.
Boskey, A. L. (1990), *Orthop. Clin. North Am.,* 21, 19.
Boskey, A. L., Rimnac C. M., Bansal, M., Federman, M., Lian, J., and Boyan, B. D. (1992), *J. Orthopaedic Res.,* 10, 774–783.
Brody et al. (1994), *JAMA,* 272, 277–283.

Bronner, F. (1994), *Am. J. Clin. Nutr.,* 60, 831–836.
Burch, R. E., Hahn, H. K. J., and Sullivan, J. F. (1975), *Clin. Chim.,* 21, 501–520.
Burnell, J. M., Liu, C., Miller, A. G., and Teubner, E. (1986), *Am. J. Physiol.,* 250, 1302–1307.
Cardenas, A., Bernard, A. M., and Lauwerys, R. R. (1991), *Toxicol. Appl. Pharmacol.,* 108, 547–558.
Carpenter, T. O., MacKowiak, S. J., Troiano, N., and Gundberg, C. M. (1992), *Am. J. Physiol.,* 263, E107–E114.
Chazan, J. A., Libbey, N. P., London, M. R., Pono, L., and Abuelo, J. G. (1991), *Clin. Nephrol.,* 35, 78–85.
Choi, H. U., Tang, L., Johnson, T. L., Pal, S., and Rosenberg, L. C. (1983), *J. Biol. Chem.,* 258, 655.
Ciancaglini, P., Pizauro, J. M., Grecchi, M. J., Curti, C., and Leone, F. A. (1989), *Cell. Mol. Biol.,* 35, 503–510.
Code of Federal Regulations, (1993), Cadmium, 29CFR 1910.1027, 182–183.
Cohen, L., Laor, A., and Kitzes, R. (1985a), *Magnesium,* 4, 148–152.
Cohen, L. (1988), *Magnes. Res. (England),* 1, 85–87.
Cohen, N. L., Keen, C. L., Hurley, L. S., and Lönnerdal, B. (1985b), *J. Nutr.,* 115, 710–725.
Coleman, J. E. and Gettins, P. (1983), *Adv. Enzymol.,* 55, 382–452.
Cournot-Witmer, G., Zingraff, J., Plachott, J. J., Escaig, F., Lefevre, R., Boumati, P., Bourdeau, A., Garabedian, M., Galle, P., Bourdon, R., Drueke, T., and Balsan, S. (1981), *Kidney Int.,* 20, 375–385.
Daniell, H. W. (1976), *Arch. Intern. Med.,* 136, 298–304.
Dollwet, H. H. A. and Sorenson, J. R. J. (1988), *Biol. Trace Elem. Res.,* 18, 39–48.
Eisenstein, R. and Kawanoue, S. (1975), *Am. J. Pathol.,* 80, 309–316.
Elinder, C.-G., Friberg, L., Lind, B., and Jawaid, M. (1983), *Environ. Res.,* 32, 332–227.
Evans, J. L. and Abraham, P. A. (1973), *J. Nutr.,* 103, 196–201.
Friberg, L. T., Piscator, M., Nordberg, G. F., and Kjellstrom, T. (1974), *Cadmium in the Environment,* 2nd ed., CRC Press, Cleveland, 137–195.
Friberg, L., Elinder, C., Kjellstrom, T., and Nordberg, G. F. (1986), *Cadmium and Health: A Toxicological and Epidemiological Appraisal, Vol. 2. Effects and Responses,* CRC Press, Boca Raton, FL.
Friedman, B. J., Freeland-Graves, J. H., Bales, C. W., Behmardi, F., Shorey-Kutschke, R. L., Willis, R. A., Crosby, J. B., Trickett, P. C., and Houston, S. D. (1987), *J. Nutr.,* 117, 133–143.
Frisancho, A. R. and Ryan, A. S. (1991), *Am. J. Clin. Nutr.,* 54, 516–519.
Furuta, H. (1978), *Experientia,* 34, 1317–1318.
Genge, B. R., Savor, G. R., Wu, L. N. Y., McLean, F. M., and Wuthier, R. E. (1988), *J. Biol. Chem.,* ??, 18513–18519.
Goodman, W. G. (1986), *Kidney Int.,* 29, S32–36.
Hamilton, J. D., O'Flaherty, E. J., Ross, R., Shukla, R., and Gartside, P. S. (1994), *Environ. Res.,* 64, 53–64.
Hammond, P. B., Aronson, A. L., and Olson, C. (1967), *J. Pharmacol. Exp. Ther.,* 157, 196–206.
Hempe, J. B., Carlson, J. M., and Cousins, R. J. (1991), *J. Nutr.,* 121, 1389–1396.
Howell, J. McC. and Gawthorne, J. M. (1985), in *Copper in Animals and Man,* Howell, J. McC. and Gawthorne, J. M., Eds., CRC Press, Boca Raton, FL.
Hruska, K. A., Rolnick, F., Duncan, R. L., Medhora, M., and Yamakawa, K. (1993), in *Cellular and Molecular Biology of Bone,* Noda, M., Ed., Academic Press, San Diego, 414–445.
Hu, H., Milder, F. L., and Burger, D. E. (1990), *Arch. Environ. Health,* 45, 335–341.
Hurley, L. S., Everson, G. J., and Geiger, J. F. (1958), *J. Nutr.,* 66, 309.
Hurley, L. S. (1980), in *Developmental Nutrition,* Prentice-Hall, Englewood Cliffs, NJ, 199.
Iguchi, H., Kasai, R., Okumura, H., Yamamuro, T., and Kagan, H. M. (1990), *Bone Mineral,* 10, 51–59.
Itokawa, Y., Nishino, K., Takashima, M., Nakuta, T., Kaito, H., Okamoto, E., Daijo, K., and Kawamura, J. (1978), *Environ. Res.,* 15, 206–217.
Iwami, K. and Moriyama, T. (1993), *Arch. Toxicol.,* 67, 352–357.
Jee, W. S. S. (1991), in *A Basic Science Primer in Orthopaedics,* Bronner, F. and Worrell, R. V. Eds.), Williams & Wilkins, Baltimore, 3–34.
Jeffery, E. H. (1994), in *Handbook of Experimental Pharmacology,* Vol. 118, pp. 139–161.
Jonas, J., Burns, J., Abel, E. W., Cresswell, M. J., Strain, J. J., and Paterson, C. R. (1993a), *Ann. Nutr. Metab.,* 37, 245–252.
Jonas, J., Burns, J., Abel, E. W., Cresswell, M. J., Strain, J. J., and Paterson, C. R. (1993b), *J. Biomech.,* 26, 271–276.
Kaji, T., Kawatani, R., Takata, M., Hoshino, T., Miyahara, T., Kozuka, H., and Koizumi, F. (1988a), *Toxicology,* 50, 303–316.
Kaji, T., Takata, M., Hoshino, T., Miyahara, T., Kozuka, H., Kurashige, Y., and Koizumi, F. (1988b), *Toxicol. Lett.,* 44, 219–227.
Kaji, T., Takata, M., Miyahara, T., Kozuka, H., and Koisumi, F. (1990), *Arch. Environ. Contam. Toxicol.,* 19, 653–656.
Kajikawa, K., Nakanishi, I., and Kuroda, K. (1981), *Exp. Mol. Pathol.,* 34, 9–24.
Kido, T., Honda, R., Tsuritani, I., Ishizaske, M., Yamada, Y., Nogawa, K., Nakagawa, H., and Dohi, Y. (1991), *J. Appl. Toxicol.,* 11, 161–166.
Kimura, M., Otaki, N., Yoshiki, S., Suzuki, M., Horiuchi, N., and Suda, T. (1974), *Archiv. Biochem. Biophys.,* 165, 340–348.
Kirchgessner, M. and Roth, H.-P. (1983), in *Metal Ions Biol. Syst.,* 15, 363–414.
Kjellstrom, T. (1986), in *Cadmium and Health: A Toxicological and Epidemiological Appraisal, Vol. 2., Effects and Responses,* Friberg, L., Elinder, C., Kjellstrom, T., and Nordberg, G. F., Eds., CRC Press, Boca Raton, FL, 111–158.
Kjellstrom, T. (1992), in *Cadmium in the Human Environment,* Nordberg, G. F., Publ. 188, International Agency for Research on Cancer, Herber, R. F. M. and Alessio, L. Eds., pp. 301–310.
Klein, R. F. and Wiren, K. M. (1993), *Endocrinology,* 132, 2531–2537.
Konz, R. P., Bhattacharyya, M. H., and Seed, T. M. (1989), *Proc. 3rd Natl. Conf. Undergrad. Res.,* 2, 411–416.

Larsson, S. E. and Piscator, M. (1971), *Isr. J. Med. Sci.,* 7, 495–498.
Lau, K.-H. W., Utrapiromsuk, S., Yoo, A., Mohan, S., Strong, D. D., and Baylink, D. J. (1993), *Arch. Biochem. Biophys.,* 303, 267–273.
Lauwerys, R. (1979), in *Chemistry, Biochemistry and Biology of Cadmium,* Webb, M., Ed., Elsevier/North-Holland Biomechanical Press, New York, 444–447.
Leach, R. M., Jr. and Muenster, A.-M. (1962), *J. Nutr.,* 78, 51–56.
Leach, R. M., Jr. (1971), *Fed. Proc.,* 30, 991–994.
Leach, R. M., Jr. (1986), in *Nutritional Bioavailability of Manganese,* Kies, C. Ed., Academic Press, Orlando, 81.
Lieberherr, M., Grosse, B., Cournot-Witmer, G., Hermann-Erlee, M. P. M., and Balsan, S. (1987), *Kidney Int.,* 31, 736–743.
Long, G. J., Rosen, J. F., and Pounds, J. G. (1990), *Toxicol. Appl. Pharmacol.,* 106, 270–277.
Long, G. J., Pounds, J. G., and Rosen, J. F. (1992), *Calcif. Tissue Int.,* 50, 451–458.
Makgoba, M. W. and Datta, H. K. (1992), *Eur. J. Clin. Invest.,* 22, 692–696.
Miyahara, T., Kato, T., Nakagawa, S., Kozuka, H., Sakai, T., Nomura, N., and Takayanagi, N. (1978), *Eisei Kagaku,* 24, 36–42.
Miyahara, T., Yamada, H., Takeuchi, M., Kozuka, H., Kato, T., and Sudo, H. (1988), *Toxicol. Appl. Pharm.,* 96, 52–59.
Miyahara, T., Takata, M., Miyata, M., Nagai, M., Sugure, A., Kozuka, H., and Kuze, S. (1991), *Bull. Environ. Contam. Toxicol.,* 47, 283–287.
Miyahara, T., Masakazu, T., Mori-Uchi, S., Miyata, M., Nagai, M., Sugure, A., Matsusta, M., Kozuka, H., and Kuze, S. (1992), *Toxicology,* 73, 93–99.
Miyahara, T., Komiyama, H., Mihanishi, A., *Calcif. Tissue Int.,* 54, 165–169.
Morgan, K. J., Stampley, G. L., Zabik, M. E., and Fischer, D. R. (1985), *J. Am. Coll. Nutr.,* 4, 195–206.
Morrissey, J., Rothstein, M., Mayor, G., and Slatopolsky, A. (1983), *Kidney Int.,* 23, 699–704.
Nicaud, P., Lafitte, A., and Gros, A. (1942), *Arch. Mal. Prof.,* 4, 192–202.
Nishizumi, M. (1972), *Arch. Environ. Health,* 24, 215–225.
Noda, M., Ed. (1993), *Cellular and Molecular Biology of Bone,* Academic Press, San Diego, 1–567.
Nogawa, K. (1981), in *Cadmium in the Environment,* Nriagu, J. O., Ed., John Wiley & Sons, New York, 1–37.
Nogowa, K., Tsuritani, I., Kido, T., Honda, R., Ishizaki, M., and Yamada, Y. (1990), *Int. Arch. Occup. Environ. Health,* 62, 189–193.
Nomiyama, K., Sato, C., and Yamamoto, S. (1973), *Toxicol. Appl. Pharmacol.,* 24, 625–636.
Nomiyama, K., Sugata, Y., Yamamoto, A., and Nomiyama, H. (1975), *Toxicol. Appl. Pharmacol.* 31, 4–12.
Nordal, K. P., Dahl, E., Halse, J., Aksnes, L., Thomassen, Y., and Flatmark, A. (1992), *J. Clin. Endocrinol. Metab.,* 74, 1140–1145.
Nusbaum, R. E., Butt, E. M., Gilmour, T. C. and DiDio, S. L. (1965), *Arch. Environ. Health,* 10, 227–232.
Nye, L. J. J. (1929), *Med. J. Austral.,* 2, 145–159.
Ogoshi, K., Moriuama, T., and Nanzai, Y. (1989), *Arch. Toxicol.,* 63, 320–324.
Ogoshi, K., Moriyama, T., and Nanzai, Y. (1990), *Toxicol. Environ. Chem.,* 27, 97–103.
Ogoshi, K., Nanzai, Y., and Moriuama, T. (1992), *Arch. Toxicol.,* 66, 315–320.
Oner, G., Bhaumick, B., and Bala, R. M. (1984), *Endocrinology,* 114, 1860–1863.
Puzas, J. E., Sickel, M. J., and Felter, M. E. (1992), *Neurotoxicology,* 13, 783–788.
Quarles, L. D., Gitelman, H. J., and Drezner, M. K. (1989), *J. Clin. Invest.,* 83, 1644–1650.
Quarles, L. D., Hartle, J. E., II, Middleton, J. P., Zhang, J., Arthur, J. M., and Raymond, J. R. (1994), *J. Cell Biochem.,* 56, 106–117.
Rabinowitz, M. (1991), *Environ. Health Perspect.,* 91 33–37.
Report of the Task Group on Reference Man, ICRP Pub. 23, Snyder, W. S., Cook, M. J., Karhausen, L. R., Nasset, E. S., Howells, G. P., and Tipton, I. H., Eds., (1975), Pergamon Press, New York.
Ruoslahti, E. (1981), *Annu. Rev. Cell Biol.,* 4, 229.
Sacco-Gibson, N., Chaudhry, S., Brock, A., Sickles, A. B., Patel, B., Hegstad, R., Johnston, S., Peterson, D., and Bhattacharyya, M. (1992), *Toxicol. Appl. Pharmacol.,* 113, 274–283.
Sauk, J. J., Smith, T., Silbergeld, E. K., Fowler, B. A., and Somerman, M. J. (1992), *Toxicol. Appl. Pharmacol.,* 116, 240–247.
Sauk, J. J. and Somerman, M. J. (1991), *Environ. Health Perspect.,* 91, 9–16.
Schultz, A., Skerfving, S., Mattson, S., Christoffersson, J. O., and Ahlgren, L. (1987), *Arch. Environ. Health,* 42, 340–346.
Severson, A. R., Haut, C. F., Firling, C. E., Thomas, E. H. (1992), *Arch. Toxicol.,* 66, 706–712.
Shukla, R., Bornschein, R. L., Dietrich, K. N., Buncher, C. R., Berger, O. G., Hammond, P. B., and Succop, P. A. (1989), *Pediatrics,* 84, 604–612.
Shukla, R., Dietrich, K. N., Bornschein, R. L., Berger, O., and Hammond, P. B. (1991), *Pediatrics,* 88, 886–892.
Silbergeld, E. K., Sauk, J., Somerman, M., Todd, A., McNeill, F., Fowler, B., Fontaine, A., and van Buren, J. (1993), *Neurotoxicology,* 14, 225–236.
Silbergeld, E. K., Schwartz, J., and Mahaffey, K. (1988), *Environ. Res.,* 47, 79–94.
Silbergeld, E. K. (1991), *Environ. Health Perspect.,* 91, 63–70.
Silbergeld, E. K. and Watson, L. (in press), *Fundam. Appl. Toxicol.,*
Slemenda, C. W., Hui, S. L., Longscope, C., and Johnson, C. C., Jr. (1989), *J. Bone Min. Res.,* 4, 737–741.

Somervaille, L. J., Chettle, D. R., Scott, M. C., Krishnan, G., Brown, C. J., Aufderheide, A. S., et al. (1987), in *In Vivo Body Composition Studies,* Ellis, K. J., Yasumura, S., and Morgan, W. D., Eds., Institute of Physical Sciences in Medicine, London (IPSM3), 325–333.
Somervaille, L. J., Chettle, D. R., Scott, M. C., Tennant, D. R., McKiernan, M. J., Skilbeck, A. et al. (1988), *Br. J. Ind. Med.,* 45, 174–181.
Spencer, H., O'Sullivan, V., and Sontag, S. J. (1992), *J. Lab. Clin. Med.,* 120, 198–199.
Stendig-Lindberg, G., Tepper, R., and Leichter, I. (1993), *Magnes. Res.,* 6, 155–163.
Strause, L. G., Hegenauer, J., Saltman, P., Cone, R., and Resnick, D. (1986), *J. Nutr.,* 116, 135–141.
Strause, L., Saltman, P., Smith, K. T., Bracker, M., and Andon, M. B. (1994), *J. Nutr.,* 124, 1060–1064.
Suzuki, Y., Morita, I., Yamane, Y., and Murota, S. (1989a), *Biochem. Biophys. Res. Commun.,* 158, 508–513.
Suzuki, Y., Morita, I., Ishizaki, Y., Yamane, Y., and Murota, S. (1989), *Biochim. Biophys. Acta,* 1012, 135.
Takeuchi, J. (1978), in *Edited Proceedings, First Int. Cadmium Conf., San Francisco, CA,* Metal Bulletin, New York, 222–224.
Tell, I., Somervaille, L. J., Nilsson, U., Bensryd, I., Schutz, A., Chettle, D. R., Scott, M. C., and Skerfving, S. (1992), *Scand. J. Work Environ. Health,* 18, 113–119.
Thompson, G. N., Robertson, E. F., and Fitzgerald, S. (1985), *Med. J. Aust.,* 143, 131.
Todd, A. C., McNeill, F. E., and Fowler, B. A. (1992a), *Environ. Res.,* 59, 326–335.
Todd, A. C., McNeil, F. E., Palethorpe, J. E., Peache, D. E., Chettle, D. R., Tobin, M. J., Strosko, S. J., and Rosen, J. C. (1992b), *Environ. Res.,* 57, 117–132.
Tranquilli, A. L., Lucino, E., Garzetti, G. G., and Romanini, C. (1994), *Gynecol. Endocrinol.,* 8, 55–58.
Tsuboi, S., Nakagaki, H., Ishiguro, K., Kondo, K., Mukai, M., Robinson, C., and Weatherell, J. A. (1994), *Calcif. Tissue Int.,* 54, 34–37.
Tsuritani, I., Honda, R., Ishizaki, M., and Yamada, Y. (1992), *J. Toxicol. Environ. Health,* 37, 519–533.
U. S. Department of Agriculture (1985), Nationwide food consumption survey: continuing survey of food intakes by individuals, (USDA Reps. 85–84 and 85–85). Human Nutrition Information Services, U. S. Department of Agriculture, Washington, D.C.
Valetas, P. (1946), *Cadmiose or Cadmium Intoxication,* doctoral thesis, School of Medicine, University of Paris.
Van Den Berg, G. J., Van Houwelingen, F., Lemmens, A. G., and Beynen, A. C. (1992), *Biol. Trace Elem. Res.,* 35, 77–79.
Wallach, S. (1988), *Magnesium,* 7, 262–270.
Wallach, S. (1990), *Magnes. Trace Elem.,* 9, 1–14.
Wallach, S. (1991), *Magnes. Trace Elem.,* 10, 281–286.
Wang, C. and Bhattacharyya, M. H. (1993), *Toxicol. Appl. Pharmacol.,* 120, 228–239.
Watanabe, M., Shiroishi, K., Nishino, H., Shinmura, T., Murase, H., Shoji, T., Naruse, Y., and Kagamimori, S. (1986), *Environ. Res.,* 40, 25–46.
Weberg, R. and Berstad, A. (1986), *Eur. J. Clin. Invest.,* 16, 428–432.
Whang, R. (1987), *Am. J. Med.,* 82, 24–29.
Wielopolski, L., Ellis, K. J., Vaswani, A. N., Cohn, S. H., Greenberg, A., Puschett, J. B., Parkinson, D. K., Fetterolf, D. E., and Landrigan, P. J. (1986), *Am. J. Indust. Med.,* 9, 221–226.
Williams, D. C. and Frolik, C. A. (1991), *Int. Rev. Cytol.,* 126, 195–291.
Wolinsky, I., Klimis-Tavantzis, D. J., and Richards, L. J. (1994), in *Manganese in Health and Disease,* Klimis-Tavantzis, D. J., Ed., CRC Press, Boca Raton, FL, 115–120.
Yamaguchi, M. and Inamoto, K. (1986), *Metabolism,* 35, 1044–1047.
Yamaguchi, M., Oishi, H., and Suketa, Y. (1987), *Biochem. Pharmacol.,* 36, 4007–4012.
Yoshiki, S., Yanagisowa, T., Kimura, M., Otaki, N., Suzuki, M., and Suda, T. (1975), *Arch. Environ. Health,* 30, 559–562.
Zaidi, M., Alam, A. S. M. T., Shankar, V. S., Bax, B. E., Bax, C. M. R., Moonga, B. S., Bevis, P. J. R., Stevens, C., Blake, D. R., Pazianas, M., and Huang, C. L. H. (1993), *Biol. Rev.,* 68, 197–264.

Section IX
Reproductive and Developmental Toxicology of Metals

Overview

An Introduction to Developmental and Reproductive Toxicology of Metals

John M. Rogers

The toxicology of metals is a fascinating field that has engaged many toxicologists around the world in recent decades, as can be appreciated by the depth and breadth of the scientific information presented in this volume. Nowhere is this more evident than among developmental toxicologists and teratologists. Events demonstrating the teratogenicity of methylmercury and the developmental toxicity of lead in humans rank among the most tragic toxic episodes, while recognition of these effects and subsequent efforts to prevent their recurrence rank among our greatest public health triumphs. The realization of the affects of these metals in humans provided impetus to the then-emerging fields of developmental and reproductive toxicology.

Mercury, cadmium, and arsenic were among the best-studied experimental teratogens of the 1960s and 1970s, as was deficiency of the essential trace metal, zinc. The developmental pathology observed, as well as maternal metabolism, transport across and toxicity to the placenta, and embryo/fetal accumulation and metabolism of metals was described during this period. Beginning during this period and continuing to the present, basic research has progressed in pursuit of the mechanisms underlying the developmental effects of metals. The physicochemical properties of the metals and their compounds define a wide range of interactions with biological molecules, leading to striking differences in metabolism, transport, and toxic mechanisms. Inorganic and organic forms of metals usually display different pharmacokinetics in the body, and the pharmacodynamics of these forms can be equally diverse in terms of both their cell and molecular targets and modes of action.

The aim of this section is to review what is known about the developmental toxicity of metals, with emphasis on mechanisms underlying toxic effects. There is evidence that many of the well-known toxic metals may be essential in ultratrace quantities, while interactions of toxic metals with essential trace elements, based on common transport mechanisms or competition for coenzyme/cofactor sites, probably represent a common mechanism of metal toxicity. These interactions make maternal nutritional status of critical importance in determining the toxic manifestations of exposure to toxic metals during pregnancy, as will be exemplified in a number of the chapters in this section. In the first chapter, these topics are reviewed, to place the subsequent discussions of metal toxicity in the context of their known or potential roles in normal biology.

The transport of metals across the placenta and metabolism of metals by the placenta are reviewed in Chapter 64. The interface between the mother and the conceptus changes as development proceeds, and the ability of some metals to reach the conceptus changes concordantly. Toxic effects of metals on the placenta can impede maternal-fetal exchange processes, and such placental insufficiency can be the ultimate cause of developmental toxicity. The isolated, dually perfused human placenta affords a unique opportunity to work with normal human tissue and examine toxicant transport and metabolism as well as placental toxicity. Dr. Miller's laboratory has utilized this system extensively to study metal toxicology, and these studies are summarized in his chapter.

Several of the most-studied metals (mercury, cadmium, arsenic) will be reviewed in depth, and metals for which less data exist but which are of contemporary concern (aluminum, manganese, uranium, vanadium) will be surveyed for recent findings. Other chapters in the section take the approach of summarizing what is known about metal toxicity to a particular organ system (placenta, developing nervous tissue) and/or in emerging areas of research pertinent to a number of the metals (ontogeny of constitutive and inducible metallothionein synthesis, membrane effects, cell adhesion molecules, actions and effects of chelators, lactational exposure to metals). Although not the subject of an individual chapter in the section, the developmental toxicity and modes of action of lead are discussed in Chapters 63, 68, and 69, as well as in chapter 39 by Rice and Silbergeld in Part I and by Needleman in Chapter 24 as well as by Davis and Elias in Chapter 4. In addition, the evidence for lead as a human teratogen has been very recently reviewed (Bellinger, 1994).

In reading these chapters, one finds that a great deal of progress has been made towards elucidation of the mechanisms through which metals can induce developmental or reproductive toxicity. Yet, it is just as clear that much work remains to be done. Metals are natural constituents of the environment as well as ubiquitous anthropogenic pollutants, and detecting and preventing their adverse developmental and reproductive effects on the world population continues to be an important and rewarding endeavor.

REFERENCE

Bellinger, D. (1994), Teratogen update: lead, *Teratology,* 50, 367–373.

Chapter 63

Teratogenic Effects of Essential Trace Metals: Deficiencies and Excesses

Carl L. Keen

I. INTRODUCTION

While a large number of trace elements have been found in animal and human tissues, relatively few are generally accepted to be essential; a list of these elements would include iron, zinc, copper, manganese, iodine, selenium, fluoride, chromium, and molybdenum. As would be predicted by the term "essential," a prolonged deficit of any of the above elements in the diet of the adult organism will result in morbidity and, potentially, death. If the deficiency occurs during prenatal or early postnatal life, it can result in profound disturbances of development, including prenatal and neonatal death, congenital abnormalities, low birth weight, and functional disturbances of systems including the neurological, cardiac, pulmonary, and immune systems. Behavioral abnormalities may also result. In this chapter, our current understanding of the mechanisms underlying the effects of essential mineral deficiencies on developmental processes will be reviewed. In addition, the effects of excessive intakes of these essential trace elements on development will be considered. Finally, the developmental effects of a potential deficiency, or toxicity, of arsenic, boron, nickel, silicon, and vanadium, will be briefly discussed, as some have argued that these trace elements are also essential in the diet of mammals. It should be noted that owing to space constraints, this review will be restricted to a discussion of the effects of these metals during the prenatal development of mammals. In addition, in many places review articles have been cited rather than primary references.

II. MECHANISMS UNDERLYING THE DEVELOPMENT OF ESSENTIAL TRACE METAL DEFICIENCIES

Prior to a discussion of the developmental effects associated with deficits of specific essential metals, it is important to recognize the multiple ways by which a metal deficiency may arise (Table 1). First, a deficiency can occur as a consequence of an inadequate dietary intake of the metal (primary deficiency). An insufficient amount of any essential nutrient in the diet (solid or liquid) will eventually result in a primary deficiency of the nutrient, and potentially the death of the animal.

A secondary, or conditioned, deficiency may occur even if the dietary content of the essential nutrient would normally be adequate (Couzy et al., 1993). Conditioned deficiencies can arise by several means. First, genetic factors may create a higher than normal requirement for a nutrient. For example, individuals with the genetic disorder acrodermatitis enteropathica require a high amount of zinc in their diet due to a genetic defect in zinc absorption (Walsh et al., 1994). In addition to single gene defects, there can be

Table 1 Causative Factors in Trace Metal Deficiencies

Primary deficiency — inadequate dietary intake
Secondary ("conditioned") deficiency
    Genetic factors
      Mutant genes
      Strain differences
    Nutritional interactions
      Dietary binding factors
      Trace metal interactions with other nutrients
      Trace metal–trace metal interactions
        Metabolic interactions
        Physiochemical interactions
    Drugs or other chemicals
      Direct metal chelation
      Altered metal metabolism
    Disease associated changes in metal metabolism

significant strain differences in the response to nutritional deficiency. These differences are normally the result of subtle variations in numerous genes. Representative of a strain variation is the influence that the breed of a sheep can have on its susceptibility to the teratogenic effects of copper deficiency (see below).

Second, nutritional interactions can produce conditioned deficiencies. These interactions can be of several types. For example, dietary binding factors, such as fiber and phytate (myoinositol hexaphosphate), can form complexes with essential metals in the gut, which limit their absorption; this type of interaction can be dramatic, as is evidenced by the occurrence of a syndrome of hypogonadal dwarfism in the Middle East (Rucker et al., 1994). This zinc-deficiency syndrome arises in part as a consequence of the consumption of diets which are rich in phytates; significantly, individuals with this syndrome can be characterized by having levels of dietary zinc which on the surface appear to be adequate (Walsh et al., 1994).

Interactions between two or more minerals can also result in a "deficiency" of one of the minerals. Mineral–mineral interactions can occur in a variety of ways. First, if one mineral is involved in the metabolism of another, a deficit of the first may influence the metabolism of the second. Illustrative of the above is the iron-refractory anemia which can occur with copper deficiency. This anemia is the result, in part, of a reduction in the activity of the copper-containing protein ceruloplasmin (ferroxidase), which facilitates the incorporation of iron into transferrin (Harris et al., 1995); thus, one consequence of copper deficiency can be a secondary iron deficiency.

A second important type of metal–metal interaction is when two elements share a common transport site or ligand. Hill and Matrone (1970) developed the theoretical basis for predicting such direct mineral interactions and provided considerable experimental data which supported the validity of the concept. According to their model, ions which have similar orbitals, size, and coordination numbers, can be predicted to be antagonistic to each other in biological systems. Table 2 shows the chemical determinants of some potential trace metal interactions. Copper, zinc, and cadmium, with similar orbitals, configurations, and coordination numbers, are predicted to interact, and indeed do. Some of the consequences of such interactions are discussed below. Similarly, one would predict that iron and manganese would interact, and as is discussed below, this is the case. It is important to note that with regard to these interactions, they can occur at numerous locations. For example, the interference can be located at the intraluminal level (due to competition for a transport ligand), within the intestinal cell (due to competition for a storage/transport ligand), within the blood pool (due to competition for a transport ligand/protein), and/or at a site of subsequent tissue uptake, storage, or excretion. In theory, such tissues can include the placenta and the yolk sac, as well as embryonic and fetal membranes.

As is discussed in the following sections, metal–metal interactions of the nature described above can significantly influence the transport of metals into the developing conceptus. Finally, it should be emphasized that, while these interactions can result in the induction of a "deficiency" of an essential trace metal, the same type of interactions can result in the excessive tissue accumulation of nonessential metals. The above can occur when there is a primary dietary deficiency of an essential trace metal which

Table 2 Chemical Determinants of Potential Interactions

Ion	Orbital	Coordination Number	Group
Cu^+	d^{10}	4	Ib
Zn^{2+}	d^{10}	4	IIb
Fe^{2+}	d^6	6	VIII
Fe^{3+}	d^5	6	VIII
Mn^{2+}	d^5	6	VIIa
Cd^{2+}	d^{10}	4	IIb
Cu^{2+}	d^9	4	Ib

has physiochemical properties similar to those of the nonessential metal. Illustrative of the above is the increased absorption of cadmium and lead which can occur with dietary zinc deficiency (Goyer, 1995).

A third mechanism by which a conditioned essential metal deficiency can arise is through an effect of drugs, or other chemicals, on the metabolism of the metal. Broadly speaking, drug–metal interactions can be separated into two categories, one composed of drugs that act via direct chelation of the metal and the other composed of drugs that alter, in an indirect fashion, the normal metabolism of the metal. Representative of the first class of interaction is the chelation of copper and zinc by triethylenetetramine (TETA) (Cohen et al., 1983; Keen et al., 1983a), D-penicillamine (DPA) (Keen et al., 1983b; Mark-Savage et al., 1983), and ethylenediaminetetraacetate (EDTA) (Swenerton and Hurley, 1971), each of which has been shown to be developmentally toxic, at least in part due to induced embryonic/fetal copper and/or zinc deficiencies.

The second category of drugs/chemicals may influence mineral metabolism by multiple routes. For example, a decreased absorption of some minerals can occur as a consequence of drug-associated intestinal cell damage or drug-induced reductions in gastrointestinal transit time. Similarly, diuretics can cause an increased urinary excretion of essential metals (Murray and Healy, 1991). Finally, as is discussed below in the section concerning zinc (Section III), drugs/chemicals can alter the metabolism of numerous essential metals via the acute phase response; in select cases, these alterations can result in transitory embryonic metal deficiencies which are teratogenic (Taubeneck et al., 1994).

A fourth and final means by which a conditioned essential metal deficiency may arise is in situations where an excessive loss of the metal occurs as a consequence of an underlying disease. Illustrative of the above would be the excessive copper and zinc losses which can be associated with chronic diarrhea, and/or diabetes-associated polyuria (Ruz and Solomon, 1990; Walter et al., 1991).

III. ZINC

A. TERATOGENESIS OF ZINC DEFICIENCY

Zinc deficiency during gestation is known to be teratogenic in many species, including rats, mice, guinea pigs, sheep, and chickens, and it is thought to be teratogenic in humans (Keen and Hurley, 1989; Goldenberg et al., 1995; Keen et al., 1995). Typical abnormalities associated with severe zinc deficiency in the rat (the most frequently used experimental animal model) include: cleft lip and palate, brain and eye malformations, and numerous abnormalities of the heart, lung, and urogenital system. The most common brain malformation observed is exencephaly. Skeletal variations including cranial, vertebral, rib, sternum, long bone, and digit aberrations are also common. Typically, the majority of fetuses in litters from dams fed a diet severely deficient in zinc (≤ 0.5 µg Zn/g compared to control diets which typically contain between 25 to 100 µg/g) throughout gestation exhibit one or more of these abnormalities. In addition to gross structural defects, functional deficits, such as behavioral and biochemical alterations including defects in pancreatic function, lung function, and immune competence, can characterize the offspring of zinc-deficient mothers (Keen and Hurley, 1989; Keen and Gershwin, 1990; Apgar, 1992; Golub et al., 1995).

While the above fetal and neonatal abnormalities are typically associated with severe maternal zinc deficiency throughout pregnancy, transitory periods of zinc deficiency can also have marked negative effects. Severe zinc deficiency for a week or more prior to mating typically results in anestrous, disrupted mating patterns, or lack of fertilization. Two-cell embryos obtained from mice fed a zinc-deficient diet for 3 or 6 days during oocyte maturation and fertilization can be characterized by altered development even when they are cultured in media with supplemental zinc (Peters et al., 1991); consistent with this, oocytes removed on day 3 of gestation from zinc-deficient rats are typically growth retarded and characterized by abnormal cell divisions (Hurley and Schrader, 1975). If the zinc deficiency is imposed during different discrete periods of organogenesis, the organ systems under development at that time will be affected (Hurley et al., 1971).

The rapidity of the effects of maternal zinc deficiency on embryonic and fetal development suggest that, on an acute basic, the plasma concentrations of this metal are not homeostatically controlled. Consistent with this, it has been shown that in the rat, within 24 hours after the introduction of a zinc-deficient diet, plasma zinc concentrations can decrease by over 40% from control plasma zinc concentrations that range from 10 to 15 μM (Hurley et al., 1982). It is important to note that high concentrations of zinc are found in bone and soft tissue. As a consequence of the above, if higher than normal amounts of tissue catabolism occur (as may be the case if the animal is anorexic), plasma zinc concentrations for short periods of time can be relatively normal, even though the animal may be consuming a zinc-deficient diet.

It is evident from the above that severe maternal zinc deficiency can result in embryonic and fetal abnormalities; however, the mechanisms underlying these defects are not well understood. It is recognized that some of the effects of zinc deficiency on the embryo/fetus are due to indirect effects of the deficiency on the metabolism of the mother. For example, one effect of zinc deficiency can be a reduction in maternal food intake; based on pair feeding studies, this reduction can contribute to the fetal growth retardation associated with zinc deficiency (Keen and Hurley, 1989). However, it is important to note that maternal food restriction alone rarely results in the occurrence of gross structural defects; thus this cannot be the cause of the multiple diverse abnormalities obtained with zinc deficiency (Keen and Hurley, 1989). On the maternal side, zinc deficiency can also result in marked changes in the production and/or action of a number of hormones including estrogen, corticosterone, insulin, vitamin A, triiodothyronine, vitamin D, and growth hormone (Bunce, 1994; Harris et al., 1995; Peters et al., 1995). Given the above, it is reasonable to suggest that some of the teratogenic effects associated with maternal zinc deficiency are due in part to the altered hormonal milieu associated with a deficit of this nutrient. However, there is a growing body of evidence which supports the idea that zinc deficiency also directly affects the embryo/fetus. This evidence is drawn in part from work using *in vitro* postimplantation rat embryo culture model systems. In a number of these studies, it has been shown that embryos grown in serum with low concentrations of zinc (≈ 5 μM) develop abnormally, displaying many of the same defects which occur in embryos exposed to zinc deficiency *in vivo*. Significantly, when embryos are grown in zinc-deficient serum supplemented with zinc to concentrations similar to those found in control serum, they typically develop normally, providing evidence that the abnormal development observed in embryos grown in the unsupplemented serum is a direct consequence of its lack of zinc (Mieden et al., 1986; Daston et al., 1994; Taubeneck et al., 1995). Given that most embryo culture studies only last 48 to 72 hours, the above results also show the rapidity with which a transitory deficiency of zinc can result in abnormal development. Jornvall et al. (1994) have reported that zinc deficiency also directly affects the development of *Xenopus laevis;* however, in this work the "deficiency" was induced through the use of 1,10-phenanthroline, which chelates available zinc in the media. Since this chelator can also bind iron, it is possible that some of the defects observed in their study (craniofacial, urogenital, and skeletal) may have been due to a combined deficiency of zinc and iron. In addition, a direct effect of the chelator on the embryo cannot be ruled out.

Given the above, it is evident that zinc deficiency represents a serious challenge to the developing embryo. Mechanistically, there are multiple developmental processes which can be affected by a deficit of this nutrient. Considerable attention has been given to the idea that one of the basic defects underlying the abnormalities observed in zinc deficiency is altered nucleic acid metabolism. DNA synthesis is significantly depressed in zinc-deficient embryos and fetuses compared with controls, and this lower rate of synthesis has been linked to low activities of DNA polymerase and thymidine kinase (Chesters, 1989; Keen and Hurley, 1989; Dreosti, 1993; Falchuk, 1993; Vallee and Falchuk, 1993). If an impairment in nucleic acid synthesis or expression is sufficient, this could result in alterations in the differential rates of cellular growth which are necessary for normal morphogenesis.

In addition to alterations in nucleic acid metabolism, several other metabolic defect may also be occurring as a consequence of cellular zinc deficiency. Oteiza et al. (1989) have shown that brain tubulin polymerization rates can be markedly reduced in zinc-deficient fetuses. The mechanism by which zinc deficiency affects tubulin polymerization has yet to be established; however, zinc is known to stabilize neurotubules *in vitro* through the formation of zinc-mercaptide bridges between tubulin dimer units (Nicholson and Veldstra, 1972). In theory, impaired tubulin polymerization could have multiple effects including alterations in normal cellular endo- and exocytosis, as well as delays in the migration of cells involved in pattern formation.

There has also been considerable interest in the idea that an early consequence of zinc deficiency may be a marked increase in apoptosis in specific regions of the embryo, and that this in turn can result in select abnormalities. Consistent with this, several investigators have reported that even transitory periods of zinc deficiency (i.e., less than 4 days), can result in significant increases in embryonic cell death (Record et al., 1985, 1986; Jankowski et al., 1995; Rogers et al., in press). Significantly, it has been reported that premigratory neural crest cells may be particularly sensitive to zinc deficiency-induced apoptosis (Rogers et al., in press); this observation could help to explain the severe dysmorphogenic effects associated with a deficit of this element. As would be predicted, the patterns of embryonic cell death which occur with zinc deficiency can be markedly affected by short-term ($\simeq$ 24 h) changes in maternal plasma zinc concentrations (Jankowski et al., 1995; Rogers et al., in press), a finding which supports the idea that the increased apoptosis observed in the deficient embryos is a very early lesion. Mechanistically, it is not yet known how zinc deficiency triggers apoptosis. Like the findings with zinc-deficient embryos, a deficiency of this metal has been shown to result in apoptosis in the lymphoid Raji and myeloid HL-60 cell lines (Martin et al., 1991). Potential targets for zinc in regulating the apoptotic cascade include calcium antagonism (Brewer, 1980), Ca^{2+}/Mg^{2+} endonuclease inhibition (Waring et al., 1990), and zinc finger proteins involved in the apoptotic cascade.

Increased oxidative damage is another mechanism by which insufficient zinc may impact developmental processes. Zinc participates in oxidant defense systems in several ways. First, it is a component of the enzyme copper-zinc superoxide dismutase. Second, as a non-redox active metal, it can occupy sites which might otherwise be occupied by redox active metals (such as iron and copper) that can participate in the Fenton reaction. It is well established that one consequence of embryonic/fetal zinc deficiency can be an increase in fetal iron concentrations (Reinstein et al., 1984; Rogers et al., 1987). Third, zinc binds avidly to, and induces the synthesis of metallothionein, a protein with oxygen radical scavenging potential (Zidenberg-Cherr and Keen, 1991). Consistent with the above, zinc deficiency is associated with increased lipid peroxidation in numerous tissues including fetal liver (Dreosti et al., 1985). Zinc deficiency is also associated with increases in protein and DNA oxidative damage (Oteiza et al., 1995). While a definitive role for excessive oxidative damage in the development of zinc deficiency-related defects has yet to be proven, it seems reasonable to assume that it is contributory.

Finally, it can be speculated that some of the biochemical and morphological lesions associated with zinc deficiency probably arise as a consequence of the critical role that zinc plays in the functioning of over 150 transcription factors. During the past decade it has been recognized that numerous transcription factors contain polypeptide sequences, approximately 30 residues in length, in which a loop is stabilized into a DNA binding domain through the coordination of zinc atoms in cysteine-rich domains. The type of binding region described above is now typically referred to as a zinc finger. A critical question in developmental biology is whether fluxes in intracellular zinc concentrations can be of sufficient magnitude to influence the binding of zinc by the finger regions (O'Halloran, 1993; Cousins, 1994). If this is the case, it would provide an additional pathway by which zinc deficiency could influence a number of developmental processes.

B. PRIMARY VS. SECONDARY ZINC DEFICIENCY TERATOGENESIS

As was discussed earlier, essential metal deficiencies can arise through a variety of means. While the majority of experimental work on zinc deficiency-induced teratogenicity has been done using primary deficiency models (i.e., the deficiency is induced by feeding zinc-deficient diets), secondary zinc deficiencies may be relatively common and contribute to the teratogenic expression of a number of other insults and conditions. For example, a number of drugs, including DPA, TETA, EDTA, and acetazolamide, can impact zinc metabolism via the chelation, increased excretion and/or decreased absorption of the metal; the teratogenicity of each of these drugs can be influenced by the animal's zinc status (Keen et al., in press). With regard to different disease states, secondary zinc deficiencies have been postulated to contribute to the developmental abnormalities associated with diabetes and alcoholism (Flynn et al., 1981; Uriu-Hare et al., 1989; Dreosti, 1993).

Zinc metabolism can also be disrupted secondary to maternal exposure to stressors, including drugs, toxicants, and other environmental challenges, which trigger an acute phase response involving the induction of the synthesis in maternal liver of the metal-binding protein metallothionein (Keen, 1992). For several agents including 6-mercaptopurine, urethane, valproic acid, α-hederin, arsenate, ethanol and tumor necrosis factor-α (TNF-α), it has been shown that the induction of metallothionein in maternal liver can be of sufficient magnitude that, as a consequence of its binding available zinc in the liver, it results in low maternal plasma zinc concentrations and a subsequent reduction in the transport of zinc into the embryo (Daston et al., 1994; Taubeneck et al., 1994, 1995). That this reduction in zinc transport to the embryo can represent a severe risk is supported by two sets of observations. First, the teratogenicity of agents which can induce this acute phase response can be modulated in part by the dietary zinc intake of the mother (Taubeneck et al., 1995). Second, it has been observed that serum collected from rats undergoing an acute phase response (as evidenced in part by low serum zinc concentrations), can be teratogenic in *in vitro* rat embryo cultures. The teratogenicity of this serum can be ameliorated by the supplementation of zinc (Daston et al., 1994; Taubeneck et al., 1995). An important conclusion which can be drawn from the above is that secondary zinc deficiencies may be very common, and as such may contribute to the teratogenicity of a diverse group of insults. Indeed, an induction of a transitory embryonic/fetal zinc deficiency may be a common mechanism which underlies the developmental complications which occur with maternal "toxicity" syndromes (Keen, 1992).

Finally, embryonic/fetal zinc deficiency may arise as a consequence of cadmium interfering with the transport of zinc into the conceptus. Indeed, it has been argued that cadmium-induced changes in zinc metabolism might be the major mechanism by which cadmium influences fetal development (Samarawicknama and Webb, 1979; Daston, 1982; Simmer et al., 1992).

C. TERATOGENESIS OF EXCESS ZINC

Although rare, incidences of acute zinc toxicity in humans resulting from high intakes of zinc have been reported. Isolated outbreaks of zinc toxicity have occurred as a result of the consumption of foods and beverages contaminated with zinc released from galvanized containers. Typical signs of acute zinc toxicosis include epigastric pain, diarrhea, nausea, and vomiting (Fosmire, 1990). Signs of toxicity typically disappear within 24 hours after the removal of the insult. The long-term consumption of zinc supplements in excess of 150 mg/d has been reported to result in low plasma and erythrocyte copper concentrations, low serum HDL concentrations, gastric erosion, and depressed immune function (Fosmire, 1990). In a brief report on four women who took 300 mg zinc per day during the third trimester of pregnancy, it was stated that three gave birth prematurely and a fourth infant was stillborn; an explanation for this high rate of pregnancy complication was not given (Kumar, 1976).

The major consequence associated with the long-term ingestion of moderately high amounts of zinc is the induction of a secondary copper deficiency. It is well documented that for humans the chronic intake of zinc supplements as low as 50 mg/d can result in a marginal copper deficiency as assessed by reductions in plasma copper concentrations and reductions in the activity of erythrocyte copper-zinc superoxide dismutase (Sandstead, 1995). Given that the 1989 Recommended Dietary Allowance for zinc during pregnancy is 15 mg/d, it is evident that the margin of safety for this metal between deficiency and toxicity is small. Mechanistically, zinc is thought to inhibit copper absorption as a consequence of copper's binding to metallothionein in intestinal cells, with the metallothionein being induced by the high concentrations of zinc in the cell (Cousins, 1994).

In experimental animals, acute zinc toxicity has not been associated with any reproducible pattern of developmental abnormalities (Ferm, 1972; Chang et al., 1977). When pregnant rats are fed high concentrations of dietary zinc (>1000 μg Zn/g diet compared to average control diet of 50 μg/g) throughout pregnancy, the primary effect is the induction of secondary maternal and fetal copper deficiencies, which can result in the occurrence of fetal abnormalities consistent with those associated with primary copper deficiency (see below; O'Dell, 1968; Ketchenson et al., 1969; Reinstein et al., 1984).

Concentrations of zinc in excess of 50 μM are toxic to preimplantation mouse embryos grown *in vitro*; the mechanisms underlying this toxicity have not been identified (Peters et al., 1991). Concentrations of zinc in excess of 40 μM have also been reported to be teratogenic to frog embryos as assessed by FETAX (frog embryo teratogenesis assay: *Xenopus*), defects of the eye, gut, notochord, and heart being observed (Luo et al., 1993).

IV. COPPER

A. TERATOGENESIS OF COPPER DEFICIENCY

The importance of copper for the prenatal development of mammals was shown by Bennetts and his co-workers (Bennetts and Beck, 1942; Bennetts et al., 1948) in their demonstration that enzootic ataxia, a disease affecting the developing fetus, could be prevented by giving the ewe additional copper during pregnancy. This disorder is characterized by spastic paralysis, especially of the hind limbs, severe incoordination, blindness in some cases, and anemia. Typically, the brains of animals with enzootic ataxia are small, and characterized by collapsed cerebral hemisphoresis, shallow convolutions, and a paucity of normal myelin (see Hurley and Keen, 1979). Similar neonatal ataxia and brain abnormalities have been reported in newborn copper-deficient goats, swine, guinea pigs, and rats (Hurley, 1981). Copper deficiency, as evidenced by low plasma copper concentrations, can be induced in most mammalian species by feeding a deficient diet (<1 μg Cu/g compared to typical control diets of 8 to 15 μg/g) for a 2 to 4 week period.

The biochemical lesions underlying the brain abnormalities associated with copper deficiency have not been firmly agreed upon. One significant lesion may be a reduction in the activity of the cuproenzyme, cytochrome oxidase. Mills and Williams (1962) have demonstrated that with copper deficiency, the activity of cytochrome oxidase is significantly reduced in the large motor neurons of the red nucleus of the brain, an area where degeneration is often particularly striking. It can be argued that if the reduction in cytochrome oxidase activity is sufficient, an anoxia will develop that will result in tissue death. Furthermore, if the production of ATP is inadequate for normal phospholipid synthesis, this could explain in part the high amount of amyelination typically observed in brains of copper-deficient fetuses and neonates (Hurley and Keen, 1979).

A second mechanism which might be contributing to the brain abnormalities associated with copper deficiency is excessive cellular oxidative damage. Brain copper-zinc superoxide dismutase activity has been reported to be low in copper deficient fetuses; however, peroxidation of brain lipids has not been reported to be particularly high (Prohaska and Wells, 1975). One explanation for the lack of increased lipid peroxidation in the copper-deficient fetal brain might be that the fatty acid profile has shifted to a less peroxidizable fatty acid composition. That the amount and type of polyunsaturated fatty acids in adult tissues can change in response to oxidative insults is well established, and presumably this shift reflects a compensatory response to the insult (Zidenberg-Cherr and Keen, 1991). Whether or not a condition of high oxidative pressure can also influence the type and amount of polyunsaturated fatty acids in embryonic and fetal tissue has yet to be established. If the fatty acid profile is changed, this in itself may contribute to abnormal development.

Hawk and co-workers (Hawk et al., 1995) have reported that the ability of gestation day 12 rat embryos to metabolize reactive oxygen species is impaired by copper deficiency. In this work, embryos from copper-sufficient and copper-deficient dams were explanted on gestation day 10, and cultured for 48 h in either copper-adequate ($\simeq$10 to 15 μM) or copper-deficient serum ($\simeq$2 to 5 μM). While copper-adequate embryos cultured in copper-adequate serum developed normally, embryos from copper-deficient dams cultured in copper-deficient serum were characterized by swollen hind brains, blood pooling, and distension of the large vessels, particularly the anterior cardinal veins. As stated above, the copper-deficient embryos were characterized by a compromised free radical defense system with a reduction in copper-zinc superoxide dismutase activity. That this compromised oxidant defense system can be a factor in the teratogenicity of copper deficiency is suggested by the observation that the addition of reactive oxygen scavengers to the culture media resulted in improved embryo development (Hawk et al., 1995). Despite an apparent copper deficiency-induced increase in reactive oxygen species, in a recent study by Jankowski et al. (1995), copper-deficient embryos were not characterized by evidence of excessive cell death. It is important to note that the work by Hawk et al. (1995) was done using an *in vitro* rat embryo culture model, while the work by Jankowski et al. (1995) was done using an *in vivo* model. The effects of *in vivo* copper deficiency on free radical metabolism in the embryo organogenesis need to be investigated. It should be noted that in addition to the gross structural defects induced by copper deficiency, a deficit of this element during perinatal development can also result in a persistent reduction in the concentration of cupro-enzymes in the brain (Prohaska and Bailey, 1993). The functional consequences of these reductions have not been defined.

In addition to brain defects, copper-deficient fetuses and neonates are typically characterized by severe connective tissue abnormalities. Cardiac hemorrhages are a frequent finding in copper-deficient sheep, rats, guinea pigs, and mice (Hurley and Keen, 1979). The walls of the internal and common carotid arteries in deficient fetuses tend to have an endothelium which is normal in appearance, but with sparse, poorly developed elastin. Cerebral arteries are also often characterized by a low elastin content. Furthermore, the elastin which is present does not have the concise fibrillar arrangements seen in control animals. This reduction in elastin content and cross-linking integrity is thought to be due primarily to a decrease in the activity of the cupro-enzyme, lysyl oxidase, which catalyzes the oxidation of certain peptidyl lysine and hydroxylysine residues to peptide aldehydes, initiating the cross-linking mechanisms required for connective tissue stability (Rucker and Tinker, 1977).

Skeletal defects can occur as a result of copper deficiency. Lambs with enzootic ataxia may have poorly developed, light, brittle bones with frequent fractures. Bone abnormalities have been found in copper-deficient calves and fowls. In dogs and swine, it was found that young born to females fed copper-deficient diets had deformed leg bones (Hurley and Keen, 1979). The lesions appeared to be associated with an impairment of osteogenesis, with the resultant thinning of the cortex and trabeculae of the long bones. Copper-deficient chicks are different, having severe hypoplasia of the long bones. Amine oxidase and cytochrome oxidase activities are low and there is a high ratio of soluble to insoluble collagen. The increased fragility of copper-deficient bones appears to result from the low number of cross-linkages present in the collagenous matrix (Rucker and Tinker, 1977).

Finally, lung abnormalities are a frequent consequence of prenatal and early postnatal copper deficiency. Lungs from neonatal rabbits born to dams fed copper-deficient diets were characterized by low concentrations of copper and lysyl oxidase activity and high proportions of poorly cross-linked elastin and collagen. The lungs were also characterized by low concentrations of surfactant phospholipids (Abdel-Mageed et al., 1994). Similar results have been reported for copper-deficient rats (Dubick et al., 1985).

B. PRIMARY VS. SECONDARY COPPER DEFICIENCY TERATOGENICITY

A condition of severe copper deficiency can be rapidly induced through the use of a number of chelating drugs, including disulfiram, DPA, TETA, and meso-2,3-dimercaptosuccinic acid (DMSA) (Salgo and Osten, 1974; Keen et al., 1983a,b; Taubeneck et al., 1992). Each of the above drugs has been shown to be teratogenic, with the abnormalities produced being reminiscent of those induced by dietary copper deficiency. While the teratogenicity of the above drugs can be modulated by the amount of copper in the mother's diet (Cohen et al., 1983; Mark-Savage et al., 1983), it is important to note that drugs which bind copper typically also bind zinc; thus, the teratogenic expression of these drugs may be due to a combination of embryonic/fetal copper and zinc deficiencies.

In contrast to zinc, drugs/chemicals which induce an acute phase response in the mother do not necessarily influence copper uptake by the embryo/fetus, and in most cases, fetal copper concentrations have been found to be unaffected by chemicals which induce transitory acute phase responses (Taubeneck et al., 1994). This is not surprising, given the fact that maternal plasma copper concentrations are increased during an acute phase response (due to the hepatic production and release of the copper enzyme, ceruloplasmin), rather than decreased, as is the case for zinc. Ceruloplasmin has been postulated to be directly involved in copper transport to the embryo/fetus (Lee et al., 1993).

There are a number of diseases which can be characterized by alterations in copper metabolism, including diarrhea, diabetes, and alcoholism (Turnlund, 1994). It has been speculated that the teratogenicity associated with maternal diabetes and maternal alcoholism may be due in part to disease-induced deficiencies of copper in the embryo/fetus (Zidenberg-Cherr et al., 1988; Uriu-Hare et al., 1989).

As was discussed earlier, a secondary copper deficiency can be induced be feeding high concentrations of zinc in the diet (see above). The interaction between zinc and copper is of practical concern as it can occur with relatively low levels of zinc supplementation (Fosmire, 1990). Copper deficiency can also be induced by the feeding of high concentrations of molybdenum which, along with sulfate, can form a complex with copper which limits its absorption. The copper deficiency induced in this manner can be sufficient to pose a developmental risk (Howell et al., 1993). Recently, Shavlovski et al. (1995) have reported that embryonic copper deficiency can also occur as a consequence of maternal silver toxicity. According to these authors, by mechanisms yet defined, silver blocks the synthesis of ceruloplasmin.

Finally, a "secondary" copper deficiency may arise as a consequence of the genetic background of the animal. There are a number of strains of mice, rats, and sheep which are characterized by abnormal copper metabolism, and for several of these strains the abnormality can influence embryonic and/or fetal development.

Interactions occurring between copper and genetic factors can be classified into two groups. The first type involves strain differences which produce a differential response to diets that are deficient or marginal in copper, and the second type involves a single mutant gene, the expression of which may resemble the signs of a deficiency or a toxicity of the element. This expression may be reduced or prevented by nutritional manipulation. These two types of phenomena can interact. Thus, the phenotypic expression of a mutant gene may be modulated by the strain background.

Examples of the first type of gene–nutrient interaction are certain mutant genes in mice. For example, the mottled (Mo) mouse is characterized by a defect in cellular copper transport which is phenotypically expressed by signs of copper deficiency (Miller, 1990; Mercer et al., 1994). Over 10 alleles at the mottled locus have been described, which range in severity from hypopigmentation of hair at birth, to death in utero. The blotchy (Mo^{blo}) mutant is typically characterized by severe connective tissue defects and neurological abnormalities, similar to those observed with severe maternal copper deficiency. While the primary genetic defect in the blotchy mouse is thought to involve a mutation in a copper-transporting ATPase gene (Das et al., 1995), the phenotypic expression is thought to be the result of the reduced activities of several cuproenzymes.

Representative of the second category of gene–nutrient interaction is the observation of the influence of the breed of sheep on the incidence of enzootic ataxia within a geographical area (Weiner et al., 1978). In this regard, it has been found that the offspring of Welsh sheep often have a lower incidence of enzootic ataxia than do offspring of Blackface sheep, even when the ewes are maintained on the same pasture. This difference in the occurrence of enzootic ataxia has been correlated with the mother's ability to absorb copper. Welsh sheep absorb 20 to 40% more copper from the diet than do Blackface sheep. From embryo transfer studies (Blackface embryos to Welsh mother's and vice versa), it was firmly established that maternal blood levels in turn dictate the copper status of the newborn.

There are at least three genetic defects of copper metabolism in humans, two of which are expressed as copper deficiency syndromes (Menkes disease and occipital horn syndrome), and one which is expressed as a condition of copper toxicity (Wilson's disease). All three disorders are thought to be due in part to defects in a copper-transporting ATPase gene (Chelly et al., 1993; Mercer et al., 1993; Petrukhin et al., 1994; Das et al., 1995). Infants with Menkes disease are characterized by progressive degeneration of the brain and spinal cord, hypothermia, connective tissue abnormalities, and failure to thrive. While this disease has been recognized to be a disorder of copper metabolism for over 20 years, the prognosis for infants with this disorder is still poor and death typically occurs before 3 years of age (Danks, 1988; Turnlund, 1994). Similar to the blotchy mouse, the developmental abnormalities associated with Menkes disease are thought to be the consequence of low activities of numerous cuproenzymes during embryonic and fetal development. In contrast to patients with Menkes disease, individuals with Wilson's disease and occipital horn syndrome, are usually phenotypically normal at birth.

C. TERATOGENESIS OF EXCESS COPPER

It is known that small amounts of copper from intrauterine devices can prevent embryogenesis by blocking implantation and blastocyst development (Hurley and Keen, 1979). However, in a number of ewes where pregnancy proceeded despite the presence of a copper-containing interuterine device, teratogenic effects of copper have not been noted (Barash et al., 1990). Concentrations of copper in excess of 2.5 μM have been reported to be teratogenic to frog embryos, resulting in defects of the eye, gut, notochord, and heart (Luo et al., 1993). However, the teratogenicity of excess copper in mammals has not been firmly established; newborn rats, hamsters, rabbits, sheep, or guinea pigs which are experimentally exposed to a uterine environment of high copper are not abnormal.

Keen et al. (1982) studied the effect of feeding diets high in copper to pregnant mice under varying conditions. Mice fed diets containing up to 500 µg Cu/g had normal litters. However, mice fed diets containing 2000 µg Cu/g during pregnancy did not carry their pregnancies to term. Feeding the diet for only 5 days of pregnancy (days 7 to 12 of gestation; with a diet containing 250 µg Cu/g being fed before and after the 5-d period) resulted in a resorption frequency of more than 50%. Surviving fetuses were not visibly malformed, and their copper content was not appreciably higher than that of fetuses from dams fed diets containing 250 µg/g of copper throughout pregnancy. The high copper diet caused a severe reduction of food intake, and this caloric deprivation, rather than a direct action of copper on the fetus, was probably the cause of resorption. It has been shown that in pregnant mice, short periods of fasting (40 h) can result in litter resorption (Runner and Miller, 1956). The importance of monitoring food intake in studies assessing the teratogenic potential of a nutrient is emphasized by these results.

Ferm and Hanlon (1974) have reported that copper injected (10 µg/kg) on day 8 of pregnancy in the hamster is teratogenic. Fetal resorption, kinked-tail, thoracic and ventral hernias, microphthalmia, cleft lip, and ectopic cordis were among the abnormalities found. In rats, injections of copper from day 7 through 10 of pregnancy resulted in a resorption frequency of 50% (Marois and Bouvet, 1972).

There are no reports in the literature of teratogenesis in humans induced by excess copper. In particular, no reports could be found of abnormalities in the offspring of mothers of untreated Wilson's disease (a genetic disorder which can be characterized by very high tissue copper concentrations). However, it should be noted that pregnancy that occurs in women with untreated Wilson's disease often ends in spontaneous abortion.

V. MANGANESE

A. TERATOGENESIS OF MANGANESE DEFICIENCY

One of the major effects of prenatal manganese deficiency is on the skeleton. In rats, the offspring of dams fed a manganese-deficient diet (<1 μg Mn/g compared to typical control diets of 10 to 50 μg/g) from weaning show a disproportionate growth of the skeleton at birth, which is characterized by severe shortening of the radius, ulna, tibia, and fibula. Additional skeletal defects seen in rats are curvature of the spine, a localized dysplasia of tibial epiphysis, and anomalous development of the inner ear (Hurley, 1981; Keen et al., 1994). Manganese deficiency-associated bone defects have also been reported in young calves born to mothers which grazed on grass low in manganese. The calves were characterized by a dwarf-like appearance, joint laxity, superior branchy gnathism, and domed foreheads. Histologically, lesions were observed in the growth plates, including irregularly aligned and short columns of chondrocytes (Staley et al., 1994).

The skeletal lesions which are found with manganese deficiency are thought to be due to abnormal cartilage and bone matrix formation secondary to a reduction in the amount of essential proteoglycans. The reduction in proteoglycans is due to a reduction in the activity of one, or more, glycosyl transferases, which are manganese-activated enzymes. The reduction(s) in glycosyl transferase activity can result in both an absolute reduction in proteoglycan content, and qualitative changes in carbohydrate content as well (Leach, 1986; Keen et al., 1994; Liu et al., 1994).

A dramatic effect of prenatal manganese deficiency can be congenital irreversible ataxia, which is characterized by a lack of equilibrium and retraction of the head. This ataxia is due to the abnormal development of the otoliths in the inner ear (Erway et al., 1986). A total or partial absence of otoliths has been reported for manganese-deficient rats, mice, guinea pigs, mink, and chicks. The abnormal development of the otoliths is thought to be due to a low glycosyltransferase activity, resulting in a reduction in the proteoglycan matrix which is essential for otolith formation (Hurley, 1981).

In addition to the skeletal and inner ear lesions, manganese deficiency can result in numerous ultrastructural abnormalities in multiple tissues. Pancreatic tissue can be markedly affected, with deficient animals exhibiting aplasia or marked hypoplasia of all cellular components. The mechanism(s) underlying this pathology have not been defined, but it is thought that one mechanism involves excessive oxidative damage secondary to a reduction in the activities of manganese superoxide dismutase (Keen et al., 1994). When manganese-deficient animals are given a glucose challenge, they often respond with diabetic-type curves (Keen et al., 1994). These animals can be characterized by significant reductions in both insulin production and in insulin responsivity (Baly et al., 1985, 1988). While reductions in insulin production can in part be ascribed to the tissue damage which occurs with manganese deficiency, insulin mRNA concentrations have been shown to be low in manganese-deficient animals, even when expressed on a per islet basis (Baly, 1985). The above observation suggests that manganese has a direct role in transcription and/or message stability. While perturbations of insulin metabolism are clearly important in the postnatal manifestations of manganese deficiency, the significance (if any) of manganese-deficiency associated changes in fetal insulin metabolism is not known.

During the last decade, several human diseases have been reported to be characterized in part by low blood manganese concentrations. These diseases include epilepsy, mseleni disease, osteoporosis, and Perths' disease (Keen et al., 1994). It remains to be determined if the low blood manganese concentrations observed in these diverse diseases reflects a cause or an effect, while manganese deficiency has been cited as a possible etiologic factor for some congenital malformations in humans (Saner et al., 1985). However, this is an issue of considerable debate (Keen et al., 1994).

B. PRIMARY VS. SECONDARY MANGANESE DEFICIENCY

Secondary manganese deficiencies are rare; there are no clearly defined diseases which result in manganese deficiency, nor are drug-induced manganese deficiencies thought to be a problem. While high levels of dietary iron can interfere with manganese absorption and presumably precipitate a condition of manganese deficiency, this is thought to be a very rare event (Keen et al., 1994).

Genetic disorders that interact with manganese have been identified. The mutant gene in the pallid mouse and the gene screwneck, in mink, produce phenotypes characterized by pale coat color and ataxia caused by missing otoliths. In a series of classic studies in the area of genetic-nutrient interactions, Erway et al. (1966, 1973) demonstrated that supplementing the diet with high amounts of manganese (>1000 µg/g) during pregnancy prevents abnormal otolith development and congenital ataxia. The biochemical lesion(s) underlying the enhanced requirement of the pallid mouse and the screwneck mink for manganese have yet to be identified. Genetic defects affecting manganese metabolism during development in humans have not been reported to date.

C. TERATOGENESIS OF EXCESS MANGANESE

While high concentrations of manganese have been reported to be embryotoxic in *in vitro* models (Webster and Valois, 1987), structural malformations have not been reported in the offspring of hamsters, rats, or mice fed high levels of manganese (1 to 2000 µg/g) throughout pregnancy (Ferm, 1972; Laskey et al., 1982; Lown et al., 1984; Webster and Valois, 1987; Sanchez et al., 1993). In contrast to the dietary studies, when high amounts of manganese are given during organogenesis by either subcutaneous or intraperitoneal injection (10 to 40 mg/kg body weight), fetotoxicity can occur, as evidenced by reduced fetal body weights, delayed and reduced ossification, and minor skeletal abnormalities (Ferm, 1972; Sanchez et al., 1993). It is important to note that in the above studies, the manganese injections resulted in pronounced maternal toxicity; thus, the observed fetotoxic effects may not have been a direct consequence of the manganese per se, but rather may have resulted secondary to the physiological changes which accompanied the maternal toxicity. Significantly, teratogenic effects of manganese toxicity have not been reported for humans, despite the fact that manganese toxicosis has been documented in tens of thousands of individuals (Keen et al., 1994).

VI. IRON

A. TERATOGENESIS OF IRON DEFICIENCY

It is widely accepted that iron deficiency is the most common single-nutrient deficiency disease in the world, affecting as much as 15% of the world population (DeMaeyer and Adiels-Tegman, 1985). Depending on the criteria used, up to 15% of women of reproductive age and up to 25% of pregnant women are iron deficient (Beard, 1994). Given the above, the paucity of studies on the effects of severe iron deficiency on embryonic and fetal development is remarkable and represents an area where additional work is urgently needed.

In three early studies where low-iron diets were fed to pregnant rats, the common finding was that of low fetal hemoglobin concentrations (Alt, 1938; O'Dell et al., 1961; Murray and Stein, 1971); in the study by O'Dell et al. (1961), an additional finding was a higher than normal incidence of eye defects (7.4% compared to 3.3% in controls). Shepard et al. (1980) also observed a trend for an increased frequency of eye abnormalities (microophthalmia) in fetuses from dams fed iron-deficient diets (5 µg Fe/g vs. typical control diets of 35 g to 120 µg/g). In addition, these investigators reported a high incidence of embryonic and fetal death in the low-iron group, with the peak mortality occurring around gestation day 12. The timing of the peak mortality correlated with both the period when there is a rapid increase in red blood cells, and the time around which aerobic oxidative phosphorylation becomes important in the rat embryo (Mackler et al., 1973). In a follow-up study, Mackler et al. (1983) reported that phenylalanine, but not tyrosine, was significantly elevated in plasma iron-deficient maternal and fetal rats at gestation day 20. The concentration of total brain 5-hydroxyindoles was lower in the iron-deficient fetuses than in controls at gestation day 20, presumably as a consequence of low tryptophan hydroxylase activities. Finally, these authors reported that at gestation day 14, mitochondrial NADH oxidase activity was markedly decreased in the iron-deficient fetuses compared to controls. This latter observation would explain in part the growth retardation and high resorption frequency characteristic of severe iron deficiency in the rat. It is significant to note that in the Mackler et al. (1983) study, there was no discussion of a higher than normal frequency of eye defects occurring in the iron-deficient group.

With respect to humans, iron-deficiency anemia is generally accepted to be a significant risk factor for prematurity and low birth weight if it is present in the first trimester. Iron-deficiency anemia in the second or third trimesters has been reported to have only a minimal effect on the above parameters (Beard, 1994; Scholl and Hediger, 1994). In a study of 50,000 consecutive pregnancies followed in the National Collaborative Perinatal Project of the National Institute of Neurologic and Communicative Disorders and Stroke, Garn et al. (1981) reported that the occurrence of maternal anemia was associated with an increased risk for fetal death, short gestation length, low birth weight, and low Apgar scores; again, it is significant to note that there was no discussion of anemia being a risk factor for an increased incidence of gross development defects. In contrast to the above studies, Roszkowski and co-workers (1966), in a study of 2313 pregnancies, reported that iron deficiency during the third trimester was associated with a 2.8-fold increase in the risk for congenital malformations. However, it is critical to note that in the above study, the test for "iron deficiency" involved the measurement of serum iron concentrations. Since serum iron can be markedly reduced as a consequence of an acute phase response, it is possible that many of the cases that they studied were not iron deficient per se, but rather were individuals suffering from some other form of stress.

An important question which needs to be asked is to what extent does maternal iron deficiency influence fetal tissue iron accumulation; and do reductions in its accumulation compromise early postnatal development? Surprisingly, in numerous studies, infants of iron-depleted mothers have typically shown little evidence of anemia or depletion of iron stores (Celada et al., 1982; Milman et al., 1987; Viteri, 1994). While the above is reassuring, it is important to note that if there is a deficit of iron during early postnatal development, it can result in significant reductions in brain ferritin and nonheme iron pools. These reductions in the brain can persist, long after iron supplements are provided (Prohaska, 1987). Given the long-term behavioral consequences which have been associated with early childhood iron deficiency anemia (Pollitt, 1995), additional studies on the postnatal consequences of embryonic and fetal iron deficiency are clearly warranted.

B. PRIMARY VS. SECONDARY IRON DEFICIENCY TERATOGENESIS

With respect to the developing embryo/fetus, the only significant cause of iron deficiency other than low maternal intake of the element is diabetes (Petry et al., 1992). While diabetes-induced fetal iron deficiency has not been reported to be a risk factor for the occurrence of diabetes-associated malformations, brain and heart iron concentrations can be significantly reduced in these infants at birth. This reduction is thought to arise as a consequence of a higher than normal proportion of iron going to blood cell synthesis. The reductions in heart and brain iron concentrations have been postulated to result in impaired tissue oxygen utilization and energy metabolism secondary to impaired synthesis of cytochrome c and iron–sulfur-containing proteins (Petry et al., 1992). The above may be of particular concern with respect to the brain, since as stated above, neurodevelopmental effects associated with postnatal iron deficiency can be refractory to iron repletion (Lozoff et al., 1991; Pollitt, 1995).

Although not directly a cause of secondary iron deficiency, it should be noted that the occurrence of iron deficiency can result in an increased sensitivity to the teratogenic effects of lead (Singh et al., 1991, 1993), as these elements can share similar transport ligands.

C. TERATOGENESIS OF EXCESS IRON

As in iron deficiency, there is no clear evidence that high levels of iron represent a teratogenic risk to the developing embryo/fetus. It is important to note that for humans, high hematologic values, similar to low hematologic values, are associated with an increased risk for pregnancy complications (Garn et al., 1981). However, while this is an interesting observation, to date there is no evidence that iron toxicity plays a role in this relationship. Considerably more likely is the possibility that the high hematologic values reflect a number of diverse maternal disorders. McElhatton et al. (1991) recently reported on the pregnancy outcomes of 49 women who took iron overdoses; while there was a high frequency of pregnancy complications including fetal abnormalities among these women, the complications could not be directly linked to the period of iron toxicosis. The report by McElhatton et al. (1991) is consistent with several smaller, previous case reports (Dugdale and Powell, 1964; Richards and Brooks, 1966; Strom et al., 1976; Olenmark et al., 1987); significantly, an absence of fetal dysmorphology was reported even in the cases where the mother ultimately died as a consequence of the overload. As with the human data, no reports could be found of iron toxicity-induced developmental defects in experimental animals.

VII. SELENIUM

A. TERATOGENESIS OF SELENIUM DEFICIENCY

Chronic selenium deficiency can result in a number of pathologies, including liver necrosis in rats and swine, white muscle disease in sheep and cattle, and exudative diathesis and pancreatic atrophy in chickens (Levander, 1986). In squirrel monkeys, chronic selenium deficiency has been reported to result in alopecia, myopathy, nephrosis, and hepatic degeneration (Muth et al., 1971); however, none of these signs were observed in rhesus monkeys fed selenium-deficient diets (Butler et al., 1988).

In humans, selenium deficiency has been reported to be a critical factor in the etiology of Keshan disease, an endemic cardiomyopathy which primarily affects young children and women in parts of China. The histopathological features of this disease include multifocal necrosis and fibrous replacement of the myocardium. Myocytolysis is also often present (Levander, 1986).

A second disease which has been associated with poor selenium status in China, Siberia, and Korea is Kashin-Beck disease, a degenerative generalized osteoarthritis that primarily affects young children. The disease is characterized by joint deformation and dwarfism (Levander, 1986). The biochemical lesions underlying the various pathologies associated with selenium deficiency have not been defined. While over 20 selenoproteins have been reported to occur in mammalian tissue, only six have been purified and studied in detail. These include seleno-glutathione peroxidase, phospholipid hydroperoxide glutathione peroxidase, selenoprotein P, selenoprotein W, and type 1 deiodinase. With the exception of type 1 deiodinase, all of the selenoenzymes studied to date are thought to be involved in the metabolism of reactive oxygen species. Thus, it is reasonable to suggest that excessive cellular oxidative damage is an essential component of selenium deficiency-related pathologies (Levander, 1986). Selenium deficiency has also been postulated to contribute to the occurrence of an iodine-refractory form of goiter. Presumably, this latter disease occurs as a consequence of a low activity of type 1 deiodinase (Berry et al. 1991; Filteau et al. 1994).

While selenium deficiency disorders have been identified in animal and human populations, there is little evidence for a deficit of this element influencing prenatal development. Of particular note is the observation that a higher than normal rate of congenital defects has not been reported to occur in the regions of China characterized by a high incidence of selenium deficiency diseases. However, Hinks et al. (1989) reported that maternal white blood cell selenium concentrations in pregnancies with fetal neural tube defects were lower than those occurring in normal pregnancies. Similar data have been provided by Guvenc et al. (1995), who reported that mothers (N = 28) who also had a newborn with a neural tube defect were characterized by low serum and hair selenium concentrations compared to women with normal newborns (N = 32). Consistent with the above, the infants with the neural tube defects were also characterized by lower than normal selenium concentrations in serum and hair.

Support for the idea that selenium deficiency may be a factor contributing to the occurrence of neural tube defects is also provided by Zimmerman and Lozzio (1989), who reported that cultured fibroblasts from fetuses with neural tube defects had low selenium concentrations compared to controls.

While the results concerning the occurrence of low selenium concentrations in infants with neural tube defects are provocative, it is important to note that all of the measurements reported to date have been done using samples collected in the third trimester, or at, or after birth. Studies focusing on the selenium status of women during the first trimester when the neural tube is closing are clearly needed.

With respect to experimental animals, there is a remarkable lack of studies on the teratogenic effects of selenium deficiency. In general, rats fed low-selenium diets (<10 µg Se/kg diet compared to typical control diets of 100 to 500 µg/kg) reproduce normally, and their offspring lack gross congenital defects (Levander, 1986). In an early study by McCoy and Weswig (1969), the offspring of severely selenium-deficient dams were reported to be characterized by slow growth rates and a loss of hair. These rats were also reported to subsequently have reproductive difficulties; however, only limited information on this problem was given in the paper. For both rats and mice, the male offspring of selenium-deficient dams can be characterized by low testis and epididymal weights and immotile sperm, with many of the sperm exhibiting breakage of the tail (Wu et al., 1979; Levander, 1986). The mechanisms by which selenium deficiency affects the morphology of sperm from rodents are unknown.

Similar to rodents, chronic selenium deficiency has been linked to infertility in sheep, cattle, and pigs; again, the mechanisms underlying the poor fertility have not been defined (Levander, 1986). A common finding in the offspring of selenium-deficient ewes and cattle is white muscle disease, a form of muscular dystrophy which can affect both skeletal and cardiac tissue. The etiology of this disease is complex, and it is thought to arise as a consequence of a generalized impairment in the animal's oxidant defense system. Consistent with this, in addition to selenium, the disorder can be treated with vitamin E as well as other antioxidants (Levander, 1986). As the disease can be present at birth, it clearly can be viewed as a potential teratogenic effect of selenium deficiency. It should be noted that while the disorder is commonly described as occurring in sheep and cattle, cases have also been reported for foals, pigs, and rabbits (Levander, 1987).

B. TERATOGENIC EFFECTS OF EXCESS SELENIUM

That excess dietary selenium can be teratogenic in avian species, crustaceans, and fish is well documented; these animals show great sensitivity to even modest increases in environmental concentrations of the element (Fan and Kizer, 1990; Schuler et al., 1990; Lemly, 1993). In aquatic birds, selenium toxicity-induced defects can be dramatic and include anophthalmia, microphthalmia, limb defects, cardiac anomalies, and defects of the gastrointestinal tract (Ohlendorf et al, 1987).

In marked contrast to the sensitivity of birds and fish, high chronic intakes of dietary selenium are not normally associated with the occurrence of birth defects in mammals (Clark et al., 1989; Fan and Kizer, 1990), although, as is discussed below, selenium can be teratogenic when it is given in high acute doses.

Ferm and co-workers (1990) showed that selenium at doses greater than 2 mg/kg body weight was embryotoxic in a hamster model and that the selenium was teratogenic regardless of the route by which it was given (by gavage or by injection). Encephalocele was the most common defect produced in this model. The authors noted that the selenium administration was associated with marked evidence of maternal toxicity and concluded that some, if not all, of the embryotoxicity could have been a consequence of the maternal toxicity, rather than due to a direct effect of the selenium. Willhite et al. (1990), in a similar study using a hamster model, also reported that severe maternal toxicity occurred prior to selenium-induced embryotoxicity; and indeed these investigators suggested that maternal lethality may be needed before teratogenic effects are seen. As in the hamster, selenium has been reported to be embryo/fetotoxic in mice and rats, but only after signs of maternal toxicity are evident (Yonemoto et al., 1983; Danielsson et al., 1990).

Like the work with rodents, selenium studies in nonhuman primates only show selenium-induced developmental toxicity, at doses which are associated with significant maternal toxicity (Tarantal et al., 1991; Choy et al., 1993; Willhite, 1993; Hawkes et al., 1994). Fetuses obtained from cynomolgus macaques given 300 µg of selenium as selenomethionine per kilogram of body weight by nasogastric intubation on gestation days 20 to 50 were reported to be normal (Hawkes et al., 1994). Based on an observation of a high occurrence of miscarriages among female laboratory technicians with exposure to selenite powder and the birth of an infant with bilateral club foot, selenium was postulated to be teratogenic in humans over 25 years ago (Robertson, 1970); however, evidence supporting this idea has not been forthcoming.

VIII. CHROMIUM

A. TERATOGENESIS OF CHROMIUM DEFICIENCY

The identification of chromium as an essential metal was based on its postulated role in the restoration of glucose tolerance in rats (Schwarz and Mertz, 1959). However, the fact that chromium is found at relatively high concentrations in the environment, coupled with an apparently very low dietary need for the metal, has resulted in considerable difficulty in the production of severe chromium deficiency in the laboratory. While chromium deficiency has been reported to result in postnatal growth retardation and impaired glucose tolerance, there is no clear evidence that a deficiency of this metal is teratogenic in experimental animal models (Stoecker, 1990). Similarly, while some investigators have postulated that chromium deficiency may be a common problem in some populations (Stoecker, 1990, 1994; Anderson, 1994), there are no reports documenting an effect of chromium deficiency on pregnancy outcome in humans.

B. TERATOGENESIS OF CHROMIUM EXCESS

In contrast to chromium deficiency, chromium toxicity is easy to demonstrate and considered to be a public health problem in areas contaminated with high concentrations of hexavalent chromium (Stoecker, 1990; Outridge and Scheuhammer, 1993; Anderson, 1994). In contrast to hexavalent chromium, trivalent chromium has few toxic effects. Concentrations of hexavalent chromium in excess of 1 μM are toxic to mouse preimplantation embryos in culture (Iijima et al., 1983). Single chromium trioxide injections at different gestational periods have been shown to have embryotoxic and fetotoxic effects in rats, mice, and hamsters (Iijima et al., 1979; Gale, 1982; Mattison et al., 1983). The embryotoxic potential of oral tri- and hexavalent chromium was studied by Danielsson et al. (1982) in mice. These investigators reported that the hexavalent form was a potent teratogen, primarily affecting bone formation. In contrast, trivalent chromium was not found to be teratogenic; Trivedi et al. (1989) reported that hexavalent chromium given in drinking water (250 ppm) to mice resulted in a high incidence of embryo deaths and skeletal abnormalities. The mechanisms(s) underlying the effects of high levels of hexavalent chromium on the developing embryo/fetus have not been identified; however, it can be speculated that it might involve oxidative damage, given the strong peroxidant properties of this metal.

Similar to the case of chromium deficiency, there are no confirmed reports of excess chromium being a teratogenic insult in humans or nonhuman primates.

IX. IODINE

A. TERATOGENESIS OF IODINE DEFICIENCY

Iodine deficiency is recognized as one of the world's major public health problems, with as many as 800 million individuals living in iodine-poor areas. Iodine is an essential component of the two thyroid hormones, thyroxine (T_4) and triiodothyronine (T_3), and this is its only known physiological function. Iodine deficiency disorders include miscarriages, still births, congenital anomalies, goiter, and hypothyroidism (Hetzel and Mano, 1989; Dunn, 1993).

Iodine deficiency during development results in cretinism, a condition in which the child is mentally and physically retarded with a pot belly, large tongue, and facies resembling those of Down syndrome. Other characteristics can include shortness of stature, delayed epiphyseal development, and skin that is coarse and myxedematous. Deaf-mutism and/or impaired speech is also a frequent finding (Dunn, 1993).

While the complex set of biochemical lesions which result in cretinism are still poorly understood, there is little question that the syndrome is due to thyroid hormone deficiency secondary to the iodine deficiency. Aspects of the syndrome have been produced in sheep, rats, and marmosets through the use of iodine-deficient diets. For an in-depth discussion of this syndrome, please see the papers by Hetzel and Mano (1989), Dunn (1993), Morreale et al. (1993), Dumont et al. (1994), and Xue-Yi et al. (1994). It should be noted that since type I iodothyronine deiodinase is a selenium-dependent enzyme, selenium deficiency may also contribute to the development of cretinism (Berry et al., 1991). Finally, copper deficiency may also result in a compromised thyroid function, as one consequence of copper deficiency is a reduction in the activity of the type I deiodinase (Olin et al., 1994). The mechanism by which copper deficiency affects this enzyme is unknown, but it is interesting to note that the activity of selenoglutathione peroxidase can also be reduced in copper-deficient animals, suggesting that copper might be involved in the metabolism of selenocysteine (Olin et al., 1994).

B. TERATOGENESIS OF EXCESS IODINE

Excessive iodine intake by the mother can pose a reproductive risk. Several cases of congenital goiter and hypothyroidism due to maternal ingestion of excess iodine (typically taken for the treatment of asthma or bronchitis, hypothyroidism, or tachycardia), have been reported (Parmelee et al., 1940; Carswell et al., 1970; Pennington, 1990). Similarly, cretinism resulting from prenatal treatment with ^{131}I has been documented (Russel et al., 1957; Green et al., 1971).

X. TERATOGENIC EFFECTS OF MOLYBDENUM DEFICIENCY AND EXCESS

Molybdenum is an essential cofactor for several enzymes including aldehyde oxidase, xanthine oxidase, and sulfite oxidase (Nielsen, 1994). A naturally occurring molybdenum dietary deficiency, uncomplicated by antagonists, has not been reported to date. Anke et al. (1985) reported the induction of molybdenum deficiency in goats by feeding purified diets which contained less than 0.07 µg Mo/g (typical control diets contain >1 µg/g). The goats were reported to be characterized by impaired reproduction, including retarded fetal growth, but details concerning these effects were not provided in the paper.

There is only one confirmed case of molybdenum deficiency in humans, and this involved an individual who was on long-term total parenteral nutrition. The individual developed signs consistent with sulfite oxidase and xanthine dehydrogenase deficiencies; these signs were resolved following the subject's supplementation with molybdenum (Abumrad et al., 1981). While there is a rare inborn error of metabolism which is reflected by very low, to absent, activities of sulfite oxidase and xanthine dehydrogenase, the genetic defect is due to a lack of the molybdenum cofactor (molybdoptein), rather than molybdenum per se (Johnson et al., 1991; Slot et al., 1993). Infants with this disorder typically die shortly after birth.

Molybdenum toxicity can be a significant problem in some species, particularly ruminants. To a significant extent this is due to the fact that molybdenum can form thio- and oxythiomolybdates, which in turn can react with copper to form poorly available complexes. While the "toxic" effects of molybdenum are to a significant extent due to the ability of the metal to interfere with copper metabolism, molybdenum at high concentrations is also thought to be able to inhibit the ATP-dependent system responsible for the synthesis of active sulfate (phosphoadenosine phosphosulfate) (Mills and Davis, 1987). In cattle, high concentrations of dietary molybdenum can result in a high incidence of early abortions, fetal growth retardation, and early neonatal deaths. While some of the above effects have been linked to the ability of molybdenum to induce an embryonic/fetal copper deficiency, other yet-to-be-defined mechanisms are also thought to be involved.

Similar to cattle, in guinea pigs high dietary concentrations of molybdenum during pregnancy can result in the induction of severe fetal copper deficiency and prenatal or early postnatal death (Howell et al., 1993).

XI. TERATOGENIC EFFECTS OF FLUORIDE DEFICIENCY AND EXCESS

The rationale for considering fluoride an essential nutrient is primarily based on its ability to provide protection against the formation of dental caries (Recommended Dietary Allowances, 1989). While there have been efforts to demonstrate other essential roles for fluoride, these have not been conclusive. Schroeder et al. (1968) and Milne and Schwarz (1974) reported that the addition of fluoride to diets resulted in a stimulation of growth rates in mice and rats. Messer et al. (1973) reported that mice fed fluoride-deficient diets (<0.3 µg/g) for two generations were characterized by a high incidence of infertility, although litters which were born were normal, and subsequently showed normal growth rates. In contrast to the above, Weber and Reid (1974) and Tao and Suttie (1976) observed no negative effects of severe fluoride deficiency on reproduction in mice in four and six generation studies, respectively. Fluoride deficiency has not been reported to be teratogenic in any species studied to date.

Fluoride, consumed in high amounts, can have considerable toxicity, including hypomineralization of tooth enamel, hypermineralization of the skeleton, particularly of the spinal column and pelvis, compromised kidney function, and possibly compromised muscle and nerve function (Krishnamachari, 1987; Kleerkoper and Balena, 1991).

In contrast to the well-documented toxicity of fluoride in growing individuals and adults, there is no clear evidence for fluoride excess being teratogenic (Krishnamachari, 1987; Schellenberg et al., 1990; Kleerkoper and Balena, 1991). However, it should be noted that in a recent study by Mullenix et al. (1995), evidence was presented which suggests that the offspring of rats injected with high amounts of fluoride (0.13 mg NaF[1] (kg body weight)) on gestation days 17 to 19 were characterized by behavioral abnormalities.

XII. TERATOGENIC EFFECTS OF A DEFICIENCY OF EXCESS OF CANDIDATE ESSENTIAL ELEMENTS

As was discussed in the introduction, several investigators have argued that, in addition to the metals considered above, mammals may also have an essential dietary requirement for vanadium, nickel, arsenic, silicon, and/or boron. The argument over the "essentiality" of the above metals is due in part to changing perceptions among health professionals as to what is intended by the word "essential." In an early discussion of this issue, Cotzias (1967) suggested that for a metal to be considered essential, it should meet the following criteria: (1) it is present in all healthy tissue of all living things; (2) its concentration from one animal to the next is fairly constant; (3) its withdrawal from the body induces reproducibly the same physiological and structural abnormalities, regardless of the species studied; (4) its addition either reverses or prevents these abnormalities; (5) the abnormalities induced by deficiencies are always accompanied by specific biochemical changes; and (6) these biological changes can be prevented or cured when the deficiency is prevented or cured.

During the past decade, many have argued that the criteria listed by Cotzias are too rigid, and indeed two metals which are routinely considered essential for humans, chromium and fluoride, would not be so if the Cozias criteria were followed. Given the above, some have argued that for a metal to be considered essential, the only critical criterion that needs to be met is that its reduction or elimination from the diet should result in a consistent and reproducible impairment of a physiological function. It is under this latter, more liberal definition, that chromium and fluoride have gained the recognition of being "essential" for mammals. The metals described below, vanadium, nickel, arsenic, silicon, and boron, are in the category of metals which in low amounts may help to provide protection against select diseases (see Nielsen, 1994a). While the teratogenic effects (if any) of a deficiency of one or more of the above metals have not been defined, these metals may represent a developmental risk if they are in the diet at high concentrations.

A. ARSENIC

That arsenic may be an essential nutrient is suggested by studies from Nielsen et al. (1975a) and Anke et al. (1976, 1986). Nielsen et al. (1975a) reported that when pregnant rats were fed diets that contained less than 30 ng As/g (compared to control diets of 4.5 µg/g), their offspring were characterized by rough, sparse coats and growth retardation. At 3 months of age, the offspring were characterized by a splenomegaly that was attributed to an elevated erythrocyte osmotic fragility.

Anke et al. (1976, 1986) has reported that when diets containing less than 50 ng As/g are fed to pregnant goats and pigs, their offspring are characterized by growth retardation and an elevated mortality rate. Mechanistically, Cornatzer et al. (1983) have reported that in the rat, arsenic deficiency results in decreased activities of liver phosphatidylethanolamine methyltransferase, phosphatidyldimethylethanolamine methyltransferase, and choline phosphotransferase. Consistent with this, Uthus et al. (1989) have reported that methionine metabolism is abnormal in arsenic-deficient animals. While the above reports are interesting, it must be stressed that to date, specific arsenic-dependent enzymes have yet to be identified.

In contrast to arsenic essentiality, the teratogenicity of this metal is well established. As the toxicity and teratogenicity of arsenic are dealt with in Chapter V.3 (see Rogers, 1996), what follows is only a brief discussion of the teratogenicity of arsenic.

Arsenic has been shown to be teratogenic in a number of animal models, including hamsters, rats, mice, and rabbits, and it is a suspected teratogen in humans (Beaudoin, 1974; Ferm and Carpenter, 1968; Hood et al., 1978, 1987, 1988; Domingo, 1994; Golub, 1994; Rogers, 1996). In the experimental animal models studied to date, arsenic teratogenicity has been characterized by multiple abnormalities including exencephaly, eye defects, urogenital defects, skeletal abnormalities, and prenatal and early postnatal death (Golub, 1994; Rogers, 1996). Investigators have consistently reported that the teratogenicity of arsenic is higher when it is given by injection than by gavage, and that arsenite is more teratogenic than arsenate (Chaineau et al., 1990; Golub, 1994; Rogers, 1996).

Significantly, in contrast to several other metals, arsenic has been shown to be teratogenic at doses below those needed to induce signs of maternal toxicity (Golub, 1994). The above is a critical point, as it strongly supports the concept that arsenic has a direct teratogenic effect on the embryo. Consistent with this idea, Muller et al. (1986), Chaineau et al. (1990), and Mirkes and Cornel (1992) have reported that arsenite and arsenate are teratogenic in *in vitro* embryo culture models. Significantly, in the study by Chaineau et al. (1990), arsenite was observed to be teratogenic at doses as low as 3 μM, which is a physiologically relevant concentration. It should be noted that in contrast to arsenite and arsenate, in a study by Morrissey et al. (1990), arsine gas was not found to be teratogenic.

A number of mechanisms have been postulated to underlie the developmental toxicity of arsenic. As discussed above, maternal toxicity is a common consequence of arsenic exposure and the physiological effects associated with this undoubtedly contribute to the developmental toxicity of the metal. In addition, arsenic tends to form strong complexes with protein sulfhydryls; the formation of these complexes can alter the activities of numerous enzymes. Finally, arsenic exposure can result in a significant induction of several of the heat shock proteins (see Golub, 1994; and Rogers, 1996); the consequences of these increases are poorly understood.

B. BORON

While boron has been recognized to be essential for plants for over 50 years, its essentiality for animals is an area of active debate. Nielsen (1994b) has reported that the response of an animal to boron deficiency is affected by numerous dietary variables, including the diet content of calcium, phosphorus, magnesium, potassium, vitamin D, and methionine. The consistent observation that is made regarding severe boron deficiency is that it can influence bone metabolism (Dupre et al., 1994; Hunt, 1994; Nielsen, 1994b); however, the mechanism(s) by which this occurs have not been firmly identified, although it is thought to involve an effect of boron on vitamin D metabolism (Hunt, 1994). Teratogenic effects of boron deficiency have not been reported.

Boron has a relatively low order of toxicity, with dietary concentrations of up to 100 µg/g being well tolerated (typical dietary boron concentrations are on the order of 1 µg/g).

Boron has been identified as a male reproductive toxicant for at least 20 years (Chapin and Ku, 1994), and it is recognized to be teratogenic at high doses. Beyer et al. (1983) reported that a single dose of boric acid (500 to 3000 mg/kg body weight) to pregnant mice on gestation day 1, results in a failure to implant; similar findings have been reported for rats (Siegel and Wason, 1986). Heindel et al. (1994) reported that boric acid is developmentally toxic for rats, mice, and rabbits when it is given at high concentrations (80 to 400 mg/kg body weight per day) throughout pregnancy, or during organogenesis alone. Developmental toxicity in all three species was reflected by growth retardation and skeletal abnormalities. In rats and rabbits, defects of the central nervous system and cardiovascular system were also described. Significantly, while teratogenic effects of boric acid were observed in the mouse and rabbit only at doses which exceeded those needed to induce signs of maternal toxicity, teratogenic effects of boric acid in the rat were observed at doses lower than those needed to induce signs of maternal toxicity. The latter observation supports the idea that boron may have direct teratogenic effects. The mechanism(s) by which boron toxicity influence embryonic/fetal development have not been defined.

C. COBALT

Among the essential metals for nonruminant mammals, cobalt is unique in that the requirement for the metal is thought to be due strictly to its being a component of vitamin B-12. As a consequence of the above, for nonruminants there is no dietary requirement for cobalt in the ionic form, but rather it needs to be consumed as vitamin B-12 (Smith, 1987). Ruminants are capable of having microbial synthesis of vitamin B-12 in the rumen; thus they have an essential requirement for the metal. For both ruminants and nonruminants, the teratogenic effects of cobalt deficiency are essentially those of vitamin B-12 deficiency (Kirke et al. 1993; Fisher, 1991; Allen, 1994).

Cobalt has a low order of toxicity in all species studied, including humans. However, at very high concentrations in the diet (200 to 500 µg/g, compared to control diet of 1 µg/g), or with cobalt injections on the order of 5 to 10 mg/kg body weight, a cobalt toxicity can be induced. Signs of cobalt toxicity include anemia (secondary to cobalt-induced reductions in iron absorption), a loss of appetite, weight loss, and occasionally, death (Smith, 1987). The teratogenicity of cobalt is very low. Ferm (1972) and Leonard and Lauwerys (1990) reported that cobalt salts were not teratogenicity in either hamsters or rats. Similarly, Paternain et al. (1988) reported no teratogenic or fetotoxicity when rats were gavaged with 100 mg Co/kg body weight per day on gestation days 6 to 15, although this dose did result in modest signs of maternal toxicity. Wide (1984) reported that when cobalt was injected i.v. (5 mM) on gestation day 8 in mice, it resulted in reduced skeletal ossification.

D. NICKEL

The essentiality of nickel for mammals has been a subject of debate for over 20 years. Nielsen et al. (1975b) reported that when rats were fed a nickel-deficient diet (2 ng Ni/g compared to control diets of 3 µg/g) for three generations, the nickel deprivation resulted in several consistent pathological findings in third generation pups. The pups were reported to be characterized by a high incidence of perinatal mortality (30%, compared to control values of 14%), a rough coat with uneven hair distribution, a liver that was pale in color, low liver cholesterol concentrations, and ultrastructural changes in the liver reflected by a reduced amount of rough endoplasmic reticulum. Nielsen et al. (1984) later reported that many of the signs of nickel deficiency could be attributed to nickel deficiency-associated changes in iron metabolism; these signs could be blunted by either high concentrations of nickel (20 µg/g) or iron (100 µg/g vs. control values of 15 to 60 µg/g) in the diet. Interestingly, when nickel was fed at a very high concentration of 100 µg/g, a secondary iron deficiency was induced. The mechanism(s) by which nickel influences iron metabolism have not been identified. Anke et al. (1984) have reported that, similar to rats, severe nickel deficiency during pregnancy can result in a high frequency of anemia and perinatal mortality in goats and pigs. Gross congenital defects were not reported in any of the above studies. Signs of nickel deficiency have not been reported for humans.

Embryotoxic and teratogenic effects of nickel have been described *in vivo* for hamsters, mice, and rats (Ferm, 1972; Sunderman et al., 1978, 1983; Lu et al., 1979; Mas et al., 1985), and *in vitro* in postimplantation rat embryo cultures (Saillenfait et al., 1991, 1993). Typical malformations associated with nickel teratogenicity (normally induced by injecting either nickel acetate or nickel chloride during organogenesis at doses in the range of 4 to 30 mg Ni/kg body weight) include severe brain defects, cleft palate, and a high incidence of skeletal abnormalities. Smith et al. (1993) reported that the incidence of perinatal death was significantly increased in litters from rats given water containing 10 µg Ni/ml or more for 10 weeks prior to and during pregnancy. Pregnancy complications, including an increased occurrence of birth defects (primarily cardiovascular and musculoskeletal), have been reported to be higher than normal in women exposed to high concentrations of nickel in industrial settings (Chashschin et al., 1994). The mechanisms underlying the teratogenic actions of nickel *in vivo* are poorly understood, although it has been speculated that they could involve nickel-induced DNA protein cross-links and/or strand breaks (Sunderman et al., 1983; Kasprzak et al., 1992). It has also been suggested that nickel toxicity can trigger a transitory teratogenic fetal hyperglycemia (Mas et al., 1985); however, the mechanisms underlying this hyperglycemia have not been defined.

E. SILICON

The precise biochemical function(s) of silicon have not been defined, but there is considerable evidence that it participates as a biological cross-linking agent which helps to give resilience to the structure of collagen and possibly elastin (Nielsen, 1994). Consistent with the above, litters from pregnant rats fed diets deficient in silicon (<5 μg Si/g compared to typical control diets of 100 μg/g or more) have been reported to be characterized by bone abnormalities and poor skeletal calcification (Carlisle, 1986). Other teratogenic signs of silicon deficiency have not been reported.

Silicon has a very low level of toxicity when it is taken orally. Very high concentrations of silicon in the diet (>2500 mg Si/kg) can result in urolithiasis in rats. No reports could be found of excess silicon being teratogenic.

F. VANADIUM

Since the early 1970s there has been increasing interest in the idea that vanadium may be essential for mammals; however, results from studies in this area have been highly contradictory, and consistent patterns of vanadium deficiency signs have not been reported. During the past few years, several investigators have argued that the early reports of vanadium essentiality demonstrated pharmacological rather than nutritional actions of vanadium.

A number of pharmacological effects of vanadium have been reported. Vanadium is a strong insulinomimetic agent, and its administration can result in rapid reductions in blood glucose concentrations. However, the above effect only occurs with very high dietary concentrations of vanadium (typical vanadium concentrations in control diets are less than 10 μg/kg, compared to high-vanadium diets, which often exceed 500 μg/g). Teratogenic effects associated with vanadium deficiency have not been established (Nielsen, 1987; 1994a).

In contrast to vanadium deficiency, the developmental toxicity of vanadium is well established (Domingo, 1994, 1996). (For a detailed review of this topic, please see the Chapter V.5 by Corbella and Domingo in this volume.) Vanadium given by i.p. injection (1 to 4 mg/kg body weight per day) to pregnant hamsters during gestation days 5 to 10 resulted in micrognathia and skeletal abnormalities (Carlton et al., 1982). Similarly, administration of 0.15 ml of 1 mM V_2O_2 to pregnant mice on gestation day 8 resulted in skeletal abnormalities and reduced skeletal ossification (Wide, 1984). Paternain et al. (1990) reported that the oral administration of vanadyl sulfate pentahydrate (>5 to 150 mg/kg body weight per day) during organogenesis resulted in maternal toxicity, fetal toxicity, and teratogenicity as evidenced by cleft palates and micrognathia. Gomez et al. (1992) reported that low doses of sodium metavanadate (4 mg/kg body weight per day) were teratogenic in mice when given during organogenesis. Zhang et al. (1993) reported that V_2O_5 given at a dose of 5 mg/kg body weight per day to pregnant rats on gestation days 9 to 12 resulted in maternal toxicity, fetal death, and numerous fetal skeletal and visceral anomalies.

The mechanisms underlying the teratogenicity of vanadium have not been well defined, but vanadium has been shown to inhibit numerous enzymes that hydrolyze phosphate esters, including ribonuclease and alkaline phosphatases. Vanadium can also inhibit Na/K-ATPase, while it can activate adenylate cyclase. Finally, it can be speculated that the strong prooxidant properties of the metal probably contribute to its embryotoxicity (Elfant and Keen, 1987; Nielsen, 1994).

XIII. SUMMARY

The list of metals which are known, or suspected, to be essential for mammalian reproduction and health has been considerably lengthened during the past 2 decades. Presumably, the list will continue to grow as improvements are gained in our ability to produce diets and living environments which are effectively free of the metal being studied. Unfortunately, while progress has been made in identifying the "essential" metals, there continues to be a poor understanding of the roles played by these metals in developmental processes.

Consistent with Paracelsus' dictum, "the dose makes the poison," an excess of an essential metal can represent a developmental risk equal to or greater than that which can occur with a deficit of the metal. Here again, however, our understanding of the mechanisms underlying the teratogenicity of high concentrations of essential metals is primitive. Clearly, in many cases, the teratogenic and embryotoxic effects associated with the toxicity of essential metals are primarily due to indirect effects associated with metal toxicity-induced maternal toxicity. However, in other cases, a direct effect of the metal on the conceptus may represent the primary challenge.

Given the remarkable strides which have been made during the past decade in understanding the chemistry and metabolism of essential metals in biological systems, coupled with the anticipated progress that should be made in the next decade, it can be predicted that a better understanding of how metals influence development is on the horizon.

ACKNOWLEDGMENT

Supported in part by HD01743, HD26777, and U.S. E.P.A. Cooperative Agreement CR-816713.

REFERENCES

Abdel-Mageed, A. B., Welti, R., Oehme, F. W., and Pickrell, J. A. (1994), *Am. J. Physiol.,* 267, L679–685.
Abumrad, N. N., Schneider, A. J., Steel, D., and Rogers, L. S. (1981), *Am. J. Clin. Nutr.,* 34, 2551–2559.
Allen, L. H. (1994), *Adv. Exp. Med. Biol.,* 352, 173–186.
Alt, H. L. (1938), *Am. J. Dis. Child.,* 56, 975–984.
Anderson, R. A. (1994), Nutritional and toxicologic aspects of chromium intake: an overview, in Risk Assessment of Essential Elements, Mertz, W., Abernathy, C. O., and Olin, S. S., Eds., International Life Sciences Institute Press, Washington, D.C., 187–196.
Anke, M. (1986), *Trace Elements in Human and Animal Nutrition,* 5th ed., Vol. 1., Mertz, W., Ed., Academic Press, Orlando, FL, 347–372.
Anke, M., Groppel, B., and Grun, M. (1985), *Trace Elements in Man and Animals — TEMA 5,* Mills, C.F., Bremner, I., and Chesters, J.K., Eds., Commonwealth Agriculture Bureaux, Slough, U.K., 154–157.
Anke, M., Groppel, B., Kronemann, H. et al. (1984), *Nickel in the Human Environment.* Sunderman, F. W. et al., Eds. International Agency Research Cancer, Lyon, 339–365.
Anke, M., Grün, M., and Partschefeld, M. (1976), *Trace Substances in Environmental Health,* Hemphill, D.D., Ed., University of Missouri, Columbia, 403–409.
Apgar, J. (1992), *J. Nutr. Biochem.,* 3, 266–278.
Baly, D. L., Lee, I., and Doshi, R. (1988), *FEBS Lett.,* 239, 55–58.
Baly, D. L., Curry, D. L., Keen, C. L., and Hurley, L. S. (1985), *Endocrinology,* 116, 1734–1740.
Barash, A., Shoham, Z., Borenstein, R., and Nebel, L. (1990), *Gynecol. Obstet. Invest.,* 29, 203–206.
Beard, J. L. (1994), *Am. J. Clin. Nutr.,* 59 (Suppl.), 502S–510S.
Beaudoin, A. R. (1974), *Teratology,* 10, 153–158.
Bennetts, H. W. and Beck, A. B. (1942), *Aust. Coun. Sci. Ind. Res. Bull.,* 147, 1–52.
Bennetts, H. W., Beck, A. B., and Harley, R. (1948), *Aust. Vet. J.,* 24, 237–244.
Berry, M. J., Banu, L., and Larsen, P. R. (1991), *Nature,* 349, 438–440.
Beyer, K. H., Bergfeld, W. F., Berndt, W. O., Boutwell, R. K., Carlton, W. W., Hoffman, D. K., and Schroeter, A. L. (1983), *J. Am. Coll. Toxicol.,* 2, 87–125.
Brewer, G. J. (1980), *Am. J. Hematol.,* 8, 213–248.
Bunce, G. E. (1994), *Nutrient Regulation during Pregnancy, Lactation, and Infant Growth.* Allen, L., King, J., and Lönnerdal, B., Eds., Plenum Press, New York, 257–264.
Butler, J. A., Whanger, P. D., and Patton, N. M. (1988), *J. Am. Coll. Nutr.,* 7, 43–56.
Carlisle, E. M. (1986), *Trace Elements in Human and Animal Nutrition,* 5th ed., Vol. 2, Mertz, W., Ed., Academic Press, Orlando, FL, 373–390.
Carlton, B. D., Beneke, M. B., and Fisher, G. L. (1982), *Environ. Res.,* 29, 256–262.
Carswell, F., Kerr, M. M., and Hutchison, J. H. (1970), *Lancet,* 1, 1241–1243.
Celada, A., Busset, R., Gutierrez, J., and Herreros, V. (1982), *Helv. Paediatr. Acta,* 37, 239–244.
Chaineau, E., Binet, S., Pol, D., Chatellier, G., and Meininger, V. (1990), *Teratology,* 41, 105–112.
Chang, C. H., Mann, D. E., and Gautieri, R. F. (1977), *J. Pharm. Sci.,* 66, 1755–1758.
Chapin, R. E. and Ku, W. W. (1994), *Environ. Health Perspect.,* 102 (Suppl. 7), 87–91.
Chashschin, V. P., Artunina, G. P., and Norseth, T. (1994), *Sci. Total Environ.,* 148, 287–291.

Chelly, J., Turner, Z., Tonnesen, T., Petterson, A., Ishikawa-Brush, Y., Tommerup, N., Horn, N. et al. (1993), *Nature Genet.,* 3, 14–19.
Chesters, J. K. (1989), *Zinc in Human Biology,* Mills, C. F., Ed., Springer-Verlag, Heidelberg, 109–118.
Choy, W. N., Henika, P. R., Willhite, C. C., and Tarantal, A. F. (1993), *Env. Mol. Mutagen.,* 21, 73–80.
Clark, D. R., Jr., Ogasawara, P. A., Smith, G. J., and Ohlendorf, H. M. (1989), *Arch. Environ. Contam. Toxicol.,* 18, 787–794.
Cohen, N. L., Keen, C. L., Lonnerdal, B., and Hurley, L. S. (1983), *Drug-Nutr. Interact.,* 2, 203–210.
Cornatzer, W. E., Uthus, E. O., Haning, J. A., and Nielsen, F. H. (1983), *Nutr. Rep. Int.,* 27, 821–829.
Corbella, J. and Domingo, J. L. (1996), *Toxicology Metals.* Vol. 2. *Target Organ Toxicology,* Chang, L. W., Ed., CRC Press, Boca Raton, FL, chap. V. 3.
Cotzias, G. C. (1967), Trace Subst. Environ. Health Proc. Univ. Mo., 1st Annu. Conf., Columbia, 5.
Cousins, R. J. (1994), *Annu. Rev. Nutr.,* 14, 449–469.
Couzy, F., Keen, C. L., Gershwin, M. E., and Mareschi, J. P. (1993), *Prog. Food Nutr. Sci.,* 17, 65–87.
Danielsson, B. R. G., Danielson, M., Khayat, A., and Wide, M. (1990), *Toxicology,* 63, 123–136.
Danielsson, B. R. G., Hassoun, E., and Dencker, L. (1982), *Arch. Toxicol.,* 51, 233–245.
Danks, D. M. (1988), *Annu. Rev. Nutr.,* 8, 235–257.
Das, S., Levinson, B., Vulpe, C., Whitney, S., Gitschier, J., and Packman, S. (1995), *Am. J. Hum. Genet.,* 56, 570–576.
Daston, G. P. (1982), *Toxicology,* 24, 55–63.
Daston, G. P., Overmann, G. J., Baines, D., Taubeneck, M. W., Lehman-McKeeman, L. D., Rogers, J. M., and Keen, C. L. (1994), *Reprod. Toxicol.,* 8, 15–24.
DeMaeyer, E. and Adiels-Tegman, M. (1985), *World Health Stat. Q.,* 38, 302–316.
Dencker, L., Danielsson, B., Khayat, A., and Lindgren, A. (1983), *Reproductive and Developmental Toxicity of Metals,* Clarkson, T. W., Nordberg, G. F., and Sager, P. R., Eds., Plenum Press, New York, 607–631.
Domingo, J. L. (1994), *J. Toxicol. Environ. Health,* 42, 123–141.
Dreosti, I. E. (1993), *Ann. N. Y. Acad. Sci.,* 678, 193–204.
Dreosti, I. E., Record, I. R., and Manuel, S. J. (1985), *Biol. Trace Elem. Res.,* 7, 103–122.
Dubick, M. A., Keen, C. L., and Rucker, R. B. (1985), *Exp. Lung Res.,* 8, 227–241.
Dugdale, A. E. and Powell, I. W. (1964), *Med. J. Aust.,* 2, 990–992.
Dumont, J. E., Corvilain, B., and Contempre, B. (1994), *Mol. Cell. Endocrin.,* 100, 163–166.
Dunn, J. T. (1993), *Ann. N.Y. Acad. Sci.,* 678, 158–168.
Dupre, J. N., Keenan, M. J., Hegsted, M., and Brudevold, A. M. (1994), *Environ. Health Perspect.,* 102 (Suppl. 7), 55–58.
Elfant, M. and Keen, C. L. (1987), *Biol. Trace Elem. Res.,* 14, 193–208.
Erway, L. C., Purichia, N. A., Netzler, E. R., D'Amore, M. A., Esses, D., and Levine, M. (1986), *Scan. Electron Microsc.,* 4, 1681–1694.
Erway, L. C. and Mitchell, S. E. (1973), *J. Hered.,* 64, 111–119.
Erway, L. C., Hurley, L. S., and Fraser, A. (1966), *Science,* 152, 1766–1768.
Falchuk, K. H. (1993), *Progr. Clin. Biol. Res.,* 380, 91–111.
Fan, A. M. and Kizer, K. W. (1990), *West. J. Med.,* 153, 160–167.
Ferm, V. H. (1972), *Adv. Teratol.,* 6, 51–75.
Ferm, V. H. and Carpenter, S. J. (1968), *J. Reprod. Fertil.,* 17, 199–201.
Ferm, V. H. and Hanlon, D. P. (1974), *Biol. Reprod.,* 11, 97–101.
Ferm, V. H., Hanlon, D. P., Willhite, C. C., Choy, W. N., and Book, S. A. (1990), *Reprod. Toxicol.,* 4, 183–190.
Filteau, S. M., Sullivan, K. R., Anwar, U. S., Anwar, Z. R., and Tomkins, A. M. (1994), *Eur. J. Clin. Nutr.,* 48, 293–302.
Fisher, G. E. (1991), *Res. Vet. Sci.,* 50, 319–327.
Flynn, A., Martier, S. S., Sokol, R. J., Miller, S. I., Golden, N. L., and Villano, B. C. (1981), *Lancet,* 1, 572–574.
Fosmire, G. J. (1990), *Am. J. Clin. Nutr.,* 51, 225–227.
Fredriksson, A., Gardlund, A. T., Bergman, K., Oskarsson, A., Ohlin, B., Danielsson, B., and Archer, T. (1993), *Pharmacol. Toxicol.,* 72, 377–382.
Gale, T. F. (1982), *Environ. Res.,* 29, 196–203.
Garn, S. M., Ridella, S. A., Petzold, A. S., and Falkner, F. (1981), *Semin. Perinatol.,* 5, 155–162.
Goldenberg, R. L., Tamura, T., Neggers, Y., Copper, R. L., Johnston, K. E., DuBard, M. B., and Hauth, J. C. (1995), *JAMA,* 274, 463–468.
Golub, M. S., Keen, C. L., Gershwin, M. E., and Hendrickx, A. G. (1995), *J. Nutr.,* 125, 2263S–2271S.
Golub, M. S. (1994), *Reprod. Toxicol.,* 8, 283–295.
Gomez, M., Sanchez, D. J., Domingo, J. L., and Corbella, J. (1992), *J. Toxicol. Environ. Health,* 37, 47–56.
Goyer, R. A. (1995), *Am. J. Clin. Nutr.,* 61 (Suppl.), 646S–650S.
Green, H. G., Gareis, F. J., Shepard, T. H., and Kelley, V. C. (1971), *Am. J. Dis. Child.,* 122, 247–249.
Guvenc, H., Karatas, F., Guvenc, M., Kunc, S., Aygun, A. D., and Bektas, S. (1995), *Pediatrics,* 95, 879–882.
Harris, Z. L., Takahashi, Y., Miyajima, H., Serizawa, M., Macgillivray, R. T., and Gitlin, J. P. (1995), *Proc. Natl. Acad. Sci., U.S.A.,* 92, 2539–2543.
Hawk, S. N., Uriu-Hare, J. Y., Daston, G. P., and Keen, C. L. (1995), *Teratology,* 51, 171–172.
Hawkes, W. C., Willhite, C. C., Omaye, S. T., Cox, D. N., Choy, W. N., and Tarantal, A. F. (1994), *Teratology,* 50, 148–159.
Heindel, J. J., Price, C. J., and Schwetz, B. A. (1994), *Environ. Health Perspect.,* 102 (Suppl 7), 107–112.
Hetzel, B. S. and Mano, M. T. (1989), *J. Nutr.,* 119, 145–151.

Hill, C. H. and Matrone, G. (1970), *Fed. Proc.,* 29, 1474–1488.
Hinks, L. J., Ogilvy-Stuart, A., Hambidge, K. M., and Walker, V. (1989), *Br. J. Obstet. Gynecol.,* 96, 61–66.
Hood, R. D., Vedel, G. C., Zaworotko, M. J., Tatum, F. M., and Meeks, R. G. (1988), *J. Toxicol. Environ. Health,* 25, 423–434.
Hood, R. D., Vedel, G. C., Zaworotko, M. J., Tatum, F. M., and Meeks, R. G. (1987), *Teratology,* 35, 19–25.
Hood, R. D., Thacker, G. T., Patterson, B. L., and Szczech, G. M. (1978), *J. Environ. Pathol. Toxicol.,* 1, 857–864.
Howell, J. M., Shunxiang, Y., and Gawthorne, J. M. (1993), *Res. Vet. Sci.,* 55, 224–230.
Hunt, C. D. (1994), *Environ. Health Perspect.,* 102 (Suppl. 7), 35–43.
Hurley, L. S., Gordon, P., Keen, C. L., and Merkhofer, L. (1982), *Proc. Soc. Exp. Biol. Med.,* 170, 48–52.
Hurley, L. S. (1981), *Physiol. Rev.,* 61, 249–295.
Hurley, L. S. and Keen, C. L. (1979), *Copper in the Environment.* Part II: *Health Effects,* Nriagu, J. O., Ed., John Wiley & Sons, New York, 33–56.
Hurley, L. S. and Shrader, R. E. (1975), *Nature,* 25, 427–429.
Hurley, L. S., Gowan, J., and Swenerton, H. (1971), *Teratology,* 4, 199–204.
Iijima, S., Shimizu, M., and Matsumoto, M. (1979), *Teratology,* 20, 152.
Iijima, S., Spindle, A., and Pedersen, R. A. (1983), *Teratology,* 27, 109–115.
Jankowski, M. A., Uriu-Hare, J. Y., Rucker, R. B., Rogers, J. M., and Keen, C. L. (1995), *Teratology,* 51, 85–93.
Johnson, J. L., Rajagopalan, K. V., Lanman, J. T., Schutgens, R. B., van Gennip, A. H., Sorensen, P., and Applegarth, D. A. (1991), *J. Inher. Metab. Dis.,* 14, 932–937.
Jornvall, H., Falchuk, K. H., Geraci, G., and Vallee, B. L. (1994), *Biochem. Biophys. Res. Commun.,* 200, 1398–1406.
Kasprzak, K. S., Diwan, B. A., Rice, J. M., Misra, M., Riggs, C. W., Olinski, R., and Dizdaroglu, M. (1992), *Chem. Res. Toxicol.,* 5, 809–815.
Keen, C. L., Taubeneck, M. W., Daston, G. P., Gershwin, M. E., Ansari, A., and Rogers, J. M. (in press), *Dev. Brain Dysfunct.*
Keen, C. L., Zidenberg-Cherr, S., and Lönnerdal, B. (1994), *Risk Assessment of Essential Elements,* Mertz, W., Abernathy, C. O., and Olin, S. S., Eds., International Life Sciences Institute Press, Washington, D. C., 221–235.
Keen, C. L. (1992), *Teratology,* 46, 15–21.
Keen, C. L. and Gershwin, M. E. (1990), *Annu. Rev. Nutr.,* 10, 415–431.
Keen, C. L. and Hurley, L. S. (1989), in *Zinc in Human Biology,* Mills, C. F., Ed., Springer-Verlag, London, 183–220.
Keen, C. L., Cohen, N. L., Lönnerdal, B., and Hurley, L. S. (1983a), *Proc. Soc. Exp. Biol. Med.,* 173, 598–605.
Keen, C. L., Mark-Savage, P., Lönnerdal, B., and Hurley, L. S. (1983b), *Drug-Nutr. Interact.,* 2, 17–34.
Keen, C. L., Lönnerdal, B., and Hurley, L. S. (1982), *Inflammatory Diseases and Copper,* Sorenson, J. R. J., Ed., Humana Press, Clifton, NJ, 109–121.
Ketchenson, M. R., Barron, G. P., and Cox, D. H. (1969), *J. Nutr.,* 98, 303–311.
Kirke, P. N., Molloy, A. M., Daly, L. E., Burke, H., Weir, D. G., and Scott, J. M. (1993), *Q. J. Med.,* 86, 703–708.
Kleerekoper, M. and Balena, R. (1991), *Annu. Rev. Nutr.,* 11, 309–324.
Krishnamachari, K. A. V. R. (1987), *Trace Elements in Human and Animal Nutrition,* 5th ed., Vol.1, Mertz, W., Ed., Academic Press, San Diego, 365–415.
Kumar, S. (1976), *Nutr. Rep. Int.,* 13, 33–36.
Laskey, J. W., Rehnberg, G. L., Hein, J. F., and Carter, S. D. (1982), *J. Toxicol. Environ. Health,* 9, 677–687.
Leach, R. M., Jr. (1986), *Manganese in Metabolism and Enzyme Function,* Schramm, V. L. and Wedler, F. C., Eds., Academic Press, Orlando, FL, 81–89.
Lee, S. H., Lancey, R., Montaser, A., Madani, N., and Linder, M. C. (1993), *Proc. Soc. Expl. Biol. Med.,* 203, 428–439.
Lemley, A. D. (1993), *Ecotoxicol. Environ. Safety,* 26, 181–204.
Leonard, A. and Lauwerys, R. (1990), *Mutat. Res.,* 239, 17–27.
Levander, O. A. (1986), *Trace Elements in Human and Animal Nutrition,* 5th ed., Vol. 2, Mertz, W., Ed., Academic Press, San Diego, 209–279.
Liu, A. C., Heinrichs, B. S., and Leach R. M., Jr. (1994), *Poultry Sci.,* 73, 663–669.
Lown, B. A., Morganti, J. B., D'Agostino, R., Stineman, C. H., and Massaro, E. J. (1984), *Neurotoxicology,* 5, 119–131.
Lozoff, B., Jimenez, A., and Wolf, A. W. (1991), *N. Engl. J. Med.,* 325, 687–694.
Lu, C. C., Matsumoto, N., and Iijima, S. (1979), *Teratology,* 19, 137–142.
Luo, S. Q., Plowman, M. C., Hopfer, S. M., and Sunderman, F. W., Jr. (1993), *Ann. Clin. Lab. Sci.,* 23, 111–120.
Mackler, B., Grace R., Person R., Shepard, T. H., and Finch, C. A. (1983), *Teratology,* 28, 103–107.
Mackler, B., Grace, R., Haynes, B., Bargman, G. J., and Shepard, T. H. (1973), *Arch. Biochem. Biophys.,* 158, 885–888.
Mark-Savage, P., Keen, C. L., and Hurley, L. S. (1983), *J. Nutr.,* 113, 501–510.
Marois, M. and Bouvet, M. (1972), *C.R. Seances Soc. Biol., Paris,* 166, 1237.
Martin, S. J., Mazdai, G., Strain, J. J., Cotter, T. G., and Hannigan, B. M. (1991), *Clin. Exp. Immunol.,* 83, 338–343.
Mas, A., Holt, D., and Webb, M. (1985), *Toxicology,* 35, 45–57.
Mattison, D. R., Gates, A. H., Leonard, A., Wide, M., Hemminki, K., and Copius Peereboom-Stegeman, J. H. J. (1983), *Reproductive and Developmental Toxicity of Metals,* Clarkson, T. W., Nordberg, G. F., and Sager, R. R., Eds., New York, Plenum Press, 43–91.
McCoy, K. E. M. and Weswig, P. H. (1969), *J. Nutr.,* 98, 383–389.
McElhatton, P. R., Roberts, J. C., and Sullivan, F. M. (1991), *Human Exp. Toxicol.,* 10, 251–259.
Mercer, J. F., Grimes, A., Ambrosini, L., Lockhart, P., Paynter, J. A., Dierick, H., and Glover, T. W. (1994), *Nature Genet.,* 6, 374–378.

Mercer, J. F., Livingston, J., Hall, B., Paynter, J. A., Begy, C., Chandrasekharappa, S., Lockhart, P., Grimes, A., Bhave, M., and Siemieniak, D. (1993), *Nature Genet.,* 3, 20–25.
Messer, H. H., Armstrong, W. D., and Singer, L. (1973), *J. Nutr.,* 103, 1319–1326.
Mieden, G. D., Keen, C. L., Hurley, L. S., and Klein, N. W. (1986), *J. Nutr.,* 116, 2424–2431.
Miller, J. (1990), *X-linked Traits: A Catalogue of Loci in Nonhuman Mammals,* Cambridge University Press, Cambridge, 115–125.
Mills, C. F. and Davis, G. K. (1987), *Trace Elements in Human and Animal Nutrition,* 5th ed., Vol. 1. Mertz, W., Ed. Academic Press, Orlando, FL, 429–463.
Mills, C. F. and Williams, R. B. (1962), *Biochem. J.,* 85, 629–632.
Milman, N., Ibsen, K. K., and Christensen, J. M. (1987), *Acta Obstet. Gynecol. Scand.,* 66, 205–211.
Mirkes, P. E. and Cornel, L. (1992), *Teratology,* 46, 251–259.
Morreale de Escobar, G., Obregon, M. J., Calvo, R., and Escobar del Rey, F. (1993), *Am. J. Clin. Nutr.,* 57, 280S–285S.
Morrissey, R. E., Fowler, B. A., Harris, M. W., Moorman, M. P., Jameson, C. W., and Schwetz, B. A. (1990), *Fundam. Appl. Toxicol.,* 15, 350–356.
Mullenix, P. J., Denbesten, P. K., Schunior, A., and Kernan, W. J. (1995), *Neurotoxicol. Teratol.,* 17, 169–177.
Muller, W. U., Streffer, C., and Fischer-Lahdo, C. (1986), *Arch. Toxicol.,* 59, 172–175.
Murray, M. J. and Stein, N. (1971), *J. Nutr.,* 101, 1583–1588.
Murray, J. J. and Healy, M. D. (1991), *J. Am. Diet. Assoc.,* 91, 66–73.
Muth, O. H., Weswig, P. H., Whanger, P. D., and Oldfield, J. E. (1971), *Am. J. Vet. Res.,* 32, 1603–1605.
Nicholson, V. J. and Veldstra, H. (1972), *FEBS Lett.,* 23, 309–313.
Nielsen, F. H. (1994a), *Modern Nutrition in Health and Disease,* 8th ed., Shils, M. E., Olson, J. A., and Shike, M., Eds., Lea & Febiger, Philadelphia, 269–286.
Nielsen, F. H. (1994b), *Environ. Health Perspect.,* 102 (Suppl. 7), 59–63.
Nielsen, F. H. (1990), *Present Knowledge in Nutrition,* 6th ed., Brown, M. L., Ed., International Life Sciences Institute, Nutrition Foundation, Washington, D.C., 294–307.
Nielsen, F. H. (1987), *Trace Elements in Human and Animal Nutrition,* 5th ed., Vol. 1, Mertz, W., Ed., Academic Press, Orlando, FL, 275–300.
Nielsen, F. H., Shuler, T. R., McLeod, T. G., and Zimmerman, T. J. (1984), *J. Nutr.,* 114, 1280–1288.
Nielsen, F. H., Givand, S. H., and Myron, D. R. (1975a), *Fed. Proc.,* 34, 923.
Nielsen, F. H., Myron, D. R., Givand, S. H., Zimmerman, T. J., and Ollerich, D. A. (1975b), *J. Nutr.,* 105, 1620–1630.
O'Dell, B. L. (1968), *Fed. Proc.,* 27, 199–204.
O'Dell, B, L., Hardwick, B. C., and Reynolds, G. (1961), *J. Nutr.,* 73, 151–157.
O'Halloran, T. V. (1993), *Science,* 261, 715–730.
Ohlendorf, H. M., Hothem, R. L., Aldrich, T. W., and Krynitsky, A. J. (1987), *Sci. Total Environ.,* 66, 169–183.
Olenmark, M., Biber B., Dottori, O., and Rybo, G. (1987), *Clin. Toxicol.,* 25, 347–359.
Olin, K. L., Walter, R. M., and Keen, C. L. (1994), *Am. J. Clin. Nutr.,* 59, 654–658.
Oteiza, P. I., Olin, K. L., Fraga, C. G., and Keen, C. L. (1995), *J. Nutr.,* 125, 823–829.
Oteiza, P. I., Cuellar, S., Lönnerdal, B., Hurley, L. S., and Keen, C. L. (1990), *Teratology,* 41, 97–104.
Outridge, P. M. and Scheuhammer, A. M. (1993), *Rev. Environ. Contam. Toxicol.,* 130, 31–77.
Parmelee, A. H., Allen, E., Stein, I. F., and Buxbaum, H. (1940), *Am. J. Obstet. Gynecol.,* 40, 145.
Paternain, J. L., Domingo, J. L., Gomez, M., Ortega, A., and Corbella, J. (1990), *J. Appl. Toxicol.,* 10, 181–186.
Paternain, J. L., Domingo, J. L., and Corbella, J. (1988), *J. Toxicol. Environ. Health,* 24, 193–200.
Pennington, J. A. (1990), *J. Am. Diet. Assoc.,* 90, 1571–1581.
Peters, J. M., Wiley, L. M., Zidenberg-Cherr, S., and Keen, C. L. (1995), *Teratogenesis Carcinog. Mutagen.,* 15, 23–31.
Peters, J. M., Wiley, L. M., Zidenberg-Cherr, S., and Keen, C. L. (1991), *Proc. Soc. Expl. Biol. Med.,* 198, 561–568.
Petrukhin, K., Lutsenko, S., Chernov, I., Ross, B. M., Kaplan, J. K., and Conrad Gilliam, T. (1994), *Hum. Mol. Genet.,* 3, 1647–1656.
Petry, C. D., Eaton, M. A., Wobken, J. D., Mills, M. M., Johnson, D. E., and Georgieff, M. K. (1992), *J. Pediatr.,* 121, 109–114.
Pollitt, E. (1995), *J. Nutr.,* 125, 2272S–2277S.
Prohaska, J. R. and Bailey, W. R. (1993), *J. Nutr.,* 123, 1226–1234.
Prohaska, J. R. (1987), *Physiol. Rev.,* 67, 858–901.
Prohaska, J. R. and Wells, W. W. (1975), *J. Neurochem.,* 25, 221–228.
Recommended Dietary Allowances (1989), 10th ed., National Academy Press, Washington, D.C.
Record, I. R., Dreosti, I. E., Tulsi, R. S., and Manuel, S. J. (1986), *Teratology,* 33, 311–317.
Record, I. R., Tulsi, R. S., Dreosti, I. E., and Fraser, F. J. (1985), *Teratology,* 32, 397–405.
Reinstein, N. H., Lönnerdal, B., Keen, C. L., and Hurley, L. S. (1984), *J. Nutr.,* 114, 1266–1279.
Richards, R. and Brooks, S. E. H. (1966), *West Indian Med. J.,* 15, 134–140.
Robertson, D. S. F. (1970), *Lancet,* 1, 518–519.
Rogers, J. M. (1996), *Toxicology of Metals,* Chang, L. W., Ed., CRC Press, Boca Raton, FL, Chap. V.3.
Rogers, J. M., Taubeneck, M. W., Daston, G. P., Sulik, K. K., Zucker, R. M., Elstein, K. H., Jankowski, M. A., and Keen, C. L. (in press), *Teratology.*
Rogers, J. M., Lönnerdal, B., Hurley, L. S., and Keen, C. L. (1987), *J. Nutr.,* 117, 1875–1882.

Roszkowski, I., Wojcicka, J., and Zaleska, K. (1966), *Obstet. Gynaecol.,* 28, 820–825.
Rucker, R. B., Lönnerdal, B., and Keen, C. L. (1994), *Physiology of the Gastrointestinal Tract,* 3rd ed., Johnson, L. R., Ed., Raven Press, New York, 2183–2202.
Rucker, R. B. and Tinker, D. (1977), *Int. Rev. Exp. Pathol.,* 17, 1–47.
Runner, M. N. and Miller, J. R. (1956), *Anat. Rec.,* 124, 437–438.
Russell, K. P., Rose, H., and Starr, P. (1957), *Surg. Gynecol. Obstet.,* 104, 560.
Ruz, M. and Solomon, N. W. (1990), *Ped. Res.,* 27, 170–175.
Saillenfait, A. M., Payan, J. P., Sabate, J. P., Langonne, I., Fabry, J. P., and Beydon, D. (1993), *Toxicol. Appl. Pharmacol.,* 123, 299–308.
Saillenfait, A. M., Sabate, J. P., Langonne, I., and De Ceaurriz, J. (1991), *Toxicol. In Vitro,* 5, 83–89.
Salgo, M. P. and Oster, G. (1974), *J. Reprod. Fert.,* 39, 375–377.
Samarawickrama, G. P. and Webb, M. (1979), *Environ. Health Perspect.,* 28, 245–249.
Sánchez, D. J., Domingo, J. L., Llobet, J. M., and Keen, C. L. (1993), *Toxicol. Lett.,* 69, 45–52.
Sandstead, H. H. (1995), *Am. J. Clin. Nutr.,* 61 (Suppl.), 621S-624S.
Saner, G., Dağoğlu, and Özden, T. (1985), *Am. J. Clin. Nutr.,* 41, 1042–1044.
Schellenberg, D., Marks, T. A., Metzler, C. M., Oostveen, J. A., and Morey, M. J. (1990), *Vet. Hum. Toxicol.,* 32, 309–314.
Scholl, T. O. and Hediger, M. L. (1994), *Am. J. Clin. Nutr.,* 59 (Suppl.), 492S–501S.
Schroeder, H. A., Mitchener, M., Balassa, J. J., Kanisawa, M., and Nason, A. P. (1968), *J. Nutr.,* 95, 95–101.
Schuler, C. A., Anthony, R. G., and Ohlendorf, H. M. (1990), *Arch. Environ. Contam. Toxicol.,* 19, 845–853.
Schwarz, K. and Mertz, W. (1959), *Arch. Biochem. Biophys.,* 85, 292–295.
Shavlovski, M. M., Chebotar, N. A., Konopistseva, L. A., Zakharova, E. T., Kachourin, A. M., Vassiliev, V. B., and Gaitskhoki, V. S. (1995), *Biometals,* 8, 122–128.
Shepard, T. H., Mackler, B., and Finch, C. A. (1980), *Teratology,* 22, 329–334.
Siegel, E. and Wason, S. (1986), *Pediatr. Clin. North Am.,* 33, 363–367.
Simmer, K., Carlsson, L., and Thompson, R. P. H. (1992), *Trace Elem. Med.,* 9, 109–112.
Singh, C., Saxena, D. K., Murthy, R. C., and Chandra, S. V. (1993), *Hum. Exp. Toxicol.* 12, 25–28.
Singh, U. S., Saxena, D. K., Singh, C., Murthy, R. C., and Chandra, S. V. (1991), *Reprod. Toxicol.,* 5, 211–217.
Slot, H. M., Overweg-Plandsoen, W. C., Bakker, H. D., Abeling, N. G., Tamminga, P., Barth, P. G., and Van Gennip, A. H. (1993), *Neuropediatrics,* 24, 139–142.
Smith, M. K., George, E. L., Stober, J. A., Feng, H. A., and Kimmel, G. L. (1993), *Environ. Res.,* 61, 200–211.
Smith, R. M. (1987), Cobalt, *Trace Elements in Human and Animal Nutrition,* 5th ed., Vol. 1, Mertz, W., Ed., Academic Press, Orlando, FL, 143–183.
Staley, G. P., van der Lugt, J. J., Axsel, G., and Loock, A. H. (1994), *J. South African Vet. Assoc.,* 65, 73–78.
Stoecker, B. J. (1994), *Risk Assessment of Essential Elements,* Mertz, W., Abernathy, C. O., and Olin, S. S., Eds., International Life Sciences Institute Press, Washington, D. C., 197–205.
Stoecker, B. J. (1990), *Present Knowledge in Nutrition,* 6th ed, Brown, M. L., Ed., International Life Sciences Institute, Nutrition Foundation, Washington, D.C., 287–293.
Strom, R. L., Schiller, P., Seeds, A. E., and Ten Bensel, R. (1976), *Minnesota Med.,* 59, 483–489.
Sunderman, F. W., Jr., Reid, M. C., Shen, S. K., and Kevorkian, C. B. (1983), *Reproductive and Developmental Toxicity of Metals,* Clarkson, T. W., Nordberg, G. F., and Sager, P. F., Eds., Plenum Press, New York, 399–416.
Sunderman, F. W., Jr., Shen, S. K., Mitchell, J. M., Allpass, P. R., and Damjanov, I. (1978), *Toxicol. Appl. Pharmacol.,* 43, 381–390.
Swenerton, H. and Hurley, L. S. (1971), *Science,* 173, 575, 583.
Tao, S. and Suttie, J. W. (1976), *J. Nutr.,* 106, 1115–1122.
Tarantal, A. F., Willhite, C. C., Lasley, B. L., Murphy, C. J., Cukierski, M. J., Book, S. A., and Hendrickx, A. G. (1991), *Fundam. Applied Toxicol.,* 16, 147–160.
Taubeneck, M. W., Daston, G. P., Rogers, J. M., Gershwin, M. E., Ansari, A., and Keen, C. L. (1995), *J. Nutr.,* 125, 908–919.
Taubeneck, M. W., Daston, G. P., Rogers, J. M., and Keen, C. L. (1994), *Reprod. Toxicol.,* 8, 25–40.
Taubeneck, M. W., Domingo, J. L., Llobet, J. M., and Keen, C. L. (1992), *Toxicology,* 72, 27–40.
Trivedi, B., Saxena, D. K., Murthy, R. C., and Chandra, S. V. (1989), *Reprod. Toxicol.,* 3, 275–278.
Turnlund, J. R. (1994), *Modern Nutrition in Health and Disease,* 8th ed., Shils, M. E., Olson, J. A., and Shike, M., Eds., Lea & Febiger, Philadelphia, 231–241.
Uriu-Hare, J. Y., Stern, J. S., and Keen, C. L. (1989), *Diabetes,* 38, 1282–1290.
Uthus, E. O., Poellot, R., and Nielsen, F. H. (1989), *Spurenelement — symposium: molybdenum, vanadium, and other trace elements,* Anke, M., Baumann, W., Braünlich, H. et al., Eds., Friedrich-Schiller-Universität, Jena, Germany, 1013–1017.
Vallee, B. L. and Falchuk, K. H. (1993), *Physiol. Rev.,* 73, 79–118.
Viteri, F. E. (1994), *Nutrient Regulation During Pregnancy, Lactation, and Infant Growth,* Allen, L., King, J., and Lönnerdal, B., Eds., Plenum Press, New York, 127–139.
Walsh, C. T., Sandstead, H. H., Prasad, A. S., Newberne, P. M., and Fraker, P. J. (1994), *Environ. Health Perspect.,* 102 (Suppl. 2), 5–46.
Walter, R. M., Uriu-Hare, J. Y., Olin, K. L., Oster, M. H., Anawalt, B. D., Critchfield, J. W., and Keen, C. L. (1991), *Diabetes Care,* 14, 1050–1056.

Waring, P., Egan, M., Braithwaite, A., Mullbacher, A., and Sjaarda, A. (1990), *Int. J. Immunopharmacol.,* 12, 445–457.
Weber, C. W. and Reid, B. L. (1974), *Trace Element Metabolism in Animals, 2,* Hoekstra, W. G., Suttie, J. W., Ganther, H. E., and Mertz, W., Eds., Univerity Park Press, Baltimore, MD, 707–709.
Webster, W. S. and Valois, A. O. (1987), *Neurotoxicology,* 8, 437–444.
Wiener, G., Wilmut, I., and Field, A. C. (1978), *Trace Element Metabolism in Man and Animals, 3,* Kirchgessner, M., Ed., Technischen Universität München, Freising-Weihenstephan, Germany, 469–472.
Wide, M. (1984), *Environ. Res.,* 33, 47–53.
Willhite, C. C. (1993), *Ann. N. Y. Acad. Sci.,* 678, 169–177.
Willhite, C. C., Ferm, V. H., and Zeise, L. (1990), *Teratology,* 42, 359–371.
Wu, A. S., Oldfield, J. E., Shull, L. R., and Cheeke, P. R. (1979), *Biol. Reprod.,* 20, 793–798.
Xue-Yi, C., Xin-Min, J., Zhi-Hong, D., Rakeman, M. A., Ming-Li, Z., O'Donnell, K., Tai, M., Amette, K., DeLong, N., and DeLong, G. R. (1994), *N. Engl. J. Med.,* 331, 1739–1744.
Yonemoto, J., Satoh, H., Himeno, S., and Suzuki, T. (1983), *Teratology,* 28, 333–340.
Zhang, T., Gou, X., and Yang, Z. (1993), *J. West China Univ. Med. Sci.,* 24, 202–205.
Zidenberg-Cherr, S. and Keen, C. L. (1991), *Trace Elements, Micronutrients and Free Radicals,* Preosti, I. E., Ed., Humana Press, Englewood Cliffs, 107–127.
Zidenberg-Cherr, S., Benak, P. A., Hurley, L. S., and Keen, C. L. (1988), *Drug Nutr. Interact.,* 5, 257–274.
Zimmerman, A. W. and Lozzio, C. B. (1989), *Zentbl. Kinderchir.,* 44 (Suppl. 1), 48–50.

Chapter 64

Placental Transport, Metabolism, and Toxicity of Metals

Carol J. Eisenmann and Richard K. Miller

I. INTRODUCTION

The placenta and its associated extraembryonic membranes have simply been considered the conduit for the transfer of molecules from mother to fetus and vice versa. This transport activity is certainly an important factor, especially when considering the transfer of toxicants, which have direct action upon the embryo/fetus, e.g., lead and methylmercury. However, the placenta may also be both a modulator of such transfer as well as a site for toxic action. A number of contributions on the reproductive and developmental toxicity of metals have already been provided in this volume as well as in other reviews (Clarkson et al., 1983; Sager et al., 1986; Domingo, 1994). Therefore, the focus of this review is the placental toxicity of metals and how placental toxicity may contribute to the known fetal effects of excess metal exposure. The evaluation of the conduit between mother and embryo/fetus, the controller of maternal physiology, and anchor for the embryo, all known as the chorioallanotic placenta and its extraembryonic membranes (including yolk sac), is limited to color and weight in developmental toxicity screening and is often only superficial in the hospital. The ability to document toxic and/or pharmacologic influences on the placenta and its associated extraembryonic membranes is further limited. In addition to this issue of toxicity, the transplacental transfer, placental localization, and metabolism of metals are also examined in the human placenta to provide insight into whether the placenta serves as a barrier for some metals. The principal metals discussed are arsenic, cadmium, lead, mercury, and selenium.

The placenta, a dynamic organ at the interface between mother and developing fetus, changes structure and plays numerous roles during development, acting as the fetal lung, gut, and kidney, as well as functioning as an endocrine gland (Benirschke and Kaufmann, 1990). Structurally, the human placenta is a hemo-monochorial type. This means that the blastocyst invades the endometrium, removing endometrial connective tissue and eroding maternal blood vessels so that a single placental syncytiotrophoblastic layer is directly in contact with maternal blood. The syncytiotrophoblast layer is covered with microvilli, which enlarge the surface area available for nutrient uptake/gas exchange. Except for primates, the structure of the human placenta differs from routinely used laboratory animals which have additional trophoblast layers separating the maternal blood and the fetus, e.g., rabbit, rat, mouse, or have a maternal endothelium preventing direct contact between maternal blood and the placenta, e.g., dog. More details concerning interspecies differences in placental structure are reviewed in Benirschke and Kaufmann (1990) and Panigel (1982). Yet of principal concern is the fact that the yolk sac placenta of the rodent and lagomorph continues to be functional throughout gestation, playing a critical role in immunoglobulin transfer, while for the human, the yolk sac does not appear to continue to be functional much beyond

8 weeks of gestation. From studies of metals, it has become apparent that the vitelline circulation and the yolk sac may be an important route of entry for metals, e.g., cadmium, until 8.5 days, but not thereafter (Dencker, 1983).

A schematic of the cut surface of the human placenta from the chorionic plate, the fetal facing surface, to the basal plate, the contact zone between fetal and maternal tissues, is shown in Figure 1. Fetal arteries and veins enter the placenta from the umbilical cord, which contains two arteries and one vein. On the fetal surface of the placenta, arteries cross over veins and divide into branches that supply individual placental cotyledons, the basic unit of the placenta. Inside the placenta, the vessels enter villi, which further divide to form terminal villi. These villi are surrounded by maternal blood in what is called the subchorial lake or lacuna.

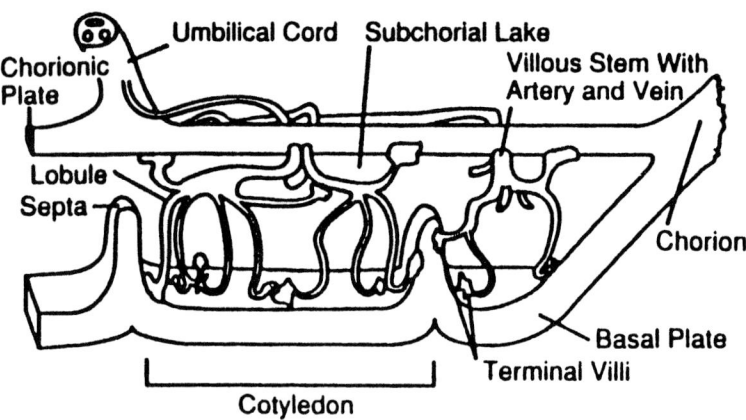

Figure 1. Schematic view of the cut surface of a human placenta. (From Novak, 1991.)

Maternal blood enters the placenta from the uterine spiral arteries, which are maximally dilated and unresponsive to vasoactive agents. Exactly how maternal blood enters the placenta and the direction of maternal blood flow through the placenta are not entirely clear (Novak, 1991). The maternal blood in the intervillous space percolates around terminal villi, allowing the transfer of oxygen and metabolites. The intervillous blood volume comprises approximately 23 to 38% of the placental volume (Benirschke and Kaufmann, 1990).

The large placental surface area in contact with a relatively large volume of maternal blood, required for normal placental function, may also make the placenta vulnerable to toxicants. This may be especially true for toxic metals which often resemble essential elements, e.g., cadmium and calcium. The toxic metals may enter the placenta and be concentrated there, or may be transferred to the fetus via mechanisms designed to transport nutrients. Because fetal demands for nutrients increase with gestation, the potential for placental intoxication with toxic metals may also increase with gestation.

It should not be overlooked that peri-implantation can also be a sensitive time for chemicals, including metals, to alter placental function, not just the outer covering, the syncytiotrophoblast, but also the cytotrophoblast and the highly invasive extravillous trophoblast (Genbacev and Miller, 1994).

To study placental function, especially in the human, numerous *in vitro* models have been applied. Unfortunately no one model can address all issues. Reviews of model systems for *in vitro* toxicity evaluations can be found in Table 1 and the following references: placental dual perfusions (Miller et al, 1994; Schneider, 1995); Explants and Cell Culture (Genbacev and Miller, 1994; Genbacev et al 1993); all technics (Wier and Miller, 1989).

To study transplacental transport, only the term "human placenta" can be currently used because the methods of delivery for first and second trimester placental tissue leave the tissue with tears in the membranes and villi. Yet the first trimester placenta can be studied in culture not only for all of the functions in Table 1, but also to examine the ability of the trophoblast to grow out of the villus and invade surrounding tissue as extravillous trophoblast. Thus, it is possible to study not only cell prolif-

Table 1 General Placental Functions which can be Studied under Different *in vitro* Conditions

Method	Hemodynamics	Transplacental Transport	Membrane/Cellular Uptake	Endocrine	Metabolism
Organ perfusion	+	+	+	+	+
Organ culture	–	–	+	+	+
Cell culture	–	–	+	+	+
Subcellular tissue Preparations	–	–	+	+	+

Modified from Wier and Miller, 1989.

eration and invasion but also cell differentiation (Genbacev and Miller, 1994; Genbacev et al., 1992, 1993, 1994). Such methods have been useful in studying the toxicity of metals, especially cadmium (Powlin et al., 1994, 1996; Eisenmann and Miller, 1994).

The placenta can also be used as a marker of environmental or occupational exposure based upon the accumulation of metals within the tissue (Miller et al., 1988). Thus, the placenta and measurements of its function and its ability to accumulate metals provide an important role as a biomarker of exposure and effect. These issues will be discussed in greater detail when examining individual metals.

II. SPECIFIC METALS

A. ARSENIC
1. General Information

There are many forms of arsenic. In the environment, arsenic is primarily in the pentavalent form (arsenate), except under reducing conditions, e.g., in deep well water, in which it may be present as the trivalent form (arsenite). Human exposure to arsenic occurs through water and food, with marine seafood containing significant amounts. Arsenic in seafood is found in organic compounds including methyl- and dimethylarsenic acids, arsenobetaine, arsenocholine, and arsonium phospholipids (WHO, 1981). Arsenic exposure can also occur through cigarette smoking, although the arsenic content of tobacco has been reduced in recent years. Occupational exposure to arsenic is a concern among smelter workers, and individuals involved in the use and production of arsenic-containing pesticides. Arsenic compounds have also been used in medicine.

Arsenic compounds are generally well absorbed from the gastrointestinal and respiratory tracts, with a majority of the dose eliminated in the urine in approximately 2 days. As a result of relatively rapid elimination of arsenic, human arsenic blood levels reflect recent exposure, and arsenic in blood rarely reaches steady state. No data are available to indicate a quantitative relationship between exposure and human blood arsenic concentrations (WHO, 1981).

Both oxidation and reduction of arsenic can occur; pentavalent arsenic can be found after treatment of animals with trivalent arsenic, and trivalent arsenic is recovered following treatment with pentavalent arsenic (Goyer, 1986). The principal form of arsenic found in the urine following inorganic arsenic exposure is dimethylarsenic acid (Goyer et al., 1986). The methylation of arsenic, which occurs via B-12 independent pathways (*S*-adenosylmethionine is not the methyl donor) (Chen et al., 1992), is considered to detoxify arsenic. The organic arsenic compounds found in fish are thought to be excreted primarily in the urine without being biotransformed (WHO, 1981).

In general, trivalent arsenic is retained in organs to a greater degree than pentavalent arsenic, and trivalent arsenite is considered to be the toxic form. Arsenic has high affinity for sulfhydryl groups, which contributes to the accumulation of arsenic in keratin-containing tissues, skin, and hair. Hair arsenic levels have been used to assess human arsenic exposure (WHO, 1981). The affinity of arsenic for sulfhydryl groups results in inhibition of numerous enzymes. In particular, arsenic interferes with enzymes of respiration and uncouples mitochondrial respiration (WHO, 1981).

The adverse effects of arsenic have been noted in numerous organ systems including the skin, respiratory, cardiovascular, gastrointestinal, nervous, and hematopoietic systems. Further details concerning the adverse effects of arsenic are summarized in WHO (1981). Arsenic is also a human carcinogen, resulting in skin cancers of low malignancy following oral ingestion, and lung cancer following inhalation exposure. Arsenic causes sister chromatid exchanges and chromosome breaks, but does not result in point mutations (Stöhrer, 1991). It induces genes in many organisms and it is this gene induction, affecting the regulation of cell growth, which is thought to indirectly result in cancer (Stöhrer, 1991).

Arsenic in drinking water is regulated on the basis of skin effects (US 50 µg/l). These effects are described as hyper- and hypopigmentation, hyperkeratosis on palms and soles, and punctate keratoses. Skin cancer (multicentric basal cell and squamous cell carcinoma) is observed only in populations in which other skin effects are observed. Based on epidemiologic studies completed in Taiwan and Bengal, (reviewed by Stöhrer, 1991), the threshold drinking water concentration for these effects appears to be 100 µg/l (400 µg/d based on tropical water intake of 4 l/d), with hyperpigmentation observed in almost everyone exposed to 200 µg/l arsenic for 1 year in the Bengal study.

2. Placental Transfer and Toxicity
a. Animal Data

Animal studies have shown that arsenic does cross the placenta, and single injection doses of arsenic are teratogenic in hamsters (Ferm and Carpenter, 1968) and mice (Hood et al., 1977). Methylated arsenic compounds are less teratogenic than inorganic arsenic (Harrison et al., 1980; Hood et al., 1982) and trivalent arsenic is more teratogenic than pentavalent arsenic. The differences in teratogenicity of trivalent and pentavalent arsenic are thought to result from differences in the toxicity of the two forms, rather than from differences in the placental pharmacokinetics (Dencker et al., 1983).

Hood and colleagues have examined the uptake, distribution, and metabolism of sodium arsenite (25 µg/kg) (1988) and sodium arsenate (40 µg/kg) (1987) given by gavage to pregnant mice on gestation day 18. Following treatment, placental arsenic concentrations peaked 4 and 2 hours after dosing with arsenite and arsenate, respectively. Fetal levels peaked 6 h after mice were treated with arsenate, while after arsenite treatment, the highest fetal arsenic concentrations were observed at the last time point, 24 h after treatment. Greater than 80% of the arsenic in the fetuses was methylated (mono and dimethyl arsenic) following treatment with either arsenic form. The source of the methylated arsenic found in fetuses has not yet been established.

Hanlon and Ferm (1987) studied the concentration and chemical state of arsenic in the placentae of hamsters treated with sodium arsenate via osmotic minipumps for 48 h. The animals were treated beginning on gestation day 6 and euthanitized on gestation day 8. The doses used ranged from minimally (106 to 114 µmol/kg) to frankly teratogenic (200 to 223 µmol/kg). Total placental arsenic concentrations were higher than maternal blood concentrations at all doses. As a percentage of total arsenic, dimethyl-arsenate accounted for about 11%. Approximately 70% of the placental arsenic was bound to macromolecules, with reversibly bound arsenic accounting for two thirds. The remaining arsenic was inorganic. Chromatography of inorganic arsenic at the high dose indicated that 40% was in the form of arsenate and 60% was arsenite.

b. Human Data

Human data have also indicated that arsenic crosses the placenta. In relatively unexposed populations, maternal and cord blood arsenic levels were similar in two studies. In a study completed in the U.S. (Kagey et al., 1977), the geometric mean of maternal blood levels was 3.0 µg/l (n = 52) and 1.17 µg/l (n = 49) in Charlotte, NC and Birmingham, AL, while cord blood concentrations were 2.23 (n = 51) and 1.69 µg/l (n = 48) in the two cities. The placental levels were 1.32 µg/kg in Charlotte, and 1.87 µg/kg in Birmingham, indicating that in persons without known occupational exposures to arsenic, the placenta does not accumulate arsenic relative to concentrations found in the blood. No difference in arsenic in maternal or cord blood was observed among smokers and nonsmokers. In a study from Taiwan (Soong et al., 1991), 82 pairs of maternal and cord blood were examined. Mean and standard deviation arsenic concentrations were 6.98 ± 5.25 and 7.89 ± 6.07 µg/l, for maternal and cord blood, respectively. A positive correlation (r = 0.57) between maternal and cord blood arsenic was noted. Placental arsenic levels were not reported in this study.

A preliminary report (Tabacova et al., 1992) suggests that placental arsenic concentrations reflect environmental exposure. Placental arsenic concentrations averaged 7.4 µg/kg in placentae from individuals living in areas of Bulgaria with no industrial sources of metal pollution, in contrast to 26.6 µg/kg in placentae from individuals living near copper smelters.

At high levels, acute exposure to arsenic can result in adverse fetal effects. In a case report (Lugo et al., 1969), a woman 30 weeks pregnant attempted suicide with arsenic (about 30 ml arsenic trioxide solution containing 1.32% arsenic). Despite treatment with the chelator, BAL, 24 hours after the ingestion, symptoms became worse during the following three days and on the third day she delivered a 1100 g live infant. The infant had progressive respiratory distress, and primarily as a result of prematurity, died at 11 hours of age. Arsenic content of fetal liver, kidney, and brain were 0.74, 0.15, and 0.022 mg

$As_2O_3/100$ g of wet tissue. Unfortunately, the placenta was discarded before it was examined, so that possible contributions of placental malfunction to the preterm delivery cannot be determined.

There is some evidence to suggest that adverse reproductive outcome can occur in humans chronically exposed to arsenic at lower levels. Nordström et al. (1979) reported an increase in multiple malformations in infants born to workers exposed to high concentrations of arsenic at a copper smelter in Sweden. These workers were also reported to have an increased frequency of chromosomal aberrations. An increase in the frequency of spontaneous abortion was observed in women living nearest to the smelter, relative to women living more than 50 km from the plant (Nordström et al., 1978). Unfortunately, exposures at and near the smelter involved a number of heavy metals as well as sulfur dioxide, so the specific cause of the malformation and spontaneous abortion excess cannot be determined (WHO, 1981).

Tabacova et al. (1992) noted a reduction in birthweight among infants born to individuals living near copper smelters in Bulgaria, relative to nonexposed individuals. The levels of lipid peroxides in the placentae increased with arsenic, while a decrease in the reduced/total glutathionine ratio was observed.

Preliminary data from a study in Hungary (Börzsönyi et al., 1992) provides additional support to the theory that arsenic exposure may increase the risk of spontaneous abortion and stillbirths. In this study, rates of spontaneous abortion and stillbirths were 69.57 and 7.68 per 1000 live births (total live births 5218) in a population that drank deep well water containing arsenic at >100 µg/l. In a similar population with low arsenic concentrations in the drinking water (not further defined), rates of spontaneous abortion and stillbirths were 51.14 and 2.84 per 1000 live births (total live births 2112). The differences between the two populations were statistically significant (chi square test) at $p = 0.007$ for spontaneous abortion and $p = 0.028$ for stillbirths. In this study the exposed population was clearly receiving relatively high levels of arsenic; there were many cases of arsenical hyperkeratosis and hyperpigmentation in both children and adults. Ongoing studies of these populations should provide further insight into the human reproductive toxicity of arsenic and the relationship to exposure.

The increase in malformations, spontaneous abortions, and stillbirths reported in arsenic-exposed individuals has been in part attributed to arsenic-induced genetic damage (Nordström et al., 1978). These effects have been noted in populations in which other arsenic effects were found, e.g., chromosomal aberrations in the smelter workers, skin effects in individuals consuming arsenic-contaminated drinking water. The possible contribution of arsenic effects on placental function to the increases in spontaneous abortions and stillbirths, including effects on placental oxidative status and enzymes of respiration, requires further investigation.

B. CADMIUM
1. Placental Transport and Toxicity
a. Animal Data

Relative to the other metals discussed in this review, the placental effects of cadmium are the most well documented. In rats and mice treated with cadmium in the drinking water (up to 100 ppm) throughout gestation cadmium has been shown to accumulate in the placenta and decrease fetal zinc levels (Webster, 1988; Sowa and Steibert, 1985). The changes in fetal zinc levels resulting from cadmium exposure may result from the induction of the metal binding metallothionein proteins in maternal tissues and the placenta (Hazelhoff Roelzema et al., 1989). Cadmium is slowly transferred from mother to fetus in the rat with concentrations appearing in the placenta in excess of those noted in either the mother or fetus (Figure 2) (Sonnawane et al. 1975; Levin et al, 1987). Metallothioneins may retain metals in maternal tissues and the placenta, reducing cadmium transport to the fetus and altering essential metal transport.

Cadmium is clearly a placental toxicant in animals given an acute subcutaneous injection. Parizek (1965) reported death (maternal and fetal), placental necrosis, with generalized visceral venous congestion and hemorrhages in the kidneys and adrenals in pregnant rats treated with a subcutaneous dose of 20 mmol Cd/kg on gestation day 18. Levin and Miller (1981) observed fetal lethality, placental necrosis, and alterations in utero-placental blood flow in pregnant rats treated with cadmium (40 mmol/kg) on gestation day 18. Microscopic examination of placentae from rats euthanatized 12 to 96 h after treatment revealed hemorrhagic necrosis, congestion of the labyrinthine portion of the placenta, and infiltration by polymorphonuclear leukocytes.

Ultrastructural studies (di Sant'Agnese et al., 1983; Cho et al., 1988) of placentae from rats treated in a similar manner and sacrificed 1 to 14 h after cadmium treatment showed that the trophoblast cell layer II (a syncytial layer) was most sensitive. Early effects noted in cell layer II included lysosmal vesiculation, nuclear chromatin clumping, nucleolar changes, and mitochondrial calcification. Later

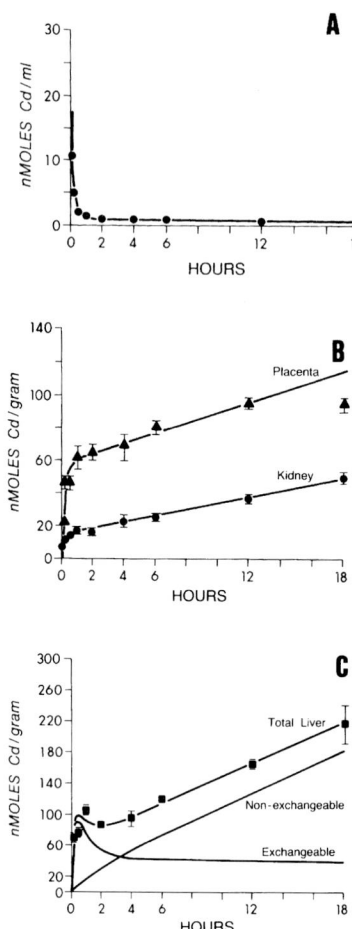

Figure 2. Toxicokinetics of cadmium chloride administered to the near-term pregnant rat. Cadmium chloride (40 μmol/kg) was administered subcutaneously in pregnant Wistar rats on day 18 of gestation. Toxicokinetic modeling was performed on a Systron Donner 10/20 analog computer using Hewlett Packard 1310A display. (A) Blood concentration of cadmium with time. The solid line represents the computer simulation of the data. This curve was used to derive all organ compartments except the fetus. (B) Cadmium concentration data for placenta and maternal kidney. The solid lines are computer simulations of the data. Note that the model for the placenta does not predict the 18 h time point due to the placental toxicity. Mean; S.E.M. (C) Cadmium concentration data for maternal liver. The lines represent computer simulations of the data. Exchangeable and nonexchangeable compartment concentrations are shown. Mean; S.E.M. (From Levin, A.A., Kilpper, R.W., and Miller, R.K., *Teratology*, 36, 163, 1987. With permission.)

changes included extracellular deposition of fibrin and platelets, congestion of the maternal blood space and fetal capillaries, and necrosis of the trophoblast layers. In contrast, fetal capillaries were often intact.

The placental toxicity was responsible for the fetal deaths observed in rats treated with cadmium on gestation day 18 (Levin and Miller, 1980). Maternal injection (subcutaneous, 40 mmol/kg) of cadmium resulting in fetal cadmium body burdens of 8.6 ± 4.4 nmol caused death of 74.9% of fetuses. In contrast, direct injection of gestation day 18 fetuses, resulting in fetal body burdens of 74 ± 34.8 nmol caused the death of 11.5% of fetuses, which was not substantially different from control injections of solvent. It should be noted that when the directly injected fetuses were examined postnatally, hydrocephalus and edema were noted (White et al., 1990). Such experiments demonstrate that the placenta is actually accumulating the cadmium and providing a protective sink preventing the movement of large amounts of cadmium to the fetus, where the cadmium would be directly damaging. It is also noted that even though the uptake of cadmium acutely by the placenta is second only to the liver on a tissue concentration basis, the placenta can become intoxicated, leading to necrosis and the eventual demise of the fetus. Interestingly, the placental toxic effects in the pregnant rats were noted at blood levels which did not produce acute gross nephrotoxicity, while cadmium-metallothioinein when injected intravenously did produce nephrotoxicity acutely in the near-term pregnant rat without any placental toxicity (Levin et al., 1983).

b. Human Data

Human cadmium exposure in the nonsmoking public is principally via food, with grains, shellfish, and liver and kidney from contaminated animals containing the highest cadmium concentrations. The total cadmium intake in the U.S. is approximately 5 to 18 µg/d, of which about 5% is absorbed (Reddy and Hayes, 1989). Cadmium blood levels in adults not excessively exposed to cadmium are usually below 1 µg/dl (0.09 nM) (Goyer, 1986).

Cigarette smoking is a major source of cadmium exposure, with approximately 15 to 30% of inhaled cadmium absorbed (Goyer, 1986). Cigarette smoking is associated with decreased infant birth weight, as well as placental changes including increased subchorionic fibrin deposits and placental calcifications (Christianson, 1979). Does cadmium contribute to the adverse developmental effects of smoking? Loiacono et al. (1992) found no association between birthweight and placental cadmium concentrations in women living near a lead smelter. Placental cadmium concentrations in these women (0.73 nmol/g dry wt) were similar to levels reported in smokers, suggesting that factors in addition to cadmium are responsible for the association between smoking and decreased birthweight.

Two case reports to PEDECS have raised the issue for the human that cadmium may also be toxic early in gestation. Two high school welding teachers, who were also cigarette smokers, had documented cadmium in their urine and increased levels of beta microglobulins. It was determined that their industrial arts classrooms did not have appropriate ventilation. Of particular interest was the additional observation that these women both had multiple miscarriages (<8 weeks of pregnancy). The association between the pregnancy losses and the high levels of cadmium are interesting. Certainly other exposures and disease processes may be involved, but were not identified for these women. Of particular interest are the observations in our laboratory that cadmium can reduce the proliferation and differentiation of cytotrophoblast cell columns in the early first trimester human placenta in explant culture (Powlin et al., 1996).

Cadmium is transferred across the human placenta with similar kinetics as noted for the rodent (Figure 3). High concentrations of cadmium are localized in the human placenta (Figure 3), which is associated with toxic action. A number of different cellular processes have been proposed as sites for the action of cadmium in producing its cellular toxicity. A number of investigations have focused on the placenta as a target site (Parizek, 1964, 1965; Levin and Miller, 1980; Levin et al., 1981, 1983; Lehman and Poisner, 1984; Wier et al., 1990; Miller et al., 1991; Torreblanca et al., 1992; Sorel and Graziano, 1990; Page et al., 1992; Eisenmann and Miller, 1995a). There are actually two major components to this placental toxicity of cadmium, (1) the direct toxicity of cadmium, and (2) the cellular defense processes available in the placenta to prevent the toxicity. Some responses by the placenta have been reported to be altered. These include

- **transport processes** — for amino acids (decrease), cyanocobalamin (decrease) (Danielsson et al, 1984), copper, iron (decrease) (Sowa and Stiebert, 1985); zinc (decrease) (Page et al., 1992; Sorel and Graziano, 1990; Wier et al., 1990) (increased) (Torreblanca et al., 1992);
- **cellular metabolism** — ATP levels (no change) (Miller et al., 1991), glucose utilization (no change) (Wier et al., 1990); lactate production (no change) (Wier et al., 1990); human chorionic gonadotropin production (decrease) (Wier et al., 1990); oxygen consumption (no change) (Wier et al., 1990); prostacyclin/thromboxane ratio (altered) (Eisenmann and Miller, 1995a);
- **enzyme activity** — succinate dehydrogenase (decrease) (Cho and Panigel, 1986); glucose-6 dehydrogenase (decrease) (Boadi et al., 1992); glutathione peroxidase (decrease) (Eisenman and Miller, 1995a);
- **morphology** — ultrastructure (Wier et al., 1990).

Cellular defense mechanisms utilized by the placenta can be important in modulating the toxicity of metals. Two such defenses will be discussed in association with cadmium: glutathione and metallothionein. Both of these molecules have an affinity for binding cadmium, and bound cadmium is not toxic to the placenta. Glutathione can be produced by the placenta, and glutathione peroxidase in both cellular and extracellular forms is present (Avissar et al., 1994). Glutathione can protect against the placental toxicity of cadmium when added exogenously to the medium (Eisenmann and Miller, 1995a). Such addition of glutathione prevents the actions of cadmium on the production of 6-keto-prostaglandin $F_{1\alpha}$, which is the metabolite of prostacyclin. Such use of glutathione maintains the tromboxane A_2/prostacyclin ratio.

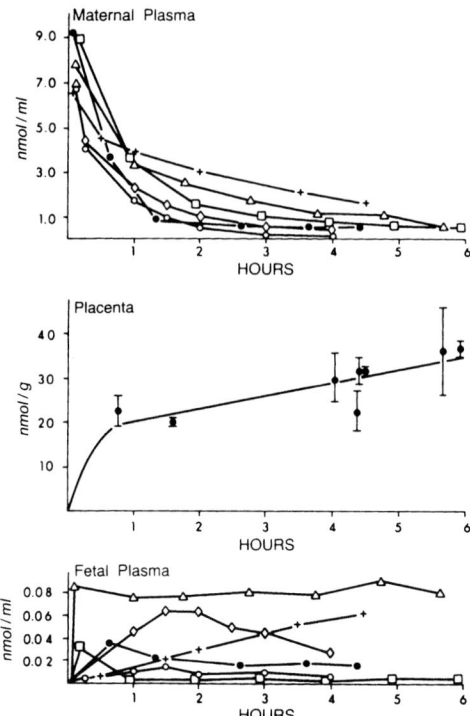

Figure 3. Cadmium concentration in maternal, placental, and fetal compartments during 6 hours of dual human placental perfusion. Cadmium was added to the maternal circulation only at 0 h of perfusion. Placental levels of cadmium (mean; S.D.) were measured at the end of perfusions lasting 0.75 to 5.9 h. The average detection limit for cadmium in the fetal perfusate was 0.03 nmole cadmium/ml. (From Wier and Miller, 1987.)

Metallothionein, a 6000-kDa cysteine-rich protein with six metal binding sites, when intracellular can protect against the toxicity of selected metals. It is well known that metallothionein when induced can protect against the cellular toxicity of metals. When metallothionein is localized to the nucleus, it can protect DNA from oxidative damage (Chubatsu et al., 1993). Metallothionein can be induced in the perfused term human placenta and in cultured human term and early trimester trophoblast cells following exposure to cadmium (Waalkes, 1984; Lehman and Poisner, 1984; Goyer et al., 1992; Boadi et al., 1991; Breen et al., 1994; 1995). In the term human placenta, cadmium-induced metallothionein is produced rapidly as noted by the accumulation of MT mRNA for MTIIa following 8 h of exposure to 20 μM of cadmium (Figure 4) (Breen et al., 1994a). Metallothionein (mRNA and protein) localizes in cells of the villous core, fetal endothelial cells, and in syncytiotrophoblast cells at term (Breen et al., 1994a). In first trimester human placental explants, metallothionein (mRNA and protein) localizes in cells of the villous core and in cytotrophoblast cells (Breen et al., 1994b). Thus, different trophoblast cells can be responding at different times during gestation to the induction of metallothionein by cadmium.

In using human trophoblast cell lines (JAr), metallothionein mRNA and protein are induced as noted for cultured term trophoblast cells (Lehman and Poisner, 1984; Wade et al., 1986). However, the JAr cells provided the opportunity to expose these human trophoblast cells chronically to low concentrations of cadmium (2 μM) for months. The human trophoblast cells do adapt to this toxic environment of cadmium by altering the expression and intracellular localization of metallothionein (Breen et al., 1995). Using conventional and confocal microscopy, the localization of metallothionein between cadmium-exposed and -unexposed trophoblast cells is markedly different. In unexposed trophoblast cells, the metallothionein was primarily perinuclear with low level, punctate expression in the cytosol. Following either chronic or 24-h exposure to cadmium (2 μM), the metallothionein protein levels increase at least 3-fold and the metallothionein localizes inside the nucleus with a lacy, cytoskeletal pattern in the cytosol. This nuclear accumulation of metallothionein is dependent upon new protein synthesis (Breen et al., 1995). The alterations, not only in the nuclear localization of metallothionein but also in the cytoskeletal

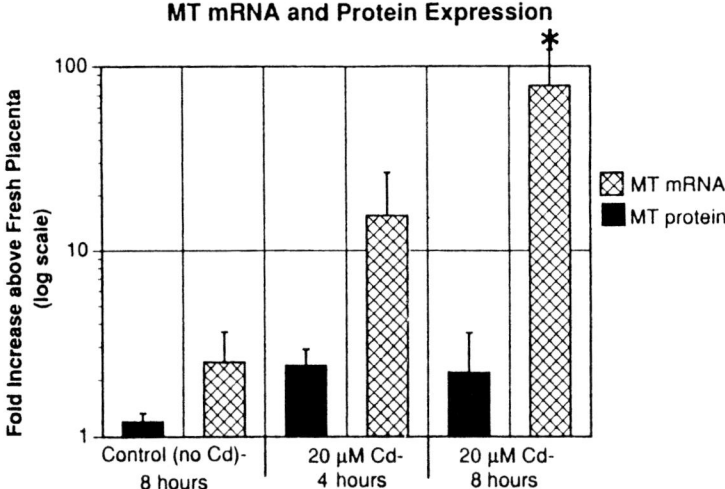

Figure 4. Metallothionein mRNA and protein levels in the human term placenta perfused with 20 μM cadmium chloride. Slot hybridization was performed using digoxigenin-UTP-labeled, antisense metallothionein IIa RNA probe and GAPDH digoxigenin-labeled DNA probe (to normalize for total amount of RNA). The ratio of MT to GAPDH signal was then used as the value for metallothionein transcript level. This value was divided by the fresh placental value to give the fold increase above fresh. Mean; SD. *significantly difference from both control and 4-h time point. (From Breen, J., Eisenmann, C., Horowitz, S., and Miller, R.K., *Reprod. Toxicol.*, 8, 297, 1994. With permission.)

localization, may play an essential part in protecting the placenta from acute higher exposures to cadmium.

Thus, cadmium can produce toxic placental responses in multiple species including the human, at term, and also peri-implanatation. Some possible sites of toxic action appear to be eliminated, e.g., carbohydrate metabolism and ATP generation, while other sites are still in question, e.g., protein synthesis (hCG, hPL), oxidative damage, enzyme inhibition (glutathione peroxidase and prostacyclin synthase), and steps in the regulation of the cycle cycle — possibly cyclins and protein kinases).

C. LEAD
1. General Information

Lead is widely distributed throughout the environment, resulting in a relatively high background exposure. Lead exposures resulting in adverse effects are associated with human activities; e.g., lead-based paints, lead in plumbing and water distribution systems, lead in air from combustion of leaded fuels and industrial activities. In countries where leaded gasoline is still used, lead emissions from cars may contribute 20% of the lead burden in adults and 35% or more in children (Krewski et al., 1989). Human lead exposure is reviewed in detail in Needleman (1992).

The absorption of lead through the gastrointestinal tract varies with age. Children absorb about 41% of ingested lead, while only 5 to 15% is absorbed in adults (Goyer, 1986). Lead that is deposited in the lungs is almost completely absorbed.

Whole blood lead is the most commonly used biological index of systemic lead exposure in human populations. Blood lead represents recent exposure, as well as a poorly defined fraction from lead previously deposited in the skeleton (Mushak et al., 1989). There appears to be at least two kinetic pools of lead; a labile soft tissue pool, and lead deposited in the skeleton which has a half life of greater than 20 years. Lead does enter the central nervous system and tends to concentrate in gray matter (Goyer, 1986). Urinary excretion is the principal route of lead elimination.

At high blood lead concentrations (>400 ng/ml), essentially all body systems, including the reproductive system, will be affected or be at risk for injury (Mushak et al., 1989). High doses of lead causes sterility, abortion, and neonatal mortality and morbidity. Further details of the reproductive effects of lead are reviewed in Miller and Bellinger (1993).

The nervous system hematopoietic system and kidneys are the principal targets of lead toxicity, while the developing nervous system is the most sensitive target of lead toxicity. Prenatal and childhood lead exposure are associated with deficits in later neurobehavioral performance, e.g., Bayley Mental Development Index. The prenatal period of exposure is thought to be the most critical (Mushak et al., 1989). Metal-analysis of retrospective studies concerning low-dose lead exposure and intellectual deficit in children support the link, as do animal studies and prospective human studies (Gatsonis and Needleman, 1992). No clear threshold for neurological deficits in children has been identified, and a blood lead concentration of 100 to 150 ng/ml is considered to be a level of concern for these effects (Mushak et al., 1989).

At the cellular level, lead can interfere with protein synthesis and inhibit membrane and mitochondrial enzymes as well as impair heme biosynthesis. These effects contribute to the observed organ system effects. For example, inhibition of heme biosynthesis not only results in anemia at high lead doses, but also affects the levels of hemoproteins in the liver, brain, and kidneys. A reduction in heme proteins in the brain may contribute to the nervous system effects, while in the kidneys, a reduction in heme production may affect the levels of 1,25-$(OH)_2$-vitamin D, which has a role in the regulation of calcium metabolism (Mushak et al., 1989). Lead, which has properties similar to calcium, may also directly compete with calcium to activate critical regulatory enzymes (Bondy, 1989). Lead can produce deficits in neurotransmission via inhibition of cholinergic function. Impairment of dopamine uptake by synaptosomes and impairment of γ-aminobutyric acid, have also been noted (Goyer, 1986).

2. Placental Transfer and Toxicity
a. Animal Data

Lead clearly crosses the placenta, and animal studies support human observations of behavioral effects following *in utero* exposure to lead (WHO, 1977; Rice, 1992). In addition to behavioral effects, lead has also been shown to decrease fetal weight. For example, fetal weight was significantly decreased in rats given lead in the drinking water throughout pregnancy at concentrations of 100 µg/l and greater (Dilts and Ahokas, 1979). A pair-feeding study showed that this effect was a result of reduced maternal food intake, as well as a direct effect on the fetus. Gerber et al. (1978) have found the treatment of mice with lead (0.5% in the diet on gestation days 8 to 18) reduced placental blood flow, an effect which may contribute to the fetal effects of lead.

Danielsson et al. (1983) examined the placental transfer of lead (^{203}lead nitrate) given to pregnant mice by intravenous injection at different stages of gestation (days 8 to 18). Lead was found in embryonic and fetal tissues at all stages of gestation, with the highest fetal uptake observed later in gestation. During early stages of gestation (days 8 to 11), lead was found predominantly in embryonic blood. Beginning on day 12, lead was taken up by the fetal liver and cartilaginous skeleton, and a large accumulation of lead was observed in calcified bone on days 14 to 18. This study was qualitative in nature, as lead transfer was assessed by autoradiography.

Nutritional status has been shown to influence the fetal uptake and toxicity of lead. Calcium deficiency increased the fetal toxicity (Jacquet and Gerber, 1979) and fetal levels of lead in mice (Leonard et al., 1983). Singh et al. (1991) examined the placental and fetal uptake of lead in iron-sufficient and -deficient rats. Rats were given lead in the drinking water at 250 to 2500 ppm on gestation days 15 to 20. Lead levels in maternal blood, placentae, and fetuses were higher in iron-deficient relative to iron-sufficient rats. Maternal blood levels showed a dose-dependent increase in lead content; this was not observed in placentae or fetuses, leading the authors to suggest that later in gestation, at high lead exposure, the placenta may serve as a partial barrier to lead transfer. Histopathological changes were observed in fetal kidneys at all dose levels. Histology of maternal tissues was not described.

b. Human Data

Human data clearly show that lead crosses the placenta. A number of studies in which lead levels were measured in maternal and cord blood as well as the placenta are summarized in Table 3. These studies show that at term, cord blood lead concentrations are generally slightly lower than maternal blood lead concentrations. Reported averages of individual cord/maternal blood lead ratios are 0.7 (Milman et al., 1988) and 0.83 (Troster and Schuartsman, 1988). The lower cord blood relative to maternal blood lead concentrations are likely a result of lead uptake by the fetus rather than the placenta acting as a barrier to lead transfer.

Placental lead levels are variable, but in general the placenta dose not appear to accumulate lead relative to maternal and cord blood. The exception is the study by Loiacona et al. (1992) in which placental lead levels (895.1 ng/g unexposed, 2871 ng/g exposed) are an order of magnitude greater than blood levels. Blood lead levels reported in this study are in the range of other studies, and no explanation for the high placental lead levels is readily apparent.

The highest cord blood levels in a nonoccupationally exposed population, 318 ng/ml, were reported by Creason et al. (1976) in a study completed in the U.S. These levels are well above the 100 to 150 ng/ml level of concern, and because no exposure was described which could account for these high levels, they may be a result of contamination or analytical and reporting errors.

In general, people living in urban areas tend to have higher blood levels than those living in rural areas. Although there are some exceptions, e.g., Toronto (Koren et al., 1990), the higher urban blood lead concentrations are reflected in Table 2. Higher blood levels, above the 100 to 150 ng/ml level of concern, were found in individuals living near a lead mine (Clark, 1977) and a lead smelter (Loiacona et al., 1992).

An interesting case report (Mayer-Popken et al., 1986) has shown that exposure early in gestation can result in fetal lead levels higher than maternal levels. In this case, conception occurred at the beginning of a 16-day period in which the woman was exposed to high concentrations of lead dust for 8 h/d at a small plant producing decorative lead plates. Production was stopped after 8 weeks, because several employees showed symptoms of acute lead intoxication and required hospitalization. Symptoms in other employees prompted blood lead measurements in the case to be completed several weeks after exposure had ended. Based on several blood lead levels determined over a 7-month period and a half-life of 20 days for the elimination of lead from blood, a maximum maternal blood lead concentration of 1200 ng/ml was estimated to have occurred near conception. At approximately 3 months of gestation, the pregnancy was terminated for unspecified medical reasons. Measurement of fetal lead content showed that lead accumulated in the skeleton (rib 1.2 µg/g dry wt) and liver (7.9 µg/g dry wt) with levels of 1.2 µg/g dry wt in the placenta. Cord blood levels, 1060 ng/ml, were more than twice that of estimated concurrent maternal blood levels (approximately 400 ng/ml).

This case suggests that following acute lead exposure early in gestation, fetal blood levels are not determined by maternal blood levels, but by liberation of fetal lead deposits in the liver and bone. This case also raises the question of the role of the placenta as a barrier to the fetal elimination of lead.

As stated earlier, the developing nervous system is the most sensitive target of lead toxicity. In addition to neurobehavioral effects, EEG profiles as well as decreased hearing acuity have been noted in children exposed to lead *in utero* (Mushak et al., 1989). The effects on the developing nervous system are likely a direct effect of lead rather than a result of placental effects.

A number of epidemiology studies reviewed by Mushak et al. (1989) and Miller and Bellinger (1993) have reported an association of lead with preterm labor and decreased birthweight. Both effects occurred at prenatal blood levels below 150 ng/ml, and it has been suggested that the length of gestation is shortened by about one half week for very 100 ng/ml increment in blood lead (Mushak et al., 1989). Effects of lead on preterm labor and birthweight were not observed in several studies summarized in Table 3 (Gersanik et al., 1974; Clark, 1977; Angell and Lavery, 1982; Ernhart et al., 1986). The possible contribution to lead effects on placental function, including placental blood flow, leading to preterm labor and decreased birthweight requires further investigation.

D. MERCURY
1. Introduction

Between 2700 to 6000 tons of mercury are released each year from the earth crust. In addition, it is estimated that 10,000 tons are generated in the mining and manufacturing processes using mercury, e.g., pulp and paper manufacturing and caustic soda manufacturing (Schardein, 1985). Additional sources are fossil fuels, combustion, production of cement, smelting of sulfur ores, and refuse incineration (WHO, 1990). Yet, it is important to distinguish the form of mercurical exposure. There are inorganic (vapor and metallic) as well as organic forms of mercury. The global cycle for mercury includes emitted mercury vapor being converted to soluble forms (Hg^{2+}) and deposited into soil and water via precipitation. Usually mercury vapor has an atmospheric residence time between 0.4 and 3 years, while the soluble mercury has a residence time of only a few weeks.

Methylmercury can be bioaccumulated many fold and represents a major source of human exposure through the food supply. Other exposures to mercury can be via gold mining/extraction and with dental amalgams. Water and air can contribute significantly to the daily intake of total mercury, depending upon the level of contamination. In most foodstuffs, mercury is in the inorganic form and below the level of detection (20 µg Hg/kg wet weight). Fish and fish products can have levels greater than 1200 µg/kg, e.g., shark, swordfish, and Mediterranean tuna (WHO, 1990). Similar levels have been noted in

Table 2 Cadmium Levels in Maternal Blood, the Placenta, and Cord Blood

Location	Population	Maternal Blood (ng/ml)	Placenta (ng/g wet wt)	Cord Blood (ng/ml)	Comment	Ref.
8 Locations, U.S.		32 (n = 177–187)	44 (n = 160–169)	28 (n = 17–187)		Creason et al., 1976
London, Canada	Nonsmokers (n = 5)		30.0 ± 4.5		Tissue perfused for 5 min before sampling	Goyer et al., 1992
Malaysia	Nonsmokers (n = 82)	3.93 ± 2.25			No difference between urban and rural	Lim et al., 1983
Nashville, TN[a]		16.86 ± 20.9 (n = 83)	17.0 ± 0.11 (n = 135)	16.2 ± 23.2 (n = 123)		Baglan et al., 1974
Augusta, GA	(n = 19)		52.8 ± 18.0		Placentae rinsed free of blood with 0.25 M sucrose	Karp and Robertson, 1977
Birmingham, AL	(n = 22)		30.3 ± 23.6			
Charlotte, NC	(n = 17)		28.1 ± 16.9			
Finland	(n = 19) 2 smokers	1.10 ± 0.90	20.2 ± 14.6 (n = 6)	0.40 ± 0.20	Amniotic fluid 1.0 ± 0.2 nmol/ml	Korpela et al., 1986
Nagoya City, Japan	Smoking status not stated	6.97 ± 4.05 (n = 106)	113.5 ± 7.9 (n = 113)	8.99 ± 5.6 (n = 97)	Placentae washed with tap water	Tsuchiya et al., 1984
Germany		1.91 ± 1.10 (n = 27)	4.95 ± 2.2 (n = 33)	1.10 ± 0.40 (n = 17)	Placental Cd 10.5 ± 5.6 nmol/g dry wt	Schramel et al., 1988
Taiwan	(n = 159)	13.0 ± 5.0		7.8 ± 5.0	No correlation between Cd and birthweight	Soong et al., 1991
Lund, Sweden	University town (n = 15)	1.10 ± 1.69	18.6 ± 18.6[b]		Cd levels of smokers and nonsmokers did not differ significantly	Fagher et al., 1993
Bailystock, Poland	Industrial town (n = 9)	1.80 ± 1.35	55.8 ± 37.2[b]			
Yugoslavia	Nonsmokers Control — Pristina (n = 55)		11.3 ± 4.3[c]		No association between placental Cd and birthweight	Loiacono et al., 1992
	Near lead smelter T. Mitrovica (n = 106)		17.1 ± 11.7[c]			
Belgium	Nonsmokers (n = 331)	1.23 ± 1.23		10.0 ± 1.4	Significant increase in maternal blood Cd in smokers	Buchet et al., 1978
	Smokers (n = 109)	2.02 ± 1.24		0.7 ± 0.9		
Cleveland, OH	Nonsmokers (n = 31)	2.25 ± 0.8	13.49 ± 6.4	1.91 ± 0.6	More small-for-dates infants born to smokers	Kuhnert et al., 1982
	Smokers (n = 41)	3.48 ± 0.17	17.98 ± 7.31	2.25 ± 0.9		
Cleveland, OH	Nonsmokers (n = 84)	1.0 ± 0.3	8.1 ± 5.0		Placentae perfused with 21 saline before analysis	Kuhnert et al., 1987
	Smokers (n = 65)	1.46 ± 0.89	12.0 ± 7.5			

Location	Group			Notes	Reference
Cleveland, OH	Nonsmokers (n = 17)	0.6 ± 0.3	5.84 ± 2.25	Placentae perfused with 2l saline before analysis; no association of Cd with preeclampsia	Lazenik et al., 1989
	Smokers (n = 9)	1.24 ± 0.79	11.69 ± 0.40		
Newark, NJ	Nonsmokers	1.91 ± 0.84 (n = 9)	1.12 ± 0.3 (n = 11)	Amniotic fluid 1.57 ± 0.51 2.14 ± 0.89 nmol/ml (n = 15) Cd in bloods significantly higher in smokers	Chatterjee et al., 1988
	Smokers	3.93 ± 0.87 (n = 11)	2.02 ± 0.11 (n = 13)		
		RBC	RBC		
Czechoslovakia	Control (n = 50)	6.07 ± 5.51	4.05 ± 3.71		Truska et al., 1989
	Urban, industrial (n = 50)	6.74 ± 4.38	3.15 ± 2.02		
			(ng/g dry wt)		
Karlstad, Sweden	Smoking: serum SCN			Placental zinc also increased with smoking	Moberg Wing et al., 1992
	<50 (n = 23)		20.2 ± 6.74		
	50–69 (n = 12)		22.5 ± 5.62		
	≥70 (n = 3)		36.0 ± 11.2		

Note: Values are mean ± SD

[a] Values presented as dry weight. Converted to wet weight according to author's directions: divided by 6.0, placenta; 5.5, maternal blood; 4.7, fetal blood.
[b] Transformed from dry weight using author's dry weight percentage 18.6 ± 2.92%.
[c] Transformed from dry weight using author's dry weight percentage 20%.

Table 3 Lead in Maternal Blood and Cord Blood

Location	Population	Maternal Blood (ng/ml)	Cord Blood (ng/ml)	Cord/Maternal Blood	Comment	Ref.
Toronto, Canada	Urban (n = 95)	29.0 ± 10.4	16.6 ± 14.5	0.57[a]		Koren et al., 1990
Roskilde, Denmark	Rural	34 (6–63)[b] (n = 78)	23 (6–50)[b] (n = 48)	0.7[c] (0.2–1.4)[b] (n = 48)		Milman et al., 1988
Germany		39 ± 14 (n = 27)	30 ± 16 (n = 17)	0.77[a]	Placental lead; 18.7 ± 7.3 ng/g wet wt; 217 ± 77 ng/g dry wt (n = 33)	Schramel et al., 1988
Finland	(n = 19)	40.4 ± 18.2	37.1 ± 13.5	0.56[a]	Placental lead, 22.6 ± 15.7 ng/g (n = 6); amniotic fluid 59.6 ± 8.3 ng/ml	Korpela et al., 1986
Cleveland, OH	Urban, disadvantaged (n = 185)	64.8 ± 18.8	58.4 ± 20.2	0.90[a]	No association of lead with decreased birthweight or anomalies	Ernhart et al., 1986
Taiwan	(n = 147)	64.8 ± 23.8	40.9 ± 15.6	0.63[a]		Soong et al., 1991
Maracaibo, Venezuela	Urban (n = 16)	66.3 ± 33.3	53.2 ± 31.8	0.80[a]		Romero et al., 1990
Nagoya City, Japan		75 ± 54 (n = 105)	84 ± 77 (n = 95)	1.1[a]	Placental lead 45 ± 34 ng/g (n = 110); placentae washed with tap water before analysis	Tsuchiya et al., 1984
Sao Paulo, Brazil	Urban, disadvantaged (n = 43)	93 ± 27	80 ± 27	0.83 ± 0.22[c]		Troster and Schuartsman, 1988
Poland	(n = 100)	98[d] (67–136)[b]	76[d] (53–109)[b]	0.8[c] (0.62–1.03)[b]		Sikorski et al., 1988
Louisville, KY	Indigent urban (n = 154)	98.5 ± 44	97.3 ± 41	0.99[a]	No relationship between lead and PROM,[e] preterm delivery, meconium, or preeclampsia	Angell and Lavery, 1982
Shreveport, LA	Indigent (n = 98)	103	101	0.98[a]	No effect of lead on birthweight	Gersanik et al., 1974
Kuala Lumpur, Malaysia	Urban (n = 114)	151.3 ± 41.4	114 ± 31.1	0.75[a]		Ong et al., 1985

Location	Subgroup			Ratio	Comment	Reference
Nashville, TN[f]		165 ± 140 (n = 84)	123.4 ± 197.8 (n=130)	0.75[a]	Placental lead 305 ± 422 ng/g (n = 234)	Baglan et al., 1974
Belgium	Smokers (n = 109)	105 ± 32	89 ± 33	0.85[a]	Smoking associated with decreased birthweight	Buchet et al., 1978
	Nonsmokers (n = 333)	100 ± 40	81 ± 35	0.81[a]		
Bialystok, Poland	Industrial town (n = 24)	37.9 ± 17.2			Lead not related to increased myometrial activity in preterm labor; placental lead: Bialystok, 0.3 ± 0.2; Lund 0.3 ± 0.1 μg/g dry μg/wt	Fagher et al., 1993
Lund, Sweden	University town (n = 6)	11.2 ± 2.9				
Sweden	Atmospheric pollution				Significantly lower blood lead in region with low atmospheric pollution	Zetterlund et al., 1977
	Low	61 ± 21 (n = 21)	44 ± 20 (n = 47)			
	Intermediate	92 ± 37 (n = 173)	80 ± 38 (n = 391)			
	High	84 ± 27 (n = 103)	73 ± 27 (n = 103)			
Yugoslavia	Near lead smelter, T. Mitrovica (n = 106)	217.6 ± 68.4	203.1 ± 76.6	0.93[a]	Placental lead: T. Mitrovica, 14.4 ± 14.8; Pristina, 4.5 ± 3.8 μg/g/g dry wt	Loiacona et al., 1992
	Nonexposed, Pristina (n = 55)	68.4 ± 47.7	55.9 ± 39.4	0.82[a]		
Kabwe, Zambia	3000 m radius of lead mine (n = 122) controls (n = 31)	412 ± 144	370 ± 153	0.9[a]	No effect of lead on birthweight	Clark, 1977
		147 ± 75	118 ± 56	0.8[a]		
Stoke-on Trent, U.K.	Pottery workers				Placental lead: lithographers (n = 16) and transferers (n = 8), 0.34 ± 0.16; painters (n = 6) 0.54 ± 0.09 μg/g/g	Khera et al., 1980
	Lithographers (n = 27)	190 ± 90				
	Transferers (n = 2)	120				
	Painters (n = 11)	240 ± 110				
Czechoslovakia	Urban, industrial (n = 50)	RBC 109.4 ± 42.4	RBC 74 ± 31.6	0.68[a]	Placental lead: urban, 42.9 ± 26.5; control 43.2 ± 29.7 ng/g	Truska et al., 1989
	Control (n = 50)	RBC 160 ± 63.1	RBC 113.4 ± 57.5	0.71[a]		
Northern Italy	Small town, rural (n = 75)	RBC 264 ± 45	RBC 254 ± 43	RBC 0.96[a]	Correlation between maternal and cord blood greater for plasma than RBCs	Cavalleri et al., 1978
		Plasma 6.6 ± 3.5	Plasma 6.2 ± 3.1	Plasma 0.94[a]		
Cleveland, OH	Urban (n = 47)	RBC 491 ± 119	RBC 329 ± 102	0.67[a]		Kuhnert et al., 1977

Note: Values are means ± SD

[a] Ratio calculated from population means.
[b] Range.
[c] Ratio is mean of individual ratios.
[d] Geometric mean.
[e] PROM = premature rupture of fetal membranes.
[f] Values were presented as dry weight. Converted to wet weight according to author's directions: divided by 6.0, placenta; 5.5, maternal blood; 4.7, fetal blood.

freshwater fish (bass, pike, and walleyes) in polluted lakes and streams. The consumption of 200 g of fish containing 500 μg Hg/kg will give an intake of 100 μg Hg as principally methylmercury. This amount of methylmercury is one half of the WHO-recommended tolerable weekly intake (WHO, 1990).

For methylmercury in the diet, practically all of the MeHg is absorbed and distributed to all tissues within 4 days. The blood to hair ratio in man is approximately 1:250. Further, cord blood levels are generally greater than maternal blood levels of MeHg. For MeHg the red blood cell to plasma distribution ratios are about 20:1 in humans, monkeys and guinea pigs, while it is 7:1 for mice and >100:1 for rats.

Methylmercury is converted to inorganic mercury in humans. The rate of mercury excretion is proportional to the body burden and fits a single compartment model with a biological half-time of 50 (39 to 70) days in fish-eaters. Interestingly, lactating females have a significantly shorter half-life for mercury.

Mean values for total mercury are whole blood, 8 μg/l; hair 2 μg/g; urine, 4 μg/l; and placenta, 10 μg/kg. In fish-eaters where consumption of mercury is 200 μg/day, the mercury blood levels can be approximately 200 μg/l with hair levels being about 50 μg/g (WHO, 1990).

When reproductive and developmental toxicity is of concern, the principal concern is organic mercurials, especially methylmercury, as a fungicide or environmental contaminant. In industry a wide array of mercurials exposures do occur in gold mining (now in South America), in electrial products, lights, switches, connectors, and in the pulp industry. The inappropriate disposal of such mercury by industry can lead to substantial environmental exposure of not only inorganic mercury but especially methylmercury.

The two largest study populations for the toxicity of methylmercury were in Minimata Bay, Japan, and in Iraq. The Japanese exposure was due to the consumption of contaminated fish from Minimata Bay, while the Iraqi exposure was due to the consumption of contaminated grain. It was especially apparent that newborns and children were more affected by the exposures, even though the adults were also affected. An exposure-dependent appearance of toxic symptoms was observed in the blood of patients exposed to methylmercury. Yet of even greater significance has been the ability to utilize hair specimens to identify the exposure spectrum. Thus, through one hair specimen one could determine when, during a pregnancy, the exposure occurred. Because of the long half-life for mercury, even if the ingestion stops, the exposure to mercury will persist.

Of particular concern is that based upon the dose-response curves, the conceptus is much more sensitive to the neurotoxic actions of methylmercury than is the mother (Clarkson, 1987). Yet equally important is the suggestion that the pregnant female may be more sensitive to methylmercury than is the normal female adult (Marsh et al., 1987; Clarkson, 1987; Coxetal 1989, 1995).

Methylmercury is fetotoxic in mice (single exposure 2.5 to 7.5 μg/kg); teratogenic in rats; and produces behavioral alterations in monkey offspring (50 to 70 μg/kg/d before and during pregnancy). Spermatogenesis in mice is affected at 1 μg/kg methylmercury. Recent quantitative and qualitative assessments in nonhuman primates have been compared with humans who have all been exposed to different concentrations of methylmercury (Burbacher et al., 1990). It is remarkable how well the behavioral and and neuropathological effects agree across species at the higher concentrations of methylmercury. Neurobehavioral functioning was similar across species at lower exposure levels for methylmercury; however, because of a lack of information in the human at lower exposure levels for the pathology, no neuropathological correlation can be established at this time (Burbacher et al., 1990).

In human adults, no adverse effects have been detected with long-term daily mercury intake (3 to 7 μg/kg body weight). The hair levels of mercury would be 50 to 125 μg/g. It should be noted that pregnant women may suffer effects at lower methylmercury exposure levels than nonpregnant adults, suggesting a greater risk for pregnant women (Marsh et al., 1987; WHO, 1990).

Severe damage to the developing central nervous system can be caused by prenatal exposure to methylmercury. For severe neurologic effects, the lowest level in maternal hair during pregnancy was 404 μg/g in the Iraqi outbreak, while the highest no-observed-effect level for severe effects was 399 μg/g (WHO, 1990). Fish-eating populations currently studied have not demonstrated such severe effects.

Psychomotor retardation in the offspring (history of seizures, abnormal reflexes, delayed achievement of developmental milestones) was noted below maternal hair levels associated with severe effects. When the data are extrapolated, motor retardation was greater than background frequency at maternal hair levels of 10 to 20 μg/g (WHO, 1990). Boys but not in girls of mothers with hair levels during pregnancy of 23.9 μg/g demonstrated abnormal muscle tone or reflexes (Canadian). Four-year-old children whose mothers had maternal hair levels from 6 to 86 µg/g (2nd highest was 19.6 µg/g) demonstrated developmental retardation according to the Denver Test.

According to the World Health Organization (WHO, 1990),

> ... the general population does not face a significant health risk from methylmercury. Certain groups with a high fish consumption may attain a blood methylmercury level (about 20 µg/litre, corresponding to 50 µg/g of hair) associated with a low (5%) risk of neurological damage to adults.
>
> The fetus is at particular risk. Recent evidence shows that at peak maternal hair mercury levels above 70 µg/g there is a high risk (more than 30%) of neurological disorder in the offspring. A prudent interpretation of the Iraqi data implies that a 5% risk may be associated with a peak mercury level of 10 to 20 µg/g in maternal hair.

Of particular concern is the change in dietary habits to reduce the daily intake of cholesterol-containing foods, e.g., meats, dairy products. Fish has been recommended as an alternative. It is essential to have a good dietary history for your patient, as well as an occupational exposure. Pregnant women are being exposed to increasing levels of methylmercury through consumption of swordfish, tuna, shark to the exclusion of other protein sources. A case report from PEDECS demonstrates that even today, families can continue to be exposed. A professional family principally eating fish as the dietary source of protein ate swordfish and tuna many times each week. The mercury levels in the blood and hair of both the father and mother exceeded the WHO levels. The mother was 35 weeks pregnant. Upon notification of the mercury levels, the mother immediately discontinued the ingestion of those types of fish. Over the next few weeks, her blood levels did fall; however, the baby's blood and hair levels were 21 µg/l and 11 ppm, while the mother's were 28 µg/l and 18 ppm.

As noted in Table 4, mercury is detected in the human placenta. The mechanism of transport of methylmercury into the brain has been associated with amino acid carriers (Clarkson, 1993). L-Cysteine accelerates methylmercury uptake into brain. The complex of L-cysteine-methylmercury is structurally similar to L-methionine, a substrate for the L-neutral amino acid transport system. Preliminary studies in the perfused human placenta have not identified a similar transport mechanism for methylmercury (Czkeridowski et al., 1993).

2. Mercury Vapor/Amalgams/Inorganic

The principal exposure to mercury vapor in the general population is via dental amalgams. Dental amalgams release mercury vapor into the mouth. When fillings are removed, an acute increase in mercury release is noted. The rate of mercury release is increased by stressing the surfaces by chewing and brushing. The released mercury from the dental amalgams is deposited in body tissues and is excreted via the kidney. Increased urinary mercury levels are noted. Estimated release rates from amalgams are consistent with mercury content in autopsy tissue in the general population (Clarkson et al., 1988b).

Occupational exposure to mercury vapor can result in renal, pulmonary, and psychomotor toxicity (cf. Clarkson, 1988). Current studies of female dentists/dental workers have not demonstrated a mercury-associated increase in birth defects or pregnancy losses (Heidam, 1984; Ericson and Kallen, 1989; Rowland et al., 1994). Unfortunately there is limited information available concerning occupational exposures to mercury (vapor/inorganic) and effects on reproduction. Menstrual disorders (hypermenorrhea/dysmenorrhea) and decreased fertility have been associated with women working in mercury plants, especially for women working longer than 3 years (Goncharuk, 1977; Rowland et al., 1994). The mercury levels in the factory fluctuated between trace and 0.08 mg/m^3. Hypermenorrhea/dysmenorrhea were reported in dental workers and women working in mercury rectifier stations. These exposures appear to be substantial, since it was reported that mercury was on the patients' hands and on the desk, tables, and floors in the work areas (Marinova et al., 1973; Mikhailova et al., 1971, as reported by Barlow and Sullivan, 1982). It is apperent from the descriptions of the working conditions in these occupational settings that substantial mercury exposure was occurring.

It has been well established that even though organic mercurials rapidly transit the placenta, inorganic mercury does not easily transit and is concentrated by the placenta itself (Table 4; cf. Miller et al., 1988). Even with such placental concentration, a recent study in sheep (Vimy et al., 1990) demonstrated the release of mercury from dental fillings with the appearance of mercury in the fetus. Further evidence was presented for the transfer to the newborn via breast milk as well. Thus, the issues of the kinetics of mercury transfer in the human require further investigation to establish method of transfer and specific

Table 4 Mercury Levels in Maternal Blood, the Placenta, and Cord Blood

Location	Population	Maternal Blood (Total ng/ml)	Placenta (Total ng/g wet wt)	Cord Blood (Total ng/ml)	Comment	Ref.
Des Moines, IA	(n = 57)	1.21 ± 1.06		1.5 ± 1.36	No difference between urban and rural	Kuntz et al., 1982
Kagoshima, Japan	(n = 38)	17.4 ± 9.8	31 ± 18.3	23.8 ± 10.4	Placentae washed before analysis; one outlier omitted: maternal blood = 966, placenta = 56, cord blood = 388	Shinkawa, 1974
Augusta, GA	(n = 19)		8 ± 4.4			Karp and Robertson, 1977
Birmingham, AL	(n = 22)		15 ± 14.1			
Charlotte, NC	(n = 17)		19 ± 12.4			
8 Locations, U.S.		10 (n = 177–187)	24 (n = 160–169)	13 (n = 177–187)	>10% of samples less than minimum detectable	Creason et al., 1976
Germany		2.7 ± 0.6 (n = 5)	4.1 ± 2.1 (n = 26)	4.6 ± 2.4 (n = 4)	In most cases blood Hg below detection limit; placental Hg 26 ± 12 ng/g dry wt	Schramel et al., 1988
Taiwan	(n = 85)	19.4 ± 13.8		28.8 ± 26.7	No association of Hg with birthweight	Soong et al., 1991
Nashville, TN[a]		87 ± 8 (n = 929) RBC 22.9 ± 11.9 Plasma 12.4 ± 7.3	21.5 ± 26.8 (n = 1061)	11.5 ± 10.2 (n = 872) RBC 30.8 ± 21.6 Plasma 11.2 ± 7.2		Baglan et al., 1974
Japan	(n = 9)		71.5 ± 27.4			Suzuki et al., 1971
Belgium	Nonsmokers	11.2 ± 6.5 (n = 331)		13.6 ± 7.9 (n = 328)	No difference in Hg between smokers and nonsmokers	Buchet et al., 1978
	Smokers	13.1 ± 7.0 (n = 109)		14.2 ± 8 (n = 108)		
		RBC/Plasma		**RBC/Plasma**		

Placental Transport, Metabolism, and Toxicity of Metals

Location	Group				Notes	Reference
Czechoslovakia	Control (n = 50)	RBC 5.5 ± 1.7	2.2 ± 1.0	RBC 4.4 ± 1.5		Truska et al., 1989
	Industrial (n = 50)	RBC 6.1 ± 1.8	2.0 ± 0.9	RBC 4.6 ± 1.9		
Oslo, Norway	Controls (n = 26)	RBC 7.96 ± 2.5 Plasma 4.62 ± 2.45	12 ± 4.74	RBC 8.85 ± 3.78 Plasma 4.31 ± 2.91	Exposed group had significantly increased Hg in placentae and membranes not in blood or amniotic fluid	Wannag and Skjaerosen, 1975
	Dental workers	RBC 8.84 ± 3.79 Plasma 4.63 ± 2.96 (n = 19)	24.47 ± 17.7 (n = 19)	RBC 10.18 ± 5.9 (n = 17) Plasma 4.06 ± 2.8 (n = 18)		
Genova, Italy	None with high seafood consumption		**Total/Organic** Total 12 ± 8 (n = 22) Organic 8 ± 7 (n = 18)		Placentae washed with deionized water	Capelli and Minganti, 1986
U.S.	Industrialized urban area (n = 3)	**Methyl/Inorganic**	**Methyl/Inorganic** Methyl: 22.3 ± 6.3 Inorganic: 16.8 ± 4.4	**Methyl/Inorganic**		Cappon and Smith, 1981
Cleveland, OH		Methyl: RBC 3 ± 2.1 (n = 29) Plasma 0.4 ± 0.2 (n = 25) Inorganic: RBC 2.5 ± 1.8 (n = 28) Plasma 1.7 ± 1.4 (n = 22)	Methyl: 1.4 ± 1.1 (n = 24) Inorganic: 5.3 ± 3.2 (n = 24)	Methyl: RBC 3.9 ± 3 (n = 29) Plasma 0.4 ± 0.3 (n = 25) Inorganic: RBC 1.4 ± 1.2 (n = 28) Plasma 2.6 ± 1.9 (n = 22)	Placentae perfused before analysis	Kuhnert et al., 1981

Note: Values, means ± SD

[a] Values were presented as dry weight. Converted to wet weight according to author's directions: divided by 6.0, placenta; 5.5, maternal blood; 4.7, fetal blood.

protein interactions within the placental cells. Yet the fact that one can accurately measure mercury and its form in the human placenta can be a useful marker of exposure for the conceptus, especially for inorganic mercury when hair analysis is not appropriate.

E. SELENIUM

Selenium is an essential element that is found in specific selenoproteins which contain selenocysteine. Mammalian selenoproteins that have been identified include the antioxidant glutathione peroxidase enzymes, selenoprotein P, and type I iodothryonine deiodinase (Stadtman, 1990; Burk et al., 1991). An additional class of selenoproteins which is not well defined is proteins that bind selenium tightly enough so that the selenium remains attached during protein purification procedures (Sunde, 1990). Selenium associated with proteins is in the form of selenide (–2). Selenide can be formed by the reduction of selenite ($-SeO_3$, +4) and selenate ($-SeO_4$, +6), two forms of selenium frequently used in experimental exposure studies. Selenite reacts with glutathione and is reduced to selenide. In this process reactive oxygen species are produced (Yan and Spallholz, 1993; Imura et al., 1994). The pathway of selenate reduction has not been well defined, but reactive oxygen species are not produced in the presence of selenate and glutathione (Yan and Spallholz, 1993).

Selenium status in animals and humans has been determined by measuring selenium concentrations and glutathione peroxidase activity in blood or plasma. During pregnancy, both plasma selenium and glutathione peroxidase activity decrease, or remain unchanged (Behne and Wolters, 1979; Swanson et al., 1983; Zachara et al., 1993).

In humans, selenium concentrations in cord blood are similar to levels in maternal blood with higher values found in the placenta (Table 5). The actual form of selenium that crosses the placenta *in vivo* is not known. Treatment of pregnant mice with selenite or selenate results in selenium transfer to fetal tissues (Danielsson et al., 1990). Selenium from selenate and selenomethionine can also cross the hamster placenta (Willhite et al., 1990). An *in vitro* study using dual perfusion of the human term placenta, in which selenite was added to the maternal circulation, identified selenite in the fetal perfusate, indicating that selenite can cross the human term placenta (Eisenmann and Miller, 1994).

Table 5 Maternal Blood, Cord Blood, and Placental Selenium Concentrations[a]

Location	Maternal Blood (μM)	Cord Blood (μM)	Placenta ($\mu M/kg$)	Ref.
Finland (21)	0.73	0.77	2.2	Korpela et al., 1984
Germany	1.0 (27)	1.0 (27)	2.4 (33)	Schramel et al., 1988
Northern Ireland (56)	0.59	0.44[b]		Wilson et al., 1991

[a] Values are means for the number of subjects indicated in parentheses.
[b] Cord blood concentrations significantly different from maternal blood concentrations, $p < 0.01$.

Although selenium is essential, it is also quite toxic, with adverse effects, including developmental effects, reported in animals following selenium intake of about 100 times larger than levels considered essential (WHO, 1987). There is little information about the effects of selenium on placental function. One study showing that an acute dose of selenite can cause abortions in mice suggests that selenite could be a reproductive toxicant, possibly acting through effects on the placenta. Yonemoto et al. (1983) studied the effect of a single subcutaneous dose of selenite given to mice on gestation day 12 or 16. Treatment on day 12 with a dose of 58.8 μmol/kg resulted in abortion and maternal deaths within 48 h, while on gestation day 16, abortions were observed at 27 μmol/kg, and abortions and maternal deaths occurred at 40 μmol/kg. Pretreatment of gestation day 12 mice with glutathione (2 or 5 mmol/kg) 20 min before selenite treatment increased the toxicity of selenite. No treatment-related malformations were observed in offspring from mice treated with selenite. Histopathological examinations of organs including the placenta were not completed. The observed increase of selenite toxicity by glutathione pretreatment is consistent with a free radical mechanism of selenite toxicity.

It is not known how a free radical mechanism of selenite toxicity may lead to abortions and the increased toxicity that was observed later in gestation in mice. One hypothesis is that oxidative stress during late-gestation pregnancy may disrupt the balance between the arachidonic acid metabolites, prostacyclin and thromboxane A2. Prostacyclin is a vasodilator that also inhibits the aggregation of platelets, while thromboxane A2 is a vasoconstrictor that stimulates platelet aggregation. Both substances are produced by the placenta, and an increase in the production of thromboxane A2 and a decrease in the production of prostacyclin have been observed in placentae from women with preeclampsia (Walsh et al., 1985).

Prostaglandin H synthase enzymes, required for the production of thromboxane A2 and prostacyclin, require small amounts of peroxides for activity, while high concentrations inhibit activity (Reddy et al., 1988). Because of the importance of peroxides to prostaglandin H synthase activity, it has been proposed that selenium in the form of the glutathione peroxidase enzymes may be important in modulating prostaglandin H synthase enzyme activity, and determining the products of the arachidonic acid cascade (Reddy et al., 1988). In addition to affecting prostaglandin H synthase, oxidizing agents can affect other enzymes of the arachidonic acid cascade. For example, prostacyclin synthase activity is also inhibited by oxidizing agents (Ham et al., 1979).

To investigate whether selenium compounds can affect the placental production of thromboxane A2 and prostacylin, an *in vitro* study was completed in which human term placental explants were exposed to selenium compounds for up to 24 h, and the production of the inactive hydrolysis products of thromboxane A2 and prostacyclin, thromboxane B2, 6-keto-$PGF_{1\alpha}$ were monitored (Eisenmann and Miller, 1995b). The selenium compounds used were selenite, which is known to produce free radicals; selenate, which does not produce free radicals; and ebselen, an organic selenium compound with glutathione peroxidase activity (Morgenstern et al., 1992). Concentrations of selenium used in these investigations did not exceed 40 μM, which is the reported blood concentration associated with selenium toxicity in humans (Yang et al., 1983).

Two 12-h exposures of human term placental explants to selenite at 20 or 40 μM significantly increased the placental production of thromboxane B2 and decreased the placental production of 6-keto-PGF_1, both changes that contributed to a significant increase in the thromboxane B2/6-keto-PGF_1 ratio (Eisenmann and Miller, in press). An increase in the thromboxane B2/6-keto-$PGF_{1\alpha}$ ratio would be conducive to vasoconstriction and blood coagulation. In contrast to selenite, two 12-h exposures to selenate at 40 μM significantly increased thromboxane B2 production with no significant effects on 6-keto-$PGF_{1\alpha}$ production or the ratio. Ebselen exposure of placental explants tended to decrease both thromboxane B2 and 6-keto-$PGF_{1\alpha}$ production with no significant changes in the thromboxane B2/6-keto-$PGF_{1\alpha}$ ratio.

Based on the results of this study (Eisenmann and Miller, 1995b), it has been proposed that selenite, which can produce oxidative stress, could lead to an imbalance between thromboxane A2 and prostacyclin. The direction of the imbalance observed following selenite exposure was the same as observed in cases of preeclampsia. How effects of selenite on thromboxane A2 and prostacyclin production contributes to toxicity of selenite during pregnancy still needs to be elucidated.

III. CONCLUSIONS

The placenta with its extraembryonic membranes is not only the anchor but the conduit and controller of the pregnancy. When considering the placenta and interactions with metals, the concern is whether the metal crosses. As noted above, all metals discussed appear in the fetal circulation; however, the amount that appears can be substantially different from what is circulating in the mother. Further, the placenta can be an accumulator of the metal, e.g., cadmium. The kinetics of transfer can elucidate how these metals and their specific charge and form can be transferred differently, e.g., organic and inorganic mercury. Most important, the metals may be directly interacting with cellular processes in the placenta, compromising placental function and eventually the developing embryo and fetus. Such considerations of not only transit of the metal but also its metabolism and direct toxic action on the placenta must be considered not only during the later trimesters, but also during peri-implantation.

ACKNOWLEDGMENTS

The authors wish to acknowledgement the support of NIH; NIEHS ES 02774, ES01247.

REFERENCES

Angell, N. F. and Lavery, J. P. (1982), *Am. J. Obstet. Gynecol.,* 142, 40.
Avissar, N., Eisenmann, C., Breen, J. G., Horowitz, S., Miller, R. K., and Cohen, H., *Am. J. Physiol.,* 267, E68.
Baglan, T. J., Brill, A. B., Schulert, A., Wilson, D., Larsen, K., Dyer, N., Mansour, M., Schaffner, W., Hoffman, L., and Davies, J. (1974), *Environ. Res.,* 8, 64.
Barlow, S. and Sullivan, F. (1982), *Reproductive Hazards of Industrial Chemicals,* Academic Press, New York.

Benirschke, K. and Kaufmann, P. (1990), Pathology of the Human Placenta, Springer-Verlag, New York.
Boadi, W., Yannai, S., Urbach, J., Brandes, J., and Sumner, K. (1991), *Arch. Toxicol.,* 65, 318.
Boadi, W., Shurtz-Swirski, R., Barnea, E., Urbach, J., Brandes, J., Philo, E., and Yannai, S. (1992), *Arch. Toxicol.,* 66, 95.
Boadi, W., Yannai, S., Urbach, J., Brandes, J., and Sumner, J. (1991), *Arch. Toxicol.,* 65, 318.
Boadi, W. Y., Urbach, J., Barnea, E. R., Brandes, J. M., and Yannai, S. (1992), *Pharm. Toxicol.,* 71, 209.
Bondy, S. (1989), *Neurotoxicol. Teratol.,* 11, 527.
Börzsönyi, M., Bereczky, A., Rudnai, P., Csanady, C., and Horvath, A. (1992), *Arch. Toxicol.,* 66, 77.
Breen, J., Eisenmann, C., Horowitz, S., and Miller, R. K. (1994a), *Reprod. Toxicol.,* 8, 297.
Breen, J., Powlin, S., and Miller, R. K. (1994b), *Toxicologist,* 14, 235.
Breen, J., Nelson, E., and Miller, R. K. (1995), *Teratology,* 51, 266.
Buchet, J. P., Roels, H., Hubermont, G., and Lauwerys, R. (1978), *Environ. Res.,* 15, 494.
Burbacher, T., Rodier, P., and Weiss, B. (1990), *Neurotoxicol. Teratol.,* 12, 191.
Burk, R. F., Hill, K. E., Read, R., and Bellew, T. (1991), *Am. J. Physiol.,* 261, E26.
Capelli, R., Minganti, V., Semino, G., and Bertarini, W. (1986), *Sci. Total Environ.,* 48, 69.
Cappon, C. J. and Crispin Smith, J. (1991), *J. Anal. Toxicol.,* 5, 90.
Cavalleri, A., Minoia, C., Polzzoli, L., Polatti, F., and Bolis, P. F. (1978), *Environ. Res.,* 17, 403.
Chatterjee, M. S., Abdel-Rahman, M., Bhandal, A., Klein, P., and Bogden, J. (1988), *J. Reprod. Med.,* 33, 417.
Chen, C. L., Brown, M., and Whanger, P. D. (1992), *FASEB J.,* 6, A1398.
Cernichiari, E., Toribara, T., Liang, L., Marsh, D., Berlin, M., Myers, G., Cox, C., Shamlaye, C., Choisy, O., Davidson, P., and Clarkson, T. (1995), *Neurotoxicology,* 16, 613.
Cho, Y. S., Panigel, M., and Wegmann, R. (1988), *Cell. Mol. Biol.,* 34, 97.
Chutbatsu, L. S. and Meneghini, R. (1993), *Biochem. J.,* 291, 193.
Clark, A. R. L. (1977), *Postgrad. Med. J.,* 53, 674.
Clarkson, T. W., Nordberg, G. F., and Sager, P. R. (1983), Eds., *Reproductive and Developmental Toxicity of Metals,* Plenum Press, New York, 607.
Clarkson, T. W. (1993), *Annu. Rev. Pharmcol. Toxicol.,* 32, 545.
Clarkson, T. W. (1987), *Environ. Health Persp.,* 74, 103.
Clarkson, T. W., Hursh, J., Sager, P., and Syverson, T. (1988b), *Biological Monitoring of Toxic Metals*, Clarkson, T., Friberg, L., Nordberg, G., and Sager, P., eds, Plenum Press, New York, 199–246.
Cox, C., Clarkson, T., Marsh, D., Amin-Zaki, L., Tikriti, S., and Myers, G. (1989), *Environ. Res.,* 49, 318.
Cox, C., Marsh, D., Myers, G., and Clarkson, T. (1995), *Neurotoxicology,* 16, 727.
Creason, J. P., Svendsgaard, D., Bumgarner, J., Pinkerton, C., and Hinners, T. (1976), *Trace Subs. Environ. Health,* 10, 53.
Czekierdowski, A., Neth-Jessee, L., and Miller, R. K. (1993), *Placenta,* 14, A13.
Danielsson, B. R. G., Dencker, L., and Lindgren, A. (1983), *Arch. Toxicol.,* 54, 97.
Danielsson, B. R. G., Danielson, M., Khayat, A., and Wide, M. (1990), *Toxicology,* 63, 123.
Dencker, L., Danielsson, B., Khayat, A., and Lingren, A. (1983), *Reproductive and Developmental Toxicity of Metals,* Clarkson, T. W., Nordberg, G. F., and Sager, P. R., Eds., Plenum Press, New York, 607.
Dilts, P. V., Jr. and Ahokas, R. A. (1979), *Am. J. Obstet. Gynecol.,* 135, 940.
di Sant'Agnese, P. A., Jensen, K., Levin A. A., Miller, R. K. (1993), *Placenta,* 4, 149.
Domingo, J. L. (1994), *J. Toxicol. Environ. Health,* 42, 123.
Eisenmann, C. J. and Miller, R. K. (1995a), *Toxicol. Appl. Pharm.,* 135, 18.
Eisenmann, C. J. and Miller, R. K. (1994), *Placenta,* 15, 883.
Eisenmann, C. J. and Miller, R. K. (in press) *Toxicol. Appl. Pharm.*
Ernhart, C. B., Wolf, A. W., Kennard, M. J., Erhard, P., Filipovich, H. F., and Sokol, R. J. (1986), *Arch. Environ. Health,* 41, 287.
Fagher, U., Laudanski, T., Schütz, A., Sipowicz, M., and Åkerlaund, M. *Int. J. Gynecol. Obstet.,* 40, 109.
Ferm, V. H. and Carpenter, S. J. (1968), *J. Reprod. Fert.,* 17, 199.
Gatsonis, C. A. and Needleman, H. L. (1992), *Human Lead Exposure,* Needleman, H. L., Ed., CRC Press, Boca Raton, chap 15.
Genbacev, O. and Miller, R. K. (1993), *Methods in Toxicology,* 3B, 205.
Genbacev, O., Schubach, S., and Miller, R. K. (1992), *Placenta,* 13, 439.
Genbacev, O., deMesy Jensen, K., Powlin, S. S., and Miller, R. K. (1993), *Placenta,* 14, 463.
Genbacev, O., White, T., and Miller, R. K. (1993), *Reproductive Toxicology,* 7, 75.
Genbacev, O., Powlin, S. S., and Miller, R. K. (1994), *Trophoblast Research,* 8, 427.
Gerber, G., Maes, J., and Deroo, J. (1978) *Arch Toxicol.,* 41, 125.
Gershanik, J. J., Brooks, G. G., and Little, J. A. (1974), *Am. J. Obstet. Gynecol.,* 119, 511.
Goyer, R. A. (1986), *Casarett and Soull's Toxicology, The Basic Science of Poisons,* 3rd ed., Klaassen, C. D., Amdur, M. O., and Doull, J, Eds., Macmillan, New York, chap 19.
Goyer, R. A., Haust, M. D., and Cherian, M. G. (1992), *Placenta,* 13, 349.
Ham, E. A., Egan, R. W., Soderman, D. D., Gale, P. H., and Kuehl, F. A., Jr. (1979), *J. Biol. Chem.,* 254, 2191.
Hanlon, D. P. and Ferm, V. H. (1987), *Environ. Res.,* 42, 546.
Harrison, W. P., Frazier, J. C., Mazzanti, E. M., and Hood, R. D. (1980), *Teratology,* 21, 43A.
Hood, R. D., Harrison, W. P., and Vedel, G. C. (submitted), *Bull. Environ. Contam. Toxicol.,* 29, 679.

Hood, R. D., Thacker, G. T., and Patterson, B. L. (1977), *Environ. Health Perspect.,* 19, 219.
Hood, R. D., Vedel, G. C., Zaworotko, M. J., and Meeks, R. G. (1988), *J. Toxicol. Environ. Health,* 25, 423.
Imura, N., Kitahra, J, Seko, Y., Utsumi, H., and Hamada, A. (1994), *Toxicologist,* 14, A1008.
Jacquet, P. and Gerber, G. B. (1979), *Biomedicine,* 30, 223.
Kagey, B. T., Bumgarner, J. E., and Creason, J. P. (1977), *Trace Substances in Environmental Health XI,* Hemphill, D. D., Ed., University of Missouri Press, Columbia.
Karp, W. B. and Robertson, A. F. (1977), *Environ. Res.,* 13, 470.
Khera, A. K., Wibberly, D. G., and Dathan, J. G. (1980), *Br. J. Indust. Med.,* 37, 394.
Koren, G., Chang, N., Gonen, R., Klein, J., Weiner, L., Demshar, H., Pizzolato, S., Radde, I., and Shime, J. (1990), *Can. Med. Assoc. J.,* 142, 1241.
Korpela, H., Loueniva, R., Yrjanheikki, E., and Kauppila, A. (1984), *Int. J. Vit. Nut. Res.,* 54, 257.
Korpela, H., Loueniva, R., Yrjänheikki, E., and Kauppila, A. (1986), *Am. J. Obstet. Gynecol.,* 155, 1086.
Krewski, D., Oxman, A., and Torrance, G. W. (1989), *The Risk Assessment of Environmental and Human Health Hazards A Textbook of Case Studies,* Paustenbach, D. J., Ed., John Wiley & Sons, New York, chap. 31.
Kuhnert, P. M., Kuhnert, B. R., Bottoms, S. F., and Erhard, P. (1982), *Am. J. Obstet. Gynecol.,* 142, 1021.
Kuhnert, P. M., Kuhnert, B. R., Erhard, P., Brashear, W. T., Groh-Wargo, S. L., and Webster, S. (1987), *Am. J. Obstet. Gynecol.,* 157, 1241.
Kuhnert, P. M., Kuhnert, B. R., and Erhard, P. (1981), *Am. J. Obstet. Gynecol.,* 139, 209.
Kuhnert, P. M., Erhard, P., and Kuhnert, B. R. (1977), *Environ. Res.,* 14, 73.
Kuntz, W. D., Pitkin, R. M., Bostrom, A. W., and Hughes, M. S. (1982), *Am. J. Obstet. Gynecol.,* 143, 440.
Lazebnik, N., Kuhnert, B. R., and Kuhnert, P. M. (1989), *Am. J. Obstet. Gynecol.,* 161, 437.
Lehman, L. and Poisner, A. (1984), *J. Toxicol. Environ. Health.,* 14, 419.
Leonard, A., Gerber, G. B., and Jacquet, P. (1983), *Reproductive and Developmental Toxicity of Metals,* Clarkson, T. W., Nordberg, G. F., and Sager, P. R., Eds., Plenum Press, New York, 357.
Levin, A. A. and Miller, R. K. (1980), *Teratology,* 22, 1–5.
Levin, A. A. and Miller, R. K. (1981), *Toxicol. Appl. Pharm.,* 58, 297.
Levin, A. A., Plautz, J. P., di Sant'Agnese, P. A., and Miller, R. K. (1981), *Placenta,* (Suppl 3), 3,303.
Levin, A. A., Miller, R. K., and di Sant'Agnese, P. A. (1983), *Reproductive and Developmental Toxicity of Metals,* Clarkson, T, Nordberg, G., and Sager, P., Eds., Plenum Press, New York, 633.
Levin, A. A., Kilpper, R. W., and Miller, R. K. (1987), *Teratology,* 36, 163.
Lim, H. H., Ong, C. N., Domala, Z., and Phoon, W. O. (1983), *Southeast Asian J. of Trop. Med. Public Health,* 14, 394.
Loiacono, N. J., Graziano, J. H., Kline, J. K., Popovac, D., Ahmedi, X., Gashi, E., Mehmeti, A., and Rajovic, B. (1992), *Arch. Environ. Health.,* 47, 250.
Lugo, G., Cassady, G., and Palmisano, P. (1969), *Am. J. Dis. Child,* 117, 328.
Marsh, D., Clarkson, T. W., Cox, C., Myers, G., Amin-Zaki, L., and Tikriti, S. (1987), *Arch. Neurol.,* 44, 1017.
Mayer-Popken, O., Denkhaus, W., and Konietzko, H. (1986), *Arch. Toxicol.,* 58, 203.
Miller, R. K. and Bellinger, D. (1993), *Occupational and Environmental Reproductive Hazards,* Paul, M., Ed., Williams & Wilkins, Baltimore, chap 17.
Miller, R. K., Malek, A., Kennedy, S., di Sant'Agnese, P. A., Mattison, D. R., Bryant, R., Panigel, M., and Neth, L. (1991), *Placenta,* 12, 420.
Miller, R. K., Mattison, D. R., and Plowchalk, D. (1988), *Biological Monitoring of Toxic Metals,* (Eds.), T. Clarkson, L. Friberg, G. Nordberg, and P. Sager, Plenum Press, New York, 567.
Milman, N., Christensen, J. M., and Ibsen, K. K. (1988), *Eur. J. Pediatr.,* 147, 71.
Moberg Wing, A., Wing, K., Tholin, K., Sjöström, R., Sandström, A., and Hallmans, G. (1992), *Eur. J. Clin. Nutr.,* 46, 585.
Morgenstern, R., Cotgreave, I. A., and Engman, L. (1992), *Chem-Biol. Interact.,* 84, 77.
Mushak, P., Davis, J. M., Crocetti, A. F., and Grant, L. D. (1989), *Environ. Res.,* 50, 11.
Needleman, H. L., Ed. (1992), *Human Lead Exposure,* CRC Press, Boca Raton.
Nordström, S., Beckman, L., and Nordenson, I. (1978), *Hereditas,* 88, 51.
Nordström, S., Beckman, L., and Nordenson, I. (1979), *Hereditas,* 90, 297.
Novak, R. F. (1991), *Arch. Pathol. Lab. Med.,* 115, 654.
Ong, C. N., Phoon, W. O., Law, H. Y., Tye, C. Y., and Lim, H. H. (1985), *Arch. Dis. Childhood,* 60, 756.
Page, K., Abramovich, D., Aggett, P., Bain, M., Chipperfield, A., Durdy, H., McLachlan, J., and Smale, A. (1992), *Placenta,* 13, 151.
Panigel, M., Les Anneses embryonnaires et le placenta des mammiferes, *Traite de Zoologie,* 16, 215–295, 1982.
Parizek, J. (1964), *J. Reprod. Fertil.,* 7, 263.
Parizek, J. (1965), *J. Reprod. Fertil.,* 9, 111.
Powlin, S., Genbacev, O., and Miller, R. K. (1994), *Toxicologist,* 14, 126.
Powlin, S. S. and Miller, R. K. (1996), *Toxicologist,* in press.
Powlin, S., Genbacev, O., and Miller, R. K. (1994), *Toxicologist,* 14, 126.
Reddy, C. S., Whelan, J., and Scholz, R. W. (1988), *Cellular Antioxidant Defense Mechanisms,* Chow, C. K., Ed., CRC Press, Boca Raton, 139.
Rice, D. C. (1992), *Human Lead Exposure,* Needleman, H. L., Ed., CRC Press, Boca Raton, chap 8.

Romero, R. A., Granadillo, V. A., Navarro, J. A., Rodriguez-Iturbe, B., Pappaterra, J., and Pierla, G. H. (1990), *J. Trace Elem. Electrolytes Health Dis.,* 4, 241.
Sager, P. R., Clarkson, T. W., and Nordberg, G. F. (1986), *Handbook on the Toxicology of Metals,* 2nd ed., Friberg, L., Nordberg, G. F., and Vouk, V., Elsevier, New York, 391.
Schramel, P., Hasse, S., and Ovcar-Pavlu, J. (1988), *Biol. Trace Elem. Res.,* 15, 111.
Shinkawa, Y. (1974), *Acta Obst. Gynaec. Jpn.,* 21, 185.
Sikorski, R., Paszkowski, T., Milart, P., Radomanski Jr., T., and Szkoda, J. (1988), *Int. J. Gynecol. Obstet.,* 26, 213.
Singh, U. S., Saxena, D. K., Singh, C., Murthy, R. C., and Chandra, S. V. (1991), *Reprod. Toxicol.,* 5, 211.
Sonnawane, B. R., Nordberg, M., Nordberg, G., and Lucier, G. (1975), *Environ. Health. Perspect.,* 28, 248.
Soong, Y.-K., Tseng, R., Liu, C., and Lin, P.-W. (1991), *J. Formosan Med. Assoc.,* 90, 59.
Stadtman, T. C. (1990), *Annu. Rev. Biochem.,* 59, 111.
Sunde, R. A. (1990), *Annu. Rev. Nutr.,* 10, 451.
Swanson, C. A., Reamer, D. C., Beillon, C., King, J. C., and Levander, O. S. (1983), *Am. J. Clin. Nutr.,* 38, 169.
Stöhrer, G. (1991), *Arch Toxicol.,* 65, 525.
Suzuki, T., Miyama, T., and Katsunuma, H. (1971), *Bull. Environ. Contam. Toxicol.,* 5, 502.
Tabacova, S., Balavaeva, L., and Petrov, I. (1992), *12th Rochester Trophoblast Conf., Program and Abstract Book,* A. 61.
Torreblanca, A., Del Ramo, J., and Bibudhendra, S. (1992), *Toxicology,* 72, 167.
Troster, E. J. and Schuartsman, S. (1988), *Biomed. Environ. Sci.,* 1, 64.
Truska, P., Rosival, L., Balazova, G., Hinst, J., Rippel, A., Palusova, O., and Grunt, J. (1989), *J. Hug. Epidem. Microbiol. Immunol.,* 33, 141.
Tsuchiya, H., Mitani, K., Kodama, K., and Nakata, T. (1984), *Arch. Environ. Health.,* 39, 11.
Wade, J., Agrawal, P., and Poisner, A. (1986), *Life Sci.,* 39, 1361.
Waalkes, M., Poisner, A., Wood, G., and Klaassen, K. (1984), *Toxicol. Appl. Pharmacol.,* 74, 179.
Walsh, S. W., Behr, M. J., and Allen, N. H. (1985), *Am. J. Obstet. Gynecol.,* 151, 110.
Wannag, A. and Skjæråsen, J. (1975), *Environ. Physiol. Biochem.,* 5, 348.
Webster, W. (1988), *J. Toxicol. Environ. Health.,* 24, 183.
White, T., Saltzman, R., di Sant'Agnese, P. A., Kung, P., Sutherland, R., and Miller, R. K. (1988), *Placenta,* 9, 583.
White, T., Baggs, R. B., and Miller, R. K. (1990), *Teratology,* 42, 7.
Willhite, C. C., Ferm, V. H., and Zeise, L. (1990), *Teratology,* 42, 359.
Wilson, D. C., Tubman, R., Bell, N., Halliday, H. L., and McMaster D. (1991), *Early Hum. Develop.,* 26, 223.
Wier, P. and Miller, R. K. (1987), *Trophoblast Research,* 2, 356.
Wier, P., Miller, R. K., Maulik, D., and di Sant'Agnese, P. A. (1990), *Toxicol. Appl. Pharm.,* 105, 156.
World Health Organization (WHO) (1990), Methylmercury, Environmental Health Criteria 101, World Health Organization, Geneva, Switzerland.
World Health Organization (WHO) (1977), *Environmental Health Criteria 3, Lead,* World Health Organization, Geneva.
World Health Organization (WHO) (1981), *Environmental Health Criteria 18, Arsenic,* World Health Organization, Geneva.
World Health Organization (WHO) (1987), *Environmental Health Criteria 58: Selenium,* World Health Organization, Finland.
Yan, L. and Spallholz, J. E. (1993), *Biochem. Pharmacol.,* 45, 429.
Yonemoto, S., Satoh, H., Himeno, S., and Suzuki, T. (1983), *Teratology,* 28, 333.
Zachara, B. A., Wardak, C., Didkowski, W., Maciag, A., and Marchaluk, E. (1993), *Gynecol. Obstet. Invest.,* 35, 12.
Zetterlund, B., Winberg, J., Lundgren, G., and Johansson, G. (1977), *Acta Paediatr. Scand.,* 66, 169.

Chapter 65

The Developmental Toxicology of Cadmium and Arsenic with Notes on Lead

John M. Rogers

Cadmium (Cd) and arsenic (As), along with mercury and lead, are among the best-studied heavy metals toxicants in terms of their developmental toxicity. Sources and patterns of human exposures are discussed elsewhere in this volume. The developmental toxicity of mercurials is reviewed by Massaro in Chapter V.4 in this section, and the developmental neurotoxicity of lead, the most critical endpoint of toxicity for this metal, is discussed in Chapters 24 and 39. The literature pertaining to the teratogenicity of lead has been recently summarized (Bellinger, 1994) and therefore will be discussed only briefly here. The aim of this chapter is to review the developmental toxicology of Cd and As, including direct effects on the conceptus and indirect effects mediated through toxicity to the mother or placenta, and to present current hypotheses concerning the biochemical mechanisms underlying the developmental toxicity of these metals. In contrast to mercury and lead, the database on human developmental toxicity of Cd or As is quite minimal, and therefore the vast majority of the data discussed in this chapter will be derived from work with experimental animals.

I. CADMIUM

Cadmium is a well-studied heavy metal toxicant in humans and animals (for recent general review, see Goering et al., 1995). The teratogenicity and developmental toxicity of Cd have been studied in a number of animal species, by multiple routes of administration, in different chemical forms, and at different developmental stages. Direct effects on the embryo, including exposures *in vitro,* have been tested, as have hypotheses of indirect developmental toxicity mediated through effects of Cd on the mother. The placental toxicity of Cd has been examined, and the effects of maternal Cd exposure on transport of various nutrients has been assessed. In this section, the developmental effects of Cd will be considered, beginning with effects on preimplantation embryos, continuing with effects on organogenesis, fetal effects, and placental toxicity. Both *in vivo* and *in vitro* studies will be reviewed, and information on putative mechanisms will be included where such data exist. Cadmium has not been demonstrated to be a human teratogen, but increased maternal environmental exposure to cadmium and higher placental Cd concentration (Loiacono et al., 1992) or fetal Cd exposure (as indicated by hair Cd concentrations) (Frery et al., 1993) have recently been associated with lower birth weights in humans.

A. EFFECTS OF CADMIUM ON PREIMPLANTATION EMBRYOS

Pedersen and Lin (1978) cultured mouse embryos from the 2-cell, 4-cell, 8-cell, or morula stage to the blastocyst stage in 10^{-8} to $10^{-5} M$ $CdCl_2$. Preimplantation development was not affected at the lower concentrations ($10^{-8} M$ to $10^{-6} M$), but cleavage-stage embryos underwent developmental arrest when exposed to $10^{-5} M$ $CdCl_2$ ($ED_{50} = 5 \times 10^{-6} M$). Cleavage arrested at the 4- to 8-cell stage, regardless of whether exposure began at the 2-, 4-, or 8-cell stage. Morula-stage embryos were less sensitive to Cd than were earlier-stage embryos, and some morulae treated with $10^{-5} M$ $CdCl_2$ developed into normal fetuses after transfer to foster mothers. Storeng and Jonsen (1980) exposed 2- and 4- to 8-cell mouse embryos in vitro to 10 to 50 μM Cd acetate for 48 h. These authors also found that earlier stage embryos were more sensitive to Cd. In 2-cell embryos, morphological alterations were observed at concentrations of 10 μM and above. When 4-cell or morula-stage mouse embryos (F_1 from C57 female × A2G male) cultured in vitro were exposed to 5 or 10 µg/ml $CdCl_2$ for 24 h, the 4-cell stage was found to be more sensitive than the morula stage. Four-cell stage embryos exposed to $CdCl_2$ developed to compacted morulae, but most then degenerated and decompacted. Exposure in vitro to 1 µg/ml $CdCl_2$ did not interfere with development to the implantation-stage mouse blastocyst, but reduced implantation in vivo was observed after embryo transfer (Yu et al., 1985). These authors proposed that Cd toxicity might be acting through general effects on cellular energy metabolism. In a subsequent study (Yu and Chan, 1986), it was found that LDH activity declined from the 4-cell stage to the late blastocyst stage of C57 female × A2G male F1 mouse embryos cultured in vitro. Exposure of 4-cell embryos to 5 or 10 µg/ml $CdCl_2$ caused arrest of development and degeneration after compaction. The LDH activity of the affected embryos at the blastocyst stage was significantly higher than that of control blastocysts, again suggesting an effect on cellular energy metabolism.

Abraham and co-workers (Abraham et al., 1984) administered $CdCl_2$ in drinking water (0.1 or 1.0%) to pregnant rabbits on days 1 to 5 of gestation and examined blastocysts by electron microscopy. Autophagic vacuoles and residual bodies were evident in the lysosomes of endodermal cells and the inner cell mass, while no such changes were noted in the trophoblast.

Yu and Chan (1987) tested the effects of cadmium ($CdCl_2$) on preimplantation as well as early postimplantation mouse embryos, including effects on their trophoblastic invasiveness. Embryos treated in vitro at the 4-cell stage for 24 h with 0.5 or 1.0 µg/ml showed normal trophoblast outgrowth. At higher concentrations of 5 or 10 µg/ml culture medium, most treated embryos underwent degeneration, and viable embryos showed variable regions of trophoblast outgrowth. When blastocysts were continuously exposed to $CdCl_2$ after attachment in vitro, trophoblast growth and the number of giant-cell nuclei were reduced at concentrations of 0.5 µg $CdCl_2$/ml culture medium and above. Subsequently, these investigators examined the effects of zinc on the toxicity of cadmium to the preimplantation mouse embryo developing in vitro (Yu and Chan, 1988a,b). It was found that 5 or 10 µg Zn/ml culture medium protected either 4-cell or morula-stage embryos from the toxic effects of exposure to up to 5 µg $CdCl_2$/ml culture medium for 24 h. The authors proposed that this protective effect of Zn was most likely due to competition between Zn and Cd for binding to and uptake by the preimplantation embryo. A similar protective effect of Zn for preimplantation mouse embryos was reported by Belmonte et al. (1989).

When rat embryos at the 8-cell, morula, or blastocyst stage were incubated for 24 h in the presence of 1 µg/ml $CdCl_2$, it was found that the development of the 8-cell and morula-stage embryos was arrested prior to blastocyst formation (Abraham, et al., 1986). Early blastocysts exposed to 1 µg $CdCl_2$/ml culture medium were also adversely affected, with cell death and debris notable in the inner cell mass.

The levels of constitutive and inducible metallothionein (MT) mRNA in mouse ova and preimplantation embryos were examined by Andrews and co-workers (Andrews et al., 1991). MT-I mRNA in ova, preimplantation embryos, and oviducts was detected by in situ hybridization, and mRNAs for MT-I and MT-II in day 1 to day 4 embryos (vaginal plug = day 1) were analyzed using RT-PCR. Low basal levels of expression were found for both MT-I and MT-II in ova and preimplantation embryos. The MT-I gene was not inducible by Zn or Cd treatment in ova, zygotes, or 2-cell embryos. Some blastomeres of some embryos showed metal-induced expression of MT-I at the 4- to 8-cell stage, while later morula and blastocyst-stage embryos exhibited metal responsiveness of this gene in all cells. Interestingly, when the stage-specificity of Zn or Cd was tested in these embryos, it was found that Cd toxicity increased with the onset of metal-responsiveness, while that of Zn decreased. In a subsequent study (De et al., 1993), 38 µmol Cd/kg maternal body weight on gestation day 2 (embryos at approximately the 2-cell stage) had little effect on pregnancy when examined on gestation day 8, but examination on gestation day 4 indicated that uterine sites of vascular permeability were absent in 62% of dams, suggesting delayed implantation. When the same dose was given on gestation day 4, pregnancy failed in all mice examined on day 8. It was determined that 2-cell embryos showed little or no accumulation of ^{109}Cd, while blastocysts exhibited rapid accumulation and efflux of Cd, suggesting that the decreased toxicity of Cd at the 2-cell stage may be due to a lack of uptake of the metal. Removal of the zona pelucida had no effect on ^{109}Cd uptake, but 500 nM nifedipine (a voltage-gated calcium channel blocker) or 100-fold molar excess of Zn significantly reduced Cd accumulation by blastocysts. In contrast to these results, most of the studies cited above found that the cleavage-stage embryo tended to be more sensitive than were later-staged embryos to the toxic effects of Cd when exposed *in vitro*.

B. EFFECTS OF CADMIUM ON THE POSTIMPLANTATION EMBRYO AND FETUS

In 1964, Pařízek reported that subcutaneous administration of cadmium chloride, acetate, or lactate to pregnant rats at a dosage of 0.04 mmol/kg body weight between the 17th and 21st day of gestation resulted in rapid, progressive placental destruction, especially in the pars fetalis. Necrotic changes could often be seen within 6 h. The placental changes were usually accompanied by hemorrhage into the uterine cavity. Complete destruction of the pars fetalis resulted in resorption or delivery of dead conceptuses. Similar effects were observed with each of the Cd salts tested.

The first demonstration of the teratogenic potential of Cd was reported by Ferm and Carpenter (1967). Single intravenous injections of 2 mg CdSO$_4$/kg body weight were administered to pregnant hamsters early on gestation day 8. Dams were killed prior to parturition and the litters were examined. There was a high incidence of resorption, and a high rate of malformation (66%) among live fetuses. Facial clefts of varying severity, exencephaly, anophthalmia, limb defects, and rib fusions were noted. Coadministration of 2 mg ZnSO$_4$/kg body weight completely ameliorated the teratogenic effects of Cd. Administration of CdSO$_4$ late on gestation day 8 induced rib and upper limb defects, and administration on day 9 produced both upper and lower limb defects (Ferm, 1971). Administration during this developmental period has widespread effects on the developing skeleton, disrupting ossification in many areas of the hamster embryo (Gale and Ferm, 1973). Gale and Layton (1980) studied the strain specificity of the developmental toxicity of Cd in one non-inbred and five inbred strains of hamster. Following a single intravenous maternal dosage of 2 mg/kg on gestation day 8, all six strains were sensitive to Cd-induced developmental toxicity, including resorption as well as external, internal, and skeletal malformations. Significant interstrain differences were observed only in the incidence of resorption, microphthalmia, and renal agenesis.

The relationship of Cd and Zn in teratogenesis was explored by varying the time of Zn administration relative to Cd administration (Ferm and Carpenter, 1968a). It was determined that i.v. ZnSO$_4$ administration concurrent with or up to 6 h following i.v. CdSO$_4$ administration markedly attenuated the teratogenic and embryolethal effects of Cd. In contrast, administration of Zn 12 or more hours after Cd administration failed to protect the embryo. Radioactive Cd administered to pregnant hamsters i.v. on gestation day 8 crossed the placenta and could be detected in the placenta and embryo 24 h later (Ferm et al., 1969).

The pathogenesis of midfacial clefts and deficiencies induced in hamsters by Cd was studied by Tassinari and Long (1982). Intravenous injection of Cd early on gestation day 8 resulted in clefts of the midface ranging from a notched lip to complete facial dysgenesis. Anophthalmia, microphthalmia, and encephalocele were also observed. Gross and histological examination of embryos on the 10th to 13th days of gestation revealed a marked mesenchymal deficiency in the medial nasal process. These authors suggested that this effect may be due to disruption of neural crest migration to this area.

The teratogenicity and developmental toxicity of cadmium has also been demonstrated in rats (Barr, 1973; Chernoff, 1973; Samarawickrama and Webb, 1979), mice (Ishizu et al., 1973; Wolkowski, 1974; Layton and Layton, 1979; Messerle and Webster, 1982), birds (Ribas and Schmidt, 1973; King and Hsu, 1977), frogs (Keino, 1973; Sunderman et al., 1991; Herkovits and Perez-Coll, 1993), and fish (Eaton 1974).

The teratogenic effects of Cd seen in rats varied with the time of administration as well as by the strain of rat used. One thing that studies demonstrating teratogenesis have in common is that high dosages (i.e., at or near maternotoxic levels) were administered parenterally. Chernoff (1973) administered 4 to 12 mg/kg/d CdCl$_2$ to pregnant Sprague-Dawley rats by subcutaneous injection on 4 consecutive days, gestation days 13 to 16, 14 to 17, 15 to 18, or 16 to 19. An increase in the incidence of dead or resorbed fetuses was observed at dosages of 6 to 12 mg/kg/d on gestation days 14 to 17, and dosages of 8 mg/kg/d for any of the dosing intervals produced increased fetal mortality and decreased fetal weight. Among the malformations observed in this study were micrognathia, cleft palate, clubfoot, and small lungs. The reduction of lung weight was over and above what would be expected given the observed fetal weight deficit, as lung/body weight ratios were significantly decreased. Barr (1973) administered 16 µmol/kg $CdCl_2$ to two strains of Wistar rats in a single intraperitoneal injection on gestation day 9, 10, or 11. Administration on day 9 caused anophthalmia or microphthalmia and attenuation of the abdominal wall in fetuses at term in both stocks, while in one stock, administration on day 10 produced limb reduction deformities, preferentially of the left forelimb. Samarawickrama and Webb (1979) treated pregnant rats of the Wistar-Porton strain with a single i.v. injection of 1.25 mg Cd^{2+}/kg body weight between days 8 and 15 of gestation and produced a variety of terata, the most frequent of which was hydrocephalus, observed in 80% of fetuses at term. Other defects observed included anophthalmia, microphthalmia, gastroschisis, and umbilical hernia. These investigators found that a dosage of 1.1 mg Cd^{2+}/kg was not teratogenic, while a dosage of 1.35 mg/kg was uniformly embryolethal.

Cadmium administered to pregnant rats by oral gavage or in the diet exhibits much less developmental toxicity than that observed with parenteral administration. Machemer and Lorke (1981) tested the developmental toxicity of Cd administered to rats by oral gavage or in the diet. During gestation day 6 to 15 rats were given 10, 30, or 100 ppm Cd^{2+} in the food or 1.8, 6.1, 18.4, or 61.3 mg Cd^{2+}/kg body weight by oral gavage. Fetal loss and malformations were observed only at dosage levels that produced moderate to severe maternal toxicity. Barański and co-workers (1982) administered $CdCl_2$ by oral gavage to pregnant rats on gestation days 7 to 16 at dosages of 2 to 40 mg/kg/d. The highest dosage elicited significant maternal toxicity, placental damage, increased resorptions, and increased fetal Cd burden. All dosage levels affected maternal gestational weight gain. Teratogenic effects were not observed, but fetal weight was affected at dosages of 8 mg/kg/d and above. In a subsequent study (Barański, et al., 1983) $CdCl_2$ was administered to female rats by oral gavage for 5 weeks prior to mating and then through gestation at dosages of 0.04, 0.4, or 4.0 mg/kg/d, and no effect on survival or fertility of females or overt fetotoxicity was found. However, male and female offspring of dams given 0.4 or 4.0 mg/kg/d exhibited reduced exploratory behavior at 2 months of age, as well as decreased performance in a rotarod test. Oral treatment of pregnant rats with 40 mg/kg/d $CdCl_2$ from day 7 to 16 of gestation produced congenital malformations including sirenomelia and amelia (Barański, 1984). In this study, other female rats were exposed to cadmium oxide by inhalation for 5 months at concentrations of 0.02 or 0.16 mg Cd/m^3, and then following mating, exposure was continued from gestation days 1 to 20. Decreased viability, lower body weight gain, prolonged latency in the negative geotaxis test, and lower locomotor activity were seen in offspring of females exposed to 0.16 mg Cd/m^3. The lower exposure level of 0.02 mg Cd/m^3 caused reduced locomotor activity and worsened consolidation of the conditioned reflex response.

The selective effect of maternal Cd exposure on the development of the lung demonstrated by Chernoff (1973) was further explored in a series of studies by Daston. The effects of Cd on growth of the fetal rat lung and development of the pulmonary surfactant system were studied following subcutaneous maternal Cd injection on gestation days 12 to 15 (Daston and Grabowski, 1979). A dosage of 8 mg/kg/d resulted in high fetal mortality and growth retardation. As observed previously by Chernoff, lung-body weight ratios were reduced by approximately 25% in fetuses of treated dams. Further, pulmonary lecithin, the principal component of surfactant, was reduced near the end of gestation. When treated dams were allowed to give birth, some of the offspring developed respiratory distress and exhibited histological evidence of alveolar hyaline membranes. Ultrastructural examination of fetal lungs revealed that surfactant-containing lamellar bodies were less numerous in alveolar epithelia of treated fetuses (Daston, 1981). Further studies demonstrated that diminished lung growth was due to hypoplasia, not hypotrophy, and that deposition of glycogen and incorporation of choline into surfactant lecithin were reduced in the fetal lung near term (Daston, 1982b). Coadministration of 12 mg/kg Zn with each Cd injection prevented all of the previously observed fetotoxic effects of Cd (Daston, 1982b). It was found that Cd did not cross the placenta, but maternal Cd treatment did decrease fetal Zn content, strongly suggesting that Cd exerted its developmental toxicity in these experiments through an indirect mechanism by inducing a fetal Zn deficiency. In further support of this idea, maternal dietary Zn deficiency can produce fetal pulmonary hypoplasia and reduced pulmonary lecithin content in rats (Vojnik and Hurley, 1977).

Chiquoine (1965) reported that a subcutaneous injection of 0.02 mmol $CdCl_2$ per kilogram body weight into pregnant mice on any single day between the 6th to 17th day of gestation resulted in loss of the litter prior to parturition. Some inbred mouse strain differences in susceptibility to the developmental toxicity of Cd appear to be due to quantitative and qualitative differences in inducibility of maternal metallothionein (MT) synthesis (Wolkowski, 1974; Wolkowski-Tyl, 1978). A dimeric form of MT was found to have more rapid and higher levels of induction in the resistant strain (NAW/Pr) compared to the sensitive strain (C57BL). Breeding experiments demonstrated a strong maternal component to this difference in sensitivity (Pierro and Haines, 1978).

Cadmium sulfate (12 or 24 µmol/kg body weight) administered to pregnant outbred CD-1 mice by i.p. injection on gestation day 9 produced malformations of the limbs, face, and trunk (Layton and Layton, 1979). The 12 µmol/kg dosage was then given to six inbred strains of mice, three that carried the gene *cdm,* known to confer resistance to Cd-induced testicular necrosis, and three that did not carry this allele. Interestingly, the three strains resistant to Cd-induced testicular damage were more sensitive to the teratogenic effects of Cd. In all of the strains, the malformations observed predominantly affected the limbs.

Messerle and Webster (1982) treated C57BL/6J mice with 4 mg/kg $CdCl_2$ i.p. on gestation day 8, 9, or 10. Embryos were examined at various times up to 3 days after dosing, or dams were killed 1 day prior to birth. Treatment on any of the experimental days produced limb defects, although early treatments were more effective. Forelimb deformities consisted primarily of postaxial reductions, while most hindlimb effects were preaxial, including reduction defects, syndactyly, and polydactyly. While C57BL/6J mice are sensitive to Cd-induced limb malformations, the SWV strain is resistant. Similarly, acetazolamide also produces postaxial ectrodactyly when administered at a critical time in mouse development, and C57BL/6J mice are sensitive, while SWV mice are relatively resistant. In addition, both Cd and acetazolamide can inhibit carbonic anhydrase, and these two agents in combination act synergistically in increasing the incidence of limb defects (Kuczuk and Scott, 1984). To investigate the mechanism(s) underlying the observed strain difference in susceptibility, Feuston and Scott (1985) measured carbonic anhydrase activity in whole embryos and embryo erythrocytes, and Zn and hemoglobin content of embryos and yolk sacs. Their principal finding was that untreated C57BL/6J (sensitive strain) embryos had much lower hemoglobin content than SWV (resistant strain) embryos, and that maternal Cd treatment reduced embryonic hemoglobin levels in the C57BL/6J embryos but not the SWV embryos. The authors proposed that Cd induced forelimb ectrodactyly by producing an acidotic embryonic environment. Using single subcutaneous doses of $CdCl_2$ to pregnant MF-1 mice on one of days 7 to 12 of gestation, Padmanabhan and Hameed (1990) examined in detail the pattern and incidence of limb malformations induced in fetuses at term and in limb buds during midgestation. Gestation days 7 to 10 were the most susceptible for the induction of multiple limb abnormalities. Postaxial ectrodactyly was more frequent in the forepaws, and no sidedness of this defect was observed. Preaxial ectrodactyly preferentially affected the left hindpaws with maternal treatment on gestation day 9. Ossification of the long bones, the carpals, tarsals, and phalanges were affected. The major effect noted in histological examination of limb buds following Cd treatment was poor organization and a decrease in the density of mesenchyme, reduced thickness of the apical ectodermal ridge, and discontinuous basement membrane.

Webster and Messerle (1980) investigated the teratogenic effects of Cd on the developing C57BL/6J mouse CNS. Injection of 4 mg/kg $CdCl_2$ on day 7, 8, 9, or 10 of gestation was followed by histologic examination of the tissues of the CNS and gross examination of embryos for up to 48 h after dosing. Treatment with Cd on gestation day 7 or 8 produced exencephaly, with cell death evident in the neural tube within 8 h after dosing. Exposure on day 9 or 10 did not cause exencephaly, but produced variable degrees of cell death in the neural tube within 24 h. In some embryos treated on gestation day 9, dorsal openings in the previously closed neural tube were observed 48 h later. Using the whole embryo culture technique, Schmid and co-workers (Schmid et al., 1985) provided further evidence that $CdCl_2$ exposure could produce neural tube defects in cultured mouse embryos by a secondary reopening of the cranial neural tube. In a study in which pregnant MF-1 mice were dosed with $CdCl_2$ (4 or 6 mg/kg) on day 7 of gestation, examination of fetuses at term revealed a high incidence of cranioschisis with exencephaly, maxillary and mandibular hypoplasia, microtia, edema, and growth retardation (Padmanabhan and Hameed, 1986). Skeletal examination showed hypoplasia of many components of the facial skeleton and basicranium, and vertebral, rib, and sternebral defects. In a follow-up study to examine abnormalities of the ear associated with Cd-induced exencephaly, Padmanabhan (1987) performed histological examination of heads from exencephalic and nonexencephalic Cd-treated fetuses in comparison to controls. Exencephalic Cd-treated embryos had multiple severe abnormalities affecting almost all components of the ear. While nonexencephalic Cd-treated embryos showed a lesser effect on the ear, there were still significant middle and inner ear abnormalities. These results suggest that exencephaly induced by Cd grossly affects the differentiation of the ear, but that Cd may also have a direct effect on ear development aside from that associated with exencephaly. Confirming earlier studies, effects of $CdCl_2$ (2 to 6 mg/kg) administered in a single dose to pregnant ICR mice on one of gestation days 8 to 14 included embryolethality, limb malformations, exencephaly, and cleft palate. A preponderance of the limb malformations were on the right side. Fetal thymus weight was reduced at the higher dosage levels when administered between the 9th and the 14th day of gestation (Soukupova and Dostal, 1991).

Treatment of pregnant Sprague-Dawley rats with 2.0 or 2.5 mg $CdCl_2$ per kilogram body weight (i.p.) on gestation days, 8, 10, 12, and 14 resulted at both dosages in significant decreases in the urinary excretion of the proximal tubular enzymes gammaglutamyl transferase, alkaline phosphatase, and N-acetyl-beta-glucosaminidase in offspring on postnatal day 3 (PND 3). The higher dosage resulted in functional deficit of the proximal tubule on PND 3, as indicated by a significant increase in urinary excretion of beta 2-microglobulin (Saillenfait et al., 1991).

When 2.5 to 5.0 mg $CdCl_2$ per kilogram body weight was administered to pregnant ICR mice on gestation day 16, postnatal immune alterations were noted in offspring at 4 weeks of age, including enhanced proliferative responses of spleen cells to concanavalin A, phytohemagglutinin, and lipopolysaccharide, and decreased delayed-type hypersensitivity to sheep red blood cells (Soukupova et al., 1991).

To test the late gestational sensitivity of the CNS to Cd, gestation d 19 rat fetuses were given direct i.p. injections of $CdCl_2$ (50, 100, 165 nmol per fetus) (White et al., 1990). All fetuses in one uterine horn were injected with Cd, while fetuses in the contralateral horn were injected with saline as a control. A dose-related increase in moderate to severe hydrocephalus was observed in fetuses collected at gestation day 21. However, direct exposure of the fetuses to Cd was required, as maternal exposure at this time does not produce hydrocephalus.

The teratogenic potential of Cd has been tested in amphibia. Perez-Coll et al. (1986) treated gastrulating *Bufo arenarum* embryos with $CdCl_2$ in concentrations ranging from 6×10^{-7} to 1.5×10^{-5} M Cd^{2+}. Defects in axial curving, microcephaly, hydrops, and abnormal tail formation were observed. The sensitivity to and types of developmental toxicity observed varied with the stage at exposure (Herkovits and Perez-Coll, 1993). For example, at the neurula stage 0.25 mg Cd/l caused 100% developmental arrest, and the neural tube stage was 16-fold more sensitive than the blastula stage. Sunderman and coworkers (Sunderman, et al., 1992) tested the teratogenicity of Cd in *Xenopus laevis* using the FETAX (frog embryo teratogenesis assay: *Xenopus*) assay. Groups of *Xenopus* embryos were grown in media containing 0.75 to 56 µmol Cd/l. In Cd-exposed groups, concentration-dependent mortality and multiple malformations were observed including gut malrotation, ocular anomalies, bent notochord, misshapen fin, facial dysplasia, cardiac anomalies, and dermal blisters. The median teratogenic concentration was 3.7 µmol Cd/l.

C. PLACENTAL TOXICITY AND PLACENTAL TRANSFER OF CADMIUM

The placenta and/or yolk sac are targets for a number of developmental toxicants, and metals are prominent among them. The placental toxicity and placental permeability of selected metals, including Cd, are reviewed by Miller and Eisenmann (Chapter 64). Only a brief review of illustrative findings will be provided here.

It appears that the developmental toxicity of Cd during mid to late gestation involves both placental toxicity (necrosis, reduced blood flow) and inhibition of nutrient transport across the placenta. The increased toxicity of Cd to the pregnant rat near term may be due to renal failure subsequent to shock from placental hemorrhage (Samarawickrama and Webb, 1981). Maternal injection of Cd during late gestation results in fetal death in rats, despite little cadmium entering the fetus (Pařízek, 1964; Levin and Miller, 1980). Fetal death occurs concomitant with reduced uteroplacental blood flow within 10 h (Levin and Miller, 1980). The authors' conclusion that fetal death was caused by placental toxicity was supported by experiments in which fetuses were directly injected with Cd. Despite fetal Cd burdens almost tenfold higher than those causing fetal death after maternal administration, only a slight increase in fetal death was observed. Alteration of uteroplacental blood flow by Cd in rats was examined using a radioactive microsphere technique (Levin and Miller, 1981). Uteroplacental blood flow was reduced 40% and 73% from control values at 12 to 16 and 18 to 24 h post dosing (40 µmol $CdCl_2$ per kilogram s.c. on gestation day 18), respectively, and placental necrosis was observed concomitantly with fetal death. It was not clear in this study whether reduced blood flow caused the placental necrosis or vice versa. These authors later proposed that the mechanism of fetal death following late-term maternal Cd administration involves trophoblastic damage which leads to a local circulatory response and a decrease in uteroplacental blood flow. These changes result in a decrease in nutrient and oxygen transport to the fetus, and this placental insufficiency ultimately leads to fetal death. An ultrastructural examination of the placenta following maternal Cd administration revealed that there is a direct toxic effect of Cd on the trophoblast, specifically on trophoblast cell layer II (di Sant' Agnese et al., 1983). Findings included lysosomal vesiculation, nuclear chromatin clumping, nucleolar alterations, and apparent mitochondrial calcification. Although necrosis first occurs in trophoblast cell layer II, necrosis of the remaining trophoblast rapidly follows. Blood flow to the chorioallantoic placenta was decreased by 35% at 16 to 18 h after maternal administration of 40 µmol $CdCl_2$ per kilogram (s.c.) on gestation day 12 in rats (Saltzman et al., 1989). Blood flow to the chorioallantoic placenta returned to control levels by 24 to 26 h, and uterine blood flow was not significantly affected at any time point.

The effects of Cd on the isolated, dually perfused (i.e., on both maternal and fetal sides) human placenta have been studied (Wier et al., 1990). Placental lobules were perfused with medium containing 0, 10, 20, or 100 nmol $CdCl_2$ per milliliter initially only on the maternal side. Every 4 h thereafter, perfusates on both sides were replaced with fresh medium without Cd. Oxygen consumption, net fetal oxygen transfer, fetal pressure, fetal volume loss, glucose uptake, lactate production, human chorionic gonadotropin (hCG), and Zn transfer were measured during perfusion. There were concentration-related alterations in the synthesis and release of hCG, volume loss from the fetal vasculature, ultrastructural changes, and necrosis at the highest concentration occurring between 5 to 8 h. In addition, the placental transfer of Zn from the maternal to the fetal circuit was inhibited by Cd.

Ferm and co-workers (Ferm et al., 1969) examined the permeability of the hamster placenta to ^{109}Cd following maternal i.v. injection on gestation day 8. Counts in maternal tissues, including the uterus, and in placenta and embryo were assayed on gestation day 9, and these tissues as well as the yolk sac were assayed on gestation day 12. Detectable counts were found in the embryo on both days examined, and concentrations in the embryo were similar to those in maternal blood on gestation day 9, but were only 2 to 5% of maternal blood concentrations by gestation day 12. On gestation day 12, ^{109}Cd concentration in the placenta and yolk sac was 50 to 100 times that in the embryo.

Hamsters and mice at different stages of pregnancy were injected with ^{109}Cd and then killed at different times afterward for whole-body autoradiography (Dencker, 1975). Cadmium administered on gestation day 8 accumulated in the embryonic gut in both species, but no ^{109}Cd was detected in embryos after day 9 in the hamster or day 11 in the mouse. These findings were attributed to uptake of Cd by the yolk sac and transfer to the primitive gut before closure of the vitelline duct but not after. High accumulation of Cd was noted in the decidua, yolk sac, ectoplacental cone, and in the placenta. In a similar study, low levels of ^{109}Cd were detected in mouse embryos following maternal i.p. injection on gestation day 9 (Christley and Webster, 1983). On a wet weight basis, Cd concentration of the embryo was highest at 1 h postdosing. Although cell damage was observed by 12 h postdose, autoradiographs demonstrated that damaged cells were not necessarily heavily labeled.

On gestation day 12, rats embryos from dams dosed with 0.1 to 0.6 mg/kg Cd showed minimal accumulations of the maternal dose, only about 1% of the amount of Cd accumulated by the placenta (Sonawane et al., 1975). Age-related changes in the uptake of Cd by the rat embryo after maternal oral administration of Cd were reported by Ahokas and Dilts (1979). Pregnant rats were given a single dose of radiolabeled $CdCl_2$ (10 to 1000 µg per rat) on one of gestation days 6, 10, 14, or 17, and maternal and embryo/fetal tissues were counted for ^{109}Cd 24 h later. Embryo accumulation of ^{109}Cd was highest prior to the formation of the chorioallantoic placenta. After establishment of the placenta, fetal levels of ^{109}Cd were decreased, while placental accumulation increased with increasing gestational age. When radiolabeled $CdCl_2$ was administered s.c. to rats on gestation day 18, minimal accumulation in the fetus was observed, while uptake by the placenta was rapid and extensive (Levin et al., 1987).

In order to investigate established maternal stores of Cd as a potential source of Cd transfer to fetuses and pups, Whelton et al. (1993) fed female mice nutrient-sufficient or -deficient diets containing 5 ppm Cd and provided drinking water with tracer amounts of ^{109}Cd. Dams were switched to water without ^{109}Cd and diets with control levels of Cd during the reproductive period. Maternal stores of Cd appeared to be a relatively minimal source of Cd transfer to the fetus, even under conditions of undernutrition. To test whether Cd exposure would increase Ca release from bone during pregnancy and lactation in relation to the etiology of itai-itai disease, virgin female mice with ^{45}Ca-prelabeled skeletons were subjected to one round of pregnancy and lactation and were fed a Ca-deficient diet containing 0, 5, or 25 ppm Cd from conception to lactation day 14 (Wang et al., 1994). While pregnancy and lactation alone produced a 40 to 75% loss of ^{45}Ca from maternal skeletal elements, concurrent exposure to Cd nearly doubled the loss of Ca from the dam's skeleton. These results suggest that Cd exposure in conjunction with Ca deficiency and pregnancy/lactation are key factors in the etiology of itai-itai disease.

Mobilization of hepatic Cd in pregnant rats was demonstrated in experiments by Chan and Cherian (1993). When female rats were Cd-loaded by 8 daily s.c. injections of 1 mg Cd/kg body weight as $CdCl_2$ and then mated, there was redistribution of hepatic Cd to maternal kidney and placenta. Maternal kidney showed a 60% increase in Cd concentration, and renal and placental MT was elevated as well compared to controls.

D. MECHANISMS OF ACTION

From the preceding discussion it is apparent that administration of Cd to pregnant animals during early gestation, prior to establishment of the placenta, can result in direct exposure of the embryo to Cd, while later exposure is likely to result in relatively little fetal Cd exposure. Most inquiries into the mechanisms of developmental toxicity of Cd have focused on the indirect effects of Cd mediated by reduced uteroplacental blood flow, reduced nutrient transport, and placental toxicity. As previously alluded to in this review, one important mechanism underlying the developmental toxicity of Cd appears to be the biochemical interactions, on several levels, between Cd and the essential trace nutrient, Zn. At the molecular level, Cd may substitute for Zn in metalloenzymes (Vallee and Ulmer, 1972; Vallee and Glades, 1974).

Cd is a transition metal similar in its physicochemical properties to Zn, and therefore Cd can often participate in Zn metabolic pathways. Cd is transported into cells through mechanisms specific for Zn uptake (Stacey and Klaassen, 1980; Waalkes and Poirier, 1985). Cadmium interferes with Zn transfer across the placenta (Samarawickrama and Webb, 1979; Ahokas, et al., 1981; Danielsson and Dencker, 1984; Sorell and Graziano, 1990), possibly via metallothionein (MT) induction in the placenta by Cd (Lehman and Poisner, 1984; De et al., 1989; Boadi et al., 1991; Goyer and Cherian, 1992). Perfusion of term human placenta with medium containing 20 μM $CdCl_2$ for 4 or 8 h results in a striking increase in MT mRNA (Breen et al., 1994). Adaptation of trophoblast cells to chronic Cd exposure (2 μM $CdCl_2$ in the medium for 6 months) results in elevated MT protein levels and increased localization of MT to the nucleus (Breen et al., 1995). Because of its high affinity for Zn, MT may sequester Zn in the placenta, impeding transfer to the conceptus. Cadmium inhibits Zn uptake by human placental microvesicles (Page et al., 1992), suggesting that Cd may also compete directly with Zn for membrane transport. Cadmium may also competitively inhibit other Zn-dependent processes in the placenta. Coadministration of Zn ameliorates the developmental toxicity of administered Cd, further indicating that interference of Cd with Zn metabolism is a key to its developmental toxicity (Ferm and Carpenter, 1967, 1968a; Garcia and Lee, 1981; Daston, 1982; Hartsfield et al., 1992). Feeding rats a Zn-deficient diet during pregnancy exacerbates the developmental toxicity of Cd (Parzyck et al., 1978), while supplemental dietary Zn can protect against Cd-induced fetal growth retardation (Ahokas et al., 1980).

Cadmium exposure can induce MT synthesis at high levels in the maternal liver in rats (Waalkes and Bell, 1980) and mice (Muñoz and Dieter, 1990), leading to sequestration of Zn to this newly synthesized MT. Sequestration of Zn in the liver can reduce maternal plasma Zn and thereby reduce Zn availability to the litter. Administration of 50 ppm $CdCl_2$ in the drinking water of pregnant rats throughout pregnancy caused a reduction in maternal serum Zn concentration, as well as reductions in Zn, Cu, Fe, and MT in fetal liver (Sowa and Steibert, 1985). Waalkes and Bell (1980) found that while maternal hepatic MT and Zn concentrations were elevated following Cd exposure, both of these parameters were decreased in fetal liver. Sasser et al. (1985) exposed pregnant rats to Cd via the drinking water throughout gestation or injected Cd into fetuses directly on gestation day 18. Fetal hepatic Zn concentrations were reduced by maternal Cd exposure, and both fetal liver and kidney MT were also reduced. Maternal hepatic MT was increased by Cd exposure, but none of the treatment regimens induced fetal MT synthesis. Barański (1987) also reported decreased fetal liver Zn concentrations following maternal Cd exposure via the drinking water throughout pregnancy. In contrast to the above, however, Waalkes et al. (1982) reported that administration of $CdCl_2$ to pregnant rabbits near term at dosages causing fetal growth retardation did induce fetal liver MT, despite relatively low transfer of Cd across the placenta. In the rabbit fetuses, liver Zn and MT-Zn concentrations were increased by maternal Cd exposure, but total fetal Zn concentration was not changed, indicating a redistribution of Zn among fetal tissues.

De and co-workers (De et al., 1990) examined MT gene expression in midgestation CD-1 mouse embryos exposed to Cd *in vivo* or in whole embryo culture. Maternal s.c. injection of a teratogenic dosage of Cd on gestation day 10 did not induce MT mRNA in the embryo, while maternal injection of Zn did. In contrast, Cd was an order of magnitude more potent than Zn for inducing MT mRNA in d 10 embryos in culture. These results indicate that the mouse embryo is able to synthesize MT mRNA, but that Cd is largely prevented from reaching the embryo at this stage.

Cadmium can inhibit pinocytosis by the rat embryo yolk sac *in vitro*, suggesting that CD may have a nonspecific effect on nutrient transport by the yolk sac (Record et al., 1982a,b). The effects of Cd on pinocytosis were markedly reduced by the addition of Zn to the culture medium. Warner and co-workers

(Warner et al., 1984) also demonstrated that Zn could ameliorate the teratogenic effects of Cd in mouse whole embryo culture. Zinc can also prevent feather malformations by Cd in chick embryos (Narbaitz et al., 1983).

Addition of $CdCl_2$ to cultures of murine embryonal carcinoma cells inhibits both proliferation and differentiation of the cells into parietal endoderm (Piersma et al., 1993). Concurrent addition of Zn counteracts these effects, and uptake of Cd by the cells is also reduced by Zn, suggesting competition between these metals for membrane transport. Metallothionein is induced by Cd, and further induced by simultaneous Zn exposure in these cells, suggesting that the ameliorative effect of Zn on Cd toxicity may also be due to detoxification by MT. Cadmium-resistant variants of Chinese hamster ovary (CHO) cells have increased MT induction capacity (Hildebrand et al., 1979; Morris and Huang, 1989). Mouse NIH/3T3 cells transfected with high copy numbers of the mouse MT-I gene were more than 10-fold more resistant to the lethal effect of Cd than were control cells (Morton et al., 1992). Metallothionein synthesis was 15-fold greater in the transfected cells and uptake of ^{109}Cd by the cells enriched in MT was 4-fold less than controls.

Glutathione (GSH) is also likely to be an important cellular defense against Cd toxicity (Singhal et al., 1987). The hepatotoxic and nephrotoxic effects of Cd are inversely related to tissue GSH levels (Dudley and Klaassen, 1984; Suzuki and Cherian, 1989). Lowered levels of GSH can exacerbate Cd toxicity even in the presence of otherwise protective levels of MT (Ochi et al., 1988; Kang et al., 1989; Suzuki and Cherian, 1989).

Cadmium has been demonstrated to induce synthesis of heat shock proteins (hsps) in a variety of cell- and embryo-based *in vitro* systems. The full set of hsps are induced by Cd in *Drosophila* embryonic cells (Bournias-Vardiabasis et al., 1990), while Honda and co-workers (Honda et al., 1991) demonstrated that hsps are induced by *in vitro* Cd exposure of human placental villus tissue or dispersed gestation day 11 mouse embryo cells. Kapron-Bras and Hales (1991, 1992) have demonstrated that exposure of cultured mouse embryos to mild hyperthermia induces tolerance to a subsequent heat exposure as well as cross-tolerance to a normally teratogenic concentration of Cd.

II. ARSENIC

Arsenic (As) has been well known as a human toxicant through the ages. Today, humans are exposed to arsenicals primarily through drinking water and seafood. Although human exposures provide evidence of the carcinogenicity and neurotoxicity of arsenic, human data pertaining to developmental toxicity are very limited. There are two reports of exposure during pregnancy resulting in fetal or neonatal death (Lugo et al., 1969; Bollinger et al., 1992). In studies correlating reproductive outcome with drinking water contaminants, some associations were noted between arsenic and adverse outcomes (Zeirler et al., 1988; Aschengrau et al., 1989). Areas around copper smelters can be subjected to considerable As (as well as other metals) contamination, and reproductive outcome in populations around smelters has been studied in Sweden and Bulgaria. In the Swedish study (Beckman, 1978), adverse reproductive outcomes (spontaneous abortions, congenital malformations, low birthweight) were studied in populations working in or residing near the Rönnskär copper smelter in northern Sweden. The study population was subdivided into workers at the plant and people residing in four regions of progressively further distance from the smelter. In addition to As, the smelter is a major source of Cu, Pb, Zn, and Hg pollution. Spontaneous abortion rates were higher and birthweights lower among workers and those living in the two closest regions. Congenital malformations were higher in occupationally exposed, but not environmentally exposed groups. The Bulgarian study, by Tabacova and co-workers (Stone, 1994; Zelikoff et al., 1995), involved 49 mother-infant pairs from hospitals near copper smelter and from an area without industrial metal contamination. Pregnancy complications and rates of mortality at birth due to malformations were higher in the smelter area than in the nonindustrial area for the 5-year period prior to the study. Placental As content was highest in areas of highest environmental contamination (Tabacova et al., 1994a). Maternal exposure to metals was associated with a decrease in reduced glutathione in the blood, suggesting that lipid peroxidation may contribute to pregnancy complications in exposed women (Tabacova et al., 1994b).

The developmental toxicity of As has been studied extensively in animals, with almost all of the work in mammals involving postimplantation exposures and evaluations. The experimental induction of terata with As was first reported in the chick (Ancel and Lallemand, 1941), and a thorough study of the effects of both trivalent (As^{III}; arsenite) and pentavalent (As^V; arsenate) As in the chick embryo was carried out by Peterkova and Puzanova (1976). The studies reviewed here primarily describe the developmental effects of As^{III} and As^V. Methylation is an important route of As metabolism and is generally, but not exclusively, considered a detoxification pathway (Styblo et al., 1995), and methylated forms are less developmentally toxic than inorganic forms (Willhite and Ferm, 1984).

A. EFFECTS OF ARSENIC ON PREIMPLANTATION MAMMALIAN EMBRYOS

In contrast to Cd, there is a paucity of data concerning the toxicity of As to the preimplantation embryo. Preimplantation embryos grown *in vitro* in the presence of As^{III} exhibited micronucleus formation at concentrations of 0.7 μM and above, inhibition of blastocyst formation at 1 μM, and immediate lethality at 100 μM (Muller et al., 1986). As^V (10 mM) was found to be a potent inhibitor of alkaline phosphatase in 8-cell mouse embryos, while dimethylarsinic acid was not inhibitory (Lepire and Ziomek, 1989).

B. EFFECTS OF MATERNAL ARSENIC EXPOSURE ON POSTIMPLANTATION EMBRYOS

The toxicity of various forms of inorganic and organic As to postimplantation development have been tested both *in vivo* and *in vitro*. Inorganic As^{III} and As^V are more developmentally toxic than the methylated arsenicals (Willhite, 1981; Hood et al., 1982). Most studies in pregnant hamsters, rats, and mice have used dosages at or near those eliciting maternal toxicity, but maternal and developmental toxicities are not necessarily well-correlated in these studies, and developmental toxicity does not appear to be secondary to maternal toxicity (Golub, 1994). Numerous studies demonstrating direct effects of As on embryos *in vitro* (see Section C, below) support this view.

The teratogenicity of As^V in a mammal was demonstrated by Ferm and Carpenter (1968b) in studies with hamsters. Intravenous injection of 20 mg sodium arsenate/per kilogram body weight on gestation day 8 resulted in 49% malformed and 84% either malformed or resorbed embryos on gestation day 13 (Holmberg and Ferm, 1969). Malformations reported include exencephaly, encephalocele, cleft lip/palate, micro/anophthalmia, and ear malformations. Coadministration of selenium (2 mg/kg sodium selenite) with arsenate reduced the frequency of malformed fetuses to 19% and the total frequency of resorbed or malformed fetuses to 39%. In a subsequent study, groups of hamsters were injected with sodium arsenate (15 to 25 mg/kg) at different times (9 a.m., 3 p.m., or 9 p.m.) on gestation day 8 (Ferm et al., 1971). Both resorption and malformation rates increased with increasing dosages. The frequency of resorption was highest at the earliest time point, while the overall frequency of malformations remained fairly uniform. The types of malformations did change with different dosing times, with anencephaly declining from 80% with dosing at 9 a.m. to 10% with dosing at 9 p.m. Conversely, the incidence of rib malformations rose from less than 1% to between 60 to 70% over the same time-period. Genitourinary malformations, including renal agenesis, were observed in 10 to 20% of fetuses at each dosing time point.

Willhite (1981) administered 20 mg/kg sodium arsenate, 2 to 10 mg/kg sodium arsenite, 20 to 100 mg/kg methylarsonic acid, or 20 to 100 mg/kg dimethylarsinic acid by i.v. injection to pregnant golden hamsters on gestation day 8. As^V administration at 20 mg/kg resulted in 44% resorptions and 86% abnormal among live fetuses. The frequency of exencephaly and encephalocele was 92% among live fetuses, and other malformations noted were anophthalmia and rib and kidney defects. As^{III} was a more potent developmental toxicant than As^V. Resorption frequency with As^{III} ranged from 4% at 2 mg/kg to 90% at 10 mg/kg, and malformed fetuses were observed at all dosage levels. The spectrum of malformations was similar to that described above for As^V. Examination of embryos 10 h after dosing with a teratogenic dosage of As^V revealed a delay in neural fold elevation and a paucity of cephalic mesoderm compared to controls. Methylarsonic acid and dimethylarsinic acid were less developmentally toxic than either As^V or As^{III}. Very low rates of resorption or malformation were seen at any dosage level and there was little evidence of a dose response. Carpenter (1987) similarly reported that retarded elevation, approximation, and fusion of the cephalic neural folds of the embryos could be seen within 24 to 48 h of maternal i.p. injection with 20 mg/kg sodium arsenate on the morning of gestation day 8.

Hood and Harrison (1982) tested the developmental toxicity of As^{III} in the hamster following administration of sodium arsenite by oral gavage at a dosage of 25 mg/kg on gestation day 8, 11, or 12, or at a dosage of 20 mg/kg on day 9 or 10. Treatment with 25 mg/kg on gestation day 8 or 12 resulted in increased fetal death, and treatment on day 12 additionally resulted in reduced fetal weight.

Ferm and Hanlon (1985) used subcutaneous osmotic minipumps to produce a constant-rate infusion of As^V to pregnant hamsters. The infusions began between days 4 and 7 of gestation. Increased resorptions and decreased fetal weight were observed with increased time or concentration of exposure, while fetal malformations increased with dose but showed no apparent effect of duration of exposure. The types of malformations observed were similar to those previously observed with intravenous injection.

The teratogenicity of As^V in rats was first demonstrated by Beaudoin (1974). Dosages of 20 to 40 mg/kg sodium arsenate were administered by i.p. injection on one of gestation days 7 to 12. Resorptions and malformations were increased in a dose-dependent manner, with rates of both of these outcomes being highest on gestation day 9. The malformations observed included exencephaly, vertebral defects, renal agenesis, rib defects, anophthalmia, and gonadal agenesis. Thus, the spectrum of malformations observed in the rat following maternal As^V exposure was similar to that observed in the hamster. The developmental origin of renal agenesis induced in rats by As^V was explored by Burk and Beaudoin (1977). A single i.p. injection of 45 mg/kg sodium arsenate was administered on gestation day 10, and embryos were harvested at various times thereafter and examined histologically. The first sign of pathological development was retardation in growth of the mesonephric duct apparent after 48 h. Absence of a ureteric bud and failure of metanephric blastema induction were subsequently observed. In the male, the shortened mesonephric duct resulted in absence of the ductus deferens, the seminal vesicle, and a portion of the epididymis. Absent uterine horns were observed in female fetuses.

Treatment of pregnant rats with arsenic acid (As^V; 30 mg/kg) on gestation day 9 was used to examine the histological and ultrastructural changes underlying As-induced cephalic dysraphism (Takeuchi, 1979). Electron micrographs in this publication show clear evidence of apoptosis in the neuroectoderm of treated embryos within 4 to 6 h after maternal exposure to As. Elevation of the neural folds is retarded by 24 h later, as is somite formation. This study suggests that As can rapidly induce apoptosis specifically in the neuroepithelium, impeding neurulation and resulting in cephalic neural tube defects. These findings are similar to those in the hamster reported by Carpenter (1987).

Fisher (1982) reported the effects of As^V on accumulation of DNA, RNA, and protein in gestation day 10 rat embryos developing *in vitro*, beginning 4 or 24 h after maternal i.p. injection of 45 mg/kg sodium arsenate. The concentrations of DNA, RNA, and protein were all lower at the beginning of culture in embryos of dams treated with As^V 24 h earlier, and these levels remained lower than controls throughout the 24 to 42 h culture period. Embryos from dams treated 24 h prior to culture also exhibited morphological abnormalities, including dorsiflexion, open anterior neuropore, facial dysmorphology, and a failure of the allantois to fuse with the chorion. In contrast, gestation day 10 embryos removed 4 h after maternal As^V treatment developed similar to controls *in vitro*.

The only report available on the developmental toxicity of maternal exposure to As^{III} in rats is an abstract (Umpierre, 1981) which reports embryotoxicity and teratogenicity similar to that described above for hamsters and below for mice.

The developmental toxicities of As^{III} and As^V have been tested in mice. A single i.p. injection of 45 mg/kg sodium arsenate on one of gestation days 6 to 12 resulted in increased fetal death (Hood and Bishop, 1972). Fetal weight was reduced, and the incidence of malformed fetuses increased with maternal exposure on any of days 6 to 11. Treatment on gestation day 11 or 12 resulted in the highest rates of resorption, 69 and 78%, respectively. Treatment on day 9 produced the highest incidence of malformed live fetuses, 63%. The specific external malformations observed depended on the day of maternal exposure, but included exencephaly, micrognathia, exophthalmia, anophthalmia, cleft lip, hydrocephalus, micromelia, and ectrodactyly. Numerous skeletal malformations were also observed. In a subsequent study, the chelating agent BAL was administered before, concurrent with, or after 45 mg/kg sodium arsenate (Hood et al., 1977). All chelator treatments reduced As^V-induced gross malformations and growth retardation, while concurrent administration diminished skeletal alterations. Hood (1972) also studied the developmental toxicity of As^{III} in mice. Swiss-Webster mice were administered single i.p. injections of sodium arsenite (10 to 12 mg As^{III}/kg maternal body weight) on one of gestation days 7 to 12. The incidence of prenatal death was elevated in all treatment groups. As^{III} administration on day 9 or 10 resulted in the highest incidences of malformations, 27 and 36%, respectively, at the high dosage. Malformations observed following day 9 exposure included exencephaly, micrognathia, open eye, rib defects, and vertebral defects. Maternal exposure on gestation day 10 produced primarily open eye, tail defects, and rib defects. Fetal weight was affected by maternal As^{III} exposure on gestation days 7 to 11.

In a comparison of the effects of oral vs. i.p. routes of administration of As^V to pregnant mice, Hood et al. (1978) found that a maternal oral dosage of 120 mg/kg sodium arsenate was less embryotoxic than an i.p. injection of 40 mg/kg when given on a single day between gestation days 7 to 15. The two treatments caused comparable maternal mortality. Embryofetal mortality was significantly higher in the i.p.-treated groups on most dose days, and the i.p. route produced 31% and 12% grossly malformed fetuses on gestation days 9 and 10, respectively, compared to only 1% and 3% in the orally treated groups. Skeletal malformations after maternal treatment on the most sensitive day, gestation day 9, occurred in 54% of the fetuses in the i.p.-treated group, compared to 14% in the orally treated group.

The developmental toxicity of orally administered As^{III} was also tested in CD-1 mice (Baxley et al., 1981). Sodium arsenite was administered by oral gavage at 20, 40, or 45 mg/kg on one of gestation days 8 to 15. The lowest dosage produced no maternal or developmental toxicity. The incidence of maternal deaths was 19% and 36% at the 40 and 45 mg/kg As^{III} dosages, respectively. With treatment at 40 mg/kg on day 8 or 9, or with 45 mg/kg on day 8, 9, or 10, a low incidence of exencephaly and open eyes was observed. This study demonstrates that, as is the case for As^V, maternal dosing with As^{III} by the oral route is less teratogenic than by the i.p. route.

Administration of BAL prior to or concurrent with maternal i.p. As^{III} administration on gestation day 9 or 12 reduced some, but not all, of the manifestations of developmental toxicity observed in mouse fetuses (Hood and Vedel-Macrander, 1984).

A standard segment II study (daily dosing on gestation days 6 to 15) and a two-generation study of arsenic acid (based on As^V), a registered herbicide, were carried out in pregnant CD-1 mice. These studies have been summarized in Golub (1994). Administration of arsenic acid was by oral gavage in the segment II study and in the diet in the two-generation study.

In the segment II study (WIL Research Laboratories, 1988a), maternal toxicity manifested as reduced weight gain after dosing, and mortality was demonstrated at the highest dosage, 64 mg/kg/d. A low incidence of malformed fetuses was observed at each dosage level, including the lowest dosage, 10 mg/kg/d, at which one fetus had exencephaly with facial cleft. In the mouse two-generation study (Hazelton Laboratories of America, 1990), severe maternal and developmental toxicity were observed at the highest dosage level (500 ppm in the diet), including increased maternal mortality and reduced maternal weight in the first generation dams, as well as smaller litter sizes and lower birth weight in this group. F_0 offspring were severely growth retarded and there was high postnatal mortality. Maternal and developmental toxicity was more severe in the second generation than in the first among the high-dose animals. At the mid-dose (100 ppm), no maternal toxicity was observed but there was mild postnatal growth retardation of F_0 offspring.

A rabbit segment II study (WIL Research Laboratories, 1988b) demonstrated that this species is quite sensitive to the acute toxicity of As^V. The maximum tolerated dose (MTD) to the doe was only 4 mg/kg/d, compared to 64 mg/kg/d in mice (WIL Research Laboratories, 1988a). The incidence of malformations was low at all dosage levels, and fetal weight was not affected. The malformations observed were rare in rabbits and similar to those induced by As^V in other species, including renal agenesis and fused ribs and sternebrae. The NOAEL for both maternal and developmental toxicity in this study was only 1 mg/kg/d. No other studies of the developmental toxicity of As in rabbits were found in the literature.

The developmental toxicity of cacodylic acid (dimethylarsinic acid), a registered herbicide, was tested in rats and mice by Rogers and co-workers (Rogers et al., 1981). The compound was administered by oral gavage on gestation days 7 to 16 (sperm-positive = day 1) in a standard Segment II study design. In mice, decreased maternal weight gain was noted at all dosage levels, including the lowest dosage of 200 mg/kg/d, and 59% maternal mortality was recorded at the highest dosage level, 600 mg/kg/d. Cleft palate and significant decreases in fetal weight were noted at 400 and 600 mg/kg/d. When these results in mice are compared to those in the Segment II study of arsenic acid presented above (in which terata were noted down to 10 mg/kg/d), it is clear that the dimethylated form is much less developmentally toxic than inorganic As when administered to the pregnant mouse. In the rat, cacodylic acid caused significant decreases in maternal weight gain at 40 mg/kg/d and above, and maternal mortality was noted at 50 mg/kg/d and above. Fetal weight was significantly lower than controls at 40 mg/kg/d and above, and abnormal fetal palatine rugae were noted at 30 mg/kg/d and above.

C. EFFECTS OF ARSENIC EXPOSURE ON EMBRYOS DEVELOPING *IN VITRO*

The direct toxicity of As^{III} and As^V to CD-1 mouse embryos developing in whole embryo culture was first reported by Chaineau and co-workers (Chaineau et al., 1990). Gestation day 8 mouse embryos (3 to 6 somites) were removed from the uterus and cultured in heat-inactivated rat serum, and sodium arsenate or sodium arsenite were added dissolved in distilled water. Embryos were cultured in the presence of As^{III} or As^V for 48 h. Viability, growth, and morphogenesis were evaluated at the end of the culture period and selected embryos were evaluated histologically. The final concentrations in the medium ranged from 10 to 400 μM for As^V and from 1 to 40 μM for As^{III}. Embryolethality was observed at concentrations above 150 μM As^V or 15 μM As^{III}, and lack of development occurred at concentrations above 40 μM As^V or 4 μM As^{III}. Both As^V and As^{III} caused prosencephalic hypoplasia in exposed embryos. In addition, As^V exposure resulted in hydropericardium, and at high concentrations both compounds caused lack of cephalic neural tube closure. These results indicate that both As^{III} and As^V are teratogens that can act directly on the developing embryo to produce effects similar to those previously reported for maternal exposure to these compounds. Also in agreement with whole animal studies, As^{III} was found to be more toxic than As^V. Mirkes and Cornel (1992) reported that 50 μM As^{III} was embryotoxic and dysmorphogenic to gestation day 10 rat embryos in culture. Growth (crown-rump length, somite number, protein content) was reduced in exposed embryos, and abnormal morphogenesis of the prosencephalon and somites was observed, as well as abnormal tail flexion. These authors also reported that As^{III} induced the embryonal synthesis of three heat shock proteins (see Section E, below).

Tabacova and Hunter (in Zelikoff et al., 1995) found that As induced dose-, time-, and valency-dependent embryotoxicity and dysmorphogenesis in ICR or CD-1 mice exposed in whole embryo culture. As^{III} was 3 to 5 times more potent than As^V with respect to specific malformations. Earlier embryos were more sensitive to As than later embryos. Exposure at progressively later stages from 4 to 20 somites revealed that the dysmorphogenic and embryolethal ED_{50} for both As^{III} and As^V increased 3 to 5 fold over this period.

As^{III} inhibited chondrogenesis in chick limb bud mesenchymal spot cultures, with an ED_{50} of about 5 to 10 μM, but As^V appeared to be without effect at concentrations up to 200 μM (Lindgren et al., 1984).

D. PHARMACOKINETICS AND PLACENTAL TRANSFER OF ARSENIC

Most of the work on pharmacokinetics of As compounds has been carried out in mice and hamsters, as rats have been shown to markedly differ from most other experimental animals and humans in the retention, excretion, and distribution of As (Styblo et al., 1995). One primary difference in the rat is the very high accumulation and retention of As in the erythrocyte compared to other species. In general, retention of As depends on the valency, the dose, and the route of administration. Methylation is a common metabolic pathway, and the extent of methylation also depends on valency, dose, and route of exposure (Vahter, 1981). In mice, dimethylarsinic acid is the primary circulating metabolite by 1 h after As administration (Vahter and Marafante, 1983).

Arsenic in the pregnant hamster is uniformly distributed between plasma and erythrocytes. Following constant rate infusion of As^V, plasma As is not bound to macromolecules, and is thus freely available to the embryo (Hanlon and Ferm, 1986a). The presence of dimethylarsinic acid and As^{III} indicate that As^V is metabolized in the pregnant hamster. Following i.p. injection of pregnant hamsters with a teratogenic dosage of As^V on gestation day 8, peak blood levels were attained approximately 30 min after dosing (Hanlon and Ferm, 1986b). Both As^{III} and dimethylarsinic acid were present in plasma at 12 min after dosing, indicating that reduction and methylation of As^V were rapid.

Hood and co-workers reported on the distribution, metabolism, and fetal uptake of maternally administered As^V (Hood et al., 1987) and As^{III} (Hood et al., 1988) in the mouse. Pregnant mice were dosed by oral gavage or i.p. injection on gestation day 18 (sperm positive = d 1) with sodium arsenate (20 mg/kg i.p.; 40 mg/kg p.o.) or sodium arsenite (8 mg/kg i.p.; 25 mg/kg p.o.). Maternal, placental, and fetal tissues were analyzed for total As at intervals up to 24 h after dosing. The results of these studies indicate that As distribution in maternal tissues is completed by 0.5 to 4 h after i.p. or p.o. administration of either As^{III} or As^V. As^V distribution in the fetus and placenta is complete at 2 h after maternal i.p. injection and at 6 h after oral administration. As^{III} distribution to the placenta occurs by 1 h after i.p. injection, but As in the fetus exhibited concentration peaks at both 4 and 24 h. After oral dosing with As^{III}, placental As concentration peaks at 4 h and fetal levels peak at 4 and 24 h. The form of As in the fetus progressively shifts from inorganic to organic forms (monomethyl and dimethyl arsenic) over time, with the methylated forms predominant by 4 to 6 h after maternal i.p. or p.o. dosing with either As^{III} or As^V. Elimination of As^V from the fetus and placenta is nearly complete by 24 h, and up to 83% of the remaining As is methylated. The half-life for As^V was approximately 10 h for either i.p. or p.o. administration. Gerber et al. (1982) reported transfer to the embryo of As^{III} administered in the diet or by i.p. injection. These investigators reported two metabolic components for i.p.-injected As, one with a half-life of about 6 h representing approximately 95% of the administered dose and another with a half-life of about 2.4 days representing about 5%.

The distribution of ^{74}As-labeled As^V and As^{III} in pregnant mice and a marmoset monkey were studied by Lindgren et al. (1984) using autoradiography and gamma counting of isolated tissues. As^V or As^{III} administered i.v. to pregnant mice on one of gestation day 7 to 18 crossed to the embryo/fetus relatively freely at all stages examined. Activity of ^{74}As in the embryo/fetus was considerably higher at 4 h compared to 24 h. Distribution of ^{74}As in the embryo/fetus showed a changing pattern depending on the stage at administration. With administration on gestation days 7 to 10, the neuroepithelium exhibited the highest concentration of all embryonic tissues. Around gestation day 13, distribution was fairly even, while near term high concentrations were observed in the skin and epithelia of the upper gastrointestinal tract. Retention time in mouse maternal tissues and placenta was approximately three times longer with As^{III} than with As^V. Affinity for the fetal ossified skeleton was exhibited by As^V but not As^{III}. The marmoset, a species known not to methylate As, exhibited a slower rate of placental transfer of ^{74}As than did mice, probably due to stronger binding and longer retention in maternal tissues.

E. MECHANISMS OF ACTION

Most of the whole animal studies of As discussed above used dosages at or above levels producing some degree of maternal toxicity. Therefore, toxic effects on the mother almost certainly play some role in the developmental toxicity of arsenicals. However, as discussed by Golub (1994), the maternal toxicity noted in studies of As does not necessarily correlate very well with the developmental toxicity observed. Placental toxicity has not been reported for As, although increased lipid peroxidation in the human placenta has been associated with exposure to As and other metals (Tabacova et al., 1994b). The ability of As to readily cross the placenta and accumulate in the conceptus, along with its demonstrated effects on embryos and cells *in vitro* suggests that direct effects of As on the embryo are of primary concern.

It is well known that As^{III} can interact with protein sulfhydryls. Free sulfhydryl groups are essential for function of a wide range of enzymes, including glutamic-oxaloacetic acid transaminase, pyruvate oxidase, monoamine oxidase, choline oxidase, glucose oxidase, urease, oxidoreductases, and kinases (Willhite and Ferm, 1984). A number of enzymes have been shown to be affected by As^{III}, including effects on mitochondrial respiration. As^V does not interact directly with sulfhydryl groups, but may due so via intracellular reduction to As^{III}. As^V can also compete with inorganic phosphate (Mitchell et al., 1971).

Stress proteins, also known as heat shock proteins (hsps) are induced in cells in response to a number of stimuli including heat and other environmental stresses, pathophysiological conditions, and physiological processes including cell cycling and embryonic development and differentiation. Metals are among the many chemical and physical inducers of stress proteins, and As is the most studied of the metals in terms of its ability to induce stress proteins. For a general review of metals and stress proteins, see Goering and Fisher (1995). The function(s) of stress proteins are not understood, but they are thought to play an important role in protecting cells after a variety of noxious environmental stimuli. It has been hypothesized, however, that changes in gene expression associated with the induction of stress protein synthesis in the embryo may be detrimental and represent a common mechanism of teratogenesis (German, 1984; German et al., 1986). Hyperthermia has been known to be teratogenic for at least 70 years, since the work of Alsop (1919) in the chick embryo.

Treatment of chick embryo fibroblasts with As^{III} induced the synthesis of four proteins of 89, 73, 35, and 27 kDa, similar to heat shock proteins (Johnston et al., 1980). Wang and Lazarides (1984) confirmed that As^{III} induced the synthesis of hsp70 in chick embryo fibroblasts, and showed that methylation of the hsp70 proteins is altered by As^{III}. German et al. (1986) demonstrated that both heat and sodium arsenite could produce the heat shock response in organogenesis-stage embryos of treated mouse dams as well as in human fetal cells treated *in vitro*. Some, but not all of the proteins induced by heat and As^{III} in the human fetal cells were similar. In particular, As^{III} induced several proteins smaller than 45 kDa that were not induced by hyperthermia. Ohtsuka et al. (1990) showed that As^{III} induced a novel 40-kDa protein in HeLa cells in addition to the higher molecular weight hsps, while Lee et al. (1991a) showed that cycloheximide treatment after As^{III} exposure eliminated the acquisition of thermotolerance in CHO cells, while similar treatment with cycloheximide did not inhibit the development of thermotolerance induced by heat. A nuclear 26-kDa protein was associated with this difference in response to As^{III} and heat. Heat treatment of As^{III}-induced thermotolerant cells reduced the level of this protein by 78%, while heat treatment of heat-induced thermotolerant cells reduced the 26-kDa protein by only 3%. Lee et al. (1991b) further showed that in CHO cells five major hsps were induced by heat (110, 87, 70, 28, and 8.5 kDa) whereas four of the major hsps (110, 87, 70, and 28 kDa) and one stress protein not induced by heat (33.3 kDa) were induced by As^{III}. Two hsp families (hsp70a,b,c and hsp28a,b,c) preferentially relocated to the nucleus after heat treatment, but only hsp70b redistributed to the nucleus after As^{III} treatment, and the extent of redistribution of hsp70b was less than that observed after heat treatment. Wiegant et al. (1994) compared the patterns of specific stress proteins induced by heat, As^{III}, Cd, dinitrophenol, and ethanol, and reported stressor-specific induction patterns for a number of these proteins, including only minimal induction of hsp100 by As^{III} compared to the other agents tested.

Expression of heat shock proteins in embryos of As^{III}-treated pregnant mice and human chorionic villus tissue treated with As^{III} *in vitro* was demonstrated by Honda et al. (1991). Mirkes and Cornel

(1992) and Mirkes et al. (1994) showed that cultured postimplantation rat embryos exposed to embryotoxic concentrations (50 μM) of As^{III} expressed three hsps, one of which was identified as heat-inducible hsp72. As^{III} treatment also induced the accumulation of heat-inducible hsp70 mRNA.

Since As^{III} is known to effectively bind protein sulfhydryls, it may be that protein denaturation caused by this binding promotes the heat shock response. Treatment of CHO cells with the sulfhydryl oxidizing agent diamide or succinimidyl propionate, a cross-linker of bifunctional amino groups, induced hsp synthesis and thermotolerance, as did As^{III}. Members of the hsp70 family interact transiently with nascent polypeptides in unstressed cells, while in As^{III}-exposed cells newly synthesized proteins bind to but are not released from hsp70 (Beckmann et al., 1992). These authors speculate that the newly synthesized proteins are unable to undergo proper folding and that binding to hsp70 may be a protective mechanism during stress, and that reduction in free hsp70 in stressed cells may trigger the stress response.

Both heat shock and As^{III} treatment activate the progesterone receptor (PR) and enhance PR-mediated gene transcription in T47D breast cancer cells. The binding of the PR with inducible hsp70 and hsp100, not observed under normal conditions, was detected after heat or As^{III} treatment (Edwards et al., 1992). In mouse L929 cells and in a CHO cell line (WCL2) overexpressing a transfected mouse glucocorticoid receptor (GR) gene, heat or As^{III} treatment results in translocation of the unliganded GR from the cytosol to the nucleus, suggesting that such stresses may provide a hormone-independent mechanism for transformation of the GR to the high affinity nuclear-binding state characteristic of the hormone-bound, transcriptionally active receptor (Sanchez, 1992). The hormone binding domain of the GR contains a vicinally spaced dithiol in a region that appears to be a contact site for hsp90, which is required for high-affinity steroid binding. Binding of the unliganded GR with As^{III} under conditions specific for reaction with vicinally spaced dithiols eliminated steroid binding, and As^{III} was shown to disrupt the dexamethasone-GR complex (Stancato et al., 1993).

Heme oxygenase (HO) is a microsomal enzyme which initiates the catabolism of heme, cleaving the tetrapyrrole ring to form biliverdin, which is subsequently converted to bilirubin. HO has been classified as a stress protein based on heat inducibility and the presence of a heat shock gene regulatory element (HSE) in the HO promotor region; numerous metals are also capable of inducing HO in a tissue-specific manner, and As^{III} is the most effective inducer in a variety of cell systems (Goering and Fisher, 1995).

Biliverdin and bilirubin, the products of HO cleavage of heme, have been demonstrated to possess antioxidant activity, thus possibly protecting the cell against free radical damage (Stocker et al., 1987a,b). Glutathione (GSH) serves as an important antioxidant protectant in the cell, and in SA7 cells, an As^{III}-resistant CHO line, resistance was correlated with cellular levels of glutathione-S-transferase (Lo et al., 1992). As^{III} has been shown to be capable of promoting free radical production and binds to GSH (Dreosti, 1991). The potential role of lipid peroxidation in the prenatal toxicity of As was examined in mouse whole embryo culture by Tabacova and Hunter (in Zelikoff et al., 1995). At teratogenic concentrations, both As^{III} and As^{V} caused a transient decrease in GSH content, and blocking GSH synthesis with buthionine sulfoximine exacerbated As teratogenicity. Excessive amounts of reactive oxygen species were observed in embryonic target sites by confocal microscopic examination of dichlorofluorescein-stained embryos. Further, the teratogenicity of As^{III}, and to a lesser extent As^{V}, was partially ameliorated by a number of exogenous antioxidants. Thus, oxidative tissue damage may play an important role in the developmental toxicity of As.

Both As^{III} and As^{V} have been observed to induce morphological transformation of Syrian hamster embryo cells, with As^{III} being more than 10-fold more potent (Lee et al., 1985; Barrett et al., 1989). Transformation and associated endoreduplication, chromosome abberations, and sister chromatid exchanges were observed to have similar dose-responses.

III. NOTES ON THE EFFECTS OF PRENATAL LEAD EXPOSURE

The effects of prenatal lead exposure on humans and experimental animals have been reviewed recently (Winder, 1993; Bellinger, 1994). High dose embryo/fetal exposures can result in malformations in experimental animals including rats, hamsters, and chicks. Malformations reported in these species include brain defects, other neural tube defects, and caudal defect of the urogenital system and tail (Winder, 1993). While high dose lead exposure is almost certainly a teratogen in humans as well, definition of the effects of low dose prenatal exposure have proved more difficult. While an increased risk of minor congenital anomalies among newborns with higher umbilical cord blood lead levels has been reported (Needleman et al., 1984), others did not observe such an association (Ernhart et al., 1986; McMichael et al., 1986). A study relating outcome of pregnancy to drinking water quality (Aschengrau et al., 1993) reported an association of drinking water lead levels with stillbirth and congenital anomalies of the ear, face, neck, and cardiovascular system, but not the central nervous system, gastrointestinal or musculoskeletal systems, genetalia, or integument. However, no individual exposure data were collected. Data on the effect of lead on the gestation length are also unclear. Lead exposure has been associated with reduced length of gestation in some studies, while others fail to find this effect or to report a positive association of gestation length with cord blood lead levels (see Bellinger, 1994 for discussion). The neurotoxicological effects of lead in humans and animal models are discussed in Chapter 39.

In addition to maternal lead exposure during pregnancy, the conceptus may be exposed via release of maternal lead stores from previous exposures. Both pregnancy and lactation (see Chapter 70) can result in mobilization of body lead, most pf which is in the skeleton, into the more bioavailable whole blood and plasma pools. Much of the evidence for this route of exposure comes from individuals with high past exposures, and the extent of mobilization will likely depend on a host of maternal physiological factors.

IV. CONCLUSIONS

Although we understand much about the effects of prenatal exposure to cadmium, arsenic, of lead, much work remains in elucidating the mechanisms underlying these effects. Epidemiological approaches to determining the effects of exposure to these metals during pregnancy have encountered numerous difficulties, especially in determiningadequate markers of exposure during critical developmental periods. While in general exposure of pregnant women to these pollutants is dropping in developed countries, many sources of potential high exposures remain. Further elucidation of the biochemical effects of these metals in pregnant laboratory animals should help guide future human epidemiological studies.

REFERENCES

Abraham, E., Ringwood, N., and Mankes, R. (1984), *J. Reprod. Fertil.,* 70, 323–325.
Abraham, R., Charles, A. K., Mankes, R., LeFevre, R., Renak, V., and Ashok, L. (1986), *Ecotoxicol. Environ. Safety,* 12, 213–219.
Ahokas, R. A. and Dilts, P. V., Jr. (1979), *Am. J. Obstet. Gynecol.,* 135, 219–222.
Ahokas, R. A., Dilts, P. V., and Lahaye, E. B. (1980), *Am. J. Obstet. Gynecol.,* 136, 216–221.
Alsop, F. M. (1919), *Anat. Rec.,* 15, 307–324.
Ancel, P. and Lallemand, S. (1941), *Arch. Phys. Biol.,* 15, 27–29.
Andrews, G. K., Huet-Hudson, Y. M., Paria, B. C., McMaster, M. T., De, S. K., and Dey, S. K. (1991), *Dev. Biol.,* 145, 13–27.
Aschengrau, A., Zeirler, S., and Cohen, A. (1989), *Arch. Environ. Health,* 44, 283–290.
Aschengrau, A., Zeirler, S., and Cohen, A. (1993), *Arch. Environ. Health,* 48, 105–113.
Barański, B. (1987), *Environ. Res.,* 42, 54–62.
Barański, B., Stetkiewicz, I., Sitarek, K., and Szymczak, W. (1983), *Arch. Toxicol.,* 54, 297–302.
Barański, B. (1984), *J. Hyg. Epidemiol. Microbiol. Immunol.,* 29, 253–262.
Barański, B., Stetkiewicz, I., Trzcinka-Ochocka, M., Sitarek, K., and Szymczak, W. (1982), *J. Appl. Toxicol.,* 2, 255–259.
Barr, M. (1973), *Teratology,* 7, 237–242.
Barrett, J. C., Lamb, P. W., Wang, T. C., and Lee, T. C. (1989), *Biol. Trace Elem. Res.,* 21, 421–429.
Baxley, M. N., Hood, R. D., Vedel, G. C., Harrison, W. P., and Szczech, G. M. (1981), *Bull. Environ. Contam. Toxicol.,* 26, 749–756.
Beaudoin, A. R. (1974), *Teratology,* 10, 153–158.
Beckman, L. (1978), *Ambio,* 7, 226–231.
Beckmann, R. P., Lovett, M., and Welch, W. J. (1992), *J. Cell Biol.,* 117, 1137–1150.
Belmonte, N. M., Rivera, O. E., and Herkovits, J. (1989), *Bull. Environ. Contam. Toxicol.,* 43, 107–110.
Bellinger, D. (1994), *Teratology,* 50, 367–373.
Boadi, W., Yannai, S., Urbach, J., Brandes, J., and Summer, K. (1991), *Arch. Toxicol.,* 65, 318–323.
Bollinger, C. T., Van Zijl, P., and Louw, J. A. (1992), *Respiration,* 59, 57–61.
Bournias-Vardiabasis, N., Buzin, C., and Flores, J. (1990), *Exp. Cell Res.,* 189, 177–182.
Breen, J. G., Eisenmann, C., Horowitz, S., and Miller, R. K. (1994), *Reprod. Toxicol.,* 8, 297–306.
Breen, J. G., Nelson, E., and Miller, R. K. (1995), *Teratology,* 51, 266–272.
Burk, D. and Beaudoin, A. R. (1977), *Teratology,* 16, 247–260.
Carpenter, S. J. (1987), *Anat. Embryol.,* 176, 345–365.
Chaineau, E., Binet, S., Pol, D., Chatellier, G., and Meininger, V. (1990), *Teratology,* 41, 105–112.
Chan, H. M. and Cherian, M. G. (1993), *Toxicol. Appl. Pharmacol.,* 120, 308–314.
Chernoff, N. (1973), *Teratology,* 8, 29–32.
Chiquoine, A. D. (1965), *J. Reprod. Fertil.,* 10, 263–265.

Christley, J. and Webster, W. S. (1983), *Teratology,* 27, 305–312.
Danielsson, B. R. and Dencker, L. (1984), *Biol. Res. Pregnancy Perinatol.,* 5, 93–101.
Darasch, S., Mosser, D. D., Bols, N. C., and Heikkila, J. J. (1988), *Biochem. Cell Biol.,* 66, 862–870.
Daston, G. P. (1982a), *J. Toxicol. Environ. Health,* 9, 51–61.
Daston, G. P. (1982b), *Toxicology,* 24, 55–63.
Daston, G. P. (1981), *Teratology,* 23, 75–84.
Daston, G. P. and Grabowski, C. T. (1979), *J. Toxicol. Environ. Health,* 5, 973–983.
De, S. K., Dey, S. K., and Andrews, G. K. (1990), *Toxicology,* 64, 89–104.
De, S. K., McMaster, M. T., Dey, S. K., Andrews, G. K. (1989), *Development,* 107, 611–621.
De, S. K., Paria, B. C., Dey, S. K., and Andrews, G. K. (1993), *Toxicology,* 80, 13–25.
Dencker, L. (1975), *J. Reprod. Fertil.,* 44, 461–471.
di Sant'Agnese, P. A., Jensen, K., Levin, A. A., and Miller, R. K. (1983), *Placenta,* 4, 149–163.
Dreosti, I. (1991), *Trace Elements, Micronutrients, and Free Radicals,* Dreosti, I., Ed., Humana Press, Clifton, NJ, 149–168.
Dudley, R. E. and Klaassen, C. D. (1984), *Toxicol. Appl. Pharmacol.,* 72, 530–538.
Eaton, J. G. (1974), *Trans. Am. Fish. Soc.,* 103, 729–735.
Edwards, D. P., Estes, P. A., Fadok, V. A., Bona, B. J., Onate, S., Nordeen, S. K., and Welch, W. J. (1992), *Biochemistry,* 31, 2482–2491.
Ernhart, C., Wolf, A., Kennard, M. Erhard, P., Filipovich, H. and Sokol, R. (1986), *Arch. Environ, Health,* 41, 287–291.
Ferm, V. H. (1971), *Biol. Neonate,* 19, 101–107.
Ferm, V. H. and Carpenter, S. J. (1967), *Nature,* 216, 1123.
Ferm, V. H. and Carpenter, S. J. (1968a), *Lab. Invest.,* 18, 429–432.
Ferm, V. H. and Carpenter, S. J. (1968b), *J. Reprod. Fertil.,* 17, 199–201.
Ferm, V. H. and Hanlon, D. P. (1985), *Environ. Res.,* 37, 425–432.
Ferm, V. H., Hanlon, D. P., and Urban, J. (1969), *J. Embryol. Exp. Morphol.,* 22, 107–113.
Ferm, V. H., Saxon, A., and Smith, B. M. (1971), *Arch. Environ. Health,* 22, 557–560.
Feuston, M. H. and Scott, W. J., Jr. (1985), *Teratology,* 32, 407–419.
Fisher, D. L. (1982), *Environ. Res.,* 28, 1–9.
Fissore, R. A., Jackson, K. V., and Kiessling, A. A. (1989), *Biol. Reprod.,* 41, 835–841.
Frery, N., Nessmann, C., Girard, F., Lafond, J., Moreau, T., Blot, P., Lellouch, J., and Huel, G. (1993), *Toxicology,* 79, 109–118.
Gale, T. F. and Ferm, V. H. (1973), *Biol. Neonate,* 23, 149–160.
Gale, T. F. and Layton, W. M. (1980), *Teratology,* 21, 181–186.
Garcia, M. and Lee, M. (1981), *Biol. Trace Element Res.,* 3, 149–156.
Gerber, G. B., Maes, J., and Eykens, B. (1982), *Arch. Toxicol.,* 49, 159–168.
German, J. (1984), *Am. J. Med.,* 76, 293–301.
German, J., Louie, E., and Banerjee, D. (1986), *Teratogenesis Carcinog. Mutagen.,* 6, 555–562.
Goering, P. L. and Fisher, B. R. (1995), *Toxicology of Metals. Biochemical Aspects,* Goyer, R. A. and Cherian, M. G., Eds., *Handbook of Experimental Pharmacology,* Vol. 115. Springer-Verlag, Berlin, 229–266.
Goering, P. L., Waalkes, M., and Klaassen, C. D. (1995), *Toxicology of Metals. Biochemical Aspects,* Goyer, R. A. and Cherian, M. G., Eds., *Handbook of Experimental Pharmacology,* Vol. 115, Springer-Verlag, Berlin, 189–214.
Golub, M. S. (1994), *Reprod. Toxicol.,* 8, 283–295.
Goyer, R. A. and Cherian, M. G. (1992), *Cadmium in the Human Environment Toxicity and Carcinogenicity,* Nordberg, G. E., Herber, R. F. M., and Alessio, L., Eds., International Agency for Research on Cancer, Lyon, 239–247.
Hanlon, D. P. and Ferm, V. H. (1986a), *Environ. Res.,* 40, 372–379.
Hanlon, D. P. and Ferm, V. H. (1986b), *Environ. Res.,* 40, 380–390.
Hartsfield, J. K., Jr., Lee, M., Morel, J. G., and Hilbelink, D. R. (1992), *Biochem. Med. Metab. Biol.,* 48, 159–173.
Hazelton Laboratories of America (1990), *Two Generation Dietary Reproduction Study with Arsenic Acid in Mice,* Hazelton Laboratories, Madison, WI, Report #HLA 6120–138.
Herkovits, J. and Perez-Coll, C. S. (1993), *Bull. Environ. Contam. Toxicol.,* 50, 608–611.
Hildebrand, C. E., Tobey, R. A., Campbell, E. W., and Enger, M. D. (1979), *Exp. Cell Res.,* 124, 237–246.
Holmberg, R. E., Jr. and Ferm, V. H. (1969), *Arch. Environ. Health,* 18, 873–877.
Honda, K., Hatayama, T., Takahashi, K., and Yukioka, M. (1991), *Teratogenesis Carcinog. Mutagen.,* 11, 235–244.
Hood, R. D. (1972), *Bull. Environ. Contam. Toxicol.,* 7, 216–222.
Hood, R. D. and Bishop, S. L. (1972), *Arch. Environ. Health,* 24, 62–65.
Hood, R. D. and Harrison, W. P. (1982), *Bull. Environ. Contam. Toxicol.,* 29, 671–678.
Hood, R. D. and Vedel-Macrander, G. C. (1984), *Toxicol. Appl. Pharmacol.,* 73, 1–7.
Hood, R. D., Thacker, G. T., and Patterson, B. L. (1977), *Environ. Health Perspect.,* 19, 219–222.
Hood, R. D., Thacker, G. T., Patterson, B. L., and Szczech, G. M. (1978), *J. Environ. Pathol. Toxicol.,* 1, 857–864.
Hood, R. D., Harrison, W. P., and Vedel, G. C. (1982), *Bull. Environ. Contam. Toxicol.,* 29, 679–687.
Hood, R. D., Vedel-Macrander, G. C., Zaworotko, M. J., Tatum, F. M., and Meeks, R. G. (1987), *Teratology,* 35, 19–25.
Hood, R. D., Vedel, G. C., Zaworotko, M. J., and Tatum, F. M. (1988), *J. Toxicol. Environ. Health,* 25, 423–434.
Ishizu, S., Minami, M., Suzuki, A., Yamada, M., Sato, M., and Yamura, K. (1973), *Ind. Health,* 11, 127–139.
Johnston, D., Oppermann, H., Jackson, J., and Levinson, W. (1980), *J. Biol. Chem.,* 255, 6975–6980.

Kang, Y.-J., Clapper, J. A., and Enger, M. D. (1989), *Cell Biol. Toxicol.,* 5, 249–260.
Kapron-Bras, C. M. and Hales, B. F. (1991), *Teratology,* 43, 83–94.
Kapron-Bras, C. M. and Hales, B. F. (1992), *Teratology,* 46, 191–200.
Keino, H. (1973), *Teratology,* 8, 96–97.
King, D. W. and Hsu, J. L. (1977), *Anat. Rec.,* 187, 770.
Kuczuk, M. H. and Scott, W. J., Jr. (1984), *Teratology,* 29, 427–435.
Layton, W. M. and Layton, M. W. (1979), *Teratology,* 19, 229–236.
Lee, T. C., Oshimura, M., and Barrett, J. C. (1985), *Carcinogenesis,* 6, 1421–1426.
Lee, K. J. and Hahn, G. M. (1988), *J. Cell Physiol.,* 136, 411–420.
Lee, Y. J., Curetty, L., and Corry, P. M. (1991a), *J. Cell Physiol.,* 149, 77–87.
Lee, Y. J., Kim, D. H., Hou, Z. Z., and Corry, P. M. (1991b), *Radiat. Res.,* 127, 325–334.
Lehman, L. D. and Poisner, A. M. (1984), *J. Toxicol. Environ. Health,* 14, 419–432.
Lepire, M. L. and Ziomek, C. A. (1989), *Biol. Reprod.,* 41, 464–473.
Levin, A. A. and Miller, R. K. (1980), *Teratology,* 22, 1–5.
Levin, A. A. and Miller, R. K. (1981), *Toxicol. Appl. Pharmacol.,* 58, 297–306.
Levin, A. A., Kilpper, R. W., and Miller, R. K. (1987), *Teratology,* 36, 163–170.
Lindgren, A., Danielsson, B. R. G., Dencker, L., and Vahter, M. (1984), *Acta Pharmacol. Toxicol.,* 54, 311–320.
Lo, J. F., Wang, H. F., Tam, M. F., and Lee, T. C. (1992), *Biochem. J.,* 288, 977–982.
Loiacono, N. J., Graziano, J. H., Kline, J. K., Popovac, D., Ahmedi, X., Gashi, E., Mehmeti, A., and Rajovic, B. (1992), *Arch. Environ. Health,* 47, 250–255.
Lugo, G., Cassady, G., and Palmisano, P. (1969), *Am. J. Dis. Child.,* 117, 328–330.
Machemer, L. and Lorke, D. (1981), *Toxicol. Appl. Pharmacol.,* 58, 438–443.
McKone, T. E. and Daniels, J. I. (1991), *Regul. Toxicol. Pharmacol.,* 13, 36–61.
McMichael, A. Vimpani, G., Robertson, E., Baghurst, P., and Clark, P. (1986), *J. Epidemiol. Commun. Health,* 40, 303–318.
Messerle, K. and Webster, W. S. (1982), *Teratology,* 25, 61–70.
Mirkes, P. E. and Cornel, L. (1992), *Teratology,* 46, 251–259.
Mirkes, P. E., Doggett, B., and Cornel, L. (1994), *Teratology,* 49, 135–142.
Mitchell, R. A., Chang, B. F., Huang, C. H., and DeMaster, E. G. (1971), *Biochemistry,* 10, 2049–2054.
Morris, S. and Huang, P. C. (1989), *Exp. Cell. Res.,* 185, 166–175.
Morton, K. A., Jones, B. J., Sohn, M. H., Schaefer, A. E., Phelps, R. C., Datz, F. L., and Lynch, R. E. (1992), *J. Biol. Chem.,* 267, 2880–2883.
Muller, W. U., Streffer, C., and Fischer-Lahdo, C. (1986), *Arch. Toxicol.,* 59, 172–175.
Muñoz, C. and Dieter, H. H. (1990), *Toxicol. Lett.,* 50, 263–274.
Narbaitz, R., Riedel, K. D., and Kacew, S. (1983), *Teratology,* 27, 207–213.
Needleman, H. Rabinowitz, M., Leviton, A., Linn, S., and Schoenbaum, S., (1984), *JAMA,* 251, 2956–2959.
Ochi, T., Otsuka, F., Takahashi, K., and Ohsawa, M. (1988), *Chem.-Biol. Interact.,* 65, 1–14.
Ohtsuka, K., Masuda, A., Nakai, A., and Nagata, K. (1990), *Biochem. Biophys. Res. Commun.,* 30, 642–647.
Padmanabhan, R. (1987), *Teratology,* 35, 9–18.
Padmanabhan, R. and Hameed, M. S. (1986), *J. Craniofac. Genet. Dev. Biol.,* 6, 245–258.
Padmanabhan, R. and Hameed, M. S. (1990), *Reprod. Toxicol.,* 4, 291–304.
Page, K., Abramovich, D., Aggett, P., Bain, M., Chipperfield, A. R., Durdy, H., MacLachlan, J., and Smale, A. (1992), *Placenta,* 13, 151–162.
Pařízek, J. (1964), *J. Reprod. Fertil.,* 7, 263–265.
Parzyck, D. C., Shaw, S. M., Kessler, W. V., Vetter, R. J., Van Sickle, D. C., and Mayes, R. A. (1978), *Bull. Environ. Contam. Toxicol.,* 19, 206–214.
Pedersen, R. A. and Lin, T. P. (1978), *Development Toxicology of Energy-Related Pollutants.* Mahlum, D., Sikov, M. R., Hackett, P. L., and Andrew, F. D., Eds., *DOE Symp. Ser.,* 47, 600–613.
Perez-Coll, C. S., Herkovits, J., and Salibian, A. (1986), *Experientia,* 42, 1174–1176.
Peterkova, R. and Puzanova, L. (1976), *Folia Morphol. Praha,* 24, 5–13.
Pierro, L. J. and Haines, J. S. (1978), *Developmental Toxicology of Energy-Related Pollutants,* Mahlum, D., Sikov, M. R., Hackett, P. L., and Andrew, F. D., Eds., *DOE Symp. Ser.,* 47, 614–626.
Piersma, A. H., Roelen, B., Roest, P., Haakmat-Hoesenie, A. S., van-Achterberg, T. A., and Mummery, C. L. (1993), *Teratology,* 48, 335–341.
Record, I. R., Dreosti, I. E., Manuel, S. J., and Buckley, R. A. (1982a), *Life Sci.,* 31, 2735–2743.
Record, I. R., Dreosti, I. E., and Manuel, S. J. (1982b), *J. Nutr.,* 112, 1994–1998.
Ribas, B. and Schmidt, W. (1973), *Gegenbaurs Morphol. Jahrb.,* 119, 358–366.
Rogers, E. H., Chernoff, N., and Kavlock, R. J. (1981), *Drug Chem. Toxicol.,* 4, 49–61.
Saillenfait, A. M., Payan, J. P., Brondeau, M. T., Zissu, D., and de Ceaurriz, J. (1991), *J. Appl. Toxicol.,* 11, 23–27.
Saltzman, R. A., Miller, R. K., and di Sant'Agnese, P. A. (1989), *Teratology,* 39, 19–30.
Samarawickrama, G. P. and Webb, M. (1979), *Environ. Health Perspect.,* 28, 245–259.
Samarawickrama, G. P. and Webb, M. (1981), *J. Appl. Toxicol.,* 1, 264–269.
Sanchez, E. R. (1992), *J. Biol. Chem.,* 267, 17–20.
Sasser, L. B., Leman, B. J., Levin, A. A., and Miller, R. K. (1985), *Toxicol. Appl. Pharmacol.,* 80, 299–307.

Schmid, B. P., Kao, J., and Goulding, E. (1985), *Experientia,* 41, 271–272.
Singhal, R. K., Andersen, M. E., and Meister, A. (1987), *FASEB J.,* 1, 220–223.
Sonawane, B. R., Nordberg, M., Nordberg, G. F., and Lucier, G. W. (1975), *Environ. Health Perspect.,* 12, 97–102.
Sorell, T. and Graziano, J. (1990), *Toxicol. Appl. Pharmacol.,* 102, 537–545.
Soukupova, D. and Dostal, M. (1991), *Funct. Dev. Morphol.,* 1, 3–9.
Soukupova, D., Dostal, M., and Piza, J. (1991), *Funct. Dev. Morphol.,* 1, 31–36.
Sowa, B. and Steibert, E. (1985), *Arch. Toxicol.,* 56, 256–262.
Stacey, N. H. and Klaassen, C. D. (1980), *Toxicol. Appl. Pharmacol.,* 55, 448–455.
Stancato, L. F., Hutchison, K. A., Chakraborti, P. K., Simons, S. S., Jr., and Pratt, W. B. (1993), *Biochemistry,* 32, 3729–3736.
Stocker, R., Glazer, A. N., and Ames, B. N. (1987a), *P.N.A.S.,* 84, 5918–5922.
Stocker, R., Yamamoto, Y., McDonagh, A. F., Glazer, A. N., and Ames, B. N. (1987b), *Science,* 235, 1043–1047.
Stone, R. (1994), *Science,* 264, 204.
Storeng, R. and Jonsen, J. (1980), *Toxicology,* 17, 183–187.
Styblo, M., Delnomdedieu, M., and Thomas, D. J. (1995), *Toxicology of Metals. Biochemical Aspects,* Goyer, R. A. and Cherian, M. G., Eds., *Handbook of Experimental Pharmacology,* Vol. 115. Springer-Verlag, Berlin, 407–433.
Sunderman, F. W. Jr., Plowman, M. C., and Hopfer, S. M. (1991), *Ann. Clin. Lab. Sci.,* 21, 381–391.
Suzuki, C. A. M. and Cherian, M. G. (1989), *Toxicol. Appl. Pharmacol.,* 98, 544–552.
Tabacova, S., Baird, D. D., Balabaeva, L., Lolova, D., and Petrov, I. (1994a), *Placenta,* 15, 873–881.
Tabacova, S., Little, R. E., Balabaeva, L., Pavlova, S., and Petrov, I. (1994b), *Reprod. Toxicol.,* 8, 217–224.
Takeuchi, I. K. (1979), *J. Toxicol. Sci.,* 4, 405–416.
Tassinari, M. S. and Long, S. Y. (1982), *Teratology,* 25, 101–113.
Umpierre, C. C. (1981), *Teratology,* 23, 66A.
Vahter, M. (1981), *Environ. Res.,* 25, 286–293.
Vahter, M. and Marafante, E. (1983), *Chem.-Biol. Interact.,* 47, 29–44.
Vallee, B. L. and Glades, A. (1974), *Adv. Enzymol.,* 56, 283–430.
Vallee, B. L. and Ulmer, D. D. (1972), *Annu. Rev. Biochem.,* 41, 91–128.
Vojnik, C. and Hurley, L. S. (1977), *J. Nutr.,* 107, 862–872.
Waalkes, M. P., Thomas, J. A., and Bell, J. U. (1982), *Toxicol. Appl. Pharmacol.,* 62, 211–218.
Waalkes, M. P. and Bell, J. U. (1980), *Toxicology,* 18, 103–110.
Wang, Z., Hou, G., and Rossman, T. G. (1994), *Environ. Health Perspect.,* 102, 97–100.
Wang, C. and Lazarides, E. (1984), *Biochem. Biophys. Res. Commun.,* 119, 735–743.
Wang, C. Brown, S., Bhattacharyya, M. H. (1994), *Toxicol. Appl. Pharmacol.,* 127, 320–330.
Warner, C. W., Sadler, T. W., Tulis, S. A., and Smith, M. K. (1984), *Teratology,* 30, 47–53.
Webster, W. S. and Messerle, K. (1980), *Teratology,* 21, 79–88.
Whelton, B. D., Toomey, J. M., and Bhattacharyya, M. H. (1993), *J. Toxicol. Environ. Health,* 49, 531–546.
White, T. E. K., Baggs, R. B., and Miller, R. K. (1990), *Teratology,* 42, 7–13.
Wiegant, F. A., Souren, J. E., van Rijn, J., and Wijk, R. (1994), *Toxicology,* 94, 143–159.
Wier, P. J., Miller, R. K., Maulik, D., and diSant'Agnese, P. A. (1990), *Toxicol. Appl. Pharmacol.,* 105, 156–171.
WIL Research Laboratories (1988a), *A teratology study in mice with arsenic acid (75%),* WIL Research Laboratories, Ashland, OH.
WIL Research Laboratories (1988b), *A teratology study in rabbits with arsenic acid (75%),* WIL Research Laboratories, Ashland, OH.
Willhite, C. C. (1981), *Exp. Mol. Pathol.,* 34, 145–158.
Willhite, C. C. and Ferm, V. H. (1984), *Adv. Exp. Med. Biol.,* 177, 205–228.
Winder, C. (1993), *Neurotoxicology,* 14, 303–318.
Wolkowski, R. M. (1974), *Teratology,* 10, 243–262.
Wolkowski-Tyl, R. M. (1978), in *Developmental Toxicology of Energy-Related Pollutants,* Mahlum, D., Sikov, M. R., Hackett, P. L. and Andrew, F. D., Eds., DOE *Symp. Ser.,* 47, 568–585.
Yu, H. S. and Chan, S. T. (1986), *Teratology,* 34, 313–319.
Yu, H. S. and Chan, S. T. (1987), *Pharmacol. Toxicol.,* 60, 129–34.
Yu, H. S. and Chan, S. T. (1988a), *Toxicology,* 48, 261–272.
Yu, H. S. and Chan, S. T. (1988b), *Teratology,* 37, 13.
Yu, H. S., Tam, P. P., and Chan, S. T. (1985), *Teratology,* 32, 347–353.
Zelikoff, J. T., Bertin, J. E., Burbacher, T. M., Hunter, E. S., Miller, R. K., Silbergeld, E. K., Tabacova, S., and Rogers, J. M. (1995), *Fundam. Appl. Toxicol.,* 25, 161–170.

Chapter 66

The Developmental Cytotoxicity of Mercurials

Edward J. Massaro

I. PREFACE

Three major categories of factors impinge upon the toxicity of a developmental toxicant: (i) the chemical/biochemical behavior of the potential toxicant in the maternal compartment, (ii) the transfer of the toxicant to the conceptus, and (iii) the chemical/biochemical behavior of the toxicant in the conceptus. To obtain a realistic view of the developmental toxicity of mercurials (or any other class of potential toxicants), the contributions from each catagory of factors must be considered. I have attempted to do so in this review.

II. HISTORICAL PERSPECTIVE

Over the last 40 years, staggering mortality and morbidity due to methylmercury (MeHg) intoxication have occurred in Japan (*Minimata Disease,* 1968, 1977), Iraq, Pakistan, Guatemala, Ghana, and the former Soviet Union, and isolated incidents have been reported elsewhere (*Mercury in the Environment,* 1972; Bakir et al., 1973; *Health Effects of Methylmercury,* 1981). In the late 1960s, following episodes of mercury (Hg) poisoning in other countries, particularly Japan and Sweden, concern arose among the few regarding agricultural and industrial practices that could lead to similar problems in the U.S. Basically, the potential for development of widespread human and wildlife intoxication arose from the gross mishandling of mercurials combined with lack of any apparent agricultural, industrial, or governmental concern about the potential consequences of such activities. The Hg problem was a blatant illustration of the public health consequences of governmental laissez faire attitude toward environmental pollution. To this day, there is general lack of knowledge/comprehension/concern by the public and sectors within governmental hierarchies regarding potential future consequences of contamination of the environment with agricultural, industrial, military, and community wastes. Indeed, even when information regarding the toxicity of methylmercury (MeHg) was publicized, it was ignored in Japan, Sweden, and the U.S., apparently in favor of unbridled industrial production. Failure to heed warnings of the hazards of exposure to MeHg resulted not only in human suffering and the destruction of wildlife, but also in a high probability of future consequences of unknown dimensions. Unfortunately, in a manner analogous to that leading to environmental contamination by MeHg, the continued apparent veneration of the seven deadly sins of modern man: overpopulation, greed, ignorance, megalomania, myopia, bureaucracy, and arrogance have allowed continued pollution of the environment with contaminants such as radioactive wastes, halogenated hydrocarbons and, no doubt, a spectrum of other known and yet-to-be-discovered or -recognized environmental toxicants. The consequences of these activities have not yet been fully realized. But, an inkling of what surprises the future may hold is reflected in the recently recognized global decrease in amphibian populations (Blaustein et al., 1994; Pechmann and Wilbur, 1994). Although there is some evidence that habitat destruction and increased UV-B radiation resulting from depletion of the ozone layer may play a role in this phenomenon, the basis of the problem has yet to be delineated. A broad spectrum of environmental factors, including heavy metals (e.g., Aks et al., 1995; Birge et al., 1979) and/or chemicals with potential estrogenic or antiestrogenic and androgenic or antiandrogenic properties (e.g., Jansen et al., 1993; Kelce et al., 1994) acting individually or in synergy (Calflisch and DuBose, 1991) could interfere with reproduction via interference with gamete production/fertilization, development, resistance to environmental stressors such as infection, etc. (e.g., *Impact of the Environment on Reproductive Health*, 1993; *Minimata Disease*, 1977). In any case, prudence dictates that a manifestation of global environmental disharmony of sufficient magnitude to result in worldwide decline of amphibian populations be heeded and the root cause(s) identified and reversed as quickly as possible.

Epidemics of MeHg poisoning occurred in the Minimata Bay area of Japan from 1953 to 1960 and in Niigata, Japan from 1964 to 1965 (*Methyl Mercury in Fish*, 1971). In the Minimata Bay area, about 1000 cases of what has become known as **Minimata disease** were recorded (Takeuchi and Eto, 1975; Tsubaki et al., 1978) and more than 3000 cases of intoxication were suspected. Of the documented victims of the disaster, 43 had died by 1968 (Takeuchi, 1972). In Niigata, a total of 520 cases of Minimata disease had been reported by the end of 1974 and the number was still increasing (*Minimata Disease*, 1968, 1977). By 1970, 6 out of 47 (1 fetal case) of the Niigata victims had died (Takeuchi, 1972). Although the Minimata incident became the benchmark of MeHg poisoning disasters, the number of cases of intoxication was small compared to the Iraqi incident of 1972 in which 459 of the 6530 people who were hospitalized died. Home-made bread prepared from the flour of seed wheat that had been treated with a MeHg-based fungicide was the culprit (Bakir et al, 1973).

Minimata disease or Hunter-Russell syndrome (Hunter et al., 1940; Hunter and Russell, 1954) is characterized by irreversible neurological disorder, especially following *in utero* exposure, that may progress to fatal consequences. The signs and symptoms of this disease have been reviewed in numerous publications (e.g., *Minimata Disease*, 1968, 1977; *Methyl Mercury in Fish*, 1971; Skerfving and Vostal, 1972; *Health Effects of Methylmercury*, 1981). Postnatally in humans, the symptoms begin with sensory disturbances in the extremities, lips, and tongue followed by tremor, ataxia, concentric constriction of the visual fields, and impaired hearing. Usually, there is a latent period between exposure and the appearance of symptoms. The symptoms are irreversible following their appearance.

A number of congenital cases of Minimata disease were born to mothers who had ingested MeHg-contaminated seafood, but exhibited no symptoms of intoxication (*Minimata Disease*, 1968, 1977; *Methyl Mercury in Fish*, 1971; Skerfving and Vostal, 1972; Bakir et al., 1973; *Health Effects of Methylmercury*, 1981). The congenital cases exhibited a nonspecific cerebral palsy with both motor disturbances and mental retardation. It is of interest to note that, during the course of the Minimata disaster, the frequency of cases of cerebral palsy in the Minimata Bay area was unusually high: 5 to 6% of the total births (up to 12% in one village, compared to the expected frequency of 0.1 to 0.6%) (Nomura, 1968).

The pathology of MeHg intoxication in humans has been reported in detail by Takeuchi (1968) and little has been added since this report. MeHg appears to affect all parts of the central nervous system (CNS), with the greatest damage apparently occurring in the cerebellar and sensorimotor cortices. The effects of MeHg intoxication appear to be dose-dependent (Kurland et al., 1960; *Methylmercury in Fish*, 1971; Dinman and Hecker, 1972). Histologically, MeHg exposure induces varying degrees of degenerative changes including disappearance of neurons, glial proliferation, hydrocephalus, perivascular edema, and demyelination (e.g. *Minimata Disease*, 1968, 1977; *Mercury in the Environment*, 1972; *Health Effects of Methylmercury*, 1981).

In 1958, the Minimata Bay area poisonings were traced to the ingestion of seafood containing high levels of Hg (up to 41 ppm wet weight). Also in 1958, Takeuchi and, independently in 1959, Tokuomi (ref. Takeuchi, 1968b), observed that alkylmercury compounds could reproduce Minimata disease-like symptoms and pathology in experimental animals. Furthermore, Uchida (1961 to 1962) had extracted a chemical he identified as methylmethylmercuric sulfide (CH_3–Hg–S–CH_3) from Minimata Bay shellfish that Takeuchi found could mimic features of Minimata disease in cats and rats (Takeuchi, 1968b). Thus, MeHg was determined to be the causal factor in Minimata disease (Kurland et al., 1960). Kurland et al. (1960) had suggested that Hg might be converted to MeHg in the sediment of Minimata Bay. However, this possibility was not immediately investigated, apparently because MeHg had been found in the chemical plant discharge. The first reports on the biological methylation of Hg were published by Jensen and Jernelöv (ref. Jernelöv, 1973) in the mid 1960s.

The effluent from a chemical plant that used large quantities of $HgCl_2$ as a catalyst in the production of vinyl chloride was the source of the Minimata Bay Hg. The effluent was discharged without treatment into Minimata Bay until 1958, when the flow was diverted to the Minimata River. The diversion produced a new outbreak of poisonings in the new discharge area. Most certainly this is a graphic example of the danger of point source pollution that still does not appear to be appreciated. It may be speculated that one reason the chemical plant may not have been implicated earlier in the course of the disaster was that the initial study of the problem was carried out by the chemical plant itself (Wood, 1971). The logic behind self-monitoring, a practice that has been and remains widespread in industrialized countries, defies comprehension.

Sweden also was greatly affected by environmental MeHg contamination. Beginning in 1950, it experienced a drastic reduction in the size of its wild bird populations apparently as the result of poisoning from the ingestion of MeHg-treated seed (Borg et al., 1966). Currently, an analogous phenomenon of much broader scope exists: the global decrease in amphibian populations (Blaustein et al., 1994; Pechmann and Wilbur, 1994). It behooves us to provide sufficient funding to identify the cause(s) of this disaster. Keep in mind that MeHg dicyandiamide had been introduced in Sweden in 1938 (under the trade name Panogen) as an effective inhibitor of the rotting of seeds after planting, unquestionably an enormous problem in Sweden's climate. It took years to recognize that Panogen had disastrous ecological side effects. Worse, after the ecological problems had been recognized, it took prolonged litigation to block the sale, manufacture, and use of the fungicide. It was not until 1966 that the agricultural use of MeHg-based compounds was banned in Sweden: a country that dotes on its social consciousness. Furthermore, it took until 1970 to ban the export of Panogen (Rissanen and Miettinen, 1972). In the age of the genome project, amphibians may not have sufficient funding priority.

During the period of MeHg use in Sweden, fish also were found to be contaminated with Hg: >0.2 ppm Hg was found in fish from the majority of Swedish fresh and coastal waters (Johnels and Westermark, 1969). In part, this appears to have resulted from the use of phenylmercury acetate (PMA) as a slimicide by pulp mills. Warning against the use of PMA had been issued in Sweden in 1947 (Vallin, 1947), but, apparently, ignored.

The U.S. was introduced to MeHg intoxication in 1969 when members of the Huckleby family (New Mexico) were poisoned by meat from hogs that had been fed MeHg dicyandiamide-treated seed. Three of the Huckelby children suffered irreversible neurological damage and a fourth child, born to the then pregnant, but asymptomatic, mother, had convulsions after birth and was mentally retarded (Curley et al., 1971; Snyder, 1971). The MeHg compound that poisoned the Huckelbys in 1969 had been banned in Sweden 3 years previously. After the Swedish ban, the U.S. continued to use approximately 200 times the amount of MeHg dicyandiamide that previously had been used in Sweden (Wallace et al., 1971). Restrictions placed on the use of MeHg in agriculture by the U.S. Department of Agriculture were reversed in 1970 by judges in Chicago and Washington. Ultimately, grassroots political pressure (i.e., common sense) overrode the lobbies (Howard, 1994).

Full realization of the potential Hg problem in the U.S. precipitated on March 24, 1970 when the Canadian government announced that 12,000 pounds of fish from Lake St. Clair had been destroyed because of Hg contamination. Subsequently, fish in many lakes and rivers in the U.S. (e.g., *Mercury in the Western Environment,* 1973; *Mercury Contamination: A Human Tragedy,* 1977) and Canada (e.g., Bligh, 1971) were found to be contaminated with up to 10 ppm Hg. In 1971, the U.S. Food and Drug Administration (FDA) recommended that consumers cease consuming swordfish because of consistent findings of high Hg levels in edible tissue (U.S. FDA, 1971).

Apparently, the Minimata disaster made no lasting impression on appropriate agencies of either the U.S. or the Canadian governments because a problem analogous to that of Minimata Bay was allowed to develop in Lake St. Clair. It was estimated that approximately 100 tons of elemental Hg had entered the lake St. Clair river system over approximately 15 years since the recognition of the health problem in the Minimata Bay area (Wood, 1971). The Hg entered the ecosystem mainly from the effluent of the chloralkali industry: the result of contamination of waste electrolysis brine by the Hg electrode. As had occurred in Minimata Bay, Hg was biotransformed into MeHg by the sediment of the lake.

Mercury remains a potential environmental problem (ref. *Mercury Pollution: Integration and Synthesis,* 1994). Humans continue to be exposed to Hg from dental amalgam, food (especially fish), water, ambient air, and the workplace (Skerfving, 1991). Indeed, recent evidence indicates that women occupationally exposed to relatively high levels of Hg vapor (e.g., dental assistants) were less fertile than similarly employed women who were not exposed (Rowland et al., 1995). Fish and other forms of aquatic life remain the most significant sources of human exposure to Hg (ref. *Mercury Pollution:*

Integration and Synthesis). Many years before the Minimata disaster, it had been observed that fish take up Hg from their environment and store it in relatively high concentration in their tissues (Stock and Cucuel, 1934; Raeder and Snekvik, 1949). Apparently, this received little attention until the Minimata disaster. Following the disaster, analyses revealed concentrations of Hg in excess of 1 ppm in the edible tissues of fish from many geographical areas including Sweden (Johnels and Westermark, 1969), Finland (Häsänen and Sjöbloom, 1968), Norway (Underdal, 1969), Denmark (Dalgaard-Mikkelsen, 1969), Italy (Ui and Kitamura, 1971; Vecchi et al., 1977), Canada (Bligh, 1971; Fimreite and Reynolds, 1973) and the U.S. (Pillay et al., 1971; *Mercury in the Western Environment,* 1973; *Mercury Pollution: Integration and Synthesis,* 1994). In 1971, Pillay et al. (1971) reported Hg concentrations up to 0.79 ppm in Lake Erie fishes.

Fishes from the Eastern basin of the lake had significantly lower Hg levels than those from the more highly industrialized Western basin that connects to Lake St. Clair via the Detroit River. In Sweden (Westöö and Norén, 1967; Johansson et al., 1970), the U.S. (Smith et al., 1971), and Canada (Bligh, 1971), practically all of the Hg present in fish was found to be in the form of MeHg. An historical perspective of Hg in the environment can be obtained from both previous reviews (e.g., *Methyl Mercury in Fish,* 1971; Holden, 1972; Vostal, 1972; *Environmental Mercury Contamination,* 1972; *Mercury in the Western Environment,* 1973) as well as more recent ones (*Health Effects of Methylmercury,* 1981; WHO, 1990; *Mercury Pollution,* 1994).

The natural concentration of Hg in fishes was reported to be not more than 0.2 ppm (Löforth, 1960). In the 1960s and 1970s, extremely high concentrations of Hg (on a wet weight basis) were reported in fish from a Swedish lake (9.8 ppm: Löforth, 1969) and the Delanger River (20 ppm: Jernelöv, 1969); fish from the Saskatchewan River (10.6 ppm: Wobeser et al., 1970); fish from Lake Wabigoon, Saskatchewan (24 ppm: Wood, 1971); fish and shellfish from Minimata Bay (1 to 36 ppm and 11.4 to 39 ppm, respectively: Takeuchi, 1972) and fish from the Agano River, Niigata, Japan (5 to 44 ppm, 1968 to 1972: *Minimata Disease,* 1977). It was reported that the Saskatchewan River fish containing 10.6 ppm exhibited no obvious ill effects. However, no rigorous evaluation or definition of "ill effects" was proffered. It is to be noted that Birke et al. (1967) estimated that the daily consumption of a meal of fish containing 5 to 6 ppm Hg might be lethal to humans.

In 1969, an international committee recommended a limit of 0.5 ppm (wet weight) for the maximum acceptable level of Hg in fish for human consumption (*International Committee Report,* 1969). The setting of such a standard was, of course, controversial since, beside human health and welfare, it would affect the economics of the fishing industry and impact the food supply of some areas. Basically, we hear similar arguments from the fishing and logging industries in the U.S. today. In any case, it was difficult to recommend banning the sale of or reducing the consumption of fish in a country such as Sweden. At the time of the limitations debate, the average daily consumption of fish in Sweden was estimated to be 50 per person with 80,000 persons consuming over 120 (Skerfving, cited in Löforth, 1969). The problem of setting of a Hg limit was compounded by the fact that very little was (and is) known about the consequences of chronic low level exposure to Hg in the context of individual diets (consider the protective affect of selenium (Ganther et al., 1972; Magos, 1991) or the potential genetic effects of such exposure (*vide infra*)).

Löforth (1969) described the controversy over establishing an acceptable Hg level for fish in Sweden. Originally, Swedish scientists recommended a limit of 0.5 ppm. However, prior to promulgation of the recommendation, it was discovered that the average Hg concentration in fish from Sweden's largest lake, Lake Vanern, exceeded the proposed maximum. Thus, the limit was raised to 1.0 ppm (wet weight). The 1.0 ppm limit was based on: (i) the victims of the Minimata disaster having consumed seafood with an average Hg concentration of 50 ppm; (ii) the assumption that reduction by a factor of 10 would prevent (in Minimata, apparently would have prevented) MeHg intoxication; and (iii) the application of a safety factor of 5.0. Löforth (1969) pointed out that the calculation was erroneous since the Minimata disaster figures were based on dry weight. On the basis of wet weight, the Swedish limit should have been set at 0.2 ppm. It would be expected that such a limit would be exceeded by fish from many geographical areas as the result of naturally occurring Hg.

Apparently, all forms of Hg can be transformed into MeHg in appropriate bottom sediment. Rate of transformation is dependant upon the level of appropriate microbial activity and the availability of carbon, nitrogen, and phosphorous (Beijer and Jernelöv, 1979; Wood et al., 1989; *Mercury Pollution: Integration and Synthesis,* 1994). If rate of elimination of MeHg from organisms occurs at a slower rate than it is ingested, it accumulates in the organism and can result in food chain/web magnification. For example, upper level predators such as the swordfish have been found to contain Hg concentrations several

III. SOURCES OF ENVIRONMENTAL MERCURY

Annual world production (1986) of Hg is approximately 6,700,000 tons (Carrico, 1986). In addition, Hg is added to the environment through the burning of coal ($\approx$3000 tons/year), oil, and lignites ($\approx$1600 tons/year), the roasting of sulfite ores ($\approx$2000 tons/year) and the production of cement ($\approx$100 per/year) (Bertine and Goldberg, 1971; Joensuu, 1971; Weiss et al., 1971; Billings and Matson, 1972).

The quantity of Hg added to the global environment by human activities may be insignificant in view of the estimated 175,000 tons per year that are released naturally into the atmosphere through continental degassing (Weiss et al., 1971). However, certain human activities such as the construction of roads and the clearing and tilling of land may increase the degassing rate. It has been reported that the Hg^{1+} concentration in the Greenland Ice Sheet has doubled since the pre-1900s (Weiss et al., 1971). It is unlikely that human activity could significantly affect the Hg concentration of the oceans since the total marine Hg content is estimated to be 10^8 tons (Peakall and Lovett, 1972) or 10^4 times annual production. However, human activity has produced potent point sources of environmental Hg contamination (e.g., Minimata Bay) that have resulted in tragic human consequences. Also, it must be kept in mind that, although the 100 tons of Hg estimated to have been deposited in the sediment of the Lake St. Clair system is, on a global scale, a relatively small quantity, the MeHg content of the fish in the system increased 100-fold in 35 years (Wood, 1971). It is predictable that point source pollution will be a significant environmental problem in "developing" countries as it is in all industrialized countries.

IV. CHEMISTRY, TOXICITY, AND MODE OF ACTION OF MERCURIALS

A. SOLUBILITY AND VOLATILITY

Tables 1 and 2 list the solubilities and vapor concentrations, respectively, of various organomercurials. These data indicate that:

1. The solubility of a mercurial salt is dependent upon its anion (e.g., MeHgOH is 10^5 times more soluble in water than MeHgCl). The solubility of MeHgCl in nonpolar solvents is greater than that in water by a factor of 10^4.
2. In general, alkyl mercurials are more volatile than aryl or alkoxy mercurials, and the anion has a large effect on volatility.
3. The anion has no effect on rate of absorption, distribution, metabolism, or toxicity of MeHg (Ulfvarson, 1962; Irukayama et al., 1965; *Methyl Mercury in Fish,* 1971; Takeuchi, 1972). Presumably, this is the result of the strong of attraction of -SH groups for $MeHg^+$.

Table 1 Solvent Effects on the Solubility of Organomercury Salts (18°C)[a,b]

Cation	Anion	Water	Ethanol	Chloroform
MethylHg-	Chloride	1.4×10^{-4}	4.2	3.7
	Bromide	0.7×10^{-4}	3.2	2.6
	Hydroxide	20		
	Dicyandiamide	3	3	
EthylHg-	Chloride	1.4×10^{-4}	0.8	2.6
	Bromide	0.6×10^{-4}	0.7	1.6
MethoxyethylHg-	Hydroxide	2		
PhenylHg-	Acetate	0.2	3.5	12.0
	Hydroxide	1.4	4.3	

[a] Grams/100 g solvent.
[b] From Ulfvarson, 1964. With permission.

B. STABILITY

Alkyl mercurials are very stable relative to aryl mercurials such as phenylmercury, which is rapidly broken down into Hg^{2+} in animals (Miller et al., 1961; Irukayama et al., 1965; Östland, 1969; Norseth, 1969; Norseth and Clarkson, 1970a,b; Norseth, 1972; Daniel et al., 1972; Doi, 1991) and the soil

Table 2 Saturated Vapor Concentration of Organomercury Salts (20°C)[a]

Cation	Anion	Concentration ($\mu g/m^3$)
MethylHg-	Chloride	94,000
	Bromide	94,000
	Iodide	90,000
	Acetate	75,000
	Hydroxide	10,000
	Dicyandiamide	300
EthylHg-	Chloride	8,000
	Bromide	7,000
	Iodide	9,000
	Dicyandiamide	400
MethoxyethylHg-	Chloride	2,600
	Acetate	2
PhenylHg-	Acetate	17
	Chloride	5
	Nitrate	1
Metallic (Hg^0)		14,000

From Ulfvarson, V., 1969.

(Andersson, 1979). Alkyl mercurials have been found to circulate unchanged in blood (Friberg, 1959) and Norseth and Clarkson (1970b) have shown that, following administration of MeHg, 98% of the total Hg in the brain is in the form of MeHg. Thus, the unique toxicologic effects of MeHg appear to be a function of the intact MeHg molecule.

C. UPTAKE AND DISTRIBUTION
1. Uptake

A spectrum of factors influences the uptake distribution and excretion of mercurials, including type of mercurial, chemical speciation, species of organism, sex, ligand availability and reactivity, nutritional status, etc. (Passow et al., 1961; Clarkson, 1972; Doi, 1991). Numerous studies of the uptake and distribution of MeHg and other mercurials in mammals have been reported (e.g., *Methyl Mercury in Fish,* 1971; Clarkson, 1972; *Mercury Contamination in Man and His Environment,* 1972; *Mercury in the Environment,* 1972; Massaro et al., 1974; *Health Effects of Methylmercury,* 1981; *Importance of Chemical "Speciation" in Environmental Processes,* 1986). From these and subsequent investigations (e.g., Salvaterra et al., 1975; Lown et al., 1977; Olson and Massaro, 1977a), it may be concluded that, in mammals, mercurials are relatively rapidly distributed to all tissues/organs, MeHg has particular affinity for the brain and RBC, and the kidney is the main target of Hg^{2+}.

In mammals, gastrointestinal absorption of MeHg has been found to be nearly complete whether it is administered as the salt or bound to protein (e.g., Clarkson, 1972; *Methylmercury in the Environment,* 1972; *Effects of Methylmercury,* 1981). At hydrogen ion concentrations of 2 M, the hydrolysis of the MeHg bond (MeHg-SR) to proteins (fish) is essentially complete with a free:bound ratio of 1800 (Westöö, 1967). Since the pH of human gastric juice is normally 1.5 to 2.0, it would appear that most ingested protein-bound MeHg would be hydrolyzed. Furthermore, MeHg is absorbed through the skin and respiratory tract with high efficiency (Hunter et al., 1940; Joselow, 1972; Swensson, 1952). Following exposure to mercury vapor (Hg^0), relatively high concentrations of Hg are rapidly accumulated in the brain (*Methylmercury in the Environment,* 1972; Clarkson, 1972; *Health Effects of Methylmercury,* 1981; *Advances in Mercury Toxicology,* 1991).

In all mammalian species investigated (e.g., mouse, rat, rabbit, monkey, pig, cat, human) and the rainbow trout (*Salmo gairdneri*), MeHg has a high affinity for components of the blood; especially the erythrocyte (RBC) (Nordberg and Skerfving, 1972; Olson et al., 1973; Giblin and Massaro, 1975; Doi, 1991). The hemoglobin (Hb) of most species contains reactive -SH groups that can bind mercurials. The total Hb concentration of human RBCs is impressively high at all stages of development. In the adult (≥ 14 years of age), it averages 35% by weight (Creighton, 1984) and ranges from 14 to 18 and 12 to 16 g/100 ml, respectively, in the blood in males and females (*Biological Data Book,* 1974). In the human fetus, the average Hb concentration has been found to be 9 g/100 ml blood at 10 weeks and 15 g/100 ml at 36 weeks of gestation (*Biological Data Book,* 1974). Because the behavior of such highly concentrated solutions is nonideal, ligand binding can be considerably greater than might occur in dilute (approaching ideal) solutions (Creighton, 1984). In addition, the potential mercurial binding capacity of the human RBC is increased by the presence of a relatively high concentration of GSH: 79 g/100 ml of blood (*Biological Data Book,* 1974). However, the RBC does not irreversibly sequester MeHg and, thereby, decrease its toxicity. On the contrary, the RBC functions as the main transporter of MeHg to the tissues (Giblin and Massaro, 1975).

Although RBCs contain a considerable number of potentially reactive -SH groups, these groups apparently are not equally available to all types of mercurials, due to differential uptake. Thus, the reactive thiols within the RBC are more readily available to mercurials like MeHg that can readily traverse the RBC membrane than to those that do not, such as charged species like Hg^{2+} (e.g., Clarkson, 1972; *Mercury in the Environment,* 1972; Doi, 1991). However, adherence to apparent "principles" of structure/activity relationships is far from absolute. Thus, the results of the investigation of Van Stevenick et al. (1965) of the interaction of Hg^{2+} and aryl mercurials with human hemoglobin and RBC stroma indicate that aryl mercurials enter the RBC more slowly than Hg^{2+} and interact with fewer binding sites. Apparently, the human RBC membrane constitutes a formidable barrier to the diffusion of aryl mercurials. In contrast, *in vivo* studies in the rat indicated that RBC uptake of phenylHg is 10 times that of Hg^{2+}; but only 1/10 that of MeHg (Ulfvarson, 1962).

Although the RBC membrane is rich in -SH groups (Jocelyn, 1972), only about 0.05% are in contact with the medium. The remainder, apparently, reside within the interior of the membrane or on its cytoplasmic face (Weed and Rothstein, 1961; Jocelyn, 1972). Lipid-soluble mercurials (nonionic forms) rapidly traverse the RBC membrane, bind to internal ligands, and primarily affect internal systems. Charged (ionic) forms primarily bind to surface ligands and affect membrane functions. They penetrate to the interior slowly (Rothstein, 1973). The binding of $^{203}Hg^{2+}$ to RBCs can be reversed by resuspension in unlabeled Hg^{2+} (Weed et al., 1962). However, once the Hg^{2+} is sequestered within the interior of the cell, it is essentially irreversibly bound (Sutherland et al., 1967).

White and Rothstein (1973) observed rapid and virtually complete uptake of MeHg by human and rat RBCs *in vitro.* Also, they reported that extracellular hemoglobin and various -SH agents were capable, *in vitro,* of eliciting release of MeHg from the cell. In effect, MeHg is not irreversibly bound within the RBC. Similar observations were made by Giblin and Massaro (1975) with the rainbow trout RBC. White and Rothstein (1973) also observed that the rat RBC released considerably less MeHg than human cells. Since almost all of the nonprotein thiol of the RBC is GSH, and the RBC contains the enzymes for GSH synthesis (Jocelyn, 1972), MeHg may be shuttled across the RBC membrane bound to GSH.

It is of interest to note that Takeda et al. (1986) reported that ethylHg administered i.v. to rats binds to -SH groups of Hb. However, Hg was not released from these cells by suspension in extracellular Hb.

The binding of Hg^{2+} to plasma proteins (*in vivo* and *in vitro*) reduces RBC uptake. Following i.v. administration of $HgCl_2$, all rabbit plasma Hg was found bound to protein and, in the rat, the highest proportion of the plasma Hg was found bound to the lipoprotein fraction (Jakubowski et al., 1970). In human and rat plasma, Hg^{2+} also has been reported to bind to albumin and gamma globulin (Suzuki et al., 1967; Cember et al., 1968).

In any case, the RBC appears to be the preeminent sink and transporter of MeHg (Birke et al., 1972; Skerfving, 1974; Massaro and Giblin, 1975; Schütz and Skerfving, 1975; Naganuma and Imura, 1979; *Health Effects of Methylmercury,* 1981; Clarkson et al., 1988); RBC transport of mercurials is not without cost. Each of the 2 titratable (reactive) -SH groups of human Hb (which has a total of 6 -SH groups) is located at position 93 of the β chains (Riggs, 1961; Riggs and Wolbach, 1956). An imidazole nitrogen of the histidine at position β-92, the proximal histidine, forms a coordinate bond with the ferrous atom of the heme prosthetic group (Cullis et al., 1962). Blocking the reactive -SH groups with mercurials results in conformational changes and perturbation of oxygen binding to Hb (Riggs, 1961; Taylor et al., 1963). For example, Noble (1971) has shown that the binding of phenylHg to the β-93 cysteine has a large effect on the reactivity of the heme group with oxygen apparently resulting from a shift in the position of the proximal histidine. In addition, human and canine Hb can be dissociated into their constituent α and β chains by excess aryl or inorganic mercurials (Bucci and Fronticelli, 1965; Enoki and Tonita, 1965). The binding of phenylHg to the reactive -SH group at β-93 results in initial dissociation of the tetramer to the dimer: $\alpha_2\beta_2 \leftrightarrow 2\alpha\beta$ (Rosemeyer and Huehns, 1967). Addition of excess mercurial results in binding to the normally unreactive -SH group at position β-112 and dissociation to the constituent α and β monomer. It does not occur in Hbs lacking an -SH at position β-112 (human fetus, horse, pig, rabbit). Dissociation and/or conformational change, resulting in subsequent exposure of additional reactive groups followed by additional ligand binding, could result in extensive molecular unfolding ("molecular unzipping") and denaturation.

2. Tissue/Organ Accumulation

Hg is ubiquitously distributed in the environment (*Toxicological Profile for Mercury*, 1989) and has a large thiol association constant ($K_{assoc} = 10^{16}$) (Jocelyn, 1972). Therefore, it is not surprising that Hg has been found in all human tissues examined (e.g., Massaro et al., 1974).

It has been estimated (*Methyl Mercury in Fish*, 1971) that, for humans exposed to MeHg, the blood Hg level above which there is onset of symptomology (i.e., the apparent threshold) is 0.2 µg/g (approximately 0.4 µg Hg/g RBCs). However, Skerfving (1972a,b,c) has pointed out uncertainties in the estimation. Furthermore, individuals with RBC total Hg levels >1.0 µg/g, resulting from exposure to MeHg through the consumption of contaminated fish, exhibited no symptoms of MeHg intoxication. In the apparently asymptomatic population studied by Massaro et al. (1974), RBC total Hg ranged from 0.003 to 0.017 µ/g, while plasma levels ranged from 0.002 to 0.011 µg/g (Table 3). The concentration of Hg in hair at onset of symptoms of MeHg intoxication has been calculated to be ≥200 µg/g and corresponds to whole blood levels ≤0.07 µHg/g (*Methyl Mercury in Fish*, 1971).

The tissue/organ distribution of MeHg is characteristically different from that of Hg^{2+} (Berlin and Ulberg, 1963a,b,c). The mammalian brain appears to be the primary target of MeHg intoxication (Suzuki, 1969). It is relatively impermeable to Hg^{2+}, which accumulates primarily in the kidney (e.g., Clarkson, 1972; *Mercury in the Environment*, 1972; Berlin et al., 1973; *Health Effects of Methylmercury*, 1981). Although the brain takes up MeHg at a slower rate than other tissues/organs, it also releases Hg at a slower rate than other tissues/organs. It has been reported that, following a single dose of Hg $(NO_3)_2$, the kidney of the rat rapidly (hours) accumulated Hg to a level 300-fold greater than that of the blood, while the brain level increased 10-fold (Swensson and Ulfvarson, 1968). Because the kidney retains Hg longer than other tissues, its proportion of the Hg body burden increases as a function of time (Rothstein and Hayes, 1960).

Åberg (1969) estimated the lowest level of MeHg in the brain capable of inducing onset of symptoms of intoxication to be 6 µg/g. Twelve victims of the Minimata disaster were reported to have brain concentrations ranging from 2.6 to 24 µg Hg/g (Takeuchi, 1968a). "Normal" brain Hg levels (neutron activation analysis) in the Eastern U.S., based on autopsy samples from 7 individuals (age range: 33 to 79 years) ranged from 0.02 to 2.0 µg/g wet weight (Glomsky et al., 1971). The range for a full-term stillborn was 0.04 to 0.05 ng/g. Analysis of similar brain areas (Table 3) from a larger sampling (30 to 100 individuals ranging in age from neonates to 91 years) from the same study area revealed a range 0.02 to 2.59 µg Hg/g (Massaro et al., 1974). A similar range has been reported more recently by Ehmann et al., (1994).

MeHg is a potent developmental toxicant. It crosses the placental barrier and accumulates in the conceptus (Berlin and Ulberg, 1963c; Suzuki et al., 1967, 1971; Childs et al., 1973; Null et al., 1973; Garcia et al., 1974; Mansour et al., 1974; Reynolds and Pitkin, 1975; Olson and Massaro, 1977a,b; Kelman and Sasser, 1977; *Health Effects of Methylmercury*, 1981). Thus, human neonates born to mothers exposed to MeHg through consumption of contaminated fish or grain were found to have higher RBC Hg levels than their mothers (Tejning, 1970; Suzuki et al., 1971). Furthermore, numerous reports indicate developmental toxicity in the absence of apparent maternal effects (Egleson and Herner, 1952; *Methyl Mercury in Fish*, 1971; Bakir et al., 1973; *Minimata Disease*, 1977; *Health Effects of Methylmercury*, 1981).

MeHg crosses the placenta and accumulates in the tissues of the fetal mouse, rat, and macaque (*Macaca mulatta*) (Berlin and Ulberg, 1963a,b,c; Childs, 1973; Null et al., 1973; Mansour et al., 1974; Garcia et al., 1974; Reynolds and Pitkin, 1975; Olson and Massaro, 1977a,b). In these studies, MeHg was administered to the pregnant female once or twice during gestation or, in the case of the macaque, directly to the fetus via catherization of the umbilical vein (Reynolds and Pitkin, 1975). Analysis of the kinetics of fetal uptake/elimination of MeHg via placental transfer in experimental animals revealed that:

- In the mouse, at a dietary MeHg content above 0.05 ppm, fetal Hg sequestration increases in direct proportion to the content of MeHg in the diet (Childs, 1973). In addition, fetal CNS Hg levels increased in proportion to maternal dose and were higher than maternal CNS levels (Null et al., 1973).
- In the rat, the net rate of MeHg transfer from the maternal circulation to fetal tissues during late gestation increases as a function of gestational age at the time of administration (Mansour et al., 1974).
- In the monkey, the rate of MeHg transfer from the maternal circulation to the fetal circulation is greater than the reverse transfer rate at equal concentration gradients (Reynolds and Pitkin, 1975).

Table 3 Human Tissue/Body Fluid Mercury Levels (Neutron Activation Analysis)

Tissue	Number of Samples	Group Range (ppm wet weight)	Mean[a]	SEM[b]
Frontal cortex[c]	100	0.03–1.69	0.336	0.038
Frontal cortex[d]	50	0.04–1.50	0.329	0.043
Cerebellar cortex[c]	100	0.08–2.59	0.358	0.047
Cerebellar cortex[d]	50	0.04–1.65	0.347	0.049
Muscle[c]	100	0.13–0.87	0.276	0.026
Medulla[e]	30	0.04–1.29	0.300	0.090
Pons[e]	30	0.04–1.66	0.341	0.091
Midbrain[e]	30	0.04–1.20	0.336	0.078
Corpus striatum[e]	30	0.03–0.61	0.135	0.038
Thalamus[e]	30	0.04–1.47	0.269	0.098
Thalamus[f]	30	0.04–0.11	0.074	0.006
Callosal white matter[e]	30	0.02–0.15	0.072	0.011
Callosal white matter[f]	30	0.04–0.11	0.070	0.010
Gyri[e]	30	0.02–0.15	0.079	0.016
Gyri[c]	30	0.04–0.11	0.077	0.013
RBC[g]	50	0.003–0.017	0.008	≫0.001
Plasma[g] + RBC washings	50	0.002–0.011	0.007	≫0.001
Urine[h]	50	0.003–0.017	0.008	≫0.001

[a]Data were normalized (logarithmic transformation) prior to statistical analysis.
[b]SEM = standard error of the mean.
[c]Frontal cortex, cerebellar cortex, and gastrocnemius muscle obtained from each of 100 adult (15 to 91 years) cadavers.
[d]Frontal cortex and cerebellar cortex obtained from each of 50 subadult (0 to 14 years) cadavers.
[e]Medulla, pons, midbrain, corpus striatum, thalamus, callosal white matter, and angular and supramarginal gyri from each of 30 adult cadavers selected at random (random numbers table) from the original 100.
[f]Thalamus, callosal white matter, and angular and supramarginal gyri from each of 30 subadult cadavers selected at random from the original 50.
[g]Blood was collected in heparin vacutainer tubes. The RBCs and plasma were separated and the cells washed 3 times (with 0.013 M NaCl). The plasma and combined washings were made 10% in bovine serum albumin (final pH 6.9 to 7.7) and lyophilized prior to activation analysis. Spiked ($^{203}HgCl_2$ or Me^{203}HgCl) samples of plasma and washings (+ BSA) exhibited no significant loss of radioactivity upon lyophilization.

The statistical analysis revealed:

a. No significant differences ($P > 0.05$) in Hg levels among the members of the following groupings: (i) Adult (15 to 91 years) and subadult (0 to 14 years) frontal and cerebellar cortex, medulla, pons, midbrain, and adult thalamus. (ii) Subadult thalamus, adult and subadult callosal white matter, and angular and supramarginal gyri. (iii) The RBCs, plasma, and urine obtained from apparently healthy volunteers. Also, there was no correlation among the Hg levels of the RBCs, plasma, and urine of any single individual.
b. A significantly higher ($P < 0.001$) Hg level in the tissues listed in a.i. (above) compared to the tissues/body fluids of a.ii. and a.iii.
c. A significantly higher ($P < 0.001$) Hg level in the tissues listed in a.ii compared to tissues/body fluid of a.iii.
d. A significantly higher Hg level in the adult gastrocnemius muscle and corpus striatum compared to the tissue/body fluids listed in a.ii ($P < 0.01$) and a.iii ($P < 0.001$); but a significantly lower Hg level ($P < 0.01$) in muscle and striatum compared to the tissues listed in a.i.
e. A significantly positive effect of age on thalamic Hg levels (adult > subadult: $P < 0.001$); but, in general, no influence of age, weight, height, sex, race, national origin, occupation, religion, cause of death, or tissue handling parameters on Hg levels.

Unfortunately, none of the times selected for administration coincided with the period of organogenesis. However, fetal sensitivity to MeHg-induced neurological damage appears to be related to enhanced CNS sequestration of the toxicant compared to maternal CNS tissue. Also, the basic tenet of cancer chemotherapeutics, that cycling cells are more sensitive to toxic insult than vegetative cells, probably also applies to the developing embryo/fetus.

To investigate fetal uptake of MeHg during organogenesis, Olson and Massaro (1977a,b) exposed pregnant Swiss-Webster CFW mice to a single s.c. dose of MeHg (5 ppm Hg: 5 mg Hg/kg maternal body weight) in phosphate-buffered saline (PBS: 0.13 M NaCl, 0.01 M NaH_2PO_4–Na_2HPO_4, pH 7.4) on day 7, 8, 9, 10, 11, 12, or 13 of gestation. Fetal Hg accumulation increased with fetal age at time of administration at least up to day 13 of gestation. MeHg exposure on days 10 to 13 of gestation produced a fetal Hg concentration higher than that of maternal blood concentrations 7 d after administration on days 10 or 11 of gestation and 5 d after administration on days 12 or 13 of gestation. Fetal Hg concentration peaked 3 d after administration, averaging 4.8 ppm in animals injected on day 13 of gestation. Placental Hg concentration was always equal to or higher than that of maternal blood. An increase in maternal net Hg elimination rate (1.3 and 3.4 µg Hg/d following administration on day 7 or 13 of gestation, respectively) was associated with increased fetal sequestration of Hg as a function of fetal age.

To determine Hg uptake via suckling, offspring of dams administered 5 ppm MeHg on day 12 of gestation and control dams were cross-fostered. The Hg concentration of the tissues of MeHgxMeHg crosslings were essentially identical with those of the dams at 10 and 21 days postpartum. The Hg concentration of the tissues of MeHgxPBS crosslings (*in utero* Hg exposure only) were 15 to 30 times greater than that of PBSxMeHg crosslings (suckling exposure only), indicative of greater Hg transfer *in utero*.

3. Subcellular Distribution

Consistent with their large thiol binding constant, mercurials tend to be present in all subcellular fractions (i.e., nuclear, mitochondrial, lysosomal, soluble) prepared by centrifugation (Berlin et al., 1973, 1975; Mehra and Choi, 1981; Norseth, 1969; Omata et al., 1980a; Syversen, 1976; Winroth et al., 1981; Yoshino et al., 1966a). However, as a function of time, Hg accumulates in the lysosomal/peroxosomal fraction (rat tissues) to a greater extent following exposure to a single dose of $HgCl_2$ compared to MeHg (Norseth, 1969). Winroth et al. (1981) observed that at 3 weeks post administration of a single subtoxic dose of Me^{203}HgOH (1 to 3 mg/kg body weight), 75% of the radioactivity in squirrel monkey (*Saimiri sciureus*) brain was associated with the particulate fraction obtained by ultracentrifugation. In the soluble fraction, 9% of the total radioactivity was associated with glutathione and the remainder with high molecular weight compounds.

Only trace amounts of Hg were found associated with low molecular weight compounds (M.W. = 100 to 300) in the kidney of rats administered $HgCl_2$ or ethylHg (e.g., Clarkson and Magos, 1966; Jakubowski et al., 1970). In equilibrium dialysis experiments, Clarkson and Magos reported (1966) that over 99% of the kidney Hg of rats administered $HgCl_2$ was present in nondiffusible form. It is generally accepted that $HgCl_2$ perturbs proximal tubule water channel function (Pratz et al., 1986). Apparently, mercurials inhibit water flow in the proximal tubule by interfering with the function of vasopressin-sensitive water channels (Hoch et al., 1989). Furthermore, it has been reported that mercurials potentiate the angiotensin II-induced increase in $[Ca^{2+}]_i$ (the second messenger for angiotensin II) in proximal tubule cells, and that this may be mediated through phospholipase C activation (Endou and Jung, 1991).

Following exposure to MeHg, it has been reported that Hg is present almost completely in the protein fraction of rat brain (e.g., Yoshino et al., 1966a; Gruenwedel et al., 1981). Only a small percentage of the total Hg was found in the lipid and nucleic acid fractions.

Mercurials have been shown to interact with phospholipid monolayers and model membranes (Rabenstein and Isab, 1986; Delnomdedieu and Allis, 1991). Hg^{2+} interacts specifically with phosphatidylserine (PS) and phosphatidylcholine (PC), phospholipids that contain a primary amine in their polar head group (Delnomdedieu et al., 1989a,b; 1990; 1992).

Mercurials also form complexes with purine and pyrimidine bases, nucleosides, nucleotides, and nucleic acids (e.g., Carty and Malone, 1979; Carrabine and Sundaralingman, 1971; Gruenwedel and Davidson, 1966; Simpson, 1964) and are mutagenic (e.g., Ramel, 1969; Mulvihill, 1972).

D. TOXICITY

The toxicologic properties of mercurials are related to their chemical speciation (e.g., *Importance of Chemical "Speciation" in Environmental Processes,* 1986). Mercurials interact with the phosphatidylserine and phosphatidylcholine of both model and biomembranes, resulting in increased membrane rigidity (Delnomdedieu, 1989a,b; 1990; 1992). By decreasing membrane fluidity and abolishing the phase transition, Hg^{2+} alters the thermotropic properties of PS model membranes and sonicated rat RBC membranes (Delnomdedieu and Allis, 1993). Amino acid nitrogens also are potential binding sites for mercurials (Eichhorn and Clark, 1963; Brownlee et al., 1978; Taylor et al., 1981; Reid and Podanyi, 1988). However, mercurials have high specificity and affinity for -SH groups (Rabenstein and Isab, 1982; Rabenstein and Reid, 1984; Banfalvi et al., 1985; Robert and Rabenstein, 1991). Hughes (1957) compared the association constant of MeHg with -SH groups to that with other ligands and concluded that, in the presence of an excess of available -SH groups, all but traces of the mercurial would bind to the -SH groups. The association constant of Hg with the –SH group is 10^{16}, which is many orders of magnitude greater than that with any other ligand (Jocelyn, 1972; Sipmson, 1961; Webb, 1966). Therefore, even a large excess of other potential ligands will not divert the binding of Hg to thiols.

The mercury ion (Hg^{2+}) can form a number of complexes with thiols including $Hg(SR)_2$, $Hg_2(SR)_2$, and $Hg_3(SR)_2$. The range of stability constants for dissociation of $Hg_x(SR)_y \leftrightarrow xHg^{2+} + yRS'$ is 10^{-44} to 10^{-40} (Jocelyn, 1972; Stricks and Kolthoff, 1953). Unlike free Hg ions, univalent organomercurials (e.g., $MeHg^+$) do not cleave disulfides except under drastic conditions (Jocelyn, 1972; Sanner and Pihl, 1962).

Although the principal reaction of MeHg in biological systems is with thiol groups, it can, theoretically, form bonds with any atom capable of donating an electron; including chloride, hydroxide, and the oxygen present in carboxyl and phosphoryl groups.

1. Postnatal Exposure

The literature on the postnatal toxicity of mercurials (including metallic Hg/vapor, inorganic salts, and alkyl, aryl, and alkoxylalkyl Hg) has been discussed/analyzed in a number of excellent historical and current reviews (e.g., Clarkson, 1972; *Mercury in the Environment,* 1972: *Health Effects of Methylmercury,* 1981; *Advances in Mercury Toxicology,* 1991). The data indicate that (at sublethal doses) the toxicity and target(s) of different types of mercurials are dependent upon route of administration. For example, the relative toxicity of metallic Hg (Hg^0) via peroral administration is low: up to 500 g have been given to humans with no apparent effects (Cantor, 1951). This is due to the fact that absorption of Hg^0 from the GI tract is limited (Bornmann et al., 1970; Suzuki and Tanaka, 1971). However, injection of 1 to 2 ml of metallic mercury can be fatal (Johnson and Koumides, 1967). Mercury vapor is a nonpolar, lipid soluble, monatomic gas, that readily crosses biomembranes (Hursh, 1985). Inhalation of Hg vapor can result in high Hg levels in brain tissue and the development of neurological symptoms (e.g., tremor) and psychological disturbances such as the "Mad Hatter's disease" (erethismus mercuralis) so graphically described in *Alice in Wonderland* (e.g., *Mercury in the Environment,* 1972; Stopford, 1979; Clarkson, 1994). Inhaled mercury vapor readily passes through the cells of the lung, dissolves in the blood, and is distributed to the tissues. From the plasma, Hg^0 enters the erythrocyte (RBC) and other cells. Oxidation of Hg^0 to Hg^{2+} occurs in the RBC and other cell types catalyzed by the enzyme catalase (Clarkson, 1994). The divalent cation is highly reactive and readily binds to thiols (Simpson, 1961; Webb, 1966). Therefore, oxidation converts Hg^0 to a more toxic species. However, charged species do not readily traverse biomembranes. In effect, oxidation should decrease the rate of transfer of the highly reactive Hg^{2+} from the RBC to other cells. Furthermore, divalent mercury binds to selenium (Kosta et al., 1975). So, even if Hg^{2+} is transferred out of the RBC into other cells, some of it apparently will be trapped in the form of highly insoluble mercuric selenide ($K_{so} = -59$: Sillen and Martell, 1964). Indeed, it has been reported that all Hg found in brain samples obtained from Idrija mercury miners at autopsy was in the form of a 1:1 complex with Se (Kosta et al., 1975). However, the ubiquitously distributed enzyme superoxide dismutase can catalyze the reduction of Hg^{2+} to Hg^0 (Ogata et al., 1987; Dunn, 1981) and it is not clear how much Hg^0 compared to Hg^{2+} reaches target tissues. The circulation should carry a considerable proportion of the regenerated Hg^0 to the lung where it would be efficiently cleared into the expired air (Magos et al., 1989).

In rats, Ikeda (cited in Brown and Kulkarni, 1967) reported the LD_{50} value for Hg^{2+} to be 104 mg/kg, approximately twice that for phenylHg (PhHg) or MeHg: 60 and 58 mg/kg, respectively. In the mouse (i.p. exposure), an LD_{50} of 5, 8, and 14 mg/kg has been reported for Hg^{2+}, PhHg, and MeHg, respectively

(Swensson, 1952). Death from Hg⁺ and PhHg primarily appears to be due to kidney damage (Flanagan and Oken, 1965; Wien, 1939), indicating biodegradation of PhHg to Hg⁺. On the other hand, MeHg accumulates in the CNS and is, primarily, a neurotoxicant. Death from MeHg exposure (at least in congenital cases of severe intoxication) results from complications of damage to the CNS apparently primarily by MeHg (*Minimata Disease,* 1968, 1977; *Methyl Mercury in Fish,* 1971; *Mercury in the Environment,* 1972; Takeuchi and Eto, 1975; Tsubaki et al, 1978; *Health Effects of Methylmercury,* 1981).

Indeed, in the rat, following the 6th subcutaneous injection of MeHg (100 μg Hg/kg) of an every other day exposure paradigm, the brain Hg level was approximately 6 times that following exposure to $Hg(NO_3)_2$ and 18 times that following exposure to PhHgOH or methoxyethylHgCl (Ulfvarson, 1962). Nevertheless, although the definitive mechanism has not been established, MeHg is biodegraded to Hg^{2+}, which accumulates in target tissues (Norseth and Clarkson, 1970a,b; Syversen, 1974; Omata et al., 1980; Hargreaves et al., 1985; Thomas et al., 1988).

A characteristic feature of human MeHg intoxication is a dose-dependent latency period prior to onset of symptoms (Takeuchi, 1968a; Röök et al., 1954). Furthermore, low-level human exposure to MeHg is difficult to diagnose. Early symptoms of intoxication include fatigue, headache, speech and hearing impairment, introversion, difficulty in following a conversation among several participants, and loss of awareness of the victim's surroundings (e.g., *Mercury in Fish,* 1971; *Minimata Disease,* 1968, 1977; Löforth, 1969). On the basis of their presentation, some victims were classified as manifesting unidentified mental illnesses and spent years in mental hospitals although, apparently, they were aware of their situation. In 1971, Herdman (cited in Åberg et al., 1969) described the case of a woman who experienced dizziness, mispronounced words, and manifested tremor of the hands and tongue and loss of memory and reading comprehension. Originally, the illness was diagnosed as psychoneurosis. However, it was learned eventually that she had consumed 0.35 kg of swordfish (approximately 1 mg Hg/kg) per day for a 21-month period prior to the onset of symptomology and had repeated the diet for periods of 3 to 6 weeks, 2 to 3 times per year for 5 years. Apparently, she had suffered MeHg intoxication. Reports such as these evoke questions regarding the interpretation of laboratory animal studies of the behavioral effects of MeHg. The question to be posed is, is the main effect of MeHg intoxication disruption of: (1) cognitive function and, therefore, the ability to acquire and process information; (2) sensorimotor systems, resulting in loss of ability to utilize (in terms of directed action) acquired information; or (3) both cognitive and sensorimotor function? The histopathological data (e.g., Takeuchi, 1968a, 1972b; Takeuchi and Eto, 1975), appear to support the latter possibility.

Compared to invalidism, a relatively high level of exposure to MeHg is required to cause death (e.g., *Methyl Mercury in Fish,* 1971, *Mercury in the Environment,* 1972). Little is known about the effects of prolonged subclinical exposure to MeHg. It appears reasonable to assume that progressive neuronal damage/loss would occur. However, regenerative processes and/or compensatory strategies might be invoked (as following limited traumatic injury) that could offset some loss of function. In any case, restorative capacity is limited, and ultimately death would ensue either as a result of loss of integrative function or damage to some other system(s). Apparently, most deaths of victims of Minimata disease were caused by infection (Kurland et al., 1960; *Minimata Disease,* 1977), implicating immune system dysfunction as a critical factor (e.g., Fiskesjö, 1970; Verschaeve et al., 1976, 1985). The increased sensitivity of mitotic cells to MeHg damage, compared to vegetative cells (*vide infra*), may be the basis for compromised immune system function.

2. Developmental

MeHg is an ideal developmental toxicant: it is insidious in that it is considerably less toxic to the dam than the conceptus (e.g., Engleson and Herner, 1952; *Methyl Mercury in Fish,* 1971; *Minimata Disease,* 1968, 1977; Bakir et al., 1973; *Mercury in the Environment,* 1972; *Health Effects of Methylmercury,* 1981), it readily crosses the placental barrier and accumulates in the conceptus (e.g., Berlin and Ulberg, 1963c; Suzuki et al., 1967, 1971; Kelman and Sasser, 1977) and it has a relatively long biological half-life (e.g., Clarkson, 1972; *Mercury in the Environment,* 1972; *Health Effects of Methylmercury,* 1981). Furthermore, cycling cells appear to be more sensitive to MeHg damage than quiescent cells (Zucker et al., 1990).

It is has been observed that offspring of pregnant women exposed to MeHg through consumption of contaminated seafood exhibited mental retardation and cerebral palsy-like symptoms in the apparent absence of maternal effects (Matsumoto et al., 1965; *Methyl Mercury in Fish,* 1971; *Minimata Disease,* 1968, 1977; Bakir et al., 1973). As in humans, fetal injury has been observed in experimental animals (e.g., mouse, rat, cat, hamster) at dose levels of MeHg that had no overt effects on the dam (Khera and

Nera, 1971; Spyker and Smithberg, 1972; Spyker et al., 1972; Khera, 1973; Khera and Tobacova, 1973; Mottet, 1974; Olson and Massaro, 1977a,b). *In utero* exposure to MeHg and other alkyl mercurials (via maternal exposure) during the period of organogenesis (days 7 through 13 in the mouse: Dagg, 1963) can result in the induction of terata (Oharazawa, 1968; Spyker, 1971; Spyker and Smithberg, 1972; Harris et al., 1972; Inouye et al., 1972; Murakami, 1972a,b; Nolen et al., 1972). For example, exposure (i.p.) of pregnant C57BL/6J, Cd or 129/SvSl mice to 4 to 8 ppm (based on maternal body weight) MeHg in a single dose (to simulate acute human exposure) on day 10 of gestation was embryocidal and/or teratogenic (anophthalmia, micrognathia, exencephaly, limb agenesis, facial malformations) (Su and Okita, 1976). Likewise, exposure of pregnant 129/SvSl mice to daily doses of 2 ppm MeHg (to simulate chronic human exposure) from days 7 to 12 of gestation was teratogenic (Su and Okita, 1976). Also, Khera (1973) reported repeated exposure of pregnant cats to MeHg (0.25 ppm/d) to be teratogenic.

Although considerable information is available on the teratogenic potential of MeHg, the mechanisms through which MeHg induces teratogenesis are unclear. However, the K_{assoc} of MeHg with thiols is 10^{16} (Hughes, 1957) and, not surprisingly, *in vivo* studies have shown MeHg to bind preferentially to sulfhydryl groups of proteins and other molecules (Simpson, 1961; Webb, 1966; Giblin and Massaro, 1975). Hypothetically, therefore, MeHg may disrupt normal development by inhibiting the function of thiol-containing enzymes and other molecules (e.g., GSH: *vide infra*) involved in developmental processes. Indeed, MeHg-induced inhibition of enzyme activity has been observed both *in vivo* (e.g., Na^+, K^+-ATPase activity in the tissues of adult mice: Salvaterra et al., 1973) and *in vitro* (e.g., membrane adenyl cyclase: Storm and Gunsalus, 1974; glutathione S-transferase: Reddy et al., 1981).

Fetal growth and development are dependent upon a continuous supply of nutrients from the circulation. These are delivered to the fetal circulation via the placenta and must be transported from the external to the internal milieu of the cell across the plasma membrane. In the case of some nutrients (e.g., amino acids), placental transport has been shown to be an energy-dependent process involving specific transport proteins (Page, 1957; Reynolds and Young, 1971; Longo et al., 1973; Schneider and Dancis, 1974). Perturbation of nutrient transport could reduce nutrient availability and MeHg might act via disruption of the function of nutrient transport systems.

Cleft palate is the most frequently noted teratogenic effect of MeHg in mice, and its incidence has been reported to increase in a dose-dependent manner (*Health Effects of Methyl Mercury*, 1981; Lee et al., 1979; Nobunaga et al., 1979; Olson and Massaro, 1977a; Su and Okita, 1976). During development of the secondary palate, the palatal shelves, consisting of mesenchyme covered by a thin epithelial layer, rotate medially and fuse (Green and Pratt, 1976). Prior to shelf rotation, components of the mesenchymal extracellular matrix, including glycosaminoglycans (e.g., Anderson and Mathiessen, 1967; Pratt et al., 1973) and collagen (e.g., Smiley, 1970; Pratt and King, 1971), are synthesized and function, at least in part, to provide rigidity to the shelf. Prior to fusion, the epithelial cells of the medial edge undergo a differentiative process involving cessation of DNA synthesis (Hudson and Shapiro, 1973; Pratt and Martin, 1975), production of lysosomal enzymes (Idoyaga-Vargas et al., 1972; Lorente et al., 1974; Mato et al., 1966, 1967) and an increase in cellular adhesiveness (Pratt and Martin, 1975). Following contact between opposing shelves, the epithelial cells autolyze and the underlying mesenchymal masses grow together. More recently, observation of the fate of MEE cells labeled *in utero* with the vital dye DiI (1,1-dioctadecyl-3,3,3′,3′-tetramethylindocarbocyanine perchlorate) indicates that the MEE undergo epithelial-mesenchymal transformation and remain as viable mesenchymal cells in the connective tissue of the palate (Shuler et al, 1992).

MeHg is a potent inducer of cleft palate in the mouse. Following administration of a single, relatively low s.c. dose of MeHg (5 mg Hg/kg maternal body weight) to pregnant Swiss-Webster CFW mice on day 12, hour 6 (12^6) of gestation, Olson and Massaro (1977a) observed a 72% incidence of cleft palate on 15^6. The incidence of cleft palate decreased, as a function of time post MeHg exposure, to 62% on 16^6 and 40% on 17^6; suggesting recovery from insult via metabolism of the toxicant. In control fetuses exposed to vehicle (PBS) only, palatal rotation had occurred by 14^{13} and fusion by approximately 14^{15}. Shelf fusion was completed by 14^{20}. Maternal administration of 7.5 ppm MeHg on 12^6 delayed the median time of fusion by 44 h (fusion increased from 0% at 14^{20} to 80% at 17^6. Administration of 10 ppm MeHg on 12^6 delayed the median time of shelf rotation by 5 h. Shelf fusion was not observed in this group at 17^6 (end of experiment).

In addition to teratogenic effects (e.g., Murakami, 1972b; Olson and Massaro, 1977a), *in utero* exposure to MeHg in laboratory animals can result in reduction of prenatal growth rate (Chen et al., 1979; Spyker and Spyker, 1977). It is conceivable that MeHg might inhibit growth rate by directly or indirectly limiting cell number via selective destruction and/or increase in the cycling time of certain cells (e.g., dividing cells: *vide infra*). Since nutritional status affects growth rate (both prenatal and postnatal), MeHg-induced increase in the cell cycle length might be mediated via reduction in nutrient availability. This is supported by the observation of Olson and Massaro (1977a) that, on 13^6, 1 d post s.c. exposure to MeHg at 5 mg Hg/kg maternal weight, total fetal protein was decreased 22% while DNA content was unaltered. Total protein was maximally decreased (28%) on 14^6 and, thereafter, returned toward the control level. Alteration of DNA content also decreased as a function of time. The maximal decrease occurred on 15^6. It was also observed that the rate of fetal protein synthesis (incorporation of 4,5 ^{3}H-isoleucine) was depressed 5% at 12^9 and the depression increased to 20 to 26% at 13^6 (end of observation). The agreement between the calculated decrease in protein synthesis (19%) and the measured decrease in protein content (22%) suggested that reduction in protein synthesis was responsible for the decreased fetal protein content. Since inadequate protein/amino acid nutritional status would adversely affect both pre- and postnatal growth rate, it seems reasonable to suggest that MeHg-induced inhibition of fetal protein synthetic rate, resulting in reduction in protein content, may be mediated through interference with amino acid availability. Indeed, Olson and Massaro (1977a) observed that fetal free amino acid concentrations at 12^{18} were generally decreased (alanine, 23.0%; valine, 9.7%; methionine, 22.6%; isoleucine, 12.0%; leucine, 18.2%). In addition, uptake of the nonmetabolizable amino acid, ^{14}C-cycloleucine, was decreased 23%. Placental blood flow and fetal water space, measured with ^{3}H-H$_2$O, at 12^{18}, were not affected by MeHg treatment. Therefore, it appears reasonable to conclude that the growth-inhibitory effects of MeHg are related, at least in part, to impaired placental/fetal transfer of amino acids.

It has been reported (Storm and Gunsalus, 1974) that MeHg is a potent inhibitor of adenosine 3′,5′-cyclic monophosphate (cAMP). At the time of palatal shelf fusion, the cAMP level of the shelves increases (Pratt and Martin, 1975); suggesting an important role of cAMP in the morphogenetic process at a critical period in palate development. Indeed, addition of the phosphodiesterase inhibitors, dibutyryl cAMP or theophylline, to prefusion palatal shelves *in vitro* precociously induces medial edge epithelial cell differentiation (Pratt and Martin, 1975). Also, as suggested in studies of cultured fibroblasts, cAMP appears to function in the control of the synthesis of mesenchymal extracellular matrix components: in response to prostaglandin E_1-induced increase in cAMP levels, fibroblasts increase synthesis of glycosaminoglycans and collagen, components of the extracellular matrix (Green and Goldberg, 1965; Green and Hammerman, 1964; Otten et al., 1972; Peery et al., 1971; Peters et al., 1974; Saito and Uzman, 1971).

Adenyl cyclase (AC) catalyzes the synthesis of cAMP from ATP and is a membrane-associated enzyme. Therefore, interaction of MeHg with the plasma membrane could interfere with cAMP metabolism via inhibition of AC either directly or indirectly. To investigate interference with cAMP metabolism as a possible mechanism of MeHg-induced perturbation of palate development, the cAMP, AC, and cAMP phosphodiesterase (PDE) levels of the palatal shelves of fetal Swiss-Webster CFW mice were quantified under conditions of normal development and following maternal exposure to the toxicant (Olson and Massaro, 1980). The cAMP, AC, and PDE levels of lung and liver also were investigated to determine if MeHg affected the palatal system in a unique manner.

It was observed that AC activity (picomoles cAMP/min/mg protein in the presence of 10 mM NaF) decreased from 13.5 ± 2.2 at 13^{22} to 2.9 ± 1.1 at 14^{10} prior to shelf rotation and increased post fusion to 13.6 ± 1.3 at 15^6. Palatal cAMP levels (pm cAMP/mg protein), quantified by radioimmunoassay, increased from <0.5 (undetectable) at 13^{22} to 25.8 ± 1.3 at 14^6. Two plateaus followed: 14^{10} to 14^2 (7.2 $\pm$ 0.7 to 8.1 $\pm$ 0.9) and 15^6 to 17^6 (5.2 $\pm$ 0.5 to 5.1 $\pm$ 0.7). From 13^{22} to 14^{22}, palatal PDE activity and cAMP levels both increased. Between 14^2 and 14^6, PDE activity decreased, while cAMP levels increased.

Although 10 ppm MeHg (maternally administered) delayed the median time of palatal shelf rotation by 5 h, AC and PDE levels were similar to those of control fetuses. Palatal cAMP levels also were similar except between 14^2 and 14^{10}. cAMP levels increased to 16.8 ± 2.8 at 14^2, decreased to 8.9 ± 3.6 at 14^6, and peaked at 18.0 ± 7.3 at 14^{10}. PDE activity covaried with cAMP levels at 14^2; but not at 14^{10}.

A transient increase in cAMP levels was noted in all control tissues at 14^2: palate (15 fold), lung (5-fold), and liver (2-fold). MeHg exposure inhibited the increase in lung; but not in palate and liver.

In addition to teratogenic effects (e.g., Murakami, 1972a,b; Olson and Massaro, 1977a) and reduction of growth rate (Chen et al., 1979; Olson and Massaro, 1977a; Spyker and Spyker, 1977), *in utero* exposure to MeHg in laboratory animals can result in postnatal behavioral perturbation (e.g., Spyker et al., 1972; Hughes and Sparber, 1978; Eccles and Annau, 1982a,b).

E. BEHAVIORAL PATHOLOGY

The most common signs and symptoms of chronic MeHg intoxication in human: (i) adults, (ii) postnatally exposed children, and (iii) prenatally exposed children have been reported to be mental disturbance, ataxia, impairment of gait, impairment of speech, impairment of chewing and swallowing, tremor (i and iii), constriction of the visual fields (i and ii), impairment of hearing (i and ii), brisk and increased tendon reflexes (ii and iii), and impairment of superficial sensation (i) (Harada, 1968a,b; Tokuomi, 1968). In the mouse, however, the dominant symptoms following adult exposure are cerebellar dysfunction and a characteristic hyperreflexia (Saito et al., 1961; Suzuki, 1969). It is to be noted that mice die without exhibiting any neurological symptoms following a heavy dose of alkylHg (Saito et al, 1961; Suzuki, 1969). During development, these effects may result from interference with mitosis (*vide infra*) (e.g., Koerker, 1980; Ramel, 1969b; Rozynkowa and Raczkiewicz, 1977; Sager, 1988; Zucker et al., 1990) and/or neuronal migration (Choi, 1991; Choi et al., 1978).

Remarkably, it has been reported that exposure to a single dose of 3 to 8 ppm MeHg on days 7 or 9 of gestation perturbs offspring active avoidance behavior, water runway and T-maze performance, and behavior in the open-field (Hughes and Annau, 1976; Spyker et al., 1972). Salvaterra et al. (1973) investigated the effects of exposure to a single dose (1, 5, 10 mg Hg/kg) of MeHg on open-field behavior (rearings and ambulations) of adult (23 to 32 g) Swiss-Webster mice. A significant dose-related decrease in rearings and ambulations was observed at both 1 and 3 h post administration (see also Diamond and Sleight, 1972), which was extinguished by 72 h. The behavioral effects correlated with neurochemical observations (Section H). However, Evans et al. (1975) argued that the rapid onset and extinction of effects and the fact that the behavioral and neurochemical alterations did not correlate with brain Hg level suggest "... a different sort of intoxication than effects observed many days after a single exposure or after a prolonged low level exposure ...", related perhaps to local irritation (e.g., Braun and Snyder, 1973). Although irritation at the site of administration was not observed by Salvaterra et al. (1973), this possibility was not systematically investigated and, therefore, could not be ruled out. In addition, hypothermia has been observed to be a reaction to toxic insult in rodents (e.g., Kobayashi et al., 1981; Watkinson et al., 1993) and may play a role in the rapid onset of effects following toxic insult. Salvaterra et al., (1973) also observed a decrease in rearings in the open-field following 6 i.p.doses of MeHg (2.5 mg Hg/kg) administered at 72-h intervals. The effect correlated significantly with the amount of ^{203}Hg present in the cerebellum and cortex, but not in the blood, fur, or whole body. Brown et al. (1972) reported impaired learning ability in the offspring of pregnant female mice chronically dosed with 2.5 ppm MeHg in the diet. The same dosing regimen had no apparent effect on young nulliparous adults (Brown et al., 1972).

Neurological effects of MeHg (e.g., on reflexes, motor coordination, sleep patterns, etc.) have been studied by numerous investigators (e.g., Arito and Takahashi; Fehling et al., 1975; Hargreaves et al., 1985; Ohi et al., 1978; Suzuki and Miyama, 1971; Zimmer and Cater, 1979). These indicators of intoxication are manifested after a variable period of latency. During the latent period, perturbation of the electrophysiological properties of peripheral neurons has been observed (Misumi, 1979; Miyama et al., 1983; Somjen et al., 1973), and morphological abnormalities of both the CNS and peripheral nervous system have been reported (Hargreaves et al., 1985; Herman et al., 1973; Klein et al., 1972; Møller-Madsen, 1990).

A variety of animal models and behavioral paradigms have been investigated to gain insight into the neurobehavioral toxicology of MeHg. For example, Berlin et al. (1973) investigated the effect of MeHg on the ability of the squirrel monkey (*Saimiri sciureus*) to perform a visual discrimination task based on form, size, and color and requiring motor ability and coordination. Clinical signs of intoxication and task performance quality correlated with rate of increase and level of Hg in the blood (a function of the dosing paradigm: subacute or prolonged). During the latent period, no adverse effects of MeHg exposure were noted. However, this may be due to the insensitivity of the detection method, since biochemical, physiological, and morphological alterations have been observed during this period (*vide supra*).

Numerous studies of the effects of MeHg on learning/performance of the rat and mouse of a variety of maze, shuttle box, etc. tasks have been reported. For example, Brown et al. (1972) investigated the effects of MeHg on maze learning ability of the mouse. Doses of 2.5 mg/kg were administered daily in the drinking water of the dams either during gestation or nursing or to the offspring, after weaning, until 45 days of age. At that time, performance of the offspring was tested in a water escape T-maze. They were retested a week later for retention. Testing revealed learning deficits in the gestational and post weaning treatment groups; but retesting indicated persistent learning deficiency only in the gestational treatment group.

Spyker et al. (1972) examined open-field and swimming behavior of the offspring of mice administered 8 mg/kg MeHg dicyandiamide (1/4–1/3 the LD_{50}) i.p. on day 7 or 9 of gestation. This dose level had no overt effect on offspring postnatal development. At 1 month of age, brain weight, protein content, and choline acetyltransferase and cholinesterase activity of the MeHg-exposed offspring were not significantly different from those of control offspring. However, certain behaviors in the open-field and the swimming behavior of the MeHg-exposed offspring differed significantly from those of control offspring.

Hughes et al. (1972) investigated the effects of exposure of the mouse to MeHg during gestation on performance in the open-field and water runway, 2-way shuttle box avoidance, and spontaneous motor activity. In this study, pregnant mice were exposed to 3 or 5 mg/kg MeHg on day 8 of gestation and the performance of their progeny was compared to that of the progeny of saline-treated control dams and progeny of the same age injected with a dose of MeHg equivalent to that given to the dams. In all cases, the performance of the gestationally exposed animals was inferior to that of the control progeny.

Post et al. (1972, 1973) gavaged 15- or 21-d-old Sprague-Dawley rats with 20 mg/kg MeHgCl. At 23 days of age, the animals were trained in a T-maze and observed in an open-field apparatus. The MeHg-exposed animals exhibited increased latencies in the start box of the open-field apparatus. However, performance in the T-maze and on several voluntary tasks in the open-field was similar between the treated and control groups.

In addition to the numerous studies on rodents, the effects of MeHg on avian behavior/learning have been investigated. Its effects on learning in the chick (e.g., Rosenthal and Sparber, 1972), and the conditioned behavior of the pigeon have been investigated (e.g., Evans and Kostyniak, 1972; Evans et al, 1975). Evans and Kostyniak (1972) examined the effects of repeated doses of MeHg (1 to 2 mg/kg administered i.p. 5 times per week for 3 to 11 weeks) on tissue Hg levels and a conditioned pecking response of the pigeon. Behavioral data were collected twice weekly, 24 h post administration. Only slight changes in the conditioned behavior were found prior to the day on which gross symptoms of intoxication appeared. Increase in the length of pause between task repetitions following food reinforcement was reported as the only reliable indicator of effects prior to onset of gross symptomology.

The physiological, biochemical, and cytostructural basis for i.e., the mechanism of the apparent learning/performance deficits induced in experimental animals by exposure to MeHg (at levels that do not result in cell death) is not clear. Conceivably, the mechanism of intoxication could involve perturbation of transmission of: (i) information (e.g., visual stimuli) for central processing, and/or (ii) the processed information into directed action. The problem may be complicated by the possibility that, in addition to the central processing, the character of the in and/or out transmission may be task specific (the mechanism through which a task is perceived and acted upon). Furthermore, the apparent specificity of MeHg lesions in the CNS (damage to the calcarine and sensorimotor cortices: *vide supra*) suggests (among other possibilities) nonuniform distribution of the target. The distribution of ion channels by type and number (and other plasma membrane components) does not appear to be uniform within or among cell types. MeHg perturbs ion channel function (e.g., Narahashi et al., 1991) and that of other plasma membrane components (e.g., Na^+, K^+-ATPase: *vide supra*) suggesting participation (perhaps a critical role) of the plasma membrane in the cascade of consequences of MeHg exposure.

F. HISTO- AND CYTOPATHOLOGY

A spectrum of animal models has been employed to investigate the pathologic effects of MeHg on the CNS, including: mouse (e.g., Berlin and Ullberg, 1963c; Inouye, 1991), rat (e.g., Chang and Hartmann, 1972a,c; Sato and Nakamura, 1991; Von Burg et al., 1980; Yoshino et al., 1966), guinea pig (e.g., Inouye, 1991), cat (e.g., Takeuchi, 1968b, 1972b; Sato and Nakamura, 1991), and various species of monkey (e.g., Berlin et al., 1973, 1975a,c; Ikeda et al., 1973; Sato and Nakamura, 1991).

Chang (1977) has published an extensive survey of the neurotoxic effects of MeHg that covers the literature through the early 1970s. As is the case for human intoxication, the predominant effects of MeHg in animals reflect damage to the central and peripheral nervous systems. Thus, in the squirrel monkey receiving daily peroral doses of $Me^{203}Hg$ for 36 d, the subcortical layer of the cerebellum and the calcarine cortex contained the highest levels of Hg (Berlin et al., 1973;1975a,c). This pattern of Hg distribution parallels the pattern of cortical lesions found in MeHg intoxicated adult humans who, in addition, also exhibit prominent lesions of the sensorimotor cortex (e.g., Takeuchi, 1968a). It is of interest to note that MeHg has been reported to induce a differential alteration of the developing cerebellar cortex between male and female mice (Sager et al., 1984). Furthermore, the pattern of cortical lesions is more broadly distributed in congenital and noncongenital infantile Minimata disease than at later stages of development. *In utero* exposure results in the broadest distribution of lesions (Takeuchi, 1968a). In a rat, a decreasing Hg concentration in the order of dorsal root ganglion neurons > calcarine cortical neurons > cerebellar Purkinje cells > ventral horn motoneurons > cerebellar granule cells has been reported (Chang and Hartmann, 1972d).

Since the classic work of Hunter et al. (1940), the neuropathological effects of alkylHg poisoning have been extensively studied and documented (e.g., Chang and Hartmann, 1972d,e); Choi, 1991; Diamond and Sleight, 1972; Inouye, 1991; Klein et al., 1972; Miyakawa and Deshimaru, 1969; Miyakawa et al., 1970; Sato and Nakamura, 1991; Takeuchi, 1968a,b, 1972b; Takeuchi and Eto, 1975). The syndrome resulting from MeHg intoxication is characterized primarily by degenerative changes in the cerebral cortex, especially the sensorimotor, calcarine, and cerebellar cortices and the sensory component of peripheral nerves. Cerebrocortical damage is characterized by loss of neurons in lamina II and III. Severe damage involves all layers except lamina I. Damage to the cerebellum is first observed to occur to neuroglia and granule cells, with Purkinje cells remaining intact until later stages of severe intoxication. The granule cell layer of the cerebellum is particularly sensitive to MeHg insult during *in utero* development. A similar pattern of destruction was observed in myelinating cultures of mouse cerebellum along with secondary degeneration of the myelin sheath (Kim, 1971). Myelin degeneration also has been observed in MeHg-intoxicated rats (Chang and Hartmann, 1972a,d).

On the subcellular level, electron microscopic observation of rat brain showed marked structural changes in cytoplasmic organelles of cerebellar granule cells with apparent sparing of the nucleus (Miyakawa et al., 1969). Similar observations were made in myelinating cultures of mouse cerebellum in which mitochondria also appeared to be spared (Kim, 1971). Chang and Hartmann reported (1972a,b) neuronal damage in the cerebellum and dorsal root ganglia. The damage in rats receiving 1.0 mg/kg MeHgCl by gavage for up to 11 weeks manifested as focal cytoplasmic degeneration and development of large cytoplasmic vacuoles in nerve cell bodies, and axonal degeneration, characterized by vacuolization and collapse.

Low doses of MeHg have been reported to perturb blood-brain barrier function. Blood-brain barrier dysfunction has been observed in the rat as early as 12 h after a single dose of MeHg at 1 mg Hg/kg (Chang and Hartmann, 1972e). In rats receiving multiple (weekly) i.p. doses of 10 mg/kg MeHg dicyandiamide, fibrinoid necrosis of cortical and hippocampal capillary walls has been observed (Diamond and Sleight, 1972). In this case, neuronal degeneration was found throughout the cerebrum, hippocampus, and cervical spinal cord. The neural lesions correlated with the vascular lesions, suggesting that vascular damage was responsible for the neuronal damage.

On the cellular level, mercurials perturb membrane permeability, ion and metabolite transport, and enzyme activity (Gutknecht, 1981; Rabenstein and Isab, 1982). Also, they can potentiate membrane deformation and osmotic fragility and affect protein solubility (Weed et al., 1962; Rothstein, 1981).

It is to be noted that cycling cells appear to be more sensitive to MeHg damage than quiescent cells (Zucker et al., 1990).

G. CELL CYCLE, CHROMOSOMAL, AND GENETIC EFFECTS

The amino group of nucleic acids is a potential binding site for Hg(II) and MeHg (Han and Li, 1970; Buncel et al., 1981; Young et al., 1982), and binding of mercurials induces DNA polymorphisms, denaturation, and strand breaks (Gruenwedel, 1985; Gruenwedel and Cruikshank, 1990; Costa et al., 1991). Amino acid nitrogens also are potential binding sites for Hg(II) and MeHg (Brownlee et al., 1978; Eichhorn and Clark, 1963; Reid and Podanyi, 1988; Taylor et al., 1981), and the binding of Hg(II) to chromatin (Rozalski and Wierzbicki, 1983) may involve interaction with both DNA and protein.

Mercurials, especially alkyl mercurials, are potent inhibitors of the cell cycle and mitosis (resulting in C-mitosis and/or aneuploidy/polyploidy) and clastogens (e.g., Costa et al., 1982; Fiskesjö, 1970; Koerker, 1980; Léonard, 1988; Léonard et al., 1983; Miura and Imura, 1991; Ramel, 1969a,b, 1971; Sager, 1988; Sharma and Talukder, 1987; Skerfving et al., 1974; Vogel et al., 1986; Zucker et al., 1990). MeHg-induced mitotic arrest, defined flow cytometrically (FCM) as the accumulation of cells in the G_2/M phase of the cell cycle (e.g., Zucker et al., 1990) results from disruption of microtubule assembly (Koerker, 1980; Miura and Imura, 1991; Sager, 1988; Vogel et al., 1986). Decreased rate of progression through the cell cycle has been attributed to lengthening of the duration of the G_1 phase as a consequence of inhibition of protein synthesis (Vogel et al., 1986). However, Costa et al. (1982) have reported that exposure to $HgCl_2$ induces an S phase-specific block in the Chinese hamster ovary cell *in vitro*. Furthermore, flow cytometric (FCM) analysis (Zucker et al., 1990) indicates that exposure of exponentially growing murine erythroleukemic (MEL) cell cultures to relatively low levels of MeHg (2.5 to 7.5 μM)

predominately inhibits progression through the S phase of the cell cycle (Figure 1). Accumulation of cells in the G_2/M phase of the cycle also occurs, but only to a minor extent — certainly to a minor extent compared to colchicine (Colcemid) exposure. Clearly, MeHg is not primarily a C-mitosis-inducing agent. The observations of Zucker et al. (1990) indicate that perturbation of DNA synthesis is a primary consequence of exposure to MeHg. However, it is unclear whether perturbation of DNA synthesis is due to direct interaction of MeHg with the DNA synthetic machinery, strand breakage, or the consequence of primary damage to some other target such as the plasma membrane (Rothstein, 1959, 1973, 1981). It appears reasonable to suggest that perturbation of the internal milieu of the cell as a consequence of plasma membrane damage could result in disruption of DNA synthesis through a spectrum of pathways. Evidence in support of the latter hypothesis has been presented by Zucker et al. (1990).

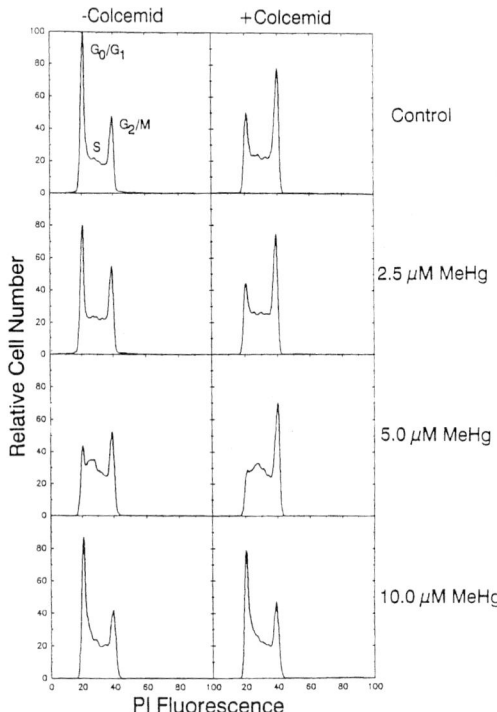

Figure 1. Effect of methylmercury exposure on the cell cycle progression (Zucker et al., 1990). The left column is comprised of representative DNA histograms of nuclei of murine erythroleukemic cells exposed to 0, 2.5, 5.0, or 10 μM MeHg for 6 h. The "+Colcemid" column is comprised of representative DNA histograms of nuclei of MEL cells exposed to Colcemid during the last 2 h of the 6-h MeHg exposure. Colcemid was added to determine rate of progression through the cell cycle. Following exposure to 5.0 μM MeHg, movement of cells through the S phase of the cycle appears to be retarded, as indicated by the decrease of cells in the G_0/G_1 phase of the cycle and increase of cells in early S phase. Exposure to 7.5 μM MeHg (data not shown) results in even greater apparent retardation as manifested by a larger percentage of cells in early S phase.

Zucker et al (1990) observed that exposure of MEL cells to MeHg concentrations >10 μM results in dose-dependent perturbation of the plasma membrane/cytoplasm complex. By FMC, this is manifested as increased 90° light scatter (refractive index: Shapiro, 1994), decreased axial light loss (apparent cell volume: Cambier and Monroe, 1983), uptake of propidium iodide (PI), a measure of plasma membrane damage (Shapiro, 1994), and resistance to nonionic detergent (NP-40) mediated cytolysis (a measure of fixation: Zucker et al., 1988). Resistance of the plasma membrane/cytoplasm complex to detergent-solubilization suggests that fixation (denaturation, cross-linking, etc.) of the proteins of the plasma membrane/cytoplasm complex may play a significant role in the mechanism of MeHg cytotoxicity.

Recently, De Flora et al. (1994) reviewed the literature on the genotoxicity of 29 mercury-containing agents including metallic mercury, mercury-containing amalgams, and compounds of mercury. They reported that the data indicate that the agents do not induce point mutations in bacteria. However, methylmercury and phenylHg have been reported to induce point mutations in *Drosophila melanogaster*

(Ramel, 1969b). In the case of phenylHg, this activity may be attributable to the release of benzene via metabolism (Daniel et al., 1972). In addition, perturbation of the pattern of synthesis of 120 individual brain protein species, obtained via reticulocyte lysate translation of polyadenylated mRNA from the brain of MeHg-exposed female Wistar rats, has been reported (Omata et al., 1991). The rats received 10 mg/kg MeHgCl per diem (s.c.) for 7 consecutive days. The number of protein species in which significant increases/decreases were observed was 25/35 on day 4 of exposure (the "early" period of intoxication), 5/67 on day 10 (latent period), and 4/99 on day 15 (symptomatic period).

Skerfving et al. (Skerfving et al., 1970) found a significant correlation between chromosome breaks and the Hg concentration of the blood of humans exposed to MeHg through consumption of contaminated fish. Light microscopic analysis of cytologic preparations of MEL cells exposed (*in vitro* in RPMI 1640 medium supplemented with 10% fetal bovine serum) to MeHg (Zucker et al., 1990) revealed a dose-dependent increase in the incidence of chromosomal aberrations, specifically condensation and pulverization (Table 4, Figure 2). Perturbation of chromosome structure was observed following exposure (6 h) to MeHg concentrations as low as 2.5 μM (Table 4). At low MeHg concentrations, condensation was the predominant chromosomal modification. Exposure to concentrations ≥ 10 μM induced formation of wreath-like chromosomal ring structures apparently involving fusion of the entire chromosomal complement (Figure 2). In human leukocytes, Fiskesjö (1970) observed chromosomes clustered into dense ring structures after *in vitro* exposure for 1 hour to 0.2 to 4 mM MeHg or methoxyethylHgCl in balanced salt solution (free of potentially available extraneous MeHg-reactive thiols). In addition, the cells exhibited increased resistance to squashing compared to control cells, which is indicative of fixation (*vide supra*). Rozynkowa and Raczkiewicz (1977) also described formation of chromosomal ring configurations, apparently resulting from "sticky" chromatin bridges, in spreads prepared from PHA-stimulated human lymphocytes exposed *in vitro* to 40 mg/ml MeHg for 10 min at 37°C in Eagle's minimal essential medium supplemented with serum.

Table 4 Cytogenetic Effects of Methylmercury

	%Viability	% Mitotics	Chromosomal Aberrations			
			% Normal	% Condensed	% Pulverized	% Rings
Control	98	3.1 ± 1.0	92 ± 6	5 ± 2	—	3 ± 4
2.5 μM	98	4.2 ± 2.0	84 ± 8	13 ± 8	2 ± 2	2 ± 2
5.0	95	5.0 ± 0.9	74 ± 22	22 ± 15	4 ± 4	—
10.0	14	4.8 ± 3.0	3 ± 4	30 ± 34	12 ± 9	55 ± 41
25.0	3	4.5 ± 0.8	—	—	—	100 ± 0
50.0	5	3.9 ± 1.8	—	—	—	100 ± 0
Colc.	97	41 ± 0	90 ± 4	10 ± 4	—	—

Note: The mitotic index and percentage of chromosomal aberrations were obtained from murine erythroleukemic cells exposed for 6 h to 0 to 50 μM MeHg or 0.2 μg/ml Colcemid. The mitotic index was based on evaluation of 500 cells. The percentage of chromosomal aberrations was based on evaluation of 200 mitotic cells. The data represent the combined results of three experiments.

Although the mechanism of chromosomal ring structure formation is unknown, direct interaction of MeHg with chromatin and/or perturbation of the plasma membrane, resulting in alteration of the intracellular environment, may be involved. Indeed, *in vitro* exposure to MeHg has been shown to affect the membranes of numerous cell lines. For example, in cultured mouse neuroblastoma cells, Koerker (1980) reported that exposure to 1 μM MeHg for 24 to 72 h at 37°C in Ham's F-12 medium supplemented with serum resulted in perturbation of the function of the plasma membrane, lysosomes, mitochondria, and endoplasmic reticulum.

The findings of Zucker et al. (1990) indicate two thresholds (at least) of cytotoxicity of MeHg. At relatively low concentrations (nominally <10 μM), MeHg perturbs cell cycle progression into, through, and out of the S and G_2 phases, and appears to be reversible. Exposure to higher concentrations, however, results in a fixation-like alteration of the plasma membrane/cytoplasm complex and induction of wreath-like chromosomal ring structures (Figure 2) and is irreversible. The fixation phenomenon (i.e., resistance to dissolution of the plasma membrane/cytoplasm complex by NP-40) is characterized (FCM) by an increase in 90° light scatter, a decrease in axial light loss (apparent cell volume), and the appearance of simultaneous CF and PI staining.

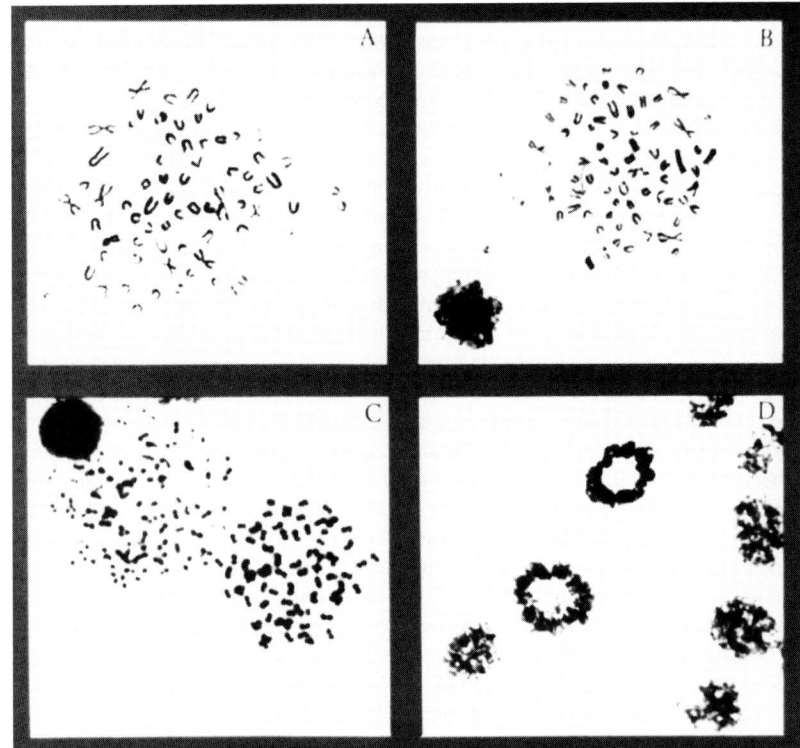

Figure 2. Effects of methylmercury on murine erythroleukemic cell chromosome morphology. Representative photomicrographs (×630) of chromosome aberrations induced by exposure to MeHg for 6 h. A. Control cell chromosomes. B. 5.0 μM MeHg. C. 10 μM MeHg. D. 25 μM MeHg. The chromosomes of cells exposed to 5.0 μM MeHg appear moderately condensed cell cycle is decreased. The percentage of cells exhibiting condensed and pulverized chromosomes (C) increases as a function of dose. Also, at 10 μM MeHg, the chromosomes of more than half of the cells occur in the form of dense ring structures (Table 4). At higher dose (D), identifiable chromosomes are found exclusively in the form of ring structures.

H. MOLECULAR PATHOLOGY

In contrast to the extensive investigation of the tissue/organ distribution, neurology and tissue/organ pathology of MeHg, knowledge of its effects at the cellular and molecular is limited. It is well known that mercurials bind to thiol groups (e.g., Hughes, 1957; Jocelyn, 1972; Webb, 1966) and are potent inhibitors of enzymes (e.g., Hellerman, 1937; Reddy et al., 1981; Rocha et al., 1993; Salvaterra et al, 1973). Tissue/organ enzyme inhibition appears to be compound specific. For example, Kuwahara (1970) observed *in vitro* inhibition of aromatic amino acid transaminase activity in rat brain by MeHg, phenylHg, and Hg^{2+}. However, a larger effect was observed for the organomercurials compared to inorganic mercury that may be primarily a function of differential distribution (*vide supra*).

In addition, MeHg inhibits amino acid transport across the placenta (e.g., Olson and Massaro, 1977a) and RNA and protein (e.g., Yoshino et al., 1966) synthesis (Chang et al., 1972a,b) in the CNS. MeHg-induced perturbation of protein synthesis has been linked to the molecular pathology of Minimata disease (*Methyl Mercury in Fish,* 1971; Brubaker et al., 1973). Apparently, inhibition of brain protein synthesis precedes neurological symptoms (Brubaker et al., 1973; Yoshino et al, 1966b).

Mecurials have the potential to perturb the metabolism of a broad spectrum of endogenous and exogenous compounds by inhibiting the function of enzymes involved in their metabolism (e.g., Lucier et al., 1971). In the rat, an acute dose of MeHg induced a biphasic alteration in the activity of the microsomal enzymes. This was manifested as an initial increase in activity 1 to 4 h post exposure, followed by a marked reduction of activity beginning about 24 h post exposure, and a return to normal levels of activity by about the 8th day post exposure (Chang and Desnoyers, 1978). Furthermore, as a consequence of MeHg-induced depression of drug metabolism via inhibition of liver microsomal oxidases and cytochrome P450 biosynthesis, Alvares et al. (1972) reported an increased hexobarbital sleeping time in the rat. In rat liver microsomes *in vitro*, Arrhenius (1967, 1969, 1969/70) observed inhibition of the detoxification of dimethylaniline, a noncarcinogenic aromatic amine, at high doses of MeHgOH. At low doses, the normal pattern of metabolites was altered. Since other aromatic amines as well as natural substances (e.g., steroid hormones) are metabolized via this same pathway, it is reasonable to surmise that their metabolism also would be perturbed by exposure to MeHg. Interference with steroid hormone metabolism alone or in concert with exposure to appropriate environmental contaminants may result in profound reproductive and/or developmental consequences. A large quantity of Hg is present naturally in the environment (*vide supra*), some of which, undoubtedly, is available for biotransformation to MeHg. However, with the exception of natural deposits, uncontaminated background Hg concentrations are relatively low. Unfortunately, certain geographical locations currently are receiving a high anthropogenically generated input of Hg (e.g., from gold-refining operations: Aks et al., 1995). It is conceivable that, either alone or in conjunction with environmental contaminants that mimic/antagonize the action of reproductive hormones (e.g., Kelce et al., 1994), MeHg may play a role in the apparently widespread alteration of sex ratios of aquatic organisms currently under intensive investigation.

Cremer (1962) observed MeHg-induced inhibition of the oxidation of glucose, pyruvate, and glutamate in rat brain slices compared to kidney slices. In addition, inhibition of succinate dehydrogenase activity was observed in rat brain cortical slices compared to control tissue in animals manifesting symptoms of MeHg intoxication; but not during the latent phase (Yoshino et al, 1966b). Interestingly, no reduction in oxygen consumption was found in the latent phase of MeHg intoxication. However, incorporation of leucine into rat brain proteins was reduced markedly in both the latent and overt stages of MeHg intoxication. Inhibition of brain protein synthesis during the latent period also has been reported by other investigators (e.g., Brubaker et al., 1973; Omata et al., 1980; Verity et al., 1977). In contrast, Brubaker et al. (1971) reported MeHg induction *in vivo* of rat liver protein synthesis characterized by an increase in ribosomes, ribosomal subunits, polysomes, and incorporation of amino acids. Also in the rat, Chang et al. (1972) observed MeHg induction of a moderate increase in the RNA content of anterior horn motoneurons and a decrease in that of spinal ganglion neurons. The Hg content of the tissues investigated in the above studies was not reported (as is the case in most studies) and may have varied significantly among them. However, the collective experience indicates that relatively high concentrations of MeHg irreversibly inhibit cellular processes and ultimately result in irreversible cellular damage (e.g., Zucker et al., 1990) while damage induced by lower concentrations apparently can be repaired.

The affinity of MeHg for thiol groups provides a basis for its toxicity via binding to enzymes, other proteins, and/or GSH. The ubiquity of thiols and their role in maintaining cellular structure, function, and viability makes it unlikely that a single molecular target is primarily responsible for MeHg cytotoxicity. Thus, *in vitro* experiments designed to determine the sensitivity of enzymes to MeHg inhibition have not been particularly useful in assessing its mechanism of neurotoxicity.

To assess MeHg-induced metabolic disruption *in vivo*, Salvaterra et al. (1973) employed an indirect approach involving investigation of the effects of MeHg on the levels of the metabolic intermediates of the glycolytic pathway and high energy phosphates in the brain. Glycolysis and the high energy phosphates were selected for investigation because of their central role in brain energy metabolism. The patterns of alteration in the levels of the intermediates and high energy phosphates were investigated in 6 to 8-week-old male Swiss-Webster CD-1 mice as a function of three factors: (i) dose of MeHg, (ii) time after dose, and (iii) brain region. The levels of the metabolic intermediates under investigation in control animals are recorded in Table 5.

Figure 3 illustrates the alterations in brain (cerebral cortex plus cerebellum) levels of glycolytic intermediates, PC and adenine nucleotides induced by MeHg at 1 and 3 h post administration. A dose-related response was indicated for all significant increases and decreases in the levels of metabolites. No significant interactions between the main effects of dose and time were observed. When data for the three dose levels at 1 and 3 h post administration were analyzed in the $4 \times 2 \times 2$ factorial design, no significant main effects of time were observed for any metabolite. However, when data for the 10 mg/kg dose (at 1, 3, or 72 h) were analyzed separately with time, significant changes were observed. Figure 4 shows the percentage change from control values as a function of time for the glycolytic intermediates, PC and adenine nucleotides. In general, the pattern was one of greater changes from control levels at 1 and 3 h than at 72 h for glycolytic intermediates, ADP and AMP. The changes in FDP, Pyr, and AMP were significant with regard to time in a one-way analysis of variance. ATP and PC levels remained essentially identical.

MeHg-induced changes in the levels of glycolytic intermediates, adenine nucleotides, and PC varied as a function of brain region. The levels of those metabolites that increased with dose (Figure 3) exhibited greater increase in the cortex than cerebellum, while those that decreased with dose, decreased more in

Table 5 Levels (μmol/g wet weight) of Glycolytic Intermediates, Phosphocreatine, and Adenine Nucleotides in the Brain (cerebrum and cerebellum) of the Swiss-Webster CD-1 Mouse

Metabolite	Control Levels (μmol/g) in Cortex and Cerebellum
Glucose-1-phosphate (G1P) [73][a]	0.006 ± 0.0007
Glucose-6-phosphate (G6P) [81]	0.041 ± 0.0010
Fructose-6-phosphate (F6P) [77]	0.008 ± 0.0002
Fructose-1, 6-bisphosphate (FDP) [71]	0.143 ± 0.0099
Dihydroxyacetone phosphate (DHAP) [77]	0.38 ± 0.0009
3-Phospho-D-glycerate (αGOP) [71]	0.271 ± 0.0107
Pyruvate (Pyr) [77]	0.118 ± 0.0038
Phosphocreatine (PC) [57]	1.219 ± 0.0311
Adenosinetriphosphate (ATP) [61]	1.452 ± 0.0347
Adenosinediphosphate (ADP) [64]	0.604 ± 0.0235
Adenosinemonophosphate (AMP) [75]	0.359 ± 0.0149

[a] Number of samples analyzed in brackets.

From Salvuterra et al., 1973. With permission.

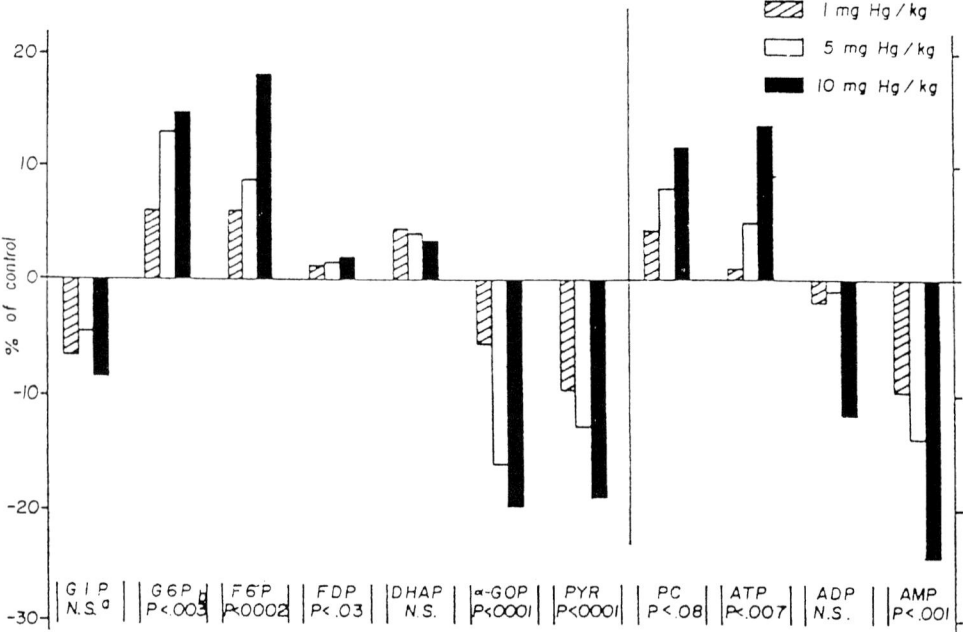

Figure 3. Methylmercury-induced changes in CD-1 mouse brain (cerebral cortex plus cerebellum) levels of glycolytic intermediates, PC, and adenine nucleotides at 1 and 3 h post intraperitoneal administration. [a]Not significantly different from the control level. [b]Level of significance from a 3-way analysis of variance. The concurrent control mean was subtracted from the individual experimental values (μmol/g wet weight) to compensate for day-to-day variations in baseline levels. An approximate 3-way analysis of variance was performed separately on the values for each metabolite using a 4 × 2 × 2 experimental design. The factors in the design were dose (0, 1, 5, or 10 mg Hg/kg body weight), time (1 and 3 h post dosing), and brain region (cortex or cerebellum). (From Salvuterra et al., 1973. With permission.)

the cerebellum. The ratio of the changes in cortical levels of αGOP, Pyr, and AMP to those in the cerebellum were less than one, indicative of greater changes in the cerebellum. They were the only statistically significant ($P < 0.02$ for αGOP, 0.002 for Pyr, and 0.04 for AMP) regional differences observed.

Results of a study of the time-dependent uptake/elimination of a 10 mg/kg dose of Me^{203}Hg by the brain and blood are shown in Figure 5. As can be seen, the rate of blood Hg decrease was faster than

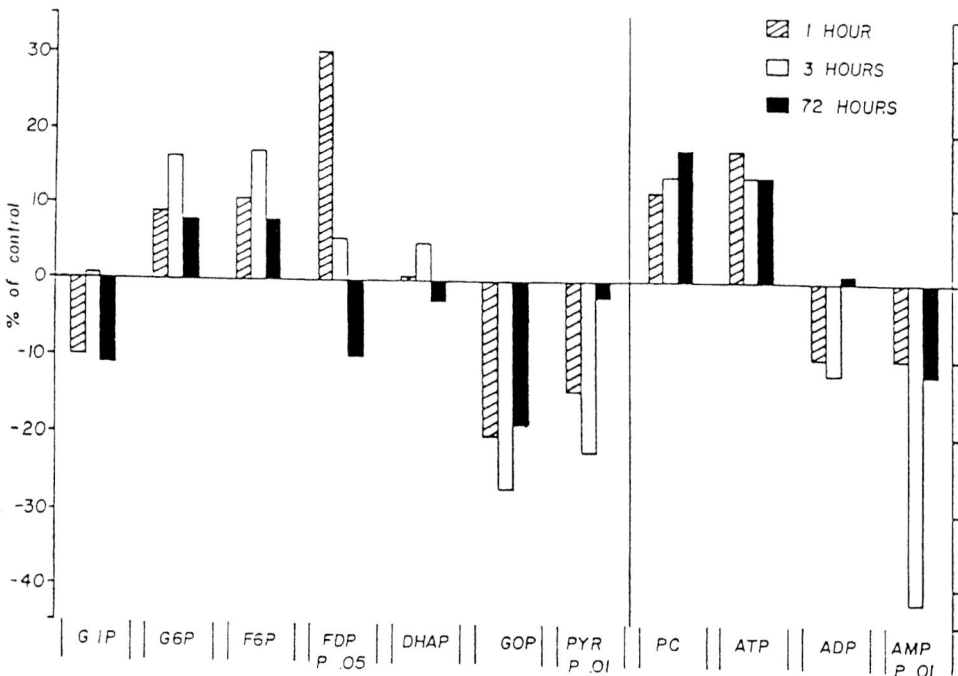

Figure 4. Changes in CD-1 mouse brain (cerebral cortex and cerebellum) levels of glycolytic intermediates, PC, and adenine nucleotides induced by 10 mg Hg/kg (body weight) MeHg at 1, 3, and 72 h post i.p. administration. [a]Level of significance from a 1-way analysis of variance (with respect to time irrespective of brain region).

the rate of brain tissue uptake. Analysis of the data (Student's t-test for paired data) for the first 4 time periods revealed an apparently significantly greater uptake of Hg by the cerebellum than cortex ($P < 0.0005$, average difference = 0.515 µg Hg/g tissue). Also, rate of apparent uptake of Hg by the two brain regions decreased with time up to 72 h, after which elimination apparently commenced. When the data for the time-dependent differences in Hg concentration between brain regions were analyzed (one-way analysis of variance), differences between the groups was highly significant ($P < 0.001$).

The brain has an absolute requirement for glucose and processes most of it by the glycolytic pathway. To fulfill its requirement for ATP, the adult human brain processes approximately 120 g of glucose per diem. As the data of Figure 3 indicate MeHg disrupts glycolysis in the CNS. The increased levels of intermediates at the beginning of the glycolytic pathway (G6P, F6P, FDP) and decreased levels of αGOP and Pyr suggest MeHg inhibition of phosphofructokinase (PFK), the limiting enzyme of glycolysis. Elevated PC levels in the CNS following MeHg exposure may function to maintain elevated levels of ATP to drive neuronal metabolism and GSH synthesis (*vide infra*) under conditions of reduced glycolytic capacity. However, high levels of ATP allosterically inhibit PFK, suggesting the potential for induction of a metabolic conundrum via activation of compensatory mechanisms. The problem may be exaggerated by mercurial-induced inhibition of Na^+, K^+-ATPase activity (Renfro et al., 1974; Singerman and Catalina, 1969). Rat brain ATPase has been shown to be inhibited by low levels of MeHg *in vitro* (Jacobson et al., 1972). Inhibition of ATPase may play a role in the increase in CNS ATP levels observed in the CD-1 mouse following exposure to a single dose (≤10 mg Hg/kg body weight) of MeHg (Salvaterra et al., 1973). The observations of Salvaterra et al. (1973) are similar to those reported for guinea pig cortex slices (Takagaki, 1968). ATPase is a good candidate for MeHg inhibition due to its location at the plasma membrane, a site at which heavy metals are likely to interact (Denomdedieu and Allis, 1993; Passow et al., 1961; Zucker et al., 1990). Mercurials have been shown to hinder the binding of ATP to binding sites associated with the plasma membrane reducing, thereby, the availability of energy for transport (Weed and Berg, 1963). Patterson and Usher (1971) also have shown dose-dependent alterations in rat brain levels of glycolytic intermediates and, less clearly, adenine nucleotides in response MeHg. The qualitative differences in the results of Salvaterra et al. (1973) and Patterson and Usher (1971) may be ascribable to differences in species, dose levels, observation times and/or brain regions investigated.

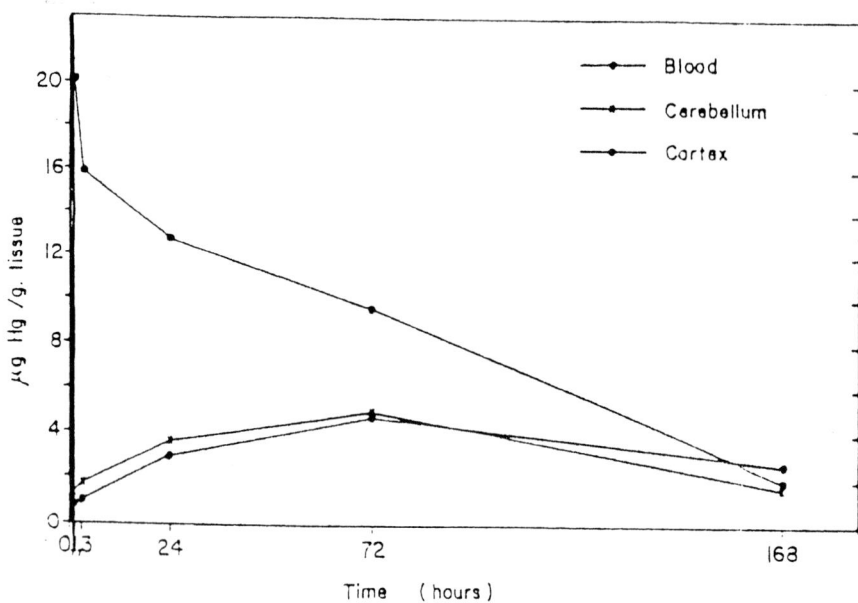

Figure 5. Time-dependent distribution of ^{203}Hg-labeled methylmercury in Cd-1 mouse blood, brain cerebral cortex, and cerebellum as a function of time post i.p. administration of 10 mg Hg/kg (body weight). Each point represents the mean of 4 animals except for that at 168 h which represents that of 2 animals. Analysis of the data for the first 4 time periods indicated significantly greater uptake of Hg by the cerebellum than cortex ($P <$ 0.0005). (From Salvuterra et al., 1973. With permission.)

MeHg and other mercurials can perturb plasma membrane structure and/or function, resulting, for example, in altered permeability and ion channel functioning (e.g., Atchison and Hare, 1994; Narahashi et al., 1991; Passow et al., 1961; Zucker et al., 1990). Hg^{2+} and aryl mercurials have been observed to inhibit RBC active transport by inhibiting Na^+, K^+-ATPase activity (Passow, 1970). Mercurials also perturb the diffusional barrier to Na^+ and K^+ movement across the membrane. The effects of mercurials on membrane transport differ with species. Thus, Hg^{2+} inhibits facilitated diffusion of glycerol across the human and rodent RBC membrane, but not the cat, dog, pig, horse, sheep, or bovine RBC membrane (Jacobi et al., 1936). Hg^{2+} (multivalent metallic cations in general) also can agglutinate washed human RBCs *in vitro*. Apparently, aggregation is facilitated initially by reduction of the negative charge on the membrane followed by the development of cross-linkages between cells (Passow, 1970). Hg^{2+}-induced RBC hemolysis occurs at concentrations considerably lower than those required to induce agglutination (Jandl and Simmons, 1956). It should be obvious that disruption of the diffusional barrier, ion channel functioning and/or development of cross-linking between cells (e.g., CNS) could have profound developmental effects.

I. ENDOGENOUS CYTOPROTECTION

Metallothionein (MT), an inducible metal-binding protein (Hamer, 1986) that protects cells from the potential damaging effects of excess levels of metals, was first isolated (from horse kidney) and characterized by Vallee and co-workers (Kägi and Vallee, 1960; Margoshes and Vallee, 1957). MT has a molecular weight of approximately 6500 daltons. It contains 61 amino acids, 20 of which are potential metal binding cysteines, and no aromatic amino acids (Pulido et al., 1966). Two distinct metal clusters have been identified in mammalian MT by nuclear magnetic and electron spin resonance spectroscopy. One of the clusters has high specificity for zinc. MT binds metals through trimercaptide bridges (Boulanger et al., 1983). Induction of MT biosynthesis has been observed in the capillary endothelium, ependymal cells, pia mater, and arachnoid of rat brain following exposure to Hg^{2+} or MeHg (Tohyama et al., 1991). However, in mammals, only Hg^{2+}, but not MeHg, appears capable of inducing MT biosynthesis. Therefore, although total Hg accumulation in rat brain following exposure to MeHg is much greater than that following comparable exposure to Hg^{2+}, MT concentration is much lower than that anticipated from a comparable intracellular level of Hg^{2+}. Apparently, MT induction in the mammalian brain is dependent upon the rate of biodegradation of MeHg and the attainable concentration of intracellular Hg^{2+}. In any case, like Hg^{2+}, MeHg apparently can bind to MT. Induction of MT in the blood-brain and cerebrospinal fluid-brain barrier systems apparently functions as a primary protective mechanism against metal-induced CNS intoxication. Furthermore, the liver and kidney of fetal and newborn mammals contain higher MT levels than at later stages of development (Bremner, 1987; Webb, 1979).

J. ELIMINATION

A large body of evidence indicates that the kinetics of mercurial excretion are influenced by the nature of the compound and the animal species under investigation (e.g., *Mercury in the Environment*, 1972). Furthermore, in all species investigated, short-chain alkyl mercurials are excreted at a slower rate than other compounds, with the half-time of MeHg excretion ranging from 8 d in the mouse to approximately 1000 d in certain species of fish and shellfish (Clarkson, 1973). In addition, effects of age and sex on retention of Hg following MeHg exposure (rat) have been reported (Thomas et al., 1982). The biologic half-life for the total body burden of Hg in humans has been reported to range from 70 to 74 d (Åberg et al., 1969) with that for the head (obtained by scanning) ranging from 64 to 95 d, and for the remainder of the body, 60 to 70 d. The rate of aryl and inorganic Hg excretion is similar to and considerably faster than that of MeHg. In the rat, it has been reported that 90% of a single i.v. dose of PhHg or inorganic Hg is eliminated in 20 d, whereas MeHg requires in excess of 150 d (Swensson and Ulfvarson, 1968).

In the liver, Hg^{2+} complexes with GSH to form GS-Hg-SG. Apparently, this complex mimics GSSG and is transported by the GSSG carrier across the specialized canalicular aspect of the liver cell plasma membrane (Ballatori and Clarkson, 1984a,b) into the bile and is eliminated in the feces (Klaassen, 1975). Less than 10% of biliary Hg is resorbed (Clarkson, 1994). In the case of the organomercurials, rate of excretion is affected by rate of biotransformation to Hg^{2+}. With time, MeHg is biodegraded to Hg^+ in tissues. However, rate of biodegradation is tissue specific. In the squirrel monkey (*Saimiri sciureus*), for example, following prolonged exposure (9 weekly doses of 0.8 mg Hg/kg $Me^{203}HgNO_3$ per os), the percentage of total Hg in inorganic form was approximately 20 for liver and 50 for kidney and ranged from 30 to 85 in bile. Less than 5% of total brain Hg was found to be in inorganic form (Berlin et al., 1975b).

V. OVERVIEW

The kinetics of uptake, distribution, retention, and elimination of mercurials are dependent on numerous interrelated factors, including species; age; sex and health status; the chemical composition of the food, which impacts the chemical composition of the tissues and may interact with (e.g., bind) toxicants/nutrients in the GI tract; the physical properties of the food, which can affect rate of emptying of the GI tract; the physical and chemical properties of the toxicant; the dose; dosing regimen; and route of administration (Daston et al., 1986; Jugo, 1976; *Mercury in the Environment*, 1972; Thomas and Smith, 1979; Thomas et al., 1982, 1986, 1988). Environmental stress also may alter the response of the subject to toxic insult (e.g., Watkinson et al., 1993; Adachi et al., 1993). The plethora of potential caveats regarding effects of exposure to a toxicant would appear to render the validity of interspecific extrapolation of toxicity data, at anything other than a superficial level, tenuous at best.

Nevertheless, generalizations have emerged from the numerous investigations of the toxicity of mercurials. For example, absorption of Hg^0 from the gastrointestinal (GI) tract is poor (0.01% of the per oral dose in the rat) while that of Hg^+ is modest (2 to 20% in the rat). On the other hand, 50 to 60% of the perorally administered dose of PhHg is absorbed (rat) while >90% of the MeHg dose is absorbed (e.g., Clarkson, 1972; *Mercury in the Environment*, 1972). In humans, complete absorption from the GI tract of a single peroral dose of MeHg, administered either in aqueous solution or bound to fish proteins, has been observed (Åberg et al., 1969; Miettinen, 1973). Complete absorption was also found in the mouse chronically fed MeHg in food (Clarkson, 1971). However, only 15% of inorganic Hg bound to liver proteins fed to human volunteers was absorbed (Miettinen, 1973). Similarly, only a fraction of inorganic mercury chronically fed to the mouse was absorbed (Clarkson, 1973). Apparently, the limited absorption (i.e., transport) of Hg^{2+} results from its bifunctional binding to proteins, dipeptides, GSH, and HS-containing amino acids (Clarkson, 1994).

MeHg also binds to proteins, dipeptides, GSH, and HS-amino acids. Indeed, the association constant of MeHg with the thiol group is 10^{15-20}, which is many orders of magnitude greater than that of other HS-ligands (Hughes, 1957; Jocelyn, 1972; Simpson, 1966; Webb, 1966). Protein digestion releases

MeHg-cysteine (MeHg-Cys), which appears to be structurally analogous to methionine (Met). It has been proposed that MeHg-Cys is transported across the gut wall into the circulation by the large neutral amino acid carrier (Clarkson, 1994; Kerper et al., 1992). If the carrier recognizes the MeHg-Cys complex as Met, does the protein-synthesizing machinery also recognize the complex as Met and incorporate it into proteins? Would such an event inhibit protein synthesis and/or interfere, at least, with achievement of optimal protein folding/conformation and the consequences thereof?

Although the mechanism(s) is/are not clear, exposure to MeHg can result in inhibition of DNA, RNA, and protein synthesis (Gruenwedel and Cruikshank, 1979; Carty and Malone, 1979). Olson and Massaro (1977a) reported that MeHg inhibits murine placental/fetal amino acid transfer and fetal protein synthesis. Inhibition of amino acid uptake/protein synthesis would appear to be analogous to protein starvation, and protein deprivation in the rat has been reported to result in rapid decrease in cellular GSH (Edwards and Westerfield, 1952), a major component of the antioxidant defense and xenobiotic metabolism systems of the cell (*Enzymatic Basis of Detoxication,* 1980; Jocelyn, 1972; Meister and Anderson, 1983). In cells with a high rate of mitosis/turnover, as in the conceptus and GI tract, even modest inhibition of protein synthesis may be deleterious. Toxicant-induced perturbation of the function/viability of the cells of the gut wall, blood-brain barrier, placenta, etc. might result in the reduction of nutrient/metabolite transport at relatively low levels of exposure and breakdown of these barriers at high levels of exposure. Limiting uptake of amino acids and other nutrients is especially deleterious during development (Olson and Massaro, 1977a) while unlimited communication between physiological compartments would be deleterious at any stage of development.

It has been reported that GSH accelerates the prophase of the cell cycle in *Amoeba proteus* (Chalkley, 1951). This suggests that decreased (for whatever reason) cellular GSH might inhibit the rate of cell cycling and adversely affect development. Zucker et al., (1990) have observed that, *in vitro,* MeHg inhibits rate of progress through the S phase of the cell cycle of the murine erythroleukemic cell. At least in part, this may be due to binding of MeHg to GSH. That MeHg binds to GSH is supported by the observation of a low molecualr weight methylmercury complex, with gel and ion exchange chromatographic elution behavior and gel electrophoretic characteristics consistent with those of MeHg-glutathione, in the brain of MeHg-exposed rats (Thomas and Smith, 1979a). The complex, which comprised 30% of total mercury in the soluble fraction prepared from the cerebrum, was reported to persist for at least 5 days following a single i.p. or femoral i.v. injection of MeHg.

Only micromolar concentrations of GSH are found in blood plasma (Meister and Anderson, 1983). Apparently, MeHg (-Cys) rapidly enters the erythrocyte (RBC) from the plasma (e.g. Clarkson, 1972; Doi, 1991; *Mercury in the Environment,* 1972). In the RBC and other cells, cysteine is incorporated into GSH by the GSH synthetic and γ-glutamyl transpeptidase pathways. It would be useful to know if MeHg-Cys is incorporated into GSH and other proteins such as metallothionein and competes with Cys for incorporation. Also, it would be useful to know if binding to MeHg interferes with the turnover rate of GSH and other target molecules. It is to be noted that Thomas and Smith (1979a) reported a relatively lengthy apparent biological half-life of MeHg apparently complexed to Cys in rat brain. In any case, unless MeHg is removed from GSH, the function of GSH would appear to be inhibited. On the other hand, binding to GSH and other HS-containing molecules would reduce the toxicity of MeHg. The intracellular concentration of GSH is high. GSH accounts for almost all of the nonprotein thiol of the RBC (e.g., Bhattacharya et al., 1955; Jocelyn, 1972; Meister and Anderson, 1983): 87 mg/100 ml blood, 79 mg in the reduced form (Beutler et al., 1963; Buell, 1935; Looney and Childs, 1934) and brain tissue: 2.2 to 3.5 μM/g wet weight rat brain (Ashwood-Smith and Smith, 1959; De Ropp and Snedecker, 1960; Martin and McIlwain, 1959) and 1.4 μM/g wet weight human brain (1.2 μM/g wet weight cerebral gray matter) (Robinson and Williams, 1965). Because of its high intracellular concentrations, GSH may function as an important intracellular sink for MeHg.

It would appear that MeHg is distributed to the tissues either as MeHg-Cys or GS-HgMe, entering the cells via the large neutral amino acid carrier (Clarkson, 1994) and γ-glutamyl transpeptidase pathway (Meister and Anderson, 1983). GSH plays a key role in the antioxidant defense mechanism of the cell (*Enzymatic Basis of Detoxication,* 1980; Massaro et al., 1988) and is present in mammalian cells in millimolar (0.5 to 10 mM concentrations (Jocelyn, 1972; Meister and Anderson, 1983). GSH is synthesized in the brain (Douglas and Mortensen, 1956; Lajtha et al., 1959). In rat brain, the concentration of protein -SH is 0.6 to 1.8 μm/g wet weight while that of nonprotein -SH is 2.2 to 3.5 μm/g wet weight (Gabay et al., 1968; Martin and McIlwain, 1959). GSH, the most abundant nonprotein thiol in brain (De Ropp and Snedecker, 1960; Martin and McIlwain, 1959), is present in highest concentration in the cortex (Ashwood-Smith and Smith, 1959). The functions of GSH (i.e., maintenance of the proper ratio of sulfhydryl: disulfide groups in proteins/other molecules and conjugation with electrophiles) may be compromised by high intracellular concentrations of MeHg or MeHg-Cys.

Perturbation of the antioxidant defense mechanism of the cell may be a significant mode of toxic action of MeHg. Support for this hypothesis derives from diverse investigations (Chang et al., 1978; Chang and Suber, 1982; Hirota et al., 1980; Kasuya, 1975, 1976; Kerper et al., 1992; Pascoe and Reed, 1989; Prasad and Ramanujam, 1980; Reddy et al., 1981; Verity and Sarafian, 1991; Welch and Soares, 1976). However, the primary site of cytotoxic interaction of MeHg (e.g., plasma membrane, cytoplasmic or nuclear proteins, etc.) is not clear and may differ with target cell type, dose, route of exposure, nontarget interactions, etc. It has been reported (Verity and Sarafian, 1991) that, as a function of dose and duration of exposure, MeHg induces both lipoperoxidation and oxygen radical generation. In addition, it has been proposed that MeHg is transported across the plasma membrane by the large neutral amino acid transporter (Kerper et al., 1992). Furthermore, employing flow cytometric methodology in conjunction with fluorescent probes, it was observed that progression of the murine erythroleukemic (MEL) cell through the S phase of the cycle was inhibited by MeHg at concentrations that had no detectable effect, *in vitro*, on the plasma membrane/cytoplasm complex (Zucker et al., 1990). At higher concentrations, MeHg interacted with the plasma membrane/cytoplasm complex in a manner akin to that of chemical fixatives (Zucker et al., 1990). Although the cytotoxicity of MeHg is manifested as interference with DNA synthesis, perhaps the primary site of toxic interaction of MeHg with the cell is at the level of the plasma membrane; but flow cytometric methodology is insufficiently sensitive to detect it. However, in light of the observations of Verity and co-workers (Sarafian et al., 1989, Verity and Sarafian, 1991), it is conceivable that MeHg-induced interference with DNA synthesis in the MEL cell may be the result of oxidative damage, mediated through interference with the antioxidant defense function of GSH.

From the above discussion, it may be concluded that MeHg cytotoxicity is the sum of multiple interactions within multiple interacting systems. In addition, evoking the basic tenet of cancer chemotherapeutic agent design, that the cycling cell is more sensitive to toxic insult than the noncycling cell, provides an explanation of the sensitivity of the developing organism to MeHg intoxication. Unfortunately, however, there are many hypotheses and few definitive answers to the majority of questions posed above.

ACKNOWLEDGMENT

This review is dedicated to all of my former students, postdoctoral fellows, and research associates. I acknowledge especially the contributions of Frank J. Giblin, Ph.D., Professor, Eye Research Institute, Oakland University, Rochester, MI; Fred C. Olson, Ph.D., M.D., Private Practice, Internal Medicine, Helena, MT, and Paul M. Salvaterra, Ph.D., Neurosciences, Beckman Research Institute, City of Hope Medical Center, Duarte, CA, who spent countless hours attempting to decifer the mechanisms of methylmercury cytotoxicity; and John B. Morganti, Ph.D. and Bradley A. Lown, Professors of Psychology, State University College of New York, Buffalo, NY, who guided me through the mazes of neurobehavioral toxicology.

DISCLAIMER

This document has been reviewed in accordance with U.S. Environmental Protection Agency policy and approved for publication. Mention of trade names of commercial products does not constitute endorsement or recommendation for use.

REFERENCES

Åberg, B., Ekman, L., Falk, R., Greitz, U., Persson, G., and Snihs, J.-O. (1969), *Arch. Environ. Health,* 19, 478–484.
Adachi, S., Kawamura, K., and Takemoto, K. (1993), *Cancer Res.,* 53, 4153–4155.
Advances in Mercury Toxicology (1991), Tsuguyoshi, T., Imura, N., and Clarkson, T. W., Eds., Plenum Press, New York.
Aks, S. E., Erickson, T. B., Branches, F. J. O., and Hryhorczuk, D. O. (1995), *Ambio,* 24, 103–105.
Alvares, A. P., Leigh, S., Cohn, J., and Kappas, A. (1972), *J. Exp. Med.,* 135, 1406–1409.
Anderson, H. and Mathiessen, M. E. (1967), *Acta Anat.,* 68, 473–508.
Andersson, A. (1979), *The Biogeochemistry of Mercury in the Environment,* Nriagu, J. O., Ed., Elsevier/North-Holland, 79–112.

Arito, H. and Takahashi, M. (1991), *Advances in Mercury Toxicology,* Suzuki, T., Imura, N., and Clarkson, T. W., Eds., Plenum Press, New York, 381–394.
Arrhenius, E. (1967), *Oikos* (Suppl.), 9, 32–35.
Arrhenius, E. (1969a), *Biochemical Aspects of Antimetabolites and of Drug Hydroxylation,* Shugar, D., Ed., Federation of European Biochemical Societies (FEBS) Symp., Vol. 16. Academic Press, 209–225.
Arrhenius, E. (1969/70), *Chem. Biol. Interact.,* 1, 381–393.
Ashwood-Smith, M. J. and Smith, A. D. (1959), *Nature,* 184, 2028–2029.
Atchison, W. D. and Hare, M. F. (1994), *FASEBJ,* 8, 622–629.
Bakir, F., Damluji, S. F., Amin-Zaki, L., Murtadha, M., Khalidi, A., Ai-Rawi, N. Y., Tikriti, S., Dhahir, H. I., Clarkson, T. W., Smith, J. C., and Doherty, R. A. (1973), *Science,* 181, 230–241.
Ballatori, N. and Clarkson, T. W. (1984a), *Biochem. Pharmacol.,* 33, 1087–1092.
Ballatori, N. and Clarkson, T. W. (1984b), *Biochem. Pharmacol.,* 33, 1093–1098.
Banfalvi, G., Bhattacharya, S., and Sarkar, N. (1985), *Anal. Biochem.,* 146, 64–70.
Beijer, K. and Jernelöv, A. (1979), in *The Biogeochemistry of Mercury in the Environment,* Nriagu, J. O., Ed., Elsevier/North-Holland, 203–230.
Berlin, M., Blomstrand, C., Grant, C. A., Hamberger, A., and Trofast, J. (1975a), *Arch. Environ. Health,* 30, 591–597.
Berlin, M., Carlson, J., and Norseth, T. (1975b), *Arch. Environ. Health,* 30, 307–313.
Berlin, M., Grant, C. A., and Helberg, J. (1975c), *Arch. Environ. Health,* 30, 340–348.
Berlin, M., Nordberg, G., and Hellberg, J. (1973), *Mercury, Mercurials and Mercaptans,* Miller, M. W. and Clarkson, T. W., Eds., Charles C Thomas, Springfield, IL, 187–208.
Berlin, M. and Ulberg, S. (1963a), *Arch. Environ. Health,* 6, 589–601.
Berlin, M. and Ulberg, S. (1963b), *Arch. Environ. Health,* 6, 602–609.
Berlin, M. and Ulberg, S. (1963c), *Arch. Environ. Health,* 6, 610–616.
Bertine, K. K. and Goldberg, E. D. (1971), *Science,* 173, 233–235.
Bhaattacharya, S. K., Robson, J. S., and Stewart, C. P. (1955), *Biochem. J.,* 60, 696–702.
Biological Data Book, Vol. 3 (1974), Federation of American Societies of Experimental Biology, Bethesda, MD.
Billings, C. E. and Matson, W. R. (1972), *Science,* 176, 1232–1233.
Birge, W. J., Black, J. A., Westerman, A. G., and Hudson, J. E. (1979), *The Biogeochemistry of Mercury in the Environment,* Nriagu, J. O., Ed., Elsevier/North-Holland, 629–655.
Birke, G., Johnels, A. G., Plantin, L.-O., Sjöstrand, B., and Westermark, T. (1967), *Läkartidningen,* 64, 3628–3637.
Blaustein, A. R., Wake, D. B., and Sousa, W. P. (1994), *Conversation Biol.,* 8, 60–71.
Bligh, E. G. (1971), Proc. Symp. on Mercury in Man's Environment, February 15–16, Royal Society of Canada, Ottawa, 73–90.
Borg, K., Wanntorp, H., Erne, K., and Hanko, E. (1966), *J. Appl. Ecol.,* 3 (Suppl.), 171–172.
Bormann, G., Henke, G., Alfes, H., and Mollmann, H. (1970), *Arch. Toxikol.,* 26, 203–209.
Boulanger, Y., Goodman, C. M., Forte, C. P., Fesik, S. W., and Armitage, I. M. (1983), *Proc. Natl. Acad. Sci. U.S.A.,* 80, 1501–1505.
Braun, J. L. and Snyder, D. R. (1973), *Bull. Psychon. Soc.,* 1, 419–420.
Bremner, I. (1987), *Experientia,* 52, 81–107.
Brown, J. R. and Kulkarni, M. V. (1967), *Med. Serv. J. Can.,* 23, 786–808.
Brown, R. V., Zenick, H., Cox, V., and Fahim, M. S. (1972), *Fed. Proc.,* 31, 552.
Brownlee, R. T. C., Canty, A. J., and McKay, M. F. (1978), *Aust. J. Chem.,* 31, 1933–1937.
Brubaker, P. E., Klein, R., Herman, S. P., Lucier, G. W., Alexander, L. T., and Long, M. D. (1973), *Exp. Mol. Pathol.,* 18, 263–280.
Brubaker, P. E., Lucier, G. W., and Klein, R. (1971), *Biochem. Biophys. Res. Commun.,* 44, 1552–1558.
Bucci, E. and Fronticelli, C., (1965), *J. Biol. Chem.,* 240, PC551–552.
Buell, M. V. (1935), *J. Biol. Chem.,* 108, 273–283.
Buncel, E., Norris, A. R., Racz, J. W., and Taylor, S. E. (1981), *Inorg. Chem.,* 20, 98–103.
Calflisch, C. R. and DuBose, T. D., Jr. (1991), *J. Toxicol. Environ. Health,* 32, 49–58.
Cambier, J. C. and Monroe, J. G. (1983), *Methods in Enzymology,* Vol. 103, Conn, P. M., Ed., Academic Press, New York, 227–245.
Cantor, M. O. (1951), *J.A.M.A.,* 146, 560–561.
Carrabine, J. A. and Sundaralingman, M. (1971), *Biochemistry,* 10, 292–299.
Carrico, L. C. (1986), *The Minerals Yearbook*. Vol. 1, Bureau of Mines, U.S. Department of Interior, Washington, D.C., 659–665.
Carty, A. J. and Malone, S. F. (1979), *The Biogeochemistry of Mercury in the Environment,* Nriagu, J. O., Ed., Elsevier/North-Holland, 433–479.
Cember, H., Gallagher, P., and Faulkner, A. (1968), *Am. Industr. Hyg. Assoc. J.,* 29, 233–237.
Chalkley, H. W. (1951), *Ann. N.Y. Acad. Sci.,* 51, 1303–1310.
Chang, L. W. (1977), *Environ. Res.,* 14, 329–373.
Chang, L. W. and Desnoyers, P. (1978), *J. Environ. Pathol. Toxicol.,* 1, 569–579.
Chang, L. W., Desnoyers, P., and Hartmann, H. A. (1972), *J. Neuropathol. Exp. Neurol.,* 31, 489–501.
Chang, L. W., Gilbert, M., and Sprecher, J. (1978), *Environ Res.,* 17, 356–366.

Chang, L. W. and Hartmann, H. A. (1972a), *Acta Neuropathol.,* 20, 122–138.
Chang, L. W. and Hartmann, H. A. (1972b), *Acta Neuropathol.,* 20, 316–334.
Chang, L. W. and Hartmann, H. A. (1972c), *Acta Neuropathol.,* 21, 179–186.
Chang, L. W. and Hartmann, H. A. (1972d), *Exp. Neurol.,* 35, 122–137.
Chang, L. W. and Hartmann, H. A. (1972e), *Acta Neuropathol.,* 21, 179–184.
Chang, L. W., Hartmann, H. A., and Desnoyers, P. (1972a), *J. Neurophathol. Exp. Neurol.,* 31, 489–501.
Chang, L. W., Martin, A., and Hartmann, H. A. (1972b), *Exp. Neurol.,* 37, 62–67.
Chang, L. W. and Suber, R. (1982), *Bull. Environ. Contam. Toxicol.,* 29, 285–289.
Chen, W.-J., Body, R. L., and Mottet, N. K. (1979), *Teratology,* 20, 31–36.
Choi, B. H. (1991), *Advances in Mercury Toxicology,* Suzuki, T., Imura, N., and Clarkson, T. W., Eds., Plenum Press, 315–337.
Choi, B. H., Lapham, L., Amin-Zaki, L., and Saleem, T. (1978), *J. Neuropathol. Exp. Neurol.,* 37, 719–733.
Childs, E. A. (1973), *Arch. Environ. Health,* 27, 50–53.
Clarkson, T. W. (1971), *Food Cosmet. Toxicol.,* 9, 229–243.
Clarkson, T. W. (1972), *Annu. Rev. Pharmacol.,* 12, 375–406.
Clarkson, T. W. (1973), *Mercury in the Western Environment,* Buhler, D. R., Ed., Continuing Education Publications, Oregon State University, Corvallis, 332–354.
Clarkson, T. W. (1994), *Mercury Pollution. Integration and Synthesis,* Watras, C. J. and Huchabee, J. W., Eds., Lewis Publishers, Boca Raton, FL, 631–641.
Clarkson, T. W. and Magos, L. (1966), *Biochem. J.,* 99, 62–70.
Costa, M., Christie, N. T., Cantoni, O., Zelikoff, J. T., Wang, X. W., and Rossman, T. G. (1991), *Advances in Mercury Toxicology,* Suzuki, T., Imura, N., and Clarkson, T. W., Eds., Plenum Press, New York, 255–273.
Creighton, T. E. (1984), *Proteins. Structure and Molecular Properties,* W. H. Freeman.
Cremer, J. (1962), *J. Neurochem.,* 9, 289–298.
Cullis, A., Muirhead, H., Perutz, M. F., Rossman, M. G., and North, A. C. T. (1962), *Proc. R. Soc. London Ser. A.,* 265, 161–187.
Curley, A., Sedlak, V. A., Girling, E. F., Hawk, R. E., Barthel, W. F., Pierce, P. E., and Likosky, W. H. (1971), *Science,* 172, 65–67.
Dagg, C. P. (1963), *The Mouse,* Green, D., Ed., McGraw Hill, New York, 309–328.
Dalgaard-Mikkelsen, S. (1964), *Nord. Hyg. T.,* 50, 34–36.
Daniel, J. W., Gage, J. C., and Lefevre, P. A. (1972), *Biochem. J.,* 129, 961–967.
Daston, G. P., Rhenberg, B. F., Hall, L. L., and Kavlock, R. J. (1986), *Toxicol. Appl. Pharmacol,* 85, 39–48.
De Flora, S., Bennicelli, C., and Bagnasco, M. (1994), *Mutat. Res.,* 317, 57–79.
Delnomdedieu, M. and Allis, J. W. (1993), *Chem.-Biol. Interact.,* 88, 71–87.
Delnomdedieu, M., Boudou, A., Desmazès, J. P., and Georgescauld, D. (1989a), *Biochim. Biophys. Acta,* 986, 191–199.
Delnomdedieu, M., Boudou, A. Desmazès, J. P., Faucon, J. F., and Georgescauld, D. (1989b), *Heavy Metals in the Environment,* Vol. 1, Vernet, J. P., Ed., CEP Consultants, Edinburgh, 578–581.
Delnomdedieu, M., Boudou, A., Georgescauld, D., and Dufourc, E. J. (1992), *Chem-Biol. Interact.,* 81, 243–269.
Delnomdedieu, M., Georgescauld, D., Boudou, A., and Dufourc, E. J. (1990), *Bull. Magn. Reson.,* 11, 420.
De Ropp, R. S. and Snedecker, E. H. (1960), *Anal. Biochem.,* 1, 424–432.
Diamond, S. S. and Sleight, S. D. (1972), *Toxicol. Appl. Pharmacol.,* 23, 197–207.
Dinman, B. D. and Hecker, L. H. (1972), *Environmental Mercury Contamination,* Ann Arbor Science, Chelsea, MI, 290–301.
Doi, R. (1991), *Advances in Mercury Toxicology,* Suzuki, T., Imura, N., and Clarkson, T. W., Eds., Plenum Press, New York, 77–98.
Douglas, G. W. and Mortensen, R. A. (1956), *J. Biol. Chem.,* 222, 581–585.
Dunn, J. D., Clarkson, T. W., and Magos, L. (1981), *Science,* 213, 1123–1125.
Eccles, C. U. and Annau, Z. (1982a), *Neurobehav. Toxicol. Teratol.,* 4, 371–376.
Eccles, J. D. and Annau, Z. (1982b), *Neurobehav. Toxicol. Teratol.,* 4, 377–382.
Edwards, S. and Westerfield, W. W. (1952), *Proc. Soc. Exp. Biol. Med.,* 79, 57–59.
Ehmann, W. D., Kasarskis, E. J., and Markesbery, W. R. (1994), *Mercury Pollution, Integration and Synthesis,* Watras, C. J. and Huckabee, J. W., Eds., Lewis Publishers, Boca Raton, FL, 651–663.
Eichhorn, G. L. and Clark, P. (1963), *J. Am. Chem. Soc.,* 85, 4020–4024.
Endou, H. and Jung, K. Y. (1991), *Advances in Mercury Toxicology,* Suzuki, T., Imura, N., and Clarkson, T. W., Eds., Plenum Press, New York, 299–314.
Engleson, G. and Herner, T. (1952), *Acta Paediatr. Scand.,* 41, 289–294.
Enoki, Y. and Tonita, S. (1965), *J. Mol. Biol.,* 11, 144–145.
Enzymatic Basis of Detoxication, Vol. 1, 2, (1980), Jacoby, W. B., Ed., Academic Press, New York.
Evans, H. L. and Kostyniak, P. S. (1967), *Fed. Proc.,* 31, 561.
Evans, H. L., Laites, V. G., and Weiss, B. (1975), *Fed. Proc.,* 34, 1858–1867.
Fehling, C., Abdulla, M., Brun, A., Dictor, M., Schultz, A., and Skerfving, S. (1975), *Toxicol. Appl. Pharmacol.,* 33, 27–37.
Fimreite, N. and Reynolds, L. M. (1973), *J. Wildlife Manag.,* 37, 62–68.
Fiskesjö, G. (1970), *Hereditas,* 64, 142–146.
Flanagan, W. J. and Oken, D. E. (1965), *J. Clin. Invest.,* 44, 449–457.

Friberg, L. (1959), *A. M. A. Arch. Industr. Health,* 20, 42–49.
Gabay, S., Cabral, A. M., and Healy, R. (1968), *Proc. Soc. Exp. Biol. Med.,* 127, 1081–1085.
Ganther, H. E., Gondie, C., Sunde, M. L., Kopecky, M. J., Wanger, P. A., Oh, S. H., and Hoekstra, W. G. (1972), *Science,* 175, 1121–1124.
Garcia, J. D., Yang, M. G., Wang, J. H. C., and Belo, P. S. (1974), *Proc. Soc. Exp. Biol. Med.,* 147, 224–231.
Giblin, F. J. and Massaro, E. J. (1975), *Toxicology,* 5, 243–254.
Glomsky, C. A., Brody, H., and Pillay, K. K. S. (1971), *Nature,* 232, 200–201.
Green, H. and Goldberg, B. (1965), *Proc. Natl. Acad. Sci. U.S.A.,* 53, 1360–1365.
Green, H. and Hamerman, D. (1964), *Nature,* 201, 710.
Greene, R. M. and Pratt, R. M. (1976), *J. Embryol. Exp. Morphol.,* 36, 225–245.
Gruenwedel, D. W. (1985), *J. Inorg. Biochem.,* 25, 109–120.
Gruenwedel, D. W. and Cruikshank, M. K. (1979), *Biochem. Pharmacol.,* 28, 651–655.
Gruenwedel, D. W. and Cruikshank, M. K. (1990), *Biochemistry,* 29, 2110–2116.
Gruenwedel, D. W. and Davidson, N. (1966), *J. Mol. Biol.,* 21, 129–144.
Gutknecht, J. (1981), *J. Membr. Biol,* 61, 61–66.
Hammond, A. L. (1971), *Science,* 172, 788–789.
Han, L. S. and Li, N. C. (1970), *J. Am. Chem. Soc.,* 92, 4823–4827.
Harada, Y. (1968a), *Minimata Disease,* Study Group of Minimata Disease, Kumamato University, Japan, 37–72.
Harada, Y. (1968b), *Minimata Disease,* Study Group of Minimata Disease, Kumamoto University, Japan, 94–117.
Hargreaves, R. J., Foster, J. R., Pelling, D., Moorhouse, S. R., Gangolli, S. D., and Rowland, I. R. (1985), *Neuropathol. Appl. Neurobiol.,* 11, 383–401.
Harris, S. B., Wilson, J. G., and Printz, R. H. (1972), *Teratology,* 6, 139–142.
Häsänen, E. and Sjöblom, V. (1967), *Suomen kalatalous (Finl. Fisk.),* 36, 5–24.
Health Effects of Methylmercury (1981), MARC Rep. No. 24, Monitoring and Assessment Research Center, Chelsea College, University of London.
Hellerman, L. (1937), *Physiol. Rev.,* 17, 454–484.
Herman, S. P., Klein, R., Talley, F. A., and Krigman, M. R. (1973), *Lab. Invest.,* 28, 104–118.
Hirota, Y., Yamaguchi, S., Shimojoh, N., and Sano, K.-I. (1980), *Toxicol. Appl. Pharmacol.,* 53, 174–176.
Hoch, B. S., Gorfien, P. C., Linzer, D., Fusco, M. J., and Levine, S. D. (1989), *Am. J. Physiol.,* 256, F948–953.
Holden, A. V. (1972), *Mercury Contamination in Man and his Environment,* Tech. Rep. Ser. No. 137, International Atomic Energy Agency, Vienna, 143–168.
Hudson, C. D. and Shapiro, B. L. (1973), *Arch. Oral. Biol.,* 18, 77–84.
Hughes, J. A. and Annau, Z. (1976), *Pharmacol. Biochem. Behav.,* 4, 385–391.
Hughes, J. A., Annau, Z., and Goldberg, A. A. (1972), *Fed. Proc.,* 31, 552.
Hughes, J. A. and Sparber, S. B. (1978), *Pharmacol. Biochem. Behav.,* 8, 365–375.
Hughes, W. L. (1957), *Ann. N.Y. Acad. Sci.,* 65, 454–460.
Hunter, D., Bromford, R. R., and Russell, D. S. (1940), *Q. J. Med.,* 33 193–219.
Hunter, D. and Russell, D. S. (1954), *J. Neurol. Neurosurg. Psychiatr.,* 17, 235–241.
Idoyaga-Vargas, V., Nasjleti, C. E., and Azcurra, J. M. (1972), *J. Embryol. Exp. Morphol.,* 27, 413–430.
Ikeda, Y., Tobe, M., Kobayashi, K., Suzuki, S., Kawasaki, Y., and Yonemaru, H. (1973), *Toxicology,* 1, 361–375.
Impact of the Environment on Reproductive Health (1993), *Environ. Health Perspect.,* 101 (Suppl. 2.), July, 1993.
Importance of Chemical "Speciation" in Environmental Processes (1986), Bernhard, M., Brinckman, F. E., and Sadler, P. J., Eds., Dehlem Workshop Reports, Vol. 33, Springer-Verlag.
Inouye, M. (1991), *Advances in Mercury Toxicology,* Suzuki, T., Imura, N., and Clarkson, T. W., Eds., Plenum Press, New York, 339–354.
Inouye, M., Hoshino, K., and Murakami, U. (1972), *Annu. Rep. Res. Inst. Environ. Med. Nagoya University,* 19, 69–74.
International Committee Report, *Arch. Environ. Health,* 19, 891–905.
Irukayama, K., Kondo, T., Ushikusa, S., Fujiki, M., and Tajima, S. (1965), *Jpn. J. Hyg.,* 20, 11–21.
Jacobi, M. H., Glassman, H. N., and Parpart, A. K. (1936), *J. Cell. Comp. Physiol.,* 7, 197–225.
Jacobson, M. S., Chen., T., and Okata, G. T. (1972), *Fed. Proc.,* 31, 561.
Jakubowski, K., Piotrowski, J., and Trojanowska, B. (1970), *Toxicol. Appl. Pharmacol.,* 16, 743–753.
Jandl, J. H. and Simmons, R. L. (1957), *Br. J. Haematol.,* 79, 19–38.
Jansen, H. T., Cooke, P. S., Porcelli, J., Liu, T.-C., and Hansen, L. G. (1993), *Reprod. Toxicol.,* 7, 237–248.
Jernelöv, A. (1973), *Mercury, Mercurials and Mercaptans,* Miller, M. W. and Clarkson, T. W., Eds., Charles C Thomas, Springfield, IL, 315–324.
Jocelyn, P. C. (1992), *Biochemistry of the SH Group,* Academic Press, New York.
Johansson, F., Ryhage, R., and Westöö, G. (1970), *Acta Chem. Scand.,* 24, 2349–2354.
Johnels, A. G. and Westermark, T. (1969), in *Chemical Fallout., Current Research on Persistent Pesticides,* Charles C Thomas, Springfield, IL, 221–239.
Johnson, H. R. M. and Koumides, O. (1967), *Br. Med. J.,* 1, 340–341.
Joselow, N. M., Louria, D. B., and Browder, A. A. (1972), *Ann. Intern. Med.,* 76, 119–130.
Jugo, S. (1976), *Health Phys.,* 30, 240–241.
Joensuu, O. I. (1971), *Science,* 172, 1027–1028.

Kägi, J. H. R. and Vallee, B. L. (1960), *J. Biol. Chem.,* 235, 3460–3465.
Kasuya, M. (1975), *Toxicol. Appl. Pharmacol.,* 32, 347–354.
Kasuya, M. (1976), *Toxicol. Appl. Pharmacol.,* 35, 11–20.
Kelce, W. R., Monosson, E., Gamcsik, M. P., Laws, S. C., and Gray, L. E., Jr. (1994), *Toxicol. Appl. Pharmacol.,* 126, 276–285.
Kelman, B. J. and Sasser, L. B. (1977), *Toxicol. Appl. Pharmacol.,* 39, 119–127.
Kerper, L., Ballatori, N., and Clarkson, T. W. (1992), *Am. J. Physiol.,* 262, R761–765.
Khera, K. S. (1973), *Teratology,* 8, 293–304.
Khera, K. S. and Nera, E. A. (1971), *Teratology,* 44, 233.
Khera K. S. and Tobacova, S. A. (1973), *Food Cosmet. Toxicol.,* 11, 245–254.
Kim, S. U. (1971), *Exp. Neurol.,* 32, 237–246.
Klaassen, C. D. (1975), *Toxicol. Appl. Pharmacol.,* 33, 356–365.
Klein, R., Herman, S. P., Brubaker, P. E., Lucier, G. W., and Krigman, M. R. (1972), *Arch Pathol.,* 93, 408–418.
Kobayashi, H., Yuyama, A., Matsusaka, N., Takeno, K., and Yanagiya, I. (1981), *Jpn. J. Pharmacol.,* 31, 711–718.
Koerker, R. L. (1980), *Toxicol. Appl. Pharmacol.,* 53, 458–469.
Kosta, L., Byrnne, A. R., and Selenko, V. (1975), *Nature,* 254, 238–239.
Kurland, L. T., Faro, S. N., and Siedler, H. (1960), *World Neurol.,* 1, 370–395.
Kuwahara, S. (1970), *J. Kumamoto Med. Soc.,* 44, 214–221.
Lajtha, A., Berl, S., and Waelsch, H. (1959), *J. Neurochem.,* 3, 322–332.
Lee, M., Chan, K. K. S., Sairenji, E., and Nickumi, T. (1979), *Environ. Res.,* 19, 39–48.
Léonard, A. (1988), *Mutat. Res.,* 198, 321–326.
Léonard, A., Jacquet, P., and Lauwerys, R. R. (1983), *Mutat. Res.,* 114, 1–18.
Löforth, G. (1969), *Ecol. Res. Comm. Bull. No. 4,* Swedish Natural Science Research Council.
Longo, L. D., Yuen, P., and Gusseck, D. J. (1973), *Nature,* 243, 531–533.
Looney, J. M. (1924), *Am. J. Psychiatr.,* 81, 29–39.
Lorernte, C. A., DePaola, D. P., Drummond, J. F., and Miller, S. A. (1974), *J. Dent. Res.,* 53, 65.
Lown, B. A., Morganti, J. B., Stineman, C. H., and Massaro, E. J. (1977), *Gen. Pharmacol.,* 8, 97–101.
Lucier, G., McDaniel, O., Brubaker, P., and Klein, R. (1971), *Chem. Biol. Interact.,* 4, 265–280.
Magos, L. (1991), *Advances in Mercury Toxicology,* Suzuki, T., Imura, N., and Clarkson, T. W., Eds., 289–297.
Magos, L., Clarkson, T. W., and Hudson, A. R. (1989), *Biochim. Biophys. Acta,* 991, 85–89.
Mansour, M. M., Dyer, N. C., Hoffman, L. H., Davies, J., and Brill, A. B. (1974), *Am. J. Obstet. Gynecol.,* 119, 557–562.
Margoshes, M. and Vallee, B. L. (1957). *J. Am. Chem. Soc.,* 79, 4813–4814.
Martin, H. and McIlwain, H. (1959), *Biochem. J.,* 71, 275–280.
Massaro, E. J., Grose, E. C., Hatch, G. E., and Slade, R. (1988), *Toxicology of the Lung,* Gardner, D. E., Crapo, J., and Massaro, E. J., Eds., Raven Press, New York, 201–218.
Massaro, E. J., Yaffe, S. J., and Thomas, C. C., Jr. (1974), *Life Sci.,* 14, 1939–1948.
Mato, M., Aikawa, E., and Katahira, M. (1966), *Gunma J. Med. Sci.,* 153, 46–56.
Matsumoto, H., Koya, G., and Takeuchi, T. (1965), *J. Neuropathol. Exp. Neurol.,* 24, 563–574.
Meister, A. and Anderson, M. E. (1983), *Annu. Rev. Biochem.,* 52, 711–760.
Mercury Contamination in Man and his Environment, Tech. Rep. Ser. No. 137, International Atomic Energy Agency, Vienna, 1972.
Mercury Contamination: A Human Tragedy (1977), D'Itri, P. A. and D'Itri, F. M., John Wiley & Sons, New York.
Mercury in the Environment (1972), Freiberg, L. and Vostal, J., Eds., CRC Press, Cleveland.
Mercury in the Western Environment (1973), Buhler, D. R., Ed., Continuing Education Publications, Oregon State University, Corvallis.
Methyl Mercury in Fish (1971), *Nord. Hyg. Tidskr.,* Suppl. 4., Stockholm.
Mercury Pollution: Integration and Synthesis (1974), Watras, C. J. and Huckabee, J. W., Eds., Lewis Publishers, Boca Raton, FL.
Miettinen, J. K. (1973), *Mercury, Mercurials and Mercaptans,* Miller, M. W. and Clarkson, T. W., Eds., Charles C Thomas, Springfield, IL, 233–243.
Miller, V. L., Klavano, P. A., Jerstad, A. C., and Csonka, E. (1961), *Toxicol. Appl. Pharmacol.,* 3, 459–468.
Minimata Disease (1968), Study Group of Minimata Disease, Kumamoto University, Japan.
Minimata Disease (1977), Tsubaki, T. and Irukayama, K., Eds., Elsevier Scientific Publishing Co. 1977.
Misumi, J. (1979), *Kumamoto Med. J.,* 32, 15–22.
Miura, K., Inokawa, M., and Imura, N. (1984), *Toxicol. Appl. Pharmacol.,* 73, 218–231.
Miura, K. and Imura, N. (1991), *Advances in Mercury Toxicology,* Suzuki, T., Imura, N., and Clarkson, T. W., Eds., Plenum Press, New York, 241–273.
Miyakawa, T. and Deshimaru, M. (1969), *Acta Neuropathol.,* 14, 126–136.
Miyakawa, T., Deshimaru, M., Sumiyoshi, S., Teraoka, A., Udo, N., and Miyakawa, K. (1969), *Psychiatr. Neurol. Jpn.,* 71, 757–763.
Miyakawa, T., Deshimaru, M., Sumiyoshi, S., Teraoka, A., Udo, N., Hatteri, E., and Tatetsu, S. (1970), *Acta Neuropathol.,* 15, 45–55.
Miyama, T., Minowa, K., Seki, H., Tamura, Y. Mizoguchi, I. Ohi, G., and Suzuki, T. (1983), *Arch. Toxicol.,* 52, 173–181.

Møller-Madsen, B. (1990), *Toxicol. Appl. Pharmacol.,* 103, 303–323.
Mottet, N. K. (1974), *Teratology,* 10, 173–181.
Mulvihill, J. (1972), *Science,* 176, 132–137.
Murakami, U. (1972a), *Annu. Rep. Inst. Environ. Med. Nagoya University,* 19, 61–68.
Murakami, V. (1972b), *Drugs and Fetal Development,* Klingberg, M. A., Abramovici, A., and Chemke, J., Eds., Plenum Press, New York, 309–336.
Naganuma, A. and Imura, N. (1979), *Toxicol. Appl. Pharmacol.,* 47, 613–616.
Narahashi, T., Arakawa, O., and Nakahiro, M. (1991), *Advances in Mercury Toxicology,* Suzuki, T., Imura, N., and Clarkson, T. W., Eds., Plenum Press, New York, 191–207.
Noble, R. W. (1971), *J. Biol. Chem.,* 246, 2972–2976.
Nobunaga, T., Satoh, S., and Suzuki, T. (1979), *Toxicol. Appl. Pharmacol.,* 47, 79–88.
Nomura, S. (1968), *Minimata Disease,* Study Group of Minimata Disease, Kumamoto University, Japan, 5–35.
Nolen, G. A., Buchler, E. V., Geil, R. G., and Goldenthal, E. I. (1972), *Toxicol. Appl. Pharmacol.,* 23, 222–237.
Nordberg, G. F. and Skerfving, S. (1972), *Mercury in the Environment,* Friberg, L. and Vostal, J., Eds., CRC Press, Cleveland, 29–90.
Norseth, T. (1969), *Chemical Fallout,* Miller, M. W. and Berg, G. G., Eds., Charles C Thomas, Springfield IL, 408–419.
Norseth, T. (1972), *Acta Pharmacol. Toxicol.,* 31, 138–148.
Norseth, T. and Clarkson. T. W. (1970a), *Biochem. Pharmacol.,* 19, 2775–2783.
Norseth, T. and Clarkson, T. W. (1970b), *Arch. Environ. Health,* 21, 717–727.
Null, D. H., Gartside, P. S., and Wei, E. (1973), *Life Sci.,* 12, 65–72.
Ogata, M., Kenmotsu, K., Hirota, N., Meguro, T., and Aikoh, H. (1987), *Arch. Environ. Health,* 42, 26–30.
Oharazawa, H. (1968), *Jpn. J. Obstet. Gynecol.,* 20, 1479–1487.
Ohi, G., Nishigaki, S., Seki, H., Tamura, Y., Mizoguchi, I., Yagyu, H., and Nagashima, K. (1978), *Environ. Res.,* 16, 353–359.
Olson, F. C. and Massaro, E. J. (1977a), *Teratology,* 16, 187–194.
Olson, F. C. and Massaro, E. J. (1977b), *Toxicol. Appl. Pharmacol.,* 39, 263–273.
Olson, F. C. and Massaro, E. J. (1980), *Teratology,* 22, 155–166.
Olson, K. R., Bergman, H. L., and Fromm, P. O. (1973), *J. Fish. Res. Bd. Can.,* 30, 1293–1299.
Omata, S., Horgome, T., Momose, Y., Kambayashi, M., Mochizuki, M., and Sugano, H. (1980b), *Toxicol. Appl. Pharmacol.,* 56, 207–215.
Omata, S., Sato, M., Sakimura, K., and Sugano, H. (1980a), *Arch. Toxicol.,* 44, 231–241.
Omata, S., Terui, Y., Kasama, H., Ichimura, T., Horigome, T., and Sugano, H. (1991), *Advances in Mercury Toxicology,* Suzuki, T., Imura, N., and Clarkson, T. W., Eds., PLenum Press, New York, 223–240.
Östlund, K. (1969), *Acta Pharmacol.,* 27, Suppl. 1, 1–132.
Otten, J., Johnson, G. S., and Pastan, I. (1972), *J. Biol. Chem.,* 247, 7082–7087.
Page, E. W. (1957), *Am. J. Obstet. Gynecol.,* 74, 705–718.
Pascoe, G. A. and Reed, D. J. (1989), *Free Radical Biol. Med.,* 6, 209–224.
Passow, H. (1970), *Effects of Metals on Cells, Subcellular Elements, and Macromolecules,* Maniloff, J., Coleman, J. R., and Miller, M. M., Eds., Charles C Thomas, Springfield, 291–344.
Passow, H., Rothstein, A., and Clarkson, T. W. (1961), *Pharmacol. Rev.,* 13, 185–224.
Paterson, R. A. and Usher, D. R. (1971), *Life Sci.,* 10, 121–128.
Peakall, D. B. and Lovett, R. J. (1972), *Bioscience,* 22, 20–25.
Pechmann, J. H. K. and Wilbur, H. M. (1994), *Hepettologica,* 50, 65–84.
Peters, H. D., Karmel, K., Padberg, D., Schoenhoefer, P. S., and Dinnendahl, V. (1974), *Polish J. Pharmacol. Pharm.,* 26, 41–49.
Pillay, K. K. S., Thomas, C. C., Sondel, J., and Hyche, C. (1971), *Anal. Chem.,* 43, 1419–1425.
Post, E. M., Yang, M. C., and King, J. A. (1972), *Fed. Proc.,* 31, 725.
Post, E. M., Yang, M. G., King, J. A., and Sanger, V. L. (1973), *Proc. Soc. Exp. Biol. Med.,* 143, 1113–1116.
Prasad, K. N. and Ramanujam, S. (1980), *Environ. Res.,* 21, 343–349.
Pratt, R. M., Goggins, J. R., Wilk, A. L., and King, C. T. G. (1973), *Dev. Biol.,* 32, 230–237.
Pratt, R. M. and King, C. T. C. (1971), *Arch. Oral. Biol.,* 16, 1181–1185.
Pratt, R. M. and Martin, G. R. (1975), *Proc. Natl. Acad. Sci. U.S.A.,* 72, 874–877.
Pratz, J., Ripoche, P., and Corman, B. (1986), *Biochim. Biophys. Acta,* 856, 259–266.
Pulido, P., Kägi, J. R. H., and Vallee, B. L. (1966), *Biochemistry,* 5, 1768–1777.
Rabenstein, D. L. and Isab, A. A. (1982), *Biochim. Biopys. Acta,* 721, 374–384.
Rabenstein, D. L. and Reid, R. S. (1984), *Inorg. Chem.,* 23, 1246–1250.
Raeder, M. G. and Snekvik, E. (1949), *K. Nor. Videnskab. Selsk. Forh.,* 21, 102–109.
Ramel, C. (1969a), *Hereditas.,* 61, 208–230.
Ramel, C. (1969b), *J. Jpn. Med. Assoc.,* 61, 1072–1077.
Ramel, C. (1971), *Mercury in the Environment,* Friberg, L. and Vostal, J., Eds., CRC Press, Cleveland, OH, 169–181.
Reddy, C. C., Scholz, R. W., and Massaro, E. J. (1981), *Toxicol. Appl. Pharmacol.,* 61, 460–468.
Reid, R. S. and Podanyi, B. (1988), *J. Inorg. Biochem.,* 61, 61–66.
Renfro, J. L., Schmudt-Nielsen, B., Miller, D., Benos, D., and Allen, J. (1974), *Pollution and Physiology of Marine Organisms,* Vernberg, F. J. and Vernberg, W. B., Eds., Academic Press, New York, 101–122.

Reynolds, M. L. and Young, M. (1971), *J. Physiol.,* 214, 583–597.
Reynolds, W. A. and Pitkin, R. M. (1975), *Proc. Soc. Exp. Biol. Med.,* 148, 523–526.
Riggs, A. (1961), *J. Biol. Chem.,* 236, 1948–1954.
Riggs, A. F. and Wolbach, R. A. (1956), *J. Gen. Physiol.,* 39, 585–605.
Rissanen, K. and Miettinen, J. K. (1972), *Mercury Contamination in Man and His Environment,* Tech. Rep. Ser. No. 137, International Atomic Energy Agency, Vienna, 5–34.
Robert, J. M. and Rabenstein, D. L. (1991), *Anal. Chem.,* 63, 2674–2679.
Robinson, N. and Williams, C. B. (1965), *Clin. Chem. Acta,* 12, 311–317.
Rocha, J. B. T., Freitas, A. J., Marqyes, M. B., Pereira, M. E., Emanuelli, T., and Souza, D. O. (1993), *Braz. J. Med. Biol. Res.,* 26, 1077–1083.
Röök, O., Lundgren, K.-D., and Swensson, Ä. (1954), *Acta Med. Scand.,* 150, 131–137.
Rosemeyer, M. A. and Huehns, E. R. (1967), *J. Mol. Biol.,* 25, 253–273.
Rosenthal, E. and Sparber, S. B. (1972), *Life Sci.,* 11, 883–892.
Rothstein, A. (1959), *Fed. Proc.,* 18, 1026–1035.
Rothstein, A. (1973), *Mercury, Mercurials and Mercaptans,* Miller, M. W. and Clarkson, T. W., Eds., Charles C Thomas, Springfield, IL, 68–95.
Rothstein, A. (1981), *Prog. Clin. Biol. Res.,* 51, 105–131.
Rothstein, A. and Hayes, A. D. (1960), *J. Pharmacol. Exp. Ther.,* 130, 166–176.
Rowland, A. S., Baird, D. D., Weinberg, C. R., Shore, D. L., Shy, C. M., and Wilcox, A. J. (1994), *Occup. Environ. Med.,* 51, 28–34.
Rozalski, M. and Wierzbicki, R. (1983), *Biochem. Pharmacol.,* 32, 2124–2126.
Rozynkowa, D. and Raczkiewicz, M. (1977), *Mutation Res.,* 56, 185–191.
Sager, P. R., Aschner, M., and Rodier, P. M. *Dev. Brain Res.,* 12, 1–11.
Sager, P. R. (1988), *Toxicol. Appl. Pharmacol.,* 94, 473–486.
Saito, M., Osono, T., Watanabe, J., Yamamoto, T., Takeuchi, M., Ohyagi, Y., and Katsunuma, H. (1961), *Jpn. J. Exp. Med.,* 31, 277–290.
Saito, H. and Uzman, B. G. (1971), *Biochem. Biophys. Res. Commun.,* 43, 728–732.
Salvaterra, P., Lown, B., Morganti, J., and Massaro, E. J. (1973), *Acta Pharmacol. Toxicol.,* 33, 177–190.
Salvaterra, P., Lown, B., Morganti, J., and Massaro, E. J. (1975), *Toxicol. Appl. Pharmacol.,* 32, 432–442.
Sanner, T. and Pihl, A. (1962), *Biochim. Biophys. Acta.,* 62, 171–172.
Sarafian, T. A., Cheung, M. K., and Verity, M. A. (1984), *Neuropathol. Appl. Neurobiol.,* 10, 85–100.
Sato, T. and Nakamura, Y. (1991), *Advances in Mercury Toxicology,* (Suzuki, T., Imura, N., and Clarkson, T. W., Eds., Plenum Press, New York, 335–365.
Schneider, H. and Dancis, J. (1974), *Am. J. Obst. Gynecol.,* 120, 1092–1098.
Shapiro, H. M. (1995), *Practical Flow Cytometry,* 3rd ed., Wiley-Liss, New York.
Sharma, A. and Talukder, G. (1987), *Mutat. Res.,* 198, 321–326.
Shuler, C. F., Halpern, D. E., Guo, Y., and Sank, A. C. (1992), *Dev. Biol.,* 154, 318–330.
Simpson, R. B. (1961), *J. Am. Chem. Soc.,* 83, 4711–4717.
Simpson, R. B. (1966), *J. Am. Chem. Soc.,* 86, 2059–2065.
Sillen, L. G. and Martell, A. E. (1964), *Stability Constants of Metal-Ion Complexes,* Spec. Pub. No. 17, 2nd ed., Chemical Society, London.
Singerman, A. and Catalina, R. L. (1971), *Proc. XVI Int. Cong. Occup. Health.,* Tokyo, 1969, 554–557.
Skerfving, S. (1972a), *Toxicity of Methylmercury with Special Reference to Exposure via Fish,* Government Publishing House, Stockholm, Appendix III (Stencils).
Skerfving, S. (1972b), *Mercury in the Environment,* Friberg, L. and Vostal, J., Eds., CRC Press, Cleveland, 109–139.
Skerfving, S. (1972c), *Mercury in the Environment,* Friberg, L. and Vostal, J., Eds., CRC Press, Cleveland, 141–168.
Skerfving, S. (1974), *Toxicology,* 2, 3–23.
Skerfving, S. (1991), *Advances in Mercury Toxicology,* Suzuki, S., Imura, N., and Clarkson, T. W., Eds., Plenum Press, New York, 411–437.
Skerfving, S., Hansson, K., and Lindsten, J. (1970), *Arch. Environ. Health,* 21, 133–139.
Skerfving, S., Hansson, K., Mangs, C., Lindsten, J., and Ryman, N. (1974), *Environ. Res.,* 7, 83–98.
Skerfving, S. and Vostal, J. (1972), *Mercury in the Environment,* Friberg, L. and Vostal, J., Eds., CRC Press, Cleveland, 93–107.
Smiley, G. R. (1970), *Arch. Oral Biol.,* 15, 287–296.
Smith, J. C., Clarkson, T. W., Greenwood, M., Von Burg, R., and Laselle, E. (1971), *Fed. Proc.* 30, 221.
Snyder, R. D. (1971), *N. Eng. J. Med.,* 18, 1014–1015.
Somjen, G. G., Herman, S. P., and Klein, R. (1973), *J. Pharmacol. Exp. Ther.,* 186, 579–592.
Spyker, J. M. and Smithberg, M. (1972), *Teratology,* 5, 181–190.
Spyker, J. M., Sparber, S. B., and Goldberg, A. M. (1972), *Science,* 177, 621–623.
Spyker, D. A. and Spyker, J. M. (1977), *Toxicol. Appl. Pharmacol.,* 40, 511–522.
Stock, A. and Cucuel, F. (1934), *Naturwissenschaften,* 22, 390–393.
Storm, D. R. and Gunsales, R. P. (1974), *Nature,* 250, 778–779.

Stopford, W. (1979), *The Biogeochemistry of Mercury in the Environment,* Nriagu, J. O., Ed., Elsevier/North-Holland, 367–397.
Stricks, W. and Kolthoff, I. M. (1953), *J. Am. Chem. Soc.,* 75, 5673–5681.
Su, M.-Q. and Okita, G. T. (1976), *Toxicol. Appl. Pharmacol.,* 38, 207–216.
Sutherland, R. M., Rothstein, A., and Weed, R. I. (1967), *J. Cell. Physiol.,* 69, 185–198.
Suzuki, T. (1969), *Chemical Fallout,* Miller, M. W. and Berg, G. G., Eds., Charles C Thomas, Springfield, IL, 245–257.
Suzuki, T., Matsumoto, N., Miyama, T., and Katsunuma, H. (1967), *Ind. Health.,* 5, 149–155.
Suzuki, T. and Miyama, T. (1971), *Indus. Health,* 9, 51–58.
Suzuki, T., Miyama, T., and Katsunuma, H. (1971), *Bull. Environ. Contam. Toxicol.,* 5, 502–508.
Suzuki, T. and Tanaka, A. (1972), *Mercury in the Environment,* Freiberg, L. and Vostal, J., Eds., CRC Press, Cleveland.
Swensson, Å. (1952), *Acta Med. Scand.,* 143, 365–384.
Swensson, Å. and Ulfvarson, U. (1968), *Acta Pharmacol.,* 26, 273–283.
Syversen, T. L. M. (1974), *Acta Pharmacol. Toxicol.,* 35, 277–283.
Syversen, T. L. M. (1976), *Biochem. Pharmacol.,* 23, 2999–3007.
Takagaki, T. (1968), *J. Neurochem.,* 15, 903–916.
Takeda, Y., Kunugi, T., Terao, T., and Ukita, T. (1968), *Toxicol. Appl. Pharmacol.,* 13, 165–173.
Takeuchi, T. (1968a), *Minimata Disease,* Study Group of Minimata Disease, Kumamoto University, Japan, 141–228.
Takeuchi, T. (1968b), *Minimata Disease,* Study Group of Minimata Disease, Kumamoto University, Japan, 229–252.
Takeuchi, T. (1972a), *Environmental Mercury Contamination,* Hartung, R. and Dinman B. D., Eds., Ann Arbor Science. Ann Arbor, MI, 79–81.
Takeuchi, T. (1972b), *Environmental Mercury Contamination,* Hartung, R. and Dinman, B. D., Eds., Ann Arbor Science, Ann Arbor, MI, 247–289.
Takeuchi, T. and Eto, K. (1975), *Studies on the Health Effects of Alkylmercury in Japan,* T. Tsubaki, Ed., Environment Agency, Japan, 28–62.
Taylor, J. F., Antonini, E., and Wyman, J. (1963), *J. Biol. Chem.,* 238, 2660–2662.
Taylor, S. E., Buncel, E., and Norris, A. R. (1981), *J. Inorg. Biochem.,* 15, 131–141.
Tejning, S. (1970), Rep. 70 05 20, Dept. Occup. Med., University Hospital, S-211 85 Lund, Stencils.
Thomas, D. A. and Smith, J. C. (1979a). *Toxicol. Appl. Pharmacol.,* 47, 547–556.
Thomas, D. A. and Smith, J. C. (1979b), *Toxicol. Appl. Pharmacol.,* 48, 43–47.
Thomas, D. J., Fisher, H. L., Hall, L. L., and Mushak, P. (1982), *Toxicol. Appl. Pharmacol.,* 62, 445–454.
Thomas, D. J., Fisher, H. L., Sumler, M. R., Marcus, A. H., Mushak, P., and Hall, L. L. (1986), *Environ. Res.,* 41, 219–234.
Thomas, D. J., Fisher, H. L., Sumler, M. R., Hall, L. L., and Mushak, P. (1988), *Environ. Res.,* 47, 59–71.
Tohyama, C., Ghaffar, A., Nakano A., Nishimura, N., and Nishimura, H. (1991), *Advances in Mercury Toxicology,* Suzuki, T., Imura, N., and Clarkson, T. W., Eds., Plenum Press, New York 155–165.
Tokuomi, H. (1968), in *Minimata Disease,* Study Group of Minimata Disease, Kumamoto University, Japan, 37–72.
Toxicological Profile for Mercury, Agency for Toxic Substances and Disease Registry, U. S. Public Health Service. 1989.
Tsubaki, T., Hirota, K., Shirakawa, K., Kondo, K., and Sato, T. (1978), *Proc. 1st Int. Congress on Toxicology,* Plaa, G. L. and Duncan, W. A. M., Eds., Academic Press, 339–357.
Ui, J. and Kitamura, S. (1971), *Mar. Pollut. Bull.,* 2, 56–58.
Ulfvarson, U. (1962), *Int. Arch. Gewerbepath.,* 19, 412–422.
Ulfvarson, U. (1969), *Fungicides, An Advanced Treatise,* Vol. 2, Torgeson, D. C., Ed., Academic Press, 303–329.
Underdal, B. (1969), *Nord. Hyg. T.,* 50, 60–63.
U.S. Food and Drug Administration, *News Release,* May 6, 1971.
Vallin, S. (1947), *Svensk Papperstidn.,* 50, 531–533.
VanStevenick, J., Weed, R. I., and Rothstein, A. (1965), *J. Gen. Physiol.,* 48, 617–632.
Vecchi, G., Bergomi, M., and Vivoli, G. (1977), *Riv. Ital. Ig.,* 37, 19–26.
Verity, M. A., Brown, W. J., Cheung, M., and Czer, G. (1977), *J. Neurochem.,* 29, 673–679.
Verity, M. A. and Sarafian, T. (1991), *Advances in Mercury Toxicology,* Suzuki, T., Imura, N., and Clarkson, T. W., Eds., Plenum Press, New York, 209–222.
Verschaeve, L., Kirsch-Volders, M., Hens, L., Susanne, C., Groetenbriel, C., Haustermans, R., Lecomte, A., and Roossels, D. (1976), *Environ. Res.,* 12, 306–316.
Verschaeve, L., Kirsch-Volders, M., and Susanne, C. (1985), *Mutation Res.,* 157, 221–226.
Von Burg, R., Northington, F. K., and Shamoo, A. (1980), *Toxicol. Appl. Pharmacol.,* 53, 285–292.
Vostal, J. (1972), *Mercury in the Environment,* Freiberg, L. and Vostal, J., Eds., CRC Press, Cleveland, 15–27.
Wallace, R. A., Fulkerson, W., Shults, W. D., and Lyon, W. S. (1971), Mercury in the Environment. The Human Element, Environmental Program, Oak Ridge National Laboratory, Oak Ridge, TN, ORNL-NSF-EP-1.
Webb, J. L., *Enzyme and Metabolic Inhibitors* (1966), Vol. 2, Academic Press.
Webb, M. (1979), *The Chemistry, Biochemistry and Biology of Cadmium,* Webb, J. L., Ed., Elsevier/North-Holland, 195.
Weed, R. and Berg, G. G. (1963), *Fed. Proc.,* 22, 213.
Weed, R. F., Eber, J., and Rothstein, A. (1962), *J. Gen. Physiol.,* 45, 395–410.
Weed, R. I. and Rothstein, A. (1961), *J. Gen. Physiol.,* 44, 301.
Wein, R. (1939), *Q. J. Pharm. Pharmacol.,* 12, 212.
Weiss, H. V., Koide, M., and Goldberg, E. D. (1971), *Science,* 174, 692–694.

Welch, S. O. and Soares, J. H., Jr. (1976), *Nutr. Rep. Int.,* 13, 43–51.
Westöö, G. (1967), *Acta Chem. Scand.,* 21, 1790–1800.
Westöö, G. and Norén, K. (1967), *Vår Föda,* 10, 138–178.
White, J. F. and Rothstein A. (1973), *Toxicol. Appl. Pharmacol.,* 26, 370–384.
WHO, *Environmental Health Criteria 101, Methylmercury* (1990), International Program on Chemical Safety, World Health Organization, Geneva.
Winroth, G., Carlstedt, I., Karlsson, H., and Berlin, M. (1981), *Acta Pharmacol. Toxicol.,* 49, 168–173.
Wobeser, G., Nielsen, N. O., Dunlop, R. H., and Atton, F. M. (1970), *J. Fish. Res. Bd. Can.,* 27, 830–838.
Wood, J. M. (1971), *Advances in Environmental Science and Technology,* Vol. 2, Pitts, J. N. and Metcalf, R. L., Eds., Wiley-Interscience, New York, 39–56.
Wood, J. M., Kennedy, F. S., and Rosen, C. G. (1968), *Nature,* 220, 173–175.
Yoshino, M., Mozai, T., and Nakao, K. (1966a), *J. Neurochem.,* 13, 397–406.
Yoshino, M., Mozai, T., and Nakao, K. (1966b), *J. Neurochem.* 13, 1223–1230.
Young, P. R., Nandi, U. S., and Kallenbach, N. R. (1982), *Biochemistry,* 21, 62–66.
Zenick, H. (1974), *Pharmacol. Biochem. Behav.,* 2, 709–713.
Zimmer, L. and Cater, D. E. (1979), *Toxicol. Appl. Pharmacol.,* 51, 29–38.
Zucker, R. M., Elstein, K. H., Easterling, R. E., Allis, J. W., Ting-Beall, H. P., and Massaro, E. J. (1988), *Toxicol. Appl. Pharmacol.,* 96, 393–403.
Zucker, R. M., Elstein, K. H., Easterling, R. E., and Massaro, E. J. (1990), *Am. J. Pathol.,* 137, 1187–1198.

Chapter 67

Developmental and Reproductive Effects of Aluminum, Manganese, Uranium, and Vanadium

Jacinto Corbella and Jose L. Domingo

I. ALUMINUM

A. DEVELOPMENTAL TOXICITY

Aluminum-containing antacids have been used for many years for the treatment of various gastrointestinal disorders, particularly peptic ulcerations, because of their acid-neutralizing and local preventive properties. However, it is now well established that, when large loads of aluminum in the form of antacids are ingested, some of this excess aluminum is absorbed with an aluminum retention rate of 0.3 to 10% (Koo and Kaplan, 1988). This fact is of special concern during the gestation periods, when aluminum containing antacids are widely used to reduce the characteristic dyspeptic symptoms (Weberg et al., 1986). Although maternal toxicity, embryo/fetal toxicity, or teratogenicity were not observed when aluminum hydroxide was given by gavage to mice and rats on gestational days 6 to 15 at high doses (66.5, 133, and 266 mg/kg/d to mice, and 192, 384, and 786 mg/kg/d to rats) (Domingo et al., 1989a; Gomez et al., 1990), it was recently found that the concurrent administration of aluminum hydroxide and some frequent organic dietary constituents caused some signs of maternal and fetal toxicity in mice (Colomina et al., 1992, 1994; Gomez et al., 1991a).

On the other hand, although oral administration of aluminum as aluminum nitrate nonahydrate (180, 360, and 720 mg/kg/d) to Sprague-Dawley rats on gestational days 6 to 14 resulted in decreased fetal body weight and increased the incidence and types of external, visceral, and skeletal malformations and variations in all the aluminum-treated groups, the number of corpora lutea, total implants, resorptions, and dead and live fetuses was not significantly different from the control group (Paternain et al., 1988).

Aluminum as aluminum chloride was found to be embryotoxic and teratogenic when given parenterally to rats (Benett et al., 1975) and mice (Cranmer et al., 1986; Wide, 1984). A high incidence of maternal death, a decrease in maternal weight gain, maternal liver damage, a significant increase in the incidence of resorptions and fetal deaths, a significant growth retardation as well as skeletal defects, were reported when aluminum chloride (100 or 200 mg/kg/d) was injected i.p. to pregnant Holtzman rats on 5 consecutive days at different stages of gestation (Benett et al., 1975), whereas i.v. administration (0.1 ml) of aluminum chloride solutions (50 or 100 mM) to pregnant NMRI mice during day 3 or early on 8 day of pregnancy caused an increased frequency of fetal internal hemorraghes and interfered with skeletal ossification (Wide, 1984). Moreover, Cranmer et al. (1986) found that the total fetal aluminum content in mice given i.p. aluminum chloride (100, 150, and 200 mg/kg/d) on days 7 to 16 of gestation was significantly increased compared to maternal aluminum content. It was also observed that exposure of mice to aluminum chloride, at levels which did not cause maternal toxicity, resulted in decreased fetal weight and increased incidence of fetal resorptions (Cranmer et al., 1986).

The importance of the route of administration in the developmental toxicity of aluminum has been clearly stressed by the above observations. Aluminum nitrate is highly water soluble, whereas in contrast, aluminum hydroxide, the most common aluminum compound given therapeutically, has a very low water solubility and consequently is poorly absorbed from the gastrointestinal tract. With regard to this question, in recent years it has been demonstrated that concurrent ingestion of aluminum compounds and a number of dietary organic factors such as citrate, ascorbate, lactate, gluconate, oxalate, etc. caused a significant increase in the gastrointestinal absorption of aluminum (Domingo et al., 1991; Partridge et al., 1989). Thus, although no significant increases in the incidence of malformations were observed when aluminum hydroxide (384 mg/kg/d) was given to rats concurrently with citric acid (62 mg/kg/d), the incidence of skeletal variations was significantly increased and the fetal body weight was reduced. Some signs of maternal toxicity were also observed (Gomez et al., 1991a). Maternal toxicity, reduced fetal body weight, and an increased number of skeletal variations were also seen when aluminum hydroxide (166 mg/kg/d) was orally administered to mice simultaneously with 570 mg/kg/d of lactic acid on gestational days 6 to 15 (Colomina et al., 1992). In contrast, no signs of maternal or developmental toxicity could be observed in Swiss mice, when aluminum hydroxide (300 mg/kg/d) was given concurrently with high doses of ascorbic acid (85 mg/kg/d) during the period of organogenesis (Colomina et al., in press).

B. NEURODEVELOPMENTAL EFFECTS

It has been demonstrated that aluminum, as aluminum lactate, administered orally to pregnant Swiss-Webster mice in a purified diet (500 or 1000 ppm) from day 0 of gestation to day 21 postpartum produced striking signs of neurotoxicity in the lactating dam including hindlimb paralysis, seizures, and death (Golub et al., 1987). Also, when aluminum lactate in a purified diet (25, 500, or 1000 µg Al/g diet) was given to Swiss-Webster mice from conception through weaning, some parameters were significantly affected by aluminum in the postweaning neurobehavioral testing: foot splay, forelimb and hindlimb grip strengths, and thermal sensitivity. However, no maternal or reproductive toxicity was detected (Donald et al., 1989). Recently, Golub et al. (1992) showed that mice are susceptible to neurodevelopmental effects of high maternal dietary aluminum intake during both gestation and lactation periods, and that high maternal intake could result in altered essential trace element metabolism in the offspring.

In studies on the developmental alterations in offspring of female rats orally intoxicated by aluminum, Bernuzzi et al. (1986, 1989a) found an increase in postnatal death rate, growth retardation, and alterations in neuromotor maturation of pups from female Wistar rats exposed to dietary aluminum chloride (100, 300, and 400 mg Al/kg/d) or aluminum lactate (100, 200, and 400 mg Al/kg/d) from day 1 to day 21 of gestation. The same investigators reported that the aluminum intoxication of the females during the first week of gestation, before the period of major organogenesis, induced long-lasting neurobehavioral deficits, while in the negative geotaxis test, the scores of pups from females treated during the 2nd and 3rd weeks of gestation were diminished (Muller et al., 1990). Also, a transitory delay of neuromotor development was observed in pups treated by gastric intubation with aluminum lactate (100, 200, and 300 mg Al/kg/d) from postnatal day 5 to 14 (Bernuzzi et al., 1989b). Recently, these authors reported a low reduction in the general activity, particularly in the radial maze test, when pregnant Wistar rats were given by gavage 200 mg Al/kg/d (as aluminum lactate) from postnatal day 5 to 14 (Cherroret et al., 1992).

In rabbits, learning behavior, as measured by the ability to acquire a classically conditioned reflex, was slightly enhanced by low aluminum doses (aluminum lactate, s.c.), but attenuated by higher aluminum doses (400 µmol) in animals exposed to the metal *in utero* (Yokel, 1985). Subcutaneous exposure to small amounts of aluminum in the milk of aluminum-exposed does did not alter the learning behavior of rabbits (Yokel, 1984) nor did aluminum injections to rabbits during the 1st or 2nd postnatal month (Yokel, 1987). Adult rabbits demonstrated more profound learning deficits after aluminum exposure than any other age group (Yokel, 1983). Measurement of memory of the learned response demonstrated significant impairment only in aluminum-exposed immature and adult rabbits. Yokel (1987) concluded that the adult rabbit demonstrates greater behavioral deficits and accumulation of aluminum in tissues than do rabbits exposed to aluminum at any earlier developmental stage. In contrast, the rabbit exposed to aluminum during the second postnatal month demonstrates the greatest aluminum-induced toxicity to the skeletal system.

C. EFFECTS ON LATE GESTATION, PARTURITION, AND LACTATION

The oral administration of aluminum nitrate nonahydrate (180, 360, and 720 mg/kg/d) to Sprague-Dawley rats from the 14th day of gestation through 21 days of lactation induced a decrease in offspring growth during the entire period of lactation at the higher doses (Domingo et al., 1987a). No adverse effects on fertility or general reproductive parameters were evident when aluminum nitrate nonahydrate (180, 360, and 720 mg/kg/d) was given orally to male Sprague-Dawley rats for 60 d prior to mating with female rats treated for 14 d prior to mating, with treatment continuing thoughout mating, gestation, parturition, and lactation. However, the survival ratio was higher for the control group, whereas a dose-dependent delay in the growth of the pups was also noted in aluminum-exposed groups (Domingo et al., 1987b).

The results of a number of studies on the developmental toxicity of aluminum are summarized in Table 1. Taken together these reports, in order to avoid possible developmental hazards, it seems that the consumption of high doses of aluminum-containing compounds should be avoided during pregnancy.

II. MANGANESE

A. INTRODUCTION

Historically, severe exposure to manganese has been restricted to adult male workers in manganese-based industries. In a study on workers exposed to various manganese oxides and salts (Lauwerys et al., 1985), it was found that the number of births occurring during the exposure period was significantly lower than that expected on the basis of the fertility experience of a matched unexposed group. A reanalysis of the data with a logistic regression model confirmed the reduced fertility in this group of manganese workers. However, in a subsequent study of the same investigators, no effect of manganese exposure on fertility of male workers from a dry alkaline battery plant was evidenced (Gennart et al., 1992). It was suggested that the workers participating in the first investigation (Lauwerys et al., 1985) absorbed more manganese than those in the later study because they were exposed to various manganese salts which are more soluble in water than manganese dioxide.

B. DEVELOPMENTAL TOXICITY

The possibility that excess exposure to manganese may occur in the vicinity of some refineries or even in the urban general environment would mean that manganese toxicity might be a serious health hazard for women and children living in high-exposure environments (Kilburn, 1987).

In a number of investigations in which laboratory animals were given large doses of manganese during pregnancy, either in their food (Jarvinen and Ahlstrom, 1975; Laskey et al., 1982), water (Konturn and Fechter, 1985), or by inhalation (Lown et al., 1984), there was no evidence of structural malformations in the offspring. Even direct i.p. exposure during the organogenic period in hamsters, although establishing an embryolethal effect, did not cause teratogenicity (Ferm, 1972). Although Webster and Valois (1987) showed in pregnant QS mice that i.p. administration of Mn^{2+} (12.5, 25, or 50 mg/kg) as manganous sulfate was embryolethal at 50 mg/kg on each day of pregnancy tested (8, 9, or 10) and teratogenic at 25 mg/kg on day 8 of gestation, these authors concluded that the results were probably of little significance in view of the high doses necessary for teratogenesis. In embryo cultures studies, these investigators showed also that mammalian embryos could tolerate manganese serum levels 50 times higher than those which normally occur in the human without causing abnormality or growth retardation. Since in previous studies, dietary intake of manganese as high as 3500 ppm did not cause malformations or fetal growth retardation in rats (Laskey et al., 1982), it was suggested that it would be unlikely that chronic exposure could elevate serum manganese to a level that would damage the embryo (Webster and Valois, 1987). However, chronic exposure to manganese during pregnancy resulted in an increased manganese content of the fetal brain, which might have some functional consequences (Konturn and Fechter, 1985). Depressed neonatal motor activity was also demonstrated in mice following prenatal exposure to manganese by inhalation (Lown et al., 1984).

Maternal and developmental toxicity of manganese was recently shown in mice when manganese chloride tetrahydrate was administered s.c. at doses of 2, 4, 8, and 16 mg/kg/d from gestation day 6 through 15 (Sanchez et al., 1993). Maternal toxicity included significant reductions in weight gain and food consumption at 8 and 16 mg/kg/d, as well as treatment-related deaths in the high-dose group. Fetotoxicity, consisting primarily of reduced fetal body weight and an increased incidence of morphological defects (wavy ribs and delayed or reduced ossification in the sternebrae, parietal, and occipital), was also observed at 8 and 16 mg/kg/d. The "no-observable-adverse-effect level" (NOAEL) for maternal

Table 1 Maternal and Developmental Toxicity of Aluminum (Al) in Laboratory Animals

Compound Tested	Species	Doses (mg/kg/d)	Dosing Period (days of gestation)	Route	Signs of Toxicity in the Mother	NOEL/NOAEL for Maternal Toxicity (mg/kg/d)	Toxic Effects in the Fetuses or in Offspring	NOEL/NOAEL for Developmental Toxicity (mg/kg/d)	Ref.
Al chloride	Holtzman rats	75, 100, and 200	9–13 or 14–18	i.p.	Deaths and decreased weight gain	Not reported	Embryolethality, growth retardation, skeletal defects	Not reported	Benett et al., 1975
Al chloride	NMRI mice	50 and 100 mM (solutions)	3 or 8	i.v.	Not reported	Not reported	Low degree of skeletal ossification, internal hemorrhages	Not reported	Wide, 1984
Al lactate	NZ rabbits	25, 100, and 400 μmol/kg/d	2–27	s.c.	Body weight loss	Not reported	Impaired weight gain and learning and memory	Not reported	Yokel, 1985
Al chloride	BALB/c mice	100, 150, 200, and 300	7–16	p.o. or i.p.	Not reported	Not reported	Fetal resorptions and decreased fetal weight	Not reported	Cranmer et al., 1986
Al chloride	Wistar rats	160 and 200	8–21	Diet	None	—	Increased preweaning mortality and delay in neuromotor development	Not reported	Bernuzzi et al., 1986
Al lactate	Swiss mice	500 or 1000 ppm	Days 0 to 21 postpartum	Diet	Signs of neurotoxicity, weight loss	Not reported	Decreases in weight gain	Not reported	Golub et al., 1987
		10, 20, and 40	3, 5, 7, 9, 12, 13, and 15	s.c.	Increased liver and spleen weight	Not reported	Decreases in fetal length	Not reported	
Al nitrate	SD rats	180, 360, and 720	14–21 (gestation); 0–21 (lactation)	p.o	Not stated	—	Decrease in offspring growth	Not reported	Domingo et al., 1987a
Al nitrate	SD rats	180, 360, and 720	Males: 60 days before mating Females: 14 days before mating	p.o.	Not stated	—	Decrease in the survival ratio and delay in offspring growth	Not reported	Domingo et al., 1987b

Compound	Strain	Dose	Days of treatment	Route	Maternal toxicity	NOAEL maternal	Fetotoxicity including teratogenicity	NOAEL fetal	Reference
Al nitrate	SD rats	180, 360, and 720	6–14	p.o.	Not stated	Not reported	Fetotoxicity including teratogenicity	<180	Paternain et al., 1988
Al lactate	Swiss-Webster mice	25, 500, and 1000 μg Al/g diet	Gestation and lactation periods	Diet	None	—	Neurobehavioral deficits in offspring	Not reported	Donald et al., 1989
Al chloride or lactate	Wistar rats	100, 300, and 400; 100, 200, and 400	1–21	Diet	None	—	Increased postnatal death rate and developmental alterations	Not reported	Bernuzzi et al., 1989a
Al hydroxide	Swiss mice	66.5, 133, and 266	6–15	p.o.	None	>266	None	>266	Domingo et al., 1989
Al lactate	Wistar rats	400	Different gestation periods	Diet	None	—	Neurobehavioral deficits in offspring	—	Muller et al., 1990
Al hydroxide	Wistar rats	192, 384, and 768	6–15	p.o.	None	>768	None	>768	Gomez et al., 1990
Al hydroxide + citric acid	SD rats	384 + 62 (citric)	6–15	p.o.	Reductions in weight gain	—	Minor skeletal variations	—	Gomez et al., 1991a
Al hydroxide + lactic acid	Swiss mice	166 + 570 (lactic)	6–15	p.o.	Reductions in weight gain and food consumption	—	Minor skeletal variations	—	Colomina et al., 1992
Al hydroxide + ascorbic acid	Swiss mice	300 + 851 (ascorbic)	6–15	p.o.	None	—	None	—	Colomina et al., in press

toxicity was established at 4 mg/kg/d, while the NOAEL for embryo/fetal toxicity was 2 mg/kg/d. The authors concluded that in case of environmental exposure to manganese via inhalation, the NOAEL might be totally different because the comparatively high pulmonary absorption rate of the metal (Sanchez et al., 1993).

III. URANIUM

A. INTRODUCTION

The mining of uranium, its subsequent widespread use in nuclear power plants, and large stores of potentially obsolescent nuclear weapons all raise serious questions concerning the ability of current technology to control the potentially toxic effects of uranium following human exposure. Although many investigations of uranium biokinetics, metabolism, and toxicity have been undertaken in recent decades, significant gaps still remain in the knowledge of the chemical toxicity of uranium in mammals. Thus, information concerning the chronic intake of uranium at low levels or the reproductive and developmental toxicity of uranium ingestion have not been available from the literature. There also was little information concerning the possible antidotes to be used therapeutically in cases of uranium poisoning.

For these and other reasons, in 1987 an extensive program was started in our laboratory to obtain an overall understanding of the toxic effects of uranium. A number of studies on the acute and semichronic toxicity of uranium in mammals (Domingo et al., 1987c; Ortega et al., 1989a), as well as on the action of several chelating agents in experimental uranium intoxication (Domingo et al., 1989b, 1990, 1992; Ortega et al., 1989b) have been performed in the last few years.

With regard to to the reproductive and developmental toxicity of uranium, Stopps and Todd (1982) reported that at the time of the initial toxicological studies of the metal carried out during World War II, two investigations were performed, one of which used high levels and the other only a brief 24-h exposure, but both showed some effects on reproduction. However, these studies were not repeated by other investigators, and so questions on the effect of uranium on reproduction remained to be answered. We here summarize some data about the reproductive and developmental toxicity of uranium.

B. EFFECTS ON REPRODUCTION

To the best of our knowledge, only a study on the effects of uranium on male reproduction in mammals has been carried out (Llobet et al., 1991). Male Swiss mice were treated with uranyl acetate dihydrate at doses of 0, 10, 20, 40, and 80 mg/kg/d given in the drinking water for 64 d. To evaluate the fertility of the uranium-treated males, mice were mated with untreated females for 4 d. There was a significant, but nondose-related decrease in the pregnancy rate of these animals. Body weights were only depressed in the 80 mg/kg/d group. Testicular function/spermatogenesis was not affected by uranium at any dose, as evidenced by normal testes and epididymis weights and normal spermatogenesis. The percentage of motile cells was similar at all doses, and the proportion of abnormal morphologic forms was also unaffected by the treatment with uranium. Histopathological examination of the testes in mice killed at the end of the treatment did not reveal any significant difference between control and uranium-treated animals at 10, 20, 40, or 80 mg/kg/d, with the exception of interstitial alterations and vacuolization of Leydig cells at 80 mg/kg/d. In addition, although not significant, tubular focal atrophy was observed in mice that received 80 mg/kg/d of uranyl acetate dihydrate.

Taken together, the reduced epididymal weights and sperm counts showed the same pattern as the reduced pregnancy rate; i.e., all groups were affected to approximately the same extent. Additionally, moderate to severe focal tubular atrophy and interstitial cell alterations occurred in all dose groups, but not in the controls. These results suggest that these changes might all have contributed to the reduction in pregnancy rate. It would also be possible that uranium exposure for 64 d might provoke some behavioral changes, including a decrease in the libido of these animals, which would contribute to the low pregnancy rate in the uranium-exposed mice. Because at the lowest dose administered in this study the pregnancy rate was significantly decreased (25% vs. 81% in the control group), the NOAEL for reproductive toxicity was established below 10 mg/kg/d.

C. PRENATAL AND POSTNATAL EFFECTS

Paternain et al. (1989) administered orally 0, 5, 10, and 25 mg/kg/d of uranyl acetate dihydrate to mature male Swiss mice (for 60 d prior to mating) and virgin female mice (for 14 d prior to mating). One half of the dams were killed on day 13 of gestation, and the remaining dams were allowed to deliver and wean their offspring. No adverse effects on fertility were evident at the doses tested in this study.

However, although the number of total implants in the uranium-exposed groups was similar to that of the controls, the number of total resorptions and dead fetuses were significantly increased at 25 mg/kg/d, whereas the number of live fetuses was also decreased in this group. Postnatal development was monitored after 0, 4, and 21 d of lactation. Significant increases in the number of dead young per litter were observed at birth and at day 4 of lactation in the 25 mg/kg/d group. Although no differences could be detected for the lactation indices between the uranium-exposed animals and the controls, the viability indices and the growth of the offspring were also significantly lower for the uranium-treated mice than for the controls.

To evaluate the developmental toxicity of the oral exposure to uranium, 5 groups of pregnant Swiss mice were given by gavage 0, 5, 10, 25, and 50 mg/kg/d of uranyl acetate dihydrate on gestational days 6 to 15 (Domingo et al., 1989c). The results of that study indicated that such exposure resulted in maternal toxicity, as evidenced by reduced weight gain and food consumption during treatment, and increased relative liver weight. There were no treatment-related effects on the number of implantation sites per dam, live fetuses per litter, percentage postimplantation loss, or fetal sex ratio. However, dose-related fetal toxicity consisting primarily of reduced fetal body weight and body length, and an increased incidence of abnormalities was observed. External malformations and variations included primarily cleft palate and hematomas (dorsal and facial areas), whereas bipartite sternebrae, delayed ossification of skull, reduced ossification of caudal, and poor ossification of some metatarsals of hindlimbs and some proximal phalanges (forelimb) were the most remarkable skeletal abnormalities. Although maternal toxicity might have an etiologic role in some of the fetal defects observed, most of these defects have also been reported to occur independent of maternal toxicity. In this study, the NOAEL for both maternal and development toxicity was below 5 mg/kg/d, as some toxic effects were observed at this dose (Domingo et al., 1989c).

Recently, the effects of multiple maternal subcutaneous injections of uranyl acetate dihydrate have been evaluated in pregnant Swiss mice. Animals were injected at doses of 0.5, 1, and 2 mg/kg/d on gestational days 6 to 15. These doses were approximately equal to 1/40, 1/20, and 1/10 of the acute s.c. LD_{50} for the chemical (Domingo et al., 1987c). This exposure resulted in maternal toxicity in all uranium-treated groups. Maternal toxicity was especially notorious at 2 mg/kg/d, where 7 animals from 25 plug-positive females died during the study. In addition, a significant decrease in weight gain during treatment and gestation periods, in maternal weight at termination, as well as in absolute liver and kidney weights was also noted. Embryofetotoxic effects in the uranium-exposed groups were evidenced by a no-dose-dependent significant increase in the number of nonviable implants per litter, as well as in the percentage postimplantation loss. Fetotoxicity was indicated by a significant decrease in fetal body weight and significant increases in the incidence of several districts unossified or with decreased ossification in the 1 and 2 mg/kg/d groups, whereas cleft palate and bipartite sternebrae were the most remarkable malformations seen at these doses. According to the results, both the maternal NOAEL and the NOAEL for embryotoxicity of uranyl acetate dihydrate were below 0.5 mg/kg/d, whereas the NOAEL for teratogenicity was 0.5 mg/kg/d (Bosque, et al., 1993a).

The effects of uranium on late gestation, parturition, lactation, and postnatal growth, and viability were assessed by Domingo and associates (1989d). Uranyl acetate dihydrate was given by gavage at doses of 0, 0.05, 0.5, 5, and 50 mg/kg/d from day 13 of pregnancy until weaning of the litters on day 21 postbirth. The viability and lactation indices, as well as the developmental milestones, pinna detachment, eye opening, and incisor eruption were monitored in all pups. Treatment with uranium had neither effects on sex ratios, mean litter size, pup body weight, or pup body length throughout lactation, nor on the developmental milestones, or on the viability and lactation indices at 0.05, 0.5, or 5 mg/kg/d. However, significant decreases in the mean litter size on postnatal day 21, and in the viability and lactation indices were seen at the 50 mg/kg/d dose level. The NOAEL for health hazards to the developing pup was established at 5 mg/kg/d of uranyl acetate dihydrate, as some anomalies were previously observed at this dose (Domingo et al., 1989c).

The results of the studies on the developmental toxicity of uranium are summarized in Table 2. Since nephrotoxicity is the primary chemically induced health effect of uranium in humans, the United Nations Scientific Committee on the Effects of Atomic Radiation (UNSCEAR) recommended a limit of uranium in drinking water of 100 µg/l in order to limit the toxic effects in the kidney (Wrenn et al., 1985). Thus, an adult of 70 kg consuming 5 l/d would not ingest more than 200 µg U/d, which would correspond to 0.005 mg/kg/d of uranyl acetate dihydrate.

Compared to the NOAEL for health hazard in the developing pup, the NOAEL for teratogenicity (<5 mg/kg/d, Domingo et al., 1989c), and the NOAEL for the effects on reproduction, gestation, and postnatal

Table 2 Maternal and Developmental Toxicity of Uranium in Mice

Compound Tested	Species	Doses (mg/kg/d)	Dosing Period (days of gestation)	Route	Signs of Toxicity in the Mother	NOEL/NOAEL for Maternal Toxicity (mg/kg/d)	Toxic Effects in the Fetuses or in Offspring	NOEL/NOAEL for Developmental Toxicity (mg/kg/d)	Ref.
Uranyl acetate dihydrate (UAD)	Swiss mice	5, 10, 25, and 50	6–15	p.o.	Reduced weight gain and food intake, increased liver weight	<5	Fetotoxicity including teratogenicity	<5	Domingo et al., 1989c
UAD	Swiss mice	0.05, 0.5, 5, and 50	13–18 (gestation) 1–21 (lactation)	p.o.	Not reported	—	Decreases in viability and lactation indices	5	Domingo et al., 1989d
UAD	Swiss mice	5, 10, and 25	Males: 60 days Females: 14 days before mating	p.o.	Not reported	Not reported	Lower viability indices and lower growth of the offspring	5	Paternain et al., 1989
UAD	Swiss mice	0.5, 1, and 2	6–15	s.c.	Deaths and decreases in weight gain	<0.5	Embryofetal toxicity and teratogenicity	<0.5	Bosque et al., 1991

survival in mice found by Paternain and associates (1989), who reported that 5 mg/kg/d of uranyl acetate dihydrate did not induce any adverse effect at this dose, a high safety factor for the intake of uranium from drinking water can be estimated. Moreover, since on the average, the percentage of uranium ingestion is about 3% from food and 97% from drinking water (Cothern and Lappenbusch, 1983), ingestion of uranium from food should not increase the risk of developmental adverse effects. With regard to the influence of chronic exposure of uranium on male reproduction, although spermatogenesis was not affected by uranium exposure, the metal produced a significant decrease in the pregnancy rate at 10 mg/kg/d (Llobet et al., 1991). Consequently, further data are required to elucidate the reason for this effect and whether it may be totally or partially reversible.

D. PREVENTION OF URANIUM-INDUCED DEVELOPMENTAL TOXICITY BY CHELATING AGENTS

Tiron (4,5-dihydroxybenzene-1,3-disulfonate), has been found to be an effective agent in increasing uranium excretion and in reducing the concentration of the metal in several tissues (Domingo et al., 1990, 1992). To evaluate whether Tiron could ameliorate the developmentally toxic effects of uranium, a series of four Tiron injections was administered i.p. to pregnant Swiss mice after a single s.c. injection of 4 mg/kg of uranyl acetate dihydrate given on day 10 of gestation and at 24, 48, and 72 h thereafter. In a preliminary study, gestational day 10 was found to be the most sensitive time for uranium-induced developmental toxicity in mice (Bosque et al., 1992). Tiron effectiveness was assessed at 500, 1000, and 1500 mg/kg/d. Although amelioration by Tiron of uranium-induced embryolethality was not noted at the two lower doses, treatment with 1500 mg Tiron/kg/d showed isolated protective effects against uranium fetotoxicity, as evidenced by a lack of differences in fetal body weight between this group and the uranium-untreated group, as well as by a decrease in the number of skeletal defects. However, according to these results, Tiron would offer only modest encouragement with regard to its possible therapeutic potential for pregnant women exposed to uranium (Bosque et al., 1993b).

IV. VANADIUM

A. INTRODUCTION

The toxicological effects of vanadium in humans and in laboratory animals are now well established (Wei et al., 1982; Jandhyala and Hom, 1983; Domingo et al., 1985; Zaporowska and Wasilewski, 1992). However, until recently, very little information on the reproductive and developmental toxicity of vanadium compounds was available. Although vanadium can exist in oxidation states from –1 to +5, the oxidation states of biological interest are V^{4+} (vanadyl) and V^{5+} (vanadate) (Nechay, 1984; Rehder, 1992). Since vanadate is more toxic than vanadyl, in most of the developmental toxicity studies of vanadium it was administered as vanadate (Table 3).

B. EFFECTS ON REPRODUCTION, GESTATION, PARTURITION, AND LACTATION

The reproductive toxicity of vanadium has been reported only in male mice (Llobet et al., 1993). Male Swiss mice were exposed to sodium metavanadate at doses of 0, 20, 40, 60, and 80 mg/kg/d given in the drinking water for 64 d prior to mating with untreated females. Although at 20 and 40 mg/kg/d fertility was not significantly different from controls (81.3%), at 60 and 80 mg/kg/d fertility dropped to 43.8 and 62.5%, respectively. The average number of total implantations, resorptions, and dead and live fetuses of half of the dams killed on days 10 to 14 of gestation was not significantly different between controls and vanadate-treated animals. Decreased body and epididymis weight was only observed in the 80 mg/kg/d group, while testicular weights were not altered by the treatment at all doses tested. Sperm count was significantly decreased at 40, 60, and 80 mg/kg/d, but the sperm motility was unaffected. Histologic study of the testis and epididymis revealed no changes at any dose of metavanadate. No gross lesions were evident in the vanadate-treated animals, whereas examination of the epididymal epithelium revealed normal cellular structures. The tubular diameters were also unaffected. In this investigation, a NOAEL of 40 mg/kg/d was established (Llobet et al., 1993). Taking as a basis a safe intake of 200 µg vanadium per person per day (10 times the average vanadium via diet) (Byrne and Kucera, 1991), as well as the high doses of vanadate that would be required to cause negative effects on the fertility, it seems clear that vanadium would not cause any adverse effect on fertility or testicular function at concentrations usually ingested by humans through the diet and/or drinking water.

Sodium metavanadate was also tested for its effects on fetal development, reproduction, gestation, and lactation in Sprague-Dawley rats. Male rats were administered sodium metavanadate by gavage at

Table 3 Maternal and Developmental Toxicity of Vanadium in Laboratory Animals

Compound Tested	Species	Doses (mg/kg/d)	Dosing Period (days of gestation)	Route	Signs of Toxicity in the Mother	NOEL/NOAEL for Maternal Toxicity (mg/kg/d)	Toxic Effects in the Fetuses or in Offspring	NOEL/NOAEL for Developmental Toxicity (mg/kg/d)	Ref.
Ammonium metavanadate	Syrian golden hamsters	0.47, 1.88, and 3.75	5–10	i.p.	<10% mortality	Not reported	Decrease in the M:F ratio, skeletal anomalies	Not stated	Carlton et al., 1982
Vanadium pentoxide	NMRI mice	1 mM (0.15 ml) (single dose)	3 or 8	i.v.	Not stated	Not stated	High frequency of "less mature" skeletons on day 8	—	Wide 1984
Sodium metavanadate	SD rats	5, 10, and 20	Males: 60 days Females: 14 days before mating	p.o.	Not reported	Not reported	Decreases in body weight and body length	<5	Domingo et al., 1986
Vanadyl sulfate pentahydrate	Swiss mice	37.5, 75, and 150	6–15	p.o.	Reductions in weight gain and liver and kidney weights	<37.5	Embryo/fetal toxicity, including teratogenicity	<37.5	Paternain et al., 1990
Sodium orthovanadate	Swiss mice	7.5, 15, 30, and 60	6–15	p.o.	Deaths and reductions in weight gain and food intake	7.5	Fetotoxicity	15	Sanchez et al., 1991
Sodium metavanadate	Swiss mice	2, 4, and 8	6–15	i.p.	Decreased weight gain	<2	Resorptions, dead fetuses, and fetal weight reductions	2	Gomez et al., 1992

doses of 0, 5, 10, and 20 mg/kg/d for 60 d before mating with females which received the same doses during the 14 d previous to mating. Plug-positive females also received 0, 5, 10, and 20 mg/kg/d of metavanadate during the periods of gestation and lactation. No significant adverse effects could be observed on the number of corpora lutea, total implants, total resorptions, and live and dead fetuses. Significant decreases in body weight and body length of the pups were observed in all the metavanadate-treated groups (Domingo et al., 1986). Elfant and Keen (1987) administered sodium metavanadate to pregnant Sprague-Dawley rats at levels of 1 (control) or 75 µg V/g diet during the gestation and lactation periods. Vanadium-fed dams had lower food intakes and weight gains than controls during pregnancy. Survival until day 21 postpartum was significantly lower in the vanadium pups compared to controls. In addition, the surviving pups gained less weight than control pups, despite similar birth weights. Vanadium-fed dams and pups had higher concentrations of hepatic vanadium than controls. The authors concluded that animals during periods of rapid growth are susceptible to vanadium toxicity, and increased lipid peroxidation might be one factor underlying this toxicity (Elfant and Keen, 1987).

C. PRENATAL AND POSTNATAL EFFECTS

In recent years, a number of investigators have assessed the developmental toxicity of vanadium in hamsters (Carlton et al., 1982), rats (Paternain et al., 1987), and mice (Gomez et al., 1992; Paternain et al., 1990; Sanchez et al., 1991; Wide, 1984). The parenteral exposure of pregnant Syrian golden hamsters to ammonium vanadate (0.47, 1.88, 3.75 mg/kg/d) from days 5 through 10 of gestation resulted in a significant decrease in the male/female sex ratio and an increase in skeletal anomalies (micrognathia, supernumerary ribs, and alterations in sternebral ossification), while embryolethality was not observed at any dosage level (Carlton et al., 1982). In contrast, the administration of sodium metavanadate to pregnant rats on gestational days 6 to 14 at sublethal doses (5, 10, and 20 mg/kg/d) was neither embryolethal nor fetotoxic at any dose (Paternain et al., 1987).

On the other hand, Wide (1984) reported that the i.v. administration of 0.15 ml of V_2O_5 1 mM injected to mice on day 8 of pregnancy caused a low degree of skeletal ossification without increasing the number of nonviable implants, while the NOAELs for maternal toxicity and for developmental toxicity were respectively, 7.5 and 15 mg/kg/d of sodium orthovanadate when this compound was given by gavage to mice on gestational days 6 to 15 at doses of 7.5, 15, 30, and 60 mg/kg/d (Sanchez et al., 1991). The administration of sodium orthovanadate at 60 mg/kg/d was extraordinarily toxic, since 17 dams (from 19 pregnant females dosed) died during the treatment period. Reduced weight gain and food consumption were also observed at 15 and 30 mg/kg/d. Embryolethality and teratogenicity were not noted at maternally toxic doses and below, but fetal toxicity was evidenced by a significant delay in the ossification process of some skeletal districts at 30 mg/kg/d. Recently, sodium metavanadate was also evaluated for developmental toxicity in mice (Gomez et al., 1992). The compound was administered i.p. on days 6 to 15 of gestation at doses of 0, 2, 4, and 8 mg/kg/d. The i.p. LD_{50} of sodium metavanadate in mice was previously reported to be 35.9 mg/kg (Llobet and Domingo, 1984). Maternal toxicity was observed at 2, 4, and 8 mg/kg/d as evidenced by decreased weight gain during treatment. Embryo/fetal toxicity was evidenced by increased resorptions and dead fetuses, increased percentage postimplantation loss, and reduced fetal body weight per litter at 4 and 8 mg/kg/d. There were no significant increases in the type or incidence of external and skeletal anomalies, but a significant increase in the incidence of cleft palate was detected at 8 mg/kg/d. The lowest-observed-adverse-effect level (LOAEL) for maternal toxicity was 2 mg/kg/d, while 2 mg/kg/d was also the NOAEL for developmental toxicity (Gomez et al., 1992).

The developmental toxicity of vanadium (+4) has also been investigated. Vanadyl sulfate pentahydrate was given by gavage to pregnant Swiss mice at doses of 0, 37.5, 75, and 150 mg/kg/d on days 6 to 15 of pregnancy. Maternal toxicity was detected in the 75 and 150 mg/kg/d groups, as evidenced by reduced weight gain, reduced body weight on gestation day 18, and decreased absolute liver and kidney weights. Embryo/fetal toxicity was evidenced by a significant increase in the number of early resorptions per litter, lower fetal weights and fetal lengths, and the presence of developmental variations (decreased ossification of supraoccipital, carpus, and tarsus). Cleft palate and micrognathia were the most relevant malformations observed at 150 mg/kg/d. The NOAEL for maternal and developmental toxicity was below 37.5 mg/kg/d (Paternain et al., 1990).

D. PREVENTION OF VANADIUM-INDUCED DEVELOPMENTAL TOXICITY BY CHELATING AGENTS

Since Tiron was reported to be effective in mobilizing vanadium following acute or repeated exposure to vanadium compounds (Gomez et al., 1991b,c), the efficacy of this chelator to ameliorate the developmentally toxic effects of vanadate was examined in mice (Domingo et al., 1993). Sodium metavanadate (25 mg/kg) was given i.p. to pregnant Swiss mice on gestation day 12. According to the results of a preliminary study, gestational day 12 is the most sensitive time for metavanadate-induced developmental toxicity in mice (Bosque et al., 1993c). Tiron was injected s.c. at 0, 24, 48, and 72 h after vanadate administration, and it was assessed at doses of 250, 500, and 1000 mg/kg. Amelioration of Tiron of metavanadate developmental toxicity was evidenced by a significant decrease in the number of resorbed fetuses, an increase in the mean fetal weight, and a reduction in the incidence of the skeletal variations caused by metavanadate. Since vanadium is capable of crossing the placental barrier and reaching the embryos (Edel and Sabbioni, 1989; Paternain et al., 1990), it would seem that the protective activity of Tiron was due to the decreased vanadium concentrations in the embryos to levels that are incapable of causing developmental toxicity in mice (Domingo et al., 1993).

REFERENCES

Benett, R. W., Persaud, T. V. N., and Moore, K. L. (1975), *Anat. Anz. Bd.,* 138(S), 365–378.
Bernuzzi, V., Desor, D., and Lehr, P. R. (1986), *Neurobehav. Toxicol. Teratol.,* 8, 115–119.
Bernuzzi, V., Desor, D., and Lehr, P. R. (1989a), *Teratology,* 40, 21–27.
Bernuzzi, V., Desor, D., and Lehr, P. R. (1989b), *Bull. Environ. Contam. Toxicol.,* 42, 451–455.
Bosque, M. A., Domingo, J. L., and Corbella, J. (1992), *Rev. Toxicol.,* 9, 107–110.
Bosque, M. A., Domingo, J. L., Llobet, J. M., and Corbella, J. (1993a), *Biol. Trace. Elem. Res.,* 36, 109–118.
Bosque, M. A., Domingo, J. L., Llobet, J. M., and Corbella, J. (1993b), *Toxicology,* 79, 149–156.
Bosque, M. A., Domingo, J. L., Llobet, J. M., and Corbella, J. (1993c), *Vet. Hum. Toxicol.,* 35, 1–3.
Byrne, A. R. and Kucera, J. (1991), Momcilovic, B., Ed., *Trace Elements in Man and Animals,* Vol. 7, IMI, Zagreb, 18.
Carlton, B. D., Beneke, M. B., and Fisher, G. L. (1982), *Environ. Res.,* 29, 256–262.
Cherroret, G., Bernuzzi, V., Desor, D., Hutin, M. F., Burnel, D., and Lehr, P. R. (1992), *Neurotoxicol. Teratol.,* 14, 259–264.
Colomina, M. T., Gomez, M., Domingo, J. L., and Corbella, J. in press, *Pharmacol Toxicol.*
Colomina, M. T., Gomez, M., Domingo, J. L., Llobet, J. M., and Corbella, J. (1992), *Res. Commun. Chem. Pathol. Pharmacol.,* 77, 95–106.
Cothern, C. R., and Lappenbusch, W. L. (1983), *Health Phys.,* 45, 89–99.
Cranmer, J. M., Wilkins, J. D., Cannon, D. J., and Smith, L. (1986), *Neurotoxicology,* 7, 601–608.
Donald, J. M., Golub, M. S., Gershwin, M. E., and Keen, C. L. (1989), *Neurotoxicol. Teratol.,* 11, 341–351.
Domingo, J. L., Llobet, J. M., Tomas, J. M., and Corbella, J. (1985), *J. Appl. Toxicol.,* 5, 418–421.
Domingo, J. L., Paternain, J. L., Llobet, J. M., and Corbella, J. (1986), *Life Sci.,* 39, 819–824.
Domingo, J. L., Paternain, J. L., Llobet, J. M., and Corbella, J. (1987a), *Res. Commun. Chem. Pathol. Pharmacol.,* 57, 129–132.
Domingo, J. L., Paternain, J. L., Llobet, J. M., and Corbella, J. (1987b), *Life Sci.,* 41, 1127–1131.
Domingo, J. L., Llobet, J. M., Tomas, J. M., and Corbella, J. (1987c), *Bull. Environ. Contam. Toxicol.,* 39, 168–174.
Domingo, J. L., Gomez, M., Bosque, M. A., and Corbella, J. (1989a), *Life Sci.,* 45, 243–247.
Domingo, J. L., Ortega, A., Llobet J. M., Paternain, J. L., and Corbella, J. (1989b), *Res. Commun. Chem. Pathol. Pharmacol.,* 64, 161–164.
Domingo, J. L., Paternain, J. L., Llobet, J. M., and Corbella, J. (1989c), *Toxicology,* 55, 143–152.
Domingo, J. L., Ortega, A., Paternain, J. L., and Corbella, J. (1989d), *Arch. Environ. Health,* 44, 395–398.
Domingo, J. L., Ortega, A., Llobet, J. M., and Corbella, J. (1990), *Fundam. Appl. Toxicol.,* 14, 88–95.
Domingo, J. L., Gomez, M., Llobet, J. M., and Corbella, J. (1991), *Kidney Int.,* 39, 598–601.
Domingo, J. L., Colomina, M. T., Llobet, J. M., Jones, J. M., Singh, P. K., and Campbell, R. A. (1992), *Fundam. Appl. Toxicol.,* 19, 350–357.
Domingo, J. L., Bosque, M. A., Luna, M., and Corbella, J. (1993), *Teratology,* 48, 133–138.
Edel, J. and Sabbioni, E. (1989), *Biol. Trace Elem. Res.,* 22, 265–275.
Elfant, M. and Keen, C. L. (1987), *Biol. Trace Elem. Res.,* 14, 193–208.
Ferm, V. H. (1972), *Adv. Teratol.,* 5, 51–75.
Gennart, J. P., Buchet, J. P., Roels, H., Ghyselen, P., Ceulemans, E., and Lauwerys, R. (1992), *Am. J. Epidemiol.,* 135, 1208–1219.
Golub, M., Gershwin, M. E., Donald, J. M., Negri, S., and Keen, C. L. (1987), *Fundam. Appl. Toxicol.,* 8, 346–357.
Golub, M., Keen, C. L., and Gershwin, M. E., (1992), *Neurotoxicol. Teratol.,* 14, 177–182.
Gomez, M., Bosque, M. A., Domingo, J. L., Llobet, J. M., and Corbella, J. (1990), *Vet. Hum. Toxicol.,* 32, 545–548.
Gomez, M., Domingo, J. L., and Llobet, J. M. (1991a), *Neurotoxicol. Teratol.,* 13, 323–328.
Gomez, M., Domingo, J. L., Llobet, J. M., and Corbella, J. (1991b), *Toxicol. Lett.,* 57, 227–234.
Gomez, M., Domingo, J. L., Llobet, J. M., and Corbella, J. (1991c), *J. Appl. Toxicol.,* 11, 195–198.
Gomez, M., Sanchez, D. J., Domingo, J. L., and Corbella, J. (1992), *J. Toxicol. Environ. Health,* 37, 47–56.
Jandhyala, B. S. and Hom, G. J. (1983), *Life Sci.,* 33, 1325–1340.
Jarvinen, R. and Ahlstrom, A. (1975), *Med. Biol.,* 53, 93–99.

Kilburn, C. J. (1987), *Neurotoxicology,* 8, 421–430.
Konturn, P. J. and Fechter, L. D. (1985), *Teratology,* 32, 1–11.
Koo, W. W. K. and Kaplan, L. A. (1988), *J. Am. Coll. Nutr.,* 7, 199–214.
Laskey, J. W., Rehnberg, G. L., Hein, J. F., and Carters, S. D. (1982), *J. Toxicol. Environ. Health,* 9, 677–687.
Lauwerys, R., Roels, H., Genet, P., Toussaint, G., Bouckaert, A., and De Cooman, S. (1985), *Am. J. Ind. Med.,* 7, 171–176.
Llobet, J. M. and Domingo, J. L. (1984), *Toxicol. Lett.,* 23, 227–231.
Llobet, J. M., Sirvent, J. J., Ortega, A., and Domingo, J. L. (1991), *Fundam. Appl. Toxicol.,* 16, 821–829.
Llobet, J. M., Colomina, M. T., Sirvent, J. J., Domingo, J. L., and Corbella, J. (1993), *Toxicology,* 80, 199–206.
Lown, B. A., Morganti, J. B., D'Agostino, R., Stineman, C. H., and Massaro, E. J. (1984), *Neurotoxicology,* 5, 119–131.
Muller, G., Bernuzzi, V., Desor, D., Hutin, M. F., Burnel, D., and Lehr, P. R. (1990), *Teratology,* 42, 253–261.
Nechay, B. R. (1984), *Annu. Rev. Pharmacol. Toxicol.,* 24, 501–524.
Ortega, A., Domingo, J. L., Llobet, J. M., Tomas, J. L., and Paternain, J. L. (1989a), *Bull. Environ. Contam. Toxicol.,* 42, 935–941.
Ortega, A., Domingo, J. L., Gomez, M., and Corbella, J. (1989b), *Pharmacol. Toxicol.,* 64, 247–251.
Partridge, N. A., Regnier, F. E., White, J. L., and Hem, S. L. (1989), *Kidney Int.,* 35, 1413–1417.
Paternain, J. L., Domingo, J. L., Llobet, J. M., and Corbella, J. (1987), *Rev. Esp. Fisiol.,* 43, 223–228.
Paternain, J. L., Domingo, J. L., Llobet, J. M., and Corbella, J. (1988), *Teratology,* 38, 253–257.
Paternain, J. L., Domingo, J. L., Ortega, A., and Llobet, J. M. (1989), *Ecotoxicol. Environ. Safety,* 17, 291–296.
Paternain, J. L., Domingo, J. L., Gomez, M., Ortega, A., and Corbella, J. (1990), *J. Appl. Toxicol.* 10, 181–186.
Rehder, D. (1992), *Biometals,* 5, 3–12.
Sanchez, D. J., Ortega, A., Domingo, J. L., and Corbella, J. (1991), *Biol. Trace Elem. Res.,* 30, 219–226.
Sanchez, D. J., Domingo, J. L., Llobet, J. M., and Keen, C. L. (1993), *Toxicol. Lett.,* 69, 45–52.
Stopps, G. J. and Todd, M. (1982), The Chemical Toxicity of Uranium with Special Reference to Effects on the Kidney and the Use of Urine for Biological Monitoring, Research Report, Atomic Energy Control Board, Ottawa, Canada.
Weberg, R., Berstad, A., Ladehaug, B., and Thomassen, Y. (1986), *Acta Pharmacol. Toxicol.,* 59(S), 63–65.
Webster, W. S. and Valois, A. A. (1987), *Neurotoxicology,* 8, 437–444.
Wei, C. I., Al Bayati, M. A., Culberston, M. R., Rossenblatt, L. S., and Hansen, L. D. (1982), *J. Toxicol. Environ. Health,* 10, 673–687.
Wide, M. (1984), *Environ. Res.,* 33, 47–53.
Wrenn, M. E., Durbin, P. W., Howard, B., Lipsztein, J., Rundo, J., Still, E., and Willis, D. L. (1985), *Health Phys.,* 13, 1321–1324.
Yokel, R. A. (1983), *Neurobehav. Toxicol. Teratol.,* 5, 41–46.
Yokel, R. A. (1984), *Toxicol. Appl. Pharmacol.,* 75, 35–43.
Yokel, R. A. (1985), *Toxicol. Appl. Pharmacol.,* 79, 121–133.
Yokel, R. A. (1987), *Fundam. Appl. Toxicol.,* 9, 795–806.
Zaporowska, H. and Wasilewski, W. (1992), *Comp. Biochem. Physiol.,* 102C, 223–231.

Chapter 68

Cell Adhesion Molecules in Metal Neurotoxicity

Kenneth R. Reuhl and P. Markus Dey

I. INTRODUCTION

Metals are the most common neurotoxicants known to man. The adverse effects of this broad class of chemicals have been recognized for more than a millennium and remain among the most important health concerns today. Despite intensive research efforts, the mechanisms by which metals perturb neural function are poorly understood. This ignorance stems in no small measure from the ability of metals to interfere with a broad spectrum of biochemical processes. The specificity of a metal's action on the cell is inversely proportional to the concentration of the metal; as levels of metal increase, progressively more cellular functions will be involved until finally, direct effects of metals are lost in the complex maze of secondary cellular responses.

A second reason for our poor understanding of metal neurotoxicity is the inherent complexity of the nervous system itself. In addition to its architectural sophistication, the brain is not a static target. Development, maturity, and aging each present unique and shifting possibilities for metal toxicity. At most, studies provide but a snapshot of toxic events which describe the effects of one metal, at one range of doses, during one specific period of life, and which can be extrapolated to other treatment conditions only with caution.

One approach to unraveling the complexities of metal toxicity in the nervous system is to focus attention on classes of molecules which, present throughout life, serve important roles in the formation and maintenance of the nervous system. Some of these molecules are constitutively expressed and unchanging throughout life, while others, often transiently expressed, are critical only to discrete cell types or regions in the brain. The morphoregulatory cell adhesion molecules are molecular systems involved in all aspects of neural formation and function. They not only play a central role in brain development but also are crucial to processes which persist throughout life, such as learning and memory. It is unlikely to be coincidental that these events are also most vulnerable to perturbation by heavy metals. That involvement of cell adhesion molecules might represent a common pathway for metal-induced injury is a tantalizing prospect for mechanistic toxicologists.

II. CELL ADHESION MOLECULES IN THE BRAIN

Development and maintenance of the nervous system is dependent upon the proper temporal and spatial expression of membrane proteins which mediate recognition, adhesion, and repulsion between cells. These glycoproteins, collectively termed cell adhesion molecules (CAMs), play essential roles in embryogenesis, neurulation, migration, and synaptogenesis (Edelman, 1985; Rutishauser and Jessell, 1988). Each CAM shows a unique pattern of expression and distribution within the nervous system.

Moreover, each shows preferential ligand binding to complimentary CAMs, often in a homophilic manner (Edelman and Chuong, 1982). During development this combination of genetically regulated expression and binding specificity assists the nervous system in the complex task of sorting neural elements into their proper cytoarchitectonic domains and helps preclude establishment of inappropriate neural connections. In the adult nervous system, CAMs mediate specialized contacts necessary to maintain metastable cytoarchitecture and participate in neural remodeling following physiologic (learning or memory) and pathologic processes (Edelman, 1984a; Rutishauser and Jessell, 1988).

Based on the enormous complexity of the nervous system, one would predict the active participation of large numbers of adhesive ligands during morphogenesis and in the mature brain. While the list of CAMs identified within the nervous system continues to expand, the number of major CAMs appears to be relatively small, perhaps only several dozen. Some of these have been characterized biochemically and their functions in the central nervous system have been partially elucidated. These CAMs have been generally divided into two families based on their dependence on calcium for their adhesive activity. The calcium-independent CAMs include the neural cell adhesion molecule (NCAM), L1 (also known as nerve growth factor inducible large extrinsic or NILE), myelin-associated glycoprotein, and P_0 (for review, see Edelman, 1986; Edelman and Crossin, 1991; Covault, 1989). Cadherins (particularly N-cadherin) are the primary calcium-dependent CAM family in the nervous system (for review, see Takeichi, 1990). A number of other CAMs have been identified whose expression is highly restricted to portions of the neural cell (e.g., axons); however, their exact functions are presently poorly understood. Multiple CAM molecules may be simultaneously expressed by neurons or glia, allowing contemporaneous operation of several discrete mechanisms of adhesion. Further, members within a family of CAMs may act cooperatively to modulate the strength of adhesion; this cooperativity has been well-documented between NCAM and L1 during neurite growth (Kadmon et al., 1990).

The role of adhesion in the formation and stabilization of the nervous system is complex. In addition to serving as recognition ligands between cells searching for compatible partners, CAMs both initiate and permit complex cellular events which accompany contact between plasma membranes. For instance, binding of CAMs is an important signal transduction pathway from the extracellular to intracellular compartments of the cell. This transduction occurs in at least two general ways: by direct signal linkage to the cytoskeleton (Pavalko and Otey, 1994; Reuhl et al., 1994) and via second messenger systems (Doherty and Walsh, 1992), including inositol phosphates (Schuch et al., 1989), protein kinases (Mackie et al., 1989), tyrosine kinases (Beggs et al., 1994), membrane channels (Doherty et al., 1991), and fibroblast growth factor receptors (Williams et al., 1994). By activating these and other signal transduction pathways, cells can rapidly affect changes in their morphology and biochemical organization in response to extracellular stimuli (Atashi et al., 1992). Reciprocal relationships are also possible; alterations of either cytoskeleton or second messenger systems will alter the expression and/or function of CAMs.

A. THE CALCIUM-INDEPENDENT CAMS-NEURAL CELL ADHESION MOLECULE (NCAM)

NCAM, the best characterized of the CAMs, is a surface glycoprotein whose structure is highly conserved across species. NCAM is a member of the immunoglobulin superfamily of molecules, and shares with other members the presence of extracellular immunoglobulin loop domains joined by disulfide bonds (Figure 1). Three major isoforms of the molecule have been resolved by SDS-PAGE; these are produced from a single gene by alternative RNA splicing and polyadenylation (Walsh, 1988). These isoforms share identical extracellular amino terminal domains and differ only in the length of the intracellular domains. NCAM 180 has the longest intracellular domain, is expressed primarily in late development and synaptogenesis, and communicates with brain spectrin, a cytoskeleton-glycoprotein linker protein (Pollerberg et al., 1987; Persohn et al., 1989). NCAM 140, seen during neurogenesis and in the adult brain, possesses a truncated cytoplasmic domain and is shorter than NCAM 180 by 267 amino acids (Barthels et al., 1988). NCAM 120 lacks a membrane-spanning domain entirely and is linked to the membrane via a phosphatidylinositol linkage (He et al., 1987). The three isoforms of NCAM are differentially expressed during development and undergo posttranslational modifications, including phosphorylation and sulfation (Hoffman and Edelman, 1983). The most striking of these modifications is the developmentally-regulated appearance of multiple chains of sialic acid polymers in an α-2,8 linkage. This heavily sialylated isoform of NCAM is referred to as *embryonic*, or eNCAM. The sialic acid residues, which may comprise 30% of the weight of the molecule, impart to the membrane surface a net negative charge which tends to repulse the negative charge of eNCAM on adjacent cells, thus inhibiting the close membrane-membrane apposition necessary for adhesion. In this manner, expression of eNCAM prevents inappropriate adhesion and permits movement of cells relative to each other

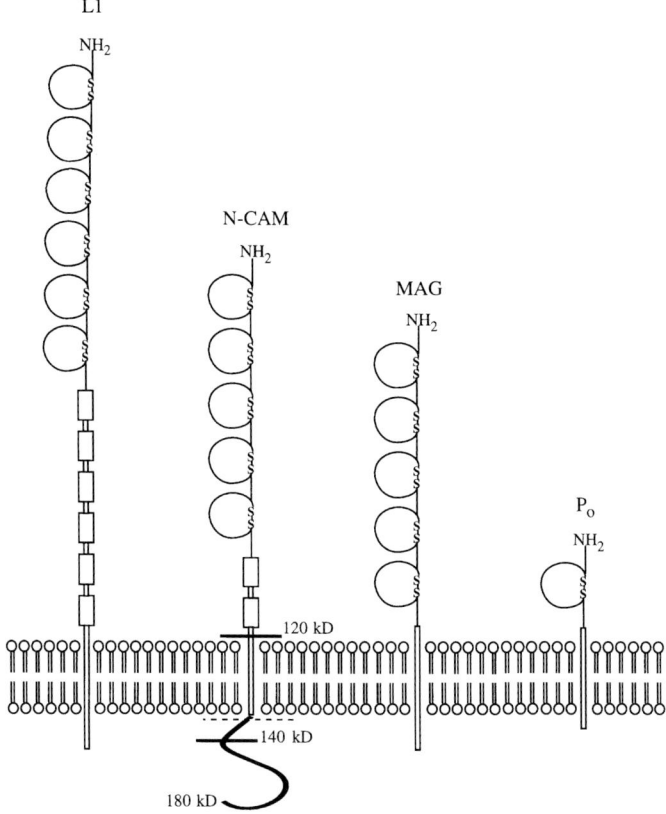

Figure 1. Schematic drawing of major members of the calcium-independent cell adhesion molecules. Multiple glycosylation sites are present on the extracellular domains of each molecule. (Adapted from Schachner, 1992.)

during migration and neurite outgrowth (Rutishauser et al., 1988; Landmesser et al., 1990). At an appropriate time (usually at the completion of neuronal migration), the amount of sialic acid decreases dramatically and the neurons express a less sialylated *adult* form, termed aNCAM. The adult form, lacking the negative charge, is more permissive of the tight membrane-membrane intimacy required for establishment of stable cytoarchitecture.

The binding rate of NCAM has been reported to be fifth order; consequently, relatively small increases or decreases in NCAM will have disproportionate impact on adhesiveness (Edelman, 1986). Cells may gain additional fine control over adhesion by varying the level of sialylation or the relative distribution of the sialylated and nonsialylated forms of NCAM on different areas of the plasma membrane, theoretically permitting adhesion of one portion of the cell and free movement of another portion of the same cell. This process is operative during differentiation, where postmigratory neurons extend axons through neuropil to establish synaptic networks. Modulation of sialic acid content on NCAM thus represents a crucial mechanism regulating the location and strength of adhesion.

While most of the brain undergoes a developmental conversion to the less sialylated adult form, several brain regions retain eNCAM into adult life (Bonfanti et al., 1992; Seki and Arai, 1993). Elements of the thalamo-neurohypophysial system (Theodosis et al., 1991), hippocampus and other limbic structures (Seki and Arai, 1991a, b; Doherty et al., 1992; Le Gal La Salle et al., 1992), and olfactory system (Miragall et al., 1988) express heavily sialylated NCAM. These brain regions are notable for their ability to remodel following injury and for their rapid synaptic plasticity (Rougon et al., 1993) under physiological or pathological stimuli. Reorganization and compensation following physical or chemical damage to neural networks recapitulates the developmental expression of CAMs in the brain (Lehman et al., 1993; Miller et al., 1994). While expression of adhesion molecules such as L1 (Poltorak et al., 1993) and the substrate adhesion molecule tenascin have been shown to be altered following lesioning, changes in NCAM expression have been most extensively investigated. It is probable that these responses are facilitated by the persistence of eNCAM which permits adjustments of the physical relationship between cells.

Disturbance of the appropriate temporal or spatial pattern of expression of the sialic acids would be expected to result in defects in normal cytoarchitectural arrangements. The premature expression of a high molecular weight heavily sialylated NCAM may underlie the neural tube defects observed in the *splotch* mutant mouse (Neale and Trasler, 1994), while brain dysmorphogenesis in *staggerer* and *reeler* mouse mutants has been associated with a failure of normal conversion of embryonal to adult NCAM (Edelman, 1986; Edelman and Chuong, 1982). Genetic deletion of NCAM180 by homologous recombination results in a phenotype characterized by abnormal neuronal migration, particularly within the olfactory bulb (Tomasiewicz et al., 1993; Cremer et al., 1994). This defect could be attributed to inappropriate loss of polysialic acid and not to the absence of the NCAM180, as a phenocopy of the defect is produced by injection of endoneuraminidase N, which removes sialic acids (Ono et al., 1994).

B. CALCIUM-DEPENDENT ADHESION MOLECULES (THE CADHERINS)

The cadherins are a family of cell transmembrane glycoproteins which mediate cell adhesion through calcium-dependent mechanisms. In recent years it has become apparent that the cadherins represent a complex, multigene family and the number of identified cadherins has grown rapidly. To date, major members of the cadherin family include N-cadherin (Takeichi, 1988), E-cadherin (also known as epithelial cadherin or uvomorulin), P-cadherin, and L-CAM (for review of cadherins, see Takeichi, 1990). Recent additions to this list are M-cadherin from mouse myoblasts, R-cadherin (retina and brain), B-cadherin (localized in choroid plexus and optic tectum), T-cadherin (a truncated form of cadherin), and several cadherins from *Xenopus* of yet unknown significance to mammalian brain development (EP-, XB-, and U-cadherin) (Grunwald, 1993). In this review, discussion will be restricted to N-cadherin, for which a role in the central nervous system is best understood.

1. N-Cadherin

N-cadherin was the first calcium-dependent adhesion molecule to be linked to nervous system development. Discovered almost simultaneously in several laboratories it was variously named N-cadherin (Takeichi et al., 1979), gp130/4.8 (Grunwald et al., 1980), and A-CAM (Geiger et al., 1990); N-cadherin has gained the widest usage. The molecule has a molecular weight of approximately 130 kDa and possesses significant homology to other cadherins, particularly E-cadherin and L-CAM. A calcium-binding site is located on the extracellular domain and imparts a specific conformation to this portion of the molecule (Figure 2). The extracellular domain is also glycosylated; the degree of glycosylation largely explains minor differences reported in the molecule's molecular weight. Homophilic binding is mediated by a highly conserved tripeptide HAV sequence located near the N-terminus (Blaschuk et al., 1990).

N-cadherin undergoes extensive posttranslational modification. Sulfation alters the net charge of the molecule and may modulate its adhesive characteristics in a manner reminiscent of sialylation of NCAM (Takeichi, 1990). Phosphorylation of N-cadherin has been demonstrated to occur in a region- and age-specific manner during brain development. During early development, N-cadherin is widely distributed in a poorly phosphorylated form, whereas later in development it is more heavily phosphorylated and restricted in distribution. The increase in phosphorylation may mediate interactions of the intracellular domain with the cytoskeleton via small linker molecules, such as the catenins (Lagunowich and Grunwald, 1991; Hirano et al., 1992).

Expression of N-cadherin is one of the first events which distinguishes ectoderm from surrounding tissues. N-cadherin participates in the early steps of neural plate development (Detrick et al., 1990) and is critical for cell migration involved in neural tube formation. Temporal and/or spatial mis-expression of N-cadherin by the developing neural tube may result in its failure to properly fuse, leading to dysraphism and possible intrauterine death. Later in development N-cadherin also functions in the formation of intercellular bonds between cells and in the guidance of growth cones during neurite elongation (Takeichi et al., 1979; Hatta and Takeichi, 1986; Matsunaga et al., 1988a,b).

The intracellular domain of N-cadherin has intimate interactions with the actin-based cytoskeleton that are mediated through cytoskeletal linker molecules, the catenins, a multigene family of cytoskeletal linker proteins (Wheelock and Knudsen, 1991). The linkage between cytoskeleton and N-cadherin appears essential for N-cadherin to display adhesive function. When a truncated N-cadherin lacking an

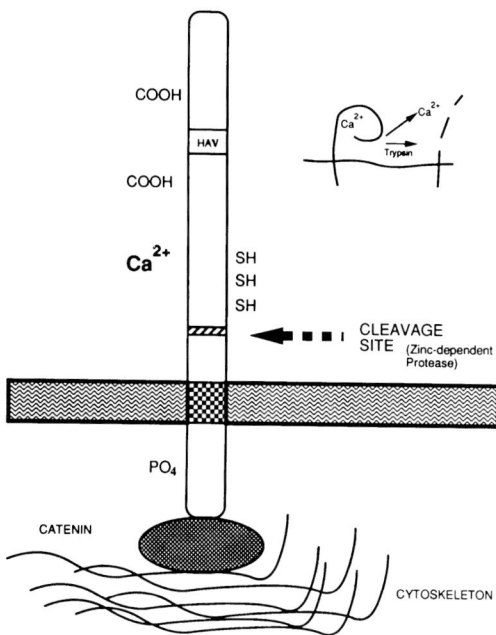

Figure 2. Schematic drawing of N-cadherin, the major calcium-dependent adhesion molecule in the brain. (From Lagunowich, L. A., Stein, A. P., and Reuhl, K. R., *Neurotoxicology*, 15, 123–132, 1994. With permission.)

intracellular domain is expressed in developing *Xenopus* embryos, development is disrupted (Kintner, 1992).

Disturbed expression of N-cadherin alters neural development (Takeichi, 1988). In late embryonic development and in adults, N-cadherin expression is down-regulated and the molecules are localized at specific sites, where they participate in the maintenance of mature neuroarchitecture as a component of some cell-cell junctions. Administration of antibodies to N-cadherin (Tomaselli et al., 1988; Bronner-Fraser et al., 1992) or toxicants (Lagunowich et al., 1994a) can block neural tube formation, neural migration, or neurite outgrowth.

III. CELL ADHESION MOLECULES AND METAL NEUROTOXICITY

Accumulating evidence implicates CAMs in neurotoxic conditions (Regan, 1993a; Lagunowich et al., 1994b; Reuhl et al., 1994). CAMs of both calcium-independent and calcium-dependent families may be perturbed by the direct action of the xenobiotic on the CAM molecule or by secondary alterations of cellular homeostasis following toxicant exposure. Metals are particularly good examples of CAM-targeting toxicants and their broad biological reactivity implies the possibility of chemical interaction with specific CAMs. Both extracellular and intracellular domains of CAMs contain regions which are potentially susceptible to interaction with metals. Metals such as mercury or lead may bind directly to specific moieties of the molecule, such as the immunoglobulin -SH loops of NCAM or the Ca^{2+} binding domain of N-cadherin, thereby disrupting the adhesive function of the molecule. Alternatively, metals may stimulate extracellular or intracellular proteases and thereby hasten CAM degradation. Effects of metals on protein synthesis, Golgi trafficking, or cytoskeleton are but a few actions which would directly or indirectly alter CAM expression and function (Reuhl et al., 1994).

To date, most studies examining metal-CAM interactions have demonstrated correlations between metals and altered CAM structure and/or function, but detailed cause-effect data are lacking. Where more detailed investigations have been performed, it has proven difficult to distinguish between a primary effect of a metal on the CAM and alterations of CAM secondary to other cellular responses. To some extent this is a distinction without difference; the consequences of CAM perturbation on cell function would be manifested regardless of how the perturbation occurred. These consequences range from disruption of the most basic morphogenetic processes such as neural tube formation and neuronal migration to subtle events such as synaptogenesis and pathway imprinting (Doherty et al., 1995). Many clinical intoxications by metals have been shown to induce this spectrum of neurotoxicity in both humans and experimental animal models. Clearly, understanding the exact molecular mechanisms is desirable in order to understand the full significance of CAM-metal interactions in neurotoxicity.

IV. CAM TOXICITY BY SPECIFIC METALS

A. LEAD

Extensive data exist documenting the neurotoxic effects of lead on nervous system morphogenesis. Lead treatment during development causes spontaneous abortion and defects in central nervous system formation. Increased frequency of anterior meningocele and hydrocephalus (Butt et al., 1952; Karnofsky and Ridgway, 1952) in animal models indicates that lead can interrupt normal neural tube closure and neuronal proliferation, events which have been shown to depend upon functional N-cadherin. Pre- and postnatal lead exposures have additionally been implicated in human psychomotor retardation and cognitive disorders (Needleman et al., 1979; Bellinger et al., 1987). Behavioral studies indicate that neurological deficits observed in experimental animals (Cory-Slechta et al., 1985; Rice, 1992) and humans may result from reduced, retarded, and/or aberrant synaptic contacts formed during development (Averill and Needleman, 1980; McCauley et al., 1982). Morphological observations of delayed synaptogenesis (Petit et al., 1983), abnormal dendritic arborization (Huttenlocher, 1974), decreased number of synaptic complexes (Press, 1977), and reductions in dendritic and synaptic structures (Lorton and Anderson, 1986; Reuhl et al., 1989) indicate that lead interrupts the temporally regulated sequence of postnatal neurogenesis and synaptic stabilization, perhaps by altering NCAM 180 function at the synapse (Pollerberg et al., 1986, 1987). Persistence of the psychomotor effects of the toxicant suggests that once the opportunity for proper synaptic patterning is missed, the brain may not be able to reconstitute the pathway at a later time, even when the toxicant is no longer present.

Since modulated expression of specific cell-cell adhesion systems is critical for normal brain development (Cunningham et al., 1987; Edelman, 1984b), any process resulting in aberrant expression and/or functional loss of these CAMs would be expected to have adverse consequences for the organism (Schachner, 1992). Studies indicate that altered expression and/or function of two specific neural adhesion molecules, the neural cell adhesion molecule (NCAM) and N-cadherin, occur following low level lead exposure (Breen and Regan, 1988b; Lagunowich et al., 1994b).

B. LEAD AND NCAM

Lead was the first toxic metal shown to adversely affect CAMs (Breen and Regan, 1988b). Treatment of rat pups with lead during early postnatal life alters posttranslational modification of NCAM at times coincident with postnatal synaptogenesis (Regan, 1993b). Chronic low-level lead administration (yielding blood lead levels of 20 to 40 µg/dl) delays conversion of NCAM from the highly sialylated embryonic form to its poorly sialylated adult form in developing rat cerebellum (Cookman et al., 1987). The mechanism by which lead delays conversion of embryonic to adult NCAM involves stimulation by lead of a Golgi-associated sialyltransferase which specifically regulates NCAM sialylation state (McCoy et al., 1985; Breen et al., 1987; Breen and Regan, 1988a). In normal development, maximal sialyltransferase activity is maintained in the developing rat pup brain until day 12, after which time enzyme activity slowly decreases to 40% maximal activity by adulthood (Regan, 1991). The decrease in sialyltransferase activity results in loss of highly-sialylated NCAM and an increase of the adult NCAM isoform, facilitating synaptic stability and formation of mature brain structure. Golgi preparations from brains of rats of increasing age were used to assay the developmental effects of lead on sialyltransferase activity. *In vitro* incubation with concentrations of lead equivalent to blood lead levels of 30 µg/dl markedly stimulates sialyltransferase activity and eNCAM expression by rats aged 16 days and older (Breen and Regan, 1988b; Regan, 1989).

The continued expression of sialylated eNCAM at times coincident with synaptogenesis may inhibit the formation of synaptic contacts during a critical period of neurodevelopment. Lead-induced injury of synaptic structure has been seen morphologically (Drew et al., 1990; Reuhl et al., 1989) and may underlie

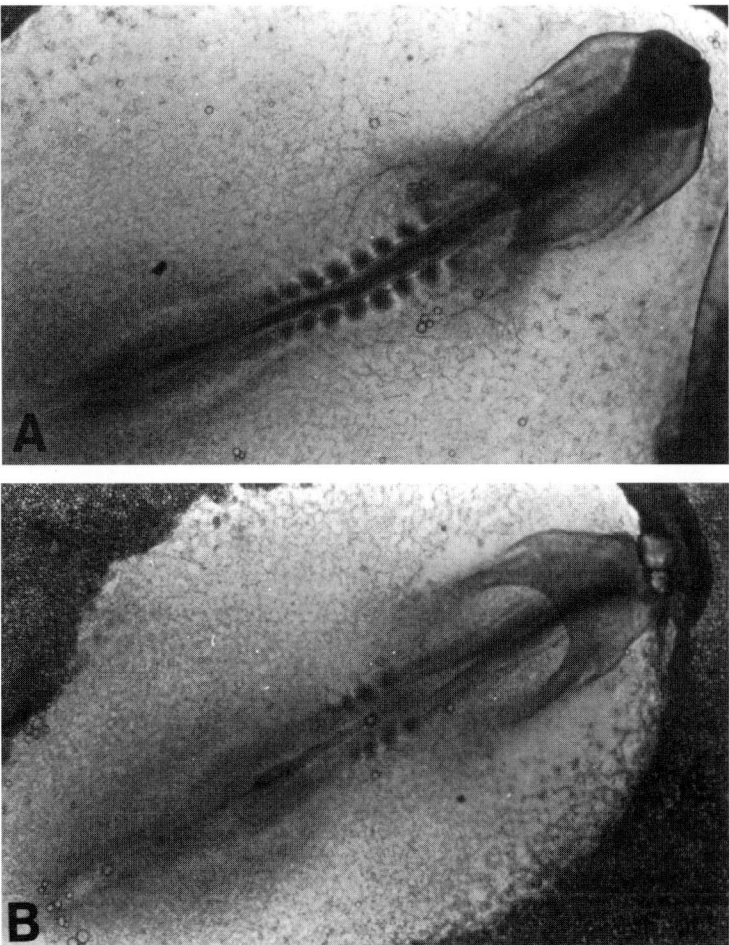

Figure 3. Chicken embryos following treatment *in ovo* with saline (A) or lead (B). Lead treatment retards growth and delays or inhibits neural tube closure, often resulting in dysraphisms or embryonic death.

some of the well-documented psychomotor and learning defects observed following neonatal lead exposure (Needleman et al., 1979).

C. LEAD AND N-CADHERIN

Functional perturbation of the calcium-dependent adhesion molecule N-cadherin by lead was reported by Lagunowich and co-workers using the chick embryo model (Lagunowich et al., 1993, 1994a). Calcium-dependent aggregation of dissociated chicken neural retina cells is inhibited in the presence of 10 to 1000 μM lead. Biochemical analysis indicated that lead-treated cells expressed normal N-cadherin levels on their cell membranes, suggesting that lead may disturb adhesion without necessarily causing proteolytic removal of the adhesion molecule. Lead exposure during neurulation *in ovo* delays neural tube closure (Lagunowich et al., 1993; Figure 3). Similar results are obtained when anti-N-cadherin antibodies are used to block binding sites. Phenocopy experiments such as these support the hypothesis that lead acts on a functional segment of the extracellular domain of N cadherin.

Effects of lead on N-cadherin function appear to result from the competition with calcium for the extracellular calcium-binding site (Lagunowich et al., 1994a,b). Lead has been shown to displace calcium from binding sites of numerous proteins and enzymes (Pounds, 1984; Goldstein, 1993), and it has been suggested that lead might replace calcium on the Ca^{2+} binding site on N-cadherin, as was observed with aforementioned retinal adhesion studies. The possibility that lead affects N-cadherin by intracellular mechanisms such as kinase activation or cytoskeletal disruption (e.g., by action on catenins) remains unexplored. Functional alterations and/or transient loss of N-cadherin induced by lead during periods

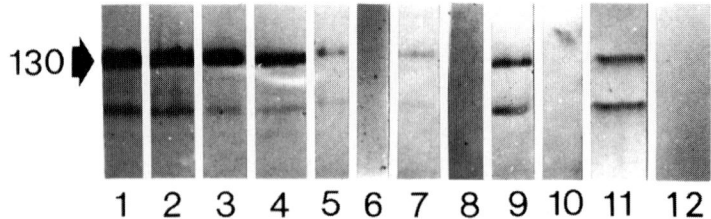

Figure 4. Expression of N-cadherin in neural retina as detected by immunoblotting; 30 min after treatment. Trypsination of cells was done in the presence (lanes 1 to 4) or absence (5 to 8) of 1 mM calcium or 10 μM lead acetate (9 to 12). Cells were allowed to reaggregate for 30 min in calcium-containing medium (1,2,5,6,9,10) or lead-containing medium (3,4,7,8,11,12) and in the presence (2,4,6,8,10,12) or absence (1,3,5,7,9,11) of cycloheximide. (From Lagunowich, L. A., Stein, A. P., and Reuhl, K. R., Neurotoxicology, 15, 123–132, 1994. With permission.)

of dynamic N-cadherin modulation in neural tube formation may help explain the dysraphisms observed following fetal exposure to this toxicant.

D. METHYLMERCURY

Methylmercury is one of the most potent and best characterized of the neurotoxic heavy metals. The severity of methylmercury neurotoxicity correlates with duration and intensity of exposure as well as the age of life during which the exposure occurs (Takeuchi, 1977; Reuhl and Chang, 1979). The most vulnerable periods are during development, particularly during the late embryonic and fetal periods (Amin-Zaki et al., 1974; Reuhl and Chang, 1979; Choi, 1986), times of prominent neuronal migration, neural stabilization, and synaptogenesis. Neuropathological findings in human infant exposures from Minimata, Japan, and Iraq include arrest of neuronal migration, deranged cortical differentiation, and marked astrocytosis (Choi et al., 1978; Takeuchi, 1977). Disturbance of cerebellar granule cell and neocortical neural migration have been described in methylmercury-treated rodents (Reuhl et al., 1981; Choi, 1986). Retardation or loss of synaptic complex formation (Reuhl et al., 1981), significant reduction in dendritic arborization (Choi, 1983), and dendritic spine dysgenesis (Stoltenburg-Didinger and Markwort, 1990) reflect subtle alterations in neural cytoarchitecture following chronic methylmercury treatment and implicate disturbance of key cellular events which mediate morphogenesis. Studies in our laboratory indicate that altered developmental expression of NCAM, but not N-cadherin, occurs following methylmercury exposure (Lagunowich et al., 1991; Graff et al., 1993).

E. METHYLMERCURY AND NCAM

Methylmercury alters NCAM expression during neural development both in vivo and in vitro. Administration of methylmercury to neonatal rodents disturbs posttranslational modifications of the NCAM molecule (Reuhl et al., 1994). Mouse pups treated with the toxicant on postnatal days 1 to 6 show delayed conversion of the molecule to its adult poorly sialylated form (Lagunowich et al., 1991). The highly-sialylated NCAM persists throughout the developmental period when conversion to the adult form is necessary for neural stabilization and synaptogenesis (Pollerberg et al., 1985, 1986; Regan, 1991). The similarity of effects on e to a conversion seen with both lead and methylmercury suggest that this event is particularly susceptible to disruption by metals.

In contrast, chronic treatment of adult mice reveals an alternative mechanism for methylmercury-induced effects on NCAM. Experiments document selective loss of the synaptic-specific isoform of NCAM (NCAM180) in certain adult brain regions following methylmercury treatment (Dey, Polunas, and Reuhl, unpublished). It is tempting to suggest that disrupted NCAM expression alters efficacy and function of synaptic contacts in these regions and that such changes underlie the memory and cognitive disturbances reported in adult, low-level MeHg poisoning.

Studies employing neurons in culture show loss of immunoreactive NCAM following exposure to methylmercury (Graff et al., 1993; Reuhl et al., 1994). NCAM staining is rapidly lost in neurons following exposure, beginning first in the periphery of the neurites and proceeding centripetally to involve the neuronal perikaryon (Figure 5). Removal of methylmercury from the medium allows slow reappearance of NCAM, with normal immunostaining regained only after several hours of recovery (Graff et al., 1992). The protracted period required for recovery suggests that methylmercury binds directly to NCAM and that its resynthesis is needed for replacement of functional NCAM.

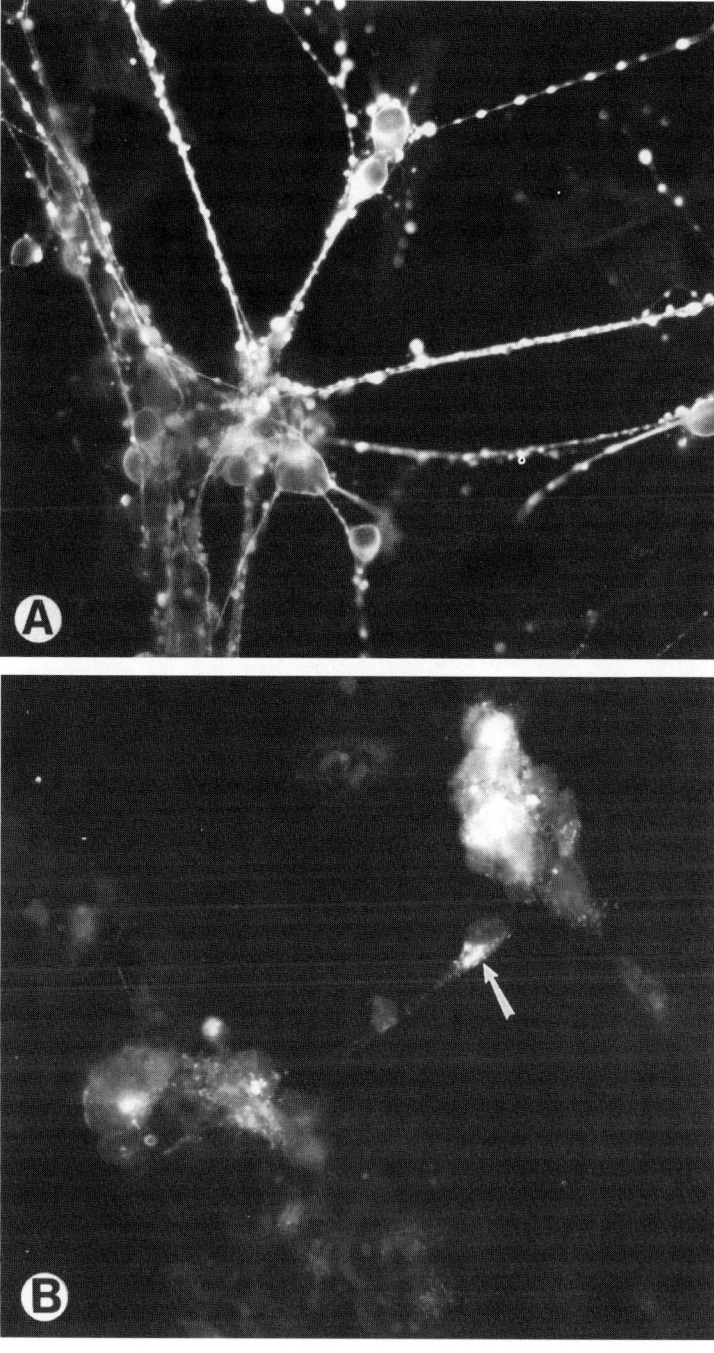

Figure 5. Photomicrographs of neurons stained with a monoclonal antibody against NCAM. Control neurons (A) display bright immunofluorescence on the perikaryon and neurites. Following exposure to methylmercury, NCAM fluorescence becomes progressively less intense (B), beginning with the peripheral neurites and proceeding toward the cell body. A focal accumulation in the perikaryon (arrow) is thought to represent NCAM in the Golgi apparatus.

F. TRIMETHYLTIN

Hallmarks of neurobehavioral disorders following acute trimethyltin toxicity include severe tremor, hyperactivity, and aggressive behavior (Brown et al., 1979; Dyer et al., 1982; McMillan and Wenger, 1985). Neuronal injury is most severe within the limbic system where hippocampus, neocortex, pyriform cortex and amygdala all exhibit varying degrees of neuronal degeneration (Aldridge et al., 1981; Bouldin et al., 1981; Chang et al., 1982). Because hippocampal injury and long-term behavioral alterations are highly reproducible consequences of trimethyltin exposure, the toxicant has been extensively used for investigations of hippocampal mechanisms of learning and memory (Loullis et al., 1985; Earley et al., 1992). Trimethyltin treatment has been shown to impair the acquisition and performance of complex spatial tasks (Walsh et al., 1982a; Swartzwelder et al., 1982) and retention of two-way and passive avoidance responses (Walsh et al., 1982b). It is believed that trimethyltin impairs task retention by disruption of the neural processes underlying learning and memory (Earley et al., 1989), but the mechanism underlying this neurotoxic outcome is unknown.

G. TRIMETHYLTIN AND NCAM

Treatment of adult mice with trimethyltin results in selective loss of the synapse-specific 180-kDa isoform of the NCAM from hippocampus and cerebellum, without affecting the 140 and 120-kDa NCAM isoforms (Dey et al., 1994). NCAM180 loss is rapid, with a 50% reduction occurring within 4 h of treatment (Figure 6). Loss does not correlate with overt neuropathology in hippocampus, as it occurs prior to the appearance of morphological damage and often persists long after hippocampal cell injury has resolved. NCAM180 in cerebellum is less severely affected and returns to normal levels within 72 h of treatment.

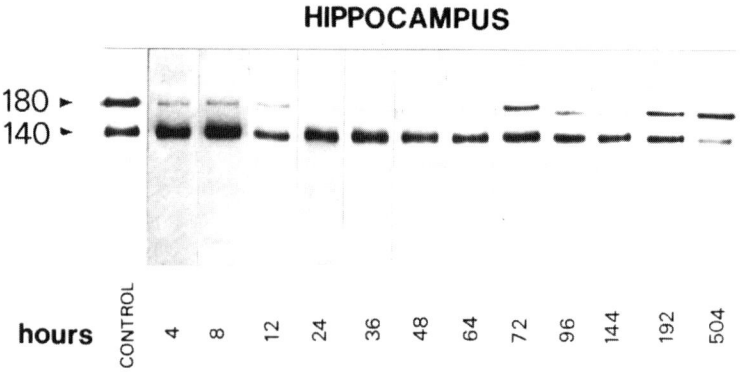

Figure 6. Immunoblots of the 140- and 180-kDa NCAM isoforms in hippocampi of mice treated with trimethyltin 4 to 504 h (3 weeks) earlier. Specific loss of the NCAM180 isoform occurs following trimethyltin in a time-dependent manner, with loss seen within 4 h, and gradual reappearance of the isoform 72 h to 3 weeks later.

The mechanisms of trimethyltin-mediated NCAM180 loss are unclear. Disruption of Golgi function by trimethyltin (Bouldin et al., 1981) may inhibit NCAM processing or insertion of the molecule into the membrane. Destabilization of cellular and synaptic contacts mediated by NCAM180 would clearly facilitate cellular plasticity and tissue remodeling, with subsequent reappearance of NCAM180 signaling the restoration of stable contacts and stable cytoarchitecture. Increases in intracellular synaptosomal calcium following organotin exposure may preferentially alter NCAM expression in synaptic regions (Komulainen and Bondy, 1987). Recent evidence implicates the unique cytoskeletal linkage and sensitivity to calcium-dependent proteolysis of NCAM180 (Sheppard et al., 1991; Covault et al., 1991) as possible excitotoxic trimethyltin-induced mechanisms (Dey et al., 1994).

V. CAMS IN LEARNING AND MEMORY

Modulation of the strong adhesive interactions between pre- and postsynaptic membranes are initial steps in formation of new synaptic contacts and in the remodeling of preexisting synapses during learning and memory (Harris and Kater, 1994; Luthl et al., 1994). Experimental models of learning and memory acquisition in hippocampus indicate that increased efficacy of synaptic contacts occurs coincident with modulation of expression of specific cell adhesion molecules (Regan, 1993a; Scholey et al., 1993; Fazeli et al., 1994). Research to date has focused on the role of the 180-kDa isoform of NCAM in hippocampal memory storage (Doyle et al., 1992b). NCAM180 is preferentially localized to synaptic membranes (Persohn et al., 1989) and is stabilized at the synapse by cytoskeletal anchoring (Pollerberg et al., 1987). The importance of NCAM in information storage was first shown with intraventricular infusions of anti-

NCAM antibodies in rodents. Administration of anti-NCAM antibodies in the early post-training period attenuates task learning and causes amnesia of the training session (Doyle et al., 1992a; Scholey et al., 1993). Doyle and co-workers (1992b) subsequently demonstrated that transient sialylation of hippocampal NCAM180 of rodents occurs during the acquisition and consolidation of task learning and hypothesized that increased sialylation of synaptic NCAM facilitates the synaptic plasticity necessary for acquisition and consolidation of a passive avoidance response. Disruption in NCAM180 sialylation occurred with administration of the amnesic agent scopolamine when the drug was given during and following the training period (Doyle et al., 1993), indicating that the memory deficits elicited by chemical agents may involve disruptions in NCAM expression. Similarly, deletion of NCAM180 using transgenic knockout mice (Cremer et al., 1994) results in persistent learning deficits.

Specific and protracted alteration of NCAM180 expression in hippocampus, such as that described above for methylmercury, lead, and trimethyltin, is likely to have important consequences for learning and memory and may partially explain the behavioral impairment which persists after administration of the toxicant. NCAM180 loss may therefore serve as a marker for changes in synaptic function and efficacy induced by metals in the absence of overt cellular changes (Jorgensen, 1995).

Many toxic metals have in common their ability to disturb behavioral parameters, particularly learning and memory. Decrement in I.Q. performance is one of the most important consequences of long-term exposure to metals and compounds such as methylmercury (Spyker et al., 1972; Rice, 1989) and trimethyltin (Walsh et al., 1982b) have been shown to induce various types of persistent behavioral deficits of developmental and adult exposure. Existing studies strongly indicate that interference with CAMs, particularly NCAM (Regan, 1993b), may represent a mechanistic common denominator underlying these observations.

VI. CONCLUSION

The last decade has seen an extraordinary expansion in our understanding of the role CAMs play in neural development and organization. Far from being merely ligands for binding, CAMs are now appreciated as critical recognition and signal transduction molecules whose proper regulation is central to nervous system integrity. Evidence is growing rapidly implicating CAM dysfunction in a wide variety of neurological disorders, including those resulting from exposure to xenobiotics, and has led to the proposal that CAMs represent a primary target of neurotoxic metals. The attractiveness of this hypothesis lies in its simplicity and testability. By implicating metal effects on disturbance of a relatively small numbers of critical molecules expressed throughout life, a heuristic model of toxicity over an animal's lifetime emerges. While this model will doubtless become more complex and challenging as new data emerge, elucidating the exact role of CAMs in neurotoxic conditions represents an exciting field of inquiry.

ACKNOWLEDGMENTS

The authors would like to thank Dr. H.E. Lowndes for his critical review and suggestions. The authors also thank Dr. Sally Pyle, Patricia Pailing, Kathleen G. Roberts, and Michele Diegmann for their assistance. Preparation of this work was supported by NIH ES 04976, ES 05022.

REFERENCES

Aldridge, W., Brown, A., Brierley, R., Verschoyle, R., and Street, B. (1981), *Lancet*, ii, 692–693.
Amin-Zaki, L., Elhassani, S., Majeed, M. A., Clarkson, T. W., Doherty, R. A., and Greenwood, M. (1974), *Pediatrics*, 54, 587–595.
Atashi, J. R., Klinz, S. G., Ingraham, C. A., Matten, W. T., Schachner, M., and Maness, P. F. (1992), *Neuron*, 8, 831–842.
Averill, D. R. and Needleman, H. L. (1980), in *Low Level Lead Exposure: The Clinical Implications of Current Research*, H. L. Needleman, Ed., Raven Press, New York, 201–210.
Barthels, D., Vopper, G., and Wille, W. (1988), *Nucl. Acids Res.*, 16, 4217–4225.
Beggs, H. E., Soriano, P., and Maness, P. F. (1994), *J. Cell Biol.*, 127, 825–833.
Bellinger, D., Leviton, A., Waternaux, C., Needleman H. L., and Rabinowitz, M. (1987), *N. Engl. J. Med.*, 316, 1037–43.
Blaschuk, O. W., Sullivan, R., David, S., and Pouliot, Y. (1990), *Dev. Biol.*, 139, 227–229.
Bonfanti, L., Olive, S., Poulain, D. A., and Theodosis, D. T. (1992), *Neuroscience*, 49, 419–436.
Bouldin, T., Goines, N., Bagnell, C., and Krigman, M. (1981), *Am. J. Pathol.*, 104, 237–249.
Breen, K. C., Kelly, P. G., and Regan, C. M. (1987), *J. Neurochem.*, 48, 1486–1493.

Breen, K. C. and Regan, C. M. (1988a), *Development,* 104, 147–154.
Breen, K. C. and Regan, C. M. (1988b), *Toxicology,* 49, 71–76.
Bronner-Fraser, M., Wolf, J. J., and Murray, B. A. (1992), *Dev. Biol.,* 153, 291–301.
Brown, A., Aldridge, W., Street, B., and Verschoyle, R. (1979), *Am. J. Pathol.,* 104, 59–81.
Butt, E. M., Pearson, H. E., and Simonsen, D. G. (1952), *PSEBM,* 79, 247–249.
Chang, L., Tiemeyer, T., Wenger, G., McMillen, D., and Reuhl, K. (1982), *Environ. Res.,* 29, 435–444.
Choi, B. H., Lapham, L. W., Amin-Zaki, L., and Saleem, T. (1978), *J. Neuropathol. Exp. Neurol.,* 37, 719–733.
Choi, B. H. (1986), *Neurotoxicology,* 7, 591–600.
Cookman, G. R., King, W., and Regan, C. M. (1987), *J. Neurochem.,* 49, 399–403.
Cory-Slecta, D. A., Weiss, B., and Cox, C. (1985), *Toxicol. Appl. Pharmacol.,* 78, 291–295.
Covault, J. (1989), *Molecular Neurobiology,* D. M. Glover and B. D. Hames, Eds., Oxford University Press, New York, 143–200.
Covault, J., Liu, Q.-Y., and El-Deeb, S. (1991), *Mol. Brain Res.,* 11, 11–16.
Cremer, H., Lange, R., Christoph, A., Plomann, M., Vopper, G., Roes, J., Brown, R., Baldwin, S., Kraemer, P., Scheff, S., Barthels, D., Rajewsky, K., and Wille, W. (1994), *Nature,* 367, 455–459.
Cunningham, B. A., Hemperly, J. J., Murray, B. A., Prediger, E. A., Brackenbury, R., and Edelman, G. M. (1987), *Science,* 236, 799.
Detrick, R. J., Dickey, D., and Kintner, C. R. (1990), *Neuron,* 4, 493–506.
Dey, P. M., Graff, R. D., Lagunowich, L. A., and Reuhl, K. R. (1994), *Toxicol. Appl. Pharmacol.,* 126, 69–74.
Doherty, P., Ashton, S. V., Moore, S. E., and Walsh, F. S. (1991), *Cell,* 67, 21–33.
Doherty, P. and Walsh, F. S. (1992), *Curr. Opinion Neurobiol.,* 2, 595–601.
Doherty, P., Skaper, S. D., Moore, S. E., Leon, A., and Walsh, F. S. (1992), *Development (Camb.),* 115, 885–892.
Doherty, P., Fazel, M., and Walsh, F. S. (1995), *Neurobiology,* 26, 437–446.
Doyle, E., Nolan, P. M., Bell, R., and Regan, C. M. (1992a), *J. Neurochem,* 59, 1570–1573.
Doyle, E., Nolan, P. M., Bell, R., and Regan, C. M. (1992b), *J. Neurosci. Res.,* 31, 513–523.
Doyle, E., Regan, C. M., and Shiotani, T. (1993), *J. Neurochem.,* 61, 266–272.
Drew, C. A., Spence, I., and Johnston, G. A. (1990), *Neurochem. Int.,* 17, 43–51.
Dyer, R., Walsh, T., Wonderlin, W., and Barcegeay, M. (1982), *Neurobehav. Toxicol. Teratol.,* 4, 127–134.
Earley, B., Burke, M., and Leonard, B. (1989), *Neuropsychobiology,* 22, 49–56, 1989.
Earley, B., Burke, M., and Leonard, B. (1992), *Neurochem. Int.,* 21, 351–366.
Edelman, G. M. (1984a), *Exp. Cell Res.,* 161, 1–16.
Edelman, G. M. (1984b), *Annu. Rev. Neurosci.,* 7, 339–377.
Edelman, G. M. (1985), *Annu. Rev. Biochem.,* 54, 135–169.
Edelman, G. M. (1986), A review, *Cell Biol.,* 2, 81–116.
Edelman, G. M. and Chuong, C. M. (1982), *Proc. Natl. Acad. Sci. U.S.A.,* 79, 7036–7040.
Edelman, G. M. and Crossin, K. L. (1991), *Annu. Rev. Biochem.,* 60, 155–190.
Fazeli, M. S., Breen, K., Errington, M. L., and Bliss, T. V. (1994), *Neurosci. Lett.,* 169, 77–80.
Geiger, B., Volberg, T., Sabanay, I., and Volk, T. (1990), *Morphoregulatory Molecules,* Edelman, G. M., Cunningham, B. A., and Thiery, J. P., Eds., John Wiley & Sons, New York, 57–79.
Goldstein, G. W. (1993), *Neurotoxicol.,* 14, 97–102.
Graff, R. D., Lagunowich, L. A., and Reuhl, K. R. (1992), *Toxicologist,* 12, 312.
Graff, R. D., Elzer, J. A., Lagunowich, L. A., and Reuhl, K. R. (1993), *Toxicologist,* 13, 168.
Grunwald, G. B. (1993), *Curr. Opinion Cell Biol.,* 5, 797–805.
Grunwald, G. B., Geller, R. L., and Lilien, J. (1980), *J. Cell Biol.,* 85, 766–776.
Harris, K. M. and Kater, S. B. (1994), *Annu Rev. Neurosci.,* 17, 341–371.
Hatta, K. and Takeichi, M. (1986), *Nature,* 320, 447–449.
He, H. T., Finne, J., and Goridis, C. (1987), *J. Cell Biol.,* 105, 2489–2500.
Hirano, S., Kimoto, N., Shimoyama, Y., Hirohashi, S., and Takeichi, M. (1992), *Cell,* 70, 293–301.
Hoffman, S. and Edelman, G. M. (1983), *Proc. Natl. Acad. Sci. U.S.A.,* 89, 5762–5766.
Huttenlocher P. R. (1974), *Neurology,* 24, 203–210.
Jorgensen, O. S. (1995), *Neurochem. Res.,* 20, 533–547.
Kadmon, G., Kowitz, A., Altevogt, P., and Schachner, M. (1990), *J. Cell Biol.,* 110, 193–208.
Karnofsky, D. A. and Ridgway, L. (1952), *J. Pharmacol. Exp. Ther.,* 104, 176–186.
Kintner C. (1992), *Cold Spring Harb. Symp. Quant. Biol.,* 57, 335–344.
Komulainen, H. and Bondy, S. (1987), *Toxicol. Appl. Pharmacol.,* 88, 77–86.
Lagunowich, L. A., Bhambhani, S., Graff, R. D., and Reuhl, K. R. (1991), *Soc. Neurosci. Abstr.,* 17, 515.
Lagunowich, L. A. and Grunwald, G. B. (1991), *Different,* 47, 19–27.
Lagunowich, L. A., Stein, A. P., and Reuhl, K. R. (1993), *Toxicologist,* 13, 168.
Lagunowich, L. A., Stein, A. P., and Reuhl, K. R. (1994a), *Neurotoxicol.,* 15, 123–132.
Lagunowich, L. A. (1994b), *Principles of Neurotoxicology,* Marcel Dekker, New York, 699–711.
Landmesser, L., Dahm, L., Tang, J., and Rutishauser, U. (1990), *Neuron,* 4, 655–667.
Le Gal La Salle, G., Rougon, G., and Valin, A. (1992), *J. Neurosci.,* 12, 872–882.

Lehman, S., Kuchler, S., Gobaille, S., Marschal, P., Badache, A., Vincenden, G. and Zanetta, J. (1992), *Brain Res. Bull.,* 30, 515–521.
Lorton D. and Anderson W. J. (1986), *Neurobehav. Toxicol. Teratol.,* 8, 51–9.
Loullis, C., Dean, R., Lippa, A., Clody, D., and Coupet, J. (1985), *Pharm. Biochem. Behav.,* 22, 147–151.
Luthl, A., Laurent, J. P., Figurov, A., Muller, D., and Schachner, M. (1994), *Nature,* 372, 777–779.
Mackie, K., Sorkin, B. C., Nairn, A. C., Greengard, P., Edelman, G. M. and Cunningham, B. A. (1989), *J. Neurosci.,* 9, 1883–1896.
Matsunaga, M., Hatta, K., Nagafuchi, A., and Takeichi, M. (1988a), *Nature (London),* 334, 62–64.
Matsunaga, M., Hatta, K., and Takeichi, M. (1988b), *Neuron,* 1, 289–295.
McCauley, P., Bull, R., Tonti, A., Lutkenhoff, S., Meister, M., Doerger, J., and Stroeber, J. (1982), *J. Toxicol. Environ. Health,* 10, 639–651.
McCoy, R. D., Vimr, E. R., and Troy, F. A. (1985), *J. Biol. Chem.,* 260, 12695–12699.
McMillan, D. and Wenger, G. (1985), *Pharmacol. Rev.,* 37, 365–379.
Miller, P. D., Styren, S. D., Lagenaur, C. F., and DeKosky, S. T. (1994), *J. Neurosci.,* 14, 4217–4225.
Miragall, F., Kadmon, G., Husmann, M., and Schachner, M. (1988), *Dev. Biol.,* 129, 516–531.
Neale, S. A. and Trasler, D. G. (1994), *Teratology,* 50, 118–124.
Needleman, H. L., Gunnoe, C., Leviton, A., Reed, R., Peresie, H., Maher, C., and Barrett, P. (1979), *N. Engl. J. Med.,* 300, 689–732.
Ono, K., Tomasiewicz, H., Magnuson, T., and Rutishauser, U. (1994), *Neuron,* 13, 595–609.
Pavalko, F. M. and Otey, C. A. (1994), *PSEBM,* 205, 282–293.
Persohn, E., Pollerberg, G. E., and Schachner, M. (1989), *J. Comp. Neurol.,* 288, 92–100.
Petit T. L., Alfano, D. P., and LeBoutillier J. C. (1983), *Neurotoxicology,* 4, 79–94.
Pollerberg, G. E., Sadoul, R., Goridis, C., and Schachner, M. (1985), *J. Cell Biol.,* 101, 1921–1929.
Pollerberg, G. E., Schachner, M., and Davoust, J. (1986), *Nature,* 324, 462–465.
Pollerberg, G. E., Burridge, K., Krebs, K. E., Goodman, S. R., and Schachner, M. (1987), *Cell Tissue Res.,* 250, 227–236.
Poltorak, M., Herranz, A. S., Williams, J., Lauretti, L., and Freed, W. J. (1993), *J. Neurosci.,* 13, 2217–2229.
Pounds, J. G. (1984), *Neurotoxicology,* 5, 295–331.
Press, M. F. (1977), *Neuropathology,* 7, 169–193.
Regan, C. M. (1989), *Neurotoxicol. Teratol.,* 11, 533–537.
Regan, C. M. (1991), *Int. J. Biochem.,* 23, 513–523.
Regan, C. M. (1993a), *Neurotoxicology,* 14, 69–74.
Regan, C. M. (1993b), *Polysialic Acid,* Roth, J., Rutishauser, U., and Troy, F. A., II, Eds., Birkhauser Verlag, Basel/Switzerland, 313–321.
Reuhl, K. R. and Chang, L. W. (1979), *Neurotoxicology,* 1, 21–55.
Reuhl, K. R., Chang, L. W., and Townsend, J. W. (1981), *Environ. Res.,* 26, 281–306.
Reuhl, K. R., Rice, D. C., Gilbert, S. G., and Mallett, J. (1989), *Toxicol. Appl. Pharmacol.,* 99, 501–509.
Reuhl, K., Lagunowich, L., and Brown, D. (1994), *Neurotoxicology,* 15, 133–146.
Rice, D. (1989), *Neurotoxicology,* 10, 645–650.
Rice, D. (1992), *Neurotox. Teratol.,* 14, 235–245.
Rougon, G., Olive, S., and Figarella-Branger, D. (1993), *Polysialic Acid,* Ruth, J., Rutishauer, A., and Troy, F. A., II., Eds., Birkhauser Verlag, Basel/Switzerland, 323–333.
Rutishauser, U. and Jessell, T. M. (1988), *Physiol. Rev.,* 68, 819–851.
Rutishauser, U., Acheson, A., Hall, A. K., Mann, D. M., and Sunshine, J. (1988), *Science,* 240, 53–57.
Schachner, M. (1992), *The Nerve Growth Cone,* Letourneau, P. C., Kater, S. B., and Macagno, E. R., Eds., Raven Press, New York, 237–254.
Scholey, A., Rose, S., Zamani, M., Bock, E., and Schachner, M. (1993), *Neuroscience,* 55, 499–509.
Schuch, U., Lohse, M. J., and Schachner, M. (1989), *Neuron,* 3, 13–20.
Schwartzwelder, H. S., Hepler, J., Holahan, W., King, S. E., Leverenz, H. A., Miller, P. A., and Myers, R. D. (1982), *Neurobehav. Toxicol. Teratol.,* 4, 169–176.
Seki, T. and Arai, Y. (1991a), *Neurosci. Res.,* 12, 503–513.
Seki, T. and Arai, Y. (1991b), *Anat. Embryol.,* 184, 395–401.
Seki, T. and Arai, Y. (1993), *Dev. Brain Res.,* 73, 141–145.
Sheppard, A., Wu, J., Rutishauser, U., and Lynch, G. (1991), *Biochim. Biophys. Acta,* 1076, 156–160.
Spyker, J., Sparber, S., and Goldberg, A. (1972), *Science,* 177, 621–623.
Stoltenburg-Didinger, G. and Markwort, S. (1990), *Neurotoxicol. Teratol.,* 12, 573–576.
Takeichi, M. (1988), *Development,* 102, 639–655.
Takeichi, M. (1990), *Annu. Rev. Biochem.,* 59, 237–252.
Takeichi, M., Ozaki, H. S., Tokunaga, K., and Okada, T. S. (1979), *Dev. Biol.,* 70, 195–205.
Takeuchi, T. (1977), *Pediatrician,* 6, 69–87.
Theodosis, D. T., Rougon, G., and Poulain, D. A. (1991), *Proc. Natl. Acad. Sci. U.S.A.,* 88, 5494–5498.
Tomaselli, K. J., Neugebauer, K. M., Bixby, J. L., Lilien, J., and Reichardt, L. F. (1988), *Neuron,* 1, 33–43.
Tomasiewicz, H., Ono, K., Yee, D., Thompson, C., Goridis, C., Rutishauser, U., and Magnuson, T. (1993), *Neuron,* 11, 1163–1174.

Walsh, T., Gallagher, M., Bostock, E., and Dyer, R. (1982a), *Neurobehav. Toxicol. Teratol.,* 4, 177–183.
Walsh, T. J., Miller, J. B., and Dyer, R. S. (1982b), *Neurobehav. Toxicol. Teratol.,* 4, 177–183.
Walsh, F. S. (1988), *Neurochem. Int.,* 12, 263–267.
Wheelock, M. J. and Knudsen, K. A. (1991), *In vivo,* 5, 505–514.
Williams, E. J., Furness, J., Walsh, F. S., and Doherty, P. (1994), *Neuron,* 13, 583–94.

Chapter 69

The Effects of Metals on Neurite Development

Teresa Audesirk and Gerald Audesirk

Proper development of the nervous system, including the elaboration of neurites (axons and dendrites), pathfinding, and establishing connections with the correct target cells, is essential for normal behavior. Relatively small changes in neurite growth may lead to significantly abnormal neuronal wiring, which may in turn lead to behavioral changes, from subtle to overt. Neuronal development is a sensitive target for neurotoxicants, including some metals. In this chapter, we will review normal neurite development and metal-induced dysgenesis, with emphasis on cellular mechanisms of normal and abnormal development.

The effects of metals on neurite development may be studied *in vivo* or *in vitro*. During *in vivo* studies, metals are administered to embryos *in utero* via the pregnant mother or postnatally via mother's milk or by injection. Some combination of anatomical, biochemical, and physiological studies is then performed on the brains of the offspring at various postnatal ages. Most *in vivo* studies to date have been anatomical, providing evidence that a given metal causes abnormal neuronal development of certain brain regions. Alternatively, *in vitro* studies assess a variety of anatomical, biochemical, or physiological parameters in neurons exposed to metals in cell culture. Most studies directed at determining the mechanisms whereby metals affect neurite development have been performed on cultured neurons.

There are many possible mechanisms whereby metals may alter neurite development; we will focus on two of these.

Interference with calcium-regulated processes: The intracellular free Ca^{2+} ion concentration ($[Ca^{2+}]_{in}$) in neurons is about 100 nM, approximately 10,000-fold lower than in the extracellular fluid, regulated by sequestration and release of Ca^{2+} by organelles, binding of Ca^{2+} to intracellular molecules, Na^+-Ca^{2+} exchange and CaATPase extrusion pumps in the plasma membrane, and permeability of neurotransmitter-gated and voltage-sensitive calcium channels (VSCC). This diversity of mechanisms reflects the multiplicity of roles of Ca^{2+} in a neuron. Several lines of evidence (summarized below) suggest that (a) $[Ca^{2+}]_{in}$ is important in regulating or modulating the initiation of neurites from the soma, the elongation of neurites, and the movement of growth cones; (b) Ca^{2+} influx through VSCC, particularly L-type channels, is an important source of $[Ca^{2+}]_{in}$ for many regulatory processes, including neurite initiation; and (c) interactions of $[Ca^{2+}]_{in}$ with second messengers such as calmodulin and protein kinases are important in mediating neuronal differentiation. Because they share certain chemical properties with calcium, a number of metal neurotoxicants affect Ca^{2+} homeostasis by altering Ca^{2+} influx through the plasma membrane, extrusion through the plasma membrane, or sequestration and/or release from intracellular stores, or by binding to Ca^{2+}-regulated molecules. Therefore, they are likely to affect neuronal development.

Interference with cytoskeletal assembly and function: Neurite development depends on intricate interactions among multiple cytoskeletal elements, including actin, neurofilaments, and microtubules, and their associated proteins (e.g., actin-binding and actin-severing proteins, microtubule-associated proteins). Several types of metals can disrupt the neuronal cytoskeleton either indirectly by acting on intracellular messengers (including $[Ca^{2+}]_{in}$, Ca^{2+}-regulated molecules, and a number of kinases) or by direct effects on cytoskeletal elements or their associated proteins.

I. NEURITE DEVELOPMENT UNDER NORMAL CONDITIONS

The control of neurite development is a subject of intense study both *in vivo* and *in vitro*. For technical reasons, including the cellular and geometric simplicity of culture systems and the relative ease of manipulation of the culture environment and cellular signalling pathways, the majority of experiments providing mechanistic data on neurite development at the cellular and subcellular levels have been performed in culture. Therefore, this section will primarily review data derived from cultured neurons.

A. OVERVIEW OF NEURITE DEVELOPMENT

The development of a neuron from a round soma to its differentiated form involves the production of neurites (neurite initiation and branching), elongation of these neurites, and movements of the growth cone that guide the neurite to its target, where synaptic connections are formed. The term "neurite" is frequently used to denote a "generic" neuronal process, encompassing not only axons and dendrites, but also processes that may not fulfill the normal biochemical and morphological criteria for either axons or dendrites (such as the processes elaborated by certain types of immortal cell lines). We will use the term "neurite" both in this sense and also to refer to mechanisms of development or toxicity for which the data suggest equal applicability to axons and dendrites. However, axons and dendrites have important differences in cytoskeleton, developmentally important proteins, and the distribution of certain organelles. Therefore, we will specify the type of process where the literature provides specific data on axonal vs. dendritic traits, or where these appear to be important for developmental neurotoxicity.

In many cultured neurons, cell adhesion to a suitable substrate in a suitable medium is followed by extension of lamellipodia (veils of plasma membrane) around the cell body. Some of these begin the process of neurite initiation; that is, they differentiate into growth cones, and direct the extension of neurites (Dotti et al., 1988). Tension, such as that exerted by an emerging growth cone, is an adequate stimulus for the initiation and elongation of a neurite (Bray, 1984; Zheng et al., 1991). The traditional view of neurite elongation is that lamellipodia add membrane at their leading edge, presumably by transport of vesicles from the Golgi apparatus into the cone, where they fuse with the lamellipodial membrane (Purves and Lichtmann, 1985), but recent studies using cultured chick sensory neurons (Zheng et al., 1991) and *Xenopus* spinal neurons (Popov et al., 1993) have demonstrated that membrane may also be added near the soma and along the neurite shaft. As the growth cone advances, microtubules and neurofilaments "consolidate" the advance by polymerizing into relatively stable structures in the neurite shaft. Neurite branching has received relatively little study, although it has been suggested that phosphorylation of microtubule-associated proteins may promote branching (Friedrich and Aszodi, 1991).

B. THE CYTOSKELETON AND NEURITE STRUCTURE

The cytoskeleton of neurons consists of a scaffolding of three types of protein fibers: microfilaments, neurofilaments, and microtubules. Excellent overviews are provided by Jacobson (1991) and Riederer (1990). Microfilaments are composed of actin polymers. Cross-linking between microfilaments by other proteins forms a gel-like matrix that underlies and is tightly linked to the plasma membrane. Actin microfilaments are the most abundant cytoskeletal component of growth cones, and interactions of microfilaments with nonmuscle myosin molecules cause the membrane movements characteristic of growth cones. Actin microfilaments are in a state of dynamic equilibrium, continuously adding actin monomers at the "plus" end and losing them at the "minus" end. An assortment of proteins, many of them regulated by $[Ca^{2+}]_{in}$, modulate actin polymerization and depolymerization, and the attachment of actin filaments to the plasma membrane.

Microtubules are polymers of tubulin. Like microfilaments, microtubules are constantly polymerizing at the "plus" end and depolymerizing at the "minus" end. They provide structural support and act as conveyers of organelles within neurons and other cells. A variety of microtubule-associated proteins

(MAPs) are found in conjunction with microtubules. Some of these MAPs promote microtubule assembly and/or stabilize microtubules by forming cross-bridges between them.

Neurofilaments are intermediate filaments that are unique to neurons. They are composed of neurofilament proteins, and unlike microfilaments and microtubules, they remain in a stable polymerized state. Neurofilaments are synthesized within the cell body and then transported into neurites. They are found distributed throughout the neuron singly or bound by cross-bridges into bundles, and oriented longitudinally. Neurofilaments presumably have a stabilizing structural role in neuronal cytoarchitecture, but their function is poorly understood.

1. The Cytoskeleton of Axons and Dendrites

Axons and dendrites share many similarities, but also differ in several important respects. Although there are many exceptions, a few generalizations can be made. Axons are usually longer than dendrites and elongate more rapidly. When the leading growth cone of an axon or axon branch contacts an appropriate target cell, it differentiates into a presynaptic terminal. Neurofilaments are more abundant in axons than in dendrites. Microtubules, while found in both types of neurites, are uniformly oriented in axons, with the plus end directed toward the periphery, but have mixed orientation in dendrites (Baas et al., 1988). Microtubule-associated proteins are differentially distributed, with MAP2 localized in dendrites, and tau localized in axons (see review by Ginzburg, 1991). Dendrites, but not axons, contain pericentriolar material (Ferreira et al., 1993), ribosomes, rough and smooth endoplasmic reticulum, and Golgi elements (Sargent, 1989). These structural differences between axons and dendrites suggest that neurotoxic metals may have differential effects on these two types of neurites.

Observations of cultured neurons undergoing differentiation have provided information as to the timing and sequence of events that leads to the differentiation of neurites into axons and dendrites (Dotti et al., 1988). Cultured hippocampal neurons initially extend several apparently identical processes. At this stage, all processes contain MAP2, which in a fully differentiated cell is confined to dendrites. After a few hours, one neurite, the presumptive axon, begins to elongate rapidly, but all neurites still express MAP2 until about 48 h in culture. Over the next few days, the axon becomes distinct from the dendrites, losing MAP2 and expressing a neurofilament subunit confined to axons (Pennypacker et al., 1991). Microtubule polarity remains uniform in all neurites until 4 to 5 days in culture, when dendritic microtubules begin to become nonuniform (Baas et al., 1989). After a week in culture, axons and dendrites are fully differentiated.

The importance of cytoskeletal elements and associated proteins in governing neurite development has been shown in a number of recent studies in which the synthesis or activity of one or more of these proteins was inhibited or enhanced. For example, antisense oligonucleotides directed against MAP1B (also called MAP5 in some papers) mRNA blocks neurite outgrowth in PC12 cells (Brugg et al., 1993). Similarly, antisense oligonucleotides directed against MAP2 mRNA inhibit neurite outgrowth in embryonal carcinoma cells treated with retinoic acid to induce neuronal differentiation (Dinsmore and Solomon, 1991). Inhibiting tau expression with antisense oligonucleotides inhibits axonal, but not dendritic, growth in cultured cerebellar macroneurons (Caceres et al., 1991) and inhibits the development of axon-like processes produced by NB2a/d1 neuroblastoma cells in response to serum deprivation (Shea et al., 1992b).

2. Regulation of the Neuronal Cytoskeleton by $[Ca^{2+}]_{in}$

Changes in $[Ca_{2+}]_{in}$ can affect the cytoskeleton, and consequently neurite growth, in many ways. Based on experiments using cytochalasin B, actin remodeling is necessary for the growth cone to guide neurite extension along appropriate pathways (Marsh and Letourneau, 1984). A Ca^{2+}-activated enzyme, gelsolin, severs actin filaments and produces a transition from a gel to a sol state that allows cytoplasmic streaming and movements of cells and growth cones. Activation of gelsolin by a brief rise in $[Ca^{2+}]_{in}$ greatly increases its affinity for actin, causing it to shut down new actin filament assembly and disrupt the existing actin network. This breakdown may be an important first step in the restructuring of the membrane cytoskeleton that is required for growth cone motility, neurite initiation, and other Ca^{2+}-influenced processes (see Forscher, 1989).

High, probably localized, $[Ca^{2+}]_{in}$ may stimulate calcium-activated proteases (calpains), which can cleave a large assortment of proteins, including neurofilaments (Johnson et al., 1991), tubulin (Billger et al., 1988), MAPs (Billger et al., 1988; Fischer et al., 1991) and several protein kinases (Melloni and Pontremoli, 1989). Calpain activity appears to be regulated through a complex set of interactions among calpain, $[Ca^{2+}]_{in}$, and a natural, Ca^{2+}-dependent inhibitor protein, calpastatin (Melloni and Pontremoli, 1989).

Calmodulin is also activated by relatively high, probably localized, $[Ca^{2+}]_{in}$. In the presence of calcium, calmodulin binds to and regulates many other proteins. For example, calmodulin binds to the 200 kDa neurofilament protein and inhibits its calpain-mediated hydrolysis (Johnson et al., 1991). Calcium–calmodulin also binds to MAP2 and inhibits microtubule assembly (Wolff, 1988). Calmodulin may also modulate the attachment of the cytoskeleton to the plasma membrane (Liu and Storm, 1990). Calcium–calmodulin-dependent protein kinase phosphorylates many cytoskeletal components, including microtubules and MAPs, and this phosphorylation helps to control microtubule bundling, the interactions between microtubules, MAPs, and neurofilaments, and the stability of microtubules (for reviews see Matus, 1988a,b; Nixon and Sihag, 1991).

A minimum concentration of Ca^{2+} is also required for activation of some isoforms of protein kinase C (PKC), which also phosphorylates cytoskeletal proteins (Matus, 1988a,b). Finally, high $[Ca^{2+}]_{in}$ activates calcineurin, a calcium-dependent phosphatase. Calcineurin dephosphorylates many of the same proteins that are phosphorylated by calcium–calmodulin-dependent protein kinase or other protein kinases (Armstrong, 1989; Liu and Storm, 1990). Therefore, depending on the exact Ca^{2+} dependence of these regulatory proteins, their spatial distribution within a neuron, and the spatial distribution of $[Ca^{2+}]_{in}$, neurites may initiate, extend, or retract.

C. GROWTH CONES

Growth cones consist of lamellipodia spanning fingerlike membrane extensions (filopodia). Both lamellipodia and filopodia are rich in actin filaments that confer motility, allowing the growth cone to direct neurite outgrowth. The growth cone adheres to the substrate (including other cells) by means of one or more of several cell adhesion molecules. As the growth cone advances, it exerts tension on the extending neurite (Lamoureux et al., 1989; Letourneau, 1975). When cultured dorsal root ganglion cells are treated and with actin-binding drug cytochalasin B, growth cone advance is halted (Yamada et al., 1971). Similarly, in cultured cerebellar granule cells, cytochalasin B induces complete growth cone collapse (Abosch and Lagenaur, 1993). Neurite elongation may persist, but even when neurites continue to elongate, they are more curved than control neurites on the same substrate (Abosch and Lagenaur, 1993; Marsh and Letourneau, 1984). These findings suggest that actin-mediated growth cone pull, rather than push from behind, mediates growth cone advance and normal neurite elongation. Growth cones appear to detect both physical and chemical environmental cues and follow them to appropriate targets for innervation (see review by Bray and Hollenbeck, 1988). Appropriate target cells may provide "stop" signals that inhibit further neurite extension (Baird et al., 1992).

1. Calcium and Growth Cones

There seems to be relatively good agreement that Ca^{2+} influx and/or fairly high $[Ca^{2+}]_{in}$ is required for growth cone motility (e.g., Anglister et al., 1982; Connor, 1986; Cohan et al., 1987; Kater et al., 1988). For example, Mattson and Kater (1987), using cultured *Helisoma* (pond snail) neurons, found that low concentrations of heavy metals, which partially inhibit Ca^{2+} influx through VSCC, cause filopodia to retract and growth cone motility to cease. At least moderate levels of Ca^{2+} influx may enhance growth cone motility in N1E-115 neuroblastoma cells (Silver et al., 1989). These investigators also found that VSCCs were clustered in "hotspots" in the growth cone but not the neurite shaft. Therefore, action potentials would result in localized Ca^{2+} influx and higher $[Ca^{2+}]_{in}$ levels in the growth cone than in the neurite shaft (Silver et al., 1990). At least in some cell types, neurite elongation seems to be favored by lower $[Ca^{2+}]_{in}$, so this distribution of VSCC may be important for normal neurite development.

D. NEURITE INITIATION

As outlined above, neurite initiation probably begins with a growth cone emerging from a cell body (or preexisting neurite, in the case of the initiation of a branch). Many studies support the hypothesis that normal levels of calcium influx are required for neurite initiation. Inhibition of Ca^{2+} influx by specific blockers of L-type VSCC (dihydropyridines, 10 μM Cd^{2+}) inhibits neurite initiation in chick retinal ganglion cells (Suarez-Isla et al., 1984), rat dorsal root ganglion neurons (Robson and Burgoyne, 1989), rat sympathetic neurons (Rogers and Hendry, 1990), chick embryo brain neurons, and N1E-115 neuroblastoma cells (Audesirk et al., 1990). Enhancing Ca^{2+} influx with the ionophore A23187 enhances neurite initiation in B50 neuroblastoma cells (Reboulleau, 1986).

E. NEURITE ELONGATION

Neurite elongation is directed and stimulated by pull from an attached growth cone (Bray, 1984; Lamoureux et al., 1989; Zheng et al., 1991; see also review by Bray and Hollenbeck, 1988) and supported by the polymerization of tubulin subunits and the addition of membrane along the extending neurite shaft. Evidence from various cell types suggests that neurite elongation may occur either at the base of the growth cone or along the length of the neurite shaft. In cultured neurons of the mollusk *Aplysia*, the growth cone lamellipodium spreads, then thickens as it is invaded by cytoplasm and organelles. As the old lamellipodium becomes an extension of the neurite shaft, new lamellipodial membrane elaborates distally (Aletta and Greene, 1988). Chick dorsal root ganglion cells in culture extend neurites by increasing the assembly of microtubules at the distal ends of the neurites (Lim et al., 1990). In cultured frog embryonic neurons, axonal elongation is associated with bundling of microtubules that extend into the growth cone (Tanaka and Kirschner, 1991). New membrane in elongating *Xenopus* neurites is added at the cell body and along the neurite shaft (Popov et al., 1993). In cultured chick sensory neurons, when neurites are artificially stimulated to elongate by applying tension, new membrane is added along the neurite shaft (Zheng et al., 1991).

1. Calcium and Neurite Elongation

The role of Ca^{2+} in neurite elongation is poorly understood, perhaps due to variability among cell types. In some cell types, neurite elongation appears to be enhanced by low $[Ca^{2+}]_{in}$ and inhibited by Ca^{2+} influx. For example, the Ca^{2+} ionophore A23187 inhibits neurite elongation in rat hippocampal neurons (Mattson et al., 1988) and *Helisoma* neurons (Mattson and Kater, 1987). Low levels of heavy metals (La^{3+}, Co^{2+}), which reduce Ca^{2+} influx through VSCC, enhance neurite elongation in *Helisoma* neurons, but higher concentrations of heavy metals inhibit elongation (Mattson and Kater, 1987). These data suggest that low, but not zero, Ca^{2+} influx and/or low $[Ca^{2+}]_{in}$ enhance neurite elongation. In contrast, neurites from *Xenopus* spinal neurons elongate in Ca^{2+}-free media with as much as 5 mM EGTA (Bixby and Spitzer, 1984), suggesting that Ca^{2+} influx is not required at all.

Intermediate levels of $[Ca^{2+}]_{in}$ favor neurite elongation in cultured chick dorsal root ganglion neurons. Neurite outgrowth is inhibited both by removing Ca^{2+} from the culture medium and by the addition of Ca^{2+} ionophores, suggesting that the permissive level of Ca^{2+} for neurite outgrowth in these cells is neither very high nor very low (Lankford and Letourneau, 1989).

Neurite elongation in some cell types is apparently enhanced by Ca^{2+} influx. Anglister et al. (1982) found that depolarization by elevated potassium or the Ca^{2+} ionophore A23187 promotes neurite elongation in N1E-115 neuroblastoma cells, suggesting that high Ca^{2+} influx stimulates elongation.

Neurite growth in still other cell types may be relatively unaffected by changes in $[Ca^{2+}]_{in}$. For example, Garyantes and Regehr (1992) electrically stimulated cultured rat superior cervical ganglion neurons and monitored a dramatic increase in $[Ca^{2+}]_{in}$, which had no influence on the rate of neurite elongation. This sampling suggests that while Ca^{2+} influx and $[Ca^{2+}]_{in}$ play important roles in neurite elongation, their roles are complex and cell-type specific.

II. METAL EFFECTS OF NEURITE DEVELOPMENT

Most metals, particularly most heavy metals, have no known essential role in the metabolism of living organisms. Even those that are essential, such as zinc or iron, are normally present in very low amounts in living organisms, and/or are regulated very precisely by specialized molecules such as ferritin. The nonessential metals, with Pb^{2+} as perhaps the best example, were practically unavailable to living organisms throughout most of evolutionary history, because they were bound in rocks and seldom mobilized in significant amounts in forms that could be assimilated by living organisms, particularly animals. Therefore, it appears that there has been little or no selective pressure favoring the evolution of macromolecules that could bind essential metals, such as zinc or calcium, while excluding nonessential heavy metals. The result is significant susceptibility of modern organisms to metals that have been mobilized by industrial processes.

Most heavy metals are polyvalent cations, and a large number of these cations can bind to proteins or other macromolecules with varying affinities. Two categories of binding sites that are probably important contributors to many aspects of the neurotoxicity of heavy metals are (1) sites that normally bind calcium ions and (2) sulfhydryl groups. Organic heavy metals, such as triethyl lead or methylmercury, may both (1) serve as carriers for metals into the brain and into the cytoplasm of neurons and glia, where they may release inorganic metals, and (2) exert significant, often enhanced, toxicity as a result of their own unique chemical configurations.

Despite decades of study, the molecular mechanisms of toxicity of most heavy metals and organoheavy metals remain incompletely understood. In this review, we will provide both descriptions of metal effects and discussions of proposed mechanisms, where these are known or hypothesized. We will not attempt to describe all effects of all metals on neural development. Rather, we will concentrate on a few, relatively well-studied metals: inorganic and organic lead, methylmercury, cadmium, and aluminum. Mechanistically, we will focus on known or hypothetical interactions of heavy metals with calcium-regulated processes and cytoskeletal assembly or function.

III. INORGANIC LEAD

As a result of human activities, lead is now so ubiquitous that even the most remote human populations probably have body burdens of lead that are orders of magnitude greater than those of our prehistoric ancestors (Settle and Patterson, 1980). Although overt lead poisoning is rare today, subtle dysfunctions can be caused by very low level exposure, especially in children (Bellinger et al., 1986; Dietrich et al., 1987, 1991; Lilienthal et al., 1990; for reviews see Bellinger et al., 1991; Marlowe, 1985; Needleman, 1987). Neurobehavioral effects of lead have been observed at blood lead concentrations of 10 to 15 µg/dl in children, and < 20 µg/dl in rodents (Davis et al., 1990).

A. *IN VIVO* EFFECTS OF INORGANIC LEAD ON NEURITE DEVELOPMENT

A large body of evidence suggests that chronic exposure to Pb^{2+} *in vivo* causes abnormal neuronal development. Exposure of fetal and neonatal rodents, usually rats, to Pb^{2+} causes a variety of changes in the fine structure of neurons and in their synaptic connections. Administration of Pb^{2+} to neonatal rats via milk of dams maintained on a diet of 4.0% $PbCO_3$ until postnatal day 25 caused changes in the morphology of pyramidal cells in the sensorimotor cerebral cortex. A reduction in the number of dendritic branches occurred at distances 80 to 100 µm away from the cell body (Petit and LeBoutillier, 1979). Lorton and Anderson (1986a) dosed rat pups with 600 mg/kg of lead acetate daily via a stomach tube for 4 days after birth. Blood lead averaged 526 µg/dl after 10 days. At 30 days of age, pyramidal cells in the motor cortex Pb^{2+}-exposed rats had a significantly decreased length of branches of both apical and basal dendrites. Cerebellar Purkinje cells from these rats showed a 40% decrease in dendritic arborization (Lorton and Anderson, 1986b). In hippocampal dentate gyrus cells, rats whose dams were fed a diet containing either 0.4% or 4.0% lead carbonate showed increased dendritic branching close to the cell body, but reduced branching at distances from the cell body of 160 µm and greater (Alfano and Petit, 1982). In 25-d old rats fed from birth by dams on a 4.0% $PbCO_3$ diet, both the length and width of the mossy fiber tract (composed of axons of hippocampal dentate gyrus neurons) were significantly reduced (Alfano et al., 1982). Kiraly and Jones (1982) investigated lead effects on the dendritic spine density of hippocampal pyramidal neurons. Newborn rats were fed by Pb^{2+}-exposed dams (given 1% lead acetate in drinking water) until day 25, then received drinking water with 1% lead acetate until day 56. At both 20 and 56 days of age, apical dendritic spine density was significantly reduced (by about 38%) in the lead-exposed pups. *In vivo* Pb^{2+} exposure has also been reported to reduce synaptogenesis in rat cortex (Averill and Needleman, 1980; Krigman et al., 1974; Petit and LeBoutillier, 1979). In a study by Reuhl et al. (1989), monkeys were dosed with Pb^{2+} from birth until 6 years of age, and their visual cortices examined. Monkeys receiving 2 mg Pb/kg/d had blood lead levels of about 50 µg/dl, while those receiving 25 µg/kg/d had levels of 20 µg/dl or lower. Dendritic arborization was decreased in pyramidal neurons of the visual cortex in the high-dose group compared to the low-dose group. Although the Pb^{2+} dosing regimes varied enormously among these studies, the predominant finding was a decrease in dendrite branching, dendritic spine number, or synapses, all of which would indicate a decrease in connectivity among neurons.

Somewhat different results were obtained when cortical pyramidal neurons were measured in guinea pig pups at 34 days of age whose dams had been dosed with several concentrations of lead acetate from gestational day 22 until birth (Legare et al., 1993). The pups showed increased apical and basal dendrite length, increased numbers of apical dendrites per cell, and increased branching of basal dendrites, although the rigorous statistical analysis applied to the data in this study showed only one significant result. Although blood lead levels in these animals were not reported, extrapolation from an earlier study by these investigators (Sierra, et al., 1989) suggests that blood lead peaked in the pups at well below 100 µg/dl.

The differences among these studies are not readily reconciled. However, it would not be unexpected that differences among species and different stages of development, routes of administration, and Pb^{2+} concentrations may produce different results. As we will describe below, *in vitro* data both from intact cells and cell-free enzyme studies suggest that Pb^{2+} may have multimodal effects. Further, as briefly described above in Section I, normal neurite development may require optimal levels of intracellular Ca^{2+} concentrations, kinase activity, etc. Departures from these optimal levels may either enhance or reduce various aspects of neurite development, and, of course, either hypo- or hypertrophy of neurites may be detrimental to brain functioning.

B. *IN VITRO* EFFECTS OF INORGANIC LEAD ON NEURITE DEVELOPMENT

The concentration of lead in cerebrospinal fluid is unknown in the experimental animals described above. In a human population not known to be exposed to lead, Cavalleri et al. (1984) found total lead concentration in CSF to be approximately 25 nM (total lead in whole blood, 21 µg/dl; total lead in plasma, 30 nM). Culture media normally contain higher protein concentrations and are somewhat higher pH than CSF, so these lead concentrations in CSF are not necessarily directly comparable to lead concentrations in culture. Nevertheless, they offer a starting point for evaluation of the likely environmental relevance of culture studies.

The effects of *in vitro* exposure of cultured neurons to Pb^{2+} vary considerably, depending on the species, cell type, and parameter of neurite development measured. The effects of Pb^{2+} exposure *in vitro* on neuronal differentiation has been studied in rat dorsal root ganglion cells or explants (Scott and Lew, 1986; Windebank, 1986), IMR32 human neuroblastoma cells (Gotti, et al., 1987), embryonic chick brain neurons (Audesirk et al., 1989), embryonic rat cortical neurons (Kern et al., 1993), embryonic rat hippocampal neurons, N1E-115 neuroblastoma cells (derived from mouse peripheral nervous system) and B-50 neuroblastoma cells (derived from rat CNS; Audesirk et al., 1991). In rat dorsal root ganglion explants, neurite elongation was inhibited by Pb^{2+} exposure for 2 to 3 days at concentrations of approximately 500 µM and above (Windebank, 1986). Myelination, however, was almost completely prevented by concentrations of 1 µM or above. In a separate study by Scott and Lew (1986), survival of adult dorsal root ganglion cells was reduced by 50% following 18 d of exposure to 35 µM Pb^{2+}. In IMR32 neuroblastoma cells, survival was reduced by 50% by exposure to 336 µM for 10 d, and 260 µM Pb^{2+} reduced neurite initiation by about 45%, although elongation remained almost normal (Gotti et al., 1987). In chick neurons, Pb^{2+} exposure for 3 to 4 d (beginning with plating) inhibited neurite initiation at concentrations of 10 µM and above, and enhanced neurite elongation at concentrations of 100 µM and above (Audesirk et al., 1989). Exposures of 2 d (beginning with plating) to Pb^{2+} concentrations between 10 nM and 500 µM had very little effect on any parameter of development in B50 neuroblastoma cells (survival, initiation rate, number of neurites per cell, elongation of neurites; Audesirk et al., 1991). For N1E-115 neuroblastoma cells, rat hippocampal neurons, and rat cortical neurons, there were complex, multimodal dose-response curves for several parameters of neuronal differentiation (Audesirk et al., 1991; Kern et al., 1993). Generally, in cortex and hippocampal neurons, mid-to-high nanomolar and mid-to-high micromolar Pb^{2+} concentrations inhibited initiation and enhanced branching, while low micromolar concentrations had little effect on any parameter of neurite development.

It is difficult to summarize such disparate data, but a few generalizations may be made. First, immortal cell lines appear to be relatively insensitive to Pb^{2+} effects, as shown by the high concentrations needed to have any significant effects (Gotti et al., 1987; Audesirk et al., 1991). Second, neuronal survival is also relatively resistant to Pb^{2+}, even in primary neurons (Audesirk et al., 1991; Kern et al., 1993). Third, myelination, neurite initiation, and perhaps branching appear to be the most sensitive parameters of neuronal differentiation, with inhibition or enhancement occurring at concentrations of 1 µM or below (Audesirk et al., 1991; Kern et al., 1993; Windebank, 1986).

Pb^{2+} also has substantial effects on glial cells *in vivo* (e.g., Cookman et al., 1988; Selvin-Testa et al., 1991) and *in vitro* (e.g., Cookman et al., 1988; Rowles et al., 1989; Sobue and Pleasure, 1985; Stark et al., 1992; for review, see Tiffany-Castiglioni et al., 1989). Pb^{2+} also affects cell adhesion molecules (Lagunowich et al., 1993; Regan, 1993). Impacts on glia or cell adhesion would also be expected to alter neuronal differentiation. A discussion of these effects is beyond the scope of this chapter.

C. CELLULAR MECHANISMS

Pb^{2+} inhibits Ca^{2+} influx through VSCCs at low micromolar concentrations in a variety of cell types, including mouse N1E-115 neuroblastoma cells (Oortgiesen et al., 1990; Audesirk and Audesirk, 1991), human neuroblastoma cells (Reuveny and Narahashi, 1991), rat dorsal root gangion cells (Evans et al.,

1991), and embryonic rat hippocampal neurons (Audesirk and Audesirk, 1993). Since normal rates of Ca^{2+} influx through VSCC promote neurite initiation (see Section I above), levels of Pb^{2+} sufficient to block VSCC would be expected to inhibit initiation. However, since these Pb^{2+} levels are considerably above those found in Pb^{2+}-exposed humans, blocking Ca^{2+} influx through VSCC is unlikely to be an important mechanism of developmental neurotoxicity of Pb^{2+}. Interactions of Pb^{2+} with VSCC may, however, still be important, because, in addition to blocking Ca^{2+} influx through VSCC, Pb^{2+} ions readily permeate through VSCC (Simons and Pocock, 1987; Tomsig and Suszkiw, 1991). Once inside a neuron, Pb^{2+} may influence a variety of Ca^{2+}-mediated intracellular processes by directly substituting for Ca^{2+}, antagonizing Ca^{2+}-transport mechanisms, or stimulating increases in $[Ca^{2+}]_{in}$ (see reviews by Bressler and Goldstein, 1991; Goldstein, 1993; Pounds, 1984; Simons, 1988, 1993). One difficulty in elucidating the mechanisms of intracellular effects of Pb^{2+} is the paucity of information on Pb^{2+} concentrations inside cells. Following short-term exposure to relatively high concentrations of Pb^{2+}, the intracellular free Pb^{2+} ion concentration has been reported to be in the range of tens of picomolar (Schanne et al., 1989a; Tomsig and Suszkiw, 1990). There are no data available on the concentration of free Pb^{2+} ions in the cytoplasm of chronically exposed cells, but if Pb^{2+} distributes across the plasma membrane similarly to Ca^{2+}, then one would expect tens to hundreds of picomolar free Pb^{2+}, even in quite highly exposed cells. Very few biochemical studies of Pb^{2+} effects on intracellular molecules have used such low Pb^{2+} concentrations.

Nevertheless, there are indications that Pb^{2+} may alter the function of a variety of intracellular proteins. For example, Pb^{2+} in picomolar concentrations may activate PKC (Markovic and Goldstein, 1988; Rosen et al., 1993), while higher levels may block PKC activation (Speizer et al., 1989; Rosen et al., 1993). PKC, in turn, has diverse and apparently variable effects on differentiation and growth in several types of cultured neurons (Cabell and Audesirk, 1993; Cambray-Deakin et al., 1990; Felipo et al., 1990; Hsu et al., 1989; Minana, et al., 1990; Tsuda et al., 1989). Pb^{2+} may also substitute for Ca^{2+} in the activation of calmodulin (Habermann et al., 1983; Goldstein and Ar, 1983). Lead may also increase $(Ca^{2+})_{in}$ (Dowd and Gupta, 1991; Rosen and Pounds, 1989; Schanne et al., 1989a,b; Schanne et al., 1990), and this increase may activate calmodulin (Goldstein, 1993). Ca^{2+}-calmodulin (or perhaps Pb^{2+}-calmodulin) in turn regulates the activity of enzymes, such as Ca^{2+}-calmodulin-dependent protein kinase II, that modulate cytoskeletal polymerization, crosslinking, and attachment to membranes (see, for example, Goedert et al., 1991; Matus, 1988b). Because the molecular mechanisms whereby the cytoskeleton and cytoskeleton-associated molecules generate normal neurite development are incompletely understood, exactly how the enhancement or inhibition of the function of these molecules by Pb^{2+} would generate the observed perturbations remains unclear.

Another potential interaction between Pb^{2+} and Ca^{2+}-mediated events occurs through Pb^{2+} effects on NMDA-type glutamate receptors and/or their associated ion channels, which are highly permeable to Ca^{2+} and Na^+. It has been reported that low micromolar concentrations of Pb^{2+} inhibit NMDA receptor-mediated currents in cultured rat hippocampal neurons (Alkondon et al., 1990). Further, chronic *in vivo* Pb^{2+} exposure of neonatal rats produces supersensitivity to the effects of NMDA (Petit et al., 1992) and an increase in binding of an NMDA-specific ligand in several brain regions (Brooks et al., 1993), both of which suggest that chronic Pb^{2+} exposure causes an upregulation of NMDA receptors. Since NMDA receptors can admit Ca^{2+} into neurons and have been implicated in neurite development in cerebellar granule cells *in vitro* (Rashid and Cambray-Deakin, 1992), it is possible that Pb^{2+} effects on NMDA receptors may be yet another pathway by which Pb^{2+} may alter neurite development *in vivo*.

Although there are very few studies available, it appears unlikely that Pb^{2+} has significant direct interactions with the cytoskeleton. For example, Roderer and Doenges (1983) found no effects on tubulin polymerization *in vitro* with Pb^{2+} at concentrations as high as 650 μM, a concentration several orders of magnitude greater than would ever occur inside a cell. However, because Pb^{2+} can substitute for Ca^{2+}, and because Ca^{2+} mediates so many of the intracellular processes governing cytoskeletal integrity (discussed above), direct or indirect interference with the neuronal cytoskeleton may be an important, but as yet undocumented, mechanism underlying lead neurotoxicity.

IV. ORGANIC LEAD

Organic lead is released into the environment primarily in the form of alkylleads. Alkylleads, being both volatile and lipid soluble, are easily absorbed through the skin and respiratory tract and readily cross the blood-brain barrier. Alkyllead compounds are potent neurotoxic agents, with particularly detrimental effects on the developing brain (see reviews by Cremer, 1984; Konat, 1984; and Verity, 1990).

A. *IN VIVO* EFFECTS OF ORGANIC LEAD ON NEURITE DEVELOPMENT

There are few detailed developmental studies of ultrastructural effects of organic lead *in vivo* that would shed light on its effects on neurite growth. Niklowitz (1975) administered massive doses (100 mg/kg) of tetraethyl lead to rabbits. Within 12 h after administration, many pyramidal cells in the frontal cortex and hippocampus exhibited neurofibrillary tangles. Yagminas et al. (1992) dosed weanling rats with triethyl lead in amounts ranging from 0.05 to 1.0 mg/kg/d, 5 d/week for 91 days. No changes were detected in the brains, but the spinal cords and lumbar/sacral nerves showed randomly distributed lesions that included a reduction of neurotubules and neurofilaments. Ferris and Cragg (1984) administered tetramethyl lead by injection to pregnant female rats at a dosage of 22 mg/kg (about 20% of the LD_{50}) on days 7, 14, and 21 of pregnancy, then injected the pups with a proportional dose at 6 d of age. Examination of brain tissue at postnatal days 13 and 28 revealed a significant reduction in the brain/body weight ratio, but no effects of tetramethyl lead treatment on a variety of histological parameters of different brain regions. Walsh et al. (1986) found that a single injection of triethyl lead (2.6 to 7.9 mg/kg) to adult rats produced damage to neurons of the hippocampus and dorsal root ganglia and that a single injection of trimethyl lead (8.8 to 26.2 mg/kg) caused damage to neurons of the spinal cord and hippocampus. Booze and Mactutus (1990) found that a single dose of triethyl lead (9.0 mg/kg) administered to 5-d-old rats caused permanent thinning of the pyramidal cell layer of the hippocampus and a decrease in pyramidal cell numbers. These studies clearly indicate that organic lead causes morphological changes in both fetal/neonatal and adult brains, but they did not address possible alterations of neurite development per se.

B. *IN VITRO* EFFECTS OF ORGANIC LEAD ON NEURITE DEVELOPMENT

Neurite production in cultured chick embryo cerebral cells was investigated by Grundt et al. (1981). Cultures aged, 1, 2, and 3 weeks were incubated in 3.16 μM triethyl lead chloride for 48 h, which caused a degeneration of neurites to about 50% of control levels. Audesirk et al. (1989) cultured chick embryo brain neurons in triethyl lead from the time of plating up to 3 to 4 days. At levels of 200 nM and higher, triethyl lead caused a significant decrease in the number of cells that produced neurites, while at concentrations of 500 nM and above, mean neurite length was significantly reduced. Recently, we exposed cultures of embryonic rat hippocampal neurons to triethyl lead chloride concentrations from 0.1 nM to 5 μM (Audesirk and Audesirk, 1992; Audesirk et al., 1995). We found a dose-dependent range of effects on various parameters of differentiation. The most sensitive target was neurite branching, which was significantly reduced at 10 nM in dendrites, and 100 nM in axons. At concentrations of 1 μM and above, the number of dendrites per cell was significantly reduced. At 2 μM and above, both axon length and neurite initiation were decreased. Survival was unaffected at concentrations below 5 μM.

C. ORGANIC LEAD AND Ca^{2+} HOMEOSTASIS

Organic lead can cause increases in $[Ca^{2+}]_{in}$, but probably only at concentrations that exceed those found with environmental exposure. Guinea pig cortical synaptosomes incubated for 10 min in triethyl lead concentrations from 5 to 30 μM showed significant $[Ca^{2+}]_{in}$ increases (50 to 170 nM), that were relatively independent of triethyl lead concentration. In addition, accumulation of $^{45}Ca^{2+}$ by synaptosomes was significantly increased by 20 μM triethyl lead, but not by lower levels (Komulainen and Bondy, 1987). This increase in $[Ca^{2+}]_{in}$ may be due to an inhibition of ATP production, impairing the ability of Ca^{2+}-ATPase to extrude Ca^{2+} (Kauppinen et al., 1988). In cultured rat hippocampal neurons, the intracellular calcium concentration of newly plated cells increased during 3 hours of exposure to triethyl lead concentrations of 1 and 5 μM, but not to lower concentrations (Audesirk et al., 1995). This suggests that increases in $[Ca^{2+}]_{in}$ could contribute to the reduced numbers of dendrites per cell seen at triethyl lead concentrations of 1 μM and higher, to the reduction of neurite-bearing cells and axon length at 2 μM, and to the decreased survival seen at 5 μM. Other mechanisms, such as interference with the integrity of the neuronal cytoskeleton (see below), may be responsible for the reduction in neurite branching at nanomolar concentrations.

D. ORGANIC LEAD AND THE CYTOSKELETON

Triethyl lead has been shown to disrupt both microtubules and neurofilaments. At concentrations as low as 1 μM, triethyl lead causes rapid depolymerization of microtubules in cultured PtK-1 cells (from kangaroo rat kidney) and human fibroblasts (Zimmerman et al., 1985; Zimmerman et al., 1987). Much higher levels (50 μM) are required to depolymerize microtubules *in vitro*, suggesting that microtubule disassembly in living cells results from the interference of triethyl lead with a more sensitive intracellular process controlling microtubule formation, perhaps elevation of $[Ca^{2+}]_{in}$, which inhibits microtubule assembly and can cause microtubule depolymerization *in vitro* both directly and through a variety of indirect mechanisms (Wolff, 1988; Yamamoto et al., 1983; Suzuki et al., 1986).

Neurofilament integrity is disrupted by far lower concentrations of triethyl lead both *in vitro* (200 nM) and in cultured Neuro-2A cells (10 nM; Zimmerman et al., 1987). As with microtubules, much higher levels of triethyl lead are required to disrupt neurofilaments *in vitro* than *in vivo*, suggesting indirect effects in cells. Neurofilaments are the most abundant cytoskeletal element in neurons and are important in maintaining complex neuronal shapes (see review by Schlaepfer, 1987). Although the mechanisms of neurite branching are unknown, the levels of triethyl lead reported by Zimmermann et al. (1987) to disrupt neurofilaments are nearly identical to those causing decreased axonal and dendritic branching in hippocampal cultures, suggesting a possible link between neurofilament integrity and branching.

V. METHYLMERCURY

Mercury enters the environment by weathering of geological deposits and from industrial sources and agriculture. After entry into aquatic ecosystems, inorganic mercury is methylated into more toxic methylmercury, which bioaccumulates in fish and shellfish. Methylmercury has devastating effects on the developing fetal nervous system where it becomes concentrated. (For reviews see Choi, 1989; O'Kusky, 1992; and Chapter 66 by E. Massaro.) Although inorganic mercury has effects on microtubules, Ca^{2+} channels and Ca^{2+} homeostasis, and therefore would be expected to impair neurite development, it appears to be less toxic than methylmercury. This review will therefore focus exclusively on methylmercury.

A. *IN VIVO* EFFECTS OF METHYLMERCURY ON NEURITE DEVELOPMENT

Stoltenburg-Didinger and Markwort (1990) exposed pregnant rats to methylmercury in doses of 0.025 to 5 mg/kg/d on days 6 to 9 of gestation. At 250 days of age, pyramidal neurons in the somatosensory cortex of the offspring of all exposed groups showed normal dendritic arborization, but both apical and basilar dendrites from the 5 mg/kg group had abnormal spines. While controls had stubby, mushroom-shaped spines, methylmercury exposed neurons had spines that were longer, thinner, more distorted, and more numerous. Since the methylmercury-exposed dendritic spines are similar to those found normally up until 30 days after birth, the authors suggest that this may be related to the developmental retardation seen in the offspring. Choi et al. (1981b) injected mice on postnatal days 3, 4, and 5 with 5 mg/kg methylmercuric chloride, and examined the morphology of Purkinje cells in the cerebellum. They found a dramatic reduction in the dendritic arborization of Purkinje cells in the exposed mice.

B. *IN VITRO* EFFECTS OF METHYLMERCURY ON NEURITE DEVELOPMENT

Neurite outgrowth and filopodial activity rapidly ceased in organotypic cultures of human fetal cerebrum exposed to 20 μM methylmercuric chloride (Choi et al., 1981a). Further, the plasma membrane became separated from the neurites, especially at the growth cones. Neurites degenerated and separated from the cell bodies within 4 h. Electron microscopic examination of cultures revealed neurotubular damage.

C. METHYLMERCURY AND CA^{2+} HOMEOSTASIS

Like Pb^{2+}, methylmercury increases $[Ca^{2+}]_{in}$ and blocks Ca^{2+} channels. In rat cerebral synaptosomes, methylmercury from 5 μM to 30 μM increases $[Ca^{2+}]_{in}$ in a dose-dependent manner. This increase is several times that induced by either triethyl lead or triethyl tin (Komulainen and Bondy, 1987). These investigators suggest that methylmercury renders the plasma membrane leaky to Ca^{2+}. Acute methylmercury exposure inhibits neuromuscular transmission (Atchison and Narahashi, 1982), blocks two types of VSCC in PC12 cells (Shafer and Atchison, 1991) and reduces ^{45}Ca uptake into synaptosomes (Shafer and Atchison, 1989), mostly in the low-to-mid micromolar concentration range. In contrast, Sarafian (1993) exposed cerebellar granule neuron cultures (1 to 3 weeks of age) to methylmercury at 1, 3, and 5 μM and found a dose-dependent increase in ^{45}Ca uptake and $[Ca^{2+}]_{in}$, suggesting that there may be some compensation of Ca^{2+} influx with chronic exposure. Since neurite initiation, at least in some cell types, is modulated by Ca^{2+} influx through VSCC, neurite production might be altered by methylmercury concentrations in the low micromolar levels.

D. METHYLMERCURY AND THE CYTOSKELETON

Methylmercury disrupts microtubules and inhibits their polymerization *in vitro* at low micromolar levels (Abe et al., 1975; Miura et al., 1984; Sager et al., 1983; Vogel et al., 1985). In cultured cells, methylmercury disrupts existing microtubules (Graff et al., 1993, and Sager et al., 1983, cultured fibroblasts; Miura et al., 1984, mouse glioma; Sager, 1988, PtK$_2$ kidney epithelial cells). Comparing data from several of these reports, it appears that, like triethyl lead, methylmercury inhibits microtubule assembly or stability at lower concentrations in intact cells than *in vitro*. This indicates that interference with other factors that influence cytoskeletal structure or function is probably more important than direct effects on the cytoskeleton itself.

Sarafian and Verity (1990) report that exposure of cerebellar granule cells in culture to methylmercury at concentrations from 200 nM to 3.0 μM stimulates phosphorylation of a variety of proteins, including cytoskeletal proteins, in a dose-dependent manner. Protein phosphorylation, in turn, influences the interactions among microtubules, MAPs, and neurofilaments, as well as the stability of microtubules (see above). Since methylmercury directly inhibits several protein kinases, this phosphorylation is probably mediated indirectly via a second messenger, such as Ca^{2+} or inositol phosphate. Methylmercury-induced increases in [Ca^{2+}]$_{in}$ could activate calmodulin, which can inhibit microtubule assembly by binding with MAPs (Wolff, 1988) and could activate Ca^{2+}-calmodulin-dependent protein kinase, which phosphorylates many cytoskeletal components. In addition, methylmercury (0.5 to 10 μM) produces a dose-dependent increase in levels of inositol phosphate in cultured cerebellar granule cells (Sarafian, 1993). This suggests that methylmercury activates phospholipase C and could stimulate protein kinase C by production of diacylglycerol. Protein kinase C in turn phosphorylates many cytoskeletal proteins.

Finally, injection of methylmercury into frog sciatic nerve impairs axonal transport (Abe et al., 1975). Neurite development is dependent on axonal transport for the delivery of many materials to the growth cone and elongating neurite shaft. Therefore, inhibiting axonal transport would be expected to inhibit neurite development.

VI. CADMIUM

Cadmium damages the kidneys, lungs, bones, ovaries, and testes. Cadmium also inhibits cell proliferation and may be carcinogenic, but its effect on the developing nervous system is largely unexplored. The effects of cadmium on cell differentiation and organogenesis are reviewed by Rogers in this volume (Chapter 65).

A. EFFECTS OF CADMIUM ON NEURITE DEVELOPMENT

There appear to be little data describing cadmium effects on neurite development *in vivo*. *In vitro*, cadmium inhibits neurite initiation in chick embryo brain neurons at concentrations of 10 μM and above, with no effect on neurite elongation (Audesirk et al., 1989). Similar effects occur in cultured embryonic rat hippocampal neurons, with initiation decreased by concentrations of 5 μM and above, while neurite elongation is not significantly affected (unpublished observations). To our knowledge, although total cadmium in intact brains of cadmium-exposed animals has been reported (e.g., about 4 μM, Vig and Nath, 1991), there have been no analyses of cadmium concentrations in cerebrospinal fluid. Therefore, it is unclear if these findings are relevant to environmental exposures.

B. CADMIUM AND CA^{2+} HOMEOSTASIS

Despite the paucity of evidence for effects of cadmium on neurite outgrowth, its well-documented disruption of calcium homeostasis leads to the prediction that cadmium should influence neurite growth via this mechanism, if concentrations in CSF reach high enough levels. For example, Cd^{2+} binds to the Ca^{2+}-binding site on CaATPase with an IC$_{50}$ of 1.6 nM, inhibiting its ability to transport Ca^{2+} out of the cell and to sequester Ca^{2+} into the endoplasmic reticulum (Verbost et al., 1987). Like Pb^{2+} and methylmercury, Cd^{2+} both blocks voltage-sensitive Ca^{2+} channels and permeates through them (Chow, 1991; Hinkle et al., 1987; Thèvenod and Jones, 1992). Further, entry through VSCC accounts for much of the cytotoxicity of Cd^{2+} to cultured cells of the pituitary cell line GH$_4$C$_1$ (Hinkle et al., 1987; see also Hinkle et al., 1992). Vig and Nath (1991) fed rats 6 mg Cd/kg/d for 4 weeks and examined the activity of calmodulin and various enzymes extracted from cerebral cortices. They found that cadmium exposure significantly reduces the activity of calmodulin, phosphodiesterase, and CaATPase, but not adenylate cyclase. These investigators suggest that Cd^{2+} may displace Ca^{2+} in calmodulin, reducing the activity of calmodulin and preventing it from activating phosphodiesterase and CaATPase. However, it has been reported that Cd^{2+} can directly activate calmodulin (Cheung, 1984; Habermann et al., 1983; Perrino and Chou, 1986, 1989), inhibit CaATPase (Hechtenberg and Beyersmann, 1991), and inhibit protein kinase C (Saijoh et al., 1988; Speizer, et al., 1989). Most of these *in vitro* effects occurred at concentrations in the micromolar range. It is not known what the intracellular cadmium concentration is in cadmium-exposed cells, although it apparently may be possible to measure cadmium concentrations with the fluorescent dye fura-2 (Hinkle et al., 1992).

C. CADMIUM AND THE CYTOSKELETON

Cadmium has deleterious effects on the cytoskeleton that could disrupt neurite outgrowth. Cadmium (10 μM for 16 h) causes microtubule disassembly and microfilament aggregation in 3T3 cells in culture (Li et al., 1993). Similar effects on microtubules have been found *in vitro*, where Cd^{2+} causes disassembly of intact microtubules (Perrino and Chou, 1986) and inhibits tubulin polymerization (Perrino and Chou, 1989), apparently through activation of calmodulin. It should be noted, however, that these *in vitro* effects were found with somewhat higher Cd^{2+} concentrations (about 50 μM), which again suggests that the effects in intact cells may not be mediated directly through actions on the cytoskeleton, but through perturbations of other intracellular signaling pathways.

VII. ALUMINUM

There is no documented neurological harm from normal levels of exposure to aluminum, but the neurotoxicity of aluminum is of some concern in areas where aluminum is in high concentration in soil and is mobilized into water by acid precipitation (see Boegman and Bates, 1984). Fatal encephalopathy has been produced by large doses of aluminum in kidney dialysis patients. Until recently, these patients were treated with dialysis solutions high in aluminum and consumed large doses of aluminum-based antacids to reduce high blood phosphate levels, creating a daily aluminum intake that could exceed 5 grams (Lione, 1985). It has also been suggested that aluminum may play a role in the development of Alzheimer's disease.

A. *IN VIVO* EFFECTS OF ALUMINUM ON NEURITE DEVELOPMENT

Some, but not all, animals are susceptible to aluminum neurotoxicity either developmentally or as adults or both. Cats, rabbits, and dogs are relatively susceptible, whereas rats and monkeys are quite resistant (Boegman and Bates, 1984). Aluminum exposure of infant or adult rabbits, usually by direct injection into the brain or spinal cord, produces neurofibrillary tangles in neurons of the hippocampus, cerebral cortex, cerebellum, and spinal cord (e.g., Forrester and Yokel, 1985; Ghetti et al., 1985; Petit et al., 1985; Wisniewski et al., 1984). There is also an ongoing controversy concerning the possible role of aluminum in the development of Alzheimer's disease or amyotrophic lateral sclerosis, particularly with respect to the paired helical filaments and neurofibrillary tangles in Alzheimer's (e.g., Good et al., 1992; Landsberg et al., 1992). In general, the animal studies involved very high doses of aluminum, and usually employed adult animals. The human Alzheimer's studies, of course, not only involved adults, but were also only correlative. Therefore, although it appears evident that large doses of aluminum *in vivo* induce abnormalities in neuronal ultrastructure, it is unclear if these studies have any bearing on developmental effects of lower doses.

B. *IN VITRO* EFFECTS OF ALUMINUM ON NEURITE DEVELOPMENT

Most studies indicate that cultured neurons are quite resistant to the effects of aluminum, with effects usually appearing at quite high concentrations. Embryonic rat hippocampal neurons in culture which were allowed to develop normally prior to aluminum administration showed no degeneration in response to exposure to aluminum levels up to 200 μM for 6 days (Mattson et al., 1993). Langui et al. (1988) found that 14-day exposure of embryonic rat cerebral neurons to 75 μM aluminum chloride resulted in some apparent cell death (although this was not quantified) and the formation of neurofibrillary tangles. There are also cell type-specific differences in susceptibility. Cultured embryonic rabbit motor neurons exposed to 1 or 10 μM aluminum chloride for 14 days developed somatal and neurite inclusions composed of phosphorylated neurofilaments; at concentrations of 25 μM and above, motor neurons did not survive (Strong and Garruto, 1991). In contrast, rabbit hippocampal neurons survived concentrations up to 100 μM, and did not develop inclusions at concentrations of 25 μM or less (Strong and Garruto, 1991). NB2a neuroblastoma cells exposed to concentrations of $AlCl_3$ ranging from 100 to 1000 μM develop normal morphology, although whorls of neurofilaments are found in the somata (Shea et al., 1989). Similar neurofibrillary abnormalities were seen in differentiated PC12 cells exposed to 500 μM aluminum chloride for 2 days (Shea and Fischer, 1991). These data tend to confirm the higher sensitivity of rabbits compared to rats, and suggest that at least some of the higher susceptibility of rabbits is directly attributable to enhanced susceptibility of neurons themselves. Further, the relative insensitivity of immortal cell lines noted with other metals (see section III) seems to be true for aluminum as well.

C. ALUMINUM AND CA^{2+} HOMEOSTASIS

Because of its chemical similarity to calcium, aluminum may impact calcium-dependent processes. For example, the calcium ionophore A23187 allows an influx of aluminum as well as Ca^{2+} (Mattson et al., 1993). Siegel and Haug (1983) found that aluminum binds to calmodulin and alters its conformation, making it unable to interact with its normal target enzymes. This observation is supported by an *in vivo* study by Farnell et al. (1985), who measured a progressive decline in the activity of calmodulin during aluminum-induced encephalopathy in rabbits. Among many other roles, calmodulin stimulates CaATPase which in turn helps maintain the large Ca^{2+} gradient across the plasma membrane. Total brain tissue Ca^{2+} levels are increased following aluminum administration in rabbits (Farnell et al. 1985) and mice (Anghileri, 1993), suggesting a disruption of normal Ca^{2+} homeostasis, which could be caused by effects on calmodulin. On the other hand, in rat forebrain synaptosomes, $AlCl_3$ concentrations of 50 μM and greater significantly inhibited voltage-dependent influx of Ca^{2+} (Koenig and Jope, 1987), which might be expected to lead to reduced $[Ca^{2+}]_{in}$. This aluminum concentration is well within the range of brain aluminum levels from humans with aluminum encephalopathy (Koenig and Jope, 1987), although probably much greater than the aluminum level in otherwise unexposed individuals.

D. ALUMINUM AND THE CYTOSKELETON

An extensive body of literature documents disruptive effects of aluminum on the neuronal cytoskeleton, suggesting that it might interfere with normal neurite development. A prominent finding in neurons exposed to aluminum *in vivo* and *in vitro* is the presence of intraneuronal tangles of neurofilament proteins (see, for example, Forrester and Yokel, 1985; Langui et al., 1988; Shea et al., 1989; Shea and Fischer, 1991; Strong and Garruto, 1991, all cited above). A possible mechanism for aluminum-induced neurofibrillary tangles is provided by Leterrier et al. (1992) who report that aluminum strongly stimulates binding between bovine spinal cord neurofilaments *in vitro*. They suggest that enhanced phosphorylation of neurofilaments by aluminum may be responsible for increasing cross-links between them. Enhanced phosphorylation of neurofilaments by aluminum has been reported in rats who consumed aluminum in drinking water (Johnson and Jope, 1988). Using an *in vivo* model in which rabbit hypoglossal nerve nuclei were directly injected with 1.5% $AlCl_3$, Bizzi et al. (1984) found accumulations of neurofilament proteins in nearly all hypoglossal neurons, and a selective reduction in the transport of neurofilaments from the cell body down the axon. This results in engorgement of proximal portions of the axons with neurofilaments. Troncoso et al. (1985), using rabbit sciatic nerve, also concluded that aluminum blocks neurofilament transport. Neurofilaments also appear to be stabilized by aluminum against degradation by calcium-dependent and independent proteases (Shea et al., 1992a).

Evidence that microtubule assembly is promoted by extremely low concentrations of aluminum (less than 10^{-10} M) was provided by Macdonald et al. (1987), who reported that the association constant for aluminum with tubulin is 10^7 times greater than that for magnesium, which normally binds to tubulin and promotes microtubule assembly. These authors also report that aluminum-bound tubulin forms microtubules that are more stable than normal to hydrolysis and to Ca^{2+}-mediated depolymerization. However, these results have not been replicated *in vivo*. Enhanced polymerization was not observed in tubulin isolated from brains of mice fed a diet high in aluminum (brain aluminum up to 165 pg/g fresh tissue; Oteiza et al., 1989). Mesco et al. (1991), using human neuroblastoma cells, found that aluminum concentrations as low as 100 μM significantly increased levels of tau microtubule-associated proteins, which are associated with microtubules in axons. Even if the total amount of microtubule-associated protein remains constant, aluminum may still alter MAP-tubulin association, through phosphorylation. For example, phosphorylation of MAP2 was significantly increased in rats given aluminum sulfate in drinking water (0.3% aluminum), although the quantity of MAP2 protein itself remained stable (Johnson and Jope, 1988).

VIII. OTHER METALS

Other metals and organometals also have the potential to alter neurite development. In general, the potential to impair neurite development exists for any metal that (1) inhibits Ca^{2+} influx through voltage-sensitive or perhaps ligand-gated ion channels; (2) alters Ca^{2+} homeostasis in other ways, such as by inhibiting Ca^{2+} extrusion by Na^+-Ca^{2+} exchange or CaATPase; or (3) alters the cytoskeleton, either directly (probably through interactions with sulfhydryl groups) or indirectly (probably through effects on phosphorylation or dephosphorylation). For example, many metals and organometals have been shown to affect cytoskeletal elements, either in intact cells or in *in vitro* assays, including arsenite (Li and Chou, 1992), chromate (Zhao and Chou, 1992), nickel (Li et al., 1993), and trialkyltins (Tan et al., 1978). Trialkyltins, which cause degeneration of adult neurons, particularly in the hippocampus and other limbic areas (Naalsund et al., 1985), also increase $[Ca^{2+}]_{in}$ (Kauppinen et al., 1988; Oyama et al., 1992), and inhibit CaATPase (Yallapragada et al., 1991). Therefore, a variety of metals and organometals, including most heavy metals, would be expected to alter neurite development if present in the embryonic and/or neonatal brain in sufficient concentrations. In most cases, whether these metals (in exposed individuals) normally approach levels that would cause neurodevelopmental toxicity is unknown.

Further, several metals, such as manganese and iron, have been suggested as causative agents in some neurodegenerative disorders, such as Parkinson's disease. These two metals may contribute to the formation of free radicals and thus lead to oxidative stress and neuronal death. Manganese also produces abnormal neurite growth in PC12 cells in culture (Tsai et al., 1993). Whether such effects might occur during neuronal development *in vivo* is unknown.

IX. SUMMARY

With a few exceptions, there is a paucity of information concerning the effects of metals on neurite development *in vivo* or *in vitro*. Studies of metal effects on neuronal development *in vivo* are quite rare, and it remains unclear in most cases whether altered development, when observed, is a direct effect on neurons and their processes or on other parts of the body with secondary deleterious effects on the developing nervous system. Even when it is known that a metal does impair neurite development *in vivo*, the search for mechanisms of action, which is often best pursued with *in vitro* studies, usually remains hampered by deficiencies in our knowledge, including the precise molecular events that control neurite initiation and elongation, growth cone movements, and target identification; interactions between neuronal and nonneuronal cells both in promoting normal development and in mediating toxicity; the concentration of toxic metals in the extracellular fluid of the developing brain; and the concentration of toxic metals in the neuronal or glial cytoplasm. For neurotoxicologists, these last three points require particular attention in order to define cellular mechanisms of action at concentrations of toxic metals that are likely to be found in humans. These caveats notwithstanding, there are a few generalizations that can be made.

1. ***Effects on Calcium Channels:*** Most heavy metals inhibit Ca^{2+} influx through voltage-sensitive and/or ligand-gated calcium channels. However, *in vitro* studies generally indicate that direct effects on channels occur at concentrations substantially higher than occur in all but the most severely exposed individuals. Nevertheless, effects on calcium channels do occur, such as the apparent up-regulation of NMDA receptors (and presumably their associated calcium-permeable ion channels) following chronic Pb^{2+} exposure (Brooks et al., 1993). It should be noted that the ultimate cause of such up-regulation is not known, but need not be an effect on the receptors or channels themselves at all; for example, altered $[Ca^{2+}]_{in}$ may alter densities of calcium channels. These facts suggest that metals may affect Ca^{2+} channels, and/or perhaps other molecules involved in Ca^{2+} homeostasis, by indirect means, such as through changes in phosphorylation or dephosphorylation. It is possible that such indirect effects might occur at very low concentrations, but would not usually be detected, for example, in the normal short-term, whole-cell, or isolated patch voltage-clamp experiment.
2. ***Effects on the Cytoskeleton:*** Many heavy metals, both organic and inorganic, alter the structure and/or function of the cytoskeleton. In virtually all cases, the metal concentration required to affect the cytoskeleton in intact cells is much lower than the concentration that inhibits polymerization of cytoskeletal elements in cell-free extracts. Therefore, it would seem to be a general rule that the primary target for cytoskeletal effects is not the cytoskeleton itself, but intracellular molecules that modulate cytoskeletal assembly and function. We should also note that, where the concentrations are known, the

metal concentration that impacts the cytoskeleton is substantially lower than the concentration that is known to affect calcium channels (at least directly; see above).
3. **Differential Sensitivity of Cell Types:** There are often substantial differences in sensitivity among cell types within a given species and also between species (see, for example, Section VII on aluminum, above). In *in vitro* studies, immortal cells, such as PC12 cells or neuroblastoma cells, are less sensitive, at least to some metals such as aluminum and lead, than are primary cultured cells.

REFERENCES

Abe, T., Haga, T., and Kurokawa, M. (1975), *Brain Res.,* 86, 504–508.
Abosch, A. and Lagenaur, C. (1993), *J. Neurobiol.,* 24, 344–355.
Aletta, J. and Green, L. A. (1988), *J. Neurosci.,* 8, 1425–1435.
Alfano, D. P., LeBoutillier, J. C., and Petit, T. L. (1982), *Exp. Neurol.,* 75, 308–319.
Alfano, D. P. and Petit, T. L. (1982), *Exp. Neurol.,* 75, 275–288.
Anghileri, L. J. (1993), *Neurotoxicology,* 13, 475–478.
Anglister, L., Farber, I. C., Shahar, A., and Grinvald, A. (1982), *Dev. Biol.,* 94, 351–365.
Alkondon, M., Costa, A. C. S., Radhakrishnan, V., Aronstam, R. S., and Albuquerque, E. X. (1990), *FEBS Lett.,* 261, 124–130.
Armstrong, D. L., (1989), *Trends Neurosci.,* 12, 117–122.
Atchison, W. D. and Narahashi, T. (1982), *Neurotoxicology,* 3, 37–50.
Audesirk, G. (1993), *Neurotoxicology,* 14, 137–148.
Audesirk, G. and Audesirk, T. (1991), *Neurotoxicology,* 12, 519–528.
Audesirk, G. and Audesirk, T. (1993), *Neurotoxicology,* 14, 259–266.
Audesirk, T. and Audesirk, G. (1992), *Neurotoxicology,* 13, 884.
Audesirk, G., Audesirk, T., Ferguson, C., Lomme, M., Shugarts, D., Rosack, J., Caracciolo, P., Gisi, T., and Nichols, P. (1990), *Dev. Brain Res.,* 55, 109–120.
Audesirk, T., Audesirk, G., Ferguson, C., and Shugarts, D. (1991), *Neurotoxicology,* 12, 529–538.
Audesirk, T., Audesirk, G., and Shugarts, D., (1995), *Cell Biol. Toxicol.,* 11, 1–10.
Audesirk, G., Shugarts, D., Nelson, G., and Przekwas J. (1989), *In Vitro Cell Dev. Biol.,* 25, 1121–1128.
Averill, D. R. and Needleman, H. L. (1980), *Low Level Lead Exposure: The Clinical Implications of Current Research,* Needleman, H. L. Ed., Raven Press, New York, 201–210.
Baird, D. H., Hatten, M. E., and Mason, C. A. (1992), *J. Neurosci.,* 12, 619–634.
Baas, P. W., Black, M. M., and Banker, G. A. (1989), *J. Cell Biol.,* 109, 3085–3094.
Baas, P. W., Deitch, J. S., Black, M. M., and Banker, G. A. (1988), *Proc. Nat. Acad. Sci. U.S.A.,* 85, 8335–8339.
Bellinger D. C., Leviton, A., Needleman, H. L., Waternaux, C., and Rabinowitz, M. (1986), *Neurobehav. Toxicol. Teratol.,* 8, 151–161.
Bellinger D. C., Sloman, J., Leviton, A., Rabinowitz, M., Needleman, H. L., and Waternaux, C. (1991), *Pediatrics,* 87, 219–227.
Billger, M., Wallin, M., and Karlsson, J. O. (1988), *Cell Calcium,* 9, 33–44.
Bixby, J. L. and Spitzer, N. C. (1984), *Dev. Biol.,* 106, 89–96.
Bizzi, A., Clark Crane, R., Autilio-Gambetti, L., and Gambetti, P. (1984), *J. Neurosci.,* 4, 722–731.
Boegman, R. J. and Bates, L. A. (1984), *Can. J. Physiol. Pharmacol.,* 62, 1010–1014.
Booze, R. M. and Mactutus., C. F. (1990), *Experientia,* 46, 292–297.
Bray, D. (1984), *Dev. Biol.,* 102, 379–389.
Bray, D. (1987), *Trends NeuroSci.,* 10, 431–438.
Bray, D. and Hollenbeck, P. J. (1988), *Annu. Rev. Cell. Biol.,* 4, 43–61.
Bressler, J. P. and Goldstein, G. W. (1991), *Biochem. Pharmacol.,* 41, 479–484.
Brooks, W. J., Petit, T. L., LeBoutillier, J. C., Nobrega, J. N., and Jarvis, M. F. (1993), *Drug. Dev. Res.,* 29, 40–47.
Brugg, B., Reddy, D., and Matus, A. (1993), *Neuroscience,* 52, 489–496.
Cabell, L. and Audesirk, L. (1993), *Int. J. Dev. Neurosci.,* 11, 357–368.
Cambray-Deakin, M. A., Adu, J., and Burgoyne, R. D. (1990), *Dev. Brain. Res.,* 53, 40–46.
Cavalleri, A., Minoia, C., Ceroni, M., and Poloni, M. (1984), *J. Appl. Toxicol.,* 4, 63–65.
Cheung, W. Y. (1984), *Fed. Proc.,* 43, 2995–2999.
Cohan, C. S., Connor, J. A., and Kater, S. B. (1987), *J. Neurosci.,* 7, 3588–3599.
Connor, J. A. (1986), *Proc. Natl. Acad. Sci. U.S.A.,* 83, 6179–6183.
Choi, B. H. (1989), *Prog. Neurobiol.,* 32, 447–470.
Choi, B. H., Cho, K. H., and Lapham, L. W. (1981a), *Environ. Res.,* 24, 61–74.
Choi, B. H., Kudo, M., and Lapham, L. W. (1981b), *Acta Neuropathol.,* 54, 233–237.
Chow, R. H. (1991), *J. Gen. Physiol.,* 98, 751–770.
Cookman, G. R., Hemmens, S. E., Keane, G. J., King, W. B., and Regna, C. M. (1988), *Neurosci. Lett.,* 86, 33–37.
Cremer, J. E. (1984), *Biological Effects of Organolead Compounds,* Grandjean, P. and Grandjean, E. C., Eds., CRC Press, Boca Raton, FL, 207–218.

Davis, J. M., Otto, D. A., Weil, D. E., and Grant, L. D. (1990), *Neurotoxicol. Teratol.,* 12, 215–229.
Devoto, S. H. (1990), *Experientia,* 46, 916–922.
Dietrich, K. N., Krafft, K. M., Bornschein, R. L., Hammond, P. B., Berger, O., Succop, P. A., and Bier, M. (1987), *Pediatrics,* 80, 721–730.
Dietrich, K. N., Succop, P. A., Berger, O. G., Hammond, P. B., and Bornschein, R. L., (1991), *Neurotoxicol. Teratol.,* 13, 203–211.
Dinsmore, J. H. and Solomon, F. (1991), *Cell,* 64, 817–826.
Dowd, T. L. and Gupta, R. K. (1991), *Biochim. Biophys. Acta,* 1092, 341–346.
Dotti, C. G., Sullivan, C. A., and Banker, G. A. (1988), *J. Neurosci.,* 8, 1454–1468.
Evans, M. L., Busselberg, D., and Carpenter, D. O. (1991), *Neurosci. Lett.,* 129, 103–106.
Farnell, B. J., Crapper McLachlan, D. R., Baimbridge, K., De Boni, U., Wong, L., and Wood, P. L. (1985), *Exp. Neurol.,* 88, 68–85.
Felipo, V., Minana, M.-D., and Grisolia, S. (1990), *J. Biol. Chem.,* 265, 9599–9601.
Ferreira, A., Palazzo, R. E., and Rebhun, L. I. (1993), *Cell Motil. Cytoskel.,* 25, 336–344.
Ferris, N. J. and Cragg, B. G. (1984), *Acta Neuropathol.,* 63, 306–312.
Fischer, I., Romano-Clarke, G., and Grynspan, F. (1991), *Neurochem. Res.,* 16, 891–898.
Forrester, T. M. and Yokel, R. A. (1985), *Neurotoxicology,* 6, 71–80.
Forscher, P. (1989), *Trends Neurosci.,* 12, 468–474.
Friedrich, P. and Aszodi, A. (1991), *FEBS Lett.,* 295, 5–9.
Garyantes, T. K. and Regehr, W. G. (1992), *J. Neurosci.,* 12, 96–103.
Ghetti, B., Musicco, M., Norton, J., and Bugiani, O. (1985), *Neuropathol. Appl. Neurobiol.,* 11, 31–53.
Ginzburg, I. (1991), *Trends Biochem. Sci.,* 16, 257–261.
Goedert, M., Crowther, R. A., and Garner, C. C. (1991), *Trends Neurosci.,* 14, 193–199.
Goering, P. L. (1993), *Neurotoxicology,* 14, 45–60.
Goldstein, G. W. (1993), *Neurotoxicology,* 14, 97–101.
Goldstein, G. W. and Ar, D. (1983), *Life Sci.,* 33, 1001–1006.
Good, P. F., Perl, D. P., Bierer, L. M., and Schmeidler, J. (1992), *Ann. Neurol.,* 31, 286–292.
Gotti, C., Cabrini, D., Sher, E., and Clementi, F. (1987), *Cell Biol. Toxicol.,* 3, 431–440.
Graff, R. D., Philbert, M. A., Lowndes, H. E., and Reuhl, K. R. (1993), *Toxicol. Appl. Pharmacol.,* 120, 20–28.
Grundt, I. K., Ammitzboll, T., and Clausen, J. (1981), *Neurochem. Res.,* 6, 193–201.
Hechtenberg, S. and Beyersmann, D. (1991), *Enzyme,* 45, 109–115.
Habermann, E., Crowell, K., and Janicki P. (1983), *Arch. Toxicol.,* 54, 61–70.
Hinkle, P. M., Kinsella, P. A., and Osterhoudt, K. C. (1987), *J. Biol. Chem.,* 262, 16333–16337.
Hinkle, P. M., Shanshala II., E. D., and Nelson, E. J. (1992), *J. Biol. Chem.,* 267, 25553–25559.
Hsu, L., Jeng, A. Y., and Chen, K. Y. (1989), *Neurosci. Lett.,* 99, 257–262.
Jacobson, Marcus (1991), *Developmental Neurobiology,* 3rd ed., Plenum Press, New York.
Johnson, G. V. W., Greenwood, J. A., Costello, A. C., and Troncoso, J. C. (1991), *Neurochem. Res.,* 16, 869–873.
Johnson, G. V. W. and Jope, R. S. (1988), *Brain Res.,* 456, 95–103.
Kater, S. B., Mattson, M. P., Cohon, C., and Connor, J. (1988), *Trends Neurosci.,* 11, 315–321.
Kauppinen, R. A., Komulainen, H., and Taipale, H. T. (1988), *J. Neurochem.,* 51, 1617–1625.
Kern, M., Audesirk, T., and Audesirk, G. (1993), *Neurotoxicology,* 14, 319–328.
Kiraly, E. and Jones D. G. (1982), *Exp. Neurol.,* 77, 236–239.
Koenig, M. L. and Jope, R. R. (1987), *J. Neurochem.,* 49, 316–320.
Komulainen, H. and Bondy, S. C. (1987), *Toxicol. Appl. Pharmacol.,* 88, 77–86.
Konat, G. (1984), *Neurotoxicology,* 5, 87–96.
Krigman, M. R., Druse, M. J., Traylor, T. D., Wilson, M. H., Newell, L. R., and Hogan, E. L. (1974), *J. Neuropathol. Exp. Neurol.,* 33, 671–687.
Lagunowich, L. A., Stein, A. P., and Reuhl, K. R. (1993), *Toxicologist,* 13, 168.
Lamoureux, P., Buxbaum, R. E., and Heidemann, S. R. (1989), *Nature,* 340, 159–162.
Landsberg, J. P., McDonald, B., and Watt, F. (1992), *Nature,* 360, 65–68.
Langui, D., Anderton, B. H., Brion, J.-P., and Ulrich, J. (1988), *Brain Res.,* 438, 67–76.
Lankford, K. L. and Letourneau, P. C. (1989), *J. Cell Biol.,* 109, 1229–1243.
Legare, M. E., Castiglioni, A. J., Rowles, T. K., Calvin, J. A., Snyder-Armstead, C., and Tiffany-Castiglioni, E. (1993), *Neurotoxicology,* 14, 77–80.
Leterrier, J. F., Langui, D., Probst, A., and Ulrich, J. (1992), *J. Neurochem.,* 58, 2060–2070.
Letourneau, P. C. (1975), *Dev. Biol.,* 44, 92–101.
Letourneau, P. C., Shattuck, T. A., and Ressler, A. H. (1987), *Cell Motil. Cytoskel.,* 8, 193–209.
Li, W. and Chou, I.-N. (1992), *Toxicol. Appl. Pharmacol.,* 114, 132–139.
Li, W., Zhao, Y., and Chou, I.-N. (1992), *Toxicol. In Vitro,* 6, 433–444.
Li, W., Zhao, Y., and Chou, I.-N. (1993), *Toxicology,* 77, 65–79.
Lione, A. (1985), *Gen. Pharmacol.,* 16, 223–228.
Lilienthal, H., Winneke, G., and Ewert, T. (1990), *Environ. Health Perspect.,* 89, 21–25.
Lim, S. S., Edson, K. L., Letourneau, P. C., Borisy, G. G. (1990), *J. Cell Biol.,* 111, 123–130.

Liu, Y. and Storm, D. R. (1990), *Trends Pharmacol. Sci.,* 11, 107–111.
Lorton, D. and Anderson, W. J. (1986a), *Neurobehav. Toxicol. Pharmacol.,* 8, 45–50.
Lorton, D. and Anderson, W. J. (1986b), *Neurobehav. Toxicol. Pharmacol.,* 8, 51–59.
Macdonald, T. L., Humphreys, W. G., and Martin, R. B. (1987), *Science,* 236, 183–186.
Markovac, J. and Goldstein, G. W. (1988), *Nature,* 334, 71–73.
Marlowe, M. (1985), *Res. Commun. Psychol. Psychiat. Behav.,* 10, 153–169.
Marsh, L. and Letourneau, P. C. (1984), *J. Cell Biol.,* 99, 2041–2047.
Mattson, M. P. and Kater, S. B. (1987), *J. Neurosci.,* 7, 4034–4043.
Mattson, M. P., Dou, P., and Kater, S. B. (1988), *J. Neurosci.,* 8, 2087–2100.
Mattson, M. P., Lovell, M. A., Ehmann, W. D., and Markesbery, W. R. (1993), *Brain Res.,* 602, 21–31.
Matus, A. (1988a), *Trends Neurosci.,* 11, 291–292.
Matus, A. (1988b), *Annu. Rev. Neurosci.,* 11, 29–44.
Melloni, E. and Pontremoli, S. (1989), *Trends Neurosci.,* 12, 438–444.
Mesco, E. R., Kachen, C., and Timiras, P. S. (1991), *Mol. Chem. Neuropathol.,* 14, 199–212.
Minana, M.-D., Felipo, V., and Grisolia, S. (1990), *Proc. Natl. Acad. Sci. U.S.A.,* 87, 4335–4339.
Miura, K., Inokawa, M., and Imura, N. (1984), *Toxicol. Appl. Pharmacol.,* 73, 218–231.
Naalsund, L. U., Allen, C. N., and Fonnum, F. (1985), *Neurotoxicology,* 6, 145–158.
Needleman, H. L. (1987), *Neurotoxicology,* 8, 389–393.
Niklowitz, W. J. (1975), *Neurology,* 25, 927–934.
Nixon, R. A. and Sihag, R. K. (1991), *Trends Neurosci.,* 14, 501–506.
O'Kusky, J. R. (1992), *The Vulnerable Brain and Environmental Risk, Vol. 2, Toxins in Food,* Isaacson, R. L. and Jensen, K. F., Eds., Plenum Press, New York, 19–34.
Oortgiesen, M., van Kleef, R. D. G. M., Bajnath, R. B., and Vijerberg, H. P. M. (1990), *Toxicol. Appl. Pharmacol.,* 103, 165–174.
Oteiza, P. I., Golub, M. S., Gershwin, M. E., Donald, J. M., and Keen, C. L. (1989), *Toxicol. Lett.,* 47, 279–285.
Oyama, Y., Chikahisa, L., Hayashi, A., Ueha, T., Sato, M., and Matoba, H. (1992), *Jpn. J. Pharmacol.,* 58, 467–471.
Pennypacker, K., Fischer, I., and Levitt, P. (1991), *Exp. Neurol.,* 111, 25–35.
Perrino, B. A. and Chou, I.-N. (1986), *Cell Biol. Intl. Rep.,* 10, 565–573.
Perrino, B. A. and Chou, I.-N., (1989), *Toxicol. in Vitro,* 3, 227–234.
Petit, T. L., Biedermann, G. B., Jonas, P., and LeBoutillier, J. C. (1985), *Exp. Neurol.,* 88, 640–651.
Petit, T. L. and LeBoutillier, J. (1979), *Exp. Neurol.,* 64, 482–492.
Popov, S., Brown, A., and Poo, M. (1993), *Science,* 259, 244–246.
Pounds, J. G. (1984), *Neurotoxicology,* 5, 295–332.
Purves, D. and Lichtman, J. W. (1985), *Principles of Neural Development,* Sinauer Associates, Sunderland, MA.
Rashid, N. A. and Cambray-Deakin, M. A. (1992), *Dev. Brain Res.,* 67, 301–308.
Reboulleau, C. P. (1986), *J. Neurochem.,* 46, 920–930.
Regan, C. M. (1993), *Neurotoxicology,* 14, 69–74.
Reuhl, K. R., Rice, D. C., Gilbert, S. G., and Mallett, J. (1989), *Toxicol. Appl. Pharmacol.,* 99, 501–509.
Reuveny, E. and Narahashi, T. (1991), *Brain Res.,* 545, 312–314.
Riederer, B. M. (1990), *Eur. J. Morphol.,* 28, 347–378.
Robson, S. J. and Burgoyne, R. D. (1989), *Neurosci. Lett.,* 104, 110–114.
Roderer, G. and Doenges, K. H. (1983), *Neurotoxicology,* 4, 171–180.
Rogers, M. and Hendry, I. (1990), *J. Neurosci. Res.,* 26, 447–454.
Rosen, J. F. and Pounds, J. G. (1989), *Toxicol. Appl. Pharmacol.,* 98, 530–543.
Rosen, J. F., Schanne, F. A., and Long, G. J. (1993), *Toxicologist,* 13, 413.
Rowles, T. K., Womac, C., Bratton, G. R., and Tiffany-Castiglioni, E. (1989), *Metab. Brain Dis.,* 4, 187–201.
Sager, P. R. (1988), *Toxicol. Appl. Pharmacol.,* 94, 473–486.
Sager, P. R., Doherty, R. A., and Olmstead, J. B. (1983), *Exp. Cell. Res.,* 146, 127–137.
Saijoh, K., Inoue, Y., Katsuyama, H., and Sumino, K. (1988), *Pharmacol. Toxicol.,* 63, 221–224.
Sarafian, T. A. (1993), *J. Neurochem.,* 61, 648–657.
Sarafian, T. and Verity, M. A. (1990), *J. Neurochem.,* 55, 913–921.
Sargent, P. B. (1989), *Trends Neurosci.,* 12, 203–205.
Schanne, F. A. X., Dowd, T. L., Gupta, R. K., and Rosen, J. F. (1989a), *Proc. Natl. Acad. Sci. U.S.A.,* 86, 5133–5135.
Schanne, F. A. X., Dowd, T. L., Gupta, R. K., and Rosen, J. F. (1989b), *Brain Res.,* 503, 308–311.
Schanne, F. A. X., Dowd, T. L., Gupta, R. K., and Rosen, J. F. (1990), *Biochim. Biophys. Acta,* 1054, 250–255.
Schlaepfer, W. W. (1987), *J. Neuropathol. Exp. Neurol.,* 46, 117–129.
Scott, B. and Lew, J. (1986), *Neurotoxicology,* 7, 57–68.
Settle, D. M. and Patterson, C. C. (1980), *Science,* 207, 1167–1176.
Selvin-Testa, A., Lopez-Costa, J. J., Nessi De Avinon, A. C., and Pecci Saavedra, J. (1991), *Glia,* 4, 384–392.
Shafer, T. J. and Atchison, W. D. (1989), *J. Pharmacol. Exp. Ther.,* 248, 696–702.
Shafer, T. J. and Atchison, W. D. (1991), *J. Pharmacol. Exp. Ther.,* 258, 149–157.
Shea, T. B., Balikian, P., and Beermann, M. L. (1992a), *FEBS* 307, 195–198.
Shea, T. B., Beermann, M. L., Nixon, R. A., and Fischer, I. (1992b), *J. Neurosci. Res.,* 32, 363–374.

Shea, T. B., Clarke, J. F., Wheelock, T. R., Paskevich, P. A., and Nixon, R. A. (1989), *Brain. Res.,* 492, 53–64.
Shea, T. B. and Fischer, I. (1991), *Neurosci. Res. Commun.,* 9, 21–26.
Siegel, N. and Haug, A. (1983), *Biochim. Biophys. Acta,* 744, 36–45.
Sierra, E. M., Rowles, T. K., Martin, J., Bratton, G. R., Womac, C., and Tiffany-Castiglioni, E. (1989), *Toxicology,* 59, 81–96.
Silver, R. A., Lamb, A. G., and Bolsover, S. R. (1990), *Nature,* 343, 751–754.
Simons, T. J. B. (1988), *Handb. Exp. Pharmacol.,* 83, 509–525.
Simons, T. J. B. (1993), *Neurotoxicology,* 14, 77–85.
Simons, T. J. B. and Pocock, G. (1987), *J. Neurochem.,* 56, 568–574.
Sobue, G. and Pleasure, D. (1985), *Ann. Neurol.,* 17, 462–468.
Speizer, L. A., Watson, M. J., Kanter, J. R., and Brunton, L. L. (1989), *J. Biol. Chem.,* 264, 5581–5585.
Stark, M., Wolff, J. E. A., and Korbmacher, A. (1992), *Neurotoxicol. Teratol.,* 14, 247–252.
Stoltenburg-Didinger, G. and Markwort, S. (1990), *Neurotoxicol. Teratol.,* 12, 573–576.
Strong, M. J. and Garruto, R. M. (1991), *Lab. Invest.,* 65, 243–249.
Suarez-Isla, B. A., Pelto, D. J., Thompson, J. M., and Rapoport, S. I. (1984), *Brain Res.,* 14, 263–270.
Suzuki T., Fujii T., and Tanaka, R. (1986), *Neurochem. Res.,* 11, 543–555.
Tan, L. P., Ng, M. L., and Kumar Das, V. G. (1978), *J. Neurochem.,* 31, 1035–1041.
Tanaka, E. M. and Kirschner, M. W. (1991), *J. Cell Biol.,* 115, 345–363.
Thèvenod, K. and Jones, S. W. (1992), *Biophys. J.,* 63, 162–168.
Tiffany-Castiglioni, E., Sierra, E. M., Wu, J.-N., and Rowles, T. K. (1989), *Neurotoxicology,* 10, 417–444.
Tomsig, J. L. and Suszkiw, J. B. (1991), *Biochim. Biophys. Acta,* 1069, 197–200.
Troncoso, J. C., Hoffman, P. N., Griffen, J. W., Hess-Kozlow, K. M., and Price, D. L. (1985), *Brain Res.,* 342, 172–175.
Tsai, S. S., Sun, A. Y., Kim, H. D., and Sun, G. Y. (1993), *Life Sci.,* 52, 1567–1575.
Tsuda, M., Ono, K., Katayama, N., Yamagata, Y., Kikuchi, K., and Tsuchiya, T. (1989), *Neurosci. Lett.,* 105, 241–245.
Verbost, P. M., Senden, M. H. M. N., and van Os, C. H. (1987), *Biochim. Biophys. Acta,* 902, 247–252.
Verity, M. A. (1990), *Environ. Health Perspect.,* 89, 43–48.
Vogel, D. G., Margolis, R. L., and Mottet, N. K. (1985), *Toxicol. Appl. Pharmacol.,* 80, 473–486.
Vig, P. J. S. and Nath, R. (1991), *Biochem. Int.,* 23, 927–934.
Walsh, T. J., McLamb, R. L., Bondy, S. C., Tilson, H. A., and Chang, L. W. (1986), *Neurotoxicology,* 7, 21–34.
Windebank, A. J. (1986), *Exp. Neurol.,* 94, 203–212.
Wisniewski, H. M., Shek, J. W., Gruca, S., and Sturman, J. A. (1984), *Acta Neuropathol.,* 63, 190–197.
Wolff, J. (1988), *Structure and Function of the Cytoskeleton,* Rousset, B. A. F., Ed., Colloque INSERM/John Libbey Eurotext, Vol 171, 477–480.
Yagminas, A. P., Little, P. B., Rousseaux, C. G., Franklin, C. A., and Villeneuve, D. C. (1992), *Fundam. Appl. Tooxicol.,* 19, 380–387.
Yallapragada, P. R., Vig, P. J. S., Kodavanti, P. R. S., and Desaiah, D. (1991), *J. Toxicol. Environ. Health,* 34, 229–237.
Yamada, K. M., Spooner, B. S., and Wessells, N. K. (1971), *J. Cell Biol.,* 49, 614–635.
Yamamoto H., Fukunaga K., Tanaka E., and Miyamoto E. (1983), *J. Neurochem.,* 41, 1119–1125.
Zheng, J., Lamoureux, P., Santiago, V., Dennerli, T., Buxbaum, R. E., and Heidemann, S. R. (1991), *J. Neuroscience,* 11, 1117–1125.
Zimmerman H.-P., Doenges K. H., and Röderer G. (1985), *Exp. Cell. Res.,* 156, 140–152.
Zimmerman H.-P., Plagens, U., and Traub P. (1987), *NeuroToxicology,* 8, 569–578.

Chapter 70

Mammary Heavy Metal Content: Contribution of Lactational Exposure to Toxicity in Suckling Infants

Sam Kacew

I. INTRODUCTION

Neonatal toxicology involves an abnormal responsiveness of newborns directly to therapeutic agents, recreational chemicals, drugs of abuse, or inadvertent exposure to environmental chemicals including heavy metals. Environmental exposure may encompass the contact of an individual with chemicals in the workplace, termed "occupational exposure," or can arise through atmospheric, water, food, or ground contamination, be it related to direct spraying of crops by farmers or an industrial accident such as in Toyama, Japan where cadmium was released into the river. An important source of environmental contamination, which has received little attention, is the excretion of natural products by humans and animals. Estrogens and testosterone are continuously excreted into the environment by cows, swine, horses, goats, and humans (Shore et al., 1993). These hormones have been identified as contaminants in drinking water but the potential toxic consequences through direct or interactive effects of these chemicals with metals needs to be addressed. It is well-known that metals such as chromium, lead, tin, and mercury exert an effect on the pituitary and hypothalamus to alter hormonal FSH and LH secretion, and thus produce reproductive abnormalities (Mattison, 1985). Since estrogen secretion by the adrenal cortex is dependent on FSH and LH secretion in a negative feedback fashion, it is possible that the heavy metals and environmental exogenous estrogens exert a synergistic adverse effect on reproduction as well as fetal and newborn development. The pregnant mother may be drinking water containing heavy metals and estrogens which can lead to a cumulative source of hormones for the fetus and may play a role in sexual dysfunction in the post-natal period. One must be cognizant that during pregnancy both mother and fetus are equally exposed to a chemical, but that the risk of adverse effects is far greater in the neonate. It is not within the scope of this chapter to examine the consequences of prenatal exposure to metals on newborn development. The focus of this chapter is on the lactational transfer of heavy metals from mothers to nursing infants and the observed potential toxic consequences.

II. IMPORTANCE OF BREAST-FEEDING

The physiological process of breast-feeding plays a critical role in human development. In poor countries the use of bottles to feed nursing infants with milk formula has resulted in enhanced morbidity and mortality (Cunningham, 1979; Cunningham et al., 1991). However, extensive studies clearly demonstrated that breast-feeding provides not only essential nutrition but also protection against infection and a variety of other immunological disorders (Lawrence, 1989; Cunningham et al., 1991). A summary of the illnesses, where it has been clearly demonstrated that breast-feeding provides protection to the nursing infant, is given in Table 1. With respect to the mother, breast-feeding is known to create a special psychological bond between infant and mother which ultimately leads to a socially healthier child (Newton and Newton, 1967). In addition, lactation enhances maternal postpartum recovery and body weight returns to prepartum levels more rapidly. The distinct advantages of breast-feeding and breast milk are widely appreciated, and it is recommended that the barriers which keep women from initiation or continuation of this physiological process be decreased (Lawrence, 1989).

Table 1 Protective Effects Attributed to Breast-feeding

Disease	Ref.
Lower respiratory tract infection	Howie et al., 1990
Otitis media	Duncan et al., 1993
Rotavirus diarrhea	Clemens et al., 1993
Shigellosis	Ahmed et al., 1992
Bacteremia and meningitis	Cochi et al., 1986
Immune system disorders	Mayer et al., 1988
	Davis et al., 1988
	Greco et al., 1988
Food allergies	Gerrard et al., 1973
Atopic dermatitis	Lucas et al., 1990
Chronic liver disease	Sveger et al., 1985

The physiological process of breast-feeding should be encouraged under most circumstances despite the presence of environmental toxins. Chemical exposure via accidents and hazardous waste sites has resulted in toxicant accumulation in breast milk. The human maternal ingestion of a fungicide, hexachlorobenzene-treated wheat, resulted in chemical accumulation in breast milk. Suckling infants subsequently developed symptoms of a disease, pembe yara, and a condition of porphyria cutanea tarda (Cam and Nigogosyan, 1963; Peters et al., 1982). Exposure to pesticides such as mirex through gestation and lactation was found to result in a dose-related incidence of irreversible cataracts in rat pups (Chernoff et al., 1979). Ingestion of polychlorinated biphenyl-contaminated rice oil by nursing mothers was determined to produce low birth weight human infants, growth retardation, and abnormal skin pigmentation as well as bone and tooth defects (Yamaguchi et al., 1971). In extensive studies in North Carolina, Rogan et al. (1986) measured the levels of polychlorinated biphenyls in human milk and found an associated hypotonicity and hyporeflexia in nursing infants. Of particular interest for this review is the Minamata Bay disaster in which nursing mothers ingesting mercury-contaminated fish contributed to production of severe neurological disorders in human infants (Matsumoto et al., 1965). The symptoms associated with aluminum during lactation in mice include reduced body weight and developmental retardation (Golub et al., 1987). Clearly, in exceptional circumstances the benefits derived from breast-feeding fail to outweigh the toxic consequences, and this process should be terminated (Kacew, 1993a,b).

A subject now recognized as an important factor for fetal growth and development is the maternal diet and nutritional status (Basu, 1988). However, with parturition the process of lactation then serves as the essential source for nutrients. In particular, maternal milk provides the essential elements zinc, iron, calcium, and vitamins for the suckling infant (Mahaffey, 1980). Few studies exist on the effects of nutritional maternal status on newborn growth and development. Vitamin D deficiency in pregnant mothers was reported to result in fetal skeletal abnormalities (Basu, 1988). In the presence of vitamin D deficiency in children, Potter et al. (1981) demonstrated that aluminum worsened hyperparathyroid bone disease. It is conceivable that in nursing mothers utilizing aluminum-containing compounds for peptic ulcer therapy, the metal content is elevated in breast milk (Weberg et al., 1986). In maternal vitamin D deficiency the suckling infant would derive proportionately more aluminum and less vitamin D, resulting in osteomalacia (Kacew, 1993a). The maternal nutritional status also plays an important role in lead-, mercury-, and cadmium-induced toxicity (Mahaffey, 1980; Kuhnert et al., 1988). In general, improvement of the nutritional status is related to a greater protective mechanism against various toxic substances. In the presence of higher mammary calcium or zinc content there is less lead or cadmium available for absorption from milk by suckling infants (Mahaffey, 1980).

The time at which maternal exposure to metals occurs and breast-feeding commences is an important factor. During prenatal exposure of mothers to metals, compounds of metals such as lead, zinc, cadmium, and iron are more water-soluble and accumulate in the colostrum, the milky fluid secreted by mammary tissue prior to and one week after parturition. Further excretion of these metals in colostrum is greater than in mature milk (Wilson et al., 1980). Thus, if metal exposure occurs only after birth, the potential for metal excretion from mother to infant is less. Knowledge that prenatal exposure of mothers to high levels of metals has taken place suggests that breast-feeding might be safer after the colostrum has been replaced by mature milk.

The presence of metal in maternal milk may be construed as a potential hazard to the infant even though only 3% of total intake is likely to be found here (Amin-Zaki et al., 1974; Wilson et al., 1980). The primary consideration is the risk to the nursing infant rather than the mere presence of toxicant in the milk. Based on the numerous advantages of breast-feeding, the benefit of this physiological process in the majority of cases far exceeds the potential risk. Although it may be inadvertent, the lactating infant derives environmental chemicals from the mother. These chemicals are excreted in breast milk and may pose a serious potential hazard to the infant. Unlike drug therapy which can be voluntarily terminated, environmental exposure may be chronic and consequently more toxic.

III. ALUMINUM

One of the most abundant elements present in the crust of the earth is aluminum. Although aluminum may be considered as an environmental agent, this metal plays a role in therapeutics. The utilization of aluminum-containing compounds in humans with renal disease, while serving to bind phosphate in dialysis therapy, has been implicated in the etiology of a number of clinical manifestations termed dialysis encephalopathy syndrome (DES) (Alfrey et al., 1976). The symptoms associated with DES in children include ataxia, loss of motor abilities, myoclonus, seizures, dementia, bulbar dysfunction, and progressive slowing in the EEG (Andreoli et al., 1984; Sedman et al., 1984). Developmental delay, microcytic anemia, and osteomalacia were also reported in children (Rotundo et al., 1982; Sedman et al., 1985; Andreoli et al., 1985). Although a direct correlation between mammary gland aluminum levels and subsequent infant disorders has not been established, it should be noted that the environment provides a continuous source of body burden of metal as shown in Table 2. To complicate matters, aluminum is an important constituent of the "over-the-counter" class of drugs used in the treatment of gastric ulcers in lactating women (Kacew, 1990). Based on their ability to neutralize acid in the stomach, aluminum-containing compounds are categorized as gastric antacids. The capacity to bind and neutralize phosphate is the rationale for addition of aluminum-containing antacids to dialysate in neonatology and this mode serves as the prime source of this element in children.

Table 2 Potential Sources for Aluminum in Suckling Infants

Source	Ref.
Municipal drinking water	King et al., 1981
Parenteral nutrients	Koo et al., 1988
Drug	Alfrey et al., 1976
	Kacew, 1993a
Breast milk	Weberg et al., 1986
Cow milk, infant formula	Pennington and Jones, 1989
Food products[a] (primarily tea, grain products, and processed cheese)	Pennington and Jones, 1989

[a] It should be noted that only minute amounts of aluminum are contributed by meat, fish, poultry, and vegetables to dietary intake.

The environment contributes to the body burden of aluminum in infants. In municipal water supplies, aluminum sulfate or alum is added as a flocculating agent in the process of water purification (King et al., 1981). Clearly, in the preparation of dialysate, the use of municipal drinking water would contain aluminum. Hence, tap water would be considered as an environmental source of aluminum (Leeming and Blair, 1979) with far-reaching implications in pediatric toxicity as exposure is not intentional. Indeed, McDermott et al. (1978) suggested that aluminum-induced encephalopathy was related to the presence of this metal in the dialysate originating from the municipal water supplies, as this syndrome did not occur in patients dialyzed with noncontaminated water. This observation is supported by the findings that individuals living in an area in Guam, where the aluminum content in soil and water was found to be high, developed a neuropathological syndrome resembling Parkinson's disease (Garruto et al., 1985). It is of interest that the concentration of aluminum was elevated in the hippocampus of these Parkinsonian-like patients.

Recently, it has been shown that in infants requiring parenteral nutrition following surgical procedure, the body burden of aluminum was elevated. Koo et al. (1988) increased the aluminum load in infants receiving parenteral nutrition and noted elevated aluminum levels in serum and bone accompanied by decreased urine metal excretion. These data demonstrated that parenteral nutrition solutions are a further source of aluminum contamination in children. In conditions of renal dysfunction, the parenteral nutritional source of aluminum can result in toxicity as assessed by aluminum accumulation in bone. In essence, the sources of aluminum for the child can be either nutritional, pharmacologic, or environmental, or a combination of both leading to increases in body burden and potential drug toxicity (Pennington and Jones, 1989).

A potential source of aluminum for the newborn which has received virtually little attention is the lactational route. During pregnancy, dyspepsia is a common complaint and mothers tend to use antacids to reduce the adverse symptoms. Weberg et al. (1986) demonstrated that low concentrations of antacids did not result in elevated serum aluminum levels in mothers and newborns. It was suggested that low-dose aluminum antacids failed to produce hyperaluminiumemia. However, the effects of high-dose antacids normally employed for peptic ulcer disease were not studied on the newborn. Despite the relative safety of low-dose antacids, Weberg and Berstad (1986) recommended against high-dose antacids during pregnancy. This is supported by recent reports that administered aluminum to pregnant mice or rats throughout gestation produced no teratogenic alterations. However, the offspring of aluminum-fed dams displayed a decrease in body weight and crown rump length accompanied by a retardation in neurobehavioral development (Golub et al., 1987; Domingo et al., 1987). It should be noted that mammary aluminum as a potential source for infant toxicity is diminished in the presence of vitamin D. In contrast, aluminum is known to decrease the bioavailability of the essential nutrients phosphate, iron, and vitamin A (Brunton, 1990). Thus, aluminum-induced toxicity can occur in a suckling infant of a mother suffering from iron-deficiency anemia. Although precise studies to determine the role of mammary aluminum as a source of toxicity in infants are not available, this element does accumulate in breast milk (Kacew, 1993a,b). Based on the knowledge that infant feeding formulas contain high levels of aluminum which can attenuate toxicity risk (Finberg et al., 1986), it was suggested that breast milk despite the presence of this element would be more beneficial to the nursing child.

IV. LEAD

The continuous emission of lead into the environment from industrial sources and automobile exhaust, the accumulation and persistence of this heavy metal in the biosphere, as well as the high affinity of lead to remain bound to mammalian tissues has resulted in extensive studies of the potential consequences of exposure to this metal (Kacew and Singhal, 1980). In particular, studies with lead have concentrated on adverse effects in the central nervous system, kidney, and hematopoietic system (Stevenson et al., 1977; Moore et al., 1980; Kimmel, 1984). Because the toxic effects of lead are age-related, a considerable amount of data exists on the pre- and/or postnatal effects of this metal on mammalian function. In the human and animal models examining lead exposure on newborn growth and development, the metal was administered (1) prenatally to pregnant dams only; (2) prenatally to pregnant dams during gestation and postnatally during lactation; (3) from the prenatal period through lactation to postnatally directly to the newborn; and (4) postnatally by direct intubation of the newborn (Grant et al., 1980; Kimmel, 1984; Dietrich, 1991). These diverse models of lead exposure have generated a considerable amount of data and the following adverse effects recorded were altered neurobehavioural, renal, reproductive, and immune responses (Luster et al., 1978; Fowler et al., 1980).

In a number of studies conducted in the U.S. and Europe the lead content in human milk ranged from .05 to .28 µg/ml (Bell and Thomas, 1980). Pharmacokinetic examination of suckling rats of metal-exposed dams revealed that mammary lead was transferred to the infant (Green and Gruener, 1974; Brown, 1975). In particular, lead was significantly increased in blood, liver, kidney, and brain in suckling rats (Millar et al., 1970; Roels et al., 1977). The observed rise in lead was associated with an increase in renal porphyrin and a decrease in delta-aminolevulinic acid dehydratase in blood, brain, liver, and kidney.

Although the contribution of only mammary lead exposure which can be attributed to the observed toxic consequences remains undefined, there is evidence that the presence of lactational metal results in newborn toxicity. Direct application of topical lead ointment on the breast was reported to produce central nervous system toxicosis in the human infant (Dillon, 1974). The lactational exposure of young animals to lead resulted in growth retardation and a delay in appearance of eye opening and body hair (Silbergeld, 1973; Kimmel, 1984). In an attempt to simulate human exposure conditions, lead was added to drinking water or to the diet either during pregnancy or from birth to weaning in animals. The decrease in birth weight observed was suggested to be due to a lead-induced deficit in the suckling response (Grant et al., 1980). With less suckling one would expect a decrease in growth but this would not account for the learning deficit noted; and thus, lead by itself may be a contributing factor to CNS dysfunction. The presence of lead in mammary tissue alters the nutritional value of milk as reflected by decreases in the essential elements copper, zinc, and iron (Bornschein et al., 1977), which are required for mammalian metabolism and CNS function. Mahaffey (1980) suggested that lead exposure may interfere with maternal metabolic pathways resulting in decreased utilization of nutrients in the diet, and thus an absence of nutritional components present in milk. This altered milk composition would consequently affect newborn development.

It is evident that lead may be exerting an adverse effect indirectly by preventing the actions of essential elements. The absorption of calcium, iron, and vitamin D is affected by lead (Mahaffey, 1980). In conditions of diets deficient in essential elements, lead absorption and toxicity is enhanced in infants. As breast milk is a source of lead for suckling infants, it is conceivable that during iron-deficiency anemia or calcium-deficient dietary intake in mothers, the bioavailability of milk lead would be increased, resulting in greater toxicity. Further, the beneficial effects of breast-feeding against immune system disorders are well-established (Davis et al., 1988; Mayer et al., 1988; Kramer, 1988). The finding that lead interferes with the immune system (Pruett et al., 1993) indicates that the presence of this metal in milk may not confer protection in lead-exposed mothers. Maternal milk serves as a source of lead for the nursing infant, but the significance of this metal source in mammalian health needs to be established.

V. CADMIUM

Although cadmium is not generally considered an essential element, the ability of this heavy metal to accumulate and persist in living organisms has aroused considerable concern over its toxicological potential (Singhal et al., 1974). The indiscriminate use of cadmium in industry and agriculture has been implicated as an etiological factor in reproductive dysfunction, prostatic cancer, as well as growth retardation (Hill et al., 1963; Schrag and Dixon, 1985). For example, in Toyama, Japan contamination of the river with cadmium effluent from industry resulted in metal accumulation in crops, ingestion of food prepared from contaminated crops, and poisoning of humans. The consequent condition, termed "itai-itai" disease, is characterized by bone pain, osteomalacia, and osteoporosis (Murata et al., 1970). The effects of cadmium during pregnancy on maternal outcome and developing progeny, on pharmacokinetics, on toxicity in the neonate as well as on the role of metallothionein, have been reviewed by Bell (1984).

Administration of radioactive cadmium to lactating rats or mice revealed the presence of metal in mammary tissue and nursing infant organs (Lucis et al., 1972; Tanaka et al., 1972). Whelton et al. (1993a) demonstrated that in mice exposed to ^{109}Cd through five successive rounds of gestation there was a progressive accumulation of metal not only in the whole animal but a 14-fold increase in mammary tissue content. It is well-known that cadmium interferes with the essential element zinc in placenta, and this may reflect the observed decrease in birth weight in metal-exposed animals (Kuhnert et al., 1988; Goyer, 1991). Thus, it is conceivable that cadmium might induce a toxic response either through an action on mammary zinc content directly or through a transfer from mammary tissue which then interferes with zinc in the nursing infant to decrease growth. It should be noted that in successive rounds of pregnancy cadmium content rises in mammary tissue, but that the highest levels occur in kidney and liver, target organs for toxicity (Whelton et al., 1993a). It is of interest that in mice fed a diet resembling female itai-itai patients where there was a deficiency of calcium, phosphorous, iron, and vitamins A, B, and D, the accumulation of mammary cadmium was higher than that seen with a normal diet (Whelton et al., 1993b). Maternal malnutrition in cadmium-exposed lactating women would clearly enhance toxicity in suckling infants.

The relationship between smoking, the presence of nicotine in breast milk, and the consequent rise in infant respiratory infections and irritation are well-known (Wilson, 1983; Riordan, 1987). Cigarette smoke is also the primary source of cadmium for humans (Lewis et al., 1972). Cadmium accumulation is approximately 6-fold greater in smokers compared to nonsmokers (Kuhnert et al., 1988). It has been suggested that a decrease in the placental zinc-cadmium ratio results in a lower newborn birth weight. However, it is also conceivable that cadmium accumulates in breast milk with multiparity is transferred to the suckling infant, and mobilized in target organs (Chan and Cherian, 1993; Whelton et al., 1993a). Evidence clearly shows that maternal milk is a source of cadmium. In light of the observed decrease in birth weight in smoking mothers, the contribution of mammary cadmium towards the deposition of this metal to infant kidney, the site of toxicity, needs to be addressed.

VI. METHYLMERCURY

The nondiscriminant use of mercury-containing compounds in industry and agriculture is known to result in environmental pollution and induce various toxic manifestations especially to the central nervous system (Chang, 1984; Clarkson, 1991). The focus of this discussion is not to negate the toxic consequences of prenatal methylmercury exposure on developing mammals, but to clarify and illustrate that this metal is present in milk and may contribute to toxicity. Thus far, few studies have concentrated on metals in breast milk and the significance of their presence to human health.

In an extensive review, Takeuchi (1968) clearly demonstrated the effects of epidemic methylmercury exposure on fetal and newborn development. Industrial release of methylmercury into Minamata Bay followed by accumulation in edible fish and ingestion by lactating females resulted in the transfer of metal to the suckling human infant (Chang, 1984). Similarly, Amin-Zaki et al. (1974) demonstrated that ingestion of homemade bread prepared from wheat treated with the fungicide methylmercury by lactating mothers produced a significant rise in human infant metal levels. In fish-eating populations in Canada, maternal ingestion of mercury-contaminated food during pregnancy and lactation resulted in abnormal muscle tone and reflexes in boys but not girls (McKeown-Eyssen et al., 1983). In a New Zealand study, Kjellstrom et al. (1989) reported developmental retardation in 4-year-old children of mothers eating mercury-contaminated fish during pregnancy and lactation. Although emphasis was placed on the consequences of prenatal exposure in the Canadian and New Zealand studies, the contribution of milk mercury to toxic outcome was neglected. Mammary transfer of methylmercury to suckling animal pups has been reported to produce neurological lesions (Chang, 1984). This finding clearly indicates a positive correlation between exposure to high concentrations of metal in mammary tissue and toxicity in suckling infants.

Although the precise contribution of mammary-derived methylmercury to the observed adverse effects on neurologic and behavioral changes in suckling pups is not known (Chang, 1984), Khera and Tabacova (1973) found that postnatal exposure directly to newborns to this metal produced ocular defects. In contrast, there was a lack of an ocular effect in fetuses of prenatal exposed dams suggesting that lactational methylmercury may in part contribute to the observed toxicity. It is well-known that methylmercury is secreted more readily in the maternal colostrum, the period at which eye defects were reported, and crosses into the suckling infant. Mercury itself decreases the suckling response in human infants (Rohyans et al., 1984). Since milk contains essential nutrients for neurological and behavioral development, it is conceivable that less feeding would contribute to the mercury-induced nervous disorders as less nutritional supply is associated with delayed growth processes.

The protective role of breast-feeding against infectious diseases is well-documented in noncontaminated mothers (Cunningham et al., 1991; Jason et al., 1984). Recently, Kosuda et al. (1994) demonstrated that methylmercury compromises the immune system by significantly reducing RT6-T lymphocytes in rats. Consequently, a renal autoimmune disease develops. Exposure of dams to methylmercury was found to result in neonatal metal accumulation in kidney and consequent nephrotoxicity (Chang and Sprecher, 1976). The relationship between mammary methylmercury ingestion and renal histopathological alterations in suckling pups is not known. However, it is conceivable that lactational mercury could serve as a potential source of metal which ultimately leads to accumulation of metal in kidney and nephrotoxicity (Fowler, 1972) through an action on the immune system. It is of interest that one of the manifestations of mercury poisoning in a human infant was a hypersensitivity and allergic reactions, an index of immune system dysfunction (Rohyans et al., 1984). It is evident that methylmercury generates autoimmune antibodies, conceivably in maternal milk, and these may be transferred to the suckling infant. Indeed, Riordan (1987) suggested that if serum levels of maternal mercury are high or infants are symptomatic, breast-feeding should be terminated.

VII. CONCLUSIONS

The physiological process of breastfeeding should be encouraged under most circumstances despite the presence of trace amounts of metals. It is essential to maintain an adequate nutritional diet to limit the potential adverse effects of lactational metals on suckling infants. In environmentally abnormal conditions where metal contamination reaches high proportions, breastfeeding should be limited especially during colostrum formation. Colostrum, unlike milk, is more water soluble and a more bioavailable source of metals for suckling infants. Clearly, studies have focused on the potential consequences of metals during pregnancy on the fetus and neonate or on the newborn directly. However, it is evident that metals accumulate in breast milk and that in certain situations the mammary source of chemicals has resulted in infant toxicity.

REFERENCES

Ahmed, F., Clemens, J. D., Rao, M. R., Sack, D. A., Khan, M. R., and Haque, E. (1992), *Pediatrics,* 90, 406–411.
Alfrey, A. C., LeGendre, G. R., and Kaehny, W. D. (1976), *N. Engl. J. Med.,* 294, 184–188.
Amin-Zaki, L., Elhassini, S., Majeed, M. A., Clarkson, T. W., Doherty, R. A., and Greenwood, M. R. (1974), *J. Pediatr.,* 85, 81–84.
Andreoli, S. P., Bergstein, J. M., and Sherrard, D. J. (1984), *N. Engl. J. Med.,* 310, 1079–1084.
Andreoli, S. P., Dunn, D., DeMyer, W., Sherrard, D. J., and Bergstein, J. M. (1985), *J. Pediatr.,* 107, 760–763.
Basu, T. K. (1988), *Toxicologic and Pharmacologic Principles in Pediatrics,* Kacew, S. and Lock, S., Eds., Hemisphere Publishing Corporation, Washington, D.C., 17–40.
Bell, J. U. (1984), *Toxicology and the Newborn,* Kacew, S., and Reasor, M. J., Eds., Elsevier Science Publishers, Amsterdam, 199–216.
Bell, J. U. and Thomas, J. A. (1980), *Lead Toxicity,* Singhal, R. L. and Thomas, J. A., Eds., Urban and Schwarzenberg, Baltimore, MD, 169–185.
Bornschein, R. L., Michaelson, I. A., Fox, D. A., and Loch, A. (1977), *Symp. Biochem. Effects of Environmental Pollutants,* U.S. Environmental Protection Agency, Cincinnati, OH, Ann Arbor Science Publishers, Ann Arbor, MI.
Brown, D. A. (1975), *Toxicol. Appl. Pharmacol.,* 32, 628–637.
Brunton, L. L. (1990), *Goodman and Gilman's The Pharmacological Basis of Therapeutics,* 8th ed., Gilman, A. G., Rall, T. W., Nies, A. S., and Taylor, P., Eds., Pergamon Press, New York, 897–913.
Cam, C. and Nigogosyan, G. (1963), *J. Am. Med. Assoc.,* 183, 88–91.
Chan, H. M. and Cherian, M. G. (1993), *Toxicol. Appl. Pharmacol.,* 120, 308–314.
Chang, L. W. (1984), *Toxicology and the Newborn,* Kacew, S. and Reasor, M. J., Eds., Elsevier Science Publishers, Amsterdam, 173–197.
Chang, L. W. and Sprecher, J. A. (1976), *Environ. Res.,* 11, 392–406.
Chernoff, N., Linder, R. E., Scotti, R. M., Rogers, E. H., Carver, B. D., and Kavlock, R. J. (1979), *Environ. Res.,* 18, 257–269.
Clarkson, T. (1991), *Fundam. Appl. Toxicol.,* 16, 20–21.
Clemens, J., Rao, M., Ahmed, F., Ward, R., Huda, S., Chakraborty, J., Yunis, M., Khan, M. R., Ali, M., Kay, B., van Loon, F., and Sack, D. (1993), *Pediatrics,* 92, 680–685.
Cochi, S. L., Fleming, D. W., Hightower, A. W., Limpakarnjanarat, K., Facklam, R. R., Smith, J. D., Sikes, R. K., and Broome, C. V. (1986), *J. Pediatr.,* 108, 887–896.
Cunningham, A. S. (1979), *J. Pediatr.,* 95, 685–689.
Cunningham, A. S., Jelliffe, D. B., and Jelliffe, E. F. P. (1991), *J. Pediatr.,* 118, 659–666.
Davis, M. K., Savitz, D. A., and Graubard, B. I. (1988), *Lancet,* 2, 365–368.
Dietrich, K. N. (1991), *Fundam. Appl. Toxicol.,* 16, 17–19.
Dillon, H. K. (1974), *Am. J. Dis. Child.,* 128, 491–492.
Domingo, J. L., Paternain, J. L., and Llobet, J. M. (1987), *Res. Commun. Chem. Pathol. Pharmacol.,* 57, 129–132.
Duncan, B., Ey, J., Holberg, C. J., Wright, A. L., Martinez, F. D., and Taussig, L. M. (1993), *Pediatrics,* 91, 867–872.
Finberg, L., Dweck, H. S., Holmes, F., Kretchmer, N., Mauer, A. M., Reynolds, J. W., and Suskind, R. M. (1986), *Pediatrics,* 78, 1150–1153.
Fowler, B. A. (1972), *Am. J. Pathol.,* 69, 163–178.
Fowler, B. A., Kimmel, C. A., Woods, J. S., McConnell, E. E., and Grant, L. D. (1980), *Toxicol. Appl. Pharmacol.,* 56, 59–77.
Garruto, R. M., Yanagihara, R., and Gadjusek, D. C. (1985), *Neurology,* 35, 193–198.

Gerrard, J. W., MacKenzie, J. W. A., Goluboff, N., Garson, J. Z., and Maningas, C. W. (1973), *Acta Paediatr. Scand.,* 234 (Suppl.), 1–21.
Golub, M. S., Gershwin, M. E., Donald, J. M., Negri, S., and Keen, C. L. (1987), *Fundam. Appl. Toxicol.,* 8, 346–357.
Goyer, R. A. (1991), *Fundam. Appl. Toxicol.,* 16, 22–23.
Grant, L. D., Kimmel, C. L., West. G. L., Martinez-Vargas, C. M., and Howard, J. L. (1980), *Toxicol. Appl. Pharmacol.,* 56, 42–58.
Greco, L., Auricchio, S., Mayer, M., and Grimaldi, M. (1988), *J. Pediatr. Gastroenterol. Nutr.,* 7, 395–399.
Green, M. and Gruener, N. (1974), *Res. Commun. Chem. Pathol. Pharmacol.,* 8, 735–738.
Hill, C. H., Matrone, G., Payne, W. L., and Barber, C. W. (1963), *J. Nutri.,* 80, 227–235.
Howie, P. W., Forsyth, J. S., Ogston, S. A., Clark, A., and Florey, C. V. (1990). *Br. Med. J.,* 300, 11–16.
Jason, J., Nieburg, P., and Marks, J. S. (1984), *Pediatrics,* 74 (Suppl.), 702–727.
Kacew, S. (1990), *Drug Toxicity and Metabolism in Pediatrics,* Kacew, S., Ed., CRC Press, Boca Raton, FL, 265–294.
Kacew, S. (1993a), *J. Clin. Pharmacol.,* 33, 213–221.
Kacew, S. (1993b), *General and Applied Toxicology,* Ballantyne, B., Marrs, T., and Turner, P., Eds., Vol. 2, Macmillan Press, London, 1047–1068.
Kacew, S. and Singhal, R. L. (1980), in *Lead Toxicity,* Singhal, R. L., and J. A. Thomas, Eds., Urban and Schwarzenberg, Baltimore, MD, 43–78.
Khera, K. S. and Tabacova, S. A. (1973), *Food Cosmet. Toxicol.,* 11, 245–254.
Kimmel, C. A. (1984), *Toxicology and the Newborn,* Kacew, S. and Reasor, M. J., Eds., Elsevier Science Publishers, Amsterdam, 217–235.
King, S. W., Savory, J., and Wills, M. R. (1981), *Crit. Rev. Clin. Lab. Sci.,* 14, 1–20.
Kjellstrom, T., Kennedy, P., Wallis, S., Stewart, A., Friberg, L., Lind, B., Wutherspoon, P., and Mantell, C. (1989), Physical and mental development of children with prenatal exposure to mercury from fish. Stage 2. Interviews and psychological tests at age 6, National Swedish Environmental Board, Solna, 112 (Rep. no. 3642).
Koo, W. W. K., Kaplan, L. A., Horn, J., Tsang, R. C., and Steichen, J. J. (1988), *J. Parenter. Enteral Nutr.,* 12, 170–173.
Kosuda, L. L., Hosseinzadeh, H., Greiner, D. L., and Bigazzi, P. E. (1994), *J. Toxicol. Environ. Health,* 42, 303–321.
Kramer, M. S. (1988), *J. Pediatr.,* 112, 181–190.
Kuhnert, D. R., Kuhnert, P. M., and Zarlingo, T. J. (1988), *Obstet. Gynecol.,* 71, 67–70.
Lawrence, R. A. (1989), *Med. Clin. N. Am.,* 73, 583–603.
Leeming, R. J. and Blair, J. A. (1979), *Lancet,* 1, 556.
Lewis, G. P., Jusko, W. J., Coughlin, L. L., and Hartz, S. (1972), *Lancet,* 2, 291–292.
Lucas, A., Brooke, O. G., Morley, R., Cole, J. T., and Bamford, M. F. (1990), *Br. Med. J.,* 300, 837–840.
Lucis, O. J., Lucis, R., and Shaikh, Z. A. (1972), *Arch. Environ. Health,* 25, 14–22.
Luster, M. I., Faith, R. E., and Kimmel, C. A. (1978), *J. Environ. Pathol. Toxicol.,* 1, 397–402.
Mahaffey, K. R. (1980), *Lead Toxicity,* Singhal, R. L. and Thomas, J. A., Eds., Urban and Schwarzenberg, Baltimore, MD, 425–460.
Matsumoto, M., Koya, G., and Takeuchi, T. (1965), *J. Neuropathol. Exp. Neurol.,* 24, 563–574.
Mattison, D. R. (1985), *Reproductive Toxicology,* Dixon, R. L., Ed., Raven Press, New York, 109–130.
Mayer, E. J., Hamman, R. F., Gay, E. C., Lezotte, D. C., Savitz, D. A., and Klingensmith, G. J. (1988), *Diabetes,* 37, 1625–1632.
McDermott, J. R., Smith, A. I., Ward, M. K., Parkinson, I. S., and Kerr, D. N. S. (1978), *Lancet,* 1, 901–904.
McKeown-Eyssen, G. E., Ruedy, J., and Neims, A. (1993), *Am. J. Epidemiol.,* 118, 470–479.
Millar, J. A., Cumming, R. L. C., Ballistini, V., Carswell, F., and Goldberg, A. (1970), *Lancet,* 2, 695.
Moore, M. R., Meredith, P. A., and Goldberg, A. (1980), *Lead Toxicity,* Singhal, R. L. and Thomas, J. A., Eds., Urban and Schwarzenberg, Baltimore, MD, 79–117.
Murata, I., Hirono, T., Saeki, Y., and Nakaga, W. S., (1970), *Bull. Soc. Int. Clin.,* 1, 34–42.
Newton, N. and Newton, M. (1967), *N. Engl. J. Med.,* 277, 1179–1188.
Pennington, J. A. T. and Jones, J. W. (1989), *Aluminum and Health,* Gitelman, H. J., Ed., Marcel Dekker, New York, 67–100.
Peters, H. A., Gocmen, A., Cripps, D. J., Bryan, G. T., and Dogramaci, I. (1982), *Arch. Neurol.,* 39, 744–749.
Potter, D. E., McDaid, T. K., McHenry, K., and Mar, H. (1981), *Trans. Am. Soc. Artif. Intern. Organs,* 27, 64–67.
Pruett, S. B., Ensley, D. K., and Crittenden, P. L. (1993), *J. Toxciol. Environ. Health,* 39, 163–192.
Riordan, J. (1987), *Problems in Pediatric Drug Therapy,* 2nd ed., Pagliaro, L. A. and Pagliaro, A. M., Eds., Drug Intelligence Publications, Hamilton, IL, 195–258.
Roels, H., Lauwerys, R., Buchet, J.-P., and Hubermont, G. (1977), *Toxicology,* 8, 107–113.
Rogan, W. J., Gladen, B. C., McKinney, J. D., Carreras, N., Hardy, P., Thullen, J., Tinglestad, J., and Tully, M. (1986), *J. Pediatr.,* 109, 335–341.
Rohyans, J., Walson, P. D., Wood, G. A., and MacDonald, W. A. (1984), *J. Pediatr.,* 104, 311–313.
Rotundo, A., Nevins, T. E., Lipton, M., Lockman, A., Mauer, S. M., and Michael, A. F. (1982), *Kidney Int.,* 21, 486–491.
Schrag, S. D. and Dixon, R. L. (1985), *Reproductive Toxicology,* Dixon, R. L., Ed., Raven Press, New York, 301–319.
Sedman, A. B., Miller, N. L., Warady, B. A., Lum, G. M., and Alfrey, A. C. (1984), *Kidney Int.,* 26, 201–204.
Sedman, A. B., Klein, G. L., Merritt, R. J., Miller, N. L., Weber, K. O., Gill, W. L., Anand, H., and Alfrey, A. C. (1985), *N. Engl. J. Med.,* 312, 1337–1343.
Shore, L. S., Gurevitz, M., and Shemesh, M. (1993), *Bull. Environ. Contam. Toxicol.,* 51, 361–366.

Silbergeld, E. K. and Goldberg, A. M. (1973), *Life Sci.,* 13, 1275–1283.
Singhal, R. L., Merali, Z., Kacew, S., and Sutherland, D. J. B. (1974), *Science,* 183, 1094–1096.
Stevenson, A. J., Kacew, S., and Singhal, R. L. (1977), *Toxicol. Appl. Pharmacol.,* 40, 161–170.
Sveger, T. and Udall, J. N. Jr. (1985), *J. Am. Med. Assoc.,* 254, 3036–3037.
Takeuchi, T. (1968), *Minimata Disease,* Kutsuna, M., Ed., Kumamoto University, Japan, 141–228.
Weberg, R. and Berstad, A. (1986), *Eur. J. Clin. Invest.,* 16, 428–432.
Weberg, R., Berstad, A., Ladehaug, B., and Thomassen, Y. (1986), *Acta Pharmacol. Toxicol.,* 59, (Suppl. 7), 63–65.
Whelton, B. D., Moretti, E. S., Peterson, D. P. and Bhattacharyya, M. H. (1993a), *J. Toxicol. Environ. Health,* 38, 115–129.
Whelton, B. D., Moretti, E. S., Peterson, D. P. and Bhattacharyya, M. H. (1993b), *J. Toxicol. Environ. Health,* 38, 131–145.
Wilson, J. T. (1983), *Nutr. Health,* 2, 191–201.
Wilson, J. T., Brown, R. D., Cherek, D. R., Dailey, J. W., Hilman, B., Jobe, P. C., Manno, B. R., Manno, J. E., Redetzki, H. M., and Stewart, J. J. (1980), *Clin. Pharmacokinet.,* 5, 1–66.
Yamaguchi, A., Yoshimura, T., and Kuratsune, M. (1971), *Fukuoka Acta Med.,* 62, 117–122.

Chapter 71

Constitutive and Induced Metallothionein Expression in Development

George P. Daston and Lois D. Lehman-McKeeman

I. INTRODUCTION

Metallothioneins (MTs) are low molecular weight metal-binding proteins originally isolated from horse kidney by Margoshes and Vallee (1957). These cysteine-rich proteins bind a variety of heavy metals with very high affinity and are induced following exposure to metals such as Cd or Zn. MTs function in the homeostasis of essential metals, such as Zn and Cu, and in the detoxication of metals such as Cd and Hg (Cousins, 1985; Klaassen and Lehman-McKeeman, 1989). These proteins are also involved in the acute phase response to stress (Karin, 1985). MTs are generally considered to be "housekeeping" proteins because they are expressed in virtually all tissues and cell types. However, tissue concentrations of MTs are significantly higher in developing animals relative to constitutive levels of MTs in adults, and the high levels of these Zn-binding proteins are believed to have an important role in supporting normal development.

MTs have been identified in developing tissues of invertebrate and vertebrate species. To date, four distinct MT gene products have been identified. MT-I and MT-II were originally isolated and characterized by Nordberg et al. (1972), and represent the major isoforms found in most species. More recently two additional gene products, MT-III and MT-IV have been isolated from neuronal (Uchida et al., 1991; Palmiter et al., 1992) and stratified squamous epithelium (Quaife et al., 1994), respectively. This chapter describes the ontogeny and inducibility of MT gene expression (MT-I and MT-II) during development and the regulation of MT expression during pregnancy and lactation, and reviews the contribution of MTs in both normal and abnormal development.

II. ONTOGENY OF MT EXPRESSION

A. PATTERN OF MT EXPRESSION DURING DEVELOPMENT

The very high levels of MT expression during development were first characterized and quantified by Wong and Klaassen (1979). This work demonstrated that the proteins expressed in neonatal rat liver were identical to those synthesized and induced in adult rat liver. Since this work, the ontogeny of MT expression has been studied in embryonic, extra-embryonic, and neonatal tissues across many species including mice (Ouellette, 1982; Kershaw et al., 1990), rats (Lehman-McKeeman et al., 1988; Acuff-Smith et al., 1994), rabbits (Waalkes and Bell, 1980; Andrews et al., 1987), and humans (Ryden and Deutsch, 1978; Riordan and Richards, 1980; Waalkes et al., 1984). Across all species, the high expression of MTs during development is a common feature, suggesting an important and conserved function for MTs during embryogenesis.

Andrews et al. (1991) described MT gene expression in mouse preimplantation embryos, representing the earliest stage of development in which MT regulation has been studied. This group established that MT-I and MT-II transcripts were present in the fertilized (single cell) egg (Figure 1). However, the genes were not induced by metals until after the third cleavage stage. These results are similar to those found in the rabbit preimplantation blastocyst (Andrews et al., 1987a). With the exception of these studies, there is essentially nothing else known about MT gene expression and regulation during preimplantation development of the mammalian embryo.

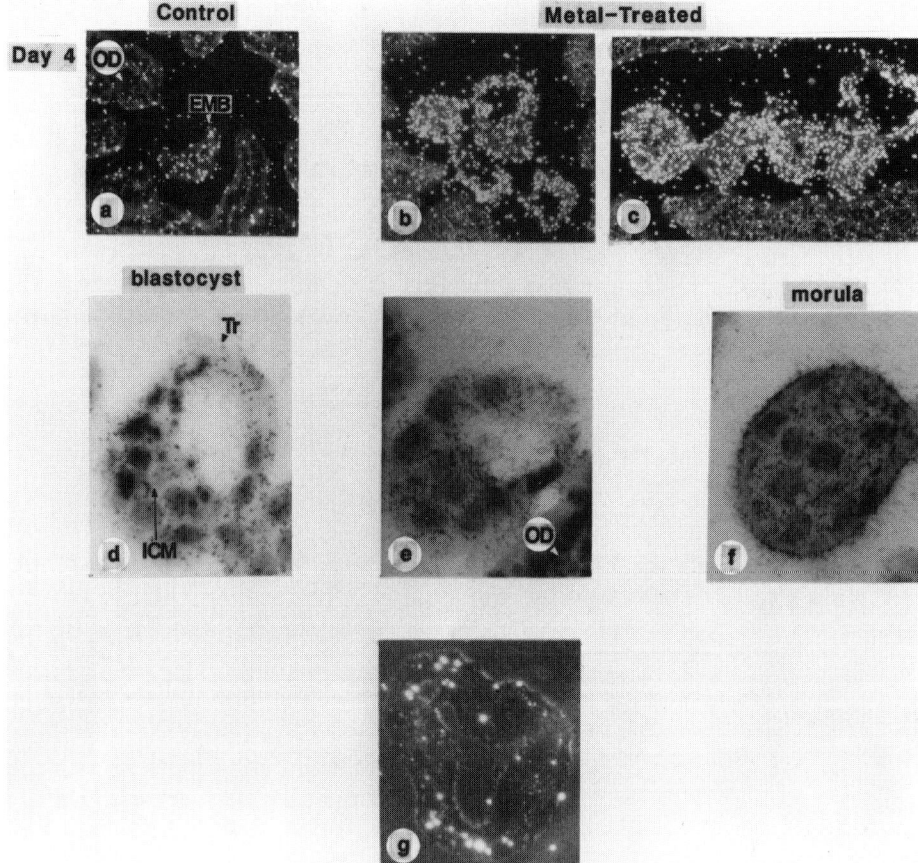

Figure 1. Detection of MT-I mRNA in preimplantation mouse embryos using *in situ* hybridization. Preimplantation mouse embryos on the indicated days of gestation were recovered and incubated for 5 h in the presence or absence of Zn (50 μ*M*) or Cd (10 μ*M*). Embryos were transferred back into the ampulla region of the oviduct, fixed in paraformaldehyde, sectioned, and mounted onto polylysine-coated slides. Sections within a given experiment were placed on the same slide. Slides were hybridized with a ^{35}S-labeled MT-I antisense strand RNA probe, and hybrids were detected by autoradiography for 5 days. A sense strand MT probe was used as a control for specificity of the hybridization. Day 4 embryos are at the late morula or blastocyst stages of development. Shown here are dark-field photomicrographs (200×) in which the autoradiographic grains appear as white dots (a to c and g), and bright-field photomicrographs (400×) in which the grains appear as black dots (d to f). Abbreviations are: ICM, inner cell mass; Tr, trophectoderm. Control embryos (a, d); Zn-treated embryos (b, e to g), Cd-treated embryos (c). (g) Zn-treated blastocyst hybridized with a sense strand MT probe as a control. (From Andrews, G. K. et al., *Dev. Biol.*, 145, 13–27, 1991. With permission.)

From implantation to late gestation, the embryo is surrounded by a variety of cells which actively synthesize MTs. MT expression in the postimplantation embryo has been most thoroughly studied in rodents. In mice, this expression is first noted in the deciduum, then the spongiotrophoblasts of the placenta (De et al., 1989) and the visceral yolk sac endoderm (Andrews et al., 1984). In the mouse visceral yolk sac. MT levels rise dramatically from day 9 to 11 of gestation, remain constant until day 15, and then decrease abruptly to detectable, but very low levels. The time course for MT gene expression in rat extra-embryonic tissues has not been as fully characterized as in the mouse. However, it has been shown that, as of gestation day 12, MT-I and MT-II mRNAs are constitutively expressed in the visceral yolk sac, decidua, and placenta (Acuff-Smith et al., 1994).

MTs have also been isolated and characterized from the term placenta in rodents (Charles-Shannon et al., 1981; Arizono et al., 1981) and humans (Waalkes et al., 1984). In human tissue, MT levels are higher in the amnionic and chorionic membranes than in the placenta proper. Whereas additional studies have shown that trophoblasts of both the human chorion and placenta directly synthesize MTs (Lehman and Poisner, 1984), it is unclear whether MTs isolated from the amnion reflect synthesis within the membrane or uptake from the extracellular environment as has been shown for other proteins (Poisner et al., 1982).

MT gene expression and regulation has been studied most extensively in the developing liver. Andrews et al. (1984) reported that MT mRNAs in mouse liver were first detected on gestation day 12. MT levels in mouse liver increase steadily from gestation day 12 through 17, but decrease abruptly in newborn mice (Ouellette, 1982; Andrews et al., 1984). In contrast, in rat liver, MT mRNAs are abundant on gestation day 15 and increase steadily throughout late gestation to maximum levels at parturition. Furthermore, unlike the mouse, rat liver MT mRNA levels remain relatively constant and at very high levels throughout the first 2 weeks of postnatal life (Andersen et al., 1983; Lehman-McKeeman et al., 1988).

Waalkes and Klaassen (1984) characterized the postnatal ontogeny of MT proteins in all major rat organs. In this work, total tissue MTs were quantified with the Cd-hemoglobin radioassay (Eaton and Toal, 1982) and the results underscored the abundance of hepatic MTs relative to any other organ. Peak MT levels approached 1000 µg/g liver between postnatal days 1 to 7. In contrast, constitutive MT levels in all other organs did not exceed 50 µg/g tissue. Since this work, additional analytical procedures have been developed which allow for the specific quantitation of MT-I and MT-II (Lehman and Klaassen, 1986). Employing the HPLC-atomic absorption spectrophotometric assay to quantify the isoproteins, Lehman-McKeeman et al. (1988a) found that at peak MT levels (postnatal days 1 to 7), total MT levels approached 1000 µg/g, and MT-II was more abundant than MT-I. In contrast, similar work in the mouse (Kershaw et al., 1990) demonstrated three major differences in the ontogeny of MT expression in rat and mouse liver (Figure 2). First, the peak levels of hepatic MT were quite different between the two species, as the total in mouse was about 300 µg/g, representing less than 30% of the maximum levels noted in the rat. Second, the temporal pattern of expression was slightly different, as mouse MTs increased between postnatal days 1 to 7 whereas rat MTs were at a maximum throughout the first week of posnatal life. Finally, the isoform composition of the total MT pool was different, as MT-II was the more abundant isoform during development of rats and the only isoform detected in adult animals. In direct contrast, MT-I was the major isoform expressed in mouse liver.

An interesting feature of MT regulation during development is the evidence for translational control of MT gene expression, particularly in rat liver. As noted above, MT protein levels increase throughout late gestation, and this increase is associated with an increase in translatable MT mRNA. However, after postnatal day 7, MT proteins decreased dramatically, whereas mRNA levels remained elevated at the levels found at parturition. This disparity between MT mRNA and protein levels is suggestive of translational control of MT gene expression. This suggestion is supported further by the finding that hepatic MT synthesis (determined *in vivo*) is highest during late gestation, but decreases steadily immediately after parturition (Piletz et al., 1983).

B. REGULATION AND FUNCTION OF MTs DURING DEVELOPMENT

In contrast to the extensive characterization of the ontogeny of MT gene expression in a variety of tissues and species, the function of these proteins during development remains to be established. That MTs are involved in development is suggested by studies showing that during the early stages of development the proteins are found in the nuclei and only move into the cytoplasm during postnatal life (Danielson et al., 1982; Panemangalore et al., 1983).

One of the most widely held theories is that MTs function to regulate Zn and Cu homeostasis during development. It is well known that these metals are cofactors for many critical enzymes and are essential for normal development (Hurley, 1976; Cousins, 1985). In this regard, it has been found that, in the developing rat liver, Zn accumulates predominantly in cytosolic fraction, whereas Cu is localized mainly in the particulate fraction (Kern et al., 1981). Furthermore, in fetal human liver, a direct correlation

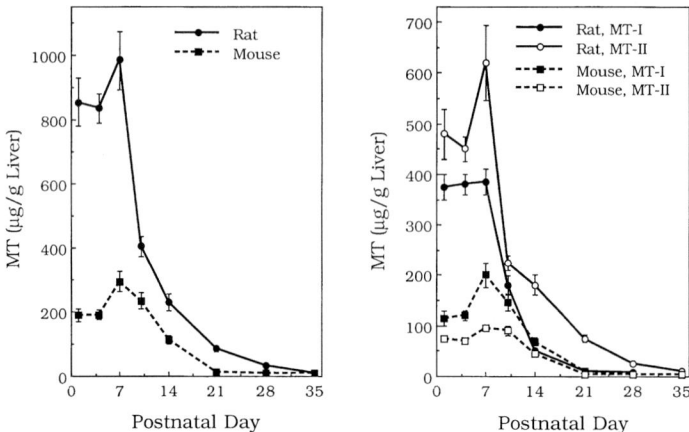

Figure 2. Concentrations of total MTs and MT-I and MT-II in rat and mouse liver during development. MT isoforms were quantified with the HPLC-AAS method described by Lehman and Klaassen (1986), and total MTs represent the sum of the isoform values. (Adapted from Lehman-McKeeman et al., 1988a and Kershaw et al., 1990.)

between cytosolic MT and Zn content has been reported, whereas no such relationship has been found between MT and Cu (Klein et al., 1991). Finally, using maternal mineral deficiency models, Gallant and Cherian (1986) reported that maternal Cu or Fe deficiency did not alter MT gene expression in rat during gestation, but maternal Zn deficiency significantly reduced hepatic MT levels in the offspring. The effects of Zn deficiency have been attributed to the lack of any increase in MT mRNA levels in fetal liver specifically during late gestation (days 18 to 20; Andrews et al., 1987). Collectively, these data suggest that Zn plays an important role in regulating MT gene expression during development and that MT functions primarily as a Zn storage protein in fetal and neonatal tissues. The importance of MT induction and embryonic Zn homeostasis is discussed more thoroughly later in this chapter (see Section IV, below).

In addition to heavy metals, a variety of physiological factors such as endogenous glucocorticoids (Karin et al., 1980a,b) and several cytokines including interleukin-1 (Durnam et al., 1984; Karin et al., 1985) and interferon-γ (Friedman and Stark, 1985) can induce MT gene expression. In this regard, corticosterone levels have been evaluated in pregnant rodents. In mice, plasma corticosterone levels in the pregnant dam increase dramatically (at least 10-fold) between gestation days 10 and 16 and then fall rapidly at parturition (Barlow et al., 1974). In contrast, fetal levels do not change during development in both mice (Dalle et al., 1978) and rats (DuPouy et al., 1975). Whereas glucocorticoids are primarily inducers of MT (Karin et al., 1980a,b; Klaassen and Lehman-McKeeman, 1989), the induction seen with these agents is quite small relative to metals (Lehman-McKeeman et al., 1988b), making it very unlikely that these endogenous hormones play an important role in regulating the high levels of MTs seen during development.

The ability of metals to induce MTs was first characterized by Terhaar et al. (1965), and since that time the inducibility of MTs during development has been extensively studied from the early blastocyst (Andrews et al., 1987a, 1991) to the developing liver (Lehman-McKeeman et al., 1988a; Kershaw et al., 1990). MTs are readily induced both *in vivo* and *in vitro* by heavy metals. As noted earlier, maternal Zn deficiency will down regulate MT expression, but injection of 20 mg Zn/kg to the Zn-deficient newborn pups can restore MT to levels seen in Zn-replete animals (Andrews et al., 1987b). Recent work has also demonstrated that maternal administration of organic compounds such as urethane can also induce MTs in embryonic tissues (Acuff-Smith et al., 1994).

Induction of MTs in developing tissues can have a protective role for the fetus. For example, induction of MT in the placenta may help to limit exposure to a toxic heavy metal such as Cd. It is known that tobacco and cigarette smoking represent sources of human exposure to Cd (Friberg et al., 1974). It has been shown that Cd accumulates in the placenta of pregnant women who smoke; however, fetal blood levels of Cd are not affected, suggesting sequestration in the placenta (Kuhnert et al., 1982). Similarly, it has been shown that neonatal (10-day-old) rats are also resistant to the hepatotoxic effects of Cd because the high constitutive levels of MTs bind up the metal and prevent it from reaching toxic concentrations at critical subcellular organelles (Goering and Klaassen, 1984).

Although many studies conducted over the past 35 years point to an important role for MTs in development, very recent work to develop mice lacking functional MT-I and MT-II genes may suggest otherwise. Two separate laboratories have developed transgenic mouse lines in which both MT genes were inactivated in embryonic stem cells (Michalska and Choo, 1993; Masters et al., 1994). In both cases, mice homozygous for the null mutant alleles were born alive and morphologically normal and were able to reproduce when maintained on standard laboratory chow. Both lines, however, were much more susceptible to Cd hepatotoxicity. The data prove that MTs do protect against Cd toxicity, but suggest that MTs may not be essential for normal development. These data also suggest that the other isoforms of MT (MT-III and MT-IV) may also contribute to normal development. These are clearly very exciting findings which will generate more research on MT regulation during development.

III. MT EXPRESSION DURING PREGNANCY AND LACTATION

The pregnant and lactating mammal, like the developing fetus and newborn, exhibits stage-specific expression of MT. Much less is known about MT expression during these life-stages than during development; in fact, the temporal pattern of expression has been evaluated only in rats and mice. Still, based on limited information in these two species it appears that there is considerable species diversity. In this section, the pattern and control of MT expression accompanying this unique life stage will be described.

A. CONSTITUTIVE HEPATIC MT EXPRESSION

In mice, MT mRNA and protein levels have been shown to increase markedly during gestation. Andrews et al. (1984) reported that the liver of the mouse on day 14 of pregnancy contained about 5 times as much MT mRNA as adult male liver (estimated from Northern blots). Studies on the time-course of changes in hepatic MT mRNA concentration during pregnancy indicate that between gestation days 8 and 10 it increases 2 to 3 fold over non-gravid levels (Daston et al., 1992), and this increase persists until just prior to parturition (Quaife et al., 1986; Daston et al., 1992). After parturition there is another increase in hepatic MT mRNA concentration which persists throughout the period of lactation. The peak postpartum MT mRNA level in maternal liver is approximately 7-fold higher than in nongravid mice and 2 to 3 fold higher than during pregnancy (Daston et al., 1992). Although the mRNA concentrations for MT-I and MT-II, are elevated, the increase in MT-II is greater, particularly postpartum. The time course for hepatic MT mRNA levels in pregnant and lactating mice is depicted in Figure 3.

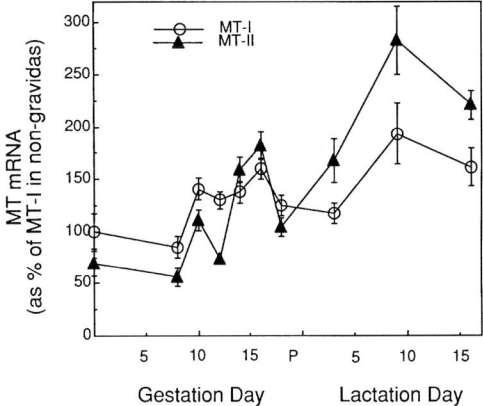

Figure 3. Concentrations of MT mRNA (MT-I and MT-II) concentrations in pregnant and lactating mouse liver mRNAs were quantitated by slot blotting. (Data from Daston et al., 1992.)

The time course for hepatic MT protein is similar to that for its mRNA, with some important differences both in pattern and magnitude of increase. There is a 10-fold increase in hepatic MT concentration between gestation days 8 and 10, which corresponds to the increase in MT mRNA, but is greater in extent. Although there is a peripartum decrease in MT concentration that corresponds with the return of MT mRNA to nonpregnant levels, MT level is still 7-fold higher than in nulliparous mice. Hepatic MT concentration remains at this level through lactation, even though mRNA levels are higher postpartum than prepartum (Daston et al., 1992). Therefore, although the patterns of increase of MT mRNA and protein concentrations coincide, the amount of protein present postpartum is somewhat less than expected given the increase in mRNA.

There are two possible explanations for the discrepancy: (1) that degradation of the protein occurs more rapidly postpartum, and (2) that there is translational control of MT expression postpartum. The former possibility makes sense in that it is likely that the lactating dam must mobilize a great deal of Zn to support the growth of her offspring. While milk from any species is rich in Zn, mice have especially high milk Zn concentrations (Reis et al., 1991). However, empirical support for this notion is lacking, as no experiments have been done to measure the MT degradation rate in the lactating mouse. On the other hand, there is experimental evidence suggesting that the translation of MT may be regulated. Specifically, it has been observed in the newborn rat than MT mRNA concentration remains high during the first 2 to 3 weeks after birth (Andersen et al., 1983; Lehman-McKeeman et al., 1988a), while MT protein synthesis declines during this period, reaching the very low level characteristic of the adult by three weeks of age (Piletz et al., 1983). It remains to be established whether a similar uncoupling of transcription and translation occurs in the lactating mouse.

The constitutive pattern of hepatic MT expression is quite different in pregnant rats. There is only a brief spike of MT around the time of parturition; otherwise, MT levels are lower than or not different from nulliparous rats (Figure 4). The relative abundance of MT mRNA, and MT protein synthesis rate have been measured in rat liver from gestation day 15 through weaning (Andersen et al., 1983; Piletz et al., 1983). Both MT mRNA concentration and [^{35}S] cysteine incorporation into MT on gestation day 15 are comparable to the nulliparous rat. This is followed by a decline in both parameters until gestation day 21, at which time there is a 10-fold increase in MT mRNA level (Andersen et al., 1983) and an estimated 2- to 3-fold increase in MT synthesis (Piletz et al., 1983). The MT mRNA level falls immediately thereafter, to a value lower than that of the nulliparous rat, and this low level persists through the suckling period (Andersen et al., 1983). The Hg-binding capacity of the TCA-soluble fraction of liver homogenate, an indirect measure of MT protein content, has been shown to decrease by about 50% between gestation days 16 and 21 (Charles-Shannon et al., 1981). Using the Cd-hemoglobin radioassay to directly quantitate MT, we have shown that hepatic MT concentration in the term pregnant rat (gestation days 19 to 21) is approximately 1 to 2 µg/g liver (Daston et al., 1994a), vs. approximately 5 to 10 µg/g on gestation day 11 (Daston et al., 1991), or in nulliparous female and male rats. The results of hepatic MT transcription, synthesis rate, and concentration are consistent and indicate that MT expression decreases during late pregnancy and through lactation in the rat, with the exception of a brief peripartum increase.

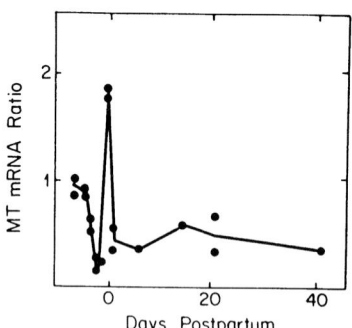

Figure 4. Concentrations of MT mRNA translation products in pregnant and postpartum rat liver. Poly(A)-rich mRNA was isolated and translated in a wheat germ cell-free system containing [^{3}H] serine. Translation products were separated by SDS-PAGE, and the film densities of autoradiograms were quantitated. The data are expressed as a ratio to a nulliparous 90-day rat. (From Andersen R. D. et al., *Eur. J. Biochem.*, 131, 497–500, 1983. With permission.)

The factors regulating MT expression during pregnancy and lactation are unknown. Quaife et al. (1986) suggested that glucocorticoids were responsible, because in mice there is an increase in circulating corticosterone concentration coincident with the rise in hepatic MT level on gestation day 10 (Dalle et al., 1978). However, the subsequent patterns of circulating corticosterone and hepatic MT mRNA are not correlated (Daston et al., 1992). The peripartum decline in MT mRNA to nongravid levels occurs despite the fact that plasma corticosterone level is still 7-fold greater than in the nonpregnant mouse; and the postpartum increase in MT mRNA concentration occurs despite the return of corticosterone concentration to the nongravid level. Furthermore, hepatic MT was not inducible by cortisol in late gestation rats, nor is the rat's peripartum spike in MT expression curtailed by adrenalectomy (Piletz et al., 1983). Another possibility is regulation by hormones of pregnancy. Andersen et al. (1983) noted that the MT mRNA expression pattern in rats is negatively related to circulating progesterone levels during late pregnancy. Progesterone inhibits MT induction by hydrocortisone in cell culture (Karin and Herschman, 1979), and it was speculated that progesterone has a similar action *in vivo*. This hypothesis makes sense, however, only if constitutive MT expression is regulated primarily by glucocorticoids, and is therefore unlikely.

It is possible that the increased Zn intake resulting from the hyperphagia that occurs during late pregnancy and lactation regulates hepatic MT. MT continues to be Zn-inducible during these stages in rats (Piletz et al., 1983) and mice (Daston et al., 1992). The failure of the two species to have similar patterns of MT expression despite similar patterns of food consumption casts doubt on this notion, however. Furthermore, changes in feed consumption are as poorly matched as corticosterone levels to changes in hepatic MT expression.

Various cytokines, including interleukins-1 and -6, and tumor necrosis factor-α are also endogenous regulators of MT expression and may be candidates for mediating MT during pregnancy and lactation. However, it is not known whether the circulating and/or hepatic levels of these inflammatory mediators change during pregnancy and lactation. In summary, at this point little can be concluded about the regulation of MT expression in the pregnant and lactating animal. The rise in circulating glucocorticoid level in the mouse may account for the initial increase in hepatic MT concentration in that species, but MT does not appear to be regulated by glucocorticoids subsequently. Regulation of MT later in gestation and during lactation may be regulated by Zn (from increased food intake), although there is no solid evidence for this.

B. Zn LEVELS DURING PREGNANCY AND LACTATION

Changes in Zn status in various tissues accompany pregnancy and lactation. In mice, the concentration of Zn in liver increases, predictably, with the increase in MT. The increase is on the order of 10 to 20% in whole liver and 25 to 30% in liver cytosol, where MT is localized (Daston et al., 1992). After parturition, Zn in whole liver decreases to nongravid levels, although cytosolic Zn does not decline. The discrepancy is accounted for by a decrease in Zn concentration of the 100,000 g pellet (microsomal fraction), which lasts through the lactational period. Coincident with the peripartum decrease in liver Zn concentrations is an increase in plasma Zn concentration (Daston et al., 1992). The likely explanation for these changes is secretion of Zn from liver into the circulation where it is available to the mammary glands.

Changes in Zn status in the pregnant rat also correlate with hepatic MT status. During midgestation (day 11), when constitutive MT level in liver is 5 to 10 µg/g, the Zn concentration in whole liver is approximately 370 nmol/g, and 175 nmol/g in 100,000 g supernatant (Daston et al., 1991). During late gestation (days 19 to 21), constitutive MT levels in liver fall to 1 to 2 µg/g, and there is a coincident drop in Zn concentration to 300 to 330 nmol/g in whole liver, and 70 to 85 nmol/g in 100,000 g supernatant (Daston et al., 1994a). Plasma Zn levels are also quite low in the term rat, averaging only 6 to 7 nmol/g, vs. 18 nmol/g during midgestation. This may be partially attributable to the increase in plasma volume and decrease in plasma protein concentration (to which Zn would be bound) in the term rat (Tam and Chan, 1977).

The circulating form(s) of Zn in the peri- and postpartum mouse is unknown. In the rat, a fraction is bound to MT. Chan et al. (1993) reported that circulating MT concentration is increased in pregnant and lactating rats beginning on gestation day 8, peaking at term at a level of approximately 600 ng/ml (vs. virtually undetectable in nongravid rats), and returning to nongravid levels between 1 and 2 weeks after parturition. The source of the MT is probably the liver, as hepatic Cd concentration decreases more rapidly in gravid than in nongravid rats (Chan and Cherian, 1993), suggesting mobilization of hepatic MT. Although this would be somewhat at odds with the observation that hepatic MT transcription and translation are not increased in the rat during pregnancy, it must be noted that the absolute quantity of plasma MT (no more than 600 ng/ml in a fluid that makes up ~5% of body weight) is only a small fraction of hepatic MT levels (5 to 10 µg/g in an organ composing ~4% of body weight). At this point, the concentration of circulating MT in the pregnant and lactating mouse has not been reported. Given the increase in plasma Zn concentration postpartum, along with the subcellular shift in Zn concentration in the liver (Daston et al., 1992), it may be speculated that Zn is being mobilized from the liver to the circulation. Whether this is true, and whether the carrier is MT, remains to be seen.

The Zn concentration of milk is considerably higher in mice than rats. Although milk Zn concentration decreases steadily during the suckling period in both species, it starts off at a much higher level in mice, approximately 40 µg/ml (Reis et al., 1991), vs. 20 µg/ml in rats (Keen et al., 1981). It may be speculated that the source of the milk Zn in mice is from the hepatic MT accumulated during pregnancy (Daston et al., 1992). It is worth noting that the Zn concentration of mature human milk is 1 µg/ml (Lonnderdal, 1989), even lower than in the rat. This suggests that the pregnant woman does not have higher than usual amounts of hepatic MT.

Why do such similar species as rats and mice have such different patterns of MT expression and tissue Zn concentration during pregnancy and lactation? Given the apparently large requirement of Zn to support growth of the offspring, it is possible that these represent different strategies for meeting the Zn requirements of the offspring. As noted in the previous section of this review, the fetuses of rats, mice, and indeed all mammalian species studied, accumulate considerable MT, particularly in liver. However, there are marked differences among species in the amount of MT accumulated. The hepatic MT concentration in rat fetuses is approximately 1000 µg/g (Wong and Klaassen, 1979; Lehman-McKeeman et al., 1988), whereas in mice it is approximately 300 µg/g (Kershaw et al., 1990). The kinetics of disappearance of the MT postnatally are also different, decreasing to adult levels after 4 weeks in the rat, vs. 2 weeks in the mouse. Despite these differences, the relative growth of these species during the first 3 to 4 weeks after birth are remarkably similar: they double in mass each week. It is likely, therefore, that the mouse neonate receives Zn from an additional source, the dam. Alternatively, the rat neonate may have more Zn than it needs.

C. INDUCIBILITY OF MT DURING PREGNANCY

Despite changes in constitutive MT expression, the inducibility of the protein by exogenous agents is not affected by pregnancy or lactation, except in the term rat (see below). In the mouse, there were no differences in the level of hepatic MT induced by 300 µmol/kg $ZnCl_2$ at any stage of pregnancy or lactation (Daston et al., 1992). Induction during midgestation in rats is comparable to male (and presumably nonpregnant female) rats after administration of urethane (Daston et al., 1991), ethanol, or arsenic (Taubeneck et al., 1994). Food deprivation may, however, induce slightly more MT in the pregnant rat. An overnight fast during midgestation produced a 7-fold increase in hepatic MT concentration (Daston et al., 1991) vs. a 6-fold induction in male rats (Bremmer and Davies, 1975; Sato and Sasaki, 1991). Hidalgo et al. (1988) also reported a greater induction of MT in the pregnant than the nulliparous rat.

The only stage in which inducibility of MT has been shown to be altered is late gestation in the rat, a time of reduced sensitivity. A 1 g/kg dosage of urethane increased hepatic MT concentration to about 140 µg/g during midgestation (Daston et al., 1991), or in the male rat (Brzeznicka et al., 1987), vs. only 40 µg/g in the term pregnant rat (Daston et al., 1994a). Even the constitutive level of MT is decreased 80 to 90% in the term rat; therefore, the relative induction (induced/control levels) is not different from earlier stages. However, the absolute level of MT is markedly lower, an effect which is associated with an increased toxicity of urethane (Daston et al., 1994a). These results suggest that there may be a stage-specific repressor of MT expression, although there is no direct evidence for this.

D. MATERNAL MT INDUCTION AND ALTERED EMBRYONIC Zn DISTRIBUTION

The metal-binding properties of MT are such that a substantial induction of the protein in liver can bring about systemic changes in Zn distribution. Numerous experiments examining the distribution of Zn after MT induction have shown an increase in Zn distribution to the liver and decreases in distribution to virtually every other tissue measured. There is a good correlation between hepatic MT concentration and Zn distribution (Taubeneck et al., 1994), confirming that MT induction is the driving force for the changes in Zn status. In the pregnant animal, maternal MT induction also elicits decreased Zn distribution to the conceptus, making it transitorily Zn deficient. Given that even short periods of dietary Zn deficiency are developmentally adverse (Hurley et al., 1971; Keen and Hurley, 1989), this had led to the formulation of the hypothesis that MT induction during pregnancy may affect the embryo by the following chain of events: (1) newly synthesized MT sequesters divalent metals, particularly Zn; (2) given a sufficiently substantial induction, this leads to a decrease in circulating Zn; (3) as plasma Zn is the proximate source for the embryo, decreased plasma Zn results in a transitory, but adverse, decrease in Zn distribution to the embryo.

This chain of events has been demonstrated to occur after treatment with a variety of MT-inducers (Table 1). For each of these, a substantial (above 100 µg/g liver) induction of MT produces a decrease in plasma Zn concentration, and a decrease in distribution of ^{65}Zn, administered after treatment with the inducer, to the conceptuses. This has been associated with an increased incidence of adverse developmental outcome, including abnormalities, embryonic death, and decreased fetal growth. The data for each of these agents fully support the hypothesis; however, these data alone do not support a causal relationship between altered Zn status and abnormal development. Therefore, additional studies have been carried out to evaluate causality.

Table 1 Metallothionein Inducers Shown to Adversely Affect Embryonic Zn Status and Development

Agent	Ref.
6-Mercaptopurine	Amemiya et al., 1986, 1989
Valproic acid	Keen et al., 1989
Urethane	Daston et al., 1991
α-Hederin	Daston et al., 1994b
Ethanol	Taubeneck et al., 1994
TNF-α	Keen et al., 1993

A number of approaches have been taken to demonstrate that the association between altered Zn distribution and abnormal development is causal. These include manipulating the diet of the dam, as well as *in vitro* procedures to isolate the Zn-related effects of treatment from possible direct embryotoxicity. The dietary manipulation approach involves feeding pregnant rats diets which are marginal, adequate, or replete in Zn, and comparing the developmental response to an MT-inducer. Amemiya et al. (1986) used diets containing 4.5 ppm (the minimum capable of sustaining normal development), 100 ppm, or 1000 ppm Zn and found that the developmental toxicity of 6-mercaptopurine was decreased by about 40% in the Zn-replete group. Leazer et al. (1992) used diets containing 12.5, 25, and 250 ppm Zn and found that the 12.5 ppm diet exacerbated the developmental toxicity of urethane given on gestation day 11 and the 250 ppm diet ameliorated it, compared to the adequate diet of 25 ppm. The rationale for these studies is the assumption that the higher levels of Zn provide a greater homeostatic buffer against MT-induced changes in Zn status. This assumption is borne out by the observation that a 1 g/kg dose of urethane decreased serum Zn concentration by more than 50% in rats fed the 12.5 ppm Zn diet, vs. only about 20% in rats fed the 25 or 250 ppm diets (Leazer, unpublished data). These dietary manipulation studies provide further support of the link between toxicant-altered Zn status and abnormal development.

Rodent whole embryo culture methods have been used to provide a more direct proof. Urethane given at a dosage of 1 g/kg i.p. to pregnant rats on gestation day 11 induced hepatic MT, altered systemic Zn distribution, and adversely affected development (Daston et al., 1991). However, urethane added directly to the culture medium of comparably staged rat embryos *in vitro* at concentrations up to 2250 µg/ml was without effect (Daston et al., 1991). Urethane distributes to total body water, so the concentrations used in culture were greater than or equal to the peak systemic concentration following a 1 g/kg i.p. dosage. No xenobiotic metabolizing system was added to these cultures; however, this would not have been expected to alter the results, as urethane is not metabolized appreciably faster by liver microsomes than by serum (Nomeir et al., 1989), and the culture media consisted of 50% rat serum. This result permits us to discount the possibility that urethane was directly embryotoxic, but does not rule out that other factors besides altered Zn homeostasis accounted for the *in vivo* result.

A more definitive proof has been provided for α-hederin. α-Hederin is a saponin isolated from species of ivy. It elicited an acute phase response when administered s.c. to pregnant rats at dosages of 30 or 300 µmol/kg, including a substantial (11 to 15 fold) induction in maternal hepatic MT, decreased serum Zn concentration, and decreased Zn distribution to the conceptuses. This was associated with increased malformations, embryolethality, and growth retardation (Daston et al., 1994b). Direct addition of α-hederin to rat embryos in culture had no effect on their development. Embryos *in vitro* were also exposed to α-hederin by culturing them in serum collected from adult rats 2 or 18 hours after treatment with a 300 µmol/kg dosage. Nothing is known about the metabolism of α-hederin, and the timing of the 2-h serum collection was chosen because it was likely that a large fraction of the s.c. injected compound would still be present, with sufficient time for active metabolites (if any) to be formed. However, this timepoint was prior to the onset of synthesis of acute phase proteins, including MT.

Serum collected 2 h after α-hederin treatment was not toxic to cultured embryos. The 18-h timepoint was selected because hepatic MT concentration was at its peak, and serum Zn at its nadir. This serum produced a high incidence of abnormal embryos. However, adding Zn to the serum to a level comparable to that in controls completely restored normal development (Daston et al., 1994b). These results convincingly demonstrate the involvement of decreased serum Zn concentration in the adverse developmental effects of maternal α-hederin administration.

These results are also relevant for the interpretation of developmental toxicology screening assays. Developmental effects produced by this indirect, maternally mediated mechanism may not be indicative of hazard at levels that are not acutely toxic. A relatively large proportion of agents are embryotoxic only in conjunction with maternal toxicity; while maternal MT induction is unlikely to be the cause of adverse embryonic effects in all these cases, it may not be uncommon given the number of toxicants that induce MT, and the relationship between tissue damage and MT induction.

The relationship between maternal toxicity and MT induction can be graphically illustrated using phenol as an example. When administered to rats by gavage at a dosage of 120 mg/kg/d on gestation days 6 to 15, phenol decreased fetal weight (Jones-Price et al., 1983). There was no maternal toxicity apparent at this dosage; however, in a pilot study dosages of 100 mg/kg/d decreased maternal weight gain and 160 mg/kg/d was lethal. On autopsy, the dead dams had substantial quantities of bedding (wood chips) in their GI tracts. This bizarre observation appears to have been a behavioral response to the caustic nature of phenol. Importantly, the pilot study employed much more concentrated dosing solutions (a dosing volume of 1 ml/kg) vs. the definitive study (5 ml/kg dosing volume). Given the different gross responses to equivalent dosages of different phenol concentrations, we were curious as to whether phenol was inducing MT, and whether MT induction would be correlated with concentration of the dose solution as well as dosage.

Female Sprague Dawley rats were gavaged with phenol at 120 mg/kg using 2.4, 6, or 12% dosing solutions, or 180 mg/kg, using 2.4 or 6% dosing solutions. Rats were sacrificed 18 h later. All of these dosing regimens induced a Cd- and Zn-binding protein in liver which was identified as MT by its heat resistance, chromatographic behavior, and characteristic UV absorbance ($A_{254} \gg A_{280}$). MT was quantitated using a Cd-hemoglobin radioassay (Eaton and Toal, 1982). The results are depicted in Figure 5. Importantly, the induction of MT is as much due to the concentration (and presumably the irritancy) of the dosing solution as it is to the dosage. Another way to illustrate the relationship between intoxication and MT induction is to plot MT induction against body weight change, an index of degree of toxic response (Figure 6). There is a good correlation between weight loss after dosing and MT induction. These data indicate that MT induction and consequent changes in Zn status may be as much a result of tissue injury and related stress as the properties of the toxicant, and suggest that altered Zn metabolism may be a widespread phenomenon in developmental toxicity screening given that the typical protocols include a maternally toxic dose level.

These results may also have relevance for human health, although data to support this notion are lacking. Still, pregnant women are exposed to dosages of agents shown to induce MT and/or alter Zn distribution in animals. The list includes the known human teratogens ethanol and valproic acid, as well as the potential teratogen hyperthermia, achieved either through fever or hot tub/sauna use. Infections provoking an acute phase response may also be in this category. If the effects of these regimens on human maternal and embryonic Zn status is comparable to animal results, then this not only provides information on the mechanism of abnormal development, but also suggests treatment strategies, such as dietary manipulation for women at risk, or more substantial intervention for cases such as severe febrile episodes or acute intoxications. Such ideas are nothing more than speculation at present, however, and await the collection of better data on mineral status and outcome in women exposed to MT inducers.

IV. CONCLUSIONS

Changes in MT expression appear to be a characteristic feature of mammalian development, both in mothers and offspring. The importance of MT in maintaining Zn and Cu homeostasis and the large demand of the growing organism for these essential metals suggest that these changes are extremely important for normal development. However, additional work must be carried out to determine the role of MT in ontogenesis, particularly in light of the apparent paradox of normal development in MT-I/MT-II null mutant mice. More optimistically, these null mutants should prove useful in understanding the relevance of elevations in constitutive MT expression during development. Whereas the high constitutive expression of MT in the perinate is suggestive of a beneficial role for the protein, excessive MT induction

Constitutive and Induced Metallothionein Expression in Development 1149

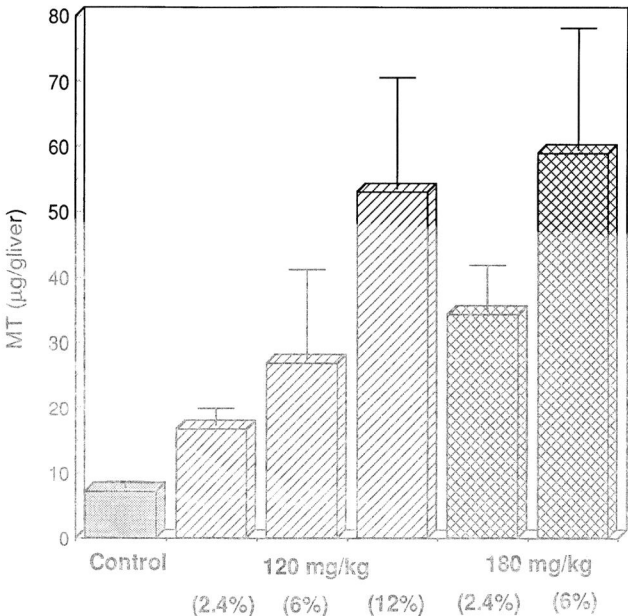

Figure 5. Hepatic MT levels of female rats 18 hours after p.o. administration of phenol at the indicated dosages and concentrations. Note that MT induction is as dependent on concentration as dosage for phenol.

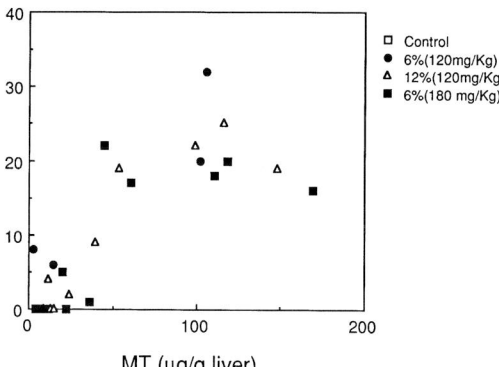

Figure 6. Correlation of severity of individual response to phenol (measured as weight loss) and hepatic MT induction.

in the pregnant mother is developmentally adverse by decreasing Zn distribution to the embryo. This phenomenon may be a fairly common mechanism contributing to developmental toxicity in animal tests. It may also contribute to adverse outcomes of human pregnancy, but this speculation remains to be proven.

REFERENCES

Acuff-Smith, K. D., Keen, C. L., Rogers, J. M., and Daston, G. P. (1994), *Teratology,* 49, 396 (abstr.).
Amemiya, K., Hurley, L. S., and Keen, C. L. (1989), *Teratology,* 39, 387–393.
Amemiya, K., Keen, C. L., and Hurley, L. S. (1986), *Teratology,* 34, 321–334.
Andersen, R. D., Piletz, J. E., Birren, B. W., and Herschman, H. R. (1983), *Eur. J. Biochem.,* 131, 497–500.
Andrews, G. K., Adamson, E. D., and Gedamu, L. (1984), *Dev. Biol.,* 103, 294–303.
Andrews, G. K., Huet, Y. M., Lehman, L. D., and Dey, S. K. (1987a), *Development,* 100, 463–469.
Andrews, G. K., Gallant, K. R., and Cherian, M. G. (1987b), *Eur. J. Biochem.,* 166, 527–531.
Andrews, G. K., Huet-Hudson, Y. M., Paria, B. C., McMaster, M. T., De, S. K., and Dey, S. K. (1991), *Dev. Biol.,* 145, 13–27.
Arizono, K., Ota, S., and Ariyoshi, T. (1981), *Bull. Environ. Contam. Toxicol.,* 27, 671–677.

Barlow, S. M., Morrison, P. J., and Sullivan, F. M. (1974), *J. Endocrinol.,* 60, 473–483.
Bremner, I. and Davies, N. T. (1975), *Biochem. J.,* 149, 733–738.
Brzeznicka, E. A., Lehman, L. D., and Klaassen, C. D. (1987), *Toxicol. Appl. Pharmacol.,* 87, 457–463.
Chan, H. M. and Cherian, M. G. (1993), *Toxicol. Appl. Pharmacol.,* 120, 308–314.
Chan, H. M., Tamura, Y., Cherian, M. G., and Goyer, R. A. (1993), *Proc. Soc. Exp. Biol. Med.,* 202, 420–427.
Charles-Shannon, V. L., Sasser, L. B., Burbank, D. K., and Kelman, B. I. (1981), *Proc. Soc. Exp. Biol. Med.,* 168, 56–61.
Cousins, R. J. (1985), *Physiol. Rev.,* 65, 238–309.
Dalle, M., Giry, J., Gay, M., and Delost, P. (1978), *J. Endocrinol,* 76, 303–309.
Danielson, K. G., Ohi, S., and Huang, P. C. (1982), *J. Histochem. Cytochem.,* 30, 1033–1039.
Daston, G. P., Overmann, G. J., Acuff-Smith, K. D., Taubeneck, M. W., and Keen, C. L. (1994a), *Toxicologist,* 14, 159 (abstr.).
Daston, G. P., Overmann, G. J., Baines, D., Taubeneck, M. W., Lehman-McKeeman, L. D., Rogers, J. M., and Keen, C. L. (1994b), *Reprod. Toxicol.,* 8, 15–24.
Daston, G. P., Overmann, G. J., Lehman-McKeeman, L. D., Taubeneck, M. W., Rogers, J. M., and Keen, C. L. (1992), *Teratology,* 45, 501–502 (abstr.).
Daston, G. P., Overmann, G. J., Taubeneck, M. W., Lehman-McKeeman, L. D., Rogers, J. M., and Keen, C. L. (1991), *Toxicol. Appl. Pharmacol.,* 110, 450–463.
De, S. K., McMaster, M. T., Dey, S. K., and Andrews, G. K. (1989), *Development,* 107, 611–621.
DuPouy, J. P., Coffigny, H., and Magre, S. (1975), *J. Endocrinol.,* 65, 347–352.
Durnam, D., Hoffman, J. S., Quaife, C. J., Benditt, E. P., Chen, H. Y., Brinster, R. L., and Palmiter, R. D. (1984), *Proc. Natl. Acad. Sci. U.S.A.,* 81, 1053–1056.
Eaton, D. L. and Toal, B. F. (1982), *Toxicol. Appl. Pharmacol.,* 66, 134–142.
Friberg, L., Piscator, M., Nordberg, G. F., and Kjellström, T. (1974), *Cadmium in the Environment,* CRC Press, Cleveland, OH.
Friedman, R. L. and Stark, G. S. (1985), *Nature,* 314, 637–639.
Gallant, K. R. and Cherian, M. G. (1986), *Biochem. Cell Biol.,* 64, 8–12.
Goering, P. L. and Klaassen, C. D. (1984), *Toxicol. Appl. Pharmacol.,* 74, 321–329.
Hidalgo, J., Giralt, M., Garvey, J. S., and Armario, A. (1988), *Biol. Neonate,* 53, 148–155.
Hurley, L. S. (1976), *Trace Elements in Human Health and Disease,* Prasad, A.S., Ed., Academic Press, New York, 301–314.
Hurley, L. S., Gowan, J., and Swenerton, H. (1971), *Teratology,* 4, 199–204.
Jones-Price, C., Ledoux, T. A., Reed, J. R., Fisher, P. W., Langhoff-Paschke, L., Marr, M. C., and Kimmel, C. A. (1983), Teratogenic evaluation of phenol (CAS No. 108–95–2) in CD rats, NTP Rep. No. PB83–247726.
Karin, M. (1985), *Cell,* 41, 9–10.
Karin, M., Anderson, R. D., Slater, E., Smith, K., and Herschman, H. R. (1980a), *Nature,* 286, 295–297.
Karin, M. and Herschman, H. R. (1979), *Science,* 204, 176–177.
Karin, M., Herschman, H. R., and Weinstein, D. (1980b), *Biochem. Biophys. Res. Commun.,* 92, 1052–1059.
Karin, M., Imbra, R. J., Heguy, A., and Wong, G. (1985), *Mol. Cell. Biol.,* 5, 2866–2869.
Keen, C. L. and Hurley, L. S. (1989), *Zinc and Human Biology,* C. F., Mills, Ed., Springer Verlag, New York, 183–220.
Keen, C. L., Lonnerdal, B., Clegg, M., and Hurley, L. S. (1981), *J. Nutr.,* 111, 226–230.
Keen, C. L., Peters, J. M., and Hurley, L. S. (1989), *J. Nutr.,* 119, 607–611.
Keen, C. L., Taubeneck, M. W., Daston, G. P., Rogers, J. M., and Gershwin, M. E. (1993), *Ann. N.Y. Acad. Sci.,* 678, 37–47.
Kern, S. R., Smith, H. A., Fontaine, D., and Bryan, S. E. (1981), *Toxicol. Appl. Pharmacol.,* 59, 346–354.
Kershaw, W. C., Lehman-McKeeman, L. D., and Klaassen, C. D. (1990), *Toxicol. Appl. Pharmacol.,* 104, 267–275.
Klaassen, C. D. and Lehman-McKeeman, L. D. (1989), *J. Am. Coll. Toxicol.,* 8, 1291–1297.
Klein, D., Scholz, P., Drasch, G. A., Muller-Hocker, J., and Summer, K. H. (1991), *Toxicol. Lett.,* 56, 61–67.
Kuhnert, P. M., Kuhnert, B. R., Bottoms, S. F., and Erhard, B. S. (1982), *Am. J. Obstet. Gynecol.,* 142, 1021–1025.
Leazer, T. M., Taubeneck, M. W., Daston, G. P., Keen, C. L., and Rogers, J. M. (1992), *Teratology,* 45, 481 (abstr.).
Lehman, L. D. and Klaassen, C. D. (1986), *Anal. Biochem.,* 153, 305–314.
Lehman, L. D. and Poisner, A. M. (1984), *J. Toxicol. Environ. Health,* 14, 419–432.
Lehman-McKeeman, L. D., Andrews, G. K., and Klaassen, C. D. (1988a), *Toxicol. Appl. Pharmacol.,* 92, 10–17.
Lehman-McKeeman, L. D., Andrews, G. K., and Klaassen, C. D. (1988b), *Biochem J.,* 249, 429.
Lonnerdal, B. (1989), *Annu. Rev. Nutr.,* 9, 109–125.
Margoshes, M. and Vallee, B. L. (1957), *J. Am. Chem. Soc.,* 79, 4813–4814.
Masters, B. A., Kelly, E. J., Quaife, C. J., Brinster, R. L., and Palmiter, R. D. (1994), *Proc. Natl. Acad. Sci. U.S.A.,* 91, 584–588.
Michalaska, A. E. and Choo, K. H. A. (1993), *Proc. Natl. Acad. Sci. U.S.A.,* 90, 8088–8092.
Nomeir, A. A., Ioannou, Y. M., Sanders, J. M., and Matthews, H. B. (1989), *Toxicol. Appl. Pharmacol.,* 97, 203–215.
Nordberg, G. F., Nordberg, M., Piscator, M., and Vesterberg, O. (1972), *Biochem. J.,* 126, 491–498.
Ouellette, A. J. (1982), *Dev. Biol.,* 92, 240–246.
Palmiter, R. D., Findley, S. D., Whitmore, T. E., and Durnam, D. M. (1992), *Proc. Natl. Acad. Sci. U.S.A.,* 89, 6333–6337.
Panemangalore, M., Banerjee, D., Onosaka, S., and Cherian, M. G. (1983), *Dev. Biol.,* 97, 95–102.
Piletz, J. E., Andersen, R. D., Birren, B. W., and Herschman, H. R. (1983), *Eur. J. Biochem.,* 131, 489–495.
Poisner, A. M., Wood, G. W., Poisner, R., and Inagami, T. (1982), *Proc. Soc. Exp. Biol. Med.,* 169, 4–6.

Quaife, C. J., Findley, S. D., Erickson, J. C., Froelick, G. J., Kelly, E. J., Zambrowicz, B. P., and Palmiter, R. D. (1994), *Biochemistry*, 33, 7250–7259.
Quaife, C., Hammer, R. E., Mottet, N. K., and Palmiter, R. D. (1986), *Dev. Biol.*, 118, 549–555.
Reis, B. L., Keen, C. L., Lonnerdal, B., and Hurley, L. S. (1991), *J. Nutr.*, 121, 687–699.
Riordan, J. R. and Richards, V. (1980), *J. Biol. Chem.*, 255, 5380–5383.
Ryden, L. and Deutsch, H. F. (1978), *J. Biol. Chem.*, 253, 519–524.
Sato, M. and Sasaki, M. (1991), *Chem.-Biol. Interact.*, 78, 143–154.
Tam, P. P. L. and Chan, S. T. H. (1977), *J. Reprod. Fert.*, 51, 41–51.
Taubeneck, M. W., Daston, G. P., Rogers, J. M., and Keen, C. L. (1994), *Reprod. Toxicol.*, 8, 25–40.
Terhaar, C. J., Vis. E., Rondabush, R. L., and Fassett, D. W. (1965), *Toxicol. Appl. Pharmacol.*, 7, 500.
Uchida, Y., Takio, K., Titani, K., Ihara, Y., and Tomonga, M. (1991), *Neuron.*, 7, 337–347.
Waalkes, M. P. and Bell, J. U. (1980), *Life Sci.*, 27, 585–593.
Waalkes, M. P. and Klaassen, C. D. (1984), *Toxicol. Appl. Pharmacol.*, 74, 314–320.
Waalkes, M. P., Poisner, A. M., Wood, G. W., and Klaassen, C. D. (1984), *Toxicol. Appl. Pharmacol.*, 74, 179–184.
Wong, K.-L. and Klaassen, C. D. (1979), *J. Biol. Chem.*, 254, 12399–12403.

Chapter 72

Developmental and Reproductive Effects of Metal Chelators

Jose L. Domingo

I. DEVELOPMENTAL TOXICITY OF METALS

Environmental contamination of heavy metals such as lead, cadmium, mercury, or chromium among others, is a serious and continuing problem throughout the world. Developmental toxicity is included among the harmful effects derived from the exposure to these elements, and embryotoxicity and teratogenicity have resulted in pregnant animals exposed to certain metals. Thus, arsenic, mercury, cadmium, and lead have been known to be developmental toxicants for many years (see review by Domingo, 1994).

Arsenate and arsenite, the main forms of inorganic arsenic in the environment, are embryotoxic and teratogenic in golden hamsters (Ferm and Carpenter, 1971; Willhite, 1981), mice (Hood and Bishop, 1972; Hood et al., 1978; Baxley et al., 1981; Morrissey and Mottet, 1983; Lindgren et al., 1984), rats (Beaudoin, 1974; Hood et al., 1977), and chicks (Peterkova and Puzanova, 1976; Chaineau et al., 1990). Inorganic mercury is also fetotoxic and may affect embryonic development and cause fetal malformations in pregnant rodents (Ferm, 1972; Chang et al., 1980; Holt and Webb, 1986), although a number of studies have shown that alkylmercury crosses the placenta more easily than inorganic mercury (Leonard et al., 1983). Methylmercury has been reported to be embryotoxic and teratogenic in rats (Fuyuta et al., 1978; Hoskins and Hupp, 1978), mice (Su and Okita, 1976; Fuyuta et al., 1978; Fujimoto et al., 1979; Yasuda et al., 1985), golden hamsters (Harris et al., 1972; Hoskins and Hupp, 1978), and cats (Khera, 1973). In animal models, cadmium has also been shown to be a developmental toxicant for hamsters, rats, mice, and chicks (Ferm, 1971; Chernoff, 1973; Dencker, 1975; Ahokas et al., 1980; Gale and Horner, 1987; Holt and Webb, 1987; Saltzman 1989). Moreover, it is well established that lead may cross the placental barrier and cause developmental toxicity in hamsters, rats, and mice (Ferm and Carpenter, 1967; McClain and Becker, 1975; Kennedy et al., 1975; Miller et al., 1982).

In humans, intrauterine intoxications by alkylmercury have been reported from several countries, the accident occurring in Japan at Minamata Bay being the best-known (Matsumoto et al., 1965; Harada, 1978; Leonard et al., 1983). With regard to arsenic, ingestion of arsenic trioxide by a pregnant woman was described to result in death of the offspring (Lugo et al., 1969). Also, it is well established that lead readily crosses the human placenta giving rise to similar maternal and fetal blood levels (Cohen and Roe, 1991). In past decades, it was found that females occupationally exposed to very high lead levels showed an increased incidence of spontaneous abortions and fetal deaths (Rom, 1976; Chang et al., 1980). Cadmium can cross the placenta, and cadmium intake by pregnant women may partly be responsible for the well-documented fact that mothers who smoke run a greater risk of having babies of low birth weight and short body length for gestational age than nonsmokers (Miller et al., 1976; Ahokas et al., 1980).

Other current metals such as lithium, vanadium, selenium, manganese, aluminum, gallium, cobalt, chromium, or nickel have also been reported to cause developmental toxicity in mammals. Lithium salts, which have been commonly used to treat manic depressive psychosis and other psychiatric disorders, provoked fetal malformations in mice (Szabo, 1970; Smithberg and Dixit, 1982) and rats (Wright et al., 1971; Marathe and Thomas, 1986), whereas children of mothers on lithium treatment during pregnancy showed malformations involving the heart and other organs (Schou, 1976).

Vanadium, as vanadium pentoxide, is an annoying pollutant produced in large amounts during natural oil combustion. Although vanadium is an element present in living organisms in trace amounts, it is toxic when introduced in excessive doses to animals and humans. Recent investigations have shown that vanadate (V^{5+}) exposure causes developmental toxicity in rats (Paternain et al., 1987) and mice (Sanchez et al., 1991; Gomez et al., 1992a), whereas vanadyl (V^{4+}) is embryotoxic and teratogenic in mice (Paternain et al., 1990a).

Selenium is an essential nutrient for animals and men. However, acute and chronic toxicity may occur in animals at relatively low dietary selenium levels or environmental concentrations (Olson, 1986; Besser et al., 1993). Teratological and embryotoxic effects of selenium compounds have also been reported (Yonemoto et al., 1983; Högberg and Alexander, 1986).

Although manganese is considered an essential trace element for all living mammals, toxic intakes of manganese (either through the air or diet) may result in severe pathologies, particularly of the CNS (Keen and Zidenberg-Cherr 1990). No evidence of structural malformations in the offspring of animals given large doses of manganese during pregnancy has been observed (Laskey et al., 1982; Webster and Valois, 1987), whereas it has been reported that the neonate may be particularly vulnerable to high levels of manganese (Webster and Valois, 1987).

Until the last decade, there was little concern about the possible toxic consequences derived from aluminum ingestion. In recent years, it has been clearly shown that individuals ingesting large amounts of aluminum compounds do absorb a definite amount of aluminum (Alfrey, 1985). Developmental toxicity of aluminum exposure was observed when aluminum compounds were administered to rabbits (Yokel, 1987), mice (Cranmer et al., 1986), and rats (Paternain et al., 1988a; Colomina et al., 1992).

Gallium, a group IIIa transition element, is particularly valuable in the field of medical oncology. Gallium nitrate administration during the period of organogenesis resulted in fetal growth retardation in mice (Gomez et al., 1992b), whereas gallium sulfate injected I.V. to pregnant hamsters showed some teratogenic effects (Ferm and Carpenter, 1970).

Cobalt is an essential trace element for mammalian nutrition, but it is toxic at high doses (Domingo, 1989a; Beyersmann and Hartwig, 1992). Teratogenicity tests of cobalt were negative in rats and hamsters (Ferm, 1972; Paternain et al., 1988b), but positive in mice (Wide, 1984) and in the frog, *Xenopus laevis* (Plowman et al., 1991).

Other elements such as indium (Ferm and Carpenter, 1970) and chromium (Matsumoto et al., 1976; Endo and Watanabe, 1988) were reported to have pronounced toxic effects upon embryonic and fetal development, being responsible for certain malformations in mammalian embryos. On the other hand, nickel, a metal which constitutes a potential hazard to our environment, due to widespread use of its chemical compounds in various industries, was found to cause embryotoxic and teratogenic effects in rats (Sunderman et al., 1978) and mice (Lu et al., 1979).

A. CHELATING AGENTS AND PREVENTION OF DEVELOPMENTAL TOXICITY OF METALS

It is well established that some agents ("antiteratogens") may act as antagonists of the developmental toxicity of various compounds. Thus, hydroxyurea, methotrexate, 2-methoxyethanol, and TCDD-induced teratogenesis was ameliorated by propyl gallate, leucovorin, ethanol, and HCB, respectively (DeSesso and Goeringer, 1990, 1991; Welsch et al., 1991; Morrissey et al., 1992). In contrast, other agents such as taurine failed to protect against the developmental toxicity of isotretinoin (Agnish et al., 1990).

With regard to the developmental toxicity of heavy metals, the antagonism of their toxic actions is generally based on the manipulation of the binding preferences of the particular toxic metal, via transfer of metal to an appropriate metal-chelate complex (Jones, 1991). Chelation for heavy metal intoxication has been practiced in various forms for approximately 50 years. In recent decades, the protective effects of a number of chelating agents against the embryotoxic and teratogenic effects of some heavy metals have been assessed by a number of investigators.

Subcutaneous treatment with 2,3-dimercaptopropanol (BAL) diminished the incidence of arsenate-induced gross malformations and growth retardation in mice (Hood and Pike, 1972), although s.c. administration of this chelator was not able to alleviate the embryotoxicity and teratogenicity of arsenite (Hood and Vedel, 1984; Domingo et al., 1992a). However, 2,3-dimercaptosuccinic acid (DMSA) and sodium 2,3-dimercaptopropane-1-sulfonate (DMPS) were effective in the prevention of both arsenite and arsenate-induced developmental toxicity in mice (Bosque et al., 1991; Domingo et al., 1991, 1992a). DMSA (Sanchez et al., 1993), DMPS (Gomez et al., 1994), and 2-mercaptopropionylglycine (Fujimoto et al., 1979) were also effective in alleviating methylmercury-induced teratogenesis in mice, whereas the administration of penicillamine to pregnant rats poisoned with methylmercury prevented morphological changes in the fetal brain (Matsumoto et al., 1967).

In contrast, methylmercury-induced teratogenicity was not influenced by the administration of trisodium nitrilotriacetate (NTA) (Nolen et al., 1972), while the incidence of malformations produced by mercury was increased by treatment with N-acetylcysteine (NAC) (Endo and Watanabe, 1988). The administration of NAC also enhanced the teratogenic effects of cadmium and chromium in mice (Endo and Watanabe, 1988). Recently, it has been demonstrated that the ability of sodium 4,5-dihydroxybenzene-1,3-disulfonate (Tiron) to protect the developing mouse embryos against uranium-induced embryo/fetal toxicity offers only modest encouragement with regard to its possible therapeutic potential (Bosque et al., 1993). However, Tiron was effective in protecting against metavanadate-induced developmental toxicity in mice (Domingo et al., 1993).

In a study on the effects of various chelating agents on the teratogenicity of lead nitrate in rats, ethylenediaminetetraacetic acid (EDTA) produced the greatest reduction in overall effect. Penicillamine, NTA, and iminodiacetic acid (IDA) exhibited only moderate protective effects (McClain and Siekierka, 1975). Nolen et al. (1972) found that NTA might protect the dam and the fetus from the toxic effects of oral cadmium, while the ability of certain dithiocarbamates (DTC) to influence the developmental toxicity of parenteral cadmium decreased within 24 h of treatment with the chelators (Hatori et al., 1990).

II. DEVELOPMENTAL TOXICITY OF CHELATING AGENTS

A number of metal binding agents have been available in the past for the treatment of heavy-metal intoxication. For example, calcium disodium ethylenediaminetetraacetic acid (EDTA) has been the mainstay of chelation therapy for lead poisoning for the past 40 years, whereas BAL-EDTA treatment continues to be standard for children with acute lead encephalopathy who are comatose or have intractable vomiting (Chisolm, 1992). D-Penicillamine has also been used in the past for treating lead toxicity as well as the toxic effects of other metals (Walshe, 1956). In recent years, a number of investigators have demonstrated that DMSA and DMPS, new oral drugs, are of potential interest for the treatment of lead poisoning (Aposhian, 1983; Aposhian and Aposhian, 1990). Both chelators are useful as arsenic antidotes, being more effective than BAL in this respect. Also, a combination of DMSA and DMPS removes mercury from most organs, with DMSA appearing to remove more organic mercury, and DMPS removing more inorganic mercury (Aposhian, 1983).

On the other hand, the polyaminocarboxylic acid diethylenetriaminepentaacetic acid (DTPA) enhances the urinary excretion of manganese (Catsch et al., 1979), zinc (Llobet et al., 1989), and thorium (Stradling et al., 1991), whereas the hydroxycarboxylic acid Tiron has been found to be effective in increasing the urinary excretion of vanadium (Gomez et al., 1991) and uranium (Domingo et al., 1992b). Although no clinical chelation treatment of cadmium intoxication has been approved, systemic cadmium poisoning can be alleviated by administration of dithiocarbamate (DTC) chelating agents (Jones and Jones, 1984). On the other hand, desferrioxamine (or deferoxamine, DFOA) is a chelator which has been used in the removal of iron (Pippard and Callender, 1983) and aluminum following acute and chronic exposure (Domingo, 1989b).

Because the above chelating agents, whose chemical structures are shown in Figure 1, may be administered to a heavy-metal-intoxicated pregnant woman, it is important to know the intrinsic developmental toxicity of these compounds. With regard to this, a number of maternal and developmental effects have been reported to occur in animal models at different doses of the chelators (Table 1). Specific developmental anomalies produced by these agents are summarized in Table 2. Studies on the developmental toxicity of the heavy metal chelators are reviewed below.

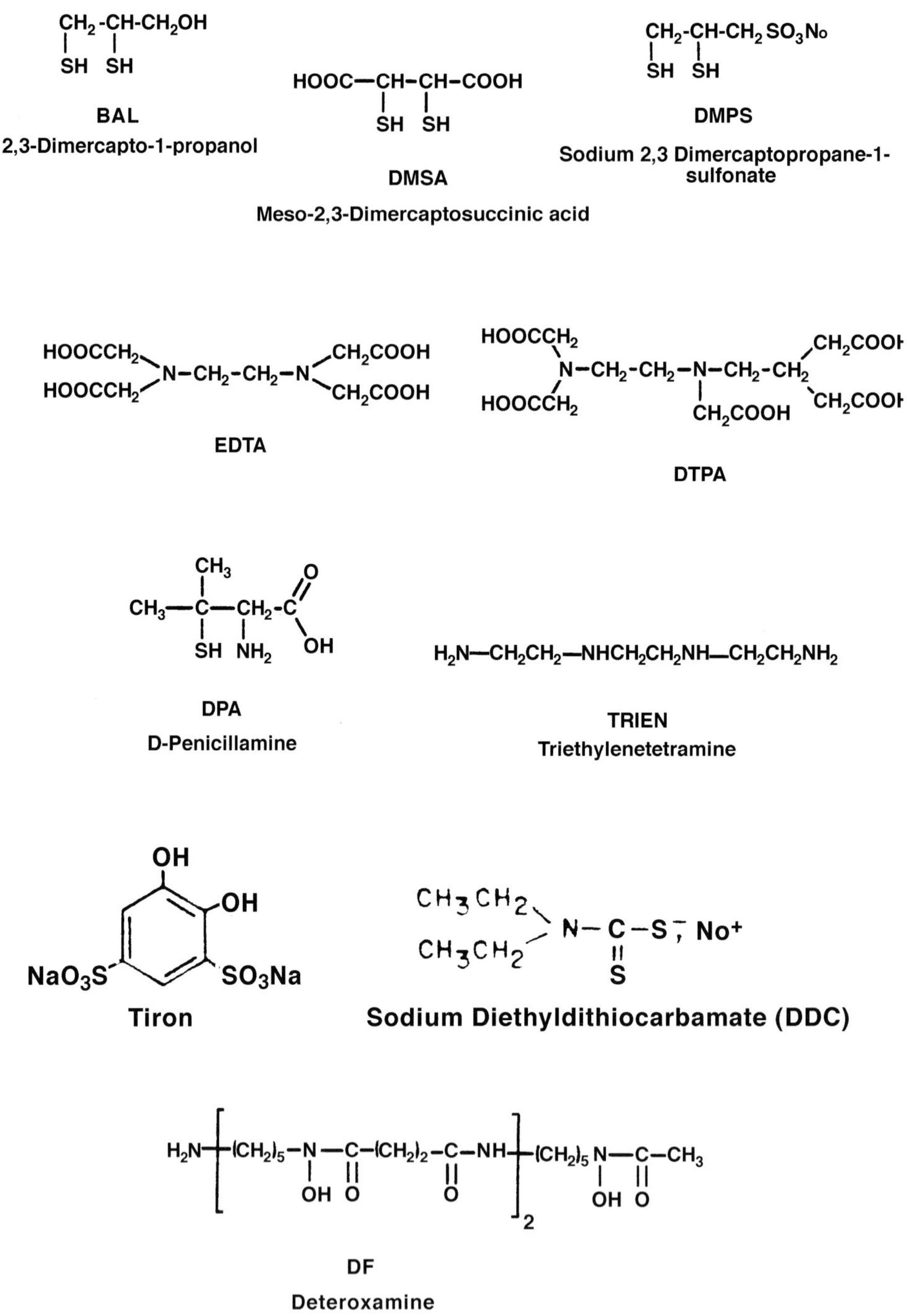

Figure 1. Chemical structures of chelating agents.

A. 2,3-DITHIOLS
1. 2,3-Dimercaptopropanol (BAL)

BAL (British anti-Lewisite) was developed as an antidote to the arsenical Lewisite (Peters et al., 1945). BAL was the first chelator successfully introduced into clinical practice as an antagonist for cases of acute and chronic metal intoxication (Jones, 1991). In spite of the wide use of this chelating agent,

Table 1 Chelating Agents and Fetal Anomalies in Animal Models

Agent	Anomaly
2,3-Dithiols	
2,3-Dimercaptopropanol (BAL)	Cleft palate
meso-2,3-Dimercaptosuccinic acid (DMSA)	Supernumerary ribs, decreased ossification in different bones, malformed sternebrae
Sodium 2,3-dimercaptopropane-1-sulfonate (DMPS)	None
Polyaminocarboxylic acids	
Diethylenetriaminepentaacetic acid (DTPA)	Exencephaly, ablepharia, spina bifida aperta, cleft palate, polydactyly
Ethylenediaminetetraacetic acid (EDTA)	Cleft palate, brain malformations, polydactyly, tail defects
Other chelators with amine groups	
D-Penicillamine (D-PA)	Cutis laxis, abdominal herniations, incomplete diaphragms, reduced ossification in phalanges and tail vertebrae
Triethylenetetramine (Trien)	None
Miscellaneous chelating agents	
Sodium 4,5-dihydroxybenzene-1,3-disulfonate (Tiron)	None
Diethyldithiocarbamates	None
Desferrioxamine (DFOA)	None

until recently only a single report on the developmental toxicity of BAL was available. BAL in 10% peanut oil solutions was injected s.c. into pregnant mice with 0.001 ml of BAL solution per gram body weight on the 9th to 12th day of gestation. Teratogenic effects were evidenced by malformations of the extremities such as digits with abnormal direction and situation, shortness of the limb or foot, brachydactyly, syndactyly, adactyly, and polydactyly. A few cases of cleft palate and one case of cerebral hernia were also recognized. Moreover, a slight growth retarding effect and a moderate effect on the mortality of embryos in early life were also found. It was concluded that the teratogenic effects of BAL might result from its action as a mitotic poison to the cells of the embryo (Nishimura and Takagaki, 1959). Since BAL density is about 1.25 g/ml, according to these results a dose of 125 mg BAL/kg body weight would be teratogenic for mice.

When BAL was administered to pregnant mice at 15, 30, and 60 mg/kg to protect against arsenite (Hood and Vedel, 1984; Domingo et al., 1992a), or methylmercury-induced developmental toxicity (Gomez et al., 1994), no embryotoxic or teratogenic effects were associated with the treatment.

2. meso-2,3-Dimercaptosuccinic acid (DMSA)

DMSA is a water-soluble chelating agent structurally similar to BAL, which is an useful antidote for the treatment of experimental and human poisoning by a number of heavy metals (Aposhian, 1983; Aposhian and Aposhian, 1990; Aposhian et al., 1992). While it is evident that DMSA can be of value in the treatment of metal intoxication, concern has been raised over the possibility that the drug may induce essential trace element deficiencies simultaneously with the removal of the target metal (Friedheim et al., 1978; Cantilena and Klaassen, 1982). It has been suggested that DMSA-induced alterations in mineral metabolism may result in pregnancy complications. Thus, when maternal and fetal toxicity of subcutaneous DMSA were evaluated in Swiss mice at 410, 820, and 1640 mg/kg/d on gestation days 6 to 15, toxic effects on fetal development were primarily seen in the 1640 mg DMSA/kg group, with the drug resulting in a significant increase in prenatal death and fetal growth retardation (lower body weight and shorter crown-rump length). Fetal body weight was also significantly lower in the 820-mg DMSA group compared with controls. Soft tissue abnormalities (small thoracic cavities, brain defects, eye defects) were observed in fetuses at 820 and 1640 mg/kg/d. Skeletal anomalies (decreased ossification of supraoccipital, tarsus and carpus, bipartite sternebrae, hypoplasia of mandible, irregular shapes of ribs) were observed at a higher than normal frequency in the 820 and 1640 mg DMSA/kg/d groups. Maternal toxicity included a significant reduction in weight gain during the exposure and postexposure periods at 1640 mg/kg/d. The developmental no-observed-effect level for DMSA under the test conditions was 410 mg/kg/d. Developmental toxicity of DMSA was thought to be a consequence of severe zinc deficiency caused by the chelator (Domingo et al., 1988).

Table 2 Chelating Agents and Maternal and Developmental Toxicity in Laboratory Animals

Chelator	Species	Doses (mg/kg/d)	Dosing Period (days of gestation)	Route	Signs of Toxicity in the Mother	NOEL for Maternal Toxicity (mg/kg/d)	Toxic Effects in the Fetuses	NOEL for Developmental Toxicity (mg/kg/d)	Ref.
BAL	Mice	125	9–12	s.c.	Not reported	Not reported	Embryotoxicity and teratogenicity	<125	Nishimura and Takagaki 1959
DMSA	Swiss mice	410, 820, or 1640	6–15	s.c.	Reductions in body weight gain	820	Prenatal deaths, reductions in body weight, soft tissue and skeletal anomalies	410	Domingo et al., 1988
	SD rats	100, 300, or 1000	6–15	p.o.	Reductions in body weight gain	<100	Resorptions and fetal body weight reductions	<100	Domingo et al., 1990a
DMPS	Swiss mice	75, 150, or 300	6–15	p.o.	None	≥300	None	≥300	Bosque et al., 1990
DTPA	C57BL mice	720–2880 (μmol CaDTPA/kg)	7–11 or 12–16	s.c. (5 inj./day)	Not stated	Not stated	Fetal mortality, gross congenital malformations	<720 μmol/kg	Fisher et al., 1976
EDTA	SD rats	2% or 3%	6–21	Diet	Diarrhea	Not reported	Embryotoxicity and teratogenicity	—	Swenerton and Hurley, 1971
	CD rats	1000	7–14	p.o.	Diarrhea, depressed activity, deaths	<1000	None	>1000	Shardein et al., 1981
D-PA	SD rats	0.17%, 0.83%, or 1.66%	0–21	Diet	Not stated	Not stated	Resorptions and malformations	<0.17%	Keen et al., 1983
Trien	Rats	0.83%	0–21	Diet	Hemorrhages	<0.83%	Malformations	<0.83%	Keen et al., 1982

Chelator	Species	Dose	Days	Route	Maternal effects	Maternal LOAEL	Developmental effects	Developmental LOAEL	Reference
Tiron	Swiss mice	750, 1500, or 3000	6–15	i.p.	Deaths and reductions in body weight gain	1,500	Resorptions reductions in fetal body weight	1,500	Ortega et al., 1991
DTC	Syrian golden hamsters	2.2 mmol/kg (four sodium DTCs)	8	i.p.	None	—	Reductions in fetal body weight	Not reported	Hatori et al., 1990
DFOA	Swiss mice	44, 88, 176, or 352	6–15	i.p.	Abortions and early deliveries, reductions in food consumption and in body weight gain	<44	Decrease in the number of live fetuses	176	Bosque et al., 1995

Note: Abbreviations: BAL, 2,3-dimercaptopropanol; DMSA, *meso*-2,3-dimercaptosuccinic acid; DMPS, sodium 2,3-dimercaptopropane-1-sulfonate; DTPA, diethylenetriaminepentaacetic acid; EDTA, ethylenedimiretetraacetic acid; D-PA, D-penicillamine; Trien, triethylentetramine; Tiron, sodium 4,5-dihydroxybenzene-1,3-disulfonate; DTC, diethyldithiocarbamates; DFOA, desferrioxamine.

Consistent with the above, maternal toxicity and fetotoxicity were observed in Sprague-Dawley rats when DMSA was administered by gavage at 100, 300, and 1000 mg/kg/d on gestational days 6 to 15. Maternal toxicity was evidenced by a significant decrease in body weight gain, whereas significant increases in the number of early resorptions and percentage postimplantation loss, as well as a significant decrease in fetal body weight were clear signs of embryo/fetal toxicity. However, DMSA did not produce teratogenicity at any dosage level. Consequently, the NOEL for maternal toxicity and for embryo/fetal toxicity was <100 mg DMSA/kg/d, while no teratogenic effects were observed at 100 mg DMSA/kg/d (Domingo et al., 1990a). In the same study, it was also found that DMSA had a pronounced effect on both maternal and fetal mineral metabolism, which was specially evidenced by the high concentrations of copper and iron in whole fetuses from dams receiving 1000 mg DMSA/kg/d. In addition, DMSA resulted in low maternal liver copper and calcium concentrations and high iron levels, whereas fetal liver calcium and copper concentrations were found to be lower in the DMSA-exposed groups than in controls (Paternain et al., 1990b).

Based on the findings that maternal DMSA administration can result in low fetal zinc concentrations (Domingo et al., 1988; Paternain et al., 1990b), and the observation that even transitory embryonic/fetal zinc deficiency can be teratogenic (Hurley and Shrader, 1975), Taubeneck et al. (1992) tested in Swiss mice the hypothesis that the developmental toxicity of DMSA is due in part to an induced zinc deficiency, and that its teratogenicity could be ameliorated by dietary zinc supplementation. DMSA was given by gavage or s.c. injections to pregnant mice at 400 and 800 mg/kg/d on gestation days 6 to 15. Mice were fed a diet containing 14 µg Zn/g diet. A subgroup of mice in the 800 mg DMSA/kg s.c. group was fed a diet containing 250 µg Zn/g. DMSA administration did not result in overt maternal toxicity. There was no effect of the drug on fetal or placental weight or on crown-rump length. However, some fetuses from DMSA-treated dams were characterized by skeletal abnormalities including supernumerary ribs, unossified anterior phalanges, and malformed sternebrae. Drug exposure was not associated with consistent changes in tissue zinc, iron, calcium, or magnesium levels, whereas fetal liver concentrations exhibited dose-dependent decreases with increasing DMSA dose. These findings agree with the results of previous studies in the rat (Domingo et al., 1990a), when changes in skeletal ossification and a dramatic dose-dependent decrease in fetal liver copper with increasing DMSA doses up to 1000 mg/kg/d were observed. A previous study by the same group did not assess mineral metabolism, but revealed an increased incidence of numerous skeletal variations with increasing DMSA doses of up to 1640 mg/kg/d (Domingo et al., 1988). Given the effects of DMSA on fetal liver copper, it has been suggested that the developmental toxicity of DMSA may be mediated, at least in part, through altered fetal copper metabolism (Paternain et al., 1990b; Taubeneck et al., 1992).

In contrast to these results, DMSA administration to mature mice from day 14 of pregnancy until postnatal day 21 did not adversely affect offspring survival and development at 200 or 400 mg/kg/d, although a significant reduction in pup body weight was noted at 800 mg DMSA/kg/d (Domingo et al., 1990b).

Although DMSA has been reported to be effective when given orally, subcutaneously, intraperitoneally, and intramuscularly (Aposhian, 1983), it is usually administered orally to humans at therapeutic levels ranging from 8 to 50 mg/kg/d (Aposhian, 1983; Ding and Liang, 1991; Graziano et al., 1992). According to the above results, DMSA at relatively high doses would not be a potent developmental toxicant in the pregnant mouse (Taubeneck et al., 1992). However, the oral embryofetal NOEL in rats was <100 mg DMSA/kg/d (Domingo et al., 1990a). Consequently, to prevent possible developmental toxicity it seems that DMSA should not be administered during organogenesis.

3. Sodium 2,3,-Dimercapto-1-Propane Sulfonate (DMPS)

DMPS is a water-soluble potent chelator which has been used in the treatment of heavy metal poisoning, in particular against arsenic, mercury, lead, and cadmium (Aposhian, 1983; Aposhian et al., 1992). No maternal toxic effects were reported when DMPS was administered by gavage to pregnant Swiss mice at 0, 75, 150, and 300 mg/kg/d on gestational days 6 to 15 (Bosque et al., 1990). In the same study, no treatment-related changes were recorded in the number of total implants, resorptions, the number of live and dead fetuses, fetal body weight, or fetal sex distribution. Also, gross external, soft tissue, and skeletal examination of the DMPS-treated fetuses did not reveal significant differences at any dose in comparison with the controls. Moreover, although it has been reported that urinary excretion of zinc and copper is increased by DMPS administration (Chisolm and Thomas, 1985) in maternal and fetal tissues of mice, Bosque et al. (1990) found only isolated statistically significant differences in calcium, magnesium, zinc, copper, and iron concentrations. The NOEL for DMPS developmental toxicity in mice was ≥300 mg/kg/d. In a previous investigation, female rats were treated p.o. with DMPS at 126 mg/kg/d before mating (14 weeks) and during pregnancy and nursing. The offspring of DMPS-treated rats, which were observed for 3 months, showed no abnormalities (Planas-Bohne, et al. 1980).

On the other hand, when DMPS was administered by gavage to pregnant Swiss mice at 0, 70, 210, and 630 mg/kg/d by two dosing schedules: gestation day 14 until birth (prenatal exposure), and gestation day 14 until postnatal day 21 (pre- and postnatal exposure), no adverse effects of the drug were observed on late gestation, parturition, lactation, or fetal and neonatal development (Domingo et al., 1990c). The NOEL for adverse effects of DMPS on the developing fetuses or pups was found to be 630 mg/kg/d. This dose is higher than the amounts usually administered in human heavy metal poisoning.

B. POLYAMINOCARBOXYLIC ACIDS
1. Diethylenetriaminepentaacetic Acid (DTPA)

Since the 1940s, DTPA has been used to remove plutonium, thorium, and other actinide elements from a number of accidentally contaminated adults (Norwood, 1962; Stradling et al., 1991). DTPA is usually injected as $CaNa_3DTPA$ (the calcium trisodium salt of DTPA), or as $ZnNa_3DTPA$ (the zinc trisodium salt of DTPA). In recent years, DTPA was also found to be effective in increasing the urinary excretion of cobalt and zinc, and in reducing the concentration of these metals accumulated in various tissues (Llobet et al., 1986, 1989). Gabard (1974) reported that calcium-DTPA impaired the synthesis of DNA, RNA, and proteins in cells, due to the removal of zinc and manganese, whereas these effects could be avoided by using zinc-DTPA. When calcium-DTPA is given, the calcium atoms in the chelate are displaced by zinc and other metals which leave body cells depleted of these elements as the chelate is excreted. It appears that the retardation of cell division due to the depletion of zinc or other essential elements may account for much of the severe toxicity caused by protracted administration of calcium-DTPA (Taylor et al., 1974).

DTPA was observed to be much more toxic to developing mice and rats than to the adult (Fisher et al., 1975; Sikov et al., 1975). Increased fetal mortality and congenital malformations (exencephaly, ablepharia, spina bifida aperta, cleft palate, and polydactyly) were induced in mice by s.c. injection of calcium-DTPA (Fisher et al., 1976). Damage was reported to occur probably as a result of the depletion by calcium-DTPA of zinc and manganese, both essential for growth and development. However, the weights of the living fetuses were not decreased compared with controls. Types of effects observed may be related to the speed with which the injected chelate is excreted from the body (Stevens et al., 1962). Extrapolating these results to humans, a high risk of fetal death or malformation was predicted if pregnant women receive calcium-DTPA at daily doses of 1 g DTPA/70 kg body weight (Mays et al., 1976).

2. Ethylenediaminetetraacetic Acid (EDTA)

EDTA has been widely used in the treatment of heavy metal intoxications, particularly lead (Chisolm, 1992). The evaluation of the teratogenic effects of EDTA in rats showed that i.m. injection of the chelator at doses of 20 and 40 mg/kg/d on days 6 or 10 of gestation induced maternal toxicity and fetal death, and polydactyly and tail defects were also observed in 9% of the offspring (Tuchmann-Duplessis and Mercier-Parot, 1956). In contrast, no adverse effects over four generations of rats were found when calcium disodium edetate was given in the diet of the animals at doses of 50, 125, and 250 mg/kg (Oser et al., 1963).

Swenerton and Hurley (1971) reported that ingestion of Na_2EDTA (the disodium salt of EDTA) by female rats during pregnancy impaired reproduction and resulted in malformed young. When Na_2EDTA was fed from days 6 to 21 of gestation, all of the full-term young had gross congenital malformations. These effects were prevented by simultaneous supplementation with 1000 ppm of dietary zinc, which suggested (although did not prove) that the severe teratogenic effects of EDTA might be due to an induced deficiency of zinc. A short-term deficiency of dietary zinc during pregnancy was previously shown to result in gross congenital malformations (Hurley and Swenerton, 1966).

The effect of the route of administration on the toxicity and teratogenicity of EDTA has also been investigated (Kimmel, 1977). On days 7 through 14 of gestation, Na_2EDTA was administered to rats in one of the following manners: (1) as 3% of the semipurified diet (954 mg/kg/d); (2) p.o. 625 mg/kg twice daily (1250 mg/kg/d), or 750 mg/kg twice daily (1500 mg/kg/d); (3) s.c., 375 mg/kg. EDTA given in the diet resulted in no maternal deaths but produced severe maternal toxicity and malformations in 71% of the offspring. EDTA/p.o. was much more toxic to the dams but produced fewer malformed young than did the slightly lower dose of EDTA administered in the diet, whereas EDTA/s.c. was lethal to 24% of the dams at a much lower dose than that given by the oral route, but did not produce a significant number of malformations in the offspring. Absorption into the circulation, interaction with essential metals, and the stress associated with the administration of the compound were reported to be possible factors involved in the differences in maternal toxicity and teratogenicity of EDTA (Kimmel, 1977).

EDTA and four of its salts, disodium, trisodium, calcium disodium, and tetrasodium edetate were also studied for teratogenic potential in rats. Equimolar doses based on 1000 mg EDTA/kg were given by gastric intubation on days 7 to 14 of gestation. In contrast to the results reported by Kimmel (1977), no teratogenic effects occurred with any of the compounds, even at maternally toxic doses (Schardein et al., 1981). The differences in outcome between these two studies were attributed to the fact that in Kimmel's study the animals were maintained on deionized water and a semipurified diet and housed in nonmetallic caging, factors not fully operative in the study by Schardein and associates. Consequently, it might be that the results in the Kimmel's study were related to the particular regimen. In summary, the results of the developmental toxicity studies indicated little or no teratogenicity for EDTA compounds in the rat when given orally.

C. OTHER CHELATORS WITH AMINE GROUPS
1. D-Penicillamine (D-PA)

D-Penicillamine, a drug used in treatment of Wilson's disease, cystinuria, rheumatoid arthritis, neoplasms, and some autoimmune diseases, has the capacity to bind divalent cations such as copper and zinc, and to increase their excretion from the body (Keen et al., 1982, 1983). The drug's chelating properties have also been exploited in the treatment of mercury, cadmium, and lead toxicoses (Lyle, 1981), and D-PA has also been used for the treatment of lead poisoning in humans to sustain the benefit when other drugs (BAL, EDTA) have been previously used (Chisolm, 1967).

The potential teratogenicity of D-PA in humans was first suggested by Mjolnerod et al. (1971). These authors reported that a young woman who had taken 2 g of D-PA daily during pregnancy for the treatment of cystinuria gave birth to a child with a generalized connective tissue defect, including lax skin, hyperflexibility of the joints, vein fragility, varicosities, and impaired wound healing. This child died of septicemia aged 7 weeks. In a second case described by Solomon et al. (1977), an abnormal child was born from a mother who had received 900 mg D-PA/day during her pregnancy for the treatment of rheumatoid arthritis. The infant had numerous congenital malformations including flattened face, low-set ears, short neck, and bilateral inguinal hernia, which were remarkably similar in appearance to those previously reported (Mjolnerod et al., 1971). The child died at 3 days of age.

Developmental toxicity studies of D-PA in rats showed that the drug was teratogenic when fed at a level of 0.83% of the diet. Malformations included loose skin, abdominal herniations, incomplete diaphragms, and reduced ossification of the phalanges and the tail vertebrae. In addition, liver and plasma copper levels were found to be lower in dams fed the drug than in controls, whereas the fetuses also had low levels of copper in whole body and liver (Keen et al., 1981, 1982). In a subsequent study, Keen et al. (1983) observed that the frequency of resorptions, and the frequency and severity of malformations observed in rats which were fed a diet containing 0.17, 0.83, or 1.66% D-PA, increased with increased levels of the drug. Moreover, maternal and fetal tissue copper levels were significantly lower in the D-PA groups than in controls, with the levels decreasing with increasing dose of D-PA.

Since the types of malformations seen after D-PA administration were similar to those observed in copper deficiency (Hurley and Keen, 1979), the strong correlation between low copper tissue concentrations and frequent fetal malformations suggested that at least one mechanism of D-PA-induced teratogenesis is through copper deficiency (Keen et al., 1982, 1983). Therefore, it was suggested that the copper status of pregnant women receiving D-PA should be carefully monitored and the copper status of infants from these mothers should be determined, even if the child has no visible anomalies, in order to ensure that sufficient copper stores exist to allow for normal postnatal growth and development.

2. Triethylenetetramine (Trien)

Trien, an example of a facultative quadridentate chelating agent, was suggested for use in Wilson's disease when treatment by D-PA is contraindicated because of a toxic response by the patient. Trien was also reported to be an effective antidote for acute nickel(II) (Horak et al., 1976) and copper (Planas-Bohne, 1979) toxicity.

When Trien was given in the diet of pregnant rats at 0.83% (level similar to that given to human patients) maternal food intake and weight gain were not affected by the treatment. However, Trien had a pronounced effect on copper in maternal and fetal tissues. Maternal plasma copper was more than 50% lower than control values, while maternal liver copper levels were about 20% lower than controls. Fetal liver copper was dramatically reduced by the drug. Maternal zinc levels were not greatly affected by Trien, although plasma zinc was slightly reduced. In contrast, fetal zinc levels were more than 2-fold higher than in controls. Although some dams fed Trien had hemorrhages, a high frequency of malformations was not found in fetuses from these dams (Keen et al., 1982).

D. MISCELLANEOUS CHELATING AGENTS
1. Sodium 4,5-Dihydroxybenzene-1,3-Disulfonate (Tiron)

Tiron, a phenolic compound widely used in analytical chemistry, has been shown to be an effective antidote in the treatment of acute uranium (Domingo et al., 1992b) and vanadium intoxication (Gomez et al., 1991). The teratogenic potential of Tiron has recently been evaluated in the mouse. The drug was administered i.p. to pregnant Swiss mice at 0, 750, 1500, and 3000 mg/kg/d on gestational days 6 to 15. Tiron caused maternal toxicity at 3000 mg/kg/d as evidenced by a high number of deaths, reduced body weight gain during gestation, and increased relative liver and kidney weights. Embryo/fetal toxicity (significant increases in the number of resorptions per litter and significant reductions in fetal body weight) was also observed at 3000 mg/kg/d, whereas Tiron was not teratogenic in mice even at maternally toxic doses. The NOEL for maternal and developmental toxicity of Tiron was 1500 mg/kg/d (Ortega et al., 1991).

2. Dithiocarbamates (DTC)

Dithiocarbamates have been used as metal complexing species for several decades, and their metal complexes have found numerous practical applications (Jones et al., 1980). Diethyldithiocarbamates and selected analogs can alleviate systemic lead (Gale et al., 1986) and cadmium (Jones and Jones, 1984) intoxication.

Nakura and associates (1984) did not find evidence for increased malformations or prenatal deaths in rats given up to 250 mg/kg/d of zinc diethyldithiocarbamate for days 7 to 15 of gestation. Recently, Hatori et al. (1990) compared the effect of a single i.p. injection (2.2 mmol/kg body weight) of four sodium dithiocarbamates: N-benzyl-D-glucamine (I), N-di(hydroxyethyl)amine (II), 4-carboxyamidopiperidine (III), and N-methyl-D-glucamine (IV) given to pregnant Syrian golden hamsters on day 8 of gestation. No signs of maternal toxicity were observed in any group. No malformations were found after dosing with DTC, I, II, or III, whereas administration of DTC IV was associated with the production of small encephaloceles in two fetuses from one litter. There were significant differences in mean fetal body weight between controls and DTC-treated groups.

Since earlier studies reported developmental toxicity with certain DTCs (Robens, 1965; Nora et al., 1977), it has been suggested that there is a structure-toxicity relationship for DTCs in mammalian teratogenesis, perhaps analogous to their structure-chelation relationship (Jones and Jones, 1984), or as a consequence of differential metabolic fate.

3. Desferrioxamine

Desferrioxamine (deferoxamine, DFOA), a trihydroxamic acid, is a by-product of the mold *Streptomyces pilosus*. DFOA has been the drug of choice for the treatment of iron poisoning (Lovejoy, 1982). In 1979, DFOA was used for the first time to remove aluminum in a patient with dialysis encephalopathy, resulting in striking improvement (Ackrill et al., 1980). This beneficial effect was subsequently corroborated by a number of investigators (see reviews by Swartz, 1985 and Domingo, 1989b). However, when DFOA is used in chronic dialysis patients with iron or aluminum overload, a number of toxic side-effects of DFOA treatment have been observed (Swartz, 1985, Domingo, 1989b).

The developmental toxicity of DFOA has recently been assessed in the mouse. DFOA was administered to pregnant Swiss mice by i.p. injection at doses of 44, 88, 176, and 352 mg/kg/d on days 6 to 15 of gestation. Maternal toxicity was evidenced by a significant reduction in food consumption, body weight at termination, and corrected body weight at all doses tested. Moreover, there was a high number of dams with abortions or early deliveries at 88, 176, and 352 mg/kg/d. However, DFOA did not produce embryolethality at 44, 88, and 176 mg/kg/d, whereas teratogenicity was not observed at any dosage level. The NOEL for maternal toxicity was <44 mg/kg/d, whereas the NOELs for embryotoxicity and teratogenicity were 176 mg/kg/d and ≥352 mg/kg/d, respectively (Bosque et al., 1995).

In humans, between 40 and 80 mg/kg are commonly given parenterally once weekly, although long-term DFOA is well tolerated at doses of 20 to 60 mg/kg, 1 to 3 times per week. Some studies on the use of DFOA therapy during iron poisoning did not find evidences of teratogenicity. In two cases, the drug was given between the 16th and 19th weeks of pregnancy to women with thalassemia (Thomas and Skalicka, 1980; Martin, 1983), whereas in another case, a pregnant woman received two i.m. DFOA injections during the 10th week (Christiansen et al., 1985). In spite of the lack of teratogenicity described in these reports, taking into account the recent results obtained in mice, it would seem evident that DFOA should not be used during pregnancy.

III. SUMMARY

The chelators BAL, EDTA, D-PA, and DFOA are currently used for the treatment of humans intoxicated by arsenic, mercury, lead, iron, or other heavy metals. In recent years, a promising development in this field has been the identification of DMSA, DMPS, Tiron, and diethyldithiocarbamates as new agents for the treatment of metal intoxication. Because these chelators may be administered to a heavy-metal intoxicated pregnant woman, the intrinsic developmental toxicity of these compounds should be taken into account. While the NOELs for developmental toxicity of chelators such as DMPS, EDTA, or Tiron were higher than the amounts usually administered in heavy metal poisoning, it has been suggested that the use of other agents such as BAL, D-PA, DTPA, or DFOA should be avoided during gestation. Developmental toxicity of chelating agents may be a consequence of severe essential trace metal deficiency caused by these agents. Among the new metal chelators, DMSA and DMPS have been reported to be effective in alleviating arsenic- and mercury-induced teratogenesis, whereas Tiron would protect against vanadium- and uranium-induced developmental toxicity.

REFERENCES

Ackrill, P., Ralston, A. J., Day, J. P., and Hodge, K. C. (1980), *Lancet,* 2, 692–693.
Agnish, N. D., Rusin, G., and Dinardo, B. (1990), *Fundam. Appl. Toxicol.,* 15, 249–257.
Ahokas, R. A., Dilts, P. V., and Lahaye, E. B. (1980), *Am. J. Obstet. Gynecol.,* 136, 216–221.
Alfrey, A. C. (1985), *Clin. Nephrol. (Suppl. 1),* 24, S84–S87.
Aposhian, H. V. (1983), *Annu. Rev. Pharmacol. Toxicol.,* 23, 193–215.
Aposhian, H. V. and Aposhian, M. M. (1990), *Annu. Rev. Pharmacol. Toxicol.,* 30, 279–306.
Aposhian, H. V., Maiorino, R. M., Rivera, M., Bruce, D. C., Dart, R. C., Hurlburt, K. M., Levine, D. J., Zheng, W., Fernando, Q., Carter, D. et al. (1992), *Clin. Toxicol.,* 30, 505–528.
Baxley, M. N., Hood, R. D., Vedel, G. C., Harrison, W. P., and Szczech, G. M. (1981), *Bull. Environ. Contam. Toxicol.,* 26, 749–756.
Beaudoin, A. R. (1974), *Teratology,* 10, 153–158.
Besser, J. M., Canfield T. J., and La Point, T. W. (1993), *Environ. Toxicol. Chem.,* 12, 57–72.
Beyersmann, D. and Hartwig, A. (1992), *Toxicol. Appl. Pharmacol.,* 115, 137–145.
Bosque, M. A., Domingo, J. L., Paternain, J. L., Llobet, J. M., and Corbella, J. (1990a), *Toxicology,* 62, 311–320.
Bosque, M. A., Domingo, J. L., Llobet, J. M., and Corbella, J. (1991), *Bull. Environ. Contam. Toxicol.,* 47, 682–688.
Bosque, M. A., Domingo, J. L., Llobet, J. M., and Corbella, J. (1993), *Toxicology,* 79, 149–156.
Bosque, M. A., Domingo, J. L., and Corbella, J. (1995), *Arch. Toxicol.,* 69, 467–471.
Cantilena, L. R., Jr. and Klaassen, C. D. (1982), *Toxicol. Appl. Pharmacol.,* 63, 344–350.
Catsch, A. and Harmuth-Hoene, A. E. (1979), *The Chelation of Heavy Metals,* Pergamon Press, Oxford, 153–155.
Chaineau, E., Binet, S., Pol, D., Chatellier, G., and Meininger, V. (1990), *Teratology,* 41, 105–112.
Chang, L. W., Wade, P. R., Pounds, J. G., and Reuhl, K. R. (1980), *Adv. Pharmacol. Chemother.,* 17, 195–231.
Chernoff, N. (1973), *Teratology,* 8, 29–32.
Chisolm, J. J., Jr. (1967), *Mod. Treat.,* 4, 710–727.
Chisolm, J. J., Jr. (1992), *Clin Toxicol.,* 30, 493–504.
Chisolm, J. J., Jr. and Thomas, D. J. (1985), *J. Pharmacol. Exp. Ther.,* 235, 665–669.
Christiansen, G. C. M., Rijksen, G., Marx, J., Hofstede, P., and Staal, G. E. J. (1985), *Arch. Gynecol.,* 237(Suppl.), 80.
Cohen, A. J. and Roe, F. J. C. (1991), *Food Chem. Toxicol.,* 29, 485–507.
Colomina, M. T., Gomez, M., Domingo, J. L., Llobet, J. M., and Corbella, J. (1992), *Res. Commun. Chem. Pathol. Pharmacol.,* 77, 95–106.
Cranmer, J. M., Wilkins, J. D., Cannon, D. J., and Smith, L. (1986), *Neurotoxicology,* 7, 601–608.
Dencker, L. (1975), *J. Reprod. Fertil.,* 44, 461–471.
DeSesso, J. M. and Goeringer, G. C. (1990), *Reprod Toxicol.,* 4, 145–152.
DeSesso, J. M. and Goeringer, G. C. (1991), *Teratology,* 43, 201–215.
Ding, G. S. and Liang, Y. Y. (1991), *J. Appl. Toxicol.,* 11, 7–14.
Domingo, J. L., Paternain, J. L., Llobet, J. M., and Corbella, J. (1988), *Fundam. Appl. Toxicol.,* 11, 715–722.
Domingo, J. L. (1989a), *Rev. Environ. Contam. Toxicol.,* 108, 105–132.

Domingo, J. L. (1989b), *Clin. Toxicol.,* 27, 355–367.
Domingo, J. L., Ortega, A., Paternain, J. L., Llobet, J. M., and Corbella, J. (1990a), *J. Toxicol. Environ. Health,* 30, 181–190.
Domingo, J. L., Bosque, M. A., and Corbella, J. (1990b), *Life Sci.,* 47, 1745–1750.
Domingo, J. L., Ortega, A., Bosque, M. A., and Corbella, J. (1990c), *Life Sci.,* 46, 1287–1292.
Domingo, J. L., Bosque, M. A., and Piera, V. (1991), *Fundam. Appl. Toxicol.,* 17, 314–320.
Domingo, J. L., Bosque, M. A., Llobet, J. M., and Corbella, J. (1992a), *Ecotoxicol. Environ. Safety,* 23, 274–281.
Domingo, J. L., Colomina, M. T., Llobet, J. M., Jones, M. M., Singh, P. K., and Campbell R. A. (1992b), *Fundam. Appl. Toxicol.,* 19, 350–357.
Domingo, J. L., Bosque, M. A., Luna, M., and Corbella J. (1993), *Teratology,* 48, 133–138.
Domingo, J. L. (1994), *J. Toxicol. Environ. Health,* 42, 123–141.
Endo, A. and Watanabe, T. (1988), *Reprod. Toxicol.,* 2, 141–144.
Ferm, V. H. and Carpenter, S. J. (1967), *Exp. Mol. Pathol.,* 7, 208–213.
Ferm, V. H. and Carpenter, S. J. (1970), *Toxicol. Appl. Pharmacol.,* 16, 166–170.
Ferm, V. H. (1971), *Biol. Neonate,* 19, 101–107.
Ferm, V. H., Saxon, A., and Smith, B. M. (1971), *Arch. Environ. Health,* 22, 557–560.
Ferm, V. H. (1972), *Adv. Teratol.,* 6, 51–75.
Fisher, D. R., Mays, C. W., and Taylor, G. N. (1975), *Health Phys.,* 29, 782–785.
Fisher, D. R., Calder, S. E., Mays, C. W., and Taylor, G. N. (1976), *Teratology,* 14, 123–128.
Friedheim, E., Graziano, J. H., Popovac, D., Dragovic, D., and Kaul, B. (1978), *Lancet,* 2, 1234–1236.
Fujimoto, T., Fuyuta, M., Kiyofuji, E., and Hirata, S. (1979), *Teratology,* 20, 297–302.
Fuyuta, M., Fujimoto, T., and Hirata, S. (1978), *Teratology,* 18, 353–366.
Gabarb, B. (1974), *Biochem. Pharmacol.,* 23, 901–909.
Gale, G. R., Atkins, L. A., Smith, A. B., and Jones, M. M. (1986), *Res. Commun. Chem. Pathol. Pharmacol.,* 52, 29–44.
Gale, T. F., and Horner, J. A. (1987), *Teratology,* 36, 379–387.
Gomez, M., Domingo, J. L., Llobet, J. M., and Corbella, J. (1991), *Toxicol. Lett.,* 57, 227–234.
Gomez, M., Sanchez, D. J., Domingo, J. L., and Corbella, J. (1992a), *J. Toxicol. Environ. Health,* 37, 47–56.
Gomez, M., Sanchez, D. J., Domingo, J. L., and Corbella, J. (1992b), *Arch. Toxicol.,* 66, 188–192.
Gomez, M., Sanchez, D. J., Colomina, M. T., Domingo, J. L., and Corbella, J. (1994), *Arch. Environ. Contam. Toxicol.,* 26, 64–68.
Graziano, J. H., LoIacono, N., Moulton, T., Mitchell, M. E., Slavkovich, V., and Zarate, C. (1992), *J. Pediatr.,* 120, 133–139.
Harada, M. (1978), *Teratology,* 18, 285–288.
Harris, S. B., Wilson, J. G., and Printz, R. H. (1972), *Teratology,* 6, 139–142.
Hatori, A., Willhite, C. C., Jones, M. M., and Sharma, R. P. (1990), *Teratology,* 42, 243–251.
Högberg, J. and Alexander, J. (1986), *Handbook on the Toxicology of Metals,* Friberg, L., Nordberg, G. F., and Vouk, V., Eds., Elsevier, Amsterdam, 482–520.
Holt, D. and Webb, M. (1986), *Arch. Toxicol.,* 58, 243–248.
Hood, R. D. and Bishop, S. L. (1972), *Arch. Environ. Health,* 24, 62–65.
Hood, R. D. and Pike, C. T. (1972), *Teratology,* 6, 235–238.
Hood, R. D., Thacker, G. T., and Patterson, B. L. (1977), *Environ. Health Perspect,* 19, 219–222.
Hood, R. D., Thacker, G. T., Patterson, B. L., and Szczech, G. M. (1978), *J. Environ. Pathol. Toxicol.,* 1, 857–864.
Hood, R. D. and Vedel, G. C. (1984), *Toxicol. Appl. Pharmacol.,* 73, 1–7.
Horak, E., Sunderman, F. W., Jr., and Sarkar, B. (1976), *Res. Commun. Chem. Pathol. Pharmacol.,* 14, 153–165.
Hoskins, B. B. and Hupp, E. W. (1978), *Environ. Res.,* 15, 5–19.
Hurley, L. S. and Shrader, R. E. (1975), *Nature,* 254, 427–429.
Hurley, L. S. and Swenerton, H. (1966), *Proc. Soc. Exp. Biol. Med.,* 123, 692–696.
Hurley, L. S. and Keen, C. L. (1979), *Copper in The Environment,* Part II. Health Effects, Nriagu, J., Ed., Wiley, New York, 33–56.
Jones, M. M. (1991), *CRC Crit. Rev. Toxicol.,* 21, 209–233.
Jones, M. M., Burka, L. T., Hunter, M. E., Basinger, M., Campo, G., and Weaver, A. D. (1980), *J. Inorg. Nucl. Chem.,* 42, 775–778.
Jones, S. G. and Jones, M. M. (1984), *Environ. Health Perspect.,* 54, 285–290.
Keen, C. L., Mark-Savage, P., Lönnerdal, B., and Hurley, L. S. (1981), *Fed. Proc.,* 40, 917.
Keen, C. L., Lönnerdal, B., and Hurley, L. S. (1982), *Inflammatory Diseases and Copper,* Sorenson, J. R. J., Ed., Humana Press, Clifton, NJ, 109–121.
Keen, C. L., Mark-Savage, P., Lonnerdal, B., and Hurley, L. S. (1983), *Drug Nutr. Interact,* 2, 17–34.
Keen, C. L. and Zidenberg-Cherr, S. (1990), *Present Knowledge in Nutrition,* Brown, M. L., Ed., ILSI, Nutrition Foundation, Washington, D. C., 279–286.
Kennedy, G. L., Arnold, D. W., and Calandra, J. C. (1975), *Food Cosmet. Toxicol.,* 13, 629–632.
Khera, K. S. (1973), *Teratology,* 8, 293–304.
Kimmel, C. A. (1977), *Toxicol. Appl. Pharmacol.,* 40, 299–306.
Laskey, J. W., Rehnberg, G. L., Hein, J. F., and Carter, S. D. (1982), *J. Toxicol. Environ. Health,* 9, 677–687.
Leonard, A., Jacquet, P., and Lauwerys, R. R. (1983), *Mutat. Res.,* 114, 1–18.
Lindgren, A., Danielsson, B. R. G., Dencker, L., and Vahter, M. (1984), *Acta Pharmacol. Toxicol.,* 54, 311–320.

Llobet, J. M., Domingo, J. L., and Corbella, J. (1986), *Arch. Toxicol.,* 58, 278–281.
Llobet, J. M., Colomina, M. T., Domingo, J. L., and Corbella, J. (1989), *Vet. Human Toxicol.,* 31, 25–28.
Lovejoy, F. H., Jr. (1982), *Clin. Toxicol.,* 19, 871–874.
Lu, C. C., Matsumoto, N., and Ijima, S. (1979), *Teratology,* 19, 137–142.
Lugo, G., Cassady, G., and Palmisano, P. (1969), *Am. J. Dis. Child.,* 117, 328–330.
Lyle, W. H. (1981), *J. Reumathol. (Suppl.),* 8, 96–99.
Marathe, M. R. and Thomas, G. P. (1986), *Toxicol. Lett.,* 34, 115–120.
Martin, K. (1983), *Aust. Paediatr. J.,* 19, 182–183.
Matsumoto, H., Koya, G., and Takeuchi, T. (1965), *J. Neuropathol. Exp. Neurol.,* 24, 563–574.
Matsumoto, H., Suzuki, A., Morita, C., Nakamura, K., and Saeki, S. (1967), *Life Sci.,* 6, 2321–2326.
Matsumoto, N., Iijima, S., and Katsunuma, H. (1976), *J. Toxicol. Sci.,* 1, 1–13.
Mays, C. W., Taylor, G. N., and Fisher, D. R. (1976), *Health Phys.,* 30, 249–250.
McClain, R. M. and Becker, B. A. (1975), *Toxicol. Appl. Pharmacol.,* 31, 72–82.
McClain, R. M. and Siekierka, J. J. (1975), *Toxicol. Appl. Pharmacol.,* 31, 434–442.
Miller, H. C., Hassanien, K., and Hensleigh, P. A. (1976), *Am. J. Obstet. Gynecol.,* 125, 55–58.
Miller, C. D., Buck, W. B., Hembrough, F. B., and Cunningham, W. L. (1982), *Vet. Human. Toxicol.,* 24, 163–166.
Mjolnerod, O. D., Rasmussen, K., Dommerud, S. A., and Gjeruldsen, S. T. (1971), *Lancet,* 1, 673–675.
Morrissey, R. E. and Mottet, N. K. (1983), *Teratology,* 28, 399–411.
Morrissey, R. E., Harris, M. W., Diliberto, J. J., and Birnbaum, L. S. (1992), *Toxicol. Lett.,* 60, 19–25.
Nakura, S., Tanaka, S., Kawashima, K., Takanaka, A., and Omori, Y. (1984), *Bull. Natl. Inst. Hyg. Sci. (Tokyo).,* 102, 55–61.
Nishimura, H. and Takagaki, S. (1959), *Anat. Rec.,* 135, 261–267.
Nolen, G. A., Bueheler, E. V., Geil, R. G., and Goldenthal, E. I. (1972), *Toxicol. Appl. Pharmacol.,* 23, 222–237.
Nora, A. H., Nora, J. J., and Blu, J. (1977), *Lancet,* 2, 664.
Norwood, W. D. (1962), *Health Phys.,* 8, 747–750.
Olson, O. E. (1986), *J. Am. Coll. Toxicol.,* 5, 45–70.
Ortega, A., Sanchez, D. J., Domingo, J. L., Llobet, J. M., and Corbella, J. (1991), *Res. Commun. Chem. Pathol. Pharmacol.,* 73, 97–106.
Oser, B. L., Oser, M., and Spencer, H. C. (1963), *Toxicol. Appl. Pharmacol.,* 5, 142–162.
Paternain, J. L., Domingo, J. L., Llobet, J. M., and Corbella, J. (1987), *Rev. Esp. Fisiol.,* 43, 223–228.
Paternain, J. L., Domingo, J. L., Llobet, J. M., and Corbella, J. (1988a), *Teratology,* 38, 253–257.
Paternain, J. L., Domingo, J. L., and Corbella, J. (1988b), *J. Toxicol. Environ. Health,* 24, 193–200.
Paternain, J. L., Domingo, J. L., Gomez, M., Ortega, A., and Corbella, J. (1990a), *J. Appl. Toxicol.,* 10, 181–186.
Paternain, J. L., Ortega, A., Domingo, J. L., Llobet, J. M., and Corbella, J. (1990b), *J. Toxicol. Environ. Health,* 30, 191–197.
Peterkova, R. and Puzanova, L. (1976), *Folia Morphol.,* 24, 5–13.
Peters, R. A., Stocken, L. A., and Thompson, R. H. S. (1945), *Nature,* 156, 616–619.
Pippard M. P. and Callender, S. T. (1983), *Br. J. Hematol.,* 54, 503–507.
Planas-Bohne, F. (1979), *Toxicol. Appl. Pharmacol.,* 50, 337–345.
Planas-Bohne, F., Gabard, B., and Schäffer, E. H. (1980), *Drug Res.,* 30, 1291–1294.
Plowman, M. C., Peracha, H., Hopfer, S. M., and Sunderman, F. W., Jr. (1991), *Teratogenesis Carcinog. Mutagen.,* 11, 83–92.
Robens, J. F. (1965), *Toxicol. Appl. Pharmacol.,* 15, 152–163.
Rom, W. N. (1976), *Mt. Sinai J. Med.,* 43, 542–552.
Saltzman, R. A., Miller, R. K., and Di Sant' Agnese, P. A. (1989), *Teratology,* 39, 19–30.
Sanchez, D. J., Ortega, A., Domingo, J. L., and Corbella, J. (1991), *Biol. Trace Elem. Res.,* 30, 219–226.
Sanchez, D. J., Gomez, M., Llobet, J. M., and Domingo, J. L. (1993), *Ecotoxicol. Environ. Safety,* 26, 33–39.
Schardein, J. L., Sakowski, R., Petrere, J., and Humphrey, R. R. (1981), *Toxicol. Appl. Pharmacol.,* 61, 423–428.
Schou, M. (1976), *Acta Psychiat. Scand.,* 54, 193–197.
Sikov, M. R., Smith, V. H., and Mahlum, D. D. (1975), Teratological effectiveness and fetal toxicity of DTPA in the rat. Pacific Northwest Laboratory Annual Report BNWL-1950, PT1, 108–110.
Smithberg, M. and Dixit, P. K. (1982), *Teratology,* 26, 239–246.
Solomon, L., Abrams, G., Dinner, M., and Berman, L. (1977), *N. Engl. J. Med.,* 296, 54–55.
Stevens, E., Rosoff, B., Weiner, M., and Spencer, H. (1962), *Proc. Soc. Exp. Biol. Med.,* 111, 235–238.
Stradling, G. N., Moody, J. C., Gray, S. A., Ellender, M., and Hodgson, A. (1991), *Human Exp. Toxicol.,* 10, 15–20.
Su, M. Q. and Okita, G. T. (1976), *Toxicol. Appl. Pharmacol.,* 38, 207–216.
Sunderman, F. W., Jr., Mitchell, J. M., Allpass P. R., and Baselt R. (1978), *Toxicol. Appl. Pharmacol.,* 43, 381–390.
Swartz, R. D. (1985), *Am. J. Kidney Dis.,* 6, 358–364.
Swenerton, H. and Hurley L. S. (1971), *Science,* 173, 62–64.
Szabo, K. T. (1970), *Nature,* 225, 73–75.
Taubeneck, M. W., Domingo J. L., Llobet J. M., and Keen C. L. (1992), *Toxicology,* 72, 27–40.
Taylor G. N., Williams J. L., Roberts, L., Atherton, D. R., and Shabestari, L. (1974), *Health Phys.,* 27, 285–288.
Thomas, R. M. and Skalicka, A. E. (1980), *Arch. Dis. Child.,* 55, 572–574.
Tuchmann-Duplessis, H. and Mercier-Parot, L. (1956), *C. R. Hebd. Seances Acad. Sci. Paris,* 243, 1064–1066.
Walshe, J. M. (1956), *Am. J. Med.,* 21, 487–495.
Webster, W. S. and Valois, A. O. (1987), *Neurotoxicology,* 8, 437–444.

Welsch, F., Sleet, R. B., and Greene J. A. (1987), *J. Biochem. Toxicol.,* 2, 225–240.
Wide, M. (1984), *Environ. Res.,* 33, 47–53.
Willhite, C. C. (1981), *Exp. Mol. Pathol.,* 34, 145–158.
Wright, T. L., Hoffman, L. H., and Davies, J. (1971), *Teratology,* 4, 151–156.
Yasuda, Y., Datu, A. R., Hirata, S., and Fujimoto, T. (1985), *Teratology,* 32, 273–286.
Yokel, R. A. (1987), *Fundam. Appl. Toxicol.,* 9, 795–806.
Yonemoto, J., Satoh, H., Himeno, S., and Suzuki, T. (1983), *Teratology,* 28, 333–340.

Epilogue

Epilogue

László Magos

After browsing through *Toxicology of Metals* you have reached the epilogue. To select Prof. Clarkson to write the foreword and me to write the epilogue is an unmistakable indication that Prof. Chang has a passion for symmetry, as both Tom and I worked most of the time on mercury toxicology. Actually, I entered into this field with Tom's guidance.

After receiving Prof. Chang's request for this epilogue, I turned to my old Webster's collegiate dictionary, brought with me in 1963 from Hungary. According to Webster, "epilogue" has a literary connotation because it is either a concluding section of a novel or a speech addressed to the audience of a play. In the context of science, the concluding section probably can be translated to general conclusions. Though only a part of the 84 metals of the periodic table have got an entry in this volume, their number is large, and I could not undertake to write an epilogue which ends the book with a "bang" (Prof. Chang's request) instead of platitudes. Naturally I could say that in every step of our evolutionary development, including *Ardiputhecus ramidus* a million years ago, mankind and his ancestors were exposed to metals. In the same spirit I could emphasize the importance of metal toxicology. I would do this with enthusiasm not because I could point with pride to work done in my late Institute on beryllium, cadmium, mercury, organolead, and organotin compounds, but because of disappointment and sorrow felt on the decision of the hierarchy of the Medical Research Council. It was decided that all the metal toxicology projects must be terminated soon after the retirement of Norman Aldridge, Michael Webb, and myself.

Instead of looking backward, congratulating the contributors, or pontificating on further research needs, I decided to be personal. The impetus was given by *Arcadia,* a play written by Tom Stoppard. This play moves on different levels in time and theme, but the scene is always a large room on the garden front of a spacious country house. Time alternates between the early 19th century and the present day; the background theme is the change in the English garden (of the aristocracy) from an artificially simple English style to an artificially romantic one. Superimposed on this background are themes of scientific interest: interpretation, confounding factors (called noise in the play), and mathematical modeling.

One of the two interpretation problems is about the participants of a duel and its outcome. The problem is caused by coincidences. Though the challenged is Septimus Hodge, the tutor of the 13-year old Thomasina Coverly, the daughter of Lord and Lady Croom, a university don, Bernard Nightingale, puts the odium on Lord Byron, who was a weekend guest of the house at the time of the duel. His case is based on circumstantial evidence and the reputation of Byron. In toxicology, reputation (I mean the reputation of a substance) has also a heavy weight. It influences the choice of subject and the interpretation of results.

Not long ago I read a paper on the relationship between estimates of exposure to mercury and the weekly frequency of sexual activity and the number of menstrual cycles without contraception before conception. All variables were collected from the participants through telephone interviews. The final conclusion was that low exposure to mercury increased fecundity and high exposure decreased it. In a situation like this there are several possibilities for interpretation. The first is to speak as little about the beneficial effect as possible, and certainly to avoid hint on causative relationship. This route was chosen by the authors of a study on the effect of prenatal exposure to methylmercury on childhood development. They found positive association between neurological defects and maternal exposure to methylmercury in boys and negative association in girls. The negative association got only a few lines and no tables. The second possibility is elaboration on the beneficial effect. This was more attractive to the authors of the fecundity study. They explained that mercury at low exposure level is beneficial because it is bound to metallothionein. However, metallothionein might neutralize some of the mercury, but cannot turn a criminal into Santa Claus. A third possibility is suggested by Valentine Coverly in the play. The number of grouses shot in the moorland of the estate was recorded for nearly 2 centuries in the game book. Valentine wanted to formulate the relationship between the yearly number of grouses shot and the grouse population. After a time his verdict was "There's just too much bloody noise!," like burning the heather, which increases the population through improving food supply, and a good year for foxes that interferes the other way. The expansion of his sharp view is that a deviation from the control level can be (or must be) taken as warning that a similar deviation in the direction of the hypothesis can also be noise related. Even when measurements are made with analytical methods, and quality assurance is given on accuracy, precision, and interlaboratory comparison, some noise is unavoidable. I suspect the noise is louder when interviews are substituted for measurements. I frequently wonder about the exchange rate between the quantitation of exposure or adverse effects by analytical methods and estimates based on questioners.

The third problem raised in Arcadia is mathematical modeling. The 13-year old Thomasina Coverly, who knew about Fermat and his theorem, followed Fermat's example when she wrote on the margin of her maths primery: "I Thomasina Coverly, have found a truly wanderful method whereby all forms of nature must give up their numerical secrets and draw themselves through number alone. This margin being too mean for my purpose, the reader must look elsewhere for the New Geometry of Irregular Forms discovered by Thomasina Coverly." Two characters of the present-day scene, Valentine and Hannah, discuss Thomasina's theorem, whether it is a sign of geniality or mischief. Hearing their discussion I could not stop thinking about mathematical models used to extrapolate from numbers of the observed range to range outside the possibility of proof or disproof. Are they genial inventions or mischiefs? There are more simple models which have their ambiguities. It must be so because there are papers which, in spite of the absence of significant differences between exposed and control groups, state that exposure and effect correlated within the exposed group. Either exposure shifted the measured parameter downward from the control level at the lower range and upward at the higher range of exposure, or the regression analysis is more liberal than tests between two groups. I think the second assumption is the correct one.

Take the following nine xy points: 10, 10; 15, 6; 15, 14; 20, 4; 20, 10; 20, 16; 25, 6; 25, 14; 30, 10. They are arranged symmetrically within an ellipsis and their plot on a graph paper looks like a cummulus. These nine points give +10 for intercept, 0.0 for slope and correlation. And now a sun from a distance from the cummulus in the form of a point of 40, 40. If you are obliged to use a ruler to draw a regression line, you would do it only under duress. Nevertheless the ten points give -5 for intercept, 0.82 for slope, and 0.683 for the correlation coefficient and therefore significance at the 0.05 level for df 8. Now plot ten other points: 9, 2; 10, 0; 12, 7; 15, 2; 17, 13; 23, 8; 25, 19; 28, 11; 30, 41; 40, 18. There are no differences in intercept, slope, and correlation coefficient between the two series; the big difference is that in the plot of the second series you visualize, without drawing it, the regression line.

Now if we assume that these points are arbitrary units of urine mercury concentration plotted against arbitrary units of occupational exposure, the negative intercept on the ordinate (often seen in urine Hg against Hg exposure plots), we face another problem. As even unexposed subjects have mercury in urine, there is something wrong with a regression line which crosses abscissa at 6, and the regression equation of $Y = 0.82 \times -5.0$ which suggests that below exposure level 6 urinary mercury concentration is negative. A hockey stick type equation, like $Y = 0.82 (X - 6) + a$ (where "a" is control U-Hg) would be nearer to the truth. However the chance to make up your mind is non-existent when the information is restricted to means and SDs and the correlation coefficient. Nowadays fewer and fewer papers give the primary data points, and with more complex statistical methods (like adjustment to several covariates in regression analysis) you get nearly always what went through the mental digestive process of the authors or their statistician. This worries me, and that is probably the true reason for writing this epilogue as it is. Reading or only browsing through the 71 chapters of this volume you may feel reassured because of hard facts, good arguments, logic, good science. And you are right. Nevertheless ...

László Magos
BIBRA Toxicology International

Index

Index

A

Acrodynia, mercury poisoning, 349–350
Actin microfilaments, 653–654
Ag. See Silver
Airborne metals
    Arctic, 25–26
    autoimmunity, 853–854
    Baltic Sea, 23–24
    long-range transport models, 17–19
    monitoring, 23–26
    North American waters, 24–25
    North Sea, 23
    overview, 9–11, 26
    source apportionment techniques, 19–23
    trace metals, 11–16
Alkyllead, 526–528
Alkyltins, 521–526
Aluminum, 96–97, 520–521, 551–557
    aluminum chelation, 395
    amyloid, 391–393
    brain aluminum concentration, 389–391
    clinical course, 395
    gene expression, 394
    histopathology, 391
    neurofibrillary degeneration, 393–394
    overview, 387, 395
    risk factors, 388
    water supply, 388–389
  biological monitoring, 97
  blood-brain barrier, 569–570, 577–579
    experimental approaches, 577–578
    mechanism of action, 578–579
    overview, 577
    permeability, 577–578
    transport systems, 578
  bone metabolism, 967
  carcinogenicity, 253–254
  cognitive functions, 554–557
  cytoskeletal toxicity, 649, 651–652
  dialysis encephalopathy, 396–397
  encephalopathy, 521
  exposure, 96
  health effects, 97
  genotoxicity, 253–254
  lactational exposure, 1131–1132
  literature reports, 398–400
  metabolism, 96–97
  motor function, 553–554
  neurite development, 1122–1123
    Ca^{2+} homeostasis and, 1123
    cytoskeleton and, 1123
    in vitro effects on, 1122–1123
    in vivo effects of, 1122
  neurodegenerative diseases, 395–396
  overview, 97, 401
  oxygen generation, neurotoxicity and, 704
  Parkinson's disease, 396
  renal disease, 396–397
  reproduction and, 1083–1085
  respiratory toxicology, 933
  in workplace, 397–400
Alzheimer's disease,
  aluminum, 387–395
    aluminum chelation, 395
    amyloid, 391–393
    brain aluminum concentration, 389–391
    chelation, 395
    clinical course, 395
    gene expression, 394
    histopathology, 391
    neurofibrillary degeneration, 393–394
    overview, 387, 395
    risk factors, 388
    water supply, 388–389
  oxygen generation, neurotoxicity and, 704
Amalgams, 481–483
  animal studies, 477
  case reports, 477
  copper amalgams, 473
  epidemiological studies, 480–481
  fetus, effects on, 482–483
  gastrointestinal absorption, 472–473
  health risk evaluation, 476–482
  immunological effects, 480
  occupational exposure, 478–480
  overview, 469, 483
  placental transport, 1019–1022

pulmonary uptake, 471–472
risk estimates, 478
tissues of amalgam bearers, 475–476
urinary excretion, mercury, 473–475
Amine groups, 1162–1163
Amyloid, 391–393
Antibodies
    cellular targets, 866–867
    exposure and, 864–866
Antigen immunogenicity, 867
Antimony, 457–458
    biokinetics, 458
    chemistry, 457
    diagnosis and therapy, 458
    mechanism, 458
    poisoning, 457–458
    therapy, 458
    use and exposure, 457
Arctic, 25–26
Arsenic, 97–99, 517–518
    acute effects, 424
        inhalation exposure, 424
        oral exposure, 424
    animal model, inadequacy of, 224
    autoimmunity, 857–858
    biological monitoring, 98
    cancer, 221–229, 291–292, 433–434
        inhalation exposure, 433
        mechanism for, 226–227
        oral exposure, 433–434
    cardiovascular effects, 424–425
        inhalation exposure, 424–425
        oral exposure, 425
    cell-mediated immunity, 786
        biological response modifiers, 786
        in vitro effects of, 786
        in vivo exposure, 786
    choroid plexus, 618
    chronic effects, 424–434
    dermal effects, 429
        dermal exposure, 429
        inhalation exposure, 429
        oral exposure, 429
    dermal exposure, 432
    developmental effects, 429–430
    developmental effects
        inhalation exposure, 429–430
        oral exposure, 430
    dose-response relationship, 223–224
    drinking water, 48–49
    in drinking water, 39–53
    endocrinological effects, 428–429
        oral exposure, 428–429
    further research, 434
    gastrointestinal effects, 427
        inhalation exposure, 427
        oral exposure, 427
    genetic effects, 226
    genotoxic effects, 432
        inhalation exposure, 432
        oral exposure, 432
    health effects, 98
    hematologic effects, 428
        inhalation exposure, 428
        oral exposure, 428
    heme metabolism effects, 430–431
        inhalation exposure, 431
        oral exposure, 430–431
    hepatic effects, 427
        inhalation exposure, 427
        oral exposure, 427
    human exposure, 221–222
    immunological effects, 431–432
        dermal exposure, 432
        inhalation exposure, 431
        oral exposure, 431
    interactions, 727
    metabolism, 98
    mutation, 224–225
    nephropathology, metals, 725
        inhalation exposure, 426
        oral exposure, 425–426
    overview, 99, 423–424
    placental transport, 1005–1007
    renal effects, 427–428
        inhalation exposure, 428
        oral exposure, 427–428
    renal gene expression, alterations in, 751–752
    reproductive effects, 430
    respiratory effects, 426–427
    respiratory toxicology, 933–934
    stressor-specific proteins and, 179
    teratogenic effects, 993
As. See Arsenic
Asthma, nickel induced, 461–468
    overview, 461–462
    study, 462, 463, 464–465
Astrocytes
    anion channels, 590
    blood-brain barrier, 593–594
    gliosis, 594–596
    homeostatic functions of, 588–591
    immune response, 594
    immunology, 591–593
    inflammatory response, 594
    ion channel, 588–590, 590
    lead, 597–599
        accumulation in astrocytes, 598–599
        neurotoxicity, 597–598
    manganese, 599–600
        effects on astrocytic function, 599–600
        uptake, in astrocytes, 600
    mercury, 600–602
        effects of, on astrocytic homeostasis, 600–601
    neurotransmitter
        receptors for, 591
        uptake systems, 590–591
    neurotrophic factors, 591–593
    overview, 587–588, 602
    pathological reactions of, 594–597
    pH carrier, 590
    swelling, 597
    as targets, for heavy metals, 597
Atmospheric transport, 16–23
Atomic absorption spectroscopy, 73–74

Atomic emission spectroscopy, 74–75
Atomic properties, 114–117
　electronegativity, 117
　geometry, 115–116
　ion size, 115
　overview, 114–115
　oxidation state, 116–117
Au. See Gold
Auditory function, methylmercury and, 547–548
Autoantibodies
　nervous system proteins, 864–867
　system-specific toxic effects, 864
Autoimmunity
　airborne metals, exposure to, 853–854
　arsenic, 857–858
　bacterial infection, susceptibility to, 856–858
　beryllium, 836
　cadmium, 840–841, 857
　chromium, 838
　cobalt, 838
　copper, 838
　gold, 841–842
　inhaled metals
　　deposition, 854–855
　　retention of, 854–855
　lead, 837–838
　lithium, 839–840
　lung tumors, host defenses against, 858–859
　manganese, 857–858
　mercury, 842–845
　nervous system, 863–864
　nickel, 856–857
　overview, 835–836, 853, 859
　pulmonary host defenses, 855–856
　viral infection, susceptibility to, 858
　zinc, 839
Axons, 1113

B

Bacterial infection, 856–858
BAL. See 2,3-dimercaptopropanol
Baltic Sea, 23–24
Beryllium
　respiratory toxicology, 929–930
Bioavailability, 126–127
Biological markers
　aluminum, 96–97
　　biological monitoring, 97
　　exposure, 96
　　health effects, 97
　　metabolism, 96–97
　　overview, 97
　arsenic, 97–99
　　biological monitoring, 98
　　exposure, 97
　　health effects, 98
　　metabolism, 98
　　overview, 99
　cadmium, 83–86, 86
　　absorption, 83
　　biological monitoring, 85–86
　　blood, levels in, 85
　　distribution, 83
　　elimination, 83–84
　　exposure, 83
　　gastrointestinal tract, 84
　　genotoxicity, 84
　　health effects, 84
　　kidney, 84
　　　cadmium concentrations in, 85–86
　　liver, cadmium concentrations in, 85–86
　　lung disease, 84
　　mechanisms of toxicity, 84
　　metabolic models, 84
　　metabolism, 83–84
　　overview, 86
　　urine, cadmium concentration in, 85
　chromium, 100
　　biological monitoring, 100
　　exposure, 100
　　health effects, 100
　　metabolism, 100
　　overview, 101
　cobalt, 99
　　biological monitoring, 99
　　exposure, 99
　　health effects, 99
　　metabolism, 99
　　overview, 99
　lead, 86, 91
　　absorption, 87
　　biological monitoring, 89–91, 92
　　blood, levels in, 89–90
　　blood and blood-forming organs, 88
　　calcified tissues, lead concentrations in, 90–91
　　cardiovascular system, 89
　　distribution, 87
　　elimination, 87
　　exposure, 86–87, 92
　　gastrointestinal tract, 88
　　genotoxicity, 89
　　health effects, 87–89, 92
　　heme metabolism, disturbances of, 91
　　inorganic lead, 86–92
　　kidneys, 88
　　mechanisms of toxicity, 87
　　metabolic model, 87
　　metabolism, 87, 92
　　mobilization tests, 91
　　nervous system, 88
　　organic, 92
　　overview, 91–92
　　reproductive effects, 89
　　urine, lead concentration in, 90
　manganese, 101–102
　　biological monitoring, 101
　　exposure, 101
　　health effects, 101
　　metabolism, 101
　　overview, 102
　mercury, 92
　　biological monitoring, 93–94, 96
　　exposure, 92–93, 95
　　health effects, 93, 95–96
　　inorganic, 92–95

metabolism, 93, 95
  organic, 95–96
  overview, 94–95, 96
 nickel, 102–103
  biological monitoring, 102–103
  exposure, 102
  health effects, 102
  metabolism, 102
  overview, 103
 overview, 81–82
Biological membranes
 cellular trapping, 141
 direct vs. indirect membrane effects, 137–138
 function of membranes, 134
 metal binding, 134–136
 metal-membrane interaction, during metal uptake, 138–141
 nature of, 134
 overview, 133–134, 141–142
Biological phenomena, 125–128
 bioavailability, 126–127
 biological residence time, 127
 compartmentalization, 125–126
 metal resistance, 127–128
 respiratory tract clearance, 126
Biological residence time, reactivity, 127
Biological response modifiers, 793
 arsenic, 786
  in vitro effects of, 786
  in vivo exposure, 786
 cadmium, 786–787
  in vitro effects, 787
  in vivo treatment, 786–787
 copper, 790–791
  in vitro effects, 791
  in vivo effects, 790
 iron, 791
  in vitro effects, 791
  in vivo effects, 791
 lead
  biological response modifiers, 788
  T cell functions, 787
  in vitro exposure, 788
 mercury, 788–789
  in vitro effects, 788–789
  in vivo exposure, 788
 nickel, 789–790
  in vitro effects, 789–790
  in vivo exposure, 789
 overview, 785–786, 794
 selenium, 791–792
  biological response modifiers, 792
  in vitro effects, 792
  in vivo effects, 791–792
 tin, 790
  in vitro effects, 790
  in vivo effects, 790
 trace metals, 790–793
 zinc, 792–793
  biological response modifiers, 793
  in vitro effects, 793
  in vivo effects, 793

Biological systems, 69–79, 75–77
 atomic absorption spectroscopy, 73–74
 atomic emission spectroscopy, 74–75
 collection of samples, 70–71
 contamination control, 71–73
 determination techniques, 73–75
 electrometric techniques, 75–76
 nuclear techniques, 76–77
 overview, 69–70, 78
 preparation of sample, 71–73
 preservation of samples, 70–71
 quality control, 77–78
 sample
  preparation, 71–73
  selection, 70–71
 storage samples, 70–71
 x-ray spectrometry techniques, 76
Bismuth toxicity
 analysis, 440
 biokinetics, 440–443
  absorption, 440–442
   enhancement of, 441
   route of, 441–442
  distribution, 442
  elimination, 443
  metabolism, 442–443
 chemistry, 440
 clinical presentation, 445–446
 diagnosis, 448–449, 449
  biological matrices, concentrations in, 448
  electroencephalography, 449
 encephalopathy, 446
 etiology, 446–447
 medical uses of Bi compounds, 439–440
 nephrotoxicity, 444
 neurotoxicity, 445–448
 osteoarthropathy, 445
 overview, 439
 pathology, 447–448
 treatment, 449–450
Blood-brain barrier
 astrocytes and, 593–594
 central nervous system, 581
  aluminum, 569–570, 577–579
   experimental approaches, 577–578
   mechanism of action, 578–579
   overview, 577
   permeability, 577–578
   transport systems, 578
  anatomy, 561–564
  barrier properties, 563–564
  blood plasma, 565
  brain fluids, 561–562
  cadmium, 580–581
  copper, 568–569
  development, 563
  experimental approaches, 571
  function of, 561–564
  high dose studies, 571–572
  iron, 569, 579–580
   overview, 579
   pathology, 579–580
   permeability, 579–580

vascular function, 580
layers, 561–562
lead, 565–566, 571–574
   low dose studies, 573
   mechanism of action, 573–574
   permeability, 572
   summary of effects, 572
   transport systems, 572
   in vitro studies, 573
manganese, 567–568
mercury, 568, 574–577
   astroglia, 576–577
   experimental approaches, 574–576
   inorganic, 575
   mechanism of action, 576–577
   membrane toxicant, 576
   organic, 575–576
   overview, 574
metals in plasma, chemical species of, 565
modulation of, 564
overview, 561, 565, 570–571
physiology, 561–564
structure, 561–564
toxicity of metals to, 570
transport, 563
transport of metals, 565–570
zinc, 566–567
Blood plasma, 565
Bonding tendencies
   covalent, 117–118
   ionic, 117–118
Bone marrow function, 939–941
Bone metabolism
   aluminum, 967
   cadmium, 964–965
   copper, 962
   lead, 965–967
   magnesium, 960–961
   manganese, 963
   zinc, 961–962

C

Cadduct formation, 293–294
Cadherin. See Calcium-dependent adhesion molecules
Cadmium
   absorption, 83
   arsenic interactions, 727
   autoimmunity, 857
   biological monitoring, 85–86
   blood
      cadmium levels, 85
      heavy metal concentrations in, 356–357
   blood findings, 353–354
   blood pressure, 363–364
   bone, radiographs of, 354–356
   bone effects, 362–363
   bone metabolism, 964–965
   carcinogenesis
      animal studies, 233–238
      epidemiology, 233
      hematopoietic tumors, 236
      human exposure, 232–233
      injection site sarcomas, 236
      metal-metal interactions, 237
      overview, 231
      pathobiologic characteristics, 232
      prostatic tumors, 234–235
      pulmonary tumors, 234
      rodents, 233–234
      sensitivity, 238
      species, 238
      strain, 238
      synergism, 237–238
      testicular tumors, 235–236
      tolerance, 238
      uses of, 232
   cell-mediated immunity, 786–787
      biological response modifiers, 787
      in vitro effects, 787
      in vivo treatment, 786–787
   cerebrovascular disease, and heart disease, 363–364
   choroid plexus, 617
   clinical picture, 353–360
   cytoskeletal toxicity, 649
   distribution, 83
   dose-response relationship, 367
   elimination, 83–84
   endoplasmic reticulum, 890
   energy metabolism, in hepatocytes, 889–890
   epidemiology, 360–361
   exposure, 83
      in endemic area, dose-response relationship, 361
   gastrointestinal tract, 84
   genotoxicity, 84, 238–240
   health effects, 84, 363–365
   heart disease, 363–364
   intracellular distribution, in hepatocytes, 889
   Japan, polluted areas in, health effects of, 361–365
   kidney, 84
      cadmium concentrations in, 85–86
   laboratory examinations, 353–360
   lactational exposure, 1133–1134
   lead, interaction, 194–195
   liver, 85–86
   lung disease, 84
   mechanisms of toxicity, 84
   metabolic models, 84
   metabolism, 83–84
   mortality, 366–367
   nephropathology, 723
   nephrotoxicity, 731–733
   neurite development, 1121–1122
      Ca^{2+} homeostasis and, 1121–1122
      cytoskeleton and, 1122
   neuropathology, 518
   overview, 86, 240–241, 353
   pathological findings, 356
   placental transport, metals, 1007–1011
   pollution, in endemic area, 360–361
   prognosis, 357–360
   renal effects, 361–362
   renal function tests, 354
   renal gene expression, alterations in, 753–754
   respiratory toxicology, 934–935
   signs, 353

stressor-specific proteins and, 179
symptoms, 353
tissues, heavy metal concentrations in, 356–357
treatment, 356
uptake by hepatocytes, 888–889
urinary findings, 354
urine
   cadmium concentration in, 85
   heavy metal concentrations in, 356–357
Cadolinium, 696
Calcium
   carcinogenicity, 290–291
   neurite development
     growth cones and, 1114
     neurite elongation and, 1115
   zinc, 195
Calcium channels, 692–695
Calculation, 36–37
Calcuria, 760
California, drinking water regulations in, 40–41
Carcinogenesis. See under specific metal
Cardiovascular system
   arsenic, 424–425
     inhalation exposure, 424–425
     oral exposure, 425
   copper, 921–928
     atherosclerosis, 922–924
     cuprotropic elements, 924–925
     depletion, 925
   overview, 925–926
Cd. See Cadmium
Cell adhesion molecules, 1097–1110
   brain, 1097–1101
   calcium-dependent, 1100–1101
   cams, 1106–1107
   lead, 1102–1103
   learning, 1106–1107
   memory, 1106–1107
   methylmercury, 1104
   N-cadherin, 1100–1101, 1103–1104
   overview, 1097, 1107
   trimethyltin, 1105–1106
Cell-mediated immunity, 785–796, 793
   arsenic, 786
     biological response modifiers, 786
     in vitro effects of, 786
     in vivo exposure, 786
   cadmium, 786–787
     biological response modifiers, 787
     in vitro effects, 787
     in vivo treatment, 786–787
   copper, 790–791
     biological response modifiers, 791
     in vitro effects, 791
     in vivo effects, 790
   iron, 791
     in vitro effects, 791
     in vivo effects, 791
   lead
     biological response modifiers, 788
     T cell functions, *in vivo* exposure, 787
     in vitro exposure, 788
   mercury, 788–789
     in vitro effects, 788–789
     in vivo exposure, 788
   nickel, 789–790
     in vitro effects, 789–790
     in vivo exposure, 789
   overview, 785–786, 794
   selenium, 791–792
     biological response modifiers, 792
     in vitro effects, 792
     in vivo effects, 791–792
   tin, 790
     in vitro effects, 790
     in vivo effects, 790
   trace metals, 790–793
   zinc, 792–793
     biological response modifiers, 793
     in vitro effects, 793
     in vivo effects, 793
Cellular oxygen-handling systems, 310–311
Cellular stress response, 167
Cellular trapping, 141
Central nervous system
   aluminum, 569–570, 577–579
     experimental approaches, 577–578
     mechanism of action, 578–579
     overview, 577
     permeability, 577–578
     transport systems, 578
   anatomy, 561–564, 562–563
   barrier properties, 563–564
   brain fluids, 561–562
   cadmium, 580–581
   copper, 568–5692
   development, 563
   experimental approaches, 571
   function of, 561–564
   high dose studies, 571–572
   iron, 569, 579–580
     overview, 579
     pathology, 579–580
     permeability, 579–580
     vascular function, 580
   layers, 561–562
   lead, 565–566, 571–574
     low dose studies, 573
     mechanism of action, 573–574
     permeability, 572
     summary of effects, 572
     transport systems, 572
     in vitro studies, 573
   manganese, 567–568
   mercury, 568, 574–577
     astroglia, 576–577
     experimental approaches, 574–576
     inorganic, 575
     mechanism of action, 576–577
     membrane toxicant, 576
     organic, 575–576
     overview, 574
   metals in plasma, chemical species of, 565
   modulation of, 564
   overview, 561, 565, 570–571

physiology, 561–564
structure, 561–564
toxicity of metals to, 570
transport of metals, 565–570
zinc, 566–567
Cerebellum, lead toxicity, 516
Cerebralo cortex, lead toxicity, 516
    aluminum, 487–493
        encephalopathy, 493
        microcytic anemia and aluminum toxicity, 487–491
        osteodystrophy, 491–492
    arsenic, 493–494
        neuropathy, 493–494
    arsine, 493
    bismuth, 494
    copper, 495
        Wilson's disease, 495
    iron, 495–496
    lead, 496–497
    manganese, 497–498
    mercury, 498–499, 499
    overview, 500–501
    radionuclides, 500
    thallium, 499–500
Chelators
    with amine groups, 1162–1163
    chelating agents, 1163–1164
    desferrioxamine, 1163–1164
    developmental toxicity, 1153–1155
    diethylenetriaminepentaacetic acid, 1161
    dimercaptopropanol, 1156–1157
    dithiocarbamates, 1163
    dithiols, 1156–1161
    ethylenediaminetetracetic acid, 1161–1162
    meso-2,3-dimercaptosuccinic acid, 1157–1160
    D-penicillamine, 1162
    polyaminocarboxylic acids, 1161–1162
    prevention, 1154–1155
    sodium 2,3-dimercapto-1-propane sulfonate, 1160–1161
    sodium 4,5-dihydroxybenzene-1,3-disulfonate, 1163
    triethylenetetramine, 1162–1163
Chemical carcinogenesis
    abnormal gene expression, 295
    adduct formation, 293–294
    arsenic, 291–292
    biochemical, 292–295
    biology, 289–292
    calcium, 290–291
    carcinogen metabolism, 293
    cobalt, 292
    DNA, oxidative damage to, 294–295
    initiation, 294
    iron, 291
    magnesium, 290
    overview, 289
    promotion, 294
    selenium, 292
    tumor formation, 289–292
    zinc, 289–290
Child. See also Pediatrics
    lead
        low dose, 405–413
            controversy, 412
            epidemiology, 410
            experimental studies, 411
            forward studies, 408
            long-term effects, 408–409
            metaanalysis, human lead studies, 410–411
            overview, 405–408
        toxicity in, 538–540
Choroid plexus
    arsenic, 618
    biochemistry, 611–612
    blood-CSF barrier, 612
    cadmium, 617
    function, 609–615, 612–615
    functional alteration, 619
    functional damage, 618–619
    gold, 618
    immunological function, 615
    iron, 618
    lead, 616–617
    manganese, 617–618
    mercury, 617
    metal binding ligand, 621
    metal ions in, 615
    metal sequestration in, 620–621
    overview, 609, 622
    pharmacology, 611–612
    secretory function, 613
    sequestration of heavy metals by, 615–618
    silver, 618
    structural damage, 618–619
    structure, 609–611, 609–615
        damage, 618–619
    as target tissue, 615–619
    tellurium, 618
    transport function, 613–615
    zinc, 618
Chromate, 727
Chromatography, 743–744
Chromium
    animal models, 208–209
    bioavailability, 209–210
    biological monitoring, 100
    carcinogenesis, 205–209, 205–219
    chromosomal effects, 213–214
    contact dermatitis, 828–829
    cytoskeletal toxicity, 649
    DNA, interactions with, 210–211
    exposure, 100
    gene expression, 213
    genotoxicity, 205–219, 209–214
    health effects, 100
    human epidemiology, 205–208
    intracellular chemistry, 209–210
    metabolism, 100
    mutagenicity, 211–213
    overview, 101, 205, 214–215
    respiratory toxicology, 930–932
    teratogenic effects, 990–991
        deficiency, 990
        excess, 990–991
Cirrhosis, 377–378
CIS-platnium, 733

Cobalt
    asthma and, 461–468
    biological monitoring, 99
    carcinogenicity, 292
        genotoxicity, 255–257
    contact dermatitis, 829
    exposure, 99
    health effects, 99
    metabolism, 99
    overview, 99
    teratogenic effects, 994
Cognitive function
    aluminum and, 554–557
    lead and, 538–543
    methylmercury and, 548–551
Compartmentalization, 125–126
Constipation, 910–911
Contact dermatitis, 827–829
    allergic, 827–829
Contact hypersensitivity, 827–833
    allergic contact dermatitis, 827–829
        chromium, 828–829
        cobalt, 829
        gold, 829
        mercury, 829
        nickel, 827–828
        palladium, 829
    animals, contact sensitization, 830–831
    molecular mechanisms, 831–832
    overview, 827, 832
Copper
    blood-brain barrier, 568–569
    bone metabolism, 962
    carcinogenicity, 291
        genotoxicity, 257–258
    cardiovascular system, 921–928
    cell-mediated immunity, 790–791
        biological response modifiers, 791
        in vitro effects, 791
        in vivo effects, 790
    drinking water, 39–53, 47–48
    hepatocytes, 893–894
    ion channel and, 686–690
    metabolism
        cirrhosis
            biliary, 378
            childhood, 377–378
        clinical manifestations, 373–374, 380–381
        copper accumulation, animal models for, 378–379
        deficiency, 382–383
        diagnosis, 374–375, 382
        genetic analysis, 372, 379
        hepatic manifestations, 373–374
        liver, 377–378
        liver transplantation, 377
        medicines, 376–377
        Menkes disease, 379–382, 382–383
        neurological symptoms, prognosis of, 377
        neuropsychiatric manifestations, 374
        occipital horn syndrome and inherited copper deficiency, 382–383
        overview, 371, 372
        pathogenesis, 372–373, 379–380
        pathology, 380
        D-penicillamine, 376
        pregnancy, 377
        prevalence, 372, 379
        treatment, 375–377, 382
        trimethylene tetramine dihydrochloride, 377
        Wilson's disease, 372–377
        zinc, 377
    oxygen generation, neurotoxicity and, 701–702
    respiratory toxicology, 935
    stressor-specific proteins and, 179
    teratogenic effects, 982–986
        deficiency, 982–984
            primary vs. secondary, 984–985
        excess, 985–986
    toxicity, from copper overload, 894–895
    zinc, 192–193
Copper amalgams, risks from, 473
Copper chelates, 873–879
Coproporphyrinogen oxidase, 949–951
Covalent bonding tendencies, 117–118
Cross-training, 305–307
Cyclic nucleotides, 630–631
Cytoskeleton
    changes, interpretation of, 640–641
    elements of, 639–641
    neurite development, effects of metals on, 1112–1114
    toxicity, 639–658
        actin microfilaments
            function of, 653–654
            structure, 653–654
        aluminum, 649, 651–652
        cadmium, 649
        chromium, 649
        cytoskeletal changes, 640–641
        cytoskeleton, 639–641
        GFAP, 652–653
        intermediate filaments, 650–653, 651
            in astrocytes, 652
            function of, 650–651
            structure, 650–651
        lead, 649–650, 652
        mercury, 647
            inorganic, 647
        methylmercury, 647–648
        microfilaments, 653–654
            disruption, 654
        microtubule, 641–645
            -associated proteins, 642–644
            alteration of, 645
            disruption of, 645–646
            function of, 641
            isotypes, 642
            metal effects on, 647–650
            proteins, 644–645
            stabilization, 642
            structure, 641
        nickel, 649
        nucleotide binding sites, 645

overview, 639
sites of metal action, 640
tubulin, 645–646
    sulfhydryl groups, 645–646

D

Dendrites, 1113
Dental amalgam
  grinding, 477–478
  mercury, 469–483
    animal studies, 477
    case reports, 477
    copper amalgams, 473
    epidemiological studies, 480–481
    fetus, effects on, 482–483
    gastrointestinal absorption, 472–473
    health risk evaluation, 476–482
    immunological effects, 480
    occupational exposure, 478–480
    overview, 469, 483
    pulmonary uptake, 471–472
    risk estimates, 478
    tissues of amalgam bearers, 475–476
    urinary excretion, mercury, 473–475
Depurination, 308
Dermal effects
  dermal exposure, 429
  inhalation exposure, 429
  oral exposure, 429
Desferrioxamine, 1163–1164
Determinants
  atomic properties, 114–117
    electronegativity, 117
    geometry, 115–116
    ion size, 115
    overview, 114–115
    oxidation state, 116–117
  biological phenomena, 125–128
    bioavailability, 126–127
    biological residence time, 127
    compartmentalization, 125–126
    metal resistance, 127–128
    respiratory tract clearance, 126
  effective ionic radii, 116
  noniosmorphic replacement, 117
  overview, 113, 128
  Pauling electronegatives, 118
  physical state, 113–114
    gases, 114
    liquids, 113–114
    solids, 113
    vapors, 114
  reactivity, 117–125
    complex formation, 120–121
    donor-atom preferences, 118–120
    ionic and covalent bonding tendencies, 117–118
    kinetic aspects, 121–122
    particle size, solid compounds, 124–125
    physicochemical properties, 125
    radical formation, 122–123
    solubility, 123–124
  table, 114

Detoxification, metal metabolism and, 179
Developmental toxicity
  aluminum, 1083–1085
  arsenic, 429–430, 1035–1041
    inhalation exposure, 429–430
    oral exposure, 430
    pharmacokinetics, 1039–1040
    placental transfer, 1039–1040
    postimplantation embryos, 1036–1038
    preimplantation embryos, 1036
  cadmium, 1027–1035
    placental toxicity, 1032–1034
    postimplantation embryo, 1029–1032
    preimplantation embryos, 1028–1029
  chelators
    with amine groups, 1162–1163
    chelating agents, 1163–1164
    desferrioxamine, 1163–1164
    diethylenetriaminepentaacetic acid, 1161
    dimercaptopropanol, 1156–1157
    dithiocarbamates, 1163
    dithiols, 1156–1161
    ethylenediaminetetracetic acid, 1161–1162
    meso-2,3-dimercaptosuccinic acid, 1157–1160
    D-penicillamine, 1162
    polyaminocarboxylic acids, 1161–1162
    sodium 2,3-dimercapto-1-propane sulfonate, 1160–1161
    sodium 4,5-dihydroxybenzene-1,3-disulfonate, 1163
    triethylenetetramine, 1162–1163
  lactational exposure, 1129–1137
    aluminum, 1131–1132
    breast-feeding, 1130–1131
    cadmium, 1133–1134
    lead, 1132–1133
    methylmercury, 1134–1135
    overview, 1129, 1135
  manganese, 1085–1088
  mercury, 1073
    behavioral pathology, 1061–1062
    cell cycle, 1063–1065
    chemistry, 1051–1071
    cytopathology, 1062–1063
    cytoprotection, 1070–1071
    developmental, 1058–1060
    distribution, 1052–1056
    elimination, 1071
    genetic effects, 1063–1065
    historical perspective, 1047–1051
    molecular pathology, 1066–1070
    overview, 1071–1073
    postnatal exposure, 1057–1058
    solubility, 1051
    sources, 1051
    stability, 1051–1052
    subcellular, 1056
    tissue/organ accumulation, 1054–1056
    toxicity, 1057–1060
    uptake, 1052–1054
    uptake and distribution, 1052–1056
    volatility, 1051
  metallothionein expression, 1139–1151
    function, during development, 1141–1143

hepatic, 1143–1145
lactation, 1143–1148
maternal induction, 1146–1148
overview, 1139, 1148–1149
pattern, 1139–1141
pregnancy, 1143–1148, 1146
regulation, during development, 1141–1143
zinc, 1145–1146
uranium, 1088–1091
vanadium, 1091–1094
in vitro developing embryos, 1038–1039
Dialysis, Minamata disease, 344
Diarrhea, 910–911
Diethylenetriaminepentaacetic acid, 1161
Diets, 36
Dimercapto-1-propane sulfonate, 1160–1161
Dimercaptopropanol, 1156–1157
Dimercaptosuccinic acid, 1157–1160
Dithiocarbamates, 1163
Dithiols, 1156–1161
DMPS. See Dimercapto-1-propane sulfonate
DMSA. See Dimercaptosuccinic acid
DNA
  chromium, 210–211
    cellular oxygen-handling systems, 310–311
    cleavage, 307–308
    cross-training, 305–307
    depurination, 308
    inflammation, stimulation of, 309–310
    lipid peroxidation enhancement, 309
    mutagenic spectrum, 308–309
    mutagenicity, 304–309
    overview, 299–301, 312–314
    oxidative DNA damage, 294–295, 304–305
      redox biochemistry, 301–304
      repair, inhibition of, 311–312
  repair
    carcinogenicity, 294
    oxidative DNA damage, 311–312
Donor-atom preferences, 118–120
Drinking water
  arsenic, 48–49
  California, regulations, 40–41
  copper, 47–48
  health effects, 44–51
  lead, 44–47
  overview, 39, 51
  risk assessment, 41–43
  selenium, 50–51
  United States, regulations, 39–41
  overview, 51
  standards, 44–51
DTC. See Dithiocarbamates
DTPA. See Diethylenetriaminepentaacetic acid

E

EAE. See Experimental aluminum encephalopathy
EDTA. See Ethylenediaminetetracetic acid
EEG. See Electroencephalography
Effective ionic radii, 116

Electrometric techniques, 75–76
Electron, 741–742
Electron x-ray microprobe, 741
Electronegativity, 117
Elongation, effects of metals on, 1115
Embryo. See Developmental toxicology
Encephalopathy, bismuth toxicity, 446
Endocrinological effects, arsenic, 428–429
  oral exposure, 428–429
Enzymes
  neurotransmitters and, 631
  urinary excretion of, 760
Erythrocytes, hematopoietic system, 941
Essential metal. See under specific metal
Ethylenediaminetetracetic acid, 1161–1162
Experimental animals, lead and, 541–542

F

Fe. See Iron
Ferrochelatase, 951–952
Fibroblasts, 326–327
Fluoride, 992

G

GABA receptors/channels, 685–686
Gallium
  nephrotoxicity, 761
  renal gene expression, alterations in, 752–753
Gallium arsenide, 725
Gastrointestinal system
  arsenic, 427
    inhalation exposure, 427
    oral exposure, 427
  bleeding, 906–910
  carcinogenesis, 913–914
  constipation, 910–911
  diarrhea, 910–911
  environmental sources, 903
  experimental investigation, 914–916
  gut, poisoning in, 904–914
  household sources, 903
  infection, 912–913
  inflammation, 912–913
  malabsorption, 911–912
  medicinal preparations, 902–903
  nausea, 905–906
  occupational sources, 903
  overview, 901, 916–917
  sources of metal exposure, 901–903
  ulceration, 906–910
  vomiting, 905–906
Gene expression, 321–329
  anchorage independence, in diploid human fibroblasts, 326–327
  arsenic, 323–326
  chromium, 323–326
  materials, 323
  methods, 323
  nickel, 323–326
  overview, 321–322
  study results, 323–327

Genotoxicity
　arsenic, 432
　　inhalation exposure, 432
　　oral exposure, 432
　beryllium, 253–284
　cadmium, 231–243
　chromium, 205–219
　nickel, 245–251
Geometry, 115–116
Germanium, 258–259
Gestation, 1088–1091. See also Developmental toxicology
　aluminum, 1083–1085
　manganese, 1085–1088
　uranium, 1088–1091
　vanadium, 1091–1094
Gliosis, 594–596
Glutamate receptor-channel complex, 690–692
Glutathione
　bile, 150–151
　chemical interactions, 146
　erythrocytes, 146–148
　　mercury sequestration, 148
　kidney, 151–161, 155–158
　liver, 148–151
　mercury
　　interorgan translocation, 146–147
　　intracellular fate, 149–150
　　renal tubular uptake, 151–152
　overview, 161
　in plasma, 146–148
　plasma
　　efflux, 150–151
　　thiols in, 147
　　uptake from, 148
　renal cellular metabolism, 158–161
　renal tubular epithelial cells, 151
Gold, 458–459
　biokinetics, 459
　chemistry, 458
　choroid plexus, 618
　contact dermatitis, 829
　diagnosis, 459
　exposure, 458
　mechanism, 459
　poisoning, 458
　therapy, 459
Grinding, 477–478
Growth cones, 1114
Gut, poisoning in, 904–905

H

Health effects, 44–51
Heavy metal. See under specific metal
Hematological effects
　arsenic, 428
　inhalation exposure, 428
　oral exposure, 428
Hematopoietic stimulation, 873
Hematopoietic system, 939–941
　bone marrow function, 939–941
　erythrocytes, 941
Hematopoietic tumors, 236

Heme biosynthetic 760–762
Heme metabolism, 942–955
　ALA dehydratase, 945–947
　ALA synthetase, regulation of, 944–945
　arsenic, 430–431
　　inhalation exposure, 431
　　oral exposure, 430–431
　biosynthesis, 944–955
　cell injury, relationship to, 954–955
　coproporphyrinogen oxidase, 949–951
　ferrochelatase, 951–952
　heme biosynthetic pathway, 942–944
　heme oxygenase, 952
　indirect effects of metals, 952–955
　pathway enzymes, 945–952
　porphobilinogen deaminase, 947–948
　porphyrinogen oxidation, 953–954
　protoporphyrinogen oxidase, 951
　subcellular organelles, structural integrity of, 952–953
　uroporphyrinogen decarboxylase, 948–949
　uroporphyrinogen III cosynthetase, 947–948
Hepatic effects
　arsenic, 427
　copper, 373–374
Heptatic effects, arsenic
　inhalation exposure, 427
　oral exposure, 427
Hg. See Mercury
Hippocampus, lead toxicity, 516
Human risk assessment, monitoring, 5–7
Hypersensitivity, contact, 827–833
　allergic contact dermatitis, 827–829
　　chromium, 828–829
　　cobalt, 829
　　gold, 829
　　mercury, 829
　　nickel, 827–828
　　palladium, 829
　animals, contact sensitization, 830–831
　molecular mechanisms, 831–832
　overview, 827, 832

I

Immune function, ion channel modulation and, 871–881
　copper chelates, 873–879
　hematopoietic stimulation, 873
　metalloelement chelates, 879–880
　metalloelements, 871–873
　overview, 871, 880
Immune response, 797–810
　arsenic, 431–432, 797–798, 812–813, 819
　　dermal exposure, 432
　　inhalation exposure, 431
　　oral exposure, 431
　astrocytes and, 594
　cadmium, 798–800, 813–814, 819–820
　copper, 800–801
　general overview, 818–819
　lead, 801–802, 814–815, 820
　macrophages, 811–818
　mercury, 802–803, 815–816, 820–821

metal-induced effects, 812–818, 819–822
natural killer cells, 818–822
nickel, 804–805, 816–817, 821
overview, 797, 811–812, 822
selenium, 805–806
tin, 806, 817–818, 822
zinc, 806–808
Immuneprivilege, 863
Immunological biomarkers, 861–869
    antibodies
        cellular targets, 866–867
        exposure and, 864–866
    antigen immunogenicity, 867
    autoantibodies
        to nervous system proteins, 864–867
        system-specific toxic effects, 864
    biological effects, markers, 862–863
    exposure, markers, 861–862
    immuneprivilege, 863
    nervous system, 863–864
    neuro-immune interactions, 863
    overview, 861, 867
    proteins, 866
    rationale, 863–864
Immunomodulation, by metals, 781–882
    overview, 783–784
Indium
    nephropathology, 725
    nephrotoxicity, 761
    renal gene expression, alterations in, 752
Infection, 912–913
Inflammation
    gastrointestinal system, 912–913
    oxidative DNA damage, 309–310
Inflammatory response, astrocytes and, 594
Inhaled metals, 854–855
Injection site sarcomas, 236
Intermediate filaments, 650–653, 651
    in astrocytes, 652
    function of, 650–651
    structure, 650–651
Involuntary movements, 342
Iodine, 991
    deficiency, 991
    excess, 991
Ion
    bonding tendencies, 117–118
    choroid plexus, normal values of, 615
    radii, 116
        metal ions, 116
    size, 115
Ion channel, 677–698
    aqueous solutions, metals in, 678
    astrocytes and, 588–590, 590
    ATP-activated cation channel, 696
    cadolinium, 696
    copper, 686–690
    GABA receptors/channels, 685–686
    glutamate receptor-channel complex, 690–692
    lanthanides, zinc and copper on GABA receptors/channels, 686–690

    lead, 682–684, 682–685, 684–685
    mercury, 678–682
        GABA-activated chloride channel, 680–682
        neuronal receptors/channels, 678–680
    modulation, immune functions and, 871–881
        copper chelates, radiation protection, recovery, 873–879
        hematopoietic stimulation, 873
        metalloelement chelates, 879–880
        metalloelements, 871–873
        overview, 871, 880
    overview, 677–678
    voltage-activated calcium channels, 692–695
    voltage-activated sodium channels, 695–696
    zinc and copper on GABA receptors/channels, 686–690
Iron, 895–897
    blood-brain barrier, 569, 579–580
        overview, 579
        pathology, 579–580
        permeability, 579–580
        vascular function, 580
    carcinogenicity, 291
        genotoxicity, 259–261
    cell-mediated immunity, 791
        in vitro effects, 791
        in vivo effects, 791
    choroid plexus, 618
    distribution in hepatocytes, 895–896
    oxygen generation, neurotoxicity and, 701–702
    stressor-specific proteins and, 180
    teratogenic effects, 987–988
        deficiency, 987–988
            primary vs. secondary, 988
        excess, 988
    toxicity, from iron overload, 896–897
Itai-itai disease, cadmium and, 353–369
    blood, 353–354
        heavy metal concentrations in, 356–357
    bone, radiographs of, 354–356
    cerebrovascular disease, and heart disease, 363–364
    clinical picture, 353–360
    epidemiology, 360–361
    heart disease, 363–364
    laboratory examinations, 353–360
    overview, 353
    pathological findings, 356
    prognosis, 357–360
    renal function tests, 354
    signs, 353
    symptoms, 353
    tissues, heavy metal concentrations in, 356–357
    treatment, 356
    urinary findings, 354

K

Kidney. See also Nephrotoxicity
    glutathione, 151–161
    liver, 145–163
Kinetics, 121–122

L

Lactational exposure, 1088–1091, 1129–1137
   aluminum, 1083–1085, 1131–1132
   breast-feeding, 1130–1131
   cadmium, 1133–1134
   lead, 1132–1133
   manganese, 1085–1088
   metallothionein expression, in development, 1143–1148
   methylmercury, 1134–1135
   overview, 1129, 1135
   uranium, 1088–1091
   vanadium, 1091–1094
Lamma, nephrotoxic metals, 740–741
Lanthanides, 686–690
Layers, 561–562
Lead
   absorption, 87
   astrocytes, 597–598, 597–599
      accumulation in astrocytes, 598–599
   biological monitoring, 89–91, 92
   blood, lead levels, 89–90
   blood-brain barrier, 565–566, 571–574
      low dose studies, 573
      mechanism of action, 573–574
      permeability, 572
      summary of effects, 572
      transport systems, 572
      in vitro studies, 573
   blood-forming organs, 88
   bone metabolism, 965–967
   calcified tissues, lead concentrations in, 90–91
   carcinogenicity, 253–284
      genotoxicity, 253–284
   cardiovascular system, 89
   cell adhesion molecules, neurotoxicity, 1102–1103
   cell-mediated immunity
      biological response modifiers, 788
      T cell functions, *in vivo* exposure, 787
      in vitro exposure, 788
   cerebellum, 516
   cerebralo cortex, 516
   child, low dose, 405–413
      controversy, 412
      epidemiology, 410
      experimental studies, 411
      forward studies, 408
      long-term effects, 408–409
      metaanalysis, 410–411
      overview, 405–408
   choroid plexus, 616–617
   cognitive function, 538–543
   cytoskeletan and, 649–650, 652
   distribution, 87
   drinking water, 39–53, 44–47
   elimination, 87
   experimental animals, 541–542
   exposure, 86–87, 92
   fixed-interval schedule controlled behavior, 543–545
   gastrointestinal tract, 88
   genotoxicity, 89
   health effects, 87–89, 92
   heme metabolism, disturbances of, 91
   hippocampus, 516
   inorganic, neurite development, 1116–1118
      in vitro effects of, 1117
      in vivo effects, 1116–1117
   ion channels, 682–685
   kidneys, 88
   lactational exposure, in suckling infants, 1132–1133
   learning deficits, 542–543
   manganese, 62–65
   mechanisms of toxicity, 87
   mechanistic considerations, 517
   metabolic model, 87
   metabolism, 87, 92
   mobilization tests, 91
   nephropathology, 721–723
      intracellular compartmentization of, 722
      mitochondria, 723
      renal gene expression, 723
      renal heme biosynthesis, 723
      renal uptake of, 722
   nephrotoxicity, 733–735
      biomarkers, 761
   neurotoxicity, 659–675
      animal literature, 663–672
      clinical research, 660–663
      complex learning, 665–667
      environmental reactivity, 671
      experimental data, 671–672
      overview, 659, 672–673
      reinforcement, schedules of, 670–671
      social behavior, 671
      spatial memory, 667–670
   occupationally-exposed populations, 540–541
   organic, 92
      neurite developmen, 1118–1120
         Ca^{2+} homeostasis and, 1119
         cytoskeleton and, 1119–1120
         in vitro effects of, 1119
         in vivo effects of, 1119
   overview, 91–92
   pediatric lead studies, 538–540
   peripheral nervous system, neuropathological effects on, 516–517
   placental transport, metals, 1011–1013
   renal gene expression, alterations in, 754–755
      organelle system effects of, 755
   reproductive effects, 89
   respiratory toxicology, 935
   risk assessment, 56 60
   in sensory-motor function, 545–546
   stressor-specific proteins and, 180
   urine, lead concentration in, 90
Learning
   cell adhesion molecules, 1106–1107
   deficits, lead and, 542–543
Linear partition coefficient, 34–36
Lipid peroxidation enhancement, 309
Liquids, 113–114

Lithium, 455–456
  biokinetics, 455–456
  chemistry, 455
  diagnosis, 456
  exposure, 455
  mechanism, 455–456
  poisoning, 455
  reatment, 456
  use, and exposure, 455
Liver
  blood, 145–163
  glutathione, 148–151
  transplantation, 377
Long-range transport models, 17–19
Lung tumors, 858–859

M

Magnesium
  bone metabolism, 960–961
  carcinogenicity, 290
  zinc, 195
Malabsorption, 911–912
Mammary heavy metal content, 1129–1137
Manganese
  astrocytes and, 599–600
    effects on astrocytic function, 599–600
    uptake, in astrocytes, 600
  biological monitoring, 101
  blood-brain barrier, central nervous system, 567–568
  carcinogenicity, genotoxicity, 262–263
  choroid plexus, 617–618
  exposure, 101
  health effects, 101
  metabolism, 101
  neurological aspects, 415–421
    absorption, 416
    course, 418
    diagnosis, 419–420
    distribution, 416
    excretion, 416
    history, 415–416
    industrial uses, 416
    laboratory examinations, 419
    pathology, 419
    prognosis, 418
    signs, 416–418
    symptoms, 416–418
    treatment, 420
  neurotransmitters and, 633–634
  overview, 102
  oxygen generation, neurotoxicity and, 702
  reproduction and, 1085–1088
  risk assessment, 60–61
  teratogenic effects, 986–987
    deficiency, 986
      primary vs. secondary, 986–987
    excess, 987
  zinc, 193–194
Manganism, 519
  mechanistic considerations, 520
  neuropathological features, 519–520
MAPs. See Microtubule-associated proteins

Memory, cell adhesion molecules, neurotoxicity, 1106–1107
Mercury
  acrodynia, 349–350
  alkoxyalkylmercury, 512
  alkylmercury, 512–513
  astrocytes and, 600–602
    effects of, on astrocytic homeostasis, 600–601
  biological monitoring, 93–94, 96
  blood-brain barrier, 568, 574–577
    astroglia, 576–577
    experimental approaches, 574–576
    inorganic, 575
    mechanism of action, 576–577
    membrane toxicant, 576
    organic, 575–576
    overview, 574
  cell-mediated immunity, 788–789
    in vitro effects, 788–789
    in vivo exposure, 788
  choroid plexus, 617
  contact dermatitis, 829
  cytoskeletan and, 647
    inorganic, 647
  dental amalgam, 469–483
    animal studies, 477
    case reports, 477
    copper amalgams, 473
    epidemiological studies, 480–481
    fetus, effects on, 482–483
    gastrointestinal absorption, 472–473
    grinding, side-effects, 477–478
    health risk evaluation, 476–482
    immunological effects, mercury, 480
    occupational exposure, 478–480
    overview, 469, 483
    pulmonary uptake, 471–472
    risk estimates, 478
    tissues of amalgam bearers, 475–476
    urinary excretion, 473–475
  developmental toxicology, 1047–1081
    behavioral pathology, 1061–1062
    cell cycle, 1063–1065
    chemistry, 1051–1071
    cytopathology, 1062–1063
    cytoprotection, 1070–1071
    developmental, 1058–1060
    distribution, 1052–1056
    elimination, 1071
    genetic effects, 1063–1065
    historical perspective, 1047–1051
    mode of action of mercurials, 1051–1071
    molecular pathology, 1066–1070
    overview, 1071–1073
    postnatal exposure, 1057–1058
    solubility, 1051
    sources, 1051
    stability, 1051–1052
    subcellular, distribution, 1056
    tissue/organ accumulation, 1054–1056
    toxicity, 1057–1060
    uptake, 1052–1056
    volatility, 1051

glutathione, of, 152–153
health effects, 93, 95–96
hepatocytes, 891–893
inorganic, 92–95
inorganic mercury salts, 512
   chronic, 348
intracellular distribution, in hepatocytes, 891
ion channels, 678–682
   GABA-activated chloride channel, 680–682
   neuronal receptors/channels, 678–680
mechanism of actions, 513
mercuric salt, 512
mercurous salt, 512
mercury salts, 349
metabolism, 93, 95
nephropathology, 723–724
   inorganic, 724
nephrotoxicity, 761
neurotransmitters and, 634–635
organic, 95–96
organomercury compounds, 512–513
overview, 94–95, 96, 346–347
placental transport, 1013–1022, 1019–1022
stressor-specific proteins and, 179
syndromes, 337–351
treatment, 348, 349, 350
uptake by hepatocytes, 891
vapor, 512
   poisoning, chronic, 347–348
vapor poisoning, 348–349
Metabolism, 630
Metal contaminants, 26, 27
Metal-metal interactions, 189–197
   cadmium, lead, 194–195
   calcium, zinc, 195
   copper, zinc, 192–193
   iron, zinc, 190–192
   lead, cadmium, 194–195
   magnesium, zinc, 195
   manganese, zinc, 193–194
   metallothionein, 194
   overview, 189–190
   sodium
      zinc, 195–196
   zinc
      calcium, 195
      copper, 192–193
      iron, 190–192
      magnesium, 195
      manganese, 193–194
      sodium, 195–196
Metal resistance, 127–128
Metal-specific cellular response, 175–179
Metal toxicology
   bioresponses, 109–197
   reactivity, 113–123
      atomic properties, 114–117
         electronegativity, 117
         geometry, 115–116
         ion size, 115
         overview, 114–115
         oxidation state, 116–117
      biological phenomena, 125–128

         bioavailability, 126–127
         biological residence time, 127
         compartmentalization, 125–126
         metal resistance, 127–128
         respiratory tract clearance, 126
      effective ionic radii, 116
      noniosmorphic replacement, 117
      overview, 113, 128
      Pauling electronegatives, 118
      physical state, 113–114
         gases, 114
         liquids, 113–114
         solids, 113
         vapors, 114
      reactivity, 117–125
         complex formation, 120–121
         donor-atom preferences, 118–120
         ionic and covalent bonding tendencies, 117–118
         kinetic aspects, 121–122
         particle size, 124–125
         physicochemical properties, 125
         radical formation, 122–123
         solubility, 123–124
      table, 114
   risk assessment, 1–504
Metalloelement chelates, 879–880
Metallothionein
   in development, 1139–1151
      function, 1141–1143
      hepatic, 1143–1145
      lactation, 1143–1148
      maternal induction, 1146–1148
      overview, 1139, 1148–1149
      pattern, 1139–1141
      pregnancy, 1143–1148, 1146
      regulation, during development, 1141–1143
      zinc, 1145–1146
   metal-induced damage, 165–187
   syntesis, 177
Methylmercury, 546–551
   auditory function, 547–548
   cell adhesion molecules, 1104
   cognitive functions, 548–551
   cytoskeletan and, 647–648
   lactational exposure, in suckling infants, 1134–1135
   mechanism of actions of, 515
   motor development, 548
   neurite development, effects of metals on, 1120–1121
      Ca^{2+} homeostasis and, 1120
      cytoskeleton and, 1121
      in vitro effects on, 1120
      in vivo effects of, 1120
   neurotransmitters and, 634–635
   sensory function, 546–548
   visual system function, 546–547
Microfilaments, 653–654
   disruption, 654
Microprobes, nephrotoxic metals, 741–742
Microtubules, 641–645
   alteration of, 645
   disruption of, 645–646

function of, 641
isotypes, 642
metal effects on, 647–650
proteins, developmental regulation of, 644–645
stabilization, 642
structure, 641
Minamata disease, 337–346
    ataxia, 341
    autonomic nerve dysfunctions, 342
    chelating agents, 344
    clinical course, 343
    clinical manifestations, 338–342
    congenital, 345–346
    delayed onset, 343
    diagnosis, 343
    dialysis, 344
    dose–effect/response, relationship, 346
    fetal, 345–346
    hearing impairment, 341–342
    involuntary movements, 342
    neurological deficit, 345
    ocular movement, 342
    overview, 337
    pathology, 342
    removal of mercury, 344–345
    sensory impairments, 339–341
    thiol resin, 345
    treatments, 344–345
    visual symptoms, 341
Mitochondria, 755
Mn. See Manganese
Molybdenum
    carcinogenicity, genotoxicity, 263–264
    teratogenic effects, 991–992
Monitoring, 23–26
Motor development, methylmercury and, 548
Motor function, aluminum and, 553–554
Mutagenic spectrum, oxidative DNA damage, carcinogenicity, damage, 308–309
Mutagenicity, 304–309

N

N-cadherin, 1100–1101, 1103–1104
Nausea, 905–906
NCAM. See Cams-neural cell adhesion molecule
Neoplastic transformation, 321–329
    arsenic, 323–326
    chromium, 323–326
    induction of anchorage independence in diploid human fibroblasts, 326–327
    materials, 323
    methods, 323
    nickel, 323–326
    overview, 321–322, 323–323, 327–328
    study results, 323–327
Nephron
    bath-to-lumen flux, 777
    calculations, 777–778
    cellular concentration, 769
        measurement, 778
    epithelium, intact, 770
    evaluation, 778–779
    harvesting kidney, 771
    leak, rate of, 778
    lumen-to-bath flux, 777
    microperfusion, 766
    micropuncture, 766
    microtechniques, 766
    overview, 765, 779
    perfusion, 775–776, 776
    perfusion chamber, 774–775
    pipette
        dissected tubular segment attachment, 775
        handcrafting, 771
    procedures, 771–779
    renal toxicology, historical development of, 766–768
    segment identification, 772–774
    solutions, 776–777
    tubule
        environment, 769–770
        transfer, to perfusion chamber, 774
Nephropathology, metals, 721–729
    arsenic, 725, 727
    cadmium, 723
        arsenic interactions, 727
    chromate, 727
    gallium arsenide, 725
    indium, 725
    lead, 721–723, 727
        intracellular compartmentization of, 722
        mitochondria, 723
        renal gene expression, 723
        renal heme biosynthesis, 723
        renal uptake of, 722
    mechanisms of, 727
    mercury, 723–724
        inorganic, 724
        organomercurials, 724
    overview, 721
    uranium, 726
Nephrotoxicity
    biomarkers, 759–763
        bismuth, 762
        calcuria, 760
        enzymes, urinary excretion of, 760
        gallium, 761
        heme biosynthetic pathway, 760–762, 762
        indium, 761
        lead, 761
        mercury, 761
        overview, 759
        proteinurias, 759–760
        renal porphyrinurias, 760–762
        urinary biomarkers, standardization of, 760
    biomolecular speciation, measurement of, 743–746
    bismuth, 444
    cadmium, 731–733
    chromatography, 743–744
    CIS-platnium, 733
    electron x-ray microprobe, 741
    exafs, 742
    fractionation, 742–743
    hepatic vs. renal toxicity, 732–733
    hybrid techniques, 745–746

lamma, 740–741
lead, 733–735
lead-binding proteins, 733–734, 734–735
microprobes, 741–742
neutron activation analysis, 738
nuclear magnetic resonance, 742
overview, 731, 735, 747
pixe, 741
radiative techniques, 740
radioisotopes, 744–745
separation of cellular components, 742–743
in situ measurement methodologies, 740–742
speciation of, 737–750
srixe, 741–742
in vitro measurement, 737–750
in vivo measurements, 738–740
x-ray fluorescence analysis, 739–740
Neural excitability, 703–704
Neurite development, effects of metals on, 1111–1128
　aluminum, 1122–1123
　　Ca^{2+} homeostasis and, 1123
　　cytoskeleton and, 1123
　　in vitro effects on, 1122–1123
　　in vivo effects of, 1122
　axons, cytoskeleton of, 1113
　cadmium, 1121–1122
　　Ca^{2+} homeostasis and, 1121–1122
　　cytoskeleton and, 1122
　calcium
　　growth cones and, 1114
　　neurite elongation and, 1115
　cellular mechanisms, 1117–1118
　cytoskeleton, neurite structure and, 1112–1114
　dendrites, cytoskeleton of, 1113
　elongation, 1115
　growth cones, 1114
　initiation, 1114
　lead
　　inorganic, 1116–1118
　　　in vitro effects of, 1117
　　　in vivo effects, 1116–1117
　　organic, 1118–1120
　　　Ca^{2+} homeostasis and, 1119
　　　cytoskeleton and, 1119–1120
　　　in vitro effects of, 1119
　　　in vivo effects of, 1119
　metal effects, 1115–1116
　methylmercury, 1120–1121
　　Ca^{2+} homeostasis and, 1120
　　cytoskeleton and, 1121
　　in vitro effects on, 1120
　　in vivo effects of, 1120
　neuronal cytoskeleton, regulation by $[Ca^{2+}]$, 1113–1114
　normal conditions, 1112–1115
　overview, 1112, 1124–1125
Neurobehavioral toxicology, 537–560
　aluminum, 551–557
　　cognitive functions, 554–557
　　motor function, 553–554

lead, 538–541, 538–546
　cognitive function, 538–543
　experimental animals, 541–542
　fixed-interval schedule controlled behavior, 543–545
　learning deficits, 542–543
　occupationally-exposed populations, 540–541
　pediatric lead studies, 538–540
　in sensory-motor function, 545–546
methylmercury, 546–551
　auditory function, 547–548
　cognitive functions, 548–551
　motor development, 548
　sensory function, 546–548
　visual system function, 546–547
overview, 537, 557–558
Neurodegenerative diseases, 395–396
Neurological deficit, 345
Neuronal cytoskeleton, 1113–1114
Neuropathology
　alkyllead, 526–528
　alkyltins, 521–526
　aluminum, 520–521
　　encephalopathy, 521
　arsenic, 425–526, 517–518
　　inhalation exposure, 426
　　oral exposure, 425–426
　cadmium, 518
　　neuropathology, 518
　lead, 513–517
　　cerebellum, 516
　　cerebralo cortex, 516
　　hippocampus, 516
　　mechanistic considerations, 517
　　peripheral nervous system, on, 516–517
　manganese, 519–520
　manganism, 519
　　mechanistic considerations, 520
　　neuropathological features, 519–520
　mercury, 511–513
　　alkoxyalkylmercury, 512
　　alkylmercury, 512–513
　　arylmercury, 512
　　inorganic mercury salts, 512
　　mechanism of actions, 513
　　mercuric salt, 512
　　mercurous salt, 512
　　mercury vapor, 512
　　organomercury compounds, 512–513
　　vapor, 512
　methylmercury, mechanism of actions of, 515
　overview, 511, 529–531
　thallium, 528–529
　triethyltin, 521–522
　trimethyltin, 523–526
Neuropsychiatric manifestations, 374
Neurotoxicity
　autoimmune responses, 863–864
　bismuth, 445–448
　cell adhesion molecules, 1097–1110
　　brain, 1097–1101
　　　calcium-dependent, 1100–1101
　　　calcium-independent, 1098–1100
　　lead, 1102–1103

learning, 1106–1107
memory, 1106–1107
   cams, 1106–1107
methylmercury, 1104
N-cadherin, 1100–1101, 1103–1104
overview, 1097, 1107
trimethyltin, 1105–1106, 1106
lead, 659–675
   animal literature, 663–672
   clinical research, 660–663
   complex learning, 665–667
   environmental reactivity, 671
   experimental data, 671–672
   overview, 659, 672–673
   reinforcement, schedules of, 670–671
   social behavior, 671
   spatial memory, 667–670
manganese, 415–421
   absorption, 416
   course, 418
   diagnosis, 419–420
   distribution, 416
   excretion, 416
   history, 415–416
   industrial uses, 416
   laboratory examinations, 419
   pathology, 419
   prognosis, 418
   signs, 416–418
   symptoms, 416–418
   treatment, 420
metals, 507–716
   overview, 509–510
   oxygen generation and, 699–706
     aluminum, 704
     Alzheimer's disease, 704
     copper, 701–702
     iron, 701–702
     manganese, 702
     neural excitability, production of, 703–704
     overview, 699–700
     selenium, high affinity for, 702–703
     sulfhydryl groups, high affinity for, 702–703
     valence flux and, 700–702
   peripheral, 516–517
   proteins, 864–867
Neurotransmitters
   brain perfusion, 631
   cyclic nucleotides, 630–631
   enzyme activity, 631
   examples, 633–635
   manganese, 633–634
   mercuric, 634–635
   metabolism, 630
   methylmercury, 634–635
   neurochemical action, 630–633
   neurotransmission, 628–630
   overview, 627, 635–636
   receptors, toxicant effects on, 632–633
   release, 631–632, 632
   second messengers, 629–630
   synapse, 628
   synthesis, 631–632
   turnover, 631–632
   uptake, 632
Neutron activation analysis, 738
Ni. See Nickel
Nickel
   asthma and, 461–468
   autoimmunity, 856–857
   biological monitoring, 102–103
   carcinogenesis, 245–251
     overview, 245–246
   cell-mediated immunity, 789–790
     in vitro effects, 789–790
     in vivo exposure, 789
   contact dermatitis, 827–828
   cytoskeletan, 649
   exposure, 102
   health effects, 102
   metabolism, 102
   overview, 103
   oxidative damage, 247–250
   porphological transformation, 323–326
   proteins, oxidative damage, 247–250
   respiratory toxicology, 932–933
   teratogenic effects, 994–995
   uptake of nickel compounds, 246–247
Noniosmorphic replacement
   essential metal ions, 117
   reactivity, 117
North American waters, 24–25
North Sea, 23
Nuclear techniques, 76–77
Nucleotides, neurotransmitters and, 630–631

O

Ocular movement, 342
Organ systems
   metal effects, 883–972
     overview, 881
   toxicology, 505–1173
Organomercurials, 724
Osteoarthropathy, 445
Oxidative DNA damage, 299–320
   cellular oxygen-handling systems, 310–311
   cleavage, 307–308
   cross-training, 305–307
   depurination, 308
   DNA base damage, 304–305
   DNA repair, inhibition of, 311–312
   inflammation, stimulation of, 309–310
   lipid peroxidation enhancement, 309
   mutagenic spectrum, 308–309
   mutagenicity, 304–309
   overview, 299–301, 312–314
   redox biochemistry, 301–304
Oxygen generation
   neurotoxicity, 699–706
     aluminum and Alzheimer's disease, 704
     Alzheimer's disease, 704
     copper, 701–702
     iron, 701–702
     manganese, 702
     neural excitability, production of, 703–704

overview, 699–700
   selenium, high affinity for, 702–703
   sulfhydryl groups, high affinity for, 702–703
   valence flux and, 700–702
neurotoxicity, 699–700

P

Palladium, 829
Parkinson's disease, 396
Particle size, 124–125
Parturition, 1088–1091, 1091. See also Developmental toxicology
   aluminum, 1083–1085
   manganese, 1085–1088
   uranium, 1088–1091
   vanadium, 1091–1094
Pauling electronegatives
   metal ions, 118
   selected metal ions, 118
Pb. See Lead
Pediatrics
   lactational exposure, in suckling infants, 1129–1137
D-penicillamine
   copper metabolism and, 376
   developmental and reproductive effects, 1162
Photon, microprobe, 741–742
Physical state, 113–114
   gases, 114
   liquids, 113–114
   solids, 113
   vapors, 114
Pixe, nephrotoxic metals, 741
Placenta, transport of metals
   amalgams, 1019–1022
   animal data, 1006, 1007–1009, 1012
   arsenic, 1005–1007
   cadmium, 1007–1011
   human data, 1006–1007, 1009–1011, 1012–1013
   lead, 1011–1013
   mercury, 1013–1022
   mercury vapor, 1019–1022
   overview, 1003–1005, 1011–1012, 1013–1019, 1023
   selenium, 1022–1023
   toxicity, 1006–1007, 1007–1011, 1012–1013
Plants, occurrence of trace elements in, 30–32
Platinum
   biokinetics, 459
   chemistry, 459
   exposure, 459
   mechanism, 459
   poisoning, 459
Polyaminocarboxylic acids, 1161–1162
Porphobilinogen deaminase, 947–948
Porphological transformation, 323–326
Porphyrinogen oxidation, 953–954
Pregnancy
   copper metabolism and, 377
   metallothionein expression, in development, 1143–1148, 1146
Preparation, of sample, biological system, metal assessment, 71–73

Preservation of samples, 70–71
Prostatic tumors, 234–235
Protein
   metal-binding, 731–736
      cadmium, 731–733
      CIS-platnium, 733
      hepatic vs. renal toxicity, 732–733
      lead, 733–735
      lead-binding proteins, 733–734, 734–735
      overview, 731, 735
   synthesis disruption, 707–716
      aluminum, 711
      description of, 707–708
      elongation, 708
      energy charge, modulation by, 708–709
      initiation, 707–708
      mercury, 712–714
         inhibition of translation due to altered GSH-GSSG ratio and/or oxidant-induced lipid peroxidation, 713
      metals and stress protein induction, 714
      modification of energy charge, 712–713
      phosphorylation and protein kinase activity, modulation of translation by, 709
      protein metabolism disruption as a mechanistic component of metal neurotoxicity, 709–710
      protein phosphorylation in neuron culture after exposure to methylmercury, 713–714
      role of glutathione and sulfhydryl group oxidation-reduction, 709
      studies on mercurial inhibition of aminoacyl-adenylation and selective aminoacyl-tRNA synthetase activity, 712–713
      termination, 708
      tin, 711–712
Proteinurias, 759–760
Proton, 741–742
Protoporphyrinogen oxidase, 951
Pulmonary host defenses, 855–856

Q

Quality assurance, 77–78
Quality control, 77–78

R

Radical formation, 122–123
Radii, ionic, effective, 116
Radioisotopes, 744–745
Reactivity, metal toxicology
   determinants, 113–123, 117–125
      atomic properties, 114–117
         electronegativity, 117
         geometry, 115–116
         ion size, 115
         overview, 114–115
         oxidation state, 116–117
      biological phenomena, 125–128
         bioavailability, 126–127
         biological residence time, 127
         compartmentalization, 125–126
         metal resistance, 127–128

respiratory tract clearance, 126
complex formation, 120–121
covalent bonding tendencies, 117–118
donor-atom preferences, 118–120
effective ionic radii, metal ions, 116
ionic, covalent bonding tendencies, 117–118
ionic bonding tendencies, covalent, 117–118
kinetic aspects, 121–122
noniosmorphic replacement, 117
overview, 113, 114–115, 128
oxidation state, 116–117
particle size, solid compounds, 124–125
Pauling electronegatives, 118
physical state, 113–114
    gases, 114
    liquids, 113–114
    solids, 113
    vapors, 114
physicochemical properties, 125
radical formation, 122–123
reactivity, 117–125
solubility, 123–124
table, 114
Receptors, 632–633
Renal cellular metabolism, 158–161
Renal disease, 396–397
Renal effects, 427–428
inhalation exposure, 428
oral exposure, 427–428
Renal gene expression, alterations in, 751–757, 755
arsenic, 751–752
cadmium, 753–754
gallium, 752–753
indium, 752
lead, 754–755
    organelle system effects of, 755
mitochondria, 755
overview, 751, 755
Renal tubular epithelial cells, 151
Renal tubular uptake, 151–152
Reproductive effects
aluminum, 1083–1085
with amine groups, 1162–1163
arsenic, 430
chelating agents, 1163–1164
desferrioxamine, 1163–1164
diethylenetriaminepentaacetic acid, 1161
dimercaptopropanol, 1156–1157
dithiocarbamates, 1163
dithiols, 1156–1161
ethylenediaminetetracetic acid, 1161–1162
manganese, 1085–1088
meso-2,3-dimercaptosuccinic acid, 1157–1160
D-penicillamine, 1162
polyaminocarboxylic acids, 1161–1162
sodium 2,3-dimercapto-1-propane sulfonate, 1160–1161
sodium 4,5-dihydroxybenzene-1,3-disulfonate, 1163
triethylenetetramine, 1162–1163
uranium, 1088–1091
vanadium, 1091–1093

Respiratory toxicology
aluminum, 933
arsenic, 426–427, 933–934
beryllium, 929–930
cadmium, 934–935
chromium, 930–932
copper, 935
lead, 935
metals affecting, 929–933
nickel, 932–933
overview, 929, 935–936
zinc, 935
Respiratory tract clearance, 126
Response modifiers, 785–796
arsenic, 786
    in vitro effects of, 786
    in vivo exposure, 786
cadmium, 786–787
    in vitro effects, 787
    in vivo treatment, 786–787
copper, 790–791
    in vitro effects, 791
    in vivo effects, 790
iron, 791
    in vitro effects, 791
    in vivo effects, 791
lead
    T cell functions, *in vivo* exposure, 787
    in vitro exposure, 788
mercury, 788–789
    in vitro effects, 788–789
    in vivo exposure, 788
nickel, 789–790
    in vitro effects, 789–790
    in vivo exposure, 789
overview, 785–786, 794
selenium, 791–792
    in vitro effects, 792
    in vivo effects, 791–792
tin, 790
    in vitro effects, 790
    in vivo effects, 790
trace metals, 790–793
zinc, 792–793
    in vitro effects, 793
    in vivo effects, 793
Risk assessment
drinking water, 41–43
metals, 55–67
    overview, 55–56

S

Sample
preparation, 71–73
selection, 70–71
storage, 70–71
Scandium, 264
Schedule controlled behavior, 543–545
Second messengers, toxicity and, 629–630
Selenium
carcinogenicity, 292
    genotoxicity, 264–266